FORD®

MECHANICAL SERVICE
2006 EDITION

THOMSON

DELMAR LEARNING

Australia • Canada • Mexico • Singapore • Spain • United Kingdom • United States

THOMSON

DELMAR LEARNING

Chilton®

Ford

Mechanical Service

2006 Edition

**Vice President,
Technology and Trades SBU:**
Alar Elken

**Executive Director,
Professional Business Unit:**
Gregory L. Clayton

**Publisher,
Professional Business Unit:**
David Koontz

Marketing Director:
Beth A. Lutz

Marketing Specialist:
Brian McGrath

Marketing Coordinator:
Marissa Maiella

Production Director:
Mary Ellen Black

Production Manager:
Larry Main

Production Editor:
Elizabeth Hough

Editorial Assistant:
Christine Wade

Chilton Publishing Coordinator:
Paula Baillie

Editors:
Terry Blomquist

Thomas A. Mellon

Richard J. Rivele

Jon Wallace

Cover Design:
Melinda Possinger

For more information contact
Thomson Delmar Learning
Executive Woods
5 Maxwell Drive, PO Box 8007,
Clifton Park, NY 12065-8007
Or find us on the World Wide Web
at
www.delmarlearning.com

ISBN: 1-4180-0601-7

ISSN: 1548-0887

NOTICE TO THE READER

Table of Contents

Model Index

USING THIS INFORMATION

Organization

To find where a particular model section or procedure is located, look in the Table of Contents. Main topics are listed with the page number on which they may be found. Following the main topics is an alphabetical listing of all of the procedures within the section and their page numbers.

Manufacturer and Model Coverage

This product covers 2002-2006 Ford Motor Company models that are produced in sufficient quantities to warrant coverage, and which have technical content available from the vehicle manufacturers before our publication date. Although this information is as complete as possible at the time of publication, some manufacturers may make changes which cannot be included here. While striving for total accuracy, the publisher cannot assume responsibility for any errors, changes, or omissions that may occur in the compilation of this data.

Part Numbers & Special Tools

Part numbers and special tools are recommended by the publisher and vehicle manufacturer to perform specific jobs. Before substituting any part or tool for the one recommended, you must be completely satisfied that neither your personal safety, nor the performance of the vehicle will be endangered.

ACKNOWLEDGEMENT

This product contains material that is reproduced and distributed under a license from Ford Motor Company. No further reproduction or distribution of the Ford Motor Company material is allowed without the express written permission of Ford Motor Company. The publisher would like to express appreciation to Ford Motor Company for its assistance in producing this product.

PRECAUTIONS

Before servicing any vehicle, please be sure to read all of the following precautions, which deal with personal safety, prevention of component damage, and important points to take into consideration when servicing a motor vehicle:

- Always wear safety glasses or goggles when drilling, cutting, grinding or prying.
- Steel-toed work shoes should be worn when working with heavy parts. Pockets should not be used for carrying tools. A slip or fall can drive a screwdriver into your body.
- Work surfaces, including tools and the floor should be kept clean of grease, oil or other slippery material.
- When working around moving parts, don't wear loose clothing. Long hair should be tied back under a hat or cap, or in a hair net.
- Always use tools only for the purpose for which they were designed. Never pry with a screwdriver.
- Keep a fire extinguisher and first aid kit handy.
- Always properly support the vehicle with approved stands or lift.
- Always have adequate ventilation when working with chemicals or hazardous material.
- Carbon monoxide is colorless, odorless and dangerous. If it is necessary to operate the engine with vehicle in a closed area such as a garage, always use an exhaust collector to vent the exhaust gases outside the closed area.
- When draining coolant, keep in mind that small children and some pets are attracted by ethylene glycol antifreeze, and are quite likely to drink any left in an open container, or in puddles on the ground. This will prove fatal in sufficient quantity. Always drain the coolant into a sealable container.
- To avoid personal injury, do not remove the coolant pressure relief cap while the engine is operating or hot. The cooling system is under pressure; steam and hot liquid can come out forcefully when the cap is loosened slightly. Failure to follow these instructions may result in personal injury. The coolant must be recovered in a suitable, clean container for reuse. If the coolant is contaminated it must be recycled or disposed of correctly.
- When carrying out maintenance on the starting system be aware that heavy gauge leads are connected directly to the battery. Make sure the protective caps are in place when maintenance is completed. Failure to follow these instructions may result in personal injury.
- Do not remove any part of the engine emission control system. Operating the engine without the engine emission control system will reduce fuel economy and engine ventilation. This will weaken engine performance and shorten engine life. It is also a violation of Federal law.
- Due to environmental concerns, when the air conditioning system is drained, the refrigerant must be collected using refrigerant recovery/recycling equipment. Federal law requires that refrigerant be recovered into appropriate recovery equipment and the process be conducted by qualified technicians who have been certified by an approved organization, such as MACS, ASI, etc. Use of a recovery machine dedicated to the appropriate refrigerant is necessary to reduce the possibility of oil and refrigerant incompatibility concerns. Refer to the instructions provided by the equipment manufacturer when removing refrigerant from or charging the air conditioning system.
- Always disconnect the battery ground when working on or around the electrical system.
- Batteries contain sulfuric acid. Avoid contact with skin, eyes, or clothing. Also, shield your eyes when working near batteries to protect against possible splashing of the acid solution. In case of acid contact with skin or eyes, flush immediately with water for a minimum of 15 minutes and get prompt medical attention. If acid is swallowed, call a physician immediately. Failure to follow these instructions may result in personal injury.
- Batteries normally produce explosive gases. Therefore, do not allow flames, sparks or lighted substances to come near the battery. When charging or working near a battery, always shield your face and protect your eyes. Always provide ventilation. Failure to follow these instructions may result in personal injury.
- When lifting a battery, excessive pressure on the end walls could cause acid to spew through the vent caps, resulting in personal injury, damage to the vehicle or battery. Lift with a battery carrier or with your hands on opposite corners. Failure to follow these instructions may result in personal injury.
- Observe all applicable safety precau-

tions when working around fuel. Whenever servicing the fuel system, always work in a well-ventilated area. Do not allow fuel spray or vapors to come in contact with a spark, open flame, or excessive heat (a hot drop light, for example). Keep a dry chemical fire extinguisher near the work area. Always keep fuel in a container specifically designed for fuel storage; also, always properly seal fuel containers to avoid the possibility of fire or explosion. Do not smoke or carry lighted tobacco or open flame of any type when working on or near any fuel-related components.

• Fuel injection systems often remain pressurized, even after the engine has been turned OFF. The fuel system pressure must be relieved before disconnecting any fuel lines. Failure to do so may result in fire and/or personal injury.

• The evaporative emissions system contains fuel vapor and condensed fuel vapor. Although not present in large quantities, it still presents the danger of explosion or fire. Disconnect the battery ground cable from the battery to minimize the possibility of an electrical spark occurring, possibly causing a fire or explosion if fuel vapor or liquid fuel is present in the area. Failure to follow these instructions can result in personal injury.

• The EPA warns that prolonged contact with used engine oil may cause a number of skin disorders, including cancer! You should make every effort to minimize your exposure to used engine oil. Protective gloves should be worn when changing oil. Wash your hands and any other exposed skin areas as soon as possible after exposure to used engine oil. Soap and water, or waterless hand cleaner should be used.

• Some vehicles are equipped with an air bag system, often referred to as a Supplemental Restraint System (SRS) or Supplemental Inflatable Restraint (SIR) system. The system must be disabled before performing service on or around system components, steering column, instrument panel components, wiring and sensors. Failure to follow safety and disabling procedures could result in accidental air bag deployment, possible personal injury and unnecessary system repairs.

• Always wear safety goggles when working with, or around, the air bag system. When carrying a non-deployed air bag, be sure the bag and trim cover are pointed away from your body. When placing a non-deployed air bag on a work surface, always face the bag and trim cover upward, away from the surface. This will reduce the motion of the module if it is accidentally deployed.

• Electronic modules are sensitive to electrical charges. The ABS module can be damaged if exposed to these charges.

• Brake pads and shoes may contain asbestos, which has been determined to be a cancer-causing agent. Never clean brake surfaces with compressed air. Avoid inhaling brake dust. Clean all brake surfaces with a commercially available brake cleaning fluid.

• When replacing brake pads, shoes, discs or drums, replace them as complete axle sets.

• When servicing drum brakes, disassemble and assemble one side at a time, leaving the remaining side intact for reference.

• Brake fluid often contains polyglycol ethers and polyglycols. Avoid contact with the eyes and wash your hands thoroughly after handling brake fluid. If you do get brake fluid in your eyes, flush your eyes with clean, running water for 15 minutes. If eye irritation persists, or if you have taken brake fluid internally, immediately seek medical assistance.

• Clean, high quality brake fluid from a sealed container is essential to the safe and proper operation of the brake system. You should always buy the correct type of brake fluid for your vehicle. If the brake fluid becomes contaminated, completely flush the system with new fluid. Never reuse any brake fluid. Any brake fluid that is removed from the system should be discarded. Also, do not allow any brake fluid to come in contact with a painted or plastic surface; it will damage the paint.

• Never operate the engine without the proper amount and type of engine oil; doing so will result in severe engine damage.

• Timing belt maintenance is extremely important! Many models utilize an interference-type, non-freewheeling engine. If the timing belt breaks, the valves in the cylinder head may strike the pistons, causing potentially serious (also time-consuming and expensive) engine damage.

• Disconnecting the negative battery cable on some vehicles may interfere with the functions of the on-board computer system(s) and may require the computer to undergo a relearning process once the negative battery cable is reconnected.

• Steering and suspension fasteners are critical parts because they affect performance of vital components and systems and their failure can result in major service expense. They must be replaced with the same grade or part number or an equivalent part if replacement is necessary. Do not use a replacement part of lesser quality or substitute design. Torque values must be used as specified during reassembly to ensure proper retention of these parts.

FORD, LINCOLN AND MERCURY

1

Aviator • Explorer • Explorer Sport • Explorer Sport-Trac • Mountaineer

SPECIFICATIONS AND MAINTENANCE CHARTS

ENGINE AND VEHICLE IDENTIFICATION

Engine								Model Year	
Code ①	Liters (cc)	Cu. In.	Cyl.	Fuel Sys.	Type	Eng. Mfg.		Code ②	Year
E	4.0 (4000)	244	6	MFI	SOHC	Ford		2	2002
H	4.6 (4601)	281	8	SFI	DOHC	Ford		3	2003
W	4.6 (4601)	281	8	MFI	SOHC	Ford		4	2004
								5	2005
								6	2006

MFI: Multi-port Fuel Injection

SFI: Sequential Multi-port Fuel Injection

DOHC: Dual Overhead Camshafts

SOHC: Single Overhead Camshaft

① 8th digit of the Vehicle Identification Number (VIN)
② 10th digit of the Vehicle Identification Number (VIN)

06017-EXPL-C01

GENERAL ENGINE SPECIFICATIONS

Year	Model	Engine Displ. Liters	Engine VIN	Net Horsepower @ rpm	Net Torque @ rpm (ft. lbs.)	Bore x Stroke (in.)	Compression Ratio	Oil Pressure @ rpm
2002	Explorer	4.0	E	210@5250	240@3000	3.81x3.39	9.0:1	40-60@2000
	Explorer	4.6	W	239@4750	282@4000	3.55x3.54	9.3:1	20-45@1500
	Mountaineer	4.0	E	210@5250	240@3000	3.81x3.39	9.0:1	40-60@2000
	Mountaineer	4.6	W	239@4750	282@4000	3.55x3.54	9.3:1	20-45@1500
	Explorer Sport	4.0	E	160@4000	225@2500	3.81x3.39	9.0:1	40-60@2000
	Explorer Sport-Trac	4.0	E	160@4000	225@2500	3.81x3.39	9.0:1	40-60@2000
2003	Explorer	4.0	E	210@5250	240@3000	3.81x3.39	9.0:1	40-60@2000
	Explorer	4.6	W	239@4750	282@4000	3.55x3.54	9.3:1	20-45@1500
	Explorer Sport	4.0	E	160@4000	225@2500	3.81x3.39	9.0:1	40-60@2000
	Explorer Sport-Trac	4.0	E	160@4000	225@2500	3.81x3.39	9.0:1	40-60@2000
	Mountaineer	4.0	E	210@5250	240@3000	3.81x3.39	9.0:1	40-60@2000
	Mountaineer	4.6	W	239@4750	282@4000	3.55x3.54	9.3:1	20-45@1500
	Aviator	4.6	H	302@5750	300@3250	3.55x3.54	10.1:1	20-45@1500
2004	Explorer	4.0	E	210@5250	240@3000	3.81x3.39	9.0:1	40-60@2000
	Explorer	4.6	W	239@4750	282@4000	3.55x3.54	9.3:1	20-45@1500
	Explorer Sport-Trac	4.0	E	160@4000	225@2500	3.81x3.39	9.0:1	40-60@2000
	Mountaineer	4.0	E	210@5250	240@3000	3.81x3.39	9.0:1	40-60@2000
	Mountaineer	4.6	W	239@4750	282@4000	3.55x3.54	9.3:1	20-45@1500
	Aviator	4.6	H	302@5750	300@3250	3.55x3.54	10.1:1	20-45@1500
2005	Explorer	4.0	E	210@5250	240@3000	3.81x3.39	9.0:1	40-60@2000
	Explorer	4.6	W	239@4750	282@4000	3.55x3.54	9.3:1	20-45@1500
	Explorer Sport-Trac	4.0	E	160@4000	225@2500	3.81x3.39	9.0:1	40-60@2000
	Mountaineer	4.0	E	210@5250	240@3000	3.81x3.39	9.0:1	40-60@2000
	Mountaineer	4.6	W	239@4750	282@4000	3.55x3.54	9.3:1	20-45@1500
	Aviator	4.6	H	302@5750	300@3250	3.55x3.54	10.1:1	20-45@1500

06017-EXPL-C02

GASOLINE ENGINE TUNE-UP SPECIFICATIONS

Year	Engine Displacement Liters	Engine VIN	Spark Plug Gap (in.)	Ignition Timing (deg.) ①		Fuel Pressure (psi) ②	Idle Speed (rpm)		Valve Clearance	
				MT	AT		MT	AT	In.	Ex.
2002	4.0	E	0.052-0.056	10B	10B	57-73	①	①	HYD	HYD
	4.6	W	0.052-0.056	—	10B	57-73	①	①	HYD	HYD
2003	4.0	E	0.052-0.056	10B	10B	57-73	①	①	HYD	HYD
	4.6	H	0.052-0.056	—	10B	25-65	①	①	HYD	HYD
	4.6	W	0.052-0.056	—	10B	57-73	①	①	HYD	HYD
2004	4.0	E	0.052-0.056	10B	10B	60-65	①	①	HYD	HYD
	4.6	W	0.052-0.056	—	10B	60-65	①	①	HYD	HYD
	4.6	H	0.052-0.056	—	10B	25-65	①	①	HYD	HYD
2005	4.0	E	0.061-0.068	10B	10B	30-40	①	①	HYD	HYD
	4.6	W	0.052-0.056	—	10B	30-40	①	①	HYD	HYD
	4.6	H	0.052-0.056	—	10B	25-65	①	①	HYD	HYD

NOTE: The Vehicle Emission Control Information label often reflects specification changes changes made during production.

The label figures must be used if they differ from those in this chart.

B: Before top dead center

HYD: Hydraulic

① Idle speed and ignition timing are electronically controlled and cannot be adjusted

② Key on; engine off

06017-EXPL-C03

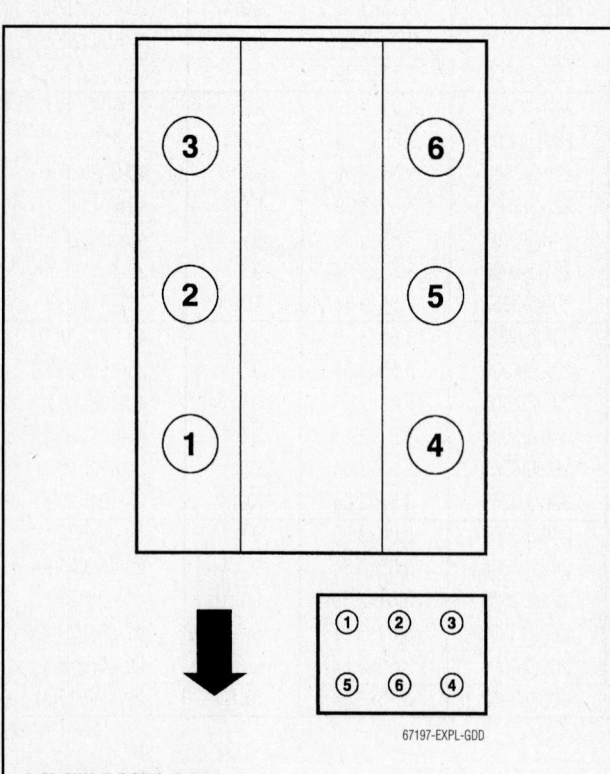

67197-EXPL-GDD

4.0L VIN E SOHC Engine
Firing order: 1–4–2–5–3–6
Distributorless ignition system

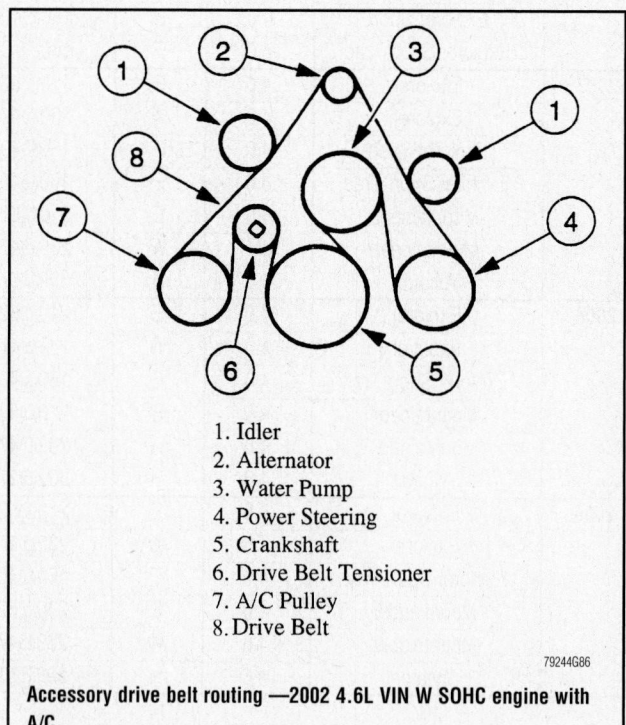

1. Idler
2. Alternator
3. Water Pump
4. Power Steering
5. Crankshaft
6. Drive Belt Tensioner
7. A/C Pulley
8. Drive Belt

79244G86

Accessory drive belt routing —2002 4.6L VIN W SOHC engine with A/C

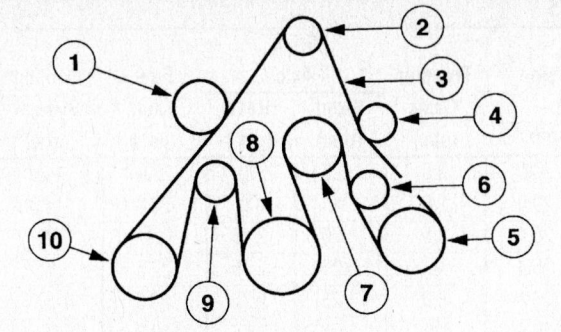

1 Belt idler pulley
2 Generator pulley
3 Drive belt
4 Belt idler pulley
5 Power steering pump pulley
6 Belt idler pulley
7 Coolant pump pulley
8 Crankshaft pulley
9 Drive belt tensioner pulley
10 A/C compressor pulley

67197-EXPL-GBB

Accessory drive belt routing —2003—05 4.6L VIN W SOHC engine

1 Belt idler pulley
2 Generator
3 Drive belt
4 Power steering pump pulley
5 Crankshaft pulley
6 Water pump pulley
7 A/C compressor
8 Drive belt tensioner pulley
9 Belt idler pulley
10 Belt idler pulley

67197-EXPL-GAA

Accessory drive belt routing —4.6L VIN H DOHC engine

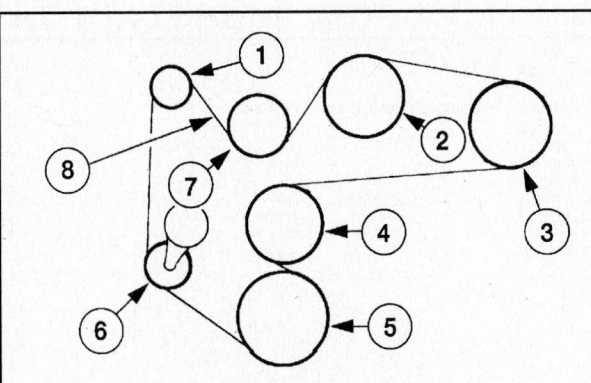

1 Generator pulley
2 Power steering pump pulley
3 A/C compressor pulley
4 Coolant pump pulley
5 Crankshaft damper
6 Drive belt tensioner pulley
7 Belt idler pulley
8 Drive belt

67197-EXPL-GCC

Accessory drive belt routing —4.0L VIN E SOHC engine

CAPACITIES

Year	Model	Engine Displ. Liters	Engine VIN	Engine Oil with Filter (qts.)	Transmission (pts.)		Transfer Case (pts.)	Drive Axle		Fuel Tank (gal.)	Cooling System (qts.)
					5-Spd	Auto. ①		Front (pts.)	Rear (pts.)		
2002	Explorer Sport	4.0	E	4.0	5.6	20.0	2.5	3.25	②	23.0	14.0
	Explorer Sport-Trac	4.0	E	4.0	5.6	20.0	2.5	3.6	②	23.0	14.0
	Explorer	4.0	E	5.0	5.0	25.4	3.0	3.25	2.75	④	⑤
	Explorer	4.6	W	6.5	—	25.4	③	3.25	2.75	22.5	⑥
	Mountaineer	4.0	E	5.0	5.0	25.4	3.0	3.25	2.75	22.5	⑤
	Mountaineer	4.6	W	6.5	—	25.4	③	3.25	2.75	22.5	⑥
2003	Explorer	4.0	E	5.0	—	25.4	3.0	2.70	3.50	④	⑤
	Explorer Sport	4.0	E	4.0	NA	⑦	2.5	3.25	②	23.0	14.0
	Explorer Sport-Trac	4.0	E	4.0	NA	⑦	2.5	3.25	②	23.0	14.0
	Explorer	4.6	W	6.5	—	25.4	③	2.70	3.50	22.5	⑥
	Mountaineer	4.0	E	5.0	—	25.4	3.0	2.70	3.50	22.5	⑤
	Mountaineer	4.6	W	6.5	—	25.4	③	2.70	3.50	22.5	⑥
	Aviator	4.6	H	6.0	—	25.4	2.6	2.70	3.50	22.5	19.6
2004	Explorer	4.0	E	5.0	—	25.4	3.0	2.70	3.50	22.5	⑤
	Explorer	4.6	W	6.5	—	25.4	③	2.70	3.50	22.5	⑥
	Explorer Sport-Trac	4.0	E	4.0	—	⑦	2.5	3.25	②	23.0	14.0
	Mountaineer	4.0	E	5.0	—	25.4	3.0	2.70	3.50	22.5	⑤
	Mountaineer	4.6	W	6.5	—	25.4	③	2.70	3.50	22.5	⑥
	Aviator	4.6	H	6.0	—	25.4	2.6	2.70	3.50	22.5	19.6
2005	Explorer	4.0	E	5.0	—	25.4	3.0	2.70	3.50	22.5	⑤
	Explorer	4.6	W	6.5	—	25.4	③	2.70	3.50	22.5	⑥
	Explorer Sport-Trac	4.0	E	4.0	—	⑦	2.5	3.25	②	23.0	14.0
	Mountaineer	4.0	E	5.0	—	25.4	3.0	2.70	3.50	22.5	⑤
	Mountaineer	4.6	W	6.5	—	25.4	③	2.70	3.50	22.5	⑥
	Aviator	4.6	H	6.0	—	25.4	2.6	2.70	3.50	22.5	19.6

NA: Information not available

NOTE: All capacities are approximate. Add fluid gradually and check to be sure a proper fluid level is obtained.

① Dry fill

② With Traction-Lok: 5.25 pts.
   Without Traction-Lok: 5.5 pts.

③ Part time: 3.0
   Full time: 2.6

④ 4-door Explorer: 22.5
   Explorer Sport: 17.5
   Explorer Sport Trac: 20.5

⑤ w/o auxiliary heater: 16.3
   w/auxiliary heater: 18.3

⑥ w/o auxiliary heater: 18.6
   w/auxiliary heater: 20.1

⑦ 4x2: 20.0 pts.
   4x4: 20.6 pts.

06017-EXPL-C04

VALVE SPECIFICATIONS

Year	Engine Displ. Liters	Engine VIN	Seat Angle (deg.)	Face Angle (deg.)	Spring Test Pressure (lbs. @ in.)	Spring Installed Height (in.)	Stem-to-Guide Clearance (in.)		Stem Diameter (in.)	
							Intake	Exhaust	Intake	Exhaust
2002	4.0	E	45	45	202-225@ 1.413-1.445	1.569-1.601	0.0010-0.0020	0.0010-0.0020	0.2740-0.2748	0.2730-0.2740
	4.6	W	44.5-45.1	45.25-45.75	162-180@ 1.134	1.675	0.0008-0.0027	0.0018-0.0037	0.2746-0.2750	0.2736-0.2740
2003	4.0	E	45	45	202-225@ 1.413-1.445	1.569-1.601	0.0010-0.0020	0.0010-0.0020	0.2740-0.2748	0.2730-0.2740
	4.6	H	45	45.5	160@1.03	1.4228	0.0008-0.0027	0.0018-0.0037	0.2746-0.2754	0.2736-0.2744
	4.6	W	44.5-45.1	45.25-45.75	162-180@ 1.134	1.675	0.0008-0.0027	0.0018-0.0037	0.2746-0.2750	0.2736-0.2740
2004	4.0	E	45	45	202-225@ 1.413-1.445	1.569-1.601	0.0010-0.0020	0.0010-0.0020	0.2740-0.2748	0.2730-0.2740
	4.6	H	45	45.5	160@1.03	1.4228	0.0008-0.0027	0.0018-0.0037	0.2746-0.2754	0.2736-0.2744
	4.6	W	44.5-45.1	45.25-45.75	162-180@ 1.134	1.675	0.0008-0.0027	0.0018-0.0037	0.2746-0.2750	0.2736-0.2740
2005	4.0	E	45	45	202-225@ 1.413-1.445	1.569-1.601	0.0010-0.0020	0.0010-0.0020	0.2740-0.2748	0.2730-0.2740
	4.6	H	45	45.5	160@1.03	1.4228	0.0008-0.0027	0.0018-0.0037	0.2746-0.2754	0.2736-0.2744
	4.6	W	44.5-45.1	45.25-45.75	162-180@ 1.134	1.675	0.0008-0.0027	0.0018-0.0037	0.2746-0.2750	0.2736-0.2740

06017-EXPL-C05

CRANKSHAFT AND CONNECTING ROD SPECIFICATIONS

All measurements are given in inches.

Year	Engine Displ. Liters	Engine VIN	Crankshaft				Connecting Rod		
			Main Brg. Journal Dia.	Main Brg. Oil Clearance	Shaft End-play	Thrust on No.	Journal Diameter	Oil Clearance	Side Clearance
2002	4.0	E	2.2430-2.2440	0.0008-0.0015	0.0020-0.0125	3	2.7252-2.7260	0.0003-0.0024	0.0036-0.0106
	4.6	W	2.6500-2.6570	0.0009-0.0018	0.0051-0.0118	3	2.0867-2.0870	0.0010-0.0027	0.0059-0.0177
2003	4.0	E	2.2430-2.2440	0.0008-0.0015	0.0020-0.0125	3	2.7252-2.7260	0.0003-0.0024	0.0036-0.0106
	4.6	H	2.6567-2.6576	0.0010-0.0018	0.0051-0.0118	3	2.0859-2.0867	0.0011-0.0027	0.0059-0.0177
	4.6	W	2.6500-2.6570	0.0009-0.0018	0.0051-0.0118	3	2.0867-2.0870	0.0010-0.0027	0.0059-0.0177
2004	4.0	E	2.2430-2.2440	0.0008-0.0015	0.0020-0.0125	3	2.7252-2.7260	0.0003-0.0024	0.0036-0.0106
	4.6	H	2.6567-2.6576	0.0010-0.0018	0.0051-0.0118	3	2.0859-2.0867	0.0011-0.0027	0.0059-0.0177
	4.6	W	2.6500-2.6570	0.0009-0.0018	0.0051-0.0118	3	2.0867-2.0870	0.0010-0.0027	0.0059-0.0177
2005	4.0	E	2.2430-2.2440	0.0008-0.0015	0.0020-0.0125	3	2.7252-2.7260	0.0003-0.0024	0.0036-0.0106
	4.6	H	2.6567-2.6576	0.0010-0.0018	0.0051-0.0118	3	2.0859-2.0867	0.0011-0.0027	0.0059-0.0177
	4.6	W	2.6500-2.6570	0.0009-0.0018	0.0051-0.0118	3	2.0867-2.0870	0.0010-0.0027	0.0059-0.0177

06017-EXPL-C07

PISTON AND RING SPECIFICATIONS

All measurements are given in inches.

Year	Engine Displ. Liters	Engine VIN	Piston Clearance	Ring Gap			Ring Side Clearance		
				Top Comp.	Bottom Comp.	Oil Control	Top Comp.	Bottom Comp.	Oil Control
2002	4.0	E	0.0008-0.0019	0.015-0.023	0.015-0.023	0.015-0.055	0.0010-0.0030	0.0010-0.0030	SNUG
	4.6	W	0.0005-0.0010	0.0394	0.0394	0.0500	0.0012-0.0031	0.0012-0.0031	SNUG
2003	4.0	E	0.0008-0.0019	0.015-0.023	0.015-0.023	0.015-0.055	0.0010-0.0030	0.0010-0.0030	SNUG
	4.6	H	0.0004	0.006-0.0120	0.012-0.0220	0.006-0.0260	0.0004-0.0009	0.0012-0.0032	SNUG
	4.6	W	0.0005-0.0010	0.0394	0.0394	0.0500	0.0012-0.0031	0.0012-0.0031	SNUG
2004	4.0	E	0.0008-0.0019	0.015-0.023	0.015-0.023	0.015-0.055	0.0010-0.0030	0.0010-0.0030	SNUG
	4.6	H	0.0004	0.006-0.0120	0.012-0.0220	0.006-0.0260	0.0004-0.0009	0.0012-0.0032	SNUG
	4.6	W	0.0005-0.0010	0.0394	0.0394	0.0500	0.0012-0.0031	0.0012-0.0031	SNUG
2005	4.0	E	0.0008-0.0019	0.015-0.023	0.015-0.023	0.015-0.055	0.0010-0.0030	0.0010-0.0030	SNUG
	4.6	H	0.0004	0.006-0.0120	0.012-0.0220	0.006-0.0260	0.0004-0.0009	0.0012-0.0032	SNUG
	4.6	W	0.0005-0.0010	0.0394	0.0394	0.0500	0.0012-0.0031	0.0012-0.0031	SNUG

06017-EXPL-C06

TORQUE SPECIFICATIONS
All readings in ft. lbs.

Engine Displ. Liters	Engine VIN	Cylinder Head Bolts	Main Bearing Bolts	Rod Bearing Bolts	Crankshaft Damper Bolts	Flywheel Bolts	Manifold Intake *	Exhaust	Spark Plugs	Oil Pan Drain Plug
2002 4.0	E	①	72	②	③	75-85	7	16	15	19
4.6	W	④	⑤	⑥	⑦	59	18	15	12	10
2003 4.0	E	①	72	②	③	75-85	7	16	15	19
4.6	H	④	⑤	⑥	⑦	59	7	15	15	10
4.6	W	④	⑤	⑥	⑦	59	18	15	12	10
2004 4.0	E	①	72	②	③	75-85	7	16	15	19
4.6	H	④	⑤	⑥	⑦	59	7	15	15	10
4.6	W	④	⑤	⑥	⑦	59	18	15	12	10
2005 4.0	E	⑧	72	②	⑨	⑩	8	16	13	19
4.6	H	④	⑤	⑥	⑦	59	7	15	15	10
4.6	W	④	⑤	⑥	⑦	59	18	15	13	10

NA: Information not available

* NOTE: Applies to Lower Manifold only.

① Step 1: 24 ft. lbs.
  Step 2: plus 90 degrees

② Step 1: 15 ft. lbs.
  Step 2: +90 degrees

③ Step 1: 37 ft. lbs.
  Step 2: plus 90 degrees

④ Step 1: 30 ft. lbs.
  Step 2: +90 degrees
  Step 3: back off 1 full turn
  Step 4: 30 ft. lbs.
  Step 5: +90 degrees
  Step 6: +90 additional degrees

⑤ Bolts 1-20:
  Step 1: 89 inch lbs.
  Step 2: 18 ft. lbs.
  Step 3: 30 ft. lbs.
  Step 4: plus 90 degrees
  Cross bolts:
  Step 1: 30 ft. lbs.
  Step 2: plus 90 degrees

⑥ Step 1: 17 ft. lbs.
  Step 2: 32 ft. lbs.
  Step 3: +90-120 degrees

⑦ Step 1: 66 ft. lbs.
  Step 2: back off 1 full turn
  Step 3: 37 ft. lbs.
  Step 4: +90 degrees
  Do not exceed 148 ft. lbs.

⑧ Step 1: 12mm bolts: 9 ft. lbs.
  Step 2: 12mm bolts: 18 ft. lbs.
  Step 3: 8mm bolts: 24 ft. lbs.
  Step 4: 12mm bolts: plus 90 degrees
  Step 5: 12mm bolts: plus an additional 90 degrees

⑨ Step 1: 33 ft. lbs.
  Step 2: plus 85 degrees

⑩ Step 1: 10 ft. lbs.
  Step 2: 52 ft. lbs.

06017-EXPL-C08

06017-EXPL-G01

Main bearing torque sequence—4.6L engines

WHEEL ALIGNMENT

Year	Model		Caster Range (+/-Deg.)	Caster Preferred Setting (Deg.)	Camber Range (+/-Deg.)	Camber Preferred Setting (Deg.)	Toe-in (in.)
2002	Explorer	F	1.00	①	0.80	-0.50	0.10+/-0.25
		R	—	—	0.80	-0.50	0.10+/-0.25
	Explorer Sport		1.00	②	0.70	-0.50	0.12+/-0.25
	Explorer Sport Trac		1.00	②	0.70	-0.50	0.12+/-0.25
	Mountaineer	F	1.00	①	0.80	-0.50	0.10+/-0.25
		R	—	—	0.80	-0.50	0.10+/-0.25
2003	Aviator	F	1.00	+5.3	0.08	-0.50	0.24+/-0.25
		R	—	—	0.80	-0.95	0.16+/-0.25
	Explorer	F	1.00	①	0.80	-0.50	0.10+/-0.25
		R	—	—	0.80	-0.50	0.10+/-0.25
	Explorer Sport		1.00	②	0.70	-0.50	0.12+/-0.25
	Explorer Sport Trac		1.00	②	0.70	-0.50	0.12+/-0.25
	Mountaineer	F	1.00	①	0.80	-0.50	0.10+/-0.25
		R	—	—	0.80	-0.50	0.10+/-0.25
2004	Aviator	F	1.00	+5.3	0.08	-0.50	0.24+/-0.25
		R	—	—	0.25	+0.80	0.16+/-0.25
	Explorer	F	1.00	L +5.1; R+5.3	0.80	-0.50	0.10+/-0.25
		R	—	—	0.80	-0.90	0.10+/-0.25
	Explorer Sport-Trac		1.00	L +3.15; R +4.05	0.70	-0.50	0.12+/-0.25
	Mountaineer	F	1.00	L +5.1; R+5.3	0.80	-0.50	0.10+/-0.25
		R	—	—	0.80	-0.90	0.10+/-0.25
2005	Aviator	F	1.00	+5.3	0.08	-0.50	0.24+/-0.25
		R	—	—	0.25	+0.80	0.16+/-0.25
	Explorer	F	1.00	L +5.1; R+5.3	0.80	-0.50	0.10+/-0.25
		R	—	—	0.80	-0.90	0.10+/-0.25
	Explorer Sport-Trac		1.00	L +3.15; R +4.05	0.70	-0.50	0.12+/-0.25
	Mountaineer	F	1.00	L +5.1; R+5.3	0.80	-0.50	0.10+/-0.25
		R	—	—	0.80	-0.90	0.10+/-0.25

① Left side: +5.1
   Right side: +5.3

② Left side: +3.95
   Right side: +4.45

06017-EXPL-C09

TIRE, WHEEL AND BALL JOINT SPECIFICATIONS

Year	Model	OEM Tires Standard	OEM Tires Optional	Tire Pressures (psi) Front	Tire Pressures (psi) Rear	Wheel Size	Ball Joint Inspection	Lugnut Toque ft. lbs.
2002	Explorer	P225/70R16	P245/70R16	②	②	7-JJ	0.030 in. ①	100
			P255/70R16	②	②	7-JJ		
			P255/70HR16	②	②	7-JJ		
	Explorer Sport	P235/70R15	P255/70R16	②	②	7-JJ	0.030 in. ①	100
	Explorer Sport Trac	P235/70R15	P255/70R16	②	②	7-JJ	0.030 in. ①	100
	Mountaineer	P225/70R16	P245/70R16	②	②	7-JJ	0.030 in. ①	100
			P255/70R16	②	②	7-JJ		
			P255/70HR16	②	②	7-JJ		
2003	Aviator	P245/65HR17	none	②	②	7.5	0.030 in. ①	100
	Explorer	P225/70R16	P245/70R16	②	②	7-JJ	0.030 in. ①	100
			P255/70R16	②	②	7-JJ		
			P255/70HR16	②	②	7-JJ		
	Explorer Sport	P235/70R15	P255/70R16	②	②	7-JJ	0.030 in. ①	100
	Explorer Sport Trac	P235/70R15	P255/70R16	②	②	7-JJ	0.030 in. ①	100
	Mountaineer	P225/70R16	P245/70R16	②	②	7-JJ	0.030 in. ①	100
			P255/70R16	②	②	7-JJ		
			P255/70HR16	②	②	7-JJ		
2004	Aviator	P245/65HR17	none	②	②	7.5	0.030 in. ①	100
	Explorer						0.030 in. ①	100
	Eddie Bauer	P245/65R17	none	②	②	NA		
	Limited	P245/65R17	none	②	②	NA		
	NBX	P245/65R17	none	②	②	NA		
	XLS	P235/70R16	none	②	②	7		
	XLS Sport	P235/70R16	none	②	②	7		
	XLT	P235/70R16	none	②	②	7		
	XLT Sport	P245/65R17	none	②	②	NA		
	Explorer Sport-Trac						0.030 in. ①	100
	Adrenalin	P255/70R16	none	②	②	7		
	XLS	P235/70R16	none	②	②	7		
	XLT	P235/70R16	none	②	②	7		
	XLT Premium	P255/70R16	none	②	②	7		
	Mountaineer						0.030 in. ①	100
	Convenience	P235/70R16	none	②	②	NA		
	Luxury	P245/65R17	none	②	②	NA		
	Premier	P245/65R17	none	②	②	NA		
2005	Aviator	P245/65HR17	none	②	②	7.5	0.030 in. ①	100
	Explorer						0.030 in. ①	100
	Eddie Bauer	P245/65R17	none	②	②	NA		
	Limited	P245/65R17	none	②	②	NA		
	XLS	P235/70R16	none	②	②	7		
	XLS Sport	P235/70R16	none	②	②	7		
	XLT	P235/70R16	none	②	②	7		
	XLT Sport	P245/65R17	none	②	②	NA		
	Explorer Sport-Trac						0.030 in. ①	100
	Adrenalin	P255/70R16	none	②	②	7		

06017-EXPL-C10

TIRE, WHEEL AND BALL JOINT SPECIFICATIONS

Year	Model	OEM Tires		Tire Pressures (psi)		Wheel Size	Ball Joint Inspection	Lugnut Toque ft. lbs.
		Standard	Optional	Front	Rear			
2005 cont.	XLS	P235/70R16	none	②	②	7		
	XLT	P235/70R16	none	②	②	7		
	XLT Premium	P255/70R16	none	②	②	7		
	Mountaineer						0.030 in. ①	100
	Convenience	P235/70R16	none	②	②	NA		
	Luxury	P245/65R17	none	②	②	NA		
	Premier	P245/65R17	none	②	②	NA		

NA: Information not available

OEM: Original Equipment Manufacturer

STD: Standard

OPT: Optional

① Both upper and lower

② See placard on vehicle

06017-EXPL-C11

BRAKE SPECIFICATIONS
All measurements in inches unless noted

| Year | Model | | Brake Disc | | | Brake Drum Diameter | | | Minimum Lining Thickness | Brake Caliper | |
			Original Thickness	Min. Thickness	Max. Runout	Original Inside Dia.	Max. Wear Limit	Max. Machine Dia.		Bracket Bolts (ft. lbs.)	Mounting Bolts (ft. lbs.)
2002	Explorer	F	1.020	0.980	0.0005	—	—	—	0.100	83	24
		R	0.480	0.440	0.0024	—	—	—	0.039	—	24
	Explorer	F	1.020	0.980	0.0016	—	—	—	0.100	83	21-26
	Sport	R	0.480	0.440	0.0024	10.00	10.09	10.06	0.039	80	20
	Explorer	F	1.020	0.980	0.0005	—	—	—	0.100	83	21-26
	Sport-Trac	R	0.480	0.440	0.0024	—	—	—	0.039	80	20
	Mountaineer	F	1.020	0.980	0.0005	—	—	—	0.100	83	21-26
		R	0.480	0.440	0.0024	—	—	—	0.039	80	24
2003	Explorer	F	1.020	0.980	0.0005	—	—	—	0.100	100	24
		R	0.480	0.440	0.0024	—	—	—	0.039	—	24
	Explorer	F	1.020	0.980	0.0016	—	—	—	0.100	83	21-26
	Sport	R	0.480	0.440	0.0024	—	—	—	0.039	80	20
	Explorer	F	1.020	0.980	0.0005	—	—	—	0.100	83	21-26
	Sport-Trac	R	0.480	0.440	0.0024	—	—	—	0.039	80	20
	Mountaineer	F	1.020	0.980	0.0005	—	—	—	0.100	100	24
		R	0.480	0.440	0.0024	—	—	—	0.039	—	24
	Aviator	F	NA	1.040	NA	—	—	—	0.118	155	32
		R	NA	0.430	NA	—	—	—	0.118	—	24
2004	Aviator	F	NA	1.040	NA	—	—	—	0.118	155	32
		R	NA	0.430	NA	—	—	—	0.118	—	24
	Explorer	F	NA	0.960	NA	—	—	—	0.118	100	24
		R	NA	0.430	NA	—	—	—	0.118	—	24
	Explorer	F	NA	0.980	NA	—	—	—	0.100	83	21-26
	Sport-Trac	R	NA	0.440	0.0024	—	—	—	0.039	80	20
	Mountaineer	F	NA	0.960	NA	—	—	—	0.118	100	24
		R	NA	0.430	NA	—	—	—	0.118	—	24
2005	Aviator	F	NA	1.040	NA	—	—	—	0.118	155	32
		R	NA	0.430	NA	—	—	—	0.118	—	24
	Explorer	F	NA	0.960	NA	—	—	—	0.118	100	24
		R	NA	0.430	NA	—	—	—	0.118	—	24
	Explorer	F	NA	0.980	NA	—	—	—	0.100	83	24
	Sport-Trac	R	NA	0.440	0.0024	—	—	—	0.039	80	20
	Mountaineer	F	NA	0.960	NA	—	—	—	0.118	100	24
		R	NA	0.430	NA	—	—	—	0.118	—	24

NA: Information not available

NOTE: Due to changes made during production, refer to manufacturer's specifications if they differ from those in this chart

06017-EXPL-C12

SCHEDULED MAINTENANCE INTERVALS

2002-03 Ford Explorer, Explorer Sport, Explorer Sport-Trac, Mercury Mountaineer
2003 Lincoln Aviator

TO BE SERVICED	TYPE OF SERVICE	VEHICLE MILEAGE INTERVAL (x1000)												
		5	10	15	20	25	30	35	40	45	50	55	60	65
Engine oil & filter	R	✓	✓	✓	✓	✓	✓	✓	✓	✓	✓	✓	✓	✓
Tires	Rotate	✓	✓	✓	✓	✓	✓	✓	✓	✓	✓	✓	✓	✓
Auto trans. fluid	I			✓			✓			✓			✓	
Brake pads/shoes	I			✓			✓			✓			✓	
Coolant hoses	S/I			✓			✓			✓			✓	
Steering linkage	I			✓			✓			✓			✓	
Suspension and driveshaft	I			✓			✓			✓			✓	
Ball joints (2wd)	I			✓			✓			✓			✓	
Wheels ①	I						✓						✓	
Cabin air filter	R		✓				✓			✓			✓	
Climate controlled seat filter	R						✓						✓	
Ball joints (2wd)	L			✓			✓			✓			✓	
Exhaust system	I						✓						✓	
Engine air filter	R						✓						✓	
Fuel filter	R						✓						✓	
Green coolant ②	R									✓				
Wheel bearings (2wd)	L												✓	
Manual trans. fluid	R												✓	
Accessory drive belts	I	every 100,000 miles												
Spark plugs	R	every 100,000 miles												
PCV valve	R	every 100,000 miles												
Yellow coolant	R	5 years or 100,000 miles, then evey 3 years or 50,000 miles												
Orange coolant	R	every 150,000 miles												
Auto trans fluid	R	every 150,000 miles												
Differential fluid	R	every 150,000 miles												
Accessory drive belts	R	every 150,000 miles if not previously done so												
Transfer case fluid	R	every 150,000 miles												

R: Replace   S: Service   I: Inspect   L: Lubricate

① Inspect wheel ends for end play and noise

① Recommended, but not required in Calif.

② Change every 30,000 miles or 36 months thereafter

Special Operating Condition Requirements

When towing a trailer or using a camper or car-top carrier:

Change engine oil and install a new oil filter every 4,800 km (3,000 miles) or 3 months.

Change transfer case fluid every 96,000 km (60,000 miles).

Change manual transmission fluid as required.

Inspect and lubricate U-joints as required.

During extensive idling and/or low speed driving for long distances, as in heavy commercial use such as delivery, taxi, patrol car or livery:

Change engine oil and install a new oil filter, lube front lower control arm and steering linkage ball joints with
zerk fittings (if equipped) every 4,800 km (3,000 miles) or 3 months.
Inspect brake system and check battery electrolyte level (Patrol cars) every 8,000 km (5,000 miles).
Install a new fuel filter every 24,000 km (15,000 miles).
Change automatic transmission fluid, lubricate 4x2 wheel bearings,
 install new grease seals and adjust bearings every 48,000 km (30,000 miles).
Install new spark plugs and change transfer case fluid every 96,000 km (60,000 miles).
Install a new cabin air filter as required.

06017-EXPL-C13

SCHEDULED MAINTENANCE INTERVALS
2002-03 Ford Explorer, Explorer Sport, Explorer Sport-Trac, Mercury Mountaineer
2003 Lincoln Aviator
Footnotes Continued

When operating in dusty conditions such as unpaved or dusty roads:

Change engine oil and install a new oil filter every 4,800 km (3,000 miles) or 3 months.

Install a new fuel filter every 24,000 km (15,000 miles).

Change automatic transmission fluid every 48,000 km (30,000 miles).

Change transfer case fluid every 96,000 km (60,000 miles).

Install a new engine air filter as required.

Install a new cabin air filter as required.

When operating in off-road conditions:

Change automatic transmission fluid every 48,000 km (30,000 miles).

Change transfer case fluid every 96,000 km (60,000 miles).

Install a new cabin air filter as required.

Inspect and lubricate U-joints.

Inspect and lubricate steering linkage ball joints with zerk fittings.

06017-EXPL-C14

SCHEDULED MAINTENANCE INTERVALS
2004-05 Lincoln Aviator, Ford Explorer, Explorer Sport-Trac, Mercury Mountaineer

TO BE SERVICED	TYPE OF SERVICE	VEHICLE MILEAGE INTERVAL (x1000)												
		5	10	15	20	25	30	35	40	45	50	55	60	65
Engine oil & filter	R	✓	✓	✓	✓	✓	✓	✓	✓	✓	✓	✓	✓	✓
Tires	Rotate	✓	✓	✓	✓	✓	✓	✓	✓	✓	✓	✓	✓	✓
Auto trans. fluid	I			✓			✓			✓			✓	
Brake pads/shoes	I			✓			✓			✓			✓	
Coolant hoses	S/I			✓			✓			✓			✓	
Steering linkage	I			✓			✓			✓			✓	
Suspension, driveshaft	I			✓			✓			✓			✓	
Cabin air filter	R			✓			✓			✓			✓	
Ball joints (2wd)	L			✓			✓			✓			✓	
Wheels ①	I						✓						✓	
Exhaust system	I						✓						✓	
Engine air filter	R						✓						✓	
Fuel filter	R						✓						✓	
Climate controlled seat filter	R						✓						✓	
Accessory drive belts	I	every 100,000 miles												
Premium Gold coolant	R	5 years or 100,000 miles, then every 3 years or 30,000 miles												
Spark plugs	R	every 100,000 miles												
PCV valve	R	every 100,000 miles												
Coolant, exc. Premium Gold	R	every 105,000 miles												
Front wheel bearings and seals (2wd)	R	every 150,000 miles if not already done so												
Manual trans. fluid	R	every 120,000 miles												
Auto trans fluid	R	every 150,000 miles												
Differential fluid	R	every 150,000 miles												
Accessory drive belts	R	every 150,000 miles if not already done so												

R: Replace   S: Service   I: Inspect   L: Lubricate

① Inspect wheel ends for end play and noise

Special Operating Condition Requirements

When towing a trailer or using a camper or car-top carrier:

Change engine oil and install a new oil filter every 4,800 km (3,000 miles) or 3 months.

Change transfer case fluid every 96,000 km (60,000 miles).

Change manual transmission fluid as required.

Inspect and lubricate U-joints as required.

During extensive idling and/or low speed driving for long distances, as in heavy commercial use such as delivery, taxi, patrol car or livery:
Change engine oil and install a new oil filter, lube front lower control arm and steering linkage ball joints with
zerk fittings (if equipped) every 4,800 km (3,000 miles) or 3 months.
Inspect brake system and check battery electrolyte level (Patrol cars) every 8,000 km (5,000 miles).
Install a new fuel filter every 24,000 km (15,000 miles).
Change automatic transmission fluid, lubricate 4x2 wheel bearings,
 install new grease seals and adjust bearings every 48,000 km (30,000 miles).
Install new spark plugs and change transfer case fluid every 96,000 km (60,000 miles).
Install a new cabin air filter as required.

When operating in dusty conditions such as unpaved or dusty roads:

Change engine oil and install a new oil filter every 4,800 km (3,000 miles) or 3 months.

Install a new fuel filter every 24,000 km (15,000 miles).

Change automatic transmission fluid every 48,000 km (30,000 miles).

Change transfer case fluid every 96,000 km (60,000 miles).

06017-EXPL-C15

SCHEDULED MAINTENANCE INTERVALS
2004-05 Lincoln Aviator, Ford Explorer, Explorer Sport-Trac, Mercury Mountaineer
Footnotes Continued

Install a new engine air filter as required.

Install a new cabin air filter as required.

When operating in off-road conditions:

Change automatic transmission fluid every 48,000 km (30,000 miles).

Change transfer case fluid every 96,000 km (60,000 miles).

Install a new cabin air filter as required.

Inspect and lubricate U-joints.

Inspect and lubricate steering linkage ball joints with zerk fittings.

06017-EXPL-C16

ENGINE REPAIR

Alternator

REMOVAL & INSTALLATION

4.0L SOHC Engine

1. Before servicing the vehicle, refer to the Precautions Section.
2. Remove or disconnect the following:

- Negative battery cable
- Air cleaner outlet tube
- Accessory drive belt
- Electrical connectors
- Wiring harness-to-generator pin-type retainer
- Stud bolt, the bolts and the alternator

3. To install, reverse the removal procedure. Torque all mounting bolts to 35 ft. lbs. (47Nm).

4.6L DOHC and SOHC Engines

1. Before servicing the vehicle, refer to the Precautions Section.
2. Remove or disconnect the following:

- Negative battery cable
- Drive belt

1 Front end accessory drive belt tensioner
2 Front end accessory drive belt
3 Generator B+ terminal nut
4 Generator B+ terminal
5 Generator electrical connectors
6 Generator bolts
7 Generator
8 Generator pulley nut
9 Generator pulley

06017-EXPL-G02

Alternator mounting—4.0L engine

1 Engine cover assembly nuts	7 Generator B+ terminal
2 Engine cover assembly	8 Generator bracket bolts
3 Front end accessory drive belt tensioner	9 Generator bracket
4 Front end accessory drive belt	10 Generator bolts
5 Generator electrical connector	11 Generator
6 Generator B+ terminal nut	12 Generator pulley nut
	13 Generator pulley

Alternator mounting—4.6L SOHC engine

06017-EXPL-G03

1 Accessory drive belt tensioner
2 Accessory drive belt
3 Generator bracket bolts (2 required)
4 Generator bracket nuts (2 required)
5 Wiring harness locators (2 required)

6 Generator B+ nut
7 Electrical connectors
8 Generator bracket
9 Generator bolts (2 required)
10 Generator
11 Generator pulley nut
12 Generator pulley

06017-EXPL-G04

Alternator mounting—4.6L DOHC engine

- Electrical connectors
- Stud bolts, the bolts and the generator bracket
- Bolts and remove the alternator

3. To install, reverse the removal procedure. Torque the mounting bolts to 18 ft. lbs. (25Nm).

Ignition Timing

ADJUSTMENT

The ignition timing is preset to 10 degrees Before Top Dead Center (BTDC) and is not adjustable.

Engine Assembly

REMOVAL & INSTALLATION

4.0L SOHC Engines

☀ CAUTION

If the fuel supply manifold is used as a leverage device, damage may occur to the supply manifold. Care must be taken when working around the fuel supply manifold.

10 RH engine support insulator lower nut
11 RH engine support insulator upper nuts (2 required)
12 RH engine support insulator

06017-EXPL-G06

Right side engine support—4.0L engine

7 LH engine support insulator nut
8 LH engine support insulator through bolt
9 LH engine support insulator

06017-EXPL-G05

Left side engine support—4.0L engine

1. Before servicing the vehicle, refer to the Precautions Section.
2. Remove or disconnect:
   - Accelerator cable from engine
   - Speed control cable from engine
   - Radiator, the fan blade, and the fan shroud
   - Accessory bracket bolts and position bracket aside
   - Alternator wiring
   - Wiring harness retainer and position generator wiring away from engine
   - Engine electrical connector
   - PCM connector
   - PCM ground wire
   - Engine ground wire
   - Brake booster vacuum hose
   - A/C high pressure switch electrical connector
   - Bolt and position the A/C lines aside

➡Heater hose will be removed with engine.

- Heater hoses
- Fuel line
- Starter motor
- Engine oil
- Oil drain plug

- Transmission portion of wiring harness
- Right and left heated oxygen sensor connectors
- Transmission control connector
- Output shaft speed sensor connector
- Digital transmission range sensor connector
- Catalyst monitor sensor electrical connector
- Transmission/transfer case portion of the wiring harness from any routing clips or pushpins. Route transmission/transfer case portion of the wiring harness to top of engine.

- Bolt, and position the transmission cooling line bracket aside
- A/C line bracket nut and position it aside
- Power steering return hose
- Power steering pressure hose
- Vapor management valve hose connector
- Eight bolts and the left and the right engine support insulator nuts

➡**The lifting eyes should be installed on the exhaust manifold studs for number three and number four cylinders.**

3. Install the lifting eyes.
4. Install the spreader bar to the lifting eyes.

5. Attach a floor crane to the spreader bar and remove the engine.

6. Installation is the reverse of removal. Observe the following torques:
- Left and right engine insulator nuts: 81 ft. lbs. (110 Nm)
- Engine mount nuts: 59 ft. lbs. (80Nm)
- Transmission-to-engine bolts: 35 ft. lbs. (47Nm)
- Torque converter nuts: 35 ft. lbs. (47Nm)

4.6L DOHC and SOHC Engines

1. Before servicing the vehicle, refer to the Precautions Section.

1 RH engine support insulator nuts (2 required)

2 LH engine support insulator through-bolt

3 RH engine support insulator nut

4 LH engine support insulator nut

5 Stabilizer bar bracket bolts (4 required)

6 Front axle bolts (3 required) (if equipped)

7 LH engine support insulator

8 RH engine support insulator

06017-EXPL-G07

Engine mounting—4.6L SOHC engines

80 Nm (59 lb-ft)

63 Nm (46 lb-ft)

1

100 Nm (74 lb-ft)

2

3

A

7

80 Nm
(59 lb-ft)

5

1 Stabilizer bar bracket bolts (4
   required)

2 Front axle bolts (3 required)
   (if equipped)

3 Flag nuts (3 required) (if
   equipped)

4 LH engine support insulator
   nut

5 RH engine support insulator
   nut

6 LH engine support insulator

7 RH engine support insulator

06017-EXPL-G08

Engine mounting—4.6L DOHC engines

2. Relieve the fuel system pressure.
3. Drain the cooling system.
4. Drain the engine oil.
5. Properly discharge the A/C system.
6. Remove or disconnect the following:
   • Battery ground cable
   • Hood
   • Accelerator cable from the throttle
     body and all clips

   • Speed control cable from the
     throttle body and all clips
   • Air cleaner and the air cleaner out-
     let pipe
   • Radiator and transmission cooler
   • A/C condenser core

   ➡**Access the heater control valve
   bracket through the wheel well.**

   • Heater control valve bracket bolt
   • Coolant hoses
   • All vacuum hoses. Position the
     heater control valve assembly
     aside.
   • A/C tubes from the A/C compres-
     sor
   • Fuel charging wiring harness con-
     nections

- Fuel charging wiring harness from the clips on the bulkhead. Release the fuel charging wiring harness pin-type retainer.
- All connections from the evaporative emissions (EVAP) canister purge solenoid
- Brake booster vacuum hose
- Fuel tube
- Alternator connections and the wiring anchors
- Wiring harness side of the left and right heated exhaust gas oxygen (HEGO) sensor connectors
- Power steering return hose. Plug the power steering reservoir.
- Power steering pressure tube from the power steering pump
- Brackets
- Oil pressure sender electrical connector
- CKP sensor electrical connector and the wiring anchor
- A/C compressor electrical connector and the wiring anchor
- Wiring harness routing clip and wiring anchor

- Starter
- On 4x4 vehicles, the bolt from the right side axle housing bushing bolt
- Right side and the left side stabilizer bar bracket nuts
- Left side lower axle housing bushing bolt
- Left side upper axle housing bushing bolt
- Right lower mount nut and washer
- Left lower mount nut and washer
- Transmission inspection cover, torque converter bolts and the transmission-to-engine block bolts

7. Install the right side Lifting Eye.
8. Install the left side Lifting Eye.
9. Install the Spreader Bar to the Lifting Eyes.
10. Attach a floor crane to the Spreader Bar and remove the engine.

➡ **If engine disassembly is to be carried out, remove the rear crankshaft oil seal slinger and the crankshaft rear oil seal.**

11. Installation is the reverse of removal. Observe the following torques:

- Lower mount nuts: 59 ft. lbs. (81Nm) for DOHC engines; 66 ft. lbs. (90 Nm) for SOHC engines
- Left side upper axle housing bushing bolt: 49 ft. lbs. (66Nm)
- Left side axle housing bushing bolt: 49 ft. lbs. (66Nm)
- Stabilizer bar bracket nuts: 30 ft. lbs. (40Nm)
- Right side axle housing bushing bolt: 49 ft. lbs. (66Nm)

Water Pump

REMOVAL & INSTALLATION

4.0L SOHC Engine

1. Before servicing the vehicle, refer to the Precautions Section.
2. Drain the cooling system.
3. Remove or disconnect the following:

- Fan shroud
- Accessory drive belt
- Idler pulley

1 Coolant pump pulley bolts (4 required)
2 Coolant pump pulley
3 Coolant pump bypass hose clamp
4 Coolant pump bypass hose
5 Heater hose clamp
6 Heater hose
7 Lower radiator hose clamp
8 Lower radiator hose
9 Coolant pump bolts (12 required)
10 Coolant pump

06017-EXPL-G09

Water pump installation—4.0L engine

1   25 Nm
    (18 lb-ft)

3   25 Nm
    (18 lb-ft)

4

5

2

1   Coolant pump pulley bolts (4
    required)

2   Coolant pump pulley

3   Coolant pump bolts (4
    required)

4   Coolant pump

5   Coolant pump O-ring

06017-EXPL-G10

Water pump installation—4.6L SOHC and DOHC engines

- Water bypass hose
- Heater hose
- Lower radiator hose
- Water pump pulley
- Water pump

✳✳ WARNING

Use care when scraping the water pump-to-engine block mating surfaces. Gouges in the aluminum could form leak paths.

4. Clean all the sealing surfaces.
5. To install, reverse the removal procedure. Torque the water pump bolts to 89 inch lbs. (10Nm). Torque the pulley bolts to 18 ft. lbs. (25Nm).

4.6L DOHC and SOHC Engines

1. Before servicing the vehicle, refer to the Precautions Section.
2. Drain the cooling system.
3. Remove or disconnect the following:
   - Engine cooling fan
   - Drive belt

- Water pump pulley
- Water pump

4. Discard the O-ring seal.
5. To install, reverse the removal procedure. Install a new O-ring seal and lubricate with engine coolant.
6. Torque the pump and pulley bolts to 18 ft. lbs. (25Nm).

Heater Core

REMOVAL & INSTALLATION

Aviator, Explorer, Explorer Sport-Trac, Mountaineer

2002

1. Before servicing the vehicle, refer to the Precautions Section.
2. Deactivate the supplemental restraints system (SRS).
3. Release the two floor console front clips (one each side).
4. On vehicles with manual transmission, remove the gearshift lever handle.

5. Remove the console finish panel mat.
6. On vehicles with automatic transmission:
   a. If equipped, remove the ashtray assembly.
   b. Disconnect the electrical connector.
   c. Remove the console finish panel screw.
7. Remove the console finish panel by lifting up at the rear and sliding it rearward. Disconnect the electrical connector(s).
8. Remove the floor console front screws.
9. If equipped, disconnect the electrical connector and release the wiring harness locators.
10. Remove the floor console center bolts.
11. Remove the two floor console access covers (one each side).
12. Remove the two floor console rear bolts (one each side).
13. Remove the floor console by lifting up at the rear of the console and sliding it rearward.
14. Remove the screws and remove the snow shield.

15. Crimp off the coolant hoses and disconnect from the heater core.

16. Remove the instrument panel center support brackets.

17. Remove the assist handle bolt covers.

18. Remove the bolts and the passenger assist handle.

19. Remove the windshield side garnish moldings.

20. Remove the defroster opening grill.

21. Remove the upper instrument panel support bolts.

22. Remove the exterior cowl grill.

23. Remove the instrument panel support bolt.

24. Remove the two instrument panel side finish panels.

25. Lower the glove compartment.

26. Loosen the left instrument panel side support bolts until half the threads are exposed.

✷✷ WARNING

Be sure the instrument panel is properly supported to avoid possible damage to the instrument panel wiring or components. To avoid damage to the instrument panel wiring or components, do not use excessive force when moving the instrument panel away from the dash panel. The instrument panel may be supported by installing threaded rods or equivalent, with the same diameter and thread pitch, in place of the instrument panel support bolts.

27. Properly support the instrument panel and remove the right instrument panel support bolts to allow the instrument panel to be pulled away from the dash panel.

28. Remove the center console floor duct.

29. Remove the screws and remove the right floor duct.

30. Remove the screws and remove the left floor duct.

31. Disconnect the electrical connectors and detach the pin-type retainers.

32. Remove the screws and remove the heater core tube cover.

33. Remove the right heater core cover screws.

34. Remove the left heater core cover screws and remove the heater core cover.

➡The instrument panel will need to be moved to facilitate removal of the heater core.

35. Remove the heater core.

36. To install, reverse the removal procedure. Lubricate the coolant hoses with coolant hose lubricant or plain water only if needed. Top-off the engine coolant level. For additional information. Reactivate the supplemental restraints system (SRS).

2003–05

FRONT

➡**If an evaporator core leak is suspected, the evaporator core must be vacuum leak tested before it is removed from the vehicle.**

➡**Installation of a new accumulator is not required when repairing the air conditioning system except when there is physical evidence of contamination from a failed A/C compressor or damage to the accumulator.**

➡**New O-ring seals, lubricated in clean mineral oil, must be installed before**

1	LH floor duct screws (2 required)	7	Heater tube cover screw (3 required)
2	LH floor duct	8	Heater tube cover
3	RH floor duct screws (2 required)	9	Heater tube seal
4	RH floor duct	10	Heater core cover screw (4 required)
5	Housing brace screw (3 required)	11	Heater core cover
6	Housing brace	12	Heater core

06017-EXPL-G11

Front heater core

reconnecting any A/C fitting which has been disconnected.

➡Lubricate the coolant hoses with plain water only if needed.

1. Recover the refrigerant.

⁂ WARNING

Always wear safety glasses when repairing an air bag supplemental restraint system (SRS) vehicle and when handling an air bag module. This will reduce the risk of injury in the event of an accidental deployment.

⁂ WARNING

Carry a live air bag module with the air bag and trim cover pointed away from your body. This will reduce the risk of injury in the event of an accidental deployment.

⁂ WARNING

Do not set a live air bag module down with the trim cover face down. This will reduce the risk of injury in the event of an accidental deployment.

⁂ WARNING

After deployment, the air bag surface can contain deposits of sodium hydroxide, a product of the gas generant combustion that is irritating to the skin. Wash your hands with soap and water afterwards.

⁂ WARNING

Never probe the connectors on the air bag module. Doing so can result in air bag deployment, which can result in personal injury.

⁂ WARNING

Air bag modules with discolored or damaged trim covers must be replaced, not repainted.

⁂ WARNING

The restraint system diagnostic tool is for restraint system service only. Remove from vehicle prior to road use. Failure to remove could result in injury and possible violation of vehicle safety standards.

⁂⁂ WARNING

To reduce the risk of personal injury, do not use any memory saver devices.

➡The air bag warning lamp illuminates when the RCM fuse is removed and the ignition switch is ON. This is normal operation and does not indicate a supplemental restraint system (SRS) fault.

➡After diagnosing or repairing an SRS, the restraint system diagnostic tools must be removed before operating the vehicle over the road.

➡The SRS must be fully operational and free of faults before releasing the vehicle to the customer.

➡Repair is made by installing a new part only. If the new part does not correct the condition, install the original part and perform the diagnostic procedure again.

2. De-power the system.
3. Remove the steering wheel back cover plugs.
4. Remove the driver air bag module bolts.
5. Remove the driver air bag module connectors.
6. Remove the horn switch connector.
7. Remove the driver air bag module.
8. Open the glove compartment.
9. Through the glove compartment opening, release the tab and disconnect the passenger air bag module electrical connector.
10. Remove the parts in the order indicated in the following illustration and table.
11. Remove the passenger air bag module nuts.
12. Reaching one hand into the glove box opening, push out on the passenger air bag module, releasing the clips at the top and remove the passenger air bag module from the instrument panel. Remove the passenger air bag module.

⁂⁂ WARNING

Do not handle the passenger air bag module by grabbing the edges of the deployment door.

13. Remove the A-pillar trim panels and right cowl side trim panel.

➡After removing the floor console, reconnect the battery to position the front seats back, then disconnect the battery.

14. Remove the floor console.
15. Remove the instrument panel side finish panel (left).
16. Remove the steering column opening cover screws.
17. Remove the steering column opening cover
18. Remove the lower steering column cover.
19. Remove the steering column electrical connector.
20. Remove the bulkhead electrical connector.
21. Remove the body harness electrical connector.
22. Remove the parking brake release handle screws.
23. Remove the instrument panel electrical connector.
24. Remove the instrument panel electrical connector.
25. Remove the pinch bolt.

⁂⁂ WARNING

To avoid damage to the clockspring, do not allow the steering column shaft to rotate while the intermediate shaft is disconnected.

26. Remove the adjustable pedal electrical connector
27. Remove the instrument panel center brace bolts.
28. Remove the instrument panel center brace nuts.
29. Remove the instrument panel center brace.
30. Remove the transmission range selector lever cable.
31. Remove the bulkhead electrical connector.
32. Remove the body harness electrical connector.
33. Remove the ground strap bolts.
34. Remove the ground straps.
35. Remove the inertia switch electrical connector.
36. Remove the instrument panel electrical connectors.
37. Remove the instrument panel center brace bolts (right).
38. Remove the instrument panel center brace nuts (right).
39. Remove the instrument panel center brace (right).
40. Remove the instrument panel bolt.

41. Remove the instrument panel defroster grille.

42. Remove the sunload sensor electrical connector.

✳✳ WARNING

To avoid damaging the sun load sensor electrical connector, remove the grille just enough to remove the connector.

43. Remove the instrument panel bolts.

44. Remove the antenna lead-in cable.

45. Remove the RCM electrical connector.

46. Remove the ground strap bolt.

47. Remove the instrument panel bolts.

✳✳ WARNING

To avoid damage to the instrument panel, an assistant is required to support the panel before carrying out this step.

✳✳ WARNING

Before removing the instrument panel make sure all electrical connector wiring is free and not hindered.

48. Remove the instrument panel.

49. Drain the engine coolant.

50. Remove the engine appearance cover.

51. Remove the EGR vacuum regulator solenoid electrical connector.

52. Remove the vacuum hoses.

53. Remove the EGR vacuum regulator solenoid mounting nuts.

54. Remove the EGR vacuum regulator solenoid.

55. Remove the wire harness.

56. Remove the heater tube bracket nut.

57. Compress the clamps and disconnect the heater hoses.

58. Remove the A/C line bracket nut at the inner fender well.

59. Remove the A/C line bracket nut at the dash panel.

60. Remove the disconnect the evaporator outlet line fitting.

61. Remove the vacuum connector.

62. Remove the heater core and evaporator core housing nut.

63. Remove the heater core and evaporator core housing.

64. Remove the floor ducts.

65. Remove the housing brace.

66. Remove the heater tube cover.

67. Remove the heater tube seal.

68. Remove the heater core cover.

69. Remove the heater core.

70. To install, reverse the removal procedure.

✳✳ WARNING

The clockspring electrical connectors are unique and cannot be reversed when connected to the driver air bag module. Match the electrical connector key to the keyway in the driver air bag module. Do not force the electrical connectors into the driver air bag module.

71. Match the electrical connector key to the keyway in the driver air bag module and connect the electrical connectors.

✳✳ WARNING

The passenger air bag module nuts must be torqued in the sequence shown.

➡**Make sure the battery negative cable is still disconnected before continuing with the installation portion of this procedure.**

72. Install the components in the order indicated in the following illustration and table.

73. Install the passenger air bag module.

74. Connect the passenger air bag module electrical connector.

75. Close the glove compartment.

76. Re-power the system.

77. Make sure the retaining clips are in place before positioning the passenger air bag module to the instrument panel.

78. Lubricate the refrigerant system with the correct amount of clean PAG oil.

79. Fill and bleed the engine cooling system.

80. Evacuate, leak test and charge the refrigerant system.

REAR

1. Before servicing the vehicle, refer to the Precautions Section.

➡**Lubricate the coolant hoses with plain water only if needed.**

2. Position the vehicle on a hoist with the gear selector in NEUTRAL.

3. Using suitable tools, clamp-off the underbody heater hoses at the floorpan bracket.

4. Remove the clamps

5. Remove the auxiliary line floorpan bracket nuts.

10	Blend door actuator electrical connectors (2 required)
11	Auxiliary blend door actuator screw (3 required)
12	Auxiliary blend door actuator
13	Auxiliary temperature blend door actuator screw (3 required)
14	Auxiliary temperature blend door actuator

15	Heater core cover screw (4 required)
16	Heater core cover
17	Heater core hose clamp
18	Heater core hose clamp
19	Auxiliary heater core

Rear heater core

06017-EXPL-G12

6. Remove the line bracket screws.

7. Remove the line bracket.

➡**The screw and line bracket are located inside the vehicle above the floorpan line bracket.**

8. Remove the auxiliary harness electrical connector.

9. Remove the auxiliary housing bolts.

10. Remove the auxiliary housing nut.

11. Remove the blend door actuator electrical connector.

12. Remove the blend door actuator screws.

13. Remove the auxiliary blend door actuator.

14. Remove the temperature blend door actuator screws.

15. Remove the auxiliary temperature blend door actuator.

16. Remove the heater core cover.

17. Remove the clamps.

18. Remove the auxiliary heater core.

19. To install, reverse the removal procedure.

20. Fill the engine cooling system.

Cylinder Head

REMOVAL & INSTALLATION

4.0L SOHC Engine

2002–04 MODELS

➡**If only one cylinder head is to be removed, only follow the procedures that apply. The following tools, or their equivalents are absolutely necessary to properly perform this procedure:**

- Cam Chain Tensioner tool T97T-6K254-A
- Cam Gear Removal tool T97T-6256-F
- Cam Gear Torque adapter T97T-6256-G
- Camshaft Gear Positioning/Holding tool T97T-6256-B
- Camshaft Gear Positioning/Holding tool adapter T97T-6256-A
- Camshaft holding tool T97T-6256-C
- Crankshaft holding tool T97T-6303-A
- Camshaft holding tool adapter T97T-6256-D

1. Before servicing the vehicle, refer to the Precautions Section.

2. Properly relieve the fuel system pressure.

3. Drain the cooling system.

4. Remove or disconnect the following:

- Negative battery cable
- Lower intake manifold
- Fan blade and shroud
- Valve cover
- Roller followers, if equipped
- Drive belt
- Upper radiator hose and tube
- Alternator electrical connectors
- Alternator mounting bracket
- Engine accessory bracket and move it aside
- Camshaft Position (CMP) electrical connector
- Crankshaft Position (CKP) sensor electrical connector
- Engine Coolant Temperature (ECT) sensor electrical connector
- Coil pack electrical connector
- Exhaust Gas Recirculation (EGR) valve electrical connector
- EGR valve bracket and move it aside
- Heater hoses
- Fuel injector electrical connectors
- Water bypass hose
- Thermostat housing
- Spark plug wires
- Fuel injection supply manifold
- Fuel injectors
- Crankcase vent separator spring
- Oil dipstick housing
- Exhaust manifold
- Hydraulic chain tensioner
- Cassette retaining bolt
- Camshaft sprocket
- Cylinder head and discard the gasket

To install:

5. Thoroughly clean all gasket mating surfaces. Remove all traces of old gasket material, oil, grease or dirt.

6. Insure that the rubber band is holding the right-hand chain to the cassette.

7. Install a new head gasket and the cylinder head.

8. Torque the new cylinder head bolts in sequence as follows:
2002

- Step 1: 24 ft. lbs. (32 Nm).
- Step 2: Plus 90 degrees.

9. Install or connect the following:

- Camshaft sprocket in the cassette and make certain that the camshaft sprocket turns freely on the camshaft
- Cassette retaining bolt. Torque the bolt to 89 inch lbs. (10 Nm).
- Exhaust manifold
- Oil level indicator tube. Torque the bolt to 18 ft. lbs. (25 Nm).
- Crankcase vent separator and spring
- Thermostat housing. Torque the bolts to 8 ft. lbs. (11 Nm).
- Water bypass hose
- Heater hoses
- EGR bracket. Torque the bolt to 89 inch lbs. (10 Nm).
- EGR tube. Torque the nut to 30 ft. lbs. (40 Nm).
- ECT sensor electrical connector
- Electrical harness retainer. Torque the bolt to 89 inch lbs. (10 Nm).
- CKP and CMP electrical connectors
- Accessory bracket. Torque the bolts to 31 ft. lbs. (42 Nm).
- Alternator mounting bracket. Torque the bolts to 31 ft. lbs. (42 Nm).
- Alternator and electrical connectors
- Drive belt

The correct cylinder head bolt loosening sequence must be used to prevent warpage–2002–04 4.0L SOHC engine

7924EG04

Cylinder head bolt torque sequence—2002–04 4.0L SOHC engine

9308EG04

- Fan shroud
- Roller followers
- Valve cover
- Lower intake manifold
- Negative battery cable

10. Change the engine oil and filter.
11. Refill the cooling system.
12. Start the engine and check for leaks, repair if necessary.

2005 MODELS

1. Before servicing the vehicle, refer to the Precautions Section.
2. Disconnect the battery ground cable.
3. With the vehicle in NEUTRAL, position it on a hoist.
4. Drain the engine cooling system.
5. Remove the roller followers.
6. Remove the accessory drive belt tensioner.
7. On the right side:
   a. Disconnect the alternator electrical connections and disconnect the push-pin.
   b. Remove the alternator mounting bracket assembly.
   c. Remove the heater hose from the thermostat housing.
   d. Disconnect the engine coolant temperature (ECT) sensor electrical connector.
   e. Disconnect the upper radiator hose.
   f. Position the coolant bypass hose clamp aside.
   g. Remove the bolts and the thermostat housing.
   h. Remove the engine wiring harness bolts and position the engine wiring harness aside.

i. Remove the engine ground strap.
8. On the left side:
   a. Disconnect the radio ignition interference capacitor electrical connector.

b. Remove the bolts.
   c. Remove the bolts and position the accessory bracket aside.
   d. Remove the bolts and the coil bracket.
   e. Remove the oil level indicator tube.

✳✳ WARNING

It is important to twist the spark plug wire boots while pulling upward to avoid possible damage to the spark plug wires.

9. Disconnect all of the spark plug wires from the spark plugs.
10. Remove the engine noise shield.
11. Remove the 4 catalytic converter-to-manifold nuts.
12. Remove both of the exhaust manifolds and gaskets.
13. Discard the gaskets and the retaining nuts.
14. On the right side:
   a. Remove the right side hydraulic chain tensioner.

Torque Wrench Extension 303-575 (T97T-6256-F) or equivalent	ST1780-A
Socket, Camshaft Sprocket Nut 303-565 (T97T-6256-G) or equivalent	ST1781-A
Holding Tool, Camshaft Sprocket 303-564 (T97T-6256-B) or equivalent	ST1777-A
Adapter for 303-564 303-578 (T97T-6256-A) or equivalent	ST1776-A
Tensioner, Timing Chain 303-571 (T97T-6K254-A) or equivalent	ST1774-A
Holding Tool, Camshaft 303-577 (T97T-6256-C) or equivalent	ST1778-A
Timing Tool, Crankshaft TDC 303-573 (T97T-6303-A) or equivalent	ST1775-A
Adapter for 303-577 303-576 (T97T-6256-D) or equivalent	ST1779-A

These special tools are necessary for cylinder head removal and installation—2005 4.0L engine

b. Install the special tools. Tighten to 10 Nm (89 inch lbs.).

c. Using the special tool with the Camshaft Sprocket Nut Socket 303-565, remove the right camshaft bolt.

d. Remove the right side cassette bolt.

✲✲ WARNING

Remove the camshaft sprocket from the timing chain to gain clearance to remove the cylinder head.

➡Hold the timing chain and cassette with a rubber band to aid in removal and to prevent the timing chain from falling into the cylinder block.

e. Remove the right camshaft sprocket from the timing chain. Install a rubber band around the cassette and the timing chain.

15. On the left side:

a. Remove the left side hydraulic chain tensioner.

b. Install the special tools. Tighten the top clamp bolts to 10 Nm (89 inch lbs.).

c. Remove the left side camshaft sprocket bolt.

d. Remove the left side cassette bolt.

✲ WARNING

Remove the camshaft sprocket from the timing chain to gain clearance to remove the cylinder head.

➡Hold the timing chain and cassette with a rubber band to aid in removal and to prevent the timing chain from falling into the cylinder block.

e. Remove the left camshaft sprocket from the timing chain. Install a rubber band around the cassette and the timing chain.

✲✲ WARNING

To avoid damage to the timing chain cassette an assistant will be required to help lift the cylinder head from the vehicle.

✲ WARNING

On the left side, when lifting the cylinder head be careful to avoid contacting the A/C tube.

➡New cylinder head bolts must be installed. They are a torque-to-yield design and cannot be reused.

Cylinder head bolt loosening sequence—2005 4.0L engine

16. Remove the cylinder heads.

17. Remove the cylinder head bolts in the sequence shown. Discard all bolts.

18. Remove and discard the head gasket.

✲✲ WARNING

Do not use metal scrapers, wire brushes, power abrasive discs or other abrasive means to clean the sealing surfaces. These tools cause scratches and gouges that make leak paths. Use a plastic scraping tool to remove all traces of the head gasket.

19. Clean and inspect the mating surfaces.

20. Inspect the cylinder head and the cylinder block for flatness

To install:

21. Position the cylinder head gasket on the block.

22. On the right side:

✲✲ WARNING

To avoid damage to the timing chain cassette an assistant will be required to help position the cylinder head in the vehicle.

✲✲ WARNING

On the right side, when lifting the cylinder head be careful to avoid contacting the A/C tube.

➡New cylinder head bolts must be installed. They are a torque-to-yield design and cannot be reused.

a. Install the cylinder head. Install 8 new 12-mm bolts and tighten in the sequence shown in 2 stages.

• Stage 1: Tighten to 12 Nm (9 ft. lbs.).

Right side cylinder head bolt torque sequence—2005 4.0L engine

Left side cylinder head bolt torque sequence—2005 4.0L engine

- Stage 2: Tighten to 25 Nm (18 ft. lbs.).

b. Install 2 new 8-mm bolts. Tighten to 32 Nm (24 ft. lbs.).

c. Tighten the 8 12-mm bolts in the sequence shown in 2 stages.

- Stage 1: Tighten the bolts 90 degrees.
- Stage 2: Tighten the bolts an additional 90 degrees.

✳✳ WARNING

The camshaft gear must turn freely on the camshaft. DO NOT tighten the bolt at this time

d. Remove the rubber band. Install the camshaft sprocket and bolt.

e. Install the right side cassette bolt. Tighten to 12 Nm (9 ft. lbs.).

23. On the left side:

✳✳ WARNING

To avoid damage to the timing chain cassette, an assistant will be required to help position the cylinder head in the vehicle.

➡**New cylinder head bolts must be installed. They are a torque-to-yield design and cannot be reused.**

a. Install the cylinder head. Install 8 new 12-mm bolts and tighten in the sequence shown in 2 stages.

- Stage 1: Tighten to 12 Nm (9 ft. lbs.).
- Stage 2: Tighten to 25 Nm (18 ft. lbs.).

b. Install 2 new 8-mm bolts. Tighten to 32 Nm (24 ft. lbs.).

c. Tighten the 8 12-mm bolts in the sequence shown in 2 stages.

- Stage 1: Tighten the bolts 90 degrees.
- Stage 2: Tighten the bolts an additional 90 degrees.

✳✳ WARNING

The camshaft gear must turn freely on the camshaft. DO NOT tighten the bolt at this time.

d. Remove the rubber band. Install the camshaft sprocket and bolt.

e. Install the left side cassette bolt. Tighten to 19 Nm (14 ft. lbs.).

24. Install both of the exhaust manifolds gaskets, the exhaust manifolds and the nuts. Tighten to 22 Nm (16 ft. lbs.).

➡**Apply dielectric compound to the inside of the spark plug wire boots.**

25. Connect all of the spark plug wires to the spark plugs.

26. Install the 4 catalytic converter-to-exhaust manifold nuts. Tighten to 40 Nm (30 ft. lbs.).

27. Install the engine noise shield.

28. On the right side:

a. Install the engine ground strap and the bolt. Tighten to 10 Nm (89 inch lbs.).

b. Position the engine wiring harness and install the bolts. Tighten to 47 Nm (35 ft. lbs.).

➡**Inspect the O-ring seal. Install a new O-ring seal if necessary.**

c. Install the thermostat housing and the bolts. Tighten to 11 Nm (8 ft. lbs.).

d. Position the coolant bypass hose clamp.

e. Install the upper radiator hose.

f. Connect the engine coolant temperature (ECT) sensor electrical connector.

g. Install the heater hose to the thermostat housing.

h. Install the alternator mounting bracket assembly and the bolts. Tighten to 42 Nm (31 ft. lbs.).

i. Connect the alternator electrical connections and install the pushpin. Tighten to 8 Nm (71 inch lbs.).

29. On the left side:

a. Install the oil level indicator tube and bolt. Tighten to 10 Nm (89 inch lbs.).

b. Install the coil bracket and the bolts. Tighten to 89 Nm (10 inch lbs.).

c. Install the accessory bracket and the bolts. Tighten to 42 Nm (31 ft. lbs.). Install the vertical bolts. Tighten to 10 Nm (89 inch lbs.).

d. Connect the radio ignition interference capacitor electrical connector.

30. Install the accessory drive belt tensioner.

Tool 303-573 installed—2005 4.0L engine

Tool 303-578 and 303 564 installed—2005 4.0L engine

Tools 303 576 and 303 577 installed—2005 4.0L engine

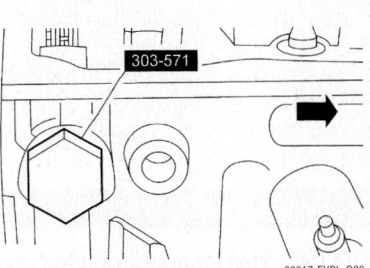

Tool 303-571 installed—2005 4.0L engine

06017-EXPL-G21

Tool 303-575 installed—2005 4.0L engine

06017-EXPL-G22

Tool 303-573 installed—2005 4.0L engine

➡ **You must retime the left and right camshafts when either camshaft is disturbed.**

31. Turn the crankshaft clockwise to position the number one cylinder at top dead center (TDC).

32. Remove the retainer and position the A/C manifold tube bracket aside.

❊❊ WARNING

Do not rotate the engine counterclockwise. Rotating the engine counterclockwise will result in incorrect timing of the engine.

➡ **The special tool must be installed on the damper and should contact the engine block, this positions the engine at TDC.**

33. Install special tool 303 573 as shown.

➡ **Leave the top 2 special tool clamp bolts loose.**

34. Install special tools 303 578 and 303 564 on the rear of the right cylinder head.

➡ **The camshaft timing slots are off-center.**

35. Position the camshaft timing slots below the centerline of the camshaft to correctly fit the special tools 303 576 and 303 577, and install the special tools on the front of the right cylinder head. Install special tool 303 571.

❊❊ WARNING

The right camshaft sprocket bolt is a left threaded bolt.

36. Tighten the bolts.

37. Tighten the special tool top clamp bolts to 10 Nm (89 inch lbs.).

38. Using the special tool with the Camshaft Sprocket Nut Socket 303-565, tighten the camshaft bolt. Tighten to 61 Nm (45 ft. lbs.).

39. Install the right camshaft tensioner. Tighten to 44 Nm (32 ft. lbs.).

➡ **Leave the top 2 special tool clamp bolts loose.**

40. Install special tools 303 564 and 303 578 on the front of the left cylinder head.

➡ **The camshaft timing slots are off-center.**

41. Position the camshaft timing slots below the centerline of the camshaft to correctly fit the special tools, and install the special tools on the rear of the left cylinder head.

42. Install special tools 303 576 and 303 577. Tighten the bolts.

43. Tighten the special tool top 2 clamp bolts to 10 Nm (89 inch lbs.).

44. Tighten the camshaft bolt. Tighten to 85 Nm (63 ft. lbs.).

45. Install the left camshaft tensioner. Tighten to 44 Nm (32 ft. lbs.).

46. Install the A/C manifold tube assembly bracket and the nut. Tighten to 20 Nm (15 ft. lbs.).

47. Install the roller followers.

48. Fill and bleed the engine cooling system.

4.6L SOHC Engine

1. Before servicing the vehicle, refer to the Precautions Section.

2. Drain the cooling system.

3. Remove or disconnect the following:

- Negative battery cable
- Intake manifold
- Timing chains
- Exhaust manifolds
- Water heater tube back and remove. Discard the O-ring seal.

➡ **The cylinder head bolts must be discarded with new bolts installed. They are tighten-to-yield designed and cannot be reused. Do not use metal scrapers, wire brushes, power abrasive discs or other abrasive means to clean the sealing surfaces. These tools cause scratches and gouges that make leak paths. Use a plastic scraping tool to remove all traces of the head gasket.**

9355EG07

Right cylinder head bolt torque sequence 4.6L SOHC Engine

9355EG01

Left cylinder head bolt torque sequence 4.6L SOHC Engine

7924EG14

Tighten valve cover bolts in the sequence shown—4.6L SOHC Engine

- Bolts and the right cylinder head
- Bolts and the left cylinder head

4. Installation is the reverse of removal. Observe the following notes and torques:

a. Install the cylinder head on the head gasket and loosely install the bolts.

b. Make sure to tighten the bolts in sequence in six stages.

- Step 1: Tighten to 40 Nm (30 ft. lbs.).
- Step 2: Tighten an additional 85–95 degrees.
- Step 3: Back out all bolts one full turn (360 degrees).
- Step 4: Tighten to 40 Nm (30 ft. lbs.).
- Step 5: Tighten an additional 85–95 degrees.
- Step 6: Tighten an additional 85–95 degrees.

c. Install a new O-ring seal and position the water heater tube forward. Lubricate the O-ring seal with premium engine coolant.

4.6L DOHC Engine

1. Before servicing the vehicle, refer to the Precautions Section.

2. Remove the engine.

3. Remove the flexplate and the spacer plate.

4. Mount the engine on a suitable engine stand.

5. Remove the two bolts and the pin-type retainer from the power steering reservoir bracket.

6. Remove the power steering reservoir and bracket.

7. Disconnect the battery cable harness ground.

8. Disconnect the electrical connector and the battery cable from the alternator.

9. Disconnect the two retainers from the engine front cover studs.

10. Disconnect the two retainers from the A/C compressor.

11. Remove the bolt and the battery cable harness from the right engine mount.

12. Disconnect the electrical connector and the vacuum hose from the fuel pressure sensor.

13. Disconnect the vacuum harness from the intake manifold and remove the harness.

14. Remove the left and right coil covers.

15. Disconnect the right and left ignition coil electrical connectors.

16. Disconnect the radio interference capacitor electrical connector.

17. Disconnect the eight fuel injector electrical connectors.

	Installer, Crankshaft Vibration Damper
	Installer, Crankshaft Front Oil Seal
	Installer, Front Cover Oil Seal
	Compressor, Intake Valve Spring
	Compressor, Exhaust Valve Spring

67197-EXPL-G01

These tools, or their equivalents, are necessary for cylinder head removal and installation on the 4.6L DOHC engine

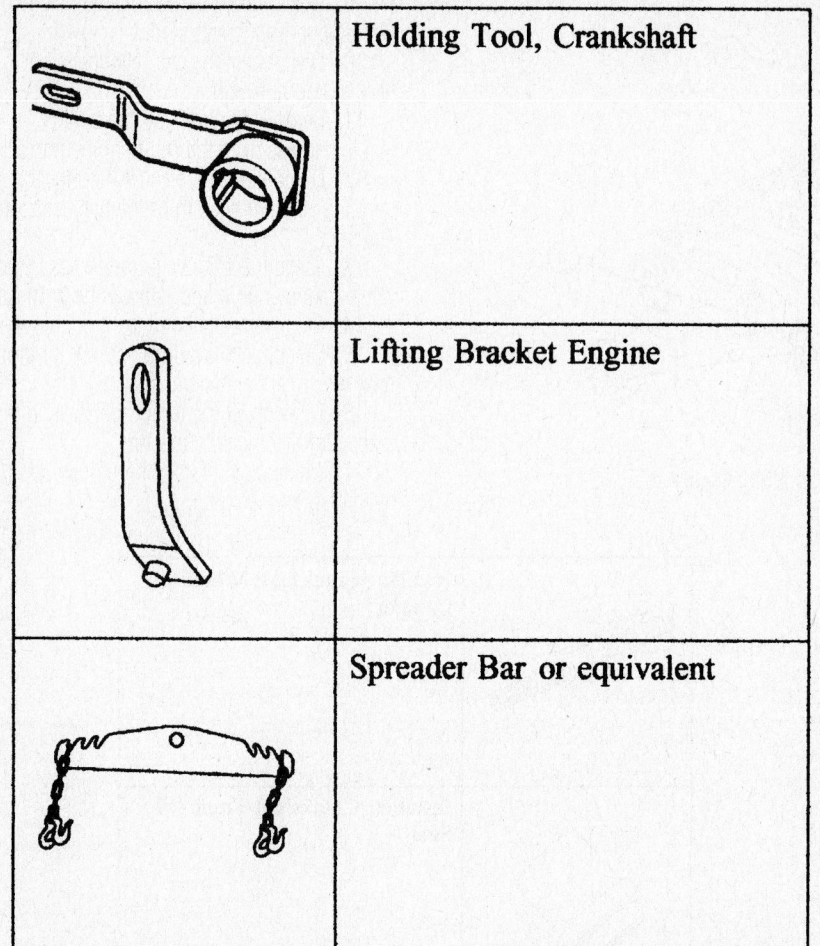

| Holding Tool, Crankshaft |
| Lifting Bracket Engine |
| Spreader Bar or equivalent |

67197-EXPL-G02

These tools, or their equivalents, are necessary for cylinder head removal and installation on the 4.6L DOHC engine

18. Separate the fuel charging wiring harness retainers from the fuel injection supply manifold.

19. Disconnect the camshaft position (CMP) sensor electrical connector and the two retainers from the valve cover.

20. Disconnect the oil pressure sensor electrical connector.

21. Disconnect the pin-type retainers from the back of the engine in the locations shown.

22. Disconnect the knock sensor (KS) electrical connector and the pin-type retainer.

23. Disconnect the wiring harness retainer from the stud on the cylinder head.

24. Disconnect the intake manifold runner control (IMRC) electrical connector.

25. Disconnect the exhaust gas recirculation (EGR) system module tube nut from the exhaust manifold.

26. Disconnect the cylinder head temperature (CHT) sensor electrical connector and the two pin-type retainers.

27. Disconnect the positive crankcase ventilation (PCV) electrical connector.

28. Remove the PCV tube.

29. Disconnect the right radio interference capacitor electrical connector and the pin-type retainer.

30. Disconnect the throttle position (TP) sensor and the idle air control (IAC) sensor electrical connectors.

31. Disconnect the two retainers from the right valve cover studs.

32. Disconnect the engine coolant temperature (ECT) sensor electrical connector.

33. Disconnect the A/C compressor and the crankshaft position (CKP) sensor electrical connectors.

34. Remove the upper radiator hose.

35. Remove the coolant bypass-to-thermostat housing hose.

36. Remove the accessory drive belt.

37. Remove the bolts and the generator support bracket.

38. Remove the bolts and the generator.

39. Remove the crossover tube.

40. Remove the eight ignition coils.

41. Remove the nut and the bolt from the oil level indicator tube and position the tube aside.

42. Remove the engine lifting eyes.

43. Remove the bolts and the left valve cover.

44. Remove the coolant hose retainers from the right valve cover.

45. Remove the bolts and the right valve cover.

46. Remove the retainers and the upper-to-lower intake bracket.

47. Remove the bolts in the sequence shown and remove the upper and lower intake.

48. Remove the bolt and the coolant bypass tube.

49. Remove the exhaust manifolds.

50. Clean and inspect the exhaust manifolds.

51. Remove the oil level indicator tube.

52. Remove the three remaining bolts and the power steering pump.

53. Install the special tool.

54. Remove the coolant pump pulley.

55. To remove the crankshaft pulley bolt, remove the upper fan shroud and electric fan.

56. Remove the pulley bolt.

57. Remove the crankshaft vibration damper.

58. Remove the crankshaft front seal.

59. Remove the bolt and the belt idler pulley.

60. Remove the four front oil pan-to-timing cover bolts.

61. Remove the engine front cover fasteners in the sequence shown.

62. Remove the crankshaft position sensor pulse wheel.

➡**Mark the roller follower locations. Reused roller followers must be installed in their original locations.**

67197-EXPL-G03

Intake manifold bolt removal sequence– 4.6L DOHC engine

1 Bolt, Hex Flange Head Pilot, M8 x 1.25 x 53

2 Bolt, Hex Flange Head Pilot, M8 x 1.25 x 53

3 Bolt, Hex Flange Head Pilot, M8 x 1.25 x 53

4 Bolt, Hex Flange Head Pilot, M8 x 1.25 x 53

5 Bolt, Hex Flange Head Pilot, M8 x 1.25 x 53

6 Stud, Hex Shldr Pilot, M8 x 1.25 x 50 —
   M6 x 1 x 10

7 Stud and Washer, Hex Head Pilot,
   M8 x 1.25 x 60 — M6 x 1 x 26

8 Bolt, Hex Flange Head Pilot, M8 x 1.25 x 53

9 Bolt, Hex Flange Head Pilot, M8 x 1.25 x 53

10 Bolt, Hex Flange Head Pilot, M8 x 1.25 x 53

11 Stud, Hex Shldr Pilot, M8 x 1.25 x 65 —
   M8 x 1.25 x 26

12 Stud, Hex Head Pilot, M8 x 1.25 x 65 —
   M8 x 1.25 x 16

13 Stud, Hex Shldr Pilot, M8 x 1.25 x 65 —
   M8 x 1.25 x 26

14 Stud, Hex Shldr Pilot, M8 x 1.25 x 65 —
   M8 x 1.25 x 26

15 Stud, Hex Shldr Pilot, M8 x 1.25 x 65 —
   M8 x 1.25 x 26

67197-EXPL-G04

Front cover bolt removal sequence—4.6L DOHC engine

303-452

67197-EXPL-G05

**Compressing the intake valve spring—4.6L
DOHC engine**

303-567

67197-EXPL-G06

**Compressing the exhaust valve spring—
4.6L DOHC engine**

63. Position the piston of the cylinder in which the roller followers are being removed at the bottom of the stroke and camshaft lobe at the base circle.

64. Compress the intake valve spring and remove the roller follower.

65. Compress the exhaust valve spring and remove the roller follower.

66. Repeat the above steps to remove all the roller followers.

67. Remove the bolts and the right and left timing chain tensioners.

68. Remove the right and left timing chain tensioner arms.

69. Remove the right and the left timing chains and the crankshaft sprocket.

70. Remove the right and left timing chain guides.

71. Remove the right and left cylinder heads. Discard the cylinder head bolts. Discard the cylinder head gaskets.

72. Clean and inspect the cylinder heads.

To install:

→Before cylinder head installation, use metal surface cleaner and a suitable plastic or wooden scraper to clean the sealing surfaces. All sealing surfaces must be clean. Make sure coolant and oil passages are clear.

All vehicles

73. Install new cylinder head gaskets.

→Lubricate the new bolt heads and threads. Use clean engine oil.

Cylinder head bolt torque sequence–4.6L DOHC engine

74. Install the left and right cylinder heads and tighten the bolts in six stages in the sequence shown.
- Stage 1: Tighten to 40 Nm (30 ft. lbs.).
- Stage 2: Tighten an additional 90 degrees.
- Stage 3: Loosen the bolts a minimum of one full turn.
- Stage 4: Tighten to 40 Nm (30 ft. lbs.).
- Stage 5: Tighten an additional 90 degrees.
- Stage 6: Tighten an additional 90 degrees.

Engines with ratcheting timing chain tensioners

✳✳ WARNING

Timing chain procedures must be followed exactly or damage to valves and pistons will result.

✳✳ WARNING

Do not compress the ratchet assembly. This will damage the ratchet assembly.

75. Compress each tensioner plunger, using an edge of a vise.
76. Using a small screwdriver or pick, push back and hold the ratchet mechanism.
77. While holding the ratchet mechanism, push the ratchet arm back into the tensioner housing.
78. Install a paper clip into the hole of each tensioner housing to hold the ratchet assembly and plunger in during installation. Remove the tensioner from the vise.

Engines with non-ratcheting timing chain tensioners

79. Compress the tensioner plunger, using a vise.
80. Install a retaining clip on the tensioner to hold the plunger in during installation.
81. Remove the tensioner from the vise.

All engines

82. If the copper links are not visible, mark one link on one end and one link on the other end, and use as timing marks.
83. Install the timing chain guides.
84. Rotate the left camshaft timing sprocket until the timing mark is approximately at the 12 o'clock position. Rotate the right camshaft timing sprocket until the timing mark is at approximately the 11 o'clock position.

✳✳ WARNING

Unless otherwise instructed, at no time when the timing chains are removed and the cylinder heads are installed is the crankshaft or camshaft to be rotated. Severe piston and valve damage will occur.

85. Using the special tool, position the crankshaft.
86. Install the crankshaft sprocket with the flange facing forward.
87. Install the left and right timing chains onto the crankshaft sprocket, aligning the one copper link on the timing chain with the slot on the crankshaft sprocket.

If the copper links are not visible, mark one link on one end and one link on the other end, and use as timing marks–4.6L DOHC engine

Using the special tool, position the crankshaft –4.6L DOHC engine

➡ **If necessary, adjust the camshaft sprocket slightly to obtain timing mark alignment.**

88. Position the left and right timing chains on the camshaft sprockets. Make sure the copper-colored links align with the camshaft sprocket timing marks.
89. Position the left and right timing chain tensioner arms on the dowel pins. Position the timing chain tensioner assemblies, and install the bolts.

Engines with ratcheting timing chain tensioners

90. Remove the retaining clip from the timing chain tensioners.

Engines with non-ratcheting timing chain tensioners

91. Remove the retaining clips from the timing chain tensioners.

All engines

92. As a post-check, verify correct alignment of all timing marks.
93. Position the crankshaft sensor ring on the crankshaft.

➡ **If the engine front cover is not secured within four minutes, the sealant must be removed and the sealing area cleaned with metal surface**

If the copper links are not visible, mark one link on one end and one link on the other end, and use as timing marks–4.6L DOHC engine

Verify timing mark alignment–4.6L DOHC engine

67197-EXPL-G11

8 mm (0.32 in)

67197-EXPL-G12

Apply silicone gasket and sealant in the locations shown –4.6L DOHC engine

cleaner. Allow to dry until there is no sign of wetness, or four minutes, whichever is longer. Failure to follow this procedure can result in future oil leakage.

94. Apply silicone gasket and sealant in the locations shown.

95. Install the engine front cover and tighten the bolts in the sequence shown.

96. Install the four oil pan bolts and tighten in the sequence shown.

- Stage 1: Tighten to 2 Nm (18 inch lbs.).
- Stage 2: Tighten to 20 Nm (15 ft. lbs.).
- Stage 3: Tighten an additional 60 degrees.

☀☀ WARNING

If reusing the roller followers, install them in their original locations.

97. Position the piston of the cylinder in which the roller followers are being installed at the bottom of the stroke and the camshaft lobe at the base circle.

98. Compress the intake valve spring and install the roller follower.

99. Compress the exhaust valve spring and install the roller follower.

100. Repeat the above steps to install all of the roller followers.

25 Nm (18 lb-ft)

1 Bolt, Hex Flange Head Pilot, M8 x 1.25 x 53

2 Bolt, Hex Flange Head Pilot, M8 x 1.25 x 53

3 Bolt, Hex Flange Head Pilot, M8 x 1.25 x 53

4 Bolt, Hex Flange Head Pilot, M8 x 1.25 x 53

5 Bolt, Hex Flange Head Pilot, M8 x 1.25 x 53

6 Stud, Hex Shldr Pilot, M8 x 1.25 x 50 — M6 x 1 x 10

7 Stud and Washer, Hex Head Pilot, M8 x 1.25 x 60 — M6 x 1 x 26

8 Bolt, Hex Flange Head Pilot, M8 x 1.25 x 53

9 Bolt, Hex Flange Head Pilot, M8 x 1.25 x 53

10 Bolt, Hex Flange Head Pilot, M8 x 1.25 x 53

11 Stud, Hex Shldr Pilot, M8 x 1.25 x 65 — M8 x 1.25 x 26

12 Stud, Hex Head Pilot, M8 x 1.25 x 65 — M8 x 1.25 x 16

13 Stud, Hex Shldr Pilot, M8 x 1.25 x 65 — M8 x 1.25 x 26

14 Stud, Hex Shldr Pilot, M8 x 1.25 x 65 — M8 x 1.25 x 26

15 Stud, Hex Shldr Pilot, M8 x 1.25 x 65 — M8 x 1.25 x 26

67197-EXPL-G13

Front cover bolt torque sequence–4.6L DOHC engine

67197-EXPL-G14

Install the four oil pan bolts and tighten in the sequence shown –4.6L DOHC engine

10 Nm (89 lb-in)

67197-EXPL-G15

Valve cover bolt torque sequence–4.6L DOHC engine

8 mm (0.32 in)

67197-EXPL-G16

Apply silicone gasket and sealant in the locations shown –4.6L DOHC engine

➡ **If the valve cover is not secured within four minutes, the sealant must be removed and the sealing area cleaned with metal surface cleaner. Allow to dry until there is no sign of wetness, or four minutes, whichever is longer. Failure to follow this procedure can result in future oil leakage.**

101. Apply silicone gasket and sealant in the locations shown.

102. Install the valve covers and tighten the bolts in the sequence shown.

103. Install the belt idler pulley.

104. Lubricate the front oil seal and the engine front cover with clean engine oil.

105. Install the front oil seal.

106. Apply silicone gasket and sealant to the Woodruff key on the crankshaft pulley.

107. Install the crankshaft pulley.

108. Install the crankshaft pulley bolt and tighten the bolt in four stages.
- Stage 1: Tighten to 90 Nm (66 ft. lbs.).
- Stage 2: Loosen the bolt one full turn.
- Stage 3: Tighten to 50 Nm (37 ft. lbs.).
- Stage 4: Tighten an additional 90 degrees.

109. Install the coolant pump pulley.

110. Install the power steering pump and the three bolts.

➡ **Install a new O-ring and lubricate with clean engine oil.**

111. Install the oil level indicator tube in the engine block.

112. Install the fasteners and tighten to 10 Nm (89 inch lbs.).

113. Install new exhaust manifold gaskets.

114. Install the exhaust manifolds.

115. Install the 16 nuts.

116. Install the bolt and the coolant bypass tube.

117. Install the upper and lower intake. Tighten all fasteners in the sequence shown to 10 Nm (89 inch lbs.).

118. Install the upper-to-lower intake bracket and the nuts.

119. Install the coolant hose retainers on the right valve cover.

120. Install the engine lifting eyes.

121. Install the eight ignition coils.

122. Install the crossover tube and the nuts. Connect the coolant hose.

123. Install the generator and the bolts.

124. Install the generator support bracket and the bolts.

125. Install the accessory drive belt.

126. Install the coolant bypass-to-thermostat housing hose.

127. Install the upper radiator hose.

128. Position the engine wiring harness on the engine.

129. Connect the A/C compressor and the crankshaft position (CKP) sensor electrical connectors.

130. Connect the engine coolant temperature (ECT) sensor electrical connector.

131. Connect the two harness retainers to the right valve cover studs.

132. Connect the throttle position (TP)

67197-EXPL-G17

Intake manifold bolt torque sequence— 4.6L DOHC engine

sensor and the idle air control (IAC) sensor electrical connectors.

133. Connect the right radio interference capacitor electrical connector and the pin-type retainer.

134. Install the positive crankcase ventilation (PCV) tube.

135. Connect the PCV electrical connector.

136. Connect the cylinder head temperature (CHT) sensor electrical connector and the two pin-type retainers.

137. Connect the exhaust gas recirculation (EGR) system module tube to the exhaust manifold.

138. Connect the intake manifold runner control (IMRC) electrical connector.

139. Connect the wiring harness retainer to the stud on the cylinder head.

140. Connect the KS electrical connector and the pin-type retainer.

141. Connect the pin-type retainers to the back of the engine in the locations shown.

142. Connect the oil pressure sensor electrical connector.

143. Connect the camshaft position (CMP) sensor electrical connector and the two retainers to the valve cover.

80 Nm (59 lb-ft)

67197-EXPL-G18

Flexplate bolt torque sequence—4.6L DOHC engine

144. Install the fuel charging wiring harness retainers to the fuel injection supply manifold.

145. Connect the eight fuel injector electrical connectors.

146. Connect the radio interference capacitor electrical connector.

147. Connect the eight ignition coil electrical connectors.

148. Install the right and left coil covers.

149. Position the vacuum harness and connect it to the intake manifold.

150. Connect the electrical connector and the vacuum hose to the fuel pressure sensor.

151. Install the battery cable harness and the bolt to the right engine mount.

152. Connect the two retainers to the A/C compressor.

153. Connect the two retainers to the engine front cover studs.

154. Connect the electrical connector and the battery cable to the alternator.

155. Connect the battery cable harness ground.

156. Position the power steering reservoir and install the two bolts and the pin-type retainer.

157. Install the remaining bolt and the hose to the power steering reservoir.

158. Remove the engine from the stand.

159. Install the spacer plate and the flexplate.

160. Install the engine.

Roller Followers

REMOVAL & INSTALLATION

4.0L SOHC Engines

➡**A special tool is required to compress the valve spring.**

1. Before servicing the vehicle, refer to the Precautions Section.

2. Disconnect the negative battery cable.

44 Nm (32 lb-ft)

1	RH hydraulic chain tensioner
2	Camshaft roller followers (6 required)
3	Hydraulic lash adjuster
4	Valve spring retainer keys (12 required)
5	Valve spring retainers (6 required)
6	Valve springs (6 required)
7	Valve seals (6 required)
8	RH camshaft sprocket bolt
9	RH camshaft sprocket
10	Camshaft bearing cap bolts (8 required)
11	Oil supply tube
12	Camshaft bearing caps (4 required)
13	RH camshaft

Right side valve train components—4.0L engine

06017-EXPL-G23

44 Nm (32 lb-ft)

14	LH hydraulic chain tensioner	20	Valve seals (6 required)
15	Camshaft roller follower (6 required)	21	LH camshaft sprocket bolt
16	Hydraulic lash adjuster	22	LH camshaft sprocket
17	Valve spring retainer keys (12 required)	23	Camshaft bearing cap bolts (8 required)
18	Valve spring retainers (6 required)	24	Oil supply tube
19	Valve springs (6 required)	25	Camshaft bearing caps (4 required)
		26	LH camshaft

06017-EXPL-G24

Left side valve train components—4.0L engine

3. Remove the valve cover.

4. Rotate the camshaft so that the base circle of the cam is against the cam follower you intend to remove.

➡ **If removing more than one cam follower, label them so they can be returned to their original position.**

5. Using special tool T97T-6565-A depress the valve spring. Slide the cam follower over the lash adjuster and out from under the camshaft.

To install:

6. Compress the valve spring and slide the roller follower into position.

7. Release the tension from the spring.

8. Install the valve cover and connect the negative battery cable.

4.6L SOHC Engine

1. Before servicing the vehicle, refer to the Precautions Section.

2. Remove the valve cover.

3. Position the piston of the cylinder being repaired at the bottom of the stroke.

4. Install the special tool between the valve spring coils to prevent valve stem seal damage.

➡ **The roller followers are positional. Mark the followers for installation in their original locations.**

- Compress the valve springs and remove the camshaft roller followers.

1	Camshaft roller followers (16 required)	8	Camshaft sprocket washer (2 required)
2	Hydraulic lash adjuster (16 required)	9	Camshaft sprocket (2 required)
3	Valve spring retainer keys (32 required)	10	Camshaft gear spacer (2 required)
4	Valve spring retainers (16 required)	11	Camshaft bearing cap assembly bolts (28 required)
5	Valve springs (16 required)	12	Camshaft bearing cap assembly (4 required)
6	Valve stem seals (16 required)	13	Camshaft (2 required)
7	Camshaft sprocket bolt (2 required)		

Valve train components—4.6L SOHC engine

Install this tool to prevent seal damage—4.6L SOHC Engine

9355EG06

303-452

67197-EXPL-G19

Compressing the intake valve spring—
4.6L DOHC engine

303-567

67197-EXPL-G20

Compressing the exhaust valve spring—
4.6L DOHC engine

➡**Mark the location of the hydraulic lash adjusters to make sure they are assembled in their original position.**

6. Remove the hydraulic lash adjusters.
7. Installation is the reverse of removal.

5. Installation is the reverse of removal.

➡**There are several kinds of bolts used in fastening the valve cover. The bolts must be installed in their original locations.**

4.6L DOHC Engine

1. Before servicing the vehicle, refer to the Precautions Section.

➡**This step is necessary for camshaft removal and installation only.**

2. Remove the engine front cover, timing gears, chain and tensioners.
3. Remove the valve covers.

➡**Position the piston of the cylinder to be serviced at the bottom of the stroke and the camshaft at base circle.**

4. Using the special tool, partially compress the intake valve spring and remove the camshaft roller follower.
5. Using the special tool, partially compress the exhaust valve spring and remove the camshaft roller follower.

1 Camshaft roller followers (intake) (16 required)
2 Hydraulic lash adjusters (intake) (16 required)
3 Camshaft roller followers (exhaust) (16 required)
4 Hydraulic lash adjusters (exhaust) (16 required)

06017-EXPL-G26

Valve train components—4.6L DOHC engine

Intake Manifold

REMOVAL & INSTALLATION

4.0L SOHC Engine

2002

1. Before servicing the vehicle, refer to the Precautions Section.
2. Remove or disconnect the following:
   • Negative battery cable
   • Splash shield
   • Air cleaner outlet pipe
   • Accelerator and speed control cables from the throttle body
   • Accelerator cable and the speed control cable from the bracket. Position the cables away from the intake manifold.
   • All the vacuum hoses
   • KS electrical connector
   • IAC valve electrical connector
   • TP sensor electrical connector
   • EGR transducer electrical connector
   • Wiring anchor
   • EGR tube from the EGR valve
   • Wiring harness pin-type retainer

- Spark plug wires
- Intake manifold retainers, then slightly reposition the manifold to access the EGR vacuum regulator (EVR) solenoid.
- EVR solenoid vacuum and electrical connections
- Intake manifold

- To install, reverse the removal procedure. Torque the bolts to 89 inch lbs. (10 Nm). There is no particular sequence.

2003–05

1. Before servicing the vehicle, refer to the Precautions Section.

2. Disconnect the battery ground cable.
3. Remove the air cleaner outlet pipe.
4. Remove the positive crankcase ventilation (PCV) tube
5. Remove the brake booster vacuum supply hose
6. Remove the engine main vacuum harness-to-intake manifold fitting

9 Nm (80 lb-in)

34 Nm (25 lb-ft)

1 Positive crankcase ventilation (PCV) tube

2 Brake booster vacuum supply hose clamp

3 Brake booster vacuum supply hose

4 Engine main vacuum harness-to-intake manifold fitting

5 Evaporative emissions (EVAP) return tube

6 EVAP tube pin-type retainer

7 Exhaust gas recirculation (EGR) system module electrical connector

8 Wiring harness pin-type retainer

9 Engine main vacuum harness-to-EGR system module fitting

10 EGR tube fitting

11 Throttle position (TP) sensor electrical connector

12 Electronic throttle body electrical connector

13 Electronic throttle body-to-intake manifold bolt

14 Intake manifold mounting bolts

15 Intake manifold

16 Intake manifold gaskets

67197-EXPL-G21

Parts to be removed prior to intake manifold removal—2003–05 4.0L SOHC engine

14 Intake manifold mounting
   bolts (8 required)

15 Intake manifold

16 Intake manifold gaskets

06017-EXPL-G27

Intake manifold removal—2003–05 4.0L SOHC engine

7. Remove the evaporative emissions (EVAP) return tube

8. Remove the EVAP tube pin-type retainer

9. Remove the exhaust gas recirculation (EGR) system module electrical connector

10. Remove the wiring harness pin-type retainer

11. Remove the engine main vacuum harness-to-EGR system module fitting

12. Remove the EGR tube fitting

13. Remove the throttle position (TP) sensor electrical connector

14. Remove the electronic throttle body electrical connector

15. Remove the electronic throttle body-to-intake manifold bolt

16. Remove the intake manifold mounting bolts

17. Remove the intake manifold

18. Remove the intake manifold gaskets

19. To install, reverse the removal procedure.

✳✳ CAUTION

Do not use metal scrapers, wire brushes, power abrasive discs or other abrasive means to clean the sealing surfaces. These tools can cause scratches and gouges which can make leak paths.

20. Clean the sealing surfaces, inspect the gaskets and install new gaskets if necessary. There is no special torque sequence.

4.6L SOHC Engine

1. Before servicing the vehicle, refer to the Precautions Section.

2. Remove or disconnect the following:

- Negative battery cable
- Throttle body
- Ignition coils
- Fuel injection supply manifold
- Upper radiator hose and the heater hose
- Alternator
- Accelerator cable bracket
- Throttle body adapter
- Thermostat housing

9355EG03

Tighten the manifold bolts in the sequence shown—4.6L SOHC Engine

23 Electronic throttle body
   adapter bolt (4 required)

24 Throttle body adapter
   assembly

25 Intake manifold bolts (9
   required)

26 Intake manifold

27 Intake manifold gasket

06017-EXPL-G28

Intake manifold and related parts—4.6L SOHC engine

- Intake manifold bolts
- Intake manifold. Tilt the manifold to slide it out from underneath the exhaust gas recirculation (EGR) tube.
- To install, reverse the removal procedure. Tighten retaining bolts is sequence shown.

4.6L DOHC Engine

UPPER

1. Before servicing the vehicle, refer to the Precautions Section.
2. Disconnect the battery ground cable.
3. Drain the cooling system.
4. Remove the air cleaner outlet pipe.

5. Remove the exhaust gas recirculation (EGR) system module.
6. Remove the positive crankcase ventilation (PCV) valve hose.
7. Remove the brake booster vacuum hose.
8. Remove the evaporative emissions (EVAP) hose.

1 Positive crankcase ventilation (PCV) valve hose	14 Accelerator cable bracket bolts
2 Brake booster hose clamp	15 Accelerator cable bracket (position aside)
3 Brake booster vacuum hose	
4 Evaporative emissions (EVAP) hose	16 Idle air control (IAC) valve electrical connector
5 Vacuum harness connector	
6 Coolant hose clamp	17 Throttle position (TP) sensor electrical connector
7 PCV coolant hose	
8 Coolant hose clamp	18 Fuel rail shield nuts
9 PCV coolant hose	19 Fuel rail shield
10 Accelerator cable	20 Radio capacitor nut
11 Speed control cable	21 Radio capacitor (position aside)
12 Accelerator cable return spring	22 Upper intake manifold bolts
13 Accelerator cable pin-type retainer	23 Upper intake manifold studs
	24 Upper intake manifold

67197-EXPL-G23

Upper intake manifold and related parts—4.6L DOHC engine

67197-EXPL-G24

Upper intake manifold bolt torque sequence—4.6L DOHC engine

9. Remove the coolant hose clamp.
10. Remove the PCV coolant hose.
11. Remove the coolant hose clamp.
12. Remove the PCV coolant hose.
13. Remove the accelerator cable.
14. Remove the speed control cable.
15. Remove the accelerator cable return spring.
16. Remove the accelerator cable pin-type retainer.
17. Remove the accelerator cable bracket.
18. Remove the idle air control (IAC) valve electrical connector.
19. Remove the throttle position (TP) sensor electrical connector.
20. Remove the fuel rail shield.
21. Remove the radio capacitor.
22. Remove the upper intake manifold bolts.
23. Remove the upper intake manifold studs.
24. Remove the upper intake manifold.
25. To install, reverse the removal procedure.

➡ **Locations 5 and 9 use studs.**

26. Tighten the bolts and studs in the sequence shown to 10 Nm (89 inch lbs.).

LOWER

1. Before servicing the vehicle, refer to the Precautions Section.
2. Remove the upper intake manifold.
3. Disconnect the spring lock couplings.
4. Remove the cooling fan.

5. Remove the accessory drive belt tensioner.
6. Remove the accessory drive belt.
7. Remove the wire harness retainer.
8. Remove the generator.
9. Remove the generator bracket studs.
10. Remove the upper coolant hose.
11. Remove the coolant bypass hose.
12. Remove the thermostat housing coolant hose.
13. Remove the engine coolant temperature (ECT) sensor electrical connector.
14. Remove the coolant crossover tube.
15. Remove the vacuum hose.
16. Remove the engine harness retainers.

17. Remove the fuel rail pressure (FRP) sensor electrical connector.
18. Remove the intake manifold runner control (IMRC) motor electrical connector.
19. Remove the fuel injector connectors.
20. Remove the lower intake manifold bolts.
21. Remove the lower intake manifold assembly.
22. Remove the intake manifold gaskets.
23. To install, reverse the removal procedure.

➡ **Hand-start all fasteners.**

24. Tighten all fasteners in the sequence shown in the illustration to 10 Nm (89 inch lbs.).

3 Wire harness retainer	13 Upper coolant hose
4 Generator bracket nuts	14 Hose clamp
5 Electrical connectors (two)	15 Coolant hose
6 Generator bracket bolts	16 Hose clamp
7 Generator bracket nuts	17 Thermostat housing coolant hose
8 Generator bracket bolts	18 Engine coolant temperature (ECT) sensor electrical connector
9 Generator bracket	19 Coolant crossover tube
10 Generator	20 Vacuum hose
11 Studs	
12 Hose clamp	

21 Engine harness retainers
22 Fuel rail pressure (FRP) sensor electrical connector
23 Intake manifold runner control (IMRC) motor electrical connector
24 Fuel injector connectors
25 Lower intake manifold bolts
26 Lower intake manifold assembly
27 Intake manifold gaskets

Lower intake manifold and related parts—4.6L DOHC engine

67197-EXPL-G25

Lower intake manifold torque sequence—4.6L DOHC engine

67197-EXPL-G26

Exhaust Manifold

REMOVAL & INSTALLATION

4.0L SOHC Engine

1. Before servicing the vehicle, refer to the Precautions Section.

2. Remove or disconnect the following:

- Negative battery cable
- Exhaust inlet pipe-to-manifold attaching bolts
- Differential Pressure Feedback EGR (DPFE) transducer hoses, left side manifold only

1 RH exhaust manifold-to-dual converter Y-pipe nuts (2 required)

2 RH exhaust manifold nuts (6 required)

3 RH exhaust manifold

4 RH exhaust manifold gasket

06017-EXPL-G30

Right side exhaust manifold—4.0L engine

1 LH exhaust manifold-to-dual converter Y-pipe nuts (2 required)

2 LH exhaust manifold nut (6 required)

3 LH exhaust manifold

4 LH exhaust manifold gasket

06017-EXPL-G29

Left side exhaust manifold—4.0L engine

- Exhaust Gas Recirculation (EGR) tube from the manifold and valve, left side manifold only
- Exhaust manifold and discard the gasket

To install:

3. Clean the gasket mating surfaces.

4. Install or connect the following:

- New gasket and the exhaust manifold. Torque the bolts to 16 ft. lbs. (22 Nm).
- EGR tube to the manifold. Torque the fastener to 30 ft. lbs. (40 Nm) left side manifold only
- DPFE transducer hoses, left side manifold only
- Exhaust inlet pipe-to-manifold attaching bolts. Torque the bolts to 30 ft. lbs. (40 Nm).
- Negative battery cable

5. Start the vehicle and check for leaks, repair if necessary.

4.6L SOHC and DOHC Engines

RIGHT SIDE

1. Before servicing the vehicle, refer to the Precautions Section.

2. Remove or disconnect the following:

- Front fender splash shield
- Heater control valve bracket bolt

4 — 10 Nm (89 lb-in)

1 — 40 Nm (30 lb-ft)

40 Nm (30 lb-ft) — 2

1   Exhaust manifold-to-exhaust gas recirculation (EGR) valve tube

2   Exhaust Y-pipe flange nuts (4 required)

3   Exhaust Y-pipe

4   Heat shield bolts (3 required)

5   Heat shield

6   Exhaust manifold nuts (8 required)

7   Exhaust manifold

8   Exhaust manifold gaskets (4 required)

06017-EXPL-G31

Right side exhaust manifold and related parts—4.6L SOHC engine

Right side exhaust manifold torque sequence—4.6L SOHC engine

06017-EXPL-G32

- Vacuum hose and heater hose. Position the heater control valve assembly aside.
- Exhaust manifold heat shield
- Exhaust gas recirculation tube from the exhaust manifold
- Eight nuts and the exhaust manifold

3. Installation is the reverse of removal. For the SOHC engine, tighten the exhaust manifold nuts in the sequence shown. There is no torque sequence given for the DOHC engine.

LEFT SIDE

1. Before servicing the vehicle, refer to the Precautions Section.

2. Remove or disconnect the following:

- Front fender splash shield
- Engine oil runoff
- Lower steering column shaft pinch bolt. Position the lower steering column shaft aside.
- Exhaust manifold heat shield
- 8 nuts and the exhaust manifold
- To install, reverse the removal pro-

20 Nm (15 lb-ft)

11 Nm (8 lb-ft)

1 Exhaust Y-pipe flange nuts (2 required)

2 Exhaust Y-pipe

3 Exhaust gas recirculation (EGR) tube

4 Heat shield bolts (3 required)

5 Heat shield

6 Exhaust manifold nuts (8 required)

7 Exhaust manifold

8 Exhaust manifold gasket

06017-EXPL-G35

Right side exhaust manifold and related parts—4.6L DOHC engine

48 Nm (35 lb-ft) — (1)

10 Nm (89 lb-in) — (4)

2 — 40 Nm (30 lb-ft)

1 Steering gear shaft

2 Exhaust Y-pipe flange nuts (4 required)

3 Exhaust Y-pipe

4 Heat shield bolts (3 required)

5 Heat shield

6 Exhaust manifold nuts (8 required)

7 Exhaust manifold

8 Exhaust manifold gaskets (4 required)

06017-EXPL-G33

Left side exhaust manifold and related parts—4.6L SOHC engine

20 Nm (15 lb-ft)

9355EG14

Left side exhaust manifold torque sequence—4.6L SOHC Engine

cedure. For the SOHC engine, tighten the nuts in the sequence shown. There is no torque sequence given for the DOHC engine.

Camshaft and Valve Lifters

➡Although Ford suggests that this component is removable while the engine is installed in the vehicle, depending on the particular options with which your truck is equipped, working clearance may be extremely tight and this procedure may be much easier to perform with the engine removed. Before commencing, read through this procedure and make certain enough clearance, or working room, exists with the engine in the vehicle; if there is not enough space, the engine should be removed.

1 Exhaust Y-pipe flange nuts (2 required)	6 Oil indicator bracket bolt
2 Exhaust Y-pipe	7 Exhaust manifold nuts (8 required)
3 Heat shield bolts (3 required)	8 Exhaust manifold
4 Heat shield	9 Exhaust manifold gasket
5 Oil indicator tube bracket nut	

06017-EXPL-G34

Left side exhaust manifold and related parts—4.6L DOHC engine

REMOVAL & INSTALLATION

4.0L SOHC Engine

1. Before servicing the vehicle, refer to the Precautions Section.

2. Remove or disconnect the following:
- Negative battery cable for safety
- Valve cover
- Hydraulic camshaft tensioner

➡ **The right-hand camshaft sprocket bolt uses left-hand threads.**

3. For the right-hand camshaft use the Cam Gear Torque Adapter tool T97T-6256-F, to remove the camshaft sprocket bolt.

4. For the left-hand camshaft, remove the sprocket bolt.

➡ **When removing the followers, label them so that they may be returned to their original positions.**

5. Using the Valve Spring Compressor tool ST1330-A, remove the camshaft roller followers.

6. Install or connect the following:
- Camshaft bearing cap bolts and the oil rail
- Camshaft

To install:

7. Lubricate all of the moving parts with SAE 50W engine oil.

8. Install camshaft onto the cylinder head.

9. Position the oil rail and install the bearing caps and bolts. Torque the bolts in 2 steps:
   a. Step 1—53.5 inch lbs. (6 Nm).
   b. Step 2—11–12.5 ft. lbs. (15–17 Nm).

10. Install or connect the following:
- Camshaft followers
- Camshaft sprocket bolt and hand tighten the bolt
- Camshaft Chain Tensioner T97T-6K254-A in the hole that the hydraulic chain tensioner was in

11. Turn the crankshaft one revolution clockwise until No. 1 piston is Top Dead Center (TDC).

12. Install or connect the following:
- Crankshaft Holding tool T97T-6303-A on the crankshaft to keep it from turning
- Position the timing slot on the rear of the camshaft to fit Camshaft Holding tool T97T-6256-C and install the holding tool on the rear of the head

- Camshaft Gear Holding tool T97T-6256-B and Camshaft Gear Holding tool T97T-6256-A on the front of the cylinder head to securely hold the camshaft gear
- Tighten the camshaft sprocket bolt to 63 ft. lbs. (85 Nm).

13. Remove the Camshaft Chain Tensioner tool and install the hydraulic chain tensioner, tighten the tensioner to 35–39 ft. lbs. (47–53 Nm).

14. Remove the special tools from the engine.

15. Install or connect the following:
- Valve cover
- Negative battery cable

16. Start the engine check for leaks and repair if necessary.

4.6L SOHC Engine

1. Before servicing the vehicle, refer to the Precautions Section.

2. Remove or disconnect the following:

※※ WARNING

At no time, when the timing chains are removed and the cylinder heads are installed may the crankshaft or camshaft be rotated. Severe piston and valve damage will occur.

- Timing chains
- Camshaft roller followers
- Camshaft sprocket

Use the proper sequence to prevent damage to the camshaft both when installing and removing the bearing caps—4.0L SOHC engine

Camshaft bearing cap torque sequence—4.0L SOHC engine

06017-EXPL-G36

10 Nm (89 lb-in)

Camshaft bearing bolt torque sequence—4.6L SOHC Engine

9355EG13

- The 13 camshaft bearing cap bolts
- Camshaft bearing caps ladders
- Camshaft from the cylinder head

➡ **The valve tappets are positional.
Mark each valve tappet for installation
in its original location.**

- Valve tappets

To install:

3. Lubricate the camshaft journals with
clean engine oil.

4. Install the camshaft onto the cylinder
head.

5. Lubricate the camshaft bearing caps
with clean engine oil.

6. Install the camshaft bearing caps and
loosely install the bolts.

7. Tighten the bolts in the sequence
shown.

8. Install the camshaft sprocket. Tighten
the sprocket bolt in two stages.

- Step 1: Tighten to 40 Nm (30 ft. lbs.)
- Step 2: Tighten an additional 90
   degrees.

9. Install the roller followers

10. Install the timing chains

4.6L DOHC Engine

1. Before servicing the vehicle, refer to
the Precautions Section.

2. Remove the engine front cover, timing gears, chain and tensioners.

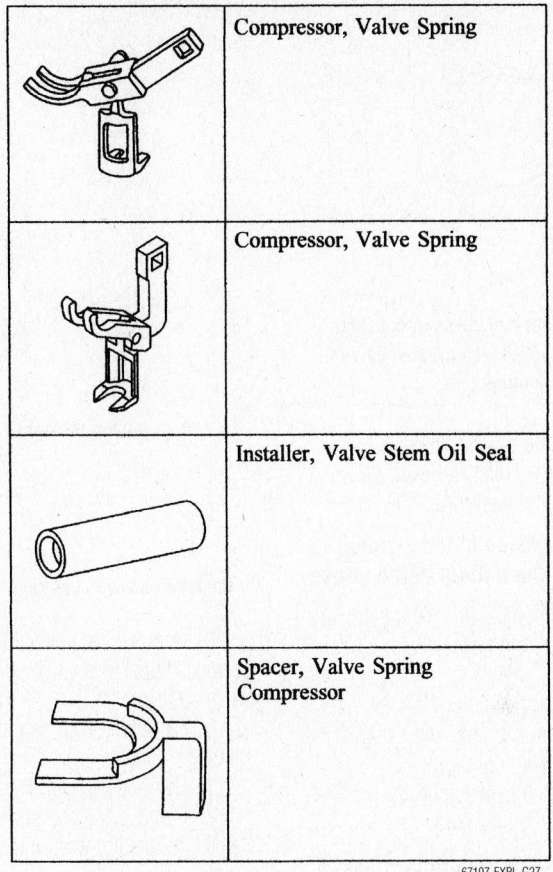

	Compressor, Valve Spring
	Compressor, Valve Spring
	Installer, Valve Stem Oil Seal
	Spacer, Valve Spring Compressor

67197-EXPL-G27

**These tools, or their equivalents, are necessary for camshaft and lifter removal—4.6L DOHC
engine**

1 Exhaust valve spring retainer keys	6 Intake valve spring retainers	10 Camshaft bearing caps (exhaust)
2 Exhaust valve spring retainers	7 Intake valve springs	11 Camshaft bearing caps (intake)
3 Exhaust valve springs	8 Intake valve stem seals	12 Camshaft (exhaust)
4 Exhaust valve stem seals	9 Camshaft bearing cap bolts (outboard row)	13 Camshaft (intake)
5 Intake valve spring retainer keys		

67197-EXPL-G28

Valve train exploded view—4.6L DOHC engine

67197-EXPL-G29

Install the special tool between the valve spring coils to protect the exhaust valve stem seal from damage

3. Remove the valve covers.
4. Remove the roller followers.
5. Remove the spark plugs.

➡**Position the piston of the cylinder to be serviced at the bottom of the stroke.**

6. Use compressed air in the cylinder to be serviced to hold both valves in position.

7. Install the special tool between the valve spring coils to protect the exhaust valve stem seal from damage.

8. Compress the valve spring. Remove the valve spring retainer keys.

➡**Valve stem seals should be visually inspected if new seals are not installed.**

9. Remove the special tool, the valve spring retainer, the valve spring and the valve stem seal.

10. Remove the camshaft bearing cap bolts.

11. Remove the camshaft bearing caps.

12. Remove the camshafts.

13. To install, reverse the removal procedure.

14. Install the camshaft bearing caps and tighten the fasteners in the sequence shown.

10 Nm (89 lb-in)

67197-EXPL-G30

Camshaft bearing cap removal and torque sequence—4.6L DOHC engine

Oil Pan

REMOVAL & INSTALLATION

4.0L SOHC Engine

➡**The 4.0L SOHC engine does not use an oil pan in the conventional sense. There is a separate access panel that unbolts from what would be considered the oil pan (which is now known as the ladder frame).**

1	Oil pan bolt	8	Bell housing-to-lower block cradle bolt	15	Lower block cradle (4x2 vehicles)
2	Oil pan	9	Bell housing-to-lower block cradle spacer	15	Lower block cradle (4x4 vehicles)
3	Oil pan gasket	10	Lower block cradle bolt	16	Lower block cradle gasket
4	Oil pump pickup tube bolt	11	Lower block cradle nut	17	Oil pump bolt
5	Oil pump pickup tube	12	Lower block cradle bolt	18	Oil pump
6	Lower block cradle bolt	13	Washer		
7	Lower block cradle bolt	14	Lower block cradle bolt		

67197-EXPL-G35

Oil pan, pump and related parts—4.0L SOHC engine

1. Before servicing the vehicle, refer to the Precautions Section.
2. Drain the engine oil.
3. Remove or disconnect the following:
   - Negative battery cable
   - Oil pan and discard the gasket

To install:

4. Install or connect the following:
   - New gasket and oil pan. Torque the bolts to 80 inch lbs. (9 Nm).
   - Negative battery cable
5. Fill the engine with clean oil.
6. Start the vehicle and check for leaks, repair if necessary.

4.6L SOHC Engine

1. Before servicing the vehicle, refer to the Precautions Section.
2. Drain the engine oil.
3. Remove or disconnect the following:
   - Front axle
   - Oil pan

➡Do not use metal scrapers, wire brushes, power abrasive discs, or other abrasive means to clean the sealing surfaces. These may cause scratches and gouges resulting in leak paths. Use a plastic scraper to clean the sealing surfaces.

4. Remove and discard the oil pan gasket. Clean the sealing surfaces with metal surface cleaner. Allow the surfaces to dry for four minutes or until there is no sign of wetness, whichever is longer. Failure to do so can cause future oil leaks.

To install:

➡If the oil pan and gasket are not secured within four minutes of sealer application, the sealant must be removed and the sealing surfaces cleaned with metal surface cleaner.

1 Oil pan bolts (16 required)
2 Oil pan
3 Oil pan gasket
4 Oil pump screen and pickup tube bolt
5 Oil pump screen and pickup tube bolt (2 required)
6 Oil pump screen and pickup tube
7 Oil pump screen and pickup tube O-ring seal
8 Oil pump bolts (3 required)
9 Oil pump

06017-EXPL-G37

Oil pan, pump and related parts—4.6L SOHC engine

Apply silicone gasket and sealant in these two places—4.6L SOHC and DOHC engines

67197-EXPL-G32

Apply silicone gasket and sealant in these two places—4.6L SOHC and DOHC engines

67197-EXPL-G33

67197-EXPL-G34

Oil pan bolt torque sequence—4.6L SOHC and DOHC engines

5. Apply silicone gasket and sealant in the two places shown.

6. Position the new oil pan gasket and the oil pan and loosely install the bolts.

7. Tighten the bolts in three stages in the sequence shown.
- Step 1: Tighten to 2 Nm (18 inch lbs.)
- Step 2: Tighten to 20 Nm (15 ft. lbs.)
- Step 3: Tighten an additional 60 degrees.

8. Install the front axle.

4.6L DOHC Engine

1. Before servicing the vehicle, refer to the Precautions Section.

2. With the vehicle in NEUTRAL position on a hoist.

3. Drain the engine oil.

4. Remove the front stabilizer bar.

➡**Carry out this step only if the oil pump is being removed.**

5. Remove the engine front cover.

6. Remove the front axle bolts.

❋❋ WARNING

Do not allow the inner halfshaft boots to contact the lower control arms. Damage to the boots can occur.

1 Front axle bolts (if equipped)	8 Oil pan
2 Crossmember bolt	9 Oil pan gasket
3 Crossmember bolts	10 Oil pump screen and pickup tube bolt
4 Crossmember	11 Oil pump screen and pickup tube bolts
5 Inspection cover bolts	12 Oil pump screen and pickup tube
6 Inspection cover	13 Oil pump bolts
7 Oil pan bolts	14 Oil pump

Oil pan, pump and related parts—4.6L DOHC engine

67197-EXPL-G31

7. Support the front axle using an appropriate jack. Lower the front axle approximately 6 inches.

8. Remove the crossmember bolts.

9. Remove the crossmember.

10. Remove the inspection cover.

11. Remove the oil pan bolts

12. Oil pan.

13. Remove the oil pan gasket.

14. To install, reverse the removal procedure.

❄❄ WARNING

Do not use metal scrapers, wire brushes, power abrasive discs, or other abrasive means to clean the sealing surfaces. These can cause scratches and gouges resulting in leak paths. Use a plastic scraper to clean the sealing surfaces.

15. Clean the sealing surfaces with metal surface cleaner.

➡**If the oil pan and gasket are not secured within four minutes of sealer application, the sealant must be removed and the sealing surfaces cleaned with metal surface cleaner.**

16. Apply silicone gasket and sealant in two places.

➡**If the oil pan and gasket are not secured within four minutes of sealer application, the sealant must be removed and the sealing surfaces cleaned with metal surface cleaner.**

17. Apply silicone gasket and sealant in two places.

18. Tighten the bolts in the sequence shown in three stages:
- Stage 1: Tighten to 2 Nm (18 inch lbs).
- Stage 2: Tighten to 20 Nm (15 ft. lbs.).
- Stage 3: Tighten an additional 60 degrees.

Oil Pump

REMOVAL & INSTALLATION

4.0L SOHC Engine

1. Before servicing the vehicle, refer to the Precautions Section.

2. With the vehicle in NEUTRAL, position it on a hoist.

3. Disconnect the negative battery cable.

4. If removing the lower block cradle, remove the starter motor.

5. Remove the air cleaner outlet tube.

6. Drain the engine oil.

7. Remove the oil pan.

8. Remove the oil pump pickup tube

9. Lower block cradle bolts

10. Remove the bell housing-to-lower block cradle bolts.

11. Remove the bell housing-to-lower block cradle spacer.

12. Remove the lower block cradle nuts

On 4x2 vehicles

13. Remove the upper air deflector.

14. Remove the weatherstrip.

➡**Not all vehicles will have screws installed in the radiator fan shroud.**

15. Remove and discard the screws. Remove the bolts and the fan shroud.

16. Remove the bolt.

17. Install the right lifting eye using the previously removed bolt.

18. Install lifting tools.

19. Remove the motor mount nuts.

20. Raise the engine.

❄❄ WARNING

Failure to back off the set screws can result in damage to the lower block cradle.

21. Back the set screws off until they are below the lower block cradle boss.

On 4x4 vehicles

22. Remove the upper air deflector.

23. Remove the weatherstrip.

➡**Not all vehicles will have screws installed in the radiator fan shroud.**

24. Remove and discard the screws. Remove the bolts and the fan shroud.

25. Remove the bolt.

26. Install the right lifting eye using the previously removed bolt.

➡**This is not a typical setup. Only the right side of the motor will be raised.**

27. Install the special tools.

28. Remove the four bolts and the crossmember.

67197-EXPL-G36

Back the set screws off until they are below the lower block cradle boss—4.0L SOHC engine

29. Remove the right motor mount nut.

30. Remove the left side through bolt and nut.

31. Raise the engine.

❄❄ WARNING

Failure to back off the set screws can result in damage to the lower block cradle.

32. Back the set screws off until they are below the lower block cradle boss.

➡**Use a floor jack to support the front axle assembly.**

33. Remove the bolt from the right side of the axle housing. Discard the nut and bolt.

34. Remove the left side lower axle housing bolt and nut. Discard the bolt and nut.

35. Remove the left side upper axle housing axle bolt and nut. Discard the bolt and nut. Lower the axle.

36. Remove the lower block cradle.

37. Lower block cradle gasket.

38. Oil pump bolt

39. Oil pump.

➡**Gasket material as well as silicone sealant may be present in the cavities in the main bearing cap. This material must be removed completely prior to assembly.**

40. Clean the sealant from the cavities on the rear main bearing cap.

To install:

➡**If the block cradle gasket is not secured within four minutes of the sealant application, the sealant must be removed and the sealing area cleaned with metal surface cleaner. Allow the silicone to dry until there is no sign of wetness, or four minutes, whichever is longer. Failure to follow this procedure can cause future oil leakage.**

41. Apply silicone in the six places shown.

42. Position the lower block cradle and gasket assembly.

43. Install and hand-tighten the outer bolts and nuts.

44. Install and hand-tighten the rear lower block cradle bolts.

45. Install the two rear lower block cradle-to-bell housing bolts.

➡**The lower block cradle to the cylinder block alignment must be within a maximum mismatch of 0.25 mm (0.01**

67197-EXPL-G37

Apply silicone in the six places shown— 4.0L SOHC engine

in) lower block cradle under-flush or 0.05 mm (0.00196 in) lower block cradle protrusion.

46. Using a straightedge, align the transmission face of the lower block cradle with the rear face of the cylinder block.

47. Tighten the outer bolts.

48. Tighten the rear lower block cradle bolts.

49. Tighten the eight inserts.

50. Install the two silver covered bolts and new washer seals. Hand-tighten them at this time.

51. Install and hand-tighten the six remaining bolts.

52. Tighten the lower block cradle bolts in two stages:

- Stage 1: Tighten to 15 Nm (11 ft. lbs.).
- Stage 2: Tighten to 34 Nm (25 ft. lbs.).

53. To install, reverse the removal procedure.

54. Fill the engine with clean engine oil.

4.6L SOHC Engines

1. Before servicing the vehicle, refer to the Precautions Section.

2. Drain the engine oil.

3. Remove or disconnect the following:

- Negative battery cable
- Timing chains
- Oil pan
- Three bolts and the oil pump screen cover and tube
- Oil pump

To install:

➡**Lubricate the new O-ring seal with clean engine oil.**

4. Clean and inspect the mating surfaces. Install a new O-ring seal.

5. Position the oil pump.

6. Loosely install the bolts.

7. Tighten the bolts to 89 inch lbs. (10Nm).

8. Install the three oil pump screen and cover bolts. Torque the bolts to 18 ft. lbs. (25Nm).

9. Install the timing chains.

10. Install the oil pan.

4.6L DOHC Engine

1. Before servicing the vehicle, refer to the Precautions Section.

2. With the vehicle in NEUTRAL position on a hoist.

3. Drain the engine oil.

4. Remove the front stabilizer bar.

➡**Carry out this step only if the oil pump is being removed.**

5. Remove the engine front cover.

6. Remove the oil pan.

7. Remove the oil pump screen and pickup tube.

8. Remove the oil pump bolts.

9. Remove the oil pump.

10. To install, reverse the removal procedure.

➡**Oil pump must be held against block until bolts are tightened.**

11. Align the inner rotor of the oil pump assembly to align with the flats on the crankshaft, and slide the oil pump toward the block until it is seated against the block.

12. Rotate the oil pump assembly to align with the bolt holes.

➡**Lubricate the new O-ring seal with clean engine oil before installation.**

13. Install O-ring seal.

Rear Main Seal

REMOVAL & INSTALLATION

4.0L SOHC Engine

1. Before servicing the vehicle, refer to the Precautions Section.

2. Remove the flexplate or flywheel.

✳✳ WARNING

Avoid scratching or damaging the oil crankshaft seal running surface during removal of the crankshaft rear oil seal.

3. Using special tool 303-409, remove the crankshaft rear oil seal.

To install:

➡**Be sure the crankshaft rear sealing surface is clean and free of any rust or corrosion. To clean the crankshaft rear sealing surface, use extra-fine emery cloth or extra-fine 0000 steel wool with metal surface cleaner.**

4. Lubricate the crankshaft rear oil seal with clean engine oil and install on the special tool.

5. Using special tool 303-S524, install the crankshaft rear oil seal.

6. Install the flexplate or flywheel.

34 Nm (25 lb-ft)

67197-EXPL-G38

Lower block cradle bolt torque sequence—4.0L SOHC engine

1 Spacer plate

2 Flexplate-to-crankshaft spacer

3 Crankshaft rear seal

06017-EXPL-G38

Rear main seal and related parts—4.0L engine

303-409

06017-EXPL-G39

Removing the seal—4.0L engine

303-S524

06017-EXPL-G40

Using special tool 303-S524, install the crankshaft rear oil seal

4.6L SOHC and DOHC Engines

1. Before servicing the vehicle, refer to the Precautions Section.

2. Remove or disconnect the following:

1 Spacer plate

2 Crankshaft rear seal slinger

3 Crankshaft rear seal

4 Crankshaft rear seal retainer plate bolts (6 required)

5 Crankshaft rear seal retainer plate

06017-EXPL-G41

Rear main seal and related parts—4.6L engine

Retainer plate removal and torque
sequence—4.6L engine

06017-EXPL-G42

- Flexplate
- Crankshaft rear oil seal slinger with
  a slide hammer
- Rear oil seal with a slide hammer

3. Installation is the reverse of removal.
Note the following:

- Lubricate the inner lip of the rear
  crankshaft seal with clean engine oil.

06017-EXPL-G43

Rear main seal installation—4.6L engine

- Use the two Crankshaft Rear Oil
  Seal Installers to install the rear oil
  seal.
- Using the two Crankshaft Rear Oil
  Seal Installers and the Crankshaft
  Rear Oil Slinger Installer, install the
  crankshaft rear oil slinger.

Front Cover and Seal

REMOVAL & INSTALLATION

4.0L Engine

1. Before servicing the vehicle, refer to
the Precautions Section.
2. Disconnect the battery ground cable
3. Drain the cooling system.
4. Remove the crankshaft front seal.
5. Remove the 2 nuts and detach the
power steering pressure hose from the front
cover.
6. Remove the 5 oil pan-to-front cover
bolts.
7. Release the clamp and disconnect
the lower radiator hose.
8. Remove the bolt and the drive belt
tensioner.

1 A/C hose bracket nuts (2 required)	6 Accessory drive belt tensioner bolt
2 A/C hose	7 Accessory drive belt tensioner
3 Oil pan-to-front cover bolts (5 required)	8 Clamp
4 Lower radiator hose clamp	9 Heater hose
5 Lower radiator hose	10 Crankshaft position sensor electrical connector

06017-EXPL-G45

Front cover removal, part 1—4.0L SOHC engine

11 Alternator wiring connections

12 Alternator bracket bolts (3 required)

13 Alternator bracket

14 Ignition coil bracket-to-LH accessory drive bracket bolts

15 LH accessory drive bracket bolts (4 required)

16 LH accessory drive bracket

17 Wiring harness routing clips

18 Engine coolant temperature (ECT) sensor electrical connector

19 Bypass hose clamp

20 Bypass hose

21 Upper radiator hose clamp

22 Upper radiator hose

23 Bypass hose clamp

24 Thermostat housing bolts (3 required)

25 Thermostat housing

06017-EXPL-G44

Front cover removal, part 2—4.0L SOHC engine

19 Nm (14 lb-ft) — (25)

06017-EXPL-G46

25 Engine front cover bolts (10 required)

26 Engine front cover

27 Gasket

Front cover removal—4.0L SOHC engine

9. Release the clamp and disconnect the heater hose from the coolant pump.

10. Disconnect the crankshaft position (CKP) sensor electrical connector.

11. Disconnect the alternator electrical connectors and the wiring harness retainer.

12. Remove the 3 alternator bracket bolts and the alternator bracket.

13. Remove the 2 coil bracket-to-left accessory drive bracket bolts.

14. Remove the 4 bolts and the left accessory drive bracket.

15. Remove and discard the CKP sensor wiring harness retainers.

16. Disconnect the engine coolant temperature (ECT) sensor electrical connector.

17. Release the clamp and disconnect the upper radiator hose from the thermostat housing.

18. Release the clamp and disconnect the heater hose from the thermostat housing.

➡**The bypass hose will be removed with the thermostat housing.**

19. Release the bypass hose clamp from the coolant pump end.

20. Remove the 3 bolts and the thermostat housing.

➡**Note the positions of the stud bolts for installation reference**

21. Remove the 10 bolts, the engine front cover and the gasket. Discard the gasket.

To install:

22. Clean and inspect the gasket mating surfaces. Use silicone gasket remover and metal surface prep and a plastic or wooden scraping tool. Follow the directions on the packaging.

23. Position the front cover gasket.

24. Apply silicone gasket and sealant to the oil pan and engine block mating surfaces.

25. Apply silicone gasket and sealant to the front cover in 2 places.

➡**Make sure the stud bolts are installed in their original positions.**

06017-EXPL-G47

Apply silicone gasket and sealant to the oil pan and engine block mating surfaces—4.0L SOHC engine

06017-EXPL-G48

Apply silicone gasket and sealant to the front cover in 2 places—4.0L SOHC engine

26. Position the front cover and install the 10 bolts. Tighten to 19 Nm (14 ft. lbs.).

➡**Make sure the coolant bypass hose is attached to the coolant pump.**

27. Position the thermostat housing and install the 3 bolts. Tighten to 11 Nm (8 ft. lbs.).

28. Position the bypass hose clamp.

29. Connect the upper radiator hose to the thermostat housing and position the clamp.

30. Connect the heater hose to the thermostat housing and position the clamp.

31. Connect the ECT sensor electrical connector.

32. Install new wiring harness retainers and position the CKP sensor wiring.

33. Connect the CKP sensor electrical connector.

34. Position the left accessory drive bracket and install the 4 bolts. Tighten to 42 Nm (31 ft. lbs.).

35. Install the 2 coil bracket-to-left accessory drive bracket bolts. Tighten to 10 Nm (89 inch lbs.).

36. Position the alternator bracket and install the 3 bolts. Tighten to 42 Nm (31 ft. lbs.).

37. Connect the alternator electrical connectors and the wiring harness retainer. Tighten to 8 Nm (71 inch lbs.).

38. Connect the heater hose to the coolant pump and position the clamp.

39. Position the drive belt tensioner and install the bolt. Tighten to 47 Nm (35 ft. lbs.).

40. Connect the lower radiator hose and position the clamp.

41. Install the 5 oil pan-to-front cover bolts. Tighten to 9 Nm (80 inch lbs.).

42. Position the power steering pressure tube and install the 2 nuts. Tighten to 18 Nm (13 ft. lbs.).

43. Install the crankshaft front seal.

44. Connect the battery ground cable.

45. Fill the engine cooling system.

4.6L SOHC Engine

SEAL ONLY

1. Before servicing the vehicle, refer to the Precautions Section.

2. Remove or disconnect the following:

- Crankshaft pulley
- Front oil seal with a puller

To install:

3. Lubricate the front cover and the front oil seal inner lip with clean engine oil.

1	Power steering pump pulley
2	Power steering pressure hose
3	Power steering pump bolts (3 required)
4	Power steering pump
5	Camshaft position (CMP) sensor electrical connector
6	CMP sensor bolt
7	CMP sensor
8	CMP sensor O-ring seal
9	Accessory drive belt idler pulley bolt
10	Accessory drive belt idler pulley
11	Accessory drive belt idler pulley bolt
12	Accessory drive belt idler pulley
13	Coolant pump pulley bolts (4 required)
14	Coolant pump pulley
15	Accessory drive belt tensioner bolts (3 required)
16	Accessory drive belt tensioner
17	Accessory drive belt idler pulley bolt
18	Accessory drive belt idler pulley
19	Crankshaft position (CKP) sensor electrical connector
20	CKP sensor bolt
21	CKP sensor
22	CKP sensor O-ring seal

06017-EXPL-G49

Remove these parts for front cover removal—4.6L SOHC engine

10 Nm
(89 lb-in) — 23

23 Oil pan bolts (4 required)
24 Engine front cover
25 Engine front cover gasket

06017-EXPL-G50

Front cover removal—4.6L SOHC engine

4. Use a seal driver to install the front oil seal into the engine front cover.
5. Install the crankshaft pulley.

FRONT COVER

1. Before servicing the vehicle, refer to the Precautions Section.
2. Drain the engine oil and coolant.
3. Remove or disconnect the following:

- Battery ground cable
- Both valve covers
- Radiator and the transmission cooler
- Drive belt
- Water pump
- CMP sensor
- Upper radiator hose bracket
- EGR vacuum regulator solenoid
- Drive belt tensioner

- Idler pulleys
- Crankshaft pulley
- Coolant hose clamps
- CKP sensor
- Oil pan to front cover bolts
- Crankshaft front oil seal
- Timing cover bolts and studs
- Front cover from the cylinder block

To install:

➡ If the engine front cover is not secured within four minutes, the sealant must be removed and the sealing area cleaned with metal surface cleaner.

4. Apply silicone gasket and sealant along the cylinder head to block surface and the oil pan to cylinder block surface.
5. Install the engine front cover on the front cover to cylinder block dowel and loosely install the bolts.

06017-EXPL-G51

Pan to cover bolt torque sequence—4.6L SOHC engine

25 Nm (18 lb-ft)

9355EG10

Front cover bolt torque sequence—4.6L SOHC Engine

6. Tighten the front cover bolts and stud bolts in the sequence shown.

7. Install a new crankshaft front oil seal.

8. Loosely install the pan-to-cover bolts, then, tighten the bolts in two stages, inner bolts first.

- Step 1: Tighten to 20 Nm (15 ft. lbs.)
- Step 2: Tighten an additional 90 degrees

9. The remainder of installation is the reverse of removal.

4.6L DOHC Engine

1. Before servicing the vehicle, refer to the Precautions Section.

2. Disconnect the battery ground cable.

3. Drain the engine oil.

4. Remove the valve covers.

5. If removing the timing chains, gears or other valve train components, remove the camshaft roller followers.

6. Remove the cooling fan.

7. Remove the accessory drive belt.

8. Remove the accessory drive belt tensioner.

9. Remove the accessory drive belt idler pulleys.

10. Remove the power steering reservoir bracket.

11. Remove the CMP sensor.

12. Remove the oil filter.

13. Remove the power steering hose bracket.

14. Remove the power steering pump.

15. Remove the A/C compressor electrical connector.

16. Remove the battery wiring harness.

17. Remove the A/C compressor.

18. Remove the transmission tube clip.

19. Remove the crankshaft position (CKP) sensor.

20. Remove the water pump pulley.

21. Remove the crankshaft pulley.

➡ **To remove the crankshaft pulley bolt, remove the upper fan shroud and electric fan.**

22. Remove the crankshaft seal.

23. Remove the engine front cover.

➡ **Remove the four oil pan-to-engine front cover bolts.**

24. Remove the engine front cover gaskets.

To install:

➡ **If the engine front cover is not secured within four minutes, the sealant must be removed and the sealing area cleaned with metal surface**

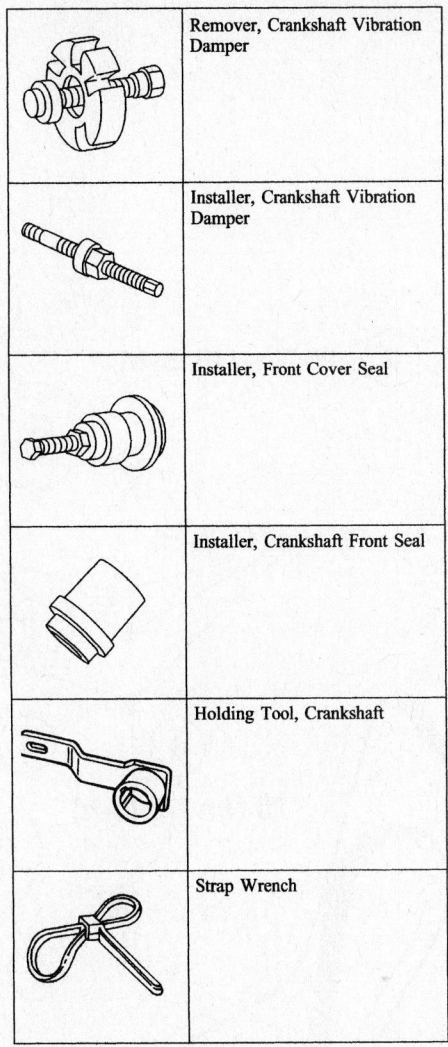

	Remover, Crankshaft Vibration Damper
	Installer, Crankshaft Vibration Damper
	Installer, Front Cover Seal
	Installer, Crankshaft Front Seal
	Holding Tool, Crankshaft
	Strap Wrench

67197-EXPL-G39

These tools, or their equivalents, are necessary for timing chain and front cover replacement—4.6L DOHC engine

	Holding Tool, Camshaft
	Remover, Crankshaft Front Seal
	Locking Tool, Flywheel

67197-EXPL-G40

These tools, or their equivalents, are necessary for timing chain and front cover replacement—4.6L DOHC engine

1 Accessory drive belt

2 Accessory drive belt idler pulley mounting bolt

3 Accessory drive belt idler pulley

4 Accessory drive belt tensioner mounting bolts

5 Accessory drive belt tensioner

6 Accessory drive belt idler
pulley mounting bolt

7 Accessory drive belt idler pulley

8 Accessory drive belt idler
pulley mounting bolt

9 Accessory drive belt idler pulley

67197-EXPL-G41

Accessory drive belt and related components—4.6L DOHC engine

10 Power steering reservoir bracket bolts

11 Power steering reservoir bracket

12 Camshaft position (CMP) sensor electrical connector

13 CMP sensor mounting bolt

14 CMP sensor

15 Oil filter

16 Power steering hose bracket mounting nut

17 Power steering hose bracket

18 Power steering pump mounting bolts

19 Power steering pump mounting bolt

20 Power steering pump mounting bolt

21 Power steering pump

67197-EXPL-G42

Power steering pump and related components—4.6L DOHC engine

10 Nm (89 lb-in)

31

22

32

30

29

28 — 25 Nm (18 lb-ft)

27

26 — 13 Nm (10 lb-ft)

23

25 — 25 Nm (18 lb-ft)

24 — 25 Nm (18 lb-ft)

22 A/C compressor electrical connector
23 Battery wiring harness
24 A/C compressor mounting stud
25 A/C compressor mounting nuts
26 A/C compressor mounting studs
27 A/C compressor (position aside)

28 Transmission tube clip mounting nut
29 Transmission tube clip
30 Crankshaft position sensor electrical connector
31 Crankshaft position sensor mounting bolt
32 Crankshaft position (CKP) sensor

67197-EXPL-G43

A/C and related components—4.6L DOHC engine

25 Nm (18 lb-ft) – (33) i

33 Water pump pulley bolts
34 Water pump pulley
35 Crankshaft pulley bolt
36 Crankshaft pulley washer

37 Crankshaft pulley
38 Crankshaft seal
39 Engine front cover
40 Engine front cover gaskets

67197-EXPL-G44

Front cover removal—4.6L DOHC engine

8 mm (0.32 in)

67197-EXPL-G58

Apply sealant in the locations shown—4.6L engine

8 mm (0.32 in)

67197-EXPL-G58

Oil pan bolt torque sequence—4.6L engine

1 Bolt, Hex Flange Head Pilot, M8 x 1.25 x 53

2 Bolt, Hex Flange Head Pilot, M8 x 1.25 x 53

3 Bolt, Hex Flange Head Pilot, M8 x 1.25 x 53

4 Bolt, Hex Flange Head Pilot, M8 x 1.25 x 53

5 Bolt, Hex Flange Head Pilot, M8 x 1.25 x 53

6 Stud, Hex Shldr Pilot, M8 x 1.25 x 50 — M6 x 1 x 10

7 Stud and Washer, Hex Head Pilot, M8 x 1.25 x 60 — M6 x 1 x 26

8 Bolt, Hex Flange Head Pilot, M8 x 1.25 x 53

9 Bolt, Hex Flange Head Pilot, M8 x 1.25 x 53

10 Bolt, Hex Flange Head Pilot, M8 x 1.25 x 53

11 Stud, Hex Shldr Pilot, M8 x 1.25 x 65 — M8 x 1.25 x 26

12 Stud, Hex Head Pilot, M8 x 1.25 x 65 — M8 x 1.25 x 16

13 Stud, Hex Shldr Pilot, M8 x 1.25 x 65 — M8 x 1.25 x 26

14 Stud, Hex Shldr Pilot, M8 x 1.25 x 65 — M8 x 1.25 x 26

15 Stud, Hex Shldr Pilot, M8 x 1.25 x 65 — M8 x 1.25 x 26

67197-EXPL-G57

Front cover bolt torque sequence—4.6L engine

cleaner. Allow to dry until there is no sign of wetness, or four minutes, whichever is longer. Failure to follow these instructions can result in future oil leakage.

25. Apply sealant in the locations shown.

26. Install the engine front cover seals and the engine front cover, and tighten the fasteners in the sequence shown.

27. Tighten the oil pan bolts in the sequence shown in three stages:
- Stage 1: Tighten to 2 Nm (18 inch lbs.).
- Stage 2: Tighten to 20 Nm (15 ft. lbs.).
- Stage 3: Tighten an additional 60 degrees.

Crankshaft Seal Installation

28. Lubricate the engine front cover and the crankshaft seal inner lip with clean engine oil.

29. Using the special tools, install the crankshaft seal.

Crankshaft Pulley Bolt Installation

➡**The crankshaft pulley must be installed within four minutes after applying the sealant.**

30. Apply sealant to the Woodruff key slot on the crankshaft pulley.

31. Using the special tool, install the crankshaft pulley.

32. Install the bolt and the washer. Tighten the bolt in four stages:
- Stage 1: Tighten the bolt to 90 Nm (66 ft. lbs.).
- Stage 2: Loosen the bolt one full turn.
- Stage 3: Tighten the bolt to 50 Nm (37 ft. lbs.).
- Stage 4: Tighten the bolt an additional 90 degrees.

Timing Chain, Sprockets, Front Cover and Seal

REMOVAL & INSTALLATION

4.0L SOHC Engine

1. Before servicing the vehicle, refer to the Precautions Section.

2. Drain the engine oil.

3. Remove or disconnect the following:
- Negative battery cable
- Engine from the vehicle
- Oil pan
- Engine front cover
- Cylinder heads

4. Lock the jackshaft tensioner by installing a pin.

Remove the jackshaft sprocket–4.0L SOHC Engine

89683G51

- Jackshaft sprocket and chain assembly
- Left front cassette retaining bolt
- Cassette chain and tensioner assembly
- Rear jackshaft plug from the engine
- Right rear cassette retaining bolt and spacer
- Right rear cassette chain and tensioner
- Timing chain (s)

To install:

5. Install or connect the following:
- Timing chain(s)
- Right rear cassette chain, tensioner and sprocket
- Jackshaft sprocket and chain on the engine and remove the tensioner pin

6. Torque the jackshaft sprocket bolt in 2 stages:
   a. 32–35 ft. lbs. (43–47 Nm).
   b. Turn an additional 65 degrees.

7. Install or connect the following:
- Cylinder heads
- Front cover
- Oil pan
- Engine to the vehicle
- Negative battery cable

8. Fill the engine with clean oil.
9. Start the vehicle, check for leaks and repair if necessary.

4.6L SOHC Engine

1. Before servicing the vehicle, refer to the Precautions Section.
2. Remove or disconnect the following:

- Front cover
- Crankshaft sensor ring from the crankshaft

3. Rotate the crankshaft until both camshaft key ways are 90 degrees from the valve cover surface. Make sure the copper links line up with the dots on the camshaft sprocket.

4. Install the special tools 303-380 and 303-413 on the camshaft.

5. Remove or disconnect the following:
- left timing chain tensioner
- right timing chain tensioner
- left and right timing chain tensioner arm from the dowel pins
- Timing chains and crankshaft sprocket
- Timing chain guides

To install:

6. Compress the tensioner plunger, using an edge of a vise.

7. Using a small screwdriver or pick, push back and hold the ratchet mechanism.

8. While holding the ratchet mechanism, push the ratchet arm back into the tensioner housing.

9. Install a paper clip into the hole in

9355EG08

Copper link alignment—4.6L SOHC Engine

Left cylinder head bolt torque sequence 4.6L SOHC Engine

9355EG01

the tensioner housing to hold the ratchet assembly and plunger in during installation.

➡**If the copper links are not visible, mark one link on one end and one link on the other end, and use as timing marks.**

10. Install or connect the following:
- Crankshaft sprocket, making sure the timing mark faces forward
- Timing chain guides
- left (inner) timing chain on the crankshaft sprocket, aligning the copper link with the dot on the crankshaft sprocket
- left (inner) timing chain on the camshaft sprocket, aligning the copper link with the dot on the camshaft sprocket
- right (outer) timing chain on the crankshaft sprocket, aligning the copper link with the dot on the crankshaft sprocket
- right (outer) timing chain on the camshaft sprocket, aligning the copper link with the dot on the camshaft sprocket

11. Make sure that the copper marks on the timing chain are lined up with the corresponding dots on the crankshaft sprockets and the camshaft sprockets.

12. Make sure that the camshaft sprocket keyway is 90 degrees from the valve cover mounting surface.

➡**The left timing chain tensioner arm has a bump near the dowel hole, for identification.**

Sprocket alignment—4.6L SOHC Engine

9355EG09

13. Position the left and right timing chain tensioner arm on the dowel pins.

14. Position the right timing chain tensioner and install the bolts.

15. Position the left timing chain tensioner and install the bolts.

16. Remove both the right and left retaining pins from the timing chain tensioner.

17. Remove the special tools from the camshaft.

18. Install the crankshaft sensor ring on the crankshaft.

19. Install the engine front cover.

4.6L DOHC Engine

1. Before servicing the vehicle, refer to the Precautions Section.

2. Remove the crankshaft sensor ignition pulse ring.

✳✳ WARNING

Since the engine is not free-wheeling, timing procedures must be followed exactly or piston and valve damage can occur.

✳✳ WARNING

Unless otherwise instructed, at no time when the timing chains are removed and the cylinder heads are installed is the crankshaft or the camshaft to be rotated. Severe piston and valve damage will occur.

3. Using special tool 303-448, position the crankshaft with the keyway at the 12 o'clock position.

4. Install special tool 303-446.

5. Remove the timing chain tensioning system from both timing chains.
   a. Remove the bolts.
   b. Remove the timing chain tensioners.
   c. Remove the timing chain tensioner arms.

6. Remove the right timing chain tensioner arm.

7. Remove the left timing chain tensioner arm.

8. Remove the right camshaft sprocket mounting bolt.

9. Remove the right camshaft sprocket washer.

10. Remove the right camshaft sprocket.

11. Remove the left camshaft sprocket mounting bolt.

12. Remove the left camshaft sprocket washer.

13. Remove the crankshaft gear.

14. Remove the left timing chain.

41 Crankshaft sensor ignition pulse ring

42 RH timing chain tensioner mounting bolts

43 RH timing chain tensioner

44 RH timing chain tensioner arm

45 LH timing chain tensioner mounting bolts

46 LH timing chain tensioner

47 LH timing chain tensioner arm

48 RH camshaft sprocket mounting bolt

49 RH camshaft sprocket washer

50 RH timing chain

51 RH camshaft sprocket

52 LH camshaft sprocket mounting bolt

53 LH camshaft sprocket washer

54 Crankshaft gear

55 LH timing chain

56 LH camshaft sprocket

67197-EXPL-G45

Timing chains and related parts—4.6L DOHC engine

10 Nm (89 lb-in)

10 Nm (89 lb-in)

10 Nm (89 lb-in)

10 Nm (89 lb-in)

10 Nm (89 lb-in)

10 Nm (89 lb-in)

10 Nm (89 lb-in)

57 LH timing chain guide upper mounting bolt

58 LH timing chain guide lower mounting bolt

59 LH timing chain guide

60 RH timing chain guide upper mounting bolt

61 RH timing chain guide lower mounting bolt

62 RH timing chain guide

63 RH intake camshaft gear mounting bolt

64 RH intake camshaft gear washer

65 RH intake camshaft gear spacer

66 LH intake camshaft gear mounting bolt

67 LH intake camshaft gear washer

68 LH intake camshaft gear spacer

69 LH secondary timing chain tensioner mounting bolts

70 LH secondary timing chain tensioner

71 RH secondary timing chain tensioner mounting bolts

72 RH secondary timing chain tensioner

73 RH exhaust secondary camshaft sprocket

74 RH secondary camshaft timing chain

75 RH intake secondary camshaft sprocket

76 LH intake secondary camshaft sprocket

77 LH exhaust secondary camshaft sprocket

78 LH secondary camshaft timing chain

67197-EXPL-G46

Secondary timing chains and related parts—4.6L DOHC engine

15. Remove the left camshaft sprocket.
16. Remove the left timing chain guide.
17. Remove the right timing chain guide.
18. Remove the right intake camshaft gear mounting bolt.
19. Remove the right intake camshaft gear washer.
20. Remove the right intake camshaft gear spacer.
21. Remove the left intake camshaft gear mounting bolt.
22. Remove the left intake camshaft gear washer.
23. Remove the left intake camshaft gear spacer.
24. Remove the left secondary timing chain tensioner.
25. Remove the right secondary timing chain tensioner.
26. Remove the right exhaust secondary camshaft sprocket.
27. Remove the right secondary camshaft timing chain.
28. Remove the right intake secondary camshaft sprocket.
29. Remove the left intake secondary camshaft sprocket.
30. Remove the left exhaust secondary camshaft sprocket.
31. Remove the left secondary camshaft timing chain.

To install:

32. Installation is the reverse of removal. Observe the following notes:

Timing Drive Components Installation

33. Compress the tensioner and install the retaining pin. Install the tensioners.

✳✳ CAUTION

Timing marks must be at the 12 o'clock position and indexed at the 6 o'clock position.

34. Install the camshaft sprockets and chain as an assembly.
35. Install special tool 303-446.

67197-EXPL-G47

Install special tool 303-446

67197-EXPL-G48

Install the camshaft sprockets and chain as an assembly

36. Install the camshaft spacer, washer and bolt, and hand-tighten the bolt.
37. Install the camshaft gear, washer and bolt, and hand-tighten the bolt.
38. Tighten the bolts in two stages:
- Stage 1: Tighten to 40 Nm (30 ft. lbs.).
- Stage 2: Tighten an additional 90 degrees.

✳✳ CAUTION

Timing chain procedures must be followed exactly or damage to valves and pistons will result.

✳✳ CAUTION

Do not compress the ratchet assembly. This will damage the ratchet assembly.

39. On engines with ratcheting timing chain tensioners only.
   a. Compress each tensioner plunger, using an edge of a vise.
   b. Using a small screwdriver or pick, push back and hold the ratchet mechanism.

67197-EXPL-G49

If the copper links are not visible, mark one link on one end and one link on the other end, and use as timing marks—4.6L DOHC engine

   c. While holding the ratchet mechanism, push the ratchet arm back into the tensioner housing.
   d. Install a paper clip into the hole of each tensioner housing to hold the ratchet assembly and plunger in during installation. Remove the tensioner from the vise.

✳✳ CAUTION

The timing chain procedures must be followed exactly or damage to valves and pistons will result.

40. On engines with non-ratcheting timing chain tensioners only.
   a. Compress each tensioner plunger, using a vise.
   b. Install a retaining clip on each tensioner to hold the plunger in during installation.
41. If the copper links are not visible, mark one link on one end and one link on the other end, and use as timing marks.
42. Install the timing chain guides.
43. Rotate the left camshaft sprocket until the timing mark is approximately at the

67197-EXPL-G50

Rotate the left camshaft sprocket until the timing mark is approximately at the 12 o'clock position. Rotate the right camshaft timing sprocket until the timing mark is approximately at the 11 o'clock position—4.6L DOHC engine

12 o'clock position. Rotate the right camshaft timing sprocket until the timing mark is approximately at the 11 o'clock position.

✸✸ CAUTION

Unless otherwise instructed, at no time when the timing chains are removed and the cylinder heads are installed is the crankshaft or the camshaft to be rotated. Severe piston and valve damage will occur.

✸✸ CAUTION

Rotate the crankshaft counterclockwise only. Do not rotate past position shown or severe piston or valve damage will occur.

➡ **The number one cylinder is at top dead center (TDC) when the stud on the engine block fits into the slot in the handle of the special tool.**

44. Using special tool 303-448, position the crankshaft so the number one cylinder is at TDC.

45. Remove the special tool.

46. Position the left (inner) timing chain on the crankshaft sprocket, aligning the copper (marked) link with the timing mark on the sprocket.

47. Install the left timing chain onto the

Using special tool 303-448, position the crankshaft so the number one cylinder is at TDC—4.6L DOHC engine

Position the left (inner) timing chain on the crankshaft sprocket, aligning the copper (marked) link with the timing mark on the sprocket—4.6L DOHC engine

67197-EXPL-G53

Install the left timing chain onto the camshaft sprocket, aligning the copper (marked) link with the timing marks on the sprocket—4.6L DOHC engine

camshaft sprocket, aligning the copper (marked) link with the timing marks on the sprocket.

➡ **The left timing chain tensioner arm has a bump near the dowel hole for identification.**

48. Position the left timing chain tensioner arms on the dowel pin and install the left timing chain tensioner.

49. On engines with ratcheting timing chain tensioners only, remove the retaining clip from the left timing chain tensioner.

50. On engines with non-ratcheting timing chain tensioners only, remove the retaining clip from the left timing chain tensioner.

51. Position the right (outer) timing chain on the crankshaft sprocket, aligning the copper (marked) link with the timing mark on the sprocket.

52. Install the right timing chain onto the camshaft sprocket, aligning the copper (marked) link with the timing marks on the sprocket.

53. Position the right timing chain tensioner arms on the dowel pin and install the right timing chain tensioner.

54. On engines with ratcheting timing chain tensioners only, remove the retaining clip from the right timing chain tensioner.

67197-EXPL-G54

Position the right (outer) timing chain on the crankshaft sprocket, aligning the copper (marked) link with the timing mark on the sprocket—4.6L DOHC engine

67197-EXPL-G55

Install the right timing chain onto the camshaft sprocket, aligning the copper (marked) link with the timing marks on the sprocket—4.6L DOHC engine

67197-EXPL-G56

Make sure that the copper (marked) chain links are lined up with the dots on the crankshaft sprockets and the camshaft sprocket—4.6L engine

55. On engines with non-ratcheting timing chain tensioners only, remove the retaining clips from the right timing chain tensioner.

➡ **Applies to all engines.**

56. Make sure that the copper (marked) chain links are lined up with the dots on the crankshaft sprockets and the camshaft sprocket.

Piston and Ring

POSITIONING

7924AG22

Piston ring positioning

Piston ring end gap spacing

Piston and connecting rod positioning on 4.0L

Piston and connecting rod positioning on 4.6L

FUEL SYSTEM

Fuel System Service Precautions

Safety is the most important factor when performing not only fuel system maintenance, but any type of maintenance. Failure to conduct maintenance and repairs in a safe manner may result in serious personal injury or death. Work on a vehicle's fuel system components can be accomplished safely and effectively by adhering to the following rules and guidelines.

• To avoid the possibility of fire and personal injury, always disconnect the negative battery cable unless the repair or test procedure requires that battery voltage be applied.

• Always relieve the fuel system pressure prior to disconnecting any fuel system component (injector, fuel rail, pressure regulator, etc.) fitting or fuel line connection. Exercise extreme caution whenever relieving fuel system pressure, to avoid exposing your skin, face and eyes to fuel spray. Please be advised that fuel under pressure may penetrate the skin or any part of the body that it contacts.

• Always place a shop towel or cloth around the fitting or connection prior to loosening to absorb any excess fuel due to spillage. Ensure that all fuel spillage is quickly remove from engine surfaces. Ensure that all fuel-soaked cloths or towels are deposited into a flame-proof waste container with a lid.

• Always keep a dry chemical (Class B) fire extinguisher near the work area.

• Do not allow fuel spray or fuel vapors to come into contact with a light bulb, spark or open flame.

• Always use a second wrench when loosening or tightening fuel line connection fittings. This will prevent unnecessary stress and torsion to fuel piping. Always follow the proper torque specifications.

• Always replace worn fuel fitting O-rings with new ones. Do not substitute fuel hose where rigid pipe is installed.

✳✳ CAUTION

Fuel in the fuel system remains under high pressure even when the engine is not running. Before repairing or disconnecting any of the fuel lines or fuel system components, the fuel system pressure must be relieved to prevent accidental spraying of fuel, causing personal injury or a fire hazard.

✳✳ WARNING

Fuel injection equipment is manufactured to very precise tolerances and fine clearances. It is therefore essential that absolute cleanliness is observed when working with these components. Always install blanking plugs to any open orifices or tubes.

✳✳ WARNING

When reusing liquid or vapor tube connectors make sure to use compressed air to remove any foreign material from the connector retaining clip area before separating from the tube.

➡During the repair or replacement of fuel-related components, all liquid and vapor fuel openings shall be capped, taped or otherwise appropriately protected to prevent the ingress of dirt or other contamination. All caps, tape and other protective materials shall be removed prior to installation.

Fuel System Pressure

RELEASE

2002–03

1. Before servicing the vehicle, refer to the Precautions Section.

All engines are equipped with a pressure relief valve located on the fuel supply manifold. Remove the fuel tank cap and attach fuel pressure gauge T80L-9974-B, to the valve to release the fuel pressure. Be sure to drain the fuel into a suitable container and to avoid gasoline spillage. If a pressure gauge is not available, disconnect the vacuum hose from the fuel pressure regulator and attach a hand-held vacuum pump. Apply about 25 in. Hg (84 kPa) of vacuum to the regulator to vent the fuel system pressure into the fuel tank through the fuel return hose. Note that this procedure will remove the fuel pressure from the lines, but not the fuel. Take precautions to avoid the risk of fire and use clean rags to soak up any spilled fuel when the lines are disconnected.

An alternate method of relieving the fuel system pressure involves disconnecting the inertia switch.

2004–05

1. Before servicing the vehicle, refer to the Precautions Section.

➡The fuel pump relay is located in the battery junction box (BJB) location C-1051.

2. Remove the fuel pump relay.

3. Start the engine and allow it to idle until it stalls.

4. After the engine stalls, crank the engine for approximately 5 seconds to make sure the fuel rail pressure has been released.

5. Turn the ignition switch to the OFF position.

6. When the fuel system service is complete, install the fuel pump relay.

➡It may take more than one key cycle to pressurize the fuel system.

7. Cycle the ignition key and wait 3 seconds to pressurize the fuel system. Check for leaks before starting the engine.

8. Start the vehicle and check the fuel system for leaks.

Fuel Line Couplings

REMOVAL & INSTALLATION

Spring Lock Couplings

Remove the safety clip, install the spring lock coupling disconnect tool

06017-EXPL-G58

Close the tool

06017-EXPL-G59

Push the tool into the open side of the cage

06017-EXPL-G60

Separate the coupler

06017-EXPL-G61

Remove the tool

06017-EXPL-G62

Lubricate the O-rings with 5W-30 oil

06017-EXPL-G63

Push the coupler together until the garter spring in engaged

06017-EXPL-G64

Pull on the coupler to make sure it's fully engaged

06017-EXPL-G65

Install the safety clip

06017-EXPL-G66

Push-Connect Fittings

Disconnect the safety clip from the male hose

06017-EXPL-G67

Install the tool and push it into the fitting

06017-EXPL-G68

Separate the fittings

06017-EXPL-G69

Align the tube and fitting and push together until you hear a click

06017-EXPL-G70

06017-EXPL-G71

Pull on the fitting to make sure it's secure, then install the safety clip

Fuel Filter

REMOVAL & INSTALLATION

1. Before servicing the vehicle, refer to the Precautions Section.

2. Properly relieve the fuel system pressure.

3. Remove or disconnect the following:
- Negative battery cable
- Fuel lines
- Fuel filter from the support

To install:

4. Install or connect the following:
- Fuel filter to the support
- Fuel lines
- Negative battery cable

5. Start the vehicle, check for leaks and repair if necessary.

06017-EXPL-G52

Fuel filter

Fuel Pump

REMOVAL & INSTALLATION

2002–03 Explorer and Mountaineer; 2004 Explorer and Mountaineer with 4.0L SOHC Early Build; 2003–05 Aviator

1. Before servicing the vehicle, refer to the Precautions Section.

❈❈ CAUTION

The fuel in the system remains under high pressure even when the engine is not running. Before repairing or disconnecting any of the fuel lines or fuel system components, the fuel

pressure must be relieved to prevent accidental spraying of fuel, causing personal injury or fire hazard.

2. With the vehicle in NEUTRAL, position it on a hoist.

3. Disconnect the battery ground cable.

4. Release the fuel system pressure.

5. Remove the rear driveshaft.

6. Disconnect the lower fill hose from the fuel intermediate steel by loosening worm gear clamp and pulling it off.

7. Insert a drain hose through the lower fill hose, past the fill tube check valve, and pump fuel out of the fuel tank.

8. Remove the fuel tank skid plate (if equipped).

9. Remove the three bolts at the outboard frame rail.

10. Remove the two nuts on the inboard side.

11. Remove the fuel tank skid plate.

12. If equipped, remove the transfer case skid plate by removing four bolts from the outer frame rail (two per side).

13. Disconnect the transmission-mounted fuel vapor line from the plastic fitting on the forward canister. This is a squeeze release connector.

14. Remove the fuel filter heat shield from the nose of the fuel tank.

15. Remove the two nuts.

16. Remove the heat shield.

17. Disconnect the transmission-mounted fuel supply line from the fuel filter. There is a redundant latch that will unhook first. Then a quick-connect fitting that will require a quick disconnect tool to remove.

18. Push the fuel line out of the retaining clip at the top rear of the fuel tank.

19. Disconnect the vapor line on the filler pipe from the tank (quick connector is on the frame rail side of tank on the top in the rear and can be accessed through the passenger rear tire opening or from the ground with the vehicle on a hoist.) Push the button on the side to release.

20. Disconnect the vapor line from the canister vent valve, which is located on the inside frame rail just rearward of the fuel tank.

21. Use a suitable jack to support the fuel tank.

22. Remove the fuel tank strap nut.

23. Remove the fuel tank strap bolt and lower tank.

24. Lower the tank slightly and disconnect the electrical connector from the pump sender to the frame loom.

25. Remove the fuel tank.

26. Clean the area around the fuel pump assembly mounting flange.

❈❈ WARNING

The fuel pump assembly must be handled carefully to avoid damage to the float arm and the filter.

27. Remove the fuel pump module lock ring with the special tool and remove the fuel pump module.

28. To install, reverse the removal procedure.

2004 Explorer and Mountaineer 4.0L SOHC Late Build and all 2004–05 Explorer and Mountaineer

1. Before servicing the vehicle, refer to the Precautions Section.

❈❈ CAUTION

Fuel in the fuel system remains under high pressure even when the engine is not running. Before repairing or disconnecting any of the fuel lines or fuel system components, the fuel system pressure must be relieved to prevent accidental spraying of fuel, causing personal injury or a fire hazard.

2. With the vehicle in NEUTRAL, position it on a hoist.

❈❈ CAUTION

Release the fuel system pressure.

3. Disconnect the battery negative cable.

4. Remove the driveshaft.

❈❈ CAUTION

Drain the fuel tank.

❈❈ CAUTION

If equipped, remove the 4 bolts and the transfer case skid plate.

5. Disconnect the filler pipe vapor tube fitting.

6. Disconnect the fuel vapor vent valve-to-fuel pressure sensor and tube assembly hose.

7. Remove the bolts and the fuel filter heat shield.

8. Remove the nuts and the fuel filter shield.

9. Disconnect the fuel filter spring lock coupling.

10. Disconnect the fuel pressure sensor and tube assembly-to-vapor tube fitting.

1   Fuel filler cap
2   Screws (3)
3   Vapor snorkel
    Removal Note
4   Fuel filler tube
    Removal Note
5   Rear axle vent hose
6   Dust box rubber hose
7   Fuel tank fill hose clamp
8   Fuel line retainer clip
    Removal Note
9   Bolts
10 Fuel tank filler pipe assembly
    Removal Note
10 Fuel tank filler pipe assembly
    Installation Note
11 Fuel supply line
    Removal Note
11 Fuel supply line
    Installation Note

12 Vapor line
    Removal Note
13 Nut
14 Front support strap
15 Bolt
16 Rear support strap
17 Fuel vapor hose
18 Fuel pump electrical
    connector
19 Fuel tank
    Removal Note
20 Quick-connect fitting
    Removal Note
21 Quick-connect fitting
    Removal Note
22 Fuel pump lock ring
    Removal Note
23 Fuel pump module
24 Fuel pump gasket
    Installation Note

06017-EXPL-G53

Fuel tank, pump and related parts—2002–03 Explorer and Mountaineer; 2004 Explorer and Mountaineer with 4.0L SOHC Early Build

06017-EXPL-G54

Fuel pump locking ring—2002–03 Explorer and Mountaineer and 2004 Explorer and Mountaineer with 4.0L SOHC Early Build; 2003–05 Aviator

✳✳ CAUTION

If equipped, remove the 3 bolts, 2 nuts and the fuel tank skid plate.

✳✳ CAUTION

Using a suitable jack, support the fuel tank.

11. Remove the bolt and the fuel tank rear support strap.
12. Remove the nut and the fuel tank front support strap.

✳✳ CAUTION

Lower the fuel tank far enough to access the fuel pump electrical connector and disconnect the connector.

13. Remove the fuel tank from the vehicle.
14. Disconnect the vapor tube fitting.

✳ CAUTION

Using the special tool, remove the fuel pump locking ring.

➡The fuel pump module must be handled carefully to avoid damage to the float arm and filter.

15. Remove the fuel pump module and the O-ring seal. Discard the O-ring seal.
To install:

➡The fuel pump module must be handled carefully to avoid damage to the float arm and filter.

➡Lubricate the O-ring seal with clean engine oil.

✳ CAUTION

Install a new O-ring seal and the fuel pump module.

1 Fuel filler cap
2 Fuel filler pipe-to-body bolts (3 required)
3 Fuel filler pipe vapor tube
4 Fuel filler pipe vapor tube fitting
5 Rear axle vent hose
6 Dust separator hose
7 Fuel filler pipe hose clamp
8 Fuel filler pipe hose
9 Fuel filler pipe bolts (2 required)
10 Fuel filler pipe assembly
11 Fuel supply tube
12 Vapor tube fitting
13 Fuel vapor vent valve
14 Pin-type retainer
15 Fuel tank skid plate bolt (3 required)
16 Fuel tank skid plate nut (2 required)
17 Fuel tank skid plate (AWD only)
18 Fuel tank support strap bolt
19 Fuel tank rear support strap
20 Fuel tank support strap nut
21 Fuel tank front support strap
22 Fuel pump electrical connector
23 Fuel tank

06017-EXPL-G55

Fuel tank and related parts—2004 Explorer and Mountaineer 4.0L SOHC Late Build and all 2004–05 Explorer and Mountaineer

➡The fuel pump module locking ring is a cam-type lock. Turn the locking ring until it stops and is locked in place.

✳✳ CAUTION

Using the special tool, install the fuel pump module locking ring.

✳✳ CAUTION

Connect the vapor tube fitting.

✳✳ CAUTION

To install the fuel tank, reverse the removal procedure. Observe the following torques:

- Fuel tank front support strap: 55 Nm (41 ft. lbs.)
- Fuel tank skid plate: 40 Nm (30 ft. lbs.).
- Fuel filter shield: 5 Nm (44 inch lbs.).
- Fuel filter heat shield: 20 Nm (15 ft. lbs.).

24 Vapor tube fitting

25 Fuel pump locking ring

26 Fuel pump module

27 Fuel pump O-ring seal

06017-EXPL-G56

Fuel pump module—2004 Explorer and Mountaineer 4.0L SOHC Late Build and all 2004–05 Explorer and Mountaineer

- Transfer case skid plate: 40 Nm (30 ft. lbs.).

2002 Explorer and Mountaineer and Sport-Trac Sport Built Before 03/2002

1. Before servicing the vehicle, refer to the Precautions Section.
2. Disconnect the battery ground cable.
3. Relieve the fuel system pressure.
4. Drain the fuel tank.
5. 4WD vehicle
   a. Remove the fuel tank skid plate bolts.
   b. Remove the fuel tank skid plate rear left bolt.
   c. Remove the fuel tank skid plate rear right bolt and lower the fuel tank skid plate.
6. All vehicles
   a. Use a suitable jack to support the fuel tank.
   b. Remove the vent hose.
   c. Remove the bolt and reposition the fuel tank support strap.
   d. Remove the support bolts and push the fuel tank back until the lip clears the support bracket.
   e. Lower the fuel tank just enough to expose the fuel lines and the fuel tank transducer connector.

   f. Disconnect the fuel line and the vapor line.
   g. Disconnect the fuel tank electrical supply connector near the frame rail. Lower the fuel tank.
7. Clean the area around the fuel pump assembly mounting flange.
8. Disconnect the electrical connector from the fuel pressure transducer.

✳✳ WARNING

The fuel pump assembly must be handled carefully to avoid damage to the float arm and the filter.

9. Remove the fuel pump mounting bolts and remove the fuel pump assembly.

To install:
10. Clean the fuel pump assembly mounting flange and the fuel tank mounting surface.
11. Position the fuel pump assembly in the fuel tank.
12. Align the arrow on the fuel pump flange with the dimple on the fuel tank.
13. Tighten the bolts in the sequence shown in three stages.
   - Stage 1: Tighten to 3 Nm (30 inch lbs.).
   - Stage 2: Tighten to 7 Nm (65 inch lbs.).

06017-EXPL-G57

Fuel pump flange bolt torque sequence— 2002 Explorer and Mountaineer and Sport-Trac Sport Built Before 03/2002

   - Stage 3: Tighten to 10 Nm (93 inch lbs.).
14. Connect the electrical connector to the fuel pressure transducer.
15. To install the tank, reverse the removal procedure. Torque the fuel tank support and strap bolts to 30 ft. lbs. (40 Nm). Torque the skid plate bolts to 89 inch lbs. (10 Nm).

2002 Explorer Sport and Sport-Trac Built After 03/2002; All 2003 Explorer Sport and Sport-Trac; 2004–05 Explorer Sport-Trac

1. Before servicing the vehicle, refer to the Precautions Section.

2. Disconnect the battery ground cable.

3. Release the fuel system pressure.

4. Drain the fuel tank.

5. On 4-wheel drive vehicles:

a. Remove the fuel tank skid plate bolts.

b. Remove the fuel tank skid plate rear left bolt.

c. Remove the fuel tank skid plate rear right bolt and lower the fuel tank skid plate.

6. On all vehicles:

a. Use a suitable jack to support the fuel tank.

b. Remove the bolt and position the fuel tank support strap aside.

c. Remove the bolt and position the fuel tank support bracket aside.

d. Remove the bolt and push the fuel tank back until the lip clears the support bracket.

e. Lower the fuel tank just enough to expose the fuel and vapor hoses.

f. Disconnect the vapor hose.

g. Disconnect the fuel tank electrical supply connector near the frame rail. Lower the fuel tank.

h. Disconnect the fuel hoses.

i. Remove the fuel tank.

7. Remove the fuel hoses from the fuel pump.

8. Clean the area around the fuel pump locking ring.

✳✳ WARNING

The fuel pump assembly must be handled carefully to avoid damage to the float arm and the filter.

9. Using special tool 310-075, remove the fuel pump locking ring and remove the fuel pump assembly.

10. To install, reverse the removal procedure. Use a new O-ring. Tighten the locking ring to 55 ft. lbs. Tighten the support strap bolts to 30 ft. lbs. (40 Nm). Tighten the skid plate bolts to 89 inch lbs. (10 Nm).

Fuel Injectors

REMOVAL & INSTALLATION

4.0L SOHC Engine

✳✳ CAUTION

Fuel in the fuel system remains under high pressure even when the engine is not running. Before working on or

disconnecting any of the fuel lines or fuel system components, the fuel system pressure must be relieved. Failure to follow these instructions may result in personal injury.

✳✳ WARNING

If used as a leverage device, the fuel rail may be damaged. Care must be taken when working around the fuel rail.

1. Before servicing the vehicle, refer to the Precautions Section.

2. Remove the intake manifold.

3. Disconnect the spring lock coupling.

4. Remove the fuel supply tube bracket bolt.

5. Disconnect the fuel injector electrical connectors.

6. Disconnect the fuel pressure and temperature sensor electrical and vacuum connectors.

7. Remove the bolts and the fuel rail and injectors as an assembly.

1	Fuel hose spring lock coupling Removal Note	8 Fuel injection supply manifold
2	Fuel hose-to-valve cover bolt	9 Fuel supply hose bolts (4 req'd)
3	Fuel injector electrical connectors (RH side)	10 Fuel supply hose
4	Wiring harness retainer	11 Fuel supply hose O-rings (2 req'd)
5	Fuel injector electrical connectors (LH side)	12 Fuel injector (6 req'd)
6	Fuel pulse damper vacuum hose	13 Fuel injector-to-intake manifold O-rings (6 req'd)
7	Fuel injection supply manifold bolts (4 req'd)	14 Fuel injector-to-supply manifold O-rings (6 req'd)

06017-EXPL-G73

Fuel rail, injectors and related parts—2002-03 4.0L SOHC engine

⁕⁕ WARNING

O-ring seals are made of special fuel-resistant material. Use of ordinary O-ring seals can cause the fuel system to leak. Do not reuse O-ring seals.

→Install new fuel injector-to-intake manifold O-ring seals and lubricate them with clean engine oil.

8. Remove the fuel injectors and the fuel injector O-ring seals.

⁕⁕ WARNING

O-ring seals are made of special fuel-resistant material. Use of ordinary O-ring seals can cause the fuel system to leak. Do not reuse O-ring seals.

→Install new fuel injector-to-fuel rail O-ring seals and lubricate them with clean engine oil.

9. To install, reverse the removal procedure Torque the fuel rail bolts to 17 ft. lbs. (23 Nm). Torque the fuel supply bracket bolt to 71 inch lbs. (8 Nm).

2004–05 4.6L SOHC Engine

⁕⁕ CAUTION

Fuel in the fuel system remains under high pressure even when the engine is not running. Before working on or disconnecting any of the fuel tubes or fuel system components, the fuel system pressure must be relieved. Failure to follow these instructions can result in personal injury.

1   Fuel tube spring lock coupling
2   Fuel supply tube bracket bolt
3   Fuel pressure and temperature sensor electrical connector
4   Fuel pressure and temperature sensor vacuum hose
5   Fuel injector electrical connectors (LH side) (3 required)
6   Fuel injector electrical connectors (RH side) (3 required)
7   Fuel rail bolts (4 required)

8   Fuel rail
9   Fuel injector (6 required)
10 Fuel injector-to-intake manifold O-ring seals (6 required)
11 Fuel injector-to-fuel rail O-ring seals (6 required)
12 Fuel supply tube bolts (4 required)
13 Fuel supply tube
14 Fuel supply tube O-ring seals (2 required)

Fuel rail, injectors and related parts—2004–05 4.0L SOHC engine

06017-EXPL-G74

19 Fuel injector electrical
   connector (8 req'd)

20 Fuel injection supply
   manifold stud/bolt (4 req'd)
   Removal Note

21 Fuel injection supply
   manifold

22 Fuel injector (8 req'd)

23 Fuel injector O-ring seals
   (16 req'd)
   Removal Note
   Installation Note

06017-EXPL-G75

Fuel rail, injectors and related parts—2004–05 4.6L SOHC engine

23 Fuel injector electrical
   connector (8 req'd)

24 Fuel injection supply
   manifold stud/bolts (4 req'd)

25 Fuel injection supply
   manifold

26 Fuel injector (8 req'd)

27 Fuel injector O-ring (16
   req'd)
   Removal Note

06017-EXPL-G76

Fuel rail, injectors and related parts—2002–03 4.6L SOHC engine

1 Throttle position (TP) sensor electrical connector
2 Idle air control (IAC) sensor electrical connector
3 Fuel pulse damper electrical connector
4 Fuel pulse damper vacuum hose
5 Fuel injector electrical connectors (8 required)
6 Exhaust gas recirculation (EGR) system module electrical connector
7 Brake booster hose clamp
8 Brake booster hose
9 Engine emissions vacuum hose

10 Evaporative emission (EVAP) canister purge valve vacuum hose
11 Upper intake manifold-to-lower intake manifold bracket nuts
12 Upper intake manifold-to-lower intake manifold bracket
13 Fuel tube spring lock coupling
14 EGR system module tube nut-to-EGR system module
15 EGR system module tube nut-to-exhaust manifold
16 EGR system module tube
17 Throttle return spring

18 Accelerator cable
19 Speed control cable
20 Accelerator control assembly bracket bolts
21 Accelerator control assembly bracket (position aside)
22 Fuel injection supply manifold bolts
23 Fuel injection supply manifold and fuel injectors
24 Fuel injectors (8 required)
25 Fuel injector O-rings

06017-EXPL-G72

Fuel rail, injectors and related parts—4.6L DOHC engine

1. Release the fuel system pressure.
2. Before servicing the vehicle, refer to the Precautions Section.
3. Disconnect the battery ground cable.
4. Remove the air cleaner outlet pipe.
5. Drain the engine cooling system.
6. Remove the parts in the order indicated in the following illustrations and tables.
7. Remove the fuel tube spring lock coupling
8. Remove the crankcase vent tube.
9. Remove the brake booster vacuum tube bracket bolt.
10. Remove the brake booster vacuum tube.
11. Disconnect the electronic throttle body electrical connector.
12. Disconnect the emission vacuum control tube.
13. Remove the all hoses and clamps in the way.
14. Disconnect the fuel injector electrical connector.
15. Remove the fuel injection supply manifold stud/bolt.
16. Remove the fuel injection supply manifold.
17. Remove the fuel injector.
18. Remove the fuel injector O-ring seals.
19. To install, reverse the removal procedure.
20. Fill and bleed the cooling system.

❊❊ WARNING

Use O-ring seals that are made of special fuel-resistant material. Use of ordinary O-rings can cause the fuel system to leak. Do not reuse the O-ring seals.

21. Install new O-ring seals.
22. Lubricate the new O-ring seals with clean engine oil before installing.

4.6L DOHC Engine

❊❊ CAUTION

Fuel in the system remains under high pressure even when the engine is not running. Before working on or disconnecting any of the fuel lines or fuel system components, the fuel system must be relieved. Failure to follow these instructions can result in personal injury.

1. Before servicing the vehicle, refer to the Precautions Section.
2. Disconnect the battery ground cable.
3. Remove the air cleaner assembly.
4. Release the fuel system pressure.
5. Disconnect the throttle position (TP) sensor electrical connector.
6. Disconnect the idle air control (IAC) sensor electrical connector.
7. Disconnect the fuel pulse damper electrical connector and vacuum hose.
8. Disconnect the 8 fuel injector electrical connectors.
9. Disconnect the exhaust gas recirculation (EGR) system module electrical connector.
10. Disconnect the brake booster hose from the intake manifold.
11. Disconnect the engine emissions vacuum hose from the intake manifold.
12. Disconnect the evaporative emission (EVAP) canister purge valve vacuum hose from the intake manifold.
13. Remove the upper intake manifold-to-lower intake manifold bracket nuts.
14. Remove the upper intake manifold-to-lower intake manifold bracket.
15. Remove the fuel tube safety clip.
16. Disconnect the fuel tube.
17. Disconnect the EGR system tube nuts and remove the tube.
18. Disconnect the throttle return spring from the throttle body.
19. Disconnect the accelerator cable and the speed control cable (if equipped) from the throttle body.
20. Remove the accelerator cable assembly bracket bolts and position the bracket and cables aside.
21. Remove the fuel rail bolts and the fuel rail.

➡Use O-ring seals that are made of special fuel-resistant material. Use of ordinary O-rings can cause the fuel system to leak. Do not reuse the O-ring seals.

22. Remove the 8 fuel injectors.
23. Remove and discard the O-ring seals.
24. To install, reverse the removal procedure. Install new O-rings and lubricate with clean engine oil. Observe the following torques:

- Fuel rail bolts: 89 inch lbs. (10 Nm)
- Cable bracket: 89 inch lbs. (10 Nm)
- EGR tube nuts: 30 ft. lbs. (40 Nm)
- Manifold bracket nuts: 89 inch lbs. (10 Nm)

DRIVE TRAIN

Manual Transmission Assembly

REMOVAL & INSTALLATION

1. Before servicing the vehicle, refer to the Precautions Section.
2. Disconnect the negative battery cable

➡Do not remove the gearshift lever knob.

3. Remove the screw and lift the console finish panel to access the gearshift lever nut. Remove the gearshift lever nut, then remove the upper gearshift lever and console finish panel as an assembly.
4. Install the nut on the front of the lever, then tighten the nut to remove the eccentric stud out of the gearshift lever.

5. Place the transmission in NEUTRAL.
6. If the transmission is being disassembled, drain the transmission fluid.
7. Remove or disconnect the following:

- Front support
- Starter
- Rear driveshaft and position it aside
- Transfer case

8. Using a suitable high lift jack, support the transmission. Securely strap the jack to the transmission.
9. Remove or disconnect the following:

- right crossmember cover, then the four bolts
- left crossmember bolts
- Heat shields from the crossmember

- Nuts, then lower the crossmember from the vehicle
- Hydraulic line from the clutch slave cylinder
- Output shaft speed (OSS) sensor electrical connector
- Reverse lamp switch electrical connector
- Wiring harness brackets from the transmission, then disconnect the heated oxygen sensor (HO2S) electrical connectors
- The two rear three way catalytic converter (TWC) bolts
- Front TWC bolts and the TWC
- The eight transmission-to-engine bolts
- Transmission.

To install:

10. Raise and position the transmission to the engine and clutch.

11. Install the eight transmission-to-engine bolts. Torque the bolts to 44 ft. lbs. (50Nm).

12. Position the front three way catalytic converter (TWC) in the vehicle and install the bolts. Do not tighten the bolts at this time.

13. Using a new gasket and nuts, install the two rear three way catalytic converter (TWC) bolts. Tighten the rear bolts, then the front bolts at this time. Tighten the bolts to 30 ft. lbs. (40Nm).

14. Install or connect the following:
- Heated oxygen sensor (HO2S) electrical connectors
- right and left crossmember bolts and the right crossmember cover. Torque the crossmember nuts to 46 ft. lbs. (63Nm).
- Transmission mount nuts. Torque the nuts to 72 ft. lbs. (98Nm).
- Heat shields to the crossmember
- Wiring harness brackets and attach the wiring harness
- Front support
- Hydraulic line to the clutch slave cylinder
- Starter
- Transfer case
- Driveshaft

15. Position the upper gearshift lever, the outer gearshift lever boot and console finish panel into the console, then install the nut.

16. Connect the battery ground cable.

17. Bleed the clutch hydraulic system.

Automatic Transmission

REMOVAL & INSTALLATION

2002–03

1. Before servicing the vehicle, refer to the Precautions Section.

2. Place the transmission in **Neutral**.

3. Remove or disconnect the following:
- Negative battery cable
- Transfer case
- Three-way catalytic converter

4. If transmission disassembly is necessary, drain the transmission fluid.

5. Support the transmission with a transmission jack.

6. Remove or disconnect the following:
- right heat shield
- left heat shield bolt
- Plastic cover from the right side of the frame rail

- Plastic shield on the right side of the crossmember near the fuel tank
- The two upper crossmember bolts (two on each side)
- right side lower crossmember bolts
- left side lower crossmember bolts
- Transmission support insulator nuts and remove the crossmember
- The two bolts and remove the transmission support insulator
- Shift cable and bracket
- Turbine shaft speed (TSS) sensor, output shaft speed (OSS) sensor and intermediate shaft speed (ISS) sensor electrical connectors

➡**Clean the area around connector to prevent contamination of the solenoid body connector.**

7. Remove or disconnect the following:
- Screw from the solenoid body connector and disconnect the connector
- Harness retainers
- Left side heated oxygen sensor (HO2S) connector from the bracket
- Digital transmission range (TR) sensor connector
- Starter

❋❋ WARNING

Do not damage the cooler tubes

- Transmission cooler tubes
- Access cover

➡**Make an identifying mark on the nut, stud, and adapter plate to allow for correct installation.**

- The 4 torque converter nuts
- The 7 engine-to-transmission retaining bolts

8. Lower the transmission from the vehicle.

To install:

9. Raise and position the transmission.

10. Align the flexplate to converter marks made at removal.

11. Install seven engine to transmission retaining bolts. Torque to 35 ft. lbs. (48Nm).

12. Install the four torque converter nuts. Torque to 28 ft. lbs. (48Nm).

13. Install or connect the following:
- Access cover
- Transmission cooler tubes
- Starter
- Digital transmission range (TR) sensor connector
- Left side heated oxygen sensor (HO2S) to the bracket
- Wire harness, and install the retainers

❋❋ WARNING

Damage will occur to the solenoid body assembly if the screw is tightened above 44 inch lbs. (5Nm). Always install new O-ring seals on vehicle harness connector. Clean the area around connector to prevent contamination of the solenoid body connector. Use petroleum jelly to lubricate the O-ring seals to aid in the installation process.

14. Install and lubricate new O-ring seals on the transmission connector and connect the connector.

15. Install or connect the following:
- Turbine shaft speed (TSS) sensor, output shaft speed (OSS) sensor, and intermediate shaft speed (ISS) sensor electrical connectors.
- Shift cable and bracket
- Three-way catalytic converter
- Transmission support insulator. Torque to 68 ft. lbs. (90Nm).
- Crossmember in place and loosely install the two nuts to hold up the crossmember.
- Crossmember bolts. Torque to 52 ft. lbs. (70Nm).
- Rear transmission support nuts. Torque to 52 ft. lbs. (70Nm).
- Heat shields
- Plastic cover to the right side of the frame rail
- Plastic shield on the right side of the crossmember near the fuel tank
- Transfer case

➡**Ford recommends, if the vehicle was not equipped with a fluid filter, install a Fluid Filter Service Kit (XC3Z-7B155-AA). Follow the instructions supplied in the kit. If the vehicle was equipped with a fluid filter, install a new filter (XC3Z-7B155-AB).**

16. Connect the battery ground cable.

17. Verify that the shift cable is correctly adjusted.

2004–05 Aviator

2-WHEEL DRIVE

1. Before servicing the vehicle, refer to the Precautions Section.

➡**When the battery has been disconnected and reconnected, some abnormal drive symptoms can occur while the vehicle relearns its adaptive strategy.**

2. Disconnect the battery ground cable.

3. Remove the fan shroud bolts. Do not remove the fan shroud.

4. Raise and support the vehicle.

5. If transmission disassembly or installation of a new transmission is necessary, drain the transmission fluid. Install the drain plug when finished.

6. Remove the 2 right side heat shield bolts and position the shield aside.

7. Disconnect the 2 heated oxygen sensor electrical connectors.

8. Disconnect the 2 catalyst monitor sensor electrical connectors.

9. Remove the right side heat shield bolts.

10. Remove the right side heat shield bolt and remove the heat shield.

11. Remove the left side heat shield bolt.

12. Remove the 2 nuts and the plastic shield on the right side of the crossmember near the fuel tank.

13. Remove the 4 upper crossmember bolts (2 on each side).

14. Remove the bolts, springs, flag nuts and separate the muffler from the converter. Discard the gasket. Discard the flag nuts.

15. Remove the 4 converter-to-manifold nuts.

16. Remove the jackstand and lower the converter assembly with the heat shield.

❊❊ CAUTION

Secure the transmission to the transmission jack with a safety chain. Failure to follow these instructions can result in personal injury.

17. Install a suitable jack under the transmission.

18. Remove the right side crossmember bolts.

19. Remove the left side crossmember bolts.

20. Remove the 2 bolts on the bracket and position aside.

21. Remove the transmission support insulator nuts and remove the crossmember.

➡ **This step may require an assistant.**

22. Remove the transmission mounted exhaust hanger and the 3-way converter assembly.

23. Remove the shift cable and bracket and position out of PARK.

24. Index-mark the location of the rear flange and pinion flange. Remove the 4 bolts. Remove the rear driveshaft pinion flange bolts.

➡ **Do not use a hammer.**

25. Remove the driveshaft flange from the pinion flange, using a pry bar.

26. Index-mark the driveshaft and the extension housing. Remove the driveshaft from the vehicle.

27. Remove the electrical connector cover.

28. Remove the starter wire and positive battery cable.

29. Remove the starter motor bolts and starter.

30. Remove the flexplate cover.

31. Remove the access cover.

➡ **Make an identifying mark on the nut, stud and flexplate to allow for correct installation.**

32. Remove the 4 torque converter nuts.

33. Lower the transmission to gain access to the sensor connectors and the transmission bolts.

34. Disconnect the right side catalyst monitor connector and the fuel lines from the bracket.

35. Disconnect the right side heated oxygen sensor from the transmission.

36. Disconnect the left side heated oxygen sensor and the harness clip from the transmission.

37. Disconnect the left side catalyst monitor connector from the transmission.

38. Move the rubber boot back to gain access to the connector. Disconnect the digital transmission range (TR) sensor connector.

39. Disconnect the turbine shaft speed (TSS) sensor, output shaft speed (OSS) sensor and intermediate shaft speed (ISS) sensor electrical connectors.

➡ **Clean the area around connector to prevent contamination of the solenoid body connector.**

40. Remove the screw from the solenoid body connector and disconnect the connector.

41. Disconnect the harness retainers.

❊❊ WARNING

Do not damage the cooler tubes.

42. Disconnect the transmission cooler tubes.

43. Remove the front crossmember.

44. Remove the 7 engine-to-transmission retaining bolts.

45. Lower the transmission assembly from the vehicle.

46. Install special tool 307-346 to hold the torque converter.

47. Carry out the transmission fluid cooler backflushing and cleaning if the transmission is being overhauled or installing a new or remanufactured transmission.

To install

❊❊ CAUTION

Secure the transmission to the transmission jack with a safety chain. Failure to follow these instructions can result in personal injury.

48. Secure the transmission to the transmission jack with a safety chain.

➡ **Rotate the torque converter so the orange or green paint daub is in the 12 o'clock position.**

49. Raise and position the transmission to the back of the engine.

50. Install the 7 engine-to-transmission retaining bolts. Tighten to 48 Nm (35 ft. lbs.).

51. Install the 4 torque converter nuts. Tighten to 38 Nm (28 ft. lbs.).

52. Install the access cover.

53. Install the flexplate cover. Tighten to 34 Nm (25 ft. lbs.).

54. Install the starter. Tighten to 24 Nm (18 ft. lbs.).

55. Install the starter motor electrical connectors. Tighten to 13 Nm (10 ft. lbs.).

56. Install the starter motor electrical connector cover.

57. Install the front crossmember. Tighten to 70 Nm (62 ft. lbs.).

❊❊ WARNING

Use care not to bend or force the cooler tubes otherwise damage to the cooler tubes and the transmission can result.

58. Install the transmission fluid cooler tubes. Tighten to 30 Nm (22 ft. lbs.).

59. Connect the digital transmission range (TR) sensor connector. Move the rubber boot back over the connector.

60. Connect the left side heated oxygen sensor (HO2S) to the transmission and install the wire harness retainer. Position the wire harness and install the retainers.

❊❊ WARNING

Damage will occur to the solenoid body assembly if the screw is tightened above the specification.

➡ **Always install new O-ring seals on the vehicle harness connector.**

➡ **Clean the area around connector to prevent contamination of the solenoid body connector.**

➡️**Use petroleum jelly to lubricate the O-ring seals to aid in the installation process.**

61. Install and lubricate new O-ring seals on the transmission connector and connect the connector. Tighten to 5 Nm (44 inch lbs.).

62. Connect the left side catalyst monitor connector to the transmission.

63. Connect the right side heated oxygen sensor (HO2S) to the transmission.

64. Connect the fuel lines and the right side catalyst monitor connector to the bracket.

65. Connect the turbine shaft speed (TSS) sensor, output shaft speed (OSS) sensor and intermediate shaft speed (ISS) sensor electrical connectors.

➡️**This step may require an assistant.**

66. Position the catalytic converter assembly with heat shield and loosely install the 4 converter-to-manifold nuts.

67. Install the rear transmission mount bolts. Tighten to 90 Nm (66 ft. lbs.).

68. Connect the 2 heated oxygen sensor electrical connectors.

69. Connect the 2 catalyst monitor sensor electrical connectors.

70. Position the crossmember in place and loosely install the 2 nuts to hold up the crossmember.

➡️**Using a transmission jack, raise the crossmember into place.**

71. Install the right side crossmember bolts. Tighten to 70 Nm (52 ft. lbs.).

72. Install the left side crossmember bolts. Tighten to 70 Nm (52 ft. lbs.).

73. Remove the transmission jack.

74. Install the 4 upper crossmember bolts (2 on each side). Tighten to 70 Nm (52 ft. lbs.).

75. Tighten the rear transmission mount nuts. Tighten to 90 Nm (66 ft. lbs.).

76. Install the cable bracket and 2 bolts. Tighten to 40 Nm (30 ft. lbs.).

77. Install the heat shield bolts. Tighten to 20 Nm (15 ft. lbs.).

78. Install the right side heat shield and 2 bolts. Tighten to 20 Nm (15 ft. lbs.).

79. Install the left side heat shield bolts. Tighten to 20 Nm (15 ft. lbs.).

80. Install the plastic shield on the right side of the crossmember near the fuel tank. Tighten to 9 Nm (80 inch lbs.).

81. Tighten the 4 converter-to-manifold nuts. Tighten to 40 Nm (30 ft. lbs.).

82. Using a new gasket install the converter-to-muffler pipe. Install the bolts, springs and flag nuts. Tighten to 40 Nm (30 ft. lbs.).

83. Align the marks made during removal. Install the front portion of the driveshaft into the transmission.

⁂ WARNING

If new bolts to retain the driveshaft to the axle are not available, coat the threads of the original bolts with Threadlock and Sealer TA-25 or equivalent meeting Ford specification WSK-M2G351-A5.

⁂ WARNING

The driveshaft flange fits tightly on the rear axle pinion flange pilot. To make sure that the driveshaft flange seats squarely on the pinion flange, tighten the bolts evenly in a cross pattern as shown.

84. Align the marks made during removal. Install the rear driveshaft and bolts. Tighten to 112 Nm (83 ft. lbs.).

85. Install the shift cable and bracket. Tighten to 40 Nm (30 ft. lbs.).

86. Use the following guidelines for installing the in-line transmission fluid filter:

   a. If the transmission was overhauled and the vehicle was equipped with an in-line fluid filter, install a new in-line fluid filter.

   b. If the transmission was overhauled and the vehicle was not equipped with an in-line fluid filter, install a new in-line fluid filter kit.

   c. If the transmission is being installed for a non-internal repair, do not install an in-line filter or filter kit.

   d. If installing a new or a Ford-authorized remanufactured transmission, install the in-line transmission fluid filter that is supplied.

   e. Prior to lowering the vehicle, install a new in-line transmission filter or a filter kit.

➡️**When the battery has been disconnected and reconnected, some abnormal drive symptoms can occur while the vehicle relearns its adaptive strategy. The customer needs to be notified that they can experience slightly different upshifts (either soft or firm) and that this is a temporary condition and will eventually return to normal operating condition.**

87. Connect the battery ground cable.

88. Position the fan shroud and install the bolts.

89. Carry out the fluid level check.

90. Verify that the shift cable is correctly adjusted.

91. Check the operation of the transmission and inspect for leaks.

4-WHEEL DRIVE

1. Before servicing the vehicle, refer to the Precautions Section.

➡️**When the battery has been disconnected and reconnected, some abnormal drive symptoms can occur while the vehicle relearns its adaptive strategy.**

2. Disconnect the battery ground cable.

3. Remove the fan shroud bolts. Do not remove the fan shroud.

4. Raise and support the vehicle.

5. If transmission disassembly or installation of new transmission is necessary, drain the transmission fluid. Install the drain plug when finished.

6. Remove the 2 right heat shield bolts and position the right heat shield aside.

7. Disconnect the 2 heated oxygen sensor electrical connectors.

8. Disconnect the 2 catalyst monitor sensor electrical connectors.

9. Remove the right heat shield bolts.

10. Remove the right heat shield bolt and remove the heat shield.

11. Remove the left heat shield bolt.

12. Remove the plastic shield on the right side of the crossmember near the fuel tank.

13. Remove the 4 upper crossmember bolts (2 on each side).

14. Remove the bolts, springs, flag nuts and separate the muffler from the converter. Discard the gasket. Discard the flag nuts.

15. Remove the 4 converter-to-manifold nuts.

16. Install a suitable jackstand under the transfer case.

17. Remove the right side crossmember bolts.

18. Remove the left side crossmember bolts.

19. Remove the transmission support insulator nuts and remove the crossmember.

➡️**This step may require an assistant.**

20. Remove the transmission mounted exhaust hanger and the 3-way converter system. Remove the bolt and converter assembly. If required, remove the 2nd bolt and mount.

21. Remove the shift cable and bracket. Position the manual lever out of park.

22. Index-mark the front axle pinion flange and the front driveshaft.

23. Index-mark the front output shaft

assembly and the front driveshaft constant velocity (CV) joint.

24. Remove and discard the bolts and washers.

25. Remove and discard the bolts and universal joint retainers.

✳✳ WARNING

Always disconnect the front driveshaft from the transfer case first. Otherwise, the weight of the driveshaft can pinch the boot between the shaft and the boot can and cause the boot to tear.

✳✳ WARNING

Tape the bearing cups to the driveshaft to prevent them from falling off of the spider.

➡**The front driveshaft CV joint is not repairable.**

26. Mark the location of the flanges. Remove the 4 bolts.

27. Remove the rear driveshaft bolts.

➡**Do not use a hammer.**

28. Remove the driveshaft from the flange using a pry bar.

29. Remove the electrical connector cover.

30. Remove the starter wire and the positive battery cable.

31. Remove the starter motor bolts and starter.

32. Remove the flexplate cover.

33. Remove the access cover.

34. Make an identifying mark on the nut, stud and flexplate to allow for correct installation. Remove the 4 torque converter nuts.

35. Lower the transmission to gain access to the sensor connectors and the transmission bolts.

36. Disconnect the right catalyst monitor connector and the fuel lines from the bracket.

37. Disconnect the right heated oxygen sensor from the transmission.

38. If equipped, disconnect the shift motor electrical connector.

39. Disconnect the left heated oxygen sensor and the harness clip from the transmission.

40. Disconnect the left catalyst monitor connector from the transmission.

41. Move the rubber boot back to gain access to the connector. Disconnect the digital transmission range (TR) sensor connector.

42. Disconnect the turbine shaft speed (TSS) sensor, output shaft speed (OSS) sensor and intermediate shaft speed (ISS) sensor electrical connectors.

➡**Clean the area around connector to prevent contamination of the solenoid body connector.**

43. Remove the screw from the solenoid body connector and disconnect the connector.

44. Disconnect the harness retainers.

✳✳ WARNING

Do not damage the cooler tubes. Hold the transmission case fittings with a wrench.

45. Disconnect the transmission cooler tubes.

✳✳ CAUTION

Secure the transmission to the transmission jack with a safety chain. Failure to follow these instructions can result in personal injury.

46. Support the transmission with a transmission jack.

47. Remove the jackstand from under the transfer case.

48. Remove the 4 front crossmember bolts and the crossmember.

49. Remove the 7 engine-to-transmission retaining bolts.

50. Lower the transmission and transfer case as an assembly from the vehicle.

51. Install special tool 307-346 to hold the converter.

➡**If the transmission is being serviced, the transfer case will need to be removed.**

52. Remove the transfer case from the back of the transmission.

53. Carry out the transmission fluid cooler backflushing and cleaning if the

307-346

06017-EXPL-G77

Torque converter holding tool

transmission is being overhauled or installing a new or remanufactured transmission.

To install:

✳✳ CAUTION

Secure the transmission to the transmission jack with a safety chain. Failure to follow these instructions can result in personal injury.

54. If removed, install the transfer case to the back of the transmission, install the bracket and connect the vent hose. Tighten to 41 Nm (30 ft. lbs.).

55. Remove the special tool.

✳✳ CAUTION

Make sure that the transmission and transfer case assembly are secured to the transmission jack with a safety chain. Failure to follow these instructions can result in personal injury.

➡**Rotate the torque converter so the orange or green paint daub is in the 12 o'clock position.**

56. Raise and position the transmission and transfer case to the back of the engine.

57. Install the 7 engine-to-transmission retaining bolts. Tighten to 48 Nm (35 ft. lbs.).

58. Install the 4 torque converter nuts. Tighten to 38 Nm (28 ft. lbs.).

59. Install the access cover.

60. Install the flexplate cover. Tighten to 34 Nm (25 ft. lbs.).

61. Install the starter. Tighten to 24 Nm (18 ft. lbs.).

62. Install the starter motor electrical connectors. Tighten to 13 Nm (10 ft. lbs.).

63. Install the starter motor electrical connector cover.

64. Install the front crossmember and the 4 bolts (2 on each side). Tighten to 70 Nm (52 ft. lbs.).

65. Install a jackstand under the transfer case.

66. Remove the transmission jack from under the transmission.

✳✳ WARNING

Use care not to bend or force the cooler tubes otherwise damage to the cooler tubes and the transmission can result.

67. Install the transmission fluid cooler tubes. Tighten to 30 Nm (22 ft. lbs.).

68. Connect the digital transmission range (TR) sensor connector. Move the rubber boot back over the connector.

69. Connect the left heated oxygen sensor (HO2S) to the transmission and install the wire harness retainer. Position the wire harness and install the retainers.

✳✳ WARNING

Damage will occur to the solenoid body assembly if the screw is tightened above the specification.

➤ **Always install new O-ring seals on the vehicle harness connector.**

➤ **Clean the area around connector to prevent contamination of the solenoid body connector.**

➤ **Use petroleum jelly to lubricate the O-ring seals to aid in the installation process.**

70. Install and lubricate new O-ring seals on the transmission connector and connect the connector. Tighten to 5 Nm (44 inch lbs.).
71. Connect the left catalyst monitor connector to the transmission.
72. Connect the right heated oxygen sensor (HO2S) to the transmission.
73. Connect the fuel lines and the right catalyst monitor connector to the bracket.
74. Connect the turbine shaft speed (TSS) sensor, output shaft speed (OSS) sensor and intermediate shaft speed (ISS) sensor electrical connectors.

➤ **This step may require an assistant.**

75. Position the catalytic converter assembly with heat shield and loosely install the 4 converter-to-manifold nuts.
76. Install the rear transmission mount bolts. Tighten to 90 Nm (66 ft. lbs.).
77. Connect the 2 heated oxygen sensor electrical connectors.
78. Connect the 2 catalyst monitor sensor electrical connectors.
79. Position the crossmember in place and loosely install the 2 nuts to hold up the crossmember.
80. Install the right side crossmember bolts. Tighten to 70 Nm (52 ft. lbs.).
81. Install the left side crossmember bolts. Tighten to 70 Nm (52 ft. lbs.).
82. Remove the jackstand.
83. Install the 4 upper crossmember bolts (2 on each side). Tighten to 70 Nm (52 ft. lbs.).
84. Tighten the rear transmission mount nuts to 90 Nm (66 ft. lbs.).
85. Install the right heat shield and bolts. Tighten to 20 Nm (15 ft. lbs.).
86. Install the left heat shield bolts. Tighten to 20 Nm (15 ft. lbs.).
87. Install the plastic shield on the right

06017-EXPL-G78

Proper can installation—2004–05 4WD models

side of the crossmember near the fuel tank. Tighten to 9 Nm (80 inch lbs.).
88. Tighten the 4 converter-to-manifold nuts. Tighten to 40 Nm (30 ft. lbs.).
89. Using a new gasket, install the converter-to-muffler pipe and the heat shield. Install the bolts, springs and flag nuts. Tighten to 40 Nm (30 ft. lbs.).
90. If equipped, connect the shift motor electrical connector.

✳✳ WARNING

The can (domed CV joint housing cover) is pressed into the CV joint housing at the factory. When housed correctly, the can will appear as shown in the illustration, top box. Do not reseat the can in the CV joint housing if the can's flange is above the CV joint housings shown in the illustration, bottom box.

91. Install a new driveshaft.
92. Verify that the can flange is housed correctly in the CV joint housing as shown in the illustration, top box. Install a new driveshaft if the can flange is not housed correctly.

✳✳ WARNING

Always connect the front driveshaft to the axle first. Otherwise, the weight of the driveshaft can pinch the boot between the shaft and the boot can and cause the boot to tear.

93. Align the index marks. Install the driveshaft.

✳✳ WARNING

Tighten the bolts evenly in a cross pattern or damage will occur to the CV joint.

➤ **Install new washers and bolts. If new bolts are not available, coat the threads of the original bolts with threadlock and sealer.**

06017-EXPL-G79

Driveshaft joint torque sequence— 2004–05 4WD Aviator models

94. Install the new washers and bolts. Tighten the bolts evenly in a cross pattern as shown. Tighten to 30 Nm (22 ft. lbs.).
95. Verify that the CV joint has seated squarely in the transfer case flange.
   a. Check that the space between the end of the CV joint cap and the end of the transfer case flange is equal from the top to the bottom of the 2 components.
   b. Rotate the driveshaft ¼ turn.
   c. Check that the space between the end of the CV joint cap and the end of the transfer case flange is equal from the top to the bottom of the 2 components.
   d. Repeat this procedure several times.
   e. If the space if not equal at any point in the attachment, remove and reinstall the driveshaft as described in this procedure.

➤ **Install new universal joint retainers and bolts. If new bolts are not available, coat the threads of the original bolts with threadlock and sealer.**

96. Align the index marks and install the new universal joint retainer and bolts. Tighten to 18 Nm (13 ft. lbs.).

✳✳ WARNING

If new bolts to retain the driveshaft to the axle are not available, coat the threads of the original bolts with Threadlock and Sealer TA-25 or equivalent meeting Ford specification WSK-M2G351-A5.

✳✳ WARNING

The driveshaft flange fits tightly on the rear axle pinion flange pilot. To make sure that the driveshaft flange seats squarely on the pinion flange, tighten the bolts evenly in a cross pattern as shown.

97. Install the rear driveshaft and the 8 bolts. Tighten to 112 Nm (83 ft. lbs.).

98. Install the shift cable and bracket. Tighten to 40 Nm (30 ft. lbs.).

99. Use the following guidelines for installing the in-line transmission fluid filter:

   a. If the transmission was overhauled and the vehicle was equipped with an in-line fluid filter, install a new in-line fluid filter.

   b. If the transmission was overhauled and the vehicle was not equipped with an in-line fluid filter, install a new in-line fluid filter kit.

   c. If the transmission is being installed for a non-internal repair, do not install an in-line filter or filter kit.

   d. If installing a new or a Ford-authorized remanufactured transmission, install the in-line transmission fluid filter that is supplied.

   e. Prior to lowering the vehicle, install a new in-line transmission filter or a filter kit.

➡ **When the battery has been disconnected and reconnected, some abnormal drive symptoms can occur while the vehicle relearns its adaptive strategy. The customer needs to be notified that they can experience slightly different upshifts (either soft or firm) and that this is a temporary condition and will eventually return to normal operating condition.**

100. Connect the battery ground cable.

101. Position the fan shroud and install the bolts.

102. Carry out the fluid level check.

103. Verify that the shift cable is correctly adjusted.

104. Check the operation of the transmission and inspect for leaks.

2004–5 Explorer and Mountaineer

2-WHEEL DRIVE

1. Before servicing the vehicle, refer to the Precautions Section.

➡ **When the battery has been disconnected and reconnected, some abnormal drive symptoms can occur while the vehicle relearns its adaptive strategy.**

2. Disconnect the battery ground cable.

3. Remove the fan shroud bolts. Do not remove the fan shroud.

4. Raise and support the vehicle.

5. If transmission disassembly or installation of new transmission is necessary, drain the transmission fluid. Install the drain plug when finished.

6. Remove the 2 right heat shield bolts and position the right heat shield aside.

7. Disconnect the 2 heated oxygen sensor electrical connectors.

8. Disconnect the 2 catalyst monitor sensor electrical connectors.

9. Remove the right heat shield bolts.

10. Remove the right heat shield bolt and remove the heat shield.

11. Remove the left heat shield bolt.

12. Remove the plastic shield on the right side of the crossmember near the fuel tank.

13. Remove the 2 upper crossmember bolts (2 on each side).

14. Remove the bolts, springs, flag nuts and separate the muffler from the converter. Discard the gasket.

15. Discard the flag nuts.

16. Remove the 4 converter-to-manifold nuts.

17. Install a suitable jackstand under the transfer case.

18. Remove the right side crossmember bolts.

19. Remove the left side crossmember bolts.

20. Remove the transmission support insulator nuts and remove the crossmember.

➡ **This step may require an assistant.**

21. Remove the transmission mounted exhaust hanger and the 3-way converter system.

22. Remove the shift cable and bracket.

23. Position the manual lever out of park.

24. Index-mark the front axle pinion flange and the front driveshaft.

25. Index-mark the front output shaft assembly and the front driveshaft constant velocity (CV) joint.

26. Remove and discard the bolts and washers.

27. Remove and discard the bolts and universal joint retainers.

✳✳ WARNING

Always disconnect the front driveshaft from the transfer case 1st. Otherwise, the weight of the driveshaft can pinch the boot between the shaft and the boot can and cause the boot to tear.

✳✳ WARNING

Tape the bearing cups to the driveshaft to prevent them from falling off of the spider.

➡ **The front driveshaft CV joint is not repairable.**

28. Remove the driveshaft.

29. Mark the location of the flanges. Remove the rear driveshaft bolts.

➡ **Do not use a hammer.**

30. Remove the driveshaft from the flange using a pry bar.

31. 4.6L engine

   a. Remove the electrical connector cover.

   b. Remove the starter wire and the positive battery cable.

   c. Remove the starter motor bolts and starter.

   d. Remove the flexplate cover.

   e. Remove the access cover.

➡ **Make an identifying mark on the nut, stud and flexplate to allow for correct installation.**

   f. Remove the 4 torque converter nuts.

32. 4.0L engine

   a. Remove the starter and position off to the side.

➡ **Make an identifying mark on the nut, stud and flexplate to allow for correct installation.**

   b. Remove the 4 torque converter nuts.

   c. Remove the lower transmission retaining bolts.

33. All vehicles

   a. Lower the transmission to gain access to the sensor connectors and the transmission bolts.

   b. Disconnect the right catalyst monitor connector and the fuel lines from the bracket.

   c. Disconnect the right heated oxygen sensor from the transmission.

   d. If equipped, disconnect the shift motor electrical connector.

   e. Disconnect the left heated oxygen sensor and the harness clip from the transmission.

   f. Disconnect the left catalyst monitor connector from the transmission.

   g. Move the rubber boot back to gain access to the connector.

   h. Disconnect the digital transmission range (TR) sensor connector.

   i. Disconnect the turbine shaft speed (TSS) sensor, output shaft speed (OSS) sensor and intermediate shaft speed (ISS) sensor electrical connectors.

➡ **Clean the area around connector to prevent contamination of the solenoid body connector.**

j. Remove the screw from the solenoid body connector and disconnect the connector.

k. Disconnect the harness retainers.

✳✳ WARNING

Do not damage the cooler tubes. Hold the transmission case fittings with a wrench.

l. Disconnect the transmission cooler tubes.

✳✳ CAUTION

Secure the transmission to the transmission jack with a safety chain. Failure to follow these instructions can result in personal injury.

m. Support the transmission with a transmission jack.

n. Remove the jackstand from under the transfer case.

34. 4.6L engine

a. Remove the 4 front crossmember bolts and the crossmember.

b. Remove the 7 engine-to-transmission retaining bolts.

35. 4.0L engine

a. Remove the 6 engine-to-transmission retaining bolts and position the fuel line bracket aside.

36. All vehicles

a. Lower the transmission and transfer case as an assembly from the vehicle.

b. Install a converter holding tool.

➡ **If the transmission is being serviced, the transfer case will need to be removed.**

c. Remove the transfer case from the back of the transmission.

d. Carry out the transmission fluid cooler backflushing and cleaning if the transmission is being overhauled or installing a new or remanufactured transmission.

To install:

✳✳ CAUTION

Secure the transmission to the transmission jack with a safety chain. Failure to follow these instructions can result in personal injury.

37. Secure the transmission to the transmission jack with a safety chain.

➡ **Rotate the torque converter so that the torque converter paint mark is in the 12 o'clock position.**

38. Raise and position the transmission to the back of the engine.

39. 4.6L engine

a. Install the 7 engine-to-transmission retaining bolts. Tighten to 48 Nm (35 ft. lbs.).

b. Install the 4 torque converter nuts. Tighten to 38 Nm (28 ft. lbs.).

c. Install the access cover.

d. Install the flexplate cover. Tighten to 34 Nm (25 ft. lbs.).

e. Install the starter. Tighten to 24 Nm (18 ft. lbs.).

f. Install the starter motor electrical connectors. Tighten the cable to 13 Nm (10 ft. lbs.). Tighten the wire to 6 Nm (53 inch lbs.).

g. Install the starter motor electrical connector cover.

h. Install the front crossmember. Tighten to 70 Nm (52 ft. lbs.).

40. 4.0L engine

a. Position the fuel line bracket in place and install the 6 retaining bolts. Tighten to 48 Nm (35 ft. lbs.).

b. Install the 4 torque converter nuts. Tighten to 38 Nm (28 ft. lbs.).

c. Install the starter motor. Tighten to 25 Nm (18 ft. lbs.).

d. Install the lower transmission retaining bolts. Tighten to 48 Nm (35 ft. lbs.).

41. All vehicles

✳✳ WARNING

Use care not to bend or force the cooler tubes otherwise damage to the cooler tubes and the transmission can result.

a. Install the transmission fluid cooler tubes. Tighten to 30 Nm (22 ft. lbs.).

b. Connect the digital transmission range (TR) sensor connector. Move the rubber boot back over the connector.

c. Connect the left heated oxygen sensor (HO2S) to the transmission and install the wire harness retainer. Position the wire harness and install the retainers.

✳✳ WARNING

Damage will occur to the solenoid body assembly if the screw is tightened above the specification.

➡ **Always install new O-ring seals on the vehicle harness connector.**

➡ **Clean the area around connector to prevent contamination of the solenoid body connector.**

➡ **Use petroleum jelly to lubricate the O-ring seals to aid in the installation process.**

d. Install and lubricate new O-ring seals on the transmission connector and connect the connector. Tighten to 5 Nm (44 inch lbs.).

e. Connect the left catalyst monitor connector to the transmission.

f. Connect the right heated oxygen sensor (HO2S) to the transmission.

g. Connect the fuel lines and the right catalyst monitor connector to the bracket.

h. Connect the turbine shaft speed (TSS) sensor, output shaft speed (OSS) sensor, and intermediate shaft speed (ISS) sensor electrical connectors.

➡ **This step may require an assistant.**

i. Position the catalytic converter assembly with heat shield and loosely install the 4 converter-to-manifold nuts.

j. Install the rear transmission mount bolts. Tighten to 90 Nm (66 ft. lbs.).

k. Connect the 2 heated oxygen sensor electrical connectors.

l. Connect the 2 catalyst monitor sensor electrical connectors.

m. Position the crossmember in place and loosely install the 2 nuts to hold up the crossmember.

➡ **Using a transmission jack, raise the crossmember into place.**

n. Install the right side crossmember bolts. Tighten to 70 Nm (52 ft. lbs.).

o. Install the left side crossmember bolts. Tighten to 70 Nm (52 ft. lbs.).

p. Remove the transmission jack.

q. Install the 2 upper crossmember bolts (2 on each side). Tighten to 70 Nm (52 ft. lbs.).

r. Tighten the rear transmission mount nuts. Tighten to 90 Nm (66 ft. lbs.).

s. Install the cable bracket and 2 bolts. Tighten to 40 Nm (30 ft. lbs.).

t. Install the heat shield bolts. Tighten to 20 Nm (15 ft. lbs.).

u. Install the right heat shield and 3 bolts. Tighten to 20 Nm (15 ft. lbs.).

v. Install the left heat shield bolts. Tighten to 20 Nm (15 ft. lbs.).

w. Install the plastic shield on the right side of the crossmember near the fuel tank. Tighten to 9 Nm (80 inch lbs.).

x. Tighten the 4 converter-to-manifold nuts. Tighten to 40 Nm (30 ft. lbs.).

y. Using a new gasket install the converter-to-muffler pipe. Install the bolts, springs and flag nuts. Tighten to 40 Nm (30 ft. lbs.).

112 Nm (83 lb-ft)

06017-EXPL-G80

Rear driveshaft flange torque sequence—2004–05 Explorer and Mountaineer

z. Align the marks made during removal. Install the front portion of the driveshaft into the transmission.

✳✳ WARNING

If new bolts to retain the driveshaft to the axle are not available, coat the threads of the original bolts with Threadlock and Sealer TA-25 or equivalent meeting Ford specification WSK-M2G351-A5.

✳✳ WARNING

The driveshaft flange fits tightly on the rear axle pinion flange pilot. To make sure that the driveshaft flange seats squarely on the pinion flange, tighten the bolts evenly in a cross pattern as shown.

aa. Align the marks made during removal. Install the rear driveshaft and bolts. Tighten to 112 Nm (83 ft. lbs.).

bb. Install the shift cable and bracket. Tighten to 40 Nm (30 ft. lbs.).

42. Use the following guidelines for installing the in-line transmission fluid filter:

a. If the transmission was overhauled and the vehicle was equipped with an in-line fluid filter, install a new in-line fluid filter.

b. If the transmission was overhauled and the vehicle was not equipped with an in-line fluid filter, install a new in-line fluid filter kit.

c. If the transmission is being installed for a non-internal repair, do not install an in-line filter or filter kit.

d. If installing a new or a Ford-authorized remanufactured transmission, install the in-line transmission fluid filter that is supplied.

e. Prior to lowering the vehicle, install a new in-line transmission filter or a filter kit.

➡When the battery has been disconnected and reconnected, some abnormal drive symptoms can occur while the vehicle relearns its adaptive strategy. The customer needs to be notified that they can experience slightly different upshifts (either soft or firm) and that this is a temporary condition and will eventually return to normal operating condition.

f. Connect the battery ground cable.

g. Position the fan shroud and install the bolts.

h. Carry out the fluid level check.

i. Verify that the shift cable is correctly adjusted. Check the operation of the transmission and inspect for leaks.

4-WHEEL DRIVE

1. Before servicing the vehicle, refer to the Precautions Section.

➡When the battery has been disconnected and reconnected, some abnormal drive symptoms can occur while the vehicle relearns its adaptive strategy.

2. Disconnect the battery ground cable.

3. Remove the fan shroud bolts. Do not remove the fan shroud.

4. Raise and support the vehicle.

5. If transmission disassembly or installation of new transmission is necessary, drain the transmission fluid. Install the drain plug when finished.

6. Remove the 2 right heat shield bolts and position the right heat shield aside.

7. Disconnect the 2 heated oxygen sensor electrical connectors.

8. Disconnect the 2 catalyst monitor sensor electrical connectors.

9. Remove the right heat shield bolts.

10. Remove the right heat shield bolt and remove the heat shield.

11. Remove the left heat shield bolt.

12. Remove the plastic shield on the right side of the crossmember near the fuel tank.

13. Remove the 2 upper crossmember bolts (2 on each side).

14. Remove the bolts, springs, flag nuts and separate the muffler from the converter. Discard the gasket. Discard the flag nuts.

15. Remove the 4 converter-to-manifold nuts.

16. Install a suitable jackstand under the transfer case.

17. Remove the right side crossmember bolts.

18. Remove the left side crossmember bolts.

19. Remove the transmission support insulator nuts and remove the crossmember.

➡This step may require an assistant.

20. Remove the transmission mounted exhaust hanger and the 3-way converter system.

21. Remove the shift cable and bracket. Position the manual lever out of park.

22. Index-mark the front axle pinion flange and the front driveshaft.

23. Index-mark the front output shaft assembly and the front driveshaft constant velocity (CV) joint.

24. Remove and discard the bolts and washers.

25. Remove and discard the bolts and universal joint retainers.

✳✳ WARNING

Always disconnect the front driveshaft from the transfer case 1st. Otherwise, the weight of the driveshaft can pinch the boot between the shaft and the boot can and cause the boot to tear.

✳✳ WARNING

Tape the bearing cups to the driveshaft to prevent them from falling off of the spider.

➡The front driveshaft CV joint is not repairable.

26. Remove the driveshaft.

➡The pinion flange is shown, transfer case flange is similar.

27. Mark the location of the flanges. Remove the rear driveshaft bolts.

➡Do not use a hammer.

28. Remove the driveshaft from the flange using a pry bar.

29. 4.6L engine

a. Remove the electrical connector cover.

b. Remove the starter wire and the positive battery cable.

c. Remove the starter motor bolts and starter.

d. Remove the flexplate cover.

e. Remove the access cover.

➡Make an identifying mark on the nut, stud and flexplate to allow for correct installation.

f. Remove the 4 torque converter nuts.

30. 4.0L engine

a. Remove the starter and position off to the side.

➡ **Make an identifying mark on the nut, stud and flexplate to allow for correct installation.**

b. Remove the 4 torque converter nuts.

c. Remove the lower transmission retaining bolts.

31. All vehicles

a. Lower the transmission to gain access to the sensor connectors and the transmission bolts.

b. Disconnect the right catalyst monitor connector and the fuel lines from the bracket.

c. Disconnect the right heated oxygen sensor from the transmission.

d. If equipped, disconnect the shift motor electrical connector.

e. Disconnect the left heated oxygen sensor and the harness clip from the transmission.

f. Disconnect the left catalyst monitor connector from the transmission. Move the rubber boot back to gain access to the connector.

g. Disconnect the digital transmission range (TR) sensor connector.

h. Disconnect the turbine shaft speed (TSS) sensor, output shaft speed (OSS) sensor and intermediate shaft speed (ISS) sensor electrical connectors.

➡ **Clean the area around connector to prevent contamination of the solenoid body connector.**

i. Remove the screw from the solenoid body connector and disconnect the connector.

j. Disconnect the harness retainers.

✳✳ **WARNING**

Do not damage the cooler tubes. Hold the transmission case fittings with a wrench.

k. Disconnect the transmission cooler tubes.

✳✳ **CAUTION**

Secure the transmission to the transmission jack with a safety chain. Failure to follow these instructions can result in personal injury.

l. Support the transmission with a transmission jack.

m. Remove the jackstand from under the transfer case.

32. 4.6L engine

a. Remove the 4 front crossmember bolts and the crossmember.

b. Remove the 7 engine-to-transmission retaining bolts.

33. 4.0L engine

a. Remove the 6 engine-to-transmission retaining bolts and position the fuel line bracket aside.

34. All vehicles

a. Lower the transmission and transfer case as an assembly from the vehicle. Install a converter holding tool.

➡ **If the transmission is being serviced, the transfer case will need to be removed.**

b. Remove the transfer case from the back of the transmission.

c. Carry out the transmission fluid cooler backflushing and cleaning if the transmission is being overhauled or installing a new or remanufactured transmission

To install:

✳✳ **CAUTION**

Secure the transmission to the transmission jack with a safety chain. Failure to follow these instructions can result in personal injury.

35. If removed, install the transfer case to the back of the transmission, install the bracket and connect the vent hose. Tighten to 41 Nm (30 ft. lbs.).

36. Remove the special tool.

✳✳ **CAUTION**

Make sure that the transmission and transfer case assembly are secured to the transmission jack with a safety chain. Failure to follow these instructions can result in personal injury.

➡ **Rotate the torque converter so that the torque converter paint mark is in the 12 o'clock position.**

37. Raise and position the transmission and transfer case to the back of the engine.

38. 4.6L engine

a. Install the 7 engine-to-transmission retaining bolts. Tighten to 48 Nm (35 ft. lbs.).

b. Install the 4 torque converter nuts. Tighten to 35 Nm (26 ft. lbs.).

c. Install the access cover.

d. Install the flexplate cover. Tighten to 34 Nm (25 ft. lbs.).

e. Install the starter. Tighten to 24 Nm (18 ft. lbs.).

f. Install the starter motor electrical connectors. Tighten the cable to 13 Nm (10 ft. lbs.). Tighten the wire to 6 Nm (53 inch lbs.).

g. Install the starter motor electrical connector cover.

➡ **right side shown, left side similar.**

h. Install the front crossmember and the 4 bolts (2 on each side). Tighten to 70 Nm (52 ft. lbs.).

39. 4.0L engine

a. Position the fuel line bracket in place and install the 6 engine-to-transmission retaining bolts. Tighten to 48 Nm (35 ft. lbs.).

b. Install the 4 torque converter nuts. Tighten to 35 Nm (26 ft. lbs.).

c. Install the starter motor. Tighten to 25 Nm (18 ft. lbs.).

d. Install the lower transmission retaining bolts. Tighten to 48 Nm (35 ft. lbs.).

40. All vehicles

a. Install a jackstand under the transfer case.

b. Remove the transmission jack from under the transmission.

✳✳ **WARNING**

Use care not to bend or force the cooler tubes otherwise damage to the cooler tubes and the transmission can result.

c. Install the transmission fluid cooler tubes. Tighten to 30 Nm (22 ft. lbs.).

d. Connect the digital transmission range (TR) sensor connector. Move the rubber boot back over the connector.

e. Connect the left heated oxygen sensor (HO2S) to the transmission and install the wire harness retainer. Position the wire harness and install the retainers.

✳✳ **WARNING**

Damage will occur to the solenoid body assembly if the screw is tightened above the specification.

➡ **Always install new O-ring seals on the vehicle harness connector.**

➡ **Clean the area around connector to prevent contamination of the solenoid body connector.**

➡ **Use petroleum jelly to lubricate the O-ring seals to aid in the installation process.**

f. Install and lubricate new O-ring seals on the transmission connector and connect the connector. Tighten to 5 Nm (44 inch lbs.).

g. Connect the left catalyst monitor connector to the transmission.

h. Connect the right heated oxygen sensor (HO2S) to the transmission.

i. Connect the fuel lines and the right catalyst monitor connector to the bracket.

j. Connect the turbine shaft speed (TSS) sensor, output shaft speed (OSS) sensor, and intermediate shaft speed (ISS) sensor electrical connectors.

➡ **This step may require an assistant.**

k. Position the catalytic converter assembly with heat shield and loosely install the 4 converter-to-manifold nuts.

l. Position the rear transmission mount and exhaust hanger. Install the rear mount bolts. Tighten to 90 Nm (66 ft. lbs.).

m. Connect the 2 heated oxygen sensor electrical connectors.

n. Connect the 2 catalyst monitor sensor electrical connectors.

o. Position the crossmember in place and loosely install the 2 nuts to hold up the crossmember.

p. Install the right side crossmember bolts. Tighten to 70 Nm (52 ft. lbs.).

q. Install the left side crossmember bolts. Tighten to 70 Nm (52 ft. lbs.).

r. Remove the jackstand.

s. Install the 2 upper crossmember bolts (2 on each side). Tighten to 70 Nm (52 ft. lbs.).

t. Tighten the rear transmission mount nuts. Tighten to 90 Nm (66 ft. lbs.).

u. Install the right heat shield and bolts. Tighten to 20 Nm (15 ft. lbs.).

v. Install the left heat shield bolts. Tighten to 20 Nm (15 ft. lbs.).

w. Install the plastic shield on the right side of the crossmember near the fuel tank. Tighten to 9 Nm (80 inch lbs.).

x. Tighten the 4 converter-to-manifold nuts. Tighten to 40 Nm (30 ft. lbs.).

y. Using a new gasket, install the converter-to-muffler pipe and the heat shield. Install the bolts, springs and flag nuts. Tighten to 40 Nm (30 ft. lbs.).

z. If equipped, connect the shift motor electrical connector.

❊❊ WARNING

The can (domed CV joint housing cover) is pressed into the CV joint housing at the factory. When housed correctly, the can will appear as shown in the cut-away illustration, top box. Do not reseat the can in the CV joint housing if the can's flange is above the CV joint housings shown in the cut-away illustration, bottom box. Install a new driveshaft.

aa. Verify that the can flange is housed correctly in the CV joint housing as shown in the cut-away illustration, top box. Install a new driveshaft if the can flange is not housed correctly.

❊❊ WARNING

Always connect the front driveshaft to the axle 1st. Otherwise, the weight of the driveshaft can pinch the boot between the shaft and the boot can and cause the boot to tear.

➡ **Align the index marks.**

bb. Install the driveshaft.

❊❊ WARNING

Tighten the bolts evenly in a cross pattern or damage will occur to the CV joint.

➡ **Install new washers and bolts. If new bolts are not available, coat the threads of the original bolts with threadlock and sealer.**

cc. Install the new washers and bolts. Tighten the bolts evenly in a cross pattern. Tighten to 30 Nm (22 ft. lbs.).

dd. Verify that the CV joint has seated squarely in the transfer case flange.

ee. Check that the space between the end of the CV joint cap and the end of the transfer case flange is equal from the top to the bottom of the 2 components.

ff. Rotate the driveshaft ¼ turn.

gg. Check that the space between the end of the CV joint cap and the end of the transfer case flange is equal from the top to the bottom of the 2 components.

hh. Repeat this procedure several times.

ii. If the space if not equal at any point in the attachment, remove and reinstall the driveshaft as described in this procedure.

➡ **Install new universal joint retainers and bolts. If new bolts are not available, coat the threads of the original bolts with threadlock and sealer.**

jj. Install the new universal joint retainer and bolts. Tighten to 18 Nm (13 ft. lbs.).

❊❊ WARNING

If new bolts to retain the driveshaft to the axle are not available, coat the threads of the original bolts with Threadlock and Sealer TA-25 or equivalent meeting Ford specification WSK-M2G351-A5.

❊❊ WARNING

The driveshaft flange fits tightly on the rear axle pinion flange pilot. To make sure that the driveshaft flange seats squarely on the pinion flange, tighten the bolts evenly in a cross pattern as shown.

kk. Install the rear driveshaft and the 8 bolts. Tighten to 112 Nm (83 ft. lbs.).

ll. Install the shift cable and bracket. Tighten to 40 Nm (30 ft. lbs.).

41. Use the following guidelines for installing the in-line transmission fluid filter:

a. If the transmission was overhauled and the vehicle was equipped with an in-line fluid filter, install a new in-line fluid filter.

b. If the transmission was overhauled and the vehicle was not equipped with an in-line fluid filter, install a new in-line fluid filter kit.

c. If the transmission is being installed for a non-internal repair, do not install an in-line filter or filter kit.

d. If installing a new or a Ford-authorized remanufactured transmission, install the in-line transmission fluid filter that is supplied.

e. Prior to lowering the vehicle, install a new in-line transmission filter or a filter kit.

➡ **When the battery has been disconnected and reconnected, some abnormal drive symptoms can occur while the vehicle relearns its adaptive strategy. The customer needs to be notified that they can experience slightly different upshifts (either soft or firm) and that this is a temporary condition and will eventually return to normal operating condition.**

f. Connect the battery ground cable.

g. Position the fan shroud and install the bolts.

h. Carry out the fluid level check.

i. Verify that the shift cable is correctly adjusted.

j. Check the operation of the transmission and inspect for leaks.

Clutch

REMOVAL & INSTALLATION

1. Before servicing the vehicle, refer to the Precautions Section.
2. Remove or disconnect the following:
   - Negative battery cable
   - Transmission

➡ **If the parts are to be reused, index-mark the clutch pressure plate to the flywheel.**

   - Bolts, clutch pressure plate and the clutch disc

To install:

3. Lubricate the transmission input shaft pilot bearing with grease.
4. Adjust the clutch pressure plate:
   a. Using a suitable press, press downward on the fingers until the adjusting ring moves freely.
   b. Rotate the adjusting ring counterclockwise to compress the tension springs. Hold the adjusting ring in this position.
   c. Release the pressure on the fingers. The adjusting ring will stay in the reset position.
5. Position the clutch disc on the flywheel.
6. Align the clutch disc and the clutch

Tighten the bolts gradually in the correct sequence to avoid warping the pressure plate

pressure plate. Install the bolts and tighten in a star pattern sequence. Torque to 21 ft. lbs. (28Nm).

➡ **An "L" is stamped by three bolt holes on the clutch pressure plate. Tighten these to specification first to ensure correct clutch alignment.**

7. Install the transmission.

ADJUSTMENT

Because the clutch is hydraulically driven, there is no adjustment required.

In the event the clutch pedal develops a squeak or uneven feel when depressing, spray the pedal bushing assembly with penetrating oil and work the pedal back-and-forth.

Hydraulic Clutch System

BLEEDING

The following procedure is recommended for bleeding the clutch hydraulic system installed on the vehicle. It is recommended that the original clutch tube, with quick-connect fitting be replaced when servicing the hydraulic system, because air can be trapped in the quick-connect fitting and prevent complete bleeding of the system. The replacement tube does not include a quick-connect fitting.

1. Before servicing the vehicle, refer to the Precautions Section.
2. Clean the dirt and grease from the dust cap.
3. Remove the cap and diaphragm and fill the reservoir to the top with approved brake fluid C6AZ-19542-AA or BA, (ESA-M6C25-A).

➡ **To keep brake fluid from entering the clutch housing, route a suitable rubber tube of appropriate inside diameter from the bleed screw to a container.**

4. Loosen the bleed screw, located in the slave cylinder body, next to the inlet connection. Fluid will now begin to move from the master cylinder down the tube to the slave cylinder.

➡ **The reservoir must be kept full at all times during the bleeding operation, to ensure no additional air enters the system.**

5. Observe the bleed screw outlet. When the slave cylinder is full, a steady stream of fluid will flow from the outlet port. Tighten the bleed screw.
6. Depress the clutch pedal to the floor and hold for 1–2 seconds. Release the pedal as rapidly as possible. The pedal must be released completely. Pause for 1–2 seconds. Repeat 10 times.
7. Check the fluid level in the reservoir. The fluid should be level with the step when the diaphragm is removed.
8. Hold the pedal to the floor, slightly open the bleed screw to allow any additional air to escape. Close the bleed screw, then release the pedal.
9. Check the fluid in the reservoir. The hydraulic system should now be fully bled, and should actuate the clutch.
10. Check the vehicle by starting, push-

Clutch disc, pressure plate and bearing assembly

ing the clutch pedal to the floor and selecting reverse gear. There should be no grating of gears. If there is, and the hydraulic system still contains air; repeat the bleeding procedure.

REMOVAL & INSTALLATION

Explorer and Mountaineer

MASTER CYLINDER

✳✳ CAUTION

Brake fluid contains polyglycol ethers and polyglycols. Avoid contact with eyes. Wash hands thoroughly after handling. If brake fluid contacts eyes, flush eyes with running water for 15 minutes. Get medical attention if irritation persists. If taken internally, drink water and induce vomiting. Get medical attention immediately.

✳✳ WARNING

Brake fluid is harmful to painted or plastic surfaces. If the brake fluid is spilled onto a painted or plastic surface, wash it immediately with water.

1. Before servicing the vehicle, refer to the Precautions Section.
2. Partially drain the brake and clutch reservoir. Drain to just below the clutch reservoir nipple.
3. Remove the hose at the brake and clutch reservoir.
4. Remove the clip and washer from the clutch pedal pin.
5. Remove the two bolts, then disconnect the clutch master cylinder from the clutch pedal.
6. Pull the clutch master cylinder forward until the line clip is accessible. Remove the clip and disconnect the line.
7. Disconnect the clutch master cylinder rod from the clutch pedal arm. Remove the clutch master cylinder from the vehicle.
8. Remove the reservoir line from the clutch master cylinder.
9. To install, reverse the removal procedure. Torque the master cylinder bolts to 84 inch lbs. (10 Nm). Install new hose clamps on the reservoir line. Refill the brake master cylinder with the specified fluid. Bleed the clutch hydraulic system.

SLAVE CYLINDER

1. Before servicing the vehicle, refer to the Precautions Section.
2. Remove the transmission.

➡**Inspect the clutch housing for traces of fluid. If brake fluid is visible, install a new clutch slave cylinder.**

3. Remove the bolts. Remove the clutch slave cylinder.
4. Inspect the clutch slave cylinder for the following:
   • Worn or damaged boot.
   • Leaking fluid.
   • Contract and return smoothly.

✳✳ CAUTION

The clutch release hub and bearing is pre-lubricated and should not be cleaned with solvent. The clutch release hub and bearing are replaced as an assembly. Do not disassemble for inspection or replacement.

5. Inspect the clutch release bearing as follows:
   • Rotate the outer race while applying pressure. If the bearing rotation is rough, install a new clutch slave cylinder.
   • Inspect for wear or damage.
   • If wear pattern is continuous (not segmented) and the wear appears to be deep in bearing face, install a new clutch slave cylinder.
6. To install, reverse the removal procedure. Position the clutch slave cylinder to the transmission and install the bolts. Torque to 9 ft. lbs. (12 Nm). Bleed the clutch hydraulic system.

Explorer Sport and Sport-Trac

MASTER CYLINDER

1. Before servicing the vehicle, refer to the Precautions Section.
2. Raise and support the vehicle.
3. Using special tool 308-182, disconnect the hydraulic line from the clutch slave cylinder.
4. Lower the vehicle.
5. Disconnect the clutch pedal position (CPP) switch electrical connector.
6. Disconnect the clutch master cylinder rod from the clutch pedal arm.
7. Remove and discard the clutch master cylinder push rod bushing.
8. Remove the CPP switch.
   a. Pinch the lock tabs and pull the clip forward.
   b. Rotate and snap the CPP switch from the clutch master cylinder rod.
9. Remove the reservoir.
10. Remove the three pushpins, four bolts, four screws and the inner fender splash shield.

11. Unclip the hydraulic line-to-frame retainer.

✳✳ CAUTION

Brake fluid is harmful to painted and plastic surfaces. If brake fluid is spilled onto a painted or plastic surface, wash the surface with water immediately.

12. Unlock and remove the clutch master cylinder assembly.
   a. Twist the clutch master cylinder clockwise 45 degrees to unlock it from the support bracket.
   b. Remove the reservoir, the hydraulic line and the clutch master cylinder as an assembly.
   c. Install a new clutch master cylinder push rod bushing.

✳✳ CAUTION

Brake fluid is harmful to painted and plastic surfaces. If brake fluid is spilled onto a painted or plastic surface, wash the surface with water immediately.

14. To install, reverse the removal procedure. Torque the master cylinder bolts to 18 inch lbs. (2.2 Nm). Bleed the clutch hydraulic system.

SLAVE CYLINDER

1. Before servicing the vehicle, refer to the Precautions Section.
2. Remove the transmission.

➡**Inspect the clutch housing for traces of fluid. If brake fluid is visible, install a new clutch slave cylinder.**

3. Remove the bolts. Remove the clutch slave cylinder.
4. Inspect the clutch slave cylinder for the following:
   a. Worn or damaged boot.
   b. Leaking fluid.
   c. Worn or damaged bearing.
5. Inspect the clutch release bearing as follows:
   a. Rotate the outer race while applying pressure. If the bearing rotation is rough, install a new clutch slave cylinder.
   b. Inspect for wear or damage.
6. To install, reverse the removal procedure. Position the clutch slave cylinder to the transmission and install the bolts. Torque to 15 ft. lbs. (20 Nm).
7. Install the transmission.
8. Depress and release the clutch pedal five full strokes.

Transfer Case Assembly

REMOVAL & INSTALLATION

Automatic Shift

1. Before servicing the vehicle, refer to the Precautions Section.
2. Place the transmission in **Neutral**.
3. Remove or disconnect the following:
   - Negative battery cable
   - Skid plate

➡ Drain the transfer case if disassembly is necessary.

➡ Match-mark the front and rear driveshaft yokes and pinion flange and the rear driveshaft yoke and rear output flange.

   - Rear driveshaft
   - Front driveshaft
   - Vent tube
   - Shift motor electrical connector

4. Using a suitable high lift jack, support the transfer case.
5. Remove or disconnect the following:

   - right crossmember cover, then the four bolts
   - The four left crossmember bolts
   - Heat shields from the crossmember
   - Transmission mount nuts
   - The seven bolts and separate the transfer case from the extension housing

6. Lower the transfer case from the vehicle.

7. Remove and discard the transfer case-to-extension housing gasket. Clean the gasket surfaces.

1 Vent hose

2 Crossmember bolt (8 required) Installation Note

3 Crossmember-to-transmission insulator nut (2 required)

4 Crossmember Removal Note

5 Transmission insulator-to-transfer case bolts (2 required)

6 Transmission insulator

7 Transfer case-to-transmission bolts (7 required)

8 Transfer case Removal Note

9 Transfer case-to-transmission gasket

06017-EXPL-G81

Transfer case mounting

8. To install, reverse the removal procedure. Use a new transfer case gasket.

9. Torque the transfer case-to-extension housing bolts to 30 ft. lbs. (40Nm).

10. Transmission mount bolts: 66 ft. lbs. (90Nm)

11. Crossmember bolts/nuts: 72 ft. lbs. (98Nm).

All Wheel Drive

1. Before servicing the vehicle, refer to the Precautions Section.

2. Place the transmission in **Neutral**.

3. Disconnect the negative battery cable.

4. Matchmark the front and rear driveshafts.

5. If transfer case disassembly is necessary, remove the drain plug and drain the fluid.

6. Remove or disconnect the following:

✳✳ WARNING

Always disconnect the front driveshaft from the transfer case first. Otherwise, the weight of the driveshaft can pinch the boot between the shaft and the boot can and cause the boot to tear.

- Front driveshaft
- Rear driveshaft
- Skid plate
- Hose from the vent
- The two bolts retaining the heat shield to the crossmember. Loosen, but do not remove, the two nuts retaining the transmission mount to the crossmember.
- The three bolts and the heat shield

7. Position a high lift jack under the transfer case.

8. Remove or disconnect the following:

- The four upper nuts and bolts retaining the crossmember to the frame
- The two lower bolts retaining the crossmember to the frame
- The two nuts retaining the transmission mount to the crossmember
- The crossmember
- The two bolts, the transmission mount, and the exhaust hanger

➡**Support the transmission.**

- The seven bolts retaining the transfer case to the transmission

9. Separate the transfer case from the extension housing and the output shaft.

10. Lower the transfer case from the vehicle.

11. Remove and discard the gasket, and clean the mating surfaces.

12. To install, reverse the removal procedure. Note the following:

- Install a new gasket on the transfer case.
- Watch for obstructions when installing the transfer case in the vehicle.
- Install the original transfer case to transmission bolts. If new bolts are required, be sure to install only the specific Ford transfer case to transmission bolts.
- Transfer case-to-extension: 30 ft. lbs. (40Nm)
- Transmission mount bolts: 66 ft. lbs. (90Nm)
- Crossmember bolts/nuts: 72 ft. lbs. (98Nm).
- Heat shield: 18 ft. lbs. (25Nm)

Driveshafts

REMOVAL & INSTALLATION

Front Shaft

1. Before servicing the vehicle, refer to the Precautions Section.

30 Nm (22 lb-ft)

18 Nm (13 lb-ft)

1 CV joint bolt
  Installation Note

2 CV joint
  Installation Note

3 U-joint bolts
  Installation Note

4 U-joint straps
  Installation Note

5 Driveshaft

6 U-joint snap ring
  Installation Note

7 U-joint spider
  Removal Note

06017-EXPL-G82

Front driveshaft exploded view

CV-joint installation—front driveshaft

06017-EXPL-G83

30 Nm (22 lb-ft)

06017-EXPL-G84

Front driveshaft CV-joint torque sequence

112 Nm (83 lb-ft)

1	U-joint flange bolt Installation Note	5	U-joint spider
2	Driveshaft flange yoke Removal Note	6	U-joint snap ring
3	Driveshaft	7	U-joint spider
4	U-joint snap ring	8	Driveshaft slip-yoke

06017-EXPL-G85

Rear driveshaft exploded view—2-wheel drive

■ **WARNING**

Always disconnect the front drive-shaft from the transfer case first. Otherwise, the weight of the driveshaft can pinch the boot between the drive-shaft and the constant velocity (CV) joint flange which can cause the boot to tear.

2. With the vehicle in NEUTRAL, position it on a hoist.

3. Index-mark the front driveshaft to maintain driveshaft balance.

4. Remove and discard the 6 CV joint bolts and CV joint washers.

5. Remove the 4 universal joint strap bolts and remove the universal joint straps. Discard the bolts.

6. Remove the front driveshaft.

■ **WARNING**

The can (domed CV joint housing cover) is pressed into the CV joint housing at the factory. When housed correctly, the can will appear as shown in the illustration. Do not reseat the can in the CV joint housing if the can's flange is above the CV joint housing as shown in the illustration. Install a new driveshaft.

❋❋ **WARNING**

Tighten the CV joint bolts evenly in a cross pattern or damage will occur to the CV joint.

➡Install new CV joint washers and CV joint bolts. If new CV joint bolts are not available, coat the threads of the original CV joint bolts with Threadlock and Sealer TA-25 meeting Ford specification WSK-M2G351-A5.

➡Install new universal joint strap bolts. If new universal joint strap bolts are not available, coat the threads of the original universal joint strap bolts with Threadlock and Sealer TA-25 meeting Ford specification WSK-M2G351-A5.

➡Verify that the universal joint bearing caps are seated in the pinion yoke flange before tightening the universal joint strap bolts.

7. To install, reverse the removal procedure. See the accompanying illustrations for torque values.

Rear Shaft

2-WHEEL DRIVE

1. Before servicing the vehicle, refer to the Precautions Section.

2. With the vehicle in NEUTRAL, position it on a hoist.

➡Make sure the index marks on the extension housing and driveshaft are aligned before separation.

➡After removing the driveshaft, place an index mark on the transmission output shaft that matches the transmission extension housing mark.

112 Nm (83 lb-ft)

06017-EXPL-G86

Rear driveshaft flange torque sequence

3. Index-mark the driveshaft and the extension housing to maintain driveshaft balance.

4. Index-mark the driveshaft flange to the pinion flange to maintain driveshaft balance.

5. Remove and discard the 4 driveshaft flange bolts.

➡The driveshaft flange yoke fits tightly on the pinion flange pilot. Never hammer on the driveshaft or any of its components to disconnect the driveshaft flange from the pinion flange.

6. Use a suitable tool and pry only in the area shown to separate the driveshaft flange yoke from the pinion flange.

❋❋ **WARNING**

The driveshaft flange fits tightly on the pinion flange pilot. To make sure that the driveshaft flange seats squarely on the pinion flange, tighten the driveshaft flange bolts evenly in a cross pattern.

06017-EXPL-G87

Pry at this point only

※※ **WARNING**

If new driveshaft flange bolts are not available, coat the threads of the original driveshaft flange bolts with Threadlock and Sealer TA-25 meeting Ford specification WSK-M2G351-A5.

7. To install, reverse the removal procedure. Tighten the flange bolts to 112 Nm (83 ft. lbs.).

4-WHEEL DRIVE

1. Before servicing the vehicle, refer to the Precautions Section.

2. With the vehicle in NEUTRAL, position it on a hoist.

3. If equipped, remove the skid plate.

4. Index-mark the driveshaft flange to the transfer case rear output flange to maintain driveshaft balance.

5. Index-mark the driveshaft flange to the pinion flange to maintain driveshaft balance.

6. Remove and discard the driveshaft flange bolts.

7. Remove and discard the transfer case rear output flange bolts.

※※ **WARNING**

The driveshaft flange fits tightly on the axle pinion flange pilot and the transfer case output flange.

→**Never hammer on the driveshaft or any of its components to disconnect the driveshaft flanges from the mating flanges. Pry only in the area shown, with a suitable tool, to disconnect the driveshaft flanges.**

8. Using a suitable tool as shown, disconnect the driveshaft flanges and remove the driveshaft.

1	Driveshaft flange bolts
2	Driveshaft flange yoke
3	Driveshaft flange bolts
4	Driveshaft flange yoke
5	Driveshaft assembly
6	U-joint snap ring
7	U-joint spider
8	U-joint snap ring
9	U-joint spider
10	Boot clamp
11	Driveshaft slip-yoke boot
12	Boot clamp

06017-EXPL-G88

Rear driveshaft exploded view—4-wheel drive

✳✳ WARNING

The driveshaft flanges fits tightly on the pinion flange pilots. To make sure that the driveshaft flanges seat squarely on the pinion flanges, tighten the driveshaft flange bolts evenly in a cross pattern.

✳✳ WARNING

If new driveshaft flange bolts are not available, coat the threads of the original driveshaft flange bolts with Threadlock and Sealer TA-25 meeting Ford specification WSK-M2G351-A5.

9. To install, reverse the removal procedure. Install new fasteners. Torque the flange bolts to 83 ft. lbs. (112 Nm).

Halfshaft

REMOVAL & INSTALLATION

Front

1. Before servicing the vehicle, refer to the Precautions Section.
2. Remove or disconnect the following:
3. Loosen the front axle wheel hub retainer.

- Wheel and tire assembly
- Hub retainer and the washer. Discard the front axle wheel hub retainer.
- The two bolts and position the disc brake caliper aside
- Tie-rod end from the knuckle. Discard the nut.

- Stabilizer bar link. Discard the nut.

✳✳ WARNING

Do not allow the knuckle to hang freely. It is possible to overextend and internally separate each inner CV joint from its housing.

- Upper ball joint from the knuckle

✳✳ WARNING

Do not use a hammer to separate the outboard CV joint from the hub. Damage to the threads and internal CV joint components may result.

4. Press the outboard CV joint until it is loose in the hub.
5. Remove the outboard CV joint from the hub.

1 Front disc brake caliper
2 Front axle wheel hub retainer
3 Tie-rod end nut
4 Stabilizer bar link nut
5 Upper ball joint nut
6 Front halfshaft assembly
7 Front axle halfshaft seal and bearing

67197-EXPL-G59

Front axle halfshaft

✻✻ WARNING

Do not damage the axle shaft oil seal or the machined sealing surface on the inboard CV joint housing.

➡**A circlip retains the inboard CV joint housing to the differential side gear in the axle.**

6. On the left side, pry the left inboard CV joint housing from the differential side gear.

7. On the right side, disengage the right inboard CV joint housing from the axle tube.

8. Pull the halfshaft and the axle shaft away from the axle tube, and separate the inboard CV joint housing from the axle shaft.

9. Remove the halfshaft assembly from the vehicle.

✻✻ WARNING

Do not damage the axle shaft oil seal, the machined sealing surface on the inboard CV joint housing, or the axle shaft splines.

10. To install, reverse the removal procedure.

11. Always install the halfshaft with a new retainer circlip and a new front axle wheel hub retainer.

12. On the right side, check the retainer circlip engagement after reseating the axle shaft and after installing the halfshaft in the axle. On the left side, check the retainer circlip engagement after installing the halfshaft in the axle. When seated, the retainer circlip will lock the axle shaft and the inboard CV joint housing to the axle.

✻✻ WARNING

Never use power tools to tighten the front axle wheel hub retainer. Torque the retainer to 184 ft. lbs. (250 Nm).

➡**It may be necessary to support the front suspension lower arm to be able to connect the upper ball joint to the knuckle.**

1 Rear disc brake caliper
2 Rear axle wheel hub retainer and washer
3 Upper ball joint pinch bolt nut
4 Toe link pinch bolt nut
5 Sway bar stud nut
6 Rear halfshaft assembly
7 Rear stub shaft housing bearing and seal

67197-EXPL-G60

Rear axle halfshaft

Rear

1. Before servicing the vehicle, refer to the Precautions Section.

2. Remove or disconnect the following:

> ### ⁂ WARNING
>
> **Do not loosen the rear axle wheel hub retainer until after the wheel and tire assembly are removed from the vehicle. Wheel bearing damage will occur if the wheel bearing is unloaded with the weight of the vehicle applied.**

- Wheel and tire assembly

➡ **Have an assistant press the brake pedal to keep the axle from rotating.**

- Hub retainer and the washer
- Caliper. Position the disc brake caliper out of the way.
- Brake disc
- Bolt retaining the parking brake cable bracket to the frame

> ### ⁂ WARNING
>
> **Using a rubber hose approximately 37.5 mm (1.5 in) long, cover the stabilizer link bolt threads and nut to prevent boot damage when removing the halfshaft assembly from the vehicle.**

> ### ⁂ WARNING
>
> **The bolt that retains the upper ball joint to the knuckle is longer and it has fewer threads than the bolt that retains the toe link to the knuckle. Switching these bolts during installation will prevent the pinch arms on the knuckle from correctly retaining the toe link to the knuckle. This may cause the toe link to separate from the knuckle during vehicle operation. Failure to follow these instructions may result in personal injury.**

- Pinch bolts and disconnect the toe link and upper ball joint from the knuckle

> ### ⁂ WARNING
>
> **Using a wood stick, approximately 450 mm (18 in) long and 25 mm (1 in) wide, support the rear suspension upper arm to prevent boot damage when removing the halfshaft assembly from the vehicle**

67197-EXPL-G61

Using a wood stick, approximately 450 mm (18 in) long and 25 mm (1 in) wide, support the rear suspension upper arm to prevent boot damage when removing the halfshaft assembly from the vehicle

➡ **Do not use a hammer to separate the outboard CV joint from the hub. Damage to the threads and internal CV joint components may result. Once the outboard CV joint separates from the hub the knuckle will continue to pivot until the brake backing plate presses against the suspension lower arm. To prevent damage to the brake backing plate, immediately after separating the outboard CV joint from the hub, rest the knuckle on a cushioned support that is tall enough to keep the backing plate from pressing against the suspension lower arm.**

3. Press the outboard CV joint until it is loose in the hub.

4. Separate the outboard CV joint from the hub.

5. Rest the knuckle on a cushioned support.

> ### ⁂ WARNING
>
> **Do not damage the axle shaft oil seal or the machined sealing surface on the inboard CV joint housing.**

➡ **A circlip retains the inboard CV joint housing to the differential side gear in the axle.**

6. Using the special tool, disengage the inboard CV joint housing from the differential side gear.

> ### ⁂ WARNING
>
> **To prevent damage to the axle shaft oil seal, install the special tool 205-461 (seal protector) before removing the inboard CV joint housing from the axle.**

7. Remove the halfshaft assembly from the vehicle.

8. To install, reverse the removal procedure.

9. To prevent damage to the axle shaft oil seal, install the seal protector before positioning the inboard CV joint housing in the axle.

10. Always install the halfshaft with a new retainer circlip and a new rear axle wheel hub retainer.

➡ **Never use power tools to tighten the rear axle wheel hub retainer. Torque the retainer to 203 ft. lbs. (275Nm).**

CV-Joints

OVERHAUL

Front Shaft

2002—04

1. Before servicing the vehicle, refer to the Precautions Section.

2. Remove or disconnect the following:
- Halfshaft
- The two inboard boot clamps
- Inboard halfshaft boot off the inboard CV joint housing

3. Separate the CV joint from the CV joint housing.

06017-EXPL-G89

Remove the stop ring—2002–04 front shaft

06017-EXPL-G90

Remove the snapring—2002–04 front shaft

Remove the joint—2002–04 front shaft

Crimping the boot clamps—2002–04 front shaft

4. Mark the shaft and the inboard CV joint to ease alignment during assembly.

5. Remove the snapring.

6. Remove the CV joint.

7. Inspect the stop ring for wear or damage. Install a new stop ring as necessary.

8. Remove the inboard halfshaft boot from the shaft assembly.

9. Remove the two outboard boot clamps.

10. Remove the outboard halfshaft boot.

➡ **If grease is contaminated, clean and inspect the joint for wear. If worn or damaged, install a new outboard CV joint and shaft assembly.**

To assemble

11. Pack the outboard CV joint with grease.

➡ **Clean the halfshaft boot mounting surfaces of access grease before positioning the halfshaft boot into place.**

12. Position the outboard halfshaft boot and outboard boot clamps.

13. Tighten the through-bolt until the installer is in the closed position.

14. Use the CV Boot Clamp Installer to install the outboard CV joint boot clamps.

15. Position the boot clamp on the halfshaft.

16. Position the inboard halfshaft boot.

17. Install a new stop ring, if necessary.

18. Install the CV joint on the halfshaft.

19. Install the snap-ring.

20. Lubricate the three CV joint needle bearings with grease.

21. Fill the inboard CV joint housing with 235 grams of grease.

22. Position the CV joint housing onto the CV joint.

➡ **Remove any excess grease from the inboard halfshaft boot mating surface before positioning into place.**

23. Position the inboard halfshaft boot and boot clamp.

24. Set the halfshaft assembled length to:

- Aviator, Explorer, Mountaineer: 24.22 inches (615.3mm) for the right side; 26.98 inches (685.3mm) for the left.
- Explorer Sport-Trac: 24.72 inches (618mm) for the left side; 21.92 inches (548mm) for the right side

25. Measure the entire assembly length. Push in or pull out on the inner joint as necessary to adjust the halfshaft assembled length. Hold the inner joint to prevent the assembled length from changing, and insert a small flat-blade screwdriver between the boot and the joint to equalize the pressure.

26. Insert a wood wand to relieve built-up air pressure in the halfshaft boot.

27. Use the CV Boot Clamp Installer to install the inboard boot clamps.

28. Install the front wheel halfshaft.

2005

1. Before servicing the vehicle, refer to the Precautions Section.

Disassembly

2. Remove the halfshaft assembly.

3. Remove and discard the boot clamps.

4. Remove the inboard CV joint housing.

5. Remove and discard the retainer circlip. Slide the boot away from the inboard CV joint.

6. Using a suitable 3-jaw puller, remove the CV joint.

7. Remove and discard the inner CV boot.

➡ **The outboard CV joint is not removable from the halfshaft. The boot must be removed or installed from the inboard CV joint side of the shaft.**

8. Remove and discard the outboard CV joint boot clamps.

9. Remove and discard the boot.

10. If necessary, press the anti-lock

Inboard CV-joint. (1) retainer clip, (2) boot—2005 front shaft and 2002–05 rear shaft

Removing the CV-joint—2005 front shaft and 2002–05 rear shaft

Tri-lobe insert—2005 front shaft and 2002–05 rear shaft

brake sensor indicator ring from the outer CV joint.

Assembly

11. If removed, install a new anti-lock brake sensor indicator ring on the outer CV joint.

06017-EXPL-G96

ABS sensor—2005 front shaft and 2002–05 rear shaft

205-343

06017-EXPL-G97

Crimping the boot clamps—2005 front shaft and 2002–05 rear shaft

205-133

06017-EXPL-G98

Installing the CV-joint—2005 front shaft and 2002–05 rear shaft

12. Lubricate the outboard CV joint. Pack the outboard CV joint with 185 grams (6.53 oz) of grease. One-third of the grease must be installed in the joint with the remainder placed in the boot. Spread the remaining grease evenly inside the boot.

13. Position the outboard halfshaft boot. Position the boot clamps.

14. Install the outboard CV boot clamps.

06017-EXPL-G99

Measuring the halfshaft length—2005 front shaft and 2002–05 rear shaft

15. Position the inboard clamp and boot.

16. Install the CV joint and snapring on the halfshaft.

17. Install the stop ring.

18. Fill the inboard CV joint housing with 210 grams (7.41 oz) of grease. One-half of the grease must be installed in the joint and the remainder placed in the boot.

19. Position the inboard halfshaft boot in the housing groove. Position the boot clamp.

20. Set the halfshaft assembled length to:

- Aviator, Explorer, Mountaineer: 24.22 inches (615.3mm) for the right side; 26.98 inches (685.3mm) for the left.
- Explorer Sport-Trac: 24.72 inches (618mm) for the left side; 21.92 inches (548mm) for the right side

21. Measure the entire assembly length. Push in or pull out on the inner joint as necessary to adjust the halfshaft assembled length. Hold the inner joint to prevent the assembled length from changing, and insert a small flat-blade screwdriver between the boot and the joint to equalize the pressure.

22. Install the boot clamps.

Rear Shaft

1. Before servicing the vehicle, refer to the Precautions Section.

2. Remove or disconnect the following:

3. Remove the halfshaft.

4. For the inboard CV joint:

    a. Remove and discard the boot clamps.

    b. Remove the inboard CV joint housing.

    c. Remove and discard the retainer circlip.

    d. Slide the boot away from the CV joint.

    e. Using a suitable 3-jaw puller, remove the CV joint.

    f. Remove and discard the tri-lobe insert and the boot.

5. For the outboard CV joint:

    a. Remove and discard the boot clamps.

    b. Remove and discard the boot.

➡ **Do not disassemble the side shaft assembly. Install a new halfshaft assembly, if the components are worn/damaged.**

6. Inspect the grease packed in the inboard CV joint and the outboard CV joint for contamination. Rub some of the grease from each joint between two fingers. Any gritty feeling indicates contamination. Wash all of the grease from the inboard CV joint, the inboard CV joint housing, the outboard CV joint, and the interconnecting shaft. Thoroughly dry all of the components, and inspect them for wear and damage. Discard the assembly, if necessary. Proceed as follows only if not discarding the assembly.

7. If necessary, remove and discard the outboard dust seal. Tap uniformly around the seal to separate it from the joint.

8. On the inboard end:

a. Remove and discard the retainer circlip.

b. If necessary, remove and discard the inboard dust seal. Tap uniformly around the seal to separate it from the housing.

To assemble:

9. For the outboard CV joint:

a. Slide the boot on the interconnecting shaft.

b. Pack the outboard CV joint with 225 grams (5.29 ounces) of grease.

c. Spread any remaining grease evenly inside the boot.

d. Clean any excess grease from the boot mounting surfaces before installing the boot.

e. Install the boot by seating it in the groove in the CV joint housing.

f. Tighten the through-bolt until the special tool is in the closed position.

g. Using the special tool, install both boot clamps.

h. If removed, use the special tools to install the dust seal.

✳✳ WARNING

Do not over-expand or twist the circlip during installation.

i. Install the retainer circlip.

10. Install the halfshaft in a soft jaw vise.

11. For the inboard CV joint:

a. Position the clamp on the interconnecting shaft.

b. Position the boot on the interconnecting shaft.

➡**The lip on the end of the tri-lobe insert must seat against the end of the boot.**

c. Install the tri-lobe insert.

➡**One side of the inboard CV joint has a chamfer cut in the edge of joint at the inner diameter near the splines. Install the inboard CV joint so that the chamfer faces the outboard end of the halfshaft.**

d. Install the CV joint.

e. Install the retainer circlip.

f. Pack the inboard CV joint housing with 225 grams (7.93 ounces) of grease.

g. Spread any remaining grease evenly inside the boot and on the CV joint.

h. Clean any excess grease from the boot mounting surfaces before installing the boot.

i. Install the inboard CV joint housing, seating the boot in the groove in the housing.

12. Set the halfshaft assembled length to:

- 2002–03 Explorer and Mountaineer: 28.44 inches (722.4mm)
- Aviator and 2004–05 Explorer and Mountaineer: right side 33.94 inches (862.1mm); left side 32.82 inches (833.7mm)

13. Hold the inner joint to prevent the assembled length from changing, and insert a wood wand between the boot and the joint to equalize the pressure.

14. Tighten the through-bolt until the special tool is in the closed position.

15. Install both boot clamps.

16. If removed, install the dust seal.

17. Install the halfshaft.

Front Axle Tube Bearing

REMOVAL & INSTALLATION

1. Before servicing the vehicle, refer to the Precautions Section.

2. Remove or disconnect the following:
- Right-hand halfshaft
- Right-hand axle shaft
- Axle seal, with a slide hammer
- Axle tube bearing, with a slide hammer

3. Clean the bearing and seal surfaces of any foreign debris.

To install:

4. Use an axle bearing replacer and the handle to replace the right axle tube bearing.

5. Check the bearing depth as shown.

6. Use an axle seal replacer and the handle to replace the axle tube seal.

➡**Care should be taken not to damage the axle seal surface.**

7. Install the axle shaft.

8. Refill the front drive axle to proper level.

9. Install the right halfshaft.

Rear Axle Shaft, Bearing and Seal

REMOVAL & INSTALLATION

Ford 8.8 inch Ring Gear Solid Axle

1. Before servicing the vehicle, refer to the Precautions Section.

2. Raise and support the vehicle.

3. Remove the wheel and tire assembly.

4. Loosen the bolts and drain the lubricant from the rear axle housing.

44.5 mm (0.875 in.)

Front axle tube bearing depth

9348EG11

5. Remove the cover bolts.
6. Remove the differential housing cover.
7. Remove the rear brake disc or brake drum and shoes.

> ❋❋ **WARNING**
>
> **Turning the differential case or an axle shaft with the differential pinion shaft removed will cause the differential pinion gears to fall out of the assembly and damage the components.**

8. Remove the differential pinion shaft lock bolt.
9. Remove the differential pinion shaft.

> ❋❋ **WARNING**
>
> **Do not damage the rubber O-ring in the U-washer groove.**

10. Push the axle shaft inboard.
11. Remove the U-washer.
12. Install the differential pinion shaft.
13. Install the differential pinion shaft lock bolt finger-tight.

> ❋❋ **WARNING**
>
> **Do not damage the wheel bearing oil seal.**

14. Remove the axle shaft.

➡ **If the wheel bearing oil seal is leaking, the axle housing vent may be plugged with foreign material.**

➡ **If only a new seal needs to be installed, use care to avoid damaging the seal bore.**

15. Using a suitable seal remover, remove the axle shaft oil seal. Discard the oil seal.
16. Inspect the rear wheel bearing and axle shaft for wear or damage.
17. If necessary, using the special tools, remove the rear wheel bearing.

Rear wheel bearing ford seal removal—8.8 inch solid axle

06017-EXPL-G100

Rear wheel bearing installation—8.8 inch solid axle

06017-EXPL-G101

Rear axle seal installation—8.8 inch solid axle

06017-EXPL-G102

To install:
18. Lubricate the new rear wheel bearing.
   a. For limited slip rear axles, use SAE 75W-140 High Performance Rear Axle Lubricant F1TZ-19580-B or equivalent meeting Ford specification WSL-M2C192-A.
   b. For conventional rear axles, use SAE 80W-90 Premium Rear Axle Lubricant XY-80W90-QL or equivalent meeting Ford specification WSP-M2C197-A.
19. Using a bearing driver, install the rear wheel bearing.
20. Lubricate the lip of the new wheel bearing oil seal. Use Premium Long-Life Grease XG-1-C or equivalent meeting Ford specification ESA-M1C75-B.
21. Using a seal driver, install the wheel bearing oil seal.
22. Lubricate the lip of the wheel bearing oil seal. Use Premium Long-Life Grease XG-1-C or equivalent meeting Ford specification ESA-M1C75-B.

> ❋❋ **WARNING**
>
> **Turning the differential case or an axle shaft with the differential pinion shaft removed will cause the differential pinion gears to fall out of the assembly and damage the components.**

23. Remove the differential pinion shaft lock bolt.
24. Remove the differential pinion shaft.

> ❋❋ **WARNING**
>
> **Do not damage the wheel bearing oil seal.**

25. Install the axle shaft.

> ❋❋ **WARNING**
>
> **Do not damage the rubber O-ring in the U-washer groove.**

26. Position the U-washer on the button end of the axle shaft.
27. Pull the axle shaft outward.

> ❋❋ **WARNING**
>
> **If a new pinion shaft lock bolt is unavailable, coat the threads with Threadlock and Sealer EOAZ-19554-AA or equivalent meeting Ford specification WSK-M2G351-A5.**

28. Align the hole in the differential pinion shaft with the differential pinion shaft lock bolt hole.
29. Install a new differential pinion shaft lock bolt. Torque to 22 ft. lbs. (30 Nm).
30. Install the rear brake disc or shoes and drum.

➡ **The machined surfaces on the differential housing and the differential housing cover must be clean and free of oil before applying the silicone sealant. Cover the inside of the rear axle prior to cleaning the machined surface to prevent contamination.**

31. Clean the cover gasket mating surfaces.

➡ **Install the differential housing cover within 15 minutes of applying the silicone, or it will be necessary to apply new sealant.**

32. Apply a continuous bead of sealant to the differential housing cover. Use Clear Silicone Rubber D6AZ-19562-AA or equivalent meeting Ford specification ESB-M4G92-A.

➡ **If possible, allow one hour before filling the axle with lubricant to allow the silicone sealant to cure.**

33. Position the differential housing cover. Install the bolts. Torque to 33 ft. lbs. (45 Nm).
34. Fill the rear axle housing with the specified lubricant type and quantity.

➡ **For Traction-Lok® axles, first fill the rear axle with 118 ml (4 oz) of Additive Friction Modifier C8AZ-19B546-A or equivalent meeting Ford specification EST-M2C118-A.**

➡ **Service refill capacities are determined by filling the rear axle to the level shown.**

35. Fill the rear axle with lubricant and install the fill plug.

a. For conventional axles, use SAE 80W-90 Premium Rear Axle Lubricant XY-80W90-QL or equivalent meeting Ford specification WSP-M2C197-A.

b. For Traction-Lok® axles, use SAE 75W-140 Synthetic Rear Axle Lubricant F1TZ-19580-B or equivalent meeting Ford specification WSL-M2C192-A.

36. Install the wheel and tire assembly.

Pinion Seal

REMOVAL & INSTALLATION

Solid Rear Axle

1. Before servicing the vehicle, refer to the Precautions Section.
2. Drain the axle housing fluid.
3. Remove or disconnect the following:
   - Negative battery cable
   - Rear wheels
   - Driveshaft
   - Brake calipers and pads or brake drum

➡ **The brake calipers and pads or brake drum must be removed so that there is no additional drag when measuring pinion bearing preload.**

4. Use an inch lb. torque wrench and measure and record the amount of torque required to maintain pinion rotation through several revolutions.
5. Remove or disconnect the following:
   - Pinion flange
   - Pinion seal
   - Pinion bearing
   - Collapsible spacer

To install:

➡ **Use a new collapsible spacer and flange nut for assembly.**

6. Install or connect the following:
   - Collapsible spacer
   - Pinion bearing

06017-EXPL-G103

Pinion flange removal—8.8 inch axle

06017-EXPL-G104

Pinion seal installation—8.8 inch axle

06017-EXPL-G105

Pinion flange installation—8.8 inch axle

   - Pinion seal
   - Pinion flange

7. Rotate the pinion flange occasionally while tightening the flange nut to make sure the pinion bearings seat correctly.
8. Take frequent bearing preload torque readings. Tighten the flange nut to achieve the preload torque readings originally recorded.

❊❊ CAUTION

Never loosen the pinion nut to reduce bearing preload. If it is necessary to reduce bearing preload, install a new collapsible spacer and pinion nut.

9. Install or connect the following:
   - Driveshaft
   - Brake calipers and pads or brake drum
   - Wheels
   - Negative battery cable
10. Fill the differential with gear lubricant and check for leaks.

Halfshaft Rear Axle

1. Before servicing the vehicle, refer to the Precautions Section.
2. Drain the axle housing fluid.
3. Remove or disconnect the following:
   - Rear wheel and tire assemblies
   - Brake caliper and support bracket from the knuckle as an assembly. Wire the caliper and support bracket assembly out of the way.

➡ **Matchmark the driveshaft flange and rear axle pinion flange to maintain initial balance during installation.**

4. Disconnect and position the driveshaft out of the way.
5. Install an inch/pound torque wrench on the nut and record the torque necessary to maintain rotation of the drive pinion gear through several revolutions.
6. Remove and discard the pinion flange nut.

➡ **Matchmark the rear axle pinion flange and drive pinion gear stem to maintain initial balance during installation.**

7. Remove the rear axle pinion flange.
8. Force up on the metal flange of the rear axle drive pinion seal. Install gripping pliers and strike with a hammer until the rear axle drive pinion seal is removed.

To install:

9. Lubricate the new rear drive pinion seal with grease.

➡ **If the rear axle drive pinion seal becomes misaligned during installation, remove the rear axle drive pinion seal and install a new seal.**

10. Drive in the rear axle drive pinion seal.
11. Inspect the rear axle pinion flange seal journal for rust, nicks and scratches prior to installing the flange. Polish the seal journal with fine crocus cloth, if necessary.
12. Lubricate the rear axle pinion flange splines.
13. Install the rear axle pinion flange.

❊❊ WARNING

Do not under any circumstance loosen the nut to reduce preload. If it is necessary to reduce preload, install a new differential drive pinion collapsible spacer and nut.

14. Rotate the pinion occasionally to make sure the pinion bearings seat correctly. Take frequent pinion bearing torque preload readings by rotating the drive pinion gear with an inch/pound torque wrench.
15. If the preload recorded prior to disassembly is lower than the specification for used bearings, then tighten the nut to specification. If the preload recorded prior to disassembly is higher than the specification for used bearings, then tighten the nut to the original reading as recorded.

Pinion bearing preload used bearings: 8–14 inch lbs. (0.9–1.16 Nm); new bearings: 16–29 inch lbs. (1.8–3.2 Nm).

16. Connect the driveshaft. Torque the bolts to 83 ft. lbs. (112 Nm).
17. Install the rear brake calipers.
18. Install the rear wheel and tire assemblies.

Front Axle

➡ **This operation disturbs the differential pinion bearing preload. Carefully reset the preload during assembly.**

�֎ CAUTION

The electrical power to the air suspension system must be shut off prior to hoisting, jacking or towing an air suspension vehicle. This can be accomplished by turning off the air suspension switch located in the rear jack storage area. Failure to do so can result in unexpected inflation or deflation of the air springs, which can result in shifting of the vehicle during these operations.

1. Before servicing the vehicle, refer to the Precautions Section.

2. Index-mark the front driveshaft and pinion flange.

3. Remove or disconnect the following:

   • Front driveshaft from the pinion flange, and position it aside

➡ **Do not allow the driveshaft to hang unsupported.**

4. Using an Nm (inch-pound) torque wrench, measure the torque required to maintain pinion rotation. Record the measurement.

5. Index-mark the pinion flange and the pinion stem.

6. Hold the pinion flange while removing the nut.

7. Place a drain pan under the differential housing.

8. Using a puller, remove the pinion flange.

9. Inspect the pinion flange for burrs and damage. Inspect the end of the pinion flange that contacts the bearing cone, the nut counterbore, and the seal surface for nicks. Discard the pinion flange as necessary.

10. Using a seal remover and impact slide hammer, remove the pinion seal.

11. Remove the front axle drive pinion shaft oil slinger and the differential pinion bearing.

12. Remove and discard the collapsible spacer.

To install:

13. Verify that the splines on the pinion stem are free of burrs. If burrs are evident, remove them with a fine crocus cloth. Work in a rotating motion to wipe the pinion clean.

14. Clean the pinion seal bore.

15. Install a new collapsible spacer.

16. Install the original differential pinion bearing and the front axle drive pinion shaft oil slinger.

17. Lubricate the pinion seal. Use Motorcraft SAE 80W90 Thermally Stable 4x4 Axle Lubricant meeting Ford specification WSP-M2C197-A.

18. Install the pinion seal.

19. Lubricate the pinion flange splines. Use Motorcraft SAE 80W90 Thermally Stable 4x4 Axle Lubricant meeting Ford specification WSP-M2C197-A.

➡ **Never use a metal hammer on the pinion flange or install the flange with power tools. If necessary, use a plastic hammer to tap on a tight fitting flange.**

   • Align the index marks and install the pinion flange.
   • Install the new nut hand-tight.

➡ **Do not loosen the nut to reduce preload. Install a new collapsible spacer and nut if preload reduction is necessary.**

20. Use the special tool to hold the pinion flange while tightening the nut to set the preload.

21. Tighten the nut, rotating the pinion occasionally to ensure the differential pinion bearings are seating correctly. Take frequent differential pinion bearing preload readings by rotating the pinion with a Nm (inch-pound) torque wrench. The final reading must be 0.56 Nm (5 inch lbs.) more than the initial reading taken during removal.

22. Align the index marks and position the front driveshaft.

23. Install the universal joint spider retainers and bolts.

24. Check the fluid level and, if necessary, fill the axle to specification. Use Motorcraft SAE 80W90 Thermally Stable 4x4 Axle Lubricant meeting Ford specification WSP-M2C197-A.

25. Lower the vehicle.

26. If so equipped, reactivate the air suspension.

Rear Axle Stub Shaft Bearing and Seal

REMOVAL & INSTALLATION

Independent Rear Suspension

1. Before servicing the vehicle, refer to the Precautions Section.

2. Remove the halfshaft assembly.

06017-EXPL-G106

Stub shaft bearing and seal removal

06017-EXPL-G107

Stub shaft bearing installation tool

06017-EXPL-G108

Stub shaft seal installation

➡ **If removing the stub shaft pilot bearing oil seal only, engage the tangs of the special tool on the stub shaft pilot bearing oil seal.**

3. Using the special tools, remove the stub shaft pilot bearing and stub shaft pilot bearing oil seal.

   a. Firmly engage the tangs of the special tool on the stub shaft pilot bearing.

   b. Remove the stub shaft pilot bearing and stub shaft pilot bearing oil seal.

4. Inspect the stub shaft pilot bearing oil seal journal for rust, nicks and scratches. Polish the seal journal surface with fine crocus cloth, if necessary.

To install:

5. Lubricate the new stub shaft pilot bearing. Use high performance rear axle lubricant.

✖ WARNING

Installation of the stub shaft pilot bearing or stub shaft pilot bearing oil seal without the correct tools can result in early stub shaft pilot bearing

or stub shaft pilot bearing oil seal failure. If the stub shaft pilot bearing becomes cocked in the bore during installation, remove it and install a new stub shaft pilot bearing.

6. Place the stub shaft pilot bearing onto the special tools.

7. Install the stub shaft pilot bearing into the rear axle housing bore.

8. Lubricate the lip of the stub shaft pilot bearing oil seal with grease and install the stub shaft pilot bearing oil seal onto the special tools. Use long-life grease.

9. Carefully align the stub shaft pilot

bearing oil seal with the housing bore and install the stub shaft pilot bearing oil seal.

⁜⁜ WARNING

Install the special tool, to avoid damaging the seal with the halfshaft.

10. Install the halfshaft assembly.

STEERING

Air Bag

PRECAUTIONS

Always wear safety glasses when servicing an air bag vehicle, and when handling an air bag.

Never attempt to service the steering wheel or steering column on an air bag-equipped vehicle without first properly disarming the air bag system. The air bag system should be properly disarmed whenever ANY service procedure in this manual indicates that you should do so.

When carrying a live air bag module, always make sure the bag and trim cover are pointed away from your body. In the unlikely event of an accidental deployment, the bag will then deploy with minimal chance of injury.

When placing a live air bag on a bench or other surface, always face the bag and trim cover up, away from the surface. This will reduce the motion of the air bag if it is accidentally deployed.

If you should come in contact with a deployed air bag, be advised that the air bag surface may contain deposits of sodium hydroxide, which is a product of the gas combustion and is irritating to the skin. Always wear gloves and safety glasses when handling a deployed air bag, and wash your hands with mild soap and water afterwards.

Always wear safety glasses when repairing an air bag supplemental restraint system (SRS) vehicle and when handling an air bag module. This will reduce the risk of injury in the event of an accidental deployment.

Carry a live air bag module with the air bag and trim cover pointed away from your body. This will reduce the risk of injury in the event of an accidental deployment.

Do not set a live air bag module down with the trim cover face down. This will reduce the risk of injury in the event of an accidental deployment.

After deployment, the air bag surface can contain deposits of sodium hydroxide, a product of the gas generant combustion that is irritating to the skin. Wash your hands with soap and water afterwards.

Never probe the connectors on the air bag module. Doing so can result in air bag deployment, which can result in personal injury.

The safety belt buckle pretensioner and safety belt retractor pretensioner are pyrotechnic devices. Never probe a pretensioner electrical connector. Doing so could result in pretensioner or air bag deployment and could result in personal injury.

The safety belt buckle pretensioner and safety belt retractor pretensioner are pyrotechnic devices. Never probe a pretensioner electrical connector. Doing so result in pretensioner or air bag deployment and could result in personal injury.

The restraint system diagnostic tool is for restraint system service only. Remove from vehicle prior to road use. Failure to remove could result in injury and possible violation of vehicle safety standards.

Never probe the connectors on the side air curtain module. Doing so can result in side air curtain deployment.

Never probe the electrical connector on a safety canopy module. Doing so can result in safety canopy deployment.

Do not handle, move or change the original horizontal mounting position of the restraints control module (RCM) while the RCM is connected and the ignition switch is ON. Failure to follow these instructions may result in the inadvertent deployment of the safety canopy and risk of personal injury.

To reduce the risk of personal injury, do not use any memory saver devices.

DISARMING THE SYSTEM

1. Before servicing the vehicle, refer to the Precautions Section.

If a seat equipped with a seat mounted side air bag and/or a safety belt pretensioner (if equipped) system is being serviced, the supplemental restraint system (SRS) must be depowered.

The air bag warning lamp illuminates when the RCM fuse is removed and the ignition switch is ON. This is normal opera-

tion and does not indicate a supplemental restraint system (SRS) fault.

All vehicles
a. Turn all vehicle accessories OFF.
b. Turn the ignition switch to OFF.

Vehicles built up to 03/2002
a. At the central junction box (CJB), located below the left side of the instrument panel, remove the restraints control module (RCM) fuse F2.28 (10A) from the CJB.

Vehicles built 03/2002 onward
a. At the central junction box (CJB), located below the left side of the instrument panel, remove the restraints control module (RCM) fuse F2.19 (10A) from the CJB.

All vehicles
a. Turn the ignition ON and visually monitor the air bag indicator for at least 30 seconds. The air bag indicator will remain lit continuously (no flashing) if the correct RCM fuse has been removed. If the air bag indicator does not remain lit continuously, remove the correct RCM fuse before proceeding.
b. Turn the ignition OFF.

⁜⁜ CAUTION

To avoid accidental deployment and possible personal injury, the backup power supply must be depleted before repairing or replacing any front or side air bag supplemental restraint system (SRS) components and before servicing, replacing, adjusting or striking components near the front or side air bag sensors or RCM, such as doors, instrument panel, console, door latches, strikers, seats and hood latches. The side impact sensors (if equipped) are located at or near the base of the B-pillars and C-pillars. To deplete the backup power supply energy, disconnect the battery ground cable and wait at least one minute. Be sure to disconnect auxiliary batteries and power supplies (if equipped). Disconnect the battery ground cable and wait at least one minute.

ARMING THE SYSTEM

✿ CAUTION

The restraint system diagnostic tool is for restraint system service only. Remove from vehicle prior to road use. Failure to remove could result in injury and possible violation of vehicle safety standards. Make sure all restraint system diagnostic tool(s) that may have been installed during the repair have been removed from the vehicle and all SRS components are connected.

1. Before servicing the vehicle, refer to the Precautions Section.
2. Turn the ignition switch from OFF to ON.
3. Install the RCM fuse to the CJB.

✿ CAUTION

Be sure that nobody is in the vehicle and that there is nothing blocking or set in front of any air bag module when the battery ground cable is connected.

4. Connect the battery ground cable.
5. Prove out the supplemental restraint system (SRS) as follows:

   a. Turn the ignition key from ON to OFF. Wait 10 seconds, then turn the key back to ON and visually monitor the air bag indicator with the air bag modules installed. The air bag indicator will light continuously for approximately 6 seconds and then turn off. If an air bag supplemental restraint system (SRS) fault is present, the air bag indicator will:

   • fail to light.
   • remain lit continuously.
   • flash.

   b. The flashing might not occur until approximately 30 seconds after the ignition switch has been turned from the OFF to the ON position. This is the time required for the restraints control module (RCM) to complete the testing of the SRS. If the air bag indicator is inoperative and a SRS fault exists, a chime will sound in a pattern of 5 sets of 5 beeps. If this occurs, the air bag indicator and any SRS fault discovered must be diagnosed and repaired.

6. Clear all continuous DTCs from the restraints control module using a diagnostic tool.

Driver's Air Bag Module, Steering Wheel, and Clockspring

REMOVAL & INSTALLATION

Aviator, Explorer and Mountaineer

✳✳ CAUTION

Always wear safety glasses when repairing an air bag supplemental restraint system (SRS) vehicle and when handling an air bag module. This will reduce the risk of injury in the event of an accidental deployment.

✳✳ CAUTION

To reduce the risk of personal injury, do not use any memory saver devices.

➡The air bag warning lamp illuminates when the RCM fuse is removed and the ignition switch is ON. This is normal operation and does not indicate a supplemental restraint system (SRS) fault.

➡The SRS must be fully operational and free of faults before releasing the vehicle to the customer.

➡Repair is made by installing a new part only. If the new part does not correct the condition, install the original part and perform the diagnostic procedure again.

1. Before servicing the vehicle, refer to the Precautions Section.
2. Depower the system.
3. Make sure the road wheels are in the straight-ahead position.
4. Remove the 2 steering wheel back cover plugs.
5. Remove the 2 driver air bag module bolts.
6. Partially remove the driver air bag module from the steering wheel.
7. Disconnect the 2 driver air bag module electrical connectors.

✳✳ WARNING

The clockspring electrical connectors are unique and cannot be reversed when connected to the driver air bag module. Match the electrical connector key to the keyway in the driver air bag module. Do not force the electri- cal connectors into the driver air bag module.

8. Disconnect the horn switch electrical connector and remove the driver air bag module.
9. Loosen the steering wheel bolt.

✳✳ WARNING

Removing the steering wheel without using a puller can damage the column bearings.

10. Using the special tool, separate the steering wheel from the steering column.

➡A new bolt must be installed.

11. Remove and discard the steering wheel bolt.
12. Remove the steering wheel.
13. Remove the 2 steering column opening cover screws.
14. Remove the steering column opening cover.
15. Remove the 3 lower steering column shroud screws.
16. Remove the lower steering column shroud.
17. Remove the upper steering column shroud. Lift where shown, releasing the retaining clips and rotate the upper steering column trim panel out of the instrument cluster finish panel. Remove the upper steering column shroud hard shell.
18. If installing the same clockspring, apply 2 strips of masking tape across the clockspring to prevent accidental rotation when the clockspring is removed.
19. Remove the multifunction switch screw.
20. While releasing the retaining tab at the top of the multifunction switch, slide the multifunction switch up and out of the way.
21. Remove the 2 clockspring screws.

➡If the clockspring is to be reinstalled, do not allow the clockspring to turn from its removal position.

22. Disconnect the 3 clockspring electrical connectors and remove the clockspring.
23. Remove the 3 clockspring mounting bracket screws and discard the clockspring mounting bracket if damaged.

To Install:

24. Position the clockspring mounting bracket, if removed.
25. Install the 3 clockspring mounting bracket screws.
26. Connect the 3 clockspring electrical connectors.

9 Nm
(80 lb-in) – ②

1 Steering wheel back cover
   plug (2 required)
2 Driver air bag module bolt (2
   required)
3 Driver air bag module
4 Driver air bag module
   electrical connectors
5 Horn switch electrical
   connector

06017-EXPL-G125

Air bag module—Aviator shown; others similar

27. If the vehicle's clockspring has rotated out of the center position, follow these steps to center the clockspring.

   a. Hold the clockspring outer housing stationary.

   b. While turning the clockspring rotor counterclockwise, carefully feel for the ribbon wire to run out of length and for a

1 Steering wheel bolt
2 Steering wheel

①– 40 Nm (30 lb-ft)

06017-EXPL-G126

Steering wheel removal—Aviator shown; others similar

1 Clockspring mounting bracket

2 Clockspring mounting bracket screws (3 required)

3 Clockspring electrical connectors (3 required)

4 Clockspring

5 Clockspring screws (2 required)

6 Retaining pin

7 Tape

8 Multi-function switch

9 Multi-function switch screw

10 Upper steering column shroud

11 Lower steering column shroud

12 Lower steering column shroud screws (3 required)

13 Lower steering column opening cover

14 Lower steering column opening cover screws (2 required)

06017-EXPL-G127

Clockspring removal—Aviator

06017-EXPL-G128

Turn the clockspring rotor clockwise approximately 3 turns

06017-EXPL-G129

Positioning the clockspring on the column

06017-EXPL-G130

Clockspring (1) and retaining clips (2)—Explorer Sport and Sport-Trac

slight resistance. Stop turning at this point.

   c. Turn the clockspring rotor clockwise approximately 3 turns. This is the center point of the clockspring.

28. Do not allow the clockspring rotor to turn from this position.

➡**Slight turning of the clockspring rotor is allowable for alignment purposes to the steering column.**

29. With the flats of the clockspring aligned to the flats of the steering column, slide the clockspring onto the steering column.

30. Install the 2 clockspring screws.

31. For vehicles receiving a new clockspring, remove the retaining pin.

32. For vehicles reusing the clockspring that was removed, remove the 2 pieces of tape from the clockspring.

33. Install the multifunction switch onto the steering column.

34. Install the multifunction switch screw.

35. Install the upper steering column shroud.

36. Install the lower steering column shroud.

37. Install the 3 lower steering column shroud screws.

38. Install the lower steering column opening cover.

39. Install the 2 lower steering column opening cover screws.

40. Install the steering wheel. Tighten to 40 Nm (30 ft. lbs.).

41. Install the driver air bag module. Tighten to 9 Nm (80 inch lbs.).

42. Repower the system

Explorer Sport and Sport-Trac

1. Before servicing the vehicle, refer to the Precautions Section.

2. Depower the system.

3. Remove the two steering wheel back cover plugs.

4. Remove the two driver air bag module retaining bolts.

5. Disconnect the driver air bag module electrical connector.

6. Remove the driver air bag module.

➡**Make sure the wheels are in the straight-ahead position.**

7. Disconnect the steering wheel accessory electrical connector.

8. Remove and discard the steering wheel bolt.

✳✳ WARNING

Removing the steering wheel without using a puller can damage the column bearings.

9. Use the 2-Jaw Puller or Steering Wheel Puller, as appropriate, to remove the steering wheel.

10. Remove the steering wheel while routing the wires from the air bag sliding contact through the steering wheel.

11. Remove the retaining screws and position the hood release handle out of the way.

12. Remove the two retaining screws from the lower steering column opening finish panel.

13. Pull out to release the retaining clips and remove the lower steering column opening finish panel.

14. Remove the bolts and the lower steering column opening finish panel reinforcement.

15. Loosen, but do not remove, the four steering column-to-instrument panel retaining bolts enough to remove the upper steering column shroud (two shown).

16. On vehicles with tilt column, remove the tilt wheel handle shank.

17. Remove the three screws. Remove the lower steering column shroud.

18. Position the ignition switch lock cylinder to the RUN position.

19. Using the suitable tool, push upward on the cylinder release tab while pulling the ignition switch lock cylinder outward.

20. Remove the upper steering column shroud.

21. Remove the passive anti-theft system (PATS) transmitter.

22. Remove the key-in-ignition warning indicator switch.

23. Remove the clockspring electrical connectors from the instrument panel and disconnect clockspring electrical connectors.

24. Remove the bolt and separate the ground wire.

25. Remove the clockspring wire harness from the two holders.

26. Release the lower clockspring retaining clip.

27. Release the remaining two clockspring retaining clips.

28. Remove the clockspring.

To install:

29. On vehicles receiving a new clockspring: A new clockspring is supplied in a centralized position and held there with a key. Remove the key from the clockspring, holding the rotor in its centralized position. Do not allow the clockspring rotor to turn.

30. On vehicles needing clockspring recentering:

❈❈ CAUTION

Incorrect centralization may result in premature component failure. If in doubt when centralizing the clockspring, repeat the centralizing procedure. Failure to follow this instruction may result in personal injury.

❈❈ WARNING

Make sure the road wheels are in the straight-ahead position.

a. Hold the clockspring outer housing (1) stationary.

Centering the clockspring—Explorer Sport and Sport-Trac

b. Depress the clockspring locking tab (2) to release the rotor.

❈❈ WARNING

Overturning will destroy the clockspring. The internal ribbon wire acts as the stop and can be broken from its internal connection.

c. While holding the clockspring locking tab in the released position, turn the rotor (3) counterclockwise, carefully feeling for the ribbon wire to run out of length, and a slight resistance is felt. Stop turning at this point.

d. While holding the clockspring locking tab in the released position, turn the rotor clockwise approximately 3 turns (4). This is the center point of the clockspring.

e. Release the clockspring locking tab. Lock the rotor in the centralized position.

31. Align the clockspring onto the steering column shaft.

32. Slide the clockspring down the steering column shaft and lock the clips into place.

33. Engage the lower clockspring retaining clip into place.

34. Connect the two clockspring electrical connectors to the vehicle harness and install the electrical connector pin-type retainers to the bracket.

35. Position the ground wire and install the bolt.

36. Route the wire harness into the two holders.

37. Install the key-in-ignition warning indicator switch to the steering column.

38. Position the PATS transmitter to the steering column.

39. Install the PATS transmitter retaining screw.

40. Position the upper steering column shroud to the steering column.

41. Position the lower steering column shroud to the steering column and the upper steering column shroud.

42. Install the screws.

43. Tighten the four steering column to instrument panel retaining bolts (two shown).

44. On vehicles with tilt column, install the tilt wheel handle and shank.

45. Align and install the ignition switch lock cylinder to the steering column. Position the ignition switch lock cylinder to the off position.

➡**Make sure the wheels are in the straight-ahead position.**

46. Position the steering wheel while routing the wires from the air bag sliding contact through the steering wheel.

47. Install a new steering wheel bolt. Torque to 25–34 ft. lbs. (34–46 Nm).

48. Connect the steering wheel accessory electrical connector.

49. Position the lower steering column opening finish panel to the instrument panel and push in at the top to seat the retaining clips.

50. Install the screws.

51. Position the hood release handle and install the screws.

52. Connect the driver air bag module electrical connector.

53. Position the driver air bag module to the steering wheel.

54. Install the two driver air bag module retaining bolts. Torque to 80 inch lbs. (9 Nm).

55. Install the two steering wheel back cover plugs.

56. Repower the system.

Power Rack and Pinion Steering Gear

REMOVAL & INSTALLATION

Explorer Sport and Sport-Trac

❈❈ WARNING

When repairing the power steering system, care should be taken to prevent the entry of contaminants or premature failure of the power steering components can result.

1. Before servicing the vehicle, refer to the Precautions Section.

2. Turn the wheel to the straight-ahead position and turn the ignition switch to the OFF position.

3. Remove the wheel and tire assemblies.

4. Remove the cooler hose clamp nuts. Loosen the clamp and disconnect the hose. Drain the fluid into a suitable container.

5. Loosen the clamp, disconnect and remove the cooler.

6. Loosen the tie rod end jam nuts.

7. Remove and discard the cotter pins and nuts.

❈❈ WARNING

Do not damage the tie rod boot when installing the special tool.

➡Remove the adapter from the ball end of the special tool. Apply a small amount of grease to the tie rod end stud and the ball of the special tool.

8. Separate the tie rod ends from the knuckles.

9. Remove the tie rod ends. Count and record the number of turns required to remove the tie rod end.

10. Remove the front stabilizer bar link nuts from the front suspension lower arms.

11. Remove the four stabilizer bar bolts and two brackets.

12. Remove the front stabilizer bar.
13. Remove the stabilizer bar insulators.

✳✳ WARNING

Do not allow the intermediate shaft to rotate while it is disconnected from the steering gear or damage to the clockspring can result. If there is evidence that the intermediate shaft has rotated, the clockspring must be removed and re-centered.

14. Remove the pinch bolt and detach the intermediate shaft from the gear.

15. Remove the fluid line bracket nut and disconnect the lines.

✳✳ WARNING

Hold the tops of the steering gear to crossmember stud bolts to avoid damaging the steering gear fluid transfer tubes.

16. Remove the steering gear mounting nuts.

17. Remove the stud bolts and washers.

18. Rotate the steering gear control valve housing toward the front of the vehicle.

19. Turn the steering gear input shaft to the right until the stop is reached.

20. Move the steering gear as far to the right side of the vehicle as possible.

21. Move the left front wheel spindle tie rod forward to clear the frame crossmember.

22. Remove the steering gear from the vehicle.

To install:

➡**Make sure the steering gear input shaft is turned to the left until the stop is reached.**

➡**Handle the steering gear with caution to avoid damage to fluid transfer tubes and to avoid dimples in the tie rod boots.**

23. Turn the steering gear input shaft to the right until the stop is reached. Note the number of turns required.

1 Stud

2 Washer

3 Insulator

4 Steering gear

5 Insulator

6 Nut

06017-EXPL-G109

Steering gear mounting—Explorer Sport and Sport-Trac

06017-EXPL-G110

Steering fluid line bracket nut—Explorer Sport and Sport-Trac

06017-EXPL-G111

Steering gear mounting nuts—Explorer Sport and Sport-Trac

1. Fluid cooler-to-crossmember bolts
2. Fluid cooler
3. Oil shield-to-crossmember bolts
4. Oil shield
5. Intermediate shaft-to-steering gear bolt
6. Intermediate shaft (detach only)
7. Clamp plate-to-steering gear nut
8. Power steering high-pressure line
9. O-ring
10. Power steering return line
11. O-ring
    Installation Note

12. Tie-rod jam nuts (loosen only)
13. Cotter pins
14. Tie-rod nuts
15. Outer tie-rods
16. Steering gear-to-crossmember nuts (2 required)
17. Steering gear-to-crossmember bolts (2 required)
18. Bracket-to-crossmember bolts (2 required)
19. Bracket
20. Steering gear
21. Outer bellows clamp
22. Inner bellows clamp
23. Steering gear bellows
24. Spindle tie-rod

06017-EXPL-G112

Steering gear and related parts—Explorer and Mountaineer

➡**Make sure the steering gear control valve housing is turned toward the front of the vehicle.**

24. Install the steering gear into the right opening of the crossmember.

✳✳ WARNING

Take care not to scuff or tear the inner tie rod boots.

25. Move the steering gear as far to the right side of the vehicle as possible.

26. Move the left front wheel spindle tie rod into the opening in the crossmember and move the steering gear into position.

27. To place the steering gear in the straight-ahead position, turn the steering gear input shaft to the left by half the number of turns previously recorded.

28. Rotate the steering gear control valve housing toward the rear of the vehicle.

29. Install the steering gear crossmember washers and stud bolts. The dished side of the washer faces downward. Install the nuts. Torque to 111 ft. lbs. (150 Nm).

✳✳ WARNING

Hold the tops of the steering gear to crossmember stud bolts to avoid damaging the steering gear fluid transfer tubes.

30. Install the fluid lines and tighten the nut. Torque to 18 ft. lbs. (25 Nm).

✳✳ WARNING

Do not allow the intermediate shaft to rotate while it is disconnected from the steering gear or damage to the clockspring can result. If there is evidence that the intermediate shaft has rotated, the clockspring must be removed and re-centered.

31. Connect the intermediate shaft to the steering gear input shaft. Install a new lower steering column pinch bolt. Torque to 35 ft. lbs. (48 Nm).

32. Install the cooler, hose and clamp. Connect and install the clamp. Install the nuts. Torque to 30 ft. lbs. (40 Nm).

➡**In the event that the self-tapping bolts cannot be installed in the frame, there is a kit available with flag nuts.**

33. Install the stabilizer bar insulators.

34. Install the front stabilizer bar brackets. Install the bolts. Torque to 25–34 ft. lbs. (34–46 Nm).

35. Install the front stabilizer bar links and the front stabilizer bar bolts.

36. Install the front stabilizer bar link nuts. Torque to 15–20 ft. lbs. (22–29 Nm).

37. Install the tie rod ends on the front wheel spindle tie rods. Rotate the tie rod ends the number of turns recorded during removal.

38. Position the tie rod ends on the steering knuckles. Install the castellated nuts and new cotter pins. Torque to 52 ft. lbs. (72 Nm). Check that the brake dust shields are not bent and are not in contact with the outer tie rod boots.

39. Tighten the tie rod end jam nuts to 59 ft. lbs. (80 Nm).

40. Install the front wheel and tire assemblies

41. Fill and leak check the system.

42. Check and, if necessary, adjust the wheel alignment.

Explorer and Mountaineer

1. Before servicing the vehicle, refer to the Precautions Section.
All vehicles

✳✳ WARNING

While repairing the power steering system, care should be taken to prevent the entry of contaminants or premature failure of the power steering components can result.

Lower arm forward nut and flag bolt

Shock absorber-to-lower arm bolt and flag nut

2. With the vehicle in NEUTRAL, position it on a hoist.

✳✳ WARNING

Do not allow the steering wheel to rotate while the lower shaft is disconnected or damage to the clockspring can result. If there is evidence that the lower shaft has rotated, the clockspring must be removed and recentered.

3. Hold the steering wheel in the straight-ahead position using a suitable device.

4. Using a suitable suction device, drain the power steering fluid reservoir.

5. Remove the 2 bolts and the oil drip shield.

➡**Install a new lower shaft-to-steering gear bolt.**

6. Remove and discard the lower shaft-to-steering gear bolt.

7. Disconnect the lower shaft from the steering gear.

8. Remove the steering line clamp plate nut.

9. Rotate the steering line clamp plate and disconnect the power steering lines.

➡**New O-rings must be installed whenever the power steering lines are disconnected.**

10. Remove and discard the 2 O-rings.

➡**New cotter pins must be installed.**

11. Remove and discard the 2 cotter pins.

➡**New tie rod end nuts must be installed.**

12. Remove and discard the 2 tie rod end nuts.

✳✳ WARNING

Do not damage the tie rod end boot when installing the special tool.

13. Using the special tool, separate the 2 tie rod ends from the wheel knuckle.

14. Remove the 2 steering gear-to-crossmember nuts and bolts.

15. Remove the 2 bolts and remove the steering gear bracket.
4WD vehicles

➡**On 4WD vehicles, the following steps (left lower arm only) must be carried out to provide clearance to remove the steering gear.**

❋❋ WARNING

Do not tighten the left lower arm inboard mounting nuts until the installation procedure is complete and the weight of the vehicle is resting on the wheel and tire

assemblies. Make sure to tighten the lower arm forward nut before tightening the lower arm-to-frame nuts.

➡ It is not necessary to disconnect the left lower ball joint.

16. Remove the lower arm forward nut and flag bolt. Discard the nut.

17. Remove and discard the 2 lower arm-to-frame nuts.

18. Remove the shock absorber-to-lower arm bolt and flag nut. Discard the flag nut.

67197-EXPL-G62

Steering gear and related parts—Aviator

19. Remove the stabilizer bar connecting link nut and disconnect the link.

All vehicles

20. Remove the steering gear from the left side of the vehicle.

21. To install, reverse the removal procedure. Install 2 new O-rings. Fill the power steering system

22. Observe the following torques:

- Oil drip shield: 15 Nm (11 ft. lbs.).
- Steering pinch bolt: 35 ft. lbs. (48 Nm)
- Fluid line clamp bolt: 18 ft. lbs. (25 Nm)
- Tie rod end nuts: 52 ft. lbs. (70 Nm)
- Steering gear-to-crossmember nuts and bolts: 200 Nm (148 ft. lbs.).
- Steering gear bracket bolts: 70 Nm (52 ft. lbs.).
- Lower arm forward nut and flag bolt: 400 Nm (295 ft. lbs.).
- Lower arm-to-frame nuts: 150 Nm (111 ft. lbs.).
- Shock absorber-to-lower arm bolt and flag nut: 350 Nm (258 ft. lbs.).
- Stabilizer bar connecting link nut: 25 Nm (18 ft. lbs.).

Aviator

1. Before servicing the vehicle, refer to the Precautions Section.

➡**New O-ring seals must be installed any time the lines are disconnected from the steering gear.**

2. With the transmission in NEUTRAL, position the vehicle on a hoist.

3. Remove the wheel and tire assembly.

※※ WARNING

Do not allow the steering column shaft to rotate while the intermediate shaft is disconnected or damage to the clockspring can result. If there is evidence that the shaft has rotated, the clockspring must be removed and recentered.

※※ WARNING

The boots and clamps are designed to provide an airtight seal and protect the internal components of the steering gear. If the seal is not airtight the vacuum generated during turning will draw water and contamination into the gear causing premature damage.

※※ WARNING

Zip ties do not produce an airtight seal and must not be used.

※※ WARNING

The inner ball joint grease is not compatible with water and contamination trapped in the grease will degrade the life of the joint.

※※ WARNING

If present, the orientation of the vent tube must be noted so the boots and vent tubes can be installed in the correct location.

4. Remove the lower arm-to-frame nuts.

5. Remove the steering gear mounting bolt

6. Remove the steering gear mounting nut.

7. Remove the steering gear mounting bracket bolt.

8. Remove the steering gear mounting bracket.

9. Remove the steering column pinch bolt.

10. Remove the variable assist power steering (VAPS) switch electrical connector.

11. Remove the power steering hose mounting plate nuts.

12. Disconnect the power steering hoses.

13. Discard the O-rings.

14. Remove the tie-rod end-to-knuckle nut. If repairing the right side it will be necessary to pull back the left inner tie rod boot to hold the steering gear.

15. Note the number of times the tie-rod ends turn for assembly reference.

16. Remove the brake disc. Rotate the front of the brake rotor outward.

17. Remove the steering gear.

18. To install, reverse the removal procedure. Final tightening of the front suspension components should be carried out at or near the ride height (curb height) setting. Install new O-ring seals onto the power steering hoses.

19. See the accompanying illustration for relevant fastener torques.

20. Fill and leak test the system.

21. Check and, if necessary, adjust the front toe.

FRONT SUSPENSION

※※ WARNING

Suspension fasteners are critical components because they affect performance of vital components and systems and their failure can result in major service expense. New fasteners must be installed with the same component number or an equivalent component if installation is necessary. Do not use a replacement component of lesser quality or substitute design. Torque values must be used as specified during reassembly to make sure of correct retention of these components. Orientation of the fasteners is also important. Make sure the fasteners are installed in the same direction as they were in when removed.

Shock Absorber

REMOVAL & INSTALLATION

Aviator, Explorer And Mountaineer

1. Before servicing the vehicle, refer to the Precautions Section.

2. Remove or disconnect the following:

- Wheel
- Upper shock mounting nuts. Discard the nuts.
- Nut and the stabilizer bar link. Discard the nut.
- Bolt, flagnut and the spring and shock absorber as an assembly. Discard the flag nut.

3. Using a suitable spring compressor, compress the spring until the tension is released from the shock absorber.

4. While holding the flats of the washer, remove the nut.

5. Remove the shock absorber. Discard the nut, remove the washer, bushing and the upper mount.

6. Remove the insulator.

7. Remove the dust shield.

8. To install, reverse the removal procedure. Observe the following torques:

- Center nut: 37 ft. lbs. (50 Nm)
- Lower shock bolt: 258 ft. lbs. (350 Nm).
- Sway bar link nut: 18 ft. lbs. (25 Nm)
- Upper shock nuts: 20 ft. lbs. (30 Nm)

① 30 Nm (20 lb-ft)

②

③ 25 Nm (18 lb-ft)

④

350 Nm (258 lb-ft)

⑤

⑥

1 Shock absorber upper mount-to frame nuts (3 required)

2 Nut and grommet

3 Stud

4 Shock absorber-to-lower arm bolt

5 Shock absorber-to-lower arm flag nut

6 Shock absorber and spring assembly

06017-EXPL-G115

Front shock absorber and spring—Aviator, Explorer and Mountaineer

Explorer Sport And Sport-Trac

➡ **Low pressure gas shocks are charged with nitrogen gas. Do not attempt to open, puncture or apply heat to them. Prior to installing a new shock absorber, hold it upright and extend it fully. Invert it and fully compress and extend it at least 3 times. This will bleed trapped air.**

1. Before servicing the vehicle, refer to the Precautions Section.
2. Remove or disconnect the following:
   • Negative battery cable
   • Upper shock-to-frame attaching nut, washer and insulator assembly
   • Lower shock-to-control arm attaching nuts

   • Slightly compress the shock absorber by hand and remove it from the vehicle

To install:
3. Install or connect the following:
   • Position the lower washer and insulator on the shock absorber rod and position the shock absorber to the upper frame bracket mount
   • Position the upper insulator and washer on the shock absorber rod and install the attaching nut loosely.
   • Position the lower shock absorber mounting studs into the control arm and install the attaching nuts loosely.

   • Torque the lower shock attaching nuts to 15–21 ft. lbs. (21–29 Nm), and the upper shock attaching bolts to 30–40 ft. lbs. (40–55 Nm).
   • Negative battery cable

Coil Spring

REMOVAL & INSTALLATION

Aviator, Explorer and Mountaineer

1. Before servicing the vehicle, refer to the Precautions Section.
2. Remove the shock absorber and spring assembly.

1 Shock absorber-to-upper mount nut

2 Shock absorber (front/rear)

3 Washer

4 Upper shock mount (front/rear)

5 Coil spring (front/rear)

06017-EXPL-G117

Front coil spring exploded view—Aviator, Explorer and Mountaineer

06017-EXPL-G118

Spring mounted in compressor—Aviator, Explorer and Mountaineer

06017-EXPL-G119

Center nut removal—Aviator, Explorer and Mountaineer

3. Using a suitable spring compressor, compress the spring until the tension is released from the shock absorber.

4. While holding the flats of the washer, remove and discard the center nut.

5. Remove the shock absorber.

6. Remove the washer.

7. Remove the upper shock absorber mount.

8. Remove the coil spring.

9. To assemble, reverse the disassembly procedure. Always install a new shock absorber-to-upper mount nut. Tighten to 56 Nm (41 ft. lbs.).

Stabilizer Bar

REMOVAL & INSTALLATION

Aviator

1. Before servicing the vehicle, refer to the Precautions Section.

2. Remove and discard the 2 stabilizer bar nuts.

3. Remove the 2 stabilizer bar links.

➥Inspect and clean the mating surfaces and the internal threads. Make sure all mating surfaces are free of foreign material and remove any thread locking compound from the internal threads.

4. Remove and discard the 4 stabilizer bar-to-frame bolts.

5. Remove the stabilizer bar.

6. To install, reverse the removal procedure. Always install new stabilizer bar-to-frame bolts. See the accompanying illustration for relevant torque values.

Explorer and Mountaineer

1. Before servicing the vehicle, refer to the Precautions Section.

2. Remove and discard the 2 stabilizer bar nut and grommets.

3. Remove the 2 stabilizer bar studs.

4. Remove the 2 stabilizer bar links.

➥Inspect and clean the mating surfaces and the internal threads. Make sure all mating surfaces are free of foreign material and remove any thread locking compound from the internal threads.

5. Remove and discard the 4 stabilizer bar-to-frame bolts.

6. Remove the 2 stabilizer bar brackets.

7. Remove the stabilizer bar.

➥Inspect the bushings for wear or damage. Install new bushings as necessary.

90 Nm (66 lb-ft)

④ 63 Nm (46 lb-ft)

① 90 Nm (66 lb-ft)

1 Stabilizer bar link-to-lower arm nut

2 Stabilizer bar link-to-stabilizer bar nut

3 Stabilizer bar link

4 Stabilizer bar-to-frame bolts

5 Stabilizer bar, bushing and bracket assembly

06017-EXPL-G120

Front stabilizer bar—Aviator

70 Nm (52 lb-ft)

25 Nm (18 lb-ft)

1 Nut and grommet (2 required)

2 Stud (2 required)

3 Stabilizer bar link (2 required)

4 Stabilizer bar-to-frame bolts (4 required)

5 Bracket (2 required)

6 Stabilizer bar

7 Bushing (2 required)

06017-EXPL-G121

Front stabilizer bar—Explorer and Mountaineer

Front stabilizer end links—Explorer Sport and Sport-Trac

Front stabilizer bar brackets—Explorer Sport and Sport-Trac

8. Remove the 2 stabilizer bar bushings.

9. To install, reverse the removal procedure. Always install new stabilizer bar-to-frame bolts and stabilizer bar nut and grommets.

Explorer Sport and Sport-Trac

1. Before servicing the vehicle, refer to the Precautions Section.

2. Raise and support the vehicle.

3. Remove the two nuts and the two stabilizer bar links.

4. Remove the four bolts on the front stabilizer bar brackets and bushings.

5. To install, reverse the removal procedure. Torque the link nuts to 18 ft. lbs. (25 Nm); the bracket bolts to 30 ft. lbs. (40 Nm).

Torsion Bar

REMOVAL & INSTALLATION

Explorer Sport and Sport-Trac

1. Before servicing the vehicle, refer to the Precautions Section.

2. Raise and support the vehicle.

Torsion bar cover plate—Explorer and Explorer Sport-Trac

Measure and record the length where indicated—Explorer and Explorer Sport-Trac

Relieving torsion bar tension—Explorer and Explorer Sport-Trac

Removing torsion bar tension bolt and nut—Explorer and Explorer Sport-Trac

3. Remove the bolts and the torsion bar cover plate.

➡Before relieving the torsion bar tension, measure and record the measurement of the torsion bar adjustment

Removing all tension from the torsion bar—Explorer and Explorer Sport-Trac

Removing the torsion bar—Explorer and Explorer Sport-Trac

bolt. This measurement will be used as the preset depth for the new torsion bar adjustment bolt during installation.

4. Make preliminary adjustment references. Measure and record the length where indicated.

5. Relieve the torsion bar tension.

a. Position the special tool and adapters.

b. Tighten the special tool until the torsion bar adjuster lifts off the adjustment bolt.

✳✳ WARNING

The torsion bar adjustment bolt is coated with dry adhesive. A new bolt must be installed if it is backed off or removed. Failure to do so can cause the adjustment bolt to loosen during operation and cause a loss of vehicle alignment.

6. Remove the torsion bar adjustment bolt and nut.

7. Loosen the special tool until the tension is removed from the torsion bar.

8. Remove the torsion bar.

a. Mark the torsion bar and the adjuster for reference during installation.

b. Remove the torsion bar insulator.

c. Grasp the torsion bar, and pull it free from the lower arm.

Measure the distance between the center line of the front suspension lower arm bushing bolt and the lift—Explorer and Explorer Sport-Trac

Measure the distance between the front wheel spindle (lowest point) and the lift—Explorer and Explorer Sport-Trac

Subtract measurement 1 from measurement 2—Explorer and Explorer Sport-Trac

9. To install, reverse the removal procedure.

10. Adjust the ride height:

a. Drive the vehicle onto a drive-on lift.

b. Jounce the vehicle's front and rear suspension to normalize the vehicle static ride height.

c. Measure the distance between the center line of the front suspension lower arm bushing bolt and the lift. Record the measurement.

d. Measure the distance between the front wheel spindle (lowest point) and the lift. Record the measurement. Take the measurement from the midpoint of the surface shown.

e. Determine ride height. Subtract measurement 1 from measurement 2. This is the ride height.

➡ The torsion bar adjusting bolt is coated with adhesive that wears off after disassembly. If the torsion bar system is ever disassembled or the torsion bar adjusting bolt is ever removed, use a new torsion bar adjusting bolt when re-assembling.

f. Adjust the torsion bars (height) as necessary by tightening or loosening the torsion bar adjusting bolt. Tighten the torsion bar adjusting bolt to increase the torque or raise the height. Loosen the torsion bar adjusting bolt to decrease the torque or lower the height.

Upper Ball Joint

REMOVAL & INSTALLATION

Aviator, Explorer Sport and Sport-Trac

The ball joint is not replaceable. If defective, replace the arm.

Explorer and Mountaineer

2-WHEEL DRIVE

1. Before servicing the vehicle, refer to the Precautions Section.

2. With the vehicle in NEUTRAL, position it on a hoist.

➡ The wheel speed sensor connectors are located in the engine compartment and are secured to the fender aprons.

3. Disconnect the wheel speed sensor connector.

4. Remove the wheel and tire assembly.

5. Detach the harness from the retainers.

✳✳ WARNING

Do not allow the disc brake caliper to hang suspended from the brake hose. Provide suitable support.

6. Remove the bolts and position the brake caliper and support bracket aside.

7. Remove the brake disc.

8. Remove and discard the cotter pin and the outer tie rod end castellated nut.

✳✳ WARNING

Do not use a hammer to separate the tie-rod from the wheel knuckle or damage to the wheel knuckle will result.

Separating the tie rod end from the knuckle

✳✳ WARNING

Do not damage the tie rod boot when installing the tool.

9. Separate the tie rod from the wheel knuckle.

10. Remove and discard the upper ball joint nut.

✳✳ WARNING

Do not use a hammer to separate the ball joint from the wheel knuckle or damage to the wheel knuckle will result.

✳✳ WARNING

Do not damage the ball joint boot when installing the special tool.

Ball joint snapring

Removing the ball joint from the control arm

11. Separate the upper ball joint from the wheel knuckle.

12. Remove and discard the snapring from the ball joint.

13. Using a suitable ball joint remover tool, remove the ball joint.

✳✳ WARNING

Do not damage the ball joint boot when installing the special tool.

➡**Clean and inspect the control arm ball joint bore for damage before installing a new ball joint.**

➡**Make sure the new ball joint snapring is fully seated.**

14. To install, reverse the removal procedure.

15. Always install new:
  • snaprings.
  • castellated nuts.
  • cotter pins.
  • upper ball joint nuts.

16. Observe the following torques:
  • Outer tie rod end castellated nut: 70 Nm (52 ft. lbs.).
  • Upper ball joint nut: 55 Nm (41 ft. lbs.).

4-WHEEL DRIVE

1. Before servicing the vehicle, refer to the Precautions Section.

2. With the vehicle in NEUTRAL, position it on a hoist.

3. Remove and discard the axle retainer nut.

➡**The wheel speed sensor connectors are located in the engine compartment and are secured to the fender aprons.**

4. Disconnect the wheel speed sensor connector.

5. Remove the wheel and tire assembly.

6. Detach the harness from the retainers.

✳✳ WARNING

Do not allow the disc brake caliper to hang suspended from the brake hose. Provide a suitable support.

7. Remove the bolts and position the brake caliper and support bracket aside.

8. Remove the brake disc.

✳✳ WARNING

Do not use a hammer to separate the outboard CV joint from the hub. Damage to the threads and internal CV joint components can result.

9. Press the outboard CV joint until it is loose in the hub.

10. Remove and discard the cotter pin and the castellated nut.

✳✳ WARNING

Do not use a hammer to separate the tie-rod from the wheel knuckle or damage to the wheel knuckle will result.

✳✳ WARNING

Do not damage the tie-rod boot when installing the special tool.

11. Separate the tie-rod from the wheel knuckle.

12. Remove and discard the upper ball joint nut.

✳✳ WARNING

Secure the front axle shaft to prevent the CV joint and boots from overextending. Failure to do so can cause damage to the CV joint and boots.

✳✳ WARNING

Do not use a hammer to separate the ball joint from the wheel knuckle or damage to the wheel knuckle will result.

✳✳ WARNING

Do not damage the ball joint when installing the special tool.

13. Separate the upper ball joint from the wheel knuckle.

14. Remove the hand-tightened lower ball joint castellated nut and remove the wheel knuckle.

15. Remove and discard the snapring from the ball joint.

16. Using a suitable ball joint remover tool, remove the ball joint.

✳✳ WARNING

Do not damage the ball joint boot when installing the special tool.

➡**Apply a thin coat of silicone sealant to the wheel hub mounting surfaces before installation.**

➡**Clean and inspect the control arm ball joint bore for damage before installing a new ball joint.**

➡**Make sure the new ball joint snapring is fully seated.**

17. To install, reverse the removal procedure.

18. Always install new:
  • castellated nuts.
  • cotter pins.
  • snaprings.
  • upper ball joint nuts.
  • axle retainer nut.

19. Observe the following torques:
  • Axle retainer nut: 250 Nm (184 ft. lbs.).
  • Tie rod end stud nut: 70 Nm (52 ft. lbs.).
  • Upper ball joint nut: 55 Nm (41 ft. lbs.).

Lower Ball Joint

REMOVAL & INSTALLATION

Aviator

The ball joints on the Aviator are not replaceable.

Explorer and Mountaineer; Explorer Sport and Sport-Trac

2-WHEEL DRIVE

1. Before servicing the vehicle, refer to the Precautions Section.

2. With the vehicle in NEUTRAL, position it on a hoist.

3. Remove the wheel and tire assembly.

✳✳ WARNING

Do not allow the disc brake caliper to hang suspended from the brake hose. Provide a suitable support.

4. Remove the caliper support bracket bolts, then position the caliper and support bracket aside.

5. Disconnect the ABS electrical connector.

6. Unclip the front ABS wire from the vehicle frame.

7. Using a suitable jack, support the front suspension lower arm.

8. Remove the tie-rod end castellated nut. Remove and discard the cotter pin and the castellated nut.

✳✳ WARNING

Do not use a hammer to separate the tie-rod from the wheel knuckle or damage to the wheel knuckle will result.

✪ WARNING

Do not damage the tie-rod boot when installing the special tool.

9. Separate the tie-rod end from the front wheel knuckle.

10. Remove the lower ball joint castellated nut.

11. Separate the front wheel knuckle from the front suspension lower arm. Then, loosely install the lower ball joint castellated nut.

12. Remove the pinch bolt and nut.

13. Remove the hand-tightened lower ball joint castellated nut, then remove the front wheel knuckle.

14. Remove the snapring from the ball joint. Discard the snapring.

15. Using a suitable ball joint remover tool, remove the ball joint.

✪ WARNING

Do not damage the ball joint boot when installing the special tool.

➡ Clean and inspect the control arm ball joint bore for damage before installing a new ball joint.

➡ Make sure the new ball joint snapring is fully seated.

16. To install, reverse the removal procedure. Always install new castellated nuts and cotter pins. Torque the ball joint nut to 98 ft. lbs. (133 Nm) for Explorer Sport and Sport-Trac models; 129 ft. lbs. (175 Nm) for Explorer and Mountaineer models. Torque the tie rod end ball stud nut to 52 ft. lbs. (70 Nm). Torque the pinch bolt to 41 ft. lbs. (55 Nm).

4-Wheel Drive

1. Before servicing the vehicle, refer to the Precautions Section.

2. With the vehicle in NEUTRAL, position it on a hoist.

3. Remove the wheel and tire assembly.

�֍ WARNING

Do not reuse the torque prevailing design hub nut and washer assembly.

4. Remove and discard the hub nut and washer assembly.

�֍ WARNING

Do not allow the disc brake caliper to hang suspended from the brake hose. Provide a suitable support.

5. Remove the caliper support bracket bolts, then position the caliper and support bracket aside.

6. Remove the brake disc.

✳ WARNING

Do not use a hammer to separate the outboard front wheel halfshaft joint from the wheel hub. Damage to the outboard CV joint stub shaft threads and internal CV joint components may result.

7. Separate the outboard front wheel halfshaft joint from the wheel hub.

8. Disconnect the ABS electrical connector.

9. Unclip the front ABS wire from the vehicle frame.

10. Using a suitable jack, support the front suspension lower arm.

✳ WARNING

Secure the front axle shaft to prevent it from overextending. Failure to do so can cause damage to the front axle shaft.

11. Support the front axle shaft with wire.

12. Remove the tie-rod end castellated nut. Discard the cotter pin. Discard the castellated nut.

✳ WARNING

Do not use a hammer to separate the tie-rod from the wheel knuckle or damage to the wheel knuckle will result.

✳ WARNING

Do not damage the tie-rod boot when installing the special tool.

13. Separate the tie-rod end from the front wheel knuckle.

14. Remove the lower ball joint castellated nut. Discard the cotter pin. Discard the castellated nut.

15. Separate the front wheel knuckle from the front suspension lower arm then, loosely install the lower ball joint castellated nut.

16. Remove the pinch bolt and nut.

17. Remove the hand-tightened lower ball joint castellated nut, then remove the front wheel knuckle.

18. Remove the snapring from the ball joint. Discard the snapring.

19. Using a suitable ball joint remover tool, remove the ball joint.

✳ WARNING

Do not damage the ball joint boot when installing the special tool.

➡ Clean and inspect the control arm ball joint bore for damage before installing a new ball joint.

➡ Make sure the new ball joint snapring is fully seated.

20. To install, reverse the removal procedure. Always install new castellated nuts and cotter pins. Observe the following torques:

- Hub nut to 162 ft. lbs. (220 Nm) for Explorer Sport and Sport-Trac models; 184 ft. lbs. (250 Nm) for Explorer and Mountaineer models.
- Tie rod end nut to 52 ft. lbs. (70 Nm)
- Ball joint nut to 98 ft. lbs. (133 Nm) for Explorer Sport and Sport-Trac models; 129 ft. lbs. (175 Nm) for Explorer and Mountaineer models
- Pinch bolt to 41 ft. lbs. (55 Nm).

Upper Control Arm

REMOVAL & INSTALLATION

Explorer Sport and Sport-Trac

1. Before servicing the vehicle, refer to the Precautions Section.

2. Remove or disconnect the following:

- Wheel
- Pinch bolt and nut from the wheel knuckle. Remove and discard the nut.
- Arm-to-frame nuts and shims. Discard the nuts.
- Upper arm

3. To install, reverse the removal procedure. Torque the upper arm nuts to 98 ft. lbs. (133 Nm). Torque the pinch bolt and nut to 41 ft. lbs. (55 Nm).

4. Check and, if necessary, align the front end.

2002 Explorer and Mountaineer

1. Before servicing the vehicle, refer to the Precautions Section.

➡ The wheel speed sensor connectors are located in the engine compartment and are secured to the fender aprons.

1 Speed sensor harness

2 Brake hose-to-wheel knuckle bolt

3 Anchor plate bolt (2 required)

4 Brake caliper, pads and anchor plate

5 Brake disc

6 Wheel hub-to-wheel knuckle bolt (3 required)

7 Wheel bearing and hub assembly

8 Cotter pins (2 required)

9 Tie-rod end-to-wheel knuckle nut

10 Lower ball joint-to-wheel knuckle nut

11 Upper ball joint-to-wheel knuckle nut

12 Wheel knuckle

06017-EXPL-G143A

Front knuckle and related parts—2-wheel drive Explorer and Mountaineer

| ST N i 12 – 52 Nm (38 lb-ft) | 4 – 112 Nm (83 lb-ft) |

13 i

15 Nm (11 lb-ft) – 3

i 2

i 5

112 Nm (83 lb-ft) – 7 N

9 N

70 Nm (52 lb-ft) – 10 ST N i

ST N i 1 – 250 Nm (184 lb-ft)

9 N

175 Nm (129 lb-ft) – 11 ST N i

8

6

1 Axle-to-wheel hub nut (4x4 only)

2 Speed sensor harness

3 Brake hose-to-wheel knuckle bolt

4 Anchor plate bolt (2 required)

5 Brake caliper, pads and anchor plate

6 Brake disc

7 Wheel hub-to-wheel knuckle bolt (3 required)

8 Wheel bearing and hub assembly

9 Cotter pins

10 Tie-rod end-to-wheel knuckle nut

11 Lower ball joint-to-wheel knuckle nut

12 Upper ball joint-to-wheel knuckle nut

13 Wheel knuckle

06017-EXPL-G144

Front knuckle and related parts—4-wheel drive Explorer and Mountaineer

14 Upper arm-to-frame nuts
15 Set shims
16 Upper arm RH/LH
17 Nut and grommet
18 Stud
19 Stabilizer bar link
20 Shock absorber upper mount-to-frame nuts

21 Shock absorber-to-lower arm bolt
22 Shock absorber-to-lower arm flag nut
23 Shock absorber and spring assembly
24 Lower arm-to-frame nut (forward mounting)
25 Lower arm-to-frame flag bolt (forward mounting)
26 Lower arm-to-frame nuts (rearward mounting)
27 Lower Arm

67197-EXPL-G68

1 Speed sensor harness

2 Brake hose-to-wheel knuckle bolt

3 Anchor plate bolt (2 required)

4 Brake caliper, pads and anchor plate

5 Brake disc

6 Wheel hub-to-wheel knuckle bolt (3 required)

7 Wheel bearing and hub assembly

8 Cotter pins (2 required)

9 Tie-rod end-to-wheel knuckle nut

10 Lower ball joint-to-wheel knuckle nut

11 Upper ball joint-to-wheel knuckle nut

12 Wheel knuckle

06017-EXPL-G145

Front knuckle and related parts—2-wheel drive Aviator

1 Axle-to-wheel hub nut

2 Speed sensor harness

3 Brake hose-to-wheel knuckle bolt

4 Anchor plate bolt (2 required)

5 Brake caliper, pads and anchor plate

6 Brake disc

7 Wheel hub-to-wheel knuckle bolt (3 required)

8 Wheel bearing and hub assembly

9 Cotter pins (2 required)

10 Tie-rod end-to-wheel knuckle nut

11 Lower ball joint-to-wheel knuckle nut

12 Upper ball joint-to-wheel knuckle nut

13 Wheel knuckle

06017-EXPL-G146

Front knuckle and related parts—4-wheel drive Aviator

14 Upper arm-to-frame nuts

15 Set shims

16 Upper arm

17 Stabilizer bar link-to-lower arm nut

18 Shock absorber-to-lower arm bolt

19 Shock absorber-to-lower arm flag nut

20 Lower arm-to-frame nut (forward mounting)

21 Lower arm-to-frame flag bolt (forward mounting)

22 Lower arm-to-frame nut (rearward mounting)

23 Lower arm-to-frame flag bolt (rearward mounting)

24 Lower arm

67197-EXPL-G69

Front suspension components—Aviator

2. Disconnect the wheel speed sensor connector.

3. Remove the wheel and tire assembly.

4. Remove and discard the nut.

❋❋ WARNING

Do not use a hammer to separate the ball joint from the wheel knuckle or damage to the wheel knuckle will result. Do not damage the ball joint boot while installing the special tool.

5. Using a puller, separate the upper ball joint from the wheel knuckle.

6. Remove the upper arm nuts and the shims. Discard the nuts.

7. Remove the upper arm.

8. To install, reverse the removal procedure. Torque the ball joint nut to 41 ft. lbs.

(55 Nm); the upper arm nuts to 111 ft. lbs. (150 Nm). Use new nuts. Check and, if necessary, align the front end.

2003–05 Aviator, Explorer and Mountaineer

1. Before servicing the vehicle, refer to the Precautions Section.

2. On 4x4 vehicles, loosen the axle retainer nut.

➡The wheel speed sensor electrical connectors are located in the engine compartment secured to the fender aprons.

3. Disconnect the wheel speed sensor.

4. Remove the wheel and tire assembly.

5. Remove the upper ball joint-to-wheel

knuckle nut. Separate the ball joint from the knuckle with a ball joint driver.

❋❋ WARNING

Do not use a hammer to separate the ball joint from the wheel knuckle or damage to the wheel knuckle can result.

6. Remove the upper arm-to-frame nuts.

7. Remove the set shims.

8. Remove the upper arm.

9. To install, reverse the removal procedure. Check and, if necessary, align the front end. Torque the ball joint nut to 41 ft. lbs. (55 Nm) for 2wd and 38 ft. lbs. (52 Nm) for 4wd; the upper arm nuts to 111 ft. lbs. (150 Nm).

UPPER CONTROL ARM BUSHING REPLACEMENT

The control arm bushings are not serviceable. If they require service, the upper or lower arm must be replaced.

Lower Control Arm

REMOVAL & INSTALLATION

Aviator, Explorer and Mountaineer

1. Before servicing the vehicle, refer to the Precautions Section.
2. Remove or disconnect the following:

➡**For reference during the installation procedure, measure the distance from the lip of the fender to the center of the wheel hub with the vehicle in a level static ground position.**

- Wheel
- Upper shock absorber mounting nuts. Discard the nuts.
- Stabilizer bar connecting link. Discard the nut.
- Bolt, flag nut and the spring and shock assembly. Discard the flag nut.
- Separate the ball joint from the wheel knuckle. Discard the nut.
- Arm-to-frame nuts. Discard the nuts.
- Arm-to-knuckle bolt. Discard the nut.

➡**On 4x4 vehicles, make sure that the crimped area of the outer CV boot clamp is not positioned downward or it will interfere with the removal of the arm.**

- Lower arm

➡**Using a suitable jack stand, raise the suspension until the distance between the lip of the fender and the center of the wheel hub is equal to the measurement taken in the removal procedure before tightening the inboard lower arm mountings. Make sure that the forward mounting is tightened first.**

3. To install, reverse the removal procedure. Observe the following torques:
- Control arm-to-frame flag bolt: 295 ft. lbs. (400 Nm)
- Control arm-to-frame nuts: 111 ft. lbs. (150 Nm)
- Ball joint nut: 129 ft. lbs. (175 Nm) for 4wd Explorer/Mountaineer; 111 ft. lbs. (150 Nm) for Aviator and 2wd Explorer/Mountaineer
- Shock lower flag bolt: 258 ft. lbs. (350 Nm)
- Sway bar link nut: 18 ft. lbs. (25 Nm)
- Upper shock mounting nuts: 22 ft. lbs. (30 Nm)
4. Check and, if necessary, align the front end.

Explorer Sport and Sport-Trac

1. Before servicing the vehicle, refer to the Precautions Section.
2. Remove the wheel and tire assembly.
3. Remove the stabilizer link nut, washer and bushing.
4. Remove the front shock absorber-to-front suspension lower arm nuts.
5. Remove the torsion bar.
6. Remove the cotter pin and the nut.

✱✱ WARNING

Do not use a hammer to separate the ball joint from the wheel knuckle or damage to the wheel knuckle will result. Do not damage the ball joint boot while installing the special tool.

7. Using a puller, separate the front suspension lower arm from the front wheel knuckle/spindle.
8. Remove the front suspension lower arm bolts and nuts.
9. Remove the front suspension lower arm.
10. To install, reverse the removal procedure. Observe the following torques:
- Stabilizer bar link: 18 ft. lbs. (25 Nm)
- Shock absorber-to-lower arm: 18 ft. lbs. (25 Nm)
- Ball joint stud nut: 98 ft. lbs. (133 Nm)
- Lower arm nuts: 148 ft. lbs. (200 Nm)
11. Check and, if necessary, adjust the ride height.
12. Check and, if necessary, align the front end.

LOWER CONTROL ARM BUSHING REPLACEMENT

The control arm bushings are not serviceable. If they require service, the upper or lower arm must be replaced.

Spindle/Knuckle

REMOVAL & INSTALLATION

Explorer Sport and Sport-Trac

2-WHEEL DRIVE

1. Before servicing the vehicle, refer to the Precautions Section.
2. Remove the brake disc and hub .
3. Remove the anti-lock sensor electrical wire from the clamp. Remove the bolt and position the sensor aside.
4. Remove the cotter pin and nut.
5. Using a puller, separate the tie-rod end from the spindle.
6. Use a suitable jack stand to support the lower arm.

✱✱ WARNING

To avoid possible damage to the front wheel spindle, secure the spindle to keep it from tilting before removing the pinch bolt and nut.

7. Remove the nut and pinch bolt.
8. Remove the cotter pin and nut.
9. Using a puller, separate the ball joint and remove the spindle.
10. To install, reverse the removal procedure.

✱✱ WARNING

Install the cotter pin into the lower ball joint from outboard to inboard with the fingers bent together at a right angle. Failure to do so can cause damage to the wheel and tire assembly.

11. Observe the following torques:
- Lower ball joint stud nut: 98 ft. lbs. (133 Nm)
- Pinch bolt: 41 ft. lbs. (55 Nm)
- Tie rod ball stud nut: 52 ft. lbs. (70 Nm)

70 Nm (52 lb-ft)

06017-EXPL-G147

Correct lower ball joint cotter pin orientation—2-wheel drive Sport and Sport-Trac

2002–04 4-WHEEL DRIVE

1. Before servicing the vehicle, refer to the Precautions Section.

2. Remove the center cap from the front wheel.

3. Loosen the hub nut.

4. Put the vehicle in NEUTRAL and position on a hoist.

5. Remove and discard the nut and washer assembly.

6. Remove the anchor plate bolts. Position aside the anchor plate, brake caliper and pads as an assembly.

7. Remove the brake disc.

8. Using a puller, separate the outboard front wheel halfshaft joint from the wheel hub.

9. Remove the brake shield.

10. Remove the bolt and the wheel speed sensor from the wheel hub.

11. Remove the three bolts and the wheel hub.

12. Remove the tie-rod end castellated nut.

✳✳ WARNING

Do not use a hammer to separate the tie-rod from the wheel knuckle or damage to the wheel knuckle will result. Do not damage the tie-rod boot when installing the special tool.

13. Using a puller, separate the tie-rod end from the front wheel knuckle.

14. Support the front suspension lower arm securely.

15. Remove the lower ball joint castellated nut.

✳✳ WARNING

Do not use a hammer to separate the ball joint from the wheel knuckle or damage to wheel knuckle will result. Do not damage the ball joint boot when installing the special tool.

16. Separate the front wheel knuckle from the front suspension lower arm.

17. Remove the pinch bolt. Remove the wheel knuckle.

18. To install, reverse the removal procedure.

19. Observe the following torques:
- Pinch bolt: 41 ft. lbs. (55 Nm)
- Lower ball joint stud nut: 98 ft. lbs. (133 Nm)
- Hub/bearing bolts: 85 ft. lbs. (115 Nm)
- Brake shield: 108 inch lbs. (12 Nm)

- Anchor plate bolts: 85 ft. lbs. (115 Nm)
- Halfshaft nut: 162 ft. lbs. (220 Nm)

2005 4-WHEEL DRIVE

1. Before servicing the vehicle, refer to the Precautions Section.

2. Remove the wheel hub.

3. Remove all tension from the torsion bar. See Torsion Bar Removal & Installation in this Chapter.

✳✳ WARNING

Secure the front axle shaft to prevent it from overextending. Failure to do so can cause damage to the front axle shaft.

4. Suspend the front axle shaft with wire.

5. Remove the tie-rod end cotter pin. Remove the tie-rod end castellated nut.

6. Separate the tie-rod end from the front wheel knuckle.

7. Remove the lower ball joint cotter pin.

8. Remove the lower ball joint castellated nut.

9. Separate the front wheel knuckle from the front suspension lower arm.

10. Remove the pinch bolt and the wheel knuckle.

11. Installation is the reverse of removal. Observe the following torques:
- Pinch bolt: 41 ft. lbs. (55 Nm)
- Lower ball joint stud nut: 98 ft. lbs. (133 Nm)
- Tie rod end stud nut: 52 ft. lbs. (70 Nm)

12. Adjust the ride height.

13. Check and if necessary align the front end.

Aviator

2-WHEEL DRIVE

➡**The wheel speed sensor electrical connectors are located in the engine compartment secured to the fender aprons.**

1. Before servicing the vehicle, refer to the Precautions Section.

2. Disconnect the wheel speed sensor.

3. Remove the wheel and tire assembly.

4. Detach the wheel speed sensor harness from the brake hose.

5. Remove the brake hose-to-wheel knuckle bolt.

6. Remove the 2 anchor plate bolts.

✳✳ WARNING

Do not allow the brake caliper to hang from the hose or damage to the hose can occur.

7. Position the caliper, pads and anchor plate aside.

8. Remove the brake disc.

9. Remove and discard the 3 wheel hub-to-wheel knuckle bolts.

10. Remove the wheel bearing and hub assembly.

11. Remove the discard the 2 cotter pins.

12. Remove and discard the tie-rod end-to-wheel knuckle castellated nut.

✳✳ WARNING

Do not use a hammer to separate the tie-rod from the wheel knuckle or damage to the wheel knuckle can result. Do not damage the tie-rod boot while installing the special tool.

13. Using a puller, separate the tie-rod from the wheel knuckle.

14. Remove and discard the lower ball joint-to-wheel knuckle castellated nut.

✳✳ WARNING

Do not use a prying device or separator fork between the upper or lower arm and the knuckle. Damage to the ball joint or ball joint dust boot can result. Do not strike the wheel knuckle with a hammer to separate the ball joint from the wheel knuckle. Damage to the wheel knuckle can result. Make sure to apply direct force to the end of the ball joint stud when separating the ball joint from the knuckle. Do not damage the boot when installing a suitable separator.

15. Using a suitable separator, separate the lower ball joint from the wheel knuckle.

16. Remove and discard the upper ball joint-to-wheel knuckle nut.

✳✳ WARNING

Do not use a hammer to separate the ball joint from the wheel knuckle or damage to the wheel knuckle can result. Do not damage the ball joint boot while installing the special tool.

17. Using the special tool, separate the upper ball joint from the wheel knuckle.

18. Remove the wheel knuckle.

19. To install, reverse the removal procedure.

20. Observe the following torques:
- Upper ball joint stud nut: 38 ft. lbs. (52 Nm)
- Lower ball joint stud nut: 111 ft. lbs. (150 Nm)
- Tie-rod end stud nut: 52 ft. lbs. (70 Nm)
- Wheel hub-to-wheel knuckle bolts: 83 ft. lbs. (112 Nm)
- Anchor plate bolts: 155 ft. lbs. (210 Nm)

21. Check and, if necessary, align the front end.
22. Always install new:
- Wheel hub-to-wheel knuckle bolts.
- Cotter pins.
- Castellated nuts.
- Upper ball joint-to-wheel knuckle nut.

4-WHEEL DRIVE

1. Before servicing the vehicle, refer to the Precautions Section.
2. Loosen the axle retainer nut.

➡ The wheel speed sensor electrical connectors are located in the engine compartment secured to the fender aprons.

3. Disconnect the wheel speed sensor.
4. Remove the wheel and tire assembly.
5. Remove and discard the axle retainer nut.
6. Separate the outboard CV joint from the wheel hub.
7. Detach the wheel speed sensor harness from the brake hose.
8. Remove the brake hose-to-wheel knuckle bolt.
9. Remove the 2 anchor plate bolts.

✳✳ WARNING

Do not allow the brake caliper to hang from the hose or damage to the hose can occur.

10. Position the caliper, pads and anchor plate aside.
11. Remove the brake disc.
12. Remove and discard the 3 wheel hub-to-wheel knuckle bolts.
13. Remove the wheel bearing and hub assembly.
14. Remove and discard the 2 cotter pins.
15. Remove and discard the tie-rod end-to-wheel knuckle castellated nut.

✳✳ WARNING

Do not use a hammer to separate the tie-rod from the wheel knuckle or

damage to the wheel knuckle can result. Do not damage the tie-rod boot while installing the special tool.

16. Separate the tie-rod from the wheel knuckle.
17. Remove and discard the lower ball joint-to-wheel knuckle castellated nut.

✳✳ WARNING

Do not use a hammer to separate the ball joint from the wheel knuckle or damage to the wheel knuckle can result. Do not damage the ball joint boot while installing the special tool.

18. Separate the lower ball joint from the wheel knuckle.
19. Remove and discard the upper ball joint-to-wheel knuckle nut.

✳✳ WARNING

Do not use a hammer to separate the ball joint from the wheel knuckle or damage to the wheel knuckle can result. Do not damage the ball joint boot while installing the special tool.

20. Separate the upper ball joint from the wheel knuckle.
21. Remove the wheel knuckle.
22. To install, reverse the removal procedure.
23. Observe the following torques:
- Upper ball joint-to-knuckle nut: 41 ft. lbs. (55 Nm)
- Lower ball joint-to-knuckle nut: 111 ft. lbs. (150 Nm)
- Tie-rod end-to-knuckle nut: 52 ft. lbs. (70 Nm)
- Wheel hub-to- knuckle bolts: 83 ft. lbs. (112 Nm)
- Anchor plate bolts: 155 ft. lbs. (210 Nm)
- Axle retainer nut: 184 ft. lbs. (250 Nm)

24. Check and, if necessary, align the front end.
25. Always install new:
- Wheel hub-to-wheel knuckle bolts.
- Cotter pins.
- Castellated nuts.
- Upper ball joint-to-wheel knuckle nut.
- Axle retainer nut.

2002 Explorer and Mountaineer

2-WHEEL DRIVE

1. Before servicing the vehicle, refer to the Precautions Section.

2. Remove the wheel hub
3. Remove and discard the cotter pin and nut.

✳✳ WARNING

Do not use a hammer to separate the tie-rod from the wheel knuckle or damage to the wheel knuckle will result. Do not damage the tie-rod boot while installing the special tool.

4. Separate the tie-rod from the wheel knuckle. Remove and discard the cotter pin and the nut.

✳✳ WARNING

Do not use a hammer to separate the ball joint from the wheel knuckle or damage to the wheel knuckle will result. Do not damage the ball joint boot while installing the special tool.

5. Separate the lower ball joint from the wheel knuckle.
6. Remove and discard the nut.

✳✳ WARNING

Do not use a hammer to separate the ball joint from the wheel knuckle or damage to the wheel knuckle will result. Do not damage the ball joint boot while installing the special tool.

7. Separate the upper ball joint from the wheel knuckle and remove the wheel knuckle.
8. To install, reverse the removal procedure. Observe the following torques:
- Upper ball joint stud nut: 41 ft. lbs. (55 Nm)
- Lower ball joint stud nut: 129 ft. lbs. (175 Nm)
- Tie-rod end stud nut: 52 ft. lbs. (70 Nm)

9. Check and, if necessary, align the front end.

4-WHEEL DRIVE

1. Before servicing the vehicle, refer to the Precautions Section.
2. Remove the wheel hub.
3. Remove and discard the cotter pin and the nut.

✳✳ WARNING

Do not use a hammer to separate the tie-rod from the wheel knuckle or damage to the wheel knuckle will result.

✳✳ WARNING

Do not damage the tie-rod boot when installing the special tool.

4. Separate the tie-rod from the wheel knuckle.

5. Remove and discard the cotter pin and the nut.

✳✳ WARNING

Do not use a hammer to separate the ball joint from the wheel knuckle or damage to the wheel knuckle will result.

✳✳ WARNING

Do not damage the ball joint boot when installing the special tool.

6. Separate the lower ball joint from the wheel knuckle.

7. Remove and discard the nut.

✳✳ WARNING

Secure the front axle shaft to prevent the CV joint and boots from overextending. Failure to do so can cause damage to the CV joint and boots.

✳✳ WARNING

Do not use a hammer to separate the ball joint from the wheel knuckle or damage to the wheel knuckle will result.

✳✳ WARNING

Do not damage the ball joint boot when installing the special tool.

8. Separate the upper ball joint from the wheel knuckle and remove the wheel knuckle.

9. To install, reverse the removal procedure. Observe the following torques:
- Upper ball joint stud nut: 41 ft. lbs. (55 Nm)
- Lower ball joint stud nut: 129 ft. lbs. (175 Nm)
- Tie-rod end stud nut: 52 ft. lbs. (70 Nm)

2003–05 Explorer and Mountaineer

2-WHEEL DRIVE

➡The wheel speed sensor electrical connectors are located in the engine compartment secured to the fender aprons.

1. Before servicing the vehicle, refer to the Precautions Section.

2. Disconnect the wheel speed sensor.

3. Remove the wheel and tire assembly.

4. Detach the wheel speed sensor harness from the brake hose.

5. Remove the brake hose-to-wheel knuckle bolt.

6. Remove the 2 anchor plate bolts.

✳✳ WARNING

Do not allow the brake caliper to hang from the hose or damage to the hose can occur.

7. Position the caliper, pads and anchor plate aside.

8. Remove the brake disc.

9. Remove and discard the 3 wheel hub-to-wheel knuckle bolts.

10. Remove the wheel bearing and hub assembly.

11. Remove the discard the 2 cotter pins.

12. Remove and discard the tie-rod end-to-wheel knuckle castellated nut.

✳✳ WARNING

Do not use a hammer to separate the tie-rod from the wheel knuckle or damage to the wheel knuckle can result. Do not damage the tie-rod boot while installing the special tool.

13. Separate the tie-rod from the wheel knuckle.

14. Remove and discard the lower ball joint-to-wheel knuckle castellated nut.

✳✳ WARNING

Do not use a hammer to separate the ball joint from the wheel knuckle or damage to the wheel knuckle can result. Do not damage the ball joint boot while installing the special tool.

15. Separate the lower ball joint from the wheel knuckle.

16. Remove and discard the upper ball joint-to-wheel knuckle nut.

✳✳ WARNING

Do not use a hammer to separate the ball joint from the wheel knuckle or damage to the wheel knuckle can result. Do not damage the ball joint boot while installing the special tool.

17. Separate the upper ball joint from the wheel knuckle.

18. Remove the wheel knuckle.

19. To install, reverse the removal procedure. Observe the following torques:
- Upper ball joint-to-knuckle nut: 38 ft. lbs. (52 Nm)
- Lower ball joint-to-knuckle nut: 111 ft. lbs. (150 Nm)
- tie-rod end-to-knuckle nut: 52 ft. lbs. (70 Nm)
- Hub-to-wheel knuckle bolts: 83 ft. lbs. (112 Nm)
- Anchor plate bolts: 83 ft. lbs. (112 Nm)

20. Check and, if necessary, align the front end.

21. Always install new:
- Wheel hub-to-wheel knuckle bolts.
- Cotter pins.
- Castellated nuts.
- Upper ball joint-to-wheel knuckle nut.

4-WHEEL DRIVE

1. Before servicing the vehicle, refer to the Precautions Section.

2. Loosen the axle retainer nut.

➡The wheel speed sensor electrical connectors are located in the engine compartment secured to the fender aprons.

3. Disconnect the wheel speed sensor.

4. Remove the wheel and tire assembly.

5. Remove and discard the axle retainer nut.

6. Using the special tool, separate the outboard CV joint from the wheel hub.

7. Detach the wheel speed sensor harness from the brake hose.

8. Remove the brake hose-to-wheel knuckle bolt.

9. Remove the 2 anchor plate bolts.

✳✳ WARNING

Do not allow the brake caliper to hang from the hose or damage to the hose can occur.

10. Position the caliper, pads and anchor plate aside.

11. Remove the brake disc.

12. Remove and discard the 3 wheel hub-to-wheel knuckle bolts.

13. Remove the wheel bearing and hub assembly.

14. Remove the discard the 2 cotter pins.

15. Remove and discard the tie-rod end-to-wheel knuckle castellated nut.

✳✳ WARNING

Do not use a hammer to separate the tie-rod from the wheel knuckle or damage to the wheel knuckle can result. Do not damage the tie-rod boot while installing the special tool.

16. Separate the tie-rod from the wheel knuckle.
17. Remove and discard the lower ball joint-to-wheel knuckle castellated nut.

✳✳ WARNING

Do not use a hammer to separate the ball joint from the wheel knuckle or damage to the wheel knuckle can result. Do not damage the ball joint boot while installing the special tool.

18. Separate the lower ball joint from the wheel knuckle.
19. Remove and discard the upper ball joint-to-wheel knuckle nut.

✳✳ WARNING

Do not use a hammer to separate the ball joint from the wheel knuckle or damage to the wheel knuckle can result. Do not damage the ball joint boot while installing the special tool.

20. Using the special tool, separate the upper ball joint from the wheel knuckle.
21. Remove the wheel knuckle.
22. To install, reverse the removal procedure.
23. Observe the following torques:
   - Upper ball joint-to-wheel knuckle nut: 38 ft. lbs. (52 Nm)
   - Lower ball joint-to-knuckle nut: 129 ft. lbs. (175 Nm)
   - Tie-rod end-to-knuckle nut: 52 ft. lbs. (70 Nm)
   - Hub-to-wheel knuckle bolts: 83 ft. lbs. (112 Nm)
   - Anchor plate bolts: 83 ft. lbs. (112 Nm)
   - Axle retainer nut: 184 ft. lbs. (250 Nm)
24. Check and, if necessary, align the front end.
25. Always install new:
   - Wheel hub-to-wheel knuckle bolts.
   - Cotter pins.
   - Castellated nuts.
   - Upper ball joint-to-wheel knuckle nut.
   - Axle retainer nut.

Front Wheel Bearings

ADJUSTMENT

Only the Explorer Sport and Sport-Trac 2-wheel drive bearings are adjustable.

2-Wheel Drive Explorer Sport and Sport-Trac

1. Before servicing the vehicle, refer to the Precautions Section.
2. Remove the front disc brake caliper anchor plate.

➡ **Match-mark the brake disc and hub.**

3. Remove the hub grease cap.
4. Remove the cotter pin.
5. Remove the nut retainer.
6. Remove the spindle nut.
7. Remove the front wheel outer bearing retainer washer.
8. Remove the outer front wheel bearing.
9. Remove the brake disc and hub.
10. Remove the wheel hub grease seal and wheel bearing.
   To install:
11. Clean and inspect the front wheel bearings and the brake disc and hub. Use Metal Brake Parts Cleaner F6AZ-2C410-AB, or equivalent.
12. Lubricate the front wheel bearings. Use Premium Long-Life Grease XG-1-C or -K or equivalent meeting Ford specification ESA-M1C75-B.
13. Install the inner front wheel bearing.
14. Install a new wheel hub grease seal.
15. Position the brake disc and hub.
16. Install the outer front wheel bearing.
17. Install the front wheel outer bearing retainer washer.
18. Install the spindle nut.
19. Tighten the spindle nut to 21 ft. lbs. (29 Nm) while rotating the brake disc and hub.
20. Loosen the spindle nut one-half turn.

67197-EXPL-G65

Front wheel bearings—2-wheel drive Explorer Sport and Sport-Trac

21. Tighten the spindle nut to 17 inch lbs. (2 Nm) while rotating the brake disc and hub.
22. Install the nut retainer.
23. Install the cotter pin.
24. Install the hub grease cap.
25. Install the front disc brake caliper anchor plate.

REMOVAL & INSTALLATION

Explorer Sport and Sport-Trac

2-WHEEL DRIVE

1. Before servicing the vehicle, refer to the Precautions Section.
2. Remove the front disc brake caliper anchor plate.

➡ **Match-mark the brake disc and hub.**

3. Remove the hub grease cap.
4. Remove the cotter pin.
5. Remove the nut retainer.
6. Remove the spindle nut.
7. Remove the front wheel outer bearing retainer washer.
8. Remove the outer front wheel bearing.
9. Remove the brake disc and hub.
10. Remove the wheel hub grease seal and inner wheel bearing.
11. Place the hub on a solid wood work surface.
12. Using a suitable drift, drive out the inner bearing race; then, in a similar fashion, drive out the outer bearing race.
   To install:
13. Clean and inspect the front wheel bearings and the brake disc and hub. Use Metal Brake Parts Cleaner F6AZ-2C410-AB, or equivalent.
14. Pack the front wheel bearings. Use Premium Long-Life Grease XG-1-C or -K or equivalent meeting Ford specification ESA-M1C75-B.
15. Lubricate the bearing race bore. Using a suitable driver, drive in a new outer bearing race until it fully seats; then, in a similar fashion, drive a new inner race into position.
16. Pack the hub bore with the same grease used to pack the bearings.
17. Install the inner front wheel bearing.
18. Install a new wheel hub grease seal.
19. Position the hub on the spindle.
20. Install the outer front wheel bearing.
21. Install the front wheel outer bearing retainer washer.
22. Install the spindle nut.
23. Tighten the spindle nut to 21 ft. lbs. (29 Nm) while rotating the brake disc and hub.

24. Loosen the spindle nut one-half turn.

25. Tighten the spindle nut to 17 inch lbs. (2 Nm) while rotating the brake disc and hub.

26. Install the nut retainer.

27. Install the cotter pin.

28. Install the hub grease cap.

29. Install the front disc brake caliper anchor plate.

4-WHEEL DRIVE

1. Before servicing the vehicle, refer to the Precautions Section.

2. Remove the brake disc.

3. Remove the three bolts and the dust shield.

✳✳ WARNING

Discard the wheel hub retainer nut and washer assembly. It is a torque prevailing design and cannot be reused.

4. Remove the wheel hub retainer nut and washer assembly.

5. Remove the bolts and position the anti-lock sensor aside.

✳✳ WARNING

Do not overextend CV joint and boots when removing the hub and bearing assembly.

➡ The CV joint is a slip fit into the wheel hub and bearing. A puller will not normally be required.

6. Remove the three bolts and the wheel hub.

7. To install, reverse the removal procedure. Observe the following torques:
- Wheel hub-to-knuckle: 85 ft. lbs. (115 Nm)
- Axle shaft nut: 162 ft. lbs. (220 Nm)
- Dust shield: 108 inch lbs. (12 Nm)

Aviator

2-WHEEL DRIVE

✳✳ WARNING

If equipped, always turn off the Automatic Ride Control (ARC) service switch before lifting the vehicle off of the ground. Failure to do so could damage the ARC system components.

➡ The wheel speed sensor electrical connectors are located in the engine

compartment secured to the fender aprons.

1. Before servicing the vehicle, refer to the Precautions Section.

2. Disconnect the wheel speed sensor.

3. Remove the wheel and tire assembly.

4. Detach the wheel speed sensor harness from the brake hose.

5. Remove the brake hose-to-wheel knuckle bolt.

6. Remove the 2 anchor plate bolts.

✳✳ WARNING

Do not allow the brake caliper to hang from the hose or damage to the hose can occur.

7. Position the caliper, pads and anchor plate aside.

8. Remove the brake disc.

9. Remove and discard the 3 wheel hub-to-wheel knuckle bolts.

10. Remove the wheel bearing and hub assembly.

11. To install, reverse the removal procedure. Always install new wheel hub-to-wheel knuckle bolts. Observe the following torques:
- Wheel hub-to-knuckle bolts: 83 ft. lbs. (112 Nm)
- Anchor plate bolts: 155 ft. lbs. (210 Nm)

4-WHEEL DRIVE

✳✳ WARNING

If equipped, always turn off the Automatic Ride Control (ARC) service switch before lifting the vehicle off of the ground. Failure to do so could damage the ARC system components.

1. Before servicing the vehicle, refer to the Precautions Section.

2. Loosen the axle retainer nut.

➡ The wheel speed sensor electrical connectors are located in the engine compartment secured to the fender aprons.

3. Disconnect the wheel speed sensor.

4. Remove the wheel and tire assembly.

5. Remove and discard the axle retainer nut.

6. Separate the outboard CV joint from the wheel hub.

7. Detach the wheel speed sensor harness from the brake hose.

8. Remove the brake hose-to-wheel knuckle bolt.

9. Remove the 2 anchor plate bolts.

✳✳ WARNING

Do not allow the brake caliper to hang from the hose or damage to the hose can occur.

10. Position the caliper, pads and anchor plate aside.

11. Remove the brake disc.

12. Remove and discard the 3 wheel hub-to-wheel knuckle bolts.

13. Remove the wheel bearing and hub assembly.

14. To install, reverse the removal procedure. Always install new wheel hub-to-wheel knuckle bolts and a new axle retainer nut.

15. Observe the following torques:
- Wheel hub-to-knuckle bolts: 83 ft. lbs. (112 Nm)
- Anchor plate bolts: 155 ft. lbs. (210 Nm)
- Axle retainer nut: 184 ft. lbs. (250 Nm)

Explorer and Mountaineer

2-WHEEL DRIVE

✳✳ WARNING

If equipped, always turn off the Automatic Ride Control (ARC) service switch before lifting the vehicle off of the ground. Failure to do so could damage the ARC system components.

1. Before servicing the vehicle, refer to the Precautions Section.

2. Remove or disconnect the following:

➡ The wheel speed sensor connectors are located in the engine compartment and are secured to the fender aprons.

- Wheel speed sensor connector
- Brake disc
- Wiring harness from the retainers
- Bolts, wheel hub and sensor as an assembly. Discard the bolts.

3. To install, reverse the removal procedure.

➡ Apply a thin coat of silicone sealant to the wheel hub mounting surfaces before installation.

4. Torque the hub-to-knuckle bolts to 83 ft. lbs. (112 Nm)

4-WHEEL DRIVE

✴ WARNING

If equipped, always turn off the Automatic Ride Control (ARC) service switch before lifting the vehicle off of the ground. Failure to do so could damage the ARC system components.

1. Before servicing the vehicle, refer to the Precautions Section.
2. Remove or disconnect the following:
   • Hub nut. Discard the nut.

➡The wheel speed sensor connectors are located in the engine compartment and are secured to the fender aprons.

   • Wheel speed sensor connector
   • Brake disc

✴✴ WARNING

Do not use a hammer to separate the outboard CV joint from the hub. Damage to the threads and internal CV joint components may result.

3. Press the outboard CV joint until it is loose in the hub.
4. Detach the harness from the retainers.

✴✴ WARNING

Do not overextend the CV joint and boots when removing the wheel hub.

5. Remove the bolts, wheel hub and sensor as an assembly. Discard the bolts.

➡Apply a thin coat of silicone sealant to the wheel hub mounting surfaces before installation.

6. To install, reverse the removal procedure. Observe the following torques:
   • Hub-to-knuckle: 83 ft. lbs. (112 Nm.
   • Hub nut: 184 ft. lbs. (250 Nm).

REAR SUSPENSION

✴✴ WARNING

Suspension fasteners are critical components because they affect performance of vital components and systems and their failure can result in major service expense. New fasteners must be installed with the same component number or an equivalent component if installation is necessary. Do not use a replacement component of lesser quality or substitute design. Torque values must be used as specified during reassembly to make sure of correct retention of these components. Orientation of the fasteners is also important. Make sure the fasteners are installed in the same direction as they were in when removed.

Shock Absorber

REMOVAL & INSTALLATION

Aviator, Explorer and Mountaineer

1. Before servicing the vehicle, refer to the Precautions Section.
2. Remove or disconnect the following:

1 Axle nut
2 Toe link-to-wheel knuckle nut
3 Toe link-to-wheel knuckle bolt
4 Upper arm-to-wheel knuckle nut
5 Upper arm-to-wheel knuckle bolt
6 Lower arm-to-wheel knuckle nut
7 Lower arm-to-wheel knuckle bolt
8 Wheel knuckle

06017-EXPL-G148

Rear hub/knuckle and related parts—Aviator, Explorer and Mountaineer

14 Toe link-to-frame nut

15 Toe link

16 Parking brake cable bracket-to-lower arm bolt

17 Stabilizer bar link-to-stabilizer bar nut and bushing

18 Stud

19 Stabilizer bar link

20 Shock absorber-to-lower arm bolt

21 Flag nut

22 Lower arm-to-frame nut (rearward mounting)

23 Set shim

24 Lower arm-to-frame flag bolt

25 Lower arm-to-frame nut (forward mounting)

26 Lower arm-to-frame flag bolt

27 Lower arm

28 Upper arm-to-frame bolt (rearward mounting) (2 required)

28 Upper arm-to-frame bolt (rearward mounting) (2 required)

29 Flag nut (2 required)

30 Upper arm bushing-to-upper arm nut

31 Upper arm bushing (rearward mounting)

32 Upper arm-to-frame bolt (forward mounting)

33 Flag nut

34 Upper arm

06017-EXPL-G149

Rear suspension—Aviator, Explorer and Mountaineer

- Wheels
- Upper shock mounting nuts. Discard the nuts.
- Nut and the stabilizer bar link. Discard the nut.
- Ball joint pinch bolt. Discard the nut.
- Bolt, flag nut and the shock absorber and spring as an assembly. Discard the flag nut.

3. Using a suitable spring compressor, compress the spring until the tension is released from the shock absorber.

4. While holding the flats of the washer, remove the nut.

5. Remove the shock absorber. Discard the nut.

6. Remove the washer, bushing and the upper mount.

7. Remove the insulator.

8. Remove the dust shield.

9. To install, reverse the removal procedure. Note the following torques:
- Center nut: 37 ft. lbs. (50 Nm)
- Lower mounting bolt: 184 ft. lbs. (250 Nm)

- Pinch bolt: 111 ft. lbs. (150 Nm)
- Sway bar nut: 18 ft. lbs. (25 Nm)
- Upper shock mounting nuts: 22 ft. lbs. (30 Nm)

Explorer Sport and Sport-Trac

➡Low pressure gas shocks are charged with nitrogen gas. Do not attempt to open, puncture or apply heat to them. Prior to installing a new shock absorber, hold it upright and extend it fully. Invert it and fully compress and extend it at least 3 times. This will bleed trapped air.

① ─ 30 Nm (22 lb-ft)

⑤ ─ 150 Nm (111 lb-ft)

250 Nm (184 lb-ft) ─ ⑦

25 Nm (18 lb-ft) ─ ④

1 Nuts (3 required)

2 Jounce bumper

3 Stabilizer link nut and grommet

4 Stud

5 Lower arm-to-wheel knuckle nut

6 Lower arm-to-wheel knuckle bolt

7 Shock absorber-to-lower arm bolt

8 Shock absorber-to-lower arm flag nut

9 Shock absorber and spring assembly

06017-EXPL-G150

Rear shock and spring assembly—Aviator, Explorer and Mountaineer

1. Before servicing the vehicle, refer to the Precautions Section.

2. Remove or disconnect the following:
   • Upper shock-to-frame attaching nut
   • Lower shock nut
   • Slightly compress the shock absorber by hand and remove it from the vehicle

To install:

3. Install or connect the following:
   • Shock absorber upper end and nut
   • Shock absorber lower end and nut
   • Torque the upper fasteners to 17 ft. lbs. (23 Nm) and lower shock attaching nuts to 46 ft. lbs. (63Nm)

56 Nm (41 lb-ft)

1 **Shock absorber-to-upper mount nut**

2 **Shock absorber (front/rear)**

3 **Washer**

4 **Upper shock mount (front/rear)**

5 **Coil spring (front/rear)**

06017-EXPL-G151

Rear spring removal—Aviator, Explorer and Mountaineer

Coil Spring

REMOVAL & INSTALLATION

Aviator, Explorer and Mountaineer

1. Before servicing the vehicle, refer to the Precautions Section.

2. Remove the shock absorber and spring assembly.

3. Using a suitable spring compressor, compress the spring until the tension is released from the shock absorber.

4. While holding the flats of the washer, remove and discard the center nut.

5. Remove the shock absorber.

6. Remove the washer.

7. Remove the upper shock absorber mount.

8. Remove the coil spring.

9. To assemble, reverse the disassembly procedure. Always install a new shock absorber-to-upper mount nut. Tighten to 56 Nm (41 ft. lbs.).

Leaf Springs

REMOVAL & INSTALLATION

1. Before servicing the vehicle, refer to the Precautions Section.

2. Turn the air suspension switch off, if equipped.

3. Remove or disconnect the following:

   • Negative battery cable
   • Rear wheels and support the rear axle
   • Separate the rear spring from the axle and position the spring plate aside
   • Rear spring

To install:

4. Install or connect the following:

   • On the 2002, the forward bolt is torqued to 129 ft. lbs. (175 Nm); the rear bolt to 85 ft. lbs. (115 Nm).
   • On the 2003–05, the forward bolt is torqued to 166 ft. lbs. (225 Nm); the rear bolt to 85 ft. lbs. (115 Nm).
   • Properly position the spring plate and install the U-bolts. Torque the bolts to 76 ft. lbs. (103 Nm).
   • Rear wheels and remove the rear axle support
   • Negative battery cable

5. Turn the air suspension switch on, if equipped.

1 Rear axle bumper
2 Rear spring
3 Shock absorber
4 Rear stabilizer bar

06017-EXPL-G116

Rear suspension—Explorer Sport and Sport-Trac

06017-EXPL-G152

Rear leaf spring U-bolts

06017-EXPL-G153

Leaf spring installation—2003–05 models

Stabilizer Bar

REMOVAL & INSTALLATION

Aviator, Explorer, Mountaineer

1. Before servicing the vehicle, refer to the Precautions Section.
2. Remove the wheel and tire assemblies.
3. Remove and discard the 2 stabilizer bar-to-link nuts.
4. Remove the 2 stabilizer bar studs.
5. Remove the 2 stabilizer bar links.
6. Remove and discard the 4 stabilizer bar-to-frame nuts.
7. Remove the 2 stabilizer bar insulators.

✳✳ WARNING

Do not damage the boot while separating the ball joint from the knuckle.

8. Remove the nut and detach the right upper ball joint.

✳✳ WARNING

Do not allow the brake caliper to hang from the hose or damage to the hose can occur.

9. Remove the bolts and position the brake caliper aside.
10. Unclip the vapor line from the fuel tank.
11. From the right side of the vehicle, remove the stabilizer bar.
12. To install, reverse the removal procedure. Always install new stabilizer bar-to-link nuts and stabilizer bar-to-frame nuts. Torque the link nuts to 18 ft. lbs (25 Nm); the bracket nuts to 41 ft. lbs. (55 Nm). Torque the ball joint nut to 66 ft. lbs. (90 Nm).

Explorer Sport and Sport-Trac

1. Before servicing the vehicle, refer to the Precautions Section.
2. Remove the wheel and tire assembly.
3. Remove the nuts, bolts, brackets and the rear stabilizer bar.
4. To install, reverse the removal proce-

4 — 55 Nm (41 lb-ft)

25 Nm (18 lb-ft) — **2**

1 Nut and grommet (2 required)

2 Stud (2 required)

3 Stabilizer bar link (2 required)

4 Stabilizer bar-to-frame nuts (4 required)

5 Bracket (2 required)

6 Stabilizer bar insulator (2 required)

7 Stabilizer bar

06017-EXPL-G124

Rear stabilizer bar—Aviator, Explorer, Mountaineer

103 Nm (76 lb-ft)

48 Nm (35 lb-ft)

06017-EXPL-G154

Rear stabilizer bar—Explorer Sport and Sport-Trac

dure. Use new fasteners. Torque the link to 76 ft. lbs. (103 Nm); the clamp to 35 ft. lbs. (48 Nm).

Upper Control Arm

REMOVAL & INSTALLATION

Aviator, Explorer and Mountaineer

1. Before servicing the vehicle, refer to the Precautions Section.
2. Remove the upper arm-to-wheel knuckle nut.

✳✳ WARNING

Do not damage the boot while separating the ball joint from the wheel knuckle.

3. Remove the bolt and separate the upper arm from the wheel knuckle.

✳✳ WARNING

Make sure that the upper arm-to-frame mounting bolts are routed through the frame and the bushing before tightening.

➡**Do not tighten until the installation procedure is complete and the weight of the vehicle is resting on the wheel and tire assemblies.**

4. Remove the 2 upper arm-to-frame bolts (rearward mounting).

5. Remove the 2 upper arm-to-frame flag nuts.

➡**Do not tighten until the installation procedure is complete and the weight of the vehicle is resting on the wheel and tire assemblies.**

6. Remove the upper arm-to-frame bolt (forward mounting).

7. Remove the upper arm-to-frame flag nut.

8. Remove the upper arm.

9. To install, reverse the removal procedure. Observe the following torques:
- Upper arm-to-frame bolt (forward mounting): 98 ft. lbs. (133 Nm)
- Upper arm-to-frame bolts (rearward mounting): 59 ft. lbs. (80 Nm)
- Upper arm-to-wheel knuckle nut: 66 ft. lbs (90 Nm).

Lower Control Arm

REMOVAL & INSTALLATION

Aviator, Explorer and Mountaineer

1. Before servicing the vehicle, refer to the Precautions Section.

2. Remove the lower arm-to-wheel knuckle nut.

3. Remove the lower arm-to-wheel knuckle bolt.

4. Remove the parking brake cable bracket-to-lower arm bolt.

5. Remove the stabilizer bar nut and the stabilizer bar link.

6. Remove the stabilizer bar stud.

➡**Do not tighten until the installation procedure is complete and the weight of the vehicle is resting on the wheel and tire assemblies.**

7. Remove the shock absorber-to-lower arm bolt. .

8. Remove and discard the flag nut.

➡**Do not tighten until the installation procedure is complete and the weight of the vehicle is resting on the wheel and tire assemblies.**

9. Remove and discard the lower arm-to-frame nut (rearward mounting).

10. Remove the set shim.

11. Remove the lower arm-to-frame flag bolt.

➡**Do not tighten until the installation procedure is complete and the weight of the vehicle is resting on the wheel and tire assemblies.**

12. Remove and discard the lower arm-to-frame nut (forward mounting).

13. Remove the lower arm-to-frame flag bolt.

14. Remove the lower arm.

15. To install, reverse the removal procedure. Observe the following torques:
- Lower arm-to-frame nut (forward mounting): 129 ft. lbs. (175Nm)
- Lower arm-to-frame nut (rearward mounting): 129 ft. lbs. (175Nm)
- Shock absorber-to-lower arm bolt: 184 ft. lbs. (250 Nm)
- Stabilizer bar stud: 18 ft. lbs. (25 Nm)
- Parking brake cable bracket-to-lower arm bolt: 108 inch lbs. (12 Nm)
- Lower arm-to-wheel knuckle nut: 111 ft. lbs. (150 Nm)

16. Always install a new:
- flag nut.
- lower arm-to-frame nut (rearward mounting).
- lower arm-to-frame nut (forward mounting).

Knuckle/Spindle

REMOVAL & INSTALLATION

⁑ WARNING

Do not loosen the axle wheel hub retainer until the wheel and tire are removed from the vehicle. Wheel bearing damage will occur if the wheel bearing is unloaded with the weight of the vehicle applied.

➡**Have an assistant press the brake pedal to keep the axle from turning.**

1. Before servicing the vehicle, refer to the Precautions Section.

2. Remove the axle-to-wheel hub retainer nut and washer. Discard the nut.

3. Remove the parking brake shoes.

⁑ WARNING

Do not use a hammer to separate the outboard CV joint from the hub. Dam-

age to the threads and internal CV joint components can result.

4. Press the outboard CV joint until it is loose from the hub.

5. Remove and discard the toe link-to-wheel knuckle nut.

⁑ WARNING

Do not damage the boot while separating the toe link from the wheel knuckle.

6. Remove the bolt and separate the toe link from the wheel knuckle.

7. Remove and discard the upper arm-to-wheel knuckle nut.

⁑ WARNING

Do not damage the boot while separating the ball joint from the wheel knuckle.

8. Remove the bolt and separate the upper arm from the wheel knuckle.

9. Remove and discard the lower arm-to-wheel knuckle nut.

10. Remove the lower arm-to-wheel knuckle bolt.

11. Remove the wheel knuckle.

12. Remove the 3 brake shield-to-wheel knuckle bolts.

13. Using a suitable press, remove and discard the wheel hub.

14. Remove the brake shield from the wheel knuckle.

15. Remove and discard the wheel bearing retainer ring.

16. Using a suitable press, remove and discard the wheel bearing.

17. To install, reverse the removal procedure. Observe the following torques:
- Brake shield-to-wheel knuckle bolts: 13 Nm (10 ft. lbs.).
- Lower arm-to-knuckle nut: 150 Nm (111 ft. lbs.).
- Upper arm-to-wheel knuckle nut: 90 Nm (66 ft. lbs.).
- Toe link-to-knuckle nut: 90 Nm (66 ft. lbs.).
- Axle-to-wheel hub retainer nut: 275 Nm (203 ft. lbs.)

18. Always install a new:
- Axle-to-wheel hub retainer nut.
- Toe link-to-wheel knuckle nut.
- Upper arm-to-wheel knuckle nut.
- Lower arm-to-wheel knuckle nut.
- Wheel hub.
- Wheel bearing retainer ring.
- Wheel bearing.

Toe Link

REMOVAL & INSTALLATION

Aviator, Explorer and Mountaineer

1. Before servicing the vehicle, refer to the Precautions Section.
2. Remove the toe link-to-wheel knuckle nut.

✳✳ WARNING

Do not damage the boot while separating the toe link from the wheel knuckle.

3. Remove the bolt and separate the toe link from the wheel knuckle.
4. Remove and discard the toe link-to-frame nut.
5. Remove the toe link.
6. To install, reverse the removal procedure. Torque the toe link-to-knuckle nut to 66 ft. lbs. (90 Nm). Torque the toe link-to-frame nut to 59 ft. lbs. (80 Nm). Always install a new toe link-to-frame nut.

Rear Wheel Bearings

REMOVAL & INSTALLATION

Aviator, Explorer and Mountaineer

✳✳ WARNING

Do not loosen the axle wheel hub retainer until the wheel and tire are removed from the vehicle. Wheel bearing damage will occur if the wheel bearing is unloaded with the weight of the vehicle applied.

➡ **Have an assistant press the brake pedal to keep the axle from turning.**

1. Before servicing the vehicle, refer to the Precautions Section.
2. Remove the axle-to-wheel hub retainer nut and washer.
3. Remove the parking brake shoes.

✳✳ WARNING

Do not use a hammer to separate the outboard CV joint from the hub. Dam- age to the threads and internal CV joint components can result.

4. Press the outboard CV joint until it is loose from the hub.
5. Remove and discard the toe link-to-wheel knuckle nut.

✳✳ WARNING

Do not damage the boot while separating the toe link from the wheel knuckle.

6. Remove the bolt and separate the toe link from the wheel knuckle.
7. Remove and discard the upper arm-to-wheel knuckle nut.

✳✳ WARNING

Do not damage the boot while separating the ball joint from the wheel knuckle.

8. Remove the bolt and separate the upper arm from the wheel knuckle.
9. Remove and discard the lower arm-to-wheel knuckle nut.

13 Nm (10 lb-ft) — 9

9   Brake shield-to-wheel knuckle bolts

10   Wheel hub

11   Brake shield

12   Retainer ring

13   Wheel bearing

06017-EXPL-G155

Rear hub/bearing Assembly—Aviator, Explorer and Mountaineer

Pressing out the bearing

06017-EXPL-G156

10. Remove the lower arm-to-wheel knuckle bolt.

11. Remove the wheel knuckle.

12. Using a suitable press, remove and discard the wheel hub.

13. Remove and discard the wheel bearing retainer ring.

14. Using a suitable press, remove the wheel bearing.

15. To install, reverse the removal procedure. Observe the following torques:
- Lower arm-to-wheel knuckle nut: 150 Nm (111 ft. lbs.).

- Upper arm-to-wheel knuckle nut: 90 Nm (66 ft. lbs.).
- Toe link-to-wheel knuckle nut: 90 Nm (66 ft. lbs.).
- Axle-to-hub retainer nut: 275 Nm (203 ft. lbs.).

16. Always install a new:
- Axle-to-wheel hub retainer nut.
- Toe link-to-wheel knuckle nut.
- Upper arm-to-wheel knuckle nut.
- Lower arm-to-wheel knuckle nut.
- Wheel hub.
- Wheel bearing retainer ring.

DISC BRAKES

Brake Caliper

REMOVAL & INSTALLATION

Aviator

FRONT

1. Before servicing the vehicle, refer to the Precautions Section.

2. Loosen the wheel lug nuts.

3. Raise and safely support the front of the vehicle. Remove the wheel.

4. Place an 8 in. (203mm) C-clamp on the caliper and tighten the clamp to bottom the caliper pistons in their bores. Remove the clamp.

5. Remove the two caliper slide pin bolts and lift the caliper from the anchor plate.

➡Use care to retain as much of the original caliper slide pin grease as possible.

6. Position the caliper on a frame member or suspend it with some wire. Do not allow the caliper to hang by the brake hose.

7. Disconnect and plug the brake hose at the caliper. Remove the caliper from the rotor.

67197-EXPL-G74

1 Bolts	6 Anchor	11 Copper washers
2 Brake caliper	7 Guide pin and boot	12 Bolts
3 Brake disc pads	8 Guide pin and boot	13 Brake disc shield (LH/RH)
4 Slippers	9 Brake disc	14 Bleed screw
5 Brake caliper anchor bracket bolt	10 Flow bolt	15 Bleed screw cap

Front brake components—Aviator

12 Nm (9 lb-ft) — 8

N 4

3

5

9

6 — 35 Nm (26 lb-ft)

1 — 32 Nm (24 lb-ft)

7

2

2

10 — 22 Nm (16 lb-ft)

1 Bolts
2 Caliper RH/LH
3 Brake disc pads
4 Slippers
5 Brake disc

6 Flow bolt
7 Copper washers
8 Bolts
9 Brake disc shield
10 Bleed screw

67197-EXPL-G76

Rear disc brake components—Aviator

1 Keeper	7 Wheel hub retaining bolts
2 Brake disc	8 Front wheel knuckle assembly (RH/LH)
3 Wheel hub	9 Brake caliper bolt
4 Brake disc shield screws	10 Brake caliper
5 Brake disc shield (RH/LH)	11 Disc caliper support bracket
6 Support bracket bolt	

67197-EXPL-G70

Front brake components—2002 Explorer and Mountaineer

1 Front brake anti-lock sensor
2 Front wheel spindle (RH/LH)
3 Brake disc shield (RH/LH)
4 Bolt
5 Grease seal
6 Front wheel bearing
7 Brake disc and hub
8 Front wheel bearing
9 Front wheel outer bearing retainer washer

10 Hub spindle nut
11 Cotter pin
12 Nut retainer
13 Hub grease cap
14 Disc brake caliper (RH/LH)
15 Front disc brake caliper anchor plate
16 Caliper anchor plate bolts
17 Disc brake caliper bolt

06017-EXPL-G157

Front brake components—2-wheel drive Explorer Sport and Sport-Trac

To install:

8. Position the caliper over the brake pads and align the slide pin mounting holes.

9. Install the slide pin bolts and tighten them to 32 ft. lbs. (44 Nm).

➡ **Tighten the bottom locator pin caliper bolt before tightening the top guide pin caliper bolt.**

10. Install the caliper brake hose using new washers. Tighten the bolt to 26 ft. lbs. (35 Nm).

11. Install the wheel and snug the lug nuts.

12. Lower the vehicle and tighten the lug nuts to 100 ft. lbs. (135 Nm).

➡ **The first couple of times you apply the brakes, the pedal may go to the floor. Continue to pump the brake pedal until it feels firm.**

13. Start the engine and apply the brakes several times to readjust the caliper pistons. Ensure that the pedal feels firm before operating the vehicle.

REAR

1. Before servicing the vehicle, refer to the Precautions Section.

2. Siphon part of the brake fluid out of the master cylinder to avoid overflow when the caliper piston is pressed into the caliper bore.

3. Raise the vehicle and support it safely. Remove the wheel and tire assembly.

4. Position an 8 in. (20cm) C-clamp on the caliper and tighten the clamp to move the caliper piston into the bore approximately ⅛ in. (3mm). Remove the clamp.

➡ **Do not pry the piston away from the rotor.**

5. Clean excess dirt from the retainer bolt area.

6. Using a Torx® socket, remove the 2 retainer bolts securing the caliper to the bracket and adapter plate.

7. Disconnect and plug the brake hose at the caliper. Remove the caliper from the rotor.

To install:

8. Make sure the caliper mounting surfaces are free of dirt. Lubricate the caliper grooves with disc brake caliper grease and install the caliper.

9. Position the caliper to the bracket and secure in place with the retainer bolts. Tighten the bolts to 24 ft. lbs. (32 Nm)

10. Install the caliper brake hose using

1 Front brake anti-lock sensor

2 Front brake hose bolt

3 Disc brake caliper

4 Pads

5 Front disc brake caliper anchor plate

6 Wheel hub

7 Brake disc

8 Front wheel knuckle assembly (RH/LH)

06017-EXPL-G158

Front brake components—4-wheel drive Explorer Sport and Sport-Trac

1 Caliper bolts	6 Anchor bracket	11 Copper washers
2 Brake caliper (RH/LH)	7 Guide pin and boot	12 Brake disc shield
3 Brake disc pads	8 Locating pin and boot	13 Brake hose bracket bolt
4 Slippers	9 Brake disc	14 Brake hose (RH/LH)
5 Anchor bracket bolt kit	10 Flow bolt	15 Brake tube

67197-EXPL-G73

Front brake components—2003–05 Explorer and Mountaineer

1 Brake disc	6 Brake caliper bolt	10 Rear disc brake caliper
2 Wheel hub	7 Flow bolt	11 Shoe slippers
3 Disc brake shield	8 Copper washer	12 Keeper nut
4 Wheel knuckle RH/LH	9 Rear wheel brake hose	13 Keeper
5 Brake pads		

67197-EXPL-G77

Rear disc brake components—2002 Explorer and Mountaineer

17 Nm
(13 lb-ft) ⑪

⑨

17 Nm
(13 lb-ft) ⑩

⑫

35 Nm
(26 lb-ft) ⑥

⑦ N

32 Nm
(24 lb-ft) ①

⑧

22 Nm
(16 lb-ft)

⑬

N ④

③

⑤

②

1	Caliper bolts	6	Flow bolt	10	Brake tube
2	Brake caliper RH/LH	7	Copper washers	11	Brake tube
3	Brake disc pads	8	Brake disc shield (RH/LH)	12	Brake hose
4	Slippers	9	Brake hose bracket bolt	13	Bleed screw
5	Brake disc				

67197-EXPL-G75

Rear disc brake components—2003–05 Explorer and Mountaineer

1 Brake caliper bolt

2 Flow bolt

3 Copper washer (2 req'd)

4 Rear wheel brake hose

5 Rear disc brake caliper

6 Brake pads

7 Adapter assy

8 Shoe slippers

9 Axle shaft

10 Wheel stud (5 req'd)

11 Brake disc

12 Keeper nut

13 Rear wheel disc brake adapter nut

14 Rear wheel disc brake adapter

15 Rear wheel disc brake adapter bolt

06017-EXPL-G159

Rear disc brake components—Explorer Sport and Sport-Trac

new washers. Tighten the bolt to 26 ft. lbs. (35 Nm).

11. Fill and bleed the brake system.

12. Install the wheel and tire assembly and lower the vehicle. Check the brake fluid level and check the brakes for proper operation.

Explorer and Mountaineer; Explorer Sport and Sport-Trac

FRONT

1. Before servicing the vehicle, refer to the Precautions Section.

2. Loosen the wheel lug nuts.

3. Raise and safely support the front of the vehicle. Remove the wheel.

4. Place an 8 in. (203mm) C-clamp on the caliper and tighten the clamp to bottom the caliper pistons in their bores. Remove the clamp.

5. Remove the two caliper slide pin bolts and lift the caliper from the anchor plate.

➡**Use care to retain as much of the original caliper slide pin grease as possible.**

6. Position the caliper on a frame member or suspend it with some wire. Do not allow the caliper to hang by the brake hose.

7. Disconnect and plug the brake hose at the caliper. Remove the caliper from the rotor.

To install:

8. Position the caliper over the brake pads and align the slide pin mounting holes.

9. Install the slide pin bolts and tighten them to 24 ft. lbs. (32 Nm).

10. Install the caliper brake hose using new washers. Tighten the bolt to 26 ft. lbs. (35 Nm).

11. Install the wheel and snug the lug nuts.

12. Lower the vehicle and tighten the lug nuts to 100 ft. lbs. (135 Nm).

➡**The first couple of times you apply the brakes, the pedal may go to the floor. Continue to pump the brake pedal until it feels firm.**

13. Start the engine and apply the brakes several times to readjust the caliper pistons. Ensure that the pedal feels firm before operating the vehicle.

REAR

1. Before servicing the vehicle, refer to the Precautions Section.

2. Siphon part of the brake fluid out of

the master cylinder to avoid overflow when the caliper piston is pressed into the caliper bore.

3. Raise the vehicle and support it safely. Remove the wheel and tire assembly.

4. Position an 8 in. (20cm) C-clamp on the caliper and tighten the clamp to move the caliper piston into the bore approximately ⅛ in. (3mm). Remove the clamp.

➡**Do not pry the piston away from the rotor.**

5. Clean excess dirt from the retainer bolt area.

6. Using a Torx® socket, remove the 2 retainer bolts securing the caliper to the bracket and adapter plate.

7. Disconnect and plug the brake hose at the caliper. Remove the caliper from the rotor.

To install:

8. Make sure the caliper mounting surfaces are free of dirt. Lubricate the caliper grooves with disc brake caliper grease and install the caliper.

9. Position the caliper to the bracket and secure in place with the retainer bolts. Tighten the bolts to 24 ft. lbs. (32 Nm).

10. Install the caliper brake hose using new washers. Tighten the bolt to 26 ft. lbs. (35 Nm).

11. Fill and bleed the brake system.

12. Install the wheel and tire assembly and lower the vehicle. Check the brake fluid level and check the brakes for proper operation.

Disc Brake Pads

REMOVAL & INSTALLATION

Aviator, Explorer and Mountaineer
FRONT

1. Before servicing the vehicle, refer to the Precautions Section.

2. Raise and safely support the front of the vehicle. Remove the wheel.

3. Place an 8 in. (203mm) C-clamp on the caliper and tighten the clamp to bottom the caliper pistons in their bores. Remove the clamp.

4. Remove the two caliper slide pin bolts and lift the caliper from the anchor plate.

➡**Use care to retain as much of the original caliper slide pin grease as possible.**

5. Position the caliper on a frame member or suspend it with some wire. Do not allow the caliper to hang by the brake hose.

6. Remove the brake pads and, if necessary, the anti-rattle clips from the anchor plate.

7. Remove the shims, if any, from the brake pads for re-use.

To install:

8. If removed, install the anti-rattle clips.

9. Install the brake pads to the anchor plate.

10. Install the caliper.

11. Install the wheel and snug the lug nuts.

12. Lower the vehicle and tighten the lug nuts to 100 ft. lbs. (135 Nm).

➡**The first couple of times you apply the brakes, the pedal may go to the floor. Continue to pump the brake pedal until it feels firm.**

13. Start the engine and apply the brakes several times to readjust the caliper pistons. Ensure that the pedal feels firm before operating the vehicle.

REAR

1. Before servicing the vehicle, refer to the Precautions Section.

2. Siphon part of the brake fluid out of the master cylinder to avoid overflow when the caliper piston is pressed into the caliper bore.

3. Raise the vehicle and support it safely. Remove the wheel and tire assembly.

4. Remove the brake caliper, but do not disconnect the brake hose. Secure the caliper aside with mechanic's wire.

5. Remove the inner and outer brake pad from the caliper.

To install:

6. Bottom out the caliper piston in the caliper bore using an 8 in. (20cm) C-clamp or equivalent and a worn out inner brake pad or block of wood to push against the piston. Do not attempt to bottom out the piston with the outer brake pad installed.

7. Position the inboard brake pad in the caliper and press the retainer spring fully into the caliper piston.

8. Start one end of the outboard brake shoe and lining on the caliper and rotate it down until the locating lugs and the retainer spring are fully seated.

9. Install new shoe slippers on the rear wheel disc brake adapter.

10. Install the caliper.

11. Install the wheel and tire assembly and lower the vehicle. Apply the brakes several times before moving the vehicle to seat the pads.

12. Check the brake fluid level. Check the brakes for proper operation.

Explorer Sport and Sport-Trac
FRONT

1. Before servicing the vehicle, refer to the Precautions Section.

2. Remove brake fluid in the master cylinder reservoir until the reservoir is half full.

3. Raise and support the vehicle.

4. Remove the wheel and tire assembly.

❊❊ WARNING

Install a new pad if worn to or past the specified thickness above the metal backing plate. Install new pads in complete axle sets.

5. Inspect the pads for wear or contamination.

6. Remove the lower caliper bolt on the right caliper and the upper caliper bolt on the left caliper first. Rotate the disc brake caliper away from the brake disc rotor.

7. Remove the pads.

8. Remove the stainless slippers.

9. Inspect the brake disc:
- If the brake disc is cracked, install a new brake disc.
- If diagnosis has revealed vibration in the steering wheel, seat, or brake pedal while braking, machine the brake disc. Heavily scored brake discs, similar to that caused by linings worn to the backing plate, must also be machined.
- Measure the brake disc for minimum thickness.
- If the brake disc is not within specification, install a new brake disc.

10. Inspect the disc brake caliper for leaks. If leaks are found, disassembly is required.

To install:

11. Use a suitable suction device to remove brake fluid from the master cylinder.

12. Clean the support bracket and slipper surfaces.

13. Install the slippers and the pads.

14. Install the slippers.

15. Install the pads.

➡**If installing new pads, the caliper pistons must be compressed using a C-clamp and a wooden block.**

16. Lower the brake fluid level in the master cylinder if required.

17. Rotate the disc brake caliper into the anchor bracket.

18. Install the caliper bolt.
19. Install the wheel and tire assembly.
20. Check the brake system for correct operation.

REAR

1. Before servicing the vehicle, refer to the Precautions Section.
2. Remove brake fluid in the master cylinder reservoir until the reservoir is half full.

❋❋ WARNING

Install new pads if worn to or past the specified thickness above the metal backing plate or rivets. Install new pads in complete axle sets.

3. Inspect the brake pads for wear or contamination. Install new pads if worn to or past specification.

➡ **It is not necessary to remove the rear wheel brake hose when carrying out this procedure.**

4. Remove the rear disc brake caliper.

❋❋ WARNING

Do not allow grease, oil, brake fluid or other contaminants to contact the brake pads.

5. Remove the brake pads.
6. Retract the caliper piston into the rear disc brake caliper.
7. Remove and discard the slippers.
8. Inspect the brake disc:
- If the brake disc is cracked, install a new brake disc.
- If diagnosis has revealed vibration in the steering wheel, seat, or brake pedal while braking, machine the brake disc. Heavily scored brake discs, similar to that caused by linings worn to the backing plate, must also be machined.
- Measure the brake disc for minimum thickness.
- If the brake disc is not within specification, install a new brake disc.

To install:

❋❋ WARNING

New stainless steel slippers must be installed when new brake pads are installed, even if the slippers appear undamaged. Make sure the slippers are correctly positioned with the slipper ends snug against the outboard end of the anchor plate rail.

9. Clean the slipper mating surface, and install the new slippers.

❋❋ WARNING

Install new brake pads in full axle sets. Do not install new brake pads on only one side of the vehicle.

10. Install the brake pads.
11. Install the rear disc brake caliper.
12. Verify correct brake operation.

Anchor Plate (Adapter, Support Bracket)

REMOVAL & INSTALLATION

Front

1. Before servicing the vehicle, refer to the Precautions Section.
2. Remove the brake pads.
3. Remove and discard the 2 brake caliper anchor plate bolts.
4. Remove the brake caliper anchor plate.
5. To install, reverse the removal procedure. Torque the anchor plate bolts to 210 Nm (155 ft. lbs.) for Aviator; 136 Nm (100 ft. lbs.) for 2003–05 Explorer and Mountaineer; 112 Nm (83 ft. lbs.) for 2002 Explorer and Mountaineer. Lubricate the brake caliper guide and locator pins with silicone grease.

Rear

EXPLORER SPORT AND SPORT-TRAC

1. Before servicing the vehicle, refer to the Precautions Section.
2. Disconnect the rear parking brake cable guide.
3. Remove the axle shaft.
4. Remove the bolts and the rear wheel disc brake adapter.
5. To install, reverse the removal procedure. Torque the adapter bolts to 80 ft. lbs. (108 Nm).

Brake Disc (Rotor)

REMOVAL & INSTALLATION

Aviator

FRONT

1. Before servicing the vehicle, refer to the Precautions Section.
2. Remove the 2 brake caliper anchor

plate bolts and position the brake caliper and anchor plate assembly aside.
3. Mark a wheel stud and mark the corresponding brake disc wheel stud hole. This minimizes front brake disc run-out during reassembly.

➡ **If the brake disc cannot be removed easily, apply rust penetrant on the brake disc-to-hub mating surfaces.**

4. Remove the brake disc.
To install:

➡ **If machining is necessary, use a hub-mounted brake lathe to machine the front brake disc.**

5. Measure the front brake disc and resurface as necessary. Install a new front brake disc if the disc thickness is beyond specifications.
6. Clean any rust or debris from the brake disc and wheel hub. Use parts cleaner to clean the front brake disc and hub surfaces.
7. Apply a thin coat of anti-seize to the hub flange.
8. Install the brake disc, aligning the marks on the disc and the wheel stud.

➡ **Tighten the bottom locator pin caliper bolt before tightening the top guide pin caliper bolt.**

9. Position the brake caliper and brake caliper anchor plate assembly and install the 2 bolts. Tighten to 210 Nm (155 ft. lbs.).

REAR

1. Before servicing the vehicle, refer to the Precautions Section.
2. Remove the brake pads.

➡ **If the brake disc binds on the parking brake shoe and lining, remove the adjustment hole knockout panel or access plug and contract the parking brake shoe and lining.**

➡ **If the brake disc cannot be removed easily, apply rust penetrant on the brake disc-to-hub mating surfaces.**

3. Remove the brake disc.
To install:

➡ **If the adjustment hole knockout panel or access plug was removed, it is necessary to install an access plug to prevent the entry of contaminants.**

4. Clean any rust or debris from brake disc and wheel hub. Use parts cleaner to clean the front brake disc and hub surfaces.

❄❄ WARNING

Do not allow the lubricant to make contact with the wheel studs, brake pads or brake disc.

5. Apply a thin coat of anti-seize lubricant to the hub flange.
6. Install the brake disc.
7. Install the brake pads.

Explorer and Mountaineer

FRONT

❄❄ WARNING

Do not allow the brake caliper to hang from the brake flexible hose. Use a suitable tool and support the brake caliper.

1. Before servicing the vehicle, refer to the Precautions Section.
2. Remove the brake caliper anchor plate bolts and position the brake caliper and anchor plate assembly aside.

➡️**If the brake disc cannot be removed easily, apply rust penetrant on the brake disc-to-hub mating surfaces.**

3. Remove the brake disc.
To install:
4. Clean any rust or debris from brake disc and wheel hub. Use parts cleaner to clean the front brake disc and hub surfaces.
5. Apply a thin coat of anti-seize lubricant to the hub flange.
6. Install the brake disc.
7. Position the brake caliper and brake caliper anchor plate assembly and install the 2 bolts. Tighten to 136 Nm (100 ft. lbs.).

REAR

1. Before servicing the vehicle, refer to the Precautions Section.
2. Remove the brake pads.

➡️**If the brake disc binds on the parking brake shoe and lining, remove the adjustment hole knockout panel or access plug and contract the parking brake shoe and lining.**

➡️**If the brake disc cannot be removed easily, apply rust penetrant on the brake disc-to-hub mating surfaces.**

3. Remove the brake disc.
To install:

➡️**If the adjustment hole knockout panel or access plug was removed, it is necessary to install an access plug to prevent the entry of contaminants.**

4. Clean any rust or debris from brake disc and wheel hub. Use parts cleaner to clean the front brake disc and hub surfaces.

❄❄ WARNING

Do not allow the lubricant to make contact with the wheel studs, brake pads or brake disc.

5. Apply a thin coat of anti-seize lubricant to the hub flange.
6. Install the brake disc.
7. Install the brake pads.

Explorer Sport and Sport-Trac

FRONT

1. Before servicing the vehicle, refer to the Precautions Section.
2. Remove the front disc brake caliper anchor plate.
3. With 4-wheel drive: Match-mark the brake disc and hub. Remove the brake disc.
4. With 2-wheel drive: Match-mark the brake disc and hub. Remove the brake disc and hub. See the procedure under Front Wheel Bearings.

❄❄ WARNING

Use a hub-mount brake lathe if necessary to machine the brake disc. Measure the brake disc and resurface as necessary.

5. Install a new brake disc if beyond specification.
6. Measure the brake disc, and resurface as necessary. Install a new brake disc if beyond specification.
To install:
7. With 4-wheel drive, position the brake disc to the wheel hub.
8. With 2-wheel drive, install and adjust the hub and bearing.
9. Use Metal Brake Parts Cleaner PM-4 or equivalent to clean the brake disc.
10. Using match-marks, mount the brake disc on the hub.
11. Apply High Temperature Nickel Anti-Seize Lubricant F6AZ-9L494-AA or equivalent meeting Ford specification ESE-M12A4-A to the hub mounting surface to prevent future corrosion.
12. Install the front disc brake caliper anchor plate.

REAR

➡️**When removing the rear brake disc in this procedure it is not necessary to disconnect the hydraulic lines.**

1. Before servicing the vehicle, refer to the Precautions Section.
2. Remove the rear disc brake caliper.

➡️**If the rear brake disc binds on the rear parking brake shoe and linings, remove the adjustment hole access plug and retract the parking brake shoe and lining.**

3. Remove the rear brake disc.

❄❄ WARNING

Use a hub-mount brake lathe if necessary to machine the rear brake disc.

4. Measure the rear brake disc, and resurface as necessary. Install a new rear brake disc if beyond specification.
5. To install, reverse the removal procedure.

Parking Brake Shoes

REMOVAL & INSTALLATION

Aviator, Explorer and Mountaineer

1. Before servicing the vehicle, refer to the Precautions Section.
2. Make sure the parking brake control is fully released.
3. Remove the left cowl side trim panel.

06017-EXPL-G160

Parking brake control—Aviator, Explorer and Mountaineer

06017-EXPL-G163

Parking brake shoes—Aviator

9   Intermediate parking brake cable

10   Rear parking brake cable (RH)

11   Rear parking brake cable bracket bolt

12   Wire form retainer bolt

13   Wire form retainer

14   Rear parking brake cable (LH)

15   Wire form retainer bolt

16   Wire form retainer

17   Brake shoe adjusting screw

18   Brake shoe adjusting screw spring

19   Brake shoe hold-down springs

20   Brake shoe hold-down pins

21   Brake shoe retracting spring

22   Parking brake shoe kit (one kit required)

23   Parking brake lever (RH/LH)

24   Parking brake lever boot

06017-EXPL-G164

Parking brake shoes and related parts—Explorer and Mountaineer

4. With an assistant, release the parking brake cable tension by pulling down on the intermediate cable at the cable-to-cable union until the parking brake control sector rotates to its stop and a 4 mm (0.15 in) x 150 mm (5.9 in) retainer pin can be inserted.

➡**Make sure the cable-to-cable union is connected to the front and rear cable before removing the brake control retaining pin, and the cable tension is reloaded slowly.**

5. Disconnect the cable at the cable-to-cable union.

6. Remove the rear brake disc.

7. Remove the parking brake shoe adjusting screw.

8. Remove the parking brake shoe adjusting screw spring.

9. Remove the 2 parking brake shoe hold-down springs and pins.

10. Remove the parking brake shoe retracting spring and the parking brake shoes.

To install:

11. Position the parking brake shoes and attach the retracting spring.

12. Install the 2 parking brake shoe hold-down pins and springs.

13. Install the parking brake shoe adjusting screw spring.

➡**Completely retract the parking brake adjusting screw before installation.**

14. Install the brake shoe adjusting screw.

15. Use a brake adjusting gauge to measure the inside diameter of the parking brake drum.

16. Adjust the parking brake shoe clearance of 1.07 mm (0.04 in) less than the inside diameter of the parking brake drum. Make sure that the parking brake shoes are correctly centered and measure across the center point of the shoes. Rotate the parking brake shoe adjuster wheel to achieve the correct parking brake shoe-to-brake disc clearance.

17. Install the rear brake disc.

18. To reload the tension on the parking brake cable, follow the release procedure in reverse.

Explorer Sport and Sport-Trac

➡**Make sure the parking brake control is fully released.**

1. Before servicing the vehicle, refer to the Precautions Section.

1	Parking brake lever (RH)
1	Parking brake lever (LH)
2	Brake shoe retracting spring
3	Brake shoe hold-down spring and pin
4	Brake adjuster screw
5	Brake shoe adjusting screw spring
6	Rear brake shoe and lining

06017-EXPL-G161

Parking brake shoes and related parts—Explorer Sport and Sport-Trac

06017-EXPL-G162

Parking brake cable (1) and control. Insert the retainer at (2)—Explorer Sport and Sport-Trac

2. Relieve the tension on the parking brake system.

   a. Pull the front parking brake cable and conduit.

   b. Insert a 4mm (5/32 inch) drill bit or equivalent retainer.

3. Raise and support the vehicle.

4. Remove the tire and wheel assembly.

5. Remove the rear brake disc.

6. Remove the brake shoe retracting spring.

7. Remove the brake shoe adjusting screw spring.

8. Remove the brake adjuster screw.

9. Remove the rear brake shoe hold-down springs.

10. Remove the rear brake shoes and linings along with the inboard brake shoe retracting spring.

11. Inspect the components for excessive wear or damage and install new as required.

To install:

➡**Lubricate the brake shoe contact point before installation of rear shoes using grease.**

12. Install the rear brake shoes and lin-

ings along with the inboard brake shoe retracting spring.

13. Install the brake shoe hold-down springs.

14. Install the brake adjuster screw.

15. Install the brake shoe adjusting screw spring.

16. Install the outboard brake shoe retracting spring.

17. Using a brake gauge, measure the inside diameter of the drum portion of the rear brake disc.

18. Using the gauge, set the rear brake shoe and lining diameter to 0.5mm (0.020

inch) less than the inside diameter of the drum portion of the rear brake disc.

19. Install the rear brake disc.

✴✴ CAUTION

Always remove any corrosion, dirt or foreign material present on the mounting surfaces of the wheel or the surface of the wheel hub, brake drum or brake disc that contacts the wheel. Installing wheels without correct metal-to-metal contact at the wheel mounting surfaces can cause the wheel nuts to loosen and come off while the vehicle is in motion, causing loss of control. Failure to follow these instructions may result in personal injury.

20. Clean the wheel hub mounting surface.
21. Install the wheel and tire assembly.
22. Apply tension to the parking brake cable system.
   a. Hold the front parking brake cable and conduit taut.
   b. Remove the retainer from the parking brake control.
23. Check operation of the parking brake.

DRUM BRAKES

Brake Drums

REMOVAL & INSTALLATION

1. Before servicing the vehicle, refer to the Precautions Section.
2. Raise and safely support the vehicle. Remove the wheel and tire assembly.
3. Remove the retaining nuts, if equipped, and remove the brake drum.
4. Inspect the brake drum surface for wear, scoring and runout. Machine or replace, as necessary.

To install:

5. Install the brake drum and secure in place with the retainer nuts, if equipped.
6. Adjust the rear brakes.
7. Install the wheel. Lower the vehicle.

Brake Shoes

REMOVAL & INSTALLATION

1. Before servicing the vehicle, refer to the Precautions Section.
2. Raise and safely support the vehicle. Remove the wheel and tire assembly and the brake drum.
3. Pull backward on the adjusting lever cable to disengage the adjusting lever from the adjusting screw. Move the outboard side of the adjusting screw upward and back off the pivot nut as far as it will go.
4. Pull the adjusting lever, cable and automatic adjuster spring down and toward the rear to unhook the pivot hook from the large hole in the secondary shoe web. Do not pry the pivot hook from the hole.
5. Remove the automatic adjuster spring and adjusting lever.
6. Remove the secondary shoe-to-anchor spring using a suitable brake spring removal/installation tool. Using the tool, remove the primary shoe-to-anchor spring and unhook the cable anchor. Remove the anchor pin plate, if equipped.
7. Remove the cable guide from the secondary shoe.

8. Remove the shoe hold-down springs, shoes, adjusting screw, pivot nut and socket. Note the color and position of each hold-down spring so they can be reassembled in the same position.
9. Remove the parking brake link and spring. Disconnect the parking brake cable from the parking brake lever.
10. Remove the secondary brake shoe. On 9 in. (22.8cm) rear brakes, remove the parking brake lever from the shoe. On 10 in. (25.4cm) rear brakes, remove the retainer clip and spring washer and remove the parking brake lever.

To install:

11. Clean the backing plate ledge pads and sand lightly. Apply a light coating of high temperature lithium grease to the points where the brake shoes touch the backing plate. Lubricate the adjusting cable eye and the anchor pin area.
12. Install the parking brake lever on the secondary shoe. On 10 in. (25.4cm) brakes, secure with the spring washer and retaining clip.
13. Position the brake shoes on the backing plate and install the hold-down spring pins, springs and cups. Install the parking brake link, spring and washer. Connect the parking brake cable to the parking brake lever.
14. Install the anchor pin plate, if equipped, and place the cable anchor over the anchor pin with the crimped side toward the backing plate.
15. Install the primary shoe-to-anchor spring using the brake spring removal/installation tool.
16. Install the cable guide on the secondary shoe with the flanged hole fitted into the hole in the secondary shoe. Thread the cable around the cable guide groove.

➡**Make sure the cable is positioned in the groove and not between the guide and shoe web.**

17. Install the secondary shoe-to-anchor (long) spring.

➡**Make sure the cable end is not cocked or binding on the anchor pin when installed. All parts should be flat on the anchor pin.**

18. Apply high temperature lithium grease to the threads and the socket end of the adjusting screw. Turn the adjusting screw into the adjusting pivot nut to the end of the threads and then loosen, ½ turn.
19. Place the adjusting socket on the screw and install the assembly between the shoe ends with the adjusting screw nearest the secondary shoe.

➡**Be sure to install the adjusting screw on the same side of the vehicle from which it came. To prevent incorrect installation, the socket end of each adjusting screw is stamped with R or L, to indicate installation on the right or left side of the vehicle. The adjusting pivot nuts have lines machined around the body of the nut, 2 lines indicating the right side nut and 1 line indicating the left side nut.**

20. Hook the cable hook into the hole in the adjusting lever from the outboard plate side. The adjusting levers are also stamped with an **R** or **L** to indicate right or left side installation.
21. Place the hooked end of the adjuster spring in the large hole in the primary shoe web and connect the loop end of the spring to the adjuster lever hole.
22. Pull the adjuster lever, cable and automatic adjuster spring down toward the rear to engage the pivot hook in the large hole in the secondary shoe web.
23. After installation, check the action of the adjuster by pulling the section of the cable between the cable guide and the adjusting lever toward the secondary shoe web far enough to lift the lever past a tooth on the adjusting screw wheel. The lever should snap into position behind the next tooth and releasing the cable should cause the adjuster spring to return the lever to its original position. This return action will turn the adjusting screw 1 tooth.

1 Wheel Cylinder-to-Backing Plate Bolt (2 Req'd)
2 Washer
3 Inspection Hole Cover
4 Brake Backing Plate
5 Lining Inspection Hole
6 Anchor Pin Guide Plate
7 Rear Wheel Cylinder
8 Wheel Cylinder Brake Shoe Link
9 Parking Brake Strut
10 Parking Brake Lever
11 Brake Shoe Adjusting Lever Cable

12 Rear Brake Shoe and Lining, Secondary
13 Washer
14 Parking Brake Lever Pin Retainer
15 Cable Guide
16 Adjusting Lever Pin
17 Adjusting Lever Return Spring
18 Brake Shoe Adjusting Lever
19 Brake Shoe Adjusting Screw Nut
20 Brake Adjuster Screw
21 Brake Shoe Adjusting Screw Spring

22 Brake Shoe Hold-Down Spring Cup
23 Brake Shoe Hold-Down Spring
24 Rear Brake Shoe and Lining, Primary
25 Brake Shoe Retracting Spring, Short
26 Parking Brake Link Spring
27 Parking Brake Spring Retainer
28 Brake Shoe Hold-Down Spring Pin
29 Brake Adjusting Hole Cover

93026G21

Exploded view of the rear brake shoes and components—Explorer Sport and Sport-Trac

24. If pulling the cable does not produce the action described previously, or if lever action is sluggish instead of positive and sharp, check the position of the lever on the adjusting screw toothed wheel. With the brake in a vertical position, anchor at the top, the lever should contact the adjusting wheel 1 tooth above the centerline of the adjusting screw. If the contact point is below the centerline, the lever will not lock on the adjusting screw wheel teeth and the screw will not turn, since the lever is actuated by the cable.

25. Adjust the brake shoes using either a brake adjustment gauge or manually with the drums installed.

26. Install the wheels, and lower the vehicle.

Wheel Cylinder

REMOVAL & INSTALLATION

1. Before servicing the vehicle, refer to the Precautions Section.

2. Remove the brake shoes and linings.

3. Disconnect the brake line fitting.

4. Remove the two rear wheel cylinder bolts and remove the rear wheel cylinder.

5. To install, reverse the removal procedure. Torque the wheel cylinder bolts to 11 ft. lbs. (15 Nm); the brake line fitting to 13 ft. lbs. (18 Nm).

6. Bleed the brake system.

SPECIFICATIONS AND MAINTENANCE CHARTS

ENGINE AND VEHICLE IDENTIFICATION

Engine							Model Year	
Code ①	Liters (cc)	Cu. In.	Cyl.	Fuel Sys.	Type	Eng. Mfg.	Code ②	Year
2	4.2 (4195)	256	6	MFI	OHV	Ford	2	2002
3	5.4 (5409)	330	8	SFI	SOHC	Ford	3	2003
5	5.4 (5409)	330	8	EFI	SOHC	Ford	4	2004
6	4.6 (4588)	280	8	MFI	SOHC	Ford	5	2005
A	5.4 (5409)	330	8	EFI	DOHC	Ford	6	2006
F	7.3 (7292)	445	8	TDI	OHV	Navistar		
D	6.8 (6802)	415	10	EFI	SOHC	Ford		
L	5.4 (5409)	330	8	EFI	SOHC	Ford		
M	5.4 (5409)	330	8	EFI ③	SOHC	Ford		
P	6.0 (5921)	365	8	TDI	OHV	Navistar		
S	6.8 (6802)	415	10	MFI	SOHC	Ford		
V ④	6.8 (6802)	415	10	MFI	SOHC	Ford		
W	4.6 (4588)	280	8	MFI	SOHC	Ford		
Z	5.4 (5409)	330	8	EFI ⑤	SOHC	Ford		

MFI: Multi-port Fuel Injection

DI: Direct Injection Turbo-Diesel

EFI: Electronic Fuel Injection

SFI: Sequential Fuel Injection

OHV: Overhead Valve

SOHC: Single Overhead Camshaft

① 8th digit of the Vehicle Identification Number (VIN)

② 10th digit of the Vehicle Identification Number (VIN)

③ Natural Gas Vehicle (NGV)

④ 3 valves per cylinder

⑤ Bi-fuel Vehicle (Natural Gas/Propane)

06017-F150-C01

GENERAL ENGINE SPECIFICATIONS

Year	Model	Engine Displ. Liters	Engine VIN	Net Horsepower @ rpm	Net Torque @ rpm (ft. lbs.)	Bore x Stroke (in.)	Compression Ratio	Oil Pressure @ rpm
2002	E-150	4.2	2	205@4400	255@3000	3.81x3.74	9.3:1	40@2500
		4.6	6	210@4400	290@3250	3.55x3.54	9.0:1	20-45@1500
		4.6	W	210@4400	290@3250	3.55x3.54	9.0:1	20-45@1500
		5.4	L	235@4250	330@3000	3.55x4.17	9.0:1	40-70@1500
	E-250	4.2	2	205@4400	255@3000	3.81x3.74	9.3:1	40@2500
		5.4	L	235@4250	330@3000	3.55x4.17	9.0:1	40-70@1500
	E-350	5.4	L	235@4250	330@3000	3.55x4.17	9.0:1	40-70@1500
		6.8	S	265@4250	410@2750	4.09x4.17	9.0:1	40-75@2000
		7.3	F	210@3000	425@2000	4.11x4.18	17.5:1	40-70@3000
	Blackwood	5.4	A	300@5000	355@2750	3.55x4.17	9.0:1	40-70@1500
	F-150	4.2	2	205@4400	255@3000	3.81x3.74	9.3:1	40@2500
		4.6	W	210@4400	290@3250	3.55x3.54	9.0:1	20-45@1500
		5.4	3	235@4250	330@3000	3.55x4.17	9.0:1	40-70@1500
		5.4	L	235@4250	330@3000	3.55x4.17	9.0:1	40-70@1500
	F-250	4.6	6	210@4400	290@3250	3.55x3.54	9.0:1	20-45@1500
		4.6	W	210@4400	290@3250	3.55x3.54	9.0:1	20-45@1500
		5.4	L	235@4250	330@3000	3.55x4.17	9.0:1	40-70@1500
	F-350	5.4	L	235@4250	330@3000	3.55x4.17	9.0:1	40-70@1500
		6.8	S	265@4250	410@2750	4.09x4.17	9.0:1	40-75@2000
		7.3	F	210@3000	425@2000	4.11x4.18	17.5:1	40-70@3000
2003	E-150	4.2	2	205@4400	255@3000	3.81x3.74	9.3:1	40@2500
		5.4	L	235@4250	330@3000	3.55x4.17	9.0:1	40-70@1500
	E-250	4.2	2	205@4400	255@3000	3.81x3.74	9.3:1	40@2500
		5.4	L	235@4250	330@3000	3.55x4.17	9.0:1	40-70@1500
	E-350	5.4	L	235@4250	330@3000	3.55x4.17	9.0:1	40-70@1500
		6.8	S	265@4250	410@2750	4.09x4.17	9.0:1	40-75@2000
		7.3	F	210@3000	425@2000	4.11x4.18	17.5:1	40-70@3000
	Blackwood	5.4	A	300@5000	355@2750	3.55x4.17	9.0:1	40-70@1500
	F-150	4.2	2	205@4400	255@3000	3.81x3.74	9.3:1	40@2500
		4.6	W	210@4400	290@3250	3.55x3.54	9.0:1	20-45@1500
		5.4	3	235@4250	330@3000	3.55x4.17	9.0:1	40-70@1500
		5.4	L	235@4250	330@3000	3.55x4.17	9.0:1	40-70@1500
	F-250	4.6	6	210@4400	290@3250	3.55x3.54	9.0:1	20-45@1500
		4.6	W	210@4400	290@3250	3.55x3.54	9.0:1	20-45@1500
		5.4	L	235@4250	330@3000	3.55x4.17	9.0:1	40-70@1500
	F-350	5.4	L	235@4250	330@3000	3.55x4.17	9.0:1	40-70@1500
		6.8	S	265@4250	410@2750	4.09x4.17	9.0:1	40-75@2000
		7.3	F	210@3000	425@2000	4.11x4.18	17.5:1	40-70@3000

06017-F150-C02

GENERAL ENGINE SPECIFICATIONS

Year	Model	Engine Displ. Liters	Engine VIN	Net Horsepower @ rpm	Net Torque @ rpm (ft. lbs.)	Bore x Stroke (in.)	Com- pression Ratio	Oil Pressure @ rpm
2004	F-150	4.2	2	202@4800	255@3400	3.81x3.74	9.3:1	40@2500
		4.6	W	210@4400	290@3250	3.55x3.54	9.0:1	20-45@1500
		5.4	5	300@5000	365@3750	3.55x4.17	9.8:1	40-60@2000
	Lightning	5.4	3	380@4750	450@3250	3.55x4.17	8.4:1	40-60@2000
	F-Super Duty	5.4	L	235@4250	330@3000	3.55x4.17	9.0:1	40-70@1500
		6.0	P	325@3300	560@2000	3.74x4.13	18.0:1	24@1200
		6.8	D	265@4250	410@2750	4.09x4.17	9.0:1	40-75@2000
		6.8	S	265@4250	410@2750	4.09x4.17	9.0:1	40-75@2000
	E-150 Wagon	4.6	W	210@4400	290@3250	3.55x3.54	9.0:1	20-45@1500
		5.4	L	235@4250	330@3000	3.55x4.17	9.0:1	40-70@1500
	E-350 Wagon	5.4	L	235@4250	330@3000	3.55x4.17	9.0:1	40-70@1500
		6.0	P	325@3300	560@2000	3.74x4.13	18.0:1	24@1200
		6.8	S	265@4250	410@2750	4.09x4.17	9.0:1	40-75@2000
	E-150 Cargo	4.6	W	210@4400	290@3250	3.55x3.54	9.0:1	20-45@1500
		5.4	L	235@4250	330@3000	3.55x4.17	9.0:1	40-70@1500
		6.0	P	325@3300	560@2000	3.74x4.13	18.0:1	24@1200
	E-250 Cargo	4.6	W	210@4400	290@3250	3.55x3.54	9.0:1	20-45@1500
		5.4	L	235@4250	330@3000	3.55x4.17	9.0:1	40-70@1500
		6.0	P	325@3300	560@2000	3.74x4.13	18.0:1	24@1200
	E-350 Cargo	5.4	L	235@4250	330@3000	3.55x4.17	9.0:1	40-70@1500
		6.0	P	325@3300	560@2000	3.74x4.13	18.0:1	24@1200
		6.8	S	265@4250	410@2750	4.09x4.17	9.0:1	40-75@2000
2005	F-150	4.2	2	202@4800	255@3400	3.81x3.74	9.3:1	40@2500
		4.6	W	210@4400	290@3250	3.55x3.54	9.0:1	20-45@1500
		5.4	5	300@5000	365@3750	3.55x4.17	9.8:1	40-60@2000
	F-Super Duty	5.4	L	235@4250	330@3000	3.55x4.17	9.0:1	40-70@1500
		6.0	P	325@3300	560@2000	3.74x4.13	18.0:1	24@1200
		6.8	V	362@4750	465@3300	4.09x4.17	9.2:1	40-75@2000
	E-150 Wagon	4.6	W	210@4400	290@3250	3.55x3.54	9.0:1	20-45@1500
		5.4	L	235@4250	330@3000	3.55x4.17	9.0:1	40-70@1500
	E-350 Wagon	5.4	L	235@4250	330@3000	3.55x4.17	9.0:1	40-70@1500
		6.0	P	325@3300	560@2000	3.74x4.13	18.0:1	24@1200
		6.8	S	265@4250	410@2750	4.09x4.17	9.0:1	40-75@2000
	E-150 Cargo	4.6	W	210@4400	290@3250	3.55x3.54	9.0:1	20-45@1500
		5.4	L	235@4250	330@3000	3.55x4.17	9.0:1	40-70@1500
		6.0	P	325@3300	560@2000	3.74x4.13	18.0:1	24@1200
	E-250 Cargo	4.6	W	210@4400	290@3250	3.55x3.54	9.0:1	20-45@1500
		5.4	L	235@4250	330@3000	3.55x4.17	9.0:1	40-70@1500
		6.0	P	325@3300	560@2000	3.74x4.13	18.0:1	24@1200
	E-350 Cargo	5.4	L	235@4250	330@3000	3.55x4.17	9.0:1	40-70@1500
		6.0	P	325@3300	560@2000	3.74x4.13	18.0:1	24@1200
		6.8	S	265@4250	410@2750	4.09x4.17	9.0:1	40-75@2000
2006	Mark LT	5.4	5	300@5000	365@3750	3.55x4.17	9.8:1	40-60@2000

06017-F150-C03

GASOLINE ENGINE TUNE-UP SPECIFICATIONS

Year	Engine Displacement Liters	Engine VIN	Spark Plug Gap (in.)	Ignition Timing (deg.) ①		Fuel Pump (psi) ②	Idle Speed (rpm)		Valve Clearance	
				MT	AT		MT	AT	In.	Ex.
2002	4.2	2	0.052-0.056	10B	10B	30-45	③	③	HYD	HYD
	4.6	6	0.052-0.056	10B	10B	30-45	③	③	HYD	HYD
	4.6	W	0.052-0.056	10B	10B	30-45	③	③	HYD	HYD
	5.4	A	0.052-0.056	10B	10B	28-45	③	③	HYD	HYD
	5.4	L	0.052-0.056	10B	10B	28-45	③	③	HYD	HYD
	5.4	3	0.052-0.056	10B	10B	28-45	③	③	HYD	HYD
	6.8	S	0.052-0.055	10B	10B	28-45	③	③	HYD	HYD
2003	4.2	2	0.052-0.056	10B	10B	30-45	③	③	HYD	HYD
	4.6	6	0.052-0.056	10B	10B	30-45	③	③	HYD	HYD
	4.6	W	0.052-0.056	10B	10B	30-45	③	③	HYD	HYD
	5.4	A	0.052-0.056	10B	10B	28-45	③	③	HYD	HYD
	5.4	L	0.052-0.056	10B	10B	28-45	③	③	HYD	HYD
	5.4	3	0.052-0.056	10B	10B	28-45	③	③	HYD	HYD
	6.8	S	0.052-0.055	10B	10B	28-45	③	③	HYD	HYD
2004	4.2	2	0.052-0.056	10B	10B	30-45	③	③	HYD	HYD
	4.6	W	0.040-0.050	10B	10B	30-45	③	③	HYD	HYD
	5.4	5	0.052-0.056	10B	10B	28-55	③	③	HYD	HYD
	5.4	3	0.052-0.056	—	10B	28-50	③	③	HYD	HYD
	5.4	L	0.052-0.056	10B	10B	28-45	③	③	HYD	HYD
	6.8	S	0.052-0.056	10B	10B	28-45	③	③	HYD	HYD
2005	4.2	2	0.052-0.056	10B	10B	30-45	③	③	HYD	HYD
	4.6	W	0.040-0.050	10B	10B	30-45	③	③	HYD	HYD
	5.4	5	0.040-0.050	10B	10B	28-55	③	③	HYD	HYD
	5.4	L	0.040-0.050	10B	10B	28-45	③	③	HYD	HYD
	6.8	S	0.052-0.056	10B	10B	28-45	③	③	HYD	HYD
	6.8	V	0.040-0.050	10B	10B	28-45	③	③	HYD	HYD
2006	5.4	5	0.040-0.050	10B	10B	28-55	③	③	HYD	HYD

NOTE: The Vehicle Emission Control Information label often reflects specification changes changes made during production. The label figures must be used
if they differ from this chart.

B: Before top dead center

HYD: Hydraulic

NA: Information not Available

① Ignition timing is preset and cannot be adjusted

② With engine running

③ Idle speed is electronically controlled and cannot be adjusted

06017-F150-C04

4.2L Engine
Firing Order: 1–4–2–5–3–6
Distributorless ignition system

79243G28

4.6L Engine
Firing Order: 1–3–7–2–6–5–4–8
Distributorless ignition system

79243G29

5.4L Engine
Firing Order: 1–3–7–2–6–5–4–8
Distributorless ignition system; one coil per cylinder

79243G58

6.8L Engine
Firing Order: 1–6–5–10–2–7–3–8–4–9
Distributorless ignition system; one coil per cylinder

79243G57

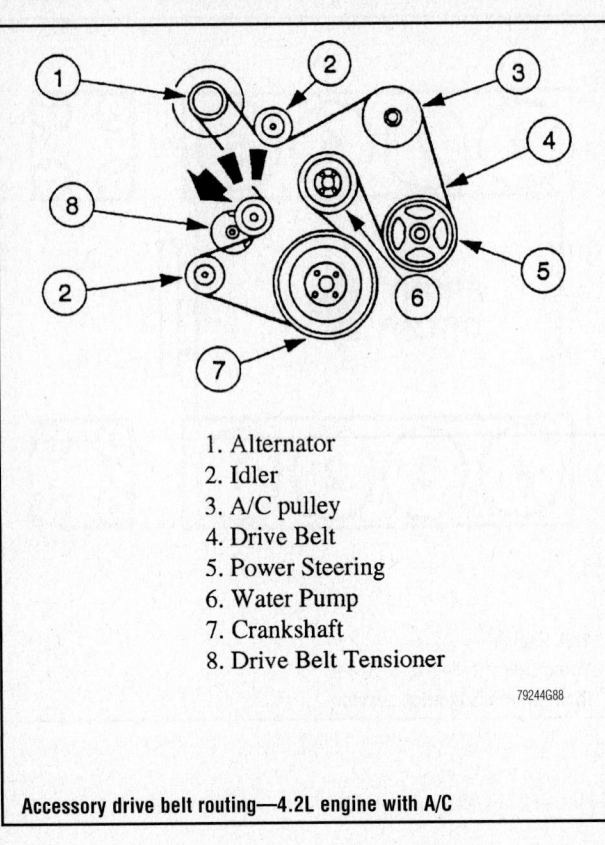

1. Alternator
2. Idler
3. A/C pulley
4. Drive Belt
5. Power Steering
6. Water Pump
7. Crankshaft
8. Drive Belt Tensioner

79244G88

Accessory drive belt routing—4.2L engine with A/C

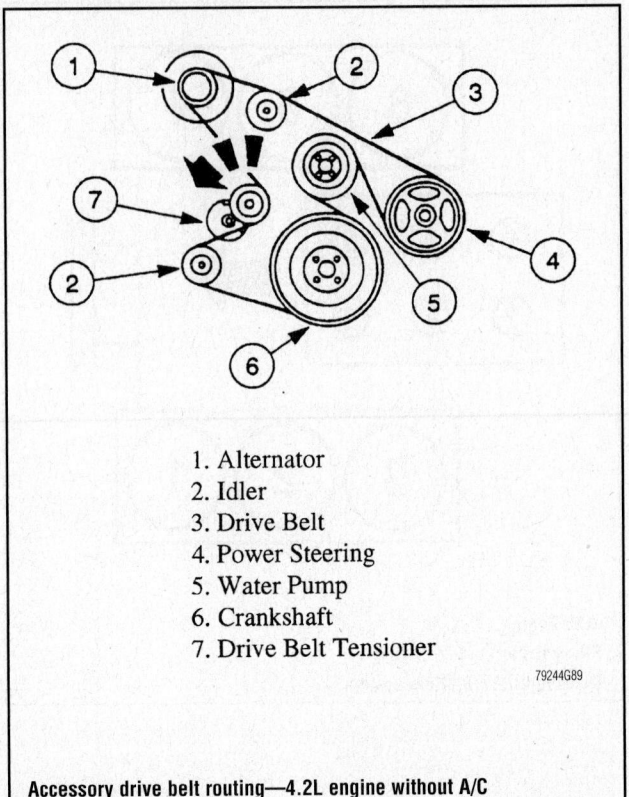

1. Alternator
2. Idler
3. Drive Belt
4. Power Steering
5. Water Pump
6. Crankshaft
7. Drive Belt Tensioner

79244G89

Accessory drive belt routing—4.2L engine without A/C

1. Idler
2. Alternator
3. Water Pump
4. Power Steering
5. Crankshaft
6. Drive Belt Tensioner
7. A/C Pulley
8. Drive Belt

79244G86

Accessory drive belt routing—4.6L, 5.4L and 6.8L engines with A/C

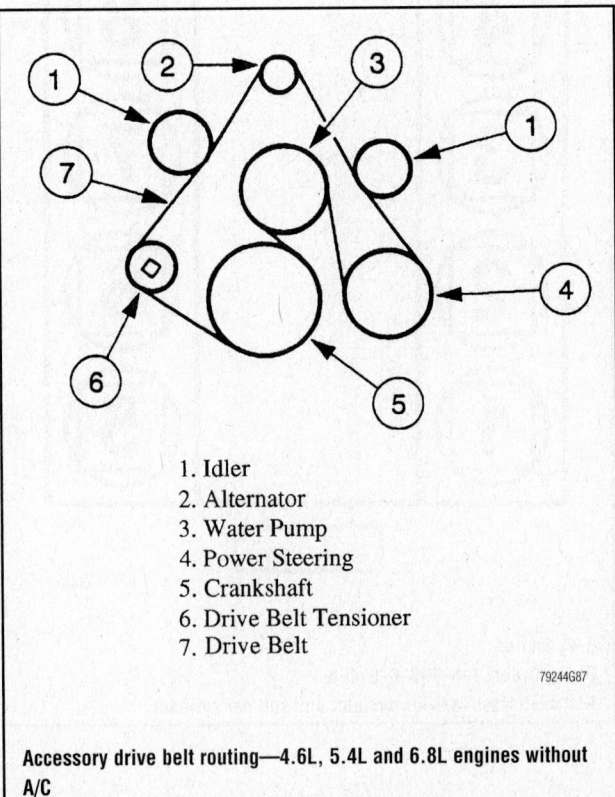

1. Idler
2. Alternator
3. Water Pump
4. Power Steering
5. Crankshaft
6. Drive Belt Tensioner
7. Drive Belt

79244G87

Accessory drive belt routing—4.6L, 5.4L and 6.8L engines without A/C

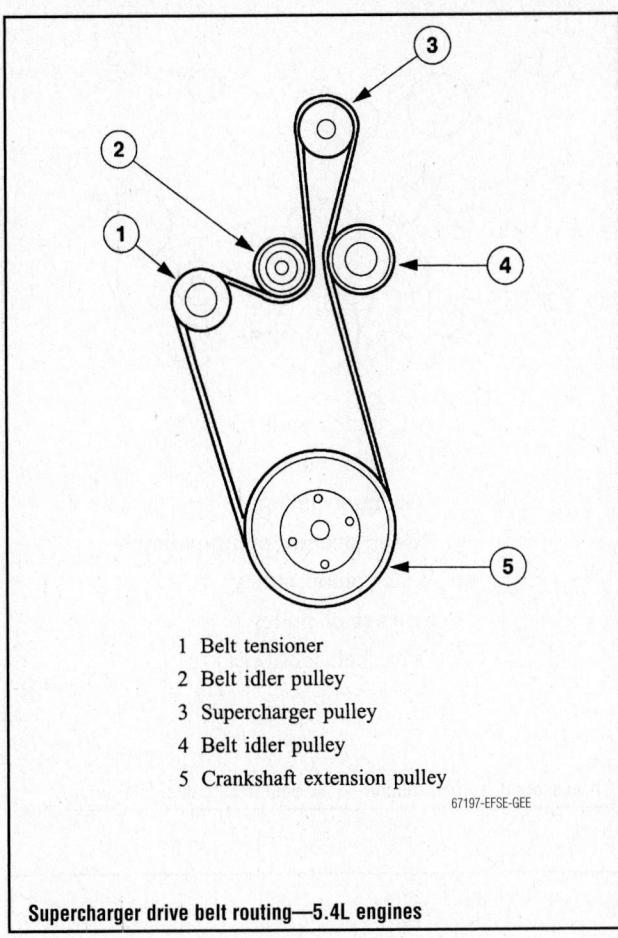

1 Belt tensioner
2 Belt idler pulley
3 Supercharger pulley
4 Belt idler pulley
5 Crankshaft extension pulley

67197-EFSE-GEE

Supercharger drive belt routing—5.4L engines

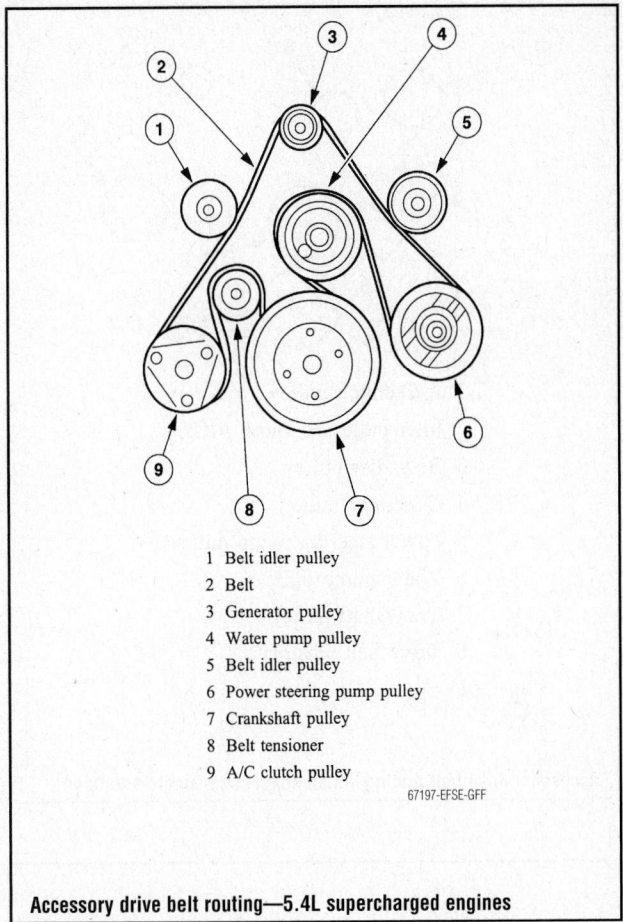

1 Belt idler pulley
2 Belt
3 Generator pulley
4 Water pump pulley
5 Belt idler pulley
6 Power steering pump pulley
7 Crankshaft pulley
8 Belt tensioner
9 A/C clutch pulley

67197-EFSE-GFF

Accessory drive belt routing—5.4L supercharged engines

DIESEL ENGINE TUNE-UP SPECIFICATIONS

Year	Engine Displ. Liters	Engine VIN	Valve Clearance		Injection Pump Setting (deg.)	Injection Nozzle Pressure (psi)		Idle Speed (rpm)	Cranking Compression Pressure (psi)
			Intake (in.)	Exhaust (in.)		New	Used		
2002	7.3	F	HYD	HYD	①	1875	1425	②	③
2003	7.3	F	HYD	HYD	①	1875	1425	②	③
2004	6.0	P	HYD	HYD	①	④	④	②	NA
2005	6.0	P	HYD	HYD	①	④	④	②	NA

NOTE: The Vehicle Emission Control Information label often reflects specification changes made during production. The label figures must be used if they differ

from those in this chart

HYD: Hydraulic

B: Before top dead center

NA: Not Available

① PCM controlled

② See underhood emission label

③ Compression pressure in the lowest cylinder must be at least 75% of the highest cylinder

   Minimum pressure: 195 psi

   Maximum pressure: 440 psi

④ Pump output pressure: 450–4,000 psi

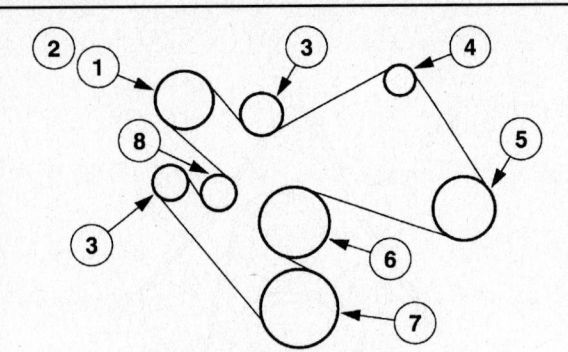

1 A/C clutch pulley
2 Idler pulley (without A/C)
3 Belt idler pulley
4 Generator pulley
5 Power steering pump pulley
6 Water pump pulley
7 Crankshaft pulley
8 Drive belt tensioner

67197-EFSE-GAA

Accessory drive belt routing—7.3L engine with single alternator

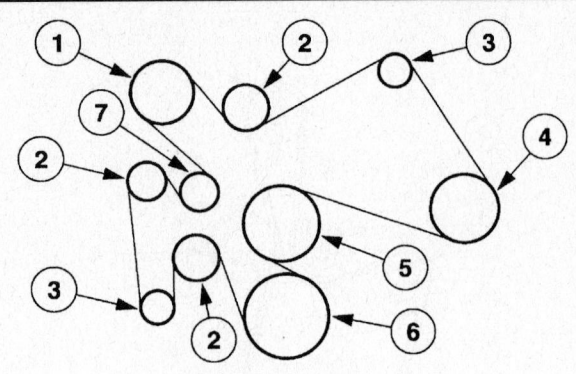

1 A/C clutch pulley
2 Belt idler pulley
3 Generator pulley
4 Power steering pump pulley
5 Water pump pulley
6 Crankshaft pulley
7 Drive belt tensioner

67197-EFSE-GBB

Accessory drive belt routing—7.3L engine with dual alternator

1 A/C clutch pulley
2 Idler pulley (without A/C)
3 Belt idler pulley
4 Generator pulley
5 Power steering pump pulley
6 Water pump pulley
7 Crankshaft pulley
8 Drive belt tensioner

67197-EFSE-GCC

Accessory drive belt routing—6.0L engine with single alternator

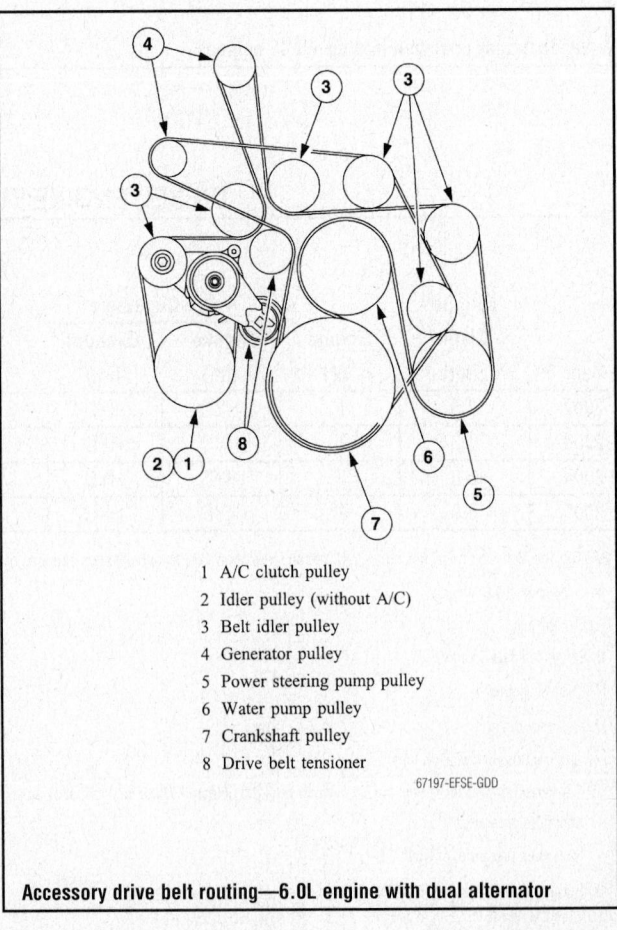

1 A/C clutch pulley
2 Idler pulley (without A/C)
3 Belt idler pulley
4 Generator pulley
5 Power steering pump pulley
6 Water pump pulley
7 Crankshaft pulley
8 Drive belt tensioner

67197-EFSE-GDD

Accessory drive belt routing—6.0L engine with dual alternator

CAPACITIES

Year	Model	Engine Displ. Liters	Engine VIN	Engine Oil with Filter (qts.)	Transmission (pts.)		Transfer Case (pts.)*	Drive Axle		Fuel Tank (gal.)	Cooling System (qts.)
					MT	Auto.		Front (pts.)	Rear (pts.)		
2002	E-150	4.2	2	6.0	—	①	—	—	6.0	35.0	②
		4.6	6	6.0	—	①	—	—	6.0	35.0	②
		4.6	W	6.0	—	①	—	—	6.0	35.0	②
		5.4	L	7.0	—	①	—	—	6.0	35.0	②
	E-250	4.2	2	6.0	—	①	—	—	6.0 ③	35.0	②
		5.4	L	7.0	—	①	—	—	6.0 ③	35.0	②
	E-350	5.4	L	7.0	—	①	—	—	6.0 ③	35.0	②
		6.8	S	6.0	—	①	—	—	6.0 ③	35.0	②
		7.3	F	14.0	—	①	—	—	6.0 ③	35.0	23.0
	Blackwood	5.4	A	6.0	—	15.9	—	—	6.0	25.0	19.2
	F-150	4.2	2	6.0	7.6	26.0	4.0	3.7	5.5	24.5 ④	②
		4.6	W	6.0	7.6	26.0	4.0	3.7	5.5	24.5 ④	②
		5.4	3	7.0	7.6	①	4.2	6.0	6.0	30.0	②
		5.4	L	7.0	7.6	①	4.2	6.0	6.0	30.0	②
	F-250	4.6	W	6.0	7.6	26.0	4.0	3.7	5.5	24.5 ④	②
		5.4	L	7.0	7.6	①	4.2	6.0	6.0	30.0	②
	F-350	5.4	L	7.0	7.6	①	4.2	6.0	6.0	30.0	②
		6.8	S	6.0	7.6	①	4.2	6.0 ⑤	6.0 ③	35.0	②
		7.3	F	14.0	7.6	①	4.2	6.0 ⑤	6.0 ③	35.0	23.0
2003	E-150	4.2	2	6.0	—	①	—	—	6.0	35.0	②
		4.6	6	6.0	—	①	—	—	6.0	35.0	②
		4.6	W	6.0	—	①	—	—	6.0	35.0	②
		5.4	L	7.0	—	①	—	—	6.0	35.0	②
	E-250	4.2	2	6.0	—	①	—	—	6.0 ③	35.0	②
		5.4	L	7.0	—	①	—	—	6.0 ③	35.0	②
	E-350	5.4	L	7.0	—	①	—	—	6.0 ③	35.0	②
		6.8	S	6.0	—	①	—	—	6.0 ③	35.0	②
		7.3	F	14.0	—	①	—	—	6.0 ③	35.0	23.0
	Blackwood	5.4	A	6.0	—	15.9	—	—	6.0	25.0	19.2
	F-150	4.2	2	6.0	7.6	26.0	4.0	3.7	5.5	24.5 ④	②
		4.6	W	6.0	7.6	26.0	4.0	3.7	5.5	24.5 ④	②
		5.4	3	7.0	7.6	①	4.2	6.0	6.0	30.0	②
		5.4	L	7.0	7.6	①	4.2	6.0	6.0	30.0	②
	F-250	4.6	W	6.0	7.6	26.0	4.0	3.7	5.5	24.5 ④	②
		5.4	L	7.0	7.6	①	4.2	6.0	6.0	30.0	②
	F-350	5.4	L	7.0	7.6	①	4.2	6.0	6.0	30.0	②
		6.8	S	6.0	7.6	①	4.2	6.0 ⑤	6.0 ③	35.0	②
		7.3	F	14.0	7.6	①	4.2	6.0 ⑤	6.0 ③	35.0	23.0

06017-F150-C06

CAPACITIES

Year	Model	Engine Displ. Liters	Engine VIN	Engine Oil with Filter (qts.)	Transmission (pts.) MT	Transmission (pts.) Auto.	Transfer Case (pts.)*	Drive Axle Front (pts.)	Drive Axle Rear (pts.)	Fuel Tank (gal.)	Cooling System (qts.)
2004	F-150	4.2	2	6.0	7.6	⑦	4.0	3.7	5.5	⑥	②
		4.6	W	6.0	—	27.8	4.0	3.6	5.5	⑥	19.8
		5.4	5	6.0	—	27.8	4.0	3.6	5.5	⑥	20.6
		5.4	3	6.0	—	34.2	—	—	NA	24.5	19.4 ⑧
	F-Super Duty	5.4	L	6.0	11.6	⑨	4.0	5.9	⑩	⑪	26.4
		6.0	P	15.0	11.6	⑨	4.0	5.9	⑩	⑪	27.5
		6.8	S	6.0	11.6	⑨	4.0	5.9	⑩	⑪	27.5
	E-150 Wagon	4.6	W	6.0	—	⑫	—	—	⑬	35.0	⑮
		5.4	L	6.0	—	⑫	—	—	⑬	35.0	⑯
	E-350 Wagon	5.4	L	6.0	—	⑫	—	—	⑬	35.0	⑯
		6.0	P	15.0	—	⑫	—	—	⑬	35.0	⑰
		6.8	S	6.0	—	⑫	—	—	⑬	35.0	⑱
	E-150 Cargo	4.6	W	6.0	—	⑫	—	—	⑬	35.0	⑮
		5.4	L	6.0	—	⑫	—	—	⑬	35.0	⑯
		6.0	P	15.0	—	⑫	—	—	⑬	35.0	⑰
	E-250 Cargo	4.6	W	6.0	—	⑫	—	—	⑬	35.0	⑮
		5.4	L	6.0	—	⑫	—	—	⑬	35.0	⑯
		6.0	P	15.0	—	⑫	—	—	⑬	35.0	⑰
	E-350 Cargo	5.4	L	6.0	—	⑫	—	—	⑬	⑭	⑯
		6.0	P	15.0	—	⑫	—	—	⑬	⑭	⑰
		6.8	S	6.0	—	⑫	—	—	⑬	⑭	⑱
2005	F-150	4.2	2	6.0	7.6	27.8	4.0	3.7	⑲	⑥	20.1
		4.6	W	6.0	—	27.8	4.0	3.7	⑲	⑥	19.8
		5.4	5	7.0	—	27.8	4.0	3.7	⑲	⑥	20.6
	F-Super Duty	5.4	L	7.0	11.6	35.0 ㉑	4.0	6.3	⑳	⑪	26.4
		6.0	P	15.0	11.6	35.0 ㉑	4.0	6.3	⑳	⑪	27.5
		6.8	V	7.0	11.6	35.0 ㉑	4.0	6.3	⑳	⑪	27.5
	E-150 Wagon	4.6	W	6.0	—	⑫	—	—	⑬	35.0	⑮
		5.4	L	6.0	—	⑫	—	—	⑬	35.0	⑯
	E-350 Wagon	5.4	L	6.0	—	⑫	—	—	⑬	35.0	⑯
		6.0	P	15.0	—	⑫	—	—	⑬	35.0	⑰
		6.8	S	6.0	—	⑫	—	—	⑬	35.0	⑱
	E-150 Cargo	4.6	W	6.0	—	⑫	—	—	⑬	35.0	⑮
		5.4	L	6.0	—	⑫	—	—	⑬	35.0	⑯
		6.0	P	15.0	—	⑫	—	—	⑬	35.0	⑰
	E-250 Cargo	4.6	W	6.0	—	⑫	—	—	⑬	35.0	⑮
		5.4	L	6.0	—	⑫	—	—	⑬	35.0	⑯
		6.0	P	15.0	—	⑫	—	—	⑬	35.0	⑰
	E-350 Cargo	5.4	L	6.0	—	⑫	—	—	⑬	⑭	⑯
		6.0	P	15.0	—	⑫	—	—	⑬	⑭	⑰
		6.8	S	6.0	—	⑫	—	—	⑬	⑭	⑱
2006	Mark LT	5.4	5	7.0	—	27.8	4.0	3.7	5.5	30	21.2

06017-F150-C07

CAPACITIES CHART FOOTNOTES

NA: Information not available

NOTE: All capacities are approximate. Add fluid gradually and check to be sure a proper fluid level is obtained.

* Overhaul

① With 4R70W: 28 pts.

    With E40D: 32.0 pts.

    With4R100: 35.4 pts.

② 4.2L. with standard cooling: 23.3 qts.

    4.2L. with aux cooling: 25.4 qts.

    4.6L. with standard cooling: 25.0 qts.

    4.6L. with aux cooling: 27.2 qts.

    5.4L. with standard cooling: 29.0 qts.

    5.4L. with aux cooling: 31.0 qts.

    6.8L. with standard cooling: 30.6 qts.

    6.8L. with aux cooling: 32.8 qts.

③ Heavy duty: 7.5 pts.

④ Also available with a 30 gallon

    tank with 8 ft. box

⑤ Without PTO: 4.2 pts.

    With PTO: 12.0 pts.

⑥ Regular cab w/126 in. wheel base and 6.5 ft. bed: 26.0

    Super cab w/132 inch wheel base and 5.5 ft. bed: 26.0

    Crew cab w/144 in. wheel base and 6.5 ft. bed: 27.0

    Regular cab w/144 in. wheel base and 8 ft. bed: 27.0

    SuperCab w/163 in. wheel base and 8 ft. bed: 27.0

    Optional for SuperCab w/144 in. wheel base and 6.5 ft. bed: 35.7

    Optional for Regular Cab w/144 in. wheel base and 8 ft. bed: 35.7

    Optional for SuperCab w/163 in. wheel base and 8 ft. bed: 35.7

    Heritage edition:

        4x4 short wheelbase Regular Cab: 24.5

        4x2 Regular Cab w/short wheelbase: 25.0

        4x2 SuperCab w/short wheelbase: 25.0

        4x4 SuperCab w/short wheelbase: 25.0

        4x2 Regular Cab w/long wheelbase: 30.0

        4x4 Regular Cab w/long wheelbase: 30.0

        4x2 SuperCab w/long wheelbase: 30.0

        4x4 SuperCab w/long wheelbase: 30.0

        Lightning: 24.5

⑦ 4R100: 34.2 pts.

    4R70W: 27.8 pts.

⑧ Plus 4.2 qts. for supercharger

⑨ 4R100: 34.2 pts.

    TorqShift: 38.4 pts.

⑩ Dana 80

    Conventional and Trutrac: 8.5 pts.

    Trac-Lok: 8.0 pts.; first fill with 0.5 pt. of friction modifier

    Dana S110: 16 pts.

    Dana S135: 24.5 pts.

    Ford 10.5 inch: 6.9 pts.

⑪ Wide frame vehicles with standard bed midship tank: 38

    Wide frame vehicles with short bed midship tank: 29

    Narrow frame chassis cab with aft of axle standard equipment: 40

    Narrow frame chassis cab with optional midship tank: 19

⑫ 4R100: 34.2 pts.

    4R70W/4R75W: 27.8 pts.

    TorqShift: 38.4 pts.

⑬ Ford 8.8 inch: 5.7 pts.

    Ford 9.5 inch: 5.7 pts.

    Traction-Lok includes 4 oz. of friction modifier

    Dana 60-1U: 6.3 pts.

    Dana 70-2U: 6.6 pts.

    Dana 70-HD: 7.4

    Limited slip includes 7 oz. of friction modifier

⑭ E-350 Cargo: 35.0

    Commercial chassis-cab: 37.0 standard; 55.0 optional

⑮ Without rear heat: 23.8 qts.

    With rear heat: 26.0 qts.

⑯ Without rear heat: 27.8 quarts

    With rear heat: 29.8 quarts

⑰ Without rear heat: 22.2 qts.

    With rear heat: 24.9 qts.

⑱ Without rear heat: 29.4 qts.

    With rear heat: 31.6 qts.

⑲ 8.8 and 9.75 inch axles: 5.5 pts.

    10.25 inch axle: 6.9 pts.

⑳ 10.5 inch: 6.9 pts.

    Dana 80: 8.5 pts.

    Dana S110: 14.0 pts.

㉑ Includes remote filter change

06017-F150-C08

VALVE SPECIFICATIONS

Year	Engine Displ. Liters	Engine VIN	Seat Angle (deg.)	Face Angle (deg.)	Spring Test Pressure (lbs. @ in.)	Spring Installed Height (in.)	Stem-to-Guide Clearance (in.)		Stem Diameter (in.)	
							Intake	Exhaust	Intake	Exhaust
2002	4.2	2	44.75	45.67	224@1.16	1.566-1.637	0.0450-0.0900	0.0015-0.0033	0.2738-0.2728	0.2751-0.2741
	4.6	6	45	45.5	132@1.100	1.570	0.0008-0.0027	0.0018-0.0037	0.2750-0.2746	0.2740-0.2736
	4.6	W	45	45.5	132@1.100	1.570	0.0008-0.0027	0.0018-0.0037	0.2750-0.2746	0.2740-0.2736
	5.4	A	44.5-45	45.25-45.75	150@1.10	1.575	0.0008-0.0027	0.0018-0.0037	0.2754-0.2746	0.2744-0.2736
	5.4	3	45	45.5	150@1.10	1.570	0.0008-0.0027	0.0018-0.0037	0.2754-0.2746	0.2744-0.2736
	5.4	L	45	45.5	150@1.10	1.570	0.0008-0.0027	0.0018-0.0037	0.275-0.2746	0.274-0.2736
	6.8	S	44.50-45.25	45.25-45.75	150@1.10	1.570	0.0008-0.0027	0.0018-0.0037	0.2754-0.2746	0.2744-0.2735
	7.3	F	①	①	200@1.38	②	0.0055	0.0055	0.3119-0.3126	0.3119-0.3126
2003	4.2	2	44.75	45.67	224@1.16	1.566-1.637	0.0450-0.0900	0.0015-0.0033	0.2738-0.2728	0.2751-0.2741
	4.6	6	45	45.5	132@1.100	1.570	0.0008-0.0027	0.0018-0.0037	0.2750-0.2746	0.2740-0.2736
	4.6	W	45	45.5	132@1.100	1.570	0.0008-0.0027	0.0018-0.0037	0.2750-0.2746	0.2740-0.2736
	5.4	A	44.5-45	45.25-45.75	150@1.10	1.575	0.0008-0.0027	0.0018-0.0037	0.2754-0.2746	0.2744-0.2736
	5.4	3	45	45.5	150@1.10	1.570	0.0008-0.0027	0.0018-0.0037	0.2754-0.2746	0.2744-0.2736
	5.4	L	45	45.5	150@1.10	1.570	0.0008-0.0027	0.0018-0.0037	0.2754-0.2746	0.2744-0.2736
	6.8	S	44.50-45.25	45.25-45.75	150@1.10	1.570	0.0008-0.0027	0.0018-0.0037	0.2754-0.2746	0.2744-0.2735
	7.3	F	①	①	200@1.38	②	0.0055	0.0055	0.3119-0.3126	0.3119-0.3126
2004	4.2	2	44.75	45.67	224@1.16	1.566-1.637	0.0450-0.0900	0.0015-0.0033	0.2738-0.2728	0.2751-0.2741
	4.6	W	45.5	45.25-45.75	132@1.103	1.563-1.586	0.0008-0.0027	0.0018-0.0037	0.2754-0.2746	0.2744-0.2736
	5.4	3	45.5	45.25-45.75	171@1.14	1.6654-1.689	0.0008-0.0027	0.0018-0.0037	0.2754-0.2746	0.2744-0.2736
	5.4	5	44.5-45	45.5	171@1.22	1.660	0.0010-0.0020	0.0030-0.0040	0.2350-0.2360	0.2340-0.2350
	5.4	L	45.5	45.25-45.75	171@1.34	1.6654-1.689	0.0008-0.0027	0.0018-0.0037	0.2754-0.2746	0.2744-0.2736
	6.0	P	①	①	191@1.51	1.820	0.0055 max.	0.0055 max.	0.2720-0.2735	0.2720-0.2735
	6.8	S	44.5-45	45.25-45.75	150@1.10	1.575	0.0008-0.0027	0.0018-0.0037	0.2754-0.2746	0.2744-0.2736

VALVE SPECIFICATIONS

Year	Engine Displ. Liters	Engine VIN	Seat Angle (deg.)	Face Angle (deg.)	Spring Test Pressure (lbs. @ in.)	Spring Installed Height (in.)	Stem-to-Guide Clearance (in.)		Stem Diameter (in.)	
							Intake	Exhaust	Intake	Exhaust
2005	4.2	2	44.75	45.67	224@1.16	1.566-1.637	0.0450-0.0900	0.0015-0.0033	0.2738-0.2728	0.2751-0.2741
	4.6	W	45.5	45.25-45.75	132@1.103	1.563-1.586	0.0008-0.0027	0.0018-0.0037	0.2754-0.2746	0.2744-0.2736
	5.4	5	44.5-45	45.5	171@1.22	1.660	0.0010-0.0020	0.0030-0.0040	0.2350-0.2360	0.2340-0.2350
	5.4	L	45.5	45.25-45.75	171@1.34	1.6654-1.689	0.0008-0.0027	0.0018-0.0037	0.2754-0.2746	0.2744-0.2736
	6.0	P	①	①	191@1.51	1.820	0.0055 max.	0.0055 max.	0.2720-0.2735	0.2720-0.2735
	6.8	S	44.5-45	45.25-45.75	150@1.10	1.575	0.0008-0.0027	0.0018-0.0037	0.2754-0.2746	0.2744-0.2736
	6.8	V	44.5-45	45.5	171@1.66	1.660	0.0010-0.0030	0.0020-0.0040	0.2350-0.2360	0.2340-0.2350
2006	5.4	5	44.5-45	45.5	171@1.22	1.660	0.0010-0.0020	0.0030-0.0040	0.2350-0.2360	0.2340-0.2350

① Intake: 30
   Exhaust: 37.5

② Intake: 1.767 in.
   Exhaust: 1.833 in.

06017-F150-C10

PISTON AND RING SPECIFICATIONS

All measurements are given in inches.

Year	Engine Displ. Liters	Engine VIN	Piston Clearance	Ring Gap			Ring Side Clearance		
				Top Compression	Bottom Compression	Oil Control	Top Compression	Bottom Compression	Oil Control
2002	4.2	2	0.0007-0.0018	0.001-0.002	0.001-0.002	0.006-0.026	0.0012-0.0031	0.0012-0.0031	SNUG
	4.6	6	0.0005-0.0010	0.010-0.020	0.010-0.020	0.006-0.026	0.0016-0.0031	0.0012-0.0031	SNUG
	4.6	W	0.0005-0.0010	0.010-0.020	0.010-0.020	0.006-0.026	0.0016-0.0031	0.0012-0.0031	SNUG
	5.4	A	0.0000-0.0010	0.005-0.011	0.010-0.016	0.006-0.026	0.0012-0.0037	0.0012-0.0037	SNUG
	5.4	3	0.0000-0.0010	0.005-0.011	0.010-0.016	0.006-0.026	0.0012-0.0037	0.0012-0.0037	SNUG
	5.4	L	0.0000-0.0010	0.005-0.011	0.010-0.016	0.006-0.026	0.0012-0.0037	0.0012-0.0037	SNUG
	6.8	S	0.0000-0.0010	0.005-0.011	0.010-0.016	0.006-0.026	0.0012-0.0037	0.0012-0.0037	SNUG
	7.3	F	0.0044-0.0057	0.014-0.024	0.062-0.072	0.012-0.024	0.0013-0.0033	0.0013-0.0033	SNUG
2003	4.2	2	0.0007-0.0018	0.001-0.002	0.001-0.002	0.006-0.026	0.0012-0.0031	0.0012-0.0031	SNUG
	4.6	6	0.0005-0.0010	0.010-0.020	0.010-0.020	0.006-0.026	0.0016-0.0031	0.0012-0.0031	SNUG
	4.6	W	0.0005-0.0010	0.010-0.020	0.010-0.020	0.006-0.026	0.0016-0.0031	0.0012-0.0031	SNUG
	5.4	A	0.0000-0.0010	0.005-0.011	0.010-0.016	0.006-0.026	0.0012-0.0037	0.0012-0.0037	SNUG
	5.4	3	0.0000-0.0010	0.005-0.011	0.010-0.016	0.006-0.026	0.0012-0.0037	0.0012-0.0037	SNUG
	5.4	L	0.0000-0.0010	0.005-0.011	0.010-0.016	0.006-0.026	0.0012-0.0037	0.0012-0.0037	SNUG
	6.8	S	0.0000-0.0010	0.005-0.011	0.010-0.016	0.006-0.026	0.0012-0.0037	0.0012-0.0037	SNUG
	7.3	F	0.0044-0.0057	0.014-0.024	0.062-0.072	0.012-0.024	0.0013-0.0033	0.0013-0.0033	SNUG
2004	4.2	2	0.0007-0.0018	0.001-0.002	0.001-0.002	0.006-0.026	0.0012-0.0031	0.0012-0.0031	SNUG
	4.6	W	0.0005-0.0010	0.010-0.020	0.010-0.020	0.006-0.026	0.0012-0.0028	0.0012-0.0028	0.0018-0.0077
	5.4	5	0.0010-0.0018	0.0059-0.0118	0.010-0.016	0.006-0.026	0.0012-0.0020	0.0012-0.0031	SNUG
	5.4	3	0.0010-0.0018	0.0059-0.0118	0.010-0.016	0.006-0.026	0.0012-0.0020	0.0012-0.0031	SNUG
	5.4	L	0.0000-0.0010	0.006-0.0150	0.010-0.020	0.006-0.026	0.0002-0.0013	0.0012-0.0031	SNUG
	6.0	P	0.0017-0.0036	0.011-0.0210	0.055-0.0650	0.009-0.0190	NA	NA	NA
	6.8	S	0.0000-0.0010	0.0059-0.0118	0.010-0.016	0.006-0.026	0.0012-0.0037	0.0012-0.0037	SNUG

PISTON AND RING SPECIFICATIONS

All measurements are given in inches.

Year	Engine Displ. Liters	Engine VIN	Piston Clearance	Ring Gap			Ring Side Clearance		
				Top Compression	Bottom Compression	Oil Control	Top Compression	Bottom Compression	Oil Control
2005	4.2	2	0.0007-0.0018	0.001-0.002	0.001-0.002	0.006-0.026	0.0012-0.0031	0.0012-0.0031	SNUG
	4.6	W	0.0005-0.0010	0.010-0.020	0.010-0.020	0.006-0.026	0.0012-0.0028	0.0012-0.0028	0.0018-0.0077
	5.4	5	0.0010-0.0018	0.005-0.011	0.010-0.016	0.006-0.026	0.0012-0.0020	0.0012-0.0031	SNUG
	5.4	L	0.0000-0.0010	0.006-0.0150	0.010-0.020	0.006-0.026	0.0002-0.0013	0.0012-0.0031	SNUG
	6.0	P	0.0017-0.0036	0.011-0.0210	0.055-0.0650	0.009-0.0190	NA	NA	NA
	6.8	S	0.0000-0.0010	0.0059-0.0118	0.010-0.016	0.006-0.026	0.0012-0.0037	0.0012-0.0037	SNUG
	6.8	V	0.0002-0.0010	0.0059-0.0118	0.010-0.0200	0.006-0.026	0.0012-0.0020	0.0012-0.0031	SNUG
2006	5.4	5	0.0010-0.0018	0.005-0.011	0.010-0.016	0.006-0.026	0.0012-0.0020	0.0012-0.0031	SNUG

NA: Information not available

06017-F150-C12

CRANKSHAFT AND CONNECTING ROD SPECIFICATIONS

All measurements are given in inches.

Year	Engine Displ. Liters	Engine VIN	Crankshaft				Connecting Rod		
			Main Brg. Journal Dia.	Main Brg. Oil Clearance	Shaft End-play	Thrust on No.	Journal Dia.	Oil Clearance	Side Clearance
2002	4.2	2	2.5190-2.5198	0.0008-0.0015	0.0080-0.0078	3	2.3103-2.3111	0.0003-0.0024	0.0047-0.0193
	4.6	6	2.6500-2.6570	0.0011-0.0026	0.0051-0.0120	5	2.0870-2.8670	0.0011-0.0026	0.0006-0.0177
	4.6	W	2.6500-2.6570	0.0011-0.0026	0.0051-0.0120	5	2.0870-2.8670	0.0011-0.0026	0.0006-0.0177
	5.4	A	2.6568-2.6576	0.0009-0.0019	0.0015-0.0030	5	2.0859-2.0867	0.0010-0.0025	0.0006-0.0177
	5.4	3	2.6568-2.6576	0.0009-0.0019	0.0015-0.0030	5	2.0859-2.0867	0.0010-0.0025	0.0006-0.0177
	5.4	L	2.6568-2.6576	0.0009-0.0019	0.0015-0.0030	5	2.0859-2.0867	0.0010-0.0025	0.0006-0.0177
	6.8	S	2.6568-2.6576	0.0009-0.0019	0.0015-0.0030	5	2.0859-2.0867	0.0010-0.0025	0.0006-0.0177
	7.3	F	3.1228-3.1236	0.0018-0.0036	0.0025-0.0085	4	2.4980-2.4990	0.0015-0.0045	0.0120-0.0240
2003	4.2	2	2.5190-2.5198	0.0008-0.0015	0.0080-0.0078	3	2.3103-2.3111	0.0003-0.0024	0.0047-0.0193
	4.6	6	2.6500-2.6570	0.0011-0.0026	0.0051-0.0120	5	2.0870-2.8670	0.0011-0.0026	0.0006-0.0177
	4.6	W	2.6500-2.6570	0.0011-0.0026	0.0051-0.0120	5	2.0870-2.8670	0.0011-0.0026	0.0006-0.0177
	5.4	A	2.6568-2.6576	0.0009-0.0019	0.0015-0.0030	5	2.0859-2.0867	0.0010-0.0025	0.0006-0.0177
	5.4	3	2.6568-2.6576	0.0009-0.0019	0.0015-0.0030	5	2.0859-2.0867	0.0010-0.0025	0.0006-0.0177
	5.4	L	2.6568-2.6576	0.0009-0.0019	0.0015-0.0030	5	2.0859-2.0867	0.0010-0.0025	0.0006-0.0177
	6.8	S	2.6568-2.6576	0.0009-0.0019	0.0015-0.0030	5	2.0859-2.0867	0.0010-0.0025	0.0006-0.0177
	7.3	F	3.1228-3.1236	0.0018-0.0036	0.0025-0.0085	4	2.4980-2.4990	0.0015-0.0045	0.0120-0.0240
2004	4.2	2	2.5190-2.5198	0.0008-0.0015	0.0080-0.0078	3	2.3103-2.3111	0.0003-0.0024	0.0047-0.0193
	4.6	W	2.6500-2.6570	0.0011-0.0026	0.0051-0.0120	5	2.0859-2.0867	0.0010-0.0027	0.0006-0.0177
	5.4	3	2.6568-2.6576	0.0009-0.0019	0.0015-0.0030	5	2.0859-2.0867	0.0010-0.0025	0.0049-0.0187
	5.4	5	2.6568-2.6576	0.0009-0.0019	0.0030-0.0148	5	2.0859-2.0867	0.0010-0.0025	0.0049-0.0187
	5.4	L	2.6568-2.6576	0.0009-0.0019	0.0030-0.0148	5	2.0859-2.0867	0.0010-0.0025	0.0049-0.0187
	6.0	P	3.1500-3.1880	NA	0.0087	NA	2.7160-2.7170	NA	0.012-0.0240

CRANKSHAFT AND CONNECTING ROD SPECIFICATIONS
All measurements are given in inches.

Year	Engine Displ. Liters	Engine VIN	Crankshaft				Connecting Rod		
			Main Brg. Journal Dia.	Main Brg. Oil Clearance	Shaft End-play	Thrust on No.	Journal Dia.	Oil Clearance	Side Clearance
2004 cont.	6.8	S	2.6568-2.6576	0.0009-0.0019	0.0015-0.0030	5	2.0859-2.0867	0.0010-0.0025	0.0006-0.0177
2005	4.2	2	2.5190-2.5198	0.0008-0.0015	0.0080-0.0078	3	2.3103-2.3111	0.0003-0.0024	0.0047-0.0193
	4.6	W	2.6500-2.6570	0.0011-0.0026	0.0051-0.0120	5	2.0859-2.0867	0.0010-0.0027	0.0006-0.0177
	5.4	5	2.6568-2.6576	0.0009-0.0019	0.0030-0.0148	5	2.0859-2.0867	0.0010-0.0025	0.0049-0.0187
	5.4	L	2.6568-2.6576	0.0009-0.0019	0.0030-0.0148	5	2.0859-2.0867	0.0010-0.0025	0.0049-0.0187
	6.0	P	3.1500-3.1880	NA	0.0087	NA	2.7160-2.7170	NA	0.012-0.0240
	6.8	S	2.6568-2.6576	0.0009-0.0019	0.0015-0.0030	5	2.0859-2.0867	0.0010-0.0025	0.0006-0.0177
	6.8	V	2.6568-2.6576	0.0009-0.0019	0.0015-0.0030	5	2.0859-2.0867	0.0010-0.0025	0.0006-0.0177
2006	5.4	5	2.6568-2.6576	0.0009-0.0019	0.0030-0.0148	5	2.0859-2.0867	0.0010-0.0025	0.0049-0.0187

NA: Information not available

06017-F150-C14

TORQUE SPECIFICATIONS
All readings in ft. lbs.

Engine Displ. Liters	Engine VIN	Cylinder Head Bolts	Main Bearing Bolts	Rod Bearing Bolts	Crankshaft Damper Bolts	Flywheel Bolts	Manifold Intake *	Manifold Exhaust	Spark Plugs	Oil Pan Drain Plug
2002 4.2	2	①	81-88	②	103-117	54-63	③	15-22	8-14	19
4.6	6	④	⑤	29-33	⑥	54-64	⑦	18	7-14	10
4.6	W	④	⑤	29-33	⑥	54-64	⑦	18	7-14	10
5.4	3	④	⑧	⑨	⑥	54-64	⑩	18	9-20	10
5.4	A	④	⑧	⑬	⑮	54-64	⑩	18	9-20	10
5.4	L	④	⑧	⑨	⑥	54-64	⑦	18	9-20	10
6.8	S	④	⑧	⑨	⑥	54-64	⑦	17-20	7-14	10
7.3	F	⑪	95	70	90	89	18	45	—	NA
2003 4.2	2	①	⑫	⑬	103-117	54-63	⑭	30	8-14	19
4.6	6	④	⑤	29-33	⑥	54-64	⑦	18	7-14	10
4.6	W	④	⑤	29-33	⑥	54-64	⑦	18	7-14	10
5.4	A	④	⑧	⑬	⑮	54-64	⑩	18	9-20	10
5.4	3	④	⑧	⑨	⑯	54-64	⑩	18	9-20	10
5.4	L	④	⑧	⑨	⑯	54-64	⑦	18	9-20	10
6.8	S	④	⑧	⑨	⑯	54-64	⑦	17-20	7-14	10
7.3	F	⑪	95	70	90	89	18	45	—	NA
2004 4.2	2	①	⑫	⑬	118	59	⑭	24	11	19
4.6	W	④	⑤	⑱	⑯	59	⑱	18	11	10
5.4	3	④	⑧	⑨	⑥	54-64	⑩	18	9-20	10
5.4	5	④	⑧	⑬	⑯	59	㉕	18	25	10
5.4	L	④	⑧	⑱	⑯	59	㉕	18	13	10
6.0	P	㉑	㉓	㉒	㉔	69	8	28 ⑳	—	18
6.8	S	④	⑧	⑱	⑯	59	⑦	18	7.5	10
2005 4.2	2	①	⑫	⑬	118	59	⑭	24	11	19
4.6	W	④	⑤	⑰	⑯	59	⑱	18	11	10
5.4	5	④	⑧	⑬	⑯	59	㉖	18	25	10
5.4	L	④	⑧	⑰	⑯	59	㉖	18	13	10
6.0	P	㉑	㉓	㉒	㉔	69	8	28 ⑳	—	18
6.8	S	④	⑧	⑰	⑯	59	㉖	18	10	10
6.8	V	④	⑧	⑰	⑯	59	㉖	18	25	17
2006 5.4	5	④	⑧	⑬	⑯	59	⑲	18	25	10

NA: Information not available

* NOTE: Applies to Lower Manifold only. For Upper Manifold, see the text.

06017-F150-C15

TORQUE SPECIFICATIONS
FOOTNOTES

① See the procedure in the text

② Step 1: 29 ft. lbs.
  Step 2: Plus 90 degrees

③ Tighten bolts to 71-101 inch lbs.

④ Step 1: 30 ft. lbs.
  Step 2: Plus 85-95 degrees
  Step 3: Plus 85-95 degrees

⑤ Vertical bolts:
  Step 1: 30 ft. lbs.
  Step 2: Plus 90 degrees
  Jack screws:
  Step 1: 44 inch lbs.
  Step 2: 89 inch lbs.
  Side bolts: 15 ft. lbs.

⑥ Step 1: 88 ft. lbs.
  Step 2: Loosen bolt
  Step 3: 39 ft. lbs.
  Step 4: Plus 90 degrees

⑦ Step 1: 18 inch lbs.
  Step 2: 18 ft. lbs.

⑧ Vertical bolts:
  Step 1: 30 ft. lbs.
  Step 2: plus 90 degrees
  Side bolts:
  Step 1: 22 ft. lbs.
  Step 2: plus 90 degrees

⑨ Step 1: 27-32 ft. lbs.
  Step 2: Plus 90 degrees

⑩ Blackwood and Lightning
  Step 1: 89 inch lbs.
  Step 2: Plus 90 degrees

⑪ Step 1: 65 ft. lbs.
  Step 2: 85 ft. lbs.
  Step 3: 105 ft. lbs.

⑫ Step 1: 37 ft. lbs.
  Step 2: plus 120 degrees

⑬ Step 1: 18 ft. lbs.
  Step 2: 33 ft. lbs.
  Step 3: plus 90-120 degrees

⑭ Step 1: 44 inch lbs.
  Step 2: 89 inch lbs.

⑮ Step 1: 66 ft. lbs.
  Step 2: Loosen 1 turn
  Step 3: 37 ft. lbs.
  Step 4: plus 90 degrees

⑯ Step 1: 66 ft. lbs.
  Step 2: loosen 1 full turn
  Step 3: 37 ft. lbs.
  Step 4: plus 90 deg. Without exceeding 148 ft. lbs.

⑰ Step 1: 32 ft. lbs.
  Step 2: 105 degrees

⑱ 89 inch lbs.

⑲ Step 1: 18 inch lbs.
  Step 2: 89 inch lbs.

⑳ Apply high temp nickel anti-seize
  to the threads

㉑ See the procedure in the text

㉒ Step 1: 33 ft. lbs.
  Step 2: 50 ft. lbs.

㉓ Step 1: 90 ft. lbs.
  Step 2: 120 ft. lbs.
  Step 3: 170 ft. lbs.
  Lower crankcase bolts: 23 ft. lbs.

㉔ Step 1: 50 ft. lbs.
  Step 2: plus 90 degrees
  See the text for torque sequence

㉕ Heritage edition:
  Step 1: 18 inch lbs.
  Step 2: 18 ft. lbs.
  Except Heritage edition: 89 inch lbs.

㉖ F-Series:
  Step 1: 18 inch lbs.
  Step 2: 89 inch lbs.
  E-Series:
  Step 1: 18 inch lbs.
  Step 2 18 ft. lbs.

06017-F150-C16

Vertical main bearing bolt torque sequence—4.6L engine

06017-F150-G01

Crankshaft jackscrew torque sequence—4.6L engine

06017-F150-G02

Crankshaft side bolt torque sequence—4.6L engine

Crankshaft vertical bolt torque sequence—5.4L engine

Crankshaft side bolt torque sequence—5.4L engine

Connecting rod bearing cap torque sequence—4.6L and 5.4L engine

06017-F150-G07

Crankshaft bearing cap torque sequence—6.8L engine

06017-F150-G08

Crankshaft side bolt torque sequence—6.8L engine

06017-F150-G09

Connecting rod cap bolt torque sequence—6.8L engine

06017-F150-G10

Crankcase (main bearing) torque sequence—6.0L diesel

Lower crankcase torque sequence—6.0L diesel

Main bearing cap torque sequence—7.3L diesel

WHEEL ALIGNMENT

| Year | Model | | Caster | | Camber | | Toe-in |
			Range (+/-Deg.)	Preferred Setting (Deg.)	Range (+/-Deg.)	Preferred Setting (Deg.)	(Deg.)
2002	E-Series	150	2.75	+4.00	0.05	0.25	0.30+/-0.25
		250-350	2.75	+4.00	0.05	0.05	0.06+/-0.25
	Blackwood		1.00	①	0.07	-0.30	0.03+/-0.25
	F-150	4x2	1.00	+6.2	0.07	0.30	0.06+/-0.25
		4x4	1.00	+4.06	0.07	0.10	0.20+/-0.25
		Lightning	—	+.007	—	0.05	0.10+/-0.25
	F-250/350	4x2	2.00	+4.00	1.00	+0.62	0.03+/-0.25
		4x4	2.00	+3.50	1.00	+0.25	0.03+/-0.25
2003	E-Series	150	2.75	+4.00	0.05	0.25	0.30+/-0.25
		250-350	2.75	+4.00	0.05	0.05	0.06+/-0.25
	Blackwood		1.00	①	0.07	-0.30	0.03+/-0.25
	F-150	4x2	1.00	①	0.07	-0.30	0.03+/-0.15
		4x4	1.00	②	0.07	-0.10	0.20+/-0.25
		Lightning	—	③	—	-0.50	-0.05+/-0.15
	F-250/350	4x2	2.00	+4.00	1.00	+0.62	0.03+/-0.25
		4x4	2.00	+3.50	1.00	+0.25	0.03+/-0.25
2004	F-150 Reg. Cab	4x2	1.00	+4.4	0.75	-0.20	0.15+/-0.25
	F-150 Super Cab	4x2	1.00	+4.6	0.75	-0.20	0.15+/-0.25
	F-150 Crew Cab	4x2	1.00	+4.6	0.75	-0.20	0.15+/-0.25
	F-150 Reg. Cab	4x4	1.00	+4.3	0.75	-0.20	0.15+/-0.25
	F-150 Super Cab	4x4	1.00	+4.6	0.75	-0.20	0.15+/-0.25
	F-150 Crew Cab	4x4	1.00	+4.6	0.75	-0.20	0.15+/-0.25
	F-150 Heritage	4x2	1.00	L +6.2 R +6.7	0.70	-0.30	0.03+/-0.15
		4x4	1.00	L +4.6 R +5.3	0.70	-0.10	0.10+/-0.15
	F-150 Lightning		NA	L +6.7 R +7.2	NA	-0.50	-0.05+/-0.15
	F-250/350	4x2	2.00	+4.00	1.00	0.62	0.03+/-0.25 ④
		4x4	2.00	+3.50	1.00	+0.25	0.03+/-0.25 ④
	E-150	Left	2.75	+4.00	0.50	+0.25	0.06+/-0.25
		Right	2.75	+4.50	—	—	—
	E-250	Left	2.75	+4.00	0.50	+0.50	0.06+/-0.25
		Right	2.75	+4.50	—	—	—
	E-350	Left	2.75	+4.00	0.50	+0.50	0.06+/-0.25
		Right	2.75	+4.50	—	—	—
	E-Super Duty	Left	2.75	+4.00	0.50	+0.50	0.06+/-0.25
		Right	2.75	+4.50	—	—	—
2005	F-150 Reg. Cab	4x2	1.00	+4.4	0.75	-0.20	0.15+/-0.25
	F-150 Super Cab	4x2	1.00	+4.6	0.75	-0.20	0.15+/-0.25
	F-150 Crew Cab	4x2	1.00	+4.6	0.75	-0.20	0.15+/-0.25
	F-150 Reg. Cab	4x4	1.00	+4.3	0.75	-0.20	0.15+/-0.25
	F-150 Super Cab	4x4	1.00	+4.6	0.75	-0.20	0.15+/-0.25
	F-150 Crew Cab	4x4	1.00	+4.6	0.75	-0.20	0.15+/-0.25
	F-250/350	4x2	1.20	+3.8	0.75	L+0.62 R+0.67	0.10+/-0.25
	F-250 Std. Susp.	4x4	1.20	+3.3	0.75	+0.15	0.10+/-0.25

WHEEL ALIGNMENT

Year	Model		Caster Range (+/-Deg.)	Caster Preferred Setting (Deg.)	Camber Range (+/-Deg.)	Camber Preferred Setting (Deg.)	Toe-in (Deg.)
2005 cont.	F-250 HD Susp.	4x4	1.30	+2.6	0.75	+0.15	0.10+/-0.25
	F-350 SRW Std. Susp.	4x4	1.20	+2.5	0.75	+0.15	0.10+/-0.25
	F-350 SRW HD Susp.	4x4	1.30	+1.8	0.75	+0.15	0.10+/-0.25
	F-350 DRW Std. Susp.	4x4	1.20	+3.0	0.75	+0.15	0.10+/-0.25
	F-350 DRW HD Susp.	4x4	1.30	+2.4	0.75	+0.15	0.10+/-0.25
	F-350 Chassis/Cab SRW Std. Susp.	4x4	1.20	+2.8	0.75	+0.15	0.10+/-0.25
	F-350 Chassis/Cab SRW HD Susp.	4x4	1.30	+2.3	0.75	+0.15	0.10+/-0.25
	F-350 Chassis/Cab DRW Std. Susp.	4x4	1.20	+2.5	0.75	+0.15	0.10+/-0.25
	F-350 Chassis/Cab DRW HD Susp.	4x4	1.30	+1.9	0.75	+0.15	0.10+/-0.25
	E-150	Left	2.75	+4.00	0.50	+0.25	0.06+/-0.25
		Right	2.75	+4.50	—	—	—
	E-250	Left	2.75	+4.00	0.50	+0.50	0.06+/-0.25
		Right	2.75	+4.50	—	—	—
	E-350	Left	2.75	+4.00	0.50	+0.50	0.06+/-0.25
		Right	2.75	+4.50	—	—	—
	E-Super Duty	Left	2.75	+4.00	0.50	+0.50	0.06+/-0.25
		Right	2.75	+4.50	—	—	—
2006	Mark LT	2wd Regular Cab	1.00	+4.4	0.75	-0.2	0.15+/-0.25
		2wd Super Cab	1.00	+4.6	0.75	-0.2	0.15+/-0.25
		2wd Crew Cab	1.00	+4.6	0.75	-0.2	0.15+/-0.25
		4wd Regular Cab	1.00	+4.3	0.75	-0.2	0.15+/-0.25
		4wd Super Cab	1.00	+4.6	0.75	-0.2	0.15+/-0.25
		4wd Crew Cab	1.00	+4.6	0.75	-0.2	0.15+/-0.25

NA: Information not available

① Left: +6.2
  Right: +6.7

② Left: +4.6
  Right: +5.3

③ Left: +6.7
  Right: +7.2

④ At curb ride height

06017-F150-C18

TIRE, WHEEL AND BALL JOINT SPECIFICATIONS

Year	Model	OEM Tires Standard	OEM Tires Optional	Tire Pressures (psi.) Front	Tire Pressures (psi.) Rear	Wheel Size	Ball Joint Inspection	Wheel Lug Torque (ft. lbs.)
2002	E-150 Club Wagon	P235/75R15XL	none	①	①	6-JJ	0.030 in. ②	③
	E-150 Van	P225/75R15SL	none	①	①	7-J	0.030 in. ②	③
	E-250	LT225/75RX16	none	①	①	7-K	0.030 in. ②	③
	E-350	LT225/75RX16	none	①	①	7-K	0.030 in. ②	③
	E-350 Super Duty	LT245/75RX16E	none	①	①	7-K	0.030 in. ②	③
	F-150	P235/70R16	P255/70R16	①	①	7-J	0.030 in. ②	③
			P265/70R17	①	①		0.030 in. ②	③
	F-250	LT245/75R16	P255/70R16	①	①	7-J	0.030 in. ②	③
	F-350	P235/85R16	none	①	①	7-J	0.030 in. ②	③
2003	E-150 Club Wagon	P235/75R15	none	①	①	6-JJ	0.030 in. ②	④
	E-350 Club Wagon	LT225/75R16E	LT245/75R16E	①	①	7JJ	0.030 in. ②	④
	E-150 cargo	P235/75R15E	none	①	①	7JJ	0.030 in. ②	④
	E-250 cargo	LT225/75R16E	none	①	①	7-K	0.030 in. ②	④
	E-350 cargo	LT245/75R16E	none	①	①	7-K	0.030 in. ②	④
	E-350 Super Duty	LT245/75RX16E	none	①	①	7-K	0.030 in. ②	④
	F-150 XL, XLT	P235/70R16	P255/70R16	①	①	7-J	0.030 in. ②	④
			P265/70R17					④
	F-150 Harley	P275/45VR20	none	①	①	9-J	0.030 in. ②	④
	F-150 King Ranch							④
	2wd	P275/60R17	none	①	①	7.5J	0.030 in. ②	④
	4wd	P265/70R17	none	①	①	7.5J		④
	F-150 Lariat							④
	2wd	P275/60R17	none	①	①	7.5J	0.030 in. ②	④
	4wd	P265/70R17	none	①	①	7.5J		④
	F-150 Lightning	P295/45ZR18	none	①	①	9.5J	0.030 in. ②	④
	F-250 Lariat	LT265/75R16E	none	①	①	7-J	0.030 in. ②	④
	F-250 XL, XLT	LT235/85R16E	LT265/75R16E	①	①	7-J	0.030 in. ②	④
	F-350	LT265/75R16E	none	①	①	7-J	0.030 in. ②	④
2004	F-150 Reg. Cab							
	XL 2wd	P235/70R17	P255/65R17	①	①	NA	0.060 in. ②	150
	XL 4wd	P235/75R17	P255/70R17	①	①	NA	0.060 in. ②	150
	XLT 2wd	P235/70R17	P255/65R17	①	①	NA	0.060 in. ②	150
	XLT 4wd	P235/75R17	P255/70R17	①	①	NA	0.060 in. ②	150
	STX 2wd	P255/65R17	none	①	①	NA	0.060 in. ②	150
	STX 4wd	P255/70R17	none	①	①	NA	0.060 in. ②	150
	FX4	P255/70R17	LT245/65R18	①	①	NA	0.060 in. ②	150
			P275/65R18	①	①	NA	0.060 in. ②	150
	F-150 SuperCab							
	XL 2wd	P235/70R17	P255/65R17	①	①	NA	0.060 in. ②	150
	XL 4wd	P235/75R17	P255/70R17	①	①	NA	0.060 in. ②	150
	XLT 2wd	P235/70R17	P255/65R17	①	①	NA	0.060 in. ②	150
	XLT 4wd	P235/75R17	P255/70R17	①	①	NA	0.060 in. ②	150

06017-F150-C19

TIRE, WHEEL AND BALL JOINT SPECIFICATIONS

Year	Model	OEM Tires Standard	OEM Tires Optional	Tire Pressures (psi.) Front	Tire Pressures (psi.) Rear	Wheel Size	Ball Joint Inspection	Wheel Lug Torque (ft. lbs.)
2004 cont.	STX 2wd	P255/65R17	none	①	①	NA	0.060 in. ②	150
	STX 4wd	P255/70R17	none	①	①	NA	0.060 in. ②	150
	Lariat 2wd	P265/60R18	LT245/65R18	①	①	NA	0.060 in. ②	150
	Lariat 4wd	P275/65R18	LT245/65R18	①	①	NA	0.060 in. ②	150
	FX4	P255/70R17	LT245/65R18	①	①	NA	0.060 in. ②	150
			P275/65R18	①	①	NA	0.060 in. ②	150
	F-150 Crew Cab							
	XLT 2wd	P255/65R17	none	①	①	NA	0.060 in. ②	150
	XLT 4wd	P255/70R17	none	①	①	NA	0.060 in. ②	150
	Lariat 2wd	P265/60R18	LT245/65R18	①	①	NA	0.060 in. ②	150
	Lariat 4wd	P275/65R18	LT245/65R18	①	①	NA	0.060 in. ②	150
	FX4	P255/70R17	LT245/65R18	①	①	NA	0.060 in. ②	150
			P275/65R18	①	①	NA	0.060 in. ②	150
	F-150 Heritage Reg. Cab							
	XL 2wd	P255/70R16	P235/70R16	①	①	NA	0.060 in. ②	④
	XL 4wd	P255/70R16	P235/70R16	①	①	NA	0.060 in. ②	④
			LT245/70R16	①	①	NA	0.060 in. ②	④
	XLT 2wd	P255/70R16	P235/70R16	①	①	NA	0.060 in. ②	④
	XLT 4wd	P255/70R16	P235/70R16	①	①	NA	0.060 in. ②	④
			LT245/70R16	①	①	NA	0.060 in. ②	④
	SuperCab							
	XL 2wd	P235/70R16	none	①	①	NA	0.060 in. ②	④
	XL 4wd	P235/70R16	LT245/75R16	①	①	NA	0.060 in. ②	④
			P235/70R16	①	①	NA	0.060 in. ②	④
	XLT 2wd	P235/70R16	none	①	①	NA	0.060 in. ②	④
	XL 4wd	P235/70R16	LT245/75R16	①	①	NA	0.060 in. ②	④
			P235/70R16	①	①	NA	0.060 in. ②	④
	F-150 SVT Lightning	P295/45ZR18	none	①	①	NA	0.060 in. ②	④
	F-250 Lariat	LT265/75R16E	none	①	①	NA	0.030 in. ②	165
	F-250 XL, XLT	LT235/85R16E	LT265/75R16E	①	①	NA	0.030 in. ②	165
	F-350 SRW	LT265/75R16E	none	①	①	NA	0.030 in. ②	165
	F-350 DRW	LT235/85R16E	none	①	①	NA	0.030 in. ②	165
	E-150 Wagon	P235/70R16	none	①	①	NA	0.030 in. ②	③
	E-350 Wagon	LT225/75R16E	none	①	①	NA	0.030 in. ②	③
	E-350 Ext. Wagon	LT245/75R16E	none	①	①	NA	0.030 in. ②	③
	E-150 Cargo	P235/70R16	none	①	①	NA	0.030 in. ②	③
	E-250 Cargo	LT245/75R16E	none	①	①	NA	0.030 in. ②	③
	E-350 cargo	LT245/75R16E	none	①	①	NA	0.030 in. ②	③
2005	F-150 Reg. Cab							
	XL 2wd	P235/70R17	P255/65R17	①	①	NA	0.060 in. ②	150
	XL 4wd	P235/75R17	P255/70R17	①	①	NA	0.060 in. ②	150
	XLT 2wd	P235/70R17	P255/65R17	①	①	NA	0.060 in. ②	150
	XLT 4wd	P235/75R17	P255/70R17	①	①	NA	0.060 in. ②	150
	STX 2wd	P255/65R17	none	①	①	NA	0.060 in. ②	150
	STX 4wd	P255/70R17	none	①	①	NA	0.060 in. ②	150

06017-F150-C20

TIRE, WHEEL AND BALL JOINT SPECIFICATIONS

Year	Model	OEM Tires Standard	OEM Tires Optional	Tire Pressures (psi.) Front	Tire Pressures (psi.) Rear	Wheel Size	Ball Joint Inspection	Wheel Lug Torque (ft. lbs.)
2005 cont.	FX4	P255/70R17	LT245/65R18	①	①	NA	0.060 in. ②	150
			P275/65R18	①	①	NA	0.060 in. ②	150
	F-150 SuperCab							
	XL 2wd	P235/70R17	P255/65R17	①	①	NA	0.060 in. ②	150
	XL 4wd	P235/75R17	P255/70R17	①	①	NA	0.060 in. ②	150
	XLT 2wd	P235/70R17	P255/65R17	①	①	NA	0.060 in. ②	150
	XLT 4wd	P235/75R17	P255/70R17	①	①	NA	0.060 in. ②	150
	STX 2wd	P255/65R17	none	①	①	NA	0.060 in. ②	150
	STX 4wd	P255/70R17	none	①	①	NA	0.060 in. ②	150
	Lariat 2wd	P265/60R18	LT245/65R18	①	①	NA	0.060 in. ②	150
	Lariat 4wd	P275/65R18	LT245/65R18	①	①	NA	0.060 in. ②	150
	FX4	P255/70R17	LT245/65R18	①	①	NA	0.060 in. ②	150
			P275/65R18	①	①	NA	0.060 in. ②	150
	F-150 Crew Cab							
	XLT 2wd	P255/65R17	none	①	①	NA	0.060 in. ②	150
	XLT 4wd	P255/70R17	none	①	①	NA	0.060 in. ②	150
	Lariat 2wd	P265/60R18	LT245/65R18	①	①	NA	0.060 in. ②	150
	Lariat 4wd	P275/65R18	LT245/65R18	①	①	NA	0.060 in. ②	150
	FX4	P255/70R17	LT245/65R18	①	①	NA	0.060 in. ②	150
			P275/65R18	①	①	NA	0.060 in. ②	150
	F-250 Lariat	LT265/75R16E	none	①	①	NA	0.030 in. ②	165
	F-250 XL, XLT	LT235/85R16E	LT265/75R16E	①	①	NA	0.030 in. ②	165
	F-350 SRW	LT265/75R16E	none	①	①	NA	0.030 in. ②	165
	F-350 DRW	LT235/85R16E	none	①	①	NA	0.030 in. ②	165
	E-150 Wagon	P235/70R16	none	①	①	NA	0.030 in. ②	③
	E-350 Wagon	LT225/75R16E	none	①	①	NA	0.030 in. ②	③
	E-350 Ext. Wagon	LT245/75R16E	none	①	①	NA	0.030 in. ②	③
	E-150 Cargo	P235/70R16	none	①	①	NA	0.030 in. ②	③
	E-250 Cargo	LT245/75R16E	none	①	①	NA	0.030 in. ②	③
	E-350 cargo	LT245/75R16E	none	①	①	NA	0.030 in. ②	③
2006	Mark LT	2wd P265/60R18	none	①	①	NA	0.060 in. ②	150
		4wd P275/65R18	none					

NA: Information not available

OEM: Original Equipment Manufacturer

PSI: Pounds Per Square Inch

STD: Standard

OPT: Optional

SRW: Single rear wheels

DRW: Dual rear wheels

① See placard on vehicle

② Both upper and lower

③ 5 lug: 74-133 ft. lbs.
   8 lug: 126-170 ft. lbs.

④ 12mm: 100 ft. lbs.
   14mm: 150 ft. lbs.

BRAKE SPECIFICATIONS

All measurements in inches unless noted

Year	Model		Brake Disc			Brake Drum Diameter			Brake Caliper	
			Original Thickness	Minimum Thickness	Maximum Runout	Original Inside Diameter	Max. Wear Limit	Maximum Machine Diameter	Bracket Bolts (ft. lbs.)	Mounting Bolts (ft. lbs.)
2002	E-150		1.160	0.960	0.0025	11.03	11.09	11.06	141-191	22-26
	E-250	F	1.300	1.100	0.0003	12.00	12.09	12.06	141-191	16-30
		R	NA	1.120	0.0015	—	—	—	128	27
	E-350	F	1.300	1.100	0.0003	12.00	12.09	12.06	141-191	16-30
		R	NA	1.120	0.0015	—	—	—	128	27
	Blackwood	F	NA	③	0.0025	—	—	—	148	20
		R	NA	0.480	0.0002	—	—	—	NA	20
	F-150	F	①	②	0.0025	11.03	11.09	11.06	136	21-26
		R	NA	0.470	NA	—	—	—	80	20
	F-250	F	①	②	0.0025	12.00	12.09	12.06	166	42
		R	NA	1.000	NA	—	—	—	128	27
	F-350	F	①	②	0.0025	12.00	12.09	12.06	166	42
		R	NA	1.000	NA	—	—	—	128	27
2003	E-150		1.160	0.960	0.0025	11.03	11.09	11.06	141-191	22-26
	E-250	F	1.300	1.100	0.0003	12.00	12.09	12.06	141-191	16-30
		R	NA	1.120	0.0015	—	—	—	128	27
	E-350	F	1.300	1.100	0.0003	12.00	12.09	12.06	141-191	16-30
		R	NA	1.120	0.0015	—	—	—	128	27
	Blackwood	F	NA	③	0.0025	—	—	—	148	20
		R	NA	0.480	0.0002	—	—	—	NA	20
	F-150	F	NA	③	0.0025	—	—	—	136	21-26
		R	NA	0.480	0.0002	NA	NA	11.29	80	20
	F-250	F	①	②	0.0025	12.00	12.09	12.06	166	42
		R	NA	1.000	NA	—	—	—	128	27
	F-350	F	①	②	0.0025	12.00	12.09	12.06	166	42
		R	NA	1.000	NA	—	—	—	128	27
2004	F-150	F	NA	1.12	NA	—	—	—	148	47
		R	NA	0.72	NA	—	—	—	60	22
	Heritage & Lightning	F	NA	HD 1.09 LD 0.965	NA	—	—	—	136	21-26
		R	NA	0.48	NA	—	11.29	—	80	20
	F-250	F	NA	1.41	NA	—	—	—	166	42
		R	NA	1.10	NA	—	—	—	④	27
	F-350	F	NA	1.41	NA	—	—	—	166	42
		R	NA	1.10	NA	—	—	—	④	27
	E-150	F	NA	1.15	NA	—	—	—	129	24
		R	NA	0.80	NA	—	—	—	129	24
	E-250	F	NA	1.10	NA	—	—	—	166	16-30
		R	NA	1.10	NA	—	—	—	⑤	27
	E-350	F	NA	1.10	NA	—	—	—	166	23
		R	NA	1.10	NA	—	—	—	⑤	27
2005	F-150	F	NA	1.12	NA	—	—	—	148	47
		R	NA	0.72	NA	—	—	—	60	22

06017-F150-C22

BRAKE SPECIFICATIONS
All measurements in inches unless noted

| Year | Model | | Brake Disc | | | Brake Drum Diameter | | | Brake Caliper | |
			Original Thickness	Minimum Thickness	Maximum Runout	Original Inside Diameter	Max. Wear Limit	Maximum Machine Diameter	Bracket Bolts (ft. lbs.)	Mounting Bolts (ft. lbs.)
2005 cont.	F-250	F	NA	1.43	NA	—	—	—	166	56
		R	NA	1.27	NA	—	—	—	⑥	⑦
	F-350	F	NA	1.43	NA	—	—	—	166	56
		R	NA	1.27	NA	—	—	—	⑥	⑦
	E-150	F	NA	1.15	NA	—	—	—	129	24
		R	NA	0.80	NA	—	—	—	129	24
	E-250	F	NA	1.10	NA	—	—	—	166	23
		R	NA	1.10	NA	—	—	—	⑤	27
	E-350	F	NA	1.10	NA	—	—	—	166	23
		R	NA	1.10	NA	—	—	—	⑤	27
2006	Mark LT	F	NA	1.12	NA	—	—	—	148	47
		R	NA	0.72	NA	—	—	—	60	22

NOTE: Due to changes made during production, refer to manufacturer's specifications if they differ from those in this chart

NA: Information not available

DRW: Dual rear wheels

SRW: Single rear wheels

① 1.020 inches for 4x2
   1.220 inches for 4x4

② 0.972 inch for 4x2
   1.090 inches for 4x4

③ LD: 0.965 inch
   HD: 1.090 inch

④ Adapter plate:
   With Dana axle: 100
   With Ford axle: 150
   Anchor plate: 128

⑤ Support bracket support bolts: 100 ft. lbs.
   Anchor plate: 129 ft. lbs.

⑥ Caliper support bolts: 101 ft. lbs.
   Anchor plate bolts: 203 ft. lbs.

⑦ F-250 and F-350 SRW: 26 ft. lbs.
   F-350 DRW: 56 ft. lbs.

06017-F150-C23

SCHEDULED MAINTENANCE INTERVALS
2002 E-150, F-150 & Blackwood

TO BE SERVICED	TYPE OF SERVICE	VEHICLE MILEAGE INTERVAL (x1000)												
		5	10	15	20	25	30	35	40	45	50	55	60	65
Engine oil & filter	R	✓	✓	✓	✓	✓	✓	✓	✓	✓	✓	✓	✓	✓
Tires	Rotate	✓	✓	✓	✓	✓	✓	✓	✓	✓	✓	✓	✓	✓
Auto trans. fluid	I			✓			✓			✓			✓	
Brake pads/shoes	I			✓			✓			✓			✓	
Coolant hoses	S/I			✓			✓			✓			✓	
Steering linkage	I			✓			✓			✓			✓	
Cabin air filter	R			✓			✓			✓			✓	
Ball joints (2wd)	L			✓			✓			✓			✓	
Exhaust system	I						✓						✓	
Engine air filter	R						✓						✓	
Fuel filter ①	R						✓						✓	
Auto trans fluid (4-speed)	R						✓						✓	
Green coolant ②	R									✓				
Wheel bearings (2wd)	L												✓	
Manual trans. fluid	R												✓	
Spark plugs	R	every 100,000 miles												
PCV valve	R	every 100,000 miles												
Orange coolant	R	every 150,000 miles												
Auto trans fluid (5-speed)	R	every 150,000 miles												
Differential fluid	R	every 150,000 miles												
Accessory drive belts	R	every 150,000 miles												
Transfer case fluid	R	every 150,000 miles												

R: Replace   S: Service   I: Inspect   L: Lubricate

① Recommended, but not required in Calif.

② Change every 30,000 miles or 36 months thereafter

FREQUENT OPERATION MAINTENANCE (SEVERE SERVICE)

If a vehicle is operated under any of the following conditions it is considered severe service:

- Towing a trailer or using a camper or car-top carrier.

- Repeated short trips of less than 5 miles in temperatures below freezing, or trips of less than 10 miles in any temperature.

- Extensive idling or low-speed driving for long distance as in heavy commercial use, such as delivery, taxi or police cars.

- Operating on rough, muddy or salt-covered roads.

- Operating on unpaved or dusty roads.

- Driving in extremely hot (over 90°) conditions.

Engine oil & filter: replace every 3000 miles.
Air cleaner filter: service or inspect every 6000 miles.
Exhaust system: check every 6000 miles.
Automatic transmission fluid & filter: change every 30,000 miles.
Transfer case fluid: change every 60,000 miles
Fule filter: change every 15,000 miles
Spark plugs: change every 60,000 miles
2wd front wheel bearings: lubricate every 30,000 miles

06017-F150-C24

SCHEDULED MAINTENANCE INTERVALS
2002 E-250/350 & F-250/350—Gasoline Engines

TO BE SERVICED	TYPE OF SERVICE	VEHICLE MILEAGE INTERVAL (x1000)																			
		5	10	15	20	25	30	35	40	45	50	55	60	65	70	75	80	85	90	95	100
Accessory drive belt	S/I																				✓
Air cleaner filter ①②	R						✓						✓						✓		
Automatic transmission fluid ③	R						✓						✓						✓		
Engine coolant ④⑤	R										✓					✓					
Engine cooling system hoses, clamps & coolant ⑥	S/I			✓			✓			✓			✓			✓			✓		
Engine oil & filter	R	✓	✓	✓	✓	✓	✓	✓	✓	✓	✓	✓	✓	✓	✓	✓	✓	✓	✓	✓	✓
Exhaust system	S/I			✓			✓			✓			✓			✓			✓		
Front wheel bearings	S/I & L																		✓		
Front/rear axle lubricant ⑦	R																				✓
Fuel filter ⑧	R						✓						✓						✓		
PCV valve	R	Every 120,000 miles																			
Rotate tires	S/I	✓	✓	✓	✓	✓	✓	✓	✓	✓	✓	✓	✓	✓	✓	✓	✓	✓	✓	✓	✓
Spark plugs	R																				✓
Steering linkage, suspension, driveshaft U joints	S/I & L	✓	✓	✓	✓	✓	✓	✓	✓	✓	✓	✓	✓	✓	✓	✓	✓	✓	✓	✓	✓

R: Replace   S/I: Service or Inspect

① Perform this at the mileage shown or every 30 months, whichever occurs first.

② 7.3L DIT Diesel engine: the air filter should be replaced when the restriction gauge is in the red zone.

③ Except the E40D transmission.

④ Drain, flush and refill the cooling system initially at 50,000 miles or 48 months, whichever occurs first, then every 30,000 miles or 30 months thereafter.

⑤ 7.3L DIT Diesel engine: add 4 pints of FW-15 each time the coolant is replaced.

⑥ 7.3L DIT Diesel engine: add 8-10 oz. of FW-15 to the engine coolant every 15,000 miles.

⑦ The axle lubricant must be replaced every 100,000 miles of if the axle has been submerged under water.

⑧ 7.3L DIT Diesel engine: the fuel filter should be replaced when the restriction lamp is illuminated.

FREQUENT OPERATION MAINTENANCE (SEVERE SERVICE)

If a vehicle is operated under any of the following conditions it is considered severe service:

- Towing a trailer or using a camper or car-top carrier.

- Repeated short trips of less than 5 miles in temperatures below freezing, or trips of less than 10 miles in any temperature.

- Extensive idling or low-speed driving for long distances as in heavy commercial use, such as delivery, taxi or police cars.

- Operating on rough, muddy or salt-covered roads.

- Operating on unpaved or dusty roads.

Engine oil & filter: replace every 3000 miles.

Tires: rotate and inspect every 6000 miles.

Steering linkage, suspension, U-joints: lubricate every 6000 miles.

Exhaust system: inspect for leaks or damage every 12,000 miles.

Fuel filter: replace every 15,000 miles.

Automatic transmission fluid: change ever 21,000 miles.

Front wheel bearings (2WD): inspect and repack every 30,000 miles.

Rear axle lubricant (E-Super Duty only): replace every 30,000 miles.

Spark plugs (except 4.2L engine): replace every 60,000 miles.

PCV valve: replace every 60,000 miles.

Accessory drive belt: inspect every 60,000 miles.

Spark plugs: replace every 99,000 miles.

SCHEDULED MAINTENANCE INTERVALS
2003 Blackwood, E-150 & F-150

TO BE SERVICED	TYPE OF SERVICE	VEHICLE MILEAGE INTERVAL (x1000)												
		5	10	15	20	25	30	35	40	45	50	55	60	65
Engine oil & filter	R	✓	✓	✓	✓	✓	✓	✓	✓	✓	✓	✓	✓	✓
Tires	Rotate	✓	✓	✓	✓	✓	✓	✓	✓	✓	✓	✓	✓	✓
Auto trans. fluid	I			✓			✓			✓			✓	
Brake pads/shoes	I			✓			✓			✓			✓	
Wheels	I ①			✓			✓			✓			✓	
Coolant hoses	S/I			✓			✓			✓			✓	
Steering linkage	I			✓			✓			✓			✓	
Suspension	I			✓			✓			✓			✓	
Cabin air filter	R			✓			✓			✓			✓	
Ball joints (2wd)	I			✓			✓			✓				
Exhaust system	I						✓						✓	
Engine air filter	R						✓						✓	
Fuel filter ②	R						✓						✓	
Auto trans fluid (4R100)	R						✓						✓	
Green coolant ③	R									✓				
Accessory drive belts	I	every 100,000 miles												
Spark plugs	R	every 100,000 miles												
Yellow coolant ⑥	R	every 5 years or 100,000 miles												
PCV valve ④	R	every 100,000 miles												
PCV valve ⑤	R	every 120,000 miles												
Auto trans fluid (exc. 4R100)	R	every 150,000 miles												
Differential fluid	R	every 150,000 miles												
Accessory drive belts	R	every 150,000 miles if not previously changed												

R: Replace   S: Service   I: Inspect   L: Lubricate

① Inspect for end play and noise

② Recommended, but not required in Calif.

③ Change every 30,000 miles or 36 months thereafter

④ Vehicle under 6,000 lbs GVW

⑤ Vehicles over 6,000 lbs GVW

⑥ Then every 3 years or 50,000 thereafter

Special Operating Condition Requirements

When towing a trailer or using a camper or car-top carrier:

Change engine oil and install a new oil filter every 4,800 km (3,000 miles), 3 months or 200 hours of engine operation (whichever occurs first).

Change transfer case fluid every 96,000 km (60,000 miles).

Change manual transmission fluid as required.

Inspect and lubricate U-joints as required.

During extensive idling and/or low speed driving for long distances, as in heavy commercial use such as delivery, taxi, patrol car or livery:

Change engine oil and install a new oil filter every 4,800 km (3,000 miles), 3 months or 200 hours of engine operation (whichever occurs first).

Lube front lower control arm and steering linkage ball joints with zerk fittings (if equipped) every 4,800 km (3,000 miles) or 3 months.

Inspect brake system and check battery electrolyte level (Patrol cars) every 8,000 km (5,000 miles).

Install a new fuel filter every 24,000 km (15,000 miles).

Change automatic transmission fluid, lubricate 4x2 wheel bearings, install new grease seals and adjust bearings every 48,000 km (30,000 miles). If equipped, change the in-line service installed transmission fluid filter.

Install new spark plugs and change transfer case fluid every 96,000 km (60,000 miles).

Install a new cabin air filter as required.

When operating in dusty conditions such as unpaved or dusty roads:

Change engine oil and install a new oil filter every 4,800 km (3,000 miles) or 3 months.

SCHEDULED MAINTENANCE INTERVALS
2003 Blackwood, E-150 & F-150
Footnotes Continued

Install a new fuel filter every 24,000 km (15,000 miles).

Change automatic transmission fluid every 48,000 km (30,000 miles). If equipped, change the in-line service installed transmission fluid filter.

Change transfer case fluid every 96,000 km (60,000 miles).

Install a new engine air filter as required.

Install a new cabin air filter as required.

When operating in off-road conditions:

Change automatic transmission fluid every 48,000 km (30,000 miles). If equipped, change the in-line service installed transmission fluid filter.

Change transfer case fluid every 96,000 km (60,000 miles).

Install a new cabin air filter as required.

Inspect and lubricate U-joints.

Inspect and lubricate steering linkage ball joints with zerk fittings.

06017-F150-C27

SCHEDULED MAINTENANCE INTERVALS
2003 E-250/350 & F-250/350—Gasoline Engines

TO BE SERVICED	TYPE OF SERVICE	5	10	15	20	25	30	35	40	45	50	55	60	65	70	75	80	85	90	95	100
Accessory drive belt	S/I																				✓
Air cleaner filter ①	R						✓						✓						✓		
Automatic transmission fluid ②	R						✓						✓						✓		
Engine coolant ③	R										✓					✓					
Engine cooling system hoses, clamps & coolant	S/I			✓			✓			✓			✓			✓			✓		
Engine oil & filter	R	✓	✓	✓	✓	✓	✓	✓	✓	✓	✓	✓	✓	✓	✓	✓	✓	✓	✓	✓	✓
Exhaust system	S/I			✓			✓			✓			✓			✓			✓		
Front wheel bearings	S/I & L																		✓		
Front/rear axle lubricant	R																				✓
Fuel filter	R						✓						✓						✓		
PCV valve	R	Every 120,000 miles																			
Rotate tires	S/I	✓	✓	✓	✓	✓	✓	✓	✓	✓	✓	✓	✓	✓	✓	✓	✓	✓	✓	✓	✓
Spark plugs	R																				✓
Steering linkage, suspension, driveshaft U joints	S/I & L	✓	✓	✓	✓	✓	✓	✓	✓	✓	✓	✓	✓	✓	✓	✓	✓	✓	✓	✓	✓

R: Replace   S/I: Service or Inspect

① Perform this at the mileage shown or every 30 months, whichever occurs first.

② Except the E40D transmission.

③ Drain, flush and refill the cooling system initially at 50,000 miles or 48 months, whichever occurs first, then every 30,000 miles or 30 months thereafter.

Special Operating Condition Requirements

When towing a trailer or using a camper or car-top carrier:

Change engine oil and install a new oil filter every 4,800 km (3,000 miles), 3 months or 200 hours of engine operation (whichever occurs first).

Change transfer case fluid every 96,000 km (60,000 miles).

Change manual transmission fluid as required.

Inspect and lubricate U-joints as required.

During extensive idling and/or low speed driving for long distances, as in heavy commercial use such as delivery, taxi, patrol car or livery:

Change engine oil and install a new oil filter every 4,800 km (3,000 miles), 3 months or 200 hours of engine operation (whichever occurs first).

Lube front lower control arm and steering linkage ball joints with zerk fittings (if equipped) every 4,800 km (3,000 miles) or 3 months.

Inspect brake system and check battery electrolyte level (Patrol cars) every 8,000 km (5,000 miles).

Install a new fuel filter every 24,000 km (15,000 miles).

Change automatic transmission fluid, lubricate 4x2 wheel bearings, install new grease seals and adjust bearings every 48,000 km (30,000 miles). If equipped, change the in-line service installed transmission fluid filter.

Install new spark plugs and change transfer case fluid every 96,000 km (60,000 miles).

Install a new cabin air filter as required.

When operating in dusty conditions such as unpaved or dusty roads:

Change engine oil and install a new oil filter every 4,800 km (3,000 miles) or 3 months.

Install a new fuel filter every 24,000 km (15,000 miles).

Change automatic transmission fluid every 48,000 km (30,000 miles). If equipped, change the in-line service installed transmission fluid filter.

Change transfer case fluid every 96,000 km (60,000 miles).

Install a new engine air filter as required.

Install a new cabin air filter as required.

06017-F150-C28

SCHEDULED MAINTENANCE INTERVALS
2003 E-250/350 & F-250/350—Gasoline Engines
Footnotes Continued

When operating in off-road conditions:

Change automatic transmission fluid every 48,000 km (30,000 miles). If equipped, change the in-line service installed transmission fluid filter.

Change transfer case fluid every 96,000 km (60,000 miles).

Install a new cabin air filter as required.

Inspect and lubricate U-joints.

Inspect and lubricate steering linkage ball joints with zerk fittings.

06017-F150-C29

SCHEDULED MAINTENANCE INTERVALS
All 2003 Vehicles with the 7.3L Diesel

TO BE SERVICED	TYPE OF SERVICE	VEHICLE MILEAGE INTERVAL (x1000)																			
		5	10	15	20	25	30	35	40	45	50	55	60	65	70	75	80	85	90	95	100
Accessory drive belt	S/I																				✓
Air cleaner filter	R	When the restriction gauge is in the red zone																			
Automatic transmission fluid ①	R						✓					✓							✓		
Engine coolant ②③	R										✓					✓					
Engine cooling system hoses, clamps & coolant ④	S/I			✓			✓			✓			✓			✓			✓		
Engine oil & filter	R	✓	✓	✓	✓	✓	✓	✓	✓	✓	✓	✓	✓	✓	✓	✓	✓	✓	✓	✓	✓
Exhaust system	S/I			✓			✓			✓			✓			✓			✓		
Front wheel bearings	S/I & L																		✓		
Front/rear axle lubricant ⑤	R																				✓
Fuel filter	R	When the retsriction lamp is illuminated																			
PCV valve	R	Every 120,000 miles																			
Rotate tires	S/I	✓	✓	✓	✓	✓	✓	✓	✓	✓	✓	✓	✓	✓	✓	✓	✓	✓	✓	✓	✓
Steering linkage, suspension, driveshaft U joints	S/I & L	✓	✓	✓	✓	✓	✓	✓	✓	✓	✓	✓	✓	✓	✓	✓	✓	✓	✓	✓	✓

R: Replace   S/I: Service or Inspect

① Except the E40D transmission.

② Drain, flush and refill the cooling system initially at 50,000 miles or 48 months, whichever occurs first, then every 30,000 miles or 30 months thereafter.

③ Add 4 pints of FW-15 each time the coolant is replaced.

④ Add 8-10 oz. of FW-15 to the engine coolant every 15,000 miles.

⑤ The axle lubricant must be replaced every 100,000 miles of if the axle has been submerged under water.

Special Operating Condition Requirements

When towing a trailer or using a camper or car-top carrier:

Change engine oil and install a new oil filter every 4,800 km (3,000 miles), 3 months or 200 hours of engine operation (whichever occurs first).

Change transfer case fluid every 96,000 km (60,000 miles).

Change manual transmission fluid as required.

Inspect and lubricate U-joints as required.

During extensive idling and/or low speed driving for long distances, as in heavy commercial use such as delivery, taxi, patrol car or livery:

Change engine oil and install a new oil filter every 4,800 km (3,000 miles), 3 months or 200 hours of engine operation (whichever occurs first).

Lube front lower control arm and steering linkage ball joints with zerk fittings (if equipped) every 4,800 km (3,000 miles) or 3 months.

Inspect brake system and check battery electrolyte level (Patrol cars) every 8,000 km (5,000 miles).

Install a new fuel filter every 24,000 km (15,000 miles).

Change automatic transmission fluid, lubricate 4x2 wheel bearings, install new grease seals and adjust bearings every 48,000 km (30,000 miles). If equipped, change the in-line service installed transmission fluid filter.

Install new spark plugs and change transfer case fluid every 96,000 km (60,000 miles).

Install a new cabin air filter as required.

When operating in dusty conditions such as unpaved or dusty roads:

Change engine oil and install a new oil filter every 4,800 km (3,000 miles) or 3 months.

Install a new fuel filter every 24,000 km (15,000 miles).

Change automatic transmission fluid every 48,000 km (30,000 miles). If equipped, change the in-line service installed transmission fluid filter.

Change transfer case fluid every 96,000 km (60,000 miles).

Install a new engine air filter as required.

Install a new cabin air filter as required.

06017-F150-C30

SCHEDULED MAINTENANCE INTERVALS
All 2003 Vehicles with the 7.3L Diesel
Footnotes Continued

When operating in off-road conditions:

Change automatic transmission fluid every 48,000 km (30,000 miles). If equipped, change the in-line service installed transmission fluid filter.

Change transfer case fluid every 96,000 km (60,000 miles).

Install a new cabin air filter as required.

Inspect and lubricate U-joints.

Inspect and lubricate steering linkage ball joints with zerk fittings.

06017-F150-C31

SCHEDULED MAINTENANCE INTERVALS
2004 F-150

TO BE SERVICED	TYPE OF SERVICE	VEHICLE MILEAGE INTERVAL (x1000)												
		5	10	15	20	25	30	35	40	45	50	55	60	65
Engine oil & filter	R	✓	✓	✓	✓	✓	✓	✓	✓	✓	✓	✓	✓	✓
Tires	Rotate	✓	✓	✓	✓	✓	✓	✓	✓	✓	✓	✓	✓	✓
Wheels	I ①			✓			✓			✓			✓	
Auto trans. fluid	I			✓			✓			✓			✓	
Brake pads/shoes	I			✓			✓			✓			✓	
Coolant hoses	S/I			✓			✓			✓			✓	
Steering linkage	I			✓			✓			✓			✓	
Suspension	I			✓			✓			✓			✓	
Driveshaft	I			✓			✓			✓			✓	
Cabin air filter	R			✓			✓			✓			✓	
Ball joints (2wd)	L			✓			✓			✓			✓	
Exhaust system	I						✓						✓	
Engine air filter	R						✓						✓	
Fuel filter	R						✓						✓	
Auto trans fluid (4R100)	R						✓						✓	
Front wheel bearings (2wd)	L/Adj												✓	
Front wheel bearings grease seal (2wd)	R												✓	
Accessory drive belts	I	every 100,000 miles												
Front wheel bearings (2wd)	R	at 150,000 miles, if not previously done so												
Spark plugs	R	every 100,000 miles												
PCV valve	R	every 120,000 miles												
Premium Gold coolant	R	every 5 years or 100,000 miles												
Auto trans fluid (all exc. 4R100)	R	every 150,000 miles												
Differential fluid	R	every 150,000 miles												
Accessory drive belts	R	every 150,000 miles, if not previously done so												

R: Replace   S: Service   I: Inspect   L: Lubricate

① Inspect for end play and noise

Special Operating Condition Requirements

When towing a trailer or using a camper or car-top carrier:

Change engine oil and install a new oil filter every 4,800 km (3,000 miles), 3 months or 200 hours of engine operation (whichever occurs first).

Change transfer case fluid every 96,000 km (60,000 miles).

Change manual transmission fluid as required.

Inspect and lubricate U-joints as required.

During extensive idling and/or low speed driving for long distances, as in heavy commercial use such as delivery, taxi, patrol car or livery:

Change engine oil and install a new oil filter every 4,800 km (3,000 miles), 3 months or 200 hours of engine operation (whichever occurs first).

Lube front lower control arm and steering linkage ball joints with zerk fittings (if equipped) every 4,800 km (3,000 miles) or 3 months.

Inspect brake system and check battery electrolyte level (Patrol cars) every 8,000 km (5,000 miles).

Install a new fuel filter every 24,000 km (15,000 miles).

Change automatic transmission fluid, lubricate 4x2 wheel bearings, install new grease seals and adjust bearings every 48,000 km (30,000 miles). If equipped, change the in-line service installed transmission fluid filter.

Install new spark plugs and change transfer case fluid every 96,000 km (60,000 miles).

Install a new cabin air filter as required.

When operating in dusty conditions such as unpaved or dusty roads:

Change engine oil and install a new oil filter every 4,800 km (3,000 miles) or 3 months.

Install a new fuel filter every 24,000 km (15,000 miles).

SCHEDULED MAINTENANCE INTERVALS
2004 F-150
Footnoes Continued

Change automatic transmission fluid every 48,000 km (30,000 miles). If equipped, change the in-line service installed transmission fluid filter.

Change transfer case fluid every 96,000 km (60,000 miles).

Install a new engine air filter as required.

Install a new cabin air filter as required.

When operating in off-road conditions:

Change automatic transmission fluid every 48,000 km (30,000 miles). If equipped, change the in-line service installed transmission fluid filter.

Change transfer case fluid every 96,000 km (60,000 miles).

Install a new cabin air filter as required.

Inspect and lubricate U-joints.

Inspect and lubricate steering linkage ball joints with zerk fittings.

Short trips in cold operating conditions:

Inspect and lubricate 4x2 ball joints and steering idler arms every 8,000 km (5,000 miles).

Change transfer case fluid every 96,000 km (60,000 miles).

06017-F150-C33

SCHEDULED MAINTENANCE INTERVALS
2004 Gasoline F-250/350 Super Duty

TO BE SERVICED	TYPE OF SERVICE	VEHICLE MILEAGE INTERVAL (x1000)												
		5	10	15	20	25	30	35	40	45	50	55	60	65
Engine oil & filter	R	✓	✓	✓	✓	✓	✓	✓	✓	✓	✓	✓	✓	✓
Tires	Rotate	✓	✓	✓	✓	✓	✓	✓	✓	✓	✓	✓	✓	✓
Wheels	I ①	✓	✓	✓	✓	✓	✓	✓	✓	✓	✓	✓	✓	✓
Auto trans. fluid	I			✓			✓			✓			✓	
Brake pads, hoses, etc.	I			✓			✓			✓			✓	
Coolant hoses	I			✓			✓			✓			✓	
Steering linkage	I/L			✓			✓			✓			✓	
Cabin air filter	R			✓			✓			✓			✓	
Ball joints (2wd)	L			✓			✓			✓			✓	
Driveshaft	I			✓			✓			✓			✓	
4x4 front axle U-joints	L			✓			✓			✓			✓	
Exhaust system	I						✓						✓	
Engine air filter	R						✓						✓	
Fuel filter	R						✓						✓	
Auto trans fluid (4R100)	R						✓						✓	
4x2 front wheel bearings	Adj.												✓	
4x2 front wheel bearing grease seals	R												✓	
Accessory drive belts	I	Every 100,000 miles												
Front wheel bearings (2wd)	R	at 150,000 miles, if not previously done so												
4x4 front hub needle bearings	L	every 120,000 miles												
Spark plugs	R	every 100,000 miles												
PCV valve	R	every 120,000 miles												
Manual trans. Fluid	R	every 120,000 miles												
Premium Gold coolant	R	every 3 years or 100,000 miles and every 50,000 miles thereafter												
Coolant (exc. Premium Gold)	R	every 135,000 miles												
Auto trans fluid (exc. 4R100)	R	every 150,000 miles												
Differential fluid	R	at 150,000 miles then every 50,000 thereafter												
Transfer case fluid	R	every 150,000 miles												
Accessory drive belts	R	every 150,000 miles, if not previously done so												

R: Replace   S: Service   I: Inspect   L: Lubricate   Adj: adjust

① Inspect for end play and noise

Special Operating Condition Requirements

When towing a trailer or using a camper or car-top carrier:

Change engine oil and install a new oil filter every 4,800 km (3,000 miles), 3 months or 200 hours of engine operation (whichever occurs first).

Change transfer case fluid every 96,000 km (60,000 miles).

Change manual transmission fluid as required.

Inspect and lubricate U-joints as required.

During extensive idling and/or low speed driving for long distances, as in heavy commercial use such as delivery, taxi, patrol car or livery:

Change engine oil and install a new oil filter every 4,800 km (3,000 miles), 3 months or 200 hours of engine operation (whichever occurs first).

Lube front lower control arm and steering linkage ball joints with zerk fittings (if equipped) every 4,800 km (3,000 miles) or 3 months.

Inspect brake system and check battery electrolyte level (Patrol cars) every 8,000 km (5,000 miles).

Install a new fuel filter every 24,000 km (15,000 miles).

Change automatic transmission fluid, lubricate 4x2 wheel bearings, install new grease seals and adjust bearings every 48,000 km (30,000 miles). If equipped, change the in-line service installed transmission fluid filter.

SCHEDULED MAINTENANCE INTERVALS
2004 Gasoline F-250/350 Super Duty
Footnotes Continued

Install new spark plugs and change transfer case fluid every 96,000 km (60,000 miles).

Install a new cabin air filter as required.

When operating in dusty conditions such as unpaved or dusty roads:

Change engine oil and install a new oil filter every 4,800 km (3,000 miles) or 3 months.

Install a new fuel filter every 24,000 km (15,000 miles).

Change automatic transmission fluid every 48,000 km (30,000 miles). If equipped, change the in-line service installed transmission fluid filter.

Change transfer case fluid every 96,000 km (60,000 miles).

Install a new engine air filter as required.

Install a new cabin air filter as required.

When operating in off-road conditions:

Change automatic transmission fluid every 48,000 km (30,000 miles). If equipped, change the in-line service installed transmission fluid filter.

Change transfer case fluid every 96,000 km (60,000 miles).

Install a new cabin air filter as required.

Inspect and lubricate U-joints.

Inspect and lubricate steering linkage ball joints with zerk fittings.

06017-F150-C35

SCHEDULED MAINTENANCE INTERVALS
2004 E-150

TO BE SERVICED	TYPE OF SERVICE	VEHICLE MILEAGE INTERVAL (x1000)												
		5	10	15	20	25	30	35	40	45	50	55	60	65
Engine oil & filter	R	✓	✓	✓	✓	✓	✓	✓	✓	✓	✓	✓	✓	✓
Tires	Rotate	✓	✓	✓	✓	✓	✓	✓	✓	✓	✓	✓	✓	✓
Wheels	I ①	✓	✓	✓	✓	✓	✓	✓	✓	✓	✓	✓	✓	✓
Auto trans. fluid	I			✓			✓			✓			✓	
Brake pads/shoes	I			✓			✓			✓			✓	
Coolant hoses	S/I			✓			✓			✓			✓	
Steering linkage	I			✓			✓			✓			✓	
Suspension	I			✓			✓			✓			✓	
Driveshaft	I			✓			✓			✓			✓	
Cabin air filter	R			✓			✓			✓			✓	
Ball joints (2wd)	I/L			✓			✓			✓			✓	
Exhaust system	I						✓						✓	
Engine air filter	R						✓						✓	
Fuel filter	R						✓						✓	
Auto trans fluid (4R100) ②	R						✓						✓	
Front wheel bearings (2wd) ③	L/Adj												✓	
Front wheel bearings grease seal (2wd)	R												✓	
Accessory drive belts	I	every 100,000 miles												
Spark plugs	R	every 100,000 miles												
Premium Gold coolant ⑤	R	at 3 years or 100,000 miles												
Coolant (exc. Premium Gold)	R	every 105,000 miles												
PCV valve ④	R	every 120,000 miles												
Front wheel bearings (2wd)	R	at 150,000 miles, if not previously done so												
Auto trans fluid (exc. 4R100) ②	R	every 150,000 miles												
Differential fluid	R	every 150,000 miles												
Accessory drive belts	R	every 150,000 miles, if not previously done so												

R: Replace   S: Service   I: Inspect   L: Lubricate   Adj: Adjust

① Inspect for end play and noise

② Change both the in-line and installed filters

③ And every 15,000 miles thereafter

④ Under 6,000 lbs. GVW, replace every 100,000 miles

⑤ And every 50,000 miles thereafter

Special Operating Condition Requirements

When towing a trailer or using a camper or car-top carrier:

Change engine oil and install a new oil filter every 4,800 km (3,000 miles), 3 months or 200 hours of engine operation (whichever occurs first).

Change transfer case fluid every 96,000 km (60,000 miles).

Change manual transmission fluid as required.

Inspect and lubricate U-joints as required.

During extensive idling and/or low speed driving for long distances, as in heavy commercial use such as delivery, taxi, patrol car or livery:

Change engine oil and install a new oil filter every 4,800 km (3,000 miles), 3 months or 200 hours of engine operation (whichever occurs first).

Lube front lower control arm and steering linkage ball joints with zerk fittings (if equipped) every 4,800 km (3,000 miles) or 3 months.

Inspect brake system and check battery electrolyte level (Patrol cars) every 8,000 km (5,000 miles).

Install a new fuel filter every 24,000 km (15,000 miles).

06017-F150-C36

SCHEDULED MAINTENANCE INTERVALS
2004 E-150
Footnotes Continued

Change automatic transmission fluid, lubricate 4x2 wheel bearings, install new grease seals and adjust bearings every

48,000 km (30,000 miles). If equipped, change the in-line service installed transmission fluid filter.

Install new spark plugs and change transfer case fluid every 96,000 km (60,000 miles).

Install a new cabin air filter as required.

When operating in dusty conditions such as unpaved or dusty roads:

Change engine oil and install a new oil filter every 4,800 km (3,000 miles) or 3 months.

Install a new fuel filter every 24,000 km (15,000 miles).

Change automatic transmission fluid every 48,000 km (30,000 miles). If equipped, change the in-line service installed transmission fluid filter.

Change transfer case fluid every 96,000 km (60,000 miles).

Install a new engine air filter as required.

Install a new cabin air filter as required.

When operating in off-road conditions:

Change automatic transmission fluid every 48,000 km (30,000 miles). If equipped, change the in-line service installed transmission fluid filter.

Change transfer case fluid every 96,000 km (60,000 miles).

Install a new cabin air filter as required.

Inspect and lubricate U-joints.

Inspect and lubricate steering linkage ball joints with zerk fittings.

06017-F150-C37

SCHEDULED MAINTENANCE INTERVALS
2004 Gasoline E-250 and 350

TO BE SERVICED	TYPE OF SERVICE	VEHICLE MILEAGE INTERVAL (x1000)												
		5	10	15	20	25	30	35	40	45	50	55	60	65
Engine oil & filter	R	✓	✓	✓	✓	✓	✓	✓	✓	✓	✓	✓	✓	✓
Tires	Rotate	✓	✓	✓	✓	✓	✓	✓	✓	✓	✓	✓	✓	✓
Wheels	I ①	✓	✓	✓	✓	✓	✓	✓	✓	✓	✓	✓	✓	✓
Auto trans. fluid	I			✓			✓			✓			✓	
Brake pads, hoses, etc.	I			✓			✓			✓			✓	
Coolant hoses	I			✓			✓			✓			✓	
Steering linkage	I/L			✓			✓			✓			✓	
Cabin air filter	R			✓			✓			✓			✓	
Ball joints (2wd)	L			✓			✓			✓			✓	
Driveshaft	I			✓			✓			✓			✓	
4x4 front axle U-joints	L			✓			✓			✓			✓	
Exhaust system	I						✓						✓	
Engine air filter	R						✓						✓	
Fuel filter	R						✓						✓	
Auto trans fluid (4R100)	R						✓						✓	
4x2 front wheel bearings	Adj.												✓	
4x2 front wheel bearing grease seals	R												✓	
Accessory drive belts	I	Every 100,000 miles												
Spark plugs	R	every 100,000 miles												
Premium Gold coolant	R	every 3 years or 100,000 miles and every 50,000 miles thereafter												
4x4 front hub needle bearings	L	every 120,000 miles												
PCV valve	R	every 120,000 miles												
Manual trans. Fluid	R	every 120,000 miles												
Coolant (exc. Premium Gold)	R	every 135,000 miles												
Front wheel bearings (2wd)	R	at 150,000 miles, if not previously done so												
Auto trans fluid (exc. 4R100)	R	every 150,000 miles												
Differential fluid	R	at 150,000 miles then every 50,000 miles thereafter												
Transfer case fluid	R	every 150,000 miles												
Accessory drive belts	R	every 150,000 miles, if not previously done so												

R: Replace   S: Service   I: Inspect   L: Lubricate   Adj: adjust

① Inspect for end play and noise

Special Operating Condition Requirements

When towing a trailer or using a camper or car-top carrier:

Change engine oil and install a new oil filter every 4,800 km (3,000 miles), 3 months or 200 hours of engine operation (whichever occurs first).

Change transfer case fluid every 96,000 km (60,000 miles).

Change manual transmission fluid as required.

Inspect and lubricate U-joints as required.

During extensive idling and/or low speed driving for long distances, as in heavy commercial use such as delivery, taxi, patrol car or livery:

Change engine oil and install a new oil filter every 4,800 km (3,000 miles), 3 months or 200 hours of engine operation (whichever occurs first).

Lube front lower control arm and steering linkage ball joints with zerk fittings (if equipped) every 4,800 km (3,000 miles) or 3 months.

Inspect brake system and check battery electrolyte level (Patrol cars) every 8,000 km (5,000 miles).

Install a new fuel filter every 24,000 km (15,000 miles).

Change automatic transmission fluid, lubricate 4x2 wheel bearings, install new grease seals and adjust bearings every 48,000 km (30,000 miles). If equipped, change the in-line service installed transmission fluid filter.

06017-F150-C38

SCHEDULED MAINTENANCE INTERVALS
2004 Gasoline E-250 and 350
Footnotes Continued

Install new spark plugs and change transfer case fluid every 96,000 km (60,000 miles).

Install a new cabin air filter as required.

When operating in dusty conditions such as unpaved or dusty roads:

Change engine oil and install a new oil filter every 4,800 km (3,000 miles) or 3 months.

Install a new fuel filter every 24,000 km (15,000 miles).

Change automatic transmission fluid every 48,000 km (30,000 miles). If equipped, change the in-line service installed transmission fluid filter.

Change transfer case fluid every 96,000 km (60,000 miles).

Install a new engine air filter as required.

Install a new cabin air filter as required.

When operating in off-road conditions:

Change automatic transmission fluid every 48,000 km (30,000 miles). If equipped, change the in-line service installed transmission fluid filter.

Change transfer case fluid every 96,000 km (60,000 miles).

Install a new cabin air filter as required.

Inspect and lubricate U-joints.

Inspect and lubricate steering linkage ball joints with zerk fittings.

06017-F150-C39

SCHEDULED MAINTENANCE INTERVALS
2005 E- and F-Series with Gasoline Engines

TO BE SERVICED	TYPE OF SERVICE	VEHICLE MILEAGE INTERVAL (x1000)												
		5	10	15	20	25	30	35	40	45	50	55	60	65
Engine oil & filter	R	✓	✓	✓	✓	✓	✓	✓	✓	✓	✓	✓	✓	✓
Tires	Rotate	✓	✓	✓	✓	✓	✓	✓	✓	✓	✓	✓	✓	✓
Wheels	I ①			✓			✓			✓			✓	
Auto trans. fluid	I			✓			✓			✓			✓	
Brake pads/shoes	I			✓			✓			✓			✓	
Coolant hoses	S/I			✓			✓			✓			✓	
Steering linkage	I			✓			✓			✓			✓	
Suspension	I			✓			✓			✓			✓	
Driveshaft	I			✓			✓			✓			✓	
Cabin air filter	R			✓			✓			✓			✓	
Ball joints (2wd)	L			✓			✓			✓			✓	
Front drive axle U-joints	I/L			✓			✓			✓			✓	
NGV fuel filter	R					✓					✓			
Exhaust system	I						✓						✓	
Engine air filter	R						✓						✓	
Fuel filter	R						✓						✓	
Auto trans fluid (4R100 and Torqshift)	R						✓						✓	
Manual trans. fluid	R												✓	
Front wheel bearings (2wd)	L/Adj												✓	
Front wheel bearings grease seal (2wd)	R												✓	
Accessory drive belts	I	every 100,000 miles												
Spark plugs	R	every 100,000 miles												
PCV valve ②	R	every 100,000 miles												
Premium Gold coolant	R	every 3 years or 100,000 miles												
Auto trans fluid (all exc. 4R100 and TorqShift)	R	every 120,000 miles												
PCV valve (5.4L 3v)	I	every 150,000 miles												
Front wheel bearings (2wd)	R	at 150,000 miles, if not previously done so												
Front drive axle needle bearings (F-Super Duty)	L	every 150,000 miles												
Fuel tank	I	every 150,000 miles												
Front drive axle fluid	R	every 150,000 miles												
Rear drive axle fluid	R	every 150,000 miles												
Transfer case fluid	R	every 150,000 miles												
Accessory drive belts	R	every 150,000 miles, if not previously done so												

R: Replace   S: Service   I: Inspect   L: Lubricate

NGV: Natural gas vehicle

① Inspect for end play and noise

② Vehicles under 6,000 lbs. GVW, exc. 5.4L 3v engines

Special Operating Condition Requirements

When towing a trailer or using a camper or car-top carrier:

Change engine oil and install a new oil filter every 4,800 km (3,000 miles), 3 months or 200 hours of engine operation (whichever occurs first).

Change transfer case fluid every 96,000 km (60,000 miles).

Change manual transmission fluid as required.

Inspect and lubricate U-joints as required.

06017-F150-C40

SCHEDULED MAINTENANCE INTERVALS
2005 E- and F-Series with Gasoline Engines
Footnotes Continued

During extensive idling and/or low speed driving for long distances, as in heavy commercial use such as delivery, taxi, patrol car or livery:

Change engine oil and install a new oil filter every 4,800 km (3,000 miles), 3 months or 200 hours of engine operation (whichever occurs first).

Lube front lower control arm and steering linkage ball joints with zerk fittings (if equipped) every 4,800 km (3,000 miles) or 3 months.

Inspect brake system and check battery electrolyte level (Patrol cars) every 8,000 km (5,000 miles).

Install a new fuel filter every 24,000 km (15,000 miles).

Change automatic transmission fluid, lubricate 4x2 wheel bearings, install new grease seals and adjust bearings every

48,000 km (30,000 miles). If equipped, change the in-line service installed transmission fluid filter.

Install new spark plugs and change transfer case fluid every 96,000 km (60,000 miles).

Install a new cabin air filter as required.

When operating in dusty conditions such as unpaved or dusty roads:

Change engine oil and install a new oil filter every 4,800 km (3,000 miles) or 3 months.

Install a new fuel filter every 24,000 km (15,000 miles).

Change automatic transmission fluid every 48,000 km (30,000 miles). If equipped, change the in-line service installed transmission fluid filter.

Change transfer case fluid every 96,000 km (60,000 miles).

Install a new engine air filter as required.

Install a new cabin air filter as required.

When operating in off-road conditions:

Change automatic transmission fluid every 48,000 km (30,000 miles). If equipped, change the in-line service installed transmission fluid filter.

Change transfer case fluid every 96,000 km (60,000 miles).

Install a new cabin air filter as required.

Inspect and lubricate U-joints.

Inspect and lubricate steering linkage ball joints with zerk fittings.

Short trips in cold operating conditions:

Inspect and lubricate 4x2 ball joints and steering idler arms every 8,000 km (5,000 miles).

Change transfer case fluid every 96,000 km (60,000 miles).

06017-F150-C41

SCHEDULED MAINTENANCE INTERVALS
All 2004 Vehicles with the 6.0L Diesel Engine

TO BE SERVICED	TYPE OF SERVICE	VEHICLE MILEAGE INTERVAL (x1000)												
		7.5	15	22.5	30	37.5	45	52.5	60	67.5	75	82.5	90	97.5
Engine oil & filter	R	✓	✓	✓	✓	✓	✓	✓	✓	✓	✓	✓	✓	✓
Tires	Rotate	✓	✓	✓	✓	✓	✓	✓	✓	✓	✓	✓	✓	✓
Air filter minder	I ①	✓	✓	✓	✓	✓	✓	✓	✓	✓	✓	✓	✓	✓
Wheels	I ②	✓	✓	✓	✓	✓	✓	✓	✓	✓	✓	✓	✓	✓
Brake pads, hoses, etc.	I		✓			✓			✓			✓		
Coolant hoses	I		✓			✓			✓			✓		
Steering linkage and suspension	I/L		✓			✓			✓			✓		
Cabin air filter	R		✓			✓			✓			✓		
Ball joints (2wd)	L		✓			✓			✓			✓		
Driveshaft	I/L		✓			✓			✓			✓		
4x4 front axle U-joints	L		✓			✓			✓			✓		
Exhaust system and heat shields	I		✓			✓			✓			✓		
Engine air filter	R			✓					✓					✓
Fuel filters ③	R		✓				✓		✓		✓		✓	
Auto trans fluid ④	R				✓				✓				✓	
4x2 front wheel bearings	L							✓						
4x2 front wheel bearing grease seals	R							✓						
Accessory drive belts	I													✓
4x4 front hub needle bearings	L							✓						
Manual trans. Fluid	R							✓						
Coolant	R										✓			✓
Rear differential fluid ⑤	R													✓
Transfer case fluid	R	every 150,000 miles												
4x2 front wheel bearings	R	every 150,000 miles												
Accessory drive belts	R	every 150,000 miles, if not previously done so												

R: Replace   S: Service   I: Inspect   L: Lubricate   Adj: adjust

① Reset after new filter is installed

② Inspect for end play and noise

③ Frame-mounted and engine

④ Including external and in-line filters

⑤ Dana axles using non-synthetic fluid only

Special Operating Condition Requirements

When towing a trailer or using a camper or car-top carrier:

Change engine oil and install a new oil filter every 4,800 km (3,000 miles), 3 months or 200 hours of engine operation (whichever occurs first).

Change transfer case fluid every 96,000 km (60,000 miles).

Change manual transmission fluid as required.

Inspect and lubricate U-joints as required.

During extensive idling and/or low speed driving for long distances, as in heavy commercial use such as delivery, taxi, patrol car or livery:

Change engine oil and install a new oil filter every 4,800 km (3,000 miles), 3 months or 200 hours of engine operation (whichever occurs first).

Lube front lower control arm and steering linkage ball joints with zerk fittings (if equipped) every 4,800 km (3,000 miles) or 3 months.

Inspect brake system and check battery electrolyte level (Patrol cars) every 8,000 km (5,000 miles).

Install a new fuel filter every 24,000 km (15,000 miles).

Change automatic transmission fluid, lubricate 4x2 wheel bearings, install new grease seals and adjust bearings every 48,000 km (30,000 miles). If equipped, change the in-line service installed transmission fluid filter.

Change transfer case fluid every 96,000 km (60,000 miles).

SCHEDULED MAINTENANCE INTERVALS
All 2004 Vehicles with the 6.0L Diesel Engine
Footnotes Continued

Install a new cabin air filter as required.

When operating in dusty conditions such as unpaved or dusty roads:

Change engine oil and install a new oil filter every 4,800 km (3,000 miles) or 3 months.

Install a new fuel filter every 24,000 km (15,000 miles).

Change automatic transmission fluid every 48,000 km (30,000 miles). If equipped, change the in-line service installed transmission fluid filter.

Change transfer case fluid every 96,000 km (60,000 miles).

Install a new engine air filter as required.

Install a new cabin air filter as required.

When operating in off-road conditions:

Change automatic transmission fluid every 48,000 km (30,000 miles). If equipped, change the in-line service installed transmission fluid filter.

Change transfer case fluid every 96,000 km (60,000 miles).

Install a new cabin air filter as required.

Inspect and lubricate U-joints.

Inspect and lubricate steering linkage ball joints with zerk fittings.

06017-F150-C43

SCHEDULED MAINTENANCE INTERVALS
All 2005 Vehicles with the 6.0L Diesel Engine

TO BE SERVICED	TYPE OF SERVICE	VEHICLE MILEAGE INTERVAL (x1000)												
		7.5	15	22.5	30	37.5	45	52.5	60	67.5	75	82.5	90	97.5
Engine oil & filter	R	✓	✓	✓	✓	✓	✓	✓	✓	✓	✓	✓	✓	✓
Tires	Rotate	✓	✓	✓	✓	✓	✓	✓	✓	✓	✓	✓	✓	✓
Air filter minder	I ①	✓	✓	✓	✓	✓	✓	✓	✓	✓	✓	✓	✓	✓
Wheels	I ②		✓		✓		✓		✓		✓		✓	
Brake pads, hoses, etc.	I		✓		✓		✓		✓		✓		✓	
Coolant hoses	I		✓		✓		✓		✓		✓		✓	
Steering linkage and suspension	I/L		✓		✓		✓		✓		✓		✓	
Cabin air filter	R					✓							✓	
Ball joints (2wd)	L		✓		✓		✓		✓		✓		✓	
Driveshaft	I/L		✓		✓		✓		✓		✓		✓	
4x4 front axle U-joints	L		✓		✓		✓		✓		✓		✓	
Exhaust system and heat shields	I		✓		✓		✓		✓		✓		✓	
Engine air filter	R		✓		✓		✓		✓		✓		✓	
Fuel filters ③	R		✓		✓		✓		✓		✓		✓	
Auto trans fluid ④	R				✓				✓				✓	
4x2 front wheel bearings	L								✓					
4x2 front wheel bearing grease seals	R								✓					
Accessory drive belts	I													✓
Manual trans. fluid	R								✓					
Rear differential fluid ⑤	R													✓
Coolant (Premium Gold)	R	every 105,000 miles												
Front drve axle fluid	R	every 150,000 miles												
Transfer case fluid	R	every 150,000 miles												
Accessory drive belts	R	every 150,000 miles, if not previously done so												

R: Replace S: Service I: Inspect L: Lubricate Adj: adjust

① Reset after new filter is installed

② Inspect for end play and noise

③ Frame-mounted and engine

④ Including external and in-line filters

⑤ Dana axles using non-synthetic fluid only

Special Operating Condition Requirements

When towing a trailer or using a camper or car-top carrier:

Change engine oil and install a new oil filter every 4,800 km (3,000 miles), 3 months or 200 hours of engine operation (whichever occurs first).

Change transfer case fluid every 96,000 km (60,000 miles).

Change manual transmission fluid as required.

Inspect and lubricate U-joints as required.

During extensive idling and/or low speed driving for long distances, as in heavy commercial use such as delivery, taxi, patrol car or livery:

Change engine oil and install a new oil filter every 4,800 km (3,000 miles), 3 months or 200 hours of engine operation (whichever occurs first).

Lube front lower control arm and steering linkage ball joints with zerk fittings (if equipped) every 4,800 km (3,000 miles) or 3 months.

Inspect brake system and check battery electrolyte level (Patrol cars) every 8,000 km (5,000 miles).

Install a new fuel filter every 24,000 km (15,000 miles).

Change automatic transmission fluid, lubricate 4x2 wheel bearings, install new grease seals and adjust bearings every 48,000 km (30,000 miles). If equipped, change the in-line service installed transmission fluid filter.

Change transfer case fluid every 96,000 km (60,000 miles).

Install a new cabin air filter as required.

06017-F150-C44

SCHEDULED MAINTENANCE INTERVALS
All 2005 Vehicles with the 6.0L Diesel Engine
Footnotes Continued

When operating in dusty conditions such as unpaved or dusty roads:

Change engine oil and install a new oil filter every 4,800 km (3,000 miles) or 3 months.

Install a new fuel filter every 24,000 km (15,000 miles).

Change automatic transmission fluid every 48,000 km (30,000 miles). If equipped, change the in-line service installed transmission fluid filter.

Change transfer case fluid every 96,000 km (60,000 miles).

Install a new engine air filter as required.

Install a new cabin air filter as required.

When operating in off-road conditions:

Change automatic transmission fluid every 48,000 km (30,000 miles). If equipped, change the in-line service installed transmission fluid filter.

Change transfer case fluid every 96,000 km (60,000 miles).

Install a new cabin air filter as required.

Inspect and lubricate U-joints.

Inspect and lubricate steering linkage ball joints with zerk fittings.

06017-F150-C45

SCHEDULED MAINTENANCE INTERVALS
2006 Lincoln Mark LT

TO BE SERVICED	TYPE OF SERVICE	VEHICLE MILEAGE INTERVAL (x1000)												
		5	10	15	20	25	30	35	40	45	50	55	60	65
Engine oil & filter	R	✓	✓	✓	✓	✓	✓	✓	✓	✓	✓	✓	✓	✓
Tires	Rotate	✓	✓	✓	✓	✓	✓	✓	✓	✓	✓	✓	✓	✓
Wheels	I ①			✓			✓			✓			✓	
Brake fluid level	S/I	✓	✓	✓	✓	✓	✓	✓	✓	✓	✓	✓	✓	✓
Tire pressure and wear	S/I	✓	✓	✓	✓	✓	✓	✓	✓	✓	✓	✓	✓	✓
Auto trans. fluid	I			✓			✓			✓			✓	
Power steering fluid level	S/I	✓	✓	✓	✓	✓	✓	✓	✓	✓	✓	✓	✓	✓
Washer fluid level	S/I	✓	✓	✓	✓	✓	✓	✓	✓	✓	✓	✓	✓	✓
Battery performance	I	✓	✓	✓	✓	✓	✓	✓	✓	✓	✓	✓	✓	✓
Exterior lights	I	✓	✓	✓	✓	✓	✓	✓	✓	✓	✓	✓	✓	✓
Brake pads/shoes	I			✓			✓			✓			✓	
Coolant hoses	S/I	✓	✓	✓	✓	✓	✓	✓	✓	✓	✓	✓	✓	✓
A/C hoses	I	✓	✓	✓	✓	✓	✓	✓	✓	✓	✓	✓	✓	✓
Engine air filter	I	✓	✓	✓	✓	✓	✓	✓	✓	✓	✓	✓	✓	✓
Steering linkage	I			✓			✓			✓			✓	
Suspension	I/L			✓			✓			✓			✓	
Halfshafts	I			✓			✓			✓			✓	
Driveshaft	I			✓			✓			✓			✓	
Cabin air filter	R			✓			✓			✓			✓	
Ball joints (2wd)	L			✓			✓			✓			✓	
Front drive axle U-joints	I/L			✓			✓			✓			✓	
Exhaust system	I	✓	✓	✓	✓	✓	✓	✓	✓	✓	✓	✓	✓	✓
Engine air filter	R						✓						✓	
Fuel filter	R						✓						✓	
Climate controlled seat filters	R						✓						✓	
Front wheel bearings (2wd)	L/Adj									✓				
Accessory drive belts	I	every 100,000 miles												
Spark plugs	R	every 100,000 miles												
Premium Gold coolant	R	every 5 years or 100,000 miles												
PCV valve	R	every 100,000 miles												
Auto trans fluid	R	every 150,000 miles												
Front wheel bearings (2wd)	R	at 150,000 miles, if not previously done so												
Fuel tank	I	every 150,000 miles												
Front drive axle fluid	R	every 150,000 miles												
Rear drive axle fluid	R	every 150,000 miles												
Transfer case fluid	R	every 150,000 miles												
Accessory drive belts	R	every 150,000 miles, if not previously done so												

R: Replace　　　S: Service　　　I: Inspect　　　L: Lubricate

NGV: Natural gas vehicle

① Inspect for end play and noise

Special Operating Condition Requirements

When towing a trailer or using a camper or car-top carrier:

Change engine oil and install a new oil filter every 4,800 km (3,000 miles), 3 months or 200 hours of engine operation (whichever occurs first).

Change transfer case fluid every 96,000 km (60,000 miles).

SCHEDULED MAINTENANCE INTERVALS
2006 Lincoln Mark LT
Footnotes Continued

Change manual transmission fluid as required.

Inspect and lubricate U-joints as required.

During extensive idling and/or low speed driving for long distances, as in heavy commercial use such as delivery, taxi, patrol car or livery:

Change engine oil and install a new oil filter every 4,800 km (3,000 miles), 3 months or 200 hours of engine operation (whichever occurs first).

Lube front lower control arm and steering linkage ball joints with zerk fittings (if equipped) every 4,800 km (3,000 miles) or 3 months.

Inspect brake system and check battery electrolyte level (Patrol cars) every 8,000 km (5,000 miles).

Install a new fuel filter every 24,000 km (15,000 miles).

Change automatic transmission fluid, lubricate 4x2 wheel bearings, install new grease seals and adjust bearings every 48,000 km (30,000 miles). If equipped, change the in-line service installed transmission fluid filter.

Install new spark plugs and change transfer case fluid every 96,000 km (60,000 miles).

Install a new cabin air filter as required.

When operating in dusty conditions such as unpaved or dusty roads:

Change engine oil and install a new oil filter every 4,800 km (3,000 miles) or 3 months.

Install a new fuel filter every 24,000 km (15,000 miles).

Change automatic transmission fluid every 48,000 km (30,000 miles). If equipped, change the in-line service installed transmission fluid filter.

Change transfer case fluid every 96,000 km (60,000 miles).

Install a new engine air filter as required.

Install a new cabin air filter as required.

When operating in off-road conditions:

Change automatic transmission fluid every 48,000 km (30,000 miles). If equipped, change the in-line service installed transmission fluid filter.

Change transfer case fluid every 96,000 km (60,000 miles).

Install a new cabin air filter as required.

Inspect and lubricate U-joints.

Inspect and lubricate steering linkage ball joints with zerk fittings.

Short trips in cold operating conditions:

Inspect and lubricate 4x2 ball joints and steering idler arms every 8,000 km (5,000 miles).

Change transfer case fluid every 96,000 km (60,000 miles).

06017-F150-C47

GASOLINE ENGINE REPAIR

➡️Disconnecting the negative battery cable on some vehicles may interfere with the functions of the on board computer systems and may require the computer to undergo a relearning process, once the negative battery cable is reconnected.

Engine Cover

REMOVAL & INSTALLATION

E-Series

1. Before servicing the vehicle, refer to the Precautions Section.

2. On the high series console, release the locking tabs and remove the storage bin.

3. Through the floor console opening, push and hold the locking clip while lifting upward.

4. Remove the instrument panel steering column cover.

5. Unfasten the engine cover latches.

6. Lift the engine cover upward from the floor, and remove from the vehicle.

7. To install, reverse the removal procedure.

➡️**When installing the floor console, be sure the locking clip is fully engaged.**

Alternator

REMOVAL & INSTALLATION

4.2L Engine

1. Before servicing the vehicle, refer to the Precautions Section.
2. Remove or disconnect the following:
   - Negative battery cable
   - Drive belt
   - Alternator electrical connectors
   - Alternator stator and voltage regulator connectors

1 Front end accessory drive belt tensioner	7 Air conditioning (A/C) line bracket
2 Front end accessory drive belt	8 Generator harness bracket nut
3 Generator electrical connector	9 Generator harness bracket
4 Generator B+ terminal nut	10 Generator stud bolts (2 required)
5 Generator B+ terminal	11 Generator bolt
6 Air conditioning (A/C) line bracket nut	12 Generator

Alternator mounting—4.2L engine

- Alternator battery cable nut and the cable
- Alternator bolts and the alternator

To install:

3. Install or connect the following:
- Alternator and tighten the bolts to 35 ft. lbs. (47 Nm)

- Alternator battery cable and the nut
- Alternator stator and voltage regulator connectors
- Alternator electrical connectors
- Drive belt
- Negative battery cable

4.6L and 5.4L Engines

1. Before servicing the vehicle, refer to the Precautions Section.

2. Remove or disconnect the following:
- Negative battery cable
- Drive belt

1	Front end accessory drive belt tensioner	6	Generator bolts (2 required)
2	Front end accessory drive belt	7	Generator electrical connector (part of 14305)
3	Generator harness locator	8	Generator B+ terminal nut
4	Generator bracket bolts (4 required)	9	Generator B+ terminal
5	Generator bracket	10	Generator

Alternator mounting—5.4L engine, exc. Lightning. 4.6L similar

06017-F150-G14

- Alternator bracket bolts
- Ignition wire from the alternator
- Alternator bolts and the alternator
- Alternator electrical connectors
- Alternator stator and voltage regulator connectors
- Alternator battery cable nut and the cable

To install:

3. Install or connect the following:

- Alternator battery cable and the nut
- Alternator stator and voltage regulator connectors
- Alternator electrical connectors
- Alternator and the bolts, tighten to 18 ft. lbs. (25 Nm)
- Ignition wire from the alternator
- Alternator bracket bolts and tighten to 89 inch lbs. (10 Nm)
- Drive belt
- Negative battery cable

Lightning Models

1. Before servicing the vehicle, refer to the Precautions Section.

2. Remove or disconnect the following:

- Negative battery cable
- Bolt retaining the degas bottle and position the bottle aside
- Supercharger drive belt
- Accessory drive belt
- Upper alternator bracket
- Lower alternator brackets
- Upper bracket bolts
- Electrical connector from the alternator bracket
- Right hand lower bolt
- Alternator bracket
- Alternator electrical connectors
- Alternator

To install:

3. Installation is the reverse of removal, please note the following torques:

   a. All alternator bracket bolts to 89 inch lbs. (10 Nm).

   b. Right hand lower bolt to 18 ft. lbs. (25 Nm).

Alternator mounting bracket installation—Lightning

67197-EFSE-G78

6.8L Engines

1. Before servicing the vehicle, refer to the Precautions Section.

2. Remove or disconnect the following:

- Negative battery cable
- Air cleaner assembly
- Mass Air Flow (MAF) sensor electrical connector
- Accessory drive belt
- Alternator electrical connectors
- Bolts retaining the upper intake plenum to the alternator bracket
- Alternator bolts and the alternator

To install:

3. Before servicing the vehicle, refer to the Precautions Section.

4. Installation is the reverse of removal, please note the following torques:

   a. Bolts retaining the upper intake plenum to the alternator bracket. Tighten to 80–106 inch lbs. (7–9 Nm).

   b. Alternator bolts and the alternator. Tighten to 16–21 ft. lbs. (22–28 Nm).

Ignition Timing

ADJUSTMENT

Distributorless Ignition System

Base timing for distributorless ignition engines is set at the factory at 10 degrees BTDC and is not adjustable.

Engine Assembly

REMOVAL & INSTALLATION

2002–04 F-150 4.2L Engine

✳✳ CAUTION

Fuel injection systems remain under pressure, even after the engine has been turned OFF. The fuel system pressure must be relieved before disconnecting any fuel lines. Failure to do so may result in fire and/or personal injury.

1. Before servicing the vehicle, refer to the Precautions Section.

2. Remove or disconnect the following:

- Both battery cables, negative cable first
- Hood
- Coolant
- Refrigerant, using approved equipment

3. Relieve the fuel system pressure as follows:

   a. Remove the fuel tank fill cap to relieve the pressure in the fuel tank.

   b. Remove the cap from the fuel pressure relief valve located on the fuel injection supply manifold.

   c. Attach a fuel pressure gauge to the relief valve and drain the fuel through the drain tube into a suitable container.

   d. After the fuel system pressure is relieved, remove the fuel pressure gauge and install the cap on the relief valve. Secure the fuel tank fill cap.

4. Remove or disconnect the following:

- Engine cooling fan, shroud and radiator
- Engine air cleaner outlet tube
- Accelerator and cruise control cables at the throttle body
- VMV hose
- Manifold vacuum connection
- Intake Manifold Runner Control (IMRC) vacuum connectors, fuel pressure regulator vacuum connector, IMRC solenoid vacuum connector and vacuum reservoir connector
- Exhaust Gas Recirculation (EGR) valve vacuum connector.
- The 3 power steering reservoir retaining bolts and position aside
- Air conditioning compressor manifold bolt and disconnect, then position the air conditioning lines aside.
- The 4 power steering pump retaining bolts and position the pump aside.
- Alternator electrical harness connectors. Remove the positive battery cable nut and disconnect the battery cable.
- Electrical harness connectors to the fuel injectors
- Wires at the spark plugs
- Both heater hoses
- Brake booster vacuum hose
- EGR Differential Pressure Feedback (DPFE) transducer hose
- Breather tube from the cylinder head cover
- Upper intake manifold
- Fuel supply and return lines, and remove the fuel injection supply manifold.
- Block heater cable
- Exhaust system from the exhaust manifolds and support with wire hung from the crossmember.
- Starter motor
- Transmission from the vehicle. If

equipped with a manual transmission, remove the clutch assembly.
- Right-hand and left-hand engine support insulator through-bolts

5. Install a suitable engine lifting bracket and connect suitable engine lifting equipment to the lifting brackets.

6. Carefully raise the engine out of the engine compartment and position on a work stand. Remove the engine lifting equipment.

To install:

7. Install the engine lifting brackets. Support the engine using a suitable floor crane installed to the lifting equipment and remove the engine from the work stand.

8. Carefully lower the engine into the engine compartment aligning the engine support insulators.

9. Remove the engine lifting equipment and brackets.

10. Raise and safely support the vehicle.

11. Install or connect the following:
- Left-hand and right-hand engine support insulator through-bolts and tighten them to 51–67 ft. lbs. (68–92 Nm).
- Transmission, and the clutch assembly
- Starter motor
- Exhaust pipes to the exhaust manifolds and tighten to 30 ft. lbs. (41 Nm).
- Block heater
- Heater cable
- Alternator electrical harness connectors and install the positive battery cable to the retaining stud. Tighten the retaining nut to 96 inch lbs. (11 Nm).
- Fuel injectors and the fuel injection supply manifold
- New upper intake gasket and install the upper intake manifold.
- Electrical harness connectors to the fuel injectors
- Power steering pump in position and install 4 retaining bolts. Tighten the bolts to 17–20 ft. lbs. (22–28 Nm).

➡**Ensure that the air conditioning manifold O-rings are in place.**

- Air conditioning manifold to the compressor and install the retaining bolt. Tighten the bolt to 14–18 ft. lbs. (18–24 Nm).
- Power steering reservoir and 3 retaining bolts. Tighten the bolts to 107 inch lbs. (12 Nm).
- VMV hose
- IMRC vacuum connectors, fuel pressure regulator vacuum connec-

tor, IMRC solenoid vacuum connector and vacuum reservoir connector.
- EGR valve vacuum connector
- Brake booster hose
- The 2 EGR DPFE transducer hoses
- Manifold vacuum connection
- Accelerator cable and speed control cable in position, if equipped. Tighten the speed control cable retaining bolt to 72 inch lbs. (8 Nm) and accelerator cable retaining bolt to 25 inch lbs. (3 Nm).
- Radiator, cooling fan and shroud
- Engine air cleaner outlet tube
- Engine oil
- Coolant; bleed the system
- Battery cables, negative cable last

12. Start the engine and allow to reach normal operating temperature while checking for leaks.

13. Check all fluid levels.

14. Properly evacuate and recharge the air conditioning system using approved equipment.

15. Install the hood, aligning the marks that were made during removal.

16. Road test the vehicle and check the engine and transmission for proper operation.

2005 F-150 4.2L Engine

1. Before servicing the vehicle, refer to the Precautions Section.

All vehicles

2. With the vehicle in NEUTRAL, position it on a hoist.

3. Release the fuel system pressure.

4. Disconnect the battery ground cable.

5. Drain the cooling system.

6. Remove the hood.

7. Remove the cowl panel grille.

8. Remove the powertrain control module (PCM).

9. Detach the electrical connector, the wiring harness retainer and the washer hose from the cowl extension panel.

Vehicles equipped with a manual transmission

10. Remove the push pins and position the clutch master cylinder reservoir aside.

11. Disconnect the fuel vapor tubes from the purge valve.

All vehicles

12. Remove the 7 bolts and the cowl extension panel.

13. Remove the 9 push pins and the radiator sight shield.

14. Remove the upper intake manifold.

15. Remove the radiator, the cooling fan and the shroud.

16. Remove the air cleaner.

17. Release the clamp and disconnect the upper radiator hose from the thermostat housing. Remove the hose from the vehicle.

18. Release the clamps, disconnect the heater hoses and position the heater hoses aside.

19. Release the clamp and disconnect the lower radiator hose from the coolant pump.

20. Disconnect the intake manifold runner control (IMRC) actuator and the fuel charging wiring harness electrical connectors.

21. Disconnect the fuel supply spring lock coupling.

22. Remove the 4 bolts and the fuel rail, the fuel injectors, the fuel charging wiring harness and the vacuum harness as an assembly.

Vehicles equipped with an automatic transmission

23. Disconnect the fuel vapor tubes from the purge valve.

All vehicles

24. Using the special tool, with a twisting motion, pull the left spark plug wires off the spark plugs.

25. Detach the left spark plug wire retainers and remove the left spark plug wires.

26. Remove the bolts and position the power steering fluid reservoir aside.

27. If equipped, remove the compressor manifold and tube assembly.

28. Rotate the accessory drive belt tensioner counterclockwise and remove the accessory drive belt.

29. Remove the 2 power steering pump upper bolts.

30. Disconnect the generator.

31. Detach the wiring harness retainer.

32. Using the special tool, with a twisting motion, pull the right spark plug wires off the spark plugs.

33. Detach the right spark plug wire retainers.

34. Disconnect the coil electrical connector.

35. Remove the bolts, the coil and the right spark plug wires as an assembly.

36. Disconnect the positive crankcase ventilation (PCV) valve hose from the PCV valve and remove the hose.

37. Disconnect the wiring harness in-line connector.

38. Remove the bolt and detach the ground wire.

39. Detach the wiring harness retainer from the rear of the right cylinder head.

40. Remove the power steering pressure hose bracket nut.

➡The front bolt cannot be removed from the power steering pump with the pressure hose in place. Remove the rear bolt first, then slide the power steering pump out while removing the second bolt.

41. Remove the bolts and position the power steering pump aside.

42. Drain the engine oil. Install the drain plug when finished. Tighten to 23 Nm (17 ft. lbs.).

43. Remove the oil filter.

Vehicles equipped with an automatic transmission

44. Detach the transmission cooler tubes from the bracket.

All vehicles

45. Remove the nut and position the starter wiring harness aside.

46. If equipped, disconnect the engine block heater electrical connector.

47. Remove the 4 3-way catalytic converter-to-exhaust manifold nuts.

48. Remove the starter motor.

Vehicles equipped with an automatic transmission

49. Remove the cylinder block opening cover.

➡Rotate the crankshaft using the crankshaft pulley bolt to access all of the torque converter nuts.

➡Mark the flexplate and one torque converter stud for installation reference.

50. Remove the 4 torque converter nuts.

51. Remove the 3 oil pan-to-transmission bolts.

52. Remove the transmission-to-engine bolts.

53. Remove the left engine support insulator through bolt.

54. Remove the right engine support insulator nuts.

55. Remove the stud bolt, the bolts and the power steering fluid reservoir bracket.

56. Remove the nuts and the right engine lifting eye.

57. Install the lifting eye on the right side of the engine.

58. Remove the nuts and the left engine lifting eye.

59. Install the lifting eye on the left side of the engine.

➡Use 2 10mm (⅜ inch) spring links to connect the Engine Lifting Bracket 303-F047 to the Lifting Bracket, Engine 303-050.

60. Install the lifting bracket.

61. Connect an engine crane to the spe-cial tool and lift the engine approximately 75 mm (3 in).

62. Remove the bolts and the left engine support insulator.

63. Remove the bolts and the right engine support insulator.

64. Support the transmission with a floor jack and a wood block.

Vehicles equipped with an automatic transmission

65. Remove the engine from the vehicle.

Vehicles equipped with a manual transmission

66. Lift the engine while pulling the engine forward until the alternator pulley is above the radiator upper support.

67. Rotate the engine 90 degrees to the left and remove the engine from the vehicle.

To install:

Vehicles equipped with an automatic transmission

➡Make sure the marks made on the torque converter stud and the flexplate are lined up.

68. Position the engine in the vehicle and mate it to the transmission.

Vehicles equipped with a manual transmission

69. Position the engine with the front of the engine facing the left side of the vehicle.

70. Lower the engine until the generator pulley is slightly higher than the radiator upper support.

71. Turn the engine 90 degrees to the right, lower the engine and mate it to the transmission.

All vehicles

72. Install the right engine support insulator and the bolts. Tighten to 63 Nm (46 ft. lbs.).

73. Install the left engine support insulator and the bolts. Tighten to 63 Nm (46 ft. lbs.).

74. Remove the floor jack and wood block from the transmission.

75. Lower the engine into position and remove the floor crane and the special tools.

76. Position the power steering fluid reservoir bracket and install the bolts and the stud bolt. Tighten to 20 Nm (15 inch lbs.).

77. Install the right engine support insulator nuts. Tighten to 175 Nm (129 ft. lbs.).

78. Install the left engine support insulator through bolt. Tighten to 175 Nm (129 ft. lbs.).

79. Install the transmission-to-engine bolts. Tighten to 47 Nm (35 ft. lbs.).

80. Install the 3 oil pan-to-transmission bolts. Tighten to 34 Nm (24 ft. lbs.).

Vehicles equipped with an automatic transmission

➡Rotate the crankshaft using the crankshaft pulley bolt to access all of the torque converter nuts.

81. Install the 4 torque converter nuts. Tighten to 36 Nm (27 ft. lbs.).

82. Install the cylinder block opening cover.

All vehicles

83. Install the starter motor.

84. Install the 4 3-way catalytic converter nuts. Tighten to 40 Nm (30 ft. lbs.).

85. If equipped, connect the engine block heater electrical connector.

86. Position the starter wiring harness bracket and install the nut. Tighten to 20 Nm (15 ft. lbs.).

Vehicles equipped with an automatic transmission

87. Attach the transmission cooler tubes to the bracket.

All vehicles

88. Install a new oil filter. Tighten to 12 Nm (9 ft. lbs.).

➡Do not tighten the bolts until the 2 upper bolts are installed.

➡Make sure the power steering pressure hose bracket is installed on the stud.

89. Position the power steering pump and install the bolts finger tight. Install the nut. Tighten to 10 Nm (89 inch lbs.).

90. Attach the wiring harness retainer to the rear of the right cylinder head.

91. Position the ground wire and install the bolt.

92. Connect the wiring harness in-line connector.

93. Connect the positive crankcase ventilation (PCV) valve hose to the PCV valve.

94. Position the coil and the right spark plug wires as an assembly and install the bolts. Tighten to 7 Nm (62 inch lbs.).

95. Connect the coil electrical connector.

96. Attach the right spark plug wire retainers.

97. Apply silicone dielectric compound to the inside of the spark plug wire boots.

98. Connect the spark plug wires to the spark plugs. Attach the wiring harness retainer.

99. Connect the generator.

➡Tighten the lower bolts at this time.

100. Install the power steering pump upper bolts. Tighten to 25 Nm (18 ft. lbs.).

101. Rotate the accessory drive belt ten-

sioner counterclockwise and install the accessory drive belt.

102. If equipped, install the compressor manifold and tube assembly.

103. Position the power steering fluid reservoir and install the bolts. Tighten to 10 Nm (89 inch lbs.).

104. Position the left spark plug wires and attach the left spark plug wire retainers.

105. Apply silicone dielectric compound to the inside of the spark plug wire boots.

106. Connect the spark plug wires to the spark plugs.

Vehicles equipped with an automatic transmission

107. Connect the fuel vapor tubes to the purge valve.

All vehicles

108. Position the fuel rail, the fuel injectors, the fuel charging wiring harness and the vacuum harness as an assembly and install the bolts. Tighten to 10 Nm (89 inch lbs.).

109. Connect the fuel supply spring lock coupling.

110. Connect the intake manifold runner control (IMRC) actuator and the fuel charging wiring harness electrical connectors.

111. Connect the lower radiator hose to the coolant pump.

112. Connect the heater hoses.

113. Connect the upper radiator hose to the thermostat housing.

114. Install the air cleaner.

115. Install the radiator and the cooling fan and shroud.

116. Install the upper intake manifold.

117. Position the radiator sight shield and install the 9 push pins.

118. Position the cowl extension panel and install the 7 bolts.

Vehicles equipped with a manual transmission

119. Connect the fuel vapor tubes to the purge valve. Position the clutch master cylinder reservoir and install the push pins.

All vehicles

120. Attach the windshield washer hose, the wiring harness retainer and the electrical connector to the cowl extension panel.

121. Install the powertrain control module (PCM).

122. Install the cowl panel grille.

123. Position the hood and install the bolts. Tighten to 30 Nm (22 ft. lbs.).

124. Connect the battery ground cable.

125. Fill the engine with clean engine oil.

126. Fill and bleed the cooling system.

127. If equipped, evacuate and charge the A/C system.

2002–03 F-Series 4.6L, 5.4L, 6.8L Engines and 2004 Heritage Edition

❋❋ CAUTION

Fuel injection systems remain under pressure, even after the engine has been turned OFF. The fuel system pressure must be relieved before disconnecting any fuel lines. Failure to do so may result in fire and/or personal injury.

1. Before servicing the vehicle, refer to the Precautions Section.

2. Remove or disconnect the following:
   - Both battery cables, negative cable first
   - Hood
   - Coolant
   - Refrigerant, using approved equipment

3. Relieve the fuel system pressure as follows:

   a. Remove the fuel tank fill cap to relieve the pressure in the fuel tank.

   b. Remove the cap from the fuel pressure relief valve located on the fuel injection supply manifold.

   c. Attach a fuel pressure gauge to the relief valve and drain the fuel through the drain tube into a suitable container.

   d. After the fuel system pressure is relieved, remove the fuel pressure gauge and install the cap on the relief valve. Secure the fuel tank fill cap.

   - Engine cooling fan, shroud and radiator
   - Accessory drive belt
   - Engine air cleaner outlet tube
   - Intake manifold assembly
   - Bulkhead connector cover and disconnect the bulkhead connector.
   - The 3 power steering reservoir bracket retaining bolts and move the reservoir aside.
   - The 2 Differential Pressure Feedback (DPFE) transducer hoses
   - Upper and lower EGR valve to exhaust manifold tube fittings and remove the tube.
   - Heater water hose
   - Ignition coil, radio capacitor and CMP sensor electrical harness connectors
   - Both ignition coils and mounting bracket bolts and remove the coil and bracket assemblies.
   - Starter motor
   - The 3 lower radiator air deflector screws. Remove the 5 clips and remove the air deflector.
   - Air conditioning compressor electrical harness connector
   - Air conditioning manifold-to-com-

pressor bolt and remove the manifold and tube assembly.
   - The 3 air conditioning compressor retaining bolts and remove the air conditioning compressor.
   - Fluid cooler hoses from the block mounted clip
   - On vehicles with automatic transmissions: the inspection cover, torque converter bolts and transmission-to-engine retaining bolts.
   - On vehicles with manual transmission: the transmission and the clutch assembly.
   - Upper and lower power steering pump bolts and move the power steering pump aside.
   - Exhaust system from the exhaust manifolds and support with wire hung from the crossmember.
   - Right-hand and left-hand engine support insulator (mount) through-bolts

4. Install a suitable engine lifting bracket and connect suitable engine lifting equipment to the lifting brackets.

5. Carefully raise the engine out of the engine compartment and place on a work stand. Remove the engine lifting equipment.

To install:

6. Install the engine lifting brackets. Support the engine using a suitable floor crane installed to the lifting equipment and remove the engine from the work stand.

7. Carefully lower the engine into the engine compartment. Start the converter pilot into the flywheel and align the paint marks on the flywheel and torque converter. Be sure the studs on the torque converter align with the holes in the flywheel.

8. Fully engage the engine to the transmission and lower onto the engine support insulators.

9. Remove the engine lifting equipment and brackets.

10. Raise and safely support the vehicle.

11. Install or connect the following:
   - If equipped with a manual transmission: the clutch and transmission assemblies.
   - The 6 engine-to-transmission retaining bolts and tighten to 30–44 ft. lbs. (40–60 Nm).
   - The engine support insulator through-bolts and tighten to 15–22 ft. lbs. (20–30 Nm).
   - The 4 torque converter retaining nuts and tighten to 22–25 ft. lbs. (20–30 Nm).
   - Transmission housing cover to the cylinder block
   - Exhaust pipes to the exhaust mani-

folds and tighten to 30 ft. lbs. (41 Nm)

- Power steering pump in position on the cylinder block and install 4 retaining nuts. Tighten to 15–20 ft. lbs. (20–30 Nm).
- Starter motor
- Transmission fluid cooler hoses into the cylinder block mounted clip
- Air conditioning compressor in position and install the 3 retaining bolts. Tighten the bolts to 15–22 ft. lbs. (20–30 Nm).
- Air conditioning manifold and tube assembly on the compressor and install the retaining bolt. Tighten the bolt to 14–18 ft. lbs. (18–24 Nm).
- Air conditioning compressor clutch electrical harness connector
- Lower radiator air deflector
- Ignition coil and bracket assemblies. Tighten the retaining nuts to 15–23 ft. lbs. (20–30 Nm).
- Ignition coil, radio capacitor and CMP sensor electrical harness connectors
- Rear heater water hose and compress and slide the clamp in position.
- EGR valve to exhaust manifold tube and tighten the upper and lower fittings to 26–33 ft. lbs. (35–45 Nm).
- DPFE transducer hoses.
- Power steering pump reservoir in position and install the 3 retaining bolts. Tighten the bolts to 71–107 inch lbs. (8–12 Nm).
- Engine bulkhead connector and install the retaining bolt. Tighten the bolt to 36–50 inch lbs. (4–6 Nm). Install the cover.
- Intake manifold assembly
- Accessory drive belt
- Radiator, cooling fan and shroud
- Engine air cleaner outlet tube
- Engine oil

12. Fill and bleed the engine cooling system.

13. Connect both battery cables, negative cable last.

14. Start the engine and allow to reach normal operating temperature while checking for leaks.

15. Check all fluid levels.

16. Properly evacuate and recharge the air conditioning system using approved equipment.

17. Install the hood, aligning the marks that were made during removal.

18. Road test the vehicle and check the

engine and transmission for proper operation.

2004 F-150 4.6L Engine, Exc. Heritage Edition

1. Before servicing the vehicle, refer to the Precautions Section.

2. With the vehicle in **Neutral**, position it on a hoist.

3. Remove the hood.

4. Remove the cowl.

5. Discharge and recover the A/C system.

6. Remove the cooling module.

7. Remove the intake manifold.

8. Remove the powertrain control module (PCM) and the support bracket.

9. Remove the starter.

10. Disconnect the electrical connector and remove the bolt and the ground strap.

11. Disconnect the electrical connectors.

12. Remove the nut and disconnect the A/C manifold and tube assembly and position aside.

13. Disconnect the heater hose.

14. If equipped, disconnect the block heater electrical connector.

15. Remove the nut and the A/C manifold and tube assembly support bracket.

16. Disconnect the degas bottle coolant hose.

17. Remove the power steering fluid reservoir and support bracket.

18. Remove the bolts.

19. Position the power steering fluid reservoir and bracket aside.

20. Remove the bolts and position the power steering pump aside.

Vehicles with early build 4.6L engines, 2WD

21. Disconnect the power steering pressure switch electrical connector.

All vehicles

22. Remove the bolts and position the A/C compressor aside.

23. Remove the nut and the transmission cooler tube support bracket.

24. Remove the bolt and the starter electrical harness support bracket.

25. Remove the bolts and the flexplate inspection cover.

26. Remove the cylinder block opening cover.

27. Remove the torque converter-to-flexplate nuts.

28. Discard the nuts.

➡**The upper two transmission-to-engine bolts will be removed later.**

29. Remove the lower five transmission-to-engine bolts.

30. Remove the nut and the transmission fluid filler tube.

31. Disconnect the heated exhaust gas oxygen sensor (HO2S) electrical connectors.

32. Remove the four exhaust manifold flange nuts.

33. Remove the RIGHT motor mount nut.

34. Remove the LEFT motor mount bolt.

35. Install the special tool.

36. Using a suitable floor crane, remove the engine assembly from the vehicle.

To install:

37. Using a suitable floor crane, position the engine assembly into the vehicle.

38. Position and install and the left engine mount bolt. Torque to 148 ft. lbs. (200 Nm).

39. Install the right motor mount nut. Torque to 148 ft. lbs. (200 Nm).

➡**Align the engine-to-transmission dowels before installing the engine-to-transmission bolts.**

40. Install the lower five transmission-to-engine bolts. Torque to 35 ft. lbs. (48 Nm).

41. Install the four exhaust manifold-to-catalytic converter nuts. Torque to 30 ft. lbs. (40 Nm).

42. Connect the heated exhaust gas oxygen sensor (HO2S) electrical connectors.

➡**Lubricate the O-ring seals with clean transmission fluid.**

43. Install the transmission fluid filler tube and nut.

44. Install the torque converter-to-flexplate nuts. Torque to 26 ft. lbs. (35 Nm).

45. Install the cylinder block opening cover.

46. Install the starter wiring harness support bracket and bolt.

47. Install the flexplate inspection cover. Torque to 25 ft. lbs. (34 Nm).

48. Install the upper two transmission-to-engine bolts. Torque to 35 ft. lbs. (48 Nm).

49. Install the transmission cooler tube support bracket and the nut.

50. Position the power steering pump and install the bolts. Torque to 18 ft. lbs. (25 Nm).

Vehicles with early build 4.6L engines, 2WD

51. Connect the power steering pressure switch electrical connector.

All vehicles

52. Position the power steering reservoir and install the two lower bolts.

53. Install the power steering reservoir support bracket and nut.

Special tool 303-F047

54. Connect the degas bottle coolant hose.

55. Position the A/C compressor and install the bolts. Torque to 18 ft. lbs. (25 Nm).

56. Position the A/C manifold and tube assembly support bracket and install the nut.

57. If equipped, connect the block heater electrical connector.

58. Connect the heater coolant hose.

59. Connect the A/C manifold and tube assembly and install the nut. Torque to 18 ft. lbs. (25 Nm).

60. Connect the electrical connectors.

61. Install the ground strap and the bolt and connect the electrical connector.

62. Remove the special tool.

63. Install the intake manifold.

64. Install the cooling module.

65. Install the starter.

66. Install the powertrain control module (PCM) and the support bracket.

67. Install the wiper cowl.

68. Install the hood and the four bolts.

69. Fill the engine with clean engine oil.

70. Evacuate and charge the A/C system.

71. Fill and bleed the engine cooling system.

2005 F-150 4.6L Engine

1. Before servicing the vehicle, refer to the Precautions Section.

2. With the vehicle in NEUTRAL, position it on a hoist.

3. Remove the hood.

4. Remove the intake manifold.

5. Remove the radiator.

6. Remove the powertrain control module (PCM) and the support bracket.

7. Remove the cowl panel grille.

8. Detach the wiring harness and the windshield washer hose from the cowl panel extension.

9. Remove the bolts and the cowl panel extension.

10. Disconnect the electrical connector and remove the bolt and the ground strap.

11. Disconnect the heater hose.

12. Remove the nut and the A/C manifold and tube assembly support bracket.

13. Disconnect the A/C compressor electrical connector.

14. Disconnect the degas bottle coolant hose.

15. Remove the bolts and position the power steering reservoir assembly aside.

16. Using the special tool, remove the power steering pump pulley.

17. Disconnect the power steering pressure tube. Drain the power steering fluid into a suitable container.

18. Remove the nut and position aside the power steering pressure tube.

19. Remove the power steering pump bolts and position the power steering pump assembly aside.

20. Remove the starter.

21. Disconnect the crankcase position (CKP) sensor electrical connector and harness retainer.

22. If equipped, remove the nut and the transmission cooler tube support bracket.

23. Remove the bolts and position the A/C compressor aside.

24. If equipped, disconnect the block heater electrical connector.

25. Detach the starter electrical harness support from the cylinder block.

26. Remove the drain plug and drain the engine oil.

27. Remove the oil filter.

28. Remove the bolts and the flexplate inspection cover.

29. Remove the cylinder block opening cover.

30. Remove the torque converter-to-flexplate nuts. Discard the nuts.

31. Disconnect the left heated exhaust gas oxygen sensor (HO$_2$S) electrical connector and detach the electrical connector retainer.

32. Disconnect the shift cable and remove the shift cable bracket.

➡**The left upper transmission-to-engine bolt will be removed later.**

33. Remove the 6 transmission-to-engine bolts.

34. Remove the bolt and position the transmission fluid filler tube aside.

35. Remove the 4 exhaust manifold flange nuts.

36. Remove the right motor mount nuts.

37. Remove the left motor mount bolt.

38. Support the transmission.

➡**On 4x4 vehicles, it may be necessary to reposition the transfer case vent hose to access the bolt.**

39. Remove the left upper transmission-to-engine bolt.

40. Install the lifting bracket.

➡**Raise the engine and position forward to disconnect the transmission wiring harness retainer at the rear of the right cylinder head and disconnect the right heated exhaust gas oxygen sensor (HO$_2$S) electrical connector and detach the electrical connector retainer.**

41. Using a suitable floor crane, remove the engine assembly from the vehicle.

To install:

✱✱ WARNING

Clean the engine support insulator-to-frame mating surfaces of any dirt or foreign material prior to engine installation.

➡**Position the engine assembly forward to connect the right heated exhaust gas oxygen sensor (HO$_2$S) electrical connector and the transmission harness retainers at the rear of the cylinder head.**

42. Using a suitable floor crane, position the engine assembly into the vehicle.

43. Apply High Strength Threadlocker to the bolt threads and install the left engine support insulator bolt.

   a. If the engine support insulator nut is missing or damaged, install a new nut using service part number W709375.

   b. If the engine support insulator nut cage is damaged or missing, install a new nut using service part number W520516-S301. Tighten to 175 Nm (129 ft. lbs.).

44. Apply High Strength Threadlocker to the bolt threads and install the right engine support insulator washer and nuts. Tighten to 175 Nm (129 ft. lbs.).

45. Install the 6 transmission-to-engine bolts. Tighten to 60 Nm (44 ft. lbs.).

46. Connect the shift cable and install the shift cable bracket and the bolts. Tighten to 25 Nm (18 ft. lbs.).

47. Connect the left heated exhaust gas oxygen sensor (HO$_2$S) electrical connector and attach the electrical connector retainer.

➡**Lubricate the O-ring seals with clean transmission fluid.**

48. Position the transmission fluid filler tube and install the bolt. Tighten to 12 Nm (9 ft. lbs.).

49. Install the 4 exhaust manifold-to-catalytic converter nuts. Tighten to 40 Nm (30 ft. lbs.).

50. Install the torque converter-to-flexplate nuts. Tighten to 35 Nm (26 ft. lbs.).

51. Install the cylinder block opening cover.

52. Install the flexplate inspection cover. Tighten to 34 Nm (25 ft. lbs.).

53. Install a new oil filter.

54. Install the drain plug. Tighten to 14 Nm (10 ft. lbs.).

55. Attach the starter electrical harness support to the cylinder block. Tighten to 10 Nm (89 inch lbs.).

56. If equipped, connect the block heater electrical connector.

57. Position the A/C compressor and install the bolts. Tighten to 25 Nm (18 ft. lbs.).

58. Install the starter.

59. Position the power steering reservoir and install the 2 lower bolts. Tighten to 23 Nm (17 ft. lbs.).

60. Position the power steering pump and install the bolts. Tighten to 25 Nm (18 ft. lbs.).

61. Using the special tool, install a new O-ring seal on the pressure line fitting.

62. Connect the power steering pressure tube. Tighten to 65 Nm (48 ft. lbs.).

63. Position the power steering pressure tube support bracket and install the nut. Tighten to 10 Nm (89 inch lbs.).

❄❄ WARNING

If the pulley has been removed and installed twice, install a new power steering pump pulley.

64. Install the power steering pump pulley. Inspect the pulley for paint marks in the web area near the hub. If there are 2 paint marks, install a new pulley. If there is one paint mark or none at all, use a pencil to mark the web area of the pulley near the hub.

65. If equipped, install the transmission cooler tube support bracket and nut. Tighten to 10 Nm (89 inch lbs.).

66. Connect the degas bottle coolant hose.

67. Connect the A/C compressor electrical connector.

68. Connect the crankcase position (CKP) sensor electrical connector and the harness retainer.

69. Position the A/C manifold and tube assembly support bracket and install the nut. Tighten to 10 Nm (89 inch lbs.).

70. Remove the lifting bracket.

71. Install the upper transmission-to-engine bolts. Tighten to 60 Nm (44 ft. lbs.).

72. Connect the heater coolant hose.

73. Install the ground strap and the bolt and connect the electrical connector. Tighten to 10 Nm (89 inch lbs.).

74. Install the cowl panel extension.

75. Attach the wiring harness and the windshield washer hose to the cowl panel extension.

76. Install the cowl panel grill.

77. Install the powertrain control module (PCM) and the support bracket.

78. Install the intake manifold.

79. Install the radiator.

80. Install the hood and the 4 bolts. Tighten to 30 Nm (22 ft. lbs.).

81. Fill the engine with clean engine oil.

82. Fill and bleed the engine cooling system.

2004 F-150 5.4L Engine, Exc. Heritage Edition

1. Before servicing the vehicle, refer to the Precautions Section.

2. With the vehicle in **Neutral**, position it on a hoist

3. Remove the hood.

4. Remove the cowl.

5. Discharge and recover the A/C system.

6. Remove the radiator.

7. Remove the intake manifold.

8. Remove the powertrain control module (PCM) and the support bracket.

9. Remove the starter.

10. Disconnect the electrical connector and remove the bolt and the ground strap.

11. Disconnect the electrical connectors.

12. Disconnect the heater hose.

13. Remove the nut and disconnect the A/C manifold and tube assembly.

14. Remove the nut and the A/C manifold and tube assembly and support bracket.

15. Remove the nut and the transmission cooler tube support bracket.

16. Remove the bolt and disconnect the A/C manifold tube assembly and the electrical connector. Discard the A/C manifold tube assembly O-ring seals.

17. Remove the bolts and the A/C compressor.

18. If equipped, disconnect the block heater electrical connector.

19. Disconnect the coolant hose.

20. Remove the power steering reservoir support bracket nut.

21. Remove the bolts and position the power steering reservoir assembly aside.

22. Remove the bolts and position the power steering pump aside.

23. Remove the bolt and the starter electrical harness support bracket.

24. Remove the bolts and the flexplate inspection cover.

25. Remove the cylinder block opening cover.

26. Remove the torque converter-to-flexplate nuts. Discard the nuts.

27. Remove the lower five transmission-to-engine bolts. The upper two transmission-to-engine bolts will be removed later.

28. Remove the nut and the transmission fluid filler tube.

29. Disconnect the heated exhaust gas oxygen sensor electrical connectors.

30. Remove the four exhaust manifold flange nuts.

31. Remove the right motor mount nut.

32. Remove the left motor mount bolt.

33. Install a suitable tool.

34. Using a suitable floor crane remove the engine assembly from the vehicle.

To install:

35. Using a suitable floor crane, position the engine assembly into the vehicle.

➤**Align the engine-to-transmission dowels before installing the engine mount bolt.**

36. Position and install and the left engine mount bolt. Torque to 148 ft. lbs. (200 Nm).

37. Install the right motor mount nut. Torque to 76 ft. lbs. (103 Nm).

38. Install the lower five transmission-to-engine bolts. Torque to 35 ft. lbs. (48 Nm).

39. Install the four exhaust manifold to catalytic converter nuts. Torque to 30 ft. lbs. (48 Nm).

40. Connect the heated exhaust gas oxygen sensor electrical connectors.

➤**Lubricate the O-ring seals with transmission fluid.**

41. Install the transmission fluid filler tube and nut.

42. Install the torque converter-to-flexplate nuts. Torque to 26 ft. lbs. (35 Nm).

43. Install the cylinder block opening cover.

44. Install the starter electrical harness support bracket and the bolt.

45. Install the flexplate inspection cover. Torque to 25 ft. lbs. (34 Nm).

46. Install the upper two transmission-to-engine bolts. Torque to 35 ft. lbs. (48 Nm).

47. Position the power steering pump and install the bolts. Torque to 18 ft. lbs. (25 Nm).

48. Position the power steering reservoir and install the two lower bolts.

49. Install the power steering reservoir support bracket nut.

50. Position the A/C compressor and install the bolts. Torque to 18 ft. lbs. (25 Nm).

➡**Install new A/C manifold tube O-ring seals.**

51. Connect the A/C compressor electrical connector and the manifold tube assembly and install the bolt.

52. Install the transmission cooler tube support bracket and the nut.

53. Position the A/C manifold and tube assembly support bracket and install the nut.

54. Connect the A/C manifold and tube assembly and install the nut.

55. If equipped, connect the block heater electrical connector.

56. Connect the heater hose.

57. Connect the coolant hose.

58. Connect the electrical connectors.

59. Install the ground strap and the bolt and connect the electrical connector.

60. Remove the lifting bracket.

61. Install the intake manifold.

62. Install the radiator.

63. Install the starter.

64. Install the powertrain control module (PCM) and the support bracket.

65. Install the cowl.

66. Install the hood. Torque to 22 ft. lbs. (30 Nm).

67. Fill the engine with clean engine oil.

68. Evacuate and charge the A/C system.

69. Fill and bleed the engine cooling system.

2005 F-150 5.4L Engine

1. Before servicing the vehicle, refer to the Precautions Section.

All vehicles

2. With the vehicle in NEUTRAL, position it on a hoist.

3. Remove the hood.

4. Remove the intake manifold.

5. Remove the accessory drive belt.

6. Recover the A/C system.

7. Remove the radiator.

8. Remove the powertrain control module (PCM) and the support bracket.

9. Remove the cowl panel grille.

10. Detach the wiring harness and the windshield washer hose retainers from the cowl panel extension.

11. Remove the bolts and the cowl extension panel.

12. Disconnect the electrical connector and remove the bolt and the ground strap.

13. Disconnect the heater hose.

14. Remove the nut, disconnect the A/C manifold and tube assembly and position aside.

15. Remove the nut and the A/C manifold and tube assembly and support bracket.

16. Disconnect the coolant hose.

17. Remove the bolt and position the power steering reservoir assembly aside.

18. Disconnect the generator wiring harness retainer from the right cylinder head.

19. Disconnect the A/C compressor and the A/C high pressure cut-off switch electrical connectors.

20. Disconnect the generator wiring harness retainer from the right cylinder head and position the harness aside.

21. Disconnect the crankshaft position (CKP) sensor wiring harness retainer from the starter motor wiring harness.

22. Remove the nut and position aside the power steering pressure hose support bracket.

23. Using a puller, remove the power steering pump pulley.

24. Disconnect the power steering pressure tube. Drain the power steering fluid into a suitable container.

25. Remove the 3 bolts and position the power steering pump aside.

Manual transmission vehicles

26. Remove the clutch.

Automatic transmission vehicles

27. Remove the nut and position aside the transmission cooler tube support bracket.

28. Remove the starter.

29. Remove the 2 bolts and the flexplate inspection cover.

30. Remove the cylinder block opening cover.

31. Remove the 4 torque converter-to-flexplate nuts. Discard the nuts.

32. Disconnect the shift cable and remove the shift cable bracket.

➡**The upper 2 transmission-to-engine bolts will be removed later.**

33. Remove the lower 5 transmission-to-engine bolts.

34. Remove the bolt and position the transmission fluid filler tube aside.

All vehicles

35. Remove the drain plug and drain the engine oil.

36. Disconnect the right heated oxygen sensor (HO2S) electrical connector and detach the wiring harness retainer.

37. Disconnect the left heated exhaust gas oxygen sensor electrical connector and detach the electrical connector retainer.

38. Remove the oil filter.

39. Disconnect the oil temperature sensor electrical connector.

40. Remove the bolts and position the A/C compressor aside.

41. If equipped, disconnect the block heater electrical connector.

42. Detach the starter electrical harness support from the cylinder block.

43. Remove the 4 exhaust manifold flange nuts.

44. Remove the right engine support insulator nuts and washer.

45. Remove the left engine support insulator bolt.

Automatic transmission vehicles

46. Support the transmission.

➡**On 4WD vehicles, it may be necessary to reposition the transfer case vent hose to access the bolts.**

47. Remove the upper 2 transmission-to-engine bolts.

All vehicles

48. Install the lifting bracket.

49. Using a suitable floor crane, remove the engine assembly from the vehicle.

To install:
All vehicles

50. Using a suitable floor crane, position the engine assembly into the vehicle.

✳✳ WARNING

Only use hand tools when installing the left engine mount bolt or damage to the engine mount can occur.

➡**Early build vehicles are equipped with an M14 through bolt. Late build vehicles are equipped with an M18 through bolt.**

51. Apply high strength threadlocker to the bolt threads and install the left engine support insulator bolt.

　a. If the M14 through bolt engine support insulator nut is missing or damaged, install a new nut using service part number W709375.

　b. If the M14 through bolt engine support insulator nut cage is damaged or missing, install a new nut using service part number W520516-S301.

　c. Early build vehicles, tighten to 220 Nm (162 ft. lbs.); Late build vehicles, tighten to 350 Nm (258 ft. lbs.).

✳✳ WARNING

Only use hand tools when installing the right engine mount nuts or damage to the engine mount can occur.

52. Apply high strength threadlocker to the stud threads and install the right engine support insulator washer and nuts. Tighten to 250 Nm (184 ft. lbs.).

53. Remove the lifting bracket.

Manual transmission vehicles

54. Install the clutch.

Automatic transmission vehicles

➡ **The upper 2 transmission-to-engine bolts will be installed later.**

55. Install the lower 5 transmission-to-engine bolts. Tighten to 60 Nm (44 ft. lbs.).

56. Position the shift cable bracket and install the bolts and connect the shift cable. Tighten to 25 Nm (18 ft. lbs.).

57. Install 4 new torque converter-to-flexplate nuts. Tighten to 36 Nm (27 ft. lbs.).

58. Install the cylinder block opening cover.

59. Install the flexplate inspection cover and the 2 bolts. Tighten to 34 Nm (25 ft. lbs.).

60. Install the starter.

61. Position the transmission cooler tube support bracket and install the bolt. Tighten to 10 Nm (89 inch lbs.).

All vehicles

62. If equipped, connect the block heater electrical connector.

63. Connect the left and right heated exhaust gas oxygen sensor (HO2S) electrical connector and attach the electrical connector retainer. •

64. Attach the starter electrical harness support bracket.

65. Connect the oil temperature sensor electrical connector.

66. Position the A/C compressor and install the 3 bolts. Tighten to 25 Nm (18 ft. lbs.).

67. Install the 4 exhaust manifold flange nuts. Tighten to 40 Nm (30 ft. lbs.).

Automatic transmission vehicles

68. Install the upper 2 transmission-to-engine bolts. Tighten to 60 Nm (44 ft. lbs.).

69. Position the transmission filler tube and install the bolt. Tighten to 20 Nm (15 ft. lbs.).

All vehicles

70. Connect the crankshaft position (CKP) sensor wiring harness retainer to the starter motor wiring harness.

71. Connect the generator wiring harness retainer to the right cylinder head.

72. Connect the A/C compressor and the A/C high pressure cut-off switch electrical connectors.

73. Connect the generator wiring harness retainer to the right cylinder head.

74. Position the power steering pump assembly and install the 3 bolts. Tighten to 25 Nm (18 ft. lbs.).

75. Install a new O-ring seal on the pressure line fitting.

76. Connect the power steering pressure tube. Tighten to 65 Nm (48 ft. lbs.).

77. Install the power steering pump pulley.

78. Position the power steering pressure hose support bracket and install the nut. Tighten to 10 Nm (89 inch lbs.).

79. Position the power steering reservoir assembly and install the bolt. Tighten to 23 Nm (17 ft. lbs.).

80. Using a new O-ring seal, connect the A/C manifold and tube assembly and install the nut. Tighten to 25 Nm (18 ft. lbs.).

81. Position the A/C manifold and tube assembly support bracket and install the nut. Tighten to 25 Nm (18 ft. lbs.).

82. Connect the heater hose.

83. Disconnect the coolant hose.

84. Connect the electrical connector and position the ground strap and install the bolt. Tighten to 10 Nm (89 inch lbs.).

85. Position the cowl extension panel and install the bolts.

86. Attach the wiring harness and the windshield washer hose retainers to the cowl panel extension.

87. Install the intake manifold.

88. Install the radiator.

89. Install the powertrain control module (PCM).

90. Install the cowl.

91. Install the hood.

92. Fill the crankcase with clean engine oil.

93. Evacuate and charge the A/C system.

94. Fill and bleed the power steering system.

2004 F-250/350 5.4L Engine

1. Before servicing the vehicle, refer to the Precautions Section.

2. Disconnect the battery ground cable.

3. Discharge and recover the air conditioning system

4. Remove the radiator grille supports.

5. Remove the A/C condenser core.

6. Remove the radiator, fan shroud and engine cooling fan.

7. Disconnect the 42-pin and the 16-pin connectors and remove the nut and the harness mounting bracket.

8. Remove the intake manifold.

9. At the oil cooler water inlet, disconnect and set aside the lower radiator hose.

10. Remove the bolts and position the power steering pump aside.

11. Disconnect the connector and position the harness aside.

12. Disconnect the A/C cycling switch, remove the harness from the studs and position aside.

13. Disconnect the suction hose at the accumulator.

14. Disconnect the heater hose.

15. Disconnect the left and right heated exhaust gas oxygen sensor connectors.

16. Remove the A/C compressor.

17. Raise and support the vehicle.

18. Remove the nuts.

Manual transmission vehicles

19. Remove the transmission and clutch.

All vehicles

20. Drain the engine oil and remove the oil bypass filter.

21. Loosen the threaded shaft and position the oil cooler aside.

Automatic transmission vehicles

22. Remove the starter motor.

23. Remove the nuts and position the transmission oil cooler lines and bracket aside.

All vehicles

24. Remove the stud bolt and position the ground strap aside.

25. Remove the nut and position the wiring bracket aside.

Automatic transmission vehicles

26. Remove the flywheel inspection cover.

27. Remove the access plug.

28. Remove and discard the four torque converter nuts.

29. Remove the four lower transmission to engine bolts.

All vehicles

30. Lower the vehicle.

31. Remove the nut retaining the ground strap, and on automatic transmission equipped vehicles remove the transmission oil filler tube.

32. Remove the two studs and the heater water tube.

Automatic transmission vehicles

33. Remove the three upper transmission to engine bolts.

All vehicles

34. Install the Modular Engine Lift Bracket and support the engine with a floor crane.

35. Remove the engine support insulator nuts.

Automatic transmission vehicles

36. Support the transmission with a jack.

All vehicles

37. Remove the engine from the vehicle.

To install:

All vehicles

38. Position the engine in the vehicle and remove the Modular Engine Lift Bracket and the jack supporting the transmission.

39. Install and tighten the nuts to 66 ft. lbs. (90 Nm).

40. On automatic transmission-equipped vehicles, install the transmission oil level tube, and the ground strap on all vehicles.

67197-EFSE-G06

**Lower transmission-to-engine bolts—2004
F-250/350 5.4L engine**

67197-EFSE-G07

**Torque converter nuts—2004 F-250/350
5.4L engine**

41. Install the heater water tube and the
two studs.

42. Raise and support the vehicle.

Automatic transmission vehicles

43. Install the lower transmission to
engine bolts. Torque to 35 ft. lbs. (48
Nm).

44. Install and tighten the four new nuts
retaining the torque converter. Torque to 26
ft. lbs. (35 Nm).

45. Install the access plug.

46. Install the flywheel inspection cover.
Torque to 25 ft. lbs. (34 Nm).

All vehicles

47. Install the wiring bracket.

Automatic transmission vehicles

48. Install the starter motor.

All vehicles

49. Install the oil cooler to the oil filter
adapter and install the oil bypass filter.
Torque to 43 ft. lbs. (58 Nm).

Manual transmission vehicles

50. Install the clutch and transmission.

All vehicles

51. Install the head pipes and torque the
nuts to 30 ft. lbs. (40 Nm).

52. Install the A/C compressor.

53. Install the ground strap.

Automatic transmission vehicles

54. Install the transmission cooler line
bracket.

55. Lower the vehicle.

56. Install the remaining transmission-
to-engine bolts and fuel line bracket. Torque
to 35 ft. lbs. (48 Nm).

57. Connect the transmission wiring harness and the left and right heated exhaust
gas oxygen sensor connectors.

58. Connect the heater hose.

59. Install the suction line to the accumulator.

60. Connect the A/C cycling switch.

61. Connect the A/C compressor clutch
connector.

62. Connect the lower radiator hose to
the oil cooler inlet.

63. Install the power steering pump.
Torque the mounting bolts to 18 ft. lbs. (25
Nm).

64. Install the intake manifold.

65. Connect the 42-pin and 16-pin connectors.

66. Install the radiator, fan shroud and
engine cooling fan.

67. Install the A/C condenser core.

68. Install the radiator grille support.

✳✳ CAUTION

**The oil pump must be primed prior to
starting the engine.**

69. Fill all fluids to the correct levels.

70. Connect the battery ground cable.

71. Start the engine and check for leaks.
Stop the engine and recheck the fluid levels.

72. Evacuate and recharge the A/C system.

2005 F-250/350 5.4L Engine

All vehicles

1. With the vehicle in NEUTRAL, position it on a hoist.

2. Remove the hood.

3. Remove the cowl vent grille.

4. Remove the radiator.

5. Remove the degas bottle.

6. Remove the intake manifold.

7. Remove the powertrain control module (PCM).

8. Drain the engine oil.

9. Disconnect the degas bottle coolant
hose.

10. Disconnect the 2 electrical connectors on the left fender splash shield.

11. Remove the nut and the starter
wiring harness support bracket.

12. Remove the 3 bolts and position the
power steering pump assembly aside.

13. Disconnect the generator wiring harness retainer from the right cylinder head.

14. Disconnect the A/C compressor and
the A/C high pressure cut-off switch electrical connectors.

15. Disconnect the generator wiring harness retainer from the right cylinder head
and position the harness aside.

16. Disconnect the crankshaft position
(CKP) sensor wiring harness retainer from
the starter motor wiring harness.

17. Disconnect the engine wiring harness retainer from the evaporative emissions (EVAP) canister purge valve
bracket.

18. Disconnect the EVAP tube quick
connect couplings and the electrical connector from EVAP canister purge valve.

19. Remove the 3 nuts and remove the
EVAP canister purge valve and bracket
assembly.

20. Disconnect the body wiring harness
retainer from the degas bottle support
bracket.

21. Remove the 3 nuts and the degas
bottle support bracket.

22. Remove the 2 nuts and the air
cleaner outlet tube support bracket.

Manual transmission vehicles

23. Remove the clutch and pressure
plate.

Automatic transmission vehicles

24. Remove the starter.

25. Remove the nut and position the
transmission oil cooler lines and bracket
aside.

26. Remove the 2 bolts and the flexplate
inspection cover.

27. Remove the cylinder block opening
cover.

28. Remove the 4 torque converter-to-
flexplate nuts. Discard the nuts.

➡**The upper 2 transmission-to-engine
bolts will be removed later.**

29. Remove the 5 transmission-to-
engine bolts.

30. Remove the nut, the ground cable
and position the transmission fluid filler
tube aside.

All vehicles

31. If equipped, disconnect the block
heater electrical connector.

32. Remove the stud bolt and position
the ground strap aside.

33. Disconnect the oil temperature sensor electrical connector.

34. Remove the 3 bolts and position the
A/C compressor aside.

35. Disconnect the 2 heated exhaust gas
oxygen sensor (HO2S) electrical connectors.

36. Remove the 4 exhaust manifold
flange nuts.

37. Remove the 6 right and left motor
mount nuts.

Automatic transmission vehicles

38. Remove the upper 2 transmission-
to-engine bolts.

All vehicles

39. Install the lifting bracket.

40. Using a suitable floor crane, remove the engine assembly from the vehicle.

To install:

All vehicles

41. Using a suitable floor crane, position the engine assembly into the vehicle.

✳✳ WARNING

Only use hand tools when installing the right engine mount nut or damage to the engine mount can occur.

➡ **Align the engine-to-transmission dowels before installing the engine mount bolt.**

42. Install the 6 engine mount nuts. Tighten to 175 Nm (129 ft. lbs.).

43. Remove the lifting bracket.

Manual transmission vehicles

44. Install the clutch.

Automatic transmission vehicles

➡ **The upper 2 transmission-to-engine bolts will be installed later.**

45. Install the lower 5 transmission-to-engine bolts. Tighten to 35 Nm (26 ft. lbs.).

46. Install 4 new torque converter-to-flexplate nuts. Tighten to 35 Nm (26 ft. lbs.).

47. Install the cylinder block opening cover.

48. Install the flexplate inspection cover and the 2 bolts. Tighten to 34 Nm (25 ft. lbs.).

49. Install the transmission fluid filler tube, the ground strap and the nut. Tighten to 12 Nm (9 ft. lbs.).

50. Install the starter.

All vehicles

51. If equipped, connect the block heater electrical connector.

52. Position the starter electrical harness support bracket and install the bolt. Tighten to 10 Nm (89 inch lbs.).

53. Position the ground strap and install the stud bolt. Tighten to 10 Nm (89 inch lbs.).

54. Connect the oil temperature sensor electrical connector.

55. Position the A/C compressor and install the 3 bolts. Tighten to 25 Nm (18 ft. lbs.).

56. Install the 4 exhaust manifold flange nuts. Tighten to 40 Nm (30 ft. lbs.).

57. Connect the 2 heated exhaust gas oxygen sensor (HO2S) electrical connectors.

Automatic transmission vehicles

58. Install the transmission cooler tube support bracket and the nut. Tighten to 10 Nm (89 inch lbs.).

59. Install the upper 2 transmission-to-engine bolts. Tighten to 48 Nm (35 ft. lbs.).

All vehicles

60. Position the degas bottle support bracket and install the 3 nuts. Tighten to 10 Nm (89 inch lbs.).

61. Connect the body wiring harness retainer to the degas bottle support bracket.

62. Position the evaporative emissions (EVAP) canister purge valve and bracket assembly, and install the 3 nuts. Tighten to 10 Nm (89 inch lbs.).

63. Connect the electrical connector and the EVAP tube quick connect couplings to the EVAP canister purge valve.

64. Connect the engine wiring harness retainer to the EVAP canister purge valve bracket.

65. Connect the crankshaft position (CKP) sensor wiring harness retainer to the starter motor wiring harness.

66. Connect the generator wiring harness retainer to the right cylinder head.

67. Connect the A/C compressor and the A/C high pressure cut-off switch electrical connectors.

68. Connect the generator wiring harness retainer to the right cylinder head.

69. Position the power steering pump and install the 3 bolts. Tighten to 25 Nm (18 ft. lbs.).

70. Position the air cleaner outlet tube support bracket and install the 2 nuts. Tighten to 25 Nm (18 ft. lbs.).

71. Connect the 2 electrical connectors on the left fender splash shield.

72. Connect the degas bottle coolant hose.

73. Install the intake manifold.

74. Install the radiator.

75. Install the degas bottle.

76. Install the powertrain control module (PCM).

77. Install the cowl vent grille.

78. Install the hood.

79. Fill the crankcase with clean engine oil.

80. Evacuate and charge the A/C system.

81. Fill and bleed the power steering system.

2004 F-250/350 6.8L Engine

1. Before servicing the vehicle, refer to the Precautions Section.

2. Disconnect the battery ground cable.

3. Discharge and recover the air conditioning system.

4. Remove the radiator grille support.

5. Remove the A/C condenser core.

6. Remove the radiator, fan shroud and engine cooling fan.

7. Disconnect the 42-pin and the 16-pin connectors and remove the nut and the harness mounting bracket.

8. Remove the intake manifold.

9. At the oil cooler water inlet, disconnect and set aside the lower radiator hose.

10. Remove the bolts and position the power steering pump aside.

11. Disconnect the connector and position the harness aside.

12. Disconnect the A/C cycling switch. Remove the harness from the studs and position aside.

13. Disconnect the suction hose at the accumulator.

14. Disconnect the left and right heated exhaust gas oxygen sensor connectors.

15. Remove the A/C compressor.

16. Remove the nuts.

Manual transmission vehicles

17. Remove the transmission and clutch.

All vehicles

18. Drain the engine oil and remove the oil bypass filter.

19. Loosen the threaded shaft and position the oil cooler aside.

Automatic transmission vehicles

20. Remove the starter motor.

21. Remove the bolts and position the transmission oil cooler lines and bracket aside.

All vehicles

22. Remove the stud bolt and position the ground strap aside.

23. Remove the nut and position the wiring bracket aside.

Automatic transmission vehicles

24. Remove the flywheel inspection cover.

25. Remove the access plug.

26. Remove and discard the four torque converter bolts.

27. Remove the four lower transmission-to-engine bolts.

All vehicles

28. Lower the vehicle.

29. Remove the nut retaining the ground strap, and on automatic transmission-equipped vehicles, remove the transmission oil filler tube.

30. Disconnect the heater hose at the rear of the engine.

31. Remove the two studs and the heater water tube.

Automatic transmission vehicles

32. Remove the three upper transmission-to-engine bolts.

All vehicles

33. Install the Modular Engine Lift Bracket and support the engine with a suitable floor crane.

34. Remove the engine support insulator nuts.

Automatic transmission vehicles

35. Support the transmission with a jack.

All vehicles

36. Remove the engine from the vehicle.

37. Remove the bolts, and remove the flywheel.

38. Using a puller set, remove the rear crankshaft oil slinger.

39. Using a puller set, remove the crankshaft rear oil seal.

40. Remove the bolts and remove the crankshaft rear oil seal retainer plate.

41. Mount the engine on an engine stand.

42. Remove the Modular Engine Lift Bracket.

To install:

43. Attach the Modular Engine Lift Bracket.

44. Remove the engine from the engine stand.

✴✴ CAUTION

Do not use metal scrapers, wire brushes, power abrasive discs or other abrasive means to clean the aluminum retainer plate. These tools cause scratches and gouges, which make leak paths. Use a plastic scraping tool to remove all traces of old sealant.

45. Clean and inspect the mating surfaces of the cylinder block and the retainer plate.

➡**If not secured within four minutes, sealant must be removed and the sealing area cleaned with metal surface cleaner. Allow to dry until there is no sign of wetness, or four minutes, whichever is longer. Failure to follow this procedure can cause future oil leakage.**

46. Apply a bead of silicone gasket and sealant around the rear oil seal retainer sealing surface.

47. Install the retainer plate. Tighten the bolts in two steps.

67197-EFSE-G02

Retainer plate installation—6.8L engine

73 - 87 Nm (54 - 64 Lb - Ft)

67197-EFSE-G03

Flywheel installation—6.8L Engine

a. Step 1: Tighten the retainer plate bolts to 10 Nm (89 inch lbs.), in the sequence shown.

b. Step 2: Tighten the oil pan bolts to 20 Nm (20 ft. lbs.), then tighten an additional 90 degrees.

48. Use the two crankshaft rear oil seal installers to install the crankshaft rear oil seal.

49. Using the two crankshaft rear oil seal installers and the crankshaft rear oil slinger installer, install the crankshaft rear oil slinger.

50. Install the flywheel. Torque to 54–64 ft. lbs. (73–87 Nm).

51. Position the engine in the vehicle and remove the Modular Engine Lift Bracket and the jack supporting the transmission.

52. Install and tighten the nuts to 66 ft. lbs. (90 Nm).

53. Install the heater water tube and the two studs. Torque to 20 ft. lbs. (25 Nm).

54. On automatic transmission-equipped vehicles, install the transmission oil level indicator tube and the ground strap on all vehicles.

55. Raise and support the vehicle.

Automatic transmission vehicles

56. Install the lower transmission to engine bolts.

57. Install the access plug.

58. Install and tighten the four new nuts retaining the torque converter. Torque to 25–34 ft. lbs. (34–46 Nm).

59. Install the flywheel inspection cover.

All vehicles

60. Install the wiring bracket.

Automatic transmission vehicles

61. Install the starter motor.

All vehicles

62. Install the oil cooler to the oil filter adapter and install the oil bypass filter.

Manual transmission vehicles

63. Install the clutch and transmission

All vehicles

64. Install the head pipes and tighten the nuts to 30 ft. lbs. (40 Nm).

65. Install the ground strap.

55-60 Nm (41-44 lb/ft)

67197-EFSE-G04

Oil bypass filter installation—6.8L Engine

Automatic transmission vehicles

66. Install the transmission cooler line bracket.

All vehicles

67. Install the A/C compressor.
68. Lower the vehicle.
69. Position back and install the heater hose at the rear of the engine.

Automatic transmission vehicles

70. Install the remaining transmission-to-engine bolts and fuel line bracket.

All vehicles

71. Connect the transmission wiring harness and the left and right heated exhaust gas oxygen sensor connectors.
72. Install the suction line to the accumulator.
73. Connect the A/C cycling switch.
74. Connect the connector and position the harness aside.
75. Install the power steering pump. Torque the mounting bolts to 20 ft. lbs. (25 Nm).
76. Position back and connect the lower radiator hose at the oil cooler water inlet.
77. Install the intake manifold.
78. Install the harness mounting bracket and connect the 42-pin and 16-pin connectors.
79. Install the radiator.
80. Install the A/C condenser core.
81. Install the radiator grille support.
82. Fill all fluids to the correct levels.
83. Connect the battery ground cable.
84. Start the engine and check for leaks. Stop the engine and recheck the fluid levels.
85. Evacuate and recharge the A/C system.

2005 F-250/350 6.8L Engine

1. Before servicing the vehicle, refer to the Precautions Section.

All vehicles

2. With the vehicle in NEUTRAL, position it on a hoist.
3. Remove the hood.
4. Remove the radiator grille support.
5. Remove the A/C condenser core.
6. Remove the cowl vent grille.
7. Remove the radiator, fan shroud and engine cooling fan.
8. Remove the intake manifold.
9. Remove the degas bottle.
10. Remove the powertrain control module (PCM).
11. Disconnect the degas bottle coolant hose.
12. Disconnect the 2 electrical connectors on the left fender splash shield.
13. If equipped, disconnect the block heater electrical connector.

20-30 Nm (15-22 lb/ft)

67197-EFSE-G05

Power steering pump installation—6.8L engine

14. Remove the 3 bolts and position the power steering pump aside.
15. Disconnect the generator wiring harness retainer from the right cylinder head.
16. Disconnect the A/C compressor and the A/C high pressure cut-off switch electrical connectors.
17. Disconnect the generator wiring harness retainer from the right cylinder head and position the harness aside.
18. Disconnect the crankshaft position (CKP) sensor wiring harness retainer from the starter motor wiring harness.
19. Disconnect the engine wiring harness retainer from the evaporative emissions (EVAP) canister purge valve bracket.
20. Disconnect EVAP tube quick connect couplings and the electrical connector from EVAP canister purge valve.
21. Remove the 3 nuts and remove EVAP canister purge valve and bracket assembly.
22. Disconnect the heater hose at the rear of the engine.
23. Disconnect the body wiring harness retainer from the degas bottle support bracket.
24. Remove the 3 nuts and the degas bottle support bracket.
25. Drain the engine oil.

26. Remove the 4 exhaust manifold-to-catalytic converter nuts.
27. Disconnect the 2 heated exhaust gas oxygen sensor (HO2S) electrical connectors.
28. Remove the nut and the ground cable.
29. Remove the 3 bolts and position the A/C compressor aside.

Manual transmission vehicles

30. Remove the clutch.

All vehicles

31. Remove and discard the oil filter.
32. Disconnect the coolant hoses from the oil cooler.
33. Loosen the threaded shaft and remove the oil cooler. Discard the oil cooler.

Automatic transmission vehicles

34. Remove the starter motor.
35. Remove the transmission filler tube.
36. Remove the bolts and position the transmission oil cooler lines and bracket aside.

All vehicles

37. Remove the stud bolt and position the ground strap aside.
38. Remove the nut and position the starter electrical harness support bracket aside.

Automatic transmission vehicles

39. Remove the 2 bolts and the flexplate inspection cover.

40. Remove the access plug.

41. Remove the 6 torque converter-to-flexplate nuts. Discard the nuts.

42. Remove the 5 lower transmission-to-engine bolts.

Automatic transmission vehicles

43. Remove the upper 2 transmission-to-engine bolts.

All vehicles

44. Install the lifting bracket and support the engine with a suitable floor crane.

> ※ **WARNING**
>
> **Only use hand tools when removing the engine mount nut or damage to the engine mount can occur.**

45. Remove the 4 engine mount nuts.

Automatic transmission vehicles

46. Support the transmission with a jack.

All vehicles

47. Using a suitable floor crane, remove the engine assembly from the vehicle.

To install:

All vehicles

48. Using a suitable floor crane, position the engine assembly into the vehicle.

> ※ **WARNING**
>
> **Only use hand tools when installing the right engine mount nut or damage to the engine mount can occur.**

➡**Align the engine-to-transmission dowels before installing the engine mount bolt.**

49. Install the 4 engine mount nuts. Tighten to 175 Nm (129 ft. lbs.).

50. Remove the lifting bracket.

Manual transmission vehicles

51. Install the clutch.

Automatic transmission vehicles

➡**The upper 2 transmission-to-engine bolts will be installed later.**

52. Install the lower 5 transmission-to-engine bolts. Tighten to 35 Nm (26 ft. lbs.).

53. Install 6 new torque converter-to-flexplate nuts. Tighten to 35 Nm (26 ft. lbs.).

54. Install the cylinder block opening cover.

55. Install the flexplate inspection cover and the 2 bolts. Tighten to 34 Nm (25 ft. lbs.).

56. Install the starter.

57. Install the transmission filler tube.

All vehicles

58. Install the ground strap and the nut. Tighten to 10 Nm (89 inch lbs.).

59. If equipped, connect the block heater electrical connector.

60. Position the starter electrical harness support bracket and install the bolt. Tighten to 10 Nm (89 inch lbs.).

61. Position the ground strap and install the stud bolt. Tighten to 10 Nm (89 inch lbs.).

62. Position the A/C compressor and install the 3 bolts. Tighten to 25 Nm (18 ft. lbs.).

63. Install the 4 exhaust manifold flange nuts. Tighten to 40 Nm (30 ft. lbs.).

64. Connect the 2 heated exhaust gas oxygen sensor (HO$_2$S) electrical connectors.

Automatic transmission vehicles

65. Install the transmission cooler tube support bracket and the nut. Tighten to 10 Nm (89 inch lbs.).

66. Install the upper 2 transmission-to-engine bolts. Tighten to 48 Nm (35 ft. lbs.).

All vehicles

67. Position the degas bottle support bracket and install the 3 nuts. Tighten to 10 Nm (89 inch lbs.).

68. Connect the heater hose at the rear of the engine.

69. Connect the body wiring harness retainer to the degas bottle support bracket.

70. Position the evaporative emissions (EVAP) canister purge valve and bracket assembly, and install the 3 nuts. Tighten to 10 Nm (89 inch lbs.).

71. Connect the electrical connector and the EVAP tube quick connect couplings to the EVAP canister purge valve.

72. Connect the engine wiring harness retainer to the EVAP canister purge valve bracket.

73. Connect the crankshaft position (CKP) sensor wiring harness retainer to the starter motor wiring harness.

74. Connect the generator wiring harness retainer to the right cylinder head.

75. Connect the A/C compressor and the A/C high pressure cut-off switch electrical connectors.

76. Connect the generator wiring harness retainer to the right cylinder head.

77. Position the power steering pump and install the 3 bolts. Tighten to 25 Nm (18 ft. lbs.).

78. Connect the 2 electrical connectors on the left fender splash shield.

79. Connect the degas bottle coolant hose.

80. Install the intake manifold.

81. Install the radiator, fan shroud and engine cooling fan.

82. Install the degas bottle.

83. Install the powertrain control module (PCM).

84. Install the cowl vent grille.

85. Install the radiator grille support.

86. Install the A/C condenser core.

87. Install the hood.

88. Fill the crankcase with clean engine oil.

E-Series 4.2L Engine

1. Before servicing the vehicle, refer to the Precautions Section.

2. Disconnect the battery ground cable.

3. Remove the engine cover.

4. Discharge the air conditioning system.

5. Remove the air cleaner assembly and the throttle body inlet tube.

6. Remove the radiator air deflector. Remove the radiator, the fan shroud and the fan.

7. Remove the radiator inner grille.

8. Remove the headlamps.

9. Remove the side marker lamps.

10. Remove the radiator grille opening panel reinforcement.

11. Remove the front end stone deflector.

12. Remove the power steering cooler.

13. Remove the auxiliary transmission cooler.

14. Remove the bolts and the transmission/power steering cooler mounting brackets.

15. Remove the A/C condenser core.

16. Set aside the hood latch.

17. Set aside the battery feed wire wiring harness to the power distribution box.

18. Remove the bolts and set the power steering reservoir aside.

19. Remove the bolts and the upper core support.

20. Set aside the headlamp dash panel junction wiring harness.

21. Remove the bolt and the lower core support.

22. Remove the engine oil level indicator. Remove the bolt and the engine oil level indicator extension tube.

23. Remove the bolt and the engine oil level indicator tube.

24. Remove the lower intake manifold.

25. Remove the accessory drive belt.

26. Set the power steering pump aside.

27. Disconnect the degas bottle supply hose.

28. Disconnect the oil cooler coolant hoses.

29. Remove the lower radiator hose.

30. Set the A/C manifold and tube aside.

31. Disconnect the A/C high-pressure line at the A/C evaporator core. Discard the O-ring.

32. Raise the vehicle.

33. Drain the engine oil.

34. Remove the starter motor.

35. Remove the torque converter access cover and the nuts. Rotate the crankshaft to access the nuts.

36. Remove the bolts on the left side.

37. Remove the bolt on the right side.

38. Remove the nuts retaining the exhaust pipe at the exhaust manifolds.

39. Remove the bolt retaining the shift cable routing bracket.

40. Remove the nuts retaining the engine mount.

41. Lower the vehicle.

42. Install the engine lifting equipment.

43. Remove the transmission dipstick tube.

44. Remove the bolts retaining the transmission to the engine.

45. Remove the engine.

To install:

46. Install the engine.

47. Install the upper bellhousing bolts. Torque to 30–41 ft. lbs. (40–55 Nm).

48. Install the transmission dipstick tube. Use a new O-ring.

49. Remove the engine lifting equipment.

50. Raise the vehicle.

51. Install the mount nuts.

52. Install the shift cable routing bracket retaining bolt. Position the wiring harness and engage the routing clips. Connect the wiring at the daytime running lamp module, if so equipped. Connect the wiring at the horn. Connect the wiring at the front air bag sensor.

53. Install the bellhousing bolts on the right side. Torque to 28–38 ft. lbs. (38–51 Nm).

54. Install the nuts retaining the exhaust pipe at the exhaust manifolds (9430).

55. Install the bellhousing bolts on the left side. Torque to 28–38 ft. lbs. (38–51 Nm).

56. Install the torque converter. Turn the crankshaft to access the nut studs.

57. Install the torque converter access cover.

58. Install the starter motor. Lower the vehicle.

59. Connect the A/C high-pressure line at the evaporator. Discard the O-ring.

60. Install the A/C manifold and tube. Use new O-rings. Torque to 14–18 ft. lbs. (18–24 Nm).

61. Install the lower radiator hose.

62. Connect the oil cooler coolant hoses.

63. Connect the degas bottle supply hose.

64. Install the power steering pump.

65. Install the accessory upper radiator hose.

66. Install the lower intake manifold.

67. Install the intake manifold spacer assembly.

68. Install the oil level indicator extension tube and oil level indicator.

69. Install the bolt retaining the routing bracket and the oil level indicator tube.

70. Install the bolt and the lower core support.

71. Install the headlamp dash panel junction wiring harness.

72. Install the bolts and the upper core support.

73. Position the power steering reservoir and install the bolts.

74. Install the battery feed wire wiring harness to the power distribution box.

75. Install the hood latch.

76. Install the air conditioning condenser.

77. Remove the bolts and the transmission/power steering cooler mounting brackets.

78. Install the auxiliary transmission cooler.

79. Install the power steering cooler.

80. Install the front end stone deflector.

81. Install the radiator grille opening panel reinforcement.

82. Install the side marker lamps.

83. Install the headlamps.

84. Install the radiator inner grille.

85. Install the radiator fan shroud and fan.

86. Install the radiator air deflector.

87. Install the air cleaner assembly and the throttle body inlet tube.

88. Charge the air conditioning system.

89. Install the engine cover.

90. Connect the battery ground cable.

91. Add the correct amount of oil to the engine and forward to the transmission.

E-Series 4.6L and 5.4L engines

1. Before servicing the vehicle, refer to the Precautions Section.

2. Recover the air conditioning system.

3. With the vehicle in **Neutral**, position it on a hoist.

4. Remove the intake manifold.

5. Remove the front bumper.

6. Disconnect the lower radiator hose from the radiator.

7. Remove the pushpin retainers and the splash shield.

8. Disconnect the transmission cooler hoses and drain the fluid into a suitable container.

9. Disconnect the exhaust system from the left and right exhaust manifolds.

10. Drain the engine oil.

11. Disconnect the oil cooler hoses and set aside.

12. Remove the oil filter.

13. Loosen the threaded insert and remove the oil cooler from the oil filter adapter.

14. Remove the starter motor solenoid terminal cover.

15. Disconnect the starter motor electrical connections.

16. Remove the starter motor.

17. Remove the flexplate inspection cover.

18. Remove the cylinder block opening cover.

19. Remove the four nuts retaining the torque converter to the flexplate. Rotate the crankshaft to access all of the torque converter nuts. Discard the torque converter nuts.

Vehicles equipped with 4R70E transmission

20. Disconnect the shift cables and the mounting bracket.

All vehicles

21. Remove the two transmission-to-engine bolts and position the shifter cable support bracket, mounting bracket and cable aside.

22. Remove the right lower transmission-to-engine bolt.

23. Remove the four engine mount nuts.

24. Disconnect the power steering reservoir hose and the power steering pressure tube at the power steering pump.

25. Remove the nut and position the transmission fluid cooler support bracket aside.

26. Remove the valance panel.

27. Disconnect the upper radiator hose and the degas hose from the radiator.

✳✳ CAUTION

The large clutch assembly nut has a right-hand thread and must be rotated counterclockwise to remove it.

28. Remove the fan blade assembly and fan clutch.

29. Disconnect the degas hose retainer from the fan shroud.

30. Remove the fan shroud, the fan, and the fan clutch.

31. Remove the bolts and the radiator support brackets.

32. Remove the radiator from the vehicle.

33. Disconnect the A/C pressure cutoff switch electrical connector.

✲✲ CAUTION

Use a wrench on each side of the fitting to prevent damage to the A/C fitting.

34. Disconnect the compressor suction tube.

✲✲ CAUTION

Use a wrench on each side of the fitting to prevent damage to the A/C fitting.

35. Disconnect the compressor discharge tube.
36. Disconnect the condenser-to-evaporator tube.
37. Remove the A/C condenser core.
38. Set the hood latch aside.
39. Remove the bolts and set the power steering reservoir aside.
40. Remove the nuts and the battery feed cable at the power distribution box.
41. Set aside the battery feed wiring harness to the power distribution box. Disengage the routing clips.
42. Remove the eight bolts and the upper radiator support.
43. Remove the pin-type retainers and the right and left air deflectors.
44. Set aside the left side of the headlamp dash panel junction wiring harness.
45. Set aside the right side of the headlamp dash panel junction wiring harness.
46. Remove the 12 bolts and the lower radiator support.
47. Disconnect the lower radiator hose from the oil filter adapter water inlet and set aside.
48. Rotate the tensioner clockwise to release belt tension and remove the drive belt.
49. Disconnect the alternator electrical connections.
50. Remove the bolts and the alternator.
51. Disconnect the electrical connectors, remove the nut, and remove the ground strap and mounting bracket.
52. Remove the engine oil filler tube support strap bolt.
53. Remove the fluid level indicator and disconnect the attachments at the transmission fluid filler tube.
54. Disconnect the evaporative emission canister purge valve electrical connector.
55. Remove the nut and the transmission fluid filler tube, and allow the fluid to drain into a suitable container.
56. Release the wiring retainers from the heater outlet tube bracket.

57. Disconnect the hose from the heater outlet tube.
58. Remove the heater outlet tube studs.
59. Remove the heater outlet tube.
60. Release the transmission wiring harness connector retainer from the bracket.
61. Disconnect the transmission wiring harness and the left and right heated oxygen sensor (HO2S) electrical connectors.
62. Remove the four upper transmission-to-engine bolts and the fuel line, and the left HO2S brackets.
63. Install a lifting tool.
64. Support the transmission with a jack.
65. Remove the engine from the vehicle.

To install:

66. Position the engine in the vehicle.
67. Remove the floor crane and the jack supporting the transmission.
68. With the vehicle in **Neutral**, position it on a hoist.

➡Align the engine-to-transmission dowels before installing the engine-to-transmission bolts.

69. Position the shifter cable support bracket and install the two transmission-to-engine bolts. Torque to 44 ft. lbs. (60 Nm).
70. Install the right lower transmission-to-engine bolt. Torque to 44 ft. lbs. (60 Nm).

Vehicles equipped with 4R70E transmission

71. Install the mounting bracket and connect the shift cable. Torque to 18 ft. lbs. (25 Nm).

All engines

72. Install four new nuts to retain the torque converter to the flexplate. Rotate the crankshaft to access all of the torque converter nuts. Torque to 26 ft. lbs. (35 Nm).
73. Install the cylinder block opening cover.
74. Install the flexplate inspection cover. Torque to 26 ft. lbs. (35 Nm).
75. Install the four engine mount nuts. Torque to 66 ft. lbs. (90 Nm).
76. Install the starter motor.
77. Connect the starter motor electrical connections.
78. Install the starter motor solenoid terminal cover.
79. Mount the exhaust system to the left and right exhaust manifolds. Torque to 30 ft. lbs. (40 Nm).

➡Make sure the tab on the oil filter adapter nests into the notch in the oil cooler.

80. Position the oil cooler on the oil filter adapter and install the threaded insert.

81. Install a new oil filter.
82. Connect the oil cooler hoses.

Romeo engine (4.6L)

83. Position the transmission fluid cooler tube support bracket and install the nut.

Windsor engine (5.4L)

84. Position the fluid cooler tube support bracket and install the nut.

All engines

85. Using a suitable tool, install a new Teflon® seal on the power steering pressure tube.
86. Connect the power steering pressure tube and the power steering reservoir hose to the power steering pump.
87. Install and tighten the oil drain plug.
88. Position the fuel line and the left heated oxygen sensor (HO2S) brackets. Install the four upper transmission-to-engine bolts. Torque to 44 ft. lbs. (60 Nm).
89. Connect the transmission wiring harness and the left and right HO2S electrical connectors. Install the transmission wiring harness connector retainer in the bracket.

➡Do not reuse the O-ring seals. Lubricate the new O-ring seals with clean engine coolant before installing the heater outlet tube.

90. Insert the heater outlet tube over the new seals.

Romeo engines (4.6L)

91. Install the heater outlet tube studs. Torque to 18 ft. lbs. (25 Nm).

➡Lubricate the O-ring seals with clean transmission fluid before installing the transmission fluid filler tube.

92. Install the transmission fluid filler tube.

Windsor engine (5.4L)

93. Install the heater outlet tube studs.

➡Lubricate the O-ring seals with clean transmission fluid before installing the transmission fluid filler tube.

94. Install the transmission fluid filler tube.

All engines

95. Connect the hose to the heater outlet tube.
96. Install the wiring retainers in the heater outlet tube bracket.
97. Connect the evaporative emission canister purge valve electrical connector.
98. Connect the attachments to the transmission fluid filler tube and install the fluid level indicator.
99. Install the engine oil filler tube support strap bolt.
100. Position the mounting bracket,

install the ground strap and nut, and connect the electrical connectors.

101. Position the alternator and install the bolts.

102. Connect the alternator electrical connections.

103. Rotate the tensioner clockwise and install the drive belt. Refer to the decal on the upper radiator air deflector for belt routing.

104. Connect the lower radiator hose to the oil filter adapter coolant inlet.

105. Position the lower radiator support and install the twelve bolts.

106. Install the right side of the headlamp dash panel junction wiring harness.

107. Install the left side of the headlamp dash panel junction wiring harness.

108. Position the right and left air deflectors and install the pin-type retainers.

109. Position the upper radiator support and install the eight bolts.

110. Route the battery feed wiring harness to the power distribution box and insert the routing clips.

111. Install the battery feed cable to the power distribution box.

112. Install the power steering reservoir.

113. Install the hood latch.

114. Install the A/C condenser core.

✷✷ CAUTION

Use a wrench on each side of the fitting to prevent damage to the A/C fitting.

115. Connect the condenser core refrigerant tubes. Torque the condenser-to-evaporator fitting to 13 ft. lbs. (18 Nm); the compressor discharge tube to 29 ft. lbs. (39 Nm).

✷✷ CAUTION

Use a wrench on each side of the fitting to prevent damage to the A/C fitting.

116. Connect the compressor suction tube. Torque to 35 ft. lbs. (47 Nm).

117. Connect the A/C pressure cutoff switch electrical connector.

118. Position the radiator in the vehicle.

119. Position the radiator support brackets and install the bolts.

120. Position the fan shroud and the fan and fan clutch in the vehicle, and install the bolts.

121. Insert the degas hose retainer in the fan shroud.

122. Install the fan blade assembly and fan clutch.

123. Connect the upper radiator hose and the degas hose to the radiator.

124. Install the valance panel.

125. Connect the transmission cooler tubes.

126. Position the splash shield and install the pushpin retainers.

127. Connect the lower radiator hose to the radiator.

128. Install the front bumper.

✷✷ CAUTION

The oil pump must be primed prior to starting the engine.

129. Fill the engine with clean engine oil.

130. Install the intake manifold.

131. Fill all fluids to the correct levels.

132. Start the engine and check for leaks. Stop the engine and recheck the fluid levels.

133. Evacuate and recharge the air conditioning system.

E-Series 6.8L Engine

1. Before servicing the vehicle, refer to the Precautions Section.

2. Discharge and recover the air conditioning system.

3. Disconnect the 42-pin and the two 16-pin connectors and remove the nut and the harness mounting bracket.

4. Remove the intake manifold.

5. Remove the radiator right and left air deflector, the radiator grille support, the valance panel and the upper and lower radiator supports.

6. Remove the radiator, fan shroud and engine cooling fan.

7. Remove the headlamps and side marker lamps.

8. Remove the A/C condenser core.

9. Remove the drive belt.

10. Disconnect the power steering reservoir hose at the power steering pump and let drain into a drain pan.

11. Disconnect the power steering pressure hose.

12. At the oil cooler water inlet, disconnect and set aside the lower radiator hose.

13. Disconnect the suction hose at the accumulator.

14. Disconnect the transmission harness and the left and right heated exhaust gas oxygen sensor connectors.

15. Remove the upper transmission-to-engine bolts and the fuel tube bracket.

16. Raise and support the vehicle.

17. Remove the nuts.

18. Drain the engine oil and remove the oil filter.

19. Disconnect the oil cooler hoses and position aside.

20. Loosen the threaded shaft and remove the oil cooler from the oil filter adapter.

21. Remove the starter motor.

22. Remove the bolt from the A/C compressor manifold and position the manifold aside.

23. Remove the bolts and position the shift cable and bracket aside.

24. Remove and discard the six nuts retaining the torque converter to the flexplate.

25. Remove the engine support insulator nuts.

26. Remove the remaining transmission to engine bolts.

27. Lower the vehicle.

28. Install the Modular Engine Lift Bracket.

29. Support the transmission with a jack.

30. Remove the engine from the vehicle.

To install:

31. Position the engine in the vehicle and remove the Modular Engine Lift Bracket and the jack supporting the transmission.

32. Raise and support the vehicle.

33. Install and tighten the nuts.

34. Install the fuel tube line bracket and the remaining transmission to engine bolts.

35. Install and tighten the six new nuts retaining the torque converter.

36. Install the starter motor.

37. Position the A/C compressor manifold assembly and install the bolt.

38. Install and tighten the nuts.

39. Connect the power steering pressure hose.

40. Install the oil cooler to the oil filter adapter and install the oil filter.

41. Position the oil cooler hoses and install the hose clamps.

42. Connect the lower radiator hose.

43. Lower the vehicle.

44. Connect the transmission wiring harness and the left and right heated exhaust gas oxygen sensor connectors.

45. Install the intake manifold.

46. Install the suction line to the accumulator.

47. Install the power steering reservoir hose at the power steering pump.

48. Connect the 16 and 42-pin connectors.

49. Install the drive belt.

50. Install the upper and lower radiator supports, valance panel, radiator grille support and the radiator right and left air deflectors.

51. Install the A/C condenser core.

52. Install the radiator.

53. Install the side marker lamps and headlamps.
54. Install the engine air cleaner assembly and tubes.
55. Fill all fluids to the correct levels.
56. Connect the battery ground cable.
57. Start the engine and check for leaks. Stop the engine and recheck the fluid levels.
58. Evacuate and recharge the A/C system.

Blackwood

1. Before servicing the vehicle, refer to the Precautions Section.
2. Remove the air cleaner and outlet tube.
3. Remove the left side push pin retainers from the hood inner liner.
4. Remove the hood light wiring harness from the hood.
5. Disconnect the windshield washer supply hose from the hood.
6. Support the hood and disconnect the hood support struts.
7. With an assistant's help, remove the hood mounting bolts and the hood.
8. Discharge the air conditioning (A/C) system.
9. Remove the lower intake manifold.
10. Remove the radiator.
11. Remove the generator.
12. Disconnect the exhaust gas recirculation (EGR) tube from the exhaust manifold.
13. Disconnect the transmission wiring harness from the rear of the right cylinder head.
14. Remove the bolts and separate the wiring harness bulkhead connector bracket.
15. Disconnect the bulkhead connectors.
16. Remove the ground strap bolt.
17. Disconnect the heater hose.
18. Remove the bolts retaining the transmission to the engine.
19. Disconnect the degas return hose.
20. Remove the nut and position the A/C muffler aside.
21. Disconnect the lower radiator hose.
22. Raise and support the vehicle.
23. Disconnect the power steering pressure (PSP) sensor.

➡**The power steering pump can be supported with mechanic's wire to keep it from hindering the removal process.**

24. Remove the bolts and position the power steering pump out of the way.
25. Remove the nut and the transmission cooler lines from the front cover.
26. Remove the starter.
27. Disconnect the crankshaft position (CKP) sensor.

28. Remove the A/C compressor.
29. Disconnect the right and the left heated oxygen sensor (HO2S).
30. Disconnect the pipes at the exhaust manifold.
31. Remove the right side exhaust manifold studs.
32. Remove the bolt from the left motor mount.
33. Remove the bolt from the right motor mount.
34. Remove the cylinder block opening cover.

➡**Gain access to the nuts by turning the crank dampener counterclockwise.**

35. Remove the four nuts from the torque converter.
36. Remove the oil bypass filter.
37. Remove the five bolts from the transmission.
38. Remove the bolt and position the transmission fluid level indicator tube out of the way.
39. Lower the vehicle.
40. Install the engine lifting bracket.
41. Support the transmission with a floor jack and a block of wood.
42. Remove the engine from the vehicle.
To install:
43. Support the transmission with a floor jack and a block of wood.
44. Install the engine in the vehicle.
45. Remove the engine lifting bracket.
46. Raise the vehicle.
47. Position the transmission level indicator tube and install the bolt.
48. Install the five bell housing lower bolts. Torque to 33 ft. lbs. (45 Nm).
49. Install the oil filter.

➡**Gain access to the nuts by turning the crank dampener counterclockwise.**

50. Install the four nuts to the torque converter. Torque to 18 ft. lbs. (25 Nm).
51. Install the cylinder block opening cover.
52. Install the bolt to the right engine motor mount. Torque to 59 ft. lbs. (80 Nm).
53. Install the bolt to the left engine motor mount. Torque to 59 ft. lbs. (80 Nm).
54. Install the right side exhaust manifold studs.
55. Connect the Y-pipe at the exhaust manifold. Torque to 30 ft. lbs. (41 Nm).
56. Connect the right and the left heated oxygen sensor (HO2S) electrical connectors.
57. Install the A/C compressor.
58. Connect the crankshaft position sensor electrical connector.
59. Install the starter.
60. Install the transmission cooler lines.

61. Position the power steering pump and install the bolts. Torque to 18 ft. lbs. (25 Nm).
62. Connect the power steering pressure sensor electrical connector.
63. Install the transmission bracket bolt.
64. Lower the vehicle.
65. Connect the lower radiator hose.
66. Position the A/C muffler and install the nut.
67. Connect the degas return hose.
68. Install the ground strap bolt.
69. Install the transmission bell housing upper bolts.
70. Connect the heater hose.
71. Connect the bulkhead connectors.
72. Install the bulkhead connector bracket bolts.
73. Install the transmission wiring harness to the rear of the right cylinder head.
74. Connect the EGR tube to the exhaust manifold. Position EGR tube and valve assembly and loosely install.
75. Install the generator. Torque to 18 ft. lbs. (25 Nm).
76. Install the lower intake manifold.
77. Install the radiator.
78. Tighten the EGR tube to 18 ft. lbs. (25 Nm).
79. Recharge the A/C system.
80. Install the air cleaner and outlet tube.
81. Fill the engine with oil.
82. Fill the cooling system.
83. With an assistant's help, install the hood. Torque to 19–25 ft. lbs. (26–34 Nm).
84. Connect the hood support struts.
85. Connect the windshield washer supply hose.
86. Install the hood light wiring harness.
87. Install the hood liner push pin retainers.

Mark LT

1. Before servicing the vehicle, refer to the Precautions Section.
2. With the vehicle in NEUTRAL, position it on a hoist.
3. Remove the hood.
4. Remove the intake manifold.
5. Remove the accessory drive belt.
6. Recover the A/C system.
7. Remove the radiator.
8. Remove the powertrain control module (PCM) and the support bracket.
9. Remove the cowl panel grille.
10. Detach the wiring harness and the windshield washer hose retainers from the cowl panel extension.
11. Remove the bolts and the cowl extension panel.

12. Disconnect the electrical connector and remove the bolt and the ground strap.

13. Disconnect the heater hose.

14. Remove the nut, disconnect the A/C manifold and tube assembly and position aside.

15. Remove the nut and the A/C manifold and tube assembly and support bracket.

16. Disconnect the coolant hose.

17. Remove the bolt and position the power steering reservoir assembly aside.

18. Disconnect the generator wiring harness retainer from the right cylinder head.

19. Disconnect the A/C compressor and the A/C high pressure cut-off switch electrical connectors.

20. Disconnect the generator wiring harness retainer from the right cylinder head and position the harness aside.

21. Disconnect the crankshaft position (CKP) sensor wiring harness retainer from the starter motor wiring harness.

22. Remove the nut and position aside the power steering pressure hose support bracket.

23. Using the special tool, remove the power steering pump pulley.

24. Disconnect the power steering pressure tube. Drain the power steering fluid into a suitable container.

25. Remove the 3 bolts and position the power steering pump aside.

26. Remove the nut and position aside the transmission cooler tube support bracket.

27. Remove the starter.

28. Remove the 2 bolts and the flexplate inspection cover.

29. Remove the cylinder block opening cover.

30. Remove the 4 torque converter-to-flexplate nuts. Discard the nuts.

31. Disconnect the shift cable and remove the shift cable bracket.

➡**The upper 2 transmission-to-engine bolts will be removed later.**

32. Remove the lower 5 transmission-to-engine bolts.

33. Remove the bolt and position the transmission fluid filler tube aside.

34. Remove the drain plug and drain the engine oil.

35. Disconnect the right heated oxygen sensor (HO2S) electrical connector and detach the wiring harness retainer.

36. Disconnect the left heated exhaust gas oxygen sensor electrical connector and detach the electrical connector retainer.

37. Remove the oil filter.

38. Disconnect the oil temperature sensor electrical connector.

39. Remove the bolts and position the A/C compressor aside.

40. If equipped, disconnect the block heater electrical connector.

41. Detach the starter electrical harness support from the cylinder block.

42. Remove the 4 exhaust manifold flange nuts.

43. Remove the right engine support insulator nuts and washer.

44. Remove the left engine support insulator bolt.

45. Support the transmission.

➡**On 4WD vehicles, it may be necessary to reposition the transfer case vent hose to access the bolts.**

46. Remove the upper 2 transmission-to-engine bolts.

47. Install the lifting bracket.

48. Using a suitable floor crane, remove the engine assembly from the vehicle.

To install:

49. Using a suitable floor crane, position the engine assembly into the vehicle.

❋❋ WARNING

Only use hand tools when installing the left engine mount bolt or damage to the engine mount can occur.

➡**Early build vehicles are equipped with an M14 through bolt. Late build vehicles are equipped with an M18 through bolt.**

50. Apply high strength threadlocker to the bolt threads and install the left engine support insulator bolt.

   a. If the M14 through bolt engine support insulator nut is missing or damaged, install a new nut using service part number W709375.

   b. If the M14 through bolt engine support insulator nut cage is damaged or missing, install a new nut using service part number W520516-S301.

   c. Early build vehicles, tighten to 220 Nm (162 ft. lbs.).

   d. Late build vehicles, tighten to 350 Nm (258 ft. lbs.).

❋❋ WARNING

Only use hand tools when installing the right engine mount nuts or damage to the engine mount can occur.

51. Apply high strength threadlocker to the stud threads and install the right engine support insulator washer and nuts. Tighten to 250 Nm (184 ft. lbs.).

52. Remove the lifting bracket.

➡**The upper 2 transmission-to-engine bolts will be installed later.**

53. Install the lower 5 transmission-to-engine bolts. Tighten to 60 Nm (44 ft. lbs.).

54. Position the shift cable bracket and install the bolts and connect the shift cable. Tighten to 25 Nm (18 ft. lbs.).

55. Install 4 new torque converter-to-flexplate nuts. Tighten to 36 Nm (27 ft. lbs.).

56. Install the cylinder block opening cover.

57. Install the flexplate inspection cover and the 2 bolts. Tighten to 34 Nm (25 ft. lbs.).

58. Install the starter.

59. Position the transmission cooler tube support bracket and install the bolt. Tighten to 10 Nm (89 inch lbs.).

60. If equipped, connect the block heater electrical connector.

61. Connect the left and right heated exhaust gas oxygen sensor (HO2S) electrical connector and attach the electrical connector retainer.

62. Attach the starter electrical harness support bracket.

63. Connect the oil temperature sensor electrical connector.

64. Position the A/C compressor and install the 3 bolts. Tighten to 25 Nm (18 ft. lbs.).

65. Install the 4 exhaust manifold flange nuts. Tighten to 40 Nm (30 ft. lbs.).

66. Install the upper 2 transmission-to-engine bolts. Tighten to 60 Nm (44 ft. lbs.).

67. Position the transmission filler tube and install the bolt. Tighten to 20 Nm (15 ft. lbs.).

68. Connect the crankshaft position (CKP) sensor wiring harness retainer to the starter motor wiring harness.

69. Connect the generator wiring harness retainer to the right cylinder head.

70. Connect the A/C compressor and the A/C high pressure cut-off switch electrical connectors.

71. Connect the generator wiring harness retainer to the right cylinder head.

72. Position the power steering pump assembly and install the 3 bolts. Tighten to 25 Nm (18 ft. lbs.).

73. Using special tool 211-D207, install a new O-ring seal on the pressure line fitting.

74. Connect the power steering pressure tube. Tighten to 65 Nm (48 ft. lbs.).

75. Install the power steering pump pulley.

76. Position the power steering pressure hose support bracket and install the nut. Tighten to 10 Nm (89 inch lbs.).

77. Position the power steering reservoir assembly and install the bolt. Tighten to 23 Nm (17 ft. lbs.).

78. Using a new O-ring seal, connect the A/C manifold and tube assembly and install the nut. Tighten to 25 Nm (18 ft. lbs.).

79. Position the A/C manifold and tube assembly support bracket and install the nut. Tighten to 25 Nm (18 ft. lbs.).

80. Connect the heater hose.

81. Disconnect the coolant hose.

82. Connect the electrical connector and position the ground strap and install the bolt. Tighten to 10 Nm (89 inch lbs.).

83. Position the cowl extension panel and install the bolts.

84. Attach the wiring harness and the windshield washer hose retainers to the cowl panel extension.

85. Install the intake manifold.

86. Install the radiator.

87. Install the powertrain control module (PCM).

88. Install the cowl.

89. Install the hood.

90. Fill the crankcase with clean engine oil.

91. Evacuate and charge the A/C system.

92. Fill and bleed the power steering system.

Water Pump

REMOVAL & INSTALLATION

Blackwood, Mark LT and F-150 4.6L and 5.4L Engines, exc. Lightning Supercharger Cooling

1. Before servicing the vehicle, refer to the Precautions Section.

➡ **The fan clutch and shroud must be removed as an assembly.**

2. Drain the engine cooling system.

3. Remove the air cleaner outlet pipe.

4. Remove the jack handle. Remove the screws. Remove the front air deflector.

5. Remove the upper radiator hose from the radiator.

6. Remove the fan blade and fan clutch. Use Fan Clutch Holding Tool to hold the water pump pulley steady. Use Fan Clutch Wrench to remove the fan blade and the fan clutch.

7. Remove the screws. Lift and remove the fan shroud and fan blade out of the vehicle together.

8. Remove the bolts. Separate the fan blade from the fan clutch.

25 Nm (18 lb-ft)

06017-F150-G15

Water pump pulley loosening/tightening sequence—4.6L, 5.4L and 6.8L engines

9. Remove the drive belt.

10. Remove the bolts in the sequence shown, and the water pump pulley.

11. Remove the bolts in the sequence shown, and the water pump.

12. Wipe the water pump mounting surface with a soft cloth if necessary.

✳✳ WARNING

Do not rotate the water pump housing once installed in the engine. Damage to the O-ring seal can occur, causing the water pump to leak.

➡ **Install a new O-ring seal and lubricate with Motorcraft Premium Engine Coolant E2FZ-19549-AA, F5FZ-19549-CC in Oregon, CXC-10 in Canada or equivalent meeting Ford specification ESE-M97B44-A (green color) or Motorcraft Premium Gold Engine Coolant VC-7-A meeting Ford specification WSS-M97B51-A1 (yellow color). Use the same coolant that was drained from the engine.**

To install:

13. To install the water pump, reverse the removal procedure. Torque the water pump bolts and the pulley bolts, in the sequence shown, to 18 ft. lbs. (25 Nm).

14. Position the fan clutch on the fan blade. Install the bolts. Torque to 13 ft. lbs. (17 Nm).

15. Position the fan shroud and the fan blade.

16. Lower the fan shroud onto the radiator clips.

17. Install the screws.

18. Position the fan blade assembly and the fan clutch on the water pump pulley.

19. Use Fan Clutch Holding Tool to hold the water pump pulley steady.

20. Use Fan Clutch Wrench to install the fan blade and fan clutch. Torque to 41 ft. lbs. (55 Nm).

21. Connect the upper radiator hose to the radiator.

25 Nm (18 lb-ft)

06017-F150-G16

Water pump bolt loosening/tightening sequence—4.6L, 5.4L and 6.8L engines

17 Nm (13 lb/ft)

Fan hub and blade assembly—4.6L, 5.4L and 6.8L engines

06017-F150-G17

55 Nm (41 lb/ft)

Tightening the fan hub—4.6L, 5.4L and 6.8L engines

06017-F150-G18

22. Install the front air deflector. Install the screws. Install the jack handle.

23. Install the air cleaner outlet pipe.

24. Fill the engine cooling system.

2002–05 E-Series and 2002–04 F-Series 4.2L Engine

1. Before servicing the vehicle, refer to the Precautions Section.

2. Drain the cooling system.

➡The fan, fan clutch and shroud must be removed and installed together due to insufficient clearance to remove them separately.

3. Drain the cooling system.

4. Remove the air cleaner assembly.

5. Disconnect the radiator upper hose from the radiator. Position the radiator hose aside.

6. Disconnect the overflow hose from the radiator.

7. Disconnect the coolant degas bottle hose from the shroud.

✳✳ WARNING

The clutch assembly nut has a right-hand thread and must be rotated counterclockwise to remove it.

8. Using the special tools, loosen the fan clutch.

9. Carefully rotate the fan and fan clutch assembly counterclockwise until the assembly is free from the water pump. Place the fan and fan clutch into the shroud opening.

10. If equipped, disconnect the retaining clips from the fan shroud.

11. Raise and support the vehicle.

12. Disconnect the lower radiator hose retaining clamp from the shroud.

13. If equipped, disconnect the underbody splash shield from the shroud.

14. Lower the vehicle.

✳✳ WARNING

Use extreme care when removing or installing the fan, fan clutch and shroud. Failure to do so will cause damage to the radiator.

15. Remove the mounting bolts, and the fan, fan clutch and shroud.

16. Remove the accessory drive belt.

17. Disconnect the engine harness from the water pump mounting stud and the water bypass tube.

18. Remove the A/C support bracket bolts and rotate the bracket out of the way.

19. Remove the bolts and the pulley from the water pump.

20. Remove the screw and the bypass tube from the water pump.

21. Raise the vehicle.

22. Remove the underbody splash shield.

23. Remove the lower radiator hose from the water pump.

24. Lower the vehicle.

25. Remove the bolts and the stud bolts from the water pump. Remove the water pump from the front cover.

26. Clean the gasket surfaces.

➡Install a new O-ring seal and lubricate with Motorcraft Premium Engine Coolant E2FZ-19549-AA, F5FZ-19549-CC in Oregon, CXC-10 in Canada or equivalent meeting Ford specification ESE-M97B44-A (green color) or Motorcraft Premium Gold Engine Coolant VC-7-A meeting Ford specification WSS-M97B51-A1 (yellow color). Use the same coolant that was drained from the engine.

27. To install, reverse the removal procedure. Tighten the water pump stud bolts, bolts, and nuts in the sequence shown to 18 ft. lbs. (25 Nm). Tighten the A/C support bracket bolts to 30–41 ft. lbs. (40–55 Nm) and the stud to 15–22 ft. lbs. (20–30 Nm). Tighten the water pump pulley bolts to 18 ft. lbs. (25 Nm). Torque the fan shroud fasteners to 53 inch lbs. (6 Nm). Torque the fan clutch to 41 ft. lbs. (55 Nm). Install an appropriately sized worm drive-type clamp in place of the constant tension clamp.

2005 F-150 4.2L Engine

1. Before servicing the vehicle, refer to the Precautions Section.

2. Drain the cooling system.

3. Remove the air cleaner outlet pipe.

25 Nm (18 lb-ft)

Water pump fastener torque sequence—2002–05 E-Series and 2002–04 F-Series 4.2L engine

06017-F150-G19

40-55 Nm (30-41 lb/ft)

20-30 Nm (15-22 lb/ft)

06017-F150-G20

A/C bracket—2002–05 E-Series and 2002–04 F-Series 4.2L engine

4. Detach the battery cable harness from the radiator shroud.

5. Using a hub wrench set, remove the cooling fan assembly and position it in the fan shroud.

6. Remove the bolts and the fan shroud and fan.

7. If necessary, remove the bolts and separate the fan and the fan clutch.

8. Remove the drive belt.

9. Remove the bolts and the coolant pump pulley.

10. Remove the bolt and disconnect the heater outlet tube assembly from the coolant pump. Clean and inspect the O-ring seal. Install a new O-ring seal if necessary.

11. Remove the A/C compressor mounting bracket support bolts.

12. Remove the nuts and the A/C compressor mounting bracket support.

13. Release the clamp and disconnect the lower radiator hose.

14. Remove the fasteners and the coolant pump.

15. Remove and discard the coolant pump gasket.

�֎֎ WARNING

Do not use metal scrapers, wire brushes, power abrasive discs, or other abrasive means to clean the sealing surfaces. These may cause scratches and gouges resulting in leak paths. Use a plastic scraper to clean the sealing surfaces.

16. Clean the coolant pump gasket mating surfaces with a plastic scraper and metal surface prep. Follow the directions on the packaging.

17. To install, reverse the removal procedure. Lubricate the O-ring seal with clean engine coolant. Observe the following torques:

- Water pump: 21 ft. lbs. (28 Nm)
- A/C compressor support bracket nuts: 21 ft. lbs. (28 Nm)
- A/C compressor support bracket bolts: 35 ft. lbs. (48 Nm)
- Heater outlet tube: 89 inch lbs. (10 Nm)
- Water pump pulley: 89 inch lbs. (10 Nm)
- Fan-to-fan clutch: 13 ft. lbs. (17 Nm)
- Fan shroud: 53 inch lbs. (6 Nm)
- Fan hub: 41 ft. lbs. (55 Nm)

E-Series 4.6L and 5.4L Engines

1. Before servicing the vehicle, refer to the Precautions Section.

6 Nm (53 lb-in) ②

① 55 Nm (41 lb-ft)

③

1 Cooling fan assembly
2 Cooling fan shroud bolts (2 required)
3 Cooling fan shroud

06017-F150-G21

Fan shroud and related parts—F-150 4.2L, 4.6L and 5.4L engines

4   Fan clutch bolts (4 required)

5   Cooling fan

6   Fan clutch

06017-F150-G22

Fan clutch assembly—F-150 4.2L, 4.6L and 5.4L engines

➡The fan, fan clutch and shroud must be removed and installed together due to insufficient clearance to remove them separately.

2. Drain the cooling system.

3. Remove the air cleaner assembly.

4. Disconnect the radiator upper hose from the radiator. Position the radiator hose aside.

5. Disconnect the overflow hose from the radiator.

6. Disconnect the coolant degas bottle hose from the shroud.

✱✱ WARNING

The clutch assembly nut has a right-hand thread and must be rotated counterclockwise to remove it.

1 Coolant pump pulley bolts (4 required)	7 A/C compressor mounting bracket support nuts (2 required)
2 Coolant pump pulley	8 A/C compressor mounting bracket support
3 Heater outlet tube assembly bolt	9 Lower radiator hose clamp
4 Heater outlet tube assembly	10 Lower radiator hose
5 Heater outlet tube assembly O-ring seal	
6 A/C compressor mounting bracket support bolts (2 required)	

06017-F150-G23

Remove the parts before water pump removal—2005 F-150 4.2L engine

11 Coolant pump bolts (3 required)

12 Coolant pump nuts (4 required)

13 Coolant pump bolt

14 Coolant pump stud bolt

15 Coolant pump

16 Coolant pump gasket

06017-F150-G24

Water pump removal/installation—2005 F-150 4.2L engine

7. Using the special tools, loosen the fan clutch.

8. Carefully rotate the fan and fan clutch assembly counterclockwise until the assembly is free from the water pump. Place the fan and fan clutch into the shroud opening.

9. If equipped, disconnect the retaining clips from the fan shroud.

10. Raise and support the vehicle.

11. Disconnect the lower radiator hose retaining clamp from the shroud.

12. If equipped, disconnect the underbody splash shield from the shroud.

13. Lower the vehicle.

❋❋ WARNING

Use extreme care when removing or installing the fan, fan clutch and shroud. Failure to do so will cause damage to the radiator.

14. Remove the mounting bolts, and the fan, fan clutch and shroud.

15. Remove the accessory drive belt.

16. Remove the bolts. Remove the coolant pump pulley.

17. Remove the bolts. Remove the coolant pump from the cylinder block.

To install:

❋❋ WARNING

Always fill the cooling system with the same coolant that is present in the system. Do not mix coolant types.

18. Install a new O-ring on the coolant pump. Coat the new O-ring with engine coolant.

❋❋ WARNING

Do not rotate the coolant pump housing once the coolant pump has been positioned in the cylinder block. Damage to the O-ring seal will occur.

19. Lubricate the new O-ring seal using engine coolant and install the O-ring seal onto the coolant pump.

20. Position the coolant pump into the engine block. Install the bolts. Torque to 18 ft. lbs. (25 Nm).

21. Position the pulley onto the coolant pump and install the bolts.

22. Install the accessory drive belt.

23. Install the fan and fan clutch.

24. To install, reverse the removal proce-

dure. Torque the fan shroud fasteners to 53 inch lbs. (6 Nm). Torque the fan clutch to 41 ft. lbs. (55 Nm). Install an appropriately sized worm drive-type clamp in place of the constant tension clamp.

E-Series 6.8L Engine

➡**The fan, fan clutch and shroud must be removed and installed together due to insufficient clearance to remove them separately.**

1. Drain the cooling system.

2. Remove the air cleaner assembly.

3. Disconnect the radiator upper hose from the radiator. Position the radiator hose aside.

4. Disconnect the overflow hose from the radiator.

5. Disconnect the coolant degas bottle hose from the shroud.

❋❋ WARNING

The clutch assembly nut has a right-hand thread and must be rotated counterclockwise to remove it.

6. Using the special tools, loosen the fan clutch.

7. Carefully rotate the fan and fan clutch assembly counterclockwise until the assembly is free from the water pump. Place the fan and fan clutch into the shroud opening.

8. If equipped, disconnect the retaining clips from the fan shroud.

9. Raise and support the vehicle.

10. Disconnect the lower radiator hose retaining clamp from the shroud.

11. If equipped, disconnect the underbody splash shield from the shroud.

12. Lower the vehicle.

✳✳ WARNING

Use extreme care when removing or installing the fan, fan clutch and shroud. Failure to do so will cause damage to the radiator.

13. Remove the mounting bolts, and the fan, fan clutch and shroud.

14. Remove the accessory drive belt.

15. Remove the bolts. Remove the coolant pump pulley.

16. Remove the bolts. Remove the coolant pump from the cylinder block.

To install:

✳✳ WARNING

Always fill the cooling system with the same coolant that is present in the system. Do not mix coolant types.

17. Install a new O-ring on the coolant pump. Coat the new O-ring with engine coolant.

✳✳ WARNING

Do not rotate the coolant pump housing once the coolant pump has been positioned in the cylinder block. Damage to the O-ring seal will occur.

18. Lubricate the new O-ring seal using engine coolant and install the O-ring seal onto the coolant pump.

19. Position the coolant pump into the engine block. Install the bolts. Torque to 18 ft. lbs. (25 Nm).

20. Position the pulley onto the coolant pump and install the bolts.

21. Install the accessory drive belt.

22. Install the fan and fan clutch.

23. To install, reverse the removal procedure. Torque the fan shroud fasteners to 53 inch lbs. (6 Nm). Torque the fan clutch to 98 ft. lbs. (133 Nm). Install an appropriately sized worm drive-type clamp in place of the constant tension clamp.

F-250/350 5.4L and 6.8L Engines

1. Before servicing the vehicle, refer to the Precautions Section.

2. Drain the cooling system.

3. Remove the air cleaner outlet tube

4. Disconnect the upper radiator hose and the degas bottle hose.

✳✳ WARNING

The large clutch assembly nut has a right-hand thread and must be rotated counterclockwise to remove it.

5. Using a hub wrench set, remove the fan and fan clutch from the coolant pump pulley.

6. Carefully position the fan and the fan clutch into the shroud.

7. Remove the 2 cooling fan shroud bolts, the shroud, fan and clutch.

8. Remove the accessory drive belt.

9. Remove the 4 bolts and the coolant pump pulley.

10. Remove the 4 bolts and the coolant pump. Discard the O-ring seal.

11. Installation is the reverse of removal. Install a new O-ring seal. Lubricate the O-ring seal with clean engine coolant prior to installation. Clean the sealing surfaces with metal surface prep. Observe the following torques:

- Water pump: 18 ft. lbs. (25 Nm)
- Water pump pulley: 18 ft. lbs. (25 Nm)
- Fan shroud: 53 inch lbs. (6 Nm)
- Fan hub nut: 98 ft. lbs. (133 Nm)

5.4L Lightning Supercharger Cooling

1. Before servicing the vehicle, refer to the Precautions Section.

2. Disconnect the battery ground cable.

3. Drain the coolant.

4. Remove the retainer and position the inner fender well aside.

10 Nm (89 lb-in)

67197-EFSE-G95

Supercharger cooling pump installation

5. Disconnect the water pump electrical connector.

6. Disconnect the coolant hoses.

7. Remove the bolts and the water pump.

8. To install, reverse the removal procedure. Torque the pump bolts to 89 inch lbs. (10 Nm).

Heater Core

REMOVAL & INSTALLATION

2002 E-Series Vans

1. Before servicing the vehicle, refer to the Precautions Section.

2. Drain the cooling system into a clean container for reuse.

3. Disconnect the battery ground cable.

4. Disconnect the quick disconnect heater hose couplings at the heater core by performing the following procedure:

✳✳ WARNING

The engine must be off, fully cool and the cooling system fully depressurized before attempting to disconnect any heater water hoses. Failure to comply with this warning can result in serious injury or burns from hot liquid escaping out of the engine cooling system.

a. Depressurize the engine cooling system.

b. Push the heater water hose toward the tube to fully expose the locking tabs.

➡ **When compressing the white coupling retainer, the quick disconnect tool must be perpendicular to and on the highest point of the coupling.**

c. Push the quick disconnect tool over the coupling retainer windows to compress the retainer locking tabs.

d. A slight twisting motion while pulling on the heater water hose may be necessary to assist in the removal.

e. Pull the heater water hose away from the heater core tube.

f. Plug the heater water hose.

g. Remove the white coupling retainer from the tube.

h. Spread the retainer tabs apart and slide the retainer off the tube.

i. Discard the retainer.

5. Plug the heater water hoses with a suitable ⅝ or ¾ inch plug.

6. Remove the engine cover.

7. Remove the instrument panel finish

panel by performing the following procedure:

> ※ **CAUTION**

Electronic modules are sensitive to static electrical charges. If exposed to these charges, damage may result.

a. Remove the driver's side air bag module by removing or disconnecting the following items:

➡ **To deplete the backup power supply energy, disconnect the battery ground cable and wait at least 1 minute. Be sure to disconnect auxiliary batteries and power supplies (if equipped).**

- Make sure the wheels are in the straight-ahead position.
- Driver's air bag module.
- Both back cover plugs.
- Both driver's air bag module screws.
- Horn electrical connector.
- Air bag sliding contact electrical connector.
- Driver's air bag module.

b. Remove the passenger's side air bag modules by removing or disconnecting the following items:

- Instrument panel finish panel and reinforcement.
- Passenger's air bag module retaining nuts.

➡ **Using a ⅜ inch x 4 inch slotted screwdriver, carefully slide the head of the screwdriver under the right bottom edge of the door and lift upward, separating the door from the clip. Separate the rest of the door from the clips by lifting the door with your hands.**

- Passenger's air bag module electrical connector and module.
- The 6 offset fasteners used for retaining the air bag module door from the instrument panel.

c. Remove the steering column.
d. Remove the engine cover.
e. Unfasten the engine cover latches and remove the engine cover.
f. Remove the instrument panel cowl top screws.
g. Remove the upper center instrument panel access panel.
h. Disconnect the climate control vacuum harness connector.
i. Remove the left-hand upper access panel.
j. Remove the instrument panel cowl top nut.

k. Remove the alternative fuel control module (AFCM).
l. Remove the instrument panel dash brace.
m. Remove the 6 dash brace bolts.
n. Remove the instrument panel dash brace.
o. From under the hood, position the coolant reservoir aside.
p. Remove the bolts.
q. Position the coolant reservoir aside.
r. Remove the main wiring assembly/lower bulkhead connector by performing the following:

- Loosen the bulkhead connector bolt.
- Disconnect the bulkhead connector clips.
- Remove the bulkhead connector.

s. Remove the right-hand A-pillar lower trim panel.
t. Disconnect the electronic crash sensor (ECS) module electrical connector by performing the following:

- Remove the locking clip.
- Disconnect the connector.

u. Disconnect the right-hand instrument panel cowl side wiring electrical connectors.
v. Disconnect the antenna cable from the radio and release the cable locators from the instrument panel.
w. Remove the right-hand instrument panel bolts.
x. Remove the left-hand instrument panel bolts.

➡ **Removing the instrument panel from the vehicle requires 2 technicians.**

y. Remove the instrument panel from the vehicle through the passenger side door.
8. Remove the heater core cover.
9. Remove the 7 screws.
10. Remove the heater core cover.
11. Remove and discard the heater core case seal.

➡ **Use care not to spill the coolant remaining in the heater core during removal.**

12. Remove the heater core.
To install:
13. Install the instrument panel finish panel by performing the following procedure:

> ※ **CAUTION**

Electronic modules are sensitive to static electrical charges. If exposed

to these charges, damage may result.

➡ **Verify that the 4-way locator on the left-hand side is properly seated.**

a. Position the instrument panel on the center dash panel.
b. Install the left-hand instrument panel bolts.
c. Install the right-hand instrument panel bolts.
d. Engage the antenna cable locators into the instrument panel and connect it to the radio.
e. Connect the right-hand instrument panel cowl side wiring electrical connectors.
f. Connect the ECS module electrical connector by performing the following:

- Connect the connector.
- Install the locking clip.

g. Install the right-hand A-pillar lower trim panel.
h. From inside the engine compartment, install the main wiring assembly/lower bulkhead connector by performing the following:

- Position the bulkhead connector.
- Secure the clips.
- Tighten the bolt.

i. Install the coolant reservoir by performing the following:

- Position the reservoir.
- Install the bolts.

j. Install the instrument panel dash brace by performing the following:

- Position the instrument panel dash brace.
- Install the 6 bolts.

k. Install the AFCM.
l. Install the instrument panel cowl top nut.
m. Install the left-hand upper access panel.
n. Connect the climate control vacuum harness connector.
o. Install the upper center instrument panel access panel.
p. Install the instrument panel cowl top screws.
q. Install the engine cover.
r. Position the engine cover in the vehicle and fasten the cover latches.
s. Install the steering column.
t. Install the driver's side air bag module by installing or connecting the following items:

- Driver's air bag module.
- Air bag sliding contact electrical connector.
- Horn electrical connector.

1. Instrument panel cluster finish panel
2. Instrument panel center finsh panel
3. passenger air bag opening cover
4. Instrument panel
5. Instrument panel dash brace
6. Instrument panel finsh panel reinforcement
7. Instrument panel finsh panel
8. Instrument panel steering column opening cover reinforcement
9. Driver side knee brace
10. Instrument panel steering column cover

93113GM4

Exploded view of the instrument panel—Ford Econoline

1. A/C evaporator core housing
2. Heater water hoses
3. Heater core housing

93113GM5

View of the heater housing and evaporator housing—Ford Econoline

- Both driver's air bag module screws.
- Both back cover plugs.

u. Install the passenger's side air bag modules by installing or connecting the following items:

- The 6 offset fasteners used for retaining the air bag module door from the instrument panel.
- Passenger's air bag module electrical connector and module.
- Passenger's air bag module retaining nuts.
- Instrument panel finish panel and reinforcement.

14. Connect the quick disconnect heater hose couplings at the heater core by performing the following procedure:

a. Clean the tubes and lubricate them with rubber insulator lube.

b. Install a new coupling retainer, spacer, and lubricated O-ring seals into the quick disconnect coupling housing.

c. Push the heater water hose with a quick disconnect coupling onto the tube.

d. Make sure the coupling is fully engaged by lightly pulling on the heater water hose.

15. Refill the cooling system.
16. Connect the battery ground cable.
17. Run the engine to normal operating temperatures; then, check the climate control operation and check for leaks.

2003–04 E-Series Primary System

1. Before servicing the vehicle, refer to the Precautions Section.

2. Clamp off and disconnect the heater hoses at the heater core.

3. Remove the instrument panel finish panel.

4. Remove the bolts and the instrument panel reinforcement.

5. Remove the bolts and the instrument panel reinforcement bracket.

6. Remove the screws and position the wire harness aside.

7. Remove the screws and the heater core cover.

➡**Use care not to spill the coolant remaining in the heater core during removal.**

8. Remove the heater core.

9. To install, reverse the removal procedure. Clean and lubricate the coolant hoses with plain water only, if needed.

10. Fill the engine cooling system.

E-Series Auxiliary System

1. Before servicing the vehicle, refer to the Precautions Section.

2. Remove the rear seats to access the quarter trim panels.

3. Remove the body side trim finish panel.

4. If equipped, disconnect the power-point electrical connector.

5. Remove the rear seat belt shoulder strap opening covers. Pry the seat belt shoulder strap covers from the upper trim panels.

6. Pull the rear door weatherstrip off at the rear door opening to access the trim panel pin-type retainers.

7. Remove the rear upper body side trim panel.

8. Remove the rear lower body side trim panel.

9. Remove the window latches, if so equipped.

10. Position the driver seat to the full forward position.

11. Remove the front seat belt guide cover.

12. Remove the nut and front safety belt guide.

13. Remove the front upper body side trim panel.

67197-EFSE-G31

Heater core cover screw locations—E-Series primary system

67197-EFSE-G32

Heater core cover screw locations—E-Series auxiliary system

14. Remove the front lower body side trim panel.

15. To install, reverse the removal procedure.

   a. To install the front upper body side trim panel, start at the lower pillar, then work to the second lower pillar.

   b. To install the rear lower body side trim panel, start with the clip to the heater housing, then snap the trim panel into the panel clips.

   c. To install the rear upper body side trim panel, start with the rear lower portion of the panel at the pillar.

16. Remove the heater core cover.

17. Clamp off and disconnect the heater hoses.

18. Remove the heater core case seal.

19. Remove the heater core.

20. To install, reverse the removal procedure. Lubricate the coolant hoses with plain water, if needed.

21. Fill the engine cooling system.

Blackwood

1. Before servicing the vehicle, refer to the Precautions Section.

➡**If a heater core leak is suspected, the heater core must be pressure leak tested before it is removed from the vehicle.**

2. Drain the radiator.

✳✳ WARNING

Electronic modules are sensitive to static electrical charges. If exposed to these charges, damage may result.

3. Move the adjustable pedals to the full forward position.

4. Position the seats as needed to remove the console.

5. Remove the console finish panel, disconnect the wiring and remove the 4 bolts.

✳✳ CAUTION

Always wear safety glasses when repairing an air bag supplemental restraint system (SRS) vehicle and when handling an air bag module. This will reduce the risk of injury in the event of an accidental deployment.

✳✳ CAUTION

Carry a live air bag module with the air bag and deployment door pointed away from your body. This will reduce the risk of injury in the event of an accidental deployment.

✳✳ CAUTION

Do not set a live air bag module down with the deployment door face down. This will reduce the risk of injury in the event of an accidental deployment.

✳✳ CAUTION

After deployment, the air bag surface can contain deposits of sodium hydroxide, a product of the gas generant combustion that is irritating to the skin. Wash your hands with soap and water afterwards.

✳✳ CAUTION

Never probe the connectors on the air bag module. Doing so can result in air bag deployment, which can result in personal injury.

✳✳ CAUTION

Air bag modules with discolored or damaged deployment doors must be installed new, not repainted.

✳✳ CAUTION

To avoid accidental deployment and possible personal injury, the backup power supply must be depleted before repairing or replacing any front or side air bag supplemental restraint system (SRS) components and before servicing, replacing, adjusting or striking components near the front or side air bag sensors, such as doors, instrument panel, console, door latches, strikers, seats and hood latches.

The side air bag sensors are located at or near the base of the B-pillar.

To deplete the backup power supply energy, disconnect the battery ground cable and wait at least one minute. Be sure to disconnect auxiliary batteries and power supplies (if equipped).

6. Disconnect the battery ground cable and wait at least one minute.

✳✳ CAUTION

To reduce the risk of serious personal injury, read and follow all warnings, cautions, notes, and instructions in the supplemental restraint system (SRS) deactivation/reactivation procedure.

7. Deactivate the supplemental restraint system (SRS). For additional information, refer to Suspension and Steering.

8. Release the passenger air bag module electrical connector pin-type retainer.

9. Remove the bolt.

10. Push in the two glove compartment door tabs and position downward.

11. Remove the two lower passenger air bag module retaining bolts.

12. Remove the passenger air bag module.

13. Remove the steering column.

   a. Remove the two back cover plugs from the steering wheel.

   b. Remove the two driver air bag module retaining bolts.

   c. Disconnect the driver air bag module electrical connector.

   d. Disconnect the horn switch electrical connector.

   e. Remove the driver air bag module.

   f. Turn the steering wheel to the straight ahead position and turn the ignition switch to the OFF position.

   g. Remove the steering wheel bolt.

✳✳ WARNING

Removing the steering wheel without using a puller can damage the column bearings.

h. Use a puller to remove the steering wheel.

i. Remove the clockspring.

✳✳ WARNING

Do not allow the steering column shaft to rotate while the intermediate shaft is disconnected or damage to the clockspring can result.

j. If there is evidence that the shaft has rotated, the clockspring must be removed and re-centered.

k. Remove the reinforcement.

l. Remove the multifunction switch.

m. Remove the ignition switch connector.

n. Remove the selector lever indicator cable.

o. Remove the intermediate shaft bolt.

p. Remove all electrical connectors.

q. Remove the column mounting nuts (4).

r. Remove the shift cable.

s. Remove the column.

14. Remove the junction block splash shield.

15. Remove the bolts and disconnect the cable ends from the starter relay.

16. Remove the bolts and position aside the junction block bracket.

17. Disconnect the heater core hose couplings.

18. Remove the A/C plenum demister adapter.

19. Remove the screws and position aside the relay bracket.

20. Disconnect the vacuum line.

21. Remove the plenum chamber top.

22. Remove the blend door assembly from the case.

23. Remove the heater core.

24. To install, reverse the removal procedure. Lubricate the coolant hoses with coolant hose lubricant or plain water only, if needed. Observe the following torques:

- Steering wheel bolt: 28 ft. lbs. (38 Nm)
- Intermediate shaft bolt: 35 ft. lbs. (48 Nm)

2002–03 F-150; 2004 Heritage Edition

1. Before servicing the vehicle, refer to the Precautions Section.

2. If equipped with power seats, move the seats fully rearward.

3. Disconnect the negative battery cable.

✳✳ CAUTION

After disconnecting the negative battery cable, wait for 1 minute for the SRS module to deplete its energy.

4. Drain the cooling system into a clean container for reuse.

5. Remove the instrument panel by performing the following procedure:

a. If equipped with automatic transmission, move the shift lever to the **1** position to ease removal.

b. At the lower driver's side instrument panel, release the electrical connector push button clip and move the connector aside.

c. Remove the fuse panel door.

d. Remove the hood latch release handle screws and move the hood release handle aside.

e. Remove the parking brake release handle screws and move the parking brake release handle aside.

f. If equipped, remove the 2 instrument panel floor duct panel push clips and release the expander clip.

g. Remove the lower steering column cover bolts and the cover.

h. Remove both front door scuff plates.

i. Remove both side cowl trim panels.

j. Disconnect the electrical connector from the Brake Pedal Position (BPP) switch.

k. Remove the radio ground and the GEM/CTM ground bolts.

l. Disconnect the left side instrument panel main wiring harness connector.

m. In the engine compartment, remove the bulkhead wiring harness connector bolts and disconnect the wiring connectors.

n. In the engine compartment, release the 6 locking tabs and remove the bulkhead electrical connector from the instrument panel.

o. Disconnect the air bag diagnostic monitor electrical connector.

p. Disconnect the inertia fuel shutoff switch electrical connector.

q. Remove the right side ground bolts.

r. Disconnect the right side instrument panel wiring harness connectors.

s. Disconnect the electronic blend door actuator electrical connector.

t. Disconnect the climate control head vacuum harness connector.

u. Remove the steering column opening cover reinforcement nuts and the cover reinforcement.

v. At the base of the steering column, disconnect the air bag sliding contact and the anti-theft sensor electrical connectors.

w. At the steering column, disconnect the remaining electrical connectors.

x. If equipped with a transmission range indicator, remove the bolt and disconnect the cable.

y. Remove the steering column-to-instrument panel nuts and lower the steering column.

z. Remove and disconnect the radio.

aa. Remove the instrument panel relay cover and disconnect the autolamp sensor electrical connector.

bb. Remove the glove box.

cc. At the passenger's air bag module, remove the screws, disconnect the electrical connector and remove the air bag module.

Place the air bag module in a safe place with the front facing upward.

a. Remove the right side assist handle screw covers, the screws and the handle.

b. At both doors, pull back the weatherstrip seals and remove the windshield garnish moldings.

c. Remove the instrument panel reinforcement bolt below the left side corner of the glove box.

d. Through the air bag module opening, remove the instrument panel bolts.

e. Remove the upper instrument panel cowl covers and bolts.

f. At the relay bracket, remove the instrument panel bolt.

g. At the lower left side of the cigar lighter, remove the instrument panel bolt.

h. At the both sides, remove the instrument panel-to-cowl side nuts.

i. At the steering column opening, remove the instrument panel bolts.

j. Remove the upper instrument panel floor brace bolt.

k. Loosen the instrument panel brace bolts and nut.

l. Using an assistant, remove the instrument panel.

6. Compress the holding tabs and disconnect the heater hoses from the heater core.

7. Remove the air conditioning plenum screw and the air conditioning plenum demister adapter.

8. Disconnect the vacuum line.

9. Remove the heater core bracket screws and the bracket.

10. Remove the 13 heater housing plenum camber cover screws and the heater housing plenum chamber cover.

11. Remove the blend door assembly from the heater housing.

12. Remove the heater core.

To install:

13. Install the heater core.

14. Install the blend door assembly to the heater housing.

15. Install the 13 heater housing plenum camber cover and the heater housing plenum chamber cover screws.

16. Install the heater core bracket and the bracket screws.

17. Connect the vacuum line.

18. Install the air conditioning plenum demister adapter and the air conditioning plenum screw.

19. Connect the heater hoses to the heater core.

20. Install the instrument panel by performing the following procedure:

a. Using an assistant, install the instrument panel.

b. Tighten the instrument panel brace bolts and nut.

c. Install the upper instrument panel floor brace bolt.

d. At the steering column opening, install the instrument panel bolts.

e. At the both sides, install the instrument panel-to-cowl side nuts.

f. At the lower left side of the cigar lighter, install the instrument panel bolt.

g. At the relay bracket, install the instrument panel bolt.

h. Install the upper instrument panel cowl bolts and covers.

i. Through the air bag module opening, install the instrument panel bolts.

j. Install the instrument panel reinforcement bolt below the left side corner of the glove box.

k. At both doors, install the windshield garnish moldings and the weatherstrip seals.

l. Install the right side assist handle, the screws and the handle screw covers.

m. At the passenger's air bag module, connect the electrical connector and install the air bag module.

n. Install the glove box.

o. Install the instrument panel relay cover and connect the autolamp sensor electrical connector.

p. Install and connect the radio.

q. Install the steering column-to-instrument panel and the steering column nuts.

1. Instrument panel finish panel
2. Instrument cluster panel
3. Instrument panel relay cover
4. Instrument panel
5. Center instrument panel finish panel
6. Instrument panel steering column cover, lower
7. Instrument panel fuse door
8. Steering column opening cover

93113GL0

Exploded view of the instrument panel components—2002–03 Ford F-150 and Blackwood; 2004 Heritage Edition

93113GM1

View of the heater core—2002–03 Ford F-150 and Blackwood; 2004 Heritage Edition

r. If equipped with a transmission range indicator, connect the cable and install the bolt.

s. At the steering column, connect the remaining electrical connectors.

t. At the base of the steering column, connect the air bag sliding contact and the anti-theft sensor electrical connectors.

u. Install the steering column opening cover reinforcement and the cover reinforcement nuts.

v. Connect the climate control head vacuum harness connector.

w. Connect the electronic blend door actuator electrical connector.

x. Connect the right side instrument panel wiring harness connectors.

y. Install the right side ground bolts.

z. Connect the inertia fuel shutoff switch electrical connector.

aa. Connect the air bag diagnostic monitor electrical connector.

bb. In the driver's compartment, install the bulkhead electrical connector to the instrument panel.

cc. In the engine compartment, install the bulkhead wiring harness connector bolts and connect the wiring connectors.

dd. Connect the left side instrument panel main wiring harness connector.

ee. Install the radio ground and the GEM/CTM ground bolts.

ff. Connect the electrical connector to the Brake Pedal Position (BPP) switch.

gg. Install both side cowl trim panels.

hh. Install both front door scuff plates.

ii. Install the lower steering column cover and the cover bolts.

jj. If equipped, install the 2 instrument panel floor duct panel push clips and the expander clip.

kk. Install the parking brake release handle and the parking brake release handle aside screws.

ll. Install the hood latch release handle screws.

mm. Install the fuse panel door.

nn. At the lower driver's side instrument panel, install the connectors.

21. Refill the cooling system.

22. Connect the negative battery cable.

23. Run the engine to normal operating temperatures; then, check the climate control operation and check for leaks.

2004 F-150

1. Before servicing the vehicle, refer to the Precautions Section.

➡**Lubricate the coolant hoses with plain water only if needed.**

✳✳ WARNING

To reduce the risk of serious personal injury, read and follow all warnings, cautions, notes and instructions in the supplemental restraint system (SRS) deactivation/reactivation procedure.

2. Remove the driver and passenger air bag modules.

3. Remove the A-pillar trim panels and right cowl side trim panel.

4. If equipped with a floor console, remove the floor console.

5. Remove the steering column opening cover.

6. Remove the steering column lower cover.

7. Remove the steering column upper cover.

8. Remove the instrument panel center finish panel

9. Remove the instrument panel finish panel.

10. Parking brake release handle.

11. Remove the hood release handle.

12. Remove the electrical connectors.

13. Remove the center brace.

14. Remove the transmission range cable.

15. Remove the pinch bolt.

16. Remove the defrost grilles.

17. Remove the instrument panel bolt.

18. Remove the instrument panel.

➡**New O-ring seals, lubricated in clean mineral oil, must be installed before reconnecting any A/C fitting which has been disconnected.**

19. Recover the refrigerant.

20. Remove the powertrain control module (PCM).

21. Remove the battery and battery tray.

22. Remove the PCM bracket.

23. Remove the heater hoses

24. Remove the heater/evaporator core housing nuts.

25. Remove the heater/evaporator core housing.

26. Discard the O-rings.

27. Remove the heater core.

28. To install, reverse the removal procedure.

29. Fill the engine cooling system.

2005 F-150 and 2006 Mark LT

1. Before servicing the vehicle, refer to the Precautions Section.

2. If equipped with adjustable pedals, move the pedals to the full forward position.

---

3. If equipped with a floor console:

a. Remove the 3 floor trim panel push pins.

b. Remove the 2 floor trim panels.

c. Remove the shifter trim panel.

d. Disconnect the shifter cable.

e. Disconnect the floor console electrical connector.

f. Remove the 4 floor console bolts.

g. Remove the floor console.

✳✳ CAUTION

To reduce the risk of serious personal injury, read and follow all warnings, cautions, notes and instructions in the supplemental restraint system (SRS) deactivation/reactivation procedure.

4. De-power the supplemental restraint system (SRS). For additional information, refer to Steering.

5. Position the left and right door weatherstrip seals aside.

6. Remove the A-pillar trim panels and right cowl side trim panel.

7. Remove the left and right instrument panel side finish panels.

8. Remove the 2 steering column opening trim screws.

9. Remove the steering column opening trim.

✳✳ WARNING

Removing the steering wheel without using a puller can damage the column bearings.

✳✳ WARNING

Do not allow the steering column shaft to rotate while the intermediate shaft is disconnected or damage to the clockspring can result. If there is evidence that the shaft has rotated, the clockspring must be removed and re-centered.

10. Place the steering wheel in the straight-ahead position and turn the ignition switch to the OFF position.

✳✳ CAUTION

Always wear safety glasses when repairing an air bag supplemental restraint system (SRS) vehicle and when handling an air bag module. This will reduce the risk of injury in the event of an accidental deployment.

9 Nm (80 lb-in)

1 Instrument panel side finish panel (RH/LH)

2 Steering column opening trim screws (2 required)

3 Steering column opening trim

4 Lower steering column cover screw

5 Lower steering column cover

6 Steering column opening panel screws (4 required)

7 Steering column opening panel

8 Upper steering column cover screws (2 required)

9 Upper steering column cover

10 Instrument cluster finish panel screws (2 required)

11 Instrument cluster finish panel

12 Instrument panel center finish panel

13 Climate control electrical connector

14 Instrument cluster center finish panel screws (2 required)

15 Instrument cluster center finish panel

Instrument panel exploded view—2005 F-150

06017-F150-G25

1 Fitting
2 O-ring (3 required)
3 Heater hoses
4 Heater core and evaporator
   core housing nuts (3 required)

5 Harness electrical connectors
6 Air inlet duct bracket nut
7 Heater core and evaporator
   core housing

06017-F150-G26

Heater/Evaporator housing—2005 F-150

� CAUTION

Carry a live air bag module with the
air bag and trim cover pointed away
from your body. This will reduce the
risk of injury in the event of an accidental deployment.

� CAUTION

Do not set a live air bag module down
with the trim cover face down. This
will reduce the risk of injury in the
event of an accidental deployment.

� CAUTION

After deployment, the air bag surface
can contain deposits of sodium
hydroxide, a product of the gas generant combustion that is irritating to
the skin. Wash your hands with soap
and water afterwards.

� CAUTION

Never probe the connectors on the air
bag module. Doing so can result in
air bag deployment, which can result
in personal injury.

1 Push pins (3 required)
2 Floor trim panel (2 required)
   (LH/RH)
3 Shifter trim panel

4 Shifter cable
5 Electrical connector
6 Floor console bolts (4
   required)
7 Floor console

06017-F150-G27

Floor console—2005 F-150

10 Nm (89 lb-in)

1 Steering wheel back cover plug (2 required)

2 Driver air bag module bolt (2 required)

3 Driver air bag module

4 Driver air bag module connectors

5 Horn switch connector

06017-F150-G28

Driver's air bag module—2005 F-150

✳✳ CAUTION

Air bag modules with discolored or damaged trim covers must be replaced, not repainted.

✳✳ CAUTION

To reduce the risk of personal injury, do not use any memory saver devices.

➡ The air bag warning lamp illuminates when the RCM fuse is removed and the ignition switch is ON. This is normal operation and does not indicate a supplemental restraint system (SRS) fault.

➡ The SRS must be fully operational and free of faults before releasing the vehicle to the customer.

➡ Repair is made by installing a new part only. If the new part does not correct the condition, install the original part and perform the diagnostic procedure again.

8 Nm (71 lb-in)

6 Nm (53 lb-in)

10 Nm (89 lb-in)

1 PCM bracket bolts (3 required)

2 PCM bracket

3 Suction accumulator outlet fitting nut

4 Suction accumulator inlet fitting

5 A/C cycling switch electrical connector

6 Suction accumulator bracket nut (2 required)

7 Suction accumulator

8 O-ring (4 required)

06017-F150-G29

Suction accumulator—2005 F-150

1 Temperature blend door
   actuator screw (3 required)

2 Temperature blend door
   actuator

3 Heater core cover screw (4
   required)

4 Heater core cover

5 Heater core

6 Dash panel seal

06017-F150-G30

Heater core removal—2005 F-150

1. Outer housing
2. Rotor
3. Clockspring

06017-F150-G31

Clockspring

11. Remove the 2 steering wheel back
cover plugs.

12. Remove the 2 driver's air bag module
bolts.

13. Partially remove the driver's air bag
module from the steering wheel.

✳✳ WARNING

**The clockspring electrical connectors
are unique and cannot be reversed
when connected to the driver's air
bag module. Match the electrical
connector key to the keyway in the
driver's air bag module. Do not force
the electrical connectors into the**

driver's air bag module during installation.

14. Disconnect the driver's air bag module electrical connectors.

15. Disconnect the horn switch electrical
connector and remove the driver's air bag
module.

➡**Do not remove the bolt until the
steering wheel has been separated
from the column.**

16. Loosen the steering wheel bolt.

17. Using a puller, separate the steering
wheel from the steering column assembly.

18. Remove the steering wheel bolt and
the steering wheel.

19. Remove the lower steering column
cover screw.

20. Remove the lower steering column
cover.

21. Remove the 2 upper steering column
cover screws.

22. Remove the upper steering column
cover.

23. Remove the instrument cluster finish
panel.

24. Remove the instrument panel center
finish panel.

25. Remove the instrument cluster center
finish panel.

26. Remove the parking brake release
handle bolt.

27. Position aside the parking brake
release handle. Remove the parking brake
release handle pin-type retainer.

28. Remove the hood release handle bolt.

29. Position aside the hood release
handle.

30. Loosen the electrical connector
screw and disconnect the bulkhead electrical connector.

31. Remove the ground bolt.

32. Disconnect the electrical connectors.

33. Position aside the front carpet.

34. Remove the 4 left center brace bolts.

35. Remove the left center brace.

36. If equipped, disconnect the transmission range indicator cable and position
aside.

✳✳ WARNING

**To avoid damage to the clockspring,
do not allow the steering column
shaft to rotate while the intermediate
shaft is disconnected.**

37. Remove the steering column pinch
bolt and disconnect the steering column
intermediate shaft.

38. Disconnect the steering column electrical connector.

39. Remove the ground bolt.
40. Disconnect the radio antenna.
41. Disconnect the 4 PDB (B1, B2, L1 and L2) electrical connectors.
42. Disconnect the ETAC electrical connector.
43. Disconnect the instrument panel electrical connectors.
44. Position aside the front carpet.
45. Remove the 4 right center brace bolts.
46. Remove the right center brace.

➡**If equipped with a floor console, remove the third center brace.**

47. If equipped, remove the 4 third center brace bolts.
48. If equipped, remove the third center brace.
49. Remove the 2 defrost grilles.
50. Disconnect the wiring harness pin-type retainer.

✳✳ WARNING

To avoid damage to the instrument panel, an assistant is required to support the panel before carrying out this step.

51. Remove the 9 instrument panel bolts.

✳✳ WARNING

Before removing the instrument panel, make sure all electrical connector wiring is free and not hindered.

52. Remove the instrument panel.

➡**Installation of a new suction accumulator is not required when repairing the air conditioning system, except when there is physical evidence of contamination from a failed A/C compressor or damage to the suction accumulator.**

53. Recover the refrigerant.
54. Remove the powertrain control module (PCM).
55. Remove the battery and battery tray.
56. Remove the PCM bracket.
57. Remove the suction accumulator outlet fitting nut and disconnect the fitting.
58. Discard the O-ring seal.
59. Disconnect the suction accumulator inlet fitting. Discard the O-ring seals.
60. Remove the 2 suction accumulator bracket nuts.
61. Remove the suction accumulator.
62. Disconnect the evaporator core

spring lock coupling. Discard the O-ring seals.
63. Clamp off and disconnect the 2 heater core quick disconnect fittings.
64. Remove the 3 heater core and evaporator core housing nuts.
65. Disconnect the 2 harness electrical connectors.
66. Remove the air inlet duct bracket nut.
67. Remove the bolt and position the junction box aside.
68. Loosen the nut and remove the heater core and evaporator core housing.
69. Remove the dash panel seal from the heater core tubes.
70. Position the temperature blend door actuator aside.
71. Remove the heater core cover.
72. Remove the heater core.

➡**The heater core seal must be correctly installed to prevent airflow from bypassing the heater core.**

✳✳ CAUTION

Incorrect centralization may result in premature component failure. If in doubt when centralizing the clockspring, repeat the centralizing procedure. Failure to follow this instruction may result in personal injury.

✳✳ WARNING

Make sure the road wheels are in the straight-ahead position.

73. If the vehicle's clockspring has rotated out of center, follow these steps to center the clockspring.
   a. Hold the clockspring outer housing stationary.

✳✳ WARNING

Overturning will destroy the clockspring. The internal ribbon wire acts as the stop and can be broken from its internal connection.

   b. While turning the rotor counterclockwise, carefully feel for the ribbon wire to run out of length and for a slight resistance. Stop turning at this point.
   c. Turn the clockspring clockwise approximately 3 turns. This is the center point of the clockspring. Do not allow the rotor to turn from this position. To prevent accidental rotation until the clockspring is installed, 2 pieces of tape may be applied to the clockspring.

➡**Slight turning of the clockspring rotor is allowable for alignment purposes to the steering column.**

   d. With the flats of the clockspring aligned to the flats of the steering column, slide the clockspring onto the steering column. For vehicles receiving a new clockspring, remove the retaining pin. For vehicles reusing the clockspring that was removed, remove the tape.
74. To install, reverse the removal procedure. Evacuate, charge and leak test the system. Install new O-ring seals. Lubricate the refrigerant system with the correct amount of clean PAG oil.
75. Observe the following torques:
   • Heater core nut: 62 inch lbs. (7 Nm)
   • Air inlet duct bracket: 62 inch lbs. (7 Nm)
   • Heater core and evaporator core housing nuts: 80 inch lbs. (9 Nm)
   • Suction accumulator bracket nuts: 53 inch lbs. (6 Nm)
   • Suction accumulator outlet fitting nut: 71 inch lbs. (8 Nm)
   • PCM bracket: 89 inch lbs. (10 Nm)
   • Instrument panel bolts: 15 ft. lbs. (20 Nm)
   • Steering column pinch bolt: 30 ft. lbs. (40 Nm)
   • Steering wheel bolt and the steering wheel: 30 ft. lbs. (40 Nm)
   • Driver's air bag module bolts: 89 inch lbs. (10 Nm)
76. Fill the engine cooling system.

2002–03 F-250

1. Before servicing the vehicle, refer to the Precautions Section.
2. If equipped with power seats, move the seats fully rearward.
3. Disconnect the negative battery cable.

✳✳ CAUTION

After disconnecting the negative battery cable, wait for 1 minute for the SRS module to deplete its energy.

4. Drain the cooling system into a clean container for reuse.
5. Remove the instrument panel by performing the following procedure:
   a. If equipped with automatic transmission, move the shift lever to the **1** position to ease removal.
   b. At the lower driver's side instrument panel, release the electrical connector push button clip and move the connector aside.

c. Remove the fuse panel door.

d. Remove the hood latch release handle screws and move the hood release handle aside.

e. Remove the parking brake release handle screws and move the parking brake release handle aside.

f. If equipped, remove the 2 instrument panel floor duct panel push clips and release the expander clip.

g. Remove the lower steering column cover bolts and the cover.

h. Remove both front door scuff plates.

i. Remove both side cowl trim panels.

j. Disconnect the electrical connector from the Brake Pedal Position (BPP) switch.

k. Remove the radio ground and the GEM/CTM ground bolts.

l. Disconnect the left side instrument panel main wiring harness connector.

m. In the engine compartment, remove the bulkhead wiring harness connector bolts and disconnect the wiring connectors.

n. In the engine compartment, release the 6 locking tabs and remove the bulkhead electrical connector from the instrument panel.

o. Disconnect the air bag diagnostic monitor electrical connector.

p. Disconnect the inertia fuel shutoff switch electrical connector.

q. Remove the right side ground bolts.

r. Disconnect the right side instrument panel wiring harness connectors.

s. Disconnect the electronic blend door actuator electrical connector.

t. Disconnect the climate control head vacuum harness connector.

u. Remove the steering column opening cover reinforcement nuts and the cover reinforcement.

v. At the base of the steering column, disconnect the air bag sliding contact and the anti-theft sensor electrical connectors.

w. At the steering column, disconnect the remaining electrical connectors.

x. If equipped with a transmission range indicator, remove the bolt and disconnect the cable.

y. Remove the steering column-to-instrument panel nuts and lower the steering column.

z. Remove and disconnect the radio.

aa. Remove the instrument panel relay cover and disconnect the autolamp sensor electrical connector.

bb. Remove the glove box.

cc. At the passenger's air bag module, remove the screws, disconnect the electrical connector and remove the air bag module.
Place the air bag module in a safe place with the front facing upward.

dd. Remove the right side assist handle screw covers, the screws and the handle.

ee. At both doors, pull back the weatherstrip seals and remove the windshield garnish moldings.

ff. Remove the instrument panel reinforcement bolt below the left side corner of the glove box.

gg. Through the air bag module opening, remove the instrument panel bolts.

hh. Remove the upper instrument panel cowl covers and bolts.

ii. At the relay bracket, remove the instrument panel bolt.

jj. At the lower left side of the cigar lighter, remove the instrument panel bolt.

kk. At the both sides, remove the instrument panel-to-cowl side nuts.

ll. At the steering column opening, remove the instrument panel bolts.

mm. Remove the upper instrument panel floor brace bolt.

nn. Loosen the instrument panel brace bolts and nut.

oo. Using an assistant, remove the instrument panel.

6. Compress the holding tabs and disconnect the heater hoses from the heater core.

7. Remove the air conditioning plenum screw and the air conditioning plenum demister adapter.

8. Disconnect the vacuum line.

9. Remove the heater core bracket screws and the bracket.

10. Remove the 13 heater housing plenum camber cover screws and the heater housing plenum chamber cover.

11. Remove the blend door assembly from the heater housing.

12. Remove the heater core.

To install:

13. Install the heater core.

14. Install the blend door assembly to the heater housing.

15. Install the 13 heater housing plenum camber cover and the heater housing plenum chamber cover screws.

16. Install the heater core bracket and the bracket screws.

17. Connect the vacuum line.

18. Install the air conditioning plenum

1. Instrument panel finish panel
2. Instrument cluster panel
3. Instrument panel relay cover
4. Instrument panel
5. Center instrument panel finish panel
6. Instrument panel steering column cover, lower
7. Instrument panel fuse door
8. Steering column opening cover

93113GL0

Exploded view of the instrument panel components—2002–03 F-250

PLENUM ASSY

HEATER CORE COVER

HEATER CORE ASSY

HEATER CORE TUBES

HEATER CORE TUBE TO COWL SPACER

84926007

Heater core removal—2002–03 F-250, 350

demister adapter and the air conditioning plenum screw.

19. Connect the heater hoses to the heater core.

20. Install the instrument panel by performing the following procedure:

a. Using an assistant, install the instrument panel.

b. Tighten the instrument panel brace bolts and nut.

c. Install the upper instrument panel floor brace bolt.

d. At the steering column opening, install the instrument panel bolts.

e. At the both sides, install the instrument panel-to-cowl side nuts.

f. At the lower left side of the cigar lighter, install the instrument panel bolt.

g. At the relay bracket, install the instrument panel bolt.

h. Install the upper instrument panel cowl bolts and covers.

i. Through the air bag module opening, install the instrument panel bolts.

j. Install the instrument panel reinforcement bolt below the left side corner of the glove box.

k. At both doors, install the windshield garnish moldings and the weatherstrip seals.

l. Install the right side assist handle, the screws and the handle screw covers.

m. At the passenger's air bag module, connect the electrical connector and install the air bag module.

n. Install the glove box.

o. Install the instrument panel relay cover and connect the autolamp sensor electrical connector.

p. Install and connect the radio.

q. Install the steering column-to-instrument panel and the steering column nuts.

r. If equipped with a transmission range indicator, connect the cable and install the bolt.

s. At the steering column, connect the remaining electrical connectors.

t. At the base of the steering column, connect the air bag sliding contact and the anti-theft sensor electrical connectors.

u. Install the steering column opening cover reinforcement and the cover reinforcement nuts.

v. Connect the climate control head vacuum harness connector.

w. Connect the electronic blend door actuator electrical connector.

x. Connect the right side instrument panel wiring harness connectors.

y. Install the right side ground bolts.

z. Connect the inertia fuel shutoff switch electrical connector.

aa. Connect the air bag diagnostic monitor electrical connector.

bb. In the driver's compartment, install the bulkhead electrical connector to the instrument panel.

cc. In the engine compartment, install the bulkhead wiring harness connector bolts and connect the wiring connectors.

dd. Connect the left side instrument panel main wiring harness connector.

ee. Install the radio ground and the GEM/CTM ground bolts.

ff. Connect the electrical connector to the Brake Pedal Position (BPP) switch.

gg. Install both side cowl trim panels.

hh. Install both front door scuff plates.

ii. Install the lower steering column cover and the cover bolts.

jj. If equipped, install the 2 instrument panel floor duct panel push clips and the expander clip.

kk. Install the parking brake release handle and the parking brake release handle aside screws.

ll. Install the hood latch release handle screws.

mm. Install the fuse panel door.

nn. At the lower driver's side instrument panel, install the connectors.

21. Refill the cooling system.

22. Connect the negative battery cable.

23. Run the engine to normal operating temperatures; then, check the climate control operation and check for leaks.

2002–03 F-350 and Super Duty

1. Before servicing the vehicle, refer to the Precautions Section.

2. If equipped with power seats, move the seats fully rearward.

3. Disconnect the negative battery cable.

❄❄ CAUTION

After disconnecting the negative battery cable, wait for 1 minute for the SRS module to deplete its energy.

4. Drain the cooling system into a clean container for reuse.

5. Remove the instrument panel by performing the following procedure:

a. If equipped with automatic transmission, move the shift lever to the **1** position to ease removal.

b. At the lower driver's side instrument panel, release the electrical connector push button clip and move the connector aside.

c. Remove the fuse panel door.

d. Remove the hood latch release handle screws and move the hood release handle aside.

e. Remove the parking brake release handle screws and move the parking brake release handle aside.

f. If equipped, remove the 2 instrument panel floor duct panel push clips and release the expander clip.

g. Remove the lower steering column cover bolts and the cover.

h. Remove both front door scuff plates.

i. Remove both side cowl trim panels.

j. Disconnect the electrical connector from the Brake Pedal Position (BPP) switch.

k. Remove the radio ground and the GEM/CTM ground bolts.

l. Disconnect the left side instrument panel main wiring harness connector.

m. In the engine compartment, remove the bulkhead wiring harness connector bolts and disconnect the wiring connectors.

n. In the engine compartment, release the 6 locking tabs and remove the bulkhead electrical connector from the instrument panel.

o. Disconnect the air bag diagnostic monitor electrical connector.

p. Disconnect the inertia fuel shutoff switch electrical connector.

q. Remove the right side ground bolts.

r. Disconnect the right side instrument panel wiring harness connectors.

s. Disconnect the electronic blend door actuator electrical connector.

t. Disconnect the climate control head vacuum harness connector.

u. Remove the steering column opening cover reinforcement nuts and the cover reinforcement.

v. At the base of the steering column, disconnect the air bag sliding contact and the anti-theft sensor electrical connectors.

w. At the steering column, disconnect the remaining electrical connectors.

x. If equipped with a transmission range indicator, remove the bolt and disconnect the cable.

y. Remove the steering column-to-instrument panel nuts and lower the steering column.

z. Remove and disconnect the radio.

aa. Remove the instrument panel relay cover and disconnect the autolamp sensor electrical connector.

bb. Remove the glove box.

cc. At the passenger's air bag module, remove the screws, disconnect the electrical connector and remove the air bag module.

Place the air bag module in a safe place with the front facing upward.

dd. Remove the right side assist handle screw covers, the screws and the handle.

ee. At both doors, pull back the weatherstrip seals and remove the windshield garnish moldings.

ff. Remove the instrument panel reinforcement bolt below the left side corner of the glove box.

gg. Through the air bag module opening, remove the instrument panel bolts.

hh. Remove the upper instrument panel cowl covers and bolts.

ii. At the relay bracket, remove the instrument panel bolt.

jj. At the lower left side of the cigar lighter, remove the instrument panel bolt.

kk. At the both sides, remove the instrument panel-to-cowl side nuts.

ll. At the steering column opening, remove the instrument panel bolts.

mm. Remove the upper instrument panel floor brace bolt.

nn. Loosen the instrument panel brace bolts and nut.

oo. Using an assistant, remove the instrument panel.

6. Compress the holding tabs and disconnect the heater hoses from the heater core.

7. Remove the air conditioning plenum screw and the air conditioning plenum demister adapter.

8. Disconnect the vacuum line.

9. Remove the heater core bracket screws and the bracket.

10. Remove the 13 heater housing plenum camber cover screws and the heater housing plenum chamber cover.

11. Remove the blend door assembly from the heater housing.

12. Remove the heater core.

To install:

13. Install the heater core.

14. Install the blend door assembly to the heater housing.

15. Install the 13 heater housing plenum camber cover and the heater housing plenum chamber cover screws.

16. Install the heater core bracket and the bracket screws.

17. Connect the vacuum line.

18. Install the air conditioning plenum demister adapter and the air conditioning plenum screw.

19. Connect the heater hoses to the heater core.

20. Install the instrument panel by performing the following procedure:

a. Using an assistant, install the instrument panel.

b. Tighten the instrument panel brace bolts and nut.

c. Install the upper instrument panel floor brace bolt.

d. At the steering column opening, install the instrument panel bolts.

e. At the both sides, install the instrument panel-to-cowl side nuts.

f. At the lower left side of the cigar lighter, install the instrument panel bolt.

g. At the relay bracket, install the instrument panel bolt.

h. Install the upper instrument panel cowl bolts and covers.

i. Through the air bag module opening, install the instrument panel bolts.

j. Install the instrument panel reinforcement bolt below the left side corner of the glove box.

k. At both doors, install the windshield garnish moldings and the weatherstrip seals.

l. Install the right side assist handle, the screws and the handle screw covers.

m. At the passenger's air bag module, connect the electrical connector and install the air bag module.

n. Install the glove box.

o. Install the instrument panel relay cover and connect the autolamp sensor electrical connector.

p. Install and connect the radio.

q. Install the steering column-to-instrument panel and the steering column nuts.

r. If equipped with a transmission range indicator, connect the cable and install the bolt.

s. At the steering column, connect the remaining electrical connectors.

t. At the base of the steering column, connect the air bag sliding contact and the anti-theft sensor electrical connectors.

u. Install the steering column opening cover reinforcement and the cover reinforcement nuts.

v. Connect the climate control head vacuum harness connector.

w. Connect the electronic blend door actuator electrical connector.

x. Connect the right side instrument panel wiring harness connectors.

y. Install the right side ground bolts.

z. Connect the inertia fuel shutoff switch electrical connector.

aa. Connect the air bag diagnostic monitor electrical connector.

bb. In the driver's compartment, install the bulkhead electrical connector to the instrument panel.

cc. In the engine compartment, install

Item	Description
1	Instrument Panel
2	Steering Column Opening Cover
3	Screw(s) 2.0-3.0 N•m (19.0-26.0 In-Lb)
4	Screw(s) 2.0-2.4 N•m (18.0-21.0 In-Lb)

Item	Description
5	Cowl Side Panel (LH)
6	Nut Insert (4 Req'd)
7	U-Nut
8	Cowl Side Panel (RH)
9	Screw(s) 2.0-2.4 N•m (18.0-21.0 In-Lb)

88280G13

Instrument panel installation—2002–03 250HD, F-350 and F-Super Duty the bulkhead wiring harness connector bolts and connect the wiring connectors.

dd. Connect the left side instrument panel main wiring harness connector.

ee. Install the radio ground and the GEM/CTM ground bolts.

ff. Connect the electrical connector to the Brake Pedal Position (BPP) switch.

gg. Install both side cowl trim panels.

hh. Install both front door scuff plates.

ii. Install the lower steering column cover and the cover bolts.

jj. If equipped, install the 2 instrument panel floor duct panel push clips and the expander clip.

kk. Install the parking brake release handle and the parking brake release handle aside screws.

ll. Install the hood latch release handle screws.

mm. Install the fuse panel door.

nn. At the lower driver's side instrument panel, install the connectors.

21. Refill the cooling system.

22. Connect the negative battery cable.

23. Run the engine to normal operating temperatures; then, check the climate control operation and check for leaks.

2004–05 F-250/350

1. Before servicing the vehicle, refer to the Precautions Section.

Remove the heater core cover—2004–05 F-250/350

Remove the heater core—2004–05 F-250/350

2. Clamp off and disconnect the heater water hoses from the heater core.

3. Disengage the stops and lower the glove compartment door.

➡**Do not attempt to bend any part of the panel/floor door lever. It is brittle and will break.**

4. Remove the panel/floor door vacuum control motor.

5. Remove the electronic blend door actuator and bracket assembly.

6. Remove the core cover screws.

➡**The heater core cover must be raised vertically before removal to avoid damage to the heater core housing.**

7. Raise the heater core cover.

8. Remove the heater core cover.

9. Remove the heater core.

➡**Position the temperature blend door manually to correctly align the actuator and the door. Do not power the actuator electrically. If it is not engaged with the temperature blend door, damage to the actuator can occur.**

➡**Add gasket between housing and cover before installing cover.**

10. To install, reverse the removal procedure.

11. Fill the engine cooling system.

Cylinder Head

REMOVAL & INSTALLATION

E- and F-Series 4.2L Engine

✳✳ CAUTION

Fuel injection systems remain under pressure, even after the engine has been turned OFF. The fuel system pressure must be relieved before disconnecting any fuel lines. Failure to do so may result in fire and/or personal injury.

1. Before servicing the vehicle, refer to the Precautions Section.

2. Remove or disconnect the following:

- Air conditioning system, using approved equipment
- Negative battery cable
- Coolant
- Upper and lower intake manifolds and related components
- Rocker arm covers
- Exhaust manifold

3. If removing the left-hand cylinder head, perform the following:

a. Position the power steering pump reservoir aside and remove the air conditioning compressor.

b. Remove and support the air conditioning compressor bracket and power steering pump aside.

4. If removing the right-hand cylinder head, perform the following:

a. Remove the alternator.

b. Remove the idler pulley.

c. Remove the alternator bracket.

➡**If the cylinder head components, such as rocker arms, valve springs, etc., are to be reinstalled, they must be installed in the same position. Mark the components for original location.**

5. Remove or disconnect the following:

- Rocker arms
- Pushrods. Be sure to label or mark the components removed for reinstallation in their original location.
- Cylinder head bolts. New cylinder head bolts must be used when the cylinder head is reinstalled.
- Cylinder head. Remove the cylinder head gasket and discard.

To install:

6. Clean and inspect the cylinder head for flatness.

7. Install a new cylinder head gasket on the cylinder block with the small hole to the front of the engine.

8. Install the cylinder head.

✳✳ WARNING

Always use new cylinder head bolts for installation. Lubricate the cylinder head bolts with clean engine oil prior to installation. Be sure to tighten the cylinder head bolts in three (3) steps.

9. Install new cylinder head bolts and torque the bolts following the proper tightening sequence in 3 steps to the following values:

a. Step 1: 14 ft. lbs. (20 Nm).

b. Step 2: 29 ft. lbs. (40 Nm).

c. Step 3: 36 ft. lbs. (50 Nm).

✳✳ WARNING

Do not loosen all of the cylinder head bolts at one time. Each cylinder head bolt must be loosened and the final tightening performed prior to loosening the next bolt in the sequence.

67197-EFSE-G29

67197-EFSE-G30

Tighten the cylinder head bolts in the following sequence—4.2L engine

10. In the same sequence as used previously, loosen the cylinder head bolt 3 turns, then tighten the cylinder head bolt to the specific value according to its length. The short bolts (A) should be tightened to 15–22 ft. lbs. (20–30 Nm) and the long bolts (B) to 30–36 ft. lbs. (40–50 Nm). Finally, tighten each cylinder head bolt, in sequence, an additional 175–185 degrees.

11. Lubricate the pushrods with clean engine oil prior to installation, then install them into their original positions.

12. Install the rocker arms. Tighten the rocker arm mounting bolts to 23–29 ft. lbs. (30–40 Nm).

13. If the valvetrain components were replaced with new components, inspect the valve clearance.

14. If installing the right-hand cylinder head, perform the following:

a. Position the alternator bracket in place, then install the 2 long bolts to 31–39 ft. lbs. (41–54 Nm). Install the short bolt and tighten to 18–22 ft. lbs. (24–31 Nm).

b. Install the idler pulley. Tighten the center retaining bolt to 35–46 ft. lbs. (47–63 Nm).

c. Install the alternator.

15. If installing the left-hand cylinder head, complete the following steps:

a. Position the air conditioning compressor bracket and power steering pump in place, then start the air conditioning compressor bracket bolt. Install the 3 compressor bracket bolts to 30–40 ft. lbs. (40–55 Nm). Then, install the 2 compressor bracket nuts to 16–21 ft. lbs. (21–29 Nm).

b. Install the air conditioning compressor.

c. Install the power steering pump reservoir. Tighten the hold-down bolts to 80–107 inch lbs. (9–12 Nm).

16. Install or connect the following:
- Exhaust manifold
- The 2 rocker arm covers. Inspect the rocker arm cover gaskets for damage prior to installation; replace them if necessary.
- Lower intake manifold and related components
- Upper intake manifold and related components

17. Drain the engine oil into a suitable container and replace the oil filter.

18. Fill the engine with the proper amount of engine oil.

19. Fill the cooling system.

20. Evacuate and recharge the air conditioning system using approved equipment.

21. Connect the negative battery cable.

22. Start the engine and check for any fluid or vacuum leaks.

2002–03 E- and F-Series, 2004 Heritage Editions, and Blackwood 4.6L, 5.4L and 6.8L Engines

✳✳ CAUTION

Fuel injection systems remain under pressure, even after the engine has been turned OFF. The fuel system pressure must be relieved before disconnecting any fuel lines. Failure to do so may result in fire and/or personal injury.

➡ **To correctly tighten the cylinder head bolts an angle torque wrench is needed.**

1. Before servicing the vehicle, refer to the Precautions Section.

2. Remove or disconnect the following:
- Air conditioning system, using approved equipment
- Negative battery cable
- Cylinder head covers
- Intake manifold
- Timing chains from the engine
- Exhaust manifolds
- The 2 heater hose retaining bolts, then compress and slide the hose clamp back to remove the heater water hose.
- Any remaining hoses, electrical connections or cables
- The cylinder head bolts, then lift the cylinder head from the engine block. Discard the cylinder head gasket and clean the engine block surface.

To install:

✳✳ WARNING

Cylinder head bolts must be replaced with new ones. They are torque-to-yield designed and cannot be reused.

3. Turn the crankshaft to position the keyway at the 12 o'clock position.

4. Clean and inspect the cylinder head for damage or warpage. Install the cylinder head gasket over the dowel pins. Then, install the cylinder head onto the engine block. Loosely install NEW cylinder head bolts.

➡ **Be sure to tighten the head bolts in 3 steps.**

RIGHT   **LEFT**

7924FG04

Tighten the cylinder head bolts using 3 steps in this sequence—4.6L and 5.4L engine

7924FG81

Tighten the cylinder head bolts using 3 steps in this sequence—6.8L engine

5. Tighten the cylinder head bolts in the correct sequence using 3 steps, as follows:

    a. Step 1—30 ft. lbs. (40 Nm).

    b. Step 2—tighten an additional 85–95 degrees.

    c. Step 3—tighten another 85–95 degrees.

6. Install or connect the following:
- Heater hose
- Exhaust manifolds
- Timing chains
- Intake manifold
- Cylinder head covers
- Electrical connections and cables

- Refrigerant
- Engine oil
- Coolant
- Negative battery cable

7. Start the engine and check for any fluid or vacuum leaks.

2004–05 E- and F-150 4.6L Engine, Exc. 2004 Heritage Edition

1. Before servicing the vehicle, refer to the Precautions Section.

2. Remove the engine.

3. Remove the bolts and the flexplate or the flywheel.

4. Install the engine onto a suitable engine stand.

    Left cylinder head

5. Remove the left exhaust manifold.

6. Remove the bolt and the oil level indicator tube.

7. Remove the engine wiring harness retainers from the left valve cover studs.

➡ **The bolts are part of the valve cover and should not be removed.**

8. Remove the bolts and the left valve cover.

    Right cylinder head

9. Remove the right exhaust manifold.

10. Remove the engine wiring harness retainers from the right valve cover studs.

➡ **The bolts are part of the valve cover and should not be removed.**

11. Remove the bolts and the right valve cover.

12. Disconnect the coolant hoses from the heater outlet tube.

13. Remove the heater outlet tube studs.

14. Remove the heater outlet tube and discard the O-ring seal.

    All cylinder heads

15. Remove the bolts, the coolant pump pulley and the accessory drive belt idler pulleys.

16. Remove the bolts and the accessory drive belt tensioner.

17. Disconnect the left and right radio ignition interference capacitor electrical connectors.

18. Remove the nuts and the two radio interference capacitors.

19. Remove the bolt and the left CMP sensor.

20. Connect the crankshaft position (CKP) sensor electrical connector.

21. Remove the bolt and washer and using a suitable tool, remove the crankshaft pulley. Discard the crankshaft bolt.

22. Using a suitable tool, remove the crankshaft seal.

23. Remove the four oil pan front bolts.

24. Remove the bolts.

✲✲ WARNING

Do not use metal scrapers, wire brushes, power abrasive discs or other abrasive means to clean the sealing surfaces. These tools cause scratches and gouges which make leak paths. Use a plastic scraping tool to remove all traces of old sealant.

25. Remove the engine front cover from the front cover to cylinder block dowel.

26. Remove the engine front cover gaskets.

27. Clean the mating surfaces with silicone gasket remover and metal surface prep. Follow the directions on the packaging.

28. Inspect the mating surfaces.

✶✶ WARNING

Use care when removing the spark plugs.

→Use compressed air to remove any foreign material from the spark plug well before removing the spark plugs.

29. Remove the eight spark plugs.

30. Install a suitable tool between the valve spring coils to prevent valve stem seal damage.

→The camshaft roller followers must be reinstalled in their original locations. Record the camshaft roller follower locations.

→Position the cam lobe away from the camshaft roller follower prior to removing each camshaft roller follower.

31. Use a suitable tool to compress the valve springs, and remove the camshaft roller followers.

32. Remove the special tool.

→The camshaft roller followers must be reinstalled in their original locations.

33. Repeat the previous three steps for each of the roller followers.

34. Remove the crankshaft sensor ring from the crankshaft.

35. Position the crankshaft with the keyway at the 12 o'clock position.

36. Remove the timing chain tensioning system from both timing chains.

37. Remove the right timing chain from the camshaft sprocket.

38. Remove the right timing chain from the crankshaft sprocket.

39. Remove the left timing chain from the camshaft sprocket.

40. Remove the left timing chain and crankshaft sprocket(s).

41. Remove both timing chain guides.

42. Install the lifting tools on both ends of the cylinder head being serviced.

✶✶ WARNING

The cylinder head must be cool before removing it from the engine. Cylinder head warpage can result if a warm or hot cylinder head is removed.

Special tool installation

67197-EFSE-G09

Special tool installation to compress the valve spring

67197-EFSE-G10

Senor ring

67197-EFSE-G11

Keyway at the 12 o'clock position

67197-EFSE-G12

✶✶ WARNING

Place clean shop towels over exposed engine cavities. Carefully remove the towels so foreign material is not dropped into the engine.

✶✶ WARNING

The cylinder head bolts must be discarded and new bolts must be

installed. They are tighten-to-yield designed and cannot be reused.

✶✶ WARNING

Do not use metal scrapers, wire brushes, power abrasive discs or other abrasive means to clean the sealing surfaces. These tools cause scratches and gouges that make leak paths. Use a plastic scraping tool to remove all traces of the head gasket.

✶✶ WARNING

Aluminum surfaces are soft and can be scratched easily. Never place the cylinder head gasket surface, unprotected, on a bench surface.

43. Remove the bolts and the cylinder head. Discard the cylinder head gasket. Discard the cylinder head bolts.

✶✶ WARNING

Do not use metal scrapers, wire brushes, power abrasive discs or other abrasive means to clean the sealing surfaces. These tools cause scratches and gouges that make leak paths. Use a plastic scraping tool to remove all traces of the head gasket.

✶✶ WARNING

Observe all warnings or cautions and follow all application directions contained on the packaging of the silicone gasket remover and the metal surface prep.

→If there is no residual gasket material present, metal surface prep can be used to clean and prepare the surfaces.

44. Clean the cylinder head-to-cylinder block mating surfaces of both the cylinder head and the cylinder block.

45. Remove any large deposits of silicone or gasket material with a plastic scraper.

46. Apply silicone gasket remover, following package directions, and allow to set for several minutes.

47. Remove the silicone gasket remover with a plastic scraper. A second application of silicone gasket remover may be required if residual traces of silicone or gasket material remain.

67197-EFSE-G13

When checking flatness, pay particular attention to these areas

48. Apply metal surface prep, following package directions, to remove any remaining traces of oil or coolant, and to prepare the surfaces to bond with the new gasket. Do not attempt to make the metal shiny. Some staining of the metal surfaces is normal.

➥Make sure all cylinder head surfaces are clear of any gasket material, RTV, oil and coolant. The cylinder head surface must be clean and dry before running a flatness check.

➥Use a straightedge that is calibrated by the manufacturer to be flat with 0.005 mm (0.0002 in.) per running foot length. For example, if the straightedge is 61 cm (24 in.) long, the machine edge must be flat with 0.010 mm (0.0004 in.) from end to end.

49. Support the cylinder head on a bench with the head gasket side up. Inspect all areas of the deck face with a straightedge, paying particular attention to the oil pressure feed area. The cylinder head must not have depressions deeper than 0.0254 mm (0.001 in.) across a 38.1 mm (1.5 in.) square area, or scratches more than 0.0254 mm (0.001 in.).

303-1040

67197-EFSE-G14

Dowel installation

To install:
All cylinder heads

☀☀ WARNING

Make sure all coolant residue and foreign material are cleaned from the block surface and cylinder bore.

☀☀ WARNING

The use of sealing aids. The gasket must be installed dry.

☀☀ WARNING

The cylinder head bolts must be discarded and new bolts installed. They are tighten-to-yield designed and cannot be reused.

➡ Do not turn the crankshaft until instructed to do so.

50. Using the lifting tools, position the cylinder head gaskets and cylinder heads over the dowels and install the cylinder head bolts loosely.

51. Tighten the bolts in the sequence shown.

Plunger held by a retaining clip

    a. Stage 1: Tighten to 40 Nm (30 ft. lbs.).
    b. Stage 2: Tighten an additional 90 degrees.
    c. Stage 3: Tighten an additional 90 degrees.

All cylinder heads

☀☀ WARNING

Timing chain procedures must be followed exactly or damage to valves and pistons will result.

52. Compress the tensioner plunger, using a vise.

53. Install a retaining clip on the tensioner to hold the plunger in during installation.

67197-EFSE-G16

If the copper links are not visible, mark one link on one end and one link on the other end, and use as timing marks

67197-EFSE-G17

Install the crankshaft sprocket, making sure the flange faces forward

Timing chain guide installation

67197-EFSE-G18

54. Remove the tensioner from the vise. If the copper links are not visible, mark one link on one end and one link on the other end, and use as timing marks.

55. Install the crankshaft sprocket, making sure the flange faces forward.

56. Position the left timing chain guide.

57. Install and tighten the left bolts.

58. Position the right timing chain guide.

59. Install and tighten the right bolts.

60. Rotate the right camshaft sprocket until the timing mark is approximately at the 11 o'clock position. Rotate the left camshaft sprocket until the timing mark is approximately at the 12 o'clock position.

➡**The number one cylinder is at top dead center (TDC) when the stud on the engine block fits into the slot in the handle of the special tool.**

61. Position the crankshaft so the number one cylinder is at TDC with the special tool.

62. Remove the Crankshaft Holding Tool.

63. Position the left (inner) timing chain on the crankshaft sprocket, aligning the copper (marked) link with the timing mark on the sprocket.

64. Install the left timing chain on the camshaft sprocket, aligning the copper (marked) link with the timing marks on the sprocket.

➡**The left timing chain tensioner arm has a bump near the dowel hole for identification.**

65. Position the left timing chain tensioner arm on the dowel pin and install the left timing chain tensioner.

66. Remove the retaining clip from the left timing chain tensioner. Torque to 18 ft. lbs. (25 Nm).

67. Position the right (outer) timing chain on the crankshaft sprocket, aligning the copper (marked) link with the timing mark on the sprocket.

68. Install the right timing chain on the camshaft sprocket, aligning the copper (marked) link with the timing marks on the sprocket.

Crankshaft alignment tool installed

67197-EFSE-G20

Position the left (inner) timing chain on the crankshaft sprocket, aligning the copper (marked) link with the timing mark on the sprocket

67197-EFSE-G21

Install the left timing chain on the camshaft sprocket, aligning the copper (marked) link with the timing marks on the sprocket

69. Position the right timing chain tensioner arm on the dowel pin and install the right timing chain tensioner. Torque to 18 ft. lbs. (25 Nm).

70. Remove the retaining clip from the right timing chain tensioner. Make sure that the copper (marked) chain links are lined up with the dots on the crankshaft sprockets and the camshaft sprocket.

71. Install the crankshaft sensor ring on the crankshaft.

❊❊ WARNING

Do not use metal scrapers, wire brushes, power abrasive discs or other abrasive means to clean the sealing surfaces. These tools cause scratches and gouges which make leak paths. Use a plastic scraping tool to remove all traces of old sealant.

Position the right (outer) timing chain on the crankshaft sprocket, aligning the copper (marked) link with the timing mark on the sprocket

67197-EFSE-G22

Install the right timing chain on the camshaft sprocket, aligning the copper (marked) link with the timing marks on the sprocket

67197-EFSE-G23

➡If the engine front cover is not secured within four minutes, the sealant must be removed and the sealing area cleaned. To clean the sealing area, use silicone gasket remover and metal surface prep. Follow the directions on the packaging. Failure to follow this procedure can cause future oil leakage.

➡Make sure that the engine front cover gasket is in place on the engine front cover before installation.

72. Apply a bead of silicone gasket and sealant along the cylinder head-to-cylinder block surface and the oil pan-to-cylinder block surface, at the locations shown.

73. Install a new engine front cover gasket on the engine front cover. Position the engine front cover. Install the fasteners finger-tight.

74. Tighten the engine front cover fasteners in sequence in three stages.

   a. Stage 1: Tighten fasteners 1 through 5 to 25 Nm (18 ft. lbs.).

   b. Stage 2: Tighten fasteners 6 and 7 to 25 Nm (18 ft. lbs.).

   c. Stage 3: Tighten fasteners 8 through 15 to 25 Nm (18 ft. lbs.).

75. Install the left camshaft position (CMP) sensor and the bolt.

76. Lubricate the new O-ring seal with clean engine oil prior to installation.

77. Lubricate the engine front cover and the crankshaft seal inner lip with clean engine oil. Use a driver to install the crankshaft seal into the engine front cover.

➡If not secured within four minutes, the sealant must be removed and the sealing area cleaned. To clean the sealing area, use silicone gasket remover and metal surface prep. Follow the directions on the packaging. Failure to follow this procedure can cause future oil leakage.

78. Apply silicone gasket and sealant to the Woodruff key slot on the crankshaft pulley. Use the special tool to install the crankshaft pulley.

79. Tighten the new crankshaft pulley bolt in four stages.

   a. Stage 1: Tighten to 90 Nm (66 ft. lbs.).

   b. Stage 2: Loosen 360 degrees.

   c. Stage 3: Tighten to 50 Nm (37 ft. lbs.).

   d. Stage 4: Tighten an additional 90 degrees.

80. Install the three accessory drive belt idler pulleys, the coolant pump pulley and the bolts.

81. Position the accessory drive belt tensioner and install the bolts.

82. Install a suitable tool between the valve spring coils to prevent valve stem seal damage.

➡The camshaft roller followers must be reinstalled in their original locations.

➡Position the cam lobe away from the valve stem prior to installing each camshaft roller follower.

83. Use a suitable tool to compress the valve springs, and install the camshaft roller follower. Remove the special tool.

➡The camshaft roller followers must be reinstalled in their original locations.

67197-EFSE-G24

Apply a bead of silicone gasket and sealant along the cylinder head-to-cylinder block surface and the oil pan-to-cylinder block surface, at the locations shown

67197-EFSE-G25

Front cover fastener tightening sequence

84. Repeat the previous four steps for each of the camshaft roller followers.

85. Install the radio frequency interference capacitors.

Left cylinder head

❋❋ WARNING

Do not use metal scrapers, wire brushes, power abrasive discs or other abrasive means to clean sealing surfaces. These tools cause scratches and gouges which make leak paths.

86. Inspect and clean the valve cover sealing surfaces with metal surface cleaner.

87. Apply instant adhesive completely around the gasket groove in the left valve cover.

88. Install the new valve cover gasket.

➡ **If not secured within four minutes, the sealant must be removed and the sealing area cleaned. To clean the sealing area, use silicone gasket remover and metal surface prep.**

89. Follow the directions on the packaging. Failure to follow this procedure can cause future oil leakage.

90. Apply silicone gasket and sealant in two places where the engine front cover meets the cylinder head.

91. Position the left valve cover and gasket on the cylinder head and install the bolts loosely.

92. Tighten the valve cover bolts in the sequence shown.

➡ **Lubricate the O-ring seal with clean engine oil.**

93. Install the oil level indicator tube.

94. Install a new O-ring seal on the oil level indicator tube.

95. Install the oil level indicator tube.

96. Install the bolt.

97. Install the left exhaust manifold and the exhaust manifold gasket. Tighten the nuts in the sequence shown.

Apply silicone gasket and sealant in two places where the engine front cover meets the cylinder head

10 Nm (89 lb-in)

67197-EFSE-G27

Tighten the valve cover bolts in the sequence shown

25 Nm (18 lb-ft)

67197-EFSE-G28

Install the left exhaust manifold nuts in the sequence shown

Right cylinder head

❋❋ WARNING

Do not use metal scrapers, wire brushes, power abrasive discs or other abrasive means to clean sealing surfaces. These tools cause scratches and gouges which make leak paths.

98. Inspect and clean the valve cover sealing surfaces with metal surface cleaner.

99. Apply instant adhesive completely around the gasket groove in the right valve cover. Install the new valve cover gasket.

➥If not secured within four minutes, the sealant must be removed and the sealing area cleaned. To clean the sealing area, use silicone gasket remover and metal surface prep. Follow the directions on the packaging. Failure to follow this procedure can cause future oil leakage.

100. Apply silicone gasket and sealant in two places where the engine front cover meets the cylinder head.

101. Position the right valve cover and gasket on the cylinder head and install the bolts loosely.

102. Tighten the valve cover bolts in the sequence shown.

103. Install the right exhaust manifold gaskets and the exhaust manifold. Tighten the nuts in the sequence shown.

104. Slide the heater outlet tube forward with a new O-ring seal into the cylinder block. Lubricate the O-ring seal with engine coolant.

105. Install the heater outlet tube studs.

106. Connect the coolant hoses to the heater outlet tube.

All cylinder heads

107. Connect the right radio ignition interference capacitor electrical connector.

108. Connect the left radio ignition interference capacitor and cylinder head temperature (CHT) sensor electrical connectors.

109. Connect the CMP sensor electrical connectors.

110. Connect the CKP sensor electrical connector.

111. Install a suitable tool.

112. Using a suitable floor crane, remove the engine from the engine stand.

113. Install the flexplate or the flywheel and bolts. Tighten the bolts in the sequence shown.

114. Install the engine.

2004–05 E- and F-150/250/350 5.4L Engine, Exc. Heritage Edition; 2006 Mark LT

1. Before servicing the vehicle, refer to the Precautions Section.

2. Remove the engine.

3. Remove the bolts and the flexplate or the flywheel.

4. Install the engine onto a suitable engine stand.

5. Remove the special tool.

Left cylinder head

6. Remove the left exhaust manifold.

7. Remove the bolt and the oil level indicator tube.

8. Remove the engine wiring harness retainers from the left valve cover studs.

✳✳ WARNING

When removing the valve cover, make sure to avoid damaging the variable camshaft timing (VCT) solenoid.

9. Remove the bolts and the left valve cover.

Right cylinder head

10. Remove the right exhaust manifold.

11. Remove the nuts.

12. Remove the right exhaust manifold.

13. Remove and discard the right exhaust manifold gasket.

14. Remove the engine wiring harness retainers from the right valve cover studs.

	Alignment Pins, Cylinder Head 303-1040 (SR-015486)
	Installer, Crankshaft Vibration Damper 303-102 (T74P-6316-B)
	Installer, Crankshaft Front Oil Seal 303-635
	Installer, Front Cover Oil Seal 303-335 (T88T-6701-A)
	Modular Engine Lift Bracket 303-F047 (014-00073) or equivalent
	Compressor, Valve Spring 303-1039

67197-EFSE-G33

Special tools needed for this procedure. These tools are referred to in the following procedure

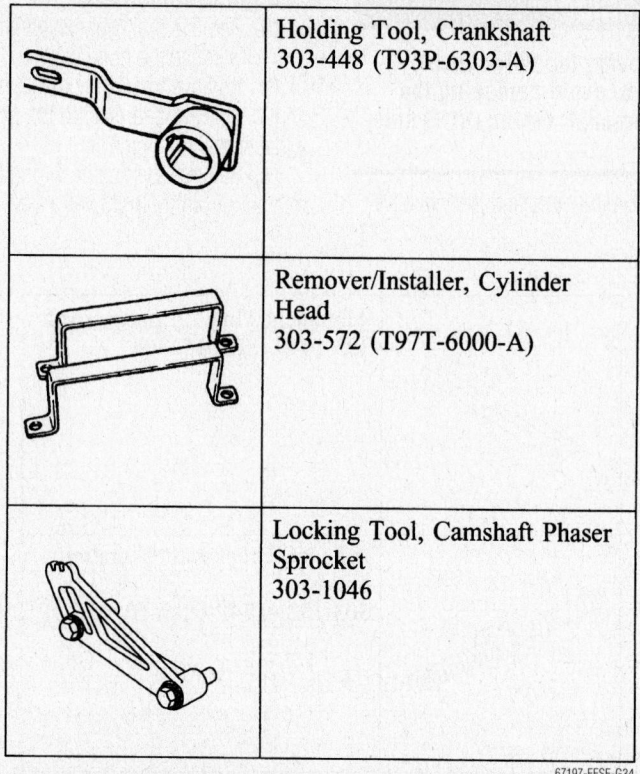

	Holding Tool, Crankshaft 303-448 (T93P-6303-A)
	Remover/Installer, Cylinder Head 303-572 (T97T-6000-A)
	Locking Tool, Camshaft Phaser Sprocket 303-1046

67197-EFSE-G34

Special tools needed for this procedure. These tools are referred to in the following procedure

※※ WARNING

When removing the valve cover, make sure to avoid damaging the variable camshaft timing (VCT) solenoid.

15. Remove the bolts and the right valve cover.

16. Remove the stud.

All cylinder heads

17. Remove the bolts, the coolant pump pulley and the accessory drive belt idler pulleys.

18. Remove the bolt and washer and using a puller set, remove the crankshaft pulley. Discard the crankshaft bolt.

19. Using a suitable tool, remove the crankshaft seal.

20. Remove the bolts and the accessory drive belt tensioner.

21. Disconnect the left and right radio ignition interference capacitor electrical connectors.

22. Remove the nuts and the two radio interference capacitors.

23. Disconnect the camshaft position (CMP) sensor electrical connectors.

24. Remove the bolt and the right CMP sensor.

25. Remove the bolt and the left CMP sensor.

26. Disconnect the crankshaft position (CKP) sensor electrical connector.

27. Remove the oil pan front bolts.

28. Remove the bolts.

※※ WARNING

Do not use metal scrapers, wire brushes, power abrasive discs or other abrasive means to clean the sealing surfaces. These tools cause scratches and gouges which make leak paths. Use a plastic scraping tool to remove all traces of old sealant.

29. Remove the engine front cover from the front cover-to-cylinder block dowels.

30. Remove the engine front cover gaskets.

31. Clean the mating surfaces with silicone gasket remover and metal surface prep. Follow the directions on the packaging.

32. Inspect the mating surfaces.

※※ WARNING

Do not allow the valve keepers to fall off of the valve or the valve can drop into the cylinder.

➡It may be necessary to push the valve down while compressing the valve spring.

➡The roller followers must be installed in their original positions.

33. Using a suitable tool, remove all of the roller followers. Record the roller follower positions.

34. Position the crankshaft keyway at the 12 o'clock position.

35. Remove the bolts, the left timing chain tensioner and tensioner arm.

36. Remove the bolts, the right timing chain tensioner and tensioner arm.

37. Remove the ignition pulse wheel from the crankshaft.

38. Remove the right timing chain from the camshaft sprocket.

39. Remove the right timing chain from the crankshaft sprocket.

40. Remove the left timing chain from the camshaft sprocket.

41. Remove the left timing chain and crankshaft sprocket.

42. Remove both timing chain guides.

※※ WARNING

Use only hand tools to remove the camshaft phaser sprocket assembly or damage can occur to the camshaft or camshaft phaser sprocket.

43. If disassembly of the cylinder head is required, using a suitable tool, loosen the camshaft phaser sprocket bolt.

44. Install a suitable tool onto the left cylinder head. Install a suitable tool onto the right cylinder head.

※※ WARNING

The cylinder head must be cool before removing it from the engine. Cylinder head warpage can result if a warm or hot cylinder head is removed.

※※ WARNING

Do not use the variable camshaft timing (VCT) phaser sprocket as a lifting point or leveraging device when removing the cylinder head or damage to the VCT phaser sprocket can occur.

※※ WARNING

Place clean shop towels over exposed engine cavities. Carefully remove the towels so foreign material is not dropped into the engine.

✳ WARNING

The cylinder head bolts must be discarded and new bolts must be installed. They are tighten-to-yield designed and cannot be reused.

✳ WARNING

Do not use metal scrapers, wire brushes, power abrasive discs or other abrasive means to clean the sealing surfaces. These tools cause scratches and gouges that make leak paths. Use a plastic scraping tool to remove all traces of the head gasket.

✳ WARNING

Aluminum surfaces are soft and can be scratched easily. Never place the cylinder head gasket surface, unprotected, on a bench surface.

45. Remove the bolts and the cylinder head. Discard the cylinder head gasket. Discard the cylinder head bolts.

✳ WARNING

Do not use metal scrapers, wire brushes, power abrasive discs or other abrasive means to clean the sealing surfaces. These tools cause scratches and gouges that make leak paths. Use a plastic scraping tool to remove all traces of the head gasket.

✳ WARNING

Observe all warnings or cautions and follow all application directions contained on the packaging of the silicone gasket remover and the metal surface prep.

➡ If there is no residual gasket material present, metal surface prep can be used to clean and prepare the surfaces.

46. Clean the cylinder head-to-cylinder block mating surfaces of both the cylinder head and the cylinder block.

47. Remove any large deposits of silicone or gasket material with a plastic scraper.

48. Apply silicone gasket remover, following package directions, and allow to set for several minutes.

49. Remove the silicone gasket remover with a plastic scraper. A second application of silicone gasket remover may be required if residual traces of silicone or gasket material remain.

50. Apply metal surface prep, following package directions, to remove any remaining traces of oil or coolant, and to prepare the surfaces to bond with the new gasket. Do not attempt to make the metal shiny. Some staining of the metal surfaces is normal.

➡ Make sure all cylinder head surfaces and engine block surfaces are clear of any gasket material, RTV, oil and coolant. The cylinder head and engine block surfaces must be clean and dry before running a flatness check.

➡ Use a straightedge that is calibrated by the manufacturer to be flat within 0.005 mm (0.0002 in) per running foot of length. For example, if the straightedge is 61 cm (24 in) long, the machined edge must be flat with 0.010 mm (0.0004 in) from end to end.

51. Support the cylinder head on a bench with the head gasket side up. Inspect all areas of the deck face with a straightedge, paying particular attention to the oil pressure feed area. The cylinder head must not have depressions deeper than 0.0254 mm (0.001 in) across a 38.1 mm (1.5 in) square area, or scratches more than 0.0254 mm (0.001 in).

To install:
All cylinder heads

✳ WARNING

Make sure all coolant residue and foreign material are cleaned from the block surface and cylinder bore.

✳ WARNING

The use of sealing aids (aviation cement, copper spray, and glue) is not permitted. The gasket must be installed dry.

✳ WARNING

The cylinder head bolts must be discarded and new bolts installed. They are tighten-to-yield designed and cannot be reused.

✳ WARNING

Do not allow the cylinder head alignment pins to contact the cylinder head gasket or cylinder head sealing surfaces or damage can occur to the cylinder head or cylinder head gasket.

➡ Do not turn the crankshaft until instructed to do so.

52. Using the cylinder head alignment pins, position the cylinder head gasket and cylinder head onto the dowels and install the cylinder head bolts loosely.

53. Tighten the bolts in the sequence shown.

   a. Stage 1: Tighten to 40 Nm (30 ft. lbs.).

   b. Stage 2: Tighten an additional 90 degrees.

   c. Stage 3: Tighten an additional 90 degrees.

54. Remove the special tool.

✳ WARNING

Timing chain procedures must be followed exactly or damage to valves and pistons will result.

55. Compress the tensioner plunger, using a vice.

56. Install a retaining clip on the tensioner to hold the plunger in during installation.

57. Remove the tensioner from the vise. If the copper links are not visible, mark two links on one end and one link on the other end, and use as timing marks.

✳ WARNING

Crankshaft keyway must be in the 12 o'clock position.

58. Install the timing chain guides.

59. Position the left timing chain guide.

60. Install the crankshaft sprocket, making sure the flange faces forward.

✳ WARNING

Only use hand tools to install the camshaft phaser sprocket assembly or damage may occur to the camshaft or camshaft phaser unit.

➡ This step is only required if cylinder head was disassembled.

61. Using a suitable tool, tighten the bolts in two stages:

   a. Stage 1: Tighten to 40 Nm (30 ft. lbs.).

   b. Stage 2: Tighten an additional 90 degrees.

62. Remove the special tool.

67197-EFSE-G35

Left cylinder head torque sequence—5.4L engine

67197-EFSE-G36

Right cylinder head torque sequence—5.4L engine

67197-EFSE-G15

Plunger held by a retaining clip

67197-EFSE-G16

If the copper links are not visible, mark one link on one end and one link on the other end, and use as timing marks

67197-EFSE-G17

Install the crankshaft sprocket, making sure the flange faces forward

63. Position the lower end of the left (inner) timing chain on the crankshaft sprocket, aligning the timing mark on the outer flange of the crankshaft sprocket with the single copper (marked) link on the chain.

➡**Make sure the upper half of the timing chain is below the tensioner arm dowel.**

➡**The left timing chain tensioner arm has a bump near the dowel hole for identification.**

64. Position the left timing chain tensioner arm on the dowel pin and install the left timing chain tensioner.

65. Remove the retaining clip from the left timing chain tensioner.

66. Position the lower end of the right

10 Nm (89 lb-in)

10 Nm
(89 lb-in)

67197-EFSE-G18

Timing chain guide installation

303-1046

67197-EFSE-G37

Camshaft phaser sprocket holding tool— 5.4L engine

(outer) timing chain on the crankshaft sprocket, aligning the timing mark on the sprocket with the single copper (marked) chain link.

➡**The lower half of the timing chain must be positioned above the tensioner arm dowel.**

67. Position the right timing chain on the camshaft sprocket. Make sure the camshaft sprocket timing mark is positioned between the two copper (marked) chain links.

68. Position the right timing chain tensioner arm on the dowel pin and install the right timing chain tensioner.

67197-EFSE-G38

Position the lower end of the left (inner) timing chain on the crankshaft sprocket, aligning the timing mark on the outer flange of the crankshaft sprocket with the single copper (marked) link on the chain—5.4L engine

69. Remove the retaining clip from the right timing chain tensioner.

➡**Both camshaft phaser sprockets are identical. Refer to the R timing mark to identify the right camshaft phaser sprocket and the L timing mark to identify the left camshaft phaser sprocket.**

70. As a post-check, verify correct alignment of all timing marks. Make sure the R

67197-EFSE-G39

Position the timing chain on the camshaft sprocket with the camshaft sprocket timing mark positioned between the two copper (marked) chain links—5.4L engine

and L timing marks on the sprockets correspond to the above note.

71. Install the crankshaft sensor ring on the crankshaft.

➡**Lubricate the camshaft roller followers using clean engine oil.**

➡**Using the mark on each camshaft roller follower, make sure it is returned to its original position.**

72. Using a suitable tool, install all of the camshaft roller followers.

67197-EFSE-G40

Position the lower end of the right (outer) timing chain on the crankshaft sprocket, aligning the timing mark on the sprocket with the single copper (marked) chain link

67197-EFSE-G41

Position the right timing chain on the camshaft sprocket. Make sure the camshaft sprocket timing mark is positioned between the two copper (marked) chain links

�֍ WARNING

Do not use metal scrapers, wire brushes, power abrasive discs or other abrasive means to clean the sealing surfaces. These tools cause scratches and gouges which make leak paths. Use a plastic scraping tool to remove all traces of old sealant.

➡**If the engine front cover is not secured within four minutes, the sealant must be removed and the sealing area cleaned. To clean the sealing area, use silicone gasket remover and metal surface prep. Follow the directions on the packaging. Failure to follow this procedure can cause future oil leakage.**

➡**Make sure that the engine front cover gasket is in place on the engine front cover before installation.**

73. Apply a bead of silicone gasket and sealant along the cylinder head-to-cylinder block surface and the oil pan-to-cylinder block surface, at the locations shown.

74. Install a new engine front cover gasket on the engine front cover. Position the engine front cover. Install the fasteners finger-tight.

75. Tighten the engine front cover fasteners in sequence in two stages.

a. Stage 1: Tighten fasteners 1 through 15 to 25 Nm (18 ft. lbs.).

b. Stage 2: Tighten fasteners 6 and 7 to 48 Nm (35 ft. lbs.).

c. Loosely install the pan-to-case bolts, then tighten the bolts in two stages, in the sequence shown.

d. Stage 1: Tighten to 20 Nm (15 ft. lbs.).

e. Stage 2: Tighten an additional 60 degrees.

76. Install the left camshaft position (CMP) sensor and the bolt.

77. Lubricate the new O-ring seal with clean engine oil prior to installation.

78. Install the right CMP sensor and the bolt.

79. Lubricate the new O-ring seal with clean engine oil prior to installation.

80. Lubricate the engine front cover and the crankshaft seal inner lip with clean engine oil.

81. Use the special tools to install the crankshaft seal into the engine front cover.

82. If not secured within four minutes, the sealant must be removed and the sealing area cleaned. To clean the sealing area, use silicone gasket remover and metal surface prep. Follow the directions on the

67197-EFSE-G42

Apply a bead of silicone gasket and sealant along the cylinder head-to-cylinder block surface and the oil pan-to-cylinder block surface, at the locations shown—5.4L engine

Front cover fastener tightening sequence

67197-EFSE-G43

Loosely install the pan-to-case bolts, then tighten the bolts in two stages, in the sequence shown

67197-EFSE-G44

packaging. Failure to follow this procedure can cause future oil leakage.

83. Apply silicone gasket and sealant to the Woodruff key slot on the crankshaft pulley. Use a suitable tool to install the crankshaft pulley.

84. Tighten the new crankshaft pulley bolt in four stages.

    a. Stage 1: Tighten to 90 Nm (66 ft. lbs.).

    b. Stage 2: Loosen 360 degrees.

    c. Stage 3: Tighten to 50 Nm (37 ft. lbs.).

    d. Stage 4: Tighten an additional 90 degrees.

85. Install the three accessory drive belt idler pulleys, the coolant pump pulley and the bolts. Torque all bolts to 18 ft. lbs. (25 Nm).

86. Position the accessory drive belt tensioner and install the bolts. Torque all bolts to 18 ft. lbs. (25 Nm).

87. Install the radio frequency interference capacitors.

Left cylinder head

✳✳ WARNING

Do not use metal scrapers, wire brushes, power abrasive discs or other abrasive means to clean sealing surfaces. These tools cause scratches and gouges which make leak paths. Use a plastic scraping tool to remove all traces of old sealant. Inspect and clean the valve cover sealing surfaces with silicone gasket remover and metal surface prep. Follow the directions on the packaging.

➡ If not secured within four minutes, the sealant must be removed and the sealing area cleaned. To clean the sealing area, use silicone gasket remover and metal surface prep. Follow the directions on the packaging. Failure to follow this procedure can cause future oil leakage.

88. Apply silicone gasket and sealant in two places where the engine front cover meets the cylinder head.

✳✳ WARNING

When installing the valve cover, make sure to avoid damaging the variable camshaft timing (VCT) solenoid.

89. Position the left valve cover and gasket on the cylinder head and install the bolts loosely. Tighten the bolts in the sequence shown.

303-635

303-102

303-335

67197-EFSE-G45

Crankshaft seal installation tools

10 Nm (89 lb-in)

67197-EFSE-G46

Left valve cover torque sequence—5.4L engine

➡**Lubricate the O-ring seal with clean engine oil.**

90. Install the oil level indicator tube.
91. Install a new O-ring seal on the oil level indicator tube.
92. Install the oil level indicator tube.
93. Install the bolt.
94. Install the left exhaust manifold and

the exhaust manifold gaskets. Tighten the nuts in the sequence shown.

Right cylinder head

✳ WARNING

Do not use metal scrapers, wire brushes, power abrasive discs or other abrasive means to clean seal-

ing surfaces. These tools cause scratches and gouges which make leak paths. Use a plastic scraping tool to remove all traces of old sealant. Inspect and clean the valve cover sealing surfaces with silicone gasket remover and metal surface prep. Follow the directions on the packaging.

➡If not secured within four minutes, the sealant must be removed and the sealing area cleaned. To clean the sealing area, use silicone gasket remover and metal surface prep. Follow the directions on the packaging. Failure to follow this procedure can cause future oil leakage.

95. Apply silicone gasket and sealant in two places where the engine front cover meets the cylinder head.

✳ WARNING

When installing the valve cover, make sure to avoid damaging the variable camshaft timing (VCT) solenoid.

96. Position the right valve cover and gasket on the cylinder head and install the bolts loosely. Tighten the bolts in the sequence shown.
97. Install the right exhaust manifold gaskets and the exhaust manifold. Tighten the nuts in the sequence shown.
98. Install the heater outlet tube stud.

All cylinder heads

99. Position the electrical harness on the valve cover and connect the engine wiring harness retainers to the valve cover studs.
100. Connect the right radio ignition interference capacitor electrical connector.
101. Connect the left radio ignition interference capacitor and cylinder head temperature (CHT) sensor electrical connectors.
102. Connect the CMP sensor electrical connectors.
103. Connect the CKP sensor electrical connector.
104. Install a suitable tool.
105. Using a suitable floor crane remove the engine from the engine stand.
106. Install the flexplate or the flywheel and bolts. Tighten the bolts in the sequence shown.
107. Install the engine.

25 Nm (18 lb-ft)

67197-EFSE-G47

Left exhaust manifold torque sequence—5.4L engine

10 Nm (89 lb-in)

67197-EFSE-G48

Right valve cover torque sequence—5.4L engine

2004–05 E- and F-250/350 6.8L Engine

1. Before servicing the vehicle, refer to the Precautions Section.
2. Remove the engine.
3. Remove the bolts and the flexplate or the flywheel.

4. Mount the engine on a suitable work stand.
5. Remove the right engine mount.
6. Remove the cylinder block drain plugs and drain the coolant into a suitable container.
7. Disconnect the left radio frequency interference capacitor and cylinder head

temperature (CHT) sensor electrical connectors.
8. Disconnect the camshaft position (CMP) electrical connector.
9. Disconnect the right radio frequency interference capacitor electrical connector.
10. Disconnect the knock sensor electrical connector.
11. Disconnect the crankshaft position (CKP) sensor electrical connector.
12. Disconnect the oil pressure switch electrical connector.
13. Remove the nuts and position the wiring harness aside.
14. Disconnect all of the harness routing clips and connector retainers. Remove the engine control sensor wiring harness.
15. Remove the nuts and the two radio interference capacitors.
16. Remove the crankcase ventilation tube from the left valve cover.
17. Remove the positive crankcase ventilation (PCV) valve and hose from the right valve cover.

➡**Do not use metal scrapers, wire brushes, power abrasive discs or other abrasive means to clean the sealing surfaces. These tools cause scratches and gouges which make leak paths. Use a plastic scraping tool to remove all traces of old sealant.**

➡**The bolts are part of the valve cover and should not be removed.**

18. Fully loosen the bolts and remove the left valve cover.
19. Clean the valve cover mating surface of the cylinder head with silicone gasket remover and metal surface prep. Follow the directions on the packaging.
20. Inspect the valve cover gasket. If the gasket is damaged, remove and discard the gasket. Clean the valve cover gasket groove with soap and water or a suitable solvent.

➡**Do not use metal scrapers, wire brushes, power abrasive discs or other abrasive means to clean the sealing surfaces. These tools cause scratches and gouges which make leak paths. Use a plastic scraping tool to remove all traces of old sealant.**

➡**The bolts are part of the valve cover and should not be removed.**

21. Fully loosen the bolts and remove the right valve cover.
22. Clean the valve cover mating surfaces of the cylinder head with silicone gasket remover and metal surface prep. Follow the directions on the packaging.

25 Nm (18 lb-ft)

67197-EFSE-G49

Right exhaust manifold torque sequence—5.4L engine

80 Nm (59 lb-ft)

67197-EFSE-G50

Flywheel torque sequence—5.4L engine

23. Inspect the valve cover gasket. If the gasket is damaged, remove and discard the gasket. Clean the valve cover gasket groove with soap and water or a suitable solvent.

24. Remove the bolt and the idler pulley.

25. Remove the coolant pump pulley.

26. Using a suitable tool remove and discard the crankshaft pulley bolt.

27. Using a suitable tool, remove the crankshaft front seal.

28. Remove the four oil pan bolts.

➡**Correct fastener location is essential for assembly procedure. Record fastener location.**

29. Remove the fasteners.

30. Remove the engine front cover from the cylinder block.

31. Remove the crankshaft sensor ring from the crankshaft.

➡**The caps must be marked for installation in their original locations or damage to the engine may occur.**

32. Remove the six bolts and remove the balance shaft bearing caps.

33. Remove the balance shaft.

➡**Use care when removing the spark plugs.**

➡**Use compressed air to remove any foreign material from the spark plug well before removing the spark plugs.**

34. Remove the 10 spark plugs.

35. Install a suitable tool between the valve spring coils to prevent valve stem seal damage.

➡**The camshaft roller followers must be reinstalled in their original locations. Record the camshaft roller follower locations.**

➡**Position the cam lobe away from the camshaft roller follower prior to removing each camshaft roller follower.**

36. Use a suitable tool to compress the valve springs and remove the camshaft roller followers.

37. Position the crankshaft with the keyway at the 12 o'clock position.

38. Remove the timing chain tensioning system from both timing chains.

39. Remove the right timing chain from the camshaft sprocket.

40. Remove the right timing chain from the crankshaft sprocket.

41. Remove the left timing chain from the camshaft sprocket.

42. Remove the left timing chain from the crankshaft sprocket.

43. Remove both timing chain guides.

Right cylinder head

44. Remove the right exhaust manifold.

Left cylinder head

45. Remove the left exhaust manifold.

46. Remove the bolt and the oil level indicator tube.

Both cylinder heads

47. Clean and inspect the exhaust manifolds.

48. Install lifting on both ends of the cylinder head.

➡**The hydraulic lash adjusters must be reinstalled in their original locations.**

Record the hydraulic lash adjuster locations.

49. Remove the hydraulic lash adjusters.

Right cylinder head

➡The cylinder head must be cool before removing it from the engine. Cylinder head warpage can result if a warm or hot cylinder head is removed.

➡Place clean shop towels over exposed engine cavities. Carefully remove the towels so foreign material is not dropped into the engine.

➡The cylinder head bolts must be discarded and new bolts installed. They are tighten-to-yield and cannot be reused.

➡Do not use metal scrapers, wire brushes, power abrasive discs or other abrasive means to clean the sealing surfaces. These tools cause scratches and gouges which make leak paths. Use a plastic scraping tool to remove all traces of the head gasket.

➡Aluminum surfaces are soft and can be scratched easily. Never place the cylinder head gasket surface, unprotected, on a bench surface.

50. Remove the bolts and the right cylinder head. Discard the cylinder head gasket. Discard the cylinder head bolts.

Left cylinder head

➡The cylinder head must be cool before removing it from the engine. Cylinder head warpage can result if a warm or hot cylinder head is removed.

➡Place clean shop towels over exposed engine cavities. Carefully remove the towels so foreign material is not dropped into the engine.

➡The cylinder head bolts must be discarded and new bolts installed. They are tighten-to-yield and cannot be reused.

➡Do not use metal scrapers, wire brushes, power abrasive discs or other abrasive means to clean the sealing surfaces. These tools cause scratches and gouges which make leak paths. Use a plastic scraping tool to remove all traces of the head gasket.

➡Aluminum surfaces are soft and can be scratched easily. Never place the cylinder head gasket surface, unprotected, on a bench surface.

51. Remove the bolts and the left cylinder head. Discard the cylinder head gasket. Discard the cylinder head bolts.

Both cylinder heads

➡Do not use metal scrapers, wire brushes, power abrasive discs or other abrasive means to clean the sealing surfaces. These tools cause scratches and gouges which make leak paths. Use a plastic scraping tool to remove all traces of the head gasket.

➡Observe all warnings or cautions and follow all application directions contained on the packaging of the silicone gasket remover and the metal surface prep.

➡If there is no residual gasket material present, metal surface prep can be used to clean and prepare surfaces.

52. Clean the cylinder head-to-cylinder block mating surfaces of both the cylinder head and the cylinder block.

53. Remove any large deposits of silicone or gasket material with a plastic scraper.

54. Apply silicone gasket remover, following package directions, and allow to set for several minutes.

55. Remove the silicone gasket remover with a plastic scraper. A second application of silicone gasket remover may be required if residual traces of silicone or gasket material remain.

56. Apply metal surface prep, following package directions, to remove any remaining traces of oil or coolant, and to prepare the surfaces to bond with the new gasket. Do not attempt to make the metal shiny. Some staining of the metal surface is normal.

➡Make sure all cylinder head surfaces are clear of any gasket material, RTV, oil and coolant. The cylinder head surface must be clean and dry before running a flatness check.

➡Use a straightedge that is calibrated by the manufacturer to be flat with 0.005 mm (0.0002 in.) per running foot length. For example, if the straightedge is 61 cm (24 in.) long, the machined edge must be flat with 0.010 mm (0.0004 in.) from end to end.

57. Support the cylinder head on a bench with the head gasket side up. Inspect all areas of the deck face with a straightedge, paying particular attention to the oil pressure feed area. The cylinder head must not have depressions deeper than 0.0254 mm (0.001 in.) across a 38.1 mm (1.5 in.) square area, or scratches more than 0.0254 mm (0.001 in.).

To install:
Both cylinder heads

➡Make sure all coolant residue and foreign material is cleaned from the block surface and the cylinder bore.

➡The use of sealing aids is not permitted. The gasket must be installed dry.

➡The new gasket has a film coating which is crucial to the gasket's ability to seal properly. Do not scratch the gasket.

58. Install the head gaskets over the dowel pins. Position the heads.

➡The new cylinder head bolts must be lightly oiled with a rag and allowed to drain for a few minutes prior to installation.

59. Loosely install cylinder head bolts.

60. Tighten the cylinder head bolts in three steps in the sequence shown.
   a. Step 1: Tighten the bolts to 40 Nm (30 ft. lbs.).
   b. Step 2: Tighten the bolts an additional 90 degrees.
   c. Step 3: Tighten the bolts an additional 90 degrees.

61. Remove the Lifting Handles from both ends of the cylinder head.

➡Lubricate the hydraulic lash adjusters with clean engine oil.

62. Install the hydraulic lash adjusters in their original locations.

Right cylinder head

63. Using a new gasket, install the right exhaust manifold. Tighten the nuts in the sequence shown.

Left cylinder head

➡Lubricate the O-ring seal with clean engine oil.

64. Install the oil level indicator tube.

65. Using a new gasket, install the left exhaust manifold. Tighten the nuts in the sequence shown.

67197-EFSE-G51

Dowel locations—6.8L engine

Left cylinder head torque sequence—6.8L engine

67197-EFSE-G52

Right cylinder head torque sequence—6.8L engine

67197-EFSE-G53

Right exhaust manifold torque sequence—6.8L engine

67197-EFSE-G54

Left exhaust manifold torque sequence—6.8L engine

67197-EFSE-G55

67197-EFSE-G56

If the copper links are not visible, mark two links on one end and one link on the other end, and use as timing marks

Both cylinder heads

➡The timing chain procedures must be followed exactly or damage to the valves and pistons will result.

66. Compress the tensioner plunger, using an edge of a vise.

67. Install a retaining clip on the tensioner to hold the plunger in during installation.

68. Remove the tensioner from the vise.

69. If the copper links are not visible, mark two links on one end and one link on the other end, and use as timing marks.

70. Install the timing chain guides.

71. Pre-position the camshafts. Rotate the left camshaft until the timing mark is approximately at 12 o'clock. Rotate the right camshaft until the timing mark is approximately at 11 o'clock.

➡Rotate the crankshaft counterclockwise only. Do not rotate past position shown or severe piston and valve damage can occur.

➡The number one piston is at top dead center (TDC) when the stud on the engine block fits into the slot in the handle of the crankshaft alignment tool 303-448, or equivalent.

72. Position the crankshaft so the number one cylinder is at TDC with a suitable tool.

73. Remove the crankshaft holding tool.

74. Install the crankshaft sprocket, making sure the flange faces forward.

75. Position the lower end of the left timing chain on the crankshaft sprocket, aligning the timing mark on the outer flange

67197-EFSE-G57

Timing chain guides installed—6.8L engine

67197-EFSE-G58

Position the lower end of the left timing chain on the crankshaft sprocket, aligning the timing mark on the outer flange of the crankshaft sprocket with the single copper (marked) link on the chain—6.8L engine

of the crankshaft sprocket with the single copper (marked) link on the chain.

➡ **Make sure the upper half of the timing chain is below the tensioner arm dowel.**

76. Position the timing chain on the camshaft sprocket with the camshaft sprocket timing mark positioned between the two copper (marked) chain links.

➡ **The left timing chain tensioner arm has a bump near the dowel hole for identification.**

67197-EFSE-G59

Position the timing chain on the camshaft sprocket with the camshaft sprocket timing mark positioned between the two copper (marked) chain links—6.8L engine

77. Position the left timing chain tensioner arm on the dowel pin and install the left timing chain tensioner.

78. Remove the retaining clip from the left timing chain tensioner.

79. Position the lower end of the right timing chain on the crankshaft sprocket, aligning the timing mark on the outer flange of the crankshaft sprocket with the single copper (marked) link on the chain.

➡**The lower half of the timing chain must be positioned above the tensioner arm dowel.**

67197-EFSE-G60

Position the lower end of the right timing chain on the crankshaft sprocket, aligning the timing mark on the outer flange of the crankshaft sprocket with the single copper (marked) link on the chain—6.8L engine

67197-EFSE-G61

Install the right timing chain on the camshaft sprocket. Make sure the camshaft sprocket timing mark is positioned between the two copper (marked) chain links—6.8L engine

67197-EFSE-G62

As a post-check, verify correct alignment of all timing marks—6.8L engine

303-382

67197-EFSE-G63

Install this tool, or equivalent, between the valve spring coils to prevent valve stem seal damage—6.8L engine

67197-EFSE-G64

Using the index mark on the balance shaft, mark the corresponding gear tooth with chalk—6.8L engine

80. Install the right timing chain on the camshaft sprocket. Make sure the camshaft sprocket timing mark is positioned between the two copper (marked) chain links.

81. Position the right timing chain tensioner arm on the dowel pin and install the right timing chain tensioner.

82. Remove the retaining clip from the right timing chain tensioner.

83. As a post-check, verify correct alignment of all timing marks.

84. Install a suitable tool between the valve spring coils to prevent valve stem seal damage.

➡**Lubricate the camshaft roller followers, using clean engine oil.**

➡**Position the cam lobe away from the camshaft roller follower location prior to installing each camshaft roller follower.**

85. Install a suitable tool. Compress the valve spring. Install the camshaft roller followers in their original locations. Remove the tool.

➡**When installing the spark plugs, use care not to exceed the recommended torque.**

86. Install the 10 spark plugs. Tighten the spark plugs to 18 Nm (13 ft. lbs.).

87. Using the index mark on the balance shaft, mark the corresponding gear tooth with chalk.

88. Position the balance shaft on the journals.

➡**It may be necessary to use an inspection mirror to see the marks. Align the chalk mark on the balance shaft with the camshaft timing mark as shown.**

89. Install the bearing caps in their original locations. Install the bolts and tighten the bolts in the sequence shown.

90. Install the crankshaft sensor ring on the crankshaft.

67197-EFSE-G65

It may be necessary to use an inspection mirror to see the marks. Align the chalk mark on the balance shaft with the camshaft timing mark as shown—6.8L engine

67197-EFSE-G66

Apply a bead of silicone gasket and sealant along the cylinder head-to-cylinder block surface and the oil pan-to-cylinder block surface, at the locations shown—6.8L engine

67197-EFSE-G67

Front cover tightening sequence—6.8L engine

➡️**If the front cover is not secured within four minutes, the sealant must be removed and the sealing area cleaned. To clean the sealing area, use silicone gasket remover and metal surface prep. Follow the directions on the packaging. Failure to follow this procedure can cause future oil leakage.**

91. Apply a bead of silicone gasket and sealant along the cylinder head-to-cylinder block surface and the oil pan-to-cylinder block surface, at the locations shown.

92. Install a new front cover gasket on the engine front cover. Position the engine front cover. Install the fasteners.

93. Tighten the engine front cover fasteners in sequence in two steps.

    a. Step 1: Tighten fasteners 1 through 7 to 25 Nm (18 ft. lbs.).

    b. Step 2: Tighten fasteners 6 through 15 to 48 Nm (35 ft. lbs.).

94. Loosely install the bolts, then tighten in two steps in the sequence shown.

    a. Step 1: Tighten to 20 Nm (15 ft. lbs.).

    b. Step 2: Tighten an additional 60 degrees.

95. Position the belt idler pulley and install the bolt. Torque to 18 ft. lbs. (25 Nm).

96. Lubricate the engine front cover and the crankshaft front seal inner lip with clean engine oil.

97. Using the special tools, install the crankshaft front seal.

➡️**If not secured within four minutes, the sealant must be removed and the sealing area cleaned. To clean the sealing area, use silicone gasket remover and metal surface prep. Follow the directions on the packaging. Failure to follow this procedure can cause future oil leakage.**

98. Apply silicone gasket and sealant to the Woodruff key slot on the crankshaft pulley. Use a suitable tool to install the crankshaft pulley.

67197-EFSE-G68

Crankshaft front seal installation—6.8L engine

➡️**Use a suitable strap wrench to hold the pulley while tightening the bolt.**

99. Tighten the new crankshaft bolt in four steps.

    a. Step 1: Tighten to 90 Nm (66 ft. lbs.).

    b. Step 2: Loosen 360 degrees.

    c. Step 3: Tighten to 50 Nm (37 ft. lbs.).

    d. Step 4: Tighten an additional 90 degrees.

100. Position the water pump pulley on the water pump and install the bolts. Torque to 18 ft. lbs. (25 Nm).

101. If a new gasket is being installed, apply instant adhesive completely around the gasket groove in the left valve cover. Install the new valve cover gasket.

➡️**If not secured within four minutes, the sealant must be removed and the sealing area cleaned. To clean the sealing area, use silicone gasket remover and metal surface prep.**

102. Follow the directions on the packaging. Failure to follow this procedure can cause future oil leakage.

103. Apply silicone gasket and sealant in

67197-EFSE-G69

Left valve cover torque sequence—6.8L engine

**10 Nm
(89 lb-in)**

67197-EFSE-G70

Left valve cover torque sequence—6.8L engine

80 Nm (59 lb-ft)

67197-EFSE-G71

Flexplate/flywheel torque sequence—6.8L engine

two places where the engine front cover meets the cylinder head.

104. Position the left valve cover and gasket on the cylinder head and install the bolts loosely. Tighten the bolts in the sequence shown. If a new gasket is being installed, apply instant adhesive completely around the gasket groove in the right valve cover. Install the new valve cover gasket.

➡**If not secured within four minutes, the sealant must be removed and the sealing area cleaned. To clean the sealing area, use silicone gasket remover and metal surface prep. Follow the directions on the packaging. Failure to follow this procedure can cause future oil leakage.**

105. Apply silicone gasket and sealant in two places where the engine front cover meets the cylinder head.

106. Position the right valve cover and gasket on the cylinder head and install the bolts loosely. Tighten the bolts in the sequence shown.

107. Install the crankcase ventilation tube on the left valve cover.

108. Install the positive crankcase ventilation (PCV) valve and hose in the right valve cover.

109. Install the radio frequency interference capacitors.

110. Position the engine control sensor wiring harness and connect the wiring harness retainers onto the valve cover studs.

111. Install the lifting bracket.

112. Connect the crankshaft position (CKP) sensor electrical connector.

113. Connect the oil pressure switch electrical connector.

114. Connect the knock sensor electrical connector.

115. Connect the right radio ignition interference capacitor electrical connector.

116. Connect the CMP electrical connector.

117. Connect the left radio ignition interference capacitor and cylinder head temperature (CHT) sensor electrical connectors.

118. Install the cylinder block drain plugs.

119. Install the right engine mount. Tighten the bolts to 63 Nm (46 ft. lbs.).

120. Install a suitable tool and remove the engine from the work stand.

121. Install the flexplate or the flywheel and the bolts in the sequence shown.

122. Install the engine.

Rocker Arms

REMOVAL & INSTALLATION

4.2L Engine

1. Before servicing the vehicle, refer to the Precautions Section.

➡ **If removing more than 1 rocker arm, mark the components for proper location.**

2. Before servicing the vehicle, refer to the Precautions Section.
3. Disconnect the negative battery cable.
4. Remove the rocker arm cover.
5. Remove the rocker arm hold-down bolt, then remove the rocker arm from the cylinder head.

To install:

6. Position the rocker arms in place, then install the hold-down bolts. Tighten the bolts in 2 passes to:
   - Step 1: 44 inch lbs. (5 Nm)
   - Step 2: 24 ft. lbs. (32 Nm)
7. Install the rocker arm cover.
8. Connect the negative battery cable.

Camshaft Roller Follower

REMOVAL & INSTALLATION

4.6L and 5.4L Engines

1. Before servicing the vehicle, refer to the Precautions Section.
2. Disconnect the negative battery cable.
3. Remove the camshaft covers.
4. Position the piston of the cylinder being serviced at the bottom of its travel.

➡ **Two different valve spring compressor tools are used for this procedure. Valve Spring Compressor (T91P-6565-A) is used on the exhaust camshaft and Valve Spring Compressor (T93P-6565-A) is used on the intake camshaft.**

5. Compress the valve spring and remove the follower.

To install:

6. Position the piston of the cylinder being serviced at the bottom of its travel.
7. Apply clean engine oil to the rocker arm, valve stem tip and tappet bore.

➡ **Valve tappet should have no more than 1/16 inch (1.5mm) of travel before installing the rocker arm.**

8. Compress the valve spring using the correct tool and install the follower.

7924FG37

Using the proper tool, compress the valve spring and remove the follower—4.6L and 5.4L engine

7924FG82

Compress the valve spring and remove the rocker arm—6.8L engine

6.8L Engine

1. Before servicing the vehicle, refer to the Precautions Section.
2. Disconnect the negative battery cable.
3. Remove the camshaft covers.

4. Position the base circle of the camshaft lobe on the rocker arm to be serviced. Also, be sure the piston is not at the top of its travel near the valve.
5. Compress the valve spring and remove the follower.

To install:

6. Position the base circle of the camshaft lobe over the place where the follower is to be installed.

7. Apply clean engine oil to the rocker arm, valve stem tip and tappet bore.

8. Compress the valve using a suitable tool and install the follower.

9. Install the rocker arm covers and the remaining components.

Supercharger

REMOVAL & INSTALLATION

1. Before servicing the vehicle, refer to the Precautions Section.

✳✳ WARNING

Before any supercharger service, clean areas around the supercharger assembly. Cover the openings of the engine and supercharger while supercharger is removed to prevent damage by foreign material.

2. Disconnect the battery ground cable.

3. Remove the air cleaner outlet tube.

4. Disconnect the throttle position (TP) sensor electrical connector.

5. Disconnect the idle air control electrical connector.

6. Disconnect the vacuum lines and the vent hose.

➡ **Pay attention to the throttle return spring orientation. It must be reinstalled in the correct orientation.**

7. Disconnect the throttle control linkages.

8. Disconnect the speed control actuator cable.

9. Disconnect the accelerator cable.

10. Remove the throttle return spring from the throttle lever.

11. Remove the three nuts and position the accelerator cable bracket aside.

12. Disconnect the exhaust gas recirculation (EGR) valve.

13. Disconnect the vacuum line.

14. Disconnect the EGR valve tube fitting.

15. Disconnect the brake booster vacuum line.

16. Remove the bracket bolt.

17. Disconnect the vacuum line.

18. Disconnect the vacuum line.

19. Remove the bolts and remove the throttle body spacer.

20. Discard the gasket.

21. Remove the accessory drive belt.

22. Drain the charge air cooling (CAC) system.

23. Loosen the CAC hose clamps and remove the hoses.

24. Remove the nuts and position the CAC water supply and return tubes aside.

25. Remove the EGR vacuum regulator bracket.

26. Remove the nuts.

27. Remove the two upper vacuum hoses from the supercharger bypass valve actuator.

28. Remove the supercharger and plenum assembly.

29. Remove the bolts.

30. Remove the stud bolts.

31. Remove the supercharger and plenum assembly.

32. Discard the gasket.

33. Remove the upper intake manifold.

34. Remove the vacuum hose.

35. Remove the CAC hoses.

36. Remove the bolts and the upper intake manifold.

37. Discard the gasket.

38. Remove the CAC.

39. Remove and discard the bolts.

40. Remove the CAC.

41. Discard the gasket.

CAC installation

67197-EFSE-G93

Supercharger fastener tightening sequence

67197-EFSE-G94

To install:

→**New bolts must be used during assembly of the charge air cooler (CAC).**

42. Install the CAC.
43. Install a new gasket.
44. Install the new bolts.
45. Install the upper intake manifold.
46. Install the upper intake manifold and the four rear bolts loosely.
47. Install the CAC hoses.
48. Install a new gasket.
49. Install the vacuum hose.
50. Install a new gasket and install the supercharger and plenum assembly, bolts, and the stud bolts. Tighten the retainers in two steps, in the sequence shown
    a. Step 1: Tighten to 2 Nm (18 inch lbs.).
    b. Step 2: Tighten 45 Nm (33 ft. lbs.).
51. Install the two upper vacuum hoses to the supercharger bypass valve actuator.
52. Install the EGR vacuum regulator bracket and install the nuts.
53. Position the bracket and install the nuts.
54. Position the CAC water supply and return tubes and install the nuts.
55. Install the CAC hoses and the clamps.
56. Fill the CAC cooling system.
57. Install the accessory drive belt.

❊❊ WARNING

The throttle body spacer sealing surfaces are soft metals. Carefully clean both sealing surfaces.

→**Inspect the throttle body spacer gasket and install a new gasket if necessary.**

58. Install a new gasket and install the throttle body spacer.
59. Tighten the bolts in two steps.
    a. Step 1: Tighten to 10 Nm (89 inch lbs.).
    b. Step 2: Tighten an additional 90 degrees.
60. Connect the vacuum lines.
61. Connect the brake booster vacuum line.
62. Connect the vacuum line.
63. Install the bracket bolt.
64. Connect the exhaust gas recirculation (EGR) valve.
65. Connect the vacuum line.
66. Connect the EGR vacuum tube fitting.
67. Install the accelerator cable bracket and install the three bolts.

→**Pay attention to the throttle return spring orientation. It must be reinstalled in the correct orientation.**

68. Connect the throttle control linkages.
69. Connect the speed control actuator cable.
70. Connect the accelerator cable.
71. Install the throttle return spring to the throttle lever.
72. Connect the vacuum lines and the vent hose.
73. Connect the idle air control electrical connector.
74. Connect the throttle position (TP) sensor electrical connector.
75. Install the air cleaner outlet tube.
76. Connect the battery ground cable.

Supercharger Drive Pulleys

REMOVAL & INSTALLATION

1. Before servicing the vehicle, refer to the Precautions Section.
2. With the vehicle in **Neutral**, position it on a hoist.
3. Remove the supercharger drive belt.
4. Remove the fan blade, clutch and shroud.
5. Remove the nut and position the shield aside.
6. Remove the transmission cooler line clip.
7. Remove the two nuts and one bolt.
8. Remove the starter motor.
9. Remove the accessory drive belt.

→**The auxiliary supercharger pulley has left-hand threads.**

10. Remove the pulley and brace assembly.
11. Remove the supercharger pulley adapter.
12. Remove and discard the bolt.
13. Using a suitable tool, remove the crankshaft pulley.

To install:

→**If the crankshaft pulley is not secured within four minutes, the sealant must be removed and the sealing area cleaned. To clean the sealing area, use silicone gasket remover and metal surface prep. Follow the directions on the packaging. Failure to follow this procedure can cause future oil leakage.**

14. Apply silicone gasket and sealant to the Woodruff key slot on the crankshaft pulley.
15. Using a suitable tool, install the crankshaft pulley.

Flywheel holding tool

Supercharger pulley adapter

Pulley and brace assembly installation

16. Install a new bolt and washer. Tighten the bolt in four steps.
    a. Step 1: Tighten to 90 Nm (66 ft. lbs.).
    b. Step 2: Loosen the bolt.
    c. Step 3: Tighten to 50 Nm (37 ft. lbs.).
    d. Step 4: Tighten an additional 90 degrees.
17. Install the supercharger pulley adapter.
18. Install the accessory drive belt.

→**Coat the threads of the supercharger pulley with anti-seize lubricant.**

→**The auxiliary supercharger pulley has left-hand threads.**

19. Install the pulley and brace assembly.
20. Remove the holding tool.
21. Install the starter motor.
22. Position the transcooler line clip and install the two nuts and one bolt.

23. Position the shield and install the nut.

24. Install the fan blade, clutch and shroud.

25. Install the supercharger drive belt.

Intake Manifold

REMOVAL & INSTALLATION

➡When the battery is disconnected and reconnected, some abnormal drive symptoms may occur while the vehicle relearns its adaptive strategy. The vehicle may need to be driven 10 miles (16 km) or more to relearn the strategy.

4.2L Engine

2002–04

✳ CAUTION

Fuel injection systems remain under pressure, even after the engine has been turned OFF. The fuel system pressure must be relieved before disconnecting any fuel lines. Failure to do so may result in fire and/or personal injury.

1. Before servicing the vehicle, refer to the Precautions Section.
2. Remove or disconnect the following:
   - Engine air cleaner outlet tube
   - Ignition coil electrical connector
   - Radio ignition interference capacitor electrical connector
   - Spark plug wires
   - Accelerator control splash shield
   - Accelerator cable end, if equipped, the speed control actuator cable end
   - Accelerator cable and actuator cable aside, after removing the hold-down bolts
   - Vapor Management Valve (VMV) hose
   - Brake booster vacuum hose
   - Manifold vacuum connection
   - Positive Crankcase Ventilation (PCV) valve from the rocker arm cover
   - Engine Vacuum Regulator (EVR) bracket aside
   - Throttle Position (TP) sensor and Idle Air Control (IAC) valve electrical connectors
   - Breather from the rocker arm cover
   - The 12 upper intake manifold retaining bolts, then lift the manifold off of the engine and discard the intake manifold upper gasket.

- The 6 fuel injector electrical connectors
- The engine coolant temperature sensor and the water temperature indicator sending unit electrical connectors.
- Exhaust Gas Recirculation (EGR) valve vacuum hose
- EGR valve tube upper fitting
- Radiator hose from the lower intake manifold
- Intake Manifold Runner Control (IMRC) actuator brackets aside
- Fuel pressure regulator vacuum line
- Fuel system pressure

- Fuel lines
- Water pump bypass hose

➡Remove the lower intake manifold with the fuel injection supply manifold and fuel injectors as one unit.

- The 6 long bolts and the 8 short bolts, then lift the lower intake manifold off of the engine.
3. Remove and discard the lower intake manifold sealing components.

To install:

4. Clean all components of dirt, grease and old gasket material.

5. Install the lower intake manifold front and rear end seals as follows:

7924FG08

Tighten the lower intake manifold bolts using this sequence—2002–04 4.2L engine

9358FG01

Tighten the upper intake manifold bolts using this sequence—2002–04 4.2L engine

a. Apply a bead of RTV sealant to the intake manifold front and rear end seal mounting points.

b. Install the lower intake manifold front and rear end seals.

6. Install new lower intake manifold gaskets onto the cylinder heads.

➡ **The lower intake manifold must be installed within 15 minutes of applying sealant.**

7. Apply a bead of RTV silicone gasket sealant to the end of the lower intake manifold end seals, where they stop on the cylinder head surface. Position the intake manifold onto the engine block and cylinder heads.

8. Install the lower intake manifold mounting bolts in the correct positions. Refer to the illustration for the correct placement of the long (A) and the short (B) mounting bolts.

➡ **Be sure to tighten the intake manifold bolts in 2 steps.**

9. Tighten the lower intake manifold mounting bolts following the tightening sequence:

a. Step 1: 44 inch lbs. (5 Nm).

b. Step 2: 71–101 inch lbs. (8.0–11.5 Nm).

10. Install or connect the following:
- Water bypass hose
- Fuel lines
- Fuel pressure regulator vacuum line
- IMRC actuators. Tighten the bolts on the brackets to 71–102 inch lbs. (8.0–11.5 Nm).
- Upper radiator hose to the lower intake manifold
- EGR valve vacuum hose
- EGR tube upper fitting to 25–34 ft. lbs. (37–47 Nm).
- Engine coolant temperature sensor and water temperature indicator sending unit electrical connectors
- The 6 fuel injector electrical connectors
- A new intake manifold upper gasket

➡ **Be sure to tighten the upper intake manifold bolts in the sequence shown.**

11. Position the upper intake manifold onto the lower intake manifold, then tighten the upper intake manifold bolts following the tightening sequence as follows:
- 53 inch lbs. (6 Nm).
- 89 inch lbs. (10 Nm).

12. Install or connect the following:
- Breather into the rocker arm cover
- TP sensor and IAC valve electrical connectors

Make certain to install the long bolts (A) and the short bolts (B) in the correct lower intake manifold holes—2002–04 4.2L engine

- EVR bracket
- Manifold vacuum connection
- PCV valve
- Brake booster vacuum hose
- Vapor Management Valve (VMV) hose
- Accelerator and speed actuator cables
- Accelerator control splash shield
- Spark plug wires
- Ignition coil and the radio ignition interference capacitor electrical connectors
- Engine air cleaner outlet tube

2005 UPPER

1. Before servicing the vehicle, refer to the Precautions Section.

2. Drain the engine cooling system.

3. Disconnect the battery ground cable.

4. Remove the air cleaner outlet pipe.

5. Remove the bolt and position the heater hoses aside.

✳✳ WARNING

It is important to twist the spark plug wire boots while pulling upward to avoid possible damage to the spark plug wire.

➡ **Spark plug wires must be connected to the correct ignition coil terminal. Mark the spark plug wires for installation reference.**

6. Disconnect the 3 left spark plug wires from the ignition coil, disconnect the 2 spark plug wire retainers from the upper

intake manifold and position the left spark plug wires aside.

7. Disconnect the exhaust gas recirculation (EGR) system module electrical and vacuum connectors.

8. Disconnect the exhaust manifold-to-EGR system module tube from the EGR system module.

9. Disconnect the electronic throttle body electrical connector.

10. Release the clamp and disconnect the fuel vapor tube from the upper intake manifold.

11. Disconnect the 5 wiring harness retainers from the upper intake manifold.

12. Release the clamp and disconnect the brake booster vacuum hose from the upper intake manifold.

13. Disconnect the vacuum hose connector from the rear of the upper intake manifold.

14. Release the clamps and disconnect the 2 heated positive crankcase ventilation (PCV) coolant hoses from the rear of the upper intake manifold.

15. Disconnect the PCV valve hose from the rear of the upper intake manifold.

16. Detach the fuel charging wiring harness-to-main engine wiring harness electrical connector retainer from the rear of the upper intake manifold.

17. Remove the 8 bolts and the upper intake manifold.

To install:

18. Inspect the 6 upper intake manifold gaskets. Install new gaskets, if necessary.

19. Position the upper intake manifold and install the bolts. Tighten the bolts in the sequence shown in 2 stages.

06017-F150-G32

Upper intake manifold torque sequence—2005 4.2L engine

- Stage 1: Tighten to 6 Nm (53 inch lbs.).
- Stage 2: Tighten to 10 Nm (89 inch lbs.).

20. Attach the fuel charging wiring harness-to-main engine wiring harness electrical connector retainer to the rear of the upper intake manifold.

21. Connect the PCV valve hose to the rear of the upper intake manifold.

22. Connect the 2 heated PCV coolant hoses to the rear of the upper intake manifold and position the clamps.

23. Connect the vacuum hose connector to the rear of the upper intake manifold.

24. Connect the brake booster vacuum hose to the upper intake manifold and position the clamp.

25. Connect the 5 wiring harness retainers to the upper intake manifold.

26. Connect the fuel vapor hose to the upper intake manifold and position the clamp.

27. Connect the electronic throttle body electrical connector.

28. Connect the exhaust manifold-to-EGR system module tube to the EGR system module. Tighten to 40 Nm (30 ft. lbs.).

29. Connect the EGR system module electrical and vacuum connectors.

➡Apply silicone brake caliper grease and dielectric compound to the inside of the spark plug wire boots.

➡Spark plug wires must be connected to the correct ignition coil terminals.

30. Position the spark plug wires, connect the 2 spark plug wire retainers to the upper intake manifold and connect the 3 spark plug wires to the ignition coil.

31. Position the heater hoses and install the bolt. Tighten to 5 Nm (44 inch lbs.).

32. Install the air cleaner outlet pipe.

33. Connect the battery ground cable.

34. Fill and bleed the engine cooling system.

2005 LOWER

1. Before servicing the vehicle, refer to the Precautions Section.

2. Relieve the fuel system pressure.

3. Disconnect the fuel supply spring lock coupling.

4. Remove the upper intake manifold.

5. Disconnect the fuel charging wiring harness-to-main engine wiring harness and the intake manifold runner control (IMRC) actuator electrical connectors.

6. Remove the bolts and the fuel rail, the fuel charging wiring harness and the engine vacuum hose harness as an assembly.

7. Release the clamp and disconnect the heater outlet hose.

8. Release the clamp and disconnect the upper radiator hose.

9. Release the clamp and disconnect the bypass hose.

10. Release the clamp and disconnect the heater inlet hose.

11. Remove the 6 long and 8 short bolts, the lower intake manifold, the lower intake manifold gaskets and the front and rear seals.

To install:

✳✳ WARNING

Do not use metal scrapers, wire brushes, power abrasive discs or other abrasive means to clean the sealing surfaces. These tools cause scratches and gouges which make leak paths. Use a plastic scraping tool to remove all traces of old sealant.

12. Using a plastic scraping tool, silicone gasket remover and metal surface prep, clean the gasket mating surfaces. Follow the directions on the packaging.

➡**If the lower intake manifold is not secured within 4 minutes, the sealant must be removed and the sealing area cleaned. To clean the sealing area, use silicone gasket remover and metal surface prep. Follow the directions on the packaging. Failure to follow this procedure can cause future oil leakage.**

13. Apply a bead of silicone gasket and sealant to the intake manifold front and rear seal mounting points as indicated.

14. Install the lower intake manifold front and rear seals.

15. Install the intake manifold gaskets.

➡**If the lower intake manifold is not secured within 4 minutes, the sealant must be removed and the sealing area cleaned. To clean the sealing area, use silicone gasket remover and metal surface prep. Follow the directions on the packaging. Failure to follow this procedure can cause future oil leakage.**

16. Apply a bead of silicone gasket and sealant to the lower intake manifold gaskets and front and rear seal junctions in 4 places as shown.

17. Position the lower intake manifold.

Intake manifold seal RTV sealer application points—4.2L engine

06017-F150-G33

Intake manifold bolt torque sequence—2005 4.2L engine

06017-F150-G34

➡**Make sure the bolts are installed in the correct locations. The long bolts are installed in positions 1, 2, 3, 4, 11 and 12.**

18. Tighten the bolts in the sequence shown in 2 stages.
   • Stage 1: Tighten to 5 Nm (44 inch lbs.).
   • Stage 2: Tighten to 10 Nm (89 inch lbs.).

19. Connect the heater inlet hose and position the clamp.

20. Connect the bypass hose and position the clamp.

21. Connect the upper radiator hose and position the clamp.

22. Connect the heater outlet hose and position the clamp.

23. Position the fuel rail, the fuel charging wiring harness and the engine vacuum hose harness as an assembly and install the bolts. Tighten to 10 Nm (89 inch lbs.).

24. Connect the IMRC actuator and the fuel charging wiring harness-to-main engine wiring harness electrical connectors.

25. Connect the fuel supply spring lock coupling.

26. Install the upper intake manifold.

2002–03 F-150 and 2004 Heritage Edition 4.6L Engine; 2002–05 E-Series 4.6L Engine

REMOVAL

1. Before servicing the vehicle, refer to the Precautions Section.

2. Disconnect the battery ground cable.

3. Remove the air cleaner outlet pipe.

4. Drain the engine cooling system.

5. Compress and slide the hose clamp and disconnect the upper radiator hose.

6. Disconnect the accelerator cable.

7. Disconnect the speed control actuator cable.

8. Remove the accelerator return spring.

9. Remove the accelerator cable bracket bolts and position the bracket and cables aside.

10. Disconnect the main vacuum harness.

11. Remove the idle air control (IAC) fresh air hose.

12. Disconnect the exhaust gas recirculation (EGR) vacuum regulator solenoid connections.

13. Disconnect the IAC and throttle position (TP) sensor electrical connectors.

14. Disconnect the EGR valve vacuum hose and the evaporative emission canister purge valve hose and position them aside.

15. Disconnect the fuel pressure regulator vacuum hose.

16. Disconnect the differential pressure feedback EGR system electrical connector.

17. Disconnect the evaporative emission canister purge valve vacuum hose.

18. Remove the nut. Disconnect the brake booster vacuum tube and position the bracket and tube aside.

19. Position the crankcase ventilation hose aside.

20. Position the exhaust manifold to EGR valve tube aside.

21. Disconnect the fuel hose spring lock couplings.

22. Disconnect the heater hose and the heated throttle body inlet hose.

23. Disconnect the positive crankcase ventilation (PCV) hose.

24. Remove the four bolts and move the throttle body adapter forward slightly.

25. Disconnect the heated throttle body hose and remove the throttle body adapter.

26. Discard the throttle body adapter-to-intake manifold gasket. Inspect the throttle body and adapter for damage.

27. Remove the power steering reservoir upper mounting bracket.

28. Disconnect and remove the eight ignition coils.

29. Disconnect the eight fuel injectors.

30. Rotate the belt tensioner clockwise and detach the drive belt from the alternator pulley.

31. Remove the bolts and the generator upper mounting bracket.

32. Remove the generator lower mounting bolts and position the generator aside.

33. Remove the bolts, thermostat housing and thermostat. Discard the O-ring seal.

34. Remove the nine bolts.

35. Lift the intake manifold. Disconnect the intake manifold tuning valve electrical connector. Remove the intake manifold and discard the gaskets. Inspect the intake manifold assembly for damage.

DISASSEMBLY

1. Before servicing the vehicle, refer to the Precautions Section.

2. Remove the fuel infection supply manifold.

3. Remove the fuel injectors from the fuel injection supply manifold.

4. Inspect the two O-ring seals from each fuel injector. Install new O-ring seals as needed.

5. Remove the engine noise shield.

6. Remove the bolts.

7. Separate the upper and lower intake manifolds and discard the lower intake manifold gasket.

❋❋ WARNING

If the engine has been internally damaged and metal or other foreign material has entered the manifold, it is recommended the manifold be discarded.

➡**If there is no residual gasket material present, metal surface prep can be used to clean and prepare the sealing surfaces.**

8. Inspect all external and internal surfaces of the upper intake manifold. Inspect for cracks or other damage. Inspect for foreign material inside of the manifold. Clean the upper intake manifold sealing surfaces with silicone gasket remover and metal surface prep. Follow the directions on the packaging.

9. Allow the upper and lower intake manifolds to dry completely.

ASSEMBLY

1. Before servicing the vehicle, refer to the Precautions Section.

2. Position a new lower intake manifold

---

gasket and the upper intake manifold on the lower intake manifold and loosely install the bolts.

3. Tighten the bolts in two stages using the sequence shown.
   • Stage 1: Tighten to 2 Nm (18 inch lbs.).
   • Stage 2: Tighten to 10 Nm (89 inch lbs.).

4. Install the engine noise shield.

➡**Lubricate the O-ring seals with clean engine oil, to aid installation.**

5. Install the fuel injectors into the fuel injection supply manifold.

6. Install the fuel injection supply manifold. Torque to 89 inch lbs. (10 Nm).

INSTALLATION

❋❋ WARNING

Do not use metal scrapers, power abrasive discs or any abrasive means to clean the sealing surfaces, these tools cause scratches and gouges that make leak paths. Use a plastic scraper only.

06017-F150-G35

Upper and lower intake manifolds—2002–03 F-150 and 2004 Heritage Edition and 2002–05 E-Series 4.6L Engine

06017-F150-G36

Upper-to-lower intake manifold torque sequence—2002–03 F-150 and 2004 Heritage Edition and 2002–05 E-Series 4.6L Engine

1. Clean all the sealing surfaces.
2. Position new gaskets and install the intake manifold assembly. Loosely install the nine bolts.
3. Connect the intake manifold tuning valve electrical connector.
4. Install the thermostat.
5. Install the thermostat.
6. Install a new O-ring seal.

➡ **The thermostat housing bolts are tightened in sequence with the intake manifold bolts.**

7. Install the thermostat housing and loosely install the bolts.
8. Tighten the 11 bolts in two stages in the sequence shown.
  • Stage 1: Tighten to 2 Nm (18 inch lbs.).
  • Stage 2: Tighten to 25 Nm (18 ft. lbs.).
9. Position the generator and install the lower mounting bolts.
10. Install the generator upper mounting bracket. Torque to 89 inch lbs. (10 Nm).
11. Rotate the belt tensioner clockwise and install the drive belt. Refer to the decal on the radiator upper air deflector for belt routing.

12. Connect the eight fuel injector electrical connectors.
13. Install and connect eight ignition coils.
14. Install the power steering reservoir upper mounting bracket. Torque the 2 upper bolts to 13 ft. lbs. (17 Nm); the one lower bolt to 96 inch lbs. (11 Nm).
15. Roughly position the throttle body adapter in the vehicle and connect the heated throttle body hose.
16. Install a new gasket, the throttle body adapter and the bolts. Torque to 89 inch lbs. (10 Nm).
17. Connect the PCV hose.
18. Connect the heater hose and the heated throttle body inlet hose.
19. Install the exhaust manifold to EGR valve tube. Hand-tighten the fittings. Tighten the upper and lower fittings to 26–33 ft. lbs. (35–45 Nm).
20. Connect the fuel hose spring lock couplings.
21. Connect the brake booster vacuum tube and bracket and install the nut.
22. Roughly position the vacuum harness and connect the evaporative emission canister purge valve vacuum hose.
23. Connect the differential pressure feedback EGR system electrical connector.
24. Connect the fuel pressure regulator vacuum hose.

06017-F150-G37

Intake manifold assembly installation torque sequence—2002–03 F-150 and 2004 Heritage Edition and 2002–05 E-Series 4.6L Engine

25. Connect the evaporative emission canister purge valve hose and the EGR valve vacuum hose.

26. Connect the TP sensor and IAC electrical connectors.

27. Connect the EGR vacuum regulator solenoid connections.

28. Install the IAC fresh air hose.

29. Connect the main vacuum harness.

30. Install the accelerator cable bracket and the bolts.

31. Connect the accelerator cable.

32. Connect the speed control actuator cable.

33. Install the throttle return spring.

34. Connect the upper radiator hose and position the clamp.

35. Install the air cleaner outlet pipe.

36. Connect the battery ground cable.

37. Fill and bleed the engine cooling system.

2002–03 F-Series and 2004 Heritage Edition 5.4L Engine, exc. Blackwood and Lightning

1. Before servicing the vehicle, refer to the Precautions Section.

2. Disconnect the battery ground cable.

3. Remove the air cleaner outlet pipe. control actuator cable.

4. Remove the accelerator return spring.

5. Remove the accelerator cable bracket bolts and position the bracket and cables aside.

6. Disconnect the PCV hose from the throttle body adapter.

7. Disconnect the heated throttle body

8. Drain the engine cooling system.

9. Compress and slide the hose clamp and disconnect the upper radiator hose.

10. Disconnect the accelerator cable.

11. Disconnect the speed outlet hose from the throttle body adapter.

12. Disconnect the PCV hose from the intake manifold.

13. Remove the PCV valve and position the harness aside.

14. Disconnect the heater hose and the heated throttle body inlet hose.

15. Disconnect the evaporative emission canister purge valve hose.

16. Disconnect the exhaust gas recirculation (EGR) valve vacuum hose and the main vacuum harness.

17. Remove the idle air control (IAC) valve fresh air tube.

18. Disconnect the IAC electrical connector.

19. Disconnect the throttle position sensor electrical connector and the EGR vac-

uum regulator solenoid electrical and vacuum connectors.

20. Disconnect the fuel pressure regulator vacuum hose.

21. Disconnect the evaporative emission canister purge valve vacuum hose and position the vacuum harness aside.

22. Disconnect the differential pressure feedback EGR system electrical connector.

23. Remove the bolt, disconnect the brake booster vacuum tube and position the bracket and tube aside.

24. Remove the nut.

25. Position the crankcase ventilation tube aside.

26. Position the exhaust manifold to EGR valve tube aside.

27. Relieve the fuel system pressure.

28. Disconnect the fuel hose spring lock couplings.

29. Remove the bolt and the differential pressure feedback EGR system bracket.

30. Remove the four bolts and remove the throttle body adapter. Discard the throttle body adapter gasket.

31. Position the power steering reservoir aside.

32. Disconnect and remove the eight ignition coils.

33. Disconnect the eight fuel injectors.

34. Rotate the belt tensioner clockwise and detach the drive belt from the alternator pulley.

35. Remove the bolts and the generator upper mounting bracket.

36. Remove the generator lower mounting bolts and position the generator aside.

37. Remove the bolts, the thermostat housing and the thermostat. Discard the O-ring seal.

38. Remove the 9 intake manifold bolts.

39. Remove the intake manifold. Discard the intake manifold gaskets.

40. Inspect the intake manifold and throttle body for damage.

To install:

✸✸ WARNING

Do not use metal scrapers, power abrasive discs or any abrasive means to clean the sealing surfaces, these tools cause scratches and gouges that make leak paths. Use a plastic scraper only.

41. Clean all the sealing surfaces.

42. Position the new intake manifold gaskets.

06017-F150-G38

Intake manifold torque sequence—2002–03 F-Series and 2004 Heritage Edition; 2002–05 E-Series 5.4L Engine, exc. Blackwood and Lightning

43. Position the intake manifold.
44. Loosely install nine bolts.
45. Install the thermostat.
46. Install a new O-ring seal.

➡ **The thermostat housing bolts are tightened in sequence with the intake manifold bolts.**

47. Install the thermostat housing and loosely install the bolts.
48. Tighten the 11 bolts in two stages in the sequence shown.
- Stage 1: Tighten to 2 Nm (18 inch lbs.).
- Stage 2: Tighten to 25 Nm (18 ft. lbs.).

49. Position the generator and install the lower mounting bolts.
50. Install the generator upper mounting bracket.
51. Rotate the belt tensioner clockwise and install the drive belt. Refer to the decal on the radiator upper air deflector for belt routing.
52. Connect the eight fuel injector electrical connectors.
53. Install and connect eight ignition coils.
54. Install the power steering reservoir upper mounting bracket. Tighten the 2 upper bolts to 13 ft. lbs. (17 Nm); the one lower bolt to 96 inch lbs. (11 Nm).
55. Install a new intake manifold-to-throttle body adapter gasket.
56. Install the throttle body adapter and the four bolts. Tighten to 96 inch lbs. (10 Nm).
57. Position the differential pressure feedback EGR system bracket and install the bolt.
58. Install the nut.
59. Connect the fuel hose spring lock couplings.
60. Install the exhaust manifold to EGR valve tube. Hand-tighten the fittings. Tighten the upper and lower fittings to 44 ft. lbs. (60 Nm). Connect the two differential pressure feedback EGR system hoses.
61. Connect the brake booster vacuum tube, position the bracket and install the bolt.
62. Connect the IAC electrical connector.
63. Connect the differential pressure feedback EGR system electrical connector.
64. Roughly position the vacuum harness and connect the evaporative emission canister purge valve vacuum hose.
65. Connect the fuel pressure regulator vacuum hose.
66. Connect the throttle position sensor electrical connector and the EGR vacuum regulator solenoid electrical and vacuum connectors.

67. Install the IAC fresh air tube.
68. Connect the main vacuum harness and the EGR valve vacuum hose.
69. Connect the evaporative emission canister purge valve hose.
70. Connect the heater hose and the heated throttle body inlet hose.
71. Install the PCV valve.
72. Connect the PCV hose to the intake manifold.
73. Connect the heated throttle body outlet hose.
74. Connect the PCV hose to the throttle body adapter.
75. Install the accelerator cable bracket and the bolts.
76. Connect the accelerator cable.
77. Connect the speed control actuator cable.
78. Install the throttle return spring.
79. Connect the upper radiator hose and position the clamp.
80. Install the air cleaner outlet pipe.
81. Connect the battery ground cable.
82. Fill and bleed the engine cooling system.

2002–04 F-250/350 6.8L Engine; 2002–05 E-Series 6.8L Engine

REMOVAL

1. Before servicing the vehicle, refer to the Precautions Section.
2. Relieve the fuel system pressure.
3. Disconnect the battery ground cable.
4. Drain the coolant system.
5. Remove the air cleaner outlet tube.
6. Remove the generator.
7. Disconnect the fuel line spring lock couplings.
8. Compress and slide the hose clamp and disconnect the water outlet hose.
9. Remove the accelerator cable splash shield.
10. Disconnect the accelerator and speed control cables and remove the return spring.
11. Position the accelerator cable bracket aside.
12. Disconnect the positive crankcase ventilation (PCV) tube and the vacuum lines.
13. Disconnect the heated PCV coolant hose.
14. Disconnect the vacuum line.
15. Disconnect the five RH fuel injector electrical connectors and five ignition coil electrical connectors.
16. Disconnect the idle air control (IAC) motor electrical connector and the bypass hose.
17. Disconnect the heater hose.
18. If equipped, disconnect the exhaust

gas recirculation (EGR) transducer and disconnect the throttle position (TP) sensor connectors.
19. If equipped, disconnect the EGR transducer vacuum lines.
20. If equipped, remove the EGR transducer bracket.
21. If equipped, position the EGR valve to exhaust manifold tube aside.
22. If equipped, disconnect the EGR valve vacuum line.

➡ **The EGR valve is shown removed for clarity.**

23. If equipped, disconnect the EGR vacuum regulator valve solenoid electrical and vacuum harness.
24. If equipped, remove the EGR vacuum regulator valve bracket.
25. Disconnect the five LH fuel injector and TP sensor electrical connectors and ignition coil connectors.
26. Remove the ignition coils.
27. Disconnect the vacuum connector from the fuel pressure regulator.
28. Remove the thermostat housing.
29. Remove the bolts, the upper intake manifold and the intake manifold gaskets.
30. Discard the intake manifold gaskets.

❊❊ WARNING

Do not use metal scrapers, wire brushes, power abrasive discs or other abrasive means to clean the aluminum retainer plate. These tools cause scratches and gouges, which make leak paths.

31. Clean all mating surfaces.

DISASSEMBLY

1. Before servicing the vehicle, refer to the Precautions Section.
2. Disconnect the heated positive crankcase valve (PCV) coolant hose.
3. Remove the coolant thermostat and the O-ring seal. Discard the O-ring seal.
4. Remove the throttle body (TB) adapter. Remove the TB adapter. Discard the TB adapter press-in-place gasket.
5. Remove the five bolts holding the fuel injection supply manifold assembly and the fuel injectors. Discard the fuel injector O-ring seals.
6. Separate the upper and lower intake manifolds. Remove the bolts. Discard the intake manifold gasket.
7. Using clean metal surface cleaner, clean and inspect the upper and lower manifold internal passages and mating surfaces.

8. Allow the upper and lower intake manifolds to dry completely.

ASSEMBLY

1. Before servicing the vehicle, refer to the Precautions Section.

2. Position a new lower intake manifold gasket and the upper intake manifold on the lower intake manifold and loosely install the bolts.

3. Tighten the bolts in two stages using the sequence shown.
- Stage 1: Tighten to 2 Nm (18 inch lbs.).
- Stage 2: Tighten to 10 Nm (89 inch lbs.).

✴✴ WARNING

Check for leakage after installation is complete.

➡**Be sure fuel injectors and fuel rail are fully seated.**

4. Lubricate the new fuel injector O-ring seals with clean engine oil prior to installation.

5. Install the fuel rail.

6. Position a new TB adapter press-in-place gasket.

7. Position the TB adapter.

8. Install the four bolts in 2 stages:
- Stage 1: Tighten to 9 Nm (80 inch lbs.).
- Stage 2: Tighten an additional 90 degrees.

➡**Thermostat must be installed as illustrated.**

9. Use a new O-ring seal to position the coolant thermostat in the upper intake manifold.

10. Connect the heated PCV hose.

INSTALLATION

1. Before servicing the vehicle, refer to the Precautions Section.

2. Position the thermostat housing.

3. Position the upper intake manifold gaskets and the intake manifold, and loosely install the bolts.

4. Tighten the bolts in two stages, in the sequence shown.
- Stage 1: Tighten to 2 Nm (18 inch lbs.).
- Stage 2: Tighten to 25 Nm (18 ft. lbs.).

5. Connect the fuel line spring lock coupling.

6. Connect the vacuum line to the fuel injection supply manifold.

7. Install the ignition coils. Torque to 53 inch lbs. (6 Nm).

06017-F150-G39

Upper and lower intake manifolds—2002–04 F-Series 6.8L engine and 2002–05 E-Series 6.8L engine

06017-F150-G40

Upper-to-lower intake manifold bolt torque sequence—2002–04 F-Series 6.8L engine and 2002–05 E-Series 6.8L engine

Intake manifold assembly installation tightening sequence—2002–04 F-Series 6.8L engine and 2002–05 E-Series 6.8L engine

8. Connect the five LH fuel injector electrical connectors and five LH ignition coil connectors.

9. If equipped, install the EGR vacuum regulator valve solenoid bracket.

➡**The EGR valve is shown removed for clarity.**

10. If equipped, connect the EGR vacuum regulator valve solenoid vacuum and electrical connectors.

11. If equipped, connect the EGR valve vacuum line.

12. If equipped, install the EGR valve-to-exhaust manifold tube. Hand-tighten the fittings. Tighten the upper and lower fittings to 40-60 Nm (30-44 ft. lbs.).

13. If equipped, install the EGR transducer bracket. Torque to 106 inch lbs. (12 Nm).

14. If equipped, connect the vacuum lines.

15. If equipped, connect the EGR transducer and throttle position sensor electrical connectors.

16. Connect the heater hose.

17. Connect the idle air control (IAC) motor bypass hose and connector.

18. Connect the five RH fuel injector electrical connectors and five RH ignition coil connectors.

19. Connect the vacuum line.

20. Connect the PCV tube and the vacuum lines.

21. Connect the heated PCV coolant hose.

22. Install the accelerator bracket.

23. Connect the accelerator and speed control cables and return spring.

24. Install the accelerator cable splash shield.

25. Connect the engine coolant outlet hose and position the hose clamp.

26. Install the generator.

27. Install the air cleaner outlet tube.

28. Connect the battery ground cable.

29. Fill the cooling system.

5.4L Blackwood DOHC

UPPER INTAKE MANIFOLD

1. Before servicing the vehicle, refer to the Precautions Section.

2. Remove or disconnect the following:
- Negative battery cable
- Air cleaner outlet tube
- Engine appearance cover
- Accelerator cable, speed control cable, and the return spring

- Bolts and position the cables and bracket out of the way
- Evaporative emission return line
- Positive crankcase ventilation tube from the upper intake manifold
- Positive Crankcase Ventilation (PCV) valve from the valve cover
- PCV valve tube from the water heated fitting and remove the tube assembly
- Coolant lines and plug the lines.
- Electrical connector from the communication valve
- Bolts and the communication valve
- Wiring harness shield bolts and 5 clips and remove the shield
- Balance tube from the engine
- Throttle Position (TP) sensor and the Idle Air Control (IAC) motor
- Vacuum lines and detach the electrical connector from the EGR vacuum regulator (EVR).
- The 2 hoses and detach the electrical connector from the differential pressure feedback EGR.
- The bolts from the power steering reservoir bracket and position it aside.
- The stud and position the oil fill tube aside.
- Brake booster vacuum line
- Vacuum line from the Exhaust Gas Recirculation (EGR) valve
- Bolts from the EGR adapter
- Retaining bolts and remove the upper intake manifold.

To install:

3. Clean and inspect the sealing surfaces.

4. Install or connect the following:
- Upper intake manifold: Tighten the bolts to 89 inch lbs. (10 Nm) and then an additional 90 degrees in the sequence shown
- EGR adapter with a new gasket
- Vacuum line to the EGR valve
- Brake booster vacuum line
- Oil fill tube and install the stud
- Power steering reservoir bracket and install the bolts.
- EGR valve at the exhaust manifold
- The 2 hoses and the electrical connector to the differential pressure feedback EGR
- Vacuum lines and the electrical connector to the EVR
- IAC valve and the TP sensor
- Balance tube on the engine
- Communication valve and the bolts
- Wiring harness shield, the bolts and 5 clips

67197-EFSE-G75

Upper intake manifold torque sequence—5.4L DOHC, exc. Lightning

- Electrical connector to the communication valve
- Tube assembly and connect the PCV valve tube to the water-heated fitting.
- Coolant hoses
- PCV valve in the valve cover
- EVAP return line
- Cables and bracket and install the bolts
- Accelerator cable, speed control cable, and the return spring
- Engine appearance cover and the 3 bolts
- Engine air cleaner and outlet tube

LOWER INTAKE MANIFOLD

❋❋ CAUTION

Fuel injection systems remain under pressure, even after the engine has been turned OFF. The fuel system pressure must be relieved before disconnecting any fuel lines. Failure to do so may result in fire and/or personal injury.

1. Before servicing the vehicle, refer to the Precautions Section.
2. Remove or disconnect the following:
   - Negative battery cable
   - Coolant
   - Upper intake manifold
   - Fuel system pressure
   - Fuel lines
   - Engine water bypass hose
   - Electrical connector from the water temperature indicator sender

- Upper radiator hose, the heater water inlet hose and the heated PCV water fitting inlet hose.
- The 4 bolts and the upper alternator support bracket
- The 8 fuel injectors
- Vacuum line from the fuel injector

pressure regulator and position out of the way.
- Lower intake manifold
- Radio ignition interference capacitors

To install:
3. Remove and inspect the gaskets, install new gaskets if necessary.
4. Clean the sealing surfaces.
5. Install or connect the following:
   - Radio ignition interference capacitors

➡**Bolts should be hand-started, positions 7–12 first then 1–6.**

- Lower intake manifold: Tighten the bolts to 89 inch lbs. (10 Nm) and then an additional 90 degrees in the sequence shown.
- Water temperature indicator sensor
- Vacuum harness and connect the vacuum line to the fuel injector pressure regulator.
- Fuel injectors
- Upper alternator support bracket and the 4 bolts
- Heater water inlet hose, the upper radiator hose and the water heated fitting inlet hose
- Engine water bypass return hose
- Fuel lines

67197-EFSE-G76

Lower intake manifold torque sequence—5.4L DOHC, exc. Lightning

- Upper intake manifold
- Coolant

5.4L Lightning

✳✳ CAUTION

Fuel injection systems remain under pressure, even after the engine has been turned OFF. The fuel system pressure must be relieved before disconnecting any fuel lines. Failure to do so may result in fire and/or personal injury.

1. Before servicing the vehicle, refer to the Precautions Section.
2. Remove or disconnect the following:
   - Negative battery cable
   - Charge air cooler
   - Fuel system pressure
   - Fuel lines
   - Coolant
   - Accelerator cable bracket retaining bolts and remove the bracket.
   - Upper radiator hose from the thermostat housing

✳✳ WARNING

Do not disconnect the PCV hose system from the intake. Installation can not be carried out with the intake in place.

- Positive Crankcase Ventilation (PCV) system
- Ground strap and both radio ignition interference capacitors
- Vacuum line
- Charge air cooler temperature sensor connector
- Heater hose
- Fuel injector electrical connectors
- Vapor management valve vacuum line
- Vacuum line near the brake vacuum booster
- Fuel injection supply manifold
- Ignition coil connectors
- Ignition coils
- Accessory drive belt
- Alternator bracket
- Intake manifold

To install:
3. Position the gaskets.

✳✳ CAUTION

If the PCV system hose becomes disconnected, the intake manifold will have to be removed to reattach the hose.

4. Make sure that the PCV system hose is securely connected.
5. Install or connect the following:
   - Intake manifold
   - Thermostat outlet
   - Intake manifold bolts. Tighten the bolts in 2 stages. Stage 1: tighten to 18 inch lbs. (2 Nm). Stage 2: tighten to plus 90 degrees.
   - Alternator bracket
   - Accessory drive belt
   - Ignition coils
   - Ignition coil connectors
   - Fuel injection supply manifold and install the bolts.
   - Vacuum line near the brake vacuum booster
   - Vapor management valve vacuum line
   - Fuel injector electrical connectors
   - Heater hose
   - Charge air cooler temperature sensor
   - Vacuum line
   - Ground strap and both radio ignition interference capacitors
   - PCV system
   - Upper radiator hose
   - Accelerator cable bracket and tighten the bolts.

- Fuel lines
- Charge air cooler

6. Fill and bleed the engine cooling system.

2004 F-150 4.6L Engine, exc. Heritage Edition

REMOVAL

1. Before servicing the vehicle, refer to the Precautions Section.
2. Release the fuel system pressure.
3. Disconnect the battery ground cable.
4. Drain the cooling system.
5. Remove the air cleaner outlet tube.
6. Remove the parts in the order indicated in the following illustrations and tables.
7. Disconnect the upper radiator coolant hose.
8. Disconnect the heater coolant hose.
9. Disconnect the vapor management valve (VMV) tube quick-release fitting.
10. Disconnect the throttle position sensor (TPS) electrical connector.
11. Power steering reservoir support bracket.
12. Disconnect the fuel pressure damper electrical connector.
13. Disconnect the fuel pressure damper vacuum connector.

67197-EFSE-G77

Intake manifold bolt torque—5.4L supercharged engine

14. Remove the nut and disconnect the brake booster vacuum tube.

15. Disconnect the exhaust gas recirculation (EGR) system module-to-exhaust manifold tube. Position the bracket and vacuum tube aside.

16. Disconnect the EGR system module vacuum connectors.

17. Disconnect the EGR system module electrical connector.

18. Disconnect the fuel tube spring lock coupling.

19. Disconnect the positive crankcase ventilation (PCV) tube quick-release fitting.

20. Remove the accessory drive belt.

21. Remove the generator.

22. Disconnect the intake manifold tuning valve (IMTV) electrical connectors.

23. Remove the ignition coils.

24. Disconnect the electronic throttle control electrical connector.

25. Remove the throttle body spacer bolts and the throttle body spacer and gasket.

26. Remove the PCV coolant return hose.

27. Remove the thermostat.

28. Detach the wiring harness from the left valve cover and position it aside.

29. Remove the 9 intake manifold bolts.

30. Remove the intake manifold.

31. Discard the intake manifold gaskets.

✷✷ WARNING

Do not use metal scrapers, wire brushes, power abrasive discs or other abrasive means to clean the sealing surfaces. These tools cause scratches and gouges which make leak paths. Use a plastic scraping tool to remove all traces of old sealant.

32. Clean the mating surfaces of the cylinder head and the intake manifold with metal surface prep and silicone gasket remover. Follow the directions on the packaging.

DISASSEMBLY

1. Before servicing the vehicle, refer to the Precautions Section.

2. Remove the coolant hose.

3. Remove the throttle body.

4. Remove the EGR system module. Discard the gasket.

5. Remove the fuel injection supply manifold.

6. Remove the fuel injectors.

7. Remove the 8 lower-to-upper intake manifold bolts.

8. Remove the lower intake manifold. Discard the gaskets.

9. Remove the Intake manifold tuning valve.

10. Remove the upper intake manifold.

ASSEMBLY

1. Before servicing the vehicle, refer to the Precautions Section.

2. To assemble, reverse the disassembly procedure. Note the following:

32	PCV coolant hose clamp
33	PCV coolant return hose
34	Thermostat housing bolts (2 req'd)
35	Thermostat housing
36	Thermostat housing O-ring seal
37	Thermostat
38	Intake manifold bolts (9 req'd)
39	Intake manifold
40	RH intake manifold gasket
41	LH intake manifold gasket

06017-F150-G44

Intake manifold assembly installation—2004 F-150 4.6L, exc. Heritage Editions

19 Lower intake manifold bolts (8 req'd)
20 Lower intake manifold
21 Lower intake manifold gasket
22 Intake manifold tuning valve (IMTV) bolts (2 req'd)
23 IMTV
24 Upper intake manifold

06017-F150-G42

Upper-to-lower intake manifold assembly—2004 F-150 4.6L engine, exc. Heritage editions

06017-F150-G43

Proper EGR system module gasket installation—2004–05 4.6L engines

a. Tighten the lower-to-upper intake manifold bolts in two stages.
- Tighten to 9 Nm (80 inch lbs.).
- Tighten an additional 90 degrees (¼ turn).

b. Tighten throttle body spacer bolts in two stages.
- Tighten to 9 Nm (80 inch lbs.).
- Tighten an additional 90 degrees (¼ turn).

c. Install a new EGR system module gasket with the side with the raised circle facing the intake manifold.

d. Tighten throttle body bolts in two stages.
- Tighten to 9 Nm (80 inch lbs.).
- Tighten an additional 90 degrees (¼ turn).

INSTALLATION

1. Before servicing the vehicle, refer to the Precautions Section.
2. Installation is the reverse of removal. Note the following:
  a. Fill and bleed the engine cooling system.

✳✳ WARNING

When reusing liquid or vapor tube connectors, make sure to use compressed air to remove any foreign material from the connector retaining.

  b. Position the new intake manifold gaskets.
  c. Position the upper intake manifold.
  d. Loosely install the nine intake manifold bolts.
  e. Tighten the intake manifold assembly bolts in the sequence shown, to 89 inch lbs. (10 Nm).
  f. Inspect and clean both fluid coupling ends. Install new O-rings and garter springs if necessary. Lubricate the O-rings with clean engine oil.
  g. Fit the male fitting into the female end and push until the garter spring snaps over the flared end of the female fitting.
  h. Make sure the coupling is engaged by pulling on the lines and install the safety clip.
3. For the VMV tube and PCV tube quick-release fitting installation, apply clean engine oil to the end of the tube before connecting the fitting. Connect the quick-release fitting. Make sure the quick-release fitting clicks into place when installing the tube. To make sure that the quick-release fitting is fully seated, pull on the tube.
  a. Install the throttle body spacer, gasket and bolts.
  b. Tighten the throttle body spacer bolts in two stages.
- Stage 1: Tighten to 9 Nm (80 inch lbs.).
- Stage 2: Tighten an additional 90 degrees (¼ turn).

  c. Connect the throttle body spacer coolant hose to the intake manifold.
  d. Connect the vacuum hose (4x4 only).

2005 F-150 4.6L Engine

REMOVAL

1. Before servicing the vehicle, refer to the Precautions Section.
2. Release the fuel system pressure.
3. Disconnect the battery ground cable.
4. Drain the cooling system.
5. Remove the air cleaner outlet tube.
6. Release the upper radiator hose clamp and position the upper radiator hose aside.
7. Release the heater hose clamp and remove and position the heater hose aside.

8. Disconnect the vapor tube quick connect fitting.

9. Disconnect the throttle position (TP) sensor electrical connector.

10. Remove the power steering reservoir support bracket bolt.

11. Remove the 2 power steering reservoir support bracket bolts and the power steering reservoir support bracket.

12. Disconnect the injection pressure (IPR) sensor electrical connector.

13. Disconnect the IPR sensor vacuum connector.

14. Remove the brake booster bracket nut and disconnect the brake booster tube and position aside the brake booster bracket and tube.

15. Remove the exhaust gas recirculation (EGR) system module-to-exhaust manifold tube.

16. Disconnect the EGR system module vacuum and electrical connectors.

17. Disconnect the fuel tube spring lock coupling.

18. Disconnect the positive crankcase ventilation (PCV) tube quick release fitting.

19. Rotate the tensioner clockwise and remove the accessory drive belt.

20. If equipped, disconnect the vacuum hose.

21. Disconnect the throttle body (TB) spacer coolant hose from the intake manifold.

22. Disconnect the generator electrical connector.

23. Remove the generator battery cable nut and the generator battery cable.

24. Release the generator wiring harness retainer.

25. Remove the 4 generator support bracket bolts and the generator support bracket.

26. Remove the 2 lower generator bolts and the generator.

27. Disconnect the 8 fuel injector electrical connectors.

28. Disconnect the 8 ignition coil electrical connectors.

29. Remove the 8 ignition coil bolts and the 8 ignition coils.

30. Disconnect the electronic throttle control electrical connector.

31. Release the PCV coolant hose clamp and remove the PCV coolant return hose.

32. Remove the 2 thermostat housing bolts and the thermostat housing and discard the O-ring seal.

33. Remove the thermostat.

34. Detach the wiring harness from the left valve cover and position aside.

35. Remove the 9 intake manifold bolts.

36. Remove the intake manifold and discard the RH and LH intake manifold gaskets.

DISASSEMBLY

1. Before servicing the vehicle, refer to the Precautions Section.

2. Release the 2 coolant hose clamps and remove the coolant hose.

3. Remove the 4 throttle body (TB) bolts and the TB. Discard the TB gasket.

4. Remove the exhaust gas recirculation (EGR) system module bolt.

5. Remove the EGR system module stud bolt and the EGR system module. Discard the EGR system module gasket.

6. Remove the 4 TB spacer bolts and the TB spacer. Discard the TB spacer gasket.

7. Remove the 4 fuel rail bolts.

8. Remove the fuel rail.

9. Remove the 8 fuel injectors Discard the 16 fuel injector O-ring seals.

10. Remove the 8 lower intake manifold bolts and the separate the upper and lower intake manifolds. Discard the lower intake manifold gasket.

ASSEMBLY

1. Before servicing the vehicle, refer to the Precautions Section.

2. Install a new lower intake manifold gasket and install the upper to the lower intake manifold.

3. Install the 8 bolts and tighten the lower intake manifold bolts in 2 stages.
- Stage 1: Tighten to 9 Nm (80 inch lbs.).
- Stage 2: Tighten an additional 90 degrees (¼ turn).

4. Install the new 16 fuel injector O-ring seals and lubricate the O-ring seals with clean engine oil and install the fuel injectors into the upper intake manifold.

19 Lower intake manifold bolts (8 required)

20 Lower intake manifold

21 Lower intake manifold gasket

22 Upper intake manifold

06017-F150-G45

Upper-to-lower intake manifold—2005 F-150 4.6L engine

5. Install the fuel rail and the 4 fuel rail bolts. Tighten to 10 Nm (89 inch lbs.).

6. Install a new throttle body spacer gasket and the throttle body spacer and tighten throttle body spacer bolts in 2 stages.

- Stage 1: Tighten to 9 Nm (80 inch lbs.).
- Stage 2: Tighten an additional 90 degrees (¼ turn).

7. Install a new EGR gasket, with the raised gasket surface facing the throttle body spacer, and the EGR system module and the stud bolt. Tighten to 25 Nm (18 ft. lbs.).

8. Install the EGR system module bolt. Tighten to 25 Nm (18 ft. lbs.).

9. Install a new throttle body gasket and the throttle body and tighten the 4 throttle body bolts in 2 stages.

- Stage 1: Tighten to 9 Nm (80 inch lbs.).

- Stage 2: Tighten an additional 90 degrees (¼ turn).

10. Install the coolant hose and the 2 coolant clamps.

INSTALLATION

1. Before servicing the vehicle, refer to the Precautions Section.

✱✱ WARNING

Do not use metal scrapers, wire brushes, power abrasive discs or other abrasive means to clean the sealing surfaces. These tools cause scratches and gouges which make leak paths. Use a plastic scraping tool to remove all traces of old sealant.

2. Clean the mating surfaces of the cylinder head and the intake manifold with metal surface prep and silicone gasket remover. Follow the directions on the packaging.

3. Position the new intake manifold gaskets.

4. Position the intake manifold.

5. Loosely install the 9 intake manifold bolts.

6. Tighten the intake manifold bolts in the sequence shown. Tighten to 10 Nm (89 inch lbs.).

7. Attach the wiring harness to the left valve cover.

8. Install the thermostat.

9. Install the thermostat housing and the thermostat housing O-ring seal. Apply clean engine coolant to the O-ring seal before installation.

10. Install the 2 thermostat housing bolts. Tighten to 25 Nm (18 ft. lbs.).

11. Connect the electronic throttle control electrical connector.

33 Thermostat housing bolts (2 required)	37 Intake manifold bolts (9 required)
34 Thermostat housing	38 Intake manifold
35 Thermostat housing O-ring seal	39 RH intake manifold gasket
36 Thermostat	40 LH intake manifold gasket

Intake manifold assembly installation—2005 F-150 4.6L engine

06017-F150-G46

06017-F150-G47

Intake manifold assembly torque sequence—2004 F-150, exc. Heritage Edition and 2005 F-150 4.6L engine

12. Install the 8 ignition coil and the 8 ignition coils bolts. Tighten to 6 Nm (53 inch lbs.).

13. Connect the 8 ignition coil electrical connectors.

14. Connect the 8 fuel injector electrical connectors.

15. Install the generator and the 2 lower generator bolts. Tighten to 25 Nm (18 ft. lbs.).

16. Install the generator support bracket and the 4 generator support bracket bolts. Tighten to 10 Nm (89 inch lbs.).

17. Attach the generator wiring harness retainer.

18. Install the generator battery cable and the generator battery cable nut. Tighten to 10 Nm (89 inch lbs.).

19. Connect the generator electrical connector.

20. Connect the TB spacer coolant hose to the intake manifold.

21. If equipped, connect the vacuum hose.

22. Rotate the tensioner clockwise and install the accessory drive belt.

23. Install the PCV coolant return hose and install the PCV coolant hose clamp.

24. Connect the PCV tube.

25. Connect the fuel tube spring lock coupling.

26. Connect the EGR system module vacuum and electrical connectors.

27. Install the EGR system module-to-exhaust manifold tube. Tighten to 40 Nm (30 ft. lbs.).

28. Connect the brake booster tube and position the brake booster bracket and install the nut. Tighten to 10 Nm (89 inch lbs.).

29. Connect the IPR sensor vacuum connector.

30. Connect the IPR sensor electrical connector.

31. Install the power steering reservoir support bracket and the 2 power steering reservoir support bracket bolts. Tighten to 17 Nm (13 ft. lbs.).

32. Install the power steering reservoir support bracket bolt. Tighten to 11 Nm (8 ft. lbs.).

33. Connect the TP sensor electrical connector.

34. Connect the vapor tube quick-connect fitting.

35. Install the heater hose and install the heater hose clamp.

36. Install the upper radiator hose and clamp.

37. Install the air cleaner outlet tube.

38. Connect the battery ground cable.

39. Fill and bleed the engine cooling system.

2004 F-150 5.4L Engine, exc. Heritage Editions and Lightning

1. Before servicing the vehicle, refer to the Precautions Section.

✳✳ WARNING

If the supply manifold is used as a leverage device, damage may occur to the supply manifold. Care must be taken when working around the supply manifold.

2. With the vehicle in NEUTRAL, position it on a hoist.

3. Relieve the fuel system pressure.

4. Drain the engine cooling system.

5. Remove the air cleaner assembly and the air cleaner inlet pipe and resonator.

6. Remove the generator.

7. Remove the parts in the order indicated in the following illustrations and tables.

8. Remove the radiator hose.

9. Remove the heater hose.

10. Disconnect the positive crankcase ventilation (PCV) tube quick-release fitting.

11. Disconnect the fuel supply tube spring lock coupling.

12. Disconnect the fuel pulse damper electrical connector.

13. Disconnect the left fuel injector electrical connectors.

14. Disconnect the electronic throttle control electrical connector.

15. Disconnect the throttle position (TP) sensor electrical connector.

16. Disconnect the right fuel injector electrical connectors.

17. Disconnect the vapor management valve (VMV) tube quick-release fitting.

18. Remove the intake manifold bolts.

➡**Position the intake manifold assembly back to access the crossover bolts.**

19. Remove the engine coolant crossover assembly. Discard the gasket.

20. Remove the engine wiring harness retainer.

21. Disconnect the charge motion control valve (CMCV) electrical connector.

22. Disconnect the cylinder head temperature (CHT) sensor electrical connector.

23. Disconnect the left and right knock sensor (KS) electrical connectors.

24. Remove the intake manifold. Discard the gaskets.

✳✳ WARNING

Do not use metal scrapers, wire brushes, power abrasive discs or other abrasive means to clean the sealing surfaces. These tools cause scratches and gouges which make leak paths. Use a plastic scraping tool to remove all traces of old sealant.

25. Clean and inspect the sealing surfaces with silicone gasket remover and metal surface prep.
26. To install, reverse the removal procedure. Note the following:

✳✳ WARNING

When reusing liquid or vapor tube connectors, make sure to use compressed air to remove any foreign material from the connector retaining clip area before separating from the tube.

a. Install new intake manifold gaskets onto the intake manifold.

➡**Electrical and vacuum harnesses must not restrict movement of the charge motion control valve (CMCV) control rods at rear of the intake manifold. Use extreme care on installation of the intake manifold to prevent any pinching of electrical and vacuum harnesses.**

b. Position the intake manifold and loosely install the bolts.
c. Tighten the bolts in two stages in the sequence shown.
- Stage 1: Tighten to 2 Nm (18 inch lbs.).
- Stage 2: Tighten to 10 Nm (89 inch lbs.).

d. Inspect and clean both fuel supply tube spring lock coupling ends. Install new O-rings and garter springs if necessary. Lubricate the O-rings with clean engine oil.
e. Fit the male fitting into the female end and push until the garter spring snaps over the flared end of the female fitting.
f. Make sure the coupling is engaged by pulling on the lines and install the safety clip.

✳✳ WARNING

Make sure the PCV tube and VMV tube quick-release fitting clicks into place when installing the tube. To make sure that the quick-release fitting is fully seated, pull on the tube.

➡**Apply clean engine oil to the end of the tube before connecting the fitting.**

27. Connect the quick-release fitting.
28. Install and connect both ends of the PCV tube.

2005 F-150 5.4L Engine and 2006 Mark LT

1. Before servicing the vehicle, refer to the Precautions Section.
2. Drain the cooling system.
3. Release the fuel system pressure.
4. Remove the generator.
5. Remove the air cleaner.
6. Disconnect the upper radiator hose from the thermostat housing.
7. Disconnect the heater coolant hose from the coolant bypass tube.
8. Disconnect the quick connect coupling and remove the evaporative emissions (EVAP) tube from the intake manifold.
9. Disconnect the quick connect coupling and remove the positive crankcase ventilation (PCV) tube.
10. Disconnect the fuel supply spring lock coupling from the fuel rail.
11. Disconnect the fuel rail pressure and temperature sensor electrical connector and vacuum connector.
12. Disconnect the 8 fuel injector electrical connectors.
13. Disconnect the throttle position (TP) sensor and electronic throttle control electrical connectors.
14. Disconnect the heated PCV element electrical connector.
15. Disconnect the brake booster vacuum hose from the intake manifold vacuum tube.
16. Remove the 10 intake manifold bolts.

✳✳ WARNING

Do not use metal scrapers, wire brushes, power abrasive discs or other abrasive means to clean the sealing surfaces. These tools cause scratches and gouges which make leak paths. Use a plastic scraping tool to remove all traces of old sealant.

17. Remove the 3 bolts, the coolant bypass tube and discard the gaskets.
18. Clean and inspect the sealing surfaces with metal surface prep. Follow the directions on the packaging.

19. Disconnect the charge motion control valve (CMCV) electrical connector.
20. Disconnect the intake manifold vacuum tube from the valve cover stud and the support bracket.
21. Disconnect the cylinder head temperature (CHT) sensor jumper harness electrical connector.
22. Disconnect the LH and RH knock sensor (KS) electrical connectors.
23. Remove the nut and disconnect the engine wiring harness retainer from the CMCV stud.

✳✳ WARNING

Do not use metal scrapers, wire brushes, power abrasive discs or other abrasive means to clean the sealing surfaces. These tools cause scratches and gouges which make leak paths. Use a plastic scraping tool to remove all traces of old sealant.

24. Remove the intake manifold and discard the gaskets.
25. Clean and inspect the sealing surfaces with metal surface prep. Follow the directions on the packaging.

To install:

➡**Electrical and vacuum harnesses must not restrict movement of the CMCV control rods at the rear of the intake manifold. Use extreme care on installation of the intake manifold to prevent any pinching of electrical and vacuum harnesses.**

26. Using new intake manifold gaskets, position the intake manifold.
27. Using new gaskets, position the coolant crossover and install the 3 bolts. Tighten to 10 Nm (89 inch lbs.).
28. Install the intake manifold bolts and tighten in 2 stages in the sequence shown.
- Stage 1: Tighten to 2 Nm (18 inch lbs.).
- Stage 2: Tighten to 10 Nm (89 inch lbs.).

29. Connect the engine wiring harness retainer to the CMCV stud and install the nut. Tighten to 10 Nm (89 inch lbs.).
30. Connect the CMCV electrical connector.
31. Connect the CHT sensor jumper harness electrical connector.
32. Connect the LH and RH KS electrical connectors.
33. Connect the intake manifold vacuum

tube to the support bracket and the valve cover stud.

34. Connect the brake booster vacuum hose to the intake manifold vacuum tube.

35. Connect the heated PCV element electrical connector.

36. Connect the TP sensor and electronic throttle control electrical connectors.

37. Connect the 8 fuel injector electrical connectors.

38. Connect the fuel rail pressure and temperature sensor electrical connector and vacuum connector.

39. Connect the fuel supply spring lock coupling to the fuel rail.

40. Position the PCV tube and connect the quick connect coupling.

41. Position the EVAP tube and connect the quick connect coupling to the intake manifold.

42. Connect the heater coolant hose to the coolant bypass.

43. Connect the upper radiator hose to the thermostat housing.

44. Install the generator.

45. Install the air cleaner.

46. Fill and bleed the engine cooling system.

2004 F-250/350 5.4L Engine

1. Before servicing the vehicle, refer to the Precautions Section.

2. Disconnect the battery ground cable.

3. Relieve the fuel system pressure.

4. Drain the cooling system.

5. Disconnect the idle air control (IAC) fresh air tube and crankcase ventilation tube from the air cleaner outlet pipe.

6. Loosen the clamp and disconnect the air cleaner outlet pipe from the throttle body.

7. Release the clamp and separate the air cleaner outlet housing from the air cleaner inlet housing.

8. Disconnect the mass air flow (MAF) sensor electrical connector and remove the air cleaner, air cleaner outlet housing and the air cleaner outlet pipe from the vehicle.

9. Compress and slide the hose clamp and disconnect the upper radiator hose.

10. On vehicles with A/C, rotate the belt tensioner clockwise and remove the belt from the generator pulley.

11. On vehicles without A/C, rotate the belt tensioner counterclockwise and remove the belt from the generator pulley.

12. Remove the harness routing clips from the generator stud and the engine front cover.

13. Remove the generator upper mounting bracket stud and bolt.

14. Remove the generator lower mounting bolts and position the generator aside.

15. Remove the accelerator control splash shield.

16. Disconnect the accelerator cable.

17. If equipped, disconnect the speed control cable.

18. Remove the throttle return spring.

19. Position the accelerator cable bracket aside.

20. Disconnect the throttle position (TP) sensor electrical connector.

21. Release the IAC fresh air tube from the routing clip and remove it from the vehicle.

22. Disconnect the evaporative emission (EVAP) canister purge valve vacuum hose from the throttle body adapter.

23. Disconnect the brake booster vacuum hose and main vacuum harness from the throttle body adapter.

24. Disconnect the IAC electrical connector.

25. On vehicles with EGR system:
   a. Disconnect the upper fitting.
   b. Loosen the lower fitting.
   c. Position the exhaust manifold to EGR valve tube aside.
   d. Position the differential pressure feedback EGR system bracket aside.
   e. Disconnect the EGR vacuum regulator solenoid connections.
   f. Disconnect the EGR valve vacuum hose.

26. Remove the throttle body adapter.

27. Remove and discard the throttle body gasket.

28. Disconnect the fuel tubes.

29. Disconnect the positive crankcase ventilation (PCV) valve tube from the intake manifold.

30. Disconnect the fuel pressure regulator vacuum hose.

31. Disconnect the eight ignition coil electrical connectors and remove the coils.

32. Disconnect the eight fuel injector electrical connectors.

33. Disconnect the hose clamp and remove the heater water hose.

34. Remove the bolts, water thermostat housing, O-ring seal and the water thermostat.

35. Discard the O-ring seal.

36. Remove the bolts.

37. Remove the intake manifold. Remove and discard the intake manifold gaskets.

✳✳ WARNING

Do not use metal scrapers, wire brushes, power abrasive discs or other abrasive means to clean the sealing surfaces. These tools cause scratches and gouges that make leak paths.

38. Clean all mating surfaces.

To install:

39. Position the intake manifold gaskets.

40. Position the intake manifold.

41. Loosely install the nine bolts.

42. Install the thermostat. Install a new O-ring seal.

➡The thermostat housing bolts are tightened in sequence with the intake manifold bolts. Do not tighten the thermostat housing bolts during thermostat installation.

43. Install the thermostat housing and loosely install the bolts.

44. Tighten the intake manifold bolts in two stages, in the sequence shown.
   • Stage 1: Tighten to 2 Nm (18 inch lbs.).
   • Stage 2: Tighten to 25 Nm (18 ft. lbs.).

45. Install the heater water hose and position the clamp.

46. Connect the eight fuel injector electrical connectors.

47. Install the eight ignition coils. Torque to 53 inch lbs. (6 Nm).

48. Connect the eight ignition coil electrical connectors.

49. Position the vacuum harness and connect the fuel pressure regulator vacuum hose.

50. Connect the PCV valve tube to the intake manifold.

51. Connect the fuel tubes and install the safety clips.

52. Install a new throttle body adapter gasket.

53. Position the throttle body adapter.

54. Install the four bolts. Torque to 89 inch lbs. (10 Nm).

55. On vehicles with EGR system:
   a. Connect the EGR valve vacuum hose.
   b. Connect the EGR vacuum regulator solenoid connections.
   c. Install the differential pressure feedback EGR system bracket.
   d. Connect the exhaust manifold to EGR valve tube upper fitting. Tighten both fittings starting at the top in three steps.
   • Hand-tighten.

- Tighten the upper fitting to 40–60 Nm (30–44 ft. lbs.).
- Tighten the lower fitting to 40–60 Nm (30–44 ft. lbs.).

56. Connect the IAC electrical connector.

57. Connect the brake booster vacuum hose and main vacuum harness to the throttle body adapter.

58. Connect the evaporative emission canister purge valve vacuum hose to the throttle body adapter.

59. Install the IAC fresh air tube and insert it in the routing clip.

60. Connect the TP sensor electrical connector.

61. Install the accelerator cable bracket.

62. Connect the accelerator cable.

63. If equipped, connect speed control cable.

64. Install the return spring.

65. Install the accelerator control splash shield.

66. Position the generator and install the lower mounting bolts. Torque to 18 ft. lbs. (25 Nm).

67. Install the generator upper mounting bracket stud and bolt. Tighten to 89 inch lbs. (10 Nm).

68. Install the harness routing clips on the generator stud and the engine front cover.

69. On vehicles without A/C, rotate the belt tensioner counterclockwise and install the belt on the generator pulley.

70. On vehicles with A/C, rotate the belt tensioner clockwise and install the belt on the generator

71. Install the upper radiator hose and reposition the clamp.

72. Position the air cleaner, air cleaner outlet housing and the air cleaner outlet pipe in the vehicle and connect the MAF sensor electrical connector.

73. Join the air cleaner outlet housing to the air cleaner inlet housing and secure the clamp.

74. Connect the air cleaner outlet pipe to the throttle body and tighten the clamp.

75. Connect the IAC fresh air tube and crankcase ventilation tube to the air cleaner outlet pipe.

76. Fill the cooling system.

77. Connect the battery ground cable.

2005 F-250/350 5.4L Engine

✳✳ WARNING

Fuel in the fuel system remains under high pressure even when the engine is not running.

1. Before servicing the vehicle, refer to the Precautions Section.

2. Disconnect the battery ground cable.

3. Relieve the fuel system pressure.

4. Drain the cooling system.

5. Disconnect the idle air control (IAC) fresh air tube and crankcase ventilation tube from the air cleaner outlet pipe.

6. Loosen the clamp and disconnect the air cleaner outlet pipe from the throttle body.

7. Release the clamp and separate the air cleaner outlet housing from the air cleaner inlet housing.

8. Disconnect the mass air flow (MAF) sensor electrical connector and remove the air cleaner, air cleaner outlet housing and the air cleaner outlet pipe from the vehicle.

9. Compress and slide the hose clamp and disconnect the upper radiator hose.

Vehicles with A/C

10. Rotate the belt tensioner clockwise and remove the belt from the alternator pulley.

Vehicles without A/C

11. Rotate the belt tensioner counterclockwise and remove the belt from the alternator pulley.

All vehicles

12. Remove the harness routing clips from the alternator stud and the engine front cover.

13. Remove the alternator upper mounting bracket stud and bolt.

14. Remove the alternator lower mounting bolts and position the alternator aside.

15. Remove the accelerator control splash shield.

16. Disconnect the throttle body cam.

17. Position the accelerator cable bracket aside.

18. Disconnect the throttle position (TP) sensor electrical connector.

19. Release the IAC fresh air tube from the routing clip and remove it from the vehicle.

20. Disconnect the evaporative emission (EVAP) canister purge valve vacuum hose from the throttle body adapter.

21. Disconnect the brake booster vacuum hose and main vacuum harness from the throttle body adapter.

22. Disconnect the IAC electrical connector.

Vehicles with EGR system

23. Position the exhaust manifold to exhaust gas recirculation (EGR) valve tube aside.

24. Position the differential pressure feedback EGR system bracket aside.

25. Disconnect the EGR vacuum regulator solenoid connections.

26. Disconnect the EGR valve vacuum hose.

All vehicles

27. Remove the throttle body adapter.

28. Disconnect the fuel tubes.

29. Disconnect the positive crankcase ventilation (PCV) valve tube from the intake manifold.

30. Disconnect the fuel pressure regulator vacuum hose.

31. Disconnect the eight ignition coil electrical connectors and remove the coils.

32. Disconnect the eight fuel injector electrical connectors.

33. Disconnect the hose clamp and remove the heater water hose.

34. Remove the bolts, water thermostat housing, O-ring seal and the water thermostat. Discard the O-ring seal.

35. Remove the bolts. Remove the intake manifold. Remove and discard the intake manifold gaskets.

✳✳ CAUTION

Do not use metal scrapers, wire brushes, power abrasive discs or other abrasive means to clean the sealing surfaces. These tools cause scratches and gouges that make leak paths.

36. Clean all mating surfaces.

To install:

All vehicles

37. Position the intake manifold gaskets.

38. Position the intake manifold. Loosely install the nine bolts.

39. Install the thermostat. Install a new O-ring seal.

➡**The thermostat housing bolts are tightened in sequence with the intake manifold bolts. Do not tighten the thermostat housing bolts during thermostat installation.**

40. Install the water thermostat housing and loosely install the bolts.

41. Tighten the bolts in two steps, in the sequence shown.
   a. Step 1: Tighten to 2 Nm (18 inch lbs.).
   b. Step 2: Tighten to 89 inch lbs.

42. Install the heater water hose and position the clamp.

43. Connect the eight fuel injector electrical connectors.

44. Install the eight ignition coils.

45. Connect the eight ignition coil electrical connectors.

46. Position the vacuum harness and

connect the fuel pressure regulator vacuum hose.

47. Connect the PCV valve tube to the intake manifold.

48. Connect the fuel tubes and install the safety clips.

49. Install a new throttle body adapter gasket.

50. Install the throttle body adapter.

Vehicles with EGR system

51. Connect the EGR valve vacuum hose.

52. Connect the EGR vacuum regulator solenoid connections.

53. Install the differential pressure feedback EGR system bracket.

54. Connect the exhaust manifold to EGR valve tube upper fitting.

55. Tighten both fittings starting at the top in three steps. Hand-tighten.

56. Tighten the upper fitting to 40–60 Nm (30–44 ft. lbs.).

57. Tighten the lower fitting to 40–60 Nm (30–44 ft. lbs.).

All vehicles

58. Connect the IAC electrical connector.

59. Connect the brake booster vacuum hose and main vacuum harness to the throttle body adapter.

60. Connect the evaporative emission canister purge valve vacuum hose to the throttle body adapter.

61. Install the IAC fresh air tube and insert it in the routing clip.

62. Connect the TP sensor electrical connector.

63. Install the accelerator cable bracket.

64. Connect the throttle body cam.

65. Install the accelerator control splash shield.

66. Position the alternator and install the lower mounting bolts.

67. Install the alternator upper mounting bracket stud and bolt.

68. Install the harness routing clips on the alternator stud and the engine front cover.

Vehicles without A/C

69. Rotate the belt tensioner counterclockwise and install the belt on the alternator pulley.

Vehicles with A/C

70. Rotate the belt tensioner clockwise and install the belt on the alternator pulley.

All vehicles

71. Install the upper radiator hose and reposition the clamp.

72. Position the air cleaner, air cleaner outlet housing and the air cleaner outlet pipe in the vehicle and connect the MAF sensor electrical connector.

73. Join the air cleaner outlet housing to

Intake manifold bolt torque sequence—2004 F-150 5.4L engine, exc. Heritage Editions and Lightning; 2005 F-250/350 5.4L engine

06017-F150-G48

the air cleaner inlet housing and secure the clamp.

74. Connect the air cleaner outlet pipe to the throttle body and tighten the clamp.

75. Connect the IAC fresh air tube and crankcase ventilation tube to the air cleaner outlet pipe.

76. Fill the cooling system.

77. Connect the battery ground cable.

2004 F-250/350 6.8L Engine

❋❋ WARNING

Fuel in the fuel system remains under high pressure, even when the engine is not running.

1. Before servicing the vehicle, refer to the Precautions Section.

2. Relieve the fuel system pressure.

3. Disconnect the battery ground cable.

4. Drain the engine cooling system.

5. Remove the air cleaner outlet tube

6. Remove the alternator.

7. Disconnect the fuel lines.

8. Compress and slide the hose clamp and disconnect the water outlet hose.

9. Remove the accelerator cable splash shield.

10. Disconnect the accelerator and speed control cables and remove the return spring.

11. Position the accelerator cable bracket aside.

12. Disconnect the positive crankcase valve (PCV) tube and the vacuum lines.

13. Disconnect the heated PCV coolant hoses.

14. Remove the bolt and position the right radio ignition interference capacitor aside.

15. Disconnect the vacuum line.

16. Disconnect the five right fuel injector electrical connectors and five ignition coil electrical connectors.

17. Disconnect the idle air control (IAC) motor electrical connector and the bypass hose.

18. Disconnect the heater hose.

19. Remove the throttle body (TB) adapter.

20. Disconnect the five left fuel injector electrical connectors and ignition coil connectors.

21. Disconnect the engine coolant temperature (ECT) sensor electrical connectors.

22. Remove the ignition coils.

23. Disconnect the vacuum connector from the fuel pressure regulator.

24. Remove the bolt and position the left ignition radio ignition interference capacitor aside.

25. Disconnect the cylinder head temperature (CHT) sensor.

26. Remove the water thermostat and housing.

27. Remove the bolts, the upper intake manifold and the intake manifold gaskets. Discard the intake manifold gaskets.

28. Separate the upper and lower intake manifolds. Remove the 10 bolts.

➡ **Do not use metal scrapers, wire brushes, power abrasive discs or other abrasive means to clean the aluminum retainer plate. These tools cause scratches and gouges, which make leak paths.**

29. Clean all mating surfaces. Clean the intake manifold mating surface of the cylinder heads with silicone gasket remover and metal surface prep. Follow the directions on the packaging.

To install:

30. Position the lower intake manifold to the upper intake manifold and loosely install the 10 bolts.

31. Tighten the bolts in the sequence shown.

   a. Step 1: Tighten to 2 Nm (18 inch lbs.).

   b. Step 2: Tighten to 10 Nm (89 inch lbs.).

32. Position the water thermostat and housing.

33. Position the upper intake manifold gaskets and the intake manifold, and loosely install the bolts.

34. Tighten the bolts in two steps, in the sequence shown.

   a. 2004–05 Step 1: Tighten to 2 Nm (18 inch lbs.).

   b. 2004 Step 2: Tighten to 25 Nm (18 ft. lbs.).

   c. 2005 Step 2: 89 inch lbs. (10 Nm).

35. Remove and discard the fuel line O-ring seals.

➡ **Lubricate the new fuel line O-ring seals with clean engine oil prior to installation.**

36. Install new fuel line O-ring seals.

37. Connect the fuel lines.

38. Connect the CHT sensor electrical connector.

39. Install the left radio ignition interference capacitor electrical connector.

40. Connect the vacuum line to the fuel injection supply manifold.

41. Connect the ECT sensor electrical connector.

67197-EFSE-G82

Lower intake manifold bolt torque sequence—2004 6.8L engine

67197-EFSE-G83

Upper-to-lower intake manifold bolt torque sequence—2004 6.8L engine

42. Install the ignition coils.
43. Connect the five left fuel injector electrical connectors and five left ignition coil connectors.
44. Position a new TB adapter press-in-place gasket.
45. Position the TB adapter.
46. Install the four bolts in two steps:
   a. Step 1: Tighten to 9 Nm (80 inch lbs.).
   b. Step 2: Tighten an additional 90 degrees.
47. Connect the coolant hose.
48. Connect the IAC motor bypass hose and connector.
49. Connect the five right fuel injector electrical connectors and five right ignition coil connectors.
50. Connect the vacuum line.
51. Install the right radio ignition interference capacitor.
52. Connect the PCV tube and the vacuum lines.
53. Connect the heated PCV coolant hoses.
54. Install the accelerator bracket.
55. Connect the accelerator and speed control cables and return spring.
56. Install the accelerator cable splash shield.
57. Connect the engine water outlet hose and position the hose clamp.
58. Install the alternator.
59. Install the air cleaner outlet tube.
60. Connect the battery ground cable.
61. Fill and bleed the cooling system.

E-Series 5.4L Engine

1. Before servicing the vehicle, refer to the Precautions Section.
2. Disconnect the battery ground cable.
3. Remove the interior engine cover.
4. Disconnect the fuel hose spring lock couplings.
5. Drain the cooling system.
6. Position the engine air cleaner (ACL).
7. Disconnect the MAF sensor electrical connector and remove the ACL.
8. Disconnect the two hoses.
9. Loosen the clamp and remove the air cleaner outlet pipe.
10. Remove the accelerator cable snow shield.
11. Disconnect the accelerator cable.
12. Disconnect the speed control actuator cable.
13. Remove the throttle return spring.
14. Compress and slide the hose clamp and disconnect the upper radiator hose.
15. Disconnect the coolant hose.

16. Disconnect the coolant hose.
17. Disconnect the throttle position (TP) sensor.
18. Disconnect the differential pressure feedback exhaust gas recirculation (EGR) system.
19. Remove the differential pressure feedback EGR system bracket.
20. Remove the bolt and the bracket.
21. Remove the bolts and the alternator upper support bracket.
22. Remove the bolt retaining the transmission fluid filler tube support bracket.
23. Remove the bolts and position the accelerator cable bracket and cables aside.
24. If equipped, disconnect the auxiliary heater hoses and position aside.
25. Disconnect and remove the idle air control (IAC) valve fresh air hose.
26. Disconnect the evaporative emission canister purge valve hose and vacuum hose.
27. Remove the two evaporative emission canister purge valve nuts and position the valve aside.
28. Disconnect the fuel pressure sensor vacuum hose and the electrical connector.
29. Disconnect the EGR vacuum regulator solenoid connections.
30. Disconnect the EGR valve vacuum hose.
31. Disconnect the IAC valve electrical connector.
32. Disconnect the brake booster and main engine vacuum harness and position aside.
33. Disconnect the differential pressure feedback EGR sensor hoses from the EGR valve.
34. Remove the exhaust manifold-to-EGR valve tube.
35. Remove the four bolts, and the throttle body and adapter as an assembly. Discard the throttle body adapter gasket.
36. Disconnect and remove the positive crankcase ventilation hose.
37. Disconnect the eight ignition coil electrical connectors.
38. Disconnect the eight fuel injector electrical connectors.
39. Remove the eight bolts and the eight ignition coils.
40. Remove the bolts, the thermostat housing and the thermostat. Disconnect the O-ring seal.
41. Remove the nine bolts retaining the intake manifold.
42. Remove the intake manifold. Discard the intake manifold gaskets.
43. Inspect the throttle body, the intake manifold, and their sealing surfaces for damage.

To install:

➡Do not use metal scrapers, wire brushes, power abrasive discs or any other abrasive means to clean the sealing surfaces. These tools cause scratches and gouges which can cause leak paths. Use a plastic scraping tool to clean these surfaces.

44. Clean and inspect all sealing surfaces.
45. Position the intake manifold gaskets and the intake manifold.
46. Loosely install the nine bolts.
47. Install the thermostat.
48. Install a new O-ring seal.

➡The thermostat housing bolts are tightened in sequence with the intake manifold bolts.

49. Loosely install the thermostat housing and bolts.
50. Tighten the bolts in two steps, in the sequence shown.
   a. Step 1: Tighten to 2 Nm (18 inch lbs.).
   b. Step 2: Tighten to 25 Nm (18 ft. lbs.).
51. Install the eight ignition coils and bolts.
52. Connect the eight fuel injector electrical connectors.
53. Connect the eight ignition coil electrical connectors.
54. Install the PCV hose and valve.
55. Install the new throttle body adapter gasket.
56. Install the throttle body and adapter assembly.
57. Install the exhaust manifold-to-EGR valve tube and tighten the fitting in two steps.
   a. Hand-tighten the fittings.
   b. Tighten the fittings to 50 Nm (37 ft. lbs.).
58. Connect the differential pressure EGR sensor hoses to the EGR valve tube.
59. Connect the brake booster and the main engine vacuum harnesses.
60. Connect the IAC electrical connector.
61. Connect the EGR vacuum hose.
62. Connect the EGR vacuum regulator solenoid connections.
63. Connect the fuel pressure sensor vacuum hose and electrical connector.
64. Install the evaporative emission canister purge valve and bracket.
65. Connect the evaporative emission canister purge valve hoses.
66. Connect the IAC fresh air hose.
67. If equipped, connect the auxiliary heater hoses.

68. Connect the fuel hose spring lock couplings.
69. Install the accelerator cable bracket.
70. Install the transmission fluid filler tube bracket and bolt.
71. Install the alternator upper support bracket.
72. Install the accelerator cable routing bracket.
73. Install the differential pressure feedback EGR system bracket.
74. Connect the differential pressure feedback EGR sensor.
75. Connect the TP sensor.
76. Connect the coolant hoses.
77. Connect the upper radiator hose.
78. Connect the accelerator cable.
79. Connect the speed control actuator cable.
80. Install the throttle return spring.
81. Install the accelerator cable snow shield.
82. Install the air cleaner outlet pipe and tighten the clamp.
83. Connect the two hoses to the air cleaner outlet pipe.
84. Connect the MAF sensor and position the ACL.
85. Install the ACL.
86. Install the interior engine cover.
87. Connect the battery ground cable.
88. Fill and bleed the engine cooling system.

E-Series 6.8L Engine

✷✷ WARNING

Fuel in the fuel system remains under high pressure, even when the engine is not running. Do not work on any fuel system component until fuel system pressure has been relieved. Failure to follow these instructions may result in personal injury.

1. Before servicing the vehicle, refer to the Precautions Section.
2. Relieve the fuel system pressure.
3. Disconnect the battery ground cable.
4. Drain the engine cooling system.
5. Remove the engine cover.
6. Remove the air cleaner and the air cleaner outlet tube.
7. Remove the alternator.
8. Disconnect the fuel tube spring lock coupling.
9. Compress and slide the hose clamp and disconnect the coolant hose.
10. Remove the accelerator cable splash shield.

11. Disconnect the accelerator and speed control cables and remove the return spring.
12. Position the accelerator cable bracket aside.
13. Disconnect the heated positive crankcase valve (PCV) tube and the vacuum hoses.
14. Disconnect the PCV electrical connector and tube.
15. Disconnect the PCV coolant hoses.
16. Disconnect the vacuum tube.
17. Disconnect the idle air control (IAC) motor electrical connector and remove the bypass hose.
18. Disconnect the throttle position (TP) sensor electrical connector.
19. Disconnect the heater hose.
20. Remove the throttle body (TB) adapter.
21. Disconnect the vacuum connector.
22. Disconnect the fuel rail pressure (FRP) sensor electrical connector.
23. Disconnect the ten fuel injector electrical connectors.
24. Disconnect the ten ignition coil electrical connectors.
25. Remove the nuts and position the right fuel charge wiring harness aside.
26. Remove the nuts and position the left fuel charge wiring harness aside.
27. Remove the ten bolts and the ten ignition coils.
28. Disconnect the engine coolant temperature (ECT) sensor electrical connector.
29. Disconnect the heater hose.
30. Remove the water thermostat and housing.
31. Remove the bolts, the upper intake manifold and the intake manifold gaskets. Discard the intake manifold gaskets.

➡Do not use metal scrapers, wire brushes, power abrasive discs or other abrasive means to clean the aluminum retainer plate. These tools cause scratches and gouges, which make leak paths.

32. Clean the intake manifold mating surface of the cylinder heads with silicone gasket remover and metal surface prep. Follow the directions on the packaging.
To install:
33. Tighten the bolts in the sequence shown.
   a. Step 1: Tighten to 2 Nm (18 inch lbs.).
   b. Step 2: Tighten to 10 Nm (89 inch lbs.).
34. Position the water thermostat and housing. Install a new O-ring seal.

35. Install the upper intake manifold. Position the new upper intake manifold gaskets and the intake manifold, and loosely install the bolts.
36. Tighten the bolts in two steps, in the sequence shown.
   a. Step 1: Tighten to 2 Nm (18 inch lbs.).
   b. Step 2: Tighten to 25 Nm (18 ft. lbs.).
37. Connect the heater hose.
38. Connect the ECT sensor electrical connector.
39. Install the ten bolts and the ten ignition coils.
40. Connect the FRP sensor electrical connector.
41. Connect the vacuum connector.
42. Install the TB adapter.
43. Install a new TB adapter gasket.
44. Position the TB adapter.
45. Install the four bolts and tighten in two steps.
   a. Step 1: Tighten the bolts to 9 Nm (80 inch lbs.).
   b. Step 2: Tighten an additional 90 degrees.
46. Connect the TP sensor electrical connector.
47. Connect the PCV coolant hoses.
48. Connect the heated PCV tube and the vacuum hoses.
49. Connect the IAC motor electrical connector and install the bypass hose.
50. Connect the vacuum tube.
51. Position the left fuel charge wiring harness and install the nuts.
52. Position the right fuel charge wiring harness and install the nuts.
53. Connect the ten ignition coil electrical connectors.
54. Connect the ten fuel injector electrical connectors.
55. Install the PVC tube and connect the electrical connector.
56. Install the accelerator bracket.
57. Connect the accelerator and speed control cables and return spring.
58. Install the accelerator cable splash shield.
59. Connect the coolant hose and position the hose clamp.
60. Connect the fuel tube spring lock coupling.
61. Install the alternator.
62. Install the air cleaner and the air cleaner outlet tube.
63. Install the engine cover.
64. Connect the battery ground cable.
65. Fill and bleed the engine cooling system.

Intake Manifold Spacer

REMOVAL & INSTALLATION

2002–03 4.2L Engines and 2004 F-150 Heritage Edition w/4.2L Engine

1. Before servicing the vehicle, refer to the Precautions Section.
2. Remove the upper intake manifold.
3. Partially drain the cooling system.
4. Disconnect the throttle position (TP) sensor.
5. Disconnect the brake booster vacuum hose.
6. Disconnect the vapor management valve (VMV) hose.
7. Disconnect the positive crankcase ventilation (PCV) hose.
8. Disconnect the heater water hoses at the intake manifold.
9. Remove the heater hose bracket at the intake manifold.
10. Disconnect the intake manifold vacuum connector.
11. Disconnect the EGR solenoid electrical and vacuum connectors.
12. Remove the eight bolts and the intake manifold spacer assembly.

To install:

13. Inspect the intake manifold spacer assembly mounting gaskets. Install new gaskets, if necessary.
14. Install the intake manifold spacer assembly. Tighten the bolts in two steps in the sequence shown.
   - Step 1: Tighten to 6 Nm (53 lb-in).
   - Step 2: Tighten to 10 Nm (89 lb-in).
15. Connect the EGR solenoid electrical and vacuum connectors.
16. Connect the intake manifold vacuum connector.
17. Install the heater hose bracket at the intake manifold.

18. Connect the heater water hoses at the intake manifold.
19. Connect the positive crankcase ventilation (PCV) hose.
20. Connect the vapor management valve (VMV) hose.
21. Connect the brake booster vacuum hose.
22. Connect the throttle position (TP) sensor.
23. Fill the cooling system.
24. Install the upper intake manifold.

06017-F150-G49

Intake manifold spacer torque sequence—2002–03 4.2L engines and 2004 4.2L Heritage Editions

Exhaust Manifold

REMOVAL & INSTALLATION

4.2L Engine

F-SERIES

1. Before servicing the vehicle, refer to the Precautions Section.
2. Remove or disconnect the following:
   - Negative battery cable

06017-F150-G50

Intake manifold spacer assembly gasket placement—2002–03 4.2L engines and 2004 4.2L Heritage Editions

20 - 30 Nm (15 - 22 Lb - Ft)

7924FG14

Tighten the left-hand exhaust manifold bolts in the order shown—4.2L engine

20-30 Nm (15-22 Lb-Ft)

7924FG15

Tighten the right-hand exhaust manifold bolts in the order shown—4.2L engine

- For the right-hand manifold: the EGR valve-to-exhaust manifold tube
- For the left-hand manifold: the oil level indicator tube bracket nut, then remove the oil level indicator tube. Remove and discard the oil level indicator tube O-ring.
- Oxygen Sensor (O_2S) electrical connector
- The 2 catalytic converter-to-exhaust manifold nuts, then disconnect the Y-pipe from the left-hand exhaust manifold.
- Exhaust manifold stud bolts, then remove the manifold mounting bolts
- Exhaust manifold. Remove and discard the exhaust manifold gasket.

To install:
3. Install or connect the following:
- New exhaust manifold gasket onto the engine, then install the exhaust manifold. Tighten the bolts and stud bolts in the sequence shown to 15–22 ft. lbs. (20–30 Nm).
- Y-pipe to the exhaust manifold, then install and tighten the catalytic converter nuts to 25–34 ft. lbs. (34–46 Nm).
- O_2S connector, then lower the vehicle.
- Left-hand exhaust manifold: a new oil level indicator tube O-ring onto the tube. Insert the tube into the engine block and tighten the

bracket retaining nut to 15–22 ft. lbs. (20–30 Nm).
- For the right-hand exhaust manifold: the EGR valve-to-exhaust manifold tube. Tighten the upper and lower fittings to 25–34 ft. lbs. (34–47 Nm).
- Negative battery cable

E-SERIES LEFT SIDE

1. Before servicing the vehicle, refer to the Precautions Section.
2. Remove the oil level indicator extension.
3. Remove the oil level indicator tube. Remove and discard the oil level indicator tube O-ring.
4. Raise and support the vehicle.
5. Remove the engine cover.
6. Disconnect the left heated oxygen sensor (HO_2S) electrical connector.
7. Remove the nuts and disconnect the Y-pipe from the left exhaust manifold.
8. Remove the left exhaust manifold and the left exhaust manifold gasket. Remove and discard the left exhaust manifold gasket.

➡ **Use a new gasket. Tighten in the sequence shown.**

9. To install, reverse the removal procedure. Observe the following torques:
- Exhaust manifold nuts: 15–22 ft. lbs. (20–30 Nm)
- Exhaust pipe-to-manifold: 30 ft. lbs. (40 Nm)
- Dipstick tube bracket: 89 inch lbs. (10 Nm)

E-SERIES RIGHT SIDE

1. Before servicing the vehicle, refer to the Precautions Section.
2. Remove the engine cover.
3. Remove the differential pressure feedback exhaust gas re-circulation (EGR) system.
4. Remove the exhaust manifold-to-EGR valve tube.
5. Disconnect the right heated oxygen sensor (HO_2S) electrical connector.
6. Raise and support the vehicle.
7. Disconnect the exhaust Y-pipe from the right exhaust manifold.
8. Remove the right exhaust manifold and the right exhaust manifold gasket. Discard the right exhaust manifold gasket.

➡ **Use a new exhaust manifold gasket.**

9. To install, reverse the removal procedure. Tighten the exhaust manifold retainers in the sequence shown to 18 ft. lbs. (25 Nm). Torque the EGR tube-to-manifold to 30 ft. lbs. (40 Nm). Torque the exhaust pipe-to-manifold to 30 ft. lbs. (40 Nm).

2002–03 Blackwood, F-150 and 2004 Heritage Edition w/4.6L or 5.4L Engine

LEFT SIDE

1. Before servicing the vehicle, refer to the Precautions Section.
2. Raise and support the vehicle.
3. On 4x4 vehicles, remove the front wheel opening molding.
4. Remove the front fender splash shield.
5. With a 4.6L engine:
   a. Remove the nut and the brake booster vacuum hose and bracket.
   b. Remove the exhaust manifold to exhaust gas re-circulation (EGR) valve tube.
6. With a 5.4L engine:
   a. Remove the bolt and the brake booster vacuum hose and bracket.
   b. Remove the exhaust manifold to EGR valve tube.
7. Remove the nuts.
8. Remove the nuts and the exhaust manifold.
9. Remove and discard the exhaust manifold gasket.
10. Clean and inspect the exhaust manifold.

To install:
11. Position the exhaust manifold gasket.
12. Position the exhaust manifold and loosely install the nuts.

Left side exhaust manifold torque sequence—2002–03 F-Series and 2004 Heritage Editions 4.6L and 5.4L engines

Right side exhaust manifold torque sequence—2002–03 F-Series and 2004 Heritage Editions 4.6L and 5.4L engines

13. Tighten the nuts in the sequence shown.

14. Install the three-way catalytic converter to exhaust manifold nuts.

15. With a 4.6L engine:

   a. Install the differential pressure feedback EGR system and connect the electrical connector.

   b. Tighten both fittings, starting at the top, in two stages.
   - Stage 1: Hand-tighten.
   - Stage 2: Tighten to 35–45 Nm (26–33 ft. lbs.).

16. Install the brake booster vacuum hose, bracket and nut.

17. With a 5.4L engine

   a. Tighten both fittings, starting at the top, in two stages.
   - Stage 1: Hand-tighten.
   - Stage 2: Tighten to 40–60 Nm (30–44 ft. lbs.).

   b. Connect the differential pressure feedback EGR system hoses.

18. Install the brake booster vacuum hose, bracket and bolt.

19. Install the front fender splash shield.

20. On 4x4 vehicles, install the front wheel opening molding.

21. Lower the vehicle.

RIGHT SIDE

1. Before servicing the vehicle, refer to the Precautions Section.

2. Raise and support the vehicle.

3. On 4x4 vehicles, remove the front wheel opening molding.

4. Remove the front fender splash shield.

5. Remove the three-way catalytic converter to exhaust manifold nuts.

6. Remove the eight nuts and the exhaust manifold.

7. Remove and discard the exhaust manifold gasket.

8. Clean and inspect the exhaust manifold.

9. To install, reverse the removal procedure. Tighten the exhaust manifold nuts in the sequence shown to 18 ft. lbs. (25 Nm). Torque the exhaust pipe-to-manifold nuts to 30 ft. lbs. (40 Nm).

2002–03 F-250/350

LEFT SIDE

1. Before servicing the vehicle, refer to the Precautions Section.

2. If equipped, remove the exhaust manifold to exhaust gas re-circulation (EGR) valve tube.

3. Raise and support the vehicle.

4. On 4x4 vehicles, remove the front wheel opening molding.

5. Remove the front fender splash shield.

6. Remove the nuts.

7. Remove the nuts and the exhaust manifold.

8. Remove the exhaust manifold gasket.

9. Clean and inspect the exhaust manifold.

10. To install, reverse the removal procedure. Tighten the exhaust manifold nuts in the sequence shown to 18 ft. lbs. (25 Nm). Tighten the exhaust pipe-to-manifold nuts to 30 ft. lbs. (40 Nm). Torque the EGR tube nuts to 30 ft. lbs. (40 Nm).

RIGHT SIDE

1. Before servicing the vehicle, refer to the Precautions Section.

✳✳ CAUTION

The electrical power to the air suspension system must be shut off prior to hoisting, jacking or towing an air suspension vehicle. This can be accomplished by turning off the air suspension switch located in the RH kick panel area. Failure to do so may result in unexpected inflation or deflation of the air springs which may result in shifting of the vehicle during these operations.

2. Raise and support the vehicle.

3. On 4x4 vehicles, remove the front wheel opening molding

4. Remove the front fender splash shield.

5. Remove the three way catalytic converter to exhaust manifold nuts.

6. Remove the eight nuts and the exhaust manifold.

7. Remove and discard the exhaust manifold gasket.

8. Clean and inspect the exhaust manifold.

9. To install, reverse the removal procedure. Tighten the exhaust manifold nuts in the sequence shown to 18 ft. lbs. (25 Nm). Torque the exhaust pipe-to-manifold nuts to 30 ft. lbs. (40 Nm).

2002–03 F-250/350 6.8L Engine

LEFT SIDE

1. Before servicing the vehicle, refer to the Precautions Section.

2. Remove the air cleaner outlet tube.

3. Raise and support the vehicle.

4. If equipped, disconnect the EGR-to-exhaust manifold tube lower fitting.

5. Remove the nuts.

6. Lower the vehicle.

7. Remove the exhaust manifold nuts and the exhaust manifold. Discard the exhaust manifold gaskets.

8. Clean and inspect the exhaust manifold.

9. To install, reverse the removal procedure. Torque the manifold nuts to 18 ft. lbs. (25 Nm). Torque the exhaust pipe-to-manifold nuts to 30 ft. lbs. (40 Nm). Torque the EGR tube nuts to 30–44 ft. lbs. (40–60 Nm).

RIGHT SIDE

1. Before servicing the vehicle, refer to the Precautions Section.

2. Disconnect the battery ground cable.

3. Raise the and support vehicle.

4. Remove the nuts.

5. Remove the right front wheel and tire.

6. Remove the right front inner fender well.

25 Nm (18 lb-ft)

06017-F150-G53

Left side exhaust manifold torque sequence—6.8L engine

23-27 Nm (17-20 lb/ft)

06017-F150-G54

Right side exhaust manifold torque sequence—6.8L engine

7. Remove the nuts, the exhaust manifold and the gasket. Discard the gasket.

8. Clean and inspect the exhaust manifold.

9. To install, reverse the removal procedure. Torque the manifold nuts to 18 ft. lbs. (25 Nm). Torque the exhaust pipe-to-manifold nuts to 30 ft. lbs. (40 Nm).

2004–05 F-150 4.6L Engine, Exc. 2004 Heritage Editions

RIGHT SIDE

1. Before servicing the vehicle, refer to the Precautions Section.

2. With the vehicle in **Neutral**, position it on a hoist.

3. Remove the starter.

4. Remove the right inner fenderwell.

5. Disconnect the Y-pipe at the manifold.

6. Remove the manifold. Discard the gasket.

➡ Do not use metal scrapers, wire brushes, power abrasive discs, or other abrasive means to clean the sealing surfaces. These may cause scratches and gouges resulting in leak paths. Use a plastic scraper to clean the sealing surfaces.

7. Clean the sealing surfaces with metal surface prep.

➡ Install a new exhaust manifold gasket.

8. Position the right exhaust manifold and tighten the nuts in the sequence shown.

9. To install, reverse the removal procedure.

LEFT SIDE

1. Before servicing the vehicle, refer to the Precautions Section.

2. With the vehicle in **Neutral**, position it on a hoist.

3. Remove the left inner fenderwell.

4. Disconnect the Y-pipe at the manifold.

5. Disconnect the EGR tube.

6. Remove the manifold. Discard the gasket.

➡ Do not use metal scrapers, wire brushes, power abrasive discs, or other abrasive means to clean the sealing surfaces. These may cause scratches and gouges resulting in leak paths. Use a plastic scraper to clean the sealing surfaces.

25 Nm (18 lb-ft)

67197-EFSE-G84

Right exhaust manifold torque sequence—2004–05 F-150 4.6L engines, exc. 2004 Heritage edition; 2006 Mark LT

25 Nm (18 lb-ft)

67197-EFSE-G85

Left exhaust manifold torque sequence—2004–05 F-150 and F-250/350 4.6L and 5.4 L engines, exc. 2004 Heritage edition; 2006 Mark LT

7. Clean the sealing surfaces with metal surface prep.

➡**Install a new exhaust manifold gasket.**

8. Position the left exhaust manifold and tighten the exhaust manifold nuts in the sequence shown.

9. To install, reverse the removal procedure.

2004–05 F-150 5.4L Engine, Exc. 2004 Heritage Editions 2006 Mark LT

LEFT SIDE

1. Before servicing the vehicle, refer to the Precautions Section.

2. With the vehicle in **Neutral**, position it on a hoist.

3. Remove the front and rear oxygen sensors.

4. Disconnect the Y-pipes from the manifolds.

5. Remove the manifold. Discard the gasket.

➡**Do not use metal scrapers, wire brushes, power abrasive discs, or other abrasive means to clean the sealing surfaces. These may cause scratches and gouges resulting in leak paths. Use a plastic scraper to clean the sealing surfaces.**

6. Clean the sealing surfaces with metal surface prep.

7. Install the manifold with a new gasket.

8. Tighten the exhaust manifold nuts in the sequence shown.

9. To install, reverse the removal procedure.

RIGHT SIDE

1. Before servicing the vehicle, refer to the Precautions Section.

2. With the vehicle in **Neutral**, position it on a hoist.

3. Remove the starter.

4. Disconnect the Y-pipes at the manifolds.

5. Remove the manifold. Discard the gasket.

➡**Do not use metal scrapers, wire brushes, power abrasive discs, or other abrasive means to clean the sealing surfaces. These may cause scratches and gouges resulting in leak paths. Use a plastic scraper to clean the sealing surfaces.**

6. Clean the sealing surfaces with metal surface prep.

7. Tighten the right exhaust manifold nuts in the sequence shown.

8. To install, reverse the removal procedure.

2004 F-250/350 5.4L Engine

RIGHT SIDE

✳✳ WARNING

The electrical power to the air suspension system must be shut off prior to hoisting, jacking or towing an air suspension vehicle. This can be accomplished by turning off the air suspension switch located in the right kick panel area. Failure to do so may result in unexpected inflation or deflation of the air springs which may result in shifting of the vehicle during these operations.

1. Before servicing the vehicle, refer to the Precautions Section.

2. Raise and support the vehicle.

3. On 4x4 vehicles, remove the front wheel opening molding.

4. Remove the front fender splash shield.

5. Remove the three way catalytic converter to exhaust manifold nuts.

6. Remove the eight nuts and the exhaust manifold.

7. Remove and discard the exhaust manifold gasket.

8. Clean and inspect the exhaust manifold.

25 Nm (18 lb-ft)

67197-EFSE-G87

Right exhaust manifold torque sequence—2004 F-250/350 w/5.4L engines

9. To install, reverse the removal procedure. Tighten the exhaust manifold nuts in the sequence shown.

LEFT SIDE

1. Before servicing the vehicle, refer to the Precautions Section.

2. If equipped, remove the exhaust manifold to exhaust gas recirculation (EGR) valve tube.

3. Raise and support the vehicle.

4. On 4x4 vehicles, remove the front wheel opening molding.

5. Remove the front fender splash shield.

6. Remove the nuts.

7. Remove the nuts and the exhaust manifold.

8. Remove the exhaust manifold gasket.

9. Clean and inspect the exhaust manifold.

10. To install, reverse the removal procedure. Tighten the exhaust manifold nuts in the sequence shown.

2005 F-250/350 5.4L Engine

RIGHT SIDE

1. Before servicing the vehicle, refer to the Precautions Section.

2. With the vehicle in NEUTRAL, position it on a hoist.

3. Remove the starter.

4. Remove the RH inner fenderwell.

5. Remove the 4 exhaust manifold-to-catalytic converter nuts.

6. Remove the 8 nuts and the exhaust manifold.

✳✳ WARNING

Do not use metal scrapers, wire brushes, power abrasive discs, or other abrasive means to clean the sealing surfaces. These may cause scratches and gouges resulting in leak paths. Use a plastic scraper to clean the sealing surfaces.

➡ **Clean the sealing surfaces with metal surface prep. Follow the directions on the packaging.**

7. Remove and discard the exhaust manifold gaskets. Clean the sealing surfaces with metal surface prep.

8. To install, reverse the removal procedure. Torque the manifold nuts to 18 ft. lbs. (25 Nm). Torque the exhaust pipe-to-manifold nuts to 30 ft. lbs. (40 Nm). Torque the heat shield nuts to 89 inch lbs. (10 Nm).

LEFT SIDE

1. Before servicing the vehicle, refer to the Precautions Section.

2. With the vehicle in NEUTRAL, position it on a hoist.

3. Remove the air cleaner housing.

4. Remove the 4 exhaust manifold-to-catalytic converter nuts.

5. Remove the 2 bolts and the exhaust manifold heat shield.

40-60 Nm (30-44 lb-ft)

40-60 Nm (30-44 lb-ft)

67197-EFSE-G88

EGR valve tube installation—2004 F-250/350 w/5.4L engine

6. With 4WD, remove the front driveshaft.

7. Remove the 8 nuts and the exhaust manifold.

✳✳ WARNING

Do not use metal scrapers, wire brushes, power abrasive discs, or other abrasive means to clean the sealing surfaces. These may cause scratches and gouges resulting in leak paths. Use a plastic scraper to clean the sealing surfaces.

➡ **Clean the sealing surfaces with metal surface prep. Follow the directions on the packaging.**

8. Remove and discard the exhaust manifold gaskets. Clean the sealing surfaces with metal surface prep.

9. To install, reverse the removal procedure. Torque the manifold nuts to 18 ft. lbs. (25 Nm). Torque the exhaust pipe-to-manifold nuts to 30 ft. lbs. (40 Nm). Torque the heat shield nuts to 89 inch lbs. (10 Nm).

2004 F-250/350 6.8L Engine

RIGHT SIDE

1. Before servicing the vehicle, refer to the Precautions Section.

2. Disconnect the battery ground cable.

3. Raise the and support vehicle.

4. Remove the nuts.

5. Remove the right front wheel and tire.

6. Remove the right front inner fender well.

7. Remove the nuts, the exhaust manifold and the gasket. Discard the gasket.

8. Clean and inspect the exhaust manifold.

9. To install, reverse the removal procedure. Torque the manifold nuts to 18 ft. lbs. (25 Nm). Torque the exhaust pipe-to-manifold nuts to 30 ft. lbs. (40 Nm). Torque the heat shield nuts to 89 inch lbs. (10 Nm).

LEFT SIDE

1. Before servicing the vehicle, refer to the Precautions Section.

2. With the vehicle in **Neutral**, position it on a hoist.

3. Remove the air cleaner outlet tube.

4. Remove the nuts.

5. Remove the left front wheel and tire.

6. Remove the left front inner fender well.

7. Remove the nuts, the exhaust manifold and the gasket. Discard the exhaust manifold gasket.

8. Clean and inspect the exhaust manifold.

9. To install, reverse the removal pro-

cedure. Torque the manifold nuts to 18 ft. lbs. (25 Nm). Torque the exhaust pipe-to-manifold nuts to 30 ft. lbs. (40 Nm). Torque the heat shield nuts to 89 inch lbs. (10 Nm).

2005 F-250/350 w/6.8L Engines

LEFT SIDE

1. Before servicing the vehicle, refer to the Precautions Section.

2. With the vehicle in NEUTRAL, position it on a hoist.

3. Remove the air cleaner housing.

4. Remove the 2 bolts and position the degas bottle aside.

5. Remove the 4 exhaust manifold-to-catalytic converter nuts.

6. Remove the 10 nuts and the exhaust manifold.

✳✳ WARNING

Do not use metal scrapers, wire brushes, power abrasive discs or other abrasive means to clean the sealing surfaces. These may cause scratches and gouges resulting in

leak paths. Use a plastic scraper to clean the sealing surfaces.

➡ **Clean the sealing surfaces with metal surface prep. Follow the directions on the packaging.**

7. Remove and discard the exhaust manifold gaskets. Clean the sealing surfaces with metal surface prep.

8. To install, reverse the removal procedure. Torque the manifold nuts to 18 ft. lbs. (25 Nm). Torque the exhaust pipe-to-manifold nuts to 30 ft. lbs. (40 Nm).

RIGHT SIDE

1. Before servicing the vehicle, refer to the Precautions Section.

2. With the vehicle in NEUTRAL, position it on a hoist.

3. If equipped, remove the transmission filler tube.

4. Remove the starter.

5. Remove the right inner fenderwell.

6. Remove the 4 exhaust manifold-to-catalytic converter nuts.

7. Remove the 10 nuts and the exhaust manifold.

06017-F150-G55

Left side exhaust manifold torque sequence—2005 F-250/350 6.8L engine

06017-F150-G56

Right side exhaust manifold torque sequence—2005 F-250/350 6.8L engine

✳✳ WARNING

Do not use metal scrapers, wire brushes, power abrasive discs or other abrasive means to clean the sealing surfaces. These may cause scratches and gouges resulting in leak paths. Use a plastic scraper to clean the sealing surfaces.

➡ Clean the sealing surfaces with metal surface prep. Follow the directions on the packaging.

8. Remove and discard the exhaust manifold gaskets. Clean the sealing surfaces with metal surface prep.

9. To install, reverse the removal procedure. Torque the manifold nuts to 18 ft. lbs. (25 Nm). Torque the exhaust pipe-to-manifold nuts to 30 ft. lbs. (40 Nm).

E-Series 4.6L and 5.4L Engines

LEFT SIDE

1. Before servicing the vehicle, refer to the Precautions Section.

All engines

2. With the vehicle in **Neutral**, position it on a hoist.

3. Remove the engine cover.

Romeo engine (4.6L)

4. Disconnect the differential pressure feedback exhaust gas recirculation (EGR) sensor hoses from the EGR valve tube.

5. Remove the EGR valve tube.

Windsor engine (5.4L)

6. Disconnect the differential pressure feedback EGR sensor hoses from the EGR valve tube.

7. Disconnect the upper and lower fittings and remove the EGR valve tube.

8. Remove the front fender splash shield.

9. Remove the nuts.

10. Remove the nuts and the exhaust manifold.

11. Remove and discard the exhaust manifold gasket.

12. Clean and inspect the exhaust manifold.

To install:

All engines

13. Position the exhaust manifold gasket.

14. Position the exhaust manifold and loosely install the nuts.

15. Tighten the nuts in the sequence shown.

16. Install the three-way catalytic converter to exhaust manifold nuts.

Windsor engine (5.4L)

17. Install the EGR valve fittings in two steps.

   a. Step 1: Hand-tighten the fittings.

   b. Step 2: Tighten the lower fitting to 50 Nm (37 ft. lbs.).

18. Connect the differential pressure feedback EGR sensor hoses to the EGR valve tube.

Romeo engine (4.6L)

19. Install the EGR valve tube in two steps.

   a. Step 1: Connect the upper and lower fittings and hand-tighten.

   b. Step 2: Tighten the fittings to 40 Nm (30 ft. lbs.).

20. Install the differential feedback EGR sensor hoses to the EGR valve tube.

All engines

21. Install the front fender splash shield.

22. Install the engine cover.

RIGHT SIDE

1. Before servicing the vehicle, refer to the Precautions Section.

2. With the vehicle in **Neutral**, position it on a hoist.

25 Nm (18 lb-ft)

67197-EFSE-G91

Left exhaust manifold torque sequence—E-Series with 4.6L or 5.4L engine

25 Nm (18 lb-ft)

67197-EFSE-G92

Right exhaust manifold torque sequence—E-Series with 4.6L or 5.4L engine

3. Remove the front fender splash shield.

4. Remove the three-way catalytic converter to exhaust manifold nuts.

5. Remove the eight nuts and the exhaust manifold.

6. Remove and discard the exhaust manifold gasket.

7. Clean and inspect the exhaust manifold.

8. To install, reverse the removal procedure. Tighten the exhaust manifold nuts in the sequence shown.

E-Series 6.8L Engine

RIGHT SIDE

1. Before servicing the vehicle, refer to the Precautions Section.

2. With the vehicle in **Neutral**, position it on a hoist.

3. Remove the engine cover.

4. Remove the nuts.

5. Remove the 10 nuts, the exhaust manifold, and the exhaust manifold gasket.

6. Discard the exhaust manifold gasket.

7. Clean and inspect the exhaust manifold.

8. To install, reverse the removal procedure.

9. Install a new exhaust manifold gasket. Tighten the exhaust manifold nuts in the sequence shown.

LEFT SIDE

1. Before servicing the vehicle, refer to the Precautions Section.

2. With the vehicle in **Neutral**, position it on a hoist.

3. Remove the engine cover.

4. Remove the nuts.

5. Remove the exhaust manifold nuts and the exhaust manifold. Discard the exhaust manifold gaskets.

6. Clean and inspect the exhaust manifold.

7. To install, reverse the removal procedure. Tighten the exhaust manifold nuts in the sequence shown.

Camshaft

REMOVAL & INSTALLATION

4.2L Engine

➡**It may be easier to remove the engine from the vehicle.**

1. Before servicing the vehicle, refer to the Precautions Section.

2. Remove or disconnect the following:

Exploded view of the camshaft retaining hardware—4.2L engine

7924FG38

- Negative battery cable
- Lower intake manifold
- Rocker arm cover
- Rocker arm hold-down bolt, then remove the rocker arm from the cylinder head
- Pushrods
- Valve lifters by pulling them up out of their bores
- Timing chain and sprockets
- Camshaft key from the end of the camshaft, then slide the engine dynamic balance shaft drive gear off the camshaft.
- The 2 camshaft thrust plate retaining bolts (1), then remove the thrust plate (2).

3. Remove the camshaft spacer (3), then slide the camshaft (4) out of the front of the engine block. Be cautious not to gouge or scratch the camshaft bearing journals.

To install:

4. Lubricate the camshaft with engine oil prior to installation.

5. Carefully slide the camshaft into the camshaft bore. Do not scratch the bearing surfaces.

6. Install the camshaft thrust plate with the spacer. Tighten the thrust plate mounting bolts to 72–120 inch lbs. (8–14 Nm).

7. Slide the engine dynamic balance shaft drive gear onto the camshaft. Install the camshaft key to the camshaft groove.

8. Install the timing chain and sprockets.

9. Install the valve lifters, pushrods, intake manifolds and rocker arm covers.

4.6L Engines

1. Before servicing the vehicle, refer to the Precautions Section.

2. Remove the camshaft roller followers.

✳✳ WARNING

At no time, when the timing chains are removed and the cylinder heads are installed may the crankshaft or camshaft be rotated. Severe piston and valve damage will occur.

3. Remove the timing chains

4. Remove the bolt. Remove the camshaft sprocket.

➡**The camshaft bearing caps must be installed in their original location. Record camshaft bearing cap location.**

5. Remove the bolts, the bearing caps and the camshaft.

6. Clean and inspect the camshaft bear-

06017-F150-G57

One of the bearing caps contains an oil flow restriction groove—4.6L engine

ing caps. One of the bearing caps contains an oil flow restriction groove. Make sure the groove is free of foreign material.

To install:

> ❄❄ **WARNING**
>
> **Do not use metal scrapers, wire brushes, power abrasive discs or other abrasive means to clean the sealing surfaces. These tools cause scratches and gouges which make leak paths. Use a plastic scraping tool to remove all traces of old sealant.**

7. Clean the valve cover mating surface of the cylinder head with silicone gasket remover and metal surface prep. Follow the directions on the packaging.

8. Lubricate the camshaft bearing caps with clean engine oil.

9. Position the camshaft bearing caps.

10. Install the bolts loosely.

11. Tighten the bolts in the sequence shown to 89 inch lbs. (10 Nm).

12. Install the camshaft sprocket.

13. Install and tighten the bolt.
- Tighten in two stages.
- Stage 1: Tighten to 40 Nm (30 ft. lbs.).
- Stage 2: Tighten an additional 90 degrees.

14. Install the timing chains.

15. Install the camshaft roller followers.

2002–F-150 5.4L Engine; 2004 F-150 Heritage Edition 5.4L Engine; 2002–04 F-250/350 5.4L Engine; 2002–05 E-Series 5.4L Engine

1. Before servicing the vehicle, refer to the Precautions Section.

2. Remove the camshaft roller followers.

> ❄❄ **WARNING**
>
> **At no time, when the timing chains are removed and the cylinder heads are installed, may the crankshaft or camshaft be rotated. Severe piston and valve damage will occur.**

3. Remove the timing chains.

➡**The camshaft bearing caps must be installed in their original location. Record camshaft bearing cap location.**

4. Remove the bolts, the bearing caps and the camshaft.

5. Clean and inspect the camshaft bearing caps. One of the bearing caps contains an oil flow restriction groove. Make sure the groove is free of foreign material.

10 Nm (89 lb-in)

06017-F150-G58

Camshaft bearing cap torque sequence—4.6L engine

To install:

> ❄❄ **WARNING**
>
> **Do not use metal scrapers, wire brushes, power abrasive discs or other abrasive means to clean the sealing surfaces. These tools cause scratches and gouges which make leak paths. Use a plastic scraping tool to clean sealing surfaces.**

6. Clean the valve cover mating surface of the cylinder head with silicone gasket remover and metal surface prep. Follow the directions on the packaging.

7. Lubricate the camshaft journals with clean engine oil.

8. Install the camshaft sprocket.

9. Install and tighten the bolt in two stages.
- Stage 1: Tighten to 40 Nm (30 ft. lbs.).
- Stage 2: Tighten an additional 90 degrees.

10. Lubricate the camshaft bearing caps with clean engine oil.

11. Position the camshaft bearing caps.

12. Install the bolts loosely.

13. Tighten the bolts in the sequence shown to 89 inch lbs. (10 Nm).

06017-F150-G59

One of the bearing caps contains an oil flow restriction groove—2002-F-150 5.4L Engine; 2004 F-150 Heritage Edition 5.4L Engine; 2002-04 F-250/350 5.4L Engine; 2002-05 E-Series 5.4L Engine

14. Install the timing chains.

15. Install the camshaft roller followers.

5.4L DOHC Engines

1. Before servicing the vehicle, refer to the Precautions Section.

2. Remove the roller followers.

3. Remove the left timing chain for the left side and both timing chains for the right side.

4. Install the special tool.

5. Remove the exhaust camshaft

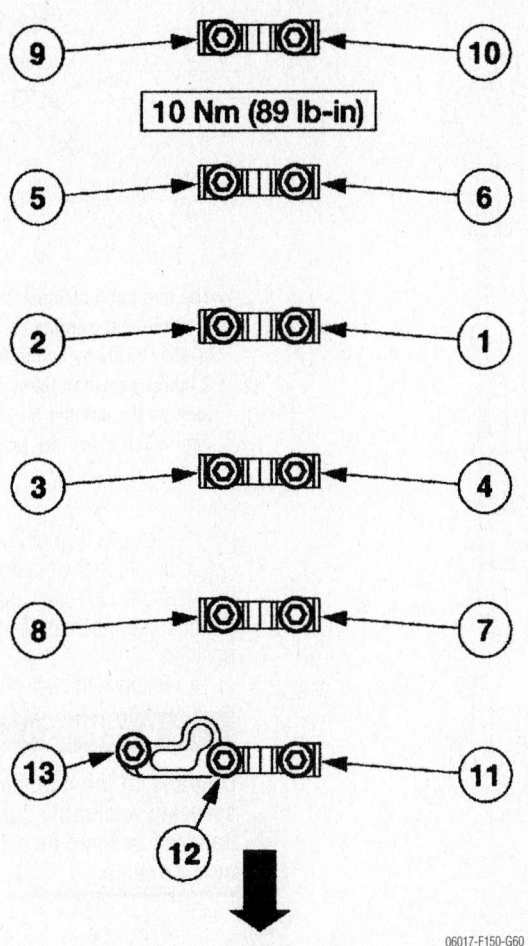

Camshaft bearing torque sequence—2002–F-150 5.4L Engine; 2004 F-150 Heritage Edition 5.4L Engine; 2002–04 F-250/350 5.4L Engine; 2002–05 E-Series 5.4L Engine

Special tool 303-446—5.4L DOHC

Compress the tensioner and install the tensioner retaining clip—5.4L DOHC

sprocket and the intake camshaft bolt, washer and spacer.

6. Remove the special tool.

7. Compress the tensioner and install the tensioner retaining clip.

8. Remove the timing chain, the sprocket, and intake camshaft sprocket spacer.

9. Remove the bolts.

✳✳ WARNING

The outer bolts on the outer cam bearing cap (exhaust) are longer and must be returned to the same location or engine damage may occur.

➡**Identify the camshaft to cylinder head location. Caps are not interchangeable.**

10. Remove the bolts and the camshaft bearing cap assemblies.

11. Remove the camshafts.

To install:

12. Install the camshafts. Lubricate the camshafts with clean engine oil.

13. Install the camshaft bearing cap assemblies. Tighten to 89 inch lbs. (10 Nm).

14. Install the bolts. Tighten to 71–106 inch lbs. (8–12 Nm).

➡**Timing marks must be at 12 o'clock and indexed at 6 o'clock.**

15. Install the camshaft sprockets and the chain as an assembly.

16. Remove the tensioner retaining clip from the chain tensioner.

17. Install the special tool.

18. Tighten the exhaust camshaft sprocket and the intake camshaft bolt washer and spacer in two stages:

  • Stage 1: Tighten to 40 Nm (30 ft. lbs.).
  • Tighten an additional 90 degrees.

19. Remove the special tool.

20. Install the left timing chain for the left side or both timing chains for the right side.

21. Install the roller followers.

Camshaft bearing torque sequence—5.4L DOHC

06017-F150-G68

The number 5 cylinder camshaft exhaust lobe must be coming up on the exhaust stroke. Verify by noting the position of the 2 intake camshaft lobes and the exhaust lobe on the number 5 cylinder—2005 5.4L engine left side F-Series and 2006 Mark LT

2. Remove the cooling fan shroud.

3. Position the crankshaft damper spoke at the 12 o'clock position and the timing mark indentation at the 1 o'clock position.

4. Remove the left valve cover.

✳✳ WARNING

Damage to the camshaft phaser sprocket assembly will occur if mis-handled or used as a lifting or lever-aging device.

2005 5.4L Engines F-Series and 2006 Mark LT

LEFT SIDE

1. Before servicing the vehicle, refer to the Precautions Section.

✳✳ WARNING

The camshaft procedure must be fol-lowed exactly or damage to the valves and pistons will result.

06017-F150-G64

Position the crankshaft damper spoke at the 12 o'clock position and the timing mark indentation at the 1 o'clock posi-tion—2005 5.4L engine F-Series and 2006 Mark LT

06017-F150-G69

Remove only these 3 roller followers at this time—2005 5.4L engine left side F-Series and 2006 Mark LT

5. Loosen and back off the left camshaft phaser bolt 1 full turn.

6. Disconnect the left camshaft position (CMP) sensor electrical connector.

7. Remove the left CMP sensor and the bolt.

❄ WARNING

If servicing both camshafts, do not rotate the crankshaft. Camshaft position has been established earlier.

➡ **If the camshaft lobes are not exactly positioned as shown, the crankshaft keyway will require 1 full additional rotation to 12 o'clock.**

8. The number 5 cylinder camshaft exhaust lobe must be coming up on the exhaust stroke. Verify by noting the position of the 2 intake camshaft lobes and the exhaust lobe on the number 5 cylinder.

9. Remove only the 3 roller followers shown in the illustration.

❄ WARNING

Do not allow the valve keepers to fall off the valve or the valve may drop into the cylinder.

➡ **The camshaft roller followers must be installed in their original locations. Record camshaft roller follower locations.**

➡ **It may be necessary to push the valve down while compressing the spring.**

10. Using special tool 303-1039, remove only the 3 designated roller followers from the previous step.

❄ WARNING

The crankshaft cannot be moved past the 6 o'clock position once set.

11. Rotate the crankshaft clockwise, as viewed from the front, positioning the crankshaft damper spoke at the 6 o'clock position and the timing mark indentation at the 7 o'clock position.

❄❄ WARNING

Engine is not freewheeling. Camshaft procedure must be followed exactly or damage to valves and pistons will result.

❄ WARNING

The Timing Chain Wedge tool must be installed square to the timing chain and the engine block.

06017-F150-G70

Using special tool 303-1039—2005 5.4L engine left side F-Series and 2006 Mark LT

➡ **Front cover removed for clarity.**

12. Install the special tools in the left timing chain as shown.

❄❄ WARNING

Do not remove the timing chain wedge tool at any time during assembly. If the special tool is removed or out of placement, the engine front cover must be removed and the engine must be retimed.

❄❄ WARNING

The timing chain must be installed in its original position onto the camshaft phaser sprocket using the scribed marks, or damage to valves and pistons will result.

13. Scribe a location mark on the timing chain and the camshaft phaser sprocket assembly.

❄❄ WARNING

Remove the front thrust camshaft bearing cap straight upward from the bearing towers, or the bearing cap may be damaged from side loading.

06017-F150-G71

Rotate the crankshaft clockwise, as viewed from the front, positioning the crankshaft damper spoke at the 6 o'clock position and the timing mark indentation at the 7 o'clock position—2005 5.4L engine F-Series and 2006 Mark LT

➡ **The camshaft bearing caps must be installed in their original locations. Record camshaft bearing cap locations.**

14. Remove the bolts in the sequence shown and remove the front camshaft bearing cap and then the remaining bearing caps.

15. Clean and inspect the left camshaft bearing caps. The camshaft front thrust bearing cap contains an oil metering groove. Make sure the groove is free of foreign material.

The Timing Chain Wedge tool must be installed square to the timing chain and the engine block—2005 5.4L engine left side F-Series and 2006 Mark LT

⁎⁎ WARNING

Damage to the camshaft phaser sprocket assembly will occur if mishandled or used as a lifting or leveraging device.

⁎⁎ WARNING

Only use hand tools to remove the camshaft phaser sprocket bolt or damage may occur to the camshaft or camshaft phaser unit.

⁎⁎ WARNING

Do not remove the timing chain wedge tool at any time during assembly. If the special tool is removed or out of placement, the engine front cover must be removed and the engine must be retimed.

16. Remove the bolt and withdraw the camshaft from the phaser sprocket assembly leaving the sprocket assembly in place. Discard the bolt and washer.

To install:

17. Lubricate the camshaft and camshaft journals with clean engine oil.

⁎⁎ WARNING

Do not remove the timing chain wedge tool at any time during assembly. If the special tool is removed or out of placement, the engine front cover must be removed and the engine must be retimed.

⁎⁎ WARNING

Damage to the camshaft phaser sprocket assembly will occur if mishandled or used as a lifting or leveraging device.

⁎⁎ WARNING

Do not allow the roller followers to move out of position when installing the camshaft.

18. Install the camshaft into the camshaft phaser sprocket assembly and onto the head.

19. Install a new camshaft phaser bolt finger tight.

⁎⁎ WARNING

Do not remove the timing chain wedge tool at any time during assembly. If the special tool is removed or out of placement, the engine front cover must be removed and the engine must be retimed.

⁎⁎ WARNING

The timing chain must be installed in its original position onto the camshaft phaser sprocket using the scribed marks, or damage to valves and pistons will result. Verify the camshaft phaser sprocket and timing chain scribe marks are still in alignment.

⁎⁎ WARNING

Do not allow the roller followers to move out of position when installing the camshaft.

20. Lubricate the camshaft bearing caps with clean engine oil.
21. Position the front camshaft bearing cap.
22. Position the remaining camshaft bearing caps.
23. Install the bolts loosely.
24. Tighten the bolts in the sequence shown. Tighten to 10 Nm (89 inch lbs.).
25. Remove the special tools.
26. Rotate the crankshaft a half turn counterclockwise and position the crankshaft damper spoke at the 12 o'clock position and the timing mark indentation at the 1 o'clock position.
27. Verify correct cam position by noting the position of the number 5 cylinder intake and exhaust camshaft lobes.
28. Using the special tool, install the 3 originally removed roller followers.
29. Install the CMP sensor and the bolt.
30. Connect the CMP electrical connector.

⁎⁎ WARNING

Only use hand tools to install the camshaft phaser sprocket assembly or damage may occur to the camshaft or camshaft phaser unit.

06017-F150-G66

Remove only these 3 roller followers at this time—2005 5.4L engine right side F-Series and 2006 Mark LT

Damage to the camshaft phaser sprocket assembly will occur if mishandled or used as a lifting or leveraging device.

31. Tighten the camshaft phaser bolt in 2 stages:
   - Stage 1: Tighten to 40 Nm (30 ft. lbs.).
   - Stage 2: Tighten an additional 90 degrees.
32. Install the left valve cover.
33. Install the cooling fan shroud.

RIGHT SIDE

1. Before servicing the vehicle, refer to the Precautions Section.

⁂ **WARNING**

The camshaft procedure must be followed exactly or damage to the valves and pistons will result.

2. Remove the cooling fan shroud.
3. Position the crankshaft damper spoke at the 12 o'clock position and the timing mark indentation at the 1 o'clock position.
4. Remove the right valve cover.

⁂⁂ **WARNING**

Damage to the camshaft phaser sprocket assembly will occur if mishandled or used as a lifting or leveraging device.

5. Loosen and back off the right camshaft phaser bolt 1 full turn.
6. Disconnect the right camshaft position (CMP) sensor electrical connector.

06017-F150-G65

The number 1 cylinder camshaft exhaust lobe must be coming up on the exhaust stroke. Verify by noting the position of the 2 intake camshaft lobes and the exhaust lobe on the number 1 cylinder—2005 5.4L engine right side F-Series and 2006 Mark LT

7. Remove the bolt and the right CMP sensor.

➡ **If the camshaft lobes are not exactly positioned as shown, the crankshaft will require 1 full additional rotation to 12 o'clock.**

8. The number 1 cylinder camshaft exhaust lobe must be coming up on the exhaust stroke. Verify by noting the position of the 2 intake camshaft lobes and the exhaust lobe on the number 1 cylinder.

9. Remove only the 3 roller followers shown in the illustration.

⁂⁂ **WARNING**

Do not allow the valve keepers to fall off the valve or the valve may drop into the cylinder.

➡ **The camshaft roller followers must be installed in their original locations. Record camshaft roller follower locations.**

➡ **It may be necessary to push the valve down while compressing the spring.**

06017-F150-G67

Using special tool 303-1039—2005 5.4L engine right side F-Series and 2006 Mark LT

10. Using special tool 303-1039, remove only the 3 designated roller followers from the previous step.

⁂⁂ **WARNING**

The crankshaft cannot be moved past the 6 o'clock position once set.

11. Rotate the crankshaft clockwise, as viewed from the front, positioning the crankshaft damper spoke at the 6 o'clock position and the timing mark indentation at the 7 o'clock position.

Timing chain wedge tool installed—2005 5.4L engine right side F-Series and 2006 Mark LT

⊗ WARNING

Engine is not freewheeling. Camshaft procedure must be followed exactly or damage to valves and pistons will result.

※ WARNING

The Timing Chain Wedge tool must be installed square to the timing chain and the engine block.

➡ Front cover removed for clarity.

12. Install the special tools in the right timing chain as shown.

⊗ WARNING

Do not remove the timing chain wedge tool at any time during assembly. If the special tool is removed or out of placement, the engine front cover must be removed and the engine must be retimed.

※ WARNING

The timing chain must be installed in its original position onto the

camshaft phaser sprocket using the scribed marks, or damage to valves and pistons will result.

13. Scribe a location mark on the timing chain and the camshaft phaser sprocket assembly.

※ WARNING

Remove the front thrust camshaft bearing cap straight upward from the bearing towers, or the bearing cap may be damaged from side loading.

➡ The camshaft bearing caps must be installed in their original locations. Record camshaft bearing cap locations.

14. Remove the bolts in the sequence shown and remove the front camshaft bearing cap and then the remaining bearing caps.

15. Clean and inspect the right camshaft bearing caps. The camshaft front thrust bearing cap contains an oil metering groove. Make sure the groove is free of foreign material.

※ WARNING

Damage to the camshaft phaser sprocket assembly will occur if mishandled or used as a lifting or leveraging device.

※ WARNING

Only use hand tools to remove the camshaft phaser sprocket bolt or damage may occur to the camshaft or camshaft phaser unit.

※ WARNING

Do not remove the timing chain wedge tool at any time during

Scribe a location mark on the timing chain and the camshaft phaser sprocket assembly—2005 5.4L engine right side F-Series and 2006 Mark LT

Camshaft bearing cap removal/installation sequence—2005 5.4L engine F-Series and 2006 Mark LT

06017-F150-G75

The camshaft front thrust bearing cap contains an oil metering groove—2005 5.4L engine F-Series and 2006 Mark LT

assembly. If the special tool is removed or out of placement, the engine front cover must be removed and the engine must be retimed.

16. Remove the bolt and withdraw the camshaft from the phaser sprocket assembly leaving the sprocket assembly in place. Discard the bolt and washer.

To install:

17. Lubricate the camshaft and camshaft journals with clean engine oil.

> ❋❋ **WARNING**
>
> Do not remove the timing chain wedge tool at any time during assembly. If the special tool is removed or out of placement, the engine front cover must be removed and the engine must be retimed.

> ❋❋ **WARNING**
>
> Damage to the camshaft phaser sprocket assembly will occur if mishandled or used as a lifting or leveraging device.

> ❋❋ **WARNING**
>
> Do not allow the roller followers to move out of position when installing the camshaft.

18. Install the camshaft into the camshaft phaser sprocket assembly and onto the head.

19. Install a new camshaft phaser bolt finger tight.

> ❋❋ **WARNING**
>
> Do not remove the timing chain wedge tool at any time during assembly. If the special tool is removed or out of placement, the engine front cover must be removed and the engine must be retimed.

> ❋❋ **WARNING**
>
> The timing chain must be installed in its original position onto the camshaft phaser sprocket using the scribed marks, or damage to valves and pistons will result.

20. Verify the camshaft phaser sprocket and timing chain scribe marks are still in alignment.

> ❋❋ **WARNING**
>
> Do not allow the roller followers to move out of position when installing the camshaft.

21. Lubricate the camshaft bearing caps with clean engine oil.

22. Position the front camshaft bearing cap.

23. Position the remaining camshaft bearing caps.

24. Install the bolts loosely.

25. Tighten the bolts in the sequence shown. Tighten to 10 Nm (89 inch lbs.).

26. Remove the special tools.

27. Rotate the crankshaft a half turn counterclockwise and position the crankshaft damper spoke at the 12 o'clock position and the timing mark indentation at the 1 o'clock position.

28. Verify correct cam position by noting the position of the number 1 cylinder intake and exhaust camshaft lobes.

29. Using the special tool, install the 3 originally removed roller followers.

30. Install the CMP sensor and the bolt.

31. Connect the CMP electrical connector.

> ❋❋ **WARNING**
>
> Only use hand tools to install the camshaft phaser sprocket assembly or damage may occur to the camshaft or camshaft phaser unit.

> ❋❋ **WARNING**
>
> Damage to the camshaft phaser sprocket assembly will occur if mishandled or used as a leveraging device.

303-567

06017-F150-G77

Use special tool 303-567 to compress the valve springs and remove the camshaft roller followers—2002–04 F-Series 6.8L engine; 2002–05 E-Series 6.8L engine

32. Tighten the camshaft phaser bolt in 2 stages:
- Stage 1: Tighten to 40 Nm (30 ft. lbs.).
- Stage 2: Tighten an additional 90 degrees.

33. Install the right valve cover.

2002–04 F-Series 6.8L Engines; 2002–05 E-Series 6.8L Engine

1. Before servicing the vehicle, refer to the Precautions Section.
2. Disconnect the battery ground cable.
3. Remove the intake manifold.
4. Remove the timing drive components.
5. Use special tool 303-567 to compress the valve springs and remove the camshaft roller followers.

✳✳ WARNING

The bearing caps must be marked for installation in their original location or engine damage can occur.

6. Remove the bolts, the camshaft bearing caps and the camshaft from the cylinder block.

To install:

7. Lubricate the camshaft journals with clean engine oil.
8. Install the camshaft and the camshaft bearing caps in their original locations onto the cylinder head. Loosely install the fifteen bolts.
9. Tighten the bolts in the sequence shown to 89 inch lbs. (10 Nm).
10. Use the special tool to compress the valve springs and install the camshaft roller followers.
11. Install the timing drive components.
12. Install the intake manifold.
13. Connect the battery ground cable.

06017-F150-G78

Camshaft and bearing caps—2002–04 F-Series 6.8L engine; 2002–05 E-Series 6.8L engine

06017-F150-G79

Camshaft bearing cap torque sequence—2002–04 F-Series 6.8L engine; 2002–05 E-Series 6.8L engine

2005 F-Series with 6.8L Engine

LEFT SIDE

1. Before servicing the vehicle, refer to the Precautions Section.

> ❋❋ **WARNING**
>
> The camshaft procedure must be followed exactly or damage to the valves and pistons will result.

2. Remove the cooling fan shroud.
3. Position the crankshaft damper spoke at the 12 o'clock position and the timing mark indentation at the 1 o'clock position.
4. Remove the left valve cover.

> ❋❋ **WARNING**
>
> Damage to the camshaft phaser sprocket assembly will occur if mishandled or used as a lifting or leveraging device.

5. Loosen and back off the left camshaft phaser bolt 1 full turn.
6. Disconnect the left camshaft position (CMP) sensor electrical connector.
7. Remove the left CMP sensor and the bolt.

> ❋❋ **WARNING**
>
> If servicing both camshafts, do not rotate the crankshaft. Camshaft position has been established earlier.

➡ If the camshaft lobes are not exactly positioned as shown, the crankshaft keyway will require 1 full additional rotation to 12 o'clock.

8. The number 5 cylinder camshaft exhaust lobe must be coming up on the exhaust stroke. Verify by noting the position

06017-F150-G64

Position the crankshaft damper spoke at the 12 o'clock position and the timing mark indentation at the 1 o'clock position—2005 F-Series 6.8L engine

06017-F150-G68

The number 1 cylinder camshaft exhaust lobe must be coming up on the exhaust stroke. Verify by noting the position of the 2 intake camshaft lobes and the exhaust lobe on the number 1 cylinder—2005 F-Series 6.8L left side[

of the 2 intake camshaft lobes and the exhaust lobe on the number 5 cylinder.

➡ The balance shaft bearing caps must be installed in their original locations. Record camshaft bearing cap locations.

9. Remove the 6 bolts in the sequence shown, the 3 bearing caps and the balance shaft.
10. Remove only the 3 roller followers shown in the illustration.

> ❋❋ **WARNING**
>
> Do not allow the valve keepers to fall off the valve or the valve may drop into the cylinder.

➡ The camshaft roller followers must be installed in their original locations. Record camshaft roller follower locations.

➡ It may be necessary to push the valve down while compressing the spring.

11. Using special tool 303-1039, remove only the 3 designated roller followers from the previous step.

> ❋❋ **WARNING**
>
> The crankshaft cannot be moved past the 6 o'clock position once set.

12. Rotate the crankshaft clockwise, as viewed from the front, positioning the crankshaft damper spoke at the 6 o'clock position and the timing mark indentation at the 7 o'clock position.

> ❋❋ **WARNING**
>
> Engine is not freewheeling. Camshaft procedure must be followed exactly or damage to valves and pistons will result.

> ❋❋ **WARNING**
>
> The Timing Chain Wedge tool must be installed square to the timing chain and the engine block.

➡ Front cover removed for clarity.

13. Install the special tools in the left timing chain as shown.

06017-F150-G81

Balance shaft bolt removal/installation sequence—2005 F-Series 6.8L engine

Remove only these 3 roller followers at this time—2005 F-Series 6.8L engine left side

Using special tool 303-1039—2005 F-Series 6.8L engine left side

Rotate the crankshaft clockwise, as viewed from the front, positioning the crankshaft damper spoke at the 6 o'clock position and the timing mark indentation at the 7 o'clock position—2005 F-Series 6.8L engine

✳✳ WARNING

Do not remove the timing chain wedge tool at any time during assembly. If the special tool is removed or out of placement, the engine front cover must be removed and the engine must be retimed.

✳✳ WARNING

The timing chain must be installed in its original position onto the camshaft phaser sprocket using the scribed marks, or damage to valves and pistons will result.

14. Scribe a location mark on the timing chain and the camshaft phaser sprocket assembly.

✳✳ WARNING

Remove the front thrust camshaft bearing cap straight upward from the bearing towers, or the bearing cap may be damaged from side loading.

➡The camshaft bearing caps must be installed in their original locations. Record camshaft bearing cap locations.

15. Remove the bolts in the sequence shown and remove the front camshaft bearing cap and then the remaining bearing caps.

16. Clean and inspect the left camshaft bearing caps. The camshaft front thrust bearing cap contains an oil metering groove. Make sure the groove is free of foreign material.

✳✳ WARNING

Damage to the camshaft phaser sprocket assembly will occur if mishandled or used as a lifting or leveraging device.

303-637

303-636

06017-F150-G84

The Timing Chain Wedge tool must be installed square to the timing chain and the engine block—2005 F-Series 6.8L engine left side

06017-F150-G85

Camshaft bearing cap removal/installation torque sequence—2005 F-Series 6.8L engine left side

❋❋ **WARNING**

Only use hand tools to remove the camshaft phaser sprocket bolt or damage may occur to the camshaft or camshaft phaser unit.

❋❋ **WARNING**

Do not remove the timing chain wedge tool at any time during assembly. If the special tool is removed or out of placement, the engine front cover must be removed and the engine must be retimed.

17. Remove the bolt and withdraw the camshaft from the phaser sprocket assembly leaving the sprocket assembly in place. Discard the bolt and washer.

To install:

18. Lubricate the camshaft and camshaft journals with clean engine oil.

❋❋ **WARNING**

Do not remove the timing chain wedge tool at any time during assembly. If the special tool is removed or out of placement, the engine front cover must be removed and the engine must be retimed.

❋❋ **WARNING**

Damage to the camshaft phaser sprocket assembly will occur if mishandled or used as a lifting or leveraging device.

❋❋ **WARNING**

Do not allow the roller followers to move out of position when installing the camshaft.

19. Install the camshaft into the camshaft phaser sprocket assembly and onto the head.

20. Install a new camshaft phaser bolt finger tight.

❋❋ **WARNING**

Do not remove the timing chain wedge tool at any time during assembly. If the special tool is removed or out of placement, the engine front cover must be removed and the engine must be retimed.

> ✳✳ **WARNING**
>
> **The timing chain must be installed in its original position onto the camshaft phaser sprocket using the scribed marks, or damage to valves and pistons will result. Verify the camshaft phaser sprocket and timing chain scribe marks are still in alignment.**

> ✳✳ **WARNING**
>
> **Do not allow the roller followers to move out of position when installing the camshaft.**

21. Lubricate the camshaft bearing caps with clean engine oil.
22. Position the front camshaft bearing cap.
23. Position the remaining camshaft bearing caps.
24. Install the bolts loosely.
25. Tighten the bolts in the sequence shown. Tighten to 10 Nm (89 inch lbs.).
26. Remove the special tools.
27. Rotate the crankshaft a half turn counterclockwise and position the crankshaft damper spoke at the 12 o'clock position and the timing mark indentation at the 1 o'clock position.
28. Verify correct cam position by noting the position of the number 5 cylinder intake and exhaust camshaft lobes.
29. Using the special tool, install the 3 originally removed roller followers.
30. Install the CMP sensor and the bolt.
31. Connect the CMP electrical connector.

> ✳✳ **WARNING**
>
> **Only use hand tools to install the camshaft phaser sprocket assembly or damage may occur to the camshaft or camshaft phaser unit.**

> ✳✳ **WARNING**
>
> **Damage to the camshaft phaser sprocket assembly will occur if mishandled or used as a lifting or leveraging device.**

32. Tighten the camshaft phaser bolt in 2 stages:
   • Stage 1: Tighten to 40 Nm (30 ft. lbs.).
   • Stage 2: Tighten an additional 90 degrees.

33. Install the left valve cover.
34. Install the cooling fan shroud.

RIGHT SIDE

1. Before servicing the vehicle, refer to the Precautions Section.

> ✳✳ **WARNING**
>
> **The camshaft procedure must be followed exactly or damage to the valves and pistons will result.**

2. Remove the cooling fan shroud.
3. Position the crankshaft damper spoke at the 12 o'clock position and the timing mark indentation at the 1 o'clock position.
4. Remove the right valve cover.

> ✳✳ **WARNING**
>
> **Damage to the camshaft phaser sprocket assembly will occur if mishandled or used as a lifting or leveraging device.**

5. Loosen and back off the right camshaft phaser bolt 1 full turn.
6. Disconnect the right camshaft position (CMP) sensor electrical connector.
7. Remove the bolt and the right CMP sensor.

06017-F150-G65

The number 1 cylinder camshaft exhaust lobe must be coming up on the exhaust stroke. Verify by noting the position of the 2 intake camshaft lobes and the exhaust lobe on the number 1 cylinder—2005 F-Series 6.8L engine

➡ **If the camshaft lobes are not exactly positioned as shown, the crankshaft will require 1 full additional rotation to 12 o'clock.**

8. The number 1 cylinder camshaft exhaust lobe must be coming up on the exhaust stroke. Verify by noting the position of the 2 intake camshaft lobes and the exhaust lobe on the number 1 cylinder.
9. Remove only the 3 roller followers shown in the illustration.

06017-F150-G80

Remove only these 3 roller followers at this time—2005 F-Series 6.8L engine right side

06017-F150-G67

Using special tool 303-1039—2005 F-Series 6.8L right side

✳✳ WARNING

Do not allow the valve keepers to fall off the valve or the valve may drop into the cylinder.

➡The camshaft roller followers must be installed in their original locations. Record camshaft roller follower locations.

➡It may be necessary to push the valve down while compressing the spring.

10. Using special tool 303-1039, remove only the 3 designated roller followers from the previous step.

06017-F150-G72

Timing chain wedge tool installed—2005 F-Series 6.8L engine right side

✳✳ WARNING

The crankshaft cannot be moved past the 6 o'clock position once set.

11. Rotate the crankshaft clockwise, as viewed from the front, positioning the crankshaft damper spoke at the 6 o'clock position and the timing mark indentation at the 7 o'clock position.

✳✳ WARNING

Engine is not freewheeling. Camshaft procedure must be followed exactly or damage to valves and pistons will result.

✳✳ WARNING

The Timing Chain Wedge tool must be installed square to the timing chain and the engine block.

➡Front cover removed for clarity.

12. Install the special tools in the right timing chain as shown.

✳✳ WARNING

Do not remove the timing chain wedge tool at any time during

06017-F150-G73

Scribe a location mark on the timing chain and the camshaft phaser sprocket assembly—2005 F-Series 6.8L engine right side

assembly. If the special tool is removed or out of placement, the engine front cover must be removed and the engine must be retimed.

✳✳ WARNING

The timing chain must be installed in its original position onto the camshaft phaser sprocket using the scribed marks, or damage to valves and pistons will result.

13. Scribe a location mark on the timing chain and the camshaft phaser sprocket assembly.

✳✳ WARNING

Remove the front thrust camshaft bearing cap straight upward from the bearing towers, or the bearing cap may be damaged from side loading.

➡The camshaft bearing caps must be installed in their original locations. Record camshaft bearing cap locations.

14. Remove the bolts in the sequence shown and remove the front camshaft bearing cap and then the remaining bearing caps.

15. Clean and inspect the right camshaft bearing caps. The camshaft front thrust bearing cap contains an oil metering groove. Make sure the groove is free of foreign material.

✳✳ WARNING

Damage to the camshaft phaser sprocket assembly will occur if mishandled or used as a lifting or leveraging device.

✳✳ WARNING

Only use hand tools to remove the camshaft phaser sprocket bolt or damage may occur to the camshaft or camshaft phaser unit.

06017-F150-G83

Camshaft bearing cap removal/installation sequence—2005 F-Series 6.8L engine right side

06017-F150-G75

The camshaft front thrust bearing cap contains an oil metering groove—2005 F-Series 6.8L engine

❋❋ WARNING

Do not remove the timing chain wedge tool at any time during assembly. If the special tool is removed or out of placement, the engine front cover must be removed and the engine must be retimed.

16. Remove the bolt and withdraw the camshaft from the phaser sprocket assembly leaving the sprocket assembly in place. Discard the bolt and washer.

To install:

17. Lubricate the camshaft and camshaft journals with clean engine oil.

❋❋ WARNING

Do not remove the timing chain wedge tool at any time during assembly. If the special tool is removed or out of placement, the engine front cover must be removed and the engine must be retimed.

❋❋ WARNING

Damage to the camshaft phaser sprocket assembly will occur if mishandled or used as a lifting or leveraging device.

❋❋ WARNING

Do not allow the roller followers to move out of position when installing the camshaft.

18. Install the camshaft into the camshaft phaser sprocket assembly and onto the head.

19. Install a new camshaft phaser bolt finger tight.

❋❋ WARNING

Do not remove the timing chain wedge tool at any time during assembly. If the special tool is removed or out of placement, the engine front cover must be removed and the engine must be retimed.

❋❋ WARNING

The timing chain must be installed in its original position onto the camshaft phaser sprocket using the scribed marks, or damage to valves and pistons will result.

20. Verify the camshaft phaser sprocket and timing chain scribe marks are still in alignment.

❋❋ WARNING

Do not allow the roller followers to move out of position when installing the camshaft.

21. Lubricate the camshaft bearing caps with clean engine oil.

22. Position the front camshaft bearing cap.

23. Position the remaining camshaft bearing caps.

24. Install the bolts loosely.

25. Tighten the bolts in the sequence shown. Tighten to 10 Nm (89 inch lbs.).

26. Remove the special tools.

27. Rotate the crankshaft a half turn counterclockwise and position the crankshaft damper spoke at the 12 o'clock position and the timing mark indentation at the 1 o'clock position.

28. Verify correct cam position by noting the position of the number 1 cylinder intake and exhaust camshaft lobes.

29. Using the special tool, install the 3 originally removed roller followers.

30. Install the CMP sensor and the bolt.
31. Connect the CMP electrical connector.

❋❋ WARNING

Only use hand tools to install the camshaft phaser sprocket assembly or damage may occur to the camshaft or camshaft phaser unit.

❋❋ WARNING

Damage to the camshaft phaser sprocket assembly will occur if mishandled or used as a lifting or leveraging device.

32. Tighten the camshaft phaser bolt in 2 stages:
   - Stage 1: Tighten to 40 Nm (30 ft. lbs.).
   - Stage 2: Tighten an additional 90 degrees.
33. Install the right valve cover.

Valve Lash

ADJUSTMENT

4.2L, 4.6L, 5.4L and 6.8L Engines

The 4.2L, 4.6L, 5.4L and 6.8L engines do not require valve lash adjusting, because they utilize hydraulic lash components in their valve actuation systems. The 4.2L engine uses hydraulic valve lifters, whereas the 4.6L, 5.4L and 6.8L engines utilize hydraulic lash adjusters, all of which automatically adjust the valve lash. No valve lash adjustment is necessary.

Starter Motor

REMOVAL & INSTALLATION

1. Before servicing the vehicle, refer to the Precautions Section.
2. Remove or disconnect the following:
   - Negative battery cable

1 Terminal cover
2 Starter solenoid B-terminal nut
3 Starter solenoid B-terminal eyelet
4 Starter solenoid S-terminal nut
5 Starter solenoid S-terminal eyelet
6 Starter motor ground cable nut
7 Starter motor ground cable eyelet
8 Starter motor stud bolt
9 Starter motor bolt
10 Starter motor

06017-F150-G86

Starter and related parts—4.2L engine

1 Terminal cover
2 Starter solenoid S-terminal nut
3 Starter solenoid S-terminal eyelet
4 Starter solenoid B-terminal nut
5 Starter solenoid B-terminal eyelet
6 Starter motor ground cable nut
7 Starter motor ground cable eyelet
8 Starter motor mounting stud bolt
9 Starter motor mounting bolt (2 required)
10 Starter motor

06017-F150-G87

Starter and related parts—4.6L, 5.4L and 6.8L engines

- Starter motor electrical connections
- Starter motor bolts and the motor

To install:

3. Installation is the reverse of removal, tighten the starter motor bolts to 18 ft. lbs. (25 Nm).

Oil Pan

REMOVAL & INSTALLATION

E-Series 4.2L Engine

1. Before servicing the vehicle, refer to the Precautions Section.
2. Disconnect the battery ground cable.
3. Remove the intake manifold spacer.
4. Remove the radiator upper hose.

� WARNING

The clutch assembly nut has a right-hand thread and must be rotated counterclockwise to remove it.

➡**Two people are needed to loosen the fan clutch pulley. From under the vehicle, have one person hold the fan pulley. The holding tool must be located at the 7 o'clock position.**

5. From the engine compartment, the second person can now loosen the fan clutch. The fan clutch nut wrench has to be located at the 12 o'clock position.
6. Using the appropriate Fan Clutch Hub Nut Wrench and Fan Pulley Holding Wrench, loosen the fan clutch.
7. Carefully rotate the fan and fan clutch assembly counterclockwise until the assembly is free from the water pump. Place the fan and fan clutch into the shroud opening.

✳✳ WARNING

Use extreme care when removing or installing the fan, fan clutch and shroud. Failure to do so will cause damage to the radiator.

8. Remove the fan, fan clutch and shroud.
9. Remove the bolt retaining the dipstick tube support bracket to the transmission converter housing.
10. Raise the vehicle.
11. Drain the engine oil.
12. Loosen the transmission-to-crossmember nuts.
13. Remove the exhaust pipe-to-manifold nuts.
14. Remove the engine mount nuts.
15. Position a screw jack and a block of wood under the oil pan.

➡**If the engine is not raised at least 330 mm (13 in), it will not be possible to remove the oil pan.**

16. Raise the engine until the distance between the crankshaft damper and the bottom of the crossmember is at least 330 mm (13 in).
17. Support the engine by placing blocks of wood between the engine mount

Position a screw jack and a block of wood under the oil pan—E-Series w/4.2L engine

330 mm (13 in)

Raise at least 330 mm (13 in)—E-Series w/4.2L engine

06017-F150-G90

Support the engine by placing blocks of wood between the engine mount bracket and the exhaust manifolds)—E-Series w/4.2L engine

bracket and the exhaust manifolds. Remove the screw jack and wood block from the oil pan.

18. Remove the bolt retaining the oil pan to the right side of the transmission converter housing.

19. Remove the bolts retaining the oil pan to the left side of the transmission converter housing.

20. Remove the remaining bolts and the oil pan. Remove and discard the oil pan seal.

21. To install, reverse the removal procedure. Clean mating surfaces on the oil pan and the cylinder block using metal surface cleaner. Apply a bead of sealant as shown in the illustration.

➡️**If the oil pan and rear oil seal are not installed within four minutes, remove sealant and reapply.**

22. Tighten the oil pan bolts in two stages in the sequence shown.
   - Stage 1: 4–5 Nm (35–44 inch lbs.)
   - Stage 2: 9–12 Nm (79–106 inch lbs.)

23. Tighten the bolts retaining the oil pan to the left side of the transmission converter housing to 32 ft. lbs. (44 Nm).

24. Tighten the bolt retaining the oil pan to the right side of the transmission converter housing to 32 ft. lbs. (44 Nm).

25. Torque the exhaust pipe-to-manifold nuts to 18 ft. lbs. (25 Nm).

26. Torque the engine mount nuts to 59 ft. lbs. (80 Nm).

27. Torque the bolt retaining the dipstick tube support bracket to the transmission converter housing to 18 ft. lbs. (25 Nm).

2002–03 F-150 and 2004 Heritage Editions 4.2L Engine

4X2 MODELS

1. Before servicing the vehicle, refer to the Precautions Section.

06017-F150-G91

Apply a bead of sealant as shown—4.2L engine

2. Remove or disconnect the following:
   - Engine
   - Dipstick tube
   - Oil filter
   - The 15 oil pan-to-cylinder block mounting bolts and the oil pan

To install:

➡If the oil pan is not installed within 15 minutes, remove the sealer and reapply.

3. Temporarily install 2 locator dowels in 2 of the oil pan-to-engine block corner mounting bolt holes.

4. Clean and apply sealant to the rear main bearing cap, the oil pan mating surface, the front cover mounting area, the front cover-to-engine block joints, then install the oil pan rear seal. Make certain to use RTV silicone gasket material for the sealant.

5. Position the oil pan, then install 13 of the oil pan mounting bolts loosely.

6. Remove the 2 locator dowels and install the remaining 2 oil pan-to-engine block bolts.

7. In sequence, tighten the 15 oil pan mounting bolts to 44 inch lbs. (5 Nm), then retighten the bolts to 89 inch lbs. (10 Nm).
   - Oil filter
   - Dipstick tube
   - Engine

Oil pan torque sequence—2002–03 F-150 4.2L engine, 2004 F-150 Heritage Edition 4.2L engine, and E-Series 4.2L engine

8. Fill the engine with the correct type and amount of engine oil.

4X4 MODELS

1. Before servicing the vehicle, refer to the Precautions Section.

2. Remove or disconnect the following:
   - Engine oil
   - The 2 front wheel driveshafts and joints, if so equipped.
   - Front differential from the vehicle
   - Front differential support
   - The 3 oil pan-to-transmission bolts
   - The 15 oil pan-to-cylinder block mounting bolts, then lower the oil pan

To install:

➡If the oil pan is not installed within 15 minutes, remove the sealer and reapply.

3. Temporarily install 2 locator dowels in 2 of the oil pan-to-engine block corner mounting bolt holes.

4. Clean and apply sealant to the rear main bearing cap, the oil pan mating surface, the front cover mounting area, the front cover-to-engine block joints, then install the oil pan rear seal. Make certain to use RTV silicone gasket material for the sealant.

5. Position the oil pan, then install 13 of the oil pan mounting bolts loosely.

6. Remove the 2 locator dowels and install the remaining 2 oil pan-to-engine block bolts.

7. In sequence, tighten the 15 oil pan mounting bolts to 44 inch lbs. (5 Nm), then retighten the bolts to 89 inch lbs. (10 Nm).

8. Install the oil pan-to-transmission bolts to 28–38 ft. lbs. (38–51 Nm).

9. Install the oil pan drain plug to 16–22 ft. lbs. (22–30 Nm).

10. Install the front differential and front differential support.

11. Install the front driveshafts and joints.

12. Lower the vehicle to the ground.

13. Fill the engine with the correct type and amount of engine oil.

2005 F-150 4.2L Engine

1. Before servicing the vehicle, refer to the Precautions Section.

✳✳ WARNING

During engine repair procedures, cleanliness is extremely important. Any foreign material, including any material created while cleaning gasket surfaces, that enters the oil passages, coolant passages or the oil pan can cause engine failure.

2. With the vehicle in NEUTRAL, position it on a hoist.

3. Remove the starter.

4. Drain the engine oil. Install the drain plug when finished.

5. Remove the 4 bolts and the crossmember.

6. Remove the 3 oil pan-to-transmission bolts.

7. Remove the 17 bolts and the oil pan.

8. Remove and discard the oil pan rear seal.

To install:

9. Clean the gasket mating surfaces using silicone gasket remover, a plastic scraping tool and metal surface prep. Follow the directions on the packaging.

➡ If the oil pan is not secured within 4 minutes, the sealant must be removed and the sealing areas cleaned. To clean the sealing area, use silicone gasket remover and metal surface prep. Follow the directions on the packaging. Failure to follow this procedure can cause future oil leakage.

10. Apply silicone gasket and sealant to the oil pan sealing areas shown and install the oil pan rear seal.

11. Apply silicone gasket and sealant to the rear main bearing cap.

12. Install the oil pan rear seal.

13. Apply a bead of silicone gasket and sealant to the oil pan mating surface.

14. Position the oil pan and install the bolts. Tighten the bolts in the sequence shown. Tighten to 10 Nm (89 inch lbs.).

15. Install the 3 oil pan-to-transmission bolts. Tighten to 34 Nm (25 ft. lbs.).

16. Install the starter.

17. Position the crossmember and install the 4 bolts. Tighten to 90 Nm (66 ft. lbs.).

2002–03 4.6L F-Series and 2004 Heritage Editions, 2-Wheel Drive

1. Before servicing the vehicle, refer to the Precautions Section.

2. Disconnect the battery ground cable.

3. Drain the engine cooling system.

4. Remove the radiator air deflector.

5. Remove the screws.

6. Remove the radiator air deflector.

7. Remove the engine air cleaner and the air cleaner outlet pipe.

8. Slide the clamp back and disconnect the upper radiator hose.

9. Disconnect the throttle body cam.

10. Disconnect the accelerator cable.

11. Disconnect the speed control actuator cable.

Oil pan torque sequence—2005 F-150 4.2L engine

06017-F150-G92

12. Replace the accelerator return spring.

13. Replace the accelerator cable bracket bolts and position the bracket and cables aside.

14. Disconnect the main vacuum harness.

15. Disconnect the throttle position (TP) sensor and exhaust gas recirculation (EGR) vacuum regulator solenoid electrical and vacuum connectors.

16. Disconnect the fuel pressure regulator vacuum hose.

17. Disconnect the evaporative emission canister purge valve vacuum hose.

18. Disconnect the differential pressure feedback EGR system electrical connector.

19. Replace the nut and disconnect the brake booster vacuum tube and bracket.

20. Disconnect the idle air control (IAC) electrical connector.

21. Position the exhaust manifold to EGR valve tube aside.

22. Disconnect the upper fitting.

23. Loosen the lower fitting. Position the tube aside.

24. Disconnect the fuel hose spring lock couplings.

25. Disconnect the heater hose and the heated throttle body hose.

26. Disconnect the positive crankcase ventilation (PCV) hose.

27. Replace the four bolts and move the throttle body adapter forward slightly.

28. Disconnect the heated throttle body hose and remove the throttle body adapter.

29. Discard the throttle body adapter-to-intake manifold gasket.

30. Inspect the throttle body and adapter for damage.

31. Position the power steering reservoir aside.

32. Replace the upper bolt.

33. Replace the lower bolts.

34. Position the reservoir aside.

35. Disconnect the eight fuel injectors.

36. Disconnect and remove the eight ignition coils.

37. Replace the alternator.

38. Position the fan shroud aside.

39. Replace the bolts.

40. Position the fan shroud aside.

41. Install a suitable lifting tool on the engine using the alternator mounting bolts.

42. Raise and support the vehicle.

43. Replace the drain plug and drain the engine oil.

44. Replace the right motor mount bolt.

45. Replace the left motor mount bolt.

46. Lower the vehicle.

47. Raise the engine using a suitable tool.

48. Raise the vehicle on the hoist.

➡Be careful when removing the oil pan gasket. The oil pan gasket is reusable if it is not damaged.

49. Replace the oil pan bolts.

50. Position the oil pan aside to gain access to the oil pump screen and pickup tube.

51. Replace the bolts and the oil pump screen and pickup tube.

To install:

➡Do not use metal scrapers, wire brushes, power abrasive discs or other abrasive means to clean the sealing surfaces. These tools cause scratches and gouges which make leak paths. Use a plastic scraping tool to remove all traces of old sealant.

52. Clean the sealing surfaces with silicone gasket remover and metal surface prep. Follow the directions on the packaging. Inspect the mating surfaces.

53. Position the oil pan gasket, the oil pan and the oil pump screen and pickup tube together to the cylinder block.

➡Make sure the O-ring is in place and not damaged. A missing or damaged O-ring can cause foam in the lubrication system, low oil pressure, and severe engine damage.

➡Install a new O-ring and lubricate with clean engine oil.

54. Position the oil pump screen and pickup tube and install the bolts.

➡If the oil pan is not secured within four minutes, the sealant must be removed and the sealing area cleaned. To clean the sealing area, use silicone gasket remover and metal surface prep. Follow the directions on the packaging. Failure to follow this procedure can cause future oil leakage.

55. Apply the silicone gasket and sealant at the engine front cover-to-cylinder block mating surface.

➡If the oil pan is not secured within four minutes, the sealant must be removed and the sealing area cleaned. To clean the sealing area, use silicone gasket remover and metal surface prep. Follow the directions on the packaging. Failure to follow this procedure can cause future oil leakage.

56. Apply the silicone gasket and sealant at the rear oil seal retainer-to-cylinder block sealing surface.

57. Position the oil pan gasket.

Apply the silicone gasket and sealant at the engine front cover-to-cylinder block mating surface—4.6L and 5.4L engines

Oil pan bolt torque sequence—2002–03 4.6L and 5.4L F-Series and 2004 Heritage Editions

58. Install the bolts and tighten in three steps, in the sequence shown.

   a. Step 1: Tighten to 2 Nm (18 inch lbs.).

   b. Step 2: Tighten to 20 Nm (15 ft. lbs.).

   c. Step 3: Tighten an additional 60 degrees.

59. Lower the vehicle.

60. Lower the engine and remove the special tool.

61. Raise the vehicle on the hoist.

62. Install the right motor mount bolt.

63. Install the left motor mount bolt.

64. Install the drain plug.

65. Install a new oil filter.

66. Lower the vehicle.

67. Replace the special tool.

68. Install the alternator.

69. Position the fan shroud and install the bolts.

70. Roughly position the throttle body adapter in the vehicle and connect the heated throttle body hose.

71. Install a new gasket and the throttle body adapter.

72. Connect the PCV hose.

73. Connect the heater hose and the heated throttle body hose.

74. Install and connect the eight ignition coils.

75. Connect the eight fuel injectors.

76. Install the power steering reservoir.

77. Position the reservoir.

78. Install the two lower bolts.

79. Install the upper bolt.

80. Install the exhaust manifold to EGR valve tube.

81. Hand-tighten the fittings.

82. Tighten the upper fitting.

83. Tighten the lower fitting.

84. Connect the fuel hose spring lock couplings.

85. Connect the IAC.

86. Connect the brake booster vacuum tube and bracket and install the nut.

87. Connect the differential pressure feedback EGR system electrical connectors.

88. Connect the evaporative emission canister purge valve vacuum hose.

89. Connect the fuel pressure regulator vacuum hose.

90. Connect the TP sensor electrical connector and the EGR vacuum regulator solenoid electrical and vacuum connectors.

91. Connect the main vacuum harness.

92. Install the accelerator cable bracket and the bolts.

93. Connect the throttle body cam.

94. Connect the accelerator cable.

95. Connect the speed control actuator cable.

96. Install the throttle return spring.

97. Connect the upper radiator hose and position the clamp.

98. Install the engine air cleaner and the air cleaner outlet pipe.

99. Install the radiator air deflector.

100. Connect the battery ground cable.

➡**The oil pump must be primed prior to starting the engine.**

101. Fill the engine with clean engine oil.

102. Fill and bleed the engine cooling system.

2002–03 5.4L Blackwood, F-Series and 2004 Heritage Editions, 2-Wheel Drive

1. Before servicing the vehicle, refer to the Precautions Section.

2. With the vehicle in **Neutral**, position it on a hoist.

3. Remove the radiator air deflector.

4. Remove the screws.

5. Remove the radiator air deflector.

6. Remove the air cleaner outlet pipe.

7. Remove the alternator.

8. Disconnect the acceleration controls from the throttle body cam.

9. Disconnect the accelerator cable.

10. Disconnect the speed control actuator cable.

11. Remove the accelerator return spring.

12. Remove the accelerator cable bracket bolts and position the bracket and cables aside.

13. Disconnect the positive crankcase ventilation (PCV) hose from the throttle body adapter.

14. Disconnect the evaporative emission (EVAP) canister purge valve hose.

15. Remove the crankcase ventilation tube and the idle air control (IAC) valve fresh air tube.

16. Disconnect the exhaust gas recirculation (EGR) valve vacuum hose and the main vacuum harness.

17. Disconnect the throttle position sensor electrical connector and the EGR vacuum regulator solenoid connections.

18. Disconnect the fuel pressure regulator vacuum hose.

19. Disconnect the EVAP canister purge valve vacuum hose.

20. Remove the bolt and disconnect the brake booster vacuum tube and bracket.

21. Remove the nut.

22. Disconnect the IAC electrical connector.

23. Disconnect the upper EGR valve tube fitting.

24. Remove the four bolts and set aside the throttle body adapter.

25. Discard the throttle body adapter gasket.

26. Position the fan shroud aside.

27. Remove the bolts.

28. Position the fan shroud aside.

29. Install a suitable tool on the engine using the alternator mounting bolts.

30. Install a suitable tool on the engine.

31. Remove the starter solenoid terminal cover.

32. Disconnect the starter motor electrical connections.

33. Remove the starter motor.

34. Remove the stud bolt.

35. Remove the two bolts.

36. Remove the starter.

37. Remove the drain plug and drain the engine oil.

38. Remove the right motor mount bolt.

39. Remove the left motor mount bolt.

40. Raise the engine using a cane.

➡**Be careful when removing the oil pan gasket. The oil pan gasket is reusable if it is not damaged.**

41. Remove the oil pan bolts.

42. Position the oil pan aside to gain access to the oil pump screen and pickup tube.

43. Remove the bolt securing the rear of the oil pump screen and pickup tube.

44. Remove the bolts securing the front of the oil pump screen and pickup tube.

45. Remove the oil pan and the oil pickup tube together.

To install:

✳✳ WARNING

Do not use metal scrapers, wire brushes, power abrasive discs or other abrasive means to clean the sealing surfaces. These tools cause scratches and gouges which make leak paths. Use a plastic scraping tool to remove all traces of old sealant.

46. Clean the sealing surfaces with silicone gasket remover and metal surface prep. Follow the directions on the packaging. Inspect the mating surfaces.

47. Position the oil pan gasket, the oil pan and the oil pump screen and pickup tube together to the cylinder block.

✳✳ WARNING

Make sure the O-ring is in place and not damaged. A missing or damaged O-ring can cause foam in the lubrication system, low oil pressure, and severe engine damage.

48. Install a new O-ring and lubricate with clean engine oil.

49. Position the oil pump screen and pickup tube and install the bolts.

➡If the oil pan is not secured within four minutes, the sealant must be removed and the sealing area cleaned. To clean the sealing area, use silicone gasket remover and metal surface prep. Follow the directions on the packaging. Failure to follow this procedure can cause future oil leakage.

50. Apply silicone gasket and sealant at the engine front cover-to-cylinder block mating surface.

➡If the oil pan is not secured within four minutes, the sealant must be removed and the sealing area cleaned. To clean the sealing area, use silicone gasket remover and metal surface prep. Follow the directions on the packaging. Failure to follow this procedure can cause future oil leakage.

51. Apply silicone gasket and sealant at the rear oil seal retainer-to-cylinder block sealing surface.

52. Position the oil pan gasket.

53. Install the bolts and tighten in three steps, in the sequence shown.

a. Step 1: Tighten to 2 Nm (18 inch lbs.).

b. Step 2: Tighten to 20 Nm (15 ft. lbs.).

c. Step 3: Tighten an additional 60 degrees.

54. Lower the engine and remove the crane.

55. Install the right motor mount bolt.

56. Install the left motor mount bolt.

57. Install the drain plug.

58. Install the starter motor.

59. Position the starter.

60. Install the two bolts.

61. Install the stud bolt.

62. Connect the starter motor electrical connections.

63. Install the starter motor solenoid terminal cover.

64. Install a new oil filter.

65. Position the fan shroud and install the bolts.

66. Install a new throttle body adapter gasket.

67. Position the throttle body adapter and install the four bolts.

68. Connect the upper EGR valve tube fitting.

69. Connect the IAC electrical connector.

70. Install the nut.

71. Connect the brake booster vacuum tube and bracket and install the bolt.

72. Connect the EVAP canister purge valve vacuum hose.

73. Connect the fuel pressure regulator vacuum hose.

74. Connect the TP sensor electrical connector and the EGR vacuum regulator solenoid electrical and vacuum connectors.

75. Connect the EGR valve vacuum hose and the main vacuum harness vacuum connections.

76. Install the crankcase ventilation tube and the IAC fresh air tube.

77. Connect the EVAP canister purge valve hose.

78. Connect the PCV hose to the throttle body adapter.

79. Install the accelerator cable bracket and the bolts.

80. Connect the accelerator controls to the throttle body cam.

81. Connect the accelerator cable.

82. Connect the speed control actuator cable.

83. Install the throttle return spring.

84. Install the alternator.

85. Install the air cleaner outlet pipe.

86. Install the radiator air deflector.

2002–03 Blackwood, F-Series and 2004 Heritage Editions, 4.6L and 5.4L Engines w/4-Wheel Drive

1. Before servicing the vehicle, refer to the Precautions Section.

2. With the vehicle in **Neutral**, position it on a hoist.

3. Drain the engine oil.

4. Support the front axle housing with a jack stand.

5. Remove front axle housing support.

6. Remove the front axle mount bolt.

7. Remove the axle support bolts.

8. Remove the axle support.

✳✳ WARNING

Use care when lowering the front axle housing, or the vacuum hoses to the axle solenoid may become disconnected or damaged.

9. Remove the right and front axle housing mount bolts and lower the axle to allow clearance for the oil pan to be removed.

10. Remove the oil pan.

✳✳ WARNING

Do not use metal scrapers, wire brushes, power abrasive discs, or other abrasive means to clean the sealing surfaces. These may cause scratches and gouges resulting in leak paths. Use a plastic scraper to clean the sealing surfaces.

11. Remove and discard the oil pan gasket. Clean the sealing surfaces with silicone gasket remover and metal surface prep. Follow the directions on the packaging. Inspect the mating surfaces.

To install:

➡If the oil pan is not secured within four minutes, the sealant must be removed and the sealing area cleaned. To clean the sealing area, use silicone gasket remover and metal surface prep. Follow the directions on the packaging. Failure to follow this procedure can cause future oil leakage.

12. Apply silicone gasket and sealant in the locations shown.

➡If the oil pan is not secured within four minutes, the sealant must be removed and the sealing area cleaned. To clean the sealing area, use silicone gasket remover and metal surface prep. Follow the directions on the packaging. Failure to follow this procedure can cause future oil leakage.

13. Apply silicone gasket and sealant in the locations shown.

14. Position the new oil pan gasket and the oil pan and loosely install the bolts.

15. Tighten the bolts in three steps in the sequence shown.

a. Step 1: Tighten to 2 Nm (18 inch lbs.).

b. Step 2: Tighten to 20 Nm (15 ft. lbs.).

c. Step 3: Tighten an additional 60 degrees.

16. Position the front axle housing and loosely install the bolts.

17. Position the axle support and install the bolts.

18. Tighten the bolts to 66 ft. lbs. (90 Nm).

19. Remove the jack stand.

20. Install a new oil filter.

✳✳ WARNING

The oil pump must be primed prior to starting the engine.

21. Fill the engine with clean engine oil.

22. Start the engine and inspect for leaks.

2004 F-150 w/4.6L or 5.4L Engine, Exc. Heritage Editions

2-WHEEL DRIVE

1. Before servicing the vehicle, refer to the Precautions Section.
2. With the vehicle in **Neutral**, position it on a hoist.
3. Drain the engine oil.
4. Remove the oil level sensor electrical connector.
5. Remove the oil pan.

> ☀ **CAUTION**
>
> **Do not use metal scrapers, wire brushes, power abrasive discs, or other abrasive means to clean the sealing surfaces. These may cause scratches and gouges resulting in leak paths. Use a plastic scraper to clean the sealing surfaces.**

6. Remove and discard the oil pan gasket. Clean the sealing surfaces with silicone gasket remover and metal surface prep. Follow the directions on the packaging. Inspect the mating surfaces.
7. To install, reverse the removal procedure.

➡ **If the oil pan is not secured within four minutes, the sealant must be removed and the sealing area cleaned. To clean the sealing area, use silicone gasket remover and metal surface prep. Follow the directions on the packaging. Failure to follow this procedure can cause future oil leakage.**

8. Apply silicone gasket and sealant in the locations shown.
9. Position the new oil pan gasket and the oil pan and loosely install the bolts.
10. Tighten the bolts in three steps in the sequence shown.
   a. Step 1: Tighten to 2 Nm (18 inch lbs.).
   b. Step 2: Tighten to 20 Nm (15 ft. lbs.).
   c. Step 3: Tighten an additional 60 degrees.

4-WHEEL DRIVE

1. With the vehicle in **Neutral**, position it on a hoist.
2. Drain the engine oil.
3. Remove the skid plate.
4. Index mark the front driveshaft flange to the front axle pinion flange.

> ☀ **CAUTION**
>
> **Do not let the driveshaft hang unsupported.**

Oil pan torque sequence—2004–05 4.6L or 5.4L engine, exc. Heritage editions

67197-EFSE-G99

1 Skid plate bolts
2 Skid plate
3 Front drive axle assembly

67197-EFSE-G100

Skid plate removal—2004 F-150 w/4.6L or 5.4L Engine, Exc. Heritage Editions

5. Disconnect and support the front driveshaft.

6. Remove the four bolts.

7. Support the front driveshaft.

8. Remove the bolts and the cross-member.

9. Position a suitable hydraulic jack under the front axle. Securely strap the jack to the axle.

➡**Rotate the steering column so the pinch bolt for the steering column coupling allows clearance for the isolator bolt.**

10. Remove the upper front axle carrier mounting bushing bolt.

11. Remove the axle shaft housing carrier bushing bolt.

12. Remove the lower front axle carrier mounting bushing bolt.

13. Lower the front axle assembly.

14. Disconnect the oil level sensor.

15. Remove the oil pan bolts and lower the pan. Discard the gasket.

✱✱ CAUTION

Do not use metal scrapers, wire brushes, power abrasive discs, or other abrasive means to clean the sealing surfaces. These may cause scratches and gouges resulting in

16. leak paths. Use a plastic scraper to clean the sealing surfaces.

17. Remove and discard the oil pan gasket. Clean the sealing surfaces with silicone gasket remover and metal surface prep. Follow the directions on the packaging. Inspect the mating surfaces.

To install:

➡**If the oil pan is not secured within four minutes, the sealant must be removed and the sealing area cleaned. To clean the sealing area, use silicone gasket remover and metal surface prep. Follow the directions on the**

Silicone sealant application—2004 F-150 w/4.6L or 5.4L Engine, Exc. Heritage Editions

67197-EFSE-G102

packaging. Failure to follow this procedure can cause future oil leakage.

18. Apply silicone gasket and sealant in the locations shown.

19. Position the new oil pan gasket and the oil pan and loosely install the bolts.

20. Tighten the bolts in three steps in the sequence shown.

   a. Step 1: Tighten to 2 Nm (18 inch lbs.).

   b. Step 2: Tighten to 20 Nm (15 ft. lbs.).

   c. Step 3: Tighten an additional 60 degrees.

21. Position the front axle.

➡**Rotate the steering column so the pinch bolt for the steering column coupling allows clearance for the isolator bolt.**

22. Install the upper front axle carrier mounting bushing bolt. Torque to 66 ft. lbs. (90 Nm).

23. Install the axle shaft housing carrier bushing nut and bolt. Torque to 66 ft. lbs. (90 Nm).

24. Install the lower front axle carrier mounting bushing bolt. Torque to 66 ft. lbs. (90 Nm).

25. Align the index marks, connect the front driveshaft and install the bolts. Torque to 82 ft. lbs. (111 Nm).

26. Install the crossmember and the bolts. Torque to 66 ft. lbs. (90 Nm).

27. Install the skid plate.

28. Add the correct amount of engine oil. Install a new filter.

2005 F-150 4.6L and 5.4L Engines and 2006 Mark LT 5.4L Engine

2-WHEEL DRIVE

1. Before servicing the vehicle, refer to the Precautions Section.

2. With the vehicle in NEUTRAL, position it on a hoist.

3. Drain the engine oil.

4. If equipped, remove the 4 skid plate bolts and the skid plate.

5. Remove the bolts and the cross-member.

✱✱ WARNING

Do not use metal scrapers, wire brushes, power abrasive discs, or other abrasive means to clean the sealing surfaces. These may cause scratches and gouges resulting in leak paths. Use a plastic scraper to clean the sealing surfaces.

Remove the bolts and the crossmember—2005 F-150 and 2006 Mark LT 5.4L Engine

06017-F150-G93

6. Remove the 16 oil pan bolts and the oil pan, discard the oil pan gasket. Clean the sealing surfaces with silicone gasket remover and metal surface prep. Follow the directions on the packaging. Inspect the mating surfaces.

To install:

➡**If the oil pan is not secured within 4 minutes, the sealant must be removed and the sealing area cleaned. To clean the sealing area, use silicone gasket remover and metal surface prep. Follow the directions on the packaging. Failure to follow this procedure can cause future oil leakage.**

7. Apply the silicone gasket and sealant at the engine front cover-to-cylinder block mating surface.

➡**If the oil pan is not secured within 4 minutes, the sealant must be removed and the sealing area cleaned. To clean the sealing area, use silicone gasket remover and metal surface prep. Follow the directions on the packaging. Failure to follow this procedure can cause future oil leakage.**

8. Apply silicone gasket and sealant in the locations shown.

9. Position the new oil pan gasket and the oil pan and loosely install the 16 bolts.

10. Tighten the bolts in 3 stages in the sequence shown.

- Stage 1: Tighten to 2 Nm (18 inch lbs.).
- Stage 2: Tighten to 20 Nm (15 ft. lbs.).
- Stage 3: Tighten an additional 60 degrees.

11. Install the crossmember and the 4 bolts. Tighten to 90 Nm (66 ft. lbs.).

12. If equipped, install the skid plate and the 4 bolts. Tighten to 25 Nm (18 ft. lbs.).

13. Install the engine oil pan drain plug and fill the engine with clean engine oil. Tighten to 14 Nm (10 ft. lbs.).

4-WHEEL DRIVE

1. With the vehicle in NEUTRAL, position it on a hoist.
2. Drain the engine oil.
3. If equipped, remove the 4 skid plate bolts and the skid plate.
4. Remove the bolts and the crossmember.
5. Position a suitable hydraulic jack under the front axle. Securely strap the jack to the axle.

➡Rotate the steering column so the pinch bolt for the steering column coupling allows clearance for the isolator bolt.

6. Remove the upper front axle carrier mounting bushing bolt.
7. Remove the axle shaft housing carrier bushing bolt.
8. Remove the lower front axle carrier mounting bushing bolt.
9. Partially lower the front axle assembly.

✳✳ WARNING

Do not use metal scrapers, wire brushes, power abrasive discs, or other abrasive means to clean the sealing surfaces. These may cause scratches and gouges resulting in leak paths. Use a plastic scraper to clean the sealing surfaces.

10. Remove the 16 oil pan bolts and the oil pan, discard the oil pan gasket. Clean the sealing surfaces with silicone gasket remover and metal surface prep. Follow the directions on the packaging. Inspect the mating surfaces.

To install:

➡If the oil pan is not secured within 4 minutes, the sealant must be removed and the sealing area cleaned. To clean the sealing area, use silicone gasket remover and metal surface prep. Follow the directions on the packaging. Failure to follow this procedure can cause future oil leakage.

11. Apply the silicone gasket and sealant at the engine front cover-to-cylinder block mating surface.

➡If the oil pan is not secured within 4 minutes, the sealant must be removed and the sealing area cleaned. To clean the sealing area, use silicone gasket remover and metal surface prep. Follow the directions on the packaging. Failure to follow this procedure can cause future oil leakage.

Position a suitable hydraulic jack under the front axle—2005 F-150 4wd

06017-F150-G94

Remove the upper front axle carrier mounting bushing bolt—2005 F-150 4wd

06017-F150-G95

Remove the axle shaft housing carrier bushing bolt—2005 F-150 4wd

06017-F150-G96

Remove the lower front axle carrier mounting bushing bolt—2005 F-150 4wd

06017-F150-G97

12. Apply silicone gasket and sealant in the locations shown.
13. Position the new oil pan gasket and the oil pan and loosely install the 16 bolts.
14. Tighten the bolts in 3 stages in the sequence shown.

Apply silicone gasket and sealant in the locations shown—2005 F-150 4.6 L engine

06017-F150-G98

- Stage 1: Tighten to 2 Nm (18 inch lbs.).
- Stage 2: Tighten to 20 Nm (15 ft. lbs.).
- Stage 3: Tighten an additional 60 degrees.

15. Position the front axle.

➡Rotate the steering column so the pinch bolt for the steering column coupling allows clearance for the isolator bolt.

16. Install the upper front axle carrier mounting bushing bolt. Tighten to 90 Nm (66 ft. lbs.).
17. Install the axle shaft housing carrier bushing nut and bolt. Tighten to 90 Nm (66 ft. lbs.).
18. Install the lower front axle carrier mounting bushing bolt. Tighten to 90 Nm (66 ft. lbs.).
19. Install the crossmember and the 4 bolts. Tighten to 90 Nm (66 ft. lbs.).
20. If equipped, install the skid plate and the 4 bolts. Tighten to 25 Nm (18 ft. lbs.).
21. Install the engine oil pan drain plug and fill the engine with clean engine oil. Tighten to 14 Nm (10 ft. lbs.).

2004 F-250/350 5.4L Engine

MANUAL TRANSMISSION

1. Before servicing the vehicle, refer to the Precautions Section.
2. With the vehicle in **Neutral**, position it on a hoist.
3. Remove the air cleaner outlet tube.
4. Remove the accelerator cable splash shield.
5. Disconnect the throttle body cable.
6. Disconnect the accelerator cable.
7. Disconnect the speed control cable.
8. Remove the throttle return spring.
9. Position the cable mounting bracket aside.
10. Remove the bolts.
11. Position the cable mounting bracket aside.

12. Disconnect the hose and electrical connector from the idle air control (IAC) valve.

13. Remove the routing clip for the IAC hose.

14. Disconnect the throttle position (TP) sensor connector.

15. Remove the two bolts retaining the vacuum transducer bracket to the intake manifold. Position the bracket and transducer out of the way.

16. If equipped, disconnect the exhaust gas recirculation (EGR) valve.

17. Loosen the EGR tube nut.

18. Disconnect the EGR tube.

19. Disconnect the vacuum line from the EGR valve.

20. If equipped, disconnect the EGR vacuum regulator (EVR) connections.

21. Disconnect electrical connector.

22. Disconnect vacuum connector.

23. Remove the four bolts. Position aside the throttle body, throttle body spacer, EGR valve, EVR and IAC valve as an assembly.

24. Hold the accelerator pedal in place and disconnect the accelerator cable from the accelerator pedal and shaft.

25. Disconnect the accelerator cable routing clip from the dash panel stud.

26. Remove the bolts and pull the accelerator cable through the bulkhead.

27. Remove the alternator.

❋❋ CAUTION

The large clutch assembly nut has a right-hand thread and must be rotated counterclockwise to remove it.

28. Remove the fan blade assembly and the fan clutch from the coolant pump pulley.

29. Use a suitable tool to hold the coolant pump pulley.

30. Using a suitable tool, remove the fan blade and the fan clutch from the coolant pump pulley.

31. Install a lifting eye on the engine using the alternator mounting holes.

32. Install a lifting bar.

33. Disconnect the heated oxygen sensor electrical connectors.

34. Remove the exhaust Y-pipe flange nuts.

35. Remove the nuts and the exhaust Y-pipe.

36. Remove the engine mount nuts.

37. Remove the stud bolt and the ground strap from the engine.

38. Remove the starter motor.

39. Remove the nut and bracket from the engine front cover stud and position the wire harness aside.

40. Remove the transmission mount nuts.

41. Drain the engine oil and remove the oil filter.

42. Remove the oil cooler threaded insert and position the oil cooler aside.

❋❋ CAUTION

Use care when raising the engine, so that the engine does not touch the body.

43. Using a crane, raise the engine.

44. Remove the bolts and remove the right engine mount from the engine.

45. Remove the bolts and partially lower the oil pan.

46. Remove the bolts from the oil pump screen and pickup tube.

47. Remove the rear bolt and the oil pump screen and pickup tube and let them drop into the oil pan.

➡ **To remove the oil pan, position the rear of the pan to the left side of the vehicle and rotate slightly to the right while pulling outward.**

48. Remove the oil pan and the oil pan gasket.

49. Clean the inside of the oil pan thoroughly and inspect the oil pan gasket.

❋❋ CAUTION

Do not use metal scrapers, wire brushes, power abrasive discs or other abrasive means to clean the sealing surfaces. These tools cause scratches and gouges, which make leak paths. Use a plastic scraping tool to remove all traces of sealant.

50. Clean the mating surfaces for the oil pan and the engine with silicone gasket remover and metal surface prep. Follow the directions on the packaging.

To install:

➡ **The oil pump screen and pickup tube must be in the oil pan when the oil pan is positioned in the vehicle.**

➡ **To install the oil pan, position the oil pan to the left rear side of the engine and rotate slightly while sliding the front of the pan toward the right side of the engine.**

51. Position the oil pan gasket and the oil pan in the vehicle from the rear of the engine.

❋❋ CAUTION

Make sure the O-ring is in place and not damaged. A missing or damaged O-ring can cause foam in the lubrication system, low oil pressure and severe engine damage.

➡ **Clean and inspect the mating surfaces and install a new O-ring. Lubricate with clean engine oil.**

52. Position the oil pump screen and pickup tube on the engine. Install the bolts. Torque the base bolts to 89 inch lbs. (10 Nm); the support bolt to 18 ft. lbs. (25 Nm).

➡ **If not secure within four minutes, the sealant must be removed and the sealing area cleaned. To clean the sealing area, follow the directions provided on the packaging of the silicone gasket remover and the metal surface prep. Failure to follow this procedure can cause future oil leakage.**

53. Apply a bead of silicone gasket and sealant at the crankshaft rear seal retainer-to-cylinder block surface.

➡ **If not secure within four minutes, the sealant must be removed and the sealing area cleaned. To clean the sealing area, follow the directions provided on the packaging of the silicone gasket remover and the metal surface prep. Failure to follow this procedure can cause future oil leakage.**

54. Apply a bead of silicone gasket and sealant at the engine front cover-to-cylinder block surface.

➡ **Be sure to tighten the bolts in two steps.**

55. Tighten the bolts in the sequence shown.

    a. Step 1: Tighten to 20 Nm (15 ft. lbs.).

    b. Step 2: Tighten an additional 90 degrees.

56. Position the right engine mount on the engine and install the bolts. Tighten the bolts to 80 Nm (59 ft. lbs.).

57. Using a crane, lower the engine and remove the crane.

58. Install the starter motor.

59. Position the oil cooler and install the threaded insert.

60. Install a new oil filter.

➡ **Make sure that both the engine and transmission mounts are correctly seated before installing the mounting nuts.**

61. Install the engine mount nuts.
62. Install the transmission mount nuts.
63. Position the wire harness back and install the bracket and nut on the engine front cover stud.
64. Install the ground strap to the engine.
65. Position the exhaust Y-pipe and install the nuts on the exhaust manifold studs.
66. Install the exhaust Y-pipe flange nuts.
67. Connect the heated oxygen sensor electrical connectors.
68. Remove the special tool.
69. Install the fan blade assembly and the fan clutch to the coolant pump pulley.
70. Install the fan blade and the fan clutch.
71. Use a suitable tool to hold the coolant pump pulley.
72. Using a suitable tool, tighten the fan clutch nut to the coolant pump pulley. Torque to 41 ft. lbs. (55 Nm).
73. Install the alternator.
74. Install the accelerator cable bracket.
75. Position the accelerator cable through the bulkhead. Install the bolts.
76. Connect the accelerator cable routing clip to the dash panel stud.
77. Connect the accelerator cable to the accelerator pedal and shaft.
78. Install a new throttle body adapter gasket.
79. Position the throttle body adapter back.
80. Position the throttle body adapter.
81. Install the four bolts. Torque to 89 inch lbs. (10 Nm).
82. Connect the EGR valve vacuum hose.
83. Connect the EGR vacuum regulator solenoid connections.
84. Install the differential pressure feedback EGR system bracket.
85. Connect the exhaust manifold to EGR valve tube upper fitting.
86. Tighten both fittings starting at the top in three steps.
   a. Hand-tighten.
   b. Tighten the upper fitting to 40–60 Nm (30–44 ft. lbs.).
   c. Tighten the lower fitting to 40–60 Nm (30–44 ft. lbs.).
87. Connect the IAC electrical connector.
88. Install the IAC fresh air tube and insert it in the routing clip.
89. Connect the TP sensor electrical connector.
90. Install the accelerator cable bracket.
91. Position the accelerator cable bracket.

92. Install the bolts.
93. Connect the throttle body cam.
94. Connect the accelerator cable.
95. If equipped, connect speed control cable.
96. Install the return spring.
97. Install the accelerator control splash shield.
98. Install the air cleaner outlet tub.

✷✷ CAUTION

The oil pump must be primed prior to starting the engine.

99. Fill the engine with clean engine oil.

AUTOMATIC TRANSMISSION

1. Before servicing the vehicle, refer to the Precautions Section.
2. Disconnect the battery ground cable.
3. Remove the air cleaner outlet pipe.
4. Remove the accelerator cable snow shield.
5. Disconnect the throttle body cam.
6. Disconnect the accelerator cable.
7. If equipped, disconnect the speed control cable.
8. Remove the throttle return spring.
9. Position the accelerator cable bracket aside.
10. Remove the bolts.
11. Position the bracket and cables aside.
12. Release the idle air control (IAC) fresh air tube from the routing clip and remove it from the vehicle.
13. Disconnect the IAC electrical connector.
14. If equipped, disconnect the differential pressure feedback exhaust gas recirculation (EGR) system electrical connector.
15. Disconnect the throttle position (TP) sensor electrical connector.
16. If equipped, disconnect the differential pressure feedback EGR system vacuum hoses.
17. If equipped, disconnect the exhaust manifold to EGR valve tube fittings and remove the tube.
18. If equipped, remove the differential pressure feedback EGR system bracket.
19. If equipped, disconnect the EGR valve vacuum hose.
20. If equipped, disconnect the EGR vacuum regulator solenoid electrical and vacuum connections.
21. Position aside the throttle body adapter.
22. Remove the four bolts.
23. Position aside the throttle body adapter.

24. Remove and discard the throttle body adapter gasket.
25. Remove the alternator.

✷✷ WARNING

The large clutch assembly nut has a right-handed thread and must be rotated counterclockwise to remove it.

26. Using a suitable tool, remove the fan and fan clutch from the coolant pump pulley.
27. Install the modular engine support bracket on the engine using the alternator mounting holes.
28. Raise and support the vehicle.
29. Remove the engine mount nuts.

✷✷ WARNING

Damage to the turbine shaft speed/output shaft speed (TSS/OSS) sensors may occur and cause the transmission or torque converter operational concerns if the transmission is raised prior to removing TSS/OSS.

✷✷ WARNING

Sensor bosses must be cleaned prior to removal and then plugged to prevent contamination from damaging internal components.

30. If the vehicle is equipped with TSS and OSS, remove the sensors and install plugs in the transmission.
31. Loosely install the nuts and the crossmember.
32. Loosely install the left crossmember nuts.
33. Loosely install the right crossmember nuts.
34. Loosely install the crossmember bolts.
35. Remove the jack from under the transmission.
36. Lower the vehicle.
37. Raise the engine using the Heavy Duty Engine Support.
38. Raise and support the vehicle.
39. Drain the engine oil and remove the oil bypass filter.
40. Disconnect the heated oxygen sensor (HO2S) electrical connectors.
41. Remove the exhaust Y-pipe flange nuts.
42. Remove the nuts and the exhaust Y-pipe.
43. Remove the flywheel inspection plate.

✳✳ WARNING

Support the transmission on the oil pan rails only or internal transmission damage can occur.

44. Support the transmission with a transmission jack.
45. Remove the transmission mount nuts.
46. Raise the transmission.
47. Remove the bolts and partially lower the oil pan.
48. Remove the bolts from the oil pump screen and pickup tube and let them drop into the oil pan.
49. Remove the rear bolt and the oil pump screen and pickup tube and let them drop into the oil pan
50. Remove the oil pan and the oil pan gasket from the rear of the engine.

Clean the oil pan thoroughly and inspect the oil pan gasket.

✳✳ WARNING

Do not use metal scrapers, wire brushes, power abrasive discs or other abrasive means to clean the sealing surface. These tools cause scratches and gouges, which make leak paths. Use a plastic scraping tool to remove all traces of sealant.

51. Clean the mating surfaces for the oil pan with silicone gasket remover and metal surface prep. Follow the directions on the packaging.

To install:

➡ **The oil pump screen and pickup tube must be in the oil pan when the oil pan is positioned in the vehicle.**

52. Position the oil pan gasket and the oil pan in the vehicle from the rear of the engine.

✳✳ WARNING

Make sure the O-ring is in place and not damaged. A missing or damaged O-ring can cause foam in the lubrication system, low oil pressure and severe engine damage.

➡ **Clean and inspect the mating surfaces and install a new O-ring. Lubricate with clean engine oil.**

53. Position the oil pump screen and pickup tube on the engine.
54. Install the bolts.

10 Nm (89 lb-in)

25 Nm (18 lb-ft)

67197-EFSE-G103

Oil pump installation—4.6L or 5.4L engine

➡ **If not secure within four minutes, the sealant must be removed and the sealing area cleaned. To clean the sealing area, follow the directions provided on the packaging of the silicone gasket remover and the metal surface prep. Failure to follow this procedure can cause future oil leakage.**

55. Apply a bead of silicone gasket and sealant at the crankshaft rear seal retainer-to-cylinder block surface.

➡ **If not secure within four minutes, the sealant must be removed and the sealing area cleaned. To clean the sealing area, follow the directions provided on the packaging of the silicone gasket remover and the metal surface prep. Failure to follow this procedure can cause future oil leakage.**

56. Apply a bead of silicone gasket and sealant at the engine front cover-to-cylinder block surface.

➡ **Be sure to tighten the bolts in two steps.**

57. Tighten the bolts in the sequence shown.
    a. Stage 1: Tighten to 20 Nm (15 ft. lbs.).
    b. Stage 2: Tighten an additional 90 degrees.
58. Lower the transmission.
59. Remove the High Lift Transmission Jack.
60. Install the flywheel inspection plate.
61. Install the oil bypass filter.
62. Lower the vehicle.
63. Align the engine mount studs and lower the engine, using the Heavy Duty Engine Support.
64. Remove the Heavy Duty Engine Support.
65. Raise and support the vehicle.
66. Install the engine mount nuts.
67. Install a suitable transmission jack and support the transmission.
68. Remove the right crossmember nuts.

69. Remove the left crossmember nuts.
70. Remove the nuts and the crossmember.
71. If the vehicle is equipped with TSS and turbine/OSS, remove the plugs and install the sensors in the transmission.
72. Position the exhaust Y-pipe and install the nuts on the exhaust manifold studs.
73. Install the exhaust Y-pipe flange nuts.
74. Connect the heated oxygen sensor electrical connectors.
75. Lower the vehicle.
76. Remove the modular engine support bracket from the engine.

✳✳ WARNING

The large clutch assembly nut has a right-handed thread and must be rotated counterclockwise to remove it.

77. Using a suitable tool, install the fan and fan clutch to the coolant pump pulley.
78. Install the alternator.
79. Install the throttle body adapter.
80. Position a new throttle body adapter gasket.
81. Position the throttle body adapter.
82. Install the four bolts. Torque to 89 inch lbs. (10 Nm).
83. If equipped, connect the EGR vacuum regulator solenoid electrical and vacuum connections.
84. If equipped, connect the EGR valve vacuum hose.
85. If equipped, install the differential pressure feedback EGR system bracket. Position the bracket. Install the bolts.
86. If equipped, install the exhaust manifold to EGR valve tube.
    a. Hand-tighten the fittings.
    b. Tighten the upper fitting to 40–60 Nm (30–44 ft. lbs.).
    c. Tighten the lower fitting to 40–60 Nm (30–44 ft. lbs.).
87. If equipped, connect the differential pressure feedback EGR system vacuum lines.

If equipped, connect the differential pressure feedback EGR system electrical connector.

88. Connect the TP sensor electrical connector.
89. Connect the IAC electrical connector.
90. Install the IAC fresh air tube and insert it in the routing clip.
91. Install the accelerator cable bracket.
92. Position the bracket and cables.
93. Install the bolts.
94. Connect the throttle body cam.

95. Connect the accelerator cable.
96. If equipped, connect the speed control cable.
97. Install the throttle return spring.
98. Install the accelerator cable snow shield.
99. Install the air cleaner outlet pipe.
100. Connect the battery cable.

✳✳ WARNING

The oil pump must be primed prior to starting the engine.

101. Fill the engine with clean engine oil.

2005 F-250/350 5.4L Engine

1. Before servicing the vehicle, refer to the Precautions Section.
2. Remove the transmission.
3. Drain the engine oil.
4. Disconnect the engine oil temperature (EOT) sensor electrical connector.
5. Remove the nut and the starter wiring harness support bracket.
6. Remove the nut and position the transmission oil cooler lines and bracket aside.
7. Remove the 16 bolts and lower the oil pan onto the crossmember.
8. Remove the 3 bolts and position the oil pump screen and pickup tube into the oil pan.

➡**Be careful when removing the oil pan gasket. It is reusable.**

9. Remove the oil pan and the oil pump screen and pickup tube.
10. Inspect the oil pan gasket for damage. If damaged, discard the oil pan gasket and the oil pan-to-oil pump gaskets. Discard the oil pump screen and pickup tube O-ring seal.
To install:

✳✳ WARNING

Do not use metal scrapers, wire brushes, power abrasive discs or other abrasive means to clean the sealing surfaces. These tools cause scratches and gouges, which make leak paths. Use a plastic scraping tool to remove all traces of old sealant.

11. Inspect the oil pan. Clean the mating surface for the oil pan with silicone gasket remover and metal surface prep. Follow the directions on the packaging.
12. Position the oil pump screen and pickup tube in the oil pan and position the oil pan into the vehicle.

✳✳ WARNING

Make sure to install a new O-ring seal. A missing or damaged O-ring seal can cause foam in the lubrication system, low oil pressure and severe engine damage.

➡**Clean and inspect the mating surfaces and install a new O-ring seal. Lubricate the O-ring seal with clean engine oil prior to installation.**

13. Position the oil pump screen and pickup tube and install the bolts.
14. Tighten the oil pump screen and pickup tube-to-oil pump bolts to 10 Nm (89 inch lbs.).
15. Tighten the oil pump screen and pickup tube-to-spacer bolt to 25 Nm (18 ft. lbs.).

➡**If not secured within 4 minutes, the sealant must be removed and the sealing area cleaned. To clean the sealing area, use silicone gasket remover and metal surface prep. Follow the directions on the packaging. Failure to follow this procedure can cause future oil leakage.**

16. Apply silicone gasket and sealant at the crankshaft rear seal retainer plate-to-cylinder block sealing surface.

➡**If not secured within 4 minutes, the sealant must be removed and the sealing area cleaned. To clean the sealing area, use silicone gasket remover and metal surface prep. Follow the directions on the packaging. Failure to follow this procedure can cause future oil leakage.**

17. Apply silicone gasket and sealant at the engine front cover-to-cylinder block sealing surface.
18. Install the oil pan gasket and the oil pan and loosely install the 16 bolts.
19. Tighten the bolts in 3 stages, in the sequence shown.
- Stage 1: Tighten to 2 Nm (18 inch lbs.).
- Stage 2: Tighten to 20 Nm (15 ft. lbs.).
- Stage 3: Tighten an additional 60 degrees.
20. Install the wire harness bracket and the bolt. Tighten to 10 Nm (89 inch lbs.).
21. Position the starter wiring harness support bracket and install the nut. Tighten to 10 Nm (89 inch lbs.).
22. Connect the engine oil temperature (EOT) sensor electrical connector.

23. Install the transmission.
24. Fill the crankcase with clean engine oil.

2002–03 F-Series 6.8L Engine

✳✳ CAUTION

Fuel injection systems remain under pressure, even after the engine has been turned OFF. The fuel system pressure must be relieved before disconnecting any fuel lines. Failure to do so may result in fire and/or personal injury.

1. Before servicing the vehicle, refer to the Precautions Section.
2. Remove or disconnect the following:
- Negative battery cable
- Fuel system pressure
- Coolant
- Upper radiator hose from the intake manifold
- Air cleaner outlet tube
- Accelerator cable from the bracket and the throttle body cam
- Speed control actuator cable from the throttle body
- All vacuum hoses, fuel lines and electrical wires from the throttle body and intake manifold
- Brake booster vacuum hose bracket
- EGR valve-to-exhaust manifold tube and disconnect the vacuum line
- Connector and vacuum line from the Engine Vacuum Regulator (EVR) solenoid
- The 4 bolts and the throttle body adapter
- Upper fan shroud mounting screws and position the shroud toward the engine.
- The alternator and install the Modular Engine Support Bracket on the engine using the mounting holes.
- Lower engine mount-to-frame nuts
- Turbine Shaft Speed (TSS) and Output Shaft Speed (OSS) sensors from the transmission. Plug the openings.
3. Lower the vehicle to the floor and raise the engine using a hoist attached to the support bracket.
4. Install an engine support fixture with a J hook to keep the engine raised, then remove the hoist.
5. Remove or disconnect the following:
- Engine oil and filter
- Dual converter Y-pipe and the flywheel inspection cover

- Driveshaft and the 2 transmission mounting nuts
- Transmission. Be sure to support the transmission along the rails of the pan to avoid damage.
- Oil pan mounting bolts and partially lower the pan.
- Oil pump pick-up tube and screen assembly and allow it to drop into the pan.
- Oil pan towards the rear of the vehicle

To install:

✳✳ WARNING

To prevent possible oil leaks, use only a plastic scraper to clean the oil pan mounting surface.

6. Clean the oil pan-to-engine mounting surface.

7. Place the oil pump pick-up tube and screen assembly in the oil pan, then position the pan and gasket near the engine.

8. Install the oil pump pick-up tube and screen assembly.

   a. Tighten the nut to 15–22 ft. lbs. (20–30 Nm).

   b. Tighten the 2 bolts to 71–106 inch lbs. (8–12 Nm).

9. Apply a bead of silicone sealant to the areas where the front cover and rear bearing cap meet the engine block.

10. Install the oil pan. Tighten the bolts in sequence using 3 steps as follows:

- 18 inch lbs. (2 Nm)
- 15 ft. lbs. (20 Nm)
- Plus an additional 90°

11. Lower the transmission and install the 2 mounting nuts. Tighten the nuts to 60–80 ft. lbs. (81–108 Nm).

12. Install the driveshaft and the TSS and OSS sensors.

13. Install the flywheel cover and dual converter Y-pipe.

14. Install the oil bypass filter.

15. Lower the vehicle and remove the engine support fixture.

16. Install the engine mounting nuts. Tighten the nuts to 66 ft. lbs. (90 Nm).

17. Remove the modular engine support bracket and install the alternator.

18. Install the fan shroud.

19. Use a new gasket and install the throttle body adapter.

20. Tighten the bolts in 2 steps:

- 71–88 inch lbs. (8–10 Nm).
- 85–95 degrees.

21. Connect the EVR solenoid harness and vacuum line.

22. Attach the vacuum line to the EGR valve.

23. Install the EGR valve-to-exhaust manifold tube. Tighten the fittings to 55 ft. lbs. (41 Nm).

24. Install the EGR transducer.

25. Install all remaining components in the reverse of the removal.

20-30 Nm (15-22 lb/ft)

8-12 Nm (71-106 lb/in)

7924FG89

Oil pump pick-up tube and screen assembly—2002–03 6.8L engine

9358FG15

To prevent leaks, tighten the oil pan bolts in the order shown—2002–03 6.8L engine

✳✳ WARNING

Operating the engine without the proper amount and type of engine oil will result in severe engine damage.

26. Fill the engine with SAE 5W30 oil.

27. Fill and bleed the cooling system.

2004 F-Series 6.8L Engine

MANUAL TRANSMISSION

1. Before servicing the vehicle, refer to the Precautions Section.

2. With the vehicle in **Neutral**, position it on a hoist.

3. Remove the air cleaner outlet tube.

4. Remove the accelerator cable splash shield.

5. Disconnect the accelerator and speed control cables and remove the return spring.

6. Position the accelerator cable bracket aside.

7. Remove the bolts.

8. Position the bracket and cables aside.

9. Disconnect the positive crankcase ventilation (PCV) tube and the vacuum hoses.

10. Disconnect the vacuum hose.

11. Disconnect the idle air control (IAC) motor electrical connector and the bypass hose.

12. Remove the routing clip for the IAC hose.

13. Disconnect the throttle position (TP) sensor electrical connector.

14. Position aside the throttle body adapter.

15. Remove the four adapter bolts.

16. Position aside the throttle body adapter.

17. Remove and discard the throttle body adapter gasket.

18. Hold the accelerator pedal in place and disconnect the accelerator cable from the accelerator pedal and shaft.

19. Disconnect the accelerator cable routing clip from the dash panel stud.

20. Remove the bolts and pull the accelerator cable through the bulkhead.

21. Remove the alternator.

> ※※ **WARNING**
>
> **The large clutch assembly nut has a right-hand thread and must be rotated counterclockwise to remove it.**

22. Using the special tools, remove the cooling fan and the fan clutch from the coolant pump pulley.

23. Install a suitable tool on the engine using the alternator mounting holes.

24. Install a suitable tool.

25. Disconnect the heated oxygen sensor (HO_2S) electrical connectors.

26. Remove the exhaust Y-pipe flange nuts.

27. Remove the nuts and the exhaust Y-pipe.

28. Remove the engine mount nuts.

29. Remove the starter motor.

30. Remove the stud bolt and the ground strap from the engine.

31. Remove the nut and bracket from the engine front cover stud and position aside the wire harness.

32. Remove the transmission mount nuts.

33. Drain the engine oil and remove the oil filter.

34. Remove the oil cooler threaded insert and position the oil cooler aside.

> ※※ **WARNING**
>
> **Use care when raising the engine, so that the engine does not touch the body.**

35. Using a suitable tool, raise the engine.

36. Remove the bolts and remove the right engine mount from the engine.

37. Remove the bolts and partially lower the oil pan.

38. Remove the oil pump screen and pickup tube and spacer and let them drop into the oil pan.

➡ **To remove the oil pan, position the rear of the pan to the left side of the vehicle and rotate slightly to the right while pulling outward.**

39. Remove the oil pan and the oil pan gasket.

40. Clean the inside of the oil pan thoroughly and inspect the oil pan gasket.

> ※※ **WARNING**
>
> **Do not use metal scrapers, wire brushes, power abrasive discs or other abrasive means to clean the sealing surfaces. These tools cause scratches and gouges, which make leak paths. Use a plastic scraping tool to remove all traces of sealant.**

41. Clean the mating surfaces for the oil pan and the engine with silicone gasket remover and metal surface prep. Follow the directions on the packaging.

To install:

➡ **The oil pump screen and pickup tube must be in the oil pan when the oil pan is positioned in the vehicle.**

➡ **To install the oil pan, position the oil pan to the left rear side of the engine and rotate slightly while sliding the front of the pan toward the right side of the engine.**

42. Position the oil pan gasket and the oil pan in the vehicle from the rear of the engine.

> ※※ **WARNING**
>
> **Make sure the O-ring is in place and not damaged. A missing or damaged O-ring can cause foam in the lubrication system, low oil pressure and severe engine damage.**

➡ **Clean and inspect the mating surfaces and install a new O-ring. Lubricate with clean engine oil.**

43. Place the oil pump screen and pickup tube in the pan.

44. Install and tighten the spacer. Torque to 18 ft. lbs. (25 Nm).

45. Position the oil pump screen and pickup tube on the engine.

46. Install the bolts. Torque the tube base bolts to 89 inch lbs. (10 Nm); the bracket bolt to 18 ft. lbs. (25 Nm).

47. Apply a bead of silicone gasket and sealant at the crankshaft rear seal retainer-to-cylinder block surface.

➡ **If not secure within four minutes, the sealant must be removed and the sealing area cleaned. To clean the sealing area, follow the directions provided on the packaging of the silicone gasket remover and the metal surface prep. Failure to follow this procedure can cause future oil leakage.**

48. Apply a bead of silicone gasket and sealant at the engine front cover-to-cylinder block surface.

➡ **Be sure to tighten the bolts in two steps.**

49. Position the oil pan. Tighten the bolts in the sequence shown.

   a. Step 1: Tighten to 20 Nm (15 ft. lbs.).

   b. Step 2: Tighten an additional 90 degrees.

50. Position the right engine mount on the engine and install the bolts. Tighten the bolts to 80 Nm (59 ft. lbs.).

51. Lower the engine and remove the lifting tool.

52. Install the starter motor.

53. Position the oil cooler and install the threaded insert. Torque to 43 ft. lbs. (58 Nm).

54. Install a new oil filter.

➡ **Make sure that both the engine and transmission mounts are correctly seated before installing the mounting nuts.**

Oil pump pick-up tube installation—2004 6.8L engine

67197-EFSE-G104

Oil pan torque sequence—2004–05 6.8L engine

67197-EFSE-G105

55. Install the engine mount nuts. Torque to 66 ft. lbs. (90 Nm).

56. Install the transmission mount nuts. Torque to 70 ft. lbs. (95 Nm).

57. Position the wire harness and install the bracket and nut on the engine front cover stud.

58. Install the ground strap to the engine.

59. Position the exhaust Y-pipe and install the nuts on the exhaust manifold studs.

60. Install the exhaust Y-pipe flange nuts.

61. Connect the heated oxygen sensor electrical connectors.

62. Remove the special tool.

63. Using the special tools, install the cooling fan and the fan clutch to the coolant pump pulley.

64. Install the alternator.

65. Install the accelerator cable bracket.

66. Position the accelerator cable through the bulkhead.

67. Install the bolts.

68. Connect the accelerator cable routing clip to the dash panel stud.

69. Connect the accelerator cable to the accelerator pedal and shaft.

70. Position back the throttle body adapter.

71. Position a new throttle body adapter gasket.

72. Position the throttle body adapter.

73. Install four bolts and tighten in two steps.

   a. Step 1: Tighten to 9 Nm (80 inch lbs.).

   b. Step 2: Tighten an additional 90 degrees.

74. Connect the TP sensor electrical connector.

75. Connect the IAC motor bypass hose and connector.

76. Install the routing clip for the IAC hose.

77. Connect the vacuum hose.

78. Connect the PCV tube and the vacuum hoses.

79. Position the accelerator bracket.

80. Install the bolts.

81. Connect the accelerator and speed control cables and return spring.

82. Install the accelerator cable splash shield.

83. Install the air cleaner outlet tube.

✳✳ WARNING

The oil pump must be primed prior to starting the engine.

84. Fill the engine with clean engine oil.

AUTOMATIC TRANSMISSION

1. Before servicing the vehicle, refer to the Precautions Section.

2. With the vehicle in **Neutral**, position it on a hoist.

3. Disconnect the battery ground cable.

4. Remove the air cleaner outlet tube.

5. Remove the accelerator cable splash shield.

6. Disconnect the accelerator and speed control cables and return spring.

7. Position the accelerator cable bracket aside.

8. Remove the bolts. Position the bracket and cables aside.

9. Disconnect the positive crankcase ventilation (PCV) tube and the vacuum hoses.

10. Disconnect the vacuum hose.
11. Disconnect the throttle positive (TP) sensor electrical connector.
12. Disconnect the idle air control (IAC) motor electrical connector and the bypass hose.
13. Remove the bypass hose routing clip.
14. Position aside the throttle body adapter.
15. Remove the four bolts.
16. Position aside the throttle body adapter.
17. Remove and discard the throttle body adapter gasket.

✳✳ WARNING

The large clutch assembly nut has a right-hand thread and must be rotated counterclockwise to remove it.

18. Using a suitable tool, remove the fan and fan clutch from the coolant pump pulley.
19. Remove the alternator.
20. Install a suitable tool on the engine, using the alternator mounting holes.
21. Raise and support the vehicle.
22. Remove the engine mount nuts.

✳✳ WARNING

Damage to the TSS/OSS may occur and cause the transmission or torque converter operational concerns if the transmission is raised prior to removing TSS/OSS.

✳✳ WARNING

Sensor bosses must be cleaned prior to removal and then plugged to prevent contamination from damaging internal components.

23. If the vehicle is equipped with turbine shaft speed (TSS) and output shaft sensors (OSS), remove the sensors and install plugs in the transmission.
24. Loosely install the nuts and the crossmember.
25. Loosely install the left crossmember nuts.
26. Loosely install the right crossmember nuts.
27. Loosely install the crossmember bolts.
28. Remove the jack assembly supporting the transmission.
29. Raise the engine, using a suitable tool.

30. Drain the engine oil and remove the oil bypass filter.
31. Remove the bolt and position the oil cooler aside.
32. Disconnect the heated oxygen sensor electrical connectors.
33. Remove the exhaust Y-pipe flange nuts.
34. Remove the nuts and the exhaust Y-pipe.
35. Remove the flywheel inspection plate.

✳✳ WARNING

Support the transmission on the oil pan rails only or internal transmission damage can occur.

36. Support the transmission with a transmission jack.
37. Remove the transmission mount nuts.
38. Using the transmission jack, raise the transmission.
39. Remove the nut and detach the transmission cooling tube bracket.
40. Remove the bolts and partially lower the oil pan.
41. Remove the bolts retaining the oil pump screen cover and pickup tube.
42. Remove the pickup tube and spacer and let them drop into the oil pan.
43. Remove the oil pan and the oil pan gasket from the rear of the engine.

✳✳ WARNING

Do not use metal scrapers, wire brushes, power abrasive discs or other abrasive means to clean the sealing surface. These tools cause scratches and gouges, which make leak paths. Use a plastic scraping tool to remove all traces of sealant.

44. Clean the mating surfaces and thoroughly clean the oil pan.
To install:

➡Clean and inspect the oil pump screen cover and tube mating surfaces and install a new O-ring. Lubricate the O-ring with clean engine oil.

➡The oil pump screen and pickup tube must be in the oil pan when the oil pan is positioned in the vehicle.

45. Position the oil pan gasket and the oil pan in the vehicle from the rear of the engine.
46. Position the oil pump screen and pickup tube.

47. Install and tighten the spacer. Torque to 18 ft. lbs. (25 Nm).
48. Insert the oil pump screen pickup tube into the oil pump and position the support bracket over the spacer.
49. Install and tighten the bolts. Torque the pump base bolts to 89 inch lbs. (10 Nm); the bracket bolt to 18 ft. lbs. (25 Nm).

➡If not secure within four minutes, the sealant must be removed and the sealing area cleaned. To clean the sealing area, follow the directions provided on the packaging of the silicone gasket remover and the metal surface prep. Failure to follow this procedure can cause future oil leakage.

50. Apply a bead of silicone gasket and sealant where the front cover and rear crankshaft seal retainer fit to the engine block.

➡Be sure to tighten the bolts in two steps.

51. Position the oil pan. Tighten the bolts in the sequence shown.
 a. Step 1: Tighten to 20 Nm (15 ft. lbs.).
 b. Step 2: Tighten an additional 90 degrees.
52. Attach the transmission cooling tube bracket and install the nut.
53. Lower the transmission.
54. Loosely install the transmission mount nuts.
55. Install the flywheel inspection plate. Torque to 25 ft. lbs. (34 Nm).
56. Position the engine oil cooler adapter and install the bolt. Torque to 43 ft. lbs. (58 Nm).
57. Install the oil bypass filter.
58. Align the engine mount studs and lower the engine.
59. Remove the lifting tool.
60. Install the engine mount nuts. Torque to 66 ft. lbs. (90 Nm).
61. Install a suitable transmission jack and support the transmission.
62. Install the right crossmember nuts.
63. Install the left crossmember nuts.
64. Install the nuts and the crossmember.
65. If the vehicle is equipped with TSS and turbine/OSS, remove the plugs and install the sensors in the transmission.
66. Position the exhaust Y-pipe and install the nuts on the exhaust manifold studs.
67. Install the exhaust Y-pipe flange nuts. Torque to 39 ft. lbs. (40 Nm).
68. Connect the heated oxygen sensor electrical connectors.

> ❉❉ **WARNING**
>
> **The large clutch assembly nut has a right-hand thread and must be rotated counterclockwise to remove it.**

69. Using a suitable tool, install the fan and fan clutch to the coolant pump pulley. Torque to 84 –112 ft. lbs. (113–153 Nm).
70. Install the alternator.
71. Position the throttle body adapter.
72. Position a new throttle body adapter gasket.
73. Position the throttle body adapter.
74. Install the four bolts and tighten in two steps.
   a. Step 1: Tighten to 9 Nm (80 inch lbs.).
   b. Step 2: Tighten an additional 90 degrees.
75. Connect the IAC electrical connector and the bypass hose.
76. Install the bypass hose routing clip.
77. Connect the vacuum hose.
78. Connect the PCV tube and the vacuum hoses.
79. Connect the TP sensor electrical connector.
80. Install the accelerator cable bracket.
81. Position the bracket and cables. Install the bolts.
82. Connect the accelerator and speed control cables and return spring.
83. Install the accelerator cable splash shield.
84. Install the air cleaner outlet tube.
85. Connect the battery cable.

> ❉❉ **WARNING**
>
> **The oil pump must be primed prior to starting the engine.**

86. Fill the engine with clean engine oil.

2005 F-Series 6.8L Engine

MANUAL TRANSMISSION

1. Before servicing the vehicle, refer to the Precautions Section.
2. Remove the flywheel.
3. Remove the nut and the starter wiring harness support bracket.
4. Remove the 18 bolts and partially lower the oil pan.
5. Remove the 3 bolts and position the oil pump screen and pickup tube into the oil pan.

➥Be careful when removing the oil pan gasket. It is reusable.

6. Remove the oil pan and the oil pump screen and pickup tube.
7. Inspect the oil pan gasket for damage. If damaged, discard the oil pan gasket and the oil pan-to-oil pump gaskets. Discard the oil pump screen and pickup tube O-ring seal.

To install:

> ❉❉ **WARNING**
>
> **Do not use metal scrapers, wire brushes, power abrasive discs or other abrasive means to clean the sealing surfaces. These tools cause scratches and gouges, which make leak paths. Use a plastic scraping tool to remove all traces of old sealant.**

8. Inspect the oil pan. Clean the mating surface for the oil pan with silicone gasket remover and metal surface prep. Follow the directions on the packaging.
9. Position the oil pump screen and pickup tube in the oil pan and position the oil pan into the vehicle.

> ❉❉ **WARNING**
>
> **Make sure to install a new O-ring seal. A missing or damaged O-ring seal can cause foam in the lubrication system, low oil pressure and severe engine damage.**

➥Clean and inspect the mating surfaces and install a new O-ring seal. Lubricate the O-ring seal with clean engine oil prior to installation.

10. Position the oil pump screen and pickup tube and install the bolts.
11. Tighten the oil pump screen and pickup tube-to-oil pump bolts to 10 Nm (89 inch lbs.).
12. Tighten the oil pump screen and pickup tube-to-spacer bolt to 25 Nm (18 ft. lbs.).

➥If not secured within 4 minutes, the sealant must be removed and the sealing area cleaned. To clean the sealing area, use silicone gasket remover and metal surface prep. Follow the directions on the packaging. Failure to follow this procedure can cause future oil leakage.

13. Apply silicone gasket and sealant at the crankshaft rear seal retainer plate-to-cylinder block sealing surface.

➥If not secured within 4 minutes, the sealant must be removed and the seal-

ing area cleaned. To clean the sealing area, use silicone gasket remover and metal surface prep. Follow the directions on the packaging. Failure to follow this procedure can cause future oil leakage.

14. Apply silicone gasket and sealant at the engine front cover-to-cylinder block sealing surface.
15. Install the oil pan gasket and the oil pan and loosely install the 18 bolts.
16. Tighten the bolts in 3 stages, in the sequence shown.
   • Stage 1: Tighten to 2 Nm (18 inch lbs.).
   • Stage 2: Tighten to 20 Nm (15 ft. lbs.).
   • Stage 3: Tighten an additional 60 degrees.
17. Install the transmission cooler tube support bracket and the nut. Tighten to 10 Nm (89 inch lbs.).
18. Install the flywheel.

AUTOMATIC TRANSMISSION

1. Before servicing the vehicle, refer to the Precautions Section.
2. With the vehicle in NEUTRAL, position it on a hoist.
3. Remove the air cleaner outlet tube.
4. Remove the generator.
5. Remove the cooling fan shroud.
6. Remove the starter.
7. Install the an engine crane and support the engine.

> ❉❉ **WARNING**
>
> **Only use hand tools when removing the transmission mount-to-crossmember nuts or damage to the transmission mount can occur.**

8. Remove the transmission mount-to-crossmember nuts.

> ❉❉ **WARNING**
>
> **Only use hand tools when installing the right engine mount nut or damage to the engine mount can occur.**

9. Remove the 6 engine mount nuts.
10. Remove the 4 exhaust manifold-to-catalytic converter nuts. To install, tighten to 40 Nm (30 ft. lbs.).
11. Remove the nut and the starter wiring harness support bracket.
12. Remove the nut and position the transmission oil cooler lines and bracket aside.

❈ WARNING

Damage to the TSS/OSS may occur and cause the transmission or torque converter operational concerns if the transmission is raised prior to removing TSS/OSS.

❈❈ WARNING

Sensor bosses must be cleaned prior to removal and then plugged to prevent contamination from damaging internal components.

13. If the vehicle is equipped with turbine shaft speed (TSS) and output shaft sensors (OSS), remove the sensors and install plugs in the transmission.

14. Using the crane, raise the engine.

❈ WARNING

Support the transmission on the oil pan rails only or internal transmission damage can occur.

15. Install a suitable transmission jack and raise the transmission.

16. Remove the 2 bolts and the flywheel inspection plate.

17. Remove the 18 bolts and partially lower the oil pan.

18. Remove the 3 bolts and position the oil pump screen and pickup tube into the oil pan.

➡**Be careful when removing the oil pan gasket. It is reusable.**

19. Remove the oil pan and the oil pump screen and pickup tube.

20. Inspect the oil pan gasket for damage. If damaged, discard the oil pan gasket and the oil pan-to-oil pump gaskets. Discard the oil pump screen and pickup tube O-ring seal.

To install:

❈❈ WARNING

Do not use metal scrapers, wire brushes, power abrasive discs or other abrasive means to clean the sealing surfaces. These tools cause scratches and gouges, which make leak paths. Use a plastic scraping tool to remove all traces of old sealant.

21. Inspect the oil pan. Clean the mating surface for the oil pan with silicone gasket remover and metal surface prep. Follow the directions on the packaging.

22. Position the oil pump screen and pickup tube in the oil pan and position the oil pan into the vehicle.

❈❈ WARNING

Make sure to install a new O-ring seal. A missing or damaged O-ring seal can cause foam in the lubrication system, low oil pressure and severe engine damage.

➡**Clean and inspect the mating surfaces and install a new O-ring seal. Lubricate the O-ring seal with clean engine oil prior to installation.**

23. Position the oil pump screen and pickup tube and install the bolts.

24. Tighten the oil pump screen and pickup tube-to-oil pump bolts to 10 Nm (89 inch lbs.).

25. Tighten the oil pump screen and pickup tube-to-spacer bolt to 25 Nm (18 ft. lbs.).

➡**If not secured within 4 minutes, the sealant must be removed and the sealing area cleaned. To clean the sealing area, use silicone gasket remover and metal surface prep. Follow the directions on the packaging. Failure to follow this procedure can cause future oil leakage.**

26. Apply silicone gasket and sealant at the crankshaft rear seal retainer plate-to-cylinder block sealing surface.

➡**If not secured within 4 minutes, the sealant must be removed and the sealing area cleaned. To clean the sealing area, use silicone gasket remover and metal surface prep. Follow the directions on the packaging. Failure to follow this procedure can cause future oil leakage.**

27. Apply silicone gasket and sealant at the engine front cover-to-cylinder block sealing surface.

28. Install the oil pan gasket and the oil pan and loosely install the 18 bolts.

29. Tighten the bolts in 3 stages, in the sequence shown.

- Stage 1: Tighten to 2 Nm (18 inch lbs.).
- Stage 2: Tighten to 20 Nm (15 ft. lbs.).
- Stage 3: Tighten an additional 60 degrees.

30. Install the flexplate inspection cover and the 2 bolts. Tighten to 34 Nm (25 ft. lbs.).

31. Lower the transmission.

32. Align the engine mount studs and lower the engine. Remove the special tool.

❈❈ WARNING

Only use hand tools when installing the right engine mount nut or damage to the engine mount can occur.

➡**Align the engine-to-transmission dowels before installing the engine mount bolt.**

33. Install the 6 engine mount nuts. Tighten to 175 Nm (129 ft. lbs.).

❈❈ WARNING

Only use hand tools when removing the transmission mount-to-crossmember nuts or damage to the transmission mount can occur.

34. Install the transmission mount-to-crossmember nuts. Tighten to 25 Nm (18 ft. lbs.).

35. Install the transmission cooler tube support bracket and the nut. Tighten to 10 Nm (89 inch lbs.).

36. Install the 4 exhaust manifold flange nuts. Tighten to 40 Nm (30 ft. lbs.).

37. If the vehicle is equipped with TSS and turbine/OSS, remove the plugs and install the sensors in the transmission.

38. Install the starter.

39. Install the generator.

40. Install the cooling fan shroud.

41. Install the air cleaner outlet tube.

42. Fill the crankcase with clean engine oil.

E-Series 4.6L and 5.4L Engines

1. Before servicing the vehicle, refer to the Precautions Section.

2. Remove the intake manifold.

3. Remove the fan shroud and the engine cooling fan.

4. Remove the retainers and the shield.

5. Remove the nut and position the transmission fluid filler tube aside.

6. Release the wiring retainers from the heater outlet tube bracket.

7. Release the knock sensor electrical connector retainer.

8. Release the engine harness routing clip retainer.

9. Disconnect the hose from the heater outlet tube.

10. Remove the heater outlet tube studs.

11. Remove the heater outlet tube.

12. Remove the two transmission-to-engine bolts.

Assemble the lifting tools, 303-F047 and 303-F694, or equivalent— E-Series 4.6L and 5.4L Engines

Install the lifting tools— E-Series 4.6L and 5.4L Engines

Install the support tools— E-Series 4.6L and 5.4L Engines

13. Assemble the lifting tools, 303-F047 and 303-F694, or equivalent.

14. Install the lifting tools.

15. Install the special tool and support the engine.

16. Raise and support the vehicle.

17. Drain the engine oil, and remove the oil filter.

18. Remove the four engine mount nuts.

19. Remove the flexplate inspection plate.

20. Lower the vehicle.

21. Using the Heavy Duty Engine Support, raise the engine 10¼ inches from the crankshaft pulley to the lower edge of the No. 1 crossmember.

22. Raise the vehicle.

23. Remove the bolts and partially lower the oil pan.

24. Remove the bolts retaining the oil pump screen cover and pickup tube and let the bolts drop into the oil pan.

25. Remove the oil pan at the rear of the engine.

26. Clean the oil pan thoroughly and inspect the oil pan gasket.

27. Clean the mating surfaces for the oil pan with silicone gasket remover and metal surface prep. Follow the directions on the packaging.

To install:
All engines

❋❋ WARNING

Make sure the O-ring is in place and not damaged. A missing or damaged O-ring can cause foam in the lubrication system, low oil pressure and severe engine damage.

➡ Clean and inspect the mating surfaces and install a new O-ring. Lubricate the O-ring with clean engine oil.

28. Install the oil pump screen and pickup tube. Torque the tube base bolts to 89 inch lbs. (10 Nm); the bracket bolt to 18 ft. lbs. (25 Nm).

➡ If not secured within four minutes, the sealant must be removed and the sealing area cleaned. To clean the sealing area, use silicone gasket remover and metal surface prep. Follow the directions on the packaging. Failure to follow this procedure can cause future oil leakage.

29. Apply a bead of silicone gasket and sealant at the crankshaft rear seal retainer-to-cylinder block surface.

30. Apply a bead of silicone gasket and sealant at the engine front cover-to-cylinder block surface.

31. Position the pan. Tighten the bolts in three steps, in the sequence shown.

    a. Stage 1: Tighten to 2 Nm (18 inch lbs.).

    b. Stage 2: Tighten to 20 Nm (15 ft. lbs.).

    c. Stage 3: Tighten an additional 90 degrees.

32. Install a new oil filter.

33. Lower the vehicle.

34. Lower the engine and remove the special tools.

35. Raise the vehicle.

36. Install the four engine mount nuts. Torque to 66 ft. lbs. (90 Nm).

37. Install the flexplate inspection plate. Torque to 25 ft. lbs. (34 Nm).

38. Lower the vehicle.

39. Install the transmission-to-engine bolts. Torque to 44 ft. lbs. (60 Nm).

Oil pan torque sequence—E-Series w/4.6L or 5.4L engine

➡ **Do not reuse the O-ring seals. Lubricate the new O-ring seals with clean engine coolant before installing the heater outlet tube.**

40. Insert the heater outlet tube over the new seals.

Romeo engines (4.6L)

41. Install the heater outlet tube studs.
42. Position the transmission fluid filler tube and install the nut.

Windsor engine (5.4L)

43. Install the heater outlet tube studs.
44. Position the transmission fluid filler tube and install the nut.

All engines

45. Connect the hose to the heater outlet tube.
46. Install the wiring retainers in the heater outlet tube bracket.
47. Install the knock sensor electrical connector retainer.
48. Install the engine harness routing clip retainer.
49. Install the retainers and the shield.
50. Install the fan shroud and the engine cooling fan.
51. Install the intake manifold.

> ❄❄ **WARNING**
>
> **The oil pump must be primed prior to starting the engine.**

52. Fill the engine with clean engine oil.
53. Start the engine and check for leaks.

E-Series 6.8L Engines

1. Before servicing the vehicle, refer to the Precautions Section.
2. With the vehicle in **Neutral**, position it on a hoist.
3. Remove the intake manifold.
4. Remove the engine cooling fan and fan shroud.
5. Remove the retainers and the shield.
6. Remove the nut attaching the transmission oil level tube.
7. Disconnect the wiring harness retainer.
8. Remove the stud bolts and position the heater outlet tube aside.
9. Remove the two transmission-to-engine bolts.
10. Disconnect the oil fill tube from the right valve cover.
11. Remove the bolt and the oil fill tube.
12. Assemble the special tools.
13. Install the special tools.
14. Install the special tool and support the engine.
15. Drain the engine oil and remove the oil filter.

16. Remove the right and left nuts retaining the front engine support insulators to the front engine support bracket.
17. Remove the flywheel inspection plate.
18. Using the special tool, raise the engine.
19. Remove the bolts and partially lower the oil pan.
20. Remove the bolts retaining the oil pump screen cover and pickup tube and allow it to drop into the oil pan. Also remove the tube and spacer and allow it to drop into the oil pan.
21. Remove the oil pan.

> ❄❄ **WARNING**
>
> **Do not use metal scrapers, wire brushes, power abrasive discs or other abrasive means to clean the sealing surfaces. These tools cause scratches and gouges, which make leak paths. Use a plastic scraping tool to remove all traces of sealant.**

22. Clean the mating surfaces for the oil pan and the engine with silicone gasket remover and metal surface prep. Follow the directions on the packaging.

To install:

➡ **The oil pump screen and pickup tube must be in the oil pan when the oil pan is positioned in the vehicle.**

23. Position the oil pan gasket and the oil pan in the vehicle from the rear of the engine.

> ❄❄ **WARNING**
>
> **Make sure the O-ring is in place and not damaged. A missing or damaged O-ring can cause foam in the lubrication system, low oil pressure and severe engine damage.**

➡ **Clean and inspect the mating surfaces and install a new O-ring. Lubricate the O-ring with clean engine oil.**

24. Position the oil pump screen cover and pickup tube.
25. Install and tighten the spacer. Torque to 18 ft. lbs. (25 Nm).
26. Insert the oil pump screen pickup tube into the oil pump and position the support bracket over the spacer.
27. Install and tighten the bolts. Torque the tube base bolts to 89 inch lbs. (10 Nm); the bracket bolt to 18 ft. lbs. (25 Nm).
28. Apply a bead of silicone gasket and sealant at the crankshaft rear seal retainer-to-cylinder block surface.

➡ **If not secure within four minutes, the sealant must be removed and the sealing area cleaned. To clean the sealing area, follow the directions provided on the packaging of the silicone gasket remover and the metal surface prep. Failure to follow this procedure can cause future oil leakage.**

29. Apply a bead of silicone gasket and sealant at the engine front cover-to-cylinder block surface.

➡ **Position the oil pan. Tighten the bolts in three steps.**

30. Tighten the bolts in the sequence shown.
   a. Step 1: Tighten to 2 Nm (18 inch lbs.).
   b. Step 2: Tighten to 20 Nm (15 ft. lbs.).
   c. Step 3: Tighten an additional 90 degrees.
31. Lower the engine and remove the special tools.
32. Install the flywheel inspection plate.
33. Tighten the engine support insulators.

➡ **Do not reuse the O-ring seal.**

34. Install the new O-ring seal and the heater outlet tube.
35. Lubricate the O-ring seal with engine coolant.
36. Hand-tighten the heater outlet tube upper stud and install the lower stud. Tighten the studs to 40 Nm (30 ft. lbs.).
37. Remove the nut attaching the transmission oil level tube.
38. Connect the wiring harness retainer.
39. Connect the oil fill tube to the right valve cover.
40. Install the oil fill tube support bracket bolt.
41. Install the transmission-to-engine bolts.
42. Install the shield and the retainers.
43. Install the intake manifold.
44. Fill the engine with clean engine oil.
45. Install the engine cooling fan and fan shroud.
46. Fill and bleed the engine cooling system.

Oil Pump

REMOVAL & INSTALLATION

4.2L Engine

2002–04

1. Before servicing the vehicle, refer to the Precautions Section.

2. Disconnect the negative battery cable.

3. Raise and support the front of the vehicle.

4. Drain the engine oil into a suitable container and dispose.

5. Remove the oil filter.

6. Remove the 6 oil pump bolts, then remove the oil pump drive gear, the oil pump driven gear, the oil pump O-ring and the oil pump itself. Discard the used oil pump O-ring.

7. Inspect the oil pump components for damage or excessive wear.

8. Check the oil pump face for warpage with a flat edge ruler. The face cannot exhibit more than 0.00157 inch (0.04mm) of distortion.

9. Remove the plug over the oil pressure relief; valve.

10. Remove the oil pressure relief valve ball and spring, then clean the parts.

To install:

➡**Lubricate the parts with clean engine oil before assembly.**

11. Assemble the oil pressure relief valve ball and spring with a new plug.

12. Install the oil pump, along with a new O-ring, the oil pump driven gear and the drive gear. Install and tighten the 6 oil pump mounting bolts to the torque value specifications indicated in the illustration.

13. Apply a film of clean engine oil to the rubber O-ring on the new filter, then install the filter onto the filter mount.

14. Install the oil pan drain plug.

15. Lower the vehicle.

16. Fill the engine with the correct amount and type of new engine oil.

17. Connect the negative battery cable.

18. Start the engine and make certain that the oil light on the instrument panel extinguishes within 6–8 seconds after the engine starts.

2005

1. Before servicing the vehicle, refer to the Precautions Section.

2. With the vehicle in NEUTRAL, position it on a hoist.

3. Drain the engine oil. Install the drain plug when finished.

4. Remove and discard the oil filter.

5. Remove the 6 bolts and the oil pump.

✳✳ WARNING

Do not reuse the oil pump seal. A new service seal must be installed.

Tighten the oil pump mounting bolts to the specifications shown—4.2L engines

Oil pump fasteners—2005 4.2L engine

6. Remove and discard the oil pump seal.

7. Remove and inspect the oil pump drive gear, the driven gear and the cover. Install new components if necessary.

8. Inspect the face of the oil pump for flatness.

To install:

9. Install a new service oil pump seal on the oil pump.

➡**Lubricate the parts with clean engine oil before assembly.**

10. Install the oil pump drive gear, the driven gear and the cover.

11. Position the oil pump and install the bolts. Tighten the bolts as shown.

• Bolts numbered 1: Tighten to 10 Nm (89 inch lbs.).

10 Nm (89 lb-in)

9358FG11

Tighten the oil pump mounting bolts in the sequence shown—2002–03 4.6L and 5.4L engines

06017-F150-G100

Pickup tube and spacer—2004–05 4.6L and 5.4L engines; 2002–05 6.8L Engine

06017-F150-G101

Oil pump fasteners—2004–05 4.6L and 5.4L engines; 2002–05 6.8L Engine

- Bolts numbered 2: Tighten to 25 Nm (18 ft. lbs.).
12. Install a new oil filter.
13. Fill the engine with clean engine oil.

2002–03 4.6L and 5.4L Engines

1. Before servicing the vehicle, refer to the Precautions Section.
2. Disconnect the negative battery cable.
3. Remove or disconnect the following:
- Crankshaft sprocket
- Oil pan
- Bolts and the oil pump
To install:
4. Clean and inspect the mating surfaces.
5. Install or connect the following:
- Oil pump, and the bolts loosely. Tighten the bolts in the sequence illustrated to 89 inch lbs. (10 Nm).
- Oil pan
- Crankshaft sprocket

2004–05 4.6L and 5.4L Engines; 2002–05 6.8L Engine

1. Before servicing the vehicle, refer to the Precautions Section.
2. With the vehicle in NEUTRAL, position it on a hoist.
3. Remove the timing drive components.
4. Remove the oil pan.
5. Remove the 3 bolts, the oil pump screen and pickup tube and the spacer.
6. Remove the 3 bolts and the oil pump.
To install:

➡**Lubricate the new O-ring seal with clean engine oil.**

7. Clean and inspect the mating surfaces and install a new O-ring seal.

✳✳ WARNING

The oil pump must be primed prior to starting the engine.

8. Install the oil pump and loosely install the 3 bolts. Tighten the bolts in the sequence shown to 10 Nm (89 inch lbs.).

✳✳ WARNING

Make sure the O-ring is in place and not damaged. A missing or damaged O-ring can cause foam in the lubrication system, low oil pressure and severe engine damage.

➡**Install a new O-ring and lubricate with clean engine oil.**

9. Install the pick up tube spacer, oil pump screen and pickup tube and the 3 bolts. Tighten bolt 1 to 25 Nm (18 ft. lbs.). Tighten bolts 2 to 10 Nm (89 inch lbs.).

10. Install the oil pan.

11. Install the timing drive components.

Rear Main Seal

REMOVAL & INSTALLATION

2002–05 4.2L Engines

2002–04 4.6L, 5.4L and 6.8L Engines without Retainer Plate

If the crankshaft rear oil seal replacement is the only operation being performed, it can be done in the vehicle as detailed in the following procedure. If the oil seal is being replaced in conjunction with a rear main bearing replacement, the engine must be removed from the vehicle and installed on a work stand.

1. Before servicing the vehicle, refer to the Precautions Section.

2. Disconnect the negative battery cable.

3. Remove the transmission from the vehicle.

4. Remove the flywheel/flexplate. If equipped, remove the crankshaft oil slinger from the crankshaft.

Rear main seal removal—2002–05 4.2L Engines; 2002–04 4.6L, 5.4L and 6.8L Engines without Retainer Plate

Assembling the rear main seal and tools 2002–05 4.2L Engines; 2002–04 4.6L, 5.4L and 6.8L Engines without Retainer Plate

Install tools 303-518 and 303-S560 on the crankshaft 2002–05 4.2L Engines; 2002–04 4.6L, 5.4L and 6.8L Engines without Retainer Plate

Installing the seal 2002–05 4.2L Engines; 2002–04 4.6L, 5.4L and 6.8L Engines without Retainer Plate

5. Use an awl to punch 2 holes in the crankshaft rear oil seal. Punch the holes on opposite sides of the crankshaft and just above the bearing cap-to-cylinder block split line.

6. Install a sheet metal screw in each hole. Use 2 small prybars to pry against both screws at the same time to remove the crankshaft rear oil seal. It may be necessary to place small blocks of wood against the cylinder block to provide a fulcrum point for the prybars. Use caution throughout this procedure to avoid scratching or otherwise damaging the crankshaft oil seal surface.

7. Clean the oil seal recess in the cylinder block and main bearing cap.

To install:

8. Clean, inspect and polish the rear oil seal rubbing surface on the crankshaft.

9. Coat the new oil seal and the crankshaft with a light film of engine oil.

10. Start the seal in the recess with the seal lip facing forward and install it with a seal driver. Keep the tool straight with the centerline of the crankshaft and install the seal until the tool contacts the cylinder block surface. Remove the tool and inspect the seal to be sure it was not damaged during installation.

11. If equipped, install the crankshaft oil slinger.

12. Position the flywheel on the crankshaft flange. Coat the threads of the flywheel attaching bolts with Loctite® and install the bolts. Tighten the bolts in sequence across from each other. See the Torque Specifications Chart.

13. Install the transmission, following the recommended procedure.

14. Install the negative battery cable.

Flywheel or flexplate torque sequence

2003–05 4.6L, 5.4L and 6.8L Engines with Retainer Plate

1. Before servicing the vehicle, refer to the Precautions Section.

2. Remove the oil pan.

3. Remove the flexplate.

4. Using a puller set, remove the crankshaft rear oil seal slinger.

5. Using a puller set, remove the crankshaft rear seal.

6. Remove the six bolts and the crankcase rear seal retainer.

To install:

✳ CAUTION

Do not use metal scrapers, wire brushes, power abrasive discs or other abrasive means to clean the sealing surfaces. These tools cause scratches and gouges which make leak paths. Use a plastic scraping tool to remove all traces of old sealant.

➡ Clean the sealing surfaces with silicone gasket remover and metal surface prep.

7. Follow the directions on the packaging. Failure to follow this procedure can cause future oil leakage.

8. Clean and inspect the mating surface.

✳ CAUTION

Do not use metal scrapers, wire brushes, power abrasive discs or other abrasive means to clean the sealing surfaces. These tools cause scratches and gouges which make leak paths. Use a plastic scraping tool to remove all traces of old sealant.

➡ If the rear crankshaft seal retaining plate is not secured within four minutes, the sealant must be removed and the sealing area cleaned. To clean the

67197-EFSE-G110

The silicone must be applied on the groove along the retainer plate—2003–05 4.6L, 5.4L and 6.8L Engines with Retainer Plate

67197-EFSE-G111

Tighten the bolts in the sequence shown—2003–05 4.6L, 5.4L and 6.8L Engines with Retainer Plate

67197-EFSE-G112

Using the special tools, install the crankshaft rear seal—2003–05 4.6L, 5.4L and 6.8L Engines with Retainer Plate

67197-EFSE-G113

Using the special tools, install the crankshaft rear oil slinger—2003–05 4.6L, 5.4L and 6.8L Engines with Retainer Plate

sealing area, follow the directions provided on the packaging of the silicone gasket remover and the metal surfacer prep. Failure to follow this procedure can cause future oil leakage.

➡ The silicone must be applied on the groove along the retainer plate.

9. Apply a 4 mm (0.16 in.) bead of silicone gasket and sealant around the rear oil seal retainer sealing surface.

10. Install the rear seal retainer and loosely install the six bolts. Tighten the bolts in the sequence shown.

➡ Lubricate the inner lip of the rear crankshaft seal with clean engine oil.

11. Using the special tools, install the crankshaft rear seal.

12. Using the special tools, install the crankshaft rear oil slinger.

13. Install the oil pan.

14. Install the flexplate.

Timing Chain, Sprockets and Front Cover

REMOVAL & INSTALLATION

4.2L Engine

1. Before servicing the vehicle, refer to the Precautions Section.

2. Remove or disconnect the following:
   - Negative battery cable
   - Accessory drive belt
   - Coolant
   - Radiator, fan blade assembly and fan shroud
   - Water pump
   - Exhaust Gas Recirculation (EGR) valve vacuum hose
   - EGR tube upper fitting
   - EGR valve and adapter assembly
   - Wiring harness from the heater water outlet tube
   - Heater water outlet bolt and position the outlet tube aside
   - Camshaft Position (CMP) sensor electrical harness connector and mark the position of the connector for proper installation.

3. Rotate the crankshaft until the Top Dead Center (TDC) timing mark lines up with the timing mark.
   - The 2 bolts retaining the CMP and remove the CMP from the camshaft synchronizer.
   - Camshaft synchronizer adjustment bolt and remove the camshaft synchronizer.

➡ The oil pump drive shaft may come out with the camshaft synchronizer.

   - Engine oil
   - Crankshaft pulley and damper
   - Engine oil pan
   - The 2 engine front cover stud bolts, front cover bolt and cap screw
   - Engine front cover and gasket off the dowels and discard the gasket.
   - CMP sensor drive gear bolt and drive gear

4. Be sure the timing marks and keyways align.

5. Compress and install a retaining pin to hold the timing chain tensioner.

After removing the CMP drive gear, be sure the timing marks (A) and keyways (B) are aligned—4.2L engine

6. Slide both sprockets and timing chain forward and remove as an assembly.

7. Remove the 3 bolts retaining the timing chain tensioner and remove the tensioner.

8. Check the timing chain and sprockets for excessive wear. Replace if necessary.

To install:

9. Before installation, clean and inspect all parts. Clean the gasket material from the engine oil pan, cylinder block and engine front cover.

10. Place the timing chain tensioner in position and install the 3 retaining bolts. Tighten the bolts to 72–120 inch lbs. (8–14 Nm).

11. Verify that the balance shaft timing gears are in correct alignment.

12. Slide both sprockets and the timing chain onto the camshaft and crankshaft with the timing marks aligned. Install the CMP sensor drive gear and bolt. Tighten the bolt to 30–36 ft. lbs. (40–50 Nm).

13. Remove the retaining pin.

14. Inspect the engine front cover seal for wear or damage and replace if necessary.

15. Install or connect the following:
- Engine front cover gasket and front cover onto the guide studs
- The 2 engine front cover stud bolts, front cover bolt and cap screw. Tighten the bolts to 15–22 ft. lbs. (20–30 Nm).
- Engine oil pan
- Crankshaft damper and pulley

✳✳ WARNING

A Synchro Positioning Tool must be used prior to installation. Failure to using this procedure will result in the fuel system being out of time, possibly causing engine damage.

16. Install Synchro Positioning Tool T89P-12200-A, or equivalent, on the camshaft synchronizer by rotating the tool until it engages the notch in the housing.

17. Install the camshaft synchronizer housing assembly so that the arrow on the tool is 54 degrees from the centerline of the engine.

18. Install the adjustment bolt and tighten to 15–22 ft. lbs. (21–30 Nm).

19. Remove the tool and position the CMP sensor. Install the 2 bolts and 40–70 inch lbs. (5–8 Nm) and install the CMP electrical harness connector.

Install the camshaft synchronizer (A) in the orientation shown (white arrow points to front of engine)—4.2L engines

20. Install or connect the following:
- Water pump
- Fan shroud, fan blade assembly and radiator
- Accessory drive belt

21. Fill the crankcase with the correct amount and type of engine oil.

22. Fill and bleed the engine cooling system.

23. Connect the negative battery cable.

24. Start the engine and check for coolant and oil leaks.

25. Road test the vehicle and check for proper engine operation.

DYNAMIC BALANCE SHAFT

1. Before servicing the vehicle, refer to the Precautions Section.

2. Remove the timing chain.

3. Remove the engine dynamic balance shaft.

4. Remove the bolts.

5. Remove the balance shaft drive gear, thrust plate and engine dynamic balance shaft as an assembly.

To install:

➡️If correctly aligned, the engine dynamic balance shaft keyway will be at 12 o'clock and the camshaft keyway will be at 6 o'clock on the camshaft.

6. Turn the camshaft so that the timing mark is at 12 o'clock and install the engine dynamic balance shaft assembly into the cylinder block. Turn the engine balance shaft driven gear so that the timing mark on the camshaft gear lines up with the timing mark on the engine balance shaft drive gear.

7. Install the bolts.

8. Install the timing chain.

**2002–05 4.6L Engine
2002–03 F-150 5.4L Engine
2004 F-150 Heritage Edition 5.4L Engine
2004 F-250/350
2002–05 E-Series 5.4L Engine**

1. Before servicing the vehicle, refer to the Precautions Section.

2. With the vehicle in NEUTRAL, position it on a hoist.

3. Disconnect the battery ground cable.

4. Drain the engine oil.

5. Remove the cooling fan.

6. Remove the right valve cover.

7. Remove the left valve cover.

8. Rotate the tensioner clockwise and remove the drive belt. Using special tool

06017-F150-G107

Using special tool 303-D055, remove the bolt and washer—4.6L engine, 2002–03 F-150 5.4L Engine, 2004 F-150 Heritage Edition 5.4L Engine, 2004 F-250/350, 2002–05 E-Series 5.4L Engine

06017-F150-G108

Using special tool 303-099, remove the crankshaft pully—all engines

303-D055, remove the bolt and washer and discard the bolt.

❄️❄️ WARNING

This bolt is torque-to-yield and cannot be reused.

9. Using special tool 303-099, remove the crankshaft pulley.

10. Using special tool 303-107, remove the crankshaft front seal.

11. Remove the 4 coolant pump pulley bolts and the coolant pump pulley.

12. Remove the accessory drive belt idler pulley bolt and the accessory drive belt idler pulley.

13. Using special tool 211-016, remove the power steering pump pulley.

14. Disconnect the power steering pressure tube. Drain the power steering fluid into a suitable container.

15. Remove the nut and position aside the power steering pressure tube.

16. Remove the power steering pump bolts and position the power steering pump assembly aside.

17. Disconnect the camshaft position (CMP) sensor electrical connector.

06017-F150-G109

Using special tool 303-107, remove the crankshaft front seal—all engines

06017-F150-G110

Using special tool 211-016, remove the power steering pump pulley—4.6L engine, 2002–03 F-150 5.4L Engine, 2004 F-150 Heritage Edition 5.4L Engine, 2004 F-250/350, 2002–05 E-Series 5.4L Engine

06017-F150-G111

Crankshaft sensor ring—all engines

18. Remove the CMP sensor bolt and the CMP sensor.

19. Remove the left radio interference capacitor nut and position the left radio interference capacitor aside.

20. Remove the nut and the upper radiator hose bracket.

21. Remove the right radio interference capacitor nut and position the right radio interference capacitor aside.

22. Disconnect the crankshaft position (CKP) sensor electrical connector.

23. Remove the 4 oil pan bolts.

24. Remove the nut and the A/C manifold and tube assembly support bracket.

25. If equipped, remove the nut and the transmission cooler tube support bracket.

26. Remove the bolts and the studs.

⁂ WARNING

Do not use metal scrapers, wire brushes, power abrasive discs or other abrasive means to clean the sealing surfaces. These tools cause scratches and gouges which make leak paths. Use a plastic scraping tool to remove all traces of old sealant.

27. Remove the engine front cover from the front cover-to-cylinder block dowel.

28. Remove the engine front cover gaskets.

29. Clean the mating surfaces with silicone gasket remover and metal surface prep. Follow the directions on the packaging.

30. Inspect the mating surfaces.

31. Remove the CKP sensor bolt and the CKP sensor.

32. Remove the crankshaft sensor ring from the crankshaft.

33. Rotate the crankshaft until the timing mark on the right camshaft sprocket is approximately at the 11 o'clock position and the timing mark on the left camshaft sprocket is approximately at the 12 o'clock position.

34. Install the special tools on the camshaft as shown.

⊕ WARNING

If one or both of the tensioner mounting bolts are loosened or removed, the tensioner-sealing bead must be inspected for seat integrity. If cracks, tears or separation from the tensioner body or permanent compression of the seal bead is observed, install a new tensioner.

35. Remove the bolts (1). Remove the timing chain tensioners (2). Remove the timing chain tensioner arms (3).

36. Remove the timing chains and crankshaft sprocket.

37. Remove the bolts (1). Remove the left timing chain guide (2).

38. Remove the bolts (3). Remove the right timing chain guide (4).

To install:

⊕ WARNING

Timing chain procedures must be followed exactly or damage to valves and pistons will result.

06017-F150-G112

Rotate the crankshaft until the timing mark on the right camshaft sprocket is approximately at the 11 o'clock position and the timing mark on the left camshaft sprocket is approximately at the 12 o'clock position—4.6L engine, 2002–03 F-150 5.4L Engine, 2004 F-150 Heritage Edition 5.4L Engine, 2004 F-250/350, 2002–05 E-Series 5.4L Engine

06017-F150-G113

Install the special tools on the camshaft—4.6L engine

06017-F150-G114

Removing the tensioners and tensioner arms—4.6L engine, 2002–03 F-150 5.4L Engine, 2004 F-150 Heritage Edition 5.4L Engine, 2004 F-250/350, 2002–05 E-Series 5.4L Engine

06017-F150-G136

Install the special tool on the camshaft—2002–03 F-150 5.4L Engine, 2004 F-150 Heritage Edition 5.4L Engine, 2004 F-250/350, 2002–05 E-Series 5.4L Engine

06017-F150-G115

Timing chain and sprocket removal—4.6L engine, 2002–03 F-150 5.4L Engine, 2004 F-150 Heritage Edition 5.4L Engine, 2004 F-250/350, 2002–05 E-Series 5.4L Engine

06017-F150-G116

Timing chain guides—4.6L engine, 2002–03 F-150 5.4L Engine, 2004 F-150 Heritage Edition 5.4L Engine, 2004 F-250/350, 2002–05 E-Series 5.4L Engine

06017-F150-G117

Compress the tensioner plunger, using a vise—all engines

06017-F150-G118

Install a retaining clip on the tensioner to hold the plunger in during installation—all engines

06017-F150-G119

Copper links—all engines

06017-F150-G120

Install the crankshaft sprocket, making sure the flange faces forward—all engines

⁂ WARNING

Prior to installation, inspect the tensioner-sealing bead for seal integrity. If cracks, tears, separation from the tensioner body or permanent compression of the seal bead is observed, install a new tensioner.

39. Compress the tensioner plunger, using a vise.

40. Install a retaining clip on the tensioner to hold the plunger in during installation.

41. If the copper links are not visible, mark one link on one end and one link on the other end, and use as timing marks.

42. Install the crankshaft sprocket, making sure the flange faces forward.

43. Position the left timing chain guide.

44. Install and tighten the left bolts. Tighten to 10 Nm (89 inch lbs.).

45. Position the right timing chain guide.

46. Install and tighten the right bolts. Tighten to 10 Nm (89 inch lbs.).

❊❊ WARNING

Unless otherwise instructed, do not rotate either the crankshaft or the camshafts, when the timing chains are removed and the cylinder heads are installed. Severe piston and valve damage will occur.

➡The number one cylinder is at top dead center (TDC) when the stud on the

06017-F150-G121

Using the special tool, position the crankshaft so the number one cylinder is at TDC—4.6L engine, 2002–03 F-150 5.4L Engine, 2004 F-150 Heritage Edition 5.4L Engine, 2004 F-250/350, 2002–05 E-Series 5.4L Engine

06017-F150-G137

Install this tool on 5.4L engines—2002–03 F-150 5.4L Engine, 2004 F-150 Heritage Edition 5.4L Engine, 2004 F-250/350, 2002–05 E-Series 5.4L Engine

06017-F150-G122

Position the left (inner) timing chain on the crankshaft sprocket, aligning the copper (marked) link with the timing mark on the sprocket—4.6L engine, 2002–03 F-150 5.4L Engine, 2004 F-150 Heritage Edition 5.4L Engine, 2004 F-250/350, 2002–05 E-Series 5.4L Engine

06017-F150-G123

Install the left timing chain on the camshaft sprocket, aligning the copper (marked) link with the timing marks on the sprocket—4.6L engine, 2002–03 F-150 5.4L Engine, 2004 F-150 Heritage Edition 5.4L Engine, 2004 F-250/350, 2002–05 E-Series 5.4L Engine

06017-F150-G124

Position the left timing chain tensioner arm on the dowel pin and install the left timing chain tensioner—4.6L engine, 2002–03 F-150 5.4L Engine, 2004 F-150 Heritage Edition 5.4L Engine, 2004 F-250/350, 2002–05 E-Series 5.4L Engine

Position the right (outer) timing chain on the crankshaft sprocket, aligning the copper (marked) link with the timing mark on the sprocket—4.6L engine, 2002–03 F-150 5.4L Engine, 2004 F-150 Heritage Edition 5.4L Engine, 2004 F-250/350, 2002–05 E-Series 5.4L Engine

Install the right timing chain on the camshaft sprocket, aligning the copper (marked) link with the timing marks on the sprocket—4.6L engine, 2002–03 F-150 5.4L Engine, 2004 F-150 Heritage Edition 5.4L Engine, 2004 F-250/350, 2002–05 E-Series 5.4L Engine

Position the right timing chain tensioner arm on the dowel pin and install the right timing chain tensioner—4.6L engine, 2002–03 F-150 5.4L Engine, 2004 F-150 Heritage Edition 5.4L Engine, 2004 F-250/350, 2002–05 E-Series 5.4L Engine

engine block fits into the slot in the handle of the special tool.

47. Using the special tool, position the crankshaft so the number one cylinder is at TDC.

48. Remove the Crankshaft Holding Tool.

49. Position the left (inner) timing chain on the crankshaft sprocket, aligning the copper (marked) link with the timing mark on the sprocket.

50. Install the left timing chain on the camshaft sprocket, aligning the copper (marked) link with the timing marks on the sprocket.

➡ **The left timing chain tensioner arm has a bump near the dowel hole for identification.**

51. Position the left timing chain tensioner arm on the dowel pin and install the left timing chain tensioner. Tighten to 25 Nm (18 ft. lbs.).

52. Remove the retaining clip from the left timing chain tensioner.

53. Position the right (outer) timing chain on the crankshaft sprocket, aligning the copper (marked) link with the timing mark on the sprocket.

55. Install the right timing chain on the camshaft sprocket, aligning the copper (marked) link with the timing marks on the sprocket.

56. Position the right timing chain tensioner arm on the dowel pin and install the right timing chain tensioner. Tighten to 25 Nm (18 ft. lbs.).

57. Remove the retaining clip from the right timing chain tensioner.

58. Make sure that the copper (marked) chain links are lined up with the dots on the crankshaft sprockets and the camshaft sprocket.

59. Remove the special tools from the camshaft.

60. Install the crankshaft sensor ring on the crankshaft.

61. Install the CKP sensor and the CKP sensor bolt. Tighten to 10 Nm (89 inch lbs.).

❊❊ WARNING

Do not use metal scrapers, wire brushes, power abrasive discs or other abrasive means to clean the sealing surfaces. These tools cause scratches and gouges which make leak paths. Use a plastic scraping tool to remove all traces of old sealant.

Make sure that the copper (marked) chain links are lined up with the dots on the crankshaft sprockets and the camshaft sprocket—4.6L engine, 2002–03 F-150 5.4L Engine, 2004 F-150 Heritage Edition 5.4L Engine, 2004 F-250/350, 2002-05 E-Series 5.4L Engine

06017-F150-G128

➡If the engine front cover is not secured within 4 minutes, the sealant must be removed and the sealing area cleaned. To clean the sealing area, use silicone gasket remover and metal surface prep. Follow the directions on the packaging. Failure to follow this procedure can cause future oil leakage.

➡Make sure that the engine front cover gasket is in place on the engine front cover before installation.

62. Apply a bead of silicone gasket and sealant along the cylinder head-to-cylinder block surface and the oil pan-to-cylinder block surface, at the locations shown.

63. Install the engine front cover with the engine front cover gasket on the front cover-to-cylinder block dowel and loosely install the bolts.

20 Crankshaft position (CKP) sensor electrical connector	26 Engine front cover upper stud bolts (4 required)
21 Oil pan bolts (4 required)	27 Engine front cover
22 Engine front cover bolt	28 Engine front cover gaskets (3 required)
23 Engine front cover lower stud bolt	29 CKP sensor bolt
24 Engine front cover lower stud bolt	30 CKP sensor
25 Engine front cover bolts (8 required)	31 CKP sensor O-ring seal

Front cover and related parts—4.6L engine, 2002–03 F-150 5.4L Engine, 2004 F-150 Heritage Edition 5.4L Engine, 2004 F-250/350, 2002-05 E-Series 5.4L Engine

06017-F150-G129

06017-F150-G130

Apply a bead of silicone gasket and sealant along the cylinder head-to-cylinder block surface and the oil pan-to-cylinder block surface, at the locations indicated—4.6L engine, 2002–03 F-150 5.4L Engine, 2004 F-150 Heritage Edition 5.4L Engine, 2004 F-250/350, 2002–05 E-Series 5.4L Engine

64. Tighten the engine front cover fasteners in the sequence shown.

65. Loosely install the oil pan-to-front cover bolts, then tighten the bolts in 2 steps, in the sequence shown.
   - Step 1: Tighten to 20 Nm (15 ft. lbs.).
   - Step 2: Tighten an additional 60 degrees.

66. If equipped, install the transmission cooler tube support bracket and nut. Tighten to 10 Nm (89 inch lbs.).

67. Install the A/C manifold and tube assembly support bracket and nut. Tighten to 10 Nm (89 inch lbs.).

68. Connect the CKP sensor electrical connector.

69. Position the right radio interference capacitor and install the nut. Tighten to 10 Nm (89 inch lbs.).

70. Position the upper radiator hose bracket and install the nut. Tighten to 10 Nm (89 inch lbs.).

71. Install the left radio interference capacitor and install the left radio interference capacitor nut. Tighten to 10 Nm (89 inch lbs.).

72. Install the CMP sensor and the bolt. Tighten to 10 Nm (89 inch lbs.).

73. Connect the CMP sensor electrical connector.

74. Position the power steering pump and install the bolts. Tighten to 25 Nm (18 ft. lbs.).

75. Using special tool 211-D207, install a new O-ring seal on the pressure line fitting.

76. Connect the power steering pressure tube. Tighten to 65 Nm (48 ft. lbs.).

77. Position the power steering pressure tube support bracket and install the nut. Tighten to 10 Nm (89 inch lbs.).

✳✳ WARNING

If the pulley has been removed and installed twice, install a new power steering pump pulley.

78. Using special tool 211-185, install the power steering pump pulley. Inspect the pulley for paint marks in the web area near the hub. If there are 2 paint marks, install a new pulley. If there is 1 paint mark or none at all, use a pencil to mark the web area of the pulley near the hub.

79. Install the accessory drive belt idler pulley and the 3 bolts. Tighten to 25 Nm (18 ft. lbs.).

80. Install the coolant pump pulley and the 4 bolts. Tighten to 25 Nm (18 ft. lbs.).

1 Bolt, Hex Flange Head Pilot, M8 x 1.25 x 53

2 Bolt, Hex Flange Head Pilot, M8 x 1.25 x 53

3 Bolt, Hex Flange Head Pilot, M8 x 1.25 x 53

4 Bolt, Hex Flange Head Pilot, M8 x 1.25 x 53

5 Bolts, Hex Flange Head Pilot, M8 x 1.25 x 53

6 Stud Hex Shoulder Pilot, M8 x 1.25 x 50— M6 x 1 x 10

7 Stud and Washer, Hex Head Pilot, M8 x 1.25 — M6 x 1 x 86.35

8 Bolt, Hex Flange Head Pilot, M8 x 1.25 x 53

9 Bolt, Hex Flange Head Pilot, M8 x 1.25 x 53

10 Bolt, Hex Flange Head Pilot, M8 x 1.25 x 53

11 Bolt, Hex Head Pilot, M8 x 1.25 x 53

12 Stud Hex Shoulder Pilot, M8 x 1.25 x 1.25 x 91.1

13 Stud Hex Shoulder Pilot, M8 x 1.25 x 1.25 x 91.1

06017-F150-G131

Front cover fastener identification—4.6L engine, 2002–03 F-150 5.4L Engine, 2004 F-150 Heritage Edition 5.4L Engine, 2004 F-250/350, 2002–05 E-Series 5.4L Engine

Oil pan-to-front cover bolt torque sequence—4.6L engine, 2002–03 F-150 5.4L Engine, 2004 F-150 Heritage Edition 5.4L Engine, 2004 F-250/350, 2002–05 E-Series 5.4L Engine

Power steering pressure line fitting connection—4.6L engine, 2002–03 F-150 5.4L Engine, 2004 F-150 Heritage Edition 5.4L Engine, 2004 F-250/350, 2002–05 E-Series 5.4L Engine

Crankshaft front seal installation—4.6L engine, 2002–03 F-150 5.4L Engine, 2004 F-150 Heritage Edition 5.4L Engine, 2004 F-250/350, 2002–05 E-Series 5.4L Engine

8 mm (0.32 in)

06017-F150-G135

Apply silicone gasket and sealant to the Woodruff key slot on the crankshaft pulley—4.6L engine, 2002–03 F-150 5.4L Engine, 2004 F-150 Heritage Edition 5.4L Engine, 2004 F-250/350, 2002–05 E-Series 5.4L Engine

81. Lubricate the engine front cover and the crankshaft front seal inner lip with clean engine oil.

82. Using the special tools shown, install the crankshaft front seal.

➡ If not secured within 4 minutes, the sealant must be removed and the sealing area cleaned. To clean the sealing area, use silicone gasket remover and metal surface prep. Follow the directions on the packaging. Failure to follow this procedure can cause future oil leakage.

83. Apply silicone gasket and sealant to the Woodruff key slot on the crankshaft pulley.

84. Using special tool 303-102, install the crankshaft pulley.

85. Using a new crankshaft pulley bolt, install the crankshaft pulley bolt and washer.

86. Using the special tool to hold the crankshaft pulley, tighten the bolt in 4 steps:

- Step 1: Tighten the bolt to 90 Nm (66 ft. lbs.).
- Step 2: Loosen the bolt one full turn.
- Step 3: Tighten the bolt to 50 Nm (37 ft. lbs.).
- Step 4: Tighten the bolt an additional 90 degrees without exceeding 200 Nm (148 ft. lbs.).

87. Rotate the tensioner clockwise and install the drive belt.

88. Install the left valve cover.

89. Install the right valve cover.

90. Install the cooling fan.

91. Connect the battery ground cable.

92. Fill the engine with clean engine oil.

2004 F-150 5.4L Engine, Exc. Heritage Edition

2005 F-150, 250, 350 5.4L Engine

2006 Mark LT 5.4L Engine

1. Before servicing the vehicle, refer to the Precautions Section.
2. With the vehicle in NEUTRAL, position it on a hoist.
3. Drain the engine oil.
4. Remove the engine cooling fan.
5. Remove the right valve cover.
6. Remove the left valve cover.
7. Rotate the tensioner clockwise and remove the accessory drive belt.
8. Remove the nut and position aside the transmission cooler tube support bracket.
9. Remove the nut and position aside the power steering pressure hose support bracket.
10. Using the special tool, remove the power steering pump pulley.
11. Disconnect the power steering pressure tube. Drain the power steering fluid into a suitable container.
12. Remove the 3 bolts and position the power steering pump aside.
13. Remove the crankshaft pulley bolt and washer. Discard the crankshaft pulley bolt.
14. Using the special tool, remove the crankshaft pulley.
15. Using the special tool, remove the crankshaft front seal.
16. Remove the 3 bolts and the 3 accessory drive idler pulleys.
17. Remove the 4 bolts and the coolant pump pulley.
18. Remove the 3 bolts and the accessory drive belt tensioner.
19. Disconnect the right camshaft position (CMP) sensor electrical connector.
20. Remove the bolt and the right CMP sensor. Discard the O-ring seal.
21. Remove the nut and the upper radiator hose support bracket.
22. Remove the nut and position the right radio ignition interference capacitor aside.
23. Disconnect the left CMP sensor electrical connector.
24. Remove the bolt and the left CMP sensor. Discard the O-ring seal.
25. Remove the nut and position the left radio ignition interference capacitor aside.
26. Disconnect the crankshaft position (CKP) sensor electrical connector.
27. Remove the 4 front oil pan bolts.
28. Remove the bolts and the studs.

06017-F150-G138

Position the crankshaft keyway at the 12 o'clock position—2004 F-150 5.4L Engine, Exc. Heritage Edition, 2005 F-150, 250, 350 5.4L Engine, 2006 Mark LT 5.4L Engine

06017-F150-G139

The number 1 cylinder camshaft exhaust lobe must be coming up on the exhaust stroke. Verify by noting the position of the 2 intake camshaft lobes and the exhaust lobe on the number 1 cylinder—2004 F-150 5.4L Engine, Exc. Heritage Edition, 2005 F-150, 250, 350 5.4L Engine, 2006 Mark LT 5.4L Engine

✳✳ WARNING

Do not use metal scrapers, wire brushes, power abrasive discs or other abrasive means to clean the sealing surfaces. These tools cause scratches and gouges which make leak paths. Use a plastic scraping tool to remove all traces of old sealant.

29. Remove the engine front cover from the front cover to cylinder block dowel.

30. Remove the engine front cover gaskets.
31. Clean the mating surfaces with silicone gasket remover and metal surface prep. Follow the directions on the packaging.
32. Inspect the mating surfaces.
33. Remove the bolt and the CKP sensor. Discard the O-ring seal.
34. Remove the crankshaft sensor ring from the crankshaft.
35. Position the crankshaft keyway at the 12 o'clock position.

Remove only these 3 roller followers from the right cylinder head—2004 F-150 5.4L Engine, Exc. Heritage Edition, 2005 F-150, 250, 350 5.4L Engine, 2006 Mark LT 5.4L Engine

➡ If the camshaft lobes are not exactly positioned as shown, the crankshaft will require one full additional rotation to 12 o'clock.

36. The number 1 cylinder camshaft exhaust lobe must be coming up on the exhaust stroke. Verify by noting the position of the 2 intake camshaft lobes and the exhaust lobe on the number 1 cylinder.

✳✳ WARNING

If the components are to be reinstalled, they must be installed in the same positions. Mark the components for installation into their original locations.

37. Remove only the 3 roller followers shown in the illustration from the right cylinder head.

✳✳ WARNING

Do not allow the valve keepers to fall off the valve or the valve may drop into the cylinder.

Using special tool 303-1039—2004 F-150 5.4L Engine, Exc. Heritage Edition, 2005 F-150, 250, 350 5.4L Engine, 2006 Mark LT 5.4L Engine

Remove only these 3 roller followers from the left cylinder head—2004 F-150 5.4L Engine, Exc. Heritage Edition, 2005 F-150, 250, 350 5.4L Engine, 2006 Mark LT 5.4L Engine

Using the special tool, remove the 3 designated roller followers from the left cylinder head—2004 F-150 5.4L Engine, Exc. Heritage Edition, 2005 F-150, 250, 350 5.4L Engine, 2006 Mark LT 5.4L Engine

06017-F150-G144

Rotate the crankshaft clockwise and position the crankshaft keyway at the 6 o'clock position—2004 F-150 5.4L Engine, Exc. Heritage Edition, 2005 F-150, 250, 350 5.4L Engine, 2006 Mark LT 5.4L Engine

06017-F150-G145

Remove the bolts, the left timing chain tensioner and tensioner arm—2004 F-150 5.4L Engine, Exc. Heritage Edition, 2005 F-150, 250, 350 5.4L Engine, 2006 Mark LT 5.4L Engine

➡It may be necessary to push the valve down while compressing the spring.

38. Using special tool 303-1039, remove the 3 designated roller followers in the previous step from the right cylinder head.

06017-F150-G146

Remove the right timing chain tensioner and tensioner arm—2004 F-150 5.4L Engine, Exc. Heritage Edition, 2005 F-150, 250, 350 5.4L Engine, 2006 Mark LT 5.4L Engine

✳✳ **WARNING**

If the components are to be reinstalled, they must be installed in the same positions.

39. Mark the components for installation into their original locations.

40. Remove only the 3 roller followers shown in the illustration from the left cylinder head.

✳✳ **WARNING**

Do not allow the valve keepers to fall off the valve or the valve may drop into the cylinder.

➡It may be necessary to push the valve down while compressing the spring.

41. Using the special tool, remove the 3 designated roller followers in the previous step from the left cylinder head.

✳✳ **WARNING**

The crankshaft cannot be moved past the 6 o'clock position once set.

42. Rotate the crankshaft clockwise and position the crankshaft keyway at the 6 o'clock position.

06017-F150-G147

Remove the right and left timing chains—2004 F-150 5.4L Engine, Exc. Heritage Edition, 2005 F-150, 250, 350 5.4L Engine, 2006 Mark LT 5.4L Engine

Remove both timing chain guides, right shown; left similar—2004 F-150 5.4L Engine, Exc. Heritage Edition, 2005 F-150, 250, 350 5.4L Engine, 2006 Mark LT 5.4L Engine

Using the special tool, remove the bolt and the right camshaft phaser sprocket assembly—2004 F-150 5.4L Engine, Exc. Heritage Edition, 2005 F-150, 250, 350 5.4L Engine, 2006 Mark LT 5.4L Engine

Using the special tool, remove the bolt and the left camshaft phaser sprocket assembly—2004 F-150 5.4L Engine, Exc. Heritage Edition, 2005 F-150, 250, 350 5.4L Engine, 2006 Mark LT 5.4L Engine

✳✳ WARNING

If one or both of the tensioner mounting bolts are loosened or removed, the tensioner-sealing bead must be inspected for seal integrity. If cracks, tears or separation from the tensioner body or permanent compres-

Camshaft bearing cap removal/installation sequence—2004 F-150 5.4L Engine, Exc. Heritage Edition, 2005 F-150, 250, 350 5.4L Engine, 2006 Mark LT 5.4L Engine

The camshaft front thrust bearing cap contains an oil metering groove—2004 F-150 5.4L Engine, Exc. Heritage Edition, 2005 F-150, 250, 350 5.4L Engine, 2006 Mark LT 5.4L Engine

sion of the seal bead is observed, install a new tensioner.

43. Remove the bolts, the left timing chain tensioner and tensioner arm.

✳✳ WARNING

If one or both of the tensioner mounting bolts are loosened or removed, the tensioner-sealing bead must be inspected for seal integrity. If cracks, tears or separation from the ten-

Position the crankshaft with the special tool—2004 F-150 5.4L Engine, Exc. Heritage Edition, 2005 F-150, 250, 350 5.4L Engine, 2006 Mark LT 5.4L Engine

sioner body or permanent compression of the seal bead is observed, install a new tensioner.

44. Remove the bolts, the right timing chain tensioner and tensioner arm.

45. Remove the right timing chain from the camshaft sprocket.

46. Remove the right timing chain from the crankshaft sprocket.

47. Remove the left timing chain from the camshaft sprocket.

48. Remove the left timing chain and crankshaft sprocket.

49. Remove the bolts. Remove both timing chain guides.

✷ WARNING

Damage to the camshaft phaser sprocket assembly will occur if mishandled or used as a lifting or leveraging device.

✷ WARNING

Only use hand tools to remove the camshaft phaser sprocket assembly or damage may occur to the camshaft or camshaft phaser unit.

50. Using the special tool, remove the bolt and the right camshaft phaser sprocket assembly. Discard the camshaft phaser sprocket bolt.

✷ WARNING

Damage to the camshaft phaser sprocket assembly will occur if mishandled or used as a lifting or leveraging device.

✷ WARNING

Only use hand tools to remove the camshaft phaser sprocket assembly or damage may occur to the camshaft or camshaft phaser unit.

51. Using the special tool, remove the bolt and the left camshaft phaser sprocket assembly. Discard the camshaft phaser sprocket bolt.

✷✷ WARNING

Remove the front thrust camshaft bearing cap straight upward from the bearing towers, or the bearing cap may be damaged from side loading.

➡The camshaft bearing caps must be installed in their original locations. Record camshaft bearing cap locations.

52. Remove the bolts in the sequence shown and remove the right cylinder head front camshaft bearing cap and then the remaining bearing caps.
53. Clean and inspect the right camshaft bearing caps. The camshaft front thrust bearing cap contains an oil metering groove. Make sure the groove is free of foreign material.
54. Remove the right camshaft.

✷✷ WARNING

Remove the front thrust camshaft bearing cap straight upward from

the bearing towers, or the bearing cap may be damaged from side loading.

➡The camshaft bearing caps must be installed in their original locations. Record camshaft bearing cap locations.

55. Remove the bolts in the sequence shown and remove the left cylinder head front camshaft bearing cap and then the remaining bearing caps.
56. Clean and inspect the left camshaft bearing caps. The camshaft front thrust bearing cap contains an oil metering groove. Make sure the groove is free of foreign material.
57. Remove the left camshaft.

✷✷ WARNING

If the components are to be reinstalled, they must be installed in the same positions.

58. Mark the components for installation into their original locations.
59. Remove all of the remaining roller followers from the cylinder heads.

To install:

60. Install the left and right camshafts. Lubricate the camshaft and camshaft journals with clean engine oil prior to installation.

Right and left timing chain guides installed—2004 F-150 5.4L Engine, Exc. Heritage Edition, 2005 F-150, 250, 350 5.4L Engine, 2006 Mark LT 5.4L Engine

61. Lubricate the camshaft bearing caps with clean engine oil.

62. Position the front camshaft bearing cap.

63. Position the remaining camshaft bearing caps.

64. Install the bolts loosely.

65. Tighten to 10 Nm (89 inch lbs.) in the sequence shown.

✳✳ WARNING

Damage to the camshaft phaser sprocket assembly will occur if mishandled or used as a lifting or leveraging device.

66. Install the camshaft phaser sprockets and new camshaft phaser bolts finger tight.

✳✳ WARNING

Damage to the camshaft phaser sprocket assembly will occur if mishandled or used as a lifting or leveraging device.

✳✳ WARNING

Only use hand tools to remove the camshaft phaser sprocket assembly or damage may occur to the camshaft or camshaft phaser unit.

67. Using the special tool, tighten the left and right camshaft phaser sprocket bolts in 2 stages.
- Stage 1: Tighten to 40 Nm (30 ft. lbs.).
- Stage 2: Tighten an additional 90 degrees.

68. Position the crankshaft with the special tool, then remove the tool.

✳✳ WARNING

Timing chain procedures must be followed exactly or damage to valves and pistons will result.

✳✳ WARNING

Prior to installation, inspect the tensioner-sealing bead for seal integrity. If cracks, tears, separation from the tensioner body or permanent compression of the seal bead is observed, install a new tensioner.

69. Compress the tensioner plunger, using a vise.

70. Install a retaining clip on the tensioner to hold the plunger in during installation.

06017-F150-G155

Position the lower end of the left (inner) timing chain on the crankshaft sprocket, aligning the timing mark on the outer flange of the crankshaft sprocket with the single copper (marked) link on the chain—2004 F-150 5.4L Engine, Exc. Heritage Edition, 2005 F-150, 250, 350 5.4L Engine, 2006 Mark LT 5.4L Engine

06017-F150-G156

Position the timing chain on the camshaft sprocket with the camshaft sprocket timing mark positioned between the 2 copper (marked) chain links—2004 F-150 5.4L Engine, Exc. Heritage Edition, 2005 F-150, 250, 350 5.4L Engine, 2006 Mark LT 5.4L Engine

06017-F150-G157

Position the lower end of the right (outer) timing chain on the crankshaft sprocket, aligning the timing mark on the sprocket with the single copper (marked) chain link—2004 F-150 5.4L Engine, Exc. Heritage Edition, 2005 F-150, 250, 350 5.4L Engine, 2006 Mark LT 5.4L Engine

71. Remove the tensioner from the vise.

72. If the copper links are not visible, mark 2 links on one end and 1 link on the other end, and use as timing marks.

73. Install the crankshaft sprocket, making sure the flange faces forward.

74. Install the 4 bolts and the left and right timing chain guides. Tighten to 10 Nm (89 inch lbs.).

75. Position the lower end of the left (inner) timing chain on the crankshaft sprocket, aligning the timing mark on the outer flange of the crankshaft sprocket with the single copper (marked) link on the chain.

➡**Make sure the upper half of the timing chain is below the tensioner arm dowel.**

76. Position the timing chain on the camshaft sprocket with the camshaft sprocket timing mark positioned between the 2 copper (marked) chain links.

➡**The left timing chain tensioner arm has a bump near the dowel hole for identification.**

77. Position the left timing chain tensioner arm on the dowel pin and install the left timing chain tensioner and bolts. Tighten to 25 Nm (18 ft. lbs.).

78. Remove the retaining clip from the left timing chain tensioner.

79. Position the lower end of the right (outer) timing chain on the crankshaft sprocket, aligning the timing mark on the sprocket with the single copper (marked) chain link.

➡**The lower half of the timing chain must be positioned above the tensioner arm dowel.**

80. Position the right timing chain on the camshaft sprocket. Make sure the camshaft sprocket timing mark is positioned between the 2 copper (marked) chain links.

81. Position the right timing chain tensioner arm on the dowel pin and install the right timing chain tensioner and bolts. Tighten to 25 Nm (18 ft. lbs.).

82. Remove the retaining clip from the right timing chain tensioner.

83. As a post-check, verify correct alignment of all timing marks.

84. Install the crankshaft sensor ring on the crankshaft.

※※ WARNING

If the components are to be reinstalled, they must be installed into their original locations.

06017-F150-G158

Position the right timing chain on the camshaft sprocket. Make sure the camshaft sprocket timing mark is positioned between the 2 copper (marked) chain links—2004 F-150 5.4L Engine, Exc. Heritage Edition, 2005 F-150, 250, 350 5.4L Engine, 2006 Mark LT 5.4L Engine

06017-F150-G159

As a post-check, verify correct alignment of all timing marks—2004 F-150 5.4L Engine, Exc. Heritage Edition, 2005 F-150, 250, 350 5.4L Engine, 2006 Mark LT 5.4L Engine

06017-F150-G161

Apply a bead of silicone gasket and sealant along the cylinder head-to-cylinder block surface and the oil pan-to-cylinder block surface—2004 F-150 5.4L Engine, Exc. Heritage Edition, 2005 F-150, 250, 350 5.4L Engine, 2006 Mark LT 5.4L Engine

85. Using the special tool, install all of the camshaft roller followers. Lubricate the roller followers with clean engine oil prior to installation.

86. Install the crankshaft position (CKP) sensor and the bolt. Tighten to 10 Nm (89 inch lbs.).

❄❄ WARNING

Do not use metal scrapers, wire brushes, power abrasive discs or other abrasive means to clean the sealing surfaces. These tools cause scratches and gouges which make leak paths. Use a plastic scraping tool to remove all traces of old sealant.

➡If the engine front cover is not secured within 4 minutes, the sealant must be removed and the sealing area cleaned. To clean the sealing area, use silicone gasket remover and metal surface prep. Follow the directions on the packaging. Failure to follow this procedure can cause future oil leakage.

➡Make sure that the engine front cover gasket is in place on the engine front cover before installation.

87. Apply a bead of silicone gasket and sealant along the cylinder head-to-cylinder block surface and the oil pan-to-cylinder block surface, at the locations shown.

88. Install a new engine front cover gasket on the engine front cover. Position the engine front cover onto the dowels. Install the fasteners finger tight.

89. Tighten the engine front cover fasteners in sequence in 2 stages.
- Stage 1: Tighten fasteners 1 through 15 to 25 Nm (18 ft. lbs.).
- Stage 2: Tighten fasteners 6 and 7 to 48 Nm (35 ft. lbs.).

90. Loosely install the oil pan-to-front cover bolts, then tighten the bolts in 2 stages, in the sequence shown.
- Stage 1: Tighten to 20 Nm (15 ft. lbs.).
- Stage 2: Tighten an additional 60 degrees.

91. Connect the CKP sensor electrical connector.

92. Position the power steering pump assembly and install the 3 bolts. Tighten to 25 Nm (18 ft. lbs.).

93. Using the special tool, install a new O-ring seal on the pressure line fitting.

94. Connect the power steering pressure tube. Tighten to 65 Nm (48 ft. lbs.).

95. Using an installer, install the power steering pump pulley.

96. Position the power steering pressure hose support bracket and install the nut. Tighten to 10 Nm (89 inch lbs.).

97. Position the transmission cooler tube support bracket and install the nut. Tighten to 10 Nm (89 inch lbs.).

➡Lubricate the O-ring seal with clean engine oil prior to installation.

98. Using a new O-ring seal, install the right camshaft position (CMP) sensor and the bolt. Tighten to 10 Nm (89 inch lbs.).

99. Connect the right CMP sensor electrical connector.

100. Install the left radio ignition interference capacitor and the nut. Tighten to 10 Nm (89 inch lbs.).

➡Lubricate the O-ring seal with clean engine oil prior to installation.

1	Bolt, Hex Flange Head Pilot, M8 x 1.25 x 50
2	Bolt, Hex Flange Head Pilot, M8 x 1.25 x 50
3	Bolt, Hex Flange Head Pilot, M8 x 1.25 x 50
4	Bolt, Hex Flange Head Pilot, M8 x 1.25 x 50
5	Bolts, Hex Flange Head Pilot, M8 x 1.25 x 50
6	Stud, Hex Head Pilot, M10 x 1.5 x 1.5 x 103
7	Stud, Hex Head Pilot, M10 x 1.5 x 1.5 x 103
8	Bolt, Hex Flange Head Pilot, M8 x 1.25 x 50
9	Bolt, Hex Flange Head Pilot, M8 x 1.25 x 50
10	Bolt, Hex Flange Head Pilot, M8 x 1.25 x 50
11	Bolt, Hex Flange Head Pilot, M8 x 1.25 x 50
12	Stud and Washer, Hex Head Pilot, M8 x 1.25 x 1.25 x 94
13	Stud and Washer, Hex Head Pilot, M8 x 1.25 x 1.25 x 94
14	Stud and Washer, Hex Head Pilot, M8 x 1.25 x 1.25 x 94
15	Bolt, Hex Head Pilot, M8 x 1.25 x 56

06017-F150-G162

Front cover fastener identification—2004 F-150 5.4L Engine, Exc. Heritage Edition, 2005 F-150, 250, 350 5.4L Engine, 2006 Mark LT 5.4L Engine

101. Using a new O-ring seal, install the left CMP sensor and the bolt. Tighten to 10 Nm (89 inch lbs.).

102. Connect the left CMP sensor electrical connector.

103. Install the right radio ignition interference capacitor and the nut. Tighten to 10 Nm (89 inch lbs.).

104. Install the accessory drive belt tensioner and the 3 bolts. Tighten to 25 Nm (18 ft. lbs.).

105. Install the coolant pump pulley and the 4 bolts. Tighten to 25 Nm (18 ft. lbs.).

106. Install the 3 accessory drive idler pulleys and the 3 bolts. Tighten to 25 Nm (18 ft. lbs.).

107. Position the upper radiator hose support bracket and install the nut. Tighten to 10 Nm (89 inch lbs.).

108. Lubricate the engine front cover and the crankshaft seal inner lip with clean engine oil.

109. Using the special tools, install the crankshaft seal into the engine front cover.

➡**If not secured within 4 minutes, the sealant must be removed and the sealing area cleaned with metal surface prep and silicone gasket remover.**

Front seal installation—2004 F-150 5.4L Engine, Exc. Heritage Edition, 2005 F-150, 250, 350 5.4L Engine, 2006 Mark LT 5.4L Engine

Crankshaft pulley installation—2004 F-150 5.4L Engine, Exc. Heritage Edition, 2005 F-150, 250, 350 5.4L Engine, 2006 Mark LT 5.4L Engine

Remove the balance shaft—2002–04 F-250/350 6.8L Engine; 2002–05 E-Series 6.8L Engine

Position the crankshaft with the keyway at the 12 o'clock position—2002–04 F-250/350 6.8L Engine; 2002–05 E-Series 6.8L Engine

Install and carefully tighten the special tool on the camshafts—2002–04 F-250/350 6.8L Engine; 2002–05 E-Series 6.8L Engine

Remove the timing chain tensioners and arms—2002–04 F-250/350 6.8L Engine; 2002–05 E-Series 6.8L Engine

Remove the timing chains—2002–04 F-250/350 6.8L Engine; 2002–05 E-Series 6.8L Engine

Allow to dry until there is no sign of wetness, or 4 minutes, whichever is longer. Failure to follow this procedure can cause future oil leakage.

110. Apply silicone gasket and sealant to the Woodruff key slot on the crankshaft pulley. Use the special tool to install the crankshaft pulley.

111. Tighten the new crankshaft pulley bolt in 4 stages.

- Stage 1: Tighten to 90 Nm (66 ft. lbs.).

Timing chain guides—2002–04 F-250/350 6.8L Engine; 2002–05 E-Series 6.8L Engine

While holding the ratchet mechanism with a pick, push the ratchet arm back into the tensioner housing—2002–04 F-250/350 6.8L Engine; 2002–05 E-Series 6.8L Engine

If copper links are not visible, mark two links on one end and one link on the other end, and use as timing marks—2002–04 F-250/350 6.8L Engine; 2002–05 E-Series 6.8L Engine

Install a paper clip on the tensioner to hold the plunger in during installation—2002–04 F-250/350 6.8L Engine; 2002–05 E-Series 6.8L Engine

- Stage 2: Loosen 360 degrees.
- Stage 3: Tighten to 50 Nm (37 ft. lbs.).
- Stage 4: Tighten an additional 90 degrees.

112. Rotate the tensioner clockwise and install the accessory drive belt.

113. Install the right valve cover.
114. Install the left valve cover.
115. Install the engine cooling fan.
116. Fill the crankcase with clean engine oil.

2002–04 F-250/350 6.8L Engine
2002–05 E-Series 6.8L Engine

1. Before servicing the vehicle, refer to the Precautions Section.
2. Disconnect the battery ground cable.
3. Remove both of the valve covers.
4. Remove the radiator.
5. Remove the water pump.
6. Raise and support the vehicle.
7. Remove the top bolts and the lower bolt and position the power steering pump aside.
8. Disconnect the crankshaft position (CKP) sensor electrical connector.
9. Remove the drain plug and drain the engine oil.
10. Remove the four front oil pan bolts.
11. Lower the vehicle.
12. Remove the crankshaft front seal.
13. Disconnect the camshaft position (CMP) sensor electrical connection.
14. Remove the bolt and the belt idler pulley.
15. Remove the bolts.

Rotate the left camshaft with the Camshaft Positioning Tool until the timing mark is approximately at 12 o'clock. Rotate the right camshaft with the Camshaft Positioning Tool until the timing mark is approximately at 11 o'clock—2002–04 F-250/350 6.8L Engine; 2002–05 E-Series 6.8L Engine

❊❊ WARNING

Do not use metal scrapers, wire brushes, power abrasive discs or other abrasive means to clean the sealing surfaces. These tools cause scratches and gouges which make leak paths. Use a plastic scraping tool to remove all traces of old sealant.

16. Remove the engine front cover from the front cover to cylinder block dowel.

17. Remove the engine front cover gasket and clean and inspect the mating surfaces.

❊❊ WARNING

Since the engine is not free-wheeling, if the crankshaft or the camshafts are moved in any manner during removal and installation the crankshaft and camshaft must be resynchronized.

18. Remove the engine front cover.

19. Remove the crankshaft sensor ring from the crankshaft.

❊❊ WARNING

The caps must be marked for installation in their original location or damage to the engine can occur.

20. Remove the six bolts and remove the balance shaft bearing caps.

21. Remove the balance shaft.

❊❊ WARNING

Unless otherwise instructed, at no time when the timing chain are removed and the cylinder heads are installed may the crankshaft or camshaft be rotated. Severe piston and valve damage will occur.

22. Position the crankshaft with the key-way at the 12 o'clock position.

23. Install and carefully tighten the special tool on the camshafts.

24. Remove the bolts. Remove the timing chain tensioners.

25. Remove the timing chain tensioner arms.

26. Remove the right timing chain from the camshaft sprocket.

27. Remove the right timing chain from the crankshaft sprocket.

28. Remove the left timing chain from the camshaft sprocket.

Position the crankshaft with the special tool—2002–04 F-250/350 6.8L Engine; 2002–05 E-Series 6.8L Engine

Install the lower end of the left (inner) timing chain, aligning the timing marks—2002–04 F-250/350 6.8L Engine; 2002–05 E-Series 6.8L Engine

Install the left timing chain on the camshaft sprocket with the two chain (marked) links and the timing marks aligned—2002–04 F-250/350 6.8L Engine; 2002–05 E-Series 6.8L Engine

29. Remove the left timing chain from the crankshaft sprocket.

30. Remove the bolts. Remove the timing chain guides.

31. On engine with bolt-on sprockets, remove the camshaft sprocket.

32. Remove the bolt.

33. Remove the camshaft sprocket.

To install:

✳✳ WARNING

Timing chain procedures must be followed exactly or damage to valves and pistons will result.

34. Compress the tensioner using a soft-jawed vice.

35. While holding the ratchet mechanism with a pick, push the ratchet arm back into the tensioner housing.

36. Install a paper clip on the tensioner to hold the plunger in during installation.

➡**There are 61 links in each timing chain.**

37. If copper links are not visible, mark two links on one end and one link on the other end, and use as timing marks.

38. On engines equipped with bolt-on sprockets, install the camshaft sprocket bolt. Tighten the bolt as follows:

    a. M10 bolt: Tighten in two stages.

    • Stage 1: Tighten to 40 Nm (30 ft. lbs.).

    • Stage 2: Tighten an additional 90 degrees.

    b. M12 bolt: Tighten to 120 Nm (90 ft. lbs.).

39. Install the timing chain guides. Torque to 89 inch lbs. (10 Nm).

➡**Slightly loosen the Camshaft Holding Tool to allow slight camshaft movement.**

40. Rotate the left camshaft with the Camshaft Positioning Tool until the timing mark is approximately at 12 o'clock.

41. Rotate the right camshaft with the Camshaft Positioning Tool until the timing mark is approximately at 11 o'clock.

42. Tighten the Camshaft Holding Tool to maintain camshaft prepositioning.

✳✳ WARNING

Rotate the crankshaft counterclockwise only. Do not rotate past the position shown or severe piston and/or valve damage can occur.

43. Position the crankshaft with the special tool, remove the tool.

Be sure the chain link and crankshaft sprocket timing marks are aligned—2002–04 F-250/350 6.8L Engine; 2002–05 E-Series 6.8L Engine

Install the right (outer) timing chain on the crankshaft sprocket—2002–04 F-250/350 6.8L Engine; 2002–05 E-Series 6.8L Engine

Position the timing chain on the camshaft sprocket. Make sure the two copper-colored (marked) links align with the camshaft sprocket timing mark—2002–04 F-250/350 6.8L Engine; 2002–05 E-Series 6.8L Engine

44. Install the crankshaft sprocket, making sure the flange faces forward.

45. Install the lower end of the left (inner) timing chain, aligning the timing marks.

➡ **Be sure the upper half of the timing chain is below the tensioner guide dowel. If necessary, use the Camshaft Positioning Tool to adjust.**

46. Install the left timing chain on the camshaft sprocket with the two chain (marked) links and the timing marks aligned.

✳ WARNING

The camshaft sprocket can jump time if the Camshaft Holding Tool is not secured.

➡ **Be sure the chain link and crankshaft sprocket timing marks are aligned.**

➡ **The lower half of the timing chain must be positioned above the dowel.**

47. Install the right (outer) timing chain on the crankshaft sprocket.

48. Position the timing chain on the camshaft sprocket. Make sure the two copper-colored (marked) links align with the camshaft sprocket timing mark.

➡ **The left timing chain tensioner arm has a bump near the dowel hole, for identification.**

49. Position the left and right timing chain tensioner arms on the dowel pins. Position the timing chain tensioners and install the bolts. Torque to 18 ft. lbs. (25 Nm).

50. Remove the retaining clips from the right and left timing chain tensioners.

51. Check for correct alignment of all timing marks.

52. Remove the special tools from the camshafts.

53. Install the crankshaft sensor ring on the crankshaft.

54. Lubricate the balance shaft journals with clean engine oil.

55. Using the index mark on the balance shaft, mark the corresponding gear tooth with chalk.

56. Position the balance shaft on the journals.

➡ **It may be necessary to use an inspection mirror to see the marks.**

57. Align the chalk mark on the balance shaft with the camshaft timing mark as shown.

Position the left and right timing chain tensioner arms on the dowel pins—2002–04 F-250/350 6.8L Engine; 2002–05 E-Series 6.8L Engine

Correct alignment of all timing marks—2002–04 F-250/350 6.8L Engine; 2002–05 E-Series 6.8L Engine

Using the index mark on the balance shaft, mark the corresponding gear tooth with chalk—2002–04 F-250/350 6.8L Engine; 2002–05 E-Series 6.8L Engine

58. Install the bearing caps in their original location. Install the bolts and tighten the bolts in the sequence shown to 89 inch lbs. (10 Nm).

✳✳ WARNING

Do not use metal scrapers, wire brushes, power abrasive discs or other abrasive means to clean the sealing surfaces. These tools cause scratches and gouges which make leak paths. Use a plastic scraping tool to remove all traces of old sealant.

➡️If the engine front cover is not secured within four minutes, the sealant must be removed and the sealing area cleaned with metal surface cleaner. Allow to dry until there is no sign of wetness, or four minutes, whichever is longer. Failure to follow this procedure can cause future oil leakage.

➡️Make sure that the engine front cover gasket is in place on the engine front cover before installation.

59. Apply silicone gasket and sealant along the cylinder head to cylinder block surface and the oil pan to cylinder block surface.

60. Install the engine front cover with engine front cover gasket on the front cover to cylinder block dowel and loosely install the bolts.

61. Tighten the engine front cover fasteners in sequence in two stages.
- Stage 1: Tighten fasteners 1 through 7 to 20–30 Nm (15–22 ft. lbs.).
- Stage 2: Tighten fasteners 6 through 15 to 40–55 Nm (30–41 ft. lbs.).

62. Connect the CMP electrical connection.

63. Install a new crankshaft front seal.

64. Raise and support the vehicle.

65. Loosely install the oil pan-to-front cover bolts, then tighten the bolts in the two stages, in the sequence shown.
- Stage 1: Tighten to 20 Nm (15 ft. lbs.).
- Stage 2: Tighten an additional 60 degrees.

66. Connect the crankshaft position sensor electrical connector.

➡️The front lower hole in the power steering pump is not used.

67. Position the power steering pump and install the bolts. Torque to 15–22 ft. lbs. (20–30 Nm).

06017-F150-G183

Align the chalk mark on the balance shaft with the camshaft timing—2002–04 F-250/350 6.8L Engine; 2002–05 E-Series 6.8L Engine

06017-F150-G184

Bearing cap torque sequence—2002–04 F-250/350 6.8L Engine; 2002–05 E-Series 6.8L Engine

06017-F150-G185

Apply silicone gasket and sealant along the cylinder head to cylinder block surface and the oil pan to cylinder block surface—2002–04 F-250/350 6.8L Engine; 2002–05 E-Series 6.8L Engine

1   Bolt, Hex
    Flange Head
    Pilot, M8 x 1.25
    x 53

2   Bolt, Hex
    Flange Head
    Pilot, M8 x 1.25
    x 53

3   Bolt, Hex
    Flange Head
    Pilot, M8 x 1.25
    x 53

4   Bolt, Hex
    Flange Head
    Pilot, M8 x 1.25
    x 53

5   Bolts, Hex
    Flange Head
    Pilot, M8 x 1.25
    x 53

6   Stud, Hex-Head
    Pilot, M10 x 1.5
    x 1.5 x 103.1

7   Stud, Hex-Head
    Pilot, M10 x 1.5
    x 1.5 x 103.1

8   Screw and
    Washer, Hex
    Pilot, M10 x 1.5
    x 57.5

9   Screw and
    Washer, Hex
    Pilot, M10 x 1.5
    x 57.5

10 Screw and
    Washer, Hex
    Pilot, M10 x 1.5
    x 57.5

11 Stud and
    Washer, Hex
    Head Pilot, M10
    x 1.5 x M8 x
    1.25 x 109.6

12 Stud and
    Washer, Hex
    Head Pilot, M10
    x 1.5 x M8 x
    1.25 x 109.6

13 Stud and
    Washer, Hex
    Head Pilot, M10
    x 1.5 x M8 x
    1.25 x 109.6

14 Stud and
    Washer, Hex
    Head Pilot, M10
    x 1.5 x M8 x
    1.25 x 109.6

15 Stud and
    Washer, Hex
    Head Pilot, M10
    x 1.5 x M8 x
    1.25 x 109.6

06017-F150-G186

Front cover fastener identification—2002–04 F-250/350 6.8L Engine; 2002–05 E-Series 6.8L Engine

Oil pan-to-front cover bolt torque sequence—2002–04 F-250/350 6.8L Engine; 2002–05 E-Series 6.8L Engine

20-30 Nm (15-22 lb/ft)

Power steering pump installation—2002–04 F-250/350 6.8L Engine; 2002–05 E-Series 6.8L Engine

Remove these 3 bolts from the A/C compressor—2005 F-250/350 6.8L Engine

68. Install the drain plug.
69. Lower the vehicle.
70. Install the valve covers.
71. Install the water pump.
72. Install the radiator.
73. Fill the engine with clean engine oil.
74. Connect the battery ground cable.

2005 F-250/350 6.8L Engine

1. Before servicing the vehicle, refer to the Precautions Section.
2. With the vehicle in NEUTRAL, position it on a hoist.
3. Drain the engine oil.
4. Remove the engine cooling fan.
5. Remove the right valve cover.
6. Remove the left valve cover.
7. Rotate the tensioner clockwise and remove the accessory drive belt.
8. Remove the 3 bolts and position the power steering pump aside.
9. Remove the nut and position the starter electrical harness support bracket aside.
10. Remove the 3 bolts and position the A/C compressor aside.
11. Remove the crankshaft pulley bolt and washer. Discard the crankshaft pulley bolt.
12. Using the special tool, remove the crankshaft pulley.
13. Using the special tool, remove the crankshaft front seal.
14. Remove the bolt and the accessory drive idler pulley.
15. Remove the 4 bolts and the coolant pump pulley.
16. Remove the 3 bolts and the accessory drive belt tensioner.
17. Disconnect the camshaft position (CMP) sensor electrical connector.
18. Remove the bolt and the CMP sensor. Discard the O-ring seal.
19. Disconnect the right and left radio ignition interference capacitor electrical connectors.
20. Remove the nut and the right and left radio ignition interference capacitors.

303-107

Front seal removal—2005 F-250/350 6.8L Engine

Balance shaft bearing cap removal/installation sequence—2005 F-250/350 6.8L Engine

Position the crankshaft keyway at the 12 o'clock position—2005 F-250/350 6.8L Engine

The number 1 cylinder camshaft exhaust lobe must be coming up on the exhaust stroke. Verify by noting the position of the 2 intake camshaft lobes and the exhaust lobe on the number 1 cylinder—2005 F-250/350 6.8L Engine

21. Disconnect the crankshaft position (CKP) sensor electrical connector.
22. Remove the 4 front oil pan bolts.
23. Remove the bolts and the studs.

✳✳ WARNING

Do not use metal scrapers, wire brushes, power abrasive discs or other abrasive means to clean the sealing surfaces. These tools cause scratches and gouges which make leak paths. Use a plastic scraping

Remove only these 3 roller followers from the right cylinder head—2005 F-250/350 6.8L Engine

tool to remove all traces of old sealant.

24. Remove the engine front cover from the front cover to cylinder block dowel.
25. Remove the engine front cover gaskets.
26. Clean the mating surfaces with silicone gasket remover and metal surface prep. Follow the directions on the packaging.
27. Inspect the mating surfaces.

Using special tool 303-1039—2005 F-250/350 6.8L Engine

28. Remove the bolt and the CKP sensor. Discard the O-ring seal.

✳✳ WARNING

Only use hand tools to loosen the camshaft sprocket bolt or damage may occur to the camshaft or camshaft sprocket.

29. Loosen and back off the right camshaft sprocket bolt 1 full turn.

✳✳ WARNING

Only use hand tools to loosen the sprocket bolt or damage may occur to the camshaft or camshaft sprocket.

30. Loosen the left camshaft sprocket bolt.

➡The balance shaft bearing caps must be installed in their original locations. Record camshaft bearing cap locations.

31. Remove the 6 bolts in the sequence shown, the 3 bearing caps and the balance shaft.
32. Remove the crankshaft sensor ring from the crankshaft.
33. Position the crankshaft keyway at the 12 o'clock position.

➡If the camshaft lobes are not exactly positioned as shown, the crankshaft will require one full additional rotation to 12 o'clock.

34. The number 1 cylinder camshaft exhaust lobe must be coming up on the exhaust stroke. Verify by noting the position of the 2 intake camshaft lobes and the exhaust lobe on the number 1 cylinder.

❋❋ WARNING

If the components are to be reinstalled, they must be installed in the same positions.

35. Mark the components for installation into the original locations.

36. Remove only the 3 roller followers shown in the illustration from the right cylinder head.

❋❋ WARNING

Do not allow the valve keepers to fall off the valve or the valve may drop into the cylinder.

➡It may be necessary to push the valve down while compressing the spring.

06017-F150-G197

Rotate the crankshaft clockwise and position the crankshaft keyway at the 6 o'clock position—2005 F-250/350 6.8L Engine

06017-F150-G198

Timing chain guide bolts—2005 F-250/350 6.8L Engine

06017-F150-G196

Remove only these 4 roller followers from the left cylinder head—2005 F-250/350 6.8L Engine

06017-F150-G199

Balance shaft drive gear—2005 F-250/350 6.8L Engine

37. Using special tool 303-1039, remove the 3 designated roller followers in the previous step from the right cylinder head.

❊❊ WARNING

If the components are to be reinstalled, they must be installed in the same positions.

38. Mark the components for installation into the original locations.

39. Remove only the 4 roller followers shown in the illustration from the left cylinder head.

❊❊ WARNING

Do not allow the valve keepers to fall off the valve or the valve may drop into the cylinder.

➡It may be necessary to push the valve down while compressing the spring.

40. Using the special tool, remove the 4 designated roller followers in the previous step from the left cylinder head.

❊❊ WARNING

The crankshaft cannot be moved past the 6 o'clock position once set.

41. Rotate the crankshaft clockwise and position the crankshaft keyway at the 6 o'clock position.

❊❊ WARNING

If one or both of the tensioner mounting bolts are loosened or removed, the tensioner-sealing bead must be inspected for seal integrity. If cracks, tears, separation from the tensioner body or permanent compression of the seal bead is observed, install a new tensioner.

42. Remove the bolts, the left timing chain tensioner and tensioner arm.

❊❊ WARNING

If one or both of the tensioner mounting bolts are loosened or removed, the tensioner-sealing bead must be inspected for seal integrity. If cracks, tears, separation from the tensioner body or permanent compression of the seal bead is observed, install a new tensioner.

43. Remove the bolts, the right timing chain tensioner and tensioner arm.

06017-F150-G200

Right side camshaft bearing cap removal/installation sequence—2005 F-250/350 6.8L Engine

44. Remove the right timing chain from the camshaft sprocket.

45. Remove the right timing chain from the crankshaft sprocket.

46. Remove the left timing chain from the camshaft sprocket.

47. Remove the left timing chain and the crankshaft sprocket.

48. Remove the bolts. Remove the timing chain guides.

49. Remove the bolts and the right and left camshaft sprockets.

50. Remove the balance shaft drive gear from the left camshaft.

❊❊ WARNING

Remove the front thrust camshaft bearing cap straight upward from the bearing towers, or the bearing cap may be damaged from side loading.

➡The camshaft bearing caps must be installed in their original locations. Record camshaft bearing cap locations.

51. Remove the bolts in the sequence shown and remove the right cylinder head front camshaft bearing cap and then the remaining bearing caps.

06017-F150-G201

The camshaft front thrust bearing cap contains an oil metering groove—2005 F-250/350 6.8L Engine

06017-F150-G202

Left side camshaft bearing cap removal/installation sequence—2005 F-250/350 6.8L Engine

52. Clean and inspect the right camshaft bearing caps. The camshaft front thrust bearing cap contains an oil metering groove. Make sure the groove is free of foreign material.

53. Remove the right camshaft.

❊❊ WARNING

Remove the front thrust camshaft bearing cap straight upward from the bearing towers, or the bearing cap may be damaged from side loading.

➡The camshaft bearing caps must be installed in their original locations. Record camshaft bearing cap locations.

54. Remove the bolts in the sequence shown and remove the left cylinder head front camshaft bearing cap and then the remaining bearing caps.

55. Clean and inspect the left camshaft bearing caps. The camshaft front thrust bearing cap contains an oil metering groove. Make sure the groove is free of foreign material.

56. Remove the left camshaft.

✳ WARNING

If the components are to be reinstalled, they must be installed in the same positions.

57. Mark the components for installation into their original locations.

58. Remove all of the remaining roller followers from the cylinder heads.

To install:

✳✳ WARNING

Timing chain procedures must be followed exactly or damage to valves and pistons will result.

✳✳ WARNING

If the components are to be reinstalled, they must be installed into their original locations.

➡Camshaft shown installed to clarify roller follower position.

59. Install only the identified camshaft roller followers onto the right cylinder head.

✳✳ WARNING

If the components are to be reinstalled, they must be installed into their original locations.

➡Camshaft shown installed to clarify camshaft roller follower position.

60. Install only the identified camshaft roller followers onto the left cylinder head.

61. Install the left and right camshafts.

62. Lubricate the camshaft and camshaft journals with clean engine oil prior to installation.

63. Lubricate the camshaft bearing caps with clean engine oil.

64. Position the front camshaft bearing cap.

65. Position the remaining camshaft bearing caps.

66. Install the bolts loosely.

67. Tighten the left camshaft bearing cap bolts in 2 steps:
- Step 1: Tighten to 8 Nm(71 inch lbs.) in the sequence shown.
- Step 2: Tighten an additional 45 degrees.

68. Tighten the right camshaft bearing cap bolts in 2 steps:
- Step 1: Tighten to 8 Nm (71 inch lbs.) in the sequence shown.
- Step 2: Tighten an additional 45 degrees.

69. Install the balance shaft drive gear onto the left camshaft.

70. Install the camshaft sprockets and camshaft sprocket bolts finger tight.

Position the crankshaft with special tool 303-448—2005 F-250/350 6.8L Engine

Install the lower end of the left timing chain, aligning the timing marks—2005 F-250/350 6.8L Engine

Install the left timing chain on the camshaft sprocket with the 2 chain (marked) links and the timing marks aligned—2005 F-250/350 6.8L Engine

Install the right (outer) timing chain on the crankshaft sprocket—2005 F-250/350 6.8L Engine

Rotate the left camshaft until the timing mark is approximately at 12 o'clock. Rotate the right camshaft until the timing mark is approximately at 11 o'clock—2005 F-250/350 6.8L Engine

06017-F150-G208

Position the timing chain on the camshaft sprocket. Make sure the 2 copper-colored (marked) links align with the camshaft sprocket timing mark—2005 F-250/350 6.8L Engine

06017-F150-G209

Position the balance shaft on the journals and align the chalk mark on the balance shaft with the camshaft timing mark— 2005 F-250/350 6.8L Engine

✳✳ WARNING

Timing chain procedures must be followed exactly or damage to valves and pistons will result.

✳✳ WARNING

Prior to installation, inspect the tensioner-sealing bead for seal integrity. If cracks, tears, separation from the tensioner body or permanent compression of the seal bead is observed, install a new tensioner.

71. Compress the tensioner plunger, using a vise.
72. Install a retaining clip on the tensioner to hold the plunger in during installation.

➡**There are 61 links in each timing chain.**

73. If copper links are not visible, mark 2 links on one end and one link on the other end, and use as timing marks.
74. Install the timing chain guides and the 4 bolts. Tighten to 10 Nm (89 inch lbs.).

75. Rotate the left camshaft until the timing mark is approximately at 12 o'clock.
76. Rotate the right camshaft until the timing mark is approximately at 11 o'clock.

✳✳ WARNING

Rotate the crankshaft counterclockwise only. Do not rotate past the position shown or severe piston and/or valve damage can occur.

77. Position the crankshaft with special tool 303-448, then remove the tool.
78. Install the crankshaft sprocket, marking sure the flange faces forward.
79. Install the lower end of the left timing chain, aligning the timing marks.

➡**Be sure the upper half of the timing chain is below the tensioner guide dowel.**

80. Install the left timing chain on the camshaft sprocket with the 2 chain (marked) links and the timing marks aligned.

➡**The left timing chain tensioner arm has a bump near the dowel hole for identification.**

81. Position the left timing chain tensioner arm on the dowel pin and install the left timing chain tensioner and bolts. Tighten to 25 Nm (18 ft. lbs.).

➡**Be sure the chain link and crankshaft sprocket timing marks are aligned.**

➡**The lower half of the timing chain must be positioned above the dowel.**

82. Install the right (outer) timing chain on the crankshaft sprocket.
83. Position the timing chain on the camshaft sprocket. Make sure the 2 copper-colored (marked) links align with the camshaft sprocket timing mark.
84. Position the right timing chain tensioner arm on the dowel pin and install the right timing chain tensioner and bolts. Tighten to 25 Nm (18 ft. lbs.).
85. Remove the retaining clips from the right and left timing chain tensioners.
86. Check for correct alignment of all timing marks.
87. Install the crankshaft sensor ring on the crankshaft.
88. Lubricate the balance shaft journals with clean engine oil.

8 mm (0.32 in)

06017-F150-G210

Apply a bead of silicone gasket and sealant along the cylinder head-to-cylinder block surface and the oil pan-to-cylinder block surface—2005 F-250/350 6.8L Engine

89. Using the index mark on the balance shaft, mark the corresponding gear tooth with chalk.

90. Position the balance shaft on the journals.

➥It may be necessary to use an inspection mirror to see the marks.

91. Position the balance shaft on the journals and align the chalk mark on the balance shaft with the camshaft timing mark as shown.

➥Install the bearing caps in their original locations.

92. Install the bolts and tighten the bolts in the sequence shown. Tighten to 10 Nm (89 inch lbs.).

✳✳ WARNING

Only use hand tools to tighten the camshaft sprocket bolt or damage may occur to the camshaft or camshaft sprocket.

93. Tighten the right camshaft sprocket bolt in 2 steps:
- Step 1: Tighten to 40 Nm (30 ft. lbs.) in the sequence shown.
- Step 2: Tighten an additional 90 degrees.

✳✳ WARNING

Only use hand tools to tighten the camshaft sprocket bolt or damage may occur to the camshaft or camshaft sprocket.

94. Tighten the left camshaft sprocket bolt in 2 steps:
- Step 1: Tighten to 40 Nm (30 ft. lbs.) in the sequence shown.
- Step 2: Tighten an additional 90 degrees.

✳✳ WARNING

If the components are to be reinstalled, they must be installed into their original locations.

95. Using the special tool, install all of the remaining camshaft roller followers.

96. Lubricate the roller followers with clean engine oil prior to installation.

✳✳ WARNING

Do not use metal scrapers, wire brushes, power abrasive discs or other abrasive means to clean the sealing surfaces. These tools cause scratches and gouges which make

leak paths. Use a plastic scraping tool to remove all traces of old sealant.

➥If the engine front cover is not secured within 4 minutes, the sealant must be removed and the sealing area cleaned. To clean the sealing area, use silicone gasket remover and metal surface prep. Follow the directions on the packaging. Failure to follow this procedure can cause future oil leakage.

➥Make sure that the engine front cover gasket is in place on the engine front cover before installation.

1	Bolt, Hex Flange Head Pilot, M8 x 1.25 x 50
2	Bolt, Hex Flange Head Pilot, M8 x 1.25 x 50
3	Bolt, Hex Flange Head Pilot, M8 x 1.25 x 50
4	Bolt, Hex Flange Head Pilot, M8 x 1.25 x 50
5	Bolts, Hex Flange Head Pilot, M8 x 1.25 x 50
6	Stud, Hex Head Pilot, M10 x 1.5 x 1.5 x 103
7	Stud, Hex Head Pilot, M10 x 1.5 x 1.5 x 103
8	Bolt, Hex Flange Head Pilot, M8 x 1.25 x 50
9	Bolt, Hex Flange Head Pilot, M8 x 1.25 x 50
10	Bolt, Hex Flange Head Pilot, M8 x 1.25 x 50
11	Stud and Washer, Hex Head Pilot, M8 x 1.25 x 1.25 x 94
12	Stud and Washer, Hex Head Pilot, M8 x 1.25 x 1.25 x 94
13	Stud and Washer, Hex Head Pilot, M8 x 1.25 x 1.25 x 94
14	Stud and Washer, Hex Head Pilot, M8 x 1.25 x 1.25 x 94
15	Stud and Washer, Hex Head Pilot, M8 x 1.25 x 56

06017-F150-G211

Front cover torque sequence—2005 F-250/350 6.8L Engine

Oil pan-to-front cover bolt torque sequence—2005 F-250/350 6.8L Engine

97. Apply a bead of silicone gasket and sealant along the cylinder head-to-cylinder block surface and the oil pan-to-cylinder block surface, at the locations shown.

98. Install a new engine front cover gasket on the engine front cover. Position the engine front cover onto the dowels. Install the fasteners finger tight.

99. Tighten the engine front cover fasteners in sequence in 2 steps.
- Step 1: Tighten fasteners 1 through 15 to 25 Nm (18 ft. lbs.).
- Step 2: Tighten fasteners 6 and 7 to 48 Nm (35 ft. lbs.).

100. Loosely install the oil pan-to-front cover bolts, then tighten the bolts in 2 steps, in the sequence shown.
- Step 1: Tighten to 20 Nm (15 ft. lbs.).
- Step 2: Tighten an additional 60 degrees.

101. Using a new O-ring seal, install the crankshaft position (CKP) sensor and the bolt.

102. Lubricate the new O-ring seal with clean engine oil prior to installation. Tighten to 10 Nm (89 inch lbs.).

103. Connect the CKP sensor electrical connector.

104. Position the A/C compressor and install the 3 bolts. Tighten to 25 Nm (18 ft. lbs.).

105. Position the power steering pump and install the 3 bolts. Tighten to 25 Nm (18 ft. lbs.).

106. Position the starter electrical harness support bracket and install the bolt. Tighten to 10 Nm (89 inch lbs.).

➡**Lubricate the O-ring seal with clean engine oil prior to installation.**

107. Using a new O-ring seal, install the camshaft position (CMP) sensor and the bolt.

108. Tighten to 10 Nm (89 inch lbs.).

109. Connect the CMP sensor electrical connector.

110. Install the left radio ignition interfer-

Front cover seal installation—2005 F-250/350 6.8L Engine

ence capacitor and the nut. Tighten to 10 Nm (89 inch lbs.).

111. Install the right radio ignition interference capacitor and the nut.
Tighten to 10 Nm (89 inch lbs.).

112. Connect the radio ignition interference capacitor electrical connectors.

113. Install the accessory drive belt tensioner and the 3 bolts. Tighten to 25 Nm (18 ft. lbs.).

114. Install the coolant pump pulley and the 4 bolts. Tighten to 25 Nm (18 ft. lbs.).

115. Install the accessory drive idler pulley and the 3 bolts. Tighten to 25 Nm (18 ft. lbs.).

116. Lubricate the engine front cover and the crankshaft seal inner lip with clean engine oil.

117. Using the special tools, install the crankshaft seal into the engine front cover.

➡**If not secured within 4 minutes, the sealant must be removed and the sealing area cleaned. To clean the sealing area, use silicone gasket remover and metal surface prep. Follow the directions on the packaging. Failure to follow this procedure can cause future oil leakage.**

118. Apply silicone gasket and sealant to the Woodruff key slot on the crankshaft pulley.

119. Use the special tool to install the crankshaft pulley.

120. Tighten the new crankshaft pulley bolt in 4 steps.
- Step 1: Tighten to 90 Nm (66 ft. lbs.).
- Step 2: Loosen 360 degrees.
- Step 3: Tighten to 50 Nm (37 ft. lbs.).
- Step 4: Tighten an additional 90 degrees.

121. Rotate the tensioner clockwise and install the accessory drive belt.

122. Install the right valve cover.

123. Install the left valve cover. Install the engine cooling fan.

124. Fill the crankcase with clean engine oil.

5.4L DOHC Engine

1. Before servicing the vehicle, refer to the Precautions Section.

2. Remove the left and right valve cover.

3. Remove the crankshaft front oil seal.

4. Remove the bolts and position the power steering pump aside.

5. Disconnect the crankshaft position (CKP) sensor electrical connector.

6. Drain the engine oil.

7. Remove the bolts.

8. Lower the vehicle.

9. Remove the water pump pulley.

10. Disconnect the camshaft position (CMP) sensor electrical connector.

Crankshaft holding tool installed—DOHC engines

Front cover fastener positions—DOHC engines

Camshaft holding tool installed—DOHC engines

11. Remove the bolt and the belt idler pulley.

✳✳ WARNING

Do not use metal scrapers, wire brushes, power abrasive discs or other abrasive means to clean the sealing surfaces. These tools cause scratches and gouges, which make leak paths. Use a plastic scraping tool to remove all traces of sealant.

12. Remove the bolts and the studs, the engine front cover and the gaskets. Clean and inspect the front cover and oil pan sealing surfaces.

✳✳ WARNING

Since the engine is not freewheeling, if the crankshaft or the camshafts are moved in any manner during removal and installation, the crankshaft and the camshafts must be re-synchronized.

13. Remove the crankshaft sensor ring from the crankshaft.

✳✳ WARNING

Unless otherwise instructed, at no time when the timing chains are removed and the cylinder heads are installed is the crankshaft or the camshaft to be rotated. Severe piston and valve damage will occur.

14. Using the special tool, position the crankshaft as shown.
15. Install the special tool.
16. Remove the two bolts, the timing chain tensioner and tensioner arm.

✳✳ WARNING

Unless otherwise instructed, at no time when the timing chains are removed and the cylinder heads are installed is the crankshaft or the camshaft to be rotated. Severe piston and valve damage will occur.

✳✳ WARNING

Do not remove the special tool from the camshafts.

17. Remove the special tool from the crankshaft.
18. Remove the timing chains from the camshaft sprockets.

19. Remove the right and left timing chains and the crankshaft sprocket from the crankshaft.

✳✳ WARNING

The bolts are different lengths and must be returned to their original location.

20. Remove the bolts and the timing chain guides.
   To install:

✳✳ WARNING

Timing chain procedures must be followed exactly or damage to the pistons or valves will result.

21. Compress the tensioner plunger, using the edge of a soft-jawed vise.
22. Install the tensioner retaining clip. Remove the tensioner from the vise.

➡**The 5.4L 4V uses a 61-link chain**

23. If the copper links are not visible, mark two links on one end and one link on the other end, and use as timing marks.
24. Install the timing chain guides. Torque to 89 inch lbs. (10 Nm).
25. Remove the special tool.
26. Rotate the left camshaft sprocket until the timing mark is approximately at the 12 o'clock position. Rotate the right camshaft timing sprocket until the timing mark is approximately in the 11 o'clock position.
27. Install the special tool.

✳✳ WARNING

Unless otherwise instructed, at no time when the timing chains are removed and the cylinder heads are installed is the crankshaft or the camshaft to be rotated. Severe piston and valve damage will occur.

Install the left timing chain onto the crankshaft sprocket, aligning the one copper link on the timing chain with the slot on the crankshaft sprocket—DOHC engines

✳✳ WARNING

Rotate the crankshaft counterclockwise only. Do not rotate past position shown or severe piston or valve damage will occur.

28. Using the special tool, position the crankshaft.
29. Remove the special tool.
30. Position the inner crankshaft sprocket with the long hub facing outward. The outer hub will face inward.
31. Install the left timing chain onto the crankshaft sprocket, aligning the one copper link on the timing chain with the slot on the crankshaft sprocket.
32. Verify the camshaft sprocket-to-copper link alignment.
33. Position the tensioner arms and tensioners, and install the bolts. Torque to 18 ft. lbs. (25 Nm).
34. Remove the tensioner retaining clip.
35. Remove the special tool.
36. Position the crankshaft sensor ring on the crankshaft.

➡**If the front cover and gasket are not secured within four minutes, the sealant must be removed and the sealing surface cleaned with metal surface cleaner.**

37. Apply silicone gasket and sealant in the locations shown.

06017-F150-G217

Verify the camshaft sprocket-to-copper link alignment—DOHC engines

06017-F150-G218

Apply silicone gasket and sealant in these locations—DOHC engines

38. Position the engine front cover and install the bolts and the studs in three steps:
- Step 1: Tighten numbers one through five to 25 Nm (18 ft. lbs.).
- Step 2: Tighten numbers six and seven to 50 Nm (37 ft. lbs.).
- Step 3: Tighten numbers 11 through 15 to 25 Nm (18 ft. lbs.).

39. Install the belt idler pulley and the bolt. Torque to 18 ft. lbs. (25 Nm).

40. Connect the CMP sensor electrical connector.

41. Install the water pump pulley and the bolts. Torque to 18 ft. lbs. (25 Nm).

42. Raise the vehicle.

43. Tighten the bolts in three steps:
- Step 1: Tighten to 2 Nm (18 inch lbs.).
- Step 2: Tighten to 20 Nm (15 ft. lbs.).
- Step 3: Tighten an additional 60 degrees.

44. Install the drain plug.

45. Connect the CKP sensor.

46. Position the power steering pump and install the bolts. Torque to 18 ft. lbs. (25 Nm).

47. Install the crankshaft front oil seal.

48. Install the left and right valve cover.

49. Fill the engine with clean engine oil.

Piston and Ring Positioning

1   Bolt, Hex Flange Head Pilot, M8 x 1.25 x 50

2   Bolt, Hex Flange Head Pilot, M8 x 1.25 x 50

3   Bolt, Hex Flange Head Pilot, M8 x 1.25 x 50

4   Bolt, Hex Flange Head Pilot, M8 x 1.25 x 50

5   Bolt, Hex Flange Head Pilot, M8 x 1.25 x 50

6   Stud, Hex-Head Pilot, M10 x 1.5 x30

7   Stud, Hex-Head Pilot, M10 x 1.5 x30

8   Bolt, Hex Flange Head Pilot, M8 x 1.25 x 50

9   Bolt,Hex Flange Head Pilot, M8 x 1.25 x 50

10   Bolt,Hex Flange Head Pilot, M8 x 1.25 x 50

11   Stud Hex Shoulder Pilot, M8 x 1.25 x 65 - M8 x 1.25 x 27

12   Stud Hex Shoulder Pilot, M8 x 1.25 x 65 - M8 x 1.25 x 27

13

14   Stud Hex Shoulder Pilot, M8 x 1.25 x 65 - M8 x 1.25 x 27

15   Bolt Hex Shoulder Pilot, M8 x 1.25 x 65

Front cover fastener torque sequence—DOHC engines

7924AG21

4.2L and 4.6L engines—piston and connecting rod front mark locations

4.2L and 4.6L engines—piston-to-engine orientation

5.4L and 6.8L engines—piston and connecting rod assembly positioning

Piston ring arrangement—all gasoline engines

DIESEL ENGINE REPAIR

Alternator

REMOVAL & INSTALLATION

F-Series w/6.0L Engine

SINGLE ALTERNATOR

1. Before servicing the vehicle, refer to the Precautions Section.
2. Disconnect the batteries.
3. Remove the cooling fan.
4. Rotate the accessory drive belt tensioner clockwise and remove the drive belt from the alternator pulley.
5. Disconnect the alternator electrical connectors.
6. Remove the bolts and the alternator.
7. To install, reverse the removal procedure. Torque the mounting bolts to 35 ft. lbs. (47 Nm).

DUAL ALTERNATORS

1. Before servicing the vehicle, refer to the Precautions Section.

➡The following applies to the secondary alternator. For primary alternator, see the Single Alternator procedure.

67197-EFSE-G114

47 Nm (35 lb-ft)

Single alternator mounting—6.0L w/single alternator; 7.3L similar

2. Disconnect the batteries.
3. Remove the coolant fan.
4. Rotate the accessory drive belt tensioner clockwise and remove the drive belt from the alternator pulley.
5. Remove the nut and position the alternator B+ cable aside.
6. Loosen the 2 clamps and remove the turbo-to-air cooler pipe.
7. Disconnect the alternator electrical connectors.
8. Remove the bolts and the alternator.
9. To install, reverse the removal procedure. Torque the mounting bolts to 35 ft. lbs. (47 Nm).

E-Series w/6.0L Engine

SINGLE ALTERNATOR

1. Before servicing the vehicle, refer to the Precautions Section.

➡This procedure also applies to the primary alternator of the dual alternator system.

2. Disconnect the dual batteries.
3. Position the cowl wire harness aside.
4. Remove the bolt and position the manifold absolute pressure (MAP) sensor aside.
5. Remove the bolt and position the ground strap aside.
6. Remove the 2 bolts and position the power steering fluid reservoir aside.
7. Remove the 3 bolts and position the cowl wire harness aside.
8. Remove the 3 bolts, 2 nuts and the intercooler tube bracket.
9. Loosen the 2 clamps and remove the intercooler tube.
10. Rotate the accessory drive belt tensioner clockwise and remove the drive belt from the alternator pulley.

11. Remove the transmission dipstick tube bracket nut from the alternator stud bolt and position the bracket aside.
12. Remove the alternator.

➡Make sure to install the engine-to-body ground strap under the front alternator stud bolt.

13. To install, reverse the removal procedure. Torque the mounting bolts to 35 ft. lbs. (47 Nm).

DUAL ALTERNATORS

1. Before servicing the vehicle, refer to the Precautions Section.

➡This procedure applies to the secondary alternator of the dual alternator system.

2. Disconnect the dual batteries.
3. Position the cowl wire harness aside.
4. Remove the bolt and position the manifold absolute pressure (MAP) sensor aside.
5. Remove the bolt and position the ground strap aside.
6. Remove the 2 bolts and position the power steering fluid reservoir aside.
7. Remove the 3 bolts and position the cowl wire harness aside.
8. Remove the 3 bolts, 2 nuts and the intercooler tube bracket.
9. Loosen the 2 clamps and remove the intercooler tube.
10. Rotate the accessory drive belt tensioner clockwise and remove the drive belt from the alternator pulley.
11. Disconnect the alternator electrical connectors.
12. Remove the bolts and the alternator.
13. To install, reverse the removal procedure. Torque the mounting bolts to 35 ft. lbs. (47 Nm).

7.3L Engine

SINGLE ALTERNATOR

1. Before servicing the vehicle, refer to the Precautions Section.
2. Remove or disconnect the following:
   - Negative battery cable
   - Air cleaner, if necessary
   - Drive belt from the alternator pulley
   - Electrical harness connectors at the alternator assembly
   - Alternator bolts and the alternator from the vehicle

To install:
3. Install or connect the following:
   - Alternator in position
   - Alternator retaining bolts and tighten to 35 ft. lbs. (47 Nm).
4. Install all remaining components in the reverse order of removal.

DUAL ALTERNATOR

1. Before servicing the vehicle, refer to the Precautions Section.
2. Remove or disconnect the following:
   - Single (upper) alternator
   - Electrical harness connectors at the alternator assembly
   - Alternator bolts and the alternator from the vehicle

To install:
3. Install or connect the following:
   - Alternator in position
   - Alternator retaining bolts and tighten to 35 ft. lbs. (47 Nm).
   - Single (upper) alternator

Engine Assembly

REMOVAL & INSTALLATION

F-Series w/6.0L Engine

1. Before servicing the vehicle, refer to the Precautions Section.
2. With the vehicle in **Neutral**, position it on a hoist.
3. Disconnect the left and right battery ground cables.

Vehicles with manual transmission

➡On vehicles equipped with manual transmissions, the transmission must be removed before the engine can be removed.

4. Remove the transmission.
5. Remove the clutch.

All vehicles
6. Remove the air cleaner assembly.
7. Remove the radiator.
8. Remove the charge air cooler.

Vehicles with A/C
9. Remove the A/C condenser assembly.

All vehicles
10. Remove the parking lamp and the headlamp assemblies.
11. Remove the radiator grille, the radiator grille opening panel, and the upper radiator core supports.
12. Remove the front bumper.
13. Disconnect the transmission cooler hoses.
14. Remove the bolts and the transmission oil cooler.
15. Remove the bolts and position the power steering cooler out of the way.
16. Remove the intake manifold.
17. Remove the accessory drive idler pulley and bolt.
18. Disconnect the heater hose at the coolant pump.
19. Remove the battery cable cover.
20. Remove the nut and position the cable out of the way.
21. Disconnect the left glow plug electrical connector and wire retainer.
22. Disconnect the camshaft position (CMP) sensor electrical connector.
23. Remove the clips and disconnect the fuel lines.
24. Remove the lower radiator hose clamp and the hose.
25. Remove the power steering upper mounting bolts.

➡Bolts need to be removed evenly.

26. Remove the bolts and position the power steering pump out of the way.
27. Remove the nut and the battery cable bracket.
28. Remove the ground stud and the ground cable.
29. If equipped, disconnect the injection control pressure (ICP) sensor electrical connector
30. Disconnect the glow plug electrical connectors.
31. Disconnect the right glow plug electrical connector and wire retainer.

Vehicles with A/C
32. Disconnect the A/C high pressure switch and the A/C clutch electrical connectors.
33. Disconnect the air conditioning manifold lines from the A/C compressor.
34. Remove the A/C compressor.

All vehicles
35. Disconnect the crankshaft position (CKP) sensor electrical connector and position the wiring aside.

Vehicles with automatic transmission
36. Remove the automatic transmission fluid indicator and tube.

All vehicles
37. Remove the ground strap at the back of the right head.
38. Remove the solenoid cap and disconnect the starter wiring.
39. Disconnect the block heater electrical connector.
40. Position the block heater and starter wiring harness out of the way.

Vehicles with automatic transmission
41. Remove the torque converter cover.
42. Remove the torque converter nuts.

All vehicles
43. Remove the bolts for the turbocharger adapter pipe.
44. Remove the motor mount nuts.
45. Loosen the nuts at the turbocharger adapter pipe flange.

Vehicles with automatic transmission
46. Remove the nine bell housing bolts.

All vehicles
47. Remove the turbocharger adapter pipe.
48. Remove the nuts and the fuel injector control module bracket.
49. Remove the left rear valve cover stud.
50. Remove the transmission cooler line bracket.
51. Secure the turbocharger outlet pipe.
52. Remove the manufacturer's lifting eye.
53. Install the engine lifting eye on the right cylinder head.
54. Install the front lifting brackets.

Vehicles with automatic transmission
55. Position a suitable jack under the transmission.

All vehicles
56. Install the Heavy Duty Floor Crane and Diesel Engine Lifting Bracket on the engine.
57. Raise the engine high enough to clear the No. 1 crossmember and pull the engine forward and clear of the vehicle.

To install:
All vehicles
58. With the vehicle in **Neutral**, position it on a hoist.
59. Raise the engine high enough to clear the No. 1 crossmember, then position the engine into the vehicle.

Vehicles with automatic transmission
60. Align the torque converter studs with the holes in the engine flywheel, then lower the engine onto the engine mount towers.

Vehicles with manual transmission

61. Lower the engine onto the engine mount towers.

All vehicles

62. Remove the Heavy Duty Floor Crane and the Diesel Engine Lifting Bracket.

Vehicles with automatic transmission

63. Remove the transmission jack.

64. Install the transmission-to-engine mounting bolts. Torque to 35 ft. lbs. (47 Nm).

65. Install the torque converter-to-flywheel retaining nuts. Torque to 26 ft. lbs. (35 Nm).

66. Install the flywheel housing cover.

All vehicles

67. Install the left and right side engine mount retaining nuts. Torque to 76 ft. lbs. (103 Nm).

68. Remove the engine lifting eye from the right side cylinder head.

69. Install the manufacturer's lifting bracket.

70. Remove the two engine lift adapters.

71. Position the turbocharger outlet pipe.

72. Install the transmission cooler tube bracket and nut.

73. Install the left rear valve cover stud.

74. Install the fuel injector control module bracket and nuts.

75. Position the turbocharger adapter pipe.

➡**Do not tighten bolts at this time.**

➡**Apply anti-seize lubricant to the threads prior to installing the bolts.**

76. Install the bolts for the turbocharger adapter pipe.

77. Position back the block heater and starter wiring.

78. Connect the block heater electrical connector.

79. Connect the starter wiring and install the solenoid cap.

80. Connect the ground strap on the right head and install the bolt.

Vehicles with automatic transmission

81. Install the automatic transmission fluid indicator and tube.

All vehicles

82. Position back the wiring and connect the crankshaft position (CKP) sensor electrical connector.

Vehicles with A/C

83. Install the A/C compressor. Torque to 18 ft. lbs. (25 Nm).

84. Position the A/C compressor mani-

fold and install the bolt. Torque to 15 ft. lbs. (21 Nm).

85. Connect the A/C high pressure switch and the clutch electrical connectors.

All vehicles

86. Connect the right glow plug electrical connector and wire retainer.

87. Connect the glow plug module electrical connectors.

88. If equipped, connect the injection control pressure (ICP) sensor electrical connector.

89. Connect the ground cable and the ground stud.

90. Install the battery cable bracket and the nut.

➡**The lower bolts need to be installed evenly.**

91. Position the power steering pump and install the lower bolts. Torque to 18 ft. lbs. (25 Nm).

92. Install the power steering pump upper bolts. Torque to 18 ft. lbs. (25 Nm).

93. Install the lower radiator hose and clamp.

94. Connect the fuel lines and install the clips.

95. Connect the camshaft position (CMP) sensor electrical connector.

96. Connect the left glow plug electrical connector and wire retainer.

97. Connect the battery crossover cable.

98. Install the battery cable cover.

99. Connect the heater hose at the coolant pump.

100. Install the accessory drive idler pulley and bolt. Torque to 35 ft. lbs. (47 Nm).

101. Install the intake manifold.

102. Tighten the turbocharger adapter pipe at the flanges. Torque to 20 ft. lbs. (27 Nm).

103. Tighten the nuts at the turbocharger adapter pipe flange. Torque to 35 ft. lbs. (47 Nm).

104. Position and install the power steering cooler and bolts.

105. Install the transmission oil cooler.

106. Connect the transmission cooler hoses.

107. Install the front bumper.

108. Install the upper radiator core support, radiator grill opening panel and the radiator grill.

109. Install the headlamp and the parking lamp assemblies.

Vehicles with A/C

110. Install the A/C condenser assembly.

All vehicles

111. Install the charge air cooler.

112. Install the radiator.

113. Install the air cleaner assembly.

Vehicles with manual transmission

114. Install the clutch assembly.

115. Install the transmission.

All vehicles

116. Fill the motor with clean engine oil.

117. Connect the left and right battery cables.

118. Fill the cooling system.

Vehicles with automatic transmission

119. Check and fill the automatic transmission.

E-Series w/6.0L Engine

1. Before servicing the vehicle, refer to the Precautions Section.

2. With the vehicle in **Neutral**, position it on a hoist.

3. Disconnect the battery ground cable.

4. Remove the A/C condenser assembly.

5. Remove the front bumper.

6. Remove the radiator grille support.

7. Remove the bolts, disconnect and remove the air bag sensor. Position the headlamp wiring harness aside.

➡**Mark the latch before removal.**

8. Remove the bolts and hood latch.

9. Remove the bolts and the radiator core supports.

10. Remove the intake manifold.

11. Remove the right fan stator stand-off. Remove the bolt and position the cable aside.

12. Remove the bolt and disconnect the A/C manifold. Disconnect the A/C electrical connector.

13. Remove and discard the O-ring seals.

14. Cap or plug the A/C openings as needed.

15. Disconnect the A/C pressure switch and remove the A/C manifold from the tee block.

16. Remove and discard the O-ring seals.

17. Cap or plug the A/C openings as needed.

18. Remove the bolts and A/C compressor.

19. Disconnect the heater hose. Position the heater hose and tube aside.

20. Remove the bolt and idler pulley.

21. Disconnect the engine coolant fill hose from the retaining clip. Loosen the clamp and remove the engine coolant fill hose.

22. Disconnect the vacuum pump hose and remove the engine coolant fill hose clip.

➡**The bolt behind the pulley is accessed thorough the pulley with an extension and a short socket.**

23. Remove the bolts and the vacuum pump.

24. Remove the power steering upper mounting bolt.

25. Remove the left fan stator stand-off.

➡️**The front bolt will remain in the power steering pump.**

26. Remove the bolts and position the power steering pump aside.

27. Remove the lower radiator hose.

28. Disconnect the fuel lines.

29. Disconnect the crankshaft position (CKP) sensor electrical connector and retaining clips.

30. Disconnect the injection control pressure (ICP) sensor electrical connector.

31. Disconnect the glow plug electrical connector and wire retainers.

32. Disconnect the camshaft position sensor electrical connector and retaining clips.

33. Disconnect the two electrical connectors and wire retainers at the transmission.

34. Disconnect the glow plug module electrical connectors.

✳✳ WARNING

Do not use power tools when removing the nut.

35. Position back the boot and disconnect the B+ wire.

36. Disconnect the exhaust back pressure sensor electrical connector and push-pin retainer.

37. Disconnect the left glow plug electrical connector.

38. Disconnect the two electrical connectors.

39. Disconnect the two powertrain control module (PCM) electrical connectors and position the engine wiring harness aside.

40. Remove the bolts and the oil filter assembly.

41. Remove the bolt at the transmission filter housing.

42. Remove the retaining nut and remove the transmission cooling tubes.

43. Remove the torque converter cover.

44. Remove and discard the torque converter nuts.

45. Disconnect the block heater electrical connector.

46. Remove the retaining nut and position the cable bracket aside.

47. Remove the starter solenoid protective cap.

48. Disconnect the starter motor electrical connections.

49. Remove the bolts and the starter motor.

50. Remove the cylinder block drain plugs and drain the coolant from the block.

➡️**Apply clean engine oil to the O-ring seal prior to installing.**

51. Install the cylinder block drain plugs.

52. Remove the motor mount nuts.

53. Remove the lower engine-to-transmission bolts.

54. Remove the mounting bolts from the lifting bracket and remove the tubes.

55. Disconnect the glow plug wire retainer. Remove the retaining nut and the oil indicator and tube.

56. Remove and discard the O-ring seal.

57. Disconnect the exhaust back pressure tube at the exhaust manifold.

58. Remove the retaining nuts and glow plug module.

➡️**Power stud removed for clarity.**

59. Remove the retaining nuts and the glow plug module mounting bracket.

➡️**Mark the location of the studs prior to removal.**

60. Remove the bolts and the left valve cover.

61. Remove the valve cover gasket.

62. Cover the cylinder head after the valve cover is removed.

63. Remove the ICP sensor.

64. Plug or cap the opening as needed.

➡️**Do not tighten the bolts at this time.**

65. Install the special tools.

66. Remove the turbocharger exhaust pipe.

67. Remove the factory installed engine lifting bracket.

68. Install the special tool.

69. Install the engine lifting attachment and lifting crane.

70. Remove the oil pan drain plug and drain the engine oil.

➡️**It may be necessary to raise the engine for oil pan removal. Disconnect the linkage from the engine lifting attachment to raise the engine.**

71. Remove the bolts and position back the oil pan until the oil pick up tube bolts are accessible.

72. Remove the bolts and let the oil pick-up tube go into the oil pan. Remove the oil pan.

73. Remove the press-in-place gasket and discard.

74. Remove and discard the oil pick-up tube O-ring seal.

75. Position a suitable transmission jack under the transmission.

76. Remove the upper engine-to-transmission bolts.

77. Raise the engine. Remove the four bolts and left motor mount.

78. Remove the bolts and the right motor mount.

➡️**The engine must be moved to the driver's side for removal.**

79. Remove the engine from the vehicle.
To install:
80. With the vehicle in **Neutral**, position it on a hoist.

➡️**The engine will need to be to the left of the center line to install.**

81. Raise the engine high enough to clear the No.1 crossmember, and position the engine into the vehicle.

82. Position the motor mounts in the vehicle.

83. Align the torque converter studs with the holes in the flywheel and push the engine in. Install the top engine-to-transmission bolts. Torque to 35 ft. lbs. (47 Nm).

84. Lower the engine until it is just above the motor mounts. Install the retaining bolts in the right motor mount. Torque to 59 ft. lbs. (80 Nm).

85. Install the four retaining bolts in the left motor mount. Torque to 59 ft. lbs. (80 Nm).

86. Remove the transmission jack.

87. Install a new O-ring seal on the oil pick-up tube and position the oil pick-up tube in the oil pan.

88. Install a new press-in-place gasket into the upper oil pan.

89. Position the oil pan in the vehicle.

90. Install the oil pick-up tube and bolts.

91. Position the oil pan and install the bolts.

92. Clean and inspect the oil pan drain plug and gasket. Install new, if necessary.

93. Install the oil pan drain plug.

94. Lower the engine. Remove the lifting crane and the engine lifting attachment.

95. Remove the special tool.

96. Install the factory engine lifting bracket.

97. Loosely install the turbocharger exhaust pipe.

98. Remove the special tools.

➡️**Remove the plug or cap as needed.**

99. Install the injection control pressure (ICP) sensor.

➡️**Remove the protective covering from the cylinder head.**

100. Install the left valve cover gasket, valve cover and bolts.

101. Install the glow plug module mounting bracket and retaining nuts.

102. Install the glow plug module and retaining nuts.

103. Connect the exhaust back pressure tube at the exhaust manifold.

➡**Install a new O-ring seal and coat with clean engine oil.**

104. Install the oil indicator and tube. Install the retaining nut. Connect the glow plug wire retainer.

105. Position the fuel tubes and install the mounting bolts for the fuel tube bracket.

106. Install the lower engine-to-transmission bolts. Torque to 35 ft. lbs. (47 Nm).

➡**The top retaining nut must be tightened first.**

107. Install the motor mount nuts. Torque to 66 ft. lbs. (90 Nm).

108. Install the starter motor and bolts.

109. Connect the starter motor electrical connections.

110. Install the starter motor solenoid protective cap.

111. Position the cable bracket and install the retaining nut.

112. Connect the block heater electrical connector.

113. Install the new torque converter nuts. Torque to 26 ft. lbs. (35 Nm).

114. Install the torque converter cover.

➡**Inspect the O-ring seals. Install new seals, if necessary.**

115. Position the transmission cooler tubes and install the retaining nut.

116. Position the tube lock and install the retaining bolt.

117. Install the oil filter assembly and bolts. Torque to 18 ft. lbs. (25 Nm).

118. Position the engine wiring harness and connect the powertrain control module (PCM) electrical connectors.

119. Connect the two electrical connectors.

120. Connect the left glow plug electrical connector.

121. Connect the push pin retainer and exhaust back pressure sensor electrical connector.

❋❋ WARNING

Do not use power tools when installing the nut. Use care not to overtighten the nut during installation.

122. Connect the B+ wire and reposition the boot.

123. Connect the two electrical connectors and wire retainers at the transmission.

124. Connect the glow plug module electrical connectors.

125. Connect the camshaft position sensor electrical connector and retaining clips.

126. Position the wiring and connect the glow plug electrical connector. Connect the ICP sensor electrical connector.

127. Connect the crankshaft position (CKP) sensor electrical connector and retaining clips.

128. Connect the fuel lines and install the retaining clips.

129. Install the lower radiator hose.

130. Position the power steering pump and install the bolts. Torque to 18 ft. lbs. (25 Nm).

131. Install the left fan stator stand-off.

132. Install the power steering upper mounting bolt. Torque to 18 ft. lbs. (25 Nm).

➡**The bolt behind the pulley is accessed through the pulley with an extension and a short socket.**

133. Install the vacuum pump and bolts. Torque to 35 ft. lbs. (47 Nm).

134. Connect the vacuum pump hose and install the engine coolant fill hose clip.

135. Install the engine coolant fill hose and tighten the clamp. Connect the engine fill hose into the retaining clip

136. Install the idler pulley and bolt. Torque to 35 ft. lbs. (47 Nm).

137. Position the heater hose back and connect.

138. Install the A/C compressor and bolts. Torque to 18 ft. lbs. (25 Nm).

➡**Remove the caps or plugs as needed. Install new O-ring seals and lubricate with PAG oil.**

139. Install the A/C manifold at the tee block. Torque to 35 ft. lbs. (47 Nm). Connect the high pressure switch electrical connector.

➡**Remove the caps or plugs as needed. Install new O-ring seals and lubricate with PAG oil.**

140. Position the A/C manifold and install the bolt. Connect the A/C clutch electrical connector.

141. Position back the cable and install the bolt. Install the right fan stator stand-off.

142. Install the intake manifold.

143. Install the radiator core supports and bolts.

144. Position the hood latch and install the bolts.

145. Position the headlamp wiring harness and connect the air bag sensor. Install the bolts for the air bag sensor.

146. Install the valance panel, radiator grille support and grille opening panel reinforcement.

147. Install the front bumper.

148. Install the A/C condenser assembly.

149. Fill the engine with clean engine oil.

150. Connect the battery ground cable.

151. Check and top off the transmission fluid level.

7.3L Engine

❋❋ CAUTION

The fuel system remains under pressure, even after the engine has been turned OFF. The fuel system pressure must be relieved before disconnecting any fuel lines. Failure to do so may result in fire and/or personal injury.

1. Before servicing the vehicle, refer to the Precautions Section.

2. Remove or disconnect the following:
   - Hood
   - Coolant
   - Engine oil
   - Negative battery cable
   - Air cleaner and intake duct assembly
   - Upper grille support bracket
   - Upper air conditioning condenser mounting bracket
   - Refrigerant, using approved equipment to remove the condenser
   - Radiator fan shroud halves
   - Fan and clutch assembly
   - Radiator hoses and the transmission cooler lines, if equipped
   - Condenser. Cap all openings immediately!
   - Radiator
   - Power steering pump and position it out of the way
   - Fuel supply line heater and alternator wires at the alternator
   - Oil pressure sending unit wire at the sending unit
   - Sender from the firewall and lay it on the engine
   - Accelerator cable and the speed control cable, if equipped, from the injection pump
   - Cable bracket with the cables attached, from the intake manifold and position it out of the way
   - Transmission kickdown rod from the injection pump, if equipped
   - Main wiring harness connector

from the right side of the engine and the ground strap from the rear of the engine
- Fuel system pressure
- Fuel return hose from the left rear of the engine
- The 2 upper transmission-to-engine attaching bolts
- Heater hoses
- Water temperature sender wire
- Overheat light switch wire and position the wire out of the way
- Battery ground cables from the front of the engine and the cables from the starter
- Fuel inlet line, and plug the fuel line at the fuel pump.
- Exhaust pipe at the exhaust manifold
- Engine insulators from the No. 1 crossmember
- Flywheel inspection plate and the 4 converter-to-flywheel attaching nuts, if equipped with automatic transmission

3. Support the transmission on a jack.
4. Remove the 4 lower transmission attaching bolts.
5. Attach an engine lifting sling and remove the engine from the vehicle.

To install:

6. Lower the engine into vehicle.
7. Align the converter to the flexplate and the engine dowels to the transmission.
8. Install the engine mount bolts and tighten them to 80 ft. lbs. (109 Nm).
9. Remove the engine lifting sling.
10. Install the 4 lower transmission attaching bolts. Tighten the bolts to 65 ft. lbs. (88 Nm).
11. Remove transmission jack.
12. Raise and support the front end.
13. Install or connect the following:
- 4 converter-to-flywheel attaching nuts, if equipped with an automatic transmission. Tighten the nuts to 34 ft. lbs. (47 Nm).
- Flywheel inspection plate. Tighten the bolts to 60–90 inch lbs. (6.7–10 Nm).
- Exhaust pipe at the exhaust manifold
- Fuel inlet line
- Battery ground cables to the front of the engine
- Starter cables at the starter
- Overheat light switch wire
- Water temperature sender wire
- Heater hoses
- The 2 upper transmission-to-engine attaching bolts. Tighten the bolts to 65 ft. lbs. (88 Nm).

- Fuel return hose at the left rear of the engine
- Main wiring harness connector at the right side of the engine and the ground strap from the rear of the engine
- Transmission kickdown rod at the injection pump, if equipped
- Accelerator cable and the speed control cable, if equipped, at the injection pump
- Cable bracket with the cables attached, to the intake manifold
- Oil pressure sending unit
- Oil pressure sending unit wire at the sending unit
- Fuel supply line heater and alternator wires at the alternator
- Power steering pump
- Radiator
- Condenser
- Radiator hoses and the transmission cooler lines, if equipped
- Fan and clutch assembly
- Radiator fan shroud halves
- Upper grille support bracket and upper air conditioning condenser mounting bracket
- Air cleaner and intake duct assembly

14. Refill the engine oil with the recommended type of engine oil and the correct amount.
15. Connect the negative battery cable.
16. If equipped with air conditioning, charge the system.
17. Fill the cooling system.
18. Install the hood.

Water Pump

REMOVAL & INSTALLATION

F-Series w/6.0L Engine

1. Before servicing the vehicle, refer to the Precautions Section.
2. With the vehicle in **Neutral**, position it on a hoist.
3. Remove the radiator and shroud assembly.
4. Disconnect the cooling fan electrical connector. Unclip and position the fan wiring aside.

➡**Use a hole in the fan hub to prevent the fan from turning.**

5. Using the hub tool, loosen the cooling fan and clutch.
6. Remove the cooling fan assembly.
7. Remove the stator bolts and the stator.

40 Nm (30 lb-ft)

67197-EFSE-G115
Remove the stator bolts and the stator

23 Nm (17 lb-ft)

67197-EFSE-G116
Water pump mounting

8. Rotate the accessory drive belt tensioner clockwise and position the drive belt aside.
9. Remove the bolts and the coolant pump pulley.
10. Remove the bolts and the coolant pump.
11. Remove and discard the O-ring seal.
12. Clean and inspect the coolant pump mounting.
13. To install, reverse the removal procedure.
14. Observe the following torques:
- Stator bolts: 30 ft. lbs. (40 Nm)
- Pulley bolts: 23 ft. lbs. (31 Nm)
- Water pump bolts: 17 ft. lbs. (23 Nm)

E-Series w/6.0L Engine

1. Before servicing the vehicle, refer to the Precautions Section.
2. Remove the radiator.
3. Disconnect the electrical connector. Release the wiring from the stator.

➡**Use a hole in the fan hub to prevent the fan from turning.**

4. Using the special tool, loosen the fan clutch by turning the wrench counterclockwise. Remove the cooling fan and clutch.
5. Remove the bolts and cooling fan stator assembly.
6. Remove the accessory drive belt.
7. Remove the bolts and the coolant pump pulley.
8. Remove the bolts and the coolant pump.

WATER PUMP

APPLY AVIATION PERMATEX®
NO. 3 OR EQUIVALENT
TO THESE BOLTS

① THESE BOLTS 2 3/4 IN. LONG
ALL OTHERS ARE 1 1/2 IN. LONG

7924FG56

Apply RTV sealant to the bolts indicated—7.3L diesel engine

67197-EFSE-G117

Apply clean engine oil to the O-rings—early build 6.0L engines

67197-EFSE-G118

Apply clean engine oil to the O-rings—late build 6.0L engines

9. Remove and discard the O-ring seal.

10. Clean and inspect the coolant pump mounting.

11. To install, reverse the removal procedure.

7.3L Engine

1. Before servicing the vehicle, refer to the Precautions Section.

2. Disconnect both battery ground cables.

3. Drain the cooling system.

4. Remove the radiator shroud and the fan clutch and fan.

➡The fan clutch bolts are right-hand thread. Remove them by turning counter-clockwise.

5. Loosen, but do not remove, the water pump pulley bolts.

6. Remove the drive belt.

7. Remove the water pump pulley bolts and remove the pulley from the pump.

8. Detach the Engine Coolant Temperature (ECT) sensor connector.

9. Remove the heater hose from the water pump.

10. Remove the bolts attaching the water pump to the front cover and lift off the pump.

To install:

11. Thoroughly clean the mating surfaces of the pump and front cover.

12. Using a new gasket, position the water pump over the dowel pins and into place on the front cover.

13. Install the attaching bolts. Tighten the bolts to 15 ft. lbs. (20 Nm).

14. Connect the heater hose to the pump.

15. Attach the ECT sensor connector.

16. Install the water pump pulley and start the water pump pulley bolts.

17. Install the drive belt.

18. Install and tighten the pulley retaining bolts to 12–18 ft. lbs. (16–24 Nm).

19. Install the fan and fan shroud assembly.

20. Fill and bleed the cooling system.

21. Connect the battery ground cables.

22. Start the engine and check for leaks.

Glow Plugs

REMOVAL & INSTALLATION

6.0L Engine

F-SERIES EARLY BUILD

1. Before servicing the vehicle, refer to the Precautions Section.

2. If servicing the right glow plugs, remove the evaporator core housing.

3. Disconnect the glow plug electrical connector.

4. Remove the glow plug buss bar.

➡If coolant residue is found on the glow plug, a new glow plug sleeve may have to be installed.

5. Remove the glow plug.

To install:

6. Install the glow plug. Torque to 14 ft. lbs. (19 Nm).

7. Clean and inspect the O-rings and install new if necessary.

8. Apply clean engine oil to the O-rings.

9. Install the glow plug buss bar.

10. Connect the glow plug electrical connector.

11. If removed, install the evaporator core housing.

F-SERIES LATE BUILD

1. Before servicing the vehicle, refer to the Precautions Section.

2. Disconnect the glow plug electrical connector.

❋❋ WARNING

Do not pull on the wiring to remove the glow plug connector or damage may occur.

3. Remove the glow plug harness.

➡If coolant residue is found on the glow plug, a new glow plug sleeve may have to be installed.

4. Remove the glow plug.

To install:

5. Install the glow plug. Torque to 14 ft. lbs. (19 Nm).

6. Clean and inspect the O-rings and install new if necessary.

7. Apply clean engine oil to the O-rings.

8. Install the glow plug harness.

9. Connect the glow plug electrical connector.

E-SERIES

1. Before servicing the vehicle, refer to the Precautions Section.

Number 2 glow plug

2. Remove the retainers and the air deflector.

3. Remove the power steering reservoir bracket retainers.

4. Remove the power steering fluid indicator and retainers. Remove the power steering reservoir mounting bracket. Install the power steering fluid indicator and position the power steering reservoir aside.

5. Disconnect the coolant hoses from the air cleaner outlet pipe.

6. Loosen the clamps and remove the air cleaner outlet pipe.

Number 1, 3-8 glow plugs

7. Remove the engine cover.

All glow plugs

❊❊ WARNING

Do not pull on the wiring to remove the glow plug connector or damage may occur.

➡**Only the number 2 glow plug is accessed from under the hood, all others are accessed from inside the cab.**

8. Remove the glow plug harness as needed.

➡**If coolant residue is found on the glow plug, a new glow plug sleeve may have to be installed.**

9. Remove the glow plug.

To install:

All glow plugs

10. Install the glow plug.

67197-EFSE-G119

Air deflector removal

19 Nm (14 lb/ft)

7924FG57

Tighten the glow plugs to 14 ft. lbs. (19 Nm) and attach the connector—7.3L engine

11. Clean and inspect the O-ring seals and install new if necessary. Apply clean engine oil to the O-ring seals.

12. Install the glow plug harness.

Number 1, 3-8 glow plug

13. Install the engine cover.

Number 2 glow plug

14. Install the air cleaner outlet pipe and tighten the clamps.

15. Connect the coolant hoses to the air cleaner outlet pipe.

16. Remove the power steering fluid indicator. Position the power steering reservoir mounting bracket. Install the retainers and power steering fluid indicator.

17. Install the power steering reservoir bracket retainers.

18. Position the air deflector and install the retainers.

7.3L Engine

❊❊ CAUTION

The red-striped wiring harness carries 115V direct current. Severe electrical shock may be received. DO NOT pierce.

1. Before servicing the vehicle, refer to the Precautions Section.

2. Disconnect the negative battery cable.

3. Remove the rocker arm cover.

4. Disconnect the glow plug electrical leads using a pair of pliers.

5. Remove the glow plugs by unscrew-

ing them from the cylinder head with a 10mm socket and wrench.

6. Inspect the tips of the plugs for any evidence of distortion or missing tip ends; replace them if necessary.

To install:

7. Install the glow plug into the cylinder head. Tighten the glow plugs to 14 ft. lbs. (19 Nm).

8. Attach the glow plug electrical connector. Be sure that the glow plug wiring is routed to avoid moving components in the engine bay.

9. Install the rocker arm cover.

➡**When the battery is disengaged and reconnected, some abnormal drive symptoms may occur while the vehicle relearns its adaptive strategy. The vehicle may need to be driven 10 miles (16 km) or more to relearn this strategy.**

10. Connect the negative battery cable.

Cylinder Head

REMOVAL & INSTALLATION

E-Series w/6.0L Engine

1. Before servicing the vehicle, refer to the Precautions Section.

2. Remove the engine.

3. Remove the bolts and flexplate.

4. Mount the engine on an engine stand.

5. Remove the rear lifting eye.

6. Remove the lifting tools.

✳✳ WARNING

Do not pull on the wiring to remove the glow plug connector or damage may occur.

7. Remove the glow plug harness.

8. Remove the eight glow plugs.

9. Prior to removing the exhaust manifolds, inspect the exhaust manifold for warpage with a feeler gauge between the manifold and cylinder head. Record the measurement and compare with the specifications.

10. Remove the bolts and the exhaust manifolds.

Left cylinder head

11. Remove the left cylinder head banjo fitting and fuel line. Discard the sealing washers.

12. Remove the bolts and turbocharger heat shield.

13. Remove the protective covering from the left cylinder head.

Right cylinder head

➡**Mark the location of the stud bolts.**

14. Remove the right valve cover.

15. Remove the stud bolts, bolts and valve cover.

16. Clean and inspect the gaskets.

17. Install a new gasket if necessary.

18. Clean and inspect the sealing surfaces.

19. Remove the right high pressure oil rail-to-valve cover gasket.

Both cylinder heads

20. Remove the bolts and the high pressure oil rail.

✳✳ WARNING

Do not attempt to apply battery voltage to the fuel injector or damage to the fuel injector will occur.

21. Using a 19 mm (0.74 in.) socket, push the fuel injector electrical connector out of the rocker arm carrier.

➡**If engine oil is found in the engine coolant or engine coolant is found in the combustion chambers, new injector sleeves may need to be installed.**

22. Remove the bolt, the fuel injector hold down and the fuel injector.

23. Remove the crankcase-to-head tube assembly.

24. Remove and discard the O-ring seals.

25. Remove and discard the 20 inner cylinder head bolts.

26. Remove the 16 bolts and the rocker arm assemblies.

27. Remove the bolts and rocker arm carrier.

28. Clean and inspect the gasket.

29. Install new gaskets if necessary.

30. Clean and inspect the sealing surfaces.

31. Mark the 16 valve bridges with a permanent marker and remove.

32. Mark and remove the 16 push rods.

33. Remove the outer cylinder head bolts.

34. Install the lifting tool.

35. Remove the cylinder head.

36. Remove and discard the cylinder head gasket and dowels.

37. Clean and inspect the gasket sealing surfaces.

To install:
Both cylinder heads

➡**Use care to avoid scratching the blue compound on the cylinder head gasket.**

38. Install a new cylinder head gasket with the part number facing upward and verify the top five holes and the head gasket push rod holes line up.

39. Install new dowels and the cylinder head gasket.

40. Using the special tools, install the cylinder head on the engine.

41. Install the lifting bracket.

42. Install the bolts.

43. Install the cylinder head.

44. Install the outer cylinder head bolts finger-tight.

➡**Higher mileage engines require push rods to be cleaned so the copper colored end of the push rod can be identified.**

45. Apply clean engine oil to each end of the push rod. Insert them into their respective positions with the copper colored end up.

➡**Coat the end of each valve stem with clean engine oil.**

46. Install the 16 valve bridges.

47. Install the rocker arm carrier.

48. Install the press-in-place gasket.

49. Install the rocker arm carrier and bolts.

✳✳ WARNING

Rotate the crankshaft until the damper locating dowel notch is in the six o'clock position or engine damage can occur.

➡**Apply clean engine oil to the top center of each valve bridge.**

50. Install the rocker arm assemblies and 16 bolts. Torque to 23 ft. lbs. (31 Nm).

✳✳ WARNING

Using too much engine oil on the threads of the cylinder head bolts can cause damage to the threads and poor sealing. Using anti-seize compounds, grease or any other lubricants other than engine oil on the cylinder head bolt threads can affect the true torque value of the bolts.

➡**Lightly lubricate the new cylinder head bolt threads and flanges with clean engine oil.**

51. Install the 20 inner cylinder head retaining bolts finger-tight.

31 Nm
(23 lb-ft)

Rocker arm carrier installation—6.0L engine

67197-EFSE-G120

Cylinder head bolt torque sequence—6.0L engine

67197-EFSE-G121

52. Tighten the head bolts in the following sequence.

a. Tighten bolts 1 through 10 to 88 Nm (65 ft. lbs.).

b. Tighten bolts 1, 3, 5, 7 and 9 to 115 Nm (85 ft. lbs.).

c. Tighten bolts in sequence 1 through 10, clockwise 90 degrees.

d. Tighten bolts in sequence 1 through 10, a second time, clockwise 90 degrees.

e. Tighten bolts in sequence 1 through 10, a third time, clockwise 90 degrees.

f. Tighten bolts 11 through 15 to 24 Nm (18 ft. lbs.).

g. Tighten bolts 11 through 15 to 31 Nm (23 ft. lbs.).

➡**Install new O-ring seals and apply clean engine oil.**

53. Install a crankcase-to-head tube assembly.

✳✳ WARNING

If the fuel injector oil inlet D-ring is damaged, a new fuel injector must be installed.

➡**Lubricate the fuel injector and O-ring seals liberally with clean engine oil.**

54. Install new O-ring seals and a copper washer on the fuel injector.

55. Install the fuel injector, the fuel injector hold-down and bolt.

✳✳ WARNING

Be sure the injector wiring is clear of all moving parts or engine damage can occur.

56. Install the fuel injector electrical connector into the rocker carrier.

➡**Apply clean engine oil to the top fuel injector O-ring seals before installing the high pressure oil rail.**

57. Position the high pressure oil rail on the injectors.

58. Place the high pressure oil rail on top of the carrier so that the four single ball tubes are engaging the fuel injector lead angle.

59. Insert three guide bolts, two on the ends of the straight side of the high pressure oil rail and one in the middle of the wavy side of the high pressure rail. Install the guide studs six to seven turns.

60. Press the high pressure oil rail into the fuel injectors.

61. Inspect that the high pressure oil rail

mounting feet are flat against the mounting surface.

62. Loosely install the six bolts.

63. Install the remaining bolts and tighten.

64. Remove the three guide bolts.

65. Loosely install the three remaining bolts.

66. Tighten the nine bolts in the sequence shown.

Right cylinder head

67. Install the right high pressure oil rail-to-valve cover gasket.

✳✳ WARNING

To prevent engine damage, do not use air powered tools when installing the valve cover.

68. Install the right valve cover. Torque to 71 inch lbs. (8 Nm).

69. Install the right exhaust manifold and the bolts.

➡**Apply anti-seize lubricant to the bolt threads prior to installing. Torque to 28 ft. lbs. (39 Nm).**

➡**Start installing the bolts with the second bolt from the rear on the top. The hole diameter is smaller, therefore allowing alignment of the remaining bolts.**

Left cylinder head

70. Cover the left cylinder head with an appropriate covering.

71. Install the turbocharger heat shield and bolts.

➡**Install new sealing washers.**

High pressure oil rail torque sequence—6.0L engine

67197-EFSE-G122

72. Install the left cylinder head fuel line. Torque to 28 ft. lbs. (39 Nm).

➡ **Start installing the bolts with the second bolt from the rear on the top. The hole diameter is smaller, therefore allowing alignment of the remaining bolts.**

➡ **Apply anti-seize lubricant to the bolt threads prior to installing.**

73. Install the left exhaust manifold and the bolts. Torque to 28 ft. lbs. (39 Nm).

Both cylinder heads

74. Install the eight glow plugs.

➡ **Clean and inspect the glow plug connector O-ring seals and install new as necessary.**

➡ **Apply clean engine oil to the O-ring seals.**

75. Install the glow plug harness.
76. Install the special tools.
77. Install the special tool.
78. Install the lifting crane and remove the engine from the engine stand.
79. Install the flexplate and the bolts.
80. Tighten the bolts in the following sequence.

   a. Tighten all bolts to 5 Nm (44 inch lbs.).

   b. Tighten all bolts to 94 Nm (69 ft. lbs.) in the sequence shown.

81. Install the engine.

F-Series w/6.0L Engine

LEFT CYLINDER HEAD

1. Before servicing the vehicle, refer to the Precautions Section.
2. With the vehicle in **Neutral**, position it on a hoist.
3. Drain the engine oil.
4. Remove the intake manifold.
5. Remove the left cylinder block drain plug.
6. Remove the fuel injector control module mounting bracket.
7. Disconnect the glow plug connector and position aside.
8. Remove the nut and position the oil level indicator and tube aside.

➡ **Mark the position of the valve cover bolts for valve cover bolt installation.**

9. Remove the eleven bolts and the valve cover. Remove the valve cover gasket.

Early build

10. Disconnect the high-pressure oil rail supply line at the high-pressure oil rail.
11. Remove the bolts and the high-pressure oil rail.
12. Disconnect and remove the high-pressure oil supply line.
13. Remove and discard the crankcase-to-head tube assembly.
14. Remove the glow plug buss bar.

Late build

➡ **Do not remove the oil rail end plugs or acoustic wave attenuator port fitting. Service parts are not available to support the components.**

15. Remove the bolts and the high-pressure oil rail.

➡ **The rings on the crankcase-to-head tube are to be used to pry the tube assembly from the branch tube assembly or the oil rail assembly.**

16. Remove the crankcase-to-head tube assembly.
17. Remove and discard the O-rings.

> ⁂ **WARNING**
>
> **Do not pull on the glow plug wire or damage may occur.**

18. Remove the glow plug harness.

All vehicles

19. Remove the four glow plugs.

> ⁂ **WARNING**
>
> **Do not attempt to apply battery voltage to the fuel injector or damage to the fuel injector will occur.**

20. Using a 19 mm socket, push the fuel injector electrical connector out of the rocker arm carrier.
21. Prior to removing the injector assembly, insert clean shop towels in the oil drain holes adjacent to each glow plug.

> ⁂ **WARNING**
>
> **Failure to account for all snaprings or pieces prior to placing the vehicle back in service can cause engine damage. A missing snapring can be ingested into the lube oil system causing severe engine damage.**

> ⁂ **WARNING**
>
> **To prevent engine damage, do not use air tools to remove the fuel injectors. The snapring that extracts the injector can dislodge and fall into the oil drain hole.**

➡ **There is no need to drain the fuel rail.**

➡ **If engine oil is found in the engine coolant or engine coolant is found in the combustion chambers, a new injector sleeve may need to be installed.**

22. Remove the bolt, fuel injector hold-down and fuel injector.

67197-EFSE-G123

Flexplate torque sequence—6.0L engine

➡️If a snapring or piece of a snapring is missing from the injector hold-down assembly, it must be located prior to removing the shop towels.

23. Remove the shop towels.
24. Remove the nuts and bolts from the turbocharger adapter pipe.
25. Remove the turbocharger adapter pipe.

➡️**Remove and discard the O-ring.**

26. Remove the oil level indicator and tube.
27. Remove the bolts and turbocharger heat shield.
28. Disconnect the exhaust backpressure tube at the exhaust manifold.
29. Remove the retaining nuts. Remove the exhaust backpressure bracket assembly.
30. Remove the retaining bolt from the fuel line bracket and position the fuel lines aside.
31. Remove the bolt and the idler pulley.
32. Disconnect the heater hose from the front cover and position aside.
33. Remove the fan shroud mounting stud.

➡️**Remove and discard the sealing washers.**

34. Remove the banjo fitting and the fuel line.
35. Remove and discard the inner cylinder head bolts.
36. Remove the eight bolts and the rocker arm assemblies.
37. Mark the eight valve bridges with a permanent marker and remove.

❈❈ WARNING

To prevent engine damage, keep the push rods in the order in which they were removed. Install all push rods back in their original positions.

38. Mark the location and remove the eight push rods.
39. Remove the outer cylinder head bolts.
40. Install the special tool and the lifting crane. With the help of an assistant, remove the cylinder from the vehicle.
41. Remove and discard the cylinder head gasket and dowels.
42. Clean and inspect the gasket sealing surfaces.
   To install:
   All vehicles, late build

❈❈ WARNING

Install new D-ring seals on the crankcase-to-head tube. It requires

several hours after installation for the D-ring seals to relax back to their original size. If the tube assembly is installed before the D-ring seals have relaxed, damage to the D-ring seals can occur.

43. Install new D-ring seals on each end of the crankcase to head tube assemblies.

➡️**Use care to avoid scratching the blue compound on the cylinder head gasket.**

44. Install the gasket with the part number facing upward and verify the five top holes and head gasket push rod holes line up.
45. Install the dowels and the cylinder head gasket.

➡️**Position the fuel lines in place before the cylinder head is installed.**

46. Using the special tools and with the help of an assistant, install the cylinder head on the engine. Remove the special tools.
47. Install the outer cylinder head bolts finger-tight.

❈❈ WARNING

To prevent engine damage, keep the push rods in the order in which they were removed. Install all push rods back in their original positions.

➡️**Higher mileage engines require push rods to be cleaned so the copper colored end of the push rod can be identified.**

48. Apply clean engine oil to each end of the push rod. Insert them into their respective positions with the copper-colored end up.

➡️**Coat the end of each valve stem with clean engine oil.**

49. Install the eight valve bridges.

❈❈ WARNING

Rotate the crankshaft until the damper locating dowel notch is in the six o'clock position or engine damage can occur.

➡️**Apply clean engine oil to the top center of each valve bridge.**

50. Install the rocker arm assemblies and eight bolts.

❈❈ WARNING

Using too much engine oil on the threads of the cylinder head bolts can

cause damage to the threads and poor sealing. Using anti-seize compounds, grease or any other lubricants other than engine oil on the cylinder head bolt threads can affect the true torque value of the bolts.

➡️**Lightly lubricate the new cylinder head bolt threads and flanges with clean engine oil.**

51. Install the 10 cylinder head retaining bolts finger-tight.
52. Tighten the head bolts in the following sequence.
   a. Tighten bolts 1 through 10 to 88 Nm (65 ft. lbs.).
   b. Tighten bolts 1, 3, 5, 7 and 9 to 115 Nm (85 ft. lbs.).
   c. Tighten bolts in sequence 1 through 10, clockwise 90 degrees.
   d. Tighten bolts in sequence 1 through 10, a second time, clockwise 90 degrees.
   e. Tighten bolts in sequence 1 through 10, a third time, clockwise 90 degrees.
   f. Tighten bolts 11 through 15 an to 24 Nm (18 ft. lbs.).
   g. Tighten bolts 11 through 15 to 31 Nm (23 ft. lbs.).

➡️**Install new sealing washers.**

53. Install the fuel line and the banjo fitting.
54. Install the fan shroud mounting stud.
55. Position back and connect the heater hose.
56. Install the idler pulley and bolt.
57. Position the exhaust backpressure bracket assembly and install the retaining nuts.
58. Tighten the exhaust backpressure tube fitting at the exhaust manifold.
59. Position the fuel lines and install the fuel line bracket retaining bolt.

➡️**Install a new O-ring and apply clean engine oil.**

60. Install the oil level indicator and tube.
61. Install the turbocharger heat shield and bolts.
62. Position the turbocharger adapter pipe.

➡️**Apply anti-seize lubricants to the bolt threads prior to installing the bolts.**

➡️**Do not tighten until after the turbocharger is installed.**

63. Install the nuts and bolts in the adapter pipe.
64. Install the four glow plugs.

⚙ WARNING

If the fuel injector oil inlet D-ring is damaged, a new fuel injector must be installed.

➡ **Lubricate the fuel injector and O-rings liberally with clean engine oil.**

65. Install new O-rings and copper washer on the fuel injector.

⚙⚙ WARNING

To prevent engine damage, do not use air tools to install the fuel injectors. The snapring that extracts the injector can dislodge and fall into the oil drain hole.

66. Install the fuel injector, fuel injector hold-down and bolt. Torque to 24 ft. lbs. (33 Nm).

⚙ WARNING

To prevent engine damage, be sure the injector wiring is clear of all moving parts.

67. Install the fuel injector electrical connector into the rocker carrier.
68. Apply engine oil to the top fuel injector O-rings.
Early build

➡ **Install a new lower O-ring.**

69. Install a new crankcase to head tube assembly. Torque to 33 ft. lbs. (45 Nm).
70. Install the high-pressure oil rail and bolts.
71. Install the high-pressure oil rail.
72. Install the bolts finger tight.
73. Tighten the bolts in the sequence shown.
74. Install the high-pressure oil line.

➡ **Clean and inspect the O-rings and install new as necessary.**

➡ **Apply clean engine oil to the O-rings before installing.**

75. Install the glow plug buss bar.
Late build

⚙⚙ WARNING

To prevent engine damage, check that the crankcase-to-head tube assemblies bottom out in the branch tube assembly. The oil rail, crankcase-to-head tube and the fuel injectors will not function correctly if the tube is not bottomed out.

11 Nm (8 lb-ft)

67197-EFSE-G124

High pressure oil rail torque installation—F-Series w/6.0L engine

➡ **Install new O-rings and .**

76. Apply clean engine oil and install the crankcase-to-head tube assembly.

➡ **Apply clean engine oil to the top fuel injector D-ring before installing the high pressure oil rail.**

77. Position the oil rail on the fuel injectors.
78. Place the oil rail on top of the carrier so that the four single ball tubes are engaging the injector lead angle.
79. Insert three guide bolts, two on the ends of the straight side of the oil rail and one in the middle of the wavy side of the oil rail. Install the guide studs six to seven turns.
80. Press the oil rail into the fuel injectors.
81. Inspect that the oil rail mounting feet are flat against the mounting surface.
82. Loosely install the six bolts.
83. Install the oil rail retaining bolts.
84. Remove the three guide bolts.
85. Loosely install the three remaining bolts.
86. Tighten the bolts in the sequence shown.

➡ **Clean and inspect the O-rings and install new as necessary.**

➡ **Apply clean engine oil to the O-rings prior to installing.**

87. Install the glow plug harness.
All vehicles

⚙⚙ WARNING

To prevent engine damage, do not use air powered tools when installing the valve cover.

➡ **Clean and inspect the valve cover gasket. Install a new gasket if necessary.**

88. Position the valve cover gasket. Install the valve cover and 11 bolts.
89. Position back the oil level indicator and install the nut.
90. Connect the glow plug connector.
91. Install the fuel injector control module mounting bracket.

➡ **Install a new oil filter.**

92. Fill the crankcase with clean engine oil.

➡ **Lightly lubricate the O-ring with clean engine oil before installing.**

93. Install the cylinder block drain plug.
94. Install the intake manifold.

RIGHT CYLINDER HEAD

1. Before servicing the vehicle, refer to the Precautions Section.
All vehicles
2. With the vehicle in **Neutral**, position it on a hoist.
3. Drain the engine oil.
4. Remove the intake manifold.
5. Remove the fuel injectors.
6. Remove the starter.
7. Remove the cylinder drain block.
8. Remove the nuts and bolts from the turbocharger adapter pipe.
9. Remove the turbocharger adapter pipe.
Early build
10. Remove and discard the crankcase to head tube assembly.
11. Remove the glow plug buss bar.

Late build

➡The rings on the crankcase-to-head tube are to be used to pry the tube assembly from the branch tube assembly or the oil rail assembly.

12. Remove the crankcase-to-head tube assembly.

13. Remove and discard the O-rings.

❄❄ WARNING

Do not pull on the wiring to remove the glow plug connector or damage may occur.

14. Remove the glow plug harness.
All vehicles

15. Remove the four glow plugs.

16. Remove and discard the 10 inner cylinder head bolts.

17. Remove the eight bolts and the rocker arm assemblies.

18. Mark the eight valve bridges with a permanent marker and remove.

❄❄ WARNING

To prevent engine damage, keep the push rods in the order in which they were removed. Install all push rods back in their original positions.

19. Mark the location and remove the eight push rods.

20. Remove the outer cylinder head bolts.

21. Install the special tool and the lifting crane. With the help of an assistant, remove the cylinder head from the vehicle.

22. Remove and discard the cylinder head gasket and dowels.

23. Clean and inspect the gasket sealing surfaces.

To install:
All vehicles, late build

❄❄ WARNING

Install new D-ring seals on the crankcase-to-head tube. It requires several hours after installation for the D-ring seals to relax back to their original size. If the tube assembly is installed before the D-ring seals have relaxed, damage to the D-ring seals can occur.

24. Install new D-ring seals on each end of the crankcase to head tube assemblies.

➡**Use care to avoid scratching the blue compound on the cylinder head gasket.**

25. Install a new cylinder head gasket with the part number facing up and verify the top five holes and the head gasket push rod holes line up.

26. Install the dowels and the cylinder head gasket.

27. Using the special tools and with the help of an assistant, install the cylinder head on the engine. Remove the special tools.

28. Install the outer cylinder head bolts finger-tight.

❄❄ WARNING

To prevent engine damage, keep the push rods in the order in which they were removed. Install all push rods back in their original positions.

➡**Higher mileage engines require push rods to be cleaned so the copper colored end of the push rod can be identified.**

29. Apply clean engine oil to each end of the push rod. Insert them into their respective positions with the copper colored end up.

➡**Coat the end of each valve stem with clean engine oil.**

30. Install the eight valve bridges.

❄❄ WARNING

Rotate the crankshaft until the damper locating dowel notch is in the six o'clock position or engine damage can occur.

➡**Apply clean engine oil to the top center of each valve bridge.**

31. Install the rocker arm assemblies and eight bolts.

❄❄ WARNING

Using too much engine oil on the threads of the cylinder head bolts can cause damage to the threads and poor sealing. Using anti-seize compounds, grease or any other lubricants other than engine oil on the cylinder head bolt threads can affect the true torque value of the bolts.

➡**Lightly lubricate the new cylinder head bolt threads and flanges with clean engine oil.**

32. Install the 10 inner cylinder head retaining bolts finger-tight.

33. Tighten the head bolts in the following sequence.

  a. Tighten bolts 1 through 10 to 88 Nm (65 ft. lbs.).

  b. Tighten bolts 1, 3, 5, 7 and 9 to 115 Nm (85 ft. lbs.).

  c. Tighten bolts in sequence 1 through 10, clockwise 90 degrees.

  d. Tighten bolts in sequence 1 through 10, a second time, clockwise 90 degrees.

  e. Tighten bolts in sequence 1 through 10, a third time, clockwise 90 degrees.

  f. Tighten bolts 11 through 15 to 24 Nm (18 ft. lbs.).

  g. Tighten bolts 11 through 15 to 31 Nm (23 ft. lbs.).

34. Install the four glow plugs.
Late build

➡**Clean and inspect the glow plug connector O-rings and install new as necessary.**

➡**Apply clean engine oil to the O-rings.**

35. Install the glow plug harness.

❄❄ WARNING

To prevent engine damage, check that the crankcase-to-head tube assemblies bottom out in the branch tube assembly. The oil rail, crankcase-to-head tube and the fuel injectors will not function correctly if the tube is not bottomed out.

36. Apply clean engine oil and install the crankcase-to-head tube assembly.
Early build

➡**Clean and inspect the glow plug buss bar O-rings and install new as necessary.**

➡**Apply clean engine oil to the O-rings before installing.**

37. Install the glow plug buss bar.

➡**Install a new lower O-ring.**

38. Install a new crankcase to head tube assembly.
All vehicles

39. Position the turbocharger adapter pipe.

➡**Apply anti-seize lubricants to the bolt threads prior to installing the bolts.**

➡**Do not tighten until after the turbocharger is installed.**

40. Install the nuts and bolts in the adapter pipe.

➡**Apply clean engine oil to the O-ring prior to installing.**

41. Install the cylinder block drain plug.

42. Install the starter.
43. Install the fuel injectors.
44. Install the intake manifold.

➡**Install a new oil filter.**

45. Fill the crankcase with clean engine oil.

F-Series w/7.3L Engine

✳✳ CAUTION

The fuel system remains under pressure, even after the engine has been turned OFF. The fuel system pressure must be relieved before disconnecting any fuel lines. Failure to do so may result in fire and/or personal injury.

1. Before servicing the vehicle, refer to the Precautions Section.
2. Remove or disconnect the following:
   • Negative battery cables
   • Bumper
   • Hood
3. Recover the refrigerant using approved equipment.
4. Drain the coolant.
5. Position a suitable container at the end of the water drain tube and move the fuel filter drain lever to drain.
6. Remove or disconnect the following:
   • Cooling module assembly
   • Cooling fan
   • Drive belt
   • Two nuts and bolts from the exhaust manifold and the exhaust adapter pipe
   • Intake Air Temperature (IAT) sensor electrical connector
   • Turbocharger inlet pipe
   • Engine air cleaner and the air cleaner intake pipe as an assembly
   • Four nuts and the engine air cleaner support brackets
   • Charge air cooler outlet pipe
   • Turbocharger outlet pipe
   • A/C pressure cutoff switch electrical connector, A/C cycling switch electrical connector, and the A/C manifold and tube
   • O-ring seals
7. Plug or cap the A/C manifold and tube, and the A/C compressor ports.
   • Wiring harness
   • Generator electrical connector and electrical lead
   • Belt tensioner
   • Bolt and the idler pulley
   • Heater water hose

• Fuel tubes
• Water-In-Fuel (WIF) sensor and fuel heater electrical connectors
• Water drain tube
• Two bolts and fuel filter/water separator
• High-pressure oil hose
• Fuel injector/glow plug nine-pin electrical connector
• Compressor manifold and turbocharger outlet pipe
• Clamp and the left half of the compressor manifold
• Right half of the compressor manifold
• Fuel tube
• Valve cover and gasket
• Electrical connectors from the fuel injectors
• Electrical leads from the glow plugs
• Rocker arms and the push rods
• Oil drain plugs
• Oil deflector from the fuel injector hold-down plate
• Outboard fuel injector hold-down bolts
• Hold-down plates from the inner hold-down bolts
• Fuel injectors
• The 4 outboard fuel injector hold-down bolts, retaining screws and 4 oil deflectors

✳✳ WARNING

Remove the oil drain plugs prior to removing the injectors or oil could enter the combustion chamber, which could result in hydrostatic lock and severe engine damage.

• Oil rail drain plugs
• Fuel injectors using Injector Remover No. T94T-9000-AH1, or equivalent. Position the tool's fulcrum beneath the fuel injector hold-down plate and over the edge of the cylinder head. Install the remover screw in the threaded hole of the fuel injector plate (see illustration). Tighten the screw to lift out the injector from its bore. Place the injector in a suitable protective sleeve such as Rotunda Injector Protective Sleeve, No. 014-00933-2 or equivalent, and set the injector in a suitable holding rack.

➡**During removal the injector tab (located above the fuel injector) must be bent completely flat and flush with the cowl and heat shield.**

8. Remove or disconnect the following:
   • The 4 glow plugs
9. Loosen the cylinder head bolts.

Cylinder head torque sequence—7.3L diesel engines

9355FG01

10. Carefully lift the cylinder head out of the engine compartment and remove the head gaskets.

➡**To prepare a good seat for the fuel injector O-rings, use a suitable injector sleeve brush to clean any debris from the bore.**

11. Carefully clean the cylinder block and head mating surfaces.

12. Install or connect the following:
- Cylinder head gasket on the engine block and carefully lower the cylinder head in place
- Cylinder head bolt and torque in 3 steps using the sequence shown in the illustration as follows:
  a. Step 1: 65 ft. lbs. (88 Nm).
  b. Step 2: 85 ft. lbs. (115 Nm).
  c. Step 3: 95 ft. lbs. (129 Nm).

➡**Lubricate the threads and the mating surfaces of the bolt heads and washers with engine oil.**

13. Install or connect the following:
- Glow plugs

✳✳ CAUTION

Remove all oil and fuel from the cylinders before installing the fuel injectors. Failure to do so can cause hydro-static lock, resulting in severe engine damage.

✳✳ CAUTION

Do not strike the top of the fuel injector to seat the injector in the cylinder head fuel injector bore. Damage to the fuel injector will occur. Use hand pressure on the top of the fuel injector until the fuel injector hold-down plate is flush with the cylinder head.

14. Install the fuel injectors using special tools as follows:
  a. Lubricate the injectors with clean engine oil. Using new copper washers, carefully push the injectors square into the bore using hand pressure only to seat the O-rings.
  b. Position the open end of Injector Replacer, No. T94T–9000–AH2, or equivalent between the fuel injector body and injector hold-down plate, while positioning the opposite end of the tool over the edge of the cylinder head.
  c. Align the hole in the tool with the threaded hole in the cylinder head and install the bolt from the tool kit. Tighten

the bolt to fully seat the injector, then remove the bolt and tool.

15. Install or connect the following:
- Outboard fuel injector hold-down bolts
- Oil deflector on the fuel injector hold-down plate, using the shoulder bolt
- Oil drain plugs
- Rocker arms and push rods
- Valve cover gasket
- Glow plug electrical leads
- Fuel injector electrical connectors
- Valve cover
- Fuel tube
- Compressor manifold and turbocharger outlet pipe
- Right half of the compressor manifold and tighten the clamps
- Left half of the compressor manifold and the clamp
- IAT sensor electrical connector
- Fuel injector/glow plug nine-pin electrical connector and high-pressure oil hose
- Fuel filter/water separator and the two bolts
- Fuel tubes
- Water drain tube and the fuel supply tube
- Fuel tubes, the WIF sensor electrical connector, and the fuel heater electrical connector
- Heater water hose
- Idler pulley and bolt
- Belt tensioner and bolt
- Generator electrical connector and electrical lead
- Wring harness
- A/C manifold and tube and the A/C cycling switch electrical connector
- Turbocharger outlet pipe
- Charge air cooler outlet pipe
- Engine air cleaner brackets and four nuts
- Engine air cleaner and the air cleaner intake pipe
- Turbocharger inlet pipe
- Two nuts and bolts in the exhaust manifold and the exhaust adapter pipe
- Drive belt
- Cooling fan
- Cooling module assembly
16. Move the fuel filter drain lever to closed.
17. Fill the engine cooling system.
- Evacuate, charge and leak check the A/C system.
- Negative battery cables.
18. Start the vehicle and check for leaks.
19. Install the hood and bumper.

E-Series w/7.3L Engine

RIGHT CYLINDER HEAD

✳✳ CAUTION

The fuel system remains under pressure, even after the engine has been turned OFF. The fuel system pressure must be relieved before disconnecting any fuel lines. Failure to do so may result in fire and/or personal injury.

1. Before servicing the vehicle, refer to the Precautions Section.
2. Remove or disconnect the following:
- Negative battery cable
- Coolant
- Engine oil
- Vacuum hose from the right intake manifold
- Both electrical harness connectors from the valve cover
- Valve cover and gasket
- Intake manifold covers
- Fuel injector electrical injectors
- Fuel lines from the heads
- Fuel supply assembly
- Heater hose
- Dipstick tube bracket retainer at the right exhaust manifold
- Rocker arms and pushrods
3. Loosen the 4 inboard fuel injector hold-down bolts.
4. Remove or disconnect the following:
- The 4 outboard fuel injector hold-down bolts, retaining screws and 4 oil deflectors

✳✳ WARNING

Remove the oil drain plugs prior to removing the injectors or oil could enter the combustion chamber, which could result in hydrostatic lock and severe engine damage.

- Oil rail drain plugs
- Fuel injectors using Injector Remover No. T94T-9000-AH1, or equivalent. Position the tool's fulcrum beneath the fuel injector hold-down plate and over the edge of the cylinder head. Install the remover screw in the threaded hole of the fuel injector plate (see illustration). Tighten the screw to lift out the injector from its bore. Place the injector in a suitable protective sleeve such as Rotunda Injector Protective Sleeve, No. 014-00933-2 or equivalent, and set the injector in a suitable holding rack.

➡️During removal the injector tab (located above the fuel injector) must be bent completely flat and flush with the cowl and heat shield.

5. Use a vacuum pump, remove the oil and fuel left over in the injector bores.

6. Remove or disconnect the following:
- The 4 glow plugs
- High pressure oil pump supply line from the cylinder head
- Grille opening reinforcement, headlamp assembly, radiator, oil reservoir and fuel filter assembly
- Exhaust back pressure line

7. Loosen the glow plug relay bracket retainers and disengage the ground wire.

8. Disconnect the fuel return line at the front of the cylinder head.

9. Loosen the cylinder head bolts.

10. Carefully lift the cylinder head out of the engine compartment and remove the head gaskets.

To install:

➡️To prepare a good seat for the fuel injector O-rings, use a suitable injector sleeve brush to clean any debris from the bore.

11. Carefully clean the cylinder block and head mating surfaces.

12. Position the cylinder head gasket on the engine block and carefully lower the cylinder head in place.

13. Install the cylinder head bolts and torque in 3 steps using the sequence shown in the illustration as follows:
   a. Step 1: 65 ft. lbs. (88 Nm).
   b. Step 2: 85 ft. lbs. (115 Nm).
   c. Step 3: 95 ft. lbs. (129 Nm).

➡️Lubricate the threads and the mating surfaces of the bolt heads and washers with engine oil.

14. Install or connect the following:
- Fuel return line to the cylinder head
- Glow plug relay bracket and ground wire, then tighten the retainers
- Exhaust back pressure line
- High pressure fuel supply line and tighten the fitting to 19 ft. lbs. (26 Nm)
- Heater hose to the cylinder head
- Manifold hoses
- Fuel supply lines to the rear of the cylinder head
- Banjo bolt through the fuel line and into the pump. Tighten the pump to 40 ft. lbs. (55 Nm).
- Glow plugs, coated with anti-seize compound. Tighten the glow plugs to 14 ft. lbs. (19 Nm).

15. Install the fuel injectors using special tools as follows:
   a. Lubricate the injector O-rings with clean engine oil. Using new copper washers, carefully push the injectors square into the bore using hand pressure only to seat the O-rings.
   b. Position the open end of Injector Replacer, No. T94T-9000-AH2, or equivalent, between the fuel injector body and injector hold-down plate, while positioning the opposite end of the tool over the edge of the cylinder head.
   c. Align the hole in the tool with the threaded hole in the cylinder head and install the bolt from the tool kit. Tighten the bolt to fully seat the injector, then remove the bolt and tool.

16. Install or connect the following:
- Oil rail drain plugs and tighten them to 53 inch lbs. (6 Nm)
- The 4 outboard fuel injector hold-down bolts, the 4 oil deflectors and retaining screws. Tighten them to 120 inch lbs. (12 Nm).
- The 4 inboard fuel injector hold-down bolts and tighten them to 120 inch lbs. (12 Nm)
- Oil deflectors

17. Turn the engine by hand until the timing mark is at the 11 o'clock position as viewed from the front.

18. Dip the pushrod ends in clean engine oil and install the pushrods with the copper colored ends toward the rocker arms, making sure the pushrods are fully seated in the tappet pushrod seats.

19. Install or connect the following:
- Rocker arms and posts in their original positions. Apply multipurpose grease to the valve stem tips. Install the rocker arm posts, bolts and tighten to 27 ft. lbs. (37 Nm).
- Valve cover gasket
- Wiring to the fuel injectors and glow plugs
- Valve cover, tightening the bolts to 97 inch lbs. (11 Nm)
- Both electrical harness connectors to the valve cover
- Vacuum hose to the right intake valve manifold cover

20. Refill the engine coolant with the recommended type and the correct amount.

21. Refill the engine oil with the recommended type of engine oil and the correct amount.

22. Install the engine in the van.

23. Connect the negative battery cable.

LEFT CYLINDER HEAD

❄️ CAUTION

Fuel system remains under pressure, even after the engine has been turned OFF. The fuel system pressure must be relieved before disconnecting any fuel lines. Failure to do so may result in fire and/or personal injury.

1. Before servicing the vehicle, refer to the Precautions Section.

2. Remove or disconnect the following:
- Negative battery cable
- Coolant
- Engine oil
- Wiring harness bracket
- Electrical connections from the valve cover gasket, then remove the valve cover
- Electrical connections from the fuel injectors and glow plugs
- Valve cover gasket
- Rocker arms and pushrods

➡️Be sure to note the location of each part prior to removal, as each reinstalled part must be returned to their original location.

- The 4 inboard fuel injector hold-down bolts

❄️ WARNING

Remove the oil drain plugs prior to removing the injectors or oil could enter the combustion chamber, which could result in hydrostatic lock and severe engine damage.

- The oil rail drain plugs

❄️ CAUTION

Be sure to retrieve the fuel injector copper washer, located at the tip of the injector during removal.

- The 4 outboard fuel injector hold-down bolts, retaining screws and 4 oil deflectors
- Fuel injectors using Injector Remover No. T94T-9000-AH1, or equivalent. Position the tool's fulcrum beneath the fuel injector hold-down plate and over the edge of the cylinder head. Install the remover screw in the threaded hole of the fuel injector plate (see illustration). Tighten the screw to lift out the injector from its bore. Place the injector in a suitable protective

sleeve such as Rotunda Injector Protective Sleeve, No. 014-00933-2 or equivalent, and set the injector in a suitable holding rack.

3. Use a vacuum pump to remove the oil and fuel left over in the injector bores.

- The 4 glow plugs
- Fuel system pressure
- Fuel supply lines from the rear of the cylinder head
- Banjo bolt from the fuel line at the pump
- Oil line from the high pressure oil pump
- Electrical connection from the injection control pressure sensor
- High pressure oil supply line from the left cylinder head

4. Loosen the fuel line nut from the intake manifold stud, then disconnect the fuel return line from the left cylinder head.

5. Loosen the fuel return line block screws at the front of the left cylinder head.

6. Remove the fuel line retaining clamp from the intake manifold cover.

7. Loosen the cylinder head bolts.

8. Remove the oil reservoir and fuel filter.

9. Remove the cylinder head and gasket.

To install:

➡**To prepare a good seat for the fuel injector O-rings, use a suitable injector sleeve brush to clean any debris from the bore.**

10. Carefully clean the cylinder block and head mating surfaces.

11. Position the cylinder head gasket on the engine block and carefully lower the cylinder head in place.

12. Install the cylinder head bolts and torque in 3 steps using the sequence shown in the illustration as follows:
   a. Step 1: 65 ft. lbs. (88 Nm).
   b. Step 2: 85 ft. lbs. (115 Nm).
   c. Step 3: 95 ft. lbs. (129 Nm).

13. Install or connect the following:
- Fuel line retaining clamp to the intake manifold cover
- Fuel return line block screws at the front of the left cylinder head
- Fuel return line to the left cylinder head
- Fuel line nut at the intake manifold stud
- High pressure oil supply line to the left cylinder head
- Electrical connection to the injection control pressure sensor
- Oil line to the high pressure oil pump

- Manifold hoses
- Fuel supply line
- Banjo bolt through the fuel line into the pump and tighten the bolt to 40 ft. lbs. (55 Nm)
- Fuel supply lines at the rear of the cylinder heads
- Glow plugs, coated with anti-seize compound. Tighten the glow plugs to 14 ft. lbs. (19 Nm).

14. Install the fuel injectors using special tools as follows:

   a. Lubricate the injector O-rings with clean engine oil. Using new copper washers, carefully push the injectors square into the bore using hand pressure only to seat the O-rings.

   b. Position the open end of Injector Replacer, No. T94T-9000-AH2 or equivalent between the fuel injector body and injector hold-down plate, while positioning the opposite end of the tool over the edge of the cylinder head.

   c. Align the hole in the tool with the threaded hole in the cylinder head and install the bolt from the tool kit. Tighten the bolt to fully seat the injector, then remove the bolt and tool.

15. Install or connect the following:
- The 4 outboard fuel injector hold-down bolts, the 4 oil deflectors and retaining screws. Tighten them to 120 inch lbs. (12 Nm).
- Oil deflectors and tighten the bolts to 120 inch lbs. (12 Nm)
- Oil rail drain plugs and tighten them to 53 inch lbs. (6 Nm)
- The 4 inboard fuel injector hold-down bolts and tighten them to 120 inch lbs. (12 Nm)

16. Turn the engine over by hand until the timing mark is at the 11 o'clock position as viewed from the front.

17. Dip the pushrod ends in clean engine oil and install the pushrods with the copper colored ends toward the rocker arms, making sure the pushrods are fully seated in the tappet pushrod seats.

18. Install the rocker arms and posts in their original positions. Apply multipurpose grease to the valve stem tips. Install the rocker arm posts, bolts and tighten to 27 ft. lbs. (37 Nm).

19. Install the valve cover gasket.

20. Connect the wiring to the fuel injectors and glow plugs.

21. Install the valve cover, tightening the bolts to 97 inch lbs. (11 Nm).

22. Connect both electrical harness connectors to the valve cover.

23. Install the engine in the van.

24. Connect the negative battery cable.

Rocker Arms

REMOVAL & INSTALLATION

6.0L Engines

EARLY BUILD

✳✳ WARNING

To prevent engine damage, always install new head bolts when installing the cylinder head, even if removing only one bolt. It will be necessary to install a new cylinder head gasket as well as all cylinder head bolts.

1. Before servicing the vehicle, refer to the Precautions Section.

2. Remove the cylinder head.

➡**Be careful when removing the rocker arm. Do not lose the rocker arm steel ball.**

06017-F150-G222

Push down on the rocker arm. Pull out on the bottom releasing the ball—6.0L early build

06017-F150-G223

Remove the steel ball from the rocker arm—6.0L early build

06017-F150-G224

Remove the rocker arm clip from the fulcrum plate—6.0L early build

3. Push down on the rocker arm. Pull out on the bottom releasing the ball.

4. Remove the steel ball from the rocker arm.

5. Remove the rocker arm clip from the fulcrum plate.

6. Clean and inspect all parts for pitting or scuffing. Install new rocker arm and valve bridges, as needed.

To install:

7. Install the rocker arm clip on the fulcrum plate.

8. Place the steel ball in the rocker arm cup and lubricate with clean engine oil.

9. Push down on the rocker arm to compress the rocker arm clip. Insert the ball and release the pressure. Verify free movement of the rocker arm on the fulcrum.

10. Install the cylinder head.

LATE BUILD

1. Before servicing the vehicle, refer to the Precautions Section.

✷✷ WARNING

To prevent engine damage, always install new head bolts when installing the cylinder head, even if removing only one bolt. It will be necessary to install a new cylinder head gasket as well as all cylinder head bolts.

2. Remove the cylinder head.

3. Disassemble the rocker arm from the fulcrum plate. With the rocker arm and fulcrum plate upside down in your palm, push down in your palm, push up on the rocker arm and with your palm push down on the fulcrum plate.

➡**Be careful when removing the rocker arm. Do not lose the rocker arm ball.**

4. Move the rocker off the ball, keeping the ball in the detent of the fulcrum plate.

With the rocker arm and fulcrum plate upside down in your palm, push down in your palm, push up on the rocker arm and with your palm push down on the fulcrum plate—6.0L late build

06017-F150-G226

Remove the rocker arm ball—6.0L late build

06017-F150-G227

Remove the rocker arm clip—6.0L late build

Item	Description
1	Snap Retaining Clip
2	Rocker Arm Pedestal
3	Rocker Arm Ball
4	Rocker Arm
5	Rocker Arm Assembly

7924FG58

Exploded view of the rocker arm assembly—7.3L diesel engines

5. Remove the rocker arm ball.

6. Remove and discard the rocker arm clip.

7. Clean all parts with a suitable solvent. Use filtered compressed air to dry the parts.

8. Inspect each rocker arm pivot foot and corresponding valve bridge for pitting and scuffing. Inspect each rocker arm ball and socket for scuffing. Install new rocker arms and valve bridges as required.

9. Inspect the rocker arm post ball socket for excessive wear. Inspect the bolts for thread damage, install new components as required.

To install:

10. Install a new rocker arm clip on the fulcrum plate.

11. Insert the ball in the detent in the fulcrum plate and lubricate with clean engine oil.

12. Hold the rocker arm upside down in your palm.

13. Press the lower part of the rocker arm against the rocker arm clip, push up with your palm and push the upper part of the rocker arm over the rocker arm ball. Check for freedom of movement of the rocker arm on the fulcrum plate.

14. Install the cylinder head.

7.3L Engines

1. Before servicing the vehicle, refer to the Precautions Section.

2. Disconnect the ground cables from both batteries.

3. Remove the valve cover attaching screws and remove both valve covers.

4. Remove the valve rocker arm post mounting bolts. Remove the rocker arms and posts in order and mark them with tape so they can be installed in their original positions.

5. If the cylinder heads are to be removed, then the pushrods can now be removed. Make a holder for the pushrods out of a piece of wood or cardboard and remove the pushrods in order. It is very important that the pushrods be reinstalled in their original order. The pushrods can remain in position if no further disassembly is required.

To install:

6. If the pushrods were removed, install them in their original locations. Make sure they are fully seated in the tappet seats.

➡**The copper colored end of the pushrod goes toward the rocker arm.**

7. Apply a polyethylene grease to the valve stem tips. Install the rocker arms and posts in their original positions.

8. Turn the engine over by hand until the valve timing mark is at the 11:00 o'clock position, as viewed from the front of the engine. Install all of the rocker arm post attaching bolts and tighten to 20 ft. lbs. (27 Nm).

9. Install new valve cover gaskets and install the valve cover.

10. Install the battery cables, start the engine and check for leaks.

Turbocharger

REMOVAL & INSTALLATION

6.0L Engines

F-SERIES

1. Before servicing the vehicle, refer to the Precautions Section.

2. Remove the air cleaner assembly.

✳✳ WARNING

To maintain the integrity of the coolant and the cooling system, add Motorcraft Premium Gold Engine Coolant VC-7-A (in Oregon VC-7-B) or equivalent meeting Ford specification WSS-M97B51-A1 (yellow color). Always fill the cooling system with the same type of coolant that was drained from the system. Do not mix coolant types.

Do not add orange-colored Motorcraft Specialty Orange Engine Coolant VC-2, or equivalent meeting WSS-M97B44-D. Mixing coolants may degrade the coolant's corrosion protection.

Do not add alcohol, methanol, brine, or any engine coolants mixed with alcohol or methanol antifreeze. These can cause engine damage from overheating or freezing.

3. Disconnect the engine vent and radiator vent hoses.

4. Remove the bolts and position the degas bottle aside.

5. Loosen the clamp at the turbocharger.

6. Remove the nuts and the turbocharger intake tube.

7. Disconnect the charge air cooler inlet pipe.

8. Remove the push pins.

9. Disconnect the two wiring harness push pins and position aside.

10. Disconnect the turbocharger variable vane hydraulic control valve electrical connector.

11. Remove the bolts for the oil supply tube.

12. Remove and discard the gasket.

13. Remove the bolt and the wire retainer.

14. Remove the marmon clamp from the turbocharger outlet.

15. Remove the marmon clamp from the turbocharger inlet.

Early build

16. Remove the rear turbocharger mounting bolt.

Late build

17. Remove the bolt and turbocharger oil supply tube.

18. Remove and discard the O-ring.

19. Remove the rear turbocharger mounting bolt.

All vehicles

20. Remove the front mounting bolts.

21. Position the turbocharger and remove the turbocharger drain tube.

22. Remove and discard the drain tube O-rings.

23. Remove the turbocharger. Remove the bolts and the turbocharger pedestal.

To install:

All vehicles

24. Position the turbocharger on the turbocharger pedestal.

➡**Install new O-rings and apply clean engine oil.**

25. Position the turbocharger and install the turbocharger drain tube.

Late build

26. Install the turbocharger and the rear mounting bolt. Torque to 28 ft. lbs. (38 Nm).

➡**Install a new O-ring and apply clean engine oil.**

27. Install the oil supply tube. Torque to 10 ft. lbs. (13 Nm).

28. Position the locking bracket and install the bolt.

Early build

29. Install the turbocharger and the rear mounting bolt. Torque to 28 ft. lbs. (38 Nm).

30. Install the oil feed tube.

All vehicles

31. Install the turbocharger front mounting bolts. Torque to 28 ft. lbs. (38 Nm).

32. Install the turbocharger inlet marmon clamp.

33. Install the turbocharger exhaust marmon clamp.

34. Position the wire retainer and install the bolt.

35. Prelubricate the oil inlet hole of the turbocharger assembly with clean engine oil and spin the compressor wheel several times to coat the bearing with oil.

36. Position the oil feed tube and install the bolts.
37. Install a new gasket.
38. Connect the turbocharger variable vane hydraulic control valve electrical connector.
39. Position the wiring harness and connect the push pins.
40. Install the push pins.
41. Connect the charge air cooler inlet pipe.
42. Install the turbocharger intake tube.

E-SERIES

1. Before servicing the vehicle, refer to the Precautions Section.
2. Remove the engine cover.
3. Remove the retainers and the air deflector.
4. Remove the power steering reservoir bracket retainers.
5. Remove the power steering fluid indicator and retainers. Remove the power steering mounting bracket. Install the power steering fluid indicator and position the power steering reservoir aside.
6. Disconnect the coolant hoses from the air cleaner outlet pipe.
7. Loosen the clamps and remove the air cleaner outlet pipe.
8. Disconnect the turbocharger intake tube breather hose.
9. Loosen the clamp at the turbocharger.
10. Remove the retaining nuts and position the heater hose bracket aside.
11. Remove the stud bolts and the turbocharger intake tube.
12. Loosen the turbocharger outlet clamp.
13. Remove the turbocharger outlet clamp.
14. Loosen the pipe at the exhaust clamp and position aside.
15. Remove the left bolts for the turbocharger adapter pipe.
16. Remove the right bolts for the turbocharger adapter pipe.
17. Remove the exhaust gas recirculation (EGR) cooler clamp.
18. Remove the turbocharger inlet clamp and turbocharger adapter pipe.
19. Disconnect the VGT actuator electrical connector and retaining clip.
20. Remove the bolts from the turbocharger oil feed tube.
21. Remove and discard the gasket.
22. Remove the bolt and turbocharger oil feed tube.
23. Remove and discard the O-ring seals.
24. Remove the right turbocharger mounting bolts.

25. Remove the left turbocharger mounting bolts.

❊❊ WARNING

Use care not to damage the turbocharger outlet hose when removing the turbocharger.

26. Position the turbocharger to remove the turbocharger drain tube. Remove the turbocharger from the vehicle.
27. Remove and discard the O-ring seals.
28. Remove the plug and drain the oil filter assembly and tubes.
29. Disconnect the wiring push-pin retainer.
30. Remove the bolt and retaining clamp at the oil filter adapter.
31. Remove the bolt and retaining clamp at the oil filter assembly.

➡**Use care not to bend the tubes, when removing the oil filter tubes.**

32. Remove the bolt and retaining clamp. Remove the oil filter tubes.
33. Remove the bolts and turbocharger pedestal.

To install:

➡**Install new drain tube O-ring seals and apply clean engine oil.**

➡**Position the turbocharger outlet flange into the outlet hose as the turbocharger is positioned on the engine.**

34. Position the turbocharger in the vehicle. Install the turbocharger drain tube and turbocharger.
35. Install the left turbocharger mounting bolts. Torque to 28 ft. lbs. (38 Nm).
36. Install the right turbocharger mounting bolts. Torque to 28 ft. lbs. (38 Nm).
37. Prelubricate the oil inlet hole of the turbocharger assembly with clean engine oil and spin the compressor wheel several times to coat the bearings with oil.

➡**Use a new gasket when installing the oil feed tube.**

38. Install the turbocharger oil feed tube, gasket and bolts. Torque to 18 ft. lbs. (25 Nm).

➡**Install a new O-ring seal and apply clean engine oil.**

39. Install the turbocharger oil feed tube. Position the clamp and install the bolt.
40. Connect the VGT actuator electrical connector and retaining clip.
41. Position the turbocharger adapter

pipe and install the turbocharger inlet clamp.
42. Install the EGR cooler clamp.
43. Install the right bolts for the turbocharger adapter pipe. Torque to 20 ft. lbs. (27 Nm).
44. Install the left bolts for the turbocharger adapter pipe.
45. Position back the turbocharger exhaust pipe. Install the turbocharger exhaust clamp.
46. Tighten the retaining nuts at the exhaust clamp.
47. Tighten the turbocharger outlet clamp.
48. Install the turbocharger intake tube.
49. Inspect the O-ring seals for the oil filter tube ends, replace as necessary.
50. Check and top off the engine oil after the engine is running

7.3L Engines

1. Before servicing the vehicle, refer to the Precautions Section.
2. Remove or disconnect the following:
   - Negative battery cable
   - The 2 air intake tube assembly bolts, clamps at the turbocharger, crankcase breather assembly, engine air cleaner and air intake tube and hoses.
   - Exhaust outlet clamp from the turbocharger
   - Engine charge exhaust pipe bolt from the transmission, if so equipped
   - Bolts and nuts from the catalytic converter-to-engine charge exhaust pipe, if so equipped
3. Loosen 2 bolts retaining the turbocharger exhaust inlet pipe to the left exhaust manifold.
4. For automatic transmissions, remove the bolts retaining the left turbocharger exhaust inlet pipe to the turbocharger exhaust inlet adapter.
5. Loosen the 2 bolts retaining the turbocharger exhaust inlet pipe to the right exhaust manifold.
6. Remove or disconnect the following:
   - Lower bolt retaining the right turbocharger exhaust inlet pipe to the turbocharger exhaust inlet adapter

➡**For automatic transmissions, the upper bolts retaining the right and left turbocharger exhaust inlet pipes to the turbocharger exhaust inlet adapter.**

   - Right engine lift hook and bolt
   - Air inlet hose clamp and hose at the turbocharger and lay the hose aside

COMPRESSOR MANIFOLD

7924FG59

After loosening the clamps, the compressor manifold can be removed—7.3L diesel engines

- Compressor manifold
- The 4 bolts retaining the turbocharger pedestal assembly to the cylinder block
- Turbocharger assembly and detach all electrical connectors from it

➡**If the turbocharger is not being removed for service, install the Fuel/Oil Turbo Protector Cap Set T94T-9395-AH or equivalent.**

- Oil gallery O-rings

To install:

7. Install or connect the following:
- Oil gallery O-rings
- Turbocharger electrical connectors
- Turbocharger assembly
- The 4 bolts retaining the turbocharger pedestal assembly to the engine block. Tighten the bolts to 18 ft. lbs. (25 Nm).
- 4 bolts retaining the right and left turbocharger exhaust inlet pipes to the turbocharger exhaust inlet adapter, loosely.
- Compressor manifold, intake manifold hoses and clamps. Be sure the compressor outlet seal is in position.
- Right engine lift hook and bolt
- The 2 right and left lower bolts (retaining the turbocharger exhaust

inlet pipes to the turbocharger exhaust inlet adapter) to 36 ft. lbs. (49 Nm)
- The 4 right and left bolts and nuts (retaining the turbocharger exhaust inlet pipes to the exhaust manifolds) to 36 ft. lbs. (49 Nm)
- Catalytic converter to the engine charge exhaust pipe bolts and nuts
- The 2 right and left upper bolts (retaining the turbocharger exhaust inlet pipes to the turbocharger exhaust inlet adapter) to 36 ft. lbs. (49 Nm)
- Exhaust outlet clamp to the turbocharger
- Air intake tube and hose assembly
- Negative battery cable

Intake Manifold

REMOVAL & INSTALLATION

6.0L Engine

E-SERIES

1. Before servicing the vehicle, refer to the Precautions Section.

All vehicles
2. Disconnect the battery ground cable.
3. Remove the cooling fan stator.
4. Remove the turbocharger pedestal.

Vehicles with dual alternator
5. Remove the accessory drive belt.
6. Remove the bolt and the accessory drive belt tensioner.
7. Remove the accessory drive belt.
8. Remove the bolts, bracket and accessory drive belt idler pulley.
9. Remove the bolts and the accessory drive belt tensioner.
10. Disconnect the alternator electrical connector and the B+ wire.
11. Remove the bolts and the alternator with mounting bracket.

Vehicles with single alternator
12. Remove the accessory drive belt.

All vehicles
13. Remove the bolt and position the ground wire aside. Disconnect the electrical connector push pin.
14. Remove the conduit and position the wiring aside.
15. Disconnect the locking tab.
16. Remove the bolts.
17. Disconnect the push pin and remove the wiring harness from the conduit.
18. Remove the bolts for the charge air cooler tube and oil fill tube.
19. Disconnect the oil fill tube at the valve cover.
20. Remove the retaining nuts, charge air cooler tube, oil fill tube and bracket.
21. Disconnect the alternator electrical connector and the B+ wire. Disconnect the wiring push pin.
22. Remove the retaining nut for the transmission fluid indicator and tube.
23. Disconnect the push pin retainer. Remove the retaining nut and position the transmission fluid indicator and tube aside.
24. Remove the three bolts, ground wire and the alternator.
25. Remove the bolts and position the heater hose tube aside.
26. Remove and discard the O-ring.

➡**It will be necessary to position back or remove the heat insulating wrap.**

27. Remove the retaining nut and disconnect the wiring retainer and the injection pressure regulator valve electrical connector.
28. Disconnect the exhaust gas recirculation (EGR) valve electrical connector.
29. Disconnect the engine oil pressure (EOP) sensor electrical connector.
30. Disconnect the engine oil temperature (EOT) sensor electrical connector.
31. Disconnect the EGR throttle position control module electrical connector.
32. Disconnect the EGR throttle position sensor electrical connector.

33. Disconnect the pin-type retainer and engine coolant temperature (ECT) sensor.

34. Remove the ECT sensor.

35. Plug or cap the opening as needed.

36. Disconnect the intake air temperature (IAT2) sensor electrical connector.

37. Remove the IAT2 sensor.

38. Plug or cap the opening as needed.

39. Disconnect the eight fuel injector electrical connectors and wiring connectors.

40. Disconnect the harness retainers. Position the engine wiring harness as needed for intake manifold removal.

41. Disconnect the manifold absolute pressure (MAP) sensor hose.

42. Disconnect the engine coolant vent hose.

43. Remove the secondary fuel filter and remove all fuel from the filter housing.

44. Disconnect the fuel tube fittings at the secondary fuel filter.

45. Remove and discard the copper sealing washers.

46. Disconnect the fuel tube at the fuel filter housing. Remove the retaining nut.

47. Remove the banjo bolt and fuel tube.

48. Discard the copper sealing washers.

49. Remove the bolts and the secondary fuel filter assembly.

50. Remove the nuts and turbocharger heat shield.

➡**Align the flat edge with the index feature located on the coolant supply port.**

51. Pull the EGR cooler clamp forward, twist and then slide the EGR cooler hose rearward to remove.

52. Remove the bolts and the intake manifold.

53. Remove the intake manifold gaskets.

54. Clean and inspect the gaskets. Install new gaskets if necessary.

55. Clean and inspect the sealing surfaces.

56. Remove and discard the front module O-ring seal.

To install:

➡**The locating tabs on the gaskets must be positioned upward and toward the center of the engine, or a leak will occur.**

57. Install the intake manifold gaskets. Install a new front module O-ring seal.

58. Install the intake manifold and bolts and tighten in the following sequence.

   a. Loosely install bolts 1–8.

   b. Tighten bolts 9–16 to 11 Nm (8 ft. lbs.).

   c. Tighten all bolts to 11 Nm (8 ft. lbs.).

Intake manifold torque sequence—6.0L engine

67197-EFSE-G125

59. Slide the EGR cooler hose forward and rotate the flat to lock.

60. Install the turbocharger heat shield and nuts.

61. Install the secondary fuel filter assembly and bolts. Torque to 18 ft. lbs. (25 Nm).

➡**Install new copper sealing washers.**

62. Install the fuel line and banjo bolt. Torque to 28 ft. lbs. (38 Nm).

63. Install the retaining nut and connect the fuel tube at the fuel filter assembly. Torque to 19 ft. lbs. (26 Nm).

➡**Install new copper sealing washers on the banjo fitting.**

64. Reposition and connect the fuel tubes at the fuel filter housing.

65. Install the secondary fuel filter and cover. Torque to 10 ft. lbs. (14 Nm).

66. Connect the engine coolant vent hose.

67. Connect the MAP sensor hose.

68. Position back the engine wiring harness as needed. Connect the harness retainers.

69. Connect the eight fuel injector electrical connectors.

70. Remove the plug or cap. Install the IAT2 sensor.

67197-EFSE-G126

Reposition and connect the fuel tubes at the fuel filter housing

71. Connect the IAT2 sensor electrical connector.

72. Remove the plug or cap. Install the ECT sensor.

73. Connect the ECT sensor electrical connector and the pin-type retainer.

74. Connect the EGR throttle position sensor electrical connector.

75. Connect the EGR throttle position control module electrical connector.

76. Connect the EOT sensor electrical connector.

77. Connect the EOP sensor electrical connector.

78. Connect the EGR valve electrical connector.

79. Connect the injector pressure regulator valve electrical connector and position back the heat insulating wrap. Install the wiring retainer and retaining nut.

➡**Install a new O-ring seal and apply clean engine coolant**

80. Install the heater tube and bolts.
81. Install the alternator, ground wire and the three bolts.
82. Position back the transmission fluid indicator and tube and install the retaining nut.
83. Connect the push pin retainer.
84. Install the retaining nut for the transmission fluid indicator and tube.
85. Connect the alternator B+ wire and electrical connector. Position back the boot. Connect the wiring push pin.
86. Position the charge air cooler tube, oil fill tube and bracket. Install the retaining nuts.
87. Connect the oil fill tube at the valve cover.
88. Install the bolts for the charge air cooler tube and oil fill tube.
89. Install the conduit and bolts.
90. Install the wiring harness into the conduit and connect the push pin.
91. Install the bolts.
92. Connect the locking tab.
93. Connect the electrical connector push pin. Position back the ground wire and install the bolt.

Vehicles with single alternator
94. Install the accessory drive belt.

Vehicles with dual alternator
95. Install the alternator with mounting bracket and bolts.
96. Connect the alternator B+ wire and electrical connector. Position the boot.
97. Install the accessory drive belt tensioner and bolts.
98. Position the accessory drive belt idler pulley. Install the bracket and bolts.
99. Install the accessory drive belt.
100. Install the accessory drive belt tensioner and bolt.
101. Install the accessory drive belt.

All vehicles
102. Install the cooling fan stator.
103. Install the turbocharger pedestal and turbocharger.
104. Connect the battery ground cable.

F-SERIES

1. Before servicing the vehicle, refer to the Precautions Section.
2. Remove the auxiliary battery.
3. Remove the cooling fan stator.
4. Remove the degas bottle.
5. Remove the upper radiator hose.

6. Remove the turbocharger-to-charge air cooler duct.

Vehicles with dual alternators
7. Remove the accessory drive belt.
8. Remove the bolt and the accessory drive belt tensioner.
9. Remove the accessory drive belt.
10. Remove the bolts, bracket and accessory drive belt idler pulley.
11. Remove the bolts and the accessory drive belt tensioner.
12. Disconnect the wire retainer, alternator electrical connector and B+ wire.
13. Remove the bolts and the alternator with mounting bracket.

Vehicles with single alternator
14. Remove the accessory drive belt.

All vehicles
15. Position the boot back and disconnect the alternator B+ wire and electrical connector.
16. Remove the three bolts and the alternator.
17. Remove the turbocharger and the turbocharger pedestal.

Early build
18. Loosen the clamps and remove the charge air cooler duct.

Late build
19. Loosen the clamps and remove the charge air cooler duct.
20. Remove the bolts and position the heater hose tube aside.
21. Remove and discard the O-ring.
22. Disconnect the manifold absolute pressure (MAP) sensor hose.
23. Disconnect the engine coolant vent hose.
24. Remove the fuel injector control module.
25. Remove or position aside the heat insulating wrap.
26. Disconnect the wiring retainer, injector pressure regulator valve and injector control pressure (ICP) sensor (if equipped) electrical connectors.
27. Disconnect the exhaust gas recirculation (EGR) valve electrical connector.
28. Disconnect the engine oil pressure (EOP) sensor electrical connector.
29. Disconnect the engine oil temperature sensor electrical connector.
30. Disconnect the exhaust gas recirculation (EGR) throttle position control module electrical connector.
31. Disconnect the EGR throttle position sensor electrical connector.
32. Disconnect the pin-type retainer and engine coolant temperature (ECT) sensor.
33. Disconnect the intake air temperature (IAT2) sensor electrical connector and wiring connector.

34. Disconnect the exhaust backpressure sensor and retaining clip.
35. Disconnect the eight fuel injectors electrical connectors. Remove the nut and the fuel injector wiring harness.

➡**It is necessary to remove the fuel filter and drain the housing.**

36. Disconnect the fuel line fittings.

➡**It is necessary to remove the oil filter and drain the housing.**

37. Remove the four bolts and the oil filter housing.
38. Remove the bolt and the oil filter return tube.
39. Remove the fuel line.
40. Remove the bolt, and the banjo fitting from the fuel line.
41. Discard the sealing washers and remove the fuel line.

➡**Align the flat edge with the index feature located on the coolant supply port.**

42. Pull the EGR cooler clamp forward, twist and then slide the EGR cooler hose rearward to remove.

Early build
43. Remove the nuts and the turbocharger heat shield.

➡**Intake removed for clarity.**

44. Remove the EGR cooler V-clamp and gasket.
45. Remove the bolts and the intake manifold.

Late build
46. Remove the nuts and the turbocharger heat shield.
47. Remove the EGR cooler V-clamp and gasket.
48. Remove the bolts and the intake manifold.

All vehicles
49. Remove the intake manifold gaskets.
50. Clean and inspect the gaskets. Install new gaskets if necessary.
51. Clean and inspect the sealing surfaces.
52. Remove and discard the front module O-ring seal.

To install:
All vehicles

➡**The locating tabs on the gaskets must be up and toward the center of the engine, or a leak will occur.**

53. Install the intake manifold gaskets and front module O-ring seal.

Late build
54. Install the intake manifold and bolts and tighten in the following sequence.

a. Loosely install bolts 1–8.

b. Tighten bolts 9–16 to 11 Nm (8 ft. lbs.).

c. Tighten all bolts to 11 Nm (8 ft. lbs.) in the sequence shown.

55. Install the gasket and the EGR cooler V-clamp.

56. Install the turbocharger heat shield and the nuts.

Early build

57. Install the intake manifold and bolts and tighten in the following sequence.

a. Loosely install bolts 1–8.

b. Tighten bolts 9–16 to 11 Nm (8 ft. lbs.).

c. Tighten all bolts to 11 Nm (8 ft. lbs.) in the sequence shown.

58. Install the gasket and the EGR cooler V-clamp.

59. Install the turbocharger heat shield and the nuts.

All vehicles

60. Slide the EGR cooler hose forward and rotate flat to lock.

61. Install the fuel line.

62. Install new sealing washers.

63. Install the fuel line.

64. Install the banjo fitting and the bolt.

65. Install the oil filter return tube and bolt.

66. On new oil filter return tubes, tighten to 6 Nm (53 inch lbs.).

67. On used oil filter return tubes, tighten to 3 Nm (27 inch lbs.).

68. Install the oil filter housing and the four bolts.

69. Clean and inspect the housing O-rings. Install new O-rings if necessary.

70. Connect the fuel line fittings.

71. Position the fuel injector harness and install the retaining nut. Connect the eight fuel injectors electrical connectors.

72. Connect the retaining clip and exhaust backpressure sensor electrical connector.

73. Connect the wiring retainer and the IAT2 sensor electrical connector.

74. Connect the engine coolant temperature (ECT) sensor and the pin-type retainer.

75. Connect the EGR throttle position sensor electrical connector.

76. Connect the EGR throttle position control module electrical connector.

77. Connect the engine oil temperature (EOT) sensor electrical connector.

78. Connect the engine oil pressure (EOP) sensor electrical connector.

79. Connect the EGR valve electrical connector.

80. Connect the wiring retainer, injector pressure regulator and ICP sensor (if equipped) electrical connectors.

81. Position back or install the heat insulating wrap.

82. Install the fuel injector control module.

83. Connect the engine coolant vent hose and clamp.

84. Connect the MAP sensor hose.

➡ **Install a new O-ring on the heater hose tube.**

85. Install the heater hose tube and bolts.

Early build

86. Install the charge air cooler duct and tighten the clamps.

Late build

87. Install the charge air cooler duct and tighten the clamp.

88. Install the turbocharger pedestal and turbocharger.

89. Install the alternator and the three bolts.

90. Connect the alternator B+ wire land electrical connector and position the boot back.

Vehicles with single alternator

91. Install the accessory drive belt.

Vehicles with dual alternators

92. Install the alternator with mounting bracket and bolts.

93. Connect the B+ wire, alternator electrical connector and wire retainer.

94. Install the accessory drive belt tensioner and bolts.

95. Position the accessory drive belt idler pulley. Install the bracket and bolts.

96. Install the accessory drive belt.

97. Install the accessory drive belt tensioner and bolt.

98. Install the accessory drive belt.

All vehicles

99. Install the turbocharger-to-charge air cooler duct.

100. Install the upper radiator hose.

101. Install the degas bottle.

102. Install the auxiliary battery.

103. Install the cooling fan stator

7.3L Engine

✳✳ CAUTION

The fuel system remains under pressure, even after the engine has been turned OFF. The fuel system pressure must be relieved before disconnecting any fuel lines. Failure to do so may result in fire and/or personal injury.

1. Before servicing the vehicle, refer to the Precautions Section.

2. Relieve the fuel system pressure.

3. Remove or disconnect the following:

- Both battery ground cables
- Air cleaner and install clean rags into the air intake of the intake manifold. It is important that no dirt or foreign objects get into the intake.
- Injection pump
- Fuel return hose from No. 7 and No. 8 rear nozzles and remove the return hose to the fuel tank.
- Engine wiring harness from the engine

➡ **The engine harness ground cables must be removed from the back of the left cylinder head.**

- Bolts attaching the intake manifold to the cylinder heads and remove the manifold.
- Crankcase Depression Regulator (CDR) valve tube grommet from the valley pan

STEP 1. TIGHTEN BOLTS TO 24 FT•LB IN NUMBERED SEQUENCE SHOWN ABOVE.
STEP 2. TIGHTEN BOLTS TO 24 FT•LB IN LINE SEQUENCE SHOWN ABOVE.

Intake manifold bolt torque sequence—7.3L diesel engines

- Bolts attaching the valley pan strap to the front of the engine block and remove the strap.
- Valley pan drain plug and remove the valley pan.

To install:

4. Apply a ⅛ inch (3mm) bead of RTV sealer to each end of the cylinder block.

➡**The RTV sealer should be applied immediately prior to the valley pan installation.**

5. Install or connect the following:
- Valley pan drain plug, CDR valve tube and new grommet into the valley pan
- New O-ring and new back-up ring on the CDR valve
- Valley pan strap on the front of the valley pan
- Intake manifold and tighten the bolts to 24 ft. lbs. (33 Nm) using the sequence illustrated
- Engine wiring harness and the engine ground wire located to the rear of the left cylinder head
- Injection pump
- No. 7 and No. 8 fuel return hoses and the fuel tank return hose
- Air cleaner
- Battery ground cables to both batteries

➡**If necessary, purge the nozzle high pressure lines of air by loosening the connector ½ to 1 turn and cranking the engine until solid stream of fuel, devoid of any bubbles, flows from the connection.**

6. Run the engine and check for oil and fuel leaks.

❋❋ CAUTION

Keep eyes and hands away from the nozzle spray. Fuel spraying from the nozzle under high pressure can penetrate the skin.

7. Check and adjust the injection pump timing.

Exhaust Manifold

REMOVAL & INSTALLATION

6.0L Engine

F-SERIES

1. Before servicing the vehicle, refer to the Precautions Section.
2. Remove the engine.

38 Nm (28 lb-ft)

Left exhaust manifold torque sequence—6.0L engine

67197-EFSE-G127

38 Nm (28 lb-ft)

Right exhaust manifold torque sequence—6.0L engine

67197-EFSE-G128

3. Mount the engine on a work stand.
4. Remove the manifold(s).
5. Installation is the reverse of removal.
6. Inspect the manifold and head surfaces of warpage.
7. Installation is the reverse of removal.

E-SERIES—LEFT SIDE

1. Before servicing the vehicle, refer to the Precautions Section.
2. With the vehicle in **Neutral**, position it on a hoist.
3. Remove the engine cover.
4. Prior to removing the exhaust manifold, inspect the exhaust manifold for warpage with a feeler gauge between the manifold and the cylinder head. Record the measurement and compare with the specifications.

5. Remove the left bolts for the turbocharger adapter pipe.
6. Disconnect the exhaust back pressure tube.
7. Remove the bolts and the left exhaust manifold.

To install:

➡**Start installing the bolts with the second bolt from the rear on the top. The hole diameter is smaller, therefore allowing alignment of the remaining bolts.**

➡**Apply anti-seize lubricant to the bolt threads prior to installing the bolts.**

8. Install the left exhaust manifold and bolts. Tighten the bolts in the sequence shown.

9. Connect the exhaust back pressure tube at the exhaust manifold.

10. Install the left bolts for the turbocharger adapter pipe.

11. Install the engine cover.

E-SERIES—RIGHT SIDE

1. Before servicing the vehicle, refer to the Precautions Section.

2. With the vehicle in **Neutral**, position it on a hoist.

3. Disconnect the battery ground cable.

4. Remove the engine cover.

5. Prior to removing the exhaust manifold, inspect the exhaust manifold for warpage with a feeler gauge between the manifold and the cylinder head. Record the measurement and compare with the specifications.

6. Remove the right bolts for the turbocharger adapter pipe.

7. Remove the bolts and the right exhaust manifold.

To install:

➡**Start installing the bolts with the second bolt from the rear on the top. The hole diameter is smaller, therefore allowing alignment of the remaining bolts.**

➡**Apply anti-seize lubricant to the bolt threads prior to installing the bolts.**

8. Install the right exhaust manifold and bolts. Tighten the bolts in the sequence shown.

9. Install the right bolts for the turbocharger adapter pipe.

10. Install the engine cover.

11. Connect the battery ground cable.

7.3L Engine

1. Before servicing the vehicle, refer to the Precautions Section.

2. Disconnect the ground cables from both batteries.

3. Raise the vehicle and safely support it.

4. Disconnect the muffler inlet pipe from the exhaust manifolds.

5. If removing the right manifold, lower the vehicle. When removing the left manifold, raise the vehicle and remove the manifold from underneath. Bend the tabs on the manifold attaching bolts, then remove the bolts and manifold.

To install:

6. Before installing, clean all mounting surfaces on the cylinder heads and the manifold. Apply an anti-seize compound on the manifold both threads and install the left manifold, using a new gasket and new locking tabs.

7. Tighten the bolts to 45 ft. lbs. (61 Nm) and bend the tabs over the flats on the bolt heads to prevent the bolts from loosening.

8. Raise the vehicle to install the right manifold. Install the right manifold Steps 5 and 6.

STEP1. TIGHTEN BOLTS TO 35 FT.LB. IN NUMBERED SEQUENCE SHOWN ABOVE
STEP2. TIGHTEN BOLTS TO 35 FT.LB. IN LINE SEQUENCE SHOWN ABOVE

7924FG26

Tighten the manifold bolts in the proper sequence—7.3L diesel engines

9. Connect the inlet pipes to the manifold and tighten. Lower the vehicle, connect the batteries and run the engine to check for exhaust leaks.

Camshaft and Valve Lifters

REMOVAL & INSTALLATION

6.0L Engine

1. Before servicing the vehicle, refer to the Precautions Section.

2. Mount the engine on an engine stand.

3. Remove the serpentine belt idler.

4. Remove the serpentine belt tensioner.

5. Remove and discard the O-rings from the oil filter base.

6. Remove the exhaust gas recirculation (EGR) cooler coolant supply port cover. Clean and inspect the gaskets. Install new gaskets if necessary. Clean and inspect the sealing surfaces.

✳✳ WARNING

In the event of a catastrophic engine failure, always install a new oil cooler cover assembly (with oil cooler). Foreign material cannot be removed from the oil cooler.

➡**The oil cooler is replaced as an assembly.**

7. Remove the oil cooler assembly. Clean and inspect the gaskets. Install a new gasket if necessary. Clean and inspect the sealing surfaces.

8. Remove the oil pump inlet strainer. Clean and inspect for tears and other damage.

9. Remove bolts and the turbocharger heat shield.

10. Remove the high-pressure oil pump cover. Use a thin gasket scraper to separate the cover from the crankcase. Clean and inspect the gaskets. Install a new gasket if necessary. Clean and inspect the sealing surfaces.

11. Remove the bolts from the high-pressure oil pump discharge pipe.

12. Disconnect and remove the high-pressure oil pump discharge pipe.

13. Remove and discard the D-ring seal.

14. Remove and discard the high-pressure pump O-ring seal.

15. Remove the bolts and the high-pressure oil pump.

16. Remove and discard the lower O-ring seal.

17. Remove the glow plug buss bar.
18. Remove the eight glow plugs.

➡ **Mark the location of the stud bolts.**

19. Remove the valve covers. Clean and inspect the gaskets. Install a new gasket if necessary. Clean and inspect the sealing surfaces.
20. Remove the bolts and the coolant pump pulley.
21. Remove the coolant pump.
22. If equipped, remove the bolts and the dual alternator pulley.
23. Prior to removing the crankshaft damper, check the crankshaft vibration damper runout.

➡ **Pry the crankshaft forward at the same point to eliminate possible error caused by crankshaft end play.**

24. Rotate the crankshaft 90 degrees. Pry the crankshaft forward. Record the measurement. Repeat every 90 degrees. If the runout exceeds 0.002 inch, install a new crankshaft vibration damper.

✳✳ WARNING

To prevent engine damage, you must always replace all four bolts when installing the vibration damper.

✳✳ CAUTION

To avoid personal injury, support the vibration damper during mounting bolt removal. The damper can slide off the nose of the crankshaft very easily.

25. Remove the bolts and the crankcase vibration damper. Discard the bolts.
26. Punch two holes in the seal. Remove the crankshaft seal with a slide-hammer.

➡ **Production engine will not have a wear sleeve. If equipped, remove the crankshaft damper wear sleeve.**

27. Remove the oil pump body. Remove and discard the O-ring seal.

➡ **Mark the front of each drive rotor for correct reassembly orientation.**

28. Remove the inner and outer oil pump drive rotors.
29. Remove the engine front cover. Clean and inspect the gaskets. Install new gaskets if necessary. Clean and inspect the sealing surfaces.
30. Using a quick-disconnect tool, disconnect the high-pressure oil rail supply line at the high-pressure oil rail.

31. Remove the bolts and the high pressure oil rail. Disconnect and remove the high-pressure oil supply line.

✳✳ WARNING

Do not attempt to put battery voltage to the fuel injector or damage to the fuel injector will occur.

32. Using a 19 mm socket, push the fuel injector electrical connector out of the rocker arm carrier.

✳✳ WARNING

To prevent engine damage, do not use air tools to remove the fuel injectors. The clip that extracts the injector can dislodge and fall into the oil drain hole.

➡ **If engine oil is found in the engine coolant or engine coolant is found in the combustion chambers, new injector sleeve may need to be installed.**

33. Remove the bolt, the fuel injector hold down and the fuel injector.
34. Remove and discard the crankcase-to-head tube assembly.
35. Remove the inner head bolts from both cylinder heads.
36. Remove the 16 bolts and the rocker arm assemblies.
37. Remove the rocker arm carrier from the cylinder head. Clean and inspect the gaskets. Install new gaskets if necessary. Clean and inspect the sealing surfaces.

➡ **Mark the location of the valve bridges before removing.**

38. Remove the 16 valve bridges.

✳✳ WARNING

To prevent engine damage, keep the push rods in the order in which they were removed. Install all push rods back in their original positions.

39. Mark the location and remove the 16 push rods.
40. Remove the 10 outer head bolts.
41. Remove the cylinder heads.
42. Remove and discard the cylinder head gasket.
43. Remove and discard the four cylinder head dowel sleeves.
44. Remove the bolts from the rear engine tube assembly.
45. Remove the bolt and the rear engine tube assembly.

✳✳ WARNING

To prevent engine damage, keep the cam followers in the order in which they were removed. Install all cam followers back in their original positions.

46. Remove the bolts and the roller follower guides. Remove the hydraulic cam followers.
47. Install the special tool and measure the camshaft gear backlash. Install a new camshaft gear if backlash is not within specification.
48. Install a dial indicator and measure the camshaft end play. Install a new camshaft thrust plate if end play is not within 0.002–0.008 inch.
49. Remove the bolt and the camshaft position (CMP) sensor.

✳✳ WARNING

Do not knick or scratch the camshaft bearings with the camshaft lobes or engine damage will occur.

50. Remove the thrust plate mounting bolts and remove the camshaft and gear.
To install:

➡ **Check alignment of the oil holes after installing the bearings.**

51. If removed, install the camshaft bearings.

✳✳ WARNING

Do not nick or scratch the camshaft bearings with the camshaft lobes or engine damage can occur.

➡ **Apply clean engine oil to the camshaft prior to installing.**

52. Using camshaft alignment tool 303-772, install the camshaft and gear assembly. Aligning it with the crankshaft. Install the thrust plate mounting bolts. Torque to 23 ft. lbs. (31 Nm).

67197-EFSE-G129

Using camshaft alignment tool 303-772, install the camshaft and gear assembly

53. The remainder of installation is the reverse of removal.

7.3L Engine

Ford recommends removing the diesel engine from the vehicle for camshaft removal.

➡ **A camshaft removal tool, Ford part No. T65L-6250-A and Adapter 14-0314 is needed to remove the diesel camshaft.**

1. Before servicing the vehicle, refer to the Precautions Section.
2. Remove the intake manifold and valley pan, if equipped.
3. Remove the rocker covers, and either remove the rocker arm shafts or loosen the rockers on their pivots and remove the pushrods. The pushrods must be reinstalled in their original positions.
4. Remove the valve lifters in sequence with a magnet. They must be replaced in their original positions.
5. Remove the timing gear cover, timing gear and sprockets.

To install:

6. Liberally coat the camshaft with oil before installing it. Slide the camshaft into the engine very carefully so as not to scratch the bearing bores with the camshaft lobes. Install the camshaft thrust plate and tighten the attaching screws to 10–12 ft. lbs. (13–16 Nm). Measure the camshaft end-play. If the end-play is more than 0.009 inch (0.228mm), replace the thrust plate. Assemble the remaining components in the reverse order of removal.
7. Install the timing gear and front cover.
8. Install the valve lifters. They must be replaced in their original positions.
9. Install the pushrods, the rocker arms and the rocker arm covers.
10. Install the intake manifold and valley pan, if equipped.

Valve Lash

ADJUSTMENT

Valve lash on the diesel engine is not adjustable.

Starter Motor

REMOVAL & INSTALLATION

6.0L Engine

1. Before servicing the vehicle, refer to the Precautions Section.

✳✳ WARNING

When performing maintenance on the starting system, be aware heavy gauge leads are connected to the battery. Make sure protective caps are in place when maintenance is complete.

2. Disconnect the battery ground cable.
3. Raise and support the vehicle.
4. Remove starter solenoid protective cap.
5. Disconnect the starter motor electrical connections.
6. Remove the bolts and the starter.
7. To install, reverse the removal procedure. Torque the bolts to 18 ft. lbs. (25Nm).

7.3L Engine

1. Before servicing the vehicle, refer to the Precautions Section.
2. Remove or disconnect the following:
   • Negative battery cable
   • Starter motor electrical connections
   • Starter motor bolts and the motor

To install:

3. Installation is the reverse of removal, tighten the starter motor bolts to 16–20 ft. lbs. (22–28 Nm).

Oil Pan

REMOVAL & INSTALLATION

F-Series w/6.0L Engine

LOWER OIL PAN

1. Before servicing the vehicle, refer to the Precautions Section.
2. With the vehicle in **Neutral**, position it on a hoist.
3. Disconnect the negative battery cable.

✳✳ WARNING

Never remove the pressure relief cap while the engine is operating or when the cooling system is hot. Failure to follow these instructions can result in damage to the cooling system or engine or result in personal injury. To avoid having scalding hot coolant or steam blow out of the degas bottle when removing the pressure relief cap, wait until the engine has cooled then wrap a thick cloth around the pressure relief cap and turn it slowly. Step back while the pressure is released from the

cooling system. When certain all the pressure has been released, (still with a cloth) turn and remove the pressure relief cap. Failure to follow these instructions can result in personal injury.

✳✳ CAUTION

The coolant must be removed in a suitable, clean container for reuse. If the coolant is contaminated, it must be recycled or disposed of correctly and the system filled with new coolant.

➡ **Less than 80% of coolant capacity can be recovered with the engine in the vehicle. Dirty, rusty, or contaminated coolant requires replacement.**

4. Place a suitable container below the radiator draincock. If equipped, disconnect the coolant return hose at the fluid cooler.
5. Remove the fill cap from the degas bottle.
6. Remove the oil pan drain plug and drain the engine oil.
7. Loosen the exhaust pipe retaining nuts.
8. Remove the motor mount retaining nuts.
9. Open the radiator draincock.
10. Disconnect the lower radiator hose.
11. Disconnect the transmission cooler tubes.
12. Remove the air cleaner assembly.
13. On vehicles with metal ducts, remove the charge air cooler duct.
14. On vehicles with blow molded ducts, loosen the clamps and remove the charge air cooler duct.
15. Remove the radiator support brackets.
16. Remove the 4 pin-type retainers and pull back the sight shield.
17. With the sight shield pulled back, remove the 3 pin-type retainers and the 2 wiring retainers and position the harness rearward out of the way.
18. Disconnect the upper radiator hose and the radiator overflow hose.
19. Remove the radiator and shroud as an assembly.
20. Disconnect the cooling fan electrical connector. Unclip and position the fan and wiring aside.

➡ **Use a hole in the fan hub to prevent the fan from turning.**

21. Using the special tool 303-591, or equivalent, loosen the fan clutch. Remove the cooling fan.

22. Remove the bolts from the stator, remove the stator.

23. On vehicles with dual alternator perform the following:

    a. Remove the secondary accessory drive belt.

    b. Remove the bolt and the secondary accessory drive belt tensioner.

    c. Remove the primary accessory drive belt.

    d. Remove the bolts, bracket and secondary accessory drive belt idler pulley.

    e. Remove the bolts and the primary accessory drive belt tensioner.

    f. Disconnect the wire retainer, secondary alternator electrical connector and B+ wire.

    g. Remove the bolts and the secondary alternator with mounting bracket.

24. On vehicles with single alternator, remove the primary accessory drive belt.

25. Remove the bolts and position the alternator back.

26. Support the hood and disconnect the hood lift assemblies.

27. Install two lifting eyes 303-D030, or equivalent, on the right hand cylinder head.

28. Remove the retaining bolt for the fuel lines.

29. Disconnect the heater hose at the coolant pump.

30. Remove the fan shroud mounting stud.

31. Install one lifting eye 303-D030 on the left hand cylinder head.

✳✳ CAUTION

Do not use the special tool to raise the engine. Damage to the special tool or vehicle may occur.

➡ **The ball studs may have to be removed.**

➡ **This procedure requires a second bolt hook assembly. The tools are available through Rotunda tools with the following numbers: bolt hook 303-F070-6, handle 303-F070-8, bracket 303-F070-7 and washer 303-F070-12004.**

32. Install the special tool.

➡ **The engine must be raised evenly.**

33. Using a lifting crane, raise the engine until the turbo charger is about to touch the cowl. Secure the engine with the special tool.

34. Remove the transmission cooler line bracket.

35. Remove the bolts and position back the oil pan until the oil pick-up tube bolts are accessible.

36. Remove the bolts and let the oil pick-up tube go into the oil pan. Remove the oil pan.

37. Remove the press-in-place gasket and discard.

38. Clean and inspect the sealing surfaces.

39. Remove and discard the oil pick-up tube O-ring.

To install:

40. Install a new O-ring on the oil pick-up tube and position the oil pick-up tube in the oil pan.

41. Install a new press-in-place gasket into the upper oil pan.

42. Position the oil pan in the vehicle.

43. Install the oil pick-up tube and bolts. Tighten to 10 ft. lbs. (13 Nm).

44. Position the oil pan and install the bolts. Install the long bolts first. Tighten to 10 ft. lbs. (13 Nm).

45. Install the transmission cooler tube bracket and nut. Tighten to 18 inch lbs. (10 Nm).

46. Using the special tool 303-F070, lower the engine.

47. Remove the lifting eyes 303-D030 and the special tool.

48. Install the fan shroud mounting stud. Tighten to 30 ft. lbs. (40 Nm).

49. Connect the heater hose at the coolant pump.

50. Install the retaining bolt for the fuel lines. Tighten to 10 ft. lbs. (10 Nm).

51. If removed, install the ball studs.

52. Connect the hood lifts and remove the hood support.

53. Position the primary alternator and install the bolts.

54. On a vehicle with a single alternator, install the primary accessory drive belt.

55. On vehicles with dual alternators, perform the following:

    a. Install the secondary alternator with mounting bracket and bolts. Tighten to 35 ft. lbs. (47 Nm).

    b. Connect the B+ wire, secondary alternator electrical connector and wire retainer.

    c. Install the primary accessory drive belt tensioner and bolts. Tighten to 18 ft. lbs. (25 Nm).

    d. Position the secondary accessory drive belt idler pulley. Install the bracket and bolts. Tighten to 18 ft. lbs. (25 Nm).

    e. Install the primary accessory drive belt.

    f. Install the secondary accessory drive belt tensioner and bolt. Tighten to 18 ft. lbs. (25 Nm).

13 Nm (10 lb-ft)

06017-F150-G228

Lower oil pan installation. (1) indicates the long bolts—6.0L engine lower oil pan

g. Install the secondary accessory drive belt.

56. Install the stator and stator bolts. Tighten to 30 ft. lbs. (40 Nm).

57. Install the cooling fan clutch. Use the special tool 303-591 to tighten the cooling fan clutch. Tighten to 98 ft. lbs. (133 Nm).

58. Position and clip the cooling fan wiring. Connect the cooling fan electrical connector.

59. Install the radiator and shroud as an assembly.

60. Connect the upper radiator hose and the radiator overflow hose.

61. Position the harness wiring, install the 2 wiring retainers and 3 pin-type retainers.

62. Position the sight shield and install the 4 pin-type retainers.

63. Install the radiator support brackets. Tighten to 22 ft. lbs. (30 Nm).

64. On vehicles with metal ducts, install the charge air cooler duct and tighten the clamps.

65. On vehicles with blow molded ducts, install the charge air cooler duct and tighten the clamps.

66. Install the air cleaner assembly.

67. Connect the transmission cooler tubes. Tighten to the specifications illustrated.

68. Close the radiator draincock. Connect the lower radiator hose.

69. Install the motor mount retaining nuts. Tighten to 76 ft. lbs. (103 Nm).

70. Tighten the exhaust pipe retaining nuts to 35 ft. lbs. (47 Nm).

71. Clean and inspect the oil pan drain plug and gasket, install new if necessary.

72. Install the oil pan drain plug and tighten to 18 ft. lbs. (25 Nm).

73. Connect the negative battery cables.

74. Fill the engine with clean engine oil.

75. Fill the coolant system.

76. Run the engine and check for leaks

UPPER OIL PAN

1. Before servicing the vehicle, refer to the Precautions Section.

2. Remove the oil pan.

3. Remove the bolts and the upper oil pan. Remove and discard the press-in-place gasket.

4. Clean and inspect the sealing surfaces.

To install:

➡**Install a new press-in-place gasket.**

5. Install the upper oil pan and bolts.

6. Install the lower oil pan.

E-Series w/6.0L Engine

LOWER PAN

1. Before servicing the vehicle, refer to the Precautions Section.

2. With the vehicle in **Neutral**, position it on a hoist.

3. Disconnect the battery ground cable.

4. Remove the engine cover.

5. Remove the cooling fan stator.

6. Remove the A/C compressor.

7. Remove the power steering pump upper mounting bolt.

8. Remove the left fan stator stand-off.

➡**The front bolt will remain in the power steering pump.**

9. Remove the bolts and position the power steering pump aside.

10. Remove the oil pan drain plug.

11. Loosen the exhaust pipe retaining nuts.

12. Remove the motor mount retaining nuts.

13. Remove the right fan stator stand-off. Remove the bolt and position the cable aside.

14. Install the special tools.

15. Install the engine lifting attachment and floor crane.

✳✷ CAUTION

Use care when raising the engine to avoid engine or body damage

---

16. Raise the engine.

17. Remove the bolts and position back the oil pan until the oil pick up tube bolts are accessible.

18. Remove the bolts and let the oil pick-up tube go into the oil pan. Remove the oil pan.

19. Remove and discard the oil pick-up tube O-ring seal.

20. Remove the press-in-place gasket and discard.

21. Clean and inspect the sealing surfaces.

To install:

22. Install a new press-in-place gasket into the upper oil pan.

23. Install a new O-ring seal on the oil pick-up tube and position the oil pick-up tube in the oil pan.

24. Position the oil pan in the vehicle.

25. Install the oil pan pick-up tube and bolts.

26. Install the oil pan and bolts.

27. Lower the engine.

28. Remove the floor crane and engine lifting attachment.

29. Remove the special tools.

30. Position back the cable and install the bolt. Install the right fan stator stand-off.

➡**Tighten the top retaining nut first.**

31. Install the motor mount retaining nuts.

32. Tighten the exhaust pipe retaining nuts.

33. Clean and inspect the oil pan drain plug and gasket. Install new, if necessary.

34. Install the oil drain plug.

35. Reposition the power steering pump and install the power steering pump bolts.

06017-F150-G229

Upper oil pan fasteners—6.0L engine

36. Install the left fan stator stand-off.

37. Install the upper power steering mounting bolt.

38. Fill the engine with clean engine oil.

39. Install the engine cover.

40. Connect the battery ground cable.

41. Install the A/C compressor.

42. Install the cooling fan stator.

43. Run the engine and check for leaks.

UPPER OIL PAN

1. Before servicing the vehicle, refer to the Precautions Section.

2. Remove the oil pan.

3. Remove the bolts and the upper oil pan. Remove and discard the press-in-place gasket.

4. Clean and inspect the sealing surfaces.

To install:

➡️**Install a new press-in-place gasket.**

5. Install the upper oil pan and bolts.

6. Install the lower oil pan.

7.3L Engine

1. Before servicing the vehicle, refer to the Precautions Section.

2. Remove the engine.

3. Remove the oil pan bolts.

4. Lower the oil pan.

➡️**The oil pan is sealed to the crankcase with RTV silicone sealant in place of a gasket. It may be necessary to separate the pan from the crankcase with a utility knife. Also, the crankshaft may have to be turned to allow the pan to clear the crankshaft throws.**

5. Clean the pan and crankcase mating surfaces thoroughly.

To install:

6. Apply a ⅛ in. (3mm) bead of RTV silicone sealant to the pan mating surfaces, and a ¼ in. (6mm) bead on the front and rear covers and in the corners; you have 15 minutes within which to install the pan!

7. Install the locating dowels into position.

8. Position the pan on the engine and install the pan bolts loosely.

9. Remove the dowels.

10. Tighten the pan bolts to:

   a. ¼ in.-20 bolts: 84 inch lbs. (10 Nm).

   b. ⁵⁄₁₆ in.-18 bolts: 14 ft. lbs. (19 Nm).

   c. ⅜ in.-16 bolts: 24 ft. lbs. (33 Nm).

11. Install the engine.

Oil Pump

REMOVAL & INSTALLATION

6.0L Engine

1. Before servicing the vehicle, refer to the Precautions Section.

2. With the vehicle in **Neutral**, position it on a hoist.

3. Disconnect the battery ground cable(s).

4. Remove the cooling fan

5. On vehicles with dual alternator, perform the following:

   a. Remove the dual alternator accessory drive belt.

   b. Remove the bolts and the dual alternator pulley.

6. Remove the accessory drive belt.

7. Check the crankshaft vibration damper runout.

8. Remove the paint from the face of the crankshaft vibration damper at four points 90 degrees apart.

9. Attach dial indicator 100-002 to the cylinder block. Position the tool on one of the unpainted surfaces.

10. Using a suitable tool, pry the crankshaft forward. Zero the dial indicator.

➡️**Pry the crankshaft forward only to eliminate possible error caused by crankshaft end play.**

11. Rotate the crankshaft 90 degrees. Pry the crankshaft forward. Record the measurement. Repeat at each unpainted surface.

12. If the runout exceeds specification, install a new crankshaft vibration damper.

✱✱ CAUTION

To avoid personal injury, support the vibration damper during mounting bolt removal. The damper can slide off the nose of the crankshaft very easily.

13. Remove the bolts and the crankshaft vibration damper.

14. Discard the bolts.

15. Punch two holes in the seal.

16. Using the special tool 303-D060, remove the crankshaft seal.

➡️**Production engine will not have a wear sleeve.**

17. If equipped, remove the crankshaft damper wear sleeve using tool 303-762.

18. Remove the bolts and the gerotor cover. Remove and discard the O-ring seal.

➡️**Mark the front of the inner and outer gerotor for correct reassembly.**

19. Remove the inner and outer gerotors.

20. Inspect the oil pump components and replace as necessary.

21. Inspect the oil pump for excessive metal particles.

22. Inspect the oil pump for gouging, cracks or deep scratches.

23. Inspect the oil pump inner and outer gear rotors for damage or excessive wear.

To install:

➡️**Install the gears with marks pointing outward.**

24. Lubricate the inner gear with lithium assembly grease and install onto the crankshaft. Lubricate the outer gear with lithium assembly grease and mesh with the inner gear rotor in the oil pump housing. Wipe off the excess assembly grease.

25. Install a new O-ring seal.

26. Install the gerotor cover and bolts. Tighten to 71 inch lbs. (8 Nm).

27. Thoroughly clean the crankshaft front seal mounting surface.

28. Apply Threadlock 262® to the outer circumference of the leading edge of the crankshaft.

➡️**New seal and wear sleeve must not be separated.**

29. Using the special tool 303-361, install the oil seal and wear sleeve assembly.

✱✱ CAUTION

To prevent engine damage, you must always install four new bolts when installing the vibration damper.

➡️**Do not use anti-seize compounds, grease or any lubricants. Lubricants have an adverse effect on the torque results.**

30. Install the crankshaft vibration damper and bolts.

31. Tighten the bolts in a criss-cross sequence as follows:

   a. Tighten the bolts to 50 ft. lbs. (68 Nm).

8 Nm (71 lb-in)

67197-EFSE-G130
Oil pump mounting—6.0L engine

BOLT

OIL PUMP
BODY PLATE

7924FG62

The oil pump is mounted on the cylinder block with 4 bolts—7.3L diesel engine

b. Tighten the bolts an additional 90 degrees.

32. Install the accessory drive belt.

33. On vehicles with dual alternator, perform the following:

a. Install the dual alternator pulley and bolts. Tighten to 35 ft. lbs. (47 Nm).

b. Install the dual alternator accessory drive belt.

34. Install the cooling fan stator

35. Connect the battery ground cables.

7.3L Engine

1. Before servicing the vehicle, refer to the Precautions Section.

2. Disconnect the negative battery cable.

3. Remove the oil pan.

4. Remove the oil pick-up tube from the pump.

5. Unbolt and remove the oil pump.

To install:

6. Assemble the pick-up tube and pump. Use a new gasket.

7. Install the oil pump and tighten the bolts to 14 ft. lbs. (19 Nm).

8. Install the oil pick up tube.

9. Install the oil pan.

10. Connect the negative battery cable.

Oil Cooler

REMOVAL & INSTALLATION

6.0L Engine

1. Before servicing the vehicle, refer to the Precautions Section.

2. Remove the intake manifold.

3. Remove the bolts and the exhaust gas recirculation (EGR) cooler supply port cover.

4. Remove and discard the press-in-place gasket. Clean and inspect the sealing surfaces.

5. Remove the bolts and the oil filter base. Remove and discard the press-in-place gasket. Clean and inspect the sealing surfaces.

❊❊ WARNING

In the event of a catastrophic engine failure, always install a new oil cooler cover assembly (with oil cooler). Foreign material cannot be removed from the oil cooler.

➡ **The oil cooler is replaced as an assembly.**

6. Remove the bolts and the oil cooler assembly. Remove and discard the press-

06017-F150-G230

Exhaust gas recirculation (EGR) cooler supply port cover—6.0L engine

22 Nm (16 lb-ft)

10 Nm (89 lb-in)

06017-F150-G231

Oil cooler bolt torque sequence—6.0L engine

in-place gasket. Clean and inspect the sealing surfaces.

7. Remove the oil pump inlet strainer. Clean and inspect the strainer for tears or other signs of damage.

To install:

8. Install the oil pump inlet strainer.

9. Install the gasket in the oil cooler assembly cover.

❊❊ WARNING

In the event of a catastrophic engine failure, always install a new oil cooler cover assembly (with oil cooler). Foreign material cannot be removed from the oil cooler.

10. Install the oil cooler assembly and the bolts. Torque the bolts as shown in the accompanying illustration.

11. Install the gasket in the oil filter base.

12. Install the oil filter base and the bolts.

13. Install the seal on the EGR cooler coolant supply port cover.

14. Install the EGR cooler coolant supply port cover and the bolts.

15. Install the intake manifold.

7.3L Engine

1. Before servicing the vehicle, refer to the Precautions Section.

06017-F150-G232

Front header bolts—7.3L engine

Rear header bolts—7.3L engine

2. Drain the radiator.
3. Raise and support the vehicle.
4. Remove the oil filter.
5. Disconnect the block heater.
6. Remove the two bolts from the oil cooler front header.
7. Remove the three bolts from the oil cooler rear header.
8. Remove the oil cooler.
9. Remove the oil cooler front and rear header gaskets from the engine.
10. Clean the gasket mating surface on the engine and the oil cooler.

To install:
11. Position the oil cooler in the vehicle.
12. Install the oil cooler front header.
13. Position the front header gasket; then place the front header against the engine.
14. Install the front header mounting bolts. Do not tighten the bolts at this time.
15. Install the oil cooler rear header.
   a. Position the rear header gasket, then place the rear header against the engine.
   b. Install the rear header mounting bolts. Torque to 18 ft. lbs. (25 Nm).
16. Tighten the oil cooler front header bolt to 18 ft. lbs. (25 Nm).
17. Connect the block heater.
18. Install the oil filter.
19. Lower the vehicle.
20. Fill the radiator.
21. Start the engine and check for leaks.

Rear Main Seal

REMOVAL & INSTALLATION

6.0L Engine

1. Before servicing the vehicle, refer to the Precautions Section.
2. Remove the transmission.
3. Remove the bolts.
4. Remove the flexplate or flywheel.

➡Use extreme care when removing the flywheel front adapter to prevent damage to the alignment dowel pin.

5. Remove the flywheel front adapter.

❊❊ CAUTION

To prevent engine damage, do not remove the rear primary crankshaft flange bolts under any circumstances. If the flange is removed and reinstalled, it will result in engine vibration and premature transmission component wear.

6. Punch two holes in the rear main seal, across from each other.
7. Using the puller tool 100-001, remove the rear main seal.

➡Production engines will not have a wear sleeve.

8. If equipped with a crankshaft wear sleeve, use the tool 303-771 to remove the crankshaft rear wear sleeve.
9. Clean and inspect the crankshaft sealing surface.

To install:

➡The crankshaft rear oil seal and wear sleeve are installed as an assembly.

Rear main seal wear sleeve removal— 6.0L engine

Rear main seal installation—6.0L engine

➡Lubricate the outer diameter of the rubber seal with a solution of dish soap and water (approximately 50/50 mix) prior to assembly. Do not use any other type of lubricant.

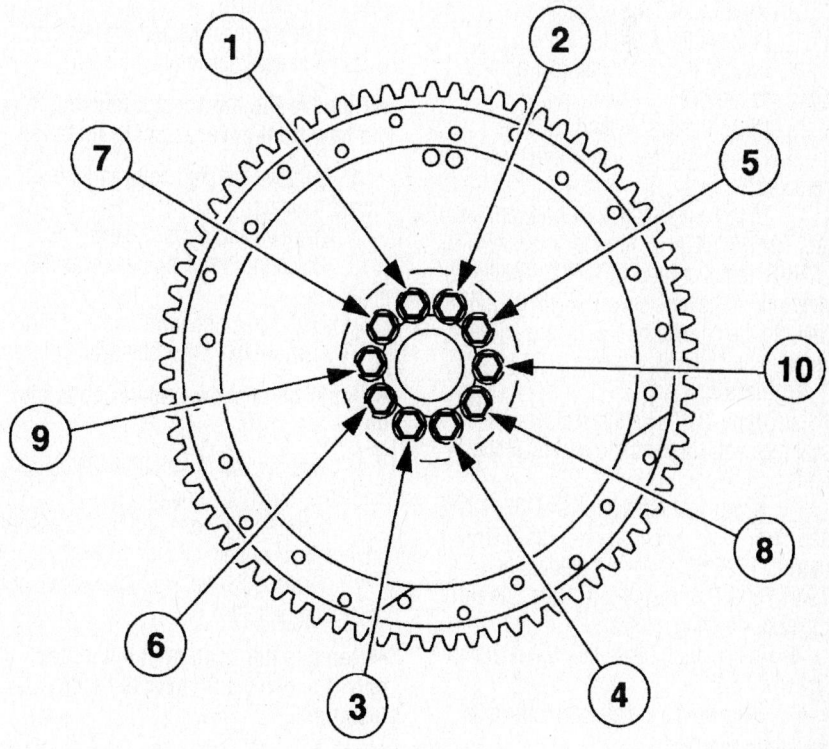

Flywheel installation torque sequence—6.0L engine

Assemble the seal and wear ring removal tools, then remove the wear ring—7.3L diesel engine

10. Apply a bead of Threadlock 262® around the circumference of the outer rear edge of the secondary crankshaft flange.

11. Using tool 303-770, install the crankshaft rear oil seal.

12. Install the flywheel front adapter.

13. Install the flexplate or flywheel.

14. Install the bolts. Snug all bolts to 44 inch lbs. (5 Nm), then tighten all bolts to 69 ft. lbs. (94 Nm) in the sequence illustrated.

15. Install the transmission.

7.3L Engine

1. Before servicing the vehicle, refer to the Precautions Section.

2. Remove the transmission.

3. Remove the flywheel.

4. Loosen the crankshaft rear oil seal bolts and remove the seal.

5. Clean the seal mating surfaces.

6. If installing the old seal, inspect it for damage.

7. Using crankshaft wear ring removal tool T94T-6701-AH1, forcing screw T84T-7025-B, remover tube T77J-7025-B and wear ring remover sleeve T94T-6701-AH2 (refer to the illustration), or their equivalents, remove the wear ring.

To install:

8. Apply RTV silicone sealant to the seal retaining ring and the seal retaining bolts.

9. Using seal replacers T94T-6701-AH3 and T94T-AH4, driver sleeve T79T-6316-A4 (part of T79T-6316-A) and guide pins T94P-7000-P or their equivalents, install the wear ring and oil seal.

10. Install and tighten the seal retaining bolts.

11. Remove the installation tools and install the flywheel.

12. Install the transmission.

Piston and Ring Positioning

ENGINE BLOCK

7.3L Diesel engines—piston-to-engine orientation

Timing Gears, Front Cover and Seal

REMOVAL & INSTALLATION

6.0L Engine

1. Before servicing the vehicle, refer to the Precautions Section.

2. Remove the intake manifold.

3. Disconnect the exhaust pressure tube at the exhaust.

4. Remove the nuts and the exhaust pressure bracket assembly.

➡**Remove the thermostat housing only if a new front cover is being installed.**

5. Remove the stud bolts and the thermostat housing.

6. Remove and discard the O-ring.

7. Disconnect the engine coolant fill hose.

8. Disconnect the lower radiator hose.

9. Remove the stator stand-off bolt.

➡**Upper bolts shown, lower bolts similar.**

10. Remove the four bolts and position aside the power steering pump.

11. Remove the bolts and the accessory drive belt tensioner.

12. Remove the bolts and the accessory drive idler pulleys.

➡**Remove the coolant pump pulley only if a new front cover is being installed.**

13. Remove the coolant pump pulley.

➡**Remove the coolant pump only if a new front cover is being installed.**

TOWARD CAMSHAFT

CAMSIDE

14. Remove the bolts and the coolant pump.

15. Remove and discard the O-ring.

16. Remove the nut and the battery cable bracket.

17. Remove the oil pump.

18. Check the crankshaft vibration damper runout.

19. Remove the paint from the face of the crankshaft vibration damper at four points 90 degrees apart.

20. Attach the special tool to the cylinder block. Position the special tool on one of the unpainted surfaces.

21. Using a suitable tool, pry the crankshaft forward. Zero the dial indicator.

➡**Pry the crankshaft forward at the same point to eliminate possible error caused by crankshaft end play.**

22. Rotate the crankshaft 90 degrees. Pry the crankshaft forward. Record the measurement. Repeat at each unpainted surface. If the runout exceeds specification, install a new crankshaft vibration damper.

※※ WARNING

To prevent engine damage, you must always install four new bolts when installing the vibration damper.

※※ WARNING

To avoid personal injury, support the vibration damper during mounting bolt removal. The damper can slide off the nose of the crankshaft very easily.

23. Remove the bolts and the crankcase vibration damper.
24. Discard the bolts.
25. Remove the bolts and the front cover.

※※ WARNING

Sealant is used where the crankcase and lower crankcase meet. Failure to cut the sealant could result in pulling the lower crankcase seal out while removing the front cover gasket.

26. Use a thin blade scraper to cut the sealant where the crankcase and the lower crankcase meet. Remove and discard the front cover gasket.
27. Clean and inspect the sealing surfaces.
28. Punch two holes in the seal.
29. Using the special tool, remove the crankshaft seal.

➡**Production engine will not have a wear sleeve.**

30. If equipped, remove the crankshaft damper wear sleeve.
31. Remove the thrust plate mounting bolts and remove the camshaft and gear.
To install:
32. Install the camshaft and gear assembly. Using the special tool, align the camshaft timing mark as shown. Install the thrust plate mounting bolts.
33. Thoroughly clean the crankshaft front seal mounting surface.
34. Apply Threadlock 262® to the outer circumference of the leading edge of the crankshaft.

➡**New seal and wear sleeve must not be separated.**

35. Using the special tool, install the oil seal and wear sleeve assembly.
36. If removed, install the front cover crankcase dowels into the cylinder block.

➡**Use guide studs to aid in installation. Studs must be fabricated locally.**

37. Install the guide studs.
38. Apply a bead of sealant at the seam where the crankcase and the lower crankcase meet.
39. Install a new engine front cover gasket.
40. Install the engine front cover and bolts.

※※ WARNING

To prevent engine damage, you must always install four new bolts when installing the vibration damper.

➡**Do not use anti-seize compounds, grease or any lubricants. Lubricants have an adverse effect on the torque results.**

41. Install the crankshaft vibration damper and bolts.
42. Tighten the bolts in the sequence shown.
  a. Tighten the bolts to 68 Nm (50 ft. lbs.).
  b. Tighten the bolts an additional 90 degrees.
43. Install the oil pump.
44. Install the battery cable bracket and nut.

➡**Install a new O-ring on the coolant pump pulley.**

45. If removed, install the coolant pump and bolts.
46. If removed, install the coolant pump pulley and bolts.
47. Install the accessory drive idler pulleys and bolts.
48. Install the accessory drive belt tensioner and bolts.
49. Position back the power steering pump and install the bolts.
50. Install the stator stand-off bolt.
51. Connect the lower radiator coolant hose.
52. Connect the engine coolant hose.

➡**Install a new O-ring.**

53. If removed, install the thermostat housing and stud bolts.

54. Install the exhaust pressure bracket assembly and retaining nuts.
55. Connect the exhaust pressure tube fitting at the exhaust manifold.
56. Install the intake manifold.

7.3L Engine

➡**The crankshaft gear sprocket is not serviced separately from the crankshaft. Do not try to remove the sprocket or you will damage the crankshaft.**

Remove the camshaft sprocket as follows:
1. Before servicing the vehicle, refer to the Precautions Section.
2. Disconnect the negative battery cable.
3. Remove the camshaft.
4. Use a press to remove the sprocket from the camshaft.
5. Remove the thrust plate and sprocket key.
6. Inspect the camshaft and related parts for wear and damage.
To install:
7. Clean the nose of the camshaft and install the thrust plate.
8. Place the key in the keyway on the camshaft.
9. Heat the sprocket in an oven to 500°F (260°C).
10. Remove the sprocket from the oven, align the sprocket keyway with the camshaft key and install the sprocket on the camshaft until it is fully seated. Allow the camshaft assembly to cool before installation
11. Install the camshaft in the engine and align the timing marks on the gears.
12. Connect the negative battery cable.

CAMSHAFT
SPROCKET

CAMSHAFT

CYLINDER
BLOCK

7924FG63

The camshaft gear is removed with the camshaft, then pressed off—7.3L diesel engines

Item	Description
1	Camshaft Sprocket
2	Crankshaft Sprocket
3	Crankshaft Sprocket Timing Mark
4	Camshaft Sprocket Timing Mark

7924FG64

Be sure the timing marks are aligned as illustrated—7.3L diesel engines

GASOLINE FUEL SYSTEM

Fuel System Service Precautions

Safety is the most important factor when performing not only fuel system maintenance, but any type of maintenance. Failure to conduct maintenance and repairs in a safe manner may result in serious personal injury or death. Maintenance and testing of the vehicle's fuel system components can be accomplished safely and effectively by adhering to the following rules and guidelines.

• To avoid the possibility of fire and personal injury, always disconnect the negative battery cable unless the repair or test procedure requires that battery voltage be applied.

• Always relieve the fuel system pressure prior to disconnecting any fuel system component (injector, fuel rail, pressure regulator, etc.), fitting or fuel line connection. Exercise extreme caution whenever relieving fuel system pressure, to avoid exposing skin, face

and eyes to fuel spray. Please be advised that fuel under pressure may penetrate the skin or any part of the body that it contacts.

• Always place a shop towel or cloth around the fitting or connection prior to loosening to absorb any excess fuel due to spillage. Ensure that all fuel spillage (should it occur) is quickly removed from engine surfaces. Ensure that all fuel soaked cloths or towels are deposited into a suitable waste container.

• Always keep a dry chemical (Class B) fire extinguisher near the work area.

• Do not allow fuel spray or fuel vapors to come into contact with a spark or open flame.

• Always use a back-up wrench when loosening and tightening fuel line connection fittings. This will prevent unnecessary stress and torsion to fuel line piping. Always follow the proper torque specifications.

• Always replace worn fuel fitting O-rings with new. Do not substitute fuel hose or equivalent where fuel pipe is installed.

✳✳ CAUTION

Do not smoke or carry lighted tobacco or open flame of any type when working on or near any fuel-related component. Highly flammable mixtures are always present and can be ignited, resulting in possible personal injury.

✳✳ CAUTION

Fuel in the fuel system remains under high pressure even when the engine is not running. Before repairing or disconnecting any of the fuel lines or fuel system components, the fuel system pressure must be relieved to prevent accidental spraying of fuel, causing personal injury or a fire hazard.

Fuel System Pressure

RELIEVING

2002–03 All Engines

2004 F-150 Heritage Edition

F-250/350

➡A fuel pressure gauge is needed to correctly perform this procedure.

✳✳ CAUTION

Fuel injection systems remain under pressure, even after the engine has been turned OFF. The fuel system pressure must be relieved before disconnecting any fuel lines. Failure to do so may result in fire and/or personal injury.

1. Before servicing the vehicle, refer to the Precautions Section.
2. Disconnect the negative battery cable and remove the fuel filler cap.
3. Remove the cap from the pressure relief valve on the fuel supply manifold. Install a fuel pressure gauge to the pressure relief valve.
4. Direct the gauge drain hose into a suitable container and depress the pressure relief button.
5. Remove the gauge and replace the cap on the pressure relief valve.

➡As an alternate method on models except F-150 and Blackwood, disconnect the inertia switch and crank the engine for 15–20 seconds until the pressure is relieved.

2002–05 E-Series

1. Before servicing the vehicle, refer to the Precautions Section.
2. Remove the fuel pump fuse from the engine compartment fuse box.
3. Start the engine and allow it to idle until it stalls.
4. After the engine stalls, crank the engine for approximately 5 seconds to make sure the fuel injection supply manifold pressure has been relieved.
5. Turn the ignition switch to the OFF position.

6. When fuel system service is complete, install the fuel pump fuse.

➡It may take more than one key cycle to pressurize the fuel system.

Cycle the ignition key and wait three seconds to pressurize the fuel system. Check for leaks before starting the engine.

7. Start the vehicle and check the fuel system for leaks.

2004 F-150 Exc. Heritage Edition

2005 F-150

2006 Mark LT

1. Before servicing the vehicle, refer to the Precautions Section.

➡The splash shield is located on the LH frame rail under the driver side door.

2. Disconnect the splash shield and position aside.
3. Disconnect the electrical connector.
4. Start the engine and allow it to idle until it stalls.
5. After the engine stalls, crank the engine for approximately 5 seconds to make sure the fuel rail pressure has been released.
6. Turn the ignition switch to the OFF position.
7. When fuel system service is complete, connect the electrical connector.
8. Reposition the splash shield and install a new push pin.

➡It may take more than one key cycle to pressurize the fuel system.

9. Cycle the ignition key and wait 3 seconds to pressurize the fuel system. Check for leaks before starting the engine.
10. Install the WDS diagnostic tool. Turn the key ON with the engine OFF. Cycle the key OFF, then ON. Select the appropriate vehicle and engine qualifier. Clear all diagnostic trouble codes (DTCs) and carry out a PCM reset.
11. Start the vehicle and check the fuel system for leaks.

2005 F-250/350

1. Before servicing the vehicle, refer to the Precautions Section.
2. With the vehicle in NEUTRAL, position it on a hoist.
3. Disconnect the fuel pump module electrical connector at the fuel tank.

4. Start the engine and allow it to idle until it stalls.
5. After the engine stalls, crank the engine for approximately 5 seconds to make sure the fuel rail pressure has been released.
6. Turn the ignition switch to the OFF position.
7. When fuel system service is complete, connect the fuel pump module electrical connector.

➡It may take more than one key cycle to pressurize the fuel system.

8. Cycle the ignition key and wait three seconds to pressurize the fuel system. Check for leaks before starting the engine.
9. Install the scan tool. Turn the key ON with the engine OFF. Cycle the key OFF, then ON. Select the appropriate vehicle and engine qualifier. Clear all diagnostic trouble codes (DTCs) and carry out a PCM reset.
10. Start the vehicle and check the fuel system for leaks.

Fuel Filter

REMOVAL & INSTALLATION

2002–03

✳✳ CAUTION

Fuel injection systems remain under pressure, even after the engine has been turned OFF. The fuel system pressure must be relieved before disconnecting any fuel lines. Failure to do so may result in fire and/or personal injury.

1. Before servicing the vehicle, refer to the Precautions Section.
2. Disconnect the negative battery cable and relieve the fuel system pressure.
3. Disconnect the fuel lines from the fuel filter. Have a drain pan handy to catch any residual fuel once the lines are separated. On newer models, disconnect the fuel lines from the filter as follows:
    a. Disconnect the safety clip from the male hose.
    b. Install and push the fuel line disconnect tool into the female fitting.
    c. Separate the male and female fittings.
    d. Inspect the fuel lines for any damage after the fuel is finished draining.
4. Remove the fuel filter from the bracket and the retainer, if equipped. Note the direction of the flow arrow so the replacement filter can be installed correctly.

Typical fuel filter mounting along an under vehicle frame rail

7924FG65

To install:

5. Position the fuel filter into the mounting bracket with the flow arrow pointing in the correct direction.

6. Install the fuel lines to the fuel filter. On newer models, align and push the male tube into the female fitting until a click is heard. Pull on the fitting to ensure that it is fully engaged, then install the safety clip.

7. Lower the vehicle to the ground.

➡When the battery has been disconnected and reconnected, some abnormal drive symptoms may occur while the PCM relearns its adaptive strategy. The vehicle may need to be driven 10 miles (16 km) or more to relearn the strategy.

8. Connect the negative battery cable.

2004–06

1. Before servicing the vehicle, refer to the Precautions Section.

2. Relieve the fuel system pressure.

3. Disconnect the fuel lines from the fuel filter.

4. Remove the fuel filter.

5. Loosen the fuel filter clamp screw.

➡Make sure that an audible click is heard when installing the fuel lines. Pull back on the fuel lines to confirm engagement.

6. To install, reverse the removal procedure.

Fuel Pump

REMOVAL & INSTALLATION

F-Series, Blackwood and Mark LT

�֎ CAUTION

Fuel injection systems remain under pressure, even after the engine has been turned OFF. The fuel system pressure must be relieved before disconnecting any fuel lines. Failure to do so may result in fire and/or personal injury.

1. Before servicing the vehicle, refer to the Precautions Section.

2. Remove or disconnect the following:

- Negative battery cable
- Fuel pressure
- Fuel tank skid plate bolts and lower the skid plate.
- Fuel
- Fuel tank filler pipe hose from the tank
- Fuel tank filler pipe vent hose from the tank
- Fuel lines from the fuel pump
- Front fuel tank connections
- Rear Evaporative Emissions (EVAP) hose clamp and the hose

- Electrical connector from the fuel pump

3. Support the fuel tank with a jack.

- Fuel tank support strap bolts and the fuel tank straps
- Fuel tank
- Fuel pump bolts, if equipped with a metal tank
- Fuel pump retaining ring, if equipped with a plastic tank

4. Squeeze the locking tabs and remove the fuel pump.

To install:

5. Install the fuel pump, using a new O-ring or gasket. With plastic tanks, make sure the alignment arrows match.

6. Installation is the reverse of removal. Note the following torques:

a. With steel tanks, tighten the fuel pump bolts to 80–107 inch lbs. (9–12 Nm), if equipped.

b. Fuel pump retaining ring if equipped with a plastic tank.

- 2002–04 F250/350: 60 ft. lbs. (82 Nm)

c. Tighten the fuel tank strap bolts to

- Blackwood: 41 ft. lbs. (55 Nm)
- 2002–03 F-150: 26 ft. lbs. (35 Nm)
- 2004 F-150 Heritage Edition: 26 ft. lbs. (35 Nm)
- 2004–05 F-150, exc. 2004 Heritage Edition: 35 ft. lbs. (47 Nm)
- 2002–04 F-250/350 mid-ships tank: 25–33 ft. lbs. (34–46 Nm)
- 2002–04 F-250/350 aft-of-axle tank: 66 ft. lbs. (90 Nm)
- 2005 F-250/350 mid-ships tank: 30 ft. lbs. (40 Nm)

7924FG67

Remove the mounting bolts (1), then lift the fuel pump assembly (2) out of the tank—E-Series

- 2005 F-250/350 aft-of-axle tank: 66 ft. lbs. (90 Nm)
- Mark LT: 35 ft. lbs. (47 Nm)
7. Tighten the skid plate bolts to:
- 2002–03 F-150: 11 ft. lbs. (15 Nm)
- 2004 F-150 Heritage Edition: 11 ft. lbs. (15 Nm)
- 2004–05 F-150, exc. 2004 Heritage Edition: 15 ft. lbs. (20 Nm)
- 2005 F-250/350 mid-ships tank shield: 15 ft. lbs. (20 Nm)
- 2005 F-250/350 aft-of-axle tank cover: 66 ft. lbs. (90 Nm)
- Mark LT: 15 ft. lbs. (20 Nm)
8. Connect the negative battery.

E-Series

> #### ✳✳ CAUTION
>
> **Fuel injection systems remain under pressure, even after the engine has been turned OFF. The fuel system pressure must be relieved before disconnecting any fuel lines. Failure to do so may result in fire and/or personal injury.**

1. Before servicing the vehicle, refer to the Precautions Section.
2. Remove or disconnect the following:
- Negative battery cable
- Fuel system pressure
- Fuel tank
- Fuel tank filler pipe vent hose and fuel tank filler pipe from the tank
3. Support the fuel tank with a jack.
- The 2 fuel tank support strap nuts
- 2 fuel tank support straps
4. Lower the fuel tank to allow access to the electrical connections
- Fuel tank connections
- Fuel and electrical connections from the fuel pump
- Fuel tank
- Fuel tank screws/nuts, fuel pump and sender

To install:
5. Install or connect the following:
- Fuel sender and fuel pump into the fuel tank. Tighten the screws/nuts to 89 inch lbs. (10 Nm).
6. Raise the fuel tank.
- Fuel and electrical connections to the fuel pump
- Fuel tank connections
- Fuel tank support straps and tighten the nuts to
- 2002–03 mid-ships tank: 13–17 ft. lbs. (17–23 Nm)

- 2002–03 aft-of-axle tank: bolts 64–88 ft. lbs. (87–119 Nm); nuts 56–76 ft. lbs. (77–103 Nm)
- 2004–05 mid-ships tank: 18 ft. lbs. (25 Nm)
- 2004–05 aft-of-axle tank: bolts 76 ft. lbs. (103 Nm); nuts 66 ft. lbs. (90 Nm)
- Fuel tank filler pipe vent hose and the fuel tank filler pipe to the tank
- Negative battery cable

Fuel Injector

REMOVAL & INSTALLATION

2002–03 4.2L Engine

> #### ✳✳ CAUTION
>
> **Fuel injection systems remain under pressure, even after the engine has been turned OFF. The fuel system pressure must be relieved before disconnecting any fuel lines. Failure to do so may result in fire and/or personal injury.**

1. Before servicing the vehicle, refer to the Precautions Section.
2. Remove or disconnect the following:
- Negative battery cable
- Engine cover, if equipped
- Air cleaner assembly
3. Relieve the fuel pressure.
- Upper intake manifold
- Fuel pressure regulator valve vacuum hose
- Fuel lines
- Fuel injector electrical connectors
- Fuel injector electrical harness from the fuel injector manifold
- Fuel injection supply manifold
- Fuel rail bolts and the rail

> #### ✳✳ CAUTION
>
> **The fuel injectors are deposit-resistant. Do not clean. Use O-rings made of special fuel resistant material. Use of ordinary O-rings can cause the fuel system to leak.**

- Fuel injector retaining clips and the fuel injectors. Inspect the fuel injector O-rings and if necessary.

To install:

➡**Make sure the injector clips must engage the upper groove on the injectors or fuel leaks may occur.**

4. Lubricate the new O-rings with clean engine oil to aid installation.
5. Install or connect the following:
- Fuel injectors and the rail. Tighten the rail bolts to 89 inch lbs. (10 Nm).
- Fuel injection supply manifold
- Fuel injector electrical harness from the fuel injector manifold
- Fuel injector electrical connectors
- Fuel lines
- Fuel pressure regulator valve vacuum hose
- Upper intake manifold
- Air cleaner assembly
- Engine cover, if equipped
- Negative battery cable

2002–03 F-150 w/4.6L Engines
2004 F-150 Heritage Editions w/4.6L Engines
2002–04 F-250/350 w/4.6L Engines
2002–05 E-Series w/4.6L Engines

1. Before servicing the vehicle, refer to the Precautions Section.
2. Disconnect the battery ground cable.
3. Partially drain the engine cooling system.
4. Remove the air cleaner outlet pipe.
5. Disconnect the exhaust gas recirculation (EGR) valve vacuum hose and the evaporative emission canister purge valve hose and position them aside.
6. Disconnect the evaporative emission canister purge valve vacuum hose.
7. Disconnect the fuel pressure regulator valve vacuum hose.
8. Remove the crankcase ventilation tube.
9. Remove the power steering reservoir upper bracket.

> #### ✳✳ WARNING
>
> **After disconnecting, plug the fuel tubes to prevent leakage.**

10. Disconnect the fuel hose spring lock couplings.
11. Disconnect the eight ignition coil electrical connectors.
12. Disconnect the eight fuel injector electrical connectors.
13. Disconnect the EGR vacuum regulator solenoid connections.
14. Disconnect the idle air control (IAC) valve and throttle position (TP) sensor electrical connectors.
15. Remove the nut and disconnect the brake booster vacuum tube and bracket.
16. Disconnect and remove the positive crankcase ventilation (PCV) hose.

1 Fuel injector electrical connector	4 Fuel rail bolts (4 required)
	5 Fuel rail
2 Fuel rail pressure sensor electrical connector	6 Fuel injector (6 required)
3 Fuel rail pressure sensor vacuum connector	7 Fuel injector O-ring (12 required)

06017-F150-G234

Fuel rail and injectors—4.2L engine

➡**The throttle body adapter is shown slightly displaced for clarity.**

17. Disconnect the heated throttle body hose and position aside.

18. Disconnect the exhaust manifold to EGR valve tube upper fitting.

19. Disconnect the differential pressure feedback EGR system electrical connector.

20. Disconnect the exhaust manifold to EGR valve tube lower fitting.

21. Disconnect the heater hose and position aside.

22. Disconnect the climate control vacuum connector.

➡**Release any wiring harnesses or vacuum hoses which are attached to the fuel injection supply manifold, before removing the manifold.**

23. Remove the bolts and lift the fuel injection supply manifold and injectors upward, out of the intake manifold.

24. Remove the fuel injectors from the fuel injection supply manifold.

25. Inspect the two O-ring seals from each fuel injector. New O-ring seals will be installed as needed.

To install:

➡**Lubricate the new O-ring seals with clean engine oil to aid installation.**

26. Install new O-ring seals, as required.

27. Install the fuel injectors into the fuel injection supply manifold.

28. Position the fuel injection supply manifold and install the four bolts. Torque to 89 inch lbs. (10 Nm).

29. Connect the climate control vacuum connector.

30. Connect the heater hose.

31. Install the exhaust manifold to EGR valve tube. Torque the fittings to 26–33 ft. lbs. (35–45 Nm).

32. Connect the differential pressure feedback EGR system electrical connector.

➡**The throttle body adapter is shown slightly displaced for clarity.**

33. Connect the heated throttle body hose.

34. Install the PCV hose.

35. Connect the brake booster vacuum tube and bracket and install the nut. Torque to 89 inch lbs. (10 Nm).

36. Connect the TP sensor and IAC valve electrical connectors.

37. Connect the EGR vacuum regulator solenoid connections.

38. Connect the eight fuel injector electrical connectors.

39. Connect the eight ignition coil electrical connectors.

40. Connect the fuel hose spring lock couplings.

41. Install the power steering reservoir upper bracket. Torque the two upper bolts to 13 ft. lbs. (17 Nm); the single lower bolt to 96 inch lbs. (11 Nm).

42. Install the crankcase ventilation tube.

43. Connect the fuel pressure regulator valve vacuum hose.

44. Connect the evaporative emission canister purge valve vacuum hose.

45. Connect the evaporative emission canister purge valve hose and the EGR valve vacuum hose.

46. Install the air cleaner outlet pipe.

47. Connect the battery ground cable.

48. Fill and bleed the engine cooling system.

2002–03 F-150 w/5.4L Engines

2004 F-150 Heritage Editions w/5.4L Engines

2002–04 F-250/350 w/5.4L Engines

2002–05 E-Series w/5.4L Engines

1. Before servicing the vehicle, refer to the Precautions Section.

2. Disconnect the battery ground cable.

3. Partially drain the engine cooling system.

4. Remove the air cleaner outlet pipe.

5. Disconnect the exhaust gas recirculation (EGR) valve vacuum hose and the evaporative emission canister purge valve hose and position them aside.

6. Disconnect the evaporative emission canister purge valve vacuum hose.

7. Disconnect the fuel pressure regulator valve vacuum hose.

8. Remove the crankcase ventilation tube.

9. Remove the power steering reservoir upper bracket.

✴✴ WARNING

After disconnecting, plug the fuel tubes to prevent leakage.

10. Disconnect the fuel hose spring lock couplings.

11. Disconnect the eight ignition coil electrical connectors.

12. Disconnect the eight fuel injector electrical connectors.

13. Remove the IAC valve fresh air tube.

14. Disconnect the IAC electrical connector.

15. Disconnect the differential pressure feedback EGR system electrical connector.

16. Remove the brake booster bracket and tube.

17. Disconnect the two differential pressure feedback EGR system hoses.

18. Disconnect the exhaust manifold to EGR valve tube upper fitting.

19. Disconnect the exhaust manifold to EGR valve tube lower fitting.

20. Disconnect the PCV hose from the throttle body adapter.

21. Disconnect the heated throttle body outlet hose from the throttle body adapter.

22. Disconnect the PCV hose from the intake manifold.

23. Remove the PCV valve and position the harness aside.

24. Disconnect the heater hose and position aside.

25. Disconnect the climate control vacuum connector.

➡**Release any wiring harnesses or vacuum hoses which are attached to the fuel injection supply manifold, before removing the manifold.**

26. Remove the bolts and lift the fuel injection supply manifold and injectors upward, out of the intake manifold.

27. Remove the fuel injectors from the fuel injection supply manifold.

28. Inspect the two O-ring seals from each fuel injector. New O-ring seals will be installed as needed.

To install:

➡**Lubricate the new O-ring seals with clean engine oil to aid installation.**

29. Install new O-ring seals, as required.

30. Install the fuel injectors into the fuel injection supply manifold.

31. Position the fuel injection supply manifold and install the four bolts. Torque to 89 inch lbs. (10 Nm).

32. Connect the climate control vacuum connector.

33. Connect the heater hose.

34. Install the PCV valve.

35. Connect the PCV hose to the intake manifold.

36. Connect the heated throttle body outlet hose.

37. Connect the PCV hose to the throttle body adapter.

38. Install the exhaust manifold to EGR valve tube and connect the differential pressure feedback EGR system hoses. Torque the fittings to 30–44 ft. lbs. (40–60 Nm).

39. Install the brake booster vacuum tube and bracket.

40. Connect the differential pressure feedback EGR system electrical connector.

41. Connect the IAC valve electrical connector.

42. Install the IAC valve fresh air tube.

43. Connect the eight fuel injector electrical connectors.

44. Connect the eight ignition coil electrical connectors.

45. Connect the fuel hose spring lock couplings.

46. Install the power steering reservoir upper bracket. Torque the two upper bolts to 13 ft. lbs. (17 Nm); the single lower bolt to 96 inch lbs. (11 Nm).

47. Install the crankcase ventilation tube.

48. Connect the fuel pressure regulator valve vacuum hose.

49. Connect the evaporative emission canister purge valve vacuum hose.

50. Connect the evaporative emission canister purge valve hose and the EGR valve vacuum hose.

51. Install the air cleaner outlet pipe.

52. Connect the battery ground cable.

53. Fill and bleed the engine cooling system.

2005 F-150 w/4.6L Engine

1. Before servicing the vehicle, refer to the Precautions Section.

2. Release the fuel system pressure.

3. Disconnect the battery ground cable.

4. Remove the air cleaner outlet pipe.

5. Disconnect the exhaust gas recirculation (EGR) system module tube upper fitting.

6. Loosen the EGR system module tube lower fitting and rotate to position aside.

7. Disconnect the brake booster vacuum hose.

8. Disconnect the throttle position (TP) sensor electrical connector.

9. Disconnect the upper vapor tube and position aside.

10. Disconnect the EGR system module vacuum and electrical connector.

11. Disconnect the crankcase vent hose quick-release fitting at the rear of the intake manifold.

12. Disconnect the crankcase vent hose quick-release fitting at the RH valve cover.

13. Disconnect the electronic throttle control electrical connector.

14. Disconnect the throttle body spacer vacuum hose.

15. Release the heated TB coolant hose clamp and remove and plug the coolant hose.

16. Release the heated TB coolant hose clamp and remove and plug the coolant hose.

17. Remove the 4 throttle body spacer bolts and the throttle body assembly and discard the gasket.

18. Disconnect the injection pressure sensor electrical connector.

19. Disconnect the 8 fuel injector electrical connectors.

20. Disconnect the injection pressure sensor vacuum connector.

18 Injection pressure sensor electrical connector	22 Fuel rail bolts (4 required)
19 Fuel injector electrical connector (8 required)	23 Fuel rail
20 Injection pressure sensor vacuum connector	24 Fuel injector (8 required)
21 Fuel tube spring lock coupling	25 Fuel injector to intake manifold O-ring seal (8 required)
	26 Fuel injector to fuel rail O-ring seal (8 required)

06017-F150-G235

Fuel rail and injectors—2005 F-150 4.6L engine

21. Disconnect the fuel tube spring lock coupling.
22. Remove the 4 fuel rail bolts.
23. Remove the fuel rail.
24. Remove the 8 fuel injectors.
25. Remove and discard the 8 upper fuel injector O-ring seals and the 8 lower fuel injector O-ring seals
26. To install, reverse the removal procedure. Torque the 4 fuel rail bolts to 10 Nm (89 inch lbs.).Torque the 4 throttle body spacer bolts to 10 Nm (89 inch lbs.). Torque the EGR system module tube fittings 40 Nm (30 ft. lbs.).

➡Install new O-ring seals on the fuel injectors. Lubricate the new fuel injector O-ring seals with clean engine oil prior to installing the fuel rail.

2005 F-Series and 2006 Mark LT w/5.4L Engine

1. Before servicing the vehicle, refer to the Precautions Section.
2. Release the fuel system pressure.
3. Disconnect the battery ground cable.
4. Remove the air cleaner and air cleaner intake pipe.
5. Disconnect the fuel vapor quick con-

nect fitting from the evaporative emission (EVAP) canister purge valve and position aside.
6. Disconnect the heated positive crankcase ventilation (PCV) valve electrical connector.
7. Disconnect the PCV hose from the intake manifold and position aside.
8. Disconnect the electronic throttle control electrical connector.
9. Disconnect the throttle position (TP) sensor electrical connector.
10. Disconnect the vacuum hose near the rear of the right valve cover.
11. Disconnect the fuel rail pressure and temperature (FRPT) sensor electrical connector.
12. Disconnect the FRPT sensor vacuum hose.
13. Disconnect the 8 fuel injector electrical connectors.
14. Disconnect the spring lock couplings at the fuel rail.
15. Remove the 4 fuel rail bolts.

✱✱ WARNING

When installing the fuel injectors, make sure that the injectors are fully engaged in the fuel injector cups.

16. Remove the fuel rail and injectors as an assembly.
17. Remove the 8 fuel injectors and discard the 8 lower and 8 upper fuel injector O-ring seals.

➡**Lubricate the O-ring seal with clean engine oil.**

18. If installing a new fuel rail, remove the 2 FRPT bolts and remove the FRPT sensor.
19. Inspect the O-ring seal and install a new seal if necessary.
20. To install, reverse the removal procedure.

➡**Install new O-ring seals on the fuel injectors. Lubricate the new fuel injector O-ring seals with clean engine oil prior to installing the fuel injectors.**

21. Torque the 2 FRPT bolts to 4 Nm (35 inch lbs.).
22. Torque the 4 fuel rail bolts to 10 Nm (89 inch lbs.).

2002–04 6.8L Engine

1. Before servicing the vehicle, refer to the Precautions Section.
2. Disconnect the battery ground cable.
3. Remove the engine air cleaner, the air cleaner outlet tube and the accelerator splash shield.
4. Disconnect both cables at the throttle bell crank.

5 — 10 Nm (89 lb-in)
10 — 4 Nm (35 lb-in)

1	Fuel rail pressure and temperature (FRPT) sensor electrical connector	7	Fuel injector (8 required)
2	FRPT sensor vacuum connector	8	Fuel injector to intake manifold O-ring seal (8 required)
3	Fuel injector electrical connector (8 required)	9	Fuel injector to fuel rail O-ring seal (8 required)
4	Fuel tube spring lock couplings	10	FRPT sensor bolts (2 required)
5	Fuel rail bolts (4 required)	11	FRPT sensor
6	Fuel rail		

06017-F150-G236

2005 F-Series 5.4L engine fuel rail and injectors

5. Disconnect both cables from the accelerator mounting bracket.

6. Remove the accelerator mounting bracket.

7. Disconnect the PCV hose.

8. If equipped, disconnect and plug or cap the coolant hoses.

9. Disconnect the power brake booster vacuum supply hose and main vacuum supply line at the intake manifold.

10. Disconnect the hose and electrical connector from the idle air control valve.

11. Disconnect the throttle position sensor electrical connector.

12. If equipped, remove the two bolts retaining the EGR transducer bracket to the intake manifold. Position the bracket and the EGR transducer out of the way.

13. Loosen the EGR tube nut.

14. Disconnect EGR tube.

15. Disconnect the vacuum line from the EGR valve.

16. Remove the four bolts from the throttle body spacer.

17. Remove the throttle body, throttle body spacer, EGR valve and idle air control valve as an assembly.

18. Disconnect the right radio capacitor assembly connector.

19. If equipped, disconnect the EGR vacuum regulator (EVR).

20. Remove the bolts and the EGR bracket assembly.

21. Disconnect the fuel pressure regulator vacuum hose.

22. Remove the fuel line tether clamps.

23. Disconnect the fuel lines.

24. Remove and discard the fuel line O-ring seals.

25. Disconnect the right fuel injector electrical connectors.

26. Disconnect the left fuel injector electrical connectors.

27. Remove the crankcase vent connector and hose.

28. Remove the bolts. Lift the fuel supply manifold. Remove the injectors.

29. Remove and discard the O-ring seals from each fuel injector.

30. Position and remove the fuel injection supply manifold.

✲✲ WARNING

Check for fuel leakage after installation is complete.

➡Be sure fuel injectors and fuel injection supply manifold are fully seated.

➡Lubricate the new O-ring seals with engine oil.

06017-F150-G237

Fuel supply manifold and injectors—2002–04 6.8L engine

31. To install, reverse the removal procedure. Torque the fuel rail bolts to 71–97 inch lbs. (8–11 Nm). Torque the EGR bracket bolts to 71–106 inch lbs. (8–12 Nm).

32. Install the four throttle body spacer bolts.

- Tighten to 8–10 Nm (71–88 inch lbs.).
- Tighten the bolts an additional 85–95 degrees.

2005 6.8L Engine

1. Before servicing the vehicle, refer to the Precautions Section.

2. Release the fuel system pressure.

3. Disconnect the battery ground cable.

4. Remove the air cleaner outlet pipe.

5. Disconnect the fuel supply tube spring lock coupling.

6. Disconnect the 2 electrical connectors from the throttle body (TB).

7. Disconnect the positive crankcase ventilation (PCV) tube from the TB spacer.

8. Disconnect the 2 vacuum lines from the TB spacer.

9. Disconnect the fuel rail pressure and temperature sensor electrical and vacuum connectors.

10. Disconnect the 10 fuel injector electrical connectors.

11. Detach the main engine wiring harness retainer from the RH valve cover.

12. Detach the 2 main engine wiring harness retainers from the intake manifold and position the harness for access to remove the fuel rail.

➡When removing the fuel rail, leave the fuel injectors in the intake manifold. This will make removal of the fuel rail easier.

13. Remove the 6 fuel rail bolts and the fuel rail.

❋❋ WARNING

Use O-ring seals that are made of special fuel-resistant material. Use of ordinary O-rings can cause the fuel system to leak. Do not reuse the O-ring seals. Lubricate the O-ring seals with clean engine oil prior to installation.

1	Fuel rail pressure and temperature sensor electrical connector
2	Fuel rail pressure and temperature sensor vacuum hose
3	Fuel rail bolts (6 required)
4	Fuel rail
5	Fuel injector electrical connector (10 required)
6	Fuel injector-to-fuel rail O-ring (10 required)
7	Fuel injector-to-intake manifold O-ring (10 required)
8	Fuel injector-to-fuel rail locking clip (10 required)
9	Fuel injector (10 required)

10 Nm (89 lb-in)

Fuel rail and injectors—2005 6.8L engine

06017-F150-G238

14. Remove the fuel injectors and the fuel injector O-ring seals. Discard the O-ring seals.

15. To install, reverse the removal procedure. Torque the 6 fuel rail bolts to 10 Nm (89 inch lbs.)

DIESEL FUEL SYSTEM

Fuel System Service Precautions

Safety is the most important factor when performing not only fuel system maintenance but any type of maintenance. Failure to conduct maintenance and repairs in a safe manner may result in serious personal injury or death. Maintenance and testing of the vehicle's fuel system components can be accomplished safely and effectively by adhering to the following rules and guidelines.

• To avoid the possibility of fire and personal injury, always disconnect the negative battery cable unless the repair or test procedure requires that battery voltage be applied.

• Always relieve the fuel system pressure prior to disconnecting any fuel system component (injector, fuel rail, pressure regulator, etc.), fitting or fuel line connection. Exercise extreme caution whenever relieving fuel system pressure, to avoid exposing skin, face and eyes to fuel spray. Please be advised that fuel under pressure may penetrate the skin or any part of the body that it contacts.

• Always place a shop towel or cloth around the fitting or connection prior to loosening to absorb any excess fuel due to spillage. Ensure that all fuel spillage (should it occur) is quickly removed from engine surfaces. Ensure that all fuel soaked cloths or towels are deposited into a suitable waste container.

• Always keep a dry chemical (Class B) fire extinguisher near the work area.

• Do not allow fuel spray or fuel vapors to come into contact with a spark or open flame.

• Always use a back-up wrench when loosening and tightening fuel line connection fittings. This will prevent unnecessary stress and torsion to fuel line piping. Always follow the proper torque specifications.

• Always replace worn fuel fitting O-rings with new. Do not substitute fuel hose or equivalent where fuel pipe is installed.

Fuel System Pressure

RELIEVING

6.0L Engine

1. Before servicing the vehicle, refer to the Precautions Section.
2. With the vehicle in NEUTRAL, position it on a hoist.
3. Disconnect both battery ground cables.
4. Open the fuel/water separator drain valve to release the fuel pressure.

7.3L Engine

> ✳✳ **CAUTION**

Before removing the fuel tank filler cap, turn the fuel tank filler cap ¼ to ¾ turn counterclockwise and wait for the tank pressure to be relieved. Personal injury may result if the fuel tank filler cap is removed without the pressure fully relieved.

1. Before servicing the vehicle, refer to the Precautions Section.
2. Remove the fuel tank filler cap to relieve any pressure in the fuel tank.
3. When servicing the fuel lines, loosen the fuel fitting to allow any residual fuel line pressure to be relieved.

Idle Speed

ADJUSTMENT

6.0L Engine

The fuel system is controlled by the PCM.

7.3L Engine

1. Before servicing the vehicle, refer to the Precautions Section.
2. Place the transmission in Neutral (manual transmissions) or **P** (automatic transmissions).
3. Bring the engine up to normal operating temperature.

➡ **Idle speed is measured with the manual transmission in Neutral or the automatic transmission in D.**

4. Ensure that the curb idle adjusting screw is against the stop. If not, correct the vehicle linkage.
5. Check curb idle speed. Curb idle speed is specified on the Vehicle Emissions Control Information (VECI) decal on the underside of the vehicle's hood. Adjust the idle speed to specification using the idle speed adjusting screw.
6. Place the transmission in Neutral (manual) or **P**. Rev the engine momentarily, then place the transmission in the specified gear and recheck the idle speed. Adjust again if necessary.
7. Remove the tachometer and close the hood.

Raise or lower the curb idle speed by turning the curb idle speed adjusting screw—7.3L diesel engines

Fuel Filter/Water Separator

DRAINING WATER

7.3L Engine

1. Before servicing the vehicle, refer to the Precautions Section.

2. Properly relieve the fuel system pressure.

3. Shut the engine **OFF**. Failure to shut the engine **OFF** before draining the separator will cause air to enter the system.

4. Unscrew the vent on the top center of the separator unit 2 ½ –3 turns.

5. Unscrew the drain screw on the bottom of the separator 1 ½ –2 turns and drain the water into an appropriate container.

6. After the water is completely drained, close the water drain finger-tight.

7. Tighten the vent until snug, then turn it an additional ¼ turn.

8. Start the engine and check the "Water in Fuel" indicator light; it should not be lit. If it is lit and continues to stay so, there is a problem somewhere else in the fuel system.

REMOVAL & INSTALLATION

7.3L Engine

> **✳✳ CAUTION**
>
> **The fuel system remains under pressure, even after the engine has been turned OFF. The fuel system pressure must be relieved before disconnecting any fuel lines. Failure to do so may result in fire and/or personal injury.**

1. Before servicing the vehicle, refer to the Precautions Section.

2. Remove or disconnect the following:
   - Negative battery cable
   - Turbocharger assembly
   - Baffle and the air inlet crossover manifold

3. Place a suitable container under the drain hose and open the filter drain.
   - The 2 capscrews securing the fuel filter base to the crankcase
   - Water drain hose from the filter
   - Fuel system pressure
   - Fuel outlet hose, located between the fuel and filter housing, and the fuel return hose from the fuel pressure regulator valve.
   - The 2 fuel supply hoses that connect the regulator block to the cylinder head fuel rails

4. Loosen the clamp at the fuel pump

Item	Description
1	Fuel Filter Cap
2	Fuel Filter Element
3	Fuel Filter Bevel Cut Gasket
4	Fuel Filter Housing and Gland
5	Fuel Filter Element and Cap Locking Tabs

7924FG68

Use a prytool to remove the fuel filter cap to gain access to the filter element-7.3L diesel fuel systems

end of the hose, which connects the fuel filter to the inlet of high pressure stage at the fuel pump.

5. Disengage the wiring harness from the right side of the filter housing.

6. Disengage the electrical connections from the Water In Fuel (WIF) sensor and the fuel heater.

7. Remove the fuel filter.

8. Use a prybar to remove the fuel filter cap and the filter element will come out with the cap.

9. Depress the element locking tabs and remove the element from the cap.

To install:

10. Clean the mating surfaces and install the filter element onto the cap, making sure the tabs engage.

11. Install the filter gap and press down firmly, but gently, to engage it.

12. Install or connect the following:
   - Wiring connections to the fuel heater and WIF sensor
   - Wiring harness to the filter housing

13. Tighten the clamp at the fuel pump end of the hose, which connects the fuel filter to the inlet of high pressure stage at the fuel pump.

14. Engage the 2 fuel supply hoses that connect the regulator block to the cylinder head fuel rails.

15. Install or connect the following:
   - Fuel outlet hose, located between the fuel and filter housing, and the fuel return hose from the fuel pressure regulator valve.
   - Water drain hose to the filter
   - The 2 capscrews securing the fuel filter base to the crankcase
   - Air inlet crossover manifold and baffle
   - Turbocharger assembly
   - Negative battery cable

Fuel Conditioning Module—6.0L Engine

The Fuel Condition Module contains the fuel pump and water separator.

DRAINING

1. Before servicing the vehicle, refer to the Precautions Section.

> **✳✳ CAUTION**
>
> **Smoking or open flame of any type must not be present when working near fuel or fuel vapor.**

2. Disconnect both battery ground cables.

3. Raise and support the vehicle.

4. Open the fuel/water separator drain valve to release the fuel pressure.

REMOVAL & INSTALLATION

1. Before servicing the vehicle, refer to the Precautions Section.

> **✳✳ CAUTION**
>
> **Smoking or open flame of any type must not be present when working near fuel or fuel vapor.**

2. Disconnect both battery ground cables.

3. Raise and support the vehicle.

67197-EFSE-G134

Fuel/water separator drain valve—6.0L engine

4. Open the fuel/water separator drain valve to release the fuel pressure.

5. Disconnect the electrical connectors.

6. Disconnect the fuel pump electrical connector.

7. Disconnect the fuel warmer electrical connector.

8. Disconnect the water-in-fuel electrical connector.

9. Disconnect the fuel hoses.

10. Remove the fuel hose retaining clips and discard. Disconnect the fuel hoses from the fuel pump.

11. Press in the retaining clips and release the fuel hoses.

12. Remove the mounting nuts and the fuel conditioning module.

13. To install, reverse the removal procedure. Torque the nuts to 11 ft. lbs. (15 Nm).

Fuel Filter—6.0L Engine

REMOVAL & INSTALLATION

1. Before servicing the vehicle, refer to the Precautions Section.

✳✳ CAUTION

Do not smoke or carry lighted tobacco or open flame of any type when working on or near any fuel related component. Highly flammable mixtures are always present and can be ignited, resulting in possible personal injury.

2. Relieve the fuel system pressure.

3. Disconnect the fuel lines from the fuel filter.

4. Remove the fuel filter.

5. Loosen the fuel filter clamp screw.

➡ **Make sure that an audible click is heard when installing the fuel lines. Pull back on the fuel lines to confirm engagement.**

6. To install, reverse the removal procedure.

67197-EFSE-G135

Fuel filter—6.0L engine

Injection Pump

REMOVAL & INSTALLATION

7.3L Engine

✳✳ CAUTION

The fuel system remains under pressure, even after the engine has been turned OFF. The fuel system pressure must be relieved before disconnecting any fuel lines. Failure to do so may result in fire and/or personal injury.

1. Before servicing the vehicle, refer to the Precautions Section.

2. Remove or disconnect the following:
   - Negative battery cable
   - Turbocharger assembly
   - Fuel system pressure
   - Fuel line banjo bolt at the pump
   - Fuel line fittings at the rear of the cylinder heads
   - Fuel lines
   - The 3 hose clamps at the injection pump fittings
   - Water drain hose at the fuel filter

   - Filter and position it forward
   - Injection pump retaining bolts, then lift the pump out of the crankcase bore
   - Injection pump tappet from the crankcase bore

To install:

3. Rotate the engine so the injection pump eccentric is on the base circle.

4. Install or connect the following:
   - Injection pump tappet in the base of the injection pump
   - O-ring on the injection pump base
   - Injection pump and tighten the bolts to 19–27 ft. lbs. (26–37 Nm)
   - Fuel filter and connect the water drain hose
   - The 3 fuel hoses at the front of the injection pump
   - The fuel line clamps
   - Fuel filter retaining bolts
   - Fuel line assembly and new seal rings at the rear of the pump
   - Fuel line fittings at the rear of the cylinder heads
   - Fuel line banjo fitting at the pump. Tighten the bolt to 18 ft. lbs. (24 Nm).
   - Turbocharger assembly
   - Negative battery cable

FUEL PUMP RETAINING BOLT

FUEL PUMP

FUEL LINE ASSEMBLY

FUEL LINE BANJO FITTING

7924FG28

Injection pump assembly components—7.3L diesel engines

Oil Pump

REMOVAL & INSTALLATION

6.0L Engine

EARLY BUILD

1. Before servicing the vehicle, refer to the Precautions Section.
2. Remove the intake manifold.
3. Remove the turbocharger heat shield.
4. Remove the bolts and the high-pressure oil pump cover. Use a thin gasket scraper to separate the cover from the crankcase.
5. Remove and discard the press-in-place gasket.
6. Position the dial indicator with bracketry onto the oil pump drive and check the oil pump drive gear backlash.

Remove the high-pressure oil pump cover

Check the oil pump drive gear backlash

Using the special tool, disconnect and remove the high-pressure oil pump discharge pipe

Install the high pressure oil pump

Clean the cover mounting surface and apply sealer at the seams

7. Remove the bolts from the high-pressure oil pump discharge pipe.
8. Using the special tool, disconnect and remove the high-pressure oil pump discharge pipe.
9. Remove and discard the D-shaped O-ring seal.
10. Remove and discard the high-pressure pump O-ring seal.
11. Remove the bolts and the high-pressure oil pump.
12. Remove and discard the lower O-ring seal.

To install:

13. Install a new lower O-ring seal.
14. Install the high-pressure oil pump and bolts. Torque to 18 ft. lbs. (24 Nm).
15. Install the high-pressure pump O-ring seal.
16. Install the oil pump discharge pipe.
17. Install the bolts for the oil discharge pipe.
18. Position the dial indicator with bracketry onto the oil pump drive and check the oil pump drive gear backlash.
19. Install a new D-ring seal on the high-pressure discharge pipe. Torque to 71 inch lbs. (8 Nm).
20. Install a new press-in-place gasket in the high-pressure pump cover.
21. Clean the cover mounting surface and apply sealer at the seams.
22. Install the high-pressure pump cover and bolts. Torque to 8 ft. lbs. (11 Nm).

23. Install the turbocharger heat shield.
24. Install the intake manifold.

LATE BUILD

1. Before servicing the vehicle, refer to the Precautions Section.
2. Remove the intake manifold.
3. Remove the turbocharger heat shield.
4. Remove the bolts and the high-pressure oil pump cover.
5. Use a thin gasket scraper to separate the cover from the crankcase.
6. Remove and discard the press-in-place gasket.
7. Position the dial indicator with bracketry onto the oil pump drive and check the oil pump drive gear backlash.
8. Remove the bolts from the high-pressure oil pump discharge pipe. Position the high-pressure discharge pipe aside.
9. Remove and discard the high-pressure pump O-ring seal.
10. Remove the bolts and the high-pressure oil pump.
11. Remove and discard the lower O-ring seal.

To install:

12. Install a new lower O-ring seal.
13. Install the high-pressure oil pump and bolts.
14. Install the high-pressure pump O-ring seal.
15. Position back the high-pressure discharge tube and install the bolts.
16. Position the dial indicator with bracketry onto the oil pump drive and check the oil pump drive gear backlash.
17. Install a new D-ring seal on the high-pressure discharge pipe.
18. Install a new press-in-place gasket in the high-pressure pump cover. Clean the cover mounting surface and apply sealer at the seams.
19. Install the high-pressure pump cover and bolts.
20. Install the turbocharger heat shield.
21. Install the intake manifold.

Electric Fuel Pump

REMOVAL & INSTALLATION

6.0L Engine

The fuel pump is located in the fuel conditioning module.

7.3L Engine

1. Before servicing the vehicle, refer to the Precautions Section.

06017-F150-G239

Electric fuel pump—7.3L engine

2. Disconnect both battery ground cables.

3. Open the fuel filter/water separator drain valve to release the fuel pressure.

4. Raise and support the vehicle.

➡ The electrical connector is located behind the fuel pump near the frame rail.

5. Disconnect the fuel pump harness.

6. Remove the fuel line retaining clip and discard, then remove the fuel line from the fuel pump.

7. Disconnect the fuel pump delivery line.

8. Unclip the brake lines from the fuel pump bracket.

9. Loosen the pinch bolt. Spread the mounting bracket. Remove the fuel pump.

10. To install, reverse the removal procedure. Torque the bolts to 13 ft. lbs. (17 Nm).

Fuel Injectors

REMOVAL & INSTALLATION

✳✳ CAUTION

Observe all applicable safety precautions when working around fuel. Whenever servicing the fuel system, always work in a well ventilated area. Do not allow fuel spray or vapors to come in contact with a spark or open flame. Keep a dry chemical fire extinguisher near the work area. Always keep fuel in a container specifically designed for fuel storage; also, always properly seal fuel containers to avoid the possibility of fire or explosion.

6.0L Engine

EARLY BUILD

1. Before servicing the vehicle, refer to the Precautions Section.

2. If removing the right fuel injectors, remove the evaporator case.

3. Remove the valve cover.

303-755

67197-EFSE-G141

Disconnect the high-pressure oil rail supply line at the high-pressure oil rail—6.0L engine

67197-EFSE-G142

Remove the bolts and the high-pressure oil rail

4. Disconnect the fuel injector electrical connector.

5. Disconnect the high-pressure oil rail supply line at the high-pressure oil rail.

6. Remove the bolts and the high-pressure oil rail.

7. Disconnect and remove the high-pressure oil supply line.

✳✳ WARNING

Do not attempt to apply battery voltage to the fuel injector or damage to the fuel injector will occur.

8. Using a 19 mm socket, push the fuel injector electrical connector out of the rocker arm carrier.

✳✳ WARNING

To prevent engine damage, do not use air tools to remove the fuel injectors. The clip that extracts the injector can dislodge and fall into the oil drain hole.

➡ There is no need to drain the fuel rail.

➡ If engine coolant is found in the combustion chambers, It may be necessary to install a new injector sleeve.

9. Remove the bolt, fuel injector hold-down clamp and fuel injector.

To install:

✳✳ WARNING

If the fuel injector oil inlet D-shaped O-ring is damaged, a new fuel injector must be installed.

10. Install new O-ring seals and copper washer on the fuel injector. Lubricate the fuel injector and O-ring seals liberally with clean engine oil.

11 Nm (8 lb-ft)

67197-EFSE-G143

High pressure oil rail torque sequence—early build 6.0L engine

❄ WARNING

To prevent engine damage, do not use air tools to install the fuel injectors. The clip that extracts the injector can dislodge and fall into the oil drain hole.

11. Install the fuel injector, fuel injector hold-down clamp and bolt. Torque to 24 ft. lbs. (33 Nm).

12. Install the fuel injector electrical connector into the rocker carrier.

13. Apply engine oil to the top fuel injector O-ring seals.

14. Install the high-pressure oil rail and bolts.

15. Install the high-pressure oil rail.

16. Install the bolts finger tight.

17. Tighten the bolts in the sequence shown.

18. Install the high-pressure oil line.

19. Connect the fuel injector electrical connector.

20. Install the valve covers.

21. If removed, install the evaporator case.

LATE BUILD

1. Before servicing the vehicle, refer to the Precautions Section.

2. Remove the valve cover.

3. Disconnect the fuel injector electrical connector.

❄ WARNING

To prevent engine damage, it is recommended that a new crankcase-to-head tube to installed each time it is removed. The D-rings are not replaceable. If you reuse the crankcase-to-head tube, you assume the risk for leakage. You must inspect each D-ring carefully for evidence of cutting, abrasion or twisting and never reuse a tube exhibiting any of these conditions.

4. Remove the crankcase-to-head tube assembly.

➡Do not remove the oil rail end plugs or acoustic wave attenuator port fitting. Service parts are not available to support the components.

5. Remove the bolts and the high-pressure oil rail.

➡Use a shop towel and brake parts cleaner to remove the oil residue prior to removal.

6. If the crankcase-to-head tube separated, use the special tool to remove the lower crankcase-to-head tube.

❄ WARNING

Do not attempt to apply battery voltage to the fuel injector or damage to the fuel injector will occur.

7. Using the special tool, push the fuel injector electrical connector out of the rocker arm carrier.

8. Prior to removing the injector assembly, insert clean shop towels in the oil drain holes adjacent to each glow plug.

❄ WARNING

Failure to account for all snaprings or pieces prior to placing the vehicle back in service can cause engine damage. A missing snapring can be ingested into the lube oil system, causing severe engine damage.

❄ WARNING

To prevent engine damage, do not use air tools to remove the fuel injectors. The snapring that extracts the injector can dislodge and fall into the oil drain hole.

➡There is no need to drain the fuel rail.

➡If engine coolant is found in the combustion chambers, it may be necessary to install a new injector sleeve.

➡Use a tool like Snap-On• SDMT440 Torx to avoid damaging the fuel injector plastic and to avoid interference with the evaporator case.

9. Remove the bolt, fuel injector hold-down clamp and fuel injector.

➡If a snapring or piece of a snapring is missing from the injector hold-down assembly, it must be located prior to removing the shop towels.

10. Remove the shop towels.

To install:

❄ WARNING

If the fuel injector oil inlet D-shaped O-ring is damaged, a new fuel injector must be installed.

11. Install new O-ring seals and copper washer on the fuel injector. Lubricate the fuel injector and O-ring seals liberally with clean engine oil.

❄ WARNING

Failure to tighten the injector correctly can lead to engine failure.

❄ WARNING

To prevent engine damage, do not use air tools to install the fuel injectors. The snapring that extracts the injector can dislodge and fall into the oil drain hole.

12. Install the fuel injector, fuel injector hold-down clamp and bolt. Torque to 24 ft. lbs. (33 Nm).

❄ WARNING

Make sure the injector wiring is clear of all moving parts or engine damage can occur.

13. Install the fuel injector electrical connector into the rocker carrier.

14. Apply engine oil to the top fuel injector O-ring seals.

➡Apply clean engine oil on the tubes prior to installing the oil manifold.

15. Position the oil rail on the fuel injectors.

a. Place the oil rail on top of the carrier so that the four single ball tubes are engaging the injector lead angle.

b. Insert three bolts, two on the ends of the straight side of the oil rail and one in the middle of the wavy side of the rail. Install the guide studs six to seven turns.

c. Press the oil rail into the fuel injectors.

d. Make sure that the oil rail mounting feet are flat against the mounting surface.

e. Loosely install the six bolts.

16. Install the oil rail retaining bolts.

a. Remove the three guide bolts.

b. Loosely install three remaining bolts.

c. Tighten the nine bolts in the sequence shown to 10 ft. lbs. (13 Nm).

33 Nm (24 lb-ft)

06017-F150-G240

Injector installation—6.0L engine

13 Nm (10 lb-ft)

06017-F150-G241

Oil rail bolt torque sequence—late build 6.0L engine

※※ WARNING

To prevent engine damage, it is recommended that a new crankcase-to-head tube to installed each time it is removed. The D-rings are not replaceable. If you reuse the crankcase-to-head tube, you assume the risk for leakage. You must inspect each D-ring carefully for evidence of cutting, abrasion or twisting and never reuse a tube exhibiting any of these conditions.

➡Apply clean engine oil to the crankcase-to-head tube prior to installation.

17. Install the crankcase-to-head tube assembly. Torque to 60 ft. lbs. (82 Nm).

06017-F150-G242

Crankcase-to-head tube assembly—late build 6.0L engine

18. Connect the fuel injector electrical connector.
19. Install the valve covers.

7.3L Engine

1. Before servicing the vehicle, refer to the Precautions Section.

※※ CAUTION

The red-striped wires on the DI Turbo carry 115 volts DC. A severe electrical shock may be given. Do not pierce the wires.

※※ WARNING

Do not pierce the wires or damage to the harness could occur.

Special tools required:
- Slide Hammer, No. T50T–100–A, or equivalent
- Injector Remover, No. T94T–9000–AH1, or equivalent
- Injector Replacer, No. T94T–9000–AH2, or equivalent
2. Remove or disconnect the following:
- Valve cover
- Fuel injector electrical connector

※※ WARNING

Remove the oil drain plugs prior to removing the injectors or oil could enter the combustion chamber, which could result in hydrostatic lock and severe engine damage.

- Oil rail drain plugs
- Retaining screw and oil deflector. The shoulder bolt on the inboard side of the fuel injector does not require removal.
- Outboard fuel injector retaining bolt
- Heater distribution box screws, nuts and clip
- Outer half of the case (to service No. 4 fuel injector only)

3. Remove the fuel injector using Injector Remover No. T94T–9000–AH1, or equivalent. Position the tool's fulcrum beneath the fuel injector hold-down plate and over the edge of the cylinder head. Install the remover screw in the threaded hole of the fuel injector plate (see illustration). Tighten the screw to lift out the injector from its bore. Place the injector in a suitable protective sleeve such as Rotunda Injector Protective Sleeve, No. 014–00933–2, and set the injector in a suitable holding rack.

4. Remove the fuel injector sleeves, if required. Insert the Injector Sleeve Tap Plug, 014–00934–3 into the injector sleeve to prevent debris from entering the combustion chamber. Insert Injector Sleeve Tap Pilot into the fuel injector sleeve and tighten 1–1 ½ turns. Attach Slide Hammer T50T–100–A to the Injector Sleeve Tap 014–00934–1 and 014–00934–2 Injector Sleeve Tap Pilot and remove the fuel injector sleeve from the bore.

5. Use Rotunda Injector Sleeve Brush 104–00934–A, or equivalent to clean the injector bore of any sealant residue. Make sure to remove any debris.
To install:
6. If removed, install the fuel injector sleeves using Rotunda Sleeve Replacer, No. 014–00934–4, or equivalent. Apply Threadlock, No. 262–E2FZ–19554–B, or equivalent to the fuel injector sleeves. Using a

1 Plug 1/8 Pipe	14 Fuel Regulator Cap	29 Fuel Injector Gasket (Copper)
2 Stud Bolt	15 Screw (2 Req'd)	30 Pipe Plug
3 Clamp Wrap	16 Fuel Return Line	31 O-Ring
4 Fitting (One-Way Flow)	17 Fuel Supply Line	32 O-Ring
5 Fuel Delivery Line (Right Cylinder Head)	18 Screw	33 Oil Pressure Sensor
6 Fitting (2 Req'd)	19 Clamp Wrap	34 Quick Connect Fitting (4 Req'd)
7 Fuel Delivery Line (Left Cylinder Head)	20 Screw	35 High-Pressure Oil Lines
8 O-Ring	21 O-Ring	36 O-Ring (Part of Kit)
9 Fuel Heater Thermostat/Water Sensor	22 Water Drain Tube	37 High-Pressure Oil Pump Flange Gasket
10 Screw (2 Req'd)	23 Oil Deflector	38 High-Pressure Oil Pump Assy
11 O-Ring	24 Fuel Injector	39 Injection Pressure Regulator
12 Hose	25 Backup O-Ring	40 Fuel Filter
13 O-Ring	26 Cushion Ring	
	27 Fuel Injector Middle Seal	
	28 Fuel Injector Lower Seal	

Fuel injection components—7.3L engine

06017-F150-G243

1   Backup Ring (Steel)
2   Cushion Ring (Black)
3   Upper Injector Seal (Black)
4   Fuel Injector
5   Lower Injector Seal (Black)
6   Injector Gasket (Copper)
7   Middle Injector Seal
    (Blue/Black)

06017-F150-G244

Fuel injector details—7.3L engine

rubber mallet, tap on the tool to seat the injector bore. Remove the tool and remove any residue sealant.

7. Clean the fuel injector sleeve using a suitable sleeve brush set. Clean any debris from the sleeve.

8. Clean the injector bore with a lint-free shop towel.

9. Install the fuel injectors using special tools as follows:

a. Lubricate the injectors with clean engine oil. Using new copper washers, carefully push the injectors square into the bore using hand pressure only to seat the O-rings.

b. Position the open end of Injector Replacer, No. T94T–9000–AH2, or equivalent between the fuel injector body and injector hold-down plate, while positioning the opposite end of the tool over the edge of the cylinder head.

c. Align the hole in the tool with the threaded hole in the cylinder head and install the bolt from the tool kit. Tighten the bolt to fully seat the injector, then remove the bolt and tool.

10. Install or connect the following:
- Outer half of the heater distribution box and retaining hardware (for No. 4 injector only)
- Oil deflector and bolt. Tighten the bolt to 108 inch lbs. (12 Nm).
- Fuel rail drain plug, tightening it to 96 inch lbs. (11 Nm)
- Oil rail drain plug, tightening it to 53 inch lbs. (6 Nm)
- Heater distribution box
- Fuel injector wiring harness
- Valve cover

DRIVE TRAIN

Automatic Transmission Assembly

REMOVAL & INSTALLATION

E-Series

4R100 TRANSMISSION

1. Before servicing the vehicle, refer to the Precautions Section.

2. Disconnecting the negative battery cable on some vehicles may interfere with the functions of the on-board computer system(s) and may require the computer to undergo a relearning process once the negative battery cable is reconnected.

3. Remove or disconnect the following:
- Air cleaner air intake duct assembly
- Transmission fluid level indicator
- Engine cover
- Vapor management valve assembly and position it aside, gas engines only
- Fluid filler tube from the short fluid inlet tube, and secure it aside
- Fluid level indicator bracket nut, on 7.3L diesel engines
- Fluid level indicator tube from the short fluid inlet tube and position it aside, on 7.3L diesel engines
- Electrical harness connectors

➡**Note the length of each bolt for correct location during installation.**

- Transmission-to-engine retaining bolts
- Driveshaft
- Transmission-mounted parking brake assembly, if equipped
- Vehicle Speed Sensor (VSS)
- Transmission Range (TR) sensor connector
- Heat shield, if equipped
- Solenoid body connector
- Shift cable at the transmission and position it aside
- Starter motor
- Cylinder block opening cover
- Torque converter-to-flexplate retaining nuts and discard

❋❋ CAUTION

Be sure not to raise the back of the transmission too high. If it makes contact with the underbody, damage to the TSS sensor can occur.

4. Install a high lift transmission jack with the special jack adapter 014-0763 to the transmission.

- Rear transmission mount nuts
- Transmission support crossmember
- Fluid cooler tubes at the Cooler Bypass Valve (CBV), plug all fittings and position the tubes aside
- Right and left hand transmission-to-engine bolts
- Flexplate inspection cover
- Transmission-to-engine bolts
- Transmission from the vehicle

To install:

5. Installation is the reverse of removal, please note the following specs:

6. Transmission-to-engine bolts to 45 ft. lbs. (61 Nm).

7. Torque converter-to-flexplate nuts to 26 ft. lbs. (35 Nm).

8. Flexplate inspection cover nuts to 25 ft. lbs. (34 Nm) on gasoline engines and 15 ft. lbs. (20 Nm) on 7.3L diesel engines.

9. Fluid filler tube into the short fluid inlet tube. Tighten the nut to 26 ft. lbs. (35 Nm).

10. Refill all transmissions with the correct amount of Motorcraft MERCON® automatic transmission fluid.

11. Connect the negative battery cable.

4R70W/4R75W TRANSMISSION

1. Before servicing the vehicle, refer to the Precautions Section.

2. Disconnecting the negative battery cable on some vehicles may interfere with the functions of the on-board computer system(s) and may require the computer to undergo a relearning process once the negative battery cable is reconnected.

3. Remove or disconnect the following:
- Air cleaner air intake duct assembly
- Transmission fluid level indicator
- Engine cover
- Fluid filler tube, on 4.6L and 5.4L engines
- Fluid filler tube from the transmission and position aside, on 4.2L engines
- Electrical harness connectors
- Torque converter housing-to-engine retaining bolts. Position the fuel and electrical harness bracket aside.
- Fluid cooler tubes and plug all fittings and position the tubes aside

4. Mark the driveshaft flange and the rear companion flange for correct alignment during assembly.
- Driveshaft
- Shift cable and bracket
- Access plugs

- Starter motor
- Fasteners and the front A/C deflector assembly, if equipped
- Four torque converter-to-flexplate retaining nuts and discard the bolts
- Catalyst monitoring sensor harness connectors, if equipped
- Right hand heat shield
- Exhaust Gas Recirculation (EGR) vacuum reservoir from the frame and position aside, if equipped

5. Attach a high lift transmission jack to the transmission.
- Crossmember

6. Support the transmission support crossmember.

- Transmission insulator and retainer-to-crossmember retaining nuts and the transmission support crossmember
- Clamp and hanger assembly, if equipped
- Transmission range sensor electrical connector
- Output Shaft Speed (OSS) sensor connector
- Solenoid body connector
- Flexplate inspection plate
- Transmission bolts

> ❈❈ **CAUTION**
>
> **Watch for clearance between the transmission and the right side heat shield retaining bracket as well as other obstructions.**

7. Separate and back the transmission away from the engine, positioning the extension housing above the exhaust crossover pipe.

8. Hold the torque converter firmly in its installed position.

> ❈❈ **CAUTION**
>
> **Watch for clearance between the transmission and the engine.**

9. Carefully tilt the front of the transmission downward.

> ❈❈ **CAUTION**
>
> **Watch for clearance between the transmission and the exhaust crossover pipe.**

10. Lower the transmission enough to clear the engine. Then, carefully pull the jack with the transmission toward the front of the vehicle, lowering the transmission as required until it is clear of the exhaust crossover pipe.

11. Tilt the front of the transmission upward until level.

12. Remove the transmission with the jack from under the vehicle.

To install:

13. Installation is the reverse of removal, please note the following specs:

14. Transmission-to-engine bolts to 45 ft. lbs. (61 Nm).

15. Torque converter-to-flexplate nuts to 26 ft. lbs. (35 Nm).

16. Flexplate inspection cover nuts to 15 ft. lbs. (20 Nm).

17. Refill all transmissions with the correct amount of Motorcraft MERCON® automatic transmission fluid.

18. Connect the negative battery cable.

TORQSHIFT

1. Before servicing the vehicle, refer to the Precautions Section.

2. With the vehicle in **Neutral**, position it on a hoist.

3. Disconnect both battery ground cables.

4. Remove the fluid level indicator.

5. Remove the driveshaft.

6. If transmission disassembly is required, drain the transmission fluid. Remove the drain plug and allow the fluid to drain

7. Install the drain plug.

8. Install a suitable high-lift transmission jack.

9. If equipped, remove the two wire harness retainers from the rear crossmember and position the wire harness out of the way.

10. If equipped, disconnect the transmission-mounted parking brake linkage.

11. Disconnect the return spring from the pin.

12. Remove the hair pin retaining clip from the pin.

13. Remove the pin from the actuating lever and move the actuating cable away from the actuating lever.

14. Compress the spring on the actuating cable and remove the cable from the cable housing.

15. Disconnect the transmission fluid cooler tubes form the filter assembly.

16. Remove the shift cable.

17. Remove the transmission fluid filter bracket nut.

18. Remove the transmission fluid filter bracket bolt, position the bracket aside.

19. Disconnect the shift cable from the manual lever.

20. Disconnect the wire harness from the shift cable bracket.

21. Remove the bolts and position the shift cable and bracket aside.

22. Disconnect the oil cooler tubes from the filter housing.

23. Remove the bolts and remove the filter housing.

24. Loosen the bolt and disconnect the solenoid body electrical connector.

25. Disconnect and position the wire harness out of the way.

26. Disconnect the output shaft speed (OSS) sensor electrical connector.

27. Disconnect the turbine shaft and intermediate shaft combination speed sensor electrical connector.

28. Disconnect the wire harness from the side of the transmission.

29. Remove the cylinder block opening cover in order to gain access to the torque converter nuts.

30. Use a suitable tool to rotate the crankshaft in order to gain access to the flexplate nuts.

31. Remove and discard the six torque nuts.

32. Remove the left crossmember nuts.

33. Remove the left crossmember bracket.

34. Remove the right crossmember nuts.

35. Remove the right crossmember bracket.

36. Remove the rear transmission mount nuts and remove the crossmember.

37. Remove the rear transmission mount from the extension housing.

38. Disconnect the wire harness from the bracket.

39. Remove the nine transmission-to-engine mounting bolts and position the wire harness bracket out of the way.

40. Slide the transmission back enough to install the special tool.

41. If equipped, remove the transmission-mounted parking brake.

42. Remove the transmission-mounted parking retaining bolts.

43. Keeping the vent in an upward position, remove the transmission-mounted parking brake assembly.

44. Remove and discard the transmission-mounted parking brake gasket.

45. Disconnect the transmission fluid cooler tubes and filter housing from the case.

46. If overhauling the transmission or installing a new transmission, carry out the transmission fluid cooler backflushing and cleaning.

To install:

47. Loosely position the fluid cooler tubes in place.

48. Loosely install the fluid filter bracket and bolt.

49. Tighten the transmission fluid cooler tube nuts and remove the bracket bolt. Torque to 30 ft. lbs. (40 Nm).

50. If equipped, install the transmission-mounted parking brake.

51. Install a new transmission-mounted parking brake gasket.

52. Keeping the vent in an upward position, install the transmission-mounted parking brake assembly.

53. Install the transmission-mounted parking retaining bolts. Torque to 41 ft. lbs. (55 Nm).

❄❄ WARNING

Prior to the installation of the assembly, the torque converter pilot hub must be correctly lubricated or damage to the torque converter or the engine crankshaft can occur.

54. Lubricate the torque converter pilot hub with multi-purpose grease.

55. If the special tool has not been installed during the assembly of the transmission, install the special tool to hold the torque converter in place while moving and positioning the transmission in place. Once the transmission is in place, prior to bolting it to the engine, remove the special tool.

56. Position the transmission in place. While raising the transmission up into the engine compartment, align the fluid filler tube with the stub tube on the transmission using the fluid level indicator as a guide.

57. While installing the transmission to the engine, align the torque converter studs with the mounting holes in the flexplate.

58. Position the electrical wiring harness bracket in place and install the eight transmission-to-engine bolts. Torque to 35 ft. lbs. (47 Nm).

59. Connect the wire harness to the bracket.

60. Using the special tool, rotate the crankshaft to gain access to the torque converter studs.

61. Install the new torque converter-to-flexplate nuts. Torque to 26 ft. lbs. (35 Nm).

62. Install the cylinder block opening cover.

63. Install the wire harness retaining clip into the transmission on the left side.

64. Connect the wire harness on the right side of the transmission.

65. Install the oil filter housing.

66. Connect the oil cooler lines.

67. Install the transmission mount. Torque to 69 ft. lbs. (94 Nm).

68. Position the crossmember to the transmission mount and loosely install the nut.

69. Install the left crossmember bracket.

70. Install the left crossmember nuts. Torque to 60 ft. lbs. (81 Nm).

71. Install the right crossmember bracket.

72. Install the right crossmember nuts. Torque to 60 ft. lbs. (81 Nm).

73. Tighten the rear insulator nuts. Torque to 69 ft. lbs. (94 Nm).

74. Connect the output shaft speed (OSS) sensor electrical connector.

75. Connect the intermediate shaft and turbine shaft combination speed sensor electrical connector.

76. Connect the solenoid body electrical connector.

77. Connect the shift cable.

78. Install the cable housing bracket with one stud bolt and one bolt.

79. Install the bolt and nut for the filter housing and bracket.

80. Connect the wire harness to the bracket.

81. Install the shift cable to the manual lever.

82. Install the fluid cooler lines to the filter housing.

83. If equipped, install the wire harness and install the two retaining clips to the rear of the crossmember.

84. If equipped, connect the transmission-mounted parking brake.

85. Install the parking cable into the bracket.

86. Position the cable end onto the lever.

87. Install the retaining pin and clip.

88. Install the return spring.

89. Install the rear driveshaft.

❄❄ WARNING

If installing a newly overhauled or a new transmission, a new transmission fluid cooler remote filter must be installed. Otherwise transmission failure can occur.

90. Install a new transmission fluid cooler remote filter.

91. Use the following guidelines for the in-line transmission fluid filter:

a. If the transmission was overhauled and the vehicle was equipped with an in-line fluid filter, install a new in-line fluid filter.

b. If the transmission was overhauled and the vehicle was not equipped with an in-line fluid filter, install a new in-line fluid filter kit.

c. If the transmission is being installed for a non-internal repair, do not install an in-line filter or filter kit.

1	Transmission fluid cooler tube nuts	13	Transmission rear support bolts
2	Starter motor bolts	14	Transmission fluid pan
3	Electrical connector nuts	15	Transmission solenoid body connector
4	Transmission assembly	16	Transmission retaining bolts
5	Rear driveshaft	17	Inspection cover shield bolts
6	Rear driveshaft retaining bolts	18	Inspection cover
7	Exhaust hanger	19	Transmission range selector lever cable bolts
8	Exhaust hanger bolts	20	Transmission range selector lever cable and bracket
9	Rear crossmember bolts and nuts	21	Transmission range selector lever cable end
10	Rear transmission mount nuts	22	Transmission electrical connectors
11	Rear crossmember	23	Torque converter nut rubber access plug
12	Transmission rear support	24	Torque converter nuts

67197-EFSE-G144

2004–05 4R70E/4R75E mounting 2-wheel drive (cont.)

d. If installing a new or a Ford authorized remanufactured transmission, install the in-line transmission fluid filter that is supplied.

92. Prior to lowering the vehicle, install a new in-line transmission filter or a filter kit.

93. Connect the battery ground cable.

94. Adjust the shift linkage. Verify that the vehicle starts in **Park** and **Neutral** and the **Reverse** lamps illuminate in **Reverse**.

95. Install the fluid level indicator.

96. With the engine running and the transmission at normal operating temperature 66–77°C (150–170°F), check and adjust the transmission fluid level, and check for any leaks. If fluid is needed, add fluid in increments of 0.24 liter (0.5 pint) until the correct level is achieved (fluid should be in the cross-hatched area of the fluid level indicator).

2002–03 F-Series and 2004 Heritage Edition

1. Before servicing the vehicle, refer to the Precautions Section.

2. Remove or disconnect the following:
- Negative battery cable
- Front driveshaft and transfer case, on 4-wheel drive vehicles
- All cables, connectors and fluid lines that may interfere with transmission removal. Tag them if helpful for installation.
- Torque converter from the flexplate and disconnect the shift linkage
- Transmission fluid

3. Position a transmission jack under the transmission and safety-chain the case to the jack.
- Driveshaft
- Transmission rear mount

- Crossmember
- The transmission-to-engine block bolts

✳✳ CAUTION

The torque converter will fall out of the transmission if it is tilted forward. Keep a hand on it while lowering the transmission out of the vehicle.

4. Roll the transmission rearward until the input shaft clears, lower the jack and remove the transmission.

To install:

5. Carefully raise the transmission to the engine or bell housing.

6. Roll the transmission forward and into position.

7. Install or connect the following:
- Tighten the bolts to 65 ft. lbs. (87 Nm) for the diesel or to 35 ft. lbs. (47 Nm) on 2002–03 gasoline engines.
- Crossmember and tighten the bolts to 59 ft. lbs. (80 Nm).
- Transmission rear insulator and lower retainer. Tighten the bolts to 60 ft. lbs. (81 Nm).

8. The rest of the installation is the reverse of removal.

9. Refill all transmissions with the correct amount of Motorcraft MERCON® automatic transmission fluid.

10. Connect the negative battery cable.

2004–05 F-150; 2006 Mark LT

2-WHEEL DRIVE

1. Before servicing the vehicle, refer to the Precautions Section.

2. Disconnect the battery ground cable.

3. With the vehicle in **Neutral**, position it on a hoist

4. Remove the rear driveshaft.

5. Drain the transmission fluid.

6. Loosen the transmission fluid pan bolts and allow the fluid to drain. After the fluid has drained, remove the bolts.

7. Remove the transmission fluid pan and transmission fluid pan gasket. Drain the rest of the fluid from the pan.

➡The transmission fluid pan gasket is reusable. Clean and inspect for damage. If not damaged, the gasket should be reused.

8. Install the transmission fluid pan and gasket.

➡If removing the transmission for a transmission related failure, it is not necessary to torque the transmission fluid pan back onto the transmission case. If removing the transmission for a non-related transmission failure it is necessary to re-torque the transmission fluid pan.

9. Install the bolts.

10. Remove the starter motor solenoid terminal cover.

11. Disconnect the starter motor electrical connectors and the ground wire.

12. Remove the starter motor.

13. Remove the rubber plug to access the nuts.

14. Remove the four torque converter nuts. Rotate either the front of the crankshaft or the flexplate to access all of the torque converter retaining nuts.

15. Remove the transmission inspection cover.

16. Remove the shift cable and bracket.

17. Disconnect the transmission electrical connectors.

18. Disconnect the solenoid body assembly electrical connector.

19. Using a suitable transmission jack, support the transmission and secure the transmission to it with a safety strap.

20. Remove the left bolt retaining the heat shield to the transmission support crossmember.

21. Remove the right bolt retaining the heat shield to the transmission support crossmember.

22. Loosen, but do not remove, the transmission mount-to-crossmember nuts.

23. Remove the two bolts for the exhaust hanger.

24. Remove the four crossmember-to-frame nuts and bolts (two on each side).

25. Remove the transmission mount-to-crossmember nuts, and remove the crossmember.

26. Remove the two bolts and the rear transmission mount.

27. Remove the exhaust hanger.

28. Disconnect the transmission fluid cooler tubes.

29. Remove the seven transmission-to-engine bolts.

30. Position the fuel lines and bracket aside.

✳✳ CAUTION

The torque converter is heavy and can result in injury if it falls out of the transmission. Secure the torque converter in the transmission. If the torque converter is dropped, a new one must be installed.

31. Slide the transmission rearward enough to install the holding tool.

✳✳ CAUTION

The transmission must be secured with a safety chain or strap.

➡ **The front of the transmission must be lowered for the transmission to clear the exhaust system during removal.**

32. Continue moving the transmission rearward, while gradually lowering the front of the transmission. As the transmission jack is being lowered, move the transmission forward and remove it from the vehicle.

33. If the transmission is to be overhauled or if installing a new transmission, carry out transmission backflushing and cleaning.

To install:

✳✳ WARNING

The torque converter cover is piloted into position to the engine crankshaft by dowels in the rear of the engine block. The torque converter must rest squarely and loosely against the flexplate. This indicates that the torque converter pilot is not binding in the engine crankshaft.

✳✳ CAUTION

The transmission must be secured with a safety chain or strap.

✳✳ CAUTION

The torque converter is heavy and can result in injury if it falls out of the transmission. Secure the torque converter in the transmission. If the torque converter is dropped, a new one must be installed.

➡ **The front of the transmission must be lowered for the transmission to clear the exhaust system during installation.**

34. Continue moving the transmission rearward, while gradually raising the front of the transmission. As the transmission jack is being raised, move the transmission back and install it in the vehicle.

35. Align the orange balancing marks between the torque converter studs and the flexplate bolt holes.

36. Remove the special tool.

➡ **While positioning the transmission to the back of the engine, use the transmission filler tube as a guide and install the transmission filler tube into the case.**

37. Install the seven transmission-to-engine bolts. Torque to 35 ft. lbs. (48 Nm).

38. Position the fuel line and bracket.

39. Install the bolts.

40. Connect the transmission fluid cooler tubes.

41. Position the rear transmission insulator and install the bolts. Torque to 59 ft. lbs. (80 Nm).

42. Install the exhaust hanger onto the exhaust crossover pipe.

43. Position the crossmember and loosely install the mount-to-crossmember nuts.

44. Install the four crossmember-to-frame nuts and bolts (two on each side). Torque to 66 ft. lbs. (90 Nm).

45. Tighten the rear transmission mount nuts. Torque to 76 ft. lbs. (103 Nm).

46. Install the bolt retaining the left heat shield to the transmission support crossmember.

47. Install the bolt retaining the right heat shield to the transmission support crossmember.

48. Install the bolts retaining the exhaust hanger to the crossmember. Torque to 30 ft. lbs. (40 Nm).

49. Connect the solenoid body assembly electrical connector.

50. Connect the transmission electrical connectors.

51. Install the shift cable bracket and connect the transmission shift linkage.

52. Install the transmission inspection cover.

53. Install the torque converter nuts. Torque to 27 ft. lbs. (36 Nm).

54. Install the rubber access plug.

55. Install the starter motor.

56. Connect the starter motor electrical connectors and the ground wire.

57. Install the starter motor solenoid terminal cover.

58. Install the rear driveshaft.

59. Use the following guidelines for the in-line transmission fluid filter:

   a. If the transmission was overhauled and the vehicle was equipped with an in-line fluid filter, install a new in-line fluid filter.

   b. If the transmission was overhauled and the vehicle was not equipped with an in-line fluid filter, install a new in-line fluid filter kit.

   c. If the transmission is being installed for a non-internal repair, do not install an in-line filter or filter kit.

   d. If installing a new or a Ford authorized remanufactured transmission, install the in-line transmission fluid filter that is supplied.

60. Prior to lowering the vehicle, install a new in-line transmission filter or a filter kit.

61. Connect the battery ground cable.

62. Fill the transmission with clean automatic transmission fluid and inspect for correct operation.

63. Check the fluid filter for any leaks.

4-WHEEL DRIVE

1. Before servicing the vehicle, refer to the Precautions Section.

2. Disconnect the battery ground cable.

3. With the vehicle in **Neutral**, position it on a hoist.

4. Drain the transmission fluid.

5. Loosen the transmission fluid pan

bolts and allow the fluid to drain. After the fluid has drained, remove the bolts.

6. Remove the transmission fluid pan and transmission fluid pan gasket. Drain the rest of the transmission fluid from the fluid pan.

➡ **The transmission fluid pan gasket is reusable. Clean and inspect for damage. If not damaged, the gasket should be reused.**

7. Install the transmission fluid pan and gasket.

8. Position the transmission fluid pan gasket.

9. Position the transmission fluid pan.

➡ **If removing the transmission for a transmission related failure, it is not necessary to torque the transmission fluid pan back onto the transmission case. If removing the transmission for a non-related transmission failure it is necessary to re-torque the transmission fluid pan.**

10. Install the bolts.

11. Using a suitable transmission jack, support the transmission and secure the transmission to it with a safety strap.

12. Index-mark the front flange of the front driveshaft.

13. Index-mark the rear flange of the front driveshaft at the transfer case.

14. Remove the front driveshaft shield.

15. Remove the transfer case.

16. Remove the front driveshaft.

17. Remove the left bolt retaining the heat shield to the transmission support crossmember.

18. Remove the right bolt retaining the heat shield to the transmission support crossmember.

19. Loosen, but do not remove, the transmission mount-to-crossmember nuts.

20. Remove the two bolts for the exhaust hanger.

21. Remove the four crossmember-to-frame nuts and bolts (two on each side).

22. Remove the transmission mount-to-crossmember nuts, and remove the crossmember.

23. Remove the two bolts and the rear transmission mount.

24. Remove the exhaust hanger.

25. Remove the starter motor solenoid terminal cover.

26. Disconnect the starter motor electrical connectors and the ground wire.

27. Remove the starter motor.

28. Remove the rubber plug to access the nuts.

29. Remove the four nuts. Rotate the crankshaft/flexplate assembly to access all the nuts.

30. Remove the transmission inspection cover.

31. Remove the shift cable and bracket.

32. Disconnect the solenoid body assembly electrical connector.

33. Disconnect the transmission electrical connectors.

34. Disconnect the transmission fluid cooler tubes.

35. Remove the seven transmission-to-engine bolts.

36. Position the fuel lines and bracket aside.

✳✳ CAUTION

The torque converter is heavy and can result in injury if it falls out of the transmission. Secure the torque converter in the transmission. If the torque converter is dropped, a new one must be installed.

37. Slide the transmission rearward enough to install a torque converter holding tool.

✳✳ CAUTION

The transmission must be secured with a safety chain or strap.

➡ **The front of the transmission must be lowered for the transmission to clear the exhaust system during removal.**

38. Continue moving the transmission rearward, while gradually lowering the front of the transmission. As the transmission jack is being lowered, move the transmission forward and remove it from the vehicle.

39. If the transmission is to be overhauled or if installing a new transmission, carry out transmission backflushing and cleaning.

To install:

✳✳ CAUTION

The torque converter cover is piloted into position to the engine crankshaft by dowels in the rear of the engine block. The torque converter must rest squarely and loosely against the flexplate. This indicates that the torque converter pilot is not binding in the engine crankshaft.

40. When installing a new transmission, install a torque converter holding tool.

✳✳ CAUTION

The transmission must be secured with a safety chain or strap.

✳✳ CAUTION

The torque converter is heavy and can result in injury if it falls out of the transmission. Secure the torque converter in the transmission. If the torque converter is dropped, a new one must be installed.

➡ **The front of the transmission must be lowered for the transmission to clear the exhaust system during installation.**

41. Continue moving the transmission rearward, while gradually raising the front of the transmission. As the transmission jack is being raised, move the transmission back and install it in the vehicle.

42. Align the orange balancing marks between the torque converter studs and the flexplate bolt holes.

43. Remove the special tool.

➡ **While positioning the transmission to the back of the engine use the transmission filler tube as a guide and install the transmission filler tube into the case.**

44. Install the seven transmission-to-engine bolts. Torque to 35 ft. lbs. (48 Nm).

45. Position the fuel line and bracket.

46. Install the bolts.

47. Connect the transmission fluid cooler tubes.

48. Install the exhaust hanger onto the exhaust crossover pipe.

49. Position the rear transmission insulator and install the bolts. Torque to 59 ft. lbs. (80 Nm).

50. Position the crossmember and loosely install the mount-to-crossmember nuts.

51. Install the four crossmember-to-frame nuts and bolts (two on each side). Torque to 66 ft. lbs. (90 Nm).

52. Tighten the rear transmission mount nuts. Torque to 76 ft. lbs. (103 Nm).

53. Install the bolt retaining the left heat shield to the transmission support crossmember.

54. Install the bolt retaining the right heat shield to the transmission support crossmember.

55. Install the bolts retaining the exhaust hanger to the crossmember.

56. Connect the solenoid body assembly electrical connector.

①

② 36 Nm (27 lb-ft)

③

31

28 Nm (21 lb-ft)

29

30

28 34 Nm (25 lb-ft)

④

⑤

⑩

47 Nm (35 lb-ft) ⑪

18 Nm (13 lb-ft) ⑥

26 Nm (19 lb-ft) ⑦

⑨

26 Nm (19 lb-ft) ⑦

23 Nm (16 lb-ft) ⑧

12 Nm (9 lb-ft) ⑧

6 Nm (53 lb-ft) ⑧

47 Nm (35 lb-ft) ⑫

⑬

⑭

16 103 Nm (76 lb-ft)

22 103 Nm (76 lb-ft)

22

21

103 Nm (76 lb-ft)

27

26 80 Nm (35 lb-ft)

20 90 Nm (66 lb-ft)

25

20

24

⑮

18 Nm (13 lb-ft) ⑰

⑱

20 90 Nm (66 lb-ft)

103 Nm (76 lb-ft) ㉓

40 Nm (66 lb-ft) ⑲

1	Torque converter nut rubber access plug	17	Front driveshaft shield nuts
2	Torque converter nuts	18	Front driveshaft shield
3	Transmission range selector lever cable end	19	Exhaust hanger bolts
4	Transmission electrical connectors	20	Rear crossmember bolts and nuts
5	Transmission solenoid body connector	21	Front driveshaft
6	Transmission fluid cooler tube nuts	22	Front driveshaft retaining bolts
7	Starter motor bolts	23	Rear transmission mount nuts
8	Starter motor electrical connector nuts	24	Exhaust hanger
9	Starter motor	25	Rear crossmember
10	Transmission assembly	26	Transmission rear support bolts
11	Transmission retaining bolts	27	Transmission rear support
12	Transfer case retaining bolts	28	Inspection cover shield bolts
13	Transfer case gasket	29	Transmission range selector lever cable bracket bolts
14	Transfer case	30	Inspection cover shield
15	Rear driveshaft	31	Transmission range selector lever cable bracket
16	Rear driveshaft retaining bolts		

67197-EFSE-G145

2004–05 4R70E/4R75E mounting—4-wheel drive

57. Connect the transmission electrical connectors.

58. Install the shift cable bracket and connect the transmission shift linkage.

59. Install the transmission inspection cover.

60. Install the torque converter nuts. Torque to 27 ft. lbs. (36 Nm).

61. Install the rubber access plug.

62. Install the starter motor.

63. Connect the starter motor electrical connectors and the ground wire.

64. Install the starter motor solenoid terminal cover.

65. Position the front driveshaft in place.

66. Install the transfer case.

67. Align the marks made during removal, install the front driveshaft flange and the four bolts.

68. Align the marks made during removal, install the rear driveshaft flange and the four bolts.

69. Install the front driveshaft shield.

70. Install the rear driveshaft.

71. Use the following guidelines for the in-line transmission fluid filter:

a. If the transmission was overhauled and the vehicle was equipped with an in-line fluid filter, install a new in-line fluid filter.

b. If the transmission was overhauled and the vehicle was not equipped with an in-line fluid filter, install a new in-line fluid filter kit.

c. If the transmission is being installed for a non-internal repair, do not install an in-line filter or filter kit.

d. If installing a new or a Ford authorized remanufactured transmission, install the in-line transmission fluid filter that is supplied.

72. Prior to lowering the vehicle, install a new in-line transmission filter or a filter kit.

73. Connect the battery ground cable.

74. Fill the transmission with clean automatic transmission fluid and inspect for correct operation.

75. Check the fluid filter for any leaks.

2004 F-250 and 350

4R100

1. Before servicing the vehicle, refer to the Precautions Section.

2. With the vehicle in **Neutral** position it on a hoist.

➡ **When the battery has been disconnected and reconnected, some abnormal drive symptoms can occur while the vehicle relearns its adaptive strategy. The customer needs to be notified that they may experience slightly different upshifts (either soft or firm) and that this is a temporary condition that will eventually return to normal operating condition.**

3. Disconnect the battery ground cable.
4x4 vehicles
4. Remove the transfer case assembly.

4x2 vehicles
5. Remove the rear driveshaft.
All vehicles
6. If transmission disassembly is required, drain the transmission fluid.

Vehicles equipped with a fluid pan drain plug
7. Remove the drain plug and allow the fluid to drain.

8. Install the drain plug.
Vehicles not equipped with a fluid pan drain plug

➡ **Do not discard the gasket unless damaged. This is a reusable gasket.**

9. Remove the transmission fluid pan and gasket.

➡ **The transmission pan gasket is reusable unless damaged.**

➡ **Apply a light coat of petroleum jelly to hold the gasket to the fluid pan.**

10. Position the gasket onto the cleaned fluid pan. Make sure the magnet is positioned over the dimple in the fluid pan.

✷✷ CAUTION

Mixing the 4x2 and the 4x4 style transmission fluid filters and transmission pan assembly components can cause transmission damage.

11. Install the correct pan with gasket for this application. Alternately tighten the bolts.

✳✳ CAUTION

Make sure securing straps or the transmission jack adapter do not touch the cooler bypass valve (CBV). Do not use the CBV as a handle. Damage to the CBV can cause a leak.

12. Install a suitable transmission jack and support the transmission.

4x4 vehicles

13. Remove the jack stand after installing a suitable transmission jack.

4x2 vehicles

14. Remove the right crossmember nuts.

15. Remove the left crossmember nuts.

16. Remove the nuts and the crossmember.

17. Remove the transmission mount.

Vehicles equipped with a transmission-mounted parking brake

18. Disconnect and lubricate.

19. Disconnect the parking brake lever return spring from the parking brake lever.

20. Apply penetrating oil to the adjusting clevis, jam nut and the threads on the front parking brake cable and conduit.

21. Disconnect the front parking brake cable and conduit.

22. Loosen the jam nut.

23. Remove the clevis locking pin.

24. Remove the clevis pin.

25. Remove the adjusting clevis from the parking brake lever.

26. Compress the retainer, and remove the front parking brake cable and conduit from the cable bracket.

All vehicles

➡ If the vehicle is equipped with a power take-off unit, all or part of the PTO unit will need to be removed.

27. Disconnect the shift cable from the transmission.

28. Disconnect the shift cable from the manual lever.

29. Remove the shift cable bracket from the transmission and position aside.

30. Disconnect the digital transmission range (TR) sensor connector and the wire loom from the shift cable bracket.

31. Disconnect the solenoid body connector.

32. Disconnect the turbine shaft speed (TSS) sensor and the output shaft speed (OSS) sensor.

33. Remove the wiring harness from the transmission and position aside.

34. Remove the flexplate inspection cover.

35. Remove the starter motor.

36. Remove the cylinder block opening cover.

37. Remove and discard the torque converter-to-flexplate nuts. 5.4L engine has four torque converter-to-flexplate nuts. 6.8L engine has six torque converter-to-flexplate nuts.

38. Position a suitable drain pan and disconnect the transmission fluid cooler tubes from the cooler bypass valve.

39. Remove the seven transmission-to-engine mounting bolts.

40. Gently rock the transmission side-to-side to disengage it from the locator dowels.

41. Move the transmission and the transmission jack rearward to clear the engine flexplate.

✳✳ WARNING

The torque converter is heavy and can result in injury if it falls out of the transmission. Secure the torque converter in the transmission. If the torque converter is dropped, a new one must be installed.

42. Hold the torque converter in place.

✳✳ CAUTION

Use care while removing the transmission to avoid obstructions.

✳✳ CAUTION

Do not use the cooler bypass valve as a handle. Damage to the cooler bypass valve assembly can occur or damage to the case can result.

✳✳ CAUTION

If a safety strap is being used to hold the transmission to the high-lift transmission jack, place the strap behind the cooler bypass valve (CBV) to prevent damage to the cooler bypass valve.

43. Lower the transmission out of the vehicle.

Vehicles equipped with a transmission-mounted parking brake

44. Remove the transmission-mounted parking brake.

45. Keep the parking brake vent in the upward position to prevent contamination of the brake shoes and linings.

46. Remove the six bolts, parking brake assembly and the gasket from the extension housing. Discard the bolts and the gasket. Clean the mating surfaces.

All vehicles

47. If the transmission is being overhauled or if installing a new or remanufactured transmission, carry out the transmission fluid cooler—backflushing and cleaning.

To install:

✳✳ CAUTION

Prior to the installation of the transmission, the fluid, cooler lines and the cooler bypass valve must be cleaned. Transmission failure can occur if this procedure is not followed.

✳✳ CAUTION

A new transmission oil-to-air (OTA) cooler must be installed if the transmission was overhauled or exchanged due to a failure of the transmission. Transmission failure can occur if this procedure is not followed.

48. Inspect the wiring harness and the connectors for damage, terminal condition, corrosion and seal integrity. Repair or install new as required.

Vehicles equipped with a transmission-mounted parking brake

49. If removed, install the parking brake. Position the parking brake assembly with a new gasket on the transmission extension housing.

50. Install six new bolts. Torque to 41 ft. lbs. (55 Nm).

All vehicles

✳✳ CAUTION

Prior to the installation of the transmission, the torque converter pilot hub must be lubricated or damage to the torque converter or the engine crankshaft can occur.

51. Lubricate the torque converter pilot hub with multi-purpose grease.

52. Raise the transmission into place.

✳✳ CAUTION

Do not use the cooler bypass valve as a handle. Damage to the cooler bypass valve assembly can occur or damage to the case can result.

✳✳ CAUTION

Be careful not to raise the transmission up too far. The sensors can

make contact with the underbody of the vehicle and cause damage to the sensors. Sensor failure or leakage can occur.

➡ **While raising the transmission up into the engine compartment, make sure to align the fluid filler tube with the stub tube on the transmission, using the dipstick as a guide.**

53. Position the transmission.
54. Remove the holding tool.

✳✳ CAUTION

Do not allow the torque converter drive flats to disengage from the pump gear. Use care not to damage the flexplate and the converter pilot. The torque converter must rest squarely against the flexplate, indicating the converter pilot is not binding in the crankshaft.

55. While installing the transmission to the engine, align the torque converter studs with the mounting holes in the flexplate.
56. Install seven transmission-to-engine bolts. Torque to 35 ft. lbs. (47 Nm).
57. Install the new torque converter-to-flexplate nuts. 5.4L engines have four torque converter nuts. 6.8L engines have six torque converter nuts. Torque to 26 ft. lbs. (35 Nm).
58. Install the cylinder block opening cover.
59. Install the flexplate inspection cover and bolts. Torque to 25 ft. lbs. (34 Nm).
60. Install the starter motor.
61. Connect the solenoid pack electrical connector.
62. Connect the digital transmission (TR) sensor connector.
63. Connect the turbine shaft speed (TSS) sensor and the output shaft speed (OSS) sensor.

➡ **If the vehicle is equipped with a power take-off (PTO) unit, all or part of the PTO unit will need to be installed.**

64. Install the cable housing bracket. Torque to 18 ft. lbs. (25 Nm).
65. Install the shift cable to the manual lever.
66. Install the transmission fluid cooler tubes to the cooler bypass valve.

Vehicle equipped with a transmission-mounted parking brake

67. Position the front parking brake cable and conduit, and press the retainer into the cable bracket until it snaps into place.

68. Set the adjusting clevis. Loosen the jam nut several turns. Position the parking brake lever in the applied position. Tighten or loosen the adjusting clevis until the adjusting clevis hole lines up with the parking brake lever hole, then loosen the adjusting clevis to the specification.
69. Install the pins, and tighten the jam nut.
70. Install the clevis pin through the adjusting clevis and the parking brake lever.
71. Install the locking pin in the clevis pin. Tighten the jam nut.
72. Install the parking brake lever return spring.

4x4 vehicles

73. Support the extension housing with a jack stand and remove the transmission jack.
74. Install the transfer case.

4x2 vehicles

75. Install the transmission mount. Torque to 70 ft. lbs. (95 Nm).
76. Position the crossmember to the transmission mount and loosely install the nuts.

All vehicles

77. Install the crossmember bolts. Torque to 52 ft. lbs. (70 Nm).
78. Install the nuts. Torque to 52 ft. lbs. (70 Nm).

4x2 vehicles

79. Remove the transmission jack. Tighten the nuts. Torque to 69 ft. lbs. (94 Nm).
80. Install the rear driveshaft.
81. Use the following guidelines for the in-line transmission fluid filter:

   a. If the transmission was overhauled and the vehicle was equipped with an in-line fluid filter, install a new in-line fluid filter.

   b. If the transmission was overhauled and the vehicle was not equipped with an in-line fluid filter, install a new in-line fluid filter kit.

   c. If the transmission is being installed for a non-internal repair, do not install an in-line filter or filter kit.

   d. If installing a new or a Ford-authorized remanufactured transmission, install the in-line transmission fluid filter that is supplied.

82. Prior to lowering the vehicle, install a new in-line transmission filter or a filter kit.

➡ **When the battery has been disconnected and reconnected, some abnormal drive symptoms can occur while the vehicle relearns its adaptive strategy. The customer needs to be notified**

that they may experience slightly different upshifts (either soft or firm) and this is a temporary condition that will eventually return to normal operating condition.

83. Connect the battery ground cable.
84. Fill the transmission to the correct level with clean automatic transmission fluid.

TORQSHIFT

1. Before servicing the vehicle, refer to the Precautions Section.
2. With the vehicle in **Neutral**, position it on a hoist.
3. Disconnect both battery ground cables.

4x4 vehicles

4. Remove the transfer case assembly.

4x2 vehicles

5. Remove the driveshaft.

All vehicles

6. If transmission disassembly is required, drain the transmission fluid. Remove the drain plug and allow the fluid to drain
7. Install the drain plug.
8. Install a suitable high-lift transmission jack.

4x4 vehicles

9. Remove the jack stand from under the extension housing after installing the transmission jack.

4x2 vehicles

10. Remove the wire harness from the rear crossmember.
11. Remove the rear transmission mount nuts.

All vehicles

12. Remove the shift cable.
13. Disconnect the shift cable from the manual lever.
14. Remove the bolts and position the shift cable and bracket out of the way.
15. Loosen the bolt and disconnect the solenoid body electrical connector.
16. Disconnect the output shaft speed (OSS) sensor electrical connector.
17. Disconnect the turbine shaft and intermediate shaft combination speed sensor electrical connector.
18. Disconnect the right and left wire harness from the side of the transmission.

4x2 vehicles

19. Remove the left crossmember bolts and nuts.
20. Loosen, but do not remove, the bolt and nut indicated.
21. Remove the right crossmember bracket bolts and nuts.
22. Remove the bracket bolts and nuts.

23. Remove the bracket.

24. Remove the right crossmember bolts and nuts.

25. Remove the left crossmember bolt, nut and the crossmember.

26. Remove the bolt and nut.

27. Remove the crossmember.

28. Remove the rear transmission mount from the extension housing.

29. If equipped with dual alternators, rotate the tensioner and remove the outer accessory drive belt from the crankshaft pulley.

All vehicles

30. Remove the cylinder block opening cover in order to gain access to the torque converter nuts.

➡**Using a suitable strap wrench, rotate the crankshaft pulley to gain access to the torque converter nuts.**

31. Remove and discard the torque converter nuts.

32. While holding the case fitting, disconnect the rear fluid cooler tube nut and tube.

33. While holding the case fitting, disconnect the front fluid cooler tube nut and tube.

34. Remove the nine transmission-to-engine mounting bolts.

35. Slide the transmission back enough to install the special tool.

36. If the transmission is being overhauled or if installing a new or remanufactured transmission, carry out the transmission fluid cooler backflushing and cleaning.

To install:

✦ CAUTION

Prior to the installation of a new or overhauled transmission, the fluid cooler lines must be cleaned. Otherwise transmission failure can occur.

✳✳ CAUTION

Prior to the installation of a new or overhauled transmission, a new transmission fluid cooler remote filter must be installed. Otherwise transmission failure can occur.

37. If necessary, install a new transmission fluid cooler remote filter.

✦ CAUTION

Prior to the installation of the assembly, the torque converter pilot hub must be correctly lubricated or dam-

age to the torque converter or the engine crankshaft can occur.

38. Lubricate the torque converter pilot hub with multi-purpose grease.

39. If the holding special tool has not been installed during the assembly of the transmission, install the tool to hold the torque converter in place while moving and positioning the transmission in place. Once the transmission is in place, prior to bolting it to the engine, remove the special tool.

40. Position the transmission in place. While raising the transmission up into the engine compartment, align the fluid filler tube with the stub tube on the transmission using the fluid level indicator as a guide.

41. While installing the transmission to the engine, align the torque converter studs with the mounting holes in the flexplate.

42. Install nine transmission-to-engine bolts. Torque to 35 ft. lbs. (47 Nm).

43. Using a suitable strap wrench to rotate the crankshaft and pulley, install the new torque converter-to-flexplate nuts. Torque to 26 ft. lbs. (35 Nm).

44. Install the cylinder block opening cover.

45. If equipped with dual alternators, rotate the tensioner and install the outer accessory drive belt onto the crankshaft pulley.

All applications

46. Install the front transmission fluid cooler tube.

47. Install the rear transmission fluid cooler tube.

4x4 applications

48. Support the extension housing with a jack stand and remove the transmission jack.

4x2 applications

49. Install the transmission mount. Torque to 69 ft. lbs. (94 Nm).

50. Position the crossmember to the transmission mount and loosely install the nut.

51. Install the left crossmember bolts. Torque to 60 ft. lbs. (81 Nm).

52. Install the right crossmember bolts. Torque to 60 ft. lbs. (81 Nm).

53. Install the bracket and loosely install the bolts.

54. Install the bracket.

55. Loosely install the bolts.

56. Tighten the right crossmember bolts. Torque to 60 ft. lbs. (81 Nm).

57. Tighten the rear insulator nuts. Torque to 69 ft. lbs. (94 Nm).

All applications

58. Connect the right and left wiring harness to the side of the transmission.

59. Connect the output shaft speed (OSS) sensor electrical connector.

60. Connect the intermediate shaft and turbine shaft combination speed sensor electrical connector.

61. Connect the solenoid body electrical connector.

➡**If the vehicle is equipped with a power take-off unit, all or part of the PTO unit will need to be installed.**

62. Connect the shift cable.

63. Install the cable housing bracket.

64. Install the shift cable to the manual lever.

4x2 applications

65. Reconnect the wire harness to the frame.

4x4 applications

66. If equipped, install the transfer case.

4x2 applications

67. Install the rear driveshaft.

All applications

➡**Use the following guidelines for the in-line transmission filter:**

a. If the transmission was overhauled and the vehicle was equipped with an in-line fluid filter, install a new in-line fluid filter.

b. If the transmission was overhauled and the vehicle was not equipped with the in-line fluid filter, install a new in-line fluid filter.

c. If the transmission is being installed for a non-internal repair, do not install an in-line filter or filter kit.

d. If installing a new or a Ford-authorized remanufactured transmission, install the in-line transmission fluid filter that is supplied.

68. Prior to lowering the vehicle, install a new in-line transmission filter or a filter kit.

69. Connect the battery ground cable.

70. Adjust the shift linkage. Verify that the vehicle starts in **Park** and **Neutral** and the **Reverse** lamps illuminate in **Reverse**.

71. With the engine running and the transmission at normal operating temperature 66–77°C (150–170°F), check and adjust the transmission fluid level, and check for any leaks. If fluid is needed, add fluid in increments of 0.24L (0.5 pint) until the correct level is achieved (fluid should be in the cross-hatched area of the fluid level indicator).

2005 F-250/350

1. Before servicing the vehicle, refer to the Precautions Section.

All vehicles

2. With the vehicle in NEUTRAL, position it on a hoist.

3. Remove the exhaust pipe between the exhaust manifolds and the exhaust flange.

4x4 vehicles

4. Remove the transfer case assembly.

4x2 vehicles

5. Remove the driveshaft.

All vehicles

6. If transmission disassembly is required, drain the transmission fluid. Remove the drain plug and allow the fluid to drain

7. Install the drain plug. Torque to 18 ft. lbs. (25 Nm).

8. Install a suitable high-lift transmission jack.

4x4 vehicles

9. Remove the jack stand from under the extension housing after installing the transmission jack.

4x2 vehicles

10. Remove the wire harness from the rear crossmember.

11. Remove the rear transmission mount nuts.

All vehicles

12. Disconnect the shift cable from the manual lever.

13. Remove the bolts and position the shift cable and bracket out of the way.

14. Loosen the bolt and disconnect the solenoid body electrical connector.

15. Disconnect the output shaft speed (OSS) sensor electrical connector.

16. Disconnect the turbine shaft and intermediate shaft combination speed sensor electrical connector.

17. Disconnect the right and left wire harness from the side of the transmission.

4x2 vehicles

18. Remove the left crossmember bolts and nuts. Loosen, but do not remove, the lower bolt.

19. Remove the right crossmember bracket bolts and nuts.

20. Remove the right crossmember bolts and nuts.

21. Remove the left crossmember bolt, nut and the crossmember.

22. Remove the rear transmission mount from the extension housing.

23. If equipped with dual generators, rotate the tensioner and remove the outer accessory drive belt from the crankshaft pulley.

All vehicles

24. Remove the cylinder block opening cover in order to gain access to the torque converter nuts.

➡**Using a suitable strap wrench, rotate the crankshaft pulley to gain access to the torque converter nuts.**

25. Remove and discard the torque converter nuts.

26. While holding the case fitting, disconnect the rear fluid cooler tube nut and tube.

27. While holding the case fitting, disconnect the front fluid cooler tube nut and tube.

28. Remove the nine transmission-to-engine mounting bolts.

29. Slide the transmission back enough to install the torque converter holding tool.

30. Carefully lower the transmission assembly.

31. If the transmission is being overhauled or if installing a new or remanufactured transmission, carry out the transmission fluid cooler backflushing and cleaning.

To install:
All vehicles

> ❊❊ **WARNING**
>
> **Prior to installation of a new or overhauled transmission, the fluid cooler lines must be cleaned. Otherwise transmission failure can occur.**

> ❊❊ **WARNING**
>
> **Prior to installation of a new or overhauled transmission, a new transmission fluid cooler remote filter element must be installed. Otherwise transmission failure can occur.**

32. If necessary, install a new transmission fluid cooler remote filter element.

> ❊❊ **WARNING**
>
> **Prior to installation of the assembly, the torque converter pilot hub must be correctly lubricated or damage to the torque converter or the engine crankshaft can occur.**

33. Lubricate the torque converter pilot hub with multi-purpose grease.

34. Rotate the torque converter so the green or orange paint daub on the converter is in the 12 o'clock position.

35. If the special tool has not been installed during the assembly of the transmission, install the special tool to hold the torque converter in place while moving and positioning the transmission in place. Once the transmission is in place, prior to bolting it to the engine, remove the special tool.

36. Position the transmission in place. While raising the transmission up into the engine compartment, align the fluid filler tube with the stub tube on the transmission using the fluid level indicator as a guide.

37. While installing the transmission to the engine, align the torque converter studs with the mounting holes in the flexplate.

38. Install nine transmission-to-engine bolts. Torque to 35 ft. lbs. (47 Nm).

39. Using a suitable strap wrench to rotate the crankshaft and pulley, install the new torque converter-to-flexplate nuts. Torque to 26 ft. lbs. (35 Nm).

40. Install the cylinder block opening cover.

41. If equipped with dual generators, rotate the tensioner and install the outer accessory drive belt onto the crankshaft pulley.

42. Install the front transmission fluid cooler tube. Torque to 30 ft. lbs. (40 Nm).

43. Install the rear transmission fluid cooler tube. Torque to 30 ft. lbs. (40 Nm).

4x4 vehicles

44. Support the extension housing with a jack stand and remove the transmission jack.

4x2 vehicles

45. Install the transmission mount. Torque to 69 ft. lbs. (94 Nm).

46. Position the crossmember to the transmission mount and loosely install the nut.

47. Install the left crossmember bolts. Torque to 60 ft. lbs. (81 Nm).

48. Install the right crossmember bolts. Torque to 60 ft. lbs. (81 Nm).

49. Install the bracket and loosely install the bolts.

50. Tighten the right crossmember bolts. Torque to 60 ft. lbs. (81 Nm).

51. Tighten the rear insulator nuts. Torque to 69 ft. lbs. (94 Nm).

All vehicles

52. Connect the right and left wiring harness to the side of the transmission.

53. Connect the output shaft speed (OSS) sensor electrical connector.

54. Connect the intermediate shaft and turbine shaft combination speed sensor electrical connector.

55. Connect the solenoid body electrical connector.

➡**If the vehicle is equipped with a power take-off (PTO) unit, all or part of the PTO unit will need to be installed.**

56. Install the shift cable housing bracket. Torque to 18 ft. lbs. (25 Nm).

57. Install the shift cable to the manual lever.

4x2 vehicles

58. Reconnect the wire harness to the frame.

59. Install the exhaust pipe between the exhaust manifold and the exhaust flange.

4x4 vehicles

60. If equipped, install the transfer case.

4x2 vehicles

61. Install the rear driveshaft.

All vehicles

➡Use the following guidelines for the in-line transmission filter:

- If the transmission was overhauled and the vehicle was equipped with an in-line fluid filter, install a new in-line fluid filter.
- If the transmission was overhauled and the vehicle was not equipped with the in-line fluid filter, install a new in-line fluid filter.
- If the transmission is being installed for a non-internal repair, do not install an in-line filter or filter kit.
- If installing a new or a Ford-authorized remanufactured transmission, install the in-line transmission fluid filter that is supplied.

62. Prior to lowering the vehicle, install a new in-line transmission filter or a filter kit.

63. Adjust the shift linkage. Verify that the vehicle starts in PARK and NEUTRAL and the REVERSE lamps illuminate in REVERSE.

64. With the engine running and the transmission at normal operating temperature 66–77°C (150–170°F), check and adjust the transmission fluid level, and check for any leaks. If fluid is needed, add fluid in increments of 0.24 liter (0.5 pint) until the correct level is achieved (fluid should be in the cross-hatched area of the fluid level indicator).

Manual Transmission

REMOVAL & INSTALLATION

2002 F-Series

1. Before servicing the vehicle, refer to the Precautions Section.

2. Remove or disconnect the following:
- Negative battery cable
- Shifter boot and lever from inside the vehicle
- Front driveshaft and transfer case, on 4-wheel drive vehicles
- All cables, connectors and fluid

lines that may interfere with transmission removal. Tag them if helpful for installation.
- Transmission fluid

3. Position a transmission jack under the transmission and safety-chain the case to the jack.
- Driveshaft
- Transmission rear mount
- Crossmember
- The transmission-to-engine block bolts
- Bolts securing the transmission to the bell housing

✳✳ CAUTION

The torque converter will fall out of the transmission if it is tilted forward. Keep a hand on it while lowering the transmission out of the vehicle.

4. Roll the transmission rearward until the input shaft clears, lower the jack and remove the transmission.

To install:

5. Carefully raise the transmission to the engine or bell housing.

6. Roll the transmission forward and into position.

7. Install or connect the following:
- Tighten the bolts to 39 ft. lbs. (54 Nm).
- Crossmember and tighten the bolts to 59 ft. lbs. (80 Nm).
- Transmission rear insulator and lower retainer. Tighten the bolts to 60 ft. lbs. (81 Nm).

8. The rest of the installation is the reverse of removal.

9. Refill all transmissions with the correct amount of Motorcraft MERCON® automatic transmission fluid.

10. Connect the negative battery cable.

2003 F-Series, 2004 F-150

1. Before servicing the vehicle, refer to the Precautions Section.

2. Disconnect the battery ground cable.

3. Remove the four screws and pull the gearshift lever upward.

4. Remove the upper gearshift lever.

5. Remove the four bolts and the lower gearshift lever boot.

6. Remove the bolts and the lower gearshift lever.

7. Raise and support the vehicle.

8. If the transmission is being disassembled, drain the transmission fluid.

9. Remove the fuel line bracket from the rear of the transmission.

10. Disconnect the wiring harness from the transmission, the transfer case (if equipped) and the heated oxygen sensors (HO2S).

11. Remove the starter.

➡Index-mark the rear driveshaft yoke and the pinion flange to maintain initial driveshaft balance during installation.

12. Remove the rear driveshaft.

Four wheel drive vehicles

13. Remove the skid plate.

➡Index-mark the front driveshaft yoke and pinion flange and at the transfer case.

14. Remove the front driveshaft.

15. Remove the torsion bars and the torsion bar crossmember.

16. Remove the transfer case.

17. Remove the transfer case heat shield.

All vehicles

18. Disconnect the clutch hydraulic line.

19. Support the transmission.

20. Securely strap the jack to the transmission.

21. Remove the two heat shield bolts from the crossmember.

22. Remove the crossmember nuts.

23. Remove the transmission mount nuts, then remove the crossmember.

24. Remove the nut.

25. Remove the bolts and the transmission mount and exhaust bracket.

Four wheel drive vehicles

26. Remove the bolt and disconnect the transfer case shift lever.

All vehicles

27. Remove the two engine oil pan-to-transmission bolts.

28. Remove the six engine-to-transmission bolts.

29. The upper bolt secures the fuel line bracket.

30. Remove the muffler from the exhaust hanger brackets.

31. Remove the tail pipe from the exhaust hanger insulator.

32. Tilt the exhaust downward and position it on top of the rear axle.

33. Remove the transmission.

34. Move the transmission rearward until the input shaft is clear of the clutch. Tilt the front of the transmission downward while lowering from the vehicle.

To install

35. Using the special tool, raise and position the transmission.

36. Install the six engine-to-transmission bolts. Tighten to 44 ft. lbs. (59 Nm).

37. Install the two engine oil pan-to-

transmission bolts. Tighten to 32 ft. lbs. (44 Nm).

38. Raise the transmission, then install the muffler and tail pipe into the exhaust hanger insulators.

39. Install the transmission mount and exhaust bracket.

40. Position the transmission mount.

41. Install the bolts. Tighten to 60–80 ft. lbs. (87–110 Nm).

42. Install the nut. Tighten to 32 ft. lbs. (44 Nm).

43. Install the crossmember.

44. Position the crossmember.

45. Install the bolts. Tighten to 58–75 ft. lbs. (78–103 Nm).

46. Install the transmission mount bolts. Tighten to 60–80 ft. lbs. (87–110 Nm).

47. Install the two heat shield bolts to the crossmember. Tighten to 10–12 ft. lbs. (13–17 Nm).

48. Remove the transmission support from the transmission.

49. Connect the clutch hydraulic line.

Four wheel drive vehicles

50. Connect the transfer case shift lever and install the bolt.

51. Install the transfer case.

52. Install the torsion bars and the torsion bar crossmember.

➡**Align the index marks made during removal.**

53. Install the front driveshaft.

54. Install the transfer case heat shield.

55. Install the skid plate.

All vehicles

➡**Align the index marks made during removal.**

56. Install the rear driveshaft.

57. Install the starter.

58. Connect the wiring harness to the transmission, the transfer case (if equipped) and the heated oxygen sensors (HO$_2$S).

59. Install the fuel line bracket.

60. Install the lower gearshift lever.

61. Install the lower gearshift lever boot.

62. Install the upper gearshift lever.

63. Install the upper gearshift lever boot.

64. Connect the battery ground cable.

65. If the transmission was disassembled, refill with transmission fluid.

66. Check the vehicle ride height. Adjust as necessary.

2004 F-250 and 350

1. Before servicing the vehicle, refer to the Precautions Section.

2. Remove the four screws and the outer shift lever boot.

3. Remove the upper gearshift lever.

4. Remove the lower shift lever boot.

5. Remove the lower gearshift and shift housing.

Vehicles with a manual shift lever

6. Shift the transfer case into 4H.

7. Remove the screws that attach the bezel and boot assembly to the floor.

8. Remove the bolt that attaches the shift lever to the transfer case control lever assembly, and remove the shift lever, and the bezel and boot assembly.

9. Slide the bezel and boot assembly upward on the shift lever.

10. Remove the bolt, the shift lever, and the bezel and boot assembly.

All vehicles

11. Raise and support the vehicle.

12. If the transmission is being disassembled, drain the transmission fluid.

13. Remove the starter.

14. Disconnect the rear driveshaft and position it aside.

15. Remove the transfer case, if equipped.

16. Remove any power take-off (PTO) equipment, if equipped.

17. Support the transmission.

18. Securely strap the jack to the transmission.

19. Remove the transmission mount nuts.

20. Remove the right crossmember nuts.

21. Remove the left crossmember bolts.

22. Disconnect the reverse lamp switch electrical connector.

23. Disconnect wiring harness from the transmission.

24. Remove the clutch slave cylinder and position it aside.

25. Push the clutch slave cylinder inward, then rotate counterclockwise 45 degrees to remove.

26. Disconnect the transmission cooling tubes.

27. Remove the dust cover bolts.

28. Remove the transmission-to-engine bolts. For vehicles equipped with 7.3L diesel engines, remove six bolts. For vehicles equipped with gasoline engines, remove seven bolts.

29. Remove the transmission. Move the transmission rearward until the input shaft is clear of the clutch, then lower from the vehicle.

To install:

30. Using the transmission jack, raise and position the transmission to the engine and clutch.

31. Install the transmission-to-engine bolts. For vehicles equipped with 7.3L diesel engines, install six bolts. For vehicles equipped with gasoline engines, install seven bolts. Torque to 46 ft. lbs. (63 Nm).

32. Install the left crossmember bolts. Torque to 52 ft. lbs. (70 Nm).

33. Install the right crossmember nuts. Torque to 52 ft. lbs. (70 Nm).

34. Install the transmission mount nuts. Torque to 60 ft. lbs. (81 Nm).

35. Install the engine plate bolts. Torque to 21 ft. lbs. (28 Nm).

36. Remove the transmission jack.

37. Connect the transmission cooler tubes.

38. Install the clutch slave cylinder. Rotate the clutch slave cylinder clockwise 45 degrees to lock in position.

39. Install the starter.

40. Install the transfer case, if equipped. If the transfer case control lever assembly was removed from the transmission, it must be correctly aligned.

41. Connect the driveshaft.

42. Install any power take-off (PTO) equipment, if equipped.

43. Connect the reverse lamp switch electrical connector.

44. Connect the wiring harness to the transmission.

45. Refill the transmission to specification.

46. Lower the vehicle.

➡**Do not use a silicone sealing compound.**

➡**Do not wait longer than ten minutes to tighten the six bolts due to the rapid cure time of the sealant.**

47. Install the lower gearshift lever and shift housing assembly. Apply gasket maker to the shift housing and the main case.

48. Install the lower shift lever boot.

49. Apply Threadlock and sealer to the gearshift lever bolts. Install the upper gearshift lever.

50. Install the screws.

Vehicles with a manual shift lever

51. Position the shift lever with the bezel and boot assembly and install the bolt.

52. Position the bezel and boot assembly and install the screws. Verify the shift sequence from 2H to 4L to 2H.

2005 F-150

1. Before servicing the vehicle, refer to the Precautions Section.

2. Disconnect the battery ground cable.

3. Unclip the console cover, then slide the cover and boot assembly up.

4. Remove the gearshift lever nut.

5. Install the gearshift lever nut on the left side of the lever, then, tighten the nut to

remove the eccentric stud out of the gearshift lever. Remove the upper gearshift lever.

6. Remove the 4 screws and the lower boot.

7. Remove the 4 bolts and the lower gearshift lever.

8. With the vehicle in NEUTRAL, position it on a hoist.

9. If transmission disassembly is required, remove the drain plug and drain the transmission fluid. Install the drain plug after draining all the fluid. Tighten to 48 Nm (35 ft. lbs.).

10. Remove the right catalytic converter.

11. Remove the 2 starter motor bolts and the starter motor. Using mechanic's wire, position the starter motor aside.

12. Disconnect the wire harness from the transmission and disconnect both of the heated oxygen (HEGO) sensor electrical connectors.

13. Disconnect the reverse lamp switch electrical connector.

14. Disconnect the vehicle speed sensor (VSS) electrical connector.

15. Remove the front stabilizer bar.

16. Remove the 2 nuts and disconnect the fuel lines and the wire harness from the rear of the transmission. Position them aside.

17. Using the special tool, disconnect the clutch hydraulic line.

18. Position a suitable jack under the transmission. Secure the transmission to the jack with a safety strap.

19. Remove the 2 exhaust hanger bolts.

20. Remove the 2 exhaust heat shield bolts.

21. Remove the 4 crossmember bolts.

22. Remove the 2 transmission mount nuts and the crossmember.

23. Lower the transmission enough to gain access to the upper transmission-to-engine bolts.

24. Remove 9 transmission-to-engine bolts.

25. Pull the transmission rearward until the input shift is clear of the pressure plate, then lower the transmission from the vehicle.

26. To install, reverse the removal procedure. Note the following:

a. Before securing the engine to the transmission, connect the hydraulic line to the clutch slave cylinder.

b. Make sure the exhaust system is correctly aligned.

c. Align the index marks when installing the rear driveshaft.

d. Check and, if necessary, fill the

transmission with the specified type and quantity of fluid.

27. Observe the following torques:
- Transmission-to-engine bolts: 60 Nm (44 ft. lbs.).
- Transmission mount nuts: 98 Nm (72 ft. lbs.).
- The 4 crossmember bolts: 90 Nm (66 ft. lbs.).
- The 2 exhaust heat shield bolts: 15 Nm (11 ft. lbs.).
- The 2 starter motor bolts: 33 Nm (24 ft. lbs.)

2005 F-250/350

1. Before servicing the vehicle, refer to the Precautions Section.

All vehicles

2. Remove the four screws and the outer shift lever boot.

3. Remove the upper gearshift lever.

4. Remove the lower shift lever boot.

5. Remove the lower gearshift and shift housing.

Vehicles with a manual shift transfer case

6. Shift the transfer case into 4H.

7. Remove the screws that attach the bezel and boot assembly to the floor.

8. Remove the bolt that attaches the shift lever to the transfer case control lever assembly, and remove the shift lever, and the bezel and boot assembly.

All vehicles

9. Raise and support the vehicle.

10. If the transmission is being disassembled, drain the transmission fluid.

11. Remove the catalytic converter.

12. Remove the starter.

➡️**Index-mark the driveshaft to the transfer case flange. Remove and discard the driveshaft bolts.**

13. Disconnect the rear driveshaft and position it aside.

14. Remove the transfer case, if equipped.

15. Disconnect the fuel lines from the transmission.

16. Remove any power take-off (PTO) equipment, if equipped.

17. Using a transmission jack, support the transmission.

✵✵ CAUTION

Securely strap the jack to the transmission.

18. Remove the right crossmember nuts.

19. Remove the left crossmember bolts.

20. Disconnect the wire harness from the crossmember.

21. Remove the transmission mount nuts and the crossmember.

22. Disconnect the reverse lamp switch electrical connector.

23. Disconnect wiring harness from the transmission.

24. Push the clutch slave cylinder inward, then rotate counterclockwise 45 degrees to remove.

25. Disconnect the transmission cooling tubes.

26. Remove the dust cover bolts. 6.0L equipped vehicles do not have a dust cover.

27. Remove the transmission-to-engine bolts.

28. Move the transmission rearward until the input shaft is clear of the clutch, then lower from the vehicle.

To install:
All vehicles

29. Using the transmission jack, raise and position the transmission to the engine and clutch.

30. Install the transmission-to-engine bolts. Torque to 46 ft. lbs. (63 Nm). Vehicles equipped with diesel engines have six bolts. Those equipped with gasoline engines, have seven bolts.

31. Position the crossmember in the vehicle. Install the transmission mount nuts. Torque to 60 ft. lbs. (81 Nm).

32. Install the left crossmember bolts. Torque to 52 ft. lbs. (70 Nm).

33. Install the right crossmember nuts. Torque to 52 ft. lbs. (70 Nm).

34. Attach the wire harness to the crossmember.

35. Install the dust cover bolts. Torque to 21 ft. lbs. (28 Nm).

➡️**6.0L equipped vehicles do not have a dust cover.**

36. Remove the transmission jack.

37. Connect the transmission cooler tubes. Torque to 20 ft. lbs. (27 Nm).

38. Install the clutch slave cylinder. Rotate the clutch slave cylinder clockwise 45 degrees to lock in position.

39. Install the starter.

40. Install the catalytic converter.

41. Install the transfer case, if equipped. If the transfer case control lever assembly was removed from the transmission, it must be correctly aligned.

42. Connect the driveshaft.

43. Install any power take-off (PTO) equipment, if equipped.

44. Connect the fuel lines to the transmission.

45. Connect the reverse lamp switch electrical connector.

46. Connect the wiring harness to the transmission.

47. Refill the transmission to specification.

48. Lower the vehicle.

✳✳ WARNING

Do not use a silicone sealing compound.

➡ **Do not wait longer than ten minutes to tighten the six bolts due to the rapid cure time of the sealant.**

49. Install the lower gearshift lever and shift housing assembly. Apply gasket maker to the shift housing and the main case. Torque to 17 ft. lbs. (23 Nm).

50. Install the lower shift lever boot.

51. Apply threadlock and sealer to the gearshift lever bolts. Install the upper gearshift lever. Torque to 21 ft. lbs. (28 Nm).

52. Install the screws.

Vehicles with a manual shift transfer case

53. Position the shift lever with the bezel and boot assembly and install the bolt. Torque to 20 ft. lbs. (27 Nm).

54. Position the bezel and boot assembly and install the screws.

55. Verify the shift sequence from 2H to 4L to 2H.

Clutch

REMOVAL & INSTALLATION

2002–04 F-150

1. Before servicing the vehicle, refer to the Precautions Section.

2. Disconnect the negative battery cable.

3. Raise and safely support the vehicle.

4. On vehicles with the externally mounted slave cylinder, remove the clutch slave cylinder. On vehicles with an internally mounted slave cylinder, disengage the quick-disconnect coupling with a spring coupling tool.

5. Remove the transmission.

6. Remove the starter. Remove the flywheel housing attaching bolts and remove the housing.

7. Mark the pressure plate and cover assembly and the flywheel so that they can be reinstalled in the same relative position.

8. Loosen the pressure plate and cover attaching bolts evenly in a staggered sequence a turn at time until the pressure

plate springs are relieved of their tension. Remove the attaching bolts.

9. Remove the pressure plate and cover assembly and the clutch disc from the flywheel.

10. Inspect the flywheel for wear, damage and flatness.

To install:

11. Position the clutch disc on the flywheel so that an aligning tool or spare transmission mainshaft can enter the clutch pilot bearing and align the disc.

12. When reinstalling the original pressure plate and cover assembly, align the assembly and flywheel according to the marks made during removal. Position the pressure plate and cover assembly on the flywheel, align the pressure plate and disc, and install the retaining bolts. Tighten the bolts in an alternating sequence a few turns at a time to 35–46 ft. lbs. (47–63 Nm).

13. Remove the tool used to align the clutch disc.

14. With the clutch fully released, apply a light coat of grease on the sides of the driving lugs.

15. Position the clutch release bearing and the bearing hub on the release lever. Install the release lever on the fulcrum in the flywheel housing. Apply a light coating

Typical clutch assembly with internal slave cylinder—2002–03

PILOT BEARING

CLUTCH DISC

BOLT
10" CLUTCH
(15-20 FT-LB)
11" CLUTCH
(20-29 FT-LB)

CLUTCH COVER

BOLT
(15-20 FT-LB)

SLAVE CYLINDER ASSEMBLY

7924FG29

of grease to the release lever fingers and the fulcrum. Fill the groove of the release bearing hub with grease.

16. If the flywheel housing has been removed, position it against the rear engine cover plate and install the attaching bolts and tighten them to 40–50 ft. lbs. (54–68 Nm).

17. Install the starter motor, if removed.

18. Install the transmission.

19. Install the slave cylinder and bleed the system.

20. Connect the negative battery cable.

2002–05 F-250/350

1. Before servicing the vehicle, refer to the Precautions Section.

2. Remove the transmission.

3. Index-mark the clutch pressure plate and the flywheel, if reinstalling these parts.

4. Remove the bolts, the clutch pressure plate, and the clutch disc.

5. Inspect the transmission input shaft pilot bearing for:

   a. Misalignment and looseness in the crankshaft (gasoline engine) or flywheel (7.3L diesel engine).

   b. Needle rollers for scoring, discoloration, wear, and broken rollers.

   c. Seal for damage and lubricant leakage.

6. Install a new transmission input shaft pilot bearing if any of these conditions are present.

To install:

➡**Sometimes, when removing the transmission, the input shaft will remove a considerable amount of lubricant from the transmission input shaft pilot bearing.**

7. Lubricate the transmission input shaft pilot bearing, as necessary. Use Krytox® High-Temperature Grease.

✳✳ WARNING

When installing the original clutch pressure plate on 5.4L, 6.0L and 6.8L applications, reset the wear indicator before installing the clutch pressure plate on the flywheel.

8. Reset the wear indicator. Using a suitable press and adapter, press downward on the fingers until the adjusting ring moves freely.

9. Rotate the adjusting ring counterclockwise to compress the tension springs. Hold the adjusting ring in this position.

10. Release the pressure on the fingers.

1 Flywheel (single mass)

2 Transmission input shaft pilot bearing

3 Clutch disc

4 Clutch pressure plate

5 Release bearing

6 Release lever

06017-F150-G245

Clutch exploded view—2002–05 F-250/350 exc. 6.0L engine

1 Pilot bearing

2 Flywheel (dual mass)

3 Clutch disc

4 Pressure plate

5 Release bearing

6 Release lever

06017-F150-G246

Clutch exploded view—F-250/350 w/6.0L engine

Rotate the adjusting ring counterclockwise to compress the tension springs—2002–05 F-250/350

06017-F150-G249

The adjusting ring will now stay in the reset position.

11. Position the clutch disc on the flywheel and the clutch alignment tool in the pilot bearing to align the clutch disc. The 5.4L/6.8L engines accept a 1¼ inch input shaft. The 6.0L engines accept a 1¼ inch input shaft with 0.98 inch pilot bearing inner diameter. The 7.3L engines accept a 1⅜ inch input shaft with 0.67 inch pilot bearing inner diameter.

➡**Align the index marks if installing the original clutch pressure plate and flywheel.**

12. Install the clutch pressure plate. Position the clutch pressure plate on the dowels.

➡**The 7.3L diesel engine flywheel has two dowels. The gasoline engine flywheel has three dowels.**

13. Using a clutch alignment tool, align the clutch disc and the pressure plate.

14. Install the bolts and tighten in a star pattern sequence to:
- 5.4L and 6.8L engines: 33 ft. lbs. (45 Nm)
- 6.0L and 7.3L engines: 21 ft. lbs. (28 Nm)

15. Remove the special tool.
16. Install the transmission.
17. Test the system for normal operation.

2005 F-150

1. Before servicing the vehicle, refer to the Precautions Section.
2. Remove the transmission.
3. If the original components are to be installed, index-mark the clutch pressure plate and the flywheel.
4. Remove the clutch pressure plate bolts, clutch pressure plate and the clutch disc.

1 Flywheel
2 Clutch disc
3 Clutch pressure plate
4 Clutch pressure plate bolts
5 Transmission

55 Nm (41 lb-ft)

06017-F150-G247

Clutch exploded view—2005 F-150

14.00 mm (0.55 in) MAX

06017-F150-G248

Clutch tension ring—2005 F-150

To install:

➡**The self-adjusting clutch pressure plate should always be adjusted before installation.**

5. Place the flywheel and the clutch pressure plate in a press. Using a suitable adapter, compress the clutch diaphragm fingers until the adjusting ring moves

6. Adjust the clutch pressure plate as follows:

a. Rotate the adjusting ring counterclockwise until the tension springs are compressed.

b. Hold the adjusting ring, then, release the pressure on the clutch pressure plate fingers.

7. Using an aligning tool, position the clutch disc on the flywheel.

➡**If installing the original clutch pressure plate, use the index marks made during removal.**

8. Position the clutch pressure plate on the dowels, install the clutch pressure plate bolts and remove the aligning tool. Apply Threadlock and Sealer TA-25 meeting Ford specification WSK-M2G351-A5 to the clutch pressure plate bolt threads. Tighten to 55 Nm (41 ft. lbs.).

9. Install the transmission. Before securing the transmission to the engine, install the hydraulic line to the clutch slave cylinder.

10. Operate the vehicle to check clutch operation.

Hydraulic Clutch System

BLEEDING

Externally Mounted Slave Cylinder

1. Before servicing the vehicle, refer to the Precautions Section.

2. Fill the clutch master cylinder reservoir to the full line with clean brake fluid.

3. Raise and support the vehicle.

4. Push the slave cylinder forward to compress the seal and at the same time twist it counterclockwise to unlock it from the transmission.

5. Disconnect the clutch hydraulic tube from the floor pan clip.

6. Position the slave cylinder and the hydraulic tube so that there are no high points that could trap air in the system.

7. Position the slave cylinder push rod downward. Route the hydraulic tube upward as straight as possible toward the master cylinder so that the air can flow freely to the fluid reservoir.

✳✳ WARNING

Do not allow the clutch master cylinder reservoir to run dry.

8. Push the push rod slowly into the slave cylinder until it bottoms out the piston. Hold the push rod in this position for five to ten seconds to allow all trapped air to rise through the system. Very slowly, so that air is not drawn back into the slave cylinder, release the push rod (the spring in the slave cylinder will force the piston outward). Wait

five to ten seconds for the air bubbles to rise. Tap the lines to assist in releasing air bubbles, which may be trapped in the line bends. Repeat this process until all air purged from the system.

9. Verify that the fluid in the reservoir is free of air bubbles.

10. Install the slave cylinder. Compress and twist the slave cylinder clockwise to lock it onto the transmission.

11. Connect the clutch hydraulic tube to the floor pan clip.

12. Lower the vehicle.

13. Slowly depress and release the clutch pedal 20 to 25 times to bleed any air still trapped in the system. Verify that the fluid in the reservoir is free of air bubbles.

14. Verify that the fluid level in the reservoir is correct, and install the cap.

Internally Mounted Slave Cylinder

EXCEPT F-150

➡**With the quick-disconnect coupling, no air should enter the system when the coupling is disengaged. However, if air should somehow enter the system, it must be bled.**

1. Before servicing the vehicle, refer to the Precautions Section.

2. Remove the reservoir cap and diaphragm. Fill the reservoir with DOT 3 brake fluid.

3. Connect a piece of rubber tubing to

the slave cylinder bleed screw. Place the other end in a container.

4. Loosen the bleed screw. Gravity will force fluid from the master cylinder to flow down to the slave cylinder, forcing air out of the bleed screw. When a steady stream—no bubbles—flows out, the system is bled. Close the bleed screw.

➡**Check periodically to be sure the master cylinder reservoir doesn't run dry.**

5. Add fluid to fill the master cylinder reservoir.

6. Fully depress the clutch pedal. Release it as quickly as possible. Pause for 2 seconds. Repeat this procedure 10 times.

7. Check the fluid level. Refill it if necessary. It should be kept full.

8. Repeat Steps 5 and 6, 5 more times.

9. Install the diaphragm and cap.

10. Have an assistant hold the pedal to the floor while you crack the bleed screw—not too far—just far enough to expel any trapped air. Close the bleed screw, then release the pedal.

11. Check, and if necessary, fill the reservoir.

F-150

➡**Be sure to keep the clutch master cylinder reservoir full of brake fluid during the bleeding process to prevent air from entering the clutch master cylinder.**

Bleed screw location for internally mounted slave cylinders

7924FG69

1. Before servicing the vehicle, refer to the Precautions Section.

2. Raise and safely support the front of the vehicle.

➡ **It is necessary to have the assistance of a helper to bleed this system.**

3. Fill the clutch system reservoir with DOT 3 brake fluid.

4. Have your assistant depress the clutch pedal rapidly for 5–10 strokes.

5. Wait 1–3 minutes.

6. Repeat Steps 3 and 4 3 more times.

7. Loosen the bleeder screw on the transmission for the slave cylinder.

8. Have the helper fully depress the clutch pedal and hold it down.

9. Tighten the bleeder screw.

10. The helper should now release the clutch pedal.

11. Apply pressure to the clutch pedal. If the clutch pedal travels more than 6–7 inches (15.3–17.7 cm), repeat the bleeding process.

Transfer Case Assembly

REMOVAL & INSTALLATION

2002

✲✲ CAUTION

The catalytic converter is located beside the transfer case. Due to the extreme high temperatures generated by the converter, be careful when removing the transfer case, or personal injury may result.

1. Before servicing the vehicle, refer to the Precautions Section.

2. Remove or disconnect the following:
- Negative battery cable
- Fluid from the transfer case
- 4WD indicator switch wire connector at the transfer case
- Skid plate from the frame, if equipped
- Front driveshaft from the front output yoke
- Rear driveshaft from the rear output shaft yoke
- Speedometer driven gear from the transfer case rear bearing retainer
- Retaining rings and shift rod from the transfer case shift lever
- Vent hose from the transfer case
- Heat shield from the frame

3. Support the transfer case with a transmission jack.

Transfer case-to-adapter bolt torque sequence—Borg-Warner model 13–45

4. Remove the bolts retaining the transfer case to the transmission adapter.

5. Lower the transfer case from the vehicle.

To install:

6. When installing place a new gasket between the transfer case and the adapter.

7. Raise the transfer case with the transmission jack so the transmission output shaft aligns with the splined transfer case input shaft. Install the bolts retaining the transfer case to the adapter.

8. Remove the transmission jack from the transfer case.

9. Install or connect the following:
- Rear driveshaft to the rear output shaft yoke. Tighten the bolts to 15 ft. lbs. (20 Nm).
- Shift lever to the transfer case and install the retaining nut
- Speedometer driven gear to the transfer case
- 4WD indicator switch wire connector at the transfer case
- Front driveshaft to the front output yoke. Tighten the bolts to 15 ft. lbs. (20 Nm).
- Heat shield to the frame crossmember and the mounting lug on the transfer case. Install and tighten the retaining bolts.
- Skid plate to the frame

10. Install the drain plug. Remove the filler plug and install 6 pts. (2.8L) of Dexron®II transmission.

11. Connect the negative battery cable.

2003 All Models and 2004 F-150 Heritage Editions

WITH 4R70W TRANSMISSION

1. Before servicing the vehicle, refer to the Precautions Section.

2. With the vehicle in **Neutral**, raise and support the vehicle.

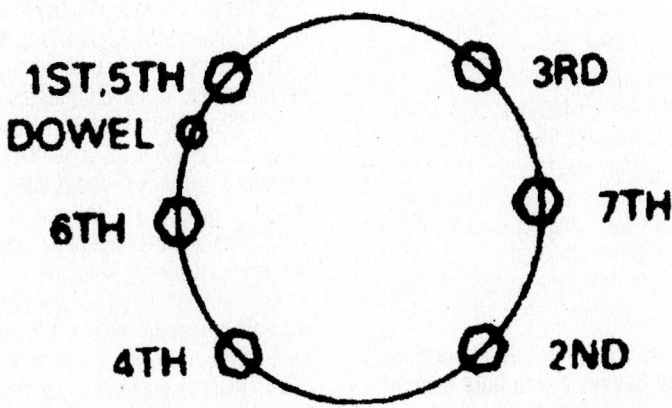

Transfer case-to-adapter bolt torque sequence—Borg-Warner 13–56 electronic and manual shift transfer case used in F-Series

➡**Drain the transfer case if disassembly is required.**

3. Remove the drain plug and drain the transfer case.

4. Remove the skid plate, if equipped.

➡**To maintain initial driveshaft balance, index-mark the rear driveshaft.**

5. Remove the rear driveshaft.

6. Remove the two nuts and the front driveshaft shield.

➡**To maintain initial driveshaft balance, index-mark the front driveshaft.**

7. Remove the front driveshaft.

8. Make preliminary adjustment references. Index-mark the torsion bars to the torsion bar crossmember. Measure and record the torsion bar adjuster bolt length.

9. Remove the bolts.

10. Using the special tools, remove the torsion bar adjustment nuts. Tighten the special tool until the torsion bar adjuster lifts off the adjuster nut.

11. Remove the six bolts, then remove the torsion bar adjusting brackets.

12. Move the torsion bar crossmember rearward. Place a hand under the torsion bar adjusting bracket. The bracket will slide off the torsion bar. Do one side at a time. Position the torsion bar crossmember against the frame support.

13. Pull the torsion bar rearward and separate it from the lower control arm. Remove the right torsion bar only.

14. Remove the torsion bar crossmember.

15. On manual shift vehicles, using the special tool, disconnect the shift rod.

16. On manual shift vehicles, disconnect the 4WD indicator switch electrical connector, if equipped.

17. Disconnect the electric shift motor electrical connector, if equipped.

18. Disconnect the vent hose.

19. Using a suitable jack, support the transfer case. Secure the transfer case to the jack with a safety strap.

20. Remove the six transfer case-to-transmission bolts and separate the transfer case from the transmission.

21. Lower the transfer case from the vehicle.

➡**Carefully clean the gasket surfaces. Nicks and gouges cause fluid leaks.**

22. Remove the transfer case-to-transmission gasket, and clean the mating surfaces.

To install:

23. To install, reverse the removal procedure.

24. Install a new transfer case-to-transmission mounting gasket.

25. Fill the transfer case.

26. Tighten the transfer case-to-transmission bolts evenly in a star pattern to 35 ft. lbs. (47 Nm).

27. Make sure the torsion bar adjuster bolt measurement is equal to the measurement taken during removal.

28. Check the ride height and adjust as necessary.

29. Check the wheel alignment and adjust as necessary.

WITH 4R100 TRANSMISSION

1. Before servicing the vehicle, refer to the Precautions Section.

2. With the vehicle in **Neutral**, raise and support the vehicle.

➡**Drain the transfer case if disassembly is required.**

3. Remove the drain plug and drain the transfer case.

4. Remove the skid plate, if equipped.

➡**To maintain initial driveshaft balance, index-mark the rear driveshaft.**

5. Remove the rear driveshaft.

6. Remove the two nuts and the front driveshaft shield.

➡**To maintain initial driveshaft balance, index-mark the front driveshaft.**

7. Remove the front driveshaft.

8. Make preliminary adjustment references. Index-mark the torsion bars to the torsion bar crossmember. Measure and record the torsion bar adjuster bolt length.

9. Remove the torsion bar adjustment nuts. Tighten the tool until the torsion bar adjuster lifts off the adjuster nut.

10. Remove the six bolts, then remove the torsion bar adjusting brackets.

11. Move the torsion bar crossmember rearward. Place a hand under the torsion bar adjusting bracket. The bracket will slide off the torsion bar. Do one side at a time.

12. Position the torsion bar crossmember against the frame support.

13. Pull the torsion bar rearward and separate it from the lower control arm. Remove the right torsion bar only.

14. Remove the torsion bar crossmember.

15. On manual shift vehicles, using the special tool, disconnect the shift rod.

16. On manual shift vehicles, disconnect the 4WD indicator switch electrical connector, if equipped.

17. Disconnect the electric shift motor electrical connector, if equipped.

18. Disconnect the vent hose.

19. Using a suitable jack, support the transfer case. Secure the transfer case to the jack with a safety strap.

20. Remove the two exhaust heat shield-to-crossmember bolts.

21. Loosen, but do not remove, the transmission mount-to-crossmember nuts.

22. Remove the six crossmember-to-frame nuts and bolts.

23. Remove the transmission mount-to-crossmember nuts and the crossmember.

24. Remove the transmission mount and the exhaust hanger.

25. Position a suitable jack stand under the extension housing.

26. Remove the six transfer case-to-transmission bolts and separate the transfer case from the transmission.

27. Lower the transfer case from the vehicle.

28. Remove the transfer case-to-transmission gasket and clean the mounting surfaces.

To install:

29. To install, reverse the removal procedure.

30. Install a new transfer case-to-transmission mounting gasket.

31. Fill the transfer case.

32. Tighten the transfer case-to-transmission bolts evenly in a star pattern to 35 ft. lbs. (47 Nm).

33. Make sure the torsion bar adjuster bolt measurement is equal to the measurement taken during removal.

34. Check the ride height and adjust as necessary.

35. Check the wheel alignment and adjust as necessary.

2004–05 F-150, Exc. 2004 Heritage Editions

MECHANICAL SHIFT

1. Before servicing the vehicle, refer to the Precautions Section.

➡**To maintain initial driveshaft balance, index-mark the rear driveshaft.**

2. Remove the rear driveshaft.

3. Remove the skid plate, if equipped.

4. Drain the fluid if the transfer case is to be disassembled.

5. Disconnect the transfer case wire harness

6. Remove the front driveshaft shield.

ⓘ ⑦ 47 Nm (35 lb-ft)

Ⓝ ⑨

⑧ ⓘ

⑥

④

⑤ ⓘ

18 Nm (13 lb-ft) ①

②

104 Nm (77 lb-ft) ③

1 Front driveshaft shield nut	4 Front driveshaft	7 Transmission-to-transfer case bolts
2 Front driveshaft shield	5 Transfer case shift linkage	8 Transfer case
3 Front driveshaft bolt	6 Transfer case vent tube	9 Transfer case-to-transmission gasket

67197-EFSE-G146

Mechanical shift transfer case—2004– F-150 exc. 2004 Heritage Editions

7. Remove the front driveshaft.
8. Remove the transfer case shift linkage.
9. Remove the vent tube.
10. Position a suitable transmission jack to the transfer case. Securely strap the transfer case to the jack.
11. Remove the transfer case.
12. Clean the mating surfaces of the transfer case and the transmission.
13. To install, reverse the removal procedure.
14. Tighten the transmission-to-transfer case bolts evenly in a star pattern to 35 ft. lbs. (47 Nm).
15. Fill the transfer case.

ELECTRONIC SHIFT

1. Before servicing the vehicle, refer to the Precautions Section.

➡**To maintain initial driveshaft balance, index-mark the rear driveshaft.**

2. Remove the rear driveshaft.
3. Remove the skid plate, if equipped.
4. Drain the fluid if the transfer case is to be disassembled.
5. Disconnect the transfer case shift motor electrical connector.
6. Remove the front driveshaft shield.
7. Remove the front driveshaft.
8. Remove the vent tube.
9. Position a suitable transmission jack to the transfer case. Securely strap the transfer case to the jack.
10. Remove the transfer case.
11. Remove the gasket.
12. Clean the mating surfaces of the transfer case and the transmission.
13. To install, reverse the removal procedure.

14. Tighten the transfer case-to-transmission bolts evenly in a star pattern to 35 ft. lbs. (47 Nm).
15. Fill the transfer case.

2004–05 F-250 and 350

1. Before servicing the vehicle, refer to the Precautions Section.

➡**Shift the transfer case to 2W HI.**

2. Raise and support the vehicle.
3. Remove the four bolts and the skid plate, if equipped.

➡**Index-mark the driveshaft to maintain driveline balance.**

4. Remove the rear driveshaft. Index-mark the front driveshaft to the transfer case flange.

6 — 47 Nm (35 lb-ft)

N **8**

7

5

2

18 Nm (13 lb-ft) — **1**

104 Nm (77 lb-ft) — **3**

4

1 Front driveshaft shield nut

2 Front driveshaft shield

3 Front driveshaft bolt

4 Front driveshaft

5 Transfer case vent tube

6 Transfer case-to-transmission bolts

7 Transfer case

8 Transfer case-to-transmission gasket

67197F150G151

Electronic shift transfer case—2004–05 F-150 Exc. 2004 Heritage Editions

➥**Support the front driveshaft with wire or a strap.**

5. Remove and discard the four bolts and position the front driveshaft aside.

Manual shift transfer case

6. Remove the manual shift linkage, if equipped.

7. Disconnect the switch electrical connector. Position the wire harness aside.

Electric shift transfer case

8. Disconnect the gear motor encoder assembly electrical connector and the gear motor electrical connector. Position the wire harness aside.

9. Disconnect the transfer case vent hose.

All transfer cases

10. If disassembly is necessary, drain the fluid into a suitable container.

11. Install the plug when finished.

12. Position a suitable high-lift jack under the transfer case and secure it with safety straps.

With automatic transmission

13. Remove the four right crossmember bolts.

14. Remove the three left crossmember bolts.

15. Remove the nuts and the crossmember.

16. Remove the transmission mount.

17. Position a suitable jack stand under the extension housing.

All vehicles

18. Remove the six transfer case-to-transmission bolts.

19. Separate the transfer case from the extension housing. Pull the transfer case rearward, then lower the transfer case from the vehicle.

➥**Carefully clean the gasket surfaces. Nicks and gouges cause fluid leaks.**

20. Remove the transfer case-to-transmission gasket. Clean the mating surfaces, using metal surface cleaner.

To install:

21. Install a new mounting gasket.

※※ CAUTION

Secure the transfer case to the high-lift jack with a safety strap.

22. Raise the transfer case into position.
23. Install the six bolts retaining the transfer case to the extension housing. Torque to 37 ft. lbs. (50 Nm).

Automatic transmission vehicles

24. Remove the jack stand from the extension housing.
25. Install the transmission mount. Torque to 70 ft. lbs. (95 Nm).
26. Position the crossmember and loosely install the two nuts.
27. Install the three left crossmember bolts. Torque to 52 ft. lbs. (70 Nm).
28. Install the right crossmember bolts. Torque to 52 ft. lbs. (70 Nm).

All vehicles

29. Remove the high-lift jack.

Automatic transmission vehicles

30. Tighten the transmission mount-to-crossmember nuts. Torque to 69 ft. lbs. (92 Nm).

All electric shift transfer cases

31. Connect the vent hose.
32. Connect the two gear motor encoder assembly electrical connectors.

All manual shift transfer cases

33. Connect the 3-position mode switch harness connector.
34. Connect the manual shift linkage.

All vehicles

➡ **Align the index-marks when installing the driveshaft.**

35. Connect the front driveshaft to the transfer case and install the four new bolts. Torque to 82 ft. lbs. (111 Nm).
36. Install the rear driveshaft.
37. If equipped, install the skid plate and the four bolts.
38. If drained, fill the transfer case.

Front Driveshaft

REMOVAL & INSTALLATION

2002–03 F-150 and 2004 F-150 Heritage Edition

1. Before servicing the vehicle, refer to the Precautions Section.
2. With the vehicle in NEUTRAL, raise and support the vehicle.
3. Remove the front driveshaft shield.
4. Index-mark the front flange and pinion flange. Repeat for the rear flange and transfer case flange.
5. Remove the four bolts retaining the rear flange to the transfer case.
6. Remove the four bolts retaining the front flange to the pinion flange.
7. Remove the front driveshaft.

※※ WARNING

If new bolts to retain the driveshaft to the axle and the driveshaft to the transfer case are not available, coat the threads of the original bolts with Threadlock and Sealer E0AZ-19554-AA or equivalent meeting Ford specification WSK-M2G351-A5.

8. To install, reverse the removal procedure. Torque the four bolts retaining the rear flange to 77 ft. lbs. (104 Nm). Torque the four bolts retaining the front flange to the pinion flange to 83 ft. lbs. (112 Nm).

2004–05 F150, exc. 2004 Heritage Edition
2006 Mark LT

1. Before servicing the vehicle, refer to the Precautions Section.
2. With the vehicle in NEUTRAL, position it on a hoist.
3. Remove the front driveshaft shield nuts and the front driveshaft shield.
4. Index-mark the front driveshaft flange to the pinion flange to maintain alignment during installation.
5. Index-mark the rear driveshaft flange to the transfer case flange to maintain alignment during installation.
6. Remove the 4 driveshaft flange bolts from the transfer case

1	Front flange	6	Spider	11	Boot groove
2	Universal joint kit	7	Splined stud shaft	12	Large clamp
3	Bearing	8	Rubber boot	13	Small clamp
4	Seal	9	Rear flange	14	Driveshaft
5	Snap ring	10	Driveshaft slip yoke	15	Front yoke

Front driveshaft—2002–03 F-150 and 2004 F-150 Heritage Edition

06017-F150-G250

1 Pinion Flange
2 Driveshaft flange
3 Driveshaft flange-to-pinion flange bolt
4 Driveshaft slip-yoke boot

5 Driveshaft flange-to-transfer case output flange bolt
6 Driveshaft flange
7 Transfer case output flange
8 Driveshaft

06017-F150-G251

Front driveshaft—2004–05 F150, exc. 2004 Heritage Edition

1 Driveshaft
2 Driveshaft centering socket yoke
3 Snap ring
4 Universal joint (U-joint)

5 Driveshaft center yoke
6 Driveshaft slip yoke boot clamp
7 Driveshaft slip yoke boot
8 Driveshaft slip yoke

06017-F150-G252

Front driveshaft—2002–05 F-250/350

7. Remove the 4 driveshaft flange bolts from the pinion flange.

8. Remove the front driveshaft.

❋❋ WARNING

If new bolts are not available, coat the threads of the original bolts with Threadlock and Sealer.

➥**Align the index marks made during removal.**

9. To install, reverse the removal procedure. Install new driveshaft flange bolts. Torque the 4 driveshaft flange bolts at the pinion flange to 103 Nm (76 ft. lbs.). Torque the 4 driveshaft flange bolts at the transfer case to 103 Nm (76 ft. lbs.). Torque the front driveshaft shield nuts to 25 Nm (18 ft. lbs.).

2002–05 F-250/350

1. Before servicing the vehicle, refer to the Precautions Section.

2. Raise and support the vehicle.

3. Index-mark the front driveshaft to the front axle flange and at the transfer case flange.

4. Remove and discard the bolts and discard the U-joint retainers.

5. Remove and discard the four flange bolts.

➥**Wrap electrical tape around the bearing cups to prevent them from falling off the U-joint spider.**

6. Compress and separate the front driveshaft from the front axle, then separate the driveshaft from the transfer case.

To install:

➥**Install the front driveshaft with the index marks aligned.**

➥**An extension must be used to tighten the driveshaft to transfer case bolts. Extensions change the calibration of the torque wrench. Use the torque wrench manual to calculate the correct settings.**

➥**Install new bolts and retainers.**

7. To install, reverse the removal procedure. Torque the flange bolts to 82 ft. lbs. (111 Nm); the U-joint retainer strap bolts to 26 ft. lbs. (35 Nm).

Rear Driveshaft

REMOVAL & INSTALLATION

F-150, E-Series, Blackwood, Mark LT One-Piece Shaft

1. Before servicing the vehicle, refer to the Precautions Section.

➥**For 2005 vehicles with automatic transmissions, when installing a new driveshaft, match the yellow dots on the driveshaft to the yellow dots on the pinion flange and the output shaft to maintain balance during installation.**

2. With the vehicle in NEUTRAL, position it on a hoist.

3. Index-mark the driveshaft flange to the pinion flange to maintain alignment during installation.

4. Index mark the driveshaft to the extension housing to maintain alignment during installation.

5. Remove and discard the 4 driveshaft flange bolts.

1 Driveshaft bolts	5 Universal joint
2 Driveshaft	6 Universal joint snap ring
3 Universal joint snap ring	7 Driveshaft slip yoke
4 Driveshaft flange yoke	8 Universal joint

06017-F150-G253

One-piece rear driveshaft—F-150, E-Series, Blackwood, Mark LT

06017-F150-

Pry here for driveshaft removal

✳✳ WARNING

The driveshaft flange fits tightly on the flange pilot. Never hammer on the driveshaft or any of its components to disconnect the driveshaft flange from the flange pilot. Pry only in the area shown with a suitable tool to disconnect the driveshaft flange from the flange pilot.

6. Using a suitable tool as shown, disconnect the driveshaft flange from the flange pilot and remove the driveshaft.

7. Plug the extension housing to prevent fluid loss.

➡The driveshaft flanges fits tightly on the pinion flange pilots. To make sure that the driveshaft flanges seat squarely on the pinion flange pilots, tighten the driveshaft flange bolts evenly in a cross pattern.

✳✳ WARNING

If new bolts are not available, coat the threads of the original driveshaft flange bolts with Threadlock and Sealer.

➡Align the index marks made during removal.

8. To install, reverse the removal procedure. Install new driveshaft flange bolts. Torque the flange bolts to:
- 2002–03 F-150, Blackwood and 2004 F-150 Heritage Editions: 83 ft. lbs. (112 Nm)
- 2002–05 E-Series: 83 ft. lbs. (112 Nm)
- 2004–05 F-150, exc. 2004 F-150 Heritage Editions: 76 ft. lbs. (103 Nm)
- 2006 Mark LT: 76 ft. lbs. (103 Nm)

F-250/350 One-Piece Shaft w/2-Wheel Drive

1. Before servicing the vehicle, refer to the Precautions Section.

2. With the vehicle in NEUTRAL, position on a hoist.

✳✳ WARNING

Do not allow the output shaft to turn.

3. On vehicles with slip yokes, index-mark the driveshaft and the transmission extension housing.

4. On vehicles with circular flange, index-mark the driveshaft yoke and the transmission flange.

5. On vehicles with split pin yoke
 a. Index-mark the driveshaft and the rear axle pinion flange.
 b. Remove and discard the four bolts and the two U-joint retainers.

6. On vehicles with circular flange:
 a. Index-mark the driveshaft yoke and the rear axle pinion flange.
 b. Remove and discard the four bolts.

➡Bolt direction is vehicle application dependent.

 c. Remove and discard the four bolts at the transmission.

7. On vehicles with split pin flange, disconnect the driveshaft from the rear axle pinion flange. For manual transmission, remove the driveshaft from the vehicle. For automatic transmission, lower the driveshaft and slide it rearward off the transmission output shaft.
 a. Tape the bearing cups to the spider to keep them from falling off the spider. Plug the automatic transmission extension housing to prevent fluid loss.

1 Rear axle pinion flange
2 Universal joint (U-joint)
3 U-joint retainer
4 Driveshaft
5 Driveshaft slip yoke (automatic transmission)
6 Driveshaft yoke (manual transmission)

06017-F150-G255

One-piece rear driveshaft 2wd with split pin yoke—F-250/350

1 Universal joint (U-joint)

2 Driveshaft

3 Driveshaft slip yoke
(automatic transmission)

4 Driveshaft yoke (manual
transmission)

06017-F150-G256

One-piece rear driveshaft 2wd with circular flange—F-250/350

❊❊ WARNING

The driveshaft yoke fits tightly on the rear axle pinion flange pilot. Never hammer on the driveshaft or any of its components to disconnect the yoke from the flange. Pry only in the area shown, with a suitable tool, to disconnect the yoke from the flange.

8. On vehicles with circular flange, using a suitable tool as shown, disconnect the driveshaft yoke from the rear axle pinion flange. For manual transmission, remove the driveshaft from the vehicle. For automatic transmission, lower the driveshaft and slide it rearward off the transmission output shaft.

9. Plug the automatic transmission extension housing to prevent fluid loss.

To install:

10. On vehicles with slip yokes:

a. Lubricate the slip-yoke spline with grease.

b. Remove the plug from the extension housing.

❊❊ WARNING

Align the index marks.

c. Slide the driveshaft slip-yoke on the transmission output shaft.

❊❊ WARNING

Align the index marks.

➡**Bolt direction is vehicle application dependent.**

11. On vehicles with circular flange, position the driveshaft yoke on the transmission flange.

12. On vehicles with split pin flange, remove the tape from the bearing cups.

❊❊ WARNING

Align the index marks.

13. Position the bearing cups in the rear axle pinion flange.

❊❊ WARNING

Align the index marks.

14. On vehicles with circular flange, position the driveshaft yoke on the rear axle pinion flange, and loosely install the four new bolts.

❊❊ WARNING

Install new bolts.

➡**Bolt direction is vehicle application dependent.**

15. Loosely install the four new bolts.

❊❊ WARNING

Install new bolts and new retainers.

16. On vehicles with split pin yoke, install the two new retainers and four new bolts. Torque to 26 ft. lbs. (35 Nm).

❊❊ WARNING

The driveshaft yoke fits tightly on the rear axle pinion flange pilot. To make sure that the yoke seats squarely on the flange, tighten the bolts evenly in a cross pattern as shown.

17. On vehicles with circular flange, tighten the four new bolts to 82 ft. lbs. (11 Nm).

18. Tighten the four transmission flange bolts to 76 ft. lbs. (103 Nm).

19. Lower the vehicle.

F-250/350 One-Piece Shaft w/4-Wheel Drive

1. Before servicing the vehicle, refer to the Precautions Section.

2. With vehicle in NEUTRAL, position on a hoist. Raise and support the vehicle.

3. Index-mark the driveshaft yoke and the transfer case flange.

4. On vehicles with split pin yoke

a. Index-mark the driveshaft and the rear axle pinion flange.

1 Driveshaft
2 Snap ring
3 Universal joint (U-joint)
4 Driveshaft slip yoke

5 Driveshaft slip yoke boot
6 Driveshaft slip yoke boot clamp

06017-F150-G257

One-piece rear driveshaft 4wd with single Cardan joint—F-250/350

1 Driveshaft
2 Driveshaft centering socket yoke
3 Snap ring
4 Universal joint (U-joint)
5 Driveshaft center yoke

6 Spring
7 Driveshaft slip yoke boot clamp
8 Driveshaft slip yoke boot
9 Driveshaft slip yoke

06017-F150-G258

One-piece rear driveshaft 4wd with double Cardan joint—F-250/350

b. Remove and discard the four bolts and the two U-joint retainers.

5. On vehicles with circular flange

a. Index-mark the driveshaft yoke and the rear axle pinion flange.

b. Remove and discard the four bolts.

6. On vehicles with split pin yoke, disconnect the driveshaft from the rear axle pinion flange, and remove the driveshaft from the vehicle. Tape the bearing cups to the spider to keep them from falling off the spider.

7. On vehicles with circular flange, using a suitable tool, disconnect the driveshaft yoke from the rear axle pinion flange, and remove the driveshaft from the vehicle.

✳✳ WARNING

The driveshaft yoke fits tightly on the rear axle pinion flange pilot. Never hammer on the driveshaft or any of its components to disconnect the yoke from the flange. Pry only in the slot on each side of the driveshaft yoke, with a suitable tool, to disconnect the yoke from the flange.

To install:

✳✳ WARNING

Align the index marks.

8. Position the driveshaft yoke on the transfer case flange.

9. On vehicles with split pin yoke:

a. Remove the tape from the bearing cups.

✳✳ WARNING

Align the index marks.

b. Position the bearing cups in the rear axle pinion flange.

10. On vehicles with circular flange, position the driveshaft yoke on the rear axle pinion flange.

✳✳ WARNING

Align the index marks.

✳✳ WARNING

Install new bolts.

➡**Bolt direction is vehicle application dependent.**

11. Loosely install the four new bolts.

12. On vehicles with split pin yoke, install the two new retainers and four new bolts. Torque to 26 ft. lbs. (35 Nm).

✳✳ WARNING

Install new bolts and new retainers.

13. On vehicles with circular flange, install the four new bolts. Torque to 76 ft. lbs. (103 Nm).

1 Transmission extension housing	6 Driveshaft flange
2 Driveshaft slip yoke	7 Pinion flange
3 Driveshaft center bearing bolt	8 Driveshaft
4 Driveshaft center bearing	9 Center bearing shim (if equipped)
5 Driveshaft flange-to-pinion flange bolt	

06017-F150-G259

F-150, Blackwood, Mark LT Two-Piece Shaft

✷✷ WARNING

Install new bolts.

✷✷ WARNING

The driveshaft yoke fits tightly on the rear axle pinion flange pilot. To make sure that the yoke seats squarely on the flange, tighten the bolts evenly in a cross pattern as shown.

14. Tighten the four transfer case flange bolts to 76 ft. lbs. (103 Nm).

15. Lower the vehicle.

F-150, Blackwood, Mark LT Two-Piece Shaft

1. Before servicing the vehicle, refer to the Precautions Section.

➡For 2005 vehicles with automatic transmissions, when installing a new driveshaft, match the yellow dots on the driveshaft to the yellow dots on the pinion flange and the output shaft to maintain balance during installation.

2. With the vehicle in NEUTRAL, position it on a hoist.

3. Index-mark the driveshaft flange to the pinion flange to maintain alignment during installation.

4. Index-mark the driveshaft to the extension housing to maintain alignment during installation.

5. Remove and discard the 4 driveshaft flange bolts.

✷✷ WARNING

The driveshaft flange fits tightly on the flange pilot. Never hammer on the driveshaft or any of its components to disconnect the driveshaft flange from the flange pilot. Pry only

1 Driveshaft slip yoke	9 Driveshaft slip yoke boot
2 Driveshaft circular flange yoke	10 Driveshaft slip yoke
3 Snap ring	11 Driveshaft slip yoke boot clamp (small)
4 Bearing cap	12 Driveshaft slip yoke boot clamp (large)
5 Spider	13 Driveshaft assembly
6 Driveshaft assembly	14 Pinion flange
7 Center bearing and insulator	15 Bearing strap
8 Driveshaft center bearing bracket	16 Bearing strap bolt

Two-piece driveshaft—F-250/350

in the area shown with a suitable tool, to disconnect the driveshaft flange from the flange pilot.

6. Using a suitable tool as shown, disconnect the driveshaft flange from the flange pilot.

7. Remove the driveshaft center bearing bolts and remove the driveshaft.

8. Remove the driveshaft center bearing bolts.

9. Lower the driveshaft and slide the driveshaft slip yoke rearward off of the output shaft.

10. Plug the extension housing to prevent fluid loss.

� WARNING

The driveshaft flanges fits tightly on the pinion flange pilots. To make sure that the driveshaft flanges seat squarely on the pinion flange pilots, tighten the driveshaft flange bolts evenly in a cross pattern.

✱✱ WARNING

If new bolts are not available, coat the threads of the original driveshaft flange bolts with Threadlock and Sealer.

➡**Align the index marks made during removal.**

11. To install, reverse the removal procedure.

12. Install new driveshaft flange bolts. Torque the bolts as follows:
- 2002–03 F-150, Blackwood and 2004 F-150 Heritage Edition: 83 ft. lbs. (112 Nm)
- 2004–05 F-150, exc. 2004 Heritage Editions: 76 ft. lbs. (103 Nm)
- 2006 Mark LT: 76 ft. lbs. (103 Nm)

13. Torque the center bearing bolts as follows:
- 2002–03 F-150, Blackwood and 2004 F-150 Heritage Edition: 46 ft. lbs. (62 Nm)
- 2004–05 F-150, exc. 2004 Heritage Editions: 35 ft. lbs. (48 Nm)
- 2006 Mark LT: 35 ft. lbs. (48 Nm)

F-250/350 w/Two-Piece Shaft

1. Before servicing the vehicle, refer to the Precautions Section.

2. Raise and support the vehicle.

3. Index-mark the driveshaft at the axle flange.

4. Disconnect the driveshaft from the axle flange. Remove and discard the bolts.

5. Remove and discard the U-joint retainers. Using mechanic's wire, support the driveshaft.

➡**Index-mark the driveshaft to transfer case or transmission flange.**

6. For four-wheel drive vehicles, or vehicles equipped with a circular flange, disconnect the driveshaft from the transfer case or transmission. Remove and discard the bolts. Using mechanic's wire, support the driveshaft.

7. For vehicles with a slip yoke at the transmission, index-mark the output shaft to the slip yoke.

8. Remove the center bearing support bolts.

9. Remove the driveshaft assembly.

10. Clean grease deposits, dirt and rust from the following:
- The driveshaft flange yoke areas.
- All driveshaft slip yoke areas.

11. Inspect the following:
- The driveshaft slip yoke boot for rips or holes. Install a new boot if necessary.
- The driveshaft center bearing support for wear or rough action. If roughness or wear is evident, install a new driveshaft center bearing support.
- The center bearing rubber insulator for evidence of hardening, cracking or deterioration. Install a new insulator if necessary.

12. Note the following:
- Inspect the transmission, transfer case and axle seals for damage.

Install new components as necessary.
- Lubricate the slip yoke splines with grease.
- Install the driveshaft so that the index marks made before removal.
- Install new bolts and retainers.
- Bolt direction is vehicle application dependent.

13. To install, reverse the removal procedure. Observe the following torques:
- Flange bolts: 76 ft. lbs. (102 Nm)
- U-joint retainer bolts: 26 ft. lbs. (35 Nm)
- Center bearing bolts: 47 ft. lbs. (63 Nm)

Halfshaft

REMOVAL & INSTALLATION

2002–04 F-150, exc. Heritage Edition

1. Before servicing the vehicle, refer to the Precautions Section.

2. Remove or disconnect the following:
- Front wheels
- Front hub cotter pin, retainer and nut

3. Using a floor hydraulic jack, support the lower suspension arm.
- Upper ball joint cotter pin and castle nut
- Knuckle from the front suspension upper arm

4. Lower the lower suspension arm and steering knuckle slightly to facilitate easier halfshaft removal.

Halfshaft-to-differential mounting bolts (3 of the 6 bolts shown)—2002–03 F-150 and 2004 F-150 Heritage Edition

7924FG70

5. Remove the 2 disc caliper mounting bolts, then lift the front disc caliper off of the front disc brake caliper anchor plate and position aside. Do not allow the caliper to hang by the brake hose; suspend it from the vehicle's frame with strong cord or wire.

6. Remove the 6 front halfshaft-to-differential bolts.

❊❊ WARNING

Use care to avoid damaging the hub seal when removing the front halfshaft.

7. Remove the inboard end of the halfshaft from the differential case or extension axle case. Separate the front halfshaft and joints from the hub, then remove the halfshaft and joints from the vehicle.

To install:

8. Slide the halfshaft outboard end into the hub, making sure that the splines engage.

9. Situate the inboard end of the halfshaft against the front differential flange and install the 6 halfshaft-to-differential bolts. Tighten the halfshaft bolts to 60 ft. lbs. (82 Nm).

10. Install the front disc brake caliper onto the rotor and anchor plate, then install and tighten the 2 caliper mounting bolts to 21–26 ft. lbs. (28–36 Nm).

11. Lift the lower suspension arm and steering knuckle up until the upper ball joint stud is inserted into the steering knuckle. Install the upper ball joint castle nut and tighten to 57–76 ft. lbs. (77–104 Nm). Install a new cotter pin.

12. Install the hub nut onto the halfshaft and tighten the hub nut to 221 ft. lbs. (300 Nm).

13. Install the hub nut retainer and a new cotter pin.

14. Install the front wheels and tighten the lug nuts in a star-shaped sequence to 83–112 ft. lbs. (113–153 Nm).

15. Lower the vehicle to the ground.

2004–05 F-150 exc. 2004 F-150 Heritage Edition
2006 Mark LT

1. Before servicing the vehicle, refer to the Precautions Section.

2. Position the vehicle on a hoist.

3. Remove the front wheel and tire assembly.

4. Remove the dust cap and axle shaft nut.

5. Disconnect the tie rod end.

6. Disconnect the upper ball joint.

7. Remove the integrated wheel end disconnect retaining bolts.

8. Remove the halfshaft flange retaining bolts.

1 Dust cap	
2 Axle halfshaft nut	
3 Tie-rod end nut	
4 Upper ball joint nut	
5 Integrated wheel end disconnect retaining bolts	
6 Front halfshaft flange retaining bolts	
7 Front axle halfshaft	
8 Integrated wheel end disconnect	

Front axle halfshaft assembly—2004 F-150, exc. Heritage Editions. 2005 and 2006 are similar

67197-EFSE-G152

✳✳ WARNING

Do not damage the hub seal.

➡**Allow the steering knuckle to swing outboard while keeping the constant velocity shaft pushed inboard.**

9. Once clearance is available, remove the constant velocity shaft joint outboard end and integrated wheel end disconnect from the steering knuckle hub bearing.

10. Separate the halfshaft assembly from the axle assembly and remove the halfshaft assembly from the vehicle.

11. Carefully remove the integrated wheel end disconnect from the outboard constant velocity joint housing to avoid damage to the vacuum chamber.

12. To install, reverse the removal procedure. Take note of the following:

➡**Maintain a clean work surface.**

13. Compress the integrated wheel end disconnect on the bench to collapse the vacuum chamber.

14. While the integrated wheel end disconnect is collapsed, install a vacuum cap on the vacuum port.

✳ WARNING

Do not install the integrated wheel end disconnect in the knuckle. It must be installed on the outer constant velocity joint housing.

✳✳ WARNING

Do not dislodge the integrated wheel end disconnect seal spring when installing the integrated wheel end on the outer constant velocity joint housing.

15. Install the integrated wheel end disconnect on the outer constant velocity joint housing.

16. Install the front axle halfshaft in the vehicle. Install the halfshaft flange retaining bolts.

✳ WARNING

Verify the spline engagement by checking for spline lash before tightening the retainers of the integrated wheel end disconnect and the front axle halfshaft retaining nut.

17. Install and tighten the integrated wheel end disconnect retaining bolts.

18. Verify the front axle lubricant level is to specifications.

19. Observe the following torques:
- Tie rod stud nut: 111 ft. lbs. (150 Nm)
- Axle halfshaft (wheel end) nut: 20 ft. lbs. (27 Nm)
- Upper ball joint stud nut: 111 ft. lbs. (150 Nm)
- Integrated wheel end disconnect bolts: 2004 models 11 ft. lbs. (15 Nm); 2005 models 108 inch lbs. (12 Nm)
- Halfshaft flange bolts: 60 ft. lbs. (82 Nm)

CV-Joint

OVERHAUL

➡**Before continuing with this procedure, make sure to have available a new CV-joint boot kit, for each CV-joint being serviced. The outer CV-joint cannot be disassembled, only the boot can be replaced.**

Inner CV-Joint And Boot

1. Before servicing the vehicle, refer to the Precautions Section.

2. Remove the halfshaft assembly from the vehicle.

3. Clamp the halfshaft in a vise equipped with jaw caps to prevent damage to machined surfaces. Do not allow the vise jaws to contact the boot or its clamp.

4. Slide 2 inboard clamp protectors off the boot clamps.

5. Carefully remove 2 boot clamps, and slide the boot off the inner CV-joint and housing.

6. Remove the CV-joint retaining ring and remove the housing.

7. Mark the inner race and the ball cage for assembly.

8. Remove 6 cage balls.

9. Remove the snapring.

10. Remove the inner race and ball cage.

11. Clean all parts in suitable parts cleaning solvent and inspect for wear.

To install:

12. Place the boot and the small boot clamp and protector on the shaft.

13. Place the ball cage on the shaft with the tapered end toward the outer CV-joint.

➡**Line up the marks made at disassembly.**

14. Position the inner race on the driveshaft in the position marked on disassembly.

15. Install the snapring.

16. Lubricate and position 6 balls with suitable CV-joint grease.

17. Place the boot protector and boot clamp on the CV-joint housing. Fill the housing with 8.29 ounces of suitable CV-joint grease.

18. Place the housing to the cage and bearings and install the retaining ring.

19. Remove any excess grease from the mating surface and position the boot and clamp.

20. Adjust the CV-joint to boot spacing to 16.43 in. (417.25mm) for 2002–03; 16.93 in. (429.92mm) for 2004–06 models.

21. After adjusting the CV-joint to boot spacing, insert a dull bladed screwdriver blade to relieve built up air pressure in the boot.

22. Use CV Boot Clamp Installer T95P-3514-A or equivalent, to install the boot clamps.

23. Place the clamp protectors over the boot clamps.

24. Install the halfshaft into the vehicle.

25. Road test the vehicle and check for proper operation.

Outer Boot

1. Before servicing the vehicle, refer to the Precautions Section.

2. Remove the inner CV-joint and housing from the halfshaft.

3. Remove the inner boot from the half-shaft.

4. Remove the outer boot clamp protectors and carefully remove the boot clamps.

5. Remove the outer boot from the half-shaft and inspect the grease for contamination.

6. If the grease is contaminated, clean and inspect the joint for wear. Replace the joint and shaft if worn or damaged.

To install:

7. Place the new CV-joint boot on the shaft.

8. Using 5.82 ounces (165 grams) of suitable CV-joint grease, pack the outer CV-joint with grease, then spread the remaining grease inside the boot.

9. Clean the boot mounting surface and position the boot in the joint grooves.

10. Place the clamps in position and use CV Boot Clamp Installer T95P-3514-A or equivalent, to install the boot clamps.

11. Place the clamp protectors over the boot clamps.

12. Install the inner boot on the half-shaft.

13. Install the inner CV-joint and housing on the halfshaft.

14. Install the halfshaft in the vehicle.

15. Road test the vehicle and check for proper operation.

Front Drive Axle Shaft and Seal

REMOVAL & INSTALLATION

See Front Hub and Bearing.

Rear Axle Drive Shaft and Seal

REMOVAL & INSTALLATION

See Rear Hub and Bearing.

Front Differential Pinion Seal

REMOVAL & INSTALLATION

2002–04 F-150

1. Before servicing the vehicle, refer to the Precautions Section.

✳✳ WARNING

This operation disturbs the pinion bearing preload. Carefully reset the preload during assembly.

➡**The front drive axle must be in Neutral before beginning this procedure.**

2. Raise and support the vehicle.
3. Remove the front differential support.
4. Match-mark the front driveshaft to the axle universal joint flange.

✳✳ WARNING

Do not allow the driveshaft to hang unsupported.

5. Disconnect and support the front driveshaft.
6. Using an Nm (inch-pound) torque wrench, measure the torque necessary to maintain pinion rotation. Record the measurement for reference during installation.

7. Match-mark the axle universal joint flange to the pinion stem.
8. Install a holding tool, and loosen, but do not remove the pinion nut.

✳✳ WARNING

Before proceeding, place a drain pan under the differential carrier.

9. With the pinion nut still engaged by a few threads, use a 2-jawed pull to separate the axle universal joint flange from the pinion gear.
10. Remove the nut and the flange.
11. Inspect the axle universal joint flange for burrs, the nut counterbore and the seal contact surface for nicks, and the bearing cone contact area for damage. Install a new flange if necessary.
12. Check the pinion stem splines for burrs. If burrs are evident, remove them with a fine crocus cloth. Working in a rotating motion, wipe the pinion clean.
13. Remove the pinion seal.
To install:
14. Clean the pinion seal bore, and use the Pinion Seal Replacer to install the pinion seal.
15. Install the front axle universal joint flange.
16. Lubricate the axle universal joint flange splines and the pinion seal. Use Motorcraft SAE 75W-90 Premium 4x4 Front Axle Lubricant XY-75W90-TQL or equivalent meeting Ford specification WSP-M2C201-A.

✳✳ WARNING

Never install the axle universal joint flange with a hammer or power tools.

➡**Disregard the scribe marks if installing a new flange.**

17. Align the index-marks and position the axle universal joint flange on the pinion shaft.

67197-EFSE-G235

Install the axle universal joint flange—Ford 8.8 inch, 9.75 inch and 10.25 inch Ring Gear Axle

➡**Rotate the pinion gear occasionally to make sure the pinion bearings seat correctly.**

18. Using the special tool, install the axle universal joint flange.

✳✳ WARNING

Do not loosen the pinion nut to reduce preload under any circumstance. If it is necessary to reduce preload, install a new drive pinion collapsible spacer and pinion nut.

19. Install the special tool, and tighten the pinion nut as follows:
  a. Rotate the pinion gear occasionally to make sure the pinion bearings are seating correctly. Take frequent pinion bearing torque preload readings by rotating the pinion gear with a Nm (inch/pound) torque wrench.
  b. If the preload recorded prior to disassembly is lower than the specification for used bearings, then tighten the pinion nut to the specification. If the preload recorded prior to disassembly is higher than the specification for used bearings, then tighten the pinion nut to the original reading as recorded.
  • Pinion bearing torque preload (used pinion bearing): 0.9–1.5 Nm (8–14 inch lbs.)
  • Pinion bearing torque preload (new pinion bearing): 1.8–3.3 Nm (16–29 inch lbs.)
20. Align the index-marks then attach the front driveshaft.
21. Install the front differential support.
22. Check and, if necessary, fill the differential. Use Motorcraft SAE 75W-90 Premium 4x4 Front Axle Lubricant XY-75W90-TQL, or equivalent, meeting Ford specification WSP-M2C201-A.
23. Lower the vehicle.

67197-EFSE-G234

Flange holding tool—Ford 8.8 inch, 9.75 inch and 10.25 inch Ring Gear Axle

67197-EFSE-G233

Installing the pinion seal—Ford 8.8 inch, 9.75 inch and 10.25 inch Ring Gear Axle

2002–04 F-250 and 350 Dana 50 and 60 Axles

1. Before servicing the vehicle, refer to the Precautions Section.

2. With the vehicle in **Neutral**, raise and support the vehicle.

3. Match-mark the front driveshaft and the front axle flange to maintain driveline balance.

4. Disconnect the front driveshaft from the front axle flange, and position it aside.

5. Rotate the pinion with a Nm (inch lb.) torque wrench. Record the torque necessary to maintain rotation of the pinion through several revolutions.

6. Remove and discard the nut and washer. Use the special tool to prevent the flange from turning while removing the nut.

➡**Match-mark the flange and the pinion shaft.**

Flange holding tool—F-250/350 Dana 50 and 60 Axles

Flange removal—F-250/350 Dana 50 and 60 Axles

Pinion seal removal—F-250/350 Dana 50 and 60 Axles

7. Using the special tool, remove the flange.

8. Using the special tools, remove the pinion seal. Discard the seal.

9. Clean and inspect the following:

   a. The seal mounting surface.

   b. The flange lugs and the flange end that contacts the bearing cone.

   c. Verify that the flange nut counterbore and the seal contact surfaces are smooth and free of nicks.

To install:

10. Using a suitable driver, install the pinion seal. Lightly coat the pinion seal lip with lubricant.

✳✳ WARNING

Never use a metal hammer on the pinion flange or install the flange with power tools. If necessary, use a plastic hammer to tap on a tight fitting flange.

➡**Align the index marks.**

11. Lightly coat the flange splines and seal mating area with lubricant, then install the flange with a new washer and nut.

✳✳ WARNING

Never back off the pinion nut to reduce preload. If preload reduction is necessary, install a new collapsible spacer and pinion nut.

12. Tighten the pinion nut as follows:

   a. Use the special tool to prevent the flange from turning while tightening the nut. Remove the special tool when taking pinion bearing torque preload readings.

   b. Take frequent pinion bearing torque preload readings.

➡**Never back off the pinion nut to reduce preload. If preload reduction is necessary, install a new collapsible spacer and pinion nut.**

   c. For new pinion bearing installation, tighten the pinion nut to a rotating torque of 1.7–3.4 Nm (15–30 inch lbs.) Pinion nut torque range is 217–678 Nm (160–500 ft. lbs.).

   d. For original pinion bearing installation, the reading must be 0.56 Nm (5 inch lbs.) more than the initial reading taken during the disassembly procedure.

13. Connect the front driveshaft to the front axle flange. Torque to 26 ft. lbs. (35 Nm).

14. Check and, if necessary, fill the axle with the specified lubricant.

15. Lower the vehicle.

Rear Differential Pinion Seal

REMOVAL & INSTALLATION

Ford 8.8 inch, 9.75 inch and 10.25 inch Ring Gear Axle

1. Before servicing the vehicle, refer to the Precautions Section.

2. Raise and support the vehicle.

3. Remove the rear wheel and tire assemblies.

✳✳ WARNING

Remove the brake drums or discs to prevent brake drag during drive pinion bearing preload adjustment.

4. Remove the brake discs or drums.

5. Mark the driveshaft flange and pinion flange for correct alignment during installation.

6. Remove the four bolts.

✳✳ WARNING

The driveshaft centering socket yoke fits tightly on the rear axle pinion flange pilot. Never hammer on the driveshaft or any of its components to disconnect the yoke from the flange.

7. Disconnect the driveshaft centering socket yoke from the rear axle pinion flange. Position the driveshaft out of the way.

8. Install a Nm (inch-pound) torque wrench on the pinion nut and record the torque required to maintain rotation of the pinion through several revolutions.

✳✳ WARNING

After removal of the pinion nut, discard it. A new nut must be used for installation.

9. Use the special tool to hold the pinion flange while removing the pinion nut.

10. Mark the driveshaft pinion flange in relation to the drive pinion stem to make sure of correct alignment during installation.

11. Using the special tool, remove the pinion flange.

To install:

12. Lubricate the pinion flange splines. Use SAE 75W-140 High Performance Rear Axle Lubricant F1TZ-19580-B or equivalent meeting Ford specification WSL-M2C192-A.

➥Disregard the scribe marks if a new pinion flange is being installed.

13. Align the pinion flange with the drive pinion shaft.

14. With the drive pinion in place in the rear axle housing, install the pinion flange using the special tool.

15. Position the new pinion nut.

✱✱ WARNING

Under no circumstances is the pinion nut to be backed off to reduce preload. If reduced preload is required, a new collapsible spacer and pinion nut must be installed.

16. Using the special tool to hold the pinion flange, tighten the pinion nut as follows:

a. Rotate the pinion occasionally to make sure the cone and roller bearings are seating correctly.

b. Install a Nm (inch-pound) torque wrench on the pinion nut.

c. Rotating the pinion through several revolutions, take frequent cone and roller bearing torque preload readings until the original recorded preload reading is obtained.

d. If the original recorded preload is lower than specifications, tighten to the appropriate specifications for used bearings. If the preload is higher than specification, tighten the nut to the original reading as recorded.

- Pinion bearing torque preload (used pinion bearing): 0.9–1.5 Nm (8–14 inch lbs.)
- Pinion bearing torque preload (new pinion bearing): 1.8–3.3 Nm (16–29 inch lbs.)

17. Position the rear driveshaft and align the marks on the pinion flange.

✱✱ WARNING

The driveshaft centering socket yoke fits tightly on the rear axle pinion flange pilot. To make sure that the yoke seats squarely on the flange, tighten the bolts evenly in a cross pattern as shown.

18. Install the bolts and tighten to 83 ft. lbs. (112 Nm)

19. Install the brake disc or drum.

20. Install the rear wheel and tire assemblies.

Ford 10.5 inch Ring Gear Axle

1. Before servicing the vehicle, refer to the Precautions Section.

➥The rear wheels and brake calipers must be removed to prevent brake drag during drive pinion bearing preload adjustment.

2. Remove the rear brake calipers.

3. Remove the driveshaft.

4. Install an Nm (inch-pound) torque wrench on the pinion nut, and record the rotational torque required to maintain rotation of the pinion through several revolutions.

➥After removal of the pinion nut, discard it. A new nut must be used for installation.

5. Use a flange holder to hold the pinion flange while removing the pinion nut.

6. Mark the pinion flange in relation to the drive pinion stem to ensure proper alignment during installation.

7. Use a 2-jaw puller to remove the pinion flange.

8. Force up on the metal flange of the rear axle drive pinion seal. Install locking pliers to the seal flange and strike with a hammer until the rear axle drive pinion seal is removed.

To install:

9. Lubricate the new pinion seal. Use Premium Long-Life Grease XG-1-C or

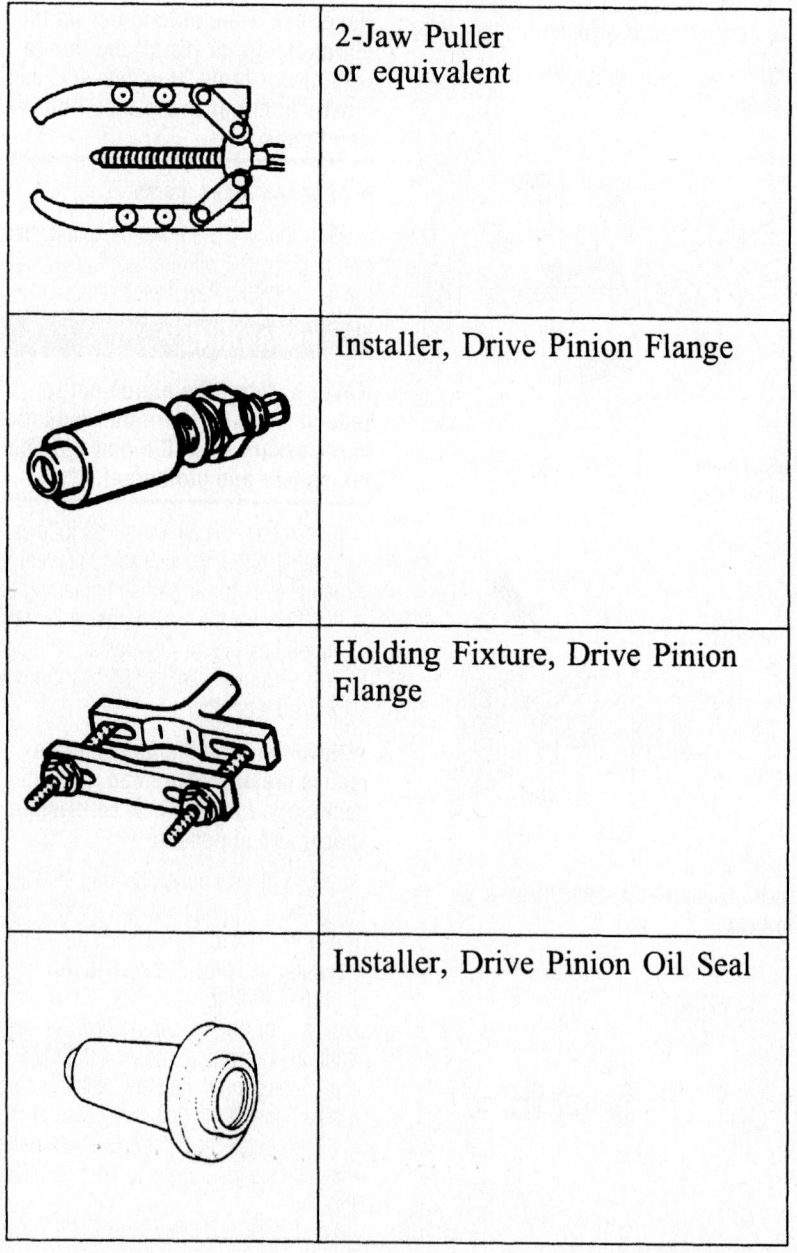

	2-Jaw Puller or equivalent
	Installer, Drive Pinion Flange
	Holding Fixture, Drive Pinion Flange
	Installer, Drive Pinion Oil Seal

Tools necessary for this job—Ford 10.5 inch ring gear axle pinion seal replacement

67197-EFSE-G240

Pinion seal removal—Ford 10.5 inch ring gear axle

Pinion seal installation—Ford 10.5 inch ring gear axle

equivalent meeting Ford specification ESA-M1C75-B.

➡**If the rear axle drive pinion seal becomes misaligned during installation, remove the rear axle drive pinion seal and replace it with a new seal.**

10. Use the Pinion Seal Replacer to install the rear axle drive pinion seal.

11. Lubricate the pinion flange splines. Use SAE 75W-140 Synthetic Rear Axle Lubricant F1TZ-19580-B or equivalent meeting Ford specification WSL-M2C192-A.

➡**Disregard the scribe marks if a new pinion flange is being installed.**

12. Align the pinion flange with the drive pinion shaft.

13. With the pinion flange in place in the rear axle housing, install the pinion flange using the Companion Flange Replacer.

14. Position the new pinion nut.

➡**Under no circumstances is the pinion nut to be backed off to reduce preload. If reduced preload is required, a new collapsible spacer and pinion nut must be installed.**

15. Use the Flange Holder to hold the pinion flange while tightening the pinion nut.

a. Tighten the pinion nut, rotating the pinion occasionally to make sure the cone and roller bearings are seating properly. Take frequent cone and roller bearing torque preload readings until the originally recorded preload reading is obtained by rotating the pinion with an Nm (inch-pound) torque wrench.

b. If the original recorded preload is lower than specifications, tighten to the appropriate specification for used bearings. If the preload is higher than specification, tighten the nut to the original reading as recorded.

c. Pinion bearing preload (used pinion bearing): 0.9–1.5Nm (8–14 inch lbs)

d. Pinion bearing preload (new pinion bearing): 1.8–3.3Nm (16–29 inch lbs.)

e. Initial minimum breakaway torque (Traction-Lok®): 27Nm (20 ft. lbs)

16. Install the driveshaft.

17. Install the brake calipers.

Dana 80 Rear Axle

1. Before servicing the vehicle, refer to the Precautions Section.

2. Raise the vehicle on a hoist or raise the rear end of the vehicle with a jack.

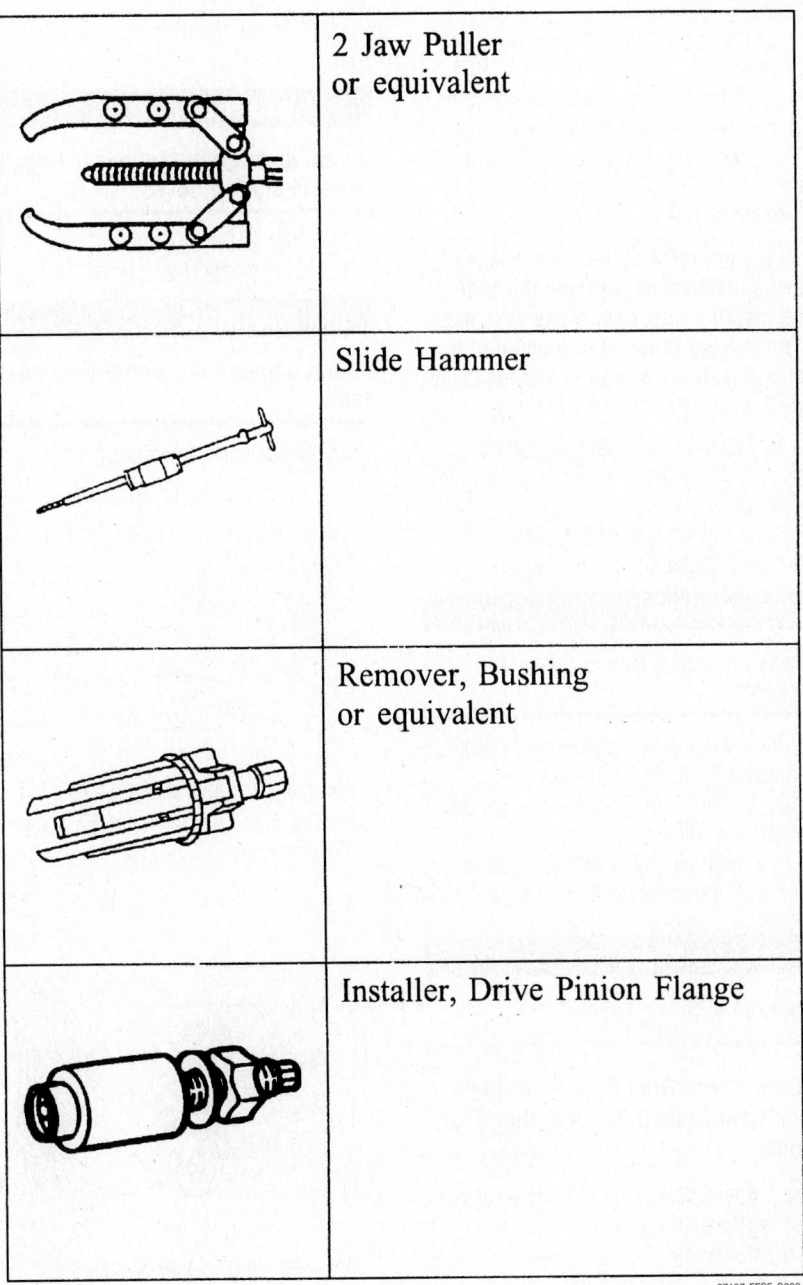

	2 Jaw Puller or equivalent
	Slide Hammer
	Remover, Bushing or equivalent
	Installer, Drive Pinion Flange

Tools necessary for this job—Dana 80 axle

Install safety stands under the frame rails and lower the jack or hoist far enough to allow the rear axle to drop into the rebound position for working clearance.

➡️To maintain driveline balance, mark the driveshaft components so they can be reinstalled in their original positions.

3. Disconnect the driveshaft at the rear axle, and position it aside.

➡️Index-mark the flange to the pinion shaft.

4. While using a flange holding tool to prevent the flange or yoke from turning, remove the pinion nut.

5. Using a 2-jaw puller, remove the flange or yoke.

6. Using a bushing remover and slide hammer, remove the pinion seal.

7. Clean the rear axle pinion seal seat.

To install:

➡️If the pinion seal becomes cocked during installation, remove the seal and install a new one. Make sure the garter spring remains in place during assembly. If the spring is dislodged, a new pinion seal must be installed.

8. Install the seal using a suitable driver. Coat the pinion seal rubber lips with lubricant.

9. Using the special tool, 205-285, or equivalent, install the pinion flange.

✳️✳️ WARNING

Always install a new washer and locknut.

10. Install the new washer and locknut. Torque to 470 ft. lbs. (637 Nm).

11. Install the driveshaft at the rear axle. Observe the following torques:
- Split pin yoke: 26 ft. lbs. (35 Nm)
- Circular flange: 82 ft. lbs. (111 Nm)

Rear Axle Bearings

REMOVAL & INSTALLATION

Ford 8.8 Inch Ring Gear, 9.75 Inch Ring Gear And 10.25 Inch Ring Gear Axles

1. Before servicing the vehicle, refer to the Precautions Section.

All Vehicles

2. Raise and support the vehicle.

3. Remove the wheel and tire assembly.

➡️Empty the lubricant into a clean container for reuse.

4. Remove the 10 differential housing cover bolts and drain the lubricant from the rear axle housing.

5. Remove the differential housing cover.

Vehicles with drum brakes

6. Remove the rear brake drums.

Vehicles with disc brakes

7. Remove the rear disc brake caliper. Wire the rear disc brake caliper aside.

8. Remove the rear brake disc.

All vehicles

9. Remove and discard the differential pinion shaft lock bolt.

10. Remove the differential pinion shaft.

✳️✳️ WARNING

Do not damage the rubber O-rings in the axle shaft grooves.

11. Push in the axle shafts.

12. Remove the U-washers.

✳️✳️ WARNING

Do not damage the wheel bearing oil seal.

13. Remove the axle shaft.

Lock bolt (1); pinion shaft (2)—Ford 8.8 inch, 9.75 inch ring gear, and 10.25 inch ring gear axles

Axle shaft (1); U-washer (2)—Ford 8.8 inch, 9.75 inch ring gear and 10.25 inch ring gear axles

Rear wheel bearing removal—Ford 8.8 inch, 9.75 inch ring gear and 10.25 inch ring gear axles

Rear wheel bearing installation—Ford 8.8 inch, 9.75 inch ring gear and 10.25 inch ring gear axles

Oil seal installation—Ford 8.8 inch, 9.75 inch ring gear and 10.25 inch ring gear axles

➡️If the wheel bearing oil seal is leaking, the axle housing vent may be plugged with foreign material.

➡️If only a new seal needs to be installed, use care to avoid damaging the seal bore.

14. Using a suitable seal remover, remove the axle shaft oil seal. Discard the oil seal.

15. Inspect the rear wheel bearing and axle shaft for wear or damage.

16. If necessary, using the special tools, remove the rear wheel bearing.

To install:

17. Lubricate the new rear wheel bearing with rear axle lubricant.

18. Using the special tools, install the rear wheel bearing.

19. Lubricate the lip of the new wheel bearing oil seal with grease.

20. Using the special tools, install the wheel bearing oil seal.
   All vehicles

✳✳ WARNING

Do not damage the wheel bearing oil seal.

21. Install the axle shaft.

✳✳ WARNING

Do not damage the rubber O-rings in the U-washer grooves.

22. Position the two U-washers on the button end of the axle shafts.
23. Pull the axle shafts outward.

➡**If a new pinion shaft lock bolt is unavailable coat the threads with Threadlock prior to installation.**

24. Align the hole in the differential pinion shaft with the case lock bolt hole.
25. Install a new differential pinion shaft lock bolt. Torque to 15–30 ft. lbs. (20–40 Nm).
   Vehicles with drum brakes
26. Install the rear brake drums.
   Vehicles with disc brakes
27. Install the rear brake disc.
28. Install the rear disc brake caliper.

➡**Clean the gasket mating surface of the rear axle and the differential housing cover.**

29. Apply a new continuous bead of sealant to the differential housing cover.

➡**The differential housing cover must be installed within 15 minutes of application of the silicone, or new sealant must be applied. If possible, allow one hour before filling with lubricant to make sure the silicone sealant has correctly cured.**

30. Install the differential housing cover.
31. Install the 10 differential housing cover bolts. Torque to 33 ft. lbs. (45 Nm).
32. Fill the rear axle housing with 2.37 liters (5 pints) with the specified lubricant.

✳✳ CAUTION

Always remove any corrosion, dirt or foreign material present on the mounting surfaces of the wheel or the surface of the wheel hub or brake drum or disc that contacts the wheel. Installing wheels without correct metal-to-metal contact at the wheel mounting surfaces can cause the lug nut to loosen and the wheel to come off while the vehicle is in motion, causing loss of control.

33. Clean the wheel hub and mounting surfaces.
34. Install the tire and wheel assembly.

Dana 60 And 70 Semi-Floating Axles

1. Before servicing the vehicle, refer to the Precautions Section.
2. Raise the vehicle on a hoist or raise the rear end of the vehicle with a jack. Install safety stands under the frame rails and lower the jack or hoist enough to allow the rear axle to drop into the rebound position for working clearance.
3. Remove the rear wheel and tire assembly.
4. Remove the brake disc.
5. Remove the differential housing cover and drain the lubricant. Clean the gasket material from the differential housing cover and the differential housing.

✳✳ WARNING

The differential assembly is equipped with either a Loctite® coated differential pinion shaft lock screw or a differential pinion shaft lock screw with torque prevailing threads. The Loctite® treated differential pinion shaft lock screw has a ⁵⁄₃₂ inch hexagram socket head. Never, under any circumstance, reuse a Loctite® coated screw after removing it. Always discard the screw and install a new one. The torque prevailing differential pinion shaft lock screw has a 12-point drive head. This type of screw is reusable for no more than four installations. When in doubt about the number of installations of a torque prevailing differential pinion shaft lock screw, discard it and install a new screw.

67197-EFSE-G219

Lock screw—Dana 60 and 70 semi-floating axle

67197-EFSE-G220

Pinion shaft—Dana 60 and 70 semi-floating axle

67197-EFSE-G221

U-washer—Dana 60 and 70 semi-floating axle

6. Remove the lock screw.
7. Remove the differential pinion shaft. The pinion shaft is a slip-fit design and is removable by hand.
8. Push the flanged end of the axle shaft toward the center of the axle and remove the U-washer.

✳✳ WARNING

Do not damage the wheel bearing oil seal when removing the axle shaft.

✳✳ WARNING

Do not rotate the differential side gears when removing the axle shaft. Rotating the differential side gears causes the differential pinion gears and differential pinion thrust washers to turn to the differential case opening and fall out of the differential case.

9. Remove the axle shaft.

✳✳ WARNING

Install the differential pinion shaft and the old lock screw (finger-tight) to prevent the differential side gears and differential pinion gears from rotating and falling out of the differential case.

10. Install the differential pinion shaft and the old lock screw in the differential case.

✳✳ WARNING

It is not necessary to remove the rear wheel bearing if only installing a new wheel bearing oil seal. To remove only the wheel bearing oil seal, pry it from the axle tube. Do not damage the seal seating surface. If removing the rear wheel bearing and the oil seal, proceed to the following step.

11. Remove the wheel bearing oil seal from the axle tube. Discard the seal.

✳✳ CAUTION

Make sure protective eyewear is in place. Failure to follow these instructions may result in personal injury.

12. Using the special tools, remove the rear wheel bearing and the wheel bearing oil seal as a unit. Discard the seal and the bearing.

✳✳ WARNING

The bearing and seal seating surfaces must be free from burrs, nicks, and spalling.

Rear wheel bearing removal—Dana 60 and 70 semi-floating axle

Rear wheel bearing installation—Dana 60 and 70 semi-floating axle

13. Clean and inspect the bore in the axle tube. Wipe the bore in the axle tube with emery cloth to smooth the surface. Clean the bore with a standard metal-cleaning solvent. Wipe the bore with a soft, lint free cloth to remove any foreign material.

To install:

14. If removed, lubricate the new rear wheel bearing with the specified axle lubricant.

✳✳ WARNING

Install the rear wheel bearing with the identification numbers facing outward.

✳✳ WARNING

Do not cock the rear wheel bearing in the axle tube.

15. Using the special tools, install the rear wheel bearing.

✳✳ WARNING

Do not cock the wheel bearing oil seal in the axle tube.

➡ The following step shows an alternate method for installing the wheel bearing oil seal.

Oil seal installation—Dana 60 and 70 semi-floating axle

Alternate oil seal installation—Dana 60 and 70 semi-floating axle

16. Using the special tools, install the new wheel bearing oil seal.

✳✳ WARNING

Do not cock the wheel bearing oil seal in the axle tube bore.

➡ This step is an alternate method for installing the wheel bearing oil seal. Carry out this step if wheel bearing oil seal installation was not done in the previous step.

17. Using the special tools, install the new wheel bearing oil seal.

18. Lubricate the cavity between the wheel bearing oil seal lips and the rear wheel bearing with grease.

19. Remove the lock screw and the differential pinion shaft.

✳✳ WARNING

Do not damage the wheel bearing oil seal when installing the axle shaft.

20. Push the axle shaft into the axle tube and engage the differential side gear with the shaft splines.

21. Push the axle shaft toward the center of the axle and install the U-washer. Pull the axle shaft outward until the U-washer locks into the differential side gear.

22. Align the differential pinion shaft lock screw hole with the hole in the differential case. Correctly position the differential pinion thrust washers. Install the differential pinion shaft.

➡ The threads in the differential case and on the lock screw must be free of dirt and oil.

23. Install the new lock screw. Torque to 20 ft. lbs. (27 Nm).

✳✳ WARNING

Clean the mounting surface on the differential housing cover and the differential housing with a suitable solvent to remove all traces of oil film.

✳✳ WARNING

The differential housing cover uses silicone rubber sealant material as a gasket. Install the differential housing cover within 15 minutes of applying the silicone or it will be necessary to apply new sealant.

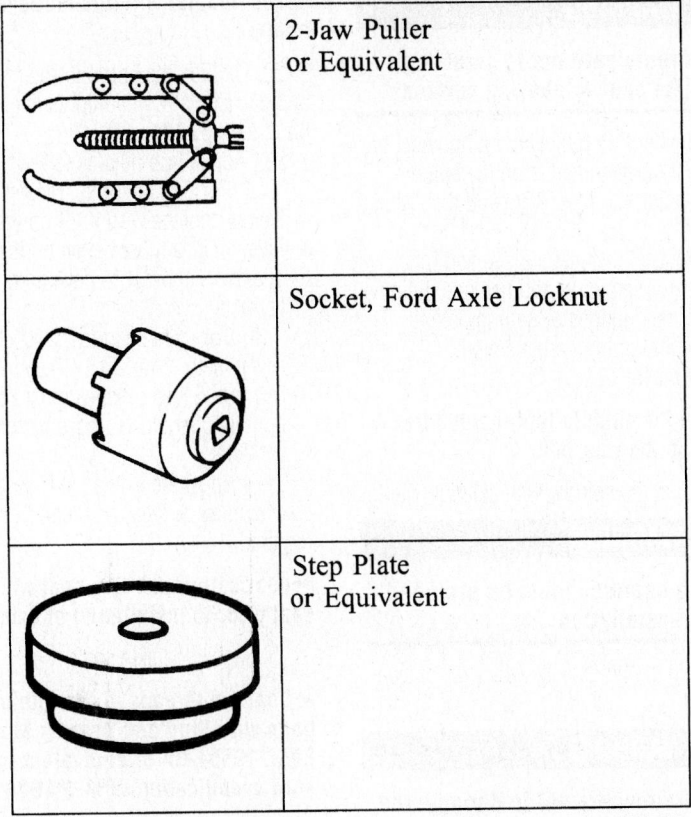

Tools needed for the following hub removal on the Ford Full-Floating axle

Axle shaft removal—Ford full-floating axle

29. Install the wheel and tire assembly.
30. Lower the vehicle.

Ford Full-Floating Axle

1. Before servicing the vehicle, refer to the Precautions Section.
2. Set the parking brake.
3. Loosen the retaining bolts.
4. Raise the vehicle to the desired working height, keeping the axle parallel with the floor.
5. Release the parking brake.
6. Remove the wheel(s).
7. Remove the brake caliper and rotor on the single rear wheel axle.
8. Remove the retaining bolts and axle shaft.

24. Apply a continuous bead of silicone sealant to the differential housing cover mounting surface.

25. Assemble two bolts into the differential housing cover at the eight o'clock and two o'clock positions. Using the two bolts as a guide, position the differential housing cover on the differential housing.

26. Install the remaining bolts. Tighten the bolts alternately and evenly. Tighten Grade 5 bolts to 47 Nm (35 ft. lbs. Tighten Grade 8 bolts to 61 Nm (45 ft. lbs.

✳✳ WARNING

Allow 1 hour cure time before filling the axle with the correct amount of specified lubricant and operating the vehicle.

✳✳ WARNING

For limited-slip axles, first fill the axle with 7 ounces of friction modifier.

27. Fill the differential housing with 3.0L (6.3 pints) of the specified rear axle lubricant. For additional information, refer to Specifications.

28. Install the brake disc.

Install the Ford Axle Locknut Socket so that the drive tangs of the tool engage the four slots in the hub nut

67197-EFSE-G213

Install the Step Plate—Ford full-floating axle

➡️**The hub nuts are right-hand thread (right hub) and left-hand thread (left hub). Each hub nut is stamped RH or LH.**

9. Install the Ford Axle Locknut Socket so that the drive tangs of the tool engage the four slots in the hub nut.

> ⚙️ **WARNING**
>
> **Discard the hub nut if the hub nut comes apart during removal.**

> ⚙️ **WARNING**
>
> **Under no circumstances are power tools to be used when performing these operations.**

➡️**The hub nut will ratchet during this operation.**

10. Remove the hub nut (counterclockwise for right-hand thread; clockwise for left-hand thread).
11. Install the Step Plate.
12. Install the 2-Jaw Puller and loosen the rear hub to the point of removal.

> ❋❋ **WARNING**
>
> **Do not drop the outer hub bearing when removing the hub.**

13. Remove the rear hub assembly.

> ⚙️ **WARNING**
>
> **Install a new hub seal each time the hub assembly is removed.**

➡️**The inner bearing is located behind the hub seal.**

14. Pack each bearing and replace the hub seals.

> ❋❋ **WARNING**
>
> **Use extreme care not to scratch or gouge the seal or bearing surfaces.**

15. If after hub removal, the hub seal or seal inner sleeve remains on the spindle, remove using the Step Plate and the 2-Jaw Puller.
16. Inspect the seal surface and inner shoulder for scratches and damage. Remove all scratches, gouges or galling damage with No. 600 or finer crocus cloth.

To install:

➡️**Clean the spindle thoroughly after removing the rear hub.**

17. Coat the spindle with axle lubricant.

> ❋❋ **WARNING**
>
> **The hub bearings must be prelubed prior to installation.**

18. Fill the hub cavity with 1 oz. of axle lubricant.

> ❋❋ **WARNING**
>
> **Use extreme care not to damage the hub seal by allowing it to contact the spindle during installation.**

➡️**Coat the spindle and hub seal inside diameter with axle lubricant.**

➡️**Installing the rear hub in this manner causes the outer bearing to act as a pilot making the installation easier.**

19. Push the rear hub and outer bearing onto the spindle as an assembly. Hold the outer bearing seated and use the bearing as a pilot.

> ❋❋ **WARNING**
>
> **Install a new hub nut if the hub nut comes apart during installation.**

➡️**Make sure the hub nut tab is located in the keyway prior to thread engagement.**

20. Install the hub nut on the spindle. Turn the hub nut clockwise for right-hand thread or counterclockwise for left-hand thread.
21. Position the Ford Axle Locknut Socket on the hub nut.

> ❋❋ **WARNING**
>
> **Under no circumstances are power tools to be used when performing these operations.**

➡️**The hub nut will ratchet as torque is applied.**

22. Tighten the hub nut, rotating the rear hub occasionally while tightening. Torque to 60 ft. lbs. (81 Nm).
23. Adjust hub nuts as follows:
   a. For new bearings, ratchet back five teeth or notches (⅛ turn) on the hub nut. Five notches must be felt during this operation in order to have performed it correctly.
   b. For used bearings, ratchet back seven teeth or notches (⅙ turn) on the hub nut. Seven notches must be felt during this operation to have performed it correctly.
24. Inspect the axle shaft O-ring seal for cracks, nicks or wear and replace it if required.

➡️**Lubricate the O-ring seal with lubricant prior to installation of axle shaft.**

25. Install the axle shaft.

➡️**Coat the threads of the retaining bolts with Stud and Bearing Mount E0AZ-19554-BA or equivalent meeting Ford specification WSK-M2G349-A1.**

26. Install and tighten the retaining bolts until they seat.

> ❋❋ **WARNING**
>
> **Remember, the last step of this procedure is to tighten the axle shaft bolts to specification, after the wheel lug nuts have been tightened.**

27. Install the brake rotor and caliper on the single rear wheel axles.
28. Install the wheels and tires but do not tighten the lug nuts to specification at this time.
29. Check the axle lubricant level.
30. Lower the vehicle.
31. Tighten the wheel lug nuts.

67197-EFSE-G243

Remove the axle shaft— Dana Axle Full-Floating Axle

205-282

67197-EFSE-G244

Using special tool 205-282, or equivalent, remove the hub nut— Dana Axle Full-Floating Axle

32. Tighten the axle shaft retaining bolts. Torque to 80 ft. lbs. (109 Nm).

Dana Axle Full-Floating Axle

1. Before servicing the vehicle, refer to the Precautions Section.
   All vehicles
2. Remove the tire and wheel assembly.
3. Remove the anchor plate.
4. Remove the axle shaft.
   Dana 70

➡ **Make sure that the drive tangs on the special tool engage the four slots of the hub nut.**

5. Using special tool 205-282, or equivalent, remove the hub nut.
   All vehicles
6. Remove the outer rear wheel bearing.
7. Remove the rear hub and brake disc assembly.
8. Remove the bolts and separate the rear hub from the rear brake disc.

9. Inspect the rear hub for the following:

   - Cracks and damage around the bolt holes.
   - Oversized holes.

   To install:

> ❋❋ **WARNING**
>
> **Install a new rear hub seal after removing the rear hub from the axle. A damaged or worn seal can permit bearing lubricant to reach the brake linings, resulting in ineffective brake operation. Failure to follow these instructions may result in personal injury.**

> ❋❋ **WARNING**
>
> **Clean and remove any dirt or foreign material in the rear hub bolt holes.**

10. Install a new rear hub seal.
11. Position the rear brake disc on the

rear hub and install the bolts. Torque to 66–88 ft. lbs. (89–119 Nm).

> ❋❋ **WARNING**
>
> **Thoroughly clean the spindle. Wrap the spindle threads with electrician's tape to prevent damage while installing the rear hub and brake disc assembly.**

> ❋❋ **WARNING**
>
> **Lightly coat the spindle and pack each rear wheel bearing with Premium Long-Life Grease XG-1-C or equivalent meeting Ford specification ESA-M1C75-B.**

12. Prepare the spindle for rear hub installation.
13. Slide the rear hub and brake disc assembly over the axle housing spindle. Remove the electrician's tape.
14. Install the outer rear wheel bearing.
15. Start the hub nut making sure that the tab aligns correctly in the keyway prior to thread engagement.

➡ **Apply inward pressure to the socket to separate the ratcheting components of the hub nut.**

16. To adjust the bearings, tighten the nut to 70 ft. lbs. (95 Nm).
17. Back off the nut 90 degrees.
18. Tighten the nut to 18 ft. lbs. (24 Nm). To verify that there is no side-to-side end play, attach a magnetically mounted dial indicator to the spindle end and place the dial indicator tip on the outboard surface of the hub. Check for side-to-side end play. Final bearing adjustment has zero end play. The maximum torque to rotate the hub is 2.3 Nm (20 inch lbs.) when end play is zero.
19. Install the axle shaft.
20. Install the anchor plate.
21. Install the tire and wheel assembly.

STEERING

Air Bag

✳✳ CAUTION

Some vehicles are equipped with an air bag system. The system must be deactivated before performing service on or around system components, steering column, instrument panel components, wiring and sensors. Failure to follow safety and disabling procedures could result in accidental air bag deployment, possible personal injury and unnecessary system repairs.

PRECAUTIONS

Several precautions must be observed when handling the inflator module to avoid accidental deployment and possible personal injury:

• Always wear safety glasses when repairing an air bag supplemental restraint system (SRS) vehicle and when handling an air bag module. This will reduce the risk of injury in the event of an accidental deployment.

• Carry a live air bag module with the air bag and trim cover pointed away from your body. This will reduce the risk of injury in the event of an accidental deployment.

• Do not set a live air bag module down with the trim cover face down. This will reduce the risk of injury in the event of an accidental deployment.

• After deployment, the air bag surface can contain deposits of sodium hydroxide, a product of the gas generant combustion that is irritating to the skin. Wash your hands with soap and water afterwards.

• Never probe the connectors on the air bag module. Doing so can result in air bag deployment, which can result in personal injury.

• Restraint system diagnostic tools are for service only. Tools must be removed prior to operating the vehicle over the road. Failure to remove restraint system diagnostic tools could result in injury and possible violation of vehicle safety standards.

• To reduce the risk of personal injury, do not use any memory saver device.

DEACTIVATION

1. Before servicing the vehicle, refer to the Precautions Section.
2. Turn all vehicle accessories OFF.

3. Turn the ignition switch to OFF.
4. At the central junction box (CJB), located below the instrument panel, remove the trim panel, the cover, and the restraints control module (RCM) fuse from the CJB. See your Owner's Manual for fuse identification.
5. Turn the ignition ON and visually monitor the air bag indicator for at least 30 seconds. The air bag indicator will remain lit continuously (no flashing) if the correct RCM fuse has been removed. If the air bag indicator does not remain lit continuously, remove the correct RCM fuse before proceeding.
6. Turn the ignition OFF.

✳✳ CAUTION

To avoid accidental deployment and possible personal injury, the backup power supply must be depleted before repairing or replacing any front or side air bag supplemental restraint system (SRS) components and before servicing, replacing, adjusting or striking components near the front or side air bag sensors, such as doors, instrument panel, console, door latches, strikers, seats and hood latches.

7. To deplete the backup power supply energy, disconnect the battery ground cable and wait at least 1 minute. Be sure to disconnect auxiliary batteries and power supplies (if equipped).
8. Disconnect the battery ground cable and wait at least 1 minute.

REACTIVATION

1. Before servicing the vehicle, refer to the Precautions Section.
2. Turn the ignition switch from OFF to ON.
3. Install the RCM fuse to the CJB and install the cover and trim panel.

✳✳ CAUTION

Be sure that nobody is in the vehicle and that there is nothing blocking or set in front of any air bag module when the battery ground cable is connected.

4. Connect the battery ground cable.
5. Prove out the supplemental restraint system (SRS) as follows:
6. Turn the ignition key from ON to OFF. Wait 10 seconds, then turn the key back to

ON and visually monitor the air bag indicator with the air bag modules installed. The air bag indicator will light continuously for approximately 6 seconds and then turn off. If an air bag supplemental restraint system (SRS) fault is present, the air bag indicator will either:

• fail to light.
• remain lit continuously.
• flash.

7. The flashing might not occur until approximately 30 seconds after the ignition switch has been turned from the OFF to the ON position. This is the time required for the restraints control module (RCM) to complete the testing of the SRS. If the air bag indicator is inoperative and a SRS fault exists, a chime will sound in a pattern of 5 sets of 5 beeps. If this occurs, the air bag indicator and any SRS fault discovered must be diagnosed and repaired.

8. Clear all continuous DTCs from the restraints control module using a diagnostic tool.

Steering Gear

REMOVAL & INSTALLATION

2002–03 F-Series and 2004 Heritage Editions

1. Before servicing the vehicle, refer to the Precautions Section.
2. Remove or disconnect the following:

• Skid plate
• Lower radiator air deflector
• Pitman arm cotter pin and castellated nut from the drag link
• Pitman arm from the drag link
• Dust cover from the steering shaft valve housing
• Intermediate shaft pinch bolt and slide the shaft off the steering gear input shaft.
• Power steering pressure hoses at the steering gear
• Steering gear-to-frame rail retaining bolts
• Steering gear

3. If replacing or servicing the steering gear, match mark the sector shaft arm to the sector shaft and remove the steering gear sector shaft arm retaining nut and lockwasher.

4. Remove the sector shaft arm.
To install:
5. If removed, install the steering gear sector shaft arm to the sector shaft aligning

the match marks made during removal. Install the retaining nut and lockwasher and tighten to 173–233 ft. lbs. (234–316 Nm) on 2002–03 models and 2004 Heritage Editions.

6. Install or connect the following:
- Steering gear into position
- 3 retaining bolts and tighten them to 50–68 ft. lbs. (68–92 Nm
- Power steering pressure hoses to the steering gear using new seals, if necessary
- Intermediate shaft on the steering gear input shaft
- Shaft pinch bolt. Tighten the pinch bolt to 30–42 ft. lbs. (41–57 Nm).
- Dust cover over the steering shaft valve housing
- Pitman arm to the drag link
- Castellated nut and tighten to 57–76 ft. lbs. (77–104 Nm). Install a new cotter pin.
- Radiator air deflector and secure with the retaining screws and push clips
- Skid plate and secure it with the retaining bolts

7. Lower the vehicle.

8. Fill and bleed the power steering system.

9. Road test the vehicle and check the steering system for proper operation.

Blackwood

1. Before servicing the vehicle, refer to the Precautions Section.

❋❋ CAUTION

The electrical power to the air suspension system must be shut off prior to hoisting or jacking an air suspension vehicle. This can accomplished by turning off the air suspension switch located below the glove box in the lower right of the passenger foot well. Failure to do so can result in unexpected inflation or deflation of the air springs, which can result in the shifting of the vehicle during these operations.

2. Turn the air suspension switch to the OFF position.

❋❋ WARNING

Do not allow the steering column shaft to rotate while the intermediate shaft is disconnected or damage to the clockspring can result. If there is evidence that the shaft has rotated, the clockspring must be removed and re-centered.

3. Remove or disconnect the following:
- Pitman arm cotter pin and castellated nut from the drag link
- Pitman arm from the drag link
- Dust cover from the steering shaft valve housing
- Intermediate shaft pinch bolt and slide the shaft off the steering gear input shaft.
- Power steering pressure hoses at the steering gear. Remove and discard the original seals.
- Steering gear-to-frame rail retaining bolts
- Steering gear and spacer.

4. If replacing or servicing the steering gear, match mark the sector shaft arm to the sector shaft and remove the steering gear sector shaft arm retaining nut and lockwasher.

5. Remove the sector shaft arm.

6. Installation is the reverse of removal. Install a new seal on the hose fittings at the power steering gear. When installing the power steering lines, use special installer tool, Installer Set, Teflon® Seal 214-D027 (D90P-3517-A), or equivalent. Stretch the seal over the special tool until it is large enough to slip over the tube nut.

7. Observe the following torques:
- Steering gear mounting bolts: 199 ft. lbs. (270 Nm)
- Intermediate shaft pinch bolt: 35 ft. lbs. (48 Nm)
- Sector shaft-to-gear: 203 ft. lbs. (275 Nm)
- Sector shaft-to-drag link: 66 ft. lbs. (90 Nm)

2004–05 F-150, Exc. 2004 F-150 Heritage Editions
2006 Mark LT

1. Before servicing the vehicle, refer to the Precautions Section.

➡**New O-ring seals must be installed any time the lines are disconnected from the steering gear.**

2. With the transmission in **Neutral**, raise and support the vehicle.

3. Remove the skid plate.

❋❋ WARNING

The boots and clamps are designed to produce an airtight seal and protect the internal components of the steering gear. If the seal is not airtight, the vacuum generated during turning will draw water and contamination into the gear, causing premature damage.

❋❋ WARNING

Zip ties do not provide an airtight seal and must not be used.

❋❋ WARNING

The inner ball joint grease is not compatible with water, and contamination trapped in the grease will degrade the life of the joint.

❋❋ WARNING

If present, the orientation of the vent tube must be noted so the boots and vent tubes can be installed in the correct location.

4. Remove the tire and wheel assembly.

5. Remove the lower shaft-to-steering gear bolt.

6. Disconnect the lower shaft.

➡**Do not allow the steering column shaft to rotate while the lower shaft is disconnected or damage to the clockspring can result. If there is evidence that the shaft has rotated, the clockspring must be removed and re-centered. Follow these steps to center the clockspring.**

a. Hold the clockspring outer housing stationary.

❋❋ WARNING

Overturning will destroy the clockspring. The internal ribbon wire acts as the stop and can be broken from its internal connection.

b. While turning the rotor counterclockwise, carefully feel for the ribbon wire to run out of length and for a slight resistance. Stop turning at this point.

c. Turn the clockspring clockwise approximately three turns. This is the center point of the clockspring.

d. Do not allow the rotor to turn from this position. Two pieces of tape may be applied to the clockspring to prevent accidental rotation, until the clockspring is installed.

7. Remove the clamp plate-to-steering gear bolt

8. Remove the power steering high pressure hose.

9. Remove the O-ring seal.

Steering gear mounting and related parts—2004–05 F-150, exc. 2004 F-150 Heritage Editions; 2006 Mark LT

1 Lower shaft-to-steering gear bolt

2 Lower shaft (disconnect only)

3 Clamp plate-to-steering gear bolt

4 Power steering high pressure hose

5 O-ring seal

6 Power steering return hose

7 O-ring seal

8 Tie-rod lock nuts (loosen only)

9 Tie-rod nuts

10 Outer tie-rods

11 Outer bellows clamp

12 Inner bellows clamp

13 Steering gear bellows

14 Inner tie-rod

15 Bracket-to-crossmember nuts

16 Bracket-to-crossmember bolts

17 Steering gear-to-crossmember nuts

18 Steering gear-to-crossmember bolts

19 Bracket

20 Steering gear

67197-EFSE-G153

Steering gear mounting and related parts—2004–05 F-150, exc. 2004 F-150 Heritage Editions; 2006 Mark LT (cont.)

10. Remove the power steering return hose

11. Remove the O-ring seal.

12. Remove the tie-rod nuts.

13. Remove the outer tie-rods. When repairing the right side, it is necessary to pull back the left inner tie-rod boot to hold the steering gear. Note the number of times the tie-rod ends turn for assembly reference. After removing the tie-rod end, remove the tie-rod end jam nut from the front wheel spindle tie-rod

14. Remove the bracket-to-crossmember nuts.

15. Remove the bracket-to-crossmember bolts.

16. Remove the steering gear-to-crossmember nuts.

17. Remove the steering gear-to-crossmember bolts.

18. Remove the bracket.

19. Remove the steering gear. Remove the steering gear through the left side wheel opening.

20. To install, reverse the removal procedure.

21. Fill and leak test the system.

22. Check and, if necessary, align the front end.

23. Install new O-ring seals onto the power steering hoses.

24. Observe the following torques:

- The 2 steering gear bracket-to-gear bolts and nuts: 150 Nm (111 ft. lbs.).
- The 2 steering gear bracket-to-crossmember bolts and nuts: 103 Nm (76 ft. lbs.).
- The 2 outer tie-rod end nuts: 150 Nm (111 ft. lbs.).
- The 2 outer tie-rod end lock nuts: 103 Nm (76 ft. lbs.).
- Power steering high pressure line clamp plate bolt: 23 Nm (17 ft. lbs.).

- Oil drip shield bolts: 10 Nm (89 inch lbs.).
- Skid plate bolts: 25 Nm (18 ft. lbs.).

2004 F-250 and 350

1. Before servicing the vehicle, refer to the Precautions Section.

➡**New O-ring seals must be installed any time the lines are disconnected from the steering gear.**

2. Remove the air cleaner assembly.

3. Disengage the steering coupling shield from the line fitting and slide upward on the steering shaft.

4. Remove the pinch bolt. Turn the steering wheel as necessary to access the bolt. Make sure the steering column is locked.

✳✳ WARNING

During installation, do not over-tighten the fittings.

5. Disconnect the lines. Discard the O-ring seals.

6. Raise the vehicle.

7. Remove the cotter pin and nut. Discard the cotter pin.

8. Using the special tool, disconnect the drag link.

9. Remove the bolts, washers and the steering gear.

10. Remove the nut.

11. Using a puller, remove the steering gear sector shaft arm.

12. To install, reverse the removal procedure.

13. Note the following torques:

- Pinch bolt: 36 ft. lbs. (48 Nm)
- Power steering lines: 26 ft. lbs. (35 Nm)

- Pitman arm-to-drag link: 66 ft. lbs. (90 Nm)
- Gear mounting bolts: 59 ft. lbs. (80 Nm)
- Pitman arm-to-gear: 199 ft. lbs. (270 Nm)

14. Install a new high pressure hose O-ring seal and a new return hose O-ring seal.

15. Fill and leak check the system.

2005 F-250/350

1. Before servicing the vehicle, refer to the Precautions Section.

➡**New Teflon® seals must be installed any time the power steering line fittings are disconnected.**

2. Remove the engine air cleaner.

3. Disconnect the steering gear coupling shield from the line fittings and slide the shield upward on the intermediate shaft.

4. Remove the intermediate shaft-to-steering gear bolt.

✳✳ WARNING

During installation, do not over-tighten the fittings.

5. Disconnect the power steering lines. Discard the Teflon® seals.

6. With the vehicle in NEUTRAL, position it on a hoist.

7. Remove the cotter pin and the drag link-to-sector shaft nut. Discard the cotter pin.

8. Using a puller, disconnect the drag link from the sector shaft arm.

➡**The sector shaft arm-to-sector shaft nut has a dry adhesive on the threads. The nut must not be reused.**

9. Remove the sector shaft arm-to-sector shaft nut. Discard the nut.

➡**The steering gear bolts have a dry adhesive on the threads. The bolts must not be reused.**

10. Remove the 3 bolts and the steering gear. Discard the bolts.

11. Using a puller, remove the sector shaft arm.

12. To install, reverse the removal procedure. Note the following:

- Using special tool 211-D027, install 2 new Teflon® seals on the power steering line fittings.
- intermediate shaft-to-steering gear bolt 36 ft. lbs. (48 Nm)
- power steering lines: 26 ft. lbs. (35 Nm)
- drag link-to-sector shaft nut: 148 ft. lbs. (200 Nm)
- sector shaft arm-to-sector shaft nut: 350 ft. lbs. (475 Nm)
- steering gear: 111 ft. lbs. (150 Nm)
- Fill the power steering system.

2002–03 E-Series

1. Before servicing the vehicle, refer to the Precautions Section.

2. Place the wheels in the straight-ahead position.

3. Remove or disconnect the following:

- Pressure and return lines. Cap the openings.
- Splash shield from the flex coupling
- Flex coupling at the gear
- Pitman arm from the sector shaft
- Steering gear mounting bolts
- Steering gear. It may be necessary to work it free of the flex coupling.

To install:

4. Install or connect the following:

- Splash shield on the steering gear lugs
- Flex coupling into place on the steering shaft. Be sure the steering wheel spokes are still horizontal.

5. Center the steering gear input shaft with the indexing flat facing downward.

- Steering gear input shaft into the flex coupling and into place on the frame side rail
- Flex coupling bolt and tighten it to 30 ft. lbs. (41 Nm)
- Gear mounting bolts and tighten them to 65 ft. lbs. (88 Nm)
- Pitman arm. Tighten the nut to 230 ft. lbs. (312 Nm).
- Pressure and return lines. Tighten the lines to 25 ft. lbs. (34 Nm).
- Flex coupling shield

6. Fill the steering reservoir.

7. Run the engine and turn the steering wheel lock-to-lock several times to expel air. Check for leaks.

2004–05 E-Series

1. Before servicing the vehicle, refer to the Precautions Section.

➡**New O-ring seals must be installed any time the lines are disconnected from the steering gear.**

2. Place the front wheels in the straight-ahead position and the ignition switch in the OFF position.

✳✳ WARNING

Do not allow the steering column shaft to rotate while the intermediate shaft is disconnected or damage to the clockspring can result. If there is evidence that the shaft has rotated, the clockspring must be removed and recentered.

---

3. Remove the bolt and detach the intermediate shaft from the gear.

4. Disconnect the lines. Discard the O-ring seals.

5. With vehicle in **Neutral**, place on a hoist.

Item	Description
1	Steering Gear
2	Lower Steering Column Shaft
3	Washer
4	Bolt 73-90 N·m (54-66 Ft-Lb)
5	Nut 230-310 N·m (170-228 Ft-Lb)
6	Washer
7	Steering Gear Sector Shaft Arm

The steering gear is mounted on the left frame rail as show—E-Series

6. Remove and discard the cotter pin and nut.

7. Using a puller, separate the drag link.

8. Remove the bolts and the steering gear.

9. Secure the steering gear in a vise and remove the nut and lockwasher.

10. Using the special tool, remove the steering gear sector shaft arm.

11. To install, reverse the removal procedure.

12. Note the following torques:
- Pinch bolt: 18 ft. lbs. (25 Nm)
- Steering gear mounting bolts: 2004 vehicles built to 1/17/04 60 ft. lbs. (81 Nm); 2004 vehicles built from 1/17/04 122 ft. lbs. 165 Nm; 2005 vehicles 60 ft. lbs. (81 Nm)

- Power steering lines: 15 ft. lbs. (20 Nm)
- Drag link-to-Pitman arm: 68 ft. lbs. (92 Nm)
- Pitman arm-to-gear: 199 ft. lbs. (270 Nm)

13. Install a new high pressure hose O-ring seal and a new return hose O-ring seal.

14. Fill and leak check the system.

FRONT SUSPENSION

✳✳ CAUTION

Suspension fasteners are critical parts because they affect performance of vital components and systems and their failure can result in major service expense. They must be replaced with the same part number or an equivalent part if replacement is necessary. Do not use a replacement part of lesser quality or substitute design. Torque values must be used as specified during reassembly to ensure proper retention of these parts.

Front Shock Absorber

REMOVAL & INSTALLATION

2002–03 F-150 and 2004 F-150 Heritage Editions

1. Before servicing the vehicle, refer to the Precautions Section.

2. If equipped with 2-wheel drive, perform the following:

a. Hold the shock absorber stem and remove the nut from the top of the shock.

b. Raise and safely support the vehicle.

c. Remove the 2 lower retaining nuts and remove the shock absorber.

3. If equipped with 4-wheel drive, perform the following:

a. Hold the shock absorber stem and remove the nut, washer and bushing from the top of the shock absorber stud.

b. Raise and support the vehicle.

c. Remove the lower retaining nut and bolt.

d. Remove the shock absorber from the vehicle.

To install:

4. On 4-wheel drive models, perform the following:

a. Install the washer and bushing to the top stem of the shock absorber.

b. Place the shock absorber up through the coil spring.

c. Install the 2 lower retaining nuts. Tighten the nuts to 19–25 ft. lbs. (26–34 Nm).

d. Lower the vehicle.

e. Install the bushing, washer and retaining nut to the top of the shock absorber stud. Tighten the nut to 34–46 ft. lbs. (47–63 Nm).

5. On 4-wheel drive models, perform the following:

a. Install the washer and bushing to the top stem of the shock absorber.

b. Place the shock absorber up through the coil spring.

c. Install the lower retaining nut and bolt. Tighten to 57–76 ft. lbs. (77–104 Nm).

d. Lower the vehicle.

e. Install the shock absorber upper bushing, washer and retaining nut. Tighten the nut to 22–29 ft. lbs. (30–40 Nm).

6. Road test the vehicle and check for proper operation.

2002–05 F-250 and 350

2-WHEEL DRIVE

1. Before servicing the vehicle, refer to the Precautions Section.

✳✳ CAUTION

All vehicles are equipped with gas-pressurized shock absorbers which will extend unassisted. Do not apply heat or flame to the shock absorbers during removal or component servicing.

✳✳ CAUTION

Suspension fasteners are critical parts because they affect performance of vital components and sys-

tems and their failure can result in major service expense. They must be replaced with the same part number or an equivalent part if replacement is necessary. Do not use a replacement part of lesser quality or substitute design. Torque values must be used as specified during reassembly to ensure proper retention of these parts.

2. Raise the hood and remove the upper shock absorber retaining nut and upper shock absorber insulator.

3. Raise and support the vehicle.

4. Remove the lower shock absorber retaining nut and remove the shock absorber.

5. Using new fasteners, follow the removal procedure in reverse order. Observe the following torques:
- Upper nut: 30 ft. lbs. (40 Nm)
- Lower nut: 60 ft. lbs. (80 Nm)

4-WHEEL DRIVE

1. Before servicing the vehicle, refer to the Precautions Section.

2. Raise and support the vehicle.

3. Disconnect the shock absorber lower mount.

4. Disconnect the shock absorber upper mount.

5. To install, reverse the removal procedure. Tighten to:
- 2002–04 upper and lower: 76 ft. lbs. (103 Nm).
- 2005 upper: 46 ft. lbs. (63 Nm)
- 2005 lower: 111 ft. lbs. (150 Nm)

E-Series

1. Before servicing the vehicle, refer to the Precautions Section.

2. Raise the vehicle and secure on support stands.

3. Remove the self-locking nut, steel washer, and rubber bushings at the upper end of the shock absorber.

1 Front Suspension Upper Arm

2 Front Suspension Upper Arm Cams

3 Shock Absorber Insulator Assy

4 Front Suspension Cam Nut

5 Front Suspension Lower Arm Bolt

6 Front Suspension Lower Arm Nut

7 Front Spring Insulator

8 Front Coil Spring

9 Front Suspension Lower Arm

10 Front Shock Absorber

11 Shock Absorber Nut

12 Ball Joint Shield

13 Front Suspension Bumper

14 Front Wheel Spindle

67197-EFSE-G154

2-wheel drive front suspension—2002–03 Blackwood, F-150 and 2004 F-150 Heritage Editions

1 Front Axle Hub Nut

2 Front Wheel Knuckle

3 Front Suspension Upper Arm

4 Front Suspension Upper Arm Shield

5 Front Suspension Upper Arm Adjusting Cam and Bolts

6 Front Suspension Upper Arm Nuts

7 Front Suspension Lower Arm Bolts

8 Front Suspension Lower Arm Nuts

9 Front Shock Absorber Lower Nut

10 Front Suspension Lower Arm

11 Front Suspension Lower Arm Shield

12 Front Shock Absorber Lower Bolt

13 Torsion Bar

14 Torsion Bar Adjuster

15 Front Shock Absorber

16 Front Wheel Driveshaft and Joint Bolt

17 Front Wheel Driveshaft and Joint

18 Brake Disc

67197-EFSE-G155

4-wheel drive front suspension—2002–03 F-150 and 2004 F-150 Heritage Editions

77 - 104 Nm (57 - 76 lb/ft)

Lower shock absorber mounting—models with torsion bar suspension

7924FG73

1 Radius arm bracket (LH/RH)
2 Shock absorber
3 Radius arm (LH/RH)
4 Front axle I-beam (LH/RH)
5 Upper ball joint adjuster

6 Lower spring seat (LH/RH)
7 Coil spring
8 Front axle jounce bumper
9 Front stabilizer bar link
10 Front stabilizer bar

06017-F150-G261

Front suspension, part 1—2002–04 F-250/350 2-wheel drive

1 Front stabilizer bar insulator

2 Front stabilizer bar retainer

3 Front stabilizer bar

4 Front spindle assembly

5 Tie rod end

6 Radius arm (LH/RH)

7 Upper ball joint adjuster

8 Front axle I-beam (LH/RH)

9 Lower spring seat (with stabilizer bar)

10 Lower spring seat (without stabilizer bar)

11 Lower spring insulator

12 Lower spring retainer

13 Coil spring

06017-F150-G262

1 Bolt, M16 x 2 x 143 x 10.9
2 Bracket assembly
3 Bolt, M12 x 1.75 x 40 x 10.9
4 Flag nut, M1.6 x 2 x 20 x 10
5 Spring shackle
6 Nut and washer, M12 x 1.75
7 Shock assembly
8 Upper shock bracket
9 Nut and retainer
10 Bolt, M12 x 1.75 x 80 x 10.9
11 Nut, M12 x 1.75 x 13 x 10
12 Tow hook
13 Nut
14 Spring hanger
15 Bolt, M18 x 2.5 x 136 x 10.9

16 Spring assembly
17 U-bolt, M14 x 2 x 92 x 201
18 Spacer
19 Nut and retainer
20 Track bar bracket
21 Bolt, M20 x 100
22 Track bar
23 Spring cap
24 Nut
25 Stabilizer bar retainer bracket
26 Drive axle, Dana Model 50
27 Drive axle, Dana Model 60
28 Nut M14 x 2 x 15 x 10
29 Stabilizer bar insulator

06017-F150-G263

Front suspension—2002–04 F-250/350 4-wheel drive

1 Radius arm bracket (LH/RH)
2 Shock absorber
3 Coil spring
4 Stabilizer bar link
5 Radius arm (LH/RH)

6 Wheel knuckle (LH/RH)
7 Track bar
8 Stabilizer bar retainer
9 Stabilizer bar
10 Jounce bumper

06017-F150-G268

Front suspension—2005 F-250/350 4-wheel drive

1 Radius arm bracket (LH/RH)

2 Shock absorber

3 Radius arm (LH/RH)

4 Front axle I-beam (LH/RH)

5 Upper ball joint adjuster

6 Lower spring seat (LH/RH)

7 Coil spring

8 Front axle jounce bumper

9 Front stabilizer bar link

10 Front stabilizer bar

11 Front suspension bar retainer

06017-F150-G264

Front suspension, part 1—2005 F-250/350 2-wheel drive

1 Radius arm bracket (LH/RH)

2 Shock absorber

3 Radius arm (LH/RH)

4 Front axle

5 Upper ball joint adjuster

6 Front spring

7 Front axle jounce bumper

8 Stabilizer bar link

9 Stabilizer bar

10 Front stabilizer bar retainer

11 Track bar mounting bracket

12 Track bar

13 Wheel knuckle (LH/RH)

06017-F150-G265

Front suspension, part 2—2005 F-250/350 2-wheel drive

E-150

E-250, E-350,
E-Super Duty

1	Shock absorber	7	Front axle (RH)
2	Front coil spring	8	Front stabilizer bar
3	Radius arm	9	Upper ball joint pinch bolt
4	Front wheel spindle	10	Caster/camber insert
5	Tie-rod end	11	Lower ball joint nut
6	Front axle (LH)	12	Tie-rod end-to-spindle nut

06017-F150-G266

Front suspension, part 1—E-Series

1 Shock absorber upper nut

2 Front spring upper retainer screw

3 Radius arm-to-front axle bolt, E150

4 Radius arm-to-front axle bolt, E250, E350, E-Super Duty

5 Shock absorber lower nut

6 Radius arm-to-rear bracket nut

7 Radius arm rear insulator washer

8 Radius arm rear insulator

9 Radius arm rear bracket

10 Radius arm spacer

11 Radius arm front insulator

12 Radius arm front insulator washer

13 Radius arm

14 Radius arm-to-front axle nut

15 Front spring insulator

16 Front spring lower retainer

17 Jounce bumper screw

06017-F150-G267

Front suspension, part 2—E-Series

4. Remove the bolt and nut at the lower end and remove the shock absorber.

To install:

5. When installing a new shock absorber, use new rubber bushings. Position the shock absorber on the mounting brackets with the stud end at the top. Install the upper bushing, steel washer and self-locking nut at the upper end, and the bolt and nut at the lower end.

6. Tighten the upper mounting studs to 30 ft. lbs. (41 Nm) and the lower mounting nuts to 59 ft. lbs. (80 Nm).

Front Strut and Spring

REMOVAL & INSTALLATION

2004–05 F-150, exc. 2004 F-150 Heritage Editions

2006 Mark LT

2-WHEEL DRIVE

1. Before servicing the vehicle, refer to the Precautions Section.
2. Remove the wheel and tire assembly.

➡Use the hex holding feature to prevent the ball studs (upper control arm, lower control arm tie-rod end, and stabilizer bar links) from turning while removing and installing the nuts.

❄❄ WARNING

Do not hammer the ball studs to separate them from the wheel knuckle or stabilizer bar. Doing so can cause damage. Lightly tap the wheel knuckle or stabilizer bar to loosen the joints.

·8 Nm (35 lb-ft)

475 Nm (351 lb-ft)

1 Shock absorber upper mount-to-frame nuts
2 Shock absorber-to-lower arm nut
3 Shock absorber-to-lower arm bolt
4 Shock absorber and spring assembly

67197-EFSE-G156

Front strut mounting—2004–05 F-150 exc. 2004 F-150 Heritage Editions; 2006 Mark LT

3. Remove the strut upper mount nuts.
4. Remove the lower mount nut/bolt.
5. Remove the nut and detach the tie-rod from the wheel knuckle. Discard the nut.
6. Remove the strut.
7. To install, reverse the removal procedure. Tighten the upper strut nuts to 35 ft. lbs. (48 Nm); the lower strut-to-arm bolts to 351 ft. lbs. (475 Nm). Torque the tie rod stud nut to 111 ft. lbs. (150 Nm).
8. Do not tighten the fasteners until the installation procedure is complete and the weight of the vehicle is resting on the wheel and tire assemblies.

4-WHEEL DRIVE

1. Before servicing the vehicle, refer to the Precautions Section.
2. Remove the wheel and tire assembly.
3. Remove and discard the 4 strut upper mount-to-frame nuts.

✳✳ WARNING

Do not hammer the ball stud to separate it from the wheel knuckle. Doing so can cause damage. Lightly tap the wheel knuckle to loosen the joint.

4. Remove and discard the lower ball joint-to-wheel knuckle nut.
5. Separate the lower ball joint from the wheel knuckle.

✳✳ WARNING

Use the hex holding feature to prevent the ball studs (upper control arm, lower control arm and tie rod end, and stabilizer bar links) from turning while removing and installing the nuts.

6. Remove and discard the tie rod-to-wheel knuckle nut.

1 Shock absorber-to-upper mount nut

2 Upper shock mount (front/rear)

3 Isolator

4 Dust tube

5 Coil spring (front)

6 Shock absorber (front/rear)

67197-EFSE-G157

Front strut disassembled

7. Separate the tie-rod from the wheel knuckle.

8. Remove and discard the strut-to-lower arm bolt.

9. Remove the strut and spring assembly from the vehicle.

➡ **Do not tighten the nuts until the installation procedure is complete and the weight of the vehicle is resting on the wheel and tire assemblies.**

10. To install, reverse the removal procedure.

   a. Always install new:

- Shock absorber upper mount-to-frame nuts.
- Lower ball joint-to-wheel knuckle nuts.
- Tie-rod-to-wheel knuckle nuts.
- Shock absorber-to-lower arm bolts.
  b. Observe the following torques:
- Strut lower bolt: 351 ft. lbs. (475 Nm)
- Tie rod stud nut: 111 ft. lbs. (150 Nm)
- Lower ball joint stud nut: 111 ft. lbs. (150 Nm)
- Upper strut nuts: 35 ft. lbs. (48 Nm)

DISASSEMBLY

1. Before servicing the vehicle, refer to the Precautions Section.

2. Remove the strut and spring assembly.

3. Using a suitable spring compressor, compress the spring until the tension is released from the shock absorber.

4. Remove the shock absorber-to-upper mount nut

5. Remove the upper shock mount

6. Remove the isolator

7. Remove the dust tube

8. Remove the coil spring

9. Remove the shock absorber

10. To assemble, reverse the disassembly procedure. Torque the nut to 22 ft. lbs. (30 Nm).

Coil Spring

REMOVAL & INSTALLATION

2002–03 E-Series

1. Before servicing the vehicle, refer to the Precautions Section.

2. Remove the stabilizer bar.

❋❋ WARNING

The axle must be supported throughout the procedure to prevent strain on the front brake hose.

➡ **The disc brake caliper may have to be removed for additional clearance.**

3. Use a suitable jack to support the affected front axle.

4. Remove the nut and disconnect the shock absorber at the lower end.

5. Remove the bolt and the upper spring retainer.

6. Lower the jack supporting the front axle and remove the front coil spring.

7. To install, reverse the removal procedure. Torque the upper spring retainer to 22 ft. lbs. (30 Nm). Torque the lower shock nut to 59 ft. lbs. (80 Nm).

2004–05 E-Series

1. Before servicing the vehicle, refer to the Precautions Section.

2. Remove the stabilizer bar.

❋❋ WARNING

The axle must be supported throughout the procedure to prevent strain on the front brake hose.

3. Using a suitable jack, support the front axle.

4. Loosen the shock absorber upper nut. Loosen the nut to the top of the threads.

5. Remove the front spring upper retainer bolt and the upper spring retainer.

➡**It may be necessary to loosen the radius arm-to-axle nut to remove the spring.**

6. Lower the jack supporting the front axle, then remove the front coil spring from the lower retainer. Using a suitable spray lubricant, spray the lower spring and spring retainer. Rotate the spring 180 degrees counterclockwise, tilt the spring outward and remove.

7. To install, reverse the removal procedure. Tighten the upper spring retainer to 22 ft. lbs. (30 Nm). Tighten the upper shock nut to 30 ft. lbs. (40 Nm).

2002–03 2-Wheel Drive Blackwood, F-150 and 2004 F-150 Heritage Editions

1. Before servicing the vehicle, refer to the Precautions Section.

2. Remove or disconnect the following:
- Wheels
- Disc brake caliper and support aside with wire
- Disc brake adapter
- Rotor
- Rotor splash shield
- Shock absorber
- Bracket supporting the brake hose
- Sway bar link retaining nut and bushing from the lower control arm. Separate the sway bar link from the lower control arm.

3. Install a coil spring compressor and compress the coil spring enough to relieve the tension of the spring between the upper and lower control arms.

4. Remove the cotter pin and castellated nut from the lower ball joint. Separate the lower ball joint from the wheel spindle.

5. Matchmark the lower control arm alignment cams for installation reference.

6. Remove the lower control arm retaining nuts and bolts.

7. Remove the lower control arm and the compressed coil spring as an assembly.

8. Loosen the coil spring compressor and remove the coil spring from the lower control arm.

9. Inspect the coil spring and replace as needed.

To install:

10. Place the coil spring correctly in the saddle of the lower control arm. Install the

Be sure the coil spring is mounted correctly in the lower control arm

7924FG75

coil spring compressor and compress the coil spring. The end of the coil spring **A**, must cover the hole designated **B** and be visible in the second hole designated as **C**, for proper installation.

11. Place the lower control arm to the frame. Install the retaining bolts, adjusting cams, and nuts.

12. Align the match marks on the adjusting cams. The forward nut must be tightened first while the control arm is held at the curb position height. Tighten the nuts to 197–241 ft. lbs. (270–330 Nm).

13. Install the lower ball joint stud into the wheel spindle. Install the castellated nut and tighten to 83–113 ft. lbs. (113–153 Nm). Install a new cotter pin.

14. Connect the sway bar link to the lower control arm. Install the bushing and retaining nut. Tighten to 15–21 ft. lbs. (21–29 Nm).

15. Remove the coil spring compressor.

16. Install the shock absorber. Tighten the lower bolts to 22 ft. lbs. (32 Nm) and the top nut to 45 ft. lbs. (61 Nm).

17. Connect the brake hose bracket.

18. Install the brake rotor splash shield. Install the 3 retaining bolts and tighten them to 90–107 inch lbs. (10–14 Nm).

19. Install the disc brake rotor and caliper assemblies.

20. Install the wheel and tire assembly.

21. Lower the vehicle.

22. Pump the brake pedal several times to position the brake pads prior to moving the vehicle.

23. Check the alignment and adjust if out of specification.

24. Road test the vehicle and check for proper operation.

2004–05 F-250 and 350

2-WHEEL DRIVE

1. Before servicing the vehicle, refer to the Precautions Section.

2. Remove the wheel and tire assembly.

3. Using a suitable jack, support the front axle assembly.

4. Remove the nut and detach the shock from the mounting stud.

5. Remove the upper spring retainer.

6. Lower the front axle until the spring is free of the upper spring seat.

7. Using an extension through the top of the spring, remove the lower spring retainer.

8. Remove the front spring.

9. Using new fasteners, reverse the removal procedure. Tighten the lower spring retainer to 99 ft. lbs. (133 Nm); the upper retainer to 26 ft. lbs. (35 Nm).

2005 F-250/350

4-WHEEL DRIVE

1. Before servicing the vehicle, refer to the Precautions Section.

2. With the vehicle in NEUTRAL, position it on a hoist.

3. Remove the wheel and tire assembly.

4. Position jack stands under the axle housing.

5. Remove the front shock absorber lower mounting bolt and nut.

6. Lower the axle and remove the spring.

7. Remove the spring insulator and retain for reinstallation.

8. To install, reverse the removal procedure.

Torsion Bar

REMOVAL & INSTALLATION

2002–03 F-150 4-Wheel Drive and 2004 Heritage Editions

1. Before servicing the vehicle, refer to the Precautions Section.

2. Raise the vehicle on a hoist.

3. Apply a liberal amount of penetrating oil to the torsion bar adjuster and to the torsion bar socket at the lower control arm.

4. Make preliminary adjustment references.

a. Mark the relation of the torsion bars to crossmember support.

b. Measure and record the torsion bar adjuster bolt lengths.

67197-EFSE-G160

Mark the relation of the torsion bars to crossmember support—2002–03 F-150 4-Wheel Drive and 2004 Heritage Editions

67197-EFSE-G159

Special tools assembled—2002–03 F-150 4-Wheel Drive and 2004 Heritage Editions

5. Relieve the torsion bar tension.

a. Remove the torsion bar adjuster bolt, 1.

b. Install the special tools T96T-5310-A and T95T-5310-A on the crossmember support and to the torsion bar adjuster, 2.

c. Tighten the special tool until the torsion bar adjuster lifts off the adjuster nut, 3.

d. Remove the torsion bar adjuster nut, 4.

e. Remove the special tool, 5. Repeat on the opposite side.

6. Remove the retainers from the torsion bar crossmember support.

7. Position the torsion bar crossmember support rearward and remove the torsion bar adjusting bracket.

8. Position the torsion bar rearward and separate it from the lower control arm. Position the torsion bar forward and remove the bar.

To install:

9. Position the torsion bar to the vehicle and install the bar to the lower control arm.

10. Reposition the torsion bar crossmember support. Install the torsion bar adjusting brackets to the torsion bars.

11. Install the crossmember support retaining bolts.

12. Load the torsion bar.

a. Install special tools T95T-5310-A and T96T-5310-A on the crossmember support and to the torsion bar adjuster.

b. Tighten the special tool to load the torsion bar.

c. Install the torsion bar adjuster nuts.

61 Nm (45 lb-ft)

47 Nm (35 lb-ft)

67197-EFSE-G161

Install the crossmember bolts—2002–03 F-150 4-Wheel Drive and 2004 Heritage Editions

d. Loosely install the torsion bar adjuster bolt.

e. Remove the special tools. Repeat on the opposite side.

13. Make an initial adjustment.

a. Turn the torsion bar adjusters until the reference marks on the torsion bars and crossmember support align.

b. Measure the torsion bar adjuster bolt lengths. The measurement must equal the measurement taken during removal.

14. Lower the vehicle.

15. Check the ride height. Adjust as necessary.

16. Check the wheel alignment. Adjust as necessary.

Radius Arm

REMOVAL & INSTALLATION

F-250 and 350

1. Before servicing the vehicle, refer to the Precautions Section.

2. Raise and support the vehicle.

3. Remove the wheel and tire assembly.

4. Remove the front disc brake caliper, front disc brake hub and rotor and front disc brake rotor shield.

5. If equipped, remove the front disc brake ABS sensor retainer bolt and the ABS sensor harness bracket bolt from the front wheel spindle. Position the ABS sensor out of the way.

6. Remove the cotter pin and the castellated nut from the tie rod end. Discard the cotter pin.

7. Using a Pitman arm puller, remove the tie rod end.

8. Remove the cotter pin and the lower ball joint castellated nut. Discard the cotter pin.

9. Remove the pinch bolt and the camber adjuster from the upper ball joint.

✳✳ WARNING

To prevent damage to the ball joint seal and the ball joint socket, do not use a pickle fork-type remover to loosen the ball joints.

10. Remove the front wheel spindle.

a. Strike the lower end of the axle to loosen the ball joint.

b. Remove the front wheel spindle.

11. Using a suitable jack, support the front axle.

12. Remove the front coil spring.

13. Remove the lower spring insulator.

14. Remove the lower spring seat.

15. Remove the radius arm-to-axle bolt and nut.

16. Remove the radius arm-to-frame bracket pivot bolt and nut. Remove the radius arm.

➡ **In order to obtain clearance for the pivot bolt removal when servicing the right front axle, it may be necessary to raise the side of the vehicle to relieve the weight on the suspension.**

17. Remove the pivot nut and bolt.

To install:

18. Inspect the pivot bushing for wear or damage. Replace as necessary.

19. Position the front axle into the pivot bracket and install a new pivot bolt and nut hand-tight.

20. Inspect the radius arm pivot bushing for wear or damage. Replace as necessary.

21. Install a new radius arm-to-axle bolt and nut hand-tight.

22. Install a new radius arm-to-frame bracket pivot bolt and nut. Tight to 185 ft. lbs. (250 Nm) for 2002–04 models; 221 ft. lbs. (300 Nm) for 2005 models.

23. Tighten the radius arm-to-axle bolt and nut to 295 ft. lbs. (400 Nm).

24. Install the lower spring seat.

25. Install the lower spring insulator.

26. Install the front coil spring.

➡ **Tighten the ball joint nut further, if necessary, in order to insert the new cotter pin.**

27. Using new fasteners, install the front wheel spindle in the front axle.

   a. Position the front wheel spindle in the front axle.

   b. Install the ball joint nut and tighten to 60 ft. lbs (80 Nm).

   c. Install the new cotter pin.

28. Install the camber adjuster and a new pinch bolt. Tighten to 60 ft. lbs (80 Nm).

29. Install the front disc brake rotor shield, front disc brake hub and rotor and the front disc brake caliper.

30. If equipped, install the front disc brake ABS sensor, retainer bolt and the ABS sensor harness bracket bolt.

31. Install the tie rod end in the front wheel spindle using a new nut and a new cotter pin. Torque to 67 ft. lbs. (90 Nm).

32. Install the tire and wheel assembly.

33. Lower the vehicle and with the vehicle weight on the suspension, tighten the axle pivot bolt and nut to 130 ft. lbs. (175 Nm).

34. Perform front end alignment.

E-Series

1. Before servicing the vehicle, refer to the Precautions Section.

2. Remove the front coil spring.

3. Remove the radius arm-to-axle nut and bolt.

4. Remove the spring retainer and insulator.

5. Remove the radius arm.

   a. Remove the nut.

   b. Remove the washer and rear insulator.

   c. Pull the radius arm from the bracket, and remove the spacer, insulator and washer.

➡ **Inspect the bushings and install new as necessary.**

6. To install, reverse the removal procedure. Torque the pivot nut to 221 ft. lbs. (300 Nm); the bracket end nut to 98 ft. lbs. (133 Nm).

Ball Joints

REMOVAL & INSTALLATION

2002–03 2-Wheel Drive F-150 and 2004 F-150 Heritage Edition

UPPER

➡ **The upper ball joints are not replaceable and are serviced with the control arm.**

LOWER

1. Before servicing the vehicle, refer to the Precautions Section.

2. With the vehicle in **Neutral**, position it on a hoist.

3. Remove the wheel and tire assembly.

✳✳ CAUTION

Do not allow the disc brake caliper to hang suspended from the brake hose. Provide a suitable support.

4. Remove the caliper support bracket bolts, then position the caliper support bracket and brake caliper aside.

5. Disconnect the front anti-lock brake sensor wire assembly.

6. Remove the tie rod castellated nut.

✳✳ CAUTION

Do not use a hammer to separate the tie rod from the wheel knuckle or damage to the wheel knuckle will result.

✳✳ CAUTION

Do not damage the tie rod boot when installing the special tool.

7. Separate the tie rod end from the front wheel knuckle.

8. Using a suitable jack, support the front suspension lower arm.

9. Remove the lower ball joint castellated nut.

10. Using the special tool, separate the front wheel knuckle from the suspension lower arm. Then, loosely reinstall the castellated nut.

11. Remove the upper ball joint castellated nut.

67197-EFSE-G163

Separate the tie rod end from the front wheel knuckle—2002–03 2-Wheel Drive F-150 and 2004 F-150 Heritage Edition

67197-EFSE-G164

Separate the knuckle from the lower arm—2002–03 2-Wheel Drive F-150 and 2004 F-150 Heritage Edition

67197-EFSE-G165

Separate the knuckle from the upper arm—2002–03 2-Wheel Drive F-150 and 2004 F-150 Heritage Edition

Press out the ball joint—2002–03 2-Wheel Drive F-150 and 2004 F-150 Heritage Edition

✳✳ CAUTION

Do not damage the ball joint boot when installing the special tool.

12. Separate the front wheel knuckle from the front suspension upper arm.

13. Remove the hand-tightened lower ball joint castellated nut, then remove the front wheel knuckle.

14. Remove the snap ring from the ball joint. Discard the snapring.

15. Press out the ball joint.

✳✳ CAUTION

Do not damage the ball joint boot when installing the special tool.

➡Clean and inspect the control arm ball joint bore for damage before installing a new ball joint.

➡Make sure the new ball joint snap ring is fully seated.

16. To install, reverse the removal procedure. Always install new castellated nuts and cotter pins. Observe the following torques:
- Caliper support bracket: 83 ft. lbs. (112 Nm)
- Tie rod end ball stud: 70 ft. lbs. (95 Nm)
- Lower ball joint stud nut: 98 ft. lbs. (133 Nm)
- Upper ball joint stud nut: 66 ft. lbs. (90 Nm)

2002–03 4-Wheel Drive F-150 and 2004 F-150 Heritage Editions

UPPER

➡The upper ball joints are not replaceable and are serviced with the control arm.

LOWER

1. Before servicing the vehicle, refer to the Precautions Section.

2. With the vehicle in **Neutral**, position it on a hoist.

3. Remove the wheel and tire assembly.

4. Remove and discard the cotter pin. Remove the retainer. Remove the hub nut.

✳✳ WARNING

Do not allow the disc brake caliper to hang suspended from the brake hose. Provide a suitable support.

5. Remove the caliper support bracket bolts, then position the caliper support bracket and caliper aside.

6. Remove the brake disc.

7. Disconnect the front anti-lock brake sensor (ABS) wire assembly.

8. Remove the tie-rod castellated nut. Discard the cotter pin and nut.

✳✳ WARNING

Do not use a hammer to separate the tie-rod from the wheel knuckle or damage to the wheel knuckle will result. Do not damage the tie-rod boot when installing the special tool.

9. Using the special tool, separate the tie-rod end from the front wheel knuckle.

10. Using a suitable jack, support the front suspension lower arm.

11. Remove the lower ball joint castellated nut. Discard the cotter pin. Discard the castellated nut.

12. Separate the front wheel knuckle from the suspension lower arm. Then, loosely install the castellated nut.

13. Remove the upper ball joint castellated nut. Discard the cotter pin. Discard the castellated nut.

14. Separate the front wheel knuckle from the front suspension upper arm.

15. Separate the halfshaft from the hub, then remove the hand-tightened lower ball joint castellated nut. Remove the front wheel knuckle. Position the halfshaft aside and support with wire.

16. Remove the snapring from the ball joint. Discard the snapring.

17. Using a suitable ball joint remover tool and receiver cup, remove the ball joint.

➡Do not damage the ball joint boot when installing the special tool.

➡Clean and inspect the control arm ball joint bore for damage before installing a new ball joint.

➡Make sure the new ball joint snapring is fully seated.

18. To install, reverse the removal procedure. Always install new castellated nuts and cotter pins. Observe the following torques:
- Upper ball joint nut: 66 ft. lbs. (90 Nm).
- Lower ball joint nut: 98 ft. lbs. (133 Nm).
- Tie rod end nut: 70 ft. lbs. (95 Nm).
- Brake caliper support bracket: 136 ft. lbs. (185 Nm).
- Hub nut: 221 ft. lbs. (300 Nm).

2002–05 2-Wheel Drive F-250 and 350

UPPER AND LOWER

1. Before servicing the vehicle, refer to the Precautions Section.

2. Raise and support the vehicle.

3. Remove the wheel and tire assembly.

4. Remove the disc brake caliper and the front disc brake hub and rotor.

5. Remove the front disc brake rotor shield.

6. If equipped, remove the ABS sensor retaining bolt, ABS sensor harness retaining bolt and the ABS sensor. Position out of the way.

7. Disconnect the tie rod end.
   a. Remove and discard the cotter pin.
   b. Remove the castellated nut.
   c. Using the Pitman Arm Puller, remove the tie rod end.

8. Remove the pinch bolt.

9. Remove the camber adjuster.

✳✳ WARNING

To prevent damage to the ball joint seal and the ball joint socket, do not use a pickle fork-type remover to loosen the ball joints.

10. Remove the front wheel spindle.
   a. Remove and discard the cotter pin.
   b. Loosen, but do not remove, the castellated nut.
   c. Strike the lower end of the front axle to loosen the ball joint.
   d. Remove the castellated nut and the front wheel spindle.

11. Position the front wheel spindle in a vise, and remove the snapring from the lower ball joint.

✳✳ WARNING

To avoid damage to the components, do not use heat to aid ball joint removal.

12. Using the ball joint press tool and suitable receiver cup, remove the lower ball joint from the front wheel spindle.

13. Using the ball joint press tool and suitable receiver cup, remove the upper ball joint.

To install:

❋❋ WARNING

To avoid damage to components, do not use heat to aid installation.

➡**Clean the wheel knuckle ball joint bores.**

➡**The lower ball joint must be installed first.**

14. Using the ball joint press with suitable receiver cups, install the lower ball joint.

15. Using the ball joint press with suitable receiver cups, install the upper ball joint.

16. Install the snapring in the groove at the bottom of the ball joint.

➡**Tighten the ball joint nut further, if necessary, in order to insert a new cotter pin.**

17. Using new fasteners, follow the removal procedure in reverse order.

18. Check the front end alignment. Observe the following torques:

- Ball joint stud nut: 99 ft. lbs. (133 Nm)
- Pinch bolt: 60 ft. lbs. (80 Nm)
- Tie rod end stud nut: 67 ft. lbs. (90 Nm)

2002–05 4-Wheel Drive F-250 and 350

UPPER

1. Before servicing the vehicle, refer to the Precautions Section.
2. Raise and support the vehicle.
3. Remove the wheel and tire assembly.
4. Remove the front brake disc.

211-003

Disconnect the tie-rod end from the wheel knuckle—2002–05 F-250 and 350

67197-EFSE-G170

67197-EFSE-G168

Removing the upper ball joint—2002–05 F-250 and 350

5. Remove the wheel hub and bearing.
6. Using a drift, drive the axle shaft main seal out of the wheel knuckle.
7. Remove the axle shaft and main seal.
8. Remove the tie-rod end castellated nut.
9. Disconnect the tie-rod end from the wheel knuckle.
10. Remove the upper ball joint castellated nut and the insert.
11. Remove the lower ball joint nut.
12. Remove the knuckle.
13. Clean and inspect the wheel knuckle ball joint bores.
14. Place the wheel knuckle into a suitable vise.

➡**Always remove the lower ball joint first.**

15. Remove the lower ball joint.
16. Using a suitable ball joint press, remove the upper ball joint.

To install:

17. Clean the wheel knuckle ball joint bores.
18. Using a suitable ball joint press, install the upper ball joint.
19. Install the lower ball joint.
20. Install the wheel knuckle.
21. Position the wheel knuckle onto the axle housing.
22. Install the nut onto the lower ball joint. Do not tighten the nut at this time.
23. Install the insert and the castellated nut onto the upper ball joint. Do not tighten the nut at this time.
24. Tighten the lower ball joint retaining nut. Pre-tighten the nut to 47 Nm (35 ft. lbs.).

➡**Do not loosen the castellated nut to install the cotter pin.**

25. Tighten the upper ball joint castellated nut. Torque to 69 ft. lbs. (94 Nm).
26. Install the cotter pin. If necessary, tighten the castellated nut until the cotter pin can be installed.
27. Tighten the lower ball joint nut to 204 Nm (150 ft. lbs.).

67197-EFSE-G171

Seat the main seal onto the axle shaft—2002–05 F-250 and 350

67197-EFSE-G172

Install the main seal into the wheel knuckle—2002–05 F-250 and 350

28. Position the tie-rod end into the wheel knuckle.
29. Install and tighten the castellated nut. Torque to 52 ft. lbs. (70 Nm).
30. Install the cotter pin.
31. Position the main seal onto the axle shaft.
32. Using the special tools and a hammer, seat the main seal onto the axle shaft.
33. Position the axle shaft into the axle housing.
34. Using the special tools and a hammer, install the main seal into the wheel knuckle.
35. Install the wheel hub and bearing.
36. Install the front brake disc.
37. Install the wheel and tire assembly.

LOWER

1. Raise and support the vehicle.
2. Remove the wheel and tire assembly.
3. Remove the front brake disc.
4. Remove the wheel hub and bearing.
5. Using a drift, drive the axle shaft main seal out of the wheel knuckle.
6. Remove the axle shaft and main seal.
7. Remove the tie-rod end castellated nut.
8. Disconnect the tie-rod end from the wheel knuckle.
9. Remove the upper ball joint castellated nut and the insert.

Disconnect the tie-rod end from the wheel knuckle—2002–05 F-250 and 350

67197-EFSE-G170

Lower ball joint removal—2002–05 F-250 and 350

67197-EFSE-G173

Lower ball joint installation—2002–05 F-250 and 350

67197-EFSE-G174

10. Remove the lower ball joint nut.

11. Remove the knuckle.

12. Clean and inspect the wheel knuckle ball joint bores.

13. Place the wheel knuckle into a suitable vise.

14. Remove the lower ball joint.

To install:

15. Clean the wheel knuckle ball joint bores.

16. Install the lower ball joint.

17. Install the wheel knuckle.

18. Position the wheel knuckle onto the axle housing.

19. Install the nut onto the lower ball joint. Do not tighten the nut at this time.

20. Install the insert and the castellated nut onto the upper ball joint. Do not tighten the nut at this time.

21. Tighten the lower ball joint retaining nut. Pre-tighten the nut to 47 Nm (35 ft. lbs.).

➡**Do not loosen the castellated nut to install the cotter pin.**

22. Tighten the upper ball joint castellated nut. Torque to 69 ft. lbs. (94 Nm).

23. Install the cotter pin. If necessary, tighten the castellated nut until the cotter pin can be installed.

24. Tighten the lower ball joint nut to 204 Nm (150 ft. lbs.).

25. Position the tie-rod end into the wheel knuckle.

26. Install and tighten the castellated nut. Torque to 52 ft. lbs. (70 Nm).

27. Install the cotter pin.

28. Position the main seal onto the axle shaft.

29. Using the special tools and a hammer, seat the main seal onto the axle shaft.

30. Position the axle shaft into the axle housing.

31. Using the special tools and a hammer, install the main seal into the wheel knuckle.

32. Install the wheel hub and bearing.

33. Install the front brake disc.

34. Install the wheel and tire assembly.

2002–05 E-Series

UPPER AND LOWER

1. Before servicing the vehicle, refer to the Precautions Section.

2. Remove the brake disc.

3. Remove the wheel speed sensor.

4. Remove and discard the cotter pin and nut.

5. Using the special tool, disconnect the tie-rod end.

6. Remove and discard the cotter pin and the nut.

7. Remove the pinch bolt adjuster.

✲✲ WARNING

To prevent damage to the ball joint seal and the ball joint socket, do not use a pickle fork-type remover to loosen the ball joints.

8. Remove the front wheel spindle.

a. Strike the lower end of the front axle to loosen the ball joint.

b. Remove the front wheel spindle.

9. Position the front wheel spindle in a vise, and remove the snapring from the lower ball joint.

✲✲ WARNING

To avoid damage to the components, do not use heat to aid ball joint removal.

10. Using the special tool and suitable receiver cup, remove the lower ball joint from the front wheel spindle.

11. Using the special tool and suitable receiver cup, remove the upper ball joint.

To install:

✲✲ WARNING

To avoid damage to components, do not use heat to aid installation.

➡**Clean the wheel knuckle ball joint bores.**

➡**The lower ball joint must be installed first.**

12. Using the special tool with suitable receiver cups, install the lower ball joint.

13. Using the special tool with suitable receiver cups, install the upper ball joint.

14. Install the snapring in the groove at the bottom of the ball joint.

15. To install, reverse the removal procedure.

16. Observe the following torques:
- Pinch bolt: 77 ft. lbs. (104 Nm)
- Lower ball stud nut: 130 ft. lbs. (176 Nm)
- Toe rod end nut: 66 ft. lbs. (90 Nm)

17. Check and, if necessary, align the front end.

2004 F-150, Exc. Heritage Editions

The upper and lower ball joints are serviced as part of the control arms and can't be replaced separately.

Leaf Spring

REMOVAL & INSTALLATION

2002–04 F-250 and 350 4-Wheel Drive

1. Before servicing the vehicle, refer to the Precautions Section.

2. Remove the wheel and tire assembly.

3. Position jack stands under the axle housing.

4. Remove the left shock absorber.

✲✲ CAUTION

Never reuse U-bolts. The U-bolts are a torque-to-yield design and cannot be retightened. Failure to use a new U-bolt can result in loose or broken springs and suspension components.

5. Remove the two U-bolts and the front spring spacer. Discard the U-bolts.

Detach the pushpin and remove the bolt and condenser bracket—2002–04 F-250 and 350 4-Wheel Drive

6. Disconnect the front driveshaft.

7. Remove the nut and bolt. Position the spring shackle away from the spring.

8. Detach the pushpin and remove the bolt and condenser bracket.

9. Remove the nut and bolt. Using a hammer, tap the spring until the spring is free from the hanger bracket.

10. Remove the spring from the vehicle.

To install:

11. Position the spring in the vehicle.

12. Position the spring into the hanger.

13. Install the bolt and nut. Do not tighten the nut and bolt at this time.

14. Install the bracket and bolt and the pushpin. Torque to 80 inch lbs. (9 Nm).

15. Position the shackle.

16. Install the bolt and nut. Do not tighten the nut and bolt at this time.

➡The U-bolts insert through the front spring cap on the left-hand side and through the axle assembly bracket on the right-hand side.

17. Position the front spring spacer and the new U-bolts.

18. Install the four nuts. Do not tighten the nuts at this time.

19. Install the shock, bolt and nut onto the spring cap. Torque to 76 ft. lbs. (103 Nm).

➡The suspension must be loaded with the weight of the vehicle before the U-bolts and the leaf spring mounting bolts can be tightened. Make sure that the locating pin is correctly aligned with the axle.

20. Lower the vehicle onto the jack stands until the front suspension is supporting the weight of the vehicle.

➡The U-bolts must be tightened in sequence.

21. Tighten the U-bolts following the sequence shown. Torque to 99 ft. lbs. (133 Nm).

22. Tighten the leaf spring rear retaining bolt and nut to 185 ft. lbs. (250 Nm).

23. Tighten the leaf spring front retaining bolt and nut to 203 ft. lbs. (275 Nm).

24. Raise the vehicle and remove the jack stands.

25. Position the driveshaft onto the pinion flange.

26. Install the two straps and the four bolts. Torque to 76 ft. lbs. (103 Nm).

27. Install the wheel and tire assembly.

Spindle

REMOVAL & INSTALLATION

E-Series

1. Before servicing the vehicle, refer to the Precautions Section.

2. Remove the brake disc.

3. Remove the wheel speed sensor.

4. Remove and discard the cotter pin and nut.

5. Disconnect the tie-rod end.

6. Remove and discard the cotter pin and the nut.

7. Remove the pinch bolt adjuster.

✳✳ WARNING

To prevent damage to the ball joint seal and the ball joint socket, do not use a pickle fork-type remover to loosen the ball joints.

8. Strike the lower end of the front axle to loosen the ball joint.

9. Remove the front wheel spindle.

10. To install, reverse the removal procedure. Observe the following torques:

• Tie rod ball stud nut: 66 ft. lbs. (90 Nm)

• Lower ball joint nut: 130 ft. lbs. (176 Nm)

• Pinch bolt adjuster bolt: 77 ft. lbs. (104 Nm)

11. Check and, if necessary, align the front end.

2002–03 F-150 2-Wheel Drive and 2004 F-150 Heritage Editions

1. Before servicing the vehicle, refer to the Precautions Section.

2. Raise the vehicle on a hoist.

3. Remove the wheel and tire assembly.

4. Remove the brake disc shield.

5. Use a suitable jack to support the front suspension lower arm.

6. Remove the upper ball joint castellated nut.

Spindle—E-Series

7. Use a pitman arm puller to separate the ball joint from the front wheel spindle.

➡Do not allow the shaft to turn while removing the nut. Hold the flats using a suitable wrench or damage to the shock can result.

8. Remove the upper shock absorber nut.

9. Remove two shock absorber lower nuts.

10. Remove the front shock absorber.

11. Remove the tie rod castellated nut.

12. Use a pitman arm puller to separate the tie rod end from the front wheel spindle.

13. Use the coil spring compressor to compress the coil spring.

14. Remove lower ball joint castellated nut.

15. Use a pitman arm puller to separate the lower ball joint from the front wheel spindle.

16. Remove the front wheel spindle.

To install:

17. Install the lower ball joint castellated nut. Torque to 83–112 ft. lbs. (113–153 Nm). Install a new cotter pin.

18. Install the upper ball joint castellated nut. Torque to 57–76 ft. lbs. (77–103 Nm). Install a new cotter pin.

19. Remove the coil spring compressor.

20. Install the tie rod castellated nut. Torque to 57–76 ft. lbs. (77–103 Nm).Install a new cotter pin.

21. Install the front shock absorber and the two shock absorber lower nuts. Torque to 22–29 ft. lbs. (30–40 Nm).

22. Install the upper shock absorber nut. Torque to 35–46 ft. lbs. (47–63 Nm).

23. Install the brake disc shield.

24. Install the tire and wheel assembly

25. Inspect and adjust the front end alignment.

2002–05 F-250 and 350 2-Wheel Drive

1. Before servicing the vehicle, refer to the Precautions Section.

2. Raise and support the vehicle.

3. Remove the wheel and tire assembly.

4. Remove the disc brake caliper and the front disc brake hub and rotor.

5. Remove the front disc brake rotor shield.

6. If equipped, remove the ABS sensor retaining bolt, ABS sensor harness retaining bolt and the ABS sensor. Position out of the way.

7. Remove and discard the cotter pin. Remove the castellated nut. Using a pitman arm puller, remove the tie rod end.

8. Remove the pinch bolt.

9. Remove the camber adjuster.

※※ WARNING

To prevent damage to the ball joint seal and the ball joint socket, do not use a pickle fork-type remover to loosen the ball joints.

10. Remove and discard the cotter pin. Loosen, but do not remove, the castellated nut. Strike the lower end of the front axle to loosen the ball joint.

11. Remove the castellated nut and the front wheel spindle.

➡**Tighten the ball joint nut further, if necessary, in order to insert a new cotter pin.**

12. Using new fasteners, follow the removal procedure in reverse order. Observe the following torques:

- Lower ball joint nut: 99 ft. lbs. (133 Nm)
- Pinch bolt: 60 ft. lbs. (80 Nm)
- Tie rod ball stud nut: 67 ft. lbs. (90 Nm)

13. Check the front end alignment.

Knuckle

REMOVAL & INSTALLATION

2002–03 F-150 4-Wheel Drive and F-150 2004 Heritage Edition

1. Before servicing the vehicle, refer to the Precautions Section.

2. Remove the wheel hub.

3. Remove the retaining bolts. Remove the torsion bar cover plate.

4. Make an alignment mark on the torsion bar and the torsion bar crossmember support. Measure and record the length.

5. Position the torsion bar tool and the adapters.

6. Tighten the torsion bar tool until the torsion bar adjuster lifts off of the adjustment bolt.

67197-EFSE-G182

Make an alignment mark on the torsion bar and the torsion bar crossmember support—2002–03 F-150 4-Wheel Drive and F-150 2004 Heritage Edition

67197-EFSE-G183

Tighten the torsion bar tool until the torsion bar adjuster lifts off of the adjustment bolt—2002–03 F-150 4-Wheel Drive and F-150 2004 Heritage Edition

67197-EFSE-G184

Remove the torsion bar adjustment bolt and nut—2002–03 F-150 4-Wheel Drive and F-150 2004 Heritage Edition

※※ CAUTION

The torsion bar adjustment bolt is coated with dry Loctite® and must be replaced if it is backed off or removed. Failure to do so can cause the adjustment bolt to loosen up during operation and a loss of vehicle alignment.

67197-EFSE-G185

Loosen the torsion bar adjustment tool until the tension is off the torsion bar—2002–03 F-150 4-Wheel Drive and F-150 2004 Heritage Edition

7. Remove the torsion bar adjustment bolt and nut.

8. Loosen the torsion bar adjustment tool until the tension is off the torsion bar.

9. If equipped with 4 wheel anti-lock brake system, safely reposition the anti-lock sensor wire.

10. Remove the upper ball joint cotter pin.

11. Remove the upper ball joint castellated nut.

12. Use a pitman arm puller to separate the front wheel knuckle from the front suspension upper arm.

13. Remove the tie rod end cotter pin.

14. Remove the tie rod end castellated nut.

15. Use a pitman arm puller to separate the tie rod end. Separate the tie rod end from the front wheel knuckle.

16. Remove the lower ball joint cotter pin.

17. Remove the lower ball joint castellated nut.

18. Use a pitman arm puller to separate the front wheel knuckle from the front suspension lower arm.

※※ WARNING

Suspend the front wheel driveshaft and joint with wire.

19. Remove the front wheel knuckle.

To install:

20. Position the front wheel knuckle.

21. Install the front suspension arm upper ball joint castellated nut. Torque to 57–75 ft. lbs. (77–103 Nm). Install a new cotter pin.

22. Position the front wheel knuckle.

23. Install the front suspension arm lower ball joint castellated nut. Torque to 83–112 ft. lbs. (113–153 Nm). Install a new cotter pin.

24. Position the tie rod end.

25. Install the tie rod end castellated nut. Torque to 57–75 ft. lbs. (77–103 Nm). Install a new cotter pin.

> ### ✳✳ CAUTION
>
> The torsion bar adjustment bolt is coated with dry Loctite® and must be replaced if it is backed off or removed. Failure to do so can cause the adjustment bolt to loosen up during operation and a loss of vehicle alignment.

26. Tighten the torsion bar tool until the new adjustment bolt and nut can be installed.

27. Turn the adjustment bolt until preliminary adjustment marks are aligned.

28. Position the torsion bar cover plate. Install the retaining bolts. Torque to 25–34 ft. lbs. (34–46 Nm).

29. Install the wheel hub.

30. If equipped with 4WABS, install the anti-lock sensor wire bracket bolt.

31. Adjust the ride height.

2002–05 F-250 and 350 4-Wheel Drive

1. Before servicing the vehicle, refer to the Precautions Section.

2. Raise and support the vehicle.

3. Remove the wheel and tire assembly.

4. Remove the front brake disc.

5. Remove the wheel hub and bearing.

6. Using a drift, drive the axle shaft main seal out of the wheel knuckle.

7. Remove the axle shaft and main seal.

8. Remove the tie-rod end castellated nut.

9. Using the special tool, disconnect the tie-rod end from the wheel knuckle.

10. Remove the upper ball joint castellated nut and the insert.

11. Remove the lower ball joint nut.

12. Remove the knuckle.

13. Clean and inspect the wheel knuckle ball joint bores.

To install:

14. Position the wheel knuckle onto the axle housing.

15. Install the nut onto the lower ball joint. Do not tighten the nut at this time.

16. Install the insert and the castellated nut onto the upper ball joint. Do not tighten the nut at this time.

17. Tighten the lower ball joint retaining nut. Pre-tighten the nut to 47 Nm (35 ft. lbs.).

➡ Do not loosen the castellated nut to install the cotter pin.

18. Tighten the upper ball joint castellated nut. Torque to 69 ft. lbs. (94 Nm).

Seat the main seal onto the axle shaft—2002–05 F-250 and 350 4-Wheel Drive

Install the main seal into the wheel knuckle—2002–05 F-250 and 350 4-Wheel Drive

19. Install the cotter pin. If necessary, tighten the castellated nut until the cotter pin can be installed.

20. Tighten the lower ball joint nut to 204 Nm (150 ft. lbs.).

21. Position the tie-rod end into the wheel knuckle.

22. Install and tighten the castellated nut. Torque to 52 ft. lbs. (70 Nm).

23. Install the cotter pin.

24. Position the main seal onto the axle shaft.

25. Using the special tools and a hammer, seat the main seal onto the axle shaft.

26. Position the axle shaft into the axle housing.

27. Using the special tools and a hammer, install the main seal into the wheel knuckle.

28. Install the wheel hub and bearing.

29. Install the front brake disc.

30. Install the wheel and tire assembly.

2004–05 F-150 2-Wheel Drive, exc. 2004 F-150 Heritage Editions

2006 Mark LT

1. Before servicing the vehicle, refer to the Precautions Section.

2. Disconnect the wheel speed sensor.

3. Remove the wheel and tire assembly.

➡ Use the hex holding feature to prevent the ball studs (upper control arm, lower control arm tie-rod end, and stabilizer bar links) from turning while removing and installing the nuts.

> ### ✳✳ WARNING
>
> Do not hammer the ball studs to separate them from the wheel knuckle or stabilizer bar. Doing so can cause damage. Lightly tap the wheel knuckle or stabilizer bar to loosen the joints.

4. Remove the anchor plate bolts.

5. Remove the brake caliper, pads and anchor plate.

> ### ✳✳ WARNING
>
> Do not allow the brake caliper to hang by the flexible brake hose.

6. Support the caliper to the vehicle.

7. Remove the axle-to-wheel hub nut, retainer and cotter pin.

8. Remove the brake disc.

9. Remove the tie-rod end-to-wheel knuckle nut.

➡ The upper ball joint, the lower ball joint, and the tie-rod end ball studs do not require special tools to be removed from the wheel knuckle.

10. Remove the lower ball joint-to-wheel knuckle nut.

11. Remove the upper ball joint-to-wheel knuckle nut.

12. To install, reverse the removal procedure. Observe the following torques:

- Anchor plate bolts: 148 ft. lbs. (200 Nm)
- Hub nut: 296 ft. lbs. (400 Nm)
- Lower ball joint nut: 111 ft. lbs. (148 Nm)
- Upper ball joint nut: 85 ft. lbs. (115 Nm)
- Tie rod end stud nut: 111 ft. lbs. (148 Nm)

13. Clean the axle (including the threaded end) of any dirt or grease using the specified solvent before installing the brake disc.

> ### ✳✳ WARNING
>
> Do not apply lubricant to the threaded part of the axle.

14. To ease installation, lubricate the inner race of the bearing assembly using the specified lubricant.

2004–05 F-150 4-Wheel Drive, exc. 2004 F-150 Heritage Edition 2006 Mark LT

1. Before servicing the vehicle, refer to the Precautions Section.
2. Loosen the axle retainer nut.

➡The wheel speed sensor electrical connectors are located in the engine compartment secured to the fender aprons.

3. Disconnect the wheel speed sensor.
4. Remove the wheel and tire assembly.

✳✳ WARNING

Use the hex holding feature to prevent the ball studs (upper control arm, lower control arm tie-rod end, and stabilizer bar links) from turning while removing and installing the nuts.

✳ WARNING

Do not hammer the ball studs to separate them from the wheel knuckle or stabilizer bar. Doing so can cause damage. Lightly tap the wheel knuckle or stabilizer bar to loosen the joints.

5. Remove the axle-to-wheel hub nut. Separate the outboard CV joint from the wheel hub.
6. Remove the wheel speed sensor harness connector.
7. Remove the anchor plate bolt.
8. Remove the brake caliper, pads and anchor plate.

✳ WARNING

Do not allow the brake caliper to hang from the hose or damage to the hose can occur.

9. Position the caliper, pads and anchor plate aside.
10. Remove the brake disc.
11. Remove the wheel hub-to-wheel knuckle bolts.
12. Remove the wheel bearing and hub assembly.
13. Remove the tie-rod end-to-wheel knuckle nut.
14. Remove the lower ball joint-to-wheel knuckle nut.
15. Remove the upper ball joint-to-wheel knuckle nut.
16. Remove the knuckle.

17. To install, reverse the removal procedure. Observe the following torques:
- Axle hub nut: 20 ft. lbs. (27 Nm)
- Anchor plate bolt: 148 ft. lbs. (200 Nm)
- Hub-to-knuckle bolts: 148 ft. lbs. (200 Nm)
- Lower ball joint nut: 111 ft. lbs. (150 Nm)
- Upper ball joint nut: 85 ft. lbs. (115 Nm)

18. Check and, if necessary, align the front end.

Upper Control Arm

REMOVAL & INSTALLATION

2002–03 F-150 and 2004 F-150 Heritage Editions

2-WHEEL DRIVE

1. Before servicing the vehicle, refer to the Precautions Section.
2. Raise the vehicle on a hoist.
3. Remove the wheel and tire assembly
4. Use a suitable jack to support the front suspension lower arm.
5. Mark the position of the front suspension upper arm cam.
6. Remove the upper ball joint castellated nut.
7. Use a Pitman Arm Puller to separate the ball joint from the front wheel spindle.
8. Remove the two nuts and the two bolts. Remove the front suspension upper arm.

To install:
9. Align the marks made during removal on the front suspension upper arm cams.

➡The forward front suspension upper arm nut must be tightened first while the control arms are held at the curb position ride height.

10. Position the front suspension upper arm. Install the two bolts and nuts. Torque to 84–112 ft. lbs. (113–153 Nm).
11. Install the upper ball joint castellated nut. Torque to 57–76 ft. lbs. (77–103 Nm). Install a new cotter pin.
12. Install the tire and wheel assembly.
13. Inspect and adjust the front end alignment.

4-WHEEL DRIVE

1. Before servicing the vehicle, refer to the Precautions Section.
2. Raise and support the vehicle.
3. Remove the wheel and tire assembly.

4. Use a suitable jack to support the lower arm.
5. If equipped with 4-wheel anti-lock brake system (4WABS), safely reposition the anti-lock sensor wire.
6. Remove the upper ball joint cotter pin.
7. Remove the upper ball joint castellated nut.
8. Use a Pitman Arm Puller to separate the front suspension upper arm from the front wheel knuckle.
9. Mark the position of the camber adjustment cam.
10. Remove the nuts. Remove the bolts. Remove the front suspension upper arm.

To install:

➡Align the marks made during removal on the camber adjustment cam before tightening the nuts.

11. Position the front suspension upper arm. Install the bolts. Install the nuts. Torque to 83–112 ft. lbs. (113–153 Nm).
12. Position the front wheel knuckle.
13. Install the upper ball joint castellated nut. Torque to 57–75 ft. lbs. (77–103 Nm). Install a new cotter pin.
14. Position the anti-lock sensor wire. Install the bolt. Torque to 60–80 inch lbs. (7–9 Nm).

✳✳ WARNING

When the wheel is installed, always remove any corrosion, dirt or foreign material present on the mounting surfaces of the wheels or the surfaces of the wheel hub, brake drum or brake disc that contact the wheel. Installing wheels without proper metal-to-metal contact at the wheel mounting surfaces can cause the wheel nut to loosen and the wheel to come off while the vehicle is in motion, causing loss of control.

15. Clean the wheel hub and mounting surfaces.
16. Install the tire and wheel assembly.
17. Lower the vehicle.
18. Check the front end alignment.

2004–05 F-150, Exc. 2004 F-150 Heritage Editions 2006 Mark LT

2-WHEEL DRIVE

1. Before servicing the vehicle, refer to the Precautions Section.
2. Disconnect the wheel speed sensor.
3. Remove the wheel and tire assembly.

200 Nm (148 lb-ft) – ① N

8

4

2 i

N ⑦ – 115 Nm (85 lb-ft)

N ⑤ – 150 Nm (111 lb-ft)

N

N ⑥ – 150 Nm (111 lb-ft)

400 Nm (296 lb-ft) – ③ N

1 Anchor plate bolt	5 Tie-rod end-to-wheel knuckle nut
2 Brake caliper, pads and anchor plate	6 Lower ball joint-to-wheel knuckle nut
3 Axle-to-wheel hub nut (retainer/cotter pin)	7 Upper ball joint-to-wheel knuckle nut
4 Brake disc	8 Wheel knuckle

67197-EFSE-G177

Spindle and hub components—2004–05 F-150 2-Wheel Drive, Exc. 2004 F-150 Heritage Editions; 2006 Mark LT

➡ Use the hex holding feature to prevent the ball studs (upper control arm, lower control arm tie-rod end, and stabilizer bar links) from turning while removing and installing the nuts.

✳✳ **WARNING**

Do not hammer the ball studs to separate them from the wheel knuckle or stabilizer bar. Doing so can cause damage. Lightly tap the wheel knuckle or stabilizer bar to loosen the joints.

4. Remove the anchor plate bolt.
5. Remove the brake caliper, pads and anchor plate.

✳✳ **WARNING**

Do not allow the brake caliper to hang by the flexible brake hose.

6. Support the caliper to the vehicle.
7. Remove the axle-to-wheel hub nut, retainer and cotter pin.
8. Remove the brake disc.
9. Remove the tie-rod end-to-wheel knuckle nut.

10. Remove the lower ball joint-to-wheel knuckle nut.
11. Remove the upper ball joint-to-wheel knuckle nut.
12. Remove the wheel knuckle.
13. Remove the strut-to-lower arm nut.
14. Remove the strut-to-lower arm bolt.
15. Remove the upper arm-to-frame nut.
16. Remove the upper arm-to-frame bolt.
17. Remove the upper arm.
18. To install, reverse the removal procedure. Observe the following torques:

10 — 115 Nm (85 lb-ft)

2

11

6

200 Nm (148 lb-ft)

200 Nm (148 lb-ft) — 3

4

8 — 150 Nm (111 lb-ft)

7

5

9 — 150 Nm (111 lb-ft)

27 Nm (20 lb-ft) — 1

1 Axle-to-wheel hub nut (4x4 only)

2 Wheel speed sensor harness connector

3 Anchor plate bolt

4 Brake caliper, pads and anchor plate

5 Brake disc

6 Wheel hub-to-wheel knuckle bolts

7 Wheel bearing and hub assembly

8 Tie-rod end-to-wheel knuckle nut

9 Lower ball joint-to-wheel knuckle nut

10 Upper ball joint-to-wheel knuckle nut

11 Wheel knuckle

67197-EFSE-G179

Spindle and hub components—2004–05 F-150 4-Wheel Drive, Exc. 2004 F-150 Heritage Editions; 2006 Mark LT

- Control arm-to-frame: 114 ft. lbs. (155 Nm)
- Upper control arm-to-knuckle: 85 ft. lbs. (115 Nm)
- Lower control arm-to-knuckle: 111 ft. lbs. (150 Nm)
- Tie rod end-to-knuckle: 111 ft. lbs. (150 Nm)
- Hub nut: 296 ft. lbs. (400 Nm)
- Anchor plate bolts: 148 ft. lbs. (200 Nm)

❊❊ CAUTION

Do not apply lubricant to the threaded part of the axle.

19. To ease installation, lubricate the inner race of the bearing assembly using the specified lubricant.

4-WHEEL DRIVE

1. Before servicing the vehicle, refer to the Precautions Section.
2. Loosen the axle retainer nut.

➡**The wheel speed sensor electrical connectors are located in the engine compartment secured to the fender aprons.**

3. Disconnect the wheel speed sensor.
4. Remove the wheel and tire assembly.

❊❊ WARNING

Use the hex holding feature to prevent the ball studs (upper control arm, lower control arm tie-rod end, and stabilizer bar links) from turning while removing and installing the nuts.

❊❊ WARNING

Do not hammer the ball studs to separate them from the wheel knuckle or stabilizer bar. Doing so can cause damage. Lightly tap the wheel knuckle or stabilizer bar to loosen the joints.

Front suspension components—2004–05 F-150 2- and 4-Wheel drive, Exc. 2004 F-150 Heritage Editions; 2006 Mark LT

12 Shock absorber-to-lower arm nut	16 Lower arm-to-frame bolt	19 Upper arm-to-frame nut
13 Shock absorber-to-lower arm bolt	17 Lower arm-to-frame bolt	20 Upper arm-to-frame bolt
14 Lower arm-to-frame nut	18 Lower arm	21 Upper arm RH/LH
15 Lower arm-to-frame nut		

67197-EFSE-G180

5. Remove the axle-to-wheel hub nut. Remove the nut and separate the outboard CV-joint from the wheel hub.

6. Remove the wheel speed sensor harness connector.

7. Remove the anchor plate bolt.

8. Remove the brake caliper, pads and anchor plate .

✳✳ WARNING

Do not allow the brake caliper to hang from the hose or damage to the hose can occur.

9. Position the caliper, pads and anchor plate aside.

10. Remove the brake disc.

11. Remove the wheel hub-to-wheel knuckle bolts.

12. Remove the wheel bearing and hub assembly.

13. Remove the tie-rod end-to-wheel knuckle nut.

14. Remove the lower ball joint-to-wheel knuckle nut.

15. Remove the upper ball joint-to-wheel knuckle nut.

16. Remove the wheel knuckle.

17. Remove the shock absorber-to-lower arm nut.

18. Remove the shock absorber-to-lower arm bolt.

19. Remove the upper arm-to-frame nut.

20. Remove the upper arm-to-frame bolt.

21. Remove the upper arm.

➡Do not tighten the control arm nuts until the installation procedure is complete and the weight of the vehicle is resting on the wheel and tire assemblies.

22. To install, reverse the removal procedure.

23. Observe the following torques:
  • Upper arm-to-frame: 114 ft. lbs. (155 Nm)
  • Strut-to-lower arm: 351 ft. lbs. (475 Nm)

- Upper arm-to-knuckle: 85 ft. lbs. (115 Nm)
- Lower arm-to-knuckle: 111 ft. lbs. (150 Nm)
- Tie rod end-to-knuckle: 111 ft. lbs. (150 Nm)
- Anchor plate: 148 ft. lbs. (200 Nm)

24. Check and, if necessary, align the front end.

Lower Control Arm

REMOVAL & INSTALLATION

2002–03 F-150 and 2004 F-150 Heritage Editions

1. Before servicing the vehicle, refer to the Precautions Section.
2. Raise the vehicle on a hoist.
3. Remove the wheel and tire assembly
4. Remove the brake disc shield.

✳✳ WARNING

Do not allow the shaft to turn while removing the nut. Hold the flats using a suitable wrench or damage to the shock can result.

5. Remove the upper shock absorber nut and washer.
6. Remove the front shock absorber.
7. Remove the brake hose bracket screw and bracket from the front suspension lower arm.
8. Remove the front stabilizer bar link nut.
9. Use the Coil Spring Compressor to compress the coil spring.
10. Remove the lower ball joint castellated nut. Discard the cotter pin.
11. Use a Pitman Arm Puller to separate the lower ball joint from the front wheel spindle.
12. Remove the front suspension lower arm nuts.
13. Remove the front suspension lower arm bolts.
14. Remove the front suspension lower arm and front coil spring.
To install:

➡**The end of the coil spring must cover the first hole and be visible in the second hole.**

15. Install the front coil spring into the front suspension lower arm.

➡**The forward front suspension lower arm nut must be tightened first while the control arm is held at the curb position ride height.**

16. Position the front suspension lower arm and front coil spring.
17. Install the bolts and nuts. Torque to 121–148 ft. lbs. (164–200 Nm).
18. Install the lower ball joint. Install the castellated nut. Torque to 83–112 ft. lbs. (113–153 Nm). Install a new cotter pin.
19. Install the front stabilizer bar link nut. Torque to 15–21 ft. lbs. (21–29 Nm).
20. Remove the Coil Spring Compressor.
21. Install the shock absorber and the shock absorber lower nuts. Torque to 22–29 ft. lbs. (30–40 Nm).
22. Install the upper shock absorber washer and nut. Torque to 35–46 ft. lbs. (47–63 Nm).
23. Install the brake hose bracket screw.
24. Install the brake disc shield.
25. Install the tire and wheel assembly.
26. Inspect and adjust the front end alignment.

2004–05 F-150, Exc. 2004 F-150 Heritage Editions
2006 Mark LT

2-WHEEL DRIVE

1. Before servicing the vehicle, refer to the Precautions Section.
2. Disconnect the wheel speed sensor.
3. Remove the wheel and tire assembly.

➡**Use the hex holding feature to prevent the ball studs (upper control arm, lower control arm tie-rod end, and stabilizer bar links) from turning while removing and installing the nuts.**

✳✳ WARNING

Do not hammer the ball studs to separate them from the wheel knuckle or stabilizer bar. Doing so can cause damage. Lightly tap the wheel knuckle or stabilizer bar to loosen the joints.

4. Remove the anchor plate bolt.
5. Remove the brake caliper, pads and anchor plate.

✳✳ WARNING

Do not allow the brake caliper to hang by the flexible brake hose.

6. Support the caliper to the vehicle.
7. Remove the axle-to-wheel hub nut, retainer and cotter pin.
8. Remove the brake disc.
9. Remove the tie-rod end-to-wheel knuckle nut.

10. Remove the lower ball joint-to-wheel knuckle nut.
11. Remove the upper ball joint-to-wheel knuckle nut.
12. Remove the wheel knuckle.
13. Remove the strut-to-lower arm nut.
14. Remove the strut-to-lower arm bolt.
15. Remove the lower arm-to-frame nut.
16. Remove the lower arm-to-frame bolt.
17. Remove the lower arm.
18. To install, reverse the removal procedure. Observe the following torques:
- Control arm-to-frame: 222 ft. lbs. (300 Nm)
- Upper control arm-to-knuckle: 85 ft. lbs. (115 Nm)
- Lower control arm-to-knuckle: 111 ft. lbs. (150 Nm)
- Tie rod end-to-knuckle: 111 ft. lbs. (150 Nm)
- Hub nut: 296 ft. lbs. (400 Nm)
- Anchor plate bolts: 148 ft. lbs. (200 Nm)

✳✳ CAUTION

Do not apply lubricant to the threaded part of the axle.

19. To ease installation, lubricate the inner race of the bearing assembly using the specified lubricant.

4-WHEEL DRIVE

1. Before servicing the vehicle, refer to the Precautions Section.
2. Loosen the axle retainer nut.

➡**The wheel speed sensor electrical connectors are located in the engine compartment secured to the fender aprons.**

3. Disconnect the wheel speed sensor.
4. Remove the wheel and tire assembly.

✳✳ WARNING

Use the hex holding feature to prevent the ball studs (upper control arm, lower control arm tie-rod end, and stabilizer bar links) from turning while removing and installing the nuts.

✳✳ WARNING

Do not hammer the ball studs to separate them from the wheel knuckle or stabilizer bar. Doing so can cause damage. Lightly tap the wheel knuckle or stabilizer bar to loosen the joints.

5. Remove the axle-to-wheel hub nut. Remove the nut and separate the outboard CV-joint from the wheel hub.

6. Remove the wheel speed sensor harness connector.

7. Remove the anchor plate bolt.

8. Remove the brake caliper, pads and anchor plate .

❋❋ WARNING

Do not allow the brake caliper to hang from the hose or damage to the hose can occur.

9. Position the caliper, pads and anchor plate aside.

10. Remove the brake disc.

11. Remove the wheel hub-to-wheel knuckle bolts.

12. Remove the wheel bearing and hub assembly.

13. Remove the tie-rod end-to-wheel knuckle nut.

14. Remove the lower ball joint-to-wheel knuckle nut.

15. Remove the upper ball joint-to-wheel knuckle nut.

16. Remove the wheel knuckle.

17. Remove the shock absorber-to-lower arm nut.

18. Remove the shock absorber-to-lower arm bolt.

19. Remove the lower arm-to-frame nut.

20. Remove the lower arm-to-frame bolt.

21. Remove the lower arm.

➡ **Do not tighten the control arm nuts until the installation procedure is complete and the weight of the vehicle is resting on the wheel and tire assemblies.**

22. To install, reverse the removal procedure.

23. Observe the following torques:
- Lower arm-to-frame: 222 ft. lbs. (300 Nm)
- Strut-to-lower arm: 351 ft. lbs. (475 Nm)
- Upper arm-to-knuckle: 85 ft. lbs. (115 Nm)
- Lower arm-to-knuckle: 111 ft. lbs. (150 Nm)

- Tie rod end-to-knuckle: 111 ft. lbs. (150 Nm)
- Anchor plate: 148 ft. lbs. (200 Nm)

24. Check and, if necessary, align the front end.

Stabilizer Bar

REMOVAL & INSTALLATION

E-Series

1. Before servicing the vehicle, refer to the Precautions Section.

2. Remove the wheel and tire assembly.

3. Remove the four bolts and the stabilizer bar.

4. To install, reverse the removal procedure. Torque the bolts to 18 ft. lbs. (25 Nm).

2002–03 Blackwood, 2- or 4-Wheel Drive F-150 and 2004 2- or 4-Wheel Drive F-150 Heritage Edition

1. Before servicing the vehicle, refer to the Precautions Section.

2. Raise the vehicle on a hoist.

26 - 35 Nm (19 - 26 Lb - Ft)

Stabilizer bar and link—2002–03 Blackwood, 2- or 4-Wheel Drive F-150 and 2004 2- or 4-Wheel Drive F-150 Heritage Edition

06017-F150-G271

90 Nm (66 lb-ft)

2

133 Nm (98 lb-ft)

1

6

5

7

6

4

48 Nm (35 lb-ft) — 4

1 Stabilizer bar link upper nut
2 Stabilizer bar link lower nut
3 Stabilizer bar link
4 Stabilizer bar bracket nuts

5 Bracket
6 Insulator (bushing)
7 Stabilizer bar

06017-F150-G269

Stabilizer bar and link—2004 2- or 4-Wheel Drive F-150, exc. Heritage Edition; Early Build 2005 2- or 4-Wheel Drive F-150

3. Remove the front stabilizer bar link nuts from the front suspension lower arms.

4. Remove the front stabilizer bar link bolts and remove the front stabilizer bar links.

5. Remove the four bracket bolts.

6. Remove the front stabilizer bar.

7. Remove the lower suspension arm stabilizer bar insulators.

8. To install, reverse the removal procedure. Torque the bracket bolts to 19–26 ft. lbs. (26–35 Nm). Torque the end links to 15–21 ft. lbs. (26–35 Nm).

2004–05 2- or 4-Wheel Drive F-150, exc. 2004 F-150 Heritage Edition 2006 Mark LT

1. Before servicing the vehicle, refer to the Precautions Section.

➡Use the hex holding feature to pre-vent the ball studs (upper ball joint, lower ball joint, tie-rod end and stabilizer bar links) from turning while removing and installing the nuts.

✲✲ WARNING

Do not hammer the ball studs to separate them from the wheel knuckle or stabilizer bar. Doing so can cause damage. Lightly tap the wheel knuckle or stabilizer bar to loosen the joints.

2. Remove the stabilizer bar link upper nuts.

3. Remove the stabilizer bar link lower nuts.

4. Remove the stabilizer bar links.

5. Remove the stabilizer bar bracket nuts, then, remove the stabilizer bar.

6. To install, reverse the removal procedure. Observe the following torques:
- Stabilizer bar bracket nuts: 48 Nm (35 ft. lbs.)
- Stabilizer bar link lower nuts: 90 Nm (66 ft. lbs.)
- Stabilizer bar link upper nuts: 133 Nm (98 ft. lbs.)

2002–05 F-250/350 2-Wheel Drive

1. Before servicing the vehicle, refer to the Precautions Section.

2. With the vehicle in NEUTRAL, position it on a hoist.

3. Remove the bolts and the retainer brackets.

90 Nm (66 lb-ft)

133 Nm (98 lb-ft)

48 Nm (35 lb-ft)

1 Stabilizer bar link-to-stabilizer
  bar nut
2 Stabilizer bar link-to-lower
  control arm nut
3 Stabilizer bar link

4 Stabilizer bar-to-frame nuts
5 Bracket
6 Insulator (bushing)
7 Stabilizer bar

06017-F150-G270

Stabilizer bar and link—Late Build 2005 2- or 4-Wheel Drive F-150; 2006 Mark LT

4. Remove the nuts, bolts, washers and the stabilizer bar and links.

5. To install, reverse the removal procedure. Torque the fasteners as follows:
- 2002–04 upper and lower link nuts: 80 ft. lbs. (109 Nm)
- 2005 upper and lower link nuts: 85 ft. lbs. (115 Nm)
- 2002–04 bracket bolts: 32 ft. lbs. (43 Nm)
- 2005 bracket bolts: 35 ft. lbs. (48 Nm)

2002–04 F-250/350 4-Wheel Drive

1. Before servicing the vehicle, refer to the Precautions Section.

2. Remove the nuts, washers, and bolts, and then disconnect the stabilizer bar links.

3. Remove the nuts and the stabilizer bar.

4. To install, reverse the removal procedure. Observe the following torques:
- Bracket nuts: 41 ft. lbs. (55 Nm)
- End link nuts to 80 ft. lbs. (109 Nm)

2005 F-250/350 4-Wheel Drive

1. Before servicing the vehicle, refer to the Precautions Section.

2. Remove the nuts and bolts and then disconnect the stabilizer bar links.

3. Remove the nuts and the stabilizer bar.

4. To install, reverse the removal procedure. Torque the bracket bolts to 35 ft. lbs. (48 Nm). Torque the link nuts to 111 ft. lbs. (150 Nm).

Track Bar

REMOVAL & INSTALLATION

2002–04 F-250/350

1. Before servicing the vehicle, refer to the Precautions Section.

2. Raise and support the vehicle.

➡**To prevent the front suspension from shifting when the track bar is removed, the front suspension must be supporting vehicle's weight.**

3. Load the front suspension with the vehicle's weight.

4. Position jack stands under the front axle housing.

5. Lower the vehicle until the front suspension is supporting the vehicle's weight.

6. Remove the nut and bolt and disconnect the track bar from the upper mounting bracket.

7. Remove the nut and bolt and remove the track bar.

8. To install, reverse the removal procedure. Observe the following torques:
- Lower end bolt/nut: 406 ft. lbs. (550 Nm)
- Upper end bolt/nut: 406 ft. lbs. (550 Nm)

2005 F-250/350

1. Before servicing the vehicle, refer to the Precautions Section.

2. With the vehicle in NEUTRAL, position it on a hoist.

➡**To prevent the front suspension from shifting when the track bar is removed, the front suspension must be supporting vehicle's weight.**

3. Position jack stands under the front axle housing.

4. Remove the nut and bolt and disconnect the track bar from the upper mounting bracket.

5. Remove the nut, then disconnect and remove the track bar.

6. To install, reverse the removal procedure. Torque the upper end nut to 406 ft. lbs. (550 Nm); the lower end nut to 183 ft. lbs. (250 Nm).

Front Wheel Bearings

ADJUSTMENT

Only the following have adjustable front wheel bearings.

E-Series, Blackwood, 2-Wheel Drive 2002–03 F-150 and 2-Wheel Drive 2004 F-150 Heritage Editions

✳✳ CAUTION

If equipped with the automatic air suspension system, the service switch near the right kick panel must be turned OFF before raising the vehicle for service.

1. Before servicing the vehicle, refer to the Precautions Section.
2. Raise and safely support the vehicle.
3. Support the front end.
4. Remove the wheel cover, if equipped.
5. Remove the grease cap.

➡**Check the wheel bearings for sufficient grease.**

6. Remove the cotter pin and retaining washer. Back off the spindle nut. Discard the cotter pin.

7. Adjust the wheel bearings as follows:
   a. Tighten the spindle nut to 30 ft. lbs. (40 Nm) while rotating the brake disc clockwise to seat the wheel bearings.
   b. Back off the nut 2 full turns.
   c. While rotating the disc counterclockwise, tighten the nut to 17–24 ft. lbs. (23–34 Nm).
   d. Back off the spindle nut about ½ turn.
   e. Tighten the spindle nut to 17 inch lbs. (2 Nm).
8. Install the retaining washer so the castellations are aligned with the cotter pin hole. Install a new cotter pin.
9. Check the wheel and tire assembly for proper rotation, then install the grease cap. If the wheel still does not rotate properly, inspect and clean or replace the wheel bearings and cups.
10. Install the wheel cover, if equipped.
11. Lower the vehicle.

1 Grease cap	5 Washer
2 Cotter pin	6 Outer wheel bearing
3 Retainer	7 Brake disc and hub assembly
4 Spindle nut	8 Grease seal

9 Inner wheel bearing
10 Inner bearing cup
11 Outer bearing cup

67197-EFSE-G191

Front hub and bearings—E-Series, Blackwood, 2-wheel drive 2002–03 F-150 and 2-wheel drive 2004 F-150 Heritage Editions; 2-wheel drive F-250/350 similar

12. Road test the vehicle and check for proper operation.

2002–05 2-Wheel Drive F-250/350

To check the wheel bearing adjustment, raise the front of the vehicle. Grasp the tire at the sides, and alternately push inward and pull outward on the tire. If any looseness is felt, adjust the front wheel bearings as follows.

1. Before servicing the vehicle, refer to the Precautions Section.
2. Remove the hub cap from the hub.
3. Remove the cotter pin and the castellated nut.
4. While rotating the wheel, tighten the adjusting nut to 21 ft. lbs. (28 Nm) to seat the bearings.
5. Back off the adjusting nut until loose (120–180 degrees).
6. While rotating the wheel, tighten the adjusting nut to 18 inch lbs. (2Nm). Torque required to rotate the hub should be 18 inch lbs. (2 Nm).
7. Install the castellated nut and insert a new cotter pin.
8. Install the hub cap.

REMOVAL & INSTALLATION

E-Series, Blackwood, 2-Wheel Drive 2002–03 F-150 and 2-Wheel Drive 2004 F-150 Heritage Editions

The hub is part of the disc brake rotor and cannot be serviced separately. The inner and outer wheel bearing and races are serviced individually. Be sure to have a new hub grease seal when servicing the wheel bearings.

1. Before servicing the vehicle, refer to the Precautions Section.
2. Remove or disconnect the following:
   - Wheels
   - Caliper
   - Brake pads and anti-rattle clips
   - Hub grease cap, cotter pin, retainer washer and the spindle nut
   - Wheel bearing retainer washer and the outer wheel bearing
   - Brake hub and rotor assembly
   - Grease seal
   - Inner wheel bearing
3. Clean and inspect the wheel bearings and races for unusual wear or damage. Replace parts as necessary.
4. Inspect the hub and brake rotor assembly. If required, the hub and brake rotor assembly must be replaced as a unit.

To install:

5. If needed, pack the wheel bearing with a suitable high temperature wheel bearing grease before assembly.
6. Install or connect the following:
   - Inner wheel bearing in the hub and brake rotor assembly
   - New grease seal
   - Hub and rotor assembly on the wheel spindle and install the outer wheel bearing.
   - Retainer washer and the spindle nut
7. Adjust the wheel bearings.
8. Install or connect the following:
   - Retaining washer, so the castellations are aligned with the cotter pin hole. Install a new cotter pin.
   - Anti-rattle clips
   - Disc brake pads
   - Caliper
   - Wheels. Tighten the lug nuts to 83–112 ft. lbs. (113–153 Nm).
9. Check the wheel and tire assembly for proper rotation, then install the grease cap.
10. Lower the vehicle.
11. Road test the vehicle and check for proper operation.

2002–03 F-150 4-Wheel Drive 2004 F-150 4-Wheel Drive Heritage Edition

The wheel bearings are of the cartridge design and are an integral part of the hub assembly. The bearings are permanently lubricated and require no maintenance or adjustments. If required, a new hub assembly must be installed.

1. Before servicing the vehicle, refer to the Precautions Section.
2. Remove the wheel and tire assembly.
3. Remove the brake disc.
4. Remove the cotter pin. Remove the retainer. Remove the front wheel hub nut.
5. If equipped with 4-wheel anti-lock brake system (4WABS), remove the brake disc shield.
6. If equipped with 4WABS, remove the front brake anti-lock sensor.

**255 - 345 Nm
(188 - 254 Lb - Ft)**

**9 - 12 Nm
(80 - 107 Lb - In)**

67197-EFSE-G192

Front hub/bearing installation—2002–03 F-150 4-Wheel Drive and 2004 F-150 4-Wheel Drive Heritage Edition

✳✳ WARNING

Do not overextend CV-joint and boots when removing the hub and bearing assembly.

➡ The CV-joint is a slip fit into the wheel hub and bearing. A puller will not normally be required.

7. Remove the bolts. Push the CV-joint inboard. Remove the wheel hub.

8. Remove the seal.

Install:

9. Install the seal.

10. Position the wheel hub on the front wheel driveshaft and joint and into the front wheel knuckle.

11. Install the bolts. Torque to 110–148 ft. lbs. (149–201 Nm).

12. If equipped with 4WABS, install the front brake anti-lock sensor. Torque to 62–80 inch lbs. (7–9 Nm).

13. Position the brake disc shield.

14. Install the bolts. Torque to 80–107 inch lbs. (9–12 Nm).

15. Install the front axle hub nut. Torque to 188–254 ft. lbs. (255–345 Nm).

16. Install the retainer.

17. Install the cotter pin.

18. Install the brake disc.

✳✳ WARNING

When the wheel is installed, always remove any corrosion, dirt or foreign material present on the mounting surfaces of the wheels or the surfaces of the wheel hub, brake drum or brake disc that contact the wheel. Installing wheels without proper metal-to-metal contact at the wheel mounting surfaces can cause the wheel nut to loosen and the wheel to come off while the vehicle is in motion, causing loss of control.

19. Clean the wheel hub and mounting surfaces.

20. Install the tire and wheel assembly.

21. Lower the vehicle.

2004–05 2-Wheel Drive F-150, exc. 2004 F-150 Heritage Editions 2006 Mark LT

1. Before servicing the vehicle, refer to the Precautions Section.

2. Disconnect the wheel speed sensor.

3. Remove the wheel and tire assembly.

4. Remove the anchor plate bolts.

5. Remove the brake caliper, pads and anchor plate. Support the caliper to the vehicle.

6. Remove the axle-to-wheel hub nut, retainer and cotter pin.

7. Remove the brake disc/hub assembly.

8. To install, reverse the removal procedure.

200 Nm (148 lb-ft) — (1) N

(8)

(4)

(2) i

(7) — 115 Nm (85 lb-ft)

(5) — 150 Nm (111 lb-ft)

(6) — 150 Nm (111 lb-ft)

400 Nm (296 lb-ft) — (3) N

1 Anchor plate bolt

2 Brake caliper, pads and anchor plate

3 Axle-to-wheel hub nut (retainer/cotter pin)

4 Brake disc

5 Tie-rod end-to-wheel knuckle nut

6 Lower ball joint-to-wheel knuckle nut

7 Upper ball joint-to-wheel knuckle nut

8 Wheel knuckle

67197-EFSE-G193

Front hub/bearing installation—2004–05 2-wheel drive F-150, exc. 2004 F-150 Heritage Editions; 2006 Mark LT

✳✳ CAUTION

Do not apply lubricant to the threaded part of the axle.

9. To ease installation, lubricate the inner race of the bearing assembly using the specified lubricant.
10. Observe the following torques:
- Hub nut: 296 ft. lbs. (400 Nm)
- Anchor plate bolts: 148 ft. lbs. (200 Nm)

2004–05 F-150 4-Wheel Drive, exc. 2004 F-150 Heritage Edition 2006 Mark LT

1. Before servicing the vehicle, refer to the Precautions Section.
2. Loosen the axle retainer nut.

➡The wheel speed sensor electrical connectors are located in the engine compartment secured to the fender aprons.

3. Disconnect the wheel speed sensor.
4. Remove the wheel and tire.
5. Remove the axle-to-wheel hub nut and, using a puller, separate the outboard CV joint from the wheel hub. assembly.
6. Remove the wheel speed sensor harness connector.
7. Remove the anchor plate bolts.
8. Remove the brake caliper, pads and anchor plate.

✳✳ CAUTION

Do not allow the brake caliper to hang from the hose or damage to the hose can occur.

9. Position the caliper, pads and anchor plate aside.
10. Remove the brake disc.
11. Remove the wheel hub-to-wheel knuckle bolts.
12. Remove the wheel bearing and hub assembly.
13. To install, reverse the removal procedure.
14. Observe the following torques:
- Anchor plate bolts: 148 ft. lbs. (200 Nm)
- Hub nut: 20 ft. lbs. (27 Nm)

2002–05 2-Wheel Drive F-250/350

1. Before servicing the vehicle, refer to the Precautions Section.
2. Raise and support the vehicle.
3. Remove the front wheel and tire assemblies.

N (10) 115 Nm (85 lb-ft)

2 i

11 i

6 N 200 Nm (148 lb-ft)

200 Nm (148 lb-ft) 3 N

7

5

4 i

N 8 150 Nm (111 lb-ft)

N 9 150 Nm (111 lb-ft)

27 Nm (20 lb-ft) 1 i

1 Axle-to-wheel hub nut (4x4 only)
2 Wheel speed sensor harness connector
3 Anchor plate bolt
4 Brake caliper, pads and anchor plate
5 Brake disc
6 Wheel hub-to-wheel knuckle bolts

7 Wheel bearing and hub assembly
8 Tie-rod end-to-wheel knuckle nut
9 Lower ball joint-to-wheel knuckle nut
10 Upper ball joint-to-wheel knuckle nut
11 Wheel knuckle

67197-EFSE-G194

Front hub/bearing assembly—2004–05 F-150 4-wheel drive, exc. 2004 F-150 Heritage Edition; 2006 Mark LT

4. Remove the front disc brake caliper and rotor, and position the caliper out of the way.

5. Remove the hub cap from the hub assembly.

6. Remove the cotter pin, adjusting nut and flat washer.

➡ **Inspect the condition of the spindle and nut threads to ensure a free turning nut when reassembling.**

7. Remove the outer bearing cone and roller assembly, and pull the hub assembly from the spindle.

8. Using care not to damage the bearing cage, use a suitable slide hammer and bearing seal remover to remove the inner bearing cone and bearing seal.

To install:

➡ **Do not spin the bearing dry with compressed air.**

➡ **Remove all traces of lubricant from the bearings, hub and axle spindle. Inspect bearings and bearing cups for pitting, spalling or unusual wear. If either bearings or bearing cups are worn or damaged, replace both bearings and bearing cups.**

➡ **It is recommended that bearings and bearing cups be replaced in sets. If cups are worn or damaged, install the inner and outer bearing cups in the hub with an appropriate bearing cup driver tool. Check for proper seating of new bearing cups by trying to insert a 0.38-mm (0.0015 inch) feeler gauge between the bottom face of the cup and wheel hub seat. You should not be able to insert the feeler gauge.**

9. Remove all burrs, nicks or scratches from the shoulder of the spindle and seal bore in the hub with emery cloth.

10. Pack the inside of the hub with lithium-base wheel bearing grease such as Motorcraft Premium Long-Life Grease XG-1-C or -K or equivalent meeting Ford specification ESA-M1C75-B. Fill the hub until the grease is flush with the inside diameters of both bearing cups.

11. Pack the bearing cone and roller assemblies with wheel bearing grease. Use a bearing packer for this operation. If a packer is not available, work as much lubricant as possible between the rollers and cages.

✳✳ WARNING

Keep the hub centered on the spindle to prevent damage to the grease seal or spindle threads.

12. Place the inner bearing cone and roller assembly in the inner cup and install the wheel bearing hub seal, using a suitable seal replacer. Make sure seal is fully seated and lubricated.

13. Install the hub assembly.

14. Install the outer bearing cone and roller assembly and the flat washer on the spindle and install the adjusting nut. Adjust the bearings. Install a new cotter pin.

15. Install the hub cap.

16. Install the front disc brake caliper and rotor.

17. Install the front wheel and tire assemblies.

18. Lower the vehicle.

2002–05 4-Wheel Drive F-250, 350

1. Before servicing the vehicle, refer to the Precautions Section.

All vehicles

2. Remove the wheel and tire assembly.

3. Remove the two front disc brake caliper anchor plate bolts.

4. Remove the front disc brake caliper anchor plate and position aside.

➡ **If excessive force must be used during brake rotor removal, the brake rotors should be checked for lateral runout prior to installation.**

5. On F-250 and F-350 4x4 SRW vehicles, remove the rotor.

6. On DRW vehicles, remove the eight hub plate nuts. Remove the hub plate. Remove the rotor.

7. Remove the retainer ring. Pull outward and remove the hub lock.

8. Remove the snapring. Remove the three thrust washers.

Vehicles equipped with ABS

➡ **Do not remove the ABS sensor from the bearing.**

9. Disconnect the ABS wheel sensor harness.

All vehicles

➡ **The wheel hub and bearing is a slip fit design and should not require a puller to remove it.**

10. Remove the four lock nuts. Remove the wheel hub and bearing.

11. If necessary, remove the brake disc shield.

Vehicles with ABS

12. If necessary, remove the bolt and the ABS sensor.

All vehicles

➡ **If necessary, position the hub in a soft-jawed vise.**

13. Install two nuts on the studs and use the inner nut to remove the studs.

14. Remove and discard the O-ring.

To install:
All vehicles

➡ **Any time the wheel hub is removed for any reason, a new O-ring seal must be installed. Failure to do so can cause a vacuum leak and loss of four wheel drive operations.**

15. Install a new O-ring.

➡ **Position the hub in a soft-jawed vise.**

16. Install two nuts on the studs and use the outer nut to install the studs.

Vehicles equipped with ABS

17. Position the ABS sensor and install the bolt. Torque to 13 ft. lbs. (18 Nm).

All vehicles

18. Position the brake disc shield.

➡ **Apply a coat of Ford High Temperature 4x4 Front Axle and Wheel Bearing Grease E8TZ-19590-A meeting Ford specification ESA-M1C198-A to the O-ring area of the wheel hub and bearing before installing the hub and bearing.**

19. Position the wheel hub and bearing. Install the four lock nuts. Torque to 133 ft. lbs. (180 Nm).

Vehicle equipped with ABS

20. Connect the ABS sensor harness.

All vehicles

✳✳ CAUTION

The non-metallic thrust washer must be installed between the two metal thrust washers. Failure to do so will cause severe wear to the non-metallic thrust washer, allowing the axle shaft to travel further in and out during torque thrust causing damage to the wheel hub and bearing, the axle shaft end seal and the axle shaft.

21. Position the three thrust washers onto the axle shaft. Install the snapring.

➡ **Any time the hub lock is removed, a new O-ring seal must be installed. Failure to do so can cause a vacuum leak and loss of four wheel drive functions.**

22. Install a new O-ring.
23. Position the hub lock.
24. Install the retainer ring.
25. Position the front disc brake rotor to the wheel hub. Make sure the wheel hub and the front disc brake rotor braking and

mounting surfaces are clean. Use brake parts cleaner to clean the front disc brake rotor.

26. For DRW vehicles, install the front wheel hub extender and nuts. Torque to 130 ft. lbs. (176 Nm).

27. Position back the front disc brake caliper and anchor plate. Install the front disc brake caliper anchor plate bolts. Torque to 166 ft. lbs. (225 Nm).
28. Install the wheel and tire assembly.
29. Test the system for normal operation.

REAR SUSPENSION

Shock Absorber

REMOVAL & INSTALATION

2002–03 F-150, 2004 F-150 Heritage Editions and 2002–05 E-Series

1. Before servicing the vehicle, refer to the Precautions Section.

2. Raise the vehicle and secure on support stands.
3. Remove the self-locking nut, steel washer, and rubber bushings at the upper end of the shock absorber.
4. Remove the bolt and nut at the lower end and remove the shock absorber. If needed, raise the rear axle assembly slightly with a jack.

To install:

5. When installing a new shock absorber, use new rubber bushings. Position the shock absorber on the mounting brackets with the stud end at the top. Install the upper bushing, steel washer and self-locking nut at the upper end, and the bolt and nut at the lower end.
6. Tighten the upper mounting studs to:

1 Brake caliper and bolts
2 Shock absorber bolts and nuts
3 Shock absorber
4 Spring front eye bolt and nut
5 Spring shackle bolt and nut (rear)
6 U-bolt nuts (light duty/heavy duty)
7 U-bolt plate
8 U-bolts
9 Rear spring assembly

Rear suspension components—2004–05 F-150, exc. 2004 F-150 Heritage Editions

67197-EFSE-G158

- E-Series: upper end 30 ft. lbs. (40 Nm); lower end 52 ft. lbs. (70 Nm)
- F-150: upper end 26 ft. lbs. (35 Nm); lower end 66 ft. lbs. (90 Nm).

2004–05 F-150, exc. 2004 F-150 Heritage Edition

1. Before servicing the vehicle, refer to the Precautions Section.
2. Raise the vehicle and secure on support stands.
3. Remove the self-locking nut, steel washer, and rubber bushings at the upper end of the shock absorber.
4. Remove the bolt and nut at the lower end and remove the shock absorber. If needed, raise the rear axle assembly slightly with a jack.

To install:
5. Installation is the reverse of removal. Tighten the upper and lower mounting nuts to 66 ft. lbs. (90 Nm).

2002–05 F-250 and 350

❋ WARNING

Suspension fasteners are critical parts because they affect performance of vital components and systems and their failure can result in major service expense. Install new parts with the same part number or an equivalent part if installation is necessary. Do not use an installation part of lesser quality or substitute design. Torque values must be used as specified during reassembly to make sure of correct retention of these parts.

1. Before servicing the vehicle, refer to the Precautions Section.

❋ WARNING

The low pressure gas shock absorbers are charged with nitrogen gas. Do not attempt to open, puncture or apply heat to shock absorbers.

2. Raise and support the vehicle.
3. Using a suitable jack, support the rear axle.
4. Remove the shock absorber lower retaining nut and bolt.
5. Remove the nut from the upper shock absorber mounting bracket and remove the shock.
6. To install, follow the removal proce-

dure in reverse order, using new fasteners. For 2002–04 models, torque the upper and lower fasteners to 46 ft. lbs. (62 Nm). For 2005 models, torque the upper bolt/nut to 46 ft. lbs. (62 Nm); the lower bolt/nut to 66 ft. lbs. (90 Nm).

Leaf Spring

REMOVAL & INSTALLATION

2002 E-Series

1. Before servicing the vehicle, refer to the Precautions Section.
2. Remove the wheel and tire assembly.
3. Support the rear axle with a suitable jack.
4. Disconnect the lower end of the rear shock absorber.
5. Remove the nuts. Remove the U-bolts. Remove the rear spring plate.
6. Remove the front shackle bolt and nut.
7. Remove the rear shackle bolt and nut.
8. Remove the rear spring.
9. Installation is the reverse of removal. Observe the following torques:
   - E-150 U-bolt nuts: 85 ft. lbs. (115 Nm)
   - E-250/350/Super Duty U-bolt nuts: 130 ft. lbs. (176 Nm)
   - E-150 spring-to-front shackle bolt/nut: 124 ft. lbs. (168 Nm)
   - E-250/350/Super Duty spring-to-front shackle bolt/nut: 265 ft. lbs. (359 Nm)
   - Spring-to-rear shackle bolt/nut: 66 ft. lbs. (90 Nm)

2003–05 E-Series

1. Before servicing the vehicle, refer to the Precautions Section.
2. Remove the wheel and tire assembly.
3. Support the rear axle with a suitable jack.
4. Disconnect the lower end of the rear shock absorber.
5. Remove the nuts. Remove the U-bolts. Remove the rear spring plate.
6. Remove the front shackle bolt and nut.
7. Remove the rear shackle bolt and nut.
8. Remove the rear spring.
9. Installation is the reverse of removal. Observe the following torques:
   - E-150 U-bolt nuts: 85 ft. lbs. (115 Nm)

- E-250/350 U-bolt nuts: 130 ft. lbs. (176 Nm)
- E-150 spring-to-front shackle bolt/nut: 148 ft. lbs. (200 Nm)
- E-250/350 spring-to-front shackle bolt/nut: 295 ft. lbs. (400 Nm)
- Spring-to-rear shackle bolt/nut, all: 85 ft. lbs. (115 Nm)
- E-150 rear shackle-to-frame: 98 ft. lbs. (133 Nm)
- E-250/350 rear shackle-to-frame: 111 ft. lbs. (150 Nm)

2002–03 F-150 and 2004 F-150 Heritage Editions, exc. Blackwood

❋❋ WARNING

Do not use heat to loosen a seized wheel nut. Heat can damage the wheel and the wheel bearings.

1. Before servicing the vehicle, refer to the Precautions Section.
2. Loosen the rear wheel nuts.
3. Remove the rear wheel center caps.
4. With the weight of the vehicle on the wheels, loosen but do not remove the wheel nuts.
5. Raise and support the vehicle.
6. Remove the wheel and tire assembly.
7. Use a jack to support the rear axle.

❋❋ WARNING

Lower the rear axle only enough to gain access to the rear spring.

8. Remove the nuts. Remove the U-bolts. Remove the rear spring plates.
9. Carefully lower the rear axle. On 4x4 vehicles, remove the rear spring spacer.
10. Remove the front and rear shackle bolts and nuts. Remove the rear spring.

To install:
11. Position the rear spring. Install the shackle nuts and bolts. Torque the front shackle bolt to 157–212 ft. lbs. (213–288 Nm); the rear shackle bolt to 73–97 ft. lbs. (98–132 Nm).
12. On 4-wheel drive vehicles, position the rear spring spacer.
13. Position the U-bolts.
14. Install the rear spring plate.
15. Install the nuts. Torque to 73–97 ft. lbs. (98–132 Nm).
16. Remove the jack.
17. Install the wheel and tire assembly.

Blackwood

1. Before servicing the vehicle, refer to the Precautions Section.

Blackwood rear suspension with air springs

1 Air spring supply line
2 Air spring retainer clip
3 Air spring
4 U-bolt nuts (4)
5 U-bolt plate
6 U-bolts (6)
7 Shock absorber lower nut
8 Shock absorber lower bolt
9 Shock absorber upper nut and insulator
10 Shock abosorber

11 Shackle-to-frame bracket bolt
12 Shackle-to-frame bracket flag nut
13 Leaf spring-to-shackle nut
14 Leaf spring-to-shackle flag bolt
15 Shackle
16 Leaf spring-to-frame bracket nut
17 Leaf spring-to-frame bracket bolt

18 Leaf spring
19 Control arm bracket bolts (4)
20 Control arm bracket flag nuts (2)
21 Control arm bracket flag nuts (2)
22 Control arm-to-bracket nut
23 Control arm-to-bracket bolt
24 Control arm bracket
25 Control arm-to-axle nut
26 Control arm-to-axle bolt
27 Control arm

67197-EFSE-G188

Blackwood rear suspension with air springs (cont.)

✳✳ CAUTION

Do not remove an air spring or solenoid valve when there is pressure in the system. Do not remove any components supporting an air spring without either venting the air or providing support for the air spring to prevent vehicle damage or personal injury.

✳ CAUTION

The electrical power to the air suspension system must be shut off prior to hoisting or jacking an air suspension vehicle. This can be accomplished by turning off the air suspension switch located below the glove box in the lower right of the passenger foot well. Failure to do so can result in unexpected inflation or deflation of the air springs, which can result in shifting of the vehicle during these operations.

2. Using a scan tool, vent the rear air springs. For additional information, see the Air Spring procedure.

✳ CAUTION

The venting procedure must be followed before proceeding!

3. Turn the air suspension switch to the OFF position.
4. Raise and support the rear of the vehicle.
5. Support the suspension with a floor jack and lower the axle to relieve tension from the springs.
6. Remove the wheel and tire assemblies.
7. Remove the U-bolt nuts.
8. Remove the U-bolt plate.

9. Remove the U-bolts.
10. Remove the shock absorber lower nut.
11. Disconnect the height sensor linkage.
12. Remove the shock absorber lower bolt.
13. Remove the lower the axle until it is no longer making contact with the leaf spring.
14. Remove the leaf spring-to-shackle flag bolt.
15. Remove the shackle.
16. Remove the leaf spring-to-frame bracket nut.
17. Remove the leaf spring-to-frame bracket.
18. Remove the leaf spring bolt.
19. To install, reverse the removal procedure. Tighten the leaf spring and shackle nuts and bolts with the vehicle at curb height. Tighten the U-bolt nuts in a criss-cross pattern.
20. Observe the following torques:
   • U-bolt nuts: 184 ft. lbs. (250 Nm)
   • Shock absorber lower nut: 66 ft. lbs. (90 Nm)
   • Leaf spring-to-shackle flag bolt: 88 ft. lbs. (115 Nm)
   • Leaf spring-to-frame bracket nut: 184 ft. lbs. (250 Nm)

2004–05 F-150, Exc. 2004 F-150 Heritage Editions

1. Before servicing the vehicle, refer to the Precautions Section.
2. Remove the wheel and tire assembly.

✳✳ WARNING

Lower the rear axle only enough to gain access to the rear spring.

3. Use the jack to support and lower the rear axle housing assembly.
4. Remove the brake caliper and bolts

✳✳ WARNING

Do not allow the brake caliper to hang from the brake hose.

5. Position the brake caliper aside using mechanic's wire.
6. Disconnect the shock absorber at the lower end.
7. Remove the spring front eye bolt and nut.

➡Lower the fuel tank to gain access to the spring shackle bolt.

8. Remove the spring shackle bolt and nut (rear)
9. Remove the U-bolt nuts
10. Remove the U-bolt plate
11. Remove the U-bolts
12. Remove the rear spring assembly
13. To install, reverse the removal procedure.

➡Tighten the U-bolt nuts in a cross pattern in 3 even steps.

14. Observe the following torques:
   • U-bolt nuts. Light duty, in 3 equal steps: 85 ft. lbs. (115 Nm)
   • U-bolt nuts. Heavy duty, in 3 equal steps: 184 ft. lbs. (250 Nm)
   • Spring front eye bolt: 222 ft. lbs. (300 Nm)
   • Spring rear shackle bolt: 98 ft. lbs. (133 Nm)
   • Shock absorber nut: (66 ft. lbs. (90 Nm)

2002–04 F-250, 350

1. Before servicing the vehicle, refer to the Precautions Section.
2. Raise and support the vehicle.
3. Remove the wheel and tire assembly.
4. Support the rear axle with a suitable jack.

1 Nut, M10-1.5 hex

2 Spacer

3 Rivet, M10 x 25

4 Rear axle bumper stop

5 Rear axle bumper assembly

6 Nut and retainer assembly,
   M10 x 1.5 x 35

7 Front auxiliary spring bracket

8 Bolt, M16-2.0 x 135 hex

9 Shackle assembly

10 Nut, M16-2.0 hex

11 Bolt, M16-2.0 x 175 x 10.9

12 Bolt, M10 x 1.5 x 33.5

13 Nut and retainer, M16-2.0 x
   20 x 10

14 Nut, M16-2.0 hex

15 Auxiliary spring

16 Rear spring assembly

17 U-bolt, M16-2.0 x 108/225

18 Rear spring spacer (4x4 only)

19 Auxiliary spring spacer

20 Rear spring plate

21 Nut and retainer, M18-2.5 x
   19 x 10

22 Bolt, M18-2.5 x 139 x 10.9

23 Auxiliary spring bumper

24 Nut and washer assembly,
   M12-1.75 hex

25 Rear shock

26 Bolt, M12-1.75 hex

27 Nut and washer assembly,
   M12-1.75 hex

28 Stabilizer bar assembly

29 Nut, M12-1.75 hex

30 Washer, M12

31 Rear stabilizer bar retainer

32 Link assembly

33 Nut, M10-1.5 hex (Ford axle
   only)

34 Lower shock bracket (Ford
   axle only)

35 Rear stabilizer bar bracket
   (Ford axle only)

36 U-bolt, M10-1.5 x 101/152.5
   (Ford axle only)

37 U-bolt, M10-1.5 x
   137.5/161.5 (Ford axle only)

38 Bolt, M10-1.5 x 33.5 (Dana
   axle only)

39 Washer, M10-15 x 33.5
   (Dana axle only)

1   Nut, M10-1.5 hex
2   Spacer
3   Rivet, M10 x 25
4   Rear axle bumper stop
5   Rear axle bumper assembly
6   Front auxiliary spring bracket
7   Auxiliary spring bracket liner (RR)
8   Bolt, M16-2.0 x 145 hex
9   Nut, M16-2.0 hex
10 Shackle assembly
11 Bolt, M16-2.0 x 175 x 10.9
12 Nut and retainer, M16-2.0 x 20 x 10
13 Nut, M16-2.0 hex
14 Auxiliary spring

15 Rear spring assembly
16 U-bolt, M16-2.0 x 108/225
17 Rear spring spacer (4x4 only)
18 Auxiliary spring spacer
19 Rear spring plate
20 Bolt, M18-2.5 x 139 x 10.9
21 Nut and retainer, M18-2.5 x 19 x 10
22 U-bolt, M10-1.50 x 101/136.7 (Ford axle only)
23 Lower shock bracket (Ford axle only)
24 Bolt, M12-1.75 x 70 hex
25 Nut, M10-1.5 hex (Ford axle only)
26 Nut and washer assembly, M12-1.75 hex

27 Nut and washer assembly, M12-1.75 hex
28 U-bolt, M10-1.75 x 137.5/161.5 (Ford axle only)
29 Rear stabilizer bar bracket (Ford axle only)
30 Nut, M12-1.75 hex
31 Washer, M12
32 Nut, M10-1.5 hex (Ford axle only)
33 Rear stabilizer bar retainer
34 Stabilizer bar assembly
35 Link assembly
36 Washer, M12
37 Nut, M12-1.75 hex
38 Bolt, M10-1.5 x 33.5 hex
39 Washer, M10-1.5 x 33.5

06017-F150-G273

F-250/350 narrow frame rear suspension—2002–04 models

1 Link assembly

2 Rear axle bumper assembly

3 U-bolt, M10-1.75 x 137.5/161.5 (Ford axle only)

4 Rear stabilizer bar bracket (Ford axle only)

5 Stabilizer bar assembly

6 Rear stabilizer bar retainer

7 Rear spring assembly

8 Rear spring plate

9 U-bolt, M16-2.0 x 108/225 (narrow frame 4 required, wide frame 6 required)

10 Rear shock absorber

11 U-bolt, M10-1.50 x 101/136.7 (Ford axle only) (LH/RH)

12 Lower shock bracket (Ford axle only)

06017-F150-G274

F-250/350 rear suspension—2005 models

5. Remove the U-bolt retaining nuts and remove the U-bolts.

6. Remove the rear spring upper plate.

7. Remove the nut and bolt from the rear spring front hanger bracket.

8. Remove the lower nut and bolt from the rear spring shackle bracket. Remove the rear spring assembly.

9. If equipped, remove the auxiliary spring and auxiliary spring spacer.

10. Using new fasteners, follow the removal procedure in reverse order. Observe the following torques:

- U-bolt nuts: 185 ft. lbs. (250 Nm)
- F-250/350 Wide Frame front and rear spring hanger/shackle bolts: 185 ft. lbs. (250 Nm)
- F-350 Narrow Frame spring-to-front shackle: 185 ft. lbs. (250 Nm)
- F-350 Narrow Frame spring-to-rear shackle: 166 ft. lbs. (225 Nm)
- Rear shackle-to-frame: Wide Frame 185 ft. lbs. (250 Nm); Narrow Frame 166 ft. lbs. (225 Nm)

2005 F-250, 350

1. Before servicing the vehicle, refer to the Precautions Section.

2. With the vehicle in NEUTRAL, position it on a hoist.

3. Remove the wheel and tire assembly.

4. Support the rear axle with a suitable jack.

5. Remove and discard the U-bolt retaining nuts and the U-bolts.

6. Remove the rear spring upper plate.

7. Remove the nut and bolt from the rear spring front hanger bracket.

➡**If the rear spring has an auxiliary spring and spacer, it is serviced as part of the rear spring assembly.**

8. Remove the lower nut and bolt from the rear spring shackle bracket. Remove the rear spring assembly.

9. To install, follow the removal procedure in reverse order, using new fasteners.

10. Use the following procedure to correctly install the U-bolts.

   a. Install the U-bolts and U-bolt retaining nuts. Do not tighten the fasteners at this time.

   b. Align the U-bolts so they are as vertical as possible.

   c. Tighten the nuts evenly in a cross-type pattern to:

   - Step 1: 37 ft. lbs. (50 Nm)
   - Step 2: 74 ft. lbs. (100 Nm)
   - Step 3: 111 ft. lbs. (150 Nm)
   - Step 4: 148 ft. lbs. (200 Nm).

11. Observe the following torques:

- Spring-to-front shackle, exc. waxed bracket: 277 ft. lbs. (375 Nm)
- Spring-to-front shackle, waxed bracket: 222 ft. lbs. (300 Nm)
- Spring-to-rear shackle: 185 ft. lbs. (250 Nm)
- Rear shackle-to-frame: 185 ft. lbs. (250 Nm)

Air Spring (Blackwood)

✳✳ CAUTION

Do not remove an air spring or solenoid valve when there is pressure in the system. Do not remove any components supporting an air spring without either venting the air or providing support for the air spring to prevent vehicle damage or personal injury.

✳✳ CAUTION

The electrical power to the air suspension system must be shut off prior to hoisting or jacking an air suspension vehicle. This can be accomplished by turning off the air suspension switch located below the glove box in the lower right of the passenger foot well. Failure to do so can result in unexpected inflation or deflation of the air springs, which can result in shifting of the vehicle during these operations.

✳✳ CAUTION

Disconnecting an air line that is connected to the air compressor can cause personal injury or damage to components as high pressure air is vented uncontrolled.

DEPRESSURIZING THE SYSTEM

1. Before servicing the vehicle, refer to the Precautions Section.

Using the scan tool, Worldwide Diagnostic System (WDS) 418-F224, New Generation STAR (NGS) Tester 418-F052 or equivalent scan tool, vent the rear air springs as follows:

2. Turn the air suspension switch on.

3. Connect the scan tool to the data link connector (DLC) and follow the directions.

4. Select Air Suspension Module under Active Command Mode:

- VENT REAR to deflate the rear down.
- LIFT REAR to inflate the rear up.

Calibration
Initialize System—Clear B2140 DTC

5. Turn the ignition key OFF and turn it back to RUN; exit the vehicle, close all doors and allow the system to vent the vehicle down to kneel height (approximately 30 seconds).

6. Connect the scan tool to the data link connector (DLC) and follow the directions.

7. Select Air Suspension Module.

➡**Do not save rear ride heights. The air suspension module has pre-calibrated values already stored. Refer to Rear Ride Height Mechanical Resetting.**

8. Select the "Save Calibration Values (Store Ride Height)" scan tool command to calibrate the air suspension module. Trigger through the warning message(s) and save "initialization" (trigger from OFF to ON).

REMOVAL & INSTALLATION

1. Before servicing the vehicle, refer to the Precautions Section.

2. Turn the air suspension switch to the OFF position.

3. Remove the wheel and tire assemblies.

4. Compress the quick connect lock ring while pulling outward on the air supply line. Remove the air spring supply line

5. Remove the air spring retainer clip

6. Remove the air spring

7. To install, reverse the removal procedure.

Ride Height Sensor (Blackwood)

REMOVAL & INSTALLATION

1. Before servicing the vehicle, refer to the Precautions Section.

✳✳ CAUTION

The electrical power to the air suspension system must be shut off prior to hoisting, jacking or towing an air suspension vehicle. This can be accomplished by turning off the air suspension switch located in the RH kick panel area. Failure to do so may result in unexpected inflation or deflation of the air springs which may result in shifting of the vehicle during these operations.

27 Nm (20 lb-ft)

67197-EFSE-G189

Height sensor installation

27 Nm (20 lb-ft)

67197-EFSE-G190

Height sensor mounting bracket installation

2. Raise and support the vehicle.
3. Disconnect the wiring.
4. Disconnect the sensor arm.
5. Remove the bolts and the mounting bracket.

6. Disconnect the sensor arm.
7. Remove the bolts and the sensor.
8. To install, reverse the removal procedure. Torque the bolts to 20 ft. lbs. (27 Nm).

Stabilizer Bar

REMOVAL & INSTALLATION

E-Series

1. Before servicing the vehicle, refer to the Precautions Section.
2. Remove the lower nut, washer and insulator.
3. Remove the upper bolt and washer.
4. Remove the rear stabilizer bar.
5. Remove the bolt and the stabilizer bar. Discard the bolts.
6. To install, reverse the removal procedure. Torque the bracket bolts to 30 ft. lbs. (40 Nm); the link-to-bar nuts to 21 ft. lbs. (28 Nm); the link-to-frame bolts to 60 ft. lbs. (81 Nm).

Lightning

1. Before servicing the vehicle, refer to the Precautions Section.
2. Remove the bar-to-end link nuts. Discard the nuts.

3. Remove the bracket nuts and U-bolts. Discard the nuts.
4. Remove the bar.
5. Installation is the reverse of removal. Torque the bar-to-link nuts to 76 ft. lbs. (103 Nm); Torque the bracket nuts to 35 ft. lbs. (48 Nm).

F-250/350

1. Before servicing the vehicle, refer to the Precautions Section.
2. With the vehicle in NEUTRAL, position it on a hoist.
3. Remove the nuts from both lower ends of the stabilizer bar links and remove the washers.
4. Remove the nuts and washers from both upper ends of the stabilizer bar links-to-frame and remove the links.
5. Remove the nuts from both stabilizer bar retainer-to-axle brackets and remove the stabilizer bar retainers, stabilizer bar mounting brackets, U-bolts and the stabilizer bar.
6. To install, reverse the removal procedure, using new fasteners. Observe the following torques:
   • Bracket U-bolt nuts: 2002–04 35 ft. lbs. (47 Nm); 2005 30 ft. lbs. (40 Nm)
   • End links top and bottom: 52 ft. lbs. (70 Nm)

FRONT DISC BRAKES

Brake Caliper

REMOVAL & INSTALLATION

2002–03 E-150

❋❋ CAUTION

Brake fluid contains polyglycol ethers and polyglycols. Avoid contact with eyes. Wash hands thoroughly after handling. If brake fluid contacts eyes, flush eyes with running water for 15 minutes. Get medical attention if irritation persists. If taken internally, drink water and induce vomiting. Get medical attention immediately. Failure to follow these instructions may result in personal injury.

1. Before servicing the vehicle, refer to the Precautions Section.
2. Raise and support the vehicle.
3. Remove the tire and wheel assembly.
4. If so equipped, unclip the speed sensor wiring from the front brake hose.

❋❋ CAUTION

To prevent contamination and reduce air entry, always plug a disconnected front brake hose.

5. Disconnect the front brake hose from the disc brake caliper.
   a. Remove the flow bolt.
   b. Disconnect the front brake hose.
   c. Remove and discard the copper washers.
6. Remove the two disc brake caliper bolts.
7. Remove the disc brake caliper from the front wheel spindle.
8. Inspect the disc brake caliper for brake fluid leakage. If the disc brake caliper is leaking, it must be rebuilt or replaced.
   To install:

❋❋ CAUTION

To prevent deterioration of the caliper sleeve boots, do not use petroleum-based lubricant.

9. Fill the rubber caliper sleeve boots with silicone brake caliper grease.
10. Install the disc brake caliper and the front brake hose.
    a. Position the disc brake caliper in the front wheel spindle.
    b. Install and tighten the disc brake caliper bolts. Torque to 22–26 ft. lbs. (30–36 Nm).
    c. Using new copper washers, install the front brake hose and the bolt. Torque to 22–30 ft. lbs. (30–40 Nm).
11. If so equipped, attach the speed sensor wiring to the front brake hose.
12. Bleed the brake system.
13. Install the wheel and tire assembly.
14. Inspect the brake system operation.

2004–05 E-150

1. Before servicing the vehicle, refer to the Precautions Section.
2. Remove the tire and wheel assembly.
3. Release the speed sensor wiring from the front brake hose.

✷✷ CAUTION

To prevent contamination and reduce air entry, always plug a disconnected front brake hose.

4. Disconnect the front brake hose from the disc brake caliper.

   a. Remove the flow bolt.

   b. Disconnect the front brake hose.

   c. Remove and discard the copper washers.

5. Remove the two disc brake caliper bolts.

6. Remove the disc brake caliper from the front disc brake caliper anchor plate.

7. Inspect the disc brake caliper for brake fluid leakage. If the disc brake caliper is leaking, it must be rebuilt or installed new.

To install:

✷✷ CAUTION

To prevent deterioration of the caliper sleeve boots, do not use petroleum-based lubricant.

8. Fill the rubber caliper sleeve boots with caliper grease.

9. Position the disc brake caliper in the front disc brake caliper anchor plate.

10. Install and tighten the disc brake caliper bolts. Torque to 24 ft. lbs. (32 Nm).

11. Using new copper washers, install the front brake hose and the bolt. Torque to 26 ft. lbs. (35 Nm).

12. Attach the speed sensor wiring to the front brake hose.

13. Bleed the brake system.

14. Install the wheel and tire assembly.

15. Inspect the brake system operation.

1 Front brake anti-lock sensor	6 Brake disc and hub	11 Nut retainer
2 Disc brake caliper bolt	7 Front wheel bearing	12 Spindle nut
3 Disc brake caliper	8 Front wheel outer bearing retainer washer	13 Front wheel bearing
4 Disc brake pad anti-rattle clip	9 Cotter pin	14 Front brake splash shield gasket
5 Grease seal	10 Hub grease cap	15 Front wheel spindle

67197-EFSE-G195

Front disc brake components—2002–03 E-150

2002–05 E-250, 350

✳✳ CAUTION

Brake fluid contains polyglycol ethers and polyglycols. Avoid contact with eyes. Wash hands thoroughly after handling. If brake fluid contacts eyes, flush eyes with running water for 15 minutes. Get medical attention if irritation persists. If taken internally, drink water and induce vomiting. Get medical attention immediately. Failure to follow these instructions may result in personal injury.

1. Before servicing the vehicle, refer to the Precautions Section.
2. Raise and support the vehicle.
3. Remove the tire and wheel assembly.
4. If so equipped, unclip the speed sensor wiring from the front brake hose.

✳✳ CAUTION

To prevent contamination and reduce air entry, always plug a disconnected front brake hose.

5. Disconnect the front brake hose from the disc brake caliper.
   a. Remove the flow bolt.
   b. Disconnect the front brake hose.
   c. Remove and discard the copper washers.

1	Front brake anti-lock sensor
2	Disc brake caliper bolt
3	Disc brake caliper
4	Disc brake pad retaining springs
5	Front disc pads
6	Disc brake pad anti-rattle clip
7	Disc brake caliper anchor bracket bolts
8	Disc brake caliper anchor bracket
9	Hub grease cap
10	Cotter pin
11	Nut retainer
12	Spindle nut
13	Front wheel outer bearing retainer washer
14	Front wheel bearing/cup
15	Brake disc and hub
16	Front wheel bearing/cup
17	Grease seal
18	Backing plate
19	Front wheel spindle
20	Backing plate screw

67197-EFSE-G198

Front disc brake components—2004–05 E-150

6. Remove the two disc brake caliper bolts.

7. Remove the disc brake caliper from the anchor plate.

8. Inspect the disc brake caliper for brake fluid leakage. If the disc brake caliper is leaking, it must be rebuilt or replaced.

To install:

❋❋ CAUTION

To prevent deterioration of the caliper sleeve boots, do not use petroleum-based lubricant.

9. Fill the rubber caliper sleeve boots with silicone brake caliper grease.

10. Install the disc brake caliper and the front brake hose.

a. Position the disc brake caliper on the anchor plate.

b. Install and tighten the disc brake caliper bolts. Torque to 16–30 ft. lbs. (22–40 Nm) for 2002–03 models; 23 ft. lbs. (31 Nm) for 2004–05 models.

c. Using new copper washers, install the front brake hose and the bolt. Torque to 22–30 ft. lbs. (30–40 Nm) for

2002–03 models; 28 ft. lbs. (35 Nm) for 2004–05 models.

11. If so equipped, attach the speed sensor wiring to the front brake hose.

12. Bleed the brake system.

13. Install the wheel and tire assembly.

14. Inspect the brake system operation.

2002–03 Blackwood, F-150 and 2004 F-150 Heritage Editions

1. Before servicing the vehicle, refer to the Precautions Section.

2. Raise and support the vehicle.

3. Remove the wheel and tire assembly.

1 Anchor bracket-to-spindle bolt	7 Front wheel bearing	13 Brake disc and hub
2 Disc brake caliper bolt	8 Cotter pin	14 Bearing cone and roller
3 Caliper bolt sleeve	9 Hub grease cap	15 Front brake anti-lock sensor
4 Disc brake caliper	10 Nut retainer	16 Front brake splash shield gasket
5 Front disc brake caliper anchor plate	11 Spindle nut	17 Disc brake pad anti-rattle clip
6 Wheel hub grease seal	12 Front wheel outer bearing retainer washer	18 Front wheel spindle

67197-EFSE-G196

Front disc brake components—E-250, 350 single rear wheel

➡Use clean, fresh Ford High Performance DOT 3 Motor Vehicle Brake Fluid C6AZ-19542-AB or equivalent meeting Ford specification ESA-M6C25A. Do not reuse brake fluid drained or bled from the system. Do not use brake fluid that has been stored in an open container. Do not use contaminated brake fluid.

4. Remove the front brake hose bolt. Disconnect the front brake hose.

5. Remove and discard the copper washers. Plug the front brake hose.

6. Remove the disc brake caliper bolts.

Lift the disc brake caliper off the front disc brake caliper anchor plate.

7. Inspect the disc brake caliper for leaks. If leaks are found, disassembly is required.

To install:

8. Install the disc brake caliper.

9. Install the disc brake caliper bolts. Torque to 21–26 ft. lbs. (28–36 Nm) for F-150 models; 20 ft. lbs. (27 Nm) for Blackwood.

10. Install the front brake hose. Use new copper washers; connect the front brake hose. Install the front brake hose bolt.

Torque to 23–29 ft. lbs. (30–40 Nm) for F-150 models; 26 ft. lbs. (35 Nm) for Blackwood.

11. Bleed the disc brake caliper.

12. Install the wheel and tire assembly.

13. Check the brake system operation.

2004–05 F-150 Exc. 2004 F-150 Heritage Editions

2006 Mark LT

1. Before servicing the vehicle, refer to the Precautions Section.

1 Anchor bracket-to-spindle bolt	11 Cotter pin
2 Disc brake caliper bolt	12 Hub grease cap
3 Caliper bolt sleeve	13 Nut retainer
4 Disc brake caliper	14 Spindle nut
5 Front disc brake caliper anchor plate	15 Front wheel bearing (outer)
6 Wheel hub grease seal	16 Front wheel bearing (inner)
7 Bearing cup, inner	17 Disc brake pad anti-rattle clip
8 Brake disc and hub (dual rear wheel)	18 Front wheel spindle
9 Bearing cup, outer	19 Front brake anti-lock sensor
10 Front wheel outer bearing retainer washer	

67197-EFSE-G197

Front disc brake components—E-250, 350 dual rear wheel

1	Bolts	8	Bolts	15	Bearing
2	Brake caliper	9	Anchor plate	16	Brake disc
3	Flow bolt	10	Grease cap	17	Bearing
4	Copper washers	11	Cotter pin	18	Seal
5	Brake hose	12	Retainer	19	Bolts
6	Slippers	13	Spindle nut Installation Note	20	Brake disc shield
7	Disc pads	14	Washer	21	Brake disc shield gasket

Front disc brake components—Blackwood

06017-F150-G275

1 Front disc brake caliper anchor plate
2 Disc brake caliper bolt
3 Pin assembly
4 Pads
5 Disc brake caliper
6 Front brake splash shield gasket
7 Brake disc shield
8 Brake disc shield bolts
9 Cotter pin
10 Hub grease cap

11 Nut retainer
12 Spindle nut
13 Front wheel outer bearing retainer washer
14 Front wheel bearing (outer)
15 Brake disc and hub
16 Front wheel bearing (inner)
17 Wheel hub grease seal
18 Front wheel spindle
19 Front brake anti-lock sensor
20 Anti-lock brake sensor bolt

67197-EFSE-G199

Front disc brake components—2002–03 2-Wheel Drive F-150 and 2004 F-150 Heritage Editions

1 Front disc brake caliper anchor plate bolt
2 Disc brake caliper bolt
3 Pin assembly
4 Front disc brake caliper anchor plate
5 Front brake hose
6 Front brake hose bolt
7 Pads
8 Disc brake caliper
9 Bolt
10 Brake disc
11 Brake disc shield
12 Front wheel knuckle

67197-EFSE-G200

Front disc brake components—2002–03 4-Wheel Drive F-150 and 2004 F-150 Heritage Editions

2. Remove the wheel and tire assembly.
3. Remove the caliper bolts
4. Remove the brake caliper

> ※※ **WARNING**
>
> **Do not allow the brake caliper to hang by the flexible brake hose.**

5. Support the caliper to the vehicle.
6. Remove the flow bolt
7. Remove the copper washers
8. Remove the brake line
9. To install, reverse the removal procedure. Tighten the bottom caliper bolt and then the top caliper bolt. Always use new copper washers.
    10. Observe the following torques:
      • Caliper mounting bolts: 47 ft. lbs. (64 Nm)
      • Brake line flow bolt: 26 ft. lbs. (35 Nm)
    11. Bleed the brake system.

2002–04 F-250, 350

1. Before servicing the vehicle, refer to the Precautions Section.
2. Raise and support the vehicle.
3. Remove the wheel and tire assembly.
4. Remove the brake hose bolt.
5. Disconnect the brake hose.
6. Remove and discard the copper washers. Plug the brake hose.
7. Remove the disc brake caliper pin bolts.
8. Lift the disc brake caliper from the disc brake caliper anchor plate.
9. Inspect the disc brake caliper for leaks. If leaks are found, disassembly is required.

To install:
10. Install the disc brake caliper.
11. Install the disc brake caliper pin bolts. Torque to 42 ft. lbs. (56 Nm).
12. Using new copper washers, connect the brake hose. Install the brake hose bolt. Torque to 26 ft. lbs. (35 Nm).
13. Bleed the brake system.
14. Install the wheel and tire assembly.
15. Fill the brake master cylinder reservoir with clean DOT 3 motor vehicle brake fluid. Install the brake master cylinder filler cap.
16. Inspect the brake system operation.

2005 F-250/350

1. Before servicing the vehicle, refer to the Precautions Section.
2. With the vehicle in NEUTRAL, position it on a hoist.
3. Remove the brake flexible hose flow bolt and position the brake flexible hose aside. Discard the 2 copper washers.

> ※※ **WARNING**
>
> **Do not pry in the caliper sight hole to retract the pistons as this can damage the pistons and boots.**

4. Remove the 2 brake caliper bolts and the brake caliper. If leaks or damaged boots are found, install a new disc brake caliper.

To install:

> ※※ **WARNING**
>
> **Tighten the bottom caliper bolt before tightening the top caliper bolt.**

> ※※ **WARNING**
>
> **Make sure the caliper pin boots are correctly seated to prevent damage to the guide pins.**

5. Position the brake caliper and install the 2 bolts. Torque to 56 ft. lbs. (76 Nm).
6. Using 2 new copper washers, position the brake flexible hose and install the brake caliper flow bolt. Torque to 35 ft. lbs. (48 Nm).

1 Caliper bolt (2 required for each side)
2 Brake caliper (RH/LH)
3 Brake disc pads (1 kit LH and RH)
4 Slippers
5 Guide pin and boot (2 required for each side)
6 Anchor bracket bolts (2 required for each side)
7 Anchor bracket
8 Cotter pin
9 Retainer
10 Spindle nut (1 each side)

11 Brake disc (heavy duty and light duty)
12 Dust shield bolts (3 required each side)
13 Dust shield (RH/LH)
14 Flow bolt
15 Copper washers (2 required each side)
16 Brake line (RH/LH)
17 Anti-lock brake sensor cable
18 Brake line bracket bolt
19 Brake line fitting
20 Retainer clip

67197-EFSE-G201

Front disc brake components—2004–05 F-150 2-Wheel Drive, exc. 2004 F-150 Heritage Edition; 2006 Mark LT

1 Caliper bolt (2 required each side)

2 Brake caliper (RH/LH)

3 Brake disc pads (1 kit LH and RH)

4 Slippers

5 Guide pin and boot (2 required each side)

6 Anchor bracket bolts (2 required each side)

7 Anchor bracket

8 Brake disc (heavy duty and light duty)

9 Dust shield bolts (3 required each side)

10 Dust shield (RH/LH)

11 Flow bolt

12 Copper washers (2 required each side)

13 Brake line (RH/LH)

14 Anti-lock brake sensor cable

15 Vacuum hose

16 Brake line bracket bolt

17 Brake line fitting

18 Retainer clip

67197-EFSE-G202

Front disc brake components—2004–05 F-150 4-Wheel Drive, exc. 2004 F-150 Heritage Edition; 2006 Mark LT

1 Front disc brake caliper anchor plate
2 Stainless steel slippers
3 Pads
4 Disc brake caliper
5 Guide pin boot
6 Guide pin
7 Caliper bolt
8 Bleeder screw and bleeder screw cap assembly
9 Front brake hose
10 Brass washer
11 Flow bolt
12 Front disc brake rotor

67197-EFSE-G203

Front disc brake—2002–04 F-250/350

7. Bleed the brake caliper.
8. Test the brake system for normal operation.

Brake Pads

REMOVAL & INSTALLATION

2002–03 E-Series

1. Before servicing the vehicle, refer to the Precautions Section.
2. Remove and discard enough brake fluid from the brake master cylinder to allow

room for the brake fluid displaced when the caliper piston is pressed to the bottom of the caliper bore.
3. Raise and support the vehicle.
4. Remove the tire and wheel assembly.
5. If so equipped, unclip the speed sensor wiring from the front brake hose.
6. Remove the two disc brake caliper bolts.

✳✳ WARNING

Do not allow the disc brake caliper to hang from the front brake hose. Use wire to support the disc brake caliper

from a convenient underbody component.

7. Remove the disc brake caliper.
8. Remove the pads.
To install:
9. If installing new pads, use a C-clamp and a pad to press the caliper piston to the bottom of the caliper bore. Install the pads.

✳✳ CAUTION

To prevent deterioration of the caliper sleeve boots, do not use petroleum-based lubricant.

10. Fill the rubber caliper sleeve boots with silicone brake caliper grease.
11. Install the disc brake caliper.
12. If so equipped, attach the speed sensor wiring to the front brake hose.
13. Install the tire and wheel assembly. Before driving, pump the brake pedal a few times to seat the pads.

2004–05 E-Series

1. Before servicing the vehicle, refer to the Precautions Section.
2. Remove and discard enough brake fluid from the brake master cylinder to allow room for the brake fluid displaced when the caliper piston is pressed to the bottom of the caliper bore.
3. Remove the tire and wheel assembly.
4. Release the speed sensor wiring from the front brake hose.
5. Remove the two disc brake caliper bolts.

✳✳ WARNING

Do not allow the disc brake caliper to hang from the front brake hose. Use wire to support the disc brake caliper from a convenient underbody component.

6. Remove and support the disc brake caliper.
7. On E-150, remove the brake pad retaining springs.
8. Remove the brake pads from the brake caliper anchor bracket.
To install:

➡**Bottoming of the caliper piston is not necessary if reusing the original pads.**

9. If installing new pads, use a suitable tool and a pad to press the caliper piston to the bottom of the caliper bore.

✳✳ CAUTION

To prevent deterioration of the caliper sleeve boots, do not use petroleum-based lubricant.

10. Fill the rubber caliper sleeve boots with caliper grease.

11. Install the brake pads.

12. On E-150, install the brake pad retaining springs.

13. Install the disc brake caliper. Install and tighten the disc brake caliper bolts.

14. Attach the speed sensor wiring to the front brake hose.

15. Install the tire and wheel assembly.

2002–03 Blackwood, F-150 and 2004 F-150 Heritage Editions

1. Before servicing the vehicle, refer to the Precautions Section.

2. Raise and support the vehicle.

3. Remove the wheel and tire assembly.

4. Inspect the pads for wear or contamination.

✳ CAUTION

When removing the front disc brake caliper, never allow it to hang from the brake hose. Provide a suitable support.

5. Remove the two disc brake caliper bolts.

6. Lift the disc brake caliper off the front disc brake caliper anchor plate.

7. Inspect the disc brake caliper for leaks. If leaks are found, disassembly is required.

8. Remove the pads.

9. Remove the disc brake anti-rattle clips.

To install:

✳ CAUTION

Do not allow grease, oil, brake fluid or other contaminants to contact the pad lining material. Do not install contaminated pads.

10. Install the disc brake anti-rattle clips.

11. Install the pads.

12. Install the disc brake caliper.

13. Install the disc brake caliper bolts. Torque to 21–26 ft. lbs. (28–36 Nm) for F-150; 20 ft. lbs. (27 Nm) for Blackwood.

14. Install the tire and wheel assembly.

15. Check brake operation.

2004–05 F-150, Exc. 2004 F-150 Heritage Editions
2006 Mark LT

1. Before servicing the vehicle, refer to the Precautions Section.

2. Remove the wheel and tire assembly.

3. Remove the caliper bolt

4. Remove the brake caliper

✳✳ CAUTION

Do not allow the brake caliper to hang by the flexible brake hose.

5. Support the caliper to the vehicle.

6. Remove the brake disc pads

7. Remove the slippers

8. To install, reverse the removal procedure. If replacing the brake pads, remove the brake fluid in the master cylinder reservoir until it is half filled.

9. If replacing the brake pads, use a suitable tool to push the brake caliper pistons into the caliper bore.

➡**Tighten the bottom caliper bolt first and then the top caliper bolt. Torque to 47 ft. lbs. (64 Nm).**

10. If the hydraulic system was opened, bleed the brake system.

2002–04 F-250, 350

1. Before servicing the vehicle, refer to the Precautions Section.

2. Raise and support the vehicle.

3. Remove the wheel and tire assembly.

4. Inspect the pads for wear or contamination.

➡**When removing the front disc brake caliper, never allow it to hang from the brake hose. Provide a suitable support.**

5. Remove the two disc brake caliper bolts.

6. Lift the disc brake caliper off the front disc brake caliper anchor plate.

7. Measure the brake disc and hub thickness. Install a new brake disc and hub if not within specification.

8. If necessary, resurface the brake disc. Ford recommends on-vehicle brake disc machining. Follow lathe manufacturer's instructions.

9. Inspect the disc brake caliper for leaks. If leaks are found, disassembly is required.

10. If necessary, remove the V-springs.

11. Remove the pads.

12. Remove the disc brake anti-rattle clips.

To install:

✳✳ CAUTION

Do not allow grease, oil, brake fluid or other contaminants to contact the pad lining material. Do not install contaminated pads.

13. Install the disc brake anti-rattle clips.

14. Install the pads.

15. Install the V-springs.

✳✳ CAUTION

Use care not to damage the bleeder screw or front wheel disc brake shield.

16. Install the disc brake caliper.

17. Install the disc brake caliper bolts.

18. Install the tire and wheel assembly.

19. Check brake operation.

2005 F-250/350

1. Before servicing the vehicle, refer to the Precautions Section.

2. Check the brake fluid level in the brake master cylinder reservoir. If required, remove the fluid until the brake master cylinder reservoir is half full.

3. With the vehicle in NEUTRAL, position it on a hoist.

✳✳ WARNING

Do not pry in the caliper sight hole to retract the pistons, as this can damage the pistons and boots.

✳✳ WARNING

Do not allow the brake caliper to hang from the brake caliper flexible hose. Use a suitable tool to support the brake caliper.

4. Remove the 2 brake caliper bolts and position the caliper aside.

5. Remove the 2 brake pads and if equipped, remove the 4 retraction clips. Discard the retraction clips.

6. Measure the front brake disc thickness. Install a new front brake disc if it is not within specification.

✳✳ WARNING

Install new brake pads if they are worn past the specified thickness above the metal backing plates. Install new brake pads in complete axle sets.

7. Inspect the brake pads for wear and contamination.

To install:

> ### ✛ WARNING
>
> **Do not allow grease, oil, brake fluid or other contaminants to contact the pad lining material. Do not install contaminated pads.**

➡ **Install all new hardware as supplied with the brake pad kit.**

8. Install the 4 new retracting clips and the 2 brake pads.

> ### ✳✳ WARNING
>
> **Protect the piston and boots when pushing the caliper piston into the caliper piston bores.**

9. Using a suitable tool, compress the disc brake caliper pistons into the caliper.

> ### ✛ WARNING
>
> **Tighten the bottom caliper bolt before tightening the top caliper bolt**

> ### ✳✳ WARNING
>
> **Make sure the caliper pin boots are correctly seated to prevent damage to the guide pins.**

10. Position the brake caliper and install the 2 bolts. Torque to 56 ft. lbs. (76 Nm).
11. Fill the brake master cylinder reservoir with clean brake fluid.
12. Test the brakes for normal operation.

Brake Rotor

REMOVAL & INSTALLATION

2002–05 E-150

1. Before servicing the vehicle, refer to the Precautions Section.
2. Remove the pads.
3. Remove the hub grease cap.
4. Remove the cotter pin.
5. Remove the nut retainer.
6. Remove the spindle nut.
7. Remove the front wheel outer bearing retainer washer.
8. Remove the outer front wheel bearing.
9. Remove the brake disc and hub.
10. Remove the wheel hub grease seal.

11. Remove the inner front wheel bearing.

To install:

12. Thoroughly clean and inspect the front wheel bearings and the brake disc and hub.
13. Lubricate the front wheel bearings.
14. Install the inner front wheel bearing.
15. Install a new wheel hub grease seal.
16. Position the brake disc and hub.
17. Install the outer front wheel bearing.
18. Install the front wheel outer bearing retainer washer.
19. Install the spindle nut.
20. Adjust the wheel bearings.
21. Install the nut retainer.
22. Install the cotter pin.
23. Install the hub grease cap.
24. Install the pads.

2002–05 E-250, 350

1. Before servicing the vehicle, refer to the Precautions Section.
2. Remove the front disc brake caliper anchor plate.
3. Remove the hub grease cap.
4. Remove the cotter pin.
5. Remove the nut retainer.
6. Remove the spindle nut.
7. Remove the front wheel outer bearing retainer washer.
8. Remove the outer front wheel bearing.
9. Remove the brake disc and hub.
10. Remove the wheel hub grease seal.
11. Remove the inner front wheel bearing.

To install:

12. Thoroughly clean and inspect the front wheel bearings and the brake disc and hub.
13. Lubricate the front wheel bearings.
14. Install the inner front wheel bearing.
15. Install a new wheel hub grease seal.
16. Position the brake disc and hub.
17. Install the outer front wheel bearing.
18. Install the front wheel outer bearing retainer washer.
19. Install the spindle nut.
20. Adjust the wheel bearings.
21. Install the nut retainer.
22. Install the cotter pin.
23. Install the hub grease cap.
24. Install the front disc brake caliper anchor plate.

2002–03 F-150 and 2004 F-150 Heritage Editions

1. Before servicing the vehicle, refer to the Precautions Section.

2. Remove the front disc brake caliper anchor plate.
3. On 4x4 vehicles, remove the brake disc.
4. On 4x2 vehicles, remove the brake disc and hub as follows:
   a. Remove the hub grease cap.
   b. Remove the cotter pin.
   c. Remove the nut retainer.
   d. Remove the spindle nut.
   e. Remove the front wheel outer bearing retainer washer.
   f. Remove the outer front wheel bearing).
   g. Remove the brake disc and hub.
5. If necessary on 4x2 vehicles, remove the front wheel bearing.
   a. Remove the wheel hub grease seal.
   b. Remove the front wheel bearing.

To install:

6. On 4x4 vehicles, position the brake disc to the wheel hub. Use Metal Brake Parts Cleaner F3AZ-19579-SA or equivalent to clean the brake disc and hub.
7. On 4x2 vehicles, thoroughly clean and inspect the front wheel bearings and the brake disc and hub. Use Metal Brake Parts Cleaner F3AZ-19579-SA or equivalent.
8. On 4x2 vehicles, lubricate the front wheel bearings. Use Premium Long-Life Grease XG-1-C or -K or equivalent meeting Ford specification ESA-M1C75-B.
9. On 4x2 vehicles, install a new wheel hub grease seal.
   a. Install the inner front wheel bearing.
   b. Install a new wheel hub grease seal.
10. On 4x2 vehicles, install the brake disc and hub.
    a. Position the brake disc and hub.
    b. Install the outer front wheel bearing.
    c. Install the front wheel outer bearing retainer washer.
    d. Install the spindle nut.
11. While rotating the brake disc and hub, tighten the spindle nut to 30 ft. lbs. (40 Nm).
12. Loosen the spindle nut two turns.
13. Tighten the spindle nut 17–24 ft. lbs. (23–34 Nm) while rotating the brake disc and hub.
14. Loosen the spindle nut ½ turn.
15. Tighten the spindle nut to 17 inch lbs. (2 Nm) while rotating the brake disc and hub.
16. On 4x2 vehicles, install the following components:

a. Install the nut retainer.
b. Install the cotter pin.
c. Install the hub grease cap.
17. Install the front disc brake caliper anchor plate.

2004–05 F-150 2-Wheel Drive Exc. 2004 F-150 Heritage Editions

1. Before servicing the vehicle, refer to the Precautions Section.
2. Remove the wheel and tire assembly.
3. Remove the anchor bracket bolts and position the caliper and bracket aside.
4. Remove the cotter pin
5. Remove the retainer
6. Remove the spindle nut
7. Remove the brake disc
8. To install, reverse the removal procedure.
9. Tighten the spindle nut to 295 ft. lbs. (400 Nm). Rotate the hub counterclockwise 5 rotations and then recheck the torque.
10. If the hydraulic system was opened, bleed the brake system.

2004–05 F-150 4-Wheel Drive Exc. 2004 F-150 Heritage Editions

1. Before servicing the vehicle, refer to the Precautions Section.

➡**Do not allow the brake caliper and anchor plate assembly to hang from the brake flexible hose. Support the brake caliper and anchor plate assembly.**

2. Remove and discard the 2 brake caliper anchor plate bolts and position the brake caliper and anchor plate assembly aside.
3. Remove the front brake disc from the wheel hub.
To install:
4. Clean any rust or debris from the brake disc and wheel hub. Use parts cleaner to clean the front brake disc and hub surfaces.
5. Apply a thin coat of anti-seize lubricant to the hub flange.
6. Install the brake disc on the wheel hub.
7. Position the brake caliper and anchor plate assembly on the wheel knuckle and install 2 new bolts. Tighten to 200 Nm (148 ft. lbs.).

2002–05 2-Wheel Drive F-250, 350

1. Before servicing the vehicle, refer to the Precautions Section.
2. Remove the front disc brake caliper anchor plate.

3. Remove the hub grease cap.
4. Remove the cotter pin.
5. Remove the nut retainer.
6. Remove the spindle nut.
7. Remove the front wheel outer bearing retainer washer.
8. Remove the outer front wheel bearing.
9. Remove the brake disc and hub.
10. Remove the wheel hub grease seal.
11. Remove the inner front wheel bearing.
To install:
12. Thoroughly clean and inspect the front wheel bearings and the brake disc and hub.
13. Lubricate the front wheel bearings.
14. Install the inner front wheel bearing.
15. Install a new wheel hub grease seal.
16. Position the brake disc and hub.
17. Install the outer front wheel bearing.
18. Install the front wheel outer bearing retainer washer.
19. Install the spindle nut.
20. Adjust the wheel bearings.
21. Install the nut retainer.
22. Install the cotter pin.
23. Install the hub grease cap.
24. Install the front disc brake caliper anchor plate.

2002–05 4-Wheel Drive F-250, 350

1. Before servicing the vehicle, refer to the Precautions Section.
2. Remove the disc brake caliper anchor plate.

➡**Perform this step for SRW vehicles.**

3. Remove the rotor.

➡**Perform this step for DRW vehicles.**

➡**If excessive force must be used during brake rotor removal, the brake rotors should be checked for lateral runout prior to installation.**

4. Remove the eight hub extender nuts, the hub plate and the rotor.
    a. Remove the eight hub plate nuts.
    b. Remove the hub plate.
    c. Remove the rotor.
To install:
5. Position the front disc brake rotor to the wheel hub. Make sure the wheel hub and the front disc brake rotor braking and mounting surfaces are clean. Use brake parts cleaner to clean the front disc brake rotor.

➡**Perform this step for DRW vehicles only.**

6. Install the front wheel hub extender and nuts. Torque to 130 ft. lbs. (176 Nm).
7. Install the front disc brake caliper anchor plate.

Anchor Plate

REMOVAL & INSTALLATION

2002–03 F-150
2004 F-150 Heritage Edition
2002–05 F-250, 350
2002–05 E-150
2002–05 E-250, 350

1. Before servicing the vehicle, refer to the Precautions Section.
2. Remove the pads.
3. Remove and discard the two front disc brake caliper anchor plate bolts.
4. Remove the front disc brake caliper anchor plate.
To install:
5. Position the front disc brake caliper anchor plate.
6. Install the front disc brake caliper anchor plate bolts. Observe the following torques:
   • F-150: 136 ft. lbs. (185 Nm)
   • F-250 and 350: 166 ft. lbs. (225 Nm)
   • 2002–03 E-Series: 141–191 ft. lbs. (191–259 Nm)
   • 2004–05 E-150: 129 ft. lbs. (175 Nm)
   • 2004–05 E-250 and 350: 166 ft. lbs. (225 Nm)
7. Install the pads.

2004–05 F-150 Exc. Heritage Editions

1. Before servicing the vehicle, refer to the Precautions Section.
2. Remove the wheel and tire assembly.
3. Remove the brake caliper.

✱✱ WARNING

Do not allow the brake caliper to hang by the flexible brake hose.

4. Support the caliper to the vehicle.
5. Remove the brake disc pads.
6. Remove the slippers.
7. Remove the guide pin and boot.
8. Remove and discard the anchor bracket bolts.
9. Remove the anchor bracket.
10. To install, reverse the removal procedure. Torque the anchor bracket bolts to 148 ft. lbs. (200 Nm).

➡**Tighten the bottom caliper bolt first and then the top caliper bolt.**

REAR DISC BRAKES

Brake Caliper

REMOVAL & INSTALLATION

2002–05 E-250, 350

1. Before servicing the vehicle, refer to the Precautions Section.
2. Raise and support the vehicle.
3. Remove the wheel and tire assembly.
4. Remove the caliper pin bolts.

5. Remove the flow bolt and discard the copper washers.
6. Remove the rear disc brake caliper.

To install:

➡ **Use new copper washers on the flow bolt.**

7. Follow the removal procedure in reverse order.
8. Torque the flow bolt to 26 ft. lbs. (35 Nm).

9. Torque the caliper pin bolts to 27 ft. lbs. (36 Nm).
10. Bleed the brake system.

2004–05 E-150

1. Before servicing the vehicle, refer to the Precautions Section.
2. Remove the wheel and tire assembly.
3. Remove the brake hose flow bolt and position the brake hose aside. Discard the copper washers.
4. Remove the caliper pin bolts.
5. Remove the rear disc brake caliper.

➡ **Use new copper washers on the brake hose flow bolt.**

6. Bleed the brake system.
7. To install, reverse the removal procedure.
8. Observe the following torques:
   • Flow bolt: 26 ft. lbs. (35 Nm)
   • Caliper pin bolts: 24 ft. lbs. (32 Nm)

Blackwood

1. Before servicing the vehicle, refer to the Precautions Section.
2. Raise and support the vehicle.
3. Remove the wheel and tire assembly.
4. Remove bolt. Disconnect the rear wheel brake hose. Remove the copper washers and plug the brake hose.

❋❋ CAUTION

Do not remove the guide pins or guide pin boots unless a problem is suspected. The guide pins are meant to be sealed for life and are not repairable. Use Silicone Brake Caliper Grease and Dielectric Compound or an equivalent silicone compound meeting Ford specifications for re-lubing the caliper slide pins. Other greases can swell the guide pin boots, resulting in contamination and accelerated corrosion or wear of the caliper slide pin mechanism.

5. Remove the brake caliper bolts.
6. Lift the rear disc brake caliper off the rear disc brake caliper anchor plate.
7. Inspect the rear disc brake caliper for leaks. If leaks are found, disassembly is required.

1 Rear brake disc and hub assembly

2 Rear disc brake caliper

3 Dana full-floating axle — Model 80

4 Axle shaft-to-rear hub bolt

5 Rear wheel gasket

6 Outer bearing cup

7 Inner bearing cup

8 Inner hub seal

9 Rear wheel bearing inner cone and roller

10 Rear wheel bearing outer cone and roller

11 Wheel bearing lock nut

12 Axle shaft

67197-EFSE-G204

Rear disc brake exploded view—2002–05 E-250, 350

1 Hub   5 Caliper

2 Brake disc   6 Brake hose

3 Caliper anchor plate   7 Stainless steel slippers

4 Caliper pin bolts

67197-EFSE-G205

Rear disc brake caliper—2002–05 E-250, 350

To install:

✷✷ CAUTION

To prevent interference with rear disc brake caliper operation, install only the correct caliper bolt.

➡**Make sure the stainless steel shoe slippers are correctly positioned. Install new slippers if worn or damaged.**

➡**When installed, the locator notch on the brake pads will be located at the upper end of the rear disc brake caliper.**

8. Install the rear disc brake caliper. Torque the caliper bolts to 20 ft. lbs. (27 Nm).

9. Connect the brake hose and install the caliper flow bolt. Use new copper washers. Torque to 26 ft. lbs. (35 Nm).

10. Bleed the disc brake caliper.

11. Install the wheel and tire assembly.

12. Verify correct brake operation.

2002–03 F-150 and 2004 Heritage Editions

1. Before servicing the vehicle, refer to the Precautions Section.

2. Raise and support the vehicle.

3. Remove the wheel and tire assembly.

4. Remove bolt. Disconnect the rear wheel brake hose. Remove the copper washers and plug the brake hose.

✷✷ CAUTION

Do not remove the guide pins or guide pin boots unless a problem is suspected. The guide pins are meant to be sealed for life and are not repairable. Use Silicone Brake Caliper Grease and Dielectric Compound D7AZ-19A331-A (Motorcraft WA-10) or an equivalent silicone compound meeting Ford specification ESE-M1C171-A for re-lubing the caliper slide pins. Other greases can swell the guide pin boots, resulting in contamination and accelerated corrosion or wear of the caliper slide pin mechanism.

5. Remove the brake caliper bolts.

6. Lift the rear disc brake caliper off the rear disc brake caliper anchor plate.

7. Inspect the rear disc brake caliper for leaks. If leaks are found, disassembly is required.

To install:

✷✷ CAUTION

To prevent interference with rear disc brake caliper operation, install only the correct caliper bolt.

➡**Make sure the stainless steel shoe slippers are correctly positioned. Install new slippers if worn or damaged.**

➡**When installed, the locator notch on the brake pads will be located at the upper end of the rear disc brake caliper.**

8. Install the rear disc brake caliper. Torque the caliper bolts to 20 ft. lbs. (27 Nm).

9. Connect the brake hose and install the caliper flow bolt. Use new copper washers. Torque to 23–29 ft. lbs. (30–40 Nm).

10. Bleed the disc brake caliper.

11. Install the wheel and tire assembly.

12. Verify correct brake operation.

2004–05 F-150, exc. 2004 F-150 Heritage Editions

1. Before servicing the vehicle, refer to the Precautions Section.

2. Remove the rear wheel and tire assembly.

3. Remove the caliper bolts
4. Remove the brake caliper

❊❊ CAUTION

Do not allow the brake caliper to hang by the flexible brake hose.

5. Support the caliper to the vehicle.
6. Remove the flow bolt
7. Remove the brass washers
8. Remove the brake hose

9. To install, reverse the removal procedure. Observe the following torques:
   • Caliper pin bolts: 22 ft. lbs. (30 Nm)
   • Flow bolt: 26 ft. lbs. (35 Nm)
10. If the hydraulic system has been opened, bleed the brake system.

2002–04 F-250, 350

1. Before servicing the vehicle, refer to the Precautions Section.

All vehicles
2. Raise and support the vehicle.
3. Remove the wheel and tire assembly.
Forward-of-axle brake calipers
4. Remove the nuts and the stone shield.
All vehicles
5. Remove the caliper pin bolts.
6. Remove the banjo bolt. Remove and discard the copper washers.
7. Remove the rear disc brake caliper.

1 Rear disc brake caliper
2 Rear brake disc and hub assembly
3 Rear disc brake caliper anchor plate
4 Brake pads
5 Rear wheel disc brake adapter

Rear disc brake and hub—2002–05 E-250, 350

67197-EFSE-G206

27 Nm (20 lb-ft)

35 Nm (26 lb-ft)

27 Nm (20 lb-ft)

1 Flow bolt
2 Copper washers
3 Bolts
4 Guide pin
5 Guide pin boot

6 Caliper RH/LH
  Removal Note
7 Disc pads
8 Slippers
9 Brake disc

06017-F150-G276

Rear disc brake components—Blackwood

1 Brake caliper bolt
2 Flow bolt
3 Copper washer
4 Rear wheel brake hose
5 Rear disc brake caliper
6 Brake pads
7 Adapter assy
8 Shoe slippers
9 Axle shaft
10 Wheel stud
11 Brake disc
12 Keeper nut
13 Rear wheel disc brake adapter nut
14 Rear wheel disc brake adapter
15 Rear wheel disc brake adapter bolt

67197-EFSE-G207

Rear disc brake components—2002–03 F-150 and 2004 F-150 Heritage Editions

1 Caliper bolts (2 required each side)

2 Brake caliper

3 Disc brake pads (1 kit)

4 Brake disc

5 Flow bolt

6 Brass washers (2 required each side)

7 Brake hose

8 Slippers (2 required each side)

9 Brake line fitting

10 Rear brake hose bracket bolt

11 Brake line fitting

12 Retainer clip

13 Brake line fittings

14 Rear axle vent tube

15 Rear brake jounce hose

16 Rear brake jounce hose bracket bolt

17 Rear brake jounce hose bracket

Rear disc brake components—2004 F-150, exc. 2004 F-150 Heritage Editions

67197-EFSE-G208

1　Rear disc brake caliper

2　Rear disc brake rotor and hub assembly

3　Rear disc brake caliper anchor plate

4　Brake pads

5　Rear wheel disc brake adapter

6　Rear wheel disc brake shield

67197-EFSE-G209

Rear disc brake components—2002–04 F-250, 350

8. To install, reverse the removal procedure. Observe the following torques:
- Stone shield nuts: 46 ft. lbs. (62 Nm)
- Caliper bolts: 27 ft. lbs. (36 Nm)
- Banjo bolt: 37 ft. lbs. (50 Nm)

➡**Prior to installing the banjo bolt, install new copper washers.**

9. Bleed the brake system.

2005 F-250/350

1. Before servicing the vehicle, refer to the Precautions Section.
2. Raise and support the vehicle.
3. Remove the brake caliper flow bolt and discard the 2 copper washers.

✸✸ WARNING

Do not pry in the caliper sight hole to retract the pistons, as this can damage the pistons and boots.

4. Remove the 2 brake caliper bolts and the caliper.
To install:
5. Position the brake caliper and install the 2 brake caliper bolts.
- For F-250, F-350 single rear wheel caliper, tighten to 35 Nm (26 ft. lbs.).
- For F-350 dual rear wheel caliper, tighten to 76 Nm (56 ft. lbs.).
6. Position the brake flexible hose, install the brake caliper flow bolt and 2 new

copper washers. Torque to 26 ft. lbs. (35 Nm).
7. Bleed the brake system.

Brake Pads

REMOVAL & INSTALLATION

2002–05 E-250, 350

1. Before servicing the vehicle, refer to the Precautions Section.
2. Remove the brake master cylinder filler cap. Check brake fluid level in brake master cylinder reservoir. Remove fluid until brake master cylinder reservoir is half full.
3. Raise and support the vehicle.
4. Remove the wheel and tire assembly.

5. Inspect the brake pads for wear or contamination. If worn, damaged or past specification, install new components.

6. Remove the caliper pin bolts.

❋❋ CAUTION

Never allow the rear disc brake caliper to hang from the brake hose. Provide suitable support.

7. Remove the rear disc brake caliper.

8. Remove the brake pads and rail clips.

9. Inspect the disc brake caliper for leaks. If leaks are found, disassembly is required.

10. Installation is the reverse of removal. Observe the following torques:
- Caliper bolts: 27 ft. lbs. (36 Nm)

2004–05 E-150

1. Before servicing the vehicle, refer to the Precautions Section.

2. Clean the area and remove the brake cylinder filler cap. Check the brake fluid level in the brake master cylinder reservoir. Remove the fluid until the brake master cylinder reservoir is half full.

3. Remove the wheel and tire assembly.

4. Inspect the brake pads for wear or contamination. If worn, damaged or past specification, install new brake pads.

5. Remove the caliper pin bolts.

❋❋ CAUTION

Never allow the rear disc brake caliper to hang from the brake hose. Provide suitable support.

6. Position aside the rear disc brake caliper aside.

7. Remove the brake pad clips and the brake pads.

8. If installing new brake pads, use a suitable tool to push the brake caliper pistons into the caliper bore.

9. Inspect the rear disc brake caliper for leaks. If leaks are found, disassembly is required.

10. To install, reverse the removal procedure. Torque the caliper pin bolts to 24 ft. lbs. (32 Nm).

Blackwood

1. Before servicing the vehicle, refer to the Precautions Section.

2. Remove brake fluid in the master cylinder reservoir until the reservoir is half full.

❋❋ CAUTION

Install new pads if worn to or past the specified thickness above the metal backing plate or rivets. Install new pads in complete axle sets.

3. Inspect the brake pads for wear or contamination. Install new pads if worn to or past specification.

➡It is not necessary to remove the rear wheel brake hose when performing this procedure.

4. Remove the rear disc brake caliper.

❋❋ CAUTION

Do not allow grease, oil, brake fluid or other contaminants to contact the brake pads.

5. Remove the brake pads.

6. Retract the caliper piston into the rear disc brake caliper.

7. Remove and discard the slippers.

To install:

❋❋ CAUTION

Stainless steel slipper replacement is mandatory with the service brake pad installation, even if the slippers appear undamaged. Make sure the slippers are correctly positioned with the slipper ends snug against the outboard end of the anchor plate rail.

8. Clean the slipper mating surface, and install the new slippers.

❋❋ WARNING

Install new brake pads in full axle sets. Do not install new brake pads on only one side of vehicle.

9. Install the brake pads.

10. Install the rear disc brake caliper.

11. Verify correct brake operation.

2002–03 F-150, 2004 F-150 Heritage Edition

1. Before servicing the vehicle, refer to the Precautions Section.

2. Remove brake fluid in the master cylinder reservoir until the reservoir is half full.

❋❋ CAUTION

Install new pads if worn to or past the specified thickness above the

metal backing plate or rivets. Install new pads in complete axle sets.

3. Inspect the brake pads for wear or contamination. Install new pads if worn to or past specification.

➡It is not necessary to remove the rear wheel brake hose when performing this procedure.

4. Remove the rear disc brake caliper.

❋❋ CAUTION

Do not allow grease, oil, brake fluid or other contaminants to contact the brake pads.

5. Remove the brake pads.

6. Retract the caliper piston into the rear disc brake caliper.

7. Remove and discard the slippers.

To install:

❋❋ CAUTION

Stainless steel slipper replacement is mandatory with the service brake pad installation, even if the slippers appear undamaged. Make sure the slippers are correctly positioned with the slipper ends snug against the outboard end of the anchor plate rail.

8. Clean the slipper mating surface, and install the new slippers.

❋❋ WARNING

Install new brake pads in full axle sets. Do not install new brake pads on only one side of vehicle.

9. Install the brake pads.

10. Install the rear disc brake caliper.

11. Verify correct brake operation.

2004–05 F-150, Exc. 2004 F-150 Heritage Editions

1. Before servicing the vehicle, refer to the Precautions Section.

2. Remove the rear wheel and tire assembly.

3. Remove the caliper bolts

4. Remove the brake caliper

❋❋ WARNING

Do not allow the brake caliper to hang by the flexible brake hose.

5. Support the caliper to the vehicle.
6. Remove the disc brake pads

❋❋ WARNING

Do not us any tools to separate the inner brake pad away from the caliper piston. Damage to the caliper piston may occur.

7. Remove the rear brake pads from the caliper.
8. To install, reverse the removal procedure.

❋❋ CAUTION

Install a new lining if worn to or past the specified thickness above the metal backing plate or if the pad is damaged. Install new pads in complete axle sets.

9. If replacing the brake pads, use a suitable suction device, remove the brake fluid in the master cylinder reservoir until it is half filled.
10. If replacing the brake pads, use a suitable tool to push the brake caliper pistons into the caliper bore.
11. If the hydraulic system has been opened, bleed the brake system.

2002–04 F-250, 350

1. Before servicing the vehicle, refer to the Precautions Section.
   All vehicles
2. Remove the brake master cylinder filler cap. Check brake fluid level in brake master cylinder reservoir. Remove fluid until brake master cylinder reservoir is half full.
3. Remove the wheel and tire assembly.
4. Inspect the brake pads for wear or contamination. For additional information, refer to Specifications.
   Forward-of-axle brake calipers
5. Remove the nuts and the stone shield.
   All vehicles
6. Remove the caliper pin bolts.

❋❋ CAUTION

Never allow the rear disc brake caliper to hang from the brake hose. Provide suitable support.

7. Remove the rear disc brake caliper.
8. Remove the brake pads and rail clips.
9. Inspect the disc brake caliper for

leaks. If leaks are found, disassembly is required.
10. To install, reverse the removal procedure. Torque the stone shield nuts to 46 ft. lbs. (62 Nm).

2005 F-250/350

1. Before servicing the vehicle, refer to the Precautions Section.
2. If necessary, remove the brake fluid until the brake master cylinder reservoir is half full.
3. Remove the wheel and tire assembly.

❋❋ WARNING

Do not pry in caliper sight hole to retract pistons as this can damage the pistons and boots.

❋❋ WARNING

When removing the brake caliper, never allow it to hang from the brake flexible hose. Provide a suitable support.

4. Remove the 2 brake caliper bolts and position the caliper aside.

➡**Install a new pads if they are worn past the specified thickness. Install new pads in complete axle sets.**

5. Remove the brake pads and retraction clips.
6. Measure the brake disc thickness. Install a new brake disc if not within specification.
7. Inspect the disc brake caliper for leaks. If leaks are found a new caliper must be installed.
8. Inspect the disc brake anchor plate assembly.
   - Check the guide pin boots for damage.
   - Check the guide pins for binding and damage.
   - Replace worn or damaged pins. Lube pins with Silicone Brake Caliper Compound.
   To install:

➡**Make sure to correctly locate the retraction clip.**

9. Install the brake pads and retraction clips.
10. Position the brake caliper and install the 2 brake caliper bolts.
   - For F-250, F-350 single rear wheel caliper, tighten to 35 Nm (26 ft. lbs.).

   - For F-350 dual rear wheel caliper, tighten to 76 Nm (56 ft. lbs.).

Brake Rotor

REMOVAL & INSTALLATION

2002–03 F-150 and 2004 Heritage Editions; 2004 E-150

1. Before servicing the vehicle, refer to the Precautions Section.
2. Remove the rear disc brake caliper.

➡**If the rear brake disc binds on the rear parking brake shoe and linings, remove the adjustment hole access plug and contract the parking brake shoe and lining.**

3. Remove the rear brake disc.
4. Measure the rear brake disc, and resurface if the disc surface has been scored. Install a new rear brake disc if below minimum thickness specification.
5. Installation is the reverse of removal.

2004 F-150, exc. F-150 Heritage Editions

1. Before servicing the vehicle, refer to the Precautions Section.
2. Remove the rear wheel and tire assembly.
3. Remove the caliper.
4. Remove the disc brake pads.
5. Remove the brake disc.
6. To install, reverse the removal procedure.

❋❋ CAUTION

Make sure to apply a thin coat of anti-seize lubricant to the hub flange only. Do not allow the lubricant to make contact with the wheel studs, brake pads or brake disc.

7. Apply a thin coat of anti-seize lubricant to the wheel hub flange.

2002–03 E-250, 350 SRW
2002–04 E-250 Single Rear Wheel
2002–04 F-250 and 350 with Ford 10.5 inch axle and Single Rear Wheel

1. Before servicing the vehicle, refer to the Precautions Section.
2. Remove the anchor plate.

3. Match-mark the brake disc and one stud.

4. Remove the brake disc.

5. Installation is the reverse of removal.

2002–05 F-250 and 350 Ford Full-Floating Axle

See the procedure under Rear Axle Bearings.

2002–05 E-350 and F-350 w/Dual Rear Wheels and Dana Axle

See the procedure under Rear Axle Bearings.

Anchor Plate

REMOVAL & INSTALLATION

2002–05 E-Series

1. Before servicing the vehicle, refer to the Precautions Section.

2. Remove the brake pads.

3. Remove the anchor plate bolts and remove the rear disc brake caliper anchor plate.

4. To install, reverse the removal procedure. Torque the anchor plate bolts to 129 ft. lbs. (175 Nm).

2002–04 F-250 and 350

1. Before servicing the vehicle, refer to the Precautions Section.

2. Remove the brake pads.

➡Forward-of-axle rear disc brake caliper anchor plates are mounted using stud bolts.

3. Remove the bolts and the rear disc brake caliper anchor plate.

To install:

4. Position the rear disc brake caliper anchor plate. Install the bolts. Torque to 128 ft. lbs. (173 Nm).

5. Install the brake pads.

2005 F-250/350

1. Before servicing the vehicle, refer to the Precautions Section.

2. Remove the brake pads.

3. Remove the bolts and the brake caliper anchor plate.

To install:

4. Position the brake caliper anchor plate and install the 2 brake caliper anchor plate bolts. Torque to 203 ft. lbs. (275 Nm).

5. Install the brake pads.

Parking Brake Shoes

REMOVAL & INSTALLATION

E-Series

2002–03

1. Before servicing the vehicle, refer to the Precautions Section.

2. Remove the wheel hub assembly.

3. Remove the brake shoe retracting spring.

4. Remove the brake shoe hold-down springs.

5. Pull the primary shoe forward and out.

6. Pull the secondary shoe rearward and remove the parking brake shoes and set on bench.

7. Remove the brake shoe inner retracting spring.

8. Remove the brake adjuster screw.

9. Remove the brake shoe adjusting screw spring.

To install:

10. Install the brake shoe adjusting screw spring.

11. Install the brake adjuster screw.

1 Parking brake lever

2 Parking brake cable and conduit

3 Brake shoe retracting spring

4 Hold down spring

5 Brake shoe adjusting screw

6 Brake shoe adjusting screw spring

7 Parking brake shoe and lining

06017-F150-G277

Parking brake shoes—2003 E-150

1 Brake shoe and lining

2 Brake shoe hold-down spring

3 Brake shoe adjusting screw
  spring

4 Brake adjuster screw

5 Brake shoe retracting spring

6 Parking brake lever

06017-F150-G278

Parking brake shoes—2002–03 E-250/350

12. Install the brake shoe inner retracting spring.

➡**Lubricate the brake shoe contact point before installation of parking brake shoes using silicone brake caliper grease.**

13. Set the primary shoe in place.

14. Pull the secondary shoe rearward and install the parking brake shoe.

15. Install the brake shoe hold-down springs.

16. Install the brake shoe retracting spring.

17. Install the wheel hub assembly.

18. Use a brake adjusting gauge to measure the inside diameter of the drum portion of the rear brake disc.

19. Use a brake adjusting gauge to set the rear brake shoe and lining diameter to 0.5 mm (0.02 in) less than the inside diameter of the drum portion of the brake disc.

20. Install the brake disc.

➡**Only apply parking brake one time.**

21. Burnish the parking brake shoe and linings, in a safe manner:

   a. Accelerate the vehicle to 40 km/h (25 mph).

b. Shift the transmission to NEUTRAL.

   c. Slowly apply the parking brake control to approximately one-half to three-quarters of its travel.

   d. Allow the vehicle to come to a complete stop.

   e. Release the parking brake.

22. Check the operation of the parking brake.

2004–05

1. Before servicing the vehicle, refer to the Precautions Section.

➡**One parking brake she kit contains the linings required for both the LH and RH side.**

   All vehicles

➡**Make sure the parking brake control is fully released.**

2. Relieve the tension on the parking brake cable.

   a. With the vehicle in NEUTRAL, position it on the hoist.

   b. Remove the left A-pillar lower trim panel.

   c. With an assistant, release the parking brake cable tension by pulling down on the intermediate cable at the cable-to-

1 Brake shoe and lining

2 Brake shoe hold-down spring

3 Brake shoe adjusting screw
  spring

4 Brake adjuster screw

5 Brake shoe retracting spring

6 Parking brake lever

06017-F150-G279

Parking brake assembly—2004–05 E-Series

cable union until the parking brake control sector rotates to its stop and a 4 mm (0.15 in) x 150 mm (5.9 in) retainer pin can be inserted.

d. Disconnect the cable at the cable-to-cable union.

E-350 with dual rear wheels (DRW)

3. Remove the wheel hub assembly.

E-150, E-250, E-350 with single rear wheels (SRW)

4. Remove the brake disc.

All vehicles

5. Remove the brake shoe adjusting screw.

6. Remove the brake shoe adjusting screw spring.

7. Remove the brake shoe hold-down clips and pins.

8. Position the brake shoes apart and remove the brake shoes and the retracting spring assembly from the axle.

9. Remove the inner and outer retraction springs from the brake shoes.

To install:

➡Lubricate the parking brake shoes where the shoe contacts the wear pad on the backing plate.

All vehicles

10. Install the inner and outer retraction springs and position the parking brake shoe assembly on the axle.

11. Position the 2 hold-down pins and install the 2 brake shoe hold-down springs.

12. Install the brake shoe adjusting screw spring.

13. Position the brake shoe adjusting screw.

14. Measure the inside diameter of the parking brake drum.

15. Adjust the parking brake shoe clearance to 0.375 mm (0.01 in) less than the inside diameter of the parking brake drum.

E-150, E-250, E-350 with single rear wheels (SRW)

16. Install the rear brake disc.

E-350 with dual rear wheels (DRW)

17. Install the wheel hub assembly.

All vehicles

18. Reload the tension on the parking brake cable.

Blackwood

1. Before servicing the vehicle, refer to the Precautions Section.

2. Remove the rear wheel and tire assembly.

➡If the brake disc binds on the parking brake shoe and lining, remove the adjustment hole access plug and contract the parking brake shoe and lining.

Parking brake assembly—Blackwood

06017-F150-G280

1 Parking brake lever
2 Parking brake rear cable and conduit
3 Brake shoe retracting spring
4 Brake shoe hold-down spring
5 Brake adjuster screw
6 Brake shoe adjusting screw spring
7 Rear brake shoe and lining

06017-F150-G281

Parking brake assembly—2002–03 F-150 and 2004 F-150 Heritage Edition

3. Remove the brake disc
4. Remove the upper spring
5. Remove the retainers
6. Remove the adjuster
7. Remove the lower spring

➡Replace the parking brake shoes if the linings have worn to within 1.0 mm (0.04 in) of the shoe.

8. Remove the parking brake shoes
9. Remove the retaining pins
10. To install, reverse the removal procedure. Apply brake lube to the points where the shoes contact the backing plate.

2002–03 F-150 and 2004 F-150 Heritage Edition

1. Before servicing the vehicle, refer to the Precautions Section.

➡Make sure the parking brake control is fully released.

2. Relieve the tension on the parking brake system.

a. Pull the front parking brake cable and conduit.
b. Insert a suitable retainer.

3. Raise and support the vehicle.

4. Remove the tire and wheel assembly.

✷✷ WARNING

Do not let the rear disc brake caliper hang from caliper hose, use wire to hang it aside.

5. Remove the rear brake disc caliper.

6. Remove the rear brake disc.

7. Remove the brake shoe retracting spring.

8. Remove the brake shoe adjusting screw spring.

9. Remove the brake adjuster screw.

10. Remove the rear brake shoe hold-down springs.

11. Remove the rear brake shoes and linings along with the inboard brake shoe retracting spring.

12. Inspect the components for excessive wear or damage and install new components as required.

To install:

➡Lubricate the brake shoe contact point before installation of rear brake shoes using Silicone Brake Caliper Grease and Dielectric Compound or an equivalent silicone compound meeting Ford specifications.

13. Install the rear brake shoes and linings along with the inboard brake shoe retracting spring.

14. Install the brake shoe hold-down springs.

15. Install the brake adjuster screw.

16. Install the brake shoe adjusting screw spring.

17. Install the outboard brake shoe retracting spring.

18. Use a brake adjusting gauge to measure the inside diameter of the drum portion of the rear brake disc.

19. Use a brake adjusting gauge to set the rear brake shoe and lining diameter to 0.5 mm (0.020 in) less than the inside diameter to the drum portion of the rear brake disc.

20. Install the rear brake disc.

21. Install the rear disc brake caliper.

22. Clean the wheel hub mounting surface.

23. Install the wheel and tire assembly.

24. Apply tension to the parking brake cable system.

　　a. Hold the front parking brake cable and conduit taut.

　　b. Remove the retainer from the parking brake control.

25. Check the operation of the parking brake.

2004–05 F-150, exc. 2004 F-150 Heritage Edition
2006 Mark LT

1. Before servicing the vehicle, refer to the Precautions Section.

➡One parking brake shoe kit contains the linings required for both the left and right side.

➡Make sure the parking brake control is fully released.

2. Relieve the tension on the parking brake cable.

　　a. With the vehicle in NEUTRAL, position it on the hoist.

　　b. Remove the left A-pillar lower trim panel.

　　c. With an assistant, release the parking brake cable tension by pulling down on the intermediate cable at the cable-to-cable union until the parking brake control sector rotates to its stop and a 4 mm (0.15 in) x 150 mm (5.9 in) retainer pin can be inserted.

　　d. Disconnect the cable at the cable-to-cable union.

3. Remove the rear brake disc.

4. Remove the brake shoe adjuster screw.

5. Remove the brake shoe adjuster screw spring.

6. Remove the 2 brake shoe hold down springs and 2 pins.

7. Remove the brake shoe retracting spring and the parking brake shoes.

To install:

➡Lubricate the parking brake shoes where the shoe contacts the wear pad on the backing plate.

8. Position the parking brake shoes and attach the retracting spring.

9. Install the 2 brake shoe hold-down pins and springs.

10. Install the brake shoe adjusting screw spring.

➡Completely retract the parking brake adjusting screw before installation.

11. Install the brake shoe adjusting screw.

12. Measure the inside diameter of the parking brake drum.

13. Adjust the parking brake shoe clearance to 0.6 mm (0.02 in) less than the inside diameter of the parking brake drum.

14. Make sure that the parking brake shoes are correctly centered by measuring across the center point of the shoes.

15. Rotate the parking brake shoe adjuster wheel to achieve the correct parking brake shoe-to-brake disc clearance.

16. Install the rear brake disc.

17. Reload the tension on the parking brake cable.

2002–04 F-250/350

1. Before servicing the vehicle, refer to the Precautions Section.

2. Remove the rear disc brake rotor.

3. Disconnect the rear parking brake cable at the parking brake intermediate cable.

4. Disconnect the parking brake cable at the parking brake lever.

5. Remove the outboard brake shoe retracting spring.

6. Remove the brake shoe adjusting screw spring.

7. Remove the brake shoe hold-down springs.

8. Remove the brake adjuster screw.

9. Remove the parking brake shoe and linings along with the inboard brake shoe retracting spring.

10. Inspect the components for excessive wear or damage and install new as required.

To install:

➡Make sure the inboard brake shoe retracting spring is attached to the parking brake shoe.

11. Install the left parking brake shoe and brake shoe hold-down spring, along with the inboard brake shoe retracting spring.

12. Connect the inboard brake shoe retracting spring to the right parking brake shoe, and install the right parking brake shoe and the brake shoe hold-down spring.

13. Install the brake adjuster screw.

14. Install the brake shoe adjusting screw spring.

➡The outboard brake shoe retracting spring mounts above the inboard brake shoe retracting spring.

15. Install the outboard brake shoe retracting spring.

16. Connect the parking brake cable at the parking brake lever.

17. Connect the rear parking brake cable at the parking brake intermediate cable.

18. Install the rear disc brake rotor.

19. Use a brake adjusting gauge to set the rear brake shoe and lining diameter to 0.76 mm (0.030 in.) less than the inside diameter of the drum portion of the rear disc brake rotor.

20. Burnish the parking brake shoe and linings, in a safe manner:

　　a. Accelerate the vehicle to 40 km/h (25 mph).

　　b. Shift the transmission to NEUTRAL.

　　c. Slowly apply the parking brake control to approximately one-half to three-quarters of its travel.

　　d. Allow the vehicle to come to a complete stop.

　　e. Release the parking brake.

2005 F-250/350

1. Before servicing the vehicle, refer to the Precautions Section.

20 Nm (15 lb-ft) – 3

30 Nm (22 lb-ft) – 1

17 Nm (13 lb-ft) – 7

1   Parking brake cable bracket bolt (RH)

2   Rear parking brake cable (RH)

3   Parking brake cable differential bracket bolt (RH)

4   Parking brake cable differential bracket

5   Double cable clamp

6   Parking brake cable (LH)

7   Parking brake cable bracket bolt (LH)

8   Parking brake cable bracket (LH)

9   Parking brake shoe return spring

10 Parking brake shoe adjuster spring

11 Parking brake shoe adjuster

12 Parking brake shoe retainer spring clip (2 required)

13 Parking brake shoe retainer pin (2 required)

14 Parking brake shoe kit (1 kit required)

15 Actuator lever kit (LH/RH)

Parking brake exploded view—2004–05 F-150, exc. 2004 F-150 Heritage Edition; 2006 Mark LT

06017-F150-G282

1 Brake shoe retracting spring
  — outboard
  (part of parking brake spring
  kit)
2 Parking brake lever
3 Parking brake shoe and
  linings
4 Brake shoe hold-down spring
  (part of parking brake spring
  kit)
5 Brake adjuster screw
6 Brake shoe adjusting screw
  spring
  (part of parking brake spring
  kit)
7 Brake shoe retracting spring
  — inboard
  (part of parking brake spring
  kit)

06017-F150-G283

Parking brake assembly—2002–04 F-250/350

2. Remove the brake disc.
3. Release the tension on the parking brake system.

  a. Remove the left lower A-pillar trim panel.

  b. With an assistant, release the parking brake cable tension by pulling down on the intermediate cable at the cable-to-cable union until the parking brake control sector rotates to its stop and a 4 mm (0.15 in) x 150 mm (5.9 in) retainer pin can be inserted.

  c. Disconnect the cable at the cable-to-cable union.

➡ **Make sure the cable-to-cable union is connected to the front and rear cable before removing the brake control retaining pin, and the cable tension is reloaded slowly.**

4. Separate the parking brake cables at the parking brake equalizer.
5. Release the parking brake cable from the left and right parking brake bracket assembly.
6. Rotate the parking brake cable 90 degrees to release it from the parking brake actuating lever.
7. Remove the 2 brake shoe hold-down clips and the 2 retaining pins.
8. Remove the brake shoe adjusting screw and the adjusting spring.
9. Remove the outboard brake shoes and the retracting spring.
10. To install, reverse the removal procedure. Inspect the components for excessive wear or damage and install new as required. To reload the tension on the parking brake cable, reverse the release procedure.

DRUM BRAKES

Drum

REMOVAL & INSTALLATION

2002–03 E-150 and 250

1. Before servicing the vehicle, refer to the Precautions Section.

✳✳ CAUTION

To reduce the possibility of uneven braking, always replace rear brake shoes and linings at both ends of an axle.

2. Remove the wheel and tire assembly.

✳✳ WARNING

Use of a brake drum puller or a torch is not recommended. Brake drum distortion can result.

➡ **If the brake drum is rusted to the axle shaft pilot diameter, tap the center of the brake drum between the wheel studs.**

3. Remove and discard the spring nut.
4. Pull the brake drum off the axle shaft.

➡ **If brake drums will not come off, follow these steps.**

  a. Using a screwdriver, move the brake shoe adjusting lever off the brake adjuster screw.

  b. Move the adjustment tool handle downward to loosen the brake shoe adjusting screw nut.

  c. Use a Brake Adjustment Tool to loosen the brake shoe adjusting screw nut.

5. Installation is the reverse of removal. Adjust the brake shoes if necessary.

2002–03 F-150 and 2004 Heritage Edition

1. Before servicing the vehicle, refer to the Precautions Section.

➡ **Make sure the parking brake control is fully released.**

2. Relieve tension on parking brake system.

a. Pull the front parking brake cable and conduit.

b. Insert a suitable retainer pin.

3. Raise and support vehicle.

4. Remove the wheel and tire assembly.

✳✳ WARNING

Use of a brake drum puller or a torch is not recommended. Brake drum distortion may result.

➡**If the brake drum is rusted to the axle shaft flange, tap the center of the brake drum.**

5. Remove the brake drum.

➡**If the brake drum will not come off, follow these steps.**

a. Disengage the brake shoe adjusting lever from the brake adjuster screw.

b. Use a brake adjustment tool to loosen the brake adjuster screw.

To install:

6. Install the brake drum.

7. If necessary, adjust the rear brakes.

✳✳ WARNING

Always remove any corrosion, dirt or foreign material present on the mounting surfaces of the wheel or

1 Wheel Cylinder-to-Backing Plate Bolt	16 Adjusting Lever Pin
2 Washer	17 Adjusting Lever Return Spring
3 Inspection Hole Cover	18 Brake Shoe Adjusting Lever
4 Brake Backing Plate	19 Brake Shoe Adjusting Screw Nut
5 Lining Inspection Hole	20 Brake Adjuster Screw
6 Anchor Pin Guide Plate	21 Brake Shoe Adjusting Screw Spring
7 Rear Wheel Cylinder	22 Brake Shoe Hold-Down Spring Cup
8 Wheel Cylinder Brake Shoe Link	23 Brake Shoe Hold-Down Spring
9 Parking Brake Strut	24 Rear Brake Shoe and Lining, Primary
10 Parking Brake Lever	25 Brake Shoe Retracting Spring, Short
11 Brake Shoe Adjusting Lever Cable	26 Parking Brake Link Spring
12 Rear Brake Shoe and Lining, Secondary	27 Parking Brake Spring Retainer
13 Washer	28 Brake Shoe Hold-Down Spring Pin
14 Parking Brake Lever Pin Retainer	29 Brake Adjusting Hole Cover
15 Cable Guide	

67197-EFSE-G226

Drum brake components—2002–03 F-150 and 2004 Heritage Editions and 2002–03 E-150

the surface of the wheel hub or brake drum that contacts the wheel. **Installing wheels without correct metal to metal contact at the wheel mounting surfaces can cause the wheel nuts to loosen and could allow the wheel to come off while the vehicle is in motion, causing loss of control.**

8. Clean the wheel hub mounting surface.

9. Install the tire and wheel assembly.

10. Lower the vehicle.

11. Remove the pin in the parking brake control and verify correct parking brake operation.

12. Check brake operation before driving.

Brake Shoes

REMOVAL & INSTALLATION

2002–03 E-150; 2002–03 F-150; 2004 Heritage Edition

1. Before servicing the vehicle, refer to the Precautions Section.

2. Remove the brake drum.

3. Inspect the rear brake assembly for the following:

- the rear wheel cylinder for excessive leakage that can contaminate brake system parts, and rebuild as necessary.
- the rear brake shoes and linings for contamination, and replace as necessary.
- the rear brake shoes and linings for minimum thickness, above the backing plate or rivets, and replace as necessary.
- the springs for heat discoloration, and replace as necessary.
- the adjusting lever contact with the brake adjuster screw.

➡**To aid installation, note the locations of the short and long brake shoe retracting springs.**

4. Remove the long brake shoe retracting spring.

5. Remove the short brake shoe retracting spring.

6. Disconnect the brake shoe adjusting lever cable from the brake shoe adjusting lever.

7. Remove the brake shoe adjusting lever cable.

8. Remove the cable guide.

9. Remove the adjusting lever return spring and the brake shoe adjusting lever.

10. Remove the brake shoe adjusting screw spring.

11. Remove the brake adjuster screw assembly.

➡**The parking brake link spring and the brake parking spring retainer will come off with the parking brake strut.**

12. Remove the brake shoe anchor pin guide plate, parking brake strut and the parking brake lever pin retainer.

13. Remove the brake shoe hold-down springs.

14. Remove the rear brake shoes and linings.

15. Remove the brake shoe hold-down spring pins.

16. Pull back the parking brake cable spring and disconnect the parking brake lever.

To install

17. Compress the parking brake cable spring and attach the parking brake lever.

18. Clean and lubricate the brake backing plate. Use Silicone Brake Caliper Grease and Dielectric Compound D7AZ-19A331-A (Motorcraft WA-10) or an equivalent silicone compound meeting Ford specification ESE-M1C171-A.

19. Attach the parking brake lever to the rear brake shoe and lining and secure the parking brake lever pin retainer.

20. Position the rear brake shoes and linings.

21. Install the brake shoe hold-down spring pins and the brake shoe hold-down springs.

22. Install the parking brake link spring and the parking brake spring retainer on the parking brake strut.

23. Install the parking brake strut.

24. Install the brake shoe anchor pin guide plate.

25. Place the brake shoe adjusting lever cable over the anchor pin, with crimped side in.

26. Install the short brake shoe retracting spring.

➡**Make sure the brake shoe adjusting lever cable is positioned in the cable guide groove.**

27. Install the cable guide.

28. Position the brake shoe adjusting lever cable in the cable guide groove.

29. Install the long brake shoe retracting spring.

➡**prevent incorrect installation, the socket end of each brake adjuster screw is stamped R or L.**

30. Apply Silicone Brake Caliper Grease and Dielectric Compound D7AZ-19A331-A (Motorcraft WA-10) or an equivalent silicone compound meeting Ford specification ESE-M1C171-A.

31. Install the brake adjuster screw into the brake shoe adjusting screw nut to the end of the threads, and then loosen one-half turn.

32. Install the brake shoe adjusting screw socket on the brake shoe adjusting screw nut.

33. Position the brake adjuster screw assembly.

34. Install the brake shoe adjusting screw spring.

35. Position the brake shoe adjusting lever cable.

36. Position the adjusting lever return spring and the brake shoe adjusting lever.

37. Hook the brake shoe adjusting lever cable to the brake shoe adjusting lever.

38. Test the operation of the automatic self-adjuster.

a. Pull the brake shoe adjusting lever cable and check that the brake shoe adjusting lever rotates the brake shoe adjuster assembly.

b. Release the brake shoe adjusting lever cable and check that the brake shoe adjusting lever advances to the next notch on the brake shoe adjusting screw nut.

39. Adjust the brakes.

40. Install the brake drum.

Wheel Cylinder

REMOVAL & INSTALLATION

1. Before servicing the vehicle, refer to the Precautions Section.

2. Remove the brake shoes and linings.

3. Disconnect the brake line fitting.

4. Remove the two rear wheel cylinder bolts.

5. Remove the rear wheel cylinder.

6. Follow the removal procedure in reverse order.

7. Torque the wheel cylinder bolts to 9–13 ft. lbs. (12–18 Nm); torque the brake line fitting to 11–14 ft. lbs. (15–20 Nm).

8. Bleed the brake system.

BRAKE SYSTEM SERVICE

Drum Brakes

BRAKE SHOE ADJUSTMENT

1. Before servicing the vehicle, refer to the Precautions Section.

Brake Drums Removed

➡ **After adjusting the rear brake shoes and linings, check the parking brake for proper operation. Make sure the parking brake cable equalizer operates freely.**

2. Measure the brake drum inside diameter. Use a brake adjustment gauge.

67197-EFSE-G228

Measuring the brake drum inside diameter

67197-EFSE-G229

Measuring overall brake shoe width

67197-EFSE-G230

Adjusting the brakes with the drum installed

3. Rotate the brake adjuster screw until the rear brake shoes and linings touch the tool.

Brake Drums Installed

4. Raise and support the vehicle.

5. At the back of the brake backing plate, remove the brake adjusting hole cover.

6. Rotate the brake shoe adjusting screw nut.

7. Using a screwdriver, move the brake shoe adjusting lever off the brake shoe adjusting screw nut.

➡ **Move the adjustment tool handle upward to tighten the brake adjuster screw.**

8. Turn the brake shoe adjusting screw nut until the brake drum begins to drag. Then loosen the brake adjuster screw until the brake drum rotates freely.

9. Replace the brake adjusting hole cover.

10. Lower the vehicle.

Disc and Drum Brakes

BRAKE BLEEDING

2002–03

1. Before servicing the vehicle, refer to the Precautions Section.

Master Cylinder Priming—In-Vehicle or Bench

➡ **When any part of the hydraulic system has been disconnected for repair or replacement, air can enter the system and cause spongy brake pedal action. This requires bleeding of the hydraulic system after it has been properly connected. The hydraulic system can be bled manually or with pressure bleeding equipment.**

➡ **When the brake master cylinder has been replaced or the system has been emptied, or partially emptied, it should be primed to prevent air from entering the system.**

2. For in-vehicle priming, disconnect the brake lines.

3. For bench priming, mount the brake master cylinder in a vise.

4. Install short brake tubes with the ends submerged in the brake master cylinder reservoir, and fill the brake master cylinder reservoir with High Performance DOT 3 Motor Vehicle Brake Fluid C6AZ-19542-AB

or equivalent DOT 3 fluid meeting Ford specification ESA-M6C25-A.

5. Have an assistant pump the brake pedal, or slowly depress the primary piston until clear fluid flows from both brake tubes, without air bubbles.

6. If the brake master cylinder is being primed at the bench, install it in the vehicle.

7. Remove the short brake tubes, and install the brake outlet tubes.

8. Bleed each brake tube at the brake master cylinder as follows:

   a. Have an assistant pump the brake pedal, and then hold firm pressure on the brake pedal.

   b. Loosen the rearmost brake tube fittings until a stream of brake fluid comes out. While the assistant maintains pressure on the brake pedal, tighten the brake tube fitting.

   c. Repeat this operation until clear, bubble-free fluid comes out.

   d. Refill the brake master cylinder reservoir as necessary. Repeat the bleeding operation at the front brake tube.

9. Tighten the brake lines at the master cylinder to 18 ft. lbs. (25 Nm) for 2002–03 models; 13 ft. lbs. (17 Nm) for 2004 models.

Four Wheel Anti-Lock Brake System (4WABS) Hydraulic Control Unit (HCU)

67197-EFSE-G231

In-vehicle bleeding the master cylinder

67197-EFSE-G232

Bleeding a caliper

→This procedure needs to be performed only if the 4-wheel anti-lock brake (4WABS) hydraulic control unit (HCU) has been replaced or if air is suspected in the HCU.

10. Clean all dirt from and remove the brake master cylinder filler cap, and fill the brake master cylinder reservoir with the specified brake fluid.

11. Connect a clear waste line to the right rear bleeder screw and the other end in a container partially filled with recommended brake fluid.

12. With the right rear bleeder screw open, cycle the brake pedal until no more air is seen in the waste line.

13. Tighten the right rear bleeder screw, and disconnect the waste line.

14. Repeat Steps 2, 3 and 4 for the left rear bleeder screw, the right front disc brake caliper bleeder screw, and the left front disc brake caliper bleeder screw, in that order.

15. Connect Worldwide Diagnostic System (WDS) 418-F224, New Generation STAR (NGS) Tester 418-F052, or equivalent diagnostic tool DCL cable adapter into the vehicle data link connector (DLC) under the dash, and follow the scan tool instructions.

16. Repeat the system bleed procedure as outlined in Steps 1 through 5.

Caliper or Wheel Cylinder

→It is not necessary to do a complete brake system bleed if only the disc brake caliper or drum brake wheel cylinder was disconnected.

17. Place a box end wrench on the disc brake caliper or wheel cylinder bleeder screw. Attach a rubber drain tube to the bleeder screw, and submerge the free end of the tube in a container partially filled with clean brake fluid.

18. Have an assistant pump the brake pedal and then hold firm pressure on the brake pedal.

19. Loosen the bleeder screw until a stream of brake fluid comes out. While the assistant maintains pressure on the brake pedal, tighten the bleeder screw. Repeat until clear, bubble-free fluid comes out. Refill the brake master cylinder reservoir as necessary.

20. Tighten the bleeder screw.

2004–06

1. Before servicing the vehicle, refer to the Precautions Section.

→When any part of the hydraulic system has been disconnected for repair or new installation, air may get into the system and cause spongy brake pedal action. This requires bleeding of the hydraulic system after it has been correctly connected. The hydraulic system can be gravity bled, manually bled or with pressure bleeding equipment.

Master Cylinder, Bench

2. Support the brake master cylinder body in a vise and fill the brake master cylinder reservoir with specified brake fluid.

→Original equipment lines are not intended to be used during this procedure.

3. Install short brake tubes with the ends submerged in the brake master cylinder reservoir.

4. Slowly press the primary piston until clear fluid flows from both brake tubes, without air bubbles.

5. Remove the short brake tubes and plug the brake tube ports.

Master Cylinder—In Vehicle

✴✴ WARNING

Do not allow the brake master cylinder reservoir to run dry during the bleeding operation. Keep the brake master cylinder reservoir filled with the specified brake fluid. Never reuse the brake fluid that has been drained from the hydraulic system.

→When a new brake master cylinder has been installed or the system has been emptied, or partially emptied, it should be primed to prevent air from getting into the system.

6. Disconnect the brake master cylinder outlet tubes.

→Original equipment lines are not intended to be used during this procedure.

7. Install short brake tubes with ends submerged in the brake master cylinder reservoir and fill the brake master cylinder reservoir with brake fluid.

8. Have an assistant pump the brake pedal until clear fluid flows from both brake tubes without air bubbles.

9. Remove the short brake tubes and install the brake outlet tubes.

10. Bleed each brake tube at the brake master cylinder as follows:

a. Have an assistant pump the brake pedal and then hold firm pressure on the brake pedal.

b. Loosen the rear brake tube fittings until a stream of brake fluid comes out. Have an assistant maintain pressure on the brake pedal while tightening the brake tube fitting.

c. Repeat this operation until clear, bubble-free fluid comes out.

d. Refill the brake master cylinder reservoir as necessary. Repeat the bleeding operation at the front brake tube.

11. While the assistant maintains pressure on the brake pedal, tighten the brake tubes.

Gravity Bleeding

✴✴ WARNING

Do not allow the brake master cylinder reservoir to run dry during the bleeding operation. Keep the brake master cylinder reservoir filled with the specified brake fluid. Never reuse the brake fluid that has been drained from the hydraulic system.

→When a new brake master cylinder has been installed or the system has been emptied, or partially emptied, it should be primed to prevent air from getting into the system.

12. Fill the brake master cylinder reservoir with brake fluid.

13. Connect a clear tube to the right rear disc brake caliper bleeder screw and the other end in a container partially filled with recommended brake fluid.

14. Open the bleeder screw and leave open until clear bubble-free brake fluid flows.

15. Refill the brake master cylinder reservoir as necessary.

16. Tighten the disc brake caliper bleeder screw.

17. Repeat Steps 1 through 4 for the three remaining brake calipers, going in order from the left rear disc brake caliper to the right front disc brake caliper ending with the left front disc brake caliper.

18. If the brake pedal feels spongy, repeat the bleed procedure.

Manual Bleeding

✴✴ WARNING

Do not allow the brake master cylinder reservoir to run dry during the bleeding operation. Keep the brake master cylinder reservoir filled with the specified brake fluid. Never reuse the brake fluid that has been drained from the hydraulic system.

19. Fill the brake master cylinder reservoir with brake fluid.

20. Connect a clear tube to the right rear disc brake caliper bleeder screw and the

other end in a container partially filled with recommended brake fluid.

21. Have an assistant pump the brake pedal and then hold firm pressure on the brake pedal.

22. Loosen the disc brake caliper bleeder screw until a stream of brake fluid comes out. Have an assistant maintain pressure on the brake pedal while tightening the disc brake caliper bleeder screw. Repeat until clear, bubble-free fluid comes out. Refill the brake master cylinder reservoir as necessary.

23. Tighten the disc brake caliper bleeder screw.

24. Repeat Steps 1 through 5 for the three remaining brake calipers, going in order from the left rear disc brake caliper to the right front disc brake caliper ending with the left front disc brake caliper.

25. If the brake pedal feels spongy, repeat the bleed procedure.

Anti-Lock Brake System Hydraulic Control Unit Bleeding

➡ **This procedure is only required when a new hydraulic control unit is installed.**

26. Connect diagnostic tool Worldwide Diagnostic System (WDS) 418-F224, New Generation STAR (NGS) Tester 418-F052, or equivalent diagnostic tool and follow the ABS system bleed instructions.

27. Use the gravity bleed or manual bleed procedure(s) to bleed the system. Begin at the right rear caliper.

FORD, LINCOLN AND MERCURY

3

Crown Victoria • Grand Marquis • Marauder • Town Car

SPECIFICATIONS AND MAINTENANCE CHARTS

ENGINE AND VEHICLE IDENTIFICATION

		Engine							Model Year	
Code ①	Liters (cc)	Cu. In.	Cyl.	Fuel Sys.	Type	Eng. Mfg.		Code ②	Year	
W	4.6 (4593)	281	8	SFI	SOHC	Ford		2	2002	
V	4.6 (4593)	281	8	SFI	DOHC	Ford		3	2003	
								4	2004	
								5	2005	
								6	2006	

SFI: Sequential Fuel Injection

SOHC: Single Overhead Camshaft

① 8th digit of the Vehicle Identification Number (VIN)

② 10th digit of the Vehicle Identification Number (VIN)

06017-CROWN-C01

GENERAL ENGINE SPECIFICATIONS

Year	Model	Engine Displacement Liters	Engine ID/VIN	Net Horsepower @ rpm	Net Torque @ rpm (ft. lbs.)	Bore x Stroke (in.)	Compression Ratio	Oil Pressure @ rpm
2002	Crown Victoria	4.6	W	①	②	3.55x3.54	③	20-45@1500
	Grand Marquis	4.6	W	①	②	3.55x3.54	9.4:0	20-45@1500
	Town Car	4.6	W	①	②	3.55x3.54	9.4:0	20-45@1500
2003	Crown Victoria	4.6	W	①	②	3.55x3.54	③	20-45@1500
	Grand Marquis	4.6	W	①	②	3.55x3.54	9.4:0	20-45@1500
	Town Car	4.6	W	①	②	3.55x3.54	9.4:0	20-45@1500
	Marauder	4.6	V	302@5750	318@4300	3.55x3.54	9.85:0	20-45@1500
2004	Crown Victoria	4.6	W	①	②	3.60x3.60	③	20-45@1500
	Grand Marquis	4.6	W	①	②	3.60x3.60	9.4:0	20-45@1500
	Town Car	4.6	W	①	②	3.60x3.60	9.4:0	20-45@1500
	Marauder	4.6	V	302@5750	318@4300	3.60x3.60	9.85:0	20-45@1500
2005	Crown Victoria	4.6	W	①	②	3.60x3.60	③	20-45@1500
	Grand Marquis	4.6	W	①	②	3.60x3.60	9.4:0	20-45@1500
	Town Car	4.6	W	①	②	3.60x3.60	9.4:0	20-45@1500

① Single exhaust: 220@4750

Dual exhaust: 235@4000

Crown Victoria with natural gas: 178@4500

② Single exhaust: 265@3250

Dual exhaust: 275@3250

Crown Victoria with natural gas: 237@3500

③ Gasoline Engine: 9.4:1

Natural Gas Engine: 10.0:1

06017-CROWN-C02

ENGINE TUNE-UP SPECIFICATIONS

Year	Engine Displacement Liters	Engine ID/VIN	Spark Plug Gap (in.)	Ignition Timing (deg.)	Fuel Pump (psi) ①	Idle Speed (rpm)	Valve Clearance	
							Intake	Exhaust
2002	4.6	W	0.054	10B	35-45	②	HYD	HYD
2003	4.6	W	0.054	10B	35-45	②	HYD	HYD
	4.6	V	0.054	10B	35-45	②	HYD	HYD
2004	4.6	W	0.054	10B	35-45	②	HYD	HYD
	4.6	V	0.054	10B	35-45	②	HYD	HYD
2005	4.6	W	0.054	10B	35-45	②	HYD	HYD

NOTE: The Vehicle Emission Control Information label often reflects specification changes made during production. The label figures must be used if they differ from those in this chart.

B: Before Top Dead Center

HYD: Hydraulic

① Fuel pressure with engine running, pressure regulator vacuum hose connected

② Refer to Vehicle Emission Control Information label

06017-CROWN-C03

4.6L (VIN W) Engine
Firing order: 1–3–7–2–6–5–4–8
Distributorless ignition system

79223G28

4.6L (VIN V) Engine
Firing order: 1–3–7–2–6–5–4–8
Distributorless ignition system—One coil per cylinder

93003G02

1. Alternator
2. Water pump
3. Power steer-
   ing pump
4. Crankshaft
5. A/C com-
   pressor
6. Drive belt
7. Tensioner
8. Idler pulley

VIEW A

FRONT OF ENGINE

VIEW A

79224G30

Accessory drive belt routing—2002–2004 4.6L (VIN W) engine

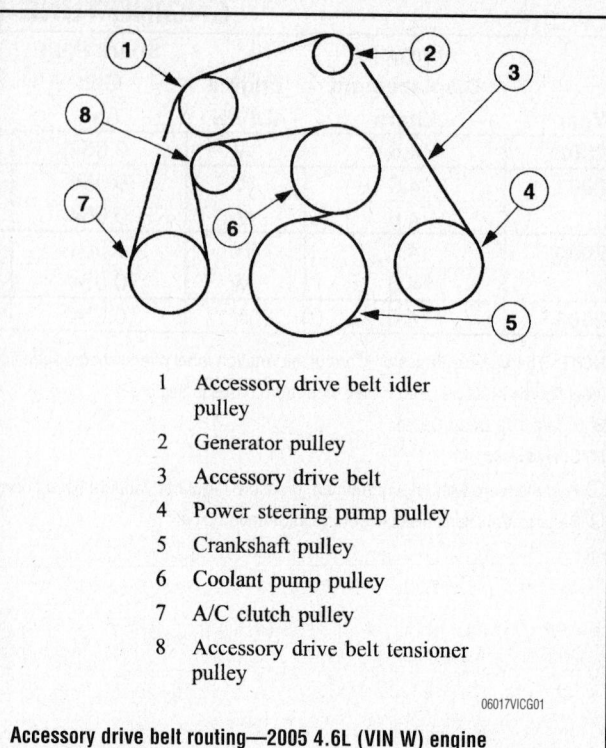

1 Accessory drive belt idler pulley
2 Generator pulley
3 Accessory drive belt
4 Power steering pump pulley
5 Crankshaft pulley
6 Coolant pump pulley
7 A/C clutch pulley
8 Accessory drive belt tensioner pulley

06017VICG01

Accessory drive belt routing—2005 4.6L (VIN W) engine

Drive Belt Routing

FRONT OF ENGINE

1. Generator
2. Water pump pulley
3. Belt idler pulley
4. Drive belt tensioner
5. Power steering pump
6. Drive belt
7. A/C compressor
8. Crankshaft pulley

79224G26

Serpentine accessory drive belt routing— 4.6L (VIN V) engine

CAPACITIES

Year	Model	Engine Displacement Liters	Engine ID/VIN	Engine Oil with Filter (qts.)	Automatic Transmission (pts.) ①	Rear Drive Axle (pts.)	Fuel Tank (gal.)	Cooling System (qts.)
2002	Crown Victoria	4.6	W	5.0	27.2	3.75	19.0	14.1
	Grand Marquis	4.6	W	5.0	28.2	3.75	19.0	15.1
	Town Car	4.6	W	5.0	28.2	3.75	19.0	15.1
2003	Crown Victoria	4.6	W	5.0	27.2	3.75	19.0	19.2
	Grand Marquis	4.6	W	5.0	28.2	3.75	19.0	19.2
	Town Car	4.6	W	5.0	28.2	3.75	19.0	19.2
	Marauder	4.6	V	6.0	27.8	3.75	19.0	18.8
2004	Crown Victoria	4.6	W	5.0	27.2	3.75	19.0	19.2
	Grand Marquis	4.6	W	5.0	28.2	3.75	19.0	19.2
	Town Car	4.6	W	5.0	28.2	3.75	19.0	19.2
	Marauder	4.6	V	6.0	27.8	3.75	19.0	18.8
2005	Crown Victoria	4.6	W	5.0	27.2	3.75	19.0	19.2
	Grand Marquis	4.6	W	5.0	28.2	3.75	19.0	19.2
	Town Car	4.6	W	5.0	28.2	3.75	19.0	19.2

NOTE: All capacities are approximate. Add fluid gradually and ensure a proper fluid level is obtained.

① Includes torque converter

06017-CROWN-C04

VALVE SPECIFICATIONS

Year	Engine Displacement Liters	Engine ID/VIN	Seat Angle (deg.)	Face Angle (deg.)	Spring Test Pressure (lbs. @ in.)	Spring Installed Height (in.)	Stem-to-Guide Clearance (in.)		Stem Diameter (in.)	
							Intake	Exhaust	Intake	Exhaust
2002	4.6	W	45	45.5	132@1.10	1.570	0.0008-0.0027	0.0018-0.0037	0.2746-0.2754	0.2736-0.2744
2003	4.6	W	45	45.5	132@1.10	1.570	0.0008-0.0027	0.0018-0.0037	0.2746-0.2754	0.2736-0.2744
	4.6	V	45	45.5	160@1.03	1.660	0.0008-0.0027	0.0018-0.0037	0.2746-0.275	0.2736-0.2744
2004	4.6	W	45	45.5	132@1.10	1.570	0.0008-0.0027	0.0018-0.0037	0.2746-0.2754	0.2736-0.2744
	4.6	V	45	45.5	160@1.03	1.660	0.0008-0.0027	0.0018-0.0037	0.2746-0.275	0.2736-0.2744
2005	4.6	W	45	45.5	132@1.10	1.570	0.0008-0.0027	0.0018-0.0037	0.2746-0.2754	0.2736-0.2744

06017-CROWN-C05

CRANKSHAFT AND CONNECTING ROD SPECIFICATIONS

All measurements are given in inches.

| Year | Engine Displacement Liters | Engine ID/VIN | Crankshaft | | | | Connecting Rod | | |
			Main Brg. Journal Dia.	Main Brg. Oil Clearance	Shaft End-play	Thrust on No.	Journal Diameter	Oil Clearance	Side Clearance
2002	4.6	W	2.6500-2.6570	0.0009-0.0026	0.0051-0.0119	5	2.0870-2.8670	0.0009-0.0026	0.0006-0.0177
2003	4.6	W	2.6500-2.6570	0.0009-0.0026	0.0051-0.0119	5	2.0870-2.8670	0.0009-0.0026	0.0006-0.0177
	4.6	V	2.6567-2.6577	0.0001-0.0018	0.0051-0.0119	5	2.0859-2.0867	0.0011-0.0027	0.0006-0.0177
2004	4.6	W	2.6500-2.6570	0.0009-0.0026	0.0051-0.0119	5	2.0870-2.8670	0.0009-0.0026	0.0006-0.0177
	4.6	V	2.6567-2.6577	0.0001-0.0018	0.0051-0.0119	5	2.0859-2.0867	0.0011-0.0027	0.0006-0.0177
2005	4.6	W	2.6500-2.6570	0.0009-0.0026	0.0051-0.0119	5	2.0870-2.8670	0.0009-0.0026	0.0006-0.0177

06017-CROWN-C06

PISTON AND RING SPECIFICATIONS

All measurements are given in inches.

| Year | Engine Displacement Liters | Engine ID/VIN | Piston Clearance ① | Ring Gap | | | Ring Side Clearance | | |
				Top Compression	Bottom Compression	Oil Control	Top Compression	Bottom Compression	Oil Control
2002	4.6	W	0.0002-0.0010	0.005-0.012	0.012-0.022	0.006-0.026	②	0.008-0.0024	0.0010-0.0077
2003	4.6	W	0.0002-0.0010	0.005-0.012	0.012-0.022	0.006-0.026	②	0.008-0.0024	0.0010-0.0077
	4.6	V	0.0000-0.0010	0.010-0.020	0.009-0.019	0.006-0.026	0.0003-0.0009	0.0012-0.0031	SNUG
2004	4.6	W	0.0002-0.0010	0.005-0.012	0.012-0.022	0.006-0.026	②	0.008-0.0024	0.0010-0.0077
	4.6	V	0.0000-0.0010	0.010-0.020	0.009-0.019	0.006-0.026	0.0003-0.0009	0.0012-0.0031	SNUG
2005	4.6	W	0.0002-0.0010	0.005-0.012	0.012-0.022	0.006-0.026	②	0.008-0.0024	0.0010-0.0077

① Measured 1.96 in. (43mm) from the top
② On 10:1 engines: 0.0012-0.0028 in.
   On 9:1 engines: 0.0008-0.0024 in.

06017-CROWN-C07

TORQUE SPECIFICATIONS

All readings in ft. lbs.

Year	Engine Displacement Liters	Engine ID/VIN	Cylinder Head Bolts	Main Bearing Bolts	Rod Bearing Bolts	Crankshaft Damper Bolts	Flywheel Bolts	Manifold		Spark Plugs	Oil Pan Drain Plug
								Intake	Exhaust		
2002	4.6	W	①	②	③	114-121	54-64	15-22	13-16	7-15	10
2003	4.6	W	①	②	③	114-121	54-64	15-22	13-16	7-15	10
	4.6	V	④	⑤	⑥	114-121	54-64	⑦	13-16	7-15	10
2004	4.6	W	①	②	③	114-121	54-64	15-22	13-16	7-15	10
	4.6	V	④	⑤	⑥	114-121	54-64	⑦	13-16	7-15	10
2005	4.6	W	①	②	③	114-121	54-64	15-22	13-16	7-15	10

NOTE: Stretch bolts are used in all procedures that require rotating the fastener a certain number of degrees. The bolts stretch and cannot be reused.

For reassembly, replace with new fasteners.

① Step 1: 22-30 ft. lbs.
 Step 2: Plus 85-95 degrees
 Step 3: Plus 85-95 degrees

② Step 1: Main bearing cap bolts: 22-25 ft. lbs.
 Step 2: Rotate each bolt 85-95 degrees
 Step 3: Main bearing cap adjusting screws: 44 inch lbs. then 80-90 inch lbs.
 Step 4: Main bearing cap side bolts: 7 ft. lbs. then 14-17 ft. lbs.

③ Step 1: 8 ft. lbs.
 Step 2: 12 ft. lbs.
 Step 3: 25-34 ft. lbs.
 Step 4: Rotate 85-95 degrees

④ Step 1: 30 ft. lbs.
 Step 2: Tighten 90 degrees
 Step 3: Loosen all bolts one (1) turn
 Step 4: 30 ft. lbs.
 Step 5: Tighten 90 degrees
 Step 6: Tighten an additional 90 degrees

⑤ Step 1: Main bearing cap bolts: 6-9 ft. lbs.
 Step 2: Main bearing cap bolts, outer: 16-21 ft. lbs.
 Step 3: Main bearing cap bolts, inner: 27-32 ft. lbs.
 Step 4: Tighten main bearing cap bolts 85-95 degrees
 Step 5: Main cap adjusting screws: 4 ft. lbs. then 7.5 ft. lbs.
 Step 6: Main cap side bolts: 7 ft. lbs. then 14-17 ft. lbs.

⑥ Step 1: 5ft. lbs.
 Step 2: 10 ft. lbs.
 Step 3: 18-25 ft. lbs.
 Step 4: Tighten 85-95 degrees

⑦ Step 1: Four inside short bolts: 9-11 ft. lbs.
 Step 2: All other bolts: 13-16 ft. lbs.
 Step 3: Tighten 85-95 degrees

06017-CROWN-C08

WHEEL ALIGNMENT

Year	Model		Caster Range (+/-Deg.)	Caster Preferred Setting (Deg.)	Camber Range (+/-Deg.)	Camber Preferred Setting (Deg.)	Toe-in (in.)
2002	Crown Victoria	F	0.75	+0.50	0.75	0	-0.25 +/- 0.25
		R	—	—	—	—	—
	Town Car	F	0.75	0	0.75	0	-0.12 +/- 0.25
		R	—	—	—	—	—
	Grand Marquis	F	0.75	+0.50	0.75	0	-0.13 +/- 0.25
		R	—	—	—	—	—
2003	Crown Victoria	F	0.75	+0.50	0.75	0	-0.25 +/- 0.25
		R	—	—	—	—	—
	Town Car	F	0.75	0	0.75	0	-0.12 +/- 0.25
		R	—	—	—	—	—
	Grand Marquis	F	0.75	+0.50	0.75	0	-0.13 +/- 0.25
		R	—	—	—	—	—
	Marauder	F	0.75	+6.10	0.75	-0.70	-0.12 +/- 0.25
		R	—	—	—	—	—
2004	Crown Victoria	F	0.75	+0.50	0.75	0	-0.25 +/- 0.25
		R	—	—	—	—	—
	Town Car	F	0.75	0	0.75	0	-0.12 +/- 0.25
		R	—	—	—	—	—
	Grand Marquis	F	0.75	+0.50	0.75	0	-0.13 +/- 0.25
		R	—	—	—	—	—
	Marauder	F	0.75	+6.10	0.75	-0.70	-0.12 +/- 0.25
		R	—	—	—	—	—
2005	Crown Victoria	F	0.75	+0.50	0.75	0	-0.25 +/- 0.25
		R	—	—	—	—	—
	Town Car	F	0.75	0	0.75	0	-0.12 +/- 0.25
		R	—	—	—	—	—
	Grand Marquis	F	0.75	+0.50	0.75	0	-0.13 +/- 0.25
		R	—	—	—	—	—

06017-CROWN-C09

TIRE, WHEEL AND BALL JOINT SPECIFICATIONS

| Year | Model | OEM Tires | | Tire Pressure | | Wheel Size | Ball Joint Inspection | Lug Nut (ft. lbs.) |
		Standard	Optional	Front	Rear			
2002	Crown Victoria	P225/60SR16	P225/60TR16	32	32	7-JJ	U① L①②	95
	Crown Victoria Police Special	P225/60SR16	P225/60VR16	35	35	7J	U① L①②	95
	Grand Marquis	P225/60SR16	None	32	32	6-JJ	U① L①②	95
	Grand Marquis w/Handling package	P225/60TR16	P225/60VR16	35	35	6-JJ	U① L①②	95
	Town Car	P225/70R15	P235/60R16	30	30	Std: 6.5-JJ Opt: 7-JJ	U① L①②	95
2003	Crown Victoria	P225/60SR16	P225/60TR16	32	32	7-JJ	U① L①②	95
	Crown Victoria Police Special	P225/60SR16	P225/60VR16	35	35	7J	U① L①②	95
	Grand Marquis	P225/60SR16	None	32	32	6-JJ	U① L①②	95
	Grand Marquis w/Handling package	P225/60TR16	P225/60VR16	35	35	6-JJ	U① L①②	95
	Town Car	P225/70R15	P235/60R16	30	30	Std: 6.5-JJ Opt: 7-JJ	U① L①②	95
	Marauder	③	None	30	30	8-JJ	U① L①②	95
2004	Crown Victoria	P225/60SR16	P225/60TR16	32	32	7-JJ	U① L①②	95
	Crown Victoria Police Special	P225/60SR16	P225/60VR16	35	35	7J	U① L①②	95
	Grand Marquis	P225/60SR16	None	32	32	6-JJ	U① L①②	95
	Grand Marquis w/Handling package	P225/60TR16	P225/60VR16	35	35	6-JJ	U① L①②	95
	Town Car	P225/70R15	P235/60R16	30	30	Std: 6.5-JJ Opt: 7-JJ	U① L①②	95
	Marauder	③	None	30	30	8-JJ	U① L①②	95
2005	Crown Victoria	P225/60SR16	P225/60TR16	32	32	7-JJ	U① L①②	95
	Crown Victoria Police Special	P225/60SR16	P225/60VR16	35	35	7J	U① L①②	95
	Grand Marquis	P225/60SR16	None	32	32	6-JJ	U① L①②	95
	Grand Marquis w/Handling package	P225/60TR16	P225/60VR16	35	35	6-JJ	U① L①②	95
	Town Car	P225/70R15	P235/60R16	30	30	Std: 6.5-JJ Opt: 7-JJ	U① L①②	95

OEM: Original Equipment Manufacturer

PSI: Pounds Per Square Inch

STD: Standard

OPT: Optional

① Replace if any measurable movement is found.

② Do not lift car. Inspect the boss into which the grease fitting is threaded. Replace if the boss is flush or receded below the surface of the ball joint.

③ Front: P235/50ZR18
  Rear: P245/55ZR18

06017-CROWN-C10

BRAKE SPECIFICATIONS
All measurements in inches unless noted

Year	Model	Front Brake Disc			Rear Brake Disc			Minimum Lining Thickness	Brake Caliper Bracket Bolts (ft. lbs.)	Mounting Bolts (ft. lbs.)
		Original Thickness	Minimum Thickness	Maximum Run-out	Original Thickness	Minimum Thickness	Maximum Run-out			
2002	Crown Victoria	1.063	1.010	0.002	NA	0.790	0.003	0.039	118	32
	Grand Marquis	1.063	1.010	0.002	NA	0.790	0.003	0.039	118	32
	Town Car	1.063	1.010	0.002	NA	0.790	0.003	0.039	118	32
2003	Crown Victoria	1.063	1.037	0.002	NA	0.790	0.003	0.039	118	32
	Grand Marquis	1.063	1.037	0.002	NA	0.790	0.003	0.039	118	32
	Town Car	1.063	1.037	0.002	NA	0.790	0.003	0.039	118	32
	Marauder	1.063	1.037	0.002	NA	0.790	0.003	0.039	118	32
2004	Crown Victoria	1.063	1.037	0.002	NA	0.790	0.003	0.039	118	32
	Grand Marquis	1.063	1.037	0.002	NA	0.790	0.003	0.039	118	32
	Town Car	1.063	1.037	0.002	NA	0.790	0.003	0.039	118	32
	Marauder	1.063	1.037	0.002	NA	0.790	0.003	0.039	118	32
2005	Crown Victoria	1.063	1.037	0.002	NA	0.790	0.003	0.039	118	32
	Grand Marquis	1.063	1.037	0.002	NA	0.790	0.003	0.039	118	32
	Town Car	1.063	1.037	0.002	NA	0.790	0.003	0.039	118	32

NOTE: Follow specifications stamped on rotor or drum if figures differ from those in this chart.

NA: Not Available

06017-CROWN-C11

SCHEDULED MAINTENANCE INTERVALS
Ford—Crown Victoria, Mercury—Grand Marquis, Marauder & Lincoln—Town Car

TO BE SERVICED	TYPE OF SERVICE	VEHICLE MILEAGE INTERVAL (x1000)												
		5	10	15	20	25	30	35	40	45	50	55	60	65
Engine oil & filter	R	✓	✓	✓	✓	✓	✓	✓	✓	✓	✓	✓	✓	✓
Rotate tires	S/I	✓		✓				✓		✓		✓		✓
Cooling system, hoses, clamps & coolant strength	S/I			✓			✓			✓			✓	
Lubricate steering linkage	S/I			✓			✓			✓			✓	
Air cleaner element	R						✓						✓	
Automatic transaxle fluid & filter	R						✓						✓	
Spark plugs ①	R													
Exhaust heat shields	S/I						✓						✓	
Fuel filter (NGV Crown Victoria) ②	R					✓					✓			
Front & rear brakes	S/I						✓						✓	
Lubricate suspension (Town Car)	S/I						✓						✓	
Engine coolant ③	R										✓			
PCV valve	R												✓	
Accessory drive belt	S/I												✓	

R: Replace   S/I: Service or Inspect

① Replace every 100,000 miles.

② Also drain coalescer assembly. Perform every 24,000 miles for severe service.

③ Change initially at 50,000 miles & thereafter every 30,000 miles.

FREQUENT OPERATION MAINTENANCE (SEVERE SERVICE)

If a vehicle is operated under any of the following conditions it is considered severe service:

- Extremely dusty areas.

- 50% or more of the vehicle operation is in 32°C (90°F) or higher temperatures, or constant operation in temperatures below 0°C (32°F).

- Prolonged idling (vehicle operation in stop and go traffic).

- Frequent short running periods (engine does not warm to normal operating temperatures).

- Police, taxi, delivery usage or trailer towing usage.

Oil & filter change: change every 3000 miles.

Rotate tires at 6000 miles & every 9000 miles thereafter.

Automatic transmission fluid & filter: change every 21,000 miles.

06017-CROWN-C12

ENGINE REPAIR

Alternator

REMOVAL

1. Remove or disconnect the following:
   - Negative battery cable
   - Engine cover
   - Pushpins and wiring harness
   - Mounting bolts
   - Alternator bracket
   - Accessory drive belt
   - Alternator

INSTALLATION

Install or connect the following:
- Alternator. Tighten the mounting bolts to 15–22 ft. lbs. (20–30 Nm).
- Alternator bracket. Tighten the bolts to 71–106 inch lbs. (8–12 Nm).
- Accessory drive belt
- Pushpins and wiring harness
- Engine cover
- Negative battery cable

Ignition Timing

ADJUSTMENT

The 4.6L engines are equipped with a Distributorless Ignition System (DIS). No adjustments are necessary.

Engine Assembly

REMOVAL & INSTALLATION

1. Drain the engine cooling system.
2. Recover the refrigerant from the air conditioning system.
3. Relieve the fuel system pressure.
4. Drain the engine oil.
5. Remove or disconnect the following:
   - Battery
   - Hood
   - Engine cooling fan, shroud and radiator
   - Windshield wiper governor and support bracket
   - Engine air cleaner outlet tube
   - Engine/transmission harness connector from the retaining bracket
   - Accelerator and cruise control cables at the throttle body
   - Electrical connector and vacuum hose from the evaporative emission canister purge valve
   - Positive battery cable from the power distribution box and harness
   - Vacuum supply hose from the throttle body adapter vacuum port
   - Both heater hoses
   - Alternator harness from the front fender apron and the power distribution box
   - Air conditioning hoses from the air conditioning compressor
   - Power steering control valve harness connector
   - Body ground strap from the dash panel
   - Exhaust system from the exhaust manifolds and support with wire hung from the crossmember
   - Retaining nut from the transmission line bracket
   - 3 bolts and 1 stud retaining the engine to the transmission knee braces
   - Starter motor
   - 4 bolts retaining the power steering pump to the cylinder block and position aside

6. Transmission housing cover from the cylinder block to access the torque converter nuts.
   Rotate the crankshaft until each of the 4 nuts is accessible and remove the nuts
7. Remove or disconnect the following:
   - 6 transmission-to-engine retaining bolts
   - Engine support insulator (mount) through-bolts
8. Support the transmission with a floor jack and a block of wood.
   - Bolt retaining the right-hand front engine support insulator to the front engine mount insulator support bracket
9. Install engine lifting bracket to the front of the left-hand cylinder head and to the rear of the right-hand cylinder head. Connect engine lifting equipment to the lifting brackets
10. Raise the engine slightly using a floor crane and carefully separate the engine from the transmission. Do not let the torque converter fall out of the transmission.
11. Carefully lift the engine out of the engine compartment and position on a workstand. Remove the engine lifting equipment.

To install:

12. Engine lifting brackets. Support the engine using a floor crane installed to the lifting equipment and remove the engine from the workstand
13. Lower the engine into the engine compartment. Start the converter pilot into the flywheel and align the paint marks on the flywheel and torque converter. Be sure the studs on the torque converter align with the holes in the flywheel.
14. Fully engage the engine to the transmission and lower onto front engine support insulators.
15. Install or connect the following:
    - Engine lifting equipment and brackets
    - Bolt retaining the right-hand front engine support insulator to the front engine mount insulator support bracket
    - 6 engine-to-transmission retaining bolts and tighten to 30–44 ft. lbs. (40–60 Nm)
    - Front engine support insulator through-bolts and tighten to 15–22 ft. lbs. (20–30 Nm)
    - 4 torque converter retaining nuts and tighten to 22–25 ft. lbs. (20–30 Nm)
    - Transmission housing cover to the cylinder block
    - Power steering pump on the cylinder block and the 4 retaining nuts. Tighten to 15–22 ft. lbs. (20–30 Nm).
    - Starter motor
    - Engine-to-transmission brace and the 3 bolts and 1 stud. Tighten the bolts and stud to 18–31 ft. lbs. (25–43 Nm).
    - Transmission line bracket to the brace stud and 1 retaining nut. Tighten to 15–22 ft. lbs. (20–30 Nm).
    - Exhaust system to the exhaust manifolds. Tighten the 4 nuts to 20–30 ft. lbs. (27–41 Nm). Be sure the exhaust system clears the No. 3 crossmember. Adjust as necessary.
    - Power steering valve harness connector
    - Ground strap to the dash panel
    - Air conditioning lines to the air conditioning compressor
    - Alternator harness at the front fender apron and the power distribution box
    - Both heater hoses
    - Vacuum supply hose to the throttle body adapter vacuum port
    - Positive battery cable to the power distribution box and harness
    - Electrical connector and vacuum hose to the evaporative emission canister purge valve

- Accelerator and cruise control cables at the throttle body
- Engine/transmission harness connector to the retaining bracket on the power brake booster
- Windshield wiper governor and support bracket
- Fuel supply and return lines
- Radiator, cooling fan and shroud
- Engine air cleaner outlet tube
- Hood
- Battery

16. Fill the crankcase to the correct level.

17. Fill the cooling system.

18. Start the engine and allow it to reach normal operating temperature.

19. Check for leaks and proper fluid levels.

20. Evacuate and recharge the air conditioning system.

21. Road test the vehicle and check the engine and transmission for proper operation.

Water Pump

REMOVAL & INSTALLATION

1. Drain the cooling system.
2. Remove or disconnect the following:
   - Negative battery cable
   - Cooling fan and shroud
   - Accessory drive belt
   - 4 water pump pulley-to-water pump bolts
   - Pulley
   - 4 water pump-to-engine bolts
   - Water pump

To install:

3. Clean the sealing surfaces of the water pump and block.

4. Install or connect the following:
   - New O-ring, lubricate it with clean antifreeze prior to installation
   - Water pump. Tighten the bolts to 15–22 ft. lbs. (20–30 Nm).
   - Water pump pulley. Tighten the bolts to 15–22 ft. lbs. (20–30 Nm).

7922RG01

Be sure to tighten the water pump mounting bolts to the specification

- Accessory drive belt
5. Fill the cooling system.
6. Operate the engine to normal operating temperatures and check for leaks.

Heater Core

REMOVAL & INSTALLATION

1. If equipped, turn the air suspension service switch to the **OFF** position before raising the vehicle.

2. Drain the cooling system.

3. Remove or disconnect the following:
   - Negative battery cable
   - Front seats
   - Carpeting
   - Rear airflow duct
   - Heater hoses

4. Remove the driver's side air bag module by removing or disconnect the following:
   - SRS module-to-steering wheel bolts, from both sides of the steering wheel.
   - SRS module and disconnect the electrical connector
   - Horn switch electrical connector

5. Remove the passenger's side SRS module by removing or disconnecting the following:
   - Open the glove box and disconnect the glove compartment isolator
   - Push inward on the 2 glove box door tabs and lower it
   - SRS module's electrical connector
   - SRS module-to-instrument panel bolts and the module

6. Remove the instrument panel by removing or disconnecting the following:
   - Speed control servo nuts and move the servo aside
   - Bolt and disconnect the left side bulkhead connector
   - Left side bulkhead connector from the dash panel
   - Windshield washer fluid reservoir screw, and position the reservoir aside
   - Blower motor electrical connector
   - Air conditioning pressure cut-off switch electrical connector
   - In-line electrical harness connector
   - Electronic Automatic Temperature Control (EATC) variable blower motor controller electrical connector
   - Right front wheel
   - Right front fender splash shield bolts and move the shield away from the cowl

- Wiring harness from the evaporator case
- Wiring harness from the cowl
- Right side instrument panel lower insulator pushpins, disconnect the power point electrical connector, remove the courtesy lamp from the socket and remove the insulator
- Unseat the wiring grommet and feed the wiring harness through the cowl
- Cowl side trim panels and the windshield side garnish moldings from both sides
- Locking clip and the Electronic Crash Sensor (ECS) module electrical connector
- Antenna connector
- Ground wire bolts from the right side
- EATC hose at the evaporator housing
- Bolt and disconnect the bulkhead connector from the right side
- Pull the door weatherstrip seals away from the instrument panel on both sides
- Tunnel brace trim cover
- Climate control head vacuum harness connector
- Instrument panel tunnel brace nuts and the brace
- Ground wire bolts from the left side
- Wiring harness connectors from the left side
- Parking brake switch electrical connector

a. Pry out the instrument panel defroster opening grille, disconnect the electrical connectors and the remove the grille.

7. Remove or disconnect the following:
- 3 upper instrument panel screws
- Instrument panel cowl side nut from both sides
- Instrument panel cowl side bolt from the left side
- Instrument panel
- Windshield wiper mounting arm and the pivot shaft
- Electrical connector and remove the upper cowl extension screw and the extension
- Vacuum hoses and the wire harness connector from the evaporative emissions canister purge valve
- Evaporative emissions purge valve nuts and the valve
- Cowl side nut
- Right side rear fender apron screws and reposition the apron

1. Instrument Panel Upper Finish Panel
2. Instrument Panel
3. Instrument Panel Upper Moulding , RH
4. Glove Compartment Lock Set
5. Glove Compartment Door Cover
6. Screw (2 Req'd)
7. Screw (10 Req'd)
8. Parking Brake Release Handle (Part of 2780)
9. Instrument Panel Steering Column Cover
10. Instrument Panel Steering Column Opening Cover Reinforcement
11. Instrument Panel Moulding , LH
12. Instrument Panel Cluster Finish Panel
A. Tighten to 2.1-2.9 N·m (19-25 Lb-In)
B. Tighten to 4-6 N·m (36-53 Lb-In)

93111GA9

View of the instrument panel finish panels

- Evaporator housing nut and screw
- Vacuum hoses and the wiring harness connectors
- Plenum chamber's lower flange nuts
- Plenum chamber's upper flange nut and the plenum chamber
- Heater core cover-to-heater plenum chamber screws and the cover
- Seal and the heater core

To install:

8. Install or connect the following:
- Seal and the heater core
- Heater core cover and the cover-to-heater plenum chamber screws
- Plenum chamber and the plenum chamber's upper flange nut
- Plenum chamber's lower flange nuts
- Vacuum hoses and the wiring harness connectors
- Evaporator housing nut and screw
- Right side rear fender apron and the apron screws
- Cowl side nut
- Evaporative emissions purge valve and the valve nuts
- Vacuum hoses and the wire harness connector to the evaporative emissions canister purge valve

- Electrical connector; then, install the upper cowl extension screw and the extension
- Windshield wiper mounting arm and the pivot shaft

9. Install the instrument panel by installing or connecting the following:
- Instrument panel
- At the left side, install the instrument panel cowl side bolt
- Instrument panel cowl side nut on both sides
- 3 upper instrument panel screws
- Electrical connectors and the install the instrument panel defroster opening grille
- Parking brake switch electrical connector
- Wiring harness connectors on the left side
- Ground wire bolts on the left side
- Instrument panel tunnel brace and the brace nuts
- Climate control head vacuum harness connector
- Install the tunnel brace trim cover.
- Door weatherstrip seals to the instrument panel on both sides
- Bulkhead connector and tighten the bolt on both sides

- EATC hose at the evaporator housing
- Ground wire bolts on the right side
- Antenna connector
- ECS module electrical connector and the locking clip
- Cowl side trim panels and the windshield side garnish moldings on both sides
- Feed the wiring harness through the cowl and seat the wiring grommet
- Right side instrument panel lower insulator, connect the power point electrical connector, install the courtesy lamp to the socket and secure the insulator with pushpins
- Wiring harness from the cowl
- Wiring harness to the evaporator case
- Right front fender splash shield and the shield bolts
- Right front wheel
- EATC variable blower motor controller electrical connector
- In-line electrical harness connector
- Air conditioning pressure cut-off switch electrical connector
- Blower motor electrical connector
- Windshield washer fluid reservoir and the reservoir screw

1 Nut (3 Req'd)
2 Rivet (2 Req'd)
3 Vehicle Identification Plate
4 Brake Pedal Support
5 J Nut
6 Stud
7 Nut
8 Instrument Panel
9 Screw (3 Req'd)
10 Steering Column Retaining Nut
11 Bolt (Part of 3F659)
12 Brake Pedal Support Steering Column Brace
13 Bolt (2 Req'd)
14 Instrument Panel Steering Column Opening Cover Reinforcement

15 Bolt (2 Req'd)
16 Nut
17 Stud
18 Bolt (3 Req'd)
A Tighten to 9-14 N·m (80-123 Lb-In)
B Tighten to 10-14 N·m (89-123 Lb-In)
C Tighten to 2-3 N·m (18-26 Lb-In)
D Tighten to 47-63 N·m (35-46 Lb-Ft)
E Tighten to 45-70 N·m (34-51 Lb-Ft)
F Tighten to 22-34 N·m (17-25 Lb-Ft)

93111GA0

Exploded view of the instrument panel—Crown Victoria and Grand Marquis

1	Suction Accumulator/Drier	9	A/C Recirculating Air Duct
2	Screw (2 Req'd)	10	Heater Air Plenum Chamber
3	Screw	11	Heater Outlet Floor Duct
4	Nut and Washer Assy	12	Nut
5	A/C Recirculating Air Duct Screen	13	A/C Evaporator Core Housing
6	Screw (2 Req'd)	14	Nut and Washer
7	Nut and Washer	A	Tighten to 2.5-3.2 N·m (23-28 Lb-In)
8	A/C Air Inlet Door Inner Seal		

93111GB1

Exploded view of the heater air plenum chamber, the evaporator housing and the air inlet duct

- Left side bulkhead connector to the dash panel
- Left side bulkhead connector and tighten the bolt
- Speed control servo and the servo nuts
- Right SRS module and torque the module-to-instrument panel bolts to 62–97 inch lbs. (7–11 Nm)
- Right SRS module electrical connector
- Glove compartment isolator
- Glove box door
- Horn switch electrical connector
- Electrical connector and install the left SRS module
- Left SRS module-to-steering wheel bolts on both sides of the wheel, and torque the bolts to 90–122 inch lbs. (10–14 Nm)
- Rear airflow duct
- Carpeting
- Front seats
- Heater hoses

10. Refill the cooling system.
11. Connect the negative battery cable.
12. Operate the engine to normal operat-

1	Screw (4 Req'd)
2	Heater Core Case Seal
3	Heater Core
4	Heater Dash Gasket
5	Heater Air Plenum Chamber
6	Heater Core Cover

93111GB2

View of the heater core

1. Instrument Panel Finish Panel
2. Instrument Panel Defroster Opening Grille
3. Passenger Side Air Bag Module
4. Instrument Panel Finish Panel
5. Glove Compartment
6. Instrument Panel Lower Insulator (RH)
7. Instrument Panel Finish Panel
8. Instrument Panel Ash Receptacle
9. Instrument Panel Lower Insulator (LH)
10. Instrument Panel Steering Column Cover
11. Instrument Panel Steering Column Opening Cover Reinforcement
12. Instrument Panel

93111GA7

Exploded view of the instrument panel—Town Car

ing temperature. Check the climate control operation and check for leaks.

Cylinder Head

REMOVAL & INSTALLATION

4.6L (VIN W) SOHC Engine

➡ **The cylinder head bolts are a torque-to-yield design and cannot be reused. Always use new cylinder head bolts for assembly.**

1. If equipped with air suspension, the air suspension switch, located on the right-hand side of the luggage compartment, must be turned to the **OFF** position before raising the vehicle.
2. Drain the engine cooling system.
3. Properly relieve the fuel system pressure.
4. Remove or disconnect the following:
   - Negative battery cable
   - Cooling fan and shroud assembly
   - Engine air cleaner outlet tube
   - Windshield wiper governor (module)
   - Accessory drive belt
   - Ignition wires from the spark plugs
   - Ignition wire brackets from the cylinder head cover studs
   - 2 ignition wire tray-to-ignition coil brackets bolts

7922RG02

Use a rubber band to support the rear cylinder head bolt to ease removal of the head

- Bolt retaining the air conditioning pressure line to the right-hand ignition coil bracket
- Wiring to both ignition coils and the Camshaft Position (CMP) sensor
- Ignition coil brackets-to-engine front cover nuts. Slide the ignition coil brackets and ignition wire assemblies off the mounting studs and from the vehicle
- Water pump pulley
- Alternator wiring harness from the junction block, fender apron and alternator
- Alternator
- Positive battery cable at the power distribution box
- Retaining bolt from the positive battery cable bracket located on the side of the right-hand cylinder head
- Vent hose from the canister purge solenoid and position the positive battery cable aside
- Positive Crankcase Ventilation (PCV) valve from the cylinder head cover
- Engine/transmission harness connector from the retaining bracket on the power brake booster
- Crankshaft Position (CKP) sensor, air conditioning compressor clutch and canister purge solenoid electrical connectors

5. Remove the bolts retaining the power steering pump to the cylinder block and engine front cover. The front lower bolt on the power steering pump will not come all the way out. Wire the power steering pump aside.

6. Remove or disconnect the following:
- Engine oil pan and oil pan gasket
- Crankshaft pulley retaining bolt
- Pulley
- Power steering control valve actuator and oil pressure sensor wiring connectors and position aside.
- Exhaust Gas Recirculation (EGR) tube from the right-hand exhaust manifold
- Exhaust pipes from the exhaust manifolds. Lower the exhaust pipes and hang with wire from the crossmember.
- Bolts retaining the starter wiring harness to the rear of the right-hand cylinder head
- Cylinder head covers to the cylinder heads
- Accelerator and cruise control cables
- Accelerator cable bracket from the intake manifold and position aside

- Vacuum hose from the throttle body elbow vacuum port
- Heated Oxygen Sensor (HO2S) and the heater water hose
- 2 bolts retaining the thermostat housing to the intake manifold and position the upper hose and thermostat housing aside

➡**The 2 thermostat housing bolts also retain the intake manifold.**

- 9 intake manifold-to-cylinder heads bolts
- Intake manifold and gaskets
- 7 stud bolts and the 4 bolts attaching the engine front cover to the engine
- Front cover
- Both timing chains
- 10 left-hand cylinder head-to-cylinder block bolts

7. Remove the cylinder head. The lower rear cylinder head bolt must stay in the cylinder head until the cylinder head is removed due to lack of clearance for removal in the vehicle. Use a rubber band to secure the cylinder head bolt in the cylinder head during removal and installation of the cylinder head and to prevent the bolt from damaging the cylinder block or head gasket

➡**The lower rear cylinder head bolt cannot be removed due to interference with the power brake booster. Use a rubber band to hold the bolt away from the cylinder block.**

8. Remove or disconnect the following:
- Ground strap, 1 stud and 1 bolt retaining the heater return line to the right-hand cylinder head
- 10 right-hand cylinder head-to-cylinder block bolts

9. Remove the cylinder head. The lower rear cylinder head bolt must stay in the cylinder head until the cylinder head is removed due to lack of clearance for removal in the vehicle. Use a rubber band to secure the cylinder head bolt in the cylinder head during removal and installation of the cylinder head and to prevent the bolt from damaging the cylinder block or head gasket.

➡**The lower rear cylinder head bolt cannot be removed due to interference with the evaporator housing. Use a rubber band to hold the bolt away from the cylinder block.**

10. Clean all gaskets mating surfaces. Check the cylinder heads and cylinder block for flatness. Check the cylinder heads for scratches near the coolant passages and combustion chambers that could provide leak paths.

To install:

11. Rotate the crankshaft counterclockwise 45 degrees. The crankshaft keyway should be at the 9 o'clock position viewed from the front of the engine. This ensures that all pistons are below the top of the engine block deck face.

12. Rotate the camshaft to a stable position where the valves do not extend below the head face.

13. Install or connect the following:
- New head gaskets on the cylinder block
- New bolts in the lower rear bolt holes on both cylinder heads and retain with rubber bands as explained during the removal procedure

14. Position the cylinder heads on the cylinder block dowels, being careful not to score the surface of the head face. Apply clean oil to the new cylinder head bolts, remove the rubber bands from the lower rear bolts and install all bolts hand-tight.

15. Tighten the new cylinder head bolts, in sequence, as follows:
    a. Step 1: 28–31 ft. lbs. (37–43 Nm).
    b. Step 2: plus 85–95 degrees.
    c. Step 3: loosen all bolts at least 1 full turn.
    d. Step 4: 27–32 ft. lbs. (37–43 Nm).
    e. Step 5: plus 85–95 degrees.
    f. Step 6: again, plus 85–95 degrees.

16. Position the heater return hose and install the 2 retaining bolts.

17. Rotate the camshafts using the flats matched at the center of the camshaft until both are in time. Install Camshaft Position-

CYLINDER HEAD TIGHTENING SEQUENCE

Cylinder head torque sequence—4.6L SOHC engine

7922RG03

ing Tools T91P-6256-A, on the flats of the camshafts to keep them from rotating.

18. Rotate the crankshaft clockwise 45 degrees to position the crankshaft at Top Dead Center (TDC) for the No. 1 cylinder.

➡**The crankshaft must only be rotated in the clockwise direction and only as far as TDC.**

19. Install or connect the following:
- Both timing chains
- New engine front cover seal and gasket. Apply silicone sealer to the lower corners of the cover where it meets the junction of the engine oil pan and cylinder block and to the points where the cover contacts the junction of the cylinder block and the cylinder heads.
- Engine front cover and the bolts. Tighten to 15–22 ft. lbs. (20–30 Nm).
- New intake manifold gaskets on the cylinder heads. Be sure the alignment tabs on the gaskets are aligned with the holes in the cylinder heads.
- Intake manifold on the cylinder heads and the retaining bolts. Tighten the bolts in sequence, to 15–22 ft. lbs. (20–30 Nm).
- Thermostat, O-ring, thermostat housing and upper hose. Tighten the 2 retaining bolts to 15–22 ft. lbs. (20–30 Nm).
- Heater water hose and both HO_2S sensors
- Vacuum hose to the throttle body adapter vacuum port
- Accelerator cable bracket on the intake manifold
- Accelerator and cruise control cables to the throttle body

20. Apply silicone sealer to both places where the engine front cover meets the cylinder heads.
- Cylinder head covers with new gasket on the cylinder heads. Tighten the bolts and stud bolts to 71–106 inch lbs. (8–12 Nm).
- Starter motor wiring harness to the right-hand cylinder head and tighten the retaining bolt
- Exhaust pipes to the exhaust manifolds. Tighten the 4 nuts to 20–30 ft. lbs. (27–41 Nm).

➡**Be sure the exhaust system clears the No. 3 crossmember. Adjust as necessary.**

- EGR tube to the right-hand exhaust manifold and tighten the line nut to 26–33 ft. lbs. (35–45 Nm).

- Power steering control valve actuator and oil pressure sensor electrical connectors

21. Apply a small amount of silicone sealer in the rear of the keyway on the crankshaft pulley.
- Pulley on the crankshaft, making sure the crankshaft key and keyway are aligned.

22. Install the crankshaft pulley and tighten the bolt as follows:
  a. Step 1: Tighten to 66 ft. lbs. (90 Nm).
  b. Step 2: Loosen one complete turn.
  c. Step 3: Tighten to 35–39 ft. lbs. (47–53 Nm).
  d. Step 4: Tighten an additional 85–95 degrees.

23. Install or connect the following:
- Engine oil pan and a new gasket
- Power steering pump in position on the cylinder block
- 4 retaining bolts. Tighten the bolts to 15–22 ft. lbs. (20–30 Nm).
- Air conditioning compressor, CKP sensor and canister purge solenoid electrical connectors
- Engine/transmission harness connector on the power brake booster
- PCV valve in the right-hand cylinder head cover and connect the canister purge solenoid vent hose
- Positive battery cable harness on the right-hand cylinder head
- Bolt retaining the cable bracket to the cylinder head
- Positive battery cable at the power distribution box and battery
- Alternator and the 2 retaining bolts. Tighten the bolts to 15–22 ft. lbs. (20–30 Nm).
- 2 bolts retaining the alternator brace to the intake manifold. Tighten to 72–96 inch lbs. (8–12 Nm).
- Water pump pulley. Tighten the bolts to 15–22 ft. lbs. (20–30 Nm).
- Ignition coil brackets and ignition wire assemblies onto the mounting studs
- 7 nuts retaining the ignition coil brackets to the engine front cover and tighten to 15–22 ft. lbs. (20–30 Nm)
- 2 bolts retaining the ignition wire tray to the ignition coil bracket and tighten to 71–106 inch lbs. (8–12 Nm)
- Ignition coil and CMP sensor harness connectors
- Air conditioning pressure line on the right-hand ignition coil bracket and tighten the retaining bolt

- Ignition wires to the spark plugs and the bracket onto the cylinder head cover studs
- Accessory drive belt and the windshield wiper governor
- Fuel supply and return lines
- Cooling fan and shroud
- Engine air cleaner outlet tube
- Negative battery cable

24. Fill the cooling system.

25. If equipped with air suspension, turn the air suspension switch to the **ON** position.

26. Refill the engine with the correct amount of oil and replace the filter.

27. Start the engine and bring to normal operating temperature while checking for leaks.

28. Road test the vehicle and check for proper engine operation.

4.6L (VIN V) DOHC Engine

1. Remove the engine from the vehicle and mount it on a suitable workstand.

2. Remove or disconnect the following:
- Fuel injectors
- Intake manifold
- Front cover
- Timing chains
- Valve covers
- Exhaust manifolds
- Cylinder heads

To install:

➡**Use new cylinder head bolts for assembly.**

3. Install the cylinder heads with new gaskets and bolts. Tighten the bolts in sequence as follows:
  a. Step 1: 28–31 ft. lbs. (37–43 Nm)
  b. Step 2: Plus 85–95 degrees
  c. Step 3: Loosen all bolts one full turn
  d. Step 4: 28–31 ft. lbs. (37–43 Nm)

Cylinder head torque sequence—4.6L DOHC engine

7922NG11

8-12 Nm (71-106 lb/in)

9346GG08

Left valve cover torque sequence—4.6L (VIN V) engine

8-12 Nm (71-106 lb/in)

9346GG09

Right valve cover torque sequence—4.6L (VIN V) engine

  e. Step 5: Plus 85–95 degrees
  f. Step 6: Plus 85–95 degrees
4. Install or connect the following:
  • Exhaust manifolds
  • Valve covers
  • Timing chains
  • Front cover
  • Intake manifold
  • Fuel injectors
5. Install the engine into the vehicle.

Rocker Arms

REMOVAL & INSTALLATION

4.6L (VIN W) SOHC Engine

1. Relieve the fuel system pressure.
2. Disconnect the negative battery cable.
3. Remove the right camshaft cover by removing or disconnecting the following:
  • Positive battery cable at the battery and at the power distribution box
  • Retaining bolt from the positive battery cable bracket located on the side of the right cylinder head
  • Crankshaft Position (CKP) sensor, air conditioning compressor clutch and canister purge solenoid connectors. Position the harness aside.

  • Vent hose from the purge solenoid and position the positive battery cable aside
  • Ignition wires from the spark plugs
  • Ignition wire brackets from the camshaft cover studs and position the wires aside
  • PCV valve from the camshaft cover grommet and position aside
  • Bolts and stud bolts and remove the camshaft cover.
4. Remove the left camshaft cover by removing or disconnecting the following:
  • Air inlet tube
  • Fuel lines
  • PSP switch and oil pressure sending unit and position the harness aside
  • 42-pin engine harness connector from the retaining bracket on the brake vacuum booster and position aside
  • Windshield wiper module
  • Ignition wires from the spark plugs
  • Ignition wire brackets from the studs and position the wires aside
  • Camshaft cover
5. Position the piston of the cylinder being serviced at the bottom of its stroke and position the camshaft lobe on the base circle.
6. Install valve spring spacer tool

VIEW A

VALVE SPRING SPACER

VALVE SPRING COMPRESSOR

VALVE SPRING SPACER

7922RG04

After installing the spring spacer, compress the valve spring and remove the rocker arm

T91P-6565-AH between the spring coils to prevent valve seal damage.

  7. Install a valve spring compressor under the camshaft and on top of the valve spring retainer.
  8. Compress the valve spring and remove the roller follower. Remove the valve spring compressor and spacer.

To install:

  9. Apply engine oil to the valve stem and tip and roller follower contact surfaces.
  10. Install valve spring spacer tool T91P-6565-AH between the spring coils. Compress the valve spring, and install the roller follower.

➡**The piston must be at the bottom of its stroke and the camshaft at the base circle.**

  11. Remove the valve spring compressor and spacer.
  12. Clean the sealing surfaces of the camshaft covers and cylinder heads. Apply silicone sealer to the places where the front cover meets the cylinder head.
  13. Position new gaskets onto the camshaft covers and install the covers. Install the bolts and stud bolts and tighten to 72–106 inch lbs. (8–12 Nm).
  14. When installing the right camshaft cover, install or connect the following:
  • PCV into the camshaft cover grommet
  • Ignition wire brackets on the studs
  • Wires to the spark plugs
  • Canister purge solenoid, air conditioning compressor clutch and CKP sensor
  • Positive battery cable harness on the right cylinder head
  • Bolt retaining the cable bracket to the cylinder head
  • Positive battery cable at the power distribution box and the battery

15. When installing the left camshaft cover, install or connect the following:
- Ignition wire brackets on the studs
- Wires to the spark plugs
- Windshield wiper module
- 42-pin and transmission harness connectors
- Retaining bracket
- PSP switch and oil pressure sending unit harness.
- Fuel lines
- Negative battery cable

16. Start the engine and check for leaks.

4.6L (VIN V) DOHC Engine

1. Remove or disconnect the following:
- Negative battery cable
- Ignition coils
- Valve covers

2. Rotate the crankshaft so that the piston on the cylinder to be serviced is at bottom dead center with the valves closed.

3. Install special tool Valve Spring Compressor T97P-6565-AH for exhaust valves, and Valve Spring Compressor T93P-6565-A for intake valves.

4. Compress the valve spring and remove the rocker arm. Repeat for each arm to be removed.

➡**If the rocker arms are to be reused, ensure that they are installed in the same position that they were removed from.**

To install:

5. Compress the valve spring and install the rocker arm. Repeat for each arm to be installed.

6. Install or connect the following:
- Valve covers
- Ignition coils
- Negative battery cable

7. Start the engine and check for proper operation.

ST1718-A

9346GG01

Rocker arm service tools—4.6L (VIN V) engine

Intake Manifold

REMOVAL & INSTALLATION

4.6L (VIN W) SOHC Engine

1. If equipped with air suspension, the air suspension switch, located on the right-hand side of the luggage compartment, must be turned to the **OFF** position before raising the vehicle.

2. Disconnect negative battery cable.

3. Drain the engine cooling system.

4. Properly relieve the fuel system pressure.

5. Remove or disconnect the following:
- Fuel supply and return lines
- Windshield wiper governor (module)
- Engine air cleaner outlet tube
- Accessory drive belt
- Ignition wires from the spark plugs
- Ignition wire brackets from the cylinder head cover studs
- Ignition coils and the Camshaft Position (CMP) sensor
- Ignition wires from both ignition coils
- 2 bolts retaining the ignition wire bracket to the ignition coil brackets
- Ignition wire assembly
- Alternator wiring harness from the junction block at the fender apron and alternator

- Bolts retaining the alternator brace to the intake manifold and the alternator to the cylinder block
- Alternator
- Oil pressure sensor and power steering control valve actuator wiring and position the wiring harness aside
- Exhaust Gas Recirculation (EGR) valve-to-exhaust manifold tube from the right-hand exhaust manifold
- Engine/transmission harness connector from the retaining bracket on the power brake booster
- Air conditioning compressor clutch, Crankshaft position (CKP) sensor and the canister purge solenoid wiring connectors
- Positive Crankcase Ventilation (PCV) valve from the cylinder head cover
- Canister purge vent hose from the PCV valve
- Accelerator and cruise control cables from the throttle body
- Accelerator cable bracket from the intake manifold and position aside
- Vacuum hose from the throttle body adapter port
- Heated Oxygen Sensor (HO2S) and the heater water hose
- 2 bolts retaining the thermostat housing to the intake manifold and

20-30 Nm (15-22 lb/ft)

7922RG06

Intake manifold torque sequence—2002-04 4.6L (VIN W) SOHC engine

25 Nm (18 lb-ft)

06017VICG02

Intake manifold torque sequence—2005
4.6L (VIN W) SOHC engine

position the upper hose and thermostat housing aside

➡The 2 thermostat housing bolts are
also used to retain the intake manifold.

- 9 bolts retaining the intake manifold to the cylinder heads
- Intake manifold and gaskets
6. If replacing the intake manifold,
swap over the necessary parts.

To install:
7. Clean all gaskets mating surfaces.
8. Position new intake manifold gaskets
on the cylinder heads. Be sure the alignment tabs on the gaskets are aligned with
the holes in the cylinder heads.
9. Install the intake manifold and the 9
retaining bolts. Hand-tighten the right-rear
bolt (viewed from the front of the engine)
before final tightening, then tighten the bolts,
in sequence, to 15–22 ft. lbs. (20–30 Nm).
10. Inspect and if necessary, replace the
O-ring seal on the thermostat housing.
Position the housing and upper hose and
install the 2 retaining bolts. Tighten to
15–22 ft. lbs. (20–30 Nm).
11. Install or connect the following:
- Heater water hose
- HO2S sensor
- Vacuum hose to the throttle body
  adapter vacuum port
- Accelerator cable bracket on the
  intake manifold
- Accelerator and cruise control
  cables to the throttle body
- PCV valve in the cylinder head
  cover
- Canister purge solenoid vent hose
- Air conditioning compressor
  clutch, CKP sensor and canister
  purge solenoid wiring connectors

- Engine/transmission harness connector the retaining bracket on the
  power brake booster
- EGR valve-to-exhaust manifold
  tube to the right-hand exhaust
  manifold. Tighten the tube nut to
  26–33 ft. lbs. (35–45 Nm).
- Power steering control valve actuator
- Oil pressure sensor wiring connectors
- Alternator. Tighten the bolts to
  15–22 ft. lbs. (20–30 Nm).
- 2 bolts retaining the alternator
  brace to the intake manifold and
  tighten to 71–106 inch lbs. (8–12
  Nm)
- Alternator wiring harness to the
  alternator, right-hand fender apron
  and junction block
- Ignition wire assembly on the
  engine
- 2 bolts retaining the ignition wire
  bracket to the ignition coil brackets.
  Tighten the bolts to 71–106 inch
  lbs. (8–12 Nm).
- Ignition wires to the ignition coils
- Ignition wires to the spark plugs
- Ignition wire brackets on the cylinder head cover studs
- Wiring connectors to both ignition
  coils and the CMP sensor
- Accessory drive belt
- Air cleaner outlet tube
- Windshield wiper governor
- Fuel supply and return lines
- Negative battery cable
12. Fill the engine cooling system.
13. If equipped with air suspension, turn
the air suspension switch to the **ON** position.
14. Start the engine and check for leaks.
15. Road test the vehicle and check for
proper operation.

4.6L (VIN V) DOHC Engine
1. Drain the cooling system.
2. Relieve the fuel system pressure.
3. Remove or disconnect the following:
- Negative battery cable
- Coolant bypass tube
- Alternator
- Throttle cable
- Cruise control cable
- Evaporative Emissions (EVAP) hose
- Throttle Position (TP) sensor con-
  nector
- Positive Crankcase Ventilation
  (PCV) hose
- Idle Air Control (IAC) valve con-
  nector

- Exhaust Gas Recirculation (EGR)
  tube
- EGR vacuum hose
- Upper intake manifold
- Fuel injection supply manifold
- Lower intake manifold
To install:
4. Install or connect the following:
- Lower intake manifold. Tighten the
  bolts in sequence to 15–22 ft. lbs.
  (20–30 Nm).
- Fuel injection supply manifold
- Upper intake manifold. Tighten the
  bolts to 15–22 ft. lbs. (20–30 Nm).
- EGR vacuum hose
- EGR tube
- IAC valve connector

7922NG21

Upper intake manifold torque sequence—
4.6L (VIN V) DOHC engine

7922NG20

Lower intake manifold torque sequence—
4.6L (VIN V) DOHC engine

- PCV hose
- TP sensor connector
- EVAP hose
- Cruise control cable
- Throttle cable
- Alternator
- Coolant bypass tube
- Negative battery cable

5. Fill the cooling system.
6. Start the engine and check for leaks.

Exhaust Manifold

REMOVAL & INSTALLATION

4.6L (VIN W) SOHC Engine

1. Drain the engine cooling system.
2. Relieve the fuel system pressure.
3. Discharge the air conditioning system
4. Remove or disconnect the following:
   - Battery cables
   - Engine air inlet tube
   - Cooling fan and shroud assembly
   - Fuel supply and return lines
   - Upper radiator hose
   - Windshield wiper governor and support bracket
   - Compressor outlet hose at the compressor and the hose assembly-to-right-hand ignition coil bracket bolt. Plug both openings.
   - Engine/transmission harness connector from the retaining bracket on the power brake booster
   - Heater hose
   - Ground strap-to-right-hand cylinder head nut
   - Upper stud and lower bolt retaining the heater hose to the right cylinder head and position aside
   - Heater blower motor switch resistor
   - Bolt retaining the right-hand front engine support insulator to the sub-frame
   - Both Heated Oxygen Sensor (HO2S)
   - Engine support insulator through-bolts
   - Exhaust Gas Recirculation (EGR) valve-to-exhaust manifold tube nut from the right-hand exhaust manifold.
   - Catalytic converter pipes from both exhaust manifolds. Lower the exhaust system and hang it from the crossmember with wire.
   - Left-hand exhaust manifold
   - Front engine support insulator from the cylinder block and the 8 nuts retaining the exhaust manifold

23-27 Nm (17-20 lb/ft)

7922RG08

Left exhaust manifold torque sequence—2002–04 4.6L (VIN W) SOHC engine

25 Nm (18 lb-ft)

06017VICG03

Left exhaust manifold torque sequence—2005 4.6L (VIN W) SOHC engine

- Left-hand exhaust manifold and the 2 manifold gaskets

5. Position an adjustable jackstand and a block of wood under the engine oil pan, rearward of the oil drain hole. Raise the engine approximately 4 inches (100mm).
6. Install or connect the following:
   - 8 exhaust manifold retaining nuts and right-hand exhaust manifold
   - Manifold and gasket

To install:

7. If the exhaust manifolds are being replaced, transfer the heated O_2 sensors and tighten to 27–33 ft. lbs. (37–45 Nm). On the right-hand exhaust manifold, transfer the EGR tube connector and tighten to 33–48 ft. lbs. (45–65 Nm).

8. Clean the mating surfaces of the exhaust manifolds and cylinder heads.

9. Install or connect the following:
   - Exhaust manifolds, using new gaskets. Tighten the bolts in sequence to 15–22 ft. lbs. (20–30 Nm).
   - EGR valve and tube assembly to the exhaust manifold. Tighten the line nut to 26–33 ft. lbs. (35–45 Nm).
   - Left-hand front engine support insulator to the cylinder block and

7922RG09

Right exhaust manifold torque sequence—4.6L (VIN W) SOHC engine tighten the bolts to 15–22 ft. lbs. (20–30 Nm).

10. Lower the engine onto the front engine support insulator and remove the jack.

11. Install or connect the following:
- Left-hand and right-hand engine support insulator through-bolts and tighten to 15–22 ft. lbs. (20–30 Nm).
- Catalytic converter pipes to both exhaust manifolds. Tighten the nuts to 20–30 ft. lbs. (27–41 Nm).

➡**Be sure the exhaust system clears the No. 3 crossmember. Adjust as necessary.**

- Both HO2S sensors
- Bolt retaining the right-hand front engine support insulator to the sub-frame. Tighten to 15–22 ft. lbs. (20–30 Nm)
- Heater blower motor switch resistor using the 2 retaining screws
- Heater hose in position
- Upper stud and lower bolt and tighten to 15–22 ft. lbs. (20–30 Nm)
- Ground strap onto the stud and tighten the nut to 15–22 ft. lbs. (20–30 Nm)
- Heater hose
- Engine/transmission harness connector
- Retaining bracket on the power brake booster
- Air conditioning compressor outlet hose to the compressor
- Bolt retaining the hose assembly to the right-hand ignition coil bracket
- Upper radiator hose
- Fuel supply and return lines
- Windshield wiper governor and retaining bracket
- Engine cooling fan blade and fan shroud
- Engine air inlet tube
- Battery cables

12. Fill the cooling system.
13. Start the engine and check for leaks.
14. Properly charge the air conditioning system.
15. Road test the vehicle and check for proper operation.

4.6L (VIN V) DOHC Engine

1. The air suspension switch, located on the left-hand side of the luggage compartment, must be turned to the **OFF** position before raising the vehicle.

18-22 Nm (14-16 lb/ft)

9346GG02

Left exhaust manifold torque sequence—4.6L (VIN V) DOHC engine

18-22 Nm (14-16 lb/ft)

9346GG03

Right exhaust manifold torque sequence—4.6L (VIN V) DOHC engine

2. Remove or disconnect the following:
- Negative battery cable
- Heated Oxygen Sensor (HO2S) connectors
- Dual converter Y-pipe
- Secondary air injection tubes
- Exhaust Gas Recirculation (EGR) tube
- Exhaust manifolds and gaskets

To install:

3. Install or connect the following:
- Exhaust manifolds with new gaskets. Tighten the nuts in sequence to 14–16 ft. lbs. (18–22 Nm).
- EGR tube
- Secondary air injection tubes
- Dual converter Y-pipe
- HO2S connectors
- Negative battery cable

4. Turn the air suspension switch to the **ON** position.

5. Start the engine and check for leaks.

Camshaft and Valve Lifters

REMOVAL & INSTALLATION

4.6L (VIN W) SOHC Engine

1. If equipped with air suspension, the air suspension switch, must be turned to the **OFF** position before raising the vehicle.

2. Properly relieve the fuel system pressure.

3. Drain the engine oil.

4. Remove or disconnect the following:
- Negative battery cable
- Fan blade and fan shroud assembly
- Fuel supply and return lines from the fuel injection supply manifold
- Windshield wiper governor (module) assembly from the vehicle
- Engine air cleaner outlet tube
- Accessory drive belt
- Ignition wires from the spark plugs
- Ignition wire brackets from the cylinder head cover studs
- 2 bolts retaining the ignition wire separator to the ignition coil brackets and the bolt retaining the air conditioning pressure line to the right-hand ignition coil bracket.
- Connectors from both ignition coils and the Crankshaft Position (CMP) sensor
- Ignition coils with brackets attached
- Electrical connector from the alternator and at the power distribution box
- Water pump pulley
- Positive battery cable at the power distribution box
- Bolt from the positive battery cable bracket located on the right-hand cylinder head
- Fuel vapor hose from the EVAP canister purge valve and position the positive battery cable aside
- Positive Crankcase Ventilation (PCV) valve from the cylinder head cover and position aside
- Engine/transmission harness connector from the bracket on the power brake booster
- Crankshaft Position (CKP) sensor and air conditioning clutch harness connectors
- Bolts retaining the power steering pump to the cylinder block and wire the pump aside

➡**The front lower bolt on the power steering pump will not come all the way out.**

- Oil pan
- Crankshaft pulley bolt and washer and crankshaft pulley
- Engine oil filter
- Power steering control valve actuator and oil pressure sensor
- Oil filter adapter
- Cylinder head covers
- Engine front cover
- Timing chains

5. Rotate the crankshaft counterclockwise no more than 45 degrees from Top Dead Center (TDC) to ensure that all pistons are below the top of the engine block deck face.

6. Install a valve spring compressor under the camshaft and on top of one of the valve spring retainers.

7. Install Valve Spring Spacer T91P-6565-AH between the valve spring coils. Be sure that the valve being compressed is on its base circle. Compress the valve spring and remove the rocker arm. Repeat the procedure until all rocker arms are removed.

8. If required, pull the lash adjusters out of their bores in the cylinder head. Note their locations, they must be installed in the same bore they were removed from.

➡ **Do not mix the camshaft bearing caps. Note the camshaft bearing cap locations for installation.**

9. To remove each camshaft, unfasten the 14 bolts retaining the camshaft bearing caps (cluster assemblies) to the cylinder head. Tap upward on the camshaft bearing

10 Nm (89 lb-in)

67197-CROWN-G03

Camshaft bearing cap torque sequence

caps at points near the upper bearing halves and gradually lift the camshaft bearing cap clusters from the cylinder head.

10. Repeat the removal procedure for the opposite cylinder head.

11. Remove the camshaft straight upward to avoid bearing damage.

12. Clean and inspect the camshafts and related components for unusual wear or damage.

To install:

13. Clean and inspect the cylinder head covers, engine front cover and cylinder head sealing surfaces.

14. Apply clean engine oil to the camshaft journals and lobes. Position the camshafts on the cylinder heads.

15. Install and seat the camshaft bearing cap cluster assemblies. Install and hand start the retaining bolts. Tighten the camshaft cluster retaining bolts in sequence to 71–106 inch lbs. (8–12 Nm). Be sure to tighten each camshaft bearing cap cluster individually.

➡ **Each camshaft bearing cap cluster assembly is tightened individually.**

16. Loosen the camshaft bearing cap cluster retaining bolts approximately 2 turns or until the heads of the bolts are free. Tighten all bolts, again in sequence, to 71–106 inch lbs. (8–12 Nm).

17. Repeat the camshaft bearing cap installation for the opposite cylinder head.

➡ **The camshafts should turn freely but with a slight drag.**

18. Check camshaft end-play as follows:
   a. Step 1: Install a dial indicator on the front of the engine. Position it so the indicator foot is resting on the camshaft sprocket bolt or the front of the camshaft.
   b. Step 2: Push the camshaft toward the rear of the engine and zero the dial indicator.
   c. Step 3: Pull the camshaft forward and release it. Specified end-play is 0.0901–0.006 inch (0.025–0.190mm).
   d. Step 4: If end-play is too tight, check for binding or foreign material in the camshaft thrust bearing. If end-play is excessive, check for worn camshaft thrust plate and replace the cylinder head, as required.
   e. Step 5: Remove the dial indicator.

19. If removed, install the lash adjusters in their original positions.

20. If necessary, install Camshaft Positioning Tools T92P-6256-A on the flats of the camshafts and install the spacers and camshaft sprockets. Install the bolts and

washers and tighten to 81–95 ft. lbs. (110–130 Nm).

21. Install a valve spring compressor under the camshaft and on top of the valve spring retainer.

22. Install Valve Spring Spacer T91P-6565-AH between the valve spring coils. Be sure that the valve being compressed is on its base circle. Compress the valve spring and install the rocker arm. Repeat the procedure until all rocker arms are installed.

23. Rotate the crankshaft clockwise 45 degrees to position the crankshaft at Top Dead Center (TDC).

➡ **The crankshaft must only be rotated in the clockwise direction and only as far as TDC.**

24. Install or connect the following:
   - Timing chains
   - Engine front cover
   - Cylinder head covers. Tighten the cylinder head cover bolts to 71–106 inch lbs. (8–12 Nm).
   - Power steering control valve actuator connector
   - Oil pressure sensor harness connector

25. Apply silicone sealer to the crankshaft keyway

26. Install the crankshaft pulley and tighten the bolt as follows:
   a. Step 1: Tighten to 66 ft. lbs. (90 Nm).
   b. Step 2: Loosen one complete turn.
   c. Step 3: Tighten to 35–39 ft. lbs. (47–53 Nm).
   d. Step 4: Tighten an additional 85–95 degrees.

27. Install or connect the following:
   - Engine oil pan
   - Power steering pump on the engine and the 4 retaining bolts. Tighten the bolts to 15–22 ft. lbs. (20–30 Nm).
   - Air conditioning clutch and CKP sensor
   - Evaporative emission canister purge valve harness connector
   - Engine/transmission harness connectors on the power brake booster retaining bracket
   - PCV valve to the right-hand cylinder head cover
   - Positive battery cable harness on the right-hand cylinder head
   - Bolt retaining the battery cable bracket to the cylinder head
   - Evaporative emission hose to the canister purge valve
   - Positive battery cable at the power distribution box

8 -12 Nm (71.4 - 107.1 Lb-In)

7922RG10

Tighten the camshaft bearing cap cluster bolts in the sequence to avoid damage to the camshaft or bearings

- Water pump pulley and tighten the bolts to 15–22 ft. lbs. (20–30 Nm)
- Ignition coil brackets and ignition wires to the engine front cover. Tighten the retaining nuts to 15–22 ft. lbs. (20–30 Nm).
- Harness connectors to the ignition coils and the CMP sensor
- Air conditioning pressure line on the right-hand ignition coil bracket and the retaining bolt
- Ignition wires to the spark plugs and the brackets onto the cylinder head cover studs
- Accessory drive belt
- Windshield wiper governor
- Fuel supply and return lines
- Fan and shroud assembly
- Negative battery cable

28. Fill the engine cooling system.
29. Fill the crankcase.
30. If equipped with air suspension, turn the air suspension switch to the **ON** position.
31. Start the engine and check for leaks.
32. Road test the vehicle and check for proper engine operation.

4.6L (VIN V) DOHC Engine

1. Drain the cooling system.
2. Remove or disconnect the following:
   - Negative battery cable
   - Windshield wiper module
   - Engine front cover

Camshafts

LEFT SIDE SHOWN
RIGHT SIDE SIMILAR

FRONT OF ENGINE

Item	Part Number	Description
1	6B277	Cap, Intake Camshaft
2	N806070	Bolt (19 Req'd Per Cylinder Head)
3	6250	Intake Camshaft
4	6049	LH Cylinder Head

Item	Part Number	Description
5	6250	Exhaust Camshaft
6	6B278	Cap, Exhaust Camshaft
7	N80757	Bolt (6 Req'd Per Cylinder Head)
A	—	Tighten to 8-12 N·m (71-106 Lb-In)

9346GG05

Exploded view of the camshaft mounting—4.6L (VIN V) engine

- Timing chains and sprockets
- Valve covers
- Rocker arms
- Hydraulic lifters
- Camshaft bearing cap assembly
- Camshafts

➡**Keep all valvetrain components in order for assembly in their original locations.**

To install:

➡**The outboard exhaust camshaft bearing cap bolts are shorter than the other bearing cap bolts.**

3. Install or connect the following:
   - Camshafts
   - Camshaft bearing cap assembly. Tighten the bolts in sequence to 71–106 inch lbs. (8–12 Nm).
   - Hydraulic lifters
   - Rocker arms
   - Valve covers
   - Timing chains and sprockets

9346GG04

Camshaft bearing cap torque sequence—4.6L (VIN V) engine

- Engine front cover
- Windshield wiper module
- Negative battery cable

4. Fill the cooling system.
5. Start the engine and check for leaks.

Valve Lash

ADJUSTMENT

The 4.6L (VIN W and V) engines are equipped with hydraulic lash adjusters. Valve clearance is not adjustable.

Starter Motor

REMOVAL & INSTALLATION

1. Remove or disconnect the following:
   - Negative battery cable
   - Red solenoid safety cap
   - Wires from solenoid
   - 2 upper bolts
   - 1 lower bolt and starter

To install:
2. Install or connect the following:
   - Starter and lower mounting bolt
   - 2 upper mounting bolts. Tighten all 3 bolts to 15–20 ft. lbs. (20–27 Nm).
   - Wires from solenoid and tighten the nut to 40–50 inch lbs.
   - Red solenoid safety cap
   - Negative battery cable

Starter motor mounting

Oil Pan

REMOVAL & INSTALLATION

1. Drain the engine cooling system.
2. Properly discharge the air conditioning system.
3. Relieve the fuel system pressure.
4. Remove or disconnect the following:
   - Negative battery cable
   - Engine air cleaner outlet tube
   - Fuel supply and return lines at the fuel injection supply manifold

- Cooling fan and fan shroud
- Upper radiator hose
- Wiper governor and support bracket
- A/C compressor outlet hose
- Engine/transmission electrical harness connector from the retaining bracket on the power brake booster
- Heater water hose
- Nut retaining the ground strap to the right-hand cylinder head
- Upper stud and loosen the lower bolt retaining the heater outlet hose to the right-hand cylinder head and position aside
- Heater blower motor switch resistor

5. Drain the engine oil and reinstall the oil pan drain plug with a new gasket. Tighten the plug to 10–12 ft. lbs. (13–16 Nm).

6. Remove or disconnect the following:
   - Bolt retaining the right-hand engine support insulator to the lower front sub-frame
   - Bolts retaining the left-hand and right-hand front engine support insulators to the engine mount supports
   - Catalytic converter pipes from both exhaust manifolds. Lower the exhaust system and support it with wire from the transmission cross-member.

7. Position a jack and a block of wood under the oil pan, rearward of the oil drain hole. Raise the engine approximately 4 inches (100mm) and insert 2 wood blocks approximately 2½–2¾ inch (60–70mm) thick under each front engine support insulator. Lower the engine onto the wood blocks and remove the jack.

8. Remove the oil pan.
9. If necessary, remove the 2 bolts retaining the oil pick-up tube to the oil pump and remove the bolt retaining the pick-up tube to the main bearing stud spacer. Remove the pick-up tube.

To install:
10. Clean the engine oil pan and inspect for damage. Clean the sealing surfaces of the front cover and cylinder block. Clean and inspect the oil pick-up tube and replace the O-ring.

11. If removed, position the oil pick-up tube on the oil pump and hand start the 2 retaining bolts. Install the bolt retaining the pick-up tube on the main bearing stud spacer, hand tight.

12. Tighten the pick-up tube-to-oil pump bolts to 72–108 inch lbs. (8–12 Nm), then tighten the pick-up tube-to-main bearing stud spacer bolt to 15–22 ft. lbs. (20–30 Nm).

13. Position a new gasket on the oil pan. Apply silicone sealer to where the front cover meets the cylinder block and the

To prevent oil leaks, tighten the oil pan bolts in the sequence shown

crankshaft rear oil seal and retainer meets the cylinder block. Position the oil pan to the engine and install the retaining bolts. Tighten the bolts in sequence, to 14 ft. lbs. (20 Nm), then rotate the oil pan retaining bolts, in sequence an additional 60 degrees within 4 minutes of applying the silicone sealer.

14. Position the jack and wood block under the engine oil pan, rearward of the oil drain hole, and raise the engine enough to remove the wood blocks. Lower the engine and remove the jack.

15. Install or connect the following:
- Left-hand and right-hand engine support insulator through-bolts and tighten to 15–22 ft. lbs. (20–30 Nm)
- Bolt retaining the right-hand engine support insulator to the lower front sub-frame. Tighten the bolt to 15–22 ft. lbs. (20–30 Nm).
- Exhaust system to the exhaust manifolds and tighten the 4 retaining nuts to 20–30 ft. lbs. (27–41 Nm). Be sure the exhaust system clears the crossmember. Adjust as necessary.
- New engine oil filter
- Heater blower motor switch resistor using the 2 retaining screws
- Heater water hose
- Upper stud and tighten the upper and lower bolts to 15–22 ft. lbs. (20–30 Nm)
- Ground strap on the stud and tighten to 15–22 ft. lbs. (20–30 Nm)
- Heater water hose
- Throttle valve cable, if equipped
- Engine/transmission electrical harness connector
- Harness connector on the power brake booster bracket
- Air conditioning compressor outlet hose to the compressor
- Bolt retaining the hose to the right-hand ignition coil bracket
- Upper radiator hose
- Fuel supply and return lines
- Wiper governor and retaining bracket
- Engine cooling fan and fan shroud
- Engine air cleaner outlet tube
- Negative battery cable

16. Fill the cooling system.

17. Fill the engine crankcase with engine oil.

18. Start the engine and check for leaks.

19. Properly evacuate and recharge the air conditioning system.

20. Road test the vehicle and check for proper engine operation.

Oil Pump

REMOVAL & INSTALLATION

1. Remove or disconnect the following:
- Negative battery cable
- Cylinder head covers
- Engine front cover
- Engine oil pan
- Timing chains
- 2 bolts retaining the oil pick-up tube to the oil pump and the bolt attaching the oil pick-up tube to the main bearing stud spacer
- Pick-up tube
- 4 bolts retaining the oil pump to the cylinder block
- Oil pump

To install:

2. Rotate the inner rotor of the oil pump to align with the flats on the crankshaft and install the oil pump flush with the cylinder block. Install the 4 retaining bolts and tighten to 72–106 inch lbs. (8–12 Nm).

3. Clean the oil pick-up tube and replace the O-ring.

4. Place the pick-up tube on the oil pump and hand start the 2 retaining bolts. Install the bolt retaining the pick-up tube to the main bearing stud spacer hand tight. Tighten the pick-up tube-to-oil pump bolts to 72–106 inch lbs. (8–12 Nm). Tighten the pick-up tube to main bearing stud spacer bolt to 15–22 ft. lbs. (20–30 Nm).

5. Install or connect the following:
- New engine oil filter
- Timing chains

The oil pump is mounted on the crankshaft at the front of the engine

- Engine oil pan
- Engine front cover
- Cylinder head covers
- Negative battery cable

6. Fill the crankcase.

7. Start the engine and check for leaks and proper engine oil pressure.

8. Road test the vehicle and check for proper engine operation.

Rear Main Seal

REMOVAL & INSTALLATION

1. Remove or disconnect the following:
- Transmission
- Flexplate or flywheel

2. With a sharp awl, carefully punch a small hole in the metal portion of the seal.

3. Remove the seal using a slide hammer with a sheet metal screw attached.

➡**If the oil leak is coming from around the seal retainer, the retainer must also be removed and resealed.**

To install:

4. If the seal retainer was removed, carefully clean the sealant from the retainer and engine block using a plastic scraper.

Apply a continuous bead of silicone sealant to the back of the seal retainer before installing it on the engine

To avoid leakage, be sure to tighten the crankshaft rear oil seal retainer bolts in the correct sequence

Remove any oil or grease residue from the sealing surfaces with a solvent.

5. Apply silicone sealant to the back of the retainer and immediately install it on the engine block. Tighten the bolts in sequence to 71–107 inch lbs. (8–12 Nm).

6. Lubricate the seal and the crankshaft with clean engine oil.

7. Install the seal with the spring side toward the engine.

8. Remove the installation tool.

9. Install or connect the following:
- Flexplate or flywheel. Tighten the bolts, in a crisscross pattern, to 54–64 ft. lbs. (73–87 Nm).
- Transmission
- Negative battery cable

10. Check the engine oil level.

11. Start the engine and check for leaks.

Timing Chain, Sprockets, Front Cover and Seal

REMOVAL & INSTALLATION

4.6L (VIN W) SOHC Engine

1. Drain the engine oil.

2. Remove or disconnect the following:
- Negative battery cable
- Cooling fan and shroud
- Accessory drive belt
- Water pump pulley
- Power steering pump
- Oil pan
- Crankshaft pulley retaining bolt and washer
- Crankshaft pulley
- Bolt retaining the air conditioning pressure line to the right-hand ignition coil bracket
- Cylinder head covers
- Wiring at both ignition coils and the Crankshaft Position (CMP) sensor

Be sure that the timing marks are aligned when the No. 1 piston is at TDC on compression—4.6L (VIN W) SOHC engine

Use the special tool to maintain camshaft position while installing the timing chains—4.6L (VIN W) SOHC engine

Install the crankshaft positioning tool to be sure the crankshaft does not turn while installing the timing chains—4.6L (VIN W) SOHC engine

- 3 bolts retaining the right-hand ignition coil bracket to the engine front cover. Position the power steering hose aside.
- 3 nuts retaining the left-hand ignition coil bracket to the engine front cover. Slide both ignition coil brackets and ignition wires off the mounting studs and lay the assembly on top of the engine.
- Bolts retaining the drive belt idler pulley and the pulley
- Wiring to the Crankshaft Position (CKP) sensor and the sensor

3. If equipped, remove the retainers for the oil cooler from the engine front cover retaining stud bolts and position the oil cooler aside.
- 9 stud bolts and the 6 standard engine front cover bolts and the cover
- Crankshaft oil seal from the cover using a seal driver
- CKP sensor pulse wheel

4. Rotate the engine to set the piston for No. 1 to Top Dead Center (TDC) on its compression stroke.

5. Install Camshaft Positioning

Adapters T92P-6256-A on the flats of both camshafts. This will prevent accidental rotation of the camshafts.

6. Remove or disconnect the following:
- 2 bolts retaining the tensioner to the right-hand cylinder head and the tensioner
- Right-hand timing chain tensioner arm
- 2 bolts retaining the right-hand timing chain guide to the cylinder head and remove the timing chain guide.
- Right-hand timing chain from the camshaft and crankshaft sprockets
- Right-hand camshaft sprocket retaining bolt, washer, sprocket and spacer, if necessary
- 2 bolts retaining the timing chain tensioner to the left-hand cylinder head
- Timing chain tensioner
- Left-hand timing chain tensioner arm
- 2 bolts retaining the timing chain guide to the left-hand cylinder head
- Timing chain guide
- Left-hand timing chain from the camshaft and crankshaft sprockets
- Left-hand camshaft sprocket retaining bolt, washer, sprocket and spacer, if necessary

7. If necessary, note the position of the crankshaft sprockets and remove the crankshaft sprockets by sliding them off the front of the crankshaft.

8. Inspect the plastic running face on the tensioner arms and chain guides. If worn or damaged, inspect the engine oil pan for contamination and thoroughly clean the oil pan. Replace the oil pick-up tube.

To install:

9. Examine the timing chains, looking for the copper links. If the copper links are not visible, lay the chain on a flat surface and pull the chain taught until the opposite sides of the chain contact one another. Mark the links at each end of the chain and use these marks in place of the copper links.

10. Be sure Camshaft Positioning Adapters T92P-6256-A are installed on the flats of the camshafts to prevent them from rotating.

11. Install or connect the following:
- Left-hand and right-hand timing chain guides and retaining bolts. Tighten the retaining bolts to 71–106 inch lbs. (8–12 Nm).
- Left-hand and right-hand camshaft spacers and sprockets, (if removed) on the camshafts, the washers and

Install the crankshaft sprockets with the tapered sides facing each other

retaining bolts but do not tighten at this time.

- Left-hand crankshaft sprocket with the tapered part of the sprocket facing away from the engine block
- Left-hand timing chain on the camshaft and crankshaft sprockets. Be sure the copper links of the timing chain line up with the timing marks on both sprockets.
- Right-hand crankshaft sprocket with the tapered part of the sprocket facing the left-hand crankshaft sprocket, if removed
- Right-hand timing chain on the camshaft and crankshaft sprockets.

Slowly compress the timing chain tensioner while holding the ratchet lock away from the stem with a suitable tool—4.6L (VIN W) SOHC engine

NOTE: Ensure ratchet stem is flush to tensioner body face prior to installing paper clip.

Install a paper clip or wire into the tensioner to hold the plunger in during assembly—4.6L (VIN W) SOHC engine

Be sure the copper links of the timing chain line up with the timing marks on both sprockets.

12. It is necessary to bleed the timing chain tensioners before installation. Proceed as follows:

   a. Step 1: position the timing chain tensioner in a soft-jawed vise.

   b. Step 2: using a small pick or similar tool, hold the ratchet lock mechanism away from the ratchet stem and slowly compress the tensioner plunger by rotating the vise handle.

✷✷ WARNING

The tensioner must be compressed slowly or damage to the internal seals will result.

   c. Step 3: once the tensioner plunger bottoms in the tensioner bore, continue to hold the ratchet lock mechanism and push down on the ratchet stem until flush with the tensioner face.

   d. Step 4: while holding the ratchet stem flush with the tensioner face, release the ratchet lock mechanism and install a paper clip or similar tool in the tensioner body to lock the tensioner in the collapsed position.

   e. Step 5: the paper clip must not be removed until the timing chain, tensioner, tensioner arm and timing chain guide are completely installed on the engine.

13. Install the right-hand and left-hand timing chain tensioners and 2 bolts on each. Tighten the bolts to 15–22 ft. lbs. (20–30 Nm).

14. Crankshaft Positioning Tool T93P-6265-A over the crankshaft and the engine front cover alignment dowel to position the crankshaft.

15. Lubricate the timing chain tensioner arm contact surfaces with clean engine oil and install the right-hand and left-hand tensioner arms on their dowel pins.

16. Position a C-clamp around the timing chain tensioner arm and timing chain guide to remove all slack from the timing chain. Use care not to bend the timing chain guide.

17. Remove the locking pins or paper clips from the timing chain tensioners and be sure that all timing marks are aligned.

18. Using Camshaft Positioning Adapters T92P-6265-A to align and hold the camshafts, tighten the camshaft sprocket retaining bolts to 81–95 ft. lbs. (110–130 Nm).

19. Position a dial indicator in the No. 1 cylinder spark plug hole to measure intake valve lift. The intake valve should be at maximum lift when the crankshaft is at 114 degrees after TDC. If the intake valve lift is not at maximum lift, loosen the camshaft sprocket bolt and repeat the steps detailing the installation of the timing chain tensioners to the tightening of the camshaft sprockets.

20. Remove the camshaft and crankshaft positioning tools.

21. Install a new crankshaft seal in the front cover. Apply engine oil to the lip of the seal.

22. Thoroughly clean the sealing surfaces of the front cover, cylinder block and oil pan. Apply silicone sealer to the points where the cylinder head meets the cylinder block.

23. Install or connect the following:
- Front cover in position using new gaskets
- Retaining bolts and studs in their proper locations. Tighten bolts 1

Timing chain front cover bolt tightening sequence—4.6L (VIN W) SOHC engine

through 5 in sequence to 18 ft. lbs. (25 Nm) and bolt 6 through 15 to 35 ft. lbs. (48 Nm) within 4 minutes of applying the silicone sealer.
- Oil cooler to the front cover retaining stud bolts, if equipped
- CKP sensor and attach the harness connector
- Drive belt idler pulley
- Ignition coil brackets and ignition wires as an assembly onto the mounting studs
- Power steering hose and the nuts retaining the coil brackets to the front cover. Tighten the nuts to 15–22 ft. lbs. (20–30 Nm).
- Wiring to both ignition coils and the CMP sensor
- Cylinder head covers
- Air conditioning pressure line on the right-hand ignition coil bracket and tighten the retaining bolt

24. Apply a small amount of silicone sealer in the rear of the keyway in the crankshaft pulley.

25. Install or connect the following:
- Pulley on the crankshaft
- Crankshaft pulley bolt and washer and tighten to 114–121 ft. lbs. (155–165 Nm)
- Oil pan
- Power steering pump on the engine. Tighten the bolts to 15–22 ft. lbs. (20–30 Nm).
- Water pump pulley. Tighten the bolts to 15–22 ft. lbs. (20–30 Nm).
- Accessory drive belt
- Engine cooling fan and shroud
- Negative battery cable

26. Fill the engine.
27. Start the engine and check for leaks.
28. Road test the vehicle and check for proper engine operation.

4.6L (VIN V) DOHC Engine

FRONT OIL SEAL

1. Remove or disconnect the following:
- Negative battery cable
- Secondary air injection diverter valve
- Accessory drive belt
- Crankshaft pulley
- Front oil seal

To install:

2. Install the front oil seal flush with the front timing cover.

3. Apply silicone sealer to the crankshaft pulley woodruff key slot.

4. Install the crankshaft pulley and tighten the bolt as follows:
  a. Step 1: 66 ft. lbs. (90 Nm)

b. Step 2: Loosen the bolt one full turn
c. Step 3: 35–39 ft. lbs. (47–53 Nm)
d. Step 4: Plus 85–90 degrees

5. Install or connect the following:
- Accessory drive belt
- Secondary air injection diverter valve. Tighten the nuts to 71–106 inch lbs. (8–12 Nm).
- Negative battery cable

FRONT COVER, TIMING CHAINS AND SPROCKETS

1. Drain the cooling system.
2. Drain the engine oil.
3. Remove or disconnect the following:
- Negative battery cable
- Water outlet hose
- Water bypass hose
- Cooling fan assembly
- Accessory drive belt
- Water pump pulley
- Secondary air injection valve
- Power steering pump harness connector
- Starter relay cable and bracket
- Camshaft Position (CMP) sensor connector
- Power steering pump and reservoir
- Crankshaft pulley
- Valve covers
- Rocker arms
- Accessory drive belt tensioner and idler pulley
- Crankshaft Position (CKP) sensor connector
- Front cover
- CKP sensor pulse wheel

4. Rotate the crankshaft so that the No. 1 cylinder is at Top Dead Center (TDC) of the compression stroke.

5. Remove or disconnect the following:
- Right primary timing chain tensioner
- Right primary timing chain tensioner arm and chain guide
- Right primary timing chain and sprockets

Secondary timing chain tensioner and locking pin—4.6L (VIN V) DOHC engine

- Left primary timing chain tensioner
- Left primary timing chain tensioner arm and chain guide
- Left primary timing chain and sprockets

6. Compress the secondary chain tensioners and install locking pins.

7. Remove the left and right secondary timing chains and sprockets.

To install:

8. Position the camshafts with the keys facing downward.

Position the camshafts for timing chain installation—4.6L (VIN V) DOHC engine

Secondary timing chain tensioning tool—4.6L (VIN V) DOHC engine

Install the camshaft positioning tool—4.6L (VIN V) DOHC engine

Install the Camshaft Holding tool to secure the camshafts and prevent damage to the camshaft positioning tool—4.6L (VIN V) DOHC engine

9. Install the secondary timing chains and sprockets. Do not tighten the sprocket bolts at this time.

10. Tension the secondary timing chains with the Secondary Timing Chain Tensioning tool T93P-6256-BH.

11. Install Camshaft Positioning tool T93P-6256-A in the rear D-slots of the camshaft.

12. Install Camshaft Holding tool T93P-6256-AH onto the camshafts to keep the camshafts from rotating and to prevent damaging the camshaft positioning tool.

13. Tighten the secondary timing chain sprocket bolts to 81–95 ft. lbs. (110–130 Nm).

14. Remove the secondary timing chain tensioner locking pins and remove the tensioner tool.

Correct alignment of the primary timing chains and sprockets—4.6L (VIN V) DOHC engine

15. Compress the primary timing chain tensioners and install locking pins.

16. Install the crankshaft timing sprockets.

17. Align the copper colored primary timing chain links with the sprocket timing marks as shown.

18. Install or connect the following:
- Left primary timing chain and sprockets
- Left primary timing chain tensioner arm and chain guide
- Left primary timing chain tensioner
- Right primary timing chain and sprockets
- Right primary timing chain tensioner arm and chain guide
- Right primary timing chain tensioner

19. Tighten the camshaft sprocket bolts to 81–95 ft. lbs. (110–130 Nm).

20. Remove the primary timing chain tensioner locking pins.

21. Remove the Camshaft Holding and Camshaft Positioning tools.

22. Install or connect the following:
- CKP sensor pulse wheel
- Front cover. Apply silicone sealant as shown. Tighten the bolts in sequence to 15–22 ft. lbs. (20–30 Nm).

Crankshaft timing sprocket installation—4.6L (VIN V) DOHC engine

Front cover torque sequence—4.6L (VIN V) DOHC engines

Sealer Location

Item	Part Number	Description
1	6049	RH Cylinder Head
2	6049	LH Cylinder Head
3	12A227	Ignition Pulse Crankshaft Sensor Ring
4	6010	Cylinder Block
5	WSE-M4G320-A2	Sealer
6	6710	Oil Pan Gasket

9306LG06

Apply sealer to these locations—4.6L (VIN V) DOHC engines

- CKP sensor connector
- Accessory drive belt tensioner and idler pulley. Tighten the bolts to 15–22 ft. lbs. (20–30 Nm).
- Rocker arms
- Valve covers

23. Apply silicone sealer to the crankshaft pulley woodruff key slot.

24. Install the crankshaft pulley and tighten the bolt as follows:
    a. Step 1: 66 ft. lbs. (90 Nm)
    b. Step 2: Loosen the bolt one full turn
    c. Step 3: 35–39 ft. lbs. (47–53 Nm)
    d. Step 4: Plus 85–90 degrees

25. Install or connect the following:
- Power steering pump and reservoir
- CMP sensor connector
- Starter relay cable and bracket
- Power steering pump harness connector
- Secondary air injection valve
- Water pump pulley. Tighten the bolts to 15–22 ft. lbs. (20–30 Nm).
- Accessory drive belt
- Cooling fan assembly

- Water bypass hose
- Water outlet hose
- Negative battery cable

26. Fill the crankcase to the correct level.
27. Fill the cooling system.
28. Turn the air suspension switch to the **ON** position.
29. Start the engine and check for leaks.

Piston and Ring

POSITIONING

7922AG22

4.6L (VIN W) SOHC engine—piston ring end-gap spacing and piston positioning

7922AG32

4.6L (VIN W) SOHC engine—piston ring positioning

4.6L (VIN V) DOHC engine—piston ring positioning

4.6L (VIN V) DOHC engine—piston ring end-gap spacing

FUEL SYSTEM

Fuel System Service Precautions

Safety is the most important factor when performing not only fuel system maintenance but any type of maintenance. Failure to conduct maintenance and repairs in a safe manner may result in serious personal injury or death. Maintenance and testing of the vehicle's fuel system components can be accomplished safely and effectively by adhering to the following rules and guidelines.

• To avoid the possibility of fire and personal injury, always disconnect the negative battery cable unless the repair or test procedure requires that battery voltage be applied.

• Always relieve the fuel system pressure prior to disconnecting any fuel system component (injector, fuel rail, pressure regulator, etc.), fitting or fuel line connection. Exercise extreme caution whenever relieving fuel system pressure, to avoid exposing skin, face and eyes to fuel spray. Please be advised that fuel under pressure may penetrate the skin or any part of the body that it contacts.

• Always place a shop towel or cloth around the fitting or connection prior to loosening to absorb any excess fuel due to spillage. Ensure that all fuel spillage (should it occur) is quickly removed from engine surfaces. Ensure that all fuel soaked cloths or towels are deposited into a waste container.

• Always keep a dry chemical (Class B) fire extinguisher near the work area.

• Do not allow fuel spray or fuel vapors to come into contact with a spark or open flame.

• Always use a back-up wrench when loosening and tightening fuel line connection fittings. This will prevent unnecessary stress and torsion to fuel line piping. Always follow the proper torque specifications.

• Always replace worn fuel fitting O-rings with new. Do not substitute fuel hose or equivalent, where fuel pipe is installed.

Fuel System Pressure

RELIEVING

Fuel supply lines on all fuel injected engines will remain pressurized for some period of time after the engine is shut **OFF**. This pressure must be relieved before servicing the fuel system. Pressure is relieved through the fuel pressure relief valve, located on the fuel rail.

To relieve the fuel system pressure, first remove the fuel tank cap to relieve pressure in the tank, then remove the cap on the fuel pressure relief valve. Attach a fuel pressure gauge and drain the system through the drain tube into a container. Remove the fuel pressure gauge and replace the cap on the relief valve.

Fuel Filter

REMOVAL & INSTALLATION

1. Disconnect the negative battery cable.
2. Relieve the fuel system pressure.
3. If equipped with air suspension, turn the air suspension switch to the **OFF** position.
4. Remove the hairpin clip push connect fittings from both ends of the fuel filter as follows:

   a. Step 1: Inspect the visible internal portion of the fitting for dirt accumulation. If more than a light coating of dust is present, clean the fitting before disassembly.

   b. Step 2: Some adhesion between the seals in the fitting and the filter will occur with time. To separate, twist the fitting on the filter, then push and pull the fitting until it moves freely on the filter.

   c. Step 3: Remove the hairpin clip from the fitting by first bending and breaking the shipping tab. Next, spread the 2 clip legs by hand about ⅛ inch each, to disengage the body and push the legs into the fitting. Lightly pull the triangular end of the clip and work it clear of the filter and fitting.

   d. Step 4: Grasp the fitting and pull in an axial direction to remove the fitting from the filter. Be careful on 90 degree elbow connectors, as excessive side loading could break the connector body.

   e. Step 5: After disassembly, inspect the inside of the fitting for any internal

parts such as O-rings and spacers that may have been dislodged from the fitting. Replace any damaged connector.

5. Remove the filter retaining clamp and remove the fuel filter. Note the direction of the flow arrow on the filter, so the replacement filter can be reinstalled in the same position.

To install:

6. Install or connect the following:
   - Fuel filter with the flow arrow facing the proper direction and tighten the filter retaining clamp
   - Rubber insulator rings on the new filter. Replace the insulator rings if the filter moves freely after the retainer is installed.
   - Filter into the retainer with the flow arrow pointing out the open end of the retainer
   - Retainer on the bracket and tighten the mounting bolts to 27–44 inch lbs. (3–5 Nm)

7. Install the hairpin clip push connect fittings at both ends of the fuel filter as follows:

   a. Step 1: Install a new connector if damage was found. Insert a new clip into any 2 adjacent openings with the triangular portion pointing away from the fitting opening. Install the clip until the legs of the clip are locked on the outside of the body. Piloting with an index finger is necessary.

   b. Step 2: Before installing the fitting on the filter, wipe the filter end with a clean cloth. Inspect the inside of the fitting to be sure it is free of dirt and/or obstructions.

   c. Step 3: Apply a light coating of engine oil to the filter end. Align the fitting and filter axially and push the fitting onto the filter end. When the fitting is engaged, a definite click will be heard. Pull on the fitting to be sure it is fully engaged.

8. If equipped with air suspension, turn the air suspension switch to the **ON** position.
9. Reconnect the negative battery cable.
10. Start the engine and check for fuel leaks and proper operation.

Fuel Pump

REMOVAL & INSTALLATION

1. Disconnect the negative battery cable.
2. Relieve the fuel system pressure.
3. Install a hose into the fuel filler pipe

and drain or siphon the fuel into a storage tank designed for fuel storage.

4. Remove any dirt that has accumulated around the fuel pump and fuel lines to prevent the entry of contaminants into the tank during fuel pump removal and installation.
5. Remove or disconnect the following:
   - Fuel supply and return line fittings at the fuel pump using fuel line disconnect tools
   - Fuel pump module electrical connector
   - 6 retaining bolts around the perimeter of the fuel pump module
   - Fuel pump module and seal from the fuel tank

To install:

6. Clean the fuel pump module mounting flange and fuel tank mounting surface.
7. Install or connect the following:
   - New seal and the fuel pump module using care not to damage the inlet filter and fuel sending unit float arm
   - 6 retaining bolts and tighten to 80–107 inch lbs. (9–12 Nm)
   - Fuel pump module electrical connector
   - Fuel supply and return lines to the fuel pump module. Pull on the fuel line fittings to verify engagement.

8. Add a minimum of 10 gallons (38L) of clean fuel to the fuel tank and check for leaks.
9. Install a Fuel pressure gauge to the Schrader valve on the fuel injection supply manifold.
10. Connect the negative battery cable.
11. Cycle the ignition switch from the **OFF** to **ON** position 5–10 times for 3 second intervals or until the fuel pressure gauge shows at least 35 psi (241 kPa).
12. Check for fuel leaks.
13. Remove the fuel pressure gauge.
14. Start the engine and recheck for fuel leaks.
15. Road test the vehicle and check for proper operation.

2-3 Nm (18-26 lb/in)

7922RG22

The fuel filter is located near the center of the vehicle on the frame rail

9-12 Nm (80-107 lb/in)

7922RG23

Tighten the fuel pump mounting bolts to 80–107 inch lbs. (9–12 Nm)

Fuel Injector

REMOVAL & INSTALLATION

4.6L (VIN W) SOHC Engine

1. Relieve the fuel system pressure.
2. Remove or disconnect the following:

- Injector supply manifold
- Wiring
- Injector by pulling it up and gently rocking it side to side
- O-rings and discard

To install:

3. Lubricate new O-rings with clean engine oil.
4. Install or connect the following:

- O-rings
- Fuel injector using a light, twisting and pushing motion
- Injector supply manifold
- Wiring

4.6L (VIN V) DOHC Engine

1. Relieve the fuel system pressure.
2. Remove or disconnect the following:

- Negative battery cable
- Air cleaner outlet tube
- Upper intake manifold
- Fuel pressure regulator vacuum line
- Fuel pulse damper vacuum line
- Brake booster vacuum line
- Fuel lines
- Fuel injector harness connectors

- Injector supply manifold
- Fuel injectors

To install:

3. Install or connect the following:

- Fuel injectors with new O-ring seals
- Injector supply manifold. Tighten the bolts to 71–106 inch lbs. (8–12 Nm).
- Fuel injector harness connectors
- Fuel lines
- Brake booster vacuum line
- Fuel pulse damper vacuum line
- Fuel pressure regulator vacuum line
- Upper intake manifold
- Air cleaner outlet tube
- Negative battery cable

4. Start the engine and check for leaks.

DRIVE TRAIN

Transmission Assembly

REMOVAL & INSTALLATION

1. Drain the transmission.
2. If equipped with air suspension, the air suspension switch, located on the right-hand side of the luggage compartment, must be turned to the **OFF** position before raising the vehicle.
3. Remove or disconnect the following:

- Negative battery cable
- Exhaust system as necessary for transmission removal
- Converter bottom access cover and adapter plate bolts
- Torque converter drain plug, to allow the converter to drain into a container, if equipped. After the converter has drained, reinstall the drain plug and tighten.
- 4 torque converter-to-flywheel retaining nuts
- Driveshaft (mark for installation), plug the transmission extension housing to prevent fluid leakage
- Vehicle Speed Sensor (VSS) or if equipped, the speedometer cable from the transmission extension housing
- Shift cable from the transmission manual control lever the throttle valve cable from the transmission throttle valve lever, if equipped
- Transmission wiring harness connectors.
- Starter motor retaining bolts and place the starter motor aside

4. Position a transmission jack under the transmission and raise it enough to allow crossmember removal.

5. Remove or disconnect the following:

- Engine rear support-to-crossmember bolts and the crossmember-to-frame side support retaining bolts
- Crossmember and transmission support insulator

6. Lower the transmission jack and allow the transmission to hang.
7. Place a jack to the front of the engine and raise the engine enough to gain access to the 2 upper transmission-to-cylinder block retaining bolts. Do not remove the bolts at this time.
8. Remove or disconnect the following:

- Transmission cooler lines at the transmission. Plug all openings to keep dirt out.
- Lower transmission-to-cylinder block retaining bolts
- Transmission fluid fill tube and plug the opening in the transmission

9. Secure the transmission to the transmission jack with a safety strap or chain.
10. Remove the 2 upper transmission-to-cylinder block bolts.
11. Carefully move the transmission rearward to disengage the bell housing from the dowel pins and the torque converter studs from the flywheel.
12. Remove or disconnect the following:

- Transmission
- Torque converter to prevent the converter from dropping out of the transmission causing possible damage or personal injury

To install:

13. Remove the safety stand and block of wood supporting the rear of the engine, if installed.
14. Install or connect the following:

- Torque converter drain plug to 21–23 ft. lbs. (28–30 Nm), if equipped
- Torque converter on the transmission and rotate into position to be sure the drive flats are fully engaged in the pump gear. When fully seated, the center of the torque converter should be about 7/16–9/16 inch (10.2–14.4mm) below the transmission mounting surface

15. Mount the transmission on a transmission jack and secure with a safely strap or chain. Raise the transmission and align with the cylinder block dowel pins.
16. Rotate the converter until the studs

DIMENSION A TO BE 10.23-14.43 mm (7/16-9/16 INCH) APPROXIMATELY

7922RG24

To prevent transmission damage, be sure that the torque converter is fully seated in the front pump of the transmission

and drain plug are in alignment with the holes in the flywheel. Align the orange balancing marks on the converter stud and flywheel bolt hole, if balancing marks are present.

17. Slide the transmission assembly forward into position, being careful not to damage the flywheel and converter pilot.

18. Install or connect the following:
- 2 transmission housing-to-cylinder block bolts at the engine dowel pin locations. Tighten the bolts to 41–50 ft. lbs. (55–68 Nm).
- Transmission housing-to-cylinder block bolts. Tighten the bolts to 41–50 ft. lbs. (55–68 Nm).

19. Remove the safety strap or chain from around the transmission.

20. Install or connect the following:
- Transmission fluid fill tube. Tighten the bolt to 28–38 ft. lbs. (38–51 Nm).
- Oil cooler lines to the transmission case. Tighten the cooler line fittings to 15–19 ft. lbs. (20–26 Nm).

21. Remove the jack supporting the front of the engine.

22. Install or connect the following:
- Crossmember using the proper jack
- Crossmember and transmission support insulators in position
- Engine rear support-to-crossmember retaining bolts and the crossmember-to-frame side support retaining bolts

23. Remove the transmission jack.

24. Install or connect the following:
- Transmission wiring harness connectors
- Starter motor and wiring
- 4 torque converter-to-flywheel retaining nuts. Tighten to 20–33 ft. lbs. (27–46 Nm).
- Torque converter access cover and cover plate bolts. Tighten the bolts to 12–16 ft. lbs. (16–22 Nm).
- Exhaust system
- VSS and the wiring, or if equipped, the speedometer cable to the transmission extension housing
- Driveshaft, aligning the marks that were made during removal
- Shift cable to the transmission manual control lever
- Throttle valve cable to the transmission throttle valve lever, if equipped

25. If equipped with air suspension, turn the air suspension switch to the **ON** position.

26. Fill the transmission.

27. Start the engine and check the transmission for leakage.

28. Road test the vehicle and check for proper transmission operation.

Axle Shaft, Bearing and Seal

REMOVAL & INSTALLATION

1. If equipped, turn the air suspension service switch to the **OFF** position before raising the vehicle.

2. Remove or disconnect the following
- Wheel
- Brake caliper and rotor
- Anti-lock brake speed sensor, if equipped

3. Clean all dirt from the area of the axle housing cover.

4. Place a drain pan under the axle housing.

5. Remove or disconnect the following:
- Axle housing cover retaining bolts and the cover, draining the axle lubricant from the housing
- Differential pinion shaft lockbolt and the differential pinion shaft

6. Push the flanged end of the axle shaft being removed toward the center of the vehicle

7. Remove or disconnect the following:
- C-lock from the button end of the axle shaft
- Axle shaft from the housing, being careful not to damage the oil seal and anti-lock brake sensor ring, if equipped.

8. Insert an axle bearing remover in the axle housing bore and position it behind the wheel bearing so the tangs on the tool engage the bearing outer race.

9. Remove the wheel bearing and seal as an assembly using an impact slide hammer attached to the bearing remover tool

DIFFERENTIAL PINION SHAFT

7922RG35

Removal of differential pinion shaft

AXLE SHAFTS

C-LOCKS

7922RG36

Removing axle shaft C-lock clips

To install:

10. Lubricate the new wheel bearing with rear axle lubricant.

11. Install the wheel bearing into the axle housing bore using a bearing installer.

12. Lubricate the lips of a new wheel bearing oil seal with wheel bearing grease.

13. Install or connect the following:
- New wheel bearing seal using a seal installer
- Axle shaft into the axle housing without damaging the bearing/seal assembly or anti-lock brake sensor ring, if equipped. Start the splines

SPEED SENSOR

CALIPER ASSY

DISC BRAKE ADAPTER ASSY

PARKING BRAKE ASSY

SPEED SENSOR RING

AXLE SHAFT ASSY

DISC BRAKE ROTOR

7922RG34

Exploded view of the rear axle shaft assembly

into the side gear and push firmly until the button end of the axle shaft can be seen in the differential case.

- C-lock on the button end of the axle shaft splines, then push the shaft outboard until the shaft splines engage and the C-lock seats in the counterbore of the differential side gear.
- Differential pinion shaft through the case and pinion gears, aligning the hole in the shaft with the lockbolt hole
- Apply a thread locking compound to the lockbolt threads and place in the case and pinion shaft. Tighten to 15–30 ft. lbs. (20–41 Nm).

14. Cover the inside of the differential case with a shop rag and clean the sealing surface of the axle housing and the axle housing cover. Remove the shop rag.

15. Apply a ⅛–³⁄₁₆ inch (3.18–4.76mm) wide bead of silicone sealer to the cover.

16. Install the axle housing and bolts and tighten in a crisscross pattern. Final torque the cover retaining bolts to 28–38 ft. lbs. (38–52 Nm).

17. Add the appropriate rear axle lubricant to the axle housing to a level ¼–⁹⁄₁₆ inch (6–14mm) below the bottom of the fill hole. If equipped with a limited slip differential, add 4 oz. (118.3 ml) of the appropriate friction modifier.

18. Install or connect the following:

- Axle housing fill plug and tighten to 15–30 ft. lbs. (20–41 Nm)
- Anti-lock brake speed sensor, if equipped. Tighten the retaining bolt to 40–60 inch lbs. (4.5–6.8 Nm).
- Brake calipers and rotors
- Wheel

19. If equipped with air suspension, turn the air suspension switch to the **ON** position.

20. Road test the vehicle and check for proper operation.

Pinion Seal

REMOVAL & INSTALLATION

1. Remove or disconnect the following:
- Driveshaft
- Rear wheels
- Rear brake calipers

→The rear brake calipers must be removed so that there is no additional drag when measuring pinion bearing preload.

2. Use an inch lb. torque wrench and measure the amount of torque required to maintain pinion rotation through several revolutions.

3. Remove the pinion flange and remove the seal.

To install:

4. Install or connect the following:

- Pinion seal and flange
- New pinion flange nut

5. Rotate the pinion flange occasionally while tightening the flange nut to make sure the pinion bearings seat correctly.

6. Take frequent bearing preload torque readings.

7. If the preload recorded prior to disassembly is **lower** than the specification for used bearings, then tighten the pinion flange nut to specification. If the preload recorder prior to disassembly is **higher** than the specification for used bearings, then tighten the pinion flange nut to the original reading as recorded.

8. The pinion bearing preload specifications are as follows:

   a. Used bearings: 8–14 inch lbs. (0.9–1.6 Nm).
   b. New bearings: 16–29 inch lbs. (1.8–3.2 Nm).

✼✼ CAUTION

Never loosen the pinion nut to reduce bearing preload. If it is necessary to reduce bearing preload, install a new collapsible spacer and pinion nut.

9. Install or connect the following:
- Driveshaft
- Brake calipers
- Rear wheels

10. Fill the differential with gear lubricant and check for leaks.

STEERING AND SUSPENSION

Air Bag

✼✼ CAUTION

Some vehicles are equipped with an air bag system. The system must be disarmed before performing service on, or around, system components, the steering column, instrument panel components, wiring and sensors. Failure to follow the safety precautions and the disarming procedure could result in accidental air bag deployment, possible injury and unnecessary system repairs.

PRECAUTIONS

Several precautions must be observed when handling the inflator module to avoid accidental deployment and possible personal injury.

- Never carry the inflator module by the

wires or connector on the underside of the module.

- When carrying a live inflator module, hold securely with both hands, and ensure that the bag and trim cover are pointed away.
- Place the inflator module on a bench or other surface with the bag and trim cover facing up.
- With the inflator module on the bench, never place anything on or close to the module which may be thrown in the event of an accidental deployment.

DISARMING

1. Before servicing the vehicle, refer to the precautions in the beginning of this section.

2. Position the vehicle with the front wheels in a straight-ahead position.

3. Disconnect both battery cables.

4. Wait at least 1 minute for the air bag back-up power supply to deplete its stored energy before continuing.

5. Proceed with the repair.
6. Once the repair is complete.
7. Reconnect both battery cables.
8. Prove out the air bag system by turning the ignition key to the **RUN** position and visually monitoring the air bag indicator lamp in the instrument cluster. The indicator lamp should illuminate for approximately 6 seconds, then turn **OFF**. If the indicator lamp does not illuminate, stays on, or flashes at any time, a fault has been detected by the air bag diagnostic monitor.

Power Steering Gear

REMOVAL & INSTALLATION

Recirculating Ball Steering Gear

1. If equipped with air suspension, the air suspension switch must be turned to the **OFF** position before raising the vehicle.

7922RG25

Remove the bolt and separate the intermediate shaft from the steering gear input shaft

7922RG26

Remove the locknut and separate the Pitman arm from the center link using the appropriate puller

2. Center the steering wheel and turn the key to the locked position.

3. Remove or disconnect the following:
- Negative battery cable
- Bolt and the intermediate shaft from the steering gear

4. On the Town Car, separate the 2 halves and remove the steering gear cover.

5. Tag the power steering pressure and return lines so they may be reassembled in their original positions.

6. Place a drain pan under the steering gear

7. Remove or disconnect the following:
- Pressure and return lines. Plug the lines and ports in the gear to prevent the entry of dirt.
- Pitman arm from the center link using a puller. It is not necessary to remove the Pitman arm from the steering gear.

8. Support the steering gear

9. Remove or disconnect the following:
- Steering gear-to-frame rail retaining bolts
- Steering gear

To install:

10. Install or connect the following:
- Steering gear on the frame rail. Tighten the steering gear-to-frame

retaining bolts to 50–67 ft. lbs. (66–90 Nm).
- Pitman arm to the center link. Tighten the retaining nut to 52–60 ft. lbs. (70–81 Nm).
- Power steering pressure and return lines to the steering gear and tighten the lines to 12–18 ft. lbs. (16–24 Nm).
- Intermediate shaft to the steering gear. Tighten the bolt to 31–41 ft. lbs. (41–55 Nm).
- Negative battery cable

11. If equipped with air suspension, turn the air suspension switch to the **ON** position.

12. Fill the reservoir with the correct power steering fluid and turn the steering wheel from stop-to-stop to distribute the fluid. Check the fluid level and add fluid, if necessary.

13. Start the engine and turn the steering wheel from left to right. Check for leaks.

14. On the Town Car, install the steering gear cover.

Rack and Pinion Steering Gear

1. If equipped with air suspension, the air suspension switch must be turned to the **OFF** position before raising the vehicle.

2. Center the steering wheel and turn the key to the locked position.

3. Remove or disconnect the following:
- Negative battery cable
- Front wheels
- Outer tie rod ends

✳✳ CAUTION

Do not allow the intermediate shaft to rotate while it is disconnected from the steering gear or damage to the clockspring can result. If there is evidence that the intermediate shaft has rotated, the clockspring must be removed and recentered.

- Intermediate shaft
- Power steering fluid pressure and return lines
- Power steering pressure switch connector
- Steering gear mounting nuts
- Steering gear mounting studs
- Steering gear

To install:

4. Installation is the reverse of the removal procedure, while using the following torque values:
- Steering gear mounting studs: 15 ft. lbs. (20 Nm)
- Steering gear mounting nuts: 76 ft. lbs. (103 Nm)

- Power steering fluid pressure and return lines: 13 ft. lbs. (18 Nm)
- Intermediate shaft pinch bolt: 20 ft. lbs. (32 Nm)
- Outer tie rod end nuts: 59 ft. lbs. (80 Nm)

Shock Absorber

REMOVAL & INSTALLATION

Front

1. If equipped with air suspension, the air suspension switch, located on the right-hand side of the luggage compartment, must be turned to the **OFF** position before raising the vehicle.

2. Remove or disconnect the following:
- Nut, washer and bushing from the upper end of the shock absorber
- 2 bolts retaining the shock absorber to the lower control arm
- Shock absorber

To install:

3. Prior to installation, prime the new shock absorber. Fully extend the shock absorber while in the right side up (installed) position. Turn the shock absorber upside down and fully compress it. Repeat the procedure at least 3 times to purge any air trapped in the shock absorber.

4. Install or connect the following:
- New bushing and washer on the stud on the top of the new shock absorber and position the unit inside the front coil spring
- 2 lower retaining bolts and tighten them to 10–12 ft. lbs. (13–17 Nm).
- New bushing and washer on the shock absorber top stud
- New retaining nut. Tighten the retaining nut to 25–34 ft. lbs. (34–46 Nm).

5. If equipped with air suspension, turn the air suspension switch to the **ON** position.

Rear

1. If equipped with air suspension, turn the air suspension service switch **OFF**.

2. Be sure the ignition switch is in the **OFF** position.

3. Support the rear axle assembly with a jack.

4. Remove or disconnect the following:
- Top retaining nut, washer and bushing
- Bottom retaining nut and washer
- Shock absorber

To install:

5. Install or connect the following:
  • Shock absorber so the upper stud enters the hole in the frame
  • Top bushing, washer and retaining nut. Tighten to 26–34 ft. lbs. (34–46 Nm).

6. Extend the shock absorber and place the lower stud through the hole in the bracket

7. Bottom retaining washer and nut. Tighten to 57–75 ft. lbs. (76–103 Nm).

8. Remove the jack from the axle assembly.

9. Turn the air suspension service switch to the **ON** position.

Coil Spring

REMOVAL & INSTALLATION

Front

1. If equipped with air suspension, turn the air suspension service switch to the **OFF** position before raising the vehicle.

2. Remove or disconnect the following:
  • Wheel
  • Shock absorber
  • Center link from the Pitman arm

3. Using a spring compressor perform the following steps:
  a. Step 1: install 1 plate with the pivot ball seat facing downward into the coils of the spring. Rotate the plate so it is flush with the upper surface of the lower arm.

```
1  Upper Ball       5  Lower Ball Nut
2  Compression Rod  6  Plate
3  Forcing Nut      7  Pin
4  Thrust Washer
```
7922RG27

Exploded view of Spring Compressor
D78P-5310-A

7922RG28

Compress the coil spring until it moves away from its seat

7922RG29

Remove the fasteners attaching the lower arm to the frame, then lower the arm and remove the spring with the compressor

b. Step 2: Install the other plate with the pivot ball seat facing upward into the coils of the spring. Insert the upper ball nut through the coils of the spring, so the nut rests in the upper plate.

c. Step 3: Insert the compression rod into the opening in the lower arm, through the upper and lower plate and upper ball nut. Insert the securing pin through the upper ball nut and compression rod.

d. Step 4: With the upper ball nut secured, turn the upper plate so it walks up the coil until it contacts the upper spring seat. Then, back off ½ turn.

e. Step 5: Install the lower ball nut and thrust washer on the compression rod and screw on the forcing nut. Tighten the forcing nut until the spring is compressed enough so it is free in its seat.

4. Remove or disconnect the following:
  • 2 lower control arm pivot bolts
  • Lower arm from the frame crossmember
  • Coil spring

5. If a new coil spring is to be installed, mark the position of the upper and lower plates on the spring with chalk. With an assistant, compress a new spring for installation and measure the compressed length

and the amount of curvature of the old spring.

6. Loosen the forcing nut to relieve the spring tension and remove the tools from the spring.

To install:

7. Assemble the spring compressor and locate in the same position as marked during disassembly.

8. Before compressing the coil spring, be sure the upper ball nut securing the pin is inserted properly.

9. Compress the coil spring until the spring height reaches the dimension measured during disassembly.

10. Position the coil spring assembly into the lower arm and position the lower arm into the frame crossmember.

11. Install both the front and rear lower control arm pivot bolts through the frame and lower arm bushings. Tighten the bolts and nuts to 109–148 ft. lbs. (148–201 Nm).

12. Remove the spring compressor from the coil spring.

13. Install or connect the following:
  • Drag link to the Pitman arm. Tighten the retaining nut to 60 ft. lbs. (80 Nm).
  • Shock absorber inside the coil spring
  • Retaining bolts
  • Wheel. Tighten the lug nuts to 85–105 ft. lbs. (115–142 Nm).

14. Place a washer and retaining nut on the shock absorber top stud. Tighten the nut to 25–34 ft. lbs. (34–46 Nm).

15. If equipped with air suspension, turn the air suspension switch to the **ON** position.

16. Check the front end alignment.

Rear

1. Place a hoist under the rear axle housing and raise and safely support the vehicle.

2. Support the frame side rails with 2 jackstands.

3. Remove or disconnect the following:
  • Rear sway bar
  • Lower studs of both rear shock absorbers from the mounting brackets on the axle tube
  • Parking brake cable from the upper arm retainer before lowering the axle housing

4. Lower the axle housing until the coil springs are released. If the axle housing is supported by the hoist, lower the hoist allowing the rear of the vehicle to rest on the jackstands. If the vehicle's axle housing is supported by the jackstands, leave the

hoist stationary and lower the jackstands or raise the hoist to release the tension on the coil springs.

   5. Remove the coil springs and insulators.

To install:

   6. Install or connect the following:

- Coil spring in the upper and lower seats with an insulator between the upper end of the spring and frame seat
- Axle housing and connect the lower studs of the shock absorbers to the mounting brackets
- Parking cable into the upper arm retainer
- Sway bar

   7. Road test the vehicle and check for proper operation.

Air Spring

REMOVAL & INSTALLATION

✳✳ CAUTION

Before servicing any air suspension component, disconnect power to the system by turning the air suspension service switch OFF or by disconnecting the negative battery cable. Do not remove an air spring under any circumstances when there is pressure in the air spring. Do not remove any components supporting an air spring without either exhausting the air or providing support for the air spring.

   1. Turn the air suspension switch to the **OFF** position.

   2. Raise and safely support the vehicle so the suspension is fully down with no load.

   3. Remove or disconnect the following:

- Heat shield, as required
- Spring retainer clip
- Air spring solenoid valve electrical connector
- Air line
- Air spring solenoid retainer

   4. Rotate the solenoid valve counterclockwise to the first stop.

   5. Pull the solenoid valve straight out slowly to the second stop to bleed air from the system.

✳✳ CAUTION

Do not fully release the solenoid until the air is completely bled from the air spring or personal injury may result.

HEAT SHIELD

SPRING RETAINER CLIP

AIR SPRING WITH INTEGRAL SOLENOID

7922RG30

Exploded view of the air spring mounting

   6. After the air is fully bled from the system, rotate the solenoid valve counterclockwise to the third stop and remove the solenoid valve from the solenoid housing. Remove the large O-ring from the solenoid housing.

   7. Remove the air spring.

To install:

   8. Check the solenoid valve O-rings for cuts or abrasions. Replace the O-rings as required. Lightly grease the O-ring area of the solenoid valve and the larger solenoid housing O-ring with silicone dielectric compound.

   9. Insert the solenoid into the air spring end cap and rotate clockwise to the third stop, push in to the second stop, then rotate clockwise to the first stop.

   10. Install or connect the following:

- Air spring solenoid retainer. Inspect the wiring harness connector and ensure the rubber gasket is in place at the bottom of the connector cavity
- Air spring into the frame (upper) spring seat, taking care to keep the solenoid air and electrical connections clean and free of damage
- Push-on ring spring retainer clip to the knob of the spring cap from the top side of the frame spring seat
- Air line and electrical connector to the solenoid

- Heat shield to the frame spring seat, if removed
- Align the air spring piston-to-axle (lower) seat. Squeeze to increase pressure and push downward on the piston, snapping the piston to the axle seat at rebound and supported by the shock absorber.
- Negative battery cable.

✳✳ WARNING

The air springs may be damaged if the suspension is allowed to compress before the spring is inflated.

   11. Refill the air spring as follows:

   a. Step 1: Turn the air suspension switch to the **ON** position. The ignition switch must be **ON** and the engine running or a battery charger must be connected to the battery to reduce battery drain.

   b. Step 2: Fold back or remove the right-hand luggage compartment trim panel and connect Super Star II Tester 007–0041-A to the air suspension DLC, which is located near the air suspension switch.

   c. Step 3: Set the tester to EEC-IV/MCU mode. Also set the tester to FAST mode. Release the tester button to the HOLD (up) position and turn the tester **ON**.

d. Step 4: Depress the tester button to TEST (down) position. A Code 10 will be displayed. Within 2 minutes a Code 13 will be displayed. After Code 13 is displayed, release the tester button to the HOLD (up) position, wait 5 seconds and depress the tester button to TEST (down) position. Ignore any codes displayed.

e. Step 5: Release the tester button to the HOLD (up) position. Wait at least 20 seconds, then depress the tester button to TEST (down) position. Within 10 seconds, the codes will be displayed in the order shown.

f. Step 6: Within 4 seconds after Code 26 is displayed, release the tester button to the HOLD (up) position. Waiting longer than 4 seconds may result in Functional Test 31 being entered. The compressor will fill the air springs with air as long as the tester button is in the HOLD (up) position. To stop filling the air springs, depress the tester button to the TEST (down) position.

g. Step 7: To exit Functional Test 26, disconnect the tester and turn the ignition switch to the **OFF** position.

12. Install the luggage trim panel, if removed.

Upper Ball Joint

REMOVAL & INSTALLATION

➡On 2005 models, the ball joint is not serviced separately. If the ball joint is faulty the upper control arm and ball joint are replaced as an assembly.

1. If equipped with air suspension, the air suspension switch to the **OFF** position before raising the vehicle.

2. Place supports under both sides of the frame just behind the lower control arms.

3. Remove the wheel.

4. Place a floor jack under the lower control arm at the lower ball joint area. The floor jack will support the spring load on the lower control arm.

5. Remove the retaining nut and pinch bolt from the upper ball joint stud.

6. Mark the position of the alignment cams. When replacing the upper ball joint this will approximate the current alignment.

7. Remove or disconnect the following:
  • 2 nuts retaining the upper ball joint to the upper control arm
  • Upper ball joint from the upper control arm and spread the slot in the wheel spindle with a prybar to remove the ball joint stud from the wheel spindle.

To install:

8. Install or connect the following:
  • Upper ball joint to the upper control arm
  • Ball stud into the wheel spindle
  • Upper ball joint pinch bolt and retaining nut. Tighten to 56–77 ft. lbs. (76–104 Nm).
  • Alignment cams to the approximate position at removal. If not marked, install in the neutral positions.
  • 2 nuts retaining the upper ball joint to the upper control arm. Hold the cams and tighten the nuts to 107–129 ft. lbs. (145–175 Nm).
  • Wheel and tire assembly. Tighten the lug nuts in a star pattern to 85–105 ft. lbs. (115–142 Nm).

9. Remove the floor jack from under the lower control arm.

10. If equipped with air suspension, turn the air suspension switch to the **ON** position.

11. Check and adjust the front wheel alignment.

Lower Ball Joint

REMOVAL & INSTALLATION

➡On 2005 models, the ball joint is not serviced separately. If the ball joint is faulty the lower control arm and ball joint are replaced as an assembly.

1. If equipped with air suspension, the air suspension service switch to the **OFF** position before raising the vehicle.

2. Place supports under both sides of the frame behind the lower control arms.

3. Remove or disconnect the following:
  • Wheel
  • Wheel spindle
  • Ball joint boot seal and discard
  • Ball joint

To install:

➡When installing a new ball joint, the protective cover should be left on to protect the ball joint seal during installation. It may be necessary to trim the cover so it can pass through the installation tool.

4. Install the ball joint.

5. Discard the protective cover and be sure the new ball joint is fully seated in the lower control arm. Ensure that the ball joint seal is not damaged.

Use a ball joint press to install the new ball joint into the lower control arm

Use a ball joint press to remove the ball joint from the lower control arm

6. Install or connect the following:
- Wheel spindle
- Wheel. Tighten the lug nuts to 85–105 ft. lbs. (115–142 Nm).

7. If equipped with air suspension, turn the air suspension service switch to the **ON** position.

8. Check the front end alignment.

Upper Control Arm

REMOVAL & INSTALLATION

1. If equipped with air suspension, the air suspension switch, located on the right-hand side of the luggage compartment, must be turned to the **OFF** position before raising the vehicle.

2. Remove or disconnect the following:
- Front wheel
- Upper ball joint. Support the lower control arm.
- Upper control arm pivot bolts
- Upper control arm

To install:

3. Install or connect the following:
- Upper control arm. Tighten the pivot bolts to 110–148 ft. lbs. (148–201 Nm).
- Upper ball joint. Tighten the pinch bolt to 56–76 ft. lbs. (76–104 Nm).
- Front wheel

4. If equipped with air suspension, turn the air suspension service switch to the **ON** position.

5. Align the vehicle.

CONTROL ARM BUSHING REPLACEMENT

The control arm bushings are serviced with the control arm as an assembly.

Lower Control Arm

REMOVAL & INSTALLATION

1. If equipped with air suspension, the air suspension switch, located on the right-hand side of the luggage compartment, must be turned to the **OFF** position before raising the vehicle.

2. Remove or disconnect the following:
- Front wheel
- Wheel speed sensor
- Brake caliper and rotor
- Sway bar link
- Shock absorber
- Drag link
- Lower ball joint
- Coil spring

- Lower control arm pivot bolts
- Lower control arm

To install:

3. Install or connect the following:
- Lower control arm. Tighten the pivot bolts to 110–148 ft. lbs. (148–201 Nm).
- Coil spring
- Lower ball joint. Tighten the nut to 110–148 ft. lbs. (148–201 Nm).
- Drag link. Tighten the nut to 35–46 ft. lbs. (47–63 Nm).
- Shock absorber
- Sway bar link
- Brake caliper and rotor
- Wheel speed sensor
- Front wheel

4. If equipped with air suspension, turn the air suspension service switch to the **ON** position.

5. Align the vehicle.

CONTROL ARM BUSHING REPLACEMENT

The control arm bushings are serviced with the control arm as an assembly.

Wheel Bearings

ADJUSTMENT

The front wheel bearings are of a hub unit design and are pre-greased, sealed and require no maintenance. The bearings are preset and cannot be adjusted.

REMOVAL & INSTALLATION

Front

1. If equipped, turn the air suspension service switch to the **OFF** position before raising the vehicle.

2. Remove or disconnect the following
- Front wheel
- Grease cap from the hub
- Disc brake caliper. Suspend the caliper with a length of wire. Do not let it hang from the brake hose. Discard the disc brake caliper mounting bolts.
- Disc brake rotor. If the factory installed push on nuts are installed, remove them first.
- Wheel hub retainer nut and discard
- Hub and bearing assembly

To install:

3. Install or connect the following:
- Hub and bearing assembly
- New wheel hub retainer nut and tighten to 189–254 ft. lbs. (255–345 Nm)
- Disc brake rotor and push on nuts, if equipped
- New grease cap seal
- Disc brake caliper using the 2 new disc brake caliper mounting bolts. Tighten the bolts to 125–170 ft. lbs. (170–230 Nm).
- Wheel. Tighten the lug nuts to 85–104 ft. lbs. (115–142 Nm).

FRONT WHEEL SPINDLE WHEEL HUB
FRONT DISC BRAKE ROTOR SHIELD

7922RG33

Front hub and bearing assembly

4. If equipped with air suspension, turn the air suspension switch to the **ON** position.

5. Pump the brake pedal several times to position the brake pads prior to moving the vehicle.

6. Check the front end alignment.

Wheel Bearings and Hub

REMOVAL & INSTALLATION

1. If equipped with air suspension, the air suspension switch, located on the right hand side of the luggage compartment, must be turned to the **OFF** position before raising the vehicle.

2. Remove or disconnect the following
- Speed sensor connector
- Wheel and tire
- Brake caliper and wire aside
- Brake rotor
- Brake hose retainers

3. Remove the 4 bolts and remove the hub and bearing assembly.

To install:

4. Install hub and bearing assembly and tighten the bolts to 74 ft. lbs (100 Nm).

5. Install or connect the following:
- Brake hose retainers
- Brake rotor
- Brake caliper and tighten to 118 ft. lbs. (160 Nm).
- Wheel and tire
- Speed sensor connector
- New grease cap seal

6. If equipped with air suspension, turn the air suspension switch to the **ON** position.

7. Pump the brake pedal several times to position the brake pads prior to moving the vehicle.

Wheel Knuckle

REMOVAL & INSTALLATION

1. If equipped with air suspension, the air suspension switch, located on the right-hand side of the luggage compartment, must be turned to the **OFF** position before raising the vehicle.

2. Remove or disconnect the following:
- Wheel bearing and hub
- Tie rod from knuckle

3. Raise the suspension arms and place a jack under them to release pressure from the stabilizer bar links.

4. Remove the upper control arm center nut.

5. Remove the stabilizer bar link nut and discard.

6. Disconnect the brake line retainer.

7. Remove the steering knuckle nuts and remove the steering knuckle.

To install:

8. Install or connect the following:
- Steering knuckle and tighten the nuts to 111 ft. lbs. (150 Nm).
- Brake line retainer.
- Stabilizer bar link using a new nut and tighten to 46 ft. lbs. (63 Nm).
- Upper control arm center nut and tighten to 46 ft. lbs. (63 Nm).
- Tie rod to knuckle and tighten to 59 ft. lbs. (80 Nm).
- Wheel bearing and hub

9. If equipped with air suspension, turn the air suspension switch to the **ON** position.

BRAKES

Brake Caliper

REMOVAL & INSTALLATION

Front

➡**Before continuing with this procedure, make sure to have available, 2 new disc brake caliper guide pin bolts and 2 banjo bolt sealing washers, per caliper. Once removed, these parts loose their torque holding ability or retention capability and must not be reused.**

1. If equipped with air suspension, the air suspension switch, located on the right-hand side of the luggage compartment, must be turned to the **OFF** position before raising the vehicle.

2. Remove or disconnect the following:
- Front wheel and tire assembly
- Banjo bolt securing the brake hose from the disc brake caliper. Plug the brake hose. Discard the sealing washers.
- 2 disc brake caliper guide pin bolts and discard. If removing both calipers, mark the right and left sides so they may be reinstalled correctly.

- Disc brake caliper off of the anchor plate

To install:

3. Retract the disc brake caliper piston fully in the piston bore, using an old brake pad or block of wood and a C-clamp.

4. Install or connect the following:
- Disc brake pads to the caliper. Make sure that the brake pad insulators are correctly attached to the brake pad plate.
- Disc brake caliper onto the anchor plate. Make sure the inner and outer pads are properly positioned and the anti-rattle spring is properly positioned. The caliper bleed screw should be positioned on top of the caliper when assembled on the vehicle.
- 2 new caliper guide pin bolts and torque to 21–26 ft. lbs. (28–26 Nm)
- Brake hose, after unplugging it, to the disc brake caliper using 2 new copper sealing washers on the banjo bolt. Torque the bolt to 30–40 ft. lbs. (41–54 Nm).

5. Bleed the brake system, filling the master cylinder as required. Only use

clean DOT 3 brake fluid from a sealed container.

6. Install the wheel and tire assembly. Torque the lug nuts in a star pattern to 85–104 ft. lbs. (115–142 Nm).

7. If equipped with air suspension, turn the air suspension switch to the **ON** position.

8. Pump the brake pedal several times to position the brake pads prior to moving the vehicle.

9. Road test the vehicle and check for proper brake system operation.

Rear

➡**Before continuing with this procedure, make sure to have available, 2 banjo bolt sealing washers, per caliper. Once removed, these parts loose their torque holding ability or retention capability and must not be reused.**

1. If equipped with air suspension, the air suspension switch, located on the right-hand side of the luggage compartment, must be turned to the **OFF** position before raising the vehicle.

2. Remove or disconnect the following:
- Rear wheel and tire assembly

- Banjo bolt securing the brake hose to the disc brake caliper. Plug the brake hose and discard both sealing washers.
- 2 disc brake caliper locating bolts. Lift the disc brake caliper off the rotor and anchor plate using a rotating motion.

To install:

3. Retract the disc brake caliper piston fully in the piston bore, using an old brake pad or block of wood and a C-clamp.

4. Install or connect the following:
- Disc brake pads on the caliper. Make sure that the pads are on the correct side.
- Caliper assembly above the rotor with the anti-rattle spring located on the lower adapter support arm. Install the caliper over the rotor with a rotating motion.

5. Clean the inner surface of the caliper bushings and locating bolts. Lubricate the caliper locating bolts with a suitable silicone dielectric compound. Install and start the locating bolts by hand only. Torque both bolts to 16–20 ft. lbs. (22–27 Nm).
- Brake hose, after unplugging it, to the disc brake caliper using 2 new copper sealing washers on the banjo bolt. Torque the bolt to 30–40 ft. lbs. (41–54 Nm).

6. Bleed the brake system, filling the master cylinder as required. Only use clean DOT 3 brake fluid from a sealed container.

7. Replace the rubber rear disc brake bleeder screw cap.

8. Install the wheel and tire assembly. Torque the lug nuts in a star pattern to 85–104 ft. lbs. (115–142 Nm).

9. If equipped with air suspension, turn the air suspension switch to the **ON** position.

10. Pump the brake pedal several times to position the brake pads prior to moving the vehicle.

11. Road test the vehicle and check for proper brake system operation.

Disc Brake Pads

REMOVAL & INSTALLATION

➡ **Before continuing with this procedure, make sure to have available, 2 new disc brake caliper anchor bracket mounting bolts, per caliper. Once removed, these parts lose their torque holding ability or retention capability and must not be reused.**

Item	Description
1	Bolt (2 Req'd)
2	Disc Brake Caliper
3	Rivet (3 Req'd)
4	Wheel Hub with ABS
5	Washer (2 Req'd)
6	Front Disc Brake Rotor
7	Wheel Hub Without ABS
8	Front Disc Brake Rotor Shield
9	Front Wheel Spindle
10	Front Disc Brake Caliper Anchor Plate Assembly

93006G83

Exploded view of the front disc brake assembly

Front

1. If equipped with air suspension, the air suspension switch, located on the right-hand side of the luggage compartment, must be turned to the **OFF** position before raising the vehicle.

2. Remove or disconnect the following:
   - ½ of the brake fluid from the brake master cylinder reservoir. Properly dispose of the used brake fluid.
   - Front wheel and tire assembly
   - 2 disc brake caliper anchor bracket mounting bolts and discard. Lift the caliper assembly from the disc brake rotor using a rotating motion. Suspend the caliper inside the fender housing with wire. Do not allow the caliper to hang from the brake hose.
   - Inner and outer disc brake pads. Inspect the rotor braking surfaces for scoring and machine as necessary. Refer to the minimum rotor thickness specification when machining. If machining is not necessary, hand-sand the glaze from the braking surfaces with medium grit sandpaper. Make sure to wear an approved respirator.

To install:

3. Use a C-clamp and an old brake pad, wood block to seat the caliper piston in its bore. Do not allow metal or sharp objects to come into direct contact with the plastic caliper piston surface or damage will result.

4. Remove all rust buildup from the inside of the caliper legs.

5. Make sure the anti-rattle spring is seated in the caliper lining inspection opening and that it is installed from the lining side.

6. Install or connect the following:
   - Inner disc brake pad to the caliper piston. Do not bend the pad clips during installation in the piston or distortion and rattles can occur. Install the outer disc brake pad. Make sure the clips are properly seated.
   - Caliper over the rotor and install 2 new anchor bracket mounting bolts. Torque the bolts to 126–169 ft. lbs. (170–230 Nm).
   - Wheel and tire assembly. Torque the lug nuts in a star pattern to 85–104 ft. lbs. (115–142 Nm).

7. If equipped with air suspension, turn the air suspension switch to the **ON** position.

8. Pump the brake pedal prior to moving the vehicle to seat the brake pads.

9. Fill the master cylinder reservoir with clean DOT 3 brake fluid from a closed container.

1 Brake disc
2 Rear wheel brake disc adapter
3 Rear wheel brake disc shield
4 Brake caliper bolt
5 Shoe slippers
6 Brake pads
7 Rear disc brake caliper

67197-CROWN-G07

Exploded view of the rear disc brake assembly

10. If the disc brake calipers were replaced or repaired be sure to bleed the system.

11. Road test the vehicle and check for proper brake system operation.

Rear

1. If equipped with air suspension, the air suspension switch, located on the right-hand side of the luggage compartment, must be turned to the **OFF** position before raising the vehicle.

2. Remove or disconnect the following:
   - ½ of the brake fluid from the brake master cylinder reservoir. Properly dispose of the used brake fluid.
   - Rear wheel and tire assembly
   - 2 disc brake caliper retaining bolts
   - Caliper off the disc brake rotor and anchor plate using a rotating motion
   - Inner and outer disc brake pads

3. Inspect the disc brake rotor for scoring and wear. Inspect the rotor braking surfaces for scoring and machine as necessary. Refer to the minimum rotor thickness specification when machining. If machining is not necessary, hand-sand the glaze from the braking surfaces with medium grit sandpaper. Make sure to wear an approved respirator.

To install:

4. Use a C-clamp and an old brake pad, wood block to seat the caliper piston in its bore. Do not allow metal or sharp objects to come into direct contact with the plastic caliper piston surface or damage will result.

5. Remove all rust buildup from the inside of the caliper legs.

6. Install or connect the following:
   - Inner disc brake pad to the caliper piston. Do not bend the pad clips during installation in the piston or distortion and rattles can occur. Install the outer disc brake pad. Make sure the clips are properly seated.

➡**Make sure the insulators are installed on the brake pads.**

   - Disc brake caliper over the disc brake rotor and install 2 caliper retaining bolts. Torque to 16–20 ft. lbs. (22–27 Nm).
   - Wheel and tire assembly. Torque the lug nuts in a star pattern to 85–104 ft. lbs. (115–142 Nm).

7. If equipped with air suspension, turn the air suspension switch to the **ON** position.

8. Pump the brake pedal prior to moving the vehicle to seat the disc brake pads.

9. Fill the master cylinder reservoir with clean DOT 3 brake fluid from a closed container.

10. If the disc brake calipers were replaced or repaired be sure to bleed the system.

11. Road test the vehicle and check for proper brake system operation.

FORD AND MERCURY

Escape • Escape Hybrid • Mariner

SPECIFICATION AND MAINTENANCE CHARTS

ENGINE AND VEHICLE IDENTIFICATION

Engine							Model Year	
Code ①	Liters	Cu. In.	Cyl.	Fuel Sys.	Engine Type	Eng. Mfg.	Code ②	Year
B	2.0	121	4	SFI	DOHC	Ford	2	2002
Z	2.3	137	4	SFI	DOHC	Ford	3	2003
H	2.3	137	4	SFI/Hybrid	DOHC	Ford	4	2004
1	3.0	182	6	SFI	DOHC	Ford	5	2005

SFI: Multi-port Fuel Injection

DOHC: Double Overhead Camshafts

① 8th digit of VIN

② 10th digit of VIN

06017-ESCA-C01

GENERAL ENGINE SPECIFICATIONS

Year	Model	Engine Displacement Liters	Engine VIN	Net Horsepower @ rpm	Net Torque @ rpm (ft. lbs.)	Bore x Stroke (in.)	Compression Ratio	Oil Pressure @ rpm
2002	Escape	2.0	B	127@5500	135@4500	3.34x3.46	9.6:1	54-80 ①
	Escape	3.0	1	200@5500	200@4500	3.50x3.13	10.0:1	45 ①
2003	Escape	2.0	B	127@5500	135@4500	3.34x3.46	9.6:1	54-80 ①
	Escape	3.0	1	200@5500	200@4500	3.50x3.13	10.0:1	45 ①
2004	Escape	2.0	B	127@5500	135@4500	3.34x3.46	9.6:1	54-80 ①
	Escape	3.0	1	200@5500	200@4500	3.50x3.13	10.0:1	45 ①
2005	Escape/Mariner	2.3	Z	153@5800	152@4250	3.44x3.70	9.7:1	29-30@2000
	Escape Hybrid	2.3	H	133@6000 ②	129@4500	3.44x3.70	12.3:1	29-30@2000
	Escape/Mariner	3.0	1	200@5500	200@4500	3.50x3.13	10.0:1	11@1500 ③

SFI: Multi-port Fuel Injection

① The manufacturer does not provide an engine speed specification for oil pump pressure.

② Combined horsepower of gasoline engine and electric motor: 155 hp

③ Minimum hot

06017-ESCA-C02

ENGINE TUNE-UP SPECIFICATIONS

Year	Engine Displacement Liters	Engine VIN	Spark Plug Gap (in.)	Ignition Timing (deg.) MT	AT	Fuel Pump (psi)	Idle Speed (rpm) MT	AT	Valve Clearance Intake	Exhaust
2002	2.0	B	0.051	10 BTDC	—	65②	①	—	HYD.	HYD.
	3.0	1	0.052-0.056	10 BTDC	10 BTDC	65②	①	①	HYD.	HYD.
2003	2.0	B	0.051	10 BTDC	—	65②	①	—	HYD.	HYD.
	3.0	1	0.052-0.056	10 BTDC	10 BTDC	65②	①	①	HYD.	HYD.
2004	2.0	B	0.051	10 BTDC	—	65②	①	—	HYD.	HYD.
	3.0	1	0.052-0.056	10 BTDC	10 BTDC	65②	①	①	HYD.	HYD.
2005	2.3	Z	0.041-0.045	10 BTDC	10 BTDC	39②	①	①	HYD.	HYD.
	2.3	H	0.049-0.053	—	NA	39②	①	①	HYD.	HYD.
	3.0	1	0.052-0.056	10 BTDC	10 BTDC	39②	①	①	HYD.	HYD.

BTDC: Before Top Dead Center

HYD: Hydraulic lash adjusters

NA: Information not available

① Refer to Vehicle Emission Control Information Label

② Key on; engine off

06017-ESCA-C03

2.0L engine
Firing order: 1–3–4–2
Distributorless ignition

79223G35

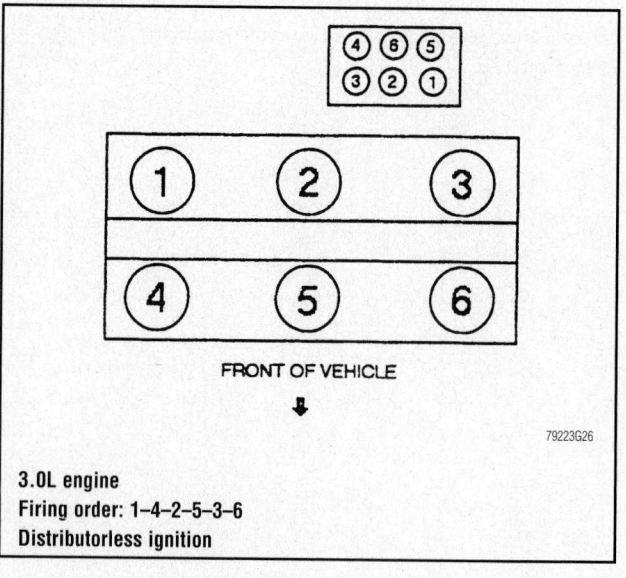

FRONT OF VEHICLE

3.0L engine
Firing order: 1–4–2–5–3–6
Distributorless ignition

79223G26

1 Ignition coil-on-plug electrical connectors
2 Ignition coil-to-valve cover bolts
3 Ignition coils
4 Spark plugs

67197-ESCA-G61

Coil and spark plug arrangement—2.3L engine

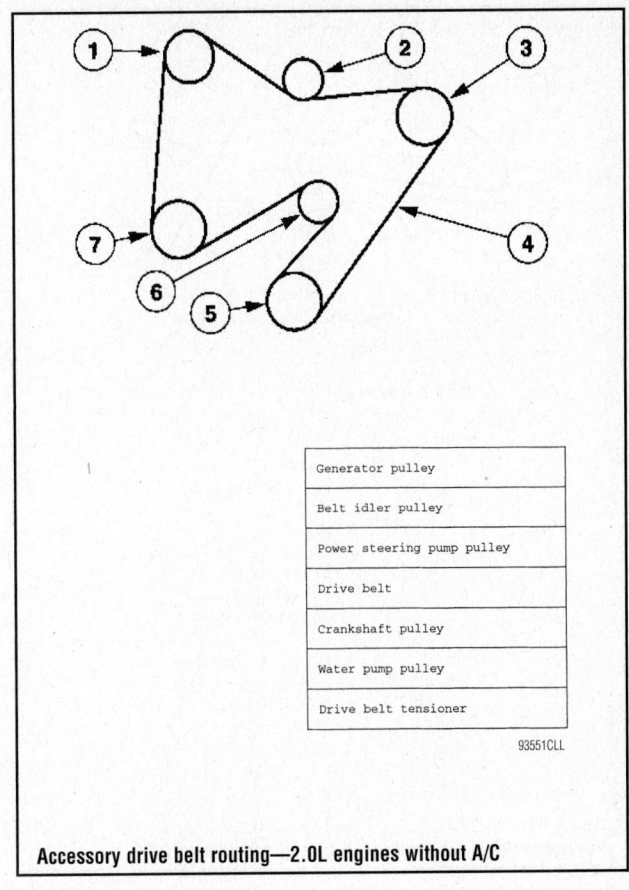

| Generator pulley |
| Belt idler pulley |
| Power steering pump pulley |
| Drive belt |
| Crankshaft pulley |
| Water pump pulley |
| Drive belt tensioner |

93551CLL

Accessory drive belt routing—2.0L engines without A/C

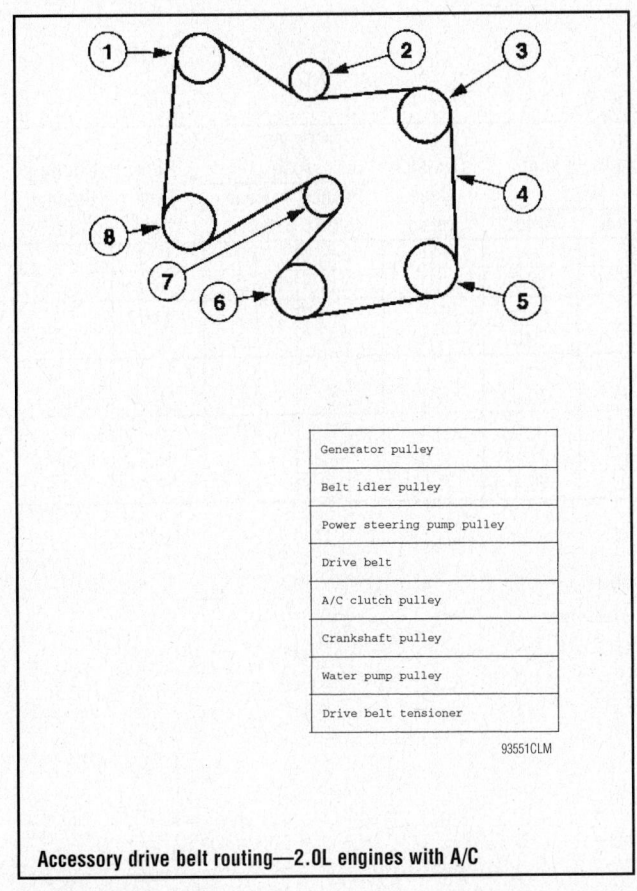

| Generator pulley |
| Belt idler pulley |
| Power steering pump pulley |
| Drive belt |
| A/C clutch pulley |
| Crankshaft pulley |
| Water pump pulley |
| Drive belt tensioner |

93551CLM

Accessory drive belt routing—2.0L engines with A/C

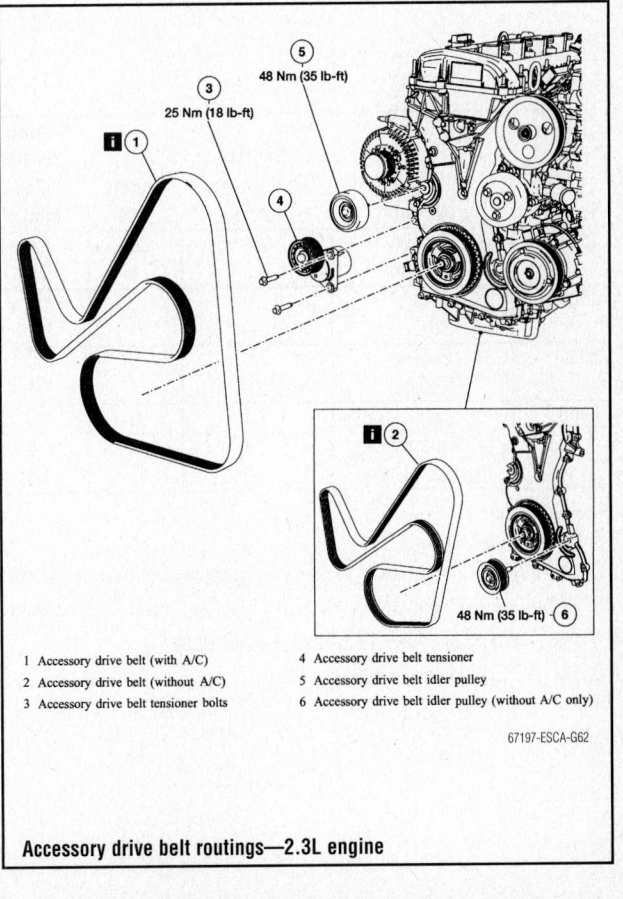

1 Accessory drive belt (with A/C)
2 Accessory drive belt (without A/C)
3 Accessory drive belt tensioner bolts
4 Accessory drive belt tensioner
5 Accessory drive belt idler pulley
6 Accessory drive belt idler pulley (without A/C only)

67197-ESCA-G62

Accessory drive belt routings—2.3L engine

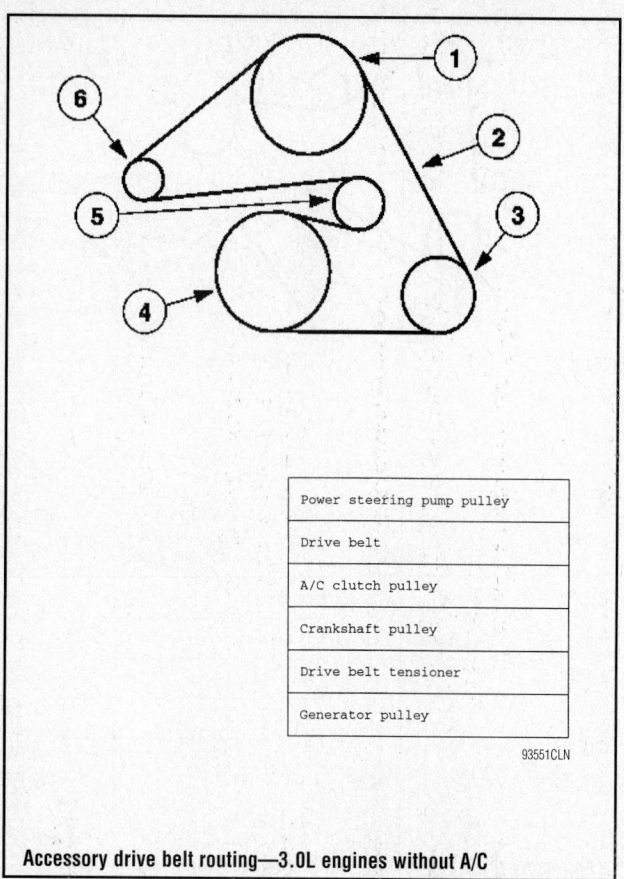

| Power steering pump pulley |
| Drive belt |
| A/C clutch pulley |
| Crankshaft pulley |
| Drive belt tensioner |
| Generator pulley |

93551CLN

Accessory drive belt routing—3.0L engines without A/C

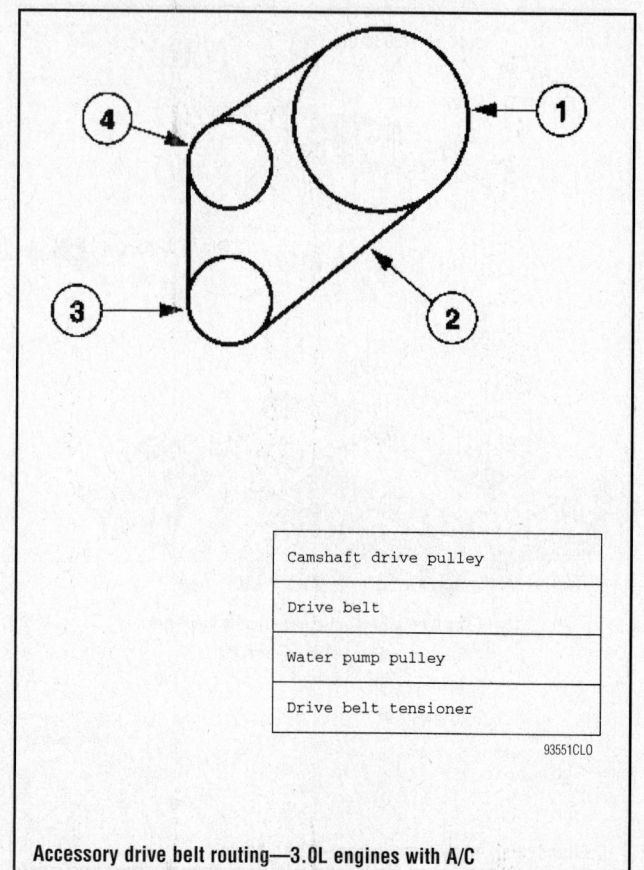

| Camshaft drive pulley |
| Drive belt |
| Water pump pulley |
| Drive belt tensioner |

93551CLO

Accessory drive belt routing—3.0L engines with A/C

CAPACITIES

Year	Model	Engine Displacement Liters	Engine VIN	Engine Oil with Filter (qts.)	Transmission (pts.) Manual	Transmission (pts.) Auto.*	Transfer Case (pts.)	Drive Axle Front (pts.)	Drive Axle Rear (pts.)	Fuel Tank (gal.)	Cooling System (qts.)
2002	Escape	2.0	B	4.5	5.7	—	①	2.6	3.0	15.0	7.0
	Escape	3.0	1	5.5	5.7	20.0	①	2.6	3.0	16.0	10.5
2003	Escape	2.0	B	4.5	5.7	—	—	—	3.0	15.0	7.0
	Escape	3.0	1	5.5	5.7	20.0	①	2.6	3.0	16.0	10.5
2004	Escape	2.0	B	4.5	5.7	—	—	—	3.0	15.0	7.0
	Escape	3.0	1	5.5	5.7	20.0	①	2.6	3.0	16.0	10.5
2005	Escape/Mariner	2.3	B	4.5	5.0	—	—	—	3.0	16.5	②
	Escape Hybrid	2.3	H	4.5	—	10.6 ③	0.75	—	3.0	15.0	8.5
	Escape/Mariner	3.0	1	5.5	5.0	20.4	0.75	2.95	3.0	16.5	10.5

*Dry fill

NA: Information not available

NOTE: All capacities are approximate. Add fluid gradually and check to be sure a proper fluid level is obtained.

① The transfer case is lubricated for life and is not to be checked unless a leak is suspected or a repair is necessary.

② With manual transaxle: 5.3 qts; with automatic transaxle: 6.3 qts.

③ Motor electronics cooling system dry fill: 7.4 pts.

06017-ESCA-C04

VALVE SPECIFICATIONS

Year	Engine Displacement Liters	Engine VIN	Seat Angle (deg.)	Face Angle (deg.)	Spring Test Pressure (lbs. @ in.)	Spring Installed Height (in.)	Stem-to-Guide Clearance (in.)		Stem Diameter (in.)	
							Intake	Exhaust	Intake	Exhaust
2002	2.0	B	45	45	①	1.346	0.0007-0.0025	0.0007-0.0025	0.2374	0.2374
	3.0	1	44.75	45.5	153@ 1.18	1.57	0.0008-0.0027	0.0018-0.0037	0.2352-0.2360	0.2343-0.2350
2003	2.0	B	45	45	①	1.346	0.0007-0.0025	0.0007-0.0025	0.2374	0.2374
	3.0	1	44.75	45.5	153@ 1.18	1.57	0.0008-0.0027	0.0018-0.0037	0.2352-0.2360	0.2343-0.2350
2004	2.0	B	45	45	①	1.346	0.0007-0.0025	0.0007-0.0025	0.2374	0.2374
	3.0	1	44.75	45.5	153@ 1.18	1.57	0.0008-0.0027	0.0018-0.0037	0.2352-0.2360	0.2343-0.2350
2005	2.3	Z	45	45	38.6@1.49	1.496	0.0009	0.0011	0.2153-0.2159	0.2151-0.2157
	2.3	H	45	45	38.6@1.49	1.496	0.0009	0.0011	0.2153-0.2159	0.2151-0.2157
	3.0	1	44.75	45.5	153@ 1.18	1.57	0.0008-0.0027	0.0018-0.0037	0.2352-0.2360	0.2343-0.2350

① Intake: 82.1@ 0.988
 Exhaust: 95@ 1.0275

06017-ESCA-C05

CRANKSHAFT AND CONNECTING ROD SPECIFICATIONS
All measurements are given in inches.

Year	Engine Displacement Liters)	Engine VIN	Crankshaft				Connecting Rod		
			Main Brg. Journal Dia.	Main Brg. Oil Clearance	Shaft End-play	Thrust on No.	Journal Diameter	Oil Clearance	Side Clearance
2002	2.0	B	2.2827-2.2835	①	0.0035-0.0102	3	1.8460-1.8468	0.0006-0.0028	0.0040-0.0110
	3.0	1	2.4791-2.4800	0.0010-0.0018	0.0043-0.0091	3	1.9673-1.9681	0.0011-0.0026	0.0039-0.0118
2003	2.0	B	2.2827-2.2835	①	0.0035-0.0102	3	1.8460-1.8468	0.0006-0.0028	0.0040-0.0110
	3.0	1	2.4791-2.4800	0.0010-0.0018	0.0043-0.0091	3	1.9673-1.9681	0.0011-0.0026	0.0039-0.0118
2004	2.0	B	2.2827-2.2835	①	0.0035-0.0102	3	1.8460-1.8468	0.0006-0.0028	0.0040-0.0110
	3.0	1	2.4791-2.4800	0.0010-0.0018	0.0043-0.0091	3	1.9673-1.9681	0.0011-0.0026	0.0039-0.0118
2005	2.3	Z	2.0460-2.0470	0.0007-0.0013	0.0080-0.0160	NA	1.9673-1.9681	0.0011-0.0026	0.0760-0.1200
	2.3	H	2.0460-2.0470	0.0007-0.0013	0.0080-0.0160	NA	1.9673-1.9681	0.0011-0.0026	0.0760-0.1200
	3.0	1	2.4791-2.4800	0.0010-0.0018	0.0043-0.0091	3	1.9673-1.9681	0.0011-0.0026	0.0039-0.0118

NA: Information not available

① Journals 1, 2 and 4: 0.0010 - 0.0017 in.
 Journal 3: 0.0012 - 0.0019 in.

06017-ESCA-C06

PISTON AND RING SPECIFICATIONS
All measurements are given in inches.

| Year | Engine Displacement Liters | Engine VIN | Piston Clearance | Ring Gap | | | Ring Side Clearance | | |
				Top Compression	Bottom Compression	Oil Control	Top Compression	Bottom Compression	Oil Control
2002	2.0	B	0.0004-0.0012	0.0100-0.0300	0.0100-0.0300	0.0160-0.0660	0.0015-0.0032	0.0015-0.0035	snug
	3.0	1	0.0005-0.0009	0.0039-0.0098	0.0106-0.0165	0.0059-0.0256	0.0016-0.0030	0.0016-0.0033	snug
2003	2.0	B	0.0004-0.0012	0.0100-0.0300	0.0100-0.0300	0.0160-0.0660	0.0015-0.0032	0.0015-0.0035	snug
	3.0	1	0.0005-0.0009	0.0039-0.0098	0.0106-0.0165	0.0059-0.0256	0.0016-0.0030	0.0016-0.0033	snug
2004	2.0	B	0.0004-0.0012	0.0100-0.0300	0.0100-0.0300	0.0160-0.0660	0.0015-0.0032	0.0015-0.0035	snug
	3.0	1	0.0005-0.0009	0.0039-0.0098	0.0106-0.0165	0.0059-0.0256	0.0016-0.0030	0.0016-0.0033	snug
2005	2.3	Z	0.0009-0.0017	0.0060-0.0120	0.0120-0.0180	0.0070-0.0270	NA	NA	NA
	2.3	H	0.0009-0.0017	0.0060-0.0120	0.0120-0.0180	0.0070-0.0270	NA	NA	NA
	3.0	1	0.0005-0.0009	0.0039-0.0098	0.0106-0.0165	0.0059-0.0256	0.0016-0.0030	0.0016-0.0033	snug

NA: Information not available

06017-ESCA-C07

TORQUE SPECIFICATIONS
All readings in ft. lbs.

| Year | Engine Displacement Liters | Engine VIN | Cylinder Head Bolts | Main Bearing Bolts | Rod Bearing Bolts | Crankshaft Damper Bolts | Flywheel Bolts | Manifold | | Spark Plugs | Oil Pan Drain Plug |
								Intake	Exhaust		
2002	2.0	B	①	②	③	80-87	83	13	12	11	18
	3.0	1	④	⑤	⑥	⑦	59	⑧	15	11	NA
2003	2.0	B	①	②	③	80-87	83	13	12	11	18
	3.0	1	④	⑤	⑥	⑦	59	⑧	15	11	NA
2004	2.0	B	①	②	③	85	83	13	12	11	18
	3.0	1	④	⑤	⑥	⑦	59	⑧	15	11	NA
2005	2.3	Z	⑨	NA	NA	⑩	⑪	13	35	11	21
	2.3	H	⑨	NA	NA	⑩	⑪	13	35	11	21
	3.0	1	④	⑤	⑥	⑫	59	⑧	15	11	NA

NA: Information not available

① Step 1: 15 ft. lbs. (20 Nm).
   Step 2: 30 ft. lbs. (40 Nm).
   Step 3: Plus an additional 90 degrees.

② Step 1: 18 ft. lbs.
   Step 2: +60 degrees

③ Step 1: 26 ft. lbs.
   Step 2: +90 degrees

④ Step 1: 30 ft. lbs. (40 Nm).
   Step 2: Tighten the bolts 90 degrees.
   Step 3: Loosen the bolts one full turn.
   Step 4: 30 ft. lbs. (40 Nm).
   Step 5: Tighten the bolts 90 degrees.
   Step 6: Tighten the bolts 90 degrees.

⑤ Step 1: Fasteners 1-8: 18 ft. lbs.
   Step 2: Fasteners 9-19: 30 ft. lbs.
   Step 3: Fasteners 1-16: +90 degrees
   Step 4: fasteners 17-22: 18 ft. lbs.

⑥ Step 1: 17 ft. lbs.
   Step 2: 32 ft. lbs.

⑦ Step 1: 89 ft. lbs.
   Step 2: Loosen 1 full turn
   Step 3: 37 ft. lbs.
   Step 4: 66 ft. lbs.

⑧ 89 inch lbs.

⑨ Step 1: 44 inch lbs.
   Step 2: 11 ft. lbs.
   Step 3: 33 ft. lbs.
   Step 4: +90 degrees
   Step 5: Plus 90 degrees

⑩ Step2: 74 ft. lbs.
   Step 2: plus 90 degrees

⑪ Step 1: 37 ft. lbs.
   Step 2: 50 ft. lbs
   Step 3: 83 ft. lbs.

⑫ Step1: 86 ft. lbs
   Step 2: Loosen 1 full turn
   Step 3: 37 ft. lbs.

06017-ESCA-C08

3.0L engine main bearing (bed plate) torque sequence

06017-ESCA-G102

WHEEL ALIGNMENT

Year	Model		Caster Range (+/-Deg.)	Caster Preferred Setting (Deg.)	Camber Range (+/-Deg.)	Camber Preferred Setting (Deg.)	Toe-in (in.)
2002	ALL	F	NA	+1.93	NA	-0.84	0.12+/-0.12
		R	NA	NA	NA	-0.04	0.09+/-0.11
2003	ALL	F	NA	+1.93	NA	-0.84	0.12+/-0.12
		R	NA	NA	NA	-0.04	0.09+/-0.11
2004①	2.3L.	F	1.00	+1.72	1.00	-0.48	-0.08+/-0.32
		R	NA	NA	1.00	+0.13	0.04+/-0.17
	3.0L	F	1.00	+1.72	1.00	-0.84	0.23+/-0.32
		R	NA	NA	1.00	+0.13	0.10+/-0.17
2005	2.3L.	F	1.00	+1.60	1.00	-1.00	-0.23+/-0.23
		R	NA	NA	0.75	②	0.03+/-0.20
	2.3L Hybrid	F	1.00	+1.60	1.00	-1.00	-0.23+/-0.23
		R	NA	NA	0.75	②	0.03+/-0.20
	3.0L	F	1.00	+1.60	1.00	-1.00	0.23+/-0.32
		R	NA	NA	0.75	②	0.03+/-0.20

Early build vehicles are those built before 8/02/04 at Kansas City, or, before 8/16/04 at Ohio.

NA: Information not available

① Assumes 8 gallons of gas

② Early build: +20 degrees
Late build: 0

06017-ESCA-C09

TIRE, WHEEL AND BALL JOINT SPECIFICATIONS

| Year | Model | OEM Tires | | Tire Pressures (psi) | | Wheel Size | Ball Joint Inspection | Lug Nuts (ft. lbs.) |
		Standard	Optional	Front	Rear			
2002	Escape	P215/70R16	P225/70SR15 P235/70R16	①	①	NA	0.030 in.	98
2003	Escape	P215/70R16	P225/70SR15 P235/70R16	①	①	NA	0.030 in.	98
2004	Escape XLS	P225/70SR15	none	①	①	6.5	0.030 in.	98
	Escape XLT	P235/70R16	none	①	①	7.5	0.030 in.	98
2005	Escape XLS Value	P225/70R15	none	①	①	6.5	0.030 in.	98
	Hybrid	P235/70R16	none	①	①	7.5	0.030 in.	98
	XLT	P235/70R16	none	①	①	7.5	0.030 in.	98
	Limited	P235/70R16	none	①	①	7.5	0.030 in.	98

OEM: Original Equipment Manufacturer

PSI: Pounds Per Square Inch

STD: Standard

OPT: Optional

06017-ESCA-C10

BRAKE SPECIFICATIONS
All measurements in inches unless noted

| Year | Model | | Brake Disc | | | Brake Drum | | Minimum Lining Thickness | Brake Caliper | |
			Original Thickness	Minimum Thickness	Maximum Run-out	Original Inside Diameter	Maximum Machine Diameter		Bracket Bolts (ft. lbs.)	Mounting Bolts (ft. lbs.)
2002	Escape		0.940	0.860	0.004	9.00	9.06	0.039	111	26
2003	Escape		0.940	0.860	0.002	9.00	9.06	0.039	111	26
2004	Escape		0.940	0.860	0.002	9.00	9.06	0.039	111	26
2005	Escape/Mariner	F	NA	①	0.004	—	—	0.118	111	②
		R	NA	0.430	0.004	NA	9.05	0.118	—	26

NA: Information not available

① Base brakes: 0.86 in.

   With 4-wheel discs: 0.95 in.

② With disc/drum: 26 ft. lbs.

   With 4-wheel disc: 33 ft. lbs.

   Escape Hybrid: 33 ft. lbs.

06017-ESCA-C11

SCHEDULED MAINTENANCE INTERVALS
2002-03 Ford Escape

TO BE SERVICED	TYPE OF SERVICE	VEHICLE MILEAGE INTERVAL (x1000)												
		5	10	15	20	25	30	35	40	45	50	55	60	65
Air cleaner filter	R						✓						✓	
Accessory drive belt	S/I	Every 100,000 miles												
Accessory drive belt	R	At 120,000 miles if not previously done so												
Auto. Trans. Fluid level	I			✓			✓			✓			✓	
Auto. Trans. Fluid	②						✓						✓	
Auto. Trans. Fluid	③	Every 150,000 miles												
Brake system ①	S/I			✓			✓			✓			✓	
Cabin air filter	R			✓			✓			✓			✓	
Camshaft belt (2.0L)	R	Every 120,000 miles												
Cooling system hoses and clamps	S/I			✓			✓			✓			✓	
Engine coolant (green coolant)	R	Every 75,000 miles and every 30,000 miles thereafter												
Engine coolant (yellow coolant)	R	At 5 years or 100,000 miles												
Engine oil & filter	R	✓	✓	✓	✓	✓	✓	✓	✓	✓	✓	✓	✓	✓
PCV valve	R	Every 100,000 miles												
Exhaust system & heat shields	S/I						✓						✓	
Fuel filter	R						✓						✓	
Rear axle lubricant	R	Every 150,000 miles												
Rotate tires	S/I	✓	✓	✓	✓	✓	✓	✓	✓	✓	✓	✓	✓	✓
Steering linkage	S/I			✓			✓			✓			✓	
Spark plugs	R	Change at 100,000 miles												
Suspension components	S/I			✓			✓			✓			✓	
Wheels for play and noise	I			✓			✓			✓			✓	

R: Replace   S/I: Inspect and service, if necessary   L: Lubricate   A: Adjust   C: Clean

① Inspect the reservoir fluid level, rotor and or drum, brake lines, hoses, calipers and or wheel cylinders

② Change automatic transmission/transaxle fluid on all vehicles equipped with AX4S, 4F50N, 4R100, 4F27E

Special Operating Condition Requirements

When towing a trailer or using a camper or car-top carrier:

Change engine oil and install a new oil filter every 4,800 km (3,000 miles) or 3 months.

Change transfer case fluid every 96,000 km (60,000 miles).

Change manual transmission fluid as required.

Inspect and lubricate U-joints as required.

During extensive idling and/or low speed driving for long distances, as in heavy commercial use such as delivery, taxi, patrol car or livery:

Change engine oil and install a new oil filter, lube front lower control arm and steering linkage ball joints with

zerk fittings (if equipped) every 4,800 km (3,000 miles) or 3 months.

Inspect brake system and check battery electrolyte level (Patrol cars) every 8,000 km (5,000 miles).

Install a new fuel filter every 24,000 km (15,000 miles).

Change automatic transmission fluid, lubricate 4x2 wheel bearings,

install new grease seals and adjust bearings every 48,000 km (30,000 miles).

Install new spark plugs and change transfer case fluid every 96,000 km (60,000 miles).

Install a new cabin air filter as required.

When operating in dusty conditions such as unpaved or dusty roads:

Change engine oil and install a new oil filter every 4,800 km (3,000 miles) or 3 months.

Install a new fuel filter every 24,000 km (15,000 miles).

Change automatic transmission fluid every 48,000 km (30,000 miles).

Change transfer case fluid every 96,000 km (60,000 miles).

Install a new engine air filter as required.

Install a new cabin air filter as required.

When operating in off-road conditions:

Change automatic transmission fluid every 48,000 km (30,000 miles).

Change transfer case fluid every 96,000 km (60,000 miles).

Install a new cabin air filter as required.

Inspect and lubricate U-joints.

Inspect and lubricate steering linkage ball joints with zerk fittings.

SCHEDULED MAINTENANCE INTERVALS
2004 Ford Escape

TO BE SERVICED	TYPE OF SERVICE	VEHICLE MILEAGE INTERVAL (x1000)												
		5	10	15	20	25	30	35	40	45	50	55	60	65
Air cleaner filter	R						✓						✓	
Accessory drive belt	S/I	Every 100,000 miles												
Accessory drive belt	R	At 150,000 miles if not previously done so												
Auto. Trans. Fluid level	I			✓			✓			✓			✓	
Auto. Trans. Fluid	②						✓						✓	
Auto. Trans. Fluid	③	Every 150,000 miles												
Ball joints (2wd)	L			✓			✓			✓			✓	
Brake system ①	S/I			✓			✓			✓			✓	
Cabin air filter	R			✓			✓			✓			✓	
Camshaft belt (2.0L)	R	Every 120,000 miles												
Cooling system hoses and clamps	S/I			✓			✓			✓			✓	
Engine coolant (exc. Premium gold)	R	Every 105,000 miles, then every 3 years or 50,000 miles												
Engine coolant (Premium gold)	R	At 5 years or 100,000 miles												
Engine oil & filter	R	✓	✓	✓	✓	✓	✓	✓	✓	✓	✓	✓	✓	✓
PCV valve (external)	R	Every 100,000 miles												
Exhaust system & heat shields	S/I						✓						✓	
Front wheel bearings and seals	R	Every 150,000 miles, if not previously done so												
Fuel filter	R						✓						✓	
Rear axle lubricant	R	Every 150,000 miles												
Rotate tires	S/I	✓	✓	✓	✓	✓	✓	✓	✓	✓	✓	✓	✓	✓
Steering linkage	S/I			✓			✓			✓			✓	
Spark plugs	R	Change at 100,000 miles												
Suspension components	S/I			✓			✓			✓			✓	
Wheels for play and noise	I			✓			✓			✓			✓	

R: Replace   S/I: Inspect and service, if necessary   L: Lubricate   A: Adjust   C: Clean

① Inspect the reservoir fluid level, rotor and or drum, brake lines, hoses, calipers and or wheel cylinders

② Change automatic transmission/transaxle fluid and filter on all vehicles equipped with AX4S, 4F50N, 4R100, 4F27E.

③ All transaxles

Special Operating Condition Requirements

When towing a trailer or using a camper or car-top carrier:

Change engine oil and install a new oil filter every 4,800 km (3,000 miles) or 3 months.

Change transfer case fluid every 96,000 km (60,000 miles).

Change manual transmission fluid as required.

Inspect and lubricate U-joints as required.

During extensive idling and/or low speed driving for long distances, as in heavy commercial use such as delivery, taxi, patrol car or livery:

Change engine oil and install a new oil filter, lube front lower control arm and steering linkage ball joints with

zerk fittings (if equipped) every 4,800 km (3,000 miles) or 3 months.

Inspect brake system and check battery electrolyte level (Patrol cars) every 8,000 km (5,000 miles).

Install a new fuel filter every 24,000 km (15,000 miles).

Change automatic transmission fluid, lubricate 4x2 wheel bearings,

install new grease seals and adjust bearings every 48,000 km (30,000 miles).

Install new spark plugs and change transfer case fluid every 96,000 km (60,000 miles).

Install a new cabin air filter as required.

When operating in dusty conditions such as unpaved or dusty roads:

Change engine oil and install a new oil filter every 4,800 km (3,000 miles) or 3 months.

Install a new fuel filter every 24,000 km (15,000 miles).

Change automatic transmission fluid every 48,000 km (30,000 miles).

Change transfer case fluid every 96,000 km (60,000 miles).

SCHEDULED MAINTENANCE INTERVALS
2004 Ford Escape
Footnotes Continued

Install a new engine air filter as required.

Install a new cabin air filter as required.

When operating in off-road conditions:

Change automatic transmission fluid every 48,000 km (30,000 miles).

Change transfer case fluid every 96,000 km (60,000 miles).

Install a new cabin air filter as required.

Inspect and lubricate U-joints.

Inspect and lubricate steering linkage ball joints with zerk fittings.

06017-ESCA-C15

SCHEDULED MAINTENANCE INTERVALS
2005 Ford Escape/Merucry Mariner, exc. Escape Hybrid

TO BE SERVICED	TYPE OF SERVICE	VEHICLE MILEAGE INTERVAL (x1000)												
		5	10	15	20	25	30	35	40	45	50	55	60	65
Air cleaner filter	R						✓						✓	
Accessory drive belt	I ⑤	Every 100,000 miles												
Auto. Trans. fluid level	I			✓			✓			✓			✓	
Auto. Trans. Fluid	③ ④						✓						✓	
Ball joints (2wd)	L			✓			✓			✓			✓	
Brake system ①	S/I			✓			✓			✓			✓	
Cabin air filter	R			✓			✓			✓			✓	
Cooling system hoses and clamps	S/I			✓			✓			✓			✓	
Driveshafts & halfshafts	S/I			✓			✓			✓			✓	
Engine coolant (Premium Gold)	R	Five years or 100,000 miles, then every 3 years or 50,000 miles												
Engine coolant (exc. Premium Gold)	R	Every 105,000 miles												
Engine oil & filter	R	✓	✓	✓	✓	✓	✓	✓	✓	✓	✓	✓	✓	✓
Front wheel bearings and seals (2wd)	R	Every 150,000 miles, if not previously done												
Fuel filter	R						✓						✓	
Man. Trans. Fluid	R	Every 120,000 miles												
PCV valve	S/I	Every 100,000 miles												
Exhaust system & heat shields	S/I						✓						✓	
Rear axle lubricant (4wd)	R	Every 150,000 miles												
Rotate tires	S/I	✓	✓	✓	✓	✓	✓	✓	✓	✓	✓	✓	✓	✓
Steering linkage	S/I			✓			✓			✓			✓	
Spark plugs	R	Change at 100,000 miles												
Suspension components	S/I			✓			✓			✓			✓	
Wheels ②	I			✓			✓			✓			✓	

R: Replace   S/I: Inspect and service, if necessary   L: Lubricate   A: Adjust   C: Clean

① Inspect the reservoir fluid level, rotor and or drum, brake lines, hoses, calipers and or wheel cylinders

② Inspect for end play and noise

③ Change automatic transmission/transaxle fluid and filter on all vehicles equipped with 4F50N, 4R100 and 4F27E.

④ Change every 150,000 miles for all transaxles

⑤ Replace at 150,000 miles, if not previously done

Special Operating Condition Requirements

When towing a trailer or using a camper or car-top carrier:

Change engine oil and install a new oil filter every 4,800 km (3,000 miles) or 3 months.

Change transfer case fluid every 96,000 km (60,000 miles).

Change manual transmission fluid as required.

Inspect and lubricate U-joints as required.

During extensive idling and/or low speed driving for long distances, as in heavy commercial use such as delivery, taxi, patrol car or livery:

Change engine oil and install a new oil filter, lube front lower control arm and steering linkage ball joints with

zerk fittings (if equipped) every 4,800 km (3,000 miles) or 3 months.

Inspect brake system and check battery electrolyte level (Patrol cars) every 8,000 km (5,000 miles).

Install a new fuel filter every 24,000 km (15,000 miles).

Change automatic transmission fluid, lubricate 4x2 wheel bearings,

install new grease seals and adjust bearings every 48,000 km (30,000 miles).

Install new spark plugs and change transfer case fluid every 96,000 km (60,000 miles).

Install a new cabin air filter as required.

06017-ESCA-C16

SCHEDULED MAINTENANCE INTERVALS
2005 Ford Escape/Merucry Mariner, exc. Escape Hybrid
Footnotes Continued

When operating in dusty conditions such as unpaved or dusty roads:

Change engine oil and install a new oil filter every 4,800 km (3,000 miles) or 3 months.

Install a new fuel filter every 24,000 km (15,000 miles).

Change automatic transmission fluid every 48,000 km (30,000 miles).

Change transfer case fluid every 96,000 km (60,000 miles).

Install a new engine air filter as required.

Install a new cabin air filter as required.

When operating in off-road conditions:

Change automatic transmission fluid every 48,000 km (30,000 miles).

Change transfer case fluid every 96,000 km (60,000 miles).

Install a new cabin air filter as required.

Inspect and lubricate U-joints.

Inspect and lubricate steering linkage ball joints with zerk fittings.

06017-ESCA-C17

SCHEDULED MAINTENANCE INTERVALS
2005 Ford Escape Hybrid

TO BE SERVICED	TYPE OF SERVICE	Vehicle mileage intervals (x1000)											
		10	20	30	40	50	60	70	80	90	100	110	120
Air cleaner filter	R			✓			✓			✓			✓
Accessory drive belt	I ①										✓		
Brake system ②	S/I		✓		✓		✓		✓		✓		✓
Cooling system hoses and clamps	S/I		✓		✓		✓		✓		✓		✓
Driveshafts & halfshafts	S/I		✓		✓		✓		✓		✓		✓
Engine coolant	R	At 5 years or 100,000 miles; then every 3 years or 50,000 miles											
Engine oil & filter	R	✓	✓	✓	✓	✓	✓	✓	✓	✓	✓	✓	✓
Exhaust system & heat shields				✓					✓				✓
Fuel filter	R										✓		
PCV valve	S/I	Every 150,000 miles											
Rear axle lubricant (4wd)	R	Every 150,000 miles											
Rear (high voltage) battery A/C filter	I	✓	✓	✓	✓	✓	✓	✓	✓	✓	✓	✓	✓
Rear (high voltage) battery A/C filter	R		✓		✓		✓		✓		✓		✓
Tires	Rotate	Every 5,000 miles											
Steering linkage	S/I		✓		✓		✓		✓		✓		✓
Spark plugs	R										✓		
Suspension components and ball joints	S/I		✓		✓		✓		✓		✓		✓

R: Replace   S/I: Inspect and service, if necessary   L: Lubricate   A: Adjust   C: Clean

① Replace at 150,000 miles, if not previously done

② Inspect the reservoir fluid level, rotor and or drum, brake lines, hoses, calipers and or wheel cylinders

Monthly Checks

Check each of the following items every month:

All interior and exterior lights

Tires for wear and correct air pressure, including spare tire

Engine oil fluid level

Windshield washer solvent fluid level

Six Month Checks

Check each of the following items at least every 6 months:

Lap/shoulder belts and seat latches for wear and function

Parking brake for correct operation

Safety warning lamps (brake, ABS, air bag, safety belt) for correct operation

Engine coolant system fluid level and correct strength

Motor/electrical cooling system fluid level and correct strength

Battery 12-volt connections. Clean if necessary

Windshield washer spray, wiper operation, clean all wiper blades

Lubricate all hinges, latches and outside locks. Inspect for correct operation

Lubricate door rubber weatherstrips. Inspect for excessive wear

Clean body and door drain holes. Inspect for clogs and obstructions

06017-ESCA-C18

ESCAPE HYBRID HIGH VOLTAGE ELECTRICAL SYSTEM

Precautions

Before working on any part of the Escape Hybrid high voltage system, observe the following precautions:

❊ CAUTION

The nominal high voltage traction battery voltage is 330 volts DC. The buffer zone must be set up and insulated rubber gloves and a face shield must be worn. Failure to follow these instructions may result in severe injury or death.

❊ CAUTION

The high voltage traction battery and charging system contains high voltage components and wiring. High voltage insulated safety gloves and a face shield must be worn when carrying out any diagnostics on this vehicle. Failure to follow these instructions may result in severe personal injury or death.

❊ CAUTION

Before carrying out any removal and installation procedures of the high voltage traction battery system, the high voltage traction battery must be de-powered. Failure to follow these instructions may result in severe personal injury or death.

❊ CAUTION

The rubber insulating gloves that are to be worn while working on the high voltage system should be of the appropriate safety and protection rating for use on the high voltage system. They must be inspected before use and must always be worn in conjunction with the leather outer gloves. Any hole in the rubber insulating glove is a potential entry point for high voltage. Failure to follow these instructions may result in severe personal injury or death.

➡The high voltage insulated safety gloves must be re-certified every 6 months to remain within Occupational Safety and Health Administration (OSHA) guidelines:

• Roll the glove up from the open end until the lower portion of the glove begins to balloon from the resulting air pressure. If the glove leaks any air, it must not be used.

• The gloves should not be used if they exhibit any signs of wear and tear.

• The leather gloves must always be worn over the rubber insulating gloves in order to protect them.

• The rubber insulating gloves must be class "00" and meet all of the American Society for Testing and Materials (ASTM) standards

❊❊ CAUTION

High voltage insulated safety gloves and a face shield must be worn when working with high voltage cables. The ignition switch must be OFF for a minimum of 5 minutes before removing high voltage cables. Failure to follow these instructions may result in severe personal injury or death.

❊❊ CAUTION

Establish a buffer zone before servicing the high voltage system. The buffer zone is required only when working with the high voltage system. See the text for buffer zone establishment. Failure to follow these instructions may result in severe personal injury or death. Do not allow any unauthorized personnel into the buffer zone during repairs involving the high voltage system. Only personnel trained for repair on the high voltage system are to be permitted in the buffer zone.

❊❊ CAUTION

De-power the high voltage traction battery (HVTB) before working on the high voltage system. See the text for the de-powering procedure. Failure to follow these instructions may result in severe personal injury or death.

Buffer Zone

1. Before servicing the vehicle, refer to the Precautions Section.

❊❊ CAUTION

Before proceeding, read and observe all of the High Voltage System Precautions.

2. Establish a buffer zone around the vehicle:
a. Position the vehicle in the repair bay.

06017-ESCA-G04

Buffer zone—Escape Hybrid

b. Position 4 orange cones at the corners of the vehicle to mark off a 1 m (3 ft.) perimeter around the vehicle.

c. Do not allow any unauthorized personnel into the buffer zone during repairs involving the high voltage system. Only personnel trained for repair on the high voltage system are to be permitted in the buffer zone.

De-powering the High Voltage Traction Battery

1. Before servicing the vehicle, refer to the Precautions Section.

❊❊ CAUTION

Before proceeding, read and observe all of the High Voltage System Precautions.

2. Establish a buffer zone. See the procedure above. Do not allow any unauthorized personnel into the buffer zone during repairs involving the high voltage system. Only personnel trained for repair on the high voltage system are to be permitted in the buffer zone.

3. De-power the high voltage traction battery (HVTB).
a. Rotate the service disconnect plug from the LOCK (1) position to the UNLOCK (2) position.
b. Remove the service disconnect plug and place in the SERVICING SHIPPING (3) position.

❊❊ WARNING

Place the service disconnect plug into the SERVICING SHIPPING position while the high voltage traction battery (HVTB) is being removed and/or while the high voltage system is having repairs carried out. If the service disconnect plug is left out and placed on the bench or toolbox, dirt or other contaminants may enter the HVTB, which can cause damage.

06017-ESCA-G05

Rotate the service disconnect plug from the LOCK (1) position to the UNLOCK (2) position—
Escape Hybrid

4. Insert the service disconnect plug into the SERVICING SHIPPING position. This disconnects the HVTB.

5. To re-power, reverse the de-power procedure.

High Voltage Traction Battery

REMOVAL & INSTALLATION

1. Before servicing the vehicle, refer to the Precautions Section.

❋❋ CAUTION

Before proceeding, read and observe all of the High Voltage System Precautions.

2. De-power the high voltage traction battery. Establish a buffer zone. See the procedure under De-power the High Voltage Traction Battery.

3. Position the carpet aside.

➡**When installing, tighten the screws on the HVTB first or an air flow loss to the HVTB may occur.**

4. Remove the 5 A/C return duct assembly screws.

5. Remove the A/C return duct assembly.

➡**Due to clearance issues, the 6-pin connector must be disconnected first during the removal process and connected last during the installation process.**

6. From the left rear door opening, fold the left rear seat backrest down and disconnect the 6-pin low voltage electrical connector.

➡**Due to clearance issues, the 40-pin connector must be disconnected last during the removal process and connected first during the installation process.**

7. Loosen the bolt and disconnect the 40-pin low voltage connector.

8. From the right rear door opening, fold the right rear seat cushion forward and remove the 2 high voltage cables shield nuts (access the shield nuts through the slotted opening in the carpet).

9. Fold the right rear seat backrest down and remove the high voltage cables shield bolt.

10. Remove the high voltage cables shield plastic rivet/screw.

11. Remove the high voltage cables shield.

12. Press the locking tab down and rotate the locking lever upward until the aligning dowels are disengaged from the locking lever to remove the high voltage cables electrical connector.

➡The attaching bolts have a conductive coating on them and are serrated under the head flange. These features ground the HVTB to the vehicle, which is required for electro-magnetic compatibility (EMC). The serration also helps the grounding effect. If a bolt(s) is lost or damaged, it must be replaced with the identical type of bolt.

13. Remove the 9 HVTB bolts.

➡The 2 front lift points are the eyelets on each front corner and the rear lift point is beneath the cap plug in the center rear of the HVTB.

➡Remove the cap plug to expose the center (rear) lifting attachment point.

➡Attach 3 M10 x 1.5 x 35 eyebolts to the 3 HVTB lift points. Obtain the eyebolts locally.

➡Make certain the HVTB does not mar or damage the interior panels during removal. There is only 6 mm (0.23 in) clearance on each side. Cover the battery mounting brackets with protective padding.

➡Do not strike the headliner with the HVTB (or floor crane) during removal.

➡Inspect the HVTB tray drain grommet. Replace it if necessary.

➡Inspect the 10 HVTB cushions for damage and proper placement. Replace them if necessary.

➡Failure to install the rear lift eye cap plug may result in noise, vibration, and harshness (NVH) issues.

14. With an assistant, attach a chain or suitable lifting device to the 3 lift points and lift the HVTB off the 2 alignment dowels using a floor crane. Remove the HVTB from the vehicle.

15. To install, reverse the removal procedure. To re-power, reverse the de-power procedure. Observe the following torques:

- The 9 HVTB bolts. Hand-start all of the bolts before tightening them to: 55 Nm (41 ft. lbs.)
- The high voltage cables shield bolt: 9 Nm (80 inch lbs.).
- The 2 high voltage cables shield nuts: 9 Nm (80 inch lbs.)
- The 40-pin low voltage connector: 5 Nm (44 inch lbs.)
- The 5 A/C return duct assembly screws: 2 Nm (18 inch lbs.)

1 A/C return duct assembly
   screws (5 required)

2 A/C return duct assembly

3 6-pin low voltage electrical
   connector

4 40-pin low voltage electrical
   connector

5 High voltage cables shield
   nuts (2 required)

6 High voltage cables shield
   bolt

7 High voltage cables shield
   plastic rivet/screw

8 High voltage cables shield

9 High voltage cables electrical
   connector

10 High voltage traction battery
    bolts (9 required)

11 High voltage traction battery

High voltage traction battery—Escape Hybrid

06017-ESCA-G06

High Voltage Cables

REMOVAL & INSTALLATION

1. Before servicing the vehicle, refer to the Precautions Section.

✳✳ CAUTION

Before proceeding, read and observe all of the High Voltage System Precautions.

2. De-power the high voltage traction battery. Establish a buffer zone. See the procedure under De-powering the High Voltage Traction Battery.

3. From the right rear door opening, fold the right rear seat cushion forward

1 High voltage cables shield nuts (2 required)

2 High voltage cables shield bolt

3 High voltage cables shield plastic rivet/screw

4 High voltage cables shield

5 High voltage cables electrical connector

6 High voltage cables floorpan grommet

7 Park brake cable bracket bolts (2 required)

8 Park brake cable bracket

9 High voltage cables conduit nuts (right side inner frame rail and floorpan) (5 required)

10 High voltage cables bracket nuts (bulkhead) (2 required)

11 High voltage cables-to-electronically controlled continuously variable transmission (eCVT) bolt

12 High voltage cables connection to eCVT

13 High voltage cables

06017-ESCA-G07

and remove the 2 high voltage cables shield nuts (access the shield nuts through the slotted opening in the carpet).

4. Fold the right rear seat backrest down and remove the high voltage cables shield bolt.

5. Remove the high voltage cables shield plastic rivet/screw.

6. Remove the high voltage cables shield.

7. Press the locking tab down and rotate the locking lever upward until the aligning dowels are disengaged from the locking lever to remove the high voltage cables electrical connector.

8. With the vehicle in NEUTRAL, position it on a hoist.

9. Remove the high voltage cables floor pan grommet.

10. Remove the 2 bolts and the park brake cable bracket.

11. Remove the 5 high voltage cables conduit nuts (right side inner frame rail and floor pan).

12. Lower the vehicle.

13. Remove the 2 high voltage cables bracket nuts (bulkhead). (Access the nuts through the engine compartment.)

14. Remove the nut and the eCVT shift cable.

15. Remove the high voltage cables-to-eCVT bolt.

16. Rotate the locking lever rearward until the alignment dowels are disengaged from the locking lever and remove the high voltage cables connection to the eCVT.

17. To install, reverse the removal procedure. Observe the following torques:
- The high voltage cables-to-eCVT bolt: 9 Nm (80 lb-in)
- The eCVT shift cable: 8 Nm (71 lb-in)
- The 2 high voltage cables bracket nuts: 8 Nm (71 lb-in)
- The 5 high voltage cables conduit nuts: 8 Nm (71 lb-in)
- The park brake cable bracket: 8 Nm (71 lb-in).
- The high voltage cables shield bolt: 9 Nm (80 lb-in).
- The 2 high voltage cables shield nuts: 9 Nm (80 lb-in)

Direct Current/Alternating Current Inverter

REMOVAL & INSTALLATION

1. Before servicing the vehicle, refer to the Precautions Section.

1 Electrical connector

2 Electrical connector

3 DC/AC inverter nuts (2 required)

4 DC/AC inverter

06017-ESCA-G08

DC/AC Inverter—Escape Hybrid

✳✳ CAUTION

Before proceeding, read and observe all of the High Voltage System Precautions.

2. Disconnect the 12 volt battery.

➡**For the Escape Hybrid, when the battery is disconnected and connected, the brake pedal needs to be calibrated. After the battery has been connected, with the vehicle in park, turn the key to the ON position. Press the brake pedal firmly, then fully release to calibrate the brake pedal.**

3. Disconnect the electrical connectors.

4. Remove the 2 DC/AC inverter nuts.

5. Remove the DC/AC inverter.

6. To install, reverse the removal procedure. Tighten the nuts to 9 Nm (80 inch lbs.).

Direct Current/Direct Current Converter

REMOVAL & INSTALLATION

1. Before servicing the vehicle, refer to the Precautions Section.

✳✳ CAUTION

Before proceeding, read and observe all of the High Voltage System Precautions.

2. De-power the high voltage traction battery. Establish a buffer zone. See the procedure under De-powering the High Voltage Traction Battery.

3. Disconnect the 12 volt battery.

1 DC/DC converter low voltage
  battery cable nuts (2 required)

2 DC/DC converter electrical
  connector

3 DC/DC converter nuts (3
  required)

4 DC/DC converter assembly

06017-ESCA-G09

DC/DC converter—Escape Hybrid

➡For the Escape Hybrid, when the battery is disconnected and connected, the brake pedal needs to be calibrated. After the battery has been connected, with the vehicle in park, turn the key to the ON position. Press the brake pedal firmly, then fully release to calibrate the brake pedal.

4. With the vehicle in NEUTRAL, position it on a hoist.

5. Drain the cooling system.

❄❄ WARNING

Always refill the motor electronics cooling system (MECS) with the same type of coolant that was drained from the system. Do not mix coolant types.

6. Drain the motor electronics cooling system:

a. With the vehicle in DRIVE, position it on a hoist.

b. Turn the ignition to the OFF position.

c. Remove the left splash shield.

d. Place a suitable container below the transaxle.

e. Loosen the hose clamps at the transaxle, then pull the hoses off to allow the coolant to drain.

7. Disconnect the motor electronics coolant vent hose.

8. Disconnect the 2 engine coolant vent hoses.

9. Remove the degas bottle bolt and nut.

10. Lift the degas bottle up and disconnect the engine coolant and motor electronics coolant return hoses.

11. Remove the degas bottle.

12. Disconnect the 2 female pin-type retainers.

➡Press the locking clip to release the connector.

13. Disconnect the DC/DC converter high voltage connector.

14. Clamp the motor/electronics coolant (MECT) hoses to prevent coolant from leaking from the hoses during the repair.

15. Loosen the hose clamps and remove the MECT hoses from the DC/DC converter.

16. Remove the 2 DC/DC converter low voltage battery cable nuts and remove the low voltage battery cables.

17. Disconnect the DC/DC converter low voltage electrical connector.

18. Remove the 3 DC/DC converter nuts.

19. Remove the DC/DC converter assembly.

20. To install, reverse the removal procedure. Re-power the high voltage battery. Observe the following torques:

• The 3 DC/DC converter nuts: 12 Nm (9 ft. lbs.)

4 — 6 Nm (53 lb-in)

6 Nm
(53 lb-in)

06017-ESCA-G10

1 Motor electronics coolant vent hose and clamp

2 Engine coolant vent hose and clamp

3 Engine coolant vent hose and clamp

4 Degas bottle-to-bracket bolt

5 Degas bottle-to-fender nut

6 Engine coolant return hose and clamp

7 Motor electronics coolant return hose and clamp

8 Degas bottle

Degas bottle

- The 2 DC/DC converter low voltage battery cable nuts: 12 Nm (9 ft. lbs.)
- The degas bottle bolt and nut: 6 Nm (53 inch lbs.)

21. Fill the Motor Electronics Cooling System as follows:

➡**Vehicle cooling systems are filled with Motorcraft Premium Gold Engine Coolant. Always fill the cooling system with the same coolant that is present**

in the system. Do not mix coolant types.

a. Connect the hoses to the transaxle and install the hose clamps.

❋❋ WARNING

Adhesives, stop-leak pellets or small debris in the Motor electronics cooling system (MECS) can cause poor performance or temporary blockage of the motor electronics pump. Only

use clean, approved coolant when filling the system.

b. Loosen the bleed screw and fill the degas bottle with coolant until it begins to flow out of the bleed hole. Then, close the bleed screw.

c. Turn the ignition to the ON position to actuate the motor electronics cooling pump and continue to fill the degas bottle to the correct level.

➡ Most of the MECS air bleeding occurs as a normal process at the degas bottle through the vent tube; very little occurs at the bleed screw.

  d. To bleed air from the system, loosen the bleed screw and allow air to escape while the M/E coolant pump is operating.

Direct Current/Direct Current Converter Wiring Harness

REMOVAL & INSTALLATION

  1. Before servicing the vehicle, refer to the Precautions Section.

✳✳ CAUTION

Before proceeding, read and observe all of the High Voltage System Precautions.

  2. De-power the high voltage traction battery. Establish a buffer zone. See the pro-

DC/DC converter wiring harness—Escape Hybrid

1 Engine cover

2 DC/DC converter high voltage electrical connector

3 Plastic loop

4 Pin-type retainers (part of 14B323) (2 required)

5 Pin-type retainers (female) (part of 14B323) (2 required)

6 DC/DC converter high voltage electrical connector (high voltage)

7 DC/DC converter wiring harness

06017-ESCA-G11

cedure under De-powering the High Voltage Traction Battery.

3. Disconnect the 12 volt battery.

➡**For the Escape Hybrid, when the battery is disconnected and connected, the brake pedal needs to be calibrated. After the battery has been connected, with the vehicle in park, turn the key to the ON position. Press the brake pedal firmly, then fully release to calibrate the brake pedal.**

4. Remove the engine cover.
5. Disconnect the DC/DC converter harness connector from the electronically controlled continuously variable transmission (ECVT).
6. Drain the cooling system.

※※ WARNING

Always refill the motor electronics cooling system (MECS) with the same type of coolant that was drained from the system. Do not mix coolant types.

7. Drain the motor electronics cooling system:
   a. With the vehicle in DRIVE, position it on a hoist.
   b. Turn the ignition to the OFF position.
   c. Remove the left splash shield.
   d. Place a suitable container below the transaxle.
   e. Loosen the hose clamps at the transaxle, then pull the hoses off to allow the coolant to drain.
8. Disconnect the motor electronics coolant vent hose.
9. Disconnect the 2 engine coolant vent hoses.
10. Remove the degas bottle bolt and nut.
11. Lift the degas bottle up and disconnect the engine coolant and motor electronics coolant return hoses.
12. Remove the degas bottle.
13. Disconnect the plastic loop.
14. Disconnect the 2 pin-type retainers.
15. Disconnect the 2 female pin-type retainers.

➡**Press the locking clip to release the connector.**

16. Disconnect the DC/DC converter high voltage electrical connector.
17. Remove the DC/DC converter wiring harness.
18. To install, reverse the removal procedure. Re-power the high voltage battery.
19. Fill the Motor Electronics Cooling System as follows:

➡**Vehicle cooling systems are filled with Motorcraft Premium Gold Engine Coolant. Always fill the cooling system with the same coolant that is present in the system. Do not mix coolant types.**

   a. Connect the hoses to the transaxle and install the hose clamps.

※※ WARNING

Adhesives, stop-leak pellets or small debris in the Motor electronics cooling system (MECS) can cause poor performance or temporary blockage of the motor electronics pump. Only use clean, approved coolant when filling the system.

   b. Loosen the bleed screw and fill the degas bottle with coolant until it begins to flow out of the bleed hole. Then, close the bleed screw.
   c. Turn the ignition to the ON position to actuate the motor electronics cooling pump and continue to fill the degas bottle to the correct level.

➡**Most of the MECS air bleeding occurs as a normal process at the degas bottle through the vent tube; very little occurs at the bleed screw.**

   d. To bleed air from the system, loosen the bleed screw and allow air to escape while the M/E coolant pump is operating.

Alternating Current Powerpoint

REMOVAL & INSTALLATION

1. Before servicing the vehicle, refer to the Precautions Section.

※※ CAUTION

Before proceeding, read and observe all of the High Voltage System Precautions.

2. Disconnect the 12 volt battery.

➡**For the Escape Hybrid, when the battery is disconnected and connected, the brake pedal needs to be calibrated. After the battery has been connected, with the vehicle in park, turn the key to the ON position. Press the brake pedal firmly, then fully release to calibrate the brake pedal.**

3. Remove the console trim insert.
4. Disconnect the electrical connector and remove the AC powerpoint.
5. To install, reverse the removal procedure.

1 Console trim insert
2 Electrical connector
3 AC powerpoint

AC powerpoint

06017-ESCA-G12

12-VOLT ELECTRICAL SYSTEM

Alternator

REMOVAL & INSTALLATION

2.0L Engine

1. Before servicing the vehicle, refer to the Precautions Section.
2. Remove or disconnect the following:
   - Negative battery cable
   - Drive belt

- Alternator electrical connectors and loosen the upper alternator bolt while moving the alternator to the rear of the engine
- Alternator. Torque the lower bolt to 35 ft. lbs. (48Nm). Torque the upper bolt to 18 ft. lbs. (25 Nm).

To install:

3. Install or connect the following:
   - Alternator with the upper bolt in the alternator before installation. Torque the bolts to 18 ft. lbs. (25 Nm).

- Alternator electrical connectors
- Drive belt
- Negative battery cable

2.3L Engine

1. Before servicing the vehicle, refer to the Precautions Section.
2. Disconnect the 12v battery.

➡ **For the Escape Hybrid, when the battery is disconnected and connected, the brake pedal needs to be calibrated.**

11 - 47 Nm (35 lb-ft)

24 Nm (18 lb-ft) - 12

9 - 20 Nm (15 lb-ft)

6 - 4 Nm (35 lb-in)

47 Nm (35 lb-ft) - 13

1 Front end accessory drive belt tensioner	8 Pin-type retainer
2 Front end accessory drive belt	9 Generator shield nut
3 Generator B+ terminal nut	10 Generator shield
4 Generator B+ terminal	11 Generator stud nut
5 Generator electrical connector	12 Generator stud
6 Generator lower air duct bolt	13 Generator bolts
7 Generator lower air duct	14 Generator

Alternator mounting—2.3L Engine

67197-ESCA-G01

After the battery has been connected, with the vehicle in park, turn the key to the ON position. Press the brake pedal firmly, then fully release to calibrate the brake pedal.

3. Remove the front end accessory drive belt tensioner. Rotate the front end accessory drive belt tensioner counter-clockwise to loosen tension on the front end accessory drive belt.

4. Remove the front end accessory drive belt.

5. Remove the alternator B+ terminal.

6. Remove the alternator electrical connector.

7. Remove the alternator lower air duct bolt.

8. Remove the alternator lower air duct. Press the locking tab to release the lower air duct from the upper air duct.

9. Remove the pin-type retainer.

10. Remove the alternator shield .

11. Remove the alternator stud nut.

12. Remove the alternator stud.

13. Remove the alternator bolts.

14. Remove the alternator.

15. To install, reverse the removal procedure. Observe the following torques:

- Alternator mounting bolts: 35 ft. lbs. (47 Nm)
- Alternator stud: 18 ft. lbs. (24 Nm)
- Stud nut: 35 ft. lbs. (47 Nm)
- Shield nut: 15 ft. lbs. (20 Nm)
- Lower air duct bolt: 35 inch lbs. (4 Nm)

3.0L Engine

1. Before servicing the vehicle, refer to the Precautions Section.

2. Remove or disconnect the following:
- Negative battery cable
- Right side intermediate axle shaft
- Right side splash shield and retainers
- Drive belt

14 4 Nm (35 lb-in)

3 8 Nm (71 lb-in)
12 7 Nm (62 lb-in)
11 47 Nm (35 lb-ft)
13 47 Nm (35 lb-ft)

47 Nm (35 lb-ft) **17**

4 Nm (35 lb-in) **6**
4 Nm (35 lb-in) **9**

1	Front end accessory drive belt tensioner	11	Generator stud nut
2	Front end accessory drive belt	12	Generator stud
3	Generator B+ terminal nut	13	Generator bolts (2 required)
4	Generator electrical connector	14	Generator lower air duct nuts (3 required)
5	Generator B+ cable	15	Generator lower air duct
6	Generator upper air duct bolt	16	Generator
7	Generator upper air duct	17	Generator bracket bolts (2 required)
8	Pin-type retainer	18	Generator bracket
9	Generator shield bolt		
10	Generator shield		

Alternator mounting—3.0L engine

06017-ESCA-G03

- Alternator electrical connectors
- Alternator. Torque the mounting and adjusting bolts to 35 ft. lbs. (48Nm).

To install:

3. Install or connect the following:

- Alternator. Torque the bolts to 35 ft. lbs. (48 Nm).
- Alternator electrical connectors
- Drive belt
- Negative battery cable

Starter Motor

REMOVAL & INSTALLATION

2.0L Engine

1. Before servicing the vehicle, refer to the Precautions Section.

2. Remove or disconnect the following:

9308TG17

Removal of the starter motor–2.0L engine

25 Nm (18 lb-ft)–⑩	20 Nm (15 lb-ft)–⑦
	⑧
⑪	25 Nm (18 lb-ft)–⑨
② 5 Nm (44 lb-in)	⑤
③	⑥
④ 12 Nm (9 lb-ft)	①

1	Starter motor solenoid terminal cover	5	Starter solenoid battery cable	9	Starter motor stud bolts
2	Starter solenoid wire nut	6	Wiring harness retainer	10	Starter motor bracket bolt
3	Starter solenoid wire	7	Ground strap nut	11	Starter motor
4	Starter solenoid battery cable nut	8	Ground strap		

67197-ESCA-G19

Starter mounting—2.3L engine

- Negative battery cable
- Starter bolts
- Exhaust system, AWD vehicles only
- Halfshaft support bracket bolts
- Starter electrical connectors
- Starter

To install:

3. Install or connect the following:
- Starter. Torque bolts to 20 ft. lbs. (27 Nm).
- Starter electrical connectors
- Halfshaft support bracket. Torque the bolts to 11 ft. lbs. (15 Nm).
- Exhaust system on AWD vehicles. Torque the bolts to 18 ft. lbs. (25 Nm).
- Negative battery cable

2.3L Engine

1. Before servicing the vehicle, refer to the Precautions Section.

✸✸ WARNING

When performing maintenance on the starting system, be aware that heavy

gauge leads are connected directly to the battery. Make sure protective caps are in place when maintenance is completed.**

2. With the vehicle in NEUTRAL, position it on a hoist
3. Disconnect the battery ground cable.
4. Starter motor solenoid terminal cover
5. Starter solenoid wire
6. Starter solenoid battery cable
7. Wiring harness retainer
8. Ground strap
9. Starter motor stud bolts
10. Starter motor bracket bolt
11. Starter motor
12. To install, reverse the removal procedure. Torque the starter and bracket bolts to 18 ft. lbs. (25 Nm).

3.0L Engine

1. Before servicing the vehicle, refer to the Precautions Section.
2. Drain the cooling system.
3. Remove or disconnect the following:
- Negative battery cable

- Air cleaner outlet tube
- Coolant hoses and move the thermostat aside
- Starter electrical connectors
- Starter

To install:

4. Install or connect the following:
- Starter. Torque bolts to 20 ft. lbs. (27 Nm).
- Starter electrical connectors and reposition the thermostat
- Connect the 4 coolant hoses
- Air cleaner outlet tube
- Negative battery cable

5. Fill the cooling system to the proper level.

6. Start the vehicle and check for leaks, repair if necessary.

Ignition Timing

ADJUSTMENT

Ignition timing is controlled by the Powertrain Control Module (PCM). No adjustment is necessary or possible.

ENGINE REPAIR

Engine Assembly

REMOVAL & INSTALLATION

2.0L Engine

MANUAL TRANSMISSION

1. Before servicing the vehicle, refer to the Precautions Section.
2. Properly recover the air conditioning system refrigerant.
3. Properly relieve the fuel system pressure.
4. Drain the cooling system.
5. Drain the engine oil.
6. Remove or disconnect the following:
- Hood
- Battery and battery tray
- Air cleaner housing
- Fuel lines
- Throttle cable and speed control cable, if equipped
- Exhaust Gas Recirculation (EGR) vacuum valve regulator
- EGR electrical connectors and vacuum hoses
- Brake booster vacuum hose
- Powertrain Control Module (PCM) wire harness and ground
- Wire harness connector

- Power distribution board electrical connectors
- Evaporative emissions (EVAP) canister vacuum lines
- Upper radiator hose
- Power steering line bracket
- Upper power steering pump bolts
- Coolant hose
- Heater hoses
- Speed control unit, if equipped
- Catalytic converter
- A/C compressor
- Both halfshafts
- Shifter linkages
- Block heater electrical connector, if equipped
- Front transmission through bolt
- Engine-to-transmission bolts
- Lower radiator hose
- Power steering pump
- Clutch slave cylinder line from the bracket and move it aside
- Rear transmission mount
- Left side transmission mount
- Lower ground cable
- Engine mount upper bracket
- Engine and transmission as an assembly by using a proper lifting device
- Alternator electrical connectors
- Knock Sensor (KS) electrical connector

- Oil pressure sender electrical connector
- Starter electrical connector
- Vehicle Speed Sensor (VSS) electrical connector
- Park Neutral Position (PNP) electrical connector
- Fuel charging wire harness electrical connector
- PCM wire harness from the bracket
- PCM ground wire
- Back up lamp switch electrical connector
- Wire harness
- Differential Pressure Feedback (DPFEE) EGR sensor

7. Separate the engine from the transmission.
8. Lock the flywheel to the engine.
9. Clutch pressure plate and disc.
10. Flywheel and rear cover plates.

To install:

11. Install or connect the following:
- Flywheel. Torque the bolts to 83 ft. lbs. (112 Nm).
- Clutch disc to the flywheel
- Pressure plate to the flywheel. Torque the bolts in the proper sequence to 18 ft. lbs. (25 Nm).
- Transmission to the engine. Torque the bolts to 33 ft. lbs. (45 Nm).

Tighten the pressure plate bolts in the proper sequence—2.0L engine

- Starter. Torque the bolts to 18 ft. lbs. (25 Nm).
- Wire harness and attach it to the powertrain assembly
- DPFEE sensor electrical connector
- Reverse lamp switch electrical connector
- Ground wire. Torque the bolt to 80 inch lbs. (9 Nm).
- PCM wire harness to the bracket
- Fuel charging wire harness electrical connector
- PNP switch electrical connector
- VSS electrical connector
- KS, Oil pressure sender and starter electrical connector. Torque the fasteners to 9 ft. lbs. (12 Nm).
- Alternator electrical connectors. Torque the fasteners to 71 inch lbs. (8 Nm).
- Powertrain assembly in the vehicle
- Left side transmission mount. Torque the side bolts to 41 ft. lbs. (55 Nm) and the center bolt to 66 ft. lbs. (90 Nm).
- Engine mount upper bracket. Torque the side bolts to 72 ft. lbs. (98 Nm) and the center bolt to 57 ft. lbs. (77 Nm).
- Ground wire. Torque the bolt to 25 ft. lbs. (34 Nm).
- Rear transmission mount. Torque the bolts to 41 ft. lbs. (55 Nm).
- Speed control unit, if equipped. Torque the bolts to 89 inch lbs. (10 Nm).
- Power steering pump and hand tighten the bolts
- Lower power steering line bracket. Torque the bolt to 89 inch lbs. (10 Nm).
- Upper power steering line bolt. Torque the bolt to 15 ft. lbs. (20 Nm).
- Slave cylinder line and clip. Torque the bolts to 16 ft. lbs. (22 Nm).
- Power steering lines. Torque the retaining bolts to 89 inch lbs. (10

Nm). Torque the power steering pump bolts to 18 ft. lbs. (25 Nm).
- Lower radiator hose
- Engine-to-transmission bolts. Torque the bolts to 33 ft. lbs. (45 Nm).
- Front transmission through bolt. Torque the bolt to 66 ft. lbs. (90 Nm).
- Block heater electrical connector
- Shifter linkages. Torque the upper bolt to 33 ft. lbs. (45 Nm) and the lower bolt to 15 ft. lbs. (20 Nm).
- Coolant hose
- Catalytic converter
- Heater hoses
- Upper radiator hose
- EVAP canister vacuum lines
- Power distribution box electrical connector. Torque the fastener to 9 ft. lbs. (12 Nm).
- Wire harness electrical connector
- Ground wires
- PCM wire harness and ground
- Brake booster vacuum supply hose to the intake manifold
- EGR vacuum regulator valve hoses and electrical connector
- Throttle cable and speed control cable, if equipped
- Fuel lines
- Battery tray and battery
- Air cleaner
- Hood

12. Fill the engine with clean oil.
13. Fill and bleed the cooling system.
14. Recharge the A/C system.
15. Start the vehicle, check for leaks and repair if necessary.

2.3L Engine, except Escape Hybrid

MANUAL TRANSAXLE

1. Before servicing the vehicle, refer to the Precautions Section.

All vehicles

2. With the vehicle in NEUTRAL, position it on a hoist.
3. Release the fuel system pressure.
4. Remove the engine air cleaner and air cleaner outlet pipe.
5. Remove the battery tray.
6. Drain the engine oil.
7. Drain the cooling system.
8. Remove the starter.
9. Remove the catalytic converter.
10. Remove the accessory drive belt.
11. Remove the bolts and the lateral support crossmember.
12. Remove the left front drive halfshaft.
13. Remove the front drive intermediate halfshaft

4x4 vehicles

14. Remove the six bolts holding the driveshaft to the transfer case.
15. Position the driveshaft aside.

All vehicles

16. If equipped, remove the bolt and ground eyelet.
17. Remove the power distribution box cover.
18. Remove the nuts and disconnect the cables.
19. Disconnect the electrical connector from the power distribution box.
20. Remove the bolt and disconnect the ground strap. Loosen the bolt and disconnect the 42-pin electrical connector.
21. Detach the wiring harness retainers from the battery tray bracket and position the wiring harness out of the way.
22. Disconnect the clutch hydraulic tube fitting. Detach the tube from the spring clip and position aside.
23. Remove the retaining clips and disconnect the transaxle control cable.
24. Remove the retaining clips and disconnect the transaxle control cable.
25. Disconnect the vehicle speed sensor electrical connector and pin-type retainer.
26. Disconnect the reversing lamp indicator switch and detach the wiring harness retainers.
27. If equipped, disconnect the block heater electrical connector. Detach all the block heater wiring harness retainers and position the wiring harness aside.
28. Disconnect the upper radiator and coolant vent hoses.
29. Remove the nuts and the coolant vent hose brackets. Position the coolant vent hose aside.
30. Detach the heater hose support strap from the stud.
31. Disconnect the heater hoses from the heater core.
32. Remove the retainers and the accelerator cable snow shield.
33. Disconnect the accelerator cable and speed control cable (if equipped).

Engine secured to lift table—2.3L engine w/manual transaxle

34. Remove the nut from the accelerator control cable bracket.

35. Remove the nut from the accelerator control cable bracket and position the accelerator control cable and bracket assembly aside.

36. Remove the nut and position the power steering tube and bracket aside.

37. Disconnect the vacuum supply tube and position aside.

38. Disconnect the fuel vapor return tube and position aside.

39. Disconnect the vacuum reservoir tube and position aside.

40. Disconnect the fuel supply tube and retainer and position aside.

41. Detach the electrical connector retainers.

42. Disconnect the powertrain control module (PCM) electrical connectors. Remove the nut and position the harness aside.

43. Remove the bolt and detach the ground wire.

44. Remove the two power steering pump bolts.

45. Disconnect the lower radiator hose from the radiator.

46. Disconnect the A/C compressor electrical connector and remove the four bolts. Position the A/C compressor aside and support the compressor with a length of mechanics wire.

47. Disconnect the power steering pressure (PSP) sensor electrical connector.

➡**The bolt under the power steering pressure tube will remain with the power steering pump.**

48. Remove the bolts and position the power steering pump aside.

49. Remove the front roll restrictor bolt and the two bolts for the engine support crossmember.

50. Remove the rear nut and the engine support crossmember.

➡**The transaxle-to-engine bolts differ in length. Mark the bolts for correct installation.**

51. Remove the two transaxle-to-engine bolts.

➡**The transaxle-to-engine bolts differ in length. Mark the bolts for correct installation.**

52. Remove the two transaxle-to-engine bolts.

53. Using the special tools, secure the engine to the lift table.

54. Remove the engine mount bracket bolt.

55. Remove the nuts and the engine mount bracket.

56. Remove the bolt from the transaxle rear mount.

57. Remove bolt from the left transaxle mount.

58. Lower the engine and transaxle from the vehicle.

59. Using the engine crane and spreader bar, remove the engine and transaxle from the lift table.

➡**The transaxle-to-engine bolts differ in length. Mark the bolts for correct installation.**

60. Remove the remaining six engine-to-transaxle bolts and separate the engine and transaxle.

To install:
All vehicles
61. Using the engine crane and spreader bar, position the engine and transaxle together. Install the six upper transaxle-to-engine bolts. Torque to 35 ft. lbs. (48 Nm).

62. Using the engine crane and spreader bar, position the engine and transaxle onto the lift table.

63. Using the special tools, secure the engine to the lift table.

64. Raise the engine and transaxle into the vehicle.

65. Install the bolt in the left transaxle mount. Torque to 76 ft. lbs. (103 Nm).

66. Install the bolt in the rear transaxle mount. Torque to 85 ft. lbs. (115 Nm).

67. Install the engine mount bracket. Torque to 66 ft. lbs. (90 Nm).

68. Install the engine mount bracket bolt. Torque to 66 ft. lbs. (90 Nm).

69. Install the 4 lower transaxle-to-engine bolts. Torque to 35 ft. lbs. (48 Nm).

70. Install the engine support crossmember and nut. Torque to 66 ft. lbs. (90 Nm).

71. Install the two bolts for the engine support crossmember. Torque to 66 ft. lbs. (90 Nm).

72. Install the front roll restrictor bolt. Torque to 85 ft. lbs. (115 Nm).

➡**The bolt under the power steering pressure tube will remain with the power steering pump.**

73. Position the power steering pump and install the bolts. Torque to 18 ft. lbs. (25 Nm).

74. Connect the power steering pressure (PSP) sensor electrical connector.

75. Install the A/C compressor and connect the A/C compressor electrical connector. Torque to 18 ft. lbs. (25 Nm).

76. Connect the lower radiator hose to the radiator.

77. Install the two lower power steering pump bolts. Torque to 18 ft. lbs. (25 Nm).

78. Install the ground wire and bolt.

79. Connect the powertrain control module (PCM) electrical connectors. Position the harness and install the nut.

80. Attach the electrical connector retainers.

81. Connect the fuel supply tube.

82. Connect the vacuum reservoir tube.

83. Connect the fuel vapor return tube and retainer.

84. Connect the vacuum supply tube.

85. Install the power steering tube and bracket.

86. Position the accelerator control cable and bracket and install the nut.

87. Install the accelerator control cable and bracket and nut.

88. Install the accelerator cable and speed control cable (if equipped).

89. Install the accelerator cable snow shield and retainers.

90. Connect the heater hoses to the heater core.

91. Attach the heater hose support strap to the stud.

92. Position the coolant vent hose and install the coolant vent hose brackets and nuts.

93. Connect the upper radiator and coolant vent hoses.

94. If equipped, route the block heater wiring harness and attach all retainers. Connect the block heater electrical connector.

95. Connect the reversing lamp indicator switch and attach the wiring harness retainers.

96. Connect the vehicle speed sensor (VSS) electrical connector and pin-type retainer.

97. Connect the transaxle control cable and install the retaining clips.

98. Connect the transaxle control cable and install the retaining clips.

99. Connect the clutch hydraulic tube fitting. Attach the tube to the spring clip.

100. Attach the wiring harness retainers to the battery tray bracket.

101. Connect the 42-pin electrical connector and tighten the bolt. Install the ground strap and bolt.

102. Connect the electrical connector to the power distribution box.

103. Connect the cables and install the nuts.

104. Install the power distribution box cover.

105. If equipped, install the ground eyelet and bolt.

4x4 vehicles

106. Install the driveshaft. Torque to 15 ft. lbs. (20 Nm).

All vehicles

107. Install the front drive intermediate halfshaft.

108. Install the left front drive halfshaft.

109. Install the lateral support crossmember. Torque to 85 ft. lbs. (115 Nm).

110. Install the accessory drive belt.

111. Install the catalytic converter.

112. Install the starter.

113. Install the battery tray and battery.

114. Install the engine air cleaner and air cleaner outlet pipe.

115. Fill the engine with clean engine oil.

116. Fill and bleed the cooling system.

117. Bleed the clutch system.

AUTOMATIC TRANSAXLE

1. Before servicing the vehicle, refer to the Precautions Section.

All vehicles

2. With the vehicle in NEUTRAL, position it on a hoist.

3. Release the fuel system pressure.

4. Remove the engine air cleaner and air cleaner outlet pipe

5. Remove the battery tray.

6. Drain the engine oil.

7. Drain the cooling system.

8. Remove the starter.

9. Remove the catalytic converter.

10. Remove the accessory drive belt.

11. Remove the left front drive halfshafts.

4wd vehicles

12. Remove the transfer case.

2wd vehicles

13. Remove the bolts and the lateral support crossmember.

14. Remove the front drive intermediate halfshaft.

All vehicles

15. If equipped, remove the bolt and ground eyelet.

16. Remove the power distribution box cover.

17. Remove the nuts and disconnect the cables.

18. Disconnect the electrical connector from the power distribution box.

19. Remove the bolt and disconnect the ground strap. Loosen the bolt and disconnect the 42-pin electrical connector.

20. Detach the wiring harness retainers from the battery tray bracket and position the wiring harness out of the way.

21. Disconnect the transaxle electrical connector.

22. Disconnect the shift cable from the transaxle manual lever.

23. Position the transaxle control cable and bracket aside.

24. Disconnect the transmission range (TR) sensor electrical connector.

25. Detach the transaxle control harness from the brackets.

26. Disconnect the fluid cooler tube.

27. Disconnect the output shaft speed (OSS) sensor electrical connector (black).

28. Disconnect the turbine shaft speed (TSS) sensor electrical connector (white connector).

29. Remove the transmission fluid cooler retaining bracket bolt.

30. Position the fluid cooler tube aside.

31. Remove the bolt and the OSS sensor.

32. Detach the transaxle control harness from the retaining clip.

33. If equipped, disconnect the block heater electrical connector. Detach all the block heater wiring harness retainers and position the wiring harness aside.

34. Disconnect the upper radiator and coolant vent hoses.

35. Remove the nuts and the coolant vent hose brackets. Position the coolant vent hose aside.

36. Detach the heater hose support strap from the stud.

37. Disconnect the heater hoses from the heater core.

38. Remove the retainers and the accelerator cable snow shield.

39. Disconnect the accelerator cable and speed control cable (if equipped).

40. Remove the nut from the accelerator control cable bracket.

41. Remove the nut from the accelerator control cable bracket and position the accelerator control cable and bracket assembly aside.

42. Remove the nut and position the power steering tube and bracket aside.

43. Disconnect the vacuum supply tube and position aside.

44. Disconnect the fuel vapor return tube and retainer and position aside.

45. Disconnect the vacuum reservoir tube and position aside.

46. Disconnect the fuel supply tube and position aside.

47. Detach the electrical connector retainers.

48. Disconnect the powertrain control module (PCM) electrical connectors. Remove the nut and position the harness aside.

49. Remove the bolt and detach the ground wire.

50. Remove the two power steering pump bolts.

51. Disconnect the lower radiator hose from the radiator.

52. Disconnect the A/C compressor electrical connector and remove the four bolts. Position the A/C compressor aside and support the compressor with a length of mechanics wire.

53. Disconnect the power steering pressure (PSP) sensor electrical connector.

➡ **The bolt under the power steering pressure tube will remain with the power steering pump.**

54. Remove the bolts and position the power steering pump aside.

55. Remove the front roll restrictor bolt and the two bolts for the engine support crossmember.

56. Remove the rear nut and the engine support crossmember.

➡ **The transaxle-to-engine bolts differ in length. Mark the bolts for correct installation.**

57. Remove the two transaxle-to-engine bolts

➡ **The transaxle-to-engine bolts differ in length. Mark the bolts for correct installation.**

58. Remove the two transaxle-to-engine bolts

59. Using the special tools, secure the engine to the lift table.

60. Remove the engine mount bracket bolt.

61. Remove the nuts and the engine mount bracket.

62. Remove the bolt from the transaxle rear mount.

63. Remove the bolt from the left transaxle mount.

64. Lower the engine and transaxle from the vehicle.

65. Using the engine crane and spreader bar remove the engine and transaxle from the lift table.

66. Remove the starter motor isolator.

67. Remove and discard the four torque converter nuts.

➡ **The transaxle-to-engine bolts differ in length. Mark the bolts for correct installation.**

68. Remove the remaining six engine-to-transaxle bolts and separate the engine and transaxle.

To install:

All vehicles

69. Using the engine crane and spreader bar, position the engine and transaxle together. Install the six upper transaxle-to-engine bolts. Torque to 35 ft. lbs. (48 Nm).

70. Install new torque converter nuts. Torque to 26 ft. lbs. (35 Nm).

71. Install the starter motor isolator.

72. Using the engine crane and spreader bar, position the engine and transaxle onto the lift table.

73. Using the special tools, secure the engine to the lift table.

74. Raise the engine and transaxle into the vehicle.

75. Install the bolt in the left transaxle mount. Torque to 76 ft. lbs. (103 Nm).

76. Install the bolt in the rear transaxle mount. Torque to 85 ft. lbs. (115 Nm).

77. Install the engine mount bracket. Torque to 66 ft. lbs. (90 Nm).

78. Install the engine mount bracket bolt. Torque to 66 ft. lbs. (90 Nm).

79. Install the 4 transaxle-to-engine bolts. Torque to 35 ft. lbs. (48 Nm).

80. Install the engine support crossmember and nut. Torque to 66 ft. lbs. (90 Nm).

81. Install the two bolts for the engine support crossmember. Torque to 66 ft. lbs. (90 Nm).

82. Install the front roll restrictor bolt . Torque to 85 ft. lbs. (115 Nm).

83. Position the power steering pump and install the upper bolts. Torque to 18 ft. lbs. (25 Nm).

84. Connect the power steering pressure (PSP) sensor electrical connector.

85. Install the A/C compressor and connect the A/C compressor electrical connector. Torque to 18 ft. lbs. (25 Nm).

86. Connect the lower radiator hose to the radiator.

87. Install the two lower power steering pump bolts. Torque to 18 ft. lbs. (25 Nm).

88. Install the ground wire and bolt.

89. Connect the powertrain control module (PCM) electrical connectors. Position the harness and install the nut.

90. Attach the electrical connector retainers.

91. Connect the fuel supply tube.

92. Connect the vacuum reservoir tube.

93. Connect the fuel vapor return tube and retainer.

94. Connect the vacuum supply tube.

95. Position the power steering tube and bracket and install the nut.

96. Position the accelerator control cable and bracket and install the nut.

97. Install the accelerator control cable and bracket and nut.

98. Install the accelerator cable and speed control cable (if equipped).

99. Install the accelerator cable snow shield and the retainers.

100. Connect the heater hoses to the heater core.

101. Attach the heater hose support strap to the stud.

102. Position the coolant vent hose and install the coolant vent hose brackets and nuts.

103. Connect the upper radiator and coolant vent hoses.

104. If equipped, route the block heater wiring harness and attach all retainers. Connect the block heater electrical connector.

105. Attach the transaxle control harness to the retaining clip.

106. Install the output shaft speed (OSS) sensor and bolt.

107. Install the fluid cooler tube.

108. Connect the transmission fluid cooler tube.

109. Attach the transaxle control harness to the brackets.

110. Connect the transmission range (TR) sensor electrical connector.

111. Install the transaxle control cable and bracket.

112. Connect the shift cable to the transaxle manual lever.

113. Connect the transaxle electrical connector.

114. Attach the wiring harness retainers to the battery tray bracket.

115. Connect the 42-pin electrical connector and tighten the bolt. Install the ground strap and bolt.

116. Connect the electrical connector to the power distribution box.

117. Connect the cables and install the nuts.

118. Install the power distribution box cover.

119. If equipped, install the ground eyelet and bolt.

2wd vehicles

120. Install the front drive intermediate halfshaft.

121. Install the lateral support crossmember. Torque to 85 ft. lbs. (115 Nm).

4wd vehicles

122. Install the transfer case.

All vehicles

123. Install the left front drive halfshaft.

124. Install the accessory drive belt.

125. Install the catalytic converter.

126. Install the starter.

127. Install the battery tray and battery.

128. Install the engine air cleaner and air cleaner outlet pipe.

129. Fill the engine with clean engine oil.

130. Fill and bleed the cooling system.

Escape Hybrid

1. Before servicing the vehicle, refer to the Precautions Section.

✳✳ CAUTION

Before proceeding, read and observe all of the High Voltage System Precautions.

All Vehicles

2. With the vehicle in NEUTRAL, position it on a hoist.

3. Release the fuel system pressure.

4. Disconnect the battery ground cable.

➡**For the Escape Hybrid, when the battery is disconnected and connected, the brake pedal needs to be calibrated. After the battery has been connected, with the vehicle in park, turn the key to the ON position. Press the brake pedal firmly, then fully release to calibrate the brake pedal.**

5. De-power the high voltage traction battery. Establish a buffer zone. See the procedure under De-powering the High Voltage Traction Battery.

6. Drain the cooling system.

✳✳ WARNING

Always refill the motor electronics cooling system (MECS) with the same type of coolant that was drained from the system. Do not mix coolant types.

7. Drain the motor electronics cooling system:

    a. With the vehicle in DRIVE, position it on a hoist.

    b. Turn the ignition to the OFF position.

    c. Remove the left splash shield.

    d. Place a suitable container below the transaxle.

    e. Loosen the hose clamps at the transaxle, then pull the hoses off to allow the coolant to drain.

8. Disconnect the motor electronics coolant vent hose.

9. Disconnect the 2 engine coolant vent hoses.

10. Remove the degas bottle bolt and nut.

11. Lift the degas bottle up and disconnect the engine coolant and motor electronics coolant return hoses.

12. Remove the degas bottle.

13. Remove the intake manifold.

14. Remove the engine air cleaner.

15. Remove the exhaust manifold/catalytic converter.

16. Remove the accessory drive belt.

17. Remove the left front drive halfshaft.

4x4 vehicles

18. Remove the transfer case.

4x2 vehicles

19. Remove the front drive intermediate halfshaft.

20. Remove the bolts and the lateral support crossmember.

All vehicles

21. Drain the engine oil.

22. Disconnect the heater hoses from the heater core.

23. Detach the heater hose retaining clip from the transaxle mount stud.

24. Disconnect the upper radiator and coolant vent hoses.

25. Disconnect the following electrical connectors:

- Secondary air injection (AIR) solenoid
- Transaxle mount control solenoid
- Purge valve

26. Disconnect the engine control harness:

a. Release the powertrain control module (PCM) connector.

b. Remove the harness retaining nut.

27. Disconnect the 2 engine control harness electrical connectors and the 2 pin-type retainers.

28. Remove the bolt and disconnect the transaxle harness electrical connector.

29. Disconnect the DC-to-DC converter electrical connector and pin-type retainer.

30. Disconnect and remove the fuel supply tube and evaporative emissions (EVAP) tube.

31. If equipped, position the block heater wiring harness aside.

a. Detach the 3 harness retainer clips.

b. Route the harness through the radiator support into the engine compartment.

32. Remove the 2 bolts and disconnect the auxiliary coolant pump electrical connector.

33. Position the auxiliary coolant pump aside.

34. Remove the bolts and the transaxle control snow shield.

35. Disconnect the transaxle control cable.

a. Release the transaxle control cable from the control lever.

b. Remove the nuts from the transaxle control cable bracket.

36. Disconnect the pin-type retainer.

37. Disconnect the pin-type retainer and position the transaxle control cable aside.

38. Disconnect the transaxle coolant

06017-ESCA-G14

Motor electronics coolant pump—Escape Hybrid

temperature sensor electrical connector and pin-type retainer.

39. Disconnect the transaxle coolant hoses.

40. Remove the nut and the ground cable.

41. Remove the 2 bolts and position the motor electronics coolant pump aside.

42. Disconnect the lower radiator hose from the radiator.

43. Disconnect the A/C compressor electrical connector and remove the 3 bolts.

44. Position the A/C compressor aside and support the compressor with a length of mechanic's wire.

45. Remove the front roll restrictor bolt.

46. Remove the rear roll restrictor bolt.

47. Remove the nut, bolts and the engine support crossmember.

➡**The transaxle-to-engine bolts differ in length. Mark the bolts for correct installation.**

48. Remove the 2 transaxle-to-engine bolts.

➡**The transaxle-to-engine bolts differ in length, mark the bolts for correct installation.**

49. Remove the 2 transaxle-to-engine bolts.

✳✳ WARNING

Due to the weight of the transaxle, special care should be taken to mount the powertrain securely to the lift table.

50. Using the special tools, secure the engine to the lift table.

51. Remove the engine mount bracket bolt.

52. Remove the nuts and the engine mount bracket.

53. Remove the right transaxle mount bracket nut.

54. Remove the bolts and the right transaxle mount bracket.

06017-ESCA-G13

Remove the bolts and the lateral support crossmember—Escape Hybrid

014-0001   014-00765

06017-ESCA-G15

Using the special tools, secure the engine to the lift table—Escape Hybrid

55. Remove the bolt, nut and the rear transaxle mount brace.

56. Remove the nuts, bolt and rear transaxle mount.

57. Lower the engine and transaxle from the vehicle.

58. Using the engine crane and spreader bar, remove the engine and transaxle from the lift table.

59. Disconnect the high voltage wiring harness electrical connector.

60. Disconnect the 2 high voltage wiring harness electrical connectors and position the harness aside.

➡**The transaxle-to-engine bolts differ in length, mark the bolts for correct installation.**

61. Remove the remaining 6 engine-to-transaxle bolts and separate the engine and transaxle.

➡**Due to packaging requirements the correct bolt must be used at the damper locations.**

06017-ESCA-G16

Transaxle damper—Escape Hybrid

➡**The damper contains a clutch which is designed to slip briefly during vehicle operation. It is essential that no grease, oil or cleaning solvents be allowed to contaminate the slip clutch. Do not use grease on transmission input shaft. Should the damper become contaminated, it must be replaced.**

62. Remove the bolts and the transaxle damper.

To install:
All vehicles

➡**Due to packaging requirements the correct bolt must be used at the damper locations.**

➡**The damper contains a clutch which is designed to slip briefly during vehicle operation. It is essential that no grease, oil or cleaning solvents be allowed to contaminate the slip clutch. Do not use grease on transmission input shaft. Should the damper become contaminated, it must be replaced.**

63. Install the transaxle damper. Tighten to 29 Nm (21 ft. lbs.).

✳✳ WARNING

When positioning the engine to the transaxle, care must be taken to maintain alignment of the damper spline with the transaxle input shaft.

64. Using the engine crane and spreader bar, position the engine and transaxle together.

65. Install the 6 transaxle-to-engine bolts. Tighten to 48 Nm (35 ft. lbs.).

66. Position the high voltage wiring harness and connect the 2 transaxle electrical connectors.

67. Connect the high voltage wiring harness electrical connector.

68. Using the engine crane and spreader bar, position the engine and transaxle onto the lift table.

✳✳ WARNING

Due to the weight of the transaxle, special care should be taken to mount the powertrain securely to the lift table.

69. Using the special tools, secure the engine to the lift table.

70. Raise the engine and transaxle into the vehicle.

71. Install the rear transaxle mount. Tighten the nuts to 115 Nm (85 ft. lbs.). Tighten the bolt to 90 Nm (66 ft. lbs.).

72. Install the rear transaxle mount brace. Tighten to 25 Nm (18 ft. lbs.).

73. Install the right transaxle mount bracket. Tighten to 90 Nm (66 ft. lbs.).

74. Install the right transaxle mount bracket nut. Tighten to 90 Nm (66 ft. lbs.).

75. Install the engine mount bracket. Tighten to 90 Nm (66 ft. lbs.).

76. Install the engine mount bracket bolt. Tighten to 90 Nm (66 ft. lbs.).

77. Install the 2 transaxle-to-engine bolts. Tighten to 48 Nm (35 ft. lbs.).

78. Install the 2 transaxle-to-engine bolts. Tighten to 48 Nm (35 ft. lbs.).

79. Install the engine support cross-member. Tighten to the bolts to 90 Nm (66 ft. lbs.). Tighten to the nut to 175 Nm (129 ft. lbs.).

80. Install the rear roll restrictor bolt. Tighten to 90 Nm (66 ft. lbs.).

81. Install the front roll restrictor bolt. Tighten to 115 Nm (85 ft. lbs.).

82. Install the A/C compressor and connect the A/C compressor electrical connector. Tighten to 25 Nm (18 ft. lbs.).

83. Connect the lower radiator hose to the radiator.

84. Install the motor electronics coolant pump. Tighten to 7 Nm (62 inch lbs.).

85. Install the ground cable and nut. Tighten to 20 Nm (15 ft. lbs.).

86. Connect the transaxle coolant hoses.

87. Connect the transaxle coolant temperature sensor electrical connector and pin-type retainer.

88. Position the transaxle control cable and connect the pin-type retainer.

89. Connect the pin-type retainer.

90. Connect the transaxle control cable:

a. Install the transaxle control cable bracket and the 2 nuts finger tight.

b. Attach the transaxle control cable to the control lever.

91. Position the control lever between the 2 casting ribs on the transaxle case.

92. Place the gear selector lever in the **D** position.

93. Tighten the transaxle control cable bracket nuts to 22 Nm (16 ft. lbs.).

94. To verify the correct cable adjustment, observe the control lever on the transaxle while an assistant shifts the gear selector lever to each range position ending in the D position.

95. Install the transaxle control snow shield. Tighten to 22 Nm (16 ft. lbs.).

96. Install the auxiliary coolant pump and connect the electrical connector. Tighten to 10 Nm (89 inch lbs.).

➡**Parts removed from view for clarity.**

97. If equipped, install the block heater wiring harness.

a.Route the harness through the radiator support.

b.Attach the 3 harness retainer clips.

98. Install the fuel supply tube and evaporative emissions (EVAP) tube.

99. Connect the DC-to-DC converter electrical connector and pin-type retainer.

100. Connect the transaxle harness electrical connector and install the bolt. Tighten to 10 Nm (89 inch lbs.).

101. Connect the 2 engine control harness electrical connectors and the 2 pin-type retainers.

102. Connect the engine control harness.

a. Engage the powertrain control module (PCM) connector.

b. Install the harness retaining nut.

c. Tighten to 6 Nm (53 inch lbs.).

103. Connect the following electrical connectors:

- Secondary air injection (AIR) solenoid
- Transaxle mount control solenoid
- Purge valve

104. Connect the upper radiator and coolant vent hoses.

105. Attach the heater hose retaining clip to the transaxle mount stud.

106. Connect the heater hoses to the heater core.

4x2 vehicles

107. Install the lateral support crossmember. Tighten to 115 Nm (85 ft. lbs.).

108. Install the front drive intermediate halfshaft.

4x4 vehicles

109. Install the transfer case.

All vehicles

110. Install the left front drive halfshaft.

111. Install the accessory drive belt.

112. Install the engine coolant degas bottle.

113. Install the catalytic converter.

114. Install the engine air cleaner.

115. Install the intake manifold.

116. Re-power the vehicle high voltage electrical system. For additional information, refer to the High Voltage Traction Battery procedure.

117. Connect the 12v battery ground cable.

118. Fill the engine with clean engine oil.

119. Fill the Motor Electronics Cooling System as follows:

➡**Vehicle cooling systems are filled with Motorcraft Premium Gold Engine Coolant. Always fill the cooling system with the same coolant that is present in the system. Do not mix coolant types.**

a. Connect the hoses to the transaxle and install the hose clamps.

✴ WARNING

Adhesives, stop-leak pellets or small debris in the Motor electronics cooling system (MECS) can cause poor performance or temporary blockage of the motor electronics pump. Only use clean, approved coolant when filling the system.

b. Loosen the bleed screw and fill the degas bottle with coolant until it begins to flow out of the bleed hole. Then, close the bleed screw.

c. Turn the ignition to the ON position to actuate the motor electronics cooling pump and continue to fill the degas bottle to the correct level.

➡**Most of the MECS air bleeding occurs as a normal process at the degas bottle through the vent tube; very little occurs at the bleed screw.**

d. To bleed air from the system, loosen the bleed screw and allow air to escape while the M/E coolant pump is operating.

120. Fill and bleed the cooling system.

3.0L Engine

1. Before servicing the vehicle, refer to the Precautions Section.

2. Properly recover the air conditioning system refrigerant.

3. Properly relieve the fuel system pressure.

4. Drain the cooling system.

5. Drain the engine oil.

6. Remove or disconnect the following:

- Hood
- Battery and battery tray
- Air cleaner outlet tube and housing
- Lower radiator air deflectors
- Fuel lines
- Water pump drive belt
- Accelerator cable and speed control cable, if equipped
- Vapor Management Valve (VMV)
- Powertrain Control Module (PCM)
- PCM ground wire
- Thermostat housing and hose assembly and move them aside
- Power distribution box electrical connector
- Power distribution box cover
- Nuts and cables from inside the power distribution box
- Transmission linkage
- Brake booster vacuum hose
- Heater hoses
- Power steering return line
- Power Steering Pressure (PSP) switch electrical connector
- Power steering supply line
- Oil level indicator
- Catalytic converter
- A/C compressor
- Both front wheels
- Intermediate drive shaft, if equipped

7. Separate both side ball joints.

8. Separate both side tie rod ends from the steering knuckles.

9. Separate both sway bar links from the strut mounts.

10. Separate the struts from the steering knuckles.

11. Remove or disconnect the following:

- Both wheel speed sensors, if equipped
- Brake calipers from the steering knuckles and properly support the struts
- Steering shaft from the rack
- Transmission line bracket bolt
- Transmission cooler lines
- Torque converter inspection cover
- Torque converter nuts
- Block heater wiring, if equipped

12. Install a powertrain lifting devise and raise the vehicle.

- Engine support bracket
- Transmission support
- 2 rear subframe bolts
- 2 subframe side bolts
- Motor mount support bolts

- Engine and transmission as an assembly
- Heated Oxygen (HO2S) sensor
- Transmission Range (TR) sensor
- Transmission harness electronic control switch
- Transmission control harness from the bracket
- Starter and wire harness
- Knock Sensor (KS) electrical connector
- Output Shaft Speed (OSS) sensor electrical connector
- HO2S sensor and Exhaust Gas Recirculation (EGR) tube from the exhaust manifold
- Alternator and electrical connectors
- Right side exhaust manifold and gasket
- Halfshaft support bracket and move it aside

13. Separate the engine from the transmission assembly

To install:

14. Install or connect the following:
- Powertrain assembly on the subframe
- Transmission-to-engine bolts. Torque the bolts to 30 ft. lbs. (40 Nm).
- Halfshaft bracket. Torque the bolts to 18 ft. lbs. (25 Nm).
- Right side exhaust manifold and new gasket. Torque the bolts to 15 ft. lbs. (25 Nm).
- Alternator. Torque the larger bolts to 18 ft. lbs. (25 Nm) and smaller bolt to 89 inch lbs. (10 Nm).
- EGR tube and HO2S sensor electrical connectors
- OSS sensor electrical connector
- KS jumper electrical connector
- Starter. Torque the bolts to 18 ft. lbs. (25 Nm).
- Transmission control harness to the bracket. Torque the bolt to 18 ft. lbs. (25 Nm).
- Transmission harness
- Transmission range sensor
- Powertrain assembly
- Motor mount support. Torque the bolts to 66 ft. lbs. (90 Nm).
- Subframe side nuts. Torque the nuts to 76 ft. lbs. (103 Nm). Raise the vehicle and support the powertrain assembly with a lifting device.
- Transmission mount. Torque the bolts to side bolts to 66 ft. lbs. (90 Nm) and the other bolts to 76 ft. lbs. (103 Nm).
- Motor mount. Torque the bolts to side bolts to 66 ft. lbs. (90 Nm)

and the other bolts to 76 ft. lbs. (103 Nm). Remove the powertrain lift.
- Block heater electrical connector, if equipped
- Torque converter. Torque the nuts to 27 ft. lbs. (37 Nm).
- Transmission cover plate and plug
- Transmission cooler lines
- Transmission cooler line bracket. Torque the bolt to 15 ft. lbs. (20 Nm).
- Steering shaft to the rack. Torque the bolt to 18 ft. lbs. (25 Nm).
- Struts to the steering knuckles. Torque the bolts to 75 ft. lbs. (102 Nm).
- Brake calipers to the steering knuckles
- Wheel speed sensors, if equipped. Torque the bolts to 89 inch lbs. (10 Nm).
- Sway bar links to the strut mount. Torque the bolts to 41 ft. lbs. (55 Nm).
- Tie rods to the steering knuckles. Torque the bolts to 41 ft. lbs. (55 Nm).
- Ball joints. Torque the bolts to 52 ft. lbs. (70 Nm).
- Intermediate drive shaft, if equipped
- Both front wheels
- A/C compressor
- Lower radiator air deflectors
- Catalytic converter
- Oil level indicator dipstick tube
- Power steering line and bracket. Torque the bolt to 13 ft. lbs. (17 Nm).
- PSP switch electrical connector
- Power steering return line
- Heater hoses
- Vacuum lines
- Transmission linkage
- Wire harness cables and nuts to the power distribution box. Torque the nuts to 89 inch lbs. (10 Nm).
- Power distribution box wire harness
- Thermostat housing and connect the hoses
- Ground wire. Torque the bolt to 89 inch lbs. (10 Nm).
- PCM electrical connector
- VMV electrical connector
- Accelerator cable and speed control cable, if equipped
- Air cleaner assembly
- Water pump drive belt
- Battery and tray

15. Fill and bleed the cooling system.
16. Fill the engine with clean oil.

17. Recharge the A/C system.
18. Inspect and top off the power steering fluid.
19. Start the vehicle, check for leaks and repair if necessary.

Water Pump

REMOVAL & INSTALLATION

2.0L Engine

1. Before servicing the vehicle, refer to the Precautions Section.
2. Drain the cooling system.
3. Remove or disconnect the following:
- Negative battery cable
- Right front wheel
- Splash shield
- Drive belt
- Water pump pulley
- Water pump

To install:
4. Install or connect the following:
- Water pump. Torque the bolts to 89 inch lbs. (10 Nm).
- Water pump pulley. Torque the bolts to 89 inch lbs. (10 Nm).
- Drive belt
- Splash shield
- Right front wheel
- Negative battery cable

5. Refill the cooling system.
6. Start the vehicle and check for leaks, repair if necessary.

2.3L Engine

1. Before servicing the vehicle, refer to the Precautions Section.
2. With the vehicle in NEUTRAL, position it on a hoist.
3. Drain the cooling system.
4. Remove the accessory drive belt.
5. Remove the coolant pump pulley bolts.
6. Remove the coolant pump pulley.

9308TG08

Exploded view of the water pump—2.0L engine

10 Nm (89 lb-in) — ③

②

④

10 Nm (89 lb-in) — ③

25 Nm (18 lb-ft) — ①

1 Coolant pump pulley bolts
2 Coolant pump pulley
3 Coolant pump bolts
4 Coolant pump
5 Coolant pump O-ring seal

67197-ESCA-G03

Water pump mounting—2.3L engine

7. Remove the coolant pump bolts.
8. Remove the coolant pump.
9. Remove the coolant pump O-ring seal.
10. To install, reverse the removal procedure. Torque the water pump bolts to 89 inch lbs. (10 Nm). Torque the pulley bolts to 18 ft. lbs. (25 Nm).
11. Fill and bleed the cooling system.

3.0L Engine

1. Before servicing the vehicle, refer to the Precautions Section.
2. Drain the cooling system.

10 Nm (89 lb-in)

67197-ESCA-G52

Water pump mounting—3.0L engine

3. Remove or disconnect the following:
• Negative battery cable
• Air cleaner outlet tube
• Water pump belt tensioner
• Coolant hoses
• Water pump
• Water pump from the housing
To install:
• Water pump to the housing. Torque the bolts to 89 inch lbs. (10 Nm).
• Water pump. Torque the bolts to 89 inch lbs. (10 Nm).
• Coolant hoses
• Water pump belt tensioner
• Air cleaner outlet tube
• Negative battery cable
4. Refill the cooling system.
5. Start the vehicle and check for leaks, repair if necessary.

Heater Core

REMOVAL & INSTALLATION

2002–04 Models

1. Before servicing the vehicle, refer to the Precautions Section.
2. Drain the engine coolant.

3. Disconnect the heater hoses from the heater core.
4. Remove the driver air bag module.
5. Remove the two front door scuff plates.
6. Remove the four pin-type retainers.
7. Remove the two front door scuff plates.
8. Remove the two A-pillar lower trim panels.
9. Remove the two pin-type retainers.
10. Remove the two A-pillar lower trim panels.
11. Disconnect the electrical connectors located by the left cowl.
12. Position the hood latch release handle aside.
13. Remove the bolts.
14. Position the hood latch release handle aside.
15. Remove the utility compartment.
16. Remove the four pin-type retainers.
17. Remove the utility compartment.
18. Disconnect the electrical connector.
19. Remove the instrument panel steering column cover.
20. Release the upper clips and rotate the cover outward to release the lower pivot retainers.

21. Remove the steering column lower cover.

22. Remove the screws.

23. Remove the steering column lower cover.

24. If equipped, disconnect the shift cable.

25. Disconnect the shift cable.

26. Disconnect the shift cable from the retaining bracket.

27. Remove the steering column coupler access cover.

28. Disconnect the steering column coupler.

29. Remove the steering column coupler bolt and nut.

30. Disconnect the steering column coupler.

31. Remove the cover panel.

32. Remove the pin-type retainer.

33. Release the retaining clip.

34. Disconnect the electrical connectors.

35. Disconnect the climate control vacuum harness connector.

36. Disconnect the in-line electrical connector.

37. Remove the four instrument panel center brace bolts.

38. Remove the passenger air bag module.

39. Disconnect the vacuum harness connector.

40. Disconnect the temperature control cable.

41. Position the locator pin.

42. Release the locking tab.

43. Disconnect the temperature control cable from the blend door shaft.

44. Close the glove compartment.

45. Press the release tabs inward while raising the glove compartment.

46. Disconnect the electrical connectors at the blower motor.

47. Disconnect the antenna cable in-line connector.

48. Open the four A-pillar passenger assist handle covers.

49. Remove the two A-pillar passenger assist handles.

50. Remove the four bolts.

51. Remove the two A-pillar passenger assist handles.

52. Remove the two windshield side garnish moldings.

53. Remove the instrument panel cowl top cover.

54. Remove the instrument panel cowl top bolt.

55. Loosen the tilt lever (if equipped) and lower steering column.

56. Position the transmission range selector lever (if equipped) down to provide

access to the instrument cluster finish panel and instrument cluster.

57. Remove the screws and the instrument cluster finish panel.

58. Remove the screws.

59. Disconnect the electrical connectors and remove the instrument cluster.

60. Through the instrument cluster opening, remove the instrument panel nut.

61. Remove the two instrument panel finish end panels.

62. Remove the four instrument panel cowl side bolts.

➡ **This step requires an assistant.**

63. Remove the instrument panel.

64. Remove the heater blending door levers.

65. Remove the screw for heater blending door.

66. Remove the levers for the blending door.

67. Remove the heater core.

68. Remove the three screws.

69. Remove the cover for the heater core and pull the heater core out of the housing.

➡ **Before installing the temperature control cable, make sure the blend door, cable and temperature switch are correctly positioned.**

70. To install, reverse the removal procedure.

✳✳ CAUTION

Electronic modules are sensitive to static electrical charges. If exposed to these charges, damage may result.

✳✳ CAUTION

Once the new module is installed, it is necessary to download the module configuration information from the scan tool into the new instrument cluster.

2005 Models

1. Before servicing the vehicle, refer to the Precautions Section.

2. Drain the engine coolant.

3. Release the 2 clamps and disconnect the 2 heater hoses from the heater core.

4. De-power the supplemental restraint system.

5. Position the seats forward and remove the 2 rear bolts.

6. Position the seats rearward and disconnect the battery.

1 Temperature blend door lever
2 Heater core cover screws (3 required)
3 Heater core cover
4 Heater core

Heater core

➥ **For the Escape Hybrid, when the battery is disconnected and connected, the brake pedal needs to be calibrated. After the battery has been connected, with the vehicle in park, turn the key to the ON position. Press the brake pedal firmly, then fully release to calibrate the brake pedal.**

7. Position the parking brake handle to the full-up position.

8. Remove the transmission selector lever bezel.

9. Release the parking brake handle boot from the floor console finish panel.

10. On the Mariner, remove the floor console top panel.

➥ **On Escape, if removing the floor console storage bin, squeeze the front and rear of the storage bin to release the retaining tabs from the floor console finish panel.**

11. Remove the floor console finish panel. Disconnect the electrical connectors.

12. Remove the 6 bolts and remove the floor console.

13. Remove the 4 pin-type retainers and the 2 front door scuff plates.

14. Remove the 2 pin-type retainers and the 2 A-pillar lower trim panels.

15. Remove the instrument panel steering column opening cover.

16. Remove the left and right instrument panel end trim panels.

17. Lower the tilt steering column to the lowest position.

18. Remove the 2 screws and the instrument cluster finish panel.

19. Remove the 4 instrument cluster screws and the instrument cluster. Disconnect the electrical connector.

20. Disconnect the main electrical connector.

21. Remove the ground wire bolt and position the ground wire aside.

22. Remove the 2 bolts and position the hood release handle aside.

23. Remove the steering column coupler access cover.

24. Remove the steering column pinch bolt and disconnect the intermediate shaft.

25. Remove the cover panel pin-type retainer and the cover panel. Release the retaining clip from the instrument panel center brace.

26. Disconnect the climate control vacuum harness connector.

27. Disconnect the restraint control module (RCM) electrical connector.

28. Disconnect the temperature control cable from the blend door shaft.

a. Align the locator holes.
b. Release the locking tab.
c. Disconnect the temperature control cable.

➥ **Automatic transmission selector lever is shown; the manual transmission selector lever is similar.**

29. Remove the 4 transmission selector lever bolts and position the transmission selector lever aside. Disconnect the electrical connectors.

30. Remove the bolts and position the parking brake control aside Disconnect the electrical connectors.

31. Open the glove compartment. Press the release tabs inward while lowering the glove compartment.

32. Disconnect the blower motor electrical connectors.

33. Disconnect the antenna cable in-line connector.

34. Remove the cover and the instrument panel cowl top bolt.

35. Remove the 4 instrument panel center brace bolts.

36. Remove the instrument panel cluster opening nut through the instrument cluster opening.

37. Remove the 4 instrument panel cowl side bolts.

➥ **This step requires an assistant.**

38. Remove the instrument panel.

39. Remove the screw and the temperature blend door lever.

40. Remove the 3 screws and the heater core cover.

41. Remove the heater core from the housing.

➥ **Make sure the temperature blend door actuator and switch are in the correct position when installing the temperature blend door lever.**

➥ **Lubricate the coolant hoses with plain water only if needed.**

42. To install, reverse the removal procedure. Observe the following torques:
- the 4 instrument panel cowl side bolts: 9 Nm (80 lb-in)
- the instrument panel cluster opening nut: 9 Nm (80 lb-in)
- The 4 instrument panel center brace bolts: 25 Nm (18 lb-ft)
- The instrument panel cowl top bolt: 6 Nm (53 lb-in)
- The parking brake control: 24 Nm (18 lb-ft)
- The 4 transmission selector lever bolts: 22 Nm (16 lb-ft) for auto-

matic transaxle; 20 Nm (15 lb-ft) for manual transaxle
- The steering column pinch bolt: 23 Nm (17 lb-ft)
- The hood release handle: 6 Nm (53 lb-in)

43. Fill and bleed the engine cooling system.

44. If equipped with automatic transmission, adjust the linkage.

45. Re-power the supplemental restraint system.

46. Following installation of the new instrument cluster, download the module configuration information from the diagnostic tool into the new module as follows:

Using the Vehicle Communication Module (VCM) When the Original Body Chassis Electrical Module is Not Available

47. Install the new module.

48. Using the VCM and the latest version of the service function card, SELECT: Programmable Module Installation.

49. Select the module being installed.

50. Follow the on-screen instructions.

51. SELECT: Retrieve Module Configuration—Old ECU and press trigger.

52. Follow the on-screen instructions.

53. The VCM attempts to retrieve the module data from the powertrain control module (PCM). If the module data is available, go to Step A. If the VCM displays: Call As-Built Data Center, go to Step B.

Step A

54. SELECT: Restore Configuration—New ECU. Press trigger.

55. The VCM completes loading the retrieved data and displays Module Download Successful.

56. Test the module for correct operation.

Step B

57. Press the trigger.

58. If the VCM asks for vehicle data, enter the vehicle data, then press store.

59. The VCM asks for module data line 1. Enter the data and press store.

60. The VCM then asks if there is an additional line of data available for that address. Select YES or NO depending on the information in the As Built Data Sheet.

61. Repeat Steps 3 and 4 until the answer is NO for Step 4.

62. The VCM should show a screen stating that the module data was stored. Press the trigger.

63. Follow the on-screen instructions.

64. SELECT: Restore Configuration—New ECU. Press the trigger.

65. The VCM completes loading the

retrieved data and displays Module Download Successful.

66. Test the module for correct operation.

Using the Vehicle Communication Module (VCM) When the Original Body Chassis Electrical Module is Available

67. With the original module still installed, using the VCM and the latest version of the service function card, SELECT: Programmable Module Installation.

68. Select the module being installed and press the trigger.

69. Follow the on-screen instructions.

70. SELECT: Retrieve Module Configuration—Old ECU. Press the trigger.

71. Follow the on-screen instructions.

72. INSTALL new module, SELECT: Restore Configuration—New ECU. Press the trigger.

73. The VCM completes loading the retrieved data and displays Module Download Successful.

74. Test the module for correct operation.

Using the Worldwide Diagnostic System (WDS) When the Original Body Chassis Electrical Module is Not Available

75. Install the new module.

76. Connect the WDS and ID the vehicle as normal.

77. From the Toolbox icon, select and highlight Module Programming. Then highlight the module that was installed and press the check mark.

78. Select and highlight Programmable Module Installation. Then highlight the module that was installed and press the check mark.

79. Follow the on-screen instructions, turn the ignition key to the OFF position and press the check mark.

80. The WDS retrieves the module data from the PCM, automatically downloads the data into the new module, and displays Module Configuration Complete.

81. If the data is not available in the PCM, the WDS displays a screen stating to contact the As-Built Data Center. Retrieve the data from WWW.FMCDEALER.COM at this time and press the check mark.

82. Enter the module data (the module address and line are displayed to the left of the 3 entry boxes) and press the check mark.

83. The WDS downloads the data into the new module and displays Operation Successful—Programming Complete.

84. Test the module for correct operation.

Using the Worldwide Diagnostic System (WDS) When the Original Body Chassis Electrical Module is Available

85. Connect the WDS and ID the vehicle as normal.

86. From the Toolbox icon, select and highlight Module Programming and press the check mark.

87. Select and highlight Programmable Module Installation.

88. Follow the on-screen instructions, turn the ignition key to the OFF position, and press the check mark.

89. Install the new module and press the check mark.

90. Follow the on-screen instructions, turn the ignition key to the ON position, and press the check mark.

91. The module configuration is complete.

92. Test the module for correct operation.

Cylinder Head

REMOVAL & INSTALLATION

2.0L Engine

1. Before servicing the vehicle, refer to the Precautions Section.

2. Properly relieve the fuel system pressure.

3. Drain the engine oil.

4. Remove or disconnect the following:
- Negative battery cable
- Ignition coil bracket
- Thermostat housing
- Positive Crankcase Ventilation (PCV) tube
- Intake manifold
- Exhaust manifold
- Power steering bracket and move it aside
- Valve tappets
- Engine mount lower bracket
- Engine mount upper bracket
- Cylinder head bolts in the proper sequence and discard the gasket

To install:

5. Install a new head gasket and the cylinder head.

6. Lubricate the cylinder head bolt threads.

7. Torque the cylinder head bolts in the proper sequence as follows:
   a. Step 1: 15 ft. lbs. (20 Nm).
   b. Step 2: 30 ft. lbs. (40 Nm).
   c. Step 3: Plus an additional 90 degrees.

8. Install or connect the following:
- Engine mount upper bracket. Torque the 2 upper bolts to 72 ft. lbs. (98 Nm) and the center bolt to 57 ft. lbs. (77 Nm).
- Engine mount lower bracket. Torque the bolts to 37 ft. lbs. (50 Nm).
- Valve tappets

Cylinder head bolt torque sequence 2.0L engine

9308TG04

- Power steering pump bracket. Torque the bolts to 20 ft. lbs. (28 Nm).
- Exhaust manifold
- Intake manifold
- PCV tube
- Thermostat housing
- Ignition coil bracket
- Negative battery cable

9. Fill the engine with clean oil and replace the filter.

10. Start the vehicle and check for leaks, repair if necessary.

2.3L Engine

1. Before servicing the vehicle, refer to the Precautions Section.

> **✳✳ WARNING**
>
> During engine repair procedures, cleanliness is extremely important. Any foreign material, including any material created while cleaning gasket surfaces that enters the oil passages, coolant passages or the oil pan can cause engine failure.

> **✳✳ WARNING**
>
> The crankshaft, the crankshaft sprocket and the pulley are fitted together by friction, using diamond washers between the flange faces on each part. For that reason, the crankshaft sprocket is also unfastened if you loosen the pulley. Therefore, the engine must be retimed each time the damper is removed. Otherwise severe engine damage can occur.

1 Radio ignition interference capacitor electrical connector
2 Exhaust gas recirculation (EGR) valve electrical connector
3 Upper radiator hose clamp
4 Upper radiator hose (position aside)
5 EGR coolant tube clamp
6 EGR coolant hose (part of heater hose) (position aside)
7 Engine coolant vent hose clamp
8 Engine coolant vent hose (position aside)

9 Heater hose clamp
10 Heater hose (position aside)
11 Bypass hose clamp
12 Bypass hose (position aside)
13 Cylinder head bolt
14 Cylinder head
15 Cylinder head gasket

Cylinder head removal—2.3L engine

67197-ESCA-G04

2. With the vehicle in NEUTRAL, position it on a hoist.

3. Remove the camshafts.

4. Remove the intake manifold.

5. Remove the catalytic converter.

6. Disconnect the radio ignition interference capacitor electrical connector

7. Disconnect the exhaust gas recirculation (EGR) valve electrical connector

8. Remove the upper radiator hose.

9. Remove the EGR coolant tube clamp.

10. Remove the EGR coolant hose.

11. Remove the engine coolant vent hose.

12. Remove the heater hose.

13. Remove the bypass hose.

14. Remove and discard the cylinder head bolts.

15. Remove the cylinder head.

16. Remove the cylinder head gasket.

17. Inspect the cylinder head for distortion.

❋❋ WARNING

Do not use metal scrapers, wire brushes, power abrasive discs or other abrasive means to clean the sealing surfaces. These tools cause scratches and gouges that make leak paths. Use a plastic scraping tool to remove all traces of the head gasket.

❋❋ WARNING

Observe all warnings or cautions and follow all application directions contained on the packaging of the silicone gasket remover and the metal surface prep.

➡ **If there is no residual gasket material present, metal surface prep can be used to clean and prepare the surfaces.**

18. Clean the cylinder head-to-cylinder block mating surface of both the cylinder head and the cylinder block.

19. Remove any large deposits of silicone or gasket material with a plastic scraper.

20. Apply silicone gasket remover, following package directions, and allow to set for several minutes.

21. Remove the silicone gasket remover with a plastic scraper. A second application of silicone gasket remover may be required if residual traces of silicone or gasket material remain.

22. Apply metal surface prep, following package directions, to remove any traces of oil or coolant, and to prepare the surfaces to bond with the new gasket. Do not attempt to make the metal shiny. Some staining of the metal surfaces is normal.

23. Apply silicone gasket and sealant to the locations shown.

24. Install a new head gasket.

➡ **The cylinder head bolts are torque-to-yield and must not be reused. New cylinder head bolts must be installed.**

➡ **Lubricate the bolts with clean engine oil prior to installation.**

25. Install new cylinder head bolts. Tighten the bolts in the sequence shown in five stages.

    a. Tighten the bolts to 5 Nm (44 inch lbs.).

    b. Tighten the bolts to 15 Nm (11 ft. lbs.).

    c. Tighten the bolts to 45 Nm (33 ft. lbs.).

    d. Turn the bolts 90 degrees.

    e. Turn the bolts an additional 90 degrees.

26. To install, reverse the removal procedure.

3.0L Engine

The procedure for the left side cylinder head and right side are similar. Changes in the procedure will be noted for either side cylinder head.

1. Before servicing the vehicle, refer to the Precautions Section.

2. Properly relieve the fuel system pressure.

• Drain the cooling system.

06017-ESCA-G18

Apply silicone gasket and sealant to the locations shown—2.3L engine

3. Remove or disconnect the following:

• Negative battery cable
• Camshaft
• Exhaust Gas Recirculation (EGR) tube, right side only
• Exhaust manifold
• Camshaft followers
• Hydraulic lash adjusters and matchmark them for proper installation
• Cylinder head bolts in sequence and discard them
• Cylinder head and discard the gasket

To install:

4. Install a new head gasket and the cylinder head.

5. Lubricate the cylinder head bolt threads.

6. Torque the cylinder head bolts in the proper sequence as follows:

    a. Step 1: 30 ft. lbs. (40 Nm).

    b. Step 2: Additional 90 degrees.

67197-ESCA-G05

Cylinder head bolt torque sequence—2.3L engine

Left side cylinder head bolt torque sequence 3.0L engine

Right side cylinder head bolt torque sequence 3.0L engine

Right side exhaust manifold bolt torque sequence–3.0L engine

c. Step 3: Loosen the bolts one full turn.

d. Step 4: 30 ft. lbs. (40 Nm).

e. Step 5: Plus an additional 90 degrees.

f. Step 6: Plus an additional 90 degrees.

7. Install or connect the following:
- Hydraulic lash adjusters
- Camshaft followers
- Camshaft
- Exhaust manifold. Torque the bolts in sequence to 15 ft. lbs. (20 Nm), right side only
- EGR tube, right side only
- Coolant bypass tube
- Negative battery cable

8. Fill the coolant to the proper level.

9. Start the vehicle and check for leaks, repair if necessary.

Intake Manifold

REMOVAL & INSTALLATION

2.0L Engine

1. Before servicing the vehicle, refer to the Precautions Section.

2. Properly relieve the fuel system pressure.

3. Remove or disconnect the following:
- Negative battery cable
- Fuel injection supply manifold

Intake manifold bolt loosening sequence—2.0L engine

18 Nm (13 lb-ft)

Tighten the intake manifold bolts in the sequence shown—2.0L engine

- Throttle Position (TP) sensor electrical connector
- Idle Air Control (IAC) electrical connector and unclip the harness from the bracket
- Main engine control sensor wiring
- Connector from the bracket
- Powertrain Control Module (PCM) wire harness from the bracket
- Brake booster vacuum hose

- 4 additional vacuum lines
- Positive Crankcase Ventilation (PCV) hose from the intake manifold
- Knock Sensor (KS) electrical connector
- Alternator
- Intake manifold and discard the gasket

4. Clean the mating surfaces.

To install:

5. Install or connect the following:
   - New gasket
   - Intake manifold. Torque the bolts, in sequence, to 13 ft. lbs. (18 Nm).
   - Alternator
   - KS electrical connector
   - PCV vacuum line
   - 4 vacuum lines
   - Brake booster vacuum supply hose

1 Vacuum tube retainer	5 Idle air control (IAC) motor electrical connector
2 Vacuum tube	6 Swirl control valve electrical connector
3 Vacuum supply hose	7 Knock sensor (KS) electrical connector
4 Fuel vapor return hose	8 Pin-type retainer

9 Temperature manifold absolute pressure (TMAP) sensor electrical connector
10 Oil pressure sender electrical connector
11 Engine control wiring harness

67197-ESCA-G06

Intake manifold and related parts—2.3L engine, exc. Hybrid

- PCM wire harness to the bracket
- Main engine control sensor wiring
- IAC valve electrical connector and attach the harness to the bracket
- TP sensor electrical connector
- Fuel injection supply manifold
- Negative battery cable

6. Start the vehicle and check for leaks, repair if necessary.

2.3L Engine

EXCEPT HYBRID

1. Before servicing the vehicle, refer to the Precautions Section.

2. With the vehicle in NEUTRAL, position it on a hoist.

3. Remove the throttle body.

4. Remove the fuel rail.

5. Remove the oil level indicator tube.

6. Remove the vacuum tube.

7. Remove the vacuum supply hose.

8. Remove the fuel vapor return hose.

9. Remove the idle air control (IAC) motor electrical connector.

10. Remove the swirl control valve electrical connector.

11. Remove the knock sensor (KS) electrical connector.

12. Remove the temperature manifold absolute pressure (TMAP) sensor electrical connector.

13. Remove the oil pressure sender electrical connector.

14. Remove the engine control wiring harness.

15. Remove the intake manifold bolts.

➡There are three different size bolts used. Mark the location of the bolts to make sure they are installed in the correct location.

16. Remove the bolts and position the intake manifold aside to access the crankcase vent hose clamp and the EGR tube.

17. Remove the crankcase vent hose.

18. Remove the exhaust gas recirculation (EGR) tube.

19. Remove the intake manifold.

20. Remove the intake manifold gasket.

21. To install, reverse the removal procedure. Torque the intake manifold bolts to 13 ft. lbs. (18 Nm).

ESCAPE HYBRID

1. Before servicing the vehicle, refer to the Precautions Section.

2. With vehicle in NEUTRAL, position it on a hoist.

3. Remove the throttle body.

4. Remove the fuel rail.

5. Loosen the clamp and disconnect the secondary air injection (AIR) tube from the AIR pump hose.

6. Remove the right and left splash shields.

7. Remove the 2 bolts and position the motor electronics coolant pump aside.

12 Intake manifold bolts

13 Crankcase vent hose clamp

14 Crankcase vent hose (position aside)

15 Exhaust gas recirculation (EGR) tube

16 Intake manifold

17 Intake manifold gasket

Intake manifold installation—2.3L engine, exc. Hybrid

67197-ESCA-G07

8. Disconnect the AIR pump electrical connector.

9. Remove the 2 bolts and the AIR pump.

10. Remove the 5 bolts and the right splash shield.

11. Remove the lower oil level indicator tube bolt.

12. Remove the oil level indictor.

13. Remove the upper bolt and the oil level indicator tube.

14. Remove and discard the oil level indicator tube O-ring seal.

15. Disconnect the oil pressure sender electrical connector.

16. Disconnect the TMAP sensor electrical connector.

17. Disconnect the swirl control valve electrical connector.

18. Disconnect the IMRC actuator electrical connector.

19. If equipped, disconnect the block heater electrical harness connector.

20. Disconnect the KS electrical connector and detach the pin-type retainer.

21. Disconnect the fuel vapor return hose from the intake manifold.

22. Disconnect the power brake booster vacuum tube. Depress the quick release

locking ring. Pull the vacuum tube out of the quick release fitting.

➡There are 3 different size bolts used. Mark the location of the bolts to make sure they are installed in the correct location.

23. Remove the 8 bolts and position the intake manifold aside to access the crankcase vent hose clamp and the EGR tube.

24. Release the clamp and disconnect the crankcase vent hose.

25. Remove the EGR tube.

1 Oil pressure sender electrical connector

2 Temperature manifold absolute pressure (TMAP) sensor electrical connector

3 Swirl control valve electrical connector

4 Intake manifold runner control (IMRC) actuator electrical connector

5 Block heater electrical harness connector (if equipped)

6 Knock sensor (KS) electrical connector

7 KS electrical connector

8 Fuel vapor return hose

9 Vacuum supply hose

06017-ESCA-G19

Intake manifold and related parts—2.3L Hybrid engine

10 Intake manifold bolts (8 required)

11 Crankcase vent hose clamp

12 Crankcase vent hose

13 Exhaust gas recirculation (EGR) tube

14 Intake manifold

15 Intake manifold gasket

06017-ESCA-G20

Intake manifold installation—2.3L Hybrid engine

26. Remove the intake manifold and gaskets.

27. To install, reverse the removal procedure. Inspect and install new intake manifold gaskets if necessary. Observe the following torques:
- EGR tube: 55 Nm (41 ft. lbs.)
- Intake manifold: 18 Nm (13 ft. lbs.)
- Upper oil level indicator tube bolt: 10 Nm (89 inch lbs.)
- Lower oil level indicator tube bolt: 47 Nm (35 ft. lbs.)
- AIR pump: 30 Nm (22 ft. lbs.)
- Motor electronics coolant pump: 7 Nm (62 inch lbs.)

3.0L Engine

UPPER

1. Before servicing the vehicle, refer to the Precautions Section.

2. Properly relieve the fuel system pressure.

3. Drain the coolant system.

4. Remove or disconnect the following:
- Negative battery cable
- Air cleaner outlet tube

1 Electronic throttle body electrical connector

2 Coolant inlet hose clamp

3 Coolant inlet hose

4 Coolant outlet hose clamp

5 Coolant outlet hose

6 Throttle body-to-upper intake manifold bolts (4 required)

7 Electronic throttle body

8 Gasket

Throttle body—2.3L Hybrid engine

06017-ESCA-G21

- Engine appearance cover
- Throttle cable
- Speed control cable, if equipped
- Throttle cable bracket
- Throttle Position (TP) sensor electrical connector
- Idle Air Control (IAC) valve electrical connector
- Exhaust Gas Recirculation (EGR) valve vacuum hose and tube
- EGR vacuum regulator valve electrical connector and hose
- Chassis vacuum hose
- Engine vacuum hose
- Positive Crankcase Ventilation (PCV) hose
- Vapor Management Valve (VMV) vacuum hose
- Electrical connectors from the left side of the upper intake manifold
- Power Steering Pressure (PSP) sensor electrical connector
- Upper intake manifold and discard the gasket

5. Clean the mating surfaces.

To install:

6. Install or connect the following:
- New gasket
- Intake manifold. Torque the bolts, in sequence, to 89 inch lbs. (10 Nm).
- PSP electrical connector
- Electrical connectors on the left side of the upper intake manifold
- VMV vacuum hose
- Chassis, engine and PCV hoses
- EGR valve vacuum regulator
- EGR valve vacuum hose and tube. Torque the nut to 30 ft. lbs. (40 Nm).
- TP sensor electrical connector
- IAC valve electrical connector
- Throttle cable and speed control cable, if equipped. Torque the bracket bolts to 89 inch lbs. (10 Nm).
- Air cleaner outlet tube
- Engine appearance cover. Torque the bolts to 53 inch lbs. (6 Nm).
- Negative battery cable

7. Fill the coolant system to the proper level.

8. Start the vehicle and check for leaks, repair if necessary.

LOWER

1. Before servicing the vehicle, refer to the Precautions Section.

2. Properly relieve the fuel system pressure.

3. Remove or disconnect the following:
- Negative battery cable
- Fuel line spring lock coupling

Tighten the upper intake manifold bolts in the sequence shown—3.0L engine

Tighten the lower intake manifold bolts in the sequence shown—3.0L engine

- Upper intake manifold
- Fuel rail
- Fuel injector electrical connectors
- Fuel pressure damper vacuum line
- Lower intake manifold
- Lower intake manifold from the fuel rail
- Fuel injectors from the manifold and discard the gasket

4. Clean the mating surfaces.

To install:

5. Inspect the fuel injector O-rings and replace if necessary.

6. Install or connect the following:
- Fuel injectors into the lower intake manifold
- Fuel rail. Torque the bolts to 89 inch lbs. (10 Nm).
- New gasket
- Intake manifold. Torque the bolts, in sequence, to 89 inch lbs. (10 Nm).
- Fuel rail electrical connectors
- Fuel injector electrical connectors
- Fuel pressure damper vacuum line
- Upper intake manifold

- Fuel line spring lock coupling
- Negative battery cable

7. Start the vehicle and check for leaks, repair if necessary.

Exhaust Manifold

REMOVAL & INSTALLATION

2.0L Engine

1. Before servicing the vehicle, refer to the Precautions Section.
2. Remove or disconnect the following:
   - Negative battery cable
   - Catalytic converter
   - Oil level indicator tube and bracket

Exhaust manifold bolt torque sequence–2.0L engine

1 Catalytic converter bracket bolts (2 required)	5 Catalyst monitor sensor
2 Heat shield bolts (6 required)	6 Heated oxygen sensor (HO2S)
3 Heat shield	7 Catalytic converter
4 Exhaust manifold nuts (7 required)	8 Exhaust manifold gasket

Catalytic converter/exhaust manifold—2.3L engine, exc. Hybrid

06017-ESCA-G22

- Exhaust manifold and discard the gasket

To install:

3. Clean the sealing surfaces of any old gasket material.

4. Install or connect the following:
    - Exhaust manifold and new gasket. Torque the bolts to 12 ft. lbs. (16 Nm).
    - Oil level indicator tube and bracket. Torque the bolt to 89 inch lbs. (10 Nm).
    - Catalytic converter
    - Negative battery cable

5. Start the vehicle and check for leaks, repair if necessary.

2.3L Engine

EXCEPT HYBRID

1. Before servicing the vehicle, refer to the Precautions Section.

> ❋❋ **WARNING**
>
> **Do not use oil or grease-based lubricants on the insulators. They may cause deterioration of the rubber.**

> ❋❋ **WARNING**
>
> **Oil or grease-based lubricants on the insulators may cause the exhaust hanger insulator to separate from the exhaust hanger bracket during vehicle operation.**

➡ Exhaust fasteners are of a torque prevailing design. Use only new fasteners with the same part number as the original. Torque values must be used as specified during reassembly to make sure of correct retention of exhaust components.

2. Remove the flex pipe nuts.
3. Remove the flex pipe gasket.
4. Remove the manifold bracket bolts.
5. Remove the heat shield.
6. Remove the exhaust manifold nuts.
7. Remove the catalyst monitor sensor.
8. Remove the heated oxygen sensor.
9. Remove the exhaust manifold.

10. To install, reverse the removal procedure. Make sure to apply anti-seize lubricant to the threads of the sensors before installation. Failure to tighten the exhaust manifold nuts to specification before installing the manifold bracket bolts will cause the manifold to develop an exhaust gas leak.

11. Observe the following torques:
    - Exhaust manifold-to-head, in the

Exhaust manifold-to-head torque sequence—2.3L engine

06017-ESCA-G23

2 - 47 Nm (35 lb-ft)
5 - 47 Nm (35 lb-ft)
7
6
47 Nm (35 lb-ft) - 4
47 Nm (35 lb-ft) - 3
25 Nm (18 lb-ft) - 1

1	Catalytic converter bracket bolts (2 required)	4	CMS (if equipped)
2	Exhaust manifold nuts (7 required)	5	Heated oxygen sensor (HO2S)
3	Catalyst monitor sensor (CMS)	6	Catalytic converter
		7	Exhaust manifold gasket

06017-ESCA-G24

Exhaust manifold/catalytic converter—2.3L Hybrid engine

sequence shown: 35 ft. lbs. (47 Nm), then recheck at 35 ft. lbs. (47 Nm)
- Flex pipe-to-manifold: 18 ft. lbs. (25 Nm)
- Heated oxygen sensor: 35 ft. lbs. (47 Nm)
- Catalyst monitor sensor: 30 ft. lbs. (40 Nm)

12. Check the exhaust system for proper alignment.

HYBRID

1. Before servicing the vehicle, refer to the Precautions Section.
2. Remove the exhaust flexible pipe.
3. Remove the 2 catalytic converter bracket bolts.

➡ **If installing a new converter, remove the heated oxygen sensor (HO2S) and the catalyst monitoring sensor (CMS) for installation in the new converter.**

4. Disconnect the HO2S and the CMS electrical connectors.
5. Remove and discard the 7 exhaust manifold nuts.
6. Remove the catalytic converter from the vehicle. Discard the exhaust manifold gasket.

To install:

> ✳ **WARNING**
>
> **If the warpage is greater than 0.76 mm (0.03 inch) replace the catalytic converter.**

7. Inspect the catalytic converter for warpage across the manifold flange area.
8. Position a new exhaust manifold gasket on the engine.

> ✳✳ **WARNING**
>
> **Failure to tighten the catalytic converter nuts to specification before installing the converter bracket bolts will cause the converter to develop an exhaust leak.**

> ✳ **WARNING**
>
> **Failure to tighten the catalytic converter nuts to specification a second time will cause the converter to develop an exhaust leak.**

9. Position the catalytic converter and tighten the 7 exhaust manifold nuts in the sequence shown.
- Step 1: Tighten to 47 Nm (35 ft. lbs.).
- Step 2: Retighten to 47 Nm (35 ft. lbs.).

➡ **Apply anti-seize lubricant to the threads of the sensors before installation.**

10. If installing a new converter, install the HO2S and the CMS. Tighten to 47 Nm (35 ft. lbs.).
11. Connect the HO2S and the CMS electrical connectors.
12. Install the 2 catalytic converter bracket bolts. Tighten to 25 Nm (18 ft. lbs.).
13. Install the exhaust flexible pipe. Tighten to 25 Nm (18 ft. lbs.)

3.0L Engine

LEFT SIDE

1. Before servicing the vehicle, refer to the Precautions Section.

2. Remove or disconnect the following:
- Negative battery cable
- Heated Oxygen (HO2S) sensor and catalyst monitor
- Splash shield
- Exhaust crossover pipe
- Drive belt
- A/C compressor and move it aside
- Exhaust manifold and discard the gasket

To install:

3. Clean the sealing surfaces of any old gasket material.
4. Install or connect the following:
- Exhaust manifold and new gasket. Torque the bolts to 15 ft. lbs. (20 Nm).
- A/C compressor. Torque the bolts to 18 ft. lbs. (20 Nm).

Left side exhaust manifold bolt torque sequence–3.0L engine

9308TG11

Right side exhaust manifold bolt torque sequence–3.0L engine

9308TG12

- Drive belt
- Exhaust crossover pipe. Torque the bolts to 30 ft. lbs. (40 Nm).
- Splash shield. Torque the bolts to 80 inch lbs. (9 Nm).
- Left side HO2S sensor and catalyst monitor
- Negative battery cable

5. Start the vehicle and check for leaks, repair if necessary.

RIGHT SIDE

1. Before servicing the vehicle, refer to the Precautions Section.
2. Remove or disconnect the following:
   - Negative battery cable
   - Exhaust Gas Recirculation (EGR) tube
   - Alternator
   - Right side Heated Oxygen (HO2S) sensor
   - Right side exhaust manifold and discard the gasket

To install:

3. Clean the sealing surfaces of any old gasket material.
4. Install or connect the following:
   - Exhaust manifold and new gasket. Torque the bolts to 15 ft. lbs. (20 Nm).
   - Right side HO2S sensor
   - Alternator
   - EGR tube
   - Negative battery cable

5. Start the vehicle and check for leaks, repair if necessary.

Front Crankshaft Seal

REMOVAL & INSTALLATION

2.0L Engine

1. Before servicing the vehicle, refer to the Precautions Section.
2. Remove or disconnect the following:
   - Negative battery cable
   - Timing belt
   - Crankshaft sprocket and timing belt guide
   - Crankshaft oil seal

➡ **Be careful not to damage the seal surface of the cover.**

To install:

3. Install or connect the following:
   - New front crankshaft oil seal
   - Timing belt guide and crankshaft sprocket
   - Timing belt
   - Negative battery cable

4. Start the vehicle and check for leaks, repair if necessary.

2.3L Engine

1. Before servicing the vehicle, refer to the Precautions Section.

✳✳ WARNING

During engine repair procedures, cleanliness is extremely important. Any foreign material, including any material created while cleaning gasket surfaces that enters the oil passages, coolant passages or the oil pan can cause engine failure.

✳✳ WARNING

The crankshaft, the crankshaft sprocket and the pulley are fitted together by friction, using diamond washers between the flange faces on each part. For that reason, the crankshaft sprocket is also unfastened if you loosen the pulley. Therefore, the engine must be retimed each time the damper is removed. Otherwise severe engine damage can occur.

	Alignment Plate, Camshaft
	Timing Peg, Crankshaft
	Holding Fixture, Drive Pinion Flange
	Adapter for 205-126
	Remover, Oil Seal
	Installer, Front Oil Seal

Tools necessary for this job—2.3L engine

67197-ESCA-G08

1 Crankshaft pulley bolt
2 Crankshaft pulley washer
3 Crankshaft pulley
4 Crankshaft front seal

67197-ESCA-G09

Crankshaft pulley and seal—2.3L engine

2. With the vehicle in NEUTRAL, position it on a hoist.

3. Remove the accessory drive belt.

4. Remove the valve cover.

✷✷ WARNING

Failure to position the No. 1 piston at top dead center (TDC) can result in damage to the engine. Turn the engine in the normal direction of rotation only.

5. Using the crankshaft pulley bolt, turn the crankshaft clockwise to position the No. 1 piston at top dead center (TDC).

✷✷ WARNING

The special tool 303-465 is for camshaft alignment only. Using this tool to prevent engine rotation can result in engine damage.

➡The camshaft timing slots are offset. If the special tool cannot be installed, rotate the crankshaft one complete revolution clockwise to correctly position the camshafts.

6. Install the special tool in the slots on the rear of both camshafts.

➡Installing the special tool in this step will prevent the engine from being rotated in the clockwise direction.

7. Install special tool 303-507.

8. Remove the crankshaft pulley bolt and washer.

9. Remove the engine plug bolt.

10. Install the crankshaft holding tools.

✷✷ WARNING

Failure to hold the crankshaft pulley in place while loosening the bolt can result in damage to the engine.

11. Remove the crankshaft pulley.

✷✷ WARNING

Use care not to damage the engine front cover or the crankshaft when removing the seal.

12. Using the special tool, remove the crankshaft front oil seal.

➡Remove the through-bolt from the special tool.

67197-ESCA-G10

Camshaft holding tool—2.3L engine

67197-ESCA-G11

Install special tool 303-507—2.3L engine

67197-ESCA-G12

Install the crankshaft holding tools—2.3L engine

67197-ESCA-G13

Using the special tool, remove the crank-shaft front oil seal—2.3L engine

To install:

➡**Lubricate the oil seal with clean engine oil.**

13. Using the a seal driver, install the crankshaft front oil seal.

➡**Do not reuse the crankshaft damper bolt.**

➡**Apply clean engine oil on the seal area before installing.**

14. Install the crankshaft pulley and hand-tighten the bolt.

✳✳ WARNING

Only hand-tighten the bolt or damage to the front cover can occur.

➡**This step will correctly align the crankshaft pulley to the crankshaft.**

15. Install a standard 6 mm x 18 mm bolt through the crankshaft pulley and thread it into the front cover. Rotate the pulley as necessary to align the bolt holes.

✳✳ WARNING

Failure to hold the crankshaft pulley in place while tightening the bolt can cause damage to the engine front cover.

16. Using the special tools to hold the crankshaft pulley in place, tighten the crankshaft pulley bolt in two stages:
   a. Stage 1: Tighten to 100 Nm (74 ft. lbs.).
   b. Stage 2: Tighten an additional 90 degrees (¼ turn).
17. Remove the 6 mm x 18 mm bolt.
18. Remove the special tools.

➡**Only turn the engine in the normal direction of rotation.**

19. Turn the engine two complete revolutions.
20. Turn the crankshaft until the No. 1 piston is at TDC.
21. Install special tool 303-507.

✳✳ WARNING

Only hand-tighten the bolt or damage to the front cover can occur.

22. Using the 6 mm x 18 mm bolt, check the position of the crankshaft pulley. If it is not possible to install the bolt, correct the engine timing.
23. Using special tool 303-465, check the position of the camshafts. If it is not possible to install the special tool, correct the engine timing.
24. Remove the 6 mm x 18 mm bolt.
25. Install the engine plug bolt.

3.0L Engine

1. Before servicing the vehicle, refer to the Precautions Section.
2. Remove or disconnect the following:
   • Negative battery cable
   • Crankshaft pulley
   • Front oil seal

To install:
3. Install or connect the following:
   • New front crankshaft oil seal
   • Crankshaft pulley. See the Torque Specifications Chart for the tightening torques.
   • Negative battery cable

4. Start the vehicle and check for leaks, repair if necessary.

Camshaft and Lifters

REMOVAL & INSTALLATION

2.0L Engine

1. Before servicing the vehicle, refer to the Precautions Section.
2. Remove or disconnect the following:
   • Negative battery cable
   • Camshaft timing sprocket and verify the valve clearance
   • Camshaft journal cap bolts by loosening them in several passes in the proper sequence
   • Camshafts
3. Inspect the camshaft for wear and discard the oil seals
To install:
4. Install or connect the following:
   • Camshaft cam followers, lubricate the bearing journals thoroughly. Torque the caps to 14 ft. lbs. (19 Nm).
   • Exhaust camshaft oil seal
   • Camshaft timing sprocket
   • Negative battery cable

9308TG13

Remove the camshaft bearing caps in sequence–2.0L engine

Camshaft bearing cap tightening sequence—2.0L engine

2.3L Engine

1. Before servicing the vehicle, refer to the Precautions Section.

※ WARNING

During engine repair procedures, cleanliness is extremely important. Any foreign material, including any material created while cleaning gasket surfaces that enters the oil passages, coolant passages or the oil pan can cause engine failure.

※ WARNING

The crankshaft, the crankshaft sprocket and the pulley are fitted together by friction, using diamond washers between the flange faces on each part. For that reason, the crankshaft sprocket is also unfastened if you loosen the pulley. Therefore, the engine must be retimed each time the damper is removed. Otherwise severe engine damage can occur.

2. With the vehicle in NEUTRAL, position it on a hoist.

➡Valve tappets are select fit and the valve clearance must be checked before removing the tappets

※ WARNING

Turn the engine clockwise only, and only use the crankshaft bolt.

➡Before removing the camshafts, measure the clearance of each valve at base circle, with the lobe pointed away from the tappet. Failure to measure all clearances prior to removing the camshafts will necessitate repeated removal and installation and wasted labor time.

3. Use a feeler gauge to measure the clearance of each valve and record its location.

➡The number on the valve tappet only reflects the digits that follow the decimal. For example, a tappet with the number 0.650 has the thickness of 3.650 mm.

➡A midrange clearance is the most desirable:

- Intake: 0.22–0.28 mm (0.008–0.011 inch)

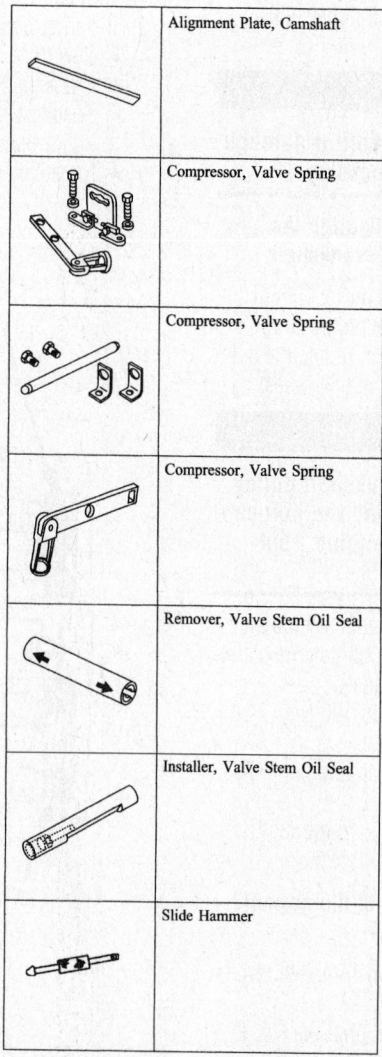

Tools necessary for camshaft and lifter removal—2.3L engine

1 Camshaft bearing cap bolt
2 Camshaft bearing cap
3 Camshaft

4 Valve tappet
5 Valve collet
6 Valve spring retainer

7 Valve spring
8 Valve seal

67197-ESCA-G16

Camshafts, lifters and related parts—2.3L engine

- Exhaust: 0.27–0.33 mm (0.010–0.013 inch)

4. Select tappets using this formula: tappet thickness = measured clearance + the base tappet thickness–most desirable thickness.

5. Select the tappets and mark the installation location.

6. If any tappets do not measure within specifications, install new tappets in these locations.

7. Remove the timing chain and sprockets.

8. Mark the position of the camshaft lobes on the No. 1 cylinder for assembly reference.

✷✷ WARNING

Failure to follow the camshaft loosening procedure can result in damage to the camshafts.

9. Loosen the camshaft bearing bolts in the sequence shown, one turn at a time. Repeat until all the tension is released.

10. Remove the camshaft bearing caps.

✷✷ WARNING

If the camshafts and valve tappets are to be reused, mark the location of the valve tappets to make sure they are assembled in their original positions.

➡The number on the valve tappets only reflects the digits that follow the decimal. For example, a tappet with the number 0.650 has the thickness of 3.650 mm.

11. Remove the camshafts.
12. Valve tappets.
13. To install, reverse the removal procedure. Coat the valve tappets with clean engine oil and insert them.

Camshaft cap removal sequence—2.3L engine

67197-ESCA-G17

67197-ESCA-G14

Valve clearance check—2.3L engine

❄ **WARNING**

Install the camshafts with the alignment slots in the camshafts lined up so the Camshaft Alignment Plate can be installed without rotating the camshafts. Make sure the lobes on the No. 1 cylinder are in the same position as noted in the removal procedure. Rotating the camshafts when the timing chain is removed, or installing the camshafts 180 degrees out of position can cause severe damage to the valves and pistons.

➡Lubricate the camshaft journals and bearing caps with clean engine oil.

14. Install the camshafts and bearing caps. Tighten the bolts in the sequence shown in three stages.

a. Stage 1: Tighten the camshaft bearing bolt caps one turn at a time until tight.

b. Stage 2: Tighten the bolts to 7 Nm (62 inch lbs.).

c. Stage 3: Tighten the bolts to 16 Nm (12 ft. lbs.).

3.0L Engine

LEFT SIDE

1. Before servicing the vehicle, refer to the Precautions Section.

2. Remove or disconnect the following:

- Negative battery cable
- Water pump belt
- Timing drive components
- Camshaft oil seal
- Camshaft oil seal retainer
- Camshaft cap bolts by loosening them in sequence
- Camshafts

To install:

3. Install or connect the following:

- Camshaft bearing caps in their original position
- Align the camshafts
- Bearing thrust caps and hand tighten the bolts. When aligned properly, torque the bolts to 89 inch lbs. (10 Nm).
- Timing drive components
- Camshaft oil seal retainer
- Crankshaft oil seal
- Water pump drive pulley
- Water pump belt
- Negative battery cable

RIGHT SIDE

1. Before servicing the vehicle, refer to the Precautions Section.

2. Remove or disconnect the following:

- Negative battery cable
- Timing drive components
- Camshaft cap bolts by loosening them in sequence
- Camshafts caps
- Camshafts

Camshaft cap torque sequence—2.3L engine

67197-ESCA-G18

Remove and install the left side camshaft bearing caps in sequence–3.0L engine

Remove and install the right side camshaft bearing caps in sequence–3.0L engine

To install:

3. Install or connect the following:
   - Camshaft bearing caps in their original position
   - Align the camshafts
   - Bearing caps and hand tighten the bolts
   - Bearing thrust caps and hand tighten the bolts. When aligned properly, torque the bolts to 89 inch lbs. (10 Nm).
   - Timing drive components
   - Negative battery cable

Valve Lash

ADJUSTMENT

2.0L Engine

1. Before servicing the vehicle, refer to the Precautions Section.
2. Remove or disconnect the following:
   - Negative battery cable
   - Timing belt
3. Measure each valve's clearance at the base circle with the lobe facing away from the tappet.
4. Use a feeler gauge to measure and record each valve's clearance
5. Remove or disconnect the following:
   - Camshafts
   - Valve tappets from the cylinder head
6. A mid range clearance is recommended as follows:
   a. Intake: 0.006 inch (0.15mm).
   b. Exhaust: 0.012 inch (0.3mm).

To install:

7. Install or connect the following:
   - Valve tappets after lubricating them with clean engine oil
   - Camshafts and verify each valve's clearance at the base circle with the lobe facing away from the tappet
   - Timing belt
   - Negative battery cable

2.3L Engine

1. Before servicing the vehicle, refer to the Precautions Section.

➡**Before removing the camshafts, measure the clearance of each valve at base circle, with the lobe pointed away from the tappet. Failure to measure all clearances prior to removing the camshafts will necessitate repeated removal and installation and wasted labor time.**

2. Use a feeler gauge to measure the clearance of each valve and record its location.

➡The number on the valve tappet only reflects the digits that follow the decimal. For example, a tappet with the number 0.650 has the thickness of 3.650 mm.

➡A midrange clearance is the most desirable:

- Intake: 0.22–0.28 mm (0.008–0.011 inch)
- Exhaust: 0.27–0.33 mm (0.010–0.013 inch)

3. Select tappets using this formula: tappet thickness = measured clearance + the base tappet thickness–most desirable thickness.

4. Select the tappets and mark the installation location.

5. If any tappets do not measure within specifications, install new tappets in these locations.

3.0L Engine

1. Before servicing the vehicle, refer to the Precautions Section.

2. Remove or disconnect the following:

- Negative battery cable
- Camshaft followers
- Hydraulic lash adjusters

➡Mark the position of the hydraulic lash adjusters to assure they are assembled in their original position

3. Inspect the adjusters for scoring or uneven wear in the bore and replace them as required.

To install:

4. Install or connect the following:

- Hydraulic lash adjusters after lubricating them with clean engine oil
- Camshaft followers
- Negative battery cable

Oil Pan

REMOVAL & INSTALLATION

2.0L Engine

1. Before servicing the vehicle, refer to the Precautions Section.

2. Drain the engine oil.

3. Support the powertrain assembly.

4. Remove or disconnect the following:

- Negative battery cable
- Catalytic converter
- Oil pan and gasket

5. Thoroughly clean the gasket mating surfaces.

Tighten the oil pan bolts in sequence–2.0L engine

To install:

6. Apply silicone sealer to the oil pan.

7. Install a new gasket on the oil pan.

8. Oil pan. Torque the bolts in sequence to:

a. Step 1: 53 inch lbs. (6 Nm).

b. Step 2: 106 in lbs. (12 Nm).

9. Install or connect the following:

- Catalytic converter
- Negative battery cable

10. Fill the engine with clean oil.

11. Start the engine and check for leaks, repair if necessary.

2.3L Engine

1. Before servicing the vehicle, refer to the Precautions Section.

2. With the vehicle in NEUTRAL, position it on a hoist.

3. Remove the oil level indicator and tube.

4. Drain the oil.

5. Remove the 4 front cover-to-oil pan bolts.

6. Remove the 4 oil pan-to-bell housing bolts.

7. Remove the 13 oil pan-to-block bolts.

To install:

✷✷ WARNING

Do not use metal scrapers, wire brushes, power abrasive discs or other abrasive means to clean the sealing surfaces. These tools cause scratches and gouges, which make leak paths. Use a plastic scraping tool to remove traces of sealant.

8. Clean and inspect all mating surfaces.

➡If the oil pan is not secured within four minutes of sealant application the sealant must be removed and the sealing area cleaned with metal surface cleaner. Allow to dry until there is no sign of wetness, or four minutes, whichever is longer. Failure to follow this procedure can cause future oil leakage.

➡The oil pan must be installed and the bolts tightened within four minutes of applying the silicone gasket and sealant.

9. Apply a 2.5 mm bead of silicone gasket and sealant to the oil pan. Install the oil pan. Install the oil pan-to-bell housing bolts. Torque to 35 ft. lbs. (48 Nm).

10. Install the front cover bolts. Torque to 89 inch lbs. (10 Nm).

11. Install the oil pan-to-bell housing bolts. Torque to 35 ft. lbs. (48 Nm).

12. Install and tighten the oil pan bolts in the sequence shown to 18 ft. lbs. (25 Nm).

13. Fill the engine with clean engine oil.

3.0L Engine

1. Before servicing the vehicle, refer to the Precautions Section.

10 Nm (89 lb-in) — 8

28 Nm (21 lb-ft)

1 Drain plug
2 Engine front cover bolt
3 Oil pan bolt
4 Oil pan-to-bell housing bolt
5 Oil pan-to-bell housing bolt

6 Oil pan bolt
7 Oil pan
8 Oil pump screen and pickup tube bolt
9 Oil pump screen and pickup tube
10 Oil pump screen and pickup tube gasket

67197-ESCA-G20

Oil pan, pump and related parts—2.3L engine

25 Nm (18 lb-ft)

67197-ESCA-G21

Oil pan bolt torque sequence—2.3L engine

2. Drain the engine oil.
3. Remove or disconnect the following:
   • Negative battery cable
   • Flexible exhaust pipe
   • Downstream catalyst monitor sensor
   • Oil pan and gasket
4. Thoroughly clean the gasket mating surfaces.
   To install:
5. Apply silicone sealer to the oil pan.
6. Install or connect the following:
   • New gasket on the oil pan
   • Oil pan. Torque the bolts in sequence to 18 ft. lbs. (25 Nm).
   • Flexible exhaust pipe
   • Downstream catalyst monitor sensor
   • Negative battery cable
7. Fill the engine with clean oil.

Oil pan torque sequence—3.0L engine

67197-ESCA-G65

Oil pump torque sequence—2.3L engine

8. Start the vehicle and check for leaks, repair if necessary.

Oil Pump

REMOVAL & INSTALLATION

2.0L Engine

1. Before servicing the vehicle, refer to the Precautions Section.
2. Drain the engine oil.
3. Remove or disconnect the following:
   • Negative battery cable
   • Oil pan
   • Oil pump screen cover and tube
   • Oil pump and discard the gasket
4. Thoroughly clean the gasket mating surfaces.

To install:

5. Install or connect the following:
   • Oil pump screen cover and tube with a new gasket. Torque the bolts to 89 inch lbs. (10 Nm).
   • Oil pump to the oil pan
   • Oil pan
   • Negative battery cable
6. Refill the engine with clean oil.
7. Start the engine and check for leaks; repair if necessary.

2.3L Engine

1. Before servicing the vehicle, refer to the Precautions Section.
2. Remove the engine from the vehicle and mount it on an engine stand.
3. Remove the oil pan.
4. Remove the oil pump pickup tube and screen.

5. Remove the front cover and the timing chain.
6. Release the tension on the tensioner spring.
7. Remove the tensioner and the shoulder bolt.
8. Remove the guide.

➡The oil pump chain sprocket must be held in place.

9. Remove the oil pump chain and sprockets.
10. Remove the oil pump assembly and gasket.

To install:

11. Install the oil pump with a new gasket. Tighten the bolts in sequence as follows:

   a. Step 1: 89 inch lbs. (10 Nm).
   b. Step 2: 17 ft. lbs. (23 Nm).
12. Install the pump chain and sprockets. Tighten the pump sprocket bolt to 18 ft. lbs. (25 Nm).
13. Install the chain guide, tensioner, and shoulder bolt. Tighten the bolts to 89 inch lbs. (10 Nm).
14. Hook the tensioner spring around the shoulder bolt.
15. Install the oil pump pickup tube and screen with a new gasket. Tighten the bolts to 89 ft. lbs. (10 Nm).
16. Install the oil pan.
17. Install the timing chain and front cover.
18. Install the engine into the vehicle.

3.0L Engine

1. Before servicing the vehicle, refer to the Precautions Section.
2. Drain the engine oil.
3. Remove or disconnect the following:
   • Negative battery cable
   • Timing drive components
   • Oil pump screen cover and tube
   • Damper bolt and crankshaft sprockets

9308TG19

Remove the oil pump bolts in the proper sequence—3.0L engine

Install the oil pump bolts in the proper sequence—3.0L engine

- Oil pump bolts in the proper sequence

4. Thoroughly clean the gasket mating surfaces.

To install:

5. Install or connect the following:
- Oil pump and bolts in the proper sequence. Torque the bolts to 89 inch lbs. (10 Nm).
- Crankshaft sprockets
- Oil pump screen cover and tube
- Timing drive components
- Negative battery cable

6. Refill the engine with clean oil.
7. Start the engine and check for leaks; repair if necessary.

Rear Main Seal

REMOVAL & INSTALLATION

2.0L Engine

1. Before servicing the vehicle, refer to the Precautions Section.
2. Remove or disconnect the following:

Rear main seal removal—2.0L engine

- Negative battery cable
- Flywheel
- Rear main seal

To install:

3. Coat the oil seal with clean engine oil.

4. Install or connect the following:
- Crankshaft rear oil seal
- Flywheel
- Negative battery cable

2.3L Engine

1. Before servicing the vehicle, refer to the Precautions Section.
2. With the vehicle in NEUTRAL, position it on a hoist.
3. If equipped, remove the automatic transaxle.

1 Flexplate or flywheel bolt	5 Oil pan bolt
2 Flexplate or flywheel	6 Oil pan
3 Engine front cover bolt	7 Crankshaft rear oil seal with retainer plate bolt
4 Oil pan bolt	8 Crankshaft rear oil seal with retainer plate

67197-ESCA-G22

Rear main seal and related parts—2.3L engine

Retainer plate torque sequence—2.3L engine

67197-ESCA-G23

Flywheel torque sequence—2.3L engine

67197-ESCA-G24

307-005

303-519

Rear main seal removal—3.0L engine

67197-ESCA-G55

303-384

303-178

Rear main seal installation—3.0L engine

67197-ESCA-G56

4. If equipped, remove the manual transaxle and clutch.

5. Remove the flexplate or flywheel.

6. Remove the oil pan.

7. Remove the crankshaft rear oil seal with retainer plate

To install:

8. Using a seal installer, position the crankshaft rear oil seal with retainer plate onto the crankshaft.

9. Install the crankshaft rear oil seal with retainer plate. Tighten the bolts in the sequence shown to 10 Nm (89 inch lbs.).

10. Install the oil pan.

➡Special bolts are used for installation. Do not use standard bolts.

11. Install the flywheel/flexplate. Tighten the bolts in the sequence shown in three stages.

    a. Stage 1: Tighten to 50 Nm (37 ft. lbs.).

    b. Stage 2: Tighten to 80 Nm (50 ft. lbs.).

    c. Stage 3: Tighten to 112 Nm (83 ft. lbs.).

3.0L Engine

1. Before servicing the vehicle, refer to the Precautions Section.

2. Remove or disconnect the following:
- Negative battery cable
- Flexplate
- Rear main oil seal

To install:

3. Coat the oil seal with clean engine oil.

4. Install or connect the following:
- Crankshaft rear oil seal

- Flywheel
- Negative battery cable

Timing Belt and Covers

REMOVAL & INSTALLATION

2.0L Engine

1. Before servicing the vehicle, refer to the Precautions Section.

2. With the vehicle in NEUTRAL, position it on a hoist.

3. Disconnect the battery ground cable.

4. Remove the air cleaner outlet pipe.

5. Remove the spark plugs.

6. Disconnect the throttle cables. If equipped, disconnect the speed control actuator cable. Disconnect the accelerator cable.

7. Disconnect the catalyst monitor and heated oxygen sensor electrical connectors.

8. Detach the electrical connectors from the bracket.

9. Remove the wiring harness anchors from the valve cover studs and position the wiring harness aside.

10. Remove the valve cover.

❈❈ WARNING

The valve cover sealing surfaces are soft materials. Do not use abrasive grinding discs to remove gasket material. Use only a plastic scraping tool. Do not scratch or gouge sealing surfaces or oil leaks can occur.

11. Clean and inspect the sealing surfaces of the valve cover and cylinder head. Both surfaces must be clean and flat.

12. Remove the heated oxygen sensor (HO_2S).

13. Disconnect the exhaust gas recirculation (EGR) tube.
- Remove the two bolts.
- Loosen the EGR tube nut.

14. Remove the catalytic converter nuts. Discard the nuts.

15. Remove the catalyst monitor sensor.

16. Remove the bolts and the catalytic converter bracket.

17. Remove the flagnuts and bolts. Discard the flagnuts, springs and bolts.

18. Remove the catalytic converter bracket nuts.

19. Remove the catalytic converter and rear support bracket. Discard the ring seal and gasket.

20. Remove the bolt and nut and position the coolant tube aside.

21. Remove the wheel and tire assembly.

06017-ESCA-G25

Rotate the crankshaft to just before top dead center (TDC) (No. 1 cylinder)—2.0L engine

06017-ESCA-G26

Remove the stud)—2.0L engine

303-574

06017-ESCA-G27

Install special too 303-574 in the stud hole)—2.0L engine

06017-ESCA-G28

Rotate the crankshaft clockwise against the peg to bring it to TDC (No. 1 cylinder)—2.0L engine

22. Remove the right lower splash shield.

23. Rotate the crankshaft to just before top dead center (TDC) (No. 1 cylinder). Remove the stud. Install special tool 303-574.

303-F072

06017-ESCA-G29

Install special tool 303-F072 to support the engine—2.0L engine

06017-ESCA-G30

Release the tension on the timing belt by disconnecting the tensioner tab from the timing cover back plate—2.0L engine

06017-ESCA-G31

(1) blanking plug, (2) exhaust camshaft pulley bolt, (3) intake camshaft pulley bolt—2.0L engine

➡**Make sure the correct (second) notch in the pulley is indexed to the lower cylinder block.**

24. Rotate the crankshaft clockwise against the peg to bring it to TDC (No. 1 cylinder).

25. Loosen the coolant pump pulley bolts.

26. Loosen the crankshaft pulley bolt.

27. Rotate the tensioner and remove the accessory drive belt.

28. Remove the bolt and the crankshaft pulley.

29. Remove the bolts and the lower timing belt cover.

30. Install special tool 303-F072 to support the engine.

31. Detach the ground strap.

32. Remove the nuts, the bolt and the engine mount upper bracket.

33. Remove the studs.

34. Detach the knock sensor electrical connector from the upper timing cover.

35. Remove the bolts and the upper timing cover.

36. Remove the coolant pump pulley.

37. Remove the accessory drive belt idler pulley.

38. Remove the bolts and the engine mount lower bracket.

39. Relieve the tension on the timing belt tensioner pulley.
- Loosen the tensioner pulley bolt.
- Release the tension on the timing belt by disconnecting the tensioner tab from the timing cover back plate.

40. Remove the timing belt. Slide the timing belt off the camshaft sprockets and the crankshaft sprocket.

41. Inspect the timing belt for wear. Install a new timing belt if required.

To install:

✲✲ WARNING

The camshaft must be held stationary at the hexagons with locking pliers. Do not use the alignment tool to hold the camshaft in position or damage to the camshaft may occur.

➡**To loosen the camshaft pulleys, hold the camshafts by the hexagon.**

42. Remove the blanking plug from the exhaust camshaft pulley.

43. Loosen the exhaust camshaft pulley bolt.

44. Loosen the intake camshaft pulley bolt.

➡**Rotate the camshafts clockwise as necessary.**

45. Install the camshaft alignment timing tool on the back of the camshafts.

303-465

06017-ESCA-G32

Install the camshaft alignment timing tool on the back of the camshafts—2.0L engine

➡Cylinder No. 1 is at top dead center (TDC) when the keyway is in the 12 o'clock position.

46. Confirm the crankshaft position is at TDC (No. 1 cylinder) by rotating it clockwise against the alignment peg.

❉❉ WARNING

Do not rotate the crankshaft; as necessary check that it is still resting against the timing pin.

➡The lug of the belt tensioner should not be hooked in the sheet metal cover during timing belt installation.

47. Position a new timing belt in place. Starting from the crankshaft timing belt pulley and working counterclockwise, position the timing belt in place while keeping it under tension.

➡Incorrect timing belt tension will cause incorrect valve timing.

48. Pre-tension the timing belt.
   a. Rotate the tensioner locating tab counterclockwise and insert the locating tab into the slot in the rear timing cover.
   b. Position the hex key slot in the tensioner adjusting washer to the 4 o'clock position.
   c. Tighten the attaching bolt enough to seat the tensioner firmly against the rear timing cover, but still allow the tensioner adjusting washer to be rotated using a 6 mm hex key.

06017-ESCA-G33

Position a new timing belt in place—2.0L engine

06017-ESCA-G34

Tensioner locating tab (1), tensioner hex slot (2), attaching bolt (3)—2.0L engine

06017-ESCA-G35

Using the hex key, rotate the adjusting washer counterclockwise until the notch in the pointer is centered over the index line on the locating tab (the pointer will move clockwise during adjustment)—2.0L engine

❉❉ WARNING

Tension the timing belt, working counterclockwise.

49. Using the hex key, rotate the adjusting washer counterclockwise until the notch in the pointer is centered over the index line on the locating tab (the pointer will move clockwise during adjustment).

50. While holding the adjusting washer in position, tighten the bolt to 25 Nm (18 ft. lbs.).

❉❉ WARNING

The camshaft must be held stationary at the hexagons with locking pliers. Do not use the alignment tool to hold the camshaft in position or damage to the camshaft may occur.

51. Tighten the bolt on the intake camshaft sprocket to 68 Nm (50 ft. lbs.).

❉❉ WARNING

The camshaft must be held stationary at the hexagons with locking pliers.

Do not use the alignment tool to hold the camshaft in position or damage to the camshaft may occur.

52. Tighten the bolt on the exhaust camshaft sprocket in three stages.
   • Stage 1: Tighten the bolt to 50 Nm (36 ft. lbs.).
   • Stage 2: Remove the TDC peg and the camshaft alignment timing tool.
   • Stage 3: Tighten the bolt to 115–125 Nm (85–92 ft. lbs.).

➡Install a new oil plug seal.

53. Screw in the new oil plug on the variable camshaft timing assembly. Torque to 37 Nm (27 ft. lbs.).

➡Turn the engine two turns in the normal direction of rotation by the crankshaft.

54. Check the valve timing by inserting the special tools and correct the alignment as necessary.
   a. Screw in special tool 303-574 and make sure that the crankshaft is resting against the special tool.
   b. Insert special tool 303-465 into the camshafts. If necessary loosen the timing pulleys and correct the camshaft alignment.
   c. Remove the special tools.

55. Install the front engine mount lower bracket. Torque to 50 Nm (37 ft. lbs.).

56. Install the accessory drive belt idler pulley. Torque to 45 Nm (35 ft. lbs.).

57. Install the coolant pump pulley. Hand-tighten the bolts.

58. Position the upper timing belt cover and install the bolts. Torque to 8 Nm (71 inch lbs.).

59. Install the studs. Torque to 34 Nm (25 ft. lbs.).

60. Connect the knock sensor electrical connector to the upper timing belt cover.

61. Install the engine mount upper bracket, the nuts and the bolt. Torque the upper nuts to 98 Nm (72 ft. lbs.); the lower bolt to 77 Nm (57 ft. lbs.).

62. Attach the ground strap and install the nut. Torque the nut to 34 Nm (25 ft. lbs.).

63. Position the lower timing belt cover and install the bolts. Torque to 7 Nm (62 inch lbs.).

64. Position the crankshaft pulley and install the bolt. Hand-tighten the bolt.

65. Rotate the tensioner and install the accessory drive belt.

66. Tighten the crankshaft pulley bolt. Torque to 115 Nm (85 ft. lbs.).

67. Tighten the coolant pump pulley bolts to 23 Nm (17 ft. lbs.).

68. Position the right lower splash shield and install the bolts.

69. Install the wheel and tire assembly.

70. Install the stud. Torque to 24 Nm (18 ft. lbs.).

71. Install the coolant tube. Torque to 20 Nm (15 ft. lbs.).

72. Position the catalytic converter into place and loosely install the two nuts. Install a new gasket and ring seal. Install new nuts.

73. Connect the converter outlet to the muffler inlet pipe. Position the rear support bracket and install the washers and nuts. Loosely install the new bolts, springs and flagnuts.

74. Lower the vehicle.

75. Install the third catalytic converter nut and tighten all the nuts. Install a new nut. Torque to 47 Nm (35 ft. lbs.).

76. Connect the EGR tube and install the bracket bolts. Torque the lower end to 63 Nm (46 ft. lbs.); the upper end to 9 Nm (80 inch lbs.).

77. Raise the vehicle.

78. Tighten the converter outlet to the muffler inlet bolts to 40 Nm (30 ft. lbs.).

79. Tighten the catalytic converter rear support bracket nuts to 47 Nm (35 ft. lbs.).

80. Install the catalytic converter bracket and bolts. Torque to 25 Nm (18 ft. lbs.).

81. Install the catalyst monitor sensor. Torque to 40 Nm (30 ft. lbs.).

82. Install the HO2S. Torque to 40 Nm (30 ft. lbs.).

83. Inspect the valve cover gasket. Install a new gasket, if necessary.

84. Position the valve cover. Install the studbolts. Install the bolts. Torque to 9 Nm) 80 inch lbs.).

85. Connect the wiring harness anchors to the valve cover studs.

86. Attach the catalyst monitor and heated oxygen sensor electrical connectors to the bracket.

87. Connect the electrical connectors.

88. Connect the accelerator cable. If equipped, connect the speed control actuator cable.

89. Install the spark plugs.

90. Install the air cleaner outlet pipe.

91. Connect the battery ground cable.

Timing Chain, Gears, Front Cover and Seal

REMOVAL & INSTALLATION

2.3L Engine

1. Before servicing the vehicle, refer to the Precautions Section.

�֎ CAUTION

During engine repair procedures, cleanliness is extremely important. Any foreign material, including any material created while cleaning gasket surfaces that enters the oil passages, coolant passages or the oil pan can cause engine failure.

✶✶ CAUTION

The crankshaft, the crankshaft sprocket and the pulley are fitted together by friction, using diamond washers between the flange faces on each part. For that reason, the crankshaft sprocket is also unfastened if you loosen the pulley. Therefore, the engine must be retimed each time the damper is removed. Otherwise severe engine damage can occur.

2. With the vehicle in NEUTRAL, position it on a hoist.

3. Remove the accessory drive belt and idler pulleys.

4. Remove the engine mount.

5. Remove the valve cover.

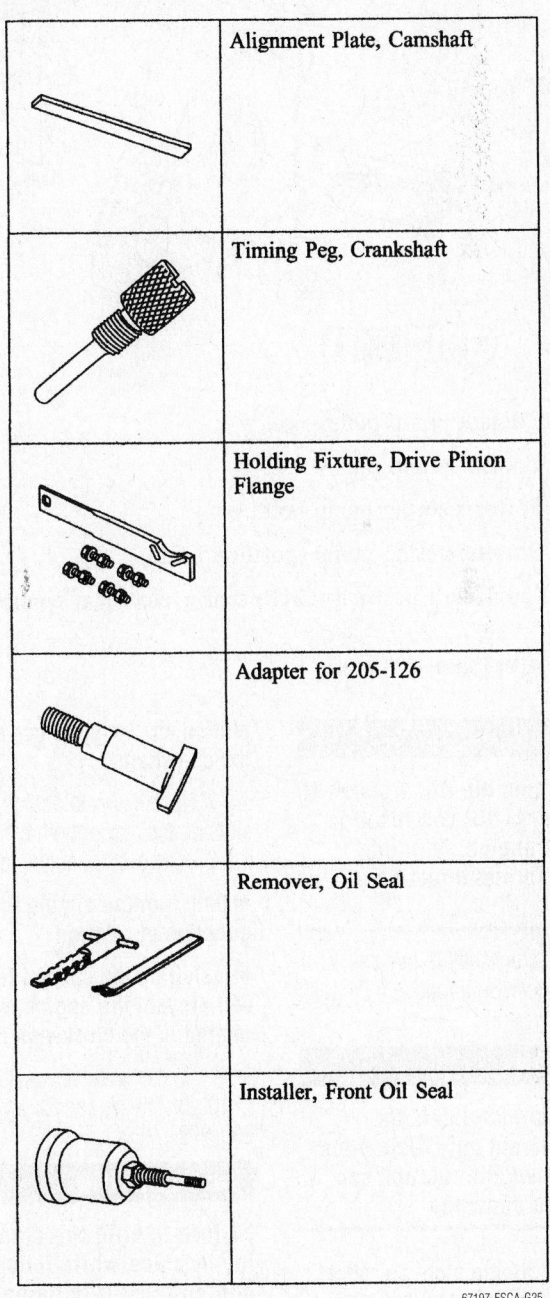

	Alignment Plate, Camshaft
	Timing Peg, Crankshaft
	Holding Fixture, Drive Pinion Flange
	Adapter for 205-126
	Remover, Oil Seal
	Installer, Front Oil Seal

67197-ESCA-G25

Tools needed for timing chain and gears replacement—2.3L engine

20 Nm (15 lb-ft)

25 Nm
(18 lb-ft)

67197-ESCA-G26

5 Coolant pump pulley bolt
6 Coolant pump pulley
7 Power steering pump bolt
8 Power steering pump (position aside)
9 Crankshaft position (CKP) sensor electrical connector

10 CKP sensor bolts
11 CKP sensor
12 Engine front cover bolt
13 Engine front cover

Front cover and related parts—2.3L engine

✳✳ CAUTION

Failure to position the No. 1 piston at top dead center (TDC) can result in damage to the engine. Turn the engine in the normal direction of rotation only.

6. Using the crankshaft pulley bolt, turn the crankshaft clockwise to position the No. 1 piston at TDC.

✳✳ CAUTION

The special tool 303-465 is for camshaft alignment only. Using this tool to prevent engine rotation can result in engine damage.

➡The camshaft timing slots are offset. If the special tool cannot be installed, rotate the crankshaft one complete rev-olution clockwise to correctly position the camshafts.

7. Install special tool 303-465 in the slots on the rear of both camshafts.
8. Remove the engine plug bolt.

➡Only turn the engine in the normal direction of rotation.

➡Installing the special tool in this step will prevent the engine from being rotated in the clockwise direction.

9. Install special tool 303-507.
10. Install the special tools 205-126 and 205-072-02.

✳✳ CAUTION

Failure to hold the crankshaft pulley in place while loosening the bolt can result in damage to the engine.

11. Remove the crankshaft pulley bolt and washer.
12. Remove the crankshaft pulley.
13. Remove the crankshaft front seal.
14. Remove the coolant pump pulley.
15. Remove the power steering pump and position it aside.

➡The bolt under the power steering pressure tube will remain with the power steering pump.

16. Remove the CKP sensor.

➡Whenever the crankshaft position (CKP) sensor is removed, a new one must be installed, using the alignment jig supplied with the new part.

17. Remove the engine front cover bolts (there are 22).
18. Remove the engine front cover.
19. Remove the timing chain tensioner.

14 Timing chain tensioner bolt

15 Timing chain tensioner

16 RH timing chain guide

17 Timing chain

18 LH timing chain guide bolt

19 LH timing chain guide

20 Camshaft sprocket bolt

21 Camshaft sprocket

67197-ESCA-G27

Timing chain and related parts—2.3L engine

67197-ESCA-G28

Install special tool 303-465 in the slots on the rear of both camshafts—2.3L engine

Compress the timing chain tensioner, and insert a paper clip into the hole to retain the tensioner.
   20. Remove the right timing chain guide.
   21. Remove the timing chain.
   22. Remove the left timing chain guide.
   23. Remove the camshaft sprocket bolts.
   24. Remove the camshaft sprockets.

67197-ESCA-G29

Install special tool 303-507—2.3L engine

⁂ **CAUTION**

Do not rely on the Camshaft Alignment Plate to prevent camshaft rotation. Damage to the tool or the camshaft can occur. Use the flats on the camshaft to prevent camshaft rotation.

To install:
   25. Installation is the reverse of removal. Note the following:

67197-ESCA-G30

Install the special tools 205-126 and 205-072-02—2.3L engine

⁂ **CAUTION**

Do not use metal scrapers, wire brushes, power abrasive disks or other abrasive means to clean sealing surfaces. These tools cause scratches and gouges which make leak paths.

   26. Clean and inspect the mounting surfaces of the engine and the front cover.

67197-ESCA-G31

Compress the timing chain tensioner, and insert a paper clip into the hole to retain the tensioner—2.3L engine

67197-ESCA-G32

Use the flats on the camshaft to prevent camshaft rotation—2.3L engine

67197-ESCA-G33

Front cover bolt torque sequence—2.3L engine

06017-ESCA-G36

Ignition pulse wheel—3.0L engine

06017-ESCA-G37

Rotate the crankshaft clockwise to position the keyway at the 11 o'clock position—3.0L engine

06017-ESCA-G38

Rotate the crankshaft clockwise 120 degrees to the 3 o'clock position to locate the right side camshafts in the neutral position—3.0L engine

06017-ESCA-G39

Verify that the right camshafts are in the neutral position—3.0L engine

➡ **The engine front cover must be installed and the bolts tightened within four minutes of applying the silicone gasket and sealant.**

27. Apply a 2.5 mm bead of silicone gasket and sealant to the cylinder head and oil pan joint areas. Apply a 2.5 mm bead of silicone gasket and sealant to the front cover.

28. Install the engine front cover. Tighten the bolts in the sequence shown, to the following specifications:

    a. Tighten the 8 mm bolts to 10 Nm (89 inch lbs.).

    b. Tighten the 13 mm bolts to 48 Nm (35 ft. lbs.).

29. Position the power steering pump and install the bolts.

➡ **Remove the through-bolt from the special tool.**

➡ **Lubricate the oil seal with clean engine oil.**

30. Using a seal driver, install the crankshaft front oil seal.

➡ **Do not reuse the crankshaft damper bolt.**

➡ **Apply clean engine oil on the seal area before installing.**

31. Install the crankshaft pulley and hand-tighten the bolt.

❋❋ CAUTION

Only hand-tighten the bolt or damage to the front cover can occur.

➡ **This step will correctly align the crankshaft pulley to the crankshaft.**

32. Install a standard 6 mm x 18 mm bolt through the crankshaft pulley and thread it into the front cover. Rotate the pulley as necessary to align the bolt holes.

❋❋ CAUTION

Failure to hold the crankshaft pulley in place while tightening the bolt can cause damage to the engine front cover.

33. Using the special tools to hold the crankshaft pulley in place, tighten the crankshaft pulley bolt in two stages:

    a. Stage 1: Tighten to 100 Nm (74 ft. lbs.).

    b. Stage 2: Tighten an additional 90 degrees (¼ turn).

34. Remove the 6 mm x 18 mm bolt.

35. Remove special tool 303-507.

36. Remove special tool 303-465.

➡ **Only turn the engine in the normal direction of rotation.**

37. Turn the engine two complete revolutions.

➡ **Only turn the engine in the normal direction of rotation.**

38. Turn the crankshaft until the No. 1 piston is at TDC.

39. Install special tool 303-507.

❋❋ CAUTION

Only hand-tighten the bolt or damage to the front cover can occur.

40. Using the 6 mm x 18 mm bolt, check the position of the crankshaft pulley. If it is not possible to install the bolt, correct engine timing.

41. Using special tool 303-465, check the position of the camshafts. If it is not possible to install the special tool, correct the engine timing.

42. Install the CKP sensor. Do not tighten the bolts at this time.

Rotate the crankshaft clockwise 1 ⅔ times to position the keyway at the 11 o'clock position—3.0L engine

Verify that the left side camshafts are in the neutral position—3.0L engine

Hold the ratchet lock mechanism away from the ratchet stem—3.0L engine

43. Adjust the CKP sensor alignment jig and tighten the bolts.
44. Remove the 6 mm x 18 mm bolt.
45. Install the engine plug bolt.

3.0L Engine

1. Before servicing the vehicle, refer to the Precautions Section.
2. Remove or disconnect the following:
   • Negative battery cable
   • Engine front cover

➡**This pulse wheel is used in several different engines. Install the pulse wheel with the keyway in the slot stamped "30" or "30RFF" (orange in color).**

   • Ignition pulse wheel and install the damper bolt
   • Spark plugs
3. Rotate the crankshaft clockwise to position the keyway at the 11 o'clock position and the camshafts in the correct positions. The No. 1 cylinder will be at Top Dead Center (TDC).
4. Rotate the crankshaft clockwise 120 degrees to the 3 o'clock position to locate the right side camshafts in the neutral position. Verify that the right camshafts are in the neutral position.
5. Remove or disconnect the following:
   • Right side timing chain and tensioner
   • Tensioner arm and timing chain guide
6. Rotate the crankshaft clockwise 1 ⅔ times to position the keyway at the 11 o'clock position. This will position the left side camshafts in the neutral position.
7. Verify that the left side camshafts are in the neutral position and mark the link position on the crankshaft sprocket.
8. Remove or disconnect the following:
   • Left side timing chain and tensioner
   • Tensioner arm and timing chain guide
   • Damper bolt and crankshaft sprockets

To install:

9. Install the crankshaft sprockets.
10. Position the timing chain tensioner in a soft jaw vise. Hold the ratchet lock mechanism away from the ratchet stem and slowly compress the timing chain tensioner. Retain the piston with a 1.5mm wire or paper clip.
11. If the timing marks on the chain are not visible, use a permanent marker to mark the left and right side timing chains. Mark the timing chains in the following sequence:

Verify that the left camshafts are correctly positioned–3.0L engine

06017-ESCA-G43

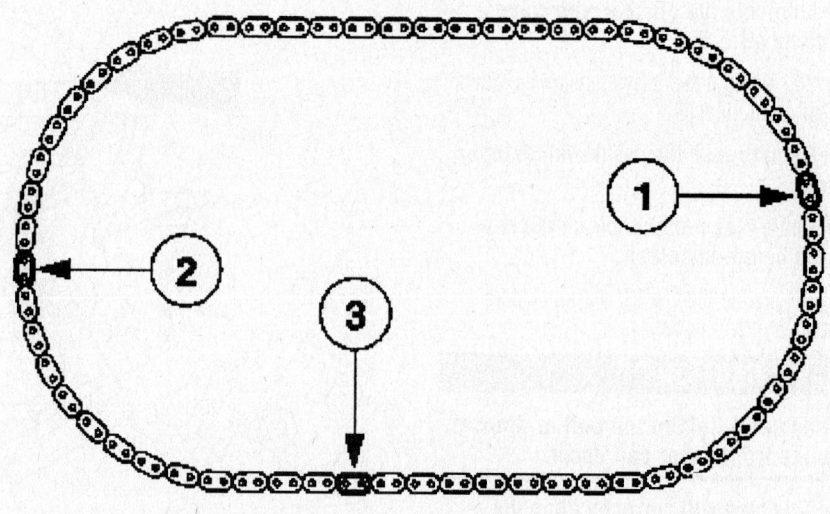

Mark the timing chain in the proper sequence–3.0L engine

9308TG21

Left side timing chain installed—3.0L engine

06017-ESCA-G44

06017-ESCA-G45

Verify that the right camshafts are correctly positioned—3.0L engine

06017-ESCA-G46

Timing chains correctly installed—3.0L engine

a. Mark any link to use as the crankshaft timing mark.

b. Count 29 links from the crankshaft timing mark and mark the link as the exhaust cam sprocket timing mark.

c. Continue counting to 42 and mark the link as the intake sprocket timing mark.

12. Verify that the left camshafts are correctly positioned.

13. Install the guide. Torque the bolts to 18 ft. lbs. (25 Nm).

14. Install the left side timing chain and align the chain in the following sequence:

a. Mark any link to use as the crankshaft timing mark.

b. Count 29 links from the crankshaft

timing mark and mark the link as the exhaust cam sprocket timing mark.

c. Continue counting to 42 and mark the link as the intake sprocket timing mark

15. Install or connect the following:
- Left side timing chain and tensioner arm. Torque the bolts to 18 ft. lbs. (25 Nm).
- Crankshaft damper bolt and rotate the keyway to the 3 o'clock position.

16. Verify that the right side camshafts are properly positioned and install the right side timing chain and guide. Torque the bolts to 18 ft. lbs. (25 Nm).

17. Make certain that the timing chain aligns with the marks on the camshaft and crankshaft sprockets

✳✳ CAUTION

Install the pulse wheel with the keyway in the slot stamped 20–25–34Y–30M (Color Blur).

18. Install or connect the following:
- Right side timing chain tensioner and arm. Torque the bolts to 18 ft. lbs. (25 Nm) and remove the damper bolt
- Ignition pulse wheel
- Spark plugs
- Engine front cover
- Negative battery cable

Piston and Ring

POSITIONING

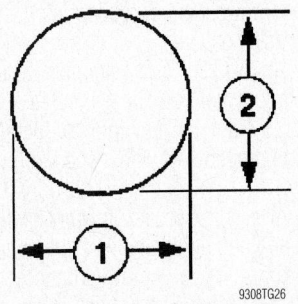

9308TG26

2.0L and 2.3L engine —piston ring end-gap spacing

9308TG25

3.0L (VIN 1) engine—piston ring end-gap spacing

FUEL SYSTEM

Fuel System Service Precautions

Safety is the most important factor when performing not only fuel system maintenance but also any type of maintenance. Failure to conduct maintenance and repairs in a safe manner may result in serious personal injury or death. Maintenance and testing of the vehicle's fuel system components can be accomplished safely and effectively by adhering to the following rules and guidelines.

1. To avoid the possibility of fire and personal injury, always disconnect the negative battery cable unless the repair or test procedure requires that battery voltage be applied.

2. Always relieve the fuel system pressure prior to disconnecting any fuel system component (injector, fuel rail, pressure regulator, etc.), fitting or fuel line connection. Exercise extreme caution whenever relieving fuel system pressure, to avoid exposing skin, face and eyes to fuel spray. Please be advised that fuel under pressure may penetrate the skin or any part of the body that it contacts.

3. Always place a shop towel or cloth around the fitting or connection prior to loosening to absorb any excess fuel due to spillage. Ensure that all fuel spillage (should it occur) is quickly removed from engine surfaces. Ensure that all fuel soaked cloths or towels are deposited into a suitable waste container.

4. Always keep a dry chemical (Class B) fire extinguisher near the work area.

5. Do not allow fuel spray or fuel vapors to come into contact with a spark or open flame.

6. Always use a backup wrench when loosening and tightening fuel line connection fittings. This will prevent unnecessary stress and torsion to fuel line piping.

7. Always replace worn fuel fitting O-rings with new. Do not substitute fuel hose or equivalent, where fuel pipe is installed.

Before servicing the vehicle, make sure to refer to the Precautions Section as well.

Fuel System Pressure

RELIEVING

2.0L Engine

1. Before servicing the vehicle, refer to the Precautions Section.
2. Remove or disconnect the following:

3. Remove the fuel pump relay and start the engine.
4. After the engine stalls, crank the engine 2 more times to be certain that all fuel pressure has been relieved.
5. Turn the ignition switch to the **OFF** position.
6. Install the fuel pump relay.

2002–04 3.0L Engine

1. Before servicing the vehicle, refer to the Precautions Section.
2. Remove or disconnect the following:
3. Remove the schrader valve cap at the end of the fuel injection supply manifold and attach a fuel pressure gauge.
4. Open the manual valve slowly and drain the fuel into a suitable container.
5. Continue draining the fuel system to relieve fuel pressure.

2.3L Engine, exc. Hybrid 2005 3.0L Engine

1. Before servicing the vehicle, refer to the Precautions Section.
2. Remove the fuel pump relay.
3. Start the engine and allow it to idle until it stalls.
4. After the engine stalls, crank the engine for approximately 5 seconds to make sure the fuel injection supply manifold pressure has been released.
5. Turn the ignition switch to the OFF position.
6. When fuel system service is complete, install the fuel pump relay.

➡ **It may take more than one key cycle to pressurize the fuel system.**

7. Cycle the ignition key and wait three seconds to pressurize the fuel system. Check for leaks before starting the engine.
8. Start the vehicle and check the fuel system for leaks.

Escape Hybrid

1. Before servicing the vehicle, refer to the Precautions Section.
2. With the vehicle in NEUTRAL, position it on a hoist.
3. Disconnect the fuel pump driver module electrical connector.
4. Start the engine and allow it to idle until it stalls.
5. After the engine stalls, crank the engine for approximately 5 seconds to make

sure the fuel injection supply manifold pressure has been released.
6. Turn the ignition switch to the OFF position.
7. When fuel system service is complete, connect the fuel pump driver module electrical connector.

➡ **It may take more than 1 key cycle to pressurize the fuel system.**

8. Cycle the ignition key and wait 3 seconds to pressurize the fuel system. Check for leaks before starting the engine.
9. Start the vehicle and check the fuel system for leaks.

Fuel Filter

REMOVAL & INSTALLATION

1. Before servicing the vehicle, refer to the Precautions Section.
2. Properly relieve the fuel system pressure.
3. Remove or disconnect the following:

   - Negative battery cable
   - Fuel line to the fuel filter

4. Loosen the clamp and remove the filter

To install:
5. Install or connect the following:
   - New clips to the fuel lines
   - Fuel filter and tighten the clamp
   - Fuel lines to the fuel filter
   - Negative battery cable
6. Start the vehicle and check for leaks, repair if necessary.

Fuel Pump

REMOVAL & INSTALLATION

2002–04

1. Before servicing the vehicle, refer to the Precautions Section.
2. Properly relieve the fuel system pressure.
3. Remove or disconnect the following:

   - Negative battery cable
   - Gas cap to relieve any additional fuel pressure
   - Left rear seat cushion and lift the access cover on the scuff plate
   - Pin type retainers and move the carpet aside

81 Nm (60 lb-ft)

310-069

67197-ESCA-G57

**Fuel pump module removal/installation—
2002–04 models**

- Screws from the fuel pump module access cover
- Fuel pump module electrical connectors
- Fuel and vapor lines from the fuel tank
- Fuel pump module and discard the gasket

To install:

4. Install or connect the following:
- New fuel pump module gasket
- Fuel pump module. Torque the module to 60 ft. lbs. (81 Nm).

- Fuel and vapor lines to the fuel tank
- Fuel pump module electrical connectors
- Fuel pump module access cover and tighten the screws securely
- Pin type retainers and reposition the carpet
- Left rear seat cushion
- Gas cap
- Negative battery cable

5. Start the engine and check for leaks, repair if necessary.

1	Fuel pump module access cover screws	5	Fuel supply tube quick connect coupling
2	Fuel pump module access cover	6	Fuel pump module lock ring
3	Fuel pump module electrical connector	7	Fuel pump module
4	Fuel vapor control tube assembly valve electrical connector	8	Fuel pump module O-ring seal

67197-ESCA-G34

Access to fuel pump module—2005 models

2005

1. Before servicing the vehicle, refer to the Precautions Section.

2. With the vehicle in NEUTRAL, position it on a hoist.

3. Release the fuel system pressure.

4. Disconnect the battery ground cable.

5. If removing the fuel tank on a four wheel drive vehicle, it is necessary to lower the exhaust system from the catalytic converter back. Support the exhaust system with a suitable stand, release the three rear exhaust hangers and carefully lower the exhaust system to allow enough clearance to remove the fuel tank.

6. If removing the fuel tank on a four wheel drive vehicle, remove the rear driveshaft..

7. Lift the left rear seat cushion, position the carpet aside and remove the screws and the fuel pump module access cover.

8. Disconnect the fuel pump module electrical connector

9. Fuel vapor control tube assembly valve electrical connector

10. Using a suitable fuel pump lock ring remover, rotate the lock ring counterclockwise and remove.

✳✳ WARNING

The fuel pump module must be handled carefully to avoid damage to the float arm and filter.

✳✳ WARNING

Some fuel will remain in the fuel pump module after draining the fuel tank. Carefully drain the fuel pump module into a suitable container.

11. Prior to completely removing the fuel pump, position it aside and using the special tool and a suitable fuel recovery system, drain the fuel tank.

12. To release the bottom-mounted fuel pump module, reach into the fuel pump module opening and squeeze the retainer tabs on the pump module housing and pull upward.

13. Remove the fuel pump module O-ring seal
To install:
14. Turn the ignition key to the ON position to pressurize the fuel system.

15. Visually inspect the fuel system for leaks.

✳✳ WARNING

Make sure the fuel tube clicks into place when installing the tube. To

make sure the tube is fully seated, pull on the tube.

➡**Apply clean engine oil to the end of the tube before inserting the tube into the connector.**

16. Install the fuel tube quick release coupling.

✳✳ WARNING

Inspect the surfaces of the fuel pump module flange and fuel tank O-ring contact surfaces. Do not polish or adjust the O-ring contact area of the fuel pump flange or fuel tank. Install a new fuel pump module or fuel tank if the O-ring contact area is bent, scratched or corroded.

✳✳ WARNING

Make sure to install a new fuel pump module O-ring and lock ring.

17. Lubricate and install a new fuel pump module O-ring seal upon installing the fuel pump module.

18. When installing the fuel pump module, make sure to align the locator tabs on the fuel tank mounting flange.

➡**Be sure the aligning tabs of the fuel pump module unit are positioned in the slot before tightening the lock ring.**

19. Holding the fuel pump module O-ring seal in place, rotate the lock ring clockwise until it stops against the retainer tabs.

➡**Make sure the collar on the fuel tube is inserted fully into the quick release coupling before the locking tab is locked.**

20. Connect the fuel supply quick connect coupling to the fuel supply manifold.

21. Press the fuel supply quick connect coupling locking tab into position.

22. Pull on the fitting to make sure it is fully engaged.

Fuel Injectors

REMOVAL & INSTALLATION

2.0L Engine

1. Before servicing the vehicle, refer to the Precautions Section.

2. Release the fuel system pressure.

3. Remove or disconnect the following:
   • Negative battery cable
   • Fuel injection supply manifold
   • Retaining clips and gently twist the fuel injector out of the manifold

4. Check the O-rings and replace if damaged.
   To install:
5. Install or connect the following:
   • Fuel injector(s) using new O-rings lubricated with clean engine oil
   • Fuel injector into the supply manifold
   • Retaining clips when the fuel injectors are seated properly
   • Fuel injection supply manifold
   • Negative battery cable

6. Start the vehicle and check for leaks, repair if necessary.

2.3L Engine

✳✳ WARNING

Do not smoke or carry lighted tobacco or open flame of any type when working on or near fuel-related components. Highly flammable vapors are always present and can ignite. Failure to follow these instructions can result in personal injury.

9308TG22

Remove the fuel injectors from the fuel supply manifold–2.0L engine

※※ WARNING

This procedure involves fuel handling. Be prepared for fuel spillage at all times and always observe fuel handling precautions. Failure to follow these instructions can result in personal injury.

1. Before servicing the vehicle, refer to the Precautions Section.
2. Disconnect the battery ground cable.
3. Release the fuel pressure.

4. Disconnect the fuel injector electrical connectors.
5. Disconnect the fuel pressure regulator electrical connector and the vacuum hose.
6. Disconnect the fuel injector harness retaining clips from the fuel injection supply manifold.
7. Disconnect the fuel tube.
8. Remove the bolts and the fuel injection supply manifold.

➡**Remove and discard the fuel injector O-rings.**

9. If necessary, remove the fuel injectors.
10. Install new O-rings and lubricate them with clean engine oil.
11. To install, reverse the removal procedure. Tighten the fuel supply manifold bolts to 18 ft. lbs. (25 Nm).

3.0L Engine

1. Before servicing the vehicle, refer to the Precautions Section.
2. Release the fuel system pressure
3. Disconnect the battery ground cable.
4. Remove the upper intake manifold.

1 Fuel tube quick release coupling (position aside)

2 Fuel rail pressure and temperature sensor electrical connector

3 Fuel rail pressure and temperature sensor vacuum tube (position aside)

4 Fuel injector electrical connectors

5 Fuel rail bolts

6 Fuel rail

7 Fuel injector clips

8 Fuel injectors

9 Fuel injector O-ring seals

67197-ESCA-G35

Fuel rail and injectors—2.3L engine

1 Fuel tube quick release coupling (position aside)

2 Fuel rail pressure and temperature sensor vacuum tube (position aside)

3 Fuel rail pressure and temperature sensor electrical connector

4 Fuel injector electrical connector

5 Fuel rail bolt

6 Fuel rail

7 Fuel injector

8 Fuel injector O-ring seals

67197-ESCA-G36

Fuel rail and injectors—3.0L engine

5. Disconnect the fuel tube quick release coupling.

6. Disconnect the fuel rail pressure and temperature sensor vacuum tube

7. Disconnect the fuel rail pressure and temperature sensor electrical connector.

8. Disconnect the fuel injector electrical connectors.

9. Remove the fuel rail bolts.

10. Remove the fuel rail.

11. Remove the fuel injector.

12. Remove the fuel injector O-ring seals.

13. To install, reverse the removal procedure. Install new fuel injector O-ring seals.

14. Lubricate the new O-ring seals with clean engine oil before installing. Torque the fuel rail bolts to 89 inch lbs. (10 Nm).

DRIVE TRAIN

Transmission Assembly, Except Escape Hybrid

REMOVAL & INSTALLATION

Manual Transmission

1. Before servicing the vehicle, refer to the Precautions Section.

2. Drain the transmission fluid.

3. Remove or disconnect the following:

- Battery cables
- Battery and tray
- Mass Air Flow (MAF) sensor electrical connector
- Accelerator cable from the air cleaner outlet tube

- Emission management tube and hose
- Crankcase ventilation hose
- Air cleaner outlet tube
- Air cleaner housing
- Back-up lamp switch electrical connector
- Front wire harness bracket and move it aside
- Front wire harness bracket spacer
- Wire harness from the rear harness bracket
- Park Neutral Position (PNP) electrical connector
- Rear wire harness bracket and move it aside
- Vehicle Speed Sensor (VSS) electrical connector

- Clutch slave cylinder line from the bracket and move it aside while properly supporting the engine
- Left side transmission support insulator and bracket
- Rear transmission support insulator
- Front transmission support insulator and bracket
- Starter and move it aside
- Top transmission flywheel housing bolts
- Front transmission flywheel housing bolts
- Transfer case, if equipped
- Left side halfshaft
- Rear transmission support insulator bracket
- Shifter linkage and stabilizer bar

- Transverse crossmember
- Front to aft crossmember
- Left side splash shield and properly support the transmission
- Remaining transmission flywheel housing bolts
- Transmission and separate the right side halfshaft from the transmission

To install:

4. Align the right side half shaft to the transmission and position the transmission to the engine.

5. Install or connect the following:

- Transmission flywheel housing bolts. Torque the bolts to 33 ft. lbs. (45 Nm) and remove the transmission support
- Left side splash shield
- Front-to-aft crossmember. Torque the bolts to 66 ft. lbs. (90 Nm).
- Transverse crossmember. Torque the bolts to 85 ft. lbs. (115 Nm).
- Shifter linkage. Torque the bolt to 15 ft. lbs. (20 Nm).
- Stabilizer bar. Torque the bolt to 30 ft. lbs. (40 Nm).
- Rear transmission support bracket. Torque the bolts to 66 ft. lbs. (90 Nm).
- Left side halfshaft
- Transfer case, if equipped
- Front transmission flywheel housing bolts. Torque the bolts to 33 ft. lbs. (45 Nm).
- Top transmission flywheel housing bolts. Torque the bolts to 33 ft. lbs. (45 Nm).
- Starter. Torque the bolts to 33 ft. lbs. (45 Nm).
- Front transmission support insulator and bracket. Torque the lower bolt to 66 ft. lbs. (90 Nm) and the 3 upper bolts to 41 ft. lbs. (55 Nm).
- Rear transmission support insulator bolt. Torque the bolt to 66 ft. lbs. (90 Nm).
- Left side transmission support insulator bracket. Torque the bolts to 66 ft. lbs. (90 Nm).
- Left side transmission support insulator. Torque the large bolt to 66 ft. lbs. (90 Nm) and the 3 remaining bolts to 41 ft. lbs. (55 Nm).
- Clutch slave cylinder. Torque the bolt to 15 ft. lbs. (20 Nm).
- Clutch slave cylinder line to the bracket and install the retaining clip
- VSS electrical connector
- Rear wire harness bracket. Torque the bolts to 80 inch lbs. (9 Nm).
- PNP switch electrical connector

- Wire harness to the rear bracket
- Front wire harness bracket spacer and bracket. Torque the bolt to 9 ft. lbs. (12 Nm).
- Back-up lamp switch electrical connector
- Air cleaner housing
- MAF sensor electrical connector
- Air cleaner outlet tube
- Crankcase ventilation hose
- Emission management tube and hose
- Accelerator cable to the air cleaner outlet tube
- Battery and tray
- Both battery cables

6. Fill the transmission to the proper level.

7. Start the vehicle and check for leaks, repair if necessary.

Automatic Transmission

1. Before servicing the vehicle, refer to the Precautions Section.

2. Remove or disconnect the following:

- Battery cables
- Battery and tray
- Breather tube
- Mass Air Flow (MAF) sensor
- Intake tube and air cleaner cover
- Air cleaner assembly
- Transmission Range (TR) sensor
- Heated Oxygen (HO2S) sensors
- Transmission harness connector and bracket
- Wire harness bracket spacer and move the bracket aside
- Shift cable
- Shift cable bracket and move the bracket aside
- Starter electrical connectors
- Starter
- Electrical connectors from the valve cover and install an engine support bar
- Upper transmission retaining bolts
- Left side upper transmission mounting plate
- Rear transmission mount
- Right side engine mount bolt and slightly raise the engine
- Both front wheels and splash shields
- Right side halfshaft and intermediate shaft assembly after matchmarking them
- Cross brace
- Center exhaust pipe and rubber hanger
- Front exhaust pipe and flange
- Rear exhaust pipe flange
- Driveshaft

- PTU vent tube
- Lower transmission bracket
- Access cover
- Flexplate nuts
- Output Shaft Speed (OSS) sensor
- Turbine Shaft Speed (TSS) sensor
- Fluid cooler tube and move it aside
- Fluid cooler line and install a transmission jack
- Bolts from the PTU unit
- Transmission with the PTU unit attached

To install:

3. Install or connect the following:

- Transmission with the PTU unit. Torque the engine-to-transmission mounting bolts to 30 ft. lbs. (40 Nm).
- Fluid cooler line. Torque the fastener to 17 ft. lbs. (23 Nm) and remove the transmission jack.
- Fluid cooler tube. Torque the bolt to 17 ft. lbs. (23 Nm).
- OSS sensor
- TSS sensor
- Flexplate nuts. Torque the nuts to 27 ft. lbs. (36 Nm).
- Access cover
- Cross brace. Torque the bolts to 96 ft. lbs. (130 Nm).
- Transmission bracket. Torque the bolts to 30 ft. lbs. (40 Nm).
- PTU vent tube
- Driveshaft. Torque the bolts to 15 ft. lbs. (20 Nm).
- Exhaust pipe and flange. Torque the bolts to 21 ft. lbs. (29 Nm).
- Exhaust pipe and rubber hanger. Torque the bolts to 21 ft. lbs. (29 Nm).
- Left side halfshaft assembly
- Right side halfshaft and intermediate shaft assembly by aligning the matchmarks
- Splash shields
- Both front wheels and lower the engine on to the right side engine mount
- Right side engine mount bolt. Torque the bolt to 89 ft. lbs. (120 Nm).
- Rear transmission mount. Torque the upper bolt to 89 ft. lbs. (120 Nm) and the lower bolts to 35 ft. lbs. (45 Nm).
- Transmission mount assemble. Torque the bolts to 30 ft. lbs. (40 Nm) and remove the engine support bar.
- Electrical connectors to the valve cover
- Starter. Torque the bolts to 20 ft. lbs. (27 Nm).

- Starter electrical connectors
- Shifter cable and bracket. Torque the bolt to 14 ft. lbs. (19 Nm) and connect the shifter cable.
- Wire harness and install the harness bracket spacer
- Wire harness bracket. Torque the bolt to 89 inch lbs. (10 Nm).
- HO2S sensor
- TR sensor and make certain it is properly aligned
- Air cleaner assembly
- Intake tube and air cleaner cover
- Breather tube
- MAF sensor
- Battery tray
- Battery and cables

4. Fill the transmission with clean fluid to the proper level.

5. Start the vehicle and check for leaks, repair if necessary

Transmission Assembly—Escape Hybrid

REMOVAL & INSTALLATION

Upper Transmission Insulator and Retainer

1. Before servicing the vehicle, refer to the Precautions Section.

2. Remove the air cleaner assembly.

3. Support the transaxle with a suitable floor jack.

4. Remove the bolts and the bracket.

5. Remove the nuts, bolt and the insulator.

6. To install, reverse the removal procedure. Torque the insulator bracket bolts to 18 ft. lbs. (25 Nm); the insulator bolts/nuts to 66 ft. lbs. (90 Nm).

Lower Transmission Insulator and Retainer

1. Before servicing the vehicle, refer to the Precautions Section.

2. With the vehicle in DRIVE, position it on a hoist.

3. Turn the ignition to the OFF position.

4. Remove the splash shield retainers and remove the left splash shield.

5. Remove the bolts and the lateral support crossmember.

6. Remove the front insulator bolt.

7. Remove the rear insulator bolt.

8. Remove the nut, bolts and the crossmember.

9. Remove the lower rear insulator bolt and remove it from the bracket.

10. To install, reverse the removal procedure. Observe the following torques:

1 Degas bottle
2 Coolant outlet hose
3 Degas bottle-to-pump hose
4 Motor electronics (M/E) coolant pump
5 Motor electronics coolant pump-to-transaxle hose
6 Transaxle-to-radiator hose
7 Motor electronics (M/E) cooler

Motor electronics cooling system—Escape Hybrid

06017-ESCA-G47

8 Motor electronics coolant pump-to-transaxle hose

9 Coolant tube (inlet/outlet)

10 Motor electronics coolant temperature sensor

11 Transaxle-to-radiator hose

06017-ESCA-G48

Motor electronics cooler hose connections—Escape Hybrid

- Lower rear insulator bolt: 66 ft. lbs. (90 Nm)
- Crossmember front bolts: 66 ft. lbs. (90 Nm)
- Crossmember rear nut: 111 ft. lbs. (150 Nm)
- Front and rear insulator bolts: 66 ft. lbs. (90 Nm)
- Lateral support crossmember bolts: 85 ft. lbs. (115 Nm)

Rear Transmission Insulator and Retainer

1. Before servicing the vehicle, refer to the Precautions Section.
2. With the vehicle in DRIVE, position it on a hoist.
3. Turn the ignition to the OFF position.
4. Disconnect the heater hose retaining clip from the transaxle mount stud.
5. Remove insulator nut.
6. Remove the nuts and the bracket.
7. Remove the nuts, heat shield, and insulator.
8. To install, reverse the removal procedure. Tighten all fasteners to 66 ft. lbs. (90 Nm).

Motor Electronics Cooler

1. Before servicing the vehicle, refer to the Precautions Section.

➡**Always refill the motor electronics cooling system (MECS) with the same type of coolant that was drained from the system. Do not mix coolant types.**

2. With the vehicle in DRIVE, position it on a hoist.
3. Turn the ignition to the OFF position.
4. Remove the left splash shield.
5. Place a suitable container below the transaxle.
6. Loosen the hose clamps at the transaxle, then, pull the hoses off to allow the coolant to drain.
7. Remove the front bumper cover.
   a. Remove the 6 pin-type retainers (3 each side).
   b. Remove the 4 fender splash shield bolts (2 each side).
   c. Remove the 4 air deflector bolts.
   d. Remove the 2 center front bumper cover pin-type retainers.
   e. Remove the 2 front bumper cover outer pin-type retainers (1 each side).
   f. If equipped, disconnect the fog lamp electrical connectors.
   g. Remove the 2 upper bolts and the front bumper cover.
8. Loosen the hose clamp and disconnect the inlet hose.
9. Loosen the hose clamp and disconnect the outlet hose.

10. Remove the motor electronics (M/E) cooler bolts.
11. To install, reverse the removal procedure. Observe the following torques:
    - Motor electronics (M/E) cooler bolts: 6 Nm (53 inch lbs.)
    - Front bumper cover bolts: 8 Nm (71 inch lbs.)
    - Air deflector bolts: 8 Nm (71 inch lbs.)
    - Fender splash shield bolts: 8 Nm (71 inch lbs.)
12. Connect the hoses to the transaxle and install the hose clamps.

❋❋ WARNING

Adhesives, stop-leak pellets or small debris in the Motor electronics cooling system (MECS) can cause poor performance or temporary blockage of the motor electronics pump. Only use clean, approved coolant when filling the system.

13. Loosen the bleed screw and fill the degas bottle with coolant until it begins to flow out of the bleed hole. Then, close the bleed screw.
14. Turn the ignition to the ON position to actuate the motor electronics cooling pump and continue to fill the degas bottle to the correct level.

➡Most of the MECS air bleeding occurs as a normal process at the degas bottle through the vent tube; very little occurs at the bleed screw.

15. To bleed air from the system, loosen the bleed screw and allow air to escape while the M/E coolant pump is operating.

Motor Electronics Pump

1. Before servicing the vehicle, refer to the Precautions Section.

➡Always refill the motor electronics cooling system (MECS) with the same type of coolant that was drained from the system. Do not mix coolant types.

2. With the vehicle in DRIVE, position it on a hoist.

3. Turn the ignition to the OFF position.

4. Remove the left splash shield.

5. Place a suitable container below the transaxle.

6. Disconnect the inlet and outlet hoses.

7. Remove the bolts.

8. To install, reverse the removal procedure. Tighten to 20 Nm (15 ft. lbs.).

⁕⁕ WARNING

Adhesives, stop-leak pellets or small debris in the Motor electronics cooling system (MECS) can cause poor performance or temporary blockage of the motor electronics pump. Only use clean, approved coolant when filling the system.

9. Loosen the bleed screw and fill the degas bottle with coolant until it begins to flow out of the bleed hole. Then, close the bleed screw.

10. Turn the ignition to the ON position to actuate the motor electronics cooling pump and continue to fill the degas bottle to the correct level.

➡Most of the MECS air bleeding occurs as a normal process at the degas bottle through the vent tube; very little occurs at the bleed screw.

11. To bleed air from the system, loosen the bleed screw and allow air to escape while the M/E coolant pump is operating.

eCVT Transmission

1. Before servicing the vehicle, refer to the Precautions Section.

⁕⁕ CAUTION

Before proceeding, read and observe all of the High Voltage System Precautions.

⁕⁕ CAUTION

To prevent exposure to high voltage, do not for any reason open the electronically controlled, continuously variable transmission (eCVT) cover. Failure to follow these instructions may result in personal injury.

All vehicles

2. With the vehicle in DRIVE, position it on a hoist.

3. Release the fuel system pressure.

4. Disconnect the battery ground cable.

⁕⁕ WARNING

Before removing the high voltage cables, the vehicle electrical system must be completely shut down for at least 10 minutes to allow for the high voltage capacitors to discharge.

5. De-power the high voltage traction battery. Establish a buffer zone. See the procedure under De-powering the High Voltage Traction Battery.

6. Drain the engine cooling system.

⁕⁕ WARNING

The coolant must be recovered in a suitable, clean container for reuse. If the coolant is contaminated, it must be recycled or disposed of correctly.

⁕⁕ WARNING

Always refill the motor electronics cooling system (MECS) with the same type of coolant that was drained from the system. Do not mix coolant types.

7. Turn the ignition to the OFF position.

8. Remove the left splash shield.

9. Place a suitable container below the transaxle.

10. Loosen the hose clamps at the transaxle, then pull the hoses off to allow the coolant to drain.

11. Remove the upper and lower intake manifold.

12. Remove the engine air cleaner.

13. Remove the exhaust manifold/catalytic converter.

14. Disconnect the motor electronics coolant vent hose.

15. Disconnect the 2 engine coolant vent hoses.

16. Remove the degas bottle bolt and nut.

17. Lift the degas bottle up and disconnect the engine coolant and motor electronics coolant return hoses.

18. Remove the degas bottle.

19. Remove the accessory drive belt.

20. Remove the left front drive halfshaft.

4WD vehicles

21. Remove the transfer case.

2WD vehicles

22. Remove the front drive intermediate halfshaft.

23. Remove the bolts and the lateral support crossmember.

All vehicles

24. Drain the engine oil.

25. Disconnect the heater hoses from the heater core.

26. Detach the heater hose retaining clip from the transaxle mount stud.

27. Disconnect the upper radiator and coolant vent hoses.

28. Disconnect the following electrical connectors:
- Secondary air injection (AIR) solenoid
- Transaxle mount control solenoid
- Purge valve

29. Disconnect the engine control harness.

30. Disconnect the 2 engine control harness electrical connectors and the 2 pin-type retainers.

31. Remove the bolt and disconnect the transaxle harness electrical connector.

32. Disconnect the DC-to-DC converter electrical connector and pin-type retainer.

33. Disconnect and remove the fuel supply tube and evaporative emissions (EVAP) tube.

34. If equipped, position the block heater wiring harness aside. Detach the 3 harness retainer clips. Route the harness through the radiator support into the engine compartment.

35. Remove the 2 bolts and disconnect the auxiliary coolant pump electrical connector.

36. Position the auxiliary coolant pump aside.

37. Remove the bolts and the transaxle control splash shield.

38. Disconnect the transaxle control cable.

39. Disconnect the pin-type retainer.

40. Disconnect the pin-type retainer and position the transaxle control cable aside.

41. Disconnect the motor electronics coolant temperature sensor electrical connector and pin-type retainer.

42. Disconnect the transaxle coolant hoses.

43. Remove the nut and the ground cable.

44. Remove the 2 bolts and position the motor electronics coolant pump aside.

45. Disconnect the lower radiator hose from the radiator.

46. Disconnect the A/C compressor electrical connector and remove the 3 bolts. Position the A/C compressor aside and support the compressor with a length of mechanic's wire.

47. Remove the front roll restrictor bolt.

48. Remove the rear roll restrictor bolt.

49. Remove the nut, bolts and the engine support crossmember.

➡ **The transaxle-to-engine bolts differ in length. Mark the bolts for correct installation.**

50. Remove the 2 transaxle-to-engine bolts.

➡ **The transaxle-to-engine bolts differ in length. Mark the bolts for correct installation.**

51. Remove the 2 transaxle-to-engine bolts.

➡ **Due to the weight of the transaxle, special care should be taken to mount the powertrain securely to the lift table.**

52. Using the special tools, secure the engine to the lift table.

53. Remove the engine mount bracket bolt.

54. Remove the nuts and the engine mount bracket.

55. Remove the right transaxle mount bracket nut.

56. Remove the bolts and the right transaxle mount bracket.

57. Remove the bolts and the rear transaxle mount brace.

58. Remove the nuts, bolt and rear transaxle mount.

59. Lower the engine and transaxle from the vehicle.

60. Using the engine crane and spreader bar, remove the engine and transaxle from the lift table.

61. Disconnect the high voltage wiring harness electrical connector.

62. Disconnect the 2 low voltage wiring harness electrical connectors and position the harness aside.

➡ **The transaxle-to-engine bolts differ in length. Mark the bolts for correct installation.**

63. Remove the remaining 6 engine-to-transaxle bolts and separate the engine and transaxle.

❋❋ **WARNING**

The damper contains a clutch which designed to slip briefly during vehicle operation. It is essential that no grease, oil or cleaning solvents be allowed to contaminate the slip clutch. Do not use grease on transmission input shaft or input shaft splines.

64. Should the damper become contaminated it must be replaced.

➡ **Due to packaging requirements, the correct bolt must be used at the damper locations.**

65. Remove the bolts and the transaxle damper.

To install:

❋❋ **CAUTION**

To prevent exposure to high voltage, do not for any reason open the electronically controlled, continuously variable transmission (eCVT) cover. Failure to follow these instructions may result in personal injury.

All vehicles

❋❋ **WARNING**

The damper contains a clutch which designed to slip briefly during vehicle operation. It is essential that no grease, oil or cleaning solvents be allowed to contaminate the slip clutch. Do not use grease on transmission input shaft or input shaft splines.

66. Should the damper become contaminated it must be replaced.

➡ **Due to packaging requirements, the correct bolt must be used at the damper locations.**

67. Install the transaxle damper. Tighten to 29 Nm (21 ft. lbs.).

❋❋ **WARNING**

When positioning the engine to the transaxle, care must be taken to maintain alignment of the damper spline with the transaxle input shaft.

➡ **Transaxle removed from view for clarity.**

68. Using the engine crane and spreader bar, position the engine and transaxle together.

69. Install the 6 transaxle-to-engine bolts. Tighten to 47 Nm (35 ft. lbs.).

70. Position the low voltage wiring harness and connect the 2 transaxle electrical connectors.

71. Connect the high voltage wiring harness electrical connector.

72. Using the engine crane and spreader bar, position the engine and transaxle onto the lift table.

❋❋ **WARNING**

Due to the weight of the transaxle, special care should be taken to mount the powertrain securely to the lift table.

73. Using the special tools, secure the engine to the lift table.

74. Raise the engine and transaxle into the vehicle.

75. Install the rear transaxle mount. Tighten to 90 Nm (66 ft. lbs.).

76. Install the rear transaxle mount bracket. Tighten to 25 Nm (18 ft. lbs.).

77. Install the right transaxle mount bracket. Tighten to 90 Nm (66 ft. lbs.).

78. Install the right transaxle mount bracket nut. Tighten to 90 Nm (66 ft. lbs.).

79. Install the engine mount bracket. Tighten to 90 Nm (66 ft. lbs.).

80. Install the engine mount bracket bolt. Tighten to 90 Nm (66 ft. lbs.).

81. Install the 2 transaxle-to-engine bolts. Tighten to 47 Nm (35 ft. lbs.).

82. Install the 2 transaxle-to-engine bolts. Tighten to 47 Nm (35 ft. lbs.).

83. Install the engine support crossmember nut and bolts. Tighten to 175 Nm (129 ft. lbs.).

84. Install the rear roll restrictor bolt. Tighten to 90 Nm (66 ft. lbs.).

85. Install the front roll restrictor bolt. Tighten to 115 Nm (85 ft. lbs.).

86. Install the A/C compressor and connect the A/C compressor electrical connector. Tighten to 25 Nm (18 ft. lbs.).

87. Connect the lower radiator hose to the radiator.

88. Install the motor electronics coolant pump. Tighten to 25 Nm (18 ft. lbs.).

89. Install the ground cable and nut. Tighten to 20 Nm (15 ft. lbs.).

90. Connect the transaxle coolant hoses.

91. Connect the motor electronics coolant temperature sensor electrical connector and pin-type retainer.

92. Position the transaxle control cable and connect the pin-type retainer.

93. Connect the pin-type retainer.

94. Connect the transaxle control cable. Install the transaxle control cable bracket and the 2 nuts finger tight. Attach the transaxle control cable to the control lever. Tighten to 22 Nm (16 ft. lbs.).

95. Position the control lever between the 2 casting ribs on the transaxle case.

96. Place the selector lever in the D position.

97. Tighten the transaxle control cable bracket nuts to 22 Nm (16 ft. lbs.).

98. To verify the correct cable adjustment, observe the control lever on the transaxle while an assistant shifts the selector lever to each range position ending in the D position.

99. Install the transaxle control splash shield. Tighten to 15 Nm (11 ft. lbs.).

100. Install the motor auxiliary coolant pump and connect the electrical connector. Tighten to 10 Nm (89 inch lbs.).

➡Parts removed from view for clarity.

101. If equipped, install the block heater wiring harness. Route the harness through the radiator support. Attach the 3 harness retainer clips.

102. Install the fuel supply tube and evaporative emissions (EVAP) tube.

103. Connect the DC-to-DC converter electrical connector and pin-type retainer.

104. Connect the transaxle harness electrical connector and install the bolt. Tighten to 25 Nm (18 ft. lbs.).

105. Connect the 2 engine control harness electrical connectors and the 2 pin-type retainers.

106. Connect the engine control harness. Engage the powertrain control module (PCM) connector. Install the harness retaining nut. Tighten to 6 Nm (53 inch lbs.).

107. Connect the following electrical connectors:
- Secondary air injection (AIR) solenoid
- Transaxle mount control solenoid
- Purge valve

108. Connect the upper radiator and coolant vent hoses.

109. Attach the heater hose retaining clip to the transaxle mount stud.

110. Connect the heater hoses to the heater core.

2WD vehicles

111. Install the lateral support crossmember. Tighten to 115 Nm (85 ft. lbs.).

112. Install the front drive intermediate halfshaft.

4WD vehicles

113. Install the transfer case.

All vehicles

114. Install the left front drive halfshaft.

115. Install the accessory drive belt.

116. Install the engine coolant degas bottle.

117. Install the exhaust manifold/catalytic converter.

118. Install the engine air cleaner.

119. Install the intake manifold.

120. Enable the vehicle high voltage electrical system. See the procedure under High Voltage Traction Battery.

121. Connect the battery ground cable.

122. Fill the engine with clean engine oil.

123. Connect the hoses to the transaxle and install the hose clamps.

❋❋ WARNING

Adhesives, stop-leak pellets or small debris in the Motor electronics cooling system (MECS) can cause poor performance or temporary blockage of the motor electronics pump. Only use clean, approved coolant when filling the system.

124. Loosen the bleed screw and fill the degas bottle with coolant until it begins to flow out of the bleed hole. Then, close the bleed screw.

125. Turn the ignition to the ON position to actuate the motor electronics cooling pump and continue to fill the degas bottle to the correct level.

➡**Most of the MECS air bleeding occurs as a normal process at the degas bottle through the vent tube; very little occurs at the bleed screw.**

126. To bleed air from the system, loosen the bleed screw and allow air to escape while the M/E coolant pump is operating. Fill and bleed the motor electronics cooling system.

127. Fill and bleed the engine cooling system.

Clutch

ADJUSTMENTS

The clutch is hydraulically driven and therefore no adjustment is required.

REMOVAL & INSTALLATION

1. Before servicing the vehicle, refer to the Precautions Section.

1. Clutch pressure plate bolts (6 required)
2. Clutch pressure plate
3. Clutch disc

Clutch components

Torque the pressure plate bolts in the proper sequence

2. Remove or disconnect the following:
   - Negative battery cable
   - Transmission and lock the flywheel to the engine with special tool 303–103
   - Pressure plate bolts by loosening them evenly
   - Clutch pressure plate and disc

3. Clean the pressure plate and inspect it for burn marks, scores, flatness or ridges, replace if damaged.

4. Inspect the pressure plate diaphragm finger for wear, replace if damaged.

5. Measure the depth of the rivet heads. Minimum depth is 0.012 inch (0.3mm).

6. Inspect the clutch disc for signs of wear and replace if needed.

7. Check the clutch disc runout. Replace the disc if not with specification: 0.027 inch (0.7mm).

To install:

8. Install or connect the following:
   - Clutch disc to the flywheel
   - Pressure plate to the flywheel. Torque the bolts in sequence to 21 ft. lbs. (29 Nm).
   - Transmission
   - Negative battery cable

9. Check the transmission fluid level and top off if necessary.

Hydraulic Clutch System

BLEEDING

The following procedure is recommended for bleeding the clutch hydraulic system installed on the vehicle. It is recommended that the original clutch tube, with quick-connect fitting be replaced when servicing the hydraulic system, because air can be trapped in the quick-connect fitting and prevent complete bleeding of the system.

1. Before servicing the vehicle, refer to the Precautions Section.

2. Clean the dirt and grease from the dust cap.

3. Remove the cap and diaphragm and fill the reservoir ¾ of the way with approved brake fluid C6AZ-19542-AB or DOT 3 equivalent fluid (ESA-M6C25-A).

4. Loosen the bleeder screw cover from the slave cylinder and attach a hose to the screw.

5. Place the hose in a container and slowly pump the clutch pedal several times.

6. With the clutch pedal depressed, loosen the bleeder screw to release the fluid and air.

7. Remove the hose and tighten the bleeder screw.

8. Repeat this procedure until all the air is removed from the hydraulic system.

REMOVAL & INSTALLATION

Master Cylinder

2002–04

1. Before servicing the vehicle, refer to the Precautions Section.

> ❋❋ **CAUTION**
>
> Note the location and orientation of the clutch push rod retainer. If any damage to the retainer occurs during removal or installation, replace the retainer clip. The retainer must be correctly installed during assembly.

2. Disconnect the clutch master cylinder rod from the clutch pedal.

3. Remove the clutch pedal bracket nut.

> ❋❋ **CAUTION**
>
> Brake fluid contains polyglycol ethers and polyglycols. Avoid contact with eyes. Wash hands thoroughly after handling. If brake fluid contacts eyes, flush eyes with running water for 15 minutes. Get medical attention if irritation persists. If taken internally, drink water and induce vomiting. Get medical attention immediately. Failure to follow these instructions may result in personal injury.

> ❋❋ **CAUTION**
>
> Brake fluid is harmful to painted and plastic surfaces. If brake fluid is spilled onto a painted or plastic surface, wash it immediately with water.

4. Disconnect the clutch master cylinder hose from the clutch master cylinder and plug it to prevent excess fluid loss.

5. Disconnect the clutch slave cylinder tube from the clutch master cylinder.

6. Remove the clutch master cylinder nut.

7. Remove the clutch master cylinder.

8. To install, reverse the removal procedure. Make sure that the push rod retainer is correctly positioned. Observe the following torques:
   - Clutch pedal bracket nut: 17 ft. lbs. (23 Nm).
   - Slave cylinder tube-to-master cylinder: 22 ft. lbs. (30 Nm).
   - Master cylinder mounting nuts: 17 ft. lbs. (23 Nm).

9. Bleed the air from the system.

Clutch pedal bracket—2002–04 models

1	Clutch master cylinder hose	5	Clutch master cylinder	9	Clutch line bracket
2	Nut	6	Clutch line	10	Clutch slave cylinder to clutch line adapter
3	Clutch master cylinder push rod	7	Fitting	11	Bolt
4	Fitting	8	Bolt	12	Clutch slave cylinder

67197-ESCA-G60

Hydraulic clutch components—2005 models shown

2005

1. Before servicing the vehicle, refer to the Precautions Section.

✴✴ WARNING

Brake fluid is harmful to painted and plastic surfaces. If brake fluid is spilled onto a painted or plastic surface, wash it immediately with water.

➡If removing the clutch line, remove the engine air cleaner.

➡If removing the clutch slave cylinder or the clutch slave cylinder-to-clutch line adapter, remove the transaxle.

2. Remove the clutch master cylinder hose

3. Remove the clutch master cylinder push rod

4. Remove the fitting at the master cylinder.

5. Remove the clutch master cylinder

✴✴ WARNING

Make sure the O-rings are properly positioned on the hydraulic line fittings or leaks may occur.

6. To install, reverse the removal procedure. Torque the mounting nuts to 17 ft. lbs. (23 Nm).

7. Fill and bleed the system.

Slave Cylinder

2002–04

1. Before servicing the vehicle, refer to the Precautions Section.

✴✴ WARNING

Brake fluid is harmful to painted and plastic surfaces. If brake fluid is spilled onto a painted or plastic surface, wash it immediately with water.

2. Disconnect the clutch slave cylinder tube and plug it to prevent excess fluid loss.

3. Remove the clutch slave cylinder bolts.

4. Remove the clutch slave cylinder.
5. To install, reverse the removal procedure. Observe the following torques:
- Slave cylinder mounting bolts: 15 ft. lbs. (20 Nm)
- Fluid line-to-slave cylinder: 18 ft. lbs. (25 Nm)

6. Bleed the air from the system.

2005

1. Before servicing the vehicle, refer to the Precautions Section.

> ✴✴ **WARNING**
>
> **Brake fluid is harmful to painted and plastic surfaces. If brake fluid is spilled onto a painted or plastic surface, wash it immediately with water.**

2. Remove the transaxle.
3. Disconnect the clutch slave cylinder-to-clutch hydraulic fluid tube adapter.
4. Remove and the clutch slave cylinder.
5. To install, reverse the removal procedure. Tighten the 3 clutch slave cylinder bolts to 12 Nm (9 ft. lbs.)
6. Bleed the air from the system.

BLEEDING

1. Before servicing the vehicle, refer to the Precautions Section.

> ✴✴ **WARNING**
>
> **To prevent damage do not spill brake fluid onto plastic or painted surfaces.**

➡ The fluid in the brake master cylinder reservoir must be maintained at the ¾ level or higher during air bleeding. Be sure to use High Performance DOT 3 Motor Vehicle Brake Fluid, or equivalent.

➡ The clutch hydraulic system must be bled whenever a clutch master cylinder, clutch slave cylinder tube or hydraulic clutch hose is disconnected.

2. Attach a rubber drain hose to the bleeder screw and submerge the free end of the hose in a container partially filled with clean brake fluid.
3. Fully press the clutch pedal to the floor and hold it.

4. With the clutch pedal held to the floor, loosen the bleeder screw until fluid and air are expelled from the system.
5. With the clutch pedal held to the floor, tighten the bleeder screw.
6. Repeat Steps 2 through 4 until no air bubbles appear in the fluid.

Transfer Case

REMOVAL & INSTALLATION

2002 w/2.0L Engine

1. Before servicing the vehicle, refer to the Precautions Section.
2. Remove the driveshaft.
3. Remove the four bolts and remove the crossmember brace.
4. Remove the two bolts and the two nuts.
5. Remove the transfer case.
6. Remove and discard the O-ring seal.

To install:
7. Install a new O-ring seal.
8. Install the transfer case.

1 Transfer case-to-transaxle bolts
2 Transfer case-to-transaxle nuts
3 Transfer case

67197-ESCA-G37

Transfer case mounting with manual transaxle—2005 model shown

9. Install the two bolts and the two nuts. Torque to 33 ft. lbs. (45 Nm).

10. Install the crossmember brace and the four bolts. Torque to 30 ft. lbs. (40 Nm).

11. Install the driveshaft.

2002–04

1. Before servicing the vehicle, refer to the Precautions Section.

2. Disconnect the battery.

3. Drain the transfer case.

4. Remove the driveshaft.

5. Remove the 4 bolts and the crossmember brace.

6. Remove the generator.

7. Disconnect the right catalyst monitor.

8. Remove and discard the 2 nuts and separate the flexible Y-pipe from the manifold.

9. Remove and discard the nuts. Discard the gasket.

10. Remove and discard the nuts. Discard the gasket.

11. Remove the flexible pipe. Disconnect the hanger.

12. Disconnect the EGR valve to exhaust manifold tube at the manifold.

13. Disconnect the right heated oxygen sensor.

14. Remove the right exhaust manifold. Discard the gasket.

15. Remove the support bracket.

16. Remove the 3 transfer case bolts.

➡ **The transfer case driven gear seal must be replaced whenever the link shaft or transfer case is removed from the vehicle.**

➡ **If necessary, replace the right differential fluid seal.**

17. Remove the bolt and the transfer case.

To install:

➡ **The transfer case driven gear seal must be replaced whenever the link shaft or transfer case is removed from the vehicle. If necessary, replace the right differential fluid seal.**

18. Install the transfer case. Install the bolts. Torque to 33 ft. lbs. (45 Nm).

19. Install the side mount bolt. Torque to 33 ft. lbs. (45 Nm).

20. Install the bracket and the 6 bolts. Torque to 30 ft. lbs. (40 Nm).

21. Install the right exhaust manifold.

22. Connect the right heated oxygen sensor.

23. Connect the EGR to exhaust manifold tube. Torque to 30 ft. lbs. (40 Nm).

24. Install the flex pipe and connect the hanger.

25. Install the gasket and the nuts. Torque to 37 ft. lbs. (50 Nm).

1 Heat shield bolt
2 Heat shield
3 Transfer case-to-transaxle bolts
4 Transfer case-to-transaxle bolts
5 Transfer case

Transfer case mounting with automatic transaxle—2005 model shown

67197-ESCA-G38

26. Install the flex pipe gasket and the nuts. Torque to 37 ft. lbs. (50 Nm).

27. Connect the flexible Y-pipe to the exhaust manifold and install the two nuts. Torque to 30 ft. lbs. (40 Nm).

28. Connect the right catalyst monitor sensor.

29. Install the generator.

30. Install the crossmember brace and the four bolts. Torque to 30 ft. lbs. (40 Nm).

31. Check the transfer case fluid level.

32. Install the driveshaft.

33. Connect the battery ground cable.

34. Check the transaxle fluid level.

2005

WITH MANUAL TRANSAXLE

1. Before servicing the vehicle, refer to the Precautions Section.

2. Remove the driveshaft.

3. Remove the 4 bolts and the cross-member brace.

4. Remove the transfer case-to-transaxle bolts.

5. Remove the transfer case-to-transaxle nut.

6. Remove the transfer case.

7. To install, reverse the removal procedure.

✳✳ WARNING

The O-ring must be properly installed before mating the transfer case to the manual transaxle. Failure to properly install the O-ring may cause the O-ring to be damaged resulting in transaxle oil leak.

8. Install a new O-ring seal. Observe the following torques:
- Crossmember bolts: 30 ft. lbs. (40 Nm).
- Transfer case mounting nuts: 33 ft. lbs. (45 Nm).
- Transfer case mounting bolts: 33 ft. lbs. (45 Nm).

WITH AUTOMATIC TRANSAXLE

1. Before servicing the vehicle, refer to the Precautions Section.

2. Disconnect the battery.

3. Drain the transfer case.

4. Remove the front right intermediate shaft.

5. Remove the driveshaft.

6. Remove the 4 bolts and the cross-member brace.

7. Remove the generator.

8. Remove the exhaust as required.

9. Remove the heat shield.

10. Remove the transfer case-to-transaxle bolts.

11. Remove the transfer case.

✳✳ WARNING

A new transfer case driven gear seal must be installed whenever the intermediate shaft or transfer case is removed from the vehicle.

➡ If necessary, replace the right differential fluid seal.

12. To install, reverse the removal procedure. Observe the following torques:
- Crossmember bolts: 30 ft. lbs. (40 Nm).
- Transfer case-to-transaxle bolts: 33 ft. lbs. (45 Nm).

Front Halfshaft

REMOVAL & INSTALLATION

1. Before servicing the vehicle, refer to the Precautions Section.

2. Place the transmission in the PARK position.

1. Front axle wheel hub nut
2. Wheel speed sensor bolt
3. Wheel speed sensor
4. Lower control arm pinch nut
5. Lower control arm pinch bolt
6. Front halfshaft

67197-ESCA-G39

Front halfshaft

3. Remove or disconnect the following:
- Negative battery cable
- Front wheel
- Front brake disc
- Front axle wheel hub nut and discard the nut
- Tie rod end and separate the lower ball from the steering knuckle
- Halfshaft from the steering knuckle
- Halfshaft

To install:

4. When seated properly, the halfshaft bearing retainer circlip will snap into the differential side gear groove.

5. Position the halfshaft and joint so that the splines align with differential side gear splines. Push the halfshaft into side gear.

6. Install or connect the following:
- Halfshaft into the steering knuckle
- Lower ball joint to steering knuckle. Torque the pinch bolt to 52 ft. lbs. (70 Nm).
- Tie rod end. Torque the nut to 41 ft. lbs. (55 Nm).
- New front axle wheel hub nut. Torque the nut to 214 ft. lbs. (290 Nm).
- Front brake disc

- Front wheel
- Negative battery cable

7. Check the fluid level and adjust as needed.

Rear Halfshaft

REMOVAL & INSTALLATION

1. Before servicing the vehicle, refer to the Precautions Section.
2. Place the selector lever in NEUTRAL.
3. Raise and support the vehicle.

1 Rear axle wheel hub retainer
2 Rear axle hub assembly
3 Lower ball joint pinch bolt nut
4 Rear axle halfshaft assembly
5 Stub shaft seal and bearing

67197-ESCA-G40

❊ WARNING

Do not loosen the rear axle wheel hub retainer until after the wheel and tire assembly are removed from the vehicle. Wheel bearing damage will occur if the wheel bearing is unloaded with the weight of the vehicle applied.

4. Remove the rear brake drum or brake disc.
5. Remove the rear coil spring.
6. Rear axle wheel hub retainer.
7. Using a puller, press the outboard CV joint until it is loose in the hub.
8. Separate the outboard CV joint from the hub.
9. Remove the rear axle hub assembly.
10. Remove the lower ball joint pinch bolt nut.

❊ WARNING

Do not use a hammer to separate the rear axle halfshaft assembly from the hub. Damage to the threads and the internal CV joint can result.

11. Using a prybar, remove the halfshaft.
12. Using a puller, remove the stub and shaft seal.
13. Using a slidehammer and adapter, remove the stub shaft pilot bearing and seal.
14. To install, reverse the removal procedure.
15. Fill the axle with the specified quantity of the specified lubricant.

➡**Lubricate the new stub shaft pilot bearing with rear axle lubricant.**

16. Using suitable drivers, install the stub shaft pilot bearing.

➡**Lubricate the new stub shaft pilot bearing housing seal with grease.**

17. Using the special tools, install the stub shaft pilot bearing housing seal.
18. Install a new circlip on the inboard CV joint.
19. Install the halfshaft end into the hub assembly.
20. Observe the following torques:
   • Lower ball joint: 85 ft. lbs. (115 Nm).
   • Halfshaft nut: 2002–04 models, 214 ft. lbs. (290 Nm); 2005 models, 221 ft. lbs. (300 Nm).

CV-Joints

1. Before servicing the vehicle, refer to the Precautions Section.

Observe the following cautionary notes:
   • Never pick up or hold the halfshaft by only the inner or outboard CV joint.
   • Handle the halfshaft by only the interconnecting shaft to avoid pull-apart and potential damage to the CV joints.
   • Do not over-angle the CV joints.
   • Damage will occur to an assembled inner CV joint if it is over-plunged outward from the joint housing.
   • Never use a hammer to remove or install the halfshafts.
   • Never use the halfshaft assembly as a lever to position other components. Always support the free-ends of the halfshaft.
   • Do not allow the boots to contact sharp edges or hot exhaust components.
   • Do not drop assembled halfshafts. The impact may cut the boots from the inside without evidence of external damage.

OVERHAUL

Front Halfshaft

1. Before servicing the vehicle, refer to the Precautions Section.
2. Remove or disconnect the following:

   • Negative battery cable
   • Halfshaft and secure it in a soft-jawed vise
   • Inboard halfshaft boot clamp
   • Boot from the inboard CV-joint housing
   • Tripod joint from the CV-joint housing and matchmark the tripod joint to the halfshaft
   • Snapring and boot from the halfshaft
   • Outboard halfshaft boot clamps
   • Outboard boot back to expose the CV-joint and matchmark the joint to the halfshaft
   • Outboard CV-joint from the halfshaft
   • Halfshaft retainer circlip and discard it
   • Boot from the halfshaft

To install:
3. Lubricate the outer CV-joint with grease.
4. Install or connect the following:
   • Outboard CV-joint and boot
   • New halfshaft bearing circlip
   • Inboard CV-joint to the halfshaft
   • Outboard halfshaft boot forward on to the outboard CV-joint
   • New outboard halfshaft boot clamps

06017-ESCA-G51

Matchmark the tripod joint to the halfshaft—front shaft

06017-ESCA-G52

Remove the snapring from the from the shaft end—front shaft

06017-ESCA-G53

Halfshaft retainer circlip—front shaft

06017-ESCA-G54

Lubricate the outer CV-joint with grease—front shaft

Installing new outer CV-joint boot clamps—front shaft

06017-ESCA-G55

Lubricate the needle bearings while filling the housing with CV-joint grease—front shaft

06017-ESCA-G56

- Inboard halfshaft boot
- Tripod joint on the halfshaft by aligning the matchmarks
- New snapring to the tripod joint and lubricate the needle bearings while filling the housing with CV-joint grease, E43Z–19590–A
- Inboard halfshaft boot with new clamps
- Halfshaft
- Negative battery cable

Rear Halfshaft

1. Before servicing the vehicle, refer to the Precautions Section.
2. Remove the halfshaft.
3. Secure the halfshaft in a vise using protective jaw covers.
4. Remove the 2 inner halfshaft boot clamps.
5. Slide the inner halfshaft boot off the inner CV-joint housing.
6. If reinstalling the original inner joint, mark the inner CV-joint and the halfshaft to be sure of correct installation.
7. Using a soft face hammer, separate the halfshaft from the inner joint housing.
8. Remove and discard the bearing retainer circlip.
9. Remove and discard the snap ring.
10. Remove the inner halfshaft boot from the halfshaft.

11. Remove the 2 outer halfshaft boot clamps.
12. Slide the outer halfshaft joint boot back out of the way exposing the outer CV-joint.
13. If reinstalling the original inner CV-joint, mark the outer CV-joint and the half-shaft to be sure of correct installation.
14. Using a soft-face hammer, separate the outer CV-joint by gently tapping it off the halfshaft.
15. Remove and discard the bearing retainer circlip.
16. Remove the snap ring from the half-shaft.
17. Slide the outer halfshaft boot off the halfshaft.

Assembly

18. Lubricate the inner and outer CV-joints with joint grease.
19. Install the outer halfshaft boot.
20. Install the snap ring on the halfshaft.
21. Install a new halfshaft bearing retainer circlip.
22. Using a soft-face hammer, install the inner CV-joint by gently tapping it onto the halfshaft.
23. Remove any excess grease on the mating surfaces and slide the outer half-shaft joint boot forward onto the outer CV-joint.
24. Remove any excess air trapped in the outer halfshaft boot using a cloth-covered screwdriver after adjusting the outer half-shaft boot spacing.
25. Using a crimping tool, install 2 new outer halfshaft boot clamps.
26. Position the inner halfshaft boot.
27. Install the snap ring.
28. Install the bearing retainer circlip.
29. Using a soft face hammer, install the halfshaft on the inner CV-joint.
30. Position the inner halfshaft boot.
31. Remove any excess air trapped in the inner halfshaft boot using a cloth-covered screwdriver after adjusting the inner half-shaft boot spacing.
32. Using a crimping tool, install 2 new inner CV-joint boot clamps.
33. Install the halfshaft.

Rear Differential Mass Damper

REMOVAL & INSTALLATION

1. Before servicing the vehicle, refer to the Precautions Section.
2. Remove the rear differential mass damper.
3. Remove the bolts.
4. Remove the rear differential mass damper.

4 Nm (35 lb-in)

06017-ESCA-G59

Tighten the rear differential mass damper bolts to 35 inch lbs. in the sequence shown—2002–04 models

54 Nm (40 lb-ft)

06017-ESCA-G60

Tighten the rear differential mass damper bolts to 40 ft. lbs. in the sequence shown—2002–04 models

To install:

✳✳ WARNING

The mass damper bolts must be installed in the sequence shown or damage to the vehicle may occur.

➡️Install the mass damper bolts by hand until finger tight.

5. Install the rear differential mass damper.
6. Position the rear differential mass damper and install the bolts in the sequence shown to 35 inch lbs. (4 Nm).
7. Final torque the mass damper bolts in the sequence shown to 40 ft. lbs. (54 Nm).

Rear Drive Axle Housing

REMOVAL & INSTALLATION

2002–04

1. Before servicing the vehicle, refer to the Precautions Section.
2. Remove the negative battery cable.
3. Remove the driveshaft.
4. Disconnect the electrical connector.

➡️Drain the differential fluid into a suitable drain pan.

5. Remove the bolts and the rear differential cover.

6. Drain the differential fluid from the housing.

7. Remove the six bolts and rotary blade coupling assembly.

8. Remove or disconnect:
- Rear halfshafts
- Axle assembly-to-front bracket bolts
- Rear axle-to-side bracket bolts
- Axle assembly

To install:

9. Install or connect the following:
- Axle assembly
- Rear axle-to-side-bearing bolts. Torque the bolts to 59 ft. lbs. (80 Nm).
- Axle assembly-to-front bracket bolts. Torque the bolts to 59 ft. lbs. (80 Nm).

❋❋ WARNING

Make sure the machined surfaces on the rear axle housing, the differential housing cover and the rotary blade coupling housing are clean and free of oil before installing the new silicone sealant. The inside of the rear axle must be covered when cleaning the machined surface to prevent contamination.

10. Clean the gasket mating surfaces of the differential housing, the differential housing cover and the rotary blade coupling housing.

➡Make sure the grommet is seated all the way in the slot.

➡The rotary blade coupling assembly must be installed within 15 minutes of application of the silicone, or new sealant must be applied. If possible, allow one hour before filling with lubricant to make sure the silicone sealant has properly cured.

11. Apply a new continuous bead of clear silicone rubber to the rotary blade coupling housing.

12. Install the rotary blade coupling assembly and six bolts. Torque to 24 ft. lbs. (32 Nm).

➡The differential housing cover must be installed within 15 minutes of application of the silicone, or new sealant must be applied. If possible, allow one hour before filling with lubricant to make sure the silicone sealant has properly cured.

Differential housing-to-front insulator bracket bolts—2002–04 models

Front insulator bracket-to-subframe bolts—2002–04 models

Differential housing-to-side insulator bracket bolts—2002–04 models

Rotary blade coupling—2002–04 models

06017-ESCA-G58

Apply a new continuous bead of clear silicone rubber to the differential housing cover as shown—2002–04 models

13. Apply a new continuous bead of clear silicone rubber to the differential housing cover as shown in the illustration.

14. Install the differential housing cover and bolts. Torque to 17 ft. lbs. (23 Nm).

15. Connect the electrical connector.

16. Install the driveshaft.

17. Fill the rear axle with 1.4 liters (2.95 pints) of rear axle lubricant. Tighten the filler plug. Torque to 20 ft. lbs. (27 Nm).

18. Connect the negative battery cable.

2005

1. Before servicing the vehicle, refer to the Precautions Section.

2. Remove the spare tire.

3. Remove the rear driveshaft assembly.

4. Remove the rear halfshafts.

5. Position a suitable transmission hydraulic jack to the axle housing. Securely strap the jack to the housing.

6. Remove the electrical connector at the axle.

7. Remove the rear axle differential housing-to-front insulator bracket bolts.

8. Remove the front insulator-to-bracket subframe bolts.

9. Remove the front insulator brackets.

10. Remove the side insulator bracket-to-subframe nut.

115 Nm (85 lb-ft) 3

80 Nm (59 lb-ft) 8

115 Nm (85 lb-ft) 5

80 Nm (59 lb-ft) 2

115 Nm (85lb-ft) 3

1 Electrical connector

2 Rear axle differential housing-to-front insulator bracket bolts

3 Front insulator-to-bracket subframe bolts

4 Front insulator brackets

5 Side insulator bracket-to-subframe nut

6 Side insulator bracket-to-subframe bolt

7 Rear axle assembly

8 Side insulator bracket-to-rear axle differential bolts

9 Side insulator bracket

67197-ESCA-G41

Rear drive axle removal—2005 models

11. Remove the side insulator bracket-to-subframe bolt.

12. Remove the rear axle assembly.

13. Remove the side insulator bracket-to-rear axle differential bolts.

14. Remove the side insulator bracket.

15. To install, reverse the removal procedure. Observe the following torques:
- Axle housing-to-front insulator bracket: 59 ft. lbs. (80 Nm)
- Front insulator-to-bracket subframe bolts: 85 ft. lbs. (115 Nm)
- Side insulator bracket-to-subframe nut: 85 ft. lbs. (115 Nm)
- Side insulator bracket-to-differential bolts: 59 ft. lbs. (80 Nm)

Differential Pinion Seal

REMOVAL & INSTALLATION

Rear Differential

1. Before servicing the vehicle, refer to the Precautions Section.

2. With the vehicle in NEUTRAL, position it on a hoist.

3. Index-mark the pinion and pinion flange to the rear of the driveshaft.

4. Remove the 4 bolts and the 2 cap straps. Disconnect and support the driveshaft.

➡**Discard the nut after removing it. Install a new nut during installation.**

5. Using the special tool, hold the pinion flange while removing the nut. Remove the nut.

6. Index-mark the location of the pinion to the yoke.

7. Using a puller, remove the pinion flange.

8. Using the special tool, remove the seal.

To install:

➡**Make sure that the mating surface is clean before installing the new seal.**

9. Using a seal driver, install the seal.

➡**Lubricate the pinion flange with premium long-life grease.**

10. Line up the index marks and position the pinion flange.

11. Using the special tool, install the pinion nut. Tighten to 244 Nm (180 ft. lbs.).

12. Line up the index marks and position the rear driveshaft.

13. Install the 2 cap straps and the 4 bolts. Tighten to 23 Nm (17 ft. lbs.).

06017-ESCA-G67

Flange holding tool

06017-ESCA-G64

Removing the pinion flange

06017-ESCA-G65

Removing the pinion seal

06017-ESCA-G66

Installing the pinion seal

Rear Driveshaft

REMOVAL & INSTALLATION

1. Before servicing the vehicle, refer to the Precautions Section.

✽✽ CAUTION

The normal operating temperature of the exhaust system is very high. Never attempt to remove any part of the system until it has cooled. Be especially careful when working around the catalytic converters. The temperature of the converter rises to a high level after only a few minutes of engine operation. Failure to follow these instructions may result in personal injury.

➡Do not swap driveshaft assembles from different vehicles. The Escape Hybrid drive shaft is longer than the driveshaft in the Escape/Mariner and is not interchangeable.

With the vehicle in NEUTRAL, position it on a hoist.

2. Remove the ground strap bolt.

✽✽ WARNING

Do not reuse the CV-joint bolts and washers. Install new bolts and washers or damage to the vehicle may occur.

3. Remove and discard the 6 front driveshaft-to-transfer case bolts and washers. Index-mark the front driveshaft to the center bearing.

1	Ground strap bolt	5	Center bearing support nuts	9	Rear driveshaft
2	Universal joint cap bolts	6	Universal joint cap bolts	10	Front driveshaft U-joint
3	Universal joint cap straps	7	Universal joint cap straps	11	Rear driveshaft U-joint
4	Front driveshaft-to-transfer case bolts	8	Front driveshaft		

67197-ESCA-G42

Rear driveshaft assembly

> ⊕ **WARNING**
>
> **Do not reuse the bolts and cap straps for the center U-joint. Install new bolts and cap straps or damage to the vehicle may occur.**

➡There is a difference in the length of the head of the replacement cap strap bolts from the production bolts. The longer head pinion bolts can be used in either location.

4. Remove and discard the 4 universal joint cap strap bolts and 2 cap straps and remove the front driveshaft.

5. Index-mark the pinion and yoke to the driveshaft.

> ⊕ **WARNING**
>
> **Do not reuse the bolts and cap straps for the rear U-joint. Install new bolts and cap straps.**

➡There is a difference in the length of the head of the replacement strap bolts from the production bolts. The longer head pinion bolts can be used in either location.

6. Remove and discard the 4 universal joint cap bolts and 2 cap straps from the rear driveshaft universal joint.

7. With the help of an assistant, remove the center bearing support nuts and the driveshaft.

8. To install, reverse the removal procedure. Observe the following torques:
- Center bearing support nuts: 48 Nm (35 ft. lbs.)
- Rear universal joint cap bolts: 23 Nm (17 ft. lbs.)
- Front universal joint cap strap bolts: 23 Nm (17 ft. lbs.)
- The 6 front driveshaft-to-transfer case bolts: 37 Nm (27 ft. lbs.)
- Ground strap bolt: 40 Nm (30 ft. lbs.)

9. If a driveshaft is installed and driveshaft vibration is encountered after installation, index the driveshaft.

   a. With the vehicle in NEUTRAL, position it on a hoist.

> ✳✳ **WARNING**
>
> **Do not reuse the CV-joint bolts and washers. Install new bolts and washers or damage to the vehicle may occur.**

10. Remove and discard the 6 front driveshaft-to-transfer case bolts and washers.

11. Rotate the flange 60 degrees.

12. Connect the front driveshaft and install the 6 new bolts and washers. Tighten to 37 Nm (27 ft. lbs.).

> ✳✳ **WARNING**
>
> **Do not reuse the bolts and cap straps for the pinion yoke. Install new bolts**

and cap straps or damage to the vehicle may occur.

13. Disconnect the rear driveshaft universal joint. Discard the 4 bolts and the 2 cap straps.

14. Rotate the rear pinion 180 degrees.

15. Connect the rear driveshaft and install 4 new bolts and 2 new cap straps. Tighten to 23 Nm (17 ft. lbs.).

16. Lower the vehicle and test drive.

17. Repeat the procedure if necessary.

Intermediate Shaft

REMOVAL & INSTALLATION

1. Before servicing the vehicle, refer to the Precautions Section.

➡If removing the intermediate shaft in order to repair a separate component, it should only be removed as an assembly with the right front halfshaft.

2. Remove the right front halfshaft.

3. Remove the inner halfshaft bearing retainer nuts

4. Remove the intermediate shaft

5. To install, reverse the removal procedure. Apply a thin coat of grease to the splines of the intermediate shaft.

6. Verify the front axle lubricant level is to specifications.

27 Nm (20 lb-ft) — 1

1 Inner halfshaft bearing retainer nuts

2 Intermediate shaft

67197-ESCA-G43

Intermediate shaft

STEERING

Air Bag

> ✳✳ **CAUTION**
>
> Some vehicles are equipped with an air bag system. The system MUST BE disabled before performing service on or around system components, steering column, instrument panel components, wiring and sensors. Failure to follow safety and disabling procedures could result in accidental air bag deployment, possible personal injury and unnecessary system repairs.

PRECAUTIONS

• Always wear safety glasses when repairing an air bag supplemental restraint system (SRS) vehicle and when handling an air bag module. This will reduce the risk of injury in the event of an accidental deployment.

• After deployment, the air bag surface can contain deposits of sodium hydroxide, a product of the gas generant combustion that is irritating to the skin. Wash your hands with soap and water afterwards.

• Never probe the connectors on the air bag module. Doing so can result in air bag deployment, which can result in personal injury.

• Never probe the connectors on the safety canopy module. Doing so can result in safety canopy deployment, which can result in personal injury.

• The safety belt pretensioner is a pyrotechnic device. Always wear safety glasses when repairing an air bag equipped vehicle and when handling a safety belt buckle pretensioner or safety belt retractor pretensioner. Never probe a pretensioner electrical connector. Doing so could result in pretensioner or air bag deployment and could result in personal injury.

• To reduce the risk of personal injury, do not use any memory saver devices.

• Never carry the inflator module by the wires or connector on the underside of the module.

• When carrying a live inflator module, hold securely with both hands and ensure that the bag and trim cover are pointed away.

• Place the inflator module on a bench or other surface with the bag and trim cover facing up.

• With the inflator module on the bench, never place anything on or close to the module, which may be thrown in the event of an accidental deployment.

DE-POWERING

1. Before servicing the vehicle, refer to the Precautions Section.
2. Turn all vehicle accessories OFF.
3. Turn the ignition switch to OFF.
4. At the central junction box (CJB), located below the left side of the instrument panel, remove the cover and the restraints control module (RCM) fuse(s) from the CJB. See the Owner's Manual.
5. Turn the ignition ON and visually monitor the air bag indicator for at least 30 seconds. The air bag indicator will remain lit continuously (no flashing) if the correct RCM fuse has been removed. If the air bag indicator does not remain lit continuously, remove the correct RCM fuse before proceeding.
6. Turn the ignition OFF.

> ✳✳ **CAUTION**
>
> To avoid accidental deployment and possible personal injury, the backup power supply must be depleted before repairing or replacing any front or side air bag supplemental restraint system (SRS) components and before servicing, replacing, adjusting or striking components near the front or side air bag sensors, such as doors, instrument panel, console, door latches, strikers, seats and hood latches.

The front impact severity sensor is located on the radiator support bracket.

The first row side impact sensors (if equipped) are located at or near the base of the B-pillars.

The second row side impact sensors (if equipped) are located on the C-pillar.

➡ To deplete the backup power supply energy, disconnect the battery ground cable and wait at least one minute. Be sure to disconnect auxiliary batteries and power supplies (if equipped).

7. Disconnect the battery ground cable and wait at least one minute.

RE-POWERING

1. Before servicing the vehicle, refer to the Precautions Section.

> ✳✳ **CAUTION**
>
> The restraint system diagnostic tool is for restraint system service only. Remove from vehicle prior to road use. Failure to remove could result in injury and possible violation of vehicle safety standards.

2. Make sure all restraint system diagnostic tool(s) that may have been installed during the repair have been removed from the vehicle and all SRS components are connected.
3. Turn the ignition switch from OFF to ON.
4. Install RCM fuse(s) to the CJB and close the cover.

> ✳✳ **CAUTION**
>
> Be sure that nobody is in the vehicle and that there is nothing blocking or set in front of any air bag module when the battery ground cable is connected.

5. Connect the battery ground cable.
6. Prove out the supplemental restraint system (SRS) as follows:

   a. Turn the ignition key from ON to OFF. Wait 10 seconds, then turn the key back to ON and visually monitor the air bag indicator with the air bag modules installed. The air bag indicator will light continuously for approximately six seconds and then turn off. If an air bag supplemental restraint system (SRS) fault is present, the air bag indicator will either:

   • fail to light.
   • remain lit continuously.
   • flash.

   b. The flashing might not occur until approximately 30 seconds after the ignition switch has been turned from the OFF to the ON position. This is the time required for the restraints control module (RCM) to complete the testing of the SRS. If the air bag indicator is inoperative and a SRS fault exists, a chime will sound in a pattern of five sets of five beeps. If this occurs, the air bag indicator and any SRS fault discovered must be diagnosed and repaired.

7. Clear all continuous DTCs from the restraints control module using a scan tool.

Steering Gear

REMOVAL & INSTALLATION

Except Escape Hybrid

2002–04 MODELS

1. Before servicing the vehicle, refer to the Precautions Section.
2. Place the steering wheel in the straight-ahead position. Lock the steering wheel in place, using a steering wheel holder.

➡**Locking the steering wheel keeps the clockspring in alignment position.**

3. Drain the power steering fluid.
4. Remove or disconnect the following:
   - Negative battery cable
   - Rear transmission insulator
   - Rear transmission insulator bracket, if equipped with an automatic transmission
   - Both front wheels
   - Rear transmission insulator bracket, if equipped with a manual transmission
   - Tie rod end cotter pin and nut
   - Tie rod end from the steering knuckle and record the number of turns required to remove the tie rod end
   - Steering gear coupling pinch bolt
   - Power steering pressure and return lines and bracket

①	47 Nm (35 lb-ft)
③	55 Nm (41 lb-ft)	⑨	25 Nm (18 lb-ft)
⑤	40 Nm (30 lb-ft)	⑬	115 Nm (85 lb-ft)
⑥ ⑦ ⑧	10 Nm (89 lb-in)	⑱	110 Nm (81 lb-ft)

1 Tie-rod jam nuts (loosen)
2 Cotter pins
3 Tie-rod end nuts
4 Tie-rod end, outer
5 Steering column coupling-to-steering gear pinch bolt
6 Power steering return line bracket-to-subframe bolt
7 Power steering pressure line bracket-to-steering gear bolt
8 Power steering return line bracket-to-steering gear stud
9 Power steering line clamp plate bolt

10 O-rings
11 Power steering pressure line
12 Power steering return line
13 Steering gear mounting bolts
14 Steering gear
15 Tie-rod boot clamp, outer
16 Tie-rod boot clamp, inner
17 Tie-rod boot
18 Tie-rod, inner

67197-ESCA-G44

Steering gear mounting—2002–04 models

1   Tie-rod end jam nuts (part of
    3280) (2 required)

2   Cotter pins (2 required)

3   Outer tie-rod end nuts (2
    required)

4   Outer tie-rod end

5   Steering column
    coupling-to-steering gear
    pinch bolt

6   Steering gear-to-fluid cooler
    return hose
    bracket-to-subframe bolt

7   Power steering pressure line
    bracket-to-steering gear bolt

8   Steering gear-to-fluid cooler
    return hose bracket-to-steering
    gear stud

9   Power steering line clamp
    plate bolt

10 Power steering pressure line

11 Steering gear-to-fluid cooler
    return hose

12 O-rings (1 each required)

13 Steering gear mounting bolts
    (2 required)

14 Steering gear

Steering gear mounting—2005 models, exc. Escape Hybrid

06017-ESCA-G68

- Steering gear mounting bolts
- Steering gear and separate the steering coupling from the steering gear shaft
- Steering gear

To install:

- Slide the steering gear rearward to connect the steering coupling to the steering gear shaft

5. Install or connect the following:

- Steering gear mounting bolts. Torque the bolts to 93 ft. lbs. (126 Nm).
- Pressure and return lines and bracket. Torque the bracket bolts to 89 inch lbs. (10 Nm).
- Power steering pressure and return lines to the steering gear. Torque the bolt to 18 ft. lbs. (25 Nm).
- Steering gear pinch bolt and reposition the boot. Torque the bolt to 18 ft. lbs. (25 Nm).
- Tie rod end to the tie rod using the number of turns required to remove the tie rod end
- Jam nuts. Torque the nuts to 35 ft. lbs. (47 Nm).
- Tie rod end to the steering knuckle. Torque the nut to 41 ft. lbs. (57 Nm) and install a new cotter pin
- Rear transmission insulator bracket. Torque the bolts to 66 ft. lbs. (90 Nm).
- Both front wheels
- Rear transmission insulator bracket. Torque the bolts to 66 ft. lbs. (90 Nm).
- Rear transmission insulator. Torque the bolts to 66 ft. lbs. (90 Nm).
- Negative battery cable

6. Fill and bleed the power steering system.

7. Start the vehicle and check for leaks, repair if necessary.

8. Check and adjust the front end alignment.

2005 MODELS

1. Before servicing the vehicle, refer to the Precautions Section.

2. With the vehicle in NEUTRAL, position it on a hoist.

3. Vehicles equipped with 3.0L engine, remove the EGR valve.

   a. Disconnect the exhaust manifold-to-exhaust gas recirculation (EGR) tube fitting from the EGR valve.

   b. Disconnect the vacuum tube fitting from the EGR valve.

➡**The EGR valve sealing surfaces are soft metals. Do not reuse the EGR valve gasket.**

   c. Remove the 2 bolts and the EGR valve. Remove and discard the EGR valve gasket.

4. Carefully clean the EGR valve sealing surfaces.

5. Remove the rear transaxle insulator bracket.

6. Remove the 3 bolts and the rear transaxle mounting plate.

➡**Do not loosen the tie-rod end nut to align the slot in the nut with the tie-rod end stud through-hole.**

7. Remove the 2 outer tie-rod end cotter pins and the 2 tie-rod end nuts.

8. Separate the tie-rod end from the knuckle.

✳✳ WARNING

Do not allow the steering wheel to rotate while the intermediate shaft is disconnected or damage to the clockspring can result. If there is evidence that the shaft has rotated, the clockspring must be removed and recentered.

9. Hold the steering wheel in a straight-ahead position using a suitable tool.

10. Remove the nuts and the steering column boot.

11. Remove the steering column coupling-to-steering gear pinch bolt.

12. Remove the steering gear-to-fluid cooler return hose bracket-to-subframe bolt.

13. Remove the power steering pressure line bracket-to-steering gear bolt.

14. Remove the steering gear-to-fluid return hose bracket-to-steering gear stud.

15. Remove the power steering line clamp plate bolt.

➡**Install a new high pressure hose O-ring seal and a new return hose O-ring seal.**

16. Remove the power steering pressure and return lines.

17. Remove the 2 steering gear mounting bolts.

➡**Remove the steering gear from the right side of the vehicle.**

18. Remove the steering gear.

19. To install, reverse the removal procedure. Observe the following torques:

- Steering gear mounting bolts: 126 Nm (93 ft. lbs.)
- Power steering line clamp plate bolt: 25 Nm (18 ft. lbs.)
- Steering gear-to-fluid return hose bracket-to-steering gear stud: 10 Nm (89 inch lbs.)

- Power steering pressure line bracket-to-steering gear bolt: 10 Nm (89 inch lbs.)
- Steering gear-to-fluid cooler return hose bracket-to-subframe bolt: 10 Nm (89 inch lbs.)
- Steering column coupling-to-steering gear pinch bolt: 40 Nm (30 ft. lbs.)
- Outer tie-rod end cotter pins and the 2 tie-rod end nuts: 55 Nm (41 ft. lbs.). If necessary, continue to tighten the tie-rod end nut until the slot in the nut aligns with the tie-rod end stud through-hole.
- Rear transaxle mounting plate: 90 Nm (66 ft. lbs.)
- Rear transaxle insulator-to-bracket through-bolt: 90 Nm (66 ft. lbs.)
- Rear transaxle insulator bracket nuts: 90 Nm (66 ft. lbs.)
- Rear transaxle insulator bracket bolt: 90 Nm (66 ft. lbs.)
- EGR valve: 25 Nm (18 ft. lbs.)
- EGR tube fitting: 40 Nm (30 ft. lbs.)

20. Fill the power steering system.

21. Check and, if necessary, align the front end.

Escape Hybrid

1. Before servicing the vehicle, refer to the Precautions Section.

✳✳ CAUTION

Before proceeding, read and observe all of the High Voltage System Precautions.

2. With the vehicle in NEUTRAL, position it on a hoist.

3. Disconnect the battery.

4. De-power the high voltage traction battery. Establish a buffer zone. See the procedure under De-power the High Voltage Traction Battery.

5. Remove the engine appearance cover.

6. Remove the exhaust flexible pipe.

7. If equipped, remove the driveshaft.

8. Remove the rear engine mount stud nut.

9. Remove the 3 rear engine mount bracket bolts.

10. Remove the rear engine mount bracket.

11. Remove the 2 rear engine mount heat shield nuts and the heat shield.

➡**Disconnect the vacuum hose from the rear engine mount.**

1 - 90 Nm (66 lb-ft)

90 Nm (66 lb-ft) - 3

5 - 15 Nm (11 lb-ft)

4

6

90 Nm (66 lb-ft) - 2

55 Nm (41 lb-ft) - 7

12

47 Nm (35 lb-ft)

8

24 Nm (18 lb-ft)

10

16 - 55 Nm (41 lb-ft)

40 Nm (30 lb-ft) - 11

9

15

12

47 Nm (35 lb-ft)

19

90 Nm (66 lb-ft)

13

15

13

21

20 - 175 Nm (129 lb-ft)

14 - 55 Nm (41 lb-ft)

18 - 115 Nm (85 lb-ft)

14 - 55 Nm (41 lb-ft)

17 - 90 Nm (66 lb-ft)

1 Rear engine mount stud nut	12 Tie-rod end jam nuts (2 required)
2 Rear engine mount bracket aft bolt	13 Cotter pin
3 Rear engine mount bracket forward bolts (2 required)	14 Tie-rod end nuts (2 required)
4 Rear engine mount bracket	15 Outer tie-rod end
5 Rear engine mount heat shield nuts (2 required)	16 Lower stabilizer bar link nuts (2 required)
6 Rear engine mount heat shield	17 Engine support crossmember bolts (2 required)
7 Rear engine mount nuts (4 required)	18 Front insulator bolt
8 Rear engine mount	19 Engine roll restrictor bracket bolts (2 required)
9 Electrical connectors	20 Engine support crossmember nut
10 Upper steering column pinch bolt	21 Engine support crossmember
11 Steering gear coupling pinch bolt	

06017-ESCA-G74

These components must be removed prior to steering gear removal—Escape Hybrid

22 Electrical connector
23 Subframe nuts (2 required)
24 Subframe bolts (2 required)
25 Ball joint pinch bolt nuts (2 required)
26 Ball joint pinch bolts (2 required)
27 Subframe
28 Steering gear bolts (2 required)
29 Steering gear

06017-ESCA-G75

Steering gear mounting—Escape Hybrid

➡When the rear engine mount nuts are removed, the rear engine mount bolts will drop through the rear engine mount bracket. Secure the rear engine mount bolts to the rear engine mount bracket.

12. Remove the 4 rear engine mount nuts and the rear engine mount.

13. Disconnect the electrical connectors at the steering gear and the power steering control module.

✳✳ WARNING

Do not allow the steering wheel to rotate while the intermediate shaft is disconnected or damage to the clockspring can result. If there is evidence that the shaft has rotated, the clockspring must be removed and recentered.

14. Hold the steering wheel in the straight-ahead position using a suitable tool.

➡Position the steering column shaft boot upward.

15. Loosen the upper steering column pinch bolt.

16. Remove the steering gear coupling pinch bolt.

17. Remove the 2 outer tie-rod end cotter pins and the 2 outer tie-rod nuts.

18. Using the special tool, separate the tie-rod end from the knuckle.

19. Remove the 2 lower stabilizer bar link nuts.

20. Remove the 2 engine support crossmember bolts.

21. Remove the front insulator bolt.

22. Remove the 2 engine roll restrictor bracket bolts.

23. Remove the engine crossmember-to-engine splash shield bolt and position aside.

24. Remove the nut and the engine support crossmember.

25. Remove the 2 subframe nuts.

➡Using a suitable support, support the subframe.

➡Completely loosen but do not remove the 2 subframe bolts from the subframe.

26. Loosen the 2 subframe bolts.

27. Remove the 2 ball joint pinch bolts and nuts.

28. Lower the front subframe while simultaneously disconnecting the front ball joints.

29. Remove the 2 steering gear bolts.

➡**Remove the steering gear from the left side of the vehicle.**

30. Remove the steering gear.

31. To install, reverse the removal procedure. Observe the following torques:

- Steering gear bolts: 115 Nm (85 ft. lbs.)
- Ball joint pinch bolts and nuts: 70 Nm (52 ft. lbs.)
- Subframe bolts: 200 Nm (148 ft. lbs.)
- Subframe nuts: 150 Nm (111 ft. lbs.)
- Engine support crossmember: 175 Nm (129 ft. lbs.)
- Engine roll restrictor bracket bolts: 90 Nm (66 ft. lbs.)
- Front insulator bolt: 115 Nm (85 ft. lbs.)
- Engine support crossmember bolts: 90 Nm (66 ft. lbs.)
- Lower stabilizer bar link nuts: 55 Nm (41 ft. lbs.)
- Outer tie-rod nuts: 55 Nm (41 ft. lbs.)
- Steering gear coupling pinch bolt: 40 Nm (30 ft. lbs.)
- Upper steering column pinch bolt: 24 Nm (18 ft. lbs.)
- Rear engine mount nuts and the rear engine mount: 55 Nm (41 ft. lbs.)
- Rear engine mount heat shield nuts and the heat shield: 15 Nm (11 ft. lbs.)
- Rear engine mount bracket bolts: 90 Nm (66 ft. lbs.)
- Rear engine mount stud nut: 90 Nm (66 ft. lbs.)

32. Check and, if necessary, align the front end.

Clockspring

REMOVAL

1. Before servicing the vehicle, refer to the Precautions Section.

2. De-power the air bag system.

3. Make sure the road wheels are in the straight-ahead position.

4. Position the steering wheel in the straight-ahead position and remove the ignition key. Rotate the steering wheel

1 Upper steering column shroud

2 Lower steering column shroud screws

3 Clockspring electrical connectors

4 Tilt column locking lever (if equipped)

5 Lower steering column shroud

6 Clockspring

7 Clockspring screws

06017-ESCA-G69

Push in where indicated, releasing the retaining tabs, and remove the upper steering column shroud

If installing the same clockspring, apply 2 strips of masking tape across the clock-spring to prevent accidental rotation when the clockspring is removed

until the steering column locks into position.

5. Open the cover on the underside of the steering wheel.

6. Remove the steering wheel pinion bolt.

7. Remove the steering wheel. Position the steering wheel rearward. Disconnect the 2 electrical connectors.

8. If equipped with tilt steering, position the steering column completely downward and lock in place.

9. Push in where indicated, releasing the retaining tabs, and remove the upper steering column shroud.

10. Release the tilt column locking lever, allowing the steering column to move upward. Do not lock the tilt column locking lever back in place.

11. Remove the 3 screws and position the lower steering column shroud aside.

12. If installing the same clockspring, apply 2 strips of masking tape across the clockspring to prevent accidental rotation when the clockspring is removed.

13. Remove the 3 clockspring screws.

➡**If the clockspring is to be reinstalled, do not allow the clockspring to turn from its removal position.**

14. Partially remove the clockspring, then disconnect the 2 electrical connectors and remove the clockspring.

Outer housing (1), rotor (2), clockspring aligning

Clockspring installation

INSTALLATION

Vehicles Needing Clockspring Recentering

1. Before servicing the vehicle, refer to the Precautions Section.

❊ CAUTION

Incorrect centralization may result in premature component failure. If in doubt when centralizing the clockspring, repeat the centralizing procedure. Failure to follow this instruction may result in personal injury.

➡ **Make sure the road wheels are in the straight-ahead position.**

2. If the vehicle's clockspring has rotated out of center, follow these steps to center the clockspring.
   a. Hold the clockspring outer housing stationary.

❊❊ WARNING

Overturning will destroy the clockspring. The internal ribbon wire acts as the stop and can be broken from its internal connection.

   b. While turning the rotor clockwise, carefully feel for the ribbon wire to run out of length and for a slight resistance. Stop turning at this point.
   c. Turn the clockspring counterclockwise until the yellow indicator shows anywhere in the window (window will be near the 1 o'clock position) and the arrow on the rotor lines up with the arrow on the top of the housing. The clockspring is now centered. Do not allow the rotor to turn from this position.

All Vehicles

1. Before servicing the vehicle, refer to the Precautions Section.
2. Connect the 2 clockspring electrical connectors to the clockspring.

➡ **Slight turning of the clockspring rotor is allowable for alignment purposes to the steering column.**

3. Align the clockspring for installation.
   a. Align the large slot to the large tab in the clockspring.
   b. Align the small slot to the small tab in the clockspring.
4. Install the 3 clockspring screws.
5. For vehicles reusing a clockspring that was removed, remove the tape. For vehicles installing a new clockspring, remove the retaining pin.
6. Install the lower steering column shroud and the 3 screws.
7. Position the steering column completely downward and lock in place.
8. Install the upper steering column shroud and engage the retaining tabs.
9. Install the steering wheel. Torque the steering wheel pinion bolt to 12 Nm (9 ft. lbs.).
10. Re-power the system.

SUSPENSION

Front Strut

REMOVAL & INSTALLATION

2002–03

1. Before servicing the vehicle, refer to the Precautions Section.
2. Install or connect the following:
   - Negative battery cable
   - Front wheel
   - Brake hose grommet from the bracket
   - Antilock Brake System (ABS) harness from the strut assembly and move the brake hose bracket aside
   - Stabilizer bar link nut and move the bar aside
   - Strut to steering knuckle bolts and support the strut assembly
   - Upper strut nuts
   - Strut and coil spring assembly

To install:
3. Install or connect the following:
   - Strut and spring assembly. Torque the upper nuts to 41 ft. lbs. (55 Nm).
   - Lower strut assembly to the steering knuckle. Torque the lower bolts to 85 ft. lbs. (115 Nm).
   - Stabilizer bar into position. Torque the bolts to 35 ft. lbs. (48 Nm).
   - Brake hose bracket. Torque the bolts to 14 ft. lbs. (18 Nm).

   - ABS harness to the strut assembly, if equipped
   - Brake hose grommet to the bracket
   - Front wheel
   - Negative battery cable

2004

1. Before servicing the vehicle, refer to the Precautions Section.
2. Install or connect the following:
   - Negative battery cable
   - Front wheel
   - Brake hose grommet from the bracket
   - Antilock Brake System (ABS) harness from the strut assembly and move the brake hose bracket aside
   - Stabilizer bar link nut and move the bar aside
   - Strut to steering knuckle bolts and support the strut assembly
   - Upper strut nuts
   - Strut and coil spring assembly

To install:
3. Install or connect the following:
   - Strut and spring assembly. Torque the upper nuts to 41 ft. lbs. (55 Nm).
   - Lower strut assembly to the steering knuckle. Torque the lower bolts to 85 ft. lbs. (115 Nm).
   - Stabilizer bar into position. Torque the bolts to 35 ft. lbs. (48 Nm).

   - Brake hose bracket. Torque the bolts to 13 ft. lbs. (18 Nm).
   - ABS harness to the strut assembly, if equipped
   - Brake hose grommet to the bracket
   - Front wheel
   - Negative battery cable

2005

1. Before servicing the vehicle, refer to the Precautions Section.

➡ **Make sure the steering wheel is in the unlocked position.**

➡ **Use the hex holding feature to prevent the ball studs from turning while removing or installing the stabilizer bar link nuts.**

2. Raise and support the vehicle.
3. Remove the brake jounce hose clip.
4. Remove the brake jounce hose. Pull the brake jounce hose downward slightly to remove the hose from the bracket.
5. Remove the ABS sensor harness bolt.
6. Remove the stabilizer bar link nut.
7. Remove the strut-to-knuckle nuts.
8. Remove the strut-to-knuckle bolts.
9. Remove the strut upper bushing nuts. Reference mark the strut mounting plate nuts.
10. Remove the strut and spring assembly.

Front strut installation—2005 model shown

1 Brake jounce hose clip
2 Brake jounce hose (LH/RH)
3 ABS sensor harness bolt
4 Stabilizer bar link nut
5 Strut-to-knuckle nuts
6 Strut-to-knuckle bolts
7 Strut upper bushing nuts
8 Strut and spring assembly

67197-ESCA-G45

✱✱ WARNING

Do not allow the axle shaft to move outboard. Over-extension of the tripod CV joint can result in separation of internal parts, causing failure of the axle shaft.

11. To install, reverse the removal procedure. See the illustration for the appropriate torque values.

12. Align the strut mounting plate nuts to the reference marks.

13. Check the front end alignment and adjust as necessary.

Shock Absorber

REMOVAL & INSTALLATION

Rear

2002–04

1. Before servicing the vehicle, refer to the Precautions Section.

2. Remove or disconnect the following:
   • Negative battery cable
   • Rear quarter trim panel
   • Upper shock absorber nut and raise the vehicle enough to relax the suspension
   • Lower shock absorber nut
   • Shock absorber

To install:
3. Install or connect the following:
   • Shock absorber. Torque the lower nut to 85 ft. lbs. (115 Nm).

18 Nm (13 lb-ft) – 1 ℹ

2

3

8

ℹ 11

175 Nm (129 lb-ft) – 4

7

ℹ 9 – 103 Nm (76 lb-ft)

10 – 103 Nm (76 lb-ft)

5

6

1 Upper shock absorber nuts
2 Washer
3 Bushing
4 Lower shock absorber nut

5 Washer
6 Lower shock absorber bolt
7 Shock absorber
8 Bushing

9 Upper ball joint nut
10 Lower ball joint nut
11 Coil spring

67197-ESCA-G46

Rear shock absorber and spring—2005 model shown

- Upper shock absorber nut. Torque the nut to 13 ft. lbs. (18 Nm).
- Rear quarter trim panel
- Negative battery cable

2005

1. Remove the wheel and tire assemblies.
2. Remove the rear quarter trim panel. Remove the upper shock absorber nut, bushing and washer.
3. Remove the lower shock absorber nut, bolt and washer.
4. Remove the shock absorber and bushing.
5. To install, reverse the removal procedure. Torque the upper nut to 13 ft. lbs.; the lower nut to 129 ft. lbs. (175 Nm).

Coil Spring

REMOVAL & INSTALLATION

Front

2002–04 MODELS

1. Before servicing the vehicle, refer to the Precautions Section.
2. Install or connect the following:
   - Negative battery cable
   - Front wheel
   - Strut and spring assembly and mount the strut assembly in a holding fixture and compress the coil spring using a suitable tool
   - Strut piston rod nut

3. Coil spring by disassembling the strut in the following sequence:
   a. Step 1: Metal sheet plate.
   b. Step 2: Upper strut mount.
   c. Step 3: Thrust bearing plate.
   d. Step 4: Thrust bearing.
   e. Step 5: Upper spring seat.
   f. Step 6: Upper spring seat isolator.
   g. Step 7: Coil spring.
   h. Step 8: Dust boot.
   i. Step 9: Rubber bump stopper.
   j. Step 10: Lower spring seat.
To install:
- Assemble the strut assembly in the reverse order of the removal procedure
4. Install or connect the following:
   - Strut piston rod nut. Torque the nut

1	Metal sheet plate
2	Upper strut mount
3	Thrust bearing plate
4	Thrust bearing
5	Upper spring seat
6	Upper spring seat isolator
7	Spring
8	Dust boot
9	Rubber bump stopper
10	Lower spring seat

9308TG23

Disassemble the strut assembly in the proper sequence—2002–04 models

to 76 ft. lbs. (103 Nm) and remove the assembly from the holding fixture
- Strut and spring assembly
- Front wheel
- Negative battery cable

2005 MODELS

1. Before servicing the vehicle, refer to the Precautions Section.

❋❋ CAUTION

Always wear safety goggles when using a spring compressor. Failure to follow these instructions may result in personal injury.

➡ **Do not use an impact wrench on the nut.**

2. Mount the strut and spring assembly in a suitable spring compressor.

3. Compress the coil spring enough to relieve the tension on the strut assembly.

4. Remove the strut piston rod-to-bushing nut.

5. Remove the strut.

6. Remove the lower coil spring insulator.

7. Remove the coil spring.

8. Remove the upper coil spring insulator.

➡ **During assembly, assemble the bearing plate to the strut so the arrow on the bearing plate points to the outboard side of the vehicle when the strut is installed.**

9. Remove the bearing plate.

10. Remove the bearing.

11. Remove the strut upper bushing.

12. Remove the dust boot and the bumper.

13. To assemble, reverse the disassembly procedure. Tighten the strut piston rod-to-bushing nut to 103 Nm (76 ft. lbs.).

Rear

2002–04

1. Before servicing the vehicle, refer to the Precautions Section.

2. Remove or disconnect the following:
- Wheel and install 1 lug nut to retain the brake drum

1	Strut piston rod-to-bushing nut
2	Strut (LH/RH)
3	Lower coil spring insulator
4	Coil spring
5	Upper coil spring insulator
6	Bearing plate
7	Bearing
8	Strut upper bushing
9	Dust boot
10	Bumper

06017-ESCA-G76

Strut/spring disassembly—2005 models

- Brake line from the wheel cylinder
- Brake line bracket
- Bolts from the Antilock Braking System (ABS) sensor bracket and move the sensor aside, if equipped
- Rear knuckle and loosen the inside upper and lower arm bolts
- Shock absorber lower nut
- Spring

To install:

3. Install or connect the following:
- Spring to the shock absorber
- Lower shock absorber nut. Torque the nut to 85 ft. lbs. (115 Nm).
- Inside upper and lower arm bolts. Torque the bolts to 85 ft. lbs. (115 Nm).
- ABS sensor bracket, if equipped. Torque the bolts to 80 inch lbs. (9 Nm).
- Brake line bracket. Torque the bolt to 15 ft. lbs. (20 Nm).
- Brake line to the wheel cylinder.

Torque the fastener to 11 ft. lbs. (15 Nm) and remove the lug nut
- Wheel
- Negative battery cable

2005

1. Before servicing the vehicle, refer to the Precautions Section.
2. Remove the wheel and tire assemblies.
3. Remove the lower shock absorber nut and washer
4. Remove the lower shock absorber bolt
5. Support the wheel knuckle.
6. Remove the upper ball joint nut.
7. Remove the upper arm inner bolt.
8. Loosen the lower arm inner bolt.

➡**Note the position of the coil spring insulator and coil spring for installation.**

9. Carefully lower the wheel knuckle support.

10. Remove the coil spring
11. To install, reverse the removal procedure.
12. Align the coil spring and coil spring insulator to the previously noted position.
13. Observe the following torques:
- Stabilizer bar link nut: 46 ft. lbs. (63 Nm)
- Strut-to-knuckle nuts: 85 ft. lbs. (115 Nm)
- Upper ball joint nut: 76 ft. lbs. (103 Nm)
- Lower shock absorber nut: 129 ft. lbs. (175 Nm)

Front Lower Control Arm

REMOVAL & INSTALLATION

2002–2003

1. Before servicing the vehicle, refer to the Precautions Section.

115 Nm (85 lb-ft)
115 Nm (85 lb-ft)
290 Nm (214 lb-ft)
70 Nm (52 lb-ft)
55 Nm (41 lb-ft)
200 Nm (148 lb-ft)

1 Wheel hub nut	6 Strut-to-knuckle nuts	11 Bearing
2 Cotter pin	7 Strut-to-knuckle bolts	12 Wheel stud
3 Tie rod end-to-knuckle nut	8 Wheel knuckle (LH/RH)	13 Lower arm bolt (front)
4 Lower ball joint pinch bolt	9 Wheel hub	14 Lower arm bolt (rear)
5 Lower ball joint pinch bolt	10 Snap ring	15 Lower arm

Front lower control arm assembly—2004–05 model shown

67197-ESCA-G48

2. Remove or disconnect the following:
- Negative battery cable
- Front wheel
- Lower ball joint from the knuckle and support the subframe
- Lower control arm

To install:

3. Install or connect the following:
- Lower control arm bolts and hand tighten them
- Pinch bolt to the wheel knuckle. Torque the nut to 52 ft. lbs. (70 Nm) and remove the subframe support
- Front wheel and jounce the vehicle

4. Torque the inner lower control arm bolt to 148 ft. lbs. (200 Nm) and outer bolt 85 ft. lbs. (115 Nm).

2004–05

1. Before servicing the vehicle, refer to the Precautions Section.
2. Remove the wheel.
3. Lift the lower arm with a floor jack until the vehicle starts to lift.
4. Record the ride height. It's measure from the center of the halfshaft to the fender lip.
5. Remove the floor jack.
6. Disconnect the ball joint from the knuckle.
7. Support the sub-frame and remove the lower arm.

To install:

8. Install the lower arm, with the bolts loose.

9. Connect the ball joint. Torque the bolt to 52 ft. lbs. (70 Nm).
10. Remove the support.
11. Position the jack under the ball joint and raise the arm to the previously recorded ride height.
12. Tighten the lower arm bolts. Horizontal 85 ft. lbs. (115 Nm); vertical 148 ft. lbs. (200 Nm).

Rear Lower Control Arm

REMOVAL & INSTALLATION

1. Before servicing the vehicle, refer to the Precautions Section.
2. Remove or disconnect the following:

		290 Nm (214 lb-ft)
5		175 Nm (129 lb-ft)
8	9	103 Nm (76 lb-ft)
11 16 18		115 Nm (85 lb-ft)

1 Wheel hub nut (4WD)	8 Upper ball joint nut	15 Wheel stud
2 Wheel hub nut (2WD)	9 Lower ball joint nut	16 Upper arm inner bolt
3 ABS sensor ring (2WD)	10 Cam nut	17 Upper arm
4 Wheel hub	11 Wheel knuckle bolt	18 Lower arm inner bolt
5 Lower shock absorber nut	12 Knuckle assembly (LH/RH)	19 Lower arm
6 Washer	13 Wheel bearing snap ring	
7 Lower shock absorber bolt	14 Wheel bearing	

67197-ESCA-G47

Rear lower control arm and related parts

- Negative battery cable
- Lower ball joint from the knuckle while holding the ball joint stud from moving
- Lower ball joint nut
- Lower control arm
- Lower control arm inner bolt

To install:

3. Install or connect the following:
   - Lower control arm inner bolt
   - Lower control arm. Torque the bolts to 85 ft. lbs. (115 Nm).
   - Lower ball joint nut
   - Lower ball joint the knuckle. Torque the ball joint nut to 76 ft. lbs. (103 Nm).
   - Rear wheel
   - Negative battery cable

Rear Upper Control Arm

REMOVAL & INSTALLATION

1. Before servicing the vehicle, refer to the Precautions Section.
2. Remove the wheel and tire.

➡**It may be necessary to hold the ball joint stud to keep it from turning while removing the nut.**

3. Separate the upper arm from the wheel knuckle. Remove the upper ball joint nut.
4. Remove the upper arm inner bolt.
5. Remove the upper arm.
6. To install, reverse the removal procedure. Observe the following torques:
   - Ball joint nut: 76 ft. lbs. (103 Nm)
   - Lower arm bolts: 85 ft. lbs. (115 Nm)

Front Knuckle

REMOVAL & INSTALLATION

1. Before servicing the vehicle, refer to the Precautions Section.

06017-ESCA-G77

Separate the outer CV-joint spindle from the wheel hub

06017-ESCA-G78

Insert the halfshaft into the wheel hub

2. Remove the brake disc.
3. Remove and discard the wheel hub nut.
4. Separate the outer CV-joint spindle from the wheel hub.
5. Remove the cotter pin and the tie-rod end-to-knuckle nut.

✳✳ WARNING

Do not use a hammer to separate the tie-rod end from the wheel knuckle or damage to the wheel knuckle can result. Do not damage the tie-rod end boot while installing the special tool.

6. Separate the tie-rod from the wheel knuckle.
7. Remove the lower ball joint pinch bolt nut and the pinch bolt.
8. Remove the anti-lock brake system (ABS) wheel speed sensor bolt and position the sensor aside.
Separate the lower ball joint from the wheel knuckle.
9. Remove the 2 strut-to-knuckle nuts, bolts and the wheel knuckle.

To install:

10. Position the wheel knuckle and install the 2 strut-to-knuckle bolts and nuts. Tighten to 115 Nm (85 ft. lbs.).
11. Position and align the ball joint stud into the wheel knuckle.
12. Install the lower ball joint pinch bolt and nut. Tighten to 70 Nm (52 ft. lbs.).
13. Install the ABS wheel speed sensor and the bolt. Tighten to 9 Nm (80 inch lbs.).
14. Position the tie rod-end into the wheel knuckle and install the tie-rod end-to-knuckle nut and a new cotter pin. Tighten to 55 Nm (41 ft. lbs.).
15. Insert the halfshaft into the wheel hub.
16. Install the wheel hub nut. Tighten to 290 Nm (214 ft. lbs.).
17. Install the brake disc.
18. Check and, if necessary, align the front end.

Rear Knuckle

REMOVAL & INSTALLATION

2002–04 2-Wheel Drive

1. Before servicing the vehicle, refer to the Precautions Section.

➡**When a new wheel knuckle is installed a new wheel bearing must also be installed.**

2. Remove the brake shoes.

➡**If equipped, remove ABS sensor ring.**

3. Remove the wheel hub nut.
4. Remove the wheel hub.

➡**This step may not be necessary if the inner wheel bearing race remains in the wheel knuckle, after removing the wheel hub.**

5. Press the inner wheel bearing race from the wheel hub.
6. Remove the snapring.
7. Vehicles with ABS: Remove the ABS sensor bolt and bracket bolts. Position the ABS sensor aside.
8. Disconnect the parking brake cable from the brake backing plate. Remove the parking brake cable bolt from the wheel knuckle.
9. Disconnect the brake line from the

06017-ESCA-G79

Remove the wheel hub

206-D002

06017-ESCA-G80

Press the inner wheel bearing race from the wheel hub

06017-ESCA-G81

Remove the snapring

wheel cylinder. Remove the brake line bracket bolt.

10. Support the wheel knuckle.

11. Remove the lower shock absorber nut.

➡**It may be necessary to hold the ball joint stud to keep it from turning while removing the nut.**

12. Remove the nut. Separate the lower ball joint from the wheel knuckle.

➡**It may be necessary to hold the ball joint stud to keep it from turning while removing the nut.**

13. Remove the nut. Separate the upper ball joint from the wheel knuckle.

➡**Note the position of the spring insulator and spring for installation.**

14. Remove the spring. Lower the support to the wheel knuckle.

15. Mark the position of the adjustment cam notch.

16. Remove the wheel knuckle cam bolt and nut.

17. Remove the wheel knuckle cam.

18. Remove the wheel knuckle.

24. Install the wheel hub nut. Torque to 214 ft. lbs. (290Nm).

25. Position the wheel knuckle and install the wheel knuckle cam.

26. Install the wheel knuckle cam bolt and nut loosely.

27. Support the wheel knuckle and install the spring.

28. Install the lower shock absorber nut. Torque to 85 ft. lbs. (115 Nm).

29. Position the upper ball joint and install the nut. Torque to 76 ft. lbs. (103 Nm).

30. Position the lower ball joint and install the nut. Torque to 76 ft. lbs. (103 Nm).

31. Tighten the wheel knuckle cam bolt and nut.

    a. Align the wheel knuckle cam.

    b. Tighten the wheel knuckle cam bolt and nut to 85 ft. lbs. (115 Nm).

06017-ESCA-G83

Adjustment cam notch (1), wheel knuckle cam bolt and nut (2)

19. Remove the brake backing plate from the wheel knuckle.

To install:

20. Install a new wheel bearing into the wheel knuckle.

21. Install the snapring into the wheel knuckle.

22. Position the brake backing plate to the wheel knuckle and install the bolts. Torque to 49 ft. lbs. (66 Nm).

23. Using the wheel hub washer and the special tool, install the wheel hub into the wheel bearing.

➡**If equipped, install the ABS sensor ring.**

06017-ESCA-G84

Remove the brake backing plate from the wheel knuckle

Install the wheel hub into the wheel bearing

32. Connect the brake line to the wheel cylinder. Install the brake line bracket bolt. Torque the brake line-to-wheel cylinder to 13 ft. lbs. (17 Nm). Torque the 16 ft. lbs. (22 Nm).

33. Install the parking brake cable through the brake backing plate. Torque the bolt to 16 ft. lbs. (22 Nm).

34. Vehicles with ABS: Position the ABS sensor and bracket bolts. Install the bolts. Torque to 80 inch lbs. (9 Nm).

35. Install the brake shoes.

36. Bleed the brake system.

37. Check and adjust the wheel alignment as necessary.

2002–04 4-Wheel Drive

1. Before servicing the vehicle, refer to the Precautions Section.

➡ **When a new wheel knuckle is installed a new wheel bearing must also be installed.**

2. Raise the vehicle on the hoist.

3. Remove the wheel and tire assembly.

⁂ WARNING

Use of a brake drum puller or a torch is not recommended. Brake drum distortion can occur.

➡ **If the brake drum is rusted to the axle shaft pilot diameter, tap the center of the brake drum between the wheel studs.**

4. Remove the brake drum.

5. If the brake drum will not come off, follow these steps.

  a. Move the brake shoe adjusting lever off the brake adjuster screw.

  b. Loosen the brake shoe adjuster screw nut by adjusting the nut upward.

6. Remove the rear halfshaft nut.

7. Loosen the halfshaft from the wheel hub.

8. Remove the wheel hub.

➡ **This step may not be necessary if the inner wheel bearing race remains in the wheel knuckle, after removing the wheel hub.**

9. Press the inner wheel bearing race from the wheel hub.

10. Remove the snapring.

11. Vehicles with ABS: Remove the ABS sensor bolt and bracket bolts. Position the ABS sensor aside.

12. Disconnect the parking brake cable. Remove the parking brake cable bolt from the wheel knuckle.

13. Remove brake line bracket bolt.

14. Support the wheel knuckle.

15. Remove the lower shock absorber nut.

➡ **It may be necessary to hold the ball joint stud to keep it from turning while removing the nut.**

16. Disconnect the lower ball joint.

➡ **It may be necessary to hold the ball joint stud to keep it from turning while removing the nut.**

17. Disconnect the upper ball joint.

➡ **Note the position of the spring insulator and spring for installation.**

18. Remove the spring. Lower the support to the wheel knuckle.

19. Mark the position of the adjustment cam notch.

20. Remove the wheel knuckle cam bolt and nut.

21. Remove the wheel knuckle cam.

22. Remove the wheel knuckle.

23. Remove the brake backing plate from the wheel knuckle.

To install:

24. Using the special tools, install a new wheel bearing into the wheel knuckle.

25. Install the snapring into the wheel knuckle.

26. Position the wheel knuckle and install the wheel knuckle cam.

27. Install the wheel knuckle cam bolt loosely.

➡ **Brake shoes shown removed for clarity.**

28. Install the brake backing plate into the wheel knuckle. Install the bolts. Torque to 49 ft. lbs. (66 Nm).

29. Install the wheel hub.

30. Position the halfshaft into the wheel hub.

31. Support the wheel knuckle and install the spring.

32. Install the lower shock absorber nut. Torque to 85 ft. lbs. (115 Nm).

33. Remove the support from the wheel knuckle.

34. Install the upper ball joint. Install the nut. Torque to 76 ft. lbs. (103 Nm).

35. Install the lower ball joint. Install the nut. Torque to 76 ft. lbs. (103 Nm).

36. Align the wheel knuckle cam.

37. Tighten the wheel knuckle cam bolt and nut to 85 ft. lbs. (115 Nm).

38. Install brake line bracket bolt. Torque to 16 ft. lbs. (22 Nm).

39. Install the parking brake cable bolt. Torque to 16 ft. lbs. (22 Nm).

40. Vehicles with ABS: Position the ABS brake sensor and bracket bolts. Install the bolts. Torque to 80 inch lbs. (9 Nm).

41. Install the halfshaft end into the hub assembly.

42. Install the halfshaft nut. Torque to 214 ft. lbs. (290 Nm).

43. Install the brake drum.

44. Install the wheel and tire assembly.

45. Check and adjust the wheel alignment as necessary.

2005 Models

1. Before servicing the vehicle, refer to the Precautions Section.

2. Drum brake vehicles:

  a. Remove the brake shoes.

  b. Disconnect the parking brake cable from the brake backing plate. Remove the parking brake cable from the brake backing plate.

  c. Disconnect the brake line from the wheel cylinder and remove the brake line bracket bolt.

3. Disc brake vehicles, remove the parking brake shoes.

4. Remove and discard the wheel hub nut.

❄❄ WARNING

Do not use a hammer to separate the outer constant velocity (CV) joint from the hub.

5. Damage to the threads and internal CV joint components can result.

6. With 4wd, separate the outer CV-joint from the wheel hub.

7. Remove the anti-lock brake system (ABS) wheel speed sensor bolt and the 2 ABS wheel speed sensor wire bolts.

8. Remove and position the wheel speed sensor and harness aside.

9. Remove the coil spring.

10. Remove the lower ball joint nut.

➡**The joint surfaces must be clean. Clean the general area of the joint to prevent debris from entering the joint. Clean using only mild liquids.**

11. Reference mark the notch on the cam nut adjustment cam.

12. Remove and discard the wheel knuckle bolt.

13. Remove and discard the cam nut.

14. Remove the wheel knuckle.

To install:

❄❄ WARNING

The joint area must be free of debris to ensure correct clamping.

➡**The joint surfaces and the bushing sleeve serrations must be clean before assembly.**

15. Clean the joint surfaces and the bushing sleeve serrations with a wire brush.

➡**Align the notch on the cam nut with the reference marks.**

16. Position the wheel knuckle and install a new wheel knuckle bolt and cam nut.

17. Using a suitable tool, hold the cam nut stationary while tightening the wheel knuckle bolt. Tighten to 150 Nm (111 ft. lbs.).

18. Position the ABS wheel speed sensor harness and the sensor.

19. Position the lower ball joint into the wheel knuckle and install the lower ball joint nut. Tighten to 103 Nm (76 ft. lbs.).

20. Install the coil spring.

21. Install the ABS wheel speed sensor bolt and the 2 ABS wheel speed sensor wire bolts. Tighten to 9 Nm (80 inch lbs.).

22. With 4wd, install the outer CV joint into the wheel hub.

23. Install a new wheel hub nut. Tighten to 290 Nm (214 ft. lbs.).

24. Install the brake shoes.

25. Connect the brake line to the wheel cylinder. Tighten to 17 Nm (13 ft. lbs.).

26. Install the brake line bracket bolt. Tighten to 22 Nm (16 ft. lbs.).

27. Connect the parking brake cable to the brake backing plate and install the parking brake cable bracket bolt. Tighten to 23 Nm (17 ft. lbs.).

28. Install the parking brake shoes.

29. Check and adjust the wheel alignment as necessary.

Front Stabilizer Bar

REMOVAL & INSTALLATION

1. Before servicing the vehicle, refer to the Precautions Section.

2. Remove the stabilizer bar bushing bracket bolts.

➡**Use the hex holding feature to prevent the ball stud from turning while removing or installing the stabilizer link nut.**

3. Remove the 2 lower stabilizer bar link nuts.

➡**Access the stabilizer bar through the left wheel opening.**

4. Remove the stabilizer bar.

5. To install, reverse the removal procedure. Observe the following torques:

- 2002–04 link nuts: 35 ft. lbs. (48 Nm)
- 2005 link nuts: 41 ft. lbs. (55 Nm)
- Bushing bracket bolts: 52 ft. lbs. (70 Nm)

Wheel Bearings

REMOVAL & INSTALLATION

Front

1. Before servicing the vehicle, refer to the Precautions Section.

➡**If removing the wheel hub, the wheel bearing must be replaced.**

2. Remove the wheel knuckle.

3. Using the special tool, press the wheel hub from the wheel bearing.

➡**This step may not be necessary if the inner wheel bearing race remains in the wheel knuckle after removing the wheel hub.**

4. Using the special tool, press the inner wheel bearing race from the wheel hub.

5. Remove the snapring.

6. Using the special tools, press the

1 Wheel hub
2 Snap ring
3 Wheel bearing
4 Wheel studs (5 required)

Front hub and bearing

06017-ESCA-G86

Using the special tool, press the wheel hub from the wheel bearing—front hub/bearing

Using the special tool, press the inner wheel bearing race from the wheel hub—front hub/bearing

Using the special tools, press the outer wheel bearing race from the wheel knuckle—front hub/bearing

Using the special tools, install the wheel bearing into the wheel knuckle—front hub/bearing

outer wheel bearing race from the wheel knuckle.

To install:

7. Position the wheel knuckle in a vise.

➥**Special Tool 205-278 is not seen in place. It is located behind the wheel knuckle.**

8. Using the special tools, install the wheel bearing into the wheel knuckle.

9. Install the snapring.

10. Using the special tool, press the wheel hub into the wheel bearing.

11. Install the wheel knuckle.

Rear

2WD VEHICLES

1. Before servicing the vehicle, refer to the Precautions Section.

2. Remove or disconnect the following:
- Negative battery cable
- Rear wheel
- Rear brake drum
- Wheel hub nut
- Wheel hub
- Inner wheel bearing race from the hub
- Snapring
- Wheel bearing outer race from the knuckle

To install:

3. Install or connect the following:
- Wheel bearing in to the knuckle
- Snapring
- Wheel hub into the wheel bearing
- Wheel hub nut. Torque the nut to 214 ft. lbs. (290 Nm).
- Brake drum
- Rear wheel
- Negative battery cable

4WD VEHICLES

1. Before servicing the vehicle, refer to the Precautions Section.

2. Remove or disconnect the following:
- Negative battery cable
- Rear wheel
- Rear brake shoes
- Rear halfshaft nut and loosen the halfshaft from the hub
- Wheel hub and place it in a vise
- Inner wheel bearing race from the hub
- Antilock Brake System (ABS) sensor bracket and move the sensor aside, if equipped
- Parking brake cable from the steering knuckle
- Brake line from the wheel cylinder and support the knuckle
- Lower shock absorber nut

205-D014

06017-ESCA-G91

Using the special tool, press the wheel hub into the wheel bearing—front hub/bearing

- Lower ball joint by holding the ball joint stud
- Upper ball joint
- Coil spring while noting the location of the insulator
- Steering knuckle cam
- Steering knuckle
- Snapring and press out the outer wheel bearing race from the knuckle

To install:

3. Install or connect the following:
- New wheel bearing into the steering knuckle

290 Nm (214 lb-ft)

1 Wheel hub nut (FWD)
2 Anti-lock brake system (ABS) wheel speed sensor ring (FWD)
3 Wheel hub
4 Wheel bearing snap ring
5 Wheel bearing
6 Wheel studs

06017-ESCA-G92

Rear hub and bearing

205-D070

100-D006

06017-ESCA-G93

Rear hub removal

205-D015

205-153

205-D002

06017-ESCA-G96

Rear wheel bearing removal—4wd

205-D002

06017-ESCA-G94

Inner wheel bearing removal—rear hub/bearing

205-278

308-153

205-014

06017-ESCA-G97

Rear wheel bearing installation

205-193

100-D006

06017-ESCA-G95

Rear wheel bearing removal—2wd

204-161

204-142

06017-ESCA-G98

Rear hub installation

- Snapring to the knuckle
- Wheel hub
- Steering knuckle cam and hand tighten the bolt
- Coil spring
- Shock absorber lower nut. Torque the nut to 85 ft. lbs. (115 Nm) for 2002–04 models; 129 ft. lbs. (175 Nm).
- Upper ball joint. Torque the nut to 76 ft. lbs. (103 Nm).

- Lower ball joint. Torque the nut to 76 ft. lbs. (103 Nm). Align the steering knuckle cam and torque the bolt to 85 ft. lbs. (115 Nm).
- Brake line to the wheel cylinder. Torque the brake line bracket bolt to 15 ft. lbs. (20 Nm) and the brake line fastener to 11 ft. lbs. (15 Nm).
- Parking brake cable to the backing plate. Torque the bolt to 16 ft. lbs. (22 Nm).

- ABS sensor bracket. Torque the bolt to 80 inch lbs. (9 Nm), if equipped
- Halfshaft nut. Torque the nut to 214 ft. lbs. (290 Nm).
- Brake shoes
- Rear wheel
- Negative battery cable
4. Fill and bleed the brake system.
5. Check and adjust the wheel alignment as needed.

DISC BRAKES

Brake Caliper

REMOVAL & INSTALLATION

Front

2002–04

1. Before servicing the vehicle, refer to the Precautions Section.
2. Remove the wheel and tire assembly.

3. Remove the brake caliper clip.
4. Remove the brake caliper bolt caps and bolts.
5. Position the caliper aside.
6. Disconnect and cap the brake line from the caliper and remove the caliper. To install, reverse the removal procedure. Torque the mounting bolts to 26 ft. lbs. (35Nm). Torque the brake line to 15 ft. lbs. (20Nm).
7. Bleed the brake system.

2005 EXCEPT ESCAPE HYBRID

1. Before servicing the vehicle, refer to the Precautions Section.
2. Remove the wheel and tire assembly.
3. Remove the brake caliper clip.
4. Remove the brake caliper dust boot caps.
5. Remove the brake caliper guide bolts.
6. Remove the brake caliper.

1. Brake caliper clip
2. Brake caliper dust boot caps
3. Brake caliper guide bolts (disc-drum system)
4. Brake caliper guide bolts (4-wheel disc brake system)
5. Brake caliper (RH/LH)
6. Disc brake pads (kit)
7. Brake caliper dust boots
8. Brake caliper anchor plate bolts
9. Brake caliper anchor plate
10. Brake disc
11. Brake line fitting nut
12. Brake caliper jounce hose retaining clip
13. Brake caliper jounce hose
14. Bleeder screw cap
15. Bleeder screw

67197-ESCA-G49

Front caliper installation—2005 model shown

✴✴ CAUTION

Do not allow the brake caliper to hang by the flexible brake hose.

7. Remove the disc brake pads.

8. Remove the brake caliper jounce hose. Loosen the jounce hose fitting prior to removing the brake caliper.

9. To install, reverse the removal procedure.

10. Torque the caliper pin bolts to 26 ft. lbs. (35 Nm) on disc/drum systems; 33 ft. lbs. (45 Nm) on 4-wheel disc systems.

11. If the hydraulic system was opened, bleed the brake system.

➡**Thread the brake caliper jounce hose onto the brake caliper before installing the brake caliper.**

➡**Make sure that the brake caliper jounce hose is not twisted.**

12. Position the brake caliper to the anchor plate and tighten the brake caliper jounce hose.

ESCAPE HYBRID

1. Before servicing the vehicle, refer to the Precautions Section.

➡**The rear brake pads will wear at approximately twice the rate of the front brake pads.**

With the vehicle in NEUTRAL, position it on a hoist.

➡**The following steps must be followed to prevent the accumulator from charging and pressurizing the brake system.**

2. Disconnect the battery.

3. Remove the battery junction box (BJB) fuses 24 (50A) and 31 (50A).

4. Remove the brake caliper clip.

5. Loosen the brake caliper jounce hose at the brake caliper.

6. Remove the 2 brake caliper dust boot caps.

7. Remove the 2 brake caliper guide bolts.

8. Unthread the brake caliper from the brake caliper jounce hose. Remove the brake caliper.

➡**Make sure that the brake caliper jounce hose is not twisted.**

9. To install, reverse the removal procedure. Observe the following torques:
- Brake caliper guide bolts: 45 Nm (33 ft. lbs.)
- Brake hose at the brake caliper: 20 Nm (15 ft. lbs.)

10. Bleed the brake system.

Rear

EXCEPT ESCAPE HYBRID

1. Before servicing the vehicle, refer to the Precautions Section.

2. Remove the wheel and tire assembly.

3. Remove the brake caliper guide bolts.

4. Remove the caliper.

✴✴ CAUTION

Do not allow the brake caliper to hang by the flexible brake hose.

17 Nm (13 lb-ft)

15 Nm (11 lb-ft)

35 Nm (26 lb-ft)

35 Nm (26 lb-ft)

16 Nm (12 lb-ft)

1 Brake caliper guide bolts	8 Brake caliper jounce hose bracket bolt
2 Caliper (RH/LH)	9 Brake line fitting nut
3 Brake disc pads	10 Brake caliper jounce hose retaining clip
4 Brake disc	11 Jounce hose (RH/LH)
5 Brake caliper guide bolt	12 Bleeder screw cap
6 Brake caliper hose flow bolt	13 Bleeder screw
7 Copper washers	

67197-ESCA-G50

Rear caliper installation—2005 model shown

5. Remove the brake disc pads.

6. Remove the brake caliper hose flow bolt.

7. Remove and discard the copper washers.

8. To install, reverse the removal procedure. Use new copper washers. Torque the caliper pin bolts to 26 ft. lbs. (35 Nm); torque the flow bolt to 26 ft. lbs. (35 Nm).

9. If the hydraulic system was opened, bleed the brake system.

ESCAPE HYBRID

1. Before servicing the vehicle, refer to the Precautions Section.

➡**The rear brake pads will wear at approximately twice the rate of the front brake pads.**

2. With the vehicle in NEUTRAL, position it on a hoist.

➡**The following steps must be followed to prevent the accumulator from charging and pressurizing the brake system.**

3. Disconnect the battery.

4. Remove the battery junction box (BJB) fuses 24 (50A) and 31 (50A).

➡**Install new copper washers.**

5. Remove the brake caliper jounce hose flow bolt and discard the 2 copper washers.

6. Remove the 2 brake caliper guide bolts.

7. Remove the brake caliper.

8. To install, reverse the removal procedure. Observe the following torques:
   - Brake caliper guide bolts: 35 Nm (26 ft. lbs.)
   - Brake caliper hose flow bolt: 35 Nm (26 ft. lbs.)

9. Bleed the brake system.

Brake Pads

REMOVAL & INSTALLATION

Front

EXCEPT ESCAPE HYBRID

1. Before servicing the vehicle, refer to the Precautions Section.

2. Remove the wheel and tire assembly.

3. Remove the brake caliper clip.

4. Position the caliper aside.

5. Remove brake caliper bolt caps and the bolts.

6. Position the caliper aside and support.

7. Remove the brake pads.

8. Remove the outer brake pad from the anchor.

9. Remove the inner brake pad from the caliper piston.

10. To install, reverse the removal procedure.

ESCAPE HYBRID

1. Before servicing the vehicle, refer to the Precautions Section.

2. With the vehicle in NEUTRAL, position it on a hoist.

➡**The following steps must be followed to prevent the accumulator from charging and pressurizing the brake system.**

3. Disconnect the battery.

4. Remove the battery junction box (BJB) fuses 24 (50A) and 31 (50A).

5. Remove the brake caliper clip.

6. Remove the 2 brake caliper dust boot caps.

7. Remove the 2 brake caliper guide bolts.

✷✷ WARNING

Do not allow the brake caliper to hang by the brake caliper jounce hose.

8. Remove the brake caliper and support the brake caliper to the vehicle.

9. Remove the inboard brake pad from the brake caliper.

10. Remove the outboard brake pad from the brake caliper anchor plate.

11. To install, reverse the removal procedure. Torque the caliper pin bolts to 33 ft. lbs. (45 Nm).

Rear

EXCEPT ESCAPE HYBRID

1. Before servicing the vehicle, refer to the Precautions Section.

2. Remove the wheel and tire assembly.

3. Remove the brake caliper guide bolts.

4. Remove the caliper.

✷✷ CAUTION

Do not allow the brake caliper to hang by the flexible brake hose.

5. Remove the brake disc pads.

6. To install, reverse the removal procedure.

7. If the hydraulic system was opened, bleed the brake system.

ESCAPE HYBRID

1. Before servicing the vehicle, refer to the Precautions Section.

➡**The rear brake pads will wear at approximately twice the rate of the front brake pads.**

2. With the vehicle in NEUTRAL, position it on a hoist.

➡**The following steps must be followed to prevent the accumulator from charging and pressurizing the brake system.**

3. Disconnect the battery.

4. Remove the battery junction box (BJB) fuses 24 (50A) and 31 (50A).

5. Remove the 2 brake caliper guide bolts.

✷✷ WARNING

Do not allow the brake caliper to hang by the brake caliper jounce hose.

6. Remove the brake caliper and secure the brake caliper to the vehicle.

7. Remove the inboard and outboard brake pads from the brake caliper. Torque the 2 brake caliper guide bolts to 35 Nm (26 ft. lbs.).

Brake Rotor

REMOVAL & INSTALLATION

Front

EXCEPT ESCAPE HYBRID

1. Before servicing the vehicle, refer to the Precautions Section.

2. Remove the brake caliper anchor plate.

3. Remove the brake disc retaining clips (if equipped) and the brake disc.

4. To install, reverse the removal procedure. Torque the anchor plate bolts to 111 ft. lbs. (150Nm).

ESCAPE HYBRID

1. Before servicing the vehicle, refer to the Precautions Section.

➡**The following steps must be followed to prevent the accumulator from charging and pressurizing the brake system.**

2. Disconnect the battery.

3. Remove the battery junction box (BJB) fuses 24 (50A) and 31 (50A).

4. Remove the brake pads.

5. Remove the 2 brake caliper anchor plate bolts.

6. Remove the brake caliper anchor plate.

7. Remove the brake caliper anchor plate.

8. Remove the brake disc.

9. To install, reverse the removal procedure. Torque the 2 brake caliper anchor plate bolts to 150 Nm (111 ft. lbs.).

Rear

EXCEPT ESCAPE HYBRID

1. Before servicing the vehicle, refer to the Precautions Section.
2. Remove the caliper.
3. Remove the rotor.
4. Installation is the reverse of removal.

ESCAPE HYBRID

1. Before servicing the vehicle, refer to the Precautions Section.

➡ **The rear brake pads will wear at approximately twice the rate of the front brake pads.**

2. With the vehicle in NEUTRAL, position it on a hoist.

➡ **The following steps must be followed to prevent the accumulator from charging and pressurizing the brake system.**

3. Disconnect the battery.
4. Remove the battery junction box (BJB) fuses 24 (50A) and 31 (50A).
5. Remove the brake pads.
6. Remove the brake disc.
7. If the brake disc binds on the parking brake shoe and lining, remove the adjustment hole access plug and retract the parking brake shoe and lining by rotating the adjuster.

8. To install, reverse the removal procedure.
9. Adjust the parking brake shoe and lining, if necessary.

Parking Brake Shoes

REMOVAL & INSTALLATION

1. Before servicing the vehicle, refer to the Precautions Section.
2. Remove the rear brake disc.
3. Remove the parking brake shoe upper return spring.
4. Remove the 2 parking brake shoe retaining pins.

1 Parking brake control boot	15 Cable bracket-to-body bolt
2 Front cable adjuster nut	16 Cable bracket-to-trailing arm bolt
3 Parking brake control bolts (4 required)	17 Cable-to-fuel tank strap clip
4 Warning indicator switch screw	18 Rear parking brake cable (LH)
5 Warning indicator switch	19 Parking brake shoe upper return spring
6 Parking brake control	20 Parking brake shoe retaining pins (2 required)
7 Parking brake equalizer bracket bolts (2 required)	21 Parking brake shoe retaining springs (2 required)
8 Front parking brake cable	22 Parking brake shoe lower return spring
9 Grommet	23 Parking brake shoe adjuster
10 Cable connector	24 Parking brake shoe (LH/RH)
11 Parking brake equalizer and bracket	25 Support plate bolts (4 required)
12 Cable bracket-to-body bolt	26 Support plate (LH/RH)
13 Cable bracket-to-trailing arm bolt	
14 Rear parking brake cable (RH)	

06017-ESCA-G99

Parking brake system—with rear disc brakes

06017-ESCA-G100

Using the special tool, measure the inside diameter of the drum portion of the rear brake disc and set the locking screw

06017-ESCA-G101

Place the special tool over the widest diameter of the parking brake shoes

5. Remove the 2 parking brake shoe retaining springs.

6. Remove the parking brake shoe lower return spring.

7. Remove the parking brake shoe adjuster.

8. Remove the parking brake shoes.

9. To install, reverse the removal procedure.

- Using anti-seize lubricant, lubricate the parking brake shoe contact points before installation of the rear parking brake shoes.
- Lubricate the adjust screw threads with anti-seize lubricant.
- Adjust the parking brake shoe and lining.
- Check the parking brake for normal operation.

SHOE ADJUSTMENT

1. Before servicing the vehicle, refer to the Precautions Section.

2. With the vehicle in NEUTRAL, position it on a hoist.

➡**Make sure the parking brake is fully released.**

3. Using the release handle, release the parking brake control.

4. Remove the rear brake disc.

5. Using the special tool, measure the inside diameter of the drum portion of the rear brake disc and set the locking screw. Record the measurement.

6. Place the special tool over the widest diameter of the parking brake shoes.

7. Adjust the parking brake shoe clearance to 1.07 mm (0.04 in) less than the inside diameter of the drum portion of the rear brake disc. Rotate the parking brake shoe adjuster to achieve the correct parking brake shoe-to-brake disc clearance.

8. Install the rear brake disc.

9. Test the parking brake for normal operation.

DRUM BRAKES

Brake Drum

REMOVAL & INSTALLATION

1. Before servicing the vehicle, refer to the Precautions Section.
2. Remove the tire and wheel assembly.

✳✳ CAUTION

Use of a brake drum puller or a torch is not recommended. Brake drum distortion can result.

➡ If the brake drum is rusted to the axle shaft pilot diameter, tap the center of the brake drum between the wheel studs.

3. Remove the brake drum.
4. If equipped, remove the brake drum retaining clips.
5. If the brake drums will not come off, follow these steps.

Drum brake exploded view

1 Plug	10 Brake shoe (kit)	19 Brake line fitting nut
2 Brake drum	11 Adjuster lever spring	20 Jounce hose bracket bolt
3 Parking brake lever clip	12 Adjuster lever (LH/RH)	21 Jounce hose retaining clips
4 Brake shoe retaining clips	13 Pivot pin (part of 2200)	22 Jounce hose bracket
5 Brake shoe retaining pins	14 Brake line fitting nut	23 Brake line fitting nut
6 Upper return spring	15 Bleeder screw cap	24 Jounce hose (LH/RH)
7 Adjuster assembly (LH/RH)	16 Bleeder screw	25 Backing plate bolts
8 Lower return spring	17 Wheel cylinder bolts	26 Backing plate
9 Parking brake actuator lever (LH/RH)	18 Wheel cylinder	

67197-ESCA-G51

6. Move the brake shoe adjusting lever off the brake adjuster screw.

7. Loosen the brake shoe adjuster screw nut by adjusting the nut upward.

8. Using the special tool, 134-R0191, measure the brake drum inside diameter.

9. Install a new brake drum if the maximum inside diameter exceeds the specification shown on the outside of the brake drum.

To install:

✳✳ WARNING

Whenever a wheel is installed, always remove any corrosion, dirt or foreign material present on the mounting surfaces of the wheel or the surface of the wheel hub, brake drum or brake disc that contacts the wheel. Installing wheels without correct metal-to-metal contact at the wheel mounting surfaces can cause the wheel nuts to loosen and the wheel to come off while the vehicle is in motion, causing loss of control. Failure to follow these instructions may result in personal injury.

10. Clean the wheel hub mounting surface and wheel pilot.

11. Install the tire and wheel assembly.

Brake Shoes

REMOVAL & INSTALLATION

1. Before servicing the vehicle, refer to the Precautions Section.

2. Remove the brake drum.

3. Use the Brake/Clutch/Service Vacuum to remove brake dust and dirt from the brake assemblies.

➡**If new rear brake shoes and linings are being installed, resurface the brake drums to remove glazing and to ensure an equal friction surface from side-to-side. Resurfacing will also correct out-of-round and bell conditions.**

4. Using the special tool, measure the braking surface diameter. If the inside diameter measures more than the maximum specification shown on the outside of the brake drum, install a new brake drum.

5. Remove the parking brake cable from the parking brake cable lever.

6. Remove the hold-down clips and pins.

7. Remove the lower spring.

8. Remove the rear brake shoes.

9. Pull the bottom of the brake shoe forward.

10. Release the upper return spring.

11. Remove both brake shoes together.

12. Remove the self adjuster lever.

13. Remove the self adjuster and spring assembly.

14. Return the self adjuster to the fully seated position.

15. Remove the parking brake lever.

16. Remove the horseshoe clip.

17. Remove the parking brake lever.

18. Inspect the rear brake shoes for minimum thickness above the backing plate, and install new as necessary.

19. To install, reverse the removal procedure.

Wheel Cylinder

REMOVAL & INSTALLATION

1. Before servicing the vehicle, refer to the Precautions Section.

2. Remove the brakes shoes.

3. Disconnect the brake line at the wheel cylinder.

4. Remove the wheel cylinder bolts.

5. Installation is the reverse of removal. Torque the bolts to 9 ft. lbs. (12 Nm).

BRAKE HYDRAULIC SYSTEM

Calipers and Wheels Cylinders

BLEEDING

• Use of any other than the approved DOT 3 brake fluid will cause permanent damage to brake components and will render the brakes inoperative. Failure to follow these instructions may result in personal injury.

• Brake fluid contains polyglycol ethers and polyglycols. Avoid contact with eyes. Wash hands thoroughly after handling. If brake fluid contacts eyes, flush eyes with running water for 15 minutes. Get medical attention if irritation persists. If taken internally, drink water and induce vomiting. Get medical attention immediately. Failure to follow these instructions may result in personal injury.

• Do not allow the brake master cylinder reservoir to run dry during the bleeding operation. Keep the master cylinder reservoir filled with the specified brake fluid. Never reuse the brake fluid that has been drained from the hydraulic system.

• Brake fluid is harmful to painted and plastic surfaces. If brake fluid is spilled onto a painted or plastic surface, immediately wash it with water.

• When any part of the hydraulic system has been disconnected or a new component is installed, air may enter the system, causing spongy brake pedal action. This requires the bleeding of the hydraulic system after it has been correctly connected.

Manual Bleeding

EXCEPT ESCAPE HYBRID

1. Before servicing the vehicle, refer to the Precautions Section.

✳✳ WARNING

Be sure to check the brake fluid level in the brake master cylinder reservoir often. Do not let it run dry.

2. Fill the brake master cylinder reservoir with the specified brake fluid.

3. Begin bleeding the system, going in order from the right rear wheel, to the left rear wheel, to the right front wheel, and ending with the left front wheel.

4. Attach a rubber drain hose to the rear bleeder screw and submerge the free end in a container partially filled with clean brake fluid.

5. Have an assistant pump the brake pedal 10 times and then hold firm pressure on the brake pedal.

6. Loosen the bleeder screw until the fluid flow stops. Maintain pressure on the brake pedal and tighten the bleeder screw.

7. Repeat Steps 4 and 5 until clear, bubble-free brake fluid flows.

8. Tighten the bleeder screw to 16 Nm (12 ft. lbs.).

9. Refill the brake master cylinder reservoir as necessary.

10. Continue bleeding the brake hydraulic system at each wheel.

11. Fill the brake master cylinder reservoir with the specified brake fluid.

ESCAPE HYBRID

1. Before servicing the vehicle, refer to the Precautions Section.

❋❋ CAUTION

Before proceeding, read and observe all of the High Voltage System Precautions.

❋❋ WARNING

Be sure to check the brake fluid level in the brake master cylinder reservoir often. Do not let it run dry.

2. Make sure that the brake fluid level is above the MIN mark.

3. Disconnect the battery.

4. De-power the high voltage traction battery. Establish a buffer zone. See the procedure under De-powering the High Voltage Traction Battery.

5. Depower the high voltage traction battery.

6. Begin bleeding the system, going in order from the right rear wheel, to the left rear wheel, to the right front wheel, and ending with the left front wheel.

7. Attach a rubber drain hose to the rear bleeder screw and submerge the free end in a container partially filled with clean brake fluid.

8. Have an assistant pump the brake pedal 10 times and then hold firm pressure on the brake pedal.

9. Loosen the bleeder screw until the fluid flow stops. Maintain pressure on the brake pedal and tighten the bleeder screw.

10. Repeat Steps 4 and 5 until clear, bubble-free brake fluid flows.

11. Tighten the bleeder screw to 16 Nm (12 ft. lbs.).

12. Refill the brake master cylinder reservoir as necessary.

13. Continue bleeding the brake hydraulic system at each wheel.

14. Fill the brake master cylinder reservoir with the specified brake fluid.

FORD

Escort ZX2

SPECIFICATION AND MAINTENANCE CHARTS

ENGINE AND VEHICLE IDENTIFICATION

		Engine							Model Year	
Code ①	Liters (cc)	Cu. In.	Cyl.	Fuel Sys.	Type	Eng. Mfg.		Code ②	Year
3	2.0 (1990)	121	4	SFI	DOHC	Ford		1	2001

Code ②	Year
1	2001
2	2002
3	2003
4	2004
5	2005

SFI: Sequential Fuel Injection
DOHC: Double Overhead Camshafts
SOHC: Single Overhead Camshaft
① 8th digit of the VIN
② 10th digit of the VIN

67197-ESCO-C01

GENERAL ENGINE SPECIFICATIONS

Year	Model	Engine Displacement Liters (VIN)	Net Horsepower @ rpm	Net Torque @ rpm (ft. lbs.)	Bore x Stroke (in.)	Compression Ratio	Oil Pressure @ rpm
2001	ZX2	2.0 (3)	125@5500	130@4000	3.34x3.46	10.0:1	35-65@2000
2002	ZX2	2.0 (3)	125@5500	130@4000	3.34x3.46	10.0:1	35-65@2000
2003	ZX2	2.0 (3)	125@5500	130@4000	3.34x3.46	10.0:1	35-65@2000

SFI: Sequential Fuel Injection

67197-ESCO-C02

ENGINE TUNE-UP SPECIFICATIONS

Year	Engine Displacement Liters (VIN)	Spark Plug Gap (in.)	Ignition Timing (deg.) MT	AT	Fuel Pump (psi)	Idle Speed (rpm) MT	AT	Valve Clearance In.	Ex.
2001	2.0 (3)	0.052-0.056	10B	10B	31-38 ①	②	②	HYD	HYD
2002	2.0 (3)	0.052-0.056	10B	10B	31-38 ①	②	②	HYD	HYD
2003	2.0 (3)	0.052-0.056	10B	10B	31-38 ①	②	②	HYD	HYD

NOTE: The Vehicle Emission Control Information label often reflects specification changes made during production. The label figures must be used if they differ from those in this chart.
B: Before Top Dead Center
HYD: Hydraulic
① Fuel pressure with engine running, pressure regulator vacuum hose connected
② Refer to Vehicle Emission Control Information label

67197-ESCO-C03

2.0L VIN 3 Engine
Firing order: 1–3–4–2
Distributorless ignition system

FRONT OF VEHICLE

79223G06

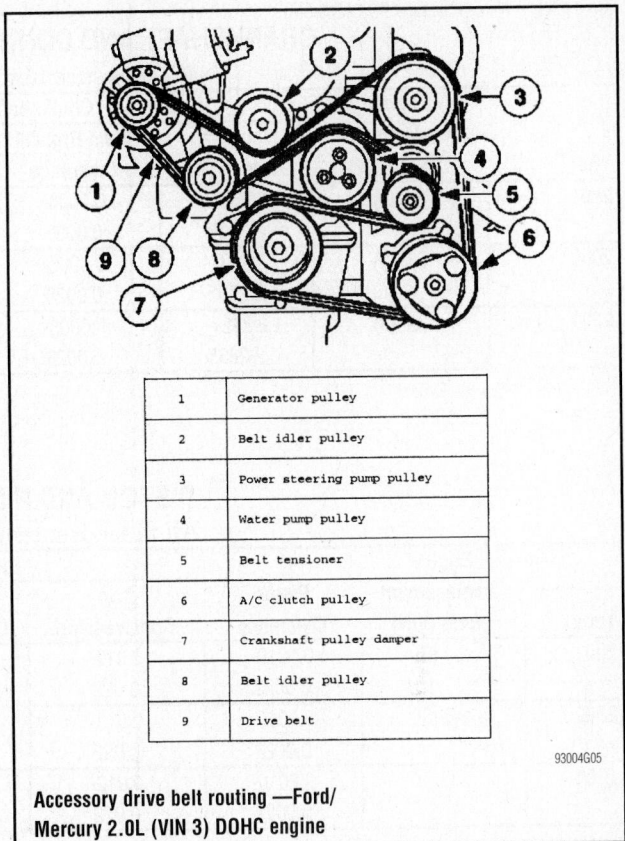

1	Generator pulley
2	Belt idler pulley
3	Power steering pump pulley
4	Water pump pulley
5	Belt tensioner
6	A/C clutch pulley
7	Crankshaft pulley damper
8	Belt idler pulley
9	Drive belt

93004G05

Accessory drive belt routing —Ford/
Mercury 2.0L (VIN 3) DOHC engine

CAPACITIES

Year	Model	Engine Displacement Liters (VIN)	Oil with Filter (qts.)	Transmission (pts.)		Front Axle (pts.)	Fuel Tank (gal.)	Cooling System (qts.)
				Manual	Auto. ①			
2001	ZX2	2.0 (3)	4.0	6.7	13.4	②	13.2	④
2002	ZX2	2.0 (3)	4.0	6.7	13.4	②	13.2	④
2003	ZX2	2.0 (3)	4.0	6.7	13.4	②	13.2	④

Note: All capacities are approximates. Add fluid gradually and ensure a proper fluid level is obtained.

① Includes torque converter

② Included in transaxle capacity

③ Manual transaxle 7.9

　Automatic transaxle 5.8

④ Manual transaxle 7.0

　Automatic transaxle 7.5

67197-ESCO-C04

VALVE SPECIFICATIONS

Year	Engine Displacement Liters (VIN)	Seat Angle (deg.)	Face Angle (deg.)	Spring Test Pressure (lbs. @ in.)	Spring Installed Height (in.)	Stem-to-Guide Clearance (in.)		Stem Diameter (in.)	
						Intake	Exhaust	Intake	Exhaust
2001	2.0 (3)	45	45	NA	1.346	0.0007-0.0025	0.0014-0.0032	0.2373-0.2379	0.2366-0.2372
2002	2.0 (3)	45	45	NA	1.346	0.0007-0.0025	0.0014-0.0032	0.2373-0.2379	0.2366-0.2372
2003	2.0 (3)	45	45	NA	1.346	0.0007-0.0025	0.0014-0.0032	0.2373-0.2379	0.2366-0.2372

67197-ESCO-C05

CRANKSHAFT AND CONNECTING ROD SPECIFICATIONS
All measurements are given in inches.

Year	Engine Displacement Liters (VIN)	Crankshaft				Connecting Rod		
		Main Brg. Journal Dia.	Main Brg. Oil Clearance	Shaft End-play	Thrust on No.	Journal Diameter	Oil Clearance	Side Clearance
2001	2.0 (3)	2.2827-2.2835	0.0008-0.0026	0.0040-0.0120	3	1.8461-1.8500	0.0006-0.0028	0.0040-0.0110
2002	2.0 (3)	2.2827-2.2835	0.0008-0.0026	0.0040-0.0120	3	1.8461-1.8500	0.0006-0.0028	0.0040-0.0110
2003	2.0 (3)	2.2827-2.2835	0.0008-0.0026	0.0040-0.0120	3	1.8461-1.8500	0.0006-0.0028	0.0040-0.0110

67197-ESCO-C06

PISTON AND RING SPECIFICATIONS
All measurements are given in inches.

Year	Engine Displacement Liters (VIN)	Piston Clearance	Ring Gap			Ring Side Clearance		
			Top Compression	Bottom Compression	Oil Control	Top Compression	Bottom Compression	Oil Control
2001	2.0 (3)	0.0010-0.0022	0.012-0.022	0.012-0.022	0.010-0.039	0.0015-0.0032	0.0015-0.0035	—
2002	2.0 (3)	0.0010-0.0022	0.012-0.022	0.012-0.022	0.010-0.039	0.0015-0.0032	0.0015-0.0035	—
2003	2.0 (3)	0.0010-0.0022	0.012-0.022	0.012-0.022	0.010-0.039	0.0015-0.0032	0.0015-0.0035	—

67197-ESCO-C07

TORQUE SPECIFICATIONS
All readings in ft. lbs.

Year	Engine Displacement Liters (VIN)	Cylinder Head Bolts	Main Bearing Bolts	Rod Bearing Bolts	Crankshaft Damper Bolts	Flywheel Bolts	Manifold		Spark Plugs	Oil Pan Drain Plug
							Intake	Exhaust		
2001	2.0 (3)	①	55-65	②	80-87	71-76	11-12	10-12	10-12	22
2002	2.0 (3)	①	55-65	②	80-87	71-76	11-12	10-12	10-12	22
2003	2.0 (3)	①	55-65	②	80-87	71-76	11-12	10-12	10-12	22

NOTE: Always follow proper torque patterns

NOTE: Stretch bolts are used in all procedures that require rotating the fastener a certain number of degrees. The bolts stretch and cannot be reused. For reassembly, replace with new fastner

① Do not reuse cylinder head bolts.
  Step 1: Tighten bolts, in sequence, to 15 ft. lbs.
  Step 2: Tighten bolts in sequence to 40 ft. lbs.
  Step 3: Turn all bolts, in sequence, plus an additional 105 degrees

② Step 1: 22-25 ft. lbs.
  Step 2: Rotate each bolt 85-95 degrees

67197-ESCO-C08

WHEEL ALIGNMENT

Year	Model		Caster Range (+/-Deg.)	Caster Preferred Setting (Deg.)	Camber Range (+/-Deg.)	Camber Preferred Setting (Deg.)	Toe-in (in.)
2001	ZX2	F	1.00	+2.20	1.00	+0.40	0.08 +/- 0.10
		R	—	—	1.00	+1.18	0.08 +/- 0.10
2002	ZX2	F	1.00	+2.20	1.00	+0.40	0.08 +/- 0.10
		R	—	—	1.00	+1.18	0.08 +/- 0.10
2003	ZX2	F	1.00	+2.20	1.00	+0.40	0.08 +/- 0.10
		R	—	—	1.00	+1.18	0.08 +/- 0.10

67197-ESCO-C09

TIRE, WHEEL AND BALL JOINT SPECIFICATIONS

Year	Model	OEM Tires Standard	OEM Tires Optional	Tire Pressures (psi) Front	Tire Pressures (psi) Rear	Wheel Size	Ball Joint Inspection	Lug Nut
2001	ZX2	P185/65R14	P185/60R15	32	32	5.5-JJ	0.030 in. ①	76
2002	ZX2	P185/65R14	P185/60R15	32	32	5.5-JJ	0.030 in. ①	76
2003	ZX2	P185/65R14	P185/60R15	32	32	5.5-JJ	0.030 in. ①	76

OEM: Original Equipment Manufacturer

PSI: Pounds Per Square Inch

STD: Standard

OPT: Optional

① Maximum radial tolerance in inches

② Replace if any measurable movement is found.

67197-ESCO-C10

BRAKE SPECIFICATIONS
All measurements in inches unless noted

Year	Model		Brake Disc Original Thickness	Brake Disc Minimum Thickness	Brake Disc Maximum Runout	Brake Drum Diameter Original Inside Diameter	Brake Drum Diameter Max. Wear Limit	Brake Drum Diameter Maximum Machine Diameter	Minimum Lining Thickness	Brake Caliper Mounting Bolts (ft. lbs.)
2001	ZX2	F	0.870	0.790	0.004	—	—	—	0.080	36-43
		R	0.350	0.280	0.004	7.87	7.95	7.91	0.040	—
2002	ZX2	F	0.870	0.790	0.004	—	—	—	0.080	36-43
		R	0.350	0.280	0.004	7.87	7.95	7.91	0.040	—
2003	ZX2	F	0.870	0.790	0.004	—	—	—	0.080	36-43
		R	0.350	0.280	0.004	7.87	7.95	7.91	0.040	—

NOTE: Follow specifications stamped on rotor or drum if figures differ from those in this chart.

NA: Not Available

F: Front

R: Rear

67197-ESCO-C11

SCHEDULED MAINTENANCE INTERVALS
Ford—Escort ZX2

TO BE SERVICED	TYPE OF SERVIC	VEHICLE MILEAGE INTERVAL (x1000)																			
		5	10	15	20	25	30	35	40	45	50	55	60	65	70	75	80	85	90	95	100
Engine oil & filter	R	✓	✓	✓	✓	✓	✓	✓	✓	✓	✓	✓	✓	✓	✓	✓	✓	✓	✓	✓	✓
Tires ①	S/I	✓		✓		✓		✓		✓		✓		✓		✓		✓		✓	
Air Cleaner	R						✓						✓						✓		
Spark Plugs	R																				✓
Drive Belts	S/I												✓								
Cooling system	S/I			✓			✓			✓			✓			✓			✓		
Engine coolant	R										✓						✓				
PCV valve	R												✓								
Exhaust heat shields	S/I						✓						✓						✓		
Brake linings & drums	S/I						✓						✓						✓		
Brake line hoses & connections	S/I						✓						✓						✓		
Front ball joints	S/I						✓						✓						✓		
Bolts & nuts on chassis body	S/I						✓						✓						✓		
Steering linkage operation	S/I						✓						✓						✓		
Brake pads & rotor	S/I						✓						✓						✓		
Clutch pedal operation	S/I						✓						✓						✓		
Halfshaft dust boots	S/I						✓						✓						✓		

R: Replace   S/I: Inspect and service, if needed
① Rotate, inspect the tire tread for wear, and adjust air pressure.

FREQUENT OPERATION MAINTENANCE (SEVERE SERVICE)

If a vehicle is operated under any of the following conditions it is considered severe service:

- Extremely dusty areas.

- 50% or more of the vehicle operation is in 32°C (90°F) or higher temperatures, or constant operation in temperatures below 0°C (32°F).

- Prolonged idling (vehicle operation in stop and go traffic).

- Frequent short running periods (engine does not warm to normal operating temperatures).

- Police, taxi, delivery usage or trailer towing usage.

Oil & filter change: change every 3000 miles.

Rotate tires at 6000 miles & every 9000 miles thereafter.

Automatic transmission fluid & filter: change every 21,000 miles.

67197-ESCO-C12

ENGINE REPAIR

➡Disconnecting the negative battery cable on some vehicles may interfere with the operation of the on-board computer system. The computer may undergo a relearning process once the negative battery cable is reconnected.

Alternator

REMOVAL

Sedan and Wagon

1. Before servicing the vehicle, refer to the Precautions Section.
2. Remove or disconnect the following:
   - Negative battery cable
   - Drive belt
   - Power steering hose bracket from the alternator bracket
   - Alternator mounting bolts
   - Alternator electrical connectors
   - Battery positive cable
   - Alternator from the vehicle

Coupe

1. Before servicing the vehicle, refer to the Precautions Section.
2. Remove or disconnect the following:
   - Negative battery cable
   - Drive belt
   - Coolant tank reservoir
   - Alternator electrical connectors
   - Alternator mounting bolts
   - Battery positive cable
   - Alternator from the vehicle

INSTALLATION

Sedan and Wagon

1. Install or connect the following:
   - Battery positive cable to the alternator. Tighten the nut to 7–9 ft. lbs. (9–12 Nm).
   - Alternator electrical connections
   - Alternator in position
   - Alternator mounting bolts. Tighten the upper mounting bolt to 29–40 ft. lbs. (40–55 Nm) and the lower mounting bolt to 15–22 ft. lbs. (20–30 Nm).
   - Power steering hose bracket to the alternator bracket
   - Drive belt
   - Negative battery cable

90982G23

Location of the alternator upper and lower mounting bolts—sedan and wagon models

90982G24

Location of the alternator upper and lower mounting bolts—coupe model

Coupe

1. Install or connect the following:
   - Battery positive cable to the alternator. Tighten the nut to 5–7 ft. lbs. (7–9 Nm).
   - Alternator in position
   - Alternator mounting bolts and tighten to 29–40 ft. lbs. (40–55 Nm)
   - Alternator electrical connections
   - Coolant tank reservoir
   - Drive belt
   - Negative battery cable

Ignition Timing

ADJUSTMENT

The Powertrain Control Module (PCM) controls ignition timing on these models. No adjustment is necessary or possible.

Engine Assembly

REMOVAL & INSTALLATION

1. Before servicing the vehicle, refer to the Precautions Section.
2. Drain the cooling system.
3. Relieve the fuel system pressure.
4. Drain the engine oil.
5. Recover the A/C refrigerant, if equipped.
6. Remove or disconnect the following:
   - Battery and tray
   - Hood
   - Air cleaner outlet tube
   - Fuel pressure regulator vacuum hose
   - Intake manifold vacuum hoses
   - Exhaust Gas Recirculation (EGR) vacuum line
   - Positive Crankcase Ventilation (PCV) valve vacuum line
   - Vacuum tree
   - Throttle cable
   - Cruise control cable, if equipped
   - Shift cable
   - Constant Control Relay Module (CCRM)
   - EGR backpressure transducer
   - Transaxle range sensor connector, if equipped
   - Transaxle solenoid connector, if equipped
   - Turbine speed sensor connector, if equipped
   - Vehicle Speed (VSS) sensor
   - Heated Oxygen (HO$_2$S) sensor connectors
   - Ground strap
   - Engine cooling fan connector
   - Power steering hose brackets
   - Fuel line
   - Exhaust manifold heat shield
   - Starter bracket
   - Clutch cylinder fluid line, if equipped
   - Transaxle cooler lines, if equipped
   - A/C line bracket, if equipped
   - A/C condenser tube, if equipped
   - A/C accumulator tube, if equipped
   - Accessory drive belt and tensioner
   - Accessory drive belt idler pulley, if equipped
   - Power steering hoses
   - Front wheels
   - Left and right splash shields
   - Front subframe crossmember
   - Catalytic converter
   - Axle halfshafts
   - Starter motor
   - Oil pressure switch connector
   - A/C compressor, if equipped
   - Radiator
   - Heater hoses
   - Alternator harness connectors
7. Attach a support fixture to the engine.
8. Remove or disconnect the following:
   - Left front engine mount
   - Transaxle mount
   - Transaxle support crossmember
   - Upper transaxle mount
   - Right engine mount
9. Attach a hoist to the engine and remove the support fixture.
10. Raise the engine and transaxle from the vehicle.

To install:

11. Installation is the reverse of removal, but please note the following important steps.
   - Front isolator nuts and bolts: 50–68 ft. lbs. (67–93 Nm)
   - Front engine support isolator through-bolt: 50–68 ft. lbs. (67–93 Nm)
   - Front roll restrictor bolts: 48–65 ft. lbs. (64–89 Nm) and the nuts to 50–69 ft. lbs. (67–93 Nm)
   - Rear roll restrictor nuts: 28–38 ft. lbs. (38–51 Nm)
   - Torque converter nuts: 27 ft. lbs. (37 Nm)
12. Fill the crankcase to the correct level.
13. Fill the cooling system.
14. Recharge the A/C system, if equipped.
15. Start the engine and check for leaks.

Water Pump

REMOVAL & INSTALLATION

1. Before servicing the vehicle, refer to the Precautions Section.

Remove the water pump mounting bolts—2.0L Zetec engine

7922MG19

1. Clutch pressure plate
2. Clutch disc
3. Support
4. Flywheel (automatic trans.)
5. Flywheel (manual trans.)
6. PVC valve
7. Oil separator
8. Engine block
9. Knock sensor
10. Oil pressure sensor
11. Oil filter

12. Throttle body
13. Fuel injection supply manifold
14. Idle Air Control (IAC) valve
15. Retainer clip
16. Fuel injector
17. Bracket
18. Water pump housing gasket
19. Water pump
20. Upper timing belt cover
21. Center timing belt cover
22. Water pump pulley

23. Crankshaft pulley
24. Crankshaft sprocket
25. Timing belt guide
26. Timing belt
27. Timing belt idler pulley
28. Oil pump
29. Oil pump housing gasket
30. Intake manifold
31. Oil pump pick-up
32. Crankshaft front seal
33. Lower timing belt cover

9300MG10

Exploded view of peripheral engine component mounting, including water pump, oil pump and intake manifold—2.0L DOHC Zetec engine

2. Drain the cooling system.
3. Remove or disconnect the following:

- Negative battery cable
- Splash shield
- Drive belt
- Air conditioning compressor
- Water pump pulley
- Water pump

To install:

4. Install or connect the following:
- Water pump. Tighten the bolts to 17 ft. lbs. (24 Nm).
- Water pump pulley. Tighten the bolts to 17 ft. lbs. (24 Nm).
- Air conditioning compressor
- Drive belt
- Splash shield
- Negative battery cable
5. Fill the cooling system.
6. Start the engine and check for leaks.

Heater Core

REMOVAL & INSTALLATION

1. Before servicing the vehicle, refer to the Precautions Section.
2. Disconnect the negative battery cable.

✳✳ CAUTION

After disconnecting the negative battery cable, wait for at least 1 minute for the SRS or air bag module to deplete its energy.

3. Drain the cooling system into a clean container for reuse.
4. Remove or disconnect the following:

- Air cleaner outlet tube, on coupe models only
- Heater hoses from the heater core
5. Place the front wheels in the straight-ahead position. Lock the steering column.
6. Remove the steering wheel by removing or disconnecting the following:
- SRS module-to-steering wheel bolts
- SRS module (carefully), and disconnect the horn switch and the SRS electrical connectors

✳✳ CAUTION

Place the SRS module in a safe place with the front facing upward.

- Steering wheel bolt and discard it
- Press the steering wheel from the steering column

Remove the screw from the center instrument panel finish panel—Escort

Unfasten the pushpins, then remove the left-hand and right-hand control box covers—Escort

Unfasten the retaining screws and position the transaxle control selector dial bezel sideways—Escort

Unfasten the PCM electrical connector bolt, unplug the connector and move it aside—Escort

Unfasten the screw and remove the instrument panel steering column cover—Escort

Remove the steering column bracket bolts and lower the column—Escort

Remove the pushpins from each cowl side trim panel, and remove the panels—Escort

Unplug the vacuum line harness connector—Escort

90980G50

Unfasten both the left-hand and right-hand center instrument panel reinforcement bolts—Escort

90980G51

Unfasten both the left-hand and right-hand lower instrument panel reinforcement bolts—Escort

90980G52

Remove the cover and unfasten the upper instrument panel reinforcement bolt—Escort

7. Remove the passenger's side SRS module by removing or disconnecting the following:
- Push inward on the 2 glove box door tabs and lower it

- SRS module's electrical connector
- SRS module-to-instrument panel bolts and the module

✴✴ CAUTION

Place the SRS module in a safe place with the front facing upward.

8. Remove the instrument panel by removing or disconnecting the following:
- Floor console
- Screw located at the center instrument panel finish panel
- Pull the center instrument panel finish panel straight out from the instrument panel reinforcement (Coupe models only)
- Power point socket electrical connectors and remove the center instrument panel finish panel, if equipped
- Control box side cover pushpins and the covers located on both sides
- Radio antenna lead-in cable, located at the instrument panel reinforcement
- Radio antenna lead-in cable
- Transmission control selector dial bezel screws; then, place the transmission control selector dial bezel sideways
- Powertrain Control (PCM) electrical connector bolt and move the PCM aside
- 2 rear side PCM bracket nuts, the 2 bolts; then, the PCM and bracket as an assembly
- Rotate the temperature control switch to the COOL position and disconnect the heater control cable
- Hood latch control handle nut; then, position the hood latch control handle and cable aside
- Instrument panel steering column cover screw and release the cover
- Light switch rheostat resistor electrical connector and remove the instrument panel steering column cover (Coupe models only)
- Steering column shroud screws and the shrouds
- Steering column bracket bolts and lower the steering column
- Pull upward on the front door scuff plates
- Cowl side trim panel pushpins and the panels, located on both sides
- Interior fuse junction panel electrical connector
- Lower instrument panel reinforce-

ment in-line electrical connector, located on the left side
- Vacuum line harness connector
- Lower instrument panel reinforcement in-line electrical connector, located on the right side
- Blower motor resistor electrical connector
- Instrument panels end panels, located on both sides
- Upper and lower instrument panel-to-chassis bolts, located on both sides
- Upper and lower instrument panel reinforcement-to-chassis bolts, located on both sides
- Cover and the upper instrument panel reinforcement bolt
- Slightly, pull the instrument panel rearward and disconnect the main electrical wiring connectors, located on the left side
- Instrument panel

9. Remove or disconnect the following:
- Antenna lead from the heater core housing
- Vacuum control motor vacuum connector
- Vacuum lines from the retainer at the evaporator housing
- Pushpins, the windshield defroster nozzle connector's screws and the connectors
- Air conditioning evaporator outlet duct clamp screw
- Lower heater core housing-to-chassis nut, the upper heater core housing-to-chassis nuts and the heater core housing
- Heater dash panel seal
- Heater core cover-to-housing screws and the cover
- Heater core from the housing

To install:
10. Install or connect the following:
- Heater core to the housing
- Heater core cover and the cover-to-housing screws
- Heater dash panel seal
- Lower heater core housing, the upper heater core housing-to-chassis nuts and the heater core housing-to-chassis nut
- Air conditioning evaporator outlet duct clamp screw
- Windshield defroster nozzle connector's screws, the connectors and the pushpins
- Vacuum lines to the retainer at the evaporator housing
- Vacuum control motor vacuum connector

1 A/C Evaporator Housing   4 Heater Core

2 Blower Motor   5 Heater Water Hoses

3 Heater Core Housing

93111GB3

Exploded view of the heater core, heater housing and related components—Escort

- Antenna lead to the heater core housing
11. Install the instrument panel by installing or connecting the following:
  - Instrument panel
  - Main electrical wiring connectors (located on the left side), and install the instrument panel
  - Upper instrument panel reinforcement bolt and the cover
  - Upper and lower instrument panel reinforcement-to-chassis bolts (both sides)
  - Upper and lower instrument panel-to-chassis bolts (both sides)
  - Instrument panels end panels (both sides)
  - Blower motor resistor electrical connector
  - Lower instrument panel reinforcement in-line electrical connector, located on the right side
  - Vacuum line harness connector.
  - Lower instrument panel reinforcement in-line electrical connector, located on the left side
  - Interior fuse junction panel electrical connector
  - Cowl side trim panels and secure the panels with the pushpins, located on both sides
  - Front door scuff plates, located on both sides
  - Steering column and the steering column bracket bolts
  - Steering column shrouds and the shroud screws
  - Light switch rheostat resistor electrical connector and install the instrument panel steering column cover (coupe models only)
  - Instrument panel steering column cover and the cover screw
  - Hood latch control handle and cable and tighten the hood latch control handle nut
  - Heater control cable
  - PCM/bracket, the 2 PCM/bracket nuts and the 2 bolts
  - PCM electrical connector and the bolt
  - Transmission control selector dial bezel screws
  - Radio antenna lead-in cable
  - Radio antenna lead-in cable
  - Control box side covers and secure with the pushpins, located at both sides
  - Power point socket electrical connectors and install the center instrument panel finish panel, if equipped

- Center instrument panel finish panel screw
- Floor console
12. Install the passenger's side SRS module by installing or connecting the following:
  - SRS module and torque the module-to-instrument panel bolts to 62–97 inch lbs. (7–11 Nm)
  - SRS module's electrical connector
  - Glove box door
13. Install the steering wheel by installing or connecting the following:
  - Steering wheel to the steering column
  - New steering wheel bolt and torque it to 26–33 ft. lbs. (34–46 Nm)
  - SRS module-to-steering wheel bolts and torque the bolts to 89–123 inch lbs. (10–14 Nm)
  - SRS module and connect the electrical connector
14. Connect the heater hoses to the heater core.
15. On the Coupe, install the air cleaner outlet tube.
16. Refill the cooling system.
17. Connect the negative battery cable.
18. Operate the engine to normal operating temperatures; then, check the climate control operation and check for leaks.

Cylinder Head

REMOVAL & INSTALLATION

✳✳ WARNING

To reduce the possibility of cylinder head warpage and/or distortion, do not remove the cylinder head while the engine is warm. Always allow the engine to cool entirely before disassembly.

1. Before servicing the vehicle, refer to the Precautions Section.
2. Drain the cooling system.
3. Relieve the fuel system pressure.
4. Remove or disconnect the following:
  - Negative battery cable
  - Air cleaner assembly outlet tube
  - Timing belt
5. Tag and disconnect the vacuum hoses from the following components:
  - Positive Crankcase Ventilation Valve (PCV) valve
  - Throttle body
  - Fuel pressure sensor
  - Intake manifold

6. Remove or disconnect the following:
  - Speed control and accelerator cables from the control lever
  - Fuel charging electrical connectors at the main engine connector
  - Crankshaft Position (CKP) sensor and Heated Oxygen (HO$_2$S) sensor electrical connections
  - Fuel line
  - Power steering pump and bracket, then set them aside with the lines still attached
  - Alternator
  - Engine oil dipstick tube
  - Splash shield bolts and the shield
  - Air conditioning compressor electrical connection
  - Air conditioning compressor retainers and set the compressor aside with the lines still attached
  - Spark plug wires and the plugs
  - Valve cover
  - Upper radiator hose from the thermostat housing water hose connection
  - Heater hose from the thermostat housing
  - Camshafts
  - Ignition coil
  - Thermostat housing
  - Cylinder head bolts
  - Cylinder head and the gasket

To install:
7. Install a new head gasket on the cylinder block.

➡**Always use new bolts when installing the cylinder head.**

8. Lubricate the new cylinder bolts when engine oil.
9. Install the cylinder head and tighten the bolts in the sequence illustrated in the following steps:
  a. Step 1: 15 ft. lbs. (20 Nm)
  b. Step 2: 40 ft. lbs. (30 Nm)
  c. Step 3: Plus 105 degrees
10. Install or connect the following:
  - Thermostat housing
  - Ignition coil and the camshafts
  - Heater and upper radiator hoses
  - Valve cover
  - Spark plugs
  - Spark plug wires
  - Air conditioning compressor and tighten the retaining bolts
  - Air conditioning compressor electrical connection
  - Splash shield and tighten its retainers
  - Engine oil dipstick tube

Item	Description
1	Water Thermostat Housing Oil Ring
2	Water Temperature Indicator Sender Unit
3	Water Thermostat Housing
4	Bolt (3 Req'd)
5	Engine Coolant Temperature Sensor
6	Water Thermostat
7	Water Thermostat Housing O-Ring
8	Water Outlet Connection
9	Bolt (3 Req'd)
10	Oil Separator Hose
11	Crankcase Ventilation Tube
12	Bolt
13	Bolt
14	Support Bracket
15	Bolt (3 Req'd)
16	Bolt (9 Req'd)

Item	Description
17	Bracket
18	Bolt (2 Req'd)
19	Bolt (6 Req'd)
20	Engine Appearance Cover
21	Ignition Wire (4 Req'd)
22	Oil Filler Cap
23	Grommet
24	Valve Cover
25	Valve Cover Gasket
26	Stud Bolt
27	Nut (9 Req'd)
28	Exhaust Manifold
29	Exhaust Manifold Gasket
30	Stud (9 Req'd)
31	Spacer
32	Cylinder Head
33	Ignition Coil Bracket
34	Radio Ignition Interference Capacitor
35	Bolt

9306MG77

Exploded view of the upper engine components—2.0L DOHC engine

Cylinder head torque sequence—2.0L
DOHC engine

Valve cover torque sequence—DOHC
Engine

- Alternator and the power steering
  reservoir and bracket
- Fuel line
- CKP sensor, HO2sensor, and 2
  main fuel charging electrical connections
- Accelerator cable and if equipped,
  the speed control cable to the control lever

11. Connect the vacuum hoses tagged
and removed to the following components:
- PCV valve
- Throttle body
- Fuel pressure sensor
- Intake manifold

12. Install or connect the following:
- Timing belt
- Air cleaner outlet tube
- Negative battery cable

13. Fill the cooling system.

14. Start the engine and check for leaks.

Rocker Arms

REMOVAL & INSTALLATION

This engine is not equipped with rocker
arms. The camshafts act directly on the
valves.

Intake Manifold

REMOVAL & INSTALLATION

1. Before servicing the vehicle, refer to
the Precautions Section.

2. Drain the cooling system.

3. Relieve the fuel system pressure.

4. Remove or disconnect the following:
- Negative battery cable
- Air cleaner outlet tube
- Throttle Position (TP) sensor elec-
  trical connection
- Bolt that secures the pipe located
  by the crankshaft pulley
- Heater hoses from the core
- Main engine control sensor
  wiring
- Connectors from the mounting
  bracket
- Vacuum hoses from the intake
  manifold by squeezing the tabs,
  twisting the hoses and pulling them
  away from the manifold
- Crankcase ventilation hose from the
  valve cover
- Drive belt
- Alternator mounting bolts and
  move the alternator aside
- Fuel lines
- Intake manifold nuts and bolts in
  the sequence illustrated
- Intake manifold and gasket

To install:

5. Clean all dirt and gasket residue from
the intake manifold mating surfaces.

6. Install or connect the following:

Intake manifold torque sequence—2.0L
DOHC engine

- New gasket and place the manifold
  in position
- Manifold bolts and nuts. Tighten in
  the sequence illustrated to 10–12
  ft. lbs. (14–17 Nm).
- Fuel lines
- Alternator
- Drive belt
- Crankcase ventilation hose to the
  valve cover
- Vacuum hoses to the manifold
  making sure the tabs are firmly
  engaged
- Connectors in the mounting bracket
- Main engine control sensor wiring
- Heater hoses to the heater core
- Bolt that secures the pipe located
  by the crankshaft pulley. Tighten
  the bolt to 71–97 inch lbs. (8–11
  Nm).
- TP sensor electrical connection
- Air cleaner outlet tube

7. Fill the cooling system.

8. Connect the negative battery cable.

Exhaust Manifold

REMOVAL & INSTALLATION

1. Before servicing the vehicle, refer to
the Precautions Section.

2. Remove or disconnect the following:
- Negative battery cable
- Fan motor electrical connection
- Fan shroud
- Heated Oxygen (HO2S) sensor elec-
  trical connection
- Catalytic converter-to-exhaust man-
  ifold bolt and nut, then disconnect
  the converter from the manifold

✳✳ WARNING

**Do not break the threadlock seal on
the engine oil dipstick tube. If the
seal is broken, relock it to prevent
any oil leaks from the block.**

Remove the intake manifold bolts and nuts in this sequence—2.0L Zetec engine

Remove the exhaust manifold nuts and bolts in the sequence illustrated—2.0L Zetec engine

- Engine oil dipstick tube bracket bolt
- Exhaust manifold heat shield bolts and nuts, then the shield
- 7 exhaust manifold bolts and 2 studs
- Manifold and gasket

To install:
3. Install or connect the following:
   - New gasket and the exhaust manifold
   - Manifold retaining bolts and nuts and tighten to 13–16 ft. lbs. (14–17 Nm)

14-17 Nm (10-12 lb/ft)

Tighten the exhaust manifold nuts and bolts in the sequence illustrated to the proper specification—2.0L Zetec engine

- Exhaust manifold heat shield and retainers. Tighten the retainers to 71–101 inch lbs. (8–11 Nm).
- Engine oil dipstick tube bracket bolt and tighten to 71–101 inch lbs. (8–11 Nm)
- Exhaust manifold-to-catalytic converter bolt and nut
- HO$_2$S sensor electrical connection
- Fan shroud and the fan motor wiring
- Negative battery cable

Front Crankshaft Seal

REMOVAL & INSTALLATION

1. Before servicing the vehicle, refer to the Precautions Section.
2. Remove the timing belt and crankshaft sprocket.

❄ WARNING

Be careful not to damage the crankshaft surface when removing the seal.

3. Using seal remover tool T92C-6700-CH, remove the crankshaft front oil seal.
   To install:
4. Use seal replacer tool T81P-6700-A, install the new seal.
5. Install the crankshaft sprocket and the timing belt.

Camshaft(s) and Valve Lifters

REMOVAL & INSTALLATION

1. Before servicing the vehicle, refer to the Precautions Section.
2. Remove or disconnect the following:
   - Timing belt
   - Valve cover and camshaft sprockets

➡**It may be necessary to rotate the oil control solenoid flange 90 degrees prior to removal.**

- Oil control solenoid flange bolts and the flange

➡**Mark the camshaft journal caps with a number to identify their location, as they must be replaced in their original position.**

- Camshaft journal caps in several passes in the sequence illustrated, then remove the bolts and caps

1	Bolt—Camshaft Bearing Cap
2	Camshaft Bearing Cap
3	Camshaft Intake
4	Camshaft Exhaust
5	Plug
6	Adjusting Shim—Valve Clearance
7	Valve Tappets
8	Valve Spring Retainer Key
9	Valve Spring Retainers
10	Valve Spring (Color Coding: Exhaust = Blue, Intake = Red)
11	Valve Stem Seal
12	Camshaft Position Sensor
13	CMP Sensor Bolt
14	Intake Valves
15	Exhaust Valves
16	Cylinder Head
17	Cylinder Head Bolts

18	Spark Plug
19	Engine Lifting Eye Bolt
20	Engine Lifting Eye
21	Camshaft Sprocket —Exhaust
22	Exhaust Camshaft Sprocket Bolt
23	Blanking Plug O-Ring
24	Blanking Plug
25	Bolt—Intake Camshaft
26	Camshaft Sprocket Intake
27	Camshaft Front Seal
28	O-Ring
29	Camshaft Front Seal
30	Oil Feed Ring
31	Oil Feed Flange Bolts
32	Oil Feed Flange
33	Guide Sleeve—Front Camshaft Bearing Cap
34	Fifth Camshaft Bearing Cap

9300MG06

Exploded view of the cylinder head, showing camshaft mounting—2.0L DOHC Zetec engine

Loosen the camshaft journal caps in several passes in this sequence—2.0L Zetec engine

7922MG58

The oil control O-ring seals must not fall out of position in the oil control solenoid flange or poor engine performance may occur.

11. Inspect the oil control solenoid flange O-rings for damage or wear and replace as necessary.
- Oil control solenoid flange and tighten the bolts to 84–92 inch lbs. (9.5–10.5 Nm) in the sequence illustrated

12. Rotate the camshafts a full turn and check for binding.
- Camshaft sprockets and timing belt
- Valve cover

9.5-10.5 Nm (84-92 lb/in)

7922MG60

Tighten the oil control solenoid flange bolts in this sequence—2.0L Zetec engine

- Camshafts from the cylinder head
- Oil control sensor and bushing

3. Inspect the camshafts for damage and wear.

To install:

4. Be sure the valve clearance is correct.

5. Install the oil control solenoid bushing and flange on the exhaust camshaft.

6. Coat the surface of the front camshaft journal cap with gasket maker E2AZ-19562-B.

7. Place the camshafts in position and lubricate the bearing surfaces with engine assembly lubricant D9AZ-19579-D.

8. Install new camshaft front oil seals.

9. Apply a thin coat of silicone gasket and sealant F6AZ-19562-AA to the sealing surface of the of the front camshaft journal bearing cap.

10. Install or connect the following:
- Caps and tighten the bolts in several 2-turn passes in the sequence illustrated to 10–12 ft. lbs. (13–17 Nm)

13-17 Nm (10-12 lb/ft)

7922MG59

Tighten the camshaft journal caps in the sequence illustrated to the proper specification—2.0L Zetec engine

Valve Lash

ADJUSTMENT

1. Before servicing the vehicle, refer to the Precautions Section.
2. Remove the valve cover.
3. Remove the timing belt.

➡**Measure each valve's clearance at the base circle before removing the camshafts. The shims are not serviceable with the camshafts in place.**

4. Measure the valve clearances, then remove the camshafts and shims.

The correct shims allow the following valve clearances:

- Intake valve clearance: 0.0043–0.0071 in. (0.11–0.18mm)
- Exhaust valve clearance: 0.0106–0.0134 in. (0.27–0.34mm)

➡**A midrange clearance is the most desirable.**

Measure each valve's clearance at the base circle before removing the camshafts—2.0L Zetec DOHC engine

7922MG62

Example of a valve lifter shim (arrow)— 2.0L Zetec DOHC engine

5. Select the shims using the following formula: shim thickness = measured clearance plus the base shim thickness minus the most desirable thickness.
6. Select the correct shims and mark their installation location.
7. Replace the shims and install the camshafts.
8. Check the new valve clearances.
9. Install the timing belt and covers.
10. Install the valve cover.
11. Start the vehicle and check for proper operation.

Starter Motor

REMOVAL & INSTALLATION

1. Before servicing the vehicle, refer to the Precautions Section.
2. Remove or disconnect the following:
   - Negative battery cable
   - Air cleaner outlet tube
   - Lower starter motor bolt
   - Two starter motor upper bolts
3. Slide the starter motor from its mounting until you can gain access to the wiring.
   - Integral connector
   - Starter motor

To install:

4. Install or connect the following:

- Integral connector and tighten the retaining nuts to 61 inch lbs. (7 Nm)
- Starter motor to its mounting
- Two starter motor upper bolts and tighten to 15–20 ft. lbs. (20–27 Nm)
- Lower starter motor bolt and tighten to 15–20 ft. lbs. (20–27 Nm)
- Air cleaner outlet tube
- Negative battery cable

Oil Pan

REMOVAL & INSTALLATION

1. Before servicing the vehicle, refer to the Precautions Section.
2. Drain the engine oil.
3. Remove or disconnect the following:
   - Catalytic converter
   - 17 oil pan bolts evenly
   - Oil pan

To install:

4. Apply a 0.1 in. (3.0mm) wide bead of silicone sealant F6AZ-19562-AA or its equivalent to the oil pan.
5. Install or connect the following:
   - Oil pan and the bolts. Tighten the oil pan bolts in the sequence illustrated to 15–22 ft. lbs. (20–30 Nm).

Tighten the oil pan bolts in the proper sequence to the correct torque specification—2.0L Zetec engine

- Oil pan drain plug
6. Fill the engine with the proper type and quantity of engine oil.
7. Start the vehicle and check for oil leaks.

Oil Pump

REMOVAL & INSTALLATION

1. Before servicing the vehicle, refer to the Precautions Section.
2. Remove or disconnect the following:
- Timing covers and belt
- Crankshaft pulley, sprocket and the timing chain guide
- Oil pan
- Oil pump cover and screen bolts
- Cover and screen
- Lower cylinder block shims, if necessary
- Lower cylinder block, the gasket and the crankshaft front oil seal, if necessary
- Oil pump retaining bolts
- Oil pump and gasket

To install:

➡**The clearance between the lower cylinder block sealing surfaces on the oil pump and the cylinder block cannot exceed 0.012–0.031 in. (0.3–0.8mm).**

3. Install or connect the following:
- New oil pump gasket and the pump
- Oil pump bolts and tighten to 88–97 inch lbs. (10–11 Nm)
- Crankshaft front oil seal, if removed
- Lower cylinder block and gasket, if removed. Tighten the lower cylinder block bolts to 15–17 ft. lbs. (20–24 Nm) in the sequence illustrated.
- Oil pump screen cover and tube. Tighten the retaining bolts to 71–97 inch lbs. (8–11 Nm).
- Oil pan, timing chain guide, crankshaft sprocket and the pulley
- Timing belt and covers

20-24 Nm
(15-17 lb/ft)

7922MG73

Tighten the lower cylinder block bolts in the sequence illustrated to the proper torque specification—2.0L Zetec engine

4. Fill the engine with the proper type and quantity of engine oil.
5. Start the vehicle and check for oil leaks.

Rear Main Seal

REMOVAL & INSTALLATION

1. Before servicing the vehicle, refer to the Precautions Section.
2. Remove the transaxle and flywheel.

✷✷ WARNING

Be careful not to damage the crankshaft surface when removing the seal.

3. Use seal replacer tool T92C-6700-Ch, to remove the old seal.
To install:
4. Coat the lip of the new seal with clean 5W30 engine oil.
5. Install the new seal using crankshaft rear seal pilot tool T88P-6701-B2 and Rear seal replacer tool T88P-6701-B1.
6. Install the flywheel and the transaxle.

Timing Belt

REMOVAL & INSTALLATION

1. Before servicing the vehicle, refer to the Precautions Section.
2. Remove or disconnect the following:
- Negative battery cable
- Spark plugs

- Catalytic converter
- Right front wheel
- Right inner splash shield
- Accessory drive belt and idler pulley

3. Rotate the crankshaft to Top Dead Center (TDC) and install Crankshaft TDC Timing Peg T97P-6000-A.
4. Rotate the crankshaft clockwise against the peg.
5. Remove or disconnect the following:
- Water pump pulley
- Valve cover
- Crankshaft pulley
- Timing belt covers

6. Install Camshaft Alignment Timing Tool T94P-6256-CH into the slots of both camshafts at the rear of the cylinder head to lock the camshafts into position.
7. Loosen the timing belt tensioner bolt and relieve the tension on the timing belt by disconnecting the tensioner tab from the timing cover back plate.
8. Remove the timing belt.

7922MG72

Remove the oil pump mounting bolts— 2.0L Zetec engine

CRANKSHAFT PULLEY

OIL PAN

ALIGN CRANKSHAFT TIMING MARKS

79225G09

Crankshaft alignment position—Ford 2.0L (VIN 3) engines

Placement of camshaft alignment timing tool T94P-6256-CH—Ford 2.0L (VIN 3) engines

Timing belt tensioner index marks—Ford 2.0L (VIN 3) engines

To install:

➡**Always replace the timing belt after loosening or removing it.**

9. Install the timing belt.
10. Engage the timing belt tensioner tab into the timing cover back plate.
11. Adjust the timing belt tensioner until the index marks are aligned. Tighten the tensioner bolt to 17–20 ft. lbs. (22–28 Nm).

12. Remove the camshaft alignment tool and the crankshaft TDC peg.
13. Install or connect the following:
- Timing belt covers
- Crankshaft pulley. Tighten the bolt to 81–89 ft. lbs. (110–120 Nm).
- Valve cover
- Water pump pulley
- Accessory drive belt and idler pulley
- Right inner splash shield
- Right front wheel
- Catalytic converter
- Spark plugs
- Negative battery cable

14. Start the engine and check for proper operation.

Piston and Ring

POSITIONING

Ford 2.0L (VIN 3) engines—piston ring positioning, end-gap spacing and piston positioning. The small directional arrow must face the front of the engine.

FUEL SYSTEM

Fuel System Service Precautions

Safety is the most important factor when performing not only fuel system maintenance but any type of maintenance. Failure to conduct maintenance and repairs in a safe manner may result in serious personal injury or death. Maintenance and testing of the vehicle's fuel system components can be accomplished safely and effectively by adhering to the following rules and guidelines.

- To avoid the possibility of fire and personal injury, always disconnect the negative battery cable unless the repair or test procedure requires that battery voltage be applied.

- Always relieve the fuel system pressure prior to disconnecting any fuel system component (injector, fuel rail, pressure regulator, etc.), fitting or fuel line connection. Exercise extreme caution whenever relieving fuel system pressure, to avoid exposing skin, face and eyes to fuel spray. Please be advised that fuel under pressure may penetrate the skin or any part of the body that it contacts.

- Always place a shop towel or cloth around the fitting or connection prior to loosening to absorb any excess fuel due to spillage. Ensure that all fuel spillage (should it occur) is quickly removed from engine surfaces. Ensure that all fuel soaked cloths or towels are deposited into a suitable waste container.

- Always keep a dry chemical (Class B) fire extinguisher near the work area.

- Do not allow fuel spray or fuel vapors to come into contact with a spark or open flame.

- Always use a back-up wrench when loosening and tightening fuel line connection fittings. This will prevent unnecessary stress and torsion to fuel line piping.

- Always replace worn fuel fitting O-rings with new. Do not substitute fuel hose or equivalent, where fuel pipe is installed.

Fuel System Pressure

RELIEVING

1. Before servicing the vehicle, refer to the Precautions Section.

2. Remove the Schrader valve cap at the end of the fuel rail and attach a fuel pressure gauge.

3. Open the manual relief valve on the fuel pressure gauge slowly to relive the fuel pressure

Fuel Filter

REMOVAL & INSTALLATION

The inline fuel filter is located in the engine compartment between the fuel tank and the fuel rail.

1. Before servicing the vehicle, refer to the Precautions Section.

2. Relieve the fuel system pressure.

3. Disconnect the negative battery cable.

4. Position a suitable container below the fuel filter to collect any excess fuel that may leak from the filter and lines.

5. Remove or disconnect the following:
- Fuel filter mounting clamp, loosen only
- Filter from the mounting bracket
- Retaining clips from the upper fuel filter hose
- Upper hose from the fuel filter and drain any excess fuel into the container. Plug the hose.
- Retaining clip from the fuel filter lower hose
- Lower hose from the fuel filter and drain any excess fuel into the container. Plug the hose.
- Fuel filter

To install:

6. Remove the plugs from the lower and upper hoses.

7. Install or connect the following:
- Lower hose to the filter
- Hose retaining clip
- Upper hose to the filter
- Hose retaining clip
- Fuel filter and tighten the filter mounting clamp
- negative battery cable

8. Start the engine and check for leaks.

Fuel Pump

The fuel pump is located inside the fuel tank.

REMOVAL & INSTALLATION

This procedure will require a new fuel pump gasket for pump installation, so be sure to have one before starting.

1. Before servicing the vehicle, refer to the Precautions Section.

2. Relieve the fuel system pressure.

3. Remove or disconnect the following:
- Negative battery cable
- Rear seat cushion
- Electrical connectors from the fuel pump
- 4 pump access cover screws and the cover
- Pump electrical connector located under the cover, if equipped
- Fuel line clip(s)
- Fuel line(s) from the pump
- Pump locking retaining ring., using a fuel pump locking ring removal tool or a brass drift and a hammer
- Fuel pump and the gasket from the tank

To install:

4. Install or connect the following:

- New gasket and place the pump into position in the tank
- Pump locking retaining ring
- Fuel line(s) to the pump
- Line retaining clip(s)
- Fuel pump electrical connection(s), if equipped
- Access cover and tighten the retainers
- Pump electrical connector(s)
- Rear seat cushion
- Negative battery cable

5. Start the vehicle and check for proper operation.

Fuel Injector

REMOVAL & INSTALLATION

❊❊ CAUTION

Fuel injection systems remain under pressure, even after the engine has been turned OFF. The fuel system pressure must be relieved before disconnecting any fuel lines. Failure to do so may result in fire and/or personal injury.

1. Before servicing the vehicle, refer to the Precautions Section.

2. Relieve the fuel system pressure.

3. Remove or disconnect the following:
- Negative battery cable
- Fuel injection fuel rail
- Injector retaining clips

4. Grasp the fuel injector body and pull up while gently rocking the fuel injector from side to side.

5. Once removed, inspect the fuel injector cap and body for signs of deterioration. Replace as required.

6. Remove the O-rings and discard. If an O-ring or end cap is missing, look in the intake manifold for the missing part.

7922MG79

Detach the upper and lower hose clips before disconnecting hoses (1) from the fuel filter

90985G07

Grasp the fuel injector's body and pull upward, while gently rocking the fuel injector from side to side

To install:

7. Install or connect the following:
  • New O-rings onto each injector and apply a small amount of clean engine oil to the O-rings

  • Injectors using a slight twisting downward motion
  • Injector retaining clips
  • Fuel injection fuel rail
  • Negative battery cable

8. Run the engine at idle for 2 minutes, then turn the engine **OFF** and check for fuel leaks and proper operation.

DRIVE TRAIN

Transaxle Assembly

REMOVAL & INSTALLATION

Manual

1. Before servicing the vehicle, refer to the Precautions Section.
2. Remove or disconnect the following:
  • Both battery cables, negative cable first
  • Battery and tray
  • Engine air cleaner outlet tube
  • Constant Control Relay Module (CCRM) module and bracket
  • Slave cylinder line from the slave cylinder hose and plug the hose
  • Slave cylinder line retaining clip
  • Slave cylinder line from the bracket
  • Heated Oxygen (HO2S) sensor electrical connection
  • Back-up light electrical connection
  • Electrical connector bracket bolt and the bracket
  • Vehicle Speed Sensor (VSS) electrical connection
  • Halfshafts
3. Install Engine Support Bar D88–6000A, and attach it to the engine lifting eyes with suitable chains or cables.
  • Insulator nuts from the left-hand engine support bracket
  • Front and rear upper transaxle-to-engine bolts
4. Drain the transaxle fluid and install the drain plug.
  • Air conditioning line from the retainer located on the engine support crossmember

Location of the left-hand engine support bracket nuts—2.0L engines

7922MG85

Loosen the transaxle gearshift bolt and remove the gearshift rod and clevis pin from the input shift shaft

  • Engine support crossmember bolts and nuts
  • Crossmember
  • Gearshift stabilizer bar-to-transaxle nut
  • Stabilizer bar and support from the transaxle
  • Transaxle gearshift rod nut
  • Transaxle gearshift bolt
  • Gearshift rod and clevis pin from the input shift shaft
  • Starter motor
  • Lower slave cylinder tube and the slave cylinder
  • Lower transaxle-to-engine bolts

5. Position and secure a suitable transaxle jack under the transaxle.
  • Catalytic converter
  • Middle transaxle-to-engine bolts
  • Transaxle from the vehicle

To install:

6. Apply a thin coating of suitable grease to the splines of the input shaft.
7. Place the transaxle onto a suitable transaxle jack. Be sure the transaxle is secure.
8. Install or connect the following:
  • Transaxle
  • Middle transaxle-to-engine bolts and tighten them to 23–38 ft. lbs. (38–51 Nm)
  • Lower engine-to-transaxle bolts and tighten them to 23–38 ft. lbs. (38–51 Nm)
  • clutch slave cylinder and tighten the nuts to 12–17 ft. lbs. (16–23 Nm)
  • Lower slave cylinder tube and tighten the fitting to 10–16 ft. lbs. (13–21 Nm)
  • Starter motor

  • Gearshift rod and clevis into position on the input shift shaft
  • Gearshift bolt and gearshift rod nut and tighten to 12–17 ft. lbs. (16–23 Nm)
  • Gearshift stabilizer bar and tighten the nut to 23–34 ft. lbs. (31–46 Nm)
  • Halfshafts and the catalytic converter
  • Engine support crossmember. Tighten the crossmember bolts and nuts to 47–65 ft. lbs. (64–89 Nm) and the insulator nuts to 28–37 ft. lbs. (38–51 Nm).
  • Front and rear upper transaxle-to-engine bolt and tighten it to 28–38 ft. lbs. (38–51 Nm)
  • Left-hand engine support insulator nuts and tighten them to 50–68 ft. lbs. (67–93 Nm)
9. Remove the engine support bar.
  • Electrical connector support bracket
  • All the electrical connectors
  • Slave cylinder line to the slave cylinder hose and attach the retaining clip. Tighten the fitting to 10–16 ft. lbs. (13–21 Nm).
10. Bleed the clutch hydraulic system.
11. Add the proper type and amount of fluid to the transaxle.
12. Install or connect the following:
  • CCRM and bracket assembly, if removed
  • CCRM retainers and attach the electrical connection
  • Engine air cleaner tube
  • Battery tray and the battery
  • Battery cables, negative cable last
13. Check for fluid leaks and proper clutch operation.
14. Road test the vehicle and check for proper transaxle operation.

Automatic

1. Before servicing the vehicle, refer to the Precautions Section.
2. Remove or disconnect the following:
  • Battery cables, negative cable first
  • Battery and tray
  • Engine air cleaner assembly
  • Computer Control Relay Module

Install the engine support crossmember and tighten the retainers to specification

(CCRM) electrical connections, relay retainers, the relay and bracket from the engine

- Shift control cable and bracket retaining nut from the manual shift lever
- Shift cable and bracket clip, then set the bracket and cable assembly aside
- All electrical connections from the transaxle
- Starter motor
- Throttle valve actuating cable bolts and disconnect the cable from the throttle cam, if equipped

3. Install Engine Support D88L-6000-A, to the engine. The engine must be properly supported for transaxle removal.

4. Remove or disconnect the following:

- Transaxle cooler lines at the transaxle
- Left-hand engine mount
- Upper transaxle housing bolts
- Left-hand splash shields, if equipped
- Transaxle plug and drain the fluid
- Halfshafts. Install 2 transaxle plugs T88C-7025-AH, into the differential side gears

✱✱ WARNING

Failure to install the transaxle plugs may cause the differential side gears to become improperly positioned. If the gears become misaligned, the differential will have to be removed from the transaxle to align them.

- Engine support crossmember and the catalytic converter
- Air conditioning line from the retainer located on the engine support crossmember
- Transaxle support crossmember bolts and nuts, the left-hand engine isolator nuts, then the crossmember
- Transaxle housing cover bolts and the cover
- 4 flywheel-to-torque converter nuts

5. Position a suitable transaxle jack under the transaxle. Secure the transaxle to the jack.

- Remaining transaxle-to-engine bolts

6. Slowly lower the transaxle out of the vehicle.

7. Inspect all components including mounts and brackets.

To install:

➡**Prior to installing the transaxle, lubricate the torque converter pilot hub with multi-purpose grease.**

8. Align the torque converter studs to the flywheel.

9. Secure the transaxle on the transaxle jack.

10. Install or connect the following:

- Transaxle. Tighten the bolts to 40–58 ft. lbs. (55–80 Nm).
- Flywheel-to-converter nuts and tighten them to 26–36 ft. lbs. (35–49 Nm)
- Transaxle housing cover and

tighten the bolts to 40–58 ft. lbs. (55–80 Nm)

- Engine support crossmember. Tighten the crossmember bolts and nuts to 47–65 ft. lbs. (64–89 Nm) and the insulator nuts to 28–37 ft. lbs. (38–51 Nm).
- Front and rear upper transaxle-to-engine bolt and tighten it to 28–38 ft. lbs. (38–51 Nm)

11. Install or connect the following:

- Halfshafts
- Crossmember to the transaxle mounts and the chassis. Tighten the crossmember-to-transaxle mount nuts to 27–38 ft. lbs. (37–52 Nm). Tighten the crossmember-to-chassis nuts and bolts to 47–66 ft. lbs. (64–89 Nm).
- Lower transaxle-to-engine oil pan bolts and tighten to 27–38 ft. lbs. (37–52 Nm)
- Engine/transaxle splash shields
- Starter motor
- All the electrical connections
- Upper transaxle-to-engine bolts and tighten to 40–58 ft. lbs. (55–80 Nm)
- Left-hand engine mount and tighten the nuts to 50–68 ft. lbs. (67–93 Nm)
- Throttle valve cable at the throttle cam, if equipped

12. Remove the engine support.

- Shift cable and bracket to the manual shift lever, the shift cable and

Connect the throttle valve cable at the throttle cam (B) and tighten the cable retaining bolts (A)

bracket clip. Tighten the nut to 12–16 ft. lbs. (16–22 Nm).
- Engine air cleaner assembly
- Battery tray and battery
- Both battery cables, negative cable last

13. Add the proper type and quantity of transaxle fluid.

14. Check the transaxle for leaks and for proper operation.

15. Check the MLP switch for proper adjustment, as follows:

a. Shift the transaxle into NEUTRAL, then align the marks on the transmission range switch and the manual control lever.

b. Install and finger-tighten the switch retaining bolts.

c. Attach an ohmmeter between terminals **A** and **B** on the switch as shown in the accompanying illustration.

d. Adjust the switch by rotating the switch housing on the manual control lever until there is no continuity between the terminals.

e. Hold the switch in place, then tighten its retaining bolts to 70–95 inch lbs. (8–11 Nm).

f. Remove the ohmmeter and attach the switch electrical connection.

Align the marks on the MLP switch and the manual control lever as shown

Attach an ohmmeter to terminals A and B on the transmission range switch

g. Place the manual control shift outer lever in position and tighten its retaining nut to 33–47 ft. lbs. (44–64 Nm).

16. Road test the vehicle and check for proper operation.

Clutch

ADJUSTMENT

Pedal Free-Play

1. Before servicing the vehicle, refer to the Precautions Section.

2. Depress the clutch pedal until resistance can be felt, and measure the distance between the upper clutch height and where the resistance is felt. The free-play should be 0.0–0.40 in. (5.0–13.9mm).

3. If an adjustment is necessary, turn the locknut and equalizer bar-to-clutch release lever rod.

4. After adjustment, measure the disengagement height from the upper surface of the clutch pedal pad to the carpet. The distance should be 2.3 in. (59mm).

Clutch pedal free play measurement

Clutch pedal free play adjustment

Pedal Height

1. Measure the distance from the upper surface of the pedal pad to the carpet. The measurement should be 8.35–8.54 in. (212–217mm).

2. If an adjustment is necessary, turn the locknut and the Clutch Pedal Position (CPP) switch until the pedal height is correct.

REMOVAL & INSTALLATION

1. Before servicing the vehicle, refer to the Precautions Section.

2. Disconnect the negative battery cable.

Clutch pedal height adjustment

3. Raise and safely support the vehicle.

4. Remove the transaxle assembly.

5. If the clutch assembly is to be reused, matchmark the pressure plate and the flywheel so they can be assembled in the same position.

6. Install a flywheel holding tool in a transaxle mounting hole on the engine and engage the tooth of the holding tool into the flywheel ring gear.

7. Install a clutch alignment tool.

8. Loosen the pressure plate-to-flywheel retaining bolts 1 turn at a time, in a crisscross pattern, until the spring tension is relieved, to prevent pressure plate damage.

❊❊ WARNING

Do not use any cleaners with a petroleum base and do not immerse the clutch pressure plate in solvent.

9. Clean the clutch pressure plate with a suitable commercial alcohol base solvent.

10. Inspect the pressure plate surface for burns, scores, flatness or ridges. Reface or replace the pressure plate.

11. Inspect the pressure plate diaphragm fingers for wear. Replace the pressure plate if necessary.

12. Using a slide caliper, measure the depth of the rivet heads. If the rivet head is within 0.012 in. (0.3mm) from the clutch surface, replace the clutch.

➡**Use emery cloth to remove minor imperfections from the clutch disc lining surface.**

13. Inspect the clutch disc for the following:

- Oil or grease saturation
- Worn or loose facings
- Warpage or loose rivets at the hub
- Loose or broken torsion dampening springs
- Wear or rust on the splines.

14. If the clutch disc shows any of these conditions, it should be replaced.

15. Use a dial indicator mounted on a metal base to measure the clutch disc run-out. If the run-out exceeds 0.0276 in. (0.700mm), replace the disc.

16. Inspect the clutch release for distortion, cracks, excessive release bearing surface wear or damaged tines and replace as necessary.

To install:

17. Clean the pressure plate and flywheel surfaces thoroughly. Position the clutch disc and pressure plate into the installed position and support them with a

Tighten the pressure plate-to-flywheel retaining bolts in the sequence illustrated to specification

clutch aligning tool. If the clutch assembly is being reused, align the matchmarks that were made during the removal procedure.

18. Install the pressure plate-to-flywheel retaining bolts. Tighten the bolts in the correct sequence to 12–24 ft. lbs. (16–32 Nm). Remove the alignment tool.

19. Install the transaxle assembly.

20. Lower the vehicle.

21. Bleed the hydraulic clutch system.

22. Adjust the clutch pedal free-play.

23. Connect the negative battery cable.

24. Road test the vehicle and check the clutch for proper operation.

Hydraulic Clutch System

BLEEDING

1. Before servicing the vehicle, refer to the Precautions Section.

2. Check that the brake master cylinder is at least ¾ full during the entire bleeding process.

3. Remove the bleeder screw cap from the clutch slave cylinder and attach a hose to the bleeder screw. Place the other end of the hose into a container to catch the fluid.

4. Have an assistant slowly pump the clutch pedal several times, then hold the clutch pedal down.

5. Loosen the bleeder screw to release the fluid and air. Tighten the bleeder screw.

6. Repeat the bleeding procedure until no more air bubbles are seen in the fluid.

7. Tighten the bleeder screw to 52–78 inch lbs. (6–9 Nm).

8. Top off the brake master cylinder to the full line.

9. Check for proper clutch system operation.

Halfshafts

REMOVAL & INSTALLATION

1. Before servicing the vehicle, refer to the Precautions Section.

2. Disconnect the negative battery cable.

3. With the vehicle sitting on the ground, carefully raise the staked portion of the halfshaft retaining nut using a suitable small chisel. Loosen the nut.

4. Remove or disconnect the following:

- Wheel and tire assembly
- Halfshaft retaining nut and discard
- Cotter pin and nut from the tie rod end, then separate the tie rod end from the steering knuckle using a suitable removal tool. Discard the cotter pin.

5. Remove or disconnect the following components to remove the stabilizer bar link:

- Stabilizer bar end nut and bolt
- Stabilizer bar end retainer
- End bushing above the stabilizer bar
- End bushing below the stabilizer bar
- Stabilizer bar spacer
- Stabilizer bar end bushings from above and below the sub-frame
- Lower stabilizer bar end retainer
- Stabilizer bar

6. Remove the ball joint bolt and nut. Carefully pry down on the lower control arm to separate the ball joint stud from the steering knuckle.

Item	Description
1	Clutch Slave Cylinder
2	Clutch Release Hub and Bearing
3	Clutch Pressure Plate
4	Clutch Disc
5	Pilot Bearing
6	Flywheel

Item	Description
7	Clutch Release Fork
8	Brake Master Cylinder Reservoir
9	Clutch Pedal
10	Clutch Pedal Position (CPP) Switch
11	Clutch Master Cylinder

9306MG78

Exploded view of the clutch system components

7. Pull outward on the steering knuckle/brake assembly. Carefully pull the halfshaft from the hub and position it aside.

➡**Removal of the left side halfshaft requires removal of the crossmember to allow access with a prybar.**

8. Support the transaxle with a suitable transaxle jack.

9. Unfasten the 4 transaxle crossmember retainers and remove the crossmember.

10. If removing the right side halfshaft, remove the right-hand shield and splash shield.

11. Position a drain pan under the transaxle.

12. Remove the left-hand halfshaft as follows:

a. Insert a prybar between the halfshaft and the transaxle case. Gently pry outward to release the halfshaft from the differential side gear. Be careful not to damage the transaxle case, oil seal, CV-joint or CV-joint boot.

b. Remove the halfshaft.

13. Remove the right-hand halfshaft as follows:

a. On models with a manual transaxle, unfasten the center support bearing bolts.

b. Lower the halfshaft and remove it from the transaxle.

c. Separate the halfshaft from the center support bearing and remove the halfshaft from the vehicle.

d. Inspect the center support bearing for damage and replace as necessary.

➡**Install suitable plugs after removing the halfshafts to prevent the differential side gears from moving out of place. Should the gears become misaligned, the differential will have to be removed from the transaxle to align the gears.**

To install:

➡**Use new locknuts, split pins and circlips for assembly.**

On models with a manual transaxle, unfasten the center support bearing bolts

---

During assembly, be sure to install a new circlip on the inner CV-joint spline—2.0L engine

14. Position a new circlip on the inner CV-joint spline so the circlip gap is at the top. Lubricate the splines lightly with a suitable grease.

15. Remove the plugs that were installed in the differential side gears.

16. Position the halfshaft so the CV-joint splines are aligned with the differential side gear splines. Push the halfshaft into the differential.

➡**When seated properly, the circlip can be felt snapping into the differential side gear groove.**

17. Install the right-hand halfshaft on models with a manual transaxle as follows:

a. Position the halfshaft and joint so that the splines line up with the splines in the halfshaft, and push the halfshaft, joint and halfshaft together with the center support bearing.

b. Install the halfshaft in the transaxle.

18. Place the center support bearing into position and tighten it retaining bolts to 32–46 ft. lbs. (46–62 Nm).

a. Install the right-hand shield and splash shield.

19. Pull outward on the steering knuckle/brake assembly and insert the halfshaft into the hub.

20. Pry downward on the lower control arm and position the lower ball joint stud in the steering knuckle.

21. Install the crossmember and the crossmember-to-frame bolts. Tighten the bolts to 69–93 ft. lbs. (94–126 Nm).

22. Remove the transaxle jack.

23. Install or connect the following:

- Steering knuckle and ball joint nut and bolt. Tighten the nut and bolt to 32–43 ft. lbs. (43–59 Nm).
- Stabilizer bar link in reverse order of removal. Tighten the bar end nut until the protruding bar end bolt length is 0.67–75 in. (17–19mm).
- Tie rod end to the steering knuckle
- Tie rod end nut and tighten to 32–41 ft. lbs. (43–56 Nm). Install a new cotter pin.

- Wheel and tire assembly
- New halfshaft retaining nut and tighten to 174–235 ft. lbs. (235–319 Nm). Stake the halfshaft retaining nut using a suitable chisel with a rounded cutting edge.

24. Check and refill the transaxle with the proper type and quantity of fluid.

25. Connect the negative battery cable.

26. Road test the vehicle and check for proper operation.

CV-Joints

OVERHAUL

Outboard Side

1. Before servicing the vehicle, refer to the Precautions Section.

2. Remove or disconnect the following:

- Halfshaft from the vehicle. Support the assembly in a vise with soft jaws.
- Front brake anti-lock sensor indicator off the halfshaft using a punch to drive it off
- Two halfshaft boot clamp and discard

3. Slide the boot back out of the way to access the outboard joint.

4. Matchmark the joint-to-halfshaft for reassembly.

5. Use a soft faced mallet to separate the outboard joint by gently tapping it off the halfshaft.

6. Remove or disconnect the following:

- Halfshaft bearing retaining circlip and discard
- snapring from the outboard side of the halfshaft

➡**Do not remove the tape until after installation of the boot.**

7. Wrap the outboard halfshaft splines with tape and then slide the boot from the shaft.

8. Clean and inspect the outboard bearings for damage, grit in the grease, pitting or cracks and replace as necessary.

If equipped with ABS, use a punch to drive the front brake anti-lock sensor indicator off the halfshaft

Clean and inspect the outboard bearings for damage

To install:

9. Lubricate the joint bearings with constant velocity grease E43Z-19590-A.

10. Install or connect the following:
- Boot and remove the tape
- Snapring on the outboard side of the halfshaft
- New bearing retaining circlip

11. Use a soft faced mallet to install the outboard joint by gently tapping it onto the halfshaft.

12. Remove any excess grease from the mating surfaces and slide the boot forward onto the joint.

13. Insert a suitable cloth covered tool between the boot and the outer bearing race to allow trapped air to escape from the boot.

14. Use boot clamp pliers to install two new boot clamps.

15. If equipped with anti-lock brakes, use sensing ring replacer tool T94P-20202-B, to install the front brake anti-lock sensor indicator onto the halfshaft.

16. Install the halfshaft.

INBOARD SIDE

1. Before servicing the vehicle, refer to the Precautions Section.

Use boot clamp pliers to install new boot clamps

Use sensing ring replacer tool T94P-20202-B to install the front brake anti-lock sensor indicator onto the halfshaft

2. Remove or disconnect the following:
- Halfshaft from the vehicle. Support the assembly in a vise with soft jaws.

➡ **The right-hand side halfshaft on models equipped with a manual transaxle do not have an inboard circlip.**

- Bearing retaining circlip from the inboard joint housing and discard
- Joint boot clamps and discard
- Joint boot from the joint housing
- Joint housing from the tripod bearing and halfshaft
- Tripod bearing snapring

3. Matchmark the tripod bearing-to-halfshaft for reassembly.

4. Remove the tripod bearing from the halfshaft.

➡ **Do not remove the tape until after installation of the boot.**

5. Wrap the halfshaft splines with tape and then slide the boot from the shaft.

6. Clean and inspect the tripod bearing assembly and outboard joint housing for damage, grit in the grease, pitting or cracks and replace as necessary.

To install:

7. Lubricate the tripod bearing and inboard joint housing with constant velocity grease E43Z-19590-A.

8. Install or connect the following:
- Boot and remove the tape
- Tripod bearing onto the halfshaft
- Tripod bearing snapring

Remove and discard the bearing retaining circlip from the inboard joint housing

Location of the tripod bearing snapring

Clean and inspect the tripod bearing assembly . . .

. . . and outboard joint housing for damage, grit in the grease, pitting or cracks, and replace as necessary

- Inboard joint housing onto the tripod bearing
- Boot onto the inboard joint housing

9. Insert a suitable cloth covered tool between the boot and the bearing to allow trapped air to escape from the boot.

10. Use boot clamp pliers to install two new boot clamps.

➡ **Measure the relaxed state of the halfshaft bearing circlip installed on the shaft. Refer to the accompanying illustration. The wrong size clip will not retain the halfshaft properly.**

11. Install a new halfshaft bearing retainer circlip of the correct size onto the outboard halfshaft joint.

12. Install the halfshaft.

A/T: 27 mm (1.063 in)
M/T: 28 mm (1.102 in)

Measure the relaxed state of the halfshaft bearing circlip installed on the shaft, and make sure it is within specification

STEERING AND SUSPENSION

Air Bag

PRECAUTIONS

Several precautions must be observed when handling the inflator module to avoid accidental deployment and possible personal injury.

- Never carry the inflator module by the wires or connector on the underside of the module.
- When carrying a live inflator module, hold securely with both hands, and ensure that the bag and trim cover are pointed away.
- Place the inflator module on a bench or other surface with the bag and trim cover facing up.
- With the inflator module on the bench, never place anything on or close to the module that may be thrown in the event of an accidental deployment.

DISARMING

1. Before servicing the vehicle, refer to the Precautions Section.
2. Disconnect both battery cables, negative cable first.
3. Wait at least 1 minute. This allows time for the back-up power supply to deplete its stored energy.
4. Remove the driver's side air bag module, then the passenger side if required.
5. Use caution when carrying live air bags. Always place the air bag with the cover up.
6. If the battery needs to be reconnected while one or both of the air bags are removed from the system, install Air Bag Simulator 105–00010, to the drivers side and/or passenger side air bag harness connectors as required. Before removing either air bag simulator, disconnect both battery cables and wait at least 1 minute before continuing.

ARMING

1. Once service is completed and the air bag modules are back in place, connect the negative battery cable and prove out the air bag system by turning the ignition key to the **RUN** position and visually monitoring the air bag indicator lamp in the instrument cluster. The indicator lamp should illuminate for approximately 6 seconds, then turn **OFF**. If the indicator lamp does not illuminate,

stays **ON**, or flashes at any time, a fault has been detected by the air bag diagnostic monitor requiring immediate attention.

Rack and Pinion Steering Gear

REMOVAL & INSTALLATION

1. Before servicing the vehicle, refer to the Precautions Section.
2. Turn the key to the **ACC** position.
3. Remove or disconnect the following:
   - Steering column tube boot nuts at the base of the column and the tube boots, from inside the passenger compartment
   - Steering column input shaft coupling-to-steering gear input shaft pinch bolt
   - Front wheel and tire assemblies
   - Separate the tie rod ends from the steering knuckles and discard the cotter pins
   - Right-hand lower splash shield
   - Crossmember
   - Pressure and return lines from the

Unfasten the steering column tube boot nuts at the base of the column and remove the tube boots

rack and pinion assembly and plug the lines
   - Strap holding the hoses to the steering gear and discard
   - Gearshift rod and clevis from the transaxle, if equipped with a manual transaxle
   - Extension bar nut and disconnect the gearshift lever stabilizer bar and support from the transaxle, if equipped with a manual transaxle
   - Retaining nuts from the rack and pinion mounting brackets
   - Pinion mounting brackets
   - Pushpin and position the right-hand boot aside
   - Rack and pinion assembly from the right-hand side of the vehicle

To install:
4. Install or connect the following:
   - Rack and pinion assembly in its mounting location
5. Align the steering column input shaft coupling and the steering gear input shaft.
   - Rack and pinion mounting brackets. Tighten the retaining nuts to 28–38 ft. lbs. (37–57 Nm).
   - Gearshift stabilizer bar, support and the gearshift rod and clevis, if equipped with a manual transaxle

➡**Install new Teflon seals on the power steering pressure hose fitting and be sure the threads are clean before connecting the hose.**

   - Plugs and connect the pressure and return lines to the rack and pinion assembly. Tighten the pressure hose to 21–25 ft. lbs. (28–33 Nm) and the return line fitting to 20–25 ft. lbs. (27.3–33.9 Nm).
   - New strap to hold the power steering lines to the rack and pinion housing

Unfasten the retaining nuts from the rack and pinion mounting brackets and remove the brackets

- Crossmember and tighten the bolts to 69–97 ft. lbs. (94–131 Nm)
- Right-hand boot shield into position
- Shield pushpin
- Right-hand splash shield
- Tie rod ends to the steering knuckles and the castellated nuts. Tighten to specification.
- New cotter pins
- Wheel and tire assemblies
- Steering column input shaft coupling-to-steering gear input shaft pinch bolt and tighten the bolt to 30–36 ft. lbs. (40–50 Nm)
- Steering column tube boots and the 5 retainers. Tighten the retainers to 18–52 inch lbs. (2–5.9 Nm).
- Negative battery cable

6. Fill and bleed the power steering system.

7. Check the alignment and adjust as required.

8. Start the engine and check for leaks.

9. Road test the vehicle and check for proper steering system operation.

Strut

REMOVAL & INSTALLATION

Front

1. Before servicing the vehicle, refer to the Precautions Section.

2. Remove or disconnect the following:
- Negative battery cable
- Front wheel and tire assembly
- Clip securing the brake hose to the strut (spring and shock) assembly.
- Anti-lock brake harness cable and clip, if equipped with anti-lock brakes
- 2 nuts and 2 bolts securing the strut assembly to the steering knuckle
- 4 upper strut retaining nuts
- Strut assembly from the vehicle

✱✱ CAUTION

Never remove the strut piston rod nut unless the coil spring is compressed. Always wear safety glasses when using a spring compressor.

3. Inspect all components and replace as needed.

To install:

4. Install or connect the following:
- Strut assembly into the wheel housing. Be sure the direction indicator on the upper mounting bracket faces inboard.
- Upper mounting bracket to the strut tower with the retaining nuts. Tighten the nuts to 22–30 ft. lbs. (29–40 Nm).
- Strut assembly to the steering knuckle and the retaining bolts and nuts. Tighten to 69–93 ft. lbs. (93–127 Nm).
- Brake hose on the strut assembly and secure it with the brake hose clip
- Anti-lock harness cable and clip, if equipped with anti-lock brakes
- Front wheel and tire assembly
- Negative battery cable

5. Check the front wheel alignment.

6. Road test the vehicle and check for proper operation.

Rear

1. Before servicing the vehicle, refer to the Precautions Section.

1	Front Shock Absorber Upper Mounting Bracket
2	Upper Spring Seat (Part of 18198)
3	Front Brake Hose
4	Disc Brake Caliper
5	Front Disc Brake Rotor
6	Front Wheel Knuckle
7	Tie Rod End
8	Front Stabilizer Bar

9	Front Suspension Lower Arm Mounting Bolt Bushing (Rear)
10	Front Wheel Spindle Tie Rod
11	Front Suspension Lower Arm
12	Front Suspension Lower Arm Mounting Bolt Bushing (Front)
13	Front Shock Absorber
14	Front Coil Spring

Identification of the front suspension components

9300MG04

Remove the 2 nuts and 2 bolts securing the strut assembly to the steering knuckle

2. Remove or disconnect the following:
   - Negative battery cable
   - Package tray trim panel, on sedan and coupe models
   - Quarter trim pane, on wagon models
   - 2 upper strut retaining nuts
   - Wheel and tire assembly
   - Clip securing the brake hose to the rear strut assembly
   - ABS sensor bolt, if equipped
   - Nuts and bolts securing the rear strut assembly to the wheel spindle
   - strut assembly from the vehicle

To install:

3. Install or connect the following:
   - Strut assembly into the vehicle wheel housing
   - Nuts and bolts securing the strut assembly to the rear wheel spindle

Location of the 2 rear strut upper retaining nuts (arrows) inside the vehicle

assembly. Tighten the lower strut bolts and nuts to 76–100 ft. lbs. (103–136 Nm).
   - ABS sensor bolt, if equipped
   - Clip securing the flexible brake hose to the rear strut assembly
   - Wheel and tire assembly
   - 2 upper strut retaining nuts and

tighten to 34–46 ft. lbs. (47–62 Nm)
   - Trim panel
   - Negative battery cable

4. Check the rear wheel alignment.
5. Road test the vehicle and check for proper operation.

Coil Spring

REMOVAL & INSTALLATION

Front

1. Before servicing the vehicle, refer to the Precautions Section.
2. Remove the strut assembly.
3. Place the strut assembly in a vise.
4. Install a suitable spring compressor tool onto the coil spring and compress the spring.
5. Unfasten the piston rod nut and remove the upper mounting bracket.
6. Remove the bound stopper, dust boot, rubber spring seat, spring, upper spring seat and thrust bearing.

To install:

7. Install the thrust bearing, upper spring seat, spring, rubber spring seat, dust boot and the bound stopper.
8. Install the piston rod nut and tighten it to 58–81 ft. lbs. (79–110 Nm).
9. After the piston rod nut has been tightened to specification, carefully remove the compressor tool from the spring while making sure the spring is properly seated in the upper and lower spring seats.

1	Rear Floor Cross Member
2	Rear Shock Absorber Insulator
3	Rear Spring
4	Rear Shock Absorber
5	Rear Wheel Spindle
6	Rear Suspension Tie Rod and Bushing
7	Rear Suspension Arm and Bushing (Rear)
8	Rear Stabilizer Bar
9	Rear Suspension Arm and Bushing (Front)
10	Rear Wheel Brake Hose

Identification of the rear suspension components

10. Install the strut assembly in the vehicle.

11. Have the vehicle alignment checked and if necessary, adjusted.

Rear

1. Remove the strut assembly.
2. Place the strut assembly in a vise.

✳✳ CAUTION

Always take the necessary precautions when using a spring compressor

3. Install a suitable spring compressor tool onto the coil spring and compress the spring.

4. Remove or disconnect the following:
- Rear strut assembly top mounting cover
- Piston rod nut and the retainer
- Strut insulator
- Spring compressor
- Rear strut dust boot and stopper seat
- Spring and the rear strut insulator

To install:

5. Place the strut assembly in a vise.
6. Install or connect the following:
- Strut insulator and spring
- Stopper seat and dust boot

7. Use the spring compressor tool to compress the strut spring.
- Strut insulator and the retainer
- Piston rod nut and tighten it to 41–49 ft. lbs. (55–67 Nm)
- Top mounting cover

8. Be sure the spring is properly aligned and carefully release the spring into the seats of the strut.

9. Remove the spring compressor from the coil spring.

10. Install the strut assembly.

Lower Ball Joint

REMOVAL & INSTALLATION

1. Before servicing the vehicle, refer to the Precautions Section.
2. Remove or disconnect the following:
- Wheel assembly
- Nut and bolt attaching the lower ball joint to the steering knuckle
- Ball joint from the steering knuckle
- Lower control arm ball joint nuts and bolt
- Ball joint

To install:

3. Install or connect the following:
- Ball joint into position on the lower control arm
- Ball joint-to-lower control arm retainers. Tighten the nuts and bolt to 69–86 ft. lbs. (93–117 Nm).

4. Apply Loctite® 290 thread locking compound to the ball joint nut and t hreads.
- Ball joint to the steering knuckle
- Ball joint nut and bolt. Tighten the nut and bolt to 32–43 ft. lbs. (43–59 Nm).
- Wheel assembly
- Check and adjust the wheel alignment as necessary

Control Arm

REMOVAL & INSTALLATION

Front

1. Before servicing the vehicle, refer to the Precautions Section.
2. Remove or disconnect the following:
- Front wheel and tire assembly
- Stabilizer bar link nuts, retainers, bushings, bolts and sleeves
- Pinch bolt and nut securing the ball joint to the steering knuckle
- Lower ball joint from the steering knuckle by prying the lower control arm down with a prybar
- Front lower control arm pivot bolt
- Three lower control arm retaining bolts and the lower control arm

3. Inspect the lower control arm, lower control arm bushings and the lower ball joint. The ball joint and bushings can be replaced individually.

To install:

4. Install or connect the following:
- Lower control arm into position at the bushings

Unfasten of the lower control arm pivot bolt (arrow)

Engage the lower ball joint to the steering knuckle and tighten the pinch bolt and nut to specification

Tighten the lower control arm retaining bolts in the sequence shown to the proper specification

- Lower ball joint to the steering knuckle and tighten the pinch bolt and nut to 32–43 ft. lbs. (43–59 Nm)

5. Use a transmission jack to raise the lower control arm so that it is parallel with the ground.
- Control arm retaining bolts and tighten them to 69–86 ft. lbs. (93–117 Nm) in the order shown in the accompanying illustration
- Control arm pivot bolt and tighten to 69–86 ft. lbs. (93–117 Nm)
- Stabilizer bolts, washers, bushings, sleeves and nuts. Tighten the stabilizer nuts so 0.67–0.75 inches (17–19mm) of thread is exposed at the end of the bolt.
- Wheel and tire assembly

6. Check the front wheel alignment.
7. Road test the vehicle and check for proper operation.

CONTROL ARM BUSHING REPLACEMENT

1. Before servicing the vehicle, refer to the Precautions Section.
2. Remove or disconnect the following:
- Control arm
- Nut and control arm mounting bushing (rear) and washer
- Front bushing from the arm using a

suitable control arm bushing tool, C-frame and clamp assembly
- Front bushing into the arm using a suitable control arm bushing tool, C-frame and clamp assembly
- Washer, control arm mounting bushing (rear) and nut. Tighten the nut to 69 ft. lbs. (93 Nm).
- Control arm

Rear

1. Before servicing the vehicle, refer to the Precautions Section.
2. Position a floor jackstand beneath the rear suspension crossmember.
3. Remove or disconnect the following:
   - Wheel and tire assembly
   - Stabilizer nuts, washers, bushings, sleeves and bolts
   - Bolts securing the stabilizer bar brackets and grommets to the rear suspension crossmember
   - Stabilizer bar
   - Bolts securing the rear suspension crossmember to the vehicle frame
4. Lower the floor jackstand to allow the rear suspension crossmember to be lowered from the vehicle frame.
   - Control arm and bushing nut
   - Control arm and bushing bolt
   - Control arm

To install:

5. Install or connect the following:
   - Control arm into position

Support the rear suspension crossmember (B) with a floor jackstand (A)

Unfasten the rear suspension crossmember bolts, then lower the floor jackstand and crossmember

- Retaining bolt and tighten it to 50–70 ft. lbs. (68–95 Nm)
- Control arm and bushing retaining nut, then tighten the nut to 64–86 ft. lbs. (87–116 Nm)
6. Raise the floor jack and place the crossmember into position.
   - Crossmember bolts and tighten them to 34–46 ft. lbs. (47–62 Nm)
   - Grommets onto the stabilizer bar and align the grommets to the positions painted on the bar
   - Stabilizer bar to the rear suspension crossmember and secure it in place with the straps and bolts. Tighten the bolts to 32–43 ft. lbs. (43–59 Nm).
   - Stabilizer bolts, washers, grommets, sleeves and nuts. Tighten the stabilizer nuts so 0.64–0.71 in. (16.2–18.2mm) of thread is exposed at the end of the bolt.
   - Wheel and tire assembly
7. Check the wheel alignment.

Wheel Bearings

ADJUSTMENT

The bearings on the front and rear wheels are a one piece cartridge design and cannot be adjusted. Wheel bearing play can be checked with a dial indicator. If wheel bearing play exceeds 0.002 in. (0.05mm) check the wheel hub retainer nut for proper torque. If the torque is correct, replacement of the wheel bearing is required.

REMOVAL & INSTALLATION

Front

1. Before servicing the vehicle, refer to the Precautions Section.
2. With the vehicle sitting on the ground, carefully raise the staked portion of the halfshaft retaining nut using a suitable small chisel. Loosen the nut.
3. Remove or disconnect the following:
   - Wheel and tire assembly
   - Brake caliper and secure it out of the way with a piece of mechanic's wire. Do not let the caliper hang on the hose.
   - Brake rotor
   - Halfshaft retaining nut and discard it
   - Cotter pin and castellated nut from the tie rod end and separate the tie rod end from the steering knuckle

using a suitable removal tool. Discard the cotter pin.
- Separate the tie rod end from the wheel knuckle
- ABS sensor bolt and the sensor, if equipped
- Strut mounting nuts and the studs which attach the strut assembly to the steering knuckle
- Strut from the steering knuckle
- Lower ball joint pinch bolt. Carefully pry down on the lower control arm to separate the ball joint stud from the steering knuckle.
- Wheel hub, knuckle and bearing assembly from the vehicle

To install:

4. Apply Loctite® 290 thread locking compound to the ball joint nut and threads.
5. Install or connect the following:
   - Wheel hub, knuckle and bearing assembly onto the ball joint and tighten the pinch bolt and nut to 32–43 ft. lbs. (43–59 Nm)
   - Tie rod end and tighten the nut to 25–33 ft. lbs. (34–46 Nm). Install a new cotter pin.
   - ABS sensor and tighten the bolt, if equipped
   - Knuckle to the strut assembly and tighten the strut mounting nuts to 69–93 ft. lbs. (93–127 Nm)
   - New halfshaft retaining nut and tighten to 174–235 ft. lbs. (235–319 Nm). Stake the retaining nut using a suitable chisel with the cutting edge rounded off.
   - Brake rotor and caliper
   - Wheel and tire assembly
6. Check the front wheel alignment.
7. Road test the vehicle and check for proper operation.

Rear

➡The wheel bearings are a cartridge design and are not serviceable. If bearing replacement is required, the bearings and hub must be replaced as an assembly.

1. Before servicing the vehicle, refer to the Precautions Section.
2. Remove or disconnect the following:
   - Wheel and tire assembly
   - Hub grease cap
   - Brake drum or disc brake caliper and rotor, as necessary
   - Wheel hub retainer nut securing the hub to the spindle, unstake the nut prior to loosening

- Hub and bearing assembly. Discard the hub retainer nut.
- Disc brake shield retaining bolts and the shield, if equipped with disc brakes
- Backing plate, if equipped with drum brakes
- Strut-to-spindle retaining bolts and nuts
- Trailing arm bolt securing the trailing arm to the spindle
- Stabilizer bar link nuts, retainers, bushings, sleeves and bolts
- Control arm nut and bolt securing both control arms to the wheel spindle
  - Spindle from the vehicle

3. Inspect all components. If the wheel spindle is damaged, replace it. Wheel bearings are sealed and must be replaced if damaged with the wheel hub.

To install:

4. Install or connect the following:
- Wheel spindle in position to the strut
- Spindle retaining bolts and nuts. Tighten to 69–93 ft. lbs. (93–127 Nm).
- Trailing arm to the spindle and tighten retaining bolt to 69–93 ft. lbs. (93–127 Nm)
- Both control arms in position to the spindle and tighten the retaining bolt and nut to 63–86 ft. lbs. (85–117 Nm)
- Stabilizer bar link nuts, retainers, bushings, sleeves and bolts
- Brake backing plate, if equipped with drum brakes

- Brake shield and the disc brake shield retaining bolts, if equipped with disc brakes
- Wheel hub and bearing assembly to the wheel spindle
- New wheel hub retainer nut and tighten to 130–174 ft. lbs. (177–235 Nm)

5. Stake the wheel hub retainer nut using a cape or round end chisel. Do not use a sharp chisel to stake the hub nut.
- Brake drum or the disc brake rotor and caliper, as equipped
- Hub grease cap
- Wheel and tire assembly

6. Pump the brake pedal several times to position the brake lining before attempting to move the vehicle.

7. Road test the vehicle and check for proper operation.

BRAKES

Brake Caliper

REMOVAL & INSTALLATION

Front

1. Before servicing the vehicle, refer to the Precautions Section.
2. Remove or disconnect the following:
- Wheels
- Disc brake pads
- Banjo bolt securing the brake hose to the brake caliper. Discard 2 copper sealing washers.
- 2 brake caliper retaining bolts and the brake caliper

To install:

3. Install or connect the following:
- Brake caliper and 2 brake caliper retaining bolts. Torque the bolts to 29–36 ft. lbs. (39–49 Nm).
- Brake hose and banjo bolt to the brake caliper with 2 new copper sealing washers. Torque the banjo bolt to 16–22 ft. lbs. (22–29 Nm).
- Disc brake pads

4. Bleed the brake system.
- Wheels

5. Pump the brake pedal several times to position the brake pads to the brake rotor.

6. Road test the vehicle and check the brake system for proper operation.

Rear

1. Before servicing the vehicle, refer to the Precautions Section.

2. Remove or disconnect the following:
- Wheels
- Disc brake pads
- Parking brake cable bracket bolt and position the bracket aside
- Parking brake cable from the operating lever
- Banjo bolt securing the brake hose to the brake caliper. Discard 2 copper sealing washers.
- Upper brake caliper retaining bolt
- Brake caliper by sliding it off the mounting bracket

To install:

3. Install or connect the following:
- Brake caliper on the mounting bracket and slide into position
- Upper brake caliper retaining bolt. Torque the bolt to 33–43 ft. lbs. (45–59 Nm).
- Brake hose to the brake caliper and secure with the banjo bolt using 2 new copper sealing washers. Torque the banjo bolt to 16–22 ft. lbs. (22–29 Nm).
- Parking brake cable to the operating lever. Position the bracket and install the bracket bolt.
- Disc brake pads

4. Bleed the brake system.
- Wheels

5. Pump the brake pedal several times to position the brake pads to the brake rotor.

6. Road test the vehicle and check the brake system for proper operation.

Disc Brake Pads

REMOVAL & INSTALLATION

Front

1. Before servicing the vehicle, refer to the Precautions Section.

2. Remove the master cylinder reservoir cap and check the fluid level in the reservoir. Remove brake fluid until the reservoir is ½ full. Discard the removed fluid.

3. Remove or disconnect the following:
- Wheels
- W-spring
- 2 disc brake pad locating pins and the M-spring
- Brake pads and shims from the brake caliper

4. Inspect the brake rotor. Resurface or replace if needed.

To install:

5. Use a suitable tool to push the piston into the brake caliper bore.

6. Apply a suitable grease between the shims and the disc brake pad guide plates.

7. Install or connect the following:
- Brake pads and shims on the brake caliper
- W-spring and 2 brake pad locating pins
- M-spring
- Wheels

8. Pump the brake pedal several times prior to moving the vehicle to position the brake pads.

1	Rear Wheel Brake Caliper Spring		7	Rear Disc Brake Adjuster Spindle
2	Rear Brake Operating Lever		8	Snap Ring
3	Brake Adjuster Screw		9	Dust Seal
4	Rear Disc Brake Caliper		10	Piston
5	Wheel Cylinder Bleeder Screw		11	O-Ring
6	Piston Seal		12	Brake Connecting Link

93006G22

Rear disc brake assembly—Escort

93006G29

Front disc brake pads—Escort

9. Check the fluid level in the master cylinder reservoir and fill as needed.

10. Road test the vehicle and check for proper brake system operation.

Rear

1. Before servicing the vehicle, refer to the Precautions Section.

2. Remove the master cylinder reservoir cap and check the fluid level in the reservoir. Remove brake fluid until the reservoir is ½ full. Discard the removed fluid.

3. Remove or disconnect the following:
   • Wheels

4. If necessary, remove the screw plug and turn the adjustment gear counterclockwise with an Allen® wrench to pull the piston fully inward.
   • Lower brake caliper bolt

5. Using a small prybar, pivot the caliper on its mounting bracket to access

the brake pads. If the upper lock bolt requires lubrication or service, remove it and suspend the caliper with mechanics wire.

- Brake pads, shims, spring and guides

6. Inspect the brake rotor. Resurface or replace if needed.

To install:

7. Apply an appropriate brake pad grease between the shims and the brake pads.

8. Pivot the caliper on its mounting bracket and position the brake pads, shims, spring and guides to the brake rotor.

9. Lubricate and install the lower caliper bolt. Tighten the bolt to 33–43 ft. lbs. (45–59 Nm).

10. Turn the adjustment gear clockwise with an Allen wrench until the brake pads just touch the rotor, then loosen the gear ⅓ of a turn. Install the screw plug and torque to 12 ft. lbs. (16 Nm).

11. Install the wheels.

12. Pump the brake pedal several times prior to moving the vehicle to position the brake pads.

13. Check the brake fluid level in the master cylinder reservoir and fill if needed.

14. Road test the vehicle and check for proper brake system operation.

Brake Drums

REMOVAL & INSTALLATION

1. Before servicing the vehicle, refer to the Precautions Section.

2. Remove or disconnect the following:
- Wheels
- 2 brake drum retaining screws
- Brake drum from the hub

3. Inspect the drum and refinish or replace, as necessary. If refinishing, check the maximum inside diameter specification.

To install:

4. If needed, adjust the brake shoes to fit the brake drum.

5. Install or connect the following:
- Brake drum on the hub
- 2 brake drum retaining screws. Torque to 89–123 inch lbs. (10–14 Nm).
- Wheels

6. Check the brake system for proper operation.

Brake Shoes

REMOVAL & INSTALLATION

1. Before servicing the vehicle, refer to the Precautions Section.

2. Remove or disconnect the following:
- Wheels
- 2 brake drum retaining screws and the brake drum
- 2 brake shoe return springs
- Right-hand anti-rattle spring
- 2 brake shoe hold-down springs

1	Parking Brake Link Spring		7	Brake Shoe Hold-Down Spring
2	Brake Shoe Hold-Down Spring Pin		8	Brake Drum Screw (2 Req'd)
3	Rear Brake Backing Plate		9	Brake Drum
4	Rear Wheel Cylinder		10	Brake Shoe Retracting Spring
5	Right Hand Anti-Rattle Spring		11	Parking Brake Lever
6	Rear Brake Shoe and Lining		12	Parking Brake Return Spring
			13	Brake Adjusting Hole Cover

93006G57

Brake shoes and related components—Escort

93006G58

Checking the adjustment of the brake shoes—Escort

• Leading and trailing shoes from the brake backing plate

To install:

3. Use a suitable high temperature grease to lightly lubricate the brake shoe contact points on the backing plate.

4. Remove or disconnect the following:
- Trailing brake shoe on the backing plate and one of the brake shoe hold-down springs
- Leading brake shoe on the backing plate and the other brake shoe hold-down spring
- Right-hand anti-rattle spring
- 2 brake shoe return springs

5. Using a brake adjusting gauge, measure the inside diameter of the brake drum.

6. Compare the brake drum measurement to the brake shoes.

7. Adjust the brake shoes by inserting a screwdriver into the knurled quadrant of the rear quad operating lever and adjust the shoes to the same measurement as the brake drum.

8. Install or connect the following:
- Brake drum
- 2 brake drum retaining screws
- Wheels

9. Complete the brake shoe adjustment by sharply applying the brakes several times while driving the vehicle alternating between forward and reverse gears.

10. Check the brake system operation by making several stops while driving forward.

FORD AND LINCOLN

Excursion • Expedition • Navigator

6

SPECIFICATION AND MAINTENANCE CHARTS

ENGINE AND VEHICLE IDENTIFICATION

Engine							Model Year		
Code ①	Liters (cc)	Cu. In.	Cyl.	Fuel Sys.	Type	Eng. Mfg.	Code ②	Year
S	6.8 (6802)	415	10	MFI	SOHC	Ford	2	2002
6	4.6 (4588)	281	8	MFI	SOHC	Ford	3	2003
F	7.3 (7292)	445	8	DI	OHV	Navistar	4	2004
R	5.4 (5409)	330	8	EFI	DOHC	Ford	5	2005
L	5.4 (5409)	330	8	EFI	SOHC	Ford	6	2006
5	5.4 (5409)	330	8	EFI	SOHC	Ford		
P	6.0 (5999)	365	8	DI ③	SOHC	Ford		
W	4.6 (4588)	281	8	MFI	SOHC	Ford		

MFI: Multi-port Fuel Injection

DI: Direct Injection Turbo-Diesel

EFI: Electronic Fuel Injection

OHV: Overhead Valve

SOHC: Single Overhead Camshaft

DOHC: Dual Overhead Camshaft

① 8th digit of the Vehicle Identification Number (VIN)

② 10th digit of the Vehicle Identification Number (VIN)

③ Turbo

06017-NAVI-C01

GENERAL ENGINE SPECIFICATIONS

Year	Model	Engine Displacement Liters (VIN)	Net Horsepower @ rpm	Net Torque @ rpm (ft. lbs.)	Bore x Stroke (in.)	Compression Ratio	Oil Pressure @ rpm
2002	Excursion	5.4 (L)	235@4250	330@3000	3.55X4.17	9.0:1	40-70@1500
	Excursion	6.8 (S)	265@4250	410@2750	4.09X4.17	9.0:1	40-70@1500
	Excursion	7.3 (F)	210@3000	425@2000	4.11x4.18	17.5:1	40-70@3000
	Expedition	4.6 (W)	210@4400	290@3250	3.55x3.54	9.0:1	20-45@1500
	Expedition	5.4 (L)	235@4250	330@3000	3.55X4.17	9.0:1	40-70@1500
	Navigator	5.4 (R)	260@4250	330@3000	3.55X4.17	9.0:1	40-60@2000
2003	Excursion	5.4 (L)	235@4250	330@3000	3.55X4.17	9.0:1	40-70@1500
	Excursion	6.8 (S)	265@4250	410@2750	4.09X4.17	9.0:1	40-70@1500
	Excursion	7.3 (F)	210@3000	425@2000	4.11x4.18	17.5:1	40-70@3000
	Excursion	6.0 (P)	325@3300	560@2000	3.74x4.13	18.0:1	24@1200
	Expedition	4.6 (W)	210@4400	290@3250	3.55x3.54	9.3:1	40-60@2000
	Expedition	5.4 (L)	235@4250	330@3000	3.55X4.17	9.0:1	40-60@2000
	Expedition	5.4 (R)	235@4250	330@3000	3.55X4.23	9.5:1	40-60@2000
	Navigator	5.4 (L)	235@4250	330@3000	3.55X4.17	9.0:1	40-60@2000
	Navigator	5.4 (R)	260@4250	330@3000	3.55X4.17	9.0:1	40-60@2000
2004	Excursion	5.4 (L)	235@4250	330@3000	3.55X4.17	9.0:1	40-70@1500
	Excursion	6.8 (S)	265@4250	410@2750	4.09X4.17	9.0:1	40-70@1500
	Excursion	6.0 (P)	325@3300	560@2000	3.74x4.13	18.0:1	24@1200
	Expedition	4.6 (W)	210@4400	290@3250	3.55x3.54	9.3:1	40-60@2000
	Expedition	5.4 (L)	235@4250	330@3000	3.55X4.17	9.0:1	40-60@2000
	Expedition	5.4 (R)	235@4250	330@3000	3.55X4.23	9.5:1	40-60@2000
	Navigator	5.4 (L)	235@4250	330@3000	3.55X4.17	9.0:1	40-60@2000
	Navigator	5.4 (R)	260@4250	330@3000	3.55X4.17	9.0:1	40-60@2000
2005	Excursion	5.4 (L)	235@4250	330@3000	3.55X4.17	9.0:1	40-70@1500
	Excursion	6.8 (S)	265@4250	410@2750	4.09X4.17	9.0:1	40-70@1500
	Excursion	6.0 (P)	325@3300	560@2000	3.74x4.13	18.0:1	24@1200
	Expedition	5.4 (5)	300@5000	365@3750	3.55X4.16	9.85:1	57@2000
	Navigator	5.4 (5)	300@5000	365@3750	3.55X4.16	9.85:1	57@2000

06017-NAVI-C02

GASOLINE ENGINE TUNE-UP SPECIFICATIONS

Year	Engine Displacement Liters (cc)	Engine ID/VIN	Spark Plug Gap (in.)	Ignition Timing (deg.) ①		Fuel Pump (psi) ①	Idle Speed (rpm)		Valve Clearance	
				MT	AT		MT	AT	In.	Ex.
2002	4.6 (4588)	W	0.052-0.056	—	10B	30-45	—	②	HYD	HYD
	4.6 (4588)	6	0.052-0.056	—	10B	30-45	—	②	HYD	HYD
	5.4 (5409)	L	0.052-0.056	—	10B	28-45	—	②	HYD	HYD
	6.8 (6802)	S	0.052-0.055	—	10B	28-45	—	②	HYD	HYD
2003	4.6 (4588)	W	0.052-0.056	—	10B	30-45	—	②	HYD	HYD
	4.6 (4588)	6	0.052-0.056	—	10B	30-45	—	②	HYD	HYD
	5.4 (5409)	L	0.052-0.056	—	10B	28-45	—	②	HYD	HYD
	6.8 (6802)	S	0.052-0.055	—	10B	28-45	—	②	HYD	HYD
2004	4.6 (4588)	W	0.052-0.056	—	10B	30-45	—	②	HYD	HYD
	5.4 (5409)	R	0.052-0.056	—	10B	28-45	—	②	HYD	HYD
	5.4 (5409)	L	0.052-0.056	—	10B	28-45	—	②	HYD	HYD
	6.8 (6802)	S	0.052-0.055	—	10B	28-45	—	②	HYD	HYD
2005	5.4 (5409)	L	0.052-0.056	—	10B	28-45	—	②	HYD	HYD
	5.4 (5409)	5	0.040-0.050	—	10B	28-45	—	②	HYD	HYD
	6.8 (6802)	S	0.052-0.055	—	10B	28-45	—	②	HYD	HYD

NOTE: The Vehicle Emission Control Information label often reflects specification changes changes made during production. The label figures must be used if they differ from those in this chart.

B: Before top dead center

HYD: Hydraulic

NA: Not Available

① With engine running

② Idle speed and timing are electronically controlled and cannot be adjusted

06017-NAVI-C03

4.6L Engines
Firing order: 1–3–7–2–6–5–4–8
Distributorless ignition system

5.4L Engines
Firing order: 1–3–7–2–6–5–4–8
Distributorless ignition system (one coil on each cylinder)

6.8L Engine
Firing order: 1–6–5–10–2–7–3–8–4–9
Distributorless ignition system (one coil on each cylinder)

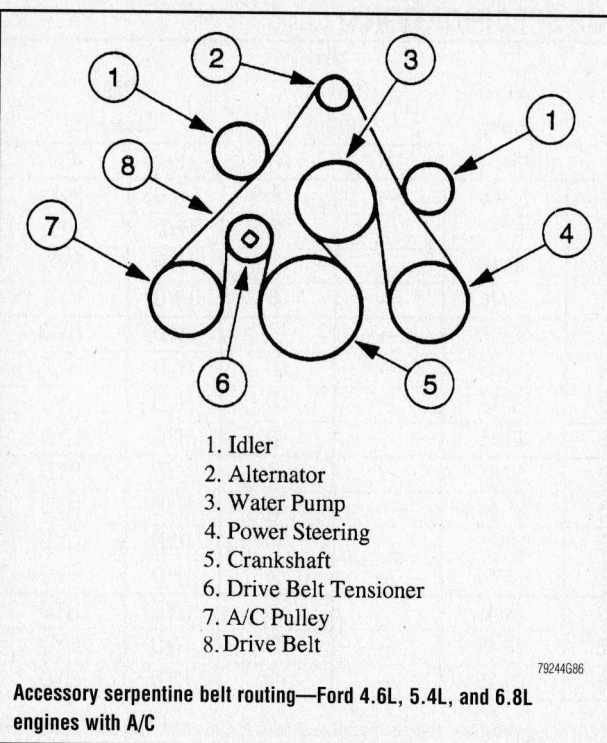

1. Idler
2. Alternator
3. Water Pump
4. Power Steering
5. Crankshaft
6. Drive Belt Tensioner
7. A/C Pulley
8. Drive Belt

79244G86

Accessory serpentine belt routing—Ford 4.6L, 5.4L, and 6.8L engines with A/C

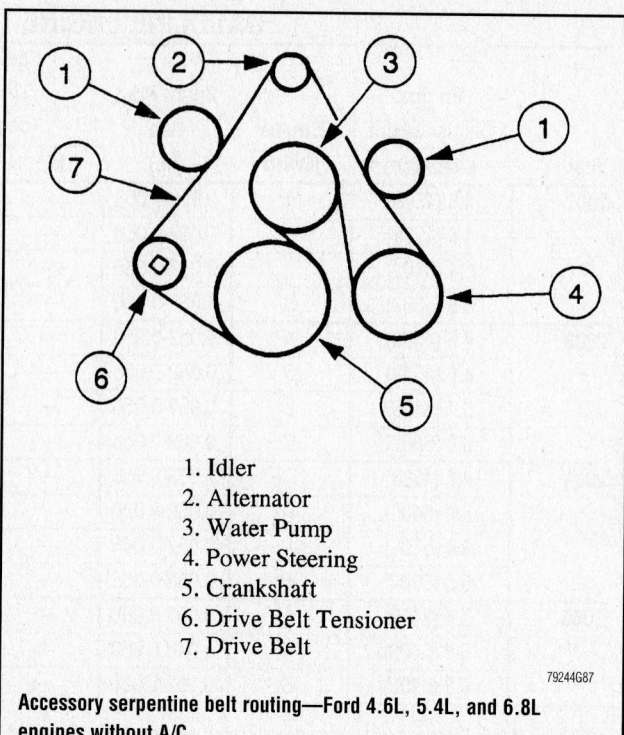

1. Idler
2. Alternator
3. Water Pump
4. Power Steering
5. Crankshaft
6. Drive Belt Tensioner
7. Drive Belt

79244G87

Accessory serpentine belt routing—Ford 4.6L, 5.4L, and 6.8L engines without A/C

79244G84

Accessory serpentine belt routing—Ford 7.3L turbo diesel engine

1 A/C clutch pulley

2 Idler pulley (without A/C)

3 Belt idler pulley

4 Generator pulley

5 Power steering pump pulley

6 Water pump pulley

7 Crankshaft pulley

8 Drive belt tensioner

67197-NAVI-G21

Accessory Drive—6.0L engine with single alternator

1 A/C clutch pulley

2 Idler pulley (without A/C)

3 Belt idler pulley

4 Generator pulley

5 Power steering pump pulley

6 Water pump pulley

7 Crankshaft pulley

8 Drive belt tensioner

67197-NAVI-G22

Accessory Drive—6.0L engine with dual alternator

DIESEL ENGINE TUNE-UP SPECIFICATIONS

Year	Engine Displacement cu. in. (VIN)	Valve Clearance Intake (in.)	Valve Clearance Exhaust (in.)	Intake Valve Opens (deg.)	Injection Pump Setting (deg.)	Injection Nozzle Pressure (psi) New	Injection Nozzle Pressure (psi) Used	Idle Speed (rpm)	Cranking Compression Pressure (psi)
2002	7.3 (F)	HYD	HYD	—	①	1875	1425	②	③
2003	7.3 (F)	HYD	HYD	—	①	1875	1425	②	③
	6.0 (P)	HYD	HYD		①	NA	NA	②	NA
2004	6.0 (P)	HYD	HYD		①	NA	NA	②	NA
2005	6.0 (P)	HYD	HYD		①	NA	NA	②	NA

NOTE: The Vehicle Emission Control Information label often reflects specification changes made during production. The label figures must be used if they differ from those in this chart.

HYD: Hydraulic

B: Before top dead center

NA: Not Available

① PCM controlled

② See underhood emission label

③ Compression pressure in the lowest cylinder must be at least 75% of the highest cylinder

　Minimum pressure: 195 psi

　Maximum pressure: 440 psi

CAPACITIES

Year	Model	Engine Displacement Liters (VIN)	Engine Oil with Filter (qts.)	Transmission (pts.) 5-Spd	Transmission (pts.) Auto.	Transfer Case (pts.)	Drive Axle Front (pts.)	Drive Axle Rear (pts.)	Fuel Tank (gal.)	Cooling System (qts.)
2002	Excursion	5.4 (L)	7.0	—	①	4.2	6.0	6.0	44.0	19.8
	Excursion	6.8 (S)	7.0	—	①	4.2	6.0 ②	6.0 ②	44.0	23.0
	Excursion	7.3 (F)	14.0	—	①	4.2	6.0 ②	6.0 ②	44.0	23.0
	Expedition	4.6 (6/W)	6.0	—	①	4.0	3.7	5.5	24.5 ③	17.9
	Expedition	5.4 (L)	7.0	—	①	4.0	3.7	5.5	24.5 ③	20.8
	Navigator	5.4 (R)	6.0	—	①	4.0	3.7	5.5	28.0	20.8
2003	Excursion	5.4 (L)	7.0	—	①	4.2	6.0	6.0	44.0	19.8
	Excursion	6.8 (S)	7.0	—	①	4.2	6.0 ②	6.0 ②	44.0	23.0
	Excursion	6.0 (P)	15.0	11.6	④	4.2	6.0 ②	6.0 ②	44.0	27.5
	Excursion	7.3 (F)	14.0	—	①	4.2	6.0 ②	6.0 ②	44.0	23.0
	Expedition	4.6 (W)	6.0	—	①	4.0	3.7	5.5	28.0	17.9
	Expedition	5.4 (L)	6.0	—	①	4.0	3.7	5.5	28.0	20.8
	Expedition	5.4 (R)	6.0	—	①	4.0	3.7	5.5	28.0	20.8
	Navigator	5.4 (L)	6.0	—	①	4.0	3.7	5.5	28.0	20.8
	Navigator	5.4 (R)	6.0	—	①	4.0	3.7	5.5	28.0	20.8
2004	Excursion	5.4 (L)	7.0	—	①	4.2	6.0	6.0	44.0	19.8
	Excursion	6.8 (S)	7.0	—	①	4.2	6.0 ②	6.0 ②	44.0	23.0
	Excursion	6.0 (P)	15.0	11.6	④	4.2	6.0 ②	6.0 ②	44.0	27.5
	Expedition	4.6 (W)	6.0	—	①	4.0	3.7	5.5	28.0	17.9
	Expedition	5.4 (L)	6.0	—	①	4.0	3.7	5.5	28.0	20.8
	Expedition	5.4 (R)	6.0	—	①	4.0	3.7	5.5	28.0	20.8
	Navigator	5.4 (L)	6.0	—	①	4.0	3.7	5.5	28.0	20.8
	Navigator	5.4 (R)	6.0	—	①	4.0	3.7	5.5	28.0	20.8
2005	Excursion	5.4 (L)	7.0	—	①	4.2	6.0	6.0	44.0	19.8
	Excursion	6.8 (S)	7.0	—	①	4.2	6.0 ②	6.0 ②	44.0	23.0
	Excursion	6.0 (P)	15.0	11.6	④	4.2	6.0 ②	6.0 ②	44.0	27.5
	Expedition	5.4 (5)	6.6	—	①	4.0	3.7	⑤	28.0	⑥
	Navigator	5.4 (5)	6.6	—	①	4.0	3.7	⑤	28.0	⑥

NOTE: All capacities are approximate. Add fluid gradually and check to be sure a proper fluid level is obtained.

① With 4R70W: 28 pts.
  With 4R100: 32.0 pts.
  With E40D: 32.0 pts.
  2005 Expedition with 4R70/75: 28 pts.
  2005 Navigator : 32 pts.
② Heavy duty: 7.5 pts.
③ Also available with a 30 gallon
  tank with 8 ft. box
④ 4R100: 34.2 pts.
  TorqShift: 38.4 pts.

⑤ 8.8 inch coventional axle: 4.0 pts.
  9.75 inch coventional axle: 4.5 pts.
  8.8 inch limited slip axle: 3.75 pts.
  9.75 inch limited slip axle: 4.25 pts.
⑥ Base radiator without aux rear heat: 19.4
  Heavy duty radiator without aux rear heat: 19.4
  Base radiator with aux rear heat: 20.7
  Heavy duty radiator with aux rear heat: 20.9

06017-NAVI-C05

VALVE SPECIFICATIONS

Year	Engine Displacement Liters (VIN)	Seat Angle (deg.)	Face Angle (deg.)	Spring Test Pressure (lbs. @ in.)	Spring Installed Height (in.)	Stem-to-Guide Clearance (in.)		Stem Diameter (in.)	
						Intake	Exhaust	Intake	Exhaust
2002	4.6 (6/W)	45	45.5	132@1.100	1.570	0.0008-0.0027	0.0018-0.0037	0.2750-0.2746	0.2740-0.2736
	5.4 (L)	45	45.5	150@1.10	1.570	0.0008-0.0027	0.0018-0.0037	0.275-0.2746	0.274-0.2736
	6.8 (S)	44.50-45.25	45.25-45.75	150@1.10	1.570	0.0008-0.0027	0.0018-0.0037	0.275-0.2746	0.274-0.2735
	7.3 (F)	①	①	200@1.38	②	0.0055	0.0055	0.3119-0.3126	0.3119-0.3126
2003	4.6 (W)	45.0-44.50	45.75-45.25	132@1.100	1.570	0.0008-0.0027	0.0018-0.0037	0.2754-0.2746	0.2744-0.2736
	5.4 (L)	45.0-44.50	45.75-45.25	150@1.10	1.570	0.0008-0.0027	0.0018-0.0037	0.2754-0.2746	0.274-0.2736
	5.4 (R)	45.0-44.50	45.75-45.25	150@1.10	1.676	0.0008-0.0027	0.0018-0.0037	0.2754-0.2746	0.2744-0.2736
	6.0 (P)	①	①	191@1.51	1.820	0.0055 max.	0.0055 max.	0.2720-0.2735	0.2720-0.2735
	6.8 (S)	44.50-45.25	45.25-45.75	150@1.10	1.570	0.0008-0.0027	0.0018-0.0037	0.275-0.2746	0.274-0.2735
	7.3 (F)	①	①	200@1.38	②	0.0055	0.0055	0.3119-0.3126	0.3119-0.3126
2004	4.6 (W)	45.0-44.50	45.75-45.25	132@1.100	1.570	0.0008-0.0027	0.0018-0.0037	0.2754-0.2746	0.2744-0.2736
	5.4 (L)	45.0-44.50	45.75-45.25	150@1.10	1.570	0.0008-0.0027	0.0018-0.0037	0.2754-0.2746	0.274-0.2736
	5.4 (R)	45.0-44.50	45.75-45.25	150@1.10	1.676	0.0008-0.0027	0.0018-0.0037	0.2754-0.2746	0.274-0.2736
	6.0 (P)	①	①	191@1.51	1.820	0.0055 max.	0.0055 max.	0.2720-0.2735	0.2720-0.2735
	6.8 (S)	44.50-45.25	45.25-45.75	150@1.10	1.570	0.0008-0.0027	0.0018-0.0037	0.275-0.2746	0.274-0.2735
2005	5.4 (L)	45.0-44.50	45.75-45.25	150@1.10	1.570	0.0008-0.0027	0.0018-0.0037	0.2754-0.2746	0.274-0.2736
	5.4 (5)	44.5-45.0	45.0	79@1.66	1.660	0.0010-0.0020	0.0030-0.0040	0.2350-0.2360	0.2340-0.2350
	6.0 (P)	①	①	191@1.51	1.820	0.0055 max.	0.0055 max.	0.2720-0.2735	0.2720-0.2735
	6.8 (S)	44.50-45.25	45.25-45.75	150@1.10	1.570	0.0008-0.0027	0.0018-0.0037	0.275-0.2746	0.274-0.2735

① Intake: 30 degrees
   Exhaust: 37.5 degrees

② Intake: 1.767 in.
   Exhaust: 1.833 in.

06017-NAVI-C06

PISTON AND RING SPECIFICATIONS

All measurements are given in inches.

Year	Engine Displacement Liters (VIN)	Piston Clearance	Ring Gap			Ring Side Clearance		
			Top Compression	Bottom Compression	Oil Control	Top Compression	Bottom Compression	Oil Control
2002	4.6 (6/W)	0.0005-0.0010	0.010-0.020	0.010-0.020	0.006-0.026	0.0016-0.0031	0.0012-0.0031	SNUG
	5.4 (L)	0.0000-0.0010	0.005-0.011	0.010-0.016	0.006-0.026	0.0012-0.0037	0.0012-0.0037	SNUG
	6.8 (S)	0.0000-0.0010	0.005-0.011	0.010-0.016	0.006-0.026	0.0012-0.0037	0.0012-0.0037	SNUG
	7.3 (F)	0.0044-0.0057	0.014-0.024	0.062-0.072	0.012-0.024	0.0013-0.0033	0.0013-0.0033	SNUG
2003	4.6 (W)	0.0005-0.0010	0.010-0.020	0.010-0.020	0.006-0.026	0.0016-0.0031	0.0012-0.0031	SNUG
	5.4 (L)	0.0000-0.0010	0.005-0.011	0.010-0.016	0.006-0.026	0.0012-0.0037	0.0012-0.0037	SNUG
	5.4 (R)	0.001-0.0020	0.005-0.011	0.0098-0.0157	0.0059-0.0256	0.0012-0.0020	0.0012-0.0031	SNUG
	6.0 (P)	0.0017-0.0036	0.011-0.0210	0.055-0.0650	0.009-0.0190	NA	NA	NA
	6.8 (S)	0.0000-0.0010	0.005-0.011	0.010-0.016	0.006-0.026	0.0012-0.0037	0.0012-0.0037	SNUG
	7.3 (F)	0.0044-0.0057	0.014-0.024	0.062-0.072	0.012-0.024	0.0013-0.0033	0.0013-0.0033	SNUG
2004	4.6 (W)	0.0005-0.0010	0.010-0.020	0.010-0.020	0.006-0.026	0.0016-0.0031	0.0012-0.0031	SNUG
	5.4 (L)	0.0000-0.0010	0.005-0.011	0.010-0.016	0.006-0.026	0.0012-0.0037	0.0012-0.0037	SNUG
	5.4 (R)	0.001-0.0020	0.005-0.011	0.0098-0.0157	0.0059-0.0256	0.0012-0.0020	0.0012-0.0031	SNUG
	6.0 (P)	0.0017-0.0036	0.011-0.0210	0.055-0.0650	0.009-0.0190	NA	NA	NA
	6.8 (S)	0.0000-0.0010	0.005-0.011	0.010-0.016	0.006-0.026	0.0012-0.0037	0.0012-0.0037	SNUG
	7.3 (F)	0.0044-0.0057	0.014-0.024	0.062-0.072	0.012-0.024	0.0013-0.0033	0.0013-0.0033	SNUG
2005	5.4 (L)	0.0000-0.0010	0.005-0.011	0.010-0.016	0.006-0.026	0.0012-0.0037	0.0012-0.0037	SNUG
	5.4 (5)	0.0010-0.0018	0.006-0.012	0.0098-0.0197	0.0059-0.0256	0.0008-0.0031	0.0012-0.0028	SNUG
	6.0 (P)	0.0017-0.0036	0.011-0.0210	0.055-0.0650	0.009-0.0190	NA	NA	NA
	6.8 (S)	0.0000-0.0010	0.005-0.011	0.010-0.016	0.006-0.026	0.0012-0.0037	0.0012-0.0037	SNUG

NA: Not Available

CRANKSHAFT AND CONNECTING ROD SPECIFICATIONS
All measurements are given in inches.

Year	Engine Displacement Liters (VIN)	Crankshaft				Connecting Rod		
		Main Brg. Journal Dia.	Main Brg. Oil Clearance	Shaft End-play	Thrust on No.	Journal Diameter	Oil Clearance	Side Clearance
2002	4.6 (6/W)	2.6500-2.6570	0.0011-0.0026	0.0051-0.0120	5	2.0870-2.8670	0.0011-0.0026	0.0006-0.0177
	5.4 (L)	2.6568-2.6576	0.0009-0.0019	0.0015-0.0030	5	2.0859-2.0867	0.0010-0.0025	0.0006-0.0177
	6.8 (S)	2.6568-2.6576	0.0009-0.0019	0.0015-0.0030	5	2.0859-2.0867	0.0010-0.0025	0.0006-0.0177
	7.3 (F)	3.1228-3.1236	0.0018-0.0036	0.0025-0.0085	4	2.4980-2.4990	0.0015-0.0045	0.0120-0.0240
2003	4.6 (W)	2.4803	0.0011-0.0026	0.0051-0.0120	5	2.0870-2.8670	0.0011-0.0026	0.0006-0.0177
	5.4 (L)	2.6568-2.6576	0.0009-0.0019	0.0051-0.0120	5	2.0859-2.0867	0.0010-0.0025	0.0006-0.0177
	5.4 (R)	2.6568-2.6576	0.0009-0.0019	0.003-0.0148	5	2.0859-2.0867	0.0010-0.0025	0.0006-0.0177
	6.0 (P)	3.1500-3.1880	NA	0.0087	NA	2.7160-2.7170	NA	0.012-0.0240
	6.8 (S)	2.6568-2.6576	0.0009-0.0019	0.0015-0.0030	5	2.0859-2.0867	0.0010-0.0025	0.0006-0.0177
	7.3 (F)	3.1228-3.1236	0.0018-0.0036	0.0025-0.0085	4	2.4980-2.4990	0.0015-0.0045	0.0120-0.0240
2004	4.6 (W)	2.4803	0.0011-0.0026	0.0051-0.0120	5	2.0870-2.8670	0.0011-0.0026	0.0006-0.0177
	5.4 (L)	2.6568-2.6576	0.0009-0.0019	0.0051-0.0120	5	2.0859-2.0867	0.0010-0.0025	0.0006-0.0177
	5.4 (R)	2.6568-2.6576	0.0009-0.0019	0.003-0.0148	5	2.0859-2.0867	0.0010-0.0025	0.0006-0.0177
	6.0 (P)	3.1500-3.1880	NA	0.0087	NA	2.7160-2.7170	NA	0.012-0.0240
	6.8 (S)	2.6568-2.6576	0.0009-0.0019	0.0015-0.0030	5	2.0859-2.0867	0.0010-0.0025	0.0006-0.0177
2005	5.4 (L)	2.6568-2.6576	0.0009-0.0019	0.0051-0.0120	5	2.0859-2.0867	0.0010-0.0025	0.0006-0.0177
	5.4 (5)	2.6567-2.6576	NA	0.0030-0.0148	5	2.0885-2.0877	NA	0.0187-0.0049
	6.0 (P)	3.1500-3.1880	NA	0.0087	NA	2.7160-2.7170	NA	0.012-0.0240
	6.8 (S)	2.6568-2.6576	0.0009-0.0019	0.0015-0.0030	5	2.0859-2.0867	0.0010-0.0025	0.0006-0.0177

06017-NAVI-C08

TORQUE SPECIFICATIONS
All readings in ft. lbs.

Engine Displacement Liters (cc)	Cylinder Head Bolts	Main Bearing Bolts	Rod Bearing Bolts	Crankshaft Damper Bolts	Flywheel Bolts	Manifold		Spark Plugs	Oil Pan Drain Plug
						Intake *	Exhaust		
2002 4.6 (6/W)	①	②	③	④	54-64	⑤	18-25	8	10
5.4 (L)	①	⑥	③	④	54-64	⑤	18-25	8	10
5.4 (R)	①	⑥	⑦	④	54-64	⑧	17-20	15	10
6.8 (S)	①	⑥	③	④	54-64	⑨	18-25	11	11
7.3 (F)	⑩	⑪	⑫	212	89	24	45	—	11
2003 4.6 (W)	①	②	③	④	54-64	⑤	18-25	8	10
5.4 (L)	①	⑥	③	④	54-64	⑤	18-25	8	10
5.4 (R)	①	⑥	⑦	④	54-64	⑧	17-20	15	10
6.0 (P)	⑬	⑭	⑮	⑯	69	8	28 ⑰	—	18
6.8 (S)	①	⑥	③	④	54-64	⑨	18-25	11	11
7.3 (F)	⑩	⑪	⑫	212	89	24	45	—	11
2004 4.6 (W)	①	⑱	⑲	④	54-64	⑤	18-25	8	10
5.4 (L)	①	⑱	⑲	④	54-64	⑤	18-25	8	10
5.4 (R)	①	⑬	⑲	④	54-64	⑫	17-20	15	10
6.0 (P)	⑬	⑭	⑮	⑯	69	8	28 ⑰	—	18
6.8 (S)	①	⑥	③	④	54-64	⑨	18-25	11	11
2005 5.4 (L)	①	⑱	⑲	④	54-64	⑤	18-25	8	10
5.4 (5)	①	⑬	⑳	④	59	⑫	18	25	10
6.0 (P)	⑬	⑭	⑮	⑯	69	8	28 ⑰	—	18
6.8 (S)	①	⑥	③	④	54-64	⑨	18-25	11	11

* NOTE: Applies to Lower Manifold only.

① Step 1: 30 ft. lbs.
Step 2: Plus 90 degrees
Step 3: Plus 90 degrees

② Vertical mounted bolts:
Step 1: 24 ft. lbs.
Step 2: Plus 90 degrees
Jack screws:
Step 1: 45 inch lbs.
Step 2: 98 inch lbs.

③ Step 1: 32 ft. lbs.
Step 2: Plus 105 degrees

④ Step 1: 66 ft. lbs.
Step 2: Loosen bolt
Step 3: 37 ft. lbs.
Step 4: Plus 90 degrees

⑤ Step 1: 18 inch lbs.
Step 2: 18 ft. lbs.

⑥ Vertical mounted bolts:
Step 1: 30 ft. lbs.
Step 2: Plus 90 degrees
Side Bolts:
Step 1: 22 ft. lbs.
Step 2: plus 90 degrees

⑦ Step 1: 18 ft. lbs.
Step 2: 30 ft. lbs.
Step 3: Plus 90 degrees

⑧ Upper manifold:
Step 1: 18 inch lbs.
Step 2: 90 inch lbs.
Lower manifold:
Step 1: 89 inch lbs.
Step 2: plus an additional 90 degrees

⑨ Lower manofold:
Step 1: 18 inch lbs.
Step 2: 89 inch lbs.
Upper Manifold:
Step 1: 18 inch lbs.
Step 2: 18 ft. lbs.

⑩ Step 1: 65. Ft. lbs.
Step 2: 85 ft. lbs.
Step 3: 95 ft. lbs.

⑪ Step 1: 76 ft. lbs.
Step 2: 96 ft. lbs.

⑫ Step 1: 18inch lbs.
Step 2: 89 inch lbs.

⑬ Vertical mounted bolts:
Step 1: 30 ft. lbs.
Step 2: Plus 90 degrees
Side Bolts:
Step 1: 22 ft. lbs.
Step 2: Plus 90 degrees

⑭ Vertical mounted bolts:
Step 1: 110 ft. lbs.
Step 2: 130 ft. lbs.
Step 3: 170 ft. lbs.
Side Bolts:
Step 1: 18 ft. lbs.

⑮ Step 1: 33 ft. lbs.
Step 2: 50 ft. lbs.

⑯ Step 1: 50 ft. lbs.
Step 2: Plus 90 degrees

⑰ Apply high temp nickel anti-seize to the threads

⑱ Cast Aluminum Block:
Vertical mounted bolts:
Step 1: 89 inch lbs.
Step 2: 18 ft. lbs.
Step 3: 30 ft. lbs.
Step 4: plus an additional 90 degrees
Side Bolts:
Step 1: 30 ft. lbs.
Step 2: plus 90 degrees
Cast Iron Block:
Step 1: 30 ft. lbs.
Step 2: Plus 90 degrees
Side Bolts:
Step 1: 44 ft. lbs.
Step 2: 89 inch lbs.

⑲ Step 1: 32 ft. lbs.
Step 2: Plus 90-120 degrees

⑳ Step 1: 32 ft. lbs.
Step 2: Plus 105 degrees

WHEEL ALIGNMENT

Year	Model		Caster Range (+/-Deg.)	Caster Preferred Setting (Deg.)	Camber Range (+/-Deg.)	Camber Preferred Setting (Deg.)	Toe-in (in.)
2002	Excursion	2WD	1.00	+0.62	+4.40	1.00	0.10+/-0.25
		4WD	1.00	+0.25	+3.63	+0.20	0.01+/-0.25
	Expedition	2WD base	1.00	+6.10	0.70	-0.30	0.06+/-0.25
		2WD air	1.00	+6.10	0.70	-0.30	0.06+/-0.25
		4WD base	1.00	+5.10	0.70	-0.20	0.30+/-0.25
		4WD air	1.00	+5.00	0.40	-0.40	0.20+/-0.25
	Navigator	2WD	1.00	+6.10	0.75	-0.33	0.06+/-0.25
		4WD	1.00	+5.10	0.40	-0.40	0.20+/-0.25
2003	Excursion	2WD	1.00	+0.62	+4.40	1.00	0.10+/-0.25
		4WD	1.00	+0.25	+3.63	+0.20	0.01+/-0.25
	Expedition	2WD base	1.00	+6.10	0.70	-0.30	0.06+/-0.25
		2WD air	1.00	+6.10	0.70	-0.30	0.06+/-0.25
		4WD base	1.00	+5.10	0.70	-0.20	0.30+/-0.25
		4WD air	1.00	+5.00	0.40	-0.40	0.20+/-0.25
	Navigator	2WD	1.00	+6.10	0.75	-0.33	0.06+/-0.25
		4WD	1.00	+5.10	0.40	-0.40	0.20+/-0.25
2004	Excursion	2WD	1.00	+0.62	+4.40	1.00	0.10+/-0.25
		4WD	1.00	+0.25	+3.63	+0.20	0.01+/-0.25
	Expedition	2WD base	1.00	+6.10	0.70	-0.30	0.06+/-0.25
		2WD air	1.00	+6.10	0.70	-0.30	0.06+/-0.25
		4WD base	1.00	+5.10	0.70	-0.20	0.30+/-0.25
		4WD air	1.00	+5.00	0.40	-0.40	0.20+/-0.25
	Navigator	2WD	1.00	+6.10	0.75	-0.33	0.06+/-0.25
		4WD	1.00	+5.10	0.40	-0.40	0.20+/-0.25
2005	Excursion	2WD	1.00	+0.62	+4.40	1.00	0.10+/-0.25
		4WD	1.00	+0.25	+3.63	+0.20	0.01+/-0.25
	Expedition	2WD base	1.00	+5.3	-0.75	-0.30	0.14+/-0.20
		2WD air	1.00	+5.5	-0.75	-0.60	0.14+/-0.20
		4WD base	1.00	+5.3	-0.75	-0.30	0.14+/-0.20
		4WD air	1.00	+5.5	-0.75	-0.60	0.14+/-0.20
	Navigator	2WD	1.00	+5.3	-0.75	-0.30	0.14+/-0.20
		4WD	1.00	+5.5	-0.75	-0.60	0.14+/-0.20

06017-NAVI-C10

TIRE, WHEEL AND BALL JOINT SPECIFICATIONS

Year	Model	OEM Tires Standard	OEM Tires Optional	Tire Pressures (psi) Front	Tire Pressures (psi) Rear	Wheel Size	Ball Joint Inspection	Lug Nut
2002	Expedition	P255/70R16	P265/70R17	35	35	7-JJ	0.030 in. ①	②
	Navigator	P245/75R16	P255/75R17	30	35	7.5	0.030 in. ①	②
2003	Expedition	P255/70R16	P265/70R17	35	35	7-JJ	0.030 in. ①	②
	Navigator	P245/75R16	P255/75R17	30	35	7.5	0.030 in. ①	②
2004	Expedition	P255/70R16	P265/70R17	35	35	7-JJ	0.030 in. ①	②
	Navigator	P245/75R16	P255/75R17	30	35	7.5	0.030 in. ①	②
2005	Expedition	P255/70R16	P265/70R17	35	35	7-JJ	0.030 in. ①	②
	Navigator	P245/75R16	P255/75R17	30	35	7.5	0.030 in. ①	②

OEM: Original Equipment Manufacturer

PSI: Pounds Per Square Inch

STD: Standard

OPT: Optional

① Both upper and lower

② 12mm nuts 100 ft. lbs.

   14mm nuts 150 ft. lbs.

06017-NAVI-C11

BRAKE SPECIFICATIONS
All measurements in inches unless noted

Year	Model		Brake Disc Original Thickness	Brake Disc Minimum Thickness	Brake Disc Maximum Runout	Brake Drum Diameter Original Inside Diameter	Brake Drum Diameter Max. Wear Limit	Brake Drum Diameter Maximum Machine Diameter	Minimum Lining Thickness	Brake Caliper Bracket Bolts (ft. lbs.)	Brake Caliper Mounting Bolts (ft. lbs.)
2002	Excursion	F	1.023	0.964	0.0025	—	—	—	0.030	125-168	41
		R	0.700	0.657	0.0250	—	—	—	0.030	120	26
	Expedition	F	1.023	0.964	0.0025	—	—	—	0.030	125-168	41
		R	0.700	0.657	0.0250	—	—	—	0.030	120	26
	Navigator	F	1.023	0.964	0.0025	—	—	—	0.030	125-168	41
		R	0.700	0.657	0.0250	—	—	—	0.030	120	26
2003	Expedition	F	1.023	0.106	0.0025	—	—	—	0.059	148	41
		R	0.700	0.075	0.0250	—	—	—	0.039	140	26
	Navigator	F	1.023	0.964	0.0025	—	—	—	0.059	148	41
		R	0.700	0.075	0.0250	—	—	—	0.039	140	26
2004	Expedition	F	1.023	0.106	0.0025	—	—	—	0.059	148	41
		R	0.700	0.075	0.0250	—	—	—	0.039	140	26
	Navigator	F	1.023	0.964	0.0025	—	—	—	0.059	148	41
		R	0.700	0.075	0.0250	—	—	—	0.039	140	26
2005-06	Expedition	F	1.023	0.106	0.0025	—	—	—	0.118	148	41
		R	0.700	0.075	0.0250	—	—	—	0.039	140	20
	Navigator	F	1.023	0.106	0.0025	—	—	—	0.118	148	41
		R	0.700	0.075	0.0250	—	—	—	0.039	140	20
	Excursion	F	1.023	1.140	0.0025	—	—	—	0.118	166	42
		R	0.700	1.10	0.0250	—	—	—	0.118	150	27

NOTE: Due to changes made during production, refer to manufacturer's specifications if they differ from those in this chart

F: Front

R: Rear

06017-NAVI-C12

SCHEDULED MAINTENANCE INTERVALS
FORD—2002—03 EXPEDITION & LINCOLN—NAVIGATOR

TO BE SERVICED	TYPE OF SERVICE	VEHICLE MILEAGE INTERVAL (x1000)																							
		5	10	15	20	25	30	35	40	45	50	55	60	65	70	75	80	85	90	95	100	105	110	115	120
Accessory drive belt	S/I												✓												✓
Air cleaner filter ①	R						✓						✓						✓			✓			✓
Automatic transmission fluid	R						✓						✓						✓			✓			✓
Automatic transmission shift linkage	S/I & L	✓	✓	✓	✓	✓	✓	✓	✓	✓	✓	✓	✓	✓	✓	✓	✓	✓	✓	✓	✓	✓	✓	✓	✓
Brake caliper, slide rails	L			✓			✓			✓			✓			✓			✓			✓			✓
Brake system, hoses & lines	S/I			✓			✓			✓			✓			✓			✓			✓			✓
Clutch reservoir fluid level	S/I	✓	✓	✓	✓	✓	✓	✓	✓	✓	✓	✓	✓	✓	✓	✓	✓	✓	✓	✓	✓	✓	✓	✓	✓
Engine coolant ②	R										✓					✓							✓		
Engine cooling system hoses, clamps & coolant	S/I			✓			✓			✓			✓			✓			✓			✓			✓
Engine oil & filter	R	✓	✓	✓	✓	✓	✓	✓	✓	✓	✓	✓	✓	✓	✓	✓	✓	✓	✓	✓	✓	✓	✓	✓	✓
Exhaust system	S/I	✓	✓	✓	✓	✓	✓	✓	✓	✓	✓	✓	✓	✓	✓	✓	✓	✓	✓	✓	✓	✓	✓	✓	✓
Front wheel bearings	S/I & L						✓						✓						✓						✓
Front/rear axle driveshaft slip yoke	L						✓						✓						✓						✓
Front/rear axle fluid ③	R																				✓				
Fuel filter	R			✓			✓			✓			✓			✓			✓			✓			✓
Manual transmission fluid	R												✓												✓
Parking brake system	S/I						✓						✓						✓						✓
PCV valve	R												✓												✓
Rotate tires	S/I	✓	✓	✓	✓	✓	✓	✓	✓	✓	✓	✓	✓	✓	✓	✓	✓	✓	✓	✓	✓	✓	✓	✓	✓
Spark plugs	R																					✓			
Steering linkage, suspension, driveshaft U joints	S/I & L	✓	✓	✓	✓	✓	✓	✓	✓	✓	✓	✓	✓	✓	✓	✓	✓	✓	✓	✓	✓	✓	✓	✓	✓

R: Replace   S/I: Service or Inspect

① Perform this at the mileage shown or every 30 months, whichever occurs first.

② Drain, flush and refill the cooling system initially at 50,000 miles or 48 months, whichever occurs first, then every 30,000 miles or 30 months thereafter.

③ The axle lubricant must be replaced every 100,000 miles or if the axle has been submerged under water. Otherwise the lube should not be checked or changed unless a repair is required.

FREQUENT OPERATION MAINTENANCE (SEVERE SERVICE)

If a vehicle is operated under any of the following conditions it is considered severe service:

- Towing a trailer or using a camper or car-top carrier.

- Repeated short trips of less than 5 miles in temperatures below freezing, or trips of less than 10 miles in any temperature.

- Extensive idling or low-speed driving for long distance as in heavy commercial use, such as delivery, taxi or police cars.

- Operating on rough, muddy or salt-covered roads.

- Operating on unpaved or dusty roads.

- Driving in extremely hot (over 90°) conditions.

Engine oil & filter: replace every 3000 miles.

Tires: rotate and inspect every 6000 miles.

Clutch reservoir fluid level: inspect every 6000 miles.

Automatic transmission shift linkage: lubricate every 6000 miles.

Steering linkage: suspension, U-joints: lubricate every 6000 miles.

Exhaust system: inspect for leaks of damage every 6000 miles.

Fuel filter: replace every 15,000 miles.

Automatic transmission fluid: change every 21,000 miles.

Crankcase emission air filter: replace every 60,000 miles.

PCV valve: replace every 60,000 miles.

Accessory drive belt: inspect every 60,000 miles.

Spark plugs: replace every 99,000 miles.

SCHEDULED MAINTENANCE INTERVALS
FORD—EXCURSION (Except 6.0L Diesel Engine)

TO BE SERVICED	TYPE OF SERVICE	VEHICLE MILEAGE INTERVAL (x1000)																			
		5	10	15	20	25	30	35	40	45	50	55	60	65	70	75	80	85	90	95	100
Accessory drive belt	S/I																				✓
Air cleaner filter ①②	R						✓						✓						✓		
Automatic transmission fluid ③	R						✓						✓						✓		
Engine coolant ④⑤	R										✓					✓					
Engine cooling system hoses, clamps & coolant ⑥	S/I			✓			✓			✓			✓			✓			✓		
Engine oil & filter	R	✓	✓	✓	✓	✓	✓	✓	✓	✓	✓	✓	✓	✓	✓	✓	✓	✓	✓	✓	✓
Exhaust system	S/I			✓			✓			✓			✓			✓			✓		
Front wheel bearings	S/I & L																✓				
Front/rear axle lubricant ⑦	R																				✓
Fuel filter ⑧	R						✓						✓						✓		
PCV valve	R	Every 120,000 miles																			
Rotate tires	S/I	✓	✓	✓	✓	✓	✓	✓	✓	✓	✓	✓	✓	✓	✓	✓	✓	✓	✓	✓	✓
Spark plugs	R																				✓
Steering linkage, suspension, driveshaft U joints	S/I & L	✓	✓	✓	✓	✓	✓	✓	✓	✓	✓	✓	✓	✓	✓	✓	✓	✓	✓	✓	✓

R: Replace   S/I: Service or Inspect

① Perform this at the mileage shown or every 30 months, whichever occurs first.

② 7.3L DIT Diesel engine: the air filter should be replaced when the restriction gauge is in the red zone.

③ Except the E40D transmission.

④ Drain, flush and refill the cooling system initially at 50,000 miles or 48 months, whichever occurs first, then every 30,000 miles or 30 months thereafter.

⑤ 7.3L DIT Diesel engine: add 4 pints of FW-15 each time the coolant is replaced.

⑥ 7.3L DIT Diesel engine: add 8-10 oz. of FW-15 to the engine coolant every 15,000 miles.

⑦ The axle lubricant must be replaced every 100,000 miles of if the axle has been submerged under water. Otherwise the lube should not be checked or changed unless a repair is required.

⑧ 7.3L DIT Diesel engine: the fuel filter should be replaced when the restriction lamp is illuminated.

FREQUENT OPERATION MAINTENANCE (SEVERE SERVICE)

If a vehicle is operated under any of the following conditions it is considered severe service:
- Towing a trailer or using a camper or car-top carrier.
- Repeated short trips of less than 5 miles in temperatures below freezing, or trips of less than 10 miles in any temperature.
- Extensive idling or low-speed driving for long distances as in heavy commercial use, such as delivery, taxi or police cars.
- Operating on rough, muddy or salt-covered roads.
- Operating on unpaved or dusty roads.

Engine oil & filter: replace every 3000 miles.
Tires: rotate and inspect every 6000 miles.
Steering linkage, suspension, U-joints: lubricate every 6000 miles.
Exhaust system: inspect for leaks or damage every 12,000 miles.
Fuel filter: replace every 15,000 miles.
Automatic transmission fluid: change ever 21,000 miles.
Front wheel bearings (2WD): inspect and repack every 30,000 miles.
Rear axle lubricant (E-Super Duty only): replace every 30,000 miles.
Spark plugs (except 4.2L engine): replace every 60,000 miles.
PCV valve: replace every 60,000 miles.
Accessory drive belt: inspect every 60,000 miles.
Spark plugs: replace every 99,000 miles.

06017-NAVI-C14

SCHEDULED MAINTENANCE INTERVALS
FORD—2004—05 EXPEDITION & LINCOLN—NAVIGATOR

TO BE SERVICED	TYPE OF SERVICE	VEHICLE MILEAGE INTERVAL (x1000)												
		5	10	15	20	25	30	35	40	45	50	55	60	65
Engine oil & filter	R	✓	✓	✓	✓	✓	✓	✓	✓	✓	✓	✓	✓	✓
Tires	Rotate	✓	✓	✓	✓	✓	✓	✓	✓	✓	✓	✓	✓	✓
Wheels	I ①	✓	✓	✓	✓	✓	✓	✓	✓	✓	✓	✓	✓	✓
Auto trans. fluid	I			✓			✓			✓			✓	
Brake pads/shoes	I			✓			✓			✓			✓	
Coolant hoses	S/I			✓			✓			✓			✓	
Steering linkage	I			✓			✓			✓			✓	
Cabin air filter	R			✓			✓			✓			✓	
Ball joints (2wd)	L			✓									✓	
Exhaust system	I						✓						✓	
Engine air filter	R						✓						✓	
Fuel filter	R						✓						✓	
Auto trans fluid (4R100)	R						✓						✓	
Front wheel bearings (2wd)	R	at 150,000 miles, if not previously done so												
Spark plugs	R	every 100,000 miles												
PCV valve	R	every 120,000 miles												
Premium Gold coolant	R	every 5 years or 100,000 miles												
Auto trans fluid (all exc. 4R100)	R	every 150,000 miles												
Differential fluid	R	every 150,000 miles												
Accessory drive belts	R	every 150,000 miles, if not previously done so												

R: Replace   S: Service   I: Inspect   L: Lubricate

① Inspect for end play and noise

Special Operating Condition Requirements

When towing a trailer or using a camper or car-top carrier:

Change engine oil and install a new oil filter every 4,800 km (3,000 miles), 3 months or 200 hours of engine operation (whichever occurs first).

Change transfer case fluid every 96,000 km (60,000 miles).

Change manual transmission fluid as required.

Inspect and lubricate U-joints as required.

During extensive idling and/or low speed driving for long distances, as in heavy commercial use such as delivery, taxi, patrol car or livery:

Change engine oil and install a new oil filter every 4,800 km (3,000 miles), 3 months or 200 hours of engine operation (whichever occurs first).

Lube front lower control arm and steering linkage ball joints with zerk fittings (if equipped) every 4,800 km (3,000 miles) or 3 months.

Inspect brake system and check battery electrolyte level (Patrol cars) every 8,000 km (5,000 miles).

Install a new fuel filter every 24,000 km (15,000 miles).

Change automatic transmission fluid, lubricate 4x2 wheel bearings, install new grease seals and adjust bearings every 48,000 km (30,000 miles). If equipped, change the in-line service installed transmission fluid filter.

Install new spark plugs and change transfer case fluid every 96,000 km (60,000 miles).

Install a new cabin air filter as required.

When operating in dusty conditions such as unpaved or dusty roads:

Change engine oil and install a new oil filter every 4,800 km (3,000 miles) or 3 months.

Install a new fuel filter every 24,000 km (15,000 miles).

Change automatic transmission fluid every 48,000 km (30,000 miles). If equipped, change the in-line service installed transmission fluid filter.

Change transfer case fluid every 96,000 km (60,000 miles).

Install a new engine air filter as required.

Install a new cabin air filter as required.

When operating in off-road conditions:

Change automatic transmission fluid every 48,000 km (30,000 miles). If equipped, change the in-line service installed transmission fluid filter.

Change transfer case fluid every 96,000 km (60,000 miles).

Install a new cabin air filter as required.

Inspect and lubricate U-joints.

Inspect and lubricate steering linkage ball joints with zerk fittings.

SCHEDULED MAINTENANCE INTERVALS
2004—05 Excursion With Gasoline Engines

TO BE SERVICED	TYPE OF SERVICE	VEHICLE MILEAGE INTERVAL (x1000)												
		5	10	15	20	25	30	35	40	45	50	55	60	65
Engine oil & filter	R	✓	✓	✓	✓	✓	✓	✓	✓	✓	✓	✓	✓	✓
Tires	Rotate	✓	✓	✓	✓	✓	✓	✓	✓	✓	✓	✓	✓	✓
Wheels	I ①	✓	✓	✓	✓	✓	✓	✓	✓	✓	✓	✓	✓	✓
Auto trans. fluid	I			✓			✓			✓			✓	
Brake pads, hoses, etc.	I			✓			✓			✓			✓	
Coolant hoses	I			✓			✓			✓			✓	
Steering linkage	I/L			✓			✓			✓			✓	
Cabin air filter	R			✓			✓			✓			✓	
Ball joints (2wd)	L			✓			✓			✓			✓	
Driveshaft	I			✓			✓			✓			✓	
4x4 front axle U-joints	L			✓			✓			✓			✓	
Exhaust system	I						✓						✓	
Engine air filter	R						✓						✓	
Fuel filter	R						✓						✓	
Auto trans fluid (4R100)	R						✓						✓	
4x2 front wheel bearings	Adj.												✓	
4x2 front wheel bearing grease seals	R												✓	
Accessory drive belts	I	Every 100,000 miles												
Front wheel bearings (2wd)	R	at 150,000 miles, if not previously done so												
4x4 front hub needle bearings	L	every 120,000 miles												
Spark plugs	R	every 100,000 miles												
PCV valve	R	every 120,000 miles												
Manual trans. Fluid	R	every 120,000 miles												
Premium Gold coolant	R	every 3 years or 100,000 miles and every 50,000 miles thereafter												
Coolant (exc. Premium Gold)	R	every 135,000 miles												
Auto trans fluid (exc. 4R100)	R	every 150,000 miles												
Differential fluid	R	at 150,000 miles then every 50,000 thereafter												
Transfer case fluid	R	every 150,000 miles												
Accessory drive belts	R	every 150,000 miles, if not previously done so												

R: Replace   S: Service   I: Inspect   L: Lubricate   Adj: adjust

① Inspect for end play and noise

Special Operating Condition Requirements

When towing a trailer or using a camper or car-top carrier:

Change engine oil and install a new oil filter every 4,800 km (3,000 miles), 3 months or 200 hours of engine operation (whichever occurs first).

Change transfer case fluid every 96,000 km (60,000 miles).

Change manual transmission fluid as required.

Inspect and lubricate U-joints as required.

During extensive idling and/or low speed driving for long distances, as in heavy commercial use such as delivery, taxi, patrol car or livery:

Change engine oil and install a new oil filter every 4,800 km (3,000 miles), 3 months or 200 hours of engine operation (whichever occurs first).

Lube front lower control arm and steering linkage ball joints with zerk fittings (if equipped) every 4,800 km (3,000 miles) or 3 months.

Inspect brake system and check battery electrolyte level (Patrol cars) every 8,000 km (5,000 miles).

Install a new fuel filter every 24,000 km (15,000 miles).

Change automatic transmission fluid, lubricate 4x2 wheel bearings, install new grease seals and adjust bearings every 48,000 km (30,000 miles). If equipped, change the in-line service installed transmission fluid filter.

Install new spark plugs and change transfer case fluid every 96,000 km (60,000 miles).

Install a new cabin air filter as required.

SCHEDULED MAINTENANCE INTERVALS
2004 Excursion With Gasoline Engines
Footnotes Continued

When operating in dusty conditions such as unpaved or dusty roads:

Change engine oil and install a new oil filter every 4,800 km (3,000 miles) or 3 months.

Install a new fuel filter every 24,000 km (15,000 miles).

Change automatic transmission fluid every 48,000 km (30,000 miles). If equipped, change the in-line service installed transmission fluid filter.

Change transfer case fluid every 96,000 km (60,000 miles).

Install a new engine air filter as required.

Install a new cabin air filter as required.

When operating in off-road conditions:

Change automatic transmission fluid every 48,000 km (30,000 miles). If equipped, change the in-line service installed transmission fluid filter.

Change transfer case fluid every 96,000 km (60,000 miles).

Install a new cabin air filter as required.

Inspect and lubricate U-joints.

Inspect and lubricate steering linkage ball joints with zerk fittings.

06017-NAVI-C17

SCHEDULED MAINTENANCE INTERVALS
6.0L Deisel Engine

TO BE SERVICED	TYPE OF SERVICE	VEHICLE MILEAGE INTERVAL (x1000)												
		7.5	15	22.5	30	37.5	45	52.5	60	67.5	75	82.5	90	97.5
Engine oil & filter	R	✓	✓	✓	✓	✓	✓	✓	✓	✓	✓	✓	✓	✓
Tires	Rotate	✓	✓	✓	✓	✓	✓	✓	✓	✓	✓	✓	✓	✓
Air filter minder	I①	✓	✓	✓	✓	✓	✓	✓	✓	✓	✓	✓	✓	✓
Wheels	I②	✓	✓	✓	✓	✓	✓	✓	✓	✓	✓	✓	✓	✓
Brake pads, hoses, etc.	I		✓		✓		✓		✓		✓		✓	
Coolant hoses	I		✓		✓		✓		✓		✓		✓	
Steering linkage and suspension	I/L		✓		✓		✓		✓		✓		✓	
Cabin air filter	R		✓		✓		✓		✓		✓		✓	
Ball joints (2wd)	L		✓		✓		✓		✓		✓		✓	
Driveshaft	I/L		✓		✓		✓		✓		✓		✓	
4x4 front axle U-joints	L		✓		✓		✓		✓		✓		✓	
Exhaust system and heat shields	I		✓		✓		✓		✓		✓		✓	
Engine air filter	R				✓				✓				✓	
Fuel filters③	R		✓		✓		✓		✓		✓		✓	
Auto trans fluid④	R				✓				✓				✓	
4x2 front wheel bearings	L								✓					
4x2 front wheel bearing grease seals	R								✓					
Accessory drive belts	I													✓
4x4 front hub needle bearings	L								✓					
Manual trans. fluid	R								✓					
Coolant	R										✓			✓
Rear differential fluid⑤	R													✓
Transfer case fluid	R	every 150,000 miles												
4x2 front wheel bearings	R	every 150,000 miles												
Accessory drive belts	R	every 150,000 miles, if not previously done so												

R: Replace S: Service I: Inspect L: Lubricate Adj: adjust

① Reset after new filter is installed

② Inspect for end play and noise

③ Frame-mounted and engine

④ Including external and in-line filters

⑤ Dana axles using non-synthetic fluid only

Special Operating Condition Requirements

When towing a trailer or using a camper or car-top carrier:

Change engine oil and install a new oil filter every 4,800 km (3,000 miles), 3 months or 200 hours of engine operation (whichever occurs first).

Change transfer case fluid every 96,000 km (60,000 miles).

Change manual transmission fluid as required.

Inspect and lubricate U-joints as required.

06017-NAVI-C18

SCHEDULED MAINTENANCE INTERVALS
6.0L Deisel Engine

Footnotes Continued

During extensive idling and/or low speed driving for long distances, as in heavy commercial use such as delivery, taxi, patrol car or livery:

Change engine oil and install a new oil filter every 4,800 km (3,000 miles), 3 months or 200 hours of engine operation (whichever occurs first).

Lube front lower control arm and steering linkage ball joints with zerk fittings (if equipped) every 4,800 km (3,000 miles) or 3 months.

Inspect brake system and check battery electrolyte level (Patrol cars) every 8,000 km (5,000 miles).

Install a new fuel filter every 24,000 km (15,000 miles).

Change automatic transmission fluid, lubricate 4x2 wheel bearings, install new grease seals and adjust bearings every

48,000 km (30,000 miles). If equipped, change the in-line service installed transmission fluid filter.

Change transfer case fluid every 96,000 km (60,000 miles).

Install a new cabin air filter as required.

When operating in dusty conditions such as unpaved or dusty roads:

Change engine oil and install a new oil filter every 4,800 km (3,000 miles) or 3 months.

Install a new fuel filter every 24,000 km (15,000 miles).

Change automatic transmission fluid every 48,000 km (30,000 miles). If equipped, change the in-line service installed transmission fluid filter.

Change transfer case fluid every 96,000 km (60,000 miles).

Install a new engine air filter as required.

Install a new cabin air filter as required.

When operating in off-road conditions:

Change automatic transmission fluid every 48,000 km (30,000 miles). If equipped, change the in-line service installed transmission fluid filter.

Change transfer case fluid every 96,000 km (60,000 miles).

Install a new cabin air filter as required.

Inspect and lubricate U-joints.

Inspect and lubricate steering linkage ball joints with zerk fittings.

06017-NAVI-C19

GASOLINE ENGINE REPAIR

➡**Disconnecting the negative battery cable on some vehicles may interfere with the functions of the on board computer systems and may require the computer to undergo a relearning process, once the negative battery cable is reconnected.**

Distributor

These vehicles are equipped with a Distributorless Ignition System (DIS).

Alternator

REMOVAL

1. Before servicing the vehicle, refer to the precautions section.
2. Remove or disconnect the following:
   • Negative battery cable
   • Air cleaner assembly, if necessary
   • Alternator bracket bolts, if necessary
   • Ignition wire pin-type retainer from the generator bracket, 4.6L engines
   • Drive belt from the alternator pulley
   • Electrical harness connectors at the alternator assembly
   • Positive battery cable, the nut and washer
   • 2 front alternator bolts
   • Rear alternator support bracket retaining bolts and the support bracket
   • Alternator from the vehicle

INSTALLATION

1. Before servicing the vehicle, refer to the precautions section.
2. Install or connect the following:
   • Alternator in position
   • 2 front alternator retaining bolts, loosely
   • Alternator bracket and 3 alternator bracket bolts. Tighten to 84 inch lbs. (10 Nm).
   • Tighten 2 front alternator retaining bolts to 19 ft. lbs. (26 Nm)
   • 2 electrical harness connectors to the alternator assembly
   • Positive battery cable, the nut and washer. Tighten the nut to 72 inch lbs. (8 Nm).
   • Drive belt on the alternator pulley
   • Ignition wire pin-type retainer from the generator bracket, 4.6L engines
   • Alternator bracket bolts, if removed

• Air cleaner assembly, if removed
• Negative battery cable
3. Start the engine and check for proper charging system operation.

Ignition Timing

ADJUSTMENT

Distributorless Ignition System

Base timing for distributorless ignition engines is set at the factory at 10 degrees Before Top Dead Center (BTDC) and is not adjustable.

Engine Assembly

REMOVAL & INSTALLATION

4.6L, 5.4L SOHC and 6.8L Engines

EXCURSION MODELS

1. Before servicing the vehicle, refer to the precautions section.
2. Disconnect the negative battery cable.
3. Remove the hood.
4. Drain the cooling system.
5. Remove the radiator grille supports.
6. Remove the radiator.
7. Recover the A/C system refrigerant.
8. Remove the condenser
9. Disconnect the two 42 pin and 16 ping connectors.
10. Remove the intake manifold.
11. Remove the lower radiator hose from the oil water cooler outlet.
12. Remove the power steering pump and set aside.
13. Disconnect the A/C cycling switch.
14. Disconnect the A/C compressor electrical connector.
15. Disconnect the suction hose at the accumulator.
16. Disconnect the heater hoses at the heater core.
17. Disconnect the Heated Oxygen Sensor (HO2S) electrical connectors.
18. Remove the bolts and position the A/C compressor aside.
19. Remove the exhaust pipe.
20. Remove the transmission and clutch, if equipped with a manual transmission.
21. Drain the oil and remove the bypass filter.
22. Loosen the threaded shaft and position the oil cooler aside.

23. Remove the starter, if equipped with an automatic transmission.
24. Remove the transmission cooler hoses from the block bracket, if equipped with an automatic transmission.
25. Remove the stud bolt and position the ground strap aside.
26. Remove the nut and position the wiring bracket aside.
27. Remove the flywheel inspection cover, if equipped with an automatic transmission.
28. Remove the access plug, if equipped with an automatic transmission.
29. Remove the torque converter nuts, if equipped with an automatic transmission.
30. Support the transmission, using a block of wood and a floor jack, if equipped with an automatic transmission.
31. Remove the four lower transmission-to-engine bolts, if equipped with an automatic transmission.
32. Remove the nut retaining the ground strap, and on automatic transmission equipped vehicles remove the transmission oil filler tube.
33. Remove the two studs and the heater water tube.
34. Remove the three upper transmission to engine bolts, if equipped with an automatic transmission.
35. Install modular engine lifting bar 303-F047 tool.
36. Remove the engine support insulator nuts.
37. Support the transmission using a transmission jack, if equipped with an automatic transmission.
38. Remove the engine from the vehicle.
   To install:
39. Install the engine in the vehicle.
40. Install the engine support insulator nuts and tighten to 66 ft. lbs. (90 Nm).
41. Remove modular engine lifting bar 303-F047 tool.
42. Install the transmission oil level tube, and the ground strap on all vehicles equipped with an automatic transmission.
43. Install the heater water tube and the two studs and tighten to 30 ft. lbs. (40 Nm) on 5.4L and 2005 6.8L engines and 15–22 ft. lbs. (20–30 Nm) on 2002–04 6.8L engines.
44. Align the transmission to the engine and install the five lower transmission-to-engine bolts. Tighten to 35 ft. lbs. (48 Nm), if equipped with an automatic transmission.
45. Install and tighten the four new nuts retaining the torque converter to 26 ft. lbs.

(35 Nm) , if equipped with an automatic transmission.

46. Install the access plug, if equipped with an automatic transmission.

47. Install the flywheel inspection cover, if equipped with an automatic transmission.

48. Install the wiring bracket and tighten to 18 ft. lbs. (25 Nm).

49. Install the starter motor, if equipped with an automatic transmission.

❋❋ CAUTION

A new oil cooler must be installed or severe damage to the engine can occur.

50. Install the oil cooler to the oil filter adapter and install the oil bypass filter. Tighten to 43 ft. lbs. (58 Nm).

51. Install the clutch and transmission, if equipped with a manual transmission

52. Tighten the exhaust pipe nuts to 30 ft. lbs. (40 Nm).

53. Install the A/C compressor

54. Install the ground strap.

55. Install the transmission cooler line bracket, if equipped with an automatic transmission and tighten to 18 ft. lbs. (25 Nm).

56. Install the remaining transmission-to-engine bolts, tighten to 35 ft. lbs. (48 Nm), if equipped with an automatic transmission.

57. Install the fuel line bracket.

58. Connect the transmission wiring harness

59. Connect the HO$_2$S electrical connectors.

60. Connect the heater hose.

61. Install the suction line to the accumulator.

62. Connect the A/C cycling switch.

63. Connect the A/C compressor clutch connector.

64. Connect the lower radiator hose to the oil cooler inlet.

65. Install the power steering pump and tighten to 18 ft. lbs. (25 Nm).

66. Install the intake manifold.

67. Connect the 42-pin and 16-pin connectors.

68. Install the condenser.

69. Install the radiator and engine cooling fan

70. Install the radiator grille supports.

71. Install the engine air cleaner.

72. Fill the engine with clean engine oil.

❋❋ CAUTION

The oil pump must be primed prior to starting the engine.

73. Connect the negative battery cable.
74. Fill the engine cooling system.
75. Fill the power steering system.
76. Recharge the A/C system.
77. Install the hood.

2002 EXPEDITION AND NAVIGATOR MODELS

1. Before servicing the vehicle, refer to the precautions section.

2. Remove the intake manifold.

3. Remove the hood.

4. Remove the engine air cleaner and reposition it to gain access to the Mass Air Flow (MAF) sensor electrical connector.

5. Disconnect the MAF sensor electrical connector and remove the air cleaner.

6. Remove the radiator.

7. Recover the A/C system refrigerant.

8. Remove the A/C condenser core.

9. Disconnect the climate control vacuum connector.

10. Remove the bolts and position the power steering reservoir aside.

11. Disconnect the exhaust manifold to Exhaust Gas Recirculation (EGR) valve tube lower fitting and remove the tube.

12. Disconnect the alternator electrical connections and remove the alternator.

13. Disconnect the 38-pin, 42-pin and 16-pin electrical connectors.

14. Disconnect the electrical connector and remove the connector from the bracket.

15. Remove the bolts and the connector bracket cover.

16. Remove the 38-pin connector from the bracket.

17. Disconnect the starter relay electrical connections and position the cables aside.

18. Remove the bolts and the connector bracket.

19. Compress the holding tabs and disconnect the heater water hoses from the heater core.

20. Remove the A/C compressor.

21. Remove the drain plug and drain the engine oil.

22. Remove the starter motor.

23. Remove the A/C line (muffler) bracket and position aside.

24. Remove the transmission fluid cooler tubes from the block-mounted clip.

25. Position the shift cable and bracket aside, by disconnecting the shift cable then remove the fasteners and position the bracket and shift cable aside.

26. Remove the bolts and the flexplate inspection cover.

27. Remove the cylinder block opening cover.

28. Remove the torque converter-to-flexplate nuts and discard the nuts.

➡**The upper two transmission-to-engine bolts will be removed later.**

29. Remove the lower five transmission-to-engine bolts.

30. Remove the nut and the transmission fluid filler tube.

31. Remove the bolts and position the power steering pump aside.

32. Disconnect the oil pressure switch electrical connection.

33. Disconnect the heated Oxygen Sensor (HO$_2$S) electrical connectors.

34. Remove the right hand exhaust manifold to three-way catalytic converter nuts.

35. Remove the right hand exhaust manifold studs.

36. Remove the left hand exhaust manifold to three-way catalytic converter nuts.

37. Position the Y-pipe aside.

38. Remove the right hand motor mount bolt.

39. Remove the left hand motor mount bolt.

40. Remove the oil bypass filter.

41. Remove the oil pressure switch.

42. Slide back the hose clamp and disconnect the lower radiator hose from the oil filter adapter.

43. Lower the vehicle.

44. Support the transmission.

45. Remove the upper two transmission-to-engine bolts and the wiring harness bracket.

46. Remove the heater water inlet tube and discard the O-ring seal.

47. Install the modular holding tool 303-F047.

48. Remove the engine.

To install:

➡**Align the engine-to-transmission dowels before installing engine to transmission bolts.**

49. Position the engine in the vehicle.

50. Remove the modular holding tool 303-F047.

51. Raise and support the vehicle.

52. Install the lower five transmission-to-engine bolts and tighten to 35 ft. lbs. (48 Nm).

53. Install the torque converter-to-flexplate nuts and tighten to 26 ft. lbs. (35 Nm).

54. Install the cylinder block opening cover.

55. Install the flexplate inspection cover and tighten to 25 ft. lbs. (34 Nm).

56. Install the bracket and shift cable, by positioning the bracket and install the fasteners, the connect the shift cable.

57. Position and install and the left hand engine mount bolt and tighten to 66 ft. lbs. (90 Nm).

58. Position and install the right hand engine mount bolt and tighten to 66 ft. lbs. (90 Nm).

59. Install the starter motor.

60. Install the right hand exhaust manifold studs.

61. Install and tighten the four nuts exhaust pipe nuts and tighten to 30 ft. lbs. (40 Nm).

62. Position the lower radiator hose on the oil filter adapter and secure the hose clamp.

➡**Only three bolts are required for installation.**

63. Install the power steering pump bolts and tighten to 18 ft. lbs. (25 Nm).

64. Install the oil bypass filter and tighten to 12 ft. lbs. (16 Nm).

65. Install the oil pressure switch.

66. Connect the oil pressure switch electrical connector.

67. Install the transmission fluid cooler tubes in the clip.

68. Install the A/C line (muffler) bracket.

69. Connect the HO2S sensor electrical connectors.

➡**Lubricate the O-ring seals with transmission fluid.**

70. Install the transmission fluid filler tube and nut.

71. Install the A/C compressor.

➡**Do not reuse the O-ring seal.**

72. On Windsor engines, slide the water heater tube forward over the new O-ring seal.

73. Lubricate the O-ring seal with engine coolant.

74. Position the transmission wiring bracket and install the upper two transmission-to-engine bolts and tighten to 35 ft. lbs. (48 Nm).

75. Connect the transmission wiring harness connectors.

76. Connect the two heater hoses at the heater core.

77. Install the connector bracket.

78. Connect the starter relay electrical connections.

79. Install the 38-pin connector in the bracket.

80. Install the connector bracket cover.

81. Connect the electrical connector and install it on the bracket.

82. Connect the 38-pin, 42-pin and 16-pin electrical connectors.

83. Position the exhaust manifold to EGR valve tube and hand-tighten the lower fitting. Then tighten the fitting to 26–33 ft. lbs. (35–45 Nm).

84. Position the power steering reservoir

and install the two lower bolts and tighten to 17 ft. lbs. (23 Nm).

85. Connect the climate control vacuum connector.

86. Connect the alternator electrical connections and install the alternator.

87. Install the A/C condenser and radiator.

88. Position the air cleaner in the vehicle and connect the MAF sensor electrical connector.

89. Install the air cleaner.

90. Install the hood.

91. Install the intake manifold.

92. Fill all fluids to the correct levels.

93. Start the engine and check for leaks. Stop the engine and recheck the fluid levels.

94. Recharge the A/C system. For

2003–04 EXPEDITION AND NAVIGATOR MODELS

1. Before servicing the vehicle, refer to the precautions section.

2. Disconnect the negative battery cable.

3. If equipped, turn the air suspension switch off.

4. Remove the intake manifold.

5. Remove the hood.

6. Remove the engine air cleaner and reposition it to gain access to the Mass Air Flow (MAF) sensor electrical connector and disconnect the sensor connector.

7. Remove the radiator.

8. Recover the A/C system refrigerant.

9. Remove the condenser.

10. Disconnect the Powertrain Control Module (PCM) electrical connectors.

11. Disconnect the engine wiring harness connector.

12. Remove the PCM mounting bolts and the PCM.

13. Remove the bolts and the PCM mounting bracket.

14. Remove the ground wire.

15. Disconnect the heater hoses at the heater core.

16. Disconnect the evaporator tubes from the manifold.

17. Remove the cable routing bracket from the heater outlet tube mounting studs.

18. On 5.4L engines disconnect the knock sensor connector.

19. Remove the wiring retainers from the heater outlet tube bracket.

20. Disconnect the hoses from the heater outlet tube.

21. Remove the heater outlet tube upper and lower studs on 4.6L engines.

22. Loosen the heater outlet tube upper and lower studs on 5.4L engines.

23. Remove the heater outlet tube and discard the O-ring.

24. Remove the wiring harness retainer from the front cover stud.

25. Remove the drive belt.

26. Remove the alternator.

27. Remove the power steering pump reservoir and set aside.

28. Remove the nuts and disconnect the transmission cooler tube and the power steering hose brackets.

29. Remove the power steering pump pulley using removal tool 211-016.

30. Disconnect the power steering pressure hose and drain the power steering fluid into a suitable container.

31. Remove the three bolts and position the power steering pump out of the way.

32. Disconnect the power steering pressure switch connector, if equipped.

33. Remove the drain plug and drain the engine oil and install the drain plug when finished.

34. Remove the oil filter.

35. Remove the lower radiator hose and degas bottle hose from the oil filter adapter.

36. Disconnect the Exhaust Gas Recirculation (EGR) valve tube lower fitting and remove the tube.

37. Disconnect the Crankshaft Position (CKP) sensor electrical connector.

38. Disconnect the A/C compressor electrical connector.

39. Remove the bolts and position the A/C compressor aside.

40. Remove the bolt and disconnect the wiring harness bracket.

41. Remove the starter.

42. Remove the transmission filler tube-to-engine block bolt which is accessed through the ride side fender well.

43. Disconnect the heated Oxygen Sensor (HO2S) electrical connectors.

44. Remove the nuts and detach the dual converter Y-pipe.

45. Remove the shift cable and bracket.

46. Remove the flexplate inspection cover.

47. Remove the cylinder block opening cover.

48. Remove the torque converter nuts.

49. Support the transmission, using a block of wood and a floor jack.

50. Remove the five lower transmission-to-engine bolts.

51. Remove the right side motor mount.

➡**If the left side engine mount through bolt nut is missing or damaged install a new nut using service part number W709375. If the left side engine mount through bolt nut cage is damaged or missing, remove the cage and install a new nut using service part number W520516-5301.**

67197-NAVI-G01

Install modular engine lifting bar 303-F047 tool

52. Remove the left side mount through bolt.

➡**On 4x4 vehicles, disconnect the transfer case vent hose from the bracket and position aside to gain access to the upper transmission-to-engine bolts.**

53. Remove the two upper transmission-to-engine bolts.

➡**On some engines it will be necessary to shim between the modular engine lifting bar 303-F047 tool mounting brackets and the cylinder block.**

54. Install modular engine lifting bar 303-F047 tool.

55. Remove the engine from the vehicle.

To install:

56. Install the engine in the vehicle

57. Remove modular engine lifting bar 303-F047 tool.

58. Remove the floor jack and wood block supporting the transmission.

59. Align the transmission to the engine and install the five lower transmission-to-engine bolts. Tighten to 35 ft. lbs. (48 Nm).

➡**If the left side engine mount through bolt nut is missing or damaged install a new nut using service part number W709375. If the left side engine mount through bolt nut cage is damaged or missing, remove the cage and install a new nut using service part number W520516-5301.**

60. Install the left side motor mount bolt and tighten to 148 ft. lbs. (200 Nm).

61. Install the right side motor mount nut and tighten to 66 ft. lbs. (90 Nm) on 2003 models, or 129 ft. lbs. (175 Nm) on 2004 models.

62. Install the four torque converter nuts and tighten to 26 ft. lbs. (35 Nm).

63. Install the cylinder block opening cover.

64. Install the flexplate inspection cover and tighten the bolts to 25 ft. lbs. (34 Nm).

65. Install the shift cable and bracket.

66. Install the dual converter Y-pipe and tighten the four nuts to 30 ft. lbs. (40 Nm).

67. Connect the HO_2S electrical connectors.

68. Install the transmission filler tube.

69. Install the starter.

70. Install the wiring harness bracket.

71. Install the A/C compressor and tighten the bolts to 18 ft. lbs. (25 Nm).

72. Connect the CKP sensor electrical connector.

73. Connect the A/C compressor electrical connector.

74. Connect the EGR valve tube and loosely install the lower fitting.

75. Install the lower radiator hose and degas bottle hose.

76. Install a new oil filter.

77. Connect the power steering pressure switch connector, if equipped.

78. Install the power steering pump and tighten the three bolts to 18 ft. lbs. (25 Nm).

79. Using the Teflon seal installer tool 211-D027, install a new O-ring seal on the power steering pressure hose fitting.

80. Connect the power steering pressure hose and tighten the fitting to 15 ft. lbs. (20 Nm).

81. Install the transmission cooler tube and the power steering hose brackets and tighten the nuts to 8 ft. lbs. (11 Nm) on 4.6L engines or 30 ft. lbs. (40 Nm) on 5.4L engines.

82. Install the power steering pump pulley using power steering pump pulley tool 211-185.

83. Install the power steering reservoir.

84. Install the alternator.

85. Install the drive belt.

86. Install the wiring harness retainer on the front cover stud.

➡**On 4x4 vehicles, connect the transfer case vent hose to the bracket after installing the upper transmission-to-engine bolts.**

211-D027

67197-NAVIG02

Use Teflon seal installer tool to install a new O-ring seal on the power steering pressure hose fitting—2003–04 Expedition and Navigator

87. Install the two upper transmission-to-engine bolts. Tighten to 35 ft. lbs. (48 Nm).

88. Install the heater outlet tube using a new O-ring.

89. Install the lower and upper heater outlet tube studs and tighten to 18 ft. lbs. (25 Nm) on 4.6L engines.

90. Tighten the lower and upper heater outlet tube studs to 30 ft. lbs. (40 Nm) on 5.4L engines.

91. Install all remaining hoses, tubes and electrical connectors.

92. Install the PCM and attach the electrical connections.

93. Install the condenser.

94. Install the radiator.

95. Connect the MAF sensor electrical connections and install the engine air cleaner.

96. Install the intake manifold.

97. Fill the engine with clean engine oil.

98. Connect the negative battery cable.

99. Fill the engine cooling system.

100. Fill the power steering system.

101. Recharge the A/C system.

102. Install the hood.

2005 EXPEDITION AND NAVIGATOR MODELS

1. Before servicing the vehicle, refer to the precautions section.

2. Disconnect the negative battery cable.

3. Remove the radiator.

4. Remove the intake manifold.

5. Remove the accessory drive belt.

6. Remove the Powertrain Control Module (PCM) and the support bracket.

7. Remove the starter.

8. Remove the hood.

9. Disconnect the heater hose.

10. Disconnect the degas bottle coolant hose.

11. If equipped, disconnect the block heater electrical connector.

12. Remove the power steering reservoir and support bracket lower bolt.

13. Remove the nut and position the power steering reservoir and support bracket aside.

14. Remove the bolts and position the power steering pump aside.

15. Disconnect the A/C compressor electrical connector.

16. Remove the bolts and position the A/C compressor aside.

17. Remove the nut and the transmission cooler tube support bracket.

18. Remove the nut and the power steering pressure hose support bracket.

19. Remove the bolt and the starter electrical harness support bracket.

20. Remove the bolts and the flexplate inspection cover.
21. Remove the cylinder block opening cover.
22. Remove the torque converter-to-flexplate nuts and discard the nuts.

➡ The upper 2 transmission-to-engine bolts will be removed later.

23. Remove the lower 5 transmission-to-engine bolts.
24. Disconnect the Heated Oxygen (HO2s) sensor connectors.
25. Remove the 4 exhaust manifold flange nuts.

✳✳ CAUTION

Only use hand tools when removing the engine mount nut or damage to the engine mount can occur.

26. Remove the right hand engine mount nut.

➡ If the left hand engine mount through bolt nut is missing or damaged install a new nut using service part number W709375. If the left hand engine mount through bolt nut cage is damaged or missing, remove the cage and install a new nut using service part number W520516-S301.

27. Remove the left hand engine mount through bolt.
28. Remove the upper 2 transmission-to-engine bolts and install the engine support tool 303-F047.
29. Using a suitable floor crane, remove the engine assembly from the vehicle.

To install:

30. Using a suitable floor crane, position the engine assembly into the vehicle.
31. Remove the engine support tool 303-F047.

✳✳ CAUTION

Only use hand tools when installing the left hand engine mount through bolt or damage to the cage nut can occur.

➡ If the left hand engine mount through bolt nut is missing or damaged install a new nut using service part number W709375. If the left hand engine mount through bolt nut cage is damaged or missing, remove the cage and install a new nut using service part number W520516-S301.

32. Install and the left and right hand engine mount through bolts and tighten to 129 ft. lbs. (175 Nm).

➡ Align the engine-to-transmission dowels before installing the engine-to-transmission bolts.

33. Install the lower 5 transmission-to-engine bolts and tighten to 35 ft. lbs. (48 Nm).
34. Install the 4 exhaust manifold-to-catalytic converter nuts and tighten to 30 ft. lbs. (40 Nm).
35. Connect the HO2s connectors.
36. Install the torque converter-to-flexplate nuts and tighten to 26 ft. lbs. (35 Nm).
37. Install the cylinder block opening cover.
38. Install the flexplate inspection cover and the 2 bolts and tighten to 26 ft. lbs. (35 Nm).
39. Install the starter wiring harness support bracket and bolt.
40. Install the upper two transmission-to-engine bolt and tighten to 35 ft. lbs. (48 Nm).
41. Install the transmission cooler tube support bracket and the nut.
42. Position the power steering pump and install the 3 bolts and tighten to 18 ft. lbs. (25 Nm).
43. Install the power steering pump support bracket and the nut.
44. Install the power steering reservoir and support bracket and install the bolt. Tighten to 18 ft. lbs. (25 Nm).
45. Install the power steering reservoir and support bracket nut. Tighten to 18 ft. lbs. (25 Nm).
46. Install the A/C compressor and install the 3 bolts. Tighten to 18 ft. lbs. (25 Nm).
47. Connect the A/C compressor electrical connector.
48. Connect the degas bottle coolant hose.
49. If equipped, connect the block heater electrical connector.
50. Connect the heater coolant hose.
51. Install the intake manifold.
52. Install the radiator.
53. Install the starter.
54. Install the PCM and the support bracket.
55. Install the hood.
56. Fill the crankcase with clean engine oil.
57. Fill and bleed the engine cooling system.
58. Start the vehicle and check for leaks.

5.4L DOHC Engine

2002 MODELS

✳✳ CAUTION

Fuel injection systems remain under pressure, even after the engine has been turned OFF. The fuel system pressure must be relieved before disconnecting any fuel lines. Failure to do so may result in fire and/or personal injury.

1. Relieve the fuel system pressure.
2. Before servicing the vehicle, refer to the precautions section.
3. Remove the hood.
4. Recover the (A/C) system.
5. Remove the lower intake manifold.
6. Remove the front air deflector.
7. Remove the radiator upper hose from the radiator.
8. Using the tools 303-240 and 303-239, remove the fan blade and fan clutch.
9. Remove the fan shroud. Lift and remove the fan shroud and fan blade, as a unit, out of the vehicle.

➡ To avoid disturbing the transmission oil cooler fittings, use a backup wrench.

10. Disconnect the transmission fluid cooler fittings.
11. Disconnect the lower radiator hose from the radiator.
12. Disconnect the degas hose from the radiator.
13. Remove the radiator.
14. Remove the accessory drive belt.
15. Disconnect the alternator electrical connectors.
16. Remove the lower bracket bolts and the alternator.
17. Disconnect the Exhaust Gas Recirculation (EGR) tube from the exhaust manifold.
18. Disconnect the transmission wiring harness retainers from the right cylinder head and position the harness aside.
19. Remove the bolts retaining the transmission to the engine.
20. Disconnect the bulkhead connectors.
21. Remove the ground strap retaining bolt.
22. Disconnect the degas return hose.
23. Disconnect the lower radiator hose.
24. Disconnect the heater hose.
25. Remove the nut and the battery-to-starter relay cable retainer.
26. Remove the nut and reposition the A/C muffler.

27. If equipped, remove the skid plate.

28. Disconnect the Power Steering Pressure (PSP) sensor.

➡**The power steering pump can be supported with mechanic's wire to keep it from hindering the removal process.**

29. Remove the bolts and position the power steering pump out of the way.

30. Remove the nut and the transmission cooler tubes from the front cover.

31. Remove the starter motor.

32. Disconnect the Crankshaft Position (CKP) sensor.

33. Remove the bolt and remove the manifold and tube assembly.

34. Disconnect the A/C compressor electrical connector.

35. Remove the bolts and the A/C compressor.

36. Disconnect the right and the left Heated Oxygen (HO2S) sensor.

37. Remove the nuts and disconnect the pipes at the exhaust manifolds.

38. Remove the right hand exhaust manifold studs.

39. Remove the bolt from the left hand motor mount.

40. Remove the bolt from the right hand motor mount.

41. Remove the bolts and the inspection cover.

42. Remove the plug to access the torque converter nuts.

➡**Gain access to the nuts by turning the crank dampener counterclockwise.**

43. Remove the four nuts from the torque converter and discard.

44. Drain the engine oil and install the drain plug when finished.

45. Remove the oil filter.

46. Remove the five bolts from the transmission.

47. Remove the bolt and position the transmission level indicator tube out of the way.

48. Lower the vehicle and install modular support too 303-Fo47.

49. Support the transmission with a floor jack and a block of wood.

50. Remove the engine from the vehicle.

To install:

51. Install the engine in the vehicle.

52. Remove the modular support tool.

53. Remove the transmission floor jack and block of wood.

54. Position the transmission level indicator tube and install the bolt. Tighten to 89 inch lbs. (10 Nm).

55. Install the five bolts to the bell housing. Tighten to 33 ft. lbs. (45 Nm).

56. Install the oil filter.

➡**Gain access to the nuts by turning the crank dampener counterclockwise.**

57. Install four new nuts to the torque converter. Tighten to 26 ft. lbs. (35 Nm).

58. Install the engine drain plug.

59. Install the inspection cover and the bolts. Tighten to 18 ft. lbs. (25 Nm).

60. Install the bolt to the right hand engine motor mount. Tighten to 59 ft. lbs. (80 Nm).

61. Install the bolt to the left hand engine motor mount. Tighten to 59 ft. lbs. (80 Nm).

62. Install the starter motor.

63. Install the right hand exhaust manifold studs.

64. Connect the right and the left HO2S sensors.

65. Connect the exhaust pipes to the exhaust manifolds and install the four nuts. Tighten to 20–30 ft. lbs. (27–41 Nm).

66. Install the A/C compressor and bolts. Tighten to 15–22 ft. lbs. (20–30 Nm).

67. Connect the A/C compressor electrical connector.

68. Install the manifold and tube assembly and bolt. Tighten to 14–18 ft. lbs. (18–24 Nm).

69. Connect the CKP sensor.

70. Install the transmission cooler line bracket. Tighten to 18 ft. lbs. (25 Nm).

71. Position the power steering pump and install the bolts. Tighten to 18 ft. lbs. (25 Nm).

72. Connect the power steering pressure sensor.

73. If equipped, install the skid plate and bolts.

74. Position the A/C muffler and install the nut. Tighten to 18 ft. lbs. (25 Nm).

75. Position the harness and install the nut. Tighten to 18 ft. lbs. (25 Nm).

76. Connect the heater hose.

77. Connect the lower radiator hose.

78. Connect the degas return hose.

79. Install the ground strap and the bolt.

80. Connect the bulkhead connectors.

81. Install the two bolts retaining the transmission to the engine. Tighten to 33 ft. lbs. (45 Nm).

82. Connect the transmission wiring harness retainers to the right cylinder head.

83. Connect the EGR tube to the exhaust manifold.

84. Position EGR tube and valve assembly and loosely install.

85. Install the alternator and the bolts. Tighten to 18 ft. lbs. (25 Nm).

86. Install the accessory drive belt.

87. Install the radiator.

88. Connect the lower radiator hoses.

➡**To avoid disturbing the transmission oil cooler fittings, use a backup wrench.**

89. Connect the transmission fluid cooler fittings. Tighten to 21 ft. lbs. (28 Nm).

90. Install the fan shroud, as a unit, in the vehicle.

91. Using the tools 303-240 and 303-239, install the fan blade and fan clutch.

92. Install the front air deflector.

93. Install the intake manifold.

94. Tighten the EGR tube to 18 ft. lbs. (25 Nm).

95. Fill the engine with oil.

96. Fill the engine cooling system.

97. Recharge the A/C system.

98. Install the hood.

2003–04 MODELS

1. Before servicing the vehicle, refer to the precautions section.

2. Disconnect the negative battery cable.

3. Remove the hood.

4. Remove the radiator.

5. Recover the A/C system refrigerant.

6. Remove the lower intake manifold.

7. Disconnect the powertrain control module (PCM) electrical connectors.

8. Remove the PCM mounting bolts and the PCM.

9. Remove the bolts and the PCM mounting bracket.

10. Disconnect the auxiliary heater hoses at the right side valve cover.

11. Disconnect the heater hoses at the heater core.

12. Disconnect the A/C hoses from the junction block.

13. Remove the safety clips and disconnect the spring lock couplers at the evaporator core.

14. Remove the drive belt.

15. Remove the alternator.

16. Remove the power steering pump pulley using power steering pump pulley tool 211-016.

17. Remove the nuts and disconnect the transmission cooler tube and the power steering hose brackets.

18. Disconnect the power steering pressure hose and drain the power steering fluid into a suitable container.

19. Remove the three bolts and position the power steering pump out of the way.

20. Disconnect the Exhaust Gas Recirculation (EGR) valve tube nut from the left side exhaust manifold and remove the tube.

21. Remove the engine noise shield.

22. Remove the positive crankcase venti-

lation (PCV) valve and hose assembly from the vehicle.

23. Remove the PVC tube.

24. Remove the bolt and disconnect the ground wire.

25. Disconnect the vacuum hose and remove the vacuum harness.

26. Disconnect the main engine harness electrical connector.

27. Remove the skid plate, if equipped.

28. Remove the transmission filler tube-to-engine block bolt which is accessed through the ride side fender well.

29. Disconnect the heated Oxygen Sensor (HO2S) electrical connectors.

30. Remove the nuts and detach the dual converter Y-pipe.

31. Remove the starter.

32. Disconnect the A/C clutch electrical connector.

33. Remove the bolts and position the A/C compressor aside.

34. Disconnect the Crankshaft Position (CKP) sensor electrical connector.

35. Remove the oil filter.

36. Remove the drain plug and drain the engine oil and install the drain plug when finished.

37. Remove the lower radiator hose.

38. Remove the inspection plate.

39. Remove the flexplate-to-torque converter nut access plug.

➡ **Mark one stud and the flexplate for assembly reference.**

40. Remove the four torque converter nuts.

41. Remove the six lower transmission-to-engine bolts.

➡ **The caged nut on the frame may break during bolt removal. If necessary, install a new nut.**

42. Remove the bolt from the left side motor mount and the nut from the right side motor mount.

43. Support the transmission, using a block of wood and a floor jack.

➡ **On 4x4 vehicles, disconnect the transfer case vent hose from the bracket and position aside to gain access to the upper transmission-to-engine bolts.**

44. Remove the two upper transmission-to-engine bolts.

➡ **The alternator bolts may be used to attach the modular engine lifting bar 303-F047 tool to the front of the engine. The two bolts used to attach the tool to the rear of the engine must**

be shorter than the transmission-to-engine mounting bolts.

45. Install modular engine lifting bar 303-F047 tool.

46. Using a floor crane, remove the engine from the vehicle.

To install:

47. Using a floor crane, install the engine in the vehicle. Make sure the marks on the flexplate and the torque converter stud made during removal are lined up.

48. Remove modular engine lifting bar 303-F047 tool.

49. Remove the floor jack and wood block supporting the transmission.

50. Install the right side motor mount nut and tighten to 66 ft. lbs. (90 Nm).

51. Install the left side motor mount bolt and tighten to 148 ft. lbs. (200 Nm).

52. Align the transmission to the engine and install the six lower transmission-to-engine bolts. Tighten to 33 ft. lbs. (45 Nm).

53. Install the four torque converter nuts and tighten to 26 ft. lbs. (35 Nm).

54. Install the flexplate-to-torque converter nut access plug.

55. Install the inspection plate and tighten the bolts to 25 ft. lbs. (34 Nm)

56. Install the lower radiator hose.

57. Install a new oil filter.

58. Connect the CKP sensor electrical connector.

59. Install the A/C compressor and tighten the bolts to 18 ft. lbs. (25 Nm).

60. Connect the A/C clutch electrical connector.

61. Install the starter.

62. Install the wire harness bracket and tighten the nut.

63. Install the dual converter Y-pipe and tighten the four nuts to 30 ft. lbs. (40 Nm).

64. Connect the HO2S electrical connectors.

65. Install the transmission filler tube.

66. Install the skid plate and the bolts, if equipped.

➡ **On 4x4 vehicles, connect the transfer case vent hose to the bracket after installing the upper transmission-to-engine bolts.**

67. Install the two upper transmission-to-engine bolts. Tighten to 33 ft. lbs. (45 Nm).

68. Connect the auxiliary heater hoses.

69. Connect the main engine harness electrical connector.

70. Connect the vacuum harness and the hose.

71. Connect the ground cable and tighten the bolt.

72. Install a new O-ring. Lubricate the O-ring with clean engine coolant. Install the PVC tube and tighten the bolt.

73. Attach the PVC hoses and install the clamps.

74. Install the engine noise shield.

75. Connect the EGR valve tube nut to the left side exhaust manifold and hand-tighten.

76. Install the power steering pump and tighten the three bolts to 18 ft. lbs. (25 Nm).

77. Using the Teflon seal installer tool 211-D027, install a new O-ring seal on the power steering pressure hose fitting.

78. Connect the power steering pressure hose and tighten the fitting to 15 ft. lbs. (20 Nm).

79. Install the transmission cooler tube and the power steering hose brackets and tighten the nuts to 18 ft. lbs. (25 Nm).

80. Install the power steering pump pulley using power steering pump pulley tool 211-185.

81. Install the alternator.

82. Install the drive belt.

83. Install all remaining hoses.

84. Install the PCM and attach the electrical connections.

85. Install the radiator and the radiator upper support brackets.

86. Install the degas hose to the radiator.

87. Install the front air deflector.

88. Install the splash shield.

89. Install the cooling fan assembly.

90. Install the lower intake manifold.

91. Tighten the EGR valve tube nut to 37 ft. lbs. (50 Nm).

92. Fill the engine with clean engine oil.

93. Connect the negative battery cable.

94. Fill the engine cooling system.

95. Fill the power steering system.

96. Recharge the A/C system.

97. Install the hood.

Water Pump

REMOVAL & INSTALLATION

1. Before servicing the vehicle, refer to the precautions section.

2. Remove or disconnect the following:
   - Negative battery cable
   - Radiator, fan blade assembly and fan shroud
   - Accessory drive belt
   - Water pump pulley
   - Heater hose from the water pump
   - Water pump bolts and nuts. Note the locations of the bolts if different lengths.
   - Water pump stud bolt, the water

20-30 Nm (15-22 lb/ft)

7924FG02

Exploded view of the water pump mounting—4.6L, 5.4L and 6.8L engines

pump and the water pump housing gasket. Discard the water pump housing gasket.

To install:

3. Before installing the water pump, be sure to completely clean the water pump mounting surfaces of all dirt, grime and old gasket material.

➥**All water pump housing bolts, nuts and studs are tightened to 15–22 ft. lbs. (20–30 Nm).**

4. Install or connect the following:
- Water pump onto the engine with a new gasket. Install the water pump stud bolt temporarily finger-tight.
- Water pump mounting nuts and bolts temporarily finger-tight, then tighten all water pump housing fasteners to 15–22 ft. lbs. (20–30 Nm).
- Water outlet tube for the heater, if equipped
- Water pump pulley and accessory drive belt
- Fan shroud, fan blade assembly and the radiator
- Coolant
- Negative battery cable

5. Start the engine and check for any fluid leaks.

6. If necessary, bleed the cooling system.

Heater Core

REMOVAL & INSTALLATION

Ford Expedition

1. Before servicing the vehicle, refer to the precautions in the beginning of this section.

2. Disconnect the negative battery cable.

✲✲ CAUTION

After disconnecting the negative bat-

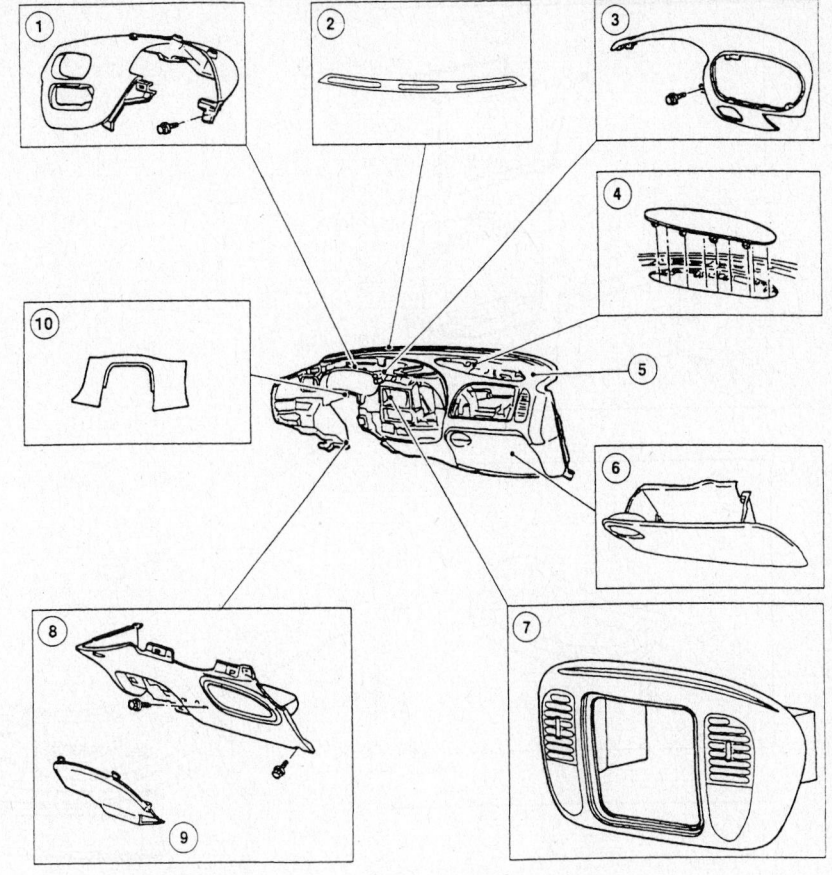

1. Instrument panel finish panel
2. Instrument panel defroster opening grille assembly
3. Instrument cluster panel
4. Instrument panel relay cover
5. Instrument panel
6. Glove compartment
7. Center instrument panel finish panel
8. Instrument panel steering column cover
9. Instrument panel fuse door
10. Steering column opening cover

93113GM2

Exploded view of the instrument panel components—Ford Expedition

tery cable, wait 1 minute for the SRS module to deplete its energy.

3. Drain the cooling system into a clean container for reuse.

4. Remove the instrument panel by performing the following procedure:

a. If equipped, remove the floor console assembly.

b. Remove the lower steering column cover bolts and the cover.

c. Remove both front door scuff plates.

d. Remove both side cowl trim panels.

e. Disconnect the electrical connector from the Brake Pedal Position (BPP) switch.

f. Remove the radio ground and the GEM/CTM ground bolts.

g. Disconnect the left side instrument panel main wiring harness connector.

h. In the engine compartment, remove the bulkhead wiring harness connector bolts and disconnect the wiring connectors.

i. In the driver's compartment, release the 6 locking tabs and remove the bulkhead electrical connector from the instrument panel.

j. Disconnect the air bag diagnostic monitor electrical connector.

k. Disconnect the inertia fuel shutoff switch electrical connector.

l. Remove the right side ground bolts.

m. Disconnect the right side instrument panel wiring harness connectors.

n. Disconnect the electronic blend door actuator electrical connector.

o. Disconnect the climate control head vacuum harness connector.

p. Remove the steering column opening cover reinforcement nuts and the cover reinforcement.

1. Digital audio compact disc player
2. Compact disc player mounting bracket
3. Compact disc player compartment trim panel
4. Radio and A/C integral control assembly
5. A/C register (upper)
6. Blower assembly
7. Center console finish panel
8. Console finish panel mat

93113GM3

Exploded view of the floor console components—Ford Expedition

91190G40

Remove the bulkhead electrical connectors from inside the engine compartment—Ford Expedition

91190G41

Remove the audio unit ground and the GEM/CTM ground bolts—Ford Expedition

91190G42

Remove the instrument panel bolts through the steering column opening—Ford Expedition

91190G43

Remove the passenger side ground bolt—Ford Expedition

91190G44

Detach the passenger side instrument panel main harness connectors—Ford Expedition

91190G45

Remove the instrument panel bolt on the relay bracket—Ford Expedition

91190G46

Remove the instrument panel bolts through the passenger side air bag module opening—Ford Expedition

91190G47

Remove the instrument panel reinforcement bolt below the driver's side corner of the glove compartment—Ford Expedition

91190G48

Position the carpet aside and loosen the instrument panel floor brace—Ford Expedition

91190G49

Remove the passenger side instrument panel cowl side nut—Ford Expedition

91190G50

On Navigator, remove the defroster grille assembly—Ford Expedition

91190G51

Remove the cowl panel mounting bolts—Ford Expedition

91190G52

Remove the instrument panel—Ford Expedition

q. At the base of the steering column, disconnect the air bag sliding contact and the anti-theft sensor electrical connectors.

r. At the steering column, disconnect the remaining electrical connectors.

s. If equipped with a transmission range indicator, remove the bolt and disconnect the cable.

t. Remove the steering column-to-instrument panel nuts and lower the steering column.

u. Remove the right side front fender splash shield screws and move the shield away from the panel.

v. Disconnect the antenna cable from the antenna base.

w. Remove the instrument panel relay cover and disconnect the autolamp sensor electrical connector and/or the sunload sensor connector.

x. Remove the glove box.

y. At the passenger's air bag module, remove the screws, disconnect the electrical connector and remove the air bag module.
Place the air bag module in a safe place with the front facing upward.

a. Remove the right side assist handle screw covers, the screws and the handle.

b. At both doors, pull back the weatherstrip seals and remove the windshield garnish moldings.

c. Remove the instrument panel reinforcement bolt below the left side corner of the glove box.

d. Through the air bag module opening, remove the instrument panel bolts.

e. Remove the upper instrument panel cowl covers and bolts.

f. At the relay bracket, remove the instrument panel bolt.

g. At the lower left side of the cigar lighter, remove the instrument panel bolt.

h. At the both sides, remove the instrument panel-to-cowl side nuts.

i. At the steering column opening, remove the instrument panel bolts.

j. Remove the upper instrument panel floor brace bolt.

k. Using an assistant, remove the instrument panel.

5. If equipped with the 5.4L 4V engine, remove the junction block splash shield.

6. If equipped with the 5.4L 4V engine, remove the bolts and disconnect the cable ends from the starter relay.

7. If equipped with the 5.4L 4V engine, remove the junction block bracket.

8. Compress the holding tabs and disconnect the heater hoses from the heater core.

9. Remove the air conditioning plenum screw and the air conditioning plenum demister adapter.

10. Disconnect the vacuum line.

11. Remove the heater core bracket screws and the bracket.

12. Remove the 13 heater housing plenum camber cover screws and the heater housing plenum chamber cover.

13. Remove the blend door assembly from the heater housing.

14. Remove the heater core.

To install:

15. Install the heater core.

16. Install the blend door assembly to the heater housing.

17. Install the 13 heater housing camber cover and the heater housing plenum chamber cover screws.

18. Install the heater core bracket and the bracket screws.

19. Connect the vacuum line.

93113GM1

View of the heater core—Ford Expedition

20. Install the air conditioning plenum demister adapter and the air conditioning plenum screw.

21. Connect the heater hoses to the heater core.

22. Install the instrument panel by performing the following procedure:

a. Using an assistant, install the instrument panel.

b. Install the upper instrument panel floor brace bolt.

c. At the steering column opening, install the instrument panel bolts.

d. At the both sides, install the instrument panel-to-cowl side nuts.

e. At the lower left side of the cigar lighter, install the instrument panel bolt.

f. At the relay bracket, install the instrument panel bolt.

g. Install the upper instrument panel cowl bolts and covers.

h. Through the air bag module opening, install the instrument panel bolts.

i. Install the instrument panel reinforcement bolt below the left side corner of the glove box.

j. At both doors, install the windshield garnish moldings and the weatherstrip seals.

k. Install the right side assist handle, the screws and the handle screw covers.

l. At the passenger's air bag module, install the air bag module, connect the electrical connector and install the air bag module screws.

m. Install the glove box.

n. Connect the autolamp sensor electrical connector and/or the sunload sensor connector; then, install the instrument panel relay cover.

o. Connect the antenna cable to the antenna base.

p. Install the right side front fender splash shield and screws.

q. Install the steering column and the steering column-to-instrument panel nuts.

r. If equipped with a transmission range indicator, connect the cable and install the bolt.

s. At the steering column, connect the remaining electrical connectors.

t. At the base of the steering column, connect the air bag sliding contact and the anti-theft sensor electrical connectors.

u. Install the steering column opening cover reinforcement and the cover reinforcement nuts.

v. Connect the climate control head vacuum harness connector.

w. Connect the electronic blend door actuator electrical connector.

x. Connect the right side instrument panel wiring harness connectors.

y. Install the right side ground bolts.

z. Connect the inertia fuel shutoff switch electrical connector.

aa. Connect the air bag diagnostic monitor electrical connector.

bb. In the driver's compartment, install the bulkhead electrical connector to the instrument panel.

cc. In the engine compartment, connect the bulkhead wiring harness connectors and the install wiring connector bolts.

dd. Connect the left side instrument panel main wiring harness connector.

ee. Install the radio ground and the GEM/CTM ground bolts.

ff. Connect the electrical connector to the Brake Pedal Position (BPP) switch.

gg. Install both side cowl trim panels.

hh. Install both front door scuff plates.

ii. Install the lower steering column cover and the cover bolts.

jj. If equipped, install the floor console assembly.

23. Refill the cooling system.

24. Connect the negative battery cable.

25. Run the engine to normal operating temperatures; then, check the climate control operation and check for leaks.

Ford Excursion

1. Before servicing the vehicle, refer to the precautions in the beginning of this section.

2. Drain and recycle the engine coolant.

✳✳ CAUTION

Never open, service or drain the radiator or cooling system when hot; serious burns can occur from the steam and hot coolant. Also, when draining engine coolant, keep in mind that cats and dogs are attracted to ethylene glycol antifreeze and could drink any that is left in an uncovered container or in puddles on the ground. This will prove fatal in sufficient quantities. Always drain coolant into a sealable container. Coolant should be reused unless it is contaminated or is several years old.

3. Remove or disconnect the following:
- Heater water hoses from the heater core.
- Stops and lower the glove compartment door
- Electronic blend door actuator and bracket assembly

✳✳ WARNING

The heater core cover must be raised vertically before removal to avoid damage to the heater core housing.

- Heater core cover screws and the cover
- Heater core from the housing

To install:

✳✳ WARNING

Position the temperature blend door manually to properly align the actuator and the door. Do not power the actuator electrically. If it is not engaged with the temperature blend door, damage to the actuator may occur.

➡ **Add gasket between housing and cover before installing cover.**

4. The installation is the reverse of the removal.

Lincoln Navigator

1. Before servicing the vehicle, refer to the precautions in the beginning of this section.

2. Disconnect the negative battery cable.

✳✳ CAUTION

After disconnecting the negative battery cable, wait 1 minute for the SRS module to deplete its energy.

3. Drain the cooling system into a clean container for reuse.

4. Remove the instrument panel by performing the following procedure:

a. If equipped, remove the floor console assembly.

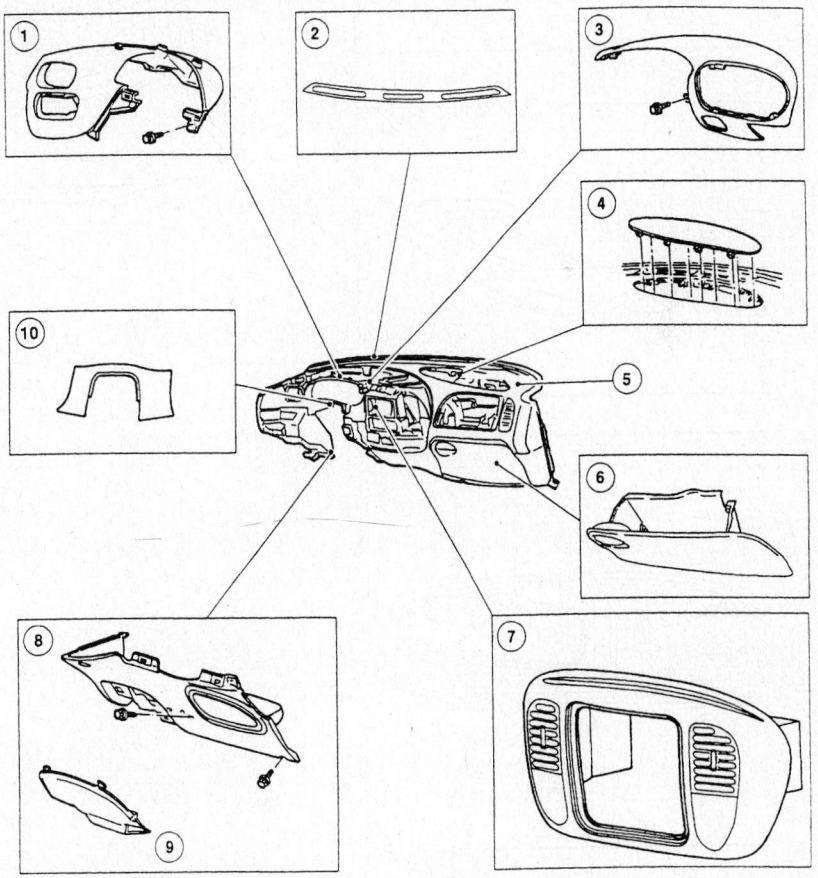

1. Instrument panel finish panel
2. Instrument panel defroster opening grille assembly
3. Instrument cluster panel
4. Instrument panel relay cover
5. Instrument panel
6. Glove compartment
7. Center instrument panel finish panel
8. Instrument panel steering column cover
9. Instrument panel fuse door
10. Steering column opening cover

93113GM2

Exploded view of the instrument panel components—Lincoln Navigator

1. Digital audio compact disc player
2. Compact disc player mounting bracket
3. Compact disc player compartment trim panel
4. Radio and A/C integral control assembly
5. A/C register (upper)
6. Blower assembly
7. Center console finish panel
8. Console finish panel mat

93113GM3

Exploded view of the floor console components—Lincoln Navigator

91190G40

Remove the bulkhead electrical connectors from inside the engine compartment—Lincoln Navigator

91190G41

Remove the audio unit ground and the GEM/CTM ground bolts—Lincoln Navigator

91190G42

Remove the instrument panel bolts through the steering column opening—Lincoln Navigator

91190G43

Remove the passenger side ground bolt—Lincoln Navigator

91190G44

Detach the passenger side instrument panel main harness connectors—Lincoln Navigator

91190G45

Remove the instrument panel bolt on the relay bracket—Lincoln Navigator

91190G46

Remove the instrument panel bolts through the passenger side air bag module opening—Lincoln Navigator

91190G47

Remove the instrument panel reinforcement bolt below the driver's side corner of the glove compartment—Lincoln Navigator

91190G48

Position the carpet aside and loosen the instrument panel floor brace—Lincoln Navigator

91190G49

Remove the passenger side instrument panel cowl side nut—Lincoln Navigator

91190G50

On Navigator, remove the defroster grille assembly—Lincoln Navigator

91190G51

Remove the cowl panel mounting bolts—Lincoln Navigator

91190G52

Remove the instrument panel—Lincoln Navigator

93113GM1

View of the heater core—Lincoln Navigator

b. Remove the lower steering column cover bolts and the cover.

c. Remove both front door scuff plates.

d. Remove both side cowl trim panels.

e. Disconnect the electrical connector from the Brake Pedal Position (BPP) switch.

f. Remove the radio ground and the GEM/CTM ground bolts.

g. Disconnect the left side instrument panel main wiring harness connector.

h. In the engine compartment, remove the bulkhead wiring harness connector bolts and disconnect the wiring connectors.

i. In the driver's compartment, release the 6 locking tabs and remove the bulkhead electrical connector from the instrument panel.

j. Disconnect the air bag diagnostic monitor electrical connector.

k. Disconnect the inertia fuel shutoff switch electrical connector.

l. Remove the right side ground bolts.

m. Disconnect the right side instrument panel wiring harness connectors.

n. Disconnect the electronic blend door actuator electrical connector.

o. Disconnect the climate control head vacuum harness connector.

p. Remove the steering column opening cover reinforcement nuts and the cover reinforcement.

q. At the base of the steering column, disconnect the air bag sliding contact and the anti-theft sensor electrical connectors.

r. At the steering column, disconnect the remaining electrical connectors.

s. If equipped with a transmission range indicator, remove the bolt and disconnect the cable.

t. Remove the steering column-to-instrument panel nuts and lower the steering column.

u. Remove the right side front fender

splash shield screws and move the shield away from the panel.

v. Disconnect the antenna cable from the antenna base.

w. Remove the instrument panel relay cover and disconnect the autolamp sensor electrical connector and/or the sun-load sensor connector.

x. Remove the glove box.

y. At the passenger's air bag module, remove the screws, disconnect the electrical connector and remove the air bag module.

Place the air bag module in a safe place with the front facing upward.

a. Remove the right side assist handle screw covers, the screws and the handle.

b. At both doors, pull back the weatherstrip seals and remove the windshield garnish moldings.

c. Remove the instrument panel reinforcement bolt below the left side corner of the glove box.

d. Through the air bag module opening, remove the instrument panel bolts.

e. Remove the instrument panel defroster grille assembly and the instrument panel cowl top bolts.

f. At the relay bracket, remove the instrument panel bolt.

g. At the lower left side of the cigar lighter, remove the instrument panel bolt.

h. At the both sides, remove the instrument panel-to-cowl side nuts.

i. At the steering column opening, remove the instrument panel bolts.

j. Remove the upper instrument panel floor brace bolt.

k. Using an assistant, remove the instrument panel.

5. If equipped with the 5.4L 4V engine, remove the junction block splash shield.

6. If equipped with the 5.4L 4V engine, remove the bolts and disconnect the cable ends from the starter relay.

7. If equipped with the 5.4L 4V engine, remove the junction block bracket.

8. Compress the holding tabs and dis-

connect the heater hoses from the heater core.

9. Remove the air conditioning plenum screw and the air conditioning plenum demister adapter.

10. Disconnect the vacuum line.

11. Remove the heater core bracket screws and the bracket.

12. Remove the 13 heater housing plenum camber cover screws and the heater housing plenum chamber cover.

13. Remove the blend door assembly from the heater housing.

14. Remove the heater core.

To install:

15. Install the heater core.

16. Install the blend door assembly to the heater housing.

17. Install the 13 heater housing camber cover and the heater housing plenum chamber cover screws.

18. Install the heater core bracket and the bracket screws.

19. Connect the vacuum line.

20. Install the air conditioning plenum demister adapter and the air conditioning plenum screw.

21. Connect the heater hoses to the heater core.

22. Install the instrument panel by performing the following procedure:

a. Using an assistant, install the instrument panel.

b. Install the upper instrument panel floor brace bolt.

c. At the steering column opening, install the instrument panel bolts.

d. At the both sides, install the instrument panel-to-cowl side nuts.

e. At the lower left side of the cigar lighter, install the instrument panel bolt.

f. At the relay bracket, install the instrument panel bolt.

g. Install the instrument panel cowl top bolts and the instrument panel defroster grille assembly.

h. Through the air bag module opening, install the instrument panel bolts.

i. Install the instrument panel reinforcement bolt below the left side corner of the glove box.

j. At both doors, install the windshield garnish moldings and the weatherstrip seals.

k. Install the right side assist handle, the screws and the handle screw covers.

l. At the passenger's air bag module, install the air bag module, connect the electrical connector and install the air bag module screws.

m. Install the glove box.

n. Connect the autolamp sensor electrical connector and/or the sunload sensor connector; then, install the instrument panel relay cover.

o. Connect the antenna cable to the antenna base.

p. Install the right side front fender splash shield and screws.

q. Install the steering column and the steering column-to-instrument panel nuts.

r. If equipped with a transmission range indicator, connect the cable and install the bolt.

s. At the steering column, connect the remaining electrical connectors.

t. At the base of the steering column, connect the air bag sliding contact and the anti-theft sensor electrical connectors.

u. Install the steering column opening cover reinforcement and the cover reinforcement nuts.

v. Connect the climate control head vacuum harness connector.

w. Connect the electronic blend door actuator electrical connector.

x. Connect the right side instrument panel wiring harness connectors.

y. Install the right side ground bolts.

z. Connect the inertia fuel shutoff switch electrical connector.

aa. Connect the air bag diagnostic monitor electrical connector.

bb. In the driver's compartment, install the bulkhead electrical connector to the instrument panel.

cc. In the engine compartment, connect the bulkhead wiring harness connectors and the install wiring connector bolts.

dd. Connect the left side instrument panel main wiring harness connector.

ee. Install the radio ground and the GEM/CTM ground bolts.

ff. Connect the electrical connector to the Brake Pedal Position (BPP) switch.

gg. Install both side cowl trim panels.

hh. Install both front door scuff plates.

ii. Install the lower steering column cover and the cover bolts.

jj. If equipped, install the floor console assembly.

23. Refill the cooling system.

24. Connect the negative battery cable.

25. Run the engine to normal operating temperatures; then, check the climate control operation and check for leaks.

Cylinder Head

REMOVAL & INSTALLATION

2002–04 Models

✳✳ CAUTION

Fuel injection systems remain under pressure, even after the engine has been turned OFF. The fuel system pressure must be relieved before disconnecting any fuel lines. Failure to do so may result in fire and/or personal injury.

➡**To correctly tighten the cylinder head bolts, an angle torque wrench is needed.**

1. Before servicing the vehicle, refer to the precautions section.

2. Remove or disconnect the following:
  • Air conditioning system, using approved equipment
  • Negative battery cable
  • Cylinder head covers
  • Intake manifold
  • Timing chains from the engine
  • Exhaust manifolds
  • The 2 heater hose retaining bolts,

then compress and slide the hose clamp back to remove the heater water hose.
  • Any remaining hoses, electrical connections or cables
  • Cylinder head bolts, then lift the cylinder head from the engine block. Discard the cylinder head gasket and clean the engine block surface.

To install:

✳✳ WARNING

Cylinder head bolts must be replaced with new ones. They are torque-to-yield designed and cannot be reused.

3. Turn the crankshaft to position the keyway at the 12 o'clock position.

4. Clean and inspect the cylinder head for damage or warpage. Install the cylinder head gasket over the dowel pins. Then, install the cylinder head onto the engine block. Loosely install NEW cylinder head bolts.

➡**Be sure to tighten the head bolts in 3 steps.**

5. Tighten the cylinder head bolts in the correct sequence using 3 steps, as follows:
  a. Step 1—30 ft. lbs. (40 Nm).
  b. Step 2—tighten an additional 85–95 degrees.
  c. Step 3—tighten another 85–95 degrees.

6. Install or connect the following:
  • Heater hose
  • Exhaust manifolds
  • Timing chains
  • Intake manifold
  • Cylinder head covers

RIGHT   LEFT

7924FG04

Tighten the cylinder head bolts using 3 steps in this sequence—4.6L and 5.4L engine

7924FG81

Tighten the cylinder head bolts using 3 steps in this sequence—6.8L engine

- Electrical connections and cables
- Refrigerant
- Engine oil
- Coolant
- Negative battery cable

7. Start the engine and check for any fluid or vacuum leaks.

2005 Excursion Models

1. Before servicing the vehicle, refer to the precautions section.
2. Remove the engine.
3. Remove the bolts and the flexplate.

✳✳ CAUTION

To prevent damage to the oil pan, use care when lowering the engine.

4. Lower the engine onto wooden blocks.
5. Remove the engine support tool.
6. Install the tools 303-D087 and 303-D088 onto the cylinder heads where illustrated.
7. Mount the engine on a suitable work stand.
8. Remove tools 303-D087 and 303-D088.
9. Remove the right hand engine mount.
10. Remove the cylinder block drain plugs, and drain the coolant in a suitable container.

06017-NAVI-G36

Install the tools 303-D087

Install the tools 303-D088 onto the cylinder heads—2005 Excursion models

06017-NAVI-G37

11. Disconnect the left hand radio frequency interference capacitor and Cylinder Head Temperature (CHT) sensor connectors.
12. Disconnect the Camshaft Position (CMP) sensor connector.
13. Disconnect the right hand radio frequency interference capacitor connector.
14. Disconnect the knock sensor connector.
15. On vehicles without A/C, disconnect the Crankshaft Position (CKP) sensor connector.
16. Disconnect the oil pressure switch connector.
17. Remove the nuts and position the wiring harness bracket aside.
18. Disconnect all of the harness routing clips and connector retainers. Remove the engine control sensor wiring harness.
19. Remove the nuts and the two radio interference capacitors.
20. Remove the crankcase ventilation tube from the left hand valve cover.
21. Remove the Positive Crankcase Ventilation (PCV) hose from the right hand valve cover.

✳✳ CAUTION

Do not use metal scrapers, wire brushes, power abrasive discs or other abrasive means to clean the sealing surfaces. These tools cause scratches and gouges which make

leak paths. Use a plastic scraping tool to remove all traces of old sealant.

➡**The bolts are part of the valve cover and should not be removed.**

22. Remove the valve covers.
23. Remove the bolt and the belt idler pulley.
24. Remove the coolant pump pulley.
25. Remove and discard the crankshaft pulley bolt.
26. Use tool 303-009 to remove the crankshaft pulley.
27. Use tool 303-107 to remove the crankshaft front oil seal.
28. Remove the front oil pan bolts.

➡**Correct fastener location is essential for front cover assembly procedure. Record fastener location.**

29. Remove the front cover fasteners.
30. Remove the engine front cover from the cylinder block.
31. Remove the crankshaft sensor ring from the crankshaft.

✳✳ CAUTION

Use care when removing the spark plugs.

➡**Use compressed air to remove any foreign material from the spark plug well before removing the spark plugs.**

32. Remove the eight spark plugs.
33. Install tool 303-382 between the valve spring coils to prevent valve stem seal damage.

➡**The camshaft roller followers must be reinstalled in their original locations. Record the camshaft roller follower locations.**

➡**Position the cam lobe away from the camshaft roller follower prior to removing each camshaft roller follower.**

34. Use tool 303-567 to compress the valve springs, and remove the camshaft roller followers.
35. Remove tool 303-382.
36. Position the crankshaft with the keyway at the 12 o'clock position.

✳✳ CAUTION

If one or both of the tensioner mounting bolts are loosened or removed, the tensioner-sealing bead must be inspected for seal integrity. If cracks, tears, separation from the tensioner body or permanent compression of the seal bead is observed, install a new tensioner.

37. Remove both timing chains and guide components.

38. Remove the nuts, exhaust manifolds and gaskets. Discard the gaskets.

39. Remove the bolt and the oil level indicator tube.

40. Install cylinder head removal/installation tools 303-572 on both ends of the cylinder head.

➡ **The hydraulic lash adjusters must be reinstalled in their original locations. Record the hydraulic lash adjuster locations.**

41. Remove the hydraulic lash adjusters.

✳✳ CAUTION

The cylinder head must be cool before removing it from the engine. Cylinder head warpage can result if a warm or hot cylinder head is removed. Place clean shop towels over exposed engine cavities. Carefully remove the towels so foreign material is not dropped into the engine.

✳✳ CAUTION

The cylinder head bolts must be discarded and new bolts installed. They are tighten-to-yield designed and cannot be reused.

✳ CAUTION

Do not use metal scrapers, wire brushes, power abrasive discs or other abrasive means to clean the sealing surfaces. These tools cause scratches and gouges that make leak paths. Use a plastic scraping tool to remove all traces of the head gasket. Aluminum surfaces are soft and can be scratched easily. Never place the cylinder head gasket surface, unprotected, on a bench surface.

42. Remove the bolts and the cylinder heads. Discard the cylinder head gasket and bolts.

To install:

✳ CAUTION

The gasket sealing surfaces on the cylinder head and cylinder block must be clean. The use of sealing aids (aviation cement, copper spray, and glue) is not permitted. The gasket must be installed dry. The new gasket has a film coating which is crucial to the gasket's ability to seal correctly. Do not scratch the gasket.

43. Install the head gasket over the dowel pins.

➡ **The new cylinder head bolts must be lightly oiled with a rag, and allowed to drain for a few minutes prior to installation.**

44. Install the cylinder head on the dowels and the head gasket. Loosely install new bolts.

➡ **Make sure to tighten the bolts in sequence in three stages.**

45. Tighten the bolts in the sequence shown:
   a. Step 1: Tighten to 40 Nm (30 lb-ft).
   b. Step 2: Tighten an additional 90 degrees.
   c. Step 3: Tighten an additional 90 degrees.

46. Remove tools 303-572 from both ends of the cylinder head.

➡ **Lubricate the hydraulic lash adjusters with clean engine oil.**

47. Install the hydraulic lash adjusters in their original locations.

48. Install the exhaust manifold gaskets and the exhaust manifold.

➡ **Lubricate the O-ring seal with clean engine oil.**

49. Install the oil level indicator tube using a new O-ring seal on the oil level indicator tube.

✳✳ CAUTION

Timing chain procedures must be followed exactly or damage to valves and pistons will result.

50. Install the timing chain assemblies.

51. Install tool 303-382 between the valve spring coils to prevent valve stem seal damage.

➡ **Lubricate the camshaft roller followers using clean engine oil. Position the cam lobe away from the camshaft roller follower location prior to installing each camshaft roller follower.**

52. Install the camshaft roller followers in their original locations. Use tool 303-567 to compress the valve spring.

53. Remove tool 303-382.

✳✳ CAUTION

When installing the spark plugs, use care not to exceed the recommended torque.

54. Install the eight spark plugs. Tighten the spark plugs to 18 Nm (13 lb-ft).

55. Install the crankshaft sensor ring on the crankshaft.

56. Install a new engine front cover gasket on the engine front cover.

57. Install the engine front cover. Install the fasteners finger-tight.

58. Loosely install the front oil bolts, then tighten the bolts in two stages, in the sequence shown:
   a. Step 1: Tighten to 20 Nm (15 lb-ft).
   b. Step 2: Tighten an additional 60 degrees.

59. Position the belt idler pulley and install the bolt. Tighten to 25 Nm (18 ft-lbs.).

60. Lubricate the engine front cover and the crankshaft front seal inner lip with clean engine oil.

61. Use the tools illustrated to install the crankshaft front seal into the engine front cover.

06017-NAVI-G37A

Install the front oil pan bolts using this sequence—2005 Excursion models

06017-NAVI-G38

Use the tools illustrated to install the crankshaft front seal into the engine front cover—2005 Excursion models

➡️**If not secured within four minutes, the sealant must be removed and the sealing area cleaned. To clean the sealing area, use silicone gasket remover and metal surface prep. Follow the directions on the packaging. Failure to follow this procedure can cause future oil leakage.**

62. Apply a 8mm (0.32 inch) bead of silicone gasket and sealant to the Woodruff key slot on the crankshaft pulley.

63. Use tool 303-102 to install the crankshaft pulley.

➡️**Use a suitable strap wrench (303-D055) to hold the pulley while tightening the bolt.**

64. Tighten the new crankshaft pulley bolt in four steps:
 a. Step 1: Tighten to 90 Nm (66 lb-ft).
 b. Step 2: Loosen 360 degrees.
 c. Step 3: Tighten to 50 Nm (37 lb-ft).
 d. Step 4: Tighten an additional 90 degrees.
65. Position the coolant pump pulley on the coolant pump and install the bolts.
66. If a new gasket is being installed, apply instant gel adhesive completely around the gasket groove in the left hand valve cover. Install the new valve cover gasket.

➡️**If not secured within four minutes, the sealant must be removed and the sealing area cleaned. To clean the sealing area, use silicone gasket remover and metal surface prep.**

67. Apply silicone gasket and sealant in two places where the engine front cover meets the left cylinder head.
68. Position the left hand valve cover and gasket on the cylinder head and install the bolts loosely.
69. Tighten the bolts in the sequence shown.
70. If a new gasket is being installed, apply instant gel adhesive completely

Apply silicone gasket and sealant in two places where the engine front cover meets the left cylinder head—2005 Excursion models

10 Nm (89 lb-in)

Tighten the left hand valve cover bolts in this sequence to the torque value shown—2005 Excursion models

around the gasket groove in the right hand valve cover. Install the new valve cover gasket.

➡️**If not secured within four minutes, the sealant must be removed and the sealing area cleaned. To clean the sealing area, use silicone gasket remover and metal surface prep.**

71. Apply silicone gasket and sealant in two places where the engine front cover meets the cylinder head.
72. Position the right hand valve cover and gasket on the cylinder head and install the bolts loosely. Tighten the bolts in the sequence shown.
73. Install the crankcase ventilation tube on the left hand valve cover.
74. Install the PCV hose on the right hand valve cover.
75. Install the radio frequency interference capacitors.

Apply silicone gasket and sealant in two places where the engine front cover meets the right cylinder head—2005 Excursion models

10 Nm (89 lb-in)

Tighten the right hand valve cover bolts in this sequence to the torque value shown—2005 Excursion models

76. Roughly position the engine control sensor wiring harness and mount it on the valve cover studs. Install the bracket and tighten to 30 Nm (22 lb-ft).
77. On vehicles without A/C, connect the CKP sensor connector.
78. Connect the oil pressure switch connector.
79. Connect the knock sensor connector.
80. Connect the right hand radio frequency interference capacitor connector.
81. Connect the CMP sensor connector.
82. Connect the left hand radio frequency interference capacitor and CHT sensor connectors.
83. Install the cylinder block drain plugs. Tighten to 24 Nm (18 lb-ft).
84. Install the right hand engine mount. Tighten the bolts to 63 Nm (46 lb-ft).
85. Install tools 303-D087 and 303-D088.
86. Install a hoist and attach to tools 303-D087 and 303-D088 and remove the engine from the work stand.
87. Lower the engine onto wooden blocks.
88. Remove the tools 303-D087 and 303-D088.
89. Install engine support tool 303-F047and raise the engine.
90. Install the flexplate and the bolts. Tighten the bolts in a star pattern to 80 Nm (59 lb-ft).
91. Install the engine.

2005 Expedition And Navigator Models

✳✳ CAUTION

Fuel injection systems remain under pressure, even after the engine has been turned OFF. The fuel system pressure must be relieved before disconnecting any fuel lines. Failure to do so may result in fire and/or personal injury.

➡ **To correctly tighten the cylinder head bolts, an angle torque wrench is needed.**

1. Before servicing the vehicle, refer to the precautions section.
2. Remove the engine.
3. Remove the bolts and the flexplate.
4. Install the engine onto a suitable engine stand.
5. Disconnect the right hand Camshaft Position (CMP) sensor connector.
6. Remove the stud bolt and the right hand radio ignition interference capacitor.
7. Disconnect the right hand Variable Camshaft Timing (VCT) solenoid connector.
8. Disconnect the 2 engine wiring harness retainers from the right hand valve cover studs.
9. Disconnect the connector retainer from the coolant tube.
10. Disconnect the 4 right hand ignition coil connectors.
11. Disconnect the Cylinder Head Temperature (CHT) sensor connector.
12. Disconnect the 2 engine wiring harness retainers from the left hand valve cover studs.
13. Disconnect the left hand VCT solenoid connector.
14. Disconnect the oil pressure sensor connector.
15. Disconnect the left hand CMP sensor connector.
16. Remove the stud bolt and the left hand radio ignition interference capacitor.
17. Disconnect the 4 left hand ignition coil connectors.
18. Remove the 8 bolts and the 8 ignition coils.
19. Disconnect the Crankshaft Position (CKP) sensor connector.
20. Remove the engine wiring harness from the engine assembly.
21. Remove the bolt and the oil level indicator tube and discard the O-ring seal.
22. Remove the bolt and the right hand CMP sensor.
23. Remove the bolt and the left hand CMP sensor.

24. Remove the bolt and the CKP sensor.

✳✳ CAUTION

Do not use metal scrapers, wire brushes, power abrasive discs or other abrasive means to clean the sealing surfaces. These tools cause scratches and gouges which make leak paths. Use a plastic scraping tool to remove all traces of old sealant.

✳✳ CAUTION

When removing the valve cover, make sure to avoid damaging the variable camshaft timing (VCT) solenoid.

➡ **The bolts are part of the valve cover and should not be removed.**

25. Remove the bolts and the valve covers.
26. Clean the valve cover mating surface of the cylinder head with silicone gasket remover and metal surface prep.
27. Inspect the valve cover gasket. If the gasket is damaged, remove and discard the gasket. Clean the valve cover gasket groove with soap and water or a suitable solvent.
28. Remove the bolts, the coolant pump pulley and the 3 accessory drive belt idler pulleys.
29. Remove the bolts and the accessory drive belt tensioner.
30. Remove and discard the crankshaft pulley bolt. Using tool 303-009, remove the crankshaft pulley.
31. Using the tool 303-107, remove the crankshaft seal.
32. Remove the front 4 oil pan bolts.

➡ **Correct fastener location is essential for assembly procedure. Record fastener location.**

33. Remove the engine front cover fasteners and the cover.
34. Remove the crankshaft sensor ring from the crankshaft.
35. Position the crankshaft keyway at the 12 o'clock position.

➡ **f the camshaft lobes are not exactly positioned at the 12 o'clock position, the crankshaft will require one full additional rotation to 12 o'clock. The number 1 cylinder camshaft exhaust lobe must be coming up on the exhaust stroke. Verify by noting the position of the 2 intake camshaft lobes and the exhaust lobe on the number 1 cylinder.**

➡ **If the components are to be reinstalled, they must be installed in the same positions. Mark the components for installation into the original locations.**

36. Remove only the 3 roller followers shown in the illustration from the right hand cylinder head.

✳✳ CAUTION

Do not allow the valve keepers to fall off the valve or the valve may drop into the cylinder.

➡ **It may be necessary to push the valve down while compressing the spring.**

37. Using the tool illustrated, remove the 3 designated roller followers in the previous step from the right hand cylinder head.
38. Remove only the 3 roller followers shown in the illustration from the left hand cylinder head.
39. Using the tool 303-1039, remove the 3 designated roller followers in the previous step from the left hand cylinder head.
40. The crankshaft cannot be moved past the 6 o'clock position once set.
41. Rotate the crankshaft clockwise and position the crankshaft keyway at the 6 o'clock position.

06017-NAVI-G01

Remove only the 3 roller followers shown from the right hand cylinder head—2005 Expedition and Navigator models

Using the tool illustrated, remove the 3 designated roller followers—2005 Expedition and Navigator models

Remove only the 3 roller followers shown from the left hand cylinder head—2005 Expedition and Navigator models

✳✳ CAUTION

If one or both of the tensioner mounting bolts are loosened or removed, the tensioner-sealing bead must be inspected for seal integrity. If cracks, tears, separation from the tensioner body or permanent compression of the seal bead is observed, install a new tensioner.

42. Remove the bolts, the left hand timing chain tensioner and tensioner arm.

43. Remove the bolts, the right hand timing chain tensioner and tensioner arm.

44. Remove the right hand and left hand timing chains and the crankshaft sprocket.

45. Remove the right hand timing chain from the camshaft sprocket.

46. Remove the right hand timing chain from the crankshaft sprocket.

47. Remove the left hand timing chain from the camshaft sprocket.

48. Remove the left hand timing chain and crankshaft sprocket.

49. Remove the left hand and right hand timing chain guides.

✳✳ CAUTION

Damage to the camshaft phaser sprocket assembly will occur if mishandled or used as a lifting or leveraging device.

✳✳ CAUTION

Only use hand tools to remove the camshaft phaser sprocket assembly or damage may occur to the camshaft or camshaft phaser unit.

✳✳ CAUTION

Damage to the camshaft phaser sprocket assembly will occur if mishandled or used as a lifting or leveraging device.

50. Using the tool illustrated, remove the bolt and the right hand camshaft phaser sprocket assembly. Discard the camshaft phaser sprocket bolt.

51. Using tool 303-1046, remove the bolt and the left hand camshaft phaser sprocket assembly. Discard the camshaft phaser sprocket bolt.

✳✳ CAUTION

When removing the front thrust camshaft bearing cap, use care as the cap may be damaged from side-loading when removing the cam unequally in height from the bearing towers.

Remove the bolt and the right hand camshaft phaser sprocket assembly—2005 Expedition and Navigator models

Right side bearing cap removal and tightening sequence, left side similar—2005 Expedition and Navigator models

➡The camshaft bearing caps must be installed in their original locations. Record camshaft bearing cap locations.

52. Remove the bolts in the sequence shown and remove the right hand cylinder head front camshaft bearing cap and then the remaining bearing caps.

53. Clean and inspect the right hand camshaft bearing caps.

54. The camshaft front thrust bearing cap contains an oil metering groove. Make sure the groove is free of foreign material.

55. Remove the right hand camshaft.

56. Remove the bolts in the sequence shown and remove the left hand cylinder head front camshaft bearing cap and then the remaining bearing caps.

57. Clean and inspect the left hand camshaft bearing caps.

58. The camshaft front thrust bearing cap contains an oil metering groove. Make sure the groove is free of foreign material.

59. Remove the left hand camshaft.

✳✳ CAUTION

If the components are to be reinstalled, they must be installed in the same positions. Mark the components for installation into the original locations.

60. Remove the all of the remaining roller followers from the cylinder heads.

61. Remove the hydraulic lash adjusters from the cylinder heads.

62. Install the tool illustrated onto the cylinder head.

06017-NAVI-G06

Install the tool illustrated onto the left or right hand cylinder head—2005 Expedition and Navigator models

63. Remove the exhaust manifold. Discard the gasket.

64. Remove the stud bolt and the coolant tube. Discard the O-ring seals.

✳✳ CAUTION

The cylinder head must be cool before removing it from the engine. Cylinder head warpage can result if a warm or hot cylinder head is removed.

✳✳ CAUTION

Place clean shop towels over exposed engine cavities. Carefully remove the towels so foreign material is not dropped into the engine.

✳ CAUTION

The cylinder head bolts must be discarded and new bolts must be installed. They are tighten-to-yield designed and cannot be reused.

✳✳ CAUTION

Do not use metal scrapers, wire brushes, power abrasive discs or other abrasive means to clean the sealing surfaces. These tools cause scratches and gouges that make leak paths. Use a plastic scraping tool to remove all traces of the head gasket.

✳ CAUTION

Aluminum surfaces are soft and can be scratched easily. Never place the cylinder head gasket surface, unprotected, on a bench surface.

65. Remove the bolts and the cylinder head.

66. Discard the cylinder head gasket and bolts.

To install:

✳✳ CAUTION

Make sure all coolant residue and foreign material are cleaned from the block surface and cylinder bore.

✳✳ CAUTION

The use of sealing aids (aviation cement, copper spray, and glue) is not permitted. The gasket must be installed dry.

➡ **Do not turn the crankshaft until instructed to do so.**

67. Position the cylinder head gaskets and cylinder heads over the dowels and install the cylinder head bolts loosely.

68. Tighten the bolts in three steps, in the sequence shown as follows:
   a. Step 1: Tighten to 30 ft. lbs. (40 Nm).
   b. Step 2: Tighten an additional 90 degrees.
   c. Step 3: Tighten an additional 90 degrees.

69. Remove the tool from the left hand cylinder head.

70. Lubricate the hydraulic lash adjusters with clean engine oil prior to installation.

71. Install the hydraulic lash adjusters into the left hand cylinder head.

72. Using a new gasket, install the exhaust manifold.

73. Install the coolant tube and the stud bolt.

74. Lubricate the camshaft and camshaft journals with clean engine oil prior to installation.

75. Install the left hand and right hand camshafts.

76. Install the left hand and right hand camshaft bearing caps in their original locations.

77. Lubricate the camshaft bearing caps with clean engine oil.

78. Install the camshaft bearing cap. Position the remaining camshaft bearing caps. Install the bolts loosely, then tighten to 18 inch lbs. (10 Nm).

✳✳ CAUTION

Damage to the camshaft phaser sprocket assembly will occur if mishandled or used as a lifting or leveraging device.

79. Install the camshaft phaser sprockets and new camshaft phaser bolts finger tight.

✳✳ CAUTION

Only use hand tools to remove the camshaft phaser sprocket assembly or damage may occur to the camshaft or camshaft phaser unit.

80. Using tool 303-1046, tighten the left hand and right hand camshaft phaser sprocket bolts in 2 steps:
   a. Step 1: Tighten to 30 ft. lbs. (40 Nm).
   b. Step 2: Tighten an additional 90 degrees.

✳✳ CAUTION

Timing chain procedures must be followed exactly or damage to valves and pistons will result.

✳✳ CAUTION

Prior to installation, inspect the tensioner-sealing bead for seal integrity. If cracks, tears, separation from the tensioner body or permanent compression of the seal bead is observed, install a new tensioner.

81. Compress the tensioner plunger, using a vise.

82. Install a retaining clip on the tensioner to hold the plunger in during installation.

83. Remove the tensioner from the vise.

84. If the copper links are not visible, mark two links on one end and one link on the other end, and use as timing marks.

85. Install the crankshaft sprocket, making sure the flange faces forward.

86. Install the 4 bolts and the left hand and right hand timing chain guides. Tighten to 89 inch lbs. (10 Nm).

87. Position the lower end of the left hand (inner) timing chain on the crankshaft sprocket, aligning the timing mark on the outer flange of the crankshaft sprocket with

06017-NAVI-G07

If the copper links are not visible, mark two links on one end and one link on the other end, and use as timing marks—2005 Expedition and Navigator models

Install the crankshaft sprocket, making sure the flange faces forward—2005 Expedition and Navigator models

Position the lower end of the left hand (inner) timing chain on the crankshaft sprocket, aligning the timing mark on the outer flange of the crankshaft sprocket with the single copper (marked) link on the chain—2005 Expedition and Navigator models

Position the timing chain on the camshaft sprocket with the camshaft sprocket timing mark positioned between the two copper (marked) chain links—2005 Expedition and Navigator models

the single copper (marked) link on the chain.

➡**Make sure the upper half of the timing chain is below the tensioner arm dowel.**

88. Position the timing chain on the camshaft sprocket with the camshaft sprocket timing mark positioned between the two copper (marked) chain links.

➡**The left hand timing chain tensioner arm has a bump near the dowel hole for identification.**

89. Position the left hand timing chain tensioner arm on the dowel pin and install the left hand timing chain tensioner and bolts. Tighten the bolts to 18 ft. lbs. (25 Nm).

90. Remove the retaining clip from the left hand timing chain tensioner.

91. Position the lower end of the right hand (outer) timing chain on the crankshaft sprocket, aligning the timing mark on the sprocket with the single copper (marked) chain link.

➡**The lower half of the timing chain must be positioned above the tensioner arm dowel.**

92. Position the right hand timing chain on the camshaft sprocket. Make sure the camshaft sprocket timing mark is positioned between the two copper (marked) chain links.

93. Position the right hand timing chain tensioner arm on the dowel pin and install

Position the lower end of the right hand (outer) timing chain on the crankshaft sprocket, aligning the timing mark on the sprocket with the single copper (marked) chain link—2005 Expedition and Navigator models

the right hand timing chain tensioner and bolts. Tighten the bolts to 18 ft. lbs. (25 Nm).

94. Remove the retaining clip from the right hand timing chain tensioner.

95. As a final-check, verify correct alignment of all timing marks.

96. Install the crankshaft sensor ring on the crankshaft.

97. Lubricate the roller followers with clean engine oil prior to installation.

98. Using tool 303-1039, install all of the camshaft roller followers.

✳✳ CAUTION

Do not use metal scrapers, wire brushes, power abrasive discs or other abrasive means to clean the sealing surfaces. These tools cause scratches and gouges which make leak paths. Use a plastic scraping tool to remove all traces of old sealant.

Position the right hand timing chain on the camshaft sprocket. Make sure the camshaft sprocket timing mark is positioned between the two copper (marked) chain links—2005 Expedition and Navigator models

Verify correct alignment of all timing marks—2005 Expedition and Navigator models

06017-NAVI-G14

Apply a bead of silicone gasket and sealant along the cylinder head-to-cylinder block surface and the oil pan-to-cylinder block surface, at the locations shown—2005 Expedition and Navigator models

1   Bolt, Hex Flange Head Pilot, M8 x 1.25 x 50

2   Bolt, Hex Flange Head Pilot, M8 x 1.25 x 50

3   Bolt, Hex Flange Head Pilot, M8 x 1.25 x 50

4   Bolt, Hex Flange Head Pilot, M8 x 1.25 x 50

5   Bolts, Hex Flange Head Pilot, M8 x 1.25 x 50

6   Stud, Hex Head Pilot, M10 x 1.5 x 1.5 x 103

7   Stud, Hex Head Pilot, M10 x 1.5 x 1.5 x 103

8   Bolt, Hex Flange Head Pilot, M8 x 1.25 x 50

9   Bolt, Hex Flange Head Pilot, M8 x 1.25 x 50

10   Bolt, Hex Flange Head Pilot, M8 x 1.25 x 50

11   Bolt, Hex Flange Head Pilot, M8 x 1.25 x 50

12   Stud and Washer, Hex Head Pilot, M8 x 1.25 x 1.25 x 94

13   Stud and Washer, Hex Head Pilot, M8 x 1.25 x 1.25 x 94

14   Stud and Washer, Hex Head Pilot, M8 x 1.25 x 1.25 x 94

15   Bolt, Hex Head Pilot, M8 x 1.25 x 56

06017-NAVI-G15

Engine front cover fastener location and torque sequence—2005 Expedition and Navigator models

➡️If the engine front cover is not secured within 4 minutes, the sealant must be removed and the sealing area cleaned. To clean the sealing area, use silicone gasket remover and metal surface prep. Failure to follow this procedure can cause future oil leakage.

➡️Make sure that the engine front cover gasket is in place on the engine front cover before installation.

99. Apply a bead of silicone gasket and sealant along the cylinder head-to-cylinder block surface and the oil pan-to-cylinder block surface, at the locations illustrated.

100. Install a new engine front cover gasket on the engine front cover. Position the engine front cover onto the dowels. Install the fasteners finger-tight.

101. Tighten the engine front cover fasteners in sequence in 2 steps:

   a. Step 1: Tighten fasteners 1 through 15 to 18 ft. lbs. (25 Nm).

   b. Step 2: Tighten fasteners 6 and 7 to 35 ft. lbs. (48 Nm).

✳✳ CAUTION

Do not use metal scrapers, wire brushes, power abrasive discs or other abrasive means to clean sealing surfaces. These tools cause scratches and gouges which make leak paths. Use a plastic scraping tool to remove all traces of old sealant.

102. Clean the valve cover mating surface with silicone gasket remover and metal surface prep.

103. Install the 4 front oil pan bolts in the sequence shown in 2 steps.

   a. Step 1: Tighten to 15 ft. lbs. (20 Nm).

   b. Step 2: Tighten an additional 60 degrees.

06017-NAVI-G16

Front oil pan bolts torque sequence—2005 Expedition and Navigator models

8 mm (0.32 in)

8 mm (0.32 in)

06017-NAVI-G17

Apply a 0.32 inch (8mm) bead of silicone gasket and sealant in 2 places where the engine front cover meets the right cylinder head—2005 Expedition and Navigator models

➡️ **If not secured within 4 minutes, the sealant must be removed and the sealing area cleaned. To clean the sealing area, use silicone gasket remover and metal surface prep. Failure to follow this procedure can cause future oil leakage.**

104. Apply a 0.32 inch (8mm) bead of silicone gasket and sealant in 2 places where the engine front cover meets the cylinder head.

❋ CAUTION

When installing the valve cover, make sure to avoid damaging the Variable Camshaft Timing (VCT) solenoid.

105. Install the right hand valve cover and gasket on the cylinder head and tighten

the bolts in the sequence shown and tighten to 89 inch lbs. (10 Nm).

➡️ **If not secured within 4 minutes, the sealant must be removed and the sealing area cleaned. To clean the sealing area, use silicone gasket remover and metal surface prep. Failure to follow this procedure can cause future oil leakage.**

106. Apply a 0.32 inch (8mm) bead of silicone gasket and sealant in 2 places where the engine front cover meets the left cylinder head.

107. Install the left hand valve cover and gasket on the cylinder head and tighten the bolts in the sequence shown and tighten to 89 inch lbs. (10 Nm).

108. Lubricate the engine front cover and the crankshaft seal inner lip with clean engine oil.

109. Use the tools illustrated to install the crankshaft seal into the engine front cover.

➡️ **If not secured within 4 minutes, the sealant must be removed and the seal-**

06017-NAVI-G20

Install the left hand valve cover and gasket on the cylinder head and tighten the bolts in sequence—2005 Expedition and Navigator models

ing area cleaned. To clean the sealing area, use silicone gasket remover and metal surface prep. Failure to follow this procedure can cause future oil leakage.

303-635

303-102

303-335

06017-NAVI-G21

Install the crankshaft seal into the engine front cover—2005 Expedition and Navigator models

06017-NAVI-G18

Install the right hand valve cover and gasket on the cylinder head and tighten the bolts in sequence—2005 Expedition and Navigator models

8 mm (0.32 in)

8 mm (0.32 in)

06017-NAVI-G19

Apply a 0.32 inch (8mm) bead of silicone gasket and sealant in 2 places where the engine front cover meets the left cylinder head—2005 Expedition and Navigator models

06017-NAVI-G22

Apply a 0.32 inch (8mm) bead of silicone gasket and sealant to the Woodruff key slot on the crankshaft pulley—2005 Expedition and Navigator models

06017-NAVI-G23

Install the crankshaft pulley—2005 Expedition and Navigator models

110. Apply a 0.32 inch (8mm) bead of silicone gasket and sealant to the Woodruff key slot on the crankshaft pulley.

111. Use the tool illustrated to install the crankshaft pulley.

112. Tighten the new crankshaft pulley bolt in 4 steps
   a. Step 1: Tighten to 66 ft. lbs. (90 Nm).
   b. Step 2: Loosen 360 degrees.
   c. Step 3: Tighten to 37 ft. lbs. (50 Nm).
   d. Step 4: Tighten an additional 90 degrees.

➡**Lubricate the new O-ring seal with clean engine oil prior to installation.**

113. Install the left hand CMP sensor and the bolt.

➡**Lubricate the new O-ring seal with clean engine oil prior to installation.**

114. Install the right hand CMP sensor and the bolt.

115. Install the CKP sensor and the bolt.

116. Install the accessory drive belt tensioner and install the 3 bolts. Tighten the bolts to 18 ft. lbs. (25 Nm).

117. Install the 3 accessory drive belt idler pulleys, the coolant pump pulley and the 7 bolts. Tighten the bolts to 18 ft. lbs. (25 Nm).

118. Position the electrical harness on the engine assembly and connect the engine wiring harness retainers to the valve cover studs.

119. Connect the CKP sensor connector.

120. Install the 8 ignition coils and the 8 bolts.

121. Connect the 4 left hand ignition coil connectors.

122. Install the left hand radio ignition interference capacitor and the stud bolt. Tighten the bolt to 18 ft. lbs. (25 Nm).

123. Connect the left hand CMP sensor connector.

124. Connect the engine oil pressure sensor connector.

125. Connect the left hand VCT solenoid connector.

126. Connect the 2 engine wiring harness retainers to the left hand valve cover studs.

127. Connect the CHT sensor connector.

128. Connect the 4 right hand ignition coil connectors.

129. Connect the connector retainer to the coolant tube.

130. Connect the 2 engine wiring harness retainers to the right hand valve cover studs.

131. Connect the right hand VCT solenoid connector.

132. Install the right hand radio ignition interference capacitor and the stud bolt. Tighten the bolt to 18 ft. lbs. (25 Nm).

133. Connect the right hand CMP sensor connector.

134. Using a suitable floor crane, remove the engine from the engine stand.

135. Install the flexplate and the 8 bolts in a star pattern. Tighten the bolts to 59 ft. lbs. (80 Nm).

136. Install the engine.

Rocker Arms

REMOVAL & INSTALLATION

4.6L and 5.4L Engines

1. Before servicing the vehicle, refer to the precautions section.
2. Disconnect the negative battery cable.
3. Remove the camshaft covers.
4. Position the piston of the cylinder being serviced at the bottom of its travel.

➡**Two different valve spring compressor tools are used for this procedure. Valve Spring Compressor (T91P-6565-A) is used on the exhaust camshaft and Valve Spring Compressor (T93P-6565-A) is used on the intake camshaft.**

5. Compress the valve spring and remove the rocker arm.

7924FG37

Using the proper tool, compress the valve spring and remove the rocker arm—4.6L and 5.4L engine

Compress the valve spring and remove the rocker arm—6.8L engine

To install:

6. Position the piston of the cylinder being serviced at the bottom of its travel.

7. Apply clean engine oil to the rocker arm, valve stem tip and tappet bore.

➡**Valve tappet should have no more than¹⁄₁₆ inch (1.5mm) of travel before installing the rocker arm.**

8. Compress the valve spring using the correct tool and install the rocker arm.

6.8L Engine

1. Before servicing the vehicle, refer to the precautions section.

2. Disconnect the negative battery cable.

3. Remove the camshaft covers.

4. Position the base circle of the camshaft lobe on the rocker arm to be serviced. Also, be sure the piston is not at the top of its travel near the valve.

5. Compress the valve spring and remove the rocker arm.

To install:

6. Position the base circle of the camshaft lobe over the place where the rocker arm is to be installed.

7. Apply clean engine oil to the rocker arm, valve stem tip and tappet bore.

8. Compress the valve using the special tool and install the rocker arm.

9. Install the rocker arm covers and the remaining components.

Intake Manifold

REMOVAL & INSTALLATION

➡**When the battery is disconnected and reconnected, some abnormal drive**

symptoms may occur while the vehicle relearns its adaptive strategy. The vehicle may need to be driven 10 miles (16 km) or more to relearn the strategy.

4.6L, 5.4L and 6.8L Engine, Except 5.4L DOHC Engines

2002–04 MODELS

> ✳✳ **CAUTION**
>
> **Fuel injection systems remain under pressure, even after the engine has been turned OFF. The fuel system pressure must be relieved before disconnecting any fuel lines. Failure to do so may result in fire and/or personal injury.**

1. Before servicing the vehicle, refer to the precautions section.

2. Remove or disconnect the following:
   - Negative battery cable
   - Engine cover, if equipped
   - Fuel system pressure
   - Coolant
   - Upper radiator hose from the intake manifold
   - Accelerator cable from the bracket and the throttle body cam
   - Speed control actuator cable from the throttle body
   - Throttle cable return spring
   - All vacuum hoses, fuel lines and electrical wires from the throttle body and intake manifold
   - Accelerator bracket from the throttle body
   - EGR valve-to-exhaust manifold tube
   - Brake booster vacuum hose bracket

   - Two evaporative emission canister purge valve nuts and position the valve aside
   - Four bolts, and the throttle body and adapter as an assembly
   - Fuel injector electrical connectors
   - Bolts and the eight ignition coils
   - On the 6.8L engine: the radio interference capacitors from the left side of the intake manifold
   - Alternator upper support bracket
   - Heater hose from the intake manifold
   - Water thermostat housing and thermostat. Discard the O-ring seal.
   - Intake manifold bolts
   - Intake manifold from the engine, then detach the Intake Manifold Tuning Valve (IMTV) electrical connector. Remove and discard the upper intake manifold gaskets.
   - Upper-to-lower intake manifold bolts, then separate the upper intake manifold from the lower intake manifold. Discard the old gasket.

To install:

3. On the 6.8L engine, position the lower intake manifold gasket and the upper intake manifold onto the lower intake manifold, then loosely install the upper-to-lower intake manifold bolts.

➡**Be sure to tighten the lower-to-upper manifold bolts in 2 steps.**

4. Tighten the 8 lower-to-upper intake manifold bolts in 2 steps following the tightening sequence as follows:
   - 18 inch lbs. (2 Nm)

Tighten the upper intake manifold bolts in 2 steps using this sequence—2002–04 4.6L shown, 5.4L SOHC engine similar

9358FGAZ

Tighten the lower intake manifold bolts in 2 steps using this sequence—2002–04 4.6L shown, 5.4L SOHC engine similar

20 - 30 Nm (15 - 22 Lb - Ft)

7924FG14

Tighten the lower intake manifold bolts in 2 steps using this sequence—6.8L engine

20-30 Nm (15-22 Lb-Ft)

7924FG15

Tighten the upper intake manifold bolts in 2 steps using this sequence—6.8L engine

- 89 inch. lbs. (10 Nm)
5. Position the 2 upper intake manifold gaskets on the cylinder heads. Set the upper intake manifold in place on the engine, then loosely install the intake manifold-to-cylinder head bolts.
6. Attach the IMTV electrical connector.

➡**Check that the thermostat housing is in the correct position before the thermostat housing is installed.**

7. Install the thermostat housing and start the 2 housing bolts.

➡**Make certain to tighten the intake manifold in 2 steps.**

8. Tighten the intake manifold bolts using the sequence shown, in 2 steps on all except 6.8L engines.
- 18 inch lbs. (2 Nm)
- 18 ft. lbs. (25 Nm)
9. Tighten the upper-to-lower intake manifold bolts using the sequence shown, in 2 steps on 6.8L engines.
- 18 inch lbs. (2 Nm)
- 18 ft. lbs. (25 Nm)
10. Install or connect the following:
- Heater water hose
- All electrical connections, fuel lines, vacuum tubes and coolant hoses to the intake manifold, fuel injectors and throttle body assembly.
- Alternator support bracket
- Ignition coils
- New throttle body adapter gasket and the throttle body adapter assembly. Tighten the bolts to 89 inch lbs. (10 Nm).

- All remaining hoses and electrical connections
- EGR valve-to-exhaust manifold tube. The tube fittings should be tightened to 26–33 ft. lbs. (35–45 Nm).
- Speed actuator cable, if equipped, and the accelerator cable to the throttle body.
- Coolant
- Negative battery cable. Start the engine and check for fuel, vacuum or coolant leaks.

2005 EXCURSION MODELS WITH 5.4L ENGINES

✳✳ CAUTION

Fuel injection systems remain under pressure, even after the engine has been turned OFF. The fuel system pressure must be relieved before disconnecting any fuel lines. Failure to do so may result in fire and/or personal injury.

1. Before servicing the vehicle, refer to the precautions section.
2. Disconnect the negative battery cable.
3. Drain and recycle the engine coolant.
4. Disconnect the Idle Air Control (IAC) fresh air tube and crankcase ventilation tube from the air cleaner outlet pipe.
5. Loosen the clamp and disconnect the air cleaner outlet pipe from the throttle body.
6. Release the clamp and separate the air cleaner outlet housing from the air cleaner inlet housing.
7. Disconnect the Mass Air Flow (MAF) sensor connector and remove the air cleaner, air cleaner outlet housing and the air cleaner outlet pipe from the vehicle.
8. Compress and slide the hose clamp and disconnect the upper radiator hose.
9. On vehicles with A/C, rotate the belt tensioner clockwise and remove the belt from the alternator pulley.
10. On vehicles without A/C, rotate the belt tensioner counterclockwise and remove the belt from the alternator pulley.
11. Remove the harness routing clips from the alternator stud and the engine front cover.
12. Remove the alternator upper mounting bracket stud and bolt.
13. Remove the alternator lower mounting bolts and position the alternator aside.
14. Remove the accelerator control splash shield.

15. Disconnect the throttle body cam.

16. Disconnect the accelerator cable.

17. If equipped, disconnect the speed control cable.

18. Remove the throttle return spring.

19. Remove accelerator cable bracket the bolts and position the bracket aside.

20. Disconnect the Throttle Position (TP) sensor connector.

21. Release the IAC fresh air tube from the routing clip and remove it from the vehicle.

22. Disconnect the Evaporative Emission (EVAP) canister purge valve vacuum hose from the throttle body adapter.

23. Disconnect the brake booster vacuum hose and main vacuum harness from the throttle body adapter.

24. Disconnect the IAC connector.

25. On vehicles with an Exhaust Gas Recirculation (EGR) system, position the exhaust manifold to EGR valve tube aside.

26. Position the differential pressure feedback EGR system bracket aside.

27. Disconnect the EGR vacuum regulator solenoid connections.

28. Disconnect the EGR valve vacuum hose.

29. Remove the throttle body adapter.

30. Remove and discard the throttle body gasket.

31. Disconnect the spring lock couplings from the fuel lines.

32. Disconnect the Positive Crankcase Ventilation (PCV) valve tube from the intake manifold.

33. Disconnect the fuel pressure regulator vacuum hose.

34. Disconnect the eight ignition coil connectors and remove the coils.

35. Disconnect the eight fuel injector connectors.

36. Disconnect the hose clamp and remove the heater water hose.

37. Remove the bolts, water thermostat housing, O-ring seal and the water thermostat. Discard the O-ring seal.

38. Remove the intake manifold bolts.

39. Remove the intake manifold.

40. Remove and discard the intake manifold gaskets.

✳✳ CAUTION

Do not use metal scrapers, wire brushes, power abrasive discs or other abrasive means to clean the sealing surfaces. These tools cause scratches and gouges that make leak paths.

41. Clean all mating surfaces.

9358FG05

Tighten the intake manifold bolts in 2 steps using this sequence—2005 Excursion models with the 5.4L SOHC engine

To install:

42. Install the intake manifold with new gaskets. Loosely install the bolts.

43. Install the water thermostat with a new O-ring seal.

➡**The thermostat housing bolts are tightened in sequence with the intake manifold bolts. Do not tighten the thermostat housing bolts during thermostat installation.**

44. Install the water thermostat housing and loosely install the bolts.

45. Tighten the bolts in two steps, in the sequence shown:
    a. Step 1: Tighten to 2 Nm (18 lb-in).
    b. Step 2: Tighten to 25 Nm (18 lb-ft).

46. Install the heater water hose and position the clamp.

47. Connect the eight fuel injector connectors.

48. Install the eight ignition coils.

49. Connect the eight ignition coil connectors.

50. Position the vacuum harness and connect the fuel pressure regulator vacuum hose.

51. Connect the PCV valve tube to the intake manifold.

52. Connect the spring lock couplings.

53. Install a new throttle body adapter gasket.

54. Install the throttle body adapter. Install the four bolts.

55. Connect the EGR valve vacuum hose.

56. Connect the EGR vacuum regulator solenoid connections.

57. Install the differential pressure feedback EGR system bracket.

58. Connect the exhaust manifold to EGR valve tube upper fitting. Tighten both fittings starting at the top in three steps: First, hand-tighten, then tighten the upper fitting to 50 Nm (37 lb-ft) and tighten the lower fitting to 50 Nm (37 lb-ft).

59. Connect the IAC connector.

60. Connect the brake booster vacuum hose and main vacuum harness to the throttle body adapter.

61. Connect the evaporative emission canister purge valve vacuum hose to the throttle body adapter.

62. Install the IAC fresh air tube and insert it in the routing clip.

63. Connect the TP sensor connector.

64. Install the accelerator cable bracket.

65. Connect the throttle body cam.

66. Connect the accelerator cable.

67. If equipped, connect speed control cable. Install the return spring.

68. Install the accelerator control splash shield.

69. Install the alternator.

70. Rotate the belt tensioner counterclockwise and install the belt on the alternator pulley, on models without A/C.

71. Rotate the belt tensioner clockwise and install the belt on the alternator pulley, on models with A/C.

72. Install the upper radiator hose and reposition the clamp.

73. Position the air cleaner, air cleaner outlet housing and the air cleaner outlet pipe in the vehicle and connect the MAF sensor connector.

74. Join the air cleaner outlet housing to the air cleaner inlet housing and secure the clamp.

75. Connect the air cleaner outlet pipe to the throttle body and tighten the clamp.

76. Connect the IAC fresh air tube and crankcase ventilation tube to the air cleaner outlet pipe.

77. Fill the cooling system.

78. Connect the negative battery cable.

2005 EXCURSION MODELS WITH 6.8L ENGINES

✳✳ CAUTION

Fuel injection systems remain under pressure, even after the engine has been turned OFF. The fuel system pressure must be relieved before disconnecting any fuel lines. Failure to do so may result in fire and/or personal injury.

1. Before servicing the vehicle, refer to the precautions section.

2. Properly relieve the fuel system pressure.

3. Disconnect the negative battery cable.

4. Drain and recycle the engine coolant.

5. Remove the air cleaner outlet tube.

6. Remove the alternator.

7. Disconnect the fuel tube spring lock couplings.

8. Compress and slide the hose clamp and disconnect the coolant outlet hose.

9. Remove the accelerator cable splash shield.

10. Disconnect the accelerator and speed control cables and remove the return spring.

11. Position the accelerator cable bracket aside.

12. Disconnect the Positive Crankcase Valve (PCV) tube and the vacuum tubes.

13. Disconnect the heated PCV coolant hose.

14. Disconnect the vacuum tube.

15. Disconnect the five right hand fuel injector connectors and five ignition coil connectors.

16. Disconnect the Idle Air Control (IAC) motor connector and the bypass hose.

17. Disconnect the heater hose.

18. Disconnect the five left hand fuel injector Throttle Position (TP) sensor connectors and ignition coil connectors.

19. Remove the ignition coils.

20. Disconnect the vacuum connector from the fuel pressure regulator.

21. Remove the bolts and the thermostat and housing.

22. Remove the bolts, the upper intake manifold and the intake manifold gaskets. Discard the intake manifold gaskets.

✳✳ CAUTION

Do not use metal scrapers, wire brushes, power abrasive discs or other abrasive means to clean the aluminum retainer plate. These tools cause scratches and gouges, which make leak paths.

23. Clean all mating surfaces. Clean the intake manifold mating surface of the cylinder heads with silicone gasket remover and metal surface prep. Follow the directions on the packaging.

To install:

24. Position the thermostat and housing and loosely install the bolts.

25. Install the upper intake manifold. Position the upper intake manifold gaskets and the intake manifold, and loosely install the bolts.

26. Tighten the bolts in two steps, in the sequence shown:

    a. Step 1: Tighten to 2 Nm (18 lb-in).

    b. Step 2: Tighten to 25 Nm (18 lb-ft).

27. Connect the fuel tube spring lock couplings .

28. Connect the vacuum tube to the fuel injection supply manifold.

29. Install the ignition coils.

30. Connect the five left hand fuel injector, TP sensor connectors and five left hand ignition coil connectors.

31. Connect the coolant hose.

32. Connect the IAC motor bypass hose and connector.

33. Connect the five right hand fuel injector connectors and five right hand ignition coil connectors.

34. Connect the vacuum tube.

35. Connect the PCV tube and the vacuum tubes.

36. Connect the heated PCV coolant hose.

37. Install the accelerator bracket.

38. Connect the accelerator and speed control cables and the return spring.

39. Install the accelerator cable splash shield.

40. Connect the engine coolant outlet hose and position the hose clamp.

41. Install the alternator.

42. Install the air cleaner outlet tube.

43. Connect the negative battery cable.

44. Fill and bleed the cooling system.

2005 EXPEDITION AND NAVIGATOR MODELS

✳✳ CAUTION

Fuel injection systems remain under pressure, even after the engine has been turned OFF. The fuel system pressure must be relieved before disconnecting any fuel lines. Failure to do so may result in fire and/or personal injury.

06017-NAVI-G43

Tighten the intake manifold bolts in this sequence—2005 Excursion models with the 6.8L engine

1. Before servicing the vehicle, refer to the precautions section.

2. Disconnect the negative battery cable.

3. Drain the cooling system.

4. Release the fuel system pressure.

5. Remove the alternator.

6. Remove the air cleaner.

7. Disconnect the upper radiator hose from the thermostat housing.

8. Disconnect the heater coolant hose from the coolant bypass tube.

9. Disconnect the Evaporative Emissions (EVAP) tube from the intake manifold.

10. Remove the Positive Crankcase Ventilation (PCV) tube.

11. Disconnect the fuel supply spring lock coupling from the fuel rail.

12. Disconnect the fuel rail pressure and temperature sensor connector and vacuum connector.

13. Disconnect the fuel injector connectors.

14. Disconnect the Throttle Position (TP) sensor and electronic acceleration control connectors.

15. Disconnect the heated PCV intake fitting connector.

16. Disconnect the brake booster vacuum hose from the intake manifold vacuum tube.

17. Remove the intake manifold bolts

✳✳ CAUTION

Do not use metal scrapers, wire brushes, power abrasive discs or other abrasive means to clean the sealing surfaces. These tools cause scratches and gouges which make leak paths. Use a plastic scraping tool to remove all traces of old sealant.

18. Remove the 3 bolts, the coolant bypass tube and discard the gaskets.

19. Clean and inspect the sealing surfaces with metal surface prep. Follow the directions on the packaging.

20. Disconnect the Charge Motion Control Valve (CMCV) connector.

21. Disconnect the manifold vacuum tube from the valve cover stud and the support bracket.

22. Disconnect the Cylinder Head Temperature (CHT) sensor jumper harness connector.

23. Disconnect the left hand and right hand Knock Sensor (KS) connectors.

24. Remove the nut and disconnect the engine wiring harness retainer from the CMCV stud.

06017-NAVI-G24

Intake manifold torque sequence—2005 Expedition and Navigator models

25. Remove the intake manifold and discard the gaskets.

26. Clean and inspect the sealing surfaces with metal surface prep.

To install:

➡**Electrical and vacuum harnesses must not restrict movement of the CMCV control rods at rear of the intake manifold. Use extreme care on installation of the intake manifold to prevent any pinching of electrical and vacuum harnesses.**

27. Using new intake manifold gaskets, position the intake manifold.

28. Using new gaskets, position the coolant crossover and install the 3 bolts. Tighten to 10 Nm (89 lb-in).

29. Install the intake manifold bolts and tighten in 2 steps in the sequence shown:
   Step 1: Tighten to 2 Nm (18 lb-in).
   Step 2: Tighten to 10 Nm (89 lb-in).

30. Connect the engine wiring harness retainer to the CMCV stud and install the nut. Tighten to 10 Nm (89 lb-in).

31. Connect the CMCV connector.

32. Connect the CHT sensor jumper harness connector.

33. Connect the left hand and right hand KS connectors.

34. Connect the manifold vacuum tube to the support bracket and the valve cover stud.

35. Connect the brake booster vacuum hose to the intake manifold vacuum tube.

36. Connect the heated PCV intake fitting connector.

37. Connect the TP sensor and electronic acceleration control connectors.

38. Connect the fuel injector connectors.

39. Connect the fuel rail pressure and temperature sensor connector and vacuum connector.

40. Connect the fuel supply spring lock coupling to the fuel rail.

41. Install the PCV tube.

42. Connect the EVAP tube to the intake manifold.

43. Connect the heater coolant hose to the coolant bypass.

44. Connect the upper radiator hose to the thermostat housing.

45. Install the alternator.

46. Install the air cleaner.

47. Connect the negative battery cable.

48. Fill and bleed the engine cooling system

5.4L DOHC Engines

UPPER INTAKE MANIFOLD

1. Before servicing the vehicle, refer to the precautions section.

2. Remove or disconnect the following:
   - Negative battery cable
   - Air cleaner outlet tube
   - Accelerator cable, speed control cable, and the return spring
   - Bolts and position the cables and bracket out of the way
   - Evaporative emission return line
   - Positive Crankcase Ventilation (PCV) tube from the upper intake manifold
   - PCV valve from the valve cover
   - PCV valve tube from the water heated fitting and remove the tube assembly
   - Coolant lines and plug the lines.
   - Electrical connector from the communication valve
   - Bolts and the communication valve
   - Wiring harness shield bolts and 5 clips and remove the shield
   - Balance tube from the engine
   - Throttle Position (TP) sensor and the Idle Air Control (IAC) motor
   - Vacuum lines and detach the electrical connector from the Exhaust Gas Recirculation (EGR) vacuum regulator (EVR)
   - The 2 hoses and detach the electrical connector from the differential pressure feedback EGR
   - The bolts from the power steering reservoir bracket and position it aside
   - The stud and position the oil fill tube aside
   - Brake booster vacuum line
   - Vacuum line from the EGR valve
   - Bolts from the EGR adapter
   - Retaining bolts and remove the upper intake manifold

Tighten the upper intake manifold bolts as shown—5.4L DOHC engines

Tighten the lower intake manifold bolts as shown—5.4L DOHC engines

To install:

3. Clean and inspect the sealing surfaces.

4. Install or connect the following:
- Upper intake manifold: Tighten the bolts to 89 inch lbs. (10 Nm) and then an additional 90 degrees in the sequence shown.
- EGR adapter with a new gasket
- Vacuum line to the EGR valve
- Brake booster vacuum line
- Oil fill tube and install the stud
- Power steering reservoir bracket and install the bolts
- EGR valve at the exhaust manifold
- The 2 hoses and the electrical connector to the differential pressure feedback EGR
- Vacuum lines and the electrical connector to the EVR
- IAC valve and the TP sensor
- Balance tube on the engine
- Communication valve and the bolts
- Wiring harness shield, the bolts and 5 clips
- Electrical connector to the communication valve
- Tube assembly and connect the PCV valve tube to the water-heated fitting
- Coolant hoses
- PCV valve in the valve cover
- EVAP return line
- Cables and bracket and install the bolts
- Accelerator cable, speed control cable, and the return spring
- Engine appearance cover and the 3 bolts
- Engine air cleaner and outlet tube

LOWER INTAKE MANIFOLD

❊❊ CAUTION

Fuel injection systems remain under pressure, even after the engine has been turned OFF. The fuel system pressure must be relieved before disconnecting any fuel lines. Failure to do so may result in fire and/or personal injury.

1. Before servicing the vehicle, refer to the precautions section.

2. Remove or disconnect the following:
- Negative battery cable
- Coolant
- Upper intake manifold
- Fuel system pressure
- Fuel lines
- Engine water bypass hose
- Electrical connector from the water temperature indicator sender
- Upper radiator hose, the heater water inlet hose and the heated PCV water fitting inlet hose
- The 4 bolts and the upper alternator support bracket
- The 8 fuel injectors
- Vacuum line from the fuel injector pressure regulator and position out of the way
- Lower intake manifold
- Radio ignition interference capacitors

To install:

3. Remove and inspect the gaskets, install new gaskets if necessary.

4. Clean the sealing surfaces.

5. Install or connect the following:
- Radio ignition interference capacitors

➡Bolts should be hand-started, positions 7–12 first then 1–6.

- Lower intake manifold: Tighten the bolts to 89 inch lbs. (10 Nm) and then an additional 90 degrees in the sequence shown.
- Water temperature indicator sensor
- Vacuum harness and connect the vacuum line to the fuel injector pressure regulator
- Fuel injectors
- Upper generator support bracket and the 4 bolts
- Heater water inlet hose, the upper radiator hose and the water heated fitting inlet hose
- Engine water bypass return hose
- Fuel lines
- Upper intake manifold
- Coolant

Exhaust Manifold

REMOVAL & INSTALLATION

Except 2005 Expedition And Navigator Models

1. Before servicing the vehicle, refer to the precautions section.

2. Remove or disconnect the following:
- Front fender splash shield
- For the left-hand exhaust manifold: the Exhaust Gas Recirculation (EGR) valve-to-exhaust manifold tube and if equipped, the DPFE gas recirculation transducer hoses.
- On the 4.6L and 5.4L engines: the catalytic converter-to-exhaust manifold bolts
- On the 6.8L engine: the front exhaust pipe from the manifold

Tighten the exhaust manifold bolts in the sequence shown—2002–04 4.6L engine shown, 5.4L engine similar

7924FG16

25 Nm (18 lb-ft)

06017-NAVI-G44

Tighten the right side exhaust manifold bolts in the sequence shown—2005 Models, except Expedition and Navigator

25 Nm (18 lb-ft)

06017-NAVI-G45

Tighten the left side exhaust manifold bolts in the sequence shown—2005 Models, except Expedition and Navigator

- The exhaust manifold mounting nuts, then remove the exhaust manifold itself. Remove and discard the old gasket.

3. Clean and inspect the exhaust manifold for damage.

23-27 Nm (17-20 lb/ft)

7924FG85

Tighten the exhaust manifold bolts in the sequence shown—right side of 6.8L engine shown

25 Nm (18 lb-ft)

06017-NAVI-G46

Tighten the exhaust manifold bolts in the sequence shown—left side of 6.8L engine shown

To install:

4. Position a new gasket and the exhaust manifold onto the engine block.

5. Install the mounting nuts and tighten following the sequence shown.

- 18 ft. lbs. (25 Nm)

6. On the 6.8L engine, tighten the exhaust manifold-to-front pipe fasteners to 27–34 ft. lbs. (34–46 Nm).

7. On the 4.6L and 5.4L engines, attach the catalytic converter to the exhaust manifold, install the catalytic converter-to-exhaust manifold bolts and tighten to 25–34 ft. lbs. (34–46 Nm)

8. For the left-hand exhaust manifold, install the DPFE transducer hoses if equipped, and the EGR valve-to-exhaust manifold tube. Tighten the upper and lower fittings to 26–33 ft. lbs. (35–45 Nm).

9. Install the front fender splash shield.

10. Lower the vehicle to the ground.

2005 Expedition And Navigator Models

RIGHT SIDE

1. Before servicing the vehicle, refer to the precautions section.

2. Remove the air cleaner housing.

3. Remove the alternator.

4. Remove the starter.

5. Remove the right hand inner fender-well.

6. Remove the upper radiator shroud.

✳✳ CAUTION

Only use hand tools when loosening the left hand engine mount through bolt or damage to the cage nut can occur.

NOTE: If the left hand engine mount through bolt nut is missing or damaged install a new nut using service part number W709375. If the left hand engine mount through bolt nut cage is damaged or missing, remove the cage and install a new nut using service part number W520516-S301.

7. Loosen the left hand engine support insulator bolt.

✳✳ CAUTION

Only use hand tools when removing the right hand engine mount nut or damage to the engine mount can occur.

8. Remove the right hand engine mount nut.

9. Remove the exhaust manifold-to-catalytic converter nuts.

✳✳ CAUTION

Only use hand tools when removing the transmission mount-to-crossmember nuts or damage to the transmission mount can occur.

10. Loosen the transmission mount-to-crossmember nuts.

11. Remove the bolts and position the sway bar down.

12. Install the and engine support tool and raise the engine.

13. Remove the nuts and the exhaust manifold.

06017-NAVI-G25

Tighten the right side exhaust manifold bolts in the sequence shown—2005 Expedition and Navigator models

✳✳ CAUTION

Do not use metal scrapers, wire brushes, power abrasive discs, or other abrasive means to clean the sealing surfaces. These may cause scratches and gouges resulting in leak paths. Use a plastic scraper to clean the sealing surfaces.

➡Clean the sealing surfaces with metal surface prep. Remove and discard the exhaust manifold gaskets. Clean the sealing surfaces with metal surface prep. To install, reverse the removal procedure.

To install:

14. Installation is the reverse of removal, please note the following torque specifications:

   a. Tighten the manifold bolts in the sequence illustrated to 25 Nm (18 ft. lbs.).

   b. Tighten the left hand engine support insulator bolt to 175 Nm (129 ft. lbs.).

   c. Tighten the right hand engine mount nut to 175 Nm (129 ft. lbs.).

   d. Tighten the exhaust manifold-to-catalytic converter nuts to 40 Nm (30 ft. lbs.).

   e. Tighten the transmission mount-to-crossmember nuts to 103 Nm (76 ft. lbs.).

   f. Tighten the sway bar bolts to 30 Nm (22 ft. lbs.).

2005 Models

LEFT SIDE

1. Before servicing the vehicle, refer to the precautions section.
2. Remove the air cleaner housing.
3. Remove the alternator.
4. Remove the left hand inner fender-well.

5. Remove the upper radiator shroud.

✳✳ CAUTION

Only use hand tools when removing the left hand engine mount through bolt or damage to the cage nut can occur.

➡If the left hand engine mount through bolt nut is missing or damaged install a new nut using service part number W709375. If the left hand engine mount through bolt nut cage is damaged or missing, remove the cage and install a new nut using service part number W520516-S301.

6. Remove the left hand engine support insulator bolt.

✳✳ CAUTION

Only use hand tools when loosening the right hand engine mount nut or damage to the engine mount can occur.

✳✳ CAUTION

Loosen the right hand engine mount nut.

7. Remove the 4 exhaust manifold-to-catalytic converter nuts.

✳✳ CAUTION

Only use hand tools when removing the transmission mount-to-cross-member nuts or damage to the transmission mount can occur.

8. Loosen the transmission mount-to-crossmember nuts.
9. Remove the bolts and the exhaust manifold heat shield.
10. Remove the front driveshaft on 4WD models.
11. Install the an engine support tool and raise the engine.
12. Remove the nuts and the exhaust manifold.

To install:

13. Installation is the reverse of removal, please note the following torque specifications:

   a. Tighten the manifold bolts in the sequence illustrated to 25 Nm (18 ft. lbs.).

   b. Tighten the left hand engine support insulator bolt to 175 Nm (129 ft. lbs.).

   c. Tighten the right hand engine mount nut to 175 Nm (129 ft. lbs.).

   d. Tighten the exhaust manifold-to-catalytic converter nuts to 40 Nm (30 ft. lbs.).

   e. Tighten the transmission mount-to-crossmember nuts to 103 Nm (76 ft. lbs.).

06017-NAVI-G26

Tighten the left side exhaust manifold bolts in the sequence shown—2005 Expedition and Navigator models

Camshaft and Valve Lifters

REMOVAL & INSTALLATION

SOHC Engines

EXCEPT 2005 EXPEDITION AND NAVIGATOR MODELS

1. Before servicing the vehicle, refer to the precautions section.
2. Remove the cylinder head covers from the engine.
3. Remove the timing chain.

✳✳ WARNING

At no time, when the timing chains are removed and the cylinder heads are installed may the crankshaft or camshaft be rotated. Severe piston and valve damage will occur.

4. On the 6.8L engine, remove the 6 bolts securing the balance shaft to the cylinder head and remove the shaft.
5. Remove the camshaft roller lifters.
6. On VIN W engines, remove the timing chain camshaft gear by removing the gear retaining bolt.

➡ Keep the bearing caps in order so they can be installed in the same position.

7. Remove the camshaft bearing cap bolts, then lift the camshaft bearing caps off of the cylinder head.
8. Lift the camshaft from the cylinder head.
9. Remove the rocker arms and pull the lash adjusters out of their bores. Keep all the parts in order. They must be installed in their original positions.

To install:

10. Install the lash adjusters and rocker arms in their original positions.
11. Lubricate the camshaft journals and bearing caps with SAE 5W30 engine oil. On the 6.8L engine, lubricate the balance shaft journals and bearing caps with the same lubricant.
12. Lower the camshaft onto the camshaft bearing journals.
13. Install the camshaft bearing caps, then loosely install the bearing cap bolts.
14. Tighten the camshaft bearing cap mounting bolts, in the sequence shown for the particular engine, to 71–107 inch lbs. (8–12 Nm).
15. On the 6.8L engine, align the timing marks and position the balance shaft on the journals, then install the bearing caps.

Tighten the bearing caps in the sequence shown—2002–04 Windsor 4.6L engine shown, 5.4L engine similar

Tighten the bearing caps in the sequence shown—2002–04 Romeo 4.6L engine shown, 5.4L engine similar

Camshaft bearing cap bolt tightening sequence—6.8L engine

Be sure to align the balance shaft timing mark with the mark on the camshaft gear—6.8L engine

Tighten the bolts in sequence to 89 inch lbs. (10 Nm).

16. On engines equipped with bolt on sprockets, install the camshaft timing chain gear by tightening the retaining bolts as follows:

    a. M10 Step 1: 30 ft. lbs. (40 Nm).
    b. M10 Step 2: Tighten 90 degrees.
    c. M12 Step 1: 90 ft. lbs. (120 Nm).

17. Install the valve lifters.

18. Install the timing chain and sprockets, if applicable.

19. Install the cylinder head covers.

2005 EXPEDITION AND NAVIGATOR MODELS

1. Before servicing the vehicle, refer to the precautions section.

✳✳ CAUTION

The camshaft procedure must be followed exactly or damage to the valves and pistons will result.

➡️ If removing both camshafts, you must remove the right hand camshaft first.

2. Position the crankshaft damper spoke at the 12 o'clock position and the timing mark indentation at the 1 o'clock position.

3. For the right hand camshaft:
    a. Remove the right hand valve cover.

✳✳ CAUTION

Damage to the camshaft phaser sprocket assembly will occur if mishandled or used as a lifting or leveraging device.

    b. Loosen and back off the right hand camshaft phaser bolt 1 full turn.
    c. Disconnect the right hand Camshaft Position (CMP) sensor connector.
    d. Remove the bolt and the right hand CMP sensor.

➡️ If the camshaft lobes are not exactly positioned as shown, the crankshaft will require 1 full additional rotation to 12 o'clock.

    e. The number 1 cylinder camshaft exhaust lobe must be coming up on the

---

Position the crankshaft damper spoke at the 12 o'clock position and the timing mark indentation at the 1 o'clock position—2005 Expedition and Navigator models

The number 1 cylinder camshaft exhaust lobe must be coming up on the exhaust stroke. Verify by noting the position of the 2 intake camshaft lobes and the exhaust lobe on the number 1 cylinder—2005 Expedition and Navigator models

---

8-12 Nm (71-106 lb/in)

Tighten the balance shaft bearing cap bolts in the sequence shown—6.8L engine

Remove only the 3 roller followers shown from the right hand cylinder head—2005 Expedition and Navigator models

exhaust stroke. Verify by noting the position of the 2 intake camshaft lobes and the exhaust lobe on the number 1 cylinder.

4. Remove only the 3 roller followers shown in the illustration.

✳✳ CAUTION

Do not allow the valve keepers to fall off the valve or the valve may drop into the cylinder.

✳✳ CAUTION

The camshaft roller followers must be installed in their original locations. Record camshaft roller follower locations.

➡ It may be necessary to push the valve down while compressing the spring.

   a. Using the tool illustrated, remove the 3 designated roller followers in the previous step from the right hand cylinder head.
   5. For the left hand camshaft:
   6. Remove the left hand valve cover.

✳✳ CAUTION

Damage to the camshaft phaser sprocket assembly will occur if mishandled or used as a lifting or leveraging device.

   a. Loosen and back off the left hand camshaft phaser bolt 1 full turn.
   b. Disconnect the left hand Camshaft Position (CMP) sensor connector.
   c. Remove the left hand CMP sensor and the bolt.

✳✳ CAUTION

If servicing both camshafts, do not rotate the crankshaft. As camshaft position has been established.

Using the tool illustrated, remove the 3 designated roller followers—2005 Expedition and Navigator models

The number 5 cylinder camshaft lobe must be coming up on the exhaust stroke. Verify by noting the position of the 2 intake camshaft lobes and the exhaust lobe on the number 5 cylinder—2005 Expedition and Navigator models

➡ If the camshaft lobes are not exactly positioned as shown, the crankshaft keyway will require one full additional rotation to 12 o'clock.

   d. The number 5 cylinder camshaft lobe must be coming up on the exhaust stroke. Verify by noting the position of the 2 intake camshaft lobes and the exhaust lobe on the number 5 cylinder.
   e. Remove only the 3 roller followers shown in the illustration from the left hand cylinder head.

✳✳ CAUTION

Do not allow the valve keepers to fall off the valve or the valve may drop into the cylinder.

➡ The camshaft roller followers must be installed in their original locations. Record camshaft roller follower locations.

➡ It may be necessary to push the valve down while compressing the spring.

   f. Using the tool illustrated, remove the 3 designated roller followers in the previous step from the left hand cylinder head.

✳✳ CAUTION

The crankshaft cannot be moved past the 6 o'clock position once set.

   7. Rotate the crankshaft clockwise, as viewed from the front, positioning the crankshaft damper spoke at the 6 o'clock position and the timing mark indentation at the 7 o'clock position.
   8. For the right hand camshaft:

Remove only the 3 roller followers shown from the left hand cylinder head—2005 Expedition and Navigator models

Rotate the crankshaft clockwise, as viewed from the front, positioning the crankshaft damper spoke at the 6 o'clock position and the timing mark indentation at the 7 o'clock position—2005 Expedition and Navigator models

✳✳ CAUTION

Engine is not freewheeling. Camshaft procedure must be followed exactly or damage to valves and pistons will result.

✳✳ CAUTION

The Timing Chain Wedge tool must be installed square to the timing chain and the engine block.

   9. Install the wedge tool illustrated in the right hand timing chain as shown.

06017-NAVI-G31

Install the wedge tool illustrated in the right hand timing chain—2005 Expedition and Navigator models

06017-NAVI-G32

Scribe a location mark on the timing chain and the camshaft phaser sprocket assembly, right side shown, left side similar—2005 Expedition and Navigator models

✳✳ CAUTION

Do not remove the timing chain wedge tool at any time during assembly. If the tool is removed or out of placement, the engine front cover must be removed and the engine must be retimed.

✳✳ CAUTION

The timing chain must be installed in its original position onto the camshaft phaser sprocket using the scribed marks, or damage to valves and pistons will result.

a. Scribe a location mark on the timing chain and the camshaft phaser sprocket assembly.

✳✳ CAUTION

When removing the front thrust camshaft bearing cap, use care as the cap may be damaged from side-loading when removing the cam unequally in height from the bearing towers.

➡The camshaft bearing caps must be installed in their original locations. Record camshaft bearing cap locations.

b. Remove the bolts in the sequence shown and remove the front camshaft bearing cap and then the remaining bearing caps.

c. Clean and inspect the right hand camshaft bearing caps.

➡The camshaft front thrust bearing cap contains an oil metering groove. Make sure the groove is free of foreign material.

✳✳ CAUTION

Damage to the camshaft phaser sprocket assembly will occur if mishandled or used as a lifting or leveraging device.

✳✳ CAUTION

Only use hand tools to remove the camshaft phaser sprocket bolt or damage may occur to the camshaft or camshaft phaser unit.

06017-NAVI-G05

Right side bearing cap removal and tightening sequence, left side similar—2005 Expedition and Navigator models

✳✳ CAUTION

Do not remove the timing chain wedge tool at any time during assembly. If the special tool is removed or out of placement, the engine front cover must be removed and the engine must be retimed.

d. Remove the bolt and withdraw the camshaft from the phaser sprocket assembly leaving the sprocket assembly in place. Discard the bolt and washer.

10. For the left hand camshaft:

✳✳ CAUTION

Engine is not freewheeling. Camshaft procedure must be followed exactly or damage to valves and pistons will result.

✳✳ CAUTION

The Timing Chain Wedge tool must be installed square to the timing chain and the engine block.

a. Install the wedge tool illustrated in the left hand timing chain as shown.

✳✳ CAUTION

Do not remove the timing chain wedge tool at any time during assembly. If the special tool is removed or out of placement, the engine front cover must be removed and the engine must be retimed.

303-637

303-636

06017-NAVI-G33

Install the wedge tool illustrated in the left hand timing chain—2005 Expedition and Navigator models

✳✳ CAUTION

The timing chain must be installed in its original position onto the camshaft phaser sprocket using the scribed marks, or damage to valves and pistons will result.

11. Scribe a location mark on the timing chain and the camshaft phaser sprocket assembly.

✳✳ CAUTION

When removing the front thrust camshaft bearing cap, use care as the cap may be damaged from side-loading when removing the cam unequally in height from the bearing towers.

➡The camshaft bearing caps must be installed in their original locations. Record camshaft bearing cap locations.

a. Remove the bolts in the sequence shown and remove the front camshaft bearing cap and then the remaining bearing caps.

b. Clean and inspect the left hand camshaft bearing caps.

➡The camshaft front thrust bearing cap contains an oil metering groove. Make

sure the groove is free of foreign material.

✳✳ CAUTION

Damage to the camshaft phaser sprocket assembly will occur if mis-handled or used as a lifting or lever-aging device.

✳✳ CAUTION

Only use hand tools to remove the camshaft phaser sprocket bolt or damage may occur to the camshaft or camshaft phaser unit.

✳✳ CAUTION

Do not remove the timing chain wedge tool at any time during assembly. If the special tool is removed or out of placement, the engine front cover must be removed and the engine must be retimed.

c. Remove the bolt and withdraw the camshaft from the phaser sprocket assembly leaving the sprocket assembly in place. Discard the bolt and washer.

To install:

12. Lubricate the camshaft and camshaft journals with clean engine oil.

13. For the right hand camshaft:

✳✳ CAUTION

Do not remove the timing chain wedge tool at any time during assembly. If the special tool is removed or out of placement, the engine front cover must be removed and the engine must be retimed.

✳✳ CAUTION

Damage to the camshaft phaser sprocket assembly will occur if mis-handled or used as a lifting or lever-aging device.

✳✳ CAUTION

Do not allow the roller followers to move out of position when installing the camshaft.

a. Install the camshaft into the camshaft phaser sprocket assembly and onto the head. Install a new camshaft phaser bolt finger tight.

✳✳ CAUTION

Do not remove the timing chain wedge tool at any time during assembly. If the special tool is removed or out of placement, the engine front cover must be removed and the engine must be retimed.

✳✳ CAUTION

The timing chain must be installed in its original position onto the camshaft phaser sprocket using the scribed marks, or damage to valves and pistons will result.

b. Verify the camshaft phaser sprocket and timing chain scribe marks are still in alignment.

✳✳ CAUTION

Do not allow the roller followers to move out of position when installing the camshaft.

c. Lubricate the camshaft bearing caps with clean engine oil.

d. Install the camshaft bearing caps in their original locations.

e. Install the bolts loosely, then

tighten the bolts in the sequence illustrated to 10 Nm (89 lb-in).

f. Remove the wedge tools.

14. For the left hand camshaft:

✳✳ CAUTION

Do not remove the timing chain wedge tool at any time during assembly. If the special tool is removed or out of placement, the engine front cover must be removed and the engine must be retimed.

✳✳ CAUTION

Damage to the camshaft phaser sprocket assembly will occur if mishandled or used as a lifting or leveraging device.

✳✳ CAUTION

Do not allow the roller followers to move out of position when installing the camshaft.

a. Install the camshaft into the camshaft phaser sprocket assembly and onto the head. Install a new camshaft phaser bolt finger tight.

✳✳ CAUTION

Do not remove the timing chain wedge tool at any time during assembly. If the special tool is removed or out of placement, the engine front cover must be removed and the engine must be retimed.

✳✳ CAUTION

The timing chain must be installed in its original position onto the camshaft phaser sprocket using the scribed marks, or damage to valves and pistons will result.

b. Verify the camshaft phaser sprocket and timing chain scribe marks are still in alignment.

✳✳ CAUTION

Do not allow the roller followers to move out of position when installing the camshaft.

c. Lubricate the camshaft bearing caps with clean engine oil.

d. Install the camshaft bearing caps in their original locations. Install the bolts loosely.

e. Tighten the bolts in the sequence illustrated to10 Nm (89 lb-in).

f. Remove the wedge tools.

15. Rotate the crankshaft a half turn counterclockwise and position the crankshaft damper spoke at the 12 o'clock position and the timing mark indentation at the 1 o'clock position.

16. For the right hand camshaft:

a. Verify correct cam position by noting the position of the number 1 cylinder intake and exhaust camshaft lobes.

b. Using the tool 303-1039, install the 3 originally removed roller followers.

c. Install the CMP sensor and the bolt.

d. Connect the CMP connector.

✳✳ CAUTION

Only use hand tools to install the camshaft phaser sprocket assembly or damage may occur to the camshaft or camshaft phaser unit.

✳✳ CAUTION

Damage to the camshaft phaser sprocket assembly will occur if mishandled or used as a lifting or leveraging device.

e. Tighten the camshaft phaser bolt in 2 steps: Tighten to 40 Nm (30 ft. lbs.), then tighten an additional 90 degrees.

f. Install the right hand valve cover.

17. For the left hand camshaft:

a. Verify correct cam position by noting the position of the number 5 cylinder intake and exhaust camshaft lobes.

b. Using the tool 303-1039, install the 3 originally removed roller followers.

c. Install the CMP sensor and the bolt.

d. Connect the CMP connector.

✳✳ CAUTION

Only use hand tools to install the camshaft phaser sprocket assembly or damage may occur to the camshaft or camshaft phaser unit.

✳✳ CAUTION

Damage to the camshaft phaser sprocket assembly will occur if mishandled or used as a lifting or leveraging device.

e. Tighten the camshaft phaser bolt in 2 steps. Tighten to 40 Nm (30 ft. lbs.), then tighten an additional 90 degrees.

f. Install the left hand valve cover.

DOHC Engines

1. Before servicing the vehicle, refer to the precautions section.

2. Remove or disconnect the following:
- Roller followers
- Left hand timing chain for the left hand side and both timing chains for the right hand side

3. Install camshaft holding tool T93P-6256-AHR.
- Exhaust camshaft sprocket and the intake washer and the spacer

4. Remove camshaft holding tool T93P-6256-AHR.

5. Compress the timing chain tensioner and install a lock pin.
- Timing chain, the sprocket, and intake camshaft sprocket spacer
- Outer cam bearing bolts shown in the accompanying illustration

✳✳ WARNING

The outer bolts on the outer cam bearing cap (exhaust) are longer and must be returned to the same location or engine damage may occur.

6. Make sure to identify the camshaft to cylinder head location. Caps are not interchangeable.
- Bolts and the camshaft bearing cap assemblies
- Camshafts

To install:

7. Make sure the timing marks are at 12 o'clock and index at 6 o'clock. Refer to the illustration for proper timing mark alignment.

8. Lubricate the camshafts. Use Super Premium SAE 5W-30.

9. Install or connect the following:
- Camshafts
- Camshaft bearing cap assemblies and bolts. Tighten the bolts in the sequence illustrated to 89 inch lbs. (10 Nm).
- Outer camshaft bearing bolts and tighten to 71–106 inch lbs. (8–12 Nm)

9359FG03

Remove the bolts shown before removing any other bearing cap bolts—DOHC models

Tighten the camshaft cap bolts as shown to specification

Make sure the timing marks are at 12 o'clock and index at 6 o'clock—DOHC models

- Timing chain, the sprocket, and intake camshaft sprocket spacer as an assembly
- Camshaft sprocket bolt, the washer and the spacer

10. Remove the lock pin.

11. Install camshaft holding tool T93P-6256-AHR.

12. Tighten the exhaust camshaft sprocket and the intake camshaft bolt washer and spacer in two steps as follows:
   a. Step 1: 30 ft. lbs. (40 Nm).
   b. Step 2: additional 90 degrees.

13. Remove camshaft holding tool T93P-6256-AHR.

- Timing chains
- Roller followers

Valve Lash

ADJUSTMENT

These engines do not require valve lash adjusting, because they utilize hydraulic

lash components in their valve actuation systems.

Starter Motor

REMOVAL & INSTALLATION

1. Before servicing the vehicle, refer to the precautions section.

2. Disconnect the negative battery cable.

3. Raise and safely support the vehicle.

4. Remove or disconnect the following:
   - Starter terminal cover
   - Terminal nut and separate the battery starter cable from the starter motor
   - Solenoid**S** terminal connector, if equipped with a starter mounted solenoid

➡**To disconnect the hard-shell connector from the solenoid S terminal, grasp the plastic shell and pull off; do not pull on the wire. Pull straight off to prevent damage to the connector and S terminal.**

5. Remove or disconnect the following:
   - Starter motor retaining bolts
   - Starter motor from the vehicle

To install:

6. Install or connect the following:
   - Starter motor and retaining bolts. Tighten the bolts to 15–20 ft. lbs. (20–27 Nm).
   - Battery starter cable and a terminal nut to the starter motor. Tighten the terminal nuts to 79 inch lbs. (9 Nm).
   - Solenoid**S** terminal connector, if equipped with a starter mounted solenoid
   - Starter solenoid safety cap, if equipped

7. Lower the vehicle.

8. Connect the negative battery cable.

9. Start the engine several times to check starter motor operation.

Oil Pan

REMOVAL & INSTALLATION

2002–04 Models—Except 6.8L Engine

4X2 MODELS WITH SOHC ENGINES

1. Before servicing the vehicle, refer to the precautions section.

2. Raise and safely support the vehicle.

3. Drain the oil and cooling system.

4. Remove or disconnect the following:
   - Negative battery cable
   - Radiator air deflector
   - Accelerator cable snow shield
   - Accelerator cable
   - Speed control actuator cable
   - Accelerator return spring
   - Accelerator cable bracket bolts and position the bracket and cables aside
   - Main vacuum harness
   - Throttle Position (TP) sensor and engine vacuum regulator connectors
   - Fuel pressure regulator vacuum line
   - Vapor management valve vacuum line
   - Differential pressure feedback Exhaust Gas Recirculation (EGR) connector
   - Brake booster vacuum line and bracket
   - Fuel lines
   - Idle Air Control (IAC) motor electrical connector
   - EGR valve-to-exhaust manifold tube aside
   - Positive Crankcase Ventilation (PCV) hose
   - Throttle body adapter
   - Upper and lower power steering reservoir bolts and position the reservoir aside
   - Fuel injector connections
   - Ignition coils
   - Alternator

5. Install an Engine Support Bracket on the engine using the alternator mounting bolts.
   - Right and left hand motor mount bolts

6. Raise the engine using the Engine Support.

➡**Be careful when removing the oil pan gasket. The oil pan gasket is reusable if it is not damaged.**

   - Oil pan bolts and pan

7. Position the oil pan aside to gain access to the oil pump screen cover and tube .
   - Bolt securing the rear of the oil pump screen cover and tube
   - Bolts securing the front of the oil pump screen cover and tube

To install:

> ✳✳ **WARNING**
>
> **Do not use metal scrapers, wire brushes, power abrasive discs or other abrasive means to clean the sealing surfaces. These tools cause scratches and gouges which make leak paths. Use a plastic scraping tool to remove all traces of old sealant.**

8. Clean and inspect the mating surfaces.

9. Install or connect the following:
- Oil pan gasket, the oil pan and the oil pump screen cover and tube together to the cylinder block
- Bolts securing the front of the oil pump screen cover and tube and tighten to 71–107 inch lbs. (8–12 Nm)

➡If the oil pan is not secured within four minutes, the sealant must be removed and the sealing area cleaned with Metal Surface Cleaner. Allow to dry until there is no sign of wetness, or four minutes, whichever is longer. Failure to follow this procedure may result in future oil leakage.

10. Apply the silicone Gasket and Sealant F7AZ-19554-EA at the rear oil seal retainer-to-cylinder block sealing surface.

11. Install the oil pan gasket and the oil pan and loosely install the 16 bolts. Tighten the bolts in three steps, in the sequence illustrated as follows:
- a. Step 1: 18 inch lbs. (2 Nm).
- b. Step 2: 15 ft. lbs. (20 Nm).
- c. Step 3: an additional 60 degrees.

12. Remove the Engine Support.
- Right and left hand motor mount bolts and tighten to 50–68 ft. lbs. (68–92 Nm)
- Alternator
- Fan shroud
- Throttle body adapter and the four bolts. Tighten the bolts to 89 inch lbs. (10 Nm).
- EGR valve-to-exhaust manifold tube
- All vacuum hoses, fuel lines and electrical connections
- Accelerator cable bracket and the bolts
- Accelerator cable, speed control actuator cable and the throttle return spring
- Accelerator cable snow shield and the bolts

- Upper radiator hose and position the clamp
- Air cleaner assembly
- Radiator air deflector
- Negative battery cable

13. Refill and bleed the cooling system.

14. Refill the engine crankcase with clean engine oil.

4X2 MODELS WITH DOHC ENGINES

1. Before servicing the vehicle, refer to the precautions section.

2. Raise and safely support the vehicle.

3. Drain the oil.

4. Remove or disconnect the following:
- Intake manifold
- Fan shroud and the engine cooling fan
- Alternator

5. Install an engine support device.
- Two bolts from the engine mounts

6. Raise the engine 2.5 inches (63.5mm).

➡Be careful when removing oil pan gasket. It is re-usable if not damaged.

- Bolts, oil pan and gasket

7. Inspect the oil pan gasket for damage.

To install:

> ✳✳ **WARNING**
>
> **Do not use metal scrapers, wire brushes, power abrasive discs or other abrasive means to clean the sealing surfaces. These tools cause scratches and gouges which make leak paths. Use a plastic scraping tool to remove all traces of old sealant.**

8. Clean the oil pan mating surfaces.

➡If the oil pan and the oil pan gasket are not secured within four minutes, the sealant must be removed and the sealing area cleaned with Metal Surface Cleaner F4AZ-19A536-RA. Allow to dry until there is no sign of wetness, or four minutes, whichever is longer. Failure to follow this procedure can cause future oil leakage.

9. Apply the silicone Gasket and Sealant F7AZ-19554-EA at the rear oil seal retainer-to-cylinder block sealing surface.

10. Install the oil pan gasket and the oil pan and loosely install the 16 bolts. Tighten the bolts in three steps, in the sequence illustrated as follows:
- a. Step 1: 18 inch lbs. (2 Nm).
- b. Step 2: 15 ft. lbs. (20 Nm).
- c. Step 3: an additional 60 degrees.

Tighten the oil pan bolts in the sequence shown

11. Lower the vehicle, then the engine.

12. Remove the engine support

13. Install or connect the following:
- Two bolts for the engine mounts and tighten to 59 ft. lbs. (80 Nm)
- Alternator
- Cooling fan
- Lower intake manifold

14. Refill the engine crankcase with clean engine oil.

4X4 MODELS WITH SOHC ENGINES

1. Before servicing the vehicle, refer to the precautions section.

2. Raise and safely support the vehicle.

3. Drain the oil.

> ✳✳ **CAUTION**
>
> **The air suspension switch must be turned off prior to raising the vehicle. Failure to do so can result in unexpected inflation or deflation of the air springs, which could result in shifting of the vehicle during the repair operation.**

4. Turn the air suspension switch off.

5. Support the front axle housing with a jackstand.

6. Remove or disconnect the following:
- Negative battery cable
- Front axle mount bolt
- Front axle support bolts and the support

> ✳✳ **CAUTION**
>
> **Use care when lowering the front axle housing, or the vacuum lines to the axle solenoid may become disconnected or damaged.**

- Right and front axle housing mount bolts and lower the axle to allow clearance for the oil pan to be removed
- Oil pan and gasket

To install:

✳✳ WARNING

Do not use metal scrapers, wire brushes, power abrasive discs or other abrasive means to clean the sealing surfaces. These tools cause scratches and gouges which make leak paths. Use a plastic scraping tool to remove all traces of old sealant.

7. Clean the oil pan mating surfaces.

➡️**If the oil pan and the oil pan gasket are not secured within four minutes, the sealant must be removed and the sealing area cleaned with Metal Surface Cleaner F4AZ-19A536-RA. Allow to dry until there is no sign of wetness, or four minutes, whichever is longer. Failure to follow this procedure can cause future oil leakage.**

8. Apply the silicone Gasket and Sealant F7AZ-19554-EA at the rear oil seal retainer-to-cylinder block sealing surface.

9. Install or connect the following:
- New oil pan gasket and the oil pan and loosely install the bolts

10. Tighten the bolts in three steps in the sequence illustrated as follows:
- a. Step 1: 18 inch lbs. (2 Nm).
- b. Step 2: 15 ft. lbs. (20 Nm).
- c. Step 3: an additional 60 degrees.

11. Install or connect the following:
- Front axle housing and loosely install the bolts
- Axle support and tighten the bolts to 56 ft. lbs. (90 Nm)

12. Refill the engine crankcase with clean engine oil.

13. Turn on the air suspension switch.

4X4 MODELS WITH DOHC ENGINES

1. Before servicing the vehicle, refer to the precautions section.
2. Raise and safely support the vehicle.
3. Drain the oil.
4. Remove or disconnect the following:
- Front drive axle assembly

➡️**Be careful when removing oil pan gasket. It is re-usable if not damaged.**

- Bolts, oil pan and gasket

5. Inspect the oil pan gasket for damage.

To install:

✳✳ WARNING

Do not use metal scrapers, wire brushes, power abrasive discs or other abrasive means to clean the sealing surfaces. These tools cause scratches and gouges which make leak paths. Use a plastic scraping tool to remove all traces of old sealant.

6. Clean the oil pan mating surfaces.

➡️**If the oil pan and the oil pan gasket are not secured within four minutes, the sealant must be removed and the sealing area cleaned with Metal Surface Cleaner F4AZ-19A536-RA. Allow to dry until there is no sign of wetness, or four minutes, whichever is longer. Failure to follow this procedure can cause future oil leakage.**

7. Apply the silicone Gasket and Sealant F7AZ-19554-EA at the rear oil seal retainer-to-cylinder block sealing surface.

8. Install the oil pan gasket and the oil pan and loosely install the 16 bolts. Tighten the bolts in three steps, in the sequence illustrated as follows:
- a. Step 1: 18 inch lbs. (2 Nm).
- b. Step 2: 15 ft. lbs. (20 Nm).
- c. Step 3: an additional 60 degrees.

9. Install the front drive axle assembly.

10. Refill the engine crankcase with clean engine oil.

6.8L Engine

✳✳ CAUTION

Fuel injection systems remain under pressure, even after the engine has been turned OFF. The fuel system pressure must be relieved before disconnecting any fuel lines. Failure to do so may result in fire and/or personal injury.

1. Before servicing the vehicle, refer to the precautions section.
2. Remove or disconnect the following:
- Negative battery cable
- Fuel system pressure
- Coolant
- Upper radiator hose from the intake manifold
- Air cleaner outlet tube
- Accelerator cable from the bracket and the throttle body cam
- Speed control actuator cable from the throttle body

- All vacuum hoses, fuel lines and electrical wires from the throttle body and intake manifold
- Brake booster vacuum hose bracket
- EGR valve-to-exhaust manifold tube and disconnect the vacuum line
- Connector and vacuum line from the Engine Vacuum Regulator (EVR) solenoid
- The 4 bolts and the throttle body adapter
- Upper fan shroud mounting screws and position the shroud toward the engine.
- The alternator and install the Modular Engine Support Bracket on the engine using the mounting holes.
- Lower engine mount-to-frame nuts
- Turbine Shaft Speed (TSS) and Output Shaft Speed (OSS) sensors from the transmission. Plug the openings.

3. Lower the vehicle to the floor and raise the engine using a hoist attached to the support bracket.

4. Install an engine support fixture with a J hook to keep the engine raised, then remove the hoist.

5. Remove or disconnect the following:
- Engine oil and filter
- Dual converter Y-pipe and the flywheel inspection cover
- Driveshaft and the 2 transmission mounting nuts
- Transmission. Be sure to support the transmission along the rails of the pan to avoid damage.
- Oil pan mounting bolts and partially lower the pan
- Oil pump pick-up tube and screen assembly and allow it to drop into the pan
- Oil pan towards the rear of the vehicle

To install:

✳✳ WARNING

To prevent possible oil leaks, use only a plastic scraper to clean the oil pan mounting surface.

6. Clean the oil pan-to-engine mounting surface.

7. Place the oil pump pick-up tube and screen assembly in the oil pan, then position the pan and gasket near the engine.

8. Install the oil pump pick-up tube and screen assembly.
- a. Tighten the nut to 15–22 ft. lbs. (20–30 Nm).

20-30 Nm (15-22 lb/ft)

8-12 Nm (71-106 lb/in)

7924FG89

Oil pump pick-up tube and screen assembly—6.8L engine

b. Tighten the 2 bolts to 71–106 inch lbs. (8–12 Nm).

9. Apply a bead of silicone sealant to the areas where the front cover and rear bearing cap meet the engine block.

10. Install the oil pan. Tighten the bolts in sequence using 3 steps as follows:
- 18 inch lbs. (2 Nm)
- 15 ft. lbs. (20 Nm)
- Plus an additional 90°

11. Lower the transmission and install the 2 mounting nuts. Tighten the nuts to 60–80 ft. lbs. (81–108 Nm).

12. Install the driveshaft and the TSS and OSS sensors.

13. Install the flywheel cover and dual converter Y-pipe.

14. Install the oil bypass filter.

15. Lower the vehicle and remove the engine support fixture.

16. Install the engine mounting nuts. Tighten the nuts to 66 ft. lbs. (90 Nm).

17. Remove the modular engine support bracket and install the alternator.

18. Install the fan shroud.

19. Use a new gasket and install the throttle body adapter.

20. Tighten the bolts in 2 steps:
- 71–88 inch lbs. (8–10 Nm)
- 85–95 degrees

21. Connect the EVR solenoid harness and vacuum line.

22. Attach the vacuum line to the EGR valve.

23. Install the EGR valve-to-exhaust manifold tube. Tighten the fittings to 41 ft. lbs. (55 Nm).

24. Install the EGR transducer.

25. Install all remaining components in the reverse of the removal.

9358FG15

To prevent leaks, tighten the oil pan bolts in the order shown—2002 6.8L engine

⁂ WARNING

Operating the engine without the proper amount and type of engine oil will result in severe engine damage.

26. Fill the engine with SAE 5W30 oil.
27. Fill and bleed the cooling system.

2005 Models

EXCURSION WITH 5.4L ENGINE

1. Before servicing the vehicle, refer to the precautions section.

2. Disconnect the negative battery cable.

3. Remove the air cleaner outlet pipe.

4. Remove the accelerator cable snow shield.

5. Disconnect the throttle body cam.

6. Disconnect the accelerator cable.

7. If equipped, disconnect the speed control cable.

8. Remove the throttle return spring.

9. Position the accelerator cable bracket aside.

10. Release the Idle Air Control (IAC) fresh air tube from the routing clip and remove it from the vehicle.

11. Disconnect the IAC electrical connector.

12. If equipped, disconnect the differential pressure feedback Exhaust Gas Recirculation (EGR) system electrical connector.

13. Disconnect the Throttle Position (TP) sensor electrical connector.

14. If equipped, disconnect the differential pressure feedback EGR system vacuum hoses.

15. If equipped, disconnect the exhaust manifold to EGR valve tube fittings and remove the tube.

16. If equipped, remove the differential pressure feedback EGR system bracket.

17. If equipped, disconnect the EGR valve vacuum hose.

18. If equipped, disconnect the EGR vacuum regulator solenoid electrical and vacuum connections.

19. Position aside the throttle body adapter. Discard the throttle body adapter gasket.

20. Remove the alternator.

⁂ CAUTION

The large clutch assembly nut has a right-handed thread and must be rotated counterclockwise to remove it.

21. Using tools shown, remove the fan and fan clutch from the coolant pump pulley.

55 Nm (41 lb-ft)
303-240

303-239

06017-NAVI-G47

Using tools shown, remove the fan and fan clutch from the coolant pump pulley— 2005 Excursion models with the 5.4L engine

22. Install the modular engine support bracket on the engine using the alternator mounting holes.
23. Remove the engine mount nuts.

※ CAUTION

Damage to the Turbine Shaft Speed/Output Shaft Speed (TSS/OSS) sensors may occur and cause the transmission or torque converter operational concerns if the transmission is raised prior to removing TSS/OSS.

※ CAUTION

Sensor bosses must be cleaned prior to removal and then plugged to prevent contamination from damaging internal components.

24. Remove the TSS or the OSS sensor and install plugs in the transmission.
25. Loosely install the transmission mount nuts and the crossmember.
26. Loosely install the crossmember bolts.
27. Loosely install the crossmember bolts.
28. Remove the jack from under the transmission and lower the vehicle.
29. Raise the engine using the a heavy duty engine support.
30. Drain the engine oil and remove the oil bypass filter.
31. Disconnect the Heated Oxygen (HO$_2$S) sensor connectors.
32. Remove the nuts and the exhaust Y-pipe.
33. Remove the flywheel inspection plate.

※ CAUTION

Support the transmission on the oil pan rails only or internal transmission damage can occur.

34. Support the transmission with a transmission jack.
35. Remove the transmission mount nuts.
36. Raise the transmission.
37. Remove the bolts and partially lower the oil pan.
38. Remove the bolts from the oil pump screen and pickup tube and let them drop into the oil pan.
39. Remove the rear bolt and the oil pump screen and pickup tube and let them drop into the oil pan.
40. Remove the oil pan and the oil pan gasket from the rear of the engine.
41. Clean the oil pan thoroughly and inspect the oil pan gasket.

※ CAUTION

Do not use metal scrapers, wire brushes, power abrasive discs or other abrasive means to clean the sealing surface. These tools cause scratches and gouges, which make leak paths. Use a plastic scraping tool to remove all traces of sealant.

42. Clean the mating surfaces for the oil pan with silicone gasket remover and metal surface prep.

To install:

➡ The oil pump screen and pickup tube must be in the oil pan when the oil pan is positioned in the vehicle.

43. Position the oil pan gasket and the oil pan in the vehicle from the rear of the engine.

※ CAUTION

Make sure the O-ring is in place and not damaged. A missing or damaged O-ring can cause foam in the lubrication system, low oil pressure and severe engine damage.

➡ Clean and inspect the mating surfaces and install a new O-ring. Lubricate with clean engine oil.

44. Install the oil pump screen and pickup tube.

➡ If not secure within four minutes, the sealant must be removed and the sealing area cleaned.

45. Apply a bead of silicone gasket and sealant at the crankshaft rear seal retainer-to-cylinder block surface, at the locations shown.

➡ If not secure within four minutes, the sealant must be removed and the sealing area cleaned.

46. Apply a bead of silicone gasket and sealant at the engine front cover-to-cylinder block surface, at the locations shown.

➡ Be sure to tighten the bolts in two stages.

47. Tighten the bolts in the sequence shown:
    a. Step 1: Tighten to 20 Nm (15 lb-ft).
    b. Step 2: Tighten an additional 90 degrees.
48. Lower the transmission.
49. Remove the transmission jack.
50. Install the flywheel inspection plate. Tighten to 34 Nm (25 lb-ft.).
51. Install the oil bypass filter.
52. Align the engine mount studs and lower the engine, using the engine support.

06017-NAVI-G48

Apply a bead of silicone gasket and sealant at the crankshaft rear seal retainer-to-cylinder block surface—2005 Excursion models with the 5.4L engine

06017-NAVI-G49

Apply a bead of silicone gasket and sealant at the engine front cover-to-cylinder block surface—2005 Excursion models with the 5.4L engine

Oil pan torque sequence—2005 Excursion models with the 5.4L engine

53. Remove the engine support.

54. Install the engine mount nuts. Tighten to 90 Nm (66 lb-ft.).

55. Install a suitable transmission jack and support the transmission.

56. Remove the crossmember bolts.

57. Remove the transmission mount nuts and the crossmember.

58. Install the TSS or the OSS sensor in the transmission.

59. Position the exhaust Y-pipe and install the nuts on the exhaust manifold studs. Tighten to 40 Nm (30 lb-ft.).

60. Install the exhaust Y-pipe flange nuts.

61. Connect the oxygen sensor electrical connectors.

62. Remove the modular engine support bracket from the engine.

✳✳ CAUTION

The large clutch assembly nut has a right-handed thread and must be rotated counterclockwise to remove it.

63. Using the tools used for removal, install the fan and fan clutch to the coolant pump pulley. Tighten to 55 Nm (41 lb-ft.).

64. Install the alternator.

65. Install the throttle body adapter using a new throttle body adapter gasket.

66. If equipped, connect the EGR vacuum regulator solenoid electrical and vacuum connections.

67. If equipped, connect the EGR valve vacuum hose.

68. If equipped, install the differential pressure feedback EGR system bracket.

69. If equipped, install the exhaust manifold to EGR valve tube as follows:

   a. Step 1: Hand-tighten the fittings.

   b. Step 2: Tighten the upper fitting to 50 Nm (37 lb-ft).

   c. Step 3: Tighten the lower fitting to 50 Nm (37 lb-ft).

70. If equipped, connect the differential pressure feedback EGR system vacuum lines.

71. If equipped, connect the differential pressure feedback EGR system electrical connector.

72. Connect the TP sensor electrical connector.

73. Connect the IAC electrical connector.

74. Install the IAC fresh air tube and insert it in the routing clip.

75. Install the accelerator cable bracket.

76. Position the bracket and cables.

77. Connect the throttle body cam.

78. Connect the accelerator cable.

79. If equipped, connect the speed control cable.

80. Install the throttle return spring.

81. Install the accelerator cable snow shield.

82. Install the air cleaner outlet pipe.

83. Connect the negative battery cable.

✳✳ CAUTION

The oil pump must be primed prior to starting the engine.

84. Fill the engine with clean engine oil.

EXCURSION WITH 6.8L ENGINE

1. Before servicing the vehicle, refer to the precautions section.

2. Disconnect the negative battery cable.

3. Remove the air cleaner outlet tube.

4. Remove the accelerator cable splash shield.

5. Disconnect the accelerator and speed control cables and return spring.

6. Position the accelerator cable bracket aside.

7. Disconnect the Positive Crankcase Ventilation (PCV) tube and the vacuum hoses.

8. Disconnect the vacuum hose.

9. Disconnect the Throttle Position (TP) sensor electrical connector.

10. Disconnect the Idle Air Control (IAC) motor electrical connector and the bypass hose.

11. Remove the bypass hose routing clip.

12. Position aside the throttle body adapter. Discard the throttle body adapter gasket.

✳✳ CAUTION

The large clutch assembly nut has a right-hand thread and must be rotated counterclockwise to remove it.

13. Using the tools shown, remove the fan and fan clutch from the coolant pump pulley.

Using tools shown, remove the fan and fan clutch from the coolant pump pulley—2005 Excursion models with the 6.8L engine

14. Remove the alternator.

15. Install the tool 303-639 on the engine, using the alternator mounting holes.

16. Remove the engine mount nuts.

✳✳ CAUTION

Damage to the TSS/OSS may occur and cause the transmission or torque converter operational concerns if the transmission is raised prior to removing TSS/OSS.

✳✳ CAUTION

Sensor bosses must be cleaned prior to removal and then plugged to prevent contamination from damaging internal components.

17. Remove the Turbine Shaft Speed (TSS) or Output Shaft Sensors (OSS), and install plugs in the transmission.

18. Loosely install the transmission mount nuts and the crossmember.

19. Loosely install the crossmember bolts.

20. Loosely install the crossmember bolts.

21. Remove the jack assembly supporting the transmission.

22. Raise the engine, using support tool 303-F070.

23. Drain the engine oil and remove the oil bypass filter.

24. Remove the threaded insert and position the oil cooler aside.

25. Disconnect the Heated Oxygen (HO2S) sensor electrical connectors.

26. Remove the exhaust Y-pipe flange nuts.

27. Remove the nuts and the exhaust Y-pipe.

28. Remove the flywheel inspection plate.

✳✳ CAUTION

Support the transmission on the oil pan rails only or internal transmission damage can occur.

29. Support the transmission with a transmission jack.

30. Remove the transmission mount nuts.

31. Using the transmission jack, raise the transmission.

32. Remove the nut and detach the transmission cooling tube bracket.

33. Remove the bolts and partially lower the oil pan.

34. Remove the bolts retaining the oil pump screen cover and pickup tube.

35. Remove the pickup tube and spacer and let them drop into the oil pan

36. Remove the oil pan and the oil pan gasket from the rear of the engine.

✳✳ CAUTION

Do not use metal scrapers, wire brushes, power abrasive discs or other abrasive means to clean the sealing surface. These tools cause scratches and gouges, which make leak paths. Use a plastic scraping tool to remove all traces of sealant.

37. Clean the mating surfaces and thoroughly clean the oil pan.

To install:

38. Clean and inspect the oil pump screen cover and tube mating surfaces and install a new O-ring. Lubricate the O-ring with clean engine oil.

➡**The oil pump screen and pickup tube must be in the oil pan when the oil pan is positioned in the vehicle.**

39. Position the oil pan gasket and the oil pan in the vehicle from the rear of the engine.

40. Install the oil pump screen and pickup tube.

➡**If not secure within four minutes, the sealant must be removed and the sealing area cleaned.**

41. Apply a bead of silicone gasket and sealant where the front cover and rear crankshaft seal retainer fit to the engine block.

➡**Be sure to tighten the bolts in two stages.**

42. Tighten the oil pan bolts in the sequence shown:
   a. Step 1: Tighten to 20 Nm (15 lb-ft).

   b. Step 2: Tighten an additional 90 degrees.

43. Attach the transmission cooling tube bracket and install the nut. Tighten to 25 Nm (18 lb-ft).

44. Lower the transmission.

45. Loosely install the transmission mount nuts.

46. Install the flywheel inspection plate. Tighten to 34 Nm (25 lb-ft).

47. Position the engine oil cooler adapter and install the threaded insert. Tighten to 58 Nm (43 lb-ft).

48. Install the oil bypass filter.

49. Align the engine mount studs and lower the engine.

50. Remove the engine support.

51. Install the engine mount nuts. Tighten to 90 Nm (66 lb-ft).

52. Install a suitable transmission jack and support the transmission.

53. Remove the crossmember bolts.

54. Remove the crossmember bolts.

55. Remove the transmission mount nuts and the crossmember.

56. Install the TSS or turbine/OSS sensors in the transmission.

57. Position the exhaust Y-pipe and install the nuts on the exhaust manifold studs. Tighten to 40 Nm (30 lb-ft).

58. Install the exhaust Y-pipe flange nuts.

59. Connect the oxygen sensor electrical connectors.

✳✳ CAUTION

The large clutch assembly nut has a right-hand thread and must be rotated clockwise to install it.

60. Using the tools used during removal, install the fan and fan clutch to the coolant pump pulley. Tighten to 133 Nm (98 lb-ft).

61. Install the alternator.

62. Position the throttle body adapter with a new gasket.

Oil pan bolt torque sequence—2005 Excursion models with the 6.8L engine

06017-NAVI-G52

63. Install the four bolts and tighten in two steps.:

    a. Step 1: Tighten to 9 Nm (80 lb-in).

    b. Step 2: Tighten an additional 90 degrees.

64. Connect the IAC electrical connector and the bypass hose.

65. Install the bypass hose routing clip.

66. Connect the vacuum hose.

67. Connect the PCV tube and the vacuum hoses.

68. Connect the TP sensor electrical connector.

69. Install the accelerator cable bracket.

70. Position the bracket and cables.

71. Connect the accelerator and speed control cables and return spring.

72. Install the accelerator cable splash shield.

73. Install the air cleaner outlet tube.

74. Connect the negative battery cable.

❋❋ CAUTION

The oil pump must be primed prior to starting the engine.

75. Fill the engine with clean engine oil.

EXPEDITION AND NAVIGATOR

1. Before servicing the vehicle, refer to the precautions section.

2. Drain the engine oil.

3. Remove the bolts and the frame crossmember.

4. Remove the bolt and detach the wire harness bracket.

5. On 4WD models:

    a. Support the front axle housing with a jack stand.

➡**Mark the bolt and bracket so that alignment can be maintained on installation.**

    b. Remove the front axle housing right hand mounting bolt.

➡**Mark the bolt and bracket so that alignment can be maintained on installation.**

    c. Remove the front axle housing left hand front mounting bolt.

➡**Mark the bolt and bracket so that alignment can be maintained on installation**

    d. Remove the front axle housing left hand rear mounting bolt.

❋❋ CAUTION

Use care when lowering the front axle housing, or the vacuum lines to

the axle solenoid may become disconnected or damaged.

    e. Lower the axle to allow clearance for the oil pan to be removed.

➡**Be careful when removing the oil pan gasket. It is reusable.**

6. Remove the 16 bolts, the oil pan and the gaskets. Inspect the oil pan gasket for damage.

7. If damaged, discard the oil pan gasket and the oil pan-to-oil pump gaskets.

To install:

❋❋ CAUTION

Do not use metal scrapers, wire brushes, power abrasive discs or other abrasive means to clean the sealing surfaces. These tools cause scratches and gouges, which make leak paths. Use a plastic scraping tool to remove all traces of old sealant.

8. Inspect the oil pan. Clean the mating surface for the oil pan with silicone gasket remover and metal surface prep.

➡**If not secured within four minutes, the sealant must be removed and the sealing area cleaned. To clean the sealing area, use silicone gasket remover and metal surface prep. Follow the directions on the packaging. Failure to follow this procedure can cause future oil leakage.**

9. Apply silicone gasket and sealant at the crankshaft rear seal retainer plate-to-cylinder block sealing surface.

10. Apply silicone gasket and sealant at the engine front cover-to-cylinder block sealing surface.

11. Install the oil pan gasket and the oil pan and loosely install the 16 bolts.

12. Tighten the bolts in 3 steps, in the sequence illustrated.

    a. Step 1: Tighten to 2 Nm (18 lb-in).

    b. Step 2: Tighten to 20 Nm (15 ft. lbs.).

    c. Step 3: Tighten an additional 60 degrees.

13. On 4WD models:

❋❋ CAUTION

Use care when positioning the front axle housing, or the vacuum lines to the axle solenoid may become disconnected or damaged.

    a. Position the front axle housing and loosely install the three bolts, aligning

the bolt location marks made during removal.

    b. Install the front axle housing right hand mounting bolt. Tighten to 89 Nm (66 ft. lbs.).

    c. Install the front axle housing left hand front mounting bolt. Tighten to 89 Nm (66 ft. lbs.).

    d. Install the front axle housing left hand rear mounting bolt. Tighten to 89 Nm (66 ft. lbs.).

14. Position the frame crossmember and the 4 bolts. Tighten to 102 Nm (75 ft. lbs.).

15. Install the wire harness bracket and the bolt. Tighten to 10 Nm (89 lb-in).

16. Fill the crankcase with clean engine oil.

Oil Pump

REMOVAL & INSTALLATION

4.6L and 5.4L Engines

2002–05 MODELS

1. Before servicing the vehicle, refer to the precautions section.

2. Disconnect the negative battery cable.

3. Remove or disconnect the following:

- Crankshaft sprocket
- Oil pan
- Bolts and the oil pump

To install:

4. Clean and inspect the mating surfaces.

5. Install or connect the following:

- Oil pump, and the bolts loosely. Tighten the bolts in the sequence illustrated to 89 inch lbs. (10 Nm).
- Oil pan
- Crankshaft sprocket

6.8L Engine

1. Before servicing the vehicle, refer to the precautions section.

2. Disconnect the negative battery cable.

9358FG11

Tighten the oil pump mounting bolts in the sequence shown—4.6L and 5.4L engines

8-12 Nm (71-106 lb/in)

7924FG91

Be sure to tighten the oil pump mounting bolts in the sequence shown—6.8L engine

3. Remove the engine front cover and crankshaft sprocket.

4. Drain the engine oil into a suitable container.

5. Remove the oil pan.

6. Remove the 3 oil pump mounting bolts, then remove the oil pump from the engine.

To install:

7. Clean and inspect the mating surfaces.

8. Install the oil pump and loosely install the oil pump mounting bolts. Tighten the bolts in the sequence shown to 71–106 inch lbs. (8–12 Nm).

9. Install the oil pan.

10. Install the crankshaft sprocket and timing chains.

11. Install the front cover.

12. Refill the engine oil with the recommended engine oil and amount.

13. Install the negative battery cable.

Rear Main Seal

REMOVAL & INSTALLATION

All 2002 Engines and 2003 Engines without Retainer Plate

If the crankshaft rear oil seal replacement is the only operation being performed, it can be done in the vehicle as detailed in the following procedure. If the oil seal is being replaced in conjunction with a rear main bearing replacement, the engine must be removed from the vehicle and installed on a work stand.

1. Before servicing the vehicle, refer to the precautions in the beginning of this section.

2. Disconnect the negative battery cable.

3. Remove the transmission from the vehicle.

4. Remove the flywheel/flexplate. If equipped, remove the crankshaft oil slinger from the crankshaft.

5. Use an awl to punch 2 holes in the

crankshaft rear oil seal. Punch the holes on opposite sides of the crankshaft and just above the bearing cap-to-cylinder block split line.

6. Install a sheet metal screw in each hole. Use 2 small prybars to pry against both screws at the same time to remove the crankshaft rear oil seal. It may be necessary to place small blocks of wood against the cylinder block to provide a fulcrum point for the prybars. Use caution throughout this procedure to avoid scratching or otherwise damaging the crankshaft oil seal surface.

7. Clean the oil seal recess in the cylinder block and main bearing cap.

To install:

8. Clean, inspect and polish the rear oil seal rubbing surface on the crankshaft.

9. Coat the new oil seal and the crankshaft with a light film of engine oil.

10. Start the seal in the recess with the seal lip facing forward and install it with a seal driver. Keep the tool straight with the centerline of the crankshaft and install the seal until the tool contacts the cylinder block surface. Remove the tool and inspect the seal to be sure it was not damaged during installation.

11. If equipped, install the crankshaft oil slinger.

12. Position the flywheel on the crankshaft flange. Coat the threads of the flywheel attaching bolts with Loctite® and install the bolts. Tighten the bolts in sequence across from each other to 75–85 ft. lbs. (102–115 Nm).

13. Install the transmission, following the recommended procedure.

14. Install the negative battery cable.

2003 Engines with Retainer Plate and All 2004–05 Engines

1. Before servicing the vehicle, refer to the precautions in the beginning of this section.

2. Remove the transmission.

3. Remove the flexplate.

4. Remove the crankshaft rear oil seal slinger.

5. Using special tools 100-001 with adapter 303-519 on all 2003–04 models or 303-514 on 2005 models, remove the crankshaft rear seal.

6. Remove the six bolts and the crankcase rear seal retainer.

To install:

✳✳ CAUTION

Do not use metal scrapers, wire brushes, power abrasive discs or other abrasive means to clean the

sealing surfaces. These tools cause scratches and gouges which make leak paths. Use a plastic scraping tool to remove all traces of old sealant.

➡ Clean the sealing surfaces with silicone gasket remover and metal surface prep.

7. Follow the directions on the packaging. Failure to follow this procedure can cause future oil leakage.

8. Clean and inspect the mating surface.

✳✳ CAUTION

Do not use metal scrapers, wire brushes, power abrasive discs or other abrasive means to clean the sealing surfaces. These tools cause scratches and gouges which make leak paths. Use a plastic scraping tool to remove all traces of old sealant.

➡ If the rear crankshaft seal retaining plate is not secured within four minutes, the sealant must be removed and the sealing area cleaned. To clean the sealing area, follow the directions provided on the packaging of the silicone gasket remover and the metal surfacer prep. Failure to follow this procedure can cause future oil leakage.

➡ The silicone must be applied on the groove along the retainer plate.

303-514

100-001

06017-NAVI-G34

Remove the rear crankshaft seal

06017-NAVI-G35

Tighten the rear seal retainer tighten the bolts in the sequence shown

9. Apply a 4 mm bead of silicone gasket and sealant around the rear oil seal retainer sealing surface.

10. Install the rear seal retainer and loosely install the six bolts.

11. On 2003–04 models, tighten the bolts in the sequence shown to 89 inch lbs. (10 Nm.).

12. On 2005 models tighten the bolts in the sequence shown to 20 Nm (15 ft. lbs.), then tighten an additional 60 degrees.

➡ **Lubricate the inner lip of the rear crankshaft seal with clean engine oil.**

13. Using the special tools used during removal, install the crankshaft rear seal.

14. Using the special tools, install the crankshaft rear oil slinger.

15. Install the oil pan.

16. Install the flexplate.

Timing Chain, Sprockets and Front Cover

REMOVAL & INSTALLATION

4.6L Engine

1. Before servicing the vehicle, refer to the precautions section.

2. Remove or disconnect the following:
- Negative battery cable
- Radiator, fan blade and fan shroud assembly
- Accessory drive belt
- Water pump pulley
- Electrical harness connectors from both ignition coils
- Both ignition coils with their brackets attached
- Left-hand and right-hand cylinder head covers
- The 2 upper power steering pump retaining bolts
- The 2 lower power steering pump retaining bolts and move the pump aside
- Crankshaft Position (CKP) sensor electrical harness connector
- CKP sensor retaining bolt and the sensor
- Engine oil
- The 4 oil pan-to-engine front cover retaining bolts
- Crankshaft damper retaining bolt and washer from the crankshaft
- Damper from the crankshaft
- Camshaft Position (CMP) sensor retaining bolt and the sensor
- Idler pulley bolt and the pulley
- The 3 belt tensioner retaining bolts and the tensioner

- The 8 engine front cover retaining bolts and the 7 nuts. Swing the top of the cover out off the dowel pins and remove the cover.
- Sensor ring from the crankshaft

3. Use Camshaft Positioning Tool T91P-6256-A and Camshaft Positioning Adapters T92P-6256-A to position the camshaft.

4. Rotate the crankshaft until both camshaft keyways are 90 degrees from the cam cover surface. Be sure the copper links line up with the dots on the camshaft sprockets.

✳✳ WARNING

At no time, when the timing chains are removed and the cylinder heads are installed may the crankshaft or the camshaft be rotated. Severe piston and valve damage will occur.

5. Remove or disconnect the following:
- The 2 left-hand and right-hand tensioner bolts and the timing chain tensioners
- The left-hand and right-hand tensioner guides off the dowel pins

- The right-hand timing chain from the camshaft sprocket
- The left-hand timing chain from the camshaft sprocket
- The left-hand and right-hand timing chain guide bolts and the timing chain guides
- Camshaft gear bolt and the camshaft gear, if necessary

To install:

6. Examine the timing chains, looking for the copper links. If the copper links are not visible, lay the chain on a flat surface and pull the chain taught until the opposite sides of the chain contact one another. Mark the links at each end of the chain and use these marks in place of the copper links.

➡ **If the engine jumped time, damage has been done to valves and possibly pistons and/or connecting rods. Any damage must be corrected before installing the timing chains.**

7. Install or connect the following:
- Camshaft gears and tighten the retaining bolt to 81–95 ft. lbs. (110–130 Nm)
- Left-hand and right-hand timing chain guides and retaining bolts.

When removing the timing chains, rotate the crankshaft so that the camshaft keyways are positioned as shown—4.6L engines

Tighten the first 5 front cover fasteners in the sequence shown—4.6L engine

Continue tightening the remaining fasteners in the sequence shown here—4.6L engine

When installing the timing chains, make certain that the copper colored links (A) are aligned with the timing marks—4.6L engine

Tighten the retaining bolts to 71–106 inch lbs. (8–12 Nm).
- Left-hand crankshaft sprocket with the tapered part of the sprocket facing away from the engine block

➡The crankshaft sprockets are identical. They may only be installed one way, with the tapered part of the sprockets facing each other. Ensure that the keyway and timing marks on the crankshaft sprockets are aligned.

- Left-hand timing chain on the camshaft and crankshaft sprockets. Be sure the copper links of the timing chain line up with the timing marks on both sprockets.
- Right-hand crankshaft sprocket with the tapered part of the sprocket facing the left-hand crankshaft sprocket
- Right-hand timing chain on the camshaft and crankshaft sprockets. Be sure the copper links of the timing chain line up with the timing marks on both sprockets.

8. It is necessary to bleed the timing chain tensioners before installation. Proceed as follows:

a. Place the timing chain tensioner in a soft-jawed vise.

b. Using a small pick or similar tool, hold the ratchet lock mechanism away from the ratchet stem and slowly compress the tensioner plunger by rotating the vise handle.

✷✷ WARNING

The tensioner must be compressed slowly or damage to the internal seals will result.

c. Once the tensioner plunger bottoms in the tensioner bore, continue to hold the ratchet lock mechanism and push down on the ratchet stem until flush with the tensioner face.

d. While holding the ratchet stem flush to the tensioner face, release the ratchet lock mechanism and install a paper clip or similar tool in the tensioner body to lock the tensioner in the collapsed position.

e. The paper clip must not be removed until the timing chain, tensioner, tensioner arm and timing chain guide are completely installed on the engine.

9. Install or connect the following:

- Left-hand and right-hand timing chain tensioner guides on the dowel pins
- Left-hand and right-hand timing chain tensioners in position and install the retaining bolts. Tighten the bolts to 15–22 ft. lbs. (20–30 Nm).

10. Remove the retaining pins from the timing chain tensioners.

11. Remove Camshaft Positioning Tool T91P-6256-A and Camshaft Positioning Adapters T92P-6256-A from the camshaft.

12. Install or connect the following:
- Crankshaft sensor ring on the crankshaft

13. Apply a bead of silicone sealer along the cylinder head-to-cylinder block and the oil pan-to-cylinder block sealing surfaces.

- Engine front cover carefully onto the dowel pins. Tighten the engine front cover bolts, in sequence, to 15–22 ft. lbs. (20–30 Nm).
- Idler pulley in position and the retaining bolt. Tighten the bolt to 15–22 ft. lbs. (20–30 Nm).
- Drive belt tensioner and the 3 retaining bolts. Tighten the bolts to 15–22 ft. lbs. (20–30 Nm).
- CMP sensor and the retaining bolt. Tighten the bolt to 106 inch lbs. (12 Nm).
- The 4 oil pan-to-engine front cover bolts and tighten, in sequence, in 2 steps: Tighten the bolts to 15 ft. lbs. (20 Nm). Rotate the bolts an additional 60 degrees.
- CKP sensor and the retaining bolt. Tighten the bolt to 106 inch lbs. (12 Nm).
- CKP sensor electrical harness connector
- Power steering pump and the 2 upper and 2 lower retaining bolts. Tighten the bolts to 15–20 ft. lbs. (20–30 Nm).
- Ignition coil and brackets on the engine front cover and the bracket bolts. Tighten the bolts to 15–22 ft. lbs. (20–30 Nm).
- Ignition coil and capacitor electrical harness connectors
- CMP electrical harness connector
- Water pump pulley and tighten the bolts to 15–22 ft. lbs. (20–30 Nm).
- Radiator, fan blade and fan shroud assembly
- Accessory drive belt

14. Refill the engine oil with the recommended type of engine oil and the correct amount.

15. Connect the negative battery cable.

16. Start the engine and check for leaks.

17. Road test the vehicle and check for proper engine operation.

2002–04 5.4L SOHC And All 6.8L Engines

1. Before servicing the vehicle, refer to the precautions section.

2. Remove or disconnect the following:
- Negative battery cable
- Radiator, fan blade and fan shroud assembly
- Accessory drive belt
- Water pump pulley
- Electrical harness connectors from both ignition coils
- Both ignition coils with their brackets attached
- The 2 upper power steering pump retaining bolts
- The 2 lower power steering pump retaining bolts and move the pump aside
- Crankshaft Position (CKP) sensor electrical harness connector
- Retaining bolt and the CKP sensor.

3. Drain the engine oil.
- The 4 oil pan-to-engine front cover retaining bolts
- The crankshaft damper retaining bolt and washer from the crankshaft
- Crankshaft damper from the crankshaft
- Camshaft Position (CMP) sensor retaining bolt and remove the CMP sensor

7924FG54

Compress the tensioner while holding the ratchet mechanism with a suitable tool— 2002–04 5.4L and all 6.8L engines

- Idler pulley bolt and remove the pulley
- The 3 belt tensioner bolts and remove the tensioner
- The 8 engine front cover retaining bolts and the 7 nuts. Swing the top of the cover out off the dowel pins and remove the cover.
- Sensor ring from the crankshaft

4. Use Crankshaft Holding Tool 303-448 (T93P-6303-A), to position the camshaft.

5. Rotate the crankshaft until both camshaft keyways are 90 degrees from the cam cover surface. Be sure the copper links line up with the dots on the camshaft sprockets.

✳✳ WARNING

At no time, when the timing chains are removed and the cylinder heads are installed may the crankshaft or the camshaft be rotated. Severe piston and valve damage will occur.

6. Install Camshaft Holding Tool T96T-6256-B or equivalent, on the camshaft.

7. Remove or disconnect the following:
- 2 left-hand and right-hand tensioner bolts and the timing chain tensioners
- Left-hand and right-hand tensioner guides off the dowel pins
- Right-hand timing chain from the camshaft sprocket
- Left-hand timing chain from the camshaft sprocket
- Left-hand and right-hand timing chain guide bolts, and the timing chain guides

To install:

8. Examine the timing chains, looking for the copper links. If the copper links are not visible, lay the chain on a flat surface and pull the chain taught until the opposite sides of the chain contact one another. Mark the links at each end of the chain and use these marks in place of the copper links.

➡**If the engine jumped time, damage has been done to valves and possibly pistons and/or connecting rods. Any damage must be corrected before installing the timing chains.**

9. Install or connect the following:
- Left-hand and right-hand timing chain guides and retaining bolts. Tighten the retaining bolts to 89 inch lbs. (10 Nm).
- Left-hand crankshaft sprocket with the tapered part of the sprocket facing away from the engine block.

7924FG53

Timing chains and related components—2002–04 5.4L and all 6.8L engines

7924FG92

If the copper links are not visible, mark one link on one end of the chain and 2 links on the opposite end of the chain—2002–04 5.4L and all 6.8L engines

➡ **The crankshaft sprockets are identical. They may only be installed one way, with the tapered part of the sprockets facing each other. Ensure that the keyway and timing marks on the crankshaft sprockets are aligned.**

- Left-hand timing chain on the camshaft and crankshaft sprockets. Be sure the 1 copper link aligns with the mark on the crankshaft sprocket and the 2 copper links align with the mark on the camshaft sprocket.
- Right-hand crankshaft sprocket with the tapered part of the sprocket facing the left-hand crankshaft sprocket
- Right-hand timing chain on the camshaft and crankshaft sprockets. Be sure the copper links of the timing chain line up with the timing marks on both sprockets.

10. It is necessary to bleed the timing chain tensioners before installation. Proceed as follows:

a. Place the timing chain tensioner in a soft-jawed vise.

b. Using a small pick or similar tool, hold the ratchet lock mechanism away from the ratchet stem and slowly compress the tensioner plunger by rotating the vise handle.

✳✳ WARNING

The tensioner must be compressed slowly or damage to the internal seals will result.

c. Once the tensioner plunger bottoms in the tensioner bore, continue to hold the ratchet lock mechanism and push down on the ratchet stem until flush with the tensioner face.

d. While holding the ratchet stem flush to the tensioner face, release the ratchet lock mechanism and install a

paper clip or similar tool in the tensioner body to lock the tensioner in the collapsed position.

e. The paper clip must not be removed until the timing chain, tensioner, tensioner arm and timing chain guide are completely installed on the engine.

11. Install the left-hand and right-hand timing chain tensioner guides on the dowel pins.

12. Place the left-hand and right-hand timing chain tensioners in position and install the retaining bolts. Tighten the bolts to 15–22 ft. lbs. (20–30 Nm).

13. Remove the retaining pins from the timing chain tensioners.

14. Remove Cam Holding Tool T96T-6256-B or equivalent, from the camshaft.

15. Install the crankshaft sensor ring on the crankshaft.

16. Apply silicone gasket along the cylinder head-to-cylinder block and engine oil pan-to-cylinder block sealing surfaces.

17. Install or connect the following:

- Engine front cover carefully onto the dowel pins
- Engine front cover bolts, in

sequence, in the following manner: bolts 1 through 7: 15–22 ft. lbs. (20–30 Nm); bolts 6 through 15: 29–40 ft. lbs. (40–55 Nm).

- Idler pulley in position and install the retaining bolt. Tighten the bolt to 15–22 ft. lbs. (20–30 Nm).
- Drive belt tensioner in position and install 3 retaining bolts. Tighten the bolts to 15–22 ft. lbs. (20–30 Nm).
- CMP sensor in position and install the retaining bolt. Tighten the bolt to 106 inch lbs. (12 Nm).
- Damper on the crankshaft. Ensure the crankshaft key and keyway are aligned.
- 4 front oil pan-to-engine front cover retaining bolts and tighten in sequence in 2 steps: Tighten the bolts to 15 ft. lbs. (20 Nm); Rotate the bolts an additional 60 degrees.
- CKP sensor in position and install the retaining bolt. Tighten the bolt to 106 inch lbs. (12 Nm).
- CKP sensor electrical harness connector
- Power steering pump with 2 upper and 2 lower retaining bolts. Tighten the bolts to 15–20 ft. lbs. (20–30 Nm).
- Ignition coil and brackets on the engine front cover and install the bracket bolts. Tighten the bolts to 15–22 ft. lbs. (20–30 Nm).
- Ignition coil and capacitor electrical harness connectors
- CMP electrical harness connector
- Water pump pulley and tighten the bolts to 15–22 ft. lbs. (20–30 Nm)
- Radiator
- Negative battery cable

18. Refill the engine oil with the recom-

9358FG16

Check the alignment of the timing marks as shown—2002–04 5.4L and all 6.8L engines

Item	Description
1	Bolt, Hex Flange Head Pilot, M8 x 1.25 x 53
2	Bolt, Hex Flange Head Pilot, M8 x 1.25 x 53
3	Bolt, Hex Flange Head Pilot, M8 x 1.25 x 53
4	Bolt, Hex Flange Head Pilot, M8 x 1.25 x 53
5	Bolt, Hex Flange Head Pilot, M8 x 1.25 x 53
6	Stud, Hex-Head Pilot, M10 x 1.5 x 1.5 x 103.1
7	Stud, Hex-Head Pilot, M10 x 1.5 x 1.5 x 103.1
8	Screw and Washer, Hex Pilot, M10 x 1.5 x 57.5
9	Screw and Washer, Hex Pilot, M10 x 1.5 x 57.5
10	Screw and Washer, Hex Pilot, M10 x 1.5 x 57.5
11	Stud and Washer, Hex-Head Pilot, M10 x 1.5 x M8 x 1.25 x 109.6
12	Stud and Washer, Hex-Head Pilot, M10 x 1.5 x M8 x 1.25 x 109.6
13	Stud and Washer, Hex-Head Pilot, M10 x 1.5 x M8 x 1.25 x 109.6
14	Stud and Washer, Hex-Head Pilot, M10 x 1.5 x M8 x 1.25 x 109.6
15	Stud and Washer, Hex-Head Pilot, M10 x 1.5 x M8 x 1.25 x 109.6

9358FG17

Install the timing cover and tighten the bolts to specification—2002–04 5.4L and all 6.8L engines

mended type of engine oil and the correct amount.

19. Start the engine and check for leaks.

20. Road test the vehicle and check for proper engine operation.

2005 Excursion Models With 5.4L And 6.8L Engines

1. Disconnect the negative battery cable.

2. Remove both valve covers.

3. Remove the radiator.

4. Remove the water pump.

5. Remove the top bolts and the lower bolt and position the power steering pump aside.

6. Disconnect the Crankshaft Position (CKP) sensor electrical connector.

7. Remove the drain plug and drain the engine oil.

8. Remove the four front oil pan bolts.

9. Remove the crankshaft front seal.

10. Disconnect the Camshaft Position (CMP) sensor electrical connection.

11. Remove the nuts and position the radio interference capacitors aside.

12. Remove the bolt and the belt idler pulley.

13. Remove the nuts and position the wiring harness bracket aside.

14. Remove the wiring harness nut.

15. Remove the engine front cover bolts.

✳✳ CAUTION

Do not use metal scrapers, wire brushes, power abrasive discs or other abrasive means to clean the sealing surfaces. These tools cause scratches and gouges which make leak paths. Use a plastic scraping tool to remove all traces of old sealant.

16. Remove the engine front cover from the front cover to cylinder block dowel.

17. Remove the engine front cover gasket.

18. Clean the mating surfaces with silicone gasket remover and metal surface prep.

19. Inspect the mating surfaces.

20. Remove the crankshaft sensor ring from the crankshaft.

✳✳ CAUTION

Unless otherwise instructed, at no time when the timing chains are removed and the cylinder heads are installed is the crankshaft or camshaft to be rotated. Severe piston and valve damage will occur.

✳✳ CAUTION

The caps must be marked for installation in their original location or damage to the engine can occur.

21. On 6.8L engines, remove the six bolts and remove the balance shaft bearing caps and the balance shaft.

22. Position the crankshaft with the keyway at the 12 o'clock position.

23. Install and fully tighten the tools shown on both camshafts.

Remove the six bolts and remove the balance shaft bearing caps and the balance shaft—2005 Excursion models with 6.8L engine

06017-NAVI-G66

06017-NAVI-G53

Position the crankshaft with the keyway at the 12 o'clock position—2005 Excursion models with 5.4L and 6.8L engines

303-557

06017-NAVI-G54

Install the tools shown on the camshafts—2005 Excursion models with 5.4L and 6.8L engines

✳✳ CAUTION

If one or both of the tensioner mounting bolts are loosened or removed, the tensioner-sealing bead must be inspected for seal integrity. If cracks, tears, separation from the tensioner body or permanent compression of the seal bead is observed, install a new tensioner.

24. Remove the timing chain tensioning system from both timing chains.

✳✳ CAUTION

Unless otherwise instructed, at no time when the timing chains are removed and the cylinders heads are installed is the crankshaft or camshaft to be rotated. Severe piston and valve damage will occur.

25. Remove the right hand and left hand timing chains and the crankshaft sprocket as follows:
    a. Remove the right hand timing chain from the camshaft sprocket.
    b. Remove the right hand timing chain and from the crankshaft sprocket.

    c. Remove the left hand timing chain from the camshaft sprocket.
    d. Remove the left hand timing chain and crankshaft sprocket.
26. Remove both timing chain guides.
27. On 6.8L engines with bolt-on sprockets, remove the camshaft sprocket.
To install:

✳✳ CAUTION

Timing chain procedures must be followed exactly or damage to valves and pistons will result.

✳✳ CAUTION

Prior to installation, inspect the tensioner-sealing bead for seal integrity. If cracks, tears, separation from the tensioner body or permanent compression of the seal bead is observed, install a new tensioner.

28. Compress the tensioner plunger, using a vise.
29. Install a retaining clip on the tensioner to hold the plunger in during installation.
30. Remove the tensioner from the vise.
31. If the copper links are not visible, mark two links on one end and one link on the other end, and use as timing marks.
32. On 6.8L engines equipped with bolt-on sprockets, install the camshaft sprocket as follows:
    a. Tighten the M10 bolt in two steps, first to 40 Nm (30 lb-ft). Then tighten an additional 90 degrees.
    b. Tighten the M12 bolt to 120 Nm (90 lb
33. Install the timing chain guides.

✳✳ CAUTION

Do not turn the engine over with the camshaft positioning tool or damage to the camshaft sprocket or the bolt can occur.

34. Install the camshaft positioning tool.

➡ Slightly loosen the camshaft position aligner to allow slight camshaft movement.

35. Pre-position the camshafts as follows:
    a. Rotate the left hand camshaft with the camshaft position aligner until the timing mark is approximately at 12 o'clock.
    b. Rotate the right hand camshaft with the camshaft position aligner until the timing mark is approximately at 11 o'clock.
    c. Tighten the camshaft position aligner to maintain camshaft pre-positioning.

✳✳ CAUTION

Unless otherwise instructed, at no time when the timing chains are removed and the cylinder heads are installed is the crankshaft or camshaft to be rotated. Severe piston and valve damage will occur.

✳✳ CAUTION

Rotate the crankshaft counterclockwise only. Do not rotate past position shown or severe piston and valve damage can occur.

➡ The number one cylinder is at Top Dead Center (TDC) when the stud on the engine block fits into the slot in the handle of the special tool.

36. Position the crankshaft so the num-

303-S568

06017-NAVI-G55

Install the camshaft positioning tool—2005 Excursion models with 5.4L and 6.8L engines

Pre-position the camshafts as shown—2005 Excursion models with 5.4L and 6.8L engines

Position the crankshaft so the number one cylinder is at TDC—2005 Excursion models with 5.4L and 6.8L engines

ber one cylinder is at TDC with the tool shown.

37. Remove the crankshaft position aligner.

38. Install the crankshaft sprocket, make sure the flange faces forward.

39. Position the lower end of the left hand (inner) timing chain on the crankshaft sprocket, aligning the timing mark on the

Position the lower end of the left hand (inner) timing chain on the crankshaft sprocket, aligning the timing mark on the outer flange of the crankshaft sprocket with the single copper (marked) link on the chain—2005 Excursion models with 5.4L and 6.8L engines

outer flange of the crankshaft sprocket with the single copper (marked) link on the chain.

➡**Make sure the upper half of the timing chain is below the tensioner arm dowel.**

➡**If necessary, use the camshaft position aligner to adjust the camshaft sprocket slightly to obtain timing mark alignment.**

40. Position the timing chain on the camshaft sprocket with the camshaft sprocket timing mark positioned between the two copper (marked) chain links.

Position the timing chain on the camshaft sprocket with the camshaft sprocket timing mark positioned between the two copper (marked) chain links—2005 Excursion models with 5.4L and 6.8L engines

➡**The left hand timing chain tensioner arm has a bump near the dowel hole, for identification.**

41. Position the left hand timing chain tensioner arm on the dowel pin and install the left hand timing chain tensioner. Tighten to 25 Nm (18 lb-ft.).

42. Remove the retaining clip from the left hand timing chain tensioner.

➡**The lower half of the timing chain must be positioned above the tensioner arm dowel.**

43. Position the lower end of the right hand (outer) timing chain on the crankshaft sprocket, aligning the timing mark on the sprocket with the single copper (marked) link on the timing chain.

Position the lower end of the right hand (outer) timing chain on the crankshaft sprocket, aligning the timing mark on the sprocket with the single copper (marked) link on the timing chain—2005 Excursion models with 5.4L and 6.8L engines

06017-NAVI-G61

Position the right hand timing chain on the camshaft sprocket. Make sure the camshaft sprocket timing mark is positioned between the two copper (marked) chain links—2005 Excursion models with 5.4L and 6.8L engines

➡ If necessary, use the camshaft position aligner to adjust the camshaft sprocket slightly to obtain timing mark alignment.

44. Position the right hand timing chain on the camshaft sprocket. Make sure the camshaft sprocket timing mark is positioned between the two copper (marked) chain links.

45. Position the right hand timing chain tensioner arm on the dowel pin and install the right hand timing chain tensioner. Tighten to 25 Nm (18 lb-ft.).

46. Remove the retaining clip from the right hand timing chain tensioner.

47. As a post-check, verify correct alignment of all timing marks.

48. Remove the camshaft holding tool.

49. Position the crankshaft sensor ring on the crankshaft.

50. On 6.8L engines perform the following:

a. Lubricate the balance shaft journals with clean engine oil.

b. Using the index mark on the balance shaft, mark the corresponding gear tooth with chalk.

c. Position the balance shaft on the journals.

06017-NAVI-G62

As a post-check, verify correct alignment of all timing marks—2005 Excursion models with 5.4L and 6.8L engines

06017-NAVI-G67

Using the index mark on the balance shaft, mark the corresponding gear tooth with chalk—2005 Excursion models with the 6.8L engine

➡ It may be necessary to use an inspection mirror to see the marks.

d. Align the chalk mark on the balance shaft with the camshaft timing mark as shown.

e. Install the bearing caps in their original locations. Install the bolts and tighten the bolts in the sequence shown.

✲✲ CAUTION

Do not use metal scrapers, wire brushes, power abrasive discs or other abrasive means to clean the sealing surfaces. These tools cause scratches and gouges which make leak paths. Use a plastic scraping tool to remove all traces of old sealant.

➡ If the engine front cover is not secured within four minutes, the

06017-NAVI-G68

Align the chalk mark on the balance shaft with the camshaft timing mark as shown—2005 Excursion models with the 6.8L engine

sealant must be removed and the sealing area cleaned.

➡ Make sure that the engine front cover gasket is in place on the engine front cover before installation.

51. Apply a bead of silicone gasket and sealant along the cylinder head-to-cylinder block surface and the oil pan-to-cylinder block surface, at the locations shown.

52. Install the engine front cover with engine front cover gasket on the front cover to cylinder block dowel and loosely install the bolts.

53. On 5.4L engines, tighten the engine front cover fasteners in sequence in three steps.

a. Step 1: Tighten fasteners 1 through 7 to 25 Nm (18 lb-ft).

b. Step 2: Tighten fasteners 6 and 7 to 48 Nm (35 lb-ft).

06017-NAVI-G69

Install the bearing caps in their original locations. Install the bolts and tighten the bolts in the sequence shown—2005 Excursion models with the 6.8L engine

Apply a bead of silicone gasket and sealant along the cylinder head-to-cylinder block surface and the oil pan-to-cylinder block surface—2005 Excursion models with the 5.4L engine

8 mm (0.32 in)

06017-NAVI-G70

Apply a bead of silicone gasket and sealant along the cylinder head-to-cylinder block surface and the oil pan-to-cylinder block surface—2005 Excursion models with the 6.8L engine

c. Step 3: Tighten fasteners 8 through 15 to 48 Nm (35 lb-ft).

54. On 6.8L engines, tighten the engine front cover fasteners in sequence in two steps.

a. Step 1: Tighten fasteners 1 through 7 to 25 Nm (18 lb-ft).

b. Step 2: Tighten fasteners 6 through 15 to 48 Nm (35 lb-ft).

55. Connect the CMP electrical connection.

56. Install the bracket and the nut. Tighten to 25 Nm (18 lb-ft).

57. Install the bracket. Tighten to 30 Nm (22 lb-ft).

58. Position the radio interference capacitors and install the nuts, if equipped.

59. Install the belt idler pulley and tighten the bolt. Tighten to 25 Nm (18 lb-ft).

60. Install a new crankshaft front seal.

61. Loosely install the front oil pan bolts, then tighten the bolts in the two steps, in the sequence shown.

a. Step 1: Tighten to 20 Nm (15 lb-ft).

b. Step 2: Tighten an additional 60 degrees.

62. Connect the CKP sensor electrical connector.

➡**The front lower hole in the power steering pump is not used.**

63. Position the power steering pump and install the bolts. Tighten to 25 Nm (18 lb-ft).

64. Install the drain plug. Tighten to 23 Nm (17 lb-ft).

1   Bolt, Hex Flange Head Pilot, M8 x 1.25 x 53

2   Bolt, Hex Flange Head Pilot, M8 x 1.25 x 53

3   Bolt, Hex Flange Head Pilot, M8 x 1.25 x 53

4   Bolt, Hex Flange Head Pilot, M8 x 1.25 x 53

5   Bolt, Hex Flange Head Pilot, M8 x 1.25 x 53

6   Stud, Hex Head Pilot, M10 x 1.5 x 59 — M10 x 1.5 x 30

7   Stud, Hex Head Pilot, M10 x 1.5 x 59 — M10 x 1.5 x 30

8   Screw and Washer, Hex Pilot, M10 x 1.5 x 57.5

9   Screw and Washer, Hex Pilot, M10 x 1.5 x 57.5

10 Screw and Washer, Hex Pilot, M10 x 1.5 x 57.5

11 Stud and Washer, Hex Head Pilot, M10 x 1.5 x 68 — M8 x 1.25 x 27

12 Stud and Washer, Hex Head Pilot, M10 x 1.5 x 68 — M8 x 1.25 x 27

13 Stud and Washer, Hex Head Pilot, M10 x 1.5 x 68 — M8 x 1.25 x 27

14 Stud and Washer, Hex Head Pilot, M10 x 1.5 x 68 — M8 x 1.25 x 27

15 Stud and Washer, Hex Head Pilot, M10 x 1.5 x 68 — M8 x 1.25 x 27

Front cover bolts size, location and torque sequence—2005 Excursion models with the 5.4L engine

06017-NAVI-G64

1	Bolt, Hex Flange Head Pilot, M8 x 1.25 x 50
2	Bolt, Hex Flange Head Pilot, M8 x 1.25 x 50
3	Bolt, Hex Flange Head Pilot, M8 x 1.25 x 50
4	Bolt, Hex Flange Head Pilot, M8 x 1.25 x 50
5	Bolts, Hex Flange Head Pilot, M8 x 1.25 x 50
6	Stud, Hex-Head Pilot, M10 x 1.5 x 59 - M10 x 1.5 x 30
7	Stud, Hex-Head Pilot, M10 x 1.5 x 59 - M10 x 1.5 x 30
8	Screw and Washer, Hex Pilot, M10 x 1.5 x 54
9	Screw and Washer, Hex Pilot, M10 x 1.5 x 54
10	Screw and Washer, Hex Pilot, M10 x 1.5 x 54
11	Stud and Washer, Hex Head Pilot, M10 x 1.5 x 68 - M8 x 1.25 x 27
12	Stud and Washer, Hex Head Pilot, M10 x 1.5 x 68 - M8 x 1.25 x 27
13	Stud and Washer, Hex Head Pilot, M10 x 1.5 x 68 - M8 x 1.25 x 27
14	Stud and Washer, Hex Head Pilot, M10 x 1.5 x 68 - M8 x 1.25 x 27
15	Stud and Washer, Hex Head Pilot, M10 x 1.5 x 68 - M8 x 1.25 x 27

06017-NAVI-G71

Front cover bolts size, location and torque sequence—2005 Excursion models with the 6.8L engine

Front oil pan bolt torque sequence—2005 Excursion models with 5.4L and 6.8L engines

65. Install the valve covers.
66. Fill the engine with clean engine oil.
67. Install the water pump.
68. Install the radiator.
69. Connect the negative battery cable.

2005 Expedition And Navigator

1. Before servicing the vehicle, refer to the precautions section.
2. Drain the engine oil.
3. Remove the engine cooling fan.
4. Remove the valve covers.
5. Remove the accessory drive belt.
6. Remove the nut and the power steering pressure hose support bracket.
7. Remove the nut and the transmission cooler tube support bracket.
8. Remove the crankshaft pulley bolt and washer. Discard the crankshaft pulley bolt.
9. Using the tool 303-009, remove the crankshaft pulley.
10. Using the tool 303-107, remove the crankshaft front seal.
11. Remove the bolts and the accessory drive idler pulleys.
12. Remove the bolts and the coolant pump pulley.
13. Remove the bolts and the accessory drive belt tensioner.
14. Disconnect the right hand Camshaft Position (CMP) sensor. Discard the O-ring seal.
15. Disconnect the radio ignition interference capacitor electrical connectors.
16. Remove the nut and the right hand radio ignition interference capacitor.
17. Disconnect the left CMP sensor. Discard the O-ring seal.

18. Remove the nut and the left hand radio ignition interference capacitor.
19. Remove the bolts and position the power steering pump assembly aside.
20. Disconnect the Crankshaft Position (CKP) sensor connector.
21. Remove the 4 front oil pan bolts.
22. Remove the bolt and the CKP sensor. Discard the O-ring seal.
23. Remove the bolts and the studs from the front cover.

✳✳ CAUTION

Do not use metal scrapers, wire brushes, power abrasive discs or other abrasive means to clean the sealing surfaces. These tools cause scratches and gouges which make leak paths. Use a plastic scraping tool to remove all traces of old sealant.

24. Remove the engine front cover from the front cover to cylinder block dowel.
25. Remove the engine front cover gaskets.
26. Clean the mating surfaces with silicone gasket remover and metal surface prep. Follow the directions on the packaging.
27. Inspect the mating surfaces.
28. Remove the crankshaft sensor ring from the crankshaft.
29. Position the crankshaft keyway at the 12 o'clock position.

➡If the camshaft lobes are not exactly positioned at the 12 o'clock position, the crankshaft will require one full additional rotation to 12 o'clock. The number 1 cylinder camshaft exhaust lobe must be coming up on the exhaust stroke. Verify by noting the position of the 2 intake camshaft lobes and the exhaust lobe on the number 1 cylinder.

➡If the components are to be reinstalled, they must be installed in the same positions. Mark the components for installation into the original locations.

30. Remove only the 3 roller followers shown in the illustration from the right hand cylinder head.

✳✳ CAUTION

Do not allow the valve keepers to fall off the valve or the valve may drop into the cylinder.

06017-NAVI-G01

Remove only the 3 roller followers shown from the right hand cylinder head—2005 Expedition and Navigator models

➡It may be necessary to push the valve down while compressing the spring.

31. Using the tool illustrated, remove the 3 designated roller followers in the previous step from the right hand cylinder head.
32. Remove only the 3 roller followers shown in the illustration from the left hand cylinder head.
33. Using the tool 303-1039, remove the 3 designated roller followers in the previous step from the left hand cylinder head.
34. The crankshaft cannot be moved past the 6 o'clock position once set.
35. Rotate the crankshaft clockwise and position the crankshaft keyway at the 6 o'clock position.

06017-NAVI-G02

Remove the 3 designated roller followers from the right hand cylinder head—2005 Expedition and Navigator models

Remove only the 3 roller followers shown from the left hand cylinder head—2005 Expedition and Navigator models

✳✳ CAUTION

If one or both of the tensioner mounting bolts are loosened or removed, the tensioner-sealing bead must be inspected for seal integrity. If cracks, tears, separation from the tensioner body or permanent compression of the seal bead is observed, install a new tensioner.

36. Remove the bolts, the left hand timing chain tensioner and tensioner arm.
37. Remove the bolts, the right hand timing chain tensioner and tensioner arm.
38. Remove the right hand and left hand timing chains and the crankshaft sprocket.
39. Remove the right hand timing chain from the camshaft sprocket.
40. Remove the right hand timing chain from the crankshaft sprocket.
41. Remove the left hand timing chain from the camshaft sprocket.
42. Remove the left hand timing chain and crankshaft sprocket.
43. Remove the left hand and right hand timing chain guides.

To install:

✳✳ CAUTION

Timing chain procedures must be followed exactly or damage to valves and pistons will result.

✳✳ CAUTION

Prior to installation, inspect the tensioner-sealing bead for seal integrity. If cracks, tears, separation from the tensioner body or permanent compression of the seal bead is observed, install a new tensioner.

44. Compress the tensioner plunger, using a vise.
45. Install a retaining clip on the tensioner to hold the plunger in during installation.
46. Remove the tensioner from the vise.
47. If the copper links are not visible, mark two links on one end and one link on the other end, and use as timing marks.
48. Install the crankshaft sprocket, making sure the flange faces forward.
49. Install the 4 bolts and the left hand and right hand timing chain guides. Tighten to 89 inch lbs. (10 Nm).
50. Position the lower end of the left hand (inner) timing chain on the crankshaft sprocket, aligning the timing mark on the outer flange of the crankshaft sprocket with the single copper (marked) link on the chain.

If the copper links are not visible, mark two links on one end and one link on the other end, and use as timing marks—2005 Expedition and Navigator models

Install the crankshaft sprocket, making sure the flange faces forward—2005 Expedition and Navigator models

Position the lower end of the left hand (inner) timing chain on the crankshaft sprocket, aligning the timing mark on the outer flange of the crankshaft sprocket with the single copper (marked) link on the chain—2005 Expedition and Navigator models

➡ **Make sure the upper half of the timing chain is below the tensioner arm dowel.**

51. Position the timing chain on the camshaft sprocket with the camshaft sprocket timing mark positioned between the two copper (marked) chain links.

➡ **The left hand timing chain tensioner arm has a bump near the dowel hole for identification.**

52. Position the left hand timing chain tensioner arm on the dowel pin and install the left hand timing chain tensioner and bolts. Tighten the bolts to 18 ft. lbs. (25 Nm).
53. Remove the retaining clip from the left hand timing chain tensioner.
54. Position the lower end of the right hand (outer) timing chain on the crankshaft sprocket, aligning the timing mark on the sprocket with the single copper (marked) chain link.

➡ **The lower half of the timing chain must be positioned above the tensioner arm dowel.**

Position the timing chain on the camshaft sprocket with the camshaft sprocket timing mark positioned between the two copper (marked) chain links—2005 Expedition and Navigator models

06017-NAVI-G11

Position the lower end of the right hand (outer) timing chain on the crankshaft sprocket, aligning the timing mark on the sprocket with the single copper (marked) chain link—2005 Expedition and Navigator models

06017-NAVI-G12

Position the right hand timing chain on the camshaft sprocket. Make sure the camshaft sprocket timing mark is positioned between the two copper (marked) chain links—2005 Expedition and Navigator models

55. Position the right hand timing chain on the camshaft sprocket. Make sure the camshaft sprocket timing mark is positioned between the two copper (marked) chain links.

56. Position the right hand timing chain tensioner arm on the dowel pin and install the right hand timing chain tensioner and bolts. Tighten the bolts to 18 ft. lbs. (25 Nm).

57. Remove the retaining clip from the right hand timing chain tensioner.

58. As a final-check, verify correct alignment of all timing marks.

59. Install the crankshaft sensor ring on the crankshaft.

60. Lubricate the roller followers with clean engine oil prior to installation.

61. Using tool 303-1039, install all of the camshaft roller followers.

✳✳ CAUTION

Do not use metal scrapers, wire brushes, power abrasive discs or other abrasive means to clean the sealing surfaces. These tools cause

06017-NAVI-G13

Verify correct alignment of all timing marks—2005 Expedition and Navigator models

scratches and gouges which make leak paths. Use a plastic scraping tool to remove all traces of old sealant.

➡ If the engine front cover is not secured within 4 minutes, the sealant must be removed and the sealing area cleaned. To clean the sealing area, use silicone gasket remover and metal surface prep. Failure to follow this procedure can cause future oil leakage.

➡ Make sure that the engine front cover gasket is in place on the engine front cover before installation.

62. Apply a bead of silicone gasket and sealant along the cylinder head-to-cylinder block surface and the oil pan-to-cylinder block surface, at the locations illustrated.

63. Install a new engine front cover gasket on the engine front cover. Position the engine front cover onto the dowels. Install the fasteners finger-tight.

64. Tighten the engine front cover fasteners in sequence in 2 steps:
 a. Step 1: Tighten fasteners 1 through 15 to 18 ft. lbs. (25 Nm).
 b. Step 2: Tighten fasteners 6 and 7 to 35 ft. lbs. (48 Nm).

65. Install the 4 front oil pan bolts in the sequence shown in 2 steps.
 a. Step 1: Tighten to 15 ft. lbs. (20 Nm).
 b. Step 2: Tighten an additional 60 degrees.

06017-NAVI-G14

Apply a bead of silicone gasket and sealant along the cylinder head-to-cylinder block surface and the oil pan-to-cylinder block surface, at the locations shown—2005 Expedition and Navigator models

1   Bolt, Hex Flange Head Pilot, M8 x 1.25 x 50

2   Bolt, Hex Flange Head Pilot, M8 x 1.25 x 50

3   Bolt, Hex Flange Head Pilot, M8 x 1.25 x 50

4   Bolt, Hex Flange Head Pilot, M8 x 1.25 x 50

5   Bolts, Hex Flange Head Pilot, M8 x 1.25 x 50

6   Stud, Hex Head Pilot, M10 x 1.5 x 1.5 x 103

7   Stud, Hex Head Pilot, M10 x 1.5 x 1.5 x 103

8   Bolt, Hex Flange Head Pilot, M8 x 1.25 x 50

9   Bolt, Hex Flange Head Pilot, M8 x 1.25 x 50

10 Bolt, Hex Flange Head Pilot, M8 x 1.25 x 50

11 Bolt, Hex Flange Head Pilot, M8 x 1.25 x 50

12 Stud and Washer, Hex Head Pilot, M8 x 1.25 x 1.25 x 94

13 Stud and Washer, Hex Head Pilot, M8 x 1.25 x 1.25 x 94

14 Stud and Washer, Hex Head Pilot, M8 x 1.25 x 1.25 x 94

15 Bolt, Hex Head Pilot, M8 x 1.25 x 56

06017-NAVI-G15

Engine front cover fastener location and torque sequence—2005 Expedition and Navigator models

Front oil pan bolts torque sequence—2005
Expedition and Navigator models

✳✳ CAUTION

Do not use metal scrapers, wire brushes, power abrasive discs or other abrasive means to clean sealing surfaces. These tools cause scratches and gouges which make leak paths. Use a plastic scraping tool to remove all traces of old sealant.

66. Connect the CKP sensor electrical connector.

67. Position the power steering pump assembly and install the bolts. Tighten to 18 ft. lbs. (25 Nm).

68. Position the power steering pressure hose support bracket and install the nut.

69. Position the transmission cooler tube support bracket and install the nut.

➡ Lubricate the O-ring seal with clean engine oil prior to installation.

70. Using a new O-ring seal, install the right hand CMP sensor and the bolt.

71. Connect the right hand CMP sensor electrical connector.

72. Install the left hand radio ignition interference capacitor and the nut.

➡ Lubricate the O-ring seal with clean engine oil prior to installation.

73. Using a new O-ring seal, install the left hand CMP sensor and the bolt.

74. Connect the left hand CMP sensor electrical connector.

75. Install the right hand radio ignition interference capacitor and the nut.

76. Connect the radio ignition interference capacitor electrical connectors.

77. Install the 3 accessory drive belt idler pulleys, the coolant pump pulley and the 7 bolts. Tighten the bolts to 18 ft. lbs. (25 Nm).

78. Lubricate the engine front cover and the crankshaft seal inner lip with clean engine oil.

Install the crankshaft seal into the engine front cover—2005 Expedition and Navigator models

79. Use the tools illustrated to install the crankshaft seal into the engine front cover.

➡ If not secured within 4 minutes, the sealant must be removed and the sealing area cleaned. To clean the sealing area, use silicone gasket remover and metal surface prep. Failure to follow this procedure can cause future oil leakage.

80. Apply a 0.32 inch (8mm) bead of silicone gasket and sealant to the Woodruff key slot on the crankshaft pulley.

Apply a 0.32 inch (8mm) bead of silicone gasket and sealant to the Woodruff key slot on the crankshaft pulley—2005 Expedition and Navigator models

Install the crankshaft pulley—2005 Expedition and Navigator models

81. Use the tool illustrated to install the crankshaft pulley.

82. Tighten the new crankshaft pulley bolt in 4 steps:

   a. Step 1: Tighten to 66 ft. lbs. (90 Nm).

   b. Step 2: Loosen 360 degrees.

   c. Step 3: Tighten to 37 ft. lbs. (50 Nm).

   d. Step 4: Tighten an additional 90 degrees.

83. Install the accessory drive belt.

84. Clean the valve cover mating surface with silicone gasket remover and metal surface prep.

➡ If not secured within 4 minutes, the sealant must be removed and the sealing area cleaned. To clean the sealing area, use silicone gasket remover and metal surface prep. Failure to follow this procedure can cause future oil leakage.

85. Apply a 0.32 inch (8mm) bead of silicone gasket and sealant in 2 places where the engine front cover meets the cylinder head.

Apply a 0.32 inch (8mm) bead of silicone gasket and sealant in 2 places where the engine front cover meets the right cylinder head—2005 Expedition and Navigator models

Install the right hand valve cover and gasket on the cylinder head and tighten the bolts in sequence—2005 Expedition and Navigator models

Install the left hand valve cover and gasket on the cylinder head and tighten the bolts in sequence—2005 Expedition and Navigator models

✳✳ CAUTION

When installing the valve cover, make sure to avoid damaging the Variable Camshaft Timing (VCT) solenoid.

86. Install the right hand valve cover and gasket on the cylinder head and tighten the bolts in the sequence shown and tighten to 89 inch lbs. (10 Nm).

NOTE: If not secured within 4 minutes, the sealant must be removed and the sealing area cleaned. To clean the sealing area, use silicone gasket remover and metal surface prep. Failure to follow this procedure can cause future oil leakage.

87. Apply a 0.32 inch (8mm) bead of silicone gasket and sealant in 2 places where

the engine front cover meets the left cylinder head.

88. Install the left hand valve cover and gasket on the cylinder head and tighten the bolts in the sequence shown and tighten to 89 inch lbs. (10 Nm).

89. Install the engine cooling fan.

90. Fill the crankcase with clean engine oil.

5.4L DOHC Engines

1. Before servicing the vehicle, refer to the precautions section.

2. Drain the oil.

3. Remove or disconnect the following:
   - Negative battery cable
   - Valve covers
   - Cooling fan
   - Water pump pulley

Apply a 0.32 inch (8mm) bead of silicone gasket and sealant in 2 places where the engine front cover meets the left cylinder head—2005 Expedition and Navigator models

- Crankshaft pulley using Crankshaft Damper Remover 303-009 (T58P-6316-D)
- Nuts and the hose support, if equipped
- Power steering pump and position aside
- Air Conditioner (A/C) muffler bolt and position out of the way
- Crankshaft Position (CKP) sensor
- Crankshaft front oil seal using Front Cover Seal Remover 303-107 (T74P-6700-A)
- Camshaft Position (CMP) sensor
- Belt idler pulley
- Engine front cover bolts, studs, cover and the gaskets
- Sensor ring from the crankshaft

4. Use Crankshaft Holding Tool 303-448 (T93P-6303-A), to position the camshaft.

5. Rotate the crankshaft until both camshaft keyways are 90 degrees from the cam cover surface. Be sure the copper links line up with the dots on the camshaft sprockets.

✳✳ WARNING

At no time, when the timing chains are removed and the cylinder heads are installed may the crankshaft or the camshaft be rotated. Severe piston and valve damage will occur.

6. Install Camshaft Holding Tool T96T-6256-B or equivalent, on the camshaft.

7. Remove or disconnect the following:
   - 2 left-hand and right-hand tensioner bolts and the timing chain tensioners
   - Left-hand and right-hand tensioner guides off the dowel pins
   - Right-hand timing chain from the camshaft sprocket
   - Left-hand timing chain from the camshaft sprocket
   - Left-hand and right-hand timing chain guide bolts, and the timing chain guides

To install:

8. Examine the timing chains, looking for the copper links. If the copper links are not visible, lay the chain on a flat surface and pull the chain taught until the opposite sides of the chain contact one another. Mark the links at each end of the chain and use these marks in place of the copper links.

➡ **If the engine jumped time, damage has been done to valves and possibly pistons and/or connecting rods. Any damage must be corrected before installing the timing chains.**

If the copper links are not visible, mark one link on one end of the chain and 2 links on the opposite end of the chain—5.4L and 6.8L engines

✳✳ WARNING

Do not compress the ratchet assembly. This will damage the ratchet assembly.

9. Compress the tensioner plunger, using the edge of a soft-jawed vise.

10. Use a small screwdriver or pick, push back and hold the ratchet mechanism.

11. While holding the ratchet mechanism, push the ratchet arm back into the tensioner housing.

12. Install a paper clip into the hole in the tensioner housing to hold the ratchet assembly and plunger in during installation.

13. Install the timing chain guides.

14. Remove tool T93P-6256-A.

15. Rotate the left hand camshaft sprocket until the timing mark is approximately at the 12 o' clock position. Rotate the right hand camshaft timing sprocket until the timing mark is approximately in the 11 o' clock position.

16. Install tool T93P-6256-A.

✳✳ WARNING

Unless otherwise instructed, at no time when the timing chains are removed and the cylinder heads are installed is the crankshaft or the camshaft to be rotated. Severe piston and valve damage will occur.

✳✳ WARNING

Rotate the crankshaft counterclockwise only. Do not rotate past position shown or severe piston or valve damage will occur.

17. Using Crankshaft Holding Tool 303-448 (T93P-6303-A), position the crankshaft, and then remove the tool.

18. Position the inner crankshaft sprocket with the long hub facing outward.

➡ The outer hub will face inward.

19. Install the left hand timing chain onto the crankshaft sprocket, aligning the one copper link on the timing chain with the slot on the crankshaft sprocket.

20. Verify the camshaft sprocket to copper link alignment.

21. Position the tensioner arms and tensioners, and install the bolts. Tighten the bolts to 18 ft. lbs. (25 Nm).

22. Remove the paper clips used to retain the ratchet assemblies.

23. Remove the camshaft holding tool.

24. Position the crankshaft sensor ring on the crankshaft.

25. Apply silicone gasket along the cylinder head-to-cylinder block and engine oil pan-to-cylinder block sealing surfaces.

26. Install or connect the following:
- Engine front cover carefully onto the dowel pins
- Engine front cover bolts, in sequence, in the following manner: bolts 1 through 5: 18 ft. lbs. (25 Nm); bolts 6 through 7: 37 ft. lbs. (50 Nm), bolts 1 through 15 to 18 ft. lbs. (25 Nm)
- Idler pulley in position and install the retaining bolt. Tighten the bolt to 18 ft. lbs. (25 Nm).
- CMP sensor in position and install the retaining bolt. Tighten the bolt to 106 inch lbs. (12 Nm).
- Damper on the crankshaft. Ensure the crankshaft key and keyway are aligned.
- 4 front oil pan-to-engine front cover retaining bolts and tighten in sequence in 3 steps: Tighten the bolts to 18 ft. lbs. (25 Nm), then to 15 ft. lbs. (20 Nm); then rotate the bolts an additional 60 degrees.
- Crankshaft front oil seal using Front Crankshaft Seal Installer 303-635 and Crankshaft Seal

Installer/Aligner 303-335 (T88T-6701-A)
- CKP sensor in position and install the retaining bolt. Tighten the bolt to 106 inch lbs. (12 Nm).
- CKP sensor electrical harness connector
- A/C muffler and tighten the nut to 18 ft. lbs. (25 Nm).
- Power steering pump with the upper and lower retaining bolts. Tighten the bolts to 18 ft. lbs. (25 Nm).
- Crankshaft pulley, bolt and washer using Crankshaft Damper Replacer 303-102 (T74P-6316-B) and tighten the bolt in 4 steps as follows:
  a. Step 1: 66 ft. lbs. (90 Nm).
  b. Step 2: Loosen one full turn.
  c. Step 3: 37 ft. lbs. (50 Nm).
  d. Step 4: an additional 85-95 degrees.
27. Install or connect the following:
- Water pump pulley and tighten the bolts to 18 ft. lbs. (25 Nm)
- Cooling fan
- Valve covers
- Negative battery cable

28. Refill the engine oil with the recommended type of engine oil and the correct amount.

29. Start the engine and check for leaks.

Check the alignment of the timing marks as shown—5.4L DOHC engines

Verify the camshaft sprocket to copper link alignment—5.4L DOHC engines

30. Road test the vehicle and check for proper engine operation.

Piston and Ring Positioning

Install the timing cover and tighten the bolts to specification—5.4L DOHC engines

Item	Description
1	Bolt, Hex Flange Head Pilot, M8 x 1.25 x 50
2	Bolt, Hex Flange Head Pilot, M8 x 1.25 x 50
3	Bolt, Hex Flange Head Pilot, M8 x 1.25 x 50
4	Bolt, Hex Flange Head Pilot, M8 x 1.25 x 50
5	Bolt, Hex Flange Head Pilot, M8 x 1.25 x 50
6	Stud, Hex-Head Pilot, M10 x 1.5 x 30
7	Stud, Hex-Head Pilot, M10 x 1.5 x 30
8	Bolt, Hex-Head Flange, M8 x 1.25 x 50
9	Bolt, Hex-Head Flange, M8 x 1.25 x 50
10	Bolt, Hex-Head Flange, M8 x 1.25 x 50
11	Stud Hex Shoulder Pilot, M8 x 1.25 x 65 - M8 x 1.25 x
12	Stud Hex Shoulder Pilot, M8 x 1.25 x 65 - M8 x 1.25 x
13	Bolt Hex Shoulder Pilot, M8 x 1.25 x 65
14	Stud Hex Shoulder Pilot, M8 x 1.25 x 65 - M8 x 1.25 x
15	Bolt Hex Shoulder Pilot, M8 x 1.25 x 65

4.6L engines—piston and connecting rod front mark locations

4.6L engines—piston-to-engine orientation

5.4L, 5.8L and 6.8L engines—piston and connecting rod assembly positioning

DIESEL ENGINE REPAIR

Alternator

REMOVAL

Single Alternator

6.0L ENGINE

1. Before servicing the vehicle, refer to the precautions in the beginning of this section.
2. Disconnect the negative battery cable.
3. Remove the cooling fan.
4. Remove the accessory drive belt.
5. Disconnect the alternator electrical connectors.
6. Remove the bolts and the alternator.

7.3L ENGINE

1. Before servicing the vehicle, refer to the precautions section.
2. Remove or disconnect the following:
   - Negative battery cable
   - Air cleaner, if necessary
   - Drive belt from the alternator pulley
   - Electrical harness connectors at the alternator assembly
   - Alternator bolts and the alternator from the vehicle

Dual Alternator

6.0L ENGINE

1. Before servicing the vehicle, refer to the precautions in the beginning of this section.
2. Disconnect the negative battery cable.
3. Remove the cooling fan.
4. Remove the accessory drive belt.
5. Remove the nut and position the generator B+ cable aside.
6. Loosen the 2 clamps and remove the turbo-to-air cooler pipe.
7. Disconnect the alternator electrical connectors.
8. Remove the bolts and the alternator.

7.3L ENGINE

1. Before servicing the vehicle, refer to the precautions section.
2. Remove or disconnect the following:
   - Single (upper) alternator
   - Electrical harness connectors at the alternator assembly
   - Alternator bolts and the alternator from the vehicle

INSTALLATION

Single Alternator

6.0L ENGINE

1. Before servicing the vehicle, refer to the precautions in the beginning of this section.
2. Install the alternator and tighten the bolts to 35 ft. lbs. (47 Nm)
3. Connect the alternator electrical connectors.
4. Install the accessory drive belt.
5. Install the cooling fan.
6. Connect the negative battery cable.

7.3L ENGINE

1. Before servicing the vehicle, refer to the precautions section.
2. Install or connect the following:
   - Alternator in position
   - Alternator retaining bolts and tighten to 35 ft. lbs. (47 Nm)
3. Install all remaining components in the reverse order of removal.

Dual Alternator

6.0L ENGINE

1. Before servicing the vehicle, refer to the precautions in the beginning of this section.
2. Install the alternator and tighten the bolts to 35 ft. lbs. (47 Nm)
3. Connect the alternator electrical connectors.
4. Install the turbo-to-air cooler pipe and tighten the 2 clamps to 9 ft. lbs. (12 Nm).
5. Connect the generator B+ cable.
6. Install the accessory drive belt.
7. Install the cooling fan.
8. Connect the negative battery cable.

7.3L ENGINE

1. Before servicing the vehicle, refer to the precautions section.
2. Install or connect the following:
   - Alternator in position
   - Alternator retaining bolts and tighten to 35 ft. lbs. (47 Nm)
   - Single (upper) alternator

Engine Assembly

REMOVAL & INSTALLATION

6.0L Engine

1. Before servicing the vehicle, refer to the precautions in the beginning of this section.
2. With the vehicle in NEUTRAL, position it on a hoist.
3. Disconnect the left hand and right hand battery ground cables.
4. On vehicles with manual transmission perform the following steps:

➡ **On vehicles equipped with manual transmissions, the transmission must be removed before the engine can be removed.**

   a. Remove the transmission.
   b. Remove the clutch.
5. Remove the air cleaner assembly.
6. Remove the radiator.
7. Remove the charge air cooler.
8. Remove the A/C condenser assembly.
9. Remove the parking lamp and the headlamp assemblies.
10. Remove the radiator grille, the radiator grille opening panel, and the upper radiator core supports.
11. Remove the front bumper.
12. Disconnect the transmission cooler hoses.
13. Remove the bolts and the transmission oil cooler.
14. Remove the bolts and position the power steering cooler out of the way.
15. Remove the intake manifold.
16. Disconnect the heater hose at the coolant pump.
17. Remove the battery cable cover.
18. Remove the nut and position the cable out of the way.
19. Remove the clips and disconnect the fuel lines.
20. Remove the lower radiator hose clamp and the hose.
21. Remove the power steering upper mounting bolts.

➡ **The power steering bolts must be removed evenly.**

22. Remove the bolts and position the power steering pump out of the way.
23. Remove the bolt and the left side ground cable.

24. Remove the nut and the battery cable bracket.

25. Remove the ground stud and the ground cable.

26. Disconnect the glow plug electrical connectors.

27. On vehicles with A/C perform the following steps:

    a. Disconnect the A/C high pressure switch and the A/C clutch electrical connectors.

    b. Disconnect the air conditioning manifold lines from the A/C compressor.

    c. Remove the A/C compressor.

28. Disconnect the Crankshaft Position (CKP) sensor electrical connector.

29. On vehicles with automatic transmission, remove the automatic transmission fluid indicator and tube.

30. Remove the ground strap at the back of the right head.

31. Remove the solenoid cap and disconnect the starter wiring.

32. Disconnect the block heater electrical connector.

33. Position the block heater and starter wiring harness out of the way.

34. On vehicles with automatic transmission perform the following steps:

    a. Remove the torque converter cover.

    b. Remove the torque converter nuts.

35. Remove the bolts for the turbocharger adapter pipe.

36. Remove the motor mount nuts.

37. Loosen the nuts at the turbocharger adapter pipe flange.

38. On vehicles with automatic transmission, remove the nine bell housing bolts.

39. Remove the turbocharger adapter pipe.

40. Remove the fuel injection control module mounting bolts for access to the module electrical connectors.

41. Disconnect the electrical connectors and remove the fuel injection control module.

42. Remove the nuts and the fuel injection control module bracket.

43. Remove the left rear valve cover stud.

44. Remove the transmission cooler line bracket.

45. Secure the turbocharger outlet pipe.

46. Remove the manufacturer's lifting eye.

47. Install the engine lifting eye 303-D043-02 on the right hand cylinder head.

48. Install the front lifting brackets 303-D043-01.

49. On vehicles with automatic transmission, position a suitable jack under the transmission.

67197-NAVI-G03

Install the engine lifting eye 303-D043-02 on the right hand cylinder head—6.0L engine

67197-NAVI-G04

Install the front lifting brackets 303-D043-01—6.0L engine

50. Install the Heavy Duty Floor Crane and Diesel Engine Lifting Bracket on the engine.

51. Raise the engine high enough to clear the No. 1 crossmember and pull the engine forward and clear of the vehicle.

To install:

52. With the vehicle in NEUTRAL, position it on a hoist.

53. Raise the engine high enough to clear the No. 1 crossmember, then position the engine into the vehicle.

54. On vehicles with automatic transmission, align the torque converter studs with the holes in the engine flywheel, then lower the engine onto the engine mount towers.

55. On vehicles with manual transmission, lower the engine onto the engine mount towers.

56. Remove the Heavy Duty Floor Crane and the Diesel Engine Lifting Bracket.

57. on vehicles with automatic transmission perform the following steps:

    a. Remove the transmission jack.

    b. Install the transmission-to-engine mounting bolts and tighten to 35 ft. lbs. (47 Nm).

    c. Install the torque converter-to-flywheel retaining nuts and tighten to 26 ft. lbs. (35 Nm).

    d. Install the flywheel housing cover.

58. Install the left hand and right hand side engine mount retaining nuts and tighten to 76 ft. lbs. (103 Nm).

59. Remove the engine lifting eye from the right side cylinder head.

60. Install the manufacturer's lifting bracket.

61. Remove the two engine lift adapters.

62. Position the turbocharger outlet pipe.

63. Install the transmission cooler tube bracket and nut and tighten to 89 inch lbs. (10 Nm).

64. Install the left rear valve cover stud and tighten to 71 inch lbs. (8 Nm).

65. Install the fuel injection control module bracket and nuts and tighten to 71 inch lbs. (8 Nm).

66. Connect the fuel injection control module electrical connectors and position the module on the bracket.

67. Install the fuel injection control module bolts and nuts and tighten to 10 ft. lbs. (13 Nm).

68. Position the turbocharger adapter pipe.

69. Install the bolts for the turbocharger adapter pipe but do not tighten bolts at this time.

70. Position back the block heater and starter wiring.

71. Connect the block heater electrical connector.

72. Connect the starter wiring and install the solenoid cap.

73. Connect the ground strap on the right head and install the bolt and tighten to 89 inch lbs. (10 Nm).

74. On vehicles with automatic transmission, install the automatic transmission fluid indicator and tube and tighten to 71 inch lbs. (8 Nm).

75. Connect the CKP sensor electrical connector.

76. On vehicles with A/C perform the following steps:

    a. Install the A/C compressor and tighten the bolts to 18 ft. lbs. (25 Nm).

    b. Position the A/C compressor manifold and install the bolt and tighten to 15 ft. lbs. (21 Nm).

    c. Connect the A/C high pressure switch and the clutch electrical connectors.

77. Connect the glow plug module electrical connectors.

78. Connect the ground cable and the ground stud and tighten to 35 ft. lbs. (47 Nm).

79. Install the battery cable bracket and the nut and tighten to 35 ft. lbs. (47 Nm).

80. Install the ground cable and bolt onto the left side of the engine block and tighten to 35 ft. lbs. (47 Nm).

81. Position the power steering pump and install the lower bolts and tighten to 18 ft. lbs. (25 Nm).

→The lower bolts need to be installed evenly.

82. Install the power steering pump upper bolts and tighten to 18 ft. lbs. (25 Nm).

83. Install the lower radiator hose and clamp.

84. Connect the fuel lines and install the clips.

85. Connect the battery crossover cable and tighten to 9 ft. lbs. (12 Nm).

86. Install the battery cable cover.

87. Connect the heater hose at the coolant pump.

88. Install the intake manifold.

89. Tighten the turbocharger adapter pipe at the flanges to 20 ft. lbs. (27 Nm).

90. Install the power steering cooler and bolts and tighten to 8 ft. lbs. (11 Nm).

91. Install the transmission oil cooler and tighten to 71 inch lbs. (8 Nm).

92. Connect the transmission cooler hoses.

93. Install the bumper.

94. Install the upper radiator core support, radiator grill opening panel and the radiator grill.

95. Install the headlamp and the parking lamp assemblies.

96. On vehicles with A/C, install the A/C condenser assembly.

97. Install the charge air cooler.

98. Install the radiator.

99. Install the air cleaner assembly.

100. On vehicles with manual transmission perform the following steps:
    a. Install the clutch assembly.
    b. Install the transmission.

101. Fill the cooling system.

102. Fill the motor with clean engine oil.

103. Connect the left and right hand battery cables.

104. Check and fill the automatic transmission, if equipped.

7.3L Engine

✱✱ CAUTION

The fuel system remains under pressure, even after the engine has been turned OFF. The fuel system pressure must be relieved before disconnecting any fuel lines. Failure to do so may result in fire and/or personal injury.

1. Before servicing the vehicle, refer to the precautions section.

2. Remove or disconnect the following:
   • Hood
   • Coolant
   • Engine oil
   • Negative battery cable
   • Air cleaner and intake duct assembly
   • Upper grille support bracket
   • Upper air conditioning condenser mounting bracket
   • Refrigerant, using approved equipment to remove the condenser
   • Radiator fan shroud halves
   • Fan and clutch assembly
   • Radiator hoses and the transmission cooler lines, if equipped
   • Condenser. Cap all openings immediately!
   • Radiator
   • Power steering pump and position it out of the way
   • Fuel supply line heater and alternator wires at the alternator
   • Oil pressure sending unit wire at the sending unit
   • Sender from the firewall and lay it on the engine
   • Accelerator cable and the speed control cable, if equipped, from the injection pump
   • Accelerator cable bracket with the cables attached, from the intake manifold and position it out of the way
   • Transmission kickdown rod from the injection pump, if equipped
   • Main wiring harness connector from the right side of the engine and the ground strap from the rear of the engine
   • Fuel system pressure
   • Fuel return hose from the left rear of the engine
   • The 2 upper transmission-to-engine attaching bolts
   • Heater hoses
   • Water temperature sender wire
   • Overheat light switch wire and position the wire out of the way
   • Battery ground cables from the front of the engine and the cables from the starter
   • Fuel inlet line, and plug the fuel line at the fuel pump
   • Exhaust pipe at the exhaust manifold
   • Engine insulators from the No. 1 crossmember
   • Flywheel inspection plate and the 4 converter-to-flywheel attaching nuts, if equipped with automatic transmission

3. Support the transmission on a jack.

4. Remove the 4 lower transmission attaching bolts.

5. Attach an engine lifting sling and remove the engine from the vehicle.

To install:

6. Lower the engine into vehicle.

7. Align the converter to the flexplate and the engine dowels to the transmission.

8. Install the engine mount bolts and tighten them to 80 ft. lbs. (109 Nm).

9. Remove the engine lifting sling.

10. Install the 4 lower transmission attaching bolts. Tighten the bolts to 65 ft. lbs. (88 Nm).

11. Remove transmission jack.

12. Raise and support the front end.

13. Install or connect the following:
   • 4 converter-to-flywheel attaching nuts, if equipped with an automatic transmission. Tighten the nuts to 34 ft. lbs. (47 Nm).
   • Flywheel inspection plate. Tighten the bolts to 60–90 inch lbs. (6.7–10 Nm).
   • Exhaust pipe at the exhaust manifold
   • Fuel inlet line
   • Battery ground cables to the front of the engine
   • Starter cables at the starter
   • Overheat light switch wire
   • Water temperature sender wire
   • Heater hoses
   • The 2 upper transmission-to-engine attaching bolts. Tighten the bolts to 65 ft. lbs. (88 Nm).
   • Fuel return hose at the left rear of the engine
   • Main wiring harness connector at the right side of the engine and the ground strap from the rear of the engine.
   • Transmission kickdown rod at the injection pump, if equipped
   • Accelerator cable and the speed control cable, if equipped, at the injection pump
   • Cable bracket with the cables attached, to the intake manifold
   • Oil pressure sending unit
   • Oil pressure sending unit wire at the sending unit
   • Fuel supply line heater and alternator wires at the alternator
   • Power steering pump
   • Radiator
   • Condenser
   • Radiator hoses and the transmission cooler lines, if equipped.
   • Fan and clutch assembly
   • Radiator fan shroud halves
   • Upper grille support bracket and upper air conditioning condenser mounting bracket

- Air cleaner and intake duct assembly

14. Refill the engine oil with the recommended type of engine oil and the correct amount.

15. Connect the negative battery cable.

16. If equipped with air conditioning, charge the system.

17. Fill the cooling system.

18. Install the hood.

Water Pump

REMOVAL & INSTALLATION

6.0L Engine

1. Before servicing the vehicle, refer to the precautions in the beginning of this section.

2. With the vehicle in NEUTRAL, position it on a hoist.

3. Remove the radiator and shroud assembly.

4. Disconnect the cooling fan electrical connector. Unclip and position the fan wiring aside.

➡**Use a hole in the fan hub to prevent the fan from turning.**

5. Using the special tool 303-591, loosen the cooling fan and clutch.

6. Remove the cooling fan assembly.

7. Remove the stator bolts and the stator.

8. Remove the accessory drive belt.

9. Remove the bolts and the water pump pulley.

10. Remove the bolts and the water pump.

11. Remove and discard the O-ring seal.

To install:

Clean and inspect the coolant pump mounting.

12. Install a new O-ring seal.

13. Install the water pump and tighten the bolts to 17 ft. lbs. (23 Nm).

14. Install the water pump pulley and tighten the bolts to 23 ft. lbs. (31 Nm).

Using the special tool 303-591, loosen the cooling fan and clutch—6.0L engine

Location of the water pump retaining bolts—6.0L engine

Location of the water pump pulley retaining bolts—6.0L engine

Location of the stator retaining bolts—6.0L engine

15. Install the accessory drive belt.

16. Install the stator and tighten the bolts to 30 ft. lbs. (40 Nm).

17. Install the cooling fan assembly.

➡**Use a hole in the fan hub to prevent the fan from turning.**

18. Using the special tool 303-591, tighten the cooling fan and clutch to 98 ft. lbs. (133 Nm)

19. Connect the cooling fan electrical connector.

20. Install the radiator and shroud assembly.

7.3L Engine

1. Before servicing the vehicle, refer to the precautions section.

2. Disconnect both battery ground cables.

3. Drain the cooling system.

4. Remove or disconnect the following:
- Radiator shroud, fan clutch and fan

➡**The fan clutch bolts are right-hand thread. Remove them by turning counter-clockwise.**

- Water pump pulley bolts, loosen only at this time
- Drive belt
- Water pump pulley bolts and the pulley
- Engine Coolant Temperature (ECT) sensor connector
- Heater hose from the water pump
- Bolts attaching the water pump to the front cover and lift off the pump

To install:

5. Thoroughly clean the mating surfaces of the pump and front cover.

6. Install or connect the following:
- New gasket
- Water pump over the dowel pins and into place on the front cover
- Water pump attaching bolts. Tighten the bolts to 15 ft. lbs. (20 Nm).
- Heater hose to the pump
- ECT sensor connector
- Water pump pulley and start the water pump pulley bolts
- Drive belt
- Tighten the pulley retaining bolts to 12–18 ft. lbs. (16–24 Nm)
- Fan and fan shroud assembly

7. Fill and bleed the cooling system.

8. Connect the battery ground cables.

9. Start the engine and check for leaks.

Apply RTV sealant to the bolts indicated—7.3L diesel engine

Glow Plugs

REMOVAL & INSTALLATION

6.0L Engine

1. Before servicing the vehicle, refer to the precautions in the beginning of this section.

2. If servicing the right hand glow plugs, remove the evaporator core housing.

3. Disconnect the glow plug electrical connector.

4. Remove the glow plug buss bar.

5. Remove the glow plug.

➡**To install new glow plug sleeves, the cylinder heads must be removed from the vehicle.**

6. If required, remove the cylinder heads from the engine.

➡**The glow plug sleeve is made of stainless steel. Lubrication of the glow plug sleeve removal tap is necessary, or damage to the tap and excessive force will be needed to get the tap started.**

7. Thread the glow plug sleeve removal tap 303-764 into the glow plug sleeve.

8. Install the glow plug sleeve removal tool 303-764 into the threaded sleeve and tighten. Use a wrench to thread the tool into the sleeve until the sleeve is extracted.

9. Clean the glow plug bore with a stiff wire brush. Use filtered compressed air to remove the debris from the glow plug recess.

To install:

10. Verify the glow plug bore is completely clean and dry.

11. Apply Threadlock 262 sealant to the glow plug sleeve in the two places illustrated.

12. Install the glow plug sleeve 303-763 into the glow plug bore until it bottoms.

67197-NAVI-G10

Apply Threadlock 262 sealant to the glow plug sleeve in the two places shown— 6.0L engine

13. Clean the glow plug sleeve after installation with a nylon brush and solvent. Make sure Threadlock 262 sealant is cleaned out before it hardens.

14. Install the cylinder heads.

15. Install the glow plug and tighten to 14 ft. lbs. (19 Nm).

16. Inspect and install new the O-rings as necessary.

17. Install the glow plug buss bar.

18. Connect the glow plug electrical connector.

19. If removed, install the evaporator core housing.

7.3L Engine

✷✷ CAUTION

The red-striped wiring harness carries 115V direct current. Severe electrical shock may be received. DO NOT pierce.

1. Before servicing the vehicle, refer to the precautions section.

2. Remove or disconnect the following:
   - Negative battery cable
   - Rocker arm cover
   - Glow plug electrical leads using a pair of pliers
   - Glow plugs by unscrewing them from the cylinder head with a 10mm socket and wrench

3. Inspect the tips of the plugs for any evidence of distortion or missing tip ends; replace them if necessary.

To install:

4. Install or connect the following:
   - Glow plug into the cylinder head. Tighten the glow plugs to 14 ft. lbs. (19 Nm).
   - Glow plug electrical connector. Be sure that the glow plug wiring is routed to avoid moving components in the engine bay.
   - Rocker arm cover

➡**When the battery is disengaged and reconnected, some abnormal drive symptoms may occur while the vehicle relearns its adaptive strategy. The vehicle may need to be driven 10 miles (16 km) or more to relearn this strategy.**

   - Negative battery cable

Cylinder Head

REMOVAL & INSTALLATION

6.0L Engine

RIGHT CYLINDER HEAD

1. Before servicing the vehicle, refer to the precautions in the beginning of this section.

303-764

67197-NAVI-G09

Insert tool 303-764 into the threaded sleeve and tighten and use a wrench to thread the tool into the sleeve until the sleeve is extracted—6.0L engine

19 Nm (14 lb/ft)

7924FG57

Tighten the glow plugs to 14 ft. lbs. (19 Nm) and attach the connector—7.3L engine

2. With the vehicle in NEUTRAL, position it on a hoist.

3. Drain the engine oil.

4. Remove the intake manifold.

5. Remove the starter.

6. Remove the cylinder drain block.

7. Remove the evaporator core housing.

8. Disconnect the wiring retainers from the stud.

➡**Mark the position of the valve cover bolts for valve cover installation.**

9. Remove the bolts and the valve cover.

10. Using the special tool 303-755, disconnect the high-pressure oil rail supply line at the high-pressure oil rail.

11. Remove the bolts and the high-pressure oil rail.

12. Using the special tool 303-625, disconnect the high-pressure oil supply line.

❈❈ CAUTION

Do not attempt to put battery voltage to the fuel injector or damage to the fuel injector will occur.

13. Using a 19 mm socket, push the fuel injector electrical connector out of the rocker arm carrier.

❈❈ CAUTION

To prevent engine damage, do not use air tools to remove the fuel injectors. The clip that extracts the fuel injector can dislodge and fall into the oil drain hole.

➡**There is no need to drain the fuel rail.**

➡**If engine oil is found in the engine coolant or engine coolant found in the combustion chambers, new injector sleeves may need to be installed.**

14. Remove the bolt, fuel injector hold-down and fuel injector.

67197-NAVI-G11

Using the special tool 303-755, disconnect the high-pressure oil rail supply line at the high-pressure oil rail—6.0L engine

67197-NAVI-G12

Using the special tool 303-625, disconnect the high-pressure oil supply line—6.0L engine

67197-NAVI-G13

Remove and discard the 10 inner cylinder head bolts—6.0L engine–right side shown, left side similar

67197-NAVI-G14

Remove the eight bolts and the rocker arm assemblies—6.0L engine–right side shown, left side similar

67197-NAVI-G15

Mark the eight valve bridges with a permanent marker and remove—6.0L engine–right side shown, left side similar

15. Remove the nuts and bolts from the turbocharger adapter pipe.

16. Remove the turbocharger adapter pipe.

17. Remove the glow plug buss bar.

18. Remove the four glow plugs.

19. Remove and discard the crankcase to head tube assembly.

20. Remove and discard the 10 inner cylinder head bolts.

21. Remove the eight bolts and the rocker arm assemblies.

22. Mark the eight valve bridges with a permanent marker and remove.

23. Mark and remove the eight push rods.

24. Remove the outer cylinder head bolts.

25. Install the special tool 303-759 and the lifting crane. With the help of an assistant, remove the cylinder head from the vehicle.

26. Remove and discard the cylinder head gasket and dowels.

27. Clean and inspect the gasket sealing surfaces.

To install:

➡**Use care to avoid scratching the blue compound on the cylinder head gasket. Install a new cylinder head gasket with the part number facing up and verify the top five holes and the head gasket push rod holes line up.**

28. Install the dowels and the cylinder head gasket.

29. Using the special tool 303-759 and with the help of an assistant, install the cylinder head on the engine. Remove the special tools.

30. Install the outer cylinder head bolts finger-tight.

➡**Higher mileage engines require push rods to be cleaned so the copper colored end of the push rod can be identified.**

31. Apply clean engine oil to each end of the push rod. Insert them into their respective positions with the copper colored end up.

➡**Coat the end of each valve stem with clean engine oil.**

32. Install the eight valve bridges.

❈❈ CAUTION

Rotate the crankshaft until the damper locating dowel notch is in the six o'clock position or engine damage can occur.

➡**Apply clean engine oil to the top center of each valve bridge.**

33. Install the rocker arm assemblies and eight bolts to 23 ft. lbs. (31 Nm).

✳✳ CAUTION

Using too much engine oil on the threads of the cylinder head bolts can cause damage to the threads and poor sealing. Using anti-seize compounds, grease or any other lubricants other than engine oil on the cylinder head bolt threads can affect the true torque value of the bolts.

➡ **Lightly lubricate the cylinder head bolt threads and flanges with clean engine oil.**

34. Install the 10 inner cylinder head retaining bolts finger-tight.

35. Tighten the head bolts in the following sequence:

a. Tighten bolts 1 through 10 to 65 ft. lbs. (88 Nm).

b. Tighten bolts 1, 3, 5, 7 and 9 to 85 ft. lbs. (115 Nm).

c. Tighten bolts in sequence 1 through 10, clockwise 90 degrees.

d. Tighten bolts in sequence 1 through 10, a second time, clockwise 90 degrees.

e. Tighten bolts in sequence 1 through 10, a third time, clockwise 90 degrees.

f. Tighten bolts 11 through 15 to 18 ft. lbs. (24 Nm).

g. Tighten bolts 11 through 15 to 23 ft. lbs. (31 Nm).

➡ **Install a new lower O-ring.**

36. Install a new crankcase to head tube assembly and tighten to 33 ft. lbs. (45 Nm).

37. Install the four glow plugs and tighten to 14 ft. lbs. (19 Nm).

➡ **Clean and inspect the glow plug buss bar O-rings and install new as necessary.**

➡ **Apply clean engine oil to the O-rings before installing.**

38. Install the glow plug buss bar.

39. Position the turbocharger adapter pipe.

➡ **Apply anti-seize lubricants to the bolt threads prior to installing the bolts.**

40. Install the nuts and bolts in the adapter pipe and tighten to 20 ft. lbs. (27 Nm).

Do not tighten until after the turbocharger is installed.

✳✳ CAUTION

If the fuel injector oil inlet D-ring is damaged, a new fuel injector must be installed.

41. Install new O-rings and copper washer on the fuel injector. Lubricate the fuel injector and O-rings liberally with clean engine oil.

✳✳ CAUTION

To prevent engine damage, do not use air tools to install the fuel injectors. The clip that extracts the injector can dislodge and fall into the oil drain hole.

42. Install the fuel injector, fuel injector hold-down and bolt and tighten to 24 ft. lbs. (33 Nm).

43. Apply clean engine oil to the top fuel injector O-rings.

44. Install the high-pressure oil rail and bolts.

45. Install the high-pressure oil rail.

46. Install the bolts finger tight. Tighten the bolts in the sequence shown to 8 ft. lbs. (11 Nm).

47. Install the high-pressure oil line.

11 Nm (8 lb-ft)

67197-NAVI-G17

Tighten the high-pressure oil rail bolts in the sequence shown—6.0L engine–right side shown, left side similar

48. Install the valve cover and eleven bolts and tighten to 71 inch lbs. (8 Nm).

49. Install the evaporator core housing.

50. Connect the wiring retainers to the studs.

51. Fill the crankcase with clean engine oil.

➡ **Apply clean engine oil to the O-ring prior to installing.**

52. Install the cylinder block drain plug and tighten to 15 ft. lbs. (20 Nm).

53. Install the starter.

54. Install a new oil filter.

55. Install the intake manifold.

LEFT CYLINDER HEAD

1. Before servicing the vehicle, refer to the precautions in the beginning of this section.

2. With the vehicle in NEUTRAL, position it on a hoist.

3. Drain the engine oil.

4. Remove the intake manifold.

5. Remove the left cylinder block drain plug.

6. Remove the fuel injector control module mounting bracket.

7. Disconnect the glow plug connector and position aside.

8. Remove the nut and position the oil level indicator and tube aside.

➡ **Mark the position of the valve cover bolts for valve cover bolt installation.**

9. Remove the eleven bolts and the valve cover. Remove the valve cover gasket.

10. Disconnect the high-pressure oil rail supply line at the high-pressure oil rail.

11. Remove the bolts and the high-pressure oil rail.

12. Disconnect and remove the high-pressure oil supply line.

67197-NAVI-G16

Tighten the head bolts in the sequence shown—6.0L engine–right side shown, left side similar

❋❋ CAUTION

Do not attempt to put battery voltage to the fuel injector or damage to the fuel injector will occur.

13. Using a 19 mm socket, push the fuel injector electrical connector out of the rocker arm carrier.

❋❋ CAUTION

To prevent engine damage, do not use air tools to remove the fuel injectors. The clip that extracts the injector can dislodge and fall into the oil drain hole.

➡There is not need to drain the fuel rail.

➡If engine oil is found in the engine coolant or engine coolant is found in the combustion chambers, new injector sleeves may need to be installed.

14. Remove the bolt, fuel injector hold-down and fuel injector.

15. Remove the nuts and bolts from the turbocharger adapter pipe.

16. Remove the turbocharger adapter pipe.

➡Remove and discard the O-ring.

17. Remove the oil level indicator and tube.

18. Remove the bolts and turbocharger heat shield.

19. Remove the glow plug buss bar.

20. Remove the four glow plugs.

21. Disconnect the exhaust back pressure tube at the exhaust manifold.

22. Remove the retaining nuts. Remove the exhaust back pressure sensor and tube assembly.

23. Remove the retaining nut from the fuel line bracket and position the fuel lines aside.

24. Remove the bolt and the idler pulley.

25. Disconnect the heater hose from the front cover and position aside.

26. Remove the fan shroud mounting stud.

➡Remove and discard the sealing washers.

27. Remove the banjo fitting and the fuel line.

28. Remove and discard the crankcase-to-head tube assembly.

29. Remove and discard the inner cylinder head bolts.

30. Remove the eight bolts and the rocker arm assemblies.

31. Mark the eight valve bridges with a permanent marker and remove.

32. Mark and remove the eight push rods.

33. Remove the outer cylinder head bolts.

34. Install the special tool 303-579 and the lifting crane. With the help of an assistant, remove the cylinder from the vehicle.

35. Remove the cylinder head gasket and dowels.

36. Clean and inspect the gasket sealing surfaces.

To install:

➡Use care to avoid scratching the blue compound on the cylinder head gasket. Install the gasket with the part number facing upward.

37. Install the dowels and the cylinder head gasket.

➡Position the fuel lines in place before the cylinder head is installed.

38. Using the special tool 303-579 and with the help of an assistant, install the cylinder head on the engine. Remove the special tools.

39. Install the outer cylinder head bolts finger-tight.

➡Higher mileage engines require push rods to be cleaned so the cooper-colored end of the push rod can be identified.

40. Apply clean engine oil to each end of the push rod. Insert them into their respective positions with the copper-colored end up.

➡Coat the end of each valve stem with clean engine oil. Install the eight valve bridges.

❋❋ CAUTION

Rotate the crankshaft until the damper locating dowel notch is in the six o'clock position or engine damage can occur. Install the rocker arm assemblies and eight bolts and tighten to 23 ft. lbs. (31 Nm).

➡Lightly lubricate the cylinder head bolt threads with clean engine oil.

41. Install the inner cylinder head retaining bolts finger-tight.

42. Tighten the head bolts in the following sequence:

a. Tighten bolts 1 through 10 to 65 ft. lbs. (88 Nm).

b. Tighten bolts 1, 3, 5, 7 and 9 to 85 ft. lbs. (115 Nm).

c. Tighten bolts in sequence 1 through 10, clockwise 90 degrees.

d. Tighten bolts in sequence 1 through 10, a second time, clockwise 90 degrees.

e. Tighten bolts in sequence 1 through 10, a third time, clockwise 90 degrees.

f. Tighten bolts 11 through 15 to 18 ft. lbs. (24 Nm).

g. Tighten bolts 11 through 15 to 23 ft. lbs. (31 Nm).

➡Install a new lower O-ring.

43. Install a new crankcase to head tube assembly and tighten to 33 ft. lbs. (45 Nm)

➡Install new sealing washers.

44. Install the fuel line and the banjo fitting and tighten to 28 ft. lbs. (38 Nm).

45. Install the fan shroud mounting stud and tighten to 30 ft. lbs. (40 Nm).

46. Connect the heater hose.

47. Install the idler pulley and bolt and tighten to 35 ft. lbs. (47 Nm).

48. Position the exhaust backpressure sensor and tube assembly and install the retaining nuts and tighten to 23 ft. lbs. (31 Nm).

49. Tighten the exhaust backpressure tube fitting at the exhaust manifold to 22 ft. lbs. (30 Nm).

50. Position the fuel lines and install the fuel line bracket retaining bolt and tighten to 10 ft. lbs. (13 Nm).

51. Install the four glow plugs and tighten to 14 ft. lbs. (19 Nm).

➡Apply clean engine oil to the O-rings before installing.

52. Install the glow plug buss bar.

➡Install a new O-ring.

53. Install the oil level indicator and tube.

54. Install the turbocharger heat shield and bolts.

55. Position the turbocharger adapter pipe.

➡Do not torque until after the turbocharger is installed.

56. Install the nuts and bolts in the adapter pipe and tighten to 20 ft. lbs. (27 Nm).

❋❋ CAUTION

If the fuel injector oil inlet D-ring is damaged, a new fuel injector must be installed.

57. Install new O-rings and copper washer on the fuel injector. Lubricate the fuel injector and O-rings liberally with clean engine oil.

✳✳ CAUTION

To prevent engine damage, do not use air tools to install the fuel injectors. The clip that extracts the injector can dislodge and fall into the oil drain hole.

58. Install the fuel injector, fuel injector hold-down and bolt and tighten to 24 ft. lbs. (33 Nm).

✳✳ CAUTION

To prevent engine damage, be sure the injector wiring is clear of all moving parts.

59. Install the fuel injector electrical connector into the rocker carrier.
60. Apply engine oil to the top fuel injector O-rings.
61. Install the high-pressure oil rail and bolts.
62. Install the high-pressure oil rail.
63. Install the bolts finger tight, then tighten the bolts in the sequence illustrated to 8 ft. lbs. (11 Nm).
64. Install the high-pressure oil line.
65. Clean and inspect the valve cover gasket. Install a new gasket if necessary.
66. Position the valve cover gasket. Install the valve cover and 11 bolts and tighten to 71 ft. lbs. (8 Nm).
67. Position back the oil level indicator and install the nut.
68. Connect the glow plug connector.
69. Install the fuel injection control module mounting bracket and tighten the retainers to 71 inch lbs. (8 Nm).
70. Fill the crankcase with clean engine oil.

➡ **Lightly lubricate the O-ring with clean engine oil before installing.**

71. Install the cylinder block drain plug and tighten to 15 ft. lbs. (20 Nm).
72. Install a new oil filter.
73. Install the intake manifold.

7.3L Engine

RIGHT CYLINDER HEAD

✳✳ CAUTION

The fuel system remains under pressure, even after the engine has been turned OFF. The fuel system pressure must be relieved before disconnecting any fuel lines. Failure to do so may result in fire and/or personal injury.

1. Before servicing the vehicle, refer to the precautions section.
2. Drain the cooling system.
3. Remove or disconnect the following:
   - Negative battery cables
   - Radiator
   - Turbocharger assembly
   - Fuel lines from the rear of both cylinder heads and the fuel pump
   - Wiring from the alternator
   - Adjusting bolts and pivot bolts from the alternator and the vacuum pump and remove both units
   - Alternator and its bracket
   - Engine oil dipstick tube
   - Manifold Absolute Pressure (MAP) sensor and position it aside
   - Valve cover
   - Connectors from the injectors and glow plugs
   - Valve cover gasket
   - High pressure oil pump supply line to the right cylinder head
   - Exhaust back pressure line
   - Three glow plug relay bracket nuts and the ground wire
   - Heater hose from the cylinder head
   - Outer half of the heater distribution box
   - Four outboard fuel injector hold-down bolts, retaining screws and four oil deflectors

✳✳ WARNING

Remove the oil drain plugs prior to removing the injectors or oil could enter the combustion chamber which could result in hydrostatic lock and severe engine damage.

   - Oil rail drain plugs
   - Fuel injectors using Injector Remover No. T94T–9000–AH1, or equivalent. Position the tool's fulcrum beneath the fuel injector hold-down plate and over the edge of the cylinder head. Install the remover screw in the threaded hole of the fuel injector plate. Tighten the screw to lift out the injector from its bore. Place the injector in a suitable protective sleeve such as Rotunda Injector Protective Sleeve, No. 014–00933–2, and set the injector in a suitable holding rack.
4. Use a suitable vacuum tool, such as Rotunda Vacuum Pump, No. 021–00037, or equivalent to remove the oil and fuel left over in the injector bores.
5. Remove or disconnect the following:
   - Rocker arms and pushrods, KEEP EVERYTHING IN ORDER
   - Four glow plugs
   - Right turbo exhaust inlet pipe
   - Ground strap from the rear of the cylinder head
   - Fuel return line at the front of the cylinder head
   - Four inboard fuel injector shoulder bolts
   - Cylinder head bolts and attach a Rotunda Cylinder Head Lifting Bracket, 014–00932–2, or equivalent
6. Carefully lift the cylinder head out of the engine compartment and remove the head gaskets.

   To install:

➡ **To prepare a good seat for the fuel injector O-rings, use a suitable injector sleeve brush to clean any debris from the bore.**

7. Carefully clean the cylinder block and head mating surfaces.
8. Install or connect the following:
   - Cylinder head gasket on the engine block and carefully lower the cylinder head in place
   - Cylinder head bolt and torque in 3 steps using the sequence shown in the illustration as follows:
   a. Step 1: 65 ft. lbs. (88 Nm).
   b. Step 2: 85 ft. lbs. (115 Nm).
   c. Step 3: 95 ft. lbs. (129 Nm).

➡ **Lubricate the threads and the mating surfaces of the bolt heads and washers with engine oil.**

   - Fuel return line to the cylinder head
   - Four inboard injector shoulder bolts. Tighten them to 9 ft. lbs. (12 Nm).
9. Install the fuel injectors using special tools as follows:
   a. Lubricate the injectors with clean engine oil. Using new copper washers, carefully push the injectors square into the bore using hand pressure only to seat the O-rings.
   b. Position the open end of Injector Replacer, No. T94T–9000–AH2, or equivalent between the fuel injector body and injector hold-down plate, while positioning the opposite end of the tool over the edge of the cylinder head.
   c. Align the hole in the tool with the threaded hole in the cylinder head and install the bolt from the tool kit. Tighten

Cylinder head torque sequence—diesel engines

the bolt to fully seat the injector, then remove the bolt and tool.

10. Install or connect the following:
   - Four outboard fuel injector hold-down bolts, the four oil deflectors and retaining screws. Tighten them to 9 ft. lbs. (12 Nm).

11. Dip the pushrod ends in clean engine oil and install the pushrods with the copper colored ends toward the rocker arms, making sure the pushrods are fully seated in the tappet pushrod seats.
   - Rocker arms and posts in their original positions. Apply Lubriplate® grease to the valve stem tips. Turn the engine over by hand until the timing mark is at the 11 o'clock position as viewed from the front.
   - Rocker arm posts, bolts and tighten to 27 ft. lbs. (37 Nm)
   - Valve covers
   - Fuel rail drain plugs, tightening them to 8 ft. lbs. (11 Nm)
   - Oil rail drain plugs, tightening them to 53 inch lbs. (6 Nm)
   - Heater distribution box
   - Heater hose to the cylinder head
   - Glow plug relay bracket and ground wire
   - Exhaust back pressure line
   - Oil supply line to the cylinder head, tightening it to 19 ft. lbs. (26 Nm)
   - Dipstick tube
   - MAP sensor and screws
   - Valve cover gasket
   - Wiring to the fuel injectors and glow plugs

   - Valve cover, tightening the bolts to 8 ft. lbs. (11 Nm)
   - Injector wiring harness to the valve cover gasket
   - Alternator, tightening the bracket bolts to 40–55 ft. lbs. (54–75 Nm)
   - Alternator wiring
   - Drive belt

12. The remainder of the installation is the reverse of the removal. Tighten the fuel pump-to-fuel line banjo bolt to 40 ft. lbs. (54 Nm).

13. Connect both negative battery cables.

14. Refill and bleed the cooling system.

15. Run the engine and check for fuel, coolant and exhaust leaks.

LEFT CYLINDER HEAD

✳✳ CAUTION

The fuel system remains under pressure, even after the engine has been turned OFF. The fuel system pressure must be relieved before disconnecting any fuel lines. Failure to do so may result in fire and/or personal injury.

1. Before servicing the vehicle, refer to the precautions section.

2. Drain the cooling system.

3. Using approved equipment, recover the refrigerant from the A/C system.

4. Remove or disconnect the following:
   - Negative battery cables
   - Radiator
   - Turbocharger assembly

   - Two crankcase breather screws and the breather
   - Wiring from the air conditioning compressor
   - Four left accessory bracket bolts
   - Vacuum hose at the brake vacuum pump
   - A/C lines from the compressor
   - Power steering lines from the pump
   - Left accessory bracket and accessories as an assembly
   - Valve cover
   - Fuel line assembly between the cylinder heads and fuel pump
   - Fuel line nut from the intake manifold stud
   - Fuel return line from the cylinder head
   - High pressure oil pump supply line from the cylinder head

5. Raise the vehicle and support it safely on jackstands.
   - Left turbo exhaust pipe from the manifold

6. Lower the vehicle.
   - Oil rail drain plugs
   - Four outboard fuel injector hold-down bolts, retaining screws and four oil deflectors

✳✳ WARNING

Remove the oil drain plugs prior to removing the injectors or oil could enter the combustion chamber which could result in hydrostatic lock and severe engine damage.

   - Oil rail drain plugs

7. Remove the fuel injectors using Injector Remover No. T94T–9000–AH1, or équivalent. Position the tool's fulcrum beneath the fuel injector hold-down plate and over the edge of the cylinder head. Install the remover screw in the threaded hole of the fuel injector plate. Tighten the screw to lift out the injector from its bore. Place the injector in a suitable protective sleeve such as Rotunda Injector Protective Sleeve, No. 014–00933–2, and set the injector in a suitable holding rack.

8. Use a suitable vacuum tool, such as Rotunda Vacuum Pump, No. 021–00037, or equivalent to remove the oil and fuel left over in the injector bores.

9. Remove or disconnect the following:
   - Rocker arms and pushrods, KEEP EVERYTHING IN ORDER
   - Four glow plugs
   - Left turbo exhaust inlet pipe
   - Main engine harness connectors in the left fender well and position the harness aside

- Four inboard fuel injector shoulder bolts
- Cylinder head bolts and attach a Rotunda Cylinder Head Lifting Bracket, 014–00932–2, or equivalent

10. Carefully lift the cylinder head out of the engine compartment and remove the head gaskets.

To install:

➡ **To prepare a good seat for the fuel injector O-rings, use a suitable injector sleeve brush to clean any debris from the bore.**

11. Carefully clean the cylinder block and head mating surfaces.

12. Install or connect the following:
- Cylinder head gasket on the engine block and carefully lower the cylinder head in place
- Cylinder head bolt and torque in 3 steps using the sequence shown in the illustration as follows:
  a. Step 1: 65 ft. lbs. (88 Nm).
  b. Step 2: 85 ft. lbs. (115 Nm).
  c. Step 3: 95 ft. lbs. (129 Nm).

➡ **Lubricate the threads and the mating surfaces of the bolt heads and washers with engine oil.**

13. Apply anti-seize paste and install the glow plugs, tightening them to 14 ft. lbs. (19 Nm).

14. Install or connect the following:
- Four outboard fuel injector hold-down bolts, the four oil deflectors and retaining screws. Tighten them to 9 ft. lbs. (12 Nm).

15. Dip the pushrod ends in clean engine oil and install the pushrods with the copper colored ends toward the rocker arms, making sure the pushrods are fully seated in the tappet pushrod seats.
- Rocker arms and posts in their original positions. Apply Lubriplate® grease to the valve stem tips. Turn the engine over by hand until the timing mark is at the 11 o'clock position as viewed from the front.
- Rocker arm posts, bolts and tighten to 27 ft. lbs. (37 Nm).
- Valve covers
- Four inboard injector shoulder bolts. Tighten them to 9 ft. lbs. (12 Nm).

16. Install the fuel injectors using special tools as follows:
  a. Lubricate the injectors with clean engine oil. Using new copper washers, carefully push the injectors square into

the bore using hand pressure only to seat the O-rings.
  b. Position the open end of Injector Replacer, No. T94T–9000–AH2, or equivalent between the fuel injector body and injector hold-down plate, while positioning the opposite end of the tool over the edge of the cylinder head.
  c. Align the hole in the tool with the threaded hole in the cylinder head and install the bolt from the tool kit. Tighten the bolt to fully seat the injector, then remove the bolt and tool.

17. Install or connect the following:
- Fuel rail drain plugs, tightening them to 8 ft. lbs. (11 Nm)
- Oil rail drain plugs, tightening them to 53 inch lbs. (6 Nm)
- Heater distribution box
- Valve cover gasket
- Wiring to the fuel injectors and glow plugs
- Valve cover, tightening the bolts to 8 ft. lbs. (11 Nm)
- Engine wiring harness

18. Raise and safely support the vehicle on jackstands.

19. Loosely install the turbo exhaust pipe to the manifold.

20. Lower the vehicle.

21. The remainder of the installation is the reverse of the removal. Tighten the fuel pump-to-fuel line banjo bolt to 40 ft. lbs. (54 Nm).

22. Connect both negative battery cables.

23. Refill and bleed the cooling system.

24. Run the engine and check for fuel, coolant and exhaust leaks.

Rocker Arms

REMOVAL & INSTALLATION

6.0L Engine

1. Before servicing the vehicle, refer to the precautions in the beginning of this section.

✳✳ CAUTION

To prevent engine damage, always install new head bolts when installing the cylinder head, even if removing only one bolt. It will be necessary to install a new cylinder head gasket as well as all cylinder head bolts.

2. Remove the cylinder head.

67197-NAVI-G18

Place the steel ball in the rocker arm cup and lubricate with clean engine oil—6.0L engine

➡ **Be careful when removing the rocker arm. Do not lose the rocker arm steel ball.**

3. Push down on the rocker arm. Pull out on the bottom releasing the ball.

4. Remove the steel ball from the rocker arm.

5. Remove the rocker arm clip from the fulcrum plate.

6. Clean and inspect all parts for pitting or scuffing. Install new rocker arm and valve bridges, as needed.

To install:

7. Install the rocker arm clip on the fulcrum plate.

8. Place the steel ball in the rocker arm cup and lubricate with clean engine oil.

9. Push down on the rocker arm to compress the rocker arm clip. Insert the ball and release the pressure. Verify free movement of the rocker arm on the fulcrum.

10. Install the cylinder head.

7.3L Engine

1. Before servicing the vehicle, refer to the precautions section.

2. Disconnect the ground cables from both batteries.

3. Remove the valve cover attaching screws and remove both valve covers.

4. Remove the valve rocker arm post mounting bolts. Remove the rocker arms and posts in order and mark them with tape so they can be installed in their original positions.

5. If the cylinder heads are to be removed, then the pushrods can now be removed. Make a holder for the pushrods out of a piece of wood or cardboard and remove the pushrods in order. It is very important that the pushrods be reinstalled in their original order. The pushrods can remain in position if no further disassembly is required.

To install:

6. If the pushrods were removed, install them in their original locations. Make sure they are fully seated in the tappet seats.

Item	Description
1	Snap Retaining Clip
2	Rocker Arm Pedestal
3	Rocker Arm Ball
4	Rocker Arm
5	Rocker Arm Assembly

7924FG58

Exploded view of the rocker arm assembly—diesel engines

➡The copper colored end of the pushrod goes toward the rocker arm.

7. Apply a polyethylene grease to the valve stem tips. Install the rocker arms and posts in their original positions.

8. Turn the engine over by hand until the valve timing mark is at the 11 o'clock position, as viewed from the front of the engine. Install all of the rocker arm post attaching bolts and tighten to 20 ft. lbs. (27 Nm).

9. Install new valve cover gaskets and install the valve cover.

10. Install the battery cables, start the engine and check for leaks.

Turbocharger

REMOVAL & INSTALLATION

6.0L Engine

1. Before servicing the vehicle, refer to the precautions in the beginning of this section.

Remove the air cleaner assembly.

✳✳ CAUTION

Never remove the pressure relief cap while the engine is operating or when the cooling system is hot. Failure to follow these instructions can result in damage to the cooling system or engine or result in personal injury. To avoid having scalding hot coolant or steam blow out of the degas bottle when removing the pressure relief cap, wait until the engine has cooled, then wrap a thick cloth around the pressure relief cap and turn it slowly. Step back while the pressure is released from the cooling system. When certain all the pressure has been released, (still with a cloth) turn and remove the pressure relief cap. Failure to follow these instructions can result in personal injury.

✳✳ CAUTION

The coolant must be recovered in a suitable, clean container for reuse. If the coolant is contaminated, it must be recycled or disposed of correctly and the system filled with new coolant.

✳✳ CAUTION

Always fill the cooling system with the same type of coolant that was drained from the system. Do not mix coolant types. Do not add orange-colored Motorcraft Specialty Orange Engine Coolant VC-2, or equivalent meeting WSS-M97B44-D. Mixing coolants may degrade the coolant's corrosion protection.

Do not add alcohol, methanol, brine, or any engine coolants mixed with alcohol or methanol antifreeze. These can cause engine damage from overheating or freezing.

2. Disconnect and plug the engine vent and radiator vent hoses.

3. Remove the bolts and position the degas bottle aside.

4. Loosen the clamp at the turbocharger.

5. Remove the bolts and remove the turbocharger intake tube.

6. Disconnect the charge air cooler inlet pipe.

7. Remove the push pins.

8. Disconnect the two wiring harness push pins and position out of the way.

9. Disconnect the turbocharger variable hydraulic control valve electrical connector.

10. Remove the bolts for the oil supply tube.

11. Remove and discard the gasket.

12. Remove the bolt and the wire retainer.

13. Using the special tool 303-755, remove the oil feed tube.

14. Remove the marmon clamp from the turbocharger outlet.

15. Remove the marmon clamp from the turbocharger inlet.

16. Remove the rear turbocharger mounting bolt.

17. Remove the front mounting bolts.

18. Position the turbocharger and remove the turbocharger drain tube.

19. Remove and discard the drain tube O-rings.

20. Remove the turbocharger.

To install:

21. Position the turbocharger on the turbocharger pedestal.

➡Install new O-rings and apply clean engine oil.

22. Position the turbocharger and install the turbocharger drain tube.

23. Install the turbocharger and the rear mounting bolt and tighten to 28 ft. lbs. (38 Nm).

24. Install the turbocharger front mounting bolts and tighten to 28 ft. lbs. (38 Nm).

25. Install the turbocharger inlet marmon clamp and tighten to 9 ft. lbs. (12 Nm).

26. Install the turbocharger exhaust marmon clamp and tighten to 89 inch lbs. (10 Nm).

27. Install the oil feed tube.

28. Position the wire retainer and install the bolt.

29. Pre-lubricate the oil inlet hole of the turbocharger assembly with clean engine oil and spin the compressor wheel several times to coat the bearing with oil.

30. Position the oil feed tube and install the bolts and tighten to 18 ft. lbs. (25 Nm).

31. Install a new gasket.

32. Connect the turbocharger variable vane hydraulic control valve electrical connector.

33. Position the wiring harness and connect the push pins.

34. Install the push pins.

35. Connect the charge air cooler inlet pipe and tighten to 9 ft. lbs. (12 Nm).

36. Install the turbocharger intake tube.

7.3L Engine

1. Before servicing the vehicle, refer to the precautions section.

2. Remove or disconnect the following:
   • Negative battery cable
   • 2 air intake tube assembly bolts
   • Clamps at the turbocharger

- Crankcase breather assembly
- Engine air cleaner, air intake tube and hoses
- Exhaust outlet clamp from the turbocharger
- Engine charge exhaust pipe bolt from the transmission, if so equipped
- Bolts and nuts from the catalytic converter-to-engine charge exhaust pipe, if so equipped
- 2 bolts retaining the turbocharger exhaust inlet pipe to the left exhaust manifold, loosen only
- Bolts retaining the left turbocharger exhaust inlet pipe to the turbocharger exhaust inlet adapter, on automatic transmissions
- 2 bolts retaining the turbocharger exhaust inlet pipe to the right exhaust manifold, loosen only
- Lower bolt retaining the right turbocharger exhaust inlet pipe to the turbocharger exhaust inlet adapter
- Upper bolts retaining the right and left turbocharger exhaust inlet pipes to the turbocharger exhaust inlet adapter, on automatic transmissions
- Right engine lift hook and bolt
- Air inlet hose clamp at the turbocharger, hose and lay aside
- Compressor manifold
- 4 bolts retaining the turbocharger pedestal assembly to the cylinder block
- Turbocharger assembly and all electrical connectors from it

➡ If the turbocharger is not being removed for service, install the

Fuel/Oil Turbo Protector Cap Set T94T-9395-AH.

- Oil gallery O-rings

To install:

3. Install or connect the following:
- Oil gallery O-rings
- Turbocharger electrical connectors and the turbocharger assembly
- 4 bolts retaining the turbocharger pedestal assembly to the engine block. Tighten the bolts to 18 ft. lbs. (25 Nm).
- 4 bolts retaining the right and left turbocharger exhaust inlet pipes to the turbocharger exhaust inlet adapter.
- Compressor manifold, intake manifold hoses and clamps. Be sure the compressor outlet seal is in position.
- Right engine lift hook and bolt
- 2 right and left lower bolts (retaining the turbocharger exhaust inlet pipes to the turbocharger exhaust inlet adapter) to 36 ft. lbs. (49 Nm)
- 4 right and left bolts and nuts (retaining the turbocharger exhaust inlet pipes to the exhaust manifolds) to 36 ft. lbs. (49 Nm)
- Catalytic converter to the engine charge exhaust pipe bolts and nuts
- 2 right and left upper bolts (retaining the turbocharger exhaust inlet pipes to the turbocharger exhaust inlet adapter) to 36 ft. lbs. (49 Nm)
- Exhaust outlet clamp to the turbocharger
- Air intake tube and hose assembly
- Negative battery cable

COMPRESSOR MANIFOLD

7924FG59

After loosening the clamps, the compressor manifold can be removed—diesel engines

Intake Manifold

REMOVAL & INSTALLATION

6.0L Engine

1. Before servicing the vehicle, refer to the precautions in the beginning of this section.
2. Remove the auxiliary battery.
3. Remove the air cleaner assembly.
4. Remove the cooling fan-blade, clutch and shroud.
5. Remove the coolant reservoir.
6. Remove the upper radiator hose.
7. Remove the turbocharger-to-charge air cooler tube.
8. Remove the charge air cooler-to-engine tube.
9. on vehicles with dual alternators, perform the following steps:
   a. Remove the accessory drive belt.
   b. Remove the bolt and the accessory drive belt tensioner.
   c. Remove the accessory drive belt.
   d. Remove the bolts, bracket and accessory drive belt idler pulley.
   e. Remove the bolts and the accessory drive belt tensioner.
   f. Disconnect the wire retainer, generator electrical connector and B+ wire.
   g. Remove the bolts and the generator with mounting bracket.
10. On vehicles with single alternator, perform the following steps:
   a. Remove the accessory drive belt.
   b. Disconnect the generator B+ wire and electrical connector.
   c. Remove the three bolts and the alternator.
   d. Remove the turbocharger and the turbocharger pedestal.
11. Remove the bolts and position the heater hose tube aside.
12. Remove and discard the O-ring.
13. Disconnect the Manifold Absolute Pressure (MAP) sensor hose.
14. Disconnect the engine coolant vent hose.
15. Remove the fuel injector control module.
16. Disconnect the wiring retainer, injector pressure regulator and Injector Control Pressure (ICP) sensor electrical connector.
17. Disconnect the Exhaust Gas Recirculation (EGR) valve electrical connector.
18. Disconnect the oil pressure sensor electrical connector.
19. Disconnect the oil temperature sensor electrical connector.

20. Disconnect the Throttle Position (TP) control module electrical connector.

21. Disconnect the TP sensor electrical connector.

22. Disconnect the pin-type retainer and water temperature sensor.

23. Disconnect the Intake Air Temperature (IAT) sensor electrical connector and wiring connector.

24. Disconnect the exhaust backpressure sensor and retaining clip.

25. Disconnect the eight fuel injectors. Remove the nut and the fuel injector wiring harness.

26. Disconnect the fuel line fittings.

➡**It is necessary to remove the filter and drain the housing.**

27. Remove the four bolts and the oil filter housing.

28. Remove the bolt and the oil filter return tube.

29. Remove the fuel line.

30. Remove the bolt, and the banjo fitting from the fuel line.

31. Discard the sealing washers and remove the fuel line.

32. Remove the nuts and the turbocharger heat shield.

➡**Align the flat edge with the index feature located on the coolant supply port.**

33. Pull the EGR cooler clamp forward, twist and then slide the EGR cooler hose rearward to remove.

34. Remove the EGR cooler V-clamp and gasket.

35. Remove the bolts and the intake manifold.

36. Remove the intake manifold gaskets.

37. Clean and inspect the gaskets. Install new gaskets if necessary.

38. Clean and inspect the sealing surfaces.

To install:

➡**Locating tabs on the gaskets must be up and toward the center of the engine, or a leak will occur.**

39. Install the intake manifold gaskets.

40. Install the intake manifold and bolts and tighten in the following sequence:

a. Loosely install the bolts 1-8.

b. Tighten bolts 9-16 to 8 ft. lbs. (11 Nm).

Tighten all bolts to 11 Nm (8 ft. lbs.) in the sequence shown.

41. Install the gasket and the EGR cooler V-clamp and tighten to 53 inch lbs. (8 Nm).

42. Slide the EGR cooler hose forward and rotate flat to lock.

43. Install the turbocharger heat shield and the nuts and tighten to 8 ft. lbs. (11 Nm).

44. Install the fuel line.

45. Install new sealing washers.

46. Install the fuel line.

47. Install the banjo fitting and the bolt. Tighten the upper bolt to 10 ft. lbs. (13 Nm) and the lower bolt to 28 ft. lbs. (38 Nm).

48. Install the oil filter return tube and bolt as follows:

a. On new oil filter return tubes, tighten to 53 inch lbs. (6 Nm).

b. On used oil filter return tubes, tighten to 27 inch lbs. (3 Nm).

49. Install the oil filter and the four bolts and tighten to 11 ft. lbs. (15 Nm).

50. Connect the fuel line fittings. Tighten the upper fitting to 19 ff. lbs. (26 Nm) and the lower fitting to 32 ft. lbs. (43 Nm).

51. Position the fuel injector harness and install the retaining nut and tighten to 8 ft. lbs. (11 Nm). Connect the eight fuel injectors electrical connectors.

52. Connect the retaining clip and exhaust backpressure sensor electrical connector.

53. Connect the wiring retainer and the IAT sensor electrical connector.

54. Connect the water temperature sensor and the pin-type retainer.

55. Connect the TP sensor electrical connector.

56. Connect the TP control module electrical connector.

57. Connect the oil temperature sensor electrical connector.

58. Connect the oil pressure sensor electrical connector.

59. Connect the EGR valve electrical connector.

60. Connect the wiring retainer, injector pressure regulator and ICP sensor electrical connectors.

61. Install the fuel injector control module.

62. Connect the engine coolant vent hose and clamp.

63. Connect the MAP sensor hose.

➡**Install a new O-ring on the heater hose tube.**

64. Install the heater hose tube and bolts and tighten to 10 ft. lbs. (13 Nm).

65. Install the turbocharger pedestal and turbocharger.

66. On models with single alternators perform the following:

Intake manifold torque sequence—6.0L engine

67197-NAVI-G19

Connect the fuel line fittings—6.0L engine

67197-NAVI-G20

a. Install the alternator and the three bolts. Tighten the bolts to 35 ft. lbs. (47 Nm).

b. Connect the alternator B+ wire land electrical connector and position the boot back.

c. Install the accessory drive belt.

67. On vehicles with dual alternators:

a. Install the alternator with mounting bracket and bolts. Tighten the bolts to 35 ft. lbs. (47 Nm).

b. Connect the B+ wire, alternator electrical connector and wire retainer.

c. Install the accessory drive belt tensioner and bolts. Tighten the bolts to 18 ft. lbs. (25 Nm).

d. Position the accessory drive belt idler pulley. Install the bracket and bolts and tighten to 18 ft. lbs. (25 Nm).

e. Install the accessory drive belt tensioner and bolt and tighten to 18 ft. lbs. (25 Nm).

f. Install the accessory drive belt.

68. Install the charge air cooler-to-engine tube and tighten the retainers to 9 ft. lbs. (12 Nm).

69. Install the turbocharger-to-charge air cooler tube and tighten the retainers to 9 ft. lbs. (12 Nm).

70. Install the upper radiator hose.

71. Install the coolant reservoir.

72. Install the auxiliary battery.

73. Install the air cleaner assembly.

74. Install the cooling fan blade, clutch and shroud.

7.3L Engine

❊❊ CAUTION

The fuel system remains under pressure, even after the engine has been turned OFF. The fuel system pressure must be relieved before disconnecting any fuel lines. Failure to do so may result in fire and/or personal injury.

1. Before servicing the vehicle, refer to the precautions section.
2. Remove or disconnect the following:
   - Both battery ground cables
   - Air cleaner and install clean rags into the air intake of the intake manifold. It is important that no dirt or foreign objects get into the intake.
   - Fuel system pressure
   - Injection pump
   - Fuel return hose from No. 7 and No. 8 rear nozzles
   - Return hose to the fuel tank

STEP 1. TIGHTEN BOLTS TO 24 FT·LB IN NUMBERED SEQUENCE SHOWN ABOVE.
STEP 2. TIGHTEN BOLTS TO 24 FT·LB IN LINE SEQUENCE SHOWN ABOVE.

7924FG25

Intake manifold bolt torque sequence—diesel engines

- Engine wiring harness from the engine

➡**The engine harness ground cables must be removed from the back of the left cylinder head.**

- Bolts attaching the intake manifold to the cylinder heads and the manifold
- Crankcase Depression Regulator (CDR) valve tube grommet from the valley pan
- Bolts attaching the valley pan strap to the front of the engine block and the strap
- Valley pan drain plug and the valley pan

To install:

3. Apply a ⅛ inch (3mm) bead of RTV sealer to each end of the cylinder block.

➡**The RTV sealer should be applied immediately prior to the valley pan installation.**

4. Install or connect the following:
   - Valley pan drain plug, CDR valve tube and new grommet into the valley pan
   - New O-ring and new back-up ring on the CDR valve
   - Valley pan strap on the front of the valley pan
   - Intake manifold and tighten the bolts to 24 ft. lbs. (33 Nm) using the tightening sequence
   - Engine wiring harness and the engine ground wire located to the rear of the left cylinder head
   - Injection pump
   - No. 7 and No. 8 fuel return hoses and the fuel tank return hose
   - Air cleaner
   - Battery ground cables to both batteries

➡**If necessary, purge the nozzle high pressure lines of air by loosening the**

connector ½ to 1 turn and cranking the engine until solid stream of fuel, devoid of any bubbles, flows from the connection.

5. Run the engine and check for oil and fuel leaks.

❊❊ CAUTION

Keep eyes and hands away from the nozzle spray. Fuel spraying from the nozzle under high pressure can penetrate the skin.

6. Check and adjust the injection pump timing.

Exhaust Manifold

REMOVAL & INSTALLATION

6.0L Engine

LEFT SIDE

1. Before servicing the vehicle, refer to the precautions in the beginning of this section.
2. If equipped, remove the exhaust manifold to Exhaust Gas Recirculation (EGR) valve tube.
3. On 4x4 vehicles, remove the front wheel opening moulding.
4. Remove the front fender splash shield.
5. Remove the pipe-to-manifold nuts.
6. Remove the nuts and the exhaust manifold.
7. Remove the exhaust manifold gasket.
8. Clean and inspect the exhaust manifold.

To install:

9. Installation is the reverse of removal. Tighten the manifold retainers in the sequence illustrated to 18 ft. lbs. (25 Nm). If equipped, install the exhaust manifold to

25 Nm (18 lb-ft)

67197-NAVI-G23

Tighten the left side exhaust manifold retainers in the sequence illustrated—6.0L engine

40-60 Nm (30-44 lb-ft)

40-60 Nm (30-44 lb-ft)

67197-NAVI-G24

If equipped, install the exhaust manifold to EGR valve tube and tighten the fasteners as specified in this illustration—6.0L engine

23-27 Nm (17-20 lb/ft)

67197-NAVI-G25

Tighten the right side exhaust manifold retainers in the sequence illustrated—6.0L engine

EGR valve tube and tighten the fasteners as shown in the illustration. Tighten the pipe-to-manifold bolts to 27–24 ft. lbs. (34–46 Nm).

RIGHT SIDE

1. Before servicing the vehicle, refer to the precautions in the beginning of this section.
2. Disconnect the battery ground cable.
3. Remove the pipe-to-manifold nuts.
4. Remove the right front wheel.
5. Remove the right front inner fender well.
6. Remove the nuts, the exhaust manifold and the gasket. Discard the gasket.
7. Clean and inspect the exhaust manifold.

To install:

8. Installation is the reverse of removal. Tighten the manifold retainers in the sequence illustrated to 18 ft. lbs. (25 Nm). Tighten the pipe-to-manifold bolts to 27–24 ft. lbs. (34–46 Nm).

7.3L Engine

1. Before servicing the vehicle, refer to the precautions section.
2. Disconnect the ground cables from both batteries.
3. Raise the vehicle and safely support it.
4. Disconnect the muffler inlet pipe from the exhaust manifolds.
5. If removing the right manifold, lower the vehicle. When removing the left manifold, raise the vehicle and remove the manifold from underneath. Bend the tabs on the manifold attaching bolts, then remove the bolts and manifold.

To install:

6. Before installing, clean all mounting surfaces on the cylinder heads and the manifold. Apply an anti-seize compound on the manifold both threads and install the left

LINE SEQUENCE START HERE (STEP NO 2)

STEP1. TIGHTEN BOLTS TO 35 FT.LB. IN NUMBERED SEQUENCE SHOWN ABOVE
STEP2. TIGHTEN BOLTS TO 35 FT.LB. IN LINE SEQUENCE SHOWN ABOVE

7924FG26

Tighten the manifold bolts in the proper sequence—diesel engines

manifold, using a new gasket and new locking tabs.

7. Tighten the bolts to 45 ft. lbs. (61 Nm) and bend the tabs over the flats on the bolt heads to prevent the bolts from loosening.

8. Raise the vehicle to install the right manifold. Install the right manifold.

9. Connect the inlet pipes to the manifold and tighten. Lower the vehicle, connect the batteries and run the engine to check for exhaust leaks.

Camshaft and Valve Lifters

REMOVAL & INSTALLATION

6.0L Engine

1. Before servicing the vehicle, refer to the precautions section.

2. Mount the engine on an engine stand.

3. Remove the serpentine belt idler.

4. Remove the serpentine belt tensioner.

5. Remove and discard the O-rings from the oil filter base.

6. Remove the Exhaust Gas Recirculation (EGR) cooler coolant supply port cover. Clean and inspect the gaskets. Install new gaskets if necessary. Clean and inspect the sealing surfaces.

> ❄ **WARNING**
>
> **In the event of a catastrophic engine failure, always install a new oil cooler cover assembly (with oil cooler). Foreign material cannot be removed from the oil cooler.**

➡ **The oil cooler is replaced as an assembly.**

7. Remove the oil cooler assembly. Clean and inspect the gaskets. Install a new gasket if necessary. Clean and inspect the sealing surfaces.

8. Remove the oil pump inlet strainer. Clean and inspect for tears and other damage.

9. Remove bolts and the turbocharger heat shield.

10. Remove the high-pressure oil pump cover. Use a thin gasket scraper to separate the cover from the crankcase. Clean and inspect the gaskets. Install a new gasket if necessary. Clean and inspect the sealing surfaces.

11. Remove the bolts from the high-pressure oil pump discharge pipe.

12. Disconnect and remove the high-pressure oil pump discharge pipe.

13. Remove and discard the D-ring seal.

14. Remove and discard the high-pressure pump O-ring seal.

15. Remove the bolts and the high-pressure oil pump.

16. Remove and discard the lower O-ring seal.

17. Remove the glow plug buss bar.

18. Remove the eight glow plugs.

➡ **Mark the location of the stud bolts.**

19. Remove the valve covers. Clean and inspect the gaskets. Install a new gasket if necessary. Clean and inspect the sealing surfaces.

20. Remove the bolts and the coolant pump pulley.

21. Remove the coolant pump.

22. If equipped, remove the bolts and the dual generator pulley.

23. Prior to removing the crankshaft damper, check the crankshaft vibration damper runout.

➡ **Pry the crankshaft forward at the same point to eliminate possible error caused by crankshaft end play.**

24. Rotate the crankshaft 90 degrees. Pry the crankshaft forward. Record the measurement. Repeat every 90 degrees. If the runout exceeds specification, install a new crankshaft vibration damper.

> ❄ **WARNING**
>
> **To prevent engine damage, you must always replace all four bolts when installing the vibration damper.**

> ❄ **CAUTION**
>
> **To avoid personal injury, support the vibration damper during mounting bolt removal. The damper can slide off the nose of the crankshaft very easily.**

25. Remove the bolts and the crankcase vibration damper. Discard the bolts.

26. Punch two holes in the seal. Remove the crankshaft seal with a slide hammer.

➡ **Production engine will not have a wear sleeve. If equipped, remove the crankshaft damper wear sleeve.**

27. Remove the oil pump body. Remove and discard the O-ring seal.

➡ **Mark the front of each drive rotor for correct reassembly orientation.**

28. Remove the inner and outer oil pump drive rotors.

29. Remove the engine front cover. Clean and inspect the gaskets. Install new gaskets if necessary. Clean and inspect the sealing surfaces.

30. Using a quick-disconnect tool, disconnect the high-pressure oil rail supply line at the high-pressure oil rail.

31. Remove the bolts and the high pressure oil rail. Disconnect and remove the high-pressure oil supply line.

> ❄❄ **WARNING**
>
> **Do not attempt to put battery voltage to the fuel injector or damage to the fuel injector will occur.**

32. Using a 19 mm socket, push the fuel injector electrical connector out of the rocker arm carrier.

> ❄❄ **WARNING**
>
> **To prevent engine damage, do not use air tools to remove the fuel injectors. The clip that extracts the injector can dislodge and fall into the oil drain hole.**

➡ **If engine oil is found in the engine coolant or engine coolant is found in the combustion chambers, new injector sleeve may need to be installed.**

33. Remove the bolt, the fuel injector hold down and the fuel injector.

34. Remove and discard the crankcase-to-head tube assembly.

35. Remove the inner head bolts from both cylinder heads.

36. Remove the 16 bolts and the rocker arm assemblies.

37. Remove the rocker arm carrier from the cylinder head. Clean and inspect the gaskets. Install new gaskets if necessary. Clean and inspect the sealing surfaces.

➡ **Mark the location of the valve bridges before removing.**

38. Remove the 16 valve bridges.

> ❄❄ **WARNING**
>
> **To prevent engine damage, keep the push rods in the order in which they were removed. Install all push rods back in their original positions.**

39. Mark the location and remove the 16 push rods.

40. Remove the 10 outer head bolts.

41. Remove the cylinder heads.

42. Remove and discard the cylinder head gasket.

43. Remove and discard the four cylinder head dowel sleeves.

44. Remove the bolts from the rear engine tube assembly.

45. Remove the bolt and the rear engine tube assembly.

❊ WARNING

To prevent engine damage, keep the cam followers in the order in which they were removed. Install all cam followers back in their original positions.

46. Remove the bolts and the roller follower guides. Remove the hydraulic cam followers.

47. Install the special tool and measure the camshaft gear backlash. Install a new camshaft gear if backlash is not within specification.

48. Install the special tool and measure the camshaft end play. Install a new camshaft thrust plate if end play is not within specification.

49. Remove the bolt and the camshaft position (CMP) sensor.

❊ WARNING

Do not knick or scratch the camshaft bearings with the camshaft lobes or engine damage will occur.

50. Remove the thrust plate mounting bolts and remove the camshaft and gear.

To install:

➡**Check alignment of the oil holes after installing the bearings.**

51. If removed, install the camshaft bearings.

❊❊ WARNING

Do not nick or scratch the camshaft bearings with the camshaft lobes or engine damage can occur.

➡**Apply clean engine oil to the camshaft prior to installing.**

52. Using camshaft alignment tool 303-772, install the camshaft and gear assembly. Aligning it with the crankshaft. Install the thrust plate mounting bolts.

53. The remainder of installation is the reverse of removal.

7.3L Engine

➡**Ford recommends removing the diesel engine from the vehicle for camshaft removal.**

1. Before servicing the vehicle, refer to the precautions section.

2. Remove or disconnect the following:
- Intake manifold and valley pan, if equipped
- Rocker covers
- Either the rocker arm shafts or loosen the rockers on their pivots and remove the pushrods. The pushrods must be reinstalled in their original positions.
- Valve lifters in sequence with a magnet. They must be replaced in their original positions.
- Timing gear cover, timing gear and sprockets

➡**A camshaft removal tool, Ford part No. T65L-6250-A and Adapter 14-0314 is needed to remove the diesel camshaft.**

- Camshafts

To install:

3. Liberally coat the camshaft with oil before installing it. Slide the camshaft into the engine very carefully so as not to scratch the bearing bores with the camshaft lobes. Install the camshaft thrust plate and tighten the attaching screws to 10–12 ft. lbs. (13–16 Nm). Measure the camshaft end-play. If the end-play is more than 0.009 inch (0.228mm), replace the thrust plate. Assemble the remaining components in the reverse order of removal.

4. Install or connect the following:
- Timing gear and front cover
- Valve lifters. They must be replaced in their original positions.
- Pushrods, the rocker arms and the rocker arm covers
- Intake manifold and valley pan, if equipped

Valve Lash

ADJUSTMENT

Valve lash on these engines is not adjustable.

Starter Motor

REMOVAL & INSTALLATION

6.0L Engine

1. Before servicing the vehicle, refer to the precautions section.

2. Disconnect the battery ground cable.

3. Remove starter solenoid protective cap.

4. Disconnect the starter motor electrical connections.

5. Remove the bolts and the starter.

To install:

6. Installation is the reverse of the removal procedure. Tighten the starter bolts to 18 ft. lbs. (25 Nm)

7.3L Engine

1. Before servicing the vehicle, refer to the precautions section.

2. Remove or disconnect the following:
- Negative battery cable

3. Raise the front of the truck and install jackstands beneath the frame. Firmly apply the parking brake and place blocks in back of the rear wheels.

4. Remove or disconnect the following:
- Wiring from the starter motor terminals
- Starter motor retaining bolts, loosen
- Starter retaining bolts while supporting the starter motor
- Starter from the vehicle

To install:

5. The installation is the reverse of removal. Tighten the starter retaining bolts to 15–20 ft. lbs. (20–27 Nm)

Oil Pan

REMOVAL & INSTALLATION

6.0L Engine

1. Before servicing the vehicle, refer to the precautions section.

2. With the vehicle in NEUTRAL, position it on a hoist.

3. Disconnect the negative battery cable.

❊❊ WARNING

Never remove the pressure relief cap while the engine is operating or when the cooling system is hot. Failure to follow these instructions can result in damage to the cooling system or engine or result in personal injury. To avoid having scalding hot coolant or steam blow out of the degas bottle when removing the pressure relief cap, wait until the engine has cooled then wrap a thick cloth around the pressure relief cap and turn it slowly. Step back while the pressure is released from the cooling system. When certain all the pressure has been released, (still with a cloth) turn and remove the pressure relief cap. Failure to follow these instructions can result in personal injury.

⁕⁕ CAUTION

The coolant must be removed in a suitable, clean container for reuse. If the coolant is contaminated, it must be recycled or disposed of correctly and the system filled with new coolant.

➡ **Less than 80% of coolant capacity can be recovered with the engine in the vehicle. Dirty, rusty, or contaminated coolant requires replacement.**

4. Place a suitable container below the radiator draincock. If equipped, disconnect the coolant return hose at the fluid cooler.

5. Remove the fill cap from the degas bottle.

6. Remove the oil pan drain plug and drain the engine oil.

7. Loosen the exhaust pipe retaining nuts.

8. Remove the motor mount retaining nuts.

9. Open the radiator draincock.

10. Disconnect the lower radiator hose.

11. Disconnect the transmission cooler tubes.

12. Remove the air cleaner assembly.

13. On vehicles with metal ducts, remove the charge air cooler duct.

➡ **Charge air cooler side shown, engine side similar.**

14. On vehicles with blow molded ducts, loosen the clamps and remove the charge air cooler duct.

15. Remove the radiator support brackets.

16. Remove the 4 pin-type retainers and pull back the sight shield.

17. With the sight shield pulled back, remove the 3 pin-type retainers and the 2 wiring retainers and position the harness rearward out of the way.

18. Disconnect the upper radiator hose and the radiator overflow hose.

19. Remove the radiator and shroud as an assembly.

20. Disconnect the cooling fan electrical connector. Unclip and position the fan and wiring aside.

➡ **Use a hole in the fan hub to prevent the fan from turning.**

21. Using the special tool 303-591, loosen the fan clutch. Remove the cooling fan.

22. Remove the bolts from the stator, remove the stator.

23. On vehicles with dual alternator perform the following:

a. Remove the secondary accessory drive belt.

b. Remove the bolt and the secondary accessory drive belt tensioner.

c. Remove the primary accessory drive belt.

d. Remove the bolts, bracket and secondary accessory drive belt idler pulley.

e. Remove the bolts and the primary accessory drive belt tensioner.

f. Disconnect the wire retainer, secondary generator electrical connector and B+ wire.

g. Remove the bolts and the secondary generator with mounting bracket.

24. On vehicles with single alternator, remove the primary accessory drive belt.

25. Remove the bolts and position the alternator back.

26. Support the hood and disconnect the hood lift assemblies.

27. Install two lifting eyes 303-D030 on the right hand cylinder head.

28. Remove the retaining bolt for the fuel lines.

29. Disconnect the heater hose at the coolant pump.

30. Remove the fan shroud mounting stud.

31. Install one lifting eye 303-D030 on the left hand cylinder head.

⁕⁕ CAUTION

Do not use the special tool to raise the engine. Damage to the special tool or vehicle may occur.

➡ **The ball studs may have to be removed.**

➡ **This procedure requires a second bolt hook assembly. The tools are available through Rotunda tools with the following numbers: bolt hook 303-F070-6, handle 303-F070-8, bracket 303-F070-7 and washer 303-F070-12004.**

32. Install the special tool.

➡ **The engine must be raised evenly.**

33. Using a lifting crane, raise the engine until the turbo charger is about to touch the cowl. Secure the engine with the special tool.

34. Remove the transmission cooler line bracket.

35. Remove the bolts and position back the oil pan until the oil pick-up tube bolts are accessible.

36. Remove the bolts and let the oil pick-up tube go into the oil pan. Remove the oil pan.

37. Remove the press-in-place gasket and discard.

38. Clean and inspect the sealing surfaces.

39. Remove and discard the oil pick-up tube O-ring.

To install:

40. Install a new O-ring on the oil pick-up tube and position the oil pick-up tube in the oil pan.

41. Install a new press-in-place gasket into the upper oil pan.

42. Position the oil pan in the vehicle.

43. Install the oil pick-up tube and bolts. Tighten to 10 ft. lbs. (13 Nm).

44. Position the oil pan and install the bolts. Tighten to 10 ft. lbs. (13 Nm).

45. Install the transmission cooler tube bracket and nut. Tighten to 18 inch lbs. (10 Nm).

46. Using the special tool 303-F070, lower the engine.

47. Remove the lifting eyes 303-D030 and the special tool.

48. Install the fan shroud mounting stud. Tighten to 30 ft. lbs. (40 Nm).

49. Connect the heater hose at the coolant pump.

50. Install the retaining bolt for the fuel lines. Tighten to 10 ft. lbs. (10 Nm).

51. If removed, install the ball studs.

52. Connect the hood lifts and remove the hood support.

53. Position the primary alternator and install the bolts.

54. On a vehicle with a single alternator, install the primary accessory drive belt.

55. On vehicles with dual alternators, perform the following:

a. Install the secondary alternator with mounting bracket and bolts. Tighten to 35 ft. lbs. (47 Nm).

b. Connect the B+ wire, secondary generator electrical connector and wire retainer.

c. Install the primary accessory drive belt tensioner and bolts. Tighten to 18 ft. lbs. (25 Nm).

d. Position the secondary accessory drive belt idler pulley. Install the bracket and bolts. Tighten to 18 ft. lbs. (25 Nm).

e. Install the primary accessory drive belt.

f. Install the secondary accessory drive belt tensioner and bolt. Tighten to 18 ft. lbs. (25 Nm).

g. Install the secondary accessory drive belt.

56. Install the stator and stator bolts. Tighten to 30 ft. lbs. (40 Nm).

57. Install the cooling fan clutch. Use the

67197-NAVI-G26

Connect the transmission cooler tubes. Tighten to the specifications illustrated— 6.0L engine

special tool 303-591 to tighten the cooling fan clutch. Tighten to 98 ft. lbs. (133 Nm).

58. Position and clip the cooling fan wiring. Connect the cooling fan electrical connector.

59. Install the radiator and shroud as an assembly.

60. Connect the upper radiator hose and the radiator overflow hose.

61. Position the harness wiring, install the 2 wiring retainers and 3 pin-type retainers.

62. Position the sight shield and install the 4 pin-type retainers.

63. Install the radiator support brackets. Tighten to 22 ft. lbs. (30 Nm).

64. On vehicles with metal ducts, install the charge air cooler duct and tighten the clamps.

65. On vehicles with blow molded ducts, install the charge air cooler duct and tighten the clamps.

66. Install the air cleaner assembly.

67. Connect the transmission cooler tubes. Tighten to the specifications illustrated.

68. Close the radiator draincock. Connect the lower radiator hose.

69. Install the motor mount retaining nuts. Tighten to 76 ft. lbs. (103 Nm).

70. Tighten the exhaust pipe retaining nuts to 35 ft. lbs. (47 Nm).

71. Clean and inspect the oil pan drain plug and gasket, install new if necessary.

72. Install the oil pan drain plug and tighten to 18 ft. lbs. (25 Nm).

73. Connect the negative battery cables.

74. Fill the engine with clean engine oil.

75. Fill the coolant system.

76. Run the engine and check for leaks

7.3L Engine

1. Before servicing the vehicle, refer to the precautions section.

2. Remove the engine.

3. Remove the oil pan bolts.

4. Lower the oil pan.

➡The oil pan is sealed to the crankcase with RTV silicone sealant in place of a gasket. It may be necessary to separate the pan from the crankcase with a utility knife. Also, the crankshaft may have to be turned to allow the pan to clear the crankshaft throws.

5. Clean the pan and crankcase mating surfaces thoroughly.

To install:

6. Apply a ⅛ in. (3mm) bead of RTV silicone sealant to the pan mating surfaces, and a ¼ in. (6mm) bead on the front and rear covers and in the corners; you have 15 minutes within which to install the pan!

7. Install the locating dowels into position.

8. Position the pan on the engine and install the pan bolts loosely.

9. Remove the dowels.

10. Tighten the pan bolts to:
   - ¼ in.-20 bolts: 84 inch lbs. (10 Nm)
   - ⁵⁄₁₆ in.-18 bolts: 14 ft. lbs. (19 Nm)
   - ⅜ in.-16 bolts: 24 ft. lbs. (33 Nm)

11. Install the engine.

Oil Pump

REMOVAL & INSTALLATION

6.0L Engine

1. Before servicing the vehicle, refer to the precautions in the beginning of this section.

2. With the vehicle in NEUTRAL, position it on a hoist.

3. Disconnect the battery ground cable(s).

4. Remove the cooling fan

5. On vehicles with dual alternator, perform the following:

   a. Remove the dual generator accessory drive belt.

   b. Remove the bolts and the dual generator pulley.

6. Remove the accessory drive belt.

7. Check the crankshaft vibration damper runout.

8. Remove the paint from the face of the crankshaft vibration damper at four points 90 degrees apart.

Attach dial indicator 100-002 to the cylinder block. Position the tool on one of the unpainted surfaces.

Using a suitable tool, pry the crankshaft forward. Zero the dial indicator.

➡Pry the crankshaft forward only to eliminate possible error caused by crankshaft end play. Rotate the crankshaft 90 degrees. Pry the crankshaft forward. Record the measurement. Repeat at each unpainted surface.

9. If the runout exceeds specification, install a new crankshaft vibration damper.

❋❋ CAUTION

To avoid personal injury, support the vibration damper during mounting bolt removal. The damper can slide off the nose of the crankshaft very easily.

10. Remove the bolts and the crankshaft vibration damper.

11. Discard the bolts.

12. Punch two holes in the seal.

13. Using the special tool 303-D060, remove the crankshaft seal.

➡Production engine will not have a wear sleeve.

14. If equipped, remove the crankshaft damper wear sleeve using tool 303-762.

15. Remove the bolts and the gerotor cover. Remove and discard the O-ring seal.

➡Mark the front of the inner and outer gerotor for correct reassembly.

16. Remove the inner and outer gerotors.

17. Inspect the oil pump components and replace as necessary.

303-D060

67197-NAVI-G27

Using the special tool 303-D060, remove the crankshaft seal—6.0L engine

303-762

67197-NAVI-G28

Remove the crankshaft damper wear sleeve using tool 303-762, if equipped— 6.0L engine

18. Inspect the oil pump for excessive metal particles.

19. Inspect the oil pump for gouging, cracks or deep scratches.

20. Inspect the oil pump inner and outer gear rotors for damage or excessive wear.

To install:

➡**Install the gears with marks pointing outward.**

21. Lubricate the inner gear with lithium assembly grease and install onto the crankshaft. Lubricate the outer gear with lithium assembly grease and mesh with the inner gear rotor in the oil pump housing. Wipe off the excess assembly grease.

22. Install a new O-ring seal.

23. Install the gerotor cover and bolts. Tighten to 71 inch lbs. (8 Nm).

24. Thoroughly clean the crankshaft front seal mounting surface.

25. Apply Threadlock 262® to the outer circumference of the leading edge of the crankshaft.

➡**New seal and wear sleeve must not be separated.**

26. Using the special tool 303-361, install the oil seal and wear sleeve assembly.

❋❋ CAUTION

To prevent engine damage, you must always install four new bolts when installing the vibration damper.

➡**Do not use anti-seize compounds, grease or any lubricants. Lubricants have an adverse effect on the torque results.**

27. Install the crankshaft vibration damper and bolts.

28. Tighten the bolts in the sequence illustrated as follows:

    a. Tighten the bolts to 50 ft. lbs. (68 Nm).

    b. Tighten the bolts an additional 90 degrees.

Use tool 303-361 to install the oil seal and wear sleeve assembly—6.0L engine

Tighten the crankshaft vibration damper bolts in this sequence—6.0L engine

29. Install the accessory drive belt.

30. On vehicles with dual alternator, perform the following:

    a. Install the dual alternator pulley and bolts. Tighten to 35 ft. lbs. (47 Nm).

    b. Install the dual generator accessory drive belt.

31. Install the cooling fan stator

32. Connect the battery ground cables.

7.3L Engine

1. Before servicing the vehicle, refer to the precautions section.

2. Remove or disconnect the following:
- Negative battery cables
- Oil pan
- Oil pick-up tube from the pump
- Bolts and the oil pump

To install:

3. Assemble the pick-up tube and pump. Use a new gasket.

4. Install or connect the following:
- Oil pump and tighten the bolts to 14 ft. lbs. (19 Nm)
- Oil pick up tube
- Oil pan
- Negative battery cables

BOLT

OIL PUMP BODY PLATE

The oil pump is mounted on the cylinder block with 4 bolts—diesel engine

Rear Main Seal

REMOVAL & INSTALLATION

6.0L Engine

1. Before servicing the vehicle, refer to the precautions section.

2. Remove the transmission.

3. Remove the bolts.

4. Remove the flexplate or flywheel.

➡**Use extreme care when removing the flywheel front adapter to prevent damage to the alignment dowel pin.**

5. Remove the flywheel front adapter.

❋❋ CAUTION

To prevent engine damage, do not remove the rear primary crankshaft flange bolts under any circumstances. If the flange is removed and reinstalled, it will result in engine vibration and premature transmission component wear.

6. Punch two holes in the rear main seal, across from each other.

7. Using the puller tool 100-001, remove the rear main seal.

➡**Production engines will not have a wear sleeve.**

8. If equipped with a crankshaft wear sleeve, use the tool 303-771 to remove the crankshaft rear wear sleeve.

9. Clean and inspect the crankshaft sealing surface.

To install:

➡**The crankshaft rear oil seal and wear sleeve are installed as an assembly.**

➡**Lubricate the outer diameter of the rubber seal with a solution of dish soap and water (approximately 50/50 mix) prior to assembly. Do not use any other type of lubricant.**

If equipped with a crankshaft wear sleeve, use tool 303-771 to remove the crankshaft rear wear sleeve—6.0L engine

67197-NAVI-G32

Use tool 303-770 to install the crankshaft rear oil seal—6.0L engine

67197-NAVI-G33

Tighten the flexplate or flywheel bolts in this sequence—6.0L engine

10. Apply a bead of Threadlock 262® around the circumference of the outer rear edge of the secondary crankshaft flange.

11. Using tool 303-770, install the crankshaft rear oil seal.

12. Install the flywheel front adapter.

13. Install the flexplate or flywheel.

14. Install the bolts. Snug all bolts to 44 inch lbs. (5 Nm), then tighten all bolts to 69 ft. lbs. (94 Nm) in the sequence illustrated.

15. Install the transmission.

7.3L Engine

1. Before servicing the vehicle, refer to the precautions section.

2. Remove or disconnect the following:
- Transmission
- Flywheel
- Crankshaft rear oil seal bolts and the seal

3. Clean the seal mating surfaces.

4. If installing the old seal, inspect it for damage.

5. Using crankshaft wear ring removal tool T94T-6701-AH1, forcing screw T84T-7025-B, remover tube T77J-7025-B and wear ring remover sleeve T94T-6701-AH2 (refer to the illustration), remove the wear ring.

To install:

6. Apply RTV silicone sealant to the seal retaining ring and the seal retaining bolts.

Item	Description
1	Crankshaft Wear Ring
2	Crankshaft Rear Wear Ring Remover
3	Forcing Screw
4	Remover Tube
5	Crankshaft Rear Wear Ring Remover Sleeve

7924FG61

Assemble the seal and wear ring removal tools, then remove the wear ring—diesel engine

7. Using seal replacers T94T-6701-AH3 and T94T-AH4, driver sleeve T79T-6316-A4 (part of T79T-6316-A) and guide pins T94P-7000-P install the wear ring and oil seal.

8. Install or connect the following:
- Seal retaining bolts
- Flywheel
- Transmission

Timing Gears, Front Cover and Seal

REMOVAL & INSTALLATION

6.0L Engine

1. Before servicing the vehicle, refer to the precautions section.

2. Remove the intake manifold.

3. Disconnect the exhaust pressure tube at the exhaust.

4. Remove the nuts and the exhaust pressure bracket assembly.

➡ **Remove the thermostat housing only if a new front cover is being installed.**

5. Remove the stud bolts and the thermostat housing.

6. Remove and discard the O-ring.

7. Disconnect the engine coolant fill hose.

8. Disconnect the lower radiator hose.

9. Remove the stator stand-off bolt.

10. Remove the four bolts and position aside the power steering pump.

11. Remove the bolts and the accessory drive belt tensioner.

12. Remove the bolts and the accessory drive idler pulleys.

➡ **Remove the coolant pump pulley only if a new front cover is being installed.**

13. Remove the coolant pump pulley.

➡ **Remove the coolant pump only if a new front cover is being installed.**

14. Remove the bolts and the coolant pump.

15. Remove and discard the O-ring.

16. Remove the nut and the battery cable bracket.

17. Remove the oil pump.

18. Check the crankshaft vibration damper runout.

19. Remove the paint from the face of the crankshaft vibration damper at four points 90 degrees apart.

20. Attach the special tool to the cylinder block. Position the special tool on one of the unpainted surfaces.

21. Using a suitable tool, pry the crankshaft forward. Zero the dial indicator.

➡ **Pry the crankshaft forward at the same point to eliminate possible error caused by crankshaft end play.**

22. Rotate the crankshaft 90 degrees. Pry the crankshaft forward. Record the measurement. Repeat at each unpainted surface. If the runout exceeds specification, install a new crankshaft vibration damper.

✳✳ WARNING

To prevent engine damage, you must always install four new bolts when installing the vibration damper.

✳✳ WARNING

To avoid personal injury, support the vibration damper during mounting bolt removal. The damper can slide off the nose of the crankshaft very easily.

23. Remove the bolts and the crankcase vibration damper.

24. Discard the bolts.

25. Remove the bolts and the front cover.

※※ WARNING

Sealant is used where the crankcase and lower crankcase meet. Failure to cut the sealant could result in pulling the lower crankcase seal out while removing the front cover gasket.

26. Use a thin blade scraper to cut the sealant where the crankcase and the lower crankcase meet. Remove and discard the front cover gasket.

27. Clean and inspect the sealing surfaces.

28. Punch two holes in the seal.

29. Using the special tool, remove the crankshaft seal.

➡**Production engine will not have a wear sleeve.**

30. If equipped, remove the crankshaft damper wear sleeve.

31. Remove the thrust plate mounting bolts and remove the camshaft and gear.

To install:

32. Install the camshaft and gear assembly. Using the special tool, align the camshaft timing mark as shown. Install the thrust plate mounting bolts.

33. Thoroughly clean the crankshaft front seal mounting surface.

34. Apply Threadlock 262® to the outer circumference of the leading edge of the crankshaft.

➡**New seal and wear sleeve must not be separated.**

35. Using the special tool, install the oil seal and wear sleeve assembly.

36. If removed, install the front cover crankcase dowels into the cylinder block.

➡**Use guide studs to aid in installation. Studs must be fabricated locally.**

37. Install the guide studs.

38. Apply a bead of sealant at the seam where the crankcase and the lower crankcase meet.

39. Install a new engine front cover gasket.

40. Install the engine front cover and bolts.

※※ WARNING

To prevent engine damage, you must always install four new bolts when installing the vibration damper.

➡**Do not use anti-seize compounds, grease or any lubricants. Lubricants have an adverse effect on the torque results.**

41. Install the crankshaft vibration damper and bolts.

42. Tighten the bolts in the sequence shown.

  a. Tighten the bolts to 68 Nm (50 ft. lbs.).

  b. Tighten the bolts an additional 90 degrees.

43. Install the oil pump.

44. Install the battery cable bracket and nut.

➡**Install a new O-ring on the coolant pump pulley.**

45. If removed, install the coolant pump and bolts.

46. If removed, install the coolant pump pulley and bolts.

47. Install the accessory drive idler pulleys and bolts.

48. Install the accessory drive belt tensioner and bolts.

49. Position back the power steering pump and install the bolts.

50. Install the stator stand-off bolt.

51. Connect the lower radiator coolant hose.

52. Connect the engine coolant hose.

➡**Install a new O-ring.**

53. If removed, install the thermostat housing and stud bolts.

54. Install the exhaust pressure bracket assembly and retaining nuts.

55. Connect the exhaust pressure tube fitting at the exhaust manifold.

56. Install the intake manifold.

7.3L Engine

➡**The crankshaft gear sprocket is not serviced separately from the crankshaft. Do not try to remove the sprocket or you will damage the crankshaft.**

1. Before servicing the vehicle, refer to the precautions section.

2. Remove or disconnect the following:
  • Negative battery cables
  • Camshaft
  • Sprocket from the camshaft using a press
  • Thrust plate and sprocket key

3. Inspect the camshaft and related parts for wear and damage.

To install:

4. Clean the nose of the camshaft.

5. Install or connect the following:
  • Thrust plate
  • Key in the keyway on the camshaft

6. Heat the sprocket in an oven to 500°F (260°C).

The camshaft gear is removed with the camshaft, then pressed off—diesel engines

Item	Description
1	Camshaft Sprocket
2	Crankshaft Sprocket
3	Crankshaft Sprocket Timing Mark
4	Camshaft Sprocket Timing Mark

7924FG64

Be sure the timing marks are aligned as illustrated—diesel engines

7. Remove the sprocket from the oven, align the sprocket keyway with the camshaft key and install the sprocket on the camshaft until it is fully seated. Allow the camshaft assembly to cool before installation
  • Camshaft in the engine and align the timing marks on the gears
  • Negative battery cables

Piston and Ring Positioning

1. Top compression ring is identified with one indentation mark and a 15 degree keystone profile.
2. The intermediate compression ring is identified with two indentation marks and a square profile.
3. Oil control ring.

67197-NAVI-G34

6.0L Diesel engines—piston ring locations

7.3L Diesel engines—piston-to-engine orientation

GASOLINE FUEL SYSTEM

Fuel System Service Precautions

Safety is the most important factor when performing not only fuel system maintenance, but any type of maintenance. Failure to conduct maintenance and repairs in a safe manner may result in serious personal injury or death. Maintenance and testing of the vehicle's fuel system components can be accomplished safely and effectively by adhering to the following rules and guidelines.

• To avoid the possibility of fire and personal injury, always disconnect the negative battery cable unless the repair or test procedure requires that battery voltage be applied.

• Always relieve the fuel system pressure prior to disconnecting any fuel system component (injector, fuel rail, pressure regulator, etc.), fitting or fuel line connection. Exercise extreme caution whenever relieving fuel system pressure, to avoid exposing skin, face and eyes to fuel spray. Please be advised that fuel under pressure may penetrate the skin or any part of the body that it contacts.

• Always place a shop towel or cloth around the fitting or connection prior to loosening to absorb any excess fuel due to spillage. Ensure that all fuel spillage (should it occur) is quickly removed from engine surfaces. Ensure that all fuel soaked cloths or towels are deposited into a suitable waste container.

• Always keep a dry chemical (Class B) fire extinguisher near the work area.

• Do not allow fuel spray or fuel vapors to come into contact with a spark or open flame.

• Always use a back-up wrench when loosening and tightening fuel line connection fittings. This will prevent unnecessary stress and torsion to fuel line piping. Always follow the proper torque specifications.

• Always replace worn fuel fitting O-rings with new. Do not substitute fuel hose or equivalent where fuel pipe is installed.

Fuel System Pressure

RELIEVING

Except 2005 Expedition And Navigator

➡A fuel pressure gauge is needed to correctly perform this procedure.

✳✳ CAUTION

Fuel injection systems remain under pressure, even after the engine has been turned OFF. The fuel system pressure must be relieved before disconnecting any fuel lines. Failure to do so may result in fire and/or personal injury.

1. Before servicing the vehicle, refer to the precautions section.
2. Disconnect the negative battery cable and remove the fuel filler cap.
3. Remove the cap from the pressure relief valve on the fuel supply manifold. Install a fuel pressure gauge to the pressure relief valve.
4. Direct the gauge drain hose into a suitable container and depress the pressure relief button.
5. Remove the gauge and replace the cap on the pressure relief valve.

➡As an alternate method, disconnect the inertia switch and crank the engine for 15–20 seconds until the pressure is relieved.

2005 Expedition And Navigator

➡The electrical connector is located on the left hand frame rail under the driver side door.

1. Disconnect the electrical connector.
2. Start the engine and allow it to idle until it stalls.
3. After the engine stalls, crank the engine for approximately 5 seconds to make sure the fuel rail pressure has been released.
4. Turn the ignition switch to the OFF position.
5. When fuel system service is complete, connect the electrical connector.

➡It may take more than one key cycle to pressurize the fuel system.

6. Cycle the ignition key and wait three seconds to pressurize the fuel system. Check for leaks before starting the engine.
7. Install the diagnostic tool. Turn the key ON with the engine OFF. Cycle the key OFF, then ON. Select the appropriate vehicle and engine qualifier. Clear all diagnostic trouble codes (DTCs) and carry out a PCM reset.
8. Start the vehicle and check the fuel system for leaks

Fuel Filter

REMOVAL & INSTALLATION

➡A fuel line disconnect tool is needed for this procedure.

7924FG65

**Typical fuel filter mounting along an
under-vehicle frame rail**

✳✳ CAUTION

**Fuel injection systems remain under
pressure, even after the engine has
been turned OFF. The fuel system
pressure must be relieved before disconnecting any fuel lines. Failure to
do so may result in fire and/or personal injury.**

1. Before servicing the vehicle, refer to
the precautions section.
2. Remove or disconnect the following:
   • Negative battery cable
3. Relieve the fuel system pressure.
   • Fuel lines from the fuel filter. Have
   a drain pan handy to catch any
   residual fuel once the lines are separated.
4. On newer models, disconnect the fuel
lines from the filter as follows:
   • Safety clip from the male hose
   • Install and push the fuel line disconnect tool into the female fitting.
5. Remove or disconnect the following:
   • Male and female fittings from the
   filter

➡**Inspect the fuel lines for any damage
after the fuel is finished draining.**

   • Fuel filter from the bracket and the
   retainer, if equipped. Note the
   direction of the flow arrow so the
   replacement filter can be installed
   correctly.

To install:
6. Install or connect the following:
   • Fuel filter into the mounting bracket
   with the flow arrow pointing in the
   correct direction
   • Fuel lines to the fuel filter. Align
   and push the male tube into the
   female fitting until a click is heard.
   Pull on the fitting to ensure that it
   is fully engaged, then install the
   safety clip.
7. Lower the vehicle to the ground.

➡**When the battery has been disconnected and reconnected, some abnor-**

mal drive symptoms may occur while
the Powertrain Control Module (PCM)
relearns its adaptive strategy. The
vehicle may need to be driven 10 miles
(16 km) or more to relearn the strategy.

8. Connect the negative battery cable.

Fuel Pump

REMOVAL & INSTALLATION

Except Excursion

2002–04 MODELS

✳✳ CAUTION

**Fuel injection systems remain under
pressure, even after the engine has
been turned OFF. The fuel system
pressure must be relieved before disconnecting any fuel lines. Failure to
do so may result in fire and/or personal injury.**

1. Before servicing the vehicle, refer to
the precautions section.
2. Remove or disconnect the following:
   • Negative battery cable
   • Fuel pressure
   • Fuel tank skid plate bolts and lower
   the skid plate
   • Fuel
   • Fuel tank filler pipe hose from the
   tank
   • Fuel tank filler pipe vent hose from
   the tank
   • Fuel lines from the fuel pump
   • Front fuel tank connections
   • Rear Evaporative Emissions (EVAP)
   hose clamp and the hose
   • Electrical connector from the fuel
   pump
3. Support the fuel tank with a jack.
   • Fuel tank support strap bolts and
   the fuel tank straps
   • Fuel tank
   • Fuel pump lock ring
   • Fuel pump
To install:
4. Install or connect the following:
   • Fuel pump
   • Fuel pump lock ring and tighten to
   66 ft. lbs. lbs. (89 Nm)
   • Fuel tank
5. Tighten the fuel tank strap bolts to 41
ft. lbs. (55 Nm).
6. Tighten the skid plate bolts to 41 ft.
lbs. (55 Nm).
7. Connect the negative battery.

2005 MODELS

✳✳ CAUTION

**Fuel injection systems remain under
pressure, even after the engine has
been turned OFF. The fuel system
pressure must be relieved before disconnecting any fuel lines. Failure to
do so may result in fire and/or personal injury.**

1. Before servicing the vehicle, refer to
the precautions section.
2. Release the fuel system pressure.
3. Drain the fuel tank.
4. If equipped, remove the four nuts
and the fuel tank skid plate.
5. Disconnect the two vapor tubes at
the rear of the fuel tank.
6. Disconnect the vapor tube and the
fuel filter outlet tube forward of the fuel tank.
7. Disconnect the vapor tube from the
Evaporative Emissions (EVAP) system dust
separator.

✳✳ CAUTION

**Do not support the fuel tank directly
beneath the fuel pump mounting
area. Damage to the fuel pump
assembly can occur.**

8. Position a suitable jack under the
fuel tank.

➡**The location of the fuel vent hose is
critical to fuel system operation. Note
the routing of the fuel vent hose at this
time.**

9. Remove the two bolts and the fuel
tank front and rear support straps. Lower the
front of the fuel tank approximately 25 mm
(1 in).
10. Disconnect the vent tube from the
fuel pump module.
11. Disconnect the fuel pump module
and fuel pressure transducer electrical connectors.
12.
13. Remove the fuel tank from the
vehicle.
Disconnect the fuel and vapor tubes from
the fuel pump module.
14. Using the lock ring removal tool
303-123, remove the fuel pump lock ring.

✳✳ CAUTION

**Note the location of the fuel pump
alignment tabs.**

Carefully remove the fuel pump assembly to avoid damaging the fuel level sensor.

15. Remove the fuel pump module assembly. Discard the O-ring seal
To install:
16. Installation is the reverse of removal. Note the following torque specifications:
 a. Fuel tank front and rear support straps bolts to 47 Nm (35 ft. lbs.).
 b. If equipped, fuel tank skid plate nuts to 20 Nm (15 ft. lbs.).

Excursion

Fuel injection systems remain under pressure, even after the engine has been turned OFF. The fuel system pressure must be relieved before disconnecting any fuel lines. Failure to do so may result in fire and/or personal injury.

1. Before servicing the vehicle, refer to the precautions section.
2. Remove or disconnect the following:
 • Negative battery cable
 • Fuel pressure
 • Fuel tank
 • Fuel pump bolts or retaining ring
 • Fuel pump assembly
3. Squeeze the locking tabs while pushing down.
 • Fuel pump mounting gasket.
To install:
4. Install or connect the following:
 • Fuel pump mounting gasket.
 • Fuel pump assembly
 • Fuel pump bolts and tighten to 80–107 inch lbs. (9–12 Nm). If equipped with a retaining ring tighten to 60 ft. lbs. (82 Nm).
 • Fuel tank
 • Negative battery cable

Fuel Injector

REMOVAL & INSTALLATION

2002–04 Models

Fuel injection systems remain under pressure, even after the engine has been turned OFF. The fuel system pressure must be relieved before dis-

connecting any fuel lines. Failure to do so may result in fire and/or personal injury.

1. Before servicing the vehicle, refer to the precautions section.
2. Remove or disconnect the following:
 • Negative battery cable
 • Engine cover, if equipped
 • Air cleaner assembly
3. Relieve the fuel pressure.
 • Upper intake manifold
 • Fuel pressure regulator valve vacuum hose
 • Fuel lines
 • Fuel injector electrical connectors
 • Fuel injector electrical harness from the fuel injector manifold
 • Fuel injection supply manifold
 • Fuel rail bolts and the rail

The fuel injectors are deposit-resistant. Do not clean. Use O-rings made of special fuel resistant material. Use of ordinary O-rings can cause the fuel system to leak.

 • Fuel injector retaining clips and the fuel injectors. Inspect the fuel injector O-rings and if necessary.
To install:
➡Make sure the injector clips must engage the upper groove on the injectors or fuel leaks may occur.

4. Lubricate the new O-rings with clean engine oil to aid installation.
5. Install or connect the following:
 • Fuel injectors and the rail. Tighten the rail bolts to 89 inch lbs. (10 Nm).
 • Fuel injection supply manifold
 • Fuel injector electrical harness from the fuel injector manifold
 • Fuel injector electrical connectors
 • Fuel lines
 • Fuel pressure regulator valve vacuum hose
 • Upper intake manifold
 • Air cleaner assembly
 • Engine cover, if equipped
 • Negative battery cable

2005 Expedition And Navigator Models

Fuel injection systems remain under pressure, even after the engine has been turned OFF. The fuel system

pressure must be relieved before disconnecting any fuel lines. Failure to do so may result in fire and/or personal injury.

1. Before servicing the vehicle, refer to the precautions section.
2. Properly relieve the fuel system pressure.
3. Disconnect the negative battery cable.
4. Remove the air cleaner inlet pipe.
5. Disconnect the Mass Air Flow (MAF) sensor connector.
6. Disconnect the brake booster vacuum hose.
7. Disconnect the Positive Crankcase Ventilation (PCV) tube.
8. Disconnect the crankcase ventilation tube.
9. Disconnect the fuel supply tube spring lock coupling.
10. Disconnect the fuel injector electrical connectors.
11. Remove the fuel injector-to-fuel rail locking clips.
12. Remove the fuel rail bolts and the fuel rail.

Use O-ring seals that are made of special fuel-resistant material. Use of ordinary O-rings can cause the fuel system to leak. Do not reuse the O-ring seals. Lubricate the O-ring seals with clean engine oil prior to installation.

13. Remove the fuel injectors and the fuel injector O-ring seals. Discard the O-ring seals.
To install:
14. Lubricate the new O-rings with clean engine oil to aid installation.
15. Connect the injectors to the fuel rail.
16. Install the fuel rail and tighten the retainers to 10 Nm (89 lb-in).
17. Install the remaining components in the reverse order of removal.

2005 Excursion Models

5.4L ENGINES

Fuel injection systems remain under pressure, even after the engine has been turned OFF. The fuel system pressure must be relieved before disconnecting any fuel lines. Failure to do so may result in fire and/or personal injury.

1. Before servicing the vehicle, refer to the precautions section.

2. Properly relieve the fuel system pressure.

3. Disconnect the negative battery cable.

4. Disconnect the Vapor Management Valve (VMV) hose.

5. Remove the four fuel rail bolts.

6. Remove the fuel injectors.

7. Remove and discard the two O-ring seals from the fuel injectors.

8. Shift and remove the fuel injection supply manifold.

✳ CAUTION

Check for fuel leakage after installation is complete.

To install:

➡**Make sure the fuel injectors and the fuel rail are fully seated.**

9. Install new O-ring seals and lubricate them with clean engine oil.

10. Attach the injector to the fuel rail.

11. Install the fuel rail and tighten the retainers to 10 Nm (89 inch-lbs).

12. Install the remaining components in the reverse of removal.

13. If equipped, connect the upper fitting to the exhaust manifold-to-EGR valve tube. Tighten the fittings to 50 Nm (37 lb-ft.).

6.8L ENGINES

✳✳ CAUTION

Fuel injection systems remain under pressure, even after the engine has been turned OFF. The fuel system pressure must be relieved before disconnecting any fuel lines. Failure to do so may result in fire and/or personal injury.

1. Before servicing the vehicle, refer to the precautions section.

2. Properly relieve the fuel system pressure.

3. Disconnect the negative battery cable.

4. Remove the throttle body spacer.

5. Disconnect the right hand radio ignition capacitor electrical connector.

6. Disconnect the fuel pressure regulator vacuum hose.

7. Remove the fuel line tether clamps.

8. Disconnect the fuel lines.

9. Remove and discard the fuel line O-ring seals.

10. Disconnect the fuel injector electrical connectors.

11. Remove the crankcase vent connector and hose.

12. Remove the four bolts, the fuel injection supply manifold assembly and fuel injectors.

13. Discard the fuel injector O-ring seals.

To install:

14. Installation is the reverse of removal, please note the

   a. Lubricate the new fuel injector O-ring seals with clean engine oil prior to installation.

   b. Be sure fuel injectors and fuel injection supply manifold are fully seated.

   c. Lubricate the new fuel line O-ring seals with clean engine oil prior to installation.

   d. Tighten the fuel rail retainers to 10 Nm (89 inch-lbs).

✳✳ CAUTION

Check for fuel leakage after installation is complete.

DIESEL FUEL SYSTEM

Fuel System Service Precautions

Safety is the most important factor when performing not only fuel system maintenance but any type of maintenance. Failure to conduct maintenance and repairs in a safe manner may result in serious personal injury or death. Maintenance and testing of the vehicle's fuel system components can be accomplished safely and effectively by adhering to the following rules and guidelines.

• To avoid the possibility of fire and personal injury, always disconnect the negative battery cable unless the repair or test procedure requires that battery voltage be applied.

• Always relieve the fuel system pressure prior to disconnecting any fuel system component (injector, fuel rail, pressure regulator, etc.), fitting or fuel line connection. Exercise extreme caution whenever relieving fuel system pressure, to avoid exposing skin, face and eyes to fuel spray. Please be advised that fuel under pressure may penetrate the skin or any part of the body that it contacts.

• Always place a shop towel or cloth around the fitting or connection prior to loosening to absorb any excess fuel due to spillage. Ensure that all fuel spillage

(should it occur) is quickly removed from engine surfaces. Ensure that all fuel soaked cloths or towels are deposited into a suitable waste container.

• Always keep a dry chemical (Class B) fire extinguisher near the work area.

• Do not allow fuel spray or fuel vapors to come into contact with a spark or open flame.

• Always use a back-up wrench when loosening and tightening fuel line connection fittings. This will prevent unnecessary stress and torsion to fuel line piping. Always follow the proper torque specifications.

• Always replace worn fuel fitting O-rings with new. Do not substitute fuel hose or equivalent where fuel pipe is installed.

Fuel System Pressure

RELIEVING

6.0L Engine

1. Before servicing the vehicle, refer to the precautions in the beginning of this section.

✳✳ CAUTION

Before removing the fuel tank filler cap, turn the fuel tank filler cap ¼ to

2. Remove the Schrader valve cap and install the fuel pressure gauge 310-012.

3. Open the manual valve slowly on the fuel pressure gauge 310-012 and relieve the fuel pressure. This will drain some fuel out of the system; place the fuel in a suitable container.

7.3L Engine

1. Before servicing the vehicle, refer to the precautions in the beginning of this section.

✳✳ CAUTION

Before removing the fuel tank filler cap, turn the fuel tank filler cap ¼ to ¾ turn counterclockwise and wait for the tank pressure to be relieved. Personal injury may result if the fuel tank filler cap is removed without the pressure fully relieved.

2. Remove the fuel tank filler cap to relieve any pressure in the fuel tank.

3. When servicing the fuel lines, loosen the fuel fitting to allow any residual fuel line pressure to be relieved.

Idle Speed

ADJUSTMENT

1. Before servicing the vehicle, refer to the precautions section.

2. Place the transmission in **P**.

3. Bring the engine up to normal operating temperature.

➡ **Idle speed is measured with the transmission in D.**

4. Ensure that the curb idle adjusting screw is against the stop. If not, correct the vehicle linkage.

5. Check curb idle speed. Curb idle speed is specified on the Vehicle Emissions Control Information (VECI) decal on the underside of the vehicle's hood. Adjust the idle speed to specification using the idle speed adjusting screw.

6. Place the transmission in **P**. Rev the engine momentarily, then place the transmission in the specified gear and recheck the idle speed. Adjust again if necessary.

7. Remove the tachometer and close the hood.

Fuel Filter/Water Separator

DRAINING WATER

7.3L Engine

➡ **Drain water from the water separator manual drain valve whenever the warning light comes ON or every 5000 miles (8000km). The "Water in Fuel" light will glow when approximately 3.5 oz. (103.5ml) of water accumulates in the separator.**

❋❋ CAUTION

The fuel system remains under pressure, even after the engine has been turned OFF. The fuel system pressure must be relieved before disconnecting any fuel lines. Failure to do so may result in fire and/or personal injury.

The diesel engines are equipped with a fuel/water separator in the fuel supply line. A "Water in Fuel" indicator light is provided on the instrument panel to alert the driver. The light should glow when the ignition switch is in the **START** position to indicate proper light and water sensor function. If the light glows continuously while the engine is running, the water must be drained from the separator as soon as possible to prevent damage to the fuel injection system.

1. Before servicing the vehicle, refer to the precautions section.

2. Properly relieve the fuel system pressure.

3. Shut the engine **OFF**. Failure to shut the engine **OFF** before draining the separator will cause air to enter the system.

4. Unscrew the vent on the top center of the separator unit 2½–3 turns.

5. Unscrew the drain screw on the bottom of the separator 1½–2 turns and drain the water into an appropriate container.

6. After the water is completely drained, close the water drain finger-tight.

7. Tighten the vent until snug, then turn it an additional ¼ turn.

8. Start the engine and check the "Water in Fuel" indicator light; it should not be lit. If it is lit and continues to stay so, there is a problem somewhere else in the fuel system.

REMOVAL & INSTALLATION

6.0L Engine

1. Before servicing the vehicle, refer to the precautions in the beginning of this section.

2. Relieve the fuel system pressure.

3. Disconnect the fuel lines from the fuel filter.

4. Remove the fuel filter.

5. Loosen the fuel filter clamp screw.

To install:

6. Installation is the reverse of removal.

➡ **Make sure that an audible click is heard when installing the fuel lines. Pull back on the fuel lines to confirm engagement.**

7.3L Engine

❋❋ CAUTION

The fuel system remains under pressure, even after the engine has been turned OFF. The fuel system pressure must be relieved before disconnecting any fuel lines. Failure to do so may result in fire and/or personal injury.

1. Before servicing the vehicle, refer to the precautions section.

2. Remove or disconnect the following:
- Negative battery cable
- Turbocharger assembly
- Baffle and the air inlet crossover manifold

3. Place a suitable container under the drain hose and open the filter drain.
- The 2 capscrews securing the fuel filter base to the crankcase
- Water drain hose from the filter

Raise or lower the curb idle speed by turning the curb idle speed adjusting screw—diesel engines

COLD START ADVANCE TERMINAL

COLD IDLE SPEED ADJUSTMENT

CURB IDLE SPEED ADJUSTMENT

7924FG27

4. Relieve the fuel system pressure.
- Fuel outlet hose, located between the fuel and filter housing, and the fuel return hose from the fuel pressure regulator valve
- The 2 fuel supply hoses that connect the regulator block to the cylinder head fuel rails
- Clamp at the fuel pump end of the hose (loosen only), which connects the fuel filter to the inlet of high pressure stage at the fuel pump.
- Wiring harness from the right side of the filter housing
- Electrical connections from the Water In Fuel (WIF) sensor and the fuel heater
- Fuel filter

5. Use a prybar to remove the fuel filter cap and the filter element will come out with the cap.

6. Depress the element locking tabs and remove the element from the cap.

To install:

7. Clean the mating surfaces and install the filter element onto the cap, making sure the tabs engage.

8. Install or connect the following:
- Filter gap and press down firmly, but gently, to engage it

Item	Description
1	Fuel Filter Cap
2	Fuel Filter Element
3	Fuel Filter Bevel Cut Gasket
4	Fuel Filter Housing and Gland
5	Fuel Filter Element and Cap Locking Tabs

7924FG68

Use a prytool to remove the fuel filter cap to gain access to the filter element-diesel fuel systems

- Wiring connections to the fuel heater and WIF sensor
- Wiring harness to the filter housing
- Tighten the clamp at the fuel pump end of the hose, which connects the fuel filter to the inlet of high pressure stage at the fuel pump
- 2 fuel supply hoses that connect the regulator block to the cylinder head fuel rails
- Fuel outlet hose, located between the fuel and filter housing, and the fuel return hose from the fuel pressure regulator valve.
- Water drain hose to the filter
- The 2 capscrews securing the fuel filter base to the crankcase
- Air inlet crossover manifold and baffle
- Turbocharger assembly
- Negative battery cable

Fuel Conditioning Module

The Fuel Condition Module contains the fuel pump and water separator.

DRAINING

6.0L Diesel Engine

> ☀☀ **CAUTION**
>
> **Smoking or open flame of any type must not be present when working near fuel or fuel vapor.**

1. Disconnect both battery ground cables.
2. Raise and support the vehicle.
3. Open the fuel/water separator drain valve to release the fuel pressure.

REMOVAL & INSTALLATION

6.0L Diesel Engine

1. Before servicing the vehicle, refer to the precautions in the beginning of this section.

> ☀☀ **CAUTION**
>
> **Smoking or open flame of any type must not be present when working near fuel or fuel vapor.**

2. Disconnect both battery ground cables.
3. Raise and support the vehicle.
4. Open the fuel/water separator drain valve to release the fuel pressure.
5. Disconnect the electrical connectors.

6. Disconnect the fuel pump electrical connector.
7. Disconnect the fuel warmer electrical connector.
8. Disconnect the water-in-fuel electrical connector.
9. Disconnect the fuel hoses.
10. Remove the fuel hose retaining clips and discard. Disconnect the fuel hoses from the fuel pump.
11. Press in the retaining clips and release the fuel hoses.
12. Remove the mounting nuts and the fuel conditioning module.
13. To install, reverse the removal procedure.

High Pressure Oil Pump

REMOVAL & INSTALLATION

6.0L Engine

EARLY BUILD MODELS

1. Before servicing the vehicle, refer to the precautions in the beginning of this section.
2. Remove the intake manifold.
3. Remove the turbocharger heat shield.
4. Remove the bolts and the high-pressure oil pump cover.
5. Use a thin gasket scraper to separate the cover from the crankcase.
6. Remove and discard the press-in-place gasket.
7. Position the dial indicator with bracketry onto the oil pump drive and check the oil pump drive gear backlash. The reading should be 0.007–0.0124 inch (0.179–0.315mm).
8. Remove the bolts from the high-pressure oil pump discharge pipe.
9. Using the tool 303-755, disconnect and remove the high-pressure oil pump discharge pipe.
10. Remove and discard the D-shaped O-ring seal.
11. Remove and discard the high-pressure pump O-ring seal.
12. Remove the bolts and the high-pressure oil pump.
13. Remove and discard the lower O-ring seal.

To install:

14. Install a new lower O-ring seal.
15. Install the high-pressure oil pump and bolts. Tighten to 18 ft. lbs. (25 Nm).
16. Install the high-pressure pump O-ring seal.
17. Install the oil pump discharge pipe.

18. Install the bolts for the oil discharge pipe. Tighten to 71 inch lbs. (8 Nm).

19. Position the dial indicator with bracketry onto the oil pump drive and check the oil pump drive gear backlash. The reading should be 0.007–0.0124 inch (0.179–0.315mm).

20. Install a new D-ring seal on the high-pressure discharge pipe.

21. Install a new press-in-place gasket in the high-pressure pump cover.

22. Clean the cover mounting surface and apply sealer at the seams.

23. Install the high-pressure pump cover and bolts. Tighten to 8 ft. lbs. (11 Nm).

24. Install the turbocharger heat shield.

25. Install the intake manifold

LATE BUILD MODELS

1. Before servicing the vehicle, refer to the precautions in the beginning of this section.

2. Remove the intake manifold.

3. Remove the turbocharger heat shield.

4. Remove the bolts and the high-pressure oil pump cover.

5. Use a thin gasket scraper to separate the cover from the crankcase.

6. Remove and discard the press-in-place gasket.

7. Position the dial indicator with bracketry onto the oil pump drive and check the oil pump drive gear backlash. The reading should be 0.007–0.0124 inch (0.179–0.315mm).

8. Remove the bolts from the high-pressure oil pump discharge pipe. Position the high-pressure discharge pipe aside.

9. Remove and discard the high-pressure pump O-ring seal.

10. Remove the bolts and the high-pressure oil pump.

11. Remove and discard the lower O-ring seal.

To install:

12. Install a new lower O-ring seal.

13. Install the high-pressure oil pump and bolts. Tighten to 18 ft. lbs. (25 Nm).

14. Install the high-pressure pump O-ring seal.

15. Position back the high-pressure discharge tube and install the bolts. Tighten to 71 inch lbs. (8 Nm).

16. Position the dial indicator with bracketry onto the oil pump drive and check the oil pump drive gear backlash. The reading should be 0.007–0.0124 inch (0.179–0.315mm).

17. Install a new D-ring seal on the high-pressure discharge pipe.

18. Install a new press-in-place gasket in the high-pressure pump cover.

19. Clean the cover mounting surface and apply sealer at the seams.

20. Install the high-pressure pump cover and bolts. Tighten to 8 ft. lbs. (11 Nm).

21. Install the turbocharger heat shield.

22. Install the intake manifold

Injection Pump

REMOVAL & INSTALLATION

7.3L Engines

❊❊ CAUTION

The fuel system remains under pressure, even after the engine has been turned OFF. The fuel system pressure must be relieved before disconnecting any fuel lines. Failure to do so may result in fire and/or personal injury.

1. Before servicing the vehicle, refer to the precautions section.

2. Remove or disconnect the following:
   - Negative battery cable
   - Turbocharger assembly
   - Fuel system pressure
   - Fuel line banjo bolt at the pump
   - Fuel line fittings at the rear of the cylinder heads
   - Fuel lines
   - The 3 hose clamps at the injection pump fittings
   - Water drain hose at the fuel filter
   - Filter and position it forward
   - Injection pump retaining bolts, then lift the pump out of the crankcase bore

   - Injection pump tappet from the crankcase bore

To install:

3. Rotate the engine so the injection pump eccentric is on the base circle.

4. Install or connect the following:
   - Injection pump tappet in the base of the injection pump
   - O-ring on the injection pump base
   - Injection pump and tighten the bolts to 19–27 ft. lbs. (26–37 Nm)
   - Fuel filter and connect the water drain hose
   - The 3 fuel hoses at the front of the injection pump
   - The fuel line clamps and the fuel filter retaining bolts
   - Fuel line assembly and new seal rings at the rear of the pump
   - Fuel line fittings at the rear of the cylinder heads
   - Fuel line banjo fitting at the pump. Tighten the bolt to 18 ft. lbs. (24 Nm).
   - Turbocharger assembly
   - Negative battery cable

Fuel Injectors

REMOVAL & INSTALLATION

6.0L Engine

EARLY BUILD MODELS

1. Before servicing the vehicle, refer to the precautions in the beginning of this section.

2. If removing the right hand fuel injectors, remove the evaporator case.

3. Remove the valve cover.

FUEL PUMP RETAINING BOLT

FUEL PUMP

FUEL LINE ASSEMBLY

FUEL LINE BANJO FITTING

7924FG28

Injection pump assembly components—diesel engines

4. Disconnect the fuel injector electrical connector.

5. Disconnect the high-pressure oil rail supply line at the high-pressure oil rail.

6. Remove the bolts and the high-pressure oil rail.

7. Disconnect and remove the high-pressure oil supply line.

✳✳ CAUTION

Do not attempt to apply battery voltage to the fuel injector or damage to the fuel injector will occur.

8. Using a 19 mm socket, push the fuel injector electrical connector out of the rocker arm carrier.

▓ CAUTION

To prevent engine damage, do not use air tools to remove/install the fuel injectors. The clip that extracts the injector can dislodge and fall into the oil drain hole.

➡ There is no need to drain the fuel rail.

➡ If engine coolant is found in the combustion chambers, It may be necessary to install a new injector sleeve.

9. Remove the bolt, fuel injector hold-down clamp and fuel injector.

To install:

▓ CAUTION

If the fuel injector oil inlet D-shaped O-ring is damaged, a new fuel injector must be installed.

10. Install new O-ring seals and copper washer on the fuel injector. Lubricate the fuel injector and O-ring seals liberally with clean engine oil.

11. Install the fuel injector, fuel injector hold-down clamp and bolt. Tighten to 24 ft. lbs. (33 Nm).

12. Install the fuel injector electrical connector into the rocker carrier.

13. Apply engine oil to the top fuel injector O-ring seals.

14. Install the high-pressure oil rail and bolts.

15. Install the high-pressure oil rail.

16. Install the bolts finger tight. Tighten the bolts in the sequence shown to 8 ft. lbs. (11 Nm).

17. Install the high-pressure oil line.

18. Connect the fuel injector electrical connector.

19. Install the valve covers.

11 Nm (8 lb-ft)

67197-NAVI-G17

On early build models, tighten the high-pressure oil rail bolts in the sequence shown—6.0L engine–right side shown, left side similar

20. If removed, install the evaporator case

LATE BUILD MODELS

1. Before servicing the vehicle, refer to the precautions in the beginning of this section.

2. If removing the right hand fuel injectors, remove the evaporator case.

3. Remove the valve cover.

4. Disconnect the fuel injector electrical connector.

5. Remove the bolts and the high-pressure oil rail.

✳✳ CAUTION

Do not attempt to apply battery voltage to the fuel injector or damage to the fuel injector will occur.

6. Using a 19 mm socket, push the fuel injector electrical connector out of the rocker arm carrier.

✳✳ CAUTION

To prevent engine damage, do not use air tools to remove/install the fuel injectors. The clip that extracts the injector can dislodge and fall into the oil drain hole.

➡ There is no need to drain the fuel rail.

➡ If engine coolant is found in the combustion chambers, it may be necessary to install a new injector sleeve.

7. Remove the bolt, fuel injector hold-down clamp and fuel injector.

To install:

✳✳ CAUTION

If the fuel injector oil inlet D-shaped O-ring is damaged, a new fuel injector must be installed.

Install new O-ring seals and copper washer on the fuel injector. Lubricate the fuel injector and O-ring seals liberally with clean engine oil.

8. Install the fuel injector, fuel injector hold-down clamp and bolt. Tighten to 24 ft. lbs. (33 Nm).

9. Install the fuel injector electrical connector into the rocker carrier.

10. Apply engine oil to the top fuel injector O-ring seals.

11. Apply clean engine oil on the crankcase-to-head tube O-ring seal.

➡ Apply clean engine oil on the tubes prior to installing the oil manifold.

12. Position the oil rail on the fuel injectors as follows:

   a. Place the oil rail on top of the carrier so that the four single ball tubes are engaging the injector lead angle.

   b. Insert three bolts, two on the ends of the straight side of the oil rail and one in the middle of the wavy side of the rail. Install the guide studs six to seven turns.

   c. Press the oil rail into the fuel injectors.

   d. Make sure that the oil rail mounting feet are flat against the mounting surface.

   e. Loosely install the six bolts.

13. Install the oil rail retaining bolts.

14. Remove the three guide bolts.

15. Install the bolts finger tight. Tighten the bolts in the sequence shown to 10 ft. lbs. (14 Nm).

16. Connect the fuel injector electrical connector.

17. Install the valve covers.

18. If removed, install the evaporator case

14 Nm (10 lb-ft)

67197-NAVI-G35

On late build models, tighten the high-pressure oil rail bolts in the sequence shown—6.0L engine–right side shown, left side similar

7.3L Engines

> ⁎ **CAUTION**
>
> Observe all applicable safety precautions when working around fuel. Whenever servicing the fuel system, always work in a well ventilated area. Do not allow fuel spray or vapors to come in contact with a spark or open flame. Keep a dry chemical fire extinguisher near the work area. Always keep fuel in a container specifically designed for fuel storage; also, always properly seal fuel containers to avoid the possibility of fire or explosion.

1. Before servicing the vehicle, refer to the precautions section.

> ⁎ **CAUTION**
>
> The red-striped wires on the DI Turbo carry 115 volts DC. A severe electrical shock may be given. Do not pierce the wires.

> ⁎⁎ **WARNING**
>
> Do not pierce the wires or damage to the harness could occur.

Special tools required:
- Slide Hammer, No. T50T–100–A, or equivalent
- Injector Remover, No. T94T–9000–AH1, or equivalent
- Injector Replacer, No. T94T–9000–AH2, or equivalent

2. Remove or disconnect the following:
- Valve cover
- Fuel injector electrical connector

> ⁎⁎ **WARNING**
>
> Remove the oil drain plugs prior to removing the injectors or oil could enter the combustion chamber which could result in hydrostatic lock and severe engine damage.

- Oil rail drain plugs
- Retaining screw and oil deflector. The shoulder bolt on the inboard side of the fuel injector does not require removal.
- Outboard fuel injector retaining bolt
- Heater distribution box screws, nuts and clip
- Outer half of the case (to service No. 4 fuel injector only)

3. Remove the fuel injector using Injector Remover No. T94T–9000–AH1, or equivalent Position the tool's fulcrum beneath the fuel injector hold-down plate and over the edge of the cylinder head. Install the remover screw in the threaded hole of the fuel injector plate. Tighten the screw to lift out the injector from its bore. Place the injector in a suitable protective sleeve such as Rotunda Injector Protective Sleeve, No. 014–00933–2, and set the injector in a suitable holding rack.

4. Remove the fuel injector sleeves, if required. Insert the Injector Sleeve Tap Plug, 014–00934–3 into the injector sleeve to prevent debris from entering the combustion chamber. Insert Injector Sleeve Tap Pilot into the fuel injector sleeve and tighten 1–1½ turns. Attach Slide Hammer T50T–100–A to the Injector Sleeve Tap 014–00934–1 and 014–00934–2 Injector Sleeve Tap Pilot and remove the fuel injector sleeve from the bore.

5. Use Rotunda Injector Sleeve Brush 104–00934–A, or equivalent to clean the injector bore of any sealant residue. Make sure to remove any debris.

To install:

6. If removed, install the fuel injector sleeves using Rotunda Sleeve Replacer, No. 014–00934–4, or equivalent. Apply Threadlock, No. 262–E2FZ–19554–B, or equivalent to the fuel injector sleeves. Using a rubber mallet, tap on the tool to seat the injector bore. Remove the tool and remove any residue sealant.

7. Clean the fuel injector sleeve using a suitable sleeve brush set. Clean any debris from the sleeve.

8. Clean the injector bore with a lint-free shop towel.

9. Install the fuel injectors using special tools as follows:

   a. Lubricate the injectors with clean engine oil. Using new copper washers, carefully push the injectors square into the bore using hand pressure only to seat the O-rings.

   b. Position the open end of Injector Replacer, No. T94T–9000–AH2, or equivalent between the fuel injector body and injector hold-down plate, while positioning the opposite end of the tool over the edge of the cylinder head.

   c. Align the hole in the tool with the threaded hole in the cylinder head and install the bolt from the tool kit. Tighten the bolt to fully seat the injector, then remove the bolt and tool.

10. Install or connect the following:
- Outer half of the heater distribution box and retaining hardware (for No. 4 injector only)
- Oil deflector and bolt. Tighten the bolt to 108 inch lbs. (12 Nm).
- Fuel rail drain plug, tightening it to 96 inch lbs. (11 Nm)
- Oil rail drain plug, tightening it to 53 inch lbs. (6 Nm)
- Heater distribution box
- Fuel injector wiring harness
- Valve cover

DRIVE TRAIN

Transmission Assembly

REMOVAL & INSTALLATION

4R100 Transmission

EXCEPT EXCURSION

1. Before servicing the vehicle, refer to the precautions section.
2. Place the transmission range selector lever in the NEUTRAL position.

➥All gasoline vehicles will have new adaptive shift strategies. Whenever the vehicle's battery has been disconnected for any type of service or repair the strategy parameters that are stored in the Keep Alive Memory (KAM) will be lost. The strategy will start to relearn once the battery is reconnected and the vehicle is driven. This is a temporary condition and will return to normal operating condition once the Powertrain Control Module (PCM) relearns all the parameters from the driving conditions. There is no set time frame for this process. If this concern is present during downshifts or converter clutch apply, it is not the fault of the shift strategy and will require diagnosis as outlined in the workshop manual. The customer needs to be notified that they may experience slightly different upshifts either soft or firm and that this is a temporary condition and will eventually return to normal operating condition.

3. Remove or disconnect the following:
- Negative battery cable
- Transmission fluid filler tube
- Transmission fluid cooler tubes at the cooler bypass valve. Catch the fluid in a suitable container.
- Transfer case assembly, if equipped

4. Mark the driveshaft yoke and axle flange so they may be installed in their original position.
- Rear driveshaft and install an appropriate plug in the transmission to prevent fluid leakage
- Shift cable from the transmission
- Shift cable from the manual lever
- Cable housing from the bracket and position aside
- Transmission Range (TR) sensor connector
- Shift cable at the transmission and position it aside
- Solenoid body connector
- Turbine Shaft Speed (TSS) sensor and the Output Shaft Speed (OSS) sensor
- Starter motor
- Cylinder block opening cover
- Torque converter-to-flexplate retaining nuts and discard

✳✳ CAUTION

Be sure not to raise the back of the transmission too high. If it makes contact with the underbody, damage to the TSS sensor can occur.

5. Install a high lift transmission jack with the special jack adapter 014-0763 to the transmission.
- Exhaust and crossmember
- Six transmission-to-engine mounting bolts

6. Gently rock the transmission side-to-side to disengage it from the locator dowels.

7. Move the transmission and the transmission jack rearward to clear the engine flexplate.
- Transmission from the vehicle

To install:

8. Installation is the reverse of removal, please note the following specs:

9. Transmission-to-engine bolts to 45 ft. lbs. (61 Nm).

10. Torque converter-to-flexplate nuts to 26 ft. lbs. (35 Nm).

11. Flexplate inspection cover nuts to 25 ft. lbs. (34 Nm) on gasoline engines and 15 ft. lbs. (20 Nm) on diesel engines.

12. Fluid filler tube into the short fluid inlet tube. Tighten the nut to 15 ft. lbs. (22 Nm).

13. Refill all transmissions with the correct amount of Motorcraft MERCON® automatic transmission fluid.

14. Connect the negative battery cable.

EXCURSION

1. Before servicing the vehicle, refer to the precautions in the beginning of this section.

2. With the vehicle in NEUTRAL position it on a hoist.

➡**When the battery has been disconnected and reconnected, some abnormal drive symptoms can occur while the vehicle relearns its adaptive strategy. The vehicle owner should to be notified that they may experience slightly different upshifts (either soft or firm) and that this is a temporary condition that will return to normal operating condition.**

3. Disconnect the negative battery cable.

4. On 4x4 models, remove the transfer case assembly.

5. On 4x2 models, remove the rear driveshaft

6. Drain the transmission fluid.
CAUTION: Mixing the 4x2 and the 4x4 style transmission fluid filters and transmission pan assembly components can cause transmission damage.
Install the correct pan with gasket for this application. Alternately tighten the bolts.

✳✳ CAUTION

Make sure securing straps or the transmission jack adapter do not touch the Cooler Bypass Valve (CBV). Do not use the CBV as a handle as damage to the CBV can cause a leak.

7. Install a suitable transmission jack and support the transmission.

8. On 4x4 models, remove the jack stand after installing a suitable transmission jack.

9. On 4x2 models, perform the following:
   a. Disconnect the wire loom from the crossmember.
   b. Remove the crossmember bolts.
   c. Remove the crossmember bolts.
   d. Remove the transmission mount nuts and the crossmember.

10. On 4x2 models, perform the following:
   a. Remove the right hand crossmember nuts.
   b. Remove the left hand crossmember nuts.
   c. Remove the nuts and the crossmember.
   d. Remove the transmission mount.

11. On models equipped with a transmission-mounted parking brake, perform the following:
   a. Disconnect the parking brake lever return spring from the parking brake lever.
   b. Apply penetrating oil to the adjusting clevis, jam nut and the threads on the front parking brake cable and conduit.
   c. Disconnect the front parking brake cable and conduit.
   d. Loosen the jam nut.
   e. Remove the clevis locking pin.
   f. Remove the clevis pin.
   g. Remove the adjusting clevis from the parking brake lever.
   h. Compress the retainer, and remove the front parking brake cable and conduit from the cable bracket.

➡**If the vehicle is equipped with a Power Take-Off (PTO) unit, all or part of the PTO unit will need to be removed.**

12. Disconnect the shift cable from the transmission.

13. Disconnect the shift cable from the manual lever.

14. Remove the shift cable bracket from the transmission and position aside.

15. Disconnect the digital Transmission Range (TR) sensor connector and the wire loom from the shift cable bracket.

16. Disconnect the solenoid body connector.

17. Disconnect the Turbine Shaft Speed (TSS) sensor and the Output Shaft Speed (OSS) sensor.

18. Remove the wiring harness from the transmission and position aside.

19. Remove the flexplate inspection cover.

20. Remove the bolts.

21. Remove the inspection plate.

22. Remove the starter motor.

23. Remove the cylinder block opening cover.

24. Remove and discard the torque converter-to-flexplate nuts.

25. Position a suitable drain pan and disconnect the transmission fluid cooler tubes from the cooler bypass valve.

26. Remove the transmission.

27. Remove the seven transmission-to-engine mounting bolts and gently rock the transmission side-to-side to disengage it from the locator dowels.

28. Move the transmission and the transmission jack rearward to clear the engine flexplate.

❋❋ WARNING

The torque converter is heavy and can result in injury if it falls out of the transmission. Secure the torque converter in the transmission. If the torque converter is dropped, a new one must be installed.

29. Use tool 307-346 to hold the torque converter in place.

❋❋ CAUTION

Use care while removing the transmission to avoid obstructions.

30. Lower the transmission out of the vehicle.
31. On models equipped with a transmission-mounted parking brake, perform the following:
  a. Remove the transmission-mounted parking brake.
  b. Keep the parking brake vent in the upward position to prevent contamination of the brake shoes and linings.
  c. Remove the six bolts, parking brake assembly and the gasket from the extension housing.
  d. Discard the bolts and the gasket.
  e. Clean the mating surfaces.

To install:

❋❋ CAUTION

Prior to the installation of the transmission, the fluid, cooler lines and the cooler bypass valve must be cleaned. A new transmission Oil-To-Air (OTA) cooler must be installed if the transmission was overhauled or exchanged due to a failure of the transmission. Transmission failure can occur if these procedures are not followed.

32. Inspect the wiring harness and the connectors for damage, terminal condition, corrosion and seal integrity. Repair or install new as required.
33. On models equipped with a transmission-mounted parking brake, perform the following:
  a. Install the parking brake, if removed.
  b. Position the parking brake assembly with a new gasket on the transmission extension housing.
  c. Install six new bolts and tighten to 41 ft. lbs. (55 Nm).

❋❋ CAUTION

Prior to the installation of the transmission, the torque converter pilot hub must be lubricated or damage to the torque converter or the engine crankshaft can occur.

34. Lubricate the torque converter pilot hub with multi-purpose grease.
35. Raise the transmission into position.

❋❋ CAUTION

Do not use the CBV as a handle. Damage to the CBV assembly can occur or damage to the case can result. Be careful not to raise the transmission up too far. The sensors can make contact with the underbody of the vehicle and cause damage to the sensors. Sensor failure or leakage can occur.

➡ While raising the transmission up into the engine compartment, make sure to align the fluid filler tube with the stub tube on the transmission, using the dipstick as a guide.

36. Position the transmission.
37. Remove tool 307-346.

❋❋ CAUTION

Do not allow the torque converter drive flats to disengage from the pump gear. Use care not to damage the flexplate and the converter pilot. The torque converter must rest squarely against the flexplate, indicating the converter pilot is not binding in the crankshaft.

38. While installing the transmission to the engine, align the torque converter studs with the mounting holes in the flexplate.
39. Install the transmission-to-engine bolts and tighten to 35 ft. lbs. (47 Nm).
40. Install the new torque converter-to-flexplate nuts and tighten to 26 ft. lbs. (35 Nm).
41. Install the cylinder block opening cover.
42. Install the flexplate inspection cover and bolts and tighten to 25 ft. lbs. (34 Nm).
43. Install the starter motor.
44. Connect the solenoid pack electrical connector.
45. Connect the digital TR sensor connector.
46. Connect the TSS and the OSS sensors.

➡ If the vehicle is equipped with a PTO unit, all or part of the PTO unit will need to be installed.

47. Connect the shift cable.
48. Install the cable housing bracket.
49. Install the shift cable to the manual lever and tighten to 18 ft. lbs. (25 Nm).
50. Install the transmission fluid cooler tubes to the cooler bypass valve and tighten to 20 ft. lbs. (27 Nm).
51. On models equipped with a transmission-mounted parking brake perform the following:
  a. Position the front parking brake cable and conduit, and press the retainer into the cable bracket until it snaps into place.
  b. Set the adjusting clevis.
  c. Loosen the jam nut several turns.
  d. Position the parking brake lever in the applied position.
  e. Tighten or loosen the adjusting clevis until the adjusting clevis hole lines up with the parking brake lever hole, then loosen the adjusting clevis to 0.5 inch (13mm).
  f. Install the pins, and tighten the jam nut.
  g. Install the clevis pin through the adjusting clevis and the parking brake lever.
  h. Install the locking pin in the clevis pin. Tighten the jam nut to 17 ft. lbs. (23 Nm).
  i. Install the parking brake lever return spring.
52. On 4x4 models, perform the following:
  a. Support the extension housing with a jack stand and remove the transmission jack.
  b. Install the transfer case.
53. On 4x2 models, perform the following:
  a. Install the transmission mount and tighten to 70 ft. lbs. (95 Nm).
  b. Position the crossmember to the transmission mount and loosely install the nuts.
  c. Install the crossmember bolts and tighten to 60 ft. lbs. (81 Nm).
  d. Reconnect the wire harness to the frame.
54. On 4x2 models, remove the transmission jack and tighten the crossmember to the transmission mount nuts 69 ft. lbs. (94 Nm).
55. Install the rear driveshaft.
56. If the transmission was overhauled and the vehicle was equipped with an in-line fluid filter, install a new in-line fluid filter.

57. If the transmission was overhauled and the vehicle was not equipped with an in-line fluid filter, install a new in-line fluid filter kit.

58. If the transmission is being installed for a non-internal repair, do not install an in-line filter or filter kit.

59. If installing a new or a Ford-authorized remanufactured transmission, install the in-line transmission fluid filter that is supplied.

60. Prior to lowering the vehicle, install a new in-line transmission filter or a filter kit.

61. Connect the negative battery cable.

62. Refill all transmissions with the correct amount of automatic transmission fluid.

Torqshift Transmission

EXCURSION

1. Before servicing the vehicle, refer to the precautions section.

2. On 4x4 vehicles, remove the transfer case assembly.

3. On 4x2 vehicles, remove the driveshaft.

4. If transmission disassembly is required, drain the transmission fluid. Remove the drain plug and allow the fluid to drain.

5. Install the drain plug. Tighten to 25 Nm (18 lb-ft).

6. Install a suitable high-lift transmission jack.

7. On 4x4 vehicles, install a suitable jack under the transmission and remove the jack stand from under the extension housing.

8. On 4x2 vehicles, remove the wire harness from the rear crossmember.

9. Disconnect the shift cable from the manual lever.

10. Remove the bolts and position the shift cable and bracket out of the way.

11. Loosen the bolt and disconnect the solenoid body electrical connector.

12. Disconnect the Output Shaft Speed (OSS) sensor electrical connector.

13. Disconnect the turbine shaft and intermediate shaft combination speed sensor electrical connector.

14. Disconnect the right hand and left hand wire harness from the side of the transmission.

15. On 4x2 vehicles perform the following:

   a. Remove the left hand crossmember bolts.

   b. Remove the right hand crossmember bolts.

   c. Remove the transmission mount nuts and the crossmember.

   d. Remove the rear transmission mount from the extension housing.

16. Remove the cylinder block opening cover in order to gain access to the torque converter nuts.

➡**Using a suitable strap wrench, rotate the crankshaft pulley to gain access to the torque converter nuts.**

17. Remove and discard the six torque nuts.

18. While holding the case fitting, disconnect the front and rear fluid cooler tube nuts and tubes.

19. Secure the transmission to a suitable transmission jack.

20. Remove the nine transmission-to-engine mounting bolts.

21. Slide the transmission back enough to install torque converter holding tool 307-346

22. Remove the transmission.

To install:

✳✳ CAUTION

Prior to the installation of the transmission, the fluid cooler lines must be cleaned. Otherwise transmission failure can occur.

23. Clean the fluid cooler lines.

✳✳ CAUTION

Prior to the installation of the transmission, a new transmission fluid cooler remote filter must be installed. Otherwise transmission failure can occur.

24. Install a new transmission fluid cooler remote filter.

✳✳ CAUTION

Prior to the installation of the assembly, the torque converter pilot hub must be correctly lubricated or damage to the torque converter or the engine crankshaft can occur.

25. Lubricate the torque converter pilot hub with multi-purpose grease.

➡**Rotate the torque converter so the orange or green paint daubs are in the 12 o'clock position.**

26. If the torque converter holding tool 307-346 has not been installed during the assembly of the transmission, install the tool to hold the torque converter in place while moving and positioning the transmission in place. Once the transmission is in place and before bolting it to the engine remove the tool.

27. Position the transmission in place. While raising the transmission into the engine compartment, align the fluid filler tube with the stub tube on the transmission using the fluid level indicator as a guide.

28. While installing the transmission to the engine, align the torque converter studs with the mounting holes in the flexplate.

29. Install nine transmission-to-engine bolts. Tighten to 47 Nm (35 lb-ft).

➡**Using a suitable strap wrench, rotate the crankshaft pulley to gain access to the torque converter nuts.**

30. Install six new torque converter-to-flexplate nuts. Tighten to 35 Nm (26 lb-ft).

31. Install the cylinder block opening cover.

32. Install the front transmission fluid cooler tube. Tighten to 40 Nm (30 lb-ft).

33. Install the rear transmission fluid cooler tube. Tighten to 40 Nm (30 lb-ft).

34. On 4x4 models, support the extension housing with a jack stand and remove the transmission jack.

35. On 4x2 models, install the transmission mount. Tighten to 94 Nm (69 lb-ft).

36. Position the crossmember to the transmission mount and loosely install the nut.

37. Install the crossmember bolts. Tighten to 81 Nm (60 lb-ft).

38. Tighten the rear insulator nuts. Tighten to 94 Nm (69 lb-ft).

39. Connect the right hand and left hand wiring harness to the side of the transmission.

40. Connect the OSS sensor electrical connector.

41. Connect the intermediate shaft and turbine shaft combination speed sensor electrical connector.

42. Connect the solenoid body electrical connector.

43. Install the cable housing bracket.

44. Install the shift cable to the manual lever. Tighten to 25 Nm (18 lb-ft).

45. On 4x2 models, reconnect the wire harness to the frame.

46. On 4x4 models, if equipped, install the transfer case.

47. On 4x2 models, install the rear driveshaft.

48. Adjust the shift linkage. Verify that the vehicle starts in PARK and NEUTRAL and the REVERSE lamps illuminate in REVERSE as follows:

   a. Remove the upper instrument panel steering column cover.

   b. Place the gearshift lever in the (D) position.

c. Rotate the gearshift lever clockwise until it bottoms out (first gear), then rotate the lever three indents counterclockwise (D).

d. Hang an three pound weight on the gearshift lever.

e. Center the pointer in the middle of the (D) position.

f. Rotate the thumbwheel located on the bottom of the steering column to adjust the pointer.

g. Remove the three pound weight.

h. Carefully move the gearshift lever from detent to detent and compare with transmission settings and readjust if necessary.

i. Install the upper instrument panel steering column cover

49. With the engine running and the transmission at normal operating temperature 66-77°C (150-170°F), check and adjust the transmission fluid level, and check for any leaks. If fluid is needed, add fluid in increments of 0.24-liter (0.5-pint) until the correct level is achieved (fluid should be in the cross-hatched area of the fluid level indicator).

50. Use the following guidelines for the in-line transmission fluid filter:

a. If the transmission was overhauled and the vehicle was equipped with an in-line fluid filter, install a new in-line fluid filter.

b. If the transmission was overhauled and the vehicle was not equipped with an in-line fluid filter, install a new in-line fluid filter kit.

c. If the transmission is being installed for non-internal repair, do not install an in-line filter or filter kit.

d. If installing a new or a Ford-authorized remanufactured transmission, install the in-line transmission fluid filter that is supplied.

e. Prior to lowering the vehicle, install a new in-line transmission filter or a filter kit.

4R70W Transmission

1. Before servicing the vehicle, refer to the precautions section.

2. Place the transmission range selector lever in the NEUTRAL position.

➡ **All gasoline vehicles will have new adaptive shift strategies. Whenever the vehicle's battery has been disconnected for any type of service or repair the strategy parameters that are stored in the Keep Alive Memory (KAM) will be lost. The strategy will start to relearn once the battery is reconnected**

and the vehicle is driven. This is a temporary condition and will return to normal operating condition once the Powertrain Control Module (PCM) relearns all the parameters from the driving conditions. There is no set time frame for this process. If this concern is present during downshifts or converter clutch apply, it is not the fault of the shift strategy and will require diagnosis as outlined in the workshop manual. The customer needs to be notified that they may experience slightly different upshifts either soft or firm and that this is a temporary condition and will eventually return to normal operating condition.

3. Remove or disconnect the following:
- Negative battery cable
- Transmission fluid filler tube
- Transmission fluid cooler tubes at the cooler bypass valve. Catch the fluid in a suitable container.
- Transfer case assembly, if equipped

4. Mark the driveshaft yoke and axle flange so they may be installed in their original position.
- Starter motor
- Torque converter-to-flexplate retaining nuts
- Shift linkage
- Transmission cooler lines
- Transmission electrical connectors
- Transmission Range (TR) sensor
- Output Shaft Speed (OSS) sensor
- Solenoid body assembly electrical connector

5. Install a high lift transmission jack with the special jack adapter 014-0763 to the transmission.

6. Remove or disconnect the following:
- Three Way Converters (TWC) and crossmember
- Fuel line bracket bolt
- Transmission fill tube bolt from the back of the right hand cylinder head
- Transmission fluid fill tube
- 6 transmission to engine bolts

❊❊ CAUTION

The torque converter is heavy and may result in injury if it falls out of the transmission. Secure the torque converter in the transmission.

7. Install a Torque Converter Holding Tool.

8. Lower the transmission from the vehicle.

To install:

9. Installation is the reverse of removal, please note the following specs:

10. Transmission-to-engine bolts to 41 ft. lbs. (55 Nm).

11. Fluid filler tube into the short fluid inlet tube. Tighten the nut to 16 ft. lbs. (23 Nm).

12. Refill all transmissions with the correct amount of Motorcraft MERCON® automatic transmission fluid.

13. Connect the negative battery cable.

6HP26 Transmission

4X4 MODELS

1. Before servicing the vehicle, refer to the precautions section.

2. Disconnect the negative battery cable.

3. Remove the transfer case.

➡ **If transmission disassembly or installation of a new transmission is necessary, the transmission fluid will need to be drained.**

4. Remove the transmission fluid pan drain plug and allow the fluid to drain.

5. Install the transmission fluid pan drain plug and tighten to 8 Nm (71 lb-in).

6. Remove the fuel line bracket bolt from the bracket. Position the bracket and lines aside.

7. Disconnect the shift cable end and remove the shift cable bracket bolts. Position the cable and bracket aside.

8. Remove the flexplate inspection cover bolts and the inspection cover.

9. Remove the bolts and the left hand engine-to-transmission support bracket.

10. Remove the rubber torque converter nut access plug.

11. Remove and discard the four flexplate-to-torque converter nuts.

12. Remove the bolts and the right hand engine-to-transmission support bracket.

13. Remove the plastic starter motor electrical connector cap.

14. Remove the starter motor electrical connectors.

15. Remove the ground wire from the stud.

16. Remove the three bolts and the starter motor.

➡ **Note the band seals and backer rings may stick into the case. Remove the band seals and backer rings if still in the case.**

17. Remove the transmission cooler tube bracket bolt and the cooler tubes.

18. Remove one transmission retaining

bolt and position the transmission fluid cooler tube retainer aside.

19. Disconnect the engine oil level sensor connector.

20. Disconnect the fluid cooler tube.

21. Remove the transmission fluid cooler tube bracket nut and position the bracket and lines aside.

22. Remove the right hand exhaust shield bolt.

23. Remove the left hand exhaust shield bolt.

✳✳ CAUTION

This new style pan is designed with the two outside ribs of the pan taller than the rest for side-to-side support. The transmission can be supported on the fluid pan as long as it is resting on the two outer ribs, if not, damage to the fluid pan will occur.

➡**Make sure that the transmission jack makes contact on the outer ribs of the fluid pan. Make sure that the transmission is securely fastened to the transmission jack.**

24. Position a suitable high-lift transmission jack under the transmission.

25. Remove the four rear crossmember bolts.

26. Remove the two rear transmission nuts and the crossmember.

27. Remove the left hand exhaust hanger bolt.

28. Remove the right hand bolt for the rear transmission support.

29. Remove the left hand bolt for the rear transmission support.

30. Remove the rear transmission support.

31. Disconnect the left and right hand Heated Oxygen (HO2s) sensor connector from the side of the transmission.

32. Disconnect the wire harness from the top of the transmission.

✳✳ CAUTION

Do not pull on the wire harness to disconnect the connector or damage to the connector will occur.

33. Disconnect the main transmission electrical harness by twisting the outer shell and pulling back on the connector. Disconnect the wire harness retainer from the top of the transmission.

34. Remove the four nuts for the front sway bar U-shaped brackets and position the sway bar aside.

➡**Note the length and the location of the bolts during removal.**

35. Remove the seven transmission-to-engine retaining bolts.

36. Slide the transmission back enough to install the torque converter retaining tool 307-346.

37. Remove the transmission from the vehicle.

To install:

38. Install the torque converter retaining tool.

✳✳ CAUTION

The converter housing is piloted into position by dowels in the rear of the engine block. The torque converter must rest squarely against the flexplate. This indicates that the converter pilot is not binding in the engine crankshaft.

39. Position and secure the transmission on the high-lift transmission jack. Raise and position the transmission into the vehicle.

➡**Make sure the torque converter is fully seated in the transmission before aligning the transmission to the engine.**

40. With the transmission in a horizontal position move it toward the engine. Align the orange balancing marks between the torque converter studs and the flexplate bolt holes.

➡**Install the transmission bolts in their correct locations noted during removal.**

41. Install the seven transmission-to-engine bolts and tighten to 48 Nm (35 ft. lbs.).

42. Install the front sway bar U-brackets and install the bolts. Tighten to 102 Nm (75 ft. lbs.).

43. Connect the wire harness retainer to the top of the transmission and connect the main transmission electrical harness by pushing it in and twisting the outer shell to lock it in place.

44. Connect the wire harness to the top of the transmission.

45. Connect the HO2s connectors to the side of the transmission.

46. Install the rear transmission mount. Install the left hand bolt for the rear transmission support. Tighten the bolt to 80 Nm (59 ft. lbs.)

47. Install the right hand bolt for the rear transmission support. Tighten the bolt to 80 Nm (59 ft. lbs.)

48. Install the exhaust hanger. Tighten the bolt to 40 Nm (30 ft. lbs.)

49. Position the rear crossmember in place and loosely install the transmission retaining nuts.

50. Install the bolts for the rear crossmember. Tighten the bolt to 90 Nm (66 ft. lbs.)

51. Tighten the two rear transmission nuts to 103 Nm (76 ft. lbs.)

52. Install the right hand exhaust heat shield bolt.

53. Install the left hand exhaust heat shield bolt.

54. Connect the transmission fluid cooler tubes.

55. Install the transmission fluid cooler tube bracket and the nut.

56. Connect the engine oil level sensor electrical connector.

57. Install the transmission fluid cooler tube retainer and the bolt. Tighten the bolt to 48 Nm (35 ft. lbs.)

➡**Inspect the case to make sure that the old band seals and backer rings are not stuck in the case.**

➡**Install new band seals and backer rings on the ends of the cooler tubes prior to installing.**

58. Install new seals on the transmission fluid cooler tubes.

➡**Inspect the case to make sure that the old band seals and backer rings are not stuck in the case.**

➡**Install new band seals and backer rings on the ends of the cooler tubes prior to installing.**

59. Position the fluid cooler tubes into the transmission, then install the bolt for the transmission cooler tube retainer. Tighten the bolt to 23 Nm (17 ft. lbs.)

60. Install the starter motor.

61. Install four new torque converter nuts.

62. Rotate the crankshaft to access all the torque converter nuts. Tighten the bolts to 35 Nm (26 ft. lbs.)

63. Install the rubber access plug.

64. Install the flexplate inspection cover. Tighten the bolts to 35 Nm (26 ft. lbs.)

65. Install the right hand engine-to-transmission support bracket and the bolts. Tighten the bolts to 48 Nm (35 ft. lbs.)

66. Install the left hand engine-to-transmission support bracket and the bolts. Tighten the bolts to 48 Nm (35 ft. lbs.)

67. Install the shift cable bracket bolts and connect the shift cable end. Tighten the bolts to 48 Nm (35 ft. lbs.)

68. Install the transfer case.

69. Install the fuel line bracket and the bolt.

70. Connect the negative battery cable.

71. Remove the fluid fill plug. The fill plug is located on the right hand rear side of the case near the transmission electrical connector.

✳✳ CAUTION

The use of any other transmission fluid than specified, can result in the transmission failing to operate correctly or transmission failure.

72. Fill the transmission. Add 4.7 liters (5 quarts) of clean automatic transmission fluid to the transmission through the fluid fill hole. Stop when the fluid runs out of the bottom of the hole.

73. Install the fluid fill plug. Tighten the bolts to 35 Nm (26 ft. lbs.)

74. Verify that the shift cable is correctly adjusted for column mounted shifters as follows:

a. Place the gearshift lever in the (D) position.

b. Place a three pound weight on the gearshift lever.

c. With the vehicle in NEUTRAL, position it on a hoist.

d. Disconnect the transmission shift cable from the manual control lever.

➡**Make sure the cable moves freely.**

e. Unlock the lock tab on the transmission shift cable.

f. Place the manual control lever in the (D) position.

g. Place the manual control lever in the first gear position.

h. Move the manual control lever two detents to the (D) position.

i. Connect the transmission shift cable to the manual control lever.

j. Lock the transmission shift cable lock tab.

k. Remove the three pound weight.

l. Carefully move the manual control lever from detent to detent and compare with transmission settings. Verify that the vehicle will start in PARK or NEUTRAL and backup lamps illuminate in REVERSE. If not, the adjustment steps must be repeated.

75. Verify that the shift cable is correctly adjusted for floor mounted shifters as follows:

a. Place the gearshift lever in the (D) position.

b. With the vehicle in NEUTRAL, position the vehicle on a hoist.

c. Disconnect the transmission shift cable from the manual control lever.

➡**Make sure the cable moves freely.**

d. Unlock the lock tab on the transmission shift cable.

e. Place the manual control lever in the (D) position.

f. Place the manual control lever in the first gear position.

g. Move the manual control lever three detents to the (D) position.

h. Connect the transmission shift cable to the manual control lever.

i. Lock the transmission shift cable lock tab.

j. Carefully move the manual control lever from detent to detent and compare with transmission settings. Verify that the vehicle will start in PARK or NEUTRAL and backup lamps illuminate in REVERSE. If not, the adjustment steps must be repeated.

76. Re-flash the transmission control module (TCM) to the latest level of software.

77. Start the engine. Move the transmission range selector lever through all the gear ranges, checking for engagements.

78. With the engine idling (600-750 rpm) in PARK and the transmission temperature at 30°C-50°C (86°F-122°F), check and adjust the transmission fluid level.

79. With the engine still running in PARK, check the fluid level. If more fluid is needed, remove the fluid fill plug on the side of the case and fill with clean automatic transmission fluid until the fluid runs out of the bottom of the hole.

80. Install the fluid fill plug. Tighten to 35 Nm (25 ft. lbs.)

4X2 MODELS

1. Before servicing the vehicle, refer to the precautions section.

2. Disconnect the negative battery cable.

➡**If transmission disassembly or installation of a new transmission is necessary, the transmission fluid will need to be drained.**

3. Remove the transmission fluid pan drain plug and allow the fluid to drain.

4. Install the transmission fluid pan drain plug. Tighten to 8 Nm (71 lb-in).

5. Mark the driveshaft flange with the insulator flange and the output shaft flange for correct alignment during installation.

6. Remove the three bolts and nuts and position the driveshaft aside.

7. Remove the fuel line bracket bolt from the bracket. Position the bracket and lines aside.

8. Disconnect the shift cable end and remove the shift cable bracket bolts. Position the cable and bracket aside.

9. Remove the flexplate inspection cover bolts and remove the inspection cover.

10. Remove the rubber torque converter nut access plug.

11. Remove and discard the four flexplate-to-torque converter nuts.

12. Remove the plastic starter motor electrical connector cap.

13. Remove the starter motor.

➡**Note the band seals and backer rings may stick into the case. Remove the band seals and backer rings if still in the case.**

14. Remove the transmission cooler tube bracket bolt and the cooler tubes.

15. Remove one transmission retaining bolt and position the transmission fluid cooler tube retainer aside.

16. Disconnect the engine oil level sensor electrical connector.

17. Disconnect the fluid cooler tube.

18. Remove the transmission fluid cooler tube bracket nut and position the bracket and lines aside.

19. Remove the right hand exhaust shield bolt.

20. Remove the left hand exhaust shield bolt.

➡**This new style pan is designed with the two outside ribs of the pan taller than the rest for side-to-side support. The transmission can be supported on the fluid pan as long as it is resting on the two outer ribs, if not, damage to the fluid pan will occur.**

➡**Make sure that the transmission jack makes contact on the outer ribs of the fluid pan. Make sure that the transmission is securely fastened to the transmission jack.**

21. Position a suitable high-lift transmission jack under the transmission.

22. Remove the four rear crossmember bolts.

23. Remove the two rear transmission nuts and the crossmember.

24. Remove the left hand exhaust hanger bolt.

25. Remove the right hand bolt for the rear transmission support.

26. Remove the left hand bolt for the rear transmission support. Remove the rear transmission support.

27. Disconnect the left and right hand

Heated Oxygen (HO2s) sensor connector from the side of the transmission.

28. Disconnect the wire harness from the top of the transmission.

✳✳ CAUTION

Do not pull on the wire harness to disconnect the connector or damage to the connector will occur.

29. Disconnect the main transmission electrical harness by twisting the outer shell and pulling back on the connector. Disconnect the wire harness retainer from the top of the transmission.

30. Remove the four nuts for the front sway bar U-shaped brackets. Position the sway bar aside.

➡️ **Note the length and the location of the bolts during removal.**

31. Remove the seven transmission-to-engine retaining bolts.

32. Slide the transmission back enough to install the torque converter retaining tool 307-346.

33. Remove the transmission from the vehicle.

To install:

34. Install the torque converter retaining tool.

✳✳ CAUTION

The converter housing is piloted into position by dowels in the rear of the engine block. The torque converter must rest squarely against the flexplate. This indicates that the converter pilot is not binding in the engine crankshaft.

35. Position and secure the transmission on the high-lift transmission jack. Raise and position the transmission into the vehicle.

➡️ **Make sure the torque converter is fully seated in the transmission before aligning the transmission to the engine.**

36. With the transmission in a horizontal position move it toward the engine. Align the orange balancing marks between the torque converter studs and the flexplate bolt holes.

➡️ **Install the transmission bolts in their correct locations noted during removal.**

37. Install the seven transmission-to-engine bolts and tighten to 48 Nm (35 ft. lbs.).

38. Install the front sway bar U-brackets

and install the bolts. Tighten to 102 Nm (75 ft. lbs.).

39. Connect the wire harness retainer to the top of the transmission and connect the main transmission electrical harness by pushing it in and twisting the outer shell to lock it in place.

40. Connect the wire harness to the top of the transmission.

41. Connect the HO2s connectors to the side of the transmission.

42. Install the rear transmission mount. Install the left hand bolt for the rear transmission support. Tighten the bolt to 80 Nm (59 ft. lbs.).

43. Install the right hand bolt for the rear transmission support. Tighten the bolt to 80 Nm (59 ft. lbs.).

44. Install the exhaust hanger. Tighten the bolt to 40 Nm (30 ft. lbs.).

45. Position the rear crossmember in place and loosely install the transmission nuts.

46. Install the bolts for the rear crossmember. Tighten the bolt to 90 Nm (66 ft. lbs.)

47. Tighten the two rear transmission nuts to 103 Nm (76 ft. lbs.).

48. Install the right hand exhaust heat shield bolt.

49. Install the left hand exhaust heat shield bolt.

50. Connect the transmission fluid cooler tubes.

51. Install the transmission fluid cooler tube bracket and install the nut.

52. Connect the engine oil level sensor electrical connector.

53. Install the transmission fluid cooler tube retainer and install the bolt. Tighten the bolt to 48 Nm (35 ft. lbs.)

➡️ **Inspect the case to make sure that the old band seals and backer rings are not stuck in the case.**

➡️ **Install new band seals and backer rings on the ends of the cooler tubes prior to installing.**

54. Install new seals on the transmission fluid cooler tubes.

➡️ **Inspect the case to make sure that the old band seals and backer rings are not stuck in the case.**

➡️ **Install new band seals and backer rings on the ends of the cooler tubes prior to installing.**

55. Position the fluid cooler tubes into the transmission, then install the bolt for the transmission cooler tube retainer.

56. Install the starter motor.

57. Install four new torque converter nuts. Rotate the crankshaft to access all the torque converter nuts. Tighten the bolt to 35 Nm (26 ft. lbs.)

58. Install the rubber access plug.

59. Install the flexplate inspection cover. Tighten the bolts to 35 Nm (26 ft. lbs.)

60. Install the shift cable bracket bolts and connect the shift cable end. Tighten the bolt to 48 Nm (35 ft. lbs.)

61. Install the fuel line bracket and install the bolt.

➡️ **To maintain initial driveshaft balance, align the index marks made during removal.**

62. Install the rear driveshaft. Tighten the bolt to 81 Nm (60 ft. lbs.)

63. Connect the negative battery cable.

64. Remove the fluid fill plug. The fill plug is located on the right hand rear side of the case near the transmission electrical connector.

✳✳ CAUTION

The use of any other transmission fluid than specified, can result in the transmission failing to operate in a normal manner or transmission failure.

65. Fill the transmission. Add 4.7 liters (5 quarts) of clean automatic transmission fluid to the transmission through the fluid fill hole. Stop when the fluid runs out of the bottom of the hole.

66. Install the fluid fill plug. Tighten to 35 Nm (25 ft. lbs.)

67. Verify that the shift cable is correctly adjusted for column mounted shifters as follows:

a. Place the gearshift lever in the (D) position.

b. Place a three pound weight on the gearshift lever.

c. With the vehicle in NEUTRAL, position it on a hoist.

d. Disconnect the transmission shift cable from the manual control lever.

➡️ **Make sure the cable moves freely.**

e. Unlock the lock tab on the transmission shift cable.

f. Place the manual control lever in the (D) position.

g. Place the manual control lever in the first gear position.

h. Move the manual control lever two detents to the (D) position.

i. Connect the transmission shift cable to the manual control lever.

j. Lock the transmission shift cable lock tab.

k. Remove the three pound weight.

l. Carefully move the manual control lever from detent to detent and compare with transmission settings. Verify that the vehicle will start in PARK or NEUTRAL and backup lamps illuminate in REVERSE. If not, the adjustment steps must be repeated.

68. Verify that the shift cable is correctly adjusted for floor mounted shifters as follows:

a. Place the gearshift lever in the (D) position.

b. With the vehicle in NEUTRAL, position the vehicle on a hoist.

c. Disconnect the transmission shift cable from the manual control lever.

➡**Make sure the cable moves freely.**

d. Unlock the lock tab on the transmission shift cable.

e. Place the manual control lever in the (D) position.

f. Place the manual control lever in the first gear position.

g. Move the manual control lever three detents to the (D) position.

h. Connect the transmission shift cable to the manual control lever.

i. Lock the transmission shift cable lock tab.

j. Carefully move the manual control lever from detent to detent and compare with transmission settings. Verify that the vehicle will start in PARK or NEUTRAL and backup lamps illuminate in REVERSE. If not, the adjustment steps must be repeated.

69. Re-flash the transmission control module (TCM) to the latest level of software.

70. Start the engine. Move the transmission range selector lever through all the gear ranges, checking for engagements.

71. With the engine idling (600-750 rpm) in PARK and the transmission temperature at 30°C-50°C (86°F-122°F), check and adjust the transmission fluid level.

72. With the engine still running in PARK, check the fluid level. If more fluid is needed, remove the fluid fill plug on the side of the case and fill with clean automatic transmission fluid until the fluid runs out of the bottom of the hole.

73. Install the fluid fill plug. Tighten to 35 Nm (25 ft. lbs.)

4R70E/4R75E Transmission

4X4 MODELS

1. Before servicing the vehicle, refer to the precautions section.

2. Disconnect the negative battery cable.

3. Remove the transfer case.

➡**Drain the transmission fluid if the transmission is disassembled, or installation of a new transmission is necessary.**

4. Loosen the transmission fluid pan bolts and allow the fluid to drain.

5. After the fluid has drained, remove the transmission fluid pan and drain the remaining fluid from the pan.

➡**It is not necessary to torque the transmission fluid pan bolts in this step.**

6. Install the transmission fluid pan.

7. Disconnect the shift cable end and remove the shift cable bracket.

8. Remove the flexplate inspection cover.

9. Remove the rubber access plug.

10. Remove and discard the 4 flexplate-to-torque converter nuts.

11. Remove the starter motor.

12. Disconnect the transmission cooler lines and position them aside.

13. Remove the bolt from the right hand cylinder head for the transmission filler tube and position the filler tube aside.

14. Remove the 2 bolts for the exhaust hanger.

15. Remove the right hand exhaust shield bolt.

16. Remove the left hand exhaust shield bolt.

➡**Make sure that the transmission is securely fastened to the transmission jack.**

17. Position a suitable high-lift transmission jack under the transmission.

18. Remove the rear crossmember bolts.

19. Remove the rear transmission retainer nuts and remove the crossmember.

20. Remove the rear transmission support.

21. Disconnect the transmission electrical connectors.

22. Disconnect the solenoid body assembly electrical connector.

➡**Note the length and the location of the bolts during removal.**

23. Remove the seven transmission-to-engine bolts.

➡**The torque converter is heavy and can result in injury if it falls out of the transmission. Secure the torque converter in the transmission.**

24. Leaving the transmission in a horizontal position, slide it back far enough to install the torque converter holding tool 307-346.

✳✳ CAUTION

To remove the transmission from the vehicle, it will be necessary to put the transmission into a nose-down position to clear the flexplate and the exhaust system.

25. Make sure that the torque converter holding tool 307-346 is in place and the transmission is securely fastened to the transmission jack before tilting the transmission.

26. Lower the transmission from the vehicle.

To install:

27. Install the torque converter holding tool 307-346 if not already done.

28. Position the vent tube on the top of the transmission.

✳✳ CAUTION

To position the transmission in the vehicle, put the transmission into a nose-down position to get the tail shaft over the exhaust system. The converter housing is piloted into position by dowels in the rear of the engine block. The torque converter must rest squarely against the flexplate. This indicates that the converter pilot is not binding in the engine crankshaft.

29. Position and secure the transmission on the high-lift transmission jack. Raise and position the transmission into the vehicle. Once the transmission is in position, place the transmission in a horizontal position.

➡**Do not remove the torque converter holding tool 307-346 until the transmission is in a horizontal position.**

30. Remove the torque converter holding tool 307-346.

➡**Make sure that the torque converter is fully seated in the transmission before aligning the transmission to the engine.**

31. With the transmission in a horizontal position move it toward the engine. Align the orange balancing marks between the

torque converter and the flexplate bolt holes.

➡️**Install the transmission retaining bolts in their correct locations made during removal. Tighten to 48 Nm (35 ft. lbs.)**

32. Install the shift cable bracket and the left hand transmission retaining bolts. Tighten to 48 Nm (35 ft. lbs.)

➡️**Install the transmission retaining bolts in their correct locations made during removal.**

33. Position the vent tube in its correct location and install the right hand transmission retaining bolts. Tighten to 48 Nm (35 ft. lbs.)

➡️**It may be necessary to disconnect the rubber vent hose in order to route the wire harness under the vent hose.**

34. Route the harness under the vent hose, connect the solenoid body connector and connect the harness retainer to the transmission.

35. Connect the transmission electrical connectors.

36. Install the rear transmission support. Tighten to 80 Nm (59 ft. lbs.)

37. Position the rear crossmember in place and install the nuts. Tighten to 103 Nm (76 ft. lbs.)

38. Position the rear crossmember in place and install the bolts and nuts. Tighten to 90 Nm (66 ft. lbs.)

39. Install the right hand exhaust bolts.

40. Install the left hand exhaust bolts.

41. Install the bolts retaining the exhaust hanger to the crossmember. Tighten to 40 Nm (30 ft. lbs.)

42. Position the transmission fluid fill tube into the transmission and install the retaining bolt on the right hand cylinder head.

43. Connect the transmission fluid cooler tubes.

44. Install the starter motor.

45. Install the 4 torque converter nuts. Tighten to 36 Nm (27 ft. lbs.)

46. Install the rubber access plug.

47. Install the flexplate inspection cover. Tighten to 34 Nm (25 ft. lbs.)

48. Install the shift cable end and the bracket. Tighten to 25 Nm (18 ft. lbs.)

49. Install the transfer case.

50. Use the following guidelines for the in-line transmission fluid filter:

   a. If the transmission was overhauled and the vehicle was equipped with an in-line fluid filter, install a new in-line fluid filter.

   b. If the transmission was overhauled and the vehicle was not equipped with an in-line fluid filter, install a new in-line fluid filter kit.

   c. If the transmission is being installed for a non-internal repair, do not install an in-line filter or filter kit.

   d. If installing a Ford-authorized remanufactured transmission, install the in-line transmission fluid filter that is supplied.

   e. Prior to lowering the vehicle, install a new in-line transmission filter or a filter kit.

51. Fill to correct level with clean automatic transmission fluid.

52. Connect the negative battery cable.

4X2 MODELS

1. Before servicing the vehicle, refer to the precautions section.

2. Disconnect the negative battery cable.

➡️**Drain the transmission fluid if the transmission is disassembled, or installation of a new transmission is required.**

3. Loosen the transmission fluid pan bolts and allow the fluid to drain.

4. After the fluid has drained, remove the transmission fluid pan and drain the remaining fluid from the pan.

➡️**It is not necessary to torque the transmission fluid pan bolts in this step.**

5. Install the transmission fluid pan.

6. Mark the driveshaft flange and rear axle pinion flange for correct alignment during installation.

7. Remove the rear driveshaft bolts and driveshaft.

8. Remove the fuel line bracket from the extension bolt.

9. Disconnect the shift cable end and remove the shift cable bracket.

10. Remove the flexplate inspection cover.

11. Remove the rubber access plug.

12. Remove and discard the 4 flexplate-to-torque converter nuts.

13. Remove the starter motor.

14. Disconnect the transmission cooler lines and position them aside.

15. Remove the bolt from the right hand cylinder head for the transmission filler tube and position the filler tube aside.

16. Remove the 2 bolts for the exhaust hanger.

17. Remove the right hand exhaust shield bolt.

18. Remove the left hand exhaust shield bolt.

➡️**Make sure that the transmission is securely fastened to the transmission jack.**

19. Position a suitable high-lift transmission jack under the transmission.

20. Remove the rear crossmember bolts.

21. Remove the rear transmission retainer nuts and remove the crossmember.

22. Remove the rear transmission support.

23. Disconnect the transmission electrical connectors.

24. Disconnect the solenoid body assembly electrical connector.

➡️**If the transmission fluid has not been drained, position a drain pan and install a suitable plug into the transmission.**

➡️**Note the length and the location of the bolts during removal.**

25. Remove the seven transmission-to-engine bolts.

> ✳✳ **CAUTION**
>
> **The torque converter is heavy and can result in injury if it falls out of the transmission. Secure the torque converter in the transmission.**

26. Leaving the transmission in a horizontal position, slide it back far enough to install the torque converter holding tool 307-346.

> ✳✳ **CAUTION**
>
> **To remove the transmission from the vehicle, it will be necessary to put the transmission into a nose-down position to clear the flexplate and the exhaust system.**

27. Make sure that the torque converter holding tool 307-346 is in place and the transmission is securely fastened to the transmission jack before tilting the transmission.

28. Lower the transmission from the vehicle.

To install:

29. Install the torque converter holding tool 307-346 if not already done.

30. Position the vent tube on the top of the transmission.

❋❋ CAUTION

To position the transmission in the vehicle, put the transmission into a nose-down position to get the tail shaft over the exhaust system. The converter housing is piloted into position by dowels in the rear of the engine block. The torque converter must rest squarely against the flexplate. This indicates that the converter pilot is not binding in the engine crankshaft.

31. Position and secure the transmission on the high-lift transmission jack. Raise and position the transmission into the vehicle. Once the transmission is in position, place the transmission in a horizontal position.

➡**Do not remove the torque converter holding tool 307-346 until the transmission is in a horizontal position.**

32. Remove the torque converter holding tool 307-346.

➡**Make sure that the torque converter is fully seated in the transmission before aligning the transmission to the engine.**

33. With the transmission in a horizontal position move it toward the engine. Align the orange balancing marks between the torque converter and the flexplate bolt holes.

➡**Install the transmission retaining bolts in their correct locations made during removal. Tighten to 48 Nm (35 ft. lbs.)**

34. Install the shift cable bracket and the left hand transmission retaining bolts. Tighten to 48 Nm (35 ft. lbs.)

➡**It may be necessary to disconnect the rubber vent hose in order to route the wire harness under the vent hose.**

35. Route the harness under the vent hose, connect the solenoid body connector and connect the harness retainer to the transmission.

36. Connect the transmission electrical connectors.

37. Install the rear transmission support. Tighten to 80 Nm (59 ft. lbs.)

38. Position the rear crossmember in place and install the nuts. Tighten to 103 Nm (76 ft. lbs.)

39. Position the rear crossmember in place and install the bolts and nuts. Tighten to 90 Nm (66 ft. lbs.)

40. Remove the transmission jack.

41. Install the right hand exhaust bolts.

42. Install the left hand exhaust bolts.

43. Install the bolts retaining the exhaust hanger to the crossmember. Tighten to 40 Nm (30 ft. lbs.)

44. Position the transmission fluid fill tube into the transmission and install the retaining bolt on the right hand cylinder head.

45. Connect the transmission fluid cooler tubes.

46. Install the starter motor.

47. Install the 4 torque converter nuts. Rotate the crankshaft to access all the torque converter nuts. Tighten to 36 Nm (27 ft. lbs.)

48. Install the rubber access plug.

49. Install the flexplate inspection cover. Tighten to 34 Nm (25 ft. lbs.)

50. Install the shift cable end and the bracket. Tighten to 25 Nm (18 ft. lbs.)

51. Install the fuel line bracket. Tighten to 25 Nm (18 ft. lbs.)

➡**The output shaft and the driveshaft are a balanced assembly.**

52. Install the driveshaft. Align the yellow dots and position the driveshaft on the transmission.

53. Position the driveshaft to the rear differential.

➡**To maintain initial driveshaft balance, align the index marks made during removal.**

54. Install the rear driveshaft. Tighten the bolts to 103 Nm (76 ft. lbs.)

55. Use the following guidelines for the in-line transmission fluid filter:

a. If the transmission was overhauled and the vehicle was equipped with an in-line fluid filter, install a new in-line fluid filter.

b. If the transmission was overhauled and the vehicle was not equipped with an in-line fluid filter, install a new in-line fluid filter kit.

c. If the transmission is being installed for a non-internal repair, do not install an in-line filter or filter kit.

d. If installing a Ford authorized remanufactured transmission, install the in-line transmission fluid filter that is supplied

e. Prior to lowering the vehicle, install a new in-line transmission filter or a filter kit.

56. Connect the negative battery cable.

57. Fill the transmission with clean automatic transmission fluid and inspect for correct operation.

Transfer Case Assembly

REMOVAL & INSTALLATION

Excursion

1. Before servicing the vehicle, refer to the precautions in the beginning of this section.

2. Shift the transfer case to the 2W HI position.

3. Remove the four bolts and the skid plate, if equipped.

4. Matchmark the driveshaft to maintain driveline balance.

5. Remove the rear driveshaft.

6. Matchmark the front driveshaft to the transfer case flange.

7. Support the front driveshaft with wire or a strap.

8. Remove and discard the four bolts and position the front driveshaft aside.

9. On models with a manual shift transfer case, perform the following:

a. Remove the manual shift linkage, if equipped.

b. Disconnect the switch electrical connector. Position the wire harness aside.

10. On models equipped with an electric shift transfer case, perform the following:

a. Disconnect the gear motor encoder assembly electrical connector and the gear motor electrical connector.

b. Position the wire harness aside.

c. Disconnect the transfer case vent hose.

11. Drain the fluid into a suitable container. Install the plug when finished.

12. Position a suitable high-lift jack under the transfer case and secure it with safety straps.

13. Detach the wire harness from the crossmember.

14. Remove the crossmember-to-frame bolts.

15. Remove the nuts and the crossmember.

16. Remove the transmission mount.

17. Position a suitable jack stand under the extension housing.

18. Remove the transfer case-to-transmission bolts.

19. Separate the transfer case from the extension housing. Pull the transfer case rearward, then lower the transfer case from the vehicle.

✳✳ CAUTION

Carefully clean the gasket surfaces as nicks and gouges can cause fluid leaks.

20. Remove the transfer case-to-transmission gasket. Clean the mating surfaces, using metal surface cleaner.

To install:

21. Install a new mounting gasket.

22. Secure the transfer case to the high-lift jack using a safety strap.

23. Raise the transfer case into position.

24. Install the bolts retaining the transfer case to the extension housing and tighten to 37 ft. lbs. (50 Nm).

25. Remove the jack stand from the extension housing.

26. Install the transmission mount and tighten to 70 ft. lbs. (95 Nm).

27. Position the crossmember and loosely install the two nuts.

28. Install the crossmember bolts and tighten to 60 ft. lbs. (81 Nm).

29. Attach the wire harness to the crossmember.

30. Remove the high-lift jack.

31. Tighten the transmission mount-to-crossmember nuts to 69 ft. lbs. (94 Nm).

32. On models with an electric shift transfer case, perform the following:

   a. Connect the vent hose.

   b. Connect the two gear motor encoder assembly electrical connectors.

33. On models with a manual shift transfer case, perform the following:

   a. Connect the 3-position mode switch harness connector.

   b. Connect the manual shift linkage.

34. Connect the front driveshaft to the transfer case, aligning the matchmarks made prior to removal and install the four new bolts. Tighten to 82 ft. lbs. (111 Nm).

35. Install the rear driveshaft.

36. Install the skid plate and the four bolts and tighten to 18 ft. lbs. (25 Nm).

   • Fill the transfer case.

Except Excursion And 2005–06 Navigator And Expedition Models

2002 4R70W AND 4R100 TRANSMISSIONS

1. Before servicing the vehicle, refer to the precautions in the beginning of this section.

2. Place the vehicle in NEUTRAL, raise and support the vehicle.

3. Drain the transfer case.

4. Remove the skid plate, if equipped.

5. Remove the rear driveshaft, match-

mark the position for installation purposes before removal.

6. Remove the two nuts and the front driveshaft shield.

7. Remove the front driveshaft, matchmark the position for installation purposes before removal.

8. Matchmark the torsion bars to the torsion bar crossmember.

9. Measure and record the torsion bar adjuster bolt length.

10. Remove the bolts.

11. Using tools 204-203 and 204-185, remove the torsion bar adjustment nuts.

12. Tighten the tool 204-185 until the torsion bar adjuster lifts off the adjuster nut.

13. Remove the torsion bar adjusting brackets.

14. Move the torsion bars forward.

15. Remove the torsion bar adjusting brackets.

16. Pull the torsion bar rearward and separate it from the lower control arm.

17. Remove the right hand torsion bar only.

18. Remove the six bolts and the torsion bar crossmember.

19. On manual shift models, use tool 307-017, disconnect the shift rod.

20. On manual shift models, disconnect the 4WD indicator switch electrical connector, if equipped.

21. Disconnect the electric shift motor electrical connector, if equipped.

22. Disconnect the vent hose.

23. Using a suitable jack, support the transfer case. Secure the transfer case to the jack with a safety strap.

24. Remove the six transfer case-to-transmission bolts and separate the transfer case from the transmission.

25. Lower the transfer case from the vehicle.

Use tools 204-203 and 204-185 to remove the torsion bar adjustment nuts—except Excursion

67197-NAVI-G37

Tighten the torsion bar crossmember bolts as shown—except Excursion

✳✳ CAUTION

Carefully clean the gasket surfaces as nicks and gouges can cause fluid leaks.

26. Remove the transfer case-to-transmission gasket.

To install:

27. Install a new transfer case-to-transmission gasket.

28. Place the transfer case in position.

29. Install the transfer case-to-transmission bolts and tighten to 35 ft. lbs. (47 Nm).

30. Remove the jack.

31. Connect the vent hose.

32. Connect the electric shift motor electrical connector, if equipped.

33. On manual shift models, Connect the 4WD indicator switch electrical connector, if equipped.

34. On manual shift models, connect the shift rod.

35. Install the torsion bar crossmember. Tighten the bolts to the specifications shown in the accompanying illustration.

36. Install the right hand torsion bar.

37. Install the torsion bar adjusting brackets.

38. Install the torsion bar adjuster bolt and tighten until the length is the same as was recorded prior to removal.

39. Install the front driveshaft, align the matchmarks marked for installation purposes before removal.

40. Install the front driveshaft shield.

41. Install the rear driveshaft, align matchmarks marked for installation purposes before removal.

42. Fill the transfer case.

2003–04 4R70W TRANSMISSION

1. Before servicing the vehicle, refer to the precautions in the beginning of this section.

2. Place the vehicle in NEUTRAL, raise and support the vehicle.

3. Drain the transfer case.

4. Remove the two nuts and the front driveshaft shield.

5. Remove the front driveshaft, match-mark the position for installation purposes before removal.

6. Remove the rear driveshaft, match-mark the position for installation purposes before removal.

7. Disconnect the electrical connector.

8. Disconnect the vent hose.

9. Remove the bolts and the damper.

10. Using a suitable jack, support the transfer case. Secure the transfer case to the jack with a safety strap.

11. Remove the six transfer case-to-transmission bolts and separate the transfer case from the transmission.

12. Lower the transfer case from the vehicle.

✳✳ CAUTION

Carefully clean the gasket surfaces as nicks and gouges can cause fluid leaks.

13. Remove the transfer case-to-transmission gasket.

To install:

14. Install a new transfer case-to-transmission gasket.

15. Place the transfer case in position.

16. Install the transfer case-to-transmission bolts and tighten to 35 ft. lbs. (47 Nm).

17. Remove the jack.

18. Connect the vent hose.

19. Connect the electrical connector.

20. Install the rear driveshaft, align matchmarks marked for installation purposes before removal.

21. Install the front driveshaft, align the matchmarks marked for installation purposes before removal.

22. Install the front driveshaft shield.

23. Fill the transfer case.

2003–04 4R100 TRANSMISSION

1. Before servicing the vehicle, refer to the precautions in the beginning of this section.

2. With the vehicle in NEUTRAL position the vehicle on a hoist.

3. Drain the transfer case fluid. Install the drain plug when finished draining.

4. Remove the transfer case-to-transmission brace.

5. Remove the bolts from the transmission.

6. Remove the nuts from the transfer case.

7. Remove the two nuts and the front driveshaft shield.

8. Remove the front driveshaft, match-mark the position for installation purposes before removal.

9. Remove the rear driveshaft, match-mark the position for installation purposes before removal.

10. Disconnect the electrical connector.

11. Disconnect the vent hose.

✳✳ CAUTION

Never support the transmission from the bottom of the transmission pan.

12. Using a suitable jack stand, support the transmission. Secure the transfer case to the jack with a safety strap.

13. Using a suitable transmission jack, support the transfer case.

14. Remove the two exhaust heat shield-to-crossmember bolts.

15. Remove the crossmember-to-frame nuts and bolts.

16. Remove the transmission mount-to-crossmember nuts, then remove the crossmember.

17. Remove the bolts and the transmission mount.

18. Remove the bolt and the exhaust hanger bracket.

19. Remove the six transfer case-to-transmission extension housing bolts.

20. Separate the transfer case from the transmission extension housing, move the transfer case rearward off the output shaft, then lower the transfer case from the vehicle.

✳✳ CAUTION

Carefully clean the gasket surfaces as nicks and gouges can cause fluid leaks.

21. Remove the transfer case-to-transmission gasket and clean the mating surfaces

To install:

22. Install a new transfer case-to-transmission gasket.

23. Place the transfer case in position.

24. Install the transfer case-to-transmission bolts and tighten to 30 ft. lbs. (40 Nm).

25. Install the exhaust hanger bracket and tighten to 30 ft. lbs. (40 Nm).

26. Install the transmission mount and tighten the bolts to 72 ft. lbs. (98 Nm).

27. Install the transmission mount-to-crossmember and tighten the nuts to 74 ft. lbs. (101 Nm).

28. Install the crossmember-to-frame nuts and bolts. Tighten to 66 ft. lbs. (90 Nm).

29. Install the two exhaust heat shield-to-crossmember bolts.

30. Remove the jack.

31. Connect the vent hose.

32. Connect the electrical connector.

33. Install the rear driveshaft, align matchmarks marked for installation purposes before removal.

34. Install the front driveshaft, align the matchmarks marked for installation purposes before removal.

35. Install the front driveshaft shield.

36. Install the transfer case-to-transmission brace. Tighten the retainers to 22 ft. lbs. (30 Nm).

37. Fill the transfer case.

2005–06 Navigator Models

1. Before servicing the vehicle, refer to the precautions in the beginning of this section.

2. Drain the fluid if the transfer case is to be disassembled.

3. Install the drain plug when finished draining.

4. Remove the transfer case–to–transmission brace.

5. Remove the bolts from the transmission.

6. Remove the nuts from the transfer case.

7. Remove the two nuts and the front driveshaft shield.

➡️Index-mark the driveshaft(s) to maintain initial driveshaft balance during installation.

8. Remove the front driveshaft.

9. Remove the rear driveshaft.

10. Disconnect the electrical connector.

11. Disconnect the vent hose.

➡️One transfer case–to–brace nut is located at the top of the transfer case input flange.

12. Remove the three transfer case–to–brace nuts.

13. Remove the stud.

14. Remove the transmission–to–brace nuts.

✳✳ CAUTION

Never support the transmission from the bottom of the transmission pan.

15. Using a suitable jack stand, support the transmission.

✳✳ CAUTION

Secure the transfer case to the jack with a safety strap.

16. Using a suitable transmission jack, support the transfer case.

17. Remove the two exhaust heat shield–to–crossmember bolts.

18. Remove the bolt and the exhaust hanger bracket.

19. Position a suitable drain pan below the transmission extension housing.

20. Remove the six transfer case–to–transmission extension housing bolts.

➡ **Watch for obstructions.**

21. Separate the transfer case from the transmission extension housing, move the transfer case rearward off the output shaft, then lower the transfer case from the vehicle.

※※ **CAUTION**

Carefully clean the gasket surfaces. Nicks and gouges cause fluid leaks.

22. Remove the transfer case–to–transmission gasket and clean the mating surfaces.

To install:

23. Installation is the reverse of removal, please not the following torque specifications:

a. Tighten the six transfer case-to-transmission extension housing bolts in a star pattern to 41 Nm (30 ft. lbs.).

b. Tighten the bolt retaining the exhaust hanger bracket to 41 Nm (30 ft. lbs.).

c. Tighten the transmission-to-brace nuts to 48 Nm (35 ft. lbs.).

d. Tighten the three transfer case-to-brace nuts to 70 Nm (52 ft. lbs.).

e. Tighten the two nuts retaining the front driveshaft shield to 18 Nm (13 ft. lbs.).

f. Tighten the transfer case-to-transmission brace bolts to 30 Nm (22 ft. lbs.) and the nuts to 28 Nm (21 ft. lbs.).

24. Fill the transfer case.

2005–06 Expedition Models

1. Before servicing the vehicle, refer to the precautions in the beginning of this section.

2. Drain the fluid if the transfer case is to be disassembled.

3. Install the drain plug when finished draining.

4. Remove the two nuts and the front driveshaft shield. Tighten to 18 Nm (13 ft. lbs.).

➡ **Index-mark the driveshaft(s) to maintain initial driveshaft balance during installation.**

5. Remove the front driveshaft.

6. Remove the rear driveshaft.

7. Disconnect the electrical connector.

8. Disconnect the vent hose.

※※ **CAUTION**

Secure the transfer case to the jack with a safety strap.

9. Using a suitable transmission jack, support the transfer case.

➡ **Position a suitable drain pan below the transmission extension housing.**

10. Remove the six transfer case-to-transmission extension housing bolts.

※※ **CAUTION**

Watch for obstructions.

11. Separate the transfer case from the transmission extension housing, move the transfer case rearward off the output shaft, then lower the transfer case from the vehicle.

※※ **CAUTION**

Carefully clean the gasket surfaces. Nicks and gouges cause fluid leaks.

12. Remove the transfer case-to-transmission gasket and clean the mating surfaces

To install:

13. Installation is the reverse of removal, please not the following torque specifications:

a. Tighten the six transfer case-to-transmission extension housing bolts in a star pattern to 47 Nm (35 ft. lbs.).

b. Tighten the two nuts retaining the front driveshaft shield to 18 Nm (13 ft. lbs.).

14. Fill the transfer case.

Halfshaft

REMOVAL & INSTALLATION

2002–04 Models

1. Before servicing the vehicle, refer to the precautions section.

2. Remove or disconnect the following:
   - Front wheels
   - Front hub cotter pin, retainer and nut

3. Using a floor hydraulic jack, support the lower suspension arm.
   - Upper ball joint cotter pin and castle nut
   - Knuckle from the front suspension upper arm

7924FG70

Halfshaft-to-differential mounting bolts (3 of the 6 bolts shown)

4. Lower the lower suspension arm and steering knuckle slightly to facilitate easier halfshaft removal.

5. Remove the 2 disc caliper mounting bolts, then lift the front disc caliper off of the front disc brake caliper anchor plate and position aside. Do not allow the caliper to hang by the brake hose; suspend it from the vehicle's frame with strong cord or wire.

6. Remove the 6 front halfshaft-to-differential bolts.

※※ **WARNING**

Use care to avoid damaging the hub seal when removing the front halfshaft.

- Inboard end of the halfshaft from the differential case or extension axle case. Separate the front halfshaft and joints from the hub, then remove the halfshaft and joints from the vehicle.

To install:

7. Slide the halfshaft outboard end into the hub, making sure that the splines engage.

8. Situate the inboard end of the halfshaft against the front differential flange and install the 6 halfshaft-to-differential bolts. Tighten the halfshaft bolts to 60 ft. lbs. (82 Nm).

9. Install the front disc brake caliper onto the rotor and anchor plate, then install and tighten the 2 caliper mounting bolts to 21–26 ft. lbs. (28–36 Nm).

10. Lift the lower suspension arm and steering knuckle up until the upper ball joint stud is inserted into the steering knuckle. Install the upper ball joint castle nut and tighten to 57–76 ft. lbs. (77–104 Nm). Install a new cotter pin.

11. Install the hub nut onto the halfshaft and tighten the hub nut to 221 ft. lbs. (300 Nm).

12. Install the hub nut retainer and a new cotter pin.

13. Install the front wheels and tighten the lug nuts in a star-shaped sequence to 83–112 ft. lbs. (113–153 Nm).

14. Lower the vehicle to the ground.

2005–06 Models

1. Before servicing the vehicle, refer to the precautions section.

⁕⁕ WARNING

The electrical power to the air suspension system must be shut off prior to hoisting, jacking or towing an air suspension vehicle. This can be accomplished by turning off the air suspension switch located in the left rear quarter trim panel. Failure to do so can result in unexpected inflation or deflation of the air springs, which can result in shifting of the vehicle during these operations.

2. Remove the front wheel and tire assembly.
3. Remove the dust cap.
4. Remove and discard the axle nut.
5. Remove the vacuum/vent line at the vacuum/vent port of the integrated wheel end disconnect.
6. Remove the three bolts retaining the integrated wheel end disconnect to the steering knuckle.
7. Remove and discard the tie rod end nut. Disconnect the tie rod end from the steering knuckle.
8. Remove and discard the upper ball joint nut. Disconnect the upper ball joint from the steering knuckle.

⁕ CAUTION

Do not damage the hub seal.

➡**Allow the steering knuckle to swing outboard while keeping the constant velocity shaft pushed inboard.**

9. Once clearance is available, remove the constant velocity shaft joint outboard end and integrated wheel end disconnect from the steering knuckle hub bearing.
10. Remove the integrated wheel end disconnect from the outboard constant velocity joint housing.
11. Remove the six bolts retaining the halfshaft to the axle.
12. Separate the halfshaft from the axle and the hub, and remove the assembly from the vehicle.

⁕ CAUTION

Verify the spline engagement by checking for spline lash before installing the halfshaft nut.

To install:

13. Installation is the reverse of removal, please not the following torque specifications:

   a. Tighten the six bolts retaining the halfshaft to the axle to 82 Nm (60 ft. lbs.).

   b. Install the ball joint and use a new nut. Tighten the nut to 150 Nm (111 ft. lbs.)

   c. Install the tie rod and use a new nut. Tighten the nut to 150 Nm (111 ft. lbs.)

   d. Install the three bolts retaining the integrated wheel and tighten to 15 Nm (11 ft. lbs.).

   e. Install a new axle nut and tighten to 27 Nm (27 ft. lbs.).

CV-Joint

OVERHAUL

➡**Before continuing with this procedure, make sure to have available a new CV-joint boot kit, for each CV-joint being serviced. The outer CV-joint cannot be disassembled, only the boot can be replaced.**

Inner CV-Joint And Boot

1. Before servicing the vehicle, refer to the precautions section.
2. Remove the halfshaft assembly from the vehicle.
3. Clamp the halfshaft in a vise equipped with jaw caps to prevent damage to machined surfaces. Do not allow the vise jaws to contact the boot or its clamp.
4. Slide 2 inboard clamp protectors off the boot clamps.
5. Carefully remove 2 boot clamps, and slide the boot off the inner CV-joint and housing.
6. Remove the CV-joint retaining ring and remove the housing.
7. Mark the inner race and the ball cage for assembly.
8. Remove 6 cage balls.
9. Remove the snapring.
10. Remove the inner race and ball cage.
11. Clean all parts in suitable parts cleaning solvent and inspect for wear.

To install:

12. Place the boot and the small boot clamp and protector on the shaft.
13. Place the ball cage on the shaft with the tapered end toward the outer CV-joint.

➡**Line up the marks made at disassembly.**

14. Position the inner race on the driveshaft in the position marked on disassembly.
15. Install the snapring.
16. Lubricate and position 6 balls with suitable CV-joint grease.
17. Place the boot protector and boot clamp on the CV-joint housing. Fill the housing with 8.29 ounces of suitable CV-joint grease.
18. Place the housing to the cage and bearings and install the retaining ring.
19. Remove any excess grease from the mating surface and position the boot and clamp.
20. Adjust the CV-joint to boot spacing to 16.43 in. (417.25mm).
21. After adjusting the CV-joint to boot spacing, insert a dull bladed screwdriver blade to relieve built up air pressure in the boot.
22. Use CV Boot Clamp Installer T95P-3514-A or equivalent, to install the boot clamps.
23. Place the clamp protectors over the boot clamps.
24. Install the halfshaft into the vehicle.
25. Road test the vehicle and check for proper operation.

Outer Boot

1. Remove the inner CV-joint and housing from the halfshaft.
2. Remove the inner boot from the halfshaft.
3. Remove the outer boot clamp protectors and carefully remove the boot clamps.
4. Remove the outer boot from the halfshaft and inspect the grease for contamination.
5. If the grease is contaminated, clean and inspect the joint for wear. Replace the joint and shaft if worn or damaged.

To install:

6. Place the new CV-joint boot on the shaft.
7. Using 5.82 ounces (165 grams) of suitable CV-joint grease, pack the outer CV-joint with grease, then spread the remaining grease inside the boot.
8. Clean the boot mounting surface and position the boot in the joint grooves.
9. Place the clamps in position and use CV Boot Clamp Installer T95P-3514-A or equivalent, to install the boot clamps.
10. Place the clamp protectors over the boot clamps.
11. Install the inner boot on the halfshaft.
12. Install the inner CV-joint and housing on the halfshaft.

13. Install the halfshaft in the vehicle.

14. Road test the vehicle and check for proper operation.

Locking Hubs

REMOVAL & INSTALLATION

1. Before servicing the vehicle, refer to the precautions section.

2. Raise and safely support the vehicle.

3. Remove or disconnect the following:
   - Tire
   - 3 screws and separate the cap from the body
   - Lockring seated in the groove of the hub assembly
   - Body assembly from the brake rotor/hub
   - Snapring from the groove in the stub-shaft
   - 3 thrust washers from the stub-shaft

4. Pull the cam assembly to remove it.

To install:

5. Align the fixed cam retaining key on the cam assembly with the keyway on the spindle. Firmly push the cam assembly on the wheel retaining nut.

6. Install or connect the following:
   - Metal, plastic, then the splined washers on the stub-shaft
   - Snapring in the groove of the stub-shaft. It may be necessary to push the stub-shaft outward from the back of the knuckle assembly.

✳✳ WARNING

Do not pack the hub assembly with grease. Too much grease will damage the hub assembly.

7. Rotate the moving cam assembly to the 1 o'clock position in relation to the fixed cam retaining key. Use any 1 of the 3 stops.

- Body assembly onto the hub by lining up the 3 legs with the 3 pockets in the cam assembly. Be sure the assembly is in far enough to see the groove in the hub.
- Large lockring in the groove on the hub. Ensure the lockring is seated completely.
- Cap using the 3 screws. Tighten the screws to 35–53 inch lbs. (4–6 Nm).
- Tire and lower the vehicle to the floor

Axle Shaft, Bearing and Seal

REMOVAL & INSTALLATION

Dana 50 Independent Front Suspension

2002–04 MODELS

1. Before servicing the vehicle, refer to the precautions section.

2. Raise and support the front end on jackstands.

3. Remove or disconnect the following:
   - Front wheels
   - Calipers
   - Hub/rotor assemblies
   - Nuts retaining the spindle to the steering knuckle. Tap the spindle with a plastic mallet to remove it from the knuckle.
   - Splash shield

4. On the left side, pull the shaft from the carrier, through the knuckle.

5. On the right side, remove and discard the keystone clamp from the shaft and joint assembly and the stub shaft. Slide the rubber boot onto the shaft and pull the shaft and joint assembly from the splines of the stub shaft.

6. Place the spindle in a soft-jawed vise clamped on the second step of the spindle.

7. Using a slide hammer and bearing puller, remove the needle bearing from the spindle.

8. Inspect all parts. If the spindle is excessively corroded or pitted it must be replaced. If the U-joints are excessively loose or don't move freely, they must be replaced. If any shaft is bent, it must be replaced.

To install:

9. Clean all dirt and grease from the spindle bearing bore. The bore must be free of nicks and burrs.

10. Insert a new spindle bearing in its bore with the printing facing outward. Drive it into place with drive T80T–4000–S for F-150 and F-250, or T80T–4000–R for the F-350, or their equivalents. Install a new bearing seal with the lip facing away from the bearing.

11. Pack the bearing and hub seal with grease. Install the hub seal with a driver.

12. Install or connect the following:
    - Thrust washer on the axle shaft
    - New slinger on the axle shaft
    - Rubber V-seal on the slinger. The seal lip should face the spindle.
    - Plastic spacer on the axle shaft. The chamfered side of the spacer should be inboard against the axle shaft.

13. Pack the thrust face of the seal in the spindle bore and the V-seal on the axle shaft with heavy duty, high temperature, waterproof wheel bearing grease.

14. On the right side, install the rubber boot and new keystone clamps on the stub shaft and slip yoke. The splines permit only one way of meshing so you'll have to properly align the missing spline in the slip yoke with the gapless male spline on the shaft. Slide the right shaft and joint assembly into the slip yoke, making sure that the splines are fully engaged. Slide the boot over the assembly and crimp the keystone clamp.

15. On the left side, slide the shaft and

Exploded view of the typical automatic locking hub assembly

7924FG93

joint assembly through the knuckle and engage the splines in the carrier.

16. Install or connect the following:
- Splash shield and spindle on the knuckle. Tighten the spindle nuts to 60 ft. lbs. (81 Nm).
- Rotor on the spindle
- Outer wheel bearing into the cup. Make sure that the grease seal lip totally encircles the spindle.
- Wheel bearing, locknut, thrust bearing, snapring and locking hubs.
- Caliper

2005 MODELS

1. Before servicing the vehicle, refer to the precautions section.
2. Remove the disc brake calipers.
3. Remove the disc brake pads.
4. Disconnect the power vacuum hub hose from the knuckle, then disconnect the ABS sensor.
5. Remove the front stabilizer bar-to-axle retaining bolts.
6. If equipped, disconnect the steering damper at the axle.
7. Disconnect the steering linkage at the steering knuckles and position it aside.

✳✳ CAUTION

Index-mark the driveshaft to the companion flange to maintain proper driveline balance.

8. Disconnect the driveshaft at the front axle and position it aside. Wrap electrical tape around the bearing cups.
9. Disconnect the vent tube and plug the fitting.

➡**It is necessary to load the suspension to remove the trackbar. Load the springs by allowing most of the front vehicle weight to rest on the axle.**

10. Support the axle with a suitable jack and lower the vehicle enough to relieve the tension on the trackbar. Then, disconnect the trackbar at the axle and position it aside.
11. Relieve the load on the suspension after disconnecting the trackbar.
12. Leave the jack supporting the axle for removal from the vehicle.
13. Remove the U-bolts securing the axle to the springs.
14. Lower and remove the axle.
To install:
15. Installation is the reverse of removal. Note the following torque specifications:
   a. Tighten the U-bolts t0 133 Nm (99 lb-ft).

b. Tighten the trackbar to axle bolts to 174 Nm (129 lb-ft).
c. Tighten the stabilizer bar-to-axle retaining bolts to 47 Nm (35 lb-ft).

Spindle and Front Axle Shaft

REMOVAL & INSTALLATION

Dana 60 Monobeam

1. Before servicing the vehicle, refer to the precautions section.
2. Raise and support the front end on jackstands.
3. Remove or disconnect the following:
- Caliper from the knuckle and wire it out of the way
- Free-running hub
- Front wheel bearings
- Hub and rotor assembly
- Spindle-to-knuckle bolts. Tap the spindle from the knuckle using a plastic mallet.
- Splash shield and caliper support
4. Pull the axle shaft out through the knuckle.
5. Using a slide hammer and bearing cup puller, remove the needle bearing from the spindle.
6. Clean the spindle bore thoroughly and make sure that it is free of nicks and burrs. If the bore is excessively pitted or scored, the spindle must be replaced.
To install:
7. Insert a new spindle bearing in its bore with the printing facing outward. Drive it into place with driver T80T–4000–R, or its equivalent. Install a new bearing seal with the lip facing away from the bearing.
8. Pack the bearing with waterproof wheel bearing grease.
9. Pack the thrust face of the seal in the spindle bore and the V-seal on the axle shaft with waterproof wheel bearing grease.
10. Carefully guide the axle shaft through the knuckle and into the housing. Align the splines and fully seat the shaft.
11. Install or connect the following:
- Bronze spacer on the shaft. The chamfered side of the spacer must be inboard.
- Splash shield and caliper support
- Spindle on the knuckle and install the bolts. Tighten the bolts to 50–60 ft. lbs. (68–81 Nm).
- Hub/rotor assembly on the spindle
- Wheel bearings
- Free-running hub

Right Side Slip Yoke and Stub Shaft, Carrier, Carrier Oil Seal and Bearing

REMOVAL & INSTALLATION

Dana 50 Independent Front Axle

1. Before servicing the vehicle, refer to the precautions section.

➡**This procedure requires the use of special tools.**

2. Raise and support the front end on jackstands.
3. Disconnect the front driveshaft from the carrier and wire it up out of the way.
4. Remove the left and right axle shafts and both spindles.
5. Support the carrier with a floor jack and unbolt the carrier from the support arm.
6. Place a drain pan under the carrier, separate the carrier from the support arm and drain the carrier.
7. Remove the carrier from the truck.
8. Place the carrier in holding fixture T57L–500–B with adapters T80T–4000–B.
9. Rotate the slip yoke and shaft assembly from the carrier.
10. Using a slide hammer/puller remove the caged needle bearing and oil seal as a unit. Discard the oil seal and bearing.
To install:
11. Clean the bearing bore thoroughly and make sure that it is free of nicks and burrs.
12. Insert a new bearing in its bore with the printing facing outward. Drive it into place with driver T83T–1244–A, or its equivalent. Install a new bearing seal with the lip facing away from the bearing. Coat the bearing and seal with waterproof wheel bearing grease.
13. Install the slip yoke and shaft assembly into the carrier so that the groove in the shaft is visible in the differential case.
14. Install the snapring in the groove in the shaft. It may be necessary to force the snapring into place with a small prybar. Don't strike the snapring!
15. Remove the carrier from the holding fixture.
16. Clean all traces of sealant from the carrier and support arm. Make sure the mating surfaces are clean. Apply a¼ in. (6mm) wide bead of RTV sealant to the mating surface of the carrier. The bead must be continuous and should not pass through or outside of the holes. Install the carrier with 5 minutes of applying the sealer.

17. Position the carrier on the jack and raise it into position using guide pins to align it if you'd like. Install and hand-tighten the bolts. Tighten the bolts in a circular pattern to 30–40 ft. lbs. (41–54 Nm).

18. Install the support arm tab bolts and tighten them to 85–100 ft. lbs. (115–136 Nm).

19. Install all other parts in reverse order of removal.

Pinion Seal

REMOVAL & INSTALLATION

Front

INDEPENDENT FRONT AXLE

1. Before servicing the vehicle, refer to the precautions section.

➡A torque wrench capable of at least 225 ft. lbs. (305 Nm) is required for pinion seal installation.

2. Raise and safely support the vehicle with jackstands under the frame rails. Allow the axle to drop to rebound position for working clearance.

3. Mark the companion flanges and U-joints for correct reinstallation position.

4. Remove the driveshaft. Use a suitable tool to hold the companion flange. Remove the pinion nut and companion flange.

5. Use a slide hammer and hook or sheet metal screw to remove the oil seal.

To install:

6. Install a new pinion seal after lubricating the sealing surfaces. Use a suitable seal driver. Install the companion flange and pinion nut. Tighten the nut to 200–220 ft. lbs. (271–298 Nm).

2002–04 MONOBEAM FRONT AXLE

1. Before servicing the vehicle, refer to the precautions section.

➡A torque wrench capable of at least 300 ft. lbs. (407 Nm) is required for pinion seal installation.

2. Raise and support the truck on jackstands.

3. Allow the axle to hang freely.

4. Matchmark and disconnect the driveshaft from the front axle.

5. Using a tool such as T75T–4851–B, or equivalent, hold the pinion flange while removing the pinion nut.

6. Using a puller, remove the pinion flange.

7. Use a puller to remove the seal, or punch the seal out using a pin punch.

To install:

8. Thoroughly clean the seal bore and make sure that it is not damaged in any way. Coat the sealing edge of the new seal with a small amount of 80W/90 oil and drive the seal into the housing using a seal driver.

9. Coat the inside of the pinion flange with clean 80W/90 oil and install the flange onto the pinion shaft.

10. Install the nut on the pinion shaft and tighten it to 250–300 ft. lbs. (339–407 Nm).

11. Connect the driveshaft.

2005 MONOBEAM FRONT AXLE

1. Before servicing the vehicle, refer to the precautions section.

2. Index-mark the front driveshaft and the front axle flange to maintain driveline balance.

3. Disconnect the front driveshaft from the front axle flange and position it aside.

4. Measure the pinion bearing torque preload. Record the reading.

5. Rotate the pinion with a Nm (inch-pound) torque wrench. Record the torque necessary to maintain rotation of the pinion through several revolutions.

6. Using tool 205-012, remove and discard the nut and washer.

NOTE: Index-mark the flange and the pinion shaft.

7. Using tool 205-018, remove the flange.

8. Using a puller and flange removal tool 205-018, remove the pinion seal. Discard the seal.

9. Clean and inspect the following:

   a. The seal mounting surface.

   b. The flange lugs and the flange end that contacts the bearing cone.

   c. Verify that the flange nut counterbore and the seal contact surfaces are smooth and free of nicks.

To install:

10. Using a suitable driver, install the pinion seal. Lightly coat the pinion seal lip with lubricant prior to installation.

✳✳ CAUTION

Never use a metal hammer on the pinion flange or install the flange with power tools. If necessary, use a plastic hammer to tap on a tight fitting flange.

➡Align the index marks.

11. Lightly coat the flange splines and seal mating area with lubricant, then install the flange with a new washer and nut.

✳✳ CAUTION

Never back off the pinion nut to reduce preload. If preload reduction is necessary, install a new collapsible spacer and pinion nut.

12. Tighten the nut as follows:

   a. Use tool 205-011 to prevent the flange from turning while tightening the nut.

   b. Remove the special tool when taking pinion bearing torque preload readings.

   c. Take frequent pinion bearing torque preload readings.

   d. The final reading must be 0.56 Nm (5 lb-in) more than the initial reading taken during removal.

13. If installing a new nut on the pinion shaft and tighten it to 250–300 ft. lbs. (339–407 Nm).

14. Connect the front driveshaft to the front axle flange. Install the two new retainers and four new bolts and tighten to 35 Nm (36 lb-ft).

15. Check and, if necessary, fill the axle with the specified lubricant.

FORD 10.50 INCH RING GEAR

➡The rear wheels and brake calipers must be removed to prevent brake drag during drive pinion bearing preload adjustment.

1. Remove the rear brake calipers.

2. Remove the driveshaft.

3. Install a Nm torque wrench on the pinion nut and record the rotational torque required to maintain rotation of the pinion through several revolutions.

✳✳ CAUTION

After removal of the pinion nut, discard it. A new nut must be used for installation.

4. Using the too 205-126l, hold the pinion flange while removing the pinion nut.

5. Mark the pinion flange in relation to the drive pinion stem to make sure alignment is correct during installation.

6. Using puller tool 205-D072, remove the pinion flange.

7. Force up on the metal flange of the rear axle drive pinion seal. Install gripping pliers to the seal flange and strike with a hammer until the rear axle drive pinion seal is removed.

To install:

8. Lubricate the new pinion seal with clean premium long-life grease.

❋ CAUTION

If the rear axle drive pinion seal becomes misaligned during installation, remove the rear axle drive pinion seal and replace it with a new seal.

9. Using seal installer tool 205-208, install a new rear axle drive pinion seal.

10. Lubricate the pinion flange splines with clean synthetic rear axle lubricant.

➡ Disregard the scribe marks if a new pinion flange is being installed.

11. Align the pinion flange with the drive pinion shaft.

12. With the pinion flange in place in the rear axle housing, use tool 205-233 to install the pinion flange.

13. Position the new pinion nut.

❋❋ CAUTION

Under no circumstances is the pinion nut to be backed off to reduce preload. If reduced preload is required, a new collapsible spacer and pinion nut must be installed.

14. Use holding tool 205-126 to hold the pinion flange while tightening the pinion nut.

15. Tighten the pinion nut, rotating the pinion occasionally to make sure the cone and roller bearings are seating correctly. Take frequent cone and roller bearing torque preload readings until the original recorded preload reading is obtained by rotating the pinion with a Nm torque wrench.

16. Install the driveshaft.

17. Install the brake calipers.

STEERING AND SUSPENSION

Air Bag

❋ CAUTION

Some vehicles are equipped with an air bag system. The system must be disabled before performing service on or around system components, steering column, instrument panel components, wiring and sensors. Failure to follow safety and disabling procedures could result in accidental air bag deployment, possible personal injury and unnecessary system repairs.

PRECAUTIONS

Several precautions must be observed when handling the inflator module to avoid accidental deployment and possible personal injury:

• Never carry the inflator module by the wires or connector on the underside of the module.

• When carrying a live inflator module, hold securely with both hands and ensure that the bag and trim cover are pointed away.

• Place the inflator module on a bench or other surface with the bag and trim cover facing up.

• With the inflator module on the bench, never place anything on or close to the module, which may be thrown in the event of an accidental deployment.

DISARMING

2002–04 Models

❋❋ CAUTION

The Supplemental Inflatable Restraint (SIR) system must be dis-

armed before performing service around SIR system components or SIR system wiring. Failure to do so may cause accidental deployment of the air bag, resulting in unnecessary SIR system repairs and/or personal injury.

The positive battery cable must be disconnected for a minimum of 1 minute before beginning any air bag work to de-energize the back-up power supply. It is a good idea to disengage both the positive and negative battery cables to ensure that the Air Bag system is definitely discharged.

2005–06 Models

❋❋ WARNING

Always wear safety glasses when repairing an air bag supplemental restraint system (SRS) vehicle and when handling an air bag module. This will reduce the risk of injury in the event of an accidental deployment.

❋❋ WARNING

Carry a live air bag module with the air bag and trim cover pointed away from your body. This will reduce the risk of injury in the event of an accidental deployment.

❋❋ WARNING

Do not set a live air bag module down with the trim cover face down. This will reduce the risk of injury in the event of an accidental deployment.

❋❋ WARNING

After deployment, the air bag surface can contain deposits of sodium hydroxide, a product of the gas generant combustion that is irritating to the skin. Wash your hands with soap and water afterwards.

❋❋ WARNING

Never probe the connectors on the air bag module. Doing so can result in air bag deployment, which can result in personal injury.

❋❋ WARNING

Never probe the connectors on the safety canopy module. Doing so can result in safety canopy module deployment.

❋❋ WARNING

To reduce the risk of personal injury, do not use any memory saver devices.

➡ The air bag warning lamp illuminates when the Restraints Control Module (RCM) fuse is removed and the ignition switch is ON. This is normal operation and does not indicate a supplemental restraint system (SRS) fault.

➡ The SRS must be fully operational and free of faults before releasing the vehicle to the customer.

1. Turn all vehicle accessories OFF.

2. Turn the ignition switch to OFF.

3. On Navigator and Expedition models, at the Central Junction Box (CJB), located below the right hand side of the instrument

panel, open the kick panel cover and remove the RCM fuse F1.19 (10A) from the CJB.

4. On Excursion models, at the central junction box CJB, located below the left hand side of the instrument panel, remove the cover and the RCM fuse F2.26 (10A) from the CJB.

5. Turn the ignition ON and visually monitor the air bag indicator for at least 30 seconds. The air bag indicator will remain lit continuously (no flashing) if the correct RCM fuse has been removed. If the air bag indicator does not remain lit continuously, remove the correct RCM fuse before proceeding.

6. Turn the ignition OFF.

✳✳ WARNING

To avoid accidental deployment and possible personal injury, the backup power supply must be depleted before repairing or replacing any front or side air bag Supplemental Restraint System (SRS) components and before servicing, replacing, adjusting or striking components near the front or side air bag sensors or RCM, such as doors, instrument panel, console, door latches, strikers, seats and hood latches.

7. To deplete the backup power supply energy, disconnect the battery ground cable and wait at least one minute. Be sure to disconnect auxiliary batteries and power supplies (if equipped).

ARMING THE SYSTEM

2002–04 Models

✳✳ WARNING

If the air bag simulators have been used, the air bag simulators must be removed and the air bags reconnected when the system is reactivated to avoid non-deployment in a collision resulting in possible personal injury.

1. Disconnect the positive battery cable.
2. Wait 1 minute, this is required for the back-up power supply in the air bag diagnostic monitor to deplete its stored energy.
3. Remove the air bag simulator from the air bag sliding contact connector at the top of the steering column. Reconnect the driver's side air bag module assembly. Position the driver's air bag module on the steering wheel and secure with the 2 bolts

and washers. Tighten the bolt and washer assembly to 8–10 ft. lbs. (10–14 Nm).

4. Connect the positive battery cable.
5. Turn the ignition switch from the **OFF** to **RUN** and visually monitor the air bag warning indicator. The light will illuminate continuously for approximately 6 seconds and then turn off. If a fault occurs, the air bag indicator will either fail to light, remain lighted continuously or flash. The flashing may not occur until approximately 30 seconds after the ignition switch has been turned from **OFF** to **RUN**. This is the time needed for the air bag diagnostic monitor to complete testing the system. If the air bag indicator is inoperative, an air bag system fault exists, a tone will sound in a pattern of 5 sets of 5 beeps. If this occurs, the air bag indicator will need to be serviced before further diagnostics can be done.

2005–06 Models

✳✳ WARNING

The restraint system diagnostic tool is for restraint system service only. Remove from vehicle prior to road use. Failure to remove could result in injury and possible violation of vehicle safety standards.

1. Make sure all restraint system diagnostic tool(s) that may have been installed during the repair have been removed from the vehicle and all SRS components are connected.
2. Turn the ignition switch from OFF to ON.
3. Install the RCM fuse to the CJB and close the cover.

✳✳ WARNING

Be sure that nobody is in the vehicle and that there is nothing blocking or set in front of any air bag module when the battery ground cable is connected.

4. Connect the battery ground cable.
5. Prove out the SRS as follows:
   a. Turn the ignition key from ON to OFF. Wait 10 seconds, then turn the key back to ON and visually monitor the air bag indicator with the air bag modules installed. The air bag indicator will light continuously for approximately six seconds and then turn off. If an air bag supplemental restraint system (SRS) fault is present, the air bag indicator will either, fail to light, remain lit continuously or flash.

b. The flashing might not occur until approximately 30 seconds after the ignition switch has been turned from the OFF to the ON position. This is the time required for the RCM to complete the testing of the SRS.

c. If the air bag indicator is inoperative and a SRS fault exists, a chime will sound in a pattern of five sets of five beeps. If this occurs, the air bag indicator and any SRS fault discovered must be diagnosed and repaired.

d. Clear all continuous DTCs from the restraints control module using a scan tool.

Steering Gear

REMOVAL & INSTALLATION

Excursion

1. Before servicing the vehicle, refer to the precautions in the beginning of this section.

➡**New O-ring seals must be installed any time the lines are disconnected from the steering gear.**

2. Remove the air cleaner assembly.
3. Disengage the steering coupling shield from the line fitting and slide upward on the steering shaft.
4. Remove the pinch bolt. Turn the steering wheel as necessary to access the bolt.
5. Make sure the steering column is locked.
6. Disconnect the lines.
7. Discard the O-ring seals.
8. Remove the tie rod cotter pin and nut. Discard the cotter pin.
9. Using tool 211-003, disconnect the drag link.
10. Remove the bolts, washers and the steering gear.
11. Remove the pitman arm nut.
12. Using tool 211-003, remove the steering gear sector shaft arm.

To install:

13. Install the steering gear sector shaft arm.
14. Install the pitman arm nut. Tighten the nut to 199 ft. lbs. (270 Nm).
15. Install steering gear, bolts and washers. Tighten to 59 ft. lbs. (80 Nm).
16. Connect the drag link.
17. Install the tie rod nut and new cotter pin. Tighten to 66 ft. lbs. (90 Nm).
18. Install a new high pressure hose O-ring seal and a new return hose O-ring seal.

19. Connect the lines. Tighten to 26 ft. lbs. (35 Nm).

20. Make sure the steering column is locked.

21. Connect the pinch bolt. Tighten to 36 ft. lbs. (48 Nm).

22. Engage the steering coupling shield to the line fitting.

23. Install the air cleaner assembly.

24. Fill and leak check the system.

Expedition And Navigator

2002 MODELS

1. Before servicing the vehicle, refer to the precautions in the beginning of this section.

✳ WARNING

The electrical power to the air suspension system must be shut off prior to hoisting, jacking or towing an air suspension vehicle. To do this, turn off the air suspension switch located in the right hand kick panel. Failure to do so can result in unexpected inflation or deflation of the air springs, which can result in shifting of the vehicle during these operations.

2. Remove the skid plate.

3. Remove the steering sector shaft arm drag link castellated nut and cotter pin.

4. Using tool 211-003, separate the steering sector shaft arm drag link.

5. Remove the dust shield.

6. Remove the pinch bolt.

✳ CAUTION

Do not rotate the steering wheel when the steering column shaft is disconnected, or damage to the air bag sliding contact will result.

7. Slide the intermediate shaft off the steering gear input shaft.

8. Turn the ignition key to the locked position.

9. Disconnect the power steering lines at the steering gear.

10. Remove the bolts and the steering gear.

To install:

11. Install the steering gear. Tighten the bolts to 173–233 ft. lbs. (234–316 Nm).

12. Install a new high pressure hose O-ring seal and a new return hose O-ring seal.

13. Connect the power steering lines to the steering gear. Tighten to 17 ft. lbs. (23 Nm).

14. Slide the intermediate shaft Into the steering gear input shaft.

✳ CAUTION

Do not rotate the steering wheel when the steering column shaft is disconnected, or damage to the air bag sliding contact will result.

15. Tighten the pinch bolt to 30–40 ft. lbs. (41–54 Nm).

16. Install the dust shield.

17. Attach the steering sector shaft arm drag link. Tighten to 173–233 ft. lbs. (234–316 Nm).

18. Install the steering sector shaft arm drag link castellated nut and new cotter pin. Tighten to 57–76 ft. lbs. (77–103 Nm).

19. Install the skid plate.

20. Fill and leak check the system

2003–04 4X2 MODELS

1. Before servicing the vehicle, refer to the precautions in the beginning of this section.

2. Place the front wheels in the straight-ahead position and the ignition switch in the OFF position.

✳ WARNING

The electrical power to the air suspension system must be shut off prior to hoisting, jacking or towing an air suspension vehicle. To do this, turn off the air suspension switch located near the jack storage area in the rear of the passenger compartment. Failure to do so can result in unexpected inflation or deflation of the air springs, which can result in shifting of the vehicle during these operations.

3. If equipped, turn the air suspension switch to the OFF position.

4. With the vehicle in NEUTRAL, position on a hoist.

➡ **The hex holding feature can be used to prevent turning of the stud while removing the nut.**

5. Remove the nuts and detach the tie-rods from the wheel knuckles.

6. Remove the bolts and the oil drip shield.

7. Detach the lines from the clip.

✳ CAUTION

Do not allow the intermediate shaft to rotate while it is disconnected

from the gear or damage to the clockspring can occur. If there is evidence that the intermediate shaft has rotated, the clockspring must be removed and recentered.

8. Remove the bolt and detach the intermediate shaft from the gear.

9. Remove the bolt and disconnect the power steering lines.

10. Discard the O-ring seals.

11. On Navigator models:

   a. Disconnect the electrical connector.

   b. Rotate the actuator until the connector housing is facing the steering gear input shaft.

12. Remove the nuts and bolts. Discard the nuts.

✳ CAUTION

Make sure that the steering lines are clear from the removal path of the gear or damage to the lines can result.

13. Remove the nuts, bolts, brackets and the steering gear. Discard the nuts.

To install:

✳ CAUTION

Make sure to tighten the bracket-to-steering gear bolts (M14) before tightening the bracket-to-crossmember bolts (M12) or damage to the steering gear can result.

14. Install the steering gear, bolts and new nuts. Tighten to 111 ft. lbs. (150 Nm).

15. Install the nuts and bolts. Tighten to 76 ft. lbs. (103 Nm).

16. On Navigator models:

   a. Rotate the actuator until the connector housing is facing the steering gear input shaft.

   b. Connect the electrical connector.

17. If equipped, turn the air suspension switch to the OFF position.

18. With the vehicle in NEUTRAL, position on a hoist.

➡ **The hex holding feature can be used to prevent turning of the stud while removing the nut.**

19. Remove the nuts and detach the tie-rods from the wheel knuckles.

20. Remove the bolts and the oil drip shield.

21. Detach the lines from the clip.

✳✳ CAUTION

Do not allow the intermediate shaft to rotate while it is disconnected from the gear or damage to the clockspring can occur. If there is evidence that the intermediate shaft has rotated, the clockspring must be removed and recentered.

22. Remove the bolt and detach the intermediate shaft from the gear.
23. Remove the bolt and disconnect the power steering lines.
24. Install a new high pressure hose O-ring seal and a new return hose O-ring seal.
25. Install the power steering lines. Tighten the bolt to 17 ft. lbs. (23 Nm).
26. Attach the intermediate shaft from the gear. Tighten the bolt to 22 ft. lbs. (30 Nm).

✳✳ CAUTION

Do not allow the intermediate shaft to rotate while it is disconnected from the gear or damage to the clockspring can occur. If there is evidence that the intermediate shaft has rotated, the clockspring must be removed and recentered.

27. Attach the lines to the clip.
28. Install the oil drip shield.
29. Install the tie-rods to the wheel knuckles and tighten the nut to 111 ft. lbs. (150 Nm).
30. If equipped, turn the air suspension switch to the ON position.
31. Fill and leak check the system

2003–04 4X4 MODELS

1. Before servicing the vehicle, refer to the precautions in the beginning of this section.

✳✳ CAUTION

While repairing the power steering system, care should be taken to prevent the entry of contaminants or premature failure of the power steering components can result.

➡ **New O-ring seals must be installed any time the lines are disconnected from the steering gear.**

2. Place the front wheels in the straight-ahead position and the ignition switch in the OFF position.

✳✳ WARNING

The electrical power to the air suspension system must be shut off prior to hoisting, jacking or towing an air suspension vehicle. This can be accomplished by turning off the air suspension switch located near the jack storage area in the rear of the passenger compartment. Failure to do so can result in unexpected inflation or deflation of the air springs, which can result in shifting of the vehicle during these operations. Failure to follow these instructions may result in personal injury.

3. Turn the air suspension switch to the OFF position, if equipped.
4. Remove the engine cooling fan.

➡ **The hex holding feature can be used to prevent turning of the stud while removing the nut.**

5. Remove the 2 nuts and disconnect the tie-rod ends from the wheel knuckles.
6. Remove the 2 bolts and the oil drip shield.
7. Disconnect the lines from the frame rail retainers.

➡ **Do not allow the intermediate shaft to rotate while it is disconnected from the gear or damage to the clockspring can occur.**

8. Remove the intermediate shaft-to-gear pinch bolt and disconnect the intermediate shaft from the gear.
9. Remove the steering line clamp plate bolt and disconnect the power steering lines.
10. Position the lines aside and discard the O-ring seals.
11. On Navigator models, disconnect the electrical connector by rotating the actuator until the connector housing is facing the steering gear input shaft.
12. On Navigator models, remove the left hand lower arm-to-frame nut and bolt and discard the nut.
13. On Navigator models, insert a block of wood between the left hand lower arm and the frame.
14. Remove the 2 steering gear bracket-to-crossmember nuts and bolts and discard the nuts.

✳✳ CAUTION

Make sure the steering lines are clear from the removal path of the gear or damage to the lines can result.

15. Remove the 2 steering gear bracket-to-gear nuts, bolts, and brackets and discard the nuts.
16. Remove the steering gear from the left hand side of the vehicle.

✳✳ CAUTION

Make sure that the wheel end vacuum line is clear of the steering gear mounting fasteners before tightening the bolts or damage to the line can result.

To install:
17. Installation is the reverse of removal, please note the following points and torque specifications:

✳✳ CAUTION

Make sure to tighten the bracket-to-steering gear bolts (M14) before tightening the bracket-to-crossmember bolts (M12) or damage to the steering gear can result.

a. Install a new high pressure line O-ring seal and a new return line O-ring seal.
b. Install the 2 new steering gear bracket-to-gear nuts, bolts and tighten to 150 Nm (111 ft. lbs.).
c. Install the 2 new steering gear bracket-to-crossmember nuts and bolts and tighten to 103 Nm (76 ft. lbs.).
d. Install the new left hand lower arm-to-frame nut and bolt and tighten to 200 Nm (148 ft. lbs.).
e. Tighten the steering line clamp bolt to 23 Nm (17 ft. lbs.).
f. Tighten the intermediate shat-to-gear pinch bolt to 30 Nm (22 ft. lbs.).
g. Tighten the new tie rod nuts to 150 Nm (111 ft. lbs.).
h. Fill and leak check the system.

2005–06 4X2 MODELS

1. Before servicing the vehicle, refer to the precautions in the beginning of this section.

✳✳ CAUTION

While repairing the power steering system, care should be taken to prevent the entry of contaminants or premature failure of the power steering components can result.

➡ **New O-ring seals must be installed any time the lines are disconnected from the steering gear.**

2. Place the front wheels in the straight-ahead position and the ignition switch in the OFF position.

✳✳ WARNING

The electrical power to the air suspension system must be shut off prior to hoisting, jacking or towing an air suspension vehicle. This can be accomplished by turning off the air suspension switch located near the jack storage area in the rear of the passenger compartment. Failure to do so can result in unexpected inflation or deflation of the air springs, which can result in shifting of the vehicle during these operations. Failure to follow these instructions may result in personal injury.

3. Turn the air suspension switch to the OFF position, if equipped.

4. With the vehicle in NEUTRAL, position the vehicle on a hoist.

➡The hex holding feature can be used to prevent turning of the stud while removing the nut.

5. Remove the 2 nuts and disconnect the tie-rod ends from the wheel knuckles.

6. Remove the 2 bolts and the oil drip shield.

7. Disconnect the lines from the frame rail retainers.

✳ CAUTION

Do not allow the intermediate shaft to rotate while it is disconnected from the gear or damage to the clockspring can occur.

8. Remove the intermediate shaft-to-gear pinch bolt and disconnect the intermediate shaft from the steering gear.

9. Remove the steering line clamp plate bolt and disconnect the power steering lines. Discard the O-ring seals.

10. On Navigator models, disconnect the electrical connector by rotating the actuator until the connector housing is facing the steering gear input shaft.

11. Remove the 2 steering gear bracket-to-crossmember nuts and bolts and discard the nuts.

✳ CAUTION

Make sure the steering lines are clear from the removal path of the gear or damage to the lines can result.

12. Remove the 2 steering gear bracket-to-gear nuts, bolts, and brackets and discard the nuts.

13. Remove the steering gear.

14. Installation is the reverse of removal, please note the following points and torque specifications:

✳✳ CAUTION

Make sure to tighten the bracket-to-steering gear bolts (M14) before tightening the bracket-to-crossmember bolts (M12) or damage to the steering gear can result.

 a. Install a new high pressure line O-ring seal and a new return line O-ring seal.

 b. Tighten the 2 new steering gear bracket-to-gear nuts and bolts to 150 Nm (111 ft. lbs.).

 c. Tighten the 2 new steering gear bracket-to-crossmember nuts and bolts to 103 Nm (76 ft. lbs.).

 d. Tighten the steering line clamp bolt to 23 Nm (17 ft. lbs.).

 e. Tighten the intermediate shat-to-gear pinch bolt to 30 Nm (22 ft. lbs.).

 f. Tighten the new tie rod nuts to 150 Nm (111 ft. lbs.).

 g. Fill the power steering system.

2005–06 4X4 MODELS

1. Before servicing the vehicle, refer to the precautions in the beginning of this section.

✳✳ CAUTION

While repairing the power steering system, care should be taken to prevent the entry of contaminants or premature failure of the power steering components can result.

➡New O-ring seals must be installed any time the lines are disconnected from the steering gear.

2. Place the front wheels in the straight-ahead position and the ignition switch in the OFF position.

✳✳ WARNING

The electrical power to the air suspension system must be shut off prior to hoisting, jacking or towing an air suspension vehicle. This can be accomplished by turning off the air suspension switch located near the jack storage area in the rear of the passenger compartment. Failure to do so can result in unexpected inflation or deflation of the air springs, which can result in shifting of the vehicle during these operations. Failure to follow these instructions may result in personal injury.

3. Turn the air suspension switch to the OFF position, if equipped.

4. Remove the engine cooling fan.

➡The hex holding feature can be used to prevent turning of the stud while removing the nut.

5. Remove the 2 nuts and disconnect the tie-rod ends from the wheel knuckles.

6. Remove the 2 bolts and the oil drip shield.

7. Disconnect the lines from the frame rail retainers.

✳✳ CAUTION

Do not allow the intermediate shaft to rotate while it is disconnected from the gear or damage to the clockspring can occur. If there is evidence that the intermediate shaft has rotated, the clockspring must be removed and recentered.

8. Remove the intermediate shaft-to-gear pinch bolt and disconnect the intermediate shaft from the gear.

9. Remove the steering line clamp plate bolt and disconnect the power steering lines. Position the lines aside and discard the O-ring seals.

10. On Navigator models:

 a. Disconnect the electrical connector. Rotate the actuator until the connector housing is facing the steering gear input shaft.

 b. Remove the left hand lower arm-to-frame nut and bolt. Discard the nut.

 c. Insert a block of wood between the left hand lower arm and the frame.

11. Remove the 2 steering gear bracket-to-crossmember nuts and bolts. Discard the nuts.

✳✳ CAUTION

Make sure the steering lines are clear from the removal path of the gear or damage to the lines can result.

12. Remove the 2 steering gear bracket-to-gear nuts, bolts, and brackets. Discard the nuts.

13. Remove the steering gear from the left hand side of the vehicle.

✳✳ CAUTION

Make sure that the wheel end vacuum line is clear of the steering gear mounting fasteners before tightening the bolts or damage to the line can result.

To install:

✳✳ CAUTION

Make sure to tighten the bracket-to-steering gear bolts (M14) before tightening the bracket-to-crossmember bolts (M12) or damage to the steering gear can result.

14. Installation is the reverse of removal, note the following torque specifications:

   a. Tighten the 2 new steering gear bracket-to-gear nuts, bolts to 150 Nm (111 lb-ft).

   b. Tighten the 2 new steering gear bracket-to-crossmember nuts and bolts to 103 Nm (76 lb-ft).

   c. Tighten the new left hand lower arm-to-frame nut and bolt to 200 Nm (148 lb-ft).

   d. Install new O-rings and tighten the steering line clamp plate bolt to 23 Nm (17 lb-ft).

   e. Tighten the intermediate shaft-to-gear pinch bolt to 30 Nm (22 lb-ft).

   f. Tighten the 2 tie-rod ends to wheel knuckle nuts to 150 Nm (111 lb-ft).

15. Fill and leak check the system.

Shock Absorber

REMOVAL & INSTALLATION

Front

EXCEPT EXCURSION AND 2005–06 MODELS

1. Before servicing the vehicle, refer to the precautions section.

2. If equipped with 2-wheel drive, perform the following:

   a. Hold the shock absorber stem and remove the nut from the top of the shock.

   b. Raise and safely support the vehicle.

   c. Remove the 2 lower retaining nuts and remove the shock absorber.

3. If equipped with 4-wheel drive, perform the following:

   a. Hold the shock absorber stem and remove the nut, washer and bushing from the top of the shock absorber stud.

   b. Raise and support the vehicle.

   c. Remove the lower retaining nut and bolt.

   d. Remove the shock absorber from the vehicle.

To install:

4. On 2-wheel drive models, perform the following:

   a. Install the washer and bushing to the top stem of the shock absorber.

   b. Place the shock absorber up through the coil spring.

   c. Install the 2 lower retaining nuts. Tighten the nuts to 22–29 ft. lbs. (30–40 Nm).

   d. Lower the vehicle.

   e. Install the bushing, washer and retaining nut to the top of the shock absorber stud. Tighten the nut to 34–46 ft. lbs. (47–63 Nm).

5. On 4-wheel drive models, perform the following:

   a. Install the washer and bushing to the top stem of the shock absorber.

   b. Place the shock absorber up through the coil spring.

   c. Install the lower retaining nut and bolt. Tighten to 57–76 ft. lbs. (77–104 Nm).

   d. Lower the vehicle.

   e. Install the shock absorber upper bushing, washer and retaining nut. Tighten the nut to 22–29 ft. lbs. (30–40 Nm).

6. Road test the vehicle and check for proper operation.

2005–06 MODELS

1. Before servicing the vehicle, refer to the precautions section.

2. Remove the wheel assembly.

3. Remove and discard the upper shock nuts. Tighten to 35 Nm (26 ft. lbs.).

4. Remove the nut and detach the tie-rod end from the wheel knuckle. Discard the nut.

5. Remove the lower nut, bolt and the shock absorber and spring assembly. Discard the nut.

To install:

6. Installation is the reverse of removal, please note the following torque specifications:

   a. New lower nut, bolt and tighten to 400 Nm (295 ft. lbs.).

   b. New tie-rod end nut and tighten to 150 Nm (111 ft. lbs.).

   c. New upper shock nuts and tighten to 35 Nm (26 ft. lbs.).

EXCURSION—4X2 MODELS

1. Before servicing the vehicle, refer to the precautions section.

2. Remove the upper shock absorber retaining nut and upper shock absorber insulator.

3. Raise and support the vehicle.

4. Remove the lower shock absorber retaining nut and remove the shock absorber.

To install:

5. Installation is the reverse of removal. Using new fasteners, tighten the lower nut to 60 ft. lbs. (80 Nm) and the upper nut to 30 ft. lbs. (40 Nm).

EXCURSION—4X4 MODELS

1. Before servicing the vehicle, refer to the precautions section.

2. Remove the lower shock absorber retaining nut and remove the shock absorber.

3. Remove the upper shock absorber retaining nut and upper shock absorber insulator.

To install:

4. Installation is the reverse of removal. Using new fasteners, tighten the lower and upper nuts to 76 ft. lbs. (103 Nm).

77 - 104 Nm (57 - 76 lb/ft)

7924FG73

Lower shock absorber mounting—late models with torsion bar suspension

Rear

2002–04 EXPEDITION AND NAVIGATOR MODELS

1. Before servicing the vehicle, refer to the precautions section.

2. Raise the vehicle and secure on support stands.

3. Remove the self-locking nut, steel washer, and rubber bushings at the upper end of the shock absorber.

4. Remove the bolt and nut at the lower end and remove the shock absorber. If needed, raise the rear axle assembly slightly with a jack.

To install:

5. When installing a new shock absorber, use new rubber bushings. Position the shock absorber on the mounting brackets with the stud end at the top. Install the upper bushing, steel washer and self-locking nut at the upper end, and the bolt and nut at the lower end.

6. Tighten the upper mounting studs to 63–84 ft. lbs. (85–115 Nm) and the lower mounting nuts to 63–84 ft. lbs. (85–115 Nm).

2005 EXPEDITION AND NAVIGATOR MODELS

1. Before servicing the vehicle, refer to the precautions section.

✻✻ WARNING

The electrical power to the air suspension system must be shut off prior to hoisting, jacking or towing an air suspension vehicle. This can be accomplished by turning off the air suspension switch located in the left rear quarter trim panel. Failure to do so can result in unexpected inflation or deflation of the air springs, which can result in shifting of the vehicle during these operations.

2. If equipped, turn the air suspension switch to the OFF position.

3. Remove the wheel assembly.

4. Remove and discard the upper nuts.

5. Remove the lower bolt and flag nut. Discard the flag nut.

6. Remove the shock absorber and spring assembly.

7. Installation is the reverse of removal, please note the following torque specifications:

   a. New lower bolt and flag nut and tighten to 475 Nm (350 ft. lbs.).

   b. New upper shock nuts and tighten to 35 Nm (26 ft. lbs.).

2005 EXCURSION MODELS

1. Before servicing the vehicle, refer to the precautions section.

✻✻ CAUTION

Suspension fasteners are critical parts because they affect performance of vital components and systems and their failure can result in major service expense. Install new parts with the same part number or an equivalent part if installation is necessary. Do not use an installation part of lesser quality or substitute design. Torque values must be used as specified during reassembly to make sure of correct retention of these parts.

✻✻ CAUTION

The low pressure gas shock absorbers are charged with nitrogen gas. Do not attempt to open, puncture or apply heat to shock absorbers.

2. Using a suitable jack, support the rear axle.

3. Remove the shock absorber lower retaining nut and bolt. Tighten to 90 Nm (66 lb-ft).

4. Remove the two shock absorber-to-frame mounting bolts. Tighten to 23 Nm (17 lb-ft).

To install:

5. Installation is the reverse of removal, please note the following torque specifications:

   a. Tighten the two shock absorber-to-frame mounting bolts to 23 Nm (17 lb-ft).

   b. Tighten the shock absorber lower retaining nut and bolt to 90 Nm (66 lb-ft).

Coil Spring

REMOVAL & INSTALLATION

Front

EXCURSION

1. Before servicing the vehicle, refer to the precautions section.

✻✻ CAUTION

Suspension fasteners are critical parts because they affect perfor-

mance of vital components and systems and their failure can result in major service expense. Install new parts with the same part number or an equivalent part if installation is necessary. Do not install a part of lesser quality or substitute design.

2. Remove the wheel and tire assembly.

3. Use a suitable jack to support the front axle assembly.

4. Remove the nut and detach the shock from the mounting stud.

5. Remove the upper spring retainer.

6. Lower the front axle until the spring is free of the upper spring seat.

7. Using an extension through the top of the spring, remove the lower spring retainer.

8. Remove the front spring.

To install:

9. Installation is the reverse of removal. Tighten the lower spring retainer to 99 ft. lbs. (133 Nm), the upper retainer to 26 ft. lbs. (35 Nm) and the lower shock bolt to 60 ft. lbs. (80 Nm).

2002 EXPEDITION AND NAVIGATOR

1. Before servicing the vehicle, refer to the precautions section.

2. Remove or disconnect the following:

- Wheels
- Disc brake caliper and support aside with wire
- Disc brake adapter
- Rotor
- Rotor splash shield
- Shock absorber
- Bracket supporting the brake hose
- Sway bar link retaining nut and bushing from the lower control arm. Separate the sway bar link from the lower control arm.

3. Install a coil spring compressor and compress the coil spring enough to relieve the tension of the spring between the upper and lower control arms.

- Cotter pin and castellated nut from the lower ball joint
- Lower ball joint from the wheel spindle

4. Matchmark the lower control arm alignment cams for installation reference.

- Lower control arm retaining nuts and bolts
- Lower control arm and the compressed coil spring as an assembly

5. Loosen the coil spring compressor and remove the coil spring from the lower control arm.

6. Inspect the coil spring and replace as needed.

7924FG75

Be sure the coil spring is mounted correctly in the lower control arm

To install:

7. Install or connect the following:
- Coil spring correctly in the saddle of the lower control arm
- Coil spring compressor and compress the coil spring. The end of the coil spring **A**, must cover the hole designated **B** and be visible in the second hole designated as **C**, for proper installation.
- Lower control arm to the frame
- Retaining bolts, adjusting cams, and nuts

8. Align the matchmarks on the adjusting cams. The forward nut must be tightened first while the control arm is held at the curb position height. Tighten the nuts to 197–241 ft. lbs. (270–330 Nm).
- Lower ball joint stud into the wheel spindle
- Castellated nut and tighten to 83–113 ft. lbs. (113–153 Nm)
- New cotter pin
- Sway bar link to the lower control arm
- Bushing and retaining nut. Tighten to 15–21 ft. lbs. (21–29 Nm).

9. Remove the coil spring compressor.
- Shock absorber. Tighten the lower bolts to 22 ft. lbs. (32 Nm) and the top nut to 45 ft. lbs. (61 Nm).
- Brake hose bracket
- Brake rotor splash shield
- 3 retaining bolts and tighten them to 90–107 inch lbs. (10–14 Nm)
- Disc brake rotor and caliper assemblies
- Wheel and tire assembly

10. Lower the vehicle.

11. Pump the brake pedal several times to position the brake pads prior to moving the vehicle.

12. Check the alignment and adjust if out of specification.

13. Road test the vehicle and check for proper operation.

2002 EXPEDITION AND NAVIGATOR

1. Before servicing the vehicle, refer to the precautions section.

2. Remove the wheel and tire assembly.

3. Remove and discard the upper shock absorber nuts.

4. Remove the nut and detach the tie-rod from the wheel knuckle. Discard the nut.

5. Remove the nut, bolt and shock absorber/spring assembly. Discard the nut.

6. For reference during assembly, index the upper mount, spring and shock absorber.

7. Using a suitable spring compressor, compress the spring until the tension is released from the shock absorber.

8. While holding the shock rod, remove the nut and washer.

9. Remove the shock absorber. Discard the nut.

10. Remove the upper mount, dust shield and insulator.

11. Installation is the reverse of removal. Tighten the shock rod nut to 22 ft. lbs. (30 Nm).Tighten the shock absorber nut and bolt to 295 ft. lbs. (400 Nm). Tighten the tie rod nut to 111 ft. lbs. (150 Nm) and the new upper shock absorber nuts to 26 ft. lbs. (35 Nm).

Rear

2002 EXPEDITION AND NAVIGATOR

1. Before servicing the vehicle, refer to the precautions section.

2. Raise and support the vehicle.
- Wheel and tire assembly
- Rear driveshaft
- Rear Anti-lock Brake System (ABS) sensor electrical connector

➡**Make sure the parking brake control is fully released.**

3. Pull the front parking brake cable and conduit.

4. Insert a suitable retainer into the parking brake control.

5. Remove or disconnect the following:
- Rear parking brake cable and conduit from the rear axle
- Rear parking brake cable and conduit from the rear axle bracket assembly
- Cable from the caliper

6. Repeat for the other side.
- Disc brake caliper bolts

7. Repeat for the other side.
- Axle vent tube
- Stabilizer bar from the rear axle

- Bolts from each stabilizer bar retainer
- Retainers

8. Use a suitable jack to support the rear axle.
- Lower shock absorber nut and bolt

9. Repeat for the other side.
- Rear axle-to-trackbar assembly bolt
- Rear suspension lower arm assembly-to-axle bolt

10. Repeat for the other side.
- Rear suspension upper arm assembly-to-rear axle bolt

11. Repeat for the other side.

12. Lower the rear axle and remove the coil spring.

To install:

➡**Do not tighten any nuts or bolts until the rear suspension is raised so that the rear suspension lower arms are parallel to the ground. Once in position, tighten all the nuts and bolts to specification.**

13. Install or connect the following:
- Rear coil spring
- Rear suspension upper arm assemblies to the axle and the bolts. Tighten to 94–127 ft. lbs. (128–172 Nm).
- Rear suspension lower arm assemblies to the axle and the bolts. Tighten to 94–127 ft. lbs. (128–172 Nm).
- Trackbar assembly to the axle and the bolts. Tighten to 125–170 ft. lbs. (170–130 Nm).
- Shock absorbers to the axle and the bolts. Tighten to 63–84 ft. lbs. (86–114 Nm).
- Stabilizer bar to the axle and the stabilizer bar retainers and bolts. Tighten to 63–84 ft. lbs. (86–114 Nm).
- Disc brake caliper and the bolts
- Parking brake cables and conduit to the axle
- Parking brake cable and conduit to the lever
- Parking brake cable and conduit to the rear axle bracket assembly
- Tire and wheel assembly
- Lug nuts and lower the vehicle
- Wheel cover.

14. Remove the pin from the parking brake control assembly.
- Rear brake anti-lock sensor (ABS) electrical connector
- Axle vent tube
- Driveshaft

15. Lower the vehicle.

2003–06 EXPEDITION AND NAVIGATOR

1. Before servicing the vehicle, refer to the precautions section.
2. Remove the wheel and tire assembly.
3. Remove and discard the upper shock absorber nuts.
4. Remove the nut, bolt and shock absorber/spring assembly. Discard the nut.
5. For reference during assembly, index the upper mount, spring and shock absorber.
6. Using a suitable spring compressor, compress the spring until the tension is released from the shock absorber.
7. While holding the shock rod, remove the nut and washer.
8. Remove the shock absorber. Discard the nut.
9. Remove the upper mount, dust shield and insulator.
10. Installation is the reverse of removal. Tighten the shock rod nut to 22 ft. lbs. (30 Nm).Tighten the shock absorber nut and bolt to 350 ft. lbs. (475 Nm). Tighten the new upper shock absorber nuts to 26 ft. lbs. (35 Nm).

2005 EXCURSION

1. Before servicing the vehicle, refer to the precautions section.
2. Remove the shock assembly.

➡**For reference during assembly, index the upper mount, spring and shock absorber.**

3. Using a suitable spring compressor, compress the spring until the tension is released from the shock absorber.
4. While holding the shock rod, remove the nut and washer.
5. Remove the shock absorber and discard the nut.
6. Remove the upper mount, dust shield and insulator.

To install:
7. Installation is the reverse of removal. Tighten the shock rod nut to 22 ft. lbs. (30 Nm).

Leaf Spring

REMOVAL & INSTALLATION

Excursion

2002–04 MODELS

1. Before servicing the vehicle, refer to the precautions section.
2. Remove or disconnect the following:
3. Raise and support the vehicle.
   • Wheel and tire assembly

4. Support the rear axle with a suitable jack.
   • U-bolt retaining nuts and the U-bolts.
   • Rear spring upper plate.
   • Nut and bolt from the rear spring front hanger bracket.
   • Lower nut and bolt from the rear spring shackle bracket
   • Rear spring assembly.
   • Nut and bolt from the rear spring shackle assembly and the rear spring shackle

To install:
5. Installation is the reverse of removal, please use new fasteners and note the following torques:
   a. Nut and bolt for the rear spring shackle assembly to 185 ft. lbs. (285 Nm).
   b. Lower nut and bolt for the rear spring shackle bracket to 185 ft. lbs. (285 Nm).
   c. Nut and bolt for the rear spring front hanger bracket to 185 ft. lbs. (285 Nm).
   d. U-bolt retaining nuts and the U-bolts to 185 ft. lbs. (285 Nm).

2005 MODELS

1. Before servicing the vehicle, refer to the precautions section.
2. Position jack stands under the axle housing.
3. Remove the shock from the front spring cap.

✳✳ CAUTION

Never reuse U-bolts. The U-bolts are a torque-to-yield design and cannot be retightened. Failure to use a new U-bolt can result in loose or broken springs and suspension components.

4. Remove the four nuts, the two U-bolts and the front spring spacer. Discard the U-bolts.
5. Disconnect the front driveshaft from the pinion flange.
6. Remove the spring shackle nut and bolt. Position the spring shackle away from the spring.
7. Detach the pushpin and remove the bolt and condenser bracket.
8. Remove the spring hanger nut and bolt.
9. Using a hammer, tap the spring until the spring is free from the hanger bracket.
10. Remove the spring from the vehicle.
To install:
11. Position the spring in the vehicle and install the spring into the spring hanger.
12. Install the bolt and nut. Do not tighten the nut and bolt at this time.

13. Install the bracket and bolt and the push pin.
14. Install the spring into the spring shackle. Position the shackle and install the bolt and nut. Do not tighten the nut and bolt at this time.

➡**The U-bolts insert through the front spring cap on the left-hand side and through the axle assembly bracket on the right-hand side.**

15. Install the upper front spring spacer and U-bolts. Install the four nuts. Do not tighten the nuts at this time.
16. Install the shock, bolt and nut onto the spring cap. Tighten the nut to 103 Nm (76 lb-ft).

➡**The suspension must be loaded with the weight of the vehicle before the U-bolts and the leaf spring mounting bolts can be tightened. Make sure that the locating pin is correctly aligned with the axle.**

17. Lower the vehicle onto the jack stands until the front suspension is supporting the weight of the vehicle.

➡**The U-bolts must be tightened in a criss-cross sequence. This will tighten the U-bolts evenly.**

18. Tighten the U-bolts following in a criss-cross pattern to 133 Nm (99 lb-ft) using several passes.
19. Tighten the leaf spring rear retaining bolt and nut to 250 Nm (185 lb-ft).
20. Tighten the leaf spring front retaining bolt and nut to 275 Nm (203 lb-ft).
21. Install the front driveshaft. Tighten the two strap bolts to 102 Nm (76 lb-ft).

Track Bar

REMOVAL & INSTALLATION

1. Before servicing the vehicle, refer to the precautions section.

✳✳ CAUTION

The electrical power to the air suspension system must be shut off prior to hoisting, jacking or towing an air suspension vehicle. This can be accomplished by turning off the air suspension switch located in the PASSENGER SIDE kick panel area. Failure to do so can result in unexpected inflation or deflation of the air springs, which can result in shifting of the vehicle during these operations.

2. Raise and support the vehicle.

> ### ✳✳ WARNING
>
> **The air suspension height sensor has a plastic harness retainer to suspension that must be unclipped prior to removal.**

3. Remove or disconnect the following:
   - Air suspension height sensor electrical connector
   - Metal retaining tabs and remove the air suspension height sensor from the ball studs
   - Passenger side trackbar bolt
   - Driver's side trackbar nut and bolt
   - Trackbar from the vehicle

To install:

4. The installation is the reverse of the removal. Tighten the upper and lower track bar bolts to 306 ft. lbs. (500 Nm) on 2002 models and 406 ft. lbs. (550 Nm) on all other models.

Torsion Bars

REMOVAL & INSTALLATION

1. Before servicing the vehicle, refer to the precautions section.
2. Raise and safely support the vehicle.
3. Matchmark and measure the length of the torsion bar being removed.

➥**Torsion bars are marked for left-hand or right-hand installation. If removing both torsion bars, make sure to reinstall in the correct position. If replacing a torsion bar, make sure that it is the correct torsion bar for the side being replaced.**

4. Install Torsion Bar Adjuster T95T-5310-A and Adapter Plates T96T-5310-A, or equivalents.
5. Tighten the adjuster tool until it touches the torsion bar adjuster.
6. Remove the torsion bar adjuster bolt and nut.
7. Remove 6 torsion bar crossmember retaining bolts.
8. Remove the crossmember support.
9. Remove the torsion bar from the vehicle.

To install:

10. Place the torsion bar in position ensuring that it is the correct torsion bar for the side being installed.
11. Install the crossmember support and 6 retaining bolts. Tighten the retaining bolts to 40–50 ft. lbs. (53–72 Nm).
12. Install the torsion bar adjuster bolt

and nut, then remove the torsion bar adjuster tool and adapters.

13. Tighten the adjusting bolt until the matchmarks are in alignment.
14. Lower the vehicle.
15. Measure the ride height and adjust if necessary.
16. Check the alignment and adjust if not within specification.
17. Road test the vehicle and check for proper operation.

ADJUSTMENT

1. Place the vehicle on a level surface.
2. Bounce the front end to normalize the ride height.
3. Measure the ride height on both sides of the vehicle from the frame to the ground.
4. Loosen the lock nut and add or delete turns until the vehicle is level and at the correct curb height.
5. Tighten the lock nut.
6. Road test the vehicle and recheck the measurements.

Ball Joint

REMOVAL & INSTALLATION

4X2 Models

EXCURSION

1. Before servicing the vehicle, refer to the precautions in the beginning of this section.
2. Remove the wheel and tire assembly.
3. Remove the disc brake caliper and the front disc brake hub and rotor.
4. Remove the front disc brake rotor shield.
5. Remove the ABS sensor retaining bolt, sensor harness retaining bolt and the sensor, if equipped and position out of the way.
6. Disconnect the tie rod end. Remove and discard the cotter pin.
7. Using a Pitman Arm Puller, remove the tie rod end.
8. Remove the pinch bolt and the camber adjuster.

> ### ✳✳ CAUTION
>
> **To prevent damage to the ball joint seal and the ball joint socket, do not use a pickle fork-type remover to loosen the ball joints.**

9. Remove and discard the cotter pin from the ball joint.

10. Loosen, but do not remove, the castellated nut.
11. Strike the lower end of the front axle to loosen the ball joint.
12. Remove the castellated nut and the front wheel spindle.
13. Position the front wheel spindle in a vise, and remove the snap ring from the lower ball joint.

> ### ✳✳ CAUTION
>
> **To avoid damage to the components, do not use heat to aid ball joint removal.**

14. Using the removal/installation tool 205-086 and suitable receiver cup, remove the lower ball joint from the front wheel spindle.
15. Using the removal/installation tool 205-086 and suitable receiver cup, remove the upper ball joint.

To install:

> ### ✳✳ CAUTION
>
> **To avoid damage to components, do not use heat to aid installation.**

16. Clean the wheel knuckle ball joint bores.

➥**The lower ball joint must be installed first.**

17. Using the removal/installation tool 205-086 with suitable receiver cups, install the lower ball joint.
18. Using the removal/installation tool 205-086 with suitable receiver cups, install the upper ball joint.
19. Install the snap ring in the groove at the bottom of the ball joint.
20. Install the front wheel spindle. Tighten the nut to 99 ft. lbs. (133 Nm). Install a new cotter pin.
21. Install the camber adjuster and pinch bolt. Tighten to 60 ft. lbs. (80 Nm).
22. Connect the tie rod end. Tighten the nut to 67 ft. lbs (90 Nm). Install a new cotter pin.
23. Install the ABS sensor, if equipped.
24. Install the front disc brake rotor shield.
25. Install the disc brake hub, rotor and caliper.
26. Install the wheel and tire assembly.

2002–04 EXPEDITION AND NAVIGATOR

1. Before servicing the vehicle, refer to the precautions in the beginning of this section.

2. Remove the wheel and tire assembly.

3. Remove the disc brake caliper and the front disc brake hub and rotor.

4. Remove the front disc brake rotor shield.

5. Remove the ABS sensor retaining bolt, sensor harness retaining bolt and the sensor, if equipped and position out of the way.

6. Disconnect the tie rod end. Remove and discard the cotter pin.

7. Using tool 211-003, remove the tie rod end.

✳✳ CAUTION

To prevent damage to the ball joint seal and the ball joint socket, do not use a pickle fork-type remover to loosen the ball joints.

8. Support the axle with a jack.

9. Remove and discard the cotter pin from the ball joint.

10. Using tool 211-003, separate the front wheel knuckle from the suspension lower arm. Then, loosely reinstall the castellated nut.

11. Remove the upper ball joint castellated nut. Discard the cotter pin and nut.

✳ CAUTION

Do not damage the ball joint boot when installing the special tool.

12. Using the too 211-003, separate the front wheel knuckle from the front suspension upper arm.

13. Remove the hand-tightened lower ball joint castellated nut, then remove the front wheel knuckle.

14. Remove the snap ring from the ball joint. Discard the snap ring.

15. Using removal/installation tool 205-086 with suitable receiver cups, remove the ball joint.

To install:

✳ CAUTION

To avoid damage to components, do not use heat to aid installation.

16. Clean the wheel knuckle ball joint bores.

17. Using the removal/installation tool 205-086 with suitable receiver cups, install the ball joint.

18. Install the snap ring in the groove at the bottom of the ball joint.

19. Install the front wheel spindle. Tighten the upper ball joint nut to 66 ft. lbs. (90 Nm) and the lower nut to 98 ft. lbs. (133 Nm). Install a new cotter pin.

20. Connect the tie rod end. Tighten the nut to 70 ft. lbs (95 Nm). Install a new cotter pin.

21. Install the ABS sensor, if equipped.

22. Install the disc brake hub, rotor and caliper.

23. Install the wheel and tire assembly.

2005–06 MODELS

The ball joints on these models are an integral part of the control am. If found to be defective the control arm must be replaced.

4X4 Models

EXCURSION

1. Before servicing the vehicle, refer to the precautions in the beginning of this section.

2. Remove the wheel and tire assembly.

3. Remove the disc brake caliper and the front disc brake hub and rotor.

4. Remove the wheel hub and bearing.

5. Using a drift, drive the axle shaft main seal out of the wheel knuckle.

6. Remove the axle shaft and main seal.

7. Remove the tie-rod end castellated nut. Discard the cotter pin.

8. Using the tool 211-003, disconnect the tie-rod end from the wheel knuckle.

9. Remove the upper ball joint castellated nut and the insert.

10. Remove the wheel knuckle.

11. Position the front wheel spindle in a vise, and remove the snap ring from the lower ball joint.

✳✳ CAUTION

To avoid damage to the components, do not use heat to aid ball joint removal.

12. Using the removal/installation tool 205-086 and suitable receiver cup, remove the lower ball joint from the front wheel spindle.

13. Using the removal/installation tool 205-086 and suitable receiver cup, remove the upper ball joint.

To install:

✳✳ CAUTION

To avoid damage to components, do not use heat to aid installation.

14. Clean the wheel knuckle ball joint bores.

➡**The lower ball joint must be installed first.**

15. Using the removal/installation tool 205-086 with suitable receiver cups, install the lower ball joint.

16. Using the removal/installation tool 205-086 with suitable receiver cups, install the upper ball joint.

17. Install the snap ring in the groove at the bottom of the ball joint.

18. Position the wheel knuckle onto the axle housing.

19. Install the nut onto the lower ball joint. Do not tighten the nut at this time.

20. Install the insert and the castellated nut onto the upper ball joint. Do not tighten the nut at this time.

21. Tighten the lower ball joint retaining nut. Pre-tighten the nut to 35 ft. lbs (47 Nm) on 2002–04 models or 47 ft. lbs (64 Nm) on 2005 models.

➡**Do not loosen the castellated nut to install the cotter pin.**

22. Install the cotter pin into the upper ball joint.

23. Tighten the upper ball joint castellated nut to 69 ft. lbs. (94 Nm). Install the cotter pin. If necessary, tighten the castellated nut until the cotter pin can be installed.

24. Tighten the lower ball joint nut to 150 ft. lbs. (204 Nm).

25. Install the tie-rod end onto the wheel knuckle.

26. Position the tie-rod end into the wheel knuckle. Tighten the castellated nut to 52 ft. lbs. (70 Nm). Install a new cotter pin.

27. Position a new main seal onto the axle shaft. Using the tools illustrated and a hammer, seat the main seal onto the axle shaft.

28. Position the axle shaft into the axle housing. Using the tools illustrated and a

67197-NAVI-G38

Using the tools shown and a hammer, seat the main seal onto the axle shaft—4X4 Excursion

67197-NAVI-G39

Using the tools shown and a hammer, install the main seal into the wheel knuckle—4X4 Excursion

hammer, install the main seal into the wheel knuckle.

29. Install the wheel hub and bearing.
30. Install the front brake disc.
31. Install the wheel and tire assembly.

2002–04 EXPEDITION AND NAVIGATOR

1. Before servicing the vehicle, refer to the precautions in the beginning of this section.
2. Remove the wheel and tire assembly.
3. Remove axle nut, cotter pin, retainer and nut.
4. Remove the brake caliper support bracket bolts and position the caliper support bracket and caliper aside.
5. Remove the brake disc.
6. Disconnect the ABS electrical connector.
7. Unclip the ABS harness routing clips.
8. Remove the tie-rod castellated nut and discard the cotter pin.

✳✳ CAUTION

Do not use a hammer to separate the tie-rod from the wheel knuckle or damage to the wheel knuckle will result.

9. Using tool 211-003, separate the tie-rod end from the front wheel knuckle.
10. Using a suitable jack, support the front suspension lower arm.
11. Remove the lower ball joint castellated nut, discard the cotter pin and nut.
12. Using tool 211-003, separate the front wheel knuckle from the suspension lower arm. Then, loosely install the castellated nut.
13. Remove the upper ball joint castellated nut discard the cotter pin and nut.
14. Using tool 211-003, separate the

front wheel knuckle from the front suspension upper arm.
15. Separate the halfshaft from the hub, then remove the hand-tightened lower ball joint castellated nut.
16. Remove the front wheel knuckle.
17. Position the halfshaft aside and support with wire.
18. Remove the snap ring from the ball joint and discard the snap ring.
19. Using a suitable ball joint remover tool and receiver cup, remove the ball joint.

To install:

✳✳ CAUTION

To avoid damage to components, do not use heat to aid installation.

20. Clean the wheel knuckle ball joint bores.

➡The lower ball joint must be installed first.

21. Using a suitable ball joint installer tool and receiver cup, install the ball joint.
22. Install a new snap ring in the groove at the bottom of the ball joint. Make sure the snap ring is fully seated.

➡Always install new castellated nuts and cotter pins.

23. Install the remaining components in the reverse of removal. Please note the following torque specifications:
   a. Upper ball joint nut: 66 ft. lbs. (90 Nm).
   b. Lower ball joint nut: 98 ft. lbs. (133 Nm).
   c. Tie rod nut: 70 ft. lbs. (95 Nm).
   d. Axle shaft nut: 221 ft. lbs. (300 Nm).

2005–05 EXPEDITION AND NAVIGATOR MODELS

The ball joints on these models are an integral part of the control am. If found to be defective the control arm must be replaced.

Upper Control Arm

REMOVAL & INSTALLATION

Navigator And Expedition

2002 4X2 MODELS

1. Before servicing the vehicle, refer to the precautions section.
2. Raise and safely support the vehicle.
3. Remove the wheel and tire assembly.
4. Support the lower control arm with a transmission jack or equivalent adjustable jack.
5. Matchmark the upper control arm adjusting cams.
6. If equipped with 4-wheel Anti-lock Brake System (ABS), remove the speed sensor harness bracket bolt and position the harness aside.
7. Remove the cotter pin and castellated nut from the upper ball joint. Discard the cotter pin.
8. Using Pitman Arm Puller T64P-3590-F or equivalent, separate the ball joint from the wheel spindle.
9. Remove 2 nuts and bolts securing the upper control arm to the body brackets.
10. Remove the upper control arm and ball joint assembly from the vehicle.

To install:
11. Align the matchmarks on the upper control arm adjusting cams.
12. Place the upper control arm in position to the body brackets and install 2 retaining bolts and nuts. Do not tighten at this time.
13. Install the ball joint stud to the wheel spindle and install the castellated nut. Tighten the nut to 56–77 ft. lbs. (76–104 Nm). Install a new cotter pin.
14. Raise the lower control arm to position the control arms at normal curb height and tighten the front upper control arm retaining nut first, to 84–112 ft. lbs. (113–153 Nm).
15. With the vehicle still at curb height, tighten the rear upper control arm retaining nut to 84–112 ft. lbs. (113–153 Nm).
16. Remove the transmission jack or equivalent from under the lower control arm.
17. If equipped with 4-wheel ABS brake system, position the speed sensor harness

and install the bracket bolt. Tighten the speed sensor bracket bolt to 62–80 inch lbs. (7–9 Nm).

18. Install the wheel and tire assembly. Torque the lug nuts to 83–112 ft. lbs. (113–153 Nm).

19. Lower the vehicle.

20. Check the alignment and adjust if not within specification.

21. Road test the vehicle and check for proper operation.

ALL 2003–06 MODELS

1. Before servicing the vehicle, refer to the precautions section.

2. Raise and safely support the vehicle.

3. Remove the wheel and tire assembly.

4. Remove the shock absorber and spring assembly.

5. On models with an air suspension, detach the height sensor from the upper arm.

➡Use the hex holding feature to prevent the stud from turning while removing the nut.

6. Remove and discard the front control arm nut. Discard the nut.

7. Remove the rear nut and bolt. Discard the nut.

8. Remove the upper arm.

To install:

NOTE: Assemble the upper arm pivot nuts and bolts. Do not tighten until the installation procedure is complete and the weight of the vehicle is resting on the wheel and tire assemblies.

9. Installation is the reverse of removal. Tighten the control arm nuts/bolts to 111 ft. lbs. (150 Nm).

Lower Control Arm

REMOVAL & INSTALLATION

Expedition and Navigator

2002 4X2 MODELS

1. Before servicing the vehicle, refer to the precautions section.

✳✳ WARNING

The electrical power to the air suspension system must be shut off prior to hoisting, jacking or towing an air suspension vehicle. To do this, turn off the air suspension switch located near the jack storage area in the rear of the passenger compartment. Failure to do so can result in unexpected

7924FG75

Be sure the coil spring is mounted correctly in the lower control arm

inflation or deflation of the air springs, which can result in shifting of the vehicle during these operations.

✳✳ CAUTION

Do not use heat to loosen a seized wheel nut. Heat can damage the wheel and the wheel bearings.

2. Remove the tire and wheel assembly.

3. Remove the brake disc shield.

4. Remove the front shock absorber.

5. Remove the brake hose bracket screw and bracket from the front suspension lower arm.

6. Remove the front stabilizer bar link nut.

7. Using the coil spring compressor, compress the front coil spring.

8. Remove the lower ball joint castle nut. Discard the cotter pin.

9. Use a Pitman arm puller to separate the lower ball joint from the front wheel spindle.

➡The front suspension lower arm nut must be installed in its original position.

10. Index the front suspension lower arm nut.

11. Remove the front suspension lower arm and front coil spring.

12. Remove the two front suspension lower arm nuts.

13. Remove the two front suspension lower arm bolts.

14. Remove the front suspension lower arm and front coil spring.

To install:

15. Install the front coil spring into the front suspension lower arm.

• Coil spring compressor and compress the coil spring. The end of the coil spring **A**, must cover the

hole designated **B** and be visible in the second hole designated as **C**, for proper installation.

➡Install the front suspension lower arm nuts in the original position.

➡The front suspension lower arm caster split adjuster is for production use only. Do not adjust.

➡The forward front suspension lower arm nut must be tightened first while the control arms are held at the curb position ride height.

16. Install the front suspension lower arm. Position the front suspension lower arm and front coil spring.

Install the bolts and nuts and tighten to 121–148 ft. lbs. (164–200 Nm).

17. Install the lower ball joint castle nut and tighten to 83–112 ft. lbs. (113–153 Nm). Install a new cotter pin.

18. Install the front stabilizer bar link nut and tighten to 21 ft. lbs. (29 Nm).

19. Remove the Coil Spring Compressor.

20. Install the front shock absorber and the shock absorber lower nuts. Tighten to 29 ft. lbs. (40 Nm).

21. Install the upper shock absorber nut and washer. Tighten to 34–46 ft. lbs. (47–63 Nm).

22. Install the brake hose bracket screw.

23. Install the brake disc shield.

➡If equipped with air suspension, reactivate the system by turning on the air suspension switch.

24. Install the tire and wheel assembly.

25. Inspect and adjust the front end alignment

2002 4X4 MODELS

1. Before servicing the vehicle, refer to the precautions section.

67197-NAVI-G40

Install the torsion bar tool 204-185 with the torsion bar tool adapters 204-203 in the B slots on the torsion bar tool—2002 Expedition and Navigator 4X4 models

✳✳ WARNING

The electrical power to the air suspension system must be shut off prior to hoisting, jacking or towing an air suspension vehicle. To do this, turn off the air suspension switch located near the jack storage area in the rear of the passenger compartment. Failure to do so can result in unexpected inflation or deflation of the air springs, which can result in shifting of the vehicle during these operations.

2. Remove the wheel hub.

3. Make an alignment mark on the torsion bar and the torsion bar crossmember support. Measure and record the length.

4. Relieve the torsion bar tension.

5. Remove the torsion bar adjuster bolt.

6. Install the torsion bar tool 204-185 with the torsion bar tool adapters 204-203 in the B slots on the torsion bar tool.

7. Tighten the torsion bar tool until the torsion bar adjuster lifts off of the adjuster nut.

8. Remove the torsion bar adjuster nut and remove the torsion bar tool.

9. Remove the torsion bar adjuster.

10. Remove the stabilizer bar link nut from the front suspension lower arm.

11. Remove the front shock absorber lower bolt and nut from the front suspension lower arm.

12. Remove the lower ball joint castle nut. Discard the cotter pin.

13. Using a Pitman arm puller, separate the front suspension lower arm from the front wheel knuckle.

14. Separate the front suspension lower arm from the front wheel knuckle.

15. Remove the two front suspension lower arm nuts.

16. Remove the two front suspension lower arm bolts.

17. Remove the front suspension lower arm and torsion bar as an assembly.

18. Separate the front suspension lower arm from the torsion bar.

To install:

19. Attach the front suspension lower arm to the torsion bar.

20. Install the front suspension lower arm and torsion bar as an assembly.

21. Install the front suspension lower arm bolts and nuts. Tighten to 121 ft. lbs. (147 Nm).

22. Install the front suspension arm lower ball joint castle nut. Tighten the nut to 83–113 ft. lbs. (113–153 Nm). Install a new cotter pin.

23. Install the front stabilizer bar link to the front suspension lower arm. Tighten to 21 ft. lbs. (29 Nm).

24. Install the front shock absorber lower bolt and nut to the front suspension lower arm. Tighten to 65–87 ft. lbs. (88–118 Nm) Nm)

25. Install the torsion bar adjuster.

26. Turn the torsion bar adjuster until the reference marks align.

27. Install the torsion bar tool 204-185 with the torsion bar tool adapters 204-203 in the B slots on the torsion bar tool.

28. Tighten the torsion bar tool to load the torsion bar.

29. Install the torsion bar adjuster nut.

30. Install the torsion bar adjuster bolt.

31. Remove torsion bar tool.

32. Install the wheel hub.

33. Adjust the ride height.

ALL 2003–06 MODELS

1. Before servicing the vehicle, refer to the precautions in the beginning of this section.

2. If equipped, turn the air suspension switch to the OFF position.

90 Nm (66 lb-ft)

67197-NAVI-G41

Remove/discard this bolt first when removing the lower control arm. . . .

270 Nm (199 lb-ft)

67197-NAVI-G42

. . . next remove/discard this bolt when removing the lower control arm

270 Nm (199 lb-ft)

67197-NAVI-G43

. . . then remove/discard this bolt when removing the lower control arm. . . .

400 Nm (295 lb-ft)

67197-NAVI-G44

. . . finally remove/discard this bolt when removing the lower control arm—All 2003–06 Expedition and Navigator models

3. Remove the wheel knuckle.

➡Use the hex holding feature to prevent the stud from turning while removing the nut.

4. Remove and discard the nuts/bolts using the illustrations below.

To install:

➡Assemble the lower arm pivot nuts and bolts. Do not tighten until the installation procedure is complete and the weight of the vehicle is resting on the wheel and tire assemblies.

5. To install, reverse the removal procedure and tighten the retainers to the specifications shown in the illustrations.

6. Check and, if necessary, align the front end.

Front Wheel Bearings

ADJUSTMENT

2002–04 Models

✳✳ WARNING

The electrical power to the air suspension system must be shut off prior

to hoisting, jacking or towing an air suspension vehicle. Failure to shut the system off may lead to an unexpected inflation or deflation of the air springs, which may result in a shift of the vehicle.

➡On 4-wheel drive vehicles, the front wheel bearings are not adjustable.

1. Before servicing the vehicle, refer to the precautions section.
2. Support the front end.
3. Remove the wheel cover, if equipped.
4. Remove the grease cap.

➡Check the wheel bearings for sufficient grease.

5. Remove the cotter pin and retaining washer. Back off the spindle nut. Discard cotter pin.

6. Adjust the wheel bearings on Navigator and Expedition models as follows:

a. Tighten the spindle nut to 30 ft. lbs. (40 Nm) while rotating the wheel and tire assembly to seat the wheel bearings.

b. Back off the spindle nut 2 turns.

c. Tighten the spindle nut to 17–24 ft. lbs. (23–34 Nm) while rotating the rotor clockwise.

d. Loosen the nut 175 degrees.

e. Tighten the nut to 17 inch lbs. (2 Nm).

7. Adjust the wheel bearings on Excursion models as follows:

a. Tighten the spindle nut to 21 ft. lbs. (28 Nm) while rotating the wheel and tire assembly to seat the wheel bearings.

b. Back off the spindle nut 120–180 degrees.

c. Tighten the nut to 17 inch lbs. (2 Nm).

8. Install the retaining washer so the castellations are aligned with the cotter pin hole. Install a new cotter pin.

9. Check the wheel and tire assembly for proper rotation, then install the grease cap. If the wheel still does not rotate properly, inspect and clean or replace the wheel bearings and cups.

10. Install the wheel cover, if equipped.

11. Lower the vehicle.

12. Road test the vehicle and check for proper operation.

2005–06 Models

The wheel bearings are not adjustable on these models.

REMOVAL & INSTALLATION

2 Wheel Drive

2002 EXPEDITION AND NAVIGATOR

✳✳ WARNING

The electrical power to the air suspension system must be shut off prior to hoisting, jacking or towing an air suspension vehicle. Failure to shut the system off may lead to an unexpected inflation or deflation of the air springs, which may result in a shift of the vehicle.

The hub is part of the disc brake rotor and cannot be serviced separately. The inner and outer wheel bearing and races are serviced individually. Be sure to have a new hub grease seal when servicing the wheel bearings.

1. Before servicing the vehicle, refer to the precautions section.

2. Remove or disconnect the following:

• Wheels
• Caliper
• Brake pads and anti-rattle clips
• Anchor plate
• Hub grease cap, cotter pin, retainer washer and the spindle nut
• Wheel bearing retainer washer and the outer wheel bearing

- Brake hub and rotor assembly
- Grease seal
- Inner wheel bearing

3. Clean and inspect the wheel bearings and races for unusual wear or damage. Replace parts as necessary.

4. Inspect the hub and brake rotor assembly. If required, the hub and brake rotor assembly must be replaced as a unit.

To install:

5. If needed, pack the wheel bearing with a suitable high temperature wheel bearing grease before assembly.

6. Install or connect the following:

- Inner wheel bearing in the hub and brake rotor assembly
- New grease seal
- Hub and rotor assembly on the wheel spindle
- Outer wheel bearing
- Retainer washer and the spindle nut

7. Adjust the wheel bearings (2WD only).

8. Install or connect the following:

- Retaining washer, so the castellations are aligned with the cotter pin hole. Install a new cotter pin.
- Anchor plate, and the 2 retaining bolts. Tighten the bolts to 125–168 ft. lbs. (170–230 Nm).
- Anti-rattle clips, and the disc brake pads
- Caliper
- Wheels. Tighten the lug nuts to 83–112 ft. lbs. (113–153 Nm).

9. Check the wheel and tire assembly for proper rotation, then install the grease cap.

10. Lower the vehicle.

11. Road test the vehicle and check for proper operation.

2003–06 EXPEDITION AND NAVIGATOR

❋❋ WARNING

The electrical power to the air suspension system must be shut off prior to hoisting, jacking or towing an air suspension vehicle. Failure to shut the system off may lead to an unexpected inflation or deflation of the air springs, which may result in a shift of the vehicle.

1. Before servicing the vehicle, refer to the precautions in the beginning of this section.

2. If equipped, turn the air suspension switch to the OFF position.

3. Disconnect the wheel speed sensor electrical connector.

4. Remove the wheel and tire assembly.

5. Remove the bolt and detach the brake line retainers.

❋❋ CAUTION

Do not allow the caliper to hang from the brake hose or damage to the hose can result.

6. Remove the caliper, pads and anchor plate and set aside.

7. Remove the brake rotor.

8. Remove the dust cap.

9. Remove and discard the axle nut.

10. Remove the bolts and the wheel bearing and hub assembly.

To install:

❋❋ CAUTION

If the original wheel bearing and hub is being reinstalled, make sure to install a new O-ring.

11. Install the remaining components in the reverse of removal. Please note the following torque specifications:

   a. Wheel bearing/hub assembly bolts: 148 ft. lbs. (200 Nm).

   b. Axle nut: 20 ft. lbs. (27 Nm).

   c. Brake anchor plate: 148 ft. lbs. (200 Nm).

EXCURSION

1. Before servicing the vehicle, refer to the precautions in the beginning of this section.

2. Remove the front wheel and tire assemblies.

3. Remove the front disc brake caliper and rotor, and position the caliper out of the way.

4. Remove the hub cap from the hub assembly.

5. Remove the cotter pin, adjusting nut and flat washer.

➡Inspect the condition of the spindle and nut threads to ensure a free turning nut when reassembling.

6. Remove the outer bearing cone and roller assembly, and pull the hub assembly from the spindle.

7. Using care not to damage the bearing cage, use a suitable slide hammer and bearing seal remover to remove the inner bearing cone and bearing seal.

To install:

❋❋ CAUTION

Do not spin the bearing dry with compressed air.

➡Remove all traces of lubricant from the bearings, hub and axle spindle. Inspect bearings and bearing cups for pitting or unusual wear. If either bearings or bearing cups are worn or damaged, replace both bearings and bearing cups.

➡It is recommended that bearings and bearing cups be replaced in sets. If cups are worn or damaged, install the inner and outer bearing cups in the hub with an appropriate bearing cup driver tool. Check for proper seating of new bearing cups by trying to insert a 0.0015 inch (0.38-mm) feeler gauge between the bottom face of the cup and wheel hub seat. You should not be able to insert the feeler gauge.

8. Remove all burrs, nicks or scratches from the shoulder of the spindle and seal bore in the hub with emery cloth.

9. Pack the inside of the hub with lithium-base wheel bearing grease. Fill the hub until the grease is flush with the inside diameters of both bearing cups.

10. Pack the bearing cone and roller assemblies with wheel bearing grease. Use a bearing packer for this operation. If a packer is not available, work as much lubricant as possible between the rollers and cages.

❋❋ CAUTION

Keep the hub centered on the spindle to prevent damage to the grease seal or spindle threads.

11. Place the inner bearing cone and roller assembly in the inner cup and install the wheel bearing hub seal, using a suitable seal replacer. Make sure seal is fully seated and lubricated.

12. Install the hub assembly.

13. Install the outer bearing cone and roller assembly and the flat washer on the spindle and install the adjusting nut. Install a new cotter pin.

14. Install the hub cap.

15. Install the front disc brake caliper and rotor.

16. Install the front wheel and tire assemblies.

4-Wheel Drive

EXCURSION

1. Before servicing the vehicle, refer to the precautions section.

2. Remove or disconnect the following:

- Front brake rotor
- Hub lock by pulling out the retainer ring. And then the hub lock

- Snap ring and the axle shaft thrust washers
- ABS wheel sensor harness and routing clips, if equipped

➡The wheel hub and bearing is a slip fit design and should not require a puller to remove it.

- Four lock nuts
- Wheel hub and bearing
- Brake rotor shield
- Bolt and the Anti-lock Brake System (ABS) sensor, if equipped

3. Place the hub in a soft-jawed vise.
4. Install two nuts on the studs and use the inner nut to remove the studs.
5. Remove and discard the O-ring.

To install:

➡Any time the wheel hub is removed for any reason, a new O-ring seal must be installed. Failure to do so can cause a vacuum leak and loss of four wheel drive operations.

6. Install a new O-ring.
7. Place the hub in a soft-jawed vise.
8. Install two nuts on the studs and use the outer nut to install the studs.
9. Install or connect the following:
   - ABS sensor and the bolt. Tighten the bolt to 13 ft. lbs. (18 Nm).
   - Brake rotor shield
10. Apply a coat of High Temperature 4x4 Front Axle and Wheel Bearing Grease E8TZ-19590-A to the O-ring area of the wheel hub and bearing before installing the hub and bearing.
    - Wheel hub and bearing
    - Four lock nuts and tighten to 133 ft. lbs. (180 Nm)
    - ABS sensor harness and routing clips, if equipped

❊❊ WARNING

The non-metallic thrust washer must be installed between the two metal thrust washers. Failure to do so will cause severe wear to the non-metallic thrust washer, allowing the axle shaft to travel further in and out during torque thrust causing damage to the wheel hub and bearing, the axle shaft end seal and the axle shaft.

- Three thrust washers onto the axle shaft
- Snap ring

➡Any time the hub lock is removed, a new O-ring seal must be installed. Failure to do so can cause a vacuum leak and loss of four wheel drive functions.

- New O-ring seal
- Hub lock and the retainer ring
- Brake rotor

11. Perform a wheel-end vacuum leak test as follows:

   a. Install the vacuum pump and gauge line on the knuckle vacuum fitting and pump to 20 in-Hg.

   b. If the vacuum drop is less than 0.5 in-Hg in 30 seconds, the 4x4 hublock is working correctly.

2002 EXPEDITION AND NAVIGATOR

❊❊ WARNING

The electrical power to the air suspension system must be shut off prior to hoisting, jacking or towing an air suspension vehicle. Failure to shut the system off may lead to an unexpected inflation or deflation of the air springs, which may result in a shift of the vehicle.

The wheel bearings are of the cartridge design and are an integral part of the hub assembly. The bearings are permanently lubricated and require no maintenance or adjustments. If required, a new hub assembly must be installed.

1. Before servicing the vehicle, refer to the precautions section.
2. Remove or disconnect the following:
   - Wheels
   - Caliper
   - Brake pads and anti-rattle clips
   - Anchor plate
   - Brake rotor
   - Rotor shield
   - Speed sensor retaining bolt and move the speed sensor and harness aside, if equipped with 4-wheel Anti-lock Brake System (ABS)
   - Hub nut cotter pin, retainer and the hub nut. Discard the cotter pin.
   - The 3 hub assembly retaining bolts from the inside of the steering knuckle
   - Hub assembly

➡If necessary, use a suitable puller to separate the hub assembly from the CV-joint. Use care not to over extend the CV-joint and boot when removing the hub assembly.

- Grease seal from the steering knuckle

To install:

3. Install or connect the following:
   - New grease seal using Bearing Cup Replacer 205-147 (T80T-4000-P),

Knuckle Seal Replacer 205-361 (T96T-1175-A) and Threaded Drawbar 204-029 (T77F-1176-A)
- Hub assembly to the steering knuckle and secure with the 3 retaining bolts. Tighten the bolts to 110–145 ft. lbs. (149–201 Nm).
- CV-joint into the hub assembly
- Speed sensor and secure with the retaining bolt, if equipped with 4-wheel ABS. Tighten the bolt to 60–84 inch lbs. (7–9 Nm).
- Brake rotor shield and the 3 retaining bolts. Tighten the bolts to 80–107 inch lbs. (9–12 Nm).
- Hub nut and tighten to 188–254 ft. lbs. (255–345 Nm)
- Hub nut retainer and a new cotter pin
- Brake rotor
- Anchor plate in position and install the 2 retaining bolts. Tighten the bolts to 125–168 ft. lbs. (170–230 Nm).
- Anti-rattle clips and the disc brake pads
- Caliper
- Wheels. Tighten the lug nuts to 83–112 ft. lbs. (113–153 Nm).

4. Lower the vehicle.
5. Pump the brake pedal several times to position the brake pads prior to moving the vehicle.
6. Road test the vehicle and check for proper operation.

2003–06 EXPEDITION AND NAVIGATOR

❊❊ WARNING

The electrical power to the air suspension system must be shut off prior to hoisting, jacking or towing an air suspension vehicle. Failure to shut the system off may lead to an unexpected inflation or deflation of the air springs, which may result in a shift of the vehicle.

1. Before servicing the vehicle, refer to the precautions in the beginning of this section.
2. If equipped, turn the air suspension switch to the OFF position.
3. Disconnect the wheel speed sensor electrical connector.
4. Remove the wheel and tire assembly.
5. Remove the bolt and detach the brake line retainers.

✳✳ CAUTION

Do not allow the caliper to hang from the brake hose or damage to the hose can result.

6. Remove the caliper, pads and anchor plate and set aside.
7. Remove the brake rotor.
8. Remove the dust cap.
9. Remove and discard the axle nut.
10. Remove the bolts and the wheel bearing and hub assembly.

To install:

✳✳ CAUTION

If the original wheel bearing and hub is being reinstalled, make sure to install a new O-ring.

11. Install the remaining components in the reverse of removal. Please note the following torque specifications:
    a. Wheel bearing/hub assembly bolts: 148 ft. lbs. (200 Nm).
    b. Axle nut: 20 ft. lbs. (27 Nm).
    c. Brake anchor plate: 148 ft. lbs. (200 Nm).

Rear Wheel Bearings

REMOVAL & INSTALLATION

2003–06 Expedition and Navigator

✳✳ WARNING

The electrical power to the air suspension system must be shut off prior to hoisting, jacking or towing an air suspension vehicle. This can be accomplished by turning off the air suspension switch located in the left hand rear quarter trim panel. Failure to do so can result in unexpected inflation or deflation of the air springs, which can result in shifting of the vehicle during these operations.

1. If equipped, turn the air suspension switch to the OFF position.
2. Remove the wheel and tire assembly.

➡ Have an assistant press the brake pedal to keep the axle from rotating.

3. Remove and discard the rear axle wheel hub retainer and the washer.
4. Using a suitable hub puller, separate the outboard CV joint from the wheel hub.

✳✳ CAUTION

Do not allow the caliper to hang from the brake hose or damage to the hose can result.

5. Remove the caliper anchor plate bolts, and position the caliper, pads and anchor plate aside.
6. Remove the rotor.
7. Disconnect the wheel speed sensor electrical connector and detach the retainer.
8. Remove the bolt and detach brake line the retainers.
9. Remove the bolts, the wheel bearing and the wheel speed sensor as an assembly.
10. Route the sensor wiring through the access hole in the brake shield and discard the bolts.
11. Installation is the reverse of removal, please note the following torque specifications:
    a. Wheel bearing bolts: tighten to 185 Nm (136 ft. lbs.).
    b. Caliper anchor plate bolts: Tighten to 190 Nm (140 ft. lbs.).
    c. New rear axle wheel hub retainer and the washer: Tighten to 300 Nm (221 ft. lbs.).

BRAKES

Brake Caliper

REMOVAL & INSTALLATION

Front

2002 MODELS

✳✳ WARNING

The electrical power to the air suspension system must be shut off prior to hoisting, jacking or towing an air suspension vehicle. Failure to shut the system off may lead to an unexpected inflation or deflation of the air springs, which may result in a shift of the vehicle.

1. Before servicing the vehicle, refer to the precautions section.
2. Remove the center cap.
3. Break the front wheel lug nuts loose.
4. Remove or disconnect the following:
- Front wheels
- Front brake hose bolt and the cop-

per washers and plug the front brake hose
- 2 front disc brake caliper slide pins, then lift the caliper off of the front caliper anchor plate

To install:
5. Install or connect the following:
- Front disc brake caliper onto the caliper anchor plate. Install the 2 slide pins. Tighten the slider pins/bolts to 42 ft. lbs. (56 Nm) on Excursion models or 21–26 ft. lbs. (28–36 Nm) on all other models.
- Front brake hose to the brake caliper, using new copper washers. Tighten the retaining bolt to 26 ft. lbs. (35 Nm) on Excursion models or 23–29 ft. lbs. (30–40 Nm) on all other models
6. Bleed the brake system.
7. Clean the wheel hub mounting surface.
8. Install the front wheels and snug the lug nuts to fully seat the wheel against the hub.
9. Lower the vehicle until some of the

vehicle's weight rests on the front tires, then tighten the lug.
10. Lower the vehicle completely.
11. Make sure that the brakes are operating correctly.

2003–06 MODELS

✳✳ WARNING

The electrical power to the air suspension system must be shut off prior to hoisting, jacking or towing an air suspension vehicle. Failure to shut the system off may lead to an unexpected inflation or deflation of the air springs, which may result in a shift of the vehicle.

1. Before servicing the vehicle, refer to the precautions section.
2. Remove or disconnect the following:
- Wheels
- Brake pads
- Front brake hose bolt and the copper washers and plug the front brake hose

- 2 front disc brake caliper bolts, then lift the caliper off of the front caliper anchor plate

To install:

3. Install or connect the following:
   - 2 front disc brake caliper bolts, tighten the bolts to 41 ft. lbs. (55 Nm)
   - Front brake hose to the brake caliper, using new copper washers. Tighten the retaining bolt to 26 ft. lbs. (35 Nm).
   - Brake pads
   - Wheels

Rear

2002 MODELS–EXCEPT EXCURSION

※※ **WARNING**

The electrical power to the air suspension system must be shut off prior to hoisting, jacking or towing an air suspension vehicle. Failure to shut the system off may lead to an unexpected inflation or deflation of the air springs, which may result in a shift of the vehicle.

1. Before servicing the vehicle, refer to the precautions section.
2. Before servicing the vehicle, refer to the precautions section.
3. Remove enough brake fluid from the brake master cylinder reservoir until it is ½ full.
4. Remove or disconnect the following:
   - Wheels
   - Banjo bolt connecting the brake hose to the disc brake caliper and plug the brake hose. Discard 2 copper sealing washers.
   - 2 brake caliper slide pins and lift the caliper off the anchor plate

To install:

5. Retract the disc brake caliper pistons fully in the piston bores using an old brake pad or block of wood and a C-clamp or equivalent.
6. Install or connect the following:
   - Disc brake caliper above the rotor and install it with a rotating motion. Make sure the inner and outer pads are properly positioned and the anti-rattle clips are correctly installed. The brake caliper bleed screw should be positioned on top of the caliper when assembled on the vehicle.
7. Lubricate the locating pins and the inside of the insulators with silicone grease.

- Locating pins through the caliper insulators and hand-start the threads into the steering knuckle attaching holes. Tighten the locating pins to 20 ft. lbs. (27 Nm).
- Brake hose to the disc brake caliper using 2 new copper sealing washers. Tighten the banjo bolt to 29 ft. lbs. (40 Nm).

8. Bleed the brake system, filling the brake master cylinder reservoir as required.
   - Wheels
9. Lower the vehicle.
10. Pump the brake pedal several times to position the brake pads prior to moving the vehicle.
11. Road-test the vehicle and check for proper brake system operation.

2002 MODELS–EXCURSION

※※ **WARNING**

The electrical power to the air suspension system must be shut off prior to hoisting, jacking or towing an air suspension vehicle. Failure to shut the system off may lead to an unexpected inflation or deflation of the air springs, which may result in a shift of the vehicle.

1. Before servicing the vehicle, refer to the precautions section.
2. Before servicing the vehicle, refer to the precautions section.
3. Remove enough brake fluid from the brake master cylinder reservoir until it is ½ full.
4. Remove or disconnect the following:
   - Wheels
   - Stone shield
   - 2 brake caliper slide pins and lift the caliper off the anchor plate
   - Banjo bolt connecting the brake hose to the disc brake caliper and plug the brake hose. Discard 2 copper sealing washers.

To install:

5. Retract the disc brake caliper pistons fully in the piston bores using an old brake pad or block of wood and a C-clamp or equivalent.
6. Install or connect the following:
   - Disc brake caliper above the rotor and install it with a rotating motion. Make sure the inner and outer pads are properly positioned and the anti-rattle clips are correctly installed. The brake caliper bleed screw should be positioned on top

of the caliper when assembled on the vehicle.
- Brake hose to the disc brake caliper using 2 new copper sealing washers. Tighten the banjo bolt to 37 ft. lbs. (50 Nm).

7. Lubricate the locating pins and the inside of the insulators with silicone grease.
   - Locating pins through the caliper insulators and hand-start the threads into the steering knuckle attaching holes. Tighten the locating pins to 27–41 ft. lbs. (36–55 Nm).
8. Bleed the brake system, filling the brake master cylinder reservoir as required.
   - Wheels
9. Lower the vehicle.
10. Pump the brake pedal several times to position the brake pads prior to moving the vehicle.
11. Road-test the vehicle and check for proper brake system operation.

2003–06 MODELS

※※ **WARNING**

The electrical power to the air suspension system must be shut off prior to hoisting, jacking or towing an air suspension vehicle. Failure to shut the system off may lead to an unexpected inflation or deflation of the air springs, which may result in a shift of the vehicle.

1. Before servicing the vehicle, refer to the precautions section.
2. Remove or disconnect the following:
   - Wheels
   - Stone shield
   - 2 rear disc brake caliper bolts, then lift the caliper off of the front caliper anchor plate
   - Rear brake hose bolt and the copper washers and plug the front brake hose

To install:

3. Install or connect the following:
   - 2 rear disc brake caliper bolts. Tighten the bolts to 26 ft. lbs. (35 Nm) on 2002–04 models, or 20 ft. lbs. (27 Nm) on 2005–06 models.
   - Rear brake hose to the brake caliper, using new copper washers. Tighten the retaining bolt to 37 ft. lbs. (50 Nm) on Excursion models or 26 ft. lbs. (35 Nm) on Expedition/Navigator models.
   - Stone shield and tighten the bolts to 46 ft. lbs. (62 Nm)

- Brake pads
- Wheels

Disc Brake Pads

REMOVAL & INSTALLATION

Front

2002 MODELS

1. Before servicing the vehicle, refer to the precautions section.

2. Break the front wheel lug nuts loose, then raise and support the front of the vehicle safely on jackstands.

3. Remove the front wheels.

4. Remove the front disc brake calipers.

5. Note the position and orientation of the brake pads and anti-rattle clip. Remove the brake shoes and lings from the front disc brake caliper anchor plate, then remove the anti-rattle clips.

To install:

6. Thoroughly clean the caliper and spindle sliding areas.

7. Place a new anti-rattle clip on the lower end of the inboard shoe. Make sure the tabs on the clip are positioned correctly and the loop-type spring is away from the rotor.

8. Place the lower end of the inner brake pad in the spindle assembly pad abutment, against the anti-rattle clip and slide the upper end of the pad into position. Be sure the clip is still in position.

9. Check and make sure the caliper piston is fully bottomed in the cylinder bore. Use a large C-clamp, bearing on a piece of wood, to bottom the piston, if necessary.

10. Position the outer brake pad on the caliper and press the pad tabs into place with your fingers. If the pad cannot be pressed into place by hand, use a C-clamp. Be careful not to damage the lining with the clamp. Bend the tabs to prevent rattling.

11. Lightly lubricate the caliper sliding grooves with caliper pin grease.

12. Install the brake caliper onto the anchor plate.

13. Bleed the brake system.

14. Clean the wheel hub mounting surface.

15. Install the front wheels and snug the lug nuts to fully seat the wheel against the hub.

16. Lower the vehicle until some of the vehicle's weight rests on the front tires, then tighten the lug nuts.

17. Lower the vehicle completely.

18. Make sure that the brakes are operating correctly.

2003–06 MODELS—EXCEPT 2005 EXCURSION

✳✳ WARNING

The electrical power to the air suspension system must be shut off prior to hoisting, jacking or towing an air suspension vehicle. Failure to shut the system off may lead to an unexpected inflation or deflation of the air springs, which may result in a shift of the vehicle.

1. Before servicing the vehicle, refer to the precautions section.

2. Using a suitable suction device, remove the brake fluid in the master cylinder reservoir until it is half filled.

3. Remove or disconnect the following:
- Wheels
- Anchor housing spring.

➡️**Ensure the anchor housing spring has one end with two tabs. If yes, the Left Hand (left hand) side anchor housing spring must be installed with the two-tabbed end in the upper brake caliper cavity.**

4. On the left hand brake caliper, release the lower portion of the anchor housing spring as follows:

a. Apply force at the center of the anchor housing spring and pull outward at the bottom of the anchor housing spring to remove it from the lower brake caliper cavity.

b. Rotate the spring upward then remove it from the brake caliper.

➡️**On the Right Hand (RH) side, the anchor housing spring must be installed with the two-tabbed end in the lower brake caliper cavity.**

5. For the RH brake caliper, release the upper portion of the anchor housing spring as follows:

a. Apply force at the center of the anchor housing spring and pull outward at the top of the anchor housing spring to remove it from the upper brake caliper cavity.

b. Rotate the spring downward then remove it from the brake caliper.

➡️**Never allow the brake caliper to hang from the brake hose.**

6. Remove the brake caliper as follows:

a. Remove the zero-drag spring.

b. Remove and discard the brake caliper-to-anchor plate bolts, guide pins and boots.

7. Remove the brake pads from the brake caliper.

8. Compress the brake caliper pistons using a C-clamp or other suitable tool.

To install:

9. Clean the inner surfaces of the brake caliper where the brake pads attach.

10. Install new brake hardware as follows:

a. Install the guide pin bushings into the caliper bores.

b. Apply grease to the inside of the guide pin bushing. Do not apply grease to the guide pin threads.

c. Push the guide pins into the bushing.

11. Install or connect the following:
- Inboard brake pad into the brake caliper.
- Outboard brake pad into the brake caliper
- Brake caliper on the brake disc. Tighten the guide pins and install the dust caps.
- Zero-drag spring

➡️**If present, the 2-tabbed end of the anchor housing spring must be installed first.**

12. Install the anchor housing spring as follows:

a. Insert tab of the anchor housing spring into the brake caliper cavity.

b. Twist tab into the brake caliper cavity (left hand side-upper brake caliper cavity/RH side-lower brake caliper cavity).

c. Rotate the anchor housing spring and position the upper portion onto the anchor plate.

d. Position the other anchor housing spring portion onto the brake caliper anchor plate.

e. Push down and inward until the upper and lower ends of the anchor housing spring are latched and seated in the brake caliper cavities.

13. Verify that the anchor housing spring is correctly latched.

14. Bleed the brake system, filling the brake master cylinder reservoir as required.

15. Install the wheels.

16. Make sure that the brakes are operating correctly.

2005 EXCURSION

⁜ WARNING

The electrical power to the air suspension system must be shut off prior to hoisting, jacking or towing an air suspension vehicle. Failure to shut the system off may lead to an unexpected inflation or deflation of the air springs, which may result in a shift of the vehicle.

1. Before servicing the vehicle, refer to the precautions section.
2. Using a suitable suction device, remove the brake fluid in the master cylinder reservoir until it is half filled.
3. Remove the wheels.
4. Remove the disc brake caliper.
5. Remove the V-springs.
6. Remove the pads and rail clips.

⁜ CAUTION

Do not allow grease, oil, brake fluid or other contaminants to contact the pad lining material. Do not install contaminated pads.

7. Install the pads.
8. Install the disc brake caliper anchor plate stainless steel rail clips.
9. If necessary, install the V-springs.
10. Compress the caliper pistons.

⁂ CAUTION

Use care not to damage the bleeder screw or front disc brake rotor shield.

11. Install the disc brake caliper.
12. Install the wheel and tire assembly.
13. Fill the brake master cylinder reservoir with DOT 3 motor vehicle brake fluid. Install the brake master cylinder filler cap.
14. Inspect brake operation.

Rear

2002 MODELS

⁂ WARNING

The electrical power to the air suspension system must be shut off prior to hoisting, jacking or towing an air suspension vehicle. Failure to shut the system off may lead to an unexpected inflation or deflation of the air springs, which may result in a shift of the vehicle.

1. Before servicing the vehicle, refer to the precautions section.

2. Remove enough brake fluid from the brake master cylinder reservoir until it is ½ full.
3. Remove or disconnect the following:
   - Wheels
   - Banjo bolt connecting the brake hose to the disc brake caliper and plug the brake hose. Discard 2 copper sealing washers.
   - 2 brake caliper slide pins and lift the caliper off the anchor plate

To install:

4. Retract the disc brake caliper pistons fully in the piston bores using an old brake pad or block of wood and a C-clamp or equivalent.
5. Install or connect the following:
   - Disc brake caliper above the rotor and install it with a rotating motion. Make sure the inner and outer pads are properly positioned and the anti-rattle clips are correctly installed. The brake caliper bleed screw should be positioned on top of the caliper when assembled on the vehicle.
6. Lubricate the locating pins and the inside of the insulators with silicone grease.
   - Locating pins through the caliper insulators and hand-start the threads into the steering knuckle attaching holes. Tighten the locating pins to 20 ft. lbs. (27 Nm).
   - Brake hose to the disc brake caliper using 2 new copper sealing washers. Tighten the banjo bolt to 29 ft. lbs. (40 Nm).
7. Bleed the brake system, filling the brake master cylinder reservoir as required.
   - Wheels
8. Lower the vehicle.
9. Pump the brake pedal several times to position the brake pads prior to moving the vehicle.
10. Road-test the vehicle and check for proper brake system operation.

2003–05 MODELS—EXCEPT 2005 EXCURSION

⁂ WARNING

The electrical power to the air suspension system must be shut off prior to hoisting, jacking or towing an air suspension vehicle. Failure to shut the system off may lead to an unexpected inflation or deflation of the air springs, which may result in a shift of the vehicle.

1. Before servicing the vehicle, refer to the precautions section.
2. Remove enough brake fluid from the brake master cylinder reservoir until it is ½ full.
3. Remove or disconnect the following:
   - Wheels
   - Anchor housing spring by squeezing at the center of the spring until it unlatches from the brake caliper at both ends, then rotate the spring to remove it from the caliper housing
   - Caps and brake caliper bolts
   - Brake caliper without disconnecting the brake hose
   - Brake pads
4. Thoroughly clean the areas of the caliper and caliper support assembly which contact each other during the sliding action of the caliper.

To install:

5. Compress the caliper piston using a C–clamp or other suitable tool.
6. Install or connect the following:
   - Brake shoes on the disc brake caliper support bracket
   - Anchor housing spring by placing the upper anchor housing spring end into the brake caliper cavity, then rotate the anchor housing spring and position the lower arm onto the anchor plate
   - Upper arm onto the anchor plate, then press down and inward until it is correctly seated and latched into the brake caliper cavities
   - Rear disc brake caliper onto the rear support bracket
7. Bleed the brake system, filling the brake master cylinder reservoir as required.
   - Front wheels
8. Make sure that the brakes are operating correctly.

2005 EXCURSION

⁂ WARNING

The electrical power to the air suspension system must be shut off prior to hoisting, jacking or towing an air suspension vehicle. Failure to shut the system off may lead to an unexpected inflation or deflation of the air springs, which may result in a shift of the vehicle.

1. Before servicing the vehicle, refer to the precautions section.
2. Remove the brake master cylinder filler cap. Check brake fluid level in brake

master cylinder reservoir. Remove fluid until brake master cylinder reservoir is half full.

3. Remove the wheel assembly.

4. Inspect the brake pads for wear or contamination.

5. Remove the nuts and the stone shield.

6. Remove the caliper pin bolts.

✳✳ CAUTION

Never allow the rear disc brake caliper to hang from the brake hose. Provide suitable support.

7. Remove the rear disc brake caliper and support with mechanics wire.

8. Remove the brake pads and rail clips.

To install:

9. Install new pads and clips.

10. Install the caliper and tighten the pin bolts to 27 ft. lbs. (36 Nm).

11. Install the stone shield and tighten to 46 ft. lbs. (62 Nm).

12. Bleed the brake system, filling the brake master cylinder reservoir as required.

13. Install the wheel assembly.

14. Make sure that the brakes are operating correctly.

FORD

Five Hundred • Montego

7

SPECIFICATIONS AND MAINTENANCE CHARTS

VEHICLE AND ENGINE IDENTIFICATION CHART

Engine							Model Year	
Code	Liters	Cu. In.	Cyl.	Fuel Sys.	Engine Type	Eng. Mfg.	Code	Year
1	3.0	183	6	SEFI	DOHC	Ford	5	2005

SEFI: Sequential Multi-port Fuel Injection

DOHC: Dual overhead cam

06017-FIVE-C01

GENERAL ENGINE SPECIFICATIONS

Year	Engine Displacement Liters	Engine VIN	Net Horsepower @ rpm	Net Torque @ rpm (ft. lbs.)	Bore x Stroke (in.)	Compression Ratio	Oil Pressure @ rpm
2005	3.0	1	203@5750	207@4500	3.50x3.13	10.0:1	11@1500

06017-FIVE-C02

GASOLINE ENGINE TUNE-UP SPECIFICATIONS

Year	Engine Displacement Liters	Engine VIN	Spark Plugs Gap (in.)	Ignition Timing (deg.) MT	AT	Fuel Pump (psi)	Idle Speed (rpm) MT	AT	Valve Clearance In.	Ex.
2005	3.0	1	0.052-0.056	—	①	②	—	①	HYD	HYD

NOTE: The Vehicle Emission Control Information label often reflects specification changes changes made during production.

The label figures must be used if they differ from those in this chart.

B: Before top dead center

HYD: Hydraulic

① Controlled by the Powertrain Control Module (PCM) and cannot be manually adjusted.

② Engine running: 40 psi

   Key On, Engine Off (KOEO): 40 psi

06017-FIVE-C03

06017-FREE-G01

3.0L engine accessory drive belt routing

CAPACITIES

Year	Model	Engine Displacement Liters	Engine ID/VIN	Engine Oil with Filter (qts.)	Transmission (pts.) 4-Spd	5-Spd	Auto.	Drive Axle Front (pts.)	Rear (pts.)	Fuel Tank (gal.)	Cooling System (qts.)
2005	Five Hundred	3.0	1	6.0	—	—	①	1.5	1.5	20	10.6
	Montego	3.0	1	6.0	—	—	①	1.5	1.5	20	10.6

NOTE: All capacities are approximate. Add fluid gradually and check to be sure a proper fluid level is obtained.

① CVT transmission: 20
  6-spd. Transmission: 15.8

06017-FIVE-C04

VALVE SPECIFICATIONS

Year	Engine VIN	Engine Displacement Liters	Seat Angle (deg.)	Face Angle (deg.)	Spring Test Pressure (lbs. @ in.)	Spring Installed Height (in.)	Stem-to-Guide Clearance (in.) Intake	Exhaust	Stem Diameter (in.) Intake	Exhaust
2005	1	3.0	44.50-45.0	45.25-45.75	156@1.18	1.562-1.577	0.0007-0.0027	0.0018-0.0037	0.2350-0.2358	0.2343-0.2350

06017-FIVE-C05

CRANKSHAFT AND CONNECTING ROD SPECIFICATIONS

All measurements are given in inches.

Year	Engine Displ. Liters	Engine VIN	Crankshaft				Connecting Rod		
			Main Brg. Journal Dia.	Main Brg. Oil Clearance	Shaft End-play	Thrust on No.	Journal Diameter	Oil Clearance	Side Clearance
2005	3.0	1	2.4790-2.4800	0.0010-0.0016	0.0050-0.0010	3	1.9670-2.9680	0.0010-0.0025	0.0039-0.0118

06017-FIVE-C06

PISTON AND RING SPECIFICATIONS

All measurements are given in inches.

Year	Engine Displ. Liters	Engine VIN	Piston Clearance	Ring Gap			Ring Side Clearance		
				Top Comp.	Bottom Comp.	Oil Control	Top Comp.	Bottom Comp.	Oil Control
2005	3.0	1	0.0007-0.0016	0.0039-0.0098	0.0106-0.0165	0.0059-0.0255	N/A	N/A	N/A

NA: Not Available

06017-FIVE-C07

TORQUE SPECIFICATIONS

All readings in ft. lbs.

Year	Engine VIN	Engine Displacement Liters	Cylinder Head Bolts	Main Bearing Bolts	Rod Bearing Bolts	Crankshaft Damper Bolts	Flywheel Bolts	Manifold		Spark Plugs	Oil Pan Drain Plug
								Intake	Exhaust		
2005	1	3.0	①	N/A	N/A	②	59	③	15	11	19

① Step 1: 30 ft. lbs.
   Step 2: plus 90 degrees
   Step 3: loosen all bolts 1 turn
   Step 4: 30 ft. lbs.
   Step 5: plus 90 degrees.
   Step 6: plus 90 degrees.

② Step 1: 86 ft. lbs.
   Step 2: loosen 1 turn
   Step 3: 37 ft. lbs.
   Step 4: plus 90 degrees

③ Upper and lower manifold: 89 INCH lbs.

06017-FIVE-C08

WHEEL ALIGNMENT

Year	Model			Caster Range (+/-Deg.)	Caster Preferred Setting (Deg.)	Camber Range (+/-Deg.)	Camber Preferred Setting (Deg.)	Toe-in (Deg.)
2005	Five Hundred	Front	AWD	0.75	+3.5	0.75	-0.40	+0.20+/-0.20
			FWD	0.75	+3.3	0.75	-0.60	+0.20+/-0.20
		Rear	AWD	—	—	0.75	-0.70	0.10+/-0.20
			FWD	—	—	0.75	-0.35	0.10+/-0.20
	Montego	Front	AWD	0.75	+3.5	0.75	-0.40	+0.20+/-0.20
			FWD	0.75	+3.3	0.75	-0.60	+0.20+/-0.20
		Rear	AWD	—	—	0.75	-0.70	0.10+/-0.20
			FWD	—	—	0.75	-0.35	0.10+/-0.20

06017-FIVE-C09

TIRE AND WHEEL SPECIFICATIONS

Year	Model	OEM Tires Standard	OEM Tires Optional	Tire Pressures (psi) Front	Tire Pressures (psi) Rear	Wheel Size	Wheel Lug Nut Torque (Ft. Lbs.)
2005	Five Hundred	P225/55R18	①	35	35	①	95
	Montego	P225/55R18	①	35	35	①	95

OEM: Original Equipment Manufacturer

PSI: Pounds Per Square Inch

STD: Standard

OPT: Optional

① Not available

06017-FIVE-C10

BRAKE SPECIFICATIONS

All measurements in inches unless noted

Year	Model		Brake Disc Original Thickness	Brake Disc Minimum Thickness	Brake Disc Maximum Runout	Minimum Lining Thickness Front	Minimum Lining Thickness Rear	Brake Caliper Bracket Bolts (ft. lbs.)	Brake Caliper Mounting Bolts (ft. lbs.)
2005	Five Hundred	F	①	1.040	0.002	0.118	—	74	44
		R	①	0.394	0.002	—	0.118	81	23
	Montego	F	①	1.040	0.002	0.118	—	74	44
		R	①	0.394	0.002	—	0.118	81	23

① Not available

06017-FIVE-C11

SCHEDULED MAINTENANCE INTERVALS
FORD FIVE HUNDRED & MERCURY MONTEGO

TO BE SERVICED	TYPE OF SERVICE	VEHICLE MILEAGE INTERVAL (x1000)																	
		5	10	15	20	25	30	35	40	45	50	55	60	65	70	75	80	85	90
Engine oil & filter ①	R																		
Rotate tires	S/I	✓	✓	✓	✓	✓	✓	✓	✓	✓	✓	✓	✓	✓	✓	✓	✓	✓	✓
Engine coolant strength hoses & clamps	S/I			✓			✓			✓			✓				✓		✓
Air cleaner filter	R						✓						✓						✓
Automatic transmission fluid & filter	R						✓						✓						✓
Engine coolant ②	R																		
PCV valve	R												✓						✓
Spark plugs ③	R																		
Drive belts	S/I						✓						✓						✓
Exhaust system & heat shields	S/I						✓						✓						✓
Front & rear brakes	S/I	✓	✓	✓	✓	✓	✓	✓	✓	✓	✓	✓	✓	✓	✓	✓	✓	✓	✓
Fuel filter	R			✓			✓			✓			✓			✓			✓

R: Replace   S/I: Service or Inspect

① Engine oil and filter: replace every 3000 miles.
② Engine coolant: change initially at 100,000 miles or 5 years.
③ Spark plugs: replace every 100,000 miles.

Special Operating Condition Requirements

During extensive idling and/or low speed driving for long distances, as in heavy commercial use such as delivery, taxi, patrol car or livery:

Lube front lower control arm and steering linkage ball joints with

Zerk fittings (if equipped) every 4,800 km (3,000 miles) or 3 months.

Inspect brake system and check battery electrolyte level (Patrol cars) every 8,000 km (5,000 miles).

Install a new cabin air filter as required.

When operating in dusty conditions such as unpaved or dusty roads:

Install a new engine air filter as required.

Install a new cabin air filter as required.

When operating in off-road conditions:

Change automatic transmission fluid every 48,000 km (30,000 miles).

Install a new cabin air filter as required.

Inspect and lubricate U-joints.

Inspect and lubricate steering linkage ball joints with zerk fittings.

06017-FIVE-C12

ENGINE REPAIR

Alternator

REMOVAL & INSTALLATION

1. Disconnect the negative battery cable.
2. Raise and support the vehicle.
3. Rotate the drive belt tensioner counterclockwise and remove the accessory drive belt.
4. Remove the right lower splash shield.
5. Disconnect the crankshaft position sensor.
6. Disconnect the wiring harness locators from the alternator splash shield.
7. Detach the alternator wiring connector.
8. Remove the positive cable nut.

9. Remove the alternator splash shield and the alternator.
   To install:
10. Position the alternator on the engine.
11. Install the alternator mounting bolts. Tighten the stud bolts to 71 INCH lbs. (8 Nm) and the bolts to 35 ft. lbs. (47 Nm).
12. Tighten the positive cable nut to 71 INCH lbs. (8 Nm).
13. Connect the wiring connector.
14. Install the wiring harness locators.
15. Connect the crankshaft position sensor.
16. Install the lower right splash shield.
17. Install and tension the accessory drive belt.
18. Connect the negative battery cable.

Engine Assembly

REMOVAL & INSTALLATION

1. Relieve the fuel system pressure.
2. Drain the engine coolant.
3. Drain the engine oil.
4. Recover the air conditioning refrigerant, into a refrigerant recovery station.
5. Place the transmission in neutral and raise the vehicle on a hoist.
6. Remove or disconnect the following:

- Both battery cables
- B+ terminal from the battery cable
- Air cleaner outlet pipe and air cleaner
- Cowl vent screen
- Left and right halfshafts
- Exhaust flex pipe
- Left side catalytic converter
- Roll restrictor cross brace-to-roll restrictor bolt and discard
- Roll restrictor cross brace
- Right and left fender ground straps
- Fuel tube from the fuel rail
- Vapor tube from the purge valve
- Powertrain Control Module (PCM) connectors and wiring conduit
- Power steering reservoir hose and retainers
- Disconnect the transmission shift cable
- Electrical connector and retainers from the shift cable bracket
- Shift cable bracket bolts and position the cable aside
- Heater and throttle body coolant hoses
- Brake booster vacuum hose
- Thermostat housing hoses
- Cooling tube retainers
- A/C high pressure switch connector
- A/C compressor manifold tube
- A/C tube retaining nuts and position the tubes aside
- Transmission cooler lines
- Power steering cooler hose
- Engine block heater connector and harness
- Torque converter lower access plug
- Torque converter attaching nuts
- 3 engine-to-transaxle mounting bolts
- Left catalytic converter bracket
- Driveshaft support bracket bolts and the driveshaft, if equipped

7 — 8 Nm (71 lb-in)

10 — 8 Nm (71 lb-in)

9 — 47 Nm (35 lb-ft)

1 Crankshaft position sensor connector	7 Generator B+ nut
2 Harness locators (2 required)	8 Generator B+ terminal
3 Engine control harness	9 Generator stud nuts (3 required)
4 Lower generator boot pin-type retainer	10 Generator studs (3 required)
5 Generator boot	11 Generator shield
6 Generator electrical connector	12 Generator

06017-FREE-G02

Alternator mounting—Five Hundred and Montego

014-00765

06017-FREE-G03

Installing subframe support tool—Five Hundred and Montego

303-D089

134-00243   134-00243

303-1140

06017-FREE-G04

Installing engine lifting tools to the engine—Five Hundred and Montego

- Stabilizer bar links from the struts
- Outer tie rod ends from the steering knuckles
- Steering column intermediate shaft bolt and separate the shaft from the steering gear

7. Position subframe support tool 014-00765 under subframe and powertrain assembly.

8. Remove the subframe bolts and rear subframe brackets.

9. Lower the powertrain and subframe assembly from the vehicle.

10. To separate the engine from the transaxle and subframe assembly, disconnect all necessary electrical connectors, fluid lines and starter wiring. Remove the starter and engine roll restrictor brackets

11. Install engine lifting eyes no. 134-00243 to the left and right cylinder heads.

12. Install lifting brackets no. 303-1140 to front of the engine.

13. Install an engine hoist and lifting bar no. 303-D089 to the engine.

14. Remove the front transaxle stabilizer bolt.

15. Remove the exhaust manifold heat shield.

16. Remove the left and right engine insulator upper nuts.

17. Remove the transaxle insulator bracket.

18. Using the hoist, lift the powertrain from the subframe.

19. Remove the transaxle-to-engine mounting bolts and separate the engine and transaxle.

To install:

20. Position the engine onto the transaxle and install the transaxle mount bolts. Torque the bolts to 30 ft. lbs. (40 Nm).

21. Install the exhaust manifold heat shield.

22. Using the hoist, position the powertrain to the subframe.

23. Install the left and right engine insulator upper nuts and tighten to 62 ft. lbs. (70 Nm).

24. Install the front transaxle stabilizer bolt and tighten to 30 ft. lbs. (40 Nm).

25. Install the transaxle insulator bracket and tighten the bolts to 41 ft. lbs. (55 Nm).

26. Install the starter and tighten the bolts to 18 ft. lbs. (25 Nm).

27. Connect the starter wiring and terminal cover.

28. Apply thread lock to the roll restrictor bolts and install the restrictor bracket and tighten the bolts to 41 ft. lbs. (55 Nm).

29. Reconnect all necessary electrical connectors, fluid lines and retainers.

30. Remove the engine hoist and lifting tools.

31. Using subframe support tool 014-00765, raise the subframe and powertrain assembly into the vehicle

32. Install the front subframe bolts and tighten the bolts to 148 ft. lbs. (200 Nm).

33. Install the rear subframe brackets. Tighten the large bolts to 111 ft. lbs. (150 Nm), and the smaller bolts to 41 ft. lbs. (55 Nm).

34. Connect the steering column intermediate shaft and tighten the pinch bolt to 18 ft. lbs. (25 Nm).

35. Install the outer tie rod ends to the steering knuckles and tighten NEW nuts to 85 ft. lbs. (115 Nm).

36. Install the stabilizer bar links to the struts using NEW nuts and tighten to 41 ft. lbs. (55 Nm).

37. Install the driveshaft and tighten the bolts to 18 ft. lbs. (25 Nm).

38. Install the driveshaft support bracket bolts and tighten the bolts to 18 ft. lbs. (25 Nm).

39. Install or connect the following:
- NEW torque converter attaching nuts and tighten to 27 ft. lbs. (36 Nm)

- Torque converter lower access plug
- Engine block heater connector and harness
- Power steering cooler hose
- Transmission cooler lines
- A/C tube and retaining nuts
- A/C compressor manifold tube
- A/C high pressure switch connector
- Cooling tube retainers
- Thermostat housing hoses
- Brake booster vacuum hose
- Heater and throttle body coolant hoses
- Shift cable and bracket bolts and tighten the bolts to 18 ft. lbs. (25 Nm)
- Electrical connector and retainers to the shift cable bracket
- Connect the transmission shift cable
- Power steering reservoir hose and retainers
- Powertrain Control Module (PCM) connectors and wiring conduit
- Vapor tube from the purge valve
- Fuel tube from the fuel rail
- Right and left fender ground straps
- Roll restrictor cross brace
- NEW roll restrictor cross brace-to-roll restrictor bolts and tighten to 41 ft. lbs. (55 Nm)
- Left catalytic converter bracket
- Left side catalytic converter
- Exhaust flex pipe
- Left and right halfshafts
- Cowl vent screen
- Air cleaner outlet pipe and air cleaner
- B+ terminal to the battery cable
- Both battery cables

40. Refill the engine, transaxle and cooling system with the correct amount of the appropriate fluids before starting the engine. Recharge the A/C system using approved recycling equipment.

Water Pump

REMOVAL & INSTALLATION

1. Remove or disconnect the following:
- Negative battery cable
- Coolant
- Accessory drive belt
- Air cleaner assembly
- Lower radiator hose
- Thermostat hoses
- Thermostat housing
- Water pump

06017-FREE-G05

Water pump mounting—Five Hundred and Montego

To install:

2. Position a new water pump housing gasket on the water pump sealing surface using gasket sealant to hold the gasket in place.

3. Install the water pump and tighten the bolts in the following sequence: tighten the center bolt to 35 inch lbs. (4 nm), loosely install the outer bolts, retighten the center bolt to 18 ft. lbs. (25 Nm), tighten the outer bolts to 89 inch lbs. (10 Nm).

4. The remainder of the installation is the reverse of removal.

5. Fill and bleed the cooling system.

6. Connect the negative battery cable.

7. Start the engine and check for leaks.

Heater Core

REMOVAL AND INSTALLATION

1. Ensure the ignition is off.

2. Remove the Smart Junction Box (SJB) cover located below the left side of the instrument panel. Remove the Restraints Control Module (RCM) fuse no. F2.21.

3. Turn the ignition on and watch the air bag indicator for 30 seconds. The indicator light will remain lit constantly if the correct fuse has been removed. If the light is not on steadily, remove the correct fuse and check the light again.

4. Turn the ignition switch off.

5. Disconnect the negative battery cable.

6. Drain the cooling system into a clean container for reuse.

7. Recover the air conditioning refrigerant, into a refrigerant recovery station.

8. Disconnect the heater hoses from the heater core inlet and outlet tubes in the engine compartment.

9. Remove the steering wheel access cover.

10. Release one side of the driver air bag module clip using needle nose pliers.

11. Lift the end of the clip over the center post.

12. Repeat the procedure on the opposite wide wire clip.

➡ **When the clips are released correctly, the clip ends will come to rest between the center two inner posts.**

13. Push on the horn at the upper part of the air bag cover and push both ends of the wire clips inward to release the driver air bag module from the steering wheel.

14. Disconnect the air bag electrical connectors, horn and accessories connectors, and remove the driver air bag module.

15. Remove and discard the steering wheel bolt.

16. Remove the steering wheel.

17. Remove the upper and lower steering column covers.

18. Disconnect the multi-function switch connectors and remove the multi-function switch.

19. Remove or disconnect the following:
- Cowl trim panels and disconnect the electrical connectors
- Lower instrument panel insulator
- Front floor console
- Floor shifter

- Instrument panel center finish panel
- Audio unit and disconnect the electrical and antenna connectors
- Instrument panel side panels
- A pillar trim panels
- Steering column cover
- Hood release handle
- Headlight switch
- Instrument cluster finish panel
- Glove box door
- Passenger air bag module connectors
- Two air bag module bolts
- Passenger air bag module
- Passenger assist handle
- Left side main wiring harness connectors
- 12 screws attaching the upper instrument panel section and remove the upper section
- Thermostatic expansion valve bolt
- Powertrain Control Module (PCM)
- Transaxle roll restrictor bolt
- Transaxle roll restrictor bar bolts
- Instrument panel cowl top bolts from both sides
- Steering column pinch bolt
- A/C floor duct
- Four floor tunnel brace bolts
- Restraints Control Module (RCM) small and large electrical connectors

20. Mark the location of the instrument panel cross beam bolts.

21. With the help of an assistant, remove the bolts from one side of the cross beam and while supporting the beam, remove the other side bolts and remove the cross beam.

22. Remove the dash panel seal from the heater core tubes.

23. Remove the heater core tube bracket.

24. Disconnect the evaporator discharge air temperature sensor connector.

25. Remove the left side blend door actuator.

26. Remove the heater core cover screws.

27. Remove the heater core cover and the heater core.

To install:

28. Install the heater core cover and the heater core.

29. Install the heater core cover screws.

30. Install the left side blend door actuator.

31. Connect the evaporator discharge air temperature sensor connector.

32. Install the heater core tube bracket.

33. Install the dash panel seal to the heater core tubes.

34. With the help of an assistant, install

1 Heater tube bracket screw

2 LH temperature blend door actuator screw

3 LH temperature blend door actuator electrical connector

4 LH temperature blend door actuator

5 Heater core cover screw

6 Heater core cover

7 Heater core

8 Evaporator discharge air temperature sensor electrical connector

06017-FREE-G06

Exploded view of heater core mounting—Five Hundred and Montego

the bolts from one side of the cross beam and while supporting the beam, install the other side bolts using the locations marked previously. Tighten the bolts to 18 ft. lbs. (25 Nm).

35. Connect the RCM small and large electrical connectors.

36. Install or connect the following:
   • Four floor tunnel brace bolts
   • A/C floor duct
   • Steering column pinch bolt and tighten to 18 ft. lbs. (25 Nm)
   • Instrument panel cowl top bolts to both sides and tighten to 18 ft. lbs. (25 Nm)
   • Transaxle roll restrictor bar bolts and tighten to 40 ft. lbs. (55 Nm)
   • Transaxle roll restrictor bolt and tighten to 35 ft. lbs. (48 Nm)
   • Powertrain Control Module (PCM)
   • Thermostatic expansion valve bolt
   • Instrument panel upper section and 12 screws
   • Left side main wiring harness connectors

   • Passenger assist handle
   • Passenger air bag module
   • Two air bag module bolts and tighten to 80 inch lbs. (9 Nm)
   • Passenger air bag module connectors
   • Glove box door
   • Instrument cluster finish panel
   • Headlight switch
   • Hood release handle
   • Steering column cover
   • A pillar trim panels
   • Instrument panel side panels
   • Audio unit and connect the electrical and antenna connectors
   • Instrument panel center finish panel
   • Floor shifter
   • Front floor console
   • Lower instrument panel insulator
   • Cowl trim panels and connect the electrical connectors
   • Multi-function switch and connectors
   • Upper and lower steering column covers

   • Steering wheel and tighten the NEW bolt to 30 ft. lbs. (40 Nm)
   • Connect the air bag electrical connectors, horn and accessories connectors, and install the driver air bag module. Make the clips are released from the center post and back in position on both sides of the air bag.
   • Steering wheel access cover
   • Heater hoses to the heater core inlet and outlet tubes

37. Recharge the air conditioning refrigerant.

38. Refill the coolant system.

39. Disconnect the negative battery cable.

40. Turn the ignition switch on.

41. Install Restraints Control Module (RCM) fuse no. F2.21 and the fuse cover.

42. Turn the ignition off.

43. Connect the negative battery cable.

44. Turn the ignition on and then off. Wait 10 seconds and turn the key back on. Watch the air bag indicator. The indicator light will remain lit constantly for 6 seconds and then go off. If the indicator does not turn on and then off, diagnose the air bag system

45. Clear all Diagnostic Trouble Codes (DTC) using a diagnostic tool.

Cylinder Head

REMOVAL & INSTALLATION

Left Side

1. Relieve the fuel system pressure.

2. Disconnect the negative battery cable.

3. Drain the engine coolant.

4. Remove or disconnect the following:
   • Accessory drive belt
   • Upper and lower intake manifolds
   • Coolant bypass tube
   • Ignition coil on plugs
   • Valve cover wiring harness
   • 6 bolts and 8 stud bolts, and remove the valve cover
   • Camshafts
   • Exhaust manifold
   • Oil dipstick and tube
   • Mark the location of the camshaft roller followers and remove them
   • Hydraulic lash adjusters
   • Cylinder head bolts in the sequence shown
   • Cylinder head from the engine block and discard the gaskets and bolts

06017-FREE-G07

Left side cylinder head bolt removal sequence—Five Hundred and Montego

06017-FREE-G09

Left side valve cover bolt installation sequence—Five Hundred and Montego

sequence using the following steps. The first step is 30 ft. lbs. (40 Nm), the second step is an additional 90 degrees, the third step is to loosen all bolts in sequence one full turn, the fourth step is to tighten the bolts to 30 ft. lbs. (40 Nm), the fifth step is to tighten and additional 90 degrees, the last step is to tighten an addition 90 degrees.

- Lash adjusters and camshaft roller followers in their correct locations. Lubricate all rocker arm components with engine assembly lubricant.
- Oil dipstick and tube
- Exhaust manifold
- Camshafts
- Place a bead of silicone gasket sealant to the front cover-to-cylinder head surface
- Valve cover. Install the 6 bolts and 8 stud bolts, and tighten in sequence to 89 inch lbs. (10 Nm).
- Valve cover wiring harness
- Ignition coil on plugs
- Coolant bypass tube
- Upper and lower intake manifolds
- Accessory drive belt

8. Fill and bleed the cooling system.

➡**Engine coolant is corrosive to engine bearing material. Replace the engine oil after removal of any coolant-carrying component to help prevent potential bearing damage.**

9. Change the engine oil and filter.
10. Connect the negative battery cable.
11. Start the engine and check for leaks.

Right Side

1. Relieve the fuel system pressure.
2. Disconnect the negative battery cable.
3. Drain the engine coolant.
4. Remove or disconnect the following:
- Accessory drive belt
- Upper and lower intake manifolds

06017-FREE-G08

Left side cylinder head bolt installation sequence—Five Hundred and Montego

To install:

5. The cylinder head should be cleaned and inspected prior to installation.

6. Clean all gasket mating surfaces thoroughly.

7. Install or connect the following:
- New head gaskets on the cylinder block

✳✳ WARNING

Always use new cylinder head bolts when installing the cylinder head or damage to the engine may occur.

- Cylinder head on the cylinder block
- Tighten the cylinder head bolts in

06017-FREE-G10

Right side cylinder head bolt removal sequence—Five Hundred and Montego

06017-FREE-G11

Right side cylinder head bolt installation sequence—Five Hundred and Montego

- Coolant bypass tube
- Ignition coil on plugs
- PCV connector and tube
- Power steering reservoir bracket and stud bolts
- Power steering pressure tube bracket
- Wiring harness and retainers from valve cover
- Valve cover bolts and stud bolts
- Valve cover
- Camshafts
- Exhaust manifold
- Mark the location of the camshaft roller followers and remove them
- Hydraulic lash adjusters
- Cylinder head bolts in the sequence shown
- Cylinder head from the engine block and discard the gaskets and bolts

To install:

5. The cylinder head should be cleaned and inspected prior to installation.
6. Clean all gasket mating surfaces thoroughly.
7. Install or connect the following:
- New head gaskets on the cylinder block

❊❊ WARNING

Always use new cylinder head bolts when installing the cylinder head or damage to the engine may occur.

06017-FREE-G12

Right side valve cover bolt installation sequence—Five Hundred and Montego

- Cylinder head on the cylinder block
- Tighten the cylinder head bolts in sequence using the following steps. The first step is 30 ft. lbs. (40 Nm), the second step is an additional 90 degrees, the third step is to loosen all bolts in sequence one full turn, the fourth step is to tighten the bolts to 30 ft. lbs. (40 Nm), the fifth step is to tighten and additional 90 degrees, the last step is to tighten an addition 90 degrees.
- Lash adjusters and camshaft roller followers in their correct locations. Lubricate all rocker arm components with engine assembly lubricant.
- Exhaust manifold
- Camshafts
- Place a bead of silicone gasket sealant to the front cover-to-cylinder head surface
- Valve cover. Install the 9 bolts and 5 stud bolts, and tighten in sequence to 89 inch lbs. (10 Nm).
- Power steering pressure tube bracket
- Power steering reservoir bracket and stud bolts
- PCV connector and tube
- Ignition coil on plugs
- Coolant bypass tube
- Upper and lower intake manifolds
- Accessory drive belt
8. Fill and bleed the cooling system.

➡**Engine coolant is corrosive to engine bearing material. Replace the engine oil after removal of any coolant-carrying component to help prevent potential bearing damage.**

9. Change the engine oil and filter
10. Connect the negative battery cable.
11. Start the engine and check for leaks.

Camshaft Roller Followers and Lash Adjusters

REMOVAL & INSTALLATION

1. Remove the valve covers. See the procedure under cylinder head.
2. Remove the spark plugs.
3. Remove the right side splash shield.
4. Rotate the crankshaft until the camshaft lobe is pointing directly away from the roller follower.
5. Using tool no. 303-473, remove the roller followers.
6. Remove the lash adjusters. If more than adjuster is to be removed, identify each adjuster location.

To install:

7. Lubricate the lash adjusters clean engine oil and install them in their original locations.
8. Reverse the removal procedure to install the remaining components.

06017-FREE-G13

Removing the camshaft roller followers using tool 303-473—Five Hundred and Montego

Intake Manifold

REMOVAL & INSTALLATION

Upper

1. Relive the fuel system pressure.
2. Drain the engine coolant.
3. Remove or disconnect the following:
- Negative battery cable
- Air cleaner outlet pipe
- Transaxle roll restrictor cross brace bolts and cross brace
- EVAP purge valve tube
- PCV, brake booster and vacuum harness tubes from the manifold
- Fuel tube retainer
- EGR connector and vacuum tube
- Throttle body and PCV coolant hoses

- Intake manifold bolts and tighten all bolts in steps in sequence to 89 inch lbs. (10 Nm)
- Transaxle roll restrictor roll brace and tighten the bolts as shown

5. Fill and bleed the engine cooling system.

6. Connect the negative battery cable.

7. Start the engine and check for leaks.

Lower

1. Remove or disconnect the following:
   - Coolant
   - Fuel system pressure
   - Upper intake manifold
   - Fuel tube from fuel rail
   - Fuel rail pressure and temperature sensor connectors and vacuum tube
   - Fuel charging wiring harness connector
   - Lower intake manifold

To install:

2. Thoroughly clean all gasket mating surfaces.

3. Install or connect the following:
   - 6 intake manifold gaskets and the intake manifold
   - Fuel charging wiring harness connector
   - Fuel rail pressure and temperature sensor connectors and vacuum tube
   - Fuel tube from fuel rail
   - Upper intake manifold
   - Fuel system pressure

4. Fill and bleed the cooling system.

5. Start the engine and check for leaks.

06017-FREE-G14

Upper intake manifold—Five Hundred and Montego

06017-FREE-G15

Upper intake manifold tightening sequence—Five Hundred and Montego

- Throttle body electrical connector
- Upper-to-lower intake manifold attaching screws
- Intake manifold

To install:

4. Installation is the reverse of the removal procedure, using the following torque specifications.

06017-FREE-G16

Transaxle roll restrictor brace bolt tightening—Five Hundred and Montego

06017-FREE-G17

Lower intake manifold and related components—Five Hundred and Montego

Exhaust Manifold

REMOVAL & INSTALLATION

➡**Spray the exhaust system fasteners with penetrating lubricant before removing them to help prevent broken studs and bolts. The use of a 6-point socket is highly recommended when removing exhaust system fasteners.**

✳✳ CAUTION

To prevent serious burns, allow the exhaust manifold to cool down before attempting to remove it.

Left Manifold

1. Disconnect the negative battery cable.
2. Remove the coolant hose retainers from the fan shroud and position aside.
3. Disconnect the cooling fan electrical connectors.
4. Remove the cooling fans.
5. Disconnect the oxygen sensor connector.
6. Remove the 2 exhaust manifold-to-catalytic converter bolts.
7. Remove 6 exhaust manifold nuts and remove the exhaust manifold.
   To install:
8. Clean all gasket mating surfaces thoroughly.

9. Install a new exhaust manifold gasket and the exhaust manifold on the cylinder head. Start 2 nuts to hold the manifold in position.
10. Install the remaining nuts. Tighten the nuts in the sequence shown to 15 ft. lbs. (20 Nm).
11. Install the 2 exhaust manifold-to-catalytic converter bolts and tighten to 30 ft. lbs. (40 Nm).

12. Connect the oxygen sensor connector.
13. Install the cooling fans.
14. Connect the cooling fan electrical connectors.
15. Install the coolant hose retainers to the fan shroud.
16. Connect the negative battery cable.
17. Start the engine and check for exhaust leaks.

Right Manifold

1. Disconnect the negative battery cable.
2. Raise the vehicle on a hoist.
3. Remove the right side catalytic converter.
4. Disconnect the oxygen sensor connector.
5. Remove the exhaust manifold heat shield.
6. Disconnect the EGR tube from the manifold.
7. Remove the exhaust manifold.
   To install:
8. Clean all gasket mating surfaces thoroughly.
9. Install a new gasket and the exhaust manifold on the cylinder head. Start 2 nuts to hold the manifold in position.
10. Install the remaining nuts. Tighten the nuts in the sequence shown to 15 ft. lbs. (20 Nm).
11. Connect the EGR tube from the manifold.
12. Install the exhaust manifold heat shield.

40 Nm (30 lb-ft)

20 Nm (15 lb-ft)

06017-FREE-G18

Exploded view of left exhaust manifold—Five Hundred and Montego

1 RH exhaust manifold heat
   shield (2 required)
2 RH exhaust manifold heat
   shield
3 Heated oxygen sensor
   (HO2S) electrical connector
4 RH exhaust manifold
5 Exhaust gas recirculation
   (EGR) system module tube
   fitting
6 RH exhaust manifold nut (6
   required)
7 RH exhaust manifold gasket

06017-FREE-G19

Exploded view of right exhaust manifold—Five Hundred and Montego

13. Connect the oxygen sensor connector.
14. Install the right side catalytic converter.
15. Lower the vehicle.
16. Connect the negative battery cable.
17. Start the engine and check for exhaust leaks.

Camshaft

REMOVAL & INSTALLATION

Right Side

1. Disconnect the negative battery cable.
2. Remove the engine front cover.
3. Remove the timing chains.
4. Loosen the camshaft bearing cap bolts evenly in the sequence shown and remove the bearing caps.

➡**The bearing caps are numbered for correct positioning.**

5. Remove the camshafts.
To install:
6. Clean and inspect all parts before installation.
7. Lubricate the camshaft lobes and journals with clean engine oil.
8. Carefully install the camshafts.
9. Align the camshaft sprocket marks as shown.
10. Lubricate the bearing surfaces of the bearing caps with clean engine oil.

06017-FREE-G20A

Right side camshaft bearing cap removal sequence—Five Hundred and Montego

06017-FREE-G21

Right side camshaft sprocket mark positioning—Five Hundred and Montego

06017-FREE-G22

Right side camshaft bearing cap tightening sequence—Five Hundred and Montego

11. Position the bearing caps in their proper locations and loosely install the bearing cap bolts.
12. Tighten the bearing caps bolts in sequence to 89 inch lbs. (10 Nm).
13. Install the timing chains.
14. Install the engine front cover.
15. Connect the negative battery cable.

Left Side

1. Disconnect the negative battery cable.
2. Remove the water pump.
3. Remove the engine front cover.
4. Remove the timing chains.

06017-FREE-G23

Left side camshaft bearing cap removal sequence—Five Hundred and Montego

5. Remove the 3 bolts and the camshaft oil seal retainer.

6. Using a press, remove the oil seal from the retainer and discard.

7. Loosen the camshaft bearing cap bolts evenly in the sequence shown and remove the bearing caps.

➡ **The bearing caps are numbered for correct positioning.**

8. Remove the camshafts.

To install:

9. Clean and inspect all parts before installation.

10. Lubricate the camshaft lobes and journals with clean engine oil.

11. Carefully install the camshafts.

12. Align the camshaft sprocket marks as shown.

13. Lubricate the bearing surfaces of the bearing caps with clean engine oil.

06017-FREE-G24

Left side camshaft sprocket mark positioning—Five Hundred and Montego

06017-FREE-G26

Left side camshaft oil seal installation—Five Hundred and Montego

14. Position the bearing caps in their proper locations and loosely install the bearing cap bolts.

15. Tighten the bearing caps bolts in sequence to 89 inch lbs. (10 Nm).

16. Install the camshaft oil seal retainer and tighten the bolts 89 inch lbs. (10 Nm).

17. Lubricate the oil seal with clean engine oil and position the seal on installer tool no. 303-463.

18. Position the installer tool and oil seal on the camshaft.

19. Using installer tools no. 303-1139, 211-185 and 303-458, install the camshaft oil seal.

20. Remove the special tools.

21. Install the timing chains.

22. Install the engine front cover.

23. Install the water pump.

24. Connect the negative battery cable.

Starter Motor

REMOVAL & INSTALLATION

Automatic Transmission Models

1. Disconnect the negative battery cable.

2. Remove the air cleaner assembly.

3. Remove the starter cover.

4. Disconnect the starter electrical connectors.

5. Remove the starter bolts.

6. Remove the starter from the vehicle.

To install:

7. Position the starter in the vehicle.

8. Install the bolts. Tighten to 18 ft. lbs. (25 Nm).

9. Connect the starter electrical connectors.

10. Install the starter cover.

11. Install the air cleaner assembly.

12. Connect the negative battery cable.

Manual Transmission Models

1. Disconnect the negative battery cable.

2. Remove the air cleaner outlet pipe.

3. Remove the transaxle roll restrictor bolts and the cross brace assembly.

4. Remove the starter cover.

5. Disconnect the starter electrical connectors.

6. Remove the starter bolts.

7. Remove the starter from the vehicle.

06017-FREE-G25

Left side camshaft bearing cap tightening sequence—Five Hundred and Montego

06017-FREE-G27

Starter mounting automatic transmission models—Five Hundred and Montego

25 Nm
(18 lb-ft)

12 Nm
(9 lb-ft)

6 Nm
(53 lb-in)

1 Transaxle roll restrictor bolt
2 Transaxle roll restrictor cross
  brace bolt
3 Transaxle roll restrictor cross
  brace
4 Starter motor solenoid
  terminal cover
5 Starter motor solenoid wire
  nut

6 Starter motor solenoid wire
7 Starter motor solenoid battery
  cable nut
8 Starter motor solenoid battery
  cable
9 Starter motor bolt
10 Starter motor

06017-FREE-G28

Starter mounting manual transmission models—Five Hundred and Montego

To install:

8. Position the starter in the vehicle.
9. Install the bolts. Tighten to 18 ft. lbs. (25 Nm).
10. Connect the starter electrical connectors.
11. Install the starter cover.
12. Install the cross brace assembly and transaxle roll restrictor bolts.
13. Install the air cleaner outlet pipe.
14. Connect the negative battery cable.

Oil Pan

✳✳ CAUTION

The EPA warns that prolonged contact with used engine oil may cause a number of skin disorders, including cancer! You should make every effort to minimize your exposure to used engine oil. Protective gloves should be worn when changing the oil. Wash your hands and any other exposed skin areas as soon as possible after exposure to used engine oil. Soap and water, or waterless hand cleaner, should be used.

REMOVAL & INSTALLATION

1. Disconnect the negative battery cable.
2. Drain the engine oil.
3. Remove the oil filter.
4. Raise the vehicle on a hoist.
5. Remove the exhaust flex pipe.
6. Remove the torque converter inspection cover.
7. Remove 2 oil pan-to-transaxle bolts.
8. Remove the oil filter adapter-to-oil pan bolt.
9. Remove the 15 retaining bolts and remove the oil pan.

To install:

10. Clean the gasket mating surfaces thoroughly.
11. Apply a drop of sealant to the locations shown.

➡**When using silicone rubber sealer, assembly must occur within 5 minutes after sealer application. After this time, the sealer may start to harden and its sealing effectiveness may be reduced.**

12. Install the oil pan using a new gasket and secure to the block with the attaching

10 mm (0.40 in)

06017-FREE-G29

Applying sealant to the oil pan mounting surface—Five Hundred and Montego

06017-FREE-G30

Oil pan bolt tightening sequence—Five Hundred and Montego

bolts. Tighten the bolts in sequence to 89 inch lbs. (10 Nm).

13. Install the oil filter adapter-to-oil pan bolt.
14. Install 2 oil pan-to-transaxle bolts.
15. Install the torque converter inspection cover.
16. Install the exhaust flex pipe.
17. Install a new oil filter.
18. Fill the engine with the proper type and amount of clean oil.
19. Connect the negative battery cable.
20. Start the engine and check for leaks.

Oil Pump

REMOVAL & INSTALLATION

1. Disconnect the negative battery cable.

06017-FREE-G31

Oil pump bolt removal sequence—Five Hundred and Montego

Oil pump bolt tightening sequence—Five Hundred and Montego

06017-FREE-G32

2. Drain the engine oil.

3. Remove the timing chain components.

4. Remove the oil pan.

5. Remove the oil pump screen and pickup tube.

6. Remove the oil pump attaching bolts in the sequence shown.

To install:

7. Place the oil pump in the proper position with a new gasket and install the retaining bolts.

8. Tighten the oil pump retaining bolts in sequence to 89 inch lbs. (10 Nm).

9. Install the oil pump screen and pickup tube.

10. Install the oil pan.

11. Install the timing chain components.

12. Fill the engine with clean oil.

13. Connect the negative battery cable.

➡**Check for proper engine oil pressure immediately after starting the engine. If engine oil pressure is not within specification a few seconds after starting the engine, stop the engine and determine the reason for the low oil pressure condition. Running an engine with low oil pressure may result in serious engine damage.**

Rear Main Seal

REMOVAL & INSTALLATION

1. Disconnect the negative battery cable.
2. Raise the vehicle on a hoist.
3. Remove the transaxle.
4. Remove the flexplate.

✳✳ WARNING

Use caution when working near the crankshaft sealing surface. If the surface becomes damaged, an oil leak may occur.

5. Screw in the threaded end of a crankshaft rear seal replacer tool, then use the tool to remove the seal.

To install:

6. Inspect the crankshaft seal area for any damage that may cause the seal to leak. If damage is evident, service or replace the crankshaft as necessary.

7. Coat the crankshaft seal area and the seal lip with engine oil.

8. Using a crankshaft seal replacer tool, install the seal. Tighten the bolts of the seal installer tool evenly so the seal is straight and seats without misalignment.

9. Install the flexplate and tighten the bolts to 59 ft. lbs. (80 Nm).

10. Install the transaxle, lower the vehicle and connect the battery.

Front Main Seal

REMOVAL & INSTALLATION

1. Disconnect the negative battery cable.
2. Raise the vehicle on a hoist.

3. Remove the accessory drive belt.

4. Remove the right side fender splash shield.

5. Hold the crankshaft pulley and remove the pulley bolt.

6. Using a puller, remove the crankshaft pulley.

7. Use a seal replacer and pull out the oil seal.

To install:

8. Coat the crankshaft seal lip with clean engine oil.

9. Using a seal installer, install the seal.

10. Lubricate the front cover and crankshaft seal inner lip with clean engine oil.

11. Apply silicone sealant to the crankshaft keyway in the pulley.

12. Using an installer, screw the crankshaft pulley on the crankshaft.

13. Hold the crankshaft pulley from turning and install the pulley bolt and washer and tighten as follows: Step 1; tighten the bolt to 86 ft. lbs. (120 Nm). Step 2; loosen the bolt one full turn. Step 3; tighten to 37 ft. lbs. (50 Nm). Step 4; tighten an additional 90 degrees.

14. Install the right side fender splash shield.

15. Install the accessory drive belt.

16. Lower the vehicle.

Timing Chains and Front Cover

REMOVAL & INSTALLATION

1. Raise the vehicle on a hoist.
2. Disconnect the negative battery cable.

14 Engine front cover stud bolt nut

15 Ground wire

16 Engine front cover stud bolt

17 Engine front cover bolt (8 required)

18 Engine front cover bolt (7 required)

19 Engine front cover

20 Engine front cover gasket (3 required)

Exploded view of engine front cover—Five Hundred and Montego

06017-FREE-G33

3. Remove the accessory drive belt.

4. Remove the oil pan.

5. Remove the crankshaft front oil seal.

6. Remove both valve covers. See the procedure in cylinder head.

7. Remove the accessory drive belt tensioner.

8. Disconnect the power steering pump and reservoir without disconnecting the lines, and position aside.

9. Disconnect the crankshaft and camshaft position sensor electrical connectors.

10. Remove the alternator-to-front cover stud nut and stud.

11. Remove the radiator splash shield.

12. Remove the A/C compressor bolts and without disconnecting the lines, position the compressor aside.

13. Remove the compressor bracket.

14. Remove the ground wire from the front cover.

15. Remove the 14 bolts, stud bolt and remove the engine front cover.

16. Remove the ignition pulse wheel.

17. Install the crankshaft damper bolt.

18. Remove the spark plugs.

19. Rotate the crankshaft clockwise to position the crankshaft keyway in the 11 o'clock position.

20. Position the camshaft sprockets timing marks in the correct position so the no.

06017-FREE-G35

Positioning right side camshaft sprockets in the neutral position—Five Hundred and Montego

06017-FREE-G36

Positioning left side camshaft sprockets in the neutral position—Five Hundred and Montego

06017-FREE-G37

Left side crankshaft and camshaft timing mark alignment—Five Hundred and Montego

06017-FREE-G38

Right side crankshaft and camshaft timing mark alignment—Five Hundred and Montego

1 cylinder is a TDC of the compression stroke.

21. Ensure all timing marks are correctly located. If not, rotate the crankshaft one additional turn and recheck the marks.

22. Rotate the crankshaft clockwise 120 degrees to the 3 o'clock position to place the right side camshafts in the neutral position.

23. Ensure the camshafts are correctly located.

24. Remove the right side timing chain tensioner arm bolts, the tensioner and the tensioner arm.

25. Remove the right side timing chain guide and timing chain.

26. Rotate the crankshaft clockwise 600 degrees to position the left side camshafts in the neutral position.

27. Remove the left side timing chain tensioner arm bolts, the tensioner and the tensioner arm.

28. Remove the left side timing chain guide and timing chain.

To install:

29. Place the timing chain tensioner in a soft-jawed vise.

30. Hold the tensioner ratchet lock away from the ratchet stem with a small pick.

06017-FREE-G34

Crankshaft and camshaft sprocket timing marks—Five Hundred and Montego

1. 12 Links
2. 27 Links
3. 30 Links

06017-FREE-G39

Correct timing chain mark alignment—Five Hundred and Montego

31. Compress the tensioner until the piston is fully bottomed in the bore and insert a 1.5mm paper clip to hold the piston. Remove the pick from the ratchet stem.

32. Repeat the procedure on the other timing chain tensioner.

33. Mark any link on the timing chain with a permanent marker to indicate the crankshaft timing mark.

34. Starting with the crankshaft timing mark, count 29 links on the chain and mark the link.

35. Continue to count to the 42nd link and mark the link. Repeat this procedure on the other timing chain.

36. Ensure the left side camshaft sprocket timing marks are still aligned.

37. Install the left side timing chain and guide and install the bolts.

38. Align the marks on the timing chain with the marks on the crankshaft and camshaft sprockets.

39. Tighten the timing chain guide bolts to 18 ft. lbs. (25 Nm).

40. Install the tensioner arm and tensioner and tighten the bolts to 18 ft. lbs. (25 Nm).

41. Rotate the crankshaft 120 degrees clockwise until the crankshaft keyway is in the 3 o'clock position.

42. Ensure the right side camshaft sprocket timing marks are still aligned.

43. Install the right side timing chain and guide and install the bolts.

44. Align the marks on the timing chain with the marks on the crankshaft and camshaft sprockets.

45. Tighten the timing chain guide bolts to 18 ft. lbs. (25 Nm).

46. Install the tensioner arm and tensioner and tighten the bolts to 18 ft. lbs. (25 Nm).

47. Remove both tensioner retaining pins and allow the piston to move.

48. Rotate the crankshaft 120 degrees counterclockwise to the TDC position.

49. Verify the correct timing as follows: There should be 12 links between the camshaft sprocket timing marks; there should be 27 links between the left side camshaft and crankshaft timing marks; there should be 30 links between the right side camshaft and crankshaft timing marks.

50. Remove the crankshaft damper bolt and install the ignition pulse wheel. Install the pulse wheel with the keyway in the slot stamped "30" or "30FF."

51. Install the spark plugs and tighten to 11 ft. lbs. (15 Nm).

52. Clean the front cover gasket mating surfaces.

53. Place silicone gasket sealant on the cylinder head and cylinder block mating surfaces as shown.

54. Install the front cover and tighten the bolts and stud bolt in the sequence shown to 18 ft. lbs. (25 Nm).

55. Install the ground wire to the front cover.

56. Install the A/C compressor bracket.

57. Install the A/C compressor.

58. Install the radiator splash shield.

0.06 mm (0.024 in)

06017-FREE-G40

Applying sealant to block and head front cover mating surfaces—Five Hundred and Montego

59. Install the alternator-to-front cover stud nut and stud.

60. Connect the crankshaft and camshaft position sensor electrical connectors.

61. Connect the power steering pump and reservoir

62. Install the accessory drive belt tensioner.

63. Install both valve covers.

64. Install the crankshaft front oil seal.

65. Install the oil pan.

66. Install the accessory drive belt.

67. Fill the crankcase with the proper type and quantity of engine oil. Fill and bleed the cooling system. Connect the negative battery cable.

68. Start the engine and check for leaks. Check the ignition timing and curb idle speed and adjust, as necessary.

06017-FREE-G41

Front cover bolt torque sequence—Five Hundred and Montego

FUEL SYSTEM

Fuel System Service Precautions

Safety is the most important factor when performing not only fuel system maintenance, but any type of maintenance. Failure to conduct maintenance and repairs in a safe manner may result in serious personal injury or death. Work on a vehicle's fuel system components can be accomplished safely and effectively by adhering to the following rules and guidelines.

• To avoid the possibility of fire and personal injury, always disconnect the negative battery cable unless the repair or test procedure requires that battery voltage by applied.

• Always relieve the fuel system pressure prior to disconnecting any fuel system component (injector, fuel rail, pressure regulator, etc.) fitting or fuel line connection. Exercise extreme caution whenever relieving fuel system pressure, to avoid exposing skin, face and eyes to fuel spray. Please be advised that fuel under pressure may penetrate the skin or any part of the body that it contacts.

• Always place a shop towel or rag around the fitting or connection prior to loosening to absorb any excess fuel due to spillage. Ensure that all fuel spillage is quickly remove from engine surfaces. Ensure that all fuel-soaked cloths or towels are deposited into a flame-proof waste container with a lid.

• Always keep a dry chemical (Class B) fire extinguisher near the work area.

• Do not allow fuel spray or fuel vapors to come into contact with a light bulb, spark or open flame.

• Always use a second wrench when loosening or tightening fuel line connections fittings. This will prevent unnecessary stress and torsion to fuel piping. Always follow the proper torque specifications.

• Always replace worn fuel fitting O-rings with new ones. Do not substitute fuel hose where rigid pipe is installed.

Fuel System Pressure

RELIEVING

Locate the fuel pump relay in the engine compartment power distribution box relay no. 53. Remove the fuel pump relay. Start the engine and allow it to idle until it stalls. Crank the engine for 5 seconds to ensure fuel supply manifold pressure is released. When vehicle service is complete, reinstall the fuel pump relay and turn the ignition on to pressurize the fuel system. Start the vehicle and check the system for leaks.

06017-FREE-G42

Fuel pump relay location—Five Hundred and Montego

Fuel Filter

REMOVAL & INSTALLATION

1. Relieve the fuel system pressure.
2. Disconnect the negative battery cable.
3. Raise and support the vehicle safely on jackstands.

4. Place a rag under the fuel filter to catch any residual fuel that may leak out when the filter is removed.
5. Remove the quick-connect fittings at both ends of the fuel filter.
6. Note the flow arrow direction for installation reference.
7. Remove the fuel filter by loosening the retaining clamp and pulling it from the bracket.

To install:

8. Install the fuel filter in its bracket, ensuring proper direction of flow as noted earlier.
9. Install the quick-connect fittings at both ends of the fuel filter. Ensure the fuel line clicks into place and is fully seated.
10. Start the engine and check the filter connections for leaks by running the tip of your finger around each connection.

9 Nm (80 lb-in)

1 Fuel supply tube quick connect coupling

2 Fuel supply tube quick connect coupling

3 Fuel filter

4 Fuel filter bracket

5 Fuel filter bracket bolt

6 Fuel tube retainer clip

7 Fuel supply tube quick connect coupling with redundant clip

06017-FREE-G43

Exploded view of fuel filter mounting—Five Hundred and Montego

11. Turn the engine off and lower the vehicle.

Fuel Pump

REMOVAL & INSTALLATION

1. Relieve the fuel system pressure.
2. Disconnect the negative battery cable.
3. Drain the fuel tank.
4. Remove the rear lower seat cushion.
5. Remove the fuel pump module cover.
6. Remove any dirt that has accumulated around the fuel pump module attaching flange to prevent it from entering the tank during service.
7. Disconnect the fuel pump electrical connectors.
8. Disconnect the fuel pump quick connect coupling from the fuel pump module.
9. Disconnect the fuel vapor tube quick connect coupling from the fuel pump module.
10. Turn the fuel pump module locking ring using a locking ring removal tool, and remove the locking ring.
11. Remove the fuel pump module and disconnect the quick connect fittings from the bottom of the fuel pump.
To install:
12. Connect the quick connect fittings to the bottom of the fuel pump. Insert the fuel

1 Fuel pump module access cover	5 Fuel pump module lock ring
2 Fuel vapor control tube assembly valve quick connect coupling	6 Fuel pump module
3 Fuel supply tube quick release coupling	7 Fuel pump module O-ring seal
4 Fuel pump module electrical connector	8 Fuel transfer supply tube quick connect coupling
	9 Fuel vapor tube quick connect coupling

06017-FREE-G44

In-tank electric fuel pump and related components—Five Hundred and Montego

pump module into the fuel tank and align the arrows on the fuel pump with the arrows on the fuel tank. Secure it in place with the locking ring.

13. Connect the fuel vapor tube quick connect coupling to the fuel pump module.

14. Connect the fuel pump quick connect coupling to the fuel pump module.

15. Connect the fuel pump electrical connectors.

16. Install the fuel pump module cover.

17. Install the rear seat latching bracket bolts and brackets and tighten to 13 ft. lbs. (17 Nm).

18. Install a minimum of 10 gallons (38L) of fuel and check for leaks.

19. Check for fuel leaks at the fittings.

20. Start the engine and check for fuel leaks.

Fuel Rail and Injectors

REMOVAL & INSTALLATION

1. Relieve the fuel system pressure.
2. Disconnect the negative battery cable.
3. Remove the upper and lower intake manifolds.
4. Disconnect the fuel rail temperature and pressure sensor connectors.
5. Remove the vacuum harness.
6. Remove the fuel rail attaching bolts.
7. Separate the fuel rail and injectors from the lower intake manifold.
8. Disconnect the fuel injector connectors.
9. Disconnect the fuel rail charging wiring harness.
10. Remove the retaining clips and separate the injectors from the supply manifold. Discard the O-rings.

To install:

11. Lubricate NEW O-rings with clean engine oil and position fuel injectors on the supply manifold. Install the retaining clips.

12. Connect the fuel rail charging wiring harness.

1 Fuel rail pressure and temperature sensor electrical connector	5 Fuel rail bolt (4 required)
	6 Fuel rail
2 Fuel injector electrical connector (6 required)	7 Fuel injector clip (6 required)
3 Fuel charging wiring harness	8 Fuel injector (6 required)
4 Vacuum harness	9 Fuel injector O-ring (12 required)

06017-FREE-G45

Fuel injectors and supply manifold—Five Hundred and Montego

13. Connect the fuel injector connectors.

14. Connect the fuel rail and injectors to the lower intake manifold.

15. Install the fuel rail attaching bolts and tighten to 89 inch lbs. (19 Nm).

16. Install the vacuum harness.

17. Connect the fuel rail temperature and pressure sensor connectors.

18. Install the upper and lower intake manifolds.

19. Connect the negative battery cable.

20. Start the engine and check for fuel leaks.

DRIVE TRAIN

Transaxle Assembly

REMOVAL & INSTALLATION

CVT Transaxle

1. Place the vehicle on a hoist.
2. Drain the transaxle fluid.
3. Remove or disconnect the following:
   - Battery and battery tray
   - Air cleaner assembly and outlet pipe
   - Cowl grille panel
   - Upper roll restrictor bolt
   - Roll restrictor brace bolts and brace
   - Shift cable end
   - Electrical connector from shift cable bracket
   - Shift able bracket and position the cable and bracket aside
   - Transaxle wiring harnesses
   - Auxiliary coolant flow pump assembly, if equipped
   - Roll restrictor bracket and roll restrictor
   - PCV hose
   - Throttle body vacuum hoses
   - Power steering line bracket
   - Starter wiring and starter
   - Transaxle fluid cooler tube bracket
   - Ground cable
   - Transaxle bulkhead electrical connector
   - 4 upper transaxle mounting bolts
   - Left and right side halfshafts
   - Front and rear exhaust pipe nuts and discard
   - Catalytic converter pipe
4. On AWD models, remove the 2 outer bolts for the driveshaft brace.
5. Index mark the rear driveshaft flange to the Power Transfer Unit (PTU).
6. Remove and the discard 6 bolts from the front driveshaft flange.
7. Remove the PTU.
8. Wire the rear driveshaft to the chassis.
9. On all models, remove or disconnect the following:
   - Engine block heater connector
   - Knock sensor heat shield and knock sensor
   - Knock sensor wiring harness from transaxle
   - Transaxle-to-engine bolt
   - Power steering hose bolt
10. Install engine lifting eyes no. 134-00243 to the left and right cylinder heads.

06017-FREE-G46

Installing engine support tools—Five Hundred and Montego

11. Install lifting brackets no. 303-1140 to front of the engine.
12. Install engine support tools to the engine as shown.
13. Remove or disconnect the following:
   - Disconnect the lower radiator hose from its retainers and position aside
   - Power steering line from subframe
   - Wiring harness from left front subframe
   - Power steering rack retainers and wire the rack to the chassis
   - Oxygen sensor connector
   - Splash shield
   - Front motor mount nut
   - Lower stabilizer mount bolts
   - Rear motor mount nut
   - Rear transaxle mount bolts
   - Rear subframe mounting bolts on both sides
   - Stabilizer bar links nuts from both sides and discard the nuts.
   - Stabilizer bar from struts
14. Using a suitabale powertrain lift, remove 4 subframe mounting bolts and remove the subframe.
15. Remove the catalytic converter.
16. Remove the torque converter access plug and remove and discard the 4 torque converter nuts.
17. Disconnect the transaxle fluid cooler tubes from the transaxle.
18. Securely fasten the transaxle to a suitable high lift transmission jack.
19. Remove the 3 engine-to-transaxle bolts, and remove the transaxle.

To install:

➡ **If a different transaxle is being installed, transfer the appropriate external components to the new transaxle.**

20. Apply multi-purpose grease to the torque converter pilot hub.
21. Place the transaxle on a suitable jack and carefully raise it into position.
22. Install 3 engine-to-transaxle mounting bolts and tighten to 30 ft. lbs. (40 Nm).
23. Install new transaxle fluid cooler tube seals and connect the tubes.
24. Install new torque converter nuts and tighten to 27 ft. lbs. (36 Nm). Install the torque converter access plug.
25. Install the catalytic converter and tighten the bolts to 30 ft. lbs. (40 Nm).
26. Using a suitabale powertrain lift, install the subframe and tighten the 4 subframe mounting bolts to 66 ft. lbs. (90 Nm).
   - Install the subframe mounting bolts on both sides and tighten to 35 ft. lbs. (48 Nm).
   - Rear transaxle mount bolts and tighten to 66 ft. lbs. (90 Nm).
   - Rear motor mount nut and tighten to 52 ft. lbs. (70 Nm).
   - Lower stabilizer mount bolts and tighten to 52 ft. lbs. (70 Nm).
   - Front motor mount nut and tighten to 52 ft. lbs. (70 Nm).
   - Splash shield
   - Oxygen sensor connector
   - Power steering rack retainers and tighten to 98 ft. lbs. (133 Nm)
   - Power steering line to subframe
   - Wiring harness to left front subframe
   - Lower radiator hose
   - Stabilizer bar to struts
   - Stabilizer bar link nuts and tighten to 41 ft. lbs. (55 Nm)
27. Remove the all the special lifting tools from the engine and cylinder heads.
28. Install or connect the following:
   - Transaxle-to-engine bolt and tighten to 30 ft. lbs. (40 Nm)
   - Power steering hose bolt
   - Knock sensor wiring harness to transaxle
   - Knock sensor and heat shield
   - Engine block heater connector
29. On AWD models, install the PTU and rear driveshaft. Tighten the bolts to 66 ft. lbs. (90 Nm).
30. Install the 2 outer bolts for the driveshaft brace and tighten to 18 ft. lbs. (25 Nm).

31. Install new bolts to the front drive-shaft flange and tighten in a criss-cross pattern to 18 ft. lbs. (25 Nm).

32. On all models, install or connect the following:

- Front and rear exhaust pipe nuts and tighten to 30 ft. lbs. (40 Nm)
- Left and right side halfshafts
- 4 upper transaxle mounting bolts and tighten to 30 ft. lbs. (40 Nm)
- Transaxle bulkhead electrical connector
- Ground cable
- Transaxle fluid cooler tube bracket
- PCV hose
- Power steering line bracket
- Throttle body vacuum hoses
- Transaxle wiring harnesses
- Starter wiring and starter and tighten to 19 ft. lbs. (26 Nm)
- Roll restrictor bracket and roll restrictor and apply treadlocker before tightening to 30 ft. lbs. (40 Nm)
- Auxiliary coolant flow pump assembly, if equipped
- Shift cable bracket and cable
- Electrical connector to shift cable bracket
- Roll restrictor brace bolts and brace and apply treadlocker before tightening to 30 ft. lbs. (40 Nm)
- Upper roll restrictor bolt and apply treadlocker before tightening to 30 ft. lbs. (40 Nm)
- Cowl grille panel
- Air cleaner assembly
- Battery tray and battery

33. Fill the transaxle with proper amount of Mercon Continuously Variable chain type transmission fluid.

34. Connect the positive, then the negative battery cable.

35. Start the engine and check for leaks.

6-Speed Transaxle

1. Place the vehicle on a hoist.
2. Drain the transaxle fluid.
3. Remove or disconnect the following:

- Battery and battery tray
- Air cleaner assembly and outlet pipe
- Cowl grille panel
- Upper roll restrictor bolt
- Roll restrictor brace bolts and brace
- Shift cable end
- Electrical connector from shift cable bracket
- Shift cable bracket and position the cable and bracket aside
- Transaxle Control Module (TCM) wiring harness
- Transaxle wiring harnesses
- Roll restrictor bracket and roll restrictor
- PCV hose
- Throttle body vacuum hoses
- Power steering line bracket
- Starter wiring and starter
- Ground cable
- 4 upper transaxle mounting bolts
- Left and right side halfshafts
- Front and rear exhaust pipe nuts and discard
- Catalytic converter pipe
- Engine block heater connector
- Knock sensor heat shield and knock sensor
- Knock sensor wiring harness from transaxle
- Transaxle-to-engine bolt
- Power steering hose bolt

4. Install engine lifting eyes no. 134-00243 to the left and right cylinder heads.

5. Install lifting brackets no. 303-1140 to front of the engine.

6. Install engine support tools to the engine as shown.

7. Remove or disconnect the following:

- Disconnect the lower radiator hose from its retainers and position aside
- Power steering line from subframe
- Wiring harness from left front subframe
- Power steering rack retainers and wire the rack to the chassis
- Oxygen sensor connector
- Splash shield
- Front motor mount nut
- Lower stabilizer mount bolts

303-290-01   303-290-01

303-290A

303-290A-12

303-290A-03A

06017-FREE-G46

Installing engine support tools—Five Hundred and Montego

- Rear motor mount nut
- Rear transaxle mount bolts
- Rear subframe mounting bolts on both sides
- Stabilizer bar links nuts from both sides and discard the nuts.
- Stabilizer bar from struts

8. Using a suitable powertrain lift, remove 4 subframe mounting bolts and remove the subframe.

9. Remove the catalytic converter.

10. Remove the torque converter access plug and remove and discard the 4 torque converter nuts.

11. Disconnect the transaxle fluid cooler tubes from the transaxle.

12. Securely fasten the transaxle to a suitable high lift transmission jack.

13. Remove the 3 engine-to-transaxle bolts, and remove the transaxle.

To install:

➡ **If a different transaxle is being installed, transfer the appropriate external components to the new transaxle.**

14. Apply multi-purpose grease to the torque converter pilot hub.

15. Place the transaxle on a suitable jack and carefully raise it into position.

16. Install 3 engine-to-transaxle mounting bolts and tighten to 30 ft. lbs. (40 Nm).

17. Install new transaxle fluid cooler tube seals and connect the tubes.

18. Install new torque converter nuts and tighten to 27 ft. lbs. (36 Nm). Install the torque converter access plug.

19. Install the catalytic converter and tighten the bolts to 30 ft. lbs. (40 Nm).

20. Using a suitable powertrain lift, install the subframe and tighten the 4 subframe mounting bolts to 66 ft. lbs. (90 Nm).

- Install the subframe mounting bolts on both sides and tighten to 35 ft. lbs. (48 Nm).
- Rear transaxle mount bolts and tighten to 66 ft. lbs. (90 Nm).
- Rear motor mount nut and tighten to 52 ft. lbs. (70 Nm).
- Lower stabilizer mount bolts and tighten to 52 ft. lbs. (70 Nm).
- Front motor mount nut and tighten to 52 ft. lbs. (70 Nm).
- Splash shield
- Oxygen sensor connector
- Power steering rack retainers and tighten to 98 ft. lbs. (133 Nm)
- Power steering line to subframe
- Wiring harness to left front subframe
- Lower radiator hose
- Stabilizer bar to struts

- Stabilizer bar link nuts and tighten to 41 ft. lbs. (55 Nm)

21. Remove the all the special lifting tools from the engine and cylinder heads.

22. Install or connect the following:
- Transaxle-to-engine bolt and tighten to 30 ft. lbs. (40 Nm)
- Power steering hose bolt
- Knock sensor wiring harness to transaxle
- Knock sensor and heat shield
- Engine block heater connector
- Front and rear exhaust pipe nuts and tighten to 30 ft. lbs. (40 Nm)
- Left and right side halfshafts
- 4 upper transaxle mounting bolts and tighten to 30 ft. lbs. (40 Nm)
- Transaxle bulkhead electrical connector
- Ground cable
- Transaxle fluid cooler tube bracket
- PCV hose
- Power steering line bracket
- Throttle body vacuum hoses
- Transaxle wiring harnesses
- Starter wiring and starter and tighten to 19 ft. lbs. (26 Nm)
- Roll restrictor bracket and roll restrictor and apply treadlocker before tightening to 30 ft. lbs. (40 Nm)
- Auxiliary coolant flow pump assembly, if equipped
- Shift cable bracket and cable
- Electrical connector to shift cable bracket
- Roll restrictor brace bolts and brace and apply treadlocker before tightening to 30 ft. lbs. (40 Nm)
- Upper roll restrictor bolt and apply treadlocker before tightening to 30 ft. lbs. (40 Nm)
- Cowl grille panel
- Air cleaner assembly
- Battery tray and battery

23. Fill the transaxle with proper amount of Motorcraft Premium automatic transmission no. XT-8-QAW.

24. Connect the positive, then the negative battery cable.

Start the engine and check for leaks.

Power Transfer Unit

REMOVAL & INSTALLATION

1. Raise the vehicle on a hoist.
2. Remove the left and right halfshafts
3. Remove the front and rear exhaust pipe nuts.

4. Remove the catalytic converter pipe.
5. Index mark the driveshaft flange to the output flange.
6. Remove the 2 outer driveshaft brace bolts.
7. Remove the 6 bolts from the driveshaft flange.
8. Remove the upper PTU bracket bolts and the bracket.
9. Remove 4 PTU mounting bolts and the PTU.
10. Remove the input shaft oil slinger and seal.
To install:
11. Using a seal installer, install the input shaft seal and oil slinger.
12. Position the PTU in place and tighten the bolts to 66 ft. lbs. (90 Nm).
13. Install the upper PTU bracket bolts and the bracket and tighten to 41 ft. lbs. (55 Nm).
14. Install the left and right halfshafts
15. Install the 2 outer driveshaft brace bolts and tighten to 18 ft. lbs. (25 Nm).
16. Install the 6 bolts to the driveshaft flange and tighten in a criss-cross pattern to 18 ft. lbs. (25 Nm).
17. Install the catalytic converter pipe, front and rear exhaust pipe nuts and tighten all fasteners to 30 ft. lbs. (40 Nm).

Output Shaft Seal and Flange

REMOVAL & INSTALLATION

1. Remove the driveshaft.
2. Remove the output flange nut and flange.
3. Remove the output shaft seal.
To install:
4. Clean the output shaft seal bore and shaft.
5. Use a seal installer and install the output shaft seal.
6. Install the output flange and tighten the nut to 74 ft. lbs. (100 Nm).
7. Install the driveshaft.

Driveshaft, Pinion Flange and Seal

REMOVAL & INSTALLATION

1. Place the vehicle on a hoist.
2. Remove the exhaust system at the flex pipe.
3. Remove the exhaust support brace.
4. Index mark the driveshaft to the rear axle pinion flange, PTU flange and center bearing bracket.

5. Disconnect the driveshaft from the rear pinion flange and discard the bolts.
6. Remove the driveshaft flange bolts from the PTU flange and discard the bolts.
7. Disconnect the driveshaft from the PTU flange.
8. Remove the center bearing bracket bolts and remove the driveshaft.
9. Remove and discard the pinion flange nut.
10. Using a puller, remove the pinion flange.
11. Remove the pinion seal.
To install:
12. Using a seal installer, install the pinion seal.
13. Install the drive pinion flange using a new nut. Tighten the nut to 74 ft. lbs. (100 Nm).
14. Install the driveshaft.
15. Install the center bearing bracket and loosely tighten the bolts.
16. Using the index marks made on removal, connect the driveshaft to the PTU flange and rear pinion flange and hand tighten the bolts.
17. Tighten the center bearing bracket bolts to 15 ft. lbs. (20 Nm).
18. Tighten the driveshaft to rear pinion flange and PTU flange bolts to 18 ft. lbs. (25 Nm).
19. Install the exhaust support brace and tighten the bolts to 22 ft. lbs. (30 Nm).
20. Install the exhaust system.
21. Lower the vehicle.

Halfshafts

REMOVAL & INSTALLATION

Front

➡**Do not begin this removal procedure unless a new wheel hub retainer nut, a new retainer circlip and a new lower ball joint-to-front wheel knuckle retaining bolt and nut are available. Once removed, these parts must not be reused during assembly. Their torque holding ability, or retention capability, is diminished during removal.**

1. Remove or disconnect the following:
- Front wheels
- ABS sensor
- Brake caliper and wire aside
- Axle hub nut and washer. Discard the nut.

2. Separate the halfshaft from the hub using tool no. 205-D070.
3. Remove and discard the lower ball joint nut.

4. Disconnect the lower control arm.

5. On the left side, on 6-speed models, use a slide hammer and remove the half-shaft from the transaxle.

6. On the left side on CVT models, use a slide hammer and halfshaft fork no. 205-241 and remove the shaft from the transaxle.

7. On the right side, remove the half-shaft bearing nuts and remove the halfshaft.

8. On both sides, inspect the halfshaft seal and replace if necessary.

To install:

✳✳ WARNING

Do not reuse the retainer circlip. A new circlip must be installed each time the inboard CV-joint stub shaft is installed into the transaxle differential.

9. Install a new retainer circlip on the inboard CV-joint stub shaft by starting one end in the groove and working the retainer circlip over the inboard shaft housing end and into the groove. This will avoid over-expanding the circlip.

➡**A non-metallic mallet may be used to aid in seating the retainer circlip into the differential side gear groove. If a mallet is necessary, tap only on the outboard CV-joint shaft.**

10. Install the halfshaft.

11. On the right side install the halfshaft bearing nuts and tighten to 20 ft. lbs. (27 Nm).

12. On both sides install the halfshaft into the hub.

13. Connect the lower control arm to the knuckle and install a new ball joint nut. Tighten the nut to 59 ft. lbs. (80 Nm).

14. Using the old axle nut and washer tighten the nut to seat the halfshaft in the hub.

15. Remove the old nut and washer.

16. Install a new axle nut and washer and in a continuous motion, tighten the nut to 184 ft. lbs. (250 Nm).

17. Install the brake caliper and tighten to 74 ft. lbs. (100 Nm).

18. Install the front wheels and lower the vehicle.

Rear

➡**Do not begin this removal procedure unless a new wheel hub retainer nut is available. Once removed, the nut must not be reused during assembly. The torque holding ability, or retention capability, is diminished during removal.**

1. Remove or disconnect the following:
   - Rear wheel
   - Brake caliper and wire aside
   - Axle hub nut and washer. Discard the nut.
   - ABS sensor

2. Using a hub remover, press the half-shaft out until it is loose in the hub.

3. Remove the brake disc.

4. Remove the 4 hub bolts and remove the hub.

5. Using a prybar, separate the half-shaft from the axle, and remove the half-shaft.

To install:

✳✳ WARNING

Do not reuse the retainer circlip. A new circlip must be installed each time.

6. Install a new retainer circlip on the CV-joint stub shaft by starting one end in the groove and working the retainer circlip over the inboard shaft housing end and into the groove. This will avoid over-expanding the circlip.

7. Install the halfshaft.

8. Install the hub and the hub bolts and tighten to 89 ft. lbs. (120 Nm).

9. Install the brake disc and tighten to 16 ft. lbs. (21 Nm).

10. Using the old axle nut and washer tighten the nut to seat the halfshaft in the hub.

11. Remove the old nut and washer.

12. Install a new axle nut and washer and in a continuous motion, tighten the nut to 148 ft. lbs. (200 Nm).

13. Install the ABS sensor.

14. Install the caliper and tighten to 81 ft. lbs. (110 Nm).

15. Install the rear wheel.

CV-Joint

OVERHAUL

➡**Overhaul procedures are not available from the manufacturer.**

STEERING AND SUSPENSION

Air Bag (Supplemental Restraint) System

The Supplemental Restraint System (SRS) is designed to work in conjunction with the standard 3-point safety belts to reduce injury in a head-on collision.

✳✳ CAUTION

The SRS can actually cause physical injury or death if the safety belts are not used, or if the manufacturer's warnings are not followed. The manufacturer's warnings can be found in your owner's manual, or, in some cases, on your sun visor.

The SRS is comprised of the following components:
- Driver's side air bag module
- Passenger's side air bag module
- Right-hand and left-hand primary crash front air bag sensors
- Air bag diagnostic monitor computer
- Electrical wiring

The SRS primary crash front air bag sensors are hard-wired to the air bag modules and determine when the air bags are deployed. During a frontal collision, the sensors quickly inflate the 2 air bags to reduce injury by cushioning the driver and front passenger from striking the dashboard, windshield, steering wheel and any other hard surfaces. The air bag inflates so quickly (in a fraction of a second) that in most cases it is fully inflated before you actually start to move during a collision.

Since the SRS is a complicated and essentially important system, its components are constantly being tested by a diag-nostic computer. The computer illuminates the air bag indicator light on the instrument cluster for approximately 6 seconds when the ignition switch is turned to the **RUN** position when the SRS is functioning properly. After being illuminated for the 6 seconds, the indicator light should then turn off.

If the air bag light does not illuminate at all, stays on continuously, or flashes at any time, a problem has been detected by the diagnostic computer.

✳✳ CAUTION

If at any time the air bag light indicates that the computer has noted a problem, immediately diagnose the problem. A faulty SRS can cause severe physical injury or death.

SERVICE PRECAUTIONS

Whenever working around, or on, the air bag supplemental restraint system, ALWAYS adhere to the following warnings and cautions.

• Always wear safety glasses when servicing an air bag vehicle and when handling an air bag module.

• Carry a live air bag module with the bag and trim cover facing away from your body, so that an accidental deployment of the air bag will have a small chance of personal injury.

• Place an air bag module on a table or other flat surface with the bag and trim cover pointing up.

• Wear gloves, a dust mask and safety glasses whenever handling a deployed air bag module. The air bag surface may contain traces of sodium hydroxide, a byproduct of the gas that inflates the air bag and which can cause skin irritation.

• Ensure to wash your hands with mild soap and water after handling a deployed air bag.

• All air bag modules with discolored or damaged cover trim must be replaced, not repainted.

• All component replacement and wiring service must be made with the negative and positive battery cables disconnected from the battery for a minimum of 1 minute prior to attempting service or replacement.

• NEVER probe the air bag electrical terminals. Doing so could result in air bag deployment, which can cause serious physical injury.

• If the vehicle is involved in a fender-bender that results in a damaged front bumper or grille, the air bag sensors should be inspected to ensure that they were not damaged.

• If at any time, the air bag light indicates that the computer has noted a problem, immediately diagnose the problem. A faulty SRS can cause severe physical injury or death.

DISARMING THE SYSTEM

1. Ensure the ignition is off.
2. Remove the Smart Junction Box (SJB) cover located below the left side of the instrument panel. Remove the Restraints Control Module (RCM) fuse no. F2.21.
3. Turn the ignition on and watch the air bag indicator for 30 seconds. The indicator light will remain lit constantly if the correct fuse has been removed. If the light is not on steadily, remove the correct fuse and check the light again.

4. Turn the ignition switch off.
5. Disconnect the negative battery cable.

ARMING THE SYSTEM

1. Turn the ignition switch on.
2. Install Restraints Control Module (RCM) fuse no. F2.21 and the fuse cover.
3. Turn the ignition off.
4. Connect the negative battery cable.
5. Turn the ignition on and then off. Wait 10 seconds and turn the key back on. Watch the air bag indicator. The indicator light will remain lit constantly for 6 seconds and then go off. If the indicator does not turn on and then off, diagnose the air bag system
6. Clear all Diagnostic Trouble Codes (DTC) using a diagnostic tool.

Rack and Pinion Steering Gear

REMOVAL & INSTALLATION

✳ WARNING

Do not allow the steering wheel to rotate when the intermediate shaft is

disconnect, or damage to the clock-spring can result. If it is suspected that the shaft has rotated, the clock-spring must be removed and re-centered.

1. Place the steering wheel in the straight ahead position and remove the ignition key.
2. Raise the vehicle on a hoist.
3. Remove or disconnect the following:

• Front wheels
• Tie rod end cotter pins and castle nuts
• Tie rod ends from the knuckles
• Power steering line clamp plate bolt

4. Place a drain pan under the vehicle and disconnect the power steering lines.
5. Release the steering shaft intermediate boot bearing clamp and release the boot from the shaft.
6. Retain the bearing boot clamp in the open position.
7. Remove the pinch bolt retaining the steering column intermediate shaft coupling.

5 — 25 Nm (18 lb-ft)
1 — 90 Nm (66 lb-ft)
3 — 20 Nm (15 lb-ft)
6 — 117 Nm (86 lb-ft)

1 Tie-rod end nuts (2 required)
2 Tie-rod ends (2 required)
3 Power steering line clamp plate bolt
4 O-ring seals (2 required)
5 Intermediate shaft-to-steering gear bolt
6 Steering gear nuts (2 required)
7 Steering gear bolts (2 required)
8 Steering gear

06017-FREE-G47

Exploded view of the rack and pinion steering gear mounting on the front subframe of the vehicle—Five Hundred and Montego

8. Remove the steering gear mounting nuts and bolts.

9. Remove the steering gear through the left side of the vehicle.

To install:

10. Install new O-rings on the power steering line fittings.

11. Place the steering gear in the vehicle through the left side.

12. Tighten the steering gear mounting bolts and nuts to 86 ft. lbs. (117 Nm).

13. Install the pinch bolt retaining the steering column intermediate shaft coupling and tighten to 18 ft. lbs. (25 Nm).

14. Position the bearing boot in the retention groove so the outer bearing race is not exposed from the groove.

15. Install boot bearing clamp and ensure the end tabs do not contact the boot.

16. Using new o-rings, install the power steering lines.

17. Install the power steering line clamp plate bolt and tighten to 15 ft. lbs. (20 Nm).

18. Connect the tie rods.

19. Install the tie rod end nuts and tighten to 66 ft. lbs. (90 Nm).

20. Install the front wheels.

21. Lower the vehicle.

22. Fill the power steering oil reservoir.

23. Start the vehicle and check for leaks.

24. Check for proper wheel alignment and steering wheel position.

Front Strut/Spring

REMOVAL & INSTALLATION

➡ Do not allow the axle shaft to move outward as damage to the CV joint may result.

❈❈ WARNING

Always use new bolts and nuts on the suspension components.

1. Turn the ignition switch **OFF** and place the steering column in the unlocked position.

2. Loosen the strut/spring upper mounting nuts.

3. Raise the vehicle on a hoist.

4. Remove or disconnect the following:
- Front wheel
- Front wheel axle hub nut and discard
- Tie rod-to-wheel knuckle nut
- Separate the tie rod from the knuckle
- Brake disc
- Wheel speed sensor and wire aside

- Lower ball joint nut using a crow's foot wrench, and discard the nut

5. Push the lower arm down until the ball joint is clear of the lower arm.

6. Install tool no. 307-102 over the ball stud.

7. Use a puller and press the halfshaft from the wheel hub and support the halfshaft level.

8. Remove and discard the strut upper mounting nuts.

9. Remove the wheel knuckle and strut as an assembly.

10. Remove and discard the knuckle-to-strut bolt and nut.

11. Separate the spring/strut assembly from the knuckle.

To install:

12. Place the strut/spring assembly in the steering knuckle.

13. Using a new bolt and nut, tighten the strut-to-knuckle bolt to 129 ft. lbs. (175 Nm).

14. Install the strut/knuckle in the vehicle and loosely tighten the new top mounting nuts.

15. Install the halfshaft in the wheel hub.

16. Remove the ball stud tool and connect the ball joint to the lower control arm

17. Install a new ball joint nut and tighten to 85 ft. lbs. (115 Nm).

18. Install the wheel speed sensor.

19. Install the brake disc.

20. Install the tie rod to the wheel knuckle and tighten the new nut to 85 ft. lbs. (115 Nm).

21. Install a new axle hub nut and tighten to 148 ft. lbs. (200 Nm).

22. Install the front wheel.

23. Lower the vehicle.

24. Tighten the upper strut nuts to 18 ft. lbs. (25 Nm).

25. Check the wheel alignment.

DISASSEMBLY

1. Using a spring compressor, compress the spring and loosen the strut rod nut.

2. Compress the spring until all tension is relieved.

3. Remove the strut rod appearance cap.

4. Remove and discard the upper strut rod nut.

5. Separate the strut, spring mount and washer.

6. Remove the spring.

7. To maintain wheel camber, position the notch and arrow on the upper strut mount outboard opposite the locator tab on the strut.

8. To adjust camber +0.5 degrees, position the notch and arrow on the upper strut

Positioning upper strut mount outboard to maintain wheel camber—Five Hundred and Montego

Positioning upper strut mount inboard to adjust wheel camber +0.5 degrees—Five Hundred and Montego

mount inboard aligned with the locator tab on the strut.

9. To reassemble, reverse the disassembly procedure. Make sure the spring is correctly seated in both spring seats. Tighten the new strut top retaining nut to 59 ft. lbs. (80 Nm).

Rear Shock Absorber/Spring

REMOVAL & INSTALLATION

❈❈ WARNING

Always use new bolts and nuts on the suspension components.

1. Measure the distance between the top of the wheel and the bottom of the wheel opening. This is the trim height.

2. Loosen the lug nuts on the rear wheels.

3. Remove the interior trim panel to access the upper shock nut, then remove and discard the nut.

4. Raise the vehicle on a hoist.

5. Remove the rear wheels.

6. Remove the brake caliper and wire aside.

83 Nm (61 lb-ft) — 22

22 Upper shock absorber nut
23 Upper shock mount
24 Bushing
25 Shock absorber/spring

06017-FREE-G50

Rear shock absorber/spring mounting—Five Hundred and Montego

7. Position a jack under the wheel knuckle at the trailing arm and knuckle attachment point.

8. Raise the wheel knuckle until the toe link is parallel to the ground.

9. Remove and discard the trailing arm-to-knuckle bolt.

10. Lower the suspension jack.

11. Remove and discard the trailing arm-to-subframe bolt.

12. Position a jack under the lower shock mount and raise it enough to compress the spring.

13. Remove and discard the lower arm-to-knuckle bolt.

14. Remove and discard the lower arm-to-subframe bolt.

15. Remove and discard the lower shock mounting bolt.

16. Hold the shock absorber and swing the lower arm down to access and remove the shock/spring assembly.

To install:

17. Tape the spring to the upper spring seat.

18. Position the spring and upper rubber seat on the shock.

19. Install a new lower arm-to-knuckle bolt in the lower arm.

20. Position a jack under the lower arm.

21. Install the shock/spring assembly into alignment with the upper mount.

22. Swing the lower arm into position under the shock/spring assembly.

23. Remove the jack.

24. Partially lower the hoist.

25. Place a floor jack under the lower arm at the lowest point.

26. Raise the jack until the top of the shock makes contact with the inside surface of the upper mount.

27. With the help of an assistant guiding the shock assembly, raise the jack until the shock seats in the upper mount.

28. Install a new top nut and tighten to 61 ft. lbs. (83 Nm).

29. Remove the floor jack.

30. Raise the vehicle.

31. Position a jack under the wheel knuckle at the trailing arm and knuckle attachment point.

32. Remove the lower arm bolt, install the shock/spring in the lower arm and install the bolt.

33. Loosen the lower arm-to-subframe bolt.

34. Raise the jack under the center of the wheel hub is at the proper trim height.

35. Tighten the lower arm-to-subframe bolt to 92 ft. lbs. (125 Nm).

36. Tighten the lower arm-to-knuckle bolt to 81 ft. lbs. (110 Nm).

37. Position the trailing arm to the knuckle and loosely install the bolt.

38. Place a jack under the lower arm and raise it until the trailing arm and subframe bushing are aligned.

39. Install a new trailing arm-to-subframe bolt and tighten to 81 ft. lbs. (100 Nm).

40. Install the trailing arm-to-knuckle bolt and tighten to 77 ft. lbs. (105 Nm).

41. Install the brake caliper.

42. Install the wheel.

43. Lower the vehicle.

Stabilizer Bar

REMOVAL & INSTALLATION

Front

✳✳ WARNING

Always use new bolts and nuts on the suspension components.

1. Place the steering wheel in the UNLOCKED position.

2. Raise and safely support the vehicle.

3. Remove the wheels.

4. Remove the stabilizer bar link nuts.

5. Remove and discard the tie rod nut and separate the tie rod from the knuckle.

6. Remove and discard the lower arm bushing-to-subframe nuts and bolts.

7. Remove and discard the lower arm-to-subframe bolt.

8. Remove and discard the rear subframe to body bolts.

9. Lower the subframe about 2 inches.

10. Remove and discard the stabilizer bar bracket bolts.

11. Remove the stabilizer bar from the right side of the vehicle.

To install:

12. Install the stabilizer bar from the right side of the vehicle.

13. Install the stabilizer bar bracket bolts and tighten to 37 ft. lbs. (50 Nm).

14. Raise the subframe about 2 inches.

15. Install the rear subframe-to-body bolts.

16. Install the lower arm-to-subframe bolt and tighten to 111 ft. lbs. (150 Nm).

17. Install the lower arm bushing-to-subframe nuts and bolts and tighten to 73 ft. lbs. (99 Nm).

18. Connect the tie rod to the knuckle and tighten the tie rod nut to 85 ft. lbs. (115 Nm).

19. Install the stabilizer bar link nuts and tighten to 41 ft. lbs. (55 Nm).

20. Install the wheels.

21. Lower the vehicle.

Rear–2WD

> ✳✳ **WARNING**
>
> **Always use new bolts and nuts on the suspension components.**

1. Separate the exhaust system from the flexpipe and support the exhaust system.
2. Remove and discard the stabilizer bar link-to-stabilizer bar nuts.
3. Remove and discard the stabilizer bar bracket-to-subframe bolts.
4. Remove the stabilizer bar.

To install:

5. Reverse the removal procedure. Tighten the stabilizer bar bracket-to-subframe bolts to 43 ft. lbs. (58 Nm).

Rear–AWD

> ✳✳ **WARNING**
>
> **Always use new bolts and nuts on the suspension components.**

1. Separate the exhaust system from the flexpipe and support the exhaust system.
2. Raise and support the vehicle.
3. Remove the rear wheels.
4. Remove the brake calipers and wire aside.
5. Remove both rear axles nuts and discard.
6. Using a hub remover, press the axle shaft from the wheel hub.
7. Remove and discard both upper arm-to-knuckle bolts.
8. Remove and discard both trailing arm-to-knuckle bolts and loosen the trailing arm-to-subframe bolt.
9. Remove and discard the lower shock-to-lower arm nuts and remove it from the lower arm.
10. Loosen both toe link-to-subframe bolts and allow the wheel knuckles to hang in a downward position.
11. Remove the rear subframe cross brace bolts and cross brace.
12. Remove and discard the stabilizer bar link-to-stabilizer bar nuts.
13. Remove and discard the stabilizer bar bracket-to-subframe bolts.
14. Remove the stabilizer bar.

To install:

15. Install the stabilizer bar.
16. Install the stabilizer bar bracket-to-subframe bolts and tighten to 43 ft. lbs. (58 Nm).
17. Install the stabilizer bar link-to-stabilizer bar nuts.

18. Install the rear subframe cross brace and bolts and tighten to 46 ft. lbs. (63 Nm).
19. Bring the wheel knuckles up and tighten both toe link-to-subframe bolts to 74 ft. lbs. (100 Nm).
20. Install the shock to the lower arm and tighten the lower shock-to-lower arm nuts to 105 ft. lbs. (142 Nm).
21. Install both trailing arm-to-knuckle bolts and tighten them and the trailing arm-to-subframe bolt to 77 ft. lbs. (110 Nm).
22. Install both upper arm-to-knuckle bolts and tighten to 77 ft. lbs. (105 Nm).
23. Install the axle shaft into the wheel hub.
24. Install new axle shaft nuts and tighten to 148 ft. lbs. (200 Nm).
25. Install the brake calipers.
26. Install the rear wheels.
27. Install the exhaust system to the flex pipe.
28. Lower the vehicle.

Lower Control Arm

REMOVAL & INSTALLATION

Front

> ✳✳ **WARNING**
>
> **Always use new bolts and nuts on the suspension components.**

1. Raise the vehicle on a hoist.
2. Remove the lower ball joint nut using a crow's foot wrench, and discard the nut
3. Push the lower arm down until the ball joint is clear of the lower arm.
4. Install tool no. 307-102 over the ball stud.
5. Remove and discard the lower arm bushing-to-subframe nuts and bolts.
6. Remove the lower control arm.

To install:

7. Install the lower control arm.
8. Install the new lower arm bushing-to-subframe bolts from the bottom of the lower arm bushing with the nuts on top.
9. Tighten the nuts to 73 ft. lbs. (99 Nm).
10. Remove the special tool.
11. Install the lower arm-to-subframe bolts and tighten to 111 ft. lbs. (150 Nm).
12. Push the lower arm upward until the ball joint is up in the arm and install a new ball joint nut and tighten to 85 ft. lbs. (115 Nm).
13. Lower the vehicle.

Rear

> ✳✳ **WARNING**
>
> **Always use new bolts and nuts on the suspension components.**

1. Measure the distance between the top of the wheel and the bottom of the wheel opening. This is the trim height.
2. Remove and discard the trailing arm-to-knuckle bolt.
3. Loosen the trailing arm-to-subframe bolt and allow it to swing down.
4. Remove and discard the knuckle-to-lower arm bolt.
5. Remove and discard the shock absorber-to-lower arm bolt.
6. Remove and discard the lower arm-to-subframe bolt.
7. Remove the lower arm.

To install:

8. Install the lower control arm.
9. Use a jack and raise the suspension to the trim height previously measured before tightening the lower arm bolts.
10. Install the lower arm-to-subframe bolt and tighten to 98 ft. lbs. (133 Nm) on AWD models and 92 ft. lbs. (125 Nm) on 2WD models.
11. Install the shock absorber-to-lower arm bolt and tighten to 105 ft. lbs. (142 Nm) on AWD models and 81 ft. lbs. (110 Nm) on 2WD models.
12. Install the knuckle-to-lower arm bolt and tighten to 66 ft. lbs. (90 Nm) on AWD models and 118 ft. lbs. (160 Nm) on 2WD models.
13. Bring the trailing arm up and tighten the trailing arm-to-subframe bolt and tighten the bolt to 81 ft. lbs. (110 Nm).
14. Install the trailing arm-to-knuckle bolt and tighten the bolt to 77 ft. lbs. (105 Nm).

Steering Knuckle

REMOVAL & INSTALLATION

Front

> ✳✳ **WARNING**
>
> **Always use new bolts and nuts on the suspension components.**

1. Raise and support the vehicle.
2. Remove the front wheel.
3. Remove and discard the axle hub nut.

4. Remove the tie rod-to-wheel nut nut and separate the tie rod from the knuckle.

5. Remove the brake disc.

6. Remove the wheel speed sensor and wire aside.

7. Remove the lower ball joint nut using a crow's foot wrench, and discard the nut

8. Push the lower arm down until the ball joint is clear of the lower arm.

9. Install tool no. 307-102 over the ball stud.

10. Use a puller and press the halfshaft from the wheel hub and support the halfshaft level.

11. Remove and discard the wheel knuckle-to-strut nut and bolt.

12. Remove the wheel knuckle.

To install:

13. Install the wheel knuckle.

14. Install and the wheel knuckle-to-strut nut and bolt and tighten to 129 ft. lbs. (175 Nm).

15. Install the halfshaft in the wheel hub.

16. Remove the ball stud tool and connect the ball joint to the lower control arm

17. Install a new ball joint nut and tighten to 85 ft. lbs. (115 Nm).

18. Install the wheel speed sensor.

19. Install the brake disc.

20. Install the tie rod to the wheel knuckle and tighten the new nut to 85 ft. lbs. (115 Nm).

21. Install a new axle hub nut and tighten to 148 ft. lbs. (200 Nm).

22. Install the front wheel.

23. Lower the vehicle.

Rear

❋❋ **WARNING**

Always use new bolts and nuts on the suspension components.

1. Measure the distance between the top of the wheel and the bottom of the wheel opening. This is the trim height.

2. Remove the wheel hub and bearing assembly.

3. Remove the brake dust shield.

4. Remove and discard the trailing arm-to-knuckle bolt.

5. Remove the toe link-to-knuckle bolt and nut.

6. Remove and discard the shock absorber-to-lower arm bolt.

7. Remove and discard the lower arm-to-subframe bolt.

8. Remove and discard the upper arm-to-subframe bolt.

9. Remove the steering knuckle.

To install:

10. Install the wheel knuckle.

11. Use a jack and raise the suspension to the trim height previously measured before tightening the upper and lower arm bolts.

12. Install the upper arm-to-subframe bolt and tighten to 81 ft. lbs. (110 Nm).

13. Install the lower arm-to-subframe bolt and tighten to 98 ft. lbs. (133 Nm) on AWD models, and 92 ft. lbs. (125 Nm) on 2WD models.

14. Install the shock absorber-to-lower arm bolt and tighten to 105 ft. lbs. (142 Nm) on AWD models, and 81 ft. lbs. (110 Nm) on 2WD models.

15. Install the toe link-to-knuckle bolt and nut and tighten to 74 ft. lbs. (100 Nm).

16. Install the trailing arm-to-knuckle bolt and tighten to 77 ft. lbs. (105 Nm).

17. Install the brake dust shield.

18. Install the wheel hub and bearing assembly.

Hub and Wheel Bearing

REMOVAL & INSTALLATION

The wheel bearing is integral with the wheel hub and can not be replaced separately. If the wheel bearing is found to be defective, the wheel hub must be replaced as an assembly.

Front

❋❋ **WARNING**

Always use new bolts and nuts on the suspension components.

1. Raise and safely support the vehicle.

2. Remove the wheel and tire assembly.

3. Remove the steering knuckle.

4. Remove and discard the wheel hub-to-knuckle bolts and separate the knuckle from the hub.

To install:

5. Clean the steering knuckle bore.

6. Install the wheel hub-to-the steering knuckle.

7. Install the wheel hub-to-knuckle bolts and tighten to 81 ft. lbs. (110 Nm).

8. Install the steering knuckle.

9. Install the wheel and tire.

Rear

1. Raise and safely support the vehicle.

2. Remove the wheel and tire assembly.

3. Remove and discard the rear axle nut.

4. Remove the brake caliper and wire it out of the way.

5. Remove the brake rotor.

6. Press the halfshaft out of the wheel hub.

7. Remove and discard the wheel hub-to-knuckle bolts and separate the hub from the knuckle.

To install:

8. Install the wheel hub-to-the steering knuckle.

9. Install the wheel hub-to-knuckle bolts and tighten to 89 ft. lbs. (120 Nm).

10. Install the halfshaft into the wheel hub.

11. Install the brake disc.

12. Install the brake caliper.

13. Install a new rear axle nut and tighten to 148 ft. lbs. (200 Nm).

14. Install the wheel.

15. Lower the vehicle.

BRAKES

Brake Caliper and Pads

REMOVAL & INSTALLATION

Front

➡The brake pads are a one time use only type. If the pads are separated from the caliper the pads must be replaced.

1. Remove and discard ½ of the brake fluid from the brake master cylinder reservoir.
2. Raise and safely support the vehicle.
3. Remove the wheel and tire assembly.
4. Mark the disc brake caliper to avoid mixing the left-hand and right-hand components.
5. Loosen the brake caliper bolts.
6. Disconnect the brake hose from the brake line fitting.

7. Remove and discard the caliper anchor bolts and position the caliper, anchor plate and pads aside. Support the steering stop when the anchor plate bolts are removed.
8. Separate the brake pads from the caliper.
9. Remove the caliper bolts from the caliper and anchor plate.
10. Remove the brake pads and spring clips.

1. Brake caliper bolts (2 required)
2. Brake caliper RH/LH
3. Brake pads
4. Spring clips (2 required)
5. Brake caliper anchor plate bolts (2 required)
6. Steering stop
7. Brake caliper anchor plate
8. Brake disc screw
9. Brake disc

10. Brake disc shield bolts (3 required)
11. Brake disc shield (RH/LH)
12. Brake caliper flow bolt
13. Copper washers (2 required)
14. Brake flexible hose RH/LH
15. Brake line fitting
16. Brake flexible hose bracket bolt
17. Bleeder screw cap
18. Bleeder screw

Exploded view of front disc brake assembly—Five Hundred and Montego

06017-FREE-G51

To install:

11. Compress the caliper pistons into the caliper bore.

12. Install the spring clips and brake pads in the caliper and anchor plate assembly.

13. Install the brake caliper anchor. Tighten the anchor bolts to 74 ft. lbs. (100 Nm).

14. Install the caliper and new guide pins and tighten the bolts to 44 ft. lbs. (60 Nm).

15. Install new copper washers in the brake flexible hoses and connect the hoses and install the flow bolt to 22 ft. lbs. (30 Nm).

16. Bleed the brake system.

17. Lower the vehicle.

18. Pump the brake pedal several times to position the brake pads before attempting to move the vehicle.

19. Check and fill the brake master cylinder as required.

20. Road-test the vehicle and check for proper brake operation.

Rear

1. Remove and discard ½ of the brake fluid from the brake master cylinder reservoir.

1   Brake caliper bolts (2 required)

2   Brake caliper (RH/LH)

3   Brake pads (kit)

4   Spring clips (4 required)

5   Brake caliper anchor plate bolts (2 bolts each side)

6   Brake caliper anchor plate (RH/LH)

7   Brake disc screw

8   Brake disc

9   Brake disc shield bolts (3 required)

10 Brake disc shield (RH/LH)

11 Brake line fitting

12 Brake flexible hose retainer

13 Brake flexible hose

14 Bleeder cap

15 Bleeder screw

Exploded view of rear disc brake assembly—Five Hundred and Montego

06017-FREE-G52

2. Raise and safely support the vehicle.

3. Remove the wheel and tire assembly.

4. Release the tension on the parking brake cable.

5. Disconnect the parking brake cable end from the caliper lever and release the cable retainer from the caliper bracket.

6. Loosen the brake caliper bolts.

7. Remove and discard the caliper anchor bolts.

8. Remove the brake pads from the caliper.

9. Remove the caliper bolts and separate the caliper from the anchor plate.

To install:

10. Compress the caliper pistons into the caliper bore.

11. Install the anchor plate and tighten the bolts to 81 ft. lbs. (110 Nm).

12. Retract the disc brake caliper piston fully into the caliper bore using tool no. 206-026.

13. Install the spring clips and brake pads in the caliper.

14. Install the caliper and tighten the guide pins to 23 ft. lbs. (31 Nm).

15. Install the parking brake cable to the caliper.

16. Reload the tension on the parking brake cable.

17. Bleed the brake system and install the rubber bleeder screw caps when complete.

18. Install the wheel and tire assembly.

19. Lower the vehicle.

20. Pump the brake pedal several times to position the brake pads before attempting to move the vehicle.

21. Check and fill the brake master cylinder as required.

22. Road-test the vehicle and check for proper brake operation.

FORD

Focus

8

SPECIFICATIONS AND MAINTENANCE CHARTS

ENGINE AND VEHICLE IDENTIFICATION

Code ①	Liters (cc)	Cu. In.	Cyl.	Fuel Sys.	Type	Eng. Mfg.	Code ②	Year
		Engine					Model Year	
3	2.0 (1988)	121	4	SFI	DOHC	Ford	2	2002
P	2.0 (1988)	121	4	SFI	SOHC	Ford	3	2003
5	2.0 (1988)	121	4	SFI	DOHC	Ford	4	2004
N	2.0 (1988)	121	4	SFI	DOHC	Ford	5	2005
Z	2.3 (2261)	138	4	SFI	DOHC	Ford	6	2006

SFI: Sequential Fuel Injection

DOHC: Double Overhead Camshafts

SOHC: Single Overhead Camshaft

① 8th digit of the VIN

② 10th digit of the VIN

06017-FOCU-C01

GENERAL ENGINE SPECIFICATIONS

Year	Model	Engine Displacement Liters	Engine VIN	Net Horsepower @ rpm	Net Torque @ rpm (ft. lbs.)	Bore x Stroke (in.)	Compression Ratio	Oil Pressure @ rpm
2002	Focus LX	2.0	P	110@5000	125@3750	3.34x3.46	9.2:1	35-65@2000
	Focus SE	2.0	P	110@5000	125@3750	3.34x3.46	9.2:1	35-65@2000
	Focus SE	2.0	3	130@5500	130@4000	3.34x3.46	10.0:1	35-65@2000
	Focus ZTS	2.0	3	130@5500	130@4000	3.34x3.46	10.0:1	35-65@2000
	Focus SVT	2.0	5	170@7000	198@5500	3.34x3.46	10.2:1	35-65@2000
	Focus ZX3	2.0	3	130@5500	130@4000	3.34x3.46	10.0:1	35-65@2000
2003	Focus LX	2.0	P	110@5000	125@3750	3.34x3.46	9.35:1	35-65@2000
	Focus SE	2.0	3	130@5500	135@4500	3.34x3.46	9.6:1	35-65@2000
	Focus ZTS	2.3	Z	144@5750	149@4200	3.44x3.70	9.9:1	35-65@2000
	Focus ZX3	2.0	3	130@5500	135@4500	3.34x3.46	9.6:1	35-65@2000
	Focus ZX5	2.0	3	130@5500	135@4500	3.34x3.46	9.6:1	35-65@2000
	Focus ZX3	2.3	Z	144@5750	149@4200	3.44x3.70	9.9:1	35-65@2000
	Focus ZX5	2.3	Z	144@5750	149@4200	3.44x3.70	9.9:1	35-65@2000
	Focus SVT	2.0	5	170@7000	198@5500	3.34x3.46	10.2:1	35-65@2000
	Focus ZTW	2.3	Z	144@5750	149@4200	3.44x3.70	9.9:1	35-65@2000
2004	Focus LX	2.0	P	110@5000	125@3750	3.34x3.46	9.35:1	35-65@2000
	Focus SE	2.0	3	130@5500	135@4500	3.34x3.46	9.6:1	35-65@2000
	Focus ZTS	2.3	Z	144@5750	149@4200	3.44x3.70	9.9:1	35-65@2000
	Focus ZX3	2.0	3	130@5500	135@4500	3.34x3.46	9.6:1	35-65@2000
	Focus ZX5	2.0	3	130@5500	135@4500	3.34x3.46	9.6:1	35-65@2000
	Focus ZX3	2.3	Z	144@5750	149@4200	3.44x3.70	9.9:1	35-65@2000
	Focus ZX5	2.3	Z	144@5750	149@4200	3.44x3.70	9.9:1	35-65@2000
	Focus SVT	2.0	5	170@7000	198@5500	3.34x3.46	10.2:1	35-65@2000
	Focus ZTW	2.3	Z	144@5750	149@4200	3.44x3.70	9.9:1	35-65@2000
2005	Focus ZX3	2.0	N	136@6000	133@4500	3.44x3.27	10.0:1	29-39@2000
	Focus ZX3	2.0	N-PZEV	130@6000	129@4000	3.44x3.27	10.0:1	29-39@2000
	Focus ZX4	2.0	N	136@6000	133@4500	3.44x3.27	10.0:1	29-39@2000
	Focus ZX4	2.0	N-PZEV	130@6000	129@4000	3.44x3.27	10.0:1	29-39@2000
	Focus ZX4 ST	2.3	Z	151@5750	154@4250	3.44x3.70	9.7:1	29-39@2000
	Focus ZX5	2.0	N	136@6000	133@4500	3.44x3.27	10.0:1	29-39@2000
	Focus ZX5	2.0	N-PZEV	130@6000	129@4000	3.44x3.27	10.0:1	29-39@2000
	Focus ZXW	2.0	N	136@6000	133@4500	3.44x3.27	10.0:1	29-39@2000
	Focus ZXW	2.0	N-PZEV	130@6000	129@4000	3.44x3.27	10.0:1	29-39@2000

SFI: Sequential Fuel Injection

PZEV: Partial Zero Emissions Vehicle

06017-FOCU-C02

ENGINE TUNE-UP SPECIFICATIONS

Year	Engine Displacement Liters	Engine VIN	Spark Plug Gap (in.)	Ignition Timing (deg.)		Fuel Pump (psi)	Idle Speed (rpm)		Valve Clearance (in.)	
				MT	AT		MT	AT	In.	Ex.
2002	2.0	3	0.052-0.056	10B	10B	31-38 ①	②	②	0.004-0.007	0.010-0.013
	2.0	5	0.041-0.045	10B	10B	31-38 ①	②	②	0.008-0.011	0.013-0.015
	2.0	P	0.052-0.056	10B	10B	31-38 ①	②	②	HYD	HYD
2003	2.0	3	0.052-0.056	10B	10B	31-38 ①	②	②	0.004-0.007	0.010-0.013
	2.0	P	0.052-0.056	10B	10B	31-38 ①	②	②	HYD	HYD
	2.0	5	0.041-0.045	10B	10B	31-38 ①	②	②	0.008-0.011	0.013-0.015
	2.3	Z	0.041-0.045	10B	10B	31-38 ①	②	②	0.008-0.011	0.010-0.013
2004	2.0	3	0.052-0.056	10B	10B	31-38 ①	②	②	0.004-0.007	0.010-0.013
	2.0	P	0.052-0.056	10B	10B	31-38 ①	②	②	HYD	HYD
	2.0	5	0.041-0.045	10B	10B	31-38 ①	②	②	0.008-0.011	0.013-0.015
	2.3	Z	0.041-0.045	10B	10B	31-38 ①	②	②	0.008-0.011	0.010-0.013
2005	2.0	N	0.049-0.053	10B	10B	NA	②	②	0.008-0.011	0.010-0.013
	2.3	Z	0.049-0.053	10B	10B	NA	②	②	0.008-0.011	0.010-0.013

NA: Information not available

NOTE: The Vehicle Emission Control Information label often reflects specification changes made during production.

The label figures must be used if they differ from those in this chart.

B: Before Top Dead Center

HYD: Hydraulic

① Fuel pressure with engine running, pressure regulator vacuum hose connected

② Refer to Vehicle Emission Control Information label

06017-FOCU-C03

2.0L (VIN P) Engine
Firing order: 1–3–4–2
Distributorless ignition system

79223G35

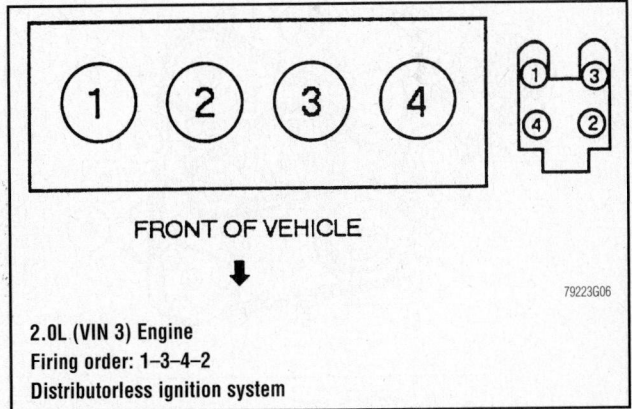

2.0L (VIN 3) Engine
Firing order: 1–3–4–2
Distributorless ignition system

79223G06

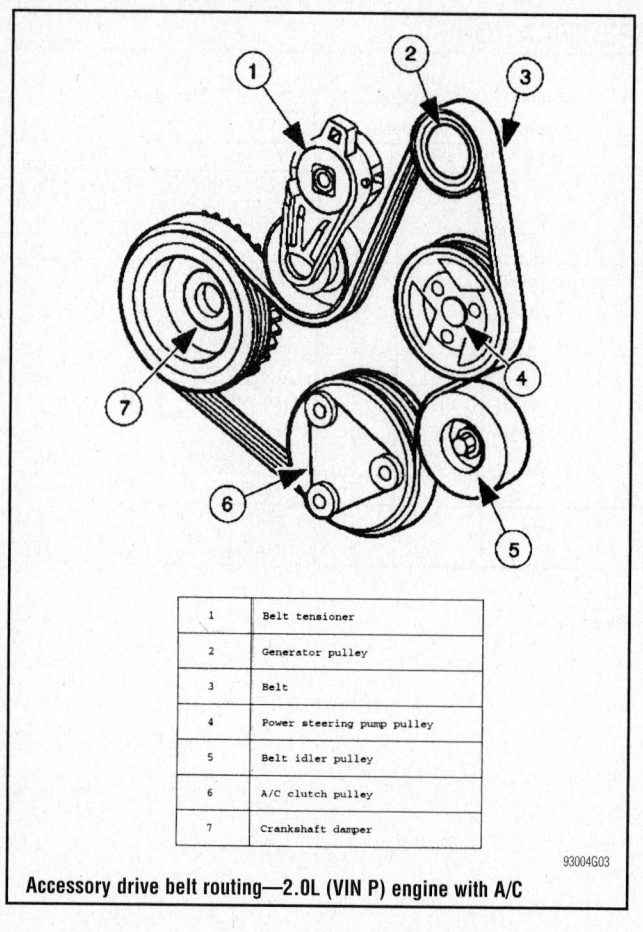

1	Belt tensioner
2	Generator pulley
3	Belt
4	Power steering pump pulley
5	Belt idler pulley
6	A/C clutch pulley
7	Crankshaft damper

93004G03

Accessory drive belt routing—2.0L (VIN P) engine with A/C

1	Belt tensioner
2	Generator pulley
3	Belt
4	Power steering pump pulley
5	Crankshaft damper

93004G04

Accessory drive belt routing—2.0L (VIN P) engine without A/C

1	Generator pulley
2	Belt idler pulley
3	Power steering pump pulley
4	Water pump pulley
5	Belt tensioner
6	A/C clutch pulley
7	Crankshaft pulley damper
8	Belt idler pulley
9	Drive belt

93004G05

Accessory drive belt routing—2.0L (VIN 3) engine

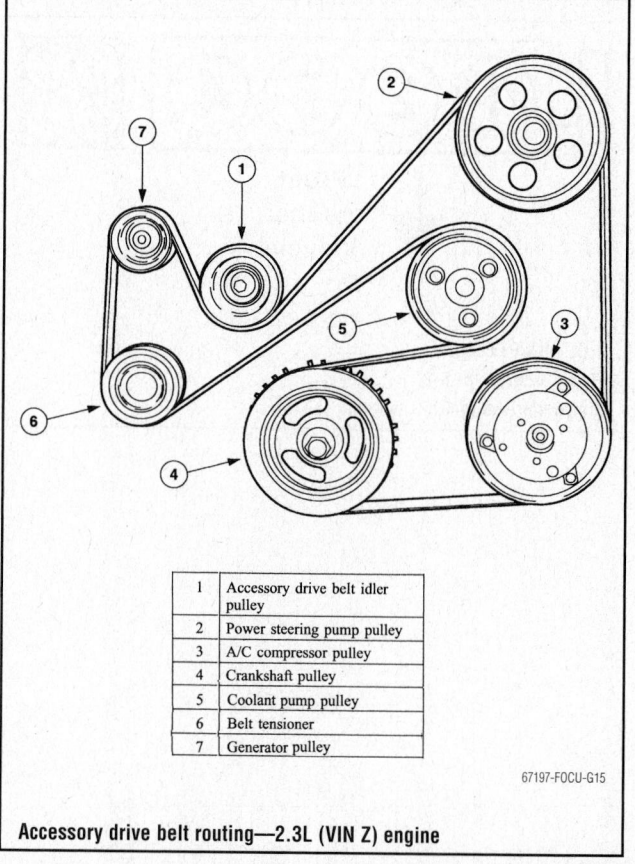

1	Accessory drive belt idler pulley
2	Power steering pump pulley
3	A/C compressor pulley
4	Crankshaft pulley
5	Coolant pump pulley
6	Belt tensioner
7	Generator pulley

67197-FOCU-G15

Accessory drive belt routing—2.3L (VIN Z) engine

CAPACITIES

Year	Model	Engine Displacement Liters	Engine VIN	Engine Oil with Filter (qts.)	Transmission (pts.)		Drive Front Axle (pts.)	Fuel Tank (gal.)	Cooling System (qts.)
					Manual	Auto.			
2002	Focus LX	2.0	P	4.0	4.8	14.0	—	13.2	①
	Focus SE	2.0	P	4.0	4.8	14.0	—	13.2	①
	Focus SE	2.0	3	4.5	4.0	14.0	—	13.2	②
	Focus ZTS	2.0	3	4.5	4.0	14.0	—	13.2	②
	Focus SVT	2.0	5	4.5	③	14.0	—	13.2	②
	Focus ZX3	2.0	3	4.5	4.0	14.0	—	13.2	②
2003	Focus LX	2.0	P	4.5	③	14.0	—	13.2	①
	Focus SE	2.0	3	4.5	③	14.0	—	13.2	②
	Focus ZTS	2.0	3	4.5	③	14.0	—	13.2	②
	Focus ZX3	2.0	3	4.5	③	14.0	—	13.2	②
	Focus ZX3	2.3	Z	4.5	③	14.0	—	14.0	②
	Focus ZX5	2.3	Z	4.5	③	14.0	—	14.0	②
	Focus SVT	2.0	5	4.5	③	14.0	—	13.2	②
	Focus ZTW	2.3	Z	4.5	③	14.0	—	14.0	②
2004	Focus LX	2.0	P	4.5	③	14.0	—	13.2	①
	Focus SE	2.0	3	4.5	③	14.0	—	13.2	②
	Focus ZTS	2.0	3	4.5	③	14.0	—	13.2	②
	Focus ZX3	2.0	3	4.5	③	14.0	—	13.2	②
	Focus ZX3	2.3	Z	4.5	③	14.0	—	14.0	②
	Focus ZX5	2.3	Z	4.5	③	14.0	—	14.0	②
	Focus SVT	2.0	5	4.5	③	14.0	—	13.2	②
	Focus ZTW	2.3	Z	4.5	③	14.0	—	14.0	②
2005	Focus ZX3	2.0	N	4.5	4.2	13.8	—	14.0	6.1
	Focus ZX4	2.0	N	4.5	4.2	13.8	—	14.0	6.1
	Focus ZX4 ST	2.3	Z	4.5	4.2	13.8	—	14.0	6.1
	Focus ZX5	2.0	N	4.5	4.2	13.8	—	14.0	6.1
	Focus ZXW	2.0	N	4.5	4.2	13.8	—	14.0	6.1

Note: All capacities are approximates. Add fluid gradually and ensure a proper fluid level is obtained.

① Manual transaxle: 7.9
  Automatic transaxle: 5.8

② Manual transaxle: 7.0
  Automatic transaxle: 7.5

③ IB5: 5 pts.
  MTX-75: 4 pts.

06017-FOCU-C04

VALVE SPECIFICATIONS

Year	Engine Displacement Liters	Engine VIN	Seat Angle (deg.)	Face Angle (deg.)	Spring Test Pressure (lbs. @ in.)	Spring Installed Height (in.)	Stem-to-Guide Clearance (in.)		Stem Diameter (in.)	
							Intake	Exhaust	Intake	Exhaust
2002	2.0	3	45	45	NA	1.346	0.0007-0.0025	0.0014-0.0032	0.2373-0.2379	0.2366-0.2372
	2.0	5	NA	NA	NA	NA	0.0007-0.0025	0.0007-0.0025	NA	NA
	2.0	P	45	45.6	200@1.09	1.420-1.540	0.0008-0.0027	0.0018-0.0037	0.3159-0.3167	0.3149-0.3156
2003	2.0	3	45	45	NA	1.346	0.0007-0.0025	0.0014-0.0032	0.2373-0.2379	0.2366-0.2372
	2.0	P	45	45.6	200@1.09	1.420-1.540	0.0008-0.0027	0.0018-0.0037	0.3159-0.3167	0.3149-0.3156
	2.0	5	NA	NA	NA	NA	0.0007-0.0025	0.0007-0.0025	NA	NA
	2.3	Z	45	45	38.667@1.492	1.492	0.0009	0.0011	0.2153-0.2159	0.2151-0.2157
2004	2.0	3	45	45	NA	1.346	0.0007-0.0025	0.0014-0.0032	0.2373-0.2379	0.2366-0.2372
	2.0	P	45	45.6	200@1.09	1.420-1.540	0.0008-0.0027	0.0018-0.0037	0.3159-0.3167	0.3149-0.3156
	2.0	5	NA	NA	NA	NA	0.0007-0.0025	0.0007-0.0025	NA	NA
	2.3	Z	45	45	38.667@1.492	1.492	0.0009	0.0011	0.2153-0.2159	0.2151-0.2157
2005	2.0	N	45	45	38.667@1.492	1.492	0.0001	0.00011	0.2153-0.2159	0.2151-0.2157
	2.3	Z	45	45	38.667@1.492	1.492	0.0009	0.0011	0.2153-0.2159	0.2151-0.2157

NA: Not Available

06017-FOCU-C05

CRANKSHAFT AND CONNECTING ROD SPECIFICATIONS
All measurements are given in inches.

Year	Engine Displacement Liters	Engine VIN	Crankshaft				Connecting Rod		
			Main Brg. Journal Dia.	Main Brg. Oil Clearance	Shaft End-play	Thrust on No.	Journal Diameter	Oil Clearance	Side Clearance
2002	2.0	3	2.2827-2.2835	0.0008-0.0026	0.0040-0.0120	3	1.8461-1.8500	0.0006-0.0028	0.0040-0.0110
	2.0	5	2.2839-2.2849	0.0004-0.0023	0.0035-0.0102	3	1.8475-1.8488	0.0006-0.0027	0.0035-0.0126
	2.0	P	2.2827-2.2835	0.0008-0.0026	0.0040-0.0120	3	1.7279-1.7287	0.0008-0.0026	0.0040-0.0110
2003	2.0	3	2.2827-2.2835	0.0008-0.0026	0.0040-0.0120	3	1.8461-1.8500	0.0006-0.0028	0.0040-0.0110
	2.0	P	2.2827-2.2835	0.0008-0.0026	0.0040-0.0120	3	1.7279-1.7287	0.0008-0.0026	0.0040-0.0110
	2.0	5	2.2839-2.2849	0.0004-0.0023	0.0035-0.0102	3	1.8475-1.8488	0.0006-0.0027	0.0035-0.0126
	2.3	Z	2.0460-2.0470	0.0007-0.0013	0.0080-0.0160	3	2.0870-2.0880	0.0010-0.0020	0.0050-0.0140
2004	2.0	3	2.2827-2.2835	0.0008-0.0026	0.0040-0.0120	3	1.8461-1.8500	0.0006-0.0028	0.0040-0.0110
	2.0	P	2.2827-2.2835	0.0008-0.0026	0.0040-0.0120	3	1.7279-1.7287	0.0008-0.0026	0.0040-0.0110
	2.0	5	2.2839-2.2849	0.0004-0.0023	0.0035-0.0102	3	1.8475-1.8488	0.0006-0.0027	0.0035-0.0126
	2.3	Z	2.0460-2.0470	0.0007-0.0013	0.0080-0.0160	3	2.0870-2.0880	0.0010-0.0020	0.0050-0.0140
2005	2.0	N	2.0460-2.0470	0.0007-0.0013	0.008-0.0160	3	2.0870-2.0880	0.0010-0.0020	0.0050-0.0140
	2.3	Z	2.0460-2.0470	0.0007-0.0013	0.0080-0.0160	3	2.0870-2.0880	0.0010-0.0020	0.0050-0.0140

06017-FOCU-C06

PISTON AND RING SPECIFICATIONS
All measurements are given in inches.

Year	Engine Displacement Liters	Engine VIN	Piston Clearance	Ring Gap			Ring Side Clearance		
				Top Compression	Bottom Compression	Oil Control	Top Compression	Bottom Compression	Oil Control
2002	2.0	3	0.0010-0.0022	0.012-0.022	0.012-0.022	0.010-0.039	0.0015-0.0032	0.0015-0.0035	—
	2.0	5	0.0004-0.0011	0.011-0.019	0.011-0.019	0.015-0.055	NA	NA	NA
	2.0	P	0.0008-0.0027	0.010-0.030	0.010-0.030	0.016-0.066	0.0015-0.0032	0.0015-0.0035	—
2003	2.0	3	0.0010-0.0022	0.012-0.022	0.012-0.022	0.010-0.039	0.0015-0.0032	0.0015-0.0035	—
	2.0	P	0.0008-0.0027	0.010-0.030	0.010-0.030	0.016-0.066	0.0015-0.0032	0.0015-0.0035	—
	2.0	5	0.0004-0.0011	0.011-0.019	0.011-0.019	0.015-0.055	NA	NA	NA
	2.3	Z	0.0009-0.0017	0.006-0.012	0.012-0.018	0.007-0.027	0.0008-0.0013	0.0004-0.0011	0.0024-0.0055
2004	2.0	3	0.0010-0.0022	0.012-0.022	0.012-0.022	0.010-0.039	0.0015-0.0032	0.0015-0.0035	—
	2.0	P	0.0008-0.0027	0.010-0.030	0.010-0.030	0.016-0.066	0.0015-0.0032	0.0015-0.0035	—
	2.0	5	0.0004-0.0011	0.011-0.019	0.011-0.019	0.015-0.055	NA	NA	NA
	2.3	Z	0.0009-0.0017	0.006-0.012	0.012-0.018	0.007-0.027	0.0008-0.0013	0.0004-0.0011	0.0024-0.0055
2005	2.0	N	0.0009-0.0017	0.006-0.0120	0.012-0.018	0.007-0.027	NA	NA	NA
	2.3	Z	0.0009-0.0017	0.006-0.012	0.012-0.018	0.007-0.027	0.0008-0.0013	0.0004-0.0011	0.0024-0.0055

NA: Not Available

06017-FOCU-C07

TORQUE SPECIFICATIONS
All readings in ft. lbs.

Year	Engine Displacement Liters	Engine VIN	Cylinder Head Bolts	Main Bearing Bolts	Rod Bearing Bolts	Crankshaft Damper Bolts	Flywheel Bolts	Manifold		Spark Plugs	Oil Pan Drain Plug
								Intake	Exhaust		
2002	2.0	P	①	66-79	26-30	80-87	71-76	15-22	15-17	12-15	18
	2.0	5	②	④	⑤	85	82	⑥	⑦	11	18
	2.0	3	②	55-65	③	80-87	71-76	11-12	10-12	10-12	18
2003	2.0	P	①	66-79	26-30	80-87	71-76	15-22	15-17	12-15	18
	2.0	3	②	55-65	③	80-87	71-76	11-12	10-12	10-12	18
	2.0	5	②	④	⑤	85	82	⑥	⑦	11	18
	2.3	Z	⑧	NA	NA	⑨	⑩	13	⑪	11	21
2004	2.0	P	①	66-79	26-30	80-87	71-76	15-22	15-17	12-15	18
	2.0	3	②	55-65	③	80-87	71-76	11-12	10-12	10-12	18
	2.0	5	②	④	⑤	85	82	⑥	⑦	11	18
	2.3	Z	⑧	NA	NA	⑨	⑩	13	⑪	11	21
2005	2.0	N	⑧	NA	NA	⑨	⑩	13	41	11	21
	2.3	Z	⑧	NA	NA	⑨	⑩	13	⑪	11	21

NA: Information Not Available

① Step 1: 37 ft. lbs.
Step 2: Loosen bolts 1/2 turn
Step 3: 37 ft. lbs.
Step 4: Plus 90 degrees
Step 5: Plus 90 degrees

② Step 1: 15 ft. lbs.
Step 2: 30 ft. lbs.
Step 3: Plus 90 degrees

③ Step 1: 22-25 ft. lbs.
Step 2: Plus 90 degrees

④ Step 1: 18 ft. lbs.
Step 2: Plus 60 degrees

⑤ Step 1: 26 ft. lbs.
Step 2: Plus 90 degrees

⑥ Studs: 44 inch lbs.
Bolts and nuts: 13 ft. lbs.

⑦ Studs: 44 inch lbs.
Nuts: 12 ft. lbs.
Bolts: 13 ft. lbs.

⑧ Step 1: 44 inch lbs.
Step 2: 11 ft. lbs.
Step 3: 33 ft. lbs.
Step 4: Plus 90 degrees
Step 5: Plus 90 degrees

⑨ Step 1: 74 ft. lbs.
Step 2: Plus 90 degrees

⑩ Step 1: 37 ft. lbs.
Step 2: 50 ft. lbs.
Step 3: 83 ft. lbs.

⑪ Studs: 13 ft. lbs.
Nuts: 40 ft. lbs.

06017-FOCU-C08

WHEEL ALIGNMENT

Year	Model		Caster		Camber		Toe-in (in.)
			Range (+/-Deg.)	Preferred Setting (Deg.)	Range (+/-Deg.)	Preferred Setting (Deg.)	
2002	Sedan	F	1.00	+2.93	1.00	-0.36	0 +/- 0.04
		R	—	—	0.30	-0.93	0.08 +/- 0.05
	Wagon	F	1.30	+2.46	1.25	-0.62	0 +/- 0.04
		R	—	—	0.75	-0.62	0.12 +/- 0.05
2003	Sedan	F	1.00	+2.93	1.00	-0.36	0 +/- 0.04
		R	—	—	0.30	-0.93	0.08 +/- 0.05
	Wagon	F	1.30	+2.46	1.25	-0.62	0 +/- 0.04
		R	—	—	0.75	-0.62	0.12 +/- 0.05
2004	Sedan	F	1.00	+2.93	1.00	-0.36	0 +/- 0.04
		R	—	—	0.30	-0.93	0.08 +/- 0.05
	Wagon	F	1.30	+2.46	1.25	-0.62	0 +/- 0.04
		R	—	—	0.75	-0.62	0.12 +/- 0.05
2005	Sedan	F	1.00	+2.75	1.25	-0.50	0 +/- 0.04
		R	—	—	0.30	-1.02	0.08 +/- 0.05
	Wagon	F	1.30	+2.75	1.25	-0.50	0 +/- 0.04
		R	—	—	0.75	-0.94	0.12 +/- 0.05

06017-FOCU-C09

TIRE, WHEEL AND BALL JOINT SPECIFICATIONS

Year	Model	OEM Tires Standard	OEM Tires Optional	Tire Pressures (psi) Front	Tire Pressures (psi) Rear	Wheel Size	Ball Joint Inspection	Lug Nut Torque (ft. lbs.)
2002	Focus LX	P185/65R14	—	32	32	5.5	①	94
	Focus SVT	P215/45ZR/17	—	32	32	5.5	①	94
	Focus ZX	P195/60/R15	P205/60/R16	32	32	5.5	①	94
2003	Focus LX	P185/65R14	—	32	32	5.5	①	94
	Focus SVT	P215/45ZR/17	—	32	32	5.5	①	94
	Focus ZX	P195/60/R15	P205/60/R16	32	32	5.5	①	94
2004	Focus LX	P185/65R14	—	32	32	5.5	①	94
	Focus SVT	P215/45ZR/17	—	32	32	5.5	①	94
	Focus ZX	P195/60/R15	P205/60/R16	32	32	5.5	①	94
2005	Focus ZX	P195/60/R15	P205/50R16	②	②	NA	①	94
	Focus ZX4 ST	P205/50R16	—	②	②	NA	①	94

OEM: Original Equipment Manufacturer

PSI: Pounds Per Square Inch

① Replace if any measurable movement is found

② Refer to the sticker on the inside of the driver's door

06017-FOCU-C10

BRAKE SPECIFICATIONS

All measurements in inches unless noted

Year	Model		Brake Disc Original Thickness	Brake Disc Minimum Thickness	Maximum Runout	Brake Drum Diameter Original Inside Diameter	Max. Wear Limit	Maximum Machine Diameter	Minimum Lining Thickness Front	Minimum Lining Thickness Rear	Brake Caliper Mounting Bolts (ft. lbs.)
2002	Focus Except SVT	F	0.870	0.790	0.002	—	—	—	0.059	—	21
		R	0.390	0.320	0.002	7.99	8.03	—	—	①	26
	Focus SVT	F	0.950	0.870	0.002	—	—	—	0.059	—	21
		R	0.390	0.350	0.002	—	—	—	—	0.059	26
2003	Focus Except SVT	F	0.870	0.790	0.002	—	—	—	0.059	—	21
		R	0.390	0.320	0.002	7.99	8.03	—	—	①	26
	Focus SVT	F	0.950	0.870	0.002	—	—	—	0.059	—	21
		R	0.390	0.350	0.002	—	—	—	—	0.059	26
2004	Focus Except SVT	F	0.870	0.790	0.002	—	—	—	0.059	—	21
		R	0.390	0.320	0.002	7.99	8.03	—	—	①	26
	Focus SVT	F	0.950	0.870	0.002	—	—	—	0.059	—	21
		R	0.390	0.350	0.002	—	—	—	—	0.059	26
2005	Focus	F	NA	0.910	0.002	—	—	—	0.118	—	21
		R	NA	0.350	0.002	NA	8.03	—	—	②	26

NA: Information Not Available

① Drum brakes: 0.039 in.
  Disc brakes: 0.059 in.

② Drum brakes: 0.14 in.
  Disc brakes: 0.118 in.

06017-FOCU-C11

SCHEDULED MAINTENANCE INTERVALS
Ford—Focus

TO BE SERVICED	TYPE OF SERVICE	VEHICLE MILEAGE INTERVAL (x1000)												
		5	10	15	20	25	30	35	40	45	50	55	60	65
Air cleaner filter	R						✓						✓	
Accessory drive belt	S/I												✓	
Brake system ①	S/I			✓			✓			✓			✓	
Clutch pedal operation	S/I						✓						✓	
Cooling fan operation	S/I		✓		✓		✓		✓		✓		✓	
Cooling system hoses and clamps	S/I			✓			✓			✓			✓	
CV-joint boots & axle seals	S/I						✓						✓	
Engine coolant	R	Ten years or 150,000 miles												
Engine oil & filter	R	✓	✓	✓	✓	✓	✓	✓	✓	✓	✓	✓	✓	✓
Exterior Lights	S/I	Check monthly												
PCV valve	S/I												✓	
Exhaust system & heat shields	S/I						✓						✓	
Parking brake system	S/I	Every 6 months												
Power steering fluid	S/I	Every 6 months												
Rotate tires	S/I	✓	✓	✓	✓	✓	✓	✓	✓	✓	✓	✓	✓	✓
Steering linkage	S/I						✓						✓	
Spark plugs	R	Change at 100,000 miles												
Suspension components	S/I						✓						✓	

R: Replace   S/I: Inspect and service, if necessary   L: Lubricate   A: Adjust   C: Clean

① Inspect the reservoir fluid level, rotor and or drum, brake lines, hoses, calipers and or wheel cylinders

FREQUENT OPERATION MAINTENANCE (SEVERE SERVICE)

If a vehicle is operated under any of the following conditions it is considered severe service:
- Extremely dusty areas.
- 50% or more of the vehicle operation is in 32°C (90°F) or higher temperatures, or constant operation in temperatures below 0°C (32°F).
- Prolonged idling (vehicle operation in stop and go traffic).
- Frequent short running periods (engine does not warm to normal operating temperatures).
- Police, taxi, delivery usage or trailer towing usage.

Oil & oil filter change: change every 3000 miles.

Air filter element: change every 15,000 miles.

Automatic transmission fluid: change every 30,000 miles

Spark plugs: change every 60,000 miles

06017-FOCU-C12

→Disconnecting the negative battery cable on some vehicles may interfere with the functions of the on board computer system. The computer may undergo a relearning process once the negative battery cable is reconnected.

Alternator

REMOVAL

2.0L (VIN P) Engine

1. Before servicing the vehicle, refer to the Precautions Section.
2. Remove or disconnect the following:
   - Negative battery cable
   - Accessory drive belt
   - Power steering pipe brackets
   - Exhaust manifold heat shield
   - Alternator harness connectors
   - Alternator

2.0L (VIN 3 and 5) Engines

1. Before servicing the vehicle, refer to the Precautions Section.
2. Remove or disconnect the following:
   - Negative battery cable
   - Accessory drive belt
   - Alternator harness connectors
   - Coolant expansion tank
   - Power steering reservoir
   - Engine wiring harness bracket
   - Ground cable
   - Evaporative Emissions (EVAP) canister purge valve
   - Alternator

2.0L (VIN N) Engine

1. Before servicing the vehicle, refer to the Precautions Section.
2. Remove or disconnect the following:
   - Negative battery cable
   - Engine heat shield
   - Upper alternator mounting nut, Non-PZEV engines only
   - Upper alternator heat shield mounting nut, PZEV engines only
   - Engine splash shield
   - Accessory drive belt
   - Alternator heat shield, PZEV engines only
   - Air intake assembly
   - Alternator B+ cable nut and cable
   - Alternator mounting bolts
   - Alternator electrical connector and upper alternator mounting nut, PZEV engines only
3. Remove the engine roll restrictor mounting bolt and position the engine forward.
4. Remove the alternator.
5. Remove the lower alternator air duct.
6. Remove the radial adapter from the alternator, non-PZEV engines only.

2.3L Engine

1. Before servicing the vehicle, refer to the Precautions Section.
2. Disconnect the negative battery cable.
3. Remove the alternator heat shield top nut.
4. Raise and support the vehicle.
5. Remove the screws and the lower splash shield.
6. Release the accessory drive belt tension and remove the belt from the generator pulley.
7. Remove the alternator heat shield lower nut and the heat shield.
8. Release the two air intake tube retainers and remove the air intake tube assembly.
9. Remove the alternator positive battery cable nut and disconnect the positive battery cable.
10. Remove the two alternator nuts and bolt.
11. Lower the vehicle.
12. Disconnect the alternator electrical connector.
13. Remove the upper alternator mounting nut.
14. Raise and support the vehicle.
15. Remove the engine roll restrictor mount bolt.
16. Position the engine forward and remove the alternator.

INSTALLATION

2.0L (VIN P) Engine

Install or connect the following:
- Alternator. Tighten the bolts to 35 ft. lbs. (45 Nm).
- Alternator harness connectors
- Exhaust manifold heat shield
- Power steering pipe brackets
- Accessory drive belt
- Negative battery cable

2.0L (VIN 3 and 5) Engines

Install or connect the following:
- Alternator. Tighten the bolts to 18 ft. lbs. (25 Nm).
- EVAP canister purge valve
- Ground cable
- Engine wiring harness bracket
- Power steering reservoir
- Coolant expansion tank
- Alternator harness connectors
- Accessory drive belt
- Negative battery cable

2.0L (VIN N) Engine

1. Install or connect the following:
   - Radial adapter to alternator, non-PZEV engines only. Tighten to 8 ft. lbs. (11 Nm).
   - Lower air duct to alternator. Tighten to 9 ft. lbs. (12 Nm).
   - Alternator. Tighten the lower mounting bolts to 35 ft. lbs. (48 Nm).
   - Engine roll restrictor bolt. Tighten to 35 ft. lbs. (48 Nm).
   - Alternator electrical connector and upper mounting nut, PZEV engines only. Tighten to 35 ft. lbs. (48 Nm).
   - Alternator positive battery cable and cable nut. Tighten to 71 inch lbs. (8 Nm).

→The cable nut must always be replaced when removed from the alternator.

- Air intake assembly
- Alternator heat shield and lower nut, PZEV engines only. Tighten to 18 ft. lbs. (25 Nm).
- Accessory drive belt
- Engine splash shield
- Alternator heat shield upper nut, PZEV engines only. Tighten to 18 ft. lbs. (25 Nm).
- Upper generator mounting nut, non-PZEV engines only. Tighten to 35 ft. lbs. (48 Nm).
- Engine heat shield
- Negative battery cable

2.3L Engine

1. Installation is the reverse of the removal procedure. Use the following torque specifications:
   - Engine roll restrictor bolt: 35 ft. lbs. (48 Nm)
   - Generator nuts and bolt: 35 ft. lbs. (47 Nm)

- Battery + terminal nut: 53 inch lbs. (6 Nm)
- Generator heat shield nuts: 18 ft. lbs. (25 Nm)

Ignition Timing

ADJUSTMENT

The engines are equipped with a Distributorless Ignition System (DIS). No adjustment is necessary.

Engine Assembly

REMOVAL & INSTALLATION

Manual Transaxle

2.0L ENGINES EXCEPT SVT AND VIN N

1. Before servicing the vehicle, refer to the Precautions Section.
2. Drain the cooling system.
3. Relieve the fuel system pressure.
4. Loosen the front strut center nuts 5 turns.
5. Remove or disconnect the following:
   - Battery, tray and cables
   - Mass Air Flow (MAF) sensor connector
   - Air cleaner housing and tubes
   - Accelerator cable
   - Cruise control cable, if equipped
   - Ignition coil connectors
   - Power steering pump pressure switch connector
   - Heated Oxygen (HO2S) sensor connectors
   - Fuel injector harness connectors
   - Powertrain Control Module (PCM) connectors
   - Vehicle Speed (VSS) sensor connector
   - Reverse lamp switch connector
   - Clutch slave cylinder fluid pipe
   - Evaporative Emissions (EVAP) canister vacuum lines
   - Power brake booster vacuum line
   - Delta Pressure Feedback Electronic (DPFE) system sensor vacuum line
   - Exhaust Gas Recirculation (EGR) vacuum line
   - Intake manifold vacuum line
   - Fuel line
   - Coolant bypass hoses
   - Radiator hoses
   - Coolant expansion tank
   - Accessory drive belt
   - Power steering pump and reservoir

   - Power steering high pressure pipe
   - Radiator cooling fan
   - Catalytic converter
   - Exhaust flex pipe
   - Drive belt cover
   - Shift cable cover
   - Shift cables
   - A/C compressor, if equipped
   - Right engine mount
   - Lower ball joints
   - Axle halfshafts
6. Support the powertrain from below
7. Remove the front and rear engine mounts.
8. Raise the vehicle away from the powertrain.

To install:

9. Lower the vehicle over the powertrain.
10. Install or connect the following:
   - Front engine mount. Tighten the nuts to 59 ft. lbs. (80 Nm) and the bolts to 35 ft. lbs. (48 Nm).
   - Rear engine mount. Tighten the outer nuts to 35 ft. lbs. (48 Nm) and the center nut to 98 ft. lbs. (133 Nm).
   - Axle halfshafts
   - Lower ball joints. Tighten the pinch bolts to 37 ft. lbs. (50 Nm).
   - Right engine mount. Tighten the bolts to 35 ft. lbs. (48 Nm).
   - A/C compressor, if equipped
   - Shift cables
   - Shift cable cover
   - Drive belt cover
   - Exhaust flex pipe
   - Catalytic converter
   - Radiator cooling fan
   - Power steering high pressure pipe
   - Power steering pump and reservoir
   - Accessory drive belt
   - Coolant expansion tank
   - Radiator hoses
   - Coolant bypass hoses
   - Fuel line
   - Intake manifold vacuum line
   - EGR vacuum line
   - DPFE system sensor vacuum line
   - Power brake booster vacuum line
   - EVAP canister vacuum lines
   - Clutch slave cylinder fluid pipe
   - Reverse lamp switch connector
   - VSS sensor connector
   - PCM connectors
   - Fuel injector harness connectors
   - HO2S sensor connectors
   - Power steering pump pressure switch connector
   - Ignition coil connectors
   - Cruise control cable, if equipped

   - Accelerator cable
   - Air cleaner housing and tubes
   - MAF sensor connector
   - Battery, tray and cables
11. Tighten the strut center nuts to 35 ft. lbs. (48 Nm).
12. Fill the cooling system.
13. Start the engine and check for leaks.

SVT

1. Before servicing the vehicle, refer to the Precautions Section.
2. Release the fuel system pressure.
3. Release the coolant system pressure.
4. Remove the battery tray.
5. Remove the air cleaner.
6. Disconnect the ground cable from the inner fender.

➡The resonator is a push fit in the bracket.

7. Remove the air cleaner intake tube and resonator.
8. Raise and support the vehicle.
9. Drain the coolant. Allow the coolant to drain into a suitable container. Install the drain plug after draining.
10. Drain the engine oil. Allow the oil to drain into a suitable container. Install the drain plug after draining.
11. Partially lower the vehicle.
12. Loosen the strut and spring assembly top mount retaining nuts by three turns on both sides.
13. Remove the splash shield.
14. Disconnect the accelerator cable and the speed control cable (if equipped) from the throttle body.
15. Disconnect the intake manifold runner control (IMRC) actuator cable from the IMRC lever.
16. Remove the electronic ignition (EI) coil cover.
17. Disconnect the block heater harness (if equipped).
18. Disconnect the starter field wire electrical connector.
19. Disconnect the fuel injector wiring harness.
20. Disconnect the ground cable from the engine lifting eye.
21. Disconnect the generator electrical connector.
22. Disconnect the hydraulic line from the clutch slave cylinder.
23. Disconnect the wiring harness electrical connector.
24. Raise and support the vehicle.
25. Disconnect the vehicle speed sensor (VSS) electrical connector.
26. Disconnect the reversing lamp switch electrical connector.

27. Partially lower the vehicle.

28. Disconnect the vacuum hoses from the intake manifold.

29. Disconnect the brake booster pipe from the intake manifold.

30. Release the quick release coupling and disconnect the brake booster pipe.

31. Disconnect the fuel lines.

32. Disconnect the ground cable.

33. Disconnect the coolant hoses.

34. Disconnect the oil cooler hose.

35. Disconnect the gearshift cables from the transaxle.

36. Disconnect the gearshift cable and the selector cable from the selector levers.

37. Disconnect the retaining bracket from the transaxle.

38. Raise and support the vehicle.

39. Remove the accessory drive belt.

40. Disconnect the air conditioning compressor and secure it to the radiator crossmember.

41. Remove the power steering pump lower retaining bolts and disconnect the radiator lower coolant hose from the water pump housing.

42. Disconnect the power steering pressure (PSP) switch electrical connector.

43. Disconnect the crankshaft position (CKP) sensor connector.

44. Disconnect the coolant hose from the oil cooler.

45. Lower the vehicle.

46. Disconnect the coolant expansion tank and position it to one side.

47. Drain the power assisted steering (PAS) reservoir.

48. Disconnect the PAS reservoir and position it to one side.

49. Disconnect the power steering line bracket from the cylinder head.

50. Disconnect the power steering pump and bracket from the engine and position it to one side.

51. Raise and support the vehicle.

52. Remove the flexible exhaust pipe.

53. Remove the crossmember.

✳✳ CAUTION

Support the halfshaft. The inner joint must not be bent at more than 18 degrees. The outer joint must not be bent at more than 45 degrees.

✳✳ CAUTION

Do not damage the halfshaft oil seal.

➡ **Plug the transaxle to prevent oil loss or dirt ingress.**

54. Remove the RH halfshaft and the intermediate shaft.

55. Allow the oil to drain into a suitable container.

56. Disconnect the LH halfshaft from the transaxle and secure it to one side. Allow the oil to drain into a suitable container. Discard the snap ring.

57. Position the workshop table with suitable wooden blocks under the engine and transmission assembly.

58. Carefully lower the vehicle until the engine and transmission assembly is positioned on the workshop table.

59. Secure the engine and transmission assembly to the assembly stand table with a retaining strap.

60. Remove the engine rear mount.

61. Remove the engine front mount.

62. Carefully raise the vehicle.

63. Pull the assembly stand forward.

64. Secure the engine and transmission assembly on an engine hoist.

✳✳ CAUTION

The catalytic converter must be supported while the support bracket is removed.

65. Remove the catalytic converter support bracket.

66. Remove the catalyst monitor sensor.

67. Disconnect the catalytic converter from the exhaust manifold. Discard the gasket and nuts.

68. Remove the heated oxygen (HO2S) sensor.

69. Remove the transaxle upper retaining bolts.

70. Disconnect the starter motor.

71. Remove the starter motor.

72. Disconnect the ground cable from the transaxle.

73. Remove the transaxle RH retaining bolts.

74. Remove the transaxle LH retaining bolts.

75. Separate the engine from the transaxle.

To install:

76. Install the transaxle. Tighten the bolts to 35 ft. lbs. (48 Nm).

77. Install the starter motor. Tighten the bolts to 26 ft. lbs. (35 Nm).

78. Attach the ground cable to the transaxle.

79. Connect the starter motor electrical connectors.

➡ **Install a new exhaust flange gasket and new nuts.**

80. Attach the catalytic converter to the exhaust manifold. Tighten the nuts to 35 ft. lbs. (47 Nm).

81. Install the heated oxygen (HO2S) sensor.

82. Install the catalytic converter bracket. Tighten the bolts to 18 ft. lbs. (25 Nm).

83. Install the catalyst monitor sensor (CMS).

84. Raise and support the vehicle.

85. Position the engine and transmission assembly under the vehicle.

86. Carefully lower the vehicle.

87. Install the engine front mount. Do not tighten the bolts and nuts at this stage.

88. Install the engine rear mount. Do not tighten the bolts and nuts at this stage.

89. Remove the retaining strap from the engine and transmission assembly, and remove the assembly stand.

90. Raise and support the vehicle.

✳✳ CAUTION

Support the halfshaft. The inner joint must not be bent more than 18 degrees. The outer joint must not be bent more than 45 degrees.

✳✳ CAUTION

Do not damage the halfshaft oil seal.

➡ **Install a new halfshaft center bearing cap and locknuts.**

91. Attach the RH halfshaft and the intermediate shaft to the transaxle.

92. Install the center bearing cap. Tighten the nuts to 18 ft. lbs. (25 Nm).

93. Attach the LH halfshaft to the transaxle.

94. Install the crossmember.

95. Install the flexible exhaust pipe.

96. Lower the vehicle.

97. Tighten the engine rear mount nuts and bolts. Tighten the bolts to 35 ft. lbs. (48 Nm), and the nuts to 59 ft. lbs. (80 Nm).

98. Tighten the engine front mount nuts. Tighten the small nuts to 35 ft. lbs. (48 Nm), and the large nut to 98 ft. lbs. (133 Nm).

99. Attach the power steering pump and the bracket to the engine. Tighten the bolts to 35 ft. lbs. (48 Nm).

100. Attach the power steering line bracket to the cylinder head. Tighten to 15 ft. lbs. (20 Nm).

101. Install the power assisted steering (PAS) reservoir.

102. Install the coolant expansion tank.

103. Raise and support the vehicle.

104. Connect the power steering pressure switch electrical connector.

105. Connect the crankshaft position sensor electrical connector.

106. Connect the radiator lower coolant hose to the coolant pump housing.

107. Connect the radiator lower coolant hose to the oil cooler.

108. Attach the air conditioning compressor. Tighten the bolts to 18 ft. lbs. (25 Nm).

109. Install the accessory drive belt.

110. Lower the vehicle.

111. Connect the oil cooler hose.

112. Attach the gearshift cables to the transaxle.

113. Attach the retaining bracket to the transaxle.

114. Attach the gearshift cable and the selector cable to the selector levers.

115. Connect the starter field wire electrical connector.

116. Attach the coolant hoses to the thermostat housing.

117. Attach the fuel pipes to the fuel rail.

118. Attach the brake servo vacuum hose to the intake manifold.

119. Attach the vacuum hoses to the throttle body.

120. Connect the reversing lamp switch electrical connector.

121. Clip on the wiring loom.

122. Connect the vehicle speed sensor (VSS) electrical connector.

123. Connect the wiring harness connectors.

124. Attach the clutch supply line to the clutch slave cylinder. Install the clip.

125. Connect the generator electrical connector.

126. Connect the fuel injector wiring harness.

127. Install the electronic ignition (EI) coil cover.

128. Attach the accelerator cable and the speed control cable (if equipped) to the throttle body.

129. Install the splash shield.

130. Connect the engine block heater (if equipped).

131. Connect the intake manifold runner control (IMRC).

132. Install the air cleaner intake tube and resonator.

133. Tighten the strut and spring assembly upper mounting retaining nuts to 18 ft. lbs. (25 Nm).

134. Attach the ground cable to the inner fender.

135. Install the air cleaner.

136. Install the battery tray.

➡**When the battery has been disconnected and reconnected, some abnormal drive symptoms may occur while the vehicle relearns its adaptive strategy. The vehicle may need to be driven 16 km (10 miles) or more to relearn the strategy.**

137. Install and connect the battery.

138. Fill the cooling system.

139. Fill the transaxle.

140. Fill the engine with engine oil.

141. Bleed the hydraulic clutch operating system.

142. Fill and bleed the PAS system.

2.0L (VIN N) ENGINE

1. Before servicing the vehicle, refer to the Precautions Section.

2. Drain the engine oil. Reinstall the drain plug.

3. Drain the cooling system.

4. Relieve the fuel system pressure.

5. Recover the A/C refrigerant, if equipped.

6. Remove or disconnect the following:

- Battery and battery tray
- Accessory drive belt splash guard
- Accessory drive belt
- Catalytic converter from the muffler assembly
- Catalytic converter support bracket
- Catalytic convert heat shield, if equipped
- Exhaust sensor electrical connectors
- Upper exhaust sensor electrical connector and retainer, if equipped
- Secondary air injection (AIR) hose, if equipped
- Catalytic converter-to-engine nuts
- Vent tube from the air cleaner outlet pipe
- Air intake resonator
- Accelerator control snow shield
- Throttle cable
- Cruise control cable, if equipped
- Fuel supply hose quick release coupling from the fuel rail. Position the fuel hose aside.
- Evaporative emissions hose
- AIR hose, if equipped
- AIR vacuum regulator electrical connector, if equipped
- Power brake booster vacuum hose
- Exhaust Gas Recirculation (EGR) valve electrical connector
- Upper radiator hose
- Heater hoses
- Coolant vent hose

- Fuel charging wiring harness
- Power distribution wiring harness electrical connector
- Wiring harness retainers
- Power distribution wiring harness eyelet
- Three main engine wiring harness electrical connectors
- Gearshift cables from transmission and retaining bracket
- Clutch slave cylinder supply hose
- Reversing lamp switch electrical connector
- Coolant expansion tank
- Power steering pump pulley, using Special Tool 211-016
- Power steering pump
- Cooling fan
- Lower radiator hose
- Alternator cooling pipe
- Starter motor electrical connectors
- Power steering pressure hose
- A/C compressor, if equipped
- Left hand brake hose support bracket
- Left hand brake caliper and support aside

7. Loosen the left hand strut and spring assembly top mounting bolts by four turns.

8. Disconnect the left hand stabilizer bar at the strut.

9. Remove the left and right tie rod end nuts.

10. Disconnect the both tie-rods from the knuckles using special tool 211-001

11. Disconnect the lower control arms from the knuckles.

12. Remove the intermediate shaft and right hand half shaft assembly from the transmission.

13. Remove the left hand halfshaft from the transmission using Special Tools 100-001 and 205-241.

➡**Support both halfshafts to one side with mechanic's wire or suitable equivalent.**

➡**Install a plug into the transmission openings.**

14. Remove or disconnect the following:
- Transmission roll restrictor
- Starter and starter isolator
- Two lower bell housing bolts
- Two oil pan-to-bell housing bolts

15. Using Special Tool 014-0000, fasten the engine to the lift table

16. Remove or disconnect the following:
- Motor mount nuts
- Transmission mount center nut

17. Lower the engine and transmission assembly from the vehicle.

06017-FOCU-G01

Removing the power steering pump pulley using Special Tool 211-016—2.0L (VIN N).

06017-FOCU-G02

Disconnecting the tie-rods using Special Tool 211-0011—2.0L (VIN N)

06017-FOCU-G03

Removing the left hand halfshaft from the transaxle using Removal Tools 100-001 and 205-241.

18. Disconnect the vehicle speed sensor (VSS) connector

19. Lower the engine on the lifting table to within a few inches of the floor. Using the engine crane, remove the engine from the lift table.

20. Remove the remaining bell housing bolts and separate the engine and transmission

To install:

21. Install the transmission onto the engine and tighten the bell housing bolts to 35 ft. lbs. (48 Nm).

22. Using engine crane, raise the engine and transmission assembly onto the lifting table.

23. Secure the engine assembly to the lift table using Special Tool 014-00001.

24. Connect the VSS connector.

25. Using the lift table, position the engine and transmission assembly into the vehicle.

26. Install or connect the following:
   - Transmission mounting center nut and tighten to 98 ft. lbs. (133 Nm)
   - Motor mount nuts and tighten to 59 ft. lbs. (80 Nm)
   - Remaining bell housing bolts and tighten to 35 ft. lbs. (48 Nm)
   - Starter motor isolator and starter motor. Tighten the bolts to 18 ft. lbs. (25 Nm).
   - Transmission roll restrictor and tighten the bolts to 35 ft. lbs. (48 Nm)
   - Left hand halfshaft
   - Right hand halfshaft and intermediate shaft. Tighten the mounting bracket to 18 ft. lbs. (25 Nm).
   - Lower control arms
   - Tie rod ends
   - Stabilizer bar at the struts

27. Tighten the left hand strut and spring assembly top mounting nuts to 18 ft. lbs. (25 Nm).

28. Install or connect the following:
   - Brake caliper
   - Brake hose support bracket
   - A/C compressor, if equipped. Tighten the bolts to 18 ft. lbs. (25 Nm).
   - PSP tube and tighten the nuts to 10 ft. lbs. (13 Nm)
   - Starter motor electrical connections
   - Alternator coolant pipe
   - Engine ground cable and tighten to 35 ft. lbs. (48 Nm)
   - Lower radiator hose
   - Cooling fan
   - Power steering pump and tighten to 13 ft. lbs. (18 Nm)

Fastening the engine to the lift table to secure the assembly—2.0L (VIN N)

06017-FOCU-G04

303-D089

303-050

06017-FOCU-G05

Using the engine lifting crane—2.0L (VIN N)

➡**Use new O-rings when connecting the PSP hose.**

- Power steering pump pulley using Special Tool 211-185
- Coolant expansion tank
- Reversing lamp electrical connector
- Clutch slave cylinder supply hose
- Transmission cables
- Three main wiring harness connectors
- Power distribution wiring harness eyelet
- Wiring harness retainers

- Power distribution wiring harness electrical connector
- Fuel charging wiring harness
- Upper radiator hose
- Heater hoses
- Coolant vent hose
- EGR electrical connector
- Power brake booster vacuum hose
- Secondary AIR electrical connector, if equipped
- AIR hose, if equipped
- Evaporative emission hose
- Fuel supply hose
- Throttle cable

- Cruise control cable, if equipped
- Accelerator control snow shield and tighten to 89 inch lbs. (10 Nm)
- Air intake resonator
- Air cleaner outlet pipe
- Catalytic converter, using new gasket and nuts. Tighten to 41 ft. lbs. (55 Nm).
- Upper exhaust sensor electrical connector, if equipped
- Exhaust sensor electrical connectors
- Catalytic converter heat shield, if equipped
- Catalytic converter support bracket
- Catalytic converter to muffler assembly using a new gasket and nuts
- Accessory drive belt
- Accessory drive belt splash shield
- Battery tray

29. Fill the engine with oil.
30. Fill and bleed the power steering system.
31. Fill the engine with coolant to the correct level.
32. Charge the A/C system.
33. Start the engine and check for leaks.

2.3L ENGINE

1. Before servicing the vehicle, refer to the Precautions Section.
2. With the vehicle in NEUTRAL, position it on a hoist.
3. Release the fuel system pressure.
4. Remove the battery tray.
5. Recover the A/C system.
6. Drain the cooling system.
7. Drain the engine oil. Install the drain plug.

8. Remove the accessory drive belt.
9. Remove the catalytic converter.
10. Remove the air intake resonator.
11. Disconnect the fuel tube.
12. Disconnect the vapor tube.
13. Disconnect the engine emissions hose and the exhaust gas recirculation (EGR) vacuum regulator solenoid electrical connector.
14. Disconnect the brake booster vacuum hose.
15. Disconnect the lower radiator hose, the heater hose and the overflow hose from the coolant bypass.
16. Disconnect the hose, remove the screw and the pin-type retainer.
17. Remove the snow shield.
18. Position aside the accelerator cable and speed control cable (if equipped).
19. Disconnect the accelerator and speed control cable (if equipped) from the throttle body.
20. Remove the three bolts and position aside the accelerator cable bracket aside.
21. Disconnect the fuel injection supply manifold wiring harness electrical connector.
22. Disconnect the EGR valve electrical connector.
23. Disconnect the gearshift cables from the transaxle.
24. Disconnect the shifter cable from the shift mast.
25. Disconnect the selector cable from the selector lever.
26. Disconnect the gearshift cables from the bracket.
27. Disconnect the shifter cable from the bracket, turning the abutment sleeves counterclockwise.
28. Disconnect the selector cable from the bracket, turning the abutment sleeves counterclockwise.
29. Remove the clutch slave cylinder supply line.
30. Remove the supply line and secure it to one side using cable tires.
31. Disconnect the reversing lamp switch electrical connector.
32. Disconnect the heater hose from the "T" fitting and position it aside.
33. Remove the coolant expansion tank.
34. Disconnect the power steering pressure (PSP) switch electrical connector.
35. Remove the power steering pump pulley.
36. Disconnect the PSP tube.
37. Disconnect the three main engine wiring harness electrical connectors.
38. Remove the wiring harness retainers from the valve cover.
39. Disconnect the cooling fan electrical connector.

40. Disconnect the upper radiator hose.
41. Remove the battery cable bracket nut.
42. Remove the generator heat shield top nut.
43. Remove the generator B+ cable nut and disconnect the B+ cable.
44. Remove the bolt and the ground cable.
45. Remove the generator cooling tube.
46. Remove the cooling fan.
47. Remove the two lower bellhousing bolts.
48. Remove the two oil pan-to-bellhousing bolts.
49. Disconnect the starter electrical connections.
50. Remove the nuts and disconnect the power steering pressure line brackets from the stud bolts.
51. Remove the three A/C compressor bolts.
52. Loosen the bolt and remove the A/C compressor.
53. Lower the A/C compressor to access the manifold bolt. Discard the O-ring seals.
54. Disconnect the field coil electrical connector.
55. Disconnect the LH brake hose from the support bracket.
56. Disconnect the LH caliper.

✳✳ CAUTION

Suspend the caliper to prevent load from being placed on the brake hose.

57. Support the brake caliper.
58. Loosen the LH strut and spring assembly top mount nuts by four turns.
59. Disconnect the LH stabilizer bar at the strut.
60. Remove the LH tie-rod end nut.
61. Remove the RH tie-rod end nut.
62. Disconnect both of the tie-rods from the knuckles.
63. Disconnect both of the lower control arms from the knuckles.
64. Remove the mounting bracket nuts, then remove the bracket from the halfshaft intermediate bearing.
65. Remove the RH front drive halfshaft. Position the halfshaft aside and support with mechanic's wire.
66. Install a plug into the transaxle opening.
67. Remove the LH front drive halfshaft from the transaxle.
68. Install a plug into the transaxle opening.
69. Support the halfshaft using mechanic's wire.

70. Remove the lower engine mount.
71. Remove the three bolts and the starter.
72. Remove the starter isolator.
73. Disconnect the vehicle speed sensor (VSS) electrical connector.
74. Fasten the engine to the lift table.
75. Remove the RH motor mount nuts.
76. Remove the LH motor mount nut.
77. Lower the engine and transaxle assembly from the vehicle.
78. Separate the transaxle from the engine.

To install:
79. Install the transaxle. Make sure the transaxle is flush to the engine.

✳✳ CAUTION

Failure to follow these steps will result in gear rattle at idle.

80. Install the two bolts at the dowels first. Tighten to 35 ft. lbs. (48 Nm).
81. Install the remaining transaxle RH flange bolts. Tighten to 35 ft. lbs. (48 Nm).
82. Connect the vehicle speed sensor (VSS) electrical connector.
83. Using the lift table, position the engine and transaxle assembly in the vehicle.
84. Install the RH motor mount nuts and tighten them to 59 ft. lbs. (80 Nm).
85. Install the LH motor mount nut and tighten it to 98 ft. lbs. (133 Nm).
86. Remove the lift table.
87. Install the lower motor mount. Tighten the bolts to 35 ft. lbs. (48 Nm).
88. Install the two lower engine-to-bellhousing bolts and tighten them to 35 ft. lbs. (48 Nm).
89. Install the remaining engine-to-bellhousing bolts and tighten them to 35 ft. lbs. (48 Nm).
90. Install the starter motor isolator.
91. Install the starter motor and tighten the bolts to 18 ft. lbs. (25 Nm).
92. Install the generator heat shield and the lower nut.

➡ **Install new O-ring seals.**

93. Position the A/C compressor to the manifold and tube assembly and tighten the bolt to 15 ft. lbs. (21 Nm).
94. Install the A/C compressor and tighten the three bolts to 18 ft. lbs. (21 Nm).
95. Connect the field coil electrical connector.
96. Attach the power steering pressure tube brackets to the stud bolts and install the nuts.
97. Connect the starter motor electrical connections.

98. Install the cooling fan.

99. Position the battery cable and install the cable bracket nut.

100. Install the generator cooling tube.

101. Install the generator heat shield top nut.

102. Install the generator B+ cable and install the cable nut.

103. Install the upper radiator hose.

104. Position the ground cable and install the bolt.

105. Connect the cooling fan electrical connector.

106. Connect the cooling fan electrical connector.

107. Connect the wiring harness retainers to the valve cover studs.

108. Position the power steering pump (PSP) and install the bolts.

109. Connect the PSP tube.

➡**Make sure the pulley is flush with the end of the power steering pump shaft.**

110. Install the power steering pump pulley.

111. Connect the PSP switch electrical connector.

112. Connect the three main engine wiring harness connectors.

113. Install the coolant expansion tank.

114. Connect the heater hose to the "T" fitting.

115. Connect the exhaust gas recirculation (EGR) valve electrical connector.

116. Connect the fuel injection supply manifold wiring harness electrical connector.

117. Position the accelerator cable bracket.

118. Install the speed control cable (if equipped) and the throttle cable.

119. Attach the hose and install the screw and the pin-type retainer.

120. Install the snow shield.

121. Install the catalytic converter.

122. Connect the lower radiator hose, the heater hose and the overflow hose to the coolant bypass.

123. Connect the engine emissions hose and EGR vacuum regulator solenoid electrical connector.

124. Install the vapor tube.

125. Connect the fuel tube.

126. Install the brake booster vacuum hose.

127. Fill the engine with clean engine oil.

➡**Install a new snap ring.**

128. Install the LH front drive halfshaft.

➡**Install new nuts and a new center bearing cap.**

129. Install the RH front halfshaft together with the intermediate shaft.

130. Install the mounting bracket for the front drive halfshaft intermediate bearing.

131. Install both of the lower control arms to the knuckles. Tighten the bolts to 37 ft. lbs. (50 Nm).

132. Install both of the tie-rod ends to the knuckles. Tighten the nuts to 35 ft. lbs. (47 Nm).

133. Connect the stabilizer bar at the strut.

134. Tighten the LH strut and spring assembly top mount nuts to 18 ft. lbs. (25 Nm).

135. Install the brake caliper. Tighten the bolts to 21 ft. lbs. (28 Nm).

136. Install the brake hose onto the support bracket.

137. Install the accessory drive belt.

138. Connect the reversing lamp switch electrical connector.

139. Connect the clutch slave cylinder supply line. Insert the clip.

➡**The shift cable abutment sleeve is colored white.**

➡**The selector cable abutment sleeve is colored black.**

140. Attach the gearshift cables to the bracket.

141. Attach the shifter cable to the bracket, turning the abutment sleeves counterclockwise to open. Position the cables into the metal holders.

142. Attach the selector cable to the bracket, turning the abutment sleeves counterclockwise to open. Position the cables into the metal holders.

143. Attach the shifter cable to the shift mast.

144. Attach the selector cable to the selector lever.

145. Check the shift cables at the metal holders. Pull outward on the cables. Cables should remain in the metal.

146. Adjust the shift cables.

147. Fill and bleed the clutch hydraulic system

148. Evacuate and recharge the A/C system.

149. Install the air intake resonator.

150. Install the battery tray.

151. Connect the battery ground cable.

152. Fill and bleed the cooling system.

153. Fill and bleed the power steering system.

Automatic Transaxle

2.0L SOHC ENGINE

1. Before servicing the vehicle, refer to the Precautions Section.

2. Drain the cooling system.

3. Relieve the fuel system pressure.

4. Loosen the front strut center nuts 5 turns.

5. Remove or disconnect the following:
- Battery, tray and cables
- Mass Air Flow (MAF) sensor connector
- Air cleaner housing and tubes
- Accelerator cable
- Cruise control cable, if equipped
- Ignition coil connectors
- Power steering pump pressure switch connector
- Heated Oxygen (HO_2S) sensor connectors
- Fuel injection wiring harness connector
- Powertrain Control Module (PCM) connectors
- Radiator cooling fan
- Evaporative Emissions (EVAP) canister vacuum lines
- Power brake booster vacuum line
- Delta Pressure Feedback Electronic (DPFE) system sensor vacuum line
- Exhaust Gas Recirculation (EGR) vacuum line
- Intake manifold vacuum line
- Fuel line
- Coolant bypass hoses
- Radiator hoses
- Coolant expansion tank
- Accessory drive belt
- Power steering pump and reservoir
- Power steering high pressure pipe
- Alternator and bracket
- Drive belt cover
- A/C compressor, if equipped
- Lower ball joints
- Exhaust front pipe
- Torque converter cover
- Torque converter
- Transaxle fluid cooler lines
- Axle halfshafts
- Intermediate shaft bracket
- Starter motor
- Crankshaft pulley
- Right engine mount
- Transaxle flange bolts. Support the transaxle
- Engine front mount
- Crankshaft Position (CKP) sensor
- Engine

To install:

6. Install or connect the following:
- Engine
- CKP sensor
- Engine front mount. Tighten the nuts to 59 ft. lbs. (80 Nm) and the bolts to 35 ft. lbs. (48 Nm).
- Transaxle flange bolts. Tighten the large bolts to 35 ft. lbs. (48 Nm) and the small bolts to 16 ft. lbs. (22 Nm).
- Right engine mount. Tighten the bolts to 35 ft. lbs. (48 Nm).
- Crankshaft pulley. Tighten the bolt to 88 ft. lbs. (120 Nm).
- Starter motor
- Intermediate shaft bracket. Tighten the bolts to 35 ft. lbs. (48 Nm).
- Axle halfshafts
- Transaxle fluid cooler lines
- Torque converter. Tighten the nuts to 27 ft. lbs. (37 Nm).
- Torque converter cover
- Exhaust front pipe
- Lower ball joints. Tighten the pinch bolts to 37 ft. lbs. (50 Nm).
- A/C compressor, if equipped
- Drive belt cover
- Alternator and bracket
- Power steering high pressure pipe
- Power steering pump and reservoir
- Accessory drive belt
- Coolant expansion tank
- Radiator hoses
- Coolant bypass hoses
- Fuel line
- Intake manifold vacuum line
- EGR vacuum line
- DPFE system sensor vacuum line
- Power brake booster vacuum line
- EVAP canister vacuum lines
- Radiator cooling fan
- PCM connectors
- HO$_2$S sensor connectors
- Fuel injection wiring harness connector
- Power steering pump pressure switch connector
- Ignition coil connectors
- Cruise control cable, if equipped
- Accelerator cable
- Air cleaner housing and tubes
- MAF sensor connector
- Battery, tray and cables

7. Tighten the strut center nuts to 35 ft. lbs. (48 Nm).

8. Fill the cooling system.

9. Start the engine and check for leaks.

2.0L DOHC ENGINE, EXCEPT VIN N

1. Before servicing the vehicle, refer to the Precautions Section.

2. Drain the cooling system.

3. Relieve the fuel system pressure.

4. Loosen the front strut center nuts 5 turns.

5. Remove or disconnect the following:
- Battery, tray and cables
- Mass Air Flow (MAF) sensor connector
- Positive Crankcase Ventilation (PCV) valve and hose
- Air cleaner assembly and tubes
- Accelerator cable
- Cruise control cable, if equipped
- Ignition coil connectors
- Heated Oxygen (HO$_2$S) sensor connectors
- Fuel injector harness connectors
- Powertrain Control Module (PCM) connectors
- Radiator cooling fan
- Evaporative Emissions (EVAP) canister vacuum lines
- Power brake booster vacuum line
- Delta Pressure Feedback Electronic (DPFE) system sensor vacuum line
- Exhaust Gas Recirculation (EGR) vacuum line
- Intake manifold vacuum line
- Fuel line
- Coolant bypass hoses
- Radiator hoses
- Coolant expansion tank
- Shift select cable and bracket
- Transaxle fluid cooler lines
- Drive belt cover
- Power steering pressure switch connector
- Accessory drive belt
- A/C compressor, if equipped
- Power steering pump
- Power steering pressure pipe
- Exhaust flex pipe
- Right engine mount
- Lower ball joints
- Axle halfshafts and intermediate shaft

6. Support the powertrain from below

7. Remove the front and rear engine mounts.

8. Raise the vehicle away from the powertrain.

To install:

9. Lower the vehicle over the powertrain.

10. Install or connect the following:
- Front engine mount. Tighten the nuts to 59 ft. lbs. (80 Nm) and the bolts to 35 ft. lbs. (48 Nm).
- Rear engine mount. Tighten the outer nuts to 35 ft. lbs. (48 Nm) and the center nut to 98 ft. lbs. (133 Nm).
- Axle halfshafts and intermediate shaft
- Lower ball joints. Tighten the pinch bolts to 37 ft. lbs. (50 Nm).
- Right engine mount. Tighten the bolts to 35 ft. lbs. (48 Nm).
- Exhaust flex pipe
- Power steering pressure pipe
- Power steering pump
- A/C compressor, if equipped
- Accessory drive belt
- Power steering pressure switch connector
- Drive belt cover
- Transaxle fluid cooler lines
- Shift select cable and bracket
- Coolant expansion tank
- Radiator hoses
- Coolant bypass hoses
- Fuel line
- Intake manifold vacuum line
- EGR vacuum line
- DPFE system sensor vacuum line
- Power brake booster vacuum line
- EVAP canister vacuum lines
- Radiator cooling fan
- PCM connectors
- Fuel injector harness connectors
- HO$_2$S sensor connectors
- Ignition coil connectors
- Cruise control cable, if equipped
- Accelerator cable
- Air cleaner assembly and tubes
- PCV valve and hose
- MAF sensor connector
- Battery, tray and cables

11. Tighten the strut center nuts to 35 ft. lbs. (48 Nm).

12. Fill the cooling system.

13. Start the engine and check for leaks.

2.0L (VIN N) ENGINE

1. Before servicing the vehicle, refer to the Precautions Section.

2. Drain the engine oil. Reinstall the drain plug.

3. Drain the cooling system.

4. Relieve the fuel system pressure.

5. Recover the A/C refrigerant, if equipped.

6. Remove or disconnect the following:
- Battery and battery tray
- Accessory drive belt splash guard
- Accessory drive belt
- Catalytic converter from the muffler assembly
- Catalytic converter support bracket
- Catalytic convert heat shield, if equipped
- Exhaust sensor electrical connectors

- Upper exhaust sensor electrical connector and retainer, if equipped
- Secondary air injection (AIR) hose, if equipped
- Catalytic converter-to-engine nuts
- Vent tube from the air cleaner outlet pipe
- Air intake resonator
- Accelerator control snow shield
- Throttle cable
- Cruise control cable, if equipped
- Fuel hose quick release coupling from the fuel rail. Position the fuel hose aside.
- Evaporative emissions hose
- AIR hose, if equipped
- AIR vacuum regulator electrical connector, if equipped
- Power brake booster vacuum hose
- Exhaust Gas Recirculation (EGR) valve electrical connector
- Upper radiator hose
- Heater hoses
- Coolant vent hose
- Fuel charging wiring harness
- Power distribution wiring harness electrical connector
- Wiring harness retainers
- Power distribution wiring harness eyelet
- Three main engine wiring harness electrical connectors
- Transmission shift cable
- Transmission cooler hoses
- Coolant expansion tank
- Power steering pump pulley, using Special Tool 211-016
- Power steering pump
- Cooling fan
- Lower radiator hose
- Alternator cooling pipe
- Starter motor electrical connectors
- Power steering pressure hose
- A/C compressor, if equipped
- Left hand brake hose support bracket
- Left hand brake caliper and support aside

7. Loosen the left hand strut and spring assembly top mounting bolts by four turns.

8. Disconnect the left hand stabilizer bar at the strut.

9. Remove the left and right tie rod end nuts.

10. Disconnect the both tie-rods from the knuckles using special tool 211-001

11. Disconnect the lower control arms from the knuckles.

12. Remove the intermediate shaft and right hand half shaft assembly from the transmission.

13. Remove the left hand halfshaft from the transmission using Special Tools 100-001 and 205-241.

➥ **Support both halfshafts to one side with mechanic's wire or suitable equivalent.**

➥ **Install a plug into the transmission openings.**

14. Remove or disconnect the following:
- Transmission roll restrictor
- Starter and starter isolator
- Two lower bell housing bolts
- Two oil pan-to-bell housing bolts

15. Using Special Tool 014-0000, fasten the engine to the lift table

16. Remove or disconnect the following:
- Motor mount nuts
- Transmission mount center nut

17. Lower the engine and transmission assembly from the vehicle.

18. Remove the four torque converter nuts

19. Disconnect the output shaft speed (OSS) connector

20. Disconnect Turbine shaft speed (TSS) connector

21. Disconnect the solenoid body and transmission range (TR) sensor connectors.

22. Lower the engine on the lifting table to within a few inches of the floor. Using the engine crane, remove the engine from the lift table.

23. Remove the remaining bell housing bolts and separate the engine and transmission

To install:

24. Install the transmission onto the engine and tighten the bell housing bolts to 35 ft. lbs. (48 Nm).

25. Using engine crane, raise the engine and transmission assembly onto the lifting table.

26. Secure the engine assembly to the lift table using Special Tool 014-00001.

27. Connect the solenoid body and transmission range (TR) sensor connectors.

28. Connect Turbine shaft speed (TSS) connector.

29. Connect the output shaft speed (OSS) connector.

30. Install the four torque converter nuts and tighten to 26 ft. lbs. (35 Nm).

31. Using the lift table, position the engine and transmission assembly into the vehicle.

32. Install or connect the following:
- Transmission mounting center nut and tighten to 98 ft. lbs. (133 Nm)
- Motor mount nuts and tighten to 59 ft. lbs. (80 Nm)
- Remaining bell housing bolts and tighten to 35 ft. lbs. (48 Nm)

- Starter motor isolator and starter motor. Tighten the bolts to 18 ft. lbs. (25 Nm).
- Transmission roll restrictor and tighten the bolts to 35 ft. lbs. (48 Nm)
- Left hand halfshaft
- Right hand halfshaft and intermediate shaft. Tighten the mounting bracket to 18 ft. lbs. (25 Nm).
- Lower control arms
- Tie rod ends
- Stabilizer bar at the struts

33. Tighten the left hand strut and spring assembly top mounting nuts to 18 ft. lbs. (25 Nm).

34. Install or connect the following:
- Brake caliper
- Brake hose support bracket
- A/C compressor, if equipped. Tighten the bolts to 18 ft. lbs. (25 Nm).
- PSP tube and tighten the nuts to 10 ft. lbs. (13 Nm)
- Starter motor electrical connections
- Alternator coolant pipe
- Engine ground cable and tighten to 35 ft. lbs. (48 Nm)
- Lower radiator hose
- Cooling fan
- Power steering pump and tighten to 13 ft. lbs. (18 Nm)

➥ **Use new O-rings when connecting the PSP hose.**

- Power steering pump pulley using Special Tool 211-185
- Coolant expansion tank
- Transmission cooler hoses
- Transmission shifter cable
- Three main wiring harness connectors
- Power distribution wiring harness eyelet
- Wiring harness retainers
- Power distribution wiring harness electrical connector
- Fuel charging wiring harness
- Upper radiator hose
- Heater hoses
- Coolant vent hose
- EGR electrical connector
- Power brake booster vacuum hose
- Secondary AIR electrical connector, if equipped
- AIR hose, if equipped
- Evaporative emission hose
- Fuel supply hose
- Throttle cable
- Cruise control cable, if equipped
- Accelerator control snow shield and tighten to 89 inch lbs. (10 Nm)

- Air intake resonator
- Air cleaner outlet pipe
- Catalytic converter, using new gasket and nuts. Tighten to 41 ft. lbs. (55 Nm).
- Upper exhaust sensor electrical connector, if equipped
- Exhaust sensor electrical connectors
- Catalytic converter heat shield, if equipped
- Catalytic converter support bracket
- Catalytic converter to muffler assembly using a new gasket and nuts
- Accessory drive belt
- Accessory drive belt splash shield
- Battery tray

35. Fill the engine with oil.
36. Fill and bleed the power steering system.
37. Fill the engine with coolant to the correct level.
38. Charge the A/C system.
39. Start the engine and check for leaks.

2.3L ENGINE

1. Before servicing the vehicle, refer to the Precautions Section.
2. With the vehicle in NEUTRAL, position it on a hoist.
3. Release the fuel system pressure.
4. Remove the battery tray.
5. Recover the A/C system.
6. Drain the cooling system.
7. Drain the engine oil. Install the drain plug.
8. Remove the accessory drive belt.
9. Remove the catalytic converter.
10. Remove the air intake resonator.
11. Disconnect the fuel tube.
12. Disconnect the vapor tube.
13. Disconnect the engine emissions hose and the exhaust gas recirculation (EGR) vacuum regulator solenoid electrical connector.
14. Disconnect the brake booster vacuum hose.
15. Disconnect the lower radiator hose, the heater hose and the overflow hose from the coolant bypass.
16. Disconnect the hose, remove the screw and the pin-type retainer.
17. Remove the snow shield.
18. Position aside the accelerator cable and speed control cable (if equipped).
19. Disconnect the accelerator and speed control cable (if equipped) from the throttle body.
20. Remove the three bolts and position aside the accelerator cable bracket aside.

21. Disconnect the fuel injection supply manifold wiring harness electrical connector.
22. Disconnect the EGR valve electrical connector.
23. Disconnect the heater hose from the "T" fitting and position it aside.
24. Remove the coolant expansion tank.
25. Disconnect the power steering pressure (PSP) switch electrical connector.
26. Remove the power steering pump pulley.
27. Disconnect the PSP tube.
28. Disconnect the three main engine wiring harness electrical connectors.
29. Remove the wiring harness retainers from the valve cover.
30. Disconnect the cooling fan electrical connector.
31. Disconnect the upper radiator hose.
32. Remove the battery cable bracket nut.
33. Remove the generator heat shield top nut.
34. Remove the generator B+ cable nut and disconnect the B+ cable.
35. Remove the bolt and the ground cable.
36. Remove the generator cooling tube.
37. Disconnect the transmission shifter cable and position it aside.
38. Disconnect the transmission cooler hoses.
39. Remove the cooling fan.
40. Remove the two lower bellhousing bolts.
41. Remove the two oil pan-to-bellhousing bolts.
42. Disconnect the starter electrical connections.
43. Remove the nuts and disconnect the power steering pressure line brackets from the stud bolts.
44. Remove the three A/C compressor bolts.
45. Loosen the bolt and remove the A/C compressor.
46. Lower the A/C compressor to access the manifold bolt. Discard the O-ring seals.
47. Disconnect the field coil electrical connector.
48. Disconnect the LH brake hose from the support bracket.
49. Disconnect the LH caliper.

✳✳ CAUTION

Suspend the caliper to prevent load from being placed on the brake hose.

50. Support the brake caliper.
51. Loosen the LH strut and spring assembly top mount nuts by four turns.

52. Disconnect the LH stabilizer bar at the strut.
53. Remove the LH tie-rod end nut.
54. Remove the RH tie-rod end nut.
55. Disconnect both of the tie-rods from the knuckles.
56. Disconnect both of the lower control arms from the knuckles.
57. Remove the mounting bracket nuts, then remove the bracket from the halfshaft intermediate bearing.
58. Remove the RH front drive halfshaft. Position the halfshaft aside and support with mechanic's wire.
59. Install a plug into the transaxle opening.
60. Remove the LH front drive halfshaft from the transaxle.
61. Install a plug into the transaxle opening.
62. Support the halfshaft using mechanic's wire.
63. Remove the lower engine mount.
64. Remove the three bolts and the starter.
65. Remove the starter isolator.

➡ **Mark one stud and the flexplate for assembly reference.**

66. Remove the four torque converter nuts.
67. Fasten the engine to the lift table.
68. Remove the RH motor mount nuts.
69. Remove the LH motor mount nut.
70. Lower the engine and transaxle assembly from the vehicle.
71. Disconnect the output shaft speed (OSS) sensor electrical connector.
72. Disconnect the turbine shaft speed (TSS) sensor electrical connector.
73. Disconnect the solenoid body and the transmission range (TR) sensor electrical connectors.
74. Separate the transaxle from the engine.

To install:

➡ **Lubricate the torque converter pilot hub with multi-purpose grease.**

75. Check the installation depth of the torque converter. Lay a steel straightedge on the automatic transaxle flange. Check the installation depth between the transaxle flange and the torque converter centering spigot for the correct clearance. The dimension should be 0.60–0.71 inches (15–18 mm).
76. Position the engine and transaxle together and install five of the six top bolts.

➡ **Make sure to note the location of the different length bolts.**

Use a straight edge (1) to check the torque converter installed depth (2)—2.3L engine

77. Install the LH flange bolts and tighten to 35 ft. lbs. (47 Nm).

78. Connect the output shaft speed (OSS) sensor electrical connector.

79. Connect the solenoid body and the transmission range (TR) sensor electrical connectors.

80. Connect the turbine shaft speed (TSS) sensor electrical connector.

81. Using the lift table, position the engine and transaxle assembly in the vehicle.

82. Install the RH motor mount nuts and tighten them to 59 ft. lbs. (80 Nm).

83. Install the LH motor mount nut and tighten it to 98 ft. lbs. (133 Nm).

84. Remove the lift table.

85. Install the lower motor mount. Tighten the bolts to 35 ft. lbs. (48 Nm).

86. Install the two lower engine-to-bell-housing bolts and tighten them to 35 ft. lbs. (48 Nm).

87. Install the remaining engine-to-bell-housing bolts and tighten them to 35 ft. lbs. (48 Nm).

➡ **If new parts are not being used, be sure to align the marks on the flexplate and the stud made during engine removal.**

88. Install the four torque converter nuts and tighten them to 27 ft. lbs. (37 Nm).

89. Install the starter motor isolator.

90. Install the starter motor and tighten the bolts to 18 ft. lbs. (25 Nm).

91. Install the generator heat shield and the lower nut.

➡ **Install new O-ring seals.**

92. Position the A/C compressor to the manifold and tube assembly and tighten the bolt to 15 ft. lbs. (21 Nm).

93. Install the A/C compressor and tighten the three bolts to 18 ft. lbs. (21 Nm).

94. Connect the field coil electrical connector.

95. Attach the power steering pressure tube brackets to the stud bolts and install the nuts.

96. Connect the starter motor electrical connections.

97. Install the cooling fan.

98. Connect the transmission cooler hoses.

99. Connect the transmission shifter cable.

100. Position the battery cable and install the cable bracket nut.

101. Install the generator cooling tube.

102. Install the generator heat shield top nut.

103. Install the generator B+ cable and install the cable nut.

104. Install the upper radiator hose.

105. Position the ground cable and install the bolt.

106. Connect the cooling fan electrical connector.

107. Connect the cooling fan electrical connector.

108. Connect the wiring harness retainers to the valve cover studs.

109. Position the power steering pump (PSP) and install the bolts.

110. Connect the PSP tube.

➡ **Make sure the pulley is flush with the end of the power steering pump shaft.**

111. Install the power steering pump pulley.

112. Connect the PSP switch electrical connector.

113. Connect the three main engine wiring harness connectors.

114. Install the coolant expansion tank.

115. Connect the heater hose to the "T" fitting.

116. Connect the exhaust gas recirculation (EGR) valve electrical connector.

117. Connect the fuel injection supply manifold wiring harness electrical connector.

118. Position the accelerator cable bracket.

119. Install the speed control cable (if equipped) and the throttle cable.

120. Attach the hose and install the screw and the pin-type retainer.

121. Install the snow shield.

122. Install the catalytic converter.

123. Connect the lower radiator hose, the heater hose and the overflow hose to the coolant bypass.

124. Connect the engine emissions hose and EGR vacuum regulator solenoid electrical connector.

125. Install the vapor tube.

126. Connect the fuel tube.

127. Install the brake booster vacuum hose.

128. Fill the engine with clean engine oil.

➡ **Install a new snap ring.**

129. Install the LH front drive halfshaft.

➡ **Install new nuts and a new center bearing cap.**

130. Install the RH front halfshaft together with the intermediate shaft.

131. Install the mounting bracket for the front drive halfshaft intermediate bearing.

132. Install both of the lower control arms to the knuckles. Tighten the bolts to 37 ft. lbs. (50 Nm).

133. Install both of the tie-rod ends to the knuckles. Tighten the nuts to 35 ft. lbs. (47 Nm).

134. Connect the stabilizer bar at the strut.

135. Tighten the LH strut and spring assembly top mount nuts to 18 ft. lbs. (25 Nm).

136. Install the brake caliper. Tighten the bolts to 21 ft. lbs. (28 Nm).

137. Install the brake hose onto the support bracket.

138. Install the accessory drive belt.

139. Evacuate and recharge the A/C system.

140. Install the air intake resonator.

141. Install the battery tray.

142. Connect the battery ground cable.

143. Fill and bleed the cooling system.

144. Fill and bleed the power steering system.

Water Pump

REMOVAL & INSTALLATION

2.0L (VIN P) Engine

1. Before servicing the vehicle, refer to the Precautions Section.

2. Drain the cooling system.

3. Remove or disconnect the following:
   - Negative battery cable
   - Accessory drive belt
   - Front cover
   - Timing belt
   - Timing belt tensioner
   - Water pump

To install:

4. Install or connect the following:
   - Water pump. Tighten the bolts to 18 ft. lbs. (25 Nm).
   - Timing belt tensioner
   - Timing belt
   - Accessory drive belt
   - Negative battery cable
5. Fill the cooling system.
6. Start the engine and check for leaks.

2.0L (VIN 3) Engines

1. Before servicing the vehicle, refer to the Precautions Section.
2. Drain the cooling system.
3. Remove or disconnect the following:
   - Negative battery cable
   - Accessory drive belt
   - Water pump pulley
   - Front cover
   - Timing belt
   - Timing belt tensioner
   - Water pump

To install:

4. Install or connect the following:
   - Water pump. Tighten the bolts to 13 ft. lbs. (18 Nm).
   - Timing belt tensioner
   - Timing belt
   - Front cover
   - Water pump pulley
   - Accessory drive belt
   - Negative battery cable
5. Fill the cooling system.
6. Start the engine and check for leaks.

2.0L (VIN N) Engine and 2.3L Engine

1. Before servicing the vehicle, refer to the Precautions Section.
2. Drain the cooling system.
3. Remove or disconnect the following:
   - Negative battery cable
   - Accessory drive belt
   - Water pump pulley
   - Water pump mounting bolts
   - Water pump

To install:

4. Install or connect the following:
   - Water pump with new O-ring. Tighten the bolts to 89 inch lbs. (10 Nm).
   - Water pump pulley. Tighten the bolts to 18 ft. lbs. (25 Nm).
   - Accessory drive belt
   - Negative battery cable
5. Fill the cooling system to the correct level.
6. Start the engine and check for leaks.

Heater Core

REMOVAL & INSTALLATION

All Models, Except 2005

1. Before servicing the vehicle, refer to the Precautions Section.
2. Drain the cooling system.
3. Recover the A/C refrigerant, if equipped.
4. Remove or disconnect the following:
   - Negative battery cable
   - Radio
   - Left and right lower panels
   - Data link electrical connector
   - Steering column upper shroud
   - Audio unit control switch, if equipped
   - Steering column lower shroud
   - Steering column
   - Instrument cluster
   - Glove compartment
   - Passenger air bag module
   - Climate control and audio trim panel
   - Ash tray
   - Heater core housing assembly panel
   - Center console
   - Instrument panel upper retaining screws
   - Instrument panel side trim panels
   - Instrument panel side bolts
   - Instrument panel bolts behind the instrument cluster, climate control and audio panel and passenger air bag module
   - Defroster tubes
   - Instrument panel
   - Heater hoses
   - A/C evaporator refrigerant lines
   - Cross vehicle beam
   - Rear footwell ventilation hoses
   - Heater housing wiring harness
   - Heater housing. Refer to the illustrations for housing fastener locations.

To install:

5. Install or connect the following:
   - Heater housing. Refer to the illustrations for housing fastener locations and torque specifications.
   - Heater housing wiring harness
   - Rear footwell ventilation hoses
   - Cross vehicle beam. Tighten the beam fasteners to 15 ft. lbs. (20 Nm) and the beam bracket bolts to 12 ft. lbs. (16 Nm).
   - A/C evaporator refrigerant lines

Heater housing fastener location

Heater housing fastener location

Heater housing fastener location

Heater housing fastener location

Location of the heater core retaining screw—2005 models

- Heater hoses
- Instrument panel
- Defroster tubes
- Instrument panel bolts behind the instrument cluster, climate control and audio panel and passenger air bag module
- Instrument panel side bolts
- Instrument panel side trim panels
- Instrument panel upper retaining screws
- Center console
- Heater core housing assembly panel
- Ash tray
- Climate control and audio trim panel
- Passenger air bag module
- Glove compartment
- Instrument cluster
- Steering column. Tighten the nuts to 13 ft. lbs. (18 Nm) and the Torx® screw to 17 ft. lbs. (23 Nm)
- Steering column lower shroud
- Audio unit control switch, if equipped
- Steering column upper shroud
- Data link electrical connector
- Left and right lower panels
- Radio
- Negative battery cable

6. Fill the cooling system.

7. Recharge the A/C system, if equipped.

8. Start the engine and check for leaks.

2005 Models

1. Before servicing the vehicle, refer to the Precautions Section.

2. Remove or disconnect the following:
- Coolant hoses from heater core
- Floor console
- Rear armrest bracket nuts, if equipped
- Rear footwell duct
- Right hand lower instrument panel insulator
- Floor duct and trim panel
- Accelerator pedal mounting nuts and position the pedal aside
- Left and right hand instrument panel brackets
- Left hand floor duct
- Right hand floor duct
- Heater core cover
- Heater core retaining screw
- Heater core

To install:

3. Install or connect the following:
- Heater core and retaining screw
- Heater core cover
- Right hand floor duct
- Left hand floor duct
- Left and right hand instrument panel brackets and tighten to 12 ft. lbs. (16 Nm)

Remove the heater core—2005 models

- Accelerator pedal mounting nuts
- Floor duct and trim panel
- Right hand lower instrument panel insulator
- Rear footwell duct
- Rear armrest bracket nuts, if equipped
- Floor console
- Coolant hoses to the heater core
- Negative battery cable

4. Fill the cooling system to the correct level.

5. Start the engine and check for leaks.

Cylinder Head

REMOVAL & INSTALLATION

2.0L (VIN P) Engine

1. Before servicing the vehicle, refer to the Precautions Section.

2. Disconnect the negative battery cable.

3. Drain the cooling system.

4. Relieve the fuel system pressure.

5. Remove or disconnect the following:
   - Timing belt
   - Engine air cleaner intake tube

6. Tag and disconnect the vacuum hoses from the following components:
   - Exhaust Gas Recirculation (EGR) valve
   - Positive Crankcase Ventilation (PCV) valve
   - Throttle body
   - Fuel pressure regulator
   - Intake manifold

7. Remove or disconnect the following:
   - Accelerator cable, throttle control lever and if equipped, the speed control cable
   - Speed control cable bracket bolt, if equipped
   - 2 fuel charging wiring electrical connections
   - Crankshaft Position (CKP) sensor and Heated Oxygen (HO2S) sensor electrical connections
   - Fuel supply line
   - Power steering pressure hose bracket bolts, then position the bracket and hose aside
   - Alternator lower bolt, loosen only
   - Alternator upper bolt and pivot the alternator forward
   - Engine oil dipstick tube bracket bolt
   - Exhaust Gas Recirculation (EGR) tube
   - Exhaust manifold heat shield

Cylinder head loosening sequence—2.0L (VIN P) engine

- Catalytic converter
- Right-hand splash shield bolts and the shield
- Engine oil dipstick tube from the cylinder block
- Air conditioning compressor
- Engine accessory drive bracket
- Valve cover
- Upper radiator hose
- Heater hose
- Cylinder head bolts. Loosen the bolts in sequence.
- Cylinder head and gasket

To install:

➡**Use new cylinder head bolts for assembly.**

8. Install the cylinder head and tighten the bolts in sequence as follows:
   a. Step 1: 37 ft. lbs. (50 Nm)
   b. Step 2: Loosen all bolts ½ turn
   c. Step 3: 37 ft. lbs. (50 Nm)
   d. Step 4: Plus 90 degrees
   e. Step 6: Plus 90 degrees

9. Install or connect the following:
   - Accessory drive bracket. Tighten the fasteners to 30–40 ft. lbs. (40–55 Nm)
   - Air conditioning line bracket and tighten the bolt to 15–18 ft. lbs. (20–25 Nm)

- Valve cover
- Air conditioning compressor
- Right-hand splash shield
- Engine oil dipstick tube
- Heater and upper radiator hoses
- Timing belt and alternator
- Power steering pressure hose
- CKP sensor and 2 main fuel charging electrical connections
- Engine oil dipstick tube bolt
- EGR valve tube
- Fuel supply line
- Catalytic converter
- Heat shield
- Speed control cable bracket, if equipped
- Throttle control lever, accelerator cable and if equipped, the speed control cable

10. Connect the vacuum hoses to the following components:
   - EGR valve
   - PCV valve
   - Throttle body
   - Fuel pressure regulator
   - Intake manifold
   - Air cleaner outlet tube
   - Negative battery cable

11. Fill the cooling system.

12. Start the engine and check for leaks.

Cylinder head torque sequence—2.0L (VIN P) engine

1. EGR valve
2. EGR valve gasket
3. Accelerator cable bracket
4. Throttle body gasket
5. Throttle Position (TP) sensor
6. Idle Air Control (IAC) valve
7. IAC valve gasket
8. Throttle body
9. Fuel injection supply manifold
10. Fuel injector
11. IMRC actuator
12. Oil filler cap
13. Valve cover
14. Cylinder head
15. EGR valve-to-exhaust manifold tube
16. Exhaust manifold gasket
17. Exhaust manifold
18. Exhaust manifold shield
19. Intake manifold gasket
20. Intake Manifold Runner Control (IMRC)
21. Intake manifold upper gasket
22. Intake manifold
23. Intake manifold tube

9300MG13

Exploded view of the cylinder head, exhaust manifold and intake manifold mounting—2.0L (VIN P) engine

2.0L (VIN 3) Engine

1. Before servicing the vehicle, refer to the Precautions Section.
2. Disconnect the negative battery cable.
3. Drain the cooling system.
4. Relieve the fuel system pressure.
5. Remove or disconnect the following:
   - Air cleaner assembly outlet tube
   - Timing belt
6. Tag and disconnect the vacuum hoses from the following components:
   - Positive Crankcase Ventilation (PCV) valve
   - Throttle body
   - Fuel pressure sensor
   - Intake manifold
7. Remove or disconnect the following:
   - Speed control and accelerator cables from the control lever
   - Fuel charging electrical connectors at the main engine connector
   - Crankshaft Position (CKP) sensor and Heated Oxygen (HO$_2$S) sensor electrical connections
   - Fuel line
   - Power steering pump and bracket
   - Alternator
   - Engine oil dipstick tube
   - Splash shield
   - Air conditioning compressor
   - Spark plug wires
   - Valve cover
   - Upper radiator hose
   - Heater hose
   - Camshafts
   - Ignition coil
   - Thermostat housing
   - Cylinder head bolts. Mark each bolt with a punch and loosen them in the sequence shown.
   - Cylinder head and the gasket

To install:

➡**The cylinder head bolts may be reused twice. Replace bolts that have two punch marks.**

8. Lubricate the cylinder bolts when engine oil.
9. Install the cylinder head and tighten the bolts in sequence as follows:
   a. Step 1: 15 ft. lbs. (20 Nm)
   b. Step 2: 30 ft. lbs. (40 Nm)
   c. Step 3: Plus 90 degrees
10. Install or connect the following:
   - Thermostat housing
   - Ignition coil and the camshafts
   - Heater and upper radiator hoses
   - Valve cover
   - Spark plugs
   - Spark plug wires
   - Air conditioning compressor

Cylinder head bolt loosening sequence—2.0L (VIN 3) engine

- Splash shield
- Engine oil dipstick tube
- Alternator and the power steering reservoir and bracket
- Fuel line
- CKP sensor, HO2sensor and 2 main fuel charging electrical connections
- Accelerator cable and if equipped, the speed control cable to the control lever

11. Connect the vacuum hoses to the following components:
- PCV valve
- Throttle body
- Fuel pressure sensor

- Intake manifold
12. Install or connect the following:
- Timing belt
- Air cleaner outlet tube
- Negative battery cable
13. Fill the cooling system.
14. Start the engine and check for leaks.

2.0L (VIN 5) Engine

1. Before servicing the vehicle, refer to the Precautions Section.
2. Release the fuel system pressure.
3. Disconnect the battery.
4. Release the coolant system pressure.

Cylinder head torque sequence—2.0L (VIN 3) engine

5. Raise and support the vehicle.
6. Drain the coolant from the radiator. Allow the coolant to drain into a suitable container. Install the radiator drain plug after draining the coolant.
7. Disconnect the brake booster pipe from the intake manifold.
8. Disconnect the oil pressure switch electrical connector.
9. Remove the intake manifold lower retaining bolt.
10. Disconnect the knock sensor electrical connector and separate it from the intake manifold.
11. Disconnect the catalyst monitor sensor electrical connector and separate it from the bracket.
12. Lower the vehicle.
13. Remove the air cleaner.
14. Remove the splash shield.
15. Disconnect the accelerator cable from the throttle body.
16. Disconnect the intake manifold runner control (IMRC) actuator cable from the IMRC lever.
17. Disconnect the vacuum hoses from the intake manifold.
18. Disconnect the fuel injector wiring harness.
19. Disconnect the camshaft position (CMP) sensor electrical connector.
20. Disconnect the fuel lines.
21. Disconnect the ground cable from the engine lifting eye.
22. Remove the exhaust manifold.
23. Disconnect the thermostat housing.
24. Disconnect the power steering pipe bracket from the cylinder head.
25. Disconnect the power steering pump bracket from the cylinder head and cylinder block.
26. Disconnect the generator.
27. Disconnect the wiring harness from the electronic ignition (EI) coil and from the engine coolant temperature (ECT) sensor.
28. Remove the upper bolt from the generator bracket.
29. Remove the camshafts.

➡ **Keep the valve tappets in order for installation.**

30. Remove the valve tappets.

✳ CAUTION

Cylinder head bolts can only be used twice, mark the bolts to indicate usage.

6-8 Nm (54-70 lb/in)

Valve cover torque sequence—2.0L (VIN 3) Engine

Locating stud—2.0L (VIN 5) engine

⚠ CAUTION

The cylinder head must be cooled to ambient temperature.

31. Remove the cylinder head bolts in the sequence shown.
32. Remove the cylinder head.
33. Using a spatula, remove gasket residue.
34. Thoroughly clean the threaded holes of the cylinder head bolts.
35. Check the cylinder head distortion.
36. Make up two locating studs as shown.
37. Install a new cylinder head gasket.
38. Insert the fabricated locating studs. Check that the locating studs are seated correctly.
39. Install the cylinder head.
40. Tighten the cylinder head bolts in sequence as follows:
   a. Step 1: 15 ft. lbs. (20 Nm)
   b. Step 2: 30 ft. lbs. (40 Nm)
   c. Step 3: Plus 90 degrees.

➡**Coat the valve tappets with clean engine oil before installation.**

41. Install the valve tappets in order.
42. Install the camshafts.

43. Check the valve clearance and, if necessary, adjust.
44. Attach the power steering pump bracket to the cylinder head and cylinder block.
45. Install the generator bracket bolt.
46. Connect the wiring harness to the electric ignition (EI) coil and to the engine coolant temperature (ECT) sensor.
47. Attach the generator.
48. Attach the power steering pipe bracket to the cylinder head.
49. Install the intake manifold lower retaining bolt.
50. Connect the catalyst monitor sensor electrical connector and attach it to the bracket.
51. Connect the knock sensor electrical connector.
52. Install the exhaust manifold.

➡**Install a new thermostat housing gasket.**

53. Attach the thermostat housing. Tighten the bolts to 15 ft. lbs. (20 Nm).
54. Attach the fuel lines.
55. Attach the ground cable to the engine lifting bracket.
56. Connect the camshaft position (CMP) sensor connector.
57. Connect the fuel injector wiring harness.
58. Attach the accelerator cable and the speed control cable (if equipped) to the throttle body.
59. Install the splash shield.
60. Attach the intake manifold runner control (IMRC) actuator cable to the IMRC lever.
61. Attach the vacuum hoses to the intake manifold.
62. Install the air cleaner.
63. Raise and support the vehicle.
64. Connect the oil pressure switch electrical connector.
65. Attach the brake booster pipe to the intake manifold.
66. Drain and fill the engine with clean engine oil.
67. Connect the battery ground cable.
68. Fill and bleed the cooling system.

2.0L (VIN N) Engine

1. Before servicing the vehicle, refer to the Precautions Section.
2. Drain the cooling system.
3. Relieve the fuel system pressure.
4. Remove or disconnect the following:

- Negative battery cable
- Battery tray
- Cooling fan
- Catalytic converter from the muffler assembly
- Catalytic converter support bracket
- Engine heat shield, if equipped
- Exhaust sensor electrical connectors
- Catalytic converter and secure it aside with mechanic's wire or equivalent
- Alternator
- Fuel rail
- Intake manifold
- Valve lifters
- Secondary air injection (AIR) vacuum regulator electrical connector and hose, if equipped
- Hoses from the AIR valve, if equipped
- Upper exhaust sensor electrical connectors, if equipped

Cylinder head bolt removal sequence—2.0L (VIN 5) engine

Cylinder head bolt torque sequence—2.0L (VIN 5) engine

- Exhaust gas recirculation (EGR) valve electrical connector
- Coolant bypass
- EGR coolant hose
- Cylinder head bolts
- Cylinder head

To install:

5. Inspect the cylinder head mating surfaces.

6. Apply silicone gasket sealant to the locations shown.

7. Install a new cylinder head gasket and cylinder head.

8. Using new bolts, tighten the cylinder bolts as follows:
   a. Step 1: 44 inch lbs. (5 Nm)
   b. Step 2: 11 ft. lbs. (15 Nm)
   c. Step 3: 33 ft. lbs. (45 Nm)
   d. Step 4: Plus 90 degrees
   e. Step 5: Plus 90 degrees again

9. Install or connect the following:
   - EGR coolant hose
   - Coolant bypass, using new gasket. Tighten to 89 inch lbs. (10 Nm).
   - EGR valve electrical connector
   - Catalytic converter, using new gasket and nuts. Tighten nuts to 41 ft. lbs. (55 Nm).
   - Exhaust sensor electrical connectors
   - Upper exhaust sensor electrical connectors, if equipped
   - Hoses to the AIR valve, if equipped
   - Secondary air injection (AIR) vacuum regulator electrical connector and hose, if equipped

- Engine heat shield, if equipped
- Catalytic converter support bracket and tighten bolts to 35 ft. lbs. (47 Nm)
- Catalytic converter to muffler assembly, using new gasket and nuts
- Valve lifters
- Intake manifold
- Fuel rail
- Alternator
- Cooling fan
- Battery tray
- Negative battery cable

10. Drain the engine oil.

11. Change the oil filter and fill the engine with oil.

12. Fill the cooling system to the correct level.

13. Start the engine and check for leaks.

2002–2004 2.3L (VIN Z) Engine

1. Before servicing the vehicle, refer to the Precautions Section.

2. Disconnect the battery ground cable.

3. Drain the cooling system.

⁕⁕ CAUTION

During engine repair procedures, cleanliness is extremely important. Any foreign material, including any material created while cleaning gasket surfaces, that enters the oil passages, coolant passages or the oil pan can cause engine failure.

⁕⁕ CAUTION

The crankshaft, the crankshaft sprocket and the pulley are fitted together by friction, using diamond washers between the flange faces on each part. For that reason, the crankshaft sprocket is also unfastened if you loosen the pulley. Therefore the engine must be retimed each time the damper is removed. Otherwise severe damage can occur.

4. Remove the camshafts.

5. Remove the catalytic converter.

6. Remove the fuel injection supply manifold.

7. Remove the intake manifold.

8. Remove the fuel injection supply manifold spacers.

9. Disconnect the exhaust gas recirculation (EGR) vacuum regulator electrical connector.

10. Remove the hoses from the air control valve.

11. Disconnect the EGR valve electrical connector.

12. Remove the coolant hoses from the coolant bypass.

13. Remove the coolant bypass bolts and the coolant bypass.

14. Disconnect the EGR coolant hose.

15. Disconnect the radio interference capacitor.

16. Remove the bolts and the cylinder head. Discard the bolts.

17. Remove and discard the head gasket.

To install:

18. Inspect the cylinder head mating surfaces.

⁕⁕ CAUTION

Do not use metal scrapers, wire brushes, power abrasive discs or other abrasive means to clean the sealing surfaces. These tools cause scratches and gouges which make leak paths.

19. Clean the cylinder head gasket surfaces with metal surface cleaner.

20. Apply silicone gasket and sealant to the locations shown.

21. Install a new head gasket.

➡The cylinder head bolts are torque-to-yield and must not be reused. New cylinder head bolts must be installed.

➡Lubricate the cylinder head bolts with clean engine oil.

22. Install the cylinder head and new

06017-FOCU-G08

Removing the optional engine heat shield from 2.0L (VIN N) engines

Apply silicone gasket sealant where indicated—2.3L (VIN Z) and 2.0L (VIN N) engines

bolts. Tighten the bolts in sequence as follows.

    a. Step 1: 44 inch lbs. (5 Nm)
    b. Step 2: 11 ft. lbs. (15 Nm)
    c. Step 3: 33 ft. lbs. (45 Nm)
    d. Step 4: Plus 90 degrees
    e. Step 5: Plus 90 degrees

23. Connect the radio interference capacitor electrical connector.

24. Install the EGR coolant hose.

25. Install the coolant bypass and tighten the bolts to 89 inch lbs. (10 Nm).

26. Install the coolant hoses onto the coolant bypass.

27. Connect the EGR valve electrical connector.

28. Install the air control valve hoses.

29. Connect the EGR vacuum regulator electrical connector.

30. Install the fuel injector supply manifold spacers.

31. Install the intake manifold.

32. Install the fuel injector supply manifold.

33. Install the catalytic converter.

34. Install the camshafts.

35. Connect the battery ground cable.

36. Fill and bleed the cooling system.

2005 2.3L (VIN Z) Engine

1. Before servicing the vehicle, refer to the Precautions Section.

2. Drain the cooling system.

3. Relieve the fuel system pressure.

4. Remove or disconnect the following:
- Negative battery cable
- Battery tray
- Cooling fan
- Catalytic converter from the muffler assembly
- Catalytic converter support bracket
- Engine heat shield
- Exhaust sensor electrical connectors
- Catalytic converter and secure it aside with mechanic's wire or equivalent
- Alternator
- Fuel rail
- Intake manifold
- Valve lifters
- Exhaust gas recirculation (EGR) valve electrical connector
- Coolant bypass
- EGR coolant hose
- Cylinder head bolts
- Cylinder head

To install:

5. Inspect the cylinder head mating surfaces.

6. Apply silicone gasket sealant to the locations shown.

7. Install a new cylinder head gasket and cylinder head.

8. Using new bolts, tighten the cylinder bolts as follows:
    a. Step 1: 44 inch lbs. (5 Nm)
    b. Step 2: 11 ft. lbs. (15 Nm)
    c. Step 3: 33 ft. lbs. (45 Nm)
    d. Step 4: Plus 90 degrees
    e. Step 5: Plus 90 degrees again

9. Install or connect the following:

- EGR coolant hose
- Coolant bypass, using new gasket. Tighten to 89 inch lbs. (10 Nm).
- EGR valve electrical connector
- Catalytic converter, using new gasket and nuts. Tighten nuts to 41 ft. lbs. (55 Nm).
- Exhaust sensor electrical connectors
- Engine heat shield
- Catalytic converter support bracket and tighten bolts to 35 ft. lbs. (47 Nm)
- Catalytic converter to muffler assembly, using new gasket and nuts
- Valve lifters
- Intake manifold
- Fuel rail
- Alternator
- Cooling fan
- Battery tray
- Negative battery cable

10. Drain the engine oil.

11. Change the oil filter and fill the engine with oil.

12. Fill the cooling system to the correct level.

13. Start the engine and check for leaks.

Rocker Arms/Shafts

REMOVAL & INSTALLATION

2.0L (VIN P) Engine

1. Before servicing the vehicle, refer to the Precautions Section.

2. Remove or disconnect the following:
- Negative battery cable
- Valve cover

➡**Keep all valve train components in order for assembly.**

- Rocker arm bolts
- Rocker arms and seats

To install:

3. Install or connect the following:
- Seats and rocker arms in their original positions
- Rocker arm bolts. Tighten to 17–22 ft. lbs. (23–30 Nm).
- Valve cover
- Negative battery cable

2.0L (VIN 3, 5 and N) and 2.3L (VIN Z) Engines

These engines are not equipped with rocker arms. The camshafts act directly on the valves.

Cylinder head bolt torque sequence—2.3L (VIN Z) and 2.0L (VIN N) engines

Intake Manifold

REMOVAL & INSTALLATION

2.0L (VIN P) Engine

1. Before servicing the vehicle, refer to the Precautions Section.
2. Drain the cooling system.
3. Relieve the fuel system pressure.
4. Remove or disconnect the following:
   - Negative battery cable
   - Air cleaner outlet tube
5. Tag and disconnect the vacuum hoses from the following components:
   - Exhaust gas Recirculation (EGR) valve
   - Positive Crankcase Ventilation (PCV) valve
   - Throttle body
   - Fuel pressure regulator
   - Intake manifold
6. Remove or disconnect the following:
   - Accelerator cable, throttle control lever and if equipped, the speed control cable
   - Speed control cable bracket bolt, if equipped
   - Idle Air Control (IAC) valve and Throttle Position (TP) sensor electrical connections
   - Engine oil dipstick tube bracket bolt that attaches the tube to the intake manifold
   - EGR manifold tube located below the EGR valve
   - Engine oil dipstick tube from the block
   - Intake manifold lower nuts
   - Intake manifold upper nuts
   - Intake manifold

To install:

7. Clean and oil the intake manifold mounting studs.
8. Install or connect the following:
   - Intake manifold with a new gasket. Tighten the nuts to 15–22 ft. lbs. (20–30 Nm).
   - Engine oil dipstick tube
   - EGR manifold tube. Tighten it to 15–20 ft. lbs. (20–28 Nm).
   - IAC valve and TP sensor electrical connections
   - Speed control cable bracket bolt, if equipped. Tighten to 71–88 inch lbs. (8–10 Nm).
   - Throttle control lever, accelerator cable and if equipped, the speed control cable
9. Connect the vacuum hoses to the following components:

- EGR valve
- PCV valve
- Throttle body
- Fuel pressure regulator
- Intake manifold
- Air cleaner outlet tube
- Negative battery cable

10. Fill the cooling system.
11. Start the engine and check for coolant leaks.

2.0L (VIN 3) Engine

1. Before servicing the vehicle, refer to the Precautions Section.
2. Drain the cooling system.
3. Relieve the fuel system pressure.
4. Remove or disconnect the following:
   - Negative battery cable
   - Air cleaner outlet tube
   - Throttle Position (TP) sensor connector
   - Bolt that secures the pipe located by the crankshaft pulley
   - Heater hoses
   - Main engine control sensor wiring
   - Connectors from the mounting bracket
   - Intake manifold vacuum lines
   - Positive Crankcase Ventilation (PCV) hose
   - Drive belt
   - Alternator
   - Fuel line
   - Intake manifold

To install:

5. Install or connect the following:
   - Intake manifold. Tighten the fasteners to 13 ft. lbs. (18 Nm).
   - Fuel line
   - Alternator
   - Drive belt
   - PCV hose
   - Intake manifold vacuum lines
   - Connectors in the mounting bracket
   - Main engine control sensor wiring
   - Heater hoses
   - Bolt that secures the pipe located by the crankshaft pulley. Tighten the bolt to 71–97 inch lbs. (8–11 Nm).
   - TP sensor electrical connection
   - Air cleaner outlet tube
   - Negative battery cable
6. Fill the cooling system.
7. Start the engine and check for leaks.

2.0L (VIN 5) Engine

1. Before servicing the vehicle, refer to the Precautions Section.

67197-FOCU-G18

Intake manifold lower retaining bolt—2.0L (VIN 5) engine

2. Remove the fuel injection supply manifold.
3. Raise and support the vehicle.
4. Remove the intake manifold lower retaining bolt.
5. Lower the vehicle.
6. Remove the engine lifting eye from the cylinder head.
7. Disconnect the idle air control (IAC) sensor electrical connector.
8. Disconnect the throttle position (TP) sensor electrical connector.
9. Remove the intake manifold runner control (IMRC) actuator cable from the IMRC lever.
10. Disconnect the IMRC actuator cable from the intake manifold assembly.
11. Disconnect the vacuum hoses from the throttle body.
12. Release the quick release coupling and pull out the brake booster pipe.
13. Separate the two sections of the intake manifold.
14. Disconnect the knock sensor (KS) electrical connector from the intake manifold inner section.
15. Remove the intake manifold inner section locknut and bolts. Discard the locknut.
16. Remove the intake manifold inner section retaining stud.
17. Remove the intake manifold sections.

To install:

➡**Clean and inspect the intake manifold gasket. Install a new intake manifold gasket if necessary.**

➡**Install a new intake manifold locknut.**

18. Install the intake manifold sections.
19. Install the intake manifold inner section retaining stud and tighten it to 88 inch lbs. (10 Nm).
20. Install the intake manifold inner section locknut and bolts. Tighten the fasteners to 13 ft. lbs. (18 Nm).
21. Connect the knock sensor (KS) electrical connector to the intake manifold inner section.

22. Connect the two sections of the intake manifold.

23. Install the brake booster pipe.

24. Connect the vacuum hoses from the throttle body.

25. Connect the IMRC actuator cable from the intake manifold assembly.

26. Install the intake manifold runner control (IMRC) actuator cable from the IMRC lever.

27. Connect the throttle position (TP) sensor electrical connector.

28. Connect the idle air control (IAC) sensor electrical connector.

29. Install the engine lifting eye from the cylinder head. Tighten the bolts to 26 ft. lbs. (35 Nm).

30. Raise the vehicle.

31. Install the intake manifold lower retaining bolt.

32. Lower the vehicle.

33. Install the fuel injection supply manifold.

2.0L (VIN N) Engine

1. Before servicing the vehicle, refer to the Precautions Section.

2. Remove or disconnect the following:
- Cooling fan motor and shroud
- Lower intake manifold bolt
- Accelerator control snow shield
- Air cleaner outlet tube
- Emissions breather tube
- Throttle cable and cruise control cable, if equipped, from the throttle linkage
- Throttle position (TP) sensor electrical connector and position the wiring harness aside
- Idle air control (IAC) valve electrical connector and position the wiring harness aside
- Evaporative emissions hose
- Power brake booster vacuum hose
- Vacuum hose from the fuel pressure regulator
- Intake manifold runner control (IMRC) actuator electrical connector
- Manifold absolute pressure (MAP) sensor electrical connector
- Secondary air injection (AIR) vacuum supply hose, if equipped
- Swirl control valve electrical connector
- Oil level indicator tube

➡**There are two different size bolts used. Mark the location of the bolts to ensure installation in the correct location.**

- Seven intake manifold bolts

Location of the intake manifold mounting bolts—2.0L (VIN N) engine.

06017-FOCU-G09

3. Raise the intake manifold enough to gain clearance to disconnect the knock sensor (KS) wiring harness.

4. Disconnect the positive crankcase ventilation (PCV) hose.

5. Remove the intake manifold.

To install:

6. Inspect and install new intake manifold gaskets, if necessary.

7. Position the intake manifold and install the PCV hose.

8. Connect the KS wiring harness.

➡**Be sure to install the bolts in the previously marked locations.**

9. Install the seven mounting bolts. Tighten the bolts to 13 ft. lbs. (18 Nm).

10. Install or connect the following:
- Oil level indicator tube
- Swirl control valve electrical connectors
- AIR vacuum supply hose, if equipped
- Lower intake bolt and tighten to 13 ft. lbs. (18 Nm)
- MAP sensor electrical connector
- IMRC actuator electrical connector
- Vacuum hose from the fuel pressure regulator
- Power brake booster vacuum hose
- Evaporative emissions hose
- IAC wiring harness and connector
- TP sensor electrical wiring harness and connector
- Throttle cable and cruise control cable, if equipped, to the throttle linkage
- Accelerator control snow shield
- Air cleaner outlet tube
- Cooling fan motor and shroud

11. Start the engine and check for leaks.

2.3L (VIN Z) Engine

1. Before servicing the vehicle, refer to the Precautions Section.

2. Remove or disconnect the following:

- Cooling fan motor and shroud
- Lower intake manifold bolt
- Accelerator control snow shield
- Air cleaner outlet tube
- Emissions breather tube
- Throttle cable and cruise control cable, if equipped, from the throttle linkage
- Throttle position (TP) sensor electrical connector and position the wiring harness aside
- Idle air control (IAC) valve electrical connector and position the wiring harness aside
- Evaporative emissions hose
- Power brake booster vacuum hose
- Vacuum hose from the fuel pressure regulator
- Intake manifold runner control (IMRC) actuator electrical connector
- Manifold absolute pressure (MAP) sensor electrical connector
- Swirl control valve electrical connectors and pin-type retainers
- Oil level indicator tube

➡**There are two different size bolts used. Mark the location of the bolts to ensure installation in the correct location.**

- Seven intake manifold bolts

3. Raise the intake manifold enough to gain clearance to disconnect the knock sensor (KS) wiring harness.

4. Disconnect the positive crankcase ventilation (PCV) hose.

5. Remove the intake manifold.

To install:

6. Inspect and install new intake manifold gaskets, if necessary.

7. Position the intake manifold and install the PCV hose.

8. Connect the KS wiring harness.

➡**Be sure to install the bolts in the previously marked locations.**

06017-FOCU-G10

Intake manifold mounting bolt locations—2.3L engine

➡**To ease the installation of the lower center mounting bolt, use a 6-inch long, 5⁄16-inch diameter hose.**

9. Install the seven mounting bolts. Tighten the bolts to 13 ft. lbs. (18 Nm).

10. Install or connect the following:
- Oil level indicator tube
- Swirl control valve electrical connectors and pin-type retainers
- Lower intake bolt and tighten to 13 ft. lbs. (18 Nm)
- MAP sensor electrical connector
- IMRC actuator electrical connector
- Vacuum hose from the fuel pressure regulator
- Power brake booster vacuum hose
- Evaporative emissions hose
- IAC wiring harness and connector
- TP sensor electrical wiring harness and connector
- Throttle cable and cruise control cable, if equipped, to the throttle linkage
- Accelerator control snow shield
- Air cleaner outlet tube
- Cooling fan motor and shroud

11. Start the engine and check for leaks.

Exhaust Manifold

REMOVAL & INSTALLATION

2.0L (VIN P) Engine

1. Before servicing the vehicle, refer to the Precautions Section.

2. Remove or disconnect the following:
- Negative battery cable
- Drive belt
- Heated Oxygen (HO2S) sensor electrical connection

- Exhaust manifold heat shield
- Exhaust Gas Recirculation (EGR) tube
- Power steering pressure hose bracket
- Alternator
- Catalytic converter
- Exhaust manifold

To install:

3. Install or connect the following:
- Exhaust manifold. Tighten the nuts to 15–18 ft. lbs. (20–24 Nm).
- Catalytic converter. Tighten the nuts to 26–34 ft. lbs. (34–48 Nm).
- Alternator
- Power steering pressure hose bracket
- EGR tube. Tighten the fasteners to 45–62 inch lbs. (5–7 Nm).
- Exhaust manifold heat shield. Tighten the nuts to 45–62 inch lbs. (5–7 Nm).
- HO2S sensor electrical connection
- Drive belt
- Negative battery cable

4. Start the engine and check for leaks.

2.0L (VIN 3) Engine

1. Before servicing the vehicle, refer to the Precautions Section.

2. Remove or disconnect the following:
- Negative battery cable
- Fan motor electrical connection
- Fan shroud
- Heated Oxygen (HO2S) sensor electrical connection
- Catalytic converter
- Engine oil dipstick tube bracket bolt
- Exhaust manifold heat shield
- Exhaust manifold

To install:

3. Install or connect the following:
- Exhaust manifold. Tighten the fasteners to 13 ft. lbs. (18 Nm).
- Exhaust manifold heat shield. Tighten the retainers to 71–101 inch lbs. (8–11 Nm).
- Engine oil dipstick tube bracket. Tighten the bolt to 71–101 inch lbs. (8–11 Nm).
- Catalytic converter
- HO2 sensor electrical connection
- Fan shroud and the fan motor wiring
- Negative battery cable

4. Start the engine and check for leaks.

2.0L (VIN 5) Engine

1. Before servicing the vehicle, refer to the Precautions Section.

2. Remove the oil level indicator and tube.

3. Remove the radiator fan and shroud.

4. Remove the exhaust manifold heat shield lower retaining bolts.

5. Detach the exhaust manifold from the catalytic converter.

6. Lower the vehicle.

7. Remove the exhaust manifold heat shield.

8. Remove the exhaust manifold. Discard the gaskets.

To install:

➡**Install new exhaust manifold gaskets.**

9. Install the exhaust manifold. Tighten the fasteners to 13 ft. lbs. (18 Nm).

10. Install the heat shield and tighten the bolts to 88 inch lbs. (10 Nm).

11. Raise and support the vehicle.

12. Attach the catalytic converter and tighten the bolts to 35 ft. lbs. (47 Nm).

13. Install the radiator fan and shroud.

14. Install the oil level indicator and tube. Tighten the bolts to 15 ft. lbs. (20 Nm).

2.0L (VIN N) Zetec and 2.3L (VIN Z) Engine

➡**The exhaust manifold is integral to the catalytic converter.**

1. Before servicing the vehicle, refer to the Precautions Section.

2. Position the vehicle on a hoist.

3. Relieve the fuel system pressure.

4. Remove the two catalytic converter-to-muffler assembly flange nuts. Discard the nuts.

5. Disconnect the catalyst monitor and

the heated oxygen sensor (HO2S) electrical connectors.

6. Remove the two catalytic converter-to-engine bracket bolts.

7. Remove the support bracket bolts.

8. Remove the two lower heat shield screws.

9. Remove and discard the lower two catalytic converter nuts.

10. Lower the vehicle.

11. Disconnect the upper HO2S electrical connector.

12. Disconnect the wiring connector, then remove the bolt and the bracket.

13. Disconnect the fuel lines.

14. Remove the two bolts and the fuel canister assembly.

15. Remove the vapor management valve (VMV).

16. Disconnect the VMV electrical connector.

17. Remove the upper fuel line.

18. Remove the lower fuel line.

19. Remove the two VMV bolts and the VMV.

20. Remove the upper three catalytic converter nuts and studs. Discard the nuts.

21. Remove the PCV hose and the secondary air hose.

22. Disconnect the heat shield from the bulkhead, then remove the heat shield.

23. Position the power steering reservoir aside.

24. Remove the catalytic converter.

➡**If installing a new converter, remove the HO2S and catalyst monitor sensors.**

➡**Make sure to apply anti-seize lubricant to the threads of the sensor before installation.**

➡**Clean the mating surfaces of the exhaust manifold and the catalytic converter mating surfaces.**

➡**Always install new fasteners and gaskets.**

➡**Do not tighten the fasteners until all components are assembled. Make sure to tighten all fasteners beginning at the front of the vehicle.**

25. To install, reverse the removal procedure.

26. Tighten the nuts to 41 ft. lbs. (55 Nm) in the sequence shown.

2.0L (VIN N) PZEV Engine

1. Before servicing the vehicle, refer to the Precautions Section.

2. Relieve the fuel system pressure.

3. Remove or disconnect the following:
- Negative battery cable
- Catalytic converter-to-muffler assembly nuts
- Catalyst monitor and heated oxygen sensor (HO2S) electrical sensors
- Catalytic converter-to-engine support bracket bolts
- Two lower heat shield bolts
- Lower four catalytic converter nuts
- Upper HO2S electrical connector
- Vapor management valve (VMV) heat shield
- VMV upper and lower fuel lines
- VMV mounting and position aside
- Upper three catalytic converter nuts
- PCV hose and secondary air hose
- Engine lifting eyelet

4. Position the power steering reservoir aside.

5. Remove the catalytic converter.

To install:

➡**If a new catalytic converter is being installed, the catalyst monitor and HO2S sensor must be transferred to the new converter.**

6. Install or connect the following:
- Catalytic converter using new nuts. Tighten bolts in sequence to 41 ft. lbs. (55 Nm).

- Power steering reservoir
- Engine lifting eyelet
- PCV hose and secondary air hose
- VMV
- VMV upper and lower fuel lines
- VMV heat shield
- Upper HO2S electrical connector
- Two lower heat shield bolts
- Catalytic converter-to-engine support bracket bolts and tighten to 35 ft. lbs. (47 Nm)
- Catalyst monitor and heated oxygen sensor (HO2S) electrical sensors
- Catalytic converter-to-muffler assembly, using new nuts and gasket. Tighten the nuts to 35 ft. lbs. (47 Nm).
- Negative battery cable

Front Crankshaft Seal

REMOVAL & INSTALLATION

2002–2004 2.0L Engines

1. Before servicing the vehicle, refer to the Precautions Section.

2. Remove the timing belt and crankshaft sprocket.

✱✱ WARNING

Be careful not to damage the crankshaft surface when removing the seal.

3. Using Seal Remover tool 303-293, remove the crankshaft front oil seal.

To install:

4. Use Seal Replacer tools 303-684 and 303-395 to install the new seal.

5. Install the crankshaft sprocket and the timing belt.

2005 2.0L and 2.3L Engines

1. Before servicing the vehicle, refer to the Precautions Section.

➡**The crankshaft, crankshaft sprocket and the pulley are fitted together by friction, using diamond washers between the flange faces on each part. For that reason, the crankshaft sprocket is also unfastened if you loosen the pulley. Therefore, the engine must be retimed each time the damper is removed.**

2. Remove the crankshaft pulley.

3. Remove the crankshaft seal using Special Tool 303-409.

67197-FOCU-G07

Exhaust manifold/catalytic converter torque sequence—2.0L (VIN N) and 2.3L (VIN Z) engines

Remove the crankshaft seal using Seal Remover 303-409—2005 models.

Installing the crankshaft seal using Special Tool 303-096—2005 models.

To install:

4. Install the front crankshaft seal using Special Tool 303-096.

5. Install the crankshaft pulley.

Camshaft and Valve Lifters

REMOVAL & INSTALLATION

2.0L (VIN P) Engine

1. Before servicing the vehicle, refer to the Precautions Section.

2. Disconnect the negative battery cable.

3. Remove the air cleaner and valve cover.

4. Remove the camshaft front seal as follows:

   a. Align the timing marks and remove the timing belt.

   b. Use cam sprocket holding/removing tool T74P-6256-B, and remove the camshaft sprocket.

   c. Unfasten the timing belt tensioner bolt and remove the tensioner.

   d. Remove the inner engine front cover.

   e. Use Seal Removal tool T92C-

1. Rocker arm
2. Rocker arm seats
3. Camshaft Position (CMP) sensor
4. Valve tappet
5. Valve tappet guide plate
6. Ignition coil
7. Radio interference capacitor
8. Ignition coil bracket
9. Cup plug
10. Camshaft
11. Water hose connection
12. Thermostat
13. Water hose connection gasket

Exploded view of the cylinder head—2.0L (VIN P) engine

Removing the camshaft front seal—2.0L (VIN P) engine

Remove the camshaft thrust plate retaining bolts (A) and the plate (B)—2.0L (VIN P) engine

Remove and discard the cup plug located at the rear of the cylinder head—2.0L (VIN P) engine

0.05-1.0 mm
(0.002-0.04 in)

Use seal Replacer T81P-6292-A, to install the camshaft front seal 0.002-0.04 in. (0.05-1.0mm) below flush with the cylinder head front face—2.0L (VIN P) engine

6700-CH, to remove the camshaft front seal.

5. Remove or disconnect the following:
   • Ignition coil and bracket
   • Rocker arms and valve tappets
   • Camshaft thrust plate bolts and the plate
   • Cup plug from the rear of the cylinder head and discard
   • Camshaft from the rear of the cylinder head

To install:

➡ **Liberally coat the cam bore in the cylinder with clean 5W-30 motor oil.**

6. Install or connect the following:
   • Camshaft through the rear of the cylinder head
   • Camshaft thrust plate and the bolts. Tighten the bolts to 71–115 inch lbs. (8–13 Nm).
   • New cup plug and the tappets
   • Rocker arms
   • Ignition coil bracket and coil

7. Install the camshaft front seal as follows:

   a. Apply a thin film of 5W-30 motor oil to the lip of the seal.

   b. Use Seal Replacer T81P-6292-A, to install the camshaft front seal.

➡ **The seal depth should be 0.002-0.04 in. (0.05-1.0mm) below flush with the cylinder head front face.**

   c. Install the inner engine front cover.

   d. Install the timing belt tensioner. Tighten the tensioner bolt to 15–22 ft. lbs. (20–30 Nm).

   e. Install the camshaft sprocket and the timing belt

8. Install the valve cover and the air cleaner assembly.

9. Connect the negative battery cable.

2.0L (VIN 3) Engine

1. Before servicing the vehicle, refer to the Precautions Section.

2. Remove or disconnect the following:
   • Timing belt
   • Valve cover and camshaft sprockets
   • Oil control solenoid flange

➡ **Mark the camshaft journal caps with a number to identify their location.**

   • Camshaft journal caps. Loosen the bolts in sequence and in several passes.
   • Camshafts
   • Oil control sensor and bushing

1	Bolt—Camshaft Bearing Cap		18	Spark Plug
2	Camshaft Bearing Cap		19	Engine Lifting Eye Bolt
3	Camshaft Intake		20	Engine Lifting Eye
4	Camshaft Exhaust		21	Camshaft Sprocket —Exhaust
5	Plug		22	Exhaust Camshaft Sprocket Bolt
6	Adjusting Shim—Valve Clearance		23	Blanking Plug O-Ring
7	Valve Tappets		24	Blanking Plug
8	Valve Spring Retainer Key		25	Bolt—Intake Camshaft
9	Valve Spring Retainers		26	Camshaft Sprocket Intake
10	Valve Spring (Color Coding: Exhaust = Blue, Intake = Red)		27	Camshaft Front Seal
			28	O-Ring
11	Valve Stem Seal		29	Camshaft Front Seal
12	Camshaft Position Sensor		30	Oil Feed Ring
13	CMP Sensor Bolt		31	Oil Feed Flange Bolts
14	Intake Valves		32	Oil Feed Flange
15	Exhaust Valves		33	Guide Sleeve—Front Camshaft Bearing Cap
16	Cylinder Head		34	Fifth Camshaft Bearing Cap
17	Cylinder Head Bolts			

9300MG06

Exploded view of the cylinder head and camshaft mounting—2.0L (VIN 3) engine

To install:

3. Install the oil control solenoid bushing and flange on the exhaust camshaft.

4. Coat the surface of the front camshaft journal cap with gasket maker E2AZ-19562-B.

5. Place the camshafts in position and lubricate the bearing surfaces with engine assembly lubricant D9AZ-19579-D.

6. Install new camshaft front oil seals.

7. Apply a thin coat of silicone gasket and sealant F6AZ-19562-AA to the sealing surface of the of the front camshaft journal bearing cap.

8. Install the caps and tighten the bolts in several 2-turn passes in the sequence illustrated to 10–12 ft. lbs. (13–17 Nm).

9. Inspect the oil control solenoid flange O-rings for damage or wear and replace, as necessary.

10. Install the oil control solenoid flange and tighten the bolts to 84–92 inch lbs. (9.5–10.5 Nm) in the sequence.

11. Rotate the camshafts a full turn and check for binding.

Camshaft journal cap loosening sequence—2.0L (VIN 3) engine

13-17 Nm (10-12 lb/ft)

7922MG59

Camshaft journal cap torque sequence—2.0L (VIN 3) engine

67197-FOCU-G19

Camshaft bolt removal sequence—2.0L (VIN 5) engine

9.5-10.5 Nm (84-92 lb/in)

7922MG60

Tighten the oil control solenoid flange bolts in this sequence—2.0L (VIN 3) engine

12. Install or connect the following:
   • Camshaft sprockets and timing belt
   • Valve cover

2.0L (VIN 5) Engine

1. Before servicing the vehicle, refer to the Precautions Section.
2. Remove the timing belt.

➡ **Hold the camshafts by the hexagon with an open ended wrench to stop them from rotating.**

3. Remove the camshaft pulleys.
4. Remove the oil feed flange. Discard the camshaft oil seals and the oil feed flange oil seal.

✷✷ CAUTION

Loosen the camshaft bearing caps in the sequence shown.

5. Working in several stages, evenly loosen each bolt two turns at a time in the indicated sequence and remove the bolts.
6. Remove the camshafts.
7. Remove the tappets.

➡ **Keep the valvetrain components in order for installation.**

To install:
8. Install the tappets in their original locations.
9. Install the camshafts.
10. Working in several stages, install the camshaft bearing caps in the indicated sequence, tighten the bolts evenly one half turn at a time.
11. Tighten the bolts in the sequence shown to 11 ft. lbs. (15 Nm).
12. Install a new oil feed flange seal.
13. Install the oil feed flange and tighten the bolts to 11 ft. lbs. (15 Nm).
14. Using the Camshaft Oil Seal Installer 303-160, install the camshaft oil seals.
15. Lubricate the camshaft and oil seal lip with clean engine oil.

Camshaft torque sequence—2.0L (VIN 5) engine

67197-FOCU-G20

67197-FOCU-G21

Apply sealant to the marked areas on the oil feed flange—2.0L (VIN 5) engine

16. Install the camshaft pulleys.

➡ **Do not tighten the camshaft pulley retaining bolts at this stage.**

17. Install the timing belt.

2.0L (VIN N) Engine

1. Before servicing the vehicle, refer to the Precautions Section.

✳✳ CAUTION

The crankshaft, the crankshaft sprocket and the pulley are fitted together by friction, using diamond washers between the flange faces on each part. For that reason, the crankshaft sprocket is also unfastened if you loosen the pulley. Therefore the engine must be retimed each time the damper is removed. Otherwise severe engine damage can occur.

➡ **Valve lifters are select fit and the valve clearance must be checked before removing the lifters.**

2. Remove the timing chain.
3. Mark the position of the camshaft lobes on the No. 1 cylinder for assembly reference.
4. Loosen the camshaft bearing bolts in the sequence shown, one turn at a time. Repeat until all tension is released.

✳✳ CAUTION

Failure to follow the camshaft loosening procedure can result in camshaft damage.

5. Remove the camshaft bearing caps.
6. Remove the camshafts.
7. Remove the valve lifters.

To install:

➡ **Lubricate the valve lifters with clean engine oil.**

8. Install the valve lifters.

➡ **Lubricate the camshaft journals and bearing caps with clean engine oil.**

9. Install the camshafts and bearing caps. Tighten the bolts as follows:
   a. Step 1: Tighten the camshaft bearing cap bolts, one turn at a time, until the cam is fully seated.
   b. Step 2: Tighten the bolts to 62 inch lbs. (7 Nm).
   c. Step 3: Tighten the bolts to 12 ft. lbs. (16 Nm).
10. Install the timing chain.

2.3L (VIN Z) Engine

✳✳ CAUTION

During engine repair procedures, cleanliness is extremely important. Any foreign material, including any material created while cleaning gasket surfaces, that enters the oil passages, coolant passages or the oil pan can cause engine failure.

06017-FOCUS-G13

Camshaft loosening sequence—2.0L (VIN N)

06017-FOCUS-G14

Camshaft tightening sequence—2.0L (VIN N)

※※ CAUTION

The crankshaft, the crankshaft sprocket and the pulley are fitted together by friction, using diamond washers between the flange faces on each part. For that reason, the crankshaft sprocket is also unfastened if you loosen the pulley. Therefore the engine must be retimed each time the damper is removed. Otherwise severe engine damage can occur.

1. Before servicing the vehicle, refer to the Precautions Section.
2. Remove the timing chain and sprockets.
3. Remove the camshaft alignment plate.

➡Mark the position of the camshaft lobes on the No. 1 cylinder for assembly reference.

※※ CAUTION

Failure to follow the camshaft loosening procedure can result in damage to the camshafts.

4. Loosen the camshaft bearing cap bolts, in sequence, one turn at a time.
5. Repeat the first step until all tension is released from the camshaft bearing caps.
6. Remove the camshaft bearing caps.
7. Remove the camshafts.

To install:

※※ CAUTION

Install the camshafts with the alignment slots in the camshafts lined up so the Camshaft Alignment Plate (T94P-6256-CH) can be installed without rotating the camshafts. Make sure the lobes on the No. 1 cylinder are in the same position as noted in the removal procedure. Rotating the camshafts when the timing chain is removed, or installing the camshafts 180 degrees out of position, can cause severe damage to the valves and pistons.

➡Lubricate the camshaft journals and bearing caps with clean engine oil.

8. Install the camshafts and bearing caps. Tighten the bolts in the sequence shown as follows:

a. Step 1: Tighten the camshaft bearing bolt caps one turn at a time until tight.
b. Step 2: Tighten the bolts to 62 inch lbs. (7 Nm)
c. Step 3: Tighten the bolts to 12 ft. lbs. (16 Nm)
9. Install the camshaft drive gears and hand-tighten the bolts.
10. Install the timing chain and sprockets.

Valve Lash

ADJUSTMENT

2.0L (VIN P) Engine

The lash adjusters are hydraulic and are not adjustable.

2.0L (VIN 3) Engine

1. Before servicing the vehicle, refer to the Precautions Section.
2. Remove the valve cover.
3. Rotate the crankshaft so that the valve is closed and measure the clearance between the lifter shim and the camshaft base circle. Intake valve clearance should be 0.004–0.007 inches. Exhaust valve clearance should be 0.010–0.013 inches.
4. Measure each valve and note the clearance.
5. If adjustment is necessary, remove the camshafts and replace the shims. To obtain the correct shim size, measure the current shim, add the measured clearance, and subtract 0.006 inches for intake valves or 0.012 inches for exhaust valves.
6. Install or connect the following:
   • Camshafts
   • Valve cover

2.0L (VIN 5) Engine

1. Before servicing the vehicle, refer to the Precautions Section.
2. Remove the valve cover.

Camshaft bolt loosening sequence—2.3L (VIN Z) engine

Camshaft bolt tightening sequence—2.3L (VIN Z) engine

Checking valve clearance—2.0L engine

3. Rotate the crankshaft so that the valve is closed and measure the clearance between the lifter shim and the camshaft base circle. Intake valve clearance should be 0.008–0.0011 inches. Exhaust valve clearance should be 0.013–0.016 inches.

4. Measure each valve and note the clearance.

5. If adjustment is necessary, remove the camshafts and replace the shims. To obtain the correct shim size, measure the current shim, add the measured clearance, and subtract 0.009 inches for intake valves or 0.014 inches for exhaust valves.

6. Install or connect the following:
- Camshafts
- Valve cover

2.0L (VIN N) and 2.3L (VIN Z) Engines

1. Before servicing the vehicle, refer to the Precautions Section.

2. Remove the valve cover.

3. Rotate the crankshaft so that the valve is closed and measure the clearance between the lifter shim and the camshaft base circle. Intake valve clearance should be 0.008–0.0011 inches. Exhaust valve clearance should be 0.010–0.013 inches.

4. Measure each valve and note the clearance.

5. If adjustment is necessary, remove the camshafts and replace the shims. To obtain the correct shim size, measure the current shim, add the measured clearance, and subtract 0.009 inches for intake valves or 0.011 inches for exhaust valves.

6. Install or connect the following:
- Camshafts
- Valve cover

Starter Motor

REMOVAL & INSTALLATION

2.0L (VIN P) Engine

1. Before servicing the vehicle, refer to the Precautions Section.

2. Remove or disconnect the following:
- Negative battery cable
- Air cleaner outlet tube
- Two top starter motor bolts
- **S** terminal wire and the **B** terminal nut and cable from the solenoid
- Lower starter motor bolt
- Starter

To install:

3. Install or connect the following:

- Starter motor assembly and tighten the lower mounting bolt to 18–20 ft. lbs. (25–27 Nm)
- **S** terminal wire, **B** terminal cable and nut. Tighten the nut to 80–120 inch lbs. (9–13.5 Nm).
- Top starter motor bolts. Tighten the bolts to 18–20 ft. lbs. (25–27 Nm).
- Air cleaner outlet tube
- Negative battery cable

2.0L (VIN 3) Engine

1. Before servicing the vehicle, refer to the Precautions Section.

2. Remove or disconnect the following:
- Negative battery cable
- Air cleaner
- Starter electrical connectors
- Starter

To install:

3. Install or connect the following:
- Starter. Tighten the bolts to 26 ft. lbs. (35 Nm).
- Starter electrical connectors
- Air cleaner
- Negative battery cable

2.0L (VIN 5) Engine

1. Before servicing the vehicle, refer to the Precautions Section.

2. Remove the intake manifold.

3. Disconnect the starter motor electrical connectors.

4. Remove the starter motor.

To install:

5. Install the starter motor and tighten the bolts to 26 ft. lbs. (35 Nm).

6. Install the starter electrical connectors.

7. Install the intake manifold.

25 Nm (18 lb-ft)

67197-FOCU-G10

Starter motor—2.0L (VIN N) and 2.3L (VIN Z) engines

2.0L (VIN N) and 2.3L (VIN Z) Engines

1. Before servicing the vehicle, refer to the Precautions Section.

2. Disconnect the battery ground cable.

3. Remove the terminal nuts and disconnect the wiring.

4. Remove the nuts and disconnect the power steering pressure line brackets from the stud bolts.

5. Remove the bolts and the starter motor.

6. To install, reverse the removal procedure. Tighten the starter bolts to 18 ft. lbs. (25 Nm).

Oil Pan

REMOVAL & INSTALLATION

2.0L (VIN P) Engine

1. Before servicing the vehicle, refer to the Precautions Section.

2. Drain the engine oil.

3. Remove or disconnect the following:
- Negative battery cable
- Catalytic converter and brackets
- Intake manifold bracket
- Intermediate shaft bracket

25 Nm

9306SG05

Oil pan torque sequence—2.0L (VIN P) engine

Oil pan torque sequence—2.0L (VIN 3) engine

67197-FOCU-G23

Install ten M6 x 20 mm studs into the blind holes indicated

- Coolant pipe bolt
- Oil pan

To install:

4. Apply silicone sealant to the oil pump and rear seal retainer joints.

5. Install or connect the following:
- Oil pan. Tighten the bolts in sequence to 18 ft. lbs. (25 Nm).
- Coolant pipe bolt
- Intermediate shaft bracket
- Intake manifold bracket
- Catalytic converter and brackets
- Negative battery cable

6. Fill the crankcase to the correct level.

7. Start the engine and check for leaks.

2.0L (VIN 3) Engine

1. Before servicing the vehicle, refer to the Precautions Section.

2. Drain the engine oil.

3. Remove or disconnect the following:
- Negative battery cable
- Catalytic converter
- Oil pan

To install:

→**Apply a bead of silicone gasket sealer to the oil pan flange.**

4. Install or connect the following:
- Oil pan. Tighten the bolts in sequence and in several passes to 18 ft. lbs. (24 Nm).
- Oil lever sensor connector
- Catalytic converter
- Negative battery cable

5. Fill the crankcase to the correct level.

6. Start the engine and check for leaks.

2.0L (VIN 5) Engine

1. Before servicing the vehicle, refer to the Precautions Section.

2. Remove the catalytic converter.

3. Drain the engine oil.

67197-FOCU-G22

Removing the oil pan with the Slide Hammer 205-047 and Adapter 303-633—2.0L (VIN 5) engine

4. Remove the oil pan retaining bolts.

✳✳ CAUTION

To prevent damage to the sealing surface, only use the Slide Hammer 205-047 and Adapter 303-633 as shown to remove the oil pan.

5. Tighten the locknut against the oil pan and tap the special tool several time to release the oil pan from the lower crankcase.

To install:

→**Do not damage the mating faces.**

6. Clean the mating faces. The mating

faces must be free from oil and sealant residue.

7. Clean all traces of oil residue and oil sludge from the oil pan.

✳✳ CAUTION

The lower crankcase may be damaged if sealant enters the blind holes.

8. Install ten M6 x 20 mm studs into the blind holes shown (engine shown removed for clarity).

→**Attach the oil pan within ten minutes of applying the sealant.**

9. Apply a 3 mm diameter bead of sealant to the mating face of the oil pan.

→**The oil pan must not be removed once it comes into contact with the lower crankcase housing.**

10. Attach the oil pan to the lower crankcase housing.

11. Remove the studs.

12. Install the oil pan retaining bolts and tighten them in sequence as follows:
   a. Step 1: 53 inch lbs. (6 Nm)
   b. Step 2: 88 inch lbs. (10 Nm)

→**Inspect the oil drain plug and gasket. Install a new oil drain plug and gasket if necessary.**

67197-FOCU-G24

Oil pan torque sequence—2.0L (VIN 5) engine

Oil pan bolt torque sequence—2.0L (VIN N) and 2.3L (VIN Z) engines

13. Install the oil pan drain plug and tighten it to 17 ft. lbs. (24 Nm).

14. Install the catalytic converter.

15. Fill the engine with clean oil.

2.0L (VIN N) and 2.3L (VIN Z) Engines

1. Before servicing the vehicle, refer to the Precautions Section.

2. Drain the engine oil.

3. Remove or disconnect the following:
- Oil level indicator tube.
- Wiring harness from the retainers and position aside
- Front cover-to-oil pan bolts
- Oil pan-to-bell housing bolts
- Oil pan mounting bolts
- Oil pan.

To install:

✳✳ CAUTION

Do not use metal scrapers, wire brushes, power abrasive discs or other abrasive means to clean the sealing surfaces. These tools cause scratches and gouges, which make leak paths. Use a plastic scraping tool to remove traces of sealant.

4. Clean and inspect all mating surfaces.

➡**If the oil pan is not secured within four minutes of sealant application the sealant must be removed and the sealing area cleaned with metal surface cleaner. Allow to dry until there is no sign of wetness, or four minutes, whichever is longer. Failure to follow this procedure can cause future oil leakage.**

➡**The oil pan must be installed and the bolts tightened within four minutes of applying the silicone gasket and sealant.**

5. Apply a 2.5 mm bead of silicone gasket and sealant to the oil pan.

6. Install or connect the following:
- Oil pan
- Oil pan-to-bell housing bolts and tighten to 35 ft. lbs. (48 Nm)
- Front cover-to-oil pan bolts and tighten to 89 inch lbs. (10 Nm)
- Wiring harness and retainers

7. Tighten the oil pan bolts in sequence to 18 ft. lbs. (25 Nm).

8. Install the engine oil level indicator tube and tighten the bolts to 89 inch lbs. (10 Nm).

9. Fill the engine with clean engine oil.

10. Start the engine and check for leaks.

Oil Pump

REMOVAL & INSTALLATION

2.0L (VIN P) Engine

1. Before servicing the vehicle, refer to the Precautions Section.

2. Remove or disconnect the following:
- Timing belt
- Oil pan
- Crankshaft Position (CKP) sensor
- Oil pump pickup tube
- Oil pump

To install:

3. Lubricate the crankshaft front oil seal lip with clean engine oil.

4. Install or connect the following:
- Oil pump. Tighten the bolts to 10–12 ft. lbs. (13–16 Nm).
- Oil pump pickup tube. Tighten the retaining bolts to 71–97 inch lbs. (8–11 Nm).
- CKP sensor
- Oil pan
- Timing belt

5. Fill the crankcase to the correct level.

6. Start the engine and check for leaks.

2.0L (VIN 3) Engine

1. Before servicing the vehicle, refer to the Precautions Section.

2. Drain the engine oil.

3. Remove or disconnect the following:
- Negative battery cable
- Timing belt
- Oil pan
- Oil pump pickup tube
- Oil pump

To install:

4. Install or connect the following:
- Oil pump. Tighten the bolts to 97 inch lbs. (11 Nm)
- Oil pump pickup tube. Tighten the fasteners to 88 inch lbs. (10 Nm).
- Oil pan
- Timing belt
- Negative battery cable

5. Fill the crankcase to the correct level.

6. Start the engine and check for leaks.

2.0L (VIN 5) Engine

1. Before servicing the vehicle, refer to the Precautions Section.

2. Remove the timing belt.

3. Remove the crankshaft timing pulley.

4. Remove the thrust washer.

5. Detach the air conditioning (A/C) compressor from the bracket and secure it on the radiator crossmember.

6. Remove the A/C compressor bracket.

7. Remove the intermediate shaft center bearing bracket.

8. Remove the oil pan.

✳✳ CAUTION

Note the location of any spacer shims between the transaxle and the lower crankcase to aid installation. The spacer shims must be installed in the same location during installation.

9. Remove the oil pick-up tube.

10. Remove the lower crankcase retaining bolts and the lower crankcase. Discard the gasket.

11. Remove the accessory drive belt tensioner.

12. Remove the oil pump. Discard the gasket.

To install:

➡**Install a new oil pump gasket**

13. Install the oil pump. Align the oil pump on both sides so that the mating face is between 0.3 mm– 0.8 mm above the lower edge of the cylinder block.

14. Tighten the oil pump retaining bolts to 97 inch lbs. (11 Nm).

15. Install the accessory drive belt tensioner and tighten the bolt to 35 ft. lbs. (48 Nm).

✳✳ CAUTION

Make sure that any spacer shims between the transaxle and the lower crankcase are installed in their original location.

16. Install the lower crankcase with a new gasket. First, tighten the transaxle bolts to 22 ft. lbs. (30 Nm). Then, tighten the rest of the lower crankcase bolts.

17. Install the oil pump pick-up tube and tighten the bolts to 88 inch lbs. (10 Nm).

18. Install the oil pan.

19. Install the intermediate shaft center bearing bracket and tighten the fasteners to 18 ft. lbs. (25 Nm).

20. Install the air conditioning (A/C) compressor bracket and tighten the fasteners to 18 ft. lbs. (25 Nm).

21. Attach the A/C compressor to the bracket and tighten the bolts to 18 ft. lbs. (25 Nm).

22. Install the thrust washer and the crankshaft timing pulley.

23. Install the timing belt.

2.3L (VIN Z) Engine

1. Before servicing the vehicle, refer to the Precautions Section.

2. Remove the engine from the vehicle and mount it on an engine stand.

3. Remove or disconnect the following:
  - Oil pan
  - Oil pump pickup tube and screen
  - Front cover and the timing chain

4. Release the tension on the tensioner spring.

5. Remove the tensioner and the shoulder bolt.

6. Remove the guide.

➡**The oil pump chain sprocket must be held in place.**

7. Remove the following:
  - Oil pump chain and sprockets
  - Oil pump assembly and gasket

To install:

8. Install the oil pump with a new gasket. Tighten the bolts in sequence as follows:
  a. Step 1: 89 inch lbs. (10 Nm).
  b. Step 2: 17 ft. lbs. (23 Nm).

9. Install the pump chain and sprock-

67197-FOCU-G12

Oil pump torque sequence—2.0L and 2.3L (VIN Z) engines

ets. Tighten the pump sprocket bolt to 18 ft. lbs. (25 Nm).

10. Install the chain guide, tensioner, and shoulder bolt. Tighten the bolts to 89 inch lbs. (10 Nm).

11. Hook the tensioner spring around the shoulder bolt.

12. Install the oil pump pickup tube and screen with a new gasket. Tighten the bolts to 89 ft. lbs. (10 Nm).

13. Install the oil pan.

14. Install the timing chain and front cover.

15. Install the engine into the vehicle.

Rear Main Seal

REMOVAL & INSTALLATION

2.0L (VIN 3, 5 and P) Engines

1. Before servicing the vehicle, refer to the Precautions Section.

2. Remove or disconnect the following:
  - Negative battery cable
  - Transaxle
  - Clutch, if equipped
  - Flexplate/flywheel
  - Oil seal

To install:

3. Install or connect the following:
  - Oil seal. Seat the seal flush with the rear of the crankshaft oil seal retainer.
  - Flexplate/flywheel. Tighten the bolts to 82 ft. lbs. (112 Nm).
  - Clutch, if equipped
  - Transaxle
  - Negative battery cable

4. Start the engine and check for leaks.

2.0L (VIN N) and 2.3L (VIN Z) Engines

1. Before servicing the vehicle, refer to the Precautions Section.

2. Remove the transaxle.

3. Remove the flywheel or flexplate.

✳✳ CAUTION

If the oil pan is not removed damage to the rear seal retainer joint can occur.

4. Remove the oil pan.

5. Remove the bolts and the crankshaft rear oil seal.

To install:

6. Install the crankshaft rear oil seal on the Crankshaft Rear Main Oil Seal Installer 303-328.

7. Install the Crankshaft Rear Main Oil Seal Installer and the crankshaft rear oil seal on the crankshaft.

8. Tighten the bolts in the sequence shown to 10 Nm (89 inch lbs.).

9. Remove the Crankshaft Rear Main Oil Seal Installer.

10. Install the oil pan.

11. Install the flywheel or flexplate. Tighten the bolts in the sequence shown as follows:.
  a. Step 1: Tighten to 37 ft. lbs. (50 Nm).
  b. Step 2: Tighten to 50 ft. lbs. (80 Nm).
  c. Step 3: Tighten to 83 ft. lbs. (112 Nm).

12. Install the transaxle.

13. Fill the engine with clean oil.

67197-FOCU-G13

Seal retainer torque sequence—2.0L (VIN N) and 2.3L (VIN Z) engines

67197-FOCU-G14

Flywheel/Flexplate torque sequence—2.3L (VIN Z) engine

Timing Belt

REMOVAL & INSTALLATION

2.0L (VIN 3) Engine

1. Remove or disconnect the following:
   - Negative battery cable
   - Spark plugs
   - Catalytic converter
   - Right front wheel
   - Right inner splash shield
   - Accessory drive belt and idler pulley

2. Rotate the crankshaft to Top Dead Center (TDC) and install Crankshaft TDC Timing Peg T97P-6000-A.

3. Rotate the crankshaft clockwise against the peg.

4. Remove or disconnect the following:
   - Water pump pulley
   - Valve cover
   - Crankshaft pulley
   - Timing belt covers

5. Install Camshaft Alignment Timing Tool T94P-6256-CH into the slots of both camshafts at the rear of the cylinder head to lock the camshafts into position.

6. Loosen the timing belt tensioner bolt

Crankshaft alignment position—Ford 2.0L (VIN 3) engines

Placement of Camshaft Alignment Timing Tool T94P-6256-CH—Ford 2.0L (VIN 3) engines

Timing belt tensioner index marks—Ford 2.0L (VIN 3) engines

and relieve the tension on the timing belt by disconnecting the tensioner tab from the timing cover back plate.

7. Remove the timing belt.

To install:

➡**Always replace the timing belt after loosening or removing it.**

8. Install the timing belt.

9. Engage the timing belt tensioner tab into the timing cover back plate.

10. Adjust the timing belt tensioner until the index marks are aligned. Tighten the tensioner bolt to 17–20 ft. lbs. (22–28 Nm).

11. Remove the camshaft alignment tool and the crankshaft TDC peg.

12. Install or connect the following:
   - Timing belt covers
   - Crankshaft pulley. Tighten the bolt to 81–89 ft. lbs. (110–120 Nm).
   - Valve cover
   - Water pump pulley
   - Accessory drive belt and idler pulley
   - Right inner splash shield
   - Right front wheel
   - Catalytic converter
   - Spark plugs
   - Negative battery cable

13. Start the engine and check for proper operation.

2.0L (VIN P) Engine

1. Remove or disconnect the following:
   - Negative battery cable
   - Accessory drive belt and tensioner
   - Timing belt cover
   - Right front wheel
   - Right inner splash shield
   - Crankshaft pulley

2. Align the timing marks as shown.

3. Refer to the illustration and remove the timing belt as follows:
   a. Loosen the timing belt tensioner bolt (1).
   b. Use an 8mm Allen wrench, and turn the tensioner (2) counterclockwise ¼ turn.

Align the timing marks before removing or installing the timing belt—2.0L (VIN P) engine

Remove the timing belt by following these 4 numbered steps (refer to the text for an explanation)—2.0L (VIN P) engine

   c. Insert a ⅛ inch drill bit in the hole (3) to lock the belt tensioner in place.
   d. Remove the timing belt (4).

To install:

4. Install the timing belt in a counterclockwise direction starting at the crankshaft.

5. Remove the drill bit to unlock the timing belt tensioner.

6. Rotate the crankshaft two complete turns and check that the timing marks align.

7. Tighten the tensioner bolt to 15–22 ft. lbs. (20–30 Nm).

8. Install or connect the following:
   - Crankshaft pulley. Tighten the bolt to 81–96 ft. lbs. (110–130 Nm).
   - Right inner splash shield
   - Right front wheel
   - Timing belt cover
   - Accessory drive belt and tensioner
   - Negative battery cable

9. Start the engine and check for proper operation.

2.0L (VIN 5) Engine

1. Before servicing the vehicle, refer to the Precautions Section.

2. Disconnect the negative battery cable.

3. Raise and support the vehicle.

4. Detach the drive belt cover.

5. Loosen the coolant pump pulley retaining bolts.

6. Remove the accessory drive belt.

7. Remove the crankshaft pulley.

8. Remove the accessory drive belt idler pulley.

9. Remove the coolant pump pulley.

10. Detach the timing belt lower cover.

11. Lower the vehicle.

12. Detach the coolant expansion tank and position it to one side.

13. Detach the power steering fluid reservoir and position it to one side.

14. Position the trolley jack with the wooden block under the oil pan and raise the engine so that the engine front mount is free from load.

➡ **Mark the position of the engine front mount.**

15. Remove the engine front mount.

16. Detach the timing belt upper cover. Leave the timing belt cover in its installed position.

17. Remove the engine front mount bracket.

18. Remove the timing belt upper cover.

19. Remove the electronic ignition (EI) coil cover.

20. Remove the engine upper cover.

21. Disconnect the variable cam timing (VCT) valve electrical connector.

➡ **Loosening sequence: from the outside to the inside, working diagonally.**

22. Disconnect the spark plug connectors.

23. Detach the positive crankcase ventilation (PCV) hose.

24. Remove the valve cover.

25. Remove the spark plugs.

26. Turn the engine to TDC on cylinder number 1.

27. Loosen the timing belt tensioner bolt four turns.

28. Position the tensioner so that the locating tab is at approximately the 4 o'clock position.

29. Line up the hexagonal key slot in the tensioner adjusting washer with the pointer that is located behind the pulley.

30. Remove the timing belt.

➡ **Use an open ended wrench to prevent the camshaft from rotating.**

31. Remove the intake camshaft pulley blanking plug.

32. Loosen the camshaft pulley retaining bolts.

To install:

➡ **The camshaft timing pulleys must be able to turn freely on the camshafts.**

33. Turn the camshafts to the ignition position on cylinder No. 1 and insert the special tool Camshaft Alignment Plate 303-574 into the camshafts.

34. Rotate the crankshaft to TDC on cylinder No. 1.

35. Remove the blanking plug and install the Crankshaft TDC Timing Peg 303-574.

36. Starting from the crankshaft timing belt pulley and working counterclockwise install the timing belt, keeping it under tension.

Camshaft Alignment Tool 303-574 installation—2.0L (VIN 5) engine

Set the crankshaft to Top Dead Center (TDC)—2.0L (VIN 5) engine

Installation of the Crankshaft TDC Timing Peg—2.0L (VIN 5) engine

⁂ CAUTION

Incorrect timing belt tension will result in incorrect valve timing.

37. Rotate the tensioner locating tab counterclockwise and insert the locating tab into the slot in the rear timing cover.

38. Position the hexagonal key slot in the tensioner adjusting washer to the 4 o'clock position.

39. Tighten the attaching bolt enough to seat the tensioner firmly against the rear timing cover, but still allow the tensioner adjusting washer to be rotated using a 6 mm hexagonal key.

⁂ CAUTION

Tension the timing belt working counterclockwise.

40. Using the hexagonal key, rotate the adjusting washer counterclockwise until the notch in the pointer is centered over the index line on the locating tab (the pointer will move clockwise during adjustment).

41. Tighten the bolt to 18 ft. lbs. (25 Nm), while holding the adjusting washer in position.

42. Use an open ended wrench to prevent the camshaft from rotating and tighten

Tensioner locating tab (1), hexagonal key slot (2) and attaching bolt (3)—2.0L (VIN 5) engine

Adjusting the timing belt tensioner—2.0L (VIN 5) engine

the camshaft pulley retaining bolts. Tighten the intake pulley bolt to 88 ft. lbs. (120 Nm) and the exhaust pulley bolt to 50 ft. lbs. (68 Nm).

43. Install the intake camshaft pulley blanking plug and tighten it to 27 ft. lbs. (37 Nm).

44. Remove the Crankshaft TDC Timing Peg.

45. Remove the Camshaft Alignment Plate from the camshafts.

46. Turn the crankshaft two revolutions in the normal direction of rotation.

47. Check the valve timing by reinstalling the Camshaft Alignment Plate and the Crankshaft TDC Timing Peg. If necessary loosen the timing pulleys and correct the camshaft alignment.

48. Remove the Camshaft Alignment Plate and the Crankshaft TDC Timing Peg.

49. Install the blanking plug.

50. Install the valve cover and tighten the bolts as follows:
    a. Step 1: 18 inch lbs. (2 Nm)
    b. Step 2: 62 inch lbs. (7 Nm)

51. Attach the PCV hose to the valve cover.

52. Connect the VCT actuator electrical connector.

53. Coat the spark plug thread with anti-seize grease, screw in the spark plugs and push in the spark plug connector until it engages.

54. Connect the VCT valve electrical connector.

55. Install the engine upper cover.

56. Install the spark plug cover.

57. Position the upper timing belt cover and the center timing belt cover.

58. Install the engine front mount bracket. Tighten the bolts to 37 ft. lbs. (50 Nm).

59. Attach the timing belt upper cover. Tighten the bolts to 88 inch lbs. (10 Nm).

60. Install the engine front mount.

61. Attach the power steering pump fluid reservoir.

62. Attach the coolant expansion tank.

63. Remove the hydraulic jack and wooden block.

64. Raise and support the vehicle.

65. Attach the lower timing belt cover. Tighten the bolts to 62 inch lbs. (7 Nm).

66. Install the crankshaft pulley. Tighten the bolt to 85 ft. lbs. (115 Nm).

67. Attach the multi-groove belt idler pulley. Tighten the bolt to 29 ft. lbs. (40 Nm).

68. Attach the coolant pump pulley. Do not tighten the bolts at this stage.

69. Install the accessory drive belt.

70. Tighten the bolts on the coolant pump pulley to 17 ft. lbs. (24 Nm).

71. Attach the drive belt cover.

72. Lower the vehicle.

➡When the battery has been disconnected and reconnected, some abnormal drive symptoms may occur while the vehicle relearns its adaptive strategy. The vehicle may need to be driven 16 km (10 miles) or more to relearn the strategy.

73. Connect the battery negative cable.

74. Install the battery cover.

75. Check fluid levels and correct if necessary.

76. Check the routing of vacuum hoses and cables and secure them with cable ties.

Timing Chain, Sprockets, Front Cover and Seal

REMOVAL & INSTALLATION

2.3L (VIN Z) Engine

✳✳ CAUTION

The crankshaft, the crankshaft sprocket and the pulley are fitted together by friction, using diamond washers between the flange faces on each part. For that reason, the crankshaft sprocket is also unfastened if you loosen the pulley. Therefore the engine must be retimed each time the damper is removed. Otherwise severe engine damage can occur.

1. Before servicing the vehicle, refer to the Precautions Section.

2. Remove or disconnect the following:
   • Negative battery cable
   • Accessory belt splash shield
   • Accessory drive belt

• Cylinder head cover

3. Using the crankshaft pulley bolt, rotate the crankshaft clockwise to the No. 1 piston at Top Dead Center (TDC).

4. Install the camshaft alignment plate 303-465 to the rear of the camshafts.

✳✳ CAUTION

The special tool Camshaft Alignment Plate (T94P-6256-CH) is for camshaft alignment only. Using this tool to prevent engine rotation can result in engine damage.

➡The camshaft timing slots are offset. If the special tool cannot be installed, rotate the crankshaft one complete revolution clockwise to correctly position the camshafts.

5. Remove the plug bolt and install Special Tool Crankshaft Timing Peg 303-507 to prevent the engine from being rotated in the clockwise direction.

6. Install the special tools Holding Fixture (205-126) and Adapter (205-072-02) to the crankshaft pulley.

7. Remove or disconnect the following
   • Crankshaft pulley bolt
   • Crankshaft pulley
   • Crankshaft Position (CKP) sensor
   • Accessory drive belt tensioner

67197-FOCU-G16

Plug bolt location—2.3L (VIN Z) engine,

06017-FOCU-G14

Using crankshaft holding fixture 205-126—2.3L engine

- Power steering pressure (PSP) switch
- PSP tube bracket
- Power steering pump pulley, using Special Tool 211-016
- PSP tube
- Power steering pump mounting bolts and position the pump aside
- Water pump pulley

8. Support the engine from above with a suitable support fixture.

9. Remove or disconnect the following:
   - Engine mounts
   - Idler pulley
   - Front cover

10. Compress the timing chain tensioner and insert a paper clip into the hole to secure the piston.

11. Remove or disconnect the following:
   - Timing chain tensioner
   - Right hand timing chain guide
   - Timing chain
   - Camshaft sprockets

Power steering pump removal tool 211-016.

1. Paper clip
2. Mounting bolts

06017-FOCU-G15

Securing the timing chain tensioner piston with a paper clip.

➡**Use a wrench on the flats of the camshaft to prevent camshaft rotation. Do not rely on the Camshaft Alignment Plate to prevent rotation. Damage to the camshaft can occur.**

   To install:
   12. Remove the Camshaft Alignment tool.

✳✳ CAUTION

Do not rotate the camshafts. Damage to the valves and pistons can occur. If the camshaft sprockets were not removed, use the flats on the camshafts to prevent camshaft rotation and loosen the sprocket bolts.

13. Install or connect the following:
   - Camshaft sprockets and the bolts. Do not tighten the bolts at this time.
   - LH timing chain guide and tighten the bolts to 89 inch lbs. (10 Nm)

- Timing chain
- RH timing chain guide
- Timing chain tensioner and tighten the bolts to 89 inch lbs. (10 Nm). Remove the paper clip to release the piston.

14. Install the Camshaft Alignment tool.

15. Using a wrench on the flats on the camshafts to prevent camshaft rotation, tighten the bolts to 48 ft. lbs. (65 Nm).

✳✳ CAUTION

Do not use metal scrapers, wire brushes, power abrasive disks or other abrasive means to clean sealing surfaces. These tools cause scratches and gouges which make leak paths.

16. Clean and inspect the mounting surfaces of the engine and the front cover.

➡**The engine front cover must be installed and the bolts tightened within four minutes of applying the silicone gasket and sealant.**

17. Apply a 0.10 inch (2.5 mm) bead of silicone gasket and sealant to the cylinder head and oil pan joint areas. Apply a 0.10 inch (2.5 mm) bead of silicone gasket and sealant to the front cover.

18. Install the engine front cover. Tighten the bolts in the sequence shown, to the following specifications:
   a. Tighten the 8 mm bolts to 89 inch lbs. (10 Nm).
   b. Tighten the 13 mm bolts to 35 ft. lbs. (48 Nm).

➡**Lubricate the oil seal with clean engine oil.**

19. If a new seal is being installed, use the special tool Front Seal Installer 303-096 to install the crankshaft front oil seal.

20. Position the power steering pump and tighten the four bolts to 18 ft. lbs. (25 Nm).

21. Using the special tool Teflon Seal Installer 211-D027, install a new O-ring on the PSP tube fitting.

22. Install or connect the following:
   - PSP tube fitting and tighten to 15 ft. lbs. (20 Nm)
   - PSP tube bracket and tighten the nut to 10 ft. lbs. (13 Nm)
   - PSP switch
   - Water pump pulley and tighten the bolts to 18 ft. lbs. (25 Nm)
   - Accessory drive belt idler pulley and tighten the bolt to 18 ft. lbs. (25 Nm)
   - Accessory drive belt tensioner and

Front cover torque sequence—2.3L (VIN Z) engine

67197-FOCU-G17

tighten the bolts to 18 ft. lbs. (25 Nm)

➡**Apply clean engine oil on seal area before installing.**

23. Install the crankshaft pulley and hand-tighten the bolt.

➡**Do not reuse the crankshaft damper bolt.**

❋❋ CAUTION

Only hand-tighten the bolt or damage to the front cover can occur.

➡**This step will correctly align the crankshaft pulley to the crankshaft.**

24. Install a standard 6 mm (0.23 in) x 18 mm (0.7 in) bolt through the crankshaft pulley and thread it into the front cover.
25. Rotate the pulley as necessary to align the bolt holes.

❋❋ CAUTION

Failure to hold the crankshaft pulley in place during bolt tightening can cause damage to the engine front cover.

26. Using the special tools to hold the crankshaft pulley in place, tighten the crankshaft pulley bolt as follows:
   a. Step 1: Tighten to 74 ft. lbs. (100 Nm).

   b. Step 2: Tighten an additional 90 degrees.
27. Remove the special tools.
28. Remove the 6 mm (0.23 in) x 18 mm (0.7 in) bolt.
29. Remove the Crankshaft Timing Peg.
30. Remove the Camshaft Alignment tool.
31. Turn the engine two complete revolutions.
32. Install the Crankshaft Timing Peg.

➡**Only turn the engine in the normal direction of rotation.**

33. Turn the crankshaft until the No. 1 piston is at top dead center (TDC).

❋❋ CAUTION

Only hand-tighten the bolt or damage to the front cover can occur.

34. Using the 6 mm (0.23 in) x 18 mm (0.7 in) bolt, check the position of the crankshaft pulley. If it is not possible to install the bolt, correct the engine timing.
35. Using the Camshaft Alignment tool, check the position of the camshafts. If it is not possible to install the special tool, correct the engine timing.
36. Remove the Camshaft Alignment tool.
37. Remove the 6 mm (0.23 in) x 18 mm (0.7 in) bolt.
38. Remove the Crankshaft Timing Peg.
39. Install or connect the following:
   • Engine plug bolt and tighten it to 15 ft. lbs. (20 Nm)
   • Engine mounts and tighten the bolts to 35 ft. lbs. (48 Nm)
   • CKP sensor
   • Cylinder head cover
   • Accessory drive belt
   • Accessory drive belt splash shield
   • Negative battery cable
40. Fill and purge the power steering fluid.
41. Fill the cooling system to the correct level.
42. Fill the engine with oil.
43. Start then engine and check for leaks.

Piston and Ring

POSITIONING

Ford 2.0L (VIN 3, 5 and P) and 2.3L (VIN Z) engines—piston ring positioning, end-gap spacing and piston positioning. The small directional arrow must face the front of the engine.

7922AG29

FUEL SYSTEM

Fuel System Service Precautions

Safety is the most important factor when performing not only fuel system maintenance but any type of maintenance. Failure to conduct maintenance and repairs in a safe manner may result in serious personal injury or death. Maintenance and testing of the vehicle's fuel system components can be accomplished safely and effectively by adhering to the following rules and guidelines.

• To avoid the possibility of fire and personal injury, always disconnect the negative battery cable unless the repair or test procedure requires that battery voltage be applied.

• Always relieve the fuel system pressure prior to disconnecting any fuel system component (injector, fuel rail, pressure regulator, etc.), fitting or fuel line connection. Exercise extreme caution whenever relieving fuel system pressure, to avoid exposing skin, face and eyes to fuel spray. Please be advised that fuel under pressure may penetrate the skin or any part of the body that it contacts.

• Always place a shop towel or cloth around the fitting or connection prior to loosening to absorb any excess fuel due to spillage. Ensure that all fuel spillage (should it occur) is quickly removed from engine surfaces. Ensure that all fuel soaked cloths or towels are deposited into a suitable waste container.

• Always keep a dry chemical (Class B) fire extinguisher near the work area.

• Do not allow fuel spray or fuel vapors to come into contact with a spark or open flame.

• Always use a backup wrench when loosening and tightening fuel line connecion fittings. This will prevent unnecessary stress and torsion to fuel line piping.

• Always replace worn fuel fitting O-rings with new. Do not substitute fuel hose or equivalent, where fuel pipe is installed.

Before servicing the vehicle, make sure to refer to the Precautions Section as well.

Fuel System Pressure

RELIEVING

1. Before servicing the vehicle, refer to the Precautions Section.

2. Remove the fuel pump fuse from the battery junction box.
3. Start the engine and allow it to idle until it stalls.
4. Crank the engine for 5 seconds to ensure that the fuel pressure has been relieved.
5. Disconnect the negative battery cable.
6. When repairs are completed, connect the negative battery cable and replace the fuel pump fuse.

Fuel Filter

REMOVAL & INSTALLATION

1. Before servicing the vehicle, refer to the Precautions Section.
2. Relieve the fuel system pressure.
3. Remove or disconnect the following:
   • Evaporative Emissions (EVAP) vapor line
   • Fuel filter lines
   • Fuel filter and bracket assembly
   • Fuel filter

To install:
4. Install or connect the following:
   • Fuel filter
   • Fuel filter and bracket assembly
   • Fuel filter lines
   • EVAP vapor line
5. Start the engine and check for leaks.

Fuel Pump

REMOVAL & INSTALLATION

1. Before servicing the vehicle, refer to the Precautions Section.
2. Relieve the fuel system pressure.
3. Drain the fuel tank.
4. Remove or disconnect the following:
   • Exhaust center muffler, if necessary
   • Heat shield
   • Fuel filler and vent hoses
   • Fuel supply line
   • Evaporative Emissions (EVAP) vapor line
   • Rollover valve line
   • Fuel tank retainer strap. Support the fuel tank.
   • Fuel pump module harness connector
   • Fuel tank pressure sensor connector
   • Fuel tank
   • Fuel pump module

To install:
5. Install or connect the following:
   • Fuel pump module. Tighten the locknut to 59 ft. lbs. (80 Nm).
   • Fuel tank
   • Fuel tank pressure sensor connector
   • Fuel pump module harness connector
   • Fuel tank retainer strap. Tighten the bolt to 18 ft. lbs. (25 Nm).
   • Rollover valve line
   • EVAP vapor line
   • Fuel supply line
   • Fuel filler and vent hoses
   • Heat shield
   • Exhaust center muffler, if removed
6. Start the engine and check for leaks.

Fuel Injector

REMOVAL & INSTALLATION

2.0L (VIN 3, 5 and P) Engines

1. Before servicing the vehicle, refer to the Precautions Section.
2. Relieve fuel system pressure.
3. Remove or disconnect the following:
   • Negative battery cable
   • Fuel line and retaining clip
   • Pressure regulator vacuum hose
   • Accelerator cable
   • Fuel injector electrical connectors
   • Fuel supply manifold with the injectors attached
   • Injectors from the supply manifold

To install:
4. Install or connect the following:
   • Fuel injectors using new O-ring seals
   • Fuel supply manifold with the injectors attached. Tighten the bolts to 88 inch lbs. (10 Nm).
   • Fuel injector electrical connectors
   • Accelerator cable
   • Pressure regulator vacuum hose
   • Fuel line and retaining clip
   • Negative battery cable

2.0L (VIN N) and 2.3L (VIN Z) Engines

1. Before servicing the vehicle, refer to the Precautions Section.
2. Relieve the fuel system pressure
3. Remove or disconnect the following:
   • Negative battery cable

- Fuel injector electrical connectors
- Fuel pressure regulator electrical connector and the vacuum hose
- Fuel injector harness retaining clips from the fuel injection supply manifold

- Fuel supply line

4. Remove the bolts and the fuel injection supply manifold.

➡**Remove and discard the fuel injector O-rings.**

5. Fuel injectors

6. Install new O-rings and lubricate them with clean engine oil.

7. To install, reverse the removal procedure. Tighten the fuel supply manifold bolts to 18 ft. lbs. (25 Nm).

DRIVE TRAIN

Transaxle Assembly

REMOVAL & INSTALLATION

Manual

EXCEPT 2.0L (VIN N) AND 2.3L (VIN Z) MODELS

1. Before servicing the vehicle, refer to the Precautions Section.

2. Attach a support fixture to the engine lifting eyes.

3. Loosen the front strut center nuts 5 turns.

4. Remove or disconnect the following:
- Battery and tray
- Mass Air Flow (MAF) sensor connector
- Air cleaner assembly
- Air intake pipe
- Clutch slave cylinder fluid line
- Shift cables
- Exhaust flex pipe
- Engine front mount
- Lower ball joints
- Axle halfshafts and intermediate shaft
- Drive belt cover
- Reverse light switch connector
- Vehicle Speed (VSS) sensor connector
- Rear engine mount and bracket
- Starter motor
- Transaxle flange bolts. Support the transaxle.
- Transaxle

To install:

5. Install or connect the following:
- Transaxle. Tighten the flange bolts to 35 ft. lbs. (48 Nm).
- Starter motor
- Rear engine mount bracket. Tighten the fastener to 59 ft. lbs. (80 Nm).
- Rear engine mount. Tighten the outer nuts to 35 ft. lbs. (48 Nm) and the center nut to 98 ft. lbs. (133 Nm).
- VSS sensor connector
- Reverse light switch connector
- Drive belt cover
- Axle halfshafts and intermediate shaft

- Lower ball joints. Tighten the pinch bolts to 37 ft. lbs. (50 Nm).
- Right engine mount. Tighten the bolts to 35 ft. lbs. (48 Nm).
- Exhaust flex pipe
- Shift cables
- Clutch slave cylinder fluid line
- Air intake pipe
- Air cleaner assembly
- MAF sensor connector
- Battery and tray

6. Tighten the strut center nuts to 35 ft. lbs. (48 Nm).

2.0L (VIN N) AND 2.3L (VIN Z) MODELS

1. Before servicing the vehicle, refer to the Precautions Section.

2. Remove or disconnect the following:
- Negative battery cable
- Steering column shaft bolt
- Battery tray
- Air intake assembly
- Resonator
- Gearshift cables from the transaxle
- Gearshift cables from the bracket

- Hydraulic clutch line from retaining bracket
- Heated Oxygen Sensors (HO2S) and Catalyst Monitor Sensor (CMS) connections

3. Support the engine using Special Tools 303-290.

4. Remove or disconnect the following:
- Rear engine mount and mount bracket
- Reverse lamp switch
- Upper transmission to engine bolts
- Tire and wheels
- Engine roll restrictor
- Tie rods from the knuckles
- Lower control arms from the wheel knuckles
- Stabilizer bar from the strut
- Power steering lines from the steering gear

➡**Matchmark the subframe to the chassis.**

5. Support the subframe with a suitable lifting device.

6. Remove the six subframe bolts and lower the subframe from the vehicle.

Supporting the engine with Engine Support 303-290.

06017-FOCU-G16

7. Remove or disconnect the following:
- Vehicle Speed Sensor (VSS)
- Right hand halfshaft from the transmission and hang securely aside
- Left hand halfshaft from the transmission and hang securely aside

8. Loosen, but do not remove, the starter mounting bolts

9. Disconnect the clutch hydraulic line.

10. Secure the transmission to a jack or suitable lifting device.

11. Remove the 10 transmission-to-engine bolts.

12. Pull the transmission rearward until the input shaft is clear of the pressure plate, then lower the transmission from the vehicle.

To install:

13. Lift the transmission into the vehicle.

14. Tighten the transmission-to-engine bolts to 35 ft. lbs. (48 Nm).

15. Connect the clutch hydraulic line.

16. Tighten the starter motor mounting bolts to 18 ft. lbs. (25 Nm).

17. Install or connect the following:
- Left hand halfshaft
- Right hand halfshaft
- VSS
- Subframe and tighten mounting bolts to 148 ft. lbs. (200 Nm)
- Power steering lines
- Stabilizer bar to the strut
- Lower control arms to the wheel knuckles
- Tie rods to the knuckles
- Engine roll restrictor
- Upper transmission-to-engine bolts and tighten to 35 ft. lbs. (48 Nm)
- Reverse lamp switch
- Rear engine mount bracket and tighten bolts to 59 ft. lbs. (80 Nm)
- Rear engine mount and tighten bolts to 35 ft. lbs. (48 Nm). Tighten the center nut to 98 ft. lbs. (133 Nm).

18. Remove the engine support tools.

19. Install or connect the following:
- HO2S and CMS sensors
- Hydraulic clutch line to the retaining bracket
- Gearshift cables to the bracket
- Gearshift cables to the transaxle
- Resonator and tighten bolts to 8 ft. lbs. (11 Nm)
- Air intake assembly
- Battery tray
- Steering column shaft bolt
- Negative battery cable

20. Fill the transmission to the correct level.

21. Adjust the shift linkage if necessary.

22. Start the engine and check for leaks.

Automatic

EXCEPT 2.0L (VIN N) ENGINE

1. Before servicing the vehicle, refer to the Precautions Section.

2. Attach a support fixture to the engine lifting eyes.

3. Loosen the front strut center nuts 5 turns.

4. Remove or disconnect the following:
- Battery and tray
- Mass Air Flow (MAF) sensor connector
- Positive Crankcase Ventilation (PCV) hose
- Air intake pipe and resonator
- Turbine Shaft Speed (TSS) sensor connector
- Transmission range sensor connector
- Transaxle harness connector
- Starter motor
- Shift cable and bracket
- Front wheels
- Exhaust front pipe
- Right engine support
- Outer tie rod ends
- Lower ball joints
- Axle halfshafts and intermediate shaft
- Output Shaft Speed (OSS) sensor connector
- Transaxle oil cooler lines
- Torque converter
- Transaxle oil dipstick tube
- Rear engine mount and bracket

➡The transaxle flange bolts vary in length. Note their locations for installation.

- Transaxle flange bolts. Support the transaxle.
- Transaxle

To install:

5. Install or connect the following:
- Transaxle. Tighten the flange bolts to 35 ft. lbs. (48 Nm).
- Rear engine mount bracket. Tighten the fastener to 59 ft. lbs. (80 Nm).
- Rear engine mount. Tighten the outer nuts to 35 ft. lbs. (48 Nm) and the center nut to 98 ft. lbs. (133 Nm).
- Transaxle oil dipstick tube
- Torque converter. Tighten the nuts to 27 ft. lbs. (37 Nm).
- Transaxle oil cooler lines
- OSS sensor connector
- Axle halfshafts and intermediate shaft
- Lower ball joints. Tighten the pinch bolts to 37 ft. lbs. (50 Nm).

- Outer tie rod ends. Tighten the nuts to 35 ft. lbs. (48 Nm).
- Right engine support. Tighten the bolts to 35 ft. lbs. (48 Nm).
- Exhaust front pipe. Tighten the fasteners to 35 ft. lbs. (48 Nm).
- Front wheels
- Shift cable and bracket
- Starter motor
- Transaxle harness connector
- Transmission range sensor connector
- TSS sensor connector
- Air intake pipe and resonator
- PCV hose
- MAF sensor connector
- Battery and tray

6. Tighten the strut center nuts to 35 ft. lbs. (48 Nm).

2.0L (VIN N) ENGINE

1. Before servicing the vehicle, refer to the Precautions Section.

2. Remove or disconnect the following:
- Negative battery cable
- Battery tray
- Air cleaner resonator
- Transmission Range (TR) sensor
- Transaxle solenoid connector
- Turbine Shaft Speed (TSS) sensor
- Upper center converter housing bolts

3. Using special tool 303-290, support the engine.

4. Remove or disconnect the following:
- Transmission fluid fill tube
- Shift cable bracket bolts
- Starter motor
- Right hand engine support insulator
- Left hand stabilizer bar at the strut
- Tie rod ends from the knuckles, using Special Tool 211-001
- Lower control arms from the knuckles
- Right hand halfshaft from the transmission and hang securely aside
- Left hand halfshaft from the transmission and hang securely aside
- Output Speed Sensor (OSS) connector
- Transmission cooler lines
- Shift cable from manual control lever

5. Remove the plastic cover and rotate the crankshaft to gain access to the converter nuts. Remove the four nuts.

6. Remove the rear engine mount and lower the engine/transaxle assembly slightly.

7. Remove the rear engine mount bracket.

8. Secure the transaxle to a jack or suitable lifting device using a safety strap.

9. Remove the converter housing bolts and separate the transaxle from the engine.

➡**Note the locations of the different bolt lengths.**

To install:

10. Position the transaxle into position onto the engine assembly.

11. Install the converter housing bolts and tighten to 35 ft. lbs. (47 Nm).

12. Using new self-locking nuts, install the torque converter to the flexplate. Tighten the nuts to 27 ft. lbs. (37 Nm).

13. Raise the engine/transaxle assembly into place and install the rear engine mount. Tighten the bracket bolts to 59 ft. lbs. (80 Nm). Tighten the outer engine mount bolts to 35 ft. lbs. (48 Nm). Tighten the center nut to 98 ft. lbs. (133 Nm).

14. Install or connect the following:
- Transmission cooler lines
- OSS connecter
- Left hand halfshaft
- Right hand halfshaft
- Lower control arms to the knuckles
- Tie rod ends to the knuckles
- Stabilizer bar to the strut
- Right hand engine support isolator and tighten to 35 ft. lbs. (48 Nm)
- Starter motor
- Transmission fluid fill tube
- Shift cable bracket bolts
- Upper center converter housing bolts and tighten to 35 ft. lbs. (47 Nm)
- TSS sensor
- TR sensor
- Transaxle solenoid connector
- Air cleaner resonator
- Battery tray
- Negative battery cable

15. Fill the transmission to the correct level.

16. Start the engine and check for leaks.

Clutch

REMOVAL & INSTALLATION

1. Before servicing the vehicle, refer to the Precautions Section.

2. Remove or disconnect the following:
- Negative battery cable
- Transaxle
- Pressure plate. Loosen the bolts evenly in ½ turn steps.
- Clutch disc

To install:

3. Install or connect the following:
- Clutch disc and pressure plate. Tighten the pressure plate bolts evenly in ½ turns to 21 ft. lbs. (29 Nm).
- Transaxle
- Negative battery cable

Hydraulic Clutch System

BLEEDING

1. Before servicing the vehicle, refer to the Precautions Section.

2. Remove or disconnect the following:
- Air cleaner assembly
- Mass Air Flow (MAF) sensor connector
- Air intake pipe

3. Remove fluid from the brake fluid reservoir until the level reaches the **MIN** mark.

4. Fill the reservoir of Special Tool 416-D002 with DOT 4 brake fluid.

5. Attach the tool to the slave cylinder bleed nipple.

6. Open the bleed nipple and pump 80 ml brake fluid into the clutch control system.

7. Tighten the bleed nipple to 88 inch lbs. (10 Nm) and remove the tool.

8. Have an assistant depress the clutch pedal 4–5 times and hold the pedal at full travel.

9. Close the bleeder before releasing the clutch pedal.

10. Repeat the procedure until no more air bubbles are seen.

11. Check the clutch for proper operation.

12. Check the brake fluid reservoir level and fill with DOT 4 brake fluid, as necessary.

13. Install or connect the following:
- Air intake pipe
- MAF sensor connector
- Air cleaner assembly

Halfshaft

REMOVAL & INSTALLATION

Left

➡**The IB5 manual transmission is used with the 2.0L (VIN P) engine and the MTX75 manual transmission is used with the 2.0L (VIN 3 and N) and 2.3L (VIN Z) engines.**

➡**The hub nut may be reused 4 times. Mark the nut at removal and only use hub nuts with 3 or fewer marks for assembly.**

1. Before servicing the vehicle, refer to the Precautions Section.

2. Loosen the strut and spring assembly mounting nuts by 4 turns.

3. Remove or disconnect the following:
- Front wheel
- Hub retainer nut. Mark the nut.
- Lower ball joint

4. Press the stub shaft out of the wheel hub.

5. Separate the inner CV-joint from the transaxle as follows:
- If equipped with the IB5 manual transaxle, use Halfshaft remover 308-256
- If equipped with the MTX75 manual transaxle or with an automatic transaxle, use Halfshaft Remover 205-241 and a slide hammer

Left halfshaft removal—IB5 manual transaxle

9306SG07

205-241

9306SG08

Left halfshaft removal—MTX75 manual transaxle and automatic transaxle

To install:

➡**Replace the circlip for assembly.**

6. Install the halfshaft inner joint so that the circlip is felt to seat. Draw the stub shaft into the wheel hub with Halfshaft Installer 205-379.

7. Install or connect the following:
- Lower ball joint. Tighten the pinch bolt to 37 ft. lbs. (50 Nm).
- Hub retainer nut. Tighten the nut to 232 ft. lbs. (316 Nm).
- Front wheel

8. Check the transaxle fluid and add, as necessary.

9. Tighten the strut center nuts to 35 ft. lbs. (48 Nm).

Right

2002-2004 MODELS

➡**The hub nut may be reused 4 times. Mark the nut at removal and only use hub nuts with 3 or fewer marks for assembly.**

1. Before servicing the vehicle, refer to the Precautions Section.

2. Loosen the strut center nut 5 turns.

3. Remove or disconnect the following:
- Front wheel
- Hub retainer nut. Mark the nut.
- Lower ball joint
- Intermediate shaft retaining clip

4. Press the stub shaft out of the wheel hub.

5. Remove the right axle halfshaft and the intermediate shaft as an assembly.

To install:

➡**Use a new retaining clip for assembly.**

6. Install or connect the following:
- Axle halfshaft and intermediate shaft assembly. Tighten the nuts to 18 ft. lbs. (25 Nm).

- Lower ball joint. Tighten the pinch bolt to 37 ft. lbs. (50 Nm).
- Hub retainer nut. Tighten the nut to 232 ft. lbs. (316 Nm).
- Front wheel

7. Check the transaxle fluid and add, as necessary.

8. Tighten the strut center nuts to 35 ft. lbs. (48 Nm).

2005 MODELS

1. Loosen the right hand strut and spring assembly by loosening the mounting bolts by five turns.

2. Loosen the hub nut.

3. Remove the right hand tire.

4. Remove the lower ball joint nut and detach the lower control arm.

✳✳ CAUTION

Do not use a prying device or separator fork between the lower ball joint and knuckle. Use a pry bar by inserting it into the lower control arm opening.

5. Install Special Tool 307-102 over the ball stud before releasing the lower control arm and knuckle into rest positions.

307-102

06017-FOCU-G17

Installing Special Tool 307-102 on the lower ball stud.

204-085

204-069

06017-FOCU-G18

Using puller 204-069 to detach the halfshaft from the wheel hub.

06017-FOCU-G19

Location of the intermediate shaft retaining nuts—2005 models

6. Unscrew the hub nut and detach the right hand halfshaft from the wheel hub using a suitable puller.

7. Remove the retaining clip nuts.

8. Pull out the right hand halfshaft and intermediate shaft as an assembly.

✳ CAUTION

Support the halfshaft. The inner half-shaft joint must not be bent by more than 18 degrees.

To install:

9. Install the halfshaft/intermediate shaft assembly.

10. Using new nuts, install the retaining clip and tighten to 18 ft. lbs. (25 Nm).

11. Install or connect the following:
- Wheel hub retaining nut
- Lower control arm
- Lower ball joint

12. Tighten the right hand strut and spring assembly mounting bolts.

CV-Joints

OVERHAUL

Inner Tripod Joint

1. Before servicing the vehicle, refer to the Precautions Section.

2. Remove the axle halfshaft from the vehicle and place it in a vise.

3. Remove or disconnect the following:
- Tripod joint boot clamps
- Tripod joint boot
- Tripod joint housing
- Snapring
- Tripod joint

To install:

➡ **Use new snaprings and boot clamps for assembly.**

4. Install or connect the following:
- Tripod joint
- Snapring
- Tripod joint housing. Fill the joint housing with grease.
- Tripod joint boot
- Tripod joint boot clamps

5. Install the halfshaft.

Outer CV-Joint

2002–04 MODELS

The outer CV-joint is serviced with the halfshaft as an assembly. The outer CV-joint boot can be serviced by removing the inner tripod joint.

2005 MODELS

1. Remove the Inner CV-Joint.

2. Remove the outer boot clamps.

3. Remove the outer boot.

To install:

4. Slide the outer boot outwards over the CV-joint.

5. Press the boot into the annular groove on the drive joint.

6. Slide a small screwdriver under the boot seat to allow the air to escape.

7. Locate the boot in position and remove the screwdriver.

8. Install the boot clamps.

9. Install the inner CV-Joint.

STEERING AND SUSPENSION

Air Bag

✳✳ CAUTION

These vehicles are equipped with an air bag system. The system must be disarmed before performing service on, or around, system components, the steering column, instrument panel components, wiring and sensors. Failure to follow the safety precautions and the disarming procedure could result in accidental air bag deployment, possible injury and unnecessary system repairs.

PRECAUTIONS

Several precautions must be observed when handling the inflator module to avoid accidental deployment and possible personal injury.

- Never carry the inflator module by the wires or connector on the underside of the module

- When carrying a live inflator module, hold securely with both hands, and ensure that the bag and trim cover are pointed away

- Place the inflator module on a bench or other surface with the bag and trim cover facing up

- With the inflator module on the bench, never place anything on or close to the module that may be thrown in the event of an accidental deployment

Before servicing the vehicle, also be sure to refer to the Precautions Section as well

DISARMING

2002–2004 Models

1. Before servicing the vehicle, refer to the Precautions Section.

2. Position the vehicle with the front wheels in a straight-ahead position.

3. Disconnect the negative battery cable.

4. Disconnect the positive battery cable.

5. Wait at least 1 minute for the air bag backup power supply to drain before continuing.

6. Proceed with the repair.

7. Once repairs are completed, connect the battery cables, negative cable last.

8. Check the functioning of the air bag system by turning the ignition key to the **RUN** position and visually monitoring the

air bag indicator lamp in the instrument cluster. The indicator lamp should illuminate for approximately 6 seconds, then turn **OFF**. If the indicator lamp does not illuminate, stays on, or flashes at any time, a fault has been detected by the air bag diagnostic monitor.

2005 Models

1. Before servicing the vehicle, refer to the Precautions Section.
2. Turn the ignition to the **OFF** position.
3. Remove the Restraints Control Module (RCM) fuse from the control junction box located below the left hand side of the instrument panel.
4. Turn the ignition to the **ON** position and visually monitor the air bag indicator. If the indicator remains solid (no flashing) the correct fuse has been removed.

Power Rack & Pinion Steering Gear

REMOVAL & INSTALLATION

1. Before servicing the vehicle, refer to the Precautions Section.
2. Center the steering wheel and turn the ignition to the **LOCK** position.
3. Remove or disconnect the following:
   - Negative battery cable
   - Lower instrument panel cover
   - Steering shaft pinch bolt
   - Front wheels
   - Outer tie rod ends
   - Stabilizer bar links
   - Power steering fluid cooler hose
   - Right engine mount
   - Steering gear heat shield
   - Power steering hose support clamp
   - Power steering hoses
   - 6 subframe bolts. Support the subframe.
   - Floor seal
   - Steering gear pinion extension
   - Steering gear

To install:
4. Install or connect the following:
   - Steering gear. Tighten the bolts to 59 ft. lbs. (80 Nm).
   - Steering gear pinion extension. Tighten the pinch bolt to 26 ft. lbs. (35 Nm).
   - Floor seal
5. Install Subframe Alignment Pins 502-002 and raise the subframe into position. Tighten the 4 rear bolts to 147 ft. lbs. (200

Nm) and the other bolts to 85 ft. lbs. (115 Nm).
6. Install or connect the following:
   - Power steering hoses
   - Power steering hose support clamp
   - Steering gear heat shield
   - Right engine mount. Tighten the bolt to 37 ft. lbs. (50 Nm).
   - Power steering fluid cooler hose
   - Stabilizer bar links. Tighten the nuts to 37 ft. lbs. (50 Nm).
   - Outer tie rod ends. Tighten the nuts to 35 ft. lbs. (48 Nm).
   - Front wheels
   - Steering shaft pinch bolt. Tighten the bolt to 21 ft. lbs. (28 Nm).
   - Lower instrument panel cover
   - Negative battery cable
7. Check the wheel alignment and adjust as necessary.

Strut

REMOVAL & INSTALLATION

2002–2004 Models

1. Before servicing the vehicle, refer to the Precautions Section.
2. Remove or disconnect the following:
   - Front wheel
   - Brake hose bracket
   - Stabilizer bar link
   - Wheel speed sensor, if equipped
   - Brake caliper and rotor
   - Outer tie rod end
   - Lower ball joint
   - Steering knuckle pinch bolt. Separate the knuckle from the strut.
   - Upper strut mount nuts
   - Strut assembly

To install:
3. Install or connect the following:
   - Strut assembly. Tighten the upper mount nuts to 18 ft. lbs. (25 Nm).
   - Steering knuckle. Tighten the pinch bolt to 66 ft. lbs. (90 Nm).
   - Lower ball joint. Tighten the pinch bolt to 37 ft. lbs. (50 Nm).
   - Outer tie rod end. Tighten the nut to 35 ft. lbs. (48 Nm).
   - Brake caliper and rotor
   - Wheel speed sensor, if equipped
   - Stabilizer bar link. Tighten the nut to 37 ft. lbs. (50 Nm).
   - Brake hose bracket
   - Front wheel
4. Check the wheel alignment and adjust as necessary.

2005 Models

1. Before servicing the vehicle, refer to the Precautions Section.
2. Remove or disconnect the following:
   - Front wheel
   - Brake hose from the support bracket
   - Stabilizer bar connecting link
   - Wheel knuckle bolt
   - Wheel knuckle
   - Strut tower nuts
   - Strut and spring assembly

To install:
3. Install or connect the following:
   - Strut and spring assembly
   - Strut tower nuts and tighten to 22 ft. lbs. (30 Nm)
   - Wheel knuckle
   - Wheel knuckle bolt and tighten to 66 ft. lbs. (90 Nm)
   - Stabilizer bar connecting link and tighten to 35 ft. lbs. (48 Nm)
   - Brake hose to the support bracket
   - Front wheel
4. Check the wheel alignment and adjust as necessary.

Shock Absorber

REMOVAL & INSTALLATION

3 and 4 Door Models

1. Before servicing the vehicle, refer to the Precautions Section.
2. Remove or disconnect the following:
   - Luggage compartment interior trim panel
   - Upper shock absorber mounting nut
   - Lower shock absorber mounting bolt
   - Shock absorber

To install:

➡ **Tighten the shock absorber mounting fasteners with the suspension at curb height and the vehicle weight supported by the wheels.**

3. Install or connect the following:
   - Shock absorber. Guide the rod into the locating hole.
   - Lower shock absorber mounting bolt. Tighten the bolt to 85 ft. lbs. (115 Nm).
   - Upper shock absorber mounting nut. Tighten the nut to 13 ft. lbs. (18 Nm).

- Luggage compartment interior trim panel

Wagon

2002–2004 MODELS

1. Before servicing the vehicle, refer to the Precautions Section.

2. Remove the upper and lower mounting bolts and remove the shock absorber.

To install:

➡️**Tighten the shock absorber mounting fasteners with the suspension at curb height and the vehicle weight supported by the wheels.**

3. Install the shock absorber and tighten the mounting bolts to 85 ft. lbs. (115 Nm).

2005 MODELS

1. Before servicing the vehicle, refer to the Precautions Section.

2. Remove the rear tire.

3. Detach the exhaust system from the rear hanger, left side only.

4. Remove the exhaust heat shield, left side only.

5. Using a transmission jack or suitable lifting device, support the lower control arm.

6. Remove the upper and lower mounting bolts.

7. Remove the shock absorber.

To install:

8. Install the shock absorber and tighten the upper mounting bolt to 85 ft. lbs. (115 Nm).

9. Lower the control arm and remove the jack.

10. Install the exhaust heat shield, if removed.

11. Attach the exhaust system to the rear hanger, if removed.

12. Install the rear tire.

13. Tighten the lower mounting bolt to 85 ft. lbs. (115 Nm).

Coil spring

REMOVAL & INSTALLATION

Front

1. Before servicing the vehicle, refer to the Precautions Section.

2. Remove the strut from the vehicle.

3. Compress the coil spring using a suitable spring compressor until the spring comes away from the seat.

Front strut assembly exploded view

48 Nm

9306SG10

4. Remove the large center nut and slowly release the spring compressor.

To install:

5. Compress the spring and install it on the strut.

6. Install the upper strut mount. Tighten the nut to 35 ft. lbs. (48 Nm).

7. Install the strut assembly in the vehicle.

Rear

EXCEPT 2005 WAGON

1. Before servicing the vehicle, refer to the Precautions Section.

2. Raise and support the vehicle.

3. Install spring compressor 204-167 with Adapters 204-215.

4. Compress the coil spring and remove it.

To install:

5. Install the coil spring and remove the compressor.

2005 WAGON

1. Before servicing the vehicle, refer to the Precautions Section.

2. Remove the rear wheel.

3. Detach the stabilizer bar from the rear lower arms.

4. Using a transmission jack or suitable lifting device, raise the rear lower arm 1.25 inches (32 mm).

5. Detach the rear lower arm from the wheel knuckle.

6. Lower the rear lower arm.

7. Remove the spring.

To install:

8. Install the spring.

9. Raise the rear lower arm using a transmission jack or suitable lifting device.

10. Attach the rear lower arm to the wheel knuckle and tighten bolt to 85 ft. lbs. (115 Nm).

204-215

204-167

9306SG09

Coil spring compressor and adapters

Control arm torque sequence

11. Attach the stabilizer bar to the rear lower arms and tighten to 11 ft. lbs. (15 Nm).

12. Install the rear wheel.

Lower Ball Joint

REMOVAL & INSTALLATION

The lower ball joint is replaced with the lower control arm as an assembly.

Lower Control Arm

REMOVAL & INSTALLATION

2002–2004 Models

1. Before servicing the vehicle, refer to the Precautions Section.

2. Remove or disconnect the following:
- Front wheel
- Lower ball joint
- Rear bracket bolts
- Front bolt
- Control arm

To install:

➡**Use new nuts, bolts and ball bearing washers for assembly.**

3. Install the control arm. Tighten the fasteners in sequence as follows:

a. Step 1: Tighten nut No. 1 to 74 ft. lbs. (100 Nm) plus 60 degrees

b. Step 2: Tighten nut No. 2 to 88 ft. lbs. (120 Nm)

c. Step 3: Tighten bolt No. 3 to 88 ft. lbs. (120 Nm) plus 90 degrees

d. Step 4: Check that bolt No. 3 is tightened to 125–169 ft. lbs. (170–230 Nm)

4. Install or connect the following:
- Lower ball joint. Tighten the pinch bolt to 37 ft. lbs. (50 Nm).
- Front wheel

5. Check the wheel alignment and adjust as necessary.

2005 Models

1. Before servicing the vehicle, refer to the Precautions Section.

2. Remove the wheel.

3. Remove the pinch bolt and separate the lower ball joint from the wheel knuckle.

4. Remove the heat shield.

5. Remove the lower control arm bolts.

6. Remove the control arm.

To install:

➡**Use new nuts, bolts and ball bearing washers for assembly.**

7. Install the control arm and tighten the bolts as follows:

a. Tighten the front pivot bolt to 103 ft. lbs. (140 Nm).

b. Tighten the rear inboard nut to 76 ft. lbs. (103 Nm).

c. Tighten the rear outboard nut to 92 ft. lbs. (125 Nm).

8. Install the lower ball joint and tighten the pinch bolt to 35 ft. lbs. (48 Nm).

9. Install the wheel.

CONTROL ARM BUSHING REPLACEMENT

The lower control arm bushings are replaced with the lower control arm as an assembly.

Wheel Bearings

ADJUSTMENT

The bearings on the front and rear wheels are a one piece cartridge design and cannot be adjusted. If wheel bearing play is excessive, check the wheel hub retainer nut for proper torque. If the torque is correct, replacement of the wheel bearing is required.

REMOVAL & REPLACEMENT

Front

➡**The hub nut may be reused 4 times. Mark the nut at removal and only use hub nuts with 3 or fewer marks for assembly.**

1. Before servicing the vehicle, refer to the Precautions Section.

2. Loosen the strut center nut 5 turns.

3. Loose the wheel hub retaining nut.

Removing hub using a suitable puller.

06017-FOCU-G21

Removing outer bearing race using puller 205-D064.

06017-FOCU-G22

Removing the bearing inner ring using Special Tool 204-180.

06017-FOCU-G23

Removing the bearing outer ring from the wheel knuckle.

4. Remove or disconnect the following:

- Front wheel
- Wheel speed sensor, if equipped
- Hub retainer nut. Mark the nut.
- Brake caliper and rotor
- Outer tie rod end
- Lower ball joint

5. Separate the wheel hub from the halfshaft using a suitable puller.

6. Remove the steering knuckle pinch bolt

7. Release the knuckle using a suitable lever and remove from the vehicle.

8. Remove the wheel hub and outer bearing race using Special Tool 205-D064

9. Remove the bearing inner ring using Special Tools 204-180 and 205-D064

10. Remove the snapring from the knuckle.

11. Reassemble the bearing cage to the bearing inner and install to the wheel knuckle.

12. Using Special Tools 205-153, 204-023 and 206-054, remove the bearing outer ring from the wheel knuckle.

To install:

13. Press the bearing into the hub using Special Tools 205-153 and 205-140.

14. Install the snapring.

15. Press the hub into the wheel bearing using Special Tools 205-153, 204-023 and 206-054.

16. Draw the stub shaft into the wheel hub with Halfshaft installer 205-379.

17. Install or connect the following:

- Steering knuckle. Tighten the pinch bolt to 66 ft. lbs. (90 Nm).
- Lower ball joint. Tighten the pinch bolt to 37 ft. lbs. (50 Nm).
- Outer tie rod end. Tighten the nut to 35 ft. lbs. (48 Nm).
- Brake caliper and rotor
- Hub retainer nut. Tighten the nut on 2002–2004 models to 232 ft. lbs. (316 Nm). Tighten 2005 Models to 199 ft. lbs. (270 Nm).
- Wheel speed sensor, if equipped
- Front wheel

18. Tighten the strut center nuts as follows:

a. 2002–2004 Models: 35 ft. lbs. (48 Nm)

b. 2005 Models: 22 ft. lbs. (30 Nm).

19. Check the wheel alignment and adjust as necessary.

Rear

1. Before servicing the vehicle, refer to the Precautions Section.

2. Remove or disconnect the following:

06017-FOCU-G24

Installing the bearing into the hub with installer 205-140.

06017-FOCU-G25

Pressing out the rear bearing.

- Rear wheel
- Dust cap
- Brake caliper and rotor, if equipped
- Brake drum, if equipped
- Hub nut
- Wheel speed sensor tone ring, if equipped with ABS
- Snapring

3. Remove the wheel bearing with a press.

To install:

➡**Use a new wheel speed sensor tone ring for assembly.**

4. Press the wheel bearing into the brake drum hub.
5. Install or connect the following:
- Snapring
- Wheel speed sensor tone ring, if equipped with ABS

- Brake drum, if equipped
- Hub nut. Tighten the hub nut to 173 ft. lbs. (235 Nm).
- Brake caliper, if equipped
- Dust cap
- Rear wheel

BRAKES

Brake Caliper

REMOVAL & INSTALLATION

Front

1. Before servicing the vehicle, refer to the Precautions Section.
2. Remove or disconnect the following:
- Front wheel
- Brake hose from the support bracket
- Outer brake pad retaining clip
- Brake caliper from the steering knuckle
- Brake pads
- Brake hose from the caliper

To install:

3. Install or connect the following:
- Brake hose to the caliper. Tighten the fitting to 11 ft. lbs. (15 Nm).
- Brake pads
- Brake caliper to the steering knuckle. Tighten the bolts to 21 ft. lbs. (28 Nm).
- Outer brake pad retaining clip
- Brake hose to the support bracket
- Front wheel

4. Bleed the brakes and check for proper operation.

Rear

1. Before servicing the vehicle, refer to the Precautions Section.
2. Remove the rear wheel and tire.
3. Detach the parking brake cable from the brake caliper.
4. Using a suitable brake hose clamp, clamp the brake hose.
5. Loosen the brake hose union.

6. Detach the brake caliper from the brake anchor plate.
7. Remove the brake caliper.
8. Disconnect the brake hose.

To install:

9. Loosely connect the brake hose to the brake caliper.

➡**When the brake caliper piston is retracted into the piston housing, brake fluid will be displaced into the brake fluid reservoir.**

06017-FOCU-G26

Retracting the caliper piston using pinion adjust and adapter.

10. Using the Rear Caliper Piston Adjuster 206-010 and Adapter 206-026, retract the brake caliper piston.

➡ **Make sure the brake caliper retaining bolt springs are correctly located.**

11. Install the caliper and tighten the bolts to 26 ft. lbs. (35 Nm).
12. Tighten the brake hose union to 11 ft. lbs. (15 Nm).
13. Remove the brake hose clamp.
14. Attach the parking brake cable to the brake caliper.
15. Install the wheel and tire.
16. Bleed the brake system.

Disc Brake Pads

REMOVAL & INSTALLATION

Front

1. Before servicing the vehicle, refer to the Precautions Section.
2. Remove or disconnect the following:
   - Front wheel
   - Brake hose from the support bracket
   - Outer brake pad retaining clip
   - Brake caliper
   - Inner and outer brake pads

To install:
3. Compress the caliper piston into the caliper bore.
4. Install or connect the following:
   - Inner and outer brake pads
   - Brake caliper. Tighten the bolts to 21 ft. lbs. (28 Nm).
   - Brake hose to the support bracket
   - Front wheel

Rear

1. Before servicing the vehicle, refer to the Precautions Section.
2. Remove the rear wheel and tire.
3. Detach the parking brake cable from the brake caliper.
4. Loosen the brake hose union.
5. Detach the brake caliper from the brake anchor plate.
6. Remove the brake caliper.
7. Remove the brake pads.

To install:

➡ **When the brake caliper piston is retracted into the piston housing, brake fluid will be displaced into the brake fluid reservoir.**

8. Using the Rear Caliper Piston Adjuster 206-010 and Adapter 206-026, retract the brake caliper piston.

➡ **Make sure the brake caliper retaining bolt springs are correctly located.**

9. Install the brake pads.
10. Install the caliper and tighten the bolts to 26 ft. lbs. (35 Nm).
11. Attach the parking brake cable to the brake caliper.
12. Install the wheel and tire.
13. Bleed the brake system.

Brake Drums

REMOVAL & INSTALLATION

1. Before servicing the vehicle, refer to the Precautions Section.
2. Remove or disconnect the following:
   - Rear wheel

- Wheel speed sensor, if equipped
- Brake drum, bearing and spindle assembly.

To install:
3. Install or connect the following:
   - Brake drum, bearing and spindle assembly. Tighten the bolts to 49 ft. lbs. (66 Nm).
   - Wheel speed sensor, if equipped
   - Rear wheel

Brake Shoes

REMOVAL & INSTALLATION

1. Before servicing the vehicle, refer to the Precautions Section.
2. Remove or disconnect the following:
   - Rear wheel
   - Wheel speed sensor, if equipped
   - Brake drum, bearing and spindle assembly
   - Brake shoe hold-down springs and pins
   - Brake shoe assembly from the wheel cylinder and anchor block
   - Parking brake cable
   - Lower return spring
   - Upper return spring
   - Primary shoe from the strut and brake shoe adjuster
   - Parking brake return spring
   - Secondary shoe from the strut support

To install:
3. Install or connect the following:
   - Secondary shoe to the strut support
   - Parking brake return spring
   - Primary shoe to the strut and brake shoe adjuster. Rotate the adjuster fully clockwise.
   - Upper return spring
   - Lower return spring
   - Parking brake cable
   - Brake shoe assembly to the wheel cylinder and anchor block
   - Brake shoe hold-down springs and pins
   - Brake drum, bearing and spindle assembly. Tighten the bolts to 49 ft. lbs. (66 Nm).
   - Wheel speed sensor, if equipped
   - Rear wheel
4. Operate the brake pedal to adjust the rear brakes.

1. Lever
2. Cable
3. Retaining clip
4. Cable

06017-FOCU-G27

Detaching the parking brake cable from the caliper.

FORD AND MERCURY

Freestar • Monterey

9

SPECIFICATIONS AND MAINTENANCE CHARTS

VEHICLE AND ENGINE IDENTIFICATION CHART

Engine							Model Year	
Code	Liters	Cu. In.	Cyl.	Fuel Sys.	Engine Type	Eng. Mfg.	Code	Year
6	3.9	238	6	SEFI	OHV	Ford	4	2004
2	4.2	256	6	SEFI	OHV	Ford	5	2005
SEFI: Sequential Multi-port Fuel Injection							6	2006

06017-FRST-C01

GENERAL ENGINE SPECIFICATIONS

Year	Engine Displacement Liters	Engine VIN	Net Horsepower @ rpm	Net Torque @ rpm (ft. lbs.)	Bore x Stroke (in.)	Com- pression Ratio	Oil Pressure @ rpm
2004	3.9	6	193@4500	240@3500	3.81x3.46	9.3:1	40-125@2500
	4.2	2	201@4250	263@3650	3.81x3.80	9.3:1	40-125@2500
2005	3.9	6	193@4500	240@3500	3.81x3.46	9.3:1	40-125@2500
	4.2	2	201@4250	263@3650	3.81x3.80	9.3:1	40-125@2500
2006	3.9	6	193@4500	240@3500	3.81x3.46	9.3:1	40-125@2500
	4.2	2	201@4250	263@3650	3.81x3.80	9.3:1	40-125@2500

06017-FRST-C02

GASOLINE ENGINE TUNE-UP SPECIFICATIONS

Year	Engine Displacement Liters	Engine VIN	Spark Plugs Gap (in.)	Ignition Timing (deg.) MT	AT	Fuel Pump (psi)	Idle Speed (rpm) MT	AT	Valve Clearance In.	Ex.
2004	3.9	6	0.052-0.056	—	①	②	—	①	HYD	HYD
	4.2	2	0.052-0.056	—	①	②	—	①	HYD	HYD
2005	3.9	6	0.052-0.056	—	①	②	—	①	HYD	HYD
	4.2	2	0.052-0.056	—	①	②	—	①	HYD	HYD
2006	3.9	6	0.052-0.056	—	①	②	—	①	HYD	HYD
	4.2	2	0.052-0.056	—	①	②	—	①	HYD	HYD

NOTE: The Vehicle Emission Control Information label often reflects specification changes changes made during production.

The label figures must be used if they differ from those in this chart.

HYD: Hydraulic

① Controlled by the Powertrain Control Module (PCM) and cannot be manually adjusted.

② Engine running: 58 psi

   Key On, Engine Off (KOEO): 67 psi

06017-FRST-C03

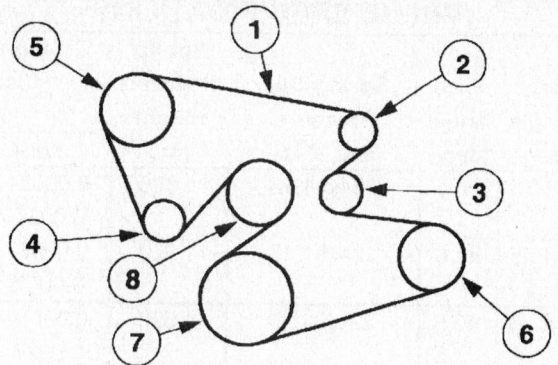

Item Description

1 Drive belt
2 Generator pulley
3 Idler pulley
4 Drive belt tensioner
5 Power steering pump pulley
6 A/C clutch pulley
7 Crankshaft vibration damper pulley
8 Coolant pump pulley

67197-FREE-G01

3.0L engine accessory drive belt routing

CAPACITIES

Year	Model	Engine Displacement Liters	Engine ID/VIN	Engine Oil with Filter (qts.)	Transmission (pts.) 4-Spd	5-Spd	Auto.	Drive Axle Front (pts.)	Rear (pts.)	Fuel Tank (gal.)	Cooling System (qts.)
2004	Freestar	3.9	6	5.0	—	—	12.25	①	—	26	②
		4.2	2	5.0	—	—	12.25	①	—	26	②
	Monterey	3.9	6	5.0	—	—	12.25	①	—	26	②
		4.2	2	5.0	—	—	12.25	①	—	26	②
2005	Freestar	3.9	6	5.0	—	—	12.25	①	—	26	②
		4.2	2	5.0	—	—	12.25	①	—	26	②
	Monterey	3.9	6	5.0	—	—	12.25	①	—	26	②
		4.2	2	5.0	—	—	12.25	①	—	26	②
2006	Freestar	3.9	6	5.0	—	—	12.25	①	—	26	②
		4.2	2	5.0	—	—	12.25	①	—	26	②
	Monterey	3.9	6	5.0	—	—	12.25	①	—	26	②
		4.2	2	5.0	—	—	12.25	①	—	26	②

NOTE: All capacities are approximate. Add fluid gradually and check to be sure a proper fluid level is obtained.

① Included in transaxle capacity
② w/auxiliary heater: 16.0
   wo/auxiliary heater: 15.0

VALVE SPECIFICATIONS

Year	Engine VIN	Engine Displacement Liters	Seat Angle (deg.)	Face Angle (deg.)	Spring Test Pressure (lbs. @ in.)	Spring Installed Height (in.)	Stem-to-Guide Clearance (in.)		Stem Diameter (in.)	
							Intake	Exhaust	Intake	Exhaust
2004	6	3.9	44.75	45.6	224@1.16	1.620	0.0008-0.0027	0.0018-0.0037	0.2746-0.2754	0.2735-0.2744
	2	4.2	44.75	45.6	224@1.16	1.620	0.0008-0.0027	0.0018-0.0037	0.2746-0.2754	0.2735-0.2744
2005	6	3.9	44.75	45.6	224@1.16	1.620	0.0008-0.0027	0.0080-0.0037	0.2746-0.2754	0.2735-0.2744
	2	4.2	44.75	45.6	224@1.16	1.620	0.0008-0.0027	0.0080-0.0037	0.2746-0.2754	0.2735-0.2744
2006	6	3.9	44.75	45.6	224@1.16	1.620	0.0008-0.0027	0.0080-0.0037	0.2746-0.2754	0.2735-0.2744
	2	4.2	44.75	45.6	224@1.16	1.620	0.0008-0.0027	0.0080-0.0037	0.2746-0.2754	0.2735-0.2744

06017-FRST-C05

CRANKSHAFT AND CONNECTING ROD SPECIFICATIONS

All measurements are given in inches.

Year	Engine Displ. Liters	Engine VIN	Crankshaft				Connecting Rod		
			Main Brg. Journal Dia.	Main Brg. Oil Clearance	Shaft End-play	Thrust on No.	Journal Diameter	Oil Clearance	Side Clearance
2004	3.9	6	2.5188-2.5196	0.0010-0.0014	0.0040-0.0080	3	2.3103-2.3111	0.0010-0.0014	0.0047-0.0193
	4.2	2	2.5188-2.5196	0.0010-0.0014	0.0040-0.0080	3	2.3103-2.3111	0.0010-0.0030	0.0047-0.0193
2005	3.9	6	2.5188-2.5196	0.0010-0.0014	0.0040-0.0080	3	2.3103-2.3111	0.0010-0.0014	0.0047-0.0193
	4.2	2	2.5188-2.5196	0.0010-0.0014	0.0040-0.0080	3	2.3103-2.3111	0.0010-0.0030	0.0047-0.0193
2006	3.9	6	2.5188-2.5196	0.0010-0.0014	0.0040-0.0080	3	2.3103-2.3111	0.0010-0.0014	0.0047-0.0193
	4.2	2	2.5188-2.5196	0.0010-0.0014	0.0040-0.0080	3	2.3103-2.3111	0.0010-0.0030	0.0047-0.0193

06017-FRST-C06

PISTON AND RING SPECIFICATIONS

All measurements are given in inches.

Year	Engine Displ. Liters	Engine VIN	Piston Clearance	Ring Gap			Ring Side Clearance		
				Top Comp.	Bottom Comp.	Oil Control	Top Comp.	Bottom Comp.	Oil Control
2004	3.9	6	0.0007-0.0017	0.0067-0.0130	0.0118-0.0217	0.0059-0.0256	0.0012-0.0026	0.0012-0.0028	Snug
	4.2	2	0.0007-0.0017	0.0066-0.0129	0.0118-0.0217	0.0059-0.0255	0.0012-0.0031	0.0012-0.0028	Snug
2005	3.9	6	0.0007-0.0017	0.0067-0.0130	0.0118-0.0217	0.0059-0.0256	0.0012-0.0026	0.0012-0.0028	Snug
	4.2	2	0.0007-0.0017	0.0066-0.0129	0.0118-0.0217	0.0059-0.0255	0.0012-0.0031	0.0012-0.0028	Snug
2006	3.9	6	0.0007-0.0017	0.0067-0.0130	0.0118-0.0217	0.0059-0.0256	0.0012-0.0026	0.0012-0.0028	Snug
	4.2	2	0.0007-0.0017	0.0066-0.0129	0.0118-0.0217	0.0059-0.0255	0.0012-0.0031	0.0012-0.0028	Snug

06017-FRST-C07

TORQUE SPECIFICATIONS

All readings in ft. lbs.

Year	Engine VIN	Engine Displacement Liters	Cylinder Head Bolts	Main Bearing Bolts	Rod Bearing Bolts	Crankshaft Damper Bolts	Flywheel Bolts	Manifold		Spark Plugs	Oil Pan Drain Plug
								Intake	Exhaust		
2004	6	3.9	①	②	③	118	59	④	18	12	19
	2	4.2	①	②	③	118	59	④	18	12	19
2005	6	3.9	①	②	③	118	59	④	18	12	19
	2	4.2	①	②	③	118	59	④	18	12	19
2006	6	3.9	①	②	③	118	59	④	18	12	19
	2	4.2	①	②	③	118	59	④	18	12	19

① Step 1: 14 ft. lbs.
Step 2: 29 ft. lbs.
Step 3: 36 ft. lbs.
Step 4: Loosen bolt no. 1 bolt 3 turns
Step 5: Tighten no. 1 bolt as listed in step 7
Step 6: Repeat steps 4 and 5 for each bolt in sequence
Step 7: Long bolts to 30 ft. lbs. plus 1/2 turn; short bolts to 18 ft. lbs. plus 1/2 turn

② Step 1: 37 ft. lbs.
Step 2: plus 120 degrees

③ Step 1: 18 ft. lbs.
Step 2: 33 ft. lbs.
Step 3: plus 105 degrees

④ Upper manifold :Step 1: 53 INCH lbs.
Step 2: 89 INCH lbs.
Lower manifold: Step 1: 44 INCH lbs.
Step 2: 89 INCH lbs.

06017-FRST-C08

WHEEL ALIGNMENT

Year	Model			Caster Range (+/-Deg.)	Caster Preferred Setting (Deg.)	Camber Range (+/-Deg.)	Camber Preferred Setting (Deg.)	Toe-in (Deg.)
2004	Freestar	Front	Left	0.75	+3.2	0.75	-0.40	-0.15+/-0.25
			Right	0.75	+3.2	0.75	-0.40	-0.15+/-0.25
		Rear		—	—	0.75	-0.30	-0.14+/-0.20
	Monterey	Front	Left	0.75	+3.2	0.75	-0.40	-0.15+/-0.25
			Right	0.75	+3.2	0.75	-0.40	-0.15+/-0.25
		Rear		—	—	0.75	-0.30	-0.14+/-0.20
2005	Freestar	Front	Left	0.75	+3.6	0.75	-0.40	-0.15+/-0.25
			Right	0.75	+3.6	0.75	-0.40	-0.15+/-0.25
		Rear		—	—	0.75	-0.30	-0.14+/-0.20
	Monterey	Front	Left	0.75	+3.6	0.75	-0.40	-0.15+/-0.25
			Right	0.75	+3.6	0.75	-0.40	-0.15+/-0.25
		Rear		—	—	0.75	-0.30	-0.14+/-0.20
2006	Freestar	Front	Left	0.75	+3.6	0.75	-0.40	-0.15+/-0.25
			Right	0.75	+3.6	0.75	-0.40	-0.15+/-0.25
		Rear		—	—	0.75	-0.30	-0.14+/-0.20
	Monterey	Front	Left	0.75	+3.5	0.75	-0.44	-0.15+/-0.25
			Right	0.75	+3.5	0.75	-0.44	-0.15+/-0.25
		Rear		—	—	0.75	-0.25	-0.14+/-0.25

06017-FRST-C09

TIRE AND WHEEL SPECIFICATIONS

Year	Model	OEM Tires Standard	OEM Tires Optional	Tire Pressures (psi) Front	Tire Pressures (psi) Rear	Wheel Size	Wheel Lug Nut Torque (Ft. Lbs.)
2004	Freestar	P225/60R16	P235/60R16	35	35	①	100
	Monterey	P225/60R16	P235/60R16	35	35	①	100
2005	Freestar	P225/60R16	P235/60R16	35	35	①	100
	Monterey	P225/60R16	P235/60R16	35	35	①	100
2006	Freestar	P225/60R16	P235/60R16	35	35	①	100
	Monterey	P225/60R16	P235/60R16	35	35	①	100

OEM: Original Equipment Manufacturer

PSI: Pounds Per Square Inch

STD: Standard

OPT: Optional

① Not available

06017-FRST-C10

BRAKE SPECIFICATIONS

All measurements in inches unless noted

Year	Model		Brake Disc Original Thickness	Brake Disc Minimum Thickness	Maximum Runout	Minimum Lining Thickness Front	Minimum Lining Thickness Rear	Brake Caliper Bracket Bolts (ft. lbs.)	Brake Caliper Mounting Bolts (ft. lbs.)
2004	Freestar	F	①	1.110	0.002	0.118	—	136	26
		R	①	0.720	0.002	—	0.118	75	24
	Monterey	F	①	1.110	0.002	0.118	—	136	26
		R	①	0.720	0.002	—	0.118	75	24
2005	Freestar	F	①	1.110	0.002	0.118	—	136	26
		R	①	0.720	0.002	—	0.118	75	24
	Monterey	F	①	1.110	0.002	0.118	—	136	26
		R	①	0.720	0.002	—	0.118	75	24
2006	Freestar	F	①	1.110	0.002	0.118	—	136	26
		R	①	0.720	0.002	—	0.118	75	24
	Monterey	F	①	1.110	0.002	0.118	—	136	26
		R	①	0.720	0.002	—	0.118	75	24

① Not available

06017-FRST-C11

SCHEDULED MAINTENANCE INTERVALS
FORD FREESTAR, MERCURY MONTEREY

TO BE SERVICED	TYPE OF SERVICE	VEHICLE MILEAGE INTERVAL (x1000)																	
		5	10	15	20	25	30	35	40	45	50	55	60	65	70	75	80	85	90
Engine oil & filter ①	R																		
Rotate tires	S/I	✓	✓	✓	✓	✓	✓	✓	✓	✓	✓	✓	✓	✓	✓	✓	✓	✓	✓
Engine coolant strength hoses & clamps	S/I			✓			✓			✓			✓				✓		✓
Air cleaner filter	R						✓						✓						✓
Automatic transmission fluid & filter	R						✓						✓						✓
Engine coolant ②	R																		
PCV valve	R												✓						✓
Spark plugs ③	R																		
Drive belts	S/I						✓						✓						✓
Exhaust system & heat shields	S/I						✓						✓						✓
Front & rear brakes	S/I	✓	✓	✓	✓	✓	✓	✓	✓	✓	✓	✓	✓	✓	✓	✓	✓	✓	✓
Fuel filter	R			✓			✓			✓			✓			✓			✓

R: Replace   S/I: Service or Inspect

① Engine oil and filter: replace every 3000 miles.

② Engine coolant: change initially at 100,000 miles or 5 years.

③ Spark plugs: replace every 100,000 miles.

Special Operating Condition Requirements

During extensive idling and/or low speed driving for long distances, as in heavy commercial use such as delivery, taxi, patrol car or livery:

Lube front lower control arm and steering linkage ball joints with

Zerk fittings (if equipped) every 4,800 km (3,000 miles) or 3 months.

Inspect brake system and check battery electrolyte level (Patrol cars) every 8,000 km (5,000 miles).

Install a new cabin air filter as required.

When operating in dusty conditions such as unpaved or dusty roads:

Install a new engine air filter as required.

Install a new cabin air filter as required.

When operating in off-road conditions:

Change automatic transmission fluid every 48,000 km (30,000 miles).

Install a new cabin air filter as required.

Inspect and lubricate U-joints.

Inspect and lubricate steering linkage ball joints with zerk fittings.

06017-FRST-C12

ENGINE REPAIR

Alternator

REMOVAL & INSTALLATION

1. Disconnect the negative battery cable.
2. Disconnect the accessory drive belt.
3. Detach the alternator wiring connector.
4. Remove the positive cable nut.
5. Remove the alternator.

To install:

6. Position the alternator on the engine.
7. Install the alternator mounting bolts. Tighten the bolts to 18 ft. lbs. (25 Nm) on 2004–05 models, or 35 ft. lbs. (47 Nm) on 2006 models.
8. Tighten the positive cable nut to 71 INCH lbs. (8 Nm).
9. Connect the wiring connector.
10. Install and tension the accessory drive belt.
11. Connect the negative battery cable.

Engine Assembly

REMOVAL & INSTALLATION

1. Relieve the fuel system pressure.
2. Drain the engine coolant.
3. Recover the air conditioning refrigerant, into a refrigerant recovery station.
4. Remove or disconnect the following:
   - Both battery cables
   - Wiper motor
   - Cowl top panel
   - Air cleaner assembly
   - Fuel supply manifold lines
   - Power steering pump
   - Starter

➡ **Do not allow the flex connector of the duel converter Y-pipe to hang unsupported or damage to the flex joint will result.**

   - Dual converter Y-pipe and support it from the body

   - Accelerator and cruise control cables from throttle assembly
   - Accelerator cable bracket
   - Upper and lower radiator hoses from the engine
   - Heater water hoses and secure to body
   - Positive terminal from Power Distribution Box (PDB)
   - PDB 68-pin connector
   - Engine wiring retainers
   - Shift cable
   - Evaporative (EVAP) return tube
   - EVAP canister connector and vacuum line
   - Brake booster vacuum line
   - Powertrain Control Module (PCM)
   - Engine grounds (3)
   - Radiator shield
   - Passenger side engine roll restrictor
   - A/C compressor and wire aside
   - Power steering cooler line
   - Transaxle cooler lines
   - Steering column pinch bolt
   - Oil pan-to-transaxle bolts
   - Torque converter nuts (4)
   - Brake calipers and wire aside
   - ABS connectors
   - Stabilizer bar links
   - Lower strut bolts on both sides
   - Position subframe support tool 014-00765 under subframe
   - Subframe bolts and lower the engine, transaxle and subframe assembly
   - Alternator brace
   - A/C compressor and wire aside
   - Driver side engine roll restrictor
   - Label and remove remaining electrical connectors
   - Transaxle fluid level tube
   - Install suitable engine lifting device
   - Bell housing bolts (4)
   - Engine mount bolts
   - Transaxle mount bolts
   - Engine

Item	Description	Item	Description
1	Generator electrical connector	4	Generator B+ cable
2	Generator B+ cable boot	5	Generator mounting bolts
3	Generator B+ cable nut	6	Generator

67197-FREE-G03

Alternator mounting—3.9L engine, 4.2L similar

67197-FREE-G04

Removing passenger side engine roll restrictor—Freestar and Monterey

Removing stabilizer bar links—Freestar and Monterey

Installing subframe support tool—Freestar and Monterey

Removing driver side engine roll restrictor—Freestar and Monterey

To install:

5. Position the engine onto the transaxle and subframe assembly.

6. Install the transaxle mount bolts. Torque the bolts to 46 ft. lbs. (63 Nm).

7. Install the oil pan-to-transaxle bolt. Torque the bolt to 33 ft. lbs. (45 Nm).

8. Install the engine mount bolts. Torque the bolts to 66 ft. lbs. (90 Nm).

9. Install the bell housing bolts. Torque the bolts to 37 ft. lbs. (50 Nm).

10. Install or connect the following:
- Transaxle fluid level tube
- Under vehicle electrical connectors
- Driver side roll restrictor bracket. Torque the bolts to 37 ft. lbs. (50 Nm).

- A/C compressor. Torque the bolts to 18 ft. lbs. (25 Nm).
- Generator brace
- Raise the engine, transaxle and subframe assembly
- Subframe bolts. Torque the bolts to 66 ft. lbs. (90 Nm).
- Lower strut bolts. Torque the bolts to 91 ft. lbs. (124 Nm).
- Stabilizer bar link bolts. Torque the bolts to 70 ft. lbs. (95 Nm).
- ABS connectors
- Brake calipers. Torque the bolts to 26 ft. lbs. (35 Nm).
- Torque converter bolts. Torque the bolts to 27 ft. lbs. (36 Nm).
- Oil pan-to-transaxle bolts. Torque the bolts to 33 ft. lbs. (45 Nm).
- Steering column pinch bolt. Torque the bolts to 30 ft. lbs. (40 Nm).

11. The remainder of the installation is the reverse of removal.

12. Please note the following torque specifications:
- Passenger side roll restrictor. Torque the bolts to 46 ft. lbs. (63 Nm).

13. Refill the engine, transaxle and cooling system with the correct amount of the appropriate fluids before starting the engine. Recharge the A/C system using approved recycling equipment.

Water Pump

REMOVAL & INSTALLATION

1. Remove or disconnect the following:
- Negative battery cable
- Coolant
- Accessory drive belt
- Lower radiator hose

- Idler pulley
- Generator bracket
- Lower engine mount nuts

➡**Install a suitable engine lifting device and raise the engine approximately 5 inches to gain access to the water pump pulley bolts.**

- Water pump pulley
- Bypass hoses
- Water pump

To install:

❋❋ WARNING

Be careful not to gouge the aluminum surfaces when scraping the old gasket material from the mating surfaces of the water pump and front cover.

2. Clean the gasket surfaces on the water pump and front cover.

3. Position a new water pump housing gasket on the water pump sealing surface using gasket sealant to hold the gasket in place.

4. Install the water pump and tighten the bolts in sequence shown to 21 ft. lbs. (28 Nm).

5. The remainder of the installation is the reverse of removal.

6. Fill and bleed the cooling system.

7. Connect the negative battery cable.

8. Start the engine and check for leaks.

Heater Core

REMOVAL & INSTALLATION

Front System

1. Disconnect the negative battery cable.

2. Drain the cooling system into a clean container for reuse.

Water pump bolt tightening sequence—Freestar and Monterey

12 Nm (9 lb-ft)

(9) ℹ️

67197-FREE-G09

Removing instrument panel support bolts—Freestar and Monterey

Support bracket

Lower air duct

Heater core

Heater core cover

67197-FREE-G10

Removing center air duct and heater core—Freestar and Monterey

3. Disconnect the heater hoses from the heater core inlet and outlet tubes in the engine compartment.

4. Remove the steering column lower covers.

5. Remove the center storage compartment

6. Remove the upper instrument panel finish cover.

7. Remove the instrument panel bolts (9) at the top and driver side of the instrument panel.

➡**Insert a long bolt or threaded rod in place of the left side instrument panel attaching bolt to support the instrument panel when removing the attaching bolts.**

8. Loosen the passenger side instrument panel bolts approximately halfway.

9. Remove the center instrument panel support brace.

10. From inside the glove box, remove the attaching bolt.

11. Pull the instrument panel outward approximately 3 inches.

12. Remove the lower center air duct.

13. Disconnect the heater core wiring connector.

14. Remove the heater core cover and heater core.

To install:

15. Installation is the reverse of removal.

16. Please note the following torque specifications:

- Upper instrument panel bolts. Torque the bolts to 44 INCH lbs. (5 Nm).
- Side and lower instrument panel bolts. Torque the bolts to 9 ft. lbs. (12 Nm).

17. Refill the cooling system.

18. Connect the negative battery cable.

19. Run the engine to normal operating temperatures; then, check the climate control operation and check for leaks.

Rear Auxiliary System

1. Disconnect the negative battery cable.

2. Drain the cooling system into a clean container for reuse.

3. Remove the left side quarter trim panel.

4. Remove the temperature blend door actuator.

Temperature blend
door activator

Rear heat core

67197-FREE-G11

Removing rear auxiliary heater core—Freestar and Monterey

5. Remove the heater hoses.
6. Remove the heater core.
To install:
7. Installation is the reverse of the removal procedure.

Cylinder Head

REMOVAL & INSTALLATION

Left Side

1. Relieve the fuel system pressure.
2. Remove or disconnect the following:
   - Negative battery cable
   - Coolant
   - Dual converter Y-pipe.
   - Exhaust manifold
   - Upper and lower intake manifolds
   - Alternator and mounting bracket
   - A/C compressor and mounting bracket
   - Valve covers

➡**Pushrods must be installed in their original positions. Note pushrod location during removal.**

   - Pushrods
   - Rotate tensioner clockwise and remove accessory drive belt

- Automatic belt tensioner assembly
- Lower exhaust manifold mounting studs
- Spark plugs
- Cylinder head bolts
- Cylinder head from the engine block and discard the gaskets

To install:
3. The cylinder head should be cleaned and inspected prior to installation.
4. Lightly oil all bolt and stud bolt threads before installation.
5. Clean all gasket mating surfaces thoroughly.

67197-FREE-G12

Left side cylinder head bolt torque sequence—3.9L and 4.2L engines

6. Install or connect the following:
- New head gaskets on the cylinder block.

✳✳ WARNING

Always use new cylinder head bolts when installing the cylinder head or damage to the engine may occur.

- Cylinder head on the cylinder block.

➡️**Long bolts go in the inside of the cylinder head and short bolts go on the outside.**

- Tighten the cylinder head bolts in steps following the proper torque sequence. The first step is 14 ft. lbs. (20 Nm), the second step is 29 ft. lbs. (40 Nm), the third step is 36 ft. lbs. (50 Nm).
- Using the torque sequence, loosen the first long bolt 3 turns, then tighten the bolt to 30 ft. lbs. (40 Nm). Tighten the bolt an additional 180 degrees. Using the torque sequence, loosen the first short bolt 3 turns, then tighten the bolt to 18 ft. lbs. (25 Nm). Tighten the bolt an additional 180 degrees. Repeat these steps for each bolt in sequence using the proper torque for long and short bolts.
- Lower exhaust manifold mounting studs
- Pushrods. Dip each end in engine assembly lubricant.
- Rocker arms, seats and retaining bolts. Lubricate all rocker arm components with engine assembly lubricant. Tighten the bolts to 44 inch lbs. (5 Nm).

➡️**The rocker arm seats must be fully seated in the cylinder head and the pushrods must be seated in the rocker arm sockets prior to the final tightening.**

- Final tighten all rocker arm retaining bolts to 24 ft. lbs. (32 Nm)
- Automatic belt tensioner assembly
- Accessory drive belt
- Valve covers
- A/C compressor and mounting bracket
- Alternator and mounting bracket
- Upper and lower intake manifolds
- Exhaust manifold
- Dual converter Y-pipe.
7. Fill and bleed the cooling system.

➡️**Engine coolant is corrosive to engine bearing material. Replace the engine oil after removal of any coolant-carrying component to help prevent potential bearing damage.**

8. Change the engine oil and filter
9. Connect the negative battery cable.
10. Start the engine and check for leaks.

Right Side

1. Remove or disconnect the following:
- Negative battery cable
- Coolant
- Dual converter Y-pipe.
- Exhaust manifold
- Upper and lower intake manifolds
- Power steering pump and mounting bracket
- Valve covers

➡️**Pushrods must be installed in their original positions. Note pushrod location during removal.**

- Pushrods
- Rotate tensioner clockwise and remove accessory drive belt
- Automatic belt tensioner assembly
- Lower exhaust manifold mounting studs
- Spark plugs
- Cylinder head bolts
- Cylinder head from the engine block and discard the gaskets

To install:

2. The cylinder head should be cleaned and inspected prior to installation.
3. Lightly oil all bolt and stud bolt threads before installation.
4. Clean all gasket mating surfaces thoroughly.
5. Install or connect the following:

- New head gaskets on the cylinder block.

✳✳ WARNING

Always use new cylinder head bolts when installing the cylinder head or damage to the engine may occur.

- Cylinder head on the cylinder block.

➡️**Long bolts go in the inside of the cylinder head and short bolts go on the outside.**

- Tighten the cylinder head bolts in steps following the proper torque sequence. The first step is 14 ft. lbs. (20 Nm), the second step is 29 ft. lbs. (40 Nm), the third step is 36 ft. lbs. (50 Nm).
- Using the torque sequence, loosen the first long bolt 3 turns, then tighten the bolt to 30 ft. lbs. (40 Nm). Tighten the bolt an additional 180 degrees. Using the torque sequence, loosen the first short bolt 3 turns, then tighten the bolt to 18 ft. lbs. (25 Nm). Tighten the bolt an additional 180 degrees. Repeat these steps for each bolt in sequence using the proper torque for long and short bolts.
- Lower exhaust manifold mounting studs
- Pushrods. Dip each end in engine assembly lubricant.
- Rocker arms, seats and retaining bolts. Lubricate all rocker arm components with engine assembly lubricant. Tighten the bolts to 44 inch lbs. (5 Nm).

Right side cylinder head bolt torque sequence—3.9L and 4.2L engines

67197-FREE-G12

➡ The rocker arm seats must be fully seated in the cylinder head and the pushrods must be seated in the rocker arm sockets prior to the final tightening.

- Final tighten all rocker arm retaining bolts to 24 ft. lbs. (32 Nm)
- Automatic belt tensioner assembly
- Accessory drive belt
- Valve covers
- Power steering pump and mounting bracket
- Upper and lower intake manifolds
- Exhaust manifold
- Dual converter Y-pipe.
6. Fill and bleed the cooling system.

➡ Engine coolant is corrosive to engine bearing material. Replace the engine oil after removal of any coolant-carrying component to help prevent potential bearing damage.

7. Change the engine oil and filter
8. Connect the negative battery cable.
9. Start the engine and check for leaks.

Rocker Arms

REMOVAL & INSTALLATION

1. Remove the valve cover.
2. Remove the rocker arm retaining bolt.

➡ Rocker the arms should be installed in their original location during assembly.

3. Remove the rocker arms. If more than 1 rocker arm is to be removed, identify each rocker arm location.

To install:
- Rocker arms, seats and retaining bolts. Lubricate all rocker arm components with engine assembly lubricant. Tighten the bolts to 44 inch lbs. (5 Nm).

➡ The rocker arm seats must be fully seated in the cylinder head and the pushrods must be seated in the rocker arm sockets prior to the final tightening.

- Final tighten all rocker arm retaining bolts to 24 ft. lbs. (32 Nm)
4. Install the valve cover.

Intake Manifold

REMOVAL & INSTALLATION

Upper

1. Remove or disconnect the following:
   - Negative battery cable

Upper intake manifold—3.9L and 4.2L engines

- Coolant
- Fuel system pressure
- Wiper motor and pivot assembly
- Cowl top panel
- Air cleaner outlet tube
- Accelerator cable
- Cruise control cable
- Cable bracket
- Evaporative (EVAP) return tube
- Brake booster vacuum hose
- Exhaust Gas Recirculation (EGR) tubes and connector
- Remaining electrical connectors
- Spark plug wires
- Coolant hoses
- Intake manifold

To install:
2. Installation is the reverse of the removal procedure, using the following torque specifications.
   - Intake manifold bolts and tighten all bolts in steps in sequence to 53 inch lbs. (6 Nm), then to 89 inch lbs. (10 Nm).
3. Fill and bleed the engine cooling system.

4. Connect the negative battery cable.
5. Start the engine and check for leaks.

Upper intake manifold tightening sequence—3.9L and 4.2L engines

Lower

1. Remove or disconnect the following:
   - Coolant
   - Fuel system pressure
   - Upper intake manifold
   - Fuel supply manifold
   - Water bypass tube from the heater water outlet tube
   - Heater hoses from manifold
   - Upper radiator hose
   - Electrical wiring harnesses

Lower intake manifold and related components—3.9L and 4.2L engines

67197-FREE-G15

Lower intake manifold bolt torque sequence—3.9L and 4.2L engines

67197-FREE-G16

- Lower intake manifold retaining bolts. Note the location of the 6 long bolts and 8 short bolts.

➡The lower intake manifold is sealed at each corner with sealer. To break the seal it may be necessary to pry on the front of the intake manifold with a pry-bar. If it is necessary, use care to prevent damage to the machined surfaces.

- Lower intake manifold

To install:

2. Thoroughly clean all gasket mating surfaces.

➡When using silicone rubber sealer, assembly must occur within 5 minutes after sealer application. After this time, the sealer may start to set up and its sealing effectiveness may be reduced.

3. Install or connect the following:
- Apply a 3mm bead of RTV silicone sealer at each corner where the cylinder head joins the engine block
- Front and rear intake manifold seals
- Lower intake manifold into position on the cylinder block, using new gaskets
- Install intake bolts in their original locations. Tighten in 2 steps in sequence to 44 inch lbs. (5 Nm), then again to 89 inch lbs. (10 Nm).
- Heater hoses
- Water bypass hose
- Fuel supply manifold
- Upper intake manifold
- Upper radiator hose

4. Fill and bleed the cooling system.
5. Start the engine and check for leaks.

Exhaust Manifold

REMOVAL & INSTALLATION

➡Spray the exhaust system fasteners with penetrating lubricant before removing them to help prevent broken studs and bolts. The use of a 6-point socket is highly recommended when removing exhaust system fasteners.

✳✳ CAUTION

To prevent serious burns, allow the exhaust manifold to cool down before attempting to remove it.

Left Manifold

1. Disconnect the negative battery cable.

2. Remove the dipstick tube bolt.
3. Disconnect the EGR tube from the manifold.
4. Raise and support the vehicle safely on jackstands.
5. Disconnect the dual converter Y-pipe from the exhaust manifold.
6. Lower the vehicle.
7. Remove the exhaust manifold.

To install:

8. Clean all gasket mating surfaces thoroughly.
9. Install a new exhaust manifold gasket and the exhaust manifold on the cylinder head. Start 2 nuts to hold the manifold in position.
10. Install the remaining nuts. Tighten the nuts in the sequence shown to 18 ft. lbs. (25 Nm).
11. Raise and support the vehicle safely.
12. Connect the dual converter Y-pipe.
13. Lower the vehicle.
14. Install the EGR tube.
15. Install the disptick bolt.
16. Connect the negative battery cable.
17. Start the engine and check for exhaust leaks.

25 Nm (18 lb-ft)

67197-FREE-G39

Left side exhaust manifold tightening sequence—3.9L and 4.2L engines

25 Nm (18 lb-ft)

67197-FREE-G44

Right side exhaust manifold tightening sequence—3.9L and 4.2L engines

Right Manifold

1. Disconnect the negative battery cable.

2. Raise and support the vehicle safely on jackstands.

3. Remove the cowl top panel.

4. Disconnect the dual converter Y-pipe from the exhaust manifold.

5. Lower the vehicle.

6. Remove the heat shield.

7. Remove the exhaust manifold.

To install:

8. Clean all gasket mating surfaces thoroughly.

9. Install a new gasket and the exhaust manifold on the cylinder head. Start 2 nuts to hold the manifold in position.

10. Install the remaining nuts. Tighten the nuts in the sequence shown to 18 ft. lbs. (25 Nm).

11. Install the heat shield.

12. Raise and support the vehicle safely on jackstands.

13. Connect the dual converter Y-pipe.

14. Lower the vehicle.

15. Connect the negative battery cable.

16. Start the engine and check for exhaust leaks.

Camshaft and Valve Lifters

REMOVAL & INSTALLATION

1. Rotate the crankshaft until the No. 1 piston is at the TDC on its compression stroke and the timing marks are aligned.

2. Remove or disconnect the following:
- Engine from the vehicle
- Valve covers
- Intake manifolds
- Pushrods
- Tappet guide plate and retainer
- Tappets
- Crankshaft pulley and damper
- Engine front cover assembly

3. Check the camshaft end-play as follows:

   a. Push the camshaft toward the rear of the engine and install a dial indicator, so the indicator point is on the camshaft sprocket attaching screw.

   b. Zero the dial indicator. Position a small prybar between the camshaft sprocket or gear and block.

   c. Pull the camshaft forward and release it. Camshaft end-play should be 0.001–0.006 in. (0.025–0.15mm).

   d. If the camshaft end-play is not within specification, replace the thrust plate upon reassembly.

Aligning balance shaft drive gear and driven gear—3.9L and 4.2L engines

67197-FREE-G17

4. Remove or disconnect the following:
- Timing chain and sprockets
- Balance shaft drive gear
- Camshaft keyway
- Balance shaft driven gear, thrust plate and balance shaft as an assembly
- Camshaft thrust plate and spacer

5. Carefully remove the camshaft by pulling it toward the front of the engine. Remove it slowly to avoid damaging the bearings, journals and lobes.

To install:

6. Clean and inspect all parts before installation.

7. Lubricate the camshaft lobes and journals with Molylube® or heavy engine oil.

8. Carefully install the camshaft, spacer and camshaft key.

➡ **If a new camshaft is being installed, recheck camshaft end-play.**

9. Lubricate the engine thrust plate with engine assembly lubricant, then install the thrust plate. Tighten the retaining bolts to 71 inch lbs. (8 Nm).

10. Install the timing chain and sprockets.

11. Install the balance shaft assembly. Tighten the bolts to 71 inch lbs. (8 Nm).

12. Install the balance shaft drive gear, ensuring that the timing marks are correctly located as shown.

➡ **Check the camshaft sprocket bolt for blockage of the drilled oil passages prior to installation, and clean if necessary.**

13. Install or connect the following:
- Engine front cover
- Crankshaft damper and pulley
- Hydraulic tappets into their original bores
- Align the valve tappet flats and install the tappet guide plate with the word **UP** facing you
- Install the intake manifold assembly
- Lubricate and install the pushrods and rocker arms
- Install the valve covers
- Install the engine assembly into the vehicle

Starter Motor

REMOVAL & INSTALLATION

1. Disconnect the negative battery cable.

2. Raise and support the vehicle safely.

3. Disconnect the starter electrical harness.

4. Remove the upper starter bolt.

5. Support the starter and remove the lower bolt.

6. Remove the starter from the vehicle.

Item	Description
1	Starter motor solenoid terminal cover
2	Starter solenoid battery cable nut
3	Starter solenoid battery cable
4	Starter solenoid wire nut
5	Starter solenoid wire

Item	Description
6	Ground strap nut
7	Ground strap
8	Starter motor stud bolt
9	Starter motor bolt
10	Starter motor

67197-FREE-G40

Starter motor mounting—3.9L and 4.2L engines

To install:

7. Position the starter in the vehicle.

8. Install the upper and lower bolts. Tighten to 21 ft. lbs. (28 Nm).

9. Connect the starter electrical harness.

10. Lower the vehicle.

11. Connect the negative battery cable.

Oil Pan

✳✳ CAUTION

The EPA warns that prolonged contact with used engine oil may cause a number of skin disorders, including cancer! You should make every effort to minimize your exposure to used engine oil. Protective gloves should be worn when changing the oil. Wash your hands and any other exposed skin areas as soon as possible after exposure to used engine oil. Soap and water, or waterless hand cleaner, should be used.

REMOVAL & INSTALLATION

1. Disconnect the negative battery cable.

2. Raise and support the vehicle safely on jackstands.

3. Drain the engine oil.

4. Remove the oil filter.

5. Remove the dual converter Y-pipe assembly.

6. Remove the starter motor.

7. Remove the engine rear plate/converter housing cover.

8. Remove the 3 oil pan-to-transaxle bolts.

9. Remove the retaining bolts and remove the oil pan.

To install:

10. Clean the gasket mating surfaces thoroughly.

11. Trial fit the oil pan to the cylinder block. Ensure that enough clearance has been provided to allow the oil pan to be installed without sealant being scraped off when pan is positioned under the engine.

12. Apply a bead of silicone sealer to the oil pan flange. Also apply a bead of sealer to the front cover/cylinder block joint and fill the grooves on both sides of the rear main seal cap.

➡️**When using silicone rubber sealer, assembly must occur within 5 minutes after sealer application. After this time, the sealer may start to harden and its sealing effectiveness may be reduced.**

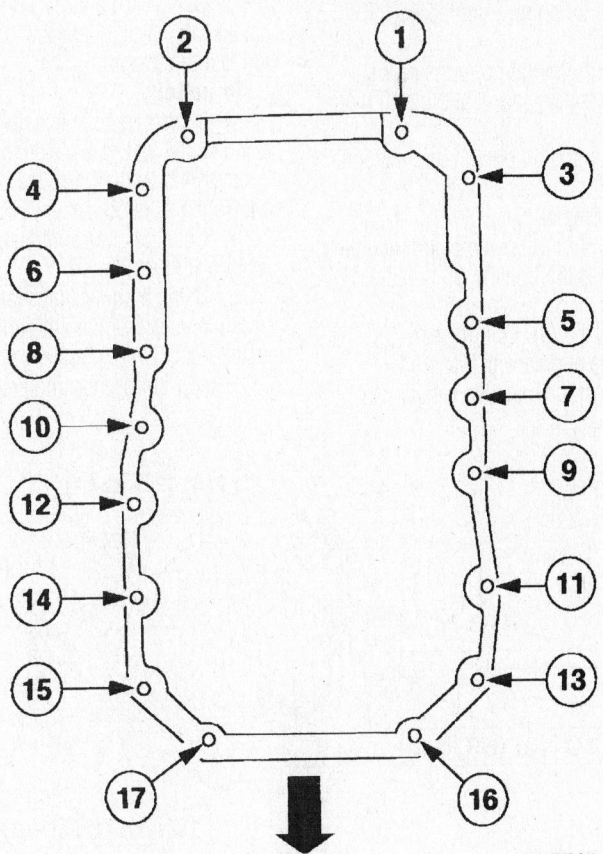

67197-FREE-G18

Oil pan bolt torque sequence—3.9L and 4.2L engines

13. Install the oil pan and secure to the block with the attaching bolts. Tighten the bolts in sequence to 89 inch lbs. (10 Nm).

14. Install the oil pan-to-transaxle bolts and tighten to 33 ft. lbs. (45 Nm).

15. Install a new oil filter.

16. Install the engine rear plate/converter housing cover.

17. Install the starter motor.

18. Install the Y-pipe converter assembly.

19. Lower the vehicle.

20. Fill the engine with the proper type and amount of clean oil.

21. Connect the negative battery cable.

22. Start the engine and check for leaks.

Oil Pump

REMOVAL & INSTALLATION

1. Disconnect the negative battery cable.

2. Drain the engine oil.

3. Remove the A/C compressor and wire aside.

4. Remove the A/C compressor mounting bracket

5. Remove the oil filter.

6. Remove the oil pump attaching bolts. Lift the oil pump from the engine.

To install:

7. Place the oil pump in the proper position with a new gasket and install the retaining bolt.

8. Tighten the oil pump retaining bolt to the proper torque as shown.

9. Install the oil filter.

10. Install the A/C compressor mounting bracket. Torque the bolts to 35 ft. lbs. 47 Nm).

11. Install the A/C compressor.

12. Fill the engine with clean oil.

13. Connect the negative battery cable.

➡ Check for proper engine oil pressure immediately after starting the engine. If engine oil pressure is not within specification a few seconds after starting the engine, stop the engine and determine the reason for the low oil pressure condition. Running an engine with low oil pressure may result in serious engine damage.

14. Start the engine and check for leaks.

Rear Main Seal

REMOVAL & INSTALLATION

1. Disconnect the negative battery cable.

2. Raise and support the vehicle safely on jackstands.

3. Remove the transaxle.

4. Remove the flywheel and the rear cover plate, if necessary.

✷✷ WARNING

Use caution when working near the crankshaft sealing surface. If the surface becomes damaged, an oil leak may occur.

5. Screw in the threaded end of a crankshaft rear seal replacer tool, then use the tool to remove the seal.

To install:

6. Inspect the crankshaft seal area for any damage that may cause the seal to leak. If damage is evident, service or replace the crankshaft as necessary.

7. Coat the crankshaft seal area and the seal lip with engine oil.

8. Using a crankshaft seal replacer tool, install the seal. Tighten the bolts of the seal installer tool evenly so the seal is straight and seats without misalignment.

67197-FREE-G20

The rear main seal must be installed with the proper tools to avoid damaging the seal or crankshaft

9. Install the flywheel.

10. Install the rear cover plate, if necessary.

11. Install the transaxle, lower the vehicle and connect the battery.

Timing Chain, Sprockets, Front Cover and Seal

REMOVAL & INSTALLATION

1. Rotate the crankshaft until No. 1 cylinder is at TDC on compression stroke.

2. Remove or disconnect the following:
   - Negative battery cable
   - Coolant
   - Air cleaner assembly and air intake duct
   - Alternator
   - Coolant hoses
   - Water pump
   - Camshaft position sensor
   - Power steering pump and position aside
   - Camshaft synchronizer
   - Alternator mounting bracket
   - Accessory drive belt and tensioner
   - Oil filter
   - Right side ABS connector
   - Place safety stand under right side of subframe
   - Remove 2 subframe bolts and lower right side of subframe about 2 inches
   - Crankshaft pulley
   - Raise subframe and reinstall bolts
   - Crankshaft front seal
   - Front oil pan bolts
   - Front cover

➡ Do not overlook the cover retaining bolt located behind the oil filter adapter. The front cover will break if pried on, and all retaining bolts are not removed.

   - Camshaft synchronizer drive gear
   - Camshaft gear

25 Nm (18 lb-in)

10 Nm (89 lb-in)

10 Nm (89 lb-in)

67197-FREE-G19

Oil pump bolt tightening torques—3.9L and 4.2L engines

Description	Description	Description
1 Camshaft position sensor (CMP) electrical connector	5 Coolant bypass hose	9 Power steering pump
2 CMP bolts	6 Coolant pump outlet tube bolts	10 Camshaft synchronizer-to-engine front cover bolt
3 CMP	7 Coolant pump outlet tube	11 Camshaft synchronizer
4 Clamp	8 Power steering pump bolts	

67197-FREE-G21

Exploded view of camshaft synchronizer assembly—3.9L and 4.2L engines

3. Ensure the crankshaft timing marks and keyways align as shown.

4. Compress and install a retaining pin to hold the timing chain tensioner.

5. Remove or disconnect the following:
- Camshaft gear, crankshaft gear and timing chain as an assembly
- Timing chain tensioner

To install:

6. Clean all the gasket mating surfaces.

7. Install the timing chain tensioner and tighten the bolt to 71 inch lbs. (11 Nm).

8. Retract the tensioner and insert a retaining pin.

9. Install or connect the following:
- Camshaft sprocket, crankshaft sprocket and timing chain. Be sure the timing marks align.
- Remove tensioner retaining pin
- Camshaft synchronizer drive gear and tighten the bolt to 33 ft. lbs. (45 Nm)
- Place silicone gasket sealant on front oil pan mating surface
- Install water pump to front cover
- Install front cover and torque bolts in sequence to 21 ft. lbs. (28 Nm)
- Front oil pan bolts

➡Front cover bolt no. 12 is tightened to 89 inch lbs. (10 Nm)

Exploded view of front crankshaft pulley—3.9L and 4.2L engines

67197-FREE-G22

Exploded view of front cover and timing components—3.9L and 4.2L engines

Aligning gear timing marks and keyway— 3.9L and 4.2L engines

- New crankshaft seal in the front cover and lubricate the seal lip with engine oil
- Remove right subframe bolts and lower subframe again about 2 inches
- Crankshaft pulley and tighten the bolt to 118 ft. lbs. (160 Nm)
- Raise subframe, install bolts and tighten to 66 ft. lbs. (90 Nm)
- Right side ABS connector

Front cover bolt torque sequence—3.9L and 4.2L engine

- Oil filter
- Accessory drive belt and tensioner
- Alternator mounting bracket
- Camshaft synchronizer

10. Coat the camshaft synchronizer gear with clean engine oil.

11. Install the special tool 303-630 onto the top of the camshaft synchronizer.

➡ **During installation the arrow on the special tool will rotate clockwise until**

Installing special tool on camshaft synchronizer—3.9L and 4.2L engines

the oil pump intermediate shaft engages the camshaft gear.

12. Install the camshaft synchronizer until the arrow on the special tool is at 54 degrees from the centerline of the engine as shown.

13. Install or connect the following:
- Power steering pump
- Camshaft position sensor
- Coolant hoses
- Alternator

- Air cleaner assembly and air intake duct

14. Fill the crankcase with the proper type and quantity of engine oil. Fill and bleed the cooling system. Connect the negative battery cable.

15. Start the engine and check for leaks. Check the ignition timing and curb idle speed and adjust, as necessary.

Piston and Ring

POSITIONING

Installing camshaft synchronizer—3.9L and 4.2L engines

67197-FREE-G28

Piston positioning in cylinder block—3.9L and 4.2L engines

FUEL SYSTEM

Fuel System Service Precautions

Safety is the most important factor when performing not only fuel system maintenance, but any type of maintenance. Failure to conduct maintenance and repairs in a safe manner may result in serious personal injury or death. Work on a vehicle's fuel system components can be accomplished safely and effectively by adhering to the following rules and guidelines.

- To avoid the possibility of fire and personal injury, always disconnect the negative battery cable unless the repair or test procedure requires that battery voltage by applied.

- Always relieve the fuel system pressure prior to disconnecting any fuel system component (injector, fuel rail, pressure regulator, etc.) fitting or fuel line connection. Exercise extreme caution whenever relieving fuel system pressure, to avoid exposing skin, face and eyes to fuel spray. Please be advised that fuel under pressure may penetrate the skin or any part of the body that it contacts.

- Always place a shop towel or rag around the fitting or connection prior to loosening to absorb any excess fuel due to spillage. Ensure that all fuel spillage is quickly remove from engine surfaces. Ensure that all fuel-soaked

cloths or towels are deposited into a flameproof waste container with a lid.

- Always keep a dry chemical (Class B) fire extinguisher near the work area.

- Do not allow fuel spray or fuel vapors to come into contact with a light bulb, spark or open flame.

- Always use a second wrench when loosening or tightening fuel line connections fittings. This will prevent unnecessary stress and torsion to fuel piping. Always follow the proper torque specifications.

- Always replace worn fuel fitting O-rings with new ones. Do not substitute fuel hose where rigid pipe is installed.

Fuel System Pressure

RELIEVING

2004–05 Models

Locate the fuel pump relay in the engine compartment fuse block at location K4/C-1051. Remove the fuel pump relay. Start the engine and allow it to idle until it stalls. Crank the engine for 5 seconds to ensure fuel supply manifold pressure is released. When vehicle service is complete, reinstall the fuel pump relay and turn

the ignition on to pressurize the fuel system. Start the vehicle and check the system for leaks.

2006 Models

Remove the Schrader valve cap from the fuel rail. Install fuel pressure gauge no. 310-012. Slowly open the manual valve to release the pressure, using a container to catch any residual fuel. When servicing is complete, remove the gauge, install the cap and start the engine. Check for leaks.

Fuel Filter

REMOVAL & INSTALLATION

1. Relieve the fuel system pressure.
2. Raise and support the vehicle safely on jackstands.
3. Place a rag under the fuel filter to catch any residual fuel that may leak out when the filter is removed.
4. Remove the quick-connect fittings at both ends of the fuel filter.
5. Install retainer clips in each fitting.
6. Note the flow arrow direction for installation reference.
7. Remove the fuel filter by pulling it from the bracket.

Fuel line quick-connect fitting removal—3.9L and 4.2L engines

67197-FREE-G41

Fuel line quick-connect fitting installation—3.9L and 4.2L engines

67197-FREE-G42

To install:

8. Install the fuel filter in its bracket, ensuring proper direction of flow as noted earlier.

9. Apply clean engine oil to end of fuel line and install the quick-connect fittings at both ends of the fuel filter. Ensure the fuel line clicks into place and is fully seated.

10. Start the engine and check the filter connections for leaks by running the tip of your finger around each connection.

11. Turn the engine off and lower the vehicle.

Fuel Pump

REMOVAL & INSTALLATION

➡️To gain access to the fuel pump, it is necessary to remove the fuel tank. The fuel tank is equipped with a composite protective shield mounted with the support straps and retainers on the right side of the tank. Make sure the shield is correctly installed when installing the tank.

1. Disconnect the negative battery cable.

2. Depressurize the fuel system and drain the fuel into a suitable container.

3. Remove the fuel filler pipe.

❄ CAUTION

Some fuel may remain in the filler pipe.

4. Support the fuel tank.

5. Disconnect the electrical connectors and fuel line fittings.

In-tank electric fuel pump and related components—Freestar and Monterey 2004–05 models

67197-FREE-G29

1	Fuel pump module lock ring	6	Fuel tube quick connect coupling
2	Fuel pump module	7	Fuel tube quick connect coupling
3	Fuel pump module O-ring	8	Fuel tank
4	Fuel level sensor		
5	Fuel pump module electrical connector		

06017-FRST-G01

In-tank electric fuel pump and related components—Freestar and Monterey 2006 models

6. Disconnect the retaining straps and lower the tank.

7. Remove any dirt that has accumulated around the fuel pump module attaching flange to prevent it from entering the tank during service.

8. Turn the fuel pump module locking ring counterclockwise using a locking ring removal tool, and remove the locking ring.

9. Remove the fuel pump module.

10. Remove the seal gasket and discard it.

To install:

11. Put a light coating of grease on a new seal ring to hold it in place during assembly. Install it in the fuel tank ring groove.

12. Insert the fuel pump module into the fuel tank, then secure it in place with the locking ring. Tighten the ring to 60 ft. lbs. (81 Nm).

13. Install the tank in the vehicle.

14. Install a minimum of 10 gallons (38L) of fuel and check for leaks.

15. Check for fuel leaks at the fittings.

16. Start the engine and check for fuel leaks.

Fuel Injectors

REMOVAL & INSTALLATION

1. Disconnect the negative battery cable.
2. Relieve the fuel system pressure.
3. Disconnect the fuel charging wiring harness.
4. Remove the fuel pressure vacuum connector and fuel supply line.
5. Remove the fuel injection supply manifold, with injectors attached.
6. Remove the retaining clips and separate the injectors from the supply manifold.

To install:

7. Inspect fuel injector O-rings for signs of deterioration and replace as required.

➡**Never use silicone grease on fuel injectors.**

8. Lubricate O-rings with clean engine oil and position fuel injectors on the supply manifold. Install the retaining clips.

9. Install the fuel injection supply manifold, making sure the injectors are seated properly. A slight rocking motion helps seat the injectors.

10. Install the fuel pressure vacuum connector and fuel supply line.

11. Install the fuel charging wiring connectors.

12. Start the engine and check for fuel leaks.

Fuel injectors and supply manifold—3.9L and 4.2L engines–2004-05

67197-FREE-G30

1	Fuel rail bolt (4 required)
2	Fuel rail
3	Fuel injector clip (6 required)
4	Fuel injector (6 required)
5	Fuel injector O-ring seal (12 required)

06017-FRST-G02

Fuel injectors and supply manifold—3.9L and 4.2L engines–2006

DRIVE TRAIN

Transaxle Assembly

REMOVAL & INSTALLATION

1. Remove or disconnect the following:
   - Battery and battery tray
   - Air cleaner assembly
   - All transaxle electrical harnesses
   - Transaxle shift cable from the lever by unsnapping the shift cable end from the lever ball stud
   - Manual control lever
   - Hood
   - Wiper arm and pivot assembly
   - Passenger side air intake box
   - Air intake tube
   - Upper radiator shield
   - Hood latch and oil filler cap

➡**Install engine lifting eyes to support the engine during transaxle removal.**

2. Install an engine support kit and suitably support the engine.
3. Remove or disconnect the following:
   - Transaxle filer tube
   - Transaxle cooler lines
   - The 2 upper transaxle-to-engine bolts
   - Raise and support the vehicle
   - Anti-roll bracket under the vehicle
   - Rear transaxle support nut
   - Transaxle fluid
   - Catalytic converter assembly
   - Front wheels
4. Disconnect the left and right tie rod ends from the knuckles.
5. Disconnect the sway bar ends from the knuckles.

Installing engine support kit—3.9L and 4.2L engines

6. Remove the left and right knuckle pinch bolts.
7. Remove or disconnect the following:
   - Steering rack bolts
   - Power steering hose retainers
   - Front and rear engine mount nuts
   - Place a jack under the subframe to support it
   - 4 subframe mounting bolts
8. Partially lower the subframe and remove the sway bar mounting bolts. Remove the sway bar.
9. Lower the subframe from the vehicle.
10. Remove or disconnect the following:
    - Wire the steering rack in place
    - Starter
    - Halfshafts from the transaxle and support them out of the way
    - Torque converter nuts and discard them
    - Transaxle electrical connectors
11. Position a transaxle jack to support the transaxle and remove the transaxle mounting nuts.
12. Remove the oil pan-to-transaxle bolt.
13. Separate the transaxle from the engine block by carefully moving the transaxle rearward until enough clearance exists to remove the transaxle from the engine compartment.

To install:

➡**If a different transaxle is being installed, transfer the heat shield to the new transaxle.**

14. Place the transaxle on a suitable jack and carefully raise it into position.
15. Install or connect the following:
    - Oil pan-to-transaxle bolt, and tighten the bolt to 46 ft. lbs. (62 Nm).
    - Engine bracket-to-transaxle bolts, and tighten the bolts to 46 ft. lbs. (62 Nm)
    - Transaxle-to-engine bolts, and tighten the bolts to 35 ft. lbs. (47 Nm)
    - Transaxle mounting bolts, and tighten the bolts to 46 ft. lbs. (62 Nm).
    - Transaxle electrical connectors
    - NEW torque converter nuts, and tighten them to 26 ft. lbs. (35 Nm)
    - Transaxle housing cover, and tighten the bolts to 89 inch lbs. (10 Nm)
    - Both halfshafts
    - Starter motor, and connect the electrical harness

- Speedometer cable
- Front subframe
- Sway bar mounting bolts, and tighten the bolts to 46 ft. lbs. (62 Nm)
- The 4 subframe bolts, and tighten them to 66 ft. lbs. (90 Nm)
- Front and rear engine mount bolts, and tighten them to 66 ft. lbs. (90 Nm)
- Power steering hose retainers
- Steering rack bolts, and tighten them to 98 ft. lbs. 133 Nm).
- Left and right knuckle pinch bolts, and tighten the bolts to 46 ft. lbs. (62 Nm).
- Sway bar end mounting nuts, and tighten the bolts to 41 ft. lbs. (55 Nm).
- Tie rod end nuts, and tighten the nuts to 41 ft. lbs. (55 Nm).
- Catalytic converter
- Rear transaxle mounting nut, and tighten to 66 ft. lbs. (90 Nm)
- Anti-roll bracket
- Upper transaxle-to-engine bolts, and tighten the bolts to 46 ft. lbs. (62 Nm)
- Fluid cooler-to-transaxle lines
- Front wheels and lower the vehicle
16. Remove the engine support kit.
17. Install or connect the following:
    - Hood latch
    - Radiator shield
    - Air intake tube and intake box
    - Wiper arm and pivot assembly
    - Hood
    - Transaxle shift cable to the manual lever ball stud
    - Transaxle electrical harnesses
    - Air cleaner assembly
    - Battery tray and battery
18. Fill the transaxle with proper amount of Mercon® V fluid.
19. Connect the positive, then the negative battery cable.

Halfshafts

REMOVAL & INSTALLATION

➡**Do not begin this removal procedure unless a new wheel hub retainer nut, a new retainer circlip and a new lower ball joint-to-front wheel knuckle retaining bolt and nut are available. Once removed, these parts must not be reused during assembly. Their torque holding ability, or retention capability, is diminished during removal.**

Washer

Halfshaft

Hub nut

Pinch bolt

67197-FREE-G31

Exploded view of front halfshaft mounting

205-D070

67197-FREE-G32

Remove the halfshaft from the transaxle using a puller

1. Remove or disconnect the following:
   - Front wheels
   - ABS sensor
   - Brake caliper and wire aside
   - Axle hub nut and washer. Discard the nut.
   - Lower control arm ball joint pinch bolt and nut and discard them

2. Separate the outboard CV-joint from the wheel hub using a front hub remover/replacer. Make sure the hub remover adapter is fully threaded onto the hub stud.

3. Separate the halfshaft from the hub, and pull the halfshaft out of the transaxle.

To install:

✴✴ WARNING

Do not reuse the retainer circlip. A new circlip must be installed each time the inboard CV-joint stub shaft is installed into the transaxle differential.

4. Install a new retainer circlip on the inboard CV-joint stub shaft by starting one end in the groove and working the retainer circlip over the inboard shaft housing end and into the groove. This will avoid over-expanding the circlip.

➡**A non-metallic mallet may be used to aid in seating the retainer circlip into the differential side gear groove. If a mallet is necessary, tap only on the outboard CV-joint shaft.**

5. Carefully align the splines of the inboard CV-joint stub shaft housing with the splines in the differential. Exerting some force, push the inboard CV-joint stub shaft

housing into the differential until the retainer circlip is felt to seat in the differential side gear. Use care to prevent damage to the inboard CV-joint stub shaft and transaxle seal.

6. Carefully align the splines of the outboard CV-joint with the splines in the wheel hub and push the shaft into the wheel hub as far as possible.

7. Temporarily fasten the front disc brake rotor to the wheel hub with washers and 2 lug nuts. Insert a steel rod into the front disc brake rotor and rotate clockwise to contact the front wheel knuckle to prevent the front disc brake rotor from turning when the nut is tightened.

➡**A new front axle wheel hub retaining nut must be installed.**

8. Manually thread the front axle wheel hub retaining nut onto the outboard CV-joint stub shaft housing as far as possible.

➡**A new bolt and nut must be used to connect the front suspension arm to the knuckle.**

9. Connect the front suspension lower arm to the front wheel knuckle. Tighten the nut and bolt to 46 ft. lbs. (63 Nm).

10. Install the front brake anti-lock sensor.

➡**Do not use power or impact tools to tighten the hub nut.**

11. Tighten front axle wheel hub retaining nut to 111ft. lbs. (150 Nm).

12. Install the front wheels and lower the vehicle.

13. Fill the transaxle to the proper level with Mercon® V automatic transmission fluid.

CV-Joint

OVERHAUL

➡**Overhaul procedures are not available from manufacturer. The assembled length of the left side halfshaft should be 24.07 inches (611.37 mm), and the assembled length of the right halfshaft should be 28 inches (711.2 mm).**

STEERING AND SUSPENSION

Air Bag (Supplemental Restraint) System

The Supplemental Restraint System (SRS) is designed to work in conjunction with the standard 3-point safety belts to reduce injury in a head-on collision.

✳✳ CAUTION

The SRS can actually cause physical injury or death if the safety belts are not used, or if the manufacturer's warnings are not followed. The manufacturer's warnings can be found in your owner's manual, or, in some cases, on your sun visor.

The SRS is comprised of the following components:

- Driver's side air bag module
- Passenger's side air bag module
- Right-hand and left-hand primary crash front air bag sensors
- Air bag diagnostic monitor computer
- Electrical wiring

The SRS primary crash front air bag sensors are hard-wired to the air bag modules and determine when the air bags are deployed. During a frontal collision, the sensors quickly inflate the 2 air bags to reduce injury by cushioning the driver and front passenger from striking the dashboard, windshield, steering wheel and any other hard surfaces. The air bag inflates so quickly (in a fraction of a second) that in most cases it is fully inflated before you actually start to move during a collision.

Since the SRS is a complicated and essentially important system, its components are constantly being tested by a diagnostic computer. The computer illuminates the air bag indicator light on the instrument cluster for approximately 6 seconds when the ignition switch is turned to the **RUN** position when the SRS is functioning properly. After being illuminated for the 6 seconds, the indicator light should then turn off.

If the air bag light does not illuminate at all, stays on continuously, or flashes at any time, a problem has been detected by the diagnostic computer.

✳✳ CAUTION

If at any time the air bag light indicates that the computer has noted a problem, immediately diagnose the problem. A faulty SRS can cause severe physical injury or death.

SERVICE PRECAUTIONS

Whenever working around, or on, the air bag supplemental restraint system, ALWAYS adhere to the following warnings and cautions.

- Always wear safety glasses when servicing an air bag vehicle and when handling an air bag module.
- Carry a live air bag module with the bag and trim cover facing away from your body, so that an accidental deployment of the air bag will have a small chance of personal injury.
- Place an air bag module on a table or other flat surface with the bag and trim cover pointing up.
- Wear gloves, a dust mask and safety glasses whenever handling a deployed air bag module. The air bag surface may contain traces of sodium hydroxide, a byproduct of the gas that inflates the air bag and which can cause skin irritation.
- Ensure to wash your hands with mild soap and water after handling a deployed air bag.
- All air bag modules with discolored or damaged cover trim must be replaced, not repainted.
- All component replacement and wiring service must be made with the negative and positive battery cables disconnected from the battery for a minimum of 1 minute prior to attempting service or replacement.
- NEVER probe the air bag electrical terminals. Doing so could result in air bag deployment, which can cause serious physical injury.
- If the vehicle is involved in a fender-bender that results in a damaged front bumper or grille, the air bag sensors should be inspected to ensure that they were not damaged.
- If at any time, the air bag light indicates that the computer has noted a problem, immediately diagnose the problem. A faulty SRS can cause severe physical injury or death.

DISARMING THE SYSTEM

1. Disconnect the negative battery cable from the battery.
2. Disconnect the positive battery cable from the battery.
3. Wait 1 minute. This time is required for the back-up power supply in the air bag diagnostic monitor to completely drain. The system is now disarmed.

ARMING THE SYSTEM

1. Connect the positive battery cable.
2. Connect the negative battery cable.
3. Stand outside the vehicle and carefully turn the ignition to the **RUN** position. Be sure that no part of your body is in front of the air bag module on the steering wheel, to prevent injury in case of an accidental air bag deployment.
4. Ensure the air bag indicator light turns off after approximately 6 seconds. If the light does not illuminate at all, does not turn off, or starts to flash, diagnose the problem. If the light does turn off after 6 seconds and does not flash, the SRS is working properly.

Rack and Pinion Steering Gear

REMOVAL & INSTALLATION

✳✳ WARNING

Do not allow the steering wheel to rotate when the intermediate shaft is disconnect, or damage to the clock-spring can result. If it is suspected that the shaft has rotated, the clock-spring must be removed and re-centered.

1. Remove or disconnect the following:
   - Front wheels
   - Tie rod end cotter pins and castle nuts
   - Tie rod ends from the knuckles
2. Position the dash opening weather seal for the steering column out of the way.
3. Remove or disconnect the following:
   - Pinch bolt retaining the steering column intermediate shaft coupling
   - Steering gear retaining nuts/bolts
4. Rotate the rack and pinion assembly to clear the bolts from the front crossmember, and pull toward the driver's side of the vehicle.
5. Place a drain pan under the vehicle and disconnect the power steering lines.
6. Remove the rack and pinion assembly through the driver's side of the vehicle.

To install:

7. Install new Teflon® O-rings on the power steering line fittings.
8. Place the rack and pinion retaining bolts in the gear housing.
9. Install or connect the following:

36 Nm
(27 lb-ft)

30 Nm
(22 lb-ft)

36 Nm (27 lb-ft)

90 Nm
(66 lb-ft)

90 Nm
(66 lb-ft)

1 — 80 Nm (59 lb-ft)

175 Nm (129 lb-ft)

80 Nm (59 lb-ft) — 1

1	Tie-rod end nuts (2 required)	6	Teflon® seals
2	Intermediate shaft-to-steering gear bolt	7	Steering gear-to-crossmember nuts (2 required)
3	Intermediate shaft	8	Stabilizer bar link nuts (2 required)
4	Power steering pressure line fitting	9	Steering gear
5	Power steering return line fitting		

06017-FRST-G03

Exploded view of the rack and pinion steering gear mounting on the front subframe of the vehicle—Freestar and Monterey

- Rack and pinion assembly through the driver's side of the vehicle
- Power steering lines on the rack and pinion assembly
- Rack and pinion assembly on the crossmember
- Tie rod ends to the knuckles. Tighten the castle nuts to 59 ft. lbs. (80 Nm) and install the cotter pins.
- Rack and pinion assembly retaining bolts, and tighten them to 129 ft. lbs. (175 Nm).
- Front wheels

10. Using a new pinch bolt, install the steering column intermediate shaft coupling on the rack input shaft. Tighten the pinch bolt to 22 ft. lbs. (30 Nm).

11. Position the steering column opening weather seal over the steering gear housing.

12. Lower the vehicle.

13. Fill the power steering oil reservoir.

14. Start the vehicle and check for leaks.

15. Check for proper wheel alignment and steering wheel position.

Strut

REMOVAL & INSTALLATION

➡**Do not allow the axle shaft to move outward as damage to the CV joint may result.**

1. Turn the ignition switch **OFF** and place the steering column in the unlocked position.

2. Remove or disconnect the following:
- Front wheel
- Stabilizer bar link nut
- Support the steering knuckle and remove the pinch bolt
- Upper strut mounting nuts

3. Push down on the steering knuckle until the strut is free of the steering knuckle and remove the strut.

To install:

4. Push down on the steering knuckle and insert the bottom of the strut into the steering knuckle.

5. Install the upper strut mounting nuts and tighten them to 26 ft. lbs. (35 Nm).

6. Install strut-to-wheel knuckle pinch bolt and tighten to 85 ft. lbs. (115 Nm).

7. Install or connect the following:
- Stabilizer bar link nut and tighten to 66 ft. lbs. (90 Nm).
- Rotor
- Caliper
- Wheels

8. Check the wheel alignment.

Stabilizer bar link nut

Steering knuckle pinch bolt

67197-FREE-G34

MacPherson strut mounting—Freestar and Monterey

DISASSEMBLY

1. Position a suitable pass through socket wrench onto the strut shaft nut.

2. Place a 10 mm six point deep socket on the strut top retaining nut.

3. Remove the strut shaft while holding the retaining nut in place.

4. Remove the shock absorber.

5. Remove the upper shock mount and bearing assembly.

6. Remove the coil spring.

7. To reassemble, reverse the disassembly procedure. Tighten the strut top retaining nut to 37 ft. lbs. (50 Nm).

Shock Absorber

REMOVAL & INSTALLATION

1. Loosen the lug nuts on the rear wheels.

2. Raise and safely support the vehicle.

3. Remove the rear wheels.

4. Position a jack under the rear axle assembly and raise it slightly to put the suspension at normal ride height.

5. Remove the lower shock absorber bolt/nut and disconnect the shock from the rear axle.

6. Lower the rear axle slightly to help aid removal of the upper shock absorber bolt/nut.

7. Remove the shock absorber.

To install:

8. Attach the shock absorber to the upper mounting bracket and install a new retaining bolt/nut.

9. Slowly raise the rear axle assembly with a jack, and guide the lower shock absorber into the bracket on the rear axle assembly. Install a new retaining bolt/nut.

10. Raise the rear suspension to normal ride height and tighten the upper shock absorber retaining bolt to 76 ft. lbs. (103 Nm), and the lower bolt to 59 ft. lbs. (80 Nm).

11. Install the wheels.

12. Lower the vehicle.

Coil Springs

REMOVAL & INSTALLATION

1. Raise and safely support the vehicle.

2. Remove the rear wheels.

3. Remove the brake caliper and wire aside.

4. Remove the brake rotor.

5. Disconnect the ABS speed sensor.

➡**The rear axle will need to be supported when the shock absorbers are removed.**

103 Nm (76 lb-ft)

133 Nm (98 lb-ft)

80 Nm (59 lb-ft)

Item	Description	Item	Description	Item	Description
5	Shock absorber-to-axle bolt	8	Upper shock absorber nut	11	Spring insulator (2 req'd)
6	Shock absorber-to-axle flagnut	9	Shock absorber	12	Trailing arm bolt (2 req'd) Removal Note
7	Upper shock absorber bolt	10	Spring	13	Trailing arm nut (2 req'd)

67197-FREE-G35

Exploded view of the rear coil spring and shock absorber mounting—Freestar and Monterey

6. Remove the wheel hub and bearing assembly.

7. Remove the shock absorber.

8. Slowly lower the rear axle assembly until the rear spring can be removed.

9. Remove the rear spring.

To install:

10. Position the rear spring insulator on the rear axle assembly and press the insulator downward into place. Verify rear spring insulator is properly seated into correct position.

11. Slowly raise the rear axle assembly with a jack, and guide the upper rear spring insulator onto the upper spring seat on the underbody.

12. Position the shock absorber on the lower rear axle assembly and tighten the upper shock absorber retaining bolt to 76 ft. lbs. (103 Nm), and the lower bolt to 59 ft. lbs. (80 Nm).

13. Install the wheel hub and bearing assembly and tighten the bolts to 85 ft. lbs. (115 Nm).

14. Connect the ABS speed sensor.

15. Install the brake rotor and caliper.

16. Install the wheels.

17. Lower the vehicle.

Stabilizer Bar

REMOVAL & INSTALLATION

1. Raise and safely support the vehicle.
2. Remove the wheels.
3. Remove the stabilizer bar link nuts.
4. Remove the stabilizer bar links.

➡**The subframe will need to be supported when the mounting bolts are removed.**

5. Remove the rear subframe mounting bolts and lower the rear of the subframe about 2-3 inches.

6. Remove the stabilizer bar brackets.

7. Remove the stabilizer bar.

To install:

8. Install or connect the following:
   - Stabilizer bar and brackets. Tighten the bracket bolts to 46 ft. lbs. (63 Nm).
   - NEW subframe bolts. Tighten the bracket bolts to 66 ft. lbs. (89 Nm).
   - Stabilizer bar links. Tighten the bracket bolts to 66 ft. lbs. (89 Nm).
   - ABS speed sensor

- Brake rotor and caliper
- Wheels
- Lower the vehicle

Lower Ball Joint

REMOVAL & INSTALLATION

The lower ball joint and seal are an integral part of the lower control arm assembly, and can not be replaced separately. If the lower ball joint or seal is found to be defective, the lower control arm must be replaced as an assembly.

Lower Control Arm, Control Arm Bushing & Steering Knuckle

REMOVAL & INSTALLATION

➡**Do not begin the removal procedure unless a new strut-to-lower arm nut, a new ball joint pinch bolt/nut and a new lower arm-to-front subframe bolt/nut are available.**

1. Remove or disconnect the following:
   - Wheel
   - Brake caliper
   - Rotor
   - Axle shaft nut
   - Wheel hub and bearing
   - Tie rod nut
   - Steering knuckle pinch bolt
   - Stabilizer bar link nut
   - Steering knuckle
   - Lower arm bushing bolts
   - Lower control arm

➡**The control arm bushing should be replaced whenever the control arm is removed.**

2. Install or connect the following:
   - Lower control arm
   - Lower arm bushing bolts. Tighten NEW bolts to 66 ft. lbs. (90 Nm).
   - Steering knuckle. Tighten NEW ball joint stud-to-knuckle bolt to 66 ft. lbs. (90 Nm).
   - Stabilizer bar link nut. Tighten NEW bolt to 66 ft. lbs. (90 Nm).
   - Steering knuckle pinch bolt. Tighten NEW bolt to 85 ft. lbs. (115 Nm).
   - Tie rod nut. Tighten NEW nut to 59 ft. lbs. (80 Nm).
   - Wheel hub and bearing. Tighten the bolts to 85 ft. lbs. (115 Nm).
   - Axle shaft nut. Tighten NEW bolt to 185 ft. lbs. (250 Nm).
   - Rotor

1	Stabilizer bar link lower nut
2	Stabilizer bar link upper nut
3	Stabilizer bar link (2 required)
4	Subframe bolt
5	Stabilizer bar bracket bolts
6	Stabilizer bar bracket
7	Stabilizer bar
8	Stabilizer bar bushing
9	Axle shaft nut
10	Wheel bearing and hub-to-wheel knuckle bolts (3 required)
11	Wheel bearing and hub assembly
12	Tie-rod end nut
13	Wheel knuckle pinch bolt
14	Wheel knuckle (LH/RH)
15	Lower ball joint pinch bolt
16	Lower arm bushing bolt (4 required)
17	Lower arm (LH/RH)
18	Wheel stud
19	Stabilizer bar link upper nut
20	Shock absorber and spring assembly
21	Shock absorber upper nut

Exploded view of the front suspension—Freestar and Monterey

06017-FRST-G04

- Brake caliper
3. Install the wheels.
4. Lower the vehicle.

Hub and Wheel Bearing

REMOVAL & INSTALLATION

The wheel bearing is integral with the wheel hub and cannot be replaced separately. If the wheel bearing is found to be defective, the wheel hub must be replaced as an assembly.

Front

1. With the vehicle on the ground, remove and discard the axle shaft nut.
2. Raise and safely support the vehicle.
3. Remove the wheel and tire assembly.

4. Remove the brake caliper and wire it out of the way.
5. Remove the brake rotor.
6. Using a puller, remove the driveshaft from the hub and bearing housing.
7. Remove 3 hub attaching bolts, and remove the hub and bearing assembly.
To install:
8. Position the dust shield over the hub and fit the hub and bearing into the knuckle and seat it fully before installing the attaching bolts. Tighten the bolts to 85 ft. lbs. (115 Nm).
9. Install or connect the following:
- Driveshaft
- Brake rotor
- Brake caliper
- Wheel and tire. Tighten the lug nuts to 100 ft. lbs. (135 Nm).
- Install the axle nut and tighten the nut to 96 ft. lbs. (130 Nm).

Rear

1. Raise and safely support the vehicle.
2. Remove the wheel and tire assembly.
3. Remove the brake caliper and wire it out of the way.
4. Remove the brake rotor.
5. Disconnect the ABS sensor connector.
6. Remove the hub and bearing assembly.
To install:
7. Install or connect the following:
- Hub and bearing assembly and tighten the bolts to 85 ft. lbs. (115 Nm).
- ABS sensor connector
- Brake rotor
- Brake caliper
- Wheel and tire. Tighten the lug nuts to 100 ft. lbs. (135 Nm).

BRAKES

Brake Caliper

REMOVAL & INSTALLATION

Front

1. Raise and safely support the vehicle.
2. Remove the wheel and tire assembly.
3. Mark the disc brake caliper to avoid mixing the left-hand and right-hand components.
4. Disconnect the brake hose from the disc brake caliper by loosening and removing the hollow retaining bolt. Discard the 2 copper sealing washers and plug the brake hose.
5. Remove the 2 brake pin retainer bolts.
6. Lift the disc brake caliper off of the disc brake rotor using a rotating motion. Do not pry against the caliper piston. Prying may damage the piston or seals.
7. Remove the disc brake caliper from the vehicle.
8. Remove the caliper anchor bracket, if necessary.
To install:
9. Install the brake caliper anchor, if removed. Tighten the anchor bolts to 136 ft. lbs. (185 Nm).
10. Retract the caliper piston fully into the caliper bore using a C-clamp and block of wood or equivalent.
11. Ensure that the disc brake pads are properly positioned and that the lining material is facing the rotor.
12. Place the disc brake caliper over the rotor and hand-start 2 brake pin retainer

bolts. Starting with the bottom bolt first, tighten the brake pin retainer bolts-to-guide pin bolts to 26 ft. lbs. (35 Nm).

➡ **If both disc brake calipers were removed, make sure that they are mounted to the proper side. The brake bleeder on the caliper when properly installed should be on top of the caliper for proper bleeding of air.**

13. Unplug and install the brake hose and hollow retaining bolt to the disc brake caliper using a new copper sealing washer on each side of the hose fitting.
14. Bleed the brake system and install the rubber bleeder screw caps when complete.
15. Install the wheel and tire assembly. Torque the lug nuts to 100 ft. lbs. (136 Nm).
16. Lower the vehicle.
17. Pump the brake pedal several times to position the brake pads before attempting to move the vehicle.
18. Check and fill the brake master cylinder as required.
19. Road-test the vehicle and check for proper brake operation.

Rear

1. Remove and discard ½ of the brake fluid from the brake master cylinder reservoir.
2. Raise and safely support the vehicle.
3. Remove the wheel and tire assembly.
4. Disconnect the brake hose from the disc brake caliper by loosening and removing the hollow retaining bolt. Discard 2 copper sealing washers and plug the brake hose.

5. Using a C-clamp or equivalent, position the clamp frame on the inboard side of the disc brake caliper housing. Place the clamp screw on the outboard disc brake pad and tighten the clamp enough to press the caliper piston into the caliper housing releasing pressure on the disc brake pads.
6. Disconnect the parking brake cable from the caliper.
7. Remove 2 disc brake caliper retaining bolts.
8. Remove the caliper by swinging out the bottom of the caliper first.
9. Remove the disc brake pads, if necessary.
10. Remove the caliper anchor bracket, if necessary.
To install:
11. If removed, install the caliper anchor bracket. Tighten the bolts to 75 ft. lbs. (102 Nm).
12. Retract the disc brake caliper piston fully into the caliper bore using a C-clamp and block of wood or equivalent.
13. Ensure that the disc brake pads are properly positioned and that the lining material is facing the rotor.
14. Install the caliper over the disc brake rotor and position on the brake adapter. Install 2 disc brake caliper retaining bolts and starting with the bottom bolt first, tighten to 24 ft. lbs. (33 Nm).
15. Install the parking brake cable to the caliper.
16. Unplug and install the brake hose and hollow retaining bolt to the disc brake caliper using a new copper sealing washer

(1)	35 Nm (26 lb-ft)	(14)	17 Nm (13 lb-ft)
(7)	185 Nm (136 lb-ft)	(17)	17 Nm (13 lb-ft)
(12)	55 Nm (41 lb-ft)	(20)	10 Nm (7 lb-ft)

Item	Description	Item	Description	Item	Description
1	Brake caliper bolt (two required for each side)	7	Brake caliper anchor bracket bolts (2 required each side)	13	Copper washers
				14	Brake line fitting
2	Brake caliper RH/LH	8	Brake caliper anchor bracket RH/LH	15	Retainer clip
3	Brake pads			16	Front brake hose RH/LH
4	Slippers (4 required each side)	9	Brake disc screws (if equipped)	17	Brake hose bracket bolt
5	Guide pin	10	Brake disc	18	Brake hose bracket
6	Locator pin	11	Dust shield (RH/LH)	19	Bleeder screw dust cover
		12	Flow bolt	20	Bleeder screw

67197-FREE-G37

Exploded view of the front disc brake caliper assembly—Freestar and Monterey

Item	Description
1	Brake caliper bolts
2	Brake caliper (RH/LH)
3	Brake pads (kit)
4	Slippers
5	Guide pin and boot

Item	Description
6	Brake caliper anchor bracket bolt kit (2 bolts each side)
7	Brake caliper anchor bracket
8	Brake disc
9	Dust shield (RH/LH)
10	Flow bolt

Item	Description
11	Copper washers
12	Brake line fitting
13	Brake hose retainer
14	Brake hose (RH/LH)
15	Brake hose bracket bolt
16	Brake hose bracket

67197-FREE-G38

Exploded view of the rear disc brake caliper assembly—Freestar and Monterey

on each side of the hose fitting. Tighten the retaining bolt to 41 ft. lbs. (55 Nm).

17. Bleed the brake system and install the rubber bleeder screw caps when complete.

18. Install the wheel and tire assembly. Torque the lug nuts to 100 ft. lbs. (136 Nm).

19. Lower the vehicle.

20. Pump the brake pedal several times to position the brake pads before attempting to move the vehicle.

21. Check and fill the brake master cylinder as required.

22. Road-test the vehicle and check for proper brake operation.

Disc Brake Pads

REMOVAL & INSTALLATION

Front

1. Remove ½ of the brake fluid from the brake master cylinder reservoir. Properly dispose of the brake fluid.

2. Raise and safely support the vehicle.

3. Remove the wheel and tire assembly.

4. Remove 2 disc brake caliper brake pin retainers. Do not remove the brake hose from the caliper.

5. Lift the disc brake caliper off of the disc brake rotor using a rotating motion. Do not pry against the caliper piston. Prying may damage the piston or seals.

6. Hang the disc brake caliper with a length of wire or equivalent to prevent damage to the brake hose.

7. Remove the inner and outer disc brake pads and the anti-rattle clip.

8. Inspect the disc brake rotor surfaces for grooves, cracks or glazing. Resurface or replace as required. If resurfacing, observe the minimum thickness specification.

To install:

9. Retract the caliper piston fully into the caliper bore using a C-clamp and wood block or equivalent. This will allow room for the new disc brake pads.

10. Install new inner and outer disc brake pads and the anti-rattle clip. Ensure that the disc brake pads are properly positioned and that the lining material is facing the rotor.

11. Place the disc brake caliper over the rotor and install 2 disc brake caliper brake pin retainers. Tighten the brake pin retainers to 26 ft. lbs. (35 Nm).

12. Install the wheel and tire assembly. Torque the lug nuts to 100 ft. lbs. (136 Nm).

13. Lower the vehicle.

14. Pump the brake pedal to position the brake pads before attempting to move the vehicle.

15. Check and fill the brake master cylinder reservoir, as required.

16. Road-test the vehicle and check for proper brake system operation.

Rear

1. Remove ½ of the brake fluid from the brake master cylinder reservoir. Properly dispose of the brake fluid.

2. Raise and safely support the vehicle.

3. Remove the wheel and tire assembly.

4. Using a C-clamp or equivalent, position the clamp frame on the inboard side of the disc brake caliper housing. Place the clamp screw on the outboard disc brake pad and tighten the clamp enough to press the caliper piston into the caliper housing releasing pressure on the disc brake pads.

5. Remove 2 disc brake caliper retaining bolts. Do not remove the disc brake caliper brake hose from the caliper.

6. Work the disc brake caliper off the brake rotor and disc brake adapter. Move the disc brake caliper aside and secure with wire or equivalent to prevent damage to the brake hose.

7. Remove the slippers from the anchor plate abutments by gently prying them off the rails and discard the slippers.

8. Remove the inner and outer disc brake pads.

9. Inspect the disc brake rotor surfaces for grooves, cracks or glazing. Resurface or replace as required. If resurfacing, observe the minimum thickness specification.

To install:

10. Retract the disc brake caliper piston fully into the caliper bore using a C-clamp and block of wood or equivalent. This will make room for the new disc brake pads.

11. Install new anti-wear slippers on the rail abutments by snapping them in place.

12. Install new inner and outer disc brake pads. Ensure that the disc brake pads are properly positioned and that the lining material is facing the rotor.

13. Install the disc brake caliper over the brake rotor and place on the brake adapter. Ensure that the notches on the upper ends of the brake pads are seated over the upper ledge of the disc brake adapter and the lower tabs are placed on the lower ledge of the disc brake adapter.

14. Lubricate 2 disc brake caliper retaining bolts with a suitable grease and install. Tighten the retaining bolts to 41ft. lbs. (55 Nm).

15. Install the wheel and tire assembly. Torque the lug nuts to 100 ft. lbs. (136 Nm).

16. Lower the vehicle.

17. Pump the brake pedal several times to position the brake pads before attempting to move the vehicle.

18. Check and fill the brake master cylinder reservoir, as required.

19. Road-test the vehicle and check for proper brake operation.

FORD

Freestyle

10

SPECIFICATIONS AND MAINTENANCE CHARTS

VEHICLE AND ENGINE IDENTIFICATION CHART

	Engine						Model Year	
Code	Liters	Cu. In.	Cyl.	Fuel Sys.	Engine Type	Eng. Mfg.	Code	Year
1	3.0	183	6	SEFI	DOHC	Ford	5	2005

SEFI: Sequential Multi-port Fuel Injection

DOHC: Dual overhead cam

06017-FREE-C01

GENERAL ENGINE SPECIFICATIONS

Year	Engine Displacement Liters	Engine VIN	Net Horsepower @ rpm	Net Torque @ rpm (ft. lbs.)	Bore x Stroke (in.)	Compression Ratio	Oil Pressure @ rpm
2005	3.0	1	203@5750	207@4500	3.50x3.13	10.0:1	11@1500

06017-FREE-C02

GASOLINE ENGINE TUNE-UP SPECIFICATIONS

Year	Engine Displacement Liters	Engine VIN	Spark Plugs Gap (in.)	Ignition Timing (deg.) MT	AT	Fuel Pump (psi)	Idle Speed (rpm) MT	AT	Valve Clearance In.	Ex.
2005	3.0	1	0.052-0.056	—	①	②	—	①	HYD	HYD

NOTE: The Vehicle Emission Control Information label often reflects specification changes changes made during production.

The label figures must be used if they differ from those in this chart.

B: Before top dead center

HYD: Hydraulic

① Controlled by the Powertrain Control Module (PCM) and cannot be manually adjusted.

② Engine running: 40 psi
   Key On, Engine Off (KOEO): 40 psi

06017-FREE-C03

06017-FREE-G01

3.0L engine accessory drive belt routing

CAPACITIES

Year	Model	Engine Displacement Liters	Engine ID/VIN	Engine Oil with Filter (qts.)	Transmission (pts.) 4-Spd	5-Spd	Auto.	Drive Axle Front (pts.)	Rear (pts.)	Fuel Tank (gal.)	Cooling System (qts.)
2005	Freesyle	3.0	1	6.0	—	—	①	1.5	1.5	19	③

NOTE: All capacities are approximate. Add fluid gradually and check to be sure a proper fluid level is obtained.

① CVT transmission: 20
   6-spd. Transmission: 15.8

② w/auxiliary heater: 12.7
   wo/auxiliary heater: 10.6

06017-FREE-C04

VALVE SPECIFICATIONS

Year	Engine VIN	Engine Displacement Liters	Seat Angle (deg.)	Face Angle (deg.)	Spring Test Pressure (lbs. @ in.)	Spring Installed Height (in.)	Stem-to-Guide Clearance (in.) Intake	Exhaust	Stem Diameter (in.) Intake	Exhaust
2005	1	3.0	44.50-45.0	45.25-45.75	156@ 1.18	1.562-1.577	0.0007-0.0027	0.0018-0.0037	0.2350-0.2358	0.2343-0.2350

06017-FREE-C05

CRANKSHAFT AND CONNECTING ROD SPECIFICATIONS

All measurements are given in inches.

Year	Engine Displ. Liters	Engine VIN	Crankshaft				Connecting Rod		
			Main Brg. Journal Dia.	Main Brg. Oil Clearance	Shaft End-play	Thrust on No.	Journal Diameter	Oil Clearance	Side Clearance
2005	3.0	1	2.4790-2.4800	0.0010-0.0016	0.0050-0.0010	3	1.9670-2.9680	0.0010-0.0025	0.0039-0.0118

06017-FREE-C06

PISTON AND RING SPECIFICATIONS

All measurements are given in inches.

Year	Engine Displ. Liters	Engine VIN	Piston Clearance	Ring Gap			Ring Side Clearance		
				Top Comp.	Bottom Comp.	Oil Control	Top Comp.	Bottom Comp.	Oil Control
2005	3.0	1	0.0007-0.0016	0.0039-0.0098	0.0106-0.0165	0.0059-0.0255	N/A	N/A	N/A

NA: Not Available

06017-FREE-C07

TORQUE SPECIFICATIONS

All readings in ft. lbs.

Year	Engine VIN	Engine Displacement Liters	Cylinder Head Bolts	Main Bearing Bolts	Rod Bearing Bolts	Crankshaft Damper Bolts	Flywheel Bolts	Manifold		Spark Plugs	Oil Pan Drain Plug
								Intake	Exhaust		
2005	1	3.0	①	N/A	N/A	②	59	③	15	11	19

① Step 1: 30 ft. lbs.
   Step 2: plus 90 degrees
   Step 3: loosen all bolts 1 turn
   Step 4: 30 ft. lbs.
   Step 5: plus 90 degrees.
   Step 6: plus 90 degrees.

② Step 1: 86 ft. lbs.
   Step 2: loosen 1 turn
   Step 3: 37 ft. lbs.
   Step 4: plus 90 degrees

③ Upper and lower manifold: 89 INCH lbs.

06017-FREE-C08

WHEEL ALIGNMENT

Year	Model			Caster Range (+/-Deg.)	Caster Preferred Setting (Deg.)	Camber Range (+/-Deg.)	Camber Preferred Setting (Deg.)	Toe-in (Deg.)
2005	Freestyle	Front	AWD	0.75	+3.0	0.75	-0.32	+0.20+/-0.20
			FWD	0.75	+3.1	0.75	-0.28	+0.20+/-0.20
		Rear	AWD	—	—	0.75	0.00	0.10+/-0.20
			FWD	—	—	0.75	-0.57	0.10+/-0.20

06017-FREE-C09

TIRE AND WHEEL SPECIFICATIONS

Year	Model	OEM Tires Standard	OEM Tires Optional	Tire Pressures (psi) Front	Tire Pressures (psi) Rear	Wheel Size	Wheel Lug Nut Torque (Ft. Lbs.)
2005	Freestyle	P215/65R17	P225/60R18	35	35	①	95

OEM: Original Equipment Manufacturer

PSI: Pounds Per Square Inch

STD: Standard

OPT: Optional

① Not available

06017-FREE-C10

BRAKE SPECIFICATIONS

All measurements in inches unless noted

Year	Model		Brake Disc Original Thickness	Brake Disc Minimum Thickness	Brake Disc Maximum Runout	Minimum Lining Thickness Front	Minimum Lining Thickness Rear	Brake Caliper Bracket Bolts (ft. lbs.)	Brake Caliper Mounting Bolts (ft. lbs.)
2005	Freestyle	F	①	1.040	0.002	0.118	—	74	44
		R	①	0.394	0.002	—	0.118	81	23

① Not available

06017-FREE-C11

SCHEDULED MAINTENANCE INTERVALS
FORD FREESTYLE

TO BE SERVICED	TYPE OF SERVICE	VEHICLE MILEAGE INTERVAL (x1000)																	
		5	10	15	20	25	30	35	40	45	50	55	60	65	70	75	80	85	90
Engine oil & filter ①	R																		
Rotate tires	S/I	✓	✓	✓	✓	✓	✓	✓	✓	✓	✓	✓	✓	✓	✓	✓	✓	✓	✓
Engine coolant strength hoses & clamps	S/I			✓			✓			✓			✓				✓		✓
Air cleaner filter	R						✓						✓						✓
Automatic transmission fluid & filter	R						✓						✓						✓
Engine coolant ②	R																		
PCV valve	R												✓						✓
Spark plugs ③	R																		
Drive belts	S/I						✓						✓						✓
Exhaust system & heat shields	S/I						✓						✓						✓
Front & rear brakes	S/I	✓	✓	✓	✓	✓	✓	✓	✓	✓	✓	✓	✓	✓	✓	✓	✓	✓	✓
Fuel filter	R			✓			✓			✓			✓			✓			✓

R: Replace   S/I: Service or Inspect

① Engine oil and filter: replace every 3000 miles.
② Engine coolant: change initially at 100,000 miles or 5 years.
③ Spark plugs: replace every 100,000 miles.

Special Operating Condition Requirements

During extensive idling and/or low speed driving for long distances, as in heavy commercial use such as delivery, taxi, patrol car or livery:

Lube front lower control arm and steering linkage ball joints with

Zerk fittings (if equipped) every 4,800 km (3,000 miles) or 3 months.

Inspect brake system and check battery electrolyte level (Patrol cars) every 8,000 km (5,000 miles).

Install a new cabin air filter as required.

When operating in dusty conditions such as unpaved or dusty roads:

Install a new engine air filter as required.

Install a new cabin air filter as required.

When operating in off-road conditions:

Change automatic transmission fluid every 48,000 km (30,000 miles).

Install a new cabin air filter as required.

Inspect and lubricate U-joints.

Inspect and lubricate steering linkage ball joints with zerk fittings.

06017-FREE-C12

ENGINE REPAIR

Alternator

REMOVAL & INSTALLATION

1. Disconnect the negative battery cable.
2. Raise and support the vehicle.
3. Rotate the drive belt tensioner counterclockwise and remove the accessory drive belt.
4. Remove the right lower splash shield.
5. Disconnect the crankshaft position sensor.
6. Disconnect the wiring harness locators from the alternator splash shield.
7. Detach the alternator wiring connector.
8. Remove the positive cable nut.
9. Remove the alternator splash shield and the alternator.

To install:

10. Position the alternator on the engine.
11. Install the alternator mounting bolts. Tighten the stud bolts to 71 INCH lbs. (8 Nm) and the bolts to 35 ft. lbs. (47 Nm).
12. Tighten the positive cable nut to 71 INCH lbs. (8 Nm).
13. Connect the wiring connector.
14. Install the wiring harness locators.
15. Connect the crankshaft position sensor.
16. Install the lower right splash shield.
17. Install and tension the accessory drive belt.
18. Connect the negative battery cable.

Engine Assembly

REMOVAL & INSTALLATION

1. Relieve the fuel system pressure.
2. Drain the engine coolant.
3. Drain the engine oil.
4. Recover the air conditioning refrigerant, into a refrigerant recovery station.
5. Place the transmission in neutral and raise the vehicle on a hoist.
6. Remove or disconnect the following:

- Both battery cables
- B+ terminal from the battery cable
- Air cleaner outlet pipe and air cleaner
- Cowl vent screen
- Left and right halfshafts
- Exhaust flex pipe
- Left side catalytic converter
- Roll restrictor cross brace-to-roll restrictor bolt and discard
- Roll restrictor cross brace
- Right and left fender ground straps
- Fuel tube from the fuel rail
- Vapor tube from the purge valve
- Powertrain Control Module (PCM) connectors and wiring conduit
- Power steering reservoir hose and retainers
- Disconnect the transmission shift cable
- Electrical connector and retainers from the shift cable bracket
- Shift cable bracket bolts and position the cable aside
- Heater and throttle body coolant hoses
- Brake booster vacuum hose
- Thermostat housing hoses
- Cooling tube retainers
- A/C high pressure switch connector
- A/C compressor manifold tube
- A/C tube retaining nuts and position the tubes aside
- Transmission cooler lines
- Power steering cooler hose
- Engine block heater connector and harness
- Torque converter lower access plug
- Torque converter attaching nuts
- 3 engine-to-transaxle mounting bolts
- Left catalytic converter bracket

8 Nm
(71 lb-in)

10 - 8 Nm (71 lb-in)

9 - 47 Nm (35 lb-ft)

1 Crankshaft position sensor connector	7 Generator B+ nut
2 Harness locators (2 required)	8 Generator B+ terminal
3 Engine control harness	9 Generator stud nuts (3 required)
4 Lower generator boot pin-type retainer	10 Generator studs (3 required)
5 Generator boot	11 Generator shield
6 Generator electrical connector	12 Generator

06017-FREE-G02

Alternator mounting—Freestyle

014-00765

06017-FREE-G03

Installing subframe support tool—Freestyle

Installing engine lifting tools to the engine—Freestyle

06017-FREE-G04

- Driveshaft support bracket bolts and the driveshaft, if equipped
- Stabilizer bar links from the struts
- Outer tie rod ends from the steering knuckles
- Steering column intermediate shaft bolt and separate the shaft from the steering gear

7. Position subframe support tool 014-00765 under subframe and powertrain assembly.

8. Remove the subframe bolts and rear subframe brackets.

9. Lower the powertrain and subframe assembly from the vehicle.

10. To separate the engine from the transaxle and subframe assembly, disconnect all necessary electrical connectors, fluid lines and starter wiring. Remove the starter and engine roll restrictor brackets

11. Install engine lifting eyes no. 134-00243 to the left and right cylinder heads.

12. Install lifting brackets no. 303-1140 to front of the engine.

13. Install an engine hoist and lifting bar no. 303-D089 to the engine.

14. Remove the front transaxle stabilizer bolt.

15. Remove the exhaust manifold heat shield.

16. Remove the left and right engine insulator upper nuts.

17. Remove the transaxle insulator bracket.

18. Using the hoist, lift the powertrain from the subframe.

19. Remove the transaxle-to-engine mounting bolts and separate the engine and transaxle.

To install:

20. Position the engine onto the transaxle and install the transaxle mount bolts. Torque the bolts to 30 ft. lbs. (40 Nm).

21. Install the exhaust manifold heat shield.

22. Using the hoist, position the powertrain to the subframe.

23. Install the left and right engine insulator upper nuts and tighten to 62 ft. lbs. (70 Nm).

24. Install the front transaxle stabilizer bolt and tighten to 30 ft. lbs. (40 Nm).

25. Install the transaxle insulator bracket and tighten the bolts to 41 ft. lbs. (55 Nm).

26. Install the starter and tighten the bolts to 18 ft. lbs. (25 Nm).

27. Connect the starter wiring and terminal cover.

28. Apply thread lock to the roll restrictor bolts and install the restrictor bracket and tighten the bolts to 41 ft. lbs. (55 Nm).

29. Reconnect all necessary electrical connectors, fluid lines and retainers.

30. Remove the engine hoist and lifting tools.

31. Using subframe support tool 014-00765, raise the subframe and powertrain assembly into the vehicle

32. Install the front subframe bolts and tighten the bolts to 148 ft. lbs. (200 Nm).

33. Install the rear subframe brackets. Tighten the large bolts to 111 ft. lbs. (150 Nm), and the smaller bolts to 41 ft. lbs. (55 Nm).

34. Connect the steering column intermediate shaft and tighten the pinch bolt to 18 ft. lbs. (25 Nm).

35. Install the outer tie rod ends to the steering knuckles and tighten NEW nuts to 85 ft. lbs. (115 Nm).

36. Install the stabilizer bar links to the struts using NEW nuts and tighten to 41 ft. lbs. (55 Nm).

37. Install the driveshaft and tighten the bolts to 18 ft. lbs. (25 Nm).

38. Install the driveshaft support bracket bolts and tighten the bolts to 18 ft. lbs. (25 Nm).

39. Install or connect the following:
- NEW torque converter attaching nuts and tighten to 27 ft. lbs. (36 Nm)
- Torque converter lower access plug
- Engine block heater connector and harness
- Power steering cooler hose
- Transmission cooler lines
- A/C tube and retaining nuts
- A/C compressor manifold tube
- A/C high pressure switch connector
- Cooling tube retainers
- Thermostat housing hoses
- Brake booster vacuum hose
- Heater and throttle body coolant hoses
- Shift cable and bracket bolts and tighten the bolts to 18 ft. (25 Nm)

- Electrical connector and retainers to the shift cable bracket
- Connect the transmission shift cable
- Power steering reservoir hose and retainers
- Powertrain Control Module (PCM) connectors and wiring conduit
- Vapor tube from the purge valve
- Fuel tube from the fuel rail
- Right and left fender ground straps
- Roll restrictor cross brace
- NEW roll restrictor cross brace-to-roll restrictor bolts and tighten to 41 ft. lbs. (55 Nm)
- Left catalytic converter bracket
- Left side catalytic converter
- Exhaust flex pipe
- Left and right halfshafts
- Cowl vent screen
- Air cleaner outlet pipe and air cleaner
- B+ terminal to the battery cable
- Both battery cables

40. Refill the engine, transaxle and cooling system with the correct amount of the appropriate fluids before starting the engine. Recharge the A/C system using approved recycling equipment.

Water Pump

REMOVAL & INSTALLATION

1. Remove or disconnect the following:
- Negative battery cable
- Coolant
- Accessory drive belt
- Air cleaner assembly
- Lower radiator hose
- Thermostat hoses
- Thermostat housing
- Water pump

To install:

2. Position a new water pump housing gasket on the water pump sealing surface using gasket sealant to hold the gasket in place.

3. Install the water pump and tighten the bolts in the following sequence: tighten the center bolt to 35 inch lbs. (4 nm), loosely install the outer bolts, retighten the center bolt to 18 ft. lbs. (25 Nm), tighten the outer bolts to 89 inch lbs. (10 Nm).

4. The remainder of the installation is the reverse of removal.

5. Fill and bleed the cooling system.

6. Connect the negative battery cable.

7. Start the engine and check for leaks.

Water pump mounting—Freestyle

Heater Core

REMOVAL AND INSTALLATION

1. Ensure the ignition is off.
2. Remove the Smart Junction Box (SJB) cover located below the left side of the instrument panel. Remove the Restraints Control Module (RCM) fuse no. F2.21.
3. Turn the ignition on and watch the air bag indicator for 30 seconds. The indicator light will remain lit constantly if the correct fuse has been removed. If the light is not on steadily, remove the correct fuse and check the light again.
4. Turn the ignition switch off.
5. Disconnect the negative battery cable.
6. Drain the cooling system into a clean container for reuse.
7. Recover the air conditioning refrigerant, into a refrigerant recovery station.
8. Disconnect the heater hoses from the heater core inlet and outlet tubes in the engine compartment.
9. Remove the steering wheel access cover.
10. Release one side of the driver air bag module clip using needle nose pliers.
11. Lift the end of the clip over the center post.
12. Repeat the procedure on the opposite wide wire clip.

➡**When the clips are released correctly, the clip ends will come to rest between the center two inner posts.**

13. Push on the horn at the upper part of the air bag cover and push both ends of the wire clips inward to release the driver air bag module from the steering wheel.
14. Disconnect the air bag electrical connectors, horn and accessories connectors, and remove the driver air bag module.

15. Remove and discard the steering wheel bolt.
16. Remove the steering wheel.
17. Remove the upper and lower steering column covers.
18. Disconnect the multi-function switch connectors and remove the multi-function switch.
19. Remove or disconnect the following:
   - Cowl trim panels and disconnect the electrical connectors
   - Lower instrument panel insulator
   - Front floor console
   - Floor shifter
   - Instrument panel center finish panel
   - Audio unit and disconnect the electrical and antenna connectors
   - Instrument panel side panels
   - A pillar trim panels
   - Steering column cover
   - Hood release handle
   - Headlight switch
   - Instrument cluster finish panel
   - Glove box door
   - Passenger air bag module connectors
   - Two air bag module bolts
   - Passenger air bag module
   - Passenger assist handle
   - Left side main wiring harness connectors

1 Heater tube bracket screw	5 Heater core cover screw
2 LH temperature blend door actuator screw	6 Heater core cover
3 LH temperature blend door actuator electrical connector	7 Heater core
4 LH temperature blend door actuator	8 Evaporator discharge air temperature sensor electrical connector

Exploded view of heater core mounting—Freestyle

06017-FREE-G06

- 12 screws attaching the upper instrument panel section and remove the upper section
- Thermostatic expansion valve bolt
- Powertrain Control Module (PCM)
- Transaxle roll restrictor bolt
- Transaxle roll restrictor bar bolts
- Instrument panel cowl top bolts from both sides
- Steering column pinch bolt
- A/C floor duct
- Four floor tunnel brace bolts
- Restraints Control Module (RCM) small and large electrical connectors

20. Mark the location of the instrument panel cross beam bolts.

21. With the help of an assistant, remove the bolts from one side of the cross beam and while supporting the beam, remove the other side bolts and remove the cross beam.

22. Remove the dash panel seal from the heater core tubes.

23. Remove the heater core tube bracket.

24. Disconnect the evaporator discharge air temperature sensor connector.

25. Remove the left side blend door actuator.

26. Remove the heater core cover screws.

27. Remove the heater core cover and the heater core.

To install:

28. Install the heater core cover and the heater core.

29. Install the heater core cover screws.

30. Install the left side blend door actuator.

31. Connect the evaporator discharge air temperature sensor connector.

32. Install the heater core tube bracket.

33. Install the dash panel seal to the heater core tubes.

34. With the help of an assistant, install the bolts from one side of the cross beam and while supporting the beam, install the other side bolts using the locations marked previously. Tighten the bolts to 18 ft. lbs. (25 Nm).

35. Connect the RCM small and large electrical connectors.

36. Install or connect the following:
- Four floor tunnel brace bolts
- A/C floor duct
- Steering column pinch bolt and tighten to 18 ft. lbs. (25 Nm)
- Instrument panel cowl top bolts to both sides and tighten to 18 ft. lbs. (25 Nm)
- Transaxle roll restrictor bar bolts and tighten to 40 ft. lbs. (55 Nm)
- Transaxle roll restrictor bolt and tighten to 35 ft. lbs. (48 Nm)
- Powertrain Control Module (PCM)

- Thermostatic expansion valve bolt
- Instrument panel upper section and 12 screws
- Left side main wiring harness connectors
- Passenger assist handle
- Passenger air bag module
- Two air bag module bolts and tighten to 80 inch lbs. (9 Nm)
- Passenger air bag module connectors
- Glove box door
- Instrument cluster finish panel
- Headlight switch
- Hood release handle
- Steering column cover
- A pillar trim panels
- Instrument panel side panels
- Audio unit and connect the electrical and antenna connectors
- Instrument panel center finish panel
- Floor shifter
- Front floor console
- Lower instrument panel insulator
- Cowl trim panels and connect the electrical connectors
- Multi-function switch and connectors
- Upper and lower steering column covers
- Steering wheel and tighten the NEW bolt to 30 ft. lbs. (40 Nm)
- Connect the air bag electrical connectors, horn and accessories connectors, and install the driver air bag

module. Make the clips are released from the center post and back in position on both sides of the air bag.
- Steering wheel access cover
- Heater hoses to the heater core inlet and outlet tubes

37. Recharge the air conditioning refrigerant.

38. Refill the coolant system.

39. Disconnect the negative battery cable.

40. Turn the ignition switch on.

41. Install Restraints Control Module (RCM) fuse no. F2.21 and the fuse cover.

42. Turn the ignition off.

43. Connect the negative battery cable.

44. Turn the ignition on and then off. Wait 10 seconds and turn the key back on. Watch the air bag indicator. The indicator light will remain lit constantly for 6 seconds and then go off. If the indicator does not turn on and then off, diagnose the air bag system

45. Clear all Diagnostic Trouble Codes (DTC) using a diagnostic tool.

Cylinder Head

REMOVAL & INSTALLATION

Left Side

1. Relieve the fuel system pressure.

2. Disconnect the negative battery cable.

06017-FREE-G07

Left side cylinder head bolt removal sequence—Freestyle

Left side cylinder head bolt installation sequence—Freestyle

06017-FREE-G08

3. Drain the engine coolant.
4. Remove or disconnect the following:
  - Accessory drive belt
  - Upper and lower intake manifolds
  - Coolant bypass tube
  - Ignition coil on plugs
  - Valve cover wiring harness
  - 6 bolts and 8 stud bolts, and remove the valve cover
  - Camshafts
  - Exhaust manifold
  - Oil dipstick and tube
  - Mark the location of the camshaft roller followers and remove them
  - Hydraulic lash adjusters
  - Cylinder head bolts in the sequence shown
  - Cylinder head from the engine block and discard the gaskets and bolts

To install:
5. The cylinder head should be cleaned and inspected prior to installation.
6. Clean all gasket mating surfaces thoroughly.
7. Install or connect the following:
  - New head gaskets on the cylinder block

✳✳ WARNING

Always use new cylinder head bolts when installing the cylinder head or damage to the engine may occur.

Left side valve cover bolt installation sequence—Freestyle

06017-FREE-G09

- Cylinder head on the cylinder block
- Tighten the cylinder head bolts in sequence using the following steps. The first step is 30 ft. lbs. (40 Nm), the second step is an additional 90 degrees, the third step is to loosen all bolts in sequence one full turn, the fourth step is to tighten the bolts to 30 ft. lbs. (40 Nm), the fifth step is to tighten and additional 90 degrees, the last step is to tighten an addition 90 degrees.
- Lash adjusters and camshaft roller followers in their correct locations. Lubricate all rocker arm components with engine assembly lubricant.
- Oil dipstick and tube
- Exhaust manifold
- Camshafts

- Place a bead of silicone gasket sealant to the front cover-to-cylinder head surface
- Valve cover. Install the 6 bolts and 8 stud bolts, and tighten in sequence to 89 inch lbs. (10 Nm).
- Valve cover wiring harness
- Ignition coil on plugs
- Coolant bypass tube
- Upper and lower intake manifolds
- Accessory drive belt
8. Fill and bleed the cooling system.

➡**Engine coolant is corrosive to engine bearing material. Replace the engine oil after removal of any coolant-carrying component to help prevent potential bearing damage.**

9. Change the engine oil and filter.
10. Connect the negative battery cable.
11. Start the engine and check for leaks.

Right Side

1. Relieve the fuel system pressure.
2. Disconnect the negative battery cable.
3. Drain the engine coolant.
4. Remove or disconnect the following:
  - Accessory drive belt
  - Upper and lower intake manifolds
  - Coolant bypass tube
  - Ignition coil on plugs
  - PCV connector and tube
  - Power steering reservoir bracket and stud bolts
  - Power steering pressure tube bracket
  - Wiring harness and retainers from valve cover
  - Valve cover bolts and stud bolts
  - Valve cover
  - Camshafts
  - Exhaust manifold
  - Mark the location of the camshaft roller followers and remove them
  - Hydraulic lash adjusters
  - Cylinder head bolts in the sequence shown

06017-FREE-G10

Right side cylinder head bolt removal sequence—Freestyle

06017-FREE-G11

Right side cylinder head bolt installation sequence—Freestyle

- Cylinder head from the engine block and discard the gaskets and bolts

To install:

5. The cylinder head should be cleaned and inspected prior to installation.

6. Clean all gasket mating surfaces thoroughly.

7. Install or connect the following:
  - New head gaskets on the cylinder block

❊❊ WARNING

Always use new cylinder head bolts when installing the cylinder head or damage to the engine may occur.

- Cylinder head on the cylinder block
- Tighten the cylinder head bolts in sequence using the following steps. The first step is 30 ft. lbs. (40 Nm), the second step is an additional 90 degrees, the third step is to loosen all bolts in sequence one full turn, the fourth step is to tighten the bolts to 30 ft. lbs. (40 Nm), the fifth step is to tighten and additional 90 degrees, the last step is to tighten an addition 90 degrees.
- Lash adjusters and camshaft roller followers in their correct locations. Lubricate all rocker arm components with engine assembly lubricant.
- Exhaust manifold
- Camshafts

06017-FREE-G12

Right side valve cover bolt installation sequence—Freestyle

- Place a bead of silicone gasket sealant to the front cover-to-cylinder head surface
- Valve cover. Install the 9 bolts and 5 stud bolts, and tighten in sequence to 89 inch lbs. (10 Nm).
- Power steering pressure tube bracket
- Power steering reservoir bracket and stud bolts
- PCV connector and tube
- Ignition coil on plugs
- Coolant bypass tube
- Upper and lower intake manifolds
- Accessory drive belt

8. Fill and bleed the cooling system.

➡**Engine coolant is corrosive to engine bearing material. Replace the engine oil after removal of any coolant-carrying component to help prevent potential bearing damage.**

9. Change the engine oil and filter
10. Connect the negative battery cable.
11. Start the engine and check for leaks.

Camshaft Roller Followers and Lash Adjusters

REMOVAL & INSTALLATION

1. Remove the valve covers. See the procedure under cylinder head.
2. Remove the spark plugs.
3. Remove the right side splash shield.
4. Rotate the crankshaft until the

303-473

06017-FREE-G13

Removing the camshaft roller followers using tool 303-473—Freestyle

camshaft lobe is pointing directly away from the roller follower.

5. Using tool no. 303-473, remove the roller followers.

6. Remove the lash adjusters. If more than adjuster is to be removed, identify each adjuster location.

To install:

7. Lubricate the lash adjusters clean engine oil and install them in their original locations.

8. Reverse the removal procedure to install the remaining components.

Intake Manifold

REMOVAL & INSTALLATION

Upper

1. Relive the fuel system pressure.
2. Drain the engine coolant.

06017-FREE-G14

Upper intake manifold—Freestyle

06017-FREE-G15

**Upper intake manifold tightening
sequence—Freestyle**

3. Remove or disconnect the following:
  - Negative battery cable
  - Air cleaner outlet pipe
  - Transaxle roll restrictor cross brace
    bolts and cross brace
  - EVAP purge valve tube
  - PCV, brake booster and vacuum
    harness tubes from the manifold
  - Fuel tube retainer
  - EGR connector and vacuum tube
  - Throttle body and PCV coolant
    hoses
  - Throttle body electrical connector
  - Upper-to-lower intake manifold
    attaching screws
  - Intake manifold

To install:

4. Installation is the reverse of the
removal procedure, using the following
torque specifications.
  - Intake manifold bolts and tighten
    all bolts in steps in sequence to 89
    inch lbs. (10 Nm)
  - Transaxle roll restrictor roll brace
    and tighten the bolts as shown

06017-FREE-G17

Lower intake manifold and related components—Freesyle

5. Fill and bleed the engine cooling system.
6. Connect the negative battery cable.
7. Start the engine and check for leaks.

Lower

1. Remove or disconnect the following:
  - Coolant
  - Fuel system pressure

- Upper intake manifold
- Fuel tube from fuel rail
- Fuel rail pressure and temperature
  sensor connectors and vacuum
  tube
- Fuel charging wiring harness connector
- Lower intake manifold

To install:

2. Thoroughly clean all gasket mating
surfaces.
3. Install or connect the following:
  - 6 intake manifold gaskets and the
    intake manifold
  - Fuel charging wiring harness connector
  - Fuel rail pressure and temperature
    sensor connectors and vacuum
    tube
  - Fuel tube from fuel rail
  - Upper intake manifold
  - Fuel system pressure
4. Fill and bleed the cooling system.
5. Start the engine and check for leaks.

Exhaust Manifold

REMOVAL & INSTALLATION

➡Spray the exhaust system fasteners
with penetrating lubricant before
removing them to help prevent broken

55 Nm (41 lb-ft)

48 Nm
(35 lb-ft)

06017-FREE-G16

Transaxle roll restrictor brace bolt tightening—Freestyle

studs and bolts. The use of a 6-point socket is highly recommended when removing exhaust system fasteners.

✳✳ CAUTION

To prevent serious burns, allow the exhaust manifold to cool down before attempting to remove it.

Left Manifold

1. Disconnect the negative battery cable.
2. Remove the coolant hose retainers from the fan shroud and position aside.
3. Disconnect the cooling fan electrical connectors.
4. Remove the cooling fans.
5. Disconnect the oxygen sensor connector.
6. Remove the 2 exhaust manifold-to-catalytic converter bolts.
7. Remove 6 exhaust manifold nuts and remove the exhaust manifold.

To install:

8. Clean all gasket mating surfaces thoroughly.
9. Install a new exhaust manifold gasket and the exhaust manifold on the cylinder head. Start 2 nuts to hold the manifold in position.
10. Install the remaining nuts. Tighten the nuts in the sequence shown to 15 ft. lbs. (20 Nm).
11. Install the 2 exhaust manifold-to-catalytic converter bolts and tighten to 30 ft. lbs. (40 Nm).
12. Connect the oxygen sensor connector.

1. RH exhaust manifold heat shield (2 required)
2. RH exhaust manifold heat shield
3. Heated oxygen sensor (HO2S) electrical connector
4. RH exhaust manifold
5. Exhaust gas recirculation (EGR) system module tube fitting
6. RH exhaust manifold nut (6 required)
7. RH exhaust manifold gasket

06017-FREE-G19

Exploded view of right exhaust manifold—Freestyle

13. Install the cooling fans.
14. Connect the cooling fan electrical connectors.
15. Install the coolant hose retainers to the fan shroud.
16. Connect the negative battery cable.
17. Start the engine and check for exhaust leaks.

Right Manifold

1. Disconnect the negative battery cable.
2. Raise the vehicle on a hoist.
3. Remove the right side catalytic converter.
4. Disconnect the oxygen sensor connector.
5. Remove the exhaust manifold heat shield.
6. Disconnect the EGR tube from the manifold.
7. Remove the exhaust manifold.

To install:

8. Clean all gasket mating surfaces thoroughly.
9. Install a new gasket and the exhaust manifold on the cylinder head. Start 2 nuts to hold the manifold in position.
10. Install the remaining nuts. Tighten the nuts in the sequence shown to 15 ft. lbs. (20 Nm).
11. Connect the EGR tube from the manifold.
12. Install the exhaust manifold heat shield.
13. Connect the oxygen sensor connector.
14. Install the right side catalytic converter.
15. Lower the vehicle.

06017-FREE-G18

Exploded view of left exhaust manifold—Freestyle

16. Connect the negative battery cable.
17. Start the engine and check for exhaust leaks.

Camshaft

REMOVAL & INSTALLATION

Right Side

1. Disconnect the negative battery cable.
2. Remove the engine front cover.
3. Remove the timing chains.
4. Loosen the camshaft bearing cap bolts evenly in the sequence shown and remove the bearing caps.

➡ **The bearing caps are numbered for correct positioning.**

5. Remove the camshafts.
To install:
6. Clean and inspect all parts before installation.
7. Lubricate the camshaft lobes and journals with clean engine oil.
8. Carefully install the camshafts.
9. Align the camshaft sprocket marks as shown.
10. Lubricate the bearing surfaces of the bearing caps with clean engine oil.
11. Position the bearing caps in their proper locations and loosely install the bearing cap bolts.
12. Tighten the bearing caps bolts in sequence to 89 inch lbs. (10 Nm).
13. Install the timing chains.
14. Install the engine front cover.
15. Connect the negative battery cable.

Left Side

1. Disconnect the negative battery cable.
2. Remove the water pump.
3. Remove the engine front cover.
4. Remove the timing chains.
5. Remove the 3 bolts and the camshaft oil seal retainer.

6. Using a press, remove the oil seal from the retainer and discard.
7. Loosen the camshaft bearing cap bolts evenly in the sequence shown and remove the bearing caps.

➡ **The bearing caps are numbered for correct positioning.**

8. Remove the camshafts.

To install:
9. Clean and inspect all parts before installation.
10. Lubricate the camshaft lobes and journals with clean engine oil.
11. Carefully install the camshafts.
12. Align the camshaft sprocket marks as shown.

06017-FREE-G20A

Right side camshaft bearing cap removal sequence—Freestyle

06017-FREE-G21

Right side camshaft sprocket mark positioning—Freestyle

06017-FREE-G22

Right side camshaft bearing cap tightening sequence—Freestyle

Left side camshaft oil seal installation—
Freestyle

211-185 and 303-458, install the camshaft oil seal.
  20. Remove the special tools.
  21. Install the timing chains.
  22. Install the engine front cover.
  23. Install the water pump.
  24. Connect the negative battery cable.

Starter Motor

REMOVAL & INSTALLATION

Automatic Transmission Models

  1. Disconnect the negative battery cable.

Left side camshaft bearing cap removal sequence—Freestyle

Left side camshaft sprocket mark positioning—Freestyle

  13. Lubricate the bearing surfaces of the bearing caps with clean engine oil.
  14. Position the bearing caps in their proper locations and loosely install the bearing cap bolts.
  15. Tighten the bearing caps bolts in sequence to 89 inch lbs. (10 Nm).
  16. Install the camshaft oil seal retainer and tighten the bolts 89 inch lbs. (10 Nm).
  17. Lubricate the oil seal with clean engine oil and position the seal on installer tool no. 303-463.
  18. Position the installer tool and oil seal on the camshaft.
  19. Using installer tools no. 303-1139,

Left side camshaft bearing cap tightening sequence—Freestyle

25 Nm (18 lb-ft)

12 Nm (9 lb-ft)

6 Nm (53 lb-in)

06017-FREE-G27

Starter mounting automatic transmission models—Freestyle

2. Remove the air cleaner assembly.
3. Remove the starter cover.
4. Disconnect the starter electrical connectors.
5. Remove the starter bolts.
6. Remove the starter from the vehicle.

To install:
7. Position the starter in the vehicle.
8. Install the bolts. Tighten to 18 ft. lbs. (25 Nm).
9. Connect the starter electrical connectors.

25 Nm (18 lb-ft)

12 Nm (9 lb-ft)

6 Nm (53 lb-in)

1 Transaxle roll restrictor bolt	6 Starter motor solenoid wire
2 Transaxle roll restrictor cross brace bolt	7 Starter motor solenoid battery cable nut
3 Transaxle roll restrictor cross brace	8 Starter motor solenoid battery cable
4 Starter motor solenoid terminal cover	9 Starter motor bolt
5 Starter motor solenoid wire nut	10 Starter motor

06017-FREE-G28

Starter mounting manual transmission models—Freestyle

10. Install the starter cover.
11. Install the air cleaner assembly.
12. Connect the negative battery cable.

Manual Transmission Models

1. Disconnect the negative battery cable.
2. Remove the air cleaner outlet pipe.
3. Remove the transaxle roll restrictor bolts and the cross brace assembly.
4. Remove the starter cover.
5. Disconnect the starter electrical connectors.
6. Remove the starter bolts.
7. Remove the starter from the vehicle.

To install:
8. Position the starter in the vehicle.
9. Install the bolts. Tighten to 18 ft. lbs. (25 Nm).
10. Connect the starter electrical connectors.
11. Install the starter cover.
12. Install the cross brace assembly and transaxle roll restrictor bolts.
13. Install the air cleaner outlet pipe.
14. Connect the negative battery cable.

Oil Pan

✳✳ CAUTION

The EPA warns that prolonged contact with used engine oil may cause a number of skin disorders, including cancer! You should make every effort to minimize your exposure to used engine oil. Protective gloves should be worn when changing the oil. Wash your hands and any other exposed skin areas as soon as possible after exposure to used engine oil. Soap and water, or waterless hand cleaner, should be used.

REMOVAL & INSTALLATION

1. Disconnect the negative battery cable.
2. Drain the engine oil.
3. Remove the oil filter.
4. Raise the vehicle on a hoist.
5. Remove the exhaust flex pipe.
6. Remove the torque converter inspection cover.
7. Remove 2 oil pan-to-transaxle bolts.
8. Remove the oil filter adapter-to-oil pan bolt.
9. Remove the 15 retaining bolts and remove the oil pan.

To install:
10. Clean the gasket mating surfaces thoroughly.

Applying sealant to the oil pan mounting surface—Freestyle

Oil pan bolt tightening sequence—Freestyle

11. Apply a drop of sealant to the locations shown.

➡When using silicone rubber sealer, assembly must occur within 5 minutes after sealer application. After this time, the sealer may start to harden and its sealing effectiveness may be reduced.

12. Install the oil pan using a new gasket and secure to the block with the attaching bolts. Tighten the bolts in sequence to 89 inch lbs. (10 Nm).
13. Install the oil filter adapter-to-oil pan bolt.
14. Install 2 oil pan-to-transaxle bolts.
15. Install the torque converter inspection cover.
16. Install the exhaust flex pipe.
17. Install a new oil filter.
18. Fill the engine with the proper type and amount of clean oil.
19. Connect the negative battery cable.
20. Start the engine and check for leaks.

Oil Pump

REMOVAL & INSTALLATION

1. Disconnect the negative battery cable.
2. Drain the engine oil.
3. Remove the timing chain components.
4. Remove the oil pan.

Oil pump bolt removal sequence—Freestyle

Oil pump bolt tightening sequence—Freestyle

5. Remove the oil pump screen and pickup tube.
6. Remove the oil pump attaching bolts in the sequence shown.
   To install:
7. Place the oil pump in the proper position with a new gasket and install the retaining bolts.
8. Tighten the oil pump retaining bolts in sequence to 89 inch lbs. (10 Nm).
9. Install the oil pump screen and pickup tube.
10. Install the oil pan.
11. Install the timing chain components.
12. Fill the engine with clean oil.
13. Connect the negative battery cable.

➡Check for proper engine oil pressure immediately after starting the engine. If engine oil pressure is not within specification a few seconds after starting the engine, stop the engine and determine the reason for the low oil pressure condition. Running an engine with low oil pressure may result in serious engine damage.

Rear Main Seal

REMOVAL & INSTALLATION

1. Disconnect the negative battery cable.
2. Raise the vehicle on a hoist.
3. Remove the transaxle.
4. Remove the flexplate.

✳✳ WARNING

Use caution when working near the crankshaft sealing surface. If the surface becomes damaged, an oil leak may occur.

5. Screw in the threaded end of a crankshaft rear seal replacer tool, then use the tool to remove the seal.
   To install:
6. Inspect the crankshaft seal area for any damage that may cause the seal to leak. If damage is evident, service or replace the crankshaft as necessary.
7. Coat the crankshaft seal area and the seal lip with engine oil.
8. Using a crankshaft seal replacer tool, install the seal. Tighten the bolts of the seal installer tool evenly so the seal is straight and seats without misalignment.
9. Install the flexplate and tighten the bolts to 59 ft. lbs. (80 Nm).
10. Install the transaxle, lower the vehicle and connect the battery.

Front Main Seal

REMOVAL & INSTALLATION

1. Disconnect the negative battery cable.
2. Raise the vehicle on a hoist.
3. Remove the accessory drive belt.
4. Remove the right side fender splash shield.
5. Hold the crankshaft pulley and remove the pulley bolt.
6. Using a puller, remove the crankshaft pulley.
7. Use a seal replacer and pull out the oil seal.
   To install:
8. Coat the crankshaft seal lip with clean engine oil.
9. Using a seal installer, install the seal.
10. Lubricate the front cover and crankshaft seal inner lip with clean engine oil.
11. Apply silicone sealant to the crankshaft keyway in the pulley.
12. Using an installer, screw the crankshaft pulley on the crankshaft.
13. Hold the crankshaft pulley from turning and install the pulley bolt and washer and tighten as follows: Step 1; tighten the bolt to 86 ft. lbs. (120 Nm). Step 2; loosen the bolt one full turn. Step 3; tighten to 37 ft. lbs. (50 Nm). Step 4; tighten an additional 90 degrees.
14. Install the right side fender splash shield.

15. Install the accessory drive belt.

16. Lower the vehicle.

Timing Chains and Front Cover

REMOVAL & INSTALLATION

1. Raise the vehicle on a hoist.

2. Disconnect the negative battery cable.

3. Remove the accessory drive belt.

4. Remove the oil pan.

5. Remove the crankshaft front oil seal.

6. Remove both valve covers. See the procedure in cylinder head.

7. Remove the accessory drive belt tensioner.

8. Disconnect the power steering pump and reservoir without disconnecting the lines, and position aside.

9. Disconnect the crankshaft and camshaft position sensor electrical connectors.

10. Remove the alternator-to-front cover stud nut and stud.

11. Remove the radiator splash shield.

12. Remove the A/C compressor bolts and without disconnecting the lines, position the compressor aside.

13. Remove the compressor bracket.

14. Remove the ground wire from the front cover.

15. Remove the 14 bolts, stud bolt and remove the engine front cover.

16. Remove the ignition pulse wheel.

17. Install the crankshaft damper bolt.

18. Remove the spark plugs.

19. Rotate the crankshaft clockwise to position the crankshaft keyway in the 11 o'clock position.

20. Position the camshaft sprockets timing marks in the correct position so the no. 1 cylinder is a TDC of the compression stroke.

21. Ensure all timing marks are correctly located. If not, rotate the crankshaft one additional turn and recheck the marks.

22. Rotate the crankshaft clockwise 120 degrees to the 3 o'clock position to place the right side camshafts in the neutral position.

23. Ensure the camshafts are correctly located.

24. Remove the right side timing chain

Crankshaft and camshaft sprocket timing marks—Freestyle

06017-FREE-G35

Positioning right side camshaft sprockets in the neutral position—Freestyle

06017-FREE-G36

Positioning left side camshaft sprockets in the neutral position—Freestyle

tensioner arm bolts, the tensioner and the tensioner arm.

25. Remove the right side timing chain guide and timing chain.

26. Rotate the crankshaft clockwise 600 degrees to position the left side camshafts in the neutral position.

27. Remove the left side timing chain tensioner arm bolts, the tensioner and the tensioner arm.

28. Remove the left side timing chain guide and timing chain.

To install:

29. Place the timing chain tensioner in a soft-jawed vise.

30. Hold the tensioner ratchet lock away from the ratchet stem with a small pick.

31. Compress the tensioner until the piston is fully bottomed in the bore and insert a 1.5mm paper clip to hold the piston. Remove the pick from the ratchet stem.

32. Repeat the procedure on the other timing chain tensioner.

33. Mark any link on the timing chain with a permanent marker to indicate the crankshaft timing mark.

34. Starting with the crankshaft timing mark, count 29 links on the chain and mark the link.

35. Continue to count to the 42nd link and mark the link. Repeat this procedure on the other timing chain.

(14)

20 Nm
(15 lb-in)

(16)

(15)

(17)

(18)

(19)

(20)

14 Engine front cover stud bolt nut

15 Ground wire

16 Engine front cover stud bolt

17 Engine front cover bolt (8 required)

18 Engine front cover bolt (7 required)

19 Engine front cover

20 Engine front cover gasket (3 required)

06017-FREE-G33

Exploded view of engine front cover—Freestyle

06017-FREE-G37

Left side crankshaft and camshaft timing mark alignment—Freestyle

06017-FREE-G38

Right side crankshaft and camshaft timing mark alignment—Freestyle

36. Ensure the left side camshaft sprocket timing marks are still aligned.

37. Install the left side timing chain and guide and install the bolts.

38. Align the marks on the timing chain with the marks on the crankshaft and camshaft sprockets.

39. Tighten the timing chain guide bolts to 18 ft. lbs. (25 Nm).

40. Install the tensioner arm and tensioner and tighten the bolts to 18 ft. lbs. (25 Nm).

41. Rotate the crankshaft 120 degrees clockwise until the crankshaft keyway is in the 3 o'clock position.

42. Ensure the right side camshaft sprocket timing marks are still aligned.

43. Install the right side timing chain and guide and install the bolts.

44. Align the marks on the timing chain with the marks on the crankshaft and camshaft sprockets.

45. Tighten the timing chain guide bolts to 18 ft. lbs. (25 Nm).

46. Install the tensioner arm and tensioner and tighten the bolts to 18 ft. lbs. (25 Nm).

47. Remove both tensioner retaining pins and allow the piston to move.

48. Rotate the crankshaft 120 degrees counterclockwise to the TDC position.

1. 12 Links
2. 27 Links
3. 30 Links

06017-FREE-G39

Correct timing chain mark alignment—Freestyle

49. Verify the correct timing as follows: There should be 12 links between the camshaft sprocket timing marks; there should be 27 links between the left side camshaft and crankshaft timing marks; there should be 30 links between the right side camshaft and crankshaft timing marks.

50. Remove the crankshaft damper bolt and install the ignition pulse wheel. Install the pulse wheel with the keyway in the slot stamped "30" or "30FF."

51. Install the spark plugs and tighten to 11 ft. lbs. (15 Nm).

52. Clean the front cover gasket mating surfaces.

53. Place silicone gasket sealant on the cylinder head and cylinder block mating surfaces as shown.

54. Install the front cover and tighten the bolts and stud bolt in the sequence shopwn to 18 ft. lbs. (25 Nm).

55. Install the ground wire to the front cover.

0.06 mm (0.024 in)

06017-FREE-G40

Applying sealant to block and head front cover mating surfaces—Freestyle

56. Install the A/C compressor bracket.

57. Install the A/C compressor.

58. Install the radiator splash shield.

59. Install the alternator-to-front cover stud nut and stud.

60. Connect the crankshaft and camshaft position sensor electrical connectors.

61. Connect the power steering pump and reservoir

62. Install the accessory drive belt tensioner.

63. Install both valve covers.

64. Install the crankshaft front oil seal.

65. Install the oil pan.

66. Install the accessory drive belt.

67. Fill the crankcase with the proper type and quantity of engine oil. Fill and bleed the cooling system. Connect the negative battery cable.

68. Start the engine and check for leaks. Check the ignition timing and curb idle speed and adjust, as necessary.

Front cover bolt torque sequence—Freestyle

06017-FREE-G41

FUEL SYSTEM

Fuel System Service Precautions

Safety is the most important factor when performing not only fuel system maintenance, but any type of maintenance. Failure to conduct maintenance and repairs in a safe manner may result in serious personal injury or death. Work on a vehicle's fuel system components can be accomplished safely and effectively by adhering to the following rules and guidelines.

• To avoid the possibility of fire and personal injury, always disconnect the negative battery cable unless the repair or test procedure requires that battery voltage by applied.

• Always relieve the fuel system pressure prior to disconnecting any fuel system component (injector, fuel rail, pressure regulator, etc.) fitting or fuel line connection. Exercise extreme caution whenever relieving

fuel system pressure, to avoid exposing skin, face and eyes to fuel spray. Please be advised that fuel under pressure may penetrate the skin or any part of the body that it contacts.

• Always place a shop towel or rag around the fitting or connection prior to loosening to absorb any excess fuel due to spillage. Ensure that all fuel spillage is quickly remove from engine surfaces. Ensure that all fuel-soaked cloths or towels are deposited into a flame-proof waste container with a lid.

• Always keep a dry chemical (Class B) fire extinguisher near the work area.

• Do not allow fuel spray or fuel vapors to come into contact with a light bulb, spark or open flame.

• Always use a second wrench when loosening or tightening fuel line connections fittings. This will prevent unnecessary

stress and torsion to fuel piping. Always follow the proper torque specifications.

• Always replace worn fuel fitting O-rings with new ones. Do not substitute fuel hose where rigid pipe is installed.

Fuel System Pressure

RELIEVING

Locate the fuel pump relay in the engine compartment power distribution box relay no. 53. Remove the fuel pump relay. Start the engine and allow it to idle until it stalls. Crank the engine for 5 seconds to ensure fuel supply manifold pressure is released. When vehicle service is complete, reinstall the fuel pump relay and turn the ignition on to pressurize the fuel system. Start the vehicle and check the system for leaks.

Fuel pump relay location—Freestyle

06017-FREE-G42

Fuel Filter

REMOVAL & INSTALLATION

1. Relieve the fuel system pressure.
2. Disconnect the negative battery cable.
3. Raise and support the vehicle safely on jackstands.

4. Place a rag under the fuel filter to catch any residual fuel that may leak out when the filter is removed.
5. Remove the quick-connect fittings at both ends of the fuel filter.
6. Note the flow arrow direction for installation reference.
7. Remove the fuel filter by loosening the retaining clamp and pulling it from the bracket.

To install:

8. Install the fuel filter in its bracket, ensuring proper direction of flow as noted earlier.
9. Install the quick-connect fittings at both ends of the fuel filter. Ensure the fuel line clicks into place and is fully seated.
10. Start the engine and check the filter connections for leaks by running the tip of your finger around each connection.

9 Nm (80 lb-in)

1 Fuel supply tube quick connect coupling

2 Fuel supply tube quick connect coupling

3 Fuel filter

4 Fuel filter bracket

5 Fuel filter bracket bolt

6 Fuel tube retainer clip

7 Fuel supply tube quick connect coupling with redundant clip

06017-FREE-G43

Exploded view of fuel filter mounting—Freestyle

11. Turn the engine off and lower the vehicle.

Fuel Pump

REMOVAL & INSTALLATION

1. Relieve the fuel system pressure.
2. Disconnect the negative battery cable.
3. Drain the fuel tank.
4. Position the second row seats in the rearward position.
5. Remove the rear seat latching bracket bolts and brackets.
6. Remove the fuel pump module cover.
7. Remove any dirt that has accumulated around the fuel pump module attaching flange to prevent it from entering the tank during service.
8. Disconnect the fuel pump electrical connectors.
9. Disconnect the fuel pump quick connect coupling from the fuel pump module.
10. Disconnect the fuel vapor tube quick connect coupling from the fuel pump module.
11. Turn the fuel pump module locking ring using a locking ring removal tool, and remove the locking ring.
12. Remove the fuel pump module and disconnect the quick connect fittings from the bottom of the fuel pump.

1 Fuel pump module access cover	5 Fuel pump module lock ring
2 Fuel vapor control tube assembly valve quick connect coupling	6 Fuel pump module
3 Fuel supply tube quick release coupling	7 Fuel pump module O-ring seal
4 Fuel pump module electrical connector	8 Fuel transfer supply tube quick connect coupling
	9 Fuel vapor tube quick connect coupling

06017-FREE-G44

In-tank electric fuel pump and related components—Freestyle

To install:

13. Connect the quick connect fittings to the bottom of the fuel pump. Insert the fuel pump module into the fuel tank and align the arrows on the fuel pump with the arrows on the fuel tank. Secure it in place with the locking ring.

14. Connect the fuel vapor tube quick connect coupling to the fuel pump module.

15. Connect the fuel pump quick connect coupling to the fuel pump module.

16. Connect the fuel pump electrical connectors.

17. Install the fuel pump module cover.

18. Install the rear seat latching bracket bolts and brackets and tighten to 13 ft. lbs. (17 Nm).

19. Install a minimum of 10 gallons (38L) of fuel and check for leaks.

20. Check for fuel leaks at the fittings.

21. Start the engine and check for fuel leaks.

Fuel Rail and Injectors

REMOVAL & INSTALLATION

1. Relieve the fuel system pressure.
2. Disconnect the negative battery cable.
3. Remove the upper and lower intake manifolds.
4. Disconnect the fuel rail temperature and pressure sensor connectors.
5. Remove the vacuum harness.
6. Remove the fuel rail attaching bolts.
7. Separate the fuel rail and injectors from the lower intake manifold.
8. Disconnect the fuel injector connectors.
9. Disconnect the fuel rail charging wiring harness.
10. Remove the retaining clips and separate the injectors from the supply manifold. Discard the O-rings.

To install:

11. Lubricate NEW O-rings with clean engine oil and position fuel injectors on the supply manifold. Install the retaining clips.

12. Connect the fuel rail charging wiring harness.

13. Connect the fuel injector connectors.

14. Connect the fuel rail and injectors to the lower intake manifold.

15. Install the fuel rail attaching bolts and tighten to 89 inch lbs. (19 Nm).

16. Install the vacuum harness.

17. Connect the fuel rail temperature and pressure sensor connectors.

18. Install the upper and lower intake manifolds.

19. Connect the negative battery cable.

20. Start the engine and check for fuel leaks.

1 Fuel rail pressure and temperature sensor electrical connector
2 Fuel injector electrical connector (6 required)
3 Fuel charging wiring harness
4 Vacuum harness
5 Fuel rail bolt (4 required)
6 Fuel rail
7 Fuel injector clip (6 required)
8 Fuel injector (6 required)
9 Fuel injector O-ring (12 required)

Fuel injectors and supply manifold—Freestyle

06017-FREE-G45

DRIVE TRAIN

Transaxle Assembly

REMOVAL & INSTALLATION

CVT Transaxle

1. Place the vehicle on a hoist.
2. Drain the transaxle fluid.
3. Remove or disconnect the following:
   - Battery and battery tray
   - Air cleaner assembly and outlet pipe
   - Cowl grille panel
   - Upper roll restrictor bolt
   - Roll restrictor brace bolts and brace
   - Shift cable end
   - Electrical connector from shift cable bracket
   - Shift cable bracket and position the cable and bracket aside
   - Transaxle wiring harnesses
   - Auxiliary coolant flow pump assembly, if equipped
   - Roll restrictor bracket and roll restrictor
   - PCV hose
   - Throttle body vacuum hoses
   - Power steering line bracket
   - Starter wiring and starter
   - Transaxle fluid cooler tube bracket
   - Ground cable
   - Transaxle bulkhead electrical connector
   - 4 upper transaxle mounting bolts
   - Left and right side halfshafts
   - Front and rear exhaust pipe nuts and discard
   - Catalytic converter pipe
4. On AWD models, remove the 2 outer bolts for the driveshaft brace.
5. Index mark the rear driveshaft flange to the Power Transfer Unit (PTU).
6. Remove and the discard 6 bolts from the front driveshaft flange.
7. Remove the PTU.
8. Wire the rear driveshaft to the chassis.
9. On all models, remove or disconnect the following:
   - Engine block heater connector
   - Knock sensor heat shield and knock sensor
   - Knock sensor wiring harness from transaxle
   - Transaxle-to-engine bolt
   - Power steering hose bolt
10. Install engine lifting eyes no. 134-00243 to the left and right cylinder heads.
11. Install lifting brackets no. 303-1140 to front of the engine.

Installing engine support tools—Freestyle

12. Install engine support tools to the engine as shown.
13. Remove or disconnect the following:
   - Disconnect the lower radiator hose from its retainers and position aside
   - Power steering line from subframe
   - Wiring harness from left front subframe
   - Power steering rack retainers and wire the rack to the chassis
   - Oxygen sensor connector
   - Splash shield
   - Front motor mount nut
   - Lower stabilizer mount bolts
   - Rear motor mount nut
   - Rear transaxle mount bolts
   - Rear subframe mounting bolts on both sides
   - Stabilizer bar links nuts from both sides and discard the nuts.
   - Stabilizer bar from struts
14. Using a suitable powertrain lift, remove 4 subframe mounting bolts and remove the subframe.
15. Remove the catalytic converter.
16. Remove the torque converter access plug and remove and discard the 4 torque converter nuts.
17. Disconnect the transaxle fluid cooler tubes from the transaxle.
18. Securely fasten the transaxle to a suitable high lift transmission jack.
19. Remove the 3 engine-to-transaxle bolts, and remove the transaxle.

To install:

➡️ **If a different transaxle is being installed, transfer the appropriate external components to the new transaxle.**

20. Apply multi-purpose grease to the torque converter pilot hub.
21. Place the transaxle on a suitable jack and carefully raise it into position.
22. Install 3 engine-to-transaxle mounting bolts and tighten to 30 ft. lbs. (40 Nm).
23. Install new transaxle fluid cooler tube seals and connect the tubes.
24. Install new torque converter nuts and tighten to 27 ft. lbs. (36 Nm). Install the torque converter access plug.
25. Install the catalytic converter and tighten the bolts to 30 ft. lbs. (40 Nm).
26. Using a suitable powertrain lift, install the subframe and tighten the 4 subframe mounting bolts to 66 ft. lbs. (90 Nm).
   - Install the subframe mounting bolts on both sides and tighten to 35 ft. lbs. (48 Nm).
   - Rear transaxle mount bolts and tighten to 66 ft. lbs. (90 Nm).
   - Rear motor mount nut and tighten to 52 ft. lbs. (70 Nm).
   - Lower stabilizer mount bolts and tighten to 52 ft. lbs. (70 Nm).
   - Front motor mount nut and tighten to 52 ft. lbs. (70 Nm).
   - Splash shield
   - Oxygen sensor connector
   - Power steering rack retainers and tighten to 98 ft. lbs. (133 Nm)
   - Power steering line to subframe
   - Wiring harness to left front subframe
   - Lower radiator hose
   - Stabilizer bar to struts
   - Stabilizer bar link nuts and tighten to 41 ft. lbs. (55 Nm)
27. Remove the all the special lifting tools from the engine and cylinder heads.
28. Install or connect the following:
   - Transaxle-to-engine bolt and tighten to 30 ft. lbs. (40 Nm)
   - Power steering hose bolt
   - Knock sensor wiring harness to transaxle
   - Knock sensor and heat shield
   - Engine block heater connector
29. On AWD models, install the PTU and rear driveshaft. Tighten the bolts to 66 ft. lbs. (90 Nm).
30. Install the 2 outer bolts for the driveshaft brace and tighten to 18 ft. lbs. (25 Nm).
31. Install new bolts to the front driveshaft flange and tighten in a criss-cross pattern to 18 ft. lbs. (25 Nm).
32. On all models, install or connect the following:

- Front and rear exhaust pipe nuts and tighten to 30 ft. lbs. (40 Nm)
- Left and right side halfshafts
- 4 upper transaxle mounting bolts and tighten to 30 ft. lbs. (40 Nm)
- Transaxle bulkhead electrical connector
- Ground cable
- Transaxle fluid cooler tube bracket
- PCV hose
- Power steering line bracket
- Throttle body vacuum hoses
- Transaxle wiring harnesses
- Starter wiring and starter and tighten to 19 ft. lbs. (26 Nm)
- Roll restrictor bracket and roll restrictor and apply treadlocker before tightening to 30 ft. lbs. (40 Nm)
- Auxiliary coolant flow pump assembly, if equipped
- Shift cable bracket and cable
- Electrical connector to shift cable bracket
- Roll restrictor brace bolts and brace and apply treadlocker before tightening to 30 ft. lbs. (40 Nm)
- Upper roll restrictor bolt and apply treadlocker before tightening to 30 ft. lbs. (40 Nm)
- Cowl grille panel
- Air cleaner assembly
- Battery tray and battery

33. Fill the transaxle with proper amount of Mercon Continuously Variable chain type transmission fluid.

34. Connect the positive, then the negative battery cable.

Start the engine and check for leaks.

6-Speed Transaxle

1. Place the vehicle on a hoist.
2. Drain the transaxle fluid.
3. Remove or disconnect the following:
   - Battery and battery tray
   - Air cleaner assembly and outlet pipe
   - Cowl grille panel
   - Upper roll restrictor bolt
   - Roll restrictor brace bolts and brace
   - Shift cable end
   - Electrical connector from shift cable bracket
   - Shift cable bracket and position the cable and bracket aside
   - Transaxle Control Module (TCM) wiring harness
   - Transaxle wiring harnesses
   - Roll restrictor bracket and roll restrictor
   - PCV hose

06017-FREE-G46

Installing engine support tools—Freestyle

- Throttle body vacuum hoses
- Power steering line bracket
- Starter wiring and starter
- Ground cable
- 4 upper transaxle mounting bolts
- Left and right side halfshafts
- Front and rear exhaust pipe nuts and discard
- Catalytic converter pipe
- Engine block heater connector
- Knock sensor heat shield and knock sensor
- Knock sensor wiring harness from transaxle
- Transaxle-to-engine bolt
- Power steering hose bolt

4. Install engine lifting eyes no. 134-00243 to the left and right cylinder heads.

5. Install lifting brackets no. 303-1140 to front of the engine.

6. Install engine support tools to the engine as shown.

7. Remove or disconnect the following:

   - Disconnect the lower radiator hose from its retainers and position aside
   - Power steering line from subframe
   - Wiring harness from left front subframe
   - Power steering rack retainers and wire the rack to the chassis
   - Oxygen sensor connector
   - Splash shield
   - Front motor mount nut
   - Lower stabilizer mount bolts
   - Rear motor mount nut
   - Rear transaxle mount bolts
   - Rear subframe mounting bolts on both sides
   - Stabilizer bar links nuts from both sides and discard the nuts
   - Stabilizer bar from struts

8. Using a suitable powertrain lift, remove 4 subframe mounting bolts and remove the subframe.

9. Remove the catalytic converter.

10. Remove the torque converter access plug and remove and discard the 4 torque converter nuts.

11. Disconnect the transaxle fluid cooler tubes from the transaxle.

12. Securely fasten the transaxle to a suitable high lift transmission jack.

13. Remove the 3 engine-to-transaxle bolts, and remove the transaxle.

To install:

➡**If a different transaxle is being installed, transfer the appropriate external components to the new transaxle.**

14. Apply multi-purpose grease to the torque converter pilot hub.

15. Place the transaxle on a suitable jack and carefully raise it into position.

16. Install 3 engine-to-transaxle mounting bolts and tighten to 30 ft. lbs. (40 Nm).

17. Install new transaxle fluid cooler tube seals and connect the tubes.

18. Install new torque converter nuts and tighten to 27 ft. lbs. (36 Nm). Install the torque converter access plug.

19. Install the catalytic converter and tighten the bolts to 30 ft. lbs. (40 Nm).

20. Using a suitable powertrain lift, install the subframe and tighten the 4 subframe mounting bolts to 66 ft. lbs. (90 Nm).

   - Install the subframe mounting bolts on both sides and tighten to 35 ft. lbs. (48 Nm).
   - Rear transaxle mount bolts and tighten to 66 ft. lbs. (90 Nm).
   - Rear motor mount nut and tighten to 52 ft. lbs. (70 Nm).
   - Lower stabilizer mount bolts and tighten to 52 ft. lbs. (70 Nm).
   - Front motor mount nut and tighten to 52 ft. lbs. (70 Nm).
   - Splash shield
   - Oxygen sensor connector
   - Power steering rack retainers and tighten to 98 ft. lbs. (133 Nm)
   - Power steering line to subframe
   - Wiring harness to left front subframe
   - Lower radiator hose
   - Stabilizer bar to struts
   - Stabilizer bar link nuts and tighten to 41 ft. lbs. (55 Nm)

21. Remove the all the special lifting tools from the engine and cylinder heads.

22. Install or connect the following:
   - Transaxle-to-engine bolt and tighten to 30 ft. lbs. (40 Nm)
   - Power steering hose bolt

- Knock sensor wiring harness to transaxle
- Knock sensor and heat shield
- Engine block heater connector
- Front and rear exhaust pipe nuts and tighten to 30 ft. lbs. (40 Nm)
- Left and right side halfshafts
- 4 upper transaxle mounting bolts and tighten to 30 ft. lbs. (40 Nm)
- Transaxle bulkhead electrical connector
- Ground cable
- Transaxle fluid cooler tube bracket
- PCV hose
- Power steering line bracket
- Throttle body vacuum hoses
- Transaxle wiring harnesses
- Starter wiring and starter and tighten to 19 ft. lbs. (26 Nm)
- Roll restrictor bracket and roll restrictor and apply treadlocker before tightening to 30 ft. lbs. (40 Nm)
- Auxiliary coolant flow pump assembly, if equipped
- Shift cable bracket and cable
- Electrical connector to shift cable bracket
- Roll restrictor brace bolts and brace and apply treadlocker before tightening to 30 ft. lbs. (40 Nm)
- Upper roll restrictor bolt and apply treadlocker before tightening to 30 ft. lbs. (40 Nm)
- Cowl grille panel
- Air cleaner assembly
- Battery tray and battery

23. Fill the transaxle with proper amount of Motorcraft Premium automatic transmission no. XT-8-QAW.

24. Connect the positive, then the negative battery cable.
Start the engine and check for leaks.

Power Transfer Unit

REMOVAL & INSTALLATION

1. Raise the vehicle on a hoist.
2. Remove the left and right halfshafts
3. Remove the front and rear exhaust pipe nuts.
4. Remove the catalytic converter pipe.
5. Index mark the driveshaft flange to the output flange.
6. Remove the 2 outer driveshaft brace bolts.
7. Remove the 6 bolts from the driveshaft flange.
8. Remove the upper PTU bracket bolts and the bracket.

9. Remove 4 PTU mounting bolts and the PTU.
10. Remove the input shaft oil slinger and seal.
To install:
11. Using a seal installer, install the input shaft seal and oil slinger.
12. Position the PTU in place and tighten the bolts to 66 ft. lbs. (90 Nm).
13. Install the upper PTU bracket bolts and the bracket and tighten to 41 ft. lbs. (55 Nm).
14. Install the left and right halfshafts
15. Install the 2 outer driveshaft brace bolts and tighten to 18 ft. lbs. (25 Nm).
16. Install the 6 bolts to the driveshaft flange and tighten in a criss-cross pattern to 18 ft. lbs. (25 Nm).
17. Install the catalytic converter pipe, front and rear exhaust pipe nuts and tighten all fasteners to 30 ft. lbs. (40 Nm).

Output Shaft Seal and Flange

REMOVAL & INSTALLATION

1. Remove the driveshaft.
2. Remove the output flange nut and flange.
3. Remove the output shaft seal.
To install:
4. Clean the output shaft seal bore and shaft.
5. Use a seal installer and install the output shaft seal.
6. Install the output flange and tighten the nut to 74 ft. lbs. (100 Nm).
7. Install the driveshaft.

Driveshaft, Pinion Flange and Seal

REMOVAL & INSTALLATION

1. Place the vehicle on a hoist.
2. Remove the exhaust system at the flex pipe.
3. Remove the exhaust support brace.
4. Index mark the driveshaft to the rear axle pinion flange, PTU flange and center bearing bracket.
5. Disconnect the driveshaft from the rear pinion flange and discard the bolts.
6. Remove the driveshaft flange bolts from the PTU flange and discard the bolts.
7. Disconnect the driveshaft from the PTU flange.
8. Remove the center bearing bracket bolts and remove the driveshaft.

9. Remove and discard the pinion flange nut.
10. Using a puller, remove the pinion flange.
11. Remove the pinion seal.
To install:
12. Using a seal installer, install the pinion seal.
13. Install the drive pinion flange using a new nut. Tighten the nut to 74 ft. lbs. (100 Nm).
14. Install the driveshaft.
15. Install the center bearing bracket and loosely tighten the bolts.
16. Using the index marks made on removal, connect the driveshaft to the PTU flange and rear pinion flange and hand tighten the bolts.
17. Tighten the center bearing bracket bolts to 15 ft. lbs. (20 Nm).
18. Tighten the driveshaft to rear pinion flange and PTU flange bolts to 18 ft. lbs. (25 Nm).
19. Install the exhaust support brace and tighten the bolts to 22 ft. lbs. (30 Nm).
20. Install the exhaust system.
21. Lower the vehicle.

Halfshafts

REMOVAL & INSTALLATION

Front

➡ **Do not begin this removal procedure unless a new wheel hub retainer nut, a new retainer circlip and a new lower ball joint-to-front wheel knuckle retaining bolt and nut are available. Once removed, these parts must not be reused during assembly. Their torque holding ability, or retention capability, is diminished during removal.**

1. Remove or disconnect the following:
- Front wheels
- ABS sensor
- Brake caliper and wire aside
- Axle hub nut and washer. Discard the nut.
2. Separate the halfshaft from the hub using tool no. 205-D070.
3. Remove and discard the lower ball joint nut.
4. Disconnect the lower control arm.
5. On the left side, on 6-speed models, use a slide hammer and remove the halfshaft from the transaxle.
6. On the left side on CVT models, use a slide hammer and halfshaft fork no. 205-241 and remove the shaft from the transaxle.

7. On the right side, remove the half-shaft bearing nuts and remove the half-shaft.

8. On both sides, inspect the halfshaft seal and replace if necessary.

To install:

❊❊ WARNING

Do not reuse the retainer circlip. A new circlip must be installed each time the inboard CV-joint stub shaft is installed into the transaxle differential.

9. Install a new retainer circlip on the inboard CV-joint stub shaft by starting one end in the groove and working the retainer circlip over the inboard shaft housing end and into the groove. This will avoid over-expanding the circlip.

➡**A non-metallic mallet may be used to aid in seating the retainer circlip into the differential side gear groove. If a mallet is necessary, tap only on the outboard CV-joint shaft.**

10. Install the halfshaft.

11. On the right side install the halfshaft bearing nuts and tighten to 20 ft. lbs. (27 Nm).

12. On both sides install the halfshaft into the hub.

13. Connect the lower control arm to the knuckle and install a new ball joint nut. Tighten the nut to 59 ft. lbs. (80 Nm).

14. Using the old axle nut and washer tighten the nut to seat the halfshaft in the hub.

15. Remove the old nut and washer.

16. Install a new axle nut and washer and in a continuous motion, tighten the nut to 184 ft. lbs. (250 Nm).

17. Install the brake caliper and tighten to 74 ft. lbs. (100 Nm).

18. Install the front wheels and lower the vehicle.

Rear

➡**Do not begin this removal procedure unless a new wheel hub retainer nut is available. Once removed, the nut must not be reused during assembly. The torque holding ability, or retention capability, is diminished during removal.**

1. Remove or disconnect the following:

- Rear wheel
- Brake caliper and wire aside
- Axle hub nut and washer. Discard the nut.
- ABS sensor

2. Using a hub remover, press the halfshaft out until it is loose in the hub.

3. Remove the brake disc.

4. Remove the 4 hub bolts and remove the hub.

5. Using a prybar, separate the halfshaft from the axle, and remove the halfshaft.

To install:

❊❊ WARNING

Do not reuse the retainer circlip. A new circlip must be installed each time.

6. Install a new retainer circlip on the CV-joint stub shaft by starting one end in the groove and working the retainer circlip over the inboard shaft housing end and into the groove. This will avoid over-expanding the circlip.

7. Install the halfshaft.

8. Install the hub and the hub bolts and tighten to 89 ft. lbs. (120 Nm).

9. Install the brake disc and tighten to 16 ft. lbs. (21 Nm).

10. Using the old axle nut and washer tighten the nut to seat the halfshaft in the hub.

11. Remove the old nut and washer.

12. Install a new axle nut and washer and in a continuous motion, tighten the nut to 148 ft. lbs. (200 Nm).

13. Install the ABS sensor.

14. Install the caliper and tighten to 81 ft. lbs. (110 Nm).

15. Install the rear wheel.

CV-Joint

OVERHAUL

➡**Overhaul procedures are not available from the manufacturer.**

STEERING AND SUSPENSION

Air Bag (Supplemental Restraint) System

The Supplemental Restraint System (SRS) is designed to work in conjunction with the standard 3-point safety belts to reduce injury in a head-on collision.

❊❊ CAUTION

The SRS can actually cause physical injury or death if the safety belts are not used, or if the manufacturer's warnings are not followed. The manufacturer's warnings can be found in your owner's manual, or, in some cases, on your sun visor.

The SRS is comprised of the following components:

- Driver's side air bag module
- Passenger's side air bag module
- Right-hand and left-hand primary crash front air bag sensors
- Air bag diagnostic monitor computer
- Electrical wiring

The SRS primary crash front air bag sensors are hard-wired to the air bag modules and determine when the air bags are deployed. During a frontal collision, the sensors quickly inflate the 2 air bags to reduce injury by cushioning the driver and front passenger from striking the dashboard, windshield, steering wheel and any other hard surfaces. The air bag inflates so quickly (in a fraction of a second) that in most cases it is fully inflated before you actually start to move during a collision.

Since the SRS is a complicated and essentially important system, its components are constantly being tested by a diagnostic computer. The computer illuminates the air bag indicator light on the instrument cluster for approximately 6 seconds when the ignition switch is turned to the **RUN** position when the SRS is functioning properly. After being illuminated for the 6 seconds, the indicator light should then turn off.

If the air bag light does not illuminate at all, stays on continuously, or flashes at any time, a problem has been detected by the diagnostic computer.

❊❊ CAUTION

If at any time the air bag light indicates that the computer has noted a problem, immediately diagnose the problem. A faulty SRS can cause severe physical injury or death.

SERVICE PRECAUTIONS

Whenever working around, or on, the air bag supplemental restraint system, ALWAYS adhere to the following warnings and cautions.

• Always wear safety glasses when servicing an air bag vehicle and when handling an air bag module.

• Carry a live air bag module with the bag and trim cover facing away from your body, so that an accidental deployment of the air bag will have a small chance of personal injury.

• Place an air bag module on a table or other flat surface with the bag and trim cover pointing up.

• Wear gloves, a dust mask and safety glasses whenever handling a deployed air bag module. The air bag surface may contain traces of sodium hydroxide, a byproduct of the gas that inflates the air bag and which can cause skin irritation.

• Ensure to wash your hands with mild soap and water after handling a deployed air bag.

• All air bag modules with discolored or damaged cover trim must be replaced, not repainted.

• All component replacement and wiring service must be made with the negative and positive battery cables disconnected from the battery for a minimum of 1 minute prior to attempting service or replacement.

• NEVER probe the air bag electrical terminals. Doing so could result in air bag deployment, which can cause serious physical injury.

• If the vehicle is involved in a fender-bender that results in a damaged front bumper or grille, the air bag sensors should be inspected to ensure that they were not damaged.

• If at any time, the air bag light indicates that the computer has noted a problem, immediately diagnose the problem. A faulty SRS can cause severe physical injury or death.

DISARMING THE SYSTEM

1. Ensure the ignition is off.
2. Remove the Smart Junction Box (SJB) cover located below the left side of the instrument panel. Remove the Restraints Control Module (RCM) fuse no. F2.21.
3. Turn the ignition on and watch the air bag indicator for 30 seconds. The indicator light will remain lit constantly if the correct fuse has been removed. If the light is not on steadily, remove the correct fuse and check the light again.
4. Turn the ignition switch off.
5. Disconnect the negative battery cable.

ARMING THE SYSTEM

1. Turn the ignition switch on.
2. Install Restraints Control Module (RCM) fuse no. F2.21 and the fuse cover.
3. Turn the ignition off.
4. Connect the negative battery cable.
5. Turn the ignition on and then off. Wait 10 seconds and turn the key back on. Watch the air bag indicator. The indicator light will remain lit constantly for 6 seconds and then go off. If the indicator does not turn on and then off, diagnose the air bag system
6. Clear all Diagnostic Trouble Codes (DTC) using a diagnostic tool.

Rack and Pinion Steering Gear

REMOVAL & INSTALLATION

❊❊ WARNING

Do not allow the steering wheel to rotate when the intermediate shaft is disconnect, or damage to the clockspring can result. If it is suspected that the shaft has rotated, the clockspring must be removed and re-centered.

5 — 25 Nm (18 lb-ft)
1 — 90 Nm (66 lb-ft)
3 — 20 Nm (15 lb-ft)
6 — 117 Nm (86 lb-ft)

1 Tie-rod end nuts (2 required)

2 Tie-rod ends (2 required)

3 Power steering line clamp plate bolt

4 O-ring seals (2 required)

5 Intermediate shaft-to-steering gear bolt

6 Steering gear nuts (2 required)

7 Steering gear bolts (2 required)

8 Steering gear

06017-FREE-G47

Exploded view of the rack and pinion steering gear mounting on the front subframe of the vehicle—Freestyle

1. Place the steering wheel in the straight ahead position and remove the ignition key.

2. Raise the vehicle on a hoist.

3. Remove or disconnect the following:
- Front wheels
- Tie rod end cotter pins and castle nuts
- Tie rod ends from the knuckles
- Power steering line clamp plate bolt

4. Place a drain pan under the vehicle and disconnect the power steering lines.

5. Release the steering shaft intermediate boot bearing clamp and release the boot from the shaft.

6. Retain the bearing boot clamp in the open position.

7. Remove the pinch bolt retaining the steering column intermediate shaft coupling.

8. Remove the steering gear mounting nuts and bolts.

9. Remove the steering gear through the left side of the vehicle.

To install:

10. Install new O-rings on the power steering line fittings.

11. Place the steering gear in the vehicle through the left side.

12. Tighten the steering gear mounting bolts and nuts to 86 ft. lbs. (117 Nm).

13. Install the pinch bolt retaining the steering column intermediate shaft coupling and tighten to 18 ft. lbs. (25 Nm).

14. Position the bearing boot in the retention groove so the outer bearing race is not exposed from the groove.

15. Install boot bearing clamp and ensure the end tabs do not contact the boot.

16. Using new o-rings, install the power steering lines.

17. Install the power steering line clamp plate bolt and tighten to 15 ft. lbs. (20 Nm).

18. Connect the tie rods.

19. Install the tie rod end nuts and tighten to 66 ft. lbs. (90 Nm).

20. Install the front wheels.

21. Lower the vehicle.

22. Fill the power steering oil reservoir.

23. Start the vehicle and check for leaks.

24. Check for proper wheel alignment and steering wheel position.

Front Strut/Spring

REMOVAL & INSTALLATION

➡**Do not allow the axle shaft to move outward as damage to the CV joint may result.**

✳✳ WARNING

Always use new bolts and nuts on the suspension components.

1. Turn the ignition switch **OFF** and place the steering column in the unlocked position.

2. Loosen the strut/spring upper mounting nuts.

3. Raise the vehicle on a hoist.

4. Remove or disconnect the following:
- Front wheel
- Front wheel axle hub nut and discard
- Tie rod-to-wheel knuckle nut
- Separate the tie rod from the knuckle
- Brake disc
- Wheel speed sensor and wire aside
- Lower ball joint nut using a crow's foot wrench, and discard the nut

5. Push the lower arm down until the ball joint is clear of the lower arm.

6. Install tool no. 307-102 over the ball stud.

7. Use a puller and press the halfshaft from the wheel hub and support the halfshaft level.

8. Remove and discard the strut upper mounting nuts.

9. Remove the wheel knuckle and strut as an assembly.

10. Remove and discard the knuckle-to-strut bolt and nut.

11. Separate the spring/strut assembly from the knuckle.

To install:

12. Place the strut/spring assembly in the steering knuckle.

13. Using a new bolt and nut, tighten the strut-to-knuckle bolt to 129 ft. lbs. (175 Nm).

14. Install the strut/knuckle in the vehicle and loosely tighten the new top mounting nuts.

15. Install the halfshaft in the wheel hub.

16. Remove the ball stud tool and connect the ball joint to the lower control arm

17. Install a new ball joint nut and tighten to 85 ft. lbs. (115 Nm).

18. Install the wheel speed sensor.

19. Install the brake disc.

20. Install the tie rod to the wheel knuckle and tighten the new nut to 85 ft. lbs. (115 Nm).

21. Install a new axle hub nut and tighten to 148 ft. lbs. (200 Nm).

22. Install the front wheel.

23. Lower the vehicle.

24. Tighten the upper strut nuts to 18 ft. lbs. (25 Nm).

25. Check the wheel alignment.

DISASSEMBLY

1. Using a spring compressor, compress the spring and loosen the strut rod nut.

2. Compress the spring until all tension is relieved.

3. Remove the strut rod appearance cap.

4. Remove and discard the upper strut rod nut.

5. Separate the strut, spring mount and washer.

6. Remove the spring.

7. To maintain wheel camber, position the notch and arrow on the upper strut mount outboard opposite the locator tab on the strut.

8. To adjust camber +0.5 degrees, position the notch and arrow on the upper strut mount inboard aligned with the locator tab on the strut.

9. To reassemble, reverse the disassembly procedure. Make sure the spring is correctly seated in both spring seats. Tighten the new strut top retaining nut to 59 ft. lbs. (80 Nm).

06017-FREE-G48

Positioning upper strut mount outboard to maintain wheel camber—Freestyle

06017-FREE-G49

Positioning upper strut mount inboard to adjust wheel camber +0.5 degrees— Freestyle

Rear Shock Absorber/Spring

REMOVAL & INSTALLATION

✳✳ WARNING

Always use new bolts and nuts on the suspension components.

1. Measure the distance between the top of the wheel and the bottom of the wheel opening. This is the trim height.
2. Loosen the lug nuts on the rear wheels.
3. Remove the interior trim panel to access the upper shock nut, then remove and discard the nut.
4. Raise the vehicle on a hoist.
5. Remove the rear wheels.
6. Remove the brake caliper and wire aside.
7. Position a jack under the wheel knuckle at the trailing arm and knuckle attachment point.
8. Raise the wheel knuckle until the toe link is parallel to the ground.
9. Remove and discard the trailing arm-to-knuckle bolt.
10. Lower the suspension jack.
11. Remove and discard the trailing arm-to-subframe bolt.
12. Position a jack under the lower shock mount and raise it enough to compress the spring.
13. Remove and discard the lower arm-to-knuckle bolt.
14. Remove and discard the lower arm-to-subframe bolt.
15. Remove and discard the lower shock mounting bolt.
16. Hold the shock absorber and swing the lower arm down to access and remove the shock/spring assembly.

To install:

17. Tape the spring to the upper spring seat.
18. Position the spring and upper rubber seat on the shock.
19. Install a new lower arm-to-knuckle bolt in the lower arm.
20. Position a jack under the lower arm.
21. Install the shock/spring assembly into alignment with the upper mount.
22. Swing the lower arm into position under the shock/spring assembly.
23. Remove the jack.
24. Partially lower the hoist.
25. Place a floor jack under the lower arm at the lowest point.
26. Raise the jack until the top of the

83 Nm (61 lb-ft) — 22

22 Upper shock absorber nut
23 Upper shock mount
24 Bushing
25 Shock absorber/spring

06017-FREE-G50

Rear shock absorber/spring mounting—Freestyle

shock makes contact with the inside surface of the upper mount.
27. With the help of an assistant guiding the shock assembly, raise the jack until the shock seats in the upper mount.
28. Install a new top nut and tighten to 61 ft. lbs. (83 Nm).
29. Remove the floor jack.
30. Raise the vehicle.
31. Position a jack under the wheel knuckle at the trailing arm and knuckle attachment point.
32. Remove the lower arm bolt, install the shock/spring in the lower arm and install the bolt.
33. Loosen the lower arm-to-subframe bolt.
34. Raise the jack under the center of the wheel hub is at the proper trim height.
35. Tighten the lower arm-to-subframe bolt to 92 ft. lbs. (125 Nm).
36. Tighten the lower arm-to-knuckle bolt to 81 ft. lbs. (110 Nm).
37. Position the trailing arm to the knuckle and loosely install the bolt.
38. Place a jack under the lower arm and raise it until the trailing arm and subframe bushing are aligned.
39. Install a new trailing arm-to-subframe bolt and tighten to 81 ft. lbs. (100 Nm).
40. Install the trailing arm-to-knuckle bolt and tighten to 77 ft. lbs. (105 Nm).

41. Install the brake caliper.
42. Install the wheel.
43. Lower the vehicle.

Stabilizer Bar

REMOVAL & INSTALLATION

Front

✳✳ WARNING

Always use new bolts and nuts on the suspension components.

1. Place the steering wheel in the UNLOCKED position.
2. Raise and safely support the vehicle.
3. Remove the wheels.
4. Remove the stabilizer bar link nuts.
5. Remove and discard the tie rod nut and separate the tie rod from the knuckle.
6. Remove and discard the lower arm bushing-to-subframe nuts and bolts.
7. Remove and discard the lower arm-to-subframe bolt.
8. Remove and discard the rear subframe to body bolts.
9. Lower the subframe about 2 inches.
10. Remove and discard the stabilizer bar bracket bolts.
11. Remove the stabilizer bar from the right side of the vehicle.

To install:

12. Install the stabilizer bar from the right side of the vehicle.

13. Install the stabilizer bar bracket bolts and tighten to 37 ft. lbs. (50 Nm).

14. Raise the subframe about 2 inches.

15. Install the rear subframe-to-body bolts.

16. Install the lower arm-to-subframe bolt and tighten to 111 ft. lbs. (150 Nm).

17. Install the lower arm bushing-to-subframe nuts and bolts and tighten to 73 ft. lbs. (99 Nm).

18. Connect the tie rod to the knuckle and tighten the tie rod nut to 85 ft. lbs. (115 Nm).

19. Install the stabilizer bar link nuts and tighten to 41 ft. lbs. (55 Nm).

20. Install the wheels.

21. Lower the vehicle.

Rear–2WD

⁕⁕ **WARNING**

Always use new bolts and nuts on the suspension components.

1. Separate the exhaust system from the flexpipe and support the exhaust system.

2. Remove and discard the stabilizer bar link-to-stabilizer bar nuts.

3. Remove and discard the stabilizer bar bracket-to-subframe bolts.

4. Remove the stabilizer bar.

To install:

5. Reverse the removal procedure. Tighten the stabilizer bar bracket-to-subframe bolts to 43 ft. lbs. (58 Nm).

Rear–AWD

⁕⁕ **WARNING**

Always use new bolts and nuts on the suspension components.

1. Separate the exhaust system from the flexpipe and support the exhaust system.

2. Raise and support the vehicle.

3. Remove the rear wheels.

4. Remove the brake calipers and wire aside.

5. Remove both rear axles nuts and discard.

6. Using a hub remover, press the axle shaft from the wheel hub.

7. Remove and discard both upper arm-to-knuckle bolts.

8. Remove and discard both trailing arm-to-knuckle bolts and loosen the trailing arm-to-subframe bolt.

9. Remove and discard the lower shock-to-lower arm nuts and remove it from the lower arm.

10. Loosen both toe link-to-subframe bolts and allow the wheel knuckles to hang in a downward position.

11. Remove the rear subframe cross brace bolts and cross brace.

12. Remove and discard the stabilizer bar link-to-stabilizer bar nuts.

13. Remove and discard the stabilizer bar bracket-to-subframe bolts.

14. Remove the stabilizer bar.

To install:

15. Install the stabilizer bar.

16. Install the stabilizer bar bracket-to-subframe bolts and tighten to 43 ft. lbs. (58 Nm).

17. Install the stabilizer bar link-to-stabilizer bar nuts.

18. Install the rear subframe cross brace and bolts and tighten to 46 ft. lbs. (63 Nm).

19. Bring the wheel knuckles up and tighten both toe link-to-subframe bolts to 74 ft. lbs. (100 Nm).

20. Install the shock to the lower arm and tighten the lower shock-to-lower arm nuts to 105 ft. lbs. (142 Nm).

21. Install both trailing arm-to-knuckle bolts and tighten them and the trailing arm-to-subframe bolt to 77 ft. lbs. (110 Nm).

22. Install both upper arm-to-knuckle bolts and tighten to 77 ft. lbs. (105 Nm).

23. Install the axle shaft into the wheel hub.

24. Install new axle shaft nuts and tighten to 148 ft. lbs. (200 Nm).

25. Install the brake calipers.

26. Install the rear wheels.

27. Install the exhaust system to the flex pipe.

28. Lower the vehicle.

Lower Control Arm

REMOVAL & INSTALLATION

Front

⁕⁕ **WARNING**

Always use new bolts and nuts on the suspension components.

1. Raise the vehicle on a hoist.

2. Remove the lower ball joint nut using a crow's foot wrench, and discard the nut

3. Push the lower arm down until the ball joint is clear of the lower arm.

4. Install tool no. 307-102 over the ball stud.

5. Remove and discard the lower arm bushing-to-subframe nuts and bolts.

6. Remove the lower control arm.

To install:

7. Install the lower control arm.

8. Install the new lower arm bushing-to-subframe bolts from the bottom of the lower arm bushing with the nuts on top.

9. Tighten the nuts to 73 ft. lbs. (99 Nm).

10. Remove the special tool.

11. Install the lower arm-to-subframe bolts and tighten to 111 ft. lbs. (150 Nm).

12. Push the lower arm upward until the ball joint is up in the arm and install a new ball joint nut and tighten to 85 ft. lbs. (115 Nm).

13. Lower the vehicle.

Rear

⁕⁕ **WARNING**

Always use new bolts and nuts on the suspension components.

1. Measure the distance between the top of the wheel and the bottom of the wheel opening. This is the trim height.

2. Remove and discard the trailing arm-to-knuckle bolt.

3. Loosen the trailing arm-to-subframe bolt and allow it to swing down.

4. Remove and discard the knuckle-to-lower arm bolt.

5. Remove and discard the shock absorber-to-lower arm bolt.

6. Remove and discard the lower arm-to-subframe bolt.

7. Remove the lower arm.

To install:

8. Install the lower control arm.

9. Use a jack and raise the suspension to the trim height previously measured before tightening the lower arm bolts.

10. Install the lower arm-to-subframe bolt and tighten to 98 ft. lbs. (133 Nm) on AWD models and 92 ft. lbs. (125 Nm) on 2WD models.

11. Install the shock absorber-to-lower arm bolt and tighten to 105 ft. lbs. (142 Nm) on AWD models and 81 ft. lbs. (110 Nm) on 2WD models.

12. Install the knuckle-to-lower arm bolt and tighten to 66 ft. lbs. (90 Nm) on AWD models and 118 ft. lbs. (160 Nm) on 2WD models.

13. Bring the trailing arm up and tighten the trailing arm-to-subframe bolt and tighten the bolt to 81 ft. lbs. (110 Nm).

14. Install the trailing arm-to-knuckle bolt and tighten the bolt to 77 ft. lbs. (105 Nm).

Steering Knuckle

REMOVAL & INSTALLATION

Front

❋❋ WARNING

Always use new bolts and nuts on the suspension components.

1. Raise and support the vehicle.
2. Remove the front wheel.
3. Remove and discard the axle hub nut.
4. Remove the tie rod-to-wheel nut nut and separate the tie rod from the knuckle.
5. Remove the brake disc.
6. Remove the wheel speed sensor and wire aside.
7. Remove the lower ball joint nut using a crow's foot wrench, and discard the nut
8. Push the lower arm down until the ball joint is clear of the lower arm.
9. Install tool no. 307-102 over the ball stud.
10. Use a puller and press the halfshaft from the wheel hub and support the halfshaft level.
11. Remove and discard the wheel knuckle-to-strut nut and bolt.
12. Remove the wheel knuckle.
To install:
13. Install the wheel knuckle.
14. Install and the wheel knuckle-to-strut nut and bolt and tighten to 129 ft. lbs. (175 Nm).
15. Install the halfshaft in the wheel hub.
16. Remove the ball stud tool and connect the ball joint to the lower control arm.
17. Install a new ball joint nut and tighten to 85 ft. lbs. (115 Nm).
18. Install the wheel speed sensor.
19. Install the brake disc.
20. Install the tie rod to the wheel knuckle and tighten the new nut to 85 ft. lbs. (115 Nm).
21. Install a new axle hub nut and tighten to 148 ft. lbs. (200 Nm).
22. Install the front wheel.
23. Lower the vehicle.

Rear

❋❋ WARNING

Always use new bolts and nuts on the suspension components.

1. Measure the distance between the top of the wheel and the bottom of the wheel opening. This is the trim height.
2. Remove the wheel hub and bearing assembly.
3. Remove the brake dust shield.
4. Remove and discard the trailing arm-to-knuckle bolt.
5. Remove the toe link-to-knuckle bolt and nut.
6. Remove and discard the shock absorber-to-lower arm bolt.
7. Remove and discard the lower arm-to-subframe bolt.
8. Remove and discard the upper arm-to-subframe bolt.
9. Remove the steering knuckle.
To install:
10. Install the wheel knuckle.
11. Use a jack and raise the suspension to the trim height previously measured before tightening the upper and lower arm bolts.
12. Install the upper arm-to-subframe bolt and tighten to 81 ft. lbs. (110 Nm).
13. Install the lower arm-to-subframe bolt and tighten to 98 ft. lbs. (133 Nm) on AWD models, and 92 ft. lbs. (125 Nm) on 2WD models.
14. Install the shock absorber-to-lower arm bolt and tighten to 105 ft. lbs. (142 Nm) on AWD models, and 81 ft. lbs. (110 Nm) on 2WD models.
15. Install the toe link-to-knuckle bolt and nut and tighten to 74 ft. lbs. (100 Nm).
16. Install the trailing arm-to-knuckle bolt and tighten to 77 ft. lbs. (105 Nm).
17. Install the brake dust shield.
18. Install the wheel hub and bearing assembly.

Hub and Wheel Bearing

REMOVAL & INSTALLATION

The wheel bearing is integral with the wheel hub and can not be replaced separately. If the wheel bearing is found to be defective, the wheel hub must be replaced as an assembly.

Front

❋❋ WARNING

Always use new bolts and nuts on the suspension components.

1. Raise and safely support the vehicle.
2. Remove the wheel and tire assembly.
3. Remove the steering knuckle.
4. Remove and discard the wheel hub-to-knuckle bolts and separate the knuckle from the hub.
To install:
5. Clean the steering knuckle bore.
6. Install the wheel hub-to-the steering knuckle.
7. Install the wheel hub-to-knuckle bolts and tighten to 81 ft. lbs. (110 Nm).
8. Install the steering knuckle.
9. Install the wheel and tire.

Rear

1. Raise and safely support the vehicle.
2. Remove the wheel and tire assembly.
3. Remove and discard the rear axle nut.
4. Remove the brake caliper and wire it out of the way.
5. Remove the brake rotor.
6. Press the halfshaft out of the wheel hub.
7. Remove and discard the wheel hub-to-knuckle bolts and separate the hub from the knuckle.
To install:
8. Install the wheel hub-to-the steering knuckle.
9. Install the wheel hub-to-knuckle bolts and tighten to 89 ft. lbs. (120 Nm).
10. Install the halfshaft into the wheel hub.
11. Install the brake disc.
12. Install the brake caliper.
13. Install a new rear axle nut and tighten to 148 ft. lbs. (200 Nm).
14. Install the wheel.
15. Lower the vehicle.

BRAKES

Brake Caliper and Pads

REMOVAL & INSTALLATION

Front

➡The brake pads are a one time use only type. If the pads are separated from the caliper the pads must be replaced.

1. Remove and discard ½ of the brake fluid from the brake master cylinder reservoir.
2. Raise and safely support the vehicle.
3. Remove the wheel and tire assembly.
4. Mark the disc brake caliper to avoid mixing the left-hand and right-hand components.
5. Loosen the brake caliper bolts.
6. Disconnect the brake hose from the brake line fitting.

7. Remove and discard the caliper anchor bolts and position the caliper, anchor plate and pads aside. Support the steering stop when the anchor plate bolts are removed.
8. Separate the brake pads from the caliper.
9. Remove the caliper bolts from the caliper and anchor plate.
10. Remove the brake pads and spring clips.

1 Brake caliper bolts (2 required)
2 Brake caliper RH/LH
3 Brake pads
4 Spring clips (2 required)
5 Brake caliper anchor plate bolts (2 required)
6 Steering stop
7 Brake caliper anchor plate
8 Brake disc screw
9 Brake disc

10 Brake disc shield bolts (3 required)
11 Brake disc shield (RH/LH)
12 Brake caliper flow bolt
13 Copper washers (2 required)
14 Brake flexible hose RH/LH
15 Brake line fitting
16 Brake flexible hose bracket bolt
17 Bleeder screw cap
18 Bleeder screw

Exploded view of front disc brake assembly—Freestyle

To install:

11. Compress the caliper pistons into the caliper bore.

12. Install the spring clips and brake pads in the caliper and anchor plate assembly.

13. Install the brake caliper anchor. Tighten the anchor bolts to 74 ft. lbs. (100 Nm).

14. Install the caliper and new guide pins and tighten the bolts to 44 ft. lbs. (60 Nm).

15. Install new copper washers in the brake flexible hoses and connect the hoses and install the flow bolt to 22 ft. lbs. (30 Nm).

16. Bleed the brake system.

17. Lower the vehicle.

18. Pump the brake pedal several times to position the brake pads before attempting to move the vehicle.

19. Check and fill the brake master cylinder as required.

20. Road-test the vehicle and check for proper brake operation.

Rear

1. Remove and discard ½ of the brake fluid from the brake master cylinder reservoir.

2. Raise and safely support the vehicle.

3. Remove the wheel and tire assembly.

4. Release the tension on the parking brake cable.

5. Disconnect the parking brake cable end from the caliper lever and release the cable retainer from the caliper bracket.

6. Loosen the brake caliper bolts.

1 Brake caliper bolts (2 required)

2 Brake caliper (RH/LH)

3 Brake pads (kit)

4 Spring clips (4 required)

5 Brake caliper anchor plate bolts (2 bolts each side)

6 Brake caliper anchor plate (RH/LH)

7 Brake disc screw

8 Brake disc

9 Brake disc shield bolts (3 required)

10 Brake disc shield (RH/LH)

11 Brake line fitting

12 Brake flexible hose retainer

13 Brake flexible hose

14 Bleeder cap

15 Bleeder screw

Exploded view of rear disc brake assembly—Freestyle

06017-FREE-G52

7. Remove and discard the caliper anchor bolts.

8. Remove the brake pads from the caliper.

9. Remove the caliper bolts and separate the caliper from the anchor plate.

To install:

10. Compress the caliper pistons into the caliper bore.

11. Install the anchor plate and tighten the bolts to 81 ft. lbs. (110 Nm).

12. Retract the disc brake caliper piston fully into the caliper bore using tool no. 206-026.

13. Install the spring clips and brake pads in the caliper.

14. Install the caliper and tighten the guide pins to 23 ft. lbs. (31 Nm).

15. Install the parking brake cable to the caliper.

16. Reload the tension on the parking brake cable.

17. Bleed the brake system and install the rubber bleeder screw caps when complete.

18. Install the wheel and tire assembly.

19. Lower the vehicle.

20. Pump the brake pedal several times to position the brake pads before attempting to move the vehicle.

21. Check and fill the brake master cylinder as required.

22. Road-test the vehicle and check for proper brake operation.

LINCOLN

LS

11

SPECIFICATION AND MAINTENANCE CHARTS

ENGINE AND VEHICLE IDENTIFICATION

Code ①	Liters (cc)	Cu. In.	Cyl.	Fuel Sys.	Type	Eng. Mfg.
S	3.0 (3049)	182	6	MFI	DOHC	Ford
A	3.9 (3947)	238	8	MFI	DOHC	Ford

MFI: Multi-Port Fuel Injection

DOHC: Double Overhead Camshaft

① 8th digit of the VIN

② 10th digit of the VIN

Code ②	Year
2	2002
3	2003
4	2004
5	2005
6	2006

06017-LS-C01

GENERAL ENGINE SPECIFICATIONS

Year	Model	Engine Displacement Liters	Engine ID/VIN	Net Horsepower @ rpm	Net Torque @ rpm (ft. lbs.)	Bore x Stroke (in.)	Compression Ratio	Oil Pressure @ rpm
2002	LS6	3.0	S	210@6500	205@4750	3.50x3.13	10.5:1	20-45@1500
	LS8	3.9	A	252@6100	267@4300	3.38x3.34	10.6:1	61-73@400
2003	LS6	3.0	S	210@6500	205@4750	3.50x3.13	10.5:1	20-45@1500
	LS8	3.9	A	280@6100	286@4300	3.38x3.34	10.75:1	61-73@400
2004	LS6	3.0	S	210@6500	205@4750	3.50x3.13	10.5:1	20-45@1500
	LS8	3.9	A	252@6100	267@4300	3.38x3.34	10.6:1	61-73@400
2005	LS6	3.0	S	210@6500	205@4750	3.50x3.13	10.5:1	20-45@1500
	LS8	3.9	A	280@6000	286@4000	3.38x3.34	10.75:1	61-73@400

NA: Not Available

06017-LS-C02

ENGINE TUNE-UP SPECIFICATIONS

Year	Engine Displacement Liters	Engine ID/VIN	Spark Plugs Gap (in.)	Ignition Timing (deg.) ① MT	Ignition Timing (deg.) ① AT	Fuel Pump (psi)	Idle Speed (rpm) ① MT	Idle Speed (rpm) ① AT	Valve Clearance In.	Valve Clearance Ex.
2002	3.0	S	0.051-0.057	12-17B	12-17B	26-45	650-750	650-750	0.007-0.009	0.013-0.015
	3.9	A	0.039-0.043	—	10-20B	43	—	650-750	0.007-0.009	0.009-0.011
2003	3.0	S	0.051-0.057	12-17B	12-17B	26-45	650-750	650-750	0.007-0.009	0.013-0.015
	3.9	A	0.039-0.043	—	10-20B	43	—	650-750	0.007-0.009	0.009-0.011
2004	3.0	S	0.051-0.057	12-17B	12-17B	26-45	650-750	650-750	0.007-0.009	0.013-0.015
	3.9	A	0.040-0.044	—	10-20B	43	—	650-750	0.007-0.009	0.009-0.011
2005	3.0	S	0.051-0.057	12-17B	12-17B	30-55	650-750	650-750	0.0069-0.0089	0.0128-0.0148
	3.9	A	0.040-0.044	—	10-20B	30-55	—	650-750	0.0069-0.0089	0.009-0.011

The underhood specifications sticker often reflects tune-up specification changes in production. Sticker figures must be used if they disagree with those in this chart.

① Controlled by the engine computer

06017-LS-C03

3.0L (VIN S) engine
Firing order:1–4–2–5–3–6
Distributorless ignition system (one coil on each cylinder)

9306TG07

3.9L (VIN A) engine
Firing order:1–5–4–2–6–3–7–8
Distributorless ignition system (one coil on each cylinder)

93003G02

1. Power steering pump
2. Water pump
3. A/C compressor
4. Drive belt
5. Crankshaft pulley
6. Drive belt tensioner
7. Generator

79224G25

Serpentine accessory drive belt routing—3.0L (VIN S) engine

93461G25

Serpentine accessory drive belt routing—3.9L (VIN A) engine

CAPACITIES

Year	Model	Engine Displacement Liters	Engine ID/VIN	Engine Oil with Filter (qts.)	Transmission (pts.)		Drive Axle Rear (pts.)	Fuel Tank (gal.)	Cooling System (qts.)
					Manual	Auto.			
2002	LS6	3.0	S	6.9	NA	23.8	3.0	18.0	10.6
	LS8	3.9	A	6.5	—	23.8	3.0	18.0	11.3
2003	LS6	3.0	S	6.9	NA	23.8	3.0	18.0	10.6
	LS8	3.9	A	6.5	—	23.8	3.0	18.0	11.3
2004	LS6	3.0	S	6.9	NA	23.8	3.0	18.0	10.6
	LS8	3.9	A	6.5	—	23.8	3.0	18.0	11.3
2005	LS6	3.0	S	6.9	NA	23.8	2.7	18.5	10.6
	LS8	3.9	A	6.5	—	23.8	2.7	18.5	11.3

N/A: Not Available

06017-LS-C04

VALVE SPECIFICATIONS

Year	Engine Displacement Liters	Engine ID/VIN	Seat Angle (deg.)	Face Angle (deg.)	Spring Test Pressure (lbs. @ in.)	Spring Free Length (in.)	Stem-to-Guide Clearance (in.)		Stem Diameter (in.)	
							Intake	Exhaust	Intake	Exhaust
2002	3.0	S	44.75	45.5	153@1.18	1.570	0.0007-0.0027	0.0017-0.0037	0.2350-0.2358	0.2343-0.2350
	3.9	A	NA	NA	NA	NA	NA	NA	NA	NA
2003	3.0	S	44.75	45.5	153@1.18	1.570	0.0007-0.0027	0.0017-0.0037	0.2350-0.2358	0.2343-0.2350
	3.9	A	NA	NA	NA	NA	NA	NA	NA	NA
2004	3.0	S	44.75	45.5	153@1.18	1.570	0.0007-0.0027	0.0017-0.0037	0.2350-0.2358	0.2343-0.2350
	3.9	A	NA	NA	NA	NA	NA	NA	NA	NA
2005	3.0	S	44.75	45.5	153@1.18	1.740	0.0008-0.0026	0.0017-0.0037	0.2160-0.2150	0.2150
	3.9	A	NA	NA	NA	NA	NA	NA	NA	NA

NA: Not Available

06017-LS-C07

CRANKSHAFT AND CONNECTING ROD SPECIFICATIONS

All measurements are given in inches.

Year	Engine Displacement Liters	Engine ID/VIN	Crankshaft				Connecting Rod		
			Main Brg. Journal Dia.	Main Brg. Oil Clearance	Shaft End-play	Thrust on No.	Journal Diameter	Oil Clearance	Side Clearance
2002	3.0	S	2.4670-2.4790	0.0009-0.0018	0.0040 0.0090	4	1.9670-1.9680	0.0010 0.0025	0.0039-0.0118
	3.9	A	NA	NA	NA	NA	NA	NA	NA
2003	3.0	S	2.4670-2.4790	0.0009-0.0018	0.0040 0.0090	4	1.9670-1.9680	0.0010 0.0025	0.0039-0.0118
	3.9	A	NA	NA	NA	NA	NA	NA	NA
2004	3.0	S	2.4670-2.4790	0.0009-0.0018	0.0040 0.0090	4	1.9670-1.9680	0.0010 0.0025	0.0039-0.0118
	3.9	A	NA	NA	NA	NA	NA	NA	NA
2005	3.0	S	2.479-2.4800	0.0009-0.0018	0.0040 0.0090	4	1.9670-1.9680	0.0010 0.0025	0.0039-0.0118
	3.9	A	NA	NA	NA	NA	NA	NA	NA

NA: Not Available

06017-LS-C05

PISTON AND RING SPECIFICATIONS

All measurements are given in inches.

Year	Engine Displacement Liters	Engine ID/VIN	Piston Clearance	Ring Gap			Ring Side Clearance		
				Top Compression	Bottom Compression	Oil Control	Top Compression	Bottom Compression	Oil Control
2002	3.0	S	0.0005-0.0009	0.004-0.010	0.011-0.017	0.005-0.026	0.0015-0.0029	0.0015-0.0033	SNUG
	3.9	A	NA	NA	NA	NA	NA	NA	NA
2003	3.0	S	0.0005-0.0009	0.004-0.010	0.011-0.017	0.005-0.026	0.0015-0.0029	0.0015-0.0033	SNUG
	3.9	A	NA	NA	NA	NA	NA	NA	NA
2004	3.0	S	0.0005-0.0009	0.004-0.010	0.011-0.017	0.005-0.026	0.0015-0.0029	0.0015-0.0033	SNUG
	3.9	A	NA	NA	NA	NA	NA	NA	NA
2005	3.0	S	0.0005-0.0009	0.0039-0.0098	0.0106-0.0165	0.0059-0.0255	0.0015-0.0029	0.0015-0.0033	SNUG
	3.9	A	NA	NA	NA	NA	NA	NA	NA

NA: Not Available

06017-LS-C06

TORQUE SPECIFICATIONS
All readings in ft. lbs.

Year	Engine Displacement Liters	Engine ID/VIN	Cylinder Head Bolts	Main Bearing Bolts	Rod Bearing Bolts	Crankshaft Damper Bolts	Flywheel Bolts	Manifold Intake	Manifold Exhaust	Spark Plugs	Oil Pan Drain Plug
2002	3.0	S	①	②	③	④	54-64	⑤	13-16	7-15	17
	3.9	A	⑥	NA	NA	⑦	⑧	18	18	19	17
2003	3.0	S	①	②	③	④	54-64	⑤	13-16	7-15	17
	3.9	A	⑥	NA	NA	⑦	⑧	18	18	19	17
2004	3.0	S	①	②	③	④	54-64	⑤	13-16	7-15	17
	3.9	A	⑥	NA	NA	⑦	⑧	18	18	19	17
2005	3.0	S	①	②	③	④	54-64	⑤	13-16	7-15	17
	3.9	A	⑥	NA	NA	⑨	⑧	15	18	20	17

NA: Not Available

① Step 1: 28-31 ft. lbs.
   Step 2: Plus 85-95 degrees
   Step 3: Loosen one turn
   Step 4: 28-31 ft. lbs.
   Step 5: Plus 85-95 degrees
   Step 6: Plus 85-95 degrees

② Step 1: Cap bolts 1-8 (outer) 17-20 ft. lbs.
   Step 2: Cap bolts 9-16 (inner) 28-31 ft. lbs.
   Step 3: Rotate bolts 1-16, 85-95 degrees
   Step 4: Bolts 17-22; 15-22 ft. lbs.

③ Step 1: 30-33 ft. lbs.
   Step 2: Plus 90-120 degrees

④ Step 1: 77-99 ft. lbs.
   Step 2: Loosen 360 degrees
   Step 3: Tighten to 35-39 ft. lbs.
   Step 4: Plus 85-95 degrees

⑤ 71-106 inch lbs.

⑥ Step 1: Tighten M10 bolts to 15 ft. lbs.
   Step 2: Tighten M10 bolts to 26 ft. lbs.
   Step 3: Tighten M10 bolts to 33 ft. lbs.
   Step 4: Tighten M10 bolts plus 90 degrees
   Step 5: Tighten M10 bolts plus 90 degrees
   Step 6: Tighten M8 bolts to 15 ft. lbs.
   Step 7: Tighten M8 bolts plus 90 degrees

⑦ Step 1: 59 ft. lbs.
   Step 2: Plus 80 degrees

⑧ Step 1: 11 ft. lbs.
   Step 2: 81 ft. lbs.

⑨ Step 1: 59 ft. lbs.
   Step 2: Loosen bolt two full turns
   Step 3: Tighten to 37 ft. lbs.
   Step 4: Tighten an additional 90 degrees

06017-LS-C08

WHEEL ALIGNMENT

Year	Model		Caster Range (+/-Deg.)	Caster Preferred Setting (Deg.)	Camber Range (+/-Deg.)	Camber Preferred Setting (Deg.)	Toe-in (in.)
2002	LS	F	0.70	0	0.70	0	0.08 +/- 0.13
		R	—	—	0.75	0	0.13 +/- 0.13
2003	LS	F	0.70	0	0.70	0	0.08 +/- 0.13
		R	—	—	0.75	0	0.13 +/- 0.13
2004	LS	F	0.70	0	0.70	0	0.08 +/- 0.13
		R	—	—	0.75	0	0.13 +/- 0.13
2005	LS	F	0.70	0	0.70	0	0.08 +/- 0.16
		R	—	—	0.75	0	0.12 +/- 0.13

06017-LS-C09

TIRE, WHEEL AND BALL JOINT SPECIFICATIONS

Year	Model	OEM Tires		Tire Pressures (psi)		Wheel Size	Ball Joint Inspection	Lug Nut (ft. lbs.)
		Standard	Optional	Front	Rear			
2002	LS6 and LS8	P215/60R16	P235/50R17	30	30	7-J/7.5J	U ① L ① ②	100
2003	LS6 and LS8	P215/60R16	P235/50R17	30	30	7-J/7.5J	U ① L ① ②	100
2004	LS6 and LS8	P215/60R16	P235/50R17	30	30	7-J/7.5J	U ① L ① ②	100
2005	LS6 and LS8	P225/55R16	P235/50R17	③	③	④	U ① L ① ②	100

OEM: Original Equipment Manufacturer

PSI: Pounds Per Square Inch

STD: Standard

OPT: Optional

L: Lower

U: Upper

① Replace if any measurable movement is found.

② Do not lift car. Inspect the boss into which the grease fitting is threaded. Replace if the boss is flush or receded below the surface of the ball joint.

③ Refer to the safety certification sticker located on driver door jamb or tire inflation placard on the passenger door jamb.

④ 16x7.5J or 7x7.5J

06017-LS-C10

BRAKE SPECIFICATIONS
All measurements in inches unless noted

Year	Model		Brake Disc			Minimum Lining Thickness	Brake Caliper	
			Original Thickness	Minimum Thickness	Maximum Runout		Bracket Bolts (ft. lbs.)	Mounting Bolts (ft. lbs.)
2002	LS6	F	1.180	1.120	0.004	0.079	76	26
		R	0.810	0.740	0.004	0.039	76	25
	LS8	F	1.180	1.120	0.004	0.079	76	26
		R	0.810	0.740	0.004	0.039	76	25
2003	LS6	F	1.180	1.120	0.004	0.079	76	26
		R	0.810	0.740	0.004	0.039	76	25
	LS8	F	1.180	1.120	0.004	0.079	76	26
		R	0.810	0.740	0.004	0.039	76	25
2004	LS6	F	1.180	1.120	0.004	0.079	76	26
		R	0.810	0.740	0.004	0.039	76	25
	LS8	F	1.180	1.120	0.004	0.079	76	26
		R	0.810	0.740	0.004	0.039	76	25
2005	LS6	F	1.180	1.120	0.004	0.118	76	26
		R	0.810	1.120	0.004	0.118	76	26
	LS8	F	1.180	1.120	0.004	0.118	76	26
		R	0.810	1.120	0.004	0.118	76	26

06017-LS-C11

SCHEDULED MAINTENANCE INTERVALS
Lincoln—LS

TO BE SERVICED	TYPE OF SERVICE	VEHICLE MILEAGE INTERVAL (x1000)												
		5	10	15	20	25	30	35	40	45	50	55	60	65
Air cleaner filter	R						✓						✓	
Accessory drive belt	S/I												✓	
Brake system ①	S/I			✓			✓			✓			✓	
Clutch pedal operation	S/I						✓						✓	
Cooling system hoses and clamps	S/I			✓			✓			✓			✓	
CV-joint boots & axle seals	S/I						✓						✓	
Engine coolant	R	Ten years or 150,000 miles												
Engine oil & filter	R	✓	✓	✓	✓	✓	✓	✓	✓	✓	✓	✓	✓	✓
Exterior Lights	S/I	Check monthly												
PCV valve	S/I												✓	
Exhaust system & heat shields	S/I						✓						✓	
Parking brake system	S/I	Every 6 months												
Power steering fluid	S/I	Every 6 months												
Rotate tires	S/I	✓		✓		✓		✓		✓		✓		✓
Steering linkage	S/I						✓						✓	
Spark plugs	R	Change at 100,000 miles												
Suspension components	S/I						✓						✓	

R: Replace   S/I: Inspect and service, if necessary   L: Lubricate   A: Adjust   C: Clean

① Inspect the reservoir fluid level, rotor and or drum, brake lines, hoses, calipers and or wheel cylinders

FREQUENT OPERATION MAINTENANCE (SEVERE SERVICE)

If a vehicle is operated under any of the following conditions it is considered severe service:

- Extremely dusty areas.
- 50% or more of the vehicle operation is in 32°C (90°F) or higher temperatures, or constant operation in temperatures below 0°C (32°F).
- Prolonged idling (vehicle operation in stop and go traffic).
- Frequent short running periods (engine does not warm to normal operating temperatures).
- Police, taxi, delivery usage or trailer towing usage.

Oil & oil filter change: change every 3000 miles.
Air filter element: change every 15,000 miles.

06017-LS-C12

ENGINE REPAIR

➡Disconnecting the negative battery cable on some vehicles may interfere with the functions of the on board computer system. The computer may undergo a relearning process once the negative battery cable is reconnected.

Alternator

REMOVAL

3.0L Engine

1. Before servicing the vehicle, refer to the Precautions Section.
2. Remove or disconnect the following:
   - Negative battery cable
   - Accessory drive belt
   - Lower splash shield
   - Alternator mounting bolts
   - Alternator harness connectors
   - Alternator

To install:

3. Before servicing the vehicle, refer to the Precautions Section.
   Install or connect the following:
   - Alternator
   - Alternator harness connectors.

1 Generator upper bolt	6 Generator B+ wire
2 Generator lower bolt	7 Generator electrical connector
3 Generator lower nut	8 Generator bracket bolts (4 required)
4 Generator	9 Generator bracket
5 Generator B+ wire nut	

06017-LS-G01

Alternator torque sequence—3.9L engine

1 Generator electrical connector	4 Generator mounting bolts
2 Generator B+ wire nut	5 Generator
3 Generator B+ wire	

67197-LILS-G01

Alternator mounting—3.0L engine

Tighten the battery cable terminal nut to 62 inch lbs. (7 Nm).
- Alternator mounting bolts. Tighten the bolts to 25 ft. lbs. (45 Nm).
- Lower splash shield
- Accessory drive belt
- Negative battery cable

3.9L Engine

1. Before servicing the vehicle, refer to the Precautions Section.

➡The A/C suction line may need to be positioned rearward to ease removal of the generator.

2. Disconnect the negative battery cable.
3. Rotate the front end accessory drive belt tensioner counterclockwise and position the belt aside.
4. Remove the front valance panel.
5. Remove the air deflector.
6. Remove the A/C suction line clip from the radiator core support.
7. Remove the left hand A/C suction line lower bracket bolt and position the bracket rearward.

8. Remove the generator upper mounting bolt.

9. Remove the generator lower mounting nut and bolt.

10. Disconnect the generator electrical connector.

11. Position the generator B+ wire protective cover aside and remove the generator B+ wire nut and cable end.

12. Remove the generator.

To install:

13. Installation is the reverse of removal. Please note the following specifications:

    a. Generator B+ wire nut to 71 inch lbs. (8 Nm).

    b. Generator lower mounting nut and bolt. Tighten to 33 ft. lbs. (45 Nm).

    c. Generator upper mounting bolt. Tighten to 15 ft. lbs. (21 Nm).

    d. Left hand A/C suction line lower bracket bolt. Tighten to 89 inch lbs. (10 Nm).

Ignition Timing

ADJUSTMENT

This vehicle is equipped with a Distributorless Ignition System (DIS). The ignition timing is not adjustable. It is controlled by the PCM.

Engine Assembly

REMOVAL & INSTALLATION

3.0L Engine

2002–03 MODELS

1. Before servicing the vehicle, refer to the Precautions Section.

2. Drain the cooling system.

3. Relieve the fuel system pressure.

4. Recover the A/C refrigerant.

5. Drain the engine oil.

6. Remove or disconnect the following:
- Negative battery cable
- Air cleaner housing and outlet tube
- Engine appearance cover
- Upper radiator shield
- Upper radiator support brackets
- A/C pressure switch connector
- Power steering reservoir
- Fuel line
- Vapor Management Valve (VMV) cover
- VMV vacuum hose
- Cowl leaf screens
- Chassis vacuum lines

- Cross vehicle support bar
- Fresh air intake housing
- Intake manifold rear main vacuum hose
- Accelerator cable and cruise control cable
- Cable bracket
- Ground strap
- Main engine wiring harness connector
- Main transmission wiring harness connector
- 2 fuel charging harness connectors
- A/C line mounting bracket
- Hydraulic cooling fan reservoir
- Left, right and center splash shields
- A/C compressor manifold and tube assembly
- Coolant hoses
- Exhaust front pipe and heat shields
- Driveshaft
- Shift cable and bracket
- Front wheels
- Front wheel speed sensor connectors
- Front brake calipers
- Lower stabilizer bar links
- Upper ball joints
- Lower strut mount bolts
- Starter motor wiring harness connectors
- Power Steering Pressure (PSP) switch connector
- Steering shaft pinch bolt
- Torque converter, if equipped

7. Support the engine, transmission, front and center crossmembers and the cooling system with a powertrain lift and transmission support bracket.

8. Support the rear of the vehicle with safety stands.

9. Remove or disconnect the following:
- Transmission crossmember bolts
- Subframe bolts
- Crossmember bolts
- Powertrain assembly

10. Attach a hoist to the engine.

11. Remove or disconnect the following:
- Wire harness retainers
- Starter motor
- Heated Oxygen (HO2S) sensor bracket
- Motor mount nuts
- Accessory drive belt
- Power steering pump
- Hydraulic cooling fan pump
- Upper radiator hose
- Transmission oil cooler lines, if equipped

12. Lift the engine and transmission out of the subframe.

13. Remove or disconnect the following:

- Oil cooler hoses
- Transmission flange bolts
- Transmission from the engine

To install:

14. Install or connect the following:
- Transmission to the engine. Tighten the flange bolts to 35 ft. lbs. (48 Nm).
- Powertrain on the subframe. Tighten the motor mount nuts to 46 ft. lbs. (63 Nm).
- Oil cooler hoses
- Transmission oil cooler lines, if equipped
- Upper radiator hose
- Hydraulic cooling fan pump. Tighten the bolts to 18 ft. lbs. (25 Nm).
- Power steering pump. Tighten the bolts to 18 ft. lbs. (25 Nm).
- Accessory drive belt
- HO2S sensor bracket. Tighten the nut to 89 inch lbs. (10 Nm).
- Starter motor. Tighten the bolts to 18 ft. lbs. (25 Nm).
- Wire harness retainers
- Powertrain assembly. Tighten the crossmember bolts to 76 ft. lbs. (103 Nm).
- Torque converter, if equipped.
- Steering shaft pinch bolt. Tighten the bolts to 18 ft. lbs. (25 Nm).
- PSP switch connector
- Starter motor wiring harness connectors
- Lower strut mount bolts. Tighten the bolts to 129 ft. lbs. (175 Nm).
- Upper ball joints. Tighten the nuts to 66 ft. lbs. (90 Nm).
- Lower stabilizer bar links. Tighten the nuts to 41 ft. lbs. (55 Nm).
- Front brake calipers
- Front wheel speed sensor connectors
- Front wheels
- Shift cable and bracket
- Driveshaft
- Exhaust front pipe and heat shields
- Coolant hoses
- A/C compressor manifold and tube assembly. Tighten the bolt to 15 ft. lbs. (21 Nm).
- Left, right and center splash shields
- Hydraulic cooling fan reservoir
- A/C line mounting bracket
- 2 fuel charging harness connectors
- Main transmission wiring harness connector. Tighten the bolt to 89 inch lbs. (10 Nm).
- Main engine wiring harness connector. Tighten the bolt to 89 inch lbs. (10 Nm).

- Ground strap. Tighten the bolt to 89 inch lbs. (10 Nm).
- Cable bracket. Tighten the bolts to 89 inch lbs. (10 Nm).
- Accelerator cable and cruise control cable
- Intake manifold rear main vacuum hose
- Fresh air intake housing
- Cross vehicle support bar. Tighten the bolts to 15 ft. lbs. (20 Nm).
- Chassis vacuum lines
- Cowl leaf screens
- VMV vacuum hose
- VMV cover
- Fuel line
- Power steering reservoir
- A/C pressure switch connector
- Upper radiator support brackets. Tighten the bolts to 89 inch lbs. (10 Nm).
- Upper radiator shield
- Engine appearance cover
- Air cleaner housing and outlet tube
- Negative battery cable

15. Fill the crankcase to the correct level.
16. Fill the cooling system.
17. Recharge the A/C system.
18. Start the engine and check for leaks.

2004–05 MODELS

1. Before servicing the vehicle, refer to the Precautions Section.
2. Drain the cooling system.
3. Relieve the fuel system pressure.
4. Recover the A/C refrigerant.
5. Drain the engine oil.
6. Disconnect the negative battery cable.
7. Remove the air cleaner outlet pipe and the air cleaner.
8. Remove the fresh air inlet duct.
9. Disconnect the fuel tube.
10. Remove the upper radiator sight shield.
11. Remove the 6 bolts and the 2 upper radiator support brackets.
12. Disconnect the A/C pressure switch electrical connector.
13. Remove the 2 power steering fluid reservoir bolts and secure the power steering fluid reservoir to the engine.
14. Disconnect the degas bottle coolant hose and the Positive Crankcase Ventilation (PCV) tube from the upper intake manifold.
15. Detach the retainer from the upper intake manifold and position the PCV tube and degas bottle coolant hose aside.
16. Disconnect the coolant hose and brake booster hose from the upper intake manifold. Detach the retainers and position the brake booster vacuum hose aside.

✳✳ CAUTION

To disconnect the quick release coupling, squeeze the tabs and pull straight out or damage to the coupling can occur.

17. Disconnect the Evaporative Emissions (EVAP) hose from the canister purge valve.
18. Remove the nut and disconnect the A/C manifold tube.
19. Remove the ground strap bolt.
20. Loosen the bolt and disconnect the main engine wiring harness connector.
21. Loosen the bolt and disconnect the main transmission wiring harness connector.
22. Disconnect the 2 engine wiring harness connectors.
23. Detach the wiring harness retainer from the bracket.
24. Disconnect the coolant valve electrical connector and detach the wiring retainer.
25. Remove the splash shields.
26. Remove the inner air deflector.
27. Remove the catalytic converters..

✳✳ CAUTION

Index-mark (color paint) the bolts, washers, nuts and the flex coupling to the transmission flange to make sure they are installed in the same location. Components that are not assembled in their original locations can cause driveshaft imbalance. Do not remove the bolts retaining the flex coupling to the driveshaft.

28. Index-mark the front driveshaft pinion flange. Remove the 3 nuts, washers and bolts.
29. Slide the front shaft assembly rearward and support it.

➡**The top heater hose has a green identifying mark to match the hose on the dual coolant flow valve. If the mark is not visible, identify the hoses for location. This will aid in the correct installation of the hose assemblies.**

30. Disconnect the 3 quick-disconnect couplings from the coolant control valve hoses.
31. Remove the nut and disconnect the A/C high-pressure tube at the right frame rail.
32. Disconnect the A/C suction tube.
33. Remove the safety clip and disconnect the A/C tube.
34. Position the right and left hand inner fender splash shields aside.

35. Remove the 3 pin-type retainers and position the inner splash shields aside.
36. Disconnect the 2 anti-lock brake sensor electrical connectors.
37. Detach the right and left hand anti-lock brake sensor harnesses from the brake hose clips.

✳✳ CAUTION

Do not allow the brake caliper to hang from the brake hose or damage to the brake hose can occur.

38. Remove the 4 bolts and position the front brake calipers aside. Support the calipers using mechanic's wire.

➡**To remove the nut, first loosen the nut, then use the hex-holding feature to prevent the stabilizer bar link ball joint from turning while removing the nut.**

39. Remove the 2 nuts and disconnect the stabilizer bar links from the lower control arms.

➡**To remove the nut, first loosen the nut, then use the hex-holding feature to prevent the upper ball joint stud from turning while removing the nut.**

40. Remove the 2 nuts and disconnect the upper ball joints from the wheel knuckles.
41. Remove the 2 strut-to-control arm bolts and nuts.
42. Disconnect the shift cable from the shifter.
43. Remove the 2 shift cable bracket-to-body bolts.
44. Detach the shift cable pin-type retainer from the exhaust heat shield.
45. Disconnect the Variable Assist Power Steering (VAPS) actuator electrical connector.
46. Detach the power steering return tube from the Power Steering Pressure (PSP) tube clip.
47. Remove the steering intermediate shaft bolt.

✳✳ CAUTION

Do not allow the steering wheel to rotate while the steering column intermediate shaft is disconnected or damage to the clockspring can result. If there is evidence that the wheel has rotated, the clockspring must be removed and recentered.

48. Remove the bolt and detach the steering column intermediate shaft from the steering gear.

Using the tools illustrated, support the engine, transmission, front and center crossmembers and the cooling system—2004–05 models

Engine lifting hooks—2004–05 models

49. Remove the nut and detach the PSP tube bracket from the oil pan stud bolt.

50. Remove the bolt and detach the PSP tube bracket from the left hand frame rail.

51. If equipped, disconnect the block heater electrical connector.

52. Remove the nut and detach the ground strap.

53. Remove the starter motor terminal cover.

54. Remove the nuts and position the starter motor cables aside.

55. Remove the nut and disconnect the B+ terminal from the generator.

56. Remove the torque converter access cover.

➡Mark one stud and the flexplate for assembly reference.

57. Remove the 4 torque converter nuts.

58. Remove the engine-to-transmission bolt.

59. Remove the 4 lower engine-to-transmission bolts.

60. Support the rear of the vehicle with suitable safety stands.

61. Using the tools in the accompanying illustration, support the engine, transmission, front and center crossmembers and the cooling system.

62. Remove the 5 transmission crossmember bolts.

63. Remove the 4 center crossmember I brace bolts.

64. Remove the 8 front and center crossmember bolts.

65. Using the tools illustrated, carefully lower the entire assembly from the vehicle.

66. Detach the 2 wire harness retainers and position the harness aside.

67. Remove the 2 bolts, the stud bolt and the starter motor.

68. Rotate the accessory drive belt tensioner counterclockwise and remove the accessory drive belt.

69. Disconnect the A/C compressor electrical connector.

70. Remove the 4 bolts and position the A/C compressor and manifold assembly aside.

71. Disconnect the Power Steering Pressure (PSP) switch electrical connector.

72. Remove the 3 bolts and position the power steering pump and reservoir aside.

73. Disconnect the coolant inlet hose.

74. Disconnect the heater hose and the lower radiator hose from the coolant pump.

75. If equipped, disconnect the oil cooler hoses.

76. Remove the nut and detach the transmission cooler tube bracket from the oil pan.

77. Install the lifting hook 303-050.

78. Using an engine crane support the engine and transmission.

79. Remove the right and left hand engine mount nut.

80. Remove the engine and transmission from the crossmember and cooling system assembly. Rest on the floor or on a bench.

81. Disconnect the transmission oil cooler tubes and plug the openings.

82. Remove the 5 transmission-to-engine bolts.

83. Using the special tools and a suitable engine crane, separate the engine from the transmission.

To install:

84. Using the special tools and a suitable engine crane, align the engine with the transmission.

85. Install the 5 transmission-to-engine bolts and tighten to 33 ft. lbs. (45 Nm).

86. Connect the transmission oil cooler tubes and tighten the fittings. Tighten to 15 ft. lbs (20 Nm).

87. Position the engine and transmission on the crossmember and cooling system assembly.

88. Install the right hand engine mount nut and tighten to 46 ft. lbs. (63 Nm).

89. Install the left hand engine mount nut and tighten to 46 ft. lbs. (63 Nm).

90. Attach the transmission cooler tube bracket to the oil pan and install the nut.

91. If equipped, connect the oil cooler hoses.

92. Connect the heater hose and the lower radiator hose to the coolant pump.

93. Connect the coolant inlet hose.

94. Position the power steering pump and install the 3 bolts. Tighten to 18 ft. lbs. (25 Nm).

95. Secure the power steering fluid reservoir to the engine.

96. Connect the PSP switch electrical connector.

97. Position the A/C compressor and manifold assembly. Install the 4 bolts and tighten to 18 ft. lbs. (25 Nm).

98. Connect the A/C compressor electrical connector.

99. Rotate the accessory drive belt tensioner counterclockwise and install the accessory drive belt.

100. Position the starter motor and install the 2 bolts and the stud bolt. Tighten to 18 ft. lbs. (25 Nm).

101. Position the wiring harness and attach the 2 retainers.

102. Carefully position the entire assembly into the vehicle.

103. Install the 8 front and center crossmember bolts. Tighten to 76 ft. lbs. (103 Nm).

104. Install the 4 center crossmember I brace bolts. Tighten to 76 ft. lbs. (103 Nm).

105. Install the 4 transmission crossmember bolts. Tighten to 41 ft. lbs. (55 Nm).

106. Install the 4 lower engine-to-transmission bolts. Tighten to 33 ft. lbs. (45 Nm).

107. Install the engine-to-transmission bolt. Tighten to 33 ft. lbs. (45 Nm).
108. Install the 4 torque converter nuts. Tighten to 28 ft. lbs. (38 Nm).
109. Install the torque converter access cover.
110. Connect the B+ terminal to the generator and install the nut.
111. Position the starter motor cables and install the nuts.
112. Install the starter motor terminal cover.
113. Position the ground strap and install the nut. Tighten to 89 inch lbs. (10 Nm).
114. If equipped, connect the block heater electrical connector.
115. Position the PSP tube bracket and install the bolt.
116. Attach the PSP tube bracket to the oil pan stud bolt and install the nut. Tighten to 71 inch lbs. (8 Nm)

✳✳ CAUTION

Do not allow the steering wheel to rotate while the steering column intermediate shaft is disconnected or damage to the clockspring can result. If there is evidence that the wheel has rotated, the clockspring must be removed and recentered.

117. Attach the steering column intermediate shaft to the steering gear and install the bolt. Tighten to 26 ft. lbs. (35 Nm).
118. Install the steering intermediate shaft bolt. Tighten to 22ft. lbs. (30 Nm).
119. Attach the power steering return tube to the PSP tube clip.
120. Connect the VAPS actuator electrical connector.
121. Attach the shift cable pin-type retainer to the exhaust heat shield.
122. Position shift cable bracket and install the bolts. Tighten to 89 inch lbs. (10 Nm).
123. Connect the shift cable from the shifter.
124. Install the 2 strut-to-control arm bolts and nuts. Tighten to 129 ft. lbs. (175 Nm).

➡**To install the nut, first loosen the nut, then use the hex-holding feature to prevent the upper ball joint stud from turning while tightening the nut.**

125. Connect the upper ball joints to the wheel knuckles and install the 2 nuts. Tighten to 66 ft. lbs. (90 Nm).
126. To install the nut, first loosen the nut, then use the hex-holding feature to prevent the stabilizer bar link ball joint from turning while tightening the nut.
127. Connect the stabilizer bar links to the lower control arms and install the 2 nuts.

✳✳ CAUTION

Do not allow the brake caliper to hang from the brake hose or damage to the brake hose can occur.

128. Position the front brake calipers and install the 4 bolts. Tighten to 76 ft. lbs. (103 Nm).
129. Attach the right and left hand anti-lock brake sensor harnesses to the brake hose clips.
130. Connect the 2 anti-lock brake sensor electrical connectors.
131. Position the right and left hand inner fender splash shields aside.
132. Position the inner splash shields.
133. Install the 3 pin-type retainers.
134. Connect the A/C low pressure hose and install the safety clip.
135. Connect the A/C high-pressure tube at the right frame rail and install the nut. Tighten to 71 inch lbs. (8 Nm).

➡**Install the hoses to the marks made during the removal procedure.**

136. Connect the three quick-disconnect couplings to the coolant control valve hoses.

✳✳ CAUTION

Install the bolts, washers and nuts in their original positions or driveshaft NVH can occur. Install the driveshaft flex coupling bolts with the head of the bolt seated against the flange and the nuts seated against the flex coupling.

➡**The bolt heads are serrated. Hold the bolt and tighten the nut. Install the bolts, washers and nuts. Coat the nut and bolt threads with sealer. Tighten to 60 ft. lbs. (81 Nm).**

137. Install the catalytic converters.
138. Install the inner air deflector.
139. Install the splash shields.
140. Connect the coolant valve electrical connector and attach the wiring retainer.
141. Attach the wiring harness retainer to the bracket.
142. Connect the 2 engine wiring harness connectors.
143. Position the main transmission wiring harness connector and tighten the bolt. Tighten to 53 inch lbs. (6 Nm).

144. Position the main engine wiring harness connector and tighten the bolt. Tighten to 53 inch lbs. (6 Nm).
145. Position the ground strap and install the bolt. Tighten to 89 inch lbs. (10 Nm).
146. Position the A/C manifold tube and install the nut. Tighten to 71 inch lbs. (8 Nm).
147. Connect EVAP hose to the canister purge valve.
148. Position the brake booster vacuum hose and attach the retainers.
149. Connect the brake booster vacuum hose to the upper intake manifold.
150. Connect the coolant hose to the upper intake manifold.
151. Position the PCV tube and degas bottle coolant hose. Attach the retainer to the upper intake manifold.
152. Connect the degas bottle coolant hose and the PCV tube to the upper intake manifold.
153. Position power steering fluid reservoir and install the 2 bolts. Tighten to 89 inch lbs. (10 Nm).
154. Connect the A/C pressure switch electrical connector.
155. Install the 2 upper radiator support brackets and the 6 bolts.
156. Install the upper radiator sight shield.
157. Connect the fuel tube.
158. Install the clean air inlet duct.
159. Install the air cleaner assembly.
160. Install the air cleaner outlet pipe and the air cleaner.
161. Fill the crankcase to the correct level.
162. Fill the cooling system.
163. Recharge the A/C system.
164. Connect the negative battery cable.
165. Start the engine and check for leaks.

3.9L Engine

2002–03 MODELS

1. Before servicing the vehicle, refer to the Precautions Section.
2. Drain the cooling system.
3. Relieve the fuel system pressure.
4. Recover the A/C refrigerant.
5. Drain the engine oil.
6. Remove or disconnect the following:

- Negative battery cable
- Air cleaner inlet tube
- Upper radiator shield
- Upper radiator support brackets
- A/C pressure switch connector
- Power steering return line clip

- Power steering reservoir
- Vapor Management Valve (VMV) cover
- VMV vacuum hose and canister purge hose
- Main vacuum supply hose
- Cowl vent screens
- Cross vehicle support bar
- Degas bottle hose
- Accelerator cable
- Cruise control cable
- Ground strap
- Fresh air filter and housing
- Powertrain harness connectors at right strut tower
- Fresh air filter panel
- Main engine wiring harness connector
- Main transmission wiring harness connector
- Heater hoses at the water control valve. Note the locations for assembly.
- Hydraulic cooling fan reservoir
- Water control valve harness connector
- Front wheels
- Inner splash shields
- Wheel speed sensor connectors and harness clips
- Brake calipers
- Lower stabilizer bar links
- Upper ball joints
- Lower strut mount bolts
- Left, right and center splash shields
- A/C suction and discharge lines
- Shift cable and bracket
- Power steering line frame rail clip
- Rack and pinion harness connectors
- Steering shaft bolt and coupling
- Starter motor harness connectors and ground cable
- Alternator harness connectors
- Lower transmission flange bolts
- Torque converter
- Inner air deflector
- Engine block heater, if equipped

7. Support the engine, transmission, front and center crossmembers and the cooling system with a powertrain lift and transmission support bracket.

8. Support the rear of the vehicle with safety stands.

9. Remove or disconnect the following:
- Transmission crossmember bolts
- Front crossmember bolts
- Center crossmember bolts
- Powertrain assembly
- A/C compressor manifold and tube assembly
- Power steering pump return hose

- Hydraulic cooling fan return hose
- Lower radiator hose
- Upper radiator hoses
- Knock Sensor (KS) connector
- Heater hose
- Transmission cooler lines and bracket
- Power steering pressure line and bracket
- Hydraulic cooling fan pressure line and bracket

10. Attach a hoist to the engine.

11. Remove the motor mount nuts and lift the powertrain out of the subframe.

12. Remove or disconnect the following:
- Wiring harness retainers
- Upper transmission flange bolts
- Transmission from the engine

To install:

13. Install or connect the following:
- Transmission to the engine. Tighten the upper flange bolts to 35 ft. lbs. (48 Nm).
- Wiring harness retainers. Tighten the nuts to 89 inch lbs. (10 Nm).
- Powertrain to the subframe. Tighten the mount nuts to 30 ft. lbs. (40 Nm).
- Hydraulic cooling fan pressure line and bracket
- Power steering pressure line and bracket
- Transmission cooler lines and bracket
- Heater hose
- Knock sensor connector
- Upper radiator hoses
- Lower radiator hose
- Hydraulic cooling fan return hose
- Power steering pump return hose
- A/C compressor manifold and tube assembly. Tighten the bolt to 15 ft. lbs. (21 Nm).
- Powertrain assembly. Tighten the front and center crossmember bolts to 76 ft. lbs. (103 Nm) and the transmission crossmember bolts to 30 ft. lbs. (40 Nm).
- Engine block heater, if equipped
- Inner air deflector
- Torque converter. Tighten the nuts to 28 ft. lbs. (38 Nm).
- Lower transmission flange bolts. Tighten the bolts to 35 ft. lbs. (47 Nm).
- Alternator harness connectors
- Starter motor harness connectors and ground cable
- Steering shaft bolt and coupling. Tighten the coupling pinch bolt to 26 ft. lbs. (35 Nm) and the shaft bolt to 22 ft. lbs. (30 Nm).

- Rack and pinion harness connectors
- Power steering line frame rail clip
- Shift cable and bracket
- A/C suction and discharge lines
- Left, right and center splash shields
- Lower strut mount bolts. Tighten the bolts to 129 ft. lbs. (175 Nm).
- Upper ball joints. Tighten the nuts to 66 ft. lbs. (90 Nm).
- Lower stabilizer bar links. Tighten the nuts to 41 ft. lbs. (55 Nm).
- Brake calipers
- Wheel speed sensor connectors and harness clips
- Inner splash shields
- Front wheels
- Water control valve harness connector
- Hydraulic cooling fan reservoir
- Heater hoses at the water control valve
- Main transmission wiring harness connector
- Main engine wiring harness connector
- Fresh air filter panel
- Powertrain harness connectors at right strut tower
- Fresh air filter and housing
- Ground strap
- Cruise control cable
- Accelerator cable
- Degas bottle hose
- Cross vehicle support bar. Tighten the bolts to 15 ft. lbs. (20 Nm).
- Cowl vent screens
- Main vacuum supply hose
- VMV vacuum hose and canister purge hose
- VMV cover
- Power steering reservoir
- Power steering return line clip
- A/C pressure switch connector
- Upper radiator support brackets
- Upper radiator shield
- Air cleaner inlet tube
- Negative battery cable

14. Fill the crankcase to the correct level.

15. Fill the cooling system.

16. Recharge the A/C system.

17. Start the engine and check for leaks.

2004–05 MODELS

1. Before servicing the vehicle, refer to the Precautions Section.

2. Drain the cooling system.

3. Relieve the fuel system pressure.

4. Recover the A/C refrigerant.

5. Drain the engine oil.

6. Disconnect the negative battery cable.

⁜ CAUTION

The engine appearance cover is equipped with a noise suppressor. If the noise suppressor becomes separated from the engine appearance cover it must be positioned on the intake manifold prior to installing the engine appearance cover.

7. Remove the engine appearance cover and the noise suppressor, if detached.

8. Remove the air cleaner assembly and outlet pipe

9. Remove the fresh air inlet duct.

10. Disconnect the fuel supply tube.

11. Remove the upper radiator sight shield.

12. Remove the 6 bolts and the 2 upper radiator support brackets.

⁜ CAUTION

To disconnect the quick release coupling, squeeze the tabs and pull straight out or damage to the coupling can occur.

13. Disconnect the Evaporative Emissions (EVAP) hose from the canister purge valve.

14. Disconnect the vacuum hose.

15. Disconnect the degas bottle hose.

16. Remove the 2 power steering reservoir bolts.

17. Remove the bolt and disconnect the ground strap.

18. Disconnect the engine wiring harness electrical connectors located on the backside of the right hand strut tower.

19. Disconnect the bulkhead electrical connectors.

20. Disconnect the coolant valve electrical connector, unclip from the radiator support and position the harness aside.

21. Disconnect the cooling fan electrical connector.

22. Disconnect the A/C pressure switch electrical connector.

23. Remove the splash shield.

24. Remove the inner air deflector.

25. Remove the catalytic converters.

⁜⁜ CAUTION

To make sure assembly will be in the exact location as when removed, index-mark the bolt, nut, washer and flex coupling to the pinion flange with paint or marker. Using a different color paint or marker, index-mark the second bolt, nut, washer and flex coupling to the pinion flange. These

fasteners are balanced weights. Always assemble them in their original positions. Failure to do so can cause driveshaft Noise, Vibration and Harshness (NVH).

⁜⁜ CAUTION

Do not remove the bolts retaining the flex coupling to the driveshaft.

26. Index-mark the front driveshaft pinion flange. Remove the 3 nuts, washers and bolts. Slide the front shaft assembly rearward and support.

➡ The top heater hose has a green identifying mark to match the hose on the dual coolant flow valve. If the mark is not visible, identify the hoses for location. This will aid in the correct installation of the hose assemblies.

27. Disconnect the 3 heater hose quick release couplings.

28. Remove the nut and disconnect the A/C high pressure hose.

29. Disconnect the low pressure A/C hose.

30. Position the right and left hand inner splash shields aside.

31. Disconnect the right and left anti-lock brake sensor electrical connectors.

32. Detach the right and left anti-lock brake sensor harnesses from the brake hose clips.

33. Remove the 4 bolts and position the front brake calipers aside. Support the calipers using mechanic's wire.

➡ To remove the nut, first loosen the nut, then use the hex-holding feature to prevent the stabilizer bar link ball joint from turning while removing the nut.

34. Remove the 2 nuts and disconnect the stabilizer bar links from the lower control arms.

➡ Use the hex-holding feature to prevent the stud from turning while removing the nut from ball joints and stabilizer bar links.

35. Remove the 2 nuts and disconnect the upper ball joints from the wheel knuckles.

36. Remove the 2 strut-to-control arm bolts and nuts.

37. Disconnect the shift cable from the shifter.

38. Remove the 2 bolts and position the shift cable bracket aside.

39. Detach the shift cable pin-type retainer from the exhaust heat shield

40. Disconnect the Variable Assist Power Steering (VAPS) actuator electrical connector.

41. Release the power steering return hose from the pressure hose clip.

42. Remove the retaining bolt for the high pressure power steering hose bracket.

43. Disconnect the block heater electrical connector, if equipped.

44. Remove the steering shaft bolt.

⁜⁜ CAUTION

Do not allow the steering wheel to rotate while the steering column intermediate shaft is disconnected or damage to the clockspring can result. If there is evidence that the wheel has rotated, the clockspring must be removed and recentered.

45. Remove the bolt and disconnect the steering coupling.

46. Remove the nut and starter motor ground cable.

47. Remove the nuts and disconnect the starter motor wiring.

48. Disconnect the generator electrical connector and pin-type retainer.

49. Remove the 6 lower transmission to engine bolts.

50. Remove the cover.

➡ Make an identifying mark on the nut, the stud and adapter plate to allow for correct installation.

51. Remove the 4 torque converter nuts.

52. Support the rear of the vehicle with suitable safety stands.

53. Support the engine, transmission, front and center crossmembers, and the cooling system with a powertrain lift and a transmission support bracket.

54. Remove the 4 transmission support bolts.

55. Remove the 4 right and left center crossmember I-brace bolts

56. Remove the 4 front crossmember and the 4 center crossmember bolts.

57. Carefully lower the entire assembly from the vehicle.

58. Remove the lower radiator hose bracket nut.

59. Disconnect the lower radiator, upper radiator and heater hose.

60. Disconnect the 2 oil cooler hoses, if equipped.

61. Rotate the tensioner and remove the accessory drive belt.

62. Disconnect the A/C compressor and Power Steering Pressure (PSP) switch electrical connectors.

63. Remove the 4 bolts and position the A/C compressor aside.

64. Remove the power steering tube bracket nut.

65. Remove the 4 bolts and position aside the power steering pump and reservoir.

66. Remove the transmission cooler tube bracket nut.

67. Disconnect the transmission cooler tubes.

68. Install the lifting hooks 303-050.

69. Using an engine crane, support the engine and transmission.

70. Remove the right and left hand upper engine mount nuts.

71. Using an engine hoist and the lifting hooks, move the engine and transmission from the crossmember and cooling system assembly to rest on the floor or on a bench.

72. Remove the nuts and position the wiring harness aside.

73. Remove the bolts and separate the engine from the transmission.

To install:

74. Align the flexplate to converter marks made during engine removal.

75. Install the engine to the transmission and install the bolts. Tighten to 35 ft. lbs. (47 Nm).

76. Position the wiring harness and install the 2 nuts.

77. Using the lifting hooks and an engine crane, move the engine and transmission onto the crossmember and cooling system assembly.

78. Install the right and left hand upper engine mount nuts. Tighten to 37 ft. lbs. (50 Nm).

79. Connect the transmission cooler tubes. Tighten to 26 ft. lbs. (35 Nm).

80. Install the transmission cooler tube bracket nut.

81. Install the power steering pump and 4 bolts. Tighten to 18 ft. lbs. (25 Nm).

82. Install the power steering tube bracket nut. Tighten to 13 ft. lbs. (18 Nm).

83. Install the A/C compressor and 4 bolts. Tighten to 18 ft. lbs. (25 Nm).

84. Connect the A/C compressor and PSP switch electrical connectors.

85. Install the accessory drive belt.

86. Connect the 2 oil cooler hoses, if equipped.

87. Connect the lower radiator, upper radiator and heater hose.

88. Install the lower radiator hose bracket nut.

89. Using the powertrain lift and the transmission support bracket, carefully raise the entire assembly into the vehicle.

90. Install the 4 front crossmember and the 4 center crossmember bolts. Tighten to 76 ft. lbs. (103 Nm).

91. Install the 4 right and left hand center crossmember I-brace bolts. Tighten to 76 ft. lbs. (103 Nm).

92. Install the 4 transmission support bolts. Tighten to 30 ft. lbs. (40 Nm).

93. Install the 4 torque converter nuts. Tighten to 28 ft. lbs. (38 Nm).

94. Install the cover.

95. Install the 6 lower transmission to engine bolts. Tighten to 35 ft. lbs. (47 Nm).

96. Connect the generator electrical connector and pin-type retainer.

97. Connect the starter motor wiring and install the nuts.

98. Install the cover.

99. Install the starter motor ground cable and nut.

✳✳ CAUTION

Do not allow the steering wheel to rotate while the steering column intermediate shaft is disconnected or damage to the clockspring can result. If there is evidence that the wheel has rotated, the clockspring must be removed and recentered.

100. Connect the steering coupling and install the bolt. Tighten to 26 ft. lbs. (35 Nm).

101. Install the steering shaft bolt. Tighten to 22 ft. lbs. (30 Nm).

102. Connect the block heater electrical connector, if equipped.

103. Install the retaining bolt for the high pressure power steering hose bracket.

104. Attach the power steering return hose to the pressure hose clip.

105. Connect the VAPS actuator electrical connector.

106. Attach the shift cable pin-type retainer to the exhaust heat shield

107. Install the shift cable bracket and 2 bolts.

108. Connect the shift cable to the shifter.

109. Install the 2 strut-to-control arm bolts and nuts. Tighten to 129 ft. lbs. (175 Nm).

110. Position the upper ball joints into the wheel knuckles and install the 2 nuts. Tighten to 66 ft. lbs. (90 Nm).

111. Position the stabilizer bar links into the lower control arms and install the 2 nuts. Tighten to 41 ft. lbs. (55 Nm).

112. Install the front brake calipers and the 4 bolts. Tighten to 26 ft. lbs. (35 Nm).

113. Attach the right and left hand anti-lock brake sensor harnesses to the brake hose clips.

114. Connect the right and left hand anti-lock brake sensor electrical connectors.

115. Install the right and left hand inner splash shields.

116. Connect the low pressure A/C hose.

117. Connect the A/C high pressure hose and install the nut. Tighten to 71 inch lbs. (8 Nm).

➡**The top heater hose has a green identifying mark to match the hose on the dual coolant flow valve. This will aid in the correct installation of the hose assemblies.**

NOTE: Hose assemblies shown with components removed for clarity.

118. Connect the 3 heater hose quick release couplings.

✳✳ CAUTION

Install the bolts, washers and nuts in their original positions or driveshaft NVH can occur. Install the driveshaft flex coupling bolts with the head of the bolt seated against the flange and the nuts seated against the flex coupling.

➡**The bolt heads are serrated. Hold the bolt and tighten the nut.**

119. Install the bolts, washers and nuts. Coat the nut and bolt threads with sealer. Tighten to 60 ft. lbs. (81 Nm).

120. Install the catalytic converters.

121. Install the inner air deflector.

122. Install the splash shields.

123. Connect the A/C pressure switch electrical connector.

124. Connect the cooling fan electrical connector.

125. Connect the coolant valve electrical connector and attach the wiring harness clip to the radiator support.

126. Connect the bulkhead electrical connectors.

127. Connect the engine wiring harness electrical connectors.

128. Position the ground strap and install the bolt. Tighten to 89 inch lbs. (10 Nm).

129. Install the power steering reservoir and 2 bolts. Tighten the top bolt to 53 inch lbs. (6 Nm) and the bottom bolt to 9 ft. lbs. (12 Nm).

130. Connect the degas bottle hose.

131. Connect the vacuum hose.

132. Connect the EVAP hose to the canister purge valve.

133. Install the 2 upper radiator support brackets and the 6 bolts.

134. Install the upper radiator sight shield.

135. Connect the fuel supply tube.

136. Install the fresh air inlet duct.

137. Install the air cleaner assembly and outlet pipe.

138. Connect the battery ground cable.

139. Fill and bleed the engine cooling system.

❄ CAUTION

The engine appearance cover is equipped with a noise suppressor. If the noise suppressor becomes separated from the engine appearance cover it must be positioned on the intake manifold prior to installing the engine appearance cover.

140. Install the noise suppressor and the engine appearance cover.

141. Fill the crankcase to the correct level.

142. Fill the cooling system.

143. Recharge the A/C system.

144. Connect the negative battery cable.

145. Start the engine and check for leaks.

Water Pump

REMOVAL & INSTALLATION

3.0L Engine

2002–03 MODELS

1. Before servicing the vehicle, refer to the Precautions Section.

2. Drain the cooling system.

3. Remove or disconnect the following:
   - Negative battery cable
   - Air cleaner outlet tube
   - Engine vent hose
   - Upper radiator hose
   - Heater supply hose
   - Water pump hose
   - Lower radiator hose
   - Water crossover assembly
   - Water inlet hose
   - Accessory drive belt and idler pulley
   - Bracket assembly
   - Water pump

   To install:
4. Install or connect the following:
   - Water pump. Tighten the bolts to 18 ft. lbs. (25 Nm).
   - Bracket assembly. Tighten the fasteners to 89 inch lbs. (10 Nm).

1 Intake manifold support bracket bolts
2 Intake manifold support bracket
3 Coolant pump-to-radiator hose clamp
4 Coolant pump-to-radiator hose
5 Coolant pump-to-coolant inlet pipe hose clamp
6 Coolant pump-to-coolant inlet pipe hose
7 Coolant pump-to-thermostat housing hose clamp
8 Coolant pump-to-thermostat housing hose

67197-LILS-G02

Water pump mounting—2002–03 3.0L engine

   - Accessory drive belt and idler pulley
   - Water inlet hose
   - Water crossover assembly
   - Lower radiator hose
   - Water pump hose
   - Heater supply hose
   - Upper radiator hose
   - Engine vent hose
   - Air cleaner outlet tube
   - Negative battery cable
5. Fill the cooling system.
6. Start the engine and check for leaks.

2004–05 MODELS

1. Before servicing the vehicle, refer to the Precautions Section.

2. Drain the cooling system.

3. Remove the cooling fan motor and shroud.

4. Remove the accessory drive belt.

5. Remove the 3 bolts and the intake manifold support bracket.

6. Disconnect the coolant pump-to-radiator hose and position aside.

7. Disconnect the coolant pump-to-coolant inlet pipe hose and position aside.

8. Disconnect the coolant pump-to-thermostat housing hose and position aside.

9. Remove the 3 bolts and the coolant pump.

To install:

10. Installation is the reverse of removal. Please note the following torque specifications:

   a. Water pump. Tighten the bolts to 18 ft. lbs. (25 Nm).

   b. Bracket assembly. Tighten the fasteners to 89 inch lbs. (10 Nm).

11. Fill the cooling system.

12. Start the engine and check for leaks.

3.9L Engine

1. Before servicing the vehicle, refer to the Precautions Section.

2. Drain the cooling system.

3. Remove or disconnect the following:
   - Thermostat housing
   - Accessory drive belt
   - Water pump pulley
   - Water pump

To install:

4. Install or connect the following:
   - Water pump. Tighten the bolts to 71 inch lbs. (8 Nm) plus 90 degrees.
   - Water pump pulley. Tighten the bolts to 89 inch lbs. (10 Nm) plus 45 degrees.
   - Accessory drive belt
   - Install a new thermostat housing O-ring seal and lubricate with clean engine coolant.

1 Coolant pump pulley bolts
2 Coolant pump pulley
3 Coolant pump bolts
4 Coolant pump
5 Coolant pump gasket

67197-LILS-G03

Exploded view of the water pump—3.9L engine

- Thermostat housing. Tighten the retainers to 89 inch lbs. (10 Nm).
5. Fill the cooling system.
6. Start the engine and check for leaks.

Heater Core

REMOVAL & INSTALLATION

1. Before servicing the vehicle, refer to the Precautions Section.
2. Drain the cooling system.
3. Recover the A/C refrigerant.
4. Remove or disconnect the following:

- Negative battery cable
- Heater hose assembly
- Cabin air filter plenum
- Thermostatic expansion valve manifold and tube assembly
- Driver's side air bag module
- Floor console and A/C duct
- Shift lever assembly, if equipped with automatic transmission
- Left and right instrument panel insulators
- Left and right door sill scuff plates
- Left and right door weatherstrips
- Left and right A pillar lower trim panels
- Left and right windshield side garnish moldings
- Instrument panel defroster opening grille assembly
- Instrument panel cowl top screws
- Instrument panel upper reinforcement bolts
- Left and right instrument panel side finish panels
- Hood release handle and cable
- Upper right bulkhead electrical connector
- Right instrument panel electrical connectors
- Passenger side tunnel electrical connector
- Steering column intermediate shaft
- Left junction box electrical connectors

- Left instrument panel electrical connectors
- Ignition shift interlock connector, if equipped
- 4 instrument panel tunnel brace bolts
- Left outer instrument panel cowl side cover and reinforcement bolt
- Left instrument panel cowl side bolt and nut
- Right instrument panel cowl side bolt and nut
- Instrument panel
- Evaporator housing electrical connector
- Cowl top attachment bolt
- Evaporator housing attachment bolt
- 3 evaporator nuts and washers in the engine compartment
- Rear seat floor ducts
- Evaporator core housing
- Evaporator core housing air inlet
- Bypass door harness connector
- Heater core

To install:

5. Install or connect the following:
- Heater core
- Bypass door harness connector
- Evaporator core housing air inlet
- Evaporator core housing
- Rear seat floor ducts
- 3 evaporator nuts and washers in the engine compartment
- Evaporator housing attachment bolt
- Cowl top attachment bolt
- Evaporator housing electrical connector
- Instrument panel
- Instrument panel cowl top screws. Tighten to 27 inch lbs. (3 Nm).
- Right instrument panel cowl side bolt and nut. Tighten the fasteners to 15 ft. lbs. (20 Nm).
- Left instrument panel cowl side bolt and nut. Tighten the fasteners to 15 ft. lbs. (20 Nm).
- Left outer instrument panel cowl side cover and reinforcement bolt. Tighten the bolt to 15 ft. lbs. (20 Nm).
- Instrument panel upper reinforcement bolts. Tighten the bolts to 15 ft. lbs. (20 Nm).
- 4 instrument panel tunnel brace bolts. Tighten the bolts to 15 ft. lbs. (20 Nm).
- Ignition shift interlock connector, if equipped
- Left instrument panel electrical connectors
- Left junction box electrical connectors

1 Heater hose clamp

2 Thermostatic expansion valve manifold bolt

3 PCM bracket bolt

4 Wire harness

5 Heater core and evaporator core housing bracket bolt

6 Heater core and evaporator core housing nut

7 Heater core and evaporator core housing

8 Thermostatic expansion valve O-ring

67197-LILS-G04

Heater core and related components (1 of 2)

- Steering column intermediate shaft. Tighten the pinch bolt to 26 ft. lbs. (35 Nm).
- Passenger side tunnel electrical connector
- Right instrument panel electrical connectors
- Upper right bulkhead electrical connector

- Hood release handle and cable
- Left and right instrument panel side finish panels
- Instrument panel defroster opening grille assembly
- Left and right windshield side garnish moldings
- Left and right A pillar lower trim panels

- Left and right door weatherstrips
- Left and right door sill scuff plates
- Left and right instrument panel insulators
- Shift lever assembly, if equipped with automatic transmission
- Floor console and A/C duct
- Driver's side air bag module

9 Wire harness

10 Heater core and evaporator core housing screw

11 Heater core and evaporator core housing screw

12 Blower motor speed control screw

13 Blower motor speed control

14 Heater core tube seal

15 Heater core tube bracket screw

67197-LILS-G05

Heater core and related components (2 of 2)

- Thermostatic expansion valve manifold and tube assembly
- Cabin air filter plenum
- Heater hose assembly
- Negative battery cable
6. Fill the cooling system.
7. Recharge the A/C system.
8. Start the engine and check for leaks.

Cylinder Head

REMOVAL & INSTALLATION

3.0L Engine

1. Before servicing the vehicle, refer to the Precautions Section.
2. Drain the cooling system.
3. Relieve the fuel system pressure.
4. Drain the engine oil.
5. Remove or disconnect the following:

- Negative battery cable
- Upper and lower intake manifolds
- Valve covers
- Accessory drive belts

Cylinder head torque sequence—3.0L engine

9306TG01

- Timing chains
- Front cover
- Oil pan
- Subframe bolts. Tighten the bolts to 76 ft. lbs. (103 Nm).
- Motor mount nuts. Tighten the nuts to 46 ft. lbs. (63 Nm).
- Accessory drive belts
- Valve covers and tighten to 89 inch lbs. (10 Nm) in the sequence illustrated.
- Upper and lower intake manifolds
- Negative battery cable

13. Fill the crankcase to the correct level.

14. Fill the cooling system.

15. Start the engine and check for leaks.

6. Install a support fixture to the engine lifting eyes.

7. Remove or disconnect the following:

- Motor mount nuts
- Subframe bolts
- Oil pan
- Front cover
- Timing chains
- Camshafts
- Exhaust manifolds
- Ground strap
- Positive Crankcase Ventilation (PCV) tube
- Ignition noise suppressor
- Coolant outlet tube
- Engine oil dipstick tube

8. Install the subframe bolts and remove the engine support fixture.

9. Remove the cylinder heads.

To install:

10. Install the cylinder heads. Tighten the bolts in sequence as follows:

    a. Step 1: 22 ft. lbs. (30 Nm)

    b. Step 2: Plus 90 degrees

    c. Step 3: Loosen all bolts one full turn

    d. Step 4: 22 ft. lbs. (30 Nm)

    e. Step 5: Plus 90 degrees

    f. Step 6: Plus 90 degrees

11. Install the engine support fixture and remove the subframe bolts.

12. Install or connect the following:

- Engine oil dipstick tube
- Coolant outlet tube
- Ignition noise suppressor
- PCV tube
- Ground strap
- Exhaust manifolds
- Camshafts

Left valve cover torque sequence—2002–03 3.0L engine

9346TG01

Right valve cover torque sequence—2002–03 3.0L engine

9346TG02

06017-LS-G04

Left valve cover torque sequence—2004–05 3.0L engine

06017-LS-G05

Right valve cover torque sequence—2004–05 3.0L engine

3.9L Engine

2002–03 MODELS

1. Before servicing the vehicle, refer to the Precautions Section.
2. Drain the cooling system.
3. Relieve the fuel system pressure.
4. Remove or disconnect the following:
   - Negative battery cable
   - Engine appearance cover
   - Intake manifold
   - Valve covers
   - Accessory drive belts
   - Front cover
   - Timing chains
   - Camshafts
   - Water outlet pipe
   - Cylinder head temperature sensor connector
   - Exhaust front pipes
   - Exhaust Gas Recirculation (EGR) tube
   - Bolts and stud bolts at the rear of the cylinder heads
   - Cylinder heads. Loosen the bolts

in reverse of the tightening sequence.

To install:

5. Install the cylinder heads. Tighten the bolts in sequence as follows:
   a. Step 1: M10 bolts to 15 ft. lbs. (20 Nm)
   b. Step 2: M10 bolts to 26 ft. lbs. (35 Nm)
   c. Step 3: M10 bolts to 33 ft. lbs. (45 Nm)
   d. Step 4: M10 bolts plus 90 degrees
   e. Step 5: M10 bolts plus 90 degrees
   f. Step 6: M8 bolts to 15 ft. lbs. (20 Nm)
   g. Step 7: M8 bolts plus 90 degrees
6. Install or connect the following:
   - Bolts and stud bolts at the rear of the cylinder heads. Tighten the bolts to 37 ft. lbs. (50 Nm).
   - EGR tube
   - Exhaust front pipes
   - Cylinder head temperature sensor connector
   - Water outlet pipe
   - Camshafts
   - Timing chains
   - Front cover
   - Accessory drive belts
   - Valve covers
   - Intake manifold
   - Engine appearance cover
   - Negative battery cable
7. Fill the cooling system.
8. Start the engine and check for leaks.

9306TG04

Right cylinder head torque sequence—2002–03 3.9L engine

Left cylinder head torque sequence—2002–03 3.9L engine

9306TG05

Left valve cover torque sequence—2002–03 3.9L engine

9346TG03

Right valve cover torque sequence—2002–03 3.9L engine

9346TG04

2004–05 MODELS

1. Before servicing the vehicle, refer to the Precautions Section.

2. Remove the engine.

3. Install the engine on a suitable engine stand.

4. Remove the engine block plug to drain the engine coolant.

5. Disconnect the oil pressure and oil temperature sensor electrical connectors.

6. Remove and discard the oil filter.

❉❉ CAUTION

A new oil cooler must be installed or severe damage to the engine can occur.

06017-LS-G10

Hold the crankshaft pulley with the tool 303-D055 while removing the bolt and washer—2004–05 3.9L engine

06017-LS-G11

Using the tool 303-009, remove and discard the front crankshaft seal—2004–05 3.9L engine

06017-LS-G12

Using the tool 303-409, remove and discard the front crankshaft seal—2004–05 3.9L engine

7. Remove the bolt and oil cooler.

8. Discard the oil cooler and gasket.

9. Remove the bolt and the accessory drive belt tensioner.

10. Remove the bolt and the accessory drive belt idler pulley.

11. Remove the 3 bolts and the coolant pump pulley.

➡Hold the crankshaft pulley with the tool 303-D055 while removing the bolt and washer.

12. Remove and discard the bolt and washer.

13. Using the puller tool 303-009, remove the crankshaft damper.

14. Using the tool 303-409, remove and discard the front crankshaft seal.

15. Remove the ignition coil covers.

16. Remove all silicone sealer and residue from the ignition coil covers.

17. Remove all silicone sealer and residue from the valve cover surface and ignition coil wiring harness.

18. Disconnect the 8 ignition coil electrical connectors.

19. Disconnect the Throttle Position (TP) sensor electrical connector.

20. Disconnect the fuel pressure sensor electrical connector.

21. Disconnect the electronic throttle body electrical connector.

22. Disconnect the Evaporative Emissions (EVAP) tube.

23. Disconnect the fuel rail temperature sensor electrical connector.

24. Disconnect the Exhaust Gas Recirculation (EGR) system module electrical connector.

25. Disconnect the coolant vent hose from the coolant outlet.

26. Remove the coolant vent hose.

27. Remove the vacuum harness.

28. Disconnect the engine wiring harness retainers from the engine appearance cover support brackets.

29. Detach the vacuum hose retainers from the left hand engine appearance cover support bracket.

30. Remove the 4 nuts and the 2 engine appearance cover brackets.

31. Disconnect the 8 fuel injector and the 2 radio interference capacitor electrical connectors

32. Disconnect the Knock Sensor (KS) and Camshaft Position (CMP) senor electrical connectors.

33. Detach the pin-type retainers.

34. Disconnect the 2 Variable Cam Timing (VCT) sensor electrical connectors.

35. Remove the 5 bolts and the power steering bracket.

36. Disconnect the wiring harness retainers from the left hand valve cover.

37. Detach the wiring harness from the engine front cover.

38. Disconnect the Crankshaft Position (CKP) sensor electrical connector and wiring harness retainers.

39. Disconnect the wiring harness retainers from the right hand valve cover.

40. Disconnect the generator electrical connector and wiring harness retainers.

41. Remove the Heated Oxygen (HO2s) sensor electrical connectors from the brackets.

42. Remove the engine wiring harness from the engine.

43. Remove the nut and the oil level indicator tube. Discard the O-ring seal.

44. Remove the 8 bolts and the 8 ignition coils.

45. Remove the crankcase vent tube.

46. Remove the 2 EGR system module bolts and loosen the EGR tube nut.

47. Remove the EGR system module and discard the gasket.

48. Remove the 2 nuts and the radio interference capacitors.

49. Disconnect the throttle body heater hoses.

50. Remove the 4 bolts and the coolant bypass housing. Discard the O-ring seals.

51. Remove the 4 bolts and the throttle body adapter. Discard the gasket.

52. Remove the intake manifold. Discard the gaskets.

06017-LS-G06

Left valve cover loosening sequence—2004–05 3.9L engine

06017-LS-G07

Right valve cover loosening sequence—2004–05 3.9L engine

✳✳ CAUTION

Do not use metal scrapers, wire brushes, power abrasive discs, or other abrasive means to clean the sealing surfaces. These tools cause scratches and gouges, which make leak paths. Use a plastic scraping tool to clean the surfaces.

53. Clean the sealing surfaces.
54. Remove the bolts in the sequence shown and remove the right valve cover. Discard the gasket.
55. Remove the bolts in the sequence shown and remove the left valve cover.

✳✳ CAUTION

Do not use metal scrapers, wire brushes, power abrasive discs, or other abrasive means to clean the sealing surfaces. These tools cause scratches and gouges, which make leak paths. Use a plastic scraping tool to clean the surfaces.

56. Remove the gaskets and all silicone sealer. Clean the mating surfaces with silicone gasket remover.
57. On the right hand side:
    a. Remove the 2 nuts and the EGR tube. Discard the gasket.
    b. Remove the right hand exhaust manifold heat shield stud bolt.
    c. Remove the 4 bolts and the right hand exhaust manifold heat shield.
58. Remove the bolts, the spacers and the right hand exhaust manifold. Discard the gasket.
59. On the left hand side:
60. Remove the left hand exhaust manifold heat shield stud bolt.
61. Remove the 4 bolts and the left hand exhaust manifold heat shield.
62. Remove the bolts, the spacers and the left exhaust manifold. Discard the gasket.
63. Remove the engine front cover bolts in the sequence shown.
64. Remove the engine front cover. Discard the gaskets.
65. Do not use metal scrapers, wire brushes, power abrasive discs or other abrasive means to clean the sealing surfaces. These tools cause scratches and gouges, which make leak paths. Use a plastic scraping tool to remove all traces of sealant.
66. Clean the sealing surfaces.
67. Remove the right and left hand Variable Cam Timing (VCT) bush carriers.

Remove the engine front cover bolts in the sequence shown—2004–05 3.9L engine

68. Remove the bolt and the CKP sensor.

➡ There is one window on the ignition pulse wheel that is unique to accept the tool 303-645.

69. Install the tool 303-645.
70. Turn the crankshaft to 45 degrees After Top Dead Center (ATDC). The crankshaft keyway will be in the 6 o'clock position.
71. Make sure the flats of the camshaft are facing upward. If not, repeat the previous step.

✳✳ CAUTION

The flats on the camshaft must completely contact the special service tool or damage to the camshafts can occur. Slowly draw the special tool down evenly until it is fully seated on the camshaft flats.

72. Install the holding tool 303-530 on the right hand cylinder head.

✳✳ CAUTION

An open-end wrench must be used on the hex area of the camshaft or damage to the head will occur.

There is one window on the ignition pulse wheel that is unique to accept the tool 303-645—2004–05 3.9L engine

Tool 303-530 is used to hold the camshafts during removal—2004–05 3.9L engine

73. Remove and discard the exhaust camshaft sprocket bolt.

74. Remove and discard the variable VCT unit bolt.

75. Remove the right hand timing chain tensioner.

76. Push the lever down and install a drill rod to lock the piston.

77. Remove the 2 bolts and the tensioner.

78. Remove the bolt and the right tensioner arm.

79. Remove the bolt and the right timing chain guide.

80. Remove the right primary timing chain.

81. Remove the holding tool from the right cylinder head.

✳✳ CAUTION

The flats on the camshaft must completely contact the special service tool or damage to the camshafts can occur. Slowly draw the special tool down evenly until it is fully seated on the camshaft flats.

82. Install the holding tool on the left cylinder head.

✳✳ CAUTION

An open-end wrench must be used on the hex area of the camshaft or damage to the cylinder head will occur.

83. Remove and discard the left VCT unit bolt and the exhaust camshaft sprocket bolt.

84. Remove the left timing chain tensioner.

85. Push the lever down and install a drill rod to lock the piston.

86. Remove the 2 bolts and the tensioner.

87. Remove the nut and the left tensioner arm.

88. Remove the bolt and the left timing chain guide.

89. Remove the left primary timing chain.

90. Remove the crankshaft sprocket.

91. Collapse the left secondary timing chain tensioner and install a pin to lock the piston.

92. Remove the left VCT unit, exhaust camshaft sprocket and timing chain as an assembly.

93. Remove the 2 bolts and the left secondary timing chain tensioner.

94. Collapse the right secondary timing chain tensioner and install a pin to lock the piston.

95. Remove the right VCT unit, exhaust camshaft sprocket and timing chain as an assembly.

96. Remove the 2 bolts and the right secondary timing chain tensioner.

97. Remove the holding tool.

98. On the left side:

✳✳ CAUTION

Record the camshaft bearing cap locations. The camshaft bearing caps are positional and must be installed in their original locations and orientation or engine damage can occur.

    a. Remove the left camshaft bearing cap bolts and caps.

    b. Remove the left hand camshafts.

✳✳ CAUTION

Do not use any means of marking the shims other than a permanent-type marker. Any scratches or paint on the shims can result in incorrect lash adjustments and severe engine damage. Record the shim and bucket tappet location. The shim and bucket tappet are positional and if installed in the incorrect location, engine damage can occur.

    c. Record the location of the valve tappets and remove the tappets.

✳✳ CAUTION

The cylinder head must be cool before removing it from the engine. Cylinder head warpage can result if a warm or hot cylinder head is removed. Place clean shop towels over exposed engine cavities. Carefully remove the towels so foreign material is not dropped into the

engine. The cylinder head bolts must be discarded and new bolts installed. They are tighten-to-yield designed and cannot be reused. Aluminum surfaces are soft and can be scratched easily. Never place the cylinder head gasket surface, unprotected, on a bench surface.

    d. Remove the left hand cylinder head end bolts and discard the bolts.

    e. Remove and discard the bolts from left cylinder head in the sequence shown.

    f. Remove the cylinder head. Discard the gasket.

99. On the right side:

✳✳ CAUTION

Record the camshaft bearing cap locations. The camshaft bearing caps are positional and must be installed in their original locations and orientation or engine damage can occur.

    a. Remove the right camshaft bearing cap bolts and caps.

    b. Remove the right camshafts.

✳✳ CAUTION

Do not use any means of marking the shims other than a permanent-type marker. Any scratches or paint on the shims can result in incorrect lash adjustments and severe engine damage. Record the shim and bucket tappet location. The shim and bucket tappet are positional and if installed in the incorrect location, engine damage can occur.

    c. Record the location of the valve tappets and remove the tappets.

06017-LS-G16

Left cylinder head bolt loosening sequence—2004–05 3.9L engine

✳✳ CAUTION

The cylinder head must be cool before removing it from the engine. Cylinder head warpage can result if a warm or hot cylinder head is removed. Place clean shop towels over exposed engine cavities. Carefully remove the towels so foreign material is not dropped into the engine. The cylinder head bolts must be discarded and new bolts installed. They are tighten-to-yield designed and cannot be reused. Aluminum surfaces are soft and can be scratched easily. Never place the cylinder head gasket surface, unprotected, on a bench surface.

    d. Remove the right cylinder head end bolts. Discard the bolts.
    e. Remove and discard the bolts from right cylinder head in the sequence shown.
    f. Remove the cylinder head. Discard the gasket.

✳✳ CAUTION

Do not use metal scrapers, wire brushes, power abrasive discs or other abrasive means to clean the sealing surfaces. These tools cause scratches and gouges, which make leak paths. Use a plastic scraping tool to remove all traces of sealant. Observe all warnings or cautions and follow all application directions contained on the packaging of the silicone gasket remover and the metal surface prep.

➡️If there is no residual gasket material present, metal surface prep can be used to clean and prepare the surfaces.

100. Clean the cylinder head-to-cylinder block mating surfaces of both the cylinder heads and the cylinder block.
101. Remove any large deposits of silicone or gasket material with a plastic scraper.
102. Apply silicone gasket remover, following package directions, and allow to set for several minutes.
103. Remove the silicone gasket remover with a plastic scraper. A second application of silicone gasket remover may be required if residual traces of silicone or gasket material remain.
104. Apply metal surface prep, following package directions, to remove any remaining traces of oil or coolant, and to prepare the surfaces to bond with the new gasket. Do not attempt to make the metal shiny. Some staining of the metal surfaces is normal.

➡️The straightedge used must be flat within 0.0002 inch (0.0051 mm) per foot of tool length.

105. Support the cylinder heads on a bench with the head gasket side up. Inspect all areas of the deck face with a straightedge, paying particular attention to the oil pressure feed area. The cylinder heads must not have depressions deeper than 0.001 inch (0.0254 mm) across a 1.5 inch (38.1 mm) square area, or scratches more than 0.001 inch (0.0254 mm).

    To install:

✳✳ CAUTION

The gasket sealing surfaces on the cylinder head and cylinder block must be clean and dry. The use of sealing aids (aviation cement, copper spray and glue) is not permitted. The gasket must be installed dry. The new gasket has a film coating which

is crucial to the gasket's ability to seal correctly. Do not scratch the gasket. Do not allow the dowels to scratch the sealing surface of the cylinder head during cylinder head installation.

106. On the right side:
    a. Lightly oil the new cylinder head bolts and allow to drain for a few minutes before installing.
    b. Install a new gasket and position the right cylinder head on the engine.

➡️The cylinder head bolts are a torque-to-yield design and must not be reused. Always use new bolts.

    c. Install and tighten the right head bolts in 4 stages in the sequence shown.
    d. Step 1: Tighten bolts 1 through 10 to 15 ft. lbs. (20 Nm).
    e. Step 2: Tighten bolts 1 through 10 to 26 ft. lbs. (35 Nm).
    f. Step 3: Tighten bolts 1 through 10 to 33 ft. lbs. (45 Nm and an additional 90 degrees.
    g. Step 4: Tighten bolts 1 through 10 an additional 90 degrees.
    h. Install the 2 end bolts and tighten to 15 ft. lbs. (20 Nm), then retighten an additional 90 degrees.
107. On the left side:
    a. Lightly oil the new cylinder head bolts and allow to drain for a few minutes before installing.
    b. Install a new gasket and position the right cylinder head on the engine.

➡️The cylinder head bolts are a torque-to-yield design and must not be reused. Always use new bolts.

    c. Install and tighten the right head bolts in 4 stages in the sequence shown.
    d. Step 1: Tighten bolts 1 through 10 to 15 ft. lbs. (20 Nm).
    e. Step 2: Tighten bolts 1 through 10 to 26 ft. lbs. (35 Nm).
    f. Step 3: Tighten bolts 1 through 10 to 33 ft. lbs. (45 Nm and an additional 90 degrees.
    g. Step 4: Tighten bolts 1 through 10 an additional 90 degrees.
    h. Install the 2 end bolts and tighten to 15 ft. lbs. (20 Nm), then retighten an additional 90 degrees.

➡️Coat the valve tappets with clean engine oil prior to installing.

108. Install the valve tappets and shims in their original locations.
109. On the right side:

06017-LS-G17

Right cylinder head bolt loosening sequence—2004–05 3.9L engine

Right cylinder head torque sequence—3.9L engine

**Right camshaft bearing cap tightening
sequence—2004–05 3.9L engine**

Left cylinder head torque sequence—03 3.9L engine

**Left camshaft bearing cap tightening
sequence—2004–05 3.9L engine**

NOTE: Apply clean engine oil to the camshaft journals, the camshaft caps and the camshaft lobes prior to installing the camshafts.

a. Position the right camshafts on the camshaft journals.

❋❋ CAUTION

Install the camshaft bearing caps. The camshaft bearing caps are positional and must be installed in their original locations and orientation or engine damage can occur.

b. Install the right camshaft bearing caps.

c. Install the bearing cap bolts. Tighten the bolts in three stages in the sequence shown.

d. Step 1: Hand-tighten.

e. Step 2: Tighten to 53 inch lbs. (6 Nm).

f. Step 3: Tighten an additional 90 degrees.

110. On the left side:

NOTE: Apply clean engine oil to the camshaft journals, the camshaft caps and the camshaft lobes prior to installing the camshafts.

a. Position the right camshafts on the camshaft journals.

❋❋ CAUTION

Install the camshaft bearing caps. The camshaft bearing caps are positional and must be installed in their original locations and orientation or engine damage can occur.

b. Install the right camshaft bearing caps.

c. Install the bearing cap bolts. Tighten the bolts in three stages in the sequence shown.

d. Step 1: Hand-tighten.

e. Step 2: Tighten to 53 inch lbs. (6 Nm).

Position all of the camshafts so the flats of the camshafts are facing upward—2004–05 3.9L engine

06017-LS-G20

f. Step 3: Tighten an additional 90 degrees.

✳✳ CAUTION

If any components are installed new, the engine must be reshimmed or engine damage can occur.

111. Using a feeler gauge, confirm that the tappet and shim clearances are within specification.

112. If tappet and shim clearances are not within specification, adjust the clearance.

113. Turn the crankshaft to 45 degrees ATDC. The crankshaft keyway will be in the 6 o'clock position.

114. Install the tool 303-645.

115. Position all of the camshafts so the flats of the camshafts are facing upward.

116. Install the left secondary timing chain tensioner and 2 bolts. Tighten to 8 ft. lbs. (11 Nm).

117. Install the left VCT unit, exhaust camshaft sprocket and timing chain as an assembly.

118. Using new bolts, install the left VCT unit and exhaust camshaft sprocket bolts finger-tight.

119. Remove the locking pin from the left secondary timing chain tensioner.

120. Install the right secondary timing chain tensioner and 2 bolts. Tighten to 8 ft. lbs. (11 Nm).

121. Install the right VCT unit, exhaust camshaft sprocket and timing chain as an assembly.

122. Using new bolts, install the right VCT unit and exhaust camshaft sprocket bolts finger-tight.

123. Remove the locking pin from the right secondary timing chain tensioner.

124. Reset the timing chain tensioners by holding the lever toward the tensioner piston. Using finger pressure, compress the tensioner. Push the lever down and install a drill rod to lock the piston.

✳✳ CAUTION

The flats on the camshaft must completely contact the special service tool or damage to the camshafts can occur. Slowly draw the special tool down evenly until it is fully seated on the camshaft flats.

125. Install the holding tool on the left head.

126. Position the left timing chain and crankshaft gear.

127. Position the timing chain over the left intake camshaft sprocket.

128. Position the crankshaft gear in the timing chain.

129. Install the left timing chain guide and bolt. Tighten to 8 ft. lbs. (11 Nm).

130. Install the left timing chain tensioner arm and nut. Tighten to 8 ft. lbs. (11 Nm).

131. Install the left timing chain tensioner.

132. Install the timing chain tensioner and bolts, and remove the drill rod.

✳✳ CAUTION

The exhaust camshaft sprocket bolt must be fully tightened before tightening the intake camshaft sprocket bolt. Using the tool, apply 89 inch lbs. (10 Nm) counterclockwise to the left exhaust camshaft sprocket.

133. Tighten the left exhaust camshaft sprocket and VCT unit bolts 26 ft. lbs. (35 Nm) then an additional 90 degrees.

134. Remove the holding tool.

✳✳ CAUTION

The flats on the camshaft must completely contact the special service tool or damage to the camshafts can occur. Slowly draw the special tool downward evenly until it is fully seated on the camshaft flats.

135. Install the holding tool on the right cylinder head.

136. Position the right timing chain over the right intake camshaft sprocket and over the crankshaft gear.

137. Install the right timing chain guide and bolt. Tighten to 8 ft. lbs. (11 Nm).

138. Install the right timing chain tensioner arm. Tighten to 8 ft. lbs. (11 Nm).

139. Install the right timing chain tensioner.

140. Install the timing chain tensioner and bolts, and remove the drill rod.

303-532

Using the tool, apply 89 inch lbs. (10 Nm) counterclockwise to the camshaft sprocket—2004–05 3.9L engine

06017-LS-G21

✳✳ CAUTION

The exhaust camshaft sprocket bolt must be fully tightened before tightening the intake camshaft sprocket bolt. Using the tool, apply 89 inch lbs. (10 Nm) counterclockwise to the left exhaust camshaft sprocket.

141. Tighten the right exhaust camshaft sprocket and VCT unit bolts 26 ft. lbs. (35 Nm) then an additional 90 degrees.

142. Remove the holding tool.

➥Lubricate the VCT bush carrier seals with clean engine oil.

143. Carefully insert the bush carrier into the VCT unit, compress ring seals with fingers as required. Fully seat the bush carrier by hand before fitting the bolts and nut. Tighten the bolts to 15 ft. lbs. (20 Nm). Tighten the nut to 89 inch lbs. (10 Nm).

144. Install new engine front cover gaskets.

➥If the front cover is not secured within 4 minutes of sealant application, the sealant must be removed and the sealing area cleaned with metal surface cleaner. Allow to dry until there is no sign of wetness, or 4 minutes, whichever is longer. Failure to follow this procedure can cause future oil leakage.

145. Apply silicone gasket and sealant in the 8 places shown.

146. Install the engine front cover. Loosely install the bolts.

147. Tighten the bolts in 2 stages in the sequence shown. Tighten to 44 inch lbs. (5 Nm) in sequence and then retighten in sequence to 89 inch lbs. (10 Nm).

➥Lubricate the front oil seal and the engine front cover mating surface with clean engine oil.

3 mm (0.12 in)

06017-LS-G22

Apply silicone gasket and sealant in the 8 places shown—2004–05 3.9L engine

06017-LS-G23

Engine front cover torque sequence—2004–05 3.9L engine

06017-LS-G24

Using the tool 303-646, install a new front crankshaft seal—2004–05 3.9L engine

06017-LS-G25

Using the tool 303-102, install the crankshaft pulley—2004–05 3.9L engine

148. Using the tool 303-646, install a new front crankshaft seal.

→**If crankshaft pulley is not secured within 4 minutes of sealant application the sealant must be removed and the sealing area cleaned with metal surface cleaner. Allow to dry until there is no sign of wetness, or 4 minutes, whichever is longer. Failure to follow this procedure can cause future oil leakage.**

149. Apply a bead of silicone gasket and sealant to the keyway.

150. Using the tool 303-102, install the crankshaft pulley.

151. Install a new bolt and using the special tool to hold the pulley, tighten as follows:.

a. Step 1: Tighten to 59 ft. lbs. (80 Nm).

b. Step 2: Loosen the bolt 2 complete turns.

c. Step 3: Tighten to 37 ft. lbs. (50 Nm).

d. Step 4: Tighten an additional 90 degrees.

1 52. On the left side:

a. Install a new gasket, left exhaust manifold, spacers and bolts.

b. Install the left exhaust manifold heat shield and 4 bolts.

c. Install the left exhaust manifold heat shield stud bolt.

153. On the right side:

06017-LS-G26

Apply 4 beads of silicone gasket and sealant as shown—2004–05 3.9L engine

a. Install a new gasket, right exhaust manifold, spacers and bolts.

b. Install the right exhaust manifold heat shield and 4 bolts.

c. Install the right exhaust manifold heat shield stud bolt.

d. Install a new gasket, EGR tube and 2 nuts. Tighten to 15 ft. lbs. (20 Nm).

154. Install new valve cover gaskets.

➡**If the valve covers are not secured within 4 minutes of sealant application the sealant must be removed and the sealing area cleaned with metal surface cleaner. Allow to dry until there is no sign of wetness, or 4 minutes, whichever is longer. Failure to follow this procedure can cause future oil leakage.**

155. Apply 4 beads of silicone gasket and sealant as shown.

156. Install the left valve cover and tighten the bolts in sequence to 89 inch lbs. (10 Nm).

157. Install the right valve cover and tighten the bolts in sequence to 89 inch lbs. (10 Nm).

158. Install new intake manifold gaskets and the manifold.

159. Install the intake manifold, the 7 bolts and the 3 studs.

160. Install a new gasket, the throttle body adapter and the 4 bolts. Tighten to 89 inch lbs. (10 Nm).

161. Install new O-ring seals, the coolant bypass housing and the 4 bolts. Tighten to 89 inch lbs. (10 Nm).

162. Connect the 2 throttle body coolant hoses.

163. Install the 2 radio interference ignition capacitors and the 2 nuts. Tighten to 89 inch lbs. (10 Nm).

164. Install a new gasket, the EGR system module and bolts. Tighten to 18 ft. lbs. (25 Nm).

165. Tighten the EGR tube nut to 30 ft. lbs. (40 Nm).

166. Install the crankcase vent tube.

06017-LS-G08

Left valve cover torque sequence—2004–05 3.9L engine

06017-LS-G09

Right valve cover torque sequence—2004–05 3.9L engine

167. Install the 8 ignition coils and bolts. Tighten to 44 inch lbs. (5 Nm).

❊❊ CAUTION

Make sure the O-ring seal remains in position and does not roll up the indicator tube during installation.

168. Install a new O-ring seal, oil level indicator tube and nut. Lubricate the O-ring seal with clean engine oil prior to installation and tighten the nut to 53 inch lbs. (6 Nm).

169. Install the CKP sensor and bolt. Tighten to 89 inch lbs. (10 Nm).

170. Position the engine wiring harness onto the engine.

171. Attach the (HO_2S) sensor electrical connectors onto the brackets.

172. Connect the generator electrical connector and wiring harness retainers.

173. Connect the wiring harness retainers to the right valve cover.

174. Connect the CKP sensor electrical connector and wiring harness retainers.

175. Attach the wiring harness onto the front cover.

176. Connect the wiring harness retainers to the left valve cover.

177. Install the power steering bracket and 5 bolts. Tighten to 18 ft. lbs. (25 Nm).

178. Connect the VCT sensor electrical connectors.

179. Connect the KS and CMP senor electrical connectors.

180. Attach the pin-type retainers.
181. Connect the 8 fuel injector and the 2 radio interference capacitor electrical connectors.
182. Install the 2 engine appearance cover brackets and the 4 nuts.
183. Attach the vacuum hose retainers onto the left engine appearance cover support bracket.
184. Connect the engine wiring harness retainers to the engine appearance cover support brackets.
185. Connect the 8 ignition coil electrical connectors.

➡**Remove all dirt and moisture from the valve cover before applying silicone sealer. Lift up the wire harness and apply a 0.5 inch (12.7 mm) bead of silicone sealer in the notch of the valve cover.**

186. Set the ignition coil harness in the bead of silicone sealer and apply enough additional sealer to surround the wire harness and notch. Tighten to 89 inch lbs. (10 Nm).
187. Install the ignition coil covers. 89 inch lbs. (10 Nm).
188. Install the vacuum harness.
189. Connect the coolant vent hose to the coolant outlet.
190. Connect the EGR system module electrical connector.
191. Connect the fuel rail temperature sensor electrical connector.
192. Connect the EVAP tube.
193. Connect the electronic throttle body electrical connector.
194. Connect the fuel pressure sensor electrical connector.
195. Connect the TP sensor electrical connector.
196. Install the coolant pump pulley and 3 bolts. Tighten to 89 inch lbs. (10 Nm).
197. Install the accessory drive belt idler pulley and bolt. Tighten to 18 ft. lbs. (25 Nm).
198. Install the accessory drive belt tensioner. Tighten to 18 ft. lbs. (25 Nm).

✳✳ CAUTION

A new oil cooler must be installed or severe damage to the engine can occur.

➡**Lubricate the gasket with clean engine oil.**

199. Position the new oil cooler, the new gasket and install the bolt.
200. Rotate the cooler clockwise until the locating pin hits the stop. Tighten to 44 ft. lbs. (60 Nm).

➡**Lubricate the gasket with clean engine oil.**

201. Install a new oil filter and tighten until the seal makes contact. Using an oil filter strap wrench, tighten filter an additional 270 degrees.
202. Connect the oil pressure and oil temperature sensor electrical connectors.
203. Install the engine block drain plug. Tighten to 37 ft. lbs. (50 Nm).
204. Install the engine in the vehicle.
205. Fill the crankcase to the correct level.
206. Fill the cooling system.
207. Recharge the A/C system.
208. Connect the negative battery cable.
209. Start the engine and check for leaks.

Rocker Arms/Shafts

REMOVAL & INSTALLATION

The vehicles covered in this section are not equipped with rocker arms/shafts. The camshaft directly actuates the valves.

Intake Manifold

REMOVAL & INSTALLATION

3.0L Engine

2002–03 MODELS

1. Before servicing the vehicle, refer to the Precautions Section.
2. Drain the cooling system.
3. Relieve the fuel system pressure.
4. Remove or disconnect the following:

- Negative battery cable
- Engine appearance cover
- Air cleaner outlet tube
- Throttle Position (TP) sensor connector
- Idle Air Control (IAC) valve connector
- Accelerator cable
- Cruise control cable
- Cable bracket
- Throttle body coolant bypass hose
- Positive Crankcase Ventilation (PCV) hose
- Evaporative Emissions (EVAP) canister purge hose
- Exhaust Gas Recirculation (EGR) vacuum line
- EGR tube
- Cowl vent screen
- Cross vehicle support bar
- Chassis vacuum supply hose
- EGR pressure transducer
- Fuel pressure sensor shield
- Upper intake manifold vacuum hose
- Intake Manifold Tuning Valve (IMTV) connector
- Exhaust Vacuum Regulator (EVR) valve connector and vacuum line
- Upper intake manifold support brackets
- Upper intake manifold
- Fuel line and bracket
- Fuel pressure sensor vacuum line
- Fuel injector harness connectors
- PCV tube
- Lower intake manifold assembly

To install:

5. Install or connect the following:
- Lower intake manifold assembly. Tighten the bolts in sequence to 89 inch lbs. (10 Nm).

10 Nm (89 lb-ft)

Lower intake manifold torque sequence—2002–03 3.0L engine

9306TG03

Upper intake manifold torque sequence—2002–03 3.0L engines

- PCV tube
- Fuel injector harness connectors
- Fuel pressure sensor vacuum line
- Fuel line and bracket. Tighten the bolt to 89 inch lbs. (10 Nm).
- Upper intake manifold. Tighten the bolts in sequence to 89 inch lbs. (10 Nm).
- Upper intake manifold support brackets. Tighten the bolts to 89 inch lbs. (10 Nm).
- EVR valve connector and vacuum line
- IMTV connector
- Upper intake manifold vacuum hose
- Fuel pressure sensor shield
- EGR pressure transducer
- Chassis vacuum supply hose
- Cross vehicle support bar. Tighten the bolts to 15 ft. lbs. (20 Nm).
- Cowl vent screen
- EGR tube
- EGR vacuum line
- EVAP canister purge hose
- PCV hose
- Throttle body coolant bypass hose
- Cable bracket
- Cruise control cable
- Accelerator cable
- IAC valve connector
- TP sensor connector
- Air cleaner outlet tube
- Engine appearance cover
- Negative battery cable

6. Fill the cooling system.
7. Start the engine and check for leaks.

2004–05 MODELS

1. Before servicing the vehicle, refer to the Precautions Section.
2. Properly relieve the fuel system pressure.
3. Drain the cooling system.
4. Disconnect the negative battery cable.
5. Remove the air cleaner outlet pipe.

6. Remove the cowl panel grill.
7. Remove the seven bolts and the cross-vehicle brace.
8. Disconnect the Throttle Position (TP) sensor and electronic throttle control electrical connectors.
9. Disconnect the brake booster vacuum hose from the upper intake manifold.
10. Disconnect the Evaporative Emissions (EVAP) and Positive Crankcase Ventilation (PCV) tubes from the upper intake manifold.
11. Disconnect and plug the two coolant hoses from the upper intake manifold.
12. Detach the coolant hose and the PCV tube retainer from the upper intake manifold.
13. Disconnect the PCV tube from the PCV valve and position the PCV tube and coolant hose aside.
14. Remove the front upper intake manifold support brace-to-upper intake manifold bolt.
15. Disconnect the two Intake Manifold Tuning (IMT) valve electrical connectors.
16. Disconnect the fuel pressure sensor electrical connector and vacuum hose.
17. Remove three bolts and the fuel pressure sensor shield.
18. Remove the right hand upper intake manifold support brace-to-upper intake manifold bolt.
19. Remove the four upper intake manifold bolts and the upper intake manifold.
20. Remove the fuel line clip.
21. Install the tool 310-D005. Close the tool and push it into the open side of the cage.
22. Separate the fitting and remove the tool.
23. Remove the fuel rail bracket bolt.
24. Disconnect the fuel pressure sensor vacuum tube.
25. Disconnect the fuel rail wiring harness electrical connector.
26. Remove the four bolts and the lower intake manifold.

Use tool 310-D005 to disconnect the fuel line coupling—2004–05 3.0L engines

Upper intake manifold torque sequence—2004–05 3.0L engines

To install:

➡**Clean and inspect all sealing surfaces. Inspect and install new gaskets as necessary.**

27. Position the upper intake manifold and loosely install the 4 bolts.
28. Position the fuel pressure sensor shield and loosely install the two upper intake manifold bolts.
29. Tighten the upper intake manifold bolts in the sequence illustrated to 89 inch lbs. (10 Nm).
30. Install the right hand upper intake manifold support brace-to-upper intake manifold bolt. Tighten to 89 inch lbs. (10 Nm).
31. Install the fuel pressure sensor shield-to-upper intake manifold bolt. Tighten to 53 inch lbs. (6 Nm).
32. Connect the fuel pressure sensor electrical connector and vacuum hose.
33. Connect the two IMT valve electrical connectors.
34. Install the front upper intake manifold support brace-to-upper intake manifold bolt. Tighten to 89 inch lbs. (10 Nm).
35. Position the PCV tube and the coolant hose and attach the PCV tube to the PCV valve.
36. Attach the coolant hose and the PCV tube retainer to the upper intake manifold.
37. Connect the two coolant hoses to the upper intake manifold.
38. Connect the EVAP and PCV tubes to the upper intake manifold.
39. Connect the brake booster vacuum hose to the upper intake manifold.
40. Connect the TP sensor and electronic throttle control electrical connectors.
41. Position the cross-vehicle brace and install the seven bolts. Tighten to 15 ft. lbs. (20 Nm).
42. Install the cowl panel grill.
43. Install the air cleaner outlet pipe.
44. Fill the cooling system.

45. Connect the negative battery cable.
46. Start the engine and check for leaks.

3.9L Engine

2002–03 MODELS

1. Before servicing the vehicle, refer to the Precautions Section.
2. Drain the cooling system.
3. Relieve the fuel system pressure.
4. Remove or disconnect the following:
- Negative battery cable
- Air cleaner outlet tube
- Cowl vent screen
- Cross vehicle support bar
- Accelerator cable
- Cruise control cable
- Intake manifold vacuum hoses
- Exhaust Gas Recirculation (EGR) valve
- Camshaft Position (CMP) sensor connector
- Evaporative Emissions (EVAP) canister purge valve line
- Fuel pressure sensor connector and vacuum line
- Fuel line
- Knock Sensor (KS) connector
- Cylinder head temperature sensor connector
- Sensor connector bracket
- Wiring harness and hose bracket
- Left bank fuel injector connectors
- Idle Air Control (IAC) valve connector
- Throttle Position (TP) sensor connector
- Positive Crankcase Ventilation (PCV) tube
- Throttle body coolant hoses
- Right bank fuel injector connectors
- Delta Pressure Feedback Electronic (DPFE) system sensor
- Intake manifold. Loosen the bolts in reverse of the tightening sequence.

To install:

5. Install or connect the following:
- Intake manifold. Tighten the bolts in sequence to 18 ft. lbs. (25 Nm).
- DPFE system sensor
- Right bank fuel injector connectors
- Throttle body coolant hoses
- PCV tube
- TP sensor connector
- IAC valve connector
- Left bank fuel injector connectors
- Wiring harness and hose bracket
- Sensor connector bracket
- Cylinder head temperature sensor connector

Intake manifold torque sequence—2002–03 3.9L engine

- KS sensor connector
- Fuel line
- Fuel pressure sensor connector and vacuum line
- EVAP canister purge valve line
- CMP sensor connector
- EGR valve
- Intake manifold vacuum hoses
- Cruise control cable
- Accelerator cable
- Cross vehicle support bar. Tighten the bolts to 15 ft. lbs. (20 Nm).
- Cowl vent screen
- Air cleaner outlet tube
- Negative battery cable
6. Fill the cooling system.
7. Start the engine and check for leaks.

2004–05 MODELS

1. Before servicing the vehicle, refer to the Precautions Section.
2. Drain the cooling system.
3. Relieve the fuel system pressure.
4. Disconnect the negative battery cable.

✳✳ CAUTION

The engine appearance cover is equipped with a noise suppressor. If the noise suppressor becomes separated from the engine appearance cover it must be positioned on the intake manifold prior to installing the engine appearance cover.

5. Remove the engine appearance cover and the noise suppressor, if detached.
6. Remove the air cleaner outlet pipe.

7. Remove the cowl panel grille.
8. Disconnect the fuel supply tube spring lock coupling from the fuel rail.
9. Remove the Exhaust Gas Recirculation (EGR) system module.
10. Remove the six bolts and the cross-vehicle brace.
11. Disconnect the Throttle Position (TP) sensor electrical connector.
12. Disconnect the fuel pressure sensor electrical connector and vacuum hose.
13. Disconnect the throttle body electrical connector.
14. Disconnect the Evaporative Emissions (EVAP) hose from the intake manifold.
15. Disconnect the fuel rail temperature sensor electrical connector.
16. Remove the crankcase ventilation tube.
17. Disconnect the vacuum supply tube.
18. Detach the vacuum harness retainers from the left hand engine appearance cover bracket.
19. Detach the engine wiring harness from the engine appearance cover brackets.
20. Remove the four nuts and the two engine appearance cover brackets.
21. Disconnect the eight fuel injector electrical connectors.
22. Disconnect the radio interference ignition capacitor electrical connector.
23. Remove the two nuts and the engine wiring harness retainer.
24. Remove the nut and the radio interference ignition capacitor.
25. Remove the four fuel rail bolts.
26. Remove the fuel rail and injectors as an assembly.

Intake manifold loosening sequence—2004–05 3.9L engines

Intake manifold tightening sequence—2004–05 3.9L engines

06017-LS-G30

27. Remove the clips and the 8 fuel injectors from the fuel rail.

28. Remove and discard the sixteen O-ring seals from the fuel injectors.

29. Disconnect and plug the 2 throttle body coolant tubes.

➡ **It is not necessary to remove the throttle body and adapter to remove the intake manifold.**

30. Remove the seven bolts and three studs in the sequence illustrated.

31. Remove the intake manifold and discard the gaskets.

To install:

❈ **CAUTION**

Do not use metal scrapers, wire brushes, power abrasive discs, or other abrasive means to clean the sealing surfaces. These tools cause scratches and gouges, which make leak paths. Use a plastic scraping tool to clean the surfaces.

32. Clean the gasket sealing surfaces.

33. Install new gaskets, the intake manifold, the seven bolts and the three studs. Tighten in the sequence illustrated to 15 ft. lbs. (20 Nm).

34. Unplug and connect the two throttle body coolant tubes.

❈ **CAUTION**

Use O-ring seals that are made of special fuel-resistant material. Use of ordinary O-rings can cause the fuel system to leak. Do not reuse the O-ring seals.

35. Install sixteen new O-ring seals onto the fuel injectors. Lubricate the O-ring seals with clean engine oil prior to installation.

36. Install the fuel injectors and clips onto the fuel rail.

37. Install the fuel rail and injectors as an assembly. Tighten the fuel rail bolts to 89 inch lbs. (10 Nm).

38. Install the radio interference ignition capacitors and nuts. Tighten to 89 inch lbs. (10 Nm).

39. Install the engine wiring harness retainer and the two nuts. Tighten to 89 inch lbs. (10 Nm).

40. Connect the radio interference ignition capacitor electrical connectors.

41. Connect the fuel injector electrical connectors.

42. Install the engine appearance cover brackets and the nuts.

43. Attach the engine wiring harness to the engine appearance cover brackets.

44. Attach the vacuum harness retainers to the left hand engine appearance cover bracket.

45. Connect the vacuum supply tube.

46. Install the crankcase ventilation tube.

47. Connect the fuel rail temperature sensor electrical connector.

48. Connect the EVAP hose to the intake manifold.

49. Connect the throttle body electrical connector.

50. Connect the fuel pressure sensor electrical connector and vacuum hose.

51. Connect the TP sensor electrical connector. Tighten to 15 ft. lbs. (20 Nm)

52. Install the six bolts and the cross-vehicle brace.

53. Install the EGR system module.

54. Install the air cleaner outlet pipe.

55. Connect the fuel supply tube spring lock coupling to the fuel rail.

❈❈ **CAUTION**

The engine appearance cover is equipped with a noise suppressor. If the noise suppressor becomes separated from the engine appearance cover it must be positioned on the intake manifold prior to installing the engine appearance cover.

56. Install the noise suppressor and the engine appearance cover.

57. Install the cowl panel grille.

58. Fill the cooling system.

59. Connect the negative battery cable.

60. Start the engine and check for leaks.

Exhaust Manifolds

REMOVAL & INSTALLATION

3.0L Engine

1. Before servicing the vehicle, refer to the Precautions Section.

2. Remove or disconnect the following:
   - Negative battery cable
   - Heat shields
   - Lower splash shields
   - Exhaust front pipes
   - Secondary air tubes
   - Exhaust Gas Recirculation (EGR) tube
   - Exhaust manifolds

To install:

3. Install the exhaust manifolds. Tighten the fasteners in sequence as follows:
   a. Step 1: 15 ft. lbs. (20 Nm)
   b. Step 2: 15 ft. lbs. (20 Nm)

4. Install or connect the following:
   - Exhaust Gas Recirculation (EGR) tube

Right exhaust manifold torque sequence—3.0L engine

9306TG09

20 Nm (15 lb-ft)

Left exhaust manifold torque sequence—3.0L engine

- Secondary air tubes
- Exhaust front pipes
- Lower splash shields
- Heat shields
- Negative battery cable

5. Start the engine and check for leaks.

3.9L Engine

RIGHT SIDE

1. Before servicing the vehicle, refer to the Precautions Section.
2. Remove the catalytic converters.
3. Remove the starter motor.
4. Remove the screws and the lower right hand splash shield.
5. Remove the two Exhaust Gas Recirculation (EGR) tube flange nuts.
6. Remove the heat shield stud bolt.
7. Remove the four bolts and the heat shield.
8. Remove the eight bolts, spacers and the exhaust manifold.
9. Discard the exhaust manifold and EGR tube gaskets.

To install:
10. Position the new exhaust manifold gasket, the new EGR tube gasket and the exhaust manifold.

11. Install the spacers and exhaust manifold bolts hand-tight.
12. Install the EGR tube nuts and tighten to 15 ft. lbs. (20 Nm).
13. Tighten the manifold bolts to 18 ft. lbs. (25 Nm) in the sequence illustrated.
14. Install the heat shield and tighten the bolts.
15. Install the heat shield stud bolt and tighten to 37 ft. lbs. (50 Nm).
16. Install the lower splash shield and screws.
17. Install the starter motor.
18. Install the catalytic converters.

LEFT SIDE

1. Before servicing the vehicle, refer to the Precautions Section.
2. Remove the catalytic converters.
3. Remove the dipstick tube.
4. Remove the bolts and position the power steering reservoir aside
5. Remove the screws and the lower left hand splash shield.
6. Remove the heat shield stud bolt.
7. Remove the four bolts and the heat shield.
8. Remove the eight bolts, spacers and the exhaust manifold.
9. Discard the exhaust manifold and EGR tube gaskets.

To install:
10. Position the new exhaust manifold gasket and the exhaust manifold.
11. Install the spacers and tighten the manifold bolts to 18 ft. lbs. (25 Nm) in the sequence illustrated.
12. Install the heat shield and tighten the bolts.
13. Install the heat shield stud bolt and tighten to 37 ft. lbs. (50 Nm).
14. Install the lower splash shield and screws.

Left side exhaust manifold tightening sequence—3.9L engines

15. Position the power steering reservoir and tighten the upper bolt to 53 inch lbs. (6 Nm) and the lower bolt to 9 ft. lbs. (12 Nm).
16. Install the dipstick tube.
17. Install the catalytic converters.

Camshaft and Valve Lifters

REMOVAL & INSTALLATION

3.0L Engine

1. Before servicing the vehicle, refer to the Precautions Section.

Right side exhaust manifold tightening sequence—3.9L engines

2. Remove or disconnect the following:
- Negative battery cable
- Valve covers
- Front cover
- Timing chains

✳✳ WARNING

The camshaft journal thrust caps must be removed before loosening the remaining camshaft journal cap bolts to ensure that the camshaft journal thrust caps are not damaged.

➡**Keep all valvetrain components in order for assembly.**

- Camshaft journal thrust caps
- Remaining camshaft journal caps
- Camshafts
- Valve tappets and shims

To install:

3. Install or connect the following:
- Valve tappets and shims in their original locations
- Camshafts
- Camshaft journal caps in their original positions. Install the thrust journal caps last. Tighten the bolts in sequence to 89 inch lbs. (10 Nm).
- Timing chains
- Front cover
- Valve covers
- Negative battery cable

4. Start the engine and check for leaks.

10 Nm (89 lb-in)

9306TG11

Camshaft journal cap torque sequence— 3.0L engine

3.9L Engine

2002–03 MODELS

1. Before servicing the vehicle, refer to the Precautions Section.
2. Remove or disconnect the following:
- Negative battery cable
- Valve covers
- Front cover
- Timing chains

➡**Keep all valvetrain components in order for assembly.**

- Camshaft journal bearing caps
- Camshafts
- Valve tappets and shims

To install:

3. Install or connect the following:
- Valve tappets and shims
- Camshafts

4. Install the camshaft journal bearing caps in their original positions. Tighten the bolts in sequence as follows:
   a. Step 1: Finger tight
   b. Step 2: 53 inch lbs. (6 Nm)
   c. Step 3: Plus 90 degrees

9306TG12

Camshaft journal bearing cap torque sequence—2002–03 3.9L engine

5. Install or connect the following:
- Timing chains
- Front cover
- Valve covers
- Negative battery cable
6. Start the engine and check for leaks.

2004–05 MODELS

1. Before servicing the vehicle, refer to the Precautions Section.

2. Disconnect the negative battery cable.
3. Remove the bolt and the accessory drive belt tensioner.
4. Remove the bolt and the accessory drive belt idler pulley.
5. Remove the 3 bolts and the coolant pump pulley.

➡**Hold the crankshaft pulley with the tool 303-D055 while removing the bolt and washer.**

6. Remove and discard the bolt and washer.
7. Using the puller tool 303-009, remove the crankshaft damper.
8. Using the tool 303-409, remove and discard the front crankshaft seal.

06017-LS-G10

Hold the crankshaft pulley with the tool 303-D055 while removing the bolt and washer—2004–05 3.9L engine

06017-LS-G11

Using the tool 303-009, remove and discard the front crankshaft seal—2004–05 3.9L engine

06017-LS-G12

Using the tool 303-409, remove and discard the front crankshaft seal—2004–05 3.9L engine

9. Remove the ignition coil covers.

10. Remove all silicone sealer and residue from the ignition coil covers.

11. Remove all silicone sealer and residue from the valve cover surface and ignition coil wiring harness.

12. Disconnect the 8 ignition coil electrical connectors.

13. Disconnect the Throttle Position (TP) sensor electrical connector.

14. Disconnect the fuel pressure sensor electrical connector.

15. Disconnect the electronic throttle body electrical connector.

16. Disconnect the Evaporative Emissions (EVAP) tube.

17. Disconnect the fuel rail temperature sensor electrical connector.

18. Disconnect the Exhaust Gas Recirculation (EGR) system module electrical connector.

19. Disconnect the coolant vent hose from the coolant outlet.

20. Remove the coolant vent hose.

21. Remove the vacuum harness.

22. Disconnect the engine wiring harness retainers from the engine appearance cover support brackets.

23. Detach the vacuum hose retainers from the left hand engine appearance cover support bracket.

24. Remove the 4 nuts and the 2 engine appearance cover brackets.

25. Disconnect the 8 fuel injector and the 2 radio interference capacitor electrical connectors

26. Disconnect the Knock Sensor (KS) and Camshaft Position (CMP) sensor electrical connectors.

27. Detach the pin-type retainers.

28. Disconnect the 2 Variable Cam Timing (VCT) sensor electrical connectors.

29. Remove the 5 bolts and the power steering bracket.

30. Disconnect the wiring harness retainers from the left hand valve cover.

31. Detach the wiring harness from the engine front cover.

32. Disconnect the Crankshaft Position (CKP) sensor electrical connector and wiring harness retainers.

33. Disconnect the wiring harness retainers from the right hand valve cover.

34. Disconnect the generator electrical connector and wiring harness retainers.

35. Remove the Heated Oxygen (HO2S) sensor electrical connectors from the brackets.

36. Remove the engine wiring harness from the engine.

37. Remove the nut and the oil level indicator tube. Discard the O-ring seal.

38. Remove the 8 bolts and the 8 ignition coils.

39. Remove the crankcase vent tube.

40. Remove the 2 EGR system module bolts and loosen the EGR tube nut.

41. Remove the EGR system module and discard the gasket.

42. Remove the 2 nuts and the radio interference capacitors.

43. Disconnect the throttle body heater hoses.

44. Remove the 4 bolts and the coolant bypass housing. Discard the O-ring seals.

45. Remove the 4 bolts and the throttle body adapter. Discard the gasket.

46. Remove the intake manifold. Discard the gaskets.

✳✳ CAUTION

Do not use metal scrapers, wire brushes, power abrasive discs, or other abrasive means to clean the sealing surfaces. These tools cause scratches and gouges, which make leak paths. Use a plastic scraping tool to clean the surfaces.

47. Clean the sealing surfaces.

48. Remove the bolts in the sequence shown and remove the right valve cover. Discard the gasket.

49. Remove the bolts in the sequence shown and remove the left valve cover.

Left valve cover loosening sequence—2004–05 3.9L engine

Right valve cover loosening sequence—2004–05 3.9L engine

Remove the engine front cover bolts in the sequence shown—2004–05 3.9L engine

06017-LS-G13

06017-LS-G14

There is one window on the ignition pulse wheel that is unique to accept the tool 303-645—2004–05 3.9L engine

303-530

06017-LS-G15

Tool 303-530 is used to hold the camshafts during removal—2004–05 3.9L engine

❋❋ CAUTION

Do not use metal scrapers, wire brushes, power abrasive discs, or other abrasive means to clean the sealing surfaces. These tools cause scratches and gouges, which make leak paths. Use a plastic scraping tool to clean the surfaces.

50. Remove the gaskets and all silicone sealer. Clean the mating surfaces with silicone gasket remover.

51. Remove the engine front cover bolts in the sequence shown.

52. Remove the engine front cover. Discard the gaskets.

53. Do not use metal scrapers, wire brushes, power abrasive discs or other abrasive means to clean the sealing surfaces. These tools cause scratches and gouges, which make leak paths. Use a plastic scraping tool to remove all traces of sealant.

54. Clean the sealing surfaces.

55. Remove the right and left hand Variable Cam Timing (VCT) bush carriers.

56. Remove the bolt and the CKP sensor.

➡There is one window on the ignition pulse wheel that is unique to accept the tool 303-645.

57. Install the tool 303-645.

58. Turn the crankshaft to 45 degrees After Top Dead Center (ATDC). The crankshaft keyway will be in the 6 o'clock position.

59. Make sure the flats of the camshaft are facing upward. If not, repeat the previous step.

❋❋ CAUTION

The flats on the camshaft must completely contact the special service tool or damage to the camshafts can occur. Slowly draw the special tool down evenly until it is fully seated on the camshaft flats.

60. Install the holding tool 303-530 on the right hand cylinder head.

❋❋ CAUTION

An open-end wrench must be used on the hex area of the camshaft or damage to the head will occur.

61. Remove and discard the exhaust camshaft sprocket bolt.

62. Remove and discard the variable VCT unit bolt.

63. Remove the right hand timing chain tensioner.

64. Push the lever down and install a drill rod to lock the piston.

65. Remove the 2 bolts and the tensioner.

66. Remove the bolt and the right tensioner arm.

67. Remove the bolt and the right timing chain guide.

68. Remove the right primary timing chain.

69. Remove the holding tool from the right cylinder head.

❋❋ CAUTION

The flats on the camshaft must completely contact the special service tool or damage to the camshafts can occur. Slowly draw the special tool down evenly until it is fully seated on the camshaft flats.

70. Install the holding tool on the left cylinder head.

❋❋ CAUTION

An open-end wrench must be used on the hex area of the camshaft or damage to the cylinder head will occur.

71. Remove and discard the left VCT unit bolt and the exhaust camshaft sprocket bolt.

72. Remove the left timing chain tensioner.

73. Push the lever down and install a drill rod to lock the piston.

74. Remove the 2 bolts and the tensioner.

75. Remove the nut and the left tensioner arm.

76. Remove the bolt and the left timing chain guide.

77. Remove the left primary timing chain.

78. Remove the crankshaft sprocket.

79. Collapse the left secondary timing chain tensioner and install a pin to lock the piston.

80. Remove the left VCT unit, exhaust camshaft sprocket and timing chain as an assembly.

81. Remove the 2 bolts and the left secondary timing chain tensioner.

82. Collapse the right secondary timing chain tensioner and install a pin to lock the piston.

83. Remove the right VCT unit, exhaust camshaft sprocket and timing chain as an assembly.

84. Remove the 2 bolts and the right secondary timing chain tensioner.

85. Remove the holding tool.

86. On the left side:

✳✳ CAUTION

Record the camshaft bearing cap locations. The camshaft bearing caps are positional and must be installed in their original locations and orientation or engine damage can occur.

    a. Remove the left camshaft bearing cap bolts and caps.

    b. Remove the left hand camshafts.

✳✳ CAUTION

Do not use any means of marking the shims other than a permanent-type marker. Any scratches or paint on the shims can result in incorrect lash adjustments and severe engine damage. Record the shim and bucket tappet location. The shim and bucket tappet are positional and if installed in the incorrect location, engine damage can occur.

    c. Record the location of the valve tappets and remove the tappets.

87. On the right side:

✳✳ CAUTION

Record the camshaft bearing cap locations. The camshaft bearing caps are positional and must be installed in their original locations and orientation or engine damage can occur.

    a. Remove the right camshaft bearing cap bolts and caps.

    b. Remove the right camshafts.

✳✳ CAUTION

Do not use any means of marking the shims other than a permanent-type marker. Any scratches or paint on the shims can result in incorrect lash adjustments and severe engine damage. Record the shim and bucket tappet location. The shim and bucket tappet are positional and if installed in the incorrect location, engine damage can occur.

To install:

88. On the right side:

NOTE: Apply clean engine oil to the camshaft journals, the camshaft caps and the camshaft lobes prior to installing the camshafts.

    a. Position the right camshafts on the camshaft journals.

✳✳ CAUTION

Install the camshaft bearing caps. The camshaft bearing caps are positional and must be installed in their original locations and orientation or engine damage can occur.

    b. Install the right camshaft bearing caps.

    c. Install the bearing cap bolts. Tighten the bolts in three stages in the sequence shown.

    d. Step 1: Hand-tighten.

Right camshaft bearing cap tightening sequence—2004–05 3.9L engine

06017-LS-G18

    e. Step 2: Tighten to 53 inch lbs. (6 Nm).

    f. Step 3: Tighten an additional 90 degrees.

89. On the left side:

NOTE: Apply clean engine oil to the camshaft journals, the camshaft caps and the camshaft lobes prior to installing the camshafts.

    a. Position the right camshafts on the camshaft journals.

✳✳ CAUTION

Install the camshaft bearing caps. The camshaft bearing caps are positional and must be installed in their original locations and orientation or engine damage can occur.

    b. Install the right camshaft bearing caps.

    c. Install the bearing cap bolts. Tighten the bolts in three stages in the sequence shown.

    d. Step 1: Hand-tighten.

    e. Step 2: Tighten to 53 inch lbs. (6 Nm).

    f. Step 3: Tighten an additional 90 degrees.

✳✳ CAUTION

If any components are installed new, the engine must be reshimmed or engine damage can occur.

90. Using a feeler gauge, confirm that the tappet and shim clearances are within specification.

Left camshaft bearing cap tightening sequence—2004–05 3.9L engine

06017-LS-G19

Position all of the camshafts so the flats of the camshafts are facing upward— 2004–05 3.9L engine

06017-LS-G20

91. If tappet and shim clearances are not within specification, adjust the clearance.

92. Turn the crankshaft to 45 degrees ATDC. The crankshaft keyway will be in the 6 o'clock position.

93. Install the tool 303-645.

94. Position all of the camshafts so the flats of the camshafts are facing upward.

95. Install the left secondary timing chain tensioner and 2 bolts. Tighten to 8 ft. lbs. (11 Nm).

96. Install the left VCT unit, exhaust camshaft sprocket and timing chain as an assembly.

97. Using new bolts, install the left VCT unit and exhaust camshaft sprocket bolts finger-tight.

98. Remove the locking pin from the left secondary timing chain tensioner.

99. Install the right secondary timing chain tensioner and 2 bolts. Tighten to 8 ft. lbs. (11 Nm).

100. Install the right VCT unit, exhaust camshaft sprocket and timing chain as an assembly.

101. Using new bolts, install the right VCT unit and exhaust camshaft sprocket bolts finger-tight.

102. Remove the locking pin from the right secondary timing chain tensioner.

103. Reset the timing chain tensioners by holding the lever toward the tensioner piston. Using finger pressure, compress the tensioner. Push the lever down and install a drill rod to lock the piston.

> ✳✳ **CAUTION**
>
> **The flats on the camshaft must completely contact the special service tool or damage to the camshafts can occur. Slowly draw the special tool down evenly until it is fully seated on the camshaft flats.**

104. Install the holding tool on the left head.

105. Position the left timing chain and crankshaft gear.

106. Position the timing chain over the left intake camshaft sprocket.

107. Position the crankshaft gear in the timing chain.

108. Install the left timing chain guide and bolt. Tighten to 8 ft. lbs. (11 Nm).

109. Install the left timing chain tensioner arm and nut. Tighten to 8 ft. lbs. (11 Nm).

110. Install the left timing chain tensioner.

111. Install the timing chain tensioner and bolts, and remove the drill rod.

> ✳✳ **CAUTION**
>
> **The exhaust camshaft sprocket bolt must be fully tightened before tightening the intake camshaft sprocket bolt. Using the tool, apply 89 inch lbs. (10 Nm) counterclockwise to the left exhaust camshaft sprocket.**

112. Tighten the left exhaust camshaft sprocket and VCT unit bolts 26 ft. lbs. (35 Nm) then an additional 90 degrees.

113. Remove the holding tool.

> ✳✳ **CAUTION**
>
> **The flats on the camshaft must completely contact the special service tool or damage to the camshafts can occur. Slowly draw the special tool downward evenly until it is fully seated on the camshaft flats.**

114. Install the holding tool on the right cylinder head.

115. Position the right timing chain over the right intake camshaft sprocket and over the crankshaft gear.

116. Install the right timing chain guide and bolt. Tighten to 8 ft. lbs. (11 Nm).

117. Install the right timing chain tensioner arm. Tighten to 8 ft. lbs. (11 Nm).

118. Install the right timing chain tensioner.

119. Install the timing chain tensioner and bolts, and remove the drill rod.

> ✳✳ **CAUTION**
>
> **The exhaust camshaft sprocket bolt must be fully tightened before tightening the intake camshaft sprocket bolt. Using the tool, apply 89 inch lbs. (10 Nm) counterclockwise to the left exhaust camshaft sprocket.**

120. Tighten the right exhaust camshaft sprocket and VCT unit bolts 26 ft. lbs. (35 Nm) then an additional 90 degrees.

Using the tool, apply 89 inch lbs. (10 Nm) counterclockwise to the camshaft sprocket—2004–05 3.9L engine

06017-LS-G21

121. Remove the holding tool.

➡ **Lubricate the VCT bush carrier seals with clean engine oil.**

122. Carefully insert the bush carrier into the VCT unit, compress ring seals with fingers as required. Fully seat the bush carrier by hand before fitting the bolts and nut. Tighten the bolts to 15 ft. lbs. (20 Nm). Tighten the nut to 89 inch lbs. (10 Nm).

123. Install new engine front cover gaskets.

➡ **If the front cover is not secured within 4 minutes of sealant application, the sealant must be removed and the sealing area cleaned with metal surface cleaner. Allow to dry until there is no sign of wetness, or 4 minutes, whichever is longer. Failure to follow this procedure can cause future oil leakage.**

124. Apply silicone gasket and sealant in the 8 places shown.

125. Install the engine front cover. Loosely install the bolts.

126. Tighten the bolts in 2 stages in the sequence shown. Tighten to 44 inch lbs. (5 Nm) in sequence and then retighten in sequence to 89 inch lbs. (10 Nm).

➡ **Lubricate the front oil seal and the engine front cover mating surface with clean engine oil.**

127. Using the tool 303-646, install a new front crankshaft seal.

➡ **If crankshaft pulley is not secured within 4 minutes of sealant application the sealant must be removed and the sealing area cleaned with metal surface cleaner. Allow to dry until there is no sign of wetness, or 4 minutes, whichever is longer. Failure to follow this procedure can cause future oil leakage.**

3 mm (0.12 in)

06017-LS-G22

Apply silicone gasket and sealant in the 8 places shown—2004–05 3.9L engine

06017-LS-G23

Engine front cover torque sequence—2004–05 3.9L engine

06017-LS-G24

Using the tool 303-646, install a new front crankshaft seal—2004–05 3.9L engine

06017-LS-G25

Using the tool 303-102, install the crankshaft pulley—2004–05 3.9L engine

128. Apply a bead of silicone gasket and sealant to the keyway.

129. Using the tool 303-102, install the crankshaft pulley.

130. Install a new bolt and using the special tool to hold the pulley, tighten as follows:.

    a. Step 1: Tighten to 59 ft. lbs. (80 Nm).

    b. Step 2: Loosen the bolt 2 complete turns.

    c. Step 3: Tighten to 37 ft. lbs. (50 Nm).

    d. Step 4: Tighten an additional 90 degrees.

131. On the left side:

    a. Install a new gasket, left exhaust manifold, spacers and bolts.

    b. Install the left exhaust manifold heat shield and 4 bolts.

    c. Install the left exhaust manifold heat shield stud bolt.

132. On the right side:

    a. Install a new gasket, right exhaust manifold, spacers and bolts.

    b. Install the right exhaust manifold heat shield and 4 bolts.

    c. Install the right exhaust manifold heat shield stud bolt.

    d. Install a new gasket, EGR tube and 2 nuts. Tighten to 15 ft. lbs. (20 Nm).

133. Install new valve cover gaskets.

➡If the valve covers are not secured within 4 minutes of sealant application the sealant must be removed and the

06017-LS-G26

Apply 4 beads of silicone gasket and sealant as shown—2004–05 3.9L engine

sealing area cleaned with metal surface cleaner. Allow to dry until there is no sign of wetness, or 4 minutes, whichever is longer. Failure to follow this procedure can cause future oil leakage.

134. Apply 4 beads of silicone gasket and sealant as shown.

135. Install the left valve cover and tighten the bolts in sequence to 89 inch lbs. (10 Nm).

136. Install the right valve cover and

Left valve cover torque sequence—2004–05 3.9L engine

06017-LS-G08

Right valve cover torque sequence—2004–05 3.9L engine

06017-LS-G09

tighten the bolts in sequence to 89 inch lbs. (10 Nm).

137. Install new intake manifold gaskets and the manifold.

138. Install the intake manifold, the 7 bolts and the 3 studs.

139. Install a new gasket, the throttle body adapter and the 4 bolts. Tighten to 89 inch lbs. (10 Nm).

140. Install new O-ring seals, the coolant bypass housing and the 4 bolts. Tighten to 89 inch lbs. (10 Nm).

141. Connect the 2 throttle body coolant hoses.

142. Install the 2 radio interference ignition capacitors and the 2 nuts. Tighten to 89 inch lbs. (10 Nm).

143. Install a new gasket, the EGR system module and bolts. Tighten to 18 ft. lbs. (25 Nm).

144. Tighten the EGR tube nut to 30 ft. lbs. (40 Nm).

145. Install the crankcase vent tube.

146. Install the 8 ignition coils and bolts. Tighten to 44 inch lbs. (5 Nm).

✳✳ CAUTION

Make sure the O-ring seal remains in position and does not roll up the indicator tube during installation.

147. Install a new O-ring seal, oil level indicator tube and nut. Lubricate the O-ring seal with clean engine oil prior to installation and tighten the nut to 53 inch lbs. (6 Nm).

148. Install the CKP sensor and bolt. Tighten to 89 inch lbs. (10 Nm).

149. Position the engine wiring harness onto the engine.

150. Attach the (HO_2S) sensor electrical connectors onto the brackets.

151. Connect the generator electrical connector and wiring harness retainers.

152. Connect the wiring harness retainers to the right valve cover.

153. Connect the CKP sensor electrical connector and wiring harness retainers.

154. Attach the wiring harness onto the front cover.

155. Connect the wiring harness retainers to the left valve cover.

156. Install the power steering bracket and 5 bolts. Tighten to 18 ft. lbs. (25 Nm).

157. Connect the VCT sensor electrical connectors.

158. Connect the KS and CMP senor electrical connectors.

159. Attach the pin-type retainers.

160. Connect the 8 fuel injector and the 2 radio interference capacitor electrical connectors.

161. Install the 2 engine appearance cover brackets and the 4 nuts.

162. Attach the vacuum hose retainers onto the left engine appearance cover support bracket.

163. Connect the engine wiring harness retainers to the engine appearance cover support brackets.

164. Connect the 8 ignition coil electrical connectors.

➡Remove all dirt and moisture from the valve cover before applying silicone sealer. Lift up the wire harness and apply a 0.5 inch (12.7 mm) bead of silicone sealer in the notch of the valve cover.

165. Set the ignition coil harness in the bead of silicone sealer and apply enough additional sealer to surround the wire harness and notch. Tighten to 89 inch lbs. (10 Nm).

166. Install the ignition coil covers. 89 inch lbs. (10 Nm).

167. Install the vacuum harness.

168. Connect the coolant vent hose to the coolant outlet.

169. Connect the EGR system module electrical connector.

170. Connect the fuel rail temperature sensor electrical connector.

171. Connect the EVAP tube.

172. Connect the electronic throttle body electrical connector.

173. Connect the fuel pressure sensor electrical connector.

174. Connect the TP sensor electrical connector.

175. Install the coolant pump pulley and 3 bolts. Tighten to 89 inch lbs. (10 Nm).

176. Install the accessory drive belt idler pulley and bolt. Tighten to 18 ft. lbs. (25 Nm).

177. Install the accessory drive belt tensioner. Tighten to 18 ft. lbs. (25 Nm).

178. Fill the crankcase to the correct level.

179. Fill the cooling system.

180. Recharge the A/C system.

181. Connect the negative battery cable.

182. Start the engine and check for leaks.

Valve Lash

ADJUSTMENT

3.0L Engine

2002–03 MODELS

1. Before servicing the vehicle, refer to the Precautions Section.

Camshaft lift tools 303-659 needed to adjust the valve lash—2002–03 3.0L engines

2. Remove or disconnect the following:
- Negative battery cable
- Engine appearance covers
- Ignition coils
- Valve covers

3. Measure the valve clearance while the camshaft lobe is pointed away from the valve shim. Rotate the crankshaft as necessary for each valve to be measured.

➡Keep all valvetrain components in order for assembly.

4. Remove the camshaft thrust cap and the rear camshaft bearing journal cap from the camshaft that requires shim adjustment.

5. Install Service Tool set 303-659 in place of the bearing journal caps, with the taller tool in place of the rear bearing cap.

6. Remove the center camshaft bearing journal caps.

7. Remove the valve shims with compressed air and replace as required to achieve the correct adjustment.

8. Valve clearance should be 0.007–0.009 in. (0.18–0.23mm) for intake valves or 0.013–0.015 in. (0.33–0.38mm) for exhaust valves.

9. Install the center bearing journal caps.

10. Remove the service tools.

11. Install or connect the following:
- Rear bearing journal cap and the thrust journal cap. Tighten the bearing journal caps in sequence to 89 inch lbs. (10 Nm).
- Valve covers
- Ignition coils
- Engine appearance covers
- Negative battery cable

2004–05 MODELS

1. Before servicing the vehicle, refer to the Precautions Section.

2. Disconnect the negative battery cable.

3. Remove the timing chains.

➡Measure each valve clearance at base circle before removing the camshafts. The shims are not repairable with the camshafts in place. Failure to measure all clearances prior to removing the camshafts will cause unnecessary repetition of the procedure.

4. Use a feeler gauge to measure each valve clearance and record its location.

5. Remove the camshafts and shims.

➡The shims are marked for thickness; example: 2.22 mm = 222 on shim.

6. The corrected shims allow the following valve clearances.
   a. Intake valve clearance: 0.00709–0.00866 inch (0.18-0.22 mm).
   b. Exhaust valve clearance: 0.0128–0.0148 inch (0.33–0.38 mm).

➡A midrange clearance of 0.00787 inch (0.20 mm) on the intake side and 0.014 inch (0.35 mm) is the most desirable:

➡Select shims using this formula: required shim thickness = measured clearance plus the base shim thickness minus most desirable clearance.

7. Select shims and mark the installation location.

8. Install new shims.

9. Install the camshafts.

10. Measure the new valve clearances.

11. Install the timing chains

12. Disconnect the negative battery cable.

3.9L Engine

2002–03 MODELS

1. Before servicing the vehicle, refer to the Precautions Section.

2. Remove or disconnect the following:
- Negative battery cable
- Engine appearance covers
- Ignition coils
- Valve covers

3. Measure the valve clearance while the camshaft lobe is pointed away from the valve shim. Rotate the crankshaft as necessary for each valve to be measured.

4. Valve clearance should be 0.007–0.009 in. (0.18–0.23mm) for intake valves or 0.009–0.011 in. (0.23–0.28mm) for exhaust valves.

5. If adjustment is necessary, compress the valves with the special tools and remove the shim with compressed air. Repeat for each valve to be adjusted.

6. Install or connect the following:
- Valve covers

Valve adjustment tool base plate—2002–03 3.9L engine

Valve adjustment tool attachment—2002–03 3.9L engine

Remove the shims with compressed air—2002–03 3.9L engine

- Ignition coils
- Engine appearance covers
- Negative battery cable

2004–05 MODELS

1. Before servicing the vehicle, refer to the Precautions Section.
2. Disconnect the negative battery cable.
3. Remove the timing chains.

➡**Measure each valve clearance at base circle before removing the camshafts. The shims are not repairable with the camshafts in place. Failure to measure all clearances prior to removing the camshafts will cause unnecessary repetition of the procedure.**

4. Use a feeler gauge to measure each valve clearance and record its location.
5. Remove the camshafts and shims.

➡**The shims are marked for thickness; example: 2.22 mm = 222 on shim.**

6. The corrected shims allow the following valve clearances.
   a. Intake valve clearance: 0.00709–0.00866 inch (0.18-0.22 mm).
   b. Exhaust valve clearance: 0.0128–0.0148 inch (0.33–0.38 mm).

➡**A midrange clearance of 0.00787 inch (0.20 mm) on the intake side and 0.014 inch (0.35 mm) is the most desirable:**

➡**Select shims using this formula: required shim thickness = measured clearance plus the base shim thickness minus most desirable clearance.**

7. Select shims and mark the installation location.
8. Install new shims.
9. Install the camshafts.
10. Measure the new valve clearances.
11. Install the timing chains.
12. Disconnect the negative battery cable.

Starter Motor

REMOVAL & INSTALLATION

1. Before servicing the vehicle, refer to the Precautions Section.
2. Remove or disconnect the following:
   - Negative battery cable
   - Starter motor wiring connectors
   - Starter motor

To install:
3. Install or connect the following:

1 Ground strap nut
2 Ground strap
3 Starter motor solenoid terminal cover
4 Starter solenoid battery cable nut
5 Starter solenoid battery cable
6 Starter solenoid wire nut
7 Starter solenoid wire
8 Starter motor bolt
9 Starter motor stud bolts
10 Starter motor

67197-LILS-G07

Starter motor mounting—3.0L engine

1 Starter motor solenoid terminal cover
2 Starter solenoid wire nut
3 Starter solenoid wire
4 Starter solenoid battery cable nut
5 Starter solenoid battery cable
6 Ground cable nut
7 Ground cable
8 Starter motor stud bolt
9 Starter motor bolt
10 Starter motor

67197-LILS-G08

Starter motor mounting—3.9L engine

- Starter motor. Tighten the bolts to 18 ft. lbs. (25 Nm).
- Starter motor wiring connectors
- Negative battery cable

Oil Pan

REMOVAL & INSTALLATION

3.0L Engine

1. Before servicing the vehicle, refer to the Precautions Section.
2. Drain the engine oil.
3. Install a support fixture to the engine lifting eyes.
4. Remove or disconnect the following:
   - Negative battery cable
   - Air cleaner outlet tube
   - Accessory drive belt
   - Alternator
   - Center and right splash shields
   - A/C compressor
   - Electronic Thermactor Air (ETA) bracket, if equipped
   - Rack and pinion steering gear
   - Lower control arm through bolts
   - Transmission cooler line bracket
   - Engine mount nuts
   - Left and right subframe bolts. Pry the subframe down for clearance.
   - Wiring harness bracket
   - Oil pan

To install:

5. Install or connect the following:
   - Oil pan. Tighten the pan bolts in sequence to 18 ft. lbs. (25 Nm) and the transmission bolts to 35 ft. lbs. (47 Nm).
   - Wiring harness bracket
   - Left and right subframe bolts. Tighten the bolts to 76 ft. lbs. (103 Nm).
   - Engine mount nuts. Tighten the nuts to 46 ft. lbs. (63 Nm).
   - Transmission cooler line bracket
   - Lower control arm through bolts. Tighten the bolts to 129 ft. lbs. (175 Nm).
   - Rack and pinion steering gear. Tighten the nuts to 76 ft. lbs. (103 Nm).
   - ETA bracket, if equipped
   - A/C compressor
   - Center and right splash shields
   - Alternator
   - Accessory drive belt
   - Air cleaner outlet tube
   - Negative battery cable

6. Fill the crankcase to the correct level.
7. Start the engine and check for leaks.

Oil pan torque sequence—3.0L engine

25 Nm (18 lb-ft)

9306TG14

Oil pan torque sequence—3.9L engine

9306TG15

3.9L Engine

1. Before servicing the vehicle, refer to the Precautions Section.
2. Drain the engine oil.
3. Remove the oil pan.

To install:

4. Install the oil pan. Tighten the bolts in sequence as follows:
   a. Step 1: 44 inch lbs. (5 Nm)
   b. Step 2: 108 inch lbs. (12 Nm)
5. Fill the crankcase to the correct level.
6. Start the engine and check for leaks.

Oil Pump

REMOVAL & INSTALLATION

3.0L Engine

2002–03 MODELS

1. Before servicing the vehicle, refer to the Precautions Section.
2. Remove or disconnect the following:

- Upper intake manifold
- Valve covers
- Accessory drive belt
- Power steering pump
- Alternator
- Water pump
- A/C compressor and bracket
- Crankshaft pulley
- Oil pan
- Oil pump screen and tube
- Front cover
- Timing chains
- Crankshaft timing gears
- Oil pump

To install:

3. Install or connect the following:
- Oil pump. Tighten the bolts to 89 inch lbs. (10 Nm).
- Crankshaft timing gears
- Timing chains
- Front cover
- Oil pump screen and tube. Tighten the bolts to 89 inch lbs. (10 Nm).
- Oil pan
- Crankshaft pulley
- A/C compressor and bracket
- Water pump
- Alternator
- Power steering pump
- Accessory drive belt
- Valve covers
- Upper intake manifold
4. Fill the crankcase.
5. Start the engine and check for leaks.

5 — 10 Nm (89 lb-in)

4

3

2

10 Nm (89 lb-in) — 1

1 Screen and pick-up tube mounting bolts
2 Screen and pick-up tube
3 Screen and pick-up tube O-ring seal

4 Oil pump mounting bolts
5 Oil pump

Exploded view of the oil pump assembly—2002–03 3.0L engine

67197-LILS-G09

2004–05 MODELS

1. Before servicing the vehicle, refer to the Precautions Section.

2. Disconnect the negative battery cable.

3. Drain the cooling system.

4. Remove the valve covers.

5. Remove the accessory drive belt tensioner and idler pulleys.

6. Remove the water pump.

7. Disconnect the upper and lower radiator hoses.

8. Remove the oil pan.

9. Remove the power steering pump.

10. Remove the crankshaft front seal.

11. Disconnect the two Camshaft Position (CMP) sensors and the Crankshaft Position (CKP) sensor electrical connectors.

12. Remove fifteen bolts, the stud bolt and the engine front cover. Discard the gaskets.

13. Remove the timing chain components.

14. Remove the two bolts and the oil pump screen and pickup tube.

15. Remove the four oil pump bolts in the sequence illustrated. Remove the pump.

To install:

16. Position the oil pump and install the bolts. Tighten in the sequence shown to 89 inch lbs. (10 Nm).

➥Inspect and install a new oil pump screen and pickup tube O-ring seal as necessary. Lubricate the oil pump

screen and pickup tube O-ring seal with clean engine oil.

17. Position the oil pump screen and pickup tube and install the bolts. Tighten to 89 inch lbs. (10 Nm).

18. Install the timing chain components.

❋❋ CAUTION

Do not use metal scrapers, wire brushes, power abrasive discs or other abrasive means to clean the sealing surfaces. These tools cause scratches and gouges which make leak paths.

19. Use a plastic scraping tool to remove all traces of sealant from the front cover. Clean all sealing surfaces with silicone gasket remover and metal surface prep.

20. Install a new gasket on the engine front cover.

➥The engine front cover must be installed and the bolts tightened within four minutes of applying sealant.

21. Apply a 0.23 inch (6 mm) diameter dot of silicone gasket and sealer to the cylinder block, the lower cylinder block and the cylinder head mating surfaces.

06017-LS-G35

Apply a 0.23 inch (6 mm) diameter dot of silicone gasket and sealer to the cylinder block, the lower cylinder block and the cylinder head mating surfaces—2004–05 3.0L engines

06017-LS-G33

Oil pump loosening sequence—2004–05 3.0L engines

06017-LS-G34

Oil pump tightening sequence—2004–05 3.0L engines

06017-LS-G36

Front cover torque sequence—2004–05 3.0L engines

➡**Number 14 is a stud and washer assembly.**

22. Position the engine front cover and install the bolts and the stud bolt. Tighten in the sequence illustrated to 18 ft. lbs (25 Nm).

23. Connect the CMP sensors and the CKP sensor electrical connectors.

24. Install the crankshaft front seal

25. Install the power steering pump.

26. Install the oil pan.

27. Connect the radiator hoses.

28. Install the water pump.

29. Install the accessory drive belt tensioner and idler pulleys.

30. Install the valve covers.

31. Fill and bleed the cooling system.

32. Fill the engine with correct type and amount of oil.

33. Connect the negative battery cable.

34. Start the engine and check for leaks.

3.9L Engine

1. Before servicing the vehicle, refer to the Precautions Section.

2. Remove or disconnect the following:
- Crankshaft pulley
- Front cover
- Primary timing chains
- Oil pump mounting bolts
- Oil pump

To install:

3. Install or connect the following:
- New gasket
- Oil pump. Tighten the bolts to 53 inch lbs. (6 Nm) plus 90 degrees.
- Primary timing chains
- Front cover
- Crankshaft pulley

Rear Main Seal

REMOVAL & INSTALLATION

3.0L Engine

1. Before servicing the vehicle, refer to the Precautions Section.

2. Attach an engine support fixture to the engine lifting eyes.

06017-LS-G37

Removing the rear crankshaft seal—3.0L engines

06017-LS-G38

Installing the rear crankshaft seal—3.0L engines

3. Remove or disconnect the following:
- Negative battery cable
- Transmission
- Clutch, if equipped
- Flywheel
- Rear crankshaft seal using the tools illustrated.

To install:

4. Install or connect the following:
- Rear main seal flush with the cylinder block surface using the tools illustrated
- Flywheel. Tighten the bolts to 59 ft. lbs. (80 Nm).
- Clutch, if equipped
- Transmission
- Negative battery cable

5. Start the engine and check for leaks.

3.9L Engine

1. Before servicing the vehicle, refer to the Precautions Section.

06017-LS-G39

Removing the rear crankshaft seal—3.9L engines

1	Oil pan bolts	5	Oil pump screen and pickup tube
2	Oil pan	6	Oil pump bolts
3	Oil pan gasket	7	Oil pump
4	Oil pump screen and pickup tube bolts	8	Oil pump gasket

67197-LILS-G10

Oil pump mounting—3.9L engine

2. Attach an engine support fixture to the engine lifting eyes.

3. Remove or disconnect the following:
- Negative battery cable
- Transmission
- Flywheel
- Rear crankshaft seal using the tools illustrated.

To install:

4. Install or connect the following:

✳✳ CAUTION

The plastic sleeve that comes with the new crankshaft rear oil seal, protects the seal while in transit and is also used as an installation tool to prevent lip seal roll when carrying out the repair. The sleeve must be used to prevent future seal concerns.

➡**Lubricate the cylinder block bore and the crankshaft with clean engine oil before installing seal.**

- Position the rear oil seal with the protective installer onto the end of the crankshaft
- Rear main seal flush with the cylinder block surface using the tools illustrated
- Flywheel

5. Tighten the flywheel bolts in a crossing pattern as follows:

06017-LS-G40

Position the rear oil seal with the protective installer onto the end of the crankshaft—3.9L engines

303-647

06017-LS-G41

Installing the rear crankshaft seal—3.9L engines

---

a. Step 1: 11 ft. lbs. (15 Nm)
b. Step 2: 81 ft. lbs. (110 Nm)

6. Install or connect the following:
- Transmission
- Negative battery cable

7. Start the engine and check for leaks.

Timing Chain, Sprockets, Front Cover and Seal

REMOVAL AND INSTALLATION

3.0L Engine

2002–03 MODELS

1. Before servicing the vehicle, refer to the Precautions Section.

2. Remove or disconnect the following:
- Upper intake manifold
- Valve covers
- Accessory drive belt
- Power steering pump

- Alternator
- Water pump
- A/C compressor and bracket
- Crankshaft pulley
- Oil pan
- Oil pump screen and tube
- Crankshaft Position (CKP) sensor connector
- Camshaft Position (CMP) sensor connector
- Front cover
- CKP sensor pulse ring

3. Rotate the crankshaft so that the keyway is at the 11 o'clock position to locate the crankshaft at Top Dead Center (TDC) for No. 1 cylinder.

4. Verify that the alignment arrows on the camshafts are aligned. If not, rotate the crankshaft 1 complete revolution and recheck.

5. Rotate the crankshaft so that the keyway is at the 3 o'clock position. This positions the right cylinder head camshafts to the neutral position.

FRONT OF ENGINE

1 RH Exhaust Camshaft	6 Timing Chain Tensioner
2 RH Intake Camshaft	7 Bolt (2 Req'd)
3 RH Timing Chain Crankshaft Sprocket	8 Timing Chain Tensioner Arm
4 RH Timing Chain	9 Timing Chain Guide
5 Bolt (2 Req'd)	10 RH Cylinder Head
	A Tighten to 20-30 N·m (15-22 Lb-Ft)

7922KG49

Exploded view of the right cylinder head timing chain and related components—2002–03 3.0L engine—left side similar

6. Remove or disconnect the following:
- Right timing chain tensioner
- Right timing chain tensioner arm
- Right timing chain and crankshaft sprocket

✳✳ WARNING

The camshaft thrust caps must be removed before loosening the remaining camshaft journal cap bolts to ensure that the thrust caps are not damaged.

➡ **The camshaft journal caps and cylinder heads are numbered to ensure that they are assembled in their original positions.**

- Camshaft thrust caps
- Camshaft journal caps. Loosen the bolts in sequence and in several passes to allow the camshaft to be raised from the cylinder head evenly.
- Right bank camshafts

7. Rotate the crankshaft 2 revolutions and locate the crankshaft keyway at the 11 o'clock position. This will position the left cylinder head camshafts to their neutral position.

8. Verify that the alignment arrows on the camshafts are aligned.

9. Remove or disconnect the following:
- Left cylinder head timing chain tensioner
- Left timing chain tensioner arm
- Left timing chain and crankshaft sprocket

✳✳ WARNING

The camshaft thrust caps must be removed before loosening the remaining camshaft journal cap bolts to ensure that the thrust caps are not damaged.

➡ **The camshaft journal caps and cylinder heads are numbered to ensure that they are assembled in their original positions.**

10. Remove or disconnect the following:
- Camshaft thrust caps
- Camshaft journal caps. Loosen the bolts in sequence and in several passes to allow the camshaft to be raised from the cylinder head evenly.
- Left bank camshafts

To install:

11. Prepare the timing chain tensioners for installation as follows:

Using a thin prytool, release and hold the timing chain tensioner ratchet/pawl mechanism—2002–03 3.0L engine

Retain the piston with a 1.5mm wire or paperclip—2002–03 3.0L engine

a. Place the left chain tensioner in a vise.

b. Using a small prytool, release and hold the timing chain tensioner ratchet/pawl mechanism through the access hole in the timing chain tensioner.

c. Slowly compress the tensioner.

d. Lock the piston with a 1.5mm wire or paperclip.

e. Repeat for the right chain tensioner.

➡ **Be sure that the crankshaft keyway is still at the 11 o'clock position.**

12. Install or connect the following:
- Left bank camshafts
- Camshaft journal caps
- Camshaft thrust caps
- Left timing chain and crankshaft sprocket. Align the colored links with the index marks on the camshaft and crankshaft sprockets.
- Left timing chain tensioner arm
- Left timing chain tensioner. Tighten the retaining bolts to 15–22 ft. lbs. (20–30 Nm).

13. Remove the retaining wire from the left timing chain tensioner.

14. Rotate the crankshaft so that the keyway is in the 3 o'clock position.

15. Install or connect the following:
- Right bank camshafts
- Camshaft journal caps
- Camshaft thrust caps
- Right timing chain and crankshaft sprocket. Align the colored links with the index marks on the camshaft and crankshaft sprockets.
- Right timing chain tensioner arm
- Right timing chain tensioner. Tighten the retaining bolts to 15–22 ft. lbs. (20–30 Nm).

16. Remove the retaining wire from the right timing chain tensioner.

17. Install the CKP sensor pulse ring.

18. Replace the crankshaft seal in the

RH CAMSHAFT TIMING INDEX MARKS

11 O'CLOCK POSITION (TDC NO. 1 CYLINDER)

CRANKSHAFT TIMING INDEX MARK

*NOTE: AT TDC NO. 1 FIRING POSITION, ARROWS ON BACK OF CAMSHAFT SPROCKETS POINT DIRECTLY AT EACH OTHER

Timing mark alignment—2002–03 3.0L engine

VIEW A

VIEW A

VIEW A

VIEW A

VIEW A

VIEW A

APPLY 3 mm (0.118 INCH)
BEAD OF SEALANT

**VIEW A
TYPICAL 6 PLACES**

7922JG29

Apply sealant to the places indicated—2002–03 3.0L engine

FRONT OF ENGINE

* LOCATION OF STUDS
FOUR PLACES

● TIGHTEN BOLTS/STUDS IN
SEQUENCE SHOWN FINGER-TIGHT

* STUD LOCATIONS

7922KG52

Front cover torque sequence—2002–03 3.0L engine

front cover with a new one. Apply clean engine oil to the seal lip.

19. Apply silicone sealer to the 6 critical areas shown in View **A**, to the cylinder block.

20. Place new front cover gaskets onto the dowel pins on the cylinder block and heads.

21. Place the front cover into position.

22. Install the 6 front cover retaining bolts and stud bolts where the silicone sealer was applied.

23. Tighten the bolts and stud bolts until the front cover contacts the cylinder block and heads and, then turn the bolts and stud bolts an additional ¼ turn.

24. Install the remaining front cover retaining bolts and stud bolts.

25. Tighten all of the front cover retaining bolts and stud bolts in sequence to 15–22 ft. lbs. (20–30 Nm).

26. Install or connect the following:
   • CMP sensor connector
   • CKP sensor connector
   • Oil pump screen and tube
   • Oil pan

27. Install the crankshaft damper and tighten the bolt as follows:
   a. Step 1: 78–99 ft. lbs. (105–135 Nm).
   b. Step 2: Loosen one full turn.
   c. Step 3: 35–39 ft. lbs. (47–53 Nm).
   d. Step 4: Plus an 85–95 degrees.

28. Install or connect the following:
   • A/C compressor and bracket
   • Water pump
   • Alternator
   • Power steering pump
   • Accessory drive belt
   • Valve covers
   • Upper intake manifold

29. Start the engine and check for leaks.

2004–05 MODELS

1. Before servicing the vehicle, refer to the Precautions Section.

2. Disconnect the negative battery cable.

3. Drain the cooling system.

4. Remove the valve covers.

5. Remove the accessory drive belt tensioner and idler pulleys.

6. Remove the water pump.

7. Disconnect the upper and lower radiator hoses.

8. Remove the oil pan.

9. Remove the power steering pump.

10. Remove the crankshaft front seal.

11. Disconnect the two Camshaft Position (CMP) sensors and the Crankshaft Position (CKP) sensor electrical connectors.

12. Remove fifteen bolts, the stud bolt

Remove the ignition pulse wheel—
2004–05 3.0L engines

Rotate the crankshaft clockwise to position
the crankshaft keyway to the 9 o'clock
position—2004–05 3.0L engines

Rotate the crankshaft clockwise to position
the crankshaft keyway to the 3 o'clock
position—2004–05 3.0L engines

Verify that the front 2 intake camshaft
lobes are pointing up—2004–05 3.0L
engines

and the engine front cover. Discard the gaskets.

13. Remove the ignition pulse wheel.
14. Install the crankshaft damper bolt
and washer.

Compress the timing chain tensioner and
install a stiff wire or paper clip—2004–05
3.0L engines

Remove the bolts and the right hand Variable Camshaft Timing (VCT) solenoid
housing assembly—2004–05 3.0L engines

15. Rotate the crankshaft clockwise to
position the crankshaft keyway to the 9
o'clock position.
16. Rotate the crankshaft clockwise to
position the crankshaft keyway to the 3
o'clock position.
17. Verify that the front 2 intake camshaft
lobes are pointing up.
18. Compress the right hand timing
chain tensioner. Release the racheting pawl.
19. Compress the timing chain tensioner
and install a stiff wire or paper clip.
20. Remove the bolts and the right hand
timing chain tensioner.

✳✳ CAUTION

**If the right hand timing chain ten-
sioner arm is to be reused, mark the
position to make sure that it is
installed on the correct side when
reassembled.**

21. Remove the right hand timing chain
tensioner arm.
22. Remove the right hand timing chain.
23. Remove the bolts and the right hand
Variable Camshaft Timing (VCT) solenoid
housing assembly.

Rotate the crankshaft counterclockwise to
position the crankshaft keyway to the 11
o'clock position—2004–05 3.0L engines

24. Rotate the crankshaft counterclockwise to position the crankshaft keyway to
the 11 o'clock position.
25. Compress the left hand timing
chain tensioner. Release the racheting
pawl.
26. Compress the timing chain tensioner
and install a stiff wire or paper clip.
27. Remove the bolts and the left hand
timing chain tensioner.

✳✳ CAUTION

**If the left hand timing chain tensioner
arm is to be reused, mark the position to make sure that it is installed
on the correct side when reassembled.**

28. Remove the left hand timing chain
tensioner arm.
29. Remove the left hand timing chain.
30. Remove the left hand VCT solenoid
housing assembly.
31. Using a wrench on the flats of the
camshaft, hold the camshaft in position and

Remove the left hand VCT solenoid housing assembly—2004–05 3.0L engines

Using a wrench on the flats of the camshaft, hold the camshaft in position and remove the bolt and the right hand VCT actuator—2004–05 3.0L engines

Using a wrench on the flats of the camshaft, hold the camshaft in position and remove the bolt and the right hand exhaust camshaft sprocket—2004–05 3.0L engines

Using a wrench on the flats of the camshaft, hold the camshaft in position and remove the bolt and the left hand VCT actuator—2004–05 3.0L engines

Using a wrench on the flats of the camshaft, hold the camshaft in position and remove the bolt and the left hand exhaust camshaft sprocket—2004–05 3.0L engines

remove the bolt and the right hand VCT actuator. Discard the bolt.

32. Using a wrench on the flats of the camshaft, hold the camshaft in position and remove the bolt and the right hand exhaust camshaft sprocket. Discard the bolt.

33. Using a wrench on the flats of the camshaft, hold the camshaft in position and remove the bolt and the left hand VCT actuator. Discard the bolt.

34. Using a wrench on the flats of the camshaft, hold the camshaft in position and remove the bolt and the left hand exhaust camshaft sprocket. Discard the bolt.

35. Remove the crankshaft sprockets.

To install:

36. Install the left hand VCT actuator and install a new bolt hand-tight. Use a wrench on the flats of the camshaft to hold it in position and tighten the bolt in 2 steps:

   a. Step 1: Tighten the VCT actuator bolt to 15 ft. lbs. (20 Nm).

   b. Step 2: Tighten an additional 90 degrees.

37. Install the left hand exhaust camshaft sprocket and install a new bolt hand-tight. Using a wrench on the flats of the camshaft, hold the camshaft in position and tighten the bolt in 2 steps:

   a. Step 1: Tighten the exhaust camshaft sprocket bolt to 15 ft. lbs. (20 Nm).

   b. Step 2: Tighten an additional 90 degrees.

38. Install the right hand variable camshaft timing (VCT) actuator and install a new bolt hand-tight. Using a wrench on the flats of the camshaft, hold the camshaft in position and tighten the bolt in 2 steps:

   a. Step 1: Tighten the VCT actuator bolt to 15 ft. lbs. (20 Nm).

   b. Step 2: Tighten an additional 90 degrees.

39. Install the left hand exhaust camshaft sprocket and install a new bolt hand-tight. Using a wrench on the flats of the camshaft, hold the camshaft in position and tighten the bolt in 2 steps:

   a. Step 1: Tighten the exhaust camshaft sprocket bolt to 15 ft. lbs. (20 Nm).

   b. Step 2: Tighten an additional 90 degrees.

40. With the 2 keyways facing forward, install the crankshaft sprockets.

41. If timing marks in the timing chains are not evident, use a permanent-type marker to mark the crankshaft and camshaft timing marks on the left and right hand timing chains.

42. Mark any link to use as the exhaust camshaft timing mark.

If timing marks in the timing chains are not evident, use a permanent-type marker to mark the crankshaft and camshaft timing marks on the left and right hand timing chains—2004–05 3.0L engines

43. Starting with the exhaust camshaft timing mark, count 16 links and mark the link as the intake camshaft sprocket timing mark.

44. Starting with the intake camshaft timing mark, count 25 links and mark the link as the crankshaft sprocket timing mark.

45. Verify that the crankshaft keyway is in the 11 o'clock position.

✳✳ CAUTION

The crankshaft keyway must remain in the 11 o'clock position until the cams are located in the NEUTRAL position, or damage to the valves can occur.

46. Rotate the left and right hand intake and exhaust camshafts to locate them in their NEUTRAL positions.

➡**The gold colored bolt must be installed in its original location.**

47. Install the left hand VCT solenoid housing assembly and the bolts. Tighten to 18 ft. lbs. (25 Nm).

48. Install the left hand timing chain, aligning the copper (or marked) link with the marks on the camshaft and crankshaft sprockets.

49. Install the left hand timing chain tensioner arm.

Verify that the crankshaft keyway is in the 11 o'clock position—2004–05 3.0L engines

Rotate the left and right hand intake and exhaust camshafts to locate them in their NEUTRAL positions—2004–05 3.0L engines

housing assembly and bolts. Tighten to 18 ft. lbs. (25 Nm).

54. Install the right hand timing chain, aligning the copper or (marked) links with the marks on the camshaft and crankshaft sprockets.

55. Verify that the copper (or marked) links on the right hand timing chain are in alignment with the timing marks on the camshaft and crankshaft sprockets.

56. Install the right hand timing chain tensioner arm.

57. Install the right hand timing chain tensioner and bolts.

58. Remove the wire or paper clip from the right hand timing chain tensioner.

Install the left hand timing chain, aligning the copper (or marked) link with the marks on the camshaft and crankshaft sprockets—2004–05 3.0L engines

➡ **Be sure to position the tensioner so that the tensioner piston is fully engaged in the tensioner arm. Install the left hand timing chain tensioner and bolts.**

50. Remove the wire or paper clip from the tensioner.

51. Rotate the crankshaft clockwise and position the keyway at the 3 o'clock position for right hand timing chain installation.

52. Verify that the copper (or marked) links on the left hand timing chain are in alignment with the timing marks on the camshaft and crankshaft sprockets.

53. Install the right hand VCT solenoid

Verify that the copper (or marked) links on the left hand timing chain are in alignment with the timing marks on the camshaft and crankshaft sprockets—2004–05 3.0L engines

Rotate the crankshaft clockwise and position the keyway at the 3 o'clock position for right hand timing chain installation—2004–05 3.0L engines

Install the right hand timing chain, aligning the copper or (marked) links with the marks on the camshaft and crankshaft sprockets—2004–05 3.0L engines

06017-LS-G62

Verify that the copper (or marked) links on the right hand timing chain are in alignment with the timing marks on the camshaft and crankshaft sprockets—2004–05 3.0L engines

06017-LS-G36

Front cover torque sequence—2004–05 3.0L engines

06017-LS-G63

Install the ignition pulse wheel using the orange marked keyway slot and the teeth facing forward—2004–05 3.0L engines

✳✳ CAUTION

This pulse wheel is used with several different engines. Install the pulse wheel with the keyway in the slot stamped "20-25-30 DAMB 30 RFF" only. This slot is also marked with an orange paint mark and aligns with the missing tooth in the trigger wheel.

59. Install the ignition pulse wheel using the orange marked keyway slot and the teeth facing forward.

60. If necessary, install new variable camshaft timing (VCT) system oil filters

✳✳ CAUTION

Do not use metal scrapers, wire brushes, power abrasive discs or other abrasive means to clean the sealing surfaces. These tools cause scratches and gouges which make leak paths.

06017-LS-G35

Apply a 0.23 inch (6 mm) diameter dot of silicone gasket and sealer to the cylinder block, the lower cylinder block and the cylinder head mating surfaces—2004–05 3.0L engines

61. Use a plastic scraping tool to remove all traces of sealant from the front cover. Clean all sealing surfaces with silicone gasket remover and metal surface prep.

62. Install a new gasket on the engine front cover.

➡ **The engine front cover must be installed and the bolts tightened within four minutes of applying sealant.**

63. Apply a 0.23 inch (6 mm) diameter dot of silicone gasket and sealer to the cylinder block, the lower cylinder block and the cylinder head mating surfaces.

➡ **Number 14 is a stud and washer assembly.**

64. Position the engine front cover and

install the bolts and the stud bolt. Tighten in the sequence illustrated to 18 ft. lbs (25 Nm).

65. Connect the CMP sensors and the CKP sensor electrical connectors.

66. Install the crankshaft front seal

67. Install the power steering pump.

68. Install the oil pan.

69. Connect the radiator hoses.

70. Install the water pump.

71. Install the accessory drive belt tensioner and idler pulleys.

72. Install the valve covers.

73. Fill and bleed the cooling system.

74. Fill the engine with correct type and amount of oil.

75. Connect the negative battery cable.

76. Start the engine and check for leaks.

3.9L Engine

2002–03 MODELS

1. Before servicing the vehicle, refer to the Precautions Section.

2. Drain the cooling system.

3. Remove or disconnect the following:

- Negative battery cable
- Engine appearance cover and brackets
- Valve covers
- Engine cooling fan assembly
- Accessory drive belt
- Water pump pulley
- Alternator
- Lower radiator hose and pipe
- Heater hose
- Idler pulleys
- Crankshaft pulley
- Front crankshaft seal
- Power steering reservoir hose
- Power steering pump and bracket
- Hydraulic cooling fan reservoir hose and bracket
- Hydraulic cooling fan pump and bracket
- Front cover wiring harness clips
- Front cover. Loosen the bolts in sequence.

- Torque converter access panel
- Crankshaft Position (CKP) sensor

4. Rotate the crankshaft to 45 degrees After Top Dead Center (ATDC). The crankshaft keyway will be in the 6 o'clock position. Check that the camshaft lobes are facing upwards. If not, rotate the crankshaft 1 full turn.

5. Install Crankshaft Holding Tool 303-645 in place of the CKP sensor.

6. Install Camshaft Locking Tool 303-530 to the right bank camshafts.

7. Loosen the exhaust and intake camshaft sprocket bolts and slide the sprockets forward on the bolts.

8. Remove or disconnect the following:

- Right bank primary timing chain tensioner and blanking plate
- Tensioner arm
- Timing chain guide
- Right bank timing chain and crankshaft timing sprocket

9. Remove the locking tool from the right bank camshafts and install it on the left bank camshafts.

10. Remove or disconnect the following:

- Left bank primary timing chain tensioner and blanking plate
- Tensioner arm
- Timing chain guide

Crankshaft Holding Tool—2002–03 3.9L engine

Camshaft Locking Tool—2002–03 3.9L engine

- Left bank timing chain and crankshaft timing sprocket
- Intake and exhaust camshaft sprocket bolts
- Intake and exhaust camshaft sprockets and secondary timing chain assemblies
- Secondary timing chain tensioners

To install:

11. Prepare the primary and secondary timing chain tensioners for installation as follows:

a. Step 1: Use a thin wire to unseat the check valve

b. Step 2: Compress the tensioner piston fully into the bore by hand

c. Step 3: Remove the wire

12. Install or connect the following:

- Secondary timing chain tensioners. Tighten the bolts to 97 inch lbs. (11 Nm).

Front cover loosening sequence—2002–03 3.9L engine

Timing chain tensioner preparation—2002–03 3.9L engine

9306TG20

Use a tie strap to remove slack from the primary timing chain—2002–03 3.9L engine

9306TG22

Crankshaft sprocket alignment—2002–03 3.9L engine

9306TG21

Timing Chain Tensioning Tool—2002–03 3.9L engine

- Intake and exhaust camshaft sprockets and secondary timing chain assemblies
- Left bank timing chain and crank-shaft timing sprocket
- Timing chain guide. Tighten the bolts to 97 inch lbs. (11 Nm).
- Tensioner arm. Tighten the bolts to 97 inch lbs. (11 Nm).
- Left bank primary timing chain ten-sioner and blanking plate. Tighten the bolts to 97 inch lbs. (11 Nm).

13. Install a tie strap to take up the slack in the timing chain.

14. Use Timing Chain Tensioning Tool 303-532 to apply tension to the left bank exhaust camshaft.

➡**Tighten the exhaust camshaft bolt first.**

15. Tighten the left bank camshaft sprocket bolts as follows:
  a. Step 1: 15 ft. lbs. (20 Nm)
  b. Step 2: Plus 90 degrees

16. Remove the locking tool from the left

3 mm (0.12 in)

9306TG23

Apply silicone sealant to the areas indicated—2002–03 3.9L engine

Front cover torque sequence—2002-03 3.9L engine

9306TG24

bank camshafts and install it on the right bank camshafts.

17. Install or connect the following:
- Right bank timing chain and crankshaft timing sprocket. Ensure that the crankshaft sprocket is installed as shown.
- Timing chain guide. Tighten the bolts to 97 inch lbs. (11 Nm).
- Tensioner arm. Tighten the bolts to 97 inch lbs. (11 Nm).
- Right bank primary timing chain tensioner and blanking plate. Tighten the bolts to 97 inch lbs. (11 Nm).

18. Install a tie strap to take up the slack in the timing chain.

19. Use Timing Chain Tensioning Tool 303-532 to apply tension to the right bank exhaust camshaft.

➡**Tighten the exhaust camshaft bolt first.**

20. Tighten the right bank camshaft sprocket bolts as follows:
- a. Step 1: 15 ft. lbs. (20 Nm)
- b. Step 2: Plus 90 degrees

21. Remove the camshaft locking tool.

22. Apply silicone sealant to the areas indicated and install the front cover.

23. Tighten the front cover bolts in sequence as follows:
- a. Step 1: 44 inch lbs. (5 Nm)
- b. Step 2: 89 inch lbs. (10 Nm)

24. Remove the crankshaft locking tool.

25. Install or connect the following:
- CKP sensor
- Torque converter access panel
- Front cover wiring harness clips
- Hydraulic cooling fan pump and bracket. Tighten the bolts to 18 ft. lbs. (25 Nm).
- Hydraulic cooling fan reservoir hose and bracket
- Power steering pump and bracket.

Tighten the bolts to 18 ft. lbs. (25 Nm).
- Power steering reservoir hose
- Front crankshaft seal

26. Install the crankshaft pulley and tighten the bolt as follows:
- a. Step 1: 59 ft. lbs. (80 Nm)
- b. Step 2: Loosen the bolt 2 complete turns
- c. Step 3: 37 ft. lbs. (50 Nm)
- d. Step 4: Plus 90 degrees

27. Install or connect the following:
- Idler pulleys. Tighten the bolts to 18 ft. lbs. (25 Nm).
- Heater hose
- Lower radiator hose and pipe
- Alternator
- Water pump pulley. Tighten the bolts to 89 inch lbs. (10 Nm) plus 45 degrees.
- Accessory drive belt
- Engine cooling fan assembly
- Valve covers
- Engine appearance cover and brackets
- Negative battery cable

28. Fill the cooling system.

29. Start the engine and check for leaks.

2004-05 MODELS

1. Before servicing the vehicle, refer to the Precautions Section.

2. Disconnect the negative battery cable.

3. Drain and recycle the engine coolant.

4. Drain the engine oil.

5. Recover the AC refrigerant using approved equipment.

6. Remove the valve covers.

7. Remove the cooling fan assembly.

8. Remove the accessory drive belt idler pulley.

9. Remove the crankshaft front seal.

10. Remove the A/C compressor..

11. Remove the lower radiator hose bracket nut.

12. Disconnect the lower radiator, upper radiator and heater hose.

13. Disconnect the two oil cooler hoses, if equipped.

14. Disconnect the power steering reservoir hose.

15. Disconnect the Power Steering Pressure (PSP) switch.

➡**One bolt is blocked by the power steering pressure hose and must be removed with the power steering pump.**

16. Remove the four bolts and position the power steering pump aside.

06017-LS-G14

There is one window on the ignition pulse wheel that is unique to accept the tool 303-645—2004–05 3.9L engine

17. Remove the five bolts and the power steering pump bracket.

18. Detach the wiring harness clips.

19. Remove the engine front cover bolts in the sequence illustrated.

20. Remove the engine front cover.

21. Remove the right and left hand Variable Cam Timing (VCT) bush carriers.

22. Remove the bolt and the Crankshaft Position (CKP) sensor.

➡There is one window on the ignition pulse wheel that is unique to accept the tool 303-645.

23. Install the tool 303-645.

24. Turn the crankshaft to 45 degrees After Top Dead Center (ATDC). The crankshaft keyway will be in the 6 o'clock position.

25. Make sure the flats of the camshaft are facing upward. If not, repeat the previous step.

❋❋ **CAUTION**

The flats on the camshaft must completely contact the special service tool or damage to the camshafts can occur. Slowly draw the special tool down evenly until it is fully seated on the camshaft flats.

06017-LS-G15

Tool 303-530 is used to hold the camshafts during removal—2004–05 3.9L engine

26. Install the holding tool 303-530 on the right hand cylinder head.

❋❋ **CAUTION**

An open-end wrench must be used on the hex area of the camshaft or damage to the head will occur.

27. Remove and discard the exhaust camshaft sprocket bolt.

28. Remove and discard the variable VCT unit bolt.

29. Remove the right hand timing chain tensioner.

30. Push the lever down and install a drill rod to lock the piston.

31. Remove the 2 bolts and the tensioner.

32. Remove the bolt and the right tensioner arm.

33. Remove the bolt and the right timing chain guide.

34. Remove the right primary timing chain.

35. Remove the holding tool from the right cylinder head.

❋❋ **CAUTION**

The flats on the camshaft must completely contact the special service tool or damage to the camshafts can occur. Slowly draw the special tool down evenly until it is fully seated on the camshaft flats.

36. Install the holding tool on the left cylinder head.

❋❋ **CAUTION**

An open-end wrench must be used on the hex area of the camshaft or damage to the cylinder head will occur.

37. Remove and discard the left VCT unit bolt and the exhaust camshaft sprocket bolt.

38. Remove the left timing chain tensioner.

39. Push the lever down and install a drill rod to lock the piston.

40. Remove the 2 bolts and the tensioner.

41. Remove the nut and the left tensioner arm.

42. Remove the bolt and the left timing chain guide.

43. Remove the left primary timing chain.

44. Remove the crankshaft sprocket.

45. Collapse the left secondary timing chain tensioner and install a pin to lock the piston.

46. Remove the left VCT unit, exhaust camshaft sprocket and timing chain as an assembly.

47. Remove the 2 bolts and the left secondary timing chain tensioner.

48. Collapse the right secondary timing chain tensioner and install a pin to lock the piston.

49. Remove the right VCT unit, exhaust camshaft sprocket and timing chain as an assembly.

50. Remove the 2 bolts and the right secondary timing chain tensioner.

To install:

51. Turn the crankshaft to 45 degrees ATDC. The crankshaft keyway will be in the 6 o'clock position.

52. Install the tool 303-645.

53. Position all of the camshafts so the flats of the camshafts are facing upward.

54. Install the left secondary timing chain tensioner and 2 bolts. Tighten to 8 ft. lbs. (11 Nm).

55. Install the left VCT unit, exhaust camshaft sprocket and timing chain as an assembly.

56. Using new bolts, install the left VCT unit and exhaust camshaft sprocket bolts finger-tight.

57. Remove the locking pin from the left secondary timing chain tensioner.

58. Install the right secondary timing chain tensioner and 2 bolts. Tighten to 8 ft. lbs. (11 Nm).

59. Install the right VCT unit, exhaust camshaft sprocket and timing chain as an assembly.

60. Using new bolts, install the right VCT unit and exhaust camshaft sprocket bolts finger-tight.

61. Remove the locking pin from the right secondary timing chain tensioner.

62. Reset the timing chain tensioners by holding the lever toward the tensioner piston. Using finger pressure, compress the

06017-LS-G20

Position all of the camshafts so the flats of the camshafts are facing upward—2004–05 3.9L engine tensioner. Push the lever down and install a drill rod to lock the piston.

✳✳ CAUTION

The flats on the camshaft must completely contact the special service tool or damage to the camshafts can occur. Slowly draw the special tool down evenly until it is fully seated on the camshaft flats.

63. Install the holding tool on the left head.
64. Position the left timing chain and crankshaft gear.
65. Position the timing chain over the left intake camshaft sprocket.
66. Position the crankshaft gear in the timing chain.
67. Install the left timing chain guide and bolt. Tighten to 8 ft. lbs. (11 Nm).
68. Install the left timing chain tensioner arm and nut. Tighten to 8 ft. lbs. (11 Nm).
69. Install the left timing chain tensioner.
70. Install the timing chain tensioner and bolts, and remove the drill rod.

✳✳ CAUTION

The exhaust camshaft sprocket bolt must be fully tightened before tightening the intake camshaft sprocket bolt. Using the tool, apply 89 inch lbs. (10 Nm) counterclockwise to the left exhaust camshaft sprocket.

71. Tighten the left exhaust camshaft sprocket and VCT unit bolts 26 ft. lbs. (35 Nm) then an additional 90 degrees.
72. Remove the holding tool.

✳✳ CAUTION

The flats on the camshaft must completely contact the special service tool or damage to the camshafts can occur. Slowly draw the special tool downward evenly until it is fully seated on the camshaft flats.

73. Install the holding tool on the right cylinder head.
74. Position the right timing chain over the right intake camshaft sprocket and over the crankshaft gear.
75. Install the right timing chain guide and bolt. Tighten to 8 ft. lbs. (11 Nm).
76. Install the right timing chain tensioner arm. Tighten to 8 ft. lbs. (11 Nm).
77. Install the right timing chain tensioner.
78. Install the timing chain tensioner and bolts, tighten to 8 ft. lbs. (11 Nm) and remove the drill rod.

✳✳ CAUTION

The exhaust camshaft sprocket bolt must be fully tightened before tightening the intake camshaft sprocket bolt. Using the tool, apply 89 inch lbs. (10 Nm) counterclockwise to the left exhaust camshaft sprocket.

79. Tighten the right exhaust camshaft sprocket and VCT unit bolts 26 ft. lbs. (35 Nm) then an additional 90 degrees.
80. Remove the holding tool.

06017-LS-G21

Using the tool, apply 89 inch lbs. (10 Nm) counterclockwise to the camshaft sprocket—2004–05 3.9L engine

➡ Lubricate the VCT bush carrier seals with clean engine oil.

81. Carefully insert the bush carrier into the VCT unit, compress ring seals with fingers as required. Fully seat the bush carrier by hand before fitting the bolts and nut. Tighten the bolts to 15 ft. lbs. (20 Nm). Tighten the nut to 89 inch lbs. (10 Nm).
82. Install new engine front cover gaskets.

➡ If the front cover is not secured within 4 minutes of sealant application, the sealant must be removed and the sealing area cleaned with metal surface cleaner. Allow to dry until there is no sign of wetness, or 4 minutes, whichever is longer. Failure to follow this procedure can cause future oil leakage.

83. Apply silicone gasket and sealant in the 8 places shown.
84. Install the engine front cover. Loosely install the bolts.
85. Tighten the bolts in 2 stages in the sequence shown. Tighten to 44 inch lbs. (5 Nm) in sequence and then retighten in sequence to 89 inch lbs. (10 Nm).

3 mm (0.12 in)

06017-LS-G22

Apply silicone gasket and sealant in the 8 places shown—2004–05 3.9L engine

Engine front cover torque sequence—3.9L engine

06017-LS-G23

3 mm (0.12 in)

06017-LS-G26

Apply 4 beads of silicone gasket and sealant as shown—2004–05 3.9L engine

→Lubricate the front oil seal and the engine front cover mating surface with clean engine oil.

86. Using the tool 303-646, install a new front crankshaft seal.

→If crankshaft pulley is not secured within 4 minutes of sealant application the sealant must be removed and the sealing area cleaned with metal surface cleaner. Allow to dry until there is no sign of wetness, or 4 minutes, whichever is longer. Failure to follow this procedure can cause future oil leakage.

87. Apply a bead of silicone gasket and sealant to the keyway.

88. Using the tool 303-102, install the crankshaft pulley.

89. Install a new bolt and using the spe-

303-646

06017-LS-G24

Using the tool 303-646, install a new front crankshaft seal—2004–05 3.9L engine

303-102

06017-LS-G25

Using the tool 303-102, install the crankshaft pulley—2004–05 3.9L engine

06017-LS-G08

Left valve cover torque sequence—2004–05 3.9L engine

Right valve cover torque sequence—2004–05 3.9L engine

06017-LS-G09

POSITIONING

7922AG16

Piston ring end-gap spacing—3.0L (VIN S) engine

cial tool to hold the pulley, tighten as follows:.

    a. Step 1: Tighten to 59 ft. lbs. (80 Nm).

    b. Step 2: Loosen the bolt 2 complete turns.

    c. Step 3: Tighten to 37 ft. lbs. (50 Nm).

    d. Step 4: Tighten an additional 90 degrees.

  90. Install new valve cover gaskets.

➡**If the valve covers are not secured within 4 minutes of sealant application the sealant must be removed and the sealing area cleaned with metal surface cleaner. Allow to dry until there is no sign of wetness, or 4 minutes, whichever is longer. Failure to follow this procedure can cause future oil leakage.**

  91. Apply 4 beads of silicone gasket and sealant as shown.

  92. Install the left valve cover and tighten the bolts in sequence to 89 inch lbs. (10 Nm).

  93. Install the right valve cover and tighten the bolts in sequence to 89 inch lbs. (10 Nm).

  94. Install the accessory drive belt idler pulley.

  95. Install the cooling fan assembly.

  96. Fill the engine with clean engine oil.

  97. Connect the negative battery cable.

  98. Fill and bleed the power steering system.

  99. Fill and bleed the engine cooling system.

  100. Evacuate and recharge the A/C system

  101. Start the engine and check for leaks.

VIEWED FROM FRONT OF ENGINE

7922AG15

Piston ring end-gap spacing—3.0L (VIN S) engine

FUEL SYSTEM

Fuel System Service Precautions

Safety is the most important factor when performing not only fuel system maintenance but any type of maintenance. Failure to conduct maintenance and repairs in a safe manner may result in serious personal injury or death. Maintenance and testing of the vehicle's fuel system components can be accomplished safely and effectively by adhering to the following rules and guidelines.

• To avoid the possibility of fire and personal injury, always disconnect the negative battery cable unless the repair or test procedure requires that battery voltage be applied.

• Always relieve the fuel system pressure prior to disconnecting any fuel system component (injector, fuel rail, pressure regulator, etc.), fitting or fuel line connection. Exercise extreme caution whenever relieving fuel system pressure, to avoid exposing skin, face and eyes to fuel spray. Please be advised that fuel under pressure may penetrate the skin or any part of the body that it contacts.

• Always place a shop towel or cloth around the fitting or connection prior to loosening to absorb any excess fuel due to spillage. Ensure that all fuel spillage (should it occur) is quickly removed from engine surfaces. Ensure that all fuel soaked cloths or towels are deposited into a suitable waste container.

• Always keep a dry chemical (Class B) fire extinguisher near the work area.

• Do not allow fuel spray or fuel vapors to come into contact with a spark or open flame.

• Always use a back-up wrench when loosening and tightening fuel line connection fittings. This will prevent unnecessary stress and torsion to fuel line piping.

• Always replace worn fuel fitting O-rings with new. Do not substitute fuel hose or equivalent where fuel pipe is installed.

Fuel System Pressure

RELIEVING

1. Remove the fuel pump relay.
2. Start the engine and allow it to idle until it stalls.
3. After the engine stalls, crank the engine for approximately 5 seconds to make sure the fuel injection supply manifold pressure has been released.

4. Turn the ignition switch to the OFF position.
5. Disconnect the negative battery cable.
6. When fuel system service is complete, install the fuel pump relay.

➡It may take more than one key cycle to pressurize the fuel system.

7. Cycle the ignition key and wait 3 seconds to pressurize the fuel system. Check for leaks before starting the engine.
8. Connect the negative battery cable.
9. Start the vehicle and check the fuel system for leaks.

Fuel Filter

REMOVAL & INSTALLATION

1. Before servicing the vehicle, refer to the Precautions Section.
2. Relieve the fuel system pressure.
3. Disconnect the negative battery cable.
4. Remove the left front wheel well rear splash shield.
5. Disconnect the fuel tube quick release fittings from both sides of the fuel filter.
6. Remove the fuel filter bracket mounting bolt, fuel filter bracket, and the fuel filter.

To install:

7. Installation is the reverse of removal. Tighten the bracket bolts to 62 inch lbs. (7 Nm).
8. Start the engine and check for leaks.

Fuel Pump

REMOVAL & INSTALLATION

2002–03 MODELS

1. Before servicing the vehicle, refer to the Precautions Section.
2. Relieve the fuel system pressure.
3. Drain the fuel tank.
4. Remove or disconnect the following:
   • Negative battery cable
   • Trunk liner
   • Trunk seal retainer
   • Rear lamp assembly interior trim finisher
   • Left and right side liners
   • Fuel feed and return lines
   • Fuel filler and vent hoses
   • Fuel tank wiring connectors

   • Fuel filler cap
   • Fuel tank retaining straps
   • Fuel tank
   • Fuel pump module

To install:

5. Install or connect the following:
   • Fuel pump module
   • Fuel tank
   • Fuel tank retaining straps
   • Fuel filler cap
   • Fuel tank wiring connectors
   • Fuel filler and vent hoses
   • Fuel feed and return lines
   • Left and right side liners
   • Rear lamp assembly interior trim finisher
   • Trunk seal retainer
   • Trunk liner
   • Negative battery cable

6. Fill the fuel tank with at least 10 gallons (38L) of fuel.
7. Start the engine and check for leaks.

2004–05 Models

1. Before servicing the vehicle, refer to the Precautions Section.
2. Remove the rear seat cushion to and any insulation padding covering the access covers.
3. Drain the fuel tank.
4. Support the tank and loosen the straps to gain access to the pump.
5. Remove the access covers.

➡The routing of the internal fuel crossover tubes is critical to correct fuel system operation. Note the routing of the internal fuel crossover hoses at this time.

6. Partially raise the fuel transfer pump and disconnect the internal fuel crossover tube quick release fittings.
7. Disconnect the fuel pump module tube quick release fitting from the fuel pump module and position aside.

❊❊ CAUTION

Drain the remaining fuel from the delivery module into the fuel tank prior to removal.

8. Before removal, attach a non-conductive material to the crossover tubes. Then remove the fuel pump module and crossover tubes as an assembly, using the non-conductive material to pull the crossover lines through the gas tank.
9. Remove and discard the fuel pump module gasket.

To install:

10. Install the fuel pump module and crossover tubes as an assembly.

11. Attaching the non-conductive material to the crossover tubes, pull the crossover tubes through the gas tank to install the fuel pump module.

12. Using tool 310-069, install the fuel pump module using a new gasket. Install the lock ring and tighten to 59 ft. lbs. (80 Nm).

➡**Make sure that the fuel crossover tubes are routed correctly.**

13. Partially raise the fuel transfer pump and connect the internal fuel crossover hose quick disconnect fittings.

14. Using the tool 310-069, install the fuel transfer pump with a new gasket and lock ring. Tighten to 59 ft. lbs. (80 Nm).

15. Connect the fuel pump module tube quick release fitting at the fuel pump module.

16. Connect the fuel pump module and fuel transfer pump electrical connectors.

17. Install the fuel pump module and fuel transfer pump access covers.

18. Install the rear seat bottom and any insulation padding covering the access covers.

19. Tighten the 2 fuel tank support strap bolts to 26 ft. lbs. (35 Nm).

20. Connect the negative battery cable.

Fuel Injector

REMOVAL & INSTALLATION

2002–03 MODELS

1. Before servicing the vehicle, refer to the Precautions Section.
2. Relieve fuel system pressure.
3. Remove or disconnect the following:

- Negative battery cable
- Engine appearance covers
- Fuel lines
- Fuel pressure regulator

1 Engine appearance cover mounting bracket nuts
2 Engine appearance cover mounting brackets
3 Fuel pressure sensor electrical connector
4 Fuel pressure sensor vacuum connector
5 Fuel injector electrical connector
6 Fuel temperature sensor electrical connector
7 Fuel tube retaining clip
8 Fuel tube spring lock coupling
9 Fuel injection supply manifold bolts
10 Fuel injection supply manifold and fuel injectors
11 Fuel injectors

Exploded view of the fuel injector mounting—3.9L engine shown

- Fuel cross over elbow
- Fuel injector connectors
- Fuel injector clamping plates
- Fuel injectors

To install:

4. Install or connect the following:
- Fuel injectors. Use new O-ring seals.
- Fuel injector clamping plates
- Fuel injector connectors
- Fuel cross over elbow
- Fuel pressure regulator
- Fuel lines
- Engine appearance covers
- Negative battery cable

5. Start the engine and check for leaks.

2004–05 MODELS

1. Before servicing the vehicle, refer to the Precautions Section.
2. Disconnect the negative battery cable.

✳✳ CAUTION

The engine appearance cover is equipped with a noise suppressor. If the noise suppressor becomes separated from the engine appearance cover it must be positioned on the intake manifold prior to installing the engine appearance cover.

3. Remove the engine appearance cover and the noise suppressor, if detached.
4. Remove the air cleaner outlet pipe.
5. Release the fuel system pressure.
6. Remove the four engine appearance cover mounting bracket nuts and the brackets.
7. Disconnect the fuel pressure sensor electrical connector from the sensor.
8. Disconnect the fuel pressure sensor vacuum connector from the sensor.
9. Disconnect the eight fuel injector electrical connectors from the injectors.
10. Disconnect the fuel temperature sensor electrical connector from the sensor.
11. Disconnect the fuel supply tube retaining clip from the fuel injection supply manifold.
12. Disconnect the fuel supply tube spring lock coupling.
13. Remove the four bolts and the fuel injection supply manifold and fuel injectors as an assembly.
14. Remove the fuel injectors from the fuel injection supply manifold.

✳✳ CAUTION

Use O-ring seals that are made of special fuel-resistant material. Use of ordinary O-ring seals can cause the fuel system to leak. Do not reuse the O-ring seals.

15. Remove and discard the 16 O-ring seals from the fuel injectors.

To install:

16. Lubricate the new fuel injector O-ring seals with clean engine oil before installing.

✳✳ CAUTION

Use O-ring seals that are made of special fuel-resistant material. Use of ordinary O-rings can cause the fuel system to leak. Do not reuse the O-ring seals.

17. Install sixteen new O-ring seals onto the fuel injectors. Lubricate the O-ring seals with clean engine oil prior to installation.
18. Install the fuel injectors and clips onto the fuel rail.
19. Install the fuel rail and injectors as an assembly. Tighten the fuel rail bolts to 89 inch lbs. (10 Nm).
20. Install the remaining components in the reverse order of removal.

DRIVE TRAIN

Transmission Assembly

REMOVAL & INSTALLATION

Automatic

2002–03 MODELS

1. Before servicing the vehicle, refer to the Precautions Section.
2. Install a support fixture to the engine lifting eyes.
3. Drain the transmission fluid.
4. Remove or disconnect the following:

- Negative battery cable
- Engine appearance covers
- Mass Air Flow (MAF) meter
- Air intake assembly
- Coolant recovery tank
- Exhaust front pipes
- Driveshaft
- Shift selector cable
- Transmission electrical connectors
- Transmission oil cooler lines
- Torque converter
- Transmission mount and bracket
- Transmission flange bolts
- Transmission

To install:

5. Install or connect the following:
- Transmission to the engine. Tighten the flange bolts to 32–42 ft. lbs. (43–57 Nm).
- Transmission mount and bracket. Tighten the mount bolts to 22–30 ft. lbs. (30–40 Nm) and the bracket bolts to 16–21 ft. lbs. (22–28 Nm).
- Torque converter. Tighten the bolts to 32–42 ft. lbs. (43–57 Nm).
- Transmission oil cooler lines
- Transmission electrical connectors
- Shift selector cable
- Driveshaft. Tighten the bolts to 55–65 ft. lbs. (75–88 Nm).
- Exhaust front pipes
- Coolant recovery tank
- Air intake assembly
- MAF meter
- Engine appearance covers
- Negative battery cable

6. Fill the transmission to the correct level with the proper fluid. Do not over-fill.
7. Start the engine and check for leaks.

75 – 88 Nm

Driveshaft flange bolts—2002–03 models

2004–05 Models

1. Before servicing the vehicle, refer to the Precautions Section.

✳✳ WARNING

Secure the transmission to the transmission jack with a safety chain. Failure to follow these instructions can result in personal injury.

➡When the battery has been disconnected and reconnected, some abnormal drive symptoms can occur while the vehicle relearns its adaptive strategy. The customer needs to be notified that they can experience slightly different upshifts either (soft or firm) and that this is a temporary condition and will eventually return to normal operating condition.

→If transmission disassembly or installation of a new transmission is necessary, drain the transmission fluid. Install the drain plug when finished. If the transmission is to be removed for an extended period of time, support the engine with a safety stand and a wood block.

2. Disconnect the negative battery cable.
3. Remove the complete exhaust system and front heat shields.
4. Remove the driveshaft.

✳✳ WARNING

Secure the transmission to the transmission jack with a safety chain. Failure to follow these instructions may result in personal injury.

5. Using a suitable transmission jack, support the transmission.
6. Disconnect the Heated Oxygen (HO2s) sensor connector and remove the sensor wire retaining clip from the right side of the transmission pan rail.
7. Remove the HO2s sensor connector from the front of the transmission pan rail.
8. Remove the transmission support center screw.
9. Remove the transmission support outer screws and remove the rear transmission support.
10. Disconnect the transmission shift cable retainer and position it aside.
11. Disconnect the transmission shift cable eyelet from the manual lever.
12. Disconnect the wire harness from the top of the transmission shift cable bracket.
13. Lower the transmission enough to gain access to the sensors. Disconnect the Turbine Shaft Speed (TSS) sensor, Output Shaft Speed (OSS) sensor and Intermediate Shaft Speed (ISS) sensor electrical connectors.
14. Remove the transmission shift cable bracket screws and bracket.

→Clean the area around connector to prevent contamination of the solenoid body connector. Remove the screw from the solenoid body connector and disconnect the connector.

15. Disconnect the wire harness retainers.
16. Remove and discard the tie strap

from the digital Transmission Range (TR) sensor connector boot.
17. Slide the rubber boot back and disconnect the digital TR sensor connector.
18. Remove the screws and pushpins, then remove the splash shield.
19. Disconnect the wire harness from the cooler line bracket.
20. Disconnect the transmission cooler line bracket.

✳✳ CAUTION

Do not damage the cooler tubes. Hold the transmission case fittings with a wrench.

21. Disconnect the transmission cooler tubes.
22. Remove the torque converter access cover.
23. Remove the torque converter nuts.
24. Remove the transmission bellhousing-to-engine bolts.
25. Carefully lower the transmission from the vehicle. Make sure the transmission fluid cooler and cooler lines are cleaned at this time.

To install:

✳✳ WARNING

Secure the transmission to the transmission jack with a safety chain. Failure to follow these instructions can result in personal injury.

26. Secure the transmission to the transmission jack with a safety chain.

→Rotate the torque converter so that the torque converter paint mark is in the 12 o'clock position.

27. While raising and positioning the transmission to the engine, install the wire harness to the top of the transmission.
28. Carefully position the transmission in alignment with the engine and install the transmission bellhousing-to-engine bolts. Refer to the accompanying illustration for component location and torque specifications.
29. Install the torque converter nuts. Refer to the accompanying illustration for component location and torque specifications.
30. Install the converter access cover.

✳✳ CAUTION

Do not damage the cooler tubes. Hold the transmission case fitting with a wrench.

31. Connect the transmission cooler tubes. Refer to the accompanying illustration for component location and torque specifications.
32. Install the cooler line bracket. Refer to the accompanying illustration for component location and torque specifications.
33. Connect the wire harness to the cooler line bracket.
34. Install the splash shield, then install the screws and pushpin.
35. Reconnect the digital TR sensor connector.
36. Slide the boot over the digital TR sensor connector and install a new tie strap.
37. Connect the wire harness retainers to the transmission.

✳✳ CAUTION

Damage will occur to the solenoid body assembly if the screw is tightened above the specification.

→Always install new O-ring seals on the vehicle harness connector. Use petroleum jelly to lubricate the O-ring seals to aid in the installation process.

38. Install and lubricate new O-ring seals on the transmission connector and connect the connector. Tighten to 44 inch lbs. (5 Nm).
39. Install the shift cable bracket screws and bracket. Refer to the accompanying illustration for component location and torque specifications.
40. Connect the TSS, OSS and ISS electrical connectors.
41. Install the wire harness to the top of the transmission shift cable bracket.
42. Connect the transmission shift cable eyelet to the manual lever.
43. Install the shift cable to the rear transmission support.
44. Position the rear transmission support and install the screws. Refer to the accompanying illustration for component location and torque specifications.
45. Install the transmission support center screw. Refer to the accompanying illustration for component location and torque specifications.
46. Install the oxygen sensor connector on the front of the transmission pan rail.
47. Install the wire in the clip on the right side of the transmission pan rail and connect the heated oxygen sensors.
48. Install the driveshaft.
49. Install the exhaust system and front heat shields.
50. Connect the negative battery cable.
51. If the transmission was overhauled

38 Nm (28 lb-ft) — ③

30 Nm (30 lb-ft) — ⑤

48 Nm (35 lb-ft) — ②

② — 48 Nm (35 lb-ft)

25 Nm (18 lb-ft)

40 Nm (30 lb-ft) — ⑰

55 Nm (41 lb-ft)

1 Transmission assembly

2 Bell housing screws

3 Torque converter nuts

4 Torque converter access plug

5 Fluid cooler tube nuts

6 Fluid cooler tube bracket nut (part of 7A030)

7 Transmission range (TR) sensor electrical connector

8 Wire harness retainers

9 Solenoid body harness connector

10 Speed sensor connectors

11 Shift cable bracket and screws

12 Wire harness retainer

13 Shift cable eyelet

14 Shift cable retainer

15 Transmission support

16 Transmission support outer screws

17 Transmission support center screw

18 Heated oxygen sensor electrical connector retainer (part of 9F472)

19 Heated oxygen sensor electrical connector (part of 9F427)

20 Transmission assembly

21 Output shaft flange coupler

22 Output shaft flange coupler screws

06017-LS-G64

Exploded view of the automatic transmission component locations and torque specifications–2004–05 models

and the vehicle was equipped with an in-line fluid filter, install a new in-line fluid filter.

52. If the transmission was overhauled and the vehicle was not equipped with an in-line fluid filter, install a new in-line fluid filter kit.

53. If the transmission is being installed for a non-internal repair, do not install an in-line filter or filter kit.

54. If installing a new or a Ford-authorized remanufactured transmission, install the in-line transmission fluid filter that is supplied.

55. Carry out the fluid level check.

56. Check the operation of the transmission and inspect for leaks.

Manual

1. Before servicing the vehicle, refer to the Precautions Section.
2. Remove or disconnect the following:
   - Negative battery cable
   - Shift linkage
   - Exhaust center section

➡ **The driveshaft flange may have additional nuts added for balance. These nuts must be replaced in the same position to maintain the driveshaft balance.**

   - Driveshaft. Matchmark the flange and note the location of the fasteners for assembly.
   - Heated Oxygen (HO2S) sensor connectors and harness
   - Vehicle Speed (VSS) sensor connector
   - Reverse light switch connector and harness
   - Transmission mount and crossmember. Support the transmission.
   - Clutch slave cylinder fluid line
   - Starter motor
   - Transmission flange bolts
   - Transmission

To install:

3. Install or connect the following:
   - Transmission. Tighten the flange bolts to 35 ft. lbs. (47 Nm).
   - Starter motor
   - Clutch slave cylinder fluid line
   - Transmission mount and crossmember. Tighten the bolts to 41 ft. lbs. (55 Nm) and the nut to 30 ft. lbs. (40 Nm).
   - Reverse light switch connector and harness
   - VSS sensor connector
   - HO2S sensor connectors and harness
   - Driveshaft. Align the matchmarks and tighten the retainers to 63 ft.

lbs. (85 Nm). Tighten any balance nuts to 18 ft. lbs. (24 Nm).
   - Exhaust center section
   - Shift linkage
   - Negative battery cable

Clutch

ADJUSTMENTS

Because the clutch system is hydraulic, the clutch pedal free-play is self-adjusting and requires no additional maintenance.

REMOVAL & INSTALLATION

1. Before servicing the vehicle, refer to the Precautions Section.
2. Remove or disconnect the following:
   - Transmission
   - Clutch pressure plate
   - Clutch disc

To install:

3. Install or connect the following:
   - Clutch disc and pressure plate. Tighten the pressure plate bolts evenly in several passes to 17 ft. lbs. (23 Nm).
   - Transmission

4. Bleed the hydraulic clutch system, if required.
5. Check the clutch system for proper operation.

Hydraulic Clutch System

BLEEDING

1. Before servicing the vehicle, refer to the Precautions Section.
2. Remove the inspection cover.
3. Connect a hose to the bleeder valve fitting on the clutch slave cylinder. Submerge the other end of the hose into a container of clean brake fluid.
4. Open the bleeder valve and have an assistant depress the clutch pedal.
5. Close the bleeder before releasing the clutch pedal.
6. Repeat the procedure until no more air bubbles are present.
7. Install the inspection cover to the bell housing.
8. Top off the clutch master cylinder fluid reservoir and install the diaphragm and cap.

Halfshaft

REMOVAL & INSTALLATION

2002–03 MODELS

1. Before servicing the vehicle, refer to the Precautions Section.
2. Remove or disconnect the following:

1 Rear wheel knuckle assembly
2 Rear halfshaft assembly
3 Halfshaft seal

67197-LILS-G14

Rear halfshaft mounting–2002–03 models

- Negative battery cable
- Rear wheel
- Brake caliper
- Wheel speed sensor
- Hub carrier pivot bolt
- Hub retaining nut

3. Press the stub shaft out of the hub and pry the inner joint out of the differential.

To install:

4. Install or connect the following:

- Halfshaft inner joint to the differential
- Halfshaft in the wheel hub by applying Loctite® 270 thread locking compound to the splines
- Hub carrier pivot bolt by aligning the bolt head matchmarks. Tighten it to 66–81 ft. lbs. (90–110 Nm).
- Hub retaining nut. Tighten the new nut to 221 ft. lbs. (300 Nm).
- Wheel speed sensor
- Brake caliper
- Rear wheel
- Negative battery cable

5. Check the wheel alignment and adjust as necessary.

➡**If the hub is removed for any reason, a new bearing assembly must be installed. Never attempt to re-use a bearing.**

2004–05 Models

1. Before servicing the vehicle, refer to the Precautions Section.

2. Disconnect the negative battery cable.

3. Remove the rear wheel.

4. Remove and discard the rear axle wheel hub retainer.

※ CAUTION

When removing the caliper, never allow it to hang from the brake hose. Provide a suitable support.

5. Remove the rear brake caliper and support bracket as an assembly.

6. Remove the rear anti-lock speed sensor from the hub.

7. Remove and discard the suspension lower arm nut and bolt.

8. Remove the toe-link nut and bolt.

9. Using the tools illustrated, press the outboard CV joint until it is loose in the hub.

※ CAUTION

Do not over-angulate the outboard CV joint or damage the boot.

10. While raising the knuckle, remove the CV joint from the hub.

11. Position the knuckle to gain clear access for the halfshaft removal.

※ CAUTION

The crown of the tool forks must face away from the axle housing. Position the removal tool correctly between the CV joint and the axle housing so as not to damage the differential seal.

12. Using the tool illustrated, exert enough pressure to overcome the axle circlip and separate the CV joint from the differential side gear.

1 Rear axle wheel hub retainer

2 Toe link nut

3 Toe link bolt

4 Suspension lower arm nut

5 Suspension lower arm bolt

6 Rear wheel knuckle assembly

7 Rear halfshaft assembly

06017-LS-G65

Exploded view of the rear halfshaft assembly mounting—2004–05 models

Using the tools shown, press the outboard CV joint until it is loose in the hub—2002–05 models

Using the tool shown, exert enough pressure to overcome the axle circlip and separate the CV joint from the differential side gear—2004–05 models

✳✳ CAUTION

To prevent damage to the axle shaft oil seal, install the tool illustrated before removing the inboard CV joint housing from the axle.

Install the tool shown to prevent axle shaft oil seal damage—2004–05 models

13. Install the tool shown to prevent axle shaft oil seal damage.

14. Install a differential plug to avoid leakage and contamination.

To install:

15. Install a new axle circlip.

16. Remove the differential plug.

17. Position the halfshaft for installation.

✳✳ CAUTION

Differential seal damage will occur if installing the halfshaft without the special tool installed.

18. Install the differential seal protector tool.

19. Seat the CV joint in the differential side gear.

20. Slide the CV joint into the axle housing until the shaft splines are past the differential seal.

21. Remove the differential seal protector tool.

22. Align the CV joint and side gear splines, and slide the joint into the gear until it seats.

23. When seated, the axle circlip will lock the CV joint in the differential side gear.

24. Check the axle circlip engagement by attempting to pull the inboard CV joint out of the differential side gear.

25. If the circlip is not seated, push the CV joint inward until the circlip is fully engaged in the differential side gear.

✳✳ CAUTION

Do not over-angulate the outboard CV joint or damage the boot.

26. While raising the knuckle, install the CV joint into the hub, then position the knuckle onto the lower control arm.

✳✳ CAUTION

Position the suspension at curb ride height before tightening the bolt and nut.

27. Install the new suspension lower arm bolt and nut. Tighten to 111 ft. lbs. (150 Nm).

28. Install the toe-link nut and bolt. Tighten to 41 ft. lbs. (55 Nm).

29. Install the rear brake caliper and bracket and the bolts. Tighten to 76 ft. lbs. (103 Nm).

30. Install the rear brake anti-lock sensor and the bolt. Tighten to 5 ft. lbs. (7 Nm).

31. Install the new rear axle wheel hub retainer. Tighten to 221 ft. lbs. (300 Nm).

32. Install the rear wheel and tire assembly

33. Connect the negative battery cable.

CV-Joint

OVERHAUL

The CV-joints are serviced with the axle halfshaft as an assembly.

STEERING AND SUSPENSION

Air Bag

✳✳ CAUTION

Some vehicles are equipped with an air bag system. The system must be disarmed before performing service on, or around, system components, the steering column, instrument panel components, wiring and sensors. Failure to follow the safety precautions and the disarming procedure could result in accidental air bag deployment, possible injury and unnecessary system repairs.

PRECAUTIONS

Several precautions must be observed when handling the inflator module to avoid accidental deployment and possible personal injury.

• Never carry the inflator module by the wires or connector on the underside of the module.

• When carrying a live inflator module, hold securely with both hands, and ensure that the bag and trim cover are pointed away.

• Place the inflator module on a bench or other surface with the bag and trim cover facing up.

• With the inflator module on the bench, never place anything on or close to the module that may be thrown in the event of an accidental deployment.

DISARMING

2002–03 MODELS

Proper SRS disarming can be obtained by disconnecting and isolating the negative battery cable. Allow the air bag system capacitor at least 2 minutes to discharge before removing any air bag system components.

2004–05 Models
DEPOWERING PROCEDURE

✲✲ WARNING

Always wear safety glasses when repairing an air bag supplemental restraint system (SRS) vehicle and when handling an air bag module. This will reduce the risk of injury in the event of an accidental deployment. Never probe the connectors on the air bag module. Doing so can result in air bag deployment, which can result in personal injury.

✲✲ WARNING

The safety belt pretensioner is a pyrotechnic device. Always wear safety glasses when repairing an air bag equipped vehicle and when handling a safety belt buckle pretensioner or safety belt retractor pretensioner. Never probe a pretensioner electrical connector. Doing so could result in pretensioner or air bag deployment and could result in personal injury.

✲✲ WARNING

To reduce the risk of personal injury, do not use any memory saver devices.

➥If a seat equipped with a seat mounted side air bag and/or a safety belt pretensioner (if equipped) system is being serviced, the Supplemental Restraint System (SRS) must be depowered.

➥The air bag warning lamp illuminates when the RCM fuse is removed and the ignition switch is ON. This is normal operation and does not indicate a SRS fault. The SRS must be fully operational and free of faults before releasing the vehicle to the customer.

1. Turn all vehicle accessories OFF.
2. Turn the ignition switch to OFF.
3. Remove the right hand kick panel cover and remove the restraints control module fuse F2.05 (10A) from the central junction box.
4. Turn the ignition ON and visually monitor the air bag indicator for at least 30 seconds. The air bag indicator will remain lit continuously (no flashing) if the correct RCM fuse has been removed. If the air bag

indicator does not remain lit continuously, remove the correct RCM fuse before proceeding.
   Turn the ignition OFF.

✲✲ WARNING

To avoid accidental deployment and possible personal injury, the backup power supply must be depleted before repairing or replacing any front or side air bag (SRS) components and before servicing, replacing, adjusting or striking components near the front or side air bag sensors, such as doors, instrument panel, console, door latches, strikers, seats and hood latches.

5. The first row side impact sensors are located at or near the base of the B-pillars.
6. The second row side impact sensors, if equipped are located at or near the base of the C-pillars.
7. To deplete the backup power supply energy, disconnect the battery ground cable and wait at least 1 minute. Be sure to disconnect auxiliary batteries and power supplies, if equipped.
8. Disconnect the negative battery cable.

REPOWERING PROCEDURE

✲✲ WARNING

The restraint system diagnostic tool is for restraint system service only. Remove from vehicle prior to road use. Failure to remove could result in injury and possible violation of vehicle safety standards.

1. Make sure all restraint system diagnostic tool(s) that may have been installed during the repair have been removed from the vehicle and all SRS components are connected.
2. Turn the ignition switch from OFF to ON.
3. Install the restraints control module fuse F2.05 (10A) to the CJB and install the right hand kick panel cover.

✲✲ WARNING

Be sure that nobody is in the vehicle and that there is nothing blocking or set in front of any air bag module when the battery ground cable is connected.

4. Connect the negative battery cable.
5. Prove out the supplemental restraint system (SRS) as follows:

a. Turn the ignition key from ON to OFF. Wait 10 seconds, then turn the key back to ON and visually monitor the air bag indicator with the air bag modules installed. The air bag indicator will light continuously for approximately 6 seconds and then turn off. If an air bag supplemental restraint system (SRS) fault is present, the air bag indicator will either, fail to light, remain lit continuously or flash.

b. The flashing might not occur until approximately 30 seconds after the ignition switch has been turned from the OFF to the ON position. This is the time required for the RCM to complete the testing of the SRS.

c. If the air bag indicator is inoperative and a SRS fault exists, a chime will sound in a pattern of 5 sets of 5 beeps. If this occurs, the air bag indicator and any SRS fault discovered must be diagnosed and repaired.

d. Clear all continuous DTCs from the restraints control module using a scan tool.

Power Rack and Pinion Steering Gear

REMOVAL & INSTALLATION

2002–03 MODELS

1. Before servicing the vehicle, refer to the Precautions Section.
2. Lock the steering wheel in the straight-ahead position.
3. Remove or disconnect the following:
   - Negative battery cable
   - Front wheels
   - Steering column intermediate shaft
   - Outer tie rod ends
   - Power steering lines
   - Steering rack and pinion gear

To install:
4. Install or connect the following:
   - Steering rack and pinion gear. Tighten the bolts to 76 ft. lbs. (103 Nm).
   - Power steering lines
   - Outer tie rod ends. Tighten the nuts to 52–63 ft. lbs. (71–85 Nm).
   - Steering column intermediate shaft
   - Front wheels
   - Negative battery cable
5. Fill the power steering fluid reservoir.
6. Start the engine and check for leaks.
7. Check the wheel alignment and adjust, as necessary.

1 Tie-rod end nuts
2 Tie-rod end jam nuts
3 Tie-rod end
4 Outer bellows clamp
5 Inner bellows clamp
6 Steering gear bellows
7 Inner tie-rod
8 Variable assist power steering (VAPS) actuator electrical connector
9 Intermediate shaft slider bolt (loosen only)
10 Intermediate shaft-to-steering gear bolt

11 Intermediate shaft
12 Steering line clamp plate bolt
13 Steering lines and clamp plate
14 O-ring (high pressure hose)
15 O-ring (return hose)
16 Steering gear-to-crossmember nuts
17 Steering gear-to-crossmember bolts
18 Steering gear
19 Variable assist power steering (VAPS) actuator bolts
20 VAPS actuator

67197-LILS-G15

Steering rack and pinion gear mounting–2002–03 models

2004–05 Models

1. Before servicing the vehicle, refer to the Precautions Section.

❋❋ CAUTION

While repairing the power steering system, care should be taken to prevent the entry of contaminants or pre- mature failure of the power steering components can result. Do not allow the steering wheel to rotate while the intermediate shaft is disconnected or damage to the clockspring can result. If there is evidence that the shaft has rotated, the clockspring must be removed and recentered.

2. Disconnect the negative battery cable.

3. Hold the steering wheel in the straight-ahead position using a suitable holding device.

4. Remove the two tie-rod end nuts and separate the 2 tie-rod ends from the wheel knuckles.

5. Remove and discard the steering line

1 Tie-rod end nuts (2 required)
2 Steering line clamp plate bolt
3 Steering lines
   (pressure/return)
4 O-ring (pressure line)
5 O-ring (return line)
6 Variable assist power steering
   (VAPS) actuator electrical
   connector

7 Intermediate shaft slider bolt
8 Intermediate shaft-to-steering
   gear bolt
9 Intermediate shaft
10 Steering gear-to-crossmember
   nuts (3 required)
11 Steering gear-to-crossmember
   bolts (3 required)
12 Steering gear

06017-LS-G69

Exploded view of the steering gear mounting—2004–05 models

clamp plate bolt and disconnect the steering lines from the steering gear.

6. Remove and discard the 2 O-ring seals.

7. Disconnect the VAPS actuator electrical connector.

8. Loosen the intermediate shaft slider bolt.

9. Remove and discard the intermediate shaft-to-steering gear bolt and disconnect

the intermediate shaft from the steering gear.

10. Remove the 3 steering gear-to-crossmember nuts and bolts.

11. Remove the steering gear.

To install:

12. Install the steering gear.

13. Install the 3 steering gear-to-crossmember nuts and bolts. Tighten to 76 ft. lbs. (103 Nm).

14. Install a new the intermediate shaft-to-steering gear bolt and connect the intermediate shaft to the steering gear. Tighten to 26 ft. lbs. (35 Nm).

15. Tighten the intermediate shaft slider bolt. Tighten to 26 ft. lbs. (35 Nm).

16. Connect the VAPS actuator electrical connector.

17. Install new O-ring seals.

18. Install a new steering line clamp

plate bolt and connect the steering lines to the steering gear. Tighten to 89 inch lbs. (10 Nm).

19. Install the two tie-rod ends to the wheel knuckles. Tighten the nuts to 59 ft. lbs. (80 Nm).

20. Connect the negative battery cable.

Strut

REMOVAL & INSTALLATION

Front

1. Before servicing the vehicle, refer to the Precautions Section.
2. Remove or disconnect the following:
- Front wheel
- Stabilizer bar link
- Lower strut mounting bolt
- Upper strut mount cover and fasteners
- Strut and spring assembly

To install:

➡**Use new fasteners for assembly.**

3. Install or connect the following:
- Strut and spring assembly. Tighten the upper mount nuts to 21 ft. lbs. (29 Nm).
- Upper strut mount cover
- Lower strut mounting bolt. Tighten the bolts to 129 ft. lbs. (175 Nm).
- Stabilizer bar link. Tighten the nut to 41 ft. lbs. (55 Nm).
- Front wheel

Rear

1. Before servicing the vehicle, refer to the Precautions Section.
2. Remove or disconnect the following:
- Trunk trim covers
- Upper strut mount nuts
- Lower strut mount bolt
- Strut and spring assembly

To install:

➡**Use new fasteners for assembly.**

3. Install or connect the following:
- Strut and spring assembly. Tighten the lower bolt to 98 ft. lbs. (133 Nm) and the upper nuts to 21 ft. lbs. (28 Nm).
- Trunk trim covers

1 Shock absorber upper mount-to-body nuts
2 Stabilizer bar link-to-lower arm nut
3 Stabilizer bar link (detach only)
4 Shock absorber-to-lower arm nut
5 Shock absorber-to-lower arm bolt
6 Shock absorber and spring assembly

67197-LILS-G16

Front strut mounting

1 Nuts
2 Shock absorber-to-lower arm bolt
3 Shock absorber and spring assembly

67197-LILS-G17

Rear strut and spring assembly mounting

Coil Spring

REMOVAL & INSTALLATION

1. Before servicing the vehicle, refer to the Precautions Section.

2. Remove the strut assembly from the vehicle.

3. Compress the coil spring and remove the piston rod nut.

4. Remove or disconnect the following:

- Upper strut mount
- Spring upper seat
- Coil spring

To install:

5. Install or connect the following:

- Coil spring
- Spring upper seat
- Upper strut mount. Tighten the piston rod nut to 37 ft. lbs. (50 Nm).

6. Remove the spring compressor and install the strut assembly to the vehicle.

Upper Ball Joint

REMOVAL & INSTALLATION

The upper ball joint is serviced with the upper control arm as an assembly.

Lower Ball Joint

REMOVAL & INSTALLATION

The lower ball joint is serviced with the lower control arm as an assembly.

Upper Control Arm

REMOVAL & INSTALLATION

2002–03 MODELS

1. Before servicing the vehicle, refer to the Precautions Section.

2. Remove or disconnect the following:

- Front wheel
- Strut and spring assembly
- Upper ball joint and tapered washer
- Inner control arm fasteners
- Upper control arm

To install:

➡**Use new fasteners for assembly.**

3. Install or connect the following:

- Upper control arm. Tighten the inner fasteners to 35 ft. lbs. (48 Nm).
- Upper ball joint and tapered washer. Tighten the nut to 66 ft. lbs. (90 Nm).
- Strut and spring assembly
- Front wheel

2004–05 Models

FRONT

1. Disconnect the negative battery cable.

2. If removing the left hand upper arm:

a. Remove and discard the power steering reservoir nut. Position the power steering reservoir aside.

b. Remove the retainers and the left upper shock mount cover, if equipped.

c. Remove the EVAP canister purge valve nuts and position the EVAP canister purge valve aside.

d. Disconnect the fluid level sensor connector.

e. Disconnect the brake master cylinder main pressure transducer electrical connector, if equipped.

f. Back out the reservoir bolt to give access to the master cylinder bolt. Tighten to 8 Nm (6 lb-ft).

g. Remove the 2 brake master cylinder nuts and position the brake master cylinder aside.

3. Remove and discard the inner and outer upper arm nuts.

✳✳ WARNING

Do not remove the center nut. This nut holds the upper spring mount in place and if this nut is removed the spring tension will be released. Failure to follow these instructions may result in personal injury.

4. Remove and discard 2 shock absorber upper mount-to-body nuts.

➡**The hex-holding feature can be used to prevent turning of the stud while removing the nut. Remove the stabilizer bar link-to-lower arm nut. While prying down on the stabilizer bar, disconnect the stabilizer bar link from the stabilizer bar.**

5. Remove and discard shock absorber lower nut and bolt, then remove the shock absorber and spring assembly.

➡**Wire the top of the wheel knuckle to the body to prevent knuckle movement. To remove the nut, first loosen the nut, then use the hex-holding feature to pre-**

vent the upper control arm ball joint from turning while removing the nut. The tapered washer on the ball joint is reused with the new nut.

6. Remove and discard the suspension upper arm nut. Disconnect the suspension upper arm from the knuckle.

7. Remove and discard the inner and outer bolts and the suspension upper arm.

8. Remove the suspension upper arm.

To install:

9. Installation is the reverse of removal, please note the following specifications:

10. New inner and outer bolts 66 ft. lbs. (90 Nm).

11. New shock absorber lower nut and bolt, tighten to 129 ft. lbs. (175 Nm).

12. Stabilizer bar link-to-lower arm nut. Tighten to 41 ft. lbs. (55 Nm).

13. New shock absorber upper mount-to-body nuts. Tighten to 21 ft. lbs. (28 Nm).

14. New inner and outer upper arm nuts. Tighten to 35 ft. lbs. (48 Nm).

15. Brake master cylinder nuts and tighten to 18 ft. lbs. (25 Nm).

16. Reservoir bolt to 6 ft. lbs. (8 Nm).

17. EVAP canister purge valve nuts and tighten to 21 ft. lbs. (28 Nm).

18. Connect the negative battery cable.

REAR

1. Before servicing the vehicle, refer to the Precautions Section.

✳✳ CAUTION

Suspension fasteners are critical parts because they affect the performance of vital components and systems and their failure can result in major service expense. A new part with the same part number must be installed if installation is necessary. Do not use a new part of lesser quality or substitute design. Torque values must be used as specified during reassembly to make sure of correct retention of these parts.

➡**The suspension upper arm bushings or ball joints are not serviced separately from the upper rear suspension arm. If the bushings or ball joints require service a new suspension upper arm must be installed.**

2. Remove the hub cap. Measure the distance from the center of the hub to the lip

of the fender with the vehicle in a level, static ground position.

3. Remove the rear wheels.

4. Unclip the wheel speed sensor wire retainer from the suspension upper arm.

5. Remove the speed sensor bolt and the sensor. Position the sensor aside.

❋ CAUTION

Use care not to damage the Anti-lock Brake System (ABS) sensor ring. A damaged sensor ring will result in incorrect ABS operation.

➡**Use the hex holding feature to prevent the ball joint from turning while removing the nut.**

6. Remove and discard the upper ball joint-to-wheel knuckle nut, then disconnect the upper ball joint from the wheel knuckle.

7. Remove the suspension upper arm-to-subframe nuts and bolts. Discard the nuts and bolts.

8. Remove the suspension upper arm.

To install:

❋ CAUTION

Do not tighten the suspension upper arm-to-subframe nuts until the suspension is at curb height. Failure to do so can cause severe damage to the bushings resulting in poor ride quality and handling.

➡**The bolts and nuts retaining the suspension upper arm are of a torque prevailing design. New bolts and nuts must be used.**

9. Install the suspension upper arm, install new upper arm-to-subframe nuts and bolts. Do not tighten at this time.

➡**Install the ball joint-to-wheel knuckle nut using the hex holding feature until snug. Final tighten the nut using a line-type crow's foot and a torque wrench.**

10. Connect the ball joint to the wheel knuckle and hub assembly. Install a new ball joint-to-wheel knuckle nut and tighten to 66 ft. lbs. (90 Nm).

11. Clip the speed sensor wire to the suspension upper arm.

12. Position a jack stand under the suspension lower arm and raise the suspension until the distance between the center of the hub and the lip of the fender is equal to the measurement taken prior to removal.

13. Tighten the suspension upper arm nuts to 66 ft. lbs. (90 Nm).

14. Lower the suspension and remove the jack stand.

CONTROL ARM BUSHING REPLACEMENT

The control arm bushings are serviced with the control arm as an assembly.

Lower Control Arm

REMOVAL & INSTALLATION

2002–03 MODELS

1. Before servicing the vehicle, refer to the Precautions Section.

2. Remove or disconnect the following:
   - Front wheel
   - Splash shield
   - Stabilizer bar link
   - Lower strut mounting bolt
   - Lower ball joint
   - Rack and pinion steering gear
   - Inner control arm mounting bolts
   - Lower control arm

To install:

➡**Use new fasteners for assembly.**

3. Install or connect the following:
   - Lower control arm. Tighten the inner mounting bolts to 129 ft. lbs. (175 Nm).
   - Rack and pinion steering gear
   - Lower ball joint. Tighten the nut to 111 ft. lbs. (150 Nm).
   - Lower strut mounting bolt. Tighten the bolt to 129 ft. lbs. (175 Nm).
   - Stabilizer bar link. Tighten the nut to 41 ft. lbs. (55 Nm).
   - Splash shield
   - Front wheel

4. Check the wheel alignment and adjust as necessary.

2004–05 Models

FRONT

1. Before servicing the vehicle, refer to the Precautions Section.

❋❋ CAUTION

Suspension fasteners are critical parts because they affect the performance of vital components and systems and their failure can result in major service expense. A new part with the same part number must be installed if installation is necessary.

Do not use a new part of lesser quality or substitute design. Torque values must be used as specified during reassembly to make sure of correct retention of these parts.

➡**New cam bolts and lock nuts must be installed whenever the lower control arm is removed.**

2. Remove the splash shield.

➡**The hex holding feature can be used to prevent turning of the stud while removing the nut.**

3. Remove and discard the stabilizer bar link-to-lower arm nut. While prying down on the stabilizer bar, disconnect the stabilizer bar link.

4. Remove and discard the shock absorber-to-lower arm nut and bolt. Disconnect the shock absorber and spring assembly from the lower arm.

➡**The hex holding feature can be used to prevent turning of the stud while removing the nut. The tapered washer is reused with the new ball joint nut.**

5. Remove and discard the lower ball joint-to-knuckle nut. Separate the front suspension lower arm from the wheel knuckle.

6. Remove and discard the front nut and bolt.

7. Remove the power steering gear nuts and bolts. Discard the nuts.

8. Position the power steering gear to access the lower control arm rear bolt.

9. Remove and discard the lower control arm rear nut and bolt. Remove the front suspension lower arm.

To install:

10. Position the front suspension lower arm.

11. Install a new caster adjustment cam bolt and a new nut.

12. The bolt must be installed from the rear, as illustrated. Install the bolt with the cam lobe down. The cam must be seated between the cam guides on the No. 1 crossmember. The nut should only be snugged up at this time to allow for wheel alignment adjustment.

13. Install a new camber adjustment cam bolt and a new nut.

14. The bolt must be installed from the rear as illustrated. Install the bolt with the cam lobe down. The cam must be seated in the groove in the No. 2 crossmember. The nut should only be snugged up at this time to allow for wheel alignment adjustment.

Position the caster adjustment cam bolt as shown

06017-LS-G70

Position the camber adjustment cam bolt as shown

06017-LS-G71

➡Make sure the tapered washer is installed on the ball joint before connecting the lower control arm to the knuckle. The hex holding feature can be used to prevent turning of the stud while tightening the nut.

15. Connect the lower control arm to the knuckle. Install a new lower ball joint-to-knuckle nut and tighten to 148 ft. lbs. (200 Nm).

➡Make sure the shock absorber lower bushing end caps are in place before installing the bolt and nut. Connect the shock absorber and spring assembly to the lower control arm.

16. Install a new shock absorber-to-lower arm bolt and nut. Tighten to 129 ft. lbs. (175 Nm).

17. While prying down on the stabilizer bar, connect the stabilizer bar link to the stabilizer bar. Install a new stabilizer bar link-to-lower arm nut. Install the nut until snug using the hex-head feature, then tighten the nut using a socket and a torque wrench to 41 ft. lbs. (55 Nm).

18. Install the splash shield.

19. Position the power steering gear in place. Install the power steering gear bolts and new nuts and tighten to 76 ft. lbs. (103 Nm).

20. Final tighten the camber and caster bolts to 129 ft. lbs. (175 Nm).

21. Check the wheel alignment. Adjust as necessary.

REAR

1. Before servicing the vehicle, refer to the Precautions Section.

❈❈ CAUTION

Suspension fasteners are critical parts because they affect the performance of vital components and systems and their failure can result in major service expense. A new part with the same part number must be installed if installation is necessary. Do not use a new part of lesser quality or substitute design. Torque values must be used as specified during reassembly to make sure of correct retention of these parts.

➡The lower arm bushings are not serviced separately from the lower suspension arm. If the bushings require service, a new lower arm must be installed.

2. With the vehicle in a static, level ground position, remove the hub cap and measure the distance from the center of the hub to the lip of the fender (curb height).

3. Remove the rear wheels.

4. Remove the rear brake rotor.

5. Disconnect the shock absorber and spring assembly.

6. Remove and discard the shock absorber-to-lower arm bolt.

7. Disconnect the shock absorber and spring assembly from the lower arm.

❈❈ CAUTION

Do not use air tools to remove the nut. Damage to the boot can result.

➡Use the hex holding feature to prevent the stabilizer bar link ball joint from turning while removing the stabilizer bar link nut.

8. Disconnect the stabilizer bar link from the suspension lower arm.

9. Remove and discard the stabilizer bar link-to-lower arm nut.

10. Disconnect the stabilizer bar link.

11. Disconnect the lower arm from the wheel knuckle.

12. Remove and discard the lower arm-to-wheel knuckle bolt. Disconnect the lower arm.

13. Remove and discard the lower arm-to-frame bolts. Remove the lower arm.

To install:

❈❈ CAUTION

Do not tighten the lower arm-to-subframe fasteners until the suspension is at curb height. Failure to do so can cause severe damage to the bushings resulting in poor ride quality and handling.

14. Install the lower arm. By positioning the suspension lower arm onto the subframe.

15. Install new lower arm-to-frame bolts and nuts. Do not tighten at this time.

16. Connect the lower arm to the knuckle. Install a new lower arm-to-wheel knuckle nut and bolt. Tighten to 111 ft. lbs. (150 Nm).

❈❈ CAUTION

Do not use air tools to install the nut. Damage to the boot can result.

➡Use the hex holding feature to prevent the stabilizer bar link ball joint from turning while installing the nut on the stabilizer bar link. Tighten the nut to specification using a socket and a torque wrench.

17. Connect the stabilizer bar link to the suspension lower arm. Install a new stabilizer bar link nut. Tighten to 35 ft. lbs. (48 Nm).

18. Position the shock absorber and spring assembly onto the suspension lower arm. Install a new shock absorber-to-lower arm bolt and tighten to 98 ft. lbs. (133 Nm).

19. Position a jack stand under the lower arm and raise the suspension until the measurement between the center of the hub and the lip of the fender is equal to the measurement taken prior to removal.

20. Tighten the lower arm-to-frame rear bolt and the lower arm-to-frame front nut to 111 ft. lbs. (150 Nm).

21. Lower the suspension and remove the jack stand.

22. Install the rear brake disc rotor and rear wheel.

CONTROL ARM BUSHING REPLACEMENT

The control arm bushings are serviced with the control arm as an assembly.

Wheel Bearings

ADJUSTMENT

The wheel bearings are not adjustable.

REMOVAL & REPLACEMENT

Front

1. Before servicing the vehicle, refer to the Precautions Section.
2. Remove or disconnect the following:
   - Front wheel
   - Brake caliper and rotor
   - Wheel speed sensor connector
   - Hub and bearing assembly

➡ The hub and bearing assembly is not pressed into the knuckle. Do not use a slide hammer or press to remove the hub and bearing assembly. Damage to the hub and bearing assembly may result.

To install:

➡ Do not remove the wheel speed sensor from the hub and bearing assembly unless it is being replaced. If installing a new hub and bearing assembly, a new wheel speed sensor must be installed.

➡ Use new fasteners for assembly.

3. Install or connect the following:
   - Hub and bearing assembly. Tighten the bolts to 66 ft. lbs. (90 Nm).

- Wheel speed sensor connector
- Brake caliper and rotor
- Front wheel

Rear

✳✳ CAUTION

Suspension fasteners are important parts because they affect performance of vital components and systems and their failure can result in major service expense. A new part with the same part number must be installed if installation is necessary. Do not use a new part of lesser quality or substitute design. Torque values must be used as specified during reassembly to make sure of correct retention of these parts.

➡ The wheel knuckle/wheel hub is serviced as an assembly. Do not attempt to separate the wheel hub from the wheel knuckle.

1. Before servicing the vehicle, refer to the Precautions Section.
2. For reference during the installation of an upper or lower arm, measure the distance between the center of the wheel hub and the lip of the fender with the weight of

Using the tools shown, separate the axle shaft from the hub

the vehicle resting on the wheel and tire assemblies.
3. Remove the rear wheel.
4. Remove the brake caliper and rotor.

✳✳ CAUTION

The axle shaft retaining nut is a one time use item and a new nut must be installed when removed. Failure to do so can cause the nut to loosen during vehicle operation resulting in loss of vehicle control.

N 1 — 103 Nm (76 lb-ft)

N 9 — 90 Nm (66 lb-ft)

i 2

N 8 — 103 Nm (76 lb-ft)

4

5

3

11

7 i

200 Nm (148 lb-ft) — 10 N

90 Nm (66 lb-ft) — 6 N

1 Anchor plate bolt
2 Brake caliper, pads and anchor plate
3 Brake disc
4 Speed sensor electrical connector (disconnect)

Front hub and bearing exploded view

5. Remove and discard the axle shaft retaining nut.

6. Disconnect the toe link from the knuckle by removing the toe link-to-wheel knuckle bolt. Discard the nut and bolt and disconnect the link.

7. Remove the ABS bolt and disconnect the ABS sensor.

✳✳ CAUTION

Support the axle shaft after removing it from the hub. Failure to do so can damage the axle shaft.

8. Do not use a hammer to separate the outboard CV joint from the hub. Damage to the threads and internal CV joint components can result.

9. Using the tools illustrated, separate the axle shaft from the hub. Support the axle shaft.

10. Disconnect the lower arm from the wheel knuckle by removing the lower arm-to-wheel knuckle bolt and disconnecting the arm. Discard the nut.

11. Remove the wheel knuckle and hub assembly.

→**Use the hex holding feature to prevent the upper arm ball joint from turning while removing the nut.**

12. Remove and discard the upper ball joint-to-wheel knuckle nut.

To install, tighten to 90 Nm (66 lb-ft).

13. Remove the wheel knuckle and hub assembly.

14. Installation is the reverse of removal, please note the following steps and torque specifications:

90 Nm (66 lb-ft)

300 Nm (221 lb-ft)

55 Nm (41 lb-ft)

150 Nm (111 lb-ft)

10 Nm (89 lb-in)

1 Axle nut
2 Toe link-to-wheel knuckle nut
3 Toe link-to-wheel knuckle bolt
4 Speed sensor-to-wheel knuckle bolt
5 Lower arm-to-wheel knuckle nut
6 Lower arm-to-wheel knuckle bolt
7 Upper ball joint-to-wheel knuckle nut
8 Wheel knuckle

67197-LILS-G21

Exploded view of the rear wheel bearing assembly

15. Before tightening the upper or lower suspension arm-to-frame fasteners, position a jack under the lower arm and raise the vehicle until the distance between the center of the hub and the lip of fender is equal to the measurement taken during the removal procedure.

a. Upper ball joint-to-wheel knuckle nut to 66 ft. lbs. (90 Nm).

b. Lower arm-to-wheel knuckle bolt to 111 ft. lbs (150 Nm).

c. ABS bolt to 89 inch lbs. (10 Nm).

d. Toe link-to-wheel knuckle bolt to 41 ft. lbs. (55 Nm).

e. Axle shaft retaining nut to 221 ft. lbs. (300 Nm).

16. Check and, if necessary, align the rear suspension

BRAKES

Brake Caliper

REMOVAL & INSTALLATION

Front

1. Before servicing the vehicle, refer to the Precautions Section.

2. Remove or disconnect the following:
   - Front wheel
   - Brake fluid hose
   - Caliper mounting bolts
   - Brake caliper

To install:

3. Install or connect the following:
   - Brake caliper. Tighten the mounting bolts to 26 ft. lbs. (35 Nm).
   - Brake fluid hose. Use new copper washers and tighten the bolt to 35 ft. lbs. (47 Nm).
   - Front wheel

4. Bleed the brake system.

5. Before attempting to move the vehicle, pump the brake pedal to seat the pads against the rotors. Make sure the vehicle has a firm brake pedal. Check the level of the brake fluid and add DOT 3 or 4 brake fluid if necessary.

Rear

1. Before servicing the vehicle, refer to the Precautions Section.

2. Remove or disconnect the following:
   - Rear wheel
   - Parking brake cable
   - Caliper mounting bolts
   - Brake fluid hose
   - Brake caliper

To install:

3. Install or connect the following:
   - Brake caliper. Tighten the mounting bolts to 26 ft. lbs. (35 Nm).
   - Brake fluid hose. Use new copper washers and tighten the bolt to 35 ft. lbs. (47 Nm).
   - Parking brake cable
   - Rear wheel

4. Bleed the brake system.

5. Before attempting to move the vehicle, pump the brake pedal to seat the pads against the rotors. Make sure the vehicle has a firm brake pedal. Check the level of the brake fluid and add DOT 3 or 4 brake fluid if necessary.

1 Bolts
2 Guide pin/boot
3 Brake caliper assembly
4 Brake pads
5 Slippers

6 Bolts
7 Brake caliper support bracket
8 Brake disc
9 Rivets
10 Brake disc shield

67197-LILS-G18

Exploded view of the front disc brake components

1 Guide pin assembly
2 Brake caliper
3 Brake pads (part of kit)
4 Bolts

5 Brake caliper support bracket
6 Brake disc
7 Rivets
8 Brake disc shield

67197-LILS-G19

Exploded view of the rear disc brake components

Disc Brake Pads

REMOVAL & INSTALLATION

Front

1. Before servicing the vehicle, refer to the Precautions Section.

2. Remove or disconnect the following:
- Master cylinder reservoir cap and check the fluid level in the reservoir
- Brake fluid until the reservoir is ½ full. Discard the removed fluid.
- Wheel and tire assembly
- Disc brake caliper locating pins. Lift the caliper assembly from the anchor plate and rotor.

3. Suspend the caliper inside the fender housing with wire. Do not allow the caliper to hang from the brake hose.
- Inner and outer brake pads. Inspect the rotor braking surfaces for scoring and machine as necessary.

To install:

4. Use a C-clamp and an old brake pad or block of wood to seat the caliper piston in its bore.

5. Remove any rust buildup from the inside of the caliper in the brake pad contact area.

6. Install or connect the following:
- Inner pad in the caliper piston
- Outer pad onto the anchor plate. Make sure the clips are properly seated.
- Disc brake caliper onto the anchor plate
- Caliper locating pins and torque to 26 ft. lbs. (35 Nm)
- Wheel and tire assembly

7. Pump the brake pedal several times prior to moving the vehicle to position the brake pads to the rotor.

8. Refill the master cylinder reservoir as necessary, using only clean DOT 3 or 4 brake fluid from a closed container.

Rear

1. Before servicing the vehicle, refer to the Precautions Section.

2. Remove or disconnect the following:
- Master cylinder reservoir cap and check the fluid level in the reservoir
- Brake fluid until the reservoir is ½ full. Discard the removed fluid.
- Wheel and tire assembly

- Disc brake caliper locating pins at the support bracket
- Caliper
- Disc brake pads

3. Inspect the rotor braking surfaces for scoring and machine as necessary.

To install:

4. Using caliper compressor tool 206-026, compress the piston until it is fully seated. Position the notch in the caliper piston up and down to align with the alignment pin on the brake pad

5. Install or connect the following:
- Brake pads in the support bracket
- Caliper assembly over the rotor into position on the support bracket. Make sure the brake pads are installed correctly.
- Disc brake caliper locating pin, and tighten to 25 ft. lbs. (33 Nm)
- Wheel and tire assembly

6. Pump the brake pedal several times prior to moving the vehicle, to position the brake pads to the rotor.

7. Refill the master cylinder reservoir if necessary, using only clean DOT 3 or 4 brake fluid from a closed container.

8. Road test the vehicle and check the brake system for proper operation.

SPECIFICATION AND MAINTENANCE CHARTS

ENGINE AND VEHICLE IDENTIFICATION

Engine							Model Year	
Code ①	Liters (cc)	Cu. In.	Cyl.	Fuel Sys.	Engine Type	Eng. Mfg.	Code ②	Year
4	3.8 (3802)	232	6	SFI	OHV	Ford	2	2002
R	4.6 (4593)	281	8	SFI	DOHC	Ford	3	2003
V	4.6 (4593)	281	8	SFI	DOHC	Ford	4	2004
X	4.6 (4593)	281	8	SFI	SOHC	Ford	5	2005
Y	4.6 (4593)	281	8	SFI	DOHC	Ford		

OHV: Overhead Valve

SOHC: Single Overhead Camshaft

DOHC: Double Overhead Camshaft

SFI: Sequential Fuel Injection

① 8th position of VIN

② 10th position of VIN

06017-MUST-C01

GENERAL ENGINE SPECIFICATIONS

All measurements are given in inches.

Year	Model	Engine Displacement Liters	Engine Series (ID/VIN)	Net Horsepower @ rpm	Net Torque @ rpm (ft. lbs.)	Bore x Stroke (in.)	Compression Ratio	Oil Pressure @ rpm
2002	Mustang	3.8	4	190@5250	220@3000	3.81x3.39	9.0:1	40-60@2500
		4.6	V	320@6000	317@4750	3.55x3.54	9.5:1	20-45@1500
		4.6	X	260@5000	302@4000	3.55x3.54	9.0:1	20-45@1500
2003	Mustang	3.8	4	190@5250	220@3000	3.81x3.39	9.0:1	40-60@2500
		4.6	V	320@6000	317@4750	3.55x3.54	9.5:1	20-45@1500
		4.6	X	260@5000	302@4000	3.55x3.54	9.0:1	20-45@1500
2004	Mustang	3.8	4	193@5500	225@2800	3.81x3.39	9.3:1	40-60@2500
		4.6	R	305@5800	320@4200	3.55x3.54	10.1:1	20-45@1500
		4.6	X	260@5250	302@4000	3.55x3.54	9.37:1	20-45@1500
		4.6	Y	390@6000	390@3500	3.55x3.54	8.5:1	20-45@1500

06017-MUST-C02

ENGINE TUNE-UP SPECIFICATIONS

Year	Engine Displacement Liters	Engine ID/VIN	Spark Plug Gap (in.)	Ignition Timing (deg.)		Fuel Pump (psi) ①	Idle Speed (rpm)		Valve Clearance	
				MT	AT		MT	AT	Intake	Exhaust
2002	3.8	4	0.054	②	②	28-54	②	②	HYD	HYD
	4.6	V	0.054	10B	—	35-45	②	—	HYD	HYD
	4.6	X	0.054	10B	10B	35-45	②	②	HYD	HYD
2003	3.8	4	0.054	②	②	28-54	②	②	HYD	HYD
	4.6	V	0.054	10B	—	35-45	②	—	HYD	HYD
	4.6	X	0.054	10B	10B	35-45	②	②	HYD	HYD
2004	3.8	4	0.052-0.056	B	B	27-37	②	②	HYD	HYD
	4.6	R	0.052-0.056	10B	10B	27-37	②	②	HYD	HYD
	4.6	X	0.052-0.056	10B	10B	27-37	②	②	HYD	HYD
	4.6	Y	0.052-0.056-	10B	10B	27-37	②	B	HYD	HYD

NOTE: The Vehicle Emission Control Information label often reflects specification changes made during production.

The label figures must be used if they differ from those in this chart.

B: Before Top Dead Center

HYD: Hydraulic

① Fuel pressure with engine running, pressure regulator vacuum hose connected

② Refer to Vehicle Emission Control Information label

06017-MUST-C03

93003G01

3.8L Engine
Firing order: 1–4–2–5–3–6
Distributorless ignition system

93003G02

4.6L Engine
Firing order: 1–3–7–2–6–5–4–8
Distributorless ignition system—One coil per cylinder

1 Tensioner
2 A/C compressor or idler pulley
3 Idler pulley
4 Alternator
5 Power steering pump
6 Crankshaft
7 Water pump

67197-MUST-G01

Accessory drive belt routing—3.8L (VIN 4) engine (2002–03 model)

1 A/C clutch
2 Drive belt
3 Water pump
4 Idler pulley
5 Alternator
6 Power steering pump
7 Vibration damper
8 Tensioner
9 Tensioner

67197-MUST-G02

Accessory drive belt routing—3.8L (VIN 4) engine (2004 model)

1 Generator pulley
2 Drive belt
3 Water pump pulley
4 Power steering pump pulley
5 Crankshaft vibration damper
6 A/C clutch pulley
7 Drive belt tensioner
8 Belt idler pulley
8 Belt idler pulley

67197-MUST-G03

Accessory drive belt routing—4.6L (VIN X and VIN R) engine

Drive Belt Routing

FRONT OF ENGINE

1. Generator
2. Water pump pulley
3. Belt idler pulley
4. Drive belt tensioner
5. Power steering pump
6. Drive belt
7. A/C compressor
8. Crankshaft pulley

79224G26

Accessory drive belt routing—4.6L (VIN V) engine

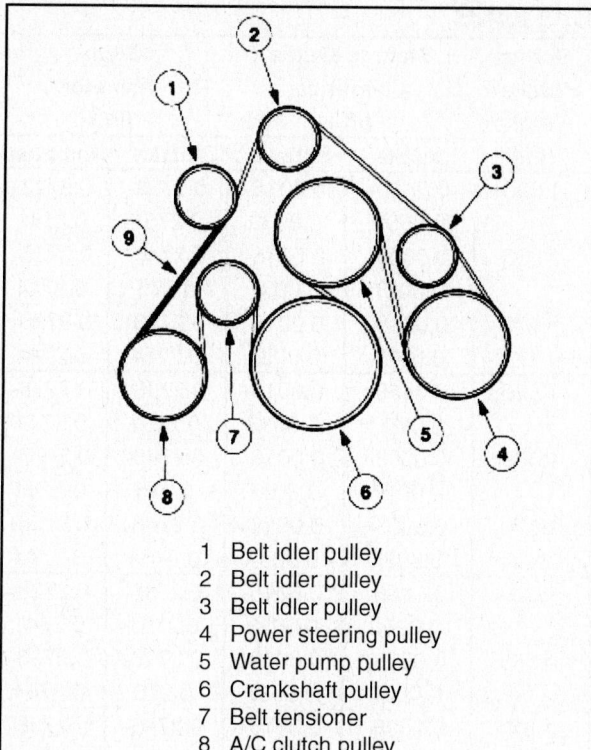

1 Belt idler pulley
2 Belt idler pulley
3 Belt idler pulley
4 Power steering pulley
5 Water pump pulley
6 Crankshaft pulley
7 Belt tensioner
8 A/C clutch pulley
9 Drive belt

67197-MUST-G04

Accessory drive belt routing—4.6L (VIN Y) engine

1 Supercharger pulley
2 Belt idler pulley
3 Alternator pulley
4 Belt idler pulley
5 Crankshaft extension pulley
6 Belt tensioner
7 Belt idler pulley
8 Drive belt

67197-MUST-G05

Supercharger accessory drive belt routing—4.6L (VIN Y) engine

CAPACITIES

Year	Model	Engine Displacement Liters	Engine ID/VIN	Engine Oil with Filter (qts.)	Transmission (pts)		Drive Axle Rear (pts.)	Fuel Tank (gal.)	Cooling System (qts.)
					Manual	Auto. ①			
2002	Mustang	3.8	4	5.0	5.6	27.8	3.50	15.7	11.8
		4.6	V	6.0	6.5	27.8	3.75	15.7	14.1
		4.6	X	②	6.5	25.6	3.75	15.7	14.1
2003	Mustang	3.8	4	5.0	5.6	27.8	3.50	15.7	11.8
		4.6	V	6.0	6.5	27.8	3.75	15.7	14.1
		4.6	X	②	6.5	25.6	3.75	15.7	14.1
2004	Mustang	3.8	4	5.0	5.6	27.8	③	15.7	11.8
		4.6	R	6.25	7.5	25.6	③	15.7	14.1
		4.6	X	6.5	7.5	25.6	③	15.7	14.1
		4.6	Y	6.25	7.5	25.6	③	15.7	④

NOTE: All capacities are approximate. Add fluid gradually and ensure a proper fluid level is obtained.

① Includes torque converter

② If equipped with an automatic transmission: 6.7 qts.
If equipped with a manual transmission: 6.4 qts.

③ If equipped with 7.5 inch ring gear: 3.25-3.50 pts.
If equipped with 8.8 inch ring gear: 3.5-3.75 pts. + 4 oz. friction modifier.
If equipped with 8.8 inch ring gear IRS: 2.6-2.9 pts. + 4 oz. friction modifier.

④ 4.6L Supercharged engine coolant capacity is 4.0 gal. in engine + 4.0 qts. In supercharger cooling system.

VALVE SPECIFICATIONS

Year	Engine Displacement Liters	Engine ID/VIN	Seat Angle (deg.)	Face Angle (deg.)	Spring Test Pressure (lbs. @ in.)	Spring Installed Height (in.)	Stem-to-Guide Clearance (in.)		Stem Diameter (in.)	
							Intake	Exhaust	Intake	Exhaust
2002	3.8	4	44.7	45.7	224@1.16	1.620	0.0450-0.0900	0.0015-0.0033	0.2738-0.2751	0.2728-0.2741
	4.6	V	45	45.5	160@1.03	1.660	0.0008-0.0027	0.0018-0.0037	0.2746-0.275	0.2736-0.2744
	4.6	X	45	45.5	161@1.03	1.570	0.0008-0.0027	0.0018-0.0037	0.2746-0.2754	0.2736-0.2744
2003	3.8	4	44.7	45.7	224@1.16	1.620	0.0450-0.0900	0.0015-0.0033	0.2738-0.2751	0.2728-0.2741
	4.6	V	45	45.5	160@1.03	1.660	0.0008-0.0027	0.0018-0.0037	0.2746-0.275	0.2736-0.2744
	4.6	X	45	45.5	161@1.03	1.570	0.0008-0.0027	0.0018-0.0037	0.2746-0.2754	0.2736-0.2744
2004	3.8	4	44.7	45.7	224@1.16	1.620	0.0450-0.0900	0.0015-0.0033	0.2738-0.2751	0.2728-0.2741
	4.6	R	45	45.5	160@1.03	1.660	0.0008-0.0027	0.0018-0.0037	0.2746-0.275	0.2736-0.2744
	4.6	X	45	45.5	162@1.13	1.67-1.690	0.0008-0.0027	0.0018-0.0037	0.2746-0.275	0.2736-0.2744
	4.6	Y	45	45.5	160@1.03	1.423	0.0008-0.0027	0.0018-0.0037	0.2746-0.2754	0.2736-0.2744

06017-MUST-C05

CRANKSHAFT AND CONNECTING ROD SPECIFICATIONS

All measurements are given in inches.

Year	Engine Displacement Liters	Engine ID/VIN	Crankshaft				Connecting Rod		
			Main Brg. Journal Dia.	Main Brg. Oil Clearance	Shaft End-play	Thrust on No.	Journal Diameter	Oil Clearance	Side Clearance
2002	3.8	4	①	0.0010-0.0014	0.0040-0.0080	3	2.3103-2.3111	0.0010-0.0014	0.0047-0.0144
	4.6	V	2.6567-2.6577	0.0001-0.0018	0.0051-0.0119	5	2.0859-2.0867	0.0011-0.0027	0.0006-0.0177
	4.6	X	2.6569-2.6576	0.0011-0.0026	0.0051-0.0119	5	2.0861-2.0867	0.0011-0.0027	0.0006-0.0177
2003	3.8	4	①	0.0010-0.0014	0.0040-0.0080	3	2.3103-2.3111	0.0010-0.0014	0.0047-0.0144
	4.6	V	2.6567-2.6577	0.0001-0.0018	0.0051-0.0119	5	2.0859-2.0867	0.0011-0.0027	0.0006-0.0177
	4.6	X	2.6569-2.6576	0.0011-0.0026	0.0051-0.0119	5	2.0861-2.0867	0.0011-0.0027	0.0006-0.0177
2004	3.8	4	①	0.0010-0.0014	0.0040-0.0080	3	2.3103-2.3111	0.0010-0.0014	0.0047-0.0144
	4.6	R	2.6567-2.6577	0.0001-0.0018	0.0051-0.0119	5	2.0859-2.0867	0.0011-0.0027	0.0006-0.0177
	4.6	X	2.6567-2.6577	0.0001-0.0018	0.0051-0.0119	5	2.0859-2.0867	0.0011-0.0027	0.0006-0.0177
	4.6	Y	2.6569-2.6576	0.0011-0.0026	0.0051-0.0119	5	2.0861-2.0867	0.0011-0.0027	0.0006-0.0177

① Journals 1, 2, 3: 2.5194-2.5186 in.
  Journal 4: 2.5100-2.5092 in.

06017-MUST-C06

PISTON AND RING SPECIFICATIONS

All measurements are given in inches.

| Year | Engine Displacement Liters | Engine ID/VIN | Piston Clearance | Ring Gap | | | Ring Side Clearance | | |
				Top Compression	Bottom Compression	Oil Control	Top Compression	Bottom Compression	Oil Control
2002	3.8	4	0.0007-0.0017	0.011-0.012	0.009-0.020	0.015-0.058	0.0016-0.0034	0.0016-0.0034	SNUG
	4.6	V	0.0000-0.0010	0.010-0.020	0.009-0.019	0.006-0.026	0.0003-0.0009	0.0012-0.0031	SNUG
	4.6	X	0.0005-0.0010	0.009-0.019	0.009-0.019	0.006-0.026	0.0016-0.0035	0.0018-0.0031	SNUG
2003	3.8	4	0.0007-0.0017	0.011-0.012	0.009-0.020	0.015-0.058	0.0016-0.0034	0.0016-0.0034	SNUG
	4.6	V	0.0000-0.0010	0.010-0.020	0.009-0.019	0.006-0.026	0.0003-0.0009	0.0012-0.0031	SNUG
	4.6	X	0.0005-0.0010	0.009-0.019	0.009-0.019	0.006-0.026	0.0016-0.0035	0.0018-0.0031	SNUG
2004	3.8	4	0.0007-0.0017	0.007-0.013	0.012-0.021	0.001-0.025	0.0012-0.0032	0.0012-0.0032	SNUG
	4.6	R	0.0000-0.0010	0.006-0.002	0.002-0.0022	0.006-0.026	0.0003-0.0009	0.0012-0.0031	SNUG
	4.6	X	0.0000-0.0010	0.005-0.011	0.0018-0.0022	0.006-0.026	0.002-0.0040	0.0012-0.0031	SNUG
	4.6	Y	0.0000-0.0010	0.0020	0.0200	0.0260	0.0014-0.0031	0.0012-0.003	0.002-0.008

06017-MUST-C07

TORQUE SPECIFICATIONS
All readings in ft. lbs.

Year	Engine Displacement Liters	Engine ID/VIN	Cylinder Head Bolts	Main Bearing Bolts	Rod Bearing Bolts	Crankshaft Damper Bolts	Flywheel Bolts	Manifold Intake	Manifold Exhaust	Spark Plugs	Oil Pan Drain Plug
2002	3.8	4	①	②	③	118	54-64	④	15-22	7-15	22-30
	4.6	V	⑤	⑥	⑦	114-121	54-64	⑧	13-16	7-15	22-30
	4.6	X	⑨	⑩	⑪	114-121	54-64	15-22	15-22	7-15	22-30
2003	3.8	4	①	②	③	118	54-64	④	15-22	7-15	22-30
	4.6	V	⑤	⑥	⑦	114-121	54-64	⑧	13-16	7-15	22-30
	4.6	X	⑨	⑩	⑪	114-121	54-64	15-22	15-22	7-15	22-30
2004	3.8	4	①	②	③	118	54-64	④	15-22	7-15	22-30
	4.6	V	⑤	⑥	⑦	114-121	54-64	⑧	13-16	7-15	22-30
	4.6	V	⑤	⑥	⑦	114-121	54-64	⑧	13-16	7-15	22-30
	4.6	X	⑨	⑩	⑪	114-121	54-64	15-22	15-22	7-15	22-30

① Do not reuse cylinder head bolts.
Step 1: 15 ft. lbs.
Step 2: 29 ft. lbs.
Step 3: 37 ft. lbs.
Step 4: Loosen bolts one at a time and retorque as follows:
Long bolts: 11-18 ft. lbs.
Short bolts: 7-15 ft. lbs.
Step 5: Tighten 85-95 degrees

② Step 1: 37 ft. lbs.
Step 2: Tighten 115-125 degrees

③ Step 1: 18 ft. lbs.
Step 2: 33 ft. lbs.
Step 3: Tighten 90-120 degrees

④ Upper intake manifold bolts:
Step 1: 8 ft. lbs.
Step 2: 15 ft. lbs.
Step 3: 24 ft. lbs
Lower intake manifold bolts:
Step 1: 13 ft. lbs.
Step 2: 16 ft. lbs.

⑤ Step 1: 30 ft. lbs.
Step 2: Tighten 90 degrees
Step 3: Loosen all bolts one (1) turn
Step 4: 30 ft. lbs.
Step 5: Tighten 90 degrees
Step 6: Tighten an additional 90 degrees

⑥ Step 1: Main bearing cap bolts: 6-9 ft. lbs.
Step 2: Main bearing cap bolts, outer: 16-21 ft. lbs.
Step 3: Main bearing cap bolts, inner: 27-32 ft. lbs.
Step 4: Tighten main bearing cap bolts 85-95 degrees
Step 5: Main cap adjusting screws: 4 ft. lbs. then 7.5 ft. lbs.
Step 6: Main cap side bolts: 7 ft. lbs. then 14-17 ft. lbs.

⑦ Step 1: 5ft. lbs.
Step 2: 10 ft. lbs.
Step 3: 18-25 ft. lbs.
Step 4: Tighten 85-95 degrees

⑧ Step 1: Four inside short bolts: 9-11 ft. lbs.
Step 2: All other bolts: 13-16 ft. lbs.
Step 3: Tighten 85-95 degrees

⑨ Do not reuse cylinder head bolts.
Step 1: 27-32 ft. lbs.
Step 2: Tighten each bolt 85-95 degrees
Step 3: Repeat Step 2

⑩ Do not reuse main cap bolts.
Step 1: Main bearing cap bolts: 22-25 ft. lbs.
Step 2: Tighten each bolt 85-95 degrees
Step 3: Main bearing cap adjust screws: 4 ft. lbs. then 6-8 ft. lbs.
Step 4: Main bearing cap side bolts: 7 ft. lbs. then 14-17 ft. lbs.

⑪ Do not reuse rod bolts.
Step 1: 12 ft. lbs.
Step 2: Tighten 85-95 degrees

06017-MUST-C08

WHEEL ALIGNMENT

Year	Model		Caster Range (+/-Deg.)	Caster Preferred Setting (Deg.)	Camber Range (+/-Deg.)	Camber Preferred Setting (Deg.)	Toe-in (in.)
2002	Mustang	F	0.75	0	0.75	0	0.13 +/- 0.13
		R	—	—	0.50	0	0.13 +/- 0.07
2003	Mustang	F	0.75	0	0.75	0	0.13 +/- 0.13
		R	—	—	0.50	0	0.13 +/- 0.07
2004	Mustang	F	0.75	0	0.75	0	0.13 +/- 0.13
		R	—	—	0.50	0	0.13 +/- 0.07
	Mustang Cobra	F	0.75	0	0.75	0	0.13 +/- 0.13
		R	—	—	0.50	0	0.13 +/- 0.07

06017-MUST-C09

TIRE, WHEEL AND BALL JOINT SPECIFICATIONS

Year	Model	OEM Tires Standard	OEM Tires Optional	Tire Pressures (psi) Front	Tire Pressures (psi) Rear	Wheel Size	Ball Joint Inspection	Wheel Lug Torque (ft. lbs.)
2002	Mustang, base	P205/65HR15	None	30	30	7-JJ	①	95
	Mustang GT	P225/55HR16	P245/45ZR17	30	30	8-JJ	①	95
2003	Mustang, base	P205/65HR15	None	30	30	7-JJ	①	95
	Mustang GT	P225/55HR16	P245/45ZR17	30	30	8-JJ	①	95
2004	Mustang, base	P225/55R16	None	30	30	-	②	95
	Mustang GT	245/45ZR17	None	30	30	-	②	95
	Mustang Cobra	245/45ZR17	None	30	30	-	②	95

OEM: Original Equipment Manufacturer

PSI: Pounds Per Square Inch

① Replace if any measurable movement is found

② Replace if movement exceeds the maximum measurement of 1/32 inches (0.8mm)

06017-MUST-C10

BRAKE SPECIFICATIONS
All measurements in inches unless noted

Year	Model		Brake Disc Original Thickness	Brake Disc Minimum Thickness	Brake Disc Maximum Run-out	Brake Drum Diameter Original Inside Diameter	Brake Drum Diameter Max. Wear Limit	Brake Drum Diameter Maximum Machine Diameter	Minimum Lining Thickness	Brake Caliper Mounting Bolts (ft. lbs.)
2002	Mustang	F	1.030	0.970	0.001	—	—	—	0.125	64
		R	0.550	0.500	0.002	—	—	—	0.123	26
	Mustang	F	1.100	1.040	0.001	—	—	—	0.125	64
	Cobra	R	0.710	0.660	0.002	—	—	—	0.123	26
2003	Mustang	F	1.030	0.970	0.001	—	—	—	0.125	64
		R	0.550	0.500	0.002	—	—	—	0.123	26
	Mustang	F	1.100	1.040	0.001	—	—	—	0.125	64
	Cobra	R	0.710	0.660	0.002	—	—	—	0.123	26
2004	Mustang	F	1.020	0.970	0.002	—	—	—	0.118	23
		R	0.590	0.500	0.004	—	—	—	0.118	25
	Mustang	F	1.100	1.030	0.002	—	—	—	0.118	23
	Cobra	R	0.590	0.500	0.004	—	—	—	0.118	25

NOTE: Follow specifications stamped on rotor or drum if figures differ from those in this chart.

NA: Not Available

F: Front

R: Rear

06017-MUST-C11

SCHEDULED MAINTENANCE INTERVALS

2002-2004 Ford Mustang

TO BE SERVICED	TYPE OF SERVICE	VEHICLE MILEAGE INTERVAL (x1000)												
		5	10	15	20	25	30	35	40	45	50	55	60	65
Engine oil & filter	R	✓	✓	✓	✓	✓	✓	✓	✓	✓	✓	✓	✓	✓
Adjust clutch pedal by lifting pedal	S/I	✓	✓	✓	✓	✓	✓	✓	✓	✓	✓	✓	✓	✓
Rotate tires	S/I	✓		✓		✓		✓		✓		✓		✓
Cooling system, hoses, clamps & coolant strength	S/I			✓			✓			✓			✓	
Air cleaner element	R						✓						✓	
Automatic transmission fluid & filter	R						✓						✓	
Engine coolant ①	R						✓						✓	
Spark plugs	R													
Accessory drive belt(s)	S/I						✓						✓	
Brake lines, hoses & connections	S/I						✓						✓	
Exhaust heat shields	S/I						✓						✓	
Front & rear brakes	S/I						✓						✓	
PCV valve	R												✓ ·	
Rear axle lubricant ②	R													

R: Replace   S/I: Service or Inspect

① Engine coolant: change engine coolant at 48,000 to 50,000 miles and thereafter every 30,000 miles.

② Replace every 100,000 miles.

FREQUENT OPERATION MAINTENANCE (SEVERE SERVICE)

If a vehicle is operated under any of the following conditions it is considered severe service:

- **Extremely dusty areas.**

- **50% or more of the vehicle operation is in 32°C (90°F) or higher temperatures, or constant operation in temperatures below 0°C (32°F).**

- **Prolonged idling (vehicle operation in stop and go traffic).**

- **Frequent short running periods (engine does not warm to normal operating temperatures).**

- **Police, taxi, delivery usage or trailer towing usage.**

Oil & filter change: change every 3000 miles.

Rotate tires at 6000 miles & every 9000 miles thereafter.

Automatic transmission fluid & filter: change every 21,000 miles.

06017-MUST-C12

ENGINE REPAIR

➡**Disconnecting the negative battery cable on some vehicles may interfere with the functions of the on board computer system. The computer may undergo a relearning process once the negative battery cable is reconnected.**

Distributor

The 3.8L and 4.6L engines use Distributorless Ignition Systems (DIS).

Alternator

REMOVAL

3.8L Engine

1. Disconnect the negative battery cable.
2. Remove the accessory drive belt.
3. Disconnect the alternator electrical connectors.
4. Remove the alternator mounting bolts.
5. Remove the alternator from the vehicle.

4.6L Engines (Non-Supercharged)

1. Disconnect the negative battery cable.
2. Remove the air intake scoop bracket, if equipped.
3. Remove the accessory drive belt.
4. Remove the upper alternator bracket.
5. Disconnect the alternator electrical connectors.
6. Remove the alternator bolts.
7. Remove the alternator from the vehicle.

4.6L Engines (Supercharged)

1. Disconnect the negative battery cable.
2. Remove the accessory drive belt.
3. Raise the vehicle.
4. Disconnect the alternator electrical connectors.
5. Remove the alternator bolts.
6. Remove the alternator from the vehicle.

INSTALLATION

3.8L Engine

1. Position the alternator onto engine. Tighten the lower bolt to 35 ft. lbs. (47 Nm) and the upper bolt to 18 ft. lbs. (25 Nm).
2. Connect the alternator electrical connectors.

3. Install the accessory drive belt.
4. Connect the negative battery cable.

4.6L Engines (Non-Supercharged)

1. Position the alternator onto engine. Tighten the mounting bolts to 18 ft. lbs. (25 Nm).
2. Connect the alternator electrical connectors.
3. Install the upper alternator bracket. Tighten the bolts to 89 inch lbs. (10 Nm).
4. Install the accessory drive belt.
5. Install the air intake scoop bracket, if equipped.
6. Connect the negative battery cable.

4.6L Engines (Supercharged)

1. Position the alternator onto engine. Tighten the lower bolt to 18 ft. lbs. (25 Nm) and the upper bolt to 37 ft. lbs. (50 Nm).
2. Connect the alternator electrical connectors.
3. Lower the vehicle.
4. Install the accessory drive belt.
5. Connect the negative battery cable.

Ignition Timing

ADJUSTMENT

The ignition timing is controlled by the Powertrain Control Module (PCM). No adjustment is necessary.

Engine Assembly

REMOVAL & INSTALLATION

3.8L Engine

1. Discharge the air conditioning system.
2. Drain the engine cooling system.
3. Drain the engine oil and remove the oil filter.
4. Disconnect the negative battery cable.
5. Relieve the fuel system pressure.
6. Remove the hood.
7. Remove cooling fan assembly.
8. Remove radiator and hoses.
9. Disconnect heater hoses.
10. Disconnect Intake Air Temperature (IAT) sensor connector.
11. Remove air intake duct.
12. Remove the accessory drive belt.
13. Disconnect alternator wiring connectors.
14. Remove coolant recovery tank.

15. Remove power steering pump and bracket.
16. Disconnect power steering pressure switch connector.
17. Remove A/C compressor and bracket.
18. Disconnect cruise control cable, if equipped.
19. Disconnect accelerator cable and remove bracket.
20. Disconnect fuel supply line.
21. Disconnect fuel return line, if equipped.
22. Disconnect fuel charging wiring harness connectors.
23. Disconnect main vacuum source hose.
24. Disconnect evaporative emissions hose.
25. Disconnect Heated Oxygen (HO2S) sensor connectors.
26. Remove dual converter Y-pipe.
27. Remove starter.
28. Remove starter wiring harness retainers.
29. Remove transmission flange bolts.
30. Disconnect torque converter, automatic transmission models only.
31. Disconnect transmission oil cooler lines, automatic transmission models only.
32. Remove left and right engine support insulators.

Install the engine lifting brackets on both sides of the engine—4.6L SOHC engine shown, other engines similar

7922NG02

33. Attach an engine hoist and remove the engine from the vehicle.

To install:

34. Lower the engine into the vehicle.
35. Install left and right engine support insulators. Tighten through bolts to 35–50 ft. lbs. (47–68 Nm).
36. Install Transmission flange bolts. Tighten bolts to 25–33 ft. lbs. (34–46 Nm).
37. Connect torque converter, automatic transmission models only. Tighten the nuts to 20–33 ft. lbs. (27–46 Nm).
38. Connect transmission oil cooler lines, automatic transmission models only.
39. Install starter wiring harness retainers.
40. Install starter.
41. Install dual converter Y-pipe.
42. Connect HO2S sensor connectors.
43. Connect evaporative emissions hose.
44. Connect main vacuum source hose.
45. Connect fuel charging wiring harness connectors.
46. Connect fuel supply line.
47. Connect fuel return line, if equipped.
48. Connect accelerator cable and install bracket.
49. Connect cruise control cable, if equipped.
50. Install A/C compressor and bracket.
51. Connect power steering pressure switch connector.
52. Install power steering pump and bracket.
53. Install coolant recovery tank.
54. Connect alternator wiring connectors.
55. Install the accessory drive belt.
56. Install air intake duct.
57. Connect IAT sensor connector.
58. Connect heater hoses.
59. Install radiator and hoses.
60. Install cooling fan assembly.
61. Install hood.

✳✳ WARNING

Be sure to check engine to see that all electrical connectors, hoses and cables are properly connected and secure.

62. Install new oil filter and fill engine with fresh oil.
63. Connect the negative battery cable.
64. Fill the cooling system.
65. Recharge A/C system.
66. Start the engine and check for leaks.

4.6L Engines (Non-Supercharged)

1. Discharge the air conditioning system.

2. Drain the engine cooling system.
3. Drain the engine oil and remove the oil filter.
4. Relieve the fuel system pressure.
5. Disconnect the negative battery cable.
6. Remove the transmission. Refer to the transmission procedure in this section.
7. Remove the hood.
8. Remove the air intake scoop bracket, if equipped.
9. Remove engine compartment brace, if necessary.
10. Disconnect Intake Air Temperature (IAT) sensor connector.
11. Disconnect Mass Airflow (MAF) sensor connector.
12. Remove air cleaner and intake duct assembly.
13. Remove coolant recovery tank assembly.
14. Remove cooling fan assembly.
15. Disconnect oil filter adapter hose, if necessary.
16. Disconnect oil pressure sensor connector.
17. Remove radiator hoses.
18. Disconnect heater hoses.
19. Disconnect cruise control cable, if equipped.
20. Disconnect accelerator cable.
21. Remove the accessory drive belt.
22. Disconnect alternator wiring connectors.
23. Disconnect fuel supply line.
24. Disconnect fuel return line, if equipped.
25. Disconnect fuel charging and engine control wiring harness connectors.
26. Separate wiring harness from cowl panel.
27. Disconnect evaporative emissions hose.
28. Disconnect vacuum hoses and label, if necessary.
29. Disconnect A/C compressor suction and discharge lines.
30. Disconnect secondary air injection switching valve, on some DOHC engines.
31. Remove power steering reservoir.
32. Disconnect power steering fluid cooler hoses.
33. Disconnect power steering pressure and return hoses.
34. Disconnect Heated Oxygen (HO2S) sensor connectors. Place wiring harness out of way.
35. Remove dual converter H-pipe.
36. Remove starter motor.
37. Position transmission cooler lines out of way, automatic transmission only.
38. Disconnect engine ground cable.

39. Remove body ground strap bolt.
40. Remove left and right engine support insulators on 2001–03 models. Remove left and right engine mount nuts on 2004 models.
41. Install left and right engine lifting brackets.
42. Attach an engine hoist and remove engine from the vehicle.

To install:

43. Lower the engine into the vehicle.
44. Install left and right engine support insulators on 2002–03 models and tighten through bolts to 15–22 ft. lbs. (20–30 Nm). Install left and right engine mount retaining nuts on 2004 models and tighten retaining nuts 111 ft. lbs. (150 Nm).
45. Remove engine lifting brackets.
46. Position body ground strap and tighten bolt.
47. Connect engine ground cable.
48. Install starter motor.
49. Install dual converter H-pipe.
50. Connect Heated Oxygen (HO2S) sensor connectors.
51. Connect power steering pressure and return hoses.
52. Connect power steering fluid cooler hoses.
53. Install power steering reservoir.
54. Connect secondary air injection switching valve, if equipped.
55. Connect A/C compressor suction and discharge lines.
56. Connect vacuum hoses.
57. Connect evaporative emissions hose.
58. Position and secure wiring harness onto cowl panel.
59. Connect fuel charging and engine control wiring harness connectors.
60. Connect fuel return line, if equipped.
61. Connect fuel supply line.
62. Connect alternator wiring connectors.
63. Install the accessory drive belt.
64. Connect accelerator cable.
65. Connect cruise control cable, if equipped.
66. Connect heater hoses.
67. Install radiator hoses.
68. Connect oil pressure sensor connector.
69. Connect oil filter adapter hose, if necessary.
70. Install cooling fan assembly.
71. Install coolant recovery tank assembly.
72. Install air cleaner and intake duct assembly.
73. Connect MAF sensor connector.
74. Connect IAT sensor connector.
75. Install engine compartment brace, if necessary.

76. Install the air intake scoop bracket, if equipped.

77. Install hood.

78. Install transmission.

✳✳ WARNING

Be sure to check engine to see that all electrical connectors, hoses and cables are properly connected and secure.

79. Install a new oil filter and fill engine with fresh oil.

80. Fill the cooling system.

81. Connect the negative battery cable.

82. Recharge the air conditioning.

83. Start the engine and check for leaks.

4.6L Engine (Supercharged)

1. Remove battery and battery tray, making sure to disconnect negative battery cable first.

2. Relieve the fuel system pressure.

3. Drain the engine cooling system.

4. Drain the engine oil and remove the oil filter.

5. Drain the supercharger cooling system.

6. Discharge the air conditioning system.

7. Remove hood and hood prop.

8. Disconnect Intake Air Temperature (IAT) sensor connector.

9. Disconnect breather hose.

10. Disconnect and label vacuum hoses.

11. Remove air cleaner assembly.

12. Disconnect Mass Airflow (MAF) connector.

13. Disconnect Throttle Position (TP) sensor and Idle Air Control (IAC) valve connectors.

14. Disconnect A/C pressure switch connector.

15. Disconnect fuel supply line.

16. Disconnect accelerator cable.

17. Disconnect cruise control cable, if equipped.

18. Remove accelerator/cruise control cable bracket and place out of the way of engine.

19. Remove throttle body and spacer assembly.

20. Disconnect 16-pin and 42-pin wiring connectors.

21. Separate wiring harness from cowl panel.

22. Remove heater hoses.

23. Remove radiator and supercharger cooling system hoses.

24. Remove A/C compressor suction and discharge lines.

25. Disconnect dual function pressure switch wiring connector.

26. Remove coolant recovery tank and support bracket.

27. Remove power steering reservoir and tube bracket.

28. Remove wiring harness support brackets.

29. Remove supercharger coolant recovery tank.

30. Remove electric cooling fan and shroud assemblies.

31. Disconnect power steering tube and bracket bolt, then remove hose from grommet.

32. Remove auxiliary crank pulley.

33. Remove clutch.

34. Disconnect Heated Oxygen (HO$_2$S) sensor connectors.

35. Disconnect transmission wiring connector, then remove harness with bracket.

36. Disconnect exhaust pipe at right and left exhaust manifolds. Move exhaust pipe out of way.

37. Disconnect oil pressure sender wiring connector.

38. Disconnect alternator wiring connectors.

39. Disconnect engine ground cable.

40. Disconnect oil filter adapter hose.

41. Remove engine mount retaining nuts.

42. Disconnect clutch cable and position out of way.

43. Remove starter motor.

44. Remove left and right ignition coil covers.

45. Install left and right engine lifting brackets.

46. Attach an engine hoist and remove the engine from the vehicle.

To install:

47. Lower the engine into the vehicle.

48. Install left and right engine mount retaining nuts. Tighten retaining nuts 111 ft. lbs. (150 Nm).

49. Remove engine lifting brackets.

50. Install ignition coil covers.

51. Install starter motor.

52. Connect clutch cable and tighten retaining screws.

53. Connect oil filter adapter hose.

54. Connect engine ground cable.

55. Connect alternator wiring connectors.

56. Connect oil pressure sender wiring connector.

57. Connect exhaust pipe to right and left exhaust manifolds and tighten manifold flange nuts.

58. Install transmission wiring harness bracket, then connect to transmission wiring connector.

59. Connect Heated Oxygen (HO$_2$S) sensor connectors.

60. Install clutch.

61. Install auxiliary crank pulley.

62. Secure power steering hose in grommet. Position power steering tube and bracket. Tighten power steering tube bracket bolt.

63. Install electric cooling fan and shroud assemblies.

64. Install supercharger coolant recovery tank.

65. Install wiring harness support brackets.

66. Install power steering reservoir and tube bracket.

67. Install coolant recovery tank and support bracket.

68. Connect dual function pressure switch wiring connector.

69. Install A/C compressor suction and discharge lines.

70. Install radiator and supercharger cooling system hoses.

71. Install heater hoses.

72. position and secure wiring harness to cowl panel.

73. Connect 16-pin and 42-pin wiring connectors.

74. Install throttle body and spacer assembly.

75. Install accelerator/cruise control cable bracket.

76. Connect cruise control cable, if equipped.

77. Connect accelerator cable.

78. Connect fuel supply line.

79. Connect A/C pressure switch connector.

80. Connect TP sensor and IAC valve connectors.

81. Connect MAF connector.

82. Install air cleaner assembly.

83. Connect vacuum hoses.

84. Connect breather hose.

85. Connect IAT sensor connector.

86. Install hood and hood prop.

✳✳ WARNING

Be sure to check engine to see that all electrical connectors, hoses and cables are properly connected and secure.

87. Install a new oil filter and fill engine with fresh oil.

88. Fill the supercharger cooling system.

89. Fill the cooling system.

90. Recharge air conditioning system.

91. Install battery tray and install battery. Connect positive battery cable first, negative battery cable second.

92. Start the engine and check for leaks.

Water Pump

REMOVAL & INSTALLATION

3.8L Engine

1. Drain the engine cooling system.
2. Remove electric cooling fan assembly.
3. Loosen water pump pulley bolts.
4. Remove the accessory drive belt.
5. Remove power steering pump pulley.
6. Remove water pump pulley.
7. Remove ignition coil and bracket, if necessary.
8. Remove power steering pump bracket.
9. Disconnect lower radiator hose from water pump.
10. Disconnect heater water outlet tube.
11. Remove water pump. Clean and inspect gasket mating surfaces.

To install:

➡ The threads of the No. 1 water pump retaining bolt must be coated with a Teflon® sealant prior to installation.

12. Install water pump. Use a new gasket and tighten the bolts and studs to 15–22 ft. lbs. (20–30 Nm). Tighten the nuts to 71–106 inch lbs. (8–12 Nm).
13. Connect lower radiator hose.
14. Connect heater water outlet tube. Use a new O-ring seal and tighten the bolts to 71–106 inch lbs. (8–12 Nm).
15. Install power steering pump bracket.
16. Install ignition coil and bracket, if necessary.
17. Install water pump pulley. Tighten the bolts to 15–21 ft. lbs. (21–29 Nm).
18. Install power steering pump pulley.
19. Install water pump pulley. Tighten the bolts to 15–21 ft. lbs. (21–29 Nm).
20. Install the accessory drive belt.
21. Install electric cooling fan assembly.
22. Fill the cooling system.
23. Start the engine and check for coolant leaks.

4.6L Engines (Non-Supercharged)

1. Remove air intake scoop, if equipped.
2. Drain the engine cooling system.
3. Remove electric cooling fan assembly.
4. Loosen water pump pulley bolts.
5. Remove the accessory drive belt.
6. Remove water pump pulley.
7. Remove water pump.

To install:

8. Install water pump. Use a new O-ring

1. Mounting stud
2. Water pump housing gasket
3. Water pump
4. Mounting nuts
5. Short mounting bolts
6. Long mounting bolt
7. Mounting stud bolt

FRONT OF ENGINE

7922NG03

Exploded view of the water pump mounting on the 3.8L engine—during removal, note the original positions of the different length bolts for reassembly

TIGHTEN IN SEQUENCE SHOWN TO 20-30 N·m (15-22 LB-FT)

REFER TO LEGEND

1. Engine front cover
2. Mounting bolts
3. Mounting nuts
4. Water pump
5. Mounting bolt
6. Mounting stud bolt
7. Engine block

Water pump torque sequence—3.8L engine

7922NG04

WATER PUMP HOUSING GASKET

BOLT

WATER PUMP

7922NG05

Exploded view of the water pump mounting—4.6L engines

seal and tighten the bolts in a crossing pattern to 15–22 ft. lbs. (20–30 Nm).

9. Install water pump pulley. Tighten the bolts to 15–21 ft. lbs. (21–29 Nm).

10. Install the accessory drive belt.

11. Install electric cooling fan assembly.

12. Install air intake scoop, if equipped.

13. Fill the cooling system.

14. Start the engine and check for coolant leaks.

4.6L Engines (Supercharged)

1. Drain the engine cooling system.

2. Drain the supercharger cooling system.

3. Remove nut and move wiring harness aside.

4. Remove stud and move coolant hose assembly aside.

5. Lower the vehicle.

6. Disconnect radiator hoses.

7. Remove top radiator shield.

8. Remove coolant recovery tank.

9. Remove supercharger drive belt cover.

10. Remove supercharger drive belt.

11. Remove supercharger cooling system hose.

12. Remove fasteners and move power steering reservoir aside.

13. Remove nut and move A/C muffler bracket aside.

14. Remove belt idler bracket support.

15. Loosen water pump pulley bolts.

16. Remove the accessory drive belt.

17. Remove water pump pulley.

18. Remove water pump.

To install:

19. Install water pump. Use a new O-ring seal and tighten the bolts in a crossing pattern to 18 ft. lbs. (25 Nm).

20. Install water pump pulley. Tighten the bolts to 18 ft. lbs. (25 Nm).

21. Install the accessory drive belt.

22. Install belt idler bracket support.

23. Place A/C muffler bracket in position and tighten mounting nut.

24. Place power steering reservoir in position and install fasteners.

25. Install supercharger cooling system hose.

26. Install supercharger drive belt.

27. Install supercharger drive belt cover.

28. Install coolant recovery tank.

29. Install top radiator shield.

30. Connect radiator hoses.

31. Place coolant hose assembly in position and tighten stud.

32. Place wiring harness in position and tighten nut.

33. Fill the cooling system.

34. Start the engine and check for coolant leaks.

Heater Core

REMOVAL & INSTALLATION

1. Drain the cooling system.

2. Recover the A/C refrigerant, if equipped.

3. Disconnect the negative battery cable.

4. Remove radio.

5. Remove CD player, if equipped.

6. Remove center console.

7. Remove left front wheel.

8. Remove left front inner fender.

9. Disconnect left bulkhead harness connector.

10. Remove left and right air bag modules.

11. Disconnect antenna cable connector.

12. Remove upper right instrument panel support bolt.

13. Disconnect climate control vacuum harness connector.

14. Remove left and right scuff plates.

15. Remove left and right A-pillar lower trim panels.

16. Remove left and right windshield side garnish moldings.

17. Remove pin-type retainers on convertible models only.

18. Position left and right door weatherstrips out of way.

19. Disconnect right main wire harness connectors.

20. Disconnect climate control wire harness connector.

21. Remove instrument panel steering column cover.

22. Remove instrument panel reinforcement.

23. Remove pinch bolt, then separate steering column shaft from intermediate shaft.

24. Disconnect left main wiring harness connectors.

25. Disconnect Generic Electronic Module (GEM) connectors.

26. Disconnect Electronic Crash Sensor (ECS) module connector.

27. Disconnect shift interlock cable, if equipped.

28. Disconnect shifter assembly harness connector, if equipped.

29. Disconnect temperature control cable.

30. Remove 4 center instrument panel support bolts.

31. Remove left instrument panel support bolt and nut.

32. Remove right instrument panel support bolt.

33. Remove instrument panel upper finish panel.

34. Remove upper instrument panel support bolts.

35. Remove instrument panel.

36. Disconnect heater hoses.

37. Disconnect vacuum supply hose.

38. Remove A/C accumulator/drier.

39. Disconnect A/C liquid line.

40. Remove evaporator housing.

41. Remove foam weather seal.

42. Remove heater core cover.

43. Remove heater core.

To install:

44. Install heater core.

45. Install heater core cover.

46. Install a new foam weather seal.

47. Install evaporator housing. Tighten the fasteners to 71 inch lbs. (8 Nm).

48. Connect A/C liquid line.

49. Install A/C accumulator/drier.

50. Connect vacuum supply hose.

51. Connect heater hoses.

52. Install instrument panel.

53. Install upper instrument panel sup-

1 Rivet (2 Req'd)
2 Instrument Panel Upper
   Finish Panel
3 Screw (4 Req'd)
4 Instrument Panel Steering
   Column Cover
5 Instrument Panel
   Reinforcement
6 Instrument Panel
7 Vehicle Identification Plate
A Tighten to 8-10 N·m (71-88
   Lb-In)

93111GA3

Exploded view of the instrument panel, cover and reinforcement—Mustang

1 U-Lock Nut (6 Req'd)
2 Screw (4 Req'd)
3 Screw (3 Req'd)
4 Stud
5 Nut
6 Instrument Panel
7 Screw (3 Req'd)
A Tighten to 8-10 N·m (71-88
   Lb-In)
B Tighten to 40-56 N·m (30-41
   Lb-Ft)

93111GA5

Exploded view of the instrument panel and related components—Mustang

port bolts. Tighten the bolts to 80 inch lbs. (9 Nm).

54. Install instrument panel upper finish panel.

55. Install right instrument panel support bolt. Tighten the bolt to 80 inch lbs. (9 Nm).

56. Install left instrument panel support bolt and nut. Tighten the bolt to 80 inch lbs. (9 Nm) and the nut to 35 ft. lbs. (48 Nm).

57. Install 4 center instrument panel support bolts. Tighten the bolts to 80 inch lbs. (9 Nm).

58. Connect temperature control cable.

59. Connect shifter assembly harness connector, if equipped.

60. Connect shift interlock cable, if equipped.

61. Connect ECS module connector.

1	A/C Evaporator Housing
2	Screw
3	Nut
4	To Vacuum Source
5	Blower Motor
6	Heater Blower Motor Switch Resistor
7	Screw
A	Tighten to 10.2-13.8 N·m (91-122 Lb-In)
B	Tighten to 1.6-2.2 N·m (15-19 Lb-In)

FRONT OF VEHICLE

93111GA4

Exploded view of the heater/air conditioning housing assembly—Mustang

9346NG05

Evaporator case interior fasteners—Mustang

9346NG06

Evaporator case screw—Mustang

9346NG07

Evaporator case under-hood nuts—Mustang

62. Connect GEM connectors.

63. Connect left main wiring harness connectors.

64. Connect steering column shaft to intermediate shaft and install pinch bolt. Tighten the pinch bolt to 35 ft. lbs. (47 Nm).

65. Install instrument panel reinforcement. Tighten the screws to 80 inch lbs. (9 Nm).

66. Install instrument panel steering column cover. Tighten the screws to 80 inch lbs. (9 Nm).

67. Connect climate control wire harness connector.

68. Connect right main wire harness connectors.

69. Place left and right door weatherstrips back in correct position.

70. Install pin-type retainers on convertible models only.

71. Install left and right windshield side garnish moldings.

72. Install left and right A-pillar lower trim panels.

73. Install left and right scuff plates.

74. Connect climate control vacuum harness connector.

75. Connect antenna cable connector.

76. Install upper right instrument panel support bolt. Tighten the bolt to 80 inch lbs. (9 Nm).

77. Install left and right air bag modules.

78. Connect left bulkhead harness connector.

79. Install left front inner fender.

80. Install left front wheel.

81. Install center console.

82. Install CD player, if equipped.

83. Install radio.

84. Connect the negative battery cable.

85. Fill the cooling system.

86. Recharge the A/C system.

87. Start the engine and check for leaks.

Cylinder Head

REMOVAL & INSTALLATION

3.8L Engine

1. Drain the cooling system.

2. Disconnect the negative battery cable.

3. Remove air intake tube.

4. Remove the accessory drive belt.

5. Remove alternator and bracket.

6. Remove power steering pump and bracket.

7. Disconnect A/C manifold and tube assembly, if necessary.

8. Remove A/C compressor and bracket.

9. Remove Positive Crankcase Ventilation (PCV) valve.

10. Remove upper intake manifold.

11. Remove spark plug wires.

12. Remove valve covers.

13. Disconnect fuel supply line.

14. Disconnect fuel return line, if equipped.

15. Disconnect fuel injector wiring connectors.

16. Remove fuel injection supply manifold.

17. Remove lower intake manifold.

18. Remove exhaust manifolds.

19. Remove exhaust manifold studs.

➡**Keep rocker arms and pushrods in order for installation.**

20. Remove rocker arms.

21. Remove pushrods.

22. Remove cylinder heads.

To install:

➡**The cylinder head bolts are a torque-to-yield design and cannot be reused.**

➡**Left and right cylinder head gaskets are not interchangeable.**

23. Install the new cylinder heads. Lubricate the new cylinder head bolt threads with clean oil and tighten them in sequence as follows:

    a. Step 1: 15 ft. lbs. (20 Nm).

    b. Step 2: 29 ft. lbs. (40 Nm).

    c. Step 3: 36 ft. lbs. (50 Nm).

➡**Loosen and tighten each bolt individually for the remainder of the procedure. Do not loosen or tighten all the bolts at the same time.**

    d. Step 4: Loosen each bolt, one at a time, 2–3 turns. Tighten the long cylinder head bolts to 29–37 ft. lbs. (40–50 Nm) plus 180 degrees. Tighten the short cylinder head bolts to 15–22 ft. lbs. (20–30 Nm) plus 180 degrees.

    e. Step 5: Repeat step 4 for the next bolt in the tightening sequence.

24. Install exhaust manifold studs.

25. Install pushrods and rocker arms in their original locations.

26. Install exhaust manifolds.

27. Install lower intake manifold.

28. Install fuel injection supply manifold.

29. Connect fuel injector wiring connectors.

30. Connect fuel supply line.

31. Connect fuel return line, if equipped.

32. Install valve covers.

33. Connect spark plug wires.

34. Install upper intake manifold.

35. Install PCV valve.

36. Install A/C compressor and bracket.

37. Connect A/C manifold and tube assembly, if necessary.

38. Install power steering pump and bracket.

39. Install alternator and bracket.

40. Install the accessory drive belt.

41. Install air intake tube.

42. Connect the negative battery cable.

43. Fill the cooling system.

44. Start the engine and check for leaks.

45. Recharge A/C system.

4.6L SOHC Engine

1. Drain the cooling system.

2. Relieve the fuel system pressure.

3. Disconnect the negative battery cable.

4. Remove electric cooling fan.

5. Disconnect fuel supply line.

6. Disconnect fuel return line, if equipped.

7. Remove air intake tube.

8. Remove windshield wiper governor, if necessary.

9. Remove the accessory drive belt.

10. Remove spark plug wires, if equipped.

11. Remove ignition coils.

12. Disconnect Camshaft Position (CMP) sensor connector.

13. Remove power steering reservoir and bracket as an assembly.

14. Remove alternator and bracket.

15. Remove water pump pulley.

16. Disconnect Mass Air Flow (MAF) sensor connector.

17. Disconnect fuel injector wiring connectors.

18. Disconnect Crankshaft Position (CKP) sensor connector.

19. Disconnect A/C compressor clutch connector.

20. Disconnect cruise control cable.

21. Disconnect accelerator cable and remove bracket.

22. Disconnect canister purge valve connector, if necessary.

23. Drain coolant from engine block.

24. Remove power steering pump.

25. Remove left and right exhaust manifolds.

26. Remove oil pan.

27. Remove crankshaft pulley.

28. Disconnect oil pressure sensor connector, if necessary.

29. Disconnect Exhaust Gas Recirculation (EGR) tube.

30. Remove PCV valve and hose.

31. Disconnect Heated Oxygen (HO2S) sensor connectors.

32. Remove dual converter H-pipe.

33. Remove starter motor wiring harness retainer, if necessary.

34. Remove valve covers.

35. Remove rocker arms.

36. Remove throttle body and adapter as an assembly.

37. Remove heater hoses.

38. Remove upper radiator hose and adapter.

39. Remove intake manifold.

40. Remove front cover.

41. Remove crankshaft sensor ring.

42. Remove timing chains.

43. Remove heater water tube.

44. Remove hydraulic lash adjusters.

45. Remove cylinder heads.

● TIGHTEN BOLTS
IN SEQUENCE

FRONT OF ENGINE

1. Engine block
2. Locating pin
3. Long cylinder head mounting bolts
4. Cylinder head
5. Short cylinder head mounting bolts
6. Alignment dowel

7922NG07

Cylinder head torque sequence—3.8L engine

CRANKSHAFT
KEYWAY

9306NG01

Place the crankshaft keyway at 9 o'clock for cylinder head installation—4.6L SOHC engine

To install:

➡ **The cylinder head bolts are a torque-to-yield design and cannot be reused.**

46. Rotate the crankshaft so that the keyway is at the 9 o'clock position.

47. Rotate each camshaft to a stable position where the valves do not extend below the cylinder head face.

48. Install the new cylinder heads. Lubricate the new cylinder head bolt threads with clean oil and tighten them in sequence as follows:

  a. Step 1: 28–31 ft. lbs. (37–43 Nm).
  b. Step 2: Plus 85–95 degrees.
  c. Step 3: Loosen all bolts one full turn.
  d. Step 4: 28–31 ft. lbs. (37–43 Nm).
  e. Step 5: Plus 85–95 degrees.
  f. Step 6: Plus 85–95 degrees.

49. Install hydraulic lash adjusters.
50. Install heater water tube.
51. Install timing chains.

52. Install crankshaft sensor ring.
53. Install front cover.
54. Install intake manifold.
55. Install upper radiator hose and adapter.
56. Install heater hoses.
57. Install throttle body and adapter as an assembly.
58. Install rocker arms.
59. Install valve covers.
60. Install starter motor wiring harness retainer.
61. Install dual converter H-pipe.
62. Connect HO_2S sensor connectors.
63. Install PCV valve and hose.
64. Connect EGR tube.
65. Connect oil pressure sensor connector.
66. Install crankshaft pulley.
67. Install oil pan.
68. Install power steering pump.
69. Connect canister purge valve connector.
70. Connect accelerator cable and install bracket.
71. Connect cruise control cable.
72. Connect A/C compressor clutch connector.
73. Connect CKP sensor connector.
74. Connect fuel injector wiring connectors.
75. Connect MAF sensor connector.
76. Install water pump pulley.
77. Install alternator and bracket.
78. Install left and right exhaust manifolds.
79. Install power steering reservoir and bracket.
80. Connect CMP sensor connector.
81. Install ignition coils.

82. Remove spark plug wires, if equipped.
83. Install the accessory drive belt.
84. Install windshield wiper governor, if necessary.
85. Install air intake tube.
86. Connect fuel supply line.
87. Connect fuel return line, if equipped.
88. Install electric cooling fan.
89. Connect the negative battery cable.
90. Fill the cooling system.
91. Start the engine and check for leaks.

10 Nm (89 lb-in)

9346NG01

Right valve cover torque sequence—4.6L SOHC Engine

9346NG02

Left valve cover torque sequence—4.6 SOHC Engine

RH CYLINDER HEAD

LH CYLINDER HEAD

FRONT OF ENGINE

7922NG08

Cylinder head torque sequence—4.6L SOHC engine

4.6L DOHC Engine (Non-Supercharged)

1. Remove the engine from the vehicle and mount it on a suitable workstand.
2. Remove accessory drive belt and tensioner.
3. Remove idler pulley.
4. Disconnect Engine Coolant Temperature (ECT) sensor connector.
5. Remove water bypass tube.
6. Remove alternator and support bracket.
7. Remove water pump pulley.
8. Remove water pump.
9. Remove power steering pump.
10. Remove crankshaft pulley.
11. Disconnect Camshaft Position (CMP) sensor.
12. Remove spark plug wires, if equipped.
13. Remove ignition coils.
14. Disconnect fuel pressure sensor connector and vacuum line.
15. Remove spark plugs.
16. Remove valve covers.
17. Remove rocker arms.
18. Remove Exhaust Gas Recirculation (EGR) vacuum regulator valve.
19. Remove EGR tube.
20. Disconnect fuel injector wiring connectors.
21. Remove intake manifold.
22. Remove left and right exhaust manifolds.
23. Remove front cover.
24. Remove Crankshaft Position (CKP) sensor pulse wheel.
25. Remove timing chains.
26. Remove hydraulic lash adjusters.
27. Remove cylinder heads.

To install:

➡ **The cylinder head bolts are a torque-to-yield design and cannot be reused.**

28. Use new gaskets and install the cylinder heads.
29. Tighten the cylinder head bolts in sequence as follows:
    a. Step 1: 28–31 ft. lbs. (37–43 Nm)
    b. Step 2: Tighten the bolts 85–95 degrees.
    c. Step 3: Loosen all bolts 1 full turn.
    d. Step 4: Tighten all bolts to 28–31 ft. lbs. (37–43 Nm).
    e. Step 5: Tighten the bolts 85–95 degrees.
    f. Step 6: Tighten the bolts 85–95 degrees.
30. Install hydraulic lash adjusters.
31. Install timing chains.
32. Install CKP sensor pulse wheel.
33. Install front cover.
34. Install left and right exhaust manifolds.
35. Install intake manifold.
36. Connect fuel injector wiring connectors.
37. Install EGR tube.
38. Install EGR vacuum regulator valve.
39. Install rocker arms.
40. Install valve covers.
41. Install spark plugs.

Cylinder head loosening sequence—4.6L DOHC engine

7922NG10

Cylinder head torque sequence—4.6L DOHC engine

7922NG11

Right valve cover torque sequence—4.6L DOHC Engine

9346NG03

10 Nm (89 lb-in)

Left valve cover torque sequence—4.6L DOHC Engine

42. Connect fuel pressure sensor connector and vacuum line.
43. Install ignition coils.
44. Install spark plug wires, if equipped.
45. Connect CMP sensor.
46. Install crankshaft pulley.
47. Install water pump.
48. Install water pump pulley.
49. Install power steering pump.
50. Install alternator and support bracket.
51. Install water bypass tube.
52. Connect ECT sensor connector.
53. Install idler pulley.
54. Install accessory drive belt and tensioner.

4.6L DOHC Engine (Supercharged)

1. Remove the engine from the vehicle and mount it on a suitable workstand.
2. Disconnect Engine Coolant Temperature (ECT) sensor connector.
3. Remove water bypass tube.
4. Disconnect Camshaft Position (CMP) sensor.
5. Disconnect fuel pressure sensor connector and vacuum line.
6. Remove ignition coils.
7. Remove spark plugs.
8. Remove valve covers.
9. Remove supercharger belt idler support bracket assembly.
10. Remove the accessory drive belt.
11. Remove water pump pulley.
12. Remove water pump.
13. Remove crankshaft pulley.
14. Remove alternator and support bracket.
15. Remove power steering pump.
16. Remove idler pulley.

17. Remove rocker arms.
18. Remove Exhaust Gas Recirculation (EGR) valve.
19. Remove EGR tube.
20. Disconnect fuel injector wiring connectors.
21. Remove intake manifold, supercharger, and fuel supply manifold as an assembly.
22. Remove left and right exhaust manifolds.
23. Remove front cover.
24. Remove Crankshaft Position (CKP) sensor pulse wheel.
25. Remove timing chains.
26. Remove hydraulic lash adjusters.
27. Remove cylinder heads.

To install:

➡**The cylinder head bolts are a torque-to-yield design and cannot be reused.**

28. Use new gaskets and install the cylinder heads.
29. Tighten the cylinder head bolts in sequence as follows:
   a. Step 1: 28–31 ft. lbs. (37–43 Nm).
   b. Step 2: Tighten the bolts 85–95 degrees.
   c. Step 3: Loosen all bolts 1 full turn.
   d. Step 4: Tighten all bolts to 28–31 ft. lbs. (37–43 Nm).
   e. Step 5: Tighten the bolts 85–95 degrees.
   f. Step 6: Tighten the bolts 85–95 degrees.
30. Install hydraulic lash adjusters.
31. Install timing chains.
32. Install CKP sensor pulse wheel.
33. Install front cover.
34. Install left and right exhaust manifolds.
35. Install intake manifold, supercharger, and fuel supply manifold as an assembly.
36. Connect fuel injector wiring connectors.
37. Install EGR valve.
38. Install EGR tube.
39. Install rocker arms.
40. Install idler pulley.
41. Install power steering pump.
42. Install alternator and support bracket.
43. Install crankshaft pulley.
44. Install water pump.
45. Install water pump pulley.
46. Install the accessory drive belt.
47. Install supercharger belt idler support bracket assembly.
48. Install valve covers.
49. Install spark plugs.
50. Install ignition coils.
51. Connect CMP sensor.

52. Connect fuel pressure sensor connector and vacuum line.
53. Install water bypass tube.

Rocker Arms

REMOVAL & INSTALLATION

3.8L Engine

1. Disconnect the negative battery cable.
2. Remove the Positive Crankcase Ventilation (PCV) valve and hose.
3. Remove valve covers.
4. Remove the rocker arms.

➡**Keep rocker arms in order for installation.**

To install:

➡**The rocker arm bolts are tightened with the valves closed. Rotate the crankshaft as necessary to position the lifter on the base circle of the camshaft lobe before tightening the corresponding rocker arm bolt.**

5. Install the rocker arms. The rocker arm bolts are tightened in two steps as follows:
   a. Step 1: 44 inch lbs. (5 Nm).
   b. Step 2: 23–29 ft. lbs. (30–40 Nm).
6. Install the valve covers.
7. Install the PCV valve and hose.
8. Connect the negative battery cable.
9. Start the engine and check for proper operation.

4.6L Engines

1. Disconnect the negative battery cable.
2. Remove the spark plug wires, if equipped.
3. Remove the ignition coils.
4. Remove the valve covers.

➡**The 4.6L DOHC engine requires special tool Valve Spring Compressor T91P-6565-A for exhaust valves, and Valve Spring Compressor T93P-6565-AR for intake valves. The 4.6L SOHC engine requires special tool Valve Spring Compressor T91P-6565-A and Valve Spring Spacer T91P-6565-AH.**

5. Rotate the crankshaft so that the piston on the cylinder to be serviced is at bottom dead center with the valves closed.
6. Compress the valve spring and remove the rocker arm. Repeat for each arm to be removed.

➡**If the rocker arms are to be reused, ensure that they are installed in the same position that they were removed from.**

Valve spring compression tool and spacer—4.6L SOHC engine

To install:

7. Compress the valve spring and install the rocker arm. Repeat for each arm to be installed.

8. Install the valve covers.

9. Install spark plug wires, if equipped.

10. Install ignition coils.

11. Connect the negative battery cable.

12. Start the engine and check for proper operation.

Supercharger

REMOVAL & INSTALLATION

1. Disconnect the negative battery cable.

2. Remove supercharger drive belt.

3. Remove PCV hoses.

4. Remove coolant supply and return manifold.

5. Remove coolant supply and return tubes and seal.

6. Remove supercharger and Charge Air Cooler (CAC) assembly.

To install:

7. Install supercharger and CAC assembly onto the engine. Use new gaskets, if necessary, and tighten the bolts in sequence as follows:

  a. Step 1: 18 inch lbs. (2 Nm).

  b. Step 2: 18 ft. lbs. (25 Nm).

8. Install coolant supply and return tubes and seals.

9. Install coolant supply and return

Supercharger assembly-to-lower intake manifold torque sequence—Cobra

manifold. Tighten the manifold fasteners as follows:

  a. Step 1: 89 inch lbs. (10 Nm).

  b. Step 2: Tighten an additional 90 degrees.

10. Install PCV hoses.

11. Install supercharger drive belt.

12. Connect the negative battery cable.

Intake Manifold

REMOVAL & INSTALLATION

3.8L Engine

1. Drain the cooling system.

2. Relieve the fuel system pressure.

3. Disconnect the negative battery cable.

4. Disconnect Intake Air Temperature (IAT) sensor.

5. Remove air intake tube.

6. Disconnect cruise control cable.

7. Disconnect accelerator cable and remove bracket.

8. Disconnect differential pressure feedback Exhaust Gas Recirculation (EGR) and vacuum regulator solenoid electrical connectors, if equipped.

9. Remove spark plug wires and ignition coil assembly.

10. Disconnect and label all area vacuum lines.

11. Remove Positive Crankcase Ventilation (PCV) valve and hose.

12. Disconnect Throttle Position (TP) sensor connector.

13. Disconnect Idle Air Control (IAC) valve connector.

14. Remove EGR valve.

15. Remove engine control sensor wiring harness retainer bracket and position harness out of the way.

16. Remove upper intake manifold.

➡ Be sure to record the locations of the long bolts and short bolts in the upper and lower intake manifolds.

17. Disconnect fuel supply line.

18. Disconnect fuel return line, if equipped.

19. Disconnect Engine Coolant Temperature (ECT) sensor connector.

20. Remove fuel supply manifold and injectors.

21. Remove upper radiator hose.

22. Remove heater hose.

23. Remove lower intake manifold.

To install:

24. Install the lower intake manifold. Use new gaskets and tighten the bolts in sequence as follows:

  a. Step 1: 45 inch lbs. (5 Nm).

  b. Step 2: 71–106 inch lbs. (8–12 Nm).

25. Install heater hose.

26. Install upper radiator hose.

27. Install fuel supply manifold and injectors.

Lower intake manifold torque sequence—
3.8L engine (2001–03 models)

Upper intake manifold torque sequence—3.8L engine (2001–03 models)

Lower intake manifold torque sequence—3.8L engine (2004 models)

Upper intake manifold torque sequence—3.8L engine (2004 models)

28. Connect ECT sensor connector.

29. Connect fuel supply line.

30. Connect fuel return line, if equipped.

31. Install the upper intake manifold and use a new gasket.

32. For 2002–03 models, tighten the bolts, in sequence, as follows:

    a. Step 1: 89 inch lbs. (10 Nm).

    b. Step 2: 15 ft. lbs. (20 Nm).

    c. Step 3: 24 ft. lbs. (32 Nm).

33. For 2004 models, tighten the bolts, in sequence, as follows:

    a. Step 1: 89 inch lbs. (10 Nm).

    b. Step 2: additional 90 degrees.

34. Install engine control sensor wiring harness retainer bracket.

35. Install EGR valve.

36. Connect IAC valve connector.

37. Connect TP sensor connector.

38. Install PCV valve and hose.

39. Connect all vacuum lines.

40. Install spark plug wires and ignition coil assembly.

41. Connect differential pressure feedback Exhaust Gas Recirculation (EGR) and vacuum regulator solenoid electrical connectors, if equipped.

42. Connect accelerator cable and install bracket.

43. Connect cruise control cable.

44. Install Air intake tube.

45. Connect IAT sensor connector.

46. Connect the negative battery cable.

47. Fill the cooling system.

48. Start the engine and check for leaks.

4.6L SOHC Engine

1. Drain the cooling system.

2. Relieve the fuel system pressure.

3. Disconnect the negative battery cable.

4. Disconnect fuel supply line.

5. Disconnect fuel return line, if equipped.

6. Disconnect Intake Air Temperature (IAT) sensor connector.

7. Remove air intake tube.

8. Remove the accessory drive belt.

9. Remove spark plug wires, if equipped.

10. Disconnect Camshaft Position (CMP) sensor connector.

11. Remove alternator and bracket.

12. Disconnect fuel injector wiring connectors.

Intake manifold torque sequence—4.6L SOHC engine

13. Disconnect oil pressure sensor connector.

14. Remove Exhaust Gas Recirculation (EGR) tube.

15. Disconnect cruise control cable.

16. Disconnect accelerator cable and remove bracket.

17. Disconnect throttle body vacuum adapter hose.

18. Remove heater hose.

19. Remove upper radiator hose adapter.

20. Remove intake manifold by removing the mounting bolts in the same sequence as the installation.

To install:

21. Install intake manifold. Use new gaskets and tighten the bolts, in sequence, to 15–22 ft. lbs. (20–30 Nm).

22. Install upper radiator hose adapter. Use a new O-ring and tighten the bolts to 15–22 ft. lbs. (20–30 Nm).

23. Install heater hose.

24. Connect throttle body vacuum adapter hose.

25. Connect accelerator cable and install bracket.

26. Connect cruise control cable.

27. Install EGR tube. Tighten the nut to 26–33 ft. lbs. (35–45 Nm).

28. Connect oil pressure sensor connector.

29. Connect fuel injector wiring connectors.

30. Install alternator and bracket.

31. Connect CMP sensor connector.

32. Install spark plug wires, if equipped.

33. Install the accessory drive belt.

34. Install air intake tube.

35. Connect IAT sensor connector.

36. Connect fuel supply line.

37. Connect fuel return line, if equipped.

38. Connect the negative battery cable.

39. Fill the engine cooling system.

40. Start the engine and check for leaks.

4.6L DOHC Engine (Non-Supercharged)

1. Drain the cooling system.

2. Relieve the fuel system pressure.

3. Disconnect the negative battery cable.

4. Remove the air intake scoop bracket, if equipped.

5. Remove engine compartment brace, if equipped.

6. Remove air intake tube.

7. Disconnect cruise control cable.

8. Disconnect accelerator cable and bracket.

9. Remove Positive Crankcase Ventilation (PCV) valve and hose.

10. Remove Exhaust Gas Recirculation (EGR) valve.

11. Remove manifold bolts, in sequence, and remove upper intake manifold.

12. Disconnect fuel supply line.

13. Disconnect fuel return line, if equipped.

14. Disconnect, and label, all intake manifold area vacuum lines.

15. Remove spark plug wires, if equipped.

16. Remove alternator and bracket.

17. Remove upper radiator hose.

18. Disconnect Throttle Position (TP) sensor connector.

19. Disconnect Engine Coolant Temperature (ECT) sensor connector.

20. Disconnect Idle Air Control (IAC) valve connector.

21. Disconnect coolant temperature gauge sender connector.

22. Remove water bypass tube.

23. Disconnect fuel injector wiring connectors.

24. Disconnect fuel pressure sensor vacuum and wiring connectors.

25. Remove manifold bolts, in sequence, and remove lower intake manifold.

To install:

26. Install lower intake manifold with new gaskets. Tighten the bolts, in sequence, to 89 inch lbs. (10 Nm).

27. Connect fuel pressure sensor vacuum and wiring connectors.

28. Connect fuel injector wiring connectors.

29. Install alternator and bracket.

30. Install water bypass tube.

31. Connect coolant temperature gauge sender connector.

32. Connect IAC valve connector.

33. Connect ECT sensor connector.

34. Connect TP sensor connector.

35. Install upper radiator hose.

36. Install spark plug wires, if equipped.

37. Install the accessory drive belt.

38. Connect fuel supply line.

39. Connect fuel return line, if equipped.

40. Install upper intake manifold and new gasket. Tighten the bolts, in sequence, to 71–106 inch lbs. (8–12 Nm).

41. Install EGR valve.

42. Install PCV valve and hose.

43. Connect accelerator cable and install bracket.

44. Connect cruise control cable.

45. Connect all intake manifold area vacuum lines.

46. Install air intake tube.

47. Install engine compartment brace, if equipped.

48. Install the air intake scoop bracket, if equipped.

Lower intake manifold loosening sequence—4.6L DOHC engine (Non-Supercharged)

Lower intake manifold torque sequence—4.6L DOHC engine (Non-Supercharged)

67197-MUST-G13

**Upper intake manifold torque sequence—
4.6L DOHC engines (2001-03 models)**

67197-MUST-G14

**Upper intake manifold torque sequence—
4.6L DOHC engines (2004 Non-Supercharged)**

49. Connect the negative battery cable.
50. Fill the engine cooling system.
51. Start the engine and check for leaks.

4.6L DOHC Engine (Supercharged)

1. Drain the supercharger cooling system.
2. Relieve the fuel system pressure.
3. Disconnect the negative battery cable.
4. Disconnect supercharger coolant hose to the overflow tank.
5. Remove supercharger belt.
6. Disconnect coolant hose.
7. Disconnect Throttle Position (TP) sensor connector.
8. Disconnect Idle Air Control (IAC) valve connector.
9. Disconnect fuel supply line.
10. Disconnect cruise control cable.
11. Disconnect accelerator cable and bracket.
12. Remove throttle body and spacer assembly.
13. Disconnect the vacuum hose and electrical connector from the fuel pulse damper.

14. Disconnect the vacuum hose and electrical connector from the EGR vacuum regulator solenoid.
15. Disconnect the vacuum hose and electrical connector from the supercharger bypass vacuum solenoid.
16. Disconnect vacuum hose and electrical connector from the differential pressure feedback EGR system.
17. Disconnect EGR valve vacuum hose and tube.
18. Disconnect Positive Crankcase Ventilation (PCV) valve hose.
19. Disconnect vacuum hoses at the back of the supercharger and position them out of the way.
20. Remove vacuum accessory bracket.
21. Disconnect Barometric (BARO) pressure sensor electrical connector.
22. Separate fuel charge wiring harness from fuel injection supply manifold and position out of the way.
23. Remove manifold bolts, in sequence, and remove lower intake manifold, supercharger, and fuel supply manifold as an assembly.

To install:

24. Install lower intake manifold with new gaskets. Tighten the bolts, in sequence, to 89 inch lbs. (10 Nm).
25. Secure fuel charge wiring harness onto the fuel injection supply manifold.
26. Connect BARO sensor electrical connector.
27. Install vacuum accessory bracket.
28. Connect vacuum hoses at the back of the supercharger.
29. Connect PCV valve hose.
30. Connect EGR valve vacuum hose and tube.
31. Connect the vacuum hose and electrical connector to the differential pressure feedback EGR system.
32. Connect the vacuum hose and electrical connector to the supercharger bypass vacuum solenoid.

33. Connect the vacuum hose and electrical connector to the EGR vacuum regulator solenoid.
34. Connect the vacuum hose and electrical connector to the fuel pulse damper.
35. Install throttle body and spacer assembly. Tighten the mounting nuts to 18 ft. lbs. (25 Nm).
36. Connect accelerator cable and install bracket.
37. Connect cruise control cable.
38. Connect fuel supply line.
39. Connect IAC valve connector.
40. Connect TP sensor connector.
41. Connect coolant hose.
42. Install supercharger belt.
43. Connect supercharger coolant hose to the overflow tank.
44. Connect the negative battery cable.
45. Fill and bleed the supercharger cooling system.
46. Start the engine and check for leaks.

Exhaust Manifold

REMOVAL & INSTALLATION

3.8L Engine

1. Disconnect the negative battery cable.
2. Disconnect, and label, spark plug wires.
3. Remove secondary air injection diverter valve, if equipped.
4. Remove secondary air injection tubes, if equipped.
5. Remove Exhaust Gas Recirculation (EGR) tube.
6. Remove oil dipstick tube.
7. Remove Heated Oxygen (HO2S) sensors.
8. Disconnect dual converter Y-pipe.
9. Remove exhaust manifolds. Note the locations of the studs and bolts for reassembly.

To install:

10. Install exhaust manifolds. Tighten the bolts in sequence to 22–26 ft. lbs. (30–36 Nm).
11. Connect dual converter Y-pipe.
12. Install HO2S sensors. Tighten the sensors to 28–33 ft. lbs. (37–45 Nm).
13. Install oil dipstick tube.
14. Install EGR tube.
15. Install secondary air injection tubes, if equipped.
16. Install secondary air injection diverter valve, if equipped.
17. Connect spark plug wires.
18. Connect the negative battery cable.

67197-MUST-G15

**Lower intake manifold torque sequence—
4.6L DOHC engine (Supercharged)**

Exhaust manifold torque sequence—3.8L engine

Exhaust manifold torque sequence—4.6L SOHC engine

Left exhaust manifold torque sequence—4.6L DOHC engine

Right exhaust manifold torque sequence—4.6L DOHC engine

4.6L Engines

1. Disconnect the negative battery cable.
2. Remove engine compartment brace, if equipped.
3. Disconnect dual converter H-pipe.
4. Remove starter.
5. Disconnect steering column intermediate shaft.
6. Remove secondary air injection tubes, if equipped.
7. Remove Exhaust Gas Recirculation (EGR) tube.
8. Remove oil dipstick tube.
9. Remove oil filter.
10. Disconnect left and right motor mounts.
11. Raise the engine about 1.5 inches for clearance and remove the exhaust manifolds.

To install:

12. Install exhaust manifolds. Use new gaskets and tighten the nuts in sequence to 15 ft. lbs. (20 Nm).
13. Connect left and right motor mounts.
14. Install oil filter.
15. Install EGR tube.
16. Install oil dipstick tube.
17. Install secondary air injection tubes, if equipped.
18. Connect steering column intermediate shaft.
19. Install starter.
20. Connect dual converter H-pipe.
21. Install engine compartment brace, if equipped.
22. Connect the negative battery cable.

Camshaft and Valve Lifters

REMOVAL & INSTALLATION

3.8L Engine

VALVE LIFTERS

➡Keep all valvetrain parts in order for installation.

1. Drain the cooling system.
2. Disconnect the negative battery cable.
3. Remove Positive Crankcase Ventilation (PCV) valve and hose.
4. Remove valve covers.
5. Remove rocker arms and pushrods.
6. Remove upper and lower intake manifolds.
7. Remove lifter guide plates and retainers.
8. Remove valve lifters.

To install:

9. Install valve lifters in their original locations.

10. Install lifter guide plates and retainers. Tighten the bolts to 88–124 inch lbs. (10–14 Nm).

11. Install upper and lower intake manifolds.

12. Install rocker arms and pushrods in their original locations.

13. Install valve covers.

14. Install PCV valve and hose.

15. Connect the negative battery cable.

16. Fill the cooling system.

17. Start the engine and check for leaks.

CAMSHAFT

1. Drain the cooling system.
2. Recover the A/C refrigerant.
3. Remove radiator and cooling fan.
4. Remove A/C condenser.
5. Remove upper and lower intake manifolds.
6. Remove valve lifters.
7. Remove front cover.
8. Remove timing chain and gears.
9. Remove camshaft thrust plate and spacer.
10. Remove camshaft.

To install:

11. Install camshaft.

12. Install camshaft thrust plate. Tighten the bolts to 71–124 inch lbs. (8–14 Nm).

13. Install timing chain and gears.

14. Install front cover.

15. Install valve lifters.

16. Install upper and lower intake manifolds.

17. Install A/C condenser.

18. Install radiator and cooling fan.

19. Fill the cooling system.

20. Recharge the A/C system.

21. Run the engine and check for leaks.

4.6L Engines

VALVE LIFTERS

➡Keep all valvetrain parts in order for installation.

1. Remove valve covers.
2. Remove rocker arms.
3. Remove hydraulic lifters.

To install:

4. Inspect each lifter. If the plunger travel exceeds 0.059 inches (1.5 mm), replace the lifter.

5. Install hydraulic lifters in their original positions.

6. Install rocker arms.

7. Install valve covers.

CAMSHAFTS

1. Remove valve covers.
2. Remove rocker arms.
3. Remove oil pan (SOHC engine).
4. Remove front cover.
5. Remove timing chains and sprockets.
6. Remove camshaft bearing caps.
7. Remove camshafts.

To install:

➡On 4.6L DOHC engines, the outboard exhaust camshaft bearing cap bolts are of a different length than the other bearing cap bolts.

8. Install the camshafts. Tighten the bearing cap bolts in sequence as follows:
   a. Step 1: 71–106 inch lbs. (8–12 Nm).
   b. Step 2: Loosen all bolts 2 turns.
   c. Step 3: 71–106 inch lbs. (8–12 Nm).

9. Install timing chains and sprockets.

10. Install front cover.

11. Install oil pan (SOHC engine).

12. Install rocker arms.

To remove the camshaft bearing caps, tap the caps with a rubber or leather mallet where shown—4.6L engines

Camshaft bearing cap torque sequence—4.6L SOHC engine

13. Install valve covers.

14. Start the engine and check for leaks.

Valve Lash

ADJUSTMENT

The 3.8L and 4.6L engines are equipped with hydraulic lash adjusters. Valve clearance is not adjustable.

Starter Motor

REMOVAL & INSTALLATION

3.8L Engine

1. Disconnect the negative battery cable.
2. Disconnect starter electrical connections.
3. Disconnect ground cable.
4. Remove starter bolts.
5. Remove starter.

Camshaft bearing cap torque sequence—4.6L DOHC engine

To install:

6. Install starter. Tighten the bolts to 17 ft. lbs. (23 Nm).

7. Connect ground cable. Tighten the bolt to 17 ft. lbs. (23 Nm).

8. Connect starter electrical connections.

9. Connect the negative battery cable.

4.6L Engines

1. Disconnect the negative battery cable.

2. Disconnect Heated Oxygen (HO$_2$S) sensor connector and bracket, if necessary.

3. Disconnect starter electrical connections.

4. Remove starter bolts.

5. Remove starter.

To install:

6. Install starter. Tighten the bolts to 17 ft. lbs. (23 Nm).

7. Connect starter electrical connections.

8. Connect HO$_2$S sensor connector and bracket, if necessary.

9. Connect the negative battery cable.

Oil Pan

REMOVAL & INSTALLATION

3.8L Engine

1. Drain the engine oil.

2. Disconnect the negative battery cable.

3. Remove air cleaner outlet pipe.

4. Remove radiator shroud cover.

5. Install engine lifting brackets and attach an engine support fixture.

6. On convertible models, remove 16 splash shield pin-type retainers and remove the cross brace support, if equipped.

7. On coupe models, remove the front subframe brace, if equipped.

8. Remove oil filter.

9. Disconnect left and right motor mounts.

10. Remove starter motor.

11. Disconnect steering column intermediate shaft.

12. Remove subframe mounting bolts, and lower subframe for clearance.

13. Remove oil pan.

To install:

14. Apply a bead of silicone sealant to the oil pan mounting flange. Install the oil pan and tighten the bolts as follows:

    a. Step 1: 44 inch lbs. (5 Nm).

    b. Step 2: 88 inch lbs. (10 Nm).

    c. Step 3: Tighten the bell housing bolts to 33 ft. lbs. (45 Nm).

Oil pan torque sequence—3.8L engine

15. Raise the subframe and tighten the upper bolts to 85 ft. lbs. (115 Nm). Tighten the lower bolts to 68 ft. lbs. (90 Nm).

16. Connect steering column intermediate shaft.

17. Install starter motor.

18. Connect left and right motor mounts.

19. Install oil filter.

20. On coupe models, install the front subframe brace, if equipped and tighten the upper bolts to 30 ft. lbs. (41 Nm).

21. On convertible models, install the cross brace support, and install 16 splash shield pin-type retainers, if equipped.

22. Install radiator shroud cover.

23. Install air cleaner outlet pipe.

24. Connect the negative battery cable.

25. Fill the engine with oil.

26. Run the engine and check for leaks.

4.6L SOHC Engine

1. Drain the engine oil.

2. Disconnect the negative battery cable.

3. Remove air cleaner outlet pipe.

4. Remove radiator shroud cover.

5. Remove engine compartment brace, if equipped.

6. Install engine lifting brackets and attach an engine support fixture.

7. Remove front subframe crossmember brace.

8. Compress the front coil springs.

9. Place a safety stand under the subframe.

10. Remove left and right motor mount nuts.

11. Remove transmission housing cover.

12. Slightly lower the front subframe.

13. Remove the oil pan.

To install:

14. Install the oil pan. Use a new gasket and tighten the bolts in sequence as follows:

    a. Step 1: 18 inch lbs. (2 Nm).

    b. Step 2: 15 ft. lbs. (20 Nm).

Oil pan torque sequence—4.6L engines

    c. Step 3: Plus 60 degrees.

15. Raise the subframe into position and tighten the bolts to 85 ft. lbs. (115 Nm).

16. Install transmission housing cover.

17. Install left and right motor mount nuts. Tighten the nuts to 95–126 ft. lbs. (128–172 Nm).

18. Install front subframe crossmember brace. Tighten the bolts to 30–40 ft. lbs. (40–55 Nm).

19. Install engine compartment brace, if necessary.

20. Install radiator shroud cover.

21. Install air cleaner outlet pipe.

22. Connect the negative battery cable.

23. Fill the engine with oil.

24. Start the engine and check for leaks.

4.6L DOHC Engine

1. Disconnect the negative battery cable.

2. Remove air intake scoop, if equipped.

3. Remove air cleaner outlet pipe.

4. Remove radiator shroud cover.

5. Drain the engine oil.

6. Remove oil filter.

7. Install engine lifting brackets and attach an engine support fixture.

8. Remove transmission.

9. Remove A/C manifold and tube assembly (accumulator to compressor).

10. Remove A/C line.

11. Remove left and right motor mount nuts.

12. Disconnect wiring from starter motor and position harness out of the way, if necessary.

13. Compress front coil springs.

14. Disconnect stabilizer bar links.

15. Disconnect steering column intermediate shaft.

16. Place a safety stand under the subframe, and loosen subframe bolts.

17. Lower subframe for clearance.

18. Remove subframe brace, if equipped.

19. Remove oil pan.

To install:

20. Apply silicone sealant to the joints where the front cover and the rear seal retainer meet the cylinder block.

21. Install the oil pan. Use a new gasket and tighten the bolts in sequence as follows:

    a. Step 1: 18 inch lbs. (2 Nm).

    b. Step 2: 15 ft. lbs. (20 Nm).

    c. Step 3: Plus 60 degrees.

22. Raise subframe into position. Tighten the large bolts to 83–113 ft. lbs. (113–153 Nm) and the small bolts to 66–97 ft. lbs. (90–132 Nm).

23. Connect steering column intermediate shaft.

24. Connect stabilizer bar links.

25. Decompress front coil springs.

26. Install left and right motor mount nuts. Tighten the nuts to 95–126 ft. lbs. (128–172 Nm).

27. Connect starter motor wiring, if necessary.

28. Install A/C line.

29. Install A/C manifold and tube assembly (accumulator to compressor).

30. Remove engine lifting brackets and attach an engine support fixture.

31. Install radiator shroud cover.

32. Install air cleaner outlet pipe.

33. Install air intake scoop, if equipped.

34. Install oil filter.

35. Install transmission.

36. Connect the negative battery cable.

37. Fill the engine with oil.

38. Start the engine and check for oil leaks.

Oil Pump

REMOVAL & INSTALLATION

3.8L Engine

1. Remove the oil filter.

2. Remove oil pump body and gears.

To install:

3. Install oil pump. Use a new O-ring seal. Tighten the large bolts to 18 ft. lbs. (25 Nm) and the small bolts to 89 inch lbs. (10 Nm).

4. Install oil filter.

5. Check for leaks and proper operation.

Exploded view of the oil pump mounting— 4.6L engines

4.6L Engines

1. Remove valve covers.

2. Remove oil pan.

3. Remove front cover.

4. Remove timing chains and sprockets.

5. Remove oil pump.

To install:

6. Install oil pump. Tighten the bolts to 71–106 inch lbs. (8–12 Nm).

7. Install timing chains and sprockets.

8. Install front cover.

9. Install oil pan.

10. Install valve covers.

11. Check for leaks and proper operation.

Rear Main Seal

REMOVAL & INSTALLATION

3.8L Engine

1. Disconnect the negative battery cable.

2. Remove transmission.

3. Remove clutch pressure plate and disc, if equipped.

4. Remove flywheel.

5. Remove engine rear plate, if equipped.

6. Remove rear main seal. Use Crankshaft Rear Seal Remover T95P-6701-EH.

To install:

7. Install rear main seal. Use Rear Crankshaft Seal Installer T82L-6701-A.

8. Install engine rear plate, if equipped.

9. Install flywheel. Tighten the bolts to 54–64 ft. lbs. (73–87 Nm).

10. Install clutch pressure plate and disc, if equipped.

11. Install transmission.

12. Connect the negative battery cable.

4.6L Engines

1. Disconnect the negative battery cable.

2. Remove transmission.

3. Remove clutch pressure plate and disc, if equipped.

4. Remove flywheel.

5. Remove oil slinger. Use Rear Crankshaft Oil Slinger Remover T95P-6701-AH.

6. Remove rear oil seal retainer.

7. Remove rear oil seal.

To install:

8. Install the rear oil seal retainer. Apply silicone sealant as shown. Tighten the bolts in sequence to 71–106 inch lbs. (8–12 Nm).

9. Install rear oil seal. Use Rear Crankshaft Seal Replacer T95P-6701-BH.

10. Install oil slinger. Use Rear Crankshaft Oil Slinger Replacer T95P-6701-CH.

11. Install flywheel. Tighten the bolts in a crossing pattern to 54–64 ft. lbs. (73–87 Nm).

Installing the rear main seal—3.8L engine

Apply silicone sealant to the engine block when installing the oil seal retainer—4.6L engines

Oil seal retainer torque sequence—4.6L engines

12. Install clutch pressure plate and disc, if equipped.
13. Install transmission.
14. Connect the negative battery cable.
15. Run the engine and check for leaks.

Timing Chain, Sprockets, Front Cover and Seal

REMOVAL & INSTALLATION

3.8L Engine

1. Drain the cooling system and the engine oil.
2. Rotate the crankshaft so that the No. 1 cylinder is at Top Dead Center (TDC) of the compression stroke.
3. Disconnect the negative battery cable.
4. Disconnect Intake Air Temperature (IAT) sensor connector.
5. Remove air cleaner and air intake tube.
6. Remove cooling fan and shroud.
7. Remove the accessory drive belt.
8. Remove power steering pump and bracket.
9. Remove A/C compressor front bracket, if equipped.
10. Remove oil filter.
11. Remove radiator hoses.
12. Remove heater water outlet tube.
13. Remove crankshaft pulley and damper.
14. Remove front crankshaft seal.

➡Note the position of the Camshaft Position (CMP) sensor connector. The installation procedure requires that the connector be located in the same position.

15. Remove Camshaft Position (CMP) sensor housing.
16. Remove Crankshaft Position (CKP) sensor.
17. Remove oil pan.
18. Remove engine front cover.
19. Remove distributor drive gear.
20. Remove timing chain and sprockets.

To install:
21. Compress the timing chain vibration damper and install a retaining pin.

POSITIONING OF TIMING MARKS AND KEYWAYS IN CAMSHAFT AND CRANKSHAFT SPROCKETS MUST BE IN LINE AS SHOWN WITH NO. 1 PISTON AT TOP DEAD CENTER FIRING.

Camshaft timing marks—3.8L engine

● TIGHTEN IN SEQUENCE SHOWN TO 20-30 N·m (15-22 LB-FT)

○ REFER TO LEGEND

1. Engine front cover
2. Mounting bolts
3. Mounting nuts
4. Water pump
5. Mounting bolt
6. Mounting stud bolt
7. Engine block

Front cover torque sequence—3.8L engine

Camshaft Position Sensor

1 Syncro Positioning Tool
2 Engine front cover
3 Hold down clamp
4 Oil pump intermediate shaft
5 Camshaft Position Sensor housing
6 Camshaft Position Sensor
7 Sensor attaching screws
A Tighten to 15-22 ft. lbs.
B Tighten to 40-69 inch lbs.

Camshaft Position Sensor housing installation—3.8L engine

22. Install timing chain and sprockets with the timing marks aligned as shown.

23. Install distributor drive gear. Tighten the bolt to 30–36 ft. lbs. (40–50 Nm) and remove the vibration damper retaining pin.

24. Install front cover. Tighten the bolts in sequence to 15–22 ft. lbs. (20–30 Nm).

25. Install oil pan.

26. Install CKP sensor.

27. Install front crankshaft seal.

28. Install crankshaft pulley and damper. Tighten the bolt to 103–132 ft. lbs. (140–180 Nm).

29. Install heater water outlet tube.

30. Install radiator hoses.

31. Install oil filter.

32. Install A/C compressor front bracket, if equipped. Tighten the bolts to 30–45 ft. lbs. (41–61 Nm).

33. Install power steering pump and bracket. Tighten the bolts to 30–45 ft. lbs. (41–61 Nm).

34. Install Synchro Positioning Tool T96T-12200-A to the CMP sensor housing and turn it clockwise until the tool boss engages the notch in the housing assembly.

35. Install the CMP sensor housing so that the CMP sensor connector is in the position noted earlier. Tighten the hold down bolt to 15–22 ft. lbs. (20–30 Nm).

36. Install the accessory drive belt.

37. Install cooling fan and shroud.

38. Install air cleaner and air intake tube.

39. Connect IAT sensor connector.

40. Connect the negative battery cable.

41. Fill the cooling system.

42. Fill the crankcase with clean engine oil.

43. Run the engine. Check for leaks and proper operation.

4.6L SOHC Engine

➡ **This is not a free wheeling engine. Do not rotate the crankshaft or camshafts with the timing chains removed.**

1. Disconnect the negative battery cable.

2. Remove cooling fan and shroud.

3. Remove the accessory drive belt.

4. Remove water pump pulley.

5. Remove power steering pump and reservoir.

6. Remove oil pan.

7. Remove crankshaft pulley.

8. Remove front crankshaft seal.

9. Remove valve covers.

10. Remove idler pulley.

11. Disconnect Camshaft Position (CMP) sensor connector.

12. Disconnect Crankshaft Position (CKP) sensor connector.

7922NG55

Crankshaft sprocket positioning—4.6L SOHC engine

13. Remove front cover.

14. Remove CKP sensor pulse wheel.

15. Rotate the crankshaft so that the No. 1 cylinder is at Top Dead Center (TDC) of the compression stroke.

16. Install Camshaft Positioning Tool Adapters T92P-6256-A and Camshaft Positioning Tool T91P-6256-A on the flats of the camshafts.

17. Remove right timing chain tensioner.

18. Remove right timing chain tensioner arm.

19. Remove right timing chain guides.

20. Remove right timing chain and sprockets.

21. Remove left timing chain tensioner.

22. Remove left timing chain tensioner arm.

23. Remove left timing chain guides.

24. Remove left timing chain and sprockets.

To install:

25. Install timing chain guides. Tighten the bolts to 71–106 inch lbs. (8–12 Nm).

26. Install camshaft sprockets. Do not tighten the bolts at this time.

27. Install crankshaft sprockets.

9306NG04

Camshaft Positioning Tool and Adapter—4.6L SOHC engine

7922NG56

Timing chain and sprocket alignment—4.6L SOHC engine

7922NG58

Timing chain tensioner bleeding procedure—4.6L SOHC engine

7922NG59

Timing chain tensioner locking procedure—4.6L SOHC engine

Crankshaft Holding Tool—4.6L SOHC engine

INSTALL TOOL OVER FRONT COVER ALIGNMENT DOWEL

CRANKSHAFT HOLDING TOOL T93P-6303-A

NOTE: ALIGN CRANKSHAFT WOODRUFF KEY WITH KEYWAY ON TOOL

9306NG05

Remove slack from the timing chains with a C-clamp—4.6L SOHC engine

9306NG06

28. Install left timing chain with the copper links aligned with the timing marks on the crankshaft and camshaft sprockets.

29. Install right timing chain with the copper links aligned with the timing marks on the crankshaft and camshaft sprockets.

30. Compress the timing chain tensioners as shown. Install locking pins.

31. Install tensioner arms and tensioners. Tighten the tensioner mounting bolts to 15–22 ft. lbs. (20–30 Nm).

32. Install crankshaft Holding Tool T93P-6303-A.

33. Use a C-clamp on the tensioner arm and chain guide to remove slack from the timing chain.

34. Remove the tensioner locking pins.

35. Tighten the camshaft sprocket bolts to 81–95 ft. lbs. (110–130 Nm).

36. Remove the C-clamps, Camshaft Positioning Tools and Adapters, and the Crankshaft Holding Tool.

37. Install CKP sensor pulse wheel.

38. Install front cover. Tighten the bolts in sequence to 15–22 ft. lbs. (20–30 Nm).

39. Connect CKP sensor connector.

40. Connect CMP sensor connector.

41. Install idler pulley.

42. Install valve covers.

APPLY SEALER

FRONT OF ENGINE

VIEW A
TYPICAL 4 PLACES

APPLY SEALER

VIEW B
TYPICAL 2 PLACES

1. Cylinder
2. Cylinder
3. Oil pan gasket
4. Gasket
5. Front cover assembly
6a. Bolts
7a. Studs
8. Dowel

7922NG54

Apply sealer when installing the front cover—4.6L SOHC engine

43. Install front crankshaft seal.

44. Install oil pan.

45. Install the crankshaft pulley and tighten the bolt as follows:
  a. Step 1: 66 ft. lbs. (90 Nm).
  b. Step 2: Loosen one complete turn.
  c. Step 3: 36 ft. lbs. (50 Nm).
  d. Step 4: Plus 90 degrees.

46. Install power steering pump and reservoir.

47. Install water pump pulley.

48. Install the accessory drive belt.

49. Install cooling fan and shroud.

50. Connect the negative battery cable.

51. Fill the crankcase with clean engine oil.

52. Start the engine. Check for leaks and proper operation.

4.6L DOHC Engine

➡ **This is not a free wheeling engine. Do not rotate the crankshaft or camshafts with the timing chains removed.**

1. Drain the cooling system.

2. Drain the supercharger cooling system, if equipped.

3. Remove supercharger drive belt, if equipped.

4. Remove the accessory drive belt.

5. Disconnect the negative battery cable.

6. Remove engine compartment brace, if equipped.

REFER TO LEGEND

TIGHTEN IN SEQUENCE SHOWN

1. Stud bolt (4 req'd)
2. Bolt (6 req'd)
3. Stud bolt (5 req'd)
4. Engine front cover

9306NG07

Front cover torque sequence and bolt identification—4.6L SOHC engine

7. Remove air intake tube.

8. Remove upper radiator hose and bypass tube.

9. Remove cooling fan and shroud.

10. Remove engine control sensor wiring support bracket.

11. Remove fuel charging wiring retainer.

12. Remove water pump pulley.

13. Remove power steering pump reservoir.

14. Remove ignition coils.

15. Remove power steering pump.

16. Remove crankshaft pulley.

17. Remove front crankshaft seal.

18. Remove valve covers.

➡**Keep rocker arms in order for installation.**

19. Remove rocker arms.

20. Disconnect Camshaft Position (CMP) sensor connector.

21. Disconnect Crankshaft Position (CKP) sensor connector.

22. Remove idler pulley.

23. Remove front cover.

24. Remove CMP sensor pulse wheel.

25. Rotate the crankshaft so that the No. 1 cylinder is at Top Dead Center (TDC) of the compression stroke.

26. Remove right primary timing chain tensioner.

27. Remove right primary timing chain tensioner arm and chain guide.

28. Remove right primary timing chain and sprockets.

29. Remove left primary timing chain tensioner.

30. Remove left primary timing chain tensioner arm and chain guide.

31. Remove left primary timing chain and sprockets.

32. Compress the secondary chain tensioners and install locking pins.

33. Remove the left and right secondary timing chains and sprockets.

To install:

34. Position the camshafts with the keys facing downward.

35. Install the secondary timing chains and sprockets. Do not tighten the sprocket bolts at this time.

36. Tension the secondary timing chains with the Secondary Timing Chain Tensioning tool T93P-6256-BH.

37. Install Camshaft Positioning tool T93P-6256-A in the rear D-slots of the camshaft.

38. Install Camshaft Holding tool T93P-6256-AH onto the camshafts to keep the camshafts from rotating and to prevent damaging the camshaft positioning tool.

39. Tighten the secondary timing chain sprocket bolts to 81–95 ft. lbs. (110–130 Nm).

40. Remove the secondary timing chain tensioner locking pins and remove the tensioner tool.

41. Compress the primary timing chain tensioners and install locking pins.

42. Install the crankshaft timing sprockets by aligning the timing marks.

43. Align the copper colored primary timing chain links with the sprocket timing marks as shown.

Install the camshaft positioning tool—4.6L DOHC engines

Install the Camshaft Holding tool to secure the camshafts and prevent damage to the camshaft positioning tool—4.6L DOHC engines

Position the camshafts for timing chain installation—4.6L DOHC engines

Secondary timing chain tensioner and locking pin—4.6L DOHC engines

Secondary timing chain tensioning tool—4.6L DOHC engines

Crankshaft timing sprocket installation—4.6L DOHC engines

Correct alignment of the primary timing chains and sprockets—4.6L DOHC engines

Front cover torque sequence—4.6L DOHC engines

44. Install left primary timing chain and sprockets.

45. Install left primary timing chain tensioner arm and chain guide.

46. Install left primary timing chain tensioner.

47. Install right primary timing chain and sprockets.

48. Install right primary timing chain tensioner arm and chain guide.

49. Install right primary timing chain tensioner.

50. Tighten the camshaft sprocket bolts to 81–95 ft. lbs. (110–130 Nm).

51. Remove the primary timing chain tensioner locking pins.

52. Remove the Camshaft Holding and Camshaft Positioning tools.

53. Install the CMP pulse wheel

54. Install the front cover. Tighten the bolts in sequence as follows:

   a. Step 1: 14 ft. lbs. (20 Nm).

   b. Step 2: Plus 60 degrees.

55. Install idler pulley.

56. Install rocker arms in their original positions.

57. Install valve covers.

58. Install front crankshaft seal.

59. Connect CKP sensor connector.

60. Connect CMP sensor connector.

61. Install Crankshaft pulley.

62. Install power steering pump.

63. Install ignition coils.

64. Install power steering pump reservoir.

65. Install water pump pulley.

66. Install fuel charging wiring retainer.

67. Install engine control sensor wiring support bracket.

68. Install cooling fan and shroud.

69. Install upper radiator hose and bypass tube.

70. Install air intake tube.

71. Install engine compartment brace, if equipped.

72. Connect the negative battery cable.

73. Install the accessory drive belt.

74. Install supercharger drive belt, if equipped.

75. Fill the cooling system.

76. Fill the supercharger cooling system, if equipped.

77. Start the engine. Check for leaks and proper operation.

Sealer Location

1 RH Cylinder Head
2 LH Cylinder Head
3 Ignition Pulse Crankshaft Sensor Ring
4 Cylinder Block
5 Sealer
6 Oil Pan Gasket

9306LG06

Apply sealer to these locations—4.6L DOHC engines

Timing chain tensioner installation—4.6L DOHC engine

7922NG62

Piston and Ring

POSITIONING

Piston and connecting rod positioning—3.8L engine

7922AG15

Piston ring end-gap spacing—3.8L engine

Piston ring positioning—4.6L SOHC engine

Piston ring end-gap spacing—4.6L SOHC engine

Piston ring positioning—4.6L DOHC engine

Piston ring end-gap spacing—4.6L DOHC engine

FUEL SYSTEM

Fuel System Service Precautions

Safety is the most important factor when performing not only fuel system maintenance, but any type of maintenance. Failure to conduct maintenance and repairs in a safe manner may result in serious personal injury or death. Work on a vehicle's fuel system components can be accomplished safely and effectively by adhering to the following rules and guidelines.

• To avoid the possibility of fire and personal injury, always disconnect the negative battery cable unless the repair or test procedure requires that battery voltage be applied.

• Always relieve the fuel system pressure prior to disconnecting any fuel system component (injector, fuel rail, pressure regulator, etc.) fitting or fuel line connection. Exercise extreme caution whenever relieving fuel system pressure to avoid exposing skin, face and eyes to fuel spray. Please be advised that fuel under pressure may penetrate the skin or any part of the body that it contacts.

• Always place a shop towel or cloth around the fitting or connection prior to loosening to absorb any excess fuel due to spillage. Ensure that all fuel spillage is quickly removed from engine surfaces. Ensure that all fuel-soaked cloths or towels are deposited into a flame-proof waste container with a lid.

• Always keep a dry chemical (Class B) fire extinguisher near the work area.

• Do not allow fuel spray or fuel vapors to come into contact with a spark or open flame.

• Always use a second wrench when loosening or tightening fuel line connection fittings. This will prevent unnecessary stress and torsion on fuel piping. Always follow the proper torque specifications.

• Always replace worn fuel fitting O-rings with new ones. Do not substitute fuel hose where rigid pipe is installed.

Fuel System Pressure

RELIEVING

Most Sequential Fuel Injection (SFI) engines , except 2004 4.6L SOHC engines, are equipped with a pressure relief valve located on the fuel supply manifold. Remove the fuel tank cap and attach fuel pressure gauge T80L-9974-B, or equivalent, to the valve on the fuel rail. Place the tube from the tool into a small container and open the valve on the tool to release the fuel pressure. Be sure to drain the fuel into a suitable container and to avoid gasoline spillage. If a pressure gauge is not available, disconnect the vacuum hose from the fuel pressure regulator and attach a hand-held vacuum pump. Apply about 25 in. Hg (84 kPa) of vacuum to the regulator to vent the fuel system pressure into the fuel tank through the fuel return hose.

For 2004 4.6L SOHC engines, remove the fuel pump fuse; located in the engine compartment fuse box. Start the engine and allow it to run at idle until it stalls. After the engine stalls, crank the engine for approximately five seconds to ensure the pressure in the fuel injection supply manifold has been released. After the fuel pressure has been fully released, turn the ignition switch to the "OFF" position.

➡This procedure will remove the fuel pressure from the lines, but not the fuel. Take precautions to avoid the risk of fire and use clean rags to soak up any spilled fuel when the lines are disconnected.

Fuel Filter

REMOVAL & INSTALLATION

➡Always replace fuel line fitting plastic clips.

1. Relieve the fuel system pressure.

Hairpin clip fuel fitting

Duckbill clip fuel fitting

7922NG44

67197-MUST-G17

Push connect fuel fitting

2. Disconnect the negative battery cable.

3. Disconnect fuel line fittings.

➡ **The filter may be secured in place with hairpin clips, duckbill clips, or push connect fittings.**

4. Disengage the filter from the fuel filter bracket clamp.

5. Remove fuel filter.

To install:

6. Install fuel filter. Note the flow direction arrow.

7. Secure the fuel filter in the bracket clamp.

8. Install fuel line fittings.

9. Connect the negative battery cable.

10. Start the engine and check for leaks.

Fuel Pump

REMOVAL & INSTALLATION

1. Relieve the fuel system pressure.
2. Disconnect the negative battery cable.
3. Drain the fuel tank.
4. Remove fuel tank filler pipe retainer.
5. Disconnect EVAP canister tube and hose, if necessary.
6. Disconnect fuel tank vent hose.
7. Remove fuel tank support straps.
8. Disconnect fuel lines.
9. Disconnect fuel pump module electrical connector.
10. Remove fuel pump module retaining bolts.
11. Raise the fuel pump module to access locking tabs, if equipped. Squeeze

1 Fuel Pump Module
2 Locking Tab (Part of 9H307)
3 O-Ring Seal
4 Connector (Part of 14405 Wiring Assy)
A Fuel Pump Must Be Snapped Into Retainer (2 Places)

9306NG08

Fuel pump module assembly

10 Nm (89 lb-in)

67197-MUST-G18

Fuel pump module torque sequence

the locking tabs together, if necessary, and remove the fuel pump module from the fuel tank.

To install:

12. Install the fuel pump module into the retainer. Use a new O-ring seal and push the module into the retainer until both locking tabs engage, if equipped.

13. Install fuel pump module retaining bolts. Tighten the bolts, in sequence, to 89 inch lbs. (10 Nm).

14. Connect fuel pump module electrical connector.

15. Connect fuel lines.

16. Install fuel tank support straps. Tighten the bolts to 22–29 ft. lbs. (30–40 Nm).

17. Connect fuel tank vent hose.

18. Connect EVAP canister tube and hose, if necessary.

19. Install fuel tank filler pipe retainer.

20. Connect the negative battery cable.

21. Add fuel (10 gallons minimum) to the tank.

22. Start the engine and check for leaks.

Fuel Injector

REMOVAL & INSTALLATION

3.8L Engine

1. Relieve the fuel system pressure.
2. Disconnect the negative battery cable.
3. Remove upper intake manifold.
4. Disconnect fuel supply line.
5. Disconnect fuel return line, if equipped.
6. Disconnect fuel injector connectors.
7. Disconnect fuel pressure sensor vacuum hose and electrical connector.
8. Remove fuel supply manifold with injectors attached.
9. Remove fuel injectors from the supply manifold.

To install:

10. Install new O-rings.
11. Install fuel injectors to the supply manifold.
12. Install fuel supply manifold with injectors. Tighten the bolts to 89 inch lbs. (10 Nm).
13. Connect fuel pressure sensor vacuum hose and electrical connector.
14. Connect fuel injector connectors.
15. Connect fuel supply line.
16. Connect fuel return line, if equipped.
17. Install upper intake manifold.
18. Connect the negative battery cable.
19. Start the engine and check for leaks.

4.6L SOHC Engine

1. Relieve the fuel system pressure.
2. Disconnect the negative battery cable.
3. Remove air intake tube.
4. Disconnect fuel supply line.
5. Disconnect fuel return line, if equipped.
6. Disconnect accelerator cable.
7. Disconnect cruise control cable, if equipped.
8. Remove throttle body.
9. Disconnect Idle Air Control (IAC) valve connector and hose.
10. Disconnect Positive Crankcase Ventilation (PCV) hose.
11. Disconnect main chassis vacuum supply hose.
12. Disconnect fuel injector connectors.
13. Disconnect fuel pressure sensor vacuum hose and electrical connector.

14. Disconnect ground wire, if necessary.

15. Disconnect Exhaust Vacuum Regulator (EVR) solenoid vacuum lines.

16. Remove Exhaust Gas Recirculation (EGR) pressure transducer bracket.

17. Disconnect EGR tube and vacuum line.

18. Remove fuel supply manifold with injectors attached.

19. Remove fuel injectors from the supply manifold.

To install:

20. Install new O-rings.

21. Install fuel injectors to the supply manifold.

22. Install fuel supply manifold with injectors attached. Tighten the bolts to 89 inch lbs. (10 Nm).

23. Install EGR tube and vacuum line.

24. Install EGR pressure transducer bracket.

25. Connect EVR solenoid vacuum lines.

26. Connect ground wire, if necessary.

27. Connect fuel pressure sensor vacuum hose and electrical connector.

28. Connect fuel injector connectors.

29. Connect main chassis vacuum supply hose.

30. Connect PCV hose.

31. Connect IAC valve connector and hose.

32. Install throttle body.

33. Connect cruise control cable, if equipped.

34. Connect accelerator cable.

35. Connect fuel supply line.

36. Connect fuel return line, if equipped.

37. Install air intake tube.

38. Connect the negative battery cable.

39. Start the engine and check for leaks.

4.6L DOHC Engine

1. Relieve the fuel system pressure.

2. Disconnect the negative battery cable.

3. Remove upper intake manifold.

4. Disconnect fuel supply line.

5. Disconnect fuel return line, if equipped.

6. Disconnect fuel injector connectors.

7. Disconnect fuel pressure sensor vacuum hose and electrical connector.

8. Remove fuel supply manifold with injectors attached.

9. Remove fuel injectors from the supply manifold.

To install:

10. Install new O-rings.

11. Install fuel injectors to the supply manifold.

12. Install fuel supply manifold with injectors attached. Tighten the bolts to 89 inch lbs. (10 Nm).

13. Connect fuel pressure sensor vacuum hose and electrical connector.

14. Connect fuel injector connectors.

15. Connect fuel supply line.

16. Connect fuel return line, if equipped.

17. Install upper intake manifold.

18. Connect the negative battery cable.

19. Start the engine and check for leaks.

DRIVE TRAIN

Transmission

REMOVAL & INSTALLATION

Manual

➡**The clutch pedal must be held in the uppermost position during clutch release cable removal and installation. Failure to properly position and support the clutch pedal can result in damage to the self-adjusting mechanism.**

1. Lift the clutch pedal and secure in place with safety wire.

2. Disconnect the negative battery cable.

3. Remove shift lever and boot.

4. Remove driveshaft.

5. Remove clutch release lever cover, if equipped.

6. Disconnect clutch cable.

7. Disconnect Heated Oxygen (HO₂S) sensor connectors.

8. Remove dual converter "Y" or "H" pipe.

9. Disconnect backup lamp switch connector.

10. Disconnect Vehicle Speed (VSS) sensor connector.

11. Remove flywheel cover.

12. Remove starter motor.

13. Support the transmission with a jack and remove the rear transmission support crossmember.

14. Lower the transmission jack and remove the transmission flange bolts.

15. Slide the transmission input shaft out of the clutch and lower the transmission away from the vehicle.

To install:

16. Install transmission. Tighten the flange bolts to 26 ft. lbs. (35 Nm) on the T56 transmission, and 55 ft. lbs. (75 Nm) on the other transmissions.

17. Install transmission crossmember. Tighten the transmission mount bolts to 43 ft. lbs. (58 Nm) and the crossmember bolts to 30 ft. lbs. (41 Nm).

18. Install starter motor. Tighten the bolts to 17 ft. lbs. (23 Nm).

19. Install flywheel cover. Tighten the bolts to 20 ft. lbs. (27 Nm).

20. Connect VSS sensor connector.

21. Connect backup lamp switch connector.

22. Install dual converter "Y" or "H" pipe.

23. Connect HO₂S sensor connectors.

24. Connect clutch cable.

25. Install driveshaft.

26. Install shift lever and boot.

27. Connect the negative battery cable.

28. Remove the safety wire from the clutch pedal and check for proper operation.

Automatic

1. Drain the transmission fluid.

2. Remove the torque converter access cover and drain the torque converter.

3. Disconnect the negative battery cable.

4. Disconnect Heated Oxygen (HO₂S) sensor connectors.

5. Remove dual converter "Y" or "H" pipe.

6. Remove torque converter nuts.

7. Remove driveshaft.

8. Disconnect shift cable.

9. Disconnect transmission wiring connectors.

10. Disconnect and plug transmission fluid cooler lines.

11. Remove starter motor.

12. Support the transmission with a jack and remove the rear transmission support crossmember.

13. Lower the transmission jack and remove the transmission flange bolts.

14. Lower the transmission away from the vehicle.

To install:

15. Install transmission. Tighten the flange bolts to 35 ft. lbs. (47 Nm).

16. Transmission crossmember. Tighten the transmission mount bolts to 59 ft. lbs. (80 Nm) and the crossmember bolts to 41 ft. lbs. (55 Nm).

17. Install starter motor. Tighten the bolts to 18 ft. lbs. (25 Nm).

18. Connect transmission fluid cooler lines

19. Connect transmission wiring connectors.

20. Connect shift cable.

21. Install driveshaft.

22. Install new torque converter nuts. Tighten the nuts to 20–33 ft. lbs. (27–46 Nm).

23. Install torque converter access cover. Tighten the bolts to 25 ft. lbs. (34 Nm).

24. Install dual converter "Y" or "H" pipe.

25. Connect HO2S sensor connectors.

26. Connect the negative battery cable.

27. Fill the transmission with fluid.

28. Start the engine. Check for leaks and proper operation.

Clutch

ADJUSTMENTS

The clutch is equipped with a self-adjusting mechanism. Pull the clutch pedal up to activate the adjuster.

REMOVAL & INSTALLATION

➡The clutch pedal must be held in the uppermost position during clutch release cable removal and installation. Failure to properly position and support the clutch pedal can result in damage to the self-adjusting mechanism.

1. Disconnect the negative battery cable.

2. Lift the clutch pedal and secure in place with safety wire.

3. Remove transmission.

4. Remove starter, if necessary.

5. Remove pressure plate bolts, loosen them evenly in several passes to avoid distortion of the pressure plate.

6. Remove pressure plate and clutch disk.

To install:

7. Install clutch disk and pressure plate. Tighten the pressure plate bolts evenly in several passes to 35 ft. lbs. (47 Nm) for 3.8L engines or to 33 ft. lbs. (45 Nm) plus an additional 60 degrees for 4.6L engines.

67197-MUST-G19

Clutch pressure plate bolts torque sequence

FRONT OF ENGINE

Clutch Release Lever Cable

Self-adjuster Quadrant

Support Bracket

Clutch Pedal

Clutch Release Lever

C-clip

HOLD PEDAL IN FULL "UP" POSITION WHILE INSTALLING CLUTCH CABLE IN RELEASE LEVER. APPROXIMATELY 44 N (10 LBS) LOAD IN DIRECTION OF ARROW

7922NG45

Clutch self-adjusting system component identification—4.6L engine—3.8L engine is similar

SPRINGS ARE NOT
TO BE BENT OR DAMAGED
DURING ASSEMBLY

✱ LUBRICATE BALL
AND POCKET

NOTE: INSTALL WITH
"FW SIDE" OR "FLYWHEEL
SIDE" STAMPED NOTATION
FACING FORWARD

✱ LUBRICATE LEVER CROWN
DO NOT DISTURB GREASE
DURING ASSEMBLY

SECTION A

✱ LUBRICATE WITHIN
63.5-165 mm (2.5-6.5 INCHES)
OF REAR SHOULDER

✱ LUBRICATE LEVER
CROWN AND SPRING
RETENTION CROWN

SPRING MUST
BE POSITIONED
WITHIN BEARING
GROOVE

SECTION A

✱ PREMIUM LONG-LIFE GREASE XG-1-C OR
XG-1-K (ESA-M1C75-B)

1. Rear face of engine block
2. Flywheel-to-clutch cover alignment dowel
3. Pilot bearing
4. Clutch disc
5. Clutch pleasure plate
6. Clutch release lever
7. Clutch release lever stud
8. Lockwasher
9. Flywheel housing
10. Bolt
11. Input shaft
12. Bolt
13. Clutch release lever dust shield
14. Bolt
15. Clutch release bearing
16. Bolt
17. Flywheel housing-to-engine block dowel

7922NG46

Exploded view of the clutch system components

8. Install starter, if necessary.
9. Install transmission.
10. Connect the negative battery cable.
11. Remove the clutch pedal safety wire and check for proper operation.

Halfshaft

REMOVAL & INSTALLATION

SVT Cobra

1. Before raising the vehicle, match-mark the rear shock absorber to indicate curb height.
2. Remove the rear wheel.
3. Remove hub retainer nut.
4. Remove disc brake caliper and rotor.

5. Remove wheel speed sensor.
6. Disconnect the tie rod link.
7. Place a support under the lower control arm.
8. Remove lower shock absorber mounting bolt.
9. Disconnect the upper control arm from the knuckle.
10. Press the outboard CV-joint from the hub. Lower the knuckle until the stub shaft clears the hub.
11. Use Halfshaft Removal Tool 205-475 to remove the halfshaft from the differential.

To install:

➡**Use new nuts, bolts, circlips, and split pins for assembly.**

12. Install Differential Seal Protector 205-461.

205-475

9306NG09

Halfshaft Removal Tool—SVT Cobra

13. Install the axle inner stub shaft into the differential until the splines are past the seal.
14. Remove the Differential Seal Protector.
15. Install the inner stub shaft into the differential until the circlip seats.

9306NG10

Differential Seal Protector—SVT Cobra

16. Install outer stub shaft into the knuckle.

17. Connect the upper control arm to the knuckle and install the lower shock absorber bolt.

18. Lower control arm to align the matchmark on the shock absorber.

19. Install lower shock absorber mounting bolt. Torque it to 98 ft. lbs. (133 Nm).

20. Install upper control arm bolt. Torque it to 66 ft. lbs. (90 Nm).

21. Remove the lower control arm support.

22. Connect the tie rod link. Tighten the nut to 35 ft. lbs. (47 Nm).

23. Install wheel speed sensor.

24. Install disc brake rotor and caliper. Tighten the caliper support bolts to 76 ft. lbs. (103 Nm).

➡**The hub retainer nut must be tightened with the brakes applied and the wheels off the ground to ensure correct bearing seating.**

25. Install hub nut. Tighten it to 240 ft. lbs. (325 Nm).

26. Install wheel.

27. Check the rear wheel alignment and adjust as necessary.

CV-Joints

REMOVAL & REPLACEMENT

Inner CV-Joint

1. Remove halfshaft assembly.

2. Remove inner CV-joint boot clamps.

3. Remove CV-joint boot, slide it away from the joint.

4. Remove snapring.

5. Remove the inner CV-joint from the halfshaft.

To install:

6. CV-joint, fill it with fresh grease and slide it onto the halfshaft.

7. Install new snapring.

8. Install new CV-joint clamps.

9. Install CV-joint boot.

10. Install halfshaft.

Outer CV-Joint

The outer CV-joint is serviced with the halfshaft as an assembly. The outer CV-joint boot can be serviced by removing the inner CV-joint.

Axle Shaft, Bearing and Seal

REMOVAL & INSTALLATION

1. Remove rear wheel.

2. Remove disc brake caliper and rotor.

3. Remove wheel speed sensor.

4. Remove axle housing cover.

5. Remove differential pinion shaft lockbolt.

6. Remove differential pinion shaft.

7. Remove axle retaining U-washer.

8. Remove axle shaft.

9. Remove bearing and seal, using a slide hammer

To install:

10. Install bearing, so that it is fully seated in the axle tube.

11. Install axle seal.

12. Install axle shaft.

13. Install axle retaining U-washer.

14. Install differential pinion shaft. Tighten the lockbolt to 15–30 ft. lbs. (20–41 Nm).

15. Install axle housing cover. Tighten the cover bolts to 18–28 ft. lbs. (24–38 Nm).

16. Install wheel speed sensor.

17. Install disc brake rotor and caliper. Tighten the caliper mounting bolts to 65–87 ft. lbs. (87–119 Nm).

18. Install rear wheel.

19. Fill the differential with gear lubricant. Tighten the filler plug to 15–30 ft. lbs. (20–41 Nm).

LOCK BOLT
TIGHTEN TO
20-40 N.m
(15-30 lb-ft)

DIFFERENTIAL
PINION SHAFT

7922NG65

Differential pinion shaft and lockbolt

Pinion Seal

REMOVAL & INSTALLATION

1. Disconnect driveshaft and position it out of the way.

2. Remove rear wheels.

3. Remove rear brake calipers.

➡**The rear brake calipers must be removed so that there is no additional drag when measuring pinion bearing preload.**

➡**Remember to index mark the driveshaft flange and pinion flange to maintain initial balance during installation.**

4. Use an inch lb. torque wrench and measure the amount of torque required to maintain pinion rotation through several revolutions.

5. Remove the pinion flange and remove the seal.

To install:

6. Install pinion seal and flange.

7. Install new pinion flange nut.

8. Rotate the pinion flange occasionally while tightening the flange nut to make sure the pinion bearings seat correctly.

9. Take frequent bearing preload torque readings.

10. If the preload recorded prior to disassembly is **lower** than the specification for used bearings, then tighten the pinion flange nut to specification. If the preload recorder prior to disassembly is **higher** than the specification for used bearings, then tighten the pinion flange nut to the original reading as recorded.

11. The pinion bearing preload specifications are as follows:

   a. Used bearings: 8–14 inch lbs. (0.9–1.5 Nm).

   b. New bearings: 16–29 inch lbs. (1.8–3.3 Nm).

✳✳ CAUTION

Never loosen the pinion nut to reduce bearing preload. If it is necessary to reduce bearing preload, install a new collapsible spacer and pinion nut.

12. Connect the driveshaft and install new bolts. Tighten the bolts to 83 ft. lbs. (112 Nm).

13. Install brake calipers.

14. Install rear wheels.

15. Fill the differential with gear lubricant and check for leaks.

STEERING AND SUSPENSION

Air Bag

❊❊ CAUTION

Some vehicles are equipped with an air bag system. The system must be disarmed before performing service on, or around, system components, the steering column, instrument panel components, wiring and sensors. Failure to follow the safety precautions and the disarming procedure could result in accidental air bag deployment, possible injury and unnecessary system repairs.

PRECAUTIONS

Several precautions must be observed when handling the inflator module to avoid accidental deployment and possible personal injury.

- Never carry the inflator module by the wires or connector on the underside of the module.
- When carrying a live inflator module, hold securely with both hands, and ensure that the bag and trim cover are pointed away.
- Place the inflator module on a bench or other surface with the bag and trim cover facing up.
- With the inflator module on the bench, never place anything on or close to the module that may be thrown in the event of an accidental deployment.

DISARMING

1. Before servicing the vehicle, refer to the Precautions Section.
2. Disconnect both battery cables from the battery, negative cable first.
3. Wait 1 minute before proceeding with the service procedure. This is the time required for the back-up power supply in the air bag diagnostic monitor to deplete its stored energy.
4. After service is completed, reconnect the battery cables, negative cable last.
5. Turn the ignition switch to the **RUN** position. The air bag indicator should light continuously for approximately 6 seconds, then turn **OFF**. If the indicator fails to light, flashes or remains lit continuously, there is a fault in the air bag system.

Power Rack and Pinion Steering Gear

REMOVAL & INSTALLATION

1. Disconnect the negative battery cable.
2. Remove front wheels.
3. Disconnect steering column intermediate shaft coupling.
4. Disconnect outer tie rod ends.
5. Remove steering gear retaining bolts.
6. Disconnect power steering pressure and return lines.
7. Remove steering gear.

To install:

8. Attach the pressure and return lines to the steering gear before attaching the steering gear to the subframe. Use new plastic seals and tighten the line fittings to 20–25 ft. lbs. (27–34 Nm).

➡**The power steering fluid lines are designed to swivel when properly tightened. Do not overtighten the fittings.**

9. Install steering gear. Tighten the bolts to 31–39 ft. lbs. (41–54 Nm).
10. Connect outer tie rod ends. Tighten the nuts to 36–46 ft. lbs. (48–63 Nm).
11. Connect steering column intermediate shaft coupling. Tighten the pinch bolt to 21–29 ft. lbs. (28–40 Nm).
12. Install front wheels.
13. Connect the negative battery cable.
14. Fill the power steering system with the proper type and quantity of fluid.

15. Check the front end alignment and adjust as necessary.

Strut

REMOVAL & INSTALLATION

Front

1. Support the front of the vehicle on jackstands placed under the control arms.
2. Remove front wheel.
3. Remove disc brake caliper.
4. Remove wheel speed sensor and bracket.
5. Remove upper strut retaining fasteners.
6. Remove wheel spindle attachment bolts.
7. Compress the strut assembly and remove it from the vehicle.

To install:

8. If replacing the strut, transfer the upper mounting bracket.
9. Install strut assembly. Tighten the spindle bolts to 141–191 ft. lbs. (190–259 Nm).
10. Install upper strut retaining fasteners. Tighten to 25–34 ft. lbs. (34–46 Nm).
11. Install wheel speed sensor and bracket.
12. Install disc brake caliper.
13. Install front wheel.
14. Remove the jackstands and check the alignment. Adjust as necessary.

1. No. 2 front crossmember
2. Mounting bolt
3. Steering gear mounting bracket
4. Retaining nut
5. Power steering rack and pinion

7922NG47

Power rack and pinion steering gear mounting on the No. 2 crossmember

Exploded view of the front strut upper mounting

Shock Absorber

REMOVAL & INSTALLATION

Rear

1. Pull aside rear compartment trim panels.

2. Remove upper shock absorber retaining nut.

3. Remove lower shock absorber bolt.

4. Remove shock absorber.

1. Insulator nut
2. Shock absorber
3. Mounting bolt
4. Mounting nut
5. Shock absorber lower mount bracket
6. Mounting nut
7. Rear axle housing

Exploded view of the rear shock absorber mounting

To install:

5. Prime the new shock absorber as follows:

    a. Step 1: With the shock absorber right side up, extend it fully.

    b. Step 2: Turn the shock absorber upside down and fully compress it.

    c. Step 3: Repeat for 3 cycles.

6. Install the shock absorber and tighten the lower bolt as follows:

    a. SVT Cobra: 98 ft. lbs. (133 Nm).

    b. All others: 57–75 ft. lbs. (76–103 Nm).

7. Install upper shock absorber nut. Tighten it to 25–33 ft. lbs. (34–46 Nm).

8. Place rear compartment trim panels back into position.

Coil Spring

REMOVAL & INSTALLATION

Front

1. Install an internal spring compressor to the spring to be serviced. Tighten the compressor to relieve spring pressure on the lower control arm.

2. Remove front wheel.

3. Disconnect outer tie rod end.

4. Disconnect ball joint stud from front wheel spindle.

5. Disconnect stabilizer bar link.

6. Loosen lower control arm mounting bolts.

7. Remove coil spring.

To install:

8. If replacing the coil spring, transfer the spring compressor to the new spring.

9. Install coil spring.

10. Connect ball joint stud to front wheel spindle. Use a jack to raise the control arm to a normal position and tighten the bolts to 109–149 ft. lbs. (148–202 Nm).

11. Connect stabilizer bar link. Tighten the nut to 11–16 ft. lbs. (16–22 Nm).

12. Tighten lower control arm mounting bolts to 148 ft. lbs. (200 Nm).

13. Connect outer tie rod end. Tighten the nut to 36–46 ft. lbs. (48–63 Nm).

14. Install front wheel.

15. Position the coil spring and remove the spring compressor.

1. Mounting bolt
2. Damper
3. Front coil spring
4. Insulator
5. Front suspension lower arm
6. Nuts

Exploded view of the front coil spring mounting

Rear

EXCEPT SVT COBRA

1. Raise and support the vehicle safely under the frame. Support the body at the rear body crossmember.

2. Disconnect wheel speed sensor wire from parking brake cable retaining bracket, on side of lower control arm.

3. Remove and discard parking brake cable bracket bolt. Separate bracket from lower control arm.

4. If equipped, remove the stabilizer bar.

5. Support the axle with a jack.

6. Place another jack under the lower arm axle pivot bolt. Remove and discard the bolt and nut. Lower the jack slowly until the coil spring load is relieved.

7. Remove the coil spring and insulator from the vehicle.

To install:

8. Place the upper spring insulator on top of the spring. Place the lower spring insulator on the lower arm.

9. Position the coil spring on the lower arm spring seat with the pigtail on the lower arm at the rear of the vehicle and pointing toward the left side of the vehicle.

10. Slowly raise the jack until the arm is in position. Insert a new rear pivot bolt and nut.

11. Raise the axle to curb height. Tighten the pivot bolt to 71–97 ft. lbs. (97–132 Nm).

12. Position parking brake cable bracket onto lower control arm. Install a new bracket bolt.

13. Connect wheel speed sensor wire to retaining bracket on side of lower control arm.

14. If equipped, install the stabilizer bar.

15. Remove the crossmember supports and lower the vehicle.

SVT COBRA

1. Support the rear subframe with a jack.

2. Remove rear wheels.

3. Remove both mufflers.

4. Remove driveshaft.

5. Disconnect parking brake cables and remove brackets.

6. Disconnect brake fluid lines.

7. Remove wheel speed sensors.

8. Disconnect tie rod links.

9. Disconnect shock absorbers at lower control arms.

10. Disconnect lower control arms.

11. Loosen the front subframe bolts and remove the rear subframe bolts.

12. Lower the rear subframe and remove the coil springs.

To install:

➡**Use new nuts, bolts and split pins for assembly.**

13. Install the coil springs and raise the rear subframe. Install new subframe bolts and tighten them to 76 ft. lbs. (103 Nm).

14. Connect lower control arms.

15. Connect shock absorbers. Tighten the bolts to 98 ft. lbs. (133 Nm).

16. Connect tie rod links.

17. Install wheel speed sensors.

18. Connect brake fluid lines.

19. Connect parking brake cables and install brackets.

20. Install driveshaft.

21. Install both mufflers.

22. Install rear wheels.

23. Check the wheel alignment and adjust as necessary.

Lower Ball Joint

REMOVAL & INSTALLATION

1. Remove front wheel.

2. Remove disc brake caliper and rotor.

3. Remove wheel speed sensor.

4. Disconnect outer tie rod end.

5. Support the lower control arm with a jackstand and remove the spindle.

6. Press the lower ball joint out of the control arm.

To install:

7. Press the lower ball joint into the control arm so that the joint is fully seated in the control arm.

8. Install spindle. Tighten the ball joint nut to 109–149 ft. lbs. (148–202 Nm). Tighten the strut bolts to 141–191 ft. lbs. (190–259 Nm).

9. Connect outer tie rod end. Tighten the nut to 36–46 ft. lbs. (48–63 Nm).

10. Install disc brake rotor and caliper. Tighten the caliper mounting bolts to 96 ft. lbs. (130 Nm).

11. Install wheel speed sensor.

12. Install front wheel.

Upper Control Arm

REMOVAL & INSTALLATION

Rear

EXCEPT SVT COBRA

1. Before raising the vehicle, matchmark the shock absorbers to indicate curb height.

2. Support the axle housing with a jack.

3. Remove the upper control arms.

To install:

➡**Use new nuts and bolts.**

4. Install the upper control arms.

5. Align the matchmarks on the shock absorbers and tighten the control arm bolts as follows:

   a. Step 1: Tighten the mounting bracket bolt to 76 ft. lbs. (103 Nm).

   b. Step 2: Tighten the axle housing bolt to 66 ft. lbs. (90 Nm).

SVT COBRA

1. Before raising the vehicle, match-mark the shock absorbers to indicate curb height.

SIDEMEMBER

C-FRAME

FRONT SUSPENSION LOWER ARM

RECEIVING CUP

7922NG51

Use the C-clamp, cup and adapters to press the ball joint out of the lower control arm

Cam bolt matchmark—SVT Cobra

2. Remove the coil springs.

3. Remove the subframe front bolts and lower the subframe from the vehicle.

4. Matchmark the upper control arm cam bolt to the knuckle.

5. Remove the upper control arm.

To install:

➡**Use new nuts and bolts.**

6. Transfer the cam bolt matchmark to the new cam bolt.

7. Install the upper control arm. Do not tighten the fasteners at this time.

8. Install the coil springs.

9. Raise the suspension to align the shock absorber matchmarks and tighten the upper control arm inner bolts to 66 ft. lbs. (90 Nm).

10. Align the cam bolt matchmark and tighten the nut to 66 ft. lbs. (90 Nm).

11. Check the wheel alignment and adjust as necessary.

CONTROL ARM BUSHING REPLACEMENT

Rear

EXCEPT SVT COBRA

The inboard control arm bushing is serviced with the control arm as an assembly.

1. Remove the upper control arm.

2. Press the bushing out of the axle housing.

3. Press a new bushing into the axle housing and install the upper control arm.

Axle bushing removal

SVT COBRA

The upper control arm bushing are serviced with the control arm as an assembly.

Lower Control Arm

REMOVAL & INSTALLATION

Front

1. Install an internal spring compressor to the coil spring. Tighten the compressor to relieve spring pressure on the lower control arm.

2. Remove front wheel.

3. Remove disc brake caliper. Hang the disc brake caliper from the body with wire. Do not let the disc brake caliper hang by the brake hose.

4. Disconnect outer tie rod end.

5. Disconnect stabilizer bar link.

6. Remove power steering gear.

7. Remove lower control arm mounting bolts.

8. Remove coil spring.

9. Disconnect lower ball joint.

10. Remove lower control arm.

To install:

11. Connect the lower ball joint to the spindle and install the nut hand tight.

12. Install coil spring.

13. Install lower control arm mounting bolts. Use a jack to raise the control arm to curb height and tighten the bolts to 141–191 ft. lbs. (191–259 Nm). Tighten the ball joint nut to 109–149 ft. lbs. (148–202 Nm).

14. Install power steering gear.

15. Connect stabilizer bar link. Tighten the nut to 11–16 ft. lbs. (16–22 Nm).

16. Connect outer tie rod end. Tighten the nut to 36–46 ft. lbs. (48–63 Nm).

17. Install front disc brake caliper.

18. Install front wheel.

19. Position the coil spring and remove the spring compressor.

20. Check the front end alignment.

Rear

EXCEPT SVT COBRA

1. Before raising the vehicle, matchmark the shock absorbers to indicate curb height.

2. Raise and support the vehicle safely under the frame. Support the body at the rear body crossmember.

3. If necessary, remove the muffler assembly.

4. If equipped, remove the stabilizer bar.

5. Support the axle with a jack.

6. Place another jack under the lower arm axle pivot bolt. Remove and discard the bolt and nut. Lower the jack slowly until the coil spring load is relieved.

7. Remove coil spring.

8. Remove lower control arm.

To install:

➡**Use new nuts and bolts.**

9. Install lower control arm to the axle housing.

10. Install coil spring.

11. Raise the control arm and install the pivot bolt.

12. Raise the axle to align the matchmarks on the shock absorbers. Tighten the lower control arm bolts to 71–97 ft. lbs. (97–132 Nm) for 2001–03 models, and 111 ft. lbs. (150 Nm) for 2004 models.

13. install the stabilizer bar, if equipped.

14. Install the muffler assembly, if necessary.

SVT COBRA

1. Before raising the vehicle, matchmark the shock absorbers to indicate curb height.

2. Remove coil springs. Refer to the coil spring procedure in this section.

3. Remove subframe front bolts and lower it from the vehicle.

4. Remove lower control arm.

To install:

➡**Use new nuts and bolts.**

5. Install lower control arm. Do not tighten the fasteners at this time.

6. Install rear subframe. Tighten the rear subframe bolts to 76 ft. lbs. (103 Nm).

7. Install coil springs.

8. Raise the suspension to align the shock absorber matchmarks.

9. Tighten the lower control arm inboard bolts to 184 ft. lbs. (250 Nm) and the knuckle bolt to 85 ft. lbs. (115 Nm).

10. Check the wheel alignment and adjust as necessary.

CONTROL ARM BUSHING REPLACEMENT

Front and Rear

ALL MODELS

The lower control arm bushings are serviced with the lower control arm as an assembly.

Wheel Bearings

ADJUSTMENT

The front wheel bearings are an integral part of the hub assembly. They require no periodic maintenance or adjustment. If the bearings are found to be defective, they must be replaced along with the hub assembly.

The rear wheel bearings are not adjustable.

REMOVAL & INSTALLATION

Front

1. Remove front wheel.
2. Remove disc brake caliper and rotor.
3. Remove grease cap.
4. Remove spindle retainer nut.
5. Remove hub and bearing assembly.

To install:

6. Install hub and bearing assembly. Tighten the spindle retainer nut to 221–295 ft. lbs. (300–400).
7. Install grease cap.
8. Install disc brake caliper and rotor.
9. Install front wheel

Rear

SVT COBRA

1. Before raising the vehicle, match-mark the shock absorbers to indicate curb height.
2. Remove rear wheel.
3. Disconnect parking brake cable.
4. Remove brake caliper and disc.
5. Remove hub retainer.
6. Disconnect shock absorber from the lower suspension arm.
7. Disconnect tie rod link from knuckle.
8. Matchmark the upper control arm cam bolt to the knuckle.
9. Remove knuckle from the vehicle.
10. Remove dust shield and press the hub out of the bearing.
11. Remove snapring and press the bearing out of the knuckle.

To install:

➡**Use new nuts, bolts, snaprings, and split pins for assembly.**

12. Install bearing so that it is fully seated in the knuckle bore.
13. Install snapring.
14. Support the bearing inner race and press the hub into the bearing.

15. Install the dust shield and tighten the bolts to 89 inch lbs. (10 Nm).
16. Transfer the cam bolt matchmark to a new cam bolt.
17. Install knuckle. Do not tighten the fasteners at this time.
18. Connect tie rod link. Tighten the nut to 35 ft. lbs. (47 Nm).
19. Connect shock absorber. Tighten the bolt to 98 ft. lbs. (133 Nm).
20. Install hub retainer. Do not tighten the nut at this time.
21. Raise the suspension to align the shock absorber matchmarks and tighten the lower control arm bolt to 85 ft. lbs. (115 Nm).
22. Align the cam bolt matchmarks and tighten the nut to 66 ft. lbs. (90 Nm).
23. Install the brake disc and caliper. Install the parking brake cable.

➡**The hub retainer nut must be tightened with the brakes applied and the wheels off the ground to ensure correct bearing seating.**

24. Tighten the hub retainer to 240 ft. lbs. (325 Nm) and install the wheel.
25. Check the wheel alignment and adjust as necessary.

Brake Caliper

REMOVAL & INSTALLATION

Front

1. Remove the wheel and tire assembly.
2. Mark the disc brake caliper to ensure that it will be installed to the same side if both calipers are being removed.
3. Remove the banjo bolt that connects the brake hose fitting to the disc brake caliper.
4. Disconnect the brake hose and plug. Discard 2 copper sealing washers.
5. Remove lower brake caliper locating pin, if equipped.
6. Remove brake caliper mounting bolts, if equipped.
7. Rotate the brake caliper approximately 90° away from the anchor assembly.
8. If equipped with locating pins, slide the brake caliper away from anchor assembly until the brake caliper disengages from the upper locating pin.
9. Remove the disc brake caliper.

To install:

10. If required, compress the caliper piston(s) using a C-clamp and an old disc brake pad or block of wood.
11. If brake caliper is equipped with locating pins, engage the upper caliper locating pin into the insulator.
12. Make sure the anti-rattle is properly installed on caliper.
13. Make sure the brake pads are properly installed.
14. Position the brake caliper assembly into place on the anchor plate.

➡**If equipped, new locating pins are recommended due to the thread locking compound on the threads.**

15. Install lower disc brake caliper locating pin, or upper and lower caliper bolts, and torque to 23 ft. lbs. (31 Nm) for dual piston calipers. Install lower disc brake caliper locating pin and torque the bolts to 65 ft. lbs. (88 Nm) for single piston calipers.
16. If equipped, install disc brake caliper locating pin and clip by pressing the caliper down to compress the bias springs and the locating pin into position and secure with the washer and clip.
17. Connect the brake hose to the disc brake caliper using 2 new copper sealing washers, 1 on each side of the fitting. Install the banjo bolt and torque to 30 ft. lbs. (40 Nm).
18. Bleed the brake system and check for leaks. Replace the bleeder screw rubber cap.
19. Install wheel and tire assembly. Torque the lug nuts in a star pattern to 85–105 ft. lbs. (115–142 Nm).
20. Pump the brake pedal several times to position the brake pads before the vehicle is moved.
21. Road test the vehicle and check for proper brake system operation.

Rear

1. Remove wheel and tire assembly.
2. Remove banjo bolt securing the brake hose to the disc brake caliper.
3. Remove brake hose from the caliper and plug the hose. Discard 2 copper sealing washers.
4. Remove retaining clip from the parking brake rear cable at the disc brake caliper. Release the tension from the parking brake cable and disengage the cable end from the parking brake lever on the brake caliper.
5. Hold the disc brake caliper locating

pin hex-heads with an open-end wrench and remove 2 brake pin retainer bolts.

6. Rotate the caliper assembly away from the disc support bracket and remove the from the vehicle.

7. Remove caliper locating pins and boots from support bracket.

To install:

8. If required, use Caliper Piston Adjuster T87P-2588-A to rotate the caliper piston into the caliper bore. Align the tool tips with the slots in the piston and rotate the tool clockwise until the piston is seated.

➡**Apply a suitable silicone dielectric compound to the inside of the locating pin boots and the caliper locating pins.**

9. Install brake caliper locating pins and boots in the disc support bracket.

10. Install disc brake pads and anti-rattle clips.

11. Apply a drop of suitable thread locking compound to the threads of the brake pin retainer bolts and install the brake pin retainers. Torque the brake pin retainer bolts to 23–26 ft. lbs. (31–35 Nm) while holding the brake caliper locating pins with an open-end wrench.

12. Connect the end of the parking brake cable to the parking brake lever on the disc brake caliper.

13. Connect the brake hose to the disc brake caliper using 2 new copper sealing washers on the fitting.

14. Install the banjo bolt through both washers and the brake line fitting and torque to 30 ft. lbs. (40 Nm).

15. Properly bleed the brake system and check for leaks.

16. Install wheel and tire assembly. Torque the lug nuts in a star pattern to 85–105 ft. lbs. (115–142 Nm).

17. Pump the brake pedal several times to position the brake pads before moving the vehicle.

18. Road test the vehicle and check the brake system for proper operation.

Disc Brake Pads

REMOVAL & INSTALLATION

Front

WITH SINGLE PISTON CALIPER

1. Remove ½ of the brake fluid from the brake master cylinder reservoir. Properly dispose of the used brake fluid.

2. Remove wheel and tire assembly.

3. Remove the front brake caliper assembly. Hang the disc brake caliper from the body with wire. Do not let the disc brake caliper hang by the brake hose.

4. Remove the outer and inner disc brake pads and the anti-rattle clip from the disc brake caliper

5. Clean any residue from the caliper anchor plate and disc brake pad contact areas.

6. Inspect the disc brake rotor for scoring and wear. Replace or machine, as necessary.

7. Inspect the piston boot and the caliper pin boots for damage. Replace as necessary.

To install:

8. Use a large C-clamp and a wood block to push the caliper piston back into its bore.

9. Remove the protective paper from the insulators on the back of the disc brake pads, if equipped.

10. Install inner and outer disc brake pads and the anti-rattle clip in the disc brake caliper.

NOTE: DO NOT REMOVE FOR CALIPER REPLACEMENT

1	Upper Caliper Brake Bolt
2	Front Caliper Sleeve
3	Insulator
4	Caliper Housing (Part of 2B119)
5	Disc Brake Pad Anti-Rattle Clip
6	Caliper Piston
7	Piston Seal
8	Dust Boot
9	Brake Shoe and Lining
10	Front Disc Brake Caliper Anchor Plate
11	Disc Brake Caliper Locating Pin
A	Tighten to 88 N·m (65 Lb-Ft)

93006G30

Disc brake pads and related components—Mustang with single piston caliper

11. Install the disc brake caliper and pads.

12. Install wheel and tire assembly and torque the lug nuts in a star pattern to 85–105 ft. lbs. (115–142 Nm).

13. Repeat the procedure for the opposite disc brake caliper assembly.

14. Pump the brake pedal prior to moving the vehicle to seat the brake pads.

15. Fill the brake master cylinder reservoir with clean DOT 3 brake fluid from a closed container.

16. If the disc brake calipers were replaced or repaired, be sure to bleed the system.

17. Road test the vehicle and check the brake system for proper operation.

WITH DUAL PISTON CALIPER

1. Remove and discard ½ of the brake fluid from the brake master cylinder reservoir. Properly dispose of the used brake fluid.

2. Remove the wheel and tire assembly.

3. Remove disc brake caliper assembly from anchor plate. Hang the disc brake caliper from the body with wire. Do not let the disc brake caliper hang by the brake hose.

4. Remove the outer and inner disc brake pads from the disc brake caliper.

5. Clean any residue from the caliper anchor plate and disc brake pad contact areas.

6. Inspect the disc brake rotor for scoring and wear. Replace or machine, as necessary.

7. Inspect the piston boots and the caliper pin boots for damage. Replace as necessary.

To install:

8. Use a large C-clamp and a wood block to push the caliper pistons back into their bores.

9. Remove the protective paper from the insulators on the back of the disc brake pads, if equipped.

10. Install the inner and outer disc brake pads in the disc brake caliper. Make sure the pads are seated properly.

11. Install the disc brake caliper and pads.

12. Install the wheel and tire assembly. Torque the lug nuts in a star pattern to 85–105 ft. lbs. (115–142 Nm).

13. Repeat the procedure for the opposite disc brake caliper assembly.

14. Pump the brake pedal prior to moving the vehicle to position the brake pads.

15. Fill the brake master cylinder reser-

voir with clean DOT 3 brake fluid from a closed container.

16. If the disc brake calipers were replaced or repaired, be sure to bleed the system.

17. Road test the vehicle and check the brake system for proper operation.

Rear

1. Before servicing the vehicle, refer to the Precautions Section.

2. Remove ½ of the brake fluid from the brake master cylinder reservoir. Properly dispose of the used brake fluid.

3. Remove wheel and tire assembly.

4. Remove screw retaining the brake hose bracket to the shock absorber bracket, if necessary.

5. Remove the rear disc brake caliper. Hang the caliper from the body with wire. Do not let the disc brake caliper hang by the brake hose.

6. Remove inner and outer disc brake pads and the anti-rattle clips from the disc support bracket.

7. Inspect the disc brake rotor for scoring and wear. Replace or machine, as necessary.

To install:

8. Using Rear Caliper Piston Adjuster T87P-2588-A, rotate the caliper piston clockwise until the piston is fully seated.

➡**Ensure that one of the 2 slots in the rear brake piston face is positioned so it will engage the nib on the back of the inner brake pad, if equipped.**

9. Install disc brake pad anti-rattle clips.

1	Caliper housing	5	Front brake locating pin	9	Washer
2	Piston seal	6	Brake shoe and lining	10	Anti-rattle clip and insulator
3	Caliper piston	7	Front disc brake caliper anchor plate	11	Wheel cylinder bleeder screw
4	Dust boot	8	Clip	12	Cover

67197-MUST-G20

Disc brake pads and related components—Mustang with dual piston caliper and locating pins

1. Disc brake caliper
2. Bleeder screw cap
3. Bleeder screw
4. Anti-rattle spring
5. Caliper piston
6. Piston seal
7. Piston dust boot
8. Pad slipper
9. Front disc brake caliper anchor plate
10. Brake pads
11. Guide pin boot
12. Guide pin
13. Caliper bolt

67197-MUST-G21

Disc brake pads and related components—Mustang with dual piston caliper and guide pins

10. Install inner and outer disc brake pads on the rear disc support bracket.

11. Install rear disc brake caliper and pads. Make sure the disc brake pads and anti-rattle clips are correctly installed.

12. Install wheel and tire assembly. Torque the lug nuts in a star pattern to 85–105 ft. lbs. (115–142 Nm).

13. Repeat the procedure for the opposite disc brake caliper assembly.

14. Pump the brake pedal prior to moving the vehicle to position the brake pads.

15. Fill the brake master cylinder reservoir with clean DOT 3 brake fluid from a closed container.

16. If the disc brake calipers were replaced or repaired, be sure to bleed the system.

17. Road test the vehicle and check the brake system for proper operation.

FORD

2005 Mustang

SPECIFICATIONS AND MAINTENANCE CHARTS

ENGINE AND VEHICLE IDENTIFICATION

| | | Engine | | | | | | Model Year | |
Code ①	Liters (cc)	Cu. In.	Cyl.	Fuel Sys.	Engine Type	Eng. Mfg.		Code ②	Year
H	4.6 (4605)	281	8	SFI	SOHC	Ford		5	2005
N	4.0 (3999)	244	6	SFI	SOHC	Ford		6	2006

SOHC: Single Overhead Camshaft

SFI: Sequential Fuel Injection

① 8th position of VIN

② 10th position of VIN

06017-TANG-C01

GENERAL ENGINE SPECIFICATIONS

All measurements are given in inches.

Year	Model	Engine Displacement Liters	Engine Series (ID/VIN)	Net Horsepower @ rpm	Net Torque @ rpm (ft. lbs.)	Bore x Stroke (in.)	Compression Ratio	Oil Pressure @ rpm
2005	Mustang	4.0	N	210@5250	250@3500	3.95x3.32	9.7:1	15@2000
		4.6	H	300@5750	320@4500	3.55x3.54	9.8:1	75@2000

06017-TANG-C02

ENGINE TUNE-UP SPECIFICATIONS

Year	Engine Displacement Liters	Engine ID/VIN	Spark Plug Gap (in.)	Ignition Timing (deg.) MT	Ignition Timing (deg.) AT	Fuel Pump (psi) ①	Idle Speed (rpm) MT	Idle Speed (rpm) AT	Valve Clearance Intake	Valve Clearance Exhaust
2005	4.0	N	0.052-0.056	N/A	N/A	27-37	②	②	HYD	HYD
	4.6	H	0.052-0.056	10B	10B	27-37	②	②	HYD	HYD

NOTE: The Vehicle Emission Control Information label often reflects specification changes made during production.

The label figures must be used if they differ from those in this chart.

N/A: Not available

B: Before Top Dead Center

HYD: Hydraulic

① Fuel pressure with engine running, pressure regulator vacuum hose connected

② Refer to Vehicle Emission Control Information label

06017-TANG-C03

FRONT

93003G01

4.0L Engine
Firing order: 1–4–2–5–3–6
Distributorless ignition system

FRONT

93003G02

4.6L Engine
Firing order: 1–3–7–2–6–5–4–8
Distributorless ignition system—One coil per cylinder

Accessory drive belt routing—4.0L (VIN N) engine

06017-TANG-G01

Accessory drive belt routing—4.6L (VIN H) engine

06017-TANG-G02

CAPACITIES

Year	Model	Engine Displacement Liters	Engine ID/VIN	Engine Oil with Filter (qts.)	Transmission (pts)		Drive Axle Rear (pts.)	Fuel Tank (gal.)	Cooling System (qts.)
					Manual	Auto. ①			
2005	Mustang	4.0	N	5.0	②	23.8	③	16.0	16.0
		4.6	H	6.5	②	28.6	③	16.0	16.0

NOTE: All capacities are approximate. Add fluid gradually and ensure a proper fluid level is obtained.

① Includes torque converter

② T50D 5-speed: 5.6 pts.
   TR3650 5-speed: 7.5 pts.

③ If equipped with 7.5 inch ring gear: 3.0 pts. + 4 oz. friction modifier.
   If equipped with 8.8 inch ring gear: 4.0 pts. + 4 oz. friction modifier.

06017-TANG-C04

VALVE SPECIFICATIONS

Year	Engine Displacement Liters	Engine ID/VIN	Seat Angle (deg.)	Face Angle (deg.)	Spring Test Pressure (lbs. @ in.)	Spring Installed Height (in.)	Stem-to-Guide Clearance (in.)		Stem Diameter (in.)	
							Intake	Exhaust	Intake	Exhaust
2005	4.0	N	45	45	①	1.569-1.609	0.0010-0.0020	0.0010-0.0030	0.2742-0.2748	0.2736-0.2742
	4.6	H	45	44.5-45	79@1.22	1.660	0.0010-0.0030	0.0020-0.0040	0.2352-0.2360	0.2742-0.2350

① 202-225@1.413-1.445

06017-TANG-C05

CRANKSHAFT AND CONNECTING ROD SPECIFICATIONS

All measurements are given in inches.

Year	Engine Displacement Liters	Engine ID/VIN	Crankshaft				Connecting Rod		
			Main Brg. Journal Dia.	Main Brg. Oil Clearance	Shaft End-play	Thrust on No.	Journal Diameter	Oil Clearance	Side Clearance
2005	4.0	N	2.2430-2.2440	0.0003-0.0024	0.0020-0.0126	5	2.125-2.1260	0.0003-0.0024	0.0036-0.0106
	4.6	H	2.6567-2.6576	0.0009-0.0019	0.0030-0.0148	5	2.0859-2.0867	0.0009-0.0026	0.0020-0.0060

06017-TANG-C06

PISTON AND RING SPECIFICATIONS

All measurements are given in inches.

Year	Engine Displacement Liters	Engine ID/VIN	Piston Clearance	Ring Gap			Ring Side Clearance		
				Top Compression	Bottom Compression	Oil Control	Top Compression	Bottom Compression	Oil Control
2005	4.0	N	0.0012-0.0020	0.0008-0.0018	0.0016-0.0024	N/A	0.0016-0.0030	0.0012-0.0026	N/A
	4.6	H	0.0007-0.0019	0.0006-0.0012	0.0098-0.0197	0.0059-0.0256	0.0008-0.0020	0.0008-0.0020	N/A

N/A: Not available

06017-TANG-C07

TORQUE SPECIFICATIONS

All readings in ft. lbs.

Year	Engine Displacement Liters	Engine ID/VIN	Cylinder Head Bolts	Main Bearing Bolts	Rod Bearing Bolts	Crankshaft Damper Bolts	Flywheel Bolts	Manifold		Spark Plugs	Oil Pan Drain Plug
								Intake	Exhaust		
2005	4.0	N	①	②	③	④	⑤	⑥	16	13	18
	4.6	H	⑦	⑧	⑨	⑩	59	⑥	18	25	19

① Do not reuse cylinder head bolts.
   M8 Bolts: 24 ft. lbs.
   M12 Bolts: Step 1: 9 ft. lbs.
   Step 2: 18 ft. lbs.
   Step 3: Tighten an additional 90 degrees
   Step 4: Tighten an additional 90 degrees

② Step 1: 35 ft. lbs.
   Step 2: Tighten an additional 57 degrees

③ Step 1: 15 ft. lbs.
   Step 2: Tighten an additional 90 degrees

④ Step 1: 33 ft. lbs.
   Step 2: Tighten an additional 85 degrees

⑤ Step 1: 10 ft. lbs.
   Step 2: 52 ft. lbs.

⑥ 89 INCH lbs.

⑦ Step 1: 30 ft. lbs.
   Step 2: Tighten an additional 90 degrees
   Step 3: Tighten an additional 90 degrees

⑧ Step 1: Bolts 1 through 20 to 89 INCH lbs.
   Step 2: Bolts 1 through 10 to 18 ft. lbs.
   Step 3: Bolts 11 through 20 to 30 ft. lbs.
   Step 4: Bolts 1 through 20 an additional 90 degrees
   Step 5: Side mounted bolts to 30 ft. lbs.
   Step 6: Side bolts tighten an additional 90 degrees

⑨ Step 1: 32 ft. lbs.
   Step 2: Tighten an additional 105 degrees

⑩ Step 1: 66 ft. lbs.
   Step 2: Loosen all bolts one turn
   Step 3: 37 ft. lbs.
   Step 4: Tighten an additional 90 degrees

06017-TANG-C08

Main bearing cap bolt tightening sequence—4.0L (VIN N) engine

06017-TANG-G03

Main bearing cap bolt tightening sequence—4.6L (VIN H) engine

06017-TANG-G04

Main bearing cap side bolt tightening sequence—4.6L (VIN H) engine

06017-TANG-G05

Connecting rod bolt tightening sequence—4.6L (VIN H) engine

06017-TANG-G06

WHEEL ALIGNMENT

Year	Model		Caster Range (+/-Deg.)	Caster Preferred Setting (Deg.)	Camber Range (+/-Deg.)	Camber Preferred Setting (Deg.)	Toe-in (in.)
2005	Mustang	F	0.70	7.1	0.70	0.75	0.10 +/- 0.20
		R	—	—	N/A	N/A	N/A

06017-TANG-C09

TIRE, WHEEL AND BALL JOINT SPECIFICATIONS

Year	Model	OEM Tires Standard	OEM Tires Optional	Tire Pressures (psi) Front	Tire Pressures (psi) Rear	Wheel Size	Ball Joint Inspection	Wheel Lug Torque (ft. lbs.)
2005	Mustang	P215/65R16	P235/55ZR17	30	30	-	②	98

OEM: Original Equipment Manufacturer

PSI: Pounds Per Square Inch

① Replace if any measurable movement is found

② Replace if movement exceeds the maximum measurement of 1/32 inches (0.8mm)

06017-TANG-C10

BRAKE SPECIFICATIONS

All measurements in inches unless noted

Year	Model		Brake Disc Original Thickness	Brake Disc Minimum Thickness	Brake Disc Maximum Run-out	Brake Drum Diameter Original Inside Diameter	Brake Drum Diameter Max. Wear Limit	Brake Drum Diameter Maximum Machine Diameter	Minimum Lining Thickness	Brake Caliper Mounting Bolts (ft. lbs.)
2005	Mustang	F	N/A	1.110	0.002	—	—	—	0.118	25
		R	N/A	0.680	0.004	—	—	—	0.118	24

NA: Not Available

F: Front

R: Rear

06017-TANG-C11

SCHEDULED MAINTENANCE INTERVALS

2005 FORD MUSTANG

TO BE SERVICED	TYPE OF SERVICE	VEHICLE MILEAGE INTERVAL (x1000)																	
		5	10	15	20	25	30	35	40	45	50	55	60	65	70	75	80	85	90
Engine oil & filter ①	R																		
Rotate tires	S/I	✓	✓	✓	✓	✓	✓	✓	✓	✓	✓	✓	✓	✓	✓	✓	✓	✓	✓
Engine coolant strength hoses & clamps	S/I			✓			✓			✓			✓			✓			✓
Air cleaner filter	R						✓						✓						✓
Automatic transmission fluid & filter	R						✓						✓						✓
Engine coolant ②	R																		
PCV valve	R												✓						✓
Spark plugs ③	R																		
Drive belts	S/I						✓						✓						✓
Exhaust system & heat shields	S/I						✓						✓						✓
Front & rear brakes	S/I	✓	✓	✓	✓	✓	✓	✓	✓	✓	✓	✓	✓	✓	✓	✓	✓	✓	✓
Fuel filter	R			✓			✓			✓			✓			✓			✓

R: Replace   S/I: Service or Inspect

① Engine oil and filter: replace every 3000 miles.

② Engine coolant: change initially at 100,000 miles or 5 years.

③ Spark plugs: replace every 100,000 miles.

Special Operating Condition Requirements

During extensive idling and/or low speed driving for long distances, as in heavy commercial use such as delivery, taxi, patrol car or livery:

Lube front lower control arm and steering linkage ball joints with

Zerk fittings (if equipped) every 4,800 km (3,000 miles) or 3 months.

Inspect brake system and check battery electrolyte level (Patrol cars) every 8,000 km (5,000 miles).

Install a new cabin air filter as required.

When operating in dusty conditions such as unpaved or dusty roads:

Install a new engine air filter as required.

Install a new cabin air filter as required.

When operating in off-road conditions:

Change automatic transmission fluid every 48,000 km (30,000 miles).

Install a new cabin air filter as required.

Inspect and lubricate U-joints.

Inspect and lubricate steering linkage ball joints with zerk fittings.

06017-TANG-C12

ENGINE REPAIR

→**Disconnecting the negative battery cable on some vehicles may interfere with the functions of the on board computer system. The computer may undergo a relearning process once the negative battery cable is reconnected.**

Distributor

The 4.0L and 4.6L engines use Distributorless Ignition Systems (DIS).

Alternator

REMOVAL

4.0L Engine

1. Disconnect the negative battery cable.
2. Remove the accessory drive belt.
3. Disconnect the alternator electrical connectors.
4. Remove the alternator shield.
5. Remove the alternator mounting bolts.
6. Remove the alternator from the vehicle.

4.6L Engine

1. Disconnect the negative battery cable.
2. Remove the crankcase vent tube from the air cleaner pipe.
3. Remove the air cleaner pipe from the throttle body.
4. Remove the accessory drive belt.
5. Remove the alternator bolts.
6. Remove the inner and outer alternator bracket.
7. Disconnect the alternator electrical connectors.
8. Remove the alternator from the vehicle.

INSTALLATION

4.0L Engine

1. Position the alternator onto engine. Tighten the bolt and the stud nuts to 35 ft. lbs. (47 Nm).
2. Install the generator shield.
3. Connect the alternator electrical connectors.
4. Install the accessory drive belt.
5. Connect the negative battery cable.

4.6L Engine

1. Position the alternator onto engine. Tighten the mounting bolts to 18 ft. lbs. (25 Nm).

2. Connect the alternator electrical connectors.
3. Install the alternator brackets. Tighten the bolts to 89 inch lbs. (10 Nm).
4. Install the accessory drive belt.
5. Install the air cleaner pipe to the throttle body.
6. Install the crankcase vent tube to the air cleaner pipe.
7. Connect the negative battery cable.

Ignition Timing

ADJUSTMENT

The ignition timing is controlled by the Powertrain Control Module (PCM). No adjustment is necessary.

Engine Assembly

REMOVAL & INSTALLATION

4.0L Engine

1. Discharge the air conditioning system.
2. Drain the engine cooling system.
3. Drain the engine oil and remove the oil filter.
4. Disconnect the battery cables.
5. Relieve the fuel system pressure.
6. Remove the hood.
7. Remove the air cleaner assembly.
8. Remove the transmission. Refer to the transmission removal procedure.
9. On manual transmission models, remove the clutch.
10. Disconnect the Heated Oxygen (HO$_2$S) sensor connectors and catalyst monitor connectors.
11. Remove 4 catalytic converter-to-exhaust manifold nuts.
12. Remove the ground strap.
13. Remove the radiator hoses.
14. Remove the power steering supply hose bracket.
15. Remove the power steering pressure tube bracket from the crossmember.
16. Remove the power steering pressure switch connector.
17. Remove the A/C pressure switch and compressor connectors.
18. Remove 2 wiring retainers from the accessory drive bracket.
19. Disconnect the fuel supply tube coupling.
20. Remove the accessory drive belt.

21. Remove the power steering pump pulley.
22. Remove the power steering pump and position aside.
23. Remove the accessory drive bracket, then remove A/C compressor without disconnecting the lines and position aside.
24. Remove the heater hose bracket from the right side valve cover.
25. Remove the coolant tube bracket bolts.
26. Remove the spark plug wires from the spark plugs.
27. Remove the ignition coil bracket.
28. Disconnect the spark plug with retainer from the intake manifold.
29. Disconnect the radio capacitor connector and the ignition coil connector.
30. Remove the upper and lower ignition coil bracket bolts, then remove the ignition coil and spark plug wires.
31. Disconnect and remove the crankcase vent tube.
32. Disconnect the brake booster vacuum hose.
33. Disconnect the PCV vapor tube.
34. Disconnect the upper and lower PCM connectors.
35. Disconnect all necessary wiring retainers.
36. Remove the power distribution box cover then remove the battery power terminal from the box.
37. Disconnect the upper and lower power distribution box covers and disconnect the 68-pin connector from the box.
38. Disconnect the ground cable from the right strut tower.
39. Attach lifting links to the cylinder head lifting eyes.
40. Remove the left and right engine mount nuts.
41. Attach an engine hoist and remove the engine from the vehicle.

To install:
42. Lower the engine into the vehicle.
43. Install left and right engine mount nuts and tighten the nuts to 37 ft. lbs. (50 Nm).
44. Connect the ground cable to the right strut tower and tighten to 89 inch lbs. (10 Nm).
45. Disconnect the upper and lower power distribution box covers and disconnect the 68-pin connector from the box.
46. Remove the power distribution box cover then remove the battery power terminal from the box.

47. Connect all necessary wiring retainers.

48. Connect the upper and lower PCM connectors.

49. Connect the PCV vapor tube.

50. Connect the brake booster vacuum hose.

51. Connect and remove the crankcase vent tube.

52. Install the upper and lower ignition coil bracket bolts, then install the ignition coil and spark plug wires.

53. Connect the radio capacitor connector and the ignition coil connector.

54. Connect the spark plug with retainer to the intake manifold.

55. Install the ignition coil bracket.

56. Install the coolant tube bracket bolts.

57. Install the heater hose bracket to the right side valve cover.

58. Install the accessory drive bracket and tighten the bolts to 35 ft. lbs. (47 Nm).

59. Install the A/C compressor.

60. Install the power steering pump and tighten the bolts to 18 ft. lbs. (25 Nm).

61. Install the power steering pump pulley and tighten the bolts to 18 ft. lbs. (25 Nm).

62. Install the accessory drive belt.

63. Connect the fuel supply tube coupling.

64. Install 2 wiring retainers from the accessory drive bracket.

65. Connect the A/C pressure switch and compressor connectors.

66. Connect the power steering pressure switch connector.

67. Install the power steering pressure tube bracket from the crossmember.

68. Install the power steering supply hose bracket.

69. Install the radiator hoses.

70. Install the ground strap.

71. Install 4 catalytic converter-to-exhaust manifold nuts and tighten to 30 ft. lbs. (40 Nm).

72. Connect the Heated Oxygen (HO2S) sensor connectors and catalyst monitor connectors.

73. On manual transmission models, install the clutch.

74. Install the transmission.

75. Install the air cleaner assembly.

76. Install hood.

⁕⁕ WARNING

Be sure to check engine to see that all electrical connectors, hoses and cables are properly connected and secure.

77. Install new oil filter and fill engine with fresh oil.

78. Connect the negative battery cable.

79. Fill the cooling system.

80. Recharge A/C system.

81. Start the engine and check for leaks.

4.6L Engine

1. Discharge the air conditioning system.

2. Drain the engine cooling system.

3. Drain the engine oil and remove the oil filter.

4. Relieve the fuel system pressure.

5. Disconnect the negative battery cable.

6. Remove the hood.

7. Remove the cowl vent screen.

8. Remove the air cleaner and outlet pipe.

9. Remove the engine coolant crossover.

10. Remove the degas bottle.

11. Remove the radiator sight shield.

12. Remove the accessory drive belt.

13. Disconnect the A/C pressure transducer connector.

14. Disconnect the coolant hose from the oil filter adapter.

15. Remove the degas hose from the cooling fan assembly.

16. Disconnect the lower radiator hose, then remove the thermostat housing and cooling hoses as an assembly.

17. Disconnect the oil temperature sensor connector.

18. Remove the ground wire.

19. Disconnect the pin retainer from the A/C compressor.

20. Disconnect the A/C clutch and crankshaft position sensor connectors.

21. Remove the A/C compressor without disconnecting the lines and position aside.

22. Remove the power steering pulley shield.

23. Disconnect the wiring harness retainer, then remove the power steering pump without disconnecting the lines, and position aside.

24. Install a lifting bracket to the threaded hole at the left side of the cylinder block.

25. Place a suitable jack to support the rear of the engine.

26. Install a 3-bar engine support to the lifting bracket on one side and the body structure near the suspension strut tower.

⁕⁕ WARNING

DO NOT place the 3-bar engine support leg on the vehicle fender.

27. On automatic transmission models, disconnect the transmission cooler tube and position aside.

28. Remove the transmission. Refer to the transmission procedure in this section.

29. On manual transmission models, remove the clutch.

30. Remove the catalytic converter flange nuts on both sides.

31. Disconnect the right side heated oxygen sensor connector.

32. Disconnect the catalyst monitor sensor connector on both sides.

33. Remove the engine-to-transmission spacer plate.

34. Disconnect the heater hoses.

35. Remove the ground strap from the cowl.

36. Disconnect the alternator jumper harness connector.

37. Remove the alternator harness and position aside.

38. Disconnect the upper and lower PCM connectors.

39. Disconnect all necessary wiring retainers.

40. Remove the power distribution box cover then remove the battery power terminal from the box.

41. Disconnect the upper and lower power distribution box covers and disconnect the 68-pin connector from the box.

42. Remove the nut, radio interference capacitor and J-bracket from the engine front cover stud bolt.

43. Remove both left and right engine insulator nuts.

44. Install an engine lifting bar on the engine.

45. Install an engine hoist to the lifting bar.

46. Remove the 3-bar support tool.

47. Remove the engine from the vehicle.

To install:

48. Install a lifting bracket to the threaded hole at the left side of the cylinder block.

49. Place a suitable jack to support the rear of the engine.

50. Lower the engine into the vehicle.

51. Install a 3-bar engine support to the lifting bracket on one side and the body structure near the suspension strut tower.

⁕⁕ WARNING

DO NOT place the 3-bar engine support leg on the vehicle fender.

52. Remove the engine hoist.

53. Install left and right engine support

insulators nuts and tighten to 46 ft. lbs. (63 Nm).

54. Install the J-bracket nut and radio interference capacitor to the engine front cover stud bolt.

55. Connect the 68-pin connector and then the upper and lower power distribution box covers.

56. Install the battery power terminal and the power distribution box cover.

57. Connect all necessary wiring retainers.

58. Connect the upper and lower PCM connectors.

59. Install the alternator harness.

60. Connect the alternator jumper harness connector.

61. Install the ground strap from the cowl.

62. Connect the heater hoses.

63. Install the engine-to-transmission spacer plate.

64. Connect the catalyst monitor sensor connector on both sides.

65. Connect the right side heated oxygen sensor connector.

66. Install the catalytic converter flange nuts on both sides and tighten to 30 ft. lbs. (40 Nm).

67. On manual transmission models, install the clutch.

68. Install the transmission. Refer to the transmission procedure in this section.

69. On automatic transmission models, connect the transmission cooler tube.

70. Remove the 3-bar engine support, chain and lifting eye.

71. Install the power steering pump and tighten the bolts to 18 ft. lbs. (25 Nm).

72. Connect the power steering tube clip and the wiring harness retainer.

73. Install the power steering pulley shield.

74. Install the A/C compressor and tighten the nuts to 18 ft. lbs. (25 Nm).

75. Connect the A/C clutch and crankshaft position sensor connectors.

76. Connect the pin retainer to the A/C compressor.

77. Install the ground wire.

78. Connect the oil temperature sensor connector.

79. Install the thermostat housing, lower radiator hose and cooling hoses as an assembly.

80. Install the degas hose to the cooling fan assembly.

81. Connect the coolant hose to the oil filter adapter.

82. Connect the A/C pressure transducer connector.

83. Install the accessory drive belt.

84. Install the radiator sight shield.

85. Install the degas bottle.

86. Install the engine coolant crossover.

87. Install the air cleaner and outlet pipe.

88. Install the cowl vent screen.

89. Install the hood.

✶✶ WARNING

Be sure to check engine to see that all electrical connectors, hoses and cables are properly connected and secure.

90. Install a new oil filter and fill engine with fresh oil.

91. Fill the cooling system.

92. Connect the negative battery cable.

93. Recharge the air conditioning.

94. Start the engine and check for leaks.

Water Pump

REMOVAL & INSTALLATION

4.0L Engine

1. Drain the engine cooling system.
2. Remove the air cleaner outlet pipe.
3. Loosen water pump pulley bolts.
4. Remove the accessory drive belt.
5. Remove the water pump pulley.

1 Coolant pump pulley bolt (4 required)
2 Coolant pump pulley
3 Hose clamp
4 Thermostat housing-to-coolant pump hose
5 Hose clamp
6 Heater hose
7 Hose clamp
8 Radiator-to-coolant pump hose
9 Coolant pump bolt (12 required)
10 Coolant pump

Exploded view of the water pump mounting—4.0L engine

06017-TANG-G07

1 Coolant pump pulley bolt (4 required)
2 Coolant pump pulley
3 Coolant pump bolt (4 required)
4 Coolant pump
5 Coolant pump O-ring seal

06017-TANG-G08

Exploded view of the water pump mounting—4.6L engine

6. Remove the upper and lower radiator hose.

7. Remove the heater hose.

8. Remove the 12 attaching bolt and remove the water pump. Clean and inspect gasket mating surfaces.

To install:

9. Install the water pump. Use a new gasket and tighten the bolts to 89 inch lbs. (10 Nm).

10. Install the heater hose.

11. Install the upper and lower radiator hose.

12. Install the water pump pulley.

13. Install the accessory drive belt.

14. Install water pump pulley bolts.

15. Install the air cleaner outlet pipe.

16. Fill the cooling system.

17. Start the engine and check for coolant leaks.

4.6L Engine

1. Drain the engine cooling system.

2. Remove the air cleaner outlet pipe.

3. Loosen water pump pulley bolts.

4. Remove the accessory drive belt.

5. Remove the water pump pulley.

6. Remove the water pump.

To install:

7. Install water pump. Use a new O-ring seal and tighten the bolts to 18 ft. lbs. (25 Nm).

8. Install the water pump pulley and tighten the bolts to 18 ft. lbs. (25 Nm).

9. Install the accessory drive belt.

10. Install air cleaner outlet pipe.

11. Fill the cooling system.

12. Start the engine and check for coolant leaks.

Heater Core

REMOVAL & INSTALLATION

➡**The heater core and evaporator core are not serviced individually. The entire housing must be replaced if either component is defective.**

1. Drain the cooling system.

2. Recover the A/C refrigerant, if equipped.

3. Disconnect the negative battery cable.

4. Disable the air bag system.

5. Remove the floor console.

6. Remove the 2 door scuff plates.

7. Remove the 2 lower A-pillar upper and lower trim panels.

8. Remove the instrument cluster.

9. Secure the steering wheel in the straight ahead position, then disconnect the steering shaft-to-steering column pinch bolt.

10. Disconnect the left side instrument panel bulkhead connector and remove the instrument panel support bolt.

11. Disconnect the antenna in-line connector and the 4 instrument panel electrical connectors from the right side.

12. Remove 2 instrument panel center support screws and 3 support bolts.

13. Through the instrument panel cluster opening, remove the instrument panel cowl top bracket nut.

14. Remove the 2 lower instrument panel support bolts.

15. Remove the 2 instrument panel end panels.

16. Remove the 4 instrument panel upper support bolts.

17. Using the help of an assistant, remove the instrument panel.

18. Clamp off and disconnect the 2 heater hoses at the heater core.

19. Disconnect the 2 evaporator core nuts and disconnect the fitting.

20. Remove the 2 heater core housing nuts at the dash panel.

21. Disconnect the antenna cable from the housing.

22. Remove the inside evaporator housing nuts.

23. Remove the heater core/evaporator housing.

To install:

24. Install the heater core/evaporator housing.

1 Bulkhead electrical connector

2 Instrument panel support bolt

3 Steering column shaft bolt

4 Antenna connector

5 Defrost mode actuator
  electrical connector

6 Bulkhead electrical connector

7 Smart junction box (SJB)
  electrical connector

8 SJB electrical connector

9 Instrument panel center
  support screws (2 required)

10 Instrument panel center
   support bolts (3 required)

11 Instrument panel lower
   support bolts (2 required)

12 Instrument panel end panels
   (2 required)

13 Instrument panel upper
   support bolts (4 required)

14 Instrument panel cowl top
   bracket nut

15 Instrument panel

Exploded view of the instrument panel—Mustang

06017-TANG-G09

1 Shift lever bezel (if equipped)
2 Floor console finish panel screws (2 required)
3 Floor console finish panel
4 Center stack wings (2 required)

5 Floor console screws (2 required)
6 Floor console mat
7 Floor console nuts (2 required)
8 Floor console

06017-TANG-G10

Removing the floor console—Mustang

25. Install the inside evaporator housing nuts and tighten to 44 inch lbs. (5 Nm).

26. Connect the antenna cable from the housing.

27. Install the 2 heater core housing nuts at the dash panel and tighten to 62 inch lbs. (7 Nm).

28. Connect the 2 evaporator core nuts and connect the fitting.

29. Connect the 2 heater hoses at the heater core.

30. Using the help of an assistant, install the instrument panel.

31. Install the 4 instrument panel upper support bolts and tighten to 15 ft. lbs. (20 Nm).

32. Install the 2 instrument panel end panels.

33. Install the 2 lower instrument panel support bolts and tighten to 80 inch lbs. (9 Nm).

34. Through the instrument panel cluster opening, install the instrument panel cowl top bracket nut and tighten to 80 inch lbs. (9 Nm).

35. Install 2 instrument panel center support screws and 3 support bolts.

36. Connect the antenna in-line connector and the 4 instrument panel electrical connectors from the right side.

37. Connect the left side instrument panel bulkhead connector and install the instrument panel support bolt and tighten to 15 ft. lbs. (20 Nm)..

38. Connect the steering shaft-to-steering column pinch bolt and tighten to 18 ft. lbs. (25 Nm).

39. Install the instrument cluster.

40. Install the 2 lower A-pillar upper and lower trim panels.

41. Install the 2 door scuff plates.

42. Install the floor console.

43. Enable the air bag system.

44. Connect the negative battery cable.

45. Fill the cooling system.

46. Recharge the A/C system.

47. Start the engine and check for leaks.

Cylinder Head

REMOVAL & INSTALLATION

4.0L Engine

1. Drain the cooling system.

2. Disconnect the negative battery cable.

3. Relieve the fuel system pressure.

4. Remove the intake manifold.

5. From the left valve cover remove or disconnect the following items:

- Ignition coil
- Fuel rail supply tube
- PCV tube
- Camshaft position sensor connector
- Fuel rail pressure and temperature sensor connector
- Fuel injector connectors
- Spark plug wire retainer
- Valve cover

5 Nm (44 lb-in)

6 Interior heater core and evaporator core housing nut
7 Antenna cable
8 Heater core and evaporator core housing

06017-TANG-G11

Evaporator/heater core mounting—Mustang

6. From the right valve cover remove or disconnect the following items:
- PCV tube
- PCV connector
- Heater hose retainer
- Wiring harness bracket
- Valve cover

7. Remove the camshaft roller followers.

8. Loosen the drive belt tensioner and remove the accessory drive belt.

9. Disconnect the alternator electrical connectors.

10. Remove the accessory drive belt tensioner.

11. Remove the alternator mounting bracket.

12. Remove the heater hose from the thermostat housing.

13. Disconnect the coolant temperature sensor connector.

14. Remove the upper radiator hose.

15. Remove the thermostat housing.

16. From the left side cylinder head, remove or disconnect the following:
- Oil level indicator tube
- Power steering supply hose bracket

17. Remove the power steering pressure hose bracket from the crossmember.

18. Remove the power steering pump pulley, then the power steering pump, without disconnecting the lines, and position aside.

19. Remove the accessory drive belt bracket and then the A/C compressor, without disconnecting the lines, and position aside.

20. Remove the starter wiring bracket retainer.

21. Remove the ground strap and the wiring harness retainer from the rear of the cylinder head.

22. Remove the spark plug wires.

23. Remove the fuel rail and injectors and discard the o-ring seals.

24. Remove the catalytic converter to exhaust manifold nuts from both sides.

25. Remove the exhaust manifolds.

26. From the right side cylinder head, remove the hydraulic chain tensioner.

27. Install camshaft locking tools no. 303-578 and 303-564 and tighten the bolts to 89 inch lbs. (10 Nm).

28. Install special tool no. 303-575 and using camshaft sprocket nut socket no. 303565, remove the right side camshaft bolt.

29. Remove the right side camshaft cassette bolt.

06017-TANG-G12

Removing hydraulic chain tensioner bolt— 4.0L engine—left side shown: right side similar

303-564
303-578

06017-TANG-G13

Installing camshaft locking tools—4.0L engine—left side shown: right side similar

06017-TANG-G14

Removing camshaft sprocket bolt—4.0L engine—left side shown: right side similar

06017-TANG-G15

Removing cassette bolt—4.0L engine— left side shown: right side similar

✳✳ CAUTION

Remove the camshaft sprocket from the timing chain to gain access to remove the cylinder head.

30. Hold the timing chain and cassette with a rubber band and remove the camshaft sprocket.

31. Remove the left side hydraulic chain tensioner.

32. Install camshaft locking tools no. 303-578 and 303-564 and tighten the bolts to 89 inch lbs. (10 Nm).

33. Remove the left side camshaft sprocket bolt.

34. Remove the left side camshaft cassette bolt.

✳✳ CAUTION

Remove the camshaft sprocket from the timing chain to gain access to remove the cylinder head.

35. Hold the timing chain and cassette with a rubber band and remove the camshaft sprocket.

36. On both sides, remove the cylinder head bolts in the sequence shown and discard the bolts and the head gasket.

37. With the aid of an assistant, remove the cylinder heads.

To install:

➡The cylinder head bolts are a torque-to-yield design and cannot be reused.

06017-TANG-G16

Cylinder head bolt removal sequence— 4.0L engine

06017-TANG-G17

Cylinder head bolt installation sequence— 4.0L engine

➡ **Left and right cylinder head gaskets are not interchangeable.**

38. Install the new cylinder heads. Lubricate the new cylinder head bolt threads with clean oil and tighten them in sequence as follows:

    a. M12 bolts, Step 1: 9 ft. lbs. (12 Nm).

    b. Step 2: 18 ft. lbs. (25 Nm).

    c. M8 bolts Step 3: 24 ft. lbs. (32 Nm).

    d. Step 4: M12 bolts an additional 90 degrees.

    e. Step 5: M12 bolts an additional 90 degrees.

39. On the right side, remove the rubber band and position the camshaft sprocket and chain in place.

40. Loosely install the sprocket bolt.

41. Install the right head cassette bolt and tighten to 9 ft. lbs. (12 Nm).

42. On the left side, remove the rubber band and position the camshaft sprocket and chain in place.

43. Loosely install the sprocket bolt.

44. Install the left head cassette bolt and tighten to 14 ft. lbs. (19 Nm).

45. Install the exhaust manifolds to both sides. See the procedure under exhaust manifolds.

46. Install the spark plug wires.

47. Install new o-rings on the fuel injectors and then install the injectors and fuel rail. Tighten the fuel rail bolts to 18 ft. lbs. (25 Nm).

48. Using a new gasket, install the thermostat housing and tighten the bolts to 8 ft. lbs. (11 Nm).

49. Install the coolant bypass hose clamp.

50. Install the upper radiator hose.

51. Connect the engine coolant temperature sensor connector.

52. Connect the heater hose to the thermostat housing.

53. Install the alternator bracket and tighten to 35 ft. lbs. (47 Nm).

54. Install the drive belt tensioner and tighten to 35 ft. lbs. (47 Nm).

55. Connect the alternator connectors.

56. Install the ground strap.

57. Connect the wiring harness retainer to the back of the left cylinder head.

58. Connect the starter wiring harness retainer.

59. Install the drive belt bracket and the A/C compressor and tighten the fasteners to 35 ft. lbs. (47 Nm).

60. Install the power steering pump and tighten the bolts to 18 ft. lbs. (25 Nm).

61. Install the power steering pump pulley and tighten the bolts to 18 ft. lbs. (25 Nm).

62. Install the power steering pressure tube bracket to the crossmember and tighten the bolt to 80 inch lbs. (10 Nm).

63. Install the power steering hose bracket.

64. Install the oil level indicator tube.

➡ **If the camshaft have been moved, the engine will require re-timing.**

65. Rotate the crankshaft until the no. 1 cylinder is at TDC.

66. Install crankshaft timing tool no. 303-573 to the crankshaft damper so it contacts the block. This positions the engine at TDC.

67. Install camshaft locking tools no. 303-578 and 303-564 to the right cylinder head.

➡ **The camshaft timing slots are off center.**

68. Position the camshaft timing slots below the centerline of the camshaft, then install tools 303-576 and 303-577 to the front of the cylinder head.

69. Tighten the bolts on tool 303-578 to 89 inch lbs. (10 Nm).

✳✳ CAUTION

The right camshaft sprocket bolt uses a left-handed thread.

70. Use camshaft sprocket nut socket 303-565 and tighten the sprocket bolt to 63 ft. lbs. (85 Nm).

71. Install the right camshaft tensioner and tighten the bolt to 32 ft. lbs. (44 Nm).

72. Remove the special tools from the right cylinder head and install them on the left cylinder head and repeat the previous procedure to tighten the left camshaft sprocket bolt.

73. Install the accessory drive belt.

74. Install the camshaft roller followers.

75. Install the intake manifold.

Install crankshaft timing tool to position the engine at TDC—4.0L engine

06017-TANG-G18

76. Install the valve covers and tighten the bolts in a criss-cross pattern to 89 inch lbs. (10 Nm).

77. Install the valve cover components and spark plug wires.

78. Connect the negative battery cable.

79. Fill the cooling system.

80. Start the engine and check for leaks.

81. Recharge the A/C system.

4.6L Engine

1. Drain the cooling system.

2. Relieve the fuel system pressure.

3. Disconnect the negative battery cable.

4. Drain the engine oil.

5. Remove the engine from the vehicle.

6. Install the engine on a suitable work stand.

7. Disconnect both camshaft position sensor connectors.

8. Disconnect both Variable Camshaft Timing (VCT) connectors.

9. Disconnect the engine wiring harness retainers.

10. Disconnect the radio interference capacitor.

11. Remove the PCV tubes from both valve covers.

12. Disconnect the ignition coil connectors.

13. Disconnect the engine wiring harness retainers from each valve cover.

14. Disconnect the cylinder head temperature sensor connector, and jumper harness connector retainer.

15. Disconnect the knock sensor connector retainer.

16. Disconnect the left side heated oxygen sensor connector.

17. Disconnect the oil pressure sender connector.

18. Remove the engine wiring harness.

19. Remove the oil filter.

20. Remove the oil level indicator tube.

21. Remove both camshaft position sensors.

22. Remove the crankshaft position sensor near the crankshaft damper.

23. Remove the ignition coils.

24. Remove both valve covers.

25. Remove the water pump pulley and the right side drive belt idler pulley.

26. Using a puller, remove the crankshaft pulley.

27. Remove the crankshaft oil seal.

28. Remove 4 engine front cover-to-oil pan bolts.

29. Remove the engine front cover.

30. Remove the timing chains. See the procedure under Timing Chain.

Installing cylinder head removal tool to
the left cylinder head—4.6L engine

06017-TANG-G19

Installing cylinder head removal tool to
the right cylinder head—4.6L engine

06017-TANG-G20

31. Remove the camshafts. See the procedure under Camshafts.

32. Remove the camshaft roller followers
and hydraulic lash adjusters.

33. Install the cylinder head removal tool
303-572 to the left cylinder head.

34. Remove the left exhaust manifold.

35. Remove the ground strap from the
left cylinder head.

36. Install the cylinder head removal tool
303-572 to the right cylinder head.

37. Remove the right exhaust manifold.

38. Remove the coolant tube from the
right cylinder head.

39. Remove the cylinder heads.

To install:

40. Inspect the cylinder head for paying
special attention to the areas around the oil
pressure feed areas. Flatness should not
exceed 0.002 inch (0.005mm) per running
foot length.

➡**The cylinder head bolts are a torque-
to-yield design and cannot be reused.**

41. Install the cylinder head gasket and
the cylinder head.

42. Tighten the cylinder head bolts in
sequence as follows:
   a. Step 1: 30 ft. lbs. (40 Nm).
   b. Step 2: Plus 90 degrees.
   c. Step 3: Plus 90 degrees.

43. Remove the special tool from the left
cylinder head.

44. Install the hydraulic lash adjusters.
After lubricating them with clean engine oil.

Cylinder head bolt tightening sequence—
4.6L engine

06017-TANG-G21

45. Install the left exhaust manifold.

46. Install the ground strap to the left
cylinder head.

47. Remove the special tool from the
right cylinder head.

48. Install the right exhaust manifold.

49. Install new o-rings on the coolant
tube and slide the tube into the cylinder
block.

50. Install the coolant tube bolt and
tighten to 89 inch lbs. (10 Nm).

51. Install the camshaft roller followers.

52. Install the camshafts. See the procedure under Camshafts.

53. Install the timing chains. See the
procedure under Timing Chain.

54. Install the engine front cover.

55. Install 4 engine front cover-to-oil
pan bolts and tighten in sequence to 15 ft.
lbs. (20 Nm), plus an additional 60 degrees.

56. Lubricate the new crankshaft oil seal
with clean engine oil and install the seal.

57. Install silicone gasket sealant to the

Oil pan-to-front cover bolt tightening
sequence—4.6L engine

06017-TANG-G22

Woodruff key slot in the crankshaft pulley
and using tool 303-102 install the crankshaft pulley and tighten as follows:
   a. Step 1: 66 ft. lbs. (90 Nm).
   b. Step 2: Loosen one full turn.
   c. Step 3: 37 ft. lbs. (50 Nm).
   d. Step 4: An additional 90 degrees.

58. Install the right side drive belt idler
pulley and tighten the bolt to 18 ft. lbs. (25
Nm).

59. Install the crankshaft position sensor
and tighten to 89 inch lbs. (10 Nm).

60. Install both camshaft position sensors after lubricating the new o-ring seal
with clean engine oil, and tighten to 89 inch
lbs. (10 Nm).

61. Apply silicone sealant to the 2 areas
where the cylinder head meets the engine
front cover on the right side.

62. Install the right side valve cover and
tighten the bolts in sequence to 89 inch lbs.
(10 Nm).

Right side valve cover bolt tightening
sequence—4.6L engine

06017-TANG-G23

63. Apply silicone sealant to the 2 areas
where the cylinder head meets the engine
front cover on the left side.

64. Install the left side valve cover and
tighten the bolts in sequence to 89 inch lbs.
(10 Nm).

65. Install the oil filter.

66. Install the ignition coils.

67. Install the engine wiring harness.

68. Connect the oil pressure sender connector.

69. Connect the left side heated oxygen
sensor connector.

70. Connect the knock sensor connector
retainer.

71. Install the oil level indicator tube.

Left side valve cover bolt tightening sequence—4.6L engine

72. Connect the left side heated oxygen sensor connector.

73. Connect the knock sensor connector retainer.

74. Connect the cylinder head temperature sensor connector, and jumper harness connector retainer.

75. Connect the engine wiring harness retainers from each valve cover.

76. Connect the ignition coil connectors.

77. Install the PCV tubes from both valve covers.

78. Connect the radio interference capacitor.

79. Connect the engine wiring harness retainers.

80. Connect both Variable Camshaft Timing (VCT) connectors.

81. Connect both camshaft position sensor connectors.

82. Install the engine.

83. Connect the negative battery cable.

84. Fill the cooling system.

85. Start the engine and check for leaks.

Camshaft Roller Followers and Lash Adjusters

REMOVAL & INSTALLATION

4.0L Engine

1. Disconnect the negative battery cable.

2. Remove the valve covers. See the procedure under Cylinder Head.

3. Rotate the crankshaft until the cylinder being serviced is at the base circle.

Removing camshaft roller follower—4.0L engine

➡ Mark the followers so they can be installed in their original positions.

4. Using Valve Spring Compressor tool no. 303-581, remove the follower.

5. Remove the lash adjusters.

To install:

6. Lubricate the lash adjusters with clean engine oil and install in the original position.

7. Lubricate the camshaft rollers with clean engine oil and install them in the original locations.

8. Install the valve covers.

9. Connect the negative battery cable.

10. Start the engine and check for proper operation.

4.6L Engine

1. Disconnect the negative battery cable.

2. Remove the valve covers. See the procedure under Cylinder Head.

3. Rotate the crankshaft so that the piston on the cylinder to be serviced is at the top of its stroke with the intake and exhaust valves closed

4. Use Valve Spring Compressor tool no. 303-1039 and remove the roller follower. Repeat for each follower o be removed.

➡ If the followers are to be reused, ensure that they are installed in the same position that they were removed from.

Removing camshaft roller follower—4.6L engine

5. To remove the lash adjusters, remove the camshaft, then the adjusters.

To install:

6. Lubricate the lash adjusters with clean engine oil and install in the original position.

7. Install the camshaft.

8. Lubricate the camshaft rollers with clean engine oil and install them in the original locations.

9. Install the valve covers.

10. Connect the negative battery cable.

11. Start the engine and check for proper operation.

Intake Manifold

REMOVAL & INSTALLATION

4.0L Engine

1. Relieve the fuel system pressure.

2. Disconnect the negative battery cable.

3. Disconnect the right side spark plug wires from the ignition coil.

4. Disconnect the right side spark plug wire retainer from the intake manifold.

5. Remove the ignition coil bracket upper and lower bolts, and remove the bracket.

6. Disconnect the Exhaust Gas Recirculation (EGR) module electrical connectors.

7. Disconnect the brake booster vacuum hose.

8. Disconnect the fuel rail temperature and pressure sensor vacuum line.

9. Disconnect the vapor tube from the manifold.

10. Remove the Positive Crankcase Ventilation (PCV) tube.

11. Disconnect the knock sensor and throttle position sensor connectors.

12. Disconnect and plug the throttle body coolant hoses.

13. Remove the intake manifold.

To install:

14. Install the intake manifold and tighten the bolts in sequence to 89 inch lbs. (10 Nm).

15. Connect the throttle body coolant hoses.

16. Connect the knock sensor and throttle position sensor connectors.

17. Install the Positive Crankcase Ventilation (PCV) tube.

18. Connect the vapor tube to the manifold.

19. Connect the fuel rail temperature and pressure sensor vacuum line.

20. Connect the brake booster vacuum hose.

Intake manifold torque sequence—4.0L engine

21. Connect the Exhaust Gas Recirculation (EGR) module electrical connectors.

22. Install the ignition coil bracket upper and lower bolts, and tighten the M8 bolt to 18 ft. lbs. (25 Nm), and the M12 bolts to 25 ft. lbs. (34 Nm).

23. Connect the right side spark plug wire retainer to the intake manifold.

24. Connect the right side spark plug wires to the ignition coil.

25. Connect the negative battery cable.

26. Fill the cooling system.

27. Start the engine and check for leaks.

4.6L Engine

1. Relieve the fuel system pressure.

2. Disconnect the negative battery cable.

3. Remove the air cleaner outlet pipe.

4. Remove the fuel rail and injectors.

5. Disconnect the EVAP tube and position aside.

6. Disconnect the PCV tube and position aside.

7. Disconnect the throttle body and throttle position sensor connectors.

8. Disconnect the charge motion control valve connector.

9. Disconnect the wiring harness from the manifold.

10. Disconnect the vacuum hose.

11. Remove the intake manifold.

To install:

12. Install the intake manifold. Use new gaskets and tighten the bolts, in sequence, to 89 inch lbs. (10 Nm).

13. Connect the vacuum hose.

14. Connect the wiring harness from the manifold.

15. Connect the charge motion control valve connector.

16. Connect the throttle body and throttle position sensor connectors.

17. Connect the PCV tube and position aside.

18. Connect the EVAP tube and position aside.

19. Install the fuel rail and injectors.

20. Install the air cleaner outlet pipe.

Intake manifold tightening sequence—4.6L engine

21. Connect the negative battery cable.

22. Start the engine and check for leaks.

Exhaust Manifold

REMOVAL & INSTALLATION

4.0L Engine

1. Disconnect the negative battery cable.

2. On the right side, remove the catalytic converter-to-exhaust manifold nuts.

3. Remove the exhaust manifold.

4. On the left side, remove the catalytic converter-to-exhaust manifold nuts.

5. Disconnect the EGR tube from the manifold.

6. Remove the exhaust manifold.

To install:

7. Install the left exhaust manifold. Tighten the bolts to 17 ft. lbs. (23 Nm).

8. Connect the EGR tube to the manifold and tighten to 29 ft. lbs. (39 Nm).

9. Install the catalytic converter-to-exhaust manifold nuts and tighten to 30 ft. lbs. (40 Nm).

10. Install the right exhaust manifold. Tighten the bolts to 17 ft. lbs. (23 Nm).

11. Install the catalytic converter-to-exhaust manifold nuts and tighten to 30 ft. lbs. (40 Nm).

4.6L Engine

1. Disconnect the negative battery cable.

2. Remove the air cleaner assembly.

3. Remove the throttle body.

4. Remove the starter.

5. On the both sides, remove the catalytic converter-to-exhaust manifold nuts.

6. Remove the radiator sight shield.

7. Remove the alternator.

8. Install an engine lifting device.

9. Remove the right and left engine support insulator nuts.

10. Raise the engine about 1-1/2 inches.

11. Remove the steering column-to-steering gear pinch bolt.

12. On automatic transmission models, disconnect the transmission cooler tube bracket.

13. Remove the exhaust manifold(s).

To install:

14. Install the exhaust manifolds. Use new gaskets and tighten the nuts in sequence to 18 ft. lbs. (25 Nm).

15. On automatic transmission models, connect the transmission cooler tube bracket.

16. Install the steering column-to-steering gear pinch bolt and tighten to 18 ft. lbs. (25 Nm).

17. Lower the engine about 1-1/2 inches.

18. Install the right and left engine support insulator nuts and tighten to 46 ft. lbs. (63 Nm).

19. Remove the engine lifting device.

20. Install the alternator.

21. Install the radiator sight shield.

22. On the both sides, install the catalytic converter-to-exhaust manifold nuts and tighten to 30 ft. lbs. (40 Nm).

23. Install the starter.

24. Install the throttle body.

25. Install the air cleaner assembly.

26. Connect the negative battery cable.

Left exhaust manifold torque sequence—4.6L engine

Right exhaust manifold torque sequence—4.6L engine

Camshaft

REMOVAL & INSTALLATION

4.0L Engine

1. Drain the cooling system.
2. Disconnect the negative battery cable.
3. Remove the valve covers.
4. Remove the camshaft roller followers.
5. Position the no. 1 cylinder at TDC.
6. Install crankshaft timing tool no. 303-573 to the crankshaft damper so it contacts the block. This positions the engine at TDC.
7. On the right side, install camshaft locking tools no. 303-578 and 303-564 and tighten the bolts to 89 inch lbs. (10 Nm).
8. Install special tool no. 303-575 and using camshaft sprocket nut socket no. 303565, remove the right side camshaft sprocket bolt.
9. On the left side, install camshaft locking tools no. 303-578 and 303-564 and tighten the bolts to 89 inch lbs. (10 Nm).
10. Remove the left side camshaft sprocket bolt.
11. Remove the camshaft sprockets from the camshaft and position aside with the timing chain in place.
12. Mark the camshaft bearing caps so they can be installed in their original positions.

303-573

Install crankshaft timing tool to position the engine at TDC—4.0L engine

06017-TANG-G31

Camshaft bearing cap bolt removal sequence—4.0L engine

06017-TANG-G32

Camshaft bearing cap bolt installation sequence—4.0L engine

13. Remove the camshaft bearing cap bolts in the sequence shown.
14. Remove the camshaft.

To install:

15. Install the camshaft.
16. Install the camshaft bearing caps in their original locations and tighten the bolts in sequence to 53 inch lbs. (6 nm), then tighten in sequence again to 12 ft. lbs. (16 Nm).
17. Install the right camshaft sprocket and loosely install the bolt.
18. Install the left camshaft sprocket and loosely install the bolt.
19. On the right side, position the camshaft timing slots below the centerline of the camshaft and install camshaft timing tools 303-577 and 303-576 to the front of the cylinder head.
20. Remove the right side camshaft tensioner bolt.
21. Install Timing Chain Tensioner tool no. 303-571 in the tensioner opening.
22. Tighten the camshaft locking tools no. 303-578 and 303-564 bolts to 89 inch lbs. (10 Nm).
23. Install special tool no. 303-575 and using camshaft sprocket nut socket no. 303-565, tighten the right side camshaft sprocket bolt to 63 ft. lbs..
24. Remove timing tensioner tool and install the right side tensioner and tighten to 32 ft. lbs. (44 Nm).
25. Remove the tools from the right cylinder head and camshaft and install them on the left cylinder head and camshaft.

303-577

303-576

06017-TANG-G33

Installing camshaft timing tools to the cylinder head—4.0L engine

26. Repeat the timing procedure on the left side.
27. Remove the special tools.
28. Install the camshaft roller followers.
29. Install the valve covers.
30. Connect the negative battery cable.

4.6L Engine

❄❄ CAUTION

The procedure must be followed exactly or damage to the valves and pistons will result.

➡**If both camshafts are being removed, the right camshaft must be removed first.**

1. Disconnect the negative battery cable.
2. Position the crankshaft damper spoke at the 12 o'clock position and the timing mark indentation at the 1 o'clock position.
3. Remove the right side valve cover.
4. Loosen and back off the camshaft phaser bolt one full turn.
5. Disconnect the camshaft position sensor connector.
6. Remove the camshaft position sensor.
7. Position the camshaft lobes as shown so the no.2 intake lobes and no. 1 exhaust lobe are in correct position..

➡**The no. 1 cylinder must be coming up on the exhaust stroke with the crankshaft keyway in the 12 o'clock position.**

8. Remove the 3 camshaft roller followers as shown, using spring compressor 303-1039.
9. Remove the left side valve cover.
10. Loosen and back off the camshaft phaser bolt one full turn.
11. Disconnect the camshaft position sensor connector.
12. Remove the camshaft position sensor.

06017-TANG-G34

Positioning camshaft lobes on right camshaft—4.6L engine

06017-TANG-G35

Removing right cylinder head camshaft roller followers—4.6L engine

06017-TANG-G37

Removing left cylinder head camshaft roller followers—4.6L engine

06017-TANG-G39

Scribing a timing mark on timing chain and camshaft phaser—4.6L engine

06017-TANG-G40

Camshaft bearing cap bolt removal sequence—4.6L engine

13. Position the camshaft lobes as shown so the no. 2 intake lobes and no. 5 exhaust lobe are in correct position.

➡ **The no. 5 cylinder must be coming up on the exhaust stroke with the crankshaft keyway in the 12 o'clock position.**

14. Remove the 3 camshaft roller followers as shown, using spring compressor 303-1039.

15. Rotate the crankshaft and position the crankshaft damper spoke at the 6 o'clock position and the timing mark indentation at the 7 o'clock position.

16. Install Timing Chain Wedge no. 303-636 and 303-637 into the right hand timing chain as shown.

➡ **The tool must be installed square to the timing chain and engine block.**

17. Scribe a mark on the right timing chain and the camshaft phaser as shown.

18. Remove the right camshaft bearing cap bolts in the sequence shown.

19. Remove the right camshaft phaser bolt and the phaser, leaving the sprocket in place.

20. Install Timing Chain Wedge no. 303-636 and 303-637 into the left hand timing chain as shown.

➡ **The tool must be installed square to the timing chain and engine block.**

21. Scribe a mark on the left timing chain and the camshaft phaser as shown.

22. Remove the left camshaft bearing cap bolts in the sequence shown.

23. Remove the left camshaft phaser bolt and the phaser, leaving the sprocket in place.

24. Remove the camshaft(s).

To install:

25. Install the camshaft(s) after lubricating the camshaft and journals with clean engine oil.

26. Install the right camshaft phaser using and new bolt. Hand tighten the bolt.

27. Ensure the phaser and timing chain marks are aligned.

28. Lubricate the bearing caps with clean engine oil.

29. Install the front bearing cap, then the remaining caps and loosely install the bolts.

30. Tighten the bearing caps in the sequence shown to 89 inch lbs. (10 Nm).

06017-TANG-G36

Positioning camshaft lobes on left camshaft—4.6L engine

303-637

303-636

06017-TANG-G38

Installing timing chain wedge in right timing chain—4.6L engine

Installing timing chain wedge in left timing chain—4.6L engine

Camshaft bearing cap bolt tightening sequence—4.6L engine

31. Remove the timing wedge tool from the right timing chain.

32. Install the left camshaft phaser using and new bolt. Hand tighten the bolt.

33. Ensure the phaser and timing chain marks are aligned.

34. Lubricate the bearing caps with clean engine oil.

35. Install the front bearing cap, then the remaining caps and loosely install the bolts.

36. Tighten the bearing caps in the sequence shown to 89 inch lbs. (10 Nm).

37. Remove the timing wedge tool from the left timing chain.

38. Rotate the crankshaft damper so the spoke is at the 12 o'clock position and the timing mark indentation is at the 1 o'clock position.

39. Verify the camshaft lobes are in the correct position.

40. Install the 3 camshaft roller followers using the spring compressor.

41. Install the camshaft position sensor and connect the connector.

42. Tighten the camshaft phaser bolt to 30 ft. lbs. (40 Nm), then tighten an additional 90 degrees.

43. Install the right valve cover.

44. Verify the left camshaft lobes are in the correct position.

45. Install the 3 camshaft roller followers using the spring compressor.

46. Install the camshaft position sensor and connect the connector.

47. Tighten the camshaft phaser bolt to 30 ft. lbs. (40 Nm), then tighten an additional 90 degrees.

48. Install the left valve cover.

49. Connect the negative battery cable.

50. Start the engine and check for leaks.

Valve Lash

ADJUSTMENT

The 4.0L and 4.6L engines are equipped with hydraulic lash adjusters. Valve clearance is not adjustable.

Starter Motor

REMOVAL & INSTALLATION

4.0L Engine

1. Disconnect the negative battery cable.
2. Disconnect starter electrical connections.
3. Remove starter bolts.
4. Remove starter.

To install:

5. Install starter. Tighten the bolts to 18 ft. lbs. (25 Nm).
6. Connect starter electrical connections.
7. Connect the negative battery cable.

4.6L Engines

1. Disconnect the negative battery cable.
2. Disconnect starter electrical connections.
3. Remove starter bolts.
4. Remove starter.

To install:

5. Install starter. Tighten the bolts to 18 ft. lbs. (25 Nm).
6. Connect starter electrical connections.
7. Connect the negative battery cable.

1	Starter solenoid terminal cover
2	Starter solenoid S-terminal nut
3	Starter solenoid S-terminal eyelet
4	Starter solenoid B+ terminal nut
5	Starter solenoid B+ terminal eyelet
6	Starter motor mounting bolt (2 required)
7	Starter motor

Starter mounting—4.0L engine

1 Starter solenoid terminal cover
2 Starter solenoid S-terminal nut
3 Starter solenoid S-terminal eyelet
4 Starter solenoid B+ terminal nut
5 Starter solenoid B+ terminal eyelet
6 Starter motor mounting bolt (3 required)
7 Starter motor

06017-TANG-G44

Starter mounting—4.6L engine

Oil Pan

REMOVAL & INSTALLATION

4.0L Engine

1. Drain the engine oil.
2. Disconnect the negative battery cable.
3. Raise the vehicle on a hoist.
4. Remove the oil pan.

To install:

5. Install a new gasket and the oil pan. Tighten the bolts to 80 inch lbs. (9 Nm).
6. Lower the vehicle.
7. Connect the negative battery cable.
8. Fill the engine with oil.
9. Run the engine and check for leaks.

4.6L Engine

1. Drain the engine oil.
2. Disconnect the negative battery cable.
3. Remove the air cleaner assembly.
4. Remove the throttle body.
5. Remove the radiator sight shield.
6. Remove the alternator.
7. Install an engine lifting device.
8. Remove the right and left engine support insulator nuts.

9. Raise the engine about 1-1/2 inches.
10. Place a suitable jack under the subframe.
11. Remove the subframe bolts and nuts.
12. Lower the subframe about 2 inches.
13. Disconnect the oil temperature sensor connector.
14. Remove the oil pan.

To install:

15. Apply a bead of sealant at the crankshaft rear seal retainer plate-to-cylinder block surface and the engine front cover-to-cylinder block surface.
16. Install the oil pan. Use a new gasket and tighten the bolts in sequence as follows:

   a. Step 1: 18 inch lbs. (2 Nm).
   b. Step 2: 15 ft. lbs. (20 Nm).
   c. Step 3: Plus 60 degrees.

17. Raise the subframe into position and tighten the bolts to 85 ft. lbs. (115 Nm).
18. Lower the engine.
19. Install the right and left engine support insulator nuts and tighten to 46 ft. lbs. (63 Nm).
20. Install an engine lifting device.
21. Install the alternator.
22. Install the radiator sight shield.
23. Install the throttle body.
24. Install the air cleaner assembly.

06017-TANG-G45

Oil pan torque sequence—4.6L engines

25. Connect the negative battery cable.
26. Fill the engine with oil.
27. Start the engine and check for leaks.

Oil Pump

REMOVAL & INSTALLATION

4.0L Engine

1. Drain the engine oil.
2. Remove the oil pan.
3. Remove the cylinder block cradle.
4. Remove oil pump.

To install:

5. Install the oil pump. Tighten the bolts to 15 ft. lbs. (20 Nm).
6. Install the cylinder block cradle.
7. Install the oil pan.
8. Fill the engine with oil.
9. Start the engine and check for leaks.

4.6L Engine

1. Remove the oil pan.
2. Remove the front cover.
3. Remove the timing chains and sprockets.
4. Remove the oil pump.

To install:

5. Install oil pump. Tighten the bolts to 79 inch lbs. (10 Nm).
6. Install timing chains and sprockets.
7. Install front cover.
8. Install oil pan.
9. Check for leaks and proper operation.

Cylinder Block Cradle

REMOVAL & INSTALLATION

4.0L Engine

1. Disconnect the negative battery cable.
2. Drain the engine oil.
3. Remove the throttle body.
4. Remove the front end accessory drive bracket bolt and install an engine lifting eye in the opening.
5. Remove the radiator sight shield.
6. Install a 3-bar engine lifting device.
7. Remove the left and right engine mount nuts.
8. Remove the oil pan.
9. Remove the oil pump screen and pickup tube.
10. Separate the steering column intermediate shaft from the steering gear.
11. Separate the steering gear from the crossmember.
12. Remove 2 cylinder block cradle rear Torx bolts.
13. Remove the 20 bolts and 2 nuts around the cylinder block cradle.

➡Note the location of 2 silver colored bolts with washer seals. Ensure they are installed in the same position.

14. Remove 8 inner cradle bolts and washer seals. Discard the seals.
15. Raise the engine using the lifting device and remove the cylinder block cradle.

To install:

16. Back off the setscrews until they are below the cylinder block cradle boss.
17. Clean all gasket mating surfaces.
18. Apply silicone to the areas shown.
19. Using a new gasket, install the cradle and hand tighten the outer bolts and nuts.
20. Install the Torx bolts and hand tighten.
21. Install the bell housing-to-cradle bolts and tighten to 35 ft. lbs. (47 Nm).
22. Tighten the cradle outer bolts and nuts to 89 inch lbs. (10 Nm).
23. Tighten the Torx bolts to 71 inch lbs. (8 Nm).
24. Tighten the 8 inserts to 27 inch lbs. (3 Nm).
25. Install the 2 silver colored bolts and new washers and hand tighten.
26. Install the 6 inner bolts and hand tighten.
27. Tighten the 8 inner bolts in sequence to 11 ft. lbs. (15 Nm), then again to 25 ft. lbs. (34 Nm).

1   Oil pan bolt (10 required)
2   Oil pan
3   Oil pan gasket
4   Oil pump pickup tube bolt
5   Oil pump pickup tube
6   Cylinder block cradle bolt
7   Cylinder block cradle bolt
8   Bell housing-to-cylinder block cradle bolt (2 required)
9   Bell housing-to-cylinder block cradle spacer (2 required)
10 Cylinder block cradle bolt (20 required)
11 Cylinder block cradle nut (2 required)
12 Cylinder block cradle bolt (2 required)
13 Washer (2 required)
14 Cylinder block cradle bolt (6 required)
15 Cylinder block cradle
16 Cylinder block cradle gasket
17 Oil pump bolt (2 required)
18 Oil pump

06017-TANG-G46

Exploded view of the engine lubrication components—4.0L engine

06017-TANG-G47

Apply sealant to the areas shown—4.0L engine

Inner cylinder block cradle bolt tightening sequence—4.0L engine

28. Install the steering gear and tighten to 85 ft. lbs. (115 Nm).

29. Install the steering column pinch bolt and tighten to 18 ft. lbs. (25 Nm).

30. Install the oil pump screen and pickup tube.

31. Install the oil pan.

32. Install the left and right engine mount nuts and tighten to 46 ft. lbs. (63 Nm).

33. Remove the engine lifting tools and install the drive bracket bolt.

34. Install the radiator sight shield.

35. Install the throttle body.

36. Connect the negative battery cable.

37. Fill the engine with oil.

Rear Main Seal

REMOVAL & INSTALLATION

4.0L Engine

1. Disconnect the negative battery cable.

2. Remove transmission.

3. Remove clutch pressure plate and disc, if equipped.

4. Remove flywheel.

5. Remove engine rear plate, if equipped.

6. Remove rear main seal. Use Crankshaft Rear Seal Remover T92C-6700-CH.

To install:

7. Place clean engine oil around the seal and install the seal. Use Rear Crankshaft Seal Installer 303-S524.

8. Install engine rear plate, if equipped.

9. Install flywheel.

10. Install clutch pressure plate and disc, if equipped.

11. Install transmission.

12. Connect the negative battery cable.

4.6L Engine

1. Disconnect the negative battery cable.

2. Remove transmission.

3. Remove clutch pressure plate and disc, if equipped.

Oil seal retainer torque sequence—4.6L engines

4. Remove flywheel.

5. Remove the spacer plate.

6. Remove the oil slinger. Use Rear Crankshaft Oil Slinger Remover 303-514.

7. Remove rear oil seal.

8. Remove the seal retainer plate.

To install:

9. Apply silicone sealer around the seal retainer plate on Essex built engines, or to the plate mating surface on the block on Romeo built engines.

10. Install the rear oil seal retainer. Tighten the bolts in sequence to 89 inch lbs. (10 Nm).

11. Install rear oil seal. Use Rear Crankshaft Seal Replacer 303-516 and 303-518.

12. Install oil slinger.

13. Install flywheel.

14. Install clutch pressure plate and disc, if equipped.

15. Install transmission.

16. Connect the negative battery cable.

17. Run the engine and check for leaks.

Timing Chain, Sprockets, Front Cover and Seal

REMOVAL & INSTALLATION

4.0L Engine

1. Drain the cooling system and the engine oil.

2. Disconnect the negative battery cable.

3. Remove the intake manifold.

4. Remove the crankshaft pulley bolt and the crankshaft damper using a puller.

5. Remove the crankshaft front oil seal using a puller.

6. Disconnect the upper and lower radiator hoses.

7. Disconnect the heater hose from the water pump.

1	Crankshaft position (CKP) sensor electrical connector	9	Block cradle-to-engine front cover bolts (5 required)
2	Wiring retainers	10	Accessory drive belt idler pulley bolt
3	Engine front cover stud bolt (4 required)	11	Accessory drive belt idler pulley
4	Clamp	12	Clamp
5	Coolant hose	13	Upper radiator hose
6	Engine front cover bolt (5 required)	14	Engine front cover
7	Clamp	15	Engine front cover gasket
8	Lower radiator hose		

Exploded view of the engine front cover—4.0L engine

8. Remove the accessory drive belt.

9. Remove the drive belt idler pulley.

10. Disconnect the crankshaft position sensor connector and retainer.

11. Disconnect the coolant bypass hose from the water pump.

12. Remove 5 engine front cover-to-engine block cradle bolts.

13. Remove 5 bolts and 4 stud bolts and remove the front cover.

14. Remove the camshaft roller followers.

15. Rotate the crankshaft so that the No. 1 cylinder is at Top Dead Center (TDC) of the compression stroke.

16. Remove the right side chain tensioner.

17. Install the special tools and remove the right camshaft sprocket bolt. See the procedure under Camshaft.

18. Hold the crankshaft from turning and remove the jackshaft sprocket bolt.

19. Remove the primary chain tensioner.

20. Remove the primary chain and sprockets as an assembly.

21. Remove the right side cassette upper and lower bolts and remove the cassette.

22. Remove the left side chain tensioner.

23. Install the special tools and remove the left camshaft sprocket bolt. See the procedure under Camshaft.

24. Remove the left side cassette upper and lower bolts and remove the cassette.

25. Remove the timing chains.

To install:

26. Install the timing chains.

27. Install the left side cassette lower bolt and tighten to 9 ft. lbs. (12 Nm).

28. Install the left side cassette upper bolt and tighten to 14 ft. lbs. (19 Nm).

29. Install the right side cassette lower bolt and tighten to 9 ft. lbs. (12 Nm).

30. Install the right side cassette upper bolt and tighten to 14 ft. lbs. (19 Nm).

31. Install the primary chain and sprockets as an assembly.

32. Install the primary chain tensioner and tighten to 80 inch lbs. (10 Nm).

33. Hold the crankshaft from turning and tighten the jackshaft sprocket bolt to 33 ft. lbs. (45 Nm), plus an additional 90 degrees.

34. Loosely install the camshaft sprocket bolts.

35. Perform the camshaft timing procedure. See the procedure under Camshaft.

36. Install the camshaft roller followers.

37. Apply silicone sealant to the front cover in 2 spots near the top center of the cover.

38. Apply thread sealant to the front cover stud bolts.

39. Install the front cover and loosely tighten the bolts and stud bolts.

40. Use front cover alignment tool 303-107 and align the front cover.

41. Tighten the front cover bolts to 14 ft. lbs. (19 Nm).

42. Install the front cover-to-engine cradle bolts and tighten to 80 inch lbs. (10 Nm).

43. Connect the coolant bypass to from the water pump.

44. Connect the crankshaft position sensor connector and retainer.

45. Install the drive belt idler pulley and tighten to 35 ft. (47 Nm).

46. Install the accessory drive belt.

47. Connect the heater hose to the water pump.

48. Connect the upper and lower radiator hoses.

49. Install the crankshaft front oil seal using a seal installer.

50. Install the crankshaft damper and the pulley bolt. Tighten the pulley bolt to 33 ft. lbs. (45 Nm), plus an additional 85 degrees.

51. Install the intake manifold.

52. Connect the negative battery cable.

53. Fill the cooling system.

54. Fill the crankcase with clean engine oil.

55. Run the engine. Check for leaks and proper operation.

4.6L Engine

1. Disconnect the negative battery cable.

2. Drain the engine oil.

3. Raise the vehicle on a hoist.

4. Remove the coolant degas bottle.

5. Remove the accessory drive belt.

6. Remove the right side idler pulley.

7. Remove the valve covers. See the procedure under Cylinder Head.

8. Disconnect the hoses from the coolant crossover.

9. Remove both left and right radio interference capacitors and position aside.

10. Remove both left and right camshaft position sensor connectors.

11. Remove the J bracket from the engine front cover stud bolt.

12. Remove the water pump pulley.

13. Remove the power steering pump pulley shield.

14. Remove the wiring harness retainer from the power steering stud bolt.

15. Remove the power steering pump and position aside.

16. Disconnect the crankshaft position sensor connector.

17. Remove the crankshaft pulley bolt and washer.

18. Using a 3-jaw puller, remove the crankshaft pulley.

19. Use a seal remover and remove the front oil seal.

20. Remove the 4 front oil pan bolts.

21. Remove the front cover stud bolts and bolts and remove the front cover.

22. Remove the crankshaft sensor ring from the crankshaft.

23. Place the crankshaft keyway at the 12 O'clock position.

24. Remove the camshaft roller followers and camshaft sprocket bolts. See the procedure under Camshaft.

25. Remove the left side timing chain tensioner and arm.

26. Remove the right side timing chain tensioner and arm.

27. Remove the right and then the left timing chain and crankshaft sprockets.

28. Remove the left and right timing chain guides.

To install:

29. Compress the timing chain tensioner plunger in a soft jaw vise and install a retaining clip to hold the plunger in place.

30. Remove the tensioner.

31. If the copper links are not visible on the timing chains, mark one link on each end of the chain.

32. Install the left and right timing chain guides and tighten the bolts to 89 inch lbs. (10 Nm).

33. Position the lower end of the left timing chain on the crankshaft sprocket, aligning the copper link with the timing mark on the outer flange of the crankshaft sprocket.

34. Position the timing chain on the left camshaft sprocket with the camshaft sprocket timing mark aligned with the copper link.

35. Position the left timing chain tensioner arm on the dowel pin and install the timing chain tensioner. Tighten the bolts to 18 ft. lbs. (25 Nm).

36. Remove the retaining clip from the tensioner plunger.

37. Position the lower end of the right timing chain on the crankshaft sprocket, aligning the copper link with the timing mark on the outer flange of the crankshaft sprocket.

38. Position the timing chain on the right camshaft sprocket with the camshaft sprocket timing mark aligned with the copper link.

39. Position the left timing chain tensioner arm on the dowel pin and install the timing chain tensioner. Tighten the bolts to 18 ft. lbs. (25 Nm).

40. Remove the retaining clip from the tensioner plunger.

24 Crankshaft position (CKP)
    sensor electrical connector

25 Oil pan bolts (4 required)

26 Engine front cover bolts (11
    required for Essex engine) (9
    required for Romeo engine)

27 Engine front cover studs (4
    required)

28 Engine front cover stud bolt
    (Romeo engine only)

29 Engine front cover stud bolt
    (Romeo engine only)

30 Engine front cover

31 Engine front cover gaskets (3
    required)

06017-TANG-G51

Exploded view of the engine front cover—4.6L engine

41. Ensure correct alignment of all timing chain links and sprocket timing marks.

42. Install the camshaft roller followers and tighten the camshaft sprocket bolts. See the procedure under Camshaft.

43. Install the crankshaft sensor ring on the crankshaft.

44. Apply sealant to the front cover mating surface as shown.

45. Install a new gasket on the front cover, then install the front cover and hand tighten the bolts.

46. Tighten the front cover bolts in the sequence shown to 18 ft. lbs. (25 Nm).

06017-TANG-G52

Timing chain link and sprocket timing mark alignment—4.6L engine

06017-TANG-G53

Sealant application to the front cover mating surface—4.6L engine

Front cover bolt tightening sequence—4.6L engine

47. Tighten the front cover-to-oil pan bolts in the sequence shown to 15 ft. lbs. (20 Nm), plus an additional 60 degrees

48. Lubricate the engine front cover and oil seal lip with clean engine oiland install the seal usinga seal installer.

49. Apply sealant to the Woodruff key slot in the crankshaft pulley, then press on the crankshaft damper.

50. Tighten the crankshaft damper as follows:

   a. Step 1: 66 ft. lbs. (90 Nm).

   b. Step 2: Loosen the bolt one full turn.

   c. Step 3: 37 ft. lbs. (50 Nm).

   d. Step 4: 90 degrees.

Front cover-to-oil pan bolt tightening sequence—4.6L engine

51. Connect the crankshaft position sensor connector.

52. Install the power steering pump and tighten the bolts to 18 ft. lbs. (25 Nm).

53. Install the wiring harness retainer to the power steering stud bolt.

54. Install the power steering pump pulley shield.

55. Install the water pump pulley.

56. Install the J bracket to the engine front cover stud bolt.

57. Install both left and right camshaft position sensor connectors.

58. Install both left and right radio interference capacitors.

59. Connect the hoses from the coolant crossover.

60. Install the valve covers. See the procedure under Cylinder Head.

61. Install the right side idler pulley.

62. Install the accessory drive belt.

63. Install the coolant degas bottle.

64. Connect the negative battery cable.

65. Fill the crankcase with clean engine oil.

66. Start the engine. Check for leaks and proper operation.

Piston and Ring

POSITIONING

A- Oil spacer
B- Oil ring
C- Compression ring

Piston ring end-gap spacing—4.0L and 4.6L engines

FUEL SYSTEM

Fuel System Service Precautions

Safety is the most important factor when performing not only fuel system maintenance, but any type of maintenance. Failure to conduct maintenance and repairs in a safe manner may result in serious personal injury or death. Work on a vehicle's fuel system components can be accomplished safely and effectively by adhering to the following rules and guidelines.

• To avoid the possibility of fire and personal injury, always disconnect the negative battery cable unless the repair or test procedure requires that battery voltage be applied.

• Always relieve the fuel system pressure prior to disconnecting any fuel system component (injector, fuel rail, pressure regulator, etc.) fitting or fuel line connection. Exercise extreme caution whenever relieving fuel system pressure to avoid exposing

skin, face and eyes to fuel spray. Please be advised that fuel under pressure may penetrate the skin or any part of the body that it contacts.

• Always place a shop towel or cloth around the fitting or connection prior to loosening to absorb any excess fuel due to spillage. Ensure that all fuel spillage is quickly removed from engine surfaces. Ensure that all fuel-soaked cloths or towels are deposited into a flame-proof waste container with a lid.

• Always keep a dry chemical (Class B) fire extinguisher near the work area.

• Do not allow fuel spray or fuel vapors to come into contact with a spark or open flame.

• Always use a second wrench when loosening or tightening fuel line connection fittings. This will prevent unnecessary stress and torsion on fuel piping. Always follow the proper torque specifications.

• Always replace worn fuel fitting O-rings with new ones. Do not substitute fuel hose where rigid pipe is installed.

Fuel System Pressure

RELIEVING

Remove the fuel pump relay; located in the bussed electrical center, relay C051. Start the engine and allow it to run at idle until it stalls. After the engine stalls, crank the engine for approximately five seconds to ensure the pressure in the fuel injection supply manifold has been released. After the fuel pressure has been fully released, turn the ignition switch to the "OFF" position. To pressurize the system, install the fuel pump relay, turn the key on for 3 seconds, then turn the key off. Check for fuel leaks.

➡This procedure will remove the fuel pressure from the lines, but not the fuel. Take precautions to avoid the risk of fire and use clean rags to soak up any spilled fuel when the lines are disconnected.

Fuel Filter

REMOVAL & INSTALLATION

➡Always replace fuel line fitting plastic clips.

1. Relieve the fuel system pressure.
2. Disconnect the negative battery cable.
3. Remove the fuel bundle shield.
4. Disconnect the fuel line fittings.
5. Remove the fuel filter bracket bolt.
6. Remove the fuel filter.

To install:

7. Install fuel filter. Note the flow direction arrow.
8. Secure the fuel filter in the bracket clamp and tighten the bolt to 44 inch lbs. (5 Nm).

9. Install fuel line fittings.
10. Install the fuel bundle shield and tighten the retainers to 62 inch lbs. (7 Nm).
11. Connect the negative battery cable.
12. Start the engine and check for leaks.

Fuel Pump

REMOVAL & INSTALLATION

1. Relieve the fuel system pressure.
2. Disconnect the negative battery cable.
3. Remove the gas cap.
4. Drain the fuel tank.
5. Remove the rear seat bottom cushion.
6. Remove the fuel pump access cover.
7. Disconnect fuel pump module electrical connector.
8. Use tool 303-D006 and remove fuel pump module locking ring.
9. Raise the fuel pump module to access quick connect fitting. Disconnect the fitting and remove the fuel pump module from the fuel tank.

To install:

10. Apply clean engine oil to the fuel pump o-ring seal.
11. Connect the quick connect fitting.
12. Install the fuel pump module into the opening and ensure the alignment arrows on the fuel pump and tank are aligned.
13. Tighten the fuel pump locking ring.
14. Connect fuel pump module electrical connector.
15. Install the fuel pump access cover.
16. Install the rear seat bottom cushion.
17. Add fuel (10 gallons minimum) to the tank.
18. Install the gas cap.
19. Connect the negative battery cable.
20. Start the engine and check for leaks.

Fuel Rail and Injector

REMOVAL & INSTALLATION

4.0L Engine

1. Relieve the fuel system pressure.
2. Disconnect the negative battery cable.
3. On the right side, remove the wiring harness retainer from the valve cover and position aside.
4. Disconnect the injector connectors.
5. Remove the 2 fuel rail supply tube bolts.
6. Remove 2 bolts and the right side fuel rail.
7. On the left side, disconnect the fuel pressure and temperature sensor vacuum hose.
8. Disconnect the fuel pressure and temperature sensor connectors.
9. Remove the wiring harness retainer from the valve cover and position aside.
10. Disconnect the injector connectors.
11. Remove the 2 fuel rail supply tube bolts.
12. Remove 2 bolts and the left side fuel rail.
13. On both sides, remove the fuel injectors from the fuel rail and discard the o-ring seals.

To install:

14. Install new O-rings on the injectors and fuel supply tube.
15. Install the injectors.
16. Install 2 bolts and the left side fuel rail. Tighten the bolts to 17 ft. lbs. (23 Nm).
17. Install the 2 fuel rail supply tube bolts and tighten to 53 inch lbs. (6 Nm).
18. Connect the injector connectors.
19. Install the wiring harness retainer to the valve cover.

8 Fuel supply tube	19 Tube spacers
9 Vapor tube	20 Fuel tube bundle pin-type retainer clip (2 required)
10 Vapor tube bracket retaining bolt	21 Fuel tube bundle retainer clip
11 Fuel tube bundle retainer clip	22 Fuel tube bundle pin-type retainer clip (2 required)
12 Fuel supply tube quick connect coupling	23 Fuel bundle shield retainer bolt (2 required)
13 Fuel filter	24 Shield retainer nuts (2 required)
14 Fuel filter bracket	25 Fuel bundle shield pin-type retainer
15 Fuel filter bracket retaining bolt	26 Fuel bundle shield retainer nuts (2 required)
16 Fuel supply tube quick connect coupling	27 Fuel bundle shield
17 Fuel supply tube quick connect coupling	
18 Fuel tube bundle retainer clip (2 required)	

06017-TANG-G57

Exploded view of fuel filter and fuel lines—Mustang

10 - 23 Nm (17 lb-ft)
6 Nm (53 lb-in)
6 Nm (53 lb-in)
6 Nm (53 lb-in)
6 Nm (53 lb-in)
10 Nm (89 lb-in) - 1

1 Lower fuel rail supply tube bracket bolt

2 Fuel rail supply tube quick connect coupling

3 Upper fuel rail supply tube bracket bolt

4 Fuel rail supply tube bolts (4 required)

5 O-ring seals (2 required)

6 Fuel rail supply tube

7 Fuel rail pressure and temperature sensor vacuum hose

8 Fuel rail pressure and temperature sensor electrical connector

9 Fuel injector electrical connector (6 required)

10 Fuel rail bolt (4 required)

11 RH fuel rail

12 LH fuel rail

13 Fuel injector (6 required)

14 O-ring seal (12 required)

06017-TANG-G58

Exploded view of fuel rail and injectors—4.0L engine

20. Connect the fuel pressure and temperature sensor connectors.

21. Connect the fuel pressure and temperature sensor vacuum hose.

22. Install 2 bolts and the right side fuel rail. Tighten the bolts to 17 ft. lbs. (23 Nm).

23. Install the 2 fuel rail supply tube bolts and tighten to 53 inch lbs. (6 Nm).

24. Install the wiring harness retainer to the valve cover.

25. Connect the negative battery cable.

26. Start the engine and check for leaks.

4.6L Engine

1. Relieve the fuel system pressure.

2. Disconnect the negative battery cable.

3. Disconnect the fuel supply line.

4. Disconnect 2 wiring harness retain-

1 Wiring harness retainer (2 required)

2 Fuel rail stud bolt (4 required)

3 Fuel rail

4 Vacuum hose

5 Fuel rail pressure and temperature sensor electrical connector

6 Fuel injector electrical connector (8 required)

7 O-ring seal (16 required)

8 Fuel injector retaining clip (8 required)

9 Fuel injector

10 Nm (89 lb-in)

06017-TANG-G59

Exploded view of fuel rail and injectors—4.6L engine

ers from the fuel rail stud bolts and position the harness aside.

5. Disconnect the fuel pressure and temperature sensor connectors and vacuum hose.

6. Disconnect the injector connectors.

7. Remove the fuel rail stud bolts.

8. Remove the fuel rail; and injectors as an assembly.

9. Remove the injectors from the fuel rail.

To install:

10. Install new O-rings.

11. Install fuel injectors to the fuel rail.

12. Install the fuel rail with injectors attached. Tighten the bolts to 89 inch lbs. (10 Nm).

13. Connect the injector connectors.

14. Connect the fuel pressure and temperature sensor connectors and vacuum hose.

15. Connect 2 wiring harness retainers from the fuel rail stud bolts and position the harness aside.

16. Connect the fuel supply line.

17. Connect the negative battery cable.

18. Start the engine and check for leaks.

DRIVE TRAIN

Transmission

REMOVAL & INSTALLATION

Manual

T50D 5-SPEED

➡**The transmission and clutch housing are removed as an assembly.**

1. Disconnect the negative battery cable.
2. Raise the vehicle on a hoist.
3. Index mark the driveshaft to the output shaft flange.
4. Remove the driveshaft from the output shaft and wire it aside.
5. Disconnect the shift linkage.
6. Disconnect all transmission electrical connectors.
7. Remove the starter.
8. Support the transmission with a transmission jack.
9. Remove 4 crossmember nuts and the transmission mount nut.
10. Move the crossmember down and slide the exhaust isolators from the brackets.
11. Remove the crossmember.
12. Separate the clutch slave cylinder line.
13. Loosen 4 exhaust manifold nuts.
14. Remove the transmission-to-engine bolts.
15. Slide the transmission rearward until it clears the input shaft, then lower and remove the transmission.

To install:

16. Reverse the removal procedure, niting the following:
    a. Tighten the transmission-to-engine bolts to 33 ft. lbs. (45 Nm).
    b. Tighten the exhaust manifold nuts to 30 ft. lbs. (40 Nm).
    c. Tighten the crossmember bolts to 46 ft. lbs. (62 Nm).
    d. Tighten the starter bolts to 18 ft. lbs. (25 Nm).
    e. Tighten the driveshaft bolts to 76 ft. lbs. (103 Nm).
17. Connect the negative battery cable.

TR3650 5-SPEED

1. Disconnect the negative battery cable.
2. Raise the vehicle on a hoist.
3. Remove the catalytic converter H pipe.
4. Index mark the driveshaft to the output shaft flange.

5. Remove the driveshaft from the output shaft and wire it aside.
6. Disconnect all transmission electrical connectors.
7. Remove the starter.
8. Support the transmission with a transmission jack.
9. Remove the crossmember-to-transmission bolt and crossmember-to-frame bolts.
10. Remove the crossmember.
11. Disconnect the shift linkage.
12. Separate the clutch slave cylinder line.
13. Remove the transmission-to-engine bolts.
14. Slide the transmission rearward until it clears the input shaft, then lower and remove the transmission.

To install:

15. Reverse the removal procedure, niting the following:
    a. Tighten the transmission-to-engine bolts to 33 ft. lbs. (45 Nm).
    b. Tighten the crossmember-to-frame bolts to 46 ft. lbs. (62 Nm).
    c. Tighten the crossmember-to-transmission bolts to 52 ft. lbs. (70 Nm).
    d. Tighten the starter bolts to 18 ft. lbs. (25 Nm).
    e. Tighten the driveshaft bolts to 76 ft. lbs. (103 Nm).
16. Connect the negative battery cable.

Automatic

1. Disconnect the negative battery cable.
2. Drain the transmission fluid.
3. Remove the driveshaft.
4. Remove the 5 transmission lower mount screws and remove the mount
5. Disconnect shift cable.
6. Position the maual lever out of "P".
7. Remove the shift cable and bracket and position aside.
8. Remove the starter motor.
9. Remove the flexplate cover.
10. Remove the torque converter access cover and remove the 4 torque converter nuts.
11. Remove the lower transmission mounting bolts.
12. Disconnect all transmission electrical connectors.
13. Disconnect and plug transmission fluid cooler lines.
14. Remove the engine-to-transmission bolts.
15. Support the transmission with a jack.

16. Lower the transmission away from the vehicle.
17. Install Torque Converter Retainer tool 307-346 to retain the torque converter.

To install:

18. Rotate the torque converter and adapter plate so the orange or green paint mark is at the 12 o'clock position.
19. Raise the transmission into and install the engine-to-transmission bolts. Tighten the bolts to 35 ft. lbs. (48 Nm).
20. Install the torque converter nuts and tighten to 28 ft. lbs. (38 Nm).
21. Install the access cover.
22. Install the flexplate cover and tighten to 25 ft. lbs. (34 Nm).
23. Install the starter and tighten to 18 ft. lbs. (25 Nm).
24. Install the lower transmission mounting bolts and tighten to 35 ft. lbs. (48 Nm).
25. Install the transmission fluid cooler lines.
26. Connect all transmission electrical connectors.
27. Install the transmission crossmember and tighten the center bolt to 30 ft. lbs. (40 Nm).
28. Install the transmission crossmember outer bolts and tighten to 41 ft. lbs. (55 Nm).
29. Remove the transmission jack.
30. Install the driveshaft.
31. Install the shift cable and bracket.
32. Connect the negative battery cable.
33. Fill the transmission with fluid.
34. Check the shift cable adjustment.
35. Start the engine. Check for leaks and proper operation.

Clutch

ADJUSTMENTS

The clutch is equipped with a self-adjusting mechanism. Pull the clutch pedal up to activate the adjuster.

REMOVAL & INSTALLATION

1. Disconnect the negative battery cable.
2. Remove the transmission.
3. Remove pressure plate bolts, loosen them evenly in several passes to avoid distortion of the pressure plate.
4. Remove pressure plate and clutch disk.

To install:

5. Install clutch disk and pressure plate.

Tighten the pressure plate bolts in a star pattern evenly to 26 ft. lbs. (35 Nm) for 4.0L engines or to 33 ft. lbs. (45 Nm) plus an additional 60 degrees for 4.6L engines.

6. Install the transmission.
7. Connect the negative battery cable.

Axle Shaft, Bearing and Seal

REMOVAL & INSTALLATION

1. Place the transmission in neutral.
2. Raise the vehicle on a hoist
3. Remove rear wheel.
4. Remove axle housing cover.
5. Remove disc brake caliper and rotor.
6. Remove wheel speed sensor.
7. Remove differential pinion shaft lockbolt.
8. Remove differential pinion shaft.
9. Remove axle retaining U-washer.
10. Remove axle shaft.
11. Remove bearing and seal, using a slide hammer

To install:

12. Coat the new bearing with axle lubricant.
13. Install bearing, so that it is fully seated in the axle tube.
14. Lubricate the seal lip with grease and install axle seal.
15. Install axle shaft.
16. Install axle retaining U-washer.
17. Install differential pinion shaft. Tighten the lockbolt to 22 ft. lbs. (30 Nm).
18. Install axle housing cover.

19. Install wheel speed sensor.
20. Install disc brake rotor and caliper. Tighten the caliper mounting bolts to 25 ft. lbs. (33 Nm).
21. Install rear wheel.
22. Fill the differential with gear lubricant.

Pinion Flange and Seal

REMOVAL & INSTALLATION

1. Place the transmission in neutral.
2. Raise the vehicle on a hoist.
3. Index mark the driveshaft, then disconnect driveshaft and position it out of the way.
4. Remove rear wheels.
5. Remove rear brake calipers.

➡**The rear brake calipers must be removed so that there is no additional drag when measuring pinion bearing preload.**

6. Use an inch lb. torque wrench and measure the amount of torque required to maintain pinion rotation through several revolutions.
7. Hold the pinion flange from turning, then remove the pinion nut.
8. Index mark the pinion flange, then use a 2-jaw puller and remove the flange.
9. Remove the pinion seal.

To install:

10. Lubricate the new seal with grease and install the seal.

11. Lubricate the pinion flange splines with grease.
12. Align the index marks and install the pinion flange.
13. Install new pinion flange nut loosely.
14. Rotate the pinion flange occasionally while tightening the flange nut to make sure the pinion bearings seat correctly.
15. Take frequent bearing preload torque readings.
16. If the preload recorded prior to disassembly is **lower** than the specification for used bearings, then tighten the pinion flange nut to specification. If the preload recorder prior to disassembly is **higher** than the specification for used bearings, then tighten the pinion flange nut to the original reading as recorded.
17. The pinion bearing preload specification is 16–29 inch lbs. (1.8–3.3 Nm).

✳✳ CAUTION

Never loosen the pinion nut to reduce bearing preload. If it is necessary to reduce bearing preload, install a new collapsible spacer and pinion nut.

18. Connect the driveshaft and install new bolts. Tighten the bolts to 76 ft. lbs. (103 Nm) on 7.5 inch gears, and 41 ft. lbs. (55 Nm) on 8.8 inch gears.
19. Install brake calipers.
20. Install rear wheels.
21. Fill the differential with gear lubricant and check for leaks.

STEERING AND SUSPENSION

Air Bag

✳✳ CAUTION

Vehicles are equipped with an air bag system. The system must be disarmed before performing service on, or around, system components, the steering column, instrument panel components, wiring and sensors. Failure to follow the safety precautions and the disarming procedure could result in accidental air bag deployment, possible injury and unnecessary system repairs.

PRECAUTIONS

Several precautions must be observed when handling the inflator module to avoid accidental deployment and possible personal injury.

• Never carry the inflator module by the wires or connector on the underside of the module.
• When carrying a live inflator module, hold securely with both hands, and ensure that the bag and trim cover are pointed away.
• Place the inflator module on a bench or other surface with the bag and trim cover facing up.
• With the inflator module on the bench, never place anything on or close to the module that may be thrown in the event of an accidental deployment.

DISARMING THE SYSTEM

1. Ensure the ignition and all accessories are off.
2. Remove the Smart Junction Box (SJB) cover located below the left side of

the instrument panel. Remove the Restraints Control Module (RCM) fuse no. F2.17.
3. Turn the ignition on and watch the air bag indicator for 30 seconds. The indicator light will remain lit constantly if the correct fuse has been removed. If the light is not on steadily, remove the correct fuse and check the light again.
4. Turn the ignition switch off.
5. Disconnect the negative battery cable.

ARMING THE SYSTEM

1. Turn the ignition switch on.
2. Install Restraints Control Module (RCM) fuse no. F2.17 and the fuse cover.
3. Turn the ignition off.
4. Connect the negative battery cable.
5. Turn the ignition on and then off. Wait 10 seconds and turn the key back on. Watch the air bag indicator. The indicator light will remain lit constantly for 6 seconds and then

go off. If the indicator does not turn on and then off, diagnose the air bag system

6. Clear all Diagnostic Trouble Codes (DTC) using a diagnostic tool.

Power Rack and Pinion Steering Gear

REMOVAL & INSTALLATION

1. Place the steering wheel in the straight ahead position.
2. Disconnect the negative battery cable.
3. Remove front wheels.
4. Disconnect the tie rod ends from the knuckle.
5. Disconnect steering column intermediate shaft coupling.
6. Disconnect power steering pressure and return lines.
7. Remove steering gear retaining bolts.
8. Remove steering gear.

To install:

9. Install steering gear. Tighten the bolts to 85 ft. lbs. (115 Nm).

Using new o-ring seals install the pressure and return lines. Tighten the pressure line to 80 inch lbs. (9 Nm), and the return line to 17 ft. lbs. (23 Nm).

10. Connect steering column intermedi-

ate shaft coupling. Tighten the pinch bolt to 18 ft. lbs. (25 Nm).

11. Connect the tie rod ends. Tighten the nuts to 59 ft. lbs. (80 Nm).
12. Install front wheels.
13. Connect the negative battery cable.
14. Fill the power steering system with the proper type and quantity of fluid.
15. Check the front end alignment and adjust as necessary.

Strut and Spring

REMOVAL & INSTALLATION

Front

1. Support the front of the vehicle on jackstands placed under the control arms.
2. Remove front wheel.
3. Remove the ABS sensor.
4. Disconnect the ABS wire from the bracket.
5. Remove the stabilizer bar upper link nut and discard.
6. Disconnect the link.
7. Place a jack under the lower control arm.
8. Index mark the lower strut mounting bolts, then remove the bolts.

9. Carefully lower the lower control arm.
10. Remove the upper strut-to-body bolts.
11. Remove the strut and spring assembly.

To install:

12. If replacing the strut, transfer the upper mounting bracket.
13. Install strut assembly. Tighten the upper nuts to 26 ft. lbs. (35 Nm).
14. Tighten the lower strut bolts to 148 ft. lbs. (200 Nm).
15. Connect the stabilizer link and tighten the new nut to 85 ft. lbs. (115 Nm).
16. Connect the ABS wire to the bracket.
17. Install the ABS sensor.
18. Install front wheel.
19. Remove the jackstands and check the alignment. Adjust as necessary.

DISASSEMBLY & REASSEMBLY

1. Use a spring compressor and compress the spring until tension is released from the strut.

1 Strut upper nut and washer
2 Strut
3 Dust boot
4 Jounce bumper
5 Upper bearing assembly
6 Spring

06017-TANG-G61

Exploded view of front strut and spring assembly—Mustang

1 Tie-rod end nut (2 required)
2 Steering column coupling-to-steering gear bolt
3 Steering column coupling
4 Power steering line clamp plate bolt
5 Power steering lines (pressure/return)
6 Pressure line O-ring seal
7 Return line O-ring seal
8 Pressure line bracket-to-crossmember bolts (2 required)
9 Steering gear bolts (2 required)
10 Steering gear

06017-TANG-G60

Exploded view of power rack and pinion steering gear mounting—Mustang

2. While holding the strut rod, remove the upper strut nut.

3. Remove the strut, dust boot, jounce bumper, washer and upper bearing.

4. Release tension on the spring and remove the spring.

5. Compress the spring.

6. Align the notch on the upper bearing with the clevis at the bottom of the strut.

7. Install the washer, jounce bumper, dust boot and strut.

8. Install a new upper strut nut and tighten to 46 ft. lbs. (62 Nm).

9. Release the tension on the spring.

Shock Absorber

REMOVAL & INSTALLATION

Rear

1. Pull aside the rear compartment carpet.

2. Support the rear axle with jack stands.

3. Remove upper shock absorber retaining nut.

4. If equipped with a stabilizer bar, remove the stabilizer bar link bolts.

5. Remove lower shock absorber bolt.

6. Remove shock absorber.

To install:

7. Install the shock absorber and tighten the lower bolt to 85 ft. lbs. (115 Nm).

8. If equipped with a stabilizer bar, install new stabilizer bar link bolts and tighten to 85 ft. lbs. (115 Nm).

9. Install upper shock absorber nut. Tighten it to 30 ft. lbs. (40 Nm).

10. Remove the jack stands.

11. Reinstall the carpet.

Coil Spring

REMOVAL & INSTALLATION

Rear

1. Raise the vehicle on a hoist.

2. Use a hi-lift jack and support the rear axle.

3. If equipped with a stabilizer bar, remove the stabilizer bar link bolts.

4. Remove lower shock absorber bolt.

5. Remove the brake hose bracket bolt.

6. Use the jack to lower the rear axle until the spring can be removed.

To install:

7. Place the upper spring insulator on top of the spring. Place the lower spring insulator on the lower arm.

| 1 | Upper nut, washer and insulator | 3 | Shock flag nut |
| 2 | Shock lower bolt | 4 | Shock |

06017-TANG-G62

Exploded view of the rear shock absorber mounting—Mustang

1	Spring (2 required)	8	Lower control arm (2 required)
2	Upper insulator (2 required)	9	Upper control arm front bolt
3	Lower insulator (2 required)	10	Upper control arm rear bolt (2 required)
4	Lower control arm rear bolt (2 required)	11	Upper control arm rear nut
5	Lower control arm rear nut (2 required)	12	Upper control arm flag bolt
6	Lower control arm front bolt (2 required)	13	Upper control arm
7	Lower control arm front nut (2 required)	14	Upper control arm bushing
		15	Rear axle assembly

06017-TANG-G63

Exploded view of the rear coil spring mounting—Mustang

8. Position the coil spring on the lower arm spring seat with the tag facing toward the rear axle.

9. Slowly raise the jack until the spring is in place.

10. Install the brake hose bracket bolt.

11. Install the lower shock absorber bolts and tighten to 85 ft. lbs. (115 Nm).

12. If equipped with a stabilizer bar, install new stabilizer bar link bolts and tighten to 85 ft. lbs. (115 Nm).

13. Remove the jack.

14. Lower the vehicle.

Stabilizer Bar

REMOVAL & INSTALLATION

Front and Rear

1. Raise the vehicle on a hoist

2. Remove the appropriate wheel wheel.

3. Remove and discard the stabilizer bar link bolts and nuts.

4. Remove and discard the stabilizer bar clamp bolts.

5. Remove the stabilizer bar.

To install:

6. Install the stabilizer bar.

7. Install new stabilizer bar clamp bolts and tighten to 52 ft. lbs. (70 Nm).

8. Install the stabilizer bar link and tighten the new bolts and nuts to 85 ft. lbs. (115 Nm).

9. Install the appropriate wheel.

10. Lower the vehicle.

Upper Control Arm

REMOVAL & INSTALLATION

Rear

1. Before raising the vehicle, match-mark the shock absorbers to indicate curb height.

2. Remove the rear seat cushion.

3. Remove and discard the upper control arm front bolt.

4. Raise the vehicle on a hoist.

5. Place a hi-lift jack under the fuel tank.

6. Remove the fuel tank rear support strap bolts and move the straps aside.

7. Partially lower the fuel tank.

8. Remove and discard the upper control arm rear bolts.

9. Remove and discard the control arm bushing flag bolt and nut.

10. Remove the control arm.

To install:

11. Install the control arm.

12. Install the new control arm bushing flag bolt and nut and tighten to 129 ft. lbs. (175 Nm).

13. Install new upper control arm rear bolts and tighten to 85 ft. lbs. (115 Nm).

14. Raise the fuel tank.

15. Attach the fuel tank straps.

16. Raise the rear axle until the shock absorbers are at curb height.

17. Install the upper control arm front bolt and tighten to 129 ft. lbs. (175 Nm).

18. Install the rear seat cushion.

CONTROL ARM BUSHING REPLACEMENT

Rear

1. Remove the upper control arm.

2. Press the bushing out of the housing using bushing remover tools 204-034 and 303-277.

3. Position the new bushing so the voids are at the top and bottom of the housing.

4. Press in the new bushing using the special tools.

5. Position the bushing so there is 0.47 inch (12mm) between the bushing flange and the axle bracket.

6. Press a new bushing into the axle housing and install the upper control arm.

7. Install the control arm.

Lower Control Arm

REMOVAL & INSTALLATION

Front

1. Remove the front wheel.

2. Raise the vehicle on a hoist.

3. Remove the pinch bolt and nut.

4. Separate the control arm from the wheel spindle.

5. Remove the steering gear mounting bolts and position the steering away enough to access the lower control arm front bolt.

6. Remove the front bolt and nut.

7. Remove the rear control arm nuts and flag bolts.

8. Remove the lower control arm and bracket.

To install:

9. Install the lower control arm and bracket.

10. Install the rear control arm nuts and flag bolts and tighten to 129 ft. lbs. (175 Nm).

11. Install the front bolt and nut and tighten to 129 ft. lbs. (175 Nm).

1	Lower control arm forward bolt
2	Lower control arm forward
3	Heat shield bolt (3 required)
4	Heat shield
5	Lower control arm rearward nut (2 required)
6	Lower control arm rearward flag bolt (2 required)
7	Lower control arm (LH/RH)

06017-TANG-G64

Exploded view of the front lower control arm—Mustang

12. Install the steering gear mounting bolts and tighten to 85 ft. lbs. (115 Nm).

13. Attach the control arm to the wheel spindle.

14. Install the pinch bolt and nut and tighten the bolt to 76 ft. lbs. (103 Nm).

15. Install front wheel.

16. Check the front end alignment.

Rear

1. Before raising the vehicle, match-mark the shock absorbers to indicate curb height.

2. Raise the vehicle on a hoist.

3. Remove the rear wheel.

4. Remove the parking brake cable from the caliper.

5. Remove the parking brake cable bracket.

6. Support the rear axle with a hi-lift jack.

7. Remove the control arm front and rear bolts and nuts.

8. Remove the control arm

To install:

➡Use new nuts and bolts.

9. Install the control arm

10. Install the control arm front and rear bolts and nuts and tighten to 129 ft. lbs. (175 Nm).

11. Install the parking brake cable bracket.
12. Install the parking brake cable from the caliper.
13. Install the rear wheel.
14. Lower the vehicle.

CONTROL ARM BUSHING REPLACEMENT

Front and Rear

The lower control arm bushings are serviced with the lower control arm as an assembly.

Wheel Bearings

ADJUSTMENT

The front wheel bearings are an integral part of the hub assembly. They require no periodic maintenance or adjustment. If the bearings are found to be defective, they must be replaced along with the hub assembly.

The rear wheel bearings are not adjustable.

REMOVAL & INSTALLATION

Front

1. Remove front wheel.
2. Remove disc brake caliper and rotor.
3. Remove grease cap.
4. Remove spindle retainer nut.
5. Remove hub and bearing assembly.

To install:
6. Install hub and bearing assembly. Tighten the spindle retainer nut to 221 ft. lbs. (300 Nm).

7. Install grease cap.
8. Install disc brake caliper and rotor.
9. Install front wheel
10. Check the wheel alignment and adjust as necessary.

Wheel Spindle

REMOVAL & INSTALLATION

Front

1. Place the transmission in neutral.
2. Raise the vehicle on a hoist.
3. Remove the front wheel.
4. Remove the front wheel hub.
5. Remove the brake disc shield.
6. Remove the ABS sensor.
7. Remove and discard the outer tie rod end nut.
8. Press the tie rod end from the wheel spindle.
9. Place a jack under the lower control arm.
10. Remove the control arm pinch bolt and nut.
11. Separate the control arm from the spindle.
12. Index mark the spindle-to-strut bolts, then remove the bolts.
13. Remove the wheel spindle.

To install:
14. Install the wheel spindle.
15. Install the spindle-to-strut bolts and tighten to 148 ft. lbs. (200 Nm).
16. Attach the control arm to the spindle.
17. Install the control arm pinch bolt and nut and tighten to 76 ft. lbs. (103 Nm).
18. Attach the tie rod end to the wheel spindle.

80 Nm (59 lb-ft)
103 Nm (76 lb-ft)
200 Nm (148 lb-ft)

1 Wheel speed sensor bolt
2 Wheel speed sensor
3 Outer tie rod end nut
4 Wheel spindle pinch bolt
5 Wheel spindle pinch nut
6 Strut-to-wheel spindle bolt (2 required)
7 Strut-to-wheel spindle flag nut (2 required)
8 Wheel spindle (RH/LH)

06017-TANG-G65

Exploded view of the front wheel spindle—Mustang

19. Install the outer tie rod end nut and tighten to 59 ft. lbs. (80 Nm).
20. Install the ABS sensor.
21. Install the brake disc shield.
22. Install the front wheel hub.
23. Install the front wheel.
24. Lower the vehicle.

BRAKES

Brake Caliper

REMOVAL & INSTALLATION

Front

1. Remove the wheel and tire assembly.
2. Mark the disc brake caliper to ensure that it will be installed to the same side if both calipers are being removed.
3. Disconnect the caliper flow bolt and discard 2 copper sealing washers.
4. Remove lower brake caliper locating pin, if equipped.
5. Remove brake caliper mounting bolts and discard.

6. Rotate the brake caliper approximately 90° away from the anchor assembly.
7. Remove the disc brake caliper.

To install:
8. If required, compress the caliper piston(s) using a C-clamp and an old disc brake pad or block of wood.
9. Position the brake caliper assembly into place on the anchor plate and tighten the bolts to 25 ft. lbs. (34 Nm).
10. Connect the caliper flow bolt using new copper sealing washers and tighten the bolt to 52 ft. lbs. (70 Nm).
11. Install the wheel and tire assembly.
12. Pump the brake pedal several times to position the brake pads before the vehicle is moved.

13. Road test the vehicle and check for proper brake system operation.

Rear

1. Remove wheel and tire assembly.
2. Remove the retaining clip and the parking brake cable and conduit from the caliper.
3. Disconnect the caliper flow bolt and discard 2 copper sealing washers.
4. Remove brake caliper mounting bolts and discard.
5. Rotate the brake caliper approximately 90° away from the anchor assembly.
6. Remove the disc brake caliper.

To install:
7. If required, compress the caliper pis-

1 Brake caliper bolts (2 required)

2 Brake caliper RH/LH

3 Brake pads kit

4 Spring clips (4 required)

5 Brake caliper anchor plate bolts (2 required)

6 Brake caliper anchor plate

7 Guide pins

8 Brake disc

9 Brake disc shield bolts (3 required)

10 Brake disc shield

11 Brake caliper flow bolt

12 Copper washers (2 required)

13 Brake line fitting

14 Brake flexible hose bracket bolts

15 Brake flexible hose RH/LH

16 Bleeder screw cap (part of 2208)

17 Bleeder screw

06017-TANG-G66

Exploded view of the front disc brake assembly—Mustang

ton(s) using a C-clamp and an old disc brake pad or block of wood.

8. Position the brake caliper assembly into place on the anchor plate and tighten the bolts to 24 ft. lbs. (33 Nm).

9. Connect the caliper flow bolt using new copper sealing washers and tighten the bolt to 52 ft. lbs. (70 Nm).

10. Install the retaining clip and the parking brake cable and conduit to the caliper.

11. Install the wheel and tire assembly.

12. Pump the brake pedal several times

to position the brake pads before the vehicle is moved.

13. Road test the vehicle and check for proper brake system operation.

Disc Brake Pads

REMOVAL & INSTALLATION

Front

1. Remove and discard ½ of the brake fluid from the brake master cylinder reser-

voir. Properly dispose of the used brake fluid.

2. Remove the wheel and tire assembly.

3. Remove disc brake caliper assembly from anchor plate. Hang the disc brake caliper from the body with wire. Do not let the disc brake caliper hang by the brake hose.

4. Remove the outer and inner disc brake pads from the disc brake caliper and discard the spring clips.

5. Clean any residue from the caliper anchor plate and disc brake pad contact areas.

1   Brake caliper bolts

2   Brake caliper (RH/LH)

3   Brake pads (kit)

4   Spring clips

5   Guide pin and boot

6   Brake caliper anchor plate bolt kit (2 bolts each side)

7   Brake caliper anchor plate

8   Brake disc

9   Brake disc shield bolts (3 required)

10   Brake disc shield

11   Brake line fitting

12   Brake caliper flow bolt

13   Copper washers (2 required)

14   Brake flexible hose (RH/LH)

15   Brake flexible hose bracket bolt

16   Bleeder screw cap

17   Bleeder screw

18   Anti-moan bracket U-bolt nuts (2 required)

19   Anti-moan bracket U-bolt and clamp

20   Anti-moan bracket

21   Brake caliper support bracket bolts (4 required)

22   Brake caliper support bracket flag nuts (4 required)

23   Brake caliper support bracket

Exploded view of the rear disc brake assembly—Mustang

06017-TANG-G67

6. Inspect the disc brake rotor for scoring and wear. Replace or machine, as necessary.

7. Inspect the piston boots and the caliper pin boots for damage. Replace as necessary.

To install:

8. Use a large C-clamp and a wood block to push the caliper pistons back into their bores.

9. Remove the protective paper from the insulators on the back of the disc brake pads, if equipped.

10. Install the inner and outer disc brake pads in the disc brake caliper. Make sure the pads are seated properly.

11. Install new spring clips.

12. Install the disc brake caliper and pads.

13. Install the wheel and tire assembly.

14. Repeat the procedure for the opposite disc brake caliper assembly.

15. Pump the brake pedal prior to moving the vehicle to position the brake pads.

16. Fill the brake master cylinder reservoir with clean DOT 3 brake fluid from a closed container.

17. If the disc brake calipers were replaced or repaired, be sure to bleed the system.

18. Road test the vehicle and check the brake system for proper operation.

Rear

1. Before servicing the vehicle, refer to the precautions at the beginning of this section.

2. Remove ½ of the brake fluid from the brake master cylinder reservoir. Properly dispose of the used brake fluid.

3. Remove wheel and tire assembly.

4. Remove the rear disc brake caliper. Hang the caliper from the body with wire. Do not let the disc brake caliper hang by the brake hose.

5. Remove inner and outer disc brake pads and the spring clips from the disc support bracket.

6. Inspect the disc brake rotor for scoring and wear. Replace or machine, as necessary.

To install:

7. Using Rear Caliper Piston Adjuster 206-026, rotate the caliper piston clockwise until the piston is fully seated.

8. Install disc brake pads and new spring clips.

9. Position the notch in the caliper piston up and down to align with the alignment pin on the brake pad.

10. Install the rear disc brake caliper and tighten to 24 ft. lbs. (33 Nm).

11. Pump the brake pedal prior to moving the vehicle to position the brake pads.

12. Fill the brake master cylinder reservoir with clean DOT 3 brake fluid from a closed container.

13. If the disc brake calipers were replaced or repaired, be sure to bleed the system.

14. Road test the vehicle and check the brake system for proper operation.

FORD

Ranger

14

SPECIFICATIONS AND MAINTENANCE CHARTS

ENGINE AND VEHICLE IDENTIFICATION

	Engine						Model Year	
Code ①	Liters (cc)	Cu. In.	Cyl.	Fuel Sys.	Type	Eng. Mfg.	Code ②	Year
D	2.3 (2261)	138	4	MFI	DOHC	Ford	2	2002
U	3.0 (2999)	183	6	MFI	OHV	Ford	3	2003
E	4.0 (4000)	244	6	MFI	SOHC	Ford	4	2004
							5	2005

MFI: Multi-port Fuel Injection

OHV: Overhead Valve

DOHC: Dual Overhead Camshafts

SOHC: Single Overhead Camshaft

① 8th digit of the Vehicle Identification Number (VIN)

② 10th digit of the Vehicle Identification Number (VIN)

06017-RANG-C01

GENERAL ENGINE SPECIFICATIONS

Year	Model	Engine Displacement Liters	Engine VIN	Net Horsepower @ rpm	Net Torque @ rpm (ft. lbs.)	Bore x Stroke (in.)	Com-pression Ratio	Oil Pressure @ rpm
2002	Ranger	2.3	D	143@5200	154@3750	3.44x3.70	9.7:1	29-39@3000
	Ranger	3.0	U	147@5000	147@5000	3.50x3.14	9.3:1	40-60@2500
	Ranger	4.0	E	160@4000	225@2500	3.81x3.39	9.0:1	40-60@2000
2003	Ranger	2.3	D	143@5200	154@3750	3.44x3.70	9.7:1	29-39@3000
	Ranger	3.0	U	147@5000	147@5000	3.50x3.14	9.3:1	40-60@2500
	Ranger	4.0	E	160@4000	225@2500	3.81x3.39	9.0:1	40-60@2000
2004	Ranger	2.3	D	143@5200	154@3750	3.44x3.70	9.7:1	29-39@3000
	Ranger	3.0	U	147@5000	147@5000	3.50x3.14	9.3:1	40-60@2500
	Ranger	4.0	E	160@4000	225@2500	3.81x3.39	9.0:1	40-60@2000
2005	Ranger	2.3	D	143@5200	154@3750	3.44x3.70	9.7:1	29-39@3000
	Ranger	3.0	U	148@4900	147@5000	3.50x3.14	9.6:1	40-60@2500
	Ranger	4.0	E	207@5250	238@3000	3.81x3.39	9.7:1	40-60@2000

06017-RANG-C02

ENGINE TUNE-UP SPECIFICATIONS

Year	Engine Displacement Liters	Engine VIN	Spark Plug Gap (in.)	Ignition Timing (deg.) ①		Fuel Pump (psi)	Idle Speed (rpm)		Valve Clearance	
				MT	AT		MT	AT	In.	Ex.
2002	2.3	D	0.041-0.045	10B	10B	64-72	①	①	HYD	HYD
	3.0	U	0.042-0.046	10B	10B	64-72	①	①	HYD	HYD
	4.0	E	0.062-0.068	10B	10B	64-72	①	①	HYD	HYD
2003	2.3	D	0.041-0.045	10B	10B	64-72	①	①	HYD	HYD
	3.0	U	0.042-0.046	10B	10B	64-72	①	①	HYD	HYD
	4.0	E	0.062-0.068	10B	10B	64-72	①	①	HYD	HYD
2004	2.3	D	0.041-0.045	10B	10B	64-72	①	①	HYD	HYD
	3.0	U	0.042-0.046	10B	10B	64-72	①	①	HYD	HYD
	4.0	E	0.062-0.068	10B	10B	64-72	①	①	HYD	HYD
2005	2.3	D	0.041-0.045	10B	10B	60-65	①	①	HYD	HYD
	3.0	U	0.042-0.046	10B	10B	60-65	①	①	HYD	HYD
	4.0	E	0.062-0.068	10B	10B	60-65	①	①	HYD	HYD

NOTE: The Vehicle Emission Control Information label often reflects specification changes changes made during production. The label figures must be used if they differ from those in this chart.

B: Before top dead center

HYD: Hydraulic

NA: Not Available

① Electronically controlled and cannot be adjusted

06017-RANG-C03

1 Fan pulley

2 Power steering pump pulley

3 Water pump pulley

4 Crankshaft pulley

5 Belt tensioner pulley

6 Generator pulley

7 A/C clutch pulley

8 Drive belt

67197-RANG-G97

Accessory drive belt routing—2.3L with A/C

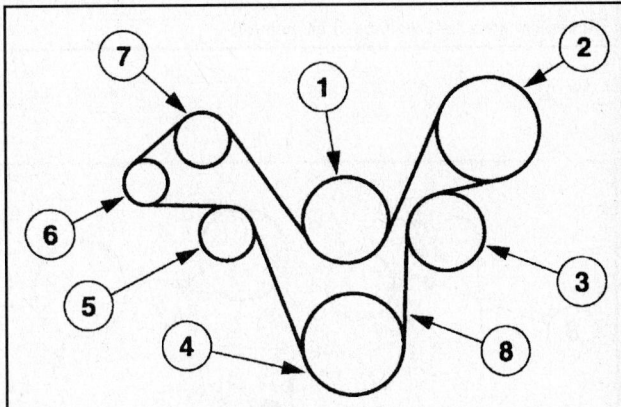

1 Fan pulley

2 Power steering pump pulley

3 Water pump pulley

4 Crankshaft pulley

5 Belt tensioner pulley

6 Generator pulley

7 Belt idler pulley

8 Drive belt

67197-RANG-G98

Accessory drive belt routing—2.3L without A/C

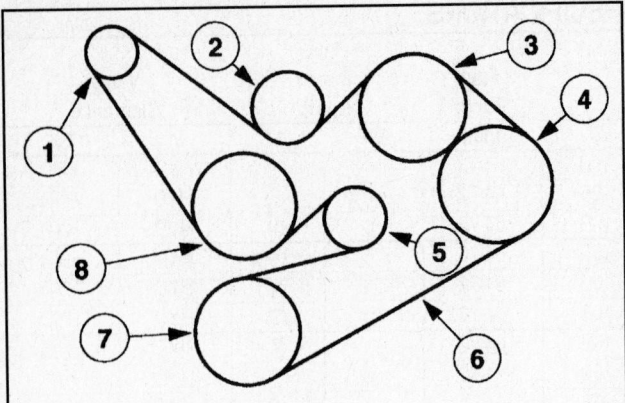

1 Generator pulley

2 Belt idler pulley

3 A/C clutch pulley

4 Power steering pump pulley

5 Belt tensioner

6 Drive belt

7 Crankshaft pulley

8 Water pump pulley

67197-RANG-G99

Accessory drive belt routing—3.0L with A/C

1	Generator pulley
2	Belt idler pulley
3	Power steering pump pulley
4	Drive belt
5	Belt tensioner
6	Crankshaft pulley
7	Water pump pulley

67197-RANG-G00

Accessory drive belt routing—3.0L without A/C

1	Generator pulley
2	Power steering pump pulley
3	A/C clutch pulley
4	Water pump pulley
5	Crankshaft pulley
6	Drive belt tensioner
7	Belt idler pulley
8	Drive belt

67197-RANG-G1A

Accessory drive belt routing—4.0L with A/C

1	Generator pulley
2	Belt idler pulley
3	Drive belt
4	Power steering pump pulley
5	Water pump pulley
6	Crankshaft pulley
7	Belt tensioner

67197-RANG-G2A

Accessory drive belt routing—4.0L without A/C

CAPACITIES

Year	Model	Engine Displacement Liters	Engine VIN	Engine Oil with Filter (qts.)	Transmission (pts.) 5-Spd	Transmission (pts.) Auto.	Transfer Case (pts.)	Drive Axle Front (pts.)	Drive Axle Rear (pts.)*	Fuel Tank (gal.)	Cooling System (qts.)
2002	Ranger	2.3	D	4.0	3.0	19.8	—	—	5.0	①	②
	Ranger	3.0	U	4.5	3.0	③	2.5	2.7	5.0	①	④
	Ranger	4.0	E	5.0	3.0	③	2.5	2.7	5.0	①	⑤
2003	Ranger	2.3	D	4.0	3.0	19.8	—	—	5.0	①	②
	Ranger	3.0	U	4.5	3.0	③	2.5	2.7	5.0	①	④
	Ranger	4.0	E	5.0	3.0	③	2.5	2.7	5.0	①	⑤
2004	Ranger	2.3	D	4.0	3.0	19.8	—	—	5.0	①	②
	Ranger	3.0	U	4.5	3.0	③	2.5	3.25	5.0	①	④
	Ranger	4.0	E	5.0	3.0	③	2.5	3.25	5.0	①	⑤
2005	Ranger	2.3	D	4.0	3.6	19.8	—	—	5.0	⑥	⑦
	Ranger	3.0	U	4.5	3.6	③	2.5	3.60	5.0	⑥	④
	Ranger	4.0	E	5.0	3.6	③	2.5	3.60	5.0	⑥	⑤

NOTE: All capacities are approximate. Add fluid gradually and check to be sure a proper fluid level is obtained.

* For limited slip axles, add 4 oz. of friction modifier, exc. for 1-ton models

① Std: 16.5
  Long Wheelbase: 20.0
  Super Cab: 19.5
② w/MT: 11.2
  w/AT: 10.9
③ 2wd: 20.0
  4wd: 20.6

④ w/MT: 15.2
  w/AT: 14.8
⑤ w/MT: 13.5
  w/AT: 13.2
⑥ Regular Cab short wheel base: 17 gal.
  Regular Cab long wheel base: 20.3 gal.
  SuperCab: 19.5 gal.

⑦ w/MT: 10.5
  w/AT: 10.2

06017-RANG-C04

VALVE SPECIFICATIONS

Year	Engine Displacement Liters	Engine VIN	Seat Angle (deg.)	Face Angle (deg.)	Spring Test Pressure (lbs. @ in.)	Spring Installed Height (in.)	Stem-to-Guide Clearance (in.) Intake	Stem-to-Guide Clearance (in.) Exhaust	Stem Diameter (in.) Intake	Stem Diameter (in.) Exhaust
2002	2.3	D	45	45	①	1.492	0.0009	0.0011	0.2153-0.2159	0.2151-0.2157
	3.0	U	45	44	185@1.16	1.580-1.610	0.0010-0.0027	0.0015-0.0032	0.3126-0.3134	0.3121-0.3129
	4.0	E	45	45	202-225@1.413-1.445	1.569-1.601	0.0010-0.0020	0.0010-0.0030	0.2742-0.2748	0.2736-0.2742
2003	2.3	D	45	45	①	1.492	0.0009	0.0011	0.2153-0.2159	0.2151-0.2157
	3.0	U	45	44	185@1.16	1.580-1.610	0.0010-0.0027	0.0015-0.0032	0.3126-0.3134	0.3121-0.3129
	4.0	E	45	45	202-225@1.413-1.445	1.569-1.601	0.0010-0.0020	0.0010-0.0030	0.2742-0.2748	0.2736-0.2742
2004	2.3	D	45	45	①	1.492	0.0009	0.0011	0.2153-0.2159	0.2151-0.2157
	3.0	U	45	44	185@1.16	1.580-1.610	0.0010-0.0027	0.0015-0.0032	0.3126-0.3134	0.3121-0.3129
	4.0	E	45	45	202-225@1.413-1.445	1.569-1.601	0.0010-0.0020	0.0010-0.0030	0.2742-0.2748	0.2736-0.2742
2005	2.3	D	45	45	①	1.492	0.0009	0.0011	0.2153-0.2159	0.2151-0.2157
	3.0	U	45	44	185@1.16	1.580-1.610	0.0010-0.0027	0.0015-0.0032	0.3126-0.3134	0.3121-0.3129
	4.0	E	45	45	202-225@1.413-1.445	1.569-1.601	0.0010-0.0020	0.0010-0.0030	0.2742-0.2748	0.2736-0.2742

① Intake: 97.0@1.201
  Exhaust: 93.3@1.201

06017-RANG-C05

CRANKSHAFT AND CONNECTING ROD SPECIFICATIONS
All measurements are given in inches.

Year	Engine Displacement Liters	Engine VIN	Crankshaft				Connecting Rod		
			Main Brg. Journal Dia.	Main Brg. Oil Clearance	Shaft End-play	Thrust on No.	Journal Diameter	Oil Clearance	Side Clearance
2002	2.3	D	2.0465-2.2059	0.0007-0.0013	0.0080-0.0160	3	1.9606-1.9685	0.0010-0.0020	0.0767-0.1200
	3.0	U	2.5190-2.2059	0.0010-0.0015	0.0040-0.0080	3	2.1253-2.0472	0.0010-0.0015	0.0060-0.0115
	4.0	E	2.2430-2.2440	0.0008-0.0015	0.0020-0.0125	3	2.7252-2.7260	0.0003-0.0024	0.0036-0.0106
2003	2.3	D	2.0465-2.2059	0.0007-0.0013	0.0080-0.0160	3	1.9606-1.9685	0.0010-0.0020	0.0767-0.1200
	3.0	U	2.5190-2.2059	0.0010-0.0015	0.0040-0.0080	3	2.1253-2.0472	0.0010-0.0015	0.0060-0.0115
	4.0	E	2.2430-2.2440	0.0008-0.0015	0.0020-0.0125	3	2.7252-2.7260	0.0003-0.0024	0.0036-0.0106
2004	2.3	D	2.0465-2.2059	0.0007-0.0013	0.0080-0.0160	3	1.9606-1.9685	0.0010-0.0020	0.0767-0.1200
	3.0	U	2.5190-2.2059	0.0010-0.0015	0.0040-0.0080	3	2.1253-2.0472	0.0010-0.0015	0.0060-0.0115
	4.0	E	2.2430-2.2440	0.0008-0.0015	0.0020-0.0125	3	2.7252-2.7260	0.0003-0.0024	0.0036-0.0106
2005	2.3	D	2.0465-2.2059	0.0007-0.0013	0.0080-0.0160	3	1.9606-1.9685	0.0010-0.0020	0.0767-0.1200
	3.0	U	2.5190-2.2059	0.0010-0.0015	0.0040-0.0080	3	2.1253-2.0472	0.0010-0.0015	0.0060-0.0115
	4.0	E	2.2430-2.2440	0.0008-0.0015	0.0020-0.0125	3	2.7252-2.7260	0.0003-0.0024	0.0036-0.0106

06017-RANG-C06

PISTON AND RING SPECIFICATIONS

All measurements are given in inches.

Year	Engine Displacement Liters	Engine VIN	Piston Clearance	Ring Gap			Ring Side Clearance		
				Top Compression	Bottom Compression	Oil Control	Top Compression	Bottom Compression	Oil Control
2002	2.3	D	0.0009-0.0017	0.006-0.0012	0.012-0.0180	0.007-0.0270	0.0008-0.0013	0.0004-0.0011	0.0025-0.0054
	3.0	U	0.0012-0.0023	0.010-0.020	0.010-0.020	0.010-0.049	0.0602-0.0612	0.0602-0.0612	SNUG
	4.0	E	0.0008-0.0019	0.015-0.023	0.015-0.023	0.015-0.055	0.0010-0.0030	0.0010-0.0030	SNUG
2003	2.3	D	0.0009-0.0017	0.006-0.0012	0.012-0.0180	0.007-0.0270	0.0008-0.0013	0.0004-0.0011	0.0025-0.0054
	3.0	U	0.0012-0.0023	0.010-0.020	0.010-0.020	0.010-0.049	0.0602-0.0612	0.0602-0.0612	SNUG
	4.0	E	0.0008-0.0019	0.015-0.023	0.015-0.023	0.015-0.055	0.0010-0.0030	0.0010-0.0030	SNUG
2004	2.3	D	0.0009-0.0017	0.006-0.0012	0.012-0.0180	0.007-0.0270	0.0008-0.0013	0.0004-0.0011	0.0025-0.0054
	3.0	U	0.0012-0.0023	0.010-0.020	0.010-0.020	0.010-0.049	0.0602-0.0612	0.0602-0.0612	SNUG
	4.0	E	0.0008-0.0019	0.015-0.023	0.015-0.023	0.015-0.055	0.0010-0.0030	0.0010-0.0030	SNUG
2005	2.3	D	0.0009-0.0017	0.006-0.0012	0.012-0.0180	0.007-0.0270	0.0008-0.0013	0.0004-0.0011	0.0025-0.0054
	3.0	U	0.0012-0.0023	0.010-0.020	0.010-0.020	0.010-0.049	0.0602-0.0612	0.0602-0.0612	SNUG
	4.0	E	0.0008-0.0019	0.015-0.023	0.015-0.023	0.015-0.055	0.0010-0.0030	0.0010-0.0030	SNUG

06017-RANG-C07

TORQUE SPECIFICATIONS
All readings in ft. lbs.

	Engine Displacement Liters	Engine VIN	Cylinder Head Bolts	Main Bearing Bolts	Rod Bearing Bolts	Crankshaft Damper Bolts	Flywheel Bolts	Manifold		Spark Plugs	Oil Pan Drain Plug
								Intake *	Exhaust		
2002	2.3	D	①	NA	NA	②	③	13	40	9	21
	3.0	U	④	60	26	107	54-64	24	⑤	8-10	NA
	4.0	E	⑥	72	⑦	⑧	75-85	7	16	13	18
2003	2.3	D	①	NA	NA	②	③	13	40	9	21
	3.0	U	④	60	26	107	54-64	24	⑤	8-10	NA
	4.0	E	⑥	72	⑦	⑧	75-85	7	16	13	18
2004	2.3	D	①	NA	NA	②	③	13	40	9	21
	3.0	U	④	60	26	107	59	21	⑤	11	NA
	4.0	E	⑥	72	⑦	⑧	⑨	7	16	13	18
2005	2.3	D	①	NA	NA	②	③	13	40	9	21
	3.0	U	⑩	59	26	107	59	21	⑤	11	NA
	4.0	E	⑥	72	⑦	⑧	⑨	7	16	13	18

NA: Information not available

* NOTE: Applies to Lower Manifold only.

① Step 1: 44 inch lbs.
   Step 2: 11 ft. lbs.
   Step 3: 33 ft. lbs.
   Step 4: +90 degrees
   Step 5: +90 degrees

② Step 1: 74 ft. lbs.
   Step 2: +90 degrees

③ Step 1: 37 ft. lbs.
   Step 2: 50 ft. lbs.
   Step 3: 83 ft. lbs.

④ Step 1: 59 ft. lbs.
   Step 2: Back off 1 full turn
   Step 3: 34-40 ft. lbs.
   Step 4: 63-73 ft. lbs.

⑤ Step 1: 89 inch lbs.
   Step 2: 17 ft. lbs.

⑥ 8mm bolts: 24 ft. lbs.
   12mm bolts: 24 ft. lbs. +80 degrees, +80 degrees more

⑦ Step 1: 15 ft. lbs.
   Step 2: +90 degrees

⑧ Step 1: 37 ft. lbs.
   Step 2: plus 90 degrees

⑨ Step 1: 10 ft.lbs.
   Step 2: 52 ft. lbs.

⑩ Step 1: 37 ft. lbs.
   Step 2: back off 1 full turn
   Step 3: 22 ft. lbs.
   Step 4: plus 90 degrees
   Step 5: plus 90 degrees more

06017-RANG-C08

80 Nm (59 lb-ft)

06017-RANG-G01

Crankshaft main bearing torque sequence—3.0L engine

WHEEL ALIGNMENT

Year	Model			Caster Range (+/-Deg.)	Caster Preferred Setting (Deg.)	Camber Range (+/-Deg.)	Camber Preferred Setting (Deg.)	Toe-in (in.)
2002	Ranger		2wd	1.0	①	0.70	-0.50	0.06+/-0.25
			4wd	1.0	②	0.70	-0.50	0.12+/-0.25
			Rear	—	—	0.75	0	0+/-0.30
2003	Ranger		2wd	1.0	①	0.70	-0.50	0.06+/-0.25
			4wd	1.0	②	0.70	-0.50	0.12+/-0.25
			Rear	—	—	0.75	0	0+/-0.30
2004	Ranger		2wd	1.0	③	0.70	④	0.06+/-0.25
			4wd	1.0	⑤	0.70	-0.50	0.12+/-0.25
			Rear	—	—	0.75	0	0+/-0.30
2005	Ranger		2wd	1.0	③	0.70	-0.50	0.06+/-0.25
			4wd	1.0	⑤	0.70	-0.50	0.12+/-0.25
			Rear	—	—	0.75	0	0+/-0.30

① Left: +4.0
  Right: +4.4

② Left: +3.9
  Right: +4.4

③ Left: +3.0
  Right: +3.4

④ Left: -0.5
  Right: -0.7

⑤ Left: +2.9
  Right: +3.6

06017-RANG-C09

TIRE, WHEEL AND BALL JOINT SPECIFICATIONS

Year	Model	OEM Tires Standard	OEM Tires Optional	Tire Pressures (psi) Front	Tire Pressures (psi) Rear	Wheel Size	Ball Joint Inspection	Lugnut Torque (ft. lbs.)
2002	2wd, exc. Tremor	P235/75R15	none	①	①	7-JJ	0.030 in. ②	100
	Tremor	P235/70R16	none	①	①	7-JJ	0.030 in. ②	100
	4wd exc. FX4	P245/75R16	none	①	①	7-JJ	0.030 in. ②	100
	FX4	31x10.5/15	none	①	①	8-JJ	0.030 in. ②	100
2003	2wd, exc. Tremor	P235/75R15	none	①	①	7-JJ	0.030 in. ②	100
	Tremor	P235/70R16	none	①	①	7-JJ	0.030 in. ②	100
	4wd exc. FX4	P245/75R16	none	①	①	7-JJ	0.030 in. ②	100
	FX4	31x10.5/15	none	①	①	8-JJ	0.030 in. ②	100
2004	2wd XL	P225/70R15	none	①	①	6-JJ	0.030 in. ②	100
	2wd XLT	P225/70R15	none	①	①	7-JJ	0.030 in. ②	100
	Edge 2wd	P235/75R15	none	①	①	7-JJ	0.030 in. ②	100
	Edge 4wd	P245/75R16	none	①	①	7-JJ	0.030 in. ②	100
	4wd XLT FX4	P245/75R16	none	①	①	7-JJ	0.030 in. ②	100
	Tremor 2wd	P235/75R16	none	①	①	7-JJ	0.030 in. ②	100
	Tremor 4wd	P245/75R16	none	①	①	7-JJ	0.030 in. ②	100
2005	Edge Reg. Cab	P235/75R15	P235/70R16	①	①	7-JJ	0.030 in. ②	100
	Edge SuperCab 2wd	P235/75R15	P235/70R16	①	①	7-JJ	0.030 in. ②	100
	FX4 Level II	NA	NA	①	①	NA	0.030 in. ②	100
	FX4 Off Road	P255/70R16	none	①	①	7-JJ	0.030 in. ②	100
	STX Reg. Cab	P225/70R15	none	①	①	6-JJ	0.030 in. ②	100
	STX SuperCab	P225/70R15	none	①	①	6-JJ	0.030 in. ②	100
	XL Reg. Cab	P235/75R15	none	①	①	6-JJ	0.030 in. ②	100
	XL SuperCab	P235/75R15	none	①	①	7-JJ	0.030 in. ②	100
	XLT Reg. Cab 2wd	P225/70R15	none	①	①	7-JJ	0.030 in. ②	100
	XLT SuperCab 2wd	P225/70R15	none	①	①	7-JJ	0.030 in. ②	100
	XLT Reg. Cab 4wd	P235/75R15	none	①	①	6-JJ	0.030 in. ②	100
	XLT SuperCab 4wd	P255/70R16	none	①	①	7-JJ	0.030 in. ②	100
	Tremor 2wd	P235/75R16	none	①	①	7-JJ	0.030 in. ②	100
	Tremor 4wd	P245/75R16	none	①	①	7-JJ	0.030 in. ②	100

NA: Information not available

OEM: Original Equipment Manufacturer

PSI: Pounds Per Square Inch

STD: Standard

OPT: Optional

① See placard on door post

② Both upper and lower

06017-RANG-C10

BRAKE SPECIFICATIONS

All measurements in inches unless noted

Year	Model		Brake Disc			Brake Drum Diameter			Minimum Lining Thickness	Brake Caliper	
			Original Thickness	Minimum Thickness	Maximum Runout	Original Inside Diameter	Max. Wear Limit	Maximum Machine Diameter		Bracket Bolts (ft. lbs.)	Mounting Bolts (ft. lbs.)
2002	Ranger	①	0.850	0.810	0.0030	9.00	9.09	9.06	0.030	85	24
		②	0.850	0.810	0.0030	10.00	10.09	10.06	0.030	85	24
		③	0.850	0.810	0.0030	10.00	10.09	10.06	0.030	85	24
2003	Ranger	①	0.850	0.810	0.0030	9.00	9.09	9.06	0.030	85	24
		②	0.850	0.810	0.0030	10.00	10.09	10.06	0.030	85	24
		③	0.850	0.810	0.0030	10.00	10.09	10.06	0.030	85	24
2004	Ranger		NA	④	NA	NA	NA	⑤	⑥	85	24
2005	Ranger		NA	0.965	NA	NA	NA	⑤	⑥	85	24

NOTE: Due to changes made during production, refer to manufacturer's specifications if they differ from those in this chart

NA: Not Available

F: Front

R: Rear

① With 9 inch brakes
② 4x2 with 10 inch brakes
③ 4x4 with 10 inch brakes
④ Molded into the disc
⑤ Molded into the drum
⑥ Front: 0.118 in.
    Rear: 0.03 in.

06017-RANG-C11

SCHEDULED MAINTENANCE INTERVALS
2002 FORD RANGER

TO BE SERVICED	TYPE OF SERVICE	VEHICLE MILEAGE INTERVAL (x1000)												
		5	10	15	20	25	30	35	40	45	50	55	60	65
Engine oil & filter	R	✓	✓	✓	✓	✓	✓	✓	✓	✓	✓	✓	✓	✓
Tires	Rotate	✓	✓	✓	✓	✓	✓	✓	✓	✓	✓	✓	✓	✓
Auto trans. fluid	I			✓			✓			✓			✓	
Brake pads/shoes	I			✓			✓			✓			✓	
Coolant hoses	S/I			✓			✓			✓			✓	
Steering linkage	I			✓			✓			✓			✓	
Cabin air filter	R			✓			✓			✓			✓	
Ball joints (2wd)	L			✓			✓			✓			✓	
Exhaust system	I						✓						✓	
Engine air filter	R						✓						✓	
Fuel filter ①	R						✓						✓	
Auto trans fluid (4-speed)	R						✓						✓	
Green coolant ②	R									✓				
Wheel bearings (2wd)	L												✓	
Manual trans. fluid	R												✓	
Spark plugs	R	every 100,000 miles												
PCV valve	R	every 100,000 miles												
Orange coolant	R	every 150,000 miles												
Auto trans fluid (5-speed)	R	every 150,000 miles												
Differential fluid	R	every 150,000 miles												
Accessory drive belts	R	every 150,000 miles												
Transfer case fluid	R	every 150,000 miles												

R: Replace   S: Service   I: Inspect   L: Lubricate

① Recommended, but not required in Calif.

② Change every 30,000 miles or 36 months thereafter

FREQUENT OPERATION MAINTENANCE (SEVERE SERVICE)

If a vehicle is operated under any of the following conditions it is considered severe service:

- Towing a trailer or using a camper or car-top carrier.

- Repeated short trips of less than 5 miles in temperatures below freezing, or trips of less than 10 miles in any temperature.

- Extensive idling or low-speed driving for long distance as in heavy commercial use, such as delivery, taxi or police cars.

- Operating on rough, muddy or salt-covered roads.

- Operating on unpaved or dusty roads.

- Driving in extremely hot (over 90°) conditions.

Engine oil & filter: replace every 3000 miles.
Air cleaner filter: service or inspect every 6000 miles.
Exhaust system: check every 6000 miles.
Automatic transmission fluid & filter: change every 30,000 miles.
Transfer case fluid: change every 60,000 miles
Fuel filter: change every 15,000 miles
Spark plugs: change every 60,000 miles
2wd front wheel bearings: lubricate every 30,000 miles

06017-RANG-C12

SCHEDULED MAINTENANCE INTERVALS
2003 FORD RANGER

TO BE SERVICED	TYPE OF SERVICE	VEHICLE MILEAGE INTERVAL (x1000)												
		5	10	15	20	25	30	35	40	45	50	55	60	65
Engine oil & filter	R	✓	✓	✓	✓	✓	✓	✓	✓	✓	✓	✓	✓	✓
Tires	Rotate	✓	✓	✓	✓	✓	✓	✓	✓	✓	✓	✓	✓	✓
Auto trans. fluid	I			✓			✓			✓			✓	
Brake pads/shoes	I			✓			✓			✓			✓	
Coolant hoses	S/I			✓			✓			✓			✓	
Steering linkage	I			✓			✓			✓			✓	
Suspension and driveshaft	I			✓			✓			✓			✓	
Cabin air filter	R			✓			✓			✓			✓	
Ball joints (2wd)	I			✓			✓			✓			✓	
Exhaust system	I						✓						✓	
Engine air filter	R						✓						✓	
Fuel filter ①	R						✓						✓	
Auto trans fluid ②	R						✓						✓	
Green coolant	R									✓				
Accessory drive belts	I	every 100,000 miles												
Spark plugs	R	every 100,000 miles												
PCV valve	R	every 100,000 miles												
Yellow coolant ③	R	first 100,000 miles or 5 years												
Auto trans fluid and filter ④	R	every 150,000 miles												
Rear axle fluid	R	every 150,000 miles												
Accessory drive belts ⑤	R	every 150,000 miles												

R: Replace   S: Service   I: Inspect   L: Lubricate

① Recommended, but not required in Calif.

② If equipped with AX4S, 4F50N, 4R100, 4F27E

③ Every 3 years or 50,000 miles thereafter

④ All except AX4S, 4F50N, 4R100, 4F27E

⑤ If not previously replaced

Special Operating Condition Requirements

When towing a trailer or using a camper or car-top carrier:
Change engine oil and install a new oil filter every 4,800 km (3,000 miles) or 3 months.
Change transfer case fluid every 96,000 km (60,000 miles).
Change manual transmission fluid as required.
Inspect and lubricate U-joints as required.

During extensive idling and/or low speed driving for long distances, as in heavy commercial use such as delivery, taxi, patrol car or livery:
Change engine oil and install a new oil filter, lube front lower control arm and steering linkage ball joints with
zerk fittings (if equipped) every 4,800 km (3,000 miles) or 3 months.
Inspect brake system and check battery electrolyte level (Patrol cars) every 8,000 km (5,000 miles).
Install a new fuel filter every 24,000 km (15,000 miles).
Change automatic transmission fluid, lubricate 4x2 wheel bearings,
install new grease seals and adjust bearings every 48,000 km (30,000 miles).
Install new spark plugs and change transfer case fluid every 96,000 km (60,000 miles).
Install a new cabin air filter as required.

When operating in dusty conditions such as unpaved or dusty roads:
Change engine oil and install a new oil filter every 4,800 km (3,000 miles) or 3 months.
Install a new fuel filter every 24,000 km (15,000 miles).
Change automatic transmission fluid every 48,000 km (30,000 miles).
Change transfer case fluid every 96,000 km (60,000 miles).
Install a new engine air filter as required.
Install a new cabin air filter as required.

When operating in off-road conditions:
Change automatic transmission fluid every 48,000 km (30,000 miles).
Change transfer case fluid every 96,000 km (60,000 miles).
Install a new cabin air filter as required.
Inspect and lubricate U-joints.
Inspect and lubricate steering linkage ball joints with zerk fittings.

06017-RANG-C13

SCHEDULED MAINTENANCE INTERVALS
2004 FORD RANGER

TO BE SERVICED	TYPE OF SERVICE	VEHICLE MILEAGE INTERVAL (x1000)												
		5	10	15	20	25	30	35	40	45	50	55	60	65
Engine oil & filter	R	✓	✓	✓	✓	✓	✓	✓	✓	✓	✓	✓	✓	✓
Tires	Rotate	✓	✓	✓	✓	✓	✓	✓	✓	✓	✓	✓	✓	✓
Auto trans. fluid	I			✓			✓			✓			✓	
Brake pads/shoes	I			✓			✓			✓			✓	
Coolant hoses	S/I			✓			✓			✓			✓	
Steering linkage	I			✓			✓			✓			✓	
Suspension and driveshaft	I			✓			✓			✓			✓	
Cabin air filter	R			✓			✓			✓			✓	
Ball joints (2wd)	I/L			✓			✓			✓			✓	
Exhaust system	I						✓						✓	
Engine air filter	R						✓						✓	
Fuel filter	R						✓						✓	
Auto trans fluid ①	R						✓						✓	
Wheel bearings (2wd)	R	at 150,000 miles, if not previously replaced, including seals												
Coolant, exc. Premium Gold	R	every 105,000 miles												
Premium Gold coolant ③	R	at 5 years or 100,000 miles												
Spark plugs	R	every 100,000 miles												
PCV valve	R	every 100,000 miles												
Auto trans fluid ④	R	every 150,000 miles												
Fuel injectors	Clean	every 100,000 miles												
Rear axle fluid	R	every 150,000 miles												
Accessory drive belts	I	every 100,000 miles												
Accessory drive belts	R	every 150,000 miles, if not previously replaced												

R: Replace   S: Service   I: Inspect   L: Lubricate

① Vehicles equipped with AX4S, 4F50N, 4R100, 4F27E

② Change every 30,000 miles or 36 months thereafter

③ After the initial change, every 3 years or 50,000 miles

④ Except vehicles equipped with AX4S, 4F50N, 4R100, 4F27E

Special Operating Condition Requirements

When towing a trailer or using a camper or car-top carrier:

Change engine oil and install a new oil filter every 4,800 km (3,000 miles), 3 months or 200 hours of engine operation (whichever occurs first).

Change transfer case fluid every 96,000 km (60,000 miles).

Change manual transmission fluid as required.

Inspect and lubricate U-joints as required.

During extensive idling and/or low speed driving for long distances, as in heavy commercial use such as delivery, taxi, patrol car or livery:

Change engine oil and install a new oil filter every 4,800 km (3,000 miles), 3 months or 200 hours of engine operation (whichever occurs first).

Lube front lower control arm and steering linkage ball joints with zerk fittings (if equipped) every 4,800 km (3,000 miles) or 3 months.

Inspect brake system and check battery electrolyte level (Patrol cars) every 8,000 km (5,000 miles).

Install a new fuel filter every 24,000 km (15,000 miles).

Change automatic transmission fluid, lubricate 4x2 wheel bearings, install new grease seals and adjust bearings every 48,000 km (30,000 miles). If equipped, change the in-line service installed transmission fluid filter.

Install new spark plugs and change transfer case fluid every 96,000 km (60,000 miles).

Install a new cabin air filter as required.

06017-RANG-C14

SCHEDULED MAINTENANCE INTERVALS
FORD RANGER
Footnotes Continued

When operating in dusty conditions such as unpaved or dusty roads:

Change engine oil and install a new oil filter every 4,800 km (3,000 miles) or 3 months.

Install a new fuel filter every 24,000 km (15,000 miles).

Change automatic transmission fluid every 48,000 km (30,000 miles). If equipped, change the in-line service installed transmission fluid filter.

Change transfer case fluid every 96,000 km (60,000 miles).

Install a new engine air filter as required.

Install a new cabin air filter as required.

When operating in off-road conditions:

Change automatic transmission fluid every 48,000 km (30,000 miles). If equipped, change the in-line service installed transmission fluid filter.

Change transfer case fluid every 96,000 km (60,000 miles).

Install a new cabin air filter as required.

Inspect and lubricate U-joints.

Inspect and lubricate steering linkage ball joints with zerk fittings.

06017-RANG-C15

SCHEDULED MAINTENANCE INTERVALS
2005 FORD RANGER

TO BE SERVICED	TYPE OF SERVICE	VEHICLE MILEAGE INTERVAL (x1000)												
		5	10	15	20	25	30	35	40	45	50	55	60	65
Engine oil & filter	R	✓	✓	✓	✓	✓	✓	✓	✓	✓	✓	✓	✓	✓
Tires	Rotate	✓	✓	✓	✓	✓	✓	✓	✓	✓	✓	✓	✓	✓
Auto trans. fluid	I									✓				
Brake pads/shoes	I			✓			✓			✓			✓	
Wheel ends ①	I			✓			✓			✓			✓	
Coolant hoses	S/I			✓			✓			✓			✓	
Steering linkage	I/L			✓			✓			✓			✓	
Suspension and driveshaft	I			✓			✓			✓			✓	
Cabin air filter	R			✓			✓			✓			✓	
Ball joints (2wd)	I/L			✓			✓			✓			✓	
Exhaust system	I						✓						✓	
Engine air filter	R						✓						✓	
Fuel filter	R						✓						✓	
Manual trans fluid	R												✓	
Wheel bearings (2wd)	R	at 150,000 miles, if not previously replaced, including seals												
Premium Gold coolant ②	R	at 5 years or 100,000 miles												
Spark plugs	R	every 100,000 miles												
PCV valve	R	every 100,000 miles												
Auto trans fluid	R	every 150,000 miles, if not previously replaced												
Rear axle fluid	R	every 150,000 miles												
Accessory drive belts	I	every 100,000 miles												
Accessory drive belts	R	every 150,000 miles, if not previously replaced												

R: Replace   S: Service   I: Inspect   L: Lubricate

① Check for play and noise

② After the initial change, every 3 years or 50,000 miles

Special Operating Condition Requirements

When towing a trailer or using a camper or car-top carrier:

Change engine oil and install a new oil filter every 4,800 km (3,000 miles), 3 months or 200 hours of engine operation (whichever occurs first).

Change transfer case fluid every 96,000 km (60,000 miles).

Change manual transmission fluid as required.

Inspect and lubricate U-joints as required.

During extensive idling and/or low speed driving for long distances, as in heavy commercial use such as delivery, taxi, patrol car or livery:

Change engine oil and install a new oil filter every 4,800 km (3,000 miles), 3 months or 200 hours of engine operation (whichever occurs first).

Lube front lower control arm and steering linkage ball joints with zerk fittings (if equipped) every 4,800 km (3,000 miles) or 3 months.

Inspect brake system and check battery electrolyte level (Patrol cars) every 8,000 km (5,000 miles).

Install a new fuel filter every 24,000 km (15,000 miles).

Change automatic transmission fluid, lubricate 4x2 wheel bearings, install new grease seals and adjust bearings every 48,000 km (30,000 miles). If equipped, change the in-line service installed transmission fluid filter.

Install new spark plugs and change transfer case fluid every 96,000 km (60,000 miles).

Install a new cabin air filter as required.

06017-RANG-C16

SCHEDULED MAINTENANCE INTERVALS
FORD RANGER
Footnotes Continued

When operating in dusty conditions such as unpaved or dusty roads:

Change engine oil and install a new oil filter every 4,800 km (3,000 miles) or 3 months.

Install a new fuel filter every 24,000 km (15,000 miles).

Change automatic transmission fluid every 48,000 km (30,000 miles). If equipped, change the in-line service installed transmission fluid filter.

Change transfer case fluid every 96,000 km (60,000 miles).

Install a new engine air filter as required.

Install a new cabin air filter as required.

When operating in off-road conditions:

Change automatic transmission fluid every 48,000 km (30,000 miles). If equipped, change the in-line service installed transmission fluid filter.

Change transfer case fluid every 96,000 km (60,000 miles).

Install a new cabin air filter as required.

Inspect and lubricate U-joints.

Inspect and lubricate steering linkage ball joints with zerk fittings.

06017-RANG-C17

ENGINE REPAIR

Alternator

REMOVAL

2.3L Engines

1. Before servicing the vehicle, refer to the Precautions Section.
2. Remove or disconnect the following:
   - Battery ground cable
   - Air cleaner outlet tube
   - Accessory drive belt
   - Mounting bolts and alternator
   - Nut and the electrical connectors.
3. To install, reverse the removal procedure. Torque all mounting bolts to 18 ft. lbs. (25Nm).

Alternator mounting bolts—2.3L engine

Alternator wiring connections—2.3L engine

3.0L Engines

1. Before servicing the vehicle, refer to the Precautions Section.
2. Remove or disconnect the following:
   - Negative battery cable
   - Air cleaner outlet tube
   - Drive belt
   - Electrical connectors from the alternator
   - Wiring harness to alternator push pin
   - Alternator

Alternator wiring connections—3.0L engine

Wiring harness pin-type retainer—3.0L engine

Alternator mounting bolts—3.0L engine

To install:

3. Install or connect the following:
   - Alternator. Torque the bolts to 40 ft. lbs. (55 Nm) for 2002–03; 35 ft. lbs (48 Nm) for 2004–05.
   - Pushpin for the alternator wiring harness
   - Electrical connectors to the alternator
   - Drive belt
   - Air cleaner outlet tube
   - Negative battery cable

4.0L Engine

1. Before servicing the vehicle, refer to the Precautions Section.
2. Remove or disconnect the following:
   - Negative battery cable

Alternator wiring connections—4.0L engine

Alternator wiring pin-type retainer—4.0L engine

Alternator mounting bolts—4.0L engine

   - Air cleaner outlet tube
   - Accessory drive belt
   - Electrical connectors
   - Wiring harness-to-generator pin-type retainer
   - Stud bolt, the bolts and the alternator

3. To install, reverse the removal procedure. Torque all mounting bolts to 35 ft. lbs. (47Nm).

Ignition Timing

ADJUSTMENT

The ignition timing is preset and is not adjustable.

Engine Assembly

REMOVAL & INSTALLATION

2.3L Engines

1. Before servicing the vehicle, refer to the Precautions Section.
2. Relieve the fuel system pressure.
3. Drain the cooling system.
4. Drain the engine oil.
5. Properly discharge the air conditioning system.
6. Remove or disconnect the following:
   - Hood
   - Accelerator control snow shield
   - Air cleaner tube
   - Upper radiator hose
   - Lower radiator hose
   - Fan and shroud
   - PCM electrical connector. Remove the retaining nut on the harness clamp. Position the harness on the engine.
   - Ground stud for the PCM
   - Heater hoses
   - All vacuum hoses
   - Coolant reservoir hoses
   - Air conditioning compressor clutch
   - MAF electrical connector
   - Air conditioning compressor manifold, plug the lines and the compressor ports
   - Accelerator and speed control cables
   - Power steering return hose
   - PSP switch electrical connector
   - High pressure power steering hose
   - Fuel supply hose
   - 42-pin electrical connector
   - VMV vacuum regulator solenoid supply hose
   - Evaporative purge hose
   - Brake booster vacuum hose and the engine ground strap
   - Positive battery cable
   - Solenoid control wire at the starter
   - Starter wiring harness clamp bolt and position it out of the way.
   - RH splash shield
   - Alternator electrical connections
   - Block heater electrical connector
   - Front heated oxygen sensor electrical connector at the bell housing
   - Oil pressure sensor electrical connector
   - Engine wiring pushpins and position the engine wiring harnesses out of the way.
   - Oil filter
   - With automatic transmission, the

bolt retaining the transmission cooling tubes to the engine. Remove the bracket.
   - Transmission dust shield
   - Starter motor
   - Heated oxygen sensor electrical connector at the rear of the transmission
   - Transmission wiring harness
   - Vehicle speed sensor, transmission range sensor, backup light switch and the transmission electrical connectors. Disconnect the pushpins and position the harness forward to the engine.
   - Oil filter adapter

➡ **Leave two side bolts in until the engine is ready to be removed.**

   - Nine of the transmission-to-engine bolts
   - With automatic transmission, the transmission fluid indicator and tube assembly
   - Starter dust shield

➡ **Mark one stud and the flexplate for assembly reference.**

   - With automatic transmission, the four torque converter nuts
7. Support the transmission with a floor jack.
8. Support the engine with a floor crane using a spreader bar.
9. Remove the two side transmission-to-engine bolts.
10. Remove the four engine support insulator.
11. Remove the engine from the vehicle.
12. Installation is the reverse of removal. Observe the following torques:
   - Torque converter bolts: 26 ft. lbs. (35Nm)
   - Nine transmission-to-engine bolts 35 ft. lbs. (48Nm)
   - Oil filter adapter: 18 ft. lbs. (25Nm)
   - Starter: 30 ft. lbs. (40Nm)
   - Engine support nuts: 75 ft. lbs. (102Nm)

67197-RANG-G09

Engine support insulator nuts, left side shown, right side similar—2.3L engine

3.0L Engines

1. Before servicing the vehicle, refer to the Precautions Section.
2. Relieve the fuel system pressure.
3. Drain the cooling system.
4. Drain the engine oil.
5. Properly discharge the air conditioning system.
6. Remove or disconnect the following:
   - Hood
   - Air cleaner outlet tube
   - Upper and the lower radiator hoses

❊❊ WARNING

The fan clutch has left-hand threads.

   - The fan clutch and blade as an assembly
   - Drive belt
   - Fan shroud
   - Radiator
   - Air conditioning manifold and tube. Remove the nut and position the line aside.
   - Air conditioning compressor wiring
   - Air conditioning compressor and the air conditioning compressor mounting bracket
   - Heater hoses
   - Ground cable
   - Fuel lines
   - Snow shield
   - Accelerator cable and the speed control actuator cable
   - All vacuum lines
   - 42-pin connector
   - Powertrain control module connector
   - Nut from the powertrain control module harness
   - Stud bolt and the powertrain control module ground strap
   - Alternator wiring and position aside
   - Both heated oxygen sensors
   - Transmission harness connectors
   - MAF sensor
   - LH heated oxygen sensor
   - Dual converter Y pipe
   - Starter motor and the starter grounding stud bolt
   - Torque converter nuts
   - 8 transmission-to-engine bolts
7. Install the lifting eyes.
8. Remove the four nuts.
9. Support the transmission.
10. Remove the engine from the vehicle.
11. Installation is the reverse of removal. Observe the following torques:
   - Engine mount nuts: 80 ft. lbs. (109Nm)
   - Transmission-to-engine bolts: 33 ft. lbs. (45Nm)

Engine support nuts, one side shown—3.0L engine

67197-RANG-G10

- Torque converter nuts: 2002–03 26 ft. lbs. (35Nm); 2004–05 35 ft. lbs. (47 Nm).

4.0L Engines

1. Before servicing the vehicle, refer to the Precautions Section.

✳✳ CAUTION

If the fuel supply manifold is used as a leverage device, damage may occur to the supply manifold. Care must be taken when working around the fuel supply manifold.

2. Remove or disconnect the following:
- Accelerator cable from engine
- Speed control cable from engine
- Radiator, the fan blade, and the fan shroud
- Accessory bracket bolts and position bracket aside
- Alternator wiring
- Wiring harness retainer and position generator wiring away from engine
- Engine electrical connector
- PCM connector
- PCM ground wire
- Engine ground wire
- Brake booster vacuum hose
- Air conditioning high pressure switch electrical connector
- Bolt and position the air conditioning lines aside

➡**Heater hose will be removed with engine.**

- Heater hoses
- Fuel line
- Starter motor
- Engine oil
- Oil drain plug
- Transmission portion of wiring harness
- RH and LH heated oxygen sensor connectors

- Transmission control connector
- Output shaft speed sensor connector
- Digital transmission range sensor connector
- Catalyst monitor sensor electrical connector
- Transmission/transfer case portion of the wiring harness from any routing clips or pushpins. Route transmission/transfer case portion of the wiring harness to top of engine.
- Bolt, and position the transmission cooling line bracket aside
- Air conditioning line bracket nut and position it aside
- Power steering return hose
- Power steering pressure hose
- Vapor management valve hose connector
- Eight bolts and the LH and the RH engine support insulator nuts

➡**The lifting eyes should be installed on the exhaust manifold studs for number three and number four cylinders.**

3. Install the lifting eyes.
4. Install the spreader bar to the lifting eyes.
5. Attach a floor crane to the spreader bar and remove the engine.
6. Installation is the reverse of removal. Observe the following torques:
- Left and right engine insulator nuts: 81 ft. lbs. (110 Nm)

Left side engine insulator nuts—4.0L engine

67197-RANG-G11

Right side engine insulator nuts—4.0L engine

67197-RANG-G12

- Engine mount nuts: 59 ft. lbs. (80Nm)
- Transmission-to-engine bolts: 35 ft. lbs. (47Nm)
- Torque converter nuts: 35 ft. lbs. (47Nm)

Water Pump

REMOVAL & INSTALLATION

2.3L Engine

1. Before servicing the vehicle, refer to the Precautions Section.
2. Drain the cooling system.
3. Remove the drive belt.
4. Remove the water pump pulley.
5. Remove the water pump.

➡**Lubricate the water pump O-ring, with MERPOL®, or equivalent.**

6. To install, reverse the removal procedure. Torque the water pump mount bolts to 89 inch lbs. (10Nm). Torque the pulley bolts to 18 ft. lbs. (25Nm).

Water pump mounting bolts—2.3L engine

67197-RANG-G13

3.0L Engines

1. Before servicing the vehicle, refer to the Precautions Section.
2. Drain the cooling system.
3. Remove or disconnect the following:
- Negative battery cable
- Air cleaner outlet tube
- Fan and radiator shroud
- Water bypass tube
- Drive belt
- Heater hose
- Water pump pulley
- Lower radiator hose
- Air conditioning compressor and bracket assembly and move them aside
- Water pump

To install:
4. Clean the mating surfaces where the water pump attaches to the engine.

Water pump mounting bolts—3.0L engine

5. Install or connect the following:
- Water pump. Torque the bolts to 89 inch lbs. (10 Nm).
- Air conditioning compressor mounting bracket. Torque the bolts to 44 ft. lbs. (61 Nm).
- Water pump pulley. Torque the bolts to 20 ft. lbs. (28 Nm) for 2002–04 models; 18 ft. lbs. (25 Nm) for 2005 models.
- Drive belt
- Heater hose
- Lower radiator hose
- Fan and shroud
- Air cleaner outlet tube
- Negative battery cable
6. Fill the cooling system.
7. Start the vehicle and check for leaks, repair if necessary.

4.0L Engine

1. Before servicing the vehicle, refer to the Precautions Section.
2. Drain the cooling system.
3. Remove or disconnect the following:
- Fan shroud
- Accessory drive belt
- Idler pulley
- Water bypass hose
- Heater hose
- Lower radiator hose
- Water pump pulley
- Water pump

> ✸✸ **WARNING**
>
> **Use care when scraping the water pump-to-engine block mating surfaces. Gouges in the aluminum could form leak paths.**

4. Clean all the sealing surfaces.
5. To install, reverse the removal procedure. Torque the water pump bolts to 89 inch lbs. (10Nm). Torque the pulley bolts to 18 ft. lbs. (25Nm).

10 Nm (89 lb-in)

Water pump mounting bolts—4.0L engine

Heater Core

REMOVAL & INSTALLATION

2002–03

1. Before servicing the vehicle, refer to the Precautions Section.
2. Disconnect the negative battery cable.

> ✸✸ **CAUTION**
>
> **After disconnecting the negative battery cable, wait for 1 minute for the SRS module to deplete its energy.**

3. Drain the cooling system into a clean container for reuse.
4. Remove the steering column by performing the following procedure:
  a. Position the front wheels in the straight-ahead direction.
  b. At the both sides of the steering wheel, remove the cover plugs, the steering wheel-to-air bag module screws, disconnect the air bag electrical connector and carefully remove the air bag module.

> ✸✸ **CAUTION**
>
> **Safely store the air bag module with the front side facing upward.**

  c. Remove the steering wheel-to-steering column nut.
  d. Using a steering wheel puller, press the steering wheel from the steering column.
  e. Remove the parking brake release handle screws and move the release handle aside.
  f. Remove the hood release screws and move the hood release aside.
  g. Remove the 2 instrument panel-to-steering column cover screws and the cover.
  h. Remove the instrument panel steering column opening reinforcement bolts and the reinforcement.

  i. Remove the ignition switch bolt and disconnect the ignition switch electrical connector.
  j. At the base of the steering column, disconnect the electrical connectors.
  k. If equipped with an automatic transmission, remove the transmission range indicator bolt and the cable.
  l. If equipped with an automatic transmission, disconnect the shift cable from the steering column shift tube lever and the steering column bracket.
  m. Disconnect the brake shift interlock solenoid electrical connector.
  n. Remove the air bag sliding contact.
  o. Remove the upper intermediate steering shaft-to-column shaft bolt and discard the bolt.
  p. Remove the lower steering column-to-instrument panel nuts and the steering column.
5. Remove the instrument panel by performing the following procedure:
  a. Remove the parking brake release handle screws and move the release handle aside.
  b. Disconnect the Brake Pedal Position (BPP) switch electrical connector.
  c. Remove both front door scuff plates.
  d. Remove the push pins and remove both cowl side trim panels.
  e. At the right side cowl panel, disconnect the electrical connectors and ground wires.
  f. Remove both sides windshield garnish moldings.
  g. Remove the instrument panel fuse door.
  h. Disconnect the power distribution box from its bracket and move it aside.
  i. In the engine compartment, loosen the bulkhead wiring harness bolts and disconnect the electrical connectors.
  j. Pull the bulkhead electrical connector handle and disconnect the wiring harness.
  k. Remove the passenger's side air bag module-to-instrument panel screws, disconnect the electrical connector and remove the air bag module.

> ✸✸ **WARNING**
>
> **Store the air bag module in a safe location with the front facing upward.**

  l. Disconnect the blend door actuator's electrical connector.
  m. Disconnect the climate control vacuum harness connector.
  n. Disconnect the radio's antenna connector.

o. Remove the glove compartment.

p. Remove the instrument panel defroster grille.

q. Remove the upper instrument panel bolts.

r. Under the steering column, remove the instrument panel brace bolt.

s. Remove both the right and left instrument panel-to-cowl bolts.

t. Pull the instrument panel away from the dash.

u. Loosen the instrument panel-to-body harness bolt and disconnect the harness.

v. Using an assistant, remove the instrument panel.

6. Remove the evaporator core by performing the following procedure:

a. Discharge and recover the air conditioning system refrigerant.

b. Remove the refrigerant lines from the evaporator core. Discard the O-rings.

c. If equipped, remove the air conditioning vacuum reservoir tank/bracket screws and reposition the tank.

d. If equipped, disconnect the speed control servo connector; then, remove the bolt and reposition the speed control servo.

e. If equipped with a 3.0L or 4.0L engine, remove the support bracket.

f. Disengage the windshield washer hose retainer and move it aside.

g. Disconnect the vacuum hose and the retainer; then, move the hose aside.

h. Remove the passenger's compartment nut.

i. At the back of the engine, remove the hose support bolts.

j. Remove the evaporator housing-to-chassis nuts.

k. Remove the air conditioning accumulator bracket screws.

l. Remove the evaporator housing cover screws, clips and the cover.

1 Instrument Panel Finish Panel
2 Instrument Panel Defroster
   Opening Grille Assembly
3 Passenger Side Air Bag Module
4 Instrument Panel Center Finish Panel
5 Instrument Panel Steering
   Column Cover

93113GL3

View of the instrument panel components—2002–03

View of the evaporator core—2002–03

93113GL4

m. Remove the evaporator core from the housing.

7. Disconnect the heater hoses from the heater core.

8. Remove the heater housing plenum chamber nuts and the plenum chamber.

9. Remove the heater core-to-heater housing screws and the cover.

10. Remove the heater core.

To install:

11. Install the heater core.

12. Install the heater core-to-heater housing cover and the cover screws.

13. Install the heater housing plenum chamber and the plenum chamber nuts.

14. Connect the heater hoses to the heater core.

93113GL6

View of the heater core cover—2002–03

93113GL5

View of the heater core—2002–03

15. Install the evaporator core by perform the following procedure:

a. Install the evaporator core to the housing.

b. Install the evaporator housing cover, clips and the cover screws.

c. Install the air conditioning accumulator bracket screws.

d. Install the evaporator housing-to-chassis nuts.

e. At the back of the engine, install the hose support bolts.

f. Install the passenger's compartment nut.

g. Connect the vacuum hose and the retainer.

h. Engage the windshield washer hose retainer.

i. If equipped with a 3.0L or 4.0L engine, install the support bracket.

j. If equipped, install the speed control servo bolt and connect the connector.

k. If equipped, install the air conditioning vacuum reservoir tank and the bracket screws.

l. Using new O-rings, install the refrigerant lines to the evaporator core.

16. Install the instrument panel by performing the following procedure:

a. Using an assistant, install the instrument panel.

b. Connect the harness and tighten the instrument panel-to-body harness bolt.

c. Push the instrument panel toward the dash.

d. Install both the right and left instrument panel-to-cowl bolts.

e. Under the steering column, install the instrument panel brace bolt.

f. Install the upper instrument panel bolts.

g. Install the instrument panel defroster grille.

h. Install the glove compartment.

i. Connect the radio's antenna connector.

j. Connect the climate control vacuum harness connector.

k. Connect the blend door actuator's electrical connector.

l. Install the passenger's side air bag module, connect the electrical connector and torque the air bag module-to-instrument panel screws to 67–92 inch lbs. (7.6–10.4 Nm).

m. Connect the bulkhead electrical connector handle wiring harness.

n. In the engine compartment, connect the electrical connectors and tighten the bulkhead wiring harness bolts.

o. Connect the power distribution box to its bracket.

p. Install the instrument panel fuse door.

q. Install both sides windshield garnish moldings.

r. At the right side cowl panel, connect the electrical connectors and ground wires.

s. Install both cowl side trim panels and the push pins.

t. Install both front door scuff plates.

u. Connect the brake pedal position (BPP) switch electrical connector.

v. Install the parking brake release handle and the release handle aside screws.

17. Install the steering column by performing the following procedure:

a. Install the lower steering column and the steering column-to-instrument panel nuts; then, torque the nuts to 10–13 ft. lbs. (13–17 Nm).

b. Using a new bolt, install the upper intermediate steering shaft-to-column shaft bolt and torque to 19–25 ft. lbs. (26–34 Nm).

c. Install the air bag sliding contact.

d. Connect the brake shift interlock solenoid electrical connector.

e. If equipped with an automatic transmission, connect the shift cable from the steering column shift tube lever and the steering column bracket.

f. If equipped with an automatic transmission, install the transmission range indicator cable and bolt.

g. At the base of the steering column, connect the electrical connectors.

h. Connect the ignition switch electrical connector and install the ignition switch bolt.

i. Install the instrument panel steering column opening reinforcement and the reinforcement bolts.

j. Install the instrument panel-to-steering column cover and the 2 cover screws.

k. Install the hood release and the hood release screws.

l. Install the parking brake release handle and the release handle screws.

m. Install the steering wheel to the steering column.

n. Install the steering wheel-to-steering column nut and torque the nut to 25–34 ft. lbs. (34–46 Nm).

o. At the both sides of the steering wheel, install the air bag module, connect the air bag electrical connector, install the steering wheel-to-air bag module screws and the cover plugs.

18. Refill the cooling system.

19. Connect the negative battery cable.

20. Evacuate and charge the air conditioning system.

21. Run the engine to normal operating temperatures; then, check the climate control operation and check for leaks.

2004–05

1. Before servicing the vehicle, refer to the Precautions Section.

2. Recover the refrigerant.

3. Remove the suction accumulator.

4. On trucks with the 2.3L engine:

a. Remove the A/C compressor.

b. Remove the engine oil indicator and tube.

5. On vehicles with the 3.0L or 4.0L engine Position the coolant reservoir and windshield washer reservoir aside.

6. Detach the speed control servo.

7. Disconnect the blower motor and blower motor resistor electrical connectors.

8. Disconnect the heater hoses. Using suitable tools, clamp-off the heater

9. Detach the pin-type retainer and position aside the windshield washer hose.

10. Disconnect and detach the heater control valve vacuum hose.

11. Disconnect the vacuum supply hose near the evaporator core housing.

12. Disconnect the condenser to evaporator line spring lock coupling from the passenger compartment evaporator core inlet.

13. Disconnect the vacuum hose connector and remove the nut.

14. Remove the nuts and the evaporator core housing.

15. Remove the driver side and passenger side air bag modules.

16. Remove the front seats.

17. Remove the screws and position the parking brake release handle aside.

18. Remove the left and right lower cowl kick panels.

19. Position the parking brake assembly aside.

20. Remove the screws and position the hood release handle aside.

21. Remove the instrument panel steering column cover.

22. Remove the instrument panel steering column opening cover reinforcement.

23. Disconnect the brake pedal position (BPP) switch electrical connector from the steering column shaft.

24. If equipped, disconnect the clutch pedal position (CPP) switch electrical connector.

25. If equipped, disconnect the shift cable from the steering column.

26. Disconnect the upper intermediate steering shaft.

27. Remove the left and right side garnish moldings.

a. Remove the bolt covers and remove the bolts.

b. Remove the assist handle.

c. Remove the windshield side garnish molding.

28. Remove the door moldings.

29. On regular cab vehicles

a. Remove the screws.

b. Remove the scuff plate.

30. Disconnect the electrical connectors and the ground wire from the RH side lower kick panel.

➡To avoid damaging the bulkhead electrical connectors, be sure the release tab is fully depressed before pulling release lever into the disconnect position.

31. Disconnect the LH side bulkhead electrical connector. Press the release tab and pull the release lever.

32. Remove the audio unit. Insert the

removal tool. Remove and support the audio unit.

33. Disconnect the audio unit electrical connector and antenna cable.

34. Lower the glove compartment. Press the release tabs inward while lowering the compartment.

35. Through the glove compartment opening, disconnect the blend door actuator electrical connector.

36. Through the glove compartment opening, disconnect the climate control vacuum harness connector.

37. Raise and secure the glove compartment. Press the release tabs inward while raising the glove compartment.

38. Remove the instrument panel defroster opening grille.

39. Remove the instrument panel cowl top bolts.

40. If equipped, remove the floor console.

41. If not equipped with the high-series floor console, remove the cup holders.

42. Release the clips and remove the restraints control module (RCM) cover.

43. If not equipped with high-series floor console, remove the consolette mat.

44. Remove the screws and remove the restraints control module (RCM) cover.

45. Remove the consolette base.

46. Remove the gearshift lever.

47. Remove the screws and remove the manual transmission consolette.

48. Disconnect the RCM electrical connector.

49. Pull the floor carpeting back.

50. On 4x2 vehicles disconnect the instrument panel main harness.

51. On 4x4 vehicles disconnect the instrument panel main harness. From underneath the vehicle, release the instrument panel main harness at the transfer case.

52. Remove the instrument panel side finish panel.

53. Disconnect the door harness electrical connector.

54. Remove the RH side instrument panel bolt. If necessary, transfer the components to the new instrument panel.

55. Remove the LH instrument panel cowl side bolts.

56. Remove the instrument panel.

57. Remove the powertrain control module (PCM).

58. Remove the PCM heat sink.

59. Remove the four nuts from the engine side of the dash panel. Position the plenum chamber on the vehicle floor.

60. Remove the heater core cover.

61. Remove the heater core.

4 Nm (35 lb-in)

67197-RANG-G18

2004–05 evaporator core housing bolts

67197-RANG-G16

2004–05 heater core cover bolts

67197-RANG-G17

2004–05 heater core removal

62. To install, reverse the removal procedure. During installation, be sure to install a new oval foam seal around the heater core inlet and outlet tubes.

63. Lubricate the refrigerant system with the correct amount of clean PAG oil or equivalent. Install new O-ring seals lubricated in clean mineral oil. Lubricate the coolant hoses with plain water only, if needed.

64. Evacuate, leak test, and charge the refrigerant system.

Cylinder Head

REMOVAL & INSTALLATION

2.3L Engines

1. Before servicing the vehicle, refer to the Precautions Section.
2. Relieve the fuel system pressure.
3. Drain the cooling system.
4. Properly discharge the air conditioning system.

5. Remove or disconnect the following:
   - Negative battery cable
   - Drive belt.
   - Engine oil level indicator assembly.
   - Engine oil level indicator.
   - Engine oil level indicator tube.
   - Water outlet tube.
   - Water outlet tube.
   - Air conditioning compressor.

➡**The generator will be removed with the accessory bracket.**

   - Accessory bracket.
   - Right motor mount.
   - Coolant hose from the thermostat.
   - Coolant hose from the EGR valve.
   - Coolant tube assembly.
   - Exhaust manifold and gasket.
   - Block heater (if so equipped).
   - Water outlet.
   - EGR valve.
   - Power steering pump and reservoir as an assembly.
   - Idle air control (IAC) valve.
   - Throttle position (TP) sensor.
   - Manifold absolute pressure (MAP) sensor.
   - Swirl control valve monitor electrical connector.
   - CKP sensor and the wiring harness pin-type retainers.
   - Knock sensor (KS).
   - Electric thermostat.
   - Swirl control valve.
   - CMP sensor electrical connector and disconnect the PCV hose from the intake manifold.
   - Engine wiring harness pin-type retainers from the intake manifold.
   - Engine wiring harness connector bracket. Position the engine wiring harness aside.
   - EGR tube.
   - Fuel supply line clip from the front of the intake manifold. Disconnect the vacuum hose from the intake manifold.
   - Intake manifold assembly.
   - Fuel injector electrical connectors. Detach the wiring harness pin-type retainers.
   - Ignition coil and the cylinder head temperature (CHT) sensor electrical connectors.
   - Engine wiring harness anchors from the valve cover studs. Remove the engine wiring harness.
   - Ignition coil.
   - Bypass hose.
   - Thermostat housing.
   - Knock sensor and the engine vent cover.

   - Left motor mount.
   - Fuel injector supply manifold with the injectors and the ground strap.
   - Water pump pulley.
   - Water pump.
   - CMP sensor.
   - CHT sensor.
   - Spark plugs.
   - Valve cover.
   - CKP sensor.
   - Crankshaft vibration damper

➡**There is one front cover bolt behind the cooling fan drive pulley. To remove this bolt, align one of the cooling fan drive pulley access holes with the bolt head to access the bolt.**

   - Front cover.
   - Timing chain tensioner.
   - Timing chain guides.
   - Timing chain assembly.

➡**Use a wrench on the flats between cylinders No. 1 and No. 2 to hold the camshaft in place.**

   - Camshaft drive sprockets.
   - Oil pump chain tensioner and guide.

➡**The oil pump chain sprocket must be held in place.**

   - Oil pump chain and sprockets.

➡**Note the position of the lobes on the No. 1 cylinder before removing the camshafts for assembly reference.**

6. Loosen the camshaft bearing cap bolts in sequence, one turn at a time. Repeat the first step until all tension is released from the camshaft bearing caps. Remove the camshaft bearing caps.

7. Remove or disconnect the following:
   - Camshafts.
   - Cylinder head bolts and the cylinder head.
   - Cylinder head gasket.

8. Installation is the reverse of removal. Apply RTV sealer to the places shown. The head must be installed within 4 minutes of application. Observe the following torques:

9348EG01

RTV sealer application—2.3L cylinder head

Head bolt torque sequence—2.3L

a. Cylinder head:
- Step 1: Tighten the bolts to 5 Nm (44 inch lbs.)
- Step 2: Tighten the bolts to 15 Nm (11 ft. lbs.)
- Step 3: tighten the bolts to 45 Nm (33 ft. lbs.)
- Step 4: Tighten the bolts an additional 90 degrees (¼ turn)
- Step 5: Tighten the bolts an additional 90 degrees (¼ turn)

b. Camshafts:

➡Install the camshafts with the alignment notches in the camshaft lined up so the camshaft alignment plate can be installed without rotating the camshafts. Make sure the lobes on the No. 1 cylinder are in the same position as noted in the disassembly procedure. Rotating the camshafts, or installing the camshafts 180 degrees out of position can cause severe damage to the valves and pistons. Lubricate the camshaft journals and bearing caps with clean engine oil. Install the camshafts and bearing caps. Tighten the bolts in the sequence shown in three stages.

- Step 1: Tighten the camshaft bearing caps one turn at a time until tight.
- Step 2: Tighten the bolts to 7 Nm (62 inch lbs.)
- Step 3: Tighten the bolts to 16 Nm (12 ft. lbs.)

c. Crankshaft vibration damper:

➡Do not reuse the crankshaft pulley bolt. Tighten the bolt in two stages.

- Step 1: Tighten the bolt to 40 Nm (30 ft. lbs.)
- Step 2: Tighten the bolt and additional 90 degrees (¼ turn).

3.0L Engine

2002–04 MODELS

➡It may be easier to remove the engine from the vehicle. If removing the engine, refer to the engine removal procedure.

1. Before servicing the vehicle, refer to the Precautions Section.
2. Evacuate the air conditioning system.
3. Drain the cooling system.
4. Drain the engine oil.
5. Remove or disconnect the following:
- Negative battery cable
- Lower intake manifold
- Air conditioning compressor
- Alternator
- Power steering pump
- Alternator mounting bracket
- Air conditioning compressor mounting bracket
- Exhaust manifolds
- Cylinder head and discard the bolts and gasket

To install:

➡The "V" in the cylinder head gasket must face the front of the engine.

6. Clean the mating surfaces where the head attaches to the engine.
7. Install a new cylinder head gasket and the cylinder head to the engine.
8. Torque the new cylinder head bolts in stages as follows:
   a. Step 1: 59 ft. lbs. (80 Nm).
   b. Step 2: Loosen the bolts one full turn.
   c. Step 3: 40 ft. lbs. (55 Nm).
   d. Step 4: 63 ft. lbs. (85 Nm).
9. Install or connect the following:
- Lower intake manifold
- Exhaust manifold
- Air conditioning compressor mounting bracket. Torque the bolts to 44 ft. lbs. (66 Nm).
- Alternator mounting bracket
- Power steering pump
- Alternator
- Air conditioning compressor
- Negative battery cable
10. Fill the engine with clean oil.
11. Fill the cooling system.
12. Recharge the air conditioning system
13. Start the vehicle and check for leaks, repair if necessary.

2005 MODELS

1. Before servicing the vehicle, refer to the Precautions Section.
2. With the vehicle in NEUTRAL, position it on a hoist.
3. Disconnect the battery ground cable.
4. Remove the lower intake manifold.

Cylinder head bolt torque sequence—2002–04 3.0L engine

⚠ WARNING

Refrigerant compressor oil (mineral oil) F73Z-19577-AA (Motorcraft YN-9-A) should be used to lubricate R-134a refrigerant system O-ring seals only and should not be added to the R-134a refrigerant system as an A/C compressor lubricant. PAG refrigerant compressor oil F7AZ-19D589-DA (Motorcraft YN-12-C) or equivalent meeting Ford specification WSH-M1C231-B only should be used as an A/C compressor lubricant.

➡ **Installation of a new suction accumulator is not required when repairing the air conditioning system except when there is physical evidence of system contamination from a failed A/C compressor or damage to the suction accumulator.**

5. If flushing of the refrigerant system has not been carried out, recover the refrigerant.

6. Position the air cleaner outlet tube aside.

7. Remove the drive belt from the A/C compressor pulley.

8. Disconnect the electrical connector.

9. Remove the clamp nut on the compressor manifold and tube assembly.

10. Loosen the bolt and disconnect the compressor manifold and tube assembly from the A/C compressor.

11. Remove the bolts and the A/C compressor. Discard the O-ring seals.

12. Remove the generator.

13. Remove the power steering pump pulley.

14. Disconnect the power steering pressure hose. Remove and discard the seal.

15. Disconnect the power steering return hose.

16. Disconnect the power steering return line hose.

17. Remove the bolts and the power steering pump.

18. Remove the generator mounting bracket.

19. Remove the rocker arms and the push rods.

20. Remove the nuts and detach the fuel tube bracket and the wiring harness bracket.

21. Remove the two studbolts from the rear of the A/C compressor mounting bracket.

22. Remove the bolts and remove the front of the A/C compressor mounting bracket.

23. Remove the exhaust manifolds.

24. Remove and discard the eight cylinder head bolts.

25. Remove the cylinder heads.

26. Remove and discard the cylinder head gaskets.

Installation

27. Clean all the sealing surfaces.

28. Check the cylinder head for flatness.

➡ **The "V" notch in the head gasket faces the front of the engine.**

29. Install new cylinder head gaskets.

30. Install the cylinder heads.

31. Install the bolts. Tighten the bolts in five steps in the sequence shown.

- Step 1: Tighten to 50 Nm (37 ft. lbs.).
- Step 2: Loosen the bolts one full turn.
- Step 3: Tighten to 30 Nm (22 ft. lbs.).
- Step 4: Tighten the bolts 90 degrees.
- Step 5: Tighten the bolts an additional 90 degrees.

32. Install the lower intake manifold.

33. Install the exhaust manifolds.

34. Install the A/C compressor mounting bracket. Torque to 34 ft. lbs. (46 Nm).

35. Install the studbolts to the A/C compressor mounting bracket. Torque to 35 ft. lbs. (48 Nm).

36. Position the wiring harness bracket and the fuel tube bracket and install the nuts.

37. Install the push rods and the rocker arms. Torque to 24 ft. lbs. (32 Nm).

38. Install the generator mounting bracket. Install the bolts. Torque to 34 ft. lbs. (46 Nm).

39. Install the power steering pump. Torque to 35 ft. lbs. (48 Nm).

40. Install the generator.

41. Install the A/C compressor. Torque to 18 ft. lbs. (25 Nm).

42. Connect the battery ground cable.

43. Drain the engine oil.

44. Fill the engine with clean engine oil.

45. Fill and bleed the cooling system.

46. Fill the power steering system.

47. Charge the A/C system.

4.0L Engine

1. Before servicing the vehicle, refer to the Precautions Section.

➡ **If only one cylinder head is to be removed, only follow the procedures that apply. The following tools, or their equivalents are absolutely necessary to properly perform this procedure:**

- Cam Chain Tensioner tool T97T-6K254-A
- Cam Gear Removal tool T97T-6256-F
- Cam Gear Torque adapter T97T-6256-G
- Camshaft Gear Positioning/Holding tool T97T-6256-B
- Camshaft Gear Positioning/Holding tool adapter T97T-6256-A
- Camshaft holding tool T97T-6256-C
- Crankshaft holding tool T97T-6303-A
- Camshaft holding tool adapter T97T-6256-D

2. Before servicing the vehicle, refer to the Precautions Section.

3. Properly relieve the fuel system pressure.

4. Drain the cooling system.

5. Remove or disconnect the following:

- Negative battery cable
- Lower intake manifold
- Fan blade and shroud
- Valve cover
- Roller followers, if equipped
- Drive belt
- Upper radiator hose and tube
- Alternator electrical connectors
- Alternator mounting bracket
- Engine accessory bracket and move it aside

06017-RANG-G02

Cylinder head bolt torque sequence—2005 3.0L engine

- Camshaft Position (CMP) electrical connector
- Crankshaft Position (CKP) sensor electrical connector
- Engine Coolant Temperature (ECT) sensor electrical connector
- Coil pack electrical connector
- Exhaust Gas Recirculation (EGR) valve electrical connector
- EGR valve bracket and move it aside
- Heater hoses
- Fuel injector electrical connectors
- Water bypass hose
- Thermostat housing
- Spark plug wires
- Fuel injection supply manifold
- Fuel injectors
- Crankcase vent separator spring
- Oil dipstick housing
- Exhaust manifold
- Hydraulic chain tensioner
- Cassette retaining bolt
- Camshaft sprocket
- Cylinder head and discard the gasket

To install:

6. Thoroughly clean all gasket mating surfaces. Remove all traces of old gasket material, oil, grease or dirt.

7. Insure that the rubber band is holding the right-hand chain to the cassette.

8. Install a new head gasket and the cylinder head.

9. Torque the new cylinder head bolts in sequence as follows:

   a. Install bolts 1, 2, 3, 4, 7, 8 (12mm) and torque, in sequence to:

- Step 1: 24 ft. lbs. (32 Nm).
- Step 2: Plus 80 degrees.
- Step 3: Plus an additional 80 degrees.
- Install bolts 5 and 6 (8mm), and torque to 24 ft. lbs. (32 Nm).

32 Nm (24 lb-ft)

67197-RANG-G9A

Cylinder head bolt torque sequence 4.0L engine

10. Install or connect the following:

- Camshaft sprocket in the cassette and make certain that the camshaft sprocket turns freely on the camshaft
- Cassette retaining bolt. Torque the bolt to 89 inch lbs. (10 Nm).
- Exhaust manifold
- Oil level indicator tube. Torque the bolt to 18 ft. lbs. (25 Nm).
- Crankcase vent separator and spring
- Thermostat housing. Torque the bolts to 8 ft. lbs. (11 Nm).
- Water bypass hose
- Heater hoses
- EGR bracket. Torque the bolt to 89 inch lbs. (10 Nm).
- EGR tube. Torque the nut to 30 ft. lbs. (40 Nm).
- ECT sensor electrical connector
- Electrical harness retainer. Torque the bolt to 89 inch lbs. (10 Nm).
- CKP and CMP electrical connectors

- Accessory bracket. Torque the bolts to 31 ft. lbs. (42 Nm).
- Alternator mounting bracket. Torque the bolts to 31 ft. lbs. (42 Nm).
- Alternator and electrical connectors
- Drive belt
- Fan shroud
- Roller followers
- Valve cover
- Lower intake manifold
- Negative battery cable

11. Change the engine oil and filter.
12. Refill the cooling system.
13. Start the engine and check for leaks, repair if necessary.

Rocker Arms/Shafts

REMOVAL & INSTALLATION

3.0L Engines

1. Before servicing the vehicle, refer to the Precautions Section.

2. Remove or disconnect the following:
- Negative battery cable
- Rocker arm covers
- Retaining bolt at each rocker arm

3. The rocker arm and pushrod may then be removed from the engine. Keep all rocker arms and pushrods in order so they may be installed in their original locations.

To install:

4. Lubricate the rocker arm assemblies with SAE 50W engine oil.

5. Ensure that the fulcrums are properly seated into the cylinder head. Torque the rocker arm fulcrum bolts to 19 ft. lbs. (26 Nm).

6. Install the rocker arm covers and connect the negative battery cable.

67197-RANG-G8A

The correct cylinder head bolt loosening sequence must be used to prevent warpage—4.0L engine

4. Rocker arm
5. Pushrod
6. Fulcrum

7. Bolt
8. Assembled rocker arm

7924EG27

Exploded view of the rocker arm assembly—3.0L engine

Intake Manifold

REMOVAL & INSTALLATION

2.3L Engine

1. Before servicing the vehicle, refer to the Precautions Section.
2. Relieve the fuel system pressure.
3. Drain the cooling system.
4. Properly discharge the air conditioning system.
5. Remove or disconnect the following:
   • Negative battery cable
   • Water outlet tube.
   • Water outlet tube.

➡**The alternator will be removed with the accessory bracket.**

   • Accessory bracket.
   • Coolant hose from the thermostat.
   • Coolant hose from the EGR valve.
   • Coolant tube assembly.
   • Block heater (if so equipped).
   • Water outlet.
   • EGR valve.
   • Idle air control (IAC) valve.
   • Throttle position (TP) sensor.
   • Manifold absolute pressure (MAP) sensor.
   • Swirl control valve monitor electrical connector.
   • Electric thermostat.
   • Swirl control valve.
   • CMP sensor electrical connector and disconnect the PCV hose from the intake manifold.

67197-RANG-G19

2.3L engine intake manifold bolts

   • Engine wiring harness pin-type retainers from the intake manifold.
   • Engine wiring harness connector bracket. Position the engine wiring harness aside.
   • EGR tube.
   • Fuel supply line clip from the front of the intake manifold. Disconnect the vacuum hose from the intake manifold.

   • Intake manifold assembly.
6. Installation is the reverse of removal. Torque the bolts to 13 ft. lbs. (18Nm). There is no special torque sequence.

3.0L Engine

UPPER MANIFOLD

1. Before servicing the vehicle, refer to the Precautions Section.
2. Disconnect the battery ground cable.
3. Remove the air cleaner outlet pipe.
4. Remove the bolt and the snow shield.
5. Disconnect the accelerator cable, speed control cable and the return spring.
6. Disconnect the accelerator cable and speed control cable from the accelerator bracket.
7. Disconnect the throttle position (TP) sensor and idle air control (IAC) valve electrical connectors.
8. Disconnect the two vacuum hoses.
9. Detach the 42-pin wiring harness retainer from the stud bolt.
10. Remove the nut and the stud bolt. Position the 42-pin wiring harness and bracket aside.
11. Remove the oil indicator tube bracket nut and the upper intake manifold support bracket stud bolt.
12. Disconnect the brake booster vacuum hose and the positive crankcase ventilation (PCV) valve tube.
13. Remove the bolts, the stud bolt and the upper intake manifold. Remove and discard the gaskets. Clean and inspect all mating surfaces.
14. To install, reverse the removal procedure. Install new gaskets and tighten the upper intake manifold bolts and stud bolt in the sequence shown, to 89 inch lbs. (10 Nm).

10 Nm
(89 lb-in)

06017-RANG-G03

Upper intake manifold torque sequence—3.0L engine

LOWER MANIFOLD

1. Before servicing the vehicle, refer to the Precautions Section.
2. Drain the engine cooling system.
3. Remove both the valve covers.
4. Disconnect the upper radiator hose, heater hose and coolant bypass hose.
5. Disconnect the engine coolant temperature (ECT) sensor and coolant temperature sender electrical connectors.
6. Disconnect the crankshaft position (CKP) sensor and detach the wiring from the stud bolt.
7. Disconnect the fuel injector electrical connectors and position the engine control wiring aside.
8. Release the fuel pressure and disconnect the fuel tube.
9. Loosen the rocker arm bolts.

➡ **Identify the location of each pushrod. Each pushrod is to be installed in the original location to prevent premature wear.**

10. Remove all the pushrods.
11. Remove the bolts from the lower intake manifold.

➡ **Gently loosen the intake manifold to separate the silicone sealant from the cylinder block.**

12. Remove the lower intake manifold.
13. Remove and discard the intake manifold gaskets and end seals.

To install:

➡ **Clean and inspect all mating surfaces.**

➡ **If the lower intake manifold is not installed within four minutes, remove the sealant and reapply.**

14. Apply a drop of silicone gasket and sealant at the four cylinder block-to-cylinder head seams.
15. Install the intake manifold gaskets and end seals.
16. Position the lower intake manifold.
17. Install the bolts and tighten in the sequence shown to 21 ft. lbs. (29 Nm).
18. Install the pushrods.
19. Tighten the rocker arms to 24 ft. lbs. (32 Nm).
20. Connect the fuel tube.
21. Connect the fuel injector electrical connectors.
22. Connect the crankshaft position (CKP) sensor and attach the wiring from the stud bolt.
23. Connect the engine coolant temperature (ECT) sensor and coolant temperature sender electrical connectors.
24. Connect the upper radiator hose, heater hose and coolant bypass hose.
25. Install both valve covers.
26. Fill and bleed the engine cooling system.

4.0L Engine

1. Before servicing the vehicle, refer to the Precautions Section.
2. Disconnect the battery ground cable.
3. Remove the bolts and the shield.
4. Remove the air cleaner outlet pipe.

5. Disconnect the idle air control (IAC) valve, throttle position (TP) sensor electrical connectors and the TP sensor wiring pin-type retainer.
6. Disconnect the MAF sensor wiring pin-type retainer.
7. Detach the accelerator and speed control cables from the throttle body.
8. Detach the accelerator and speed control cables from the bracket, and position the cables aside.
9. Disconnect the exhaust gas recirculation (EGR) valve vacuum hose and tube fitting.
10. Disconnect the EGR vacuum regulator solenoid valve electrical connector and vacuum hose.
11. Disconnect the hose.
12. Loosen the clamp and disconnect the brake booster vacuum hose.

❊❊ CAUTION

It is important to twist the spark plug wire boots while pulling upward to avoid possible damage to the spark plug wire.

➡ **Mark the spark plug wire locations before removing them.**

13. Disconnect the RH spark plug wires from the coil. Remove the spark plug wire routing clip pin-type retainer and position the wires aside.
14. Remove the wiring harness bracket retainer, then position the wiring harness aside.
15. Remove the accelerator cable routing clip pin-type retainer and the wiring harness pin-type retainer.
16. Remove the bolts.
17. Remove the bolts and position the coil and bracket aside.
18. Disconnect the vacuum hoses.
19. Remove the nut.
20. Disconnect the powertrain control module (PCM) electrical connector.
21. Remove the retainer and position the ground wires aside.
22. Detach the electrical connector retainer.
23. Remove the intake manifold bolts and lift up the intake manifold.
24. Remove the heated positive crankcase ventilation (PCV) hose retainers and remove the heated PCV fitting.
25. Remove the intake manifold.
26. To install, reverse the removal procedure. Torque the fasteners to 89 inch lbs. (10 Nm).

29 Nm (21 lb-ft)

Lower intake manifold torque sequence—3.0L engine

10 Nm (89 lb-in)

67197-RANG-G20

Intake manifold installation—4.0L engine

Exhaust Manifold

REMOVAL & INSTALLATION

2.3L Engine

1. Before servicing the vehicle, refer to the Precautions Section.
2. Remove or disconnect the following:
   - Negative battery cable
   - Exhaust flange nuts
   - Drive belt
   - Coolant
   - Upper radiator hose and the engine reservoir hose
   - Air conditioning compressor
   - Heater hose
   - Oil indicator and the upper bolt for the tube assembly
   - Lower bolt and remove the oil indicator tube assembly
   - Front radiator tube
3. Remove the pushpins and position the right inner fender splash shield out of the way.
4. Remove or disconnect the following:
   - Alternator electrical connectors
   - Lower front end accessory drive (FEAD) mounting bolts
   - Upper mounting bolt and the FEAD assembly
   - Two nuts and position the coolant tube out of the way
   - Exhaust manifold
   - Exhaust manifold gasket

To install:

5. Install or connect the following:
   - Exhaust manifold gasket
   - Exhaust manifold and the nuts
   - Coolant tube and the nuts
   - FEAD assembly and the upper mounting bolts, finger tight.
   - Lower FEAD mounting bolts, finger tight. Then, torque all FEAD bolts, in the sequence shown, to 35 ft. lbs. (47 Nm).
   - Alternator electrical connectors
   - Right inner splash shield and pushpins
   - Upper radiator tube and install the bolts
   - Oil indicator tube assembly and the lower bolt
   - Oil indicator tube upper bolt and the oil indicator
   - Heater water hose
   - Air conditioning compressor
   - Upper radiator hose and the engine reservoir hose
6. Fill the cooling system.

67197-RANG-G21

Exhaust manifold retaining nuts—2.3L engine

06017-RANG-G04

FEAD bracket—2.3L engine

7. Install the serpentine drive belt.
8. Install the exhaust flange nuts.
9. Connect the battery ground cable.

3.0L Engine

LEFT SIDE

1. Before servicing the vehicle, refer to the Precautions Section.
2. Install or connect the following:
   - Negative battery cable
   - Exhaust flange nuts
   - Exhaust Gas Recirculation (EGR) valve from the exhaust manifold tube
   - Oil lever indicator and bracket
   - Exhaust manifold and discard the gasket

To install:

3. Clean the mating surfaces for the exhaust manifold and cylinder head.
4. Install a new gasket and the exhaust manifold. Torque the bolts in sequence to:
   a. 89 inch lbs. (10 Nm).
   b. 15 ft. lbs. (20 Nm).
5. Install or connect the following:
   - Oil lever indicator tube and bracket. Torque the bolt to 12 ft. lbs. (16 Nm).
   - EGR valve to the exhaust manifold tube. Torque the fastener to 26 ft. lbs. (35 Nm).
   - Exhaust flange. Torque the nuts to 25 ft. lbs. (34 Nm).
   - Negative battery cable
6. Start the vehicle and check for leaks, repair if necessary.

RIGHT SIDE

1. Before servicing the vehicle, refer to the Precautions Section.

Tighten the right side exhaust manifold bolts in the proper sequence—3.0L

2. Remove or disconnect the following:
   - Negative battery cable
   - Exhaust manifold flange
   - Ignition coil support bracket
   - Exhaust manifold and discard the gasket

To install:

3. Clean the mating surfaces for the exhaust manifold and cylinder head
4. Install a new gasket and the exhaust manifold. Torque the bolts is sequence to:
   a. 89 inch lbs. (10 Nm).
   b. 18 ft. lbs. (25 Nm).
5. Install or connect the following:
   - Ignition coil support bracket. Torque the bolts to 15 ft. lbs. (20 Nm).
   - Exhaust flange nuts. Torque the nuts to 33 ft. lbs. (46 Nm).
   - Negative battery cable
6. Start the vehicle and check for leaks, repair if necessary.

4.0L Engine

1. Before servicing the vehicle, refer to the Precautions Section.
2. Remove or disconnect the following:
   - Negative battery cable
   - Exhaust inlet pipe-to-manifold attaching bolts
   - Differential Pressure Feedback EGR (DPFE) transducer hoses, left side manifold only
   - Exhaust Gas Recirculation (EGR) tube from the manifold and valve, left side manifold only
   - Exhaust manifold and discard the gasket

To install:

3. Clean the gasket mating surfaces.
4. Install or connect the following:
   - New gasket and the exhaust manifold. Torque the bolts to 16 ft. lbs. (22 Nm).

Tighten the exhaust manifold bolts in sequence—3.0L left side

67197-RANG-G22

Left side exhaust manifold retaining nuts—4.0L engine

67197-RANG-G23

Right side exhaust manifold retaining nuts—4.0L engine

- EGR tube to the manifold. Torque the fastener to 30 ft. lbs. (40 Nm) left side manifold only
- DPFE transducer hoses, left side manifold only
- Exhaust inlet pipe-to-manifold attaching bolts. Torque the bolts to 30 ft. lbs. (40 Nm).
- Negative battery cable
5. Start the vehicle and check for leaks, repair if necessary.

Front Crankshaft Seal

REMOVAL & INSTALLATION

2.3L Engines

1. Before servicing the vehicle, refer to the Precautions Section.
2. Remove or disconnect the following:
   - Negative battery cable
   - Crankshaft pulley

✳ WARNING

Use care not to damage the engine front cover or the crankshaft when removing the seal.

- Crankshaft front oil seal by prying the seal out of the front cover

To install:
3. Using the special tool, install the crankshaft front oil seal.
4. Install the crankshaft pulley.

67197-RANG-G24

Front crankshaft seal—2.3L engine

5. Tighten the crankshaft damper in two stages:
   - Step 1: Tighten to 74 ft. lbs. (100 Nm)
   - Step 2: Tighten an additional 90 degrees

3.0L Engine

1. Before servicing the vehicle, refer to the Precautions Section.
2. Remove the crankshaft vibration damper.
3. Remove the crankshaft front seal.

303-096

67197-RANG-G25

Front crankshaft seal installation—2.3L engine

67197-RANG-G26

Front crankshaft seal—3.0L engine

To install:
4. Install the crankshaft front seal.
   a. Lubricate the seal lip with clean engine oil.
   b. Using a seal installer, install the crankshaft front seal.
5. Install the crankshaft damper. Torque to 107 ft. lbs. (145 Nm).

4.0L Engine

1. Before servicing the vehicle, refer to the Precautions Section.
2. Remove or disconnect the following:
   - Negative battery cable
   - Crankshaft pulley
3. Using a seal remover, remove the crankshaft front oil seal.
To install:
4. Lubricate the seal lip with clean engine oil.
5. Using a seal driver, install the crankshaft front oil seal.
6. Install the crankshaft pulley. Torque to 37 ft. lbs. (50 Nm), then 90 degrees more.

303-107

67197-RANG-G28

Front crankshaft seal removal—4.0L engine

303-102

303-093

67197-RANG-G29

Front crankshaft seal installation—4.0L engine

Camshaft and Valve Lifters

➡ **Although Ford suggests that this component is removable while the engine is installed in the vehicle, depending on the particular options with which your truck is equipped, working clearance may be extremely tight and this procedure may be much easier to perform with the engine removed. Before commencing, read through this procedure and make certain enough clearance, or working room, exists with the engine in the vehicle; if there is not enough space, the engine should be removed.**

REMOVAL & INSTALLATION

2.3L Engine

1. Before servicing the vehicle, refer to the Precautions Section.
2. Relieve the fuel system pressure.
3. Drain the cooling system.
4. Properly discharge the air conditioning system.
5. Remove or disconnect the following:
   - Negative battery cable
   - Drive belt.
   - Engine oil level indicator assembly.
   - Engine oil level indicator.
   - Engine oil level indicator tube.
   - Water outlet tube.
   - Water outlet tube.
   - Air conditioning compressor.

➡ **The generator will be removed with the accessory bracket.**

   - Accessory bracket.
   - Right motor mount.
   - Coolant hose from the thermostat.
   - Coolant hose from the EGR valve.
   - Coolant tube assembly.
   - Exhaust manifold and gasket.
   - Block heater (if so equipped).

   - Water outlet.
   - EGR valve.
   - Power steering pump and reservoir as an assembly.
   - Idle air control (IAC) valve.
   - Throttle position (TP) sensor.
   - Manifold absolute pressure (MAP) sensor.
   - Swirl control valve monitor electrical connector.
   - CKP sensor and the wiring harness pin-type retainers.
   - Knock sensor (KS).
   - Electric thermostat.
   - Swirl control valve.
   - CMP sensor electrical connector and disconnect the PCV hose from the intake manifold.
   - Engine wiring harness pin-type retainers from the intake manifold.
   - Engine wiring harness connector bracket. Position the engine wiring harness aside.
   - EGR tube.
   - Fuel supply line clip from the front of the intake manifold. Disconnect the vacuum hose from the intake manifold.
   - Intake manifold assembly.
   - Fuel injector electrical connectors. Detach the wiring harness pin-type retainers.

   - Ignition coil and the cylinder head temperature (CHT) sensor electrical connectors.
   - Engine wiring harness anchors from the valve cover studs. Remove the engine wiring harness.
   - Ignition coil.
   - Bypass hose.
   - Thermostat housing.
   - Knock sensor and the engine vent cover.
   - Left motor mount.
   - Fuel injector supply manifold with the injectors and the ground strap.
   - Water pump pulley.
   - Water pump.
   - CMP sensor.
   - CHT sensor.
   - Spark plugs.
   - Valve cover.
   - CKP sensor.
   - Crankshaft vibration damper

➡ **There is one front cover bolt behind the cooling fan drive pulley. To remove this bolt, align one of the cooling fan drive pulley access holes with the bolt head to access the bolt.**

   - Front cover.
   - Timing chain tensioner.
   - Timing chain guides.
   - Timing chain assembly.

➡ **Use a wrench on the flats between cylinders No. 1 and No. 2 to hold the camshaft in place.**

   - Camshaft drive sprockets.
   - Oil pump chain tensioner and guide.

➡ **The oil pump chain sprocket must be held in place.**

   - Oil pump chain and sprockets.

➡ **Note the position of the lobes on the No. 1 cylinder before removing the camshafts for assembly reference.**

6. Loosen the camshaft bearing cap bolts in sequence, one turn at a time.

67197-RANG-G5A

Camshaft cap loosening sequence—2.3L

67197-RANG-G6A

Camshaft cap torque sequence—2.3L

Repeat the first step until all tension is released from the camshaft bearing caps. Remove the camshaft bearing caps.

7. Remove the camshafts.
8. Installation is the reverse of removal.

➡Install the camshafts with the alignment notches in the camshaft lined up so the camshaft alignment plate can be installed without rotating the camshafts. Make sure the lobes on the No. 1 cylinder are in the same position as noted in the disassembly procedure. Rotating the camshafts, or installing the camshafts 180 degrees out of position can cause severe damage to the valves and pistons. Lubricate the camshaft journals and bearing caps with clean engine oil. Install the camshafts and bearing caps. Tighten the bolts in the sequence shown in three stages.

- Step 1: Tighten the camshaft bearing caps one turn at a time until tight.
- Step 2: Tighten the bolts to 7 Nm (62 inch lbs.)
- Step 3: Tighten the bolts to 16 Nm (12 ft. lbs.)
  a. Crankshaft vibration damper:

➡Do not reuse the crankshaft pulley bolt. Tighten the bolt in two stages.

- Step 1: Tighten the bolt to 40 Nm (30 ft. lbs.)
- Step 2: Tighten the bolt and additional 90 degrees (1/4 turn).

3.0L Engine

1. Before servicing the vehicle, refer to the Precautions Section.
2. Properly relieve the fuel system pressure.
3. Drain the cooling system.
4. Drain the engine oil.
5. Evacuate the air conditioning system.

6. Remove or disconnect the following:
- Negative battery cable
- Air cleaner hoses
- Fan, spacer and shroud
- Radiator

7. Rotate the crankshaft so that No. 1 piston is at Top Dead Center (TDC) on the compression stroke.
- Air conditioning condenser
- Fuel lines from the fuel supply manifold
- Vacuum hoses
- Electrical wiring
- Engine front cover
- Water pump
- Alternator
- Power steering pump. Do not disconnect the hoses
- Air conditioning compressor. Do not disconnect the hoses
- Throttle body
- Fuel injection wire harness

8. Turn the engine by hand to TDC of the power stroke on No. 1 cylinder.
- Spark plug wires from the plugs
- Distributor cap with the spark plug wires as an assembly, if equipped

9. Matchmark the rotor, distributor body and engine. Disconnect the distributor wiring harness and remove the distributor, if equipped.
- Rocker arm covers
- Intake manifold
- Loosen the rocker arm bolts enough to pivot the rocker arms out of the way and remove the pushrods. Identify them for installation
- Lifters and identify them for installation
- Crankshaft pulley/damper
- Starter
- Oil pan
- Camshaft gear attaching bolt and washer, then slide the gear off the camshaft

- Camshaft thrust plate

10. Carefully slide the camshaft out of the engine block, using caution to avoid any damage to the camshaft bearings.

To install:

11. Oil the camshaft journals and cam lobes with heavy SJ engine oil (50W). Install the spacer ring with the chamfered side toward the camshaft, then insert the camshaft key.

12. Install or connect the following:
- Camshaft using caution to avoid any damage to the camshaft bearings
- Thrust plate. Torque the screws to 84 inch lbs. (10 Nm).

13. Rotate the camshaft and crankshaft as necessary to align the timing marks. Install the camshaft gear and chain. Torque the bolt to 46 ft. lbs. (62 Nm).

14. Coat the tappets with 50W engine oil and place them in their original locations.

15. Apply 50W engine oil to both ends of the pushrods. Install the pushrods in their original locations.

16. Pivot the rocker arms into position. Torque the fulcrum bolts to 96 inch lbs. (11 Nm).

17. Rotate the engine until both timing marks are at the top of their sprockets and aligned. Torque the following fulcrum bolts to 18 ft. lbs. (24 Nm):
   a. No.1 intake.
   b. No.2 exhaust.
   c. No.4 intake.
   d. No.5 exhaust.

18. Rotate the engine until the camshaft timing mark is at the bottom of the sprocket and the crankshaft timing mark is at the top of the sprocket, and both are aligned. Torque the following fulcrum bolts to 18 ft. lbs. (24 Nm):
   a. No.1 exhaust.
   b. No.2 intake.
   c. No.3 intake and exhaust.
   d. No.4 exhaust.
   e. No.5 intake.
   f. No.6 intake and exhaust.

19. Torque all the bolts to 24 ft. lbs. (33 Nm).

20. Turn the engine by hand to 0 degrees Before Top Dead center (BTDC) of the power stroke on No. 1 cylinder.

21. Install or connect the following:
- Engine front cover and water pump assembly
- Oil pan
- Crankshaft damper/pulley and tighten the retaining bolt to 107 ft. lbs. (145 Nm).
- Intake manifold
- Starter

- Crankshaft pulley and damper
- Rocker arm covers
- Rotor and distributor cap, if equipped
- Spark plug wires
- Fuel lines to the fuel supply manifold
- Fuel injection wire harness
- Throttle body
- Air conditioning compressor
- Power steering pump
- Alternator
- Water pump
- Engine front cover
- All electrical connectors and vacuum lines
- Air conditioning condenser
- Radiator
- Fan, spacer and shroud
- Air cleaner hoses
- Negative battery cable

22. Recharge the air conditioning system.
23. Refill the cooling system.
24. Replace the oil filter and refill the engine with the specified amount of engine oil.
25. Start the engine and check the ignition timing and idle speed. Adjust if necessary. Run the engine at fast idle and check for coolant, fuel, vacuum or oil leaks.

4.0L Engine

1. Before servicing the vehicle, refer to the Precautions Section.
2. Remove or disconnect the following:
   - Negative battery cable for safety
   - Valve cover
   - Hydraulic camshaft tensioner

➡**The right-hand camshaft sprocket bolt uses left-hand threads.**

3. For the right-hand camshaft use the Cam Gear Torque Adapter tool T97T-6256-F, to remove the camshaft sprocket bolt.
4. For the left-hand camshaft, remove the sprocket bolt.

Camshaft bolt tightening sequence—4.0L engine

➡**When removing the followers, label them so that they may be returned to their original positions.**

5. Using the Valve Spring Compressor tool ST1330-A, remove the camshaft roller followers.
6. Remove or disconnect the following:
   - Camshaft bearing cap bolts and the oil rail
   - Camshaft

To install:

7. Lubricate all of the moving parts with SAE 50W engine oil.
8. Install camshaft onto the cylinder head.
9. Position the oil rail and install the bearing caps and bolts. Torque the bolts in 2 steps:
   a. Step 1—53.5 inch lbs. (6 Nm).
   b. Step 2—11–12.5 ft. lbs. (15–17 Nm).
10. Install or connect the following:
    - Camshaft followers
    - Camshaft sprocket bolt and hand tighten the bolt
    - Camshaft Chain Tensioner T97T-6K254-A in the hole that the hydraulic chain tensioner was in
11. Turn the crankshaft one revolution clockwise until No. 1 piston is Top Dead Center (TDC).

12. Install or connect the following:
    - Crankshaft Holding tool T97T-6303-A on the crankshaft to keep it from turning
    - Position the timing slot on the rear of the camshaft to fit Camshaft Holding tool T97T-6256-C and install the holding tool on the rear of the head
    - Camshaft Gear Holding tool T97T-6256-B and Camshaft Gear Holding tool T97T-6256-A on the front of the cylinder head to securely hold the camshaft gear
    - Tighten the camshaft sprocket bolt to 63 ft. lbs. (85 Nm).
13. Remove the Camshaft Chain Tensioner tool and install the hydraulic chain tensioner, tighten the tensioner to 35–39 ft. lbs. (47–53 Nm).
14. Remove the special tools from the engine.
15. Install or connect the following:
    - Valve cover
    - Negative battery cable
16. Start the engine check for leaks and repair if necessary.

Oil Pan

REMOVAL & INSTALLATION

2.3L Engine

1. Before servicing the vehicle, refer to the Precautions Section.
2. Drain the engine oil.
3. Remove or disconnect the following:
   - Engine from the vehicle
   - Engine oil level indicator assembly
   - Engine oil pan bolts and oil pan

To install:

4. Clean and inspect all mating surfaces.

➡**The oil pan must be installed and the bolts tightened with four minutes of applying the silicone gasket and sealant.**

Camshaft bolt removal sequence—4.0L engine

25 Nm (18 lb-ft)

Oil pan torque sequence—2.3L

67197-RANG-G7A

4-6 mm (0.16-0.23 in)

Oil pan sealer application—3.0L engine

67197-RANG-G41

06017-RANG-G05

Oil pan bolt torque sequence—3.0L engine with 2-wheel drive

5. Apply a 2.5 mm bead of silicone gasket and sealant to the oil pan. Install the oil pan. Tighten the oil pan in the sequence shown.

6. Lubricate the O-ring with clean engine oil and install the engine oil level indicator assembly.

7. Install the engine into the vehicle.

3.0L Engine

2WD

1. Before servicing the vehicle, refer to the Precautions Section.

2. Drain the engine oil.

3. Remove or disconnect the following:
- Negative battery cable
- Oil level dipstick tube
- Fan shroud. Leave the fan shroud over the fan assembly
- Motor mount nuts from the frame

※※ WARNING

On models equipped with distributor ignition, failure to remove the distributor will damage or break it when the engine is lifted.

- Starter
- Transmission inspection cover
- Right hand axle I-beam. The brake caliper must be removed and secured out of the way.
- Oil pan attaching bolts, using a suitable lifting device, raise the engine about 2 in. (5cm)
- Oil pan and discard the gasket

➡ **The oil pan fits tightly between the transmission spacer plate and oil pump pick-up tube. Use care when removing the oil pan from the engine.**

4. Clean all gasket surfaces on the engine and oil pan. Remove all traces of old gasket and/or sealer.

To install:

5. Apply a ⅛ (4mm) bead of RTV sealer to the junctions of the rear main bearing cap and block, and the front cover and block. The sealer sets in 15 minutes, so work quickly!

6. Apply adhesive to the gasket surfaces and install the oil pan gasket.

7. Install or connect the following:
- Oil pan on the engine block. Torque the bolts EVENLY to 9 ft. lbs. (12 Nm) working from the center to the end position on the oil pan.
- Right hand axle I-beam
- Brake caliper
- Transmission inspection cover
- Starter

Oil pan bolts—3.0L engine with 4-wheel drive

9 Nm (80 lb-in)

Oil pan installation—4.0L engine

- Fan shroud
- Motor mount retaining nuts
- Oil level dipstick tube
- Negative battery cable

8. Fill the engine with clean oil.

9. Start the vehicle and check for leaks, repair if necessary.

4WD

1. Before servicing the vehicle, refer to the Precautions Section.

2. Drain the engine oil.

3. Remove or disconnect the following:
- Negative battery cable
- Engine from the vehicle and place it on a suitable engine stand
- Oil pan and discard the gasket

To install:

4. Install or connect the following:
- New oil pan gasket and secure the gasket with trim adhesive
- Oil pan. Torque the bolts to 9 ft. lbs. (12 Nm).
- Engine
- Negative battery cable

5. Fill the engine with clean oil.

6. Start the vehicle and check for leaks, repair if necessary.

4.0L Engine

➡**The 4.0L engine does not use an oil pan in the conventional sense. There is a separate access panel that unbolts from what would be considered the oil pan (which is now known as the ladder frame).**

1. Before servicing the vehicle, refer to the Precautions Section.

2. Drain the engine oil.

3. Remove or disconnect the following:
- Negative battery cable
- Oil pan and discard the gasket

To install:

4. Install or connect the following:
- New gasket and oil pan. Torque the bolts to 80 inch lbs. (9 Nm).
- Negative battery cable

5. Fill the engine with clean oil.

6. Start the vehicle and check for leaks, repair if necessary.

Oil Pump

REMOVAL & INSTALLATION

2.3L Engine

➡**The oil pump is located on the front of the engine and is turned by the timing belt.**

1. Before servicing the vehicle, refer to the Precautions Section.

2. Remove or disconnect the following:
- Negative battery cable
- Timing chain
- Oil pump chain and sprockets
- Oil pan
- Oil pump pickup tube and gasket
- Oil pump assembly and gasket

To install:

3. Turn the crankshaft clockwise to position the No. 1 piston.

4. Remove the plug bolt.

5. Install the Engine Timing Peg 303-507.

➡**Clean the gasket surface with metal surface cleaner.**

6. Install a new gasket and the oil pump assembly. Tighten the bolts in the sequence shown in two stages.
- Step 1: Tighten the bolts to 10 Nm (80 inch lbs.)
- Step 2: Tight the bolts to 23 Nm (17 ft. lbs.)

7. Install a new oil pump pickup tube

Oil pump torque sequence—2.3L

10 Nm (89 lb-in)

Oil pump pickup tube torque sequence—2.3L

gasket and the pickup tube. Tighten the bolts in the sequence shown

3.0 L Engines

1. Before servicing the vehicle, refer to the Precautions Section.
2. Drain the engine oil.
3. Remove or disconnect the following:
   - Negative battery cable
   - Oil pan
   - Oil pick-up and tube assembly from the pump
   - Oil pump retainer bolts and the oil pump

To install:

4. Prime the oil pump with clean engine oil by filling either the inlet or outlet port. Rotate the pump shaft to distribute the oil within the pump body.
5. Install the oil pump and tighten the mounting bolts to 30–40 ft. lbs. (41–54 Nm).

✳✳ WARNING

Do not force the oil pump if it does not seat readily. The oil pump drive-shaft may be misaligned with the distributor or shaft assembly. If the pump is tightened down with the driveshaft misaligned, damage to the pump could occur. To align, rotate the intermediate driveshaft into a new position.

40-55 Nm
(30-40 lb/ft)

67197-RANG-G46

Oil pump installed—3.0L engine

6. Install or connect the following:
   - Oil pick-up and tube assembly
   - Oil pan
7. Fill the engine with clean oil.
8. Start the vehicle and check for leaks, repair if necessary.

4.0L Engines

➡**The oil pump cannot be removed with the engine in the vehicle.**

1. Before servicing the vehicle, refer to the Precautions Section.
2. Drain the engine oil.
3. Remove or disconnect the following:
   - Engine from the vehicle
   - Oil pan
   - Unbolt the oil pick-up tube
   - The 8 ladder frame bolts that were under the oil pan
   - The 2 rear outer ladder frame bolts
   - The 7 left-hand and the 8 right-hand ladder frame bolts
   - The ladder frame from the engine
   - The 2 oil pump attaching bolts and the pump.

To install:

4. Submerge the pump in clean engine oil to prime it.
5. Install or connect the following:

10 Nm (89 lb-ft)

67197-RANG-G47

Oil pump installation—4.0L engine

   - The ladder frame on the engine
   - The 8 right-hand and 7 left-hand ladder frame bolts
   - The 2 rear outer and the 8 frame bolts under the pan
   - The oil pump. Torque the bolts to 13–15 ft. lbs. (17–21 Nm).
   - Oil pick-up tube
   - Oil pan
   - Engine to the vehicle
   - Negative battery cable
6. Fill the engine with clean oil.
7. Start the vehicle and check for leaks, repair if necessary.

Rear Main Seal

REMOVAL & INSTALLATION

2.3L Engine

1. Before servicing the vehicle, refer to the Precautions Section.
2. Remove or disconnect the following:
   - Flywheel or flexplate
   - Bolts and the crankshaft rear oil seal

To install

3. Install or connect the following:
   - Rear oil seal on the Crankshaft Rear Main Oil Seal Installer
   - Crankshaft Rear Main Oil Seal Installer and the crankshaft rear oil seal on the crankshaft

9348EG08

Rear main seal torque sequence—2.3L

4. Tighten the bolts in the sequence shown to 10 Nm (89 inch lbs.)

5. Remove the Crankshaft Rear Main Oil Seal Installer.

6. Install the flywheel or flexplate.

3.0L Engines

1. Before servicing the vehicle, refer to the Precautions Section.

2. Remove the flexplate or flywheel.

※※ WARNING

Use care to avoid scratching or damaging the oil seal surface or leakage may occur.

3. Using a sharp awl, punch one hole into the crankshaft rear oil seal metal surface between the seal lip and the cylinder block.

4. Screw the threaded end of the special tool into the oil seal. Use the special tool to remove the crankshaft rear oil seal.

To install:

5. Lubricate the outer lips and the inner seal on the crankshaft rear oil seal with clean engine oil.

6. Using the special tool, install the crankshaft rear oil seal. Alternate bolt tightening to correctly seat the crankshaft rear oil seal.

7. Install the flexplate or flywheel.

4.0L Engine

1. Before servicing the vehicle, refer to the Precautions Section.

Rear main seal removal—3.0L engine

Rear main seal installation—3.0L engine

Rear main seal removal—4.0L engine

Front part of installation tool—4.0L engine

Rear main seal installation—4.0L engine

2. Remove the flexplate or flywheel.

※※ WARNING

Avoid scratching or damaging the oil crankshaft seal running surface during removal of the crankshaft rear oil seal.

3. Using the special tool, remove the crankshaft rear oil seal.

To install:

➡ **Be sure the crankshaft rear sealing surface is clean and free of any rust or corrosion. To clean the crankshaft rear sealing surface, use extra-fine emery cloth or extra-fine 0000 steel wool with metal surface cleaner.**

4. Lubricate the crankshaft rear oil seal with clean engine oil and install on the special tool.

5. Using the special tool, install the crankshaft rear oil seal.

6. Install the flexplate or flywheel.

Timing Chain, Sprockets, Front Cover and Seal

REMOVAL & INSTALLATION

2.3L Engine

1. Before servicing the vehicle, refer to the Precautions Section.

2. Remove or disconnect the following:
   - Negative battery cable
   - Fan and shroud
   - Drive belt
   - Valve cover

3. Set No. 1 piston to TDC and install the Camshaft Alignment Plate 303-376, or equivalent.

4. Remove the plug for the crankshaft timing peg.

5. Install the Crankshaft Timing Peg 303-507, or equivalent.

6. Install an M6 bolt into the crankshaft pulley to verify the engine timing.

7. Remove or disconnect the following:
   - Camshaft pulley
   - Crankshaft position sensor
   - Belt tensioner
   - Water pump pulley
   - Power steering high pressure hose. Remove the nylon O-ring.
   - Power steering return hose
   - Power steering pump

➡ **This step is needed only if a new front cover is being installed.**

8. Using a three-jaw puller, remove the fan drive pulley.

➡ **There is one bolt behind the cooling fan drive pulley. This bolt can be accessed by lining up one of the holes in the pulley with the bolt.**

9. Remove the bolts and the engine front cover.

10. Compress the timing chain tensioner and remove the tensioner.

11. Remove the right-hand timing chain guide.

12. Remove the timing chain.

13. Remove the bolts and the left-hand timing chain guide.

※※ WARNING

Do not rely on the Camshaft Alignment Plate to prevent camshaft rotation. Damage to the tool or the camshaft can occur.

14. If necessary, remove the bolts and the camshaft sprockets. Use the flats on the camshaft to prevent camshaft rotation.

Timing chain tensioner removal; 1-paper clip, 2-bolts—2.3L engine

Camshaft alignment plate installed—2.3L engine

Right side timing chain guide—2.3L engine

Timing chain removal—2.3L engine

Left timing chain guide—2.3L engine

To install:

15. Remove the special tool.

> ✳✳ **WARNING**

Do not rotate the camshafts. Damage to the valves and pistons can occur.

If the camshaft sprockets were not removed, use the flats on the camshafts to prevent camshaft rotation and loosen the sprocket bolts.

16. If removed, install the camshaft sprockets and the bolts. Do not tighten the bolts at this time.

17. Install or connect the following:
- Left-hand timing chain guide and bolts
- Timing chain
- Right-hand timing chain guide
- Timing chain tensioner and release the piston
- Timing chain tensioner and the bolts

18. Remove the drill rod to release the piston.

19. Install the special tool.

> ✳✳ **WARNING**

Do not rely on the Camshaft Alignment Plate to prevent camshaft rotation. Damage to the tool or the camshafts can result. Using the flats on the camshafts to prevent camshaft rotation, tighten the bolts.

➡**This step is needed only if a new front cover is being installed.**

20. Install the fan drive pulley using a nut and bolt with flat washers.

21. Clean and inspect the mounting surfaces of the engine and the front cover.

➡**The engine front cover must be installed and the bolts tightened within four minutes of applying the silicone gasket and sealant.**

22. Apply a 2.5 mm bead of silicone gasket and sealant to the cylinder head and oil pan joint areas. Apply a 2.5 mm bead of silicone gasket and sealant to the front cover.

23. Install the front cover. Tighten the bolts in the sequence shown, to the following specifications:
- Step 1: 8 mm bolts to 10 Nm (89 inch lbs.)
- Step 2: 10 mm bolts to 25 Nm (18 ft. lbs.)
- Step 3: 13 mm bolts to 48 Nm (35 ft. lbs.)

24. Install or connect the following:

Front cover torque sequence—2.3L

06017-RANG-G06

- Power steering pump and lower retaining bolt
- Power steering return hose
- New nylon O-ring and install the high pressure line.
- Water pump pulley
- Belt tensioner

➡Do not reuse the crankshaft damper bolt.

- Crankshaft pulley and hand-tighten the bolt

25. Install an M6 bolt in the crankshaft pulley. Tighten the crankshaft retaining bolt in two stages.
- Step 1: 40 Nm (30 ft. lbs.)
- Step 2: Rotate the bolt an additional 90 degrees.

26. Install the crankshaft position sensor, do not tighten the bolts at this time.

27. Adjust the crankshaft position sensor with the Alignment Tool, and tighten the mounting bolts.

28. Connect the crankshaft position sensor electrical connector.

29. Remove the M6 bolt from the crankshaft pulley.

30. Remove the Crankshaft Timing Peg 303-507.

31. Install the plug.

32. Remove the Camshaft Alignment Plate 303-376.

33. Install the valve cover.

34. Install the drive belt.

35. Install the fan and shroud.

36. Connect the battery ground cable.

3.0L Engines

1. Before servicing the vehicle, refer to the Precautions Section.

2. Remove or disconnect the following:
- Negative battery cable
- Engine front cover
- Rotate the crankshaft and align the timing marks
- Timing chain tensioner, 4.0L engine only
- Sprocket bolt

67197-RANG-G58

Timing mark alignment—3.0L engine

- Timing chain, camshaft sprocket and crankshaft sprocket as an assembly

To install:

3. Install or connect the following:
- Timing chain, camshaft and crankshaft sprockets as an assembly

4. Align the timing marks.

5. Install or connect the following:
- Timing chain tensioner
- Sprocket bolt. Torque the bolt to 51 ft. lbs. (70 Nm).
- Engine front cover
- Negative battery cable

4.0L Engine

TIMING DRIVE COMPONENTS

1. Before servicing the vehicle, refer to the Precautions Section.

2. With the vehicle in neutral, position it on a hoist.

3. Remove the intake manifold.

4. Remove the fuel supply manifold.

5. Remove the accessory drive belt.

6. Remove the thermostat housing.

7. Remove the roller followers.

➡You must retime the LH and RH camshafts when either camshaft is disturbed. Turn the crankshaft clockwise to position the number one cylinder at top dead center (TDC).

➡The special tool must be installed on the damper and should contact the engine block to position the engine at TDC.

8. Install the special tool.

➡The right-hand camshaft sprocket bolt is a left-hand threaded bolt.

➡If necessary, use camshaft gear torque adapter to loosen the camshaft sprocket bolt.

9. Using the special tool, loosen the RH camshaft sprocket bolt.

➡The camshaft timing slots are off-center.

10. Position the camshaft timing slots below the centerline of the camshaft to correctly fit the special tools. Install the special tools on the front of the RH cylinder head.

11. Remove the RH lower splash shield.

12. Remove the RH camshaft tensioner.

➡Leave the top two special tool clamp bolts loose.

13. Install the special tools on the rear of the RH cylinder head.

14. Install the special tool.

TDC positioning tool installed—4.0L engine

Loosening the right side camshaft sprocket bolt—4.0L engine

Camshaft holding tool installed—4.0L engine

Camshaft gear holding tool and adapter—4.0L engine

➡️**The right-hand camshaft sprocket bolt is a left-hand threaded bolt.**

➡️**If necessary, use camshaft gear torque adapter to tighten the camshaft sprocket bolt.**

15. Tighten the bolts.
16. Tighten the special tool top two clamp bolts to 10 Nm (89 inch lbs.).
17. Tighten the camshaft bolt.
18. Install the RH camshaft tensioner.
19. Install the RH lower splash shield.
20. Remove the LH camshaft tensioner.
21. Install the special tools on the front of the LH cylinder head and tighten the top two clamp bolts to 10 Nm (89 inch lbs.).
22. Loosen the LH camshaft sprocket bolt.
23. Loosen the top two clamp bolts on the special tool to allow the camshaft sprocket to rotate freely.

➡️**The camshaft timing slots are off-center.**

24. Position the camshaft timing slots below the centerline of the camshaft to correctly fit the special tools. Install the special tools on the rear of the LH cylinder head.
25. Install the special tool.
26. Tighten the bolts.
27. Tighten the special tool top two clamp bolts to 10 Nm (89 inch lbs.).
28. Tighten the camshaft bolt.
29. Install the LH camshaft tensioner.
30. Install the roller followers.
31. Install the thermostat housing.
32. Install the accessory drive belt.
33. Install the fuel supply manifold.
34. Install the intake manifold.
35. Install the RH valve cover.
36. Install the LH valve cover.

Piston and Ring

POSITIONING

Piston ring positioning

Piston ring end gap spacing

Piston and connecting rod positioning on 3.0L engines

Piston and connecting rod positioning on 4.0L engines

FUEL SYSTEM

Fuel System Service Precautions

Safety is the most important factor when performing not only fuel system maintenance, but any type of maintenance. Failure to conduct maintenance and repairs in a safe manner may result in serious personal injury or death. Work on a vehicle's fuel system components can be accomplished safely and effectively by adhering to the following rules and guidelines.

• To avoid the possibility of fire and personal injury, always disconnect the negative battery cable unless the repair or test procedure requires that battery voltage be applied.

• Always relieve the fuel system pressure prior to disconnecting any fuel system component (injector, fuel rail, pressure regulator, etc.) fitting or fuel line connection. Exercise extreme caution whenever relieving fuel system pressure, to avoid exposing your skin, face and eyes to fuel spray. Please be advised that fuel under pressure may penetrate the skin or any part of the body that it contacts.

• Always place a shop towel or cloth around the fitting or connection prior to loosening to absorb any excess fuel due to spillage. Ensure that all fuel spillage is quickly remove from engine surfaces. Ensure that all fuel-soaked cloths or towels are deposited into a flame-proof waste container with a lid.

• Always keep a dry chemical (Class B) fire extinguisher near the work area.

• Do not allow fuel spray or fuel vapors to come into contact with a light bulb, spark or open flame.

• Always use a second wrench when loosening or tightening fuel line connection fittings. This will prevent unnecessary stress and torsion to fuel piping. Always follow the proper torque specifications.

• Always replace worn fuel fitting O-rings with new ones. Do not substitute fuel hose where rigid pipe is installed.

Relieving Fuel System Pressure

All Sequential Fuel Injection (SFI) fuel injected engines are equipped with a pressure relief valve located on the fuel supply manifold. Remove the fuel tank cap and attach fuel pressure gauge T80L-9974-B, or equivalent, to the valve to release the fuel pressure. Be sure to drain the fuel into a suitable container and to avoid gasoline spillage. If a pressure gauge is not available, disconnect the vacuum hose from the fuel pressure regulator and attach a hand-held vacuum pump. Apply about 25 in. Hg (84 kPa) of vacuum to the regulator to vent the fuel system pressure into the fuel tank through the fuel return hose. Note that this procedure will remove the fuel pressure from the lines, but not the fuel. Take precautions to avoid the risk of fire and use clean rags to soak up any spilled fuel when the lines are disconnected.

An alternate method of relieving the fuel system pressure involves disconnecting the inertia switch.

Fuel Filter

REMOVAL & INSTALLATION

1. Before servicing the vehicle, refer to the Precautions Section.
2. Properly relieve the fuel system pressure.
3. Remove or disconnect the following:
   • Negative battery cable
   • Push connect and R-clip fittings from the fuel filter
   • Fuel filter

To install:
4. Install or connect the following:
   • Fuel filter. Torque the nut to 17 ft. lbs. (23 Nm).
   • R-clip and push connect fittings

Fuel filter connections

- Negative battery cable

5. Start the vehicle, check for leaks and repair if necessary.

Fuel Pump

REMOVAL & INSTALLATION

2002–03

1. Before servicing the vehicle, refer to the Precautions Section.
2. Properly relieve the fuel system pressure.
3. Remove or disconnect the following:
   - Negative battery cable
   - Fuel tank
   - Fuel tank pump locking retainer ring
   - Fuel pump mounting gasket and discard the gasket
   - Fuel pump

To install:

4. Install or connect the following:
   - Fuel pump and a new mounting gasket
   - Fuel tank pump locking retainer ring. Torque the ring to 66 ft. lbs. (90 Nm).
   - Fuel tank
   - Negative battery cable

5. Start the vehicle, check for leaks and repair if necessary.

2004

1. Before servicing the vehicle, refer to the Precautions Section.
2. Remove the fuel tank.
3. Clean the area around the fuel pump mounting flange.
4. Using the special tool, remove the fuel tank pump assembly locking retainer ring.

✳✳ WARNING

The fuel pump assembly must be removed and handled carefully to avoid damage to the float arm and filter.

67197-RANG-G27

Lock ring wrench—2002–03 models

67197-RANG-G64

Lock ring tool—2004–05 models

5. Remove the fuel pump assembly.
6. Remove and discard the fuel pump mounting gasket.

To install:

7. Clean the fuel pump mounting flange and the fuel tank mounting surface.
8. Install a new fuel pump mounting gasket.
9. Install the fuel pump and sender assembly with the float toward the rear of the tank. Align the arrows molded into the tank and flange.
10. Install the locking ring while compressing the pump assembly into the tank.
11. Using the special tool, tighten the fuel pump assembly locking ring retainer ring until it locks in place.
12. Install the fuel tank.

Fuel Injectors

REMOVAL & INSTALLATION

2.3L Engine

1. Before servicing the vehicle, refer to the Precautions Section.
2. Properly relieve the fuel system pressure.
3. Remove or disconnect the following:
   - Negative battery cable
   - Upper intake manifold
   - Fuel injector connectors
   - Fuel injector harness from the fuel injector supply manifold
   - Fuel line spring lock
   - Fuel line
   - Fuel injection supply manifold
   - Fuel injector retaining clip
   - Fuel injector

✳✳ WARNING

Use O-ring seals that are made of special fuel-resistant material. Use of ordinary O-ring seals can cause the fuel system to leak. Do not reuse the O-ring seals.

67197-RANG-G65

Fuel injector-to-fuel rail installation. 1-retaining clip; 2-injector—2.3L engine

67197-RANG-G66

Fuel injector O-rings—2.3L engine

4. Installation is the reverse of removal. Install new O-rings. Lubricate the O-rings with clean engine oil. Torque the supply manifold bolts to 18 ft. lbs. (25 Nm).

3.0L and 4.0L Engines

1. Before servicing the vehicle, refer to the Precautions Section.
2. Properly relieve the fuel system pressure.
3. Remove or disconnect the following:
   - Negative battery cable
   - Upper intake manifold
   - Engine control sensor wiring from the fuel injectors
   - Fuel lines
   - Fuel injection supply manifold and injectors as an assembly
   - Vacuum line
   - Fuel injectors from the supply manifold
   - Inspect the O-rings and replace them as needed

67197-RANG-G67

Fuel injector wiring connectors—3.0L engine

Fuel rail bolts—3.0L engine

Fuel rail and injectors—3.0L engine

Fuel rail and injectors—4.0L engine

Fuel injector O-rings—4.0L shown; 3.0L similar

To install:

4. Install or connect the following:
- Fuel injectors
- Vacuum line
- Fuel injection supply manifold.

Torque the bolts to 89 inch lbs. (10 Nm).
- Fuel line
- Engine control sensor wiring to the fuel injectors

- Upper intake manifold
- Negative battery cable

5. Start the vehicle, check for leaks and repair if necessary.

DRIVE TRAIN

Manual Transmission Assembly

REMOVAL & INSTALLATION

2002–03

1. Before servicing the vehicle, refer to the Precautions Section.

2. Remove or disconnect the following:
- Negative battery cable
- Upper gearshift lever and the outer gearshift lever boot and console assembly as an assembly

3. If transmission disassembly is necessary, remove the drain plug, and drain the transmission fluid. Install the drain plug after draining all of the fluid.

4. Remove or disconnect the following:
- Electrical connector from the reverse lamp switch
- Electrical connector from the vehicle speed sensor (VSS)

- Heated oxygen sensor (HO$_2$S) electrical connector from the bracket
- Wiring harness from the bracket
- Electrical connectors from the heated oxygen sensors (HO$_2$S)
- Starter motor

➡The driveshaft centering socket yoke fits tightly on the rear axle pinion flange pilot. Never hammer on the driveshaft or any of its components to disconnect the yoke from the flange. Pry only in the area shown, with a suitable tool, to disconnect the yoke from the flange.

➡If equipped, always disconnect the front driveshaft from the transfer case first. Otherwise, the weight of the driveshaft can cause the boot to tear.

- Rear driveshaft, and the front driveshaft, if so equipped
- Bolts retaining the exhaust inlet

Pry here for driveshaft removal

crossover pipe to the exhaust manifold
- Bolts retaining the catalytic converter to the muffler. Discard the exhaust converter outlet gasket.
- Exhaust hanger from the insulator. Position the exhaust assembly aside.
- On 4-wheel drive vehicles, the transfer case
- Clutch hydraulic line from the clutch slave cylinder

5. Using a suitable transmission jack, support the transmission. Secure the transmission to the jack with a suitable safety strap.

6. Loosen, but do not remove the nuts retaining the transmission insulator to the crossmember.

7. Remove the six bolts retaining the crossmember to the frame.

8. Remove the nuts and the crossmember.

➡ **Lower the transmission enough to gain access to the upper bolts retaining the transmission to the engine.**

9. Remove the nine bolts retaining the transmission to the engine.

10. Remove the transmission from the vehicle.

11. Installation is the reverse of removal. Observe the following torques:

- Transmission-to-engine bolts: 44 ft. lbs. (60Nm)
- Crossmember-to-frame: 46 ft. lbs. (63Nm)
- Transmission insulator-to-crossmember: 72 ft. lbs. (98Nm)

2004–05

2-WHEEL DRIVE

1. Before servicing the vehicle, refer to the Precautions Section.

2. Remove the upper gearshift lever, the outer gearshift lever boot and the console as an assembly.

3. With the vehicle in NEUTRAL, position it on a hoist.

4. If transmission disassembly is required, remove the drain plug and drain the transmission fluid. Install the drain plug after draining all the fluid.

5. To maintain initial driveshaft balance, index-mark the driveshaft yoke to the axle flange, so they can be installed in their original positions. Remove the rear driveshaft.

6. Disconnect the wire harness from the crossmember.

7. Place a suitable jack under the transmission. Secure the transmission to the jack with a safety strap.

8. Remove the six crossmember bolts.

9. Remove the transmission mount nuts and the crossmember.

10. Remove the heated oxygen senor

(HO$_2$S) bracket nut and the bracket from the extension housing.

11. Remove the transmission mount bolts and the transmission mount.

12. For 3.0L and 4.0L engines, remove the catalytic converter Y-pipe.

13. Disconnect the vehicle speed sensor (VSS) electrical connector and the reverse lamp switch electrical connector. Then unclip the wiring harness from the transmission.

14. Remove the starter motor. Using mechanics wire, position the starter aside.

15. Using the special tool, disconnect the clutch hydraulic line.

16. Using a suitable jack, support the engine.

17. Lower the transmission enough to gain access to the upper transmission-to-engine bolts. Remove the nine transmission-to-engine bolts.

18. Pull the transmission rearward until the input shaft is clear of the pressure plate, then lower the transmission from the vehicle.

To install:

19. Before securing the engine to the transmission, connect the hydraulic line to the clutch slave cylinder.

20. Install the transfer case with a new gasket.

21. Tighten the bolts that retain the transfer case to the extension housing in a clockwise direction beginning with the upper LH bolt.

22. Install the front driveshaft with new bolts and washers and the rear driveshaft with new bolts.

23. Align the index marks when installing the front and rear driveshafts.

24. Check and, if necessary, fill the transmission with the specified type and quantity of fluid.

25. To install, reverse the removal procedure. Observe the following torques:

- Drain plug: 36 ft. lbs. (48 Nm)
- Crossmember bolts: 46 ft. lbs. (62 Nm)
- Transmission mount nuts: 72 ft. lbs. (98 Nm)
- Oxygen sensor bracket nut: 29 ft. lbs. (39 Nm)
- Transmission mount bolts: 73 ft. lbs. (99 Nm)
- Transmission-to-engine bolts: 44 ft. lbs. (60 Nm)

4-WHEEL DRIVE

1. Before servicing the vehicle, refer to the Precautions Section.

2. Remove the upper gearshift lever, the outer gearshift lever boot and the console as an assembly.

3. With the vehicle in NEUTRAL, position it on a hoist.

4. Remove the skid plate, if equipped.

5. If transmission disassembly is required, remove the drain plug and drain the transmission fluid. Install the drain plug after draining all the fluid.

6. To maintain initial driveshaft balance, index-mark the front output shaft and the front driveshaft constant velocity (CV) joint. Index-mark the rear driveshaft yokes to the axle flange and on the transfer case flange.

7. Remove the transfer case.

8. Disconnect the wire harness from the crossmember.

9. Position a suitable jack under the transmission. Secure the transmission to the jack with a safety strap.

10. Remove the six crossmember bolts.

11. Remove the transmission mount nuts and the crossmember

12. Remove the heated oxygen senor (HO$_2$S) bracket nut and the bracket from the extension housing.

13. Remove the catalytic converter Y-pipe.

14. Disconnect the vehicle speed sensor (VSS) electrical connector and the reverse lamp switch electrical connector. Then unclip the wiring harness from the transmission.

15. Remove the starter motor. Using mechanics wire, position the starter aside.

16. Using the special tool, disconnect the clutch hydraulic line.

17. Using a suitable jack, support the engine.

18. Lower the transmission enough to gain access to the upper transmission-to-engine bolts. Remove the nine transmission-to-engine bolts.

19. Pull the transmission rearward until the input shaft is clear of the pressure plate, then lower the transmission from the vehicle.

To install:

20. Before securing the engine to the transmission, connect the hydraulic line to the clutch slave cylinder.

21. Install the transfer case with a new gasket.

22. Tighten the bolts that retain the transfer case to the extension housing in a clockwise direction beginning with the upper LH bolt.

23. Install the front driveshaft with new bolts and washers and the rear driveshaft with new bolts.

24. Align the index marks when installing the front and rear driveshafts.

25. Check and, if necessary, fill the transmission with the specified type and quantity of fluid.

26. To install, reverse the removal procedure. Observe the following torques:
- Skid plate: 18 ft. lbs. (24 Nm)
- Drain plug: 36 ft. lbs. (48 Nm)
- Crossmember bolts: 46 ft. lbs. (63 Nm)
- Transmission mount nuts: 72 ft. lbs. (98 Nm)
- Oxygen sensor bracket: 29 ft. lbs. (39 Nm)
- Transmission-to-engine bolts: 44 ft. lbs. (60Nm)

Automatic Transmission

REMOVAL & INSTALLATION

2002–03

1. Before servicing the vehicle, refer to the Precautions Section.
2. Place the selector lever in NEUTRAL position.
3. Remove or disconnect the following:
- Negative battery cable
- Fluid level indicator
- The two bolts retaining the fan shroud to the radiator.

➡ **If transmission disassembly is required, drain the transmission fluid.**

- With 4wd, the transfer case

➡ **Mark the driveshaft yoke and axle flange, so they may be installed in their original alignment.**

- Rear driveshaft
- Starter motor
- Torque converter access cover

➡ **Mark the torque converter and the flexplate for correct alignment at reinstallation.**

- The four converter nuts
- Shift cable
- Transmission wiring harness
- Three way catalytic converter
- Left HO2S sensor
- Front exhaust crossover pipe
- Transmission cooler lines

4. Position a High-Lift Jack under the transmission. Raise and support the transmission.
5. Remove or disconnect the following:
- Crossmember.
- Transmission mount
- Transmission upper fill tube

➡ **Lower the High-Lift Transmission Jack to gain access to screws.**

- On 4x4 models, the vent tube assembly

✳✳ WARNING

Install the Torque Converter Holding Tool before lowering the transmission from the vehicle. Secure the transmission to the transmission jack with a safety chain. Failure to follow these instructions can result in personal injury.

6. Lower the transmission.
7. Installation is the reverse of removal. Observe the following torques:
- Transmission-to-engine bolts: 41 ft. lbs. (55Nm)
- Exhaust bracket bolts: 81 ft. lbs. (110Nm)
- Crossmember-to-frame: 87 ft. lbs. (118Nm)
- Transmission mount-to-crossmember: 81 ft. lbs. (110Nm)
- Converter-to-flexplate: 30 ft. lbs. (40Nm)
- Rear driveshaft-to-flange bolts: 95 ft. lbs. (129Nm)

2004–05

1. Before servicing the vehicle, refer to the Precautions Section.

➡ **If the transmission is to be removed for a period of time, support the engine with a safety stand and a wood block.**

2. With the vehicle in NEUTRAL, position it on a hoist.

➡ **When the battery has been disconnected and reconnected, some abnormal drive symptoms can occur while the vehicle relearns its adaptive strategy.**

3. Disconnect the battery ground cable
4. On 4.0L vehicles, remove the fluid level indicator tube bolt and remove the tube and indicator.
5. If transmission disassembly is required, drain the transmission fluid.
6. On 4x4 vehicles, remove the transfer case.
7. To maintain initial driveshaft balance, mark the driveshaft yoke and axle flange so they can be installed in their original alignment.
8. Remove the rear driveshaft.
   a. Remove the four bolts.
   b. Remove the driveshaft.
9. Remove the starter motor.
10. With a 2.3L engine:
   a. When removing the torque converter nuts, the crankshaft must be rotated only in the clockwise direction, otherwise engine damage can occur. The crankshaft, crankshaft sprocket and the

---

pulley are fitted together by friction between the flange faces on each part. For that reason, the crankshaft sprocket can also be moved when the crankshaft pulley is turned in the counterclockwise direction.
   b. It may be necessary to gain access to the flexplate nuts through the wheel well.
   c. Mark the torque converter and the flexplate for correct alignment at reinstallation.
   d. Remove and discard the four torque converter nuts. Rotate the flexplate to access to all the nuts.
11. With 3.0L and 4.0L engines:
   a. Mark the torque converter and the flexplate for correct alignment at reinstallation. Remove the four nuts. Rotate the flexplate to access to all the nuts.
   b. Disconnect the shift cable.
   c. Disconnect the transmission wiring harness from the case.
12. Disconnect the transmission wiring harness. Remove the three way catalytic converter.
13. With a 2.3L engine, remove the rear engine cover plate.

➡ **Care should be taken not to bend or damage the cooler lines.**

14. Hold the case fitting and remove the transmission cooler lines.
15. Remove the nuts.
16. Position a transmission jack under the transmission. Raise and support the transmission.
17. Remove the crossmember.
18. Remove the transmission mount.
19. With a 2.3L engine, remove the rear vibration damper.
20. 2.3L and 3.0L engines, remove the transmission upper fill tube.
21. Lower the jack to gain access to screws. Remove the transmission-to-engine bolts.
22. With a 2.3L engine, remove the lower screws.
23. Remove the HO2S connector bracket from the transmission.
24. On 4x4 vehicles, remove the vent tube assembly.

✳✳ CAUTION

The torque converter is heavy and may result in injury if it falls out of the transmission. Secure the torque converter in the transmission. Failure to follow these instructions may result in personal injury. Install a converter locking tool before lowering the transmission from the vehicle.

❄❄ CAUTION

Secure the transmission to the transmission jack with a safety chain. Failure to follow these instructions may result in personal injury.

25. Lower the transmission.
26. If the transmission is being overhauled or if installing a new or remanufactured transmission, carry out transmission fluid cooler backflushing and cleaning.

To install:

27. On 4x4 vehicles, install the vent tube assembly.

❄❄ CAUTION

Secure the transmission to the transmission jack with a safety chain. Failure to follow these instructions may result in personal injury.

28. Raise and position the transmission.
29. Remove the holding tool.
30. With the 4.0L engine:
   a. Align the flexplate to the converter marks made at removal.
   b. Install the transmission-to-engine screws. Torque to 35 ft. lbs. (48 Nm).
31. With the 2.3L and 3.0L engines
   a. Align the flexplate to the converter marks made at removal.
   b. Install the transmission-to-engine screws. Torque to 35 ft. lbs. (48 Nm).
   c. Install the upper fluid filler tube and bracket screw.
32. With the 2.3L engine:
   a. Install the lower transmission-to-engine screws. Torque to 35 ft. lbs. (48 Nm).
   b. Install the HO2S connector bracket.
   c. Install the rear engine cover plate.
33. On 4x4 vehicles, install the transfer case.
34. Install the exhaust bracket. Torque the bolts to 73 ft. lbs. (99 Nm).
35. Install the crossmember. Tighten the bolts to 74 ft. lbs. (101 Nm).
36. Install the transmission mount into the crossmember and torque the nuts to 73 ft. lbs. (99 Nm).
37. With 3.0L engines, install the rear vibration damper. Torque the nuts to 22 ft. lbs. (30 Nm).
38. With 3.0L and 4.0L engines:

➡**Prior to installing the cooler lines to the case, inspect the O-rings. If damaged new O-rings will need to be installed.**

   a. Hold the case fitting and install the transmission cooler lines. Torque to 19 ft. lbs. (28 Nm).
   b. Install four new torque converter nuts. Rotate the crankshaft as needed to gain access to all the nuts. Torque to 26 ft. lbs. (35 Nm).
39. With the 2.3L vehicles: Install four new torque converter nuts. Rotate the crankshaft as needed to gain access to all the nuts. Torque to 26 ft. lbs. (35 Nm).

➡**When installing the torque converter nuts, the crankshaft must be rotated only in the clockwise direction, otherwise engine damage can occur. The crankshaft, the crankshaft sprocket and the pulley are fitted together by friction between the flange faces on each part. For that reason, the crankshaft sprocket can also be moved when the crankshaft pulley is turned in the counterclockwise direction.**

40. Install the starter motor.
41. Install the catalytic converter assembly.
42. Position the transmission wiring harness in place.
43. Connect the transmission wiring harness.
44. Install the shift cable.
45. Align the driveshaft yoke and the axle shaft marks made at removal to maintain driveline balance. Install the rear driveshaft. Install the driveshaft bolts. Torque to 83 ft. lbs. (112 Nm).
46. Use the following guidelines for installing the in-line transmission fluid filter:
   a. If the transmission was overhauled and the vehicle was equipped with an in-line fluid filter, install a new in-line fluid filter.
   b. If the transmission was overhauled and the vehicle was not equipped with an in-line fluid filter, install a new in-line fluid filter kit.
   c. If the transmission is being installed for a non-internal repair, do not install an in-line filter or filter kit.
   d. If installing a new or re-manufactured transmission, install the in-line transmission fluid filter that is supplied.
   e. Prior to lowering the vehicle, install a new in-line transmission filter or a filter kit.
47. With the 4.0L engine, install the transmission fill tube and indicator as an assembly.

➡**When the battery has been disconnected and reconnected, some abnormal drive symptoms can occur while the vehicle relearns its adaptive strategy.**

48. Connect the battery ground cable.
49. Fill the transmission with clean automatic transmission fluid to the specified level.
50. Check the transmission for correct operation.
51. Verify that the shift cable is correctly adjusted.

Clutch

REMOVAL & INSTALLATION

1. Before servicing the vehicle, refer to the Precautions Section.
2. Remove or disconnect the following:
   - Negative battery cable
   - Transmission

➡**If the clutch disc and pressure plate are to be reinstalled, bolts must be removed evenly or permanent damage to the diaphragm spring will occur resulting in complete clutch release.**

Clutch disc, pressure plate and bearing assembly

7924EG18

Tighten the bolts gradually in the correct sequence to avoid warping the pressure plate

- Bolts, clutch pressure plate and the clutch disc.

➡️**If the parts are to be reused, index-mark the clutch pressure plate to the flywheel.**

To install:

3. Lubricate the transmission input shaft pilot bearing with front axle grease.

4. Using a suitable press, press downward on the pressure plate fingers until the adjusting ring moves freely.

5. Rotate the adjusting ring counterclockwise to compress the tension springs. Hold the adjusting ring in this position.

6. Release the pressure on the fingers. The adjusting ring will stay in the reset position.

7. Position the clutch disc on the flywheel.

➡️**If reusing the clutch pressure plate and flywheel, align the marks made during removal.**

8. Align the clutch disc and the clutch pressure plate. Install the bolts and tighten in a star pattern sequence to 24 ft. lbs. (35Nm) for 2002; 20 ft. lbs. (27 Nm) for 2003–05 models.

- Install the transmission.

ADJUSTMENT

1. Before servicing the vehicle, refer to the Precautions Section.

Because the clutch is hydraulically driven, there is no adjustment required.

In the event the clutch pedal develops a squeak or uneven feel when depressing, spray the pedal bushing assembly with penetrating oil and work the pedal back-and-forth.

Hydraulic Clutch System

BLEEDING

The following procedure is recommended for bleeding the clutch hydraulic system installed on the vehicle. It is recommended that the original clutch tube, with quick-connect fitting be replaced when servicing the hydraulic system, because air can be trapped in the quick-connect fitting and prevent complete bleeding of the system. The replacement tube does not include a quick-connect fitting.

1. Before servicing the vehicle, refer to the Precautions Section.

2. Clean the dirt and grease from the dust cap.

3. Remove the cap and diaphragm and fill the reservoir to the top with approved brake fluid C6AZ-19542-AA or BA, (ESA-M6C25-A).

➡️**To keep brake fluid from entering the clutch housing, route a suitable rubber tube of appropriate inside diameter from the bleed screw to a container.**

4. Loosen the bleed screw, located in the slave cylinder body, next to the inlet connection. Fluid will now begin to move from the master cylinder down the tube to the slave cylinder.

➡️**The reservoir must be kept full at all times during the bleeding operation, to ensure no additional air enters the system.**

5. Observe the bleed screw outlet. When the slave cylinder is full, a steady stream of fluid will flow from the outlet port. Tighten the bleed screw.

6. Depress the clutch pedal to the floor and hold for 1–2 seconds. Release the pedal as rapidly as possible. The pedal must be released completely. Pause for 1–2 seconds. Repeat 10 times.

7. Check the fluid level in the reservoir. The fluid should be level with the step when the diaphragm is removed.

8. Hold the pedal to the floor, slightly open the bleed screw to allow any additional air to escape. Close the bleed screw, then release the pedal.

9. Check the fluid in the reservoir. The hydraulic system should now be fully bled, and should actuate the clutch.

10. Check the vehicle by starting, pushing the clutch pedal to the floor and selecting reverse gear. There should be no grating of gears. If there is, and the hydraulic system still contains air; repeat the bleeding procedure.

Transfer Case Assembly

REMOVAL & INSTALLATION

2002–03

1. Before servicing the vehicle, refer to the Precautions Section.

2. Place the transmission in neutral.

3. Remove or disconnect the following:
   - Skid plate
   - Damper
   - Transfer case harness connector and position it aside

4. If transfer case disassembly is necessary, remove the drain plug and drain the fluid.

➡️**Index-mark the front output shaft assembly and the front driveshaft constant velocity (CV) joint. Always disconnect the front driveshaft from the transfer case first. Otherwise, the weight of the driveshaft can pinch the boot between the shaft and the boot can and cause the boot to tear.**

- Front driveshaft from the transfer case and position the driveshaft aside. Remove and discard the bolts and washers.

➡️**Index-mark the front flange on the rear driveshaft and the flange on the transfer case.**

- Rear driveshaft

➡️**Secure the transfer case to the jack with safety straps.**

5. Position a high lift jack under the transfer case.

6. Remove or disconnect the following:
   - Five bolts retaining the transfer case to the extension housing
   - Transfer case rearward and off of the transmission output shaft

7. Remove and discard the front extension housing gasket and clean the mating surfaces.

To install:

8. Installation is the reverse of removal. Take note of the following:
   - Install the transfer case with a new gasket.
   - Tighten the bolts that retain the transfer case to the extension housing in a clockwise direction beginning with the upper LH bolt.
   - Install the front and the rear driveshafts with new bolts. If new bolts are not available, coat the threads of the original bolts with Thread-

lock and Sealer E0AZ-19554-AA or equivalent meeting Ford specification WSK-M2G351-A5.

- When installing the front driveshaft, always connect it to the axle first and then connect it to the transfer case.
- Align the index marks when installing the front and rear driveshafts.

9. Observe the following torques:

- Nut retaining the flange to the rear output shaft: 262 ft. lbs. (355Nm)
- Bolt retaining the rear driveshaft to the flange: 82 ft. lbs. (111Nm)
- Bolt retaining the motor assembly and connector to the transfer case cover: 89 inch lbs. (10Nm)
- Bolt retaining the skid plate to the frame: 18 ft. lbs. (24Nm)
- Bolt retaining the damper to the transfer case: 30 ft. lbs. (40Nm)
- Bolt retaining the driveshaft CV joint to the front output shaft assembly: 22 ft. lbs. (30Nm)
- Bolt retaining the transfer case to the extension housing: 40 ft. lbs. (54Nm)
- Bolt retaining the front adapter to the transfer case: 30 ft. lbs. (40Nm)
- Bolt retaining the transfer case to the transfer case cover: 27 ft. lbs. (36Nm)
- Drain plug: 18 ft. lbs. (24Nm)
- Fill plug: 18 ft. lbs. (24 Nm)

2004–05

1. Before servicing the vehicle, refer to the Precautions Section.

2. With the vehicle in NEUTRAL, raise and support the vehicle.
3. Remove the skid plate.
4. Remove the damper, if so equipped.
5. Disconnect the transfer case harness connector and position it aside.
6. If transfer case disassembly is necessary, remove the drain plug and drain the fluid. Install the drain plug when all of the fluid has drained.

➡ **Index-mark the front output shaft assembly and the front driveshaft constant velocity (CV) joint.**

❊❊ WARNING

Always disconnect the front driveshaft from the transfer case first. Otherwise, the weight of the driveshaft can pinch the boot between the shaft and the boot can and cause the boot to tear.

7. Index-mark the front output shaft assembly and the front driveshaft constant velocity (CV) joint.
8. Remove and discard the bolts and washers.
9. Disconnect the front driveshaft from the transfer case and position the driveshaft aside.

➡ **Index-mark the front flange on the rear driveshaft and the flange on the transfer case.**

10. Remove the rear driveshaft.

❊❊ CAUTION

Secure the transfer case to the jack with safety straps.

11. Position a high lift jack under the transfer case.
12. Remove the five bolts retaining the transfer case to the extension housing.
13. Slide the transfer case rearward and off of the transmission output shaft.
14. Remove and discard the front extension housing gasket, and clean the mating surfaces.

To install:

15. Install the transfer case with a new gasket.
16. Tighten the bolts that retain the transfer case to the extension housing in a clockwise direction beginning with the upper LH bolt. Torque to 40 ft. lbs. (54 Nm).
17. Install the front driveshaft with new bolts and washers and the rear driveshaft with new bolts. If new bolts are not available, coat the threads of the original bolts with Threadlock and Sealer E0AZ-19554-AA, or equivalent.

➡ **When installing the front driveshaft, always connect it to the axle first and then connect it to the transfer case.**

➡ **Align the index marks when installing the front and rear driveshafts.**

18. The remainder of installation is the reverse of the removal procedure.
19. Check and, if necessary, fill the transfer case with the specified type and quantity of fluid.

TRANSMISSION EXTENSION

TRANSFER CASE TO TRANSMISSION EXTENSION GASKET

TRANSFER CASE

DISCONNECT SPEEDOMETER DRIVEN GEAR FROM THIS POINT

FRAME SIDE RAIL

TRANSFER CASE RETAINING BOLTS

INSTALL VENT HOSE WITH WHITE MARK AT THIS LOCATION

SKID SHIELD

SKID SHIELD NUTS, BOLTS, AND RETAINER PLATES

7924EG20

Exploded view of the 13-54 electronic shift transfer case-to-transmission mounting

Halfshaft

REMOVAL & INSTALLATION

1. Before servicing the vehicle, refer to the Precautions Section.

2. With the vehicle in NEUTRAL, raise and support the vehicle.

3. Remove the front wheel and tire assembly.

➡ Do not reuse the torque prevailing design hub nut and washer assembly.

4. Remove and discard the hub nut and washer assembly.

✳✳ WARNING

Do not allow the disc brake caliper to hang suspended from the brake hose. Provide a suitable support.

5. Remove the front disc brake caliper, anchor plate, and pads as an assembly, and position the assembly aside.

6. Remove the brake disc.

✳✳ WARNING

Do not use a hammer to separate the outboard front wheel halfshaft joint from the wheel hub. Damage to the outboard CV joint stub shaft threads and internal CV joint components may result.

7. Using the special tool, separate the outboard front wheel halfshaft joint from the wheel hub. Remove the special tool.

8. Support the front suspension lower arm.

9. Remove the nut and bolt retaining the upper ball joint to the front wheel knuckle.

10. Rotate the front wheel knuckle.

11. Compress the outboard front wheel halfshaft joint.

12. Remove the outboard front wheel halfshaft joint from the wheel hub.

67197-RANG-G72

Halfshaft removal tool

67197-RANG-G73

Separating the halfshaft from the axle housing

13. Using the special tools, 205-241 and 100-001, or equivalent, separate the inboard front wheel halfshaft joint from the front axle housing.

14. Remove the halfshaft assembly from the vehicle with both hands. Do not damage the axle seal.

To install:

✳✳ WARNING

Install the halfshaft with a new hub nut and washer assembly. Do not use power or impact tools to tighten the hub nut and washer assembly.

➡ Install a new retainer circlip in the groove in the LH inboard CV joint housing stub shaft before installing the halfshaft in the vehicle. To prevent the new retainer circlip from over-expanding when installing it, start one end in the groove and work the circlip over the shaft and into the groove.

15. To install, reverse the removal procedure. Observe the following torques:
- Hub nut: 162 ft. lbs. (220 Nm)
- Upper ball joint-to-knuckle nut: 41 ft. lbs. (55 Nm)

67197-RANG-G74

Circlip installed

Driveshaft

REMOVAL & INSTALLATION

Front Driveshaft

1. Before servicing the vehicle, refer to the Precautions Section.

2. With the vehicle in NEUTRAL, raise and support the vehicle.

3. Index the front axle pinion flange to the front driveshaft.

4. Index the transfer case pinion flange to the front driveshaft.

5. Remove the front driveshaft-to-transfer case bolts and retainers.

6. Remove the front driveshaft-to-front axle bolts.

7. Remove the front driveshaft.

8. To install, reverse the removal procedure. Torque the bolts at the axle end to 15 ft. lbs. (20 Nm); the transfer case end bolts to 22 ft. lbs. (30 Nm).

Rear Driveshaft

2-WHEEL DRIVE

1. Before servicing the vehicle, refer to the Precautions Section.

2. With the vehicle in NEUTRAL, raise and support the vehicle.

3. Index-mark the driveshaft flange and rear axle pinion flange.

4. Index-mark the driveshaft and the extension housing.

5. Remove the four bolts.

✳✳ WARNING

The driveshaft flange fits tightly on the rear axle pinion flange pilot. Never hammer on the driveshaft or any of its components to disconnect the driveshaft flange from the pinion flange. Pry only in the area shown, with a suitable tool, to disconnect the driveshaft flange from the pinion flange.

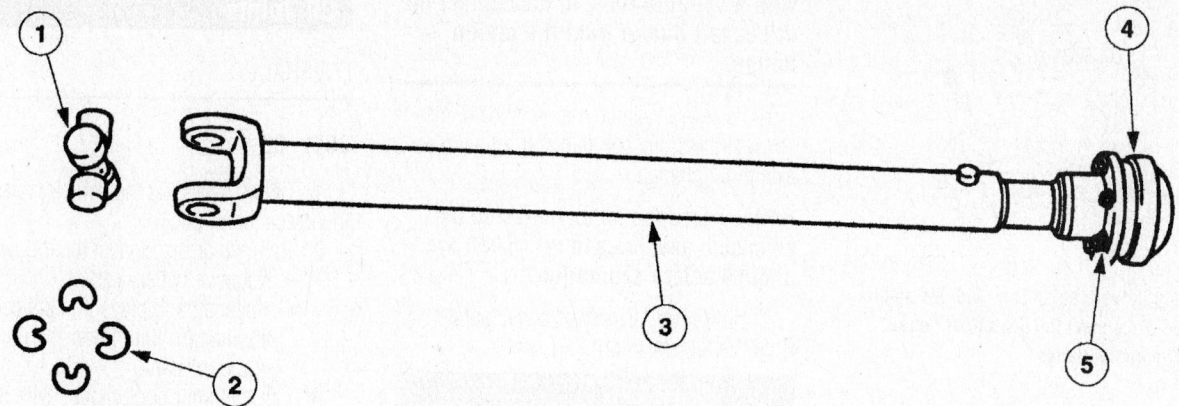

1 Universal joint

2 Snap rings

3 Front driveshaft

4 Constant velocity (CV) joint

5 Constant velocity (CV) joint
bolts and washers

06017-RANG-G08

Front driveshaft

6. Using a suitable tool as shown, disconnect the driveshaft flange from the rear axle pinion flange.

7. Lower the driveshaft and slide it off the output shaft.

8. Plug the extension housing to prevent fluid loss.

❊❊ WARNING

If new bolts to retain the driveshaft to the axle are not available, coat the threads of the original bolts with Threadlock and Sealer EOAZ-19554-AA or equivalent meeting Ford specification WSK-M2G351-A5.

❊❊ WARNING

The driveshaft flange fits tightly on the rear axle pinion flange pilot. To make sure that the driveshaft flange seats squarely on the pinion flange, tighten the bolts evenly in a cross pattern as shown.

1 Flange

2 Universal joint

3 Snap ring

4 Bearing

5 Seal

6 Spider

7 Driveshaft

8 Driveshaft slip yoke

06017-RANG-G09

Rear driveshaft—2-wheel drive

06017-RANG-G11

Pry only in the area shown, with a suitable tool, to disconnect the driveshaft flange from the pinion flange

9. To install, reverse the removal procedure. Torque the bolts to 83 ft. lbs. (112 Nm).

4-WHEEL DRIVE

1. Before servicing the vehicle, refer to the Precautions Section.
2. With the vehicle in NEUTRAL, raise and support the vehicle.
3. Index-mark the driveshaft flange to the rear axle pinion flange.
4. Remove the four rear axle flange bolts.
5. Index-mark the driveshaft and the extension housing.

�֎ WARNING

The driveshaft flange fits tightly on the rear axle pinion flange pilot. Never hammer on the driveshaft or any of its components to disconnect the driveshaft flange from the pinion flange. Pry only in the area shown,

with a suitable tool, to disconnect the driveshaft flange from the pinion flange.

6. Using a suitable tool as shown, disconnect the driveshaft flange from the rear axle pinion flange.

➡**Make sure the index marks on the extension housing and driveshaft are aligned before separation.**

7. Lower the driveshaft and separate it from the transmission.

✖✖ WARNING

If new bolts to retain the driveshaft to the axle are not available, coat the threads of the original bolts with Threadlock and Sealer E0AZ-19554-AA or equivalent meeting Ford specification WSK-M2G351-A5.

✖✖ WARNING

The driveshaft flange fits tightly on the rear axle pinion flange pilot. To make sure that the driveshaft flange seats squarely on the pinion flange, tighten the bolts evenly in a cross pattern as shown.

8. To install, reverse the removal procedure. Torque the bolts at each end to 83 ft. lbs. (112 Nm).

CV-Joints

OVERHAUL

2002–03

1. Before servicing the vehicle, refer to the Precautions Section.
2. Remove or disconnect the following:
   - Negative battery cable
   - Halfshaft and place it in a vice with the inboard joint lower than the outboard joint
3. Cut the inner boot clamps with side cutters and remove the clamp from the boot.
   - Larger boot end off the joint
   - Inboard CV-joint bolts and separate the spacer and grease cap
   - Snap-ring retaining the interconnecting shaft end to the CV-joint cage
   - CV-joint and discard the washer

➡**The outboard CV-joint is non-serviceable other than to replace the boot.**

   To install:
4. Install or connect the following:
   - Slide the boot over the shaft
5. Fill the CV-joint area with grease.
   - Assemble the outer boot to the outboard CV-joint and interconnecting shaft. Make certain that the boot is seated in the grooves on the outer race and on the shaft
   - New clamps to the boot
   - New inner boot to the shaft
   - New washer to the end of the shaft

1 Rear flange
2 Spider
3 Bearing cup
4 Snap ring
5 Driveshaft
6 Driveshaft slip-yoke boot clamp
7 Universal joint slip-yoke boot
8 Driveshaft slip-yoke
9 Front flange

06017-RANG-G10

Rear driveshaft—4-wheel drive

- Assemble the inboard CV-joint to the interconnecting shaft spline until it rests on the washer
- Snap-ring
6. Fill the CV-joint area with grease.
- Boot into position and make certain that it is seated in the grooves on the boot adapter and the shaft
- New clamps and tighten the clamps with crimping pliers
- Spacer to the CV-joint end pilot. Torque the bolts to 25 ft. lbs. (34 Nm).
- Halfshaft
- Negative battery cable

2004–05

1. Before servicing the vehicle, refer to the Precautions Section.
2. Remove the front wheel halfshaft. Do not damage the halfshaft boot.
3. Remove the two inboard boot clamps.
4. Slide the inboard halfshaft boot off the inboard CV joint housing.
5. Separate the CV joint from the CV joint housing.
6. Index-mark the shaft and the inboard CV joint for correct alignment during assembly.
7. Remove the snap ring.
8. Remove the CV joint.
9. Remove the inboard halfshaft boot from the shaft assembly.
10. Remove the front wheel excluder seal, if necessary. Discard the seal. Tap uniformly around the seal to separate it from the joint.
11. Remove the two outboard boot clamps.
12. Remove the outboard halfshaft boot.
13. If the grease is contaminated, clean and inspect the joint for wear. Install a new outboard CV joint and shaft assembly if worn/damaged.
14. Inspect the assembly for contaminated grease.

67197-RANG-G76

Separating the joint from the housing

67197-RANG-G77

Make an alignment mark for reassembly

205-343

67197-RANG-G78

Boot clamp crimping tool

To assemble

15. Pack the outboard CV joint with grease. Use Ford High Temp Constant Velocity Joint Grease E43Z-19590-A or equivalent meeting Ford specification ESP-M1C207-A. Spread any remaining grease

67197-RANG-G75

Inboard boot clamps

from the service kit evenly inside the outboard halfshaft boot.
16. Clean the halfshaft boot mounting surfaces of excess grease before positioning the halfshaft boot into place.
17. Position the outboard halfshaft boot.
18. Position the boot clamps on the outboard halfshaft boot.
19. Tighten the through-bolt until the installer is in the closed position.
20. Install the outboard CV joint boot clamps. There are special tools made for this procedure.
21. Position the boot clamp on the halfshaft.
22. Position the inboard halfshaft boot.
23. Align the index marks on the halfshaft and the CV joint.
24. Install the CV joint on the halfshaft.
25. Install the snap ring.
26. Lubricate the three CV joint needle bearings. Use Ford High Temp Constant Velocity Joint Grease E43Z-19590-A or equivalent meeting Ford specification ESP-M1C207-A.
27. Fill the inboard CV joint housing with 235 grams (8.3 oz.) of grease. Use Ford High Temp Constant Velocity Joint Grease E43Z-19590-A or equivalent meeting Ford specification ESP-M1C207-A.
28. Position the CV joint housing onto the CV joint.
29. Remove any excess grease from the inboard halfshaft boot mating surface before positioning it into place.
30. Position the inboard halfshaft boot into place.
31. Position the boot clamp.
32. Insert a dulled screwdriver blade to relieve built-up air pressure in the halfshaft boot.
33. Using the special tool, install the inboard boot clamps.
34. Using the special tool, install the new front wheel excluder seal, if removed. Seat the metal ring at the seal's inner diameter flat against the CV joint housing.
35. Install the front wheel halfshaft.

Front Axle Tube Bearing

REMOVAL & INSTALLATION

1. Before servicing the vehicle, refer to the Precautions Section.
2. Remove or disconnect the following:
- Right-hand halfshaft
- Right-hand axle shaft
- Axle seal, with a slide hammer
- Axle tube bearing, with a slide hammer

67197-RANG-G83

Right side axle shaft removal

67197-RANG-G84

Axle seal removal

67197-RANG-G85

Axle tube bearing removal

6197RANG-G79

Axle tube bearing installation

44.5 mm (0.875 in.)

67197-RANG-G80

Axle tube bearing depth

67197-RANG-G81

Axle tube seal installation

3. Clean the bearing and seal surfaces of any foreign debris.

To install:

4. Use an axle bearing replacer and the handle to replace the RH axle tube bearing.

5. Check the bearing depth as shown.

6. Use an axle seal replacer and the handle to replace the axle tube seal.

➡ **Care should be taken not to damage the axle seal surface.**

7. Install the axle shaft.

8. Refill the front drive axle to proper level using SAE 80W90.

9. Install the RH halfshaft.

67197-RANG-G82

Axle shaft installation

Rear Axle Shaft, Bearing and Seal

REMOVAL & INSTALLATION

2002–03

1. Before servicing the vehicle, refer to the Precautions Section.
2. Drain the axle housing fluid.
3. Remove or disconnect the following:
   - Negative battery cable
   - Rear wheel
   - Brake drum
   - Wheel speed sensor, if equipped
   - Axle housing cover
   - Bearing retainer nuts
   - Axle shaft and bearing
   - Axle shaft inner oil seal
4. If equipped with ABS, grind a flat spot on the wheel speed sensor tone ring, then split the ring with a chisel.
5. Press the wheel bearing off the axle shaft.
6. Remove the bearing retainer and the outer oil seal.

To install:

7. Install or connect the following:
   - Outer oil seal to the bearing retainer
   - Bearing retainer to the axle shaft

Differential pinion shaft removal. 1-lock bolt; 2-pinion shaft

67197-RANG-G90

Removing the U-washers. 1-axle shaft; 2-U-washer

67197-RANG-G91

- Bearing and retainer ring pressed onto the axle shaft
- Wheel speed sensor tone ring pressed onto the axle shaft, if equipped
- Axle shaft inner oil seal
- Axle shaft and bearing
- Bearing retainer nuts. Tighten them to 17 ft. lbs. (23 Nm).
- Wheel speed sensor, if equipped
- Brake drum
- Rear wheel
- Negative battery cable

8. Fill the rear differential to the correct level.

2004–05

FORD 7½ INCH RING GEAR

1. Before servicing the vehicle, refer to the Precautions Section.
2. Raise and support the vehicle.
3. Remove the wheel and tire assembly.
4. Remove the 10 differential housing cover bolts and drain the lubricant from the rear axle housing.
5. Remove the differential housing cover.
6. Remove the rear brake drums.
7. Remove and discard the differential pinion shaft lock bolt.
8. Remove the differential pinion shaft.

➡ **Do not damage the rubber O-rings in the axle shaft grooves.**

9. Push the axle shafts inboard.
10. Remove the U-washers.

➡ **Do not damage the wheel bearing oil seal.**

11. Remove the two axle shafts.

➡ **If only a new seal needs to be installed, use care to avoid damaging the seal bore.**

12. Using a suitable seal remover, remove the axle shaft oil seal. Discard the oil seal.
13. Inspect the rear wheel bearing and axle shaft for wear or damage.
14. If necessary, using a slidehammer, remove the rear wheel bearing.

To install:

15. Lubricate the new rear wheel bearing with lubricant.
16. Using a driver, install the rear wheel bearing.
17. Lubricate the lip of the new wheel bearing oil seal with grease.
18. Using a driver, install the wheel bearing oil seal.

➡ **Make sure the machined surfaces on both the rear axle housing and the differential housing cover are clean and free of oil before installing the new silicone sealant. The inside of the rear axle must be covered when cleaning the machined surface to prevent contamination.**

19. Clean the gasket mating surface of the rear axle and the differential housing cover.
20. Lubricate the lip of the wheel bearing oil seal with grease.

➡ **Do not damage the wheel bearing oil seal.**

21. Install the axle shafts.

➡ **Do not damage the rubber O-rings in the axle shaft grooves.**

22. Position the two U-washers on the button end of the axle shafts.
23. Pull the axle shafts outward.

➡ **If a new pinion shaft lock bolt is unavailable, coat the threads with threadlock and sealer prior to installation.**

24. Install the differential pinion shaft.
   a. Align the hole in the differential pinion shaft with the case lock bolt hole.
   b. Install a new differential pinion

shaft lock bolt. Torque to 20 ft. lbs. (33 Nm).

25. Install the rear brake drums.

26. Apply a new continuous bead of sealant of the specified thickness to the differential housing cover.

➡ **The differential housing cover must be installed within 15 minutes of application of the silicone, or new sealant must be applied. If possible, allow one hour before filling with lubricant to make sure the silicone sealant has correctly cured.**

27. Install the different housing cover.

28. Install the 10 differential housing cover bolts. Torque to 33 ft. lbs. (45Nm).

29. Fill the rear axle housing with 2.4 liters (5 pints) of lubricant.

30. Install the wheels and tires.

31. Lower the vehicle.

FORD 8.8 INCH RING GEAR

1. Before servicing the vehicle, refer to the Precautions Section.

2. Raise and support the vehicle.

3. Remove the rear wheel and tire assembly.

4. Remove the brake drum.

5. Remove the differential housing cover and drain the lubricant.

6. Remove and discard the pinion shaft bolt.

7. Remove the differential pinion shaft.

➡ **Do not damage the rubber O-ring in the U-washer groove.**

8. Push the axle shaft inboard.

9. Remove the U-washer.

10. Do not damage the wheel bearing oil seal.

11. Remove the axle shaft.

➡ **If only a new seal needs to be installed, use care to avoid damaging the seal bore. If the wheel bearing oil seal is leaking, the differential housing vent may be plugged with foreign material.**

12. Using a suitable seal remover, remove the axle shaft oil seal. Discard the oil seal.

13. Inspect the rear wheel bearing and axle shaft for wear or damage.

14. Using the special tools, remove the rear wheel bearing.

To install:

15. Lubricate the new rear wheel bearing with rear axle lubricant.

16. Using the special tools, install the rear wheel bearing.

17. Lubricate the lip of the new wheel bearing oil seal with grease.

18. Using the special tools, install the wheel bearing oil seal.

19. Install the axle shaft.

20. Lubricate the lip of the wheel bearing oil seal with grease.

➡ **Do not damage the wheel bearing oil seal.**

21. Install the axle shaft.

22. Do not damage the rubber O-ring in the U-washer groove.

23. Position the U-washer on the button end of the axle shaft.

24. Pull the axle shaft outward.

➡ **If a new bolt is unavailable, coat the bolt threads with threadlock prior to installation.**

25. Align the bolt hole in the differential pinion shaft with the bolt hole in the case.

26. Install the new bolt. Torque to 22 ft. lbs. (33 Nm).

27. Install the brake drum.

28. Install the differential housing cover and fill the differential housing with the specified lubricant.

29. Install the rear wheel and tire assembly.

30. Lower the vehicle.

Front Pinion Seal

REMOVAL & INSTALLATION

➡ **This operation disturbs the differential pinion bearing preload. Carefully reset the preload during assembly.**

✳✳ CAUTION

The electrical power to the air suspension system must be shut off prior to hoisting, jacking or towing an air suspension vehicle. This can be accomplished by turning off the air suspension switch located in the rear jack storage area. Failure to do so can result in unexpected inflation or deflation of the air springs, which can result in shifting of the vehicle during these operations.

1. Before servicing the vehicle, refer to the Precautions Section.

2. Index-mark the front driveshaft and pinion flange.

3. Remove or disconnect the following:

- Front driveshaft from the pinion flange, and position it aside

➡ **Do not allow the driveshaft to hang unsupported.**

4. Using a Nm (inch-pound) torque wrench, measure the torque required to maintain pinion rotation. Record the measurement.

5. Index-mark the pinion flange and the pinion stem.

6. Hold the pinion flange while removing the nut.

7. Place a drain pan under the differential housing.

8. Using a puller, remove the pinion flange.

9. Inspect the pinion flange for burrs and damage. Inspect the end of the pinion flange that contacts the bearing cone, the nut counterbore, and the seal surface for nicks. Discard the pinion flange as necessary.

10. Using a seal remover and impact slide hammer, remove the pinion seal.

11. Remove the front axle drive pinion shaft oil slinger and the differential pinion bearing.

12. Remove and discard the collapsible spacer.

To install:

13. Verify that the splines on the pinion stem are free of burrs. If burrs are evident, remove them with a fine crocus cloth. Work in a rotating motion to wipe the pinion clean.

14. Clean the pinion seal bore.

15. Install a new collapsible spacer.

16. Install the original differential pinion bearing and the front axle drive pinion shaft oil slinger.

17. Lubricate the pinion seal. Use Motorcraft SAE 80W90 Thermally Stable 4x4 Axle Lubricant meeting Ford specification WSP-M2C197-A.

18. Install the pinion seal.

19. Lubricate the pinion flange splines. Use Motorcraft SAE 80W90 Thermally Stable 4x4 Axle Lubricant meeting Ford specification WSP-M2C197-A.

➡ **Never use a metal hammer on the pinion flange or install the flange with power tools. If necessary, use a plastic hammer to tap on a tight fitting flange.**

- Align the index marks and install the pinion flange.
- Install the new nut hand-tight.

➡ **Do not loosen the nut to reduce preload. Install a new collapsible spacer and nut if preload reduction is necessary.**

Holding the front axle pinion flange

Removing the front axle pinion flange

Removing the front axle pinion seal

Installing the front axle pinion seal

20. Use the special tool to hold the pinion flange while tightening the nut to set the preload.

21. Tighten the nut, rotating the pinion occasionally to ensure the differential pinion bearings are seating correctly. Take frequent differential pinion bearing preload readings by rotating the pinion with a Nm (inch-pound) torque wrench. The final reading must be 0.56 Nm (5 inch lbs.) more than the initial reading taken during removal.

22. Align the index marks and position the front driveshaft.

23. Install the universal joint spider retainers and bolts.

24. Check the fluid level and, if necessary, fill the axle to specification. Use Motorcraft SAE 80W90 Thermally Stable 4x4 Axle Lubricant meeting Ford specification WSP-M2C197-A.

25. Lower the vehicle.

26. If so equipped, reactivate the air suspension.

Rear Pinion Seal

REMOVAL & INSTALLATION

1. Before servicing the vehicle, refer to the Precautions Section.

2. Drain the axle housing fluid.

3. Remove or disconnect the following:
   • Negative battery cable
   • Rear wheels
   • Driveshaft
   • Brake calipers and pads or brake drum

➡ **The brake calipers and pads or brake drum must be removed so that there is no additional drag when measuring pinion bearing preload.**

4. Use an inch lb. torque wrench and measure and record the amount of torque required to maintain pinion rotation through several revolutions.

5. Remove or disconnect the following:
   • Pinion flange
   • Pinion seal
   • Pinion bearing
   • Collapsible spacer

To install:

➡ **Use a new collapsible spacer and flange nut for assembly.**

6. Install or connect the following:
   • Collapsible spacer
   • Pinion bearing
   • Pinion seal
   • Pinion flange

7. Rotate the pinion flange occasionally while tightening the flange nut to make sure the pinion bearings seat correctly.

8. Take frequent bearing preload torque readings. Tighten the flange nut to achieve the preload torque readings originally recorded.

✳✳ CAUTION

Never loosen the pinion nut to reduce bearing preload. If it is necessary to reduce bearing preload, install a new collapsible spacer and pinion nut.

9. Install or connect the following:
- Driveshaft
- Brake calipers and pads or brake drum
- Wheels
- Negative battery cable

10. Fill the differential with gear lubricant and check for leaks.

STEERING

Air Bag

PRECAUTIONS

- Always wear safety glasses when servicing an air bag vehicle, and when handling an air bag.
- Never attempt to service the steering wheel or steering column on an air bag-equipped vehicle without first properly disarming the air bag system. The air bag system should be properly disarmed whenever ANY service procedure in this manual indicates that you should do so.
- When carrying a live air bag module, always make sure the bag and trim cover are pointed away from your body. In the unlikely event of an accidental deployment, the bag will then deploy with minimal chance of injury.
- When placing a live air bag on a bench or other surface, always face the bag and trim cover up, away from the surface. This will reduce the motion of the air bag if it is accidentally deployed.
- If you should come in contact with a deployed air bag, be advised that the air bag surface may contain deposits of sodium hydroxide, which is a product of the gas combustion and is irritating to the skin. Always wear gloves and safety glasses when handling a deployed air bag, and wash your hands with mild soap and water afterwards.

DE-POWERING

1. Before servicing the vehicle, refer to the Precautions Section.
2. Turn all vehicle accessories OFF.
3. Turn the ignition switch to OFF.
4. At the smart junction box (SJB), located below the RH side of the instrument panel, remove the RH lower cowl trim panel and remove the restraints control module (RCM) fuse from the SJB. For additional information, refer to the Owner's Manual.
5. Turn the ignition ON and visually monitor the air bag indicator for at least 30

seconds. the air bag indicator will remain lit continuously (no flashing) if the correct RCM fuse has been removed. If the air bag indicator does not remain lit continuously, remove the correct RCM fuse before proceeding.

6. Turn the ignition OFF.

✳✳ CAUTION

To avoid accidental deployment and possible personal injury, the backup power supply must be depleted before repairing or replacing any front or side air bag supplemental restraint system (SRS) components and before servicing, replacing, adjusting or striking components near the front or side air bag sensors or RCM, such as doors, instrument panel, console, door latches, strikers, seats and hood latches.

7. To deplete the backup power supply energy, disconnect the battery ground cable and wait at least one minute. Be sure to disconnect auxiliary batteries and power supplies (if equipped).

8. Disconnect the battery ground cable (14301) and wait at least one minute.

REPOWERING

1. Before servicing the vehicle, refer to the Precautions Section.

✳✳ CAUTION

The restraint system diagnostic tool is for restraint system service only. Remove from vehicle prior to road use. Failure to remove could result in injury and possible violation of vehicle safety standards. Make sure all restraint system diagnostic tool(s) that may have been installed during the repair have been removed from the vehicle and all SRS components are connected.

2. Turn the ignition switch from OFF to ON.

3. Install the RCM fuse to the SJB and install the RH lower cowl trim panel.

✳✳ CAUTION

Be sure that nobody is in the vehicle and that there is nothing blocking or set in front of any air bag module when the battery ground cable is connected.

4. Connect the battery ground cable.
5. Prove out the supplemental restraint system (SRS) as follows:
   a. Turn the ignition key from ON to OFF. Wait 10 seconds, then turn the key back to ON and visually monitor the air bag indicator with the air bag modules installed. The air bag indicator will light continuously for approximately six seconds and then turn off. If an air bag supplemental restraint system (SRS) fault is present, the air bag indicator will either:
   - fail to light.
   - remain lit continuously.
   - flash.
   b. The flashing might not occur until approximately 30 seconds after the ignition switch has been turned from the OFF to the ON position. This is the time required for the restraints control module (RCM) to complete the testing of the SRS. If the air bag indicator is inoperative and a SRS fault exists, a chime will sound in a pattern of five sets of five beeps. If this occurs, the air bag indicator and any SRS fault discovered must be diagnosed and repaired.
   c. Clear all continuous DTCs from the restraints control module using a scan tool.

Steering Gear

REMOVAL & INSTALLATION

2-Wheel Drive

1. Before servicing the vehicle, refer to the Precautions Section.

2. Turn the wheel to the straight-ahead position and turn the ignition switch to the OFF position.

3. Remove the front wheel and tire assemblies.

4. Remove the fluid cooler.

5. Remove and discard the cotter pins and nuts.

➡**Do not damage the tie-rod boot when installing the special tool.**

6. Using special tool, separate the tie-rod ends from the wheel knuckles.

➡**Do not allow the intermediate shaft to rotate while it is disconnected from the steering gear or damage to the clockspring can result. If there is evidence that the intermediate shaft has rotated, the clockspring must be removed and recentered.**

7. Remove the pinch bolt and detach the intermediate shaft from the gear. Discard the bolt.

8. Remove the nut and disconnect the lines.

9. Plug or cap the power steering return hose, power steering pressure hose, and the steering gear ports to prevent the entry of dirt.

➡**Hold the tops of the steering gear to crossmember stud bolts to avoid damaging the steering gear fluid transfer tubes.**

10. Remove the rack mounting nuts.

11. Remove the mounting stud, nut, washer and stop assemblies.

12. Remove the steering gear. Clean the mounting surfaces.

13. To install, reverse the removal procedure.

   a. Install new seals on the power

1	Stud	5	Insulator
2	Washer	6	Crossmember
3	Insulator	7	Tie-rod end — outer
4	Steering gear	8	Nut

Steering gear mounting

67197-RANG-G91A

steering return hose and power steering pressure hose.

b. The dished side of the washer faces downward.

c. Install a new intermediate shaft pinch bolt.

14. Observe the following torques:
- Pinch bolt: 35 ft. lbs. (48 Nm)
- Pressure line bracket nut: 18 ft. lbs. (25 Nm)
- Rack retaining nuts: 111 ft. lbs. (150 Nm)

15. Fill and leak check the power steering system.

16. Check and, if necessary, adjust the wheel alignment.

4-Wheel Drive

1. Before servicing the vehicle, refer to the Precautions Section.

2. Turn the wheel to the straight-ahead position and turn the ignition switch to the OFF position.

3. Remove the wheel and tire assemblies.

4. Remove the fluid cooler.

5. Remove the four air deflector retaining screws.

6. Pull downward on the air deflector to disengage the retaining pins.

7. Loosen the LH tie-rod end jam nut.

8. Remove and discard the cotter pins and nuts.

➡ **Do not damage the tie-rod boot when installing the special tool.**

9. Using a separator, separate the tie-rod ends from the wheel knuckles

10. Remove the LH tie-rod end. Count and record the number of turns required to remove the tie-rod end.

11. Remove the front stabilizer bar. Note or mark the driver side end of the sway bar for correct installation.

➡ **Do not allow the intermediate shaft to rotate while it is disconnected from the steering gear or damage to the clockspring can result. If there is evidence that the intermediate shaft has rotated, the clockspring must be removed and recentered.**

12. Remove the pinch bolt and detach the intermediate shaft from the gear.

13. Remove the nut and disconnect the lines.

14. Plug the ends of all fluid lines removed and ports in the steering gear to prevent entry of dirt.

➡ **Hold the tops of the steering gear to crossmember stud bolts to avoid dam-**

aging the steering gear fluid transfer tubes.

15. Remove the steering rack nuts.

16. Remove the stud bolts and washers.

17. Remove the steering gear to crossmember insulator bushings.

18. Rotate the steering gear control valve housing toward the front of the vehicle.

19. Turn the steering gear input shaft to the right until the stop is reached.

20. Move the steering gear as far to the RH side of the vehicle as possible.

21. Move the LH front wheel spindle tie-rod forward to clear the frame crossmember.

22. Remove the steering gear from the vehicle.

To install:

23. Using special tool 211-027, install new seals on the power steering return hose and power steering pressure hose.

➡ **Make sure the steering gear input shaft is turned to the left until the stop is reached.**

➡ **Handle the steering gear with caution to avoid damage to fluid transfer tubes and to avoid dimples in tie-rod boots.**

24. Turn the steering gear input shaft to the right until the stop is reached. Note the number of turns required.

➡ **Make sure the steering gear control valve housing is turned toward the front of the vehicle.**

25. Install the steering gear into the RH opening of the crossmember.

26. Move the steering gear as far to the RH side of the vehicle as possible.

27. Move the LH front wheel spindle tie-rod into the opening in the crossmember and move the steering gear into position.

28. To place the steering gear in the straight ahead position, turn the steering gear input shaft to the left by half the number of turns recorded previously.

29. Rotate the steering gear control valve housing toward the rear of the vehicle.

30. Install the steering gear to crossmember insulator bushings.

a. The large end of the metal sleeve must be positioned downward.

b. Check that the mounting surfaces on the crossmember are clean and free of foreign material.

31. Install the steering gear to crossmember washers and stud bolts. The dished side of the washer faces downward.

➡ **Hold the tops of the steering gear to crossmember stud bolts to avoid dam-**

aging the steering gear fluid transfer tubes.

32. Install the rack mounting nuts. Torque to 111 ft. lbs. (150 Nm).

33. Install the lines and tighten the pressure line bracket nut to 18 ft. lbs. (25 Nm).

➡ **Do not allow the intermediate shaft to rotate while it is disconnected from the steering gear or damage to the clockspring can result. If there is evidence that the intermediate shaft has rotated, the clockspring must be removed and recentered.**

34. Connect the intermediate shaft to the steering gear input shaft. Install a new lower steering column pinch bolt. Torque to 35 ft. lbs. (48 Nm).

35. Install the power steering fluid cooler.

36. Install the front stabilizer bar. Orient the front stabilizer bar as noted during removal.

37. Install the LH tie-rod end on the front wheel spindle tie-rod. Rotate the tie-rod end the number of turns recorded during removal.

38. Position the tie-rod ends on the steering knuckles. Install the castellated nuts and new cotter pins. Check that the brake dust shields are not bent and are not in contact with the outer tie-rod boot seals. Torque to 52 ft. lbs. (70 Nm).

39. Tighten the tie-rod end jam nut. Torque to 59 ft. lbs. (80 Nm).

40. Position the air deflector, and install the retaining screws.

41. Install the front wheel and tire assemblies.

42. Fill and leak check the system.

43. Check and, if necessary, adjust the wheel alignment.

Power Steering Pump

REMOVAL & INSTALLATION

2.3L Engine

1. Before servicing the vehicle, refer to the Precautions Section.

2. Remove the air cleaner outlet pipe.

3. Rotate the tensioner counterclockwise and remove the belt from the power steering pump pulley.

✷✷ WARNING

Replacement of the power steering pump pulley is necessary after being removed and installed two times.

06017-RANG-G15

Power steering pump-to-bracket bolts—2.3L engine

06017-RANG-G12

Power steering pump pulley removal—2.3L and 3.0L engines

06017-RANG-G14

Power steering pump pulley installation—2.3L engine

4. Using the special tool, remove the power steering pump pulley.

5. Inspect the pulley for paint marks in the web area near the hub. If there are two paint marks, discard the pulley. If there is no paint mark or one paint mark, use a paint pencil to mark the web area of the pulley near the hub.

6. Loosen the clamp and remove the hose.

7. Disconnect the pressure line. Remove and discard the seal.

8. Remove the bolts and the bracket.

9. Remove the bolt and the power steering pump.

10. To install, reverse the removal procedure. Rotate the tensioner clockwise, and install the belt to the power steering pump pulley. Observe the following torques:
- Mounting bolts: 13 ft. lbs. (18 Nm)
- Pressure line: 48 ft. lbs. (65 Nm)

11. Using the special tool, install a new seal on the pressure line.

12. Fill and leak test the system.

3.0L Engine

1. Before servicing the vehicle, refer to the Precautions Section.

2. Rotate the tensioner, and remove the belt from the power steering pump pulley.

✳✳ WARNING

Replacement of the power steering pump pulley is necessary after being removed and installed two times.

3. Using the special tool, remove the power steering pump pulley.

4. Inspect the pulley for paint marks in the web area near the hub. If there are two paint marks, discard the pulley. If there is no paint mark or one paint mark, use a paint pencil to mark the web area of the pulley near the hub.

5. Remove the pump pulley.

6. Disconnect the power steering pressure hose. Remove and discard the seal.

7. Disconnect the power steering return hose.

8. Disconnect the power steering return line hose.

9. Remove the bolts and the power steering pump.

10. To install, reverse the removal procedure. Using the special tool, install the power steering pump pulley. Torque the mounting bolts and pressure line to 35 ft. lbs. (48 Nm).

11. Using the special tool, install a new seal on the power steering pressure hose.

06017-RANG-G13

Power steering pump pulley installation—3.0L engine

06017-RANG-G16

Seal installation

12. Fill and leak test the system.

4.0L Engine

1. Before servicing the vehicle, refer to the Precautions Section.

2. Remove the engine cooling fan.

3. Loosen the power steering pump pulley bolts.

4. Rotate the tensioner counterclockwise and remove the belt from the power steering pump pulley.

5. Remove the bolts. Remove the power steering pump pulley.

6. Disconnect the power steering return line hose at the power steering fluid reservoir. Allow the system to drain.

7. Disconnect the power steering pressure hose from the power steering pump.

8. Compress and move the power steering fluid reservoir outlet hose clamp.

9. Disconnect the power steering fluid reservoir outlet hose from the power steering pump.

10. Remove the bolts. Remove the power steering pump.

To install:

11. Position the power steering pump.

12. Install the bolts. Torque to 18 ft. lbs. (25 Nm).

13. Connect the power steering fluid reservoir outlet hose to the power steering pump.

14. Connect the power steering fluid reservoir outlet hose to the power steering pump.

15. Compress and move the power steering fluid reservoir outlet hose clamp into place.

16. Using the special tool, install a new seal on the power steering pressure hose.

17. Connect the power steering pressure hose to the power steering pump. Torque to 48 ft. lbs. (65 Nm).

18. Connect the power steering return line hose to the power steering fluid reservoir.

19. Position the power steering pump pulley.

20. Install the bolts.

21. Install the drive belt.

22. Tighten the power steering pump pulley bolts to 18 ft. lbs. (25 Nm).

23. Install the engine cooling fan.

24. Fill and leak check the system.

06017-RANG-G17

Power steering pump mounting—4.0L engine

SUSPENSION

Shock Absorber

REMOVAL & INSTALLATION

Front

➡**Low pressure gas shocks are charged with nitrogen gas. Do not attempt to open, puncture or apply heat to them. Prior to installing a new shock absorber, hold it upright and extend it fully. Invert it and fully compress and extend it at least 3 times. This will bleed trapped air.**

1. Before servicing the vehicle, refer to the Precautions Section.

2. Remove or disconnect the following:
   - Negative battery cable
   - Upper shock-to-frame attaching nut, washer and insulator assembly
   - Lower shock-to-control arm attaching nuts
   - Slightly compress the shock absorber by hand and remove it from the vehicle

To install:

3. Install or connect the following:
   - Position the lower washer and insulator on the shock absorber rod and position the shock absorber to the upper frame bracket mount
   - Position the upper insulator and washer on the shock absorber rod and install the attaching nut loosely.
   - Position the lower shock absorber mounting studs into the control arm and install the attaching nuts loosely.
   - Torque the lower shock attaching nuts to 15–21 ft. lbs. (21–29 Nm), and the upper shock attaching bolts to 30–40 ft. lbs. (40–55 Nm).
   - Negative battery cable

Rear

➡**Low pressure gas shocks are charged with nitrogen gas. Do not attempt to open, puncture or apply heat to them. Prior to installing a new shock absorber, hold it upright and extend it fully. Invert it and fully compress and extend it at least 3 times. This will bleed trapped air.**

1. Before servicing the vehicle, refer to the Precautions Section.

2. Remove or disconnect the following:
   - Upper shock-to-frame attaching nut
   - Lower shock nut
   - Slightly compress the shock absorber by hand and remove it from the vehicle

To install:

3. Install or connect the following:
   - Shock absorber upper end and nut
   - Shock absorber lower end and nut
   - Torque the upper and lower shock attaching nuts to 53 ft. lbs. (72Nm)

1 Front suspension upper arm cam bolt (4 req'd)

2 Front suspension upper arm cam assy (2 req'd)

3 Front suspension upper arm cam assy nut (2 req'd)

4 Front shock absorber upper nut/washer assy (2 req'd)

5 Front shock absorber upper bushing (2 req'd)

6 Front suspension upper arm

6 Front suspension upper arm

7 Front wheel spindle pinch bolt (2 req'd)

8 Front wheel spindle pinch bolt nut

9   Front wheel spindle
10 Tie-rod end
11 Cotter pin
12 Lower ball joint castellated nut (2 req'd)
13 Tie-rod end castellated nut (2 req'd)
14 Cotter pin
15 Front shock absorber
16 Front coil spring
17 Front spring insulator
18 Front suspension lower arm
18 Front suspension lower arm

19 Front stabilizer bar link
20 Front stabilizer bar stud and bushing assy (2 req'd)
21 Front shock absorber lower nut (4 req'd)
22 Front stabilizer bar nut and washer assy (2 req'd)
23 Front stabilizer bar mounting bolts (4 req'd)
24 Stabilizer bar bracket (2 req'd)
25 Front stabilizer bar bushing assy (2 req'd)
26 Front stabilizer bar

67197-RANG-G92

2-wheel drive coil spring front suspension

1. Upper arm, bushing and joint assembly (LH)
2. Upper arm, bushing and joint assembly (RH)
3. Shock assembly
4. Torsion bar adjuster plate
5. Torsion bar adjuster
6. Torsion bar (LH)

7. Lower arm, bushing and joint assembly (LH)
8. Knuckle assembly
9. Protection shield (RH)
10. Protection shield (LH)
11. Screw (self-tapping)
12. Oil seal

67197-RANG-G93

4-wheel drive front suspension

Coil Spring

REMOVAL & INSTALLATION

1. Before servicing the vehicle, refer to the Precautions Section.
2. Remove or disconnect the following:
   • Wheel and tire assembly
   • Shock absorber
   • Front stabilizer bar link nut
3. Use a coil spring compressor to compress the coil spring.
4. Remove the cotter pin and castellated nut.
5. Separate the lower ball joint from the front wheel spindle.
6. Position the front wheel spindle out of the way and remove the coil spring.

To install:

➡ **The end of the coil spring must cover the first hole and should not be visible in the second hole.**

7. Install the coil spring in the lower arm.

✳ WARNING

Always install the cotter pin into the lower ball joint castellated nut from outboard to inboard. Failure to do so will result in damage to the wheel and tire assembly.

8. Install the lower ball joint.
9. Install the front stabilizer bar link nut.
10. Remove the Coil Spring Compressor.
11. Install the front shock absorber and the two lower nuts.
12. Install the upper shock absorber bushing and nut/washer assembly.
13. Install the wheel and tire assembly.

Leaf Springs

REMOVAL & INSTALLATION

1. Before servicing the vehicle, refer to the Precautions Section.
2. Remove or disconnect the following:
   • Negative battery cable
   • Rear wheels
   • U-bolts from the rear spring plate
   • Hardware from the spring to bracket at the front of the rear spring
   • Upper and lower shackle bolts at the rear of the spring
   • Spring and shackle from the bracket

To install:

3. Install or connect the following:
   • Spring and shackle to the bracket
   • Upper and lower shackle bolts at the rear of the spring. Torque the nuts to 85 ft. lbs. (115 Nm).
   • U-bolts to the spring plate. Torque the nuts 76 ft. lbs. (103 Nm).
   • Rear wheels
   • Negative battery cable

Torsion Bar

REMOVAL & INSTALLATION

✳ CAUTION

The electrical power to the air suspension system must be shut off prior to hoisting, jacking or towing an air suspension vehicle. This can be accomplished by turning off the air suspension switch located in the rear jack storage area. Failure to do so can result in unexpected inflation or deflation of the air springs or shocks, which can result in shifting of the vehicle during these operations.

1. Before servicing the vehicle, refer to the Precautions Section.
2. Remove the torsion bar cover plate.

➡ **Before relieving the torsion bar tension, measure and record the measurement of the torsion bar adjustment bolt. This measurement will be used as the preset depth for the new torsion bar adjustment bolt during installation.**

3. Relieve the torsion bar tension.
   a. Position the Torsion Bar Tool and adapters.
   b. Tighten the Torsion Bar Tool until the torsion bar adjuster lifts off the adjustment bolt.

06017-RANG-G18

Measure and record the measurement of the torsion bar adjustment bolt

06017-RANG-G19

Torsion Bar Tool

06017-RANG-G20

Torsion bar adjuster

✳ CAUTION

The torsion bar adjustment bolt is coated with dry adhesive; and must be replaced if it is backed off or removed. Failure to do so can cause the adjustment bolt to loosen during operation and cause a loss of vehicle alignment.

   c. Remove the torsion bar adjustment bolt and nut.
   d. Loosen the Torsion Bar Tool until the tension is removed from the torsion bar.
4. Mark the torsion bar and the adjuster for proper installation.
5. Remove the torsion bar insulator.
6. Grasp the torsion bar, and pull it free from the front suspension lower arm.

To install:

7. Position the torsion bar and the torsion bar adjuster.
8. Align the marks on the torsion bar and the torsion bar adjuster, then install the torsion bar adjuster.
9. Position the torsion bar insulator.
10. Install the Torsion Bar Tool and the adapters.
11. Tighten the Torsion Bar Tool until the new adjustment bolt and nut can be installed.
12. Turn the adjustment bolt until the preliminary adjustment measurement (recorded length of the old adjustment bolt) is reached.

13. Install the torsion bar cover plate. Torque the bolts to 46 ft. lbs. (63Nm).

14. If equipped with air suspension, reactivate the system by turning on the air suspension switch.

15. Lower the vehicle.

16. Adjust the ride height.

17. Check the alignment.

Upper Ball Joint

REMOVAL & INSTALLATION

The ball joints are integral with the control arm. If the ball joint is defective, the entire control arm must be replaced.

Lower Ball Joint

REMOVAL & INSTALLATION

1. Before servicing the vehicle, refer to the Precautions Section.

2. With the vehicle in NEUTRAL, position it on a hoist.

3. Remove the wheel and tire assembly.

✳✳ WARNING

Do not allow the disc brake caliper to hang suspended from the brake hose. Provide a suitable support.

4. Remove the caliper support bracket bolts, then position the caliper and support bracket aside.

5. Disconnect the front anti-lock brake sensor (ABS) wire from the vehicle frame.

6. Using a suitable jack, support the front suspension lower arm.

7. Remove the tie-rod end castellated nut. Remove and discard the cotter pin and the castellated nut.

✳✳ WARNING

Do not use a hammer to separate the tie-rod from the wheel knuckle or damage to the wheel knuckle will result. Do not damage the tie-rod boot when installing the special tool.

8. Using the special tool, separate the tie-rod end from the front wheel knuckle.

9. Remove the lower ball joint castellated nut. Remove and discard the cotter pin and the castellated nut.

10. Separate the front wheel knuckle from the front suspension lower arm. Then, loosely install the lower ball joint castellated nut.

11. Remove the pinch bolt and nut.

12. Remove the hand-tightened lower ball joint castellated nut, then remove the front wheel knuckle.

13. Remove the snapring from the ball joint. Discard the snap ring.

14. Using a suitable ball joint remover tool, remove the ball joint.

✳✳ WARNING

Do not damage the ball joint boot when installing the special tool.

➡Clean and inspect the control arm ball joint bore for damage before installing a new ball joint.

➡Make sure the new ball joint snap ring is fully seated.

15. To install, reverse the removal procedure. Always install new castellated nuts and cotter pins. Observe the following torques:

- Pinch bolt and nut: 41 ft. lbs. (55 Nm)
- Lower ball stud nut: 98 ft. lbs. (133 Nm)
- Tie rod stud nut: 52 ft. lbs. (70 Nm)
- Caliper support bracket bolts: 83 ft. lbs. (112 Nm)

Upper Control Arm

REMOVAL & INSTALLATION

Coil Spring Suspension

1. Before servicing the vehicle, refer to the Precautions Section.

2. Remove or disconnect the following:
- Wheel and tire assembly
- Brake disc shield

3. Use a jack to support the front suspension lower arm.

4. Mark the position of the front suspension upper arm adjustment cams.

5. Remove the upper ball joint retaining nut and pinch bolt.

6. Separate the ball joint from the front wheel spindle.

7. Remove the front suspension upper arm.

8. Installation is the reverse of removal. Align the marks made during removal on the front suspension upper arm adjustment cam. The forward front suspension upper arm nut must be tightened first while the arm is held at the curb position ride height. Observe the following torques:

- Control arm attaching nuts: 98 ft. lbs. (133Nm)
- Pinch bolt: 46 ft. lbs. (63Nm)

Torsion Bar Suspension

2-WHEEL DRIVE

✳✳ WARNING

The electrical power to the air suspension system must be shut off prior to hoisting, jacking or towing an air suspension vehicle. This can be accomplished by turning off the air suspension switch located in the rear jack storage area. Failure to do so can result in unexpected inflation or deflation of the air springs or shocks, which can result in shifting of the vehicle during these operations.

1. Before servicing the vehicle, refer to the Precautions Section.

2. Raise the vehicle on a hoist.

3. Remove the wheel and tire assembly

4. Use a suitable jack stand to support the front suspension lower arm.

➡To avoid possible damage to the front wheel spindle, secure the spindle to keep it from tilting before removing the pinch bolt and nut.

5. Remove the pinch bolt and nut from the front wheel spindle.

6. Remove the front suspension upper arm:
   a. Remove the two nuts.
   b. Remove the two bolts.
   c. Remove the front suspension upper arm.

To install:

7. Position the front suspension upper arm.

8. Install the two bolts and two nuts. Torque to 83–112 ft. lbs. (113–153 Nm).

9. Position the front wheel spindle.

10. Install the pinch bolt and nut. Torque to 35–46 ft. lbs. (47–63 Nm).

11. Remove the jack stand from under the front suspension lower arm.

12. Install the tire and wheel assembly

➡If equipped with air suspension, reactivate the system by turning on the air suspension switch.

13. Lower the vehicle.

14. Inspect the front end ride height

15. Inspect and adjust the front end alignment

4-WHEEL DRIVE

1. Before servicing the vehicle, refer to the Precautions Section.

2. Raise the vehicle on a hoist.

3. Remove the wheel and tire assembly.

Upper control arm—2-wheel drive torsion bar suspension; 1-nuts; 2-bolts; 3-arm

67197-RANG-G94

4. Use a suitable jack stand to support the front suspension lower arm.

5. Remove the pinch bolt.

✳✳ WARNING

Before separating the front suspension upper arm from the front wheel knuckle, secure the front wheel knuckle to prevent it from tilting outward. Failure to do so can cause damage to the front axle shaft.

6. Separate the front suspension upper arm from the front wheel knuckle.

7. Remove the front suspension upper arm:

a. Remove the two nuts and alignment plates.

b. Remove the two bolts and the cams.

c. Remove the front suspension upper arm.

To Install:

➡**When installing the front suspension upper arm, replace the alignment plates with new alignment cams.**

8. Install the front suspension upper arm.

a. Position the front suspension arm bushing joint.

b. Install the two bolts, four cams and two nuts. Torque to 83–112 ft. lbs. (113–153 Nm).

9. Install the pinch bolt and nut.

a. Position the upper arm into the front wheel knuckle.

b. Install the pinch bolt and nut. Torque to 41 ft. lbs. (55 Nm).

10. Remove the jack stand from under the front suspension lower arm.

11. Install the wheel and tire assembly.

12. Lower the vehicle.

13. Check the wheel alignment.

67197-RANG-G95

4-wheel drive upper control arm

UPPER CONTROL ARM BUSHING REPLACEMENT

The control arm bushings are not serviceable. If they require service, the upper or lower arm must be replaced.

Lower Control Arm

REMOVAL & INSTALLATION

Coil Spring Suspension

1. Before servicing the vehicle, refer to the Precautions Section.

2. Remove or disconnect the following:
- Negative battery cable
- Front wheel
- Brake rotor shield
- Shock absorber
- Stabilizer bar link hardware

3. Using a spring compressor tool, compress the coil spring.
- Lower ball joint from the spindle
- Lower control arm bolts
- Lower control arm and coil spring

To install:

4. Install or connect the following:
- Coil spring to the lower control arm

➡**The end of the coil spring must cover the first hole and should not be visible in the second hole.**

- Lower arm and front coil spring
- The two front suspension lower arm bolts and nuts. Do not tighten the nuts at this time.

➡**On the RH front suspension lower arm, install the rear bolt adjustment cam, and nut in the center of the frame slot.**

✳✳ CAUTION

Always install the cotter pin into the lower ball joint castellated nut from outboard to inboard, with the fingers bent together at a right angle. Failure to do so will result in damage to the wheel and tire assembly.

- Lower ball joint. Torque the nut to 113 ft. lbs. (153Nm).

5. Remove the Coil Spring Compressor.

6. Install or connect the following:
- Front stabilizer bar link nut. Torque the nut to 21 ft. lbs. (29Nm).
- Shock absorber and the two lower nuts
- Upper shock absorber bushing and nut/washer assembly

7. Support the lower control arm with a jackstand. Torque the bolts to 129 ft. lbs. (175 Nm).
- Brake disc shield
- Wheel and tire assembly

8. Inspect and adjust the front end alignment.

Torsion Bar Suspension

1. Before servicing the vehicle, refer to the Precautions Section.
2. Raise the vehicle on a hoist.
3. Remove the wheel and tire assembly.
4. Remove the stabilizer link nut, washer and bushing.
5. Remove the front shock absorber-to-front suspension lower arm nuts.
6. Remove the torsion bar
7. Remove the lower ball joint castellated nut.

➡ **Do not use a hammer to separate the ball joint from the wheel knuckle or damage to the wheel knuckle will result. Do not damage the ball joint boot while installing the special tool.**

8. Using the special tool, separate the front suspension lower arm from the front wheel knuckle/spindle.
9. Remove the front suspension lower arm bolts and nuts.
10. Remove the front suspension lower arm.

To install:

➡ **Tighten the front suspension lower arm pivot bolts and nuts until snug. Do not tighten to specification until the installation procedure is complete.**

11. Position the front suspension lower arm to the front suspension crossmember.
12. Install the pivot bolts and nuts and tighten until snug.

✳ WARNING

Install the cotter pin into the lower ball joint from outboard to inboard with the fingers bent together at a right angle. Failure to do so will cause damage to the wheel and tire assembly.

13. Position the lower ball joint into the front wheel knuckle/spindle.
14. Install the new castellated nut. Torque to 83–112 ft. lbs. (113–153 Nm).
15. Install a new cotter pin.
16. Install the front shock absorber-to-front suspension lower arm nuts. Torque to 15–21 ft. lbs. (21–29 Nm).
17. Install the stabilizer link bushing, washer, and nut. Torque to 15–21 ft. lbs. (21–29 Nm).

➡ **Whenever the torsion bar or torsion bar adjuster is removed, the vehicle ride height must be checked.**

18. Install the torsion bar.
19. Install the tire and wheel assembly.
20. Lower the vehicle.
21. Tighten the front suspension lower arm nuts. Torque to 111–148 ft. lbs. (150–200 Nm).
22. Inspect and adjust the front end alignment.

LOWER CONTROL ARM BUSHING REPLACEMENT

The control arm bushings are not serviceable. If they require service, the upper or lower arm must be replaced.

Wheel Bearings

ADJUSTMENT

2-Wheel Drive Vehicles

1. Before servicing the vehicle, refer to the Precautions Section.
2. Remove the grease cap from the hub and wipe the excess grease from the end of the spindle. Remove the cotter pin and retainer. Discard the cotter pin.
3. Loosen the adjusting nut 3 turns.

✳ WARNING

Obtain running clearance between the disc brake rotor surface and shoe linings by rocking the entire wheel assembly in and out several times in order to push the caliper and brake pads away from the rotor. An alter- nate method to obtain proper running clearance is to tap lightly on the caliper housing. Be sure not to tap on any other area that may damage the disc brake rotor or the brake lining surfaces. Do not pry on the phenolic caliper piston. The running clearance must be maintained throughout the adjustment procedure. If proper clearance cannot be maintained, the caliper must be removed from its mounting.**

4. While rotating the wheel assembly, tighten the adjusting nut to 17–25 ft. lbs. (23–34 Nm) in order to seat the bearings. Loosen the adjusting nut a half turn. Retighten the adjusting nut 18–20 inch lbs. (2.0–2.2 Nm).
5. Place the retainer on the adjusting nut. The castellations on the retainer must be in alignment with the cotter pin holes in the spindle. Once this is accomplished install a new cotter pin and bend the ends to insure its being locked in place.
6. Check for proper wheel rotation. If correct, install the grease cap.
7. Lower the vehicle and tighten the lug nuts to 100 ft. lbs., (136 Nm) if the wheel was removed. Before driving the vehicle, pump the brake pedal several times to restore normal brake pedal travel.

✳ CAUTION

If the wheel was removed, retighten the wheel lug nuts to specification after about 500 miles (804km) of driving. Failure to do this could result in the wheel coming off while the vehicle is in motion causing loss of vehicle control or collision.

ADJUSTING NUT

7924EG34

Loosen the adjusting nut 3 turns, then rock the entire wheel assembly in-and-out to spread the brake pads before attempting to adjust the bearing—2wd vehicles

REMOVAL & INSTALLATION

2-Wheel Drive

1. Before servicing the vehicle, refer to the Precautions Section.
2. Remove or disconnect the following:
   - Disc brake caliper anchor plate
   - Hub grease cap
   - Cotter pin
   - Nut retainer
   - Spindle nut
   - Wheel outer bearing retainer washer
   - Outer front wheel bearing
   - Brake disc and hub
   - Hub grease seal
   - Inner wheel bearing

To install:

3. Thoroughly clean and inspect the front wheel bearings and the brake disc and hub.
4. Lubricate the front wheel bearings.
5. Install the inner front wheel bearing.
6. Install a new wheel hub grease seal.
7. Position the brake disc and hub.
8. Assemble all parts and adjust the bearings.

4-Wheel Drive

1. Before servicing the vehicle, refer to the Precautions Section.
2. Remove the brake disc.
3. Remove and discard the nut and washer assembly.

✳✳ WARNING

Do not use a hammer to separate the outboard CV joint from the hub. Damage to the outboard CV threads and to internal components may result.

4. Using the special tool, separate the outboard CV joint from the hub.
5. Remove the dust shield.
6. Remove the bolt and detach the anti-lock sensor from the wheel hub.

✳✳ WARNING

Do not overextend the CV joint and boots when removing the hub and bearing assembly.

7. Remove the bolts and the wheel hub.
8. To install, reverse the removal procedure. Observe the following torques:
   - Hub/bearing bolts: 85 ft. lbs. (115 Nm)
   - Dust shield: 9 ft. lbs. (12 Nm)
   - Halfshaft nut: 162 ft. lbs. (220 Nm)

Stabilizer Bar

REMOVAL & INSTALLATION

Front

1. Before servicing the vehicle, refer to the Precautions Section.
2. Raise the vehicle on a hoist.
3. Remove the wheel and tire assembly.
4. Remove the front stabilizer bar link nuts from the front suspension lower arms.
5. Remove the front stabilizer bar link bolts and the front stabilizer bar links (5K483).
6. Remove the four bolts and two brackets.
7. Remove the front stabilizer bar.
8. Remove the stabilizer bar insulator.
9. Installation is the reverse of removal. Observe the following torques:
   - End link nuts: 18 ft. lbs. (25 Nm)

- Bracket bolts: 30 ft. lbs. (40 Nm)

➡ **In the event the self-tapping bolts cannot be installed in the frame, there is a kit available with flag nuts.**

Rear

1. Before servicing the vehicle, refer to the Precautions Section.
2. Raise the vehicle and install safety stands.
3. Remove the wheel and tire assembly.
4. Remove the rear stabilizer bar link nuts and bolt.
5. Remove the rear stabilizer bar link.
6. Remove the rear stabilizer bar mounting bracket bolts and remove the brackets.
7. Remove the rear stabilizer bar.
8. Installation is the reverse of removal. Observe the following torques:
   - End link nuts: 59 ft. lbs. (80 Nm)
   - Bracket bolts: 34 ft. lbs. (46 Nm)

34-46 Nm (25-34 lb/ft)

60-80 Nm (44-59 lb/ft)

Rear stabilizer bar

06017-RANG-G21

BRAKES

Brake Caliper

REMOVAL & INSTALLATION

1. Before servicing the vehicle, refer to the Precautions Section.
2. Loosen the wheel lug nuts.
3. Raise and safely support the front of the vehicle. Remove the wheel.

4. Place an 8 in. (203mm) C-clamp on the caliper and tighten the clamp to bottom the caliper pistons in their bores. Remove the clamp.
5. Remove the two caliper slide pin bolts and lift the caliper from the anchor plate.

➡ **Use care to retain as much of the original caliper slide pin grease as possible.**

6. Position the caliper on a frame member or suspend it with some wire. Do not allow the caliper to hang by the brake hose.
7. Disconnect and plug the brake hose at the caliper. Remove the caliper from the rotor.

To install:

8. Position the caliper over the brake pads and align the slide pin mounting holes.

1 Front Brake Anti-Lock Sensor	7 Front Disc Brake Hub and Rotor
2 Front Wheel Spindle	8 Front Wheel Bearing
3 Front Disc Brake Rotor Shield	9 Front Wheel Outer Bearing Retainer Washer
4 Rotor Shield Bolt	10 Hub Spindle Nut
5 Grease Seal	11 Cotter Pin
6 Front Wheel Bearing	12 Nut Retainer
	13 Hub Grease Cap
	14 Disc Brake Caliper
	15 Front Disc Brake Caliper Anchor Plate
	16 Caliper Anchor Plate Bolts
	17 Disc Brake Caliper Bolt

Exploded view of the 2WD front disc brake assembly

93026G22

9. Install the slide pin bolts and tighten them to 24 ft. lbs. (32 Nm).

10. Install the caliper brake hose using new washers. Tighten the bolt to 25 ft. lbs. (34 Nm)

11. Install the wheel and snug the lug nuts.

12. Lower the vehicle and tighten the lug nuts to 100 ft. lbs. (135 Nm).

➡**The first couple of times you apply the brakes, the pedal may go to the floor. Continue to pump the brake pedal until it feels firm.**

13. Start the engine and apply the brakes several times to readjust the caliper pistons. Ensure that the pedal feels firm before operating the vehicle.

Disc Brake Pads

REMOVAL & INSTALLATION

1. Before servicing the vehicle, refer to the Precautions Section.

2. Raise and safely support the front of the vehicle. Remove the wheel.

3. Place an 8 in. (203mm) C-clamp on the caliper and tighten the clamp to bottom the caliper pistons in their bores. Remove the clamp.

4. Remove the two caliper slide pin bolts and lift the caliper from the anchor plate.

➡**Use care to retain as much of the original caliper slide pin grease as possible.**

5. Position the caliper on a frame member or suspend it with some wire. Do not allow the caliper to hang by the brake hose.

6. Remove the brake pads and, if necessary, the anti-rattle clips from the anchor plate.

7. Remove the shims, if any, from the brake pads for re-use.

To install:

8. If removed, install the anti-rattle clips.

9. Install the brake pads to the anchor plate.

10. Position the caliper over the brake pads and align the slide pin mounting holes.

11. Install the slide pin bolts and tighten them to 24 ft. lbs. (32 Nm).

12. Install the wheel and snug the lug nuts.

13. Lower the vehicle and tighten the lug nuts to 100 ft. lbs. (135 Nm).

1 Front Disc Brake Caliper Anchor Plate Bolt (2 Req'd)

2 Front Brake Hose Bolt

3 Disc Brake Caliper

4 Pads

5 Front Disc Brake Caliper Anchor Plate

6 Front Disc Brake Rotor

7 Front Axle Wheel Hub Retainer

8 Front Disc Brake Rotor Shield Bolt (3 Req'd)

9 Front Disc Brake Rotor Shield

10 Front Wheel Knuckle

93026G23

Exploded view of the 4WD front disc brake assembly

1 stainless slippers
2 pads

93026G24

Position of the front disc brake components

93026G25

View of the front disc brake anti-rattle spring

➡ **The first couple of times you apply the brakes, the pedal may go to the floor. Continue to pump the brake pedal until it feels firm.**

14. Start the engine and apply the brakes several times to readjust the caliper pistons. Ensure that the pedal feels firm before operating the vehicle.

Brake Drums

REMOVAL & INSTALLATION

1. Before servicing the vehicle, refer to the Precautions Section.

2. Raise and safely support the vehicle. Remove the wheel and tire assembly.

3. Remove the retaining nuts, if equipped, and remove the brake drum.

4. Inspect the brake drum surface for wear, scoring and runout. Machine or replace, as necessary.

To install:

5. Install the brake drum and secure in place with the retainer nuts, if equipped.

6. Adjust the rear brakes.

7. Install the wheel. Lower the vehicle.

Brake Shoes

REMOVAL & INSTALLATION

1. Before servicing the vehicle, refer to the Precautions Section.

2. Raise and safely support the vehicle. Remove the wheel and tire assembly and the brake drum.

3. Pull backward on the adjusting lever cable to disengage the adjusting lever from the adjusting screw. Move the outboard side of the adjusting screw upward and back off the pivot nut as far as it will go.

4. Pull the adjusting lever, cable and automatic adjuster spring down and toward the rear to unhook the pivot hook from the large hole in the secondary shoe web. Do not pry the pivot hook from the hole.

5. Remove the automatic adjuster spring and adjusting lever.

6. Remove the secondary shoe-to-anchor spring using a suitable brake spring removal/installation tool. Using the tool, remove the primary shoe-to-anchor spring and unhook the cable anchor. Remove the anchor pin plate, if equipped.

7. Remove the cable guide from the secondary shoe.

8. Remove the shoe hold-down springs, shoes, adjusting screw, pivot nut and socket. Note the color and position of each hold-down spring so they can be reassembled in the same position.

9. Remove the parking brake link and spring. Disconnect the parking brake cable from the parking brake lever.

10. Remove the secondary brake shoe. On 9 in. (22.8cm) rear brakes, remove the parking brake lever from the shoe. On 10 in. (25.4cm) rear brakes, remove the retainer clip and spring washer and remove the parking brake lever.

To install:

11. Clean the backing plate ledge pads and sand lightly. Apply a light coating of high temperature lithium grease to the points where the brake shoes touch the backing plate. Lubricate the adjusting cable eye and the anchor pin area.

12. Install the parking brake lever on the secondary shoe. On 10 in. (25.4cm) brakes, secure with the spring washer and retaining clip.

13. Position the brake shoes on the backing plate and install the hold-down spring pins, springs and cups. Install the parking brake link, spring and washer. Connect the parking brake cable to the parking brake lever.

14. Install the anchor pin plate, if equipped, and place the cable anchor over the anchor pin with the crimped side toward the backing plate.

15. Install the primary shoe-to-anchor spring using the brake spring removal/installation tool.

16. Install the cable guide on the secondary shoe with the flanged hole fitted into the hole in the secondary shoe. Thread the cable around the cable guide groove.

➡ **Make sure the cable is positioned in the groove and not between the guide and shoe web.**

17. Install the secondary shoe-to-anchor (long) spring.

➡ **Make sure the cable end is not cocked or binding on the anchor pin when installed. All parts should be flat on the anchor pin.**

18. Apply high temperature lithium grease to the threads and the socket end of the adjusting screw. Turn the adjusting screw into the adjusting pivot nut to the end of the threads and then loosen, ½ turn.

19. Place the adjusting socket on the screw and install the assembly between the shoe ends with the adjusting screw nearest the secondary shoe.

➡ **Be sure to install the adjusting screw on the same side of the vehicle from which it came. To prevent incorrect installation, the socket end of each adjusting screw is stamped with R or L, to indicate installation on the right or left side of the vehicle. The adjusting pivot nuts have lines machined around the body of the nut, 2 lines indicating the right side nut and 1 line indicating the left side nut.**

20. Hook the cable hook into the hole in the adjusting lever from the outboard plate side. The adjusting levers are also stamped with an **R** or **L** to indicate right or left side installation.

21. Place the hooked end of the adjuster spring in the large hole in the primary shoe web and connect the loop end of the spring to the adjuster lever hole.

22. Pull the adjuster lever, cable and automatic adjuster spring down toward the

1 Wheel Cylinder-to-Backing Plate Bolt (2 Req'd)
2 Washer
3 Inspection Hole Cover
4 Brake Backing Plate
5 Lining Inspection Hole
6 Anchor Pin Guide Plate
7 Rear Wheel Cylinder
8 Wheel Cylinder Brake Shoe Link
9 Parking Brake Strut
10 Parking Brake Lever
11 Brake Shoe Adjusting Lever Cable

12 Rear Brake Shoe and Lining, Secondary
13 Washer
14 Parking Brake Lever Pin Retainer
15 Cable Guide
16 Adjusting Lever Pin
17 Adjusting Lever Return Spring
18 Brake Shoe Adjusting Lever
19 Brake Shoe Adjusting Screw Nut
20 Brake Adjuster Screw
21 Brake Shoe Adjusting Screw Spring

22 Brake Shoe Hold-Down Spring Cup
23 Brake Shoe Hold-Down Spring
24 Rear Brake Shoe and Lining, Primary
25 Brake Shoe Retracting Spring, Short
26 Parking Brake Link Spring
27 Parking Brake Spring Retainer
28 Brake Shoe Hold-Down Spring Pin
29 Brake Adjusting Hole Cover

93026G21

Exploded view of the rear brake shoes and components

rear to engage the pivot hook in the large hole in the secondary shoe web.

23. After installation, check the action of the adjuster by pulling the section of the cable between the cable guide and the adjusting lever toward the secondary shoe web far enough to lift the lever past a tooth on the adjusting screw wheel. The lever should snap into position behind the next tooth and releasing the cable should cause the adjuster spring to return the lever to its

original position. This return action will turn the adjusting screw 1 tooth.

24. If pulling the cable does not produce the action described previously, or if lever action is sluggish instead of positive and sharp, check the position of the lever on the adjusting screw toothed wheel. With the brake in a vertical position, anchor at the top, the lever should contact the adjusting wheel 1 tooth above the centerline of the adjusting screw. If the contact point is

below the centerline, the lever will not lock on the adjusting screw wheel teeth and the screw will not turn, since the lever is actuated by the cable.

25. Adjust the brake shoes using either a brake adjustment gauge or manually with the drums installed.

26. Install the wheels, and lower the vehicle.

FORD AND MERCURY

15

Sable • Taurus

SPECIFICATION AND MAINTENANCE CHARTS

ENGINE AND VEHICLE IDENTIFICATION

	Engine							Model Year	
Code ①	Liters (cc)	Cu. In.	Cyl.	Fuel Sys.	Engine Type	Eng. Mfg.		Code ②	Year
S	3.0 (3049)	182	6	EFI	DOHC	Ford		2	2002
U	3.0 (2982)	181	6	EFI	OHV	Ford		3	2003
								4	2004
								5	2005
								6	2006

OHV: Overhead Valves

DOHC: Double Overhead Camshafts

EFI: Electronic Fuel Injection

① 8th digit of the Vehicle Identification Number (VIN)

② 10th digit of the Vehicle Identification Number (VIN)

06017-TAUR-C01

GENERAL ENGINE SPECIFICATIONS

Year	Model	Engine Displacement Liters	Engine ID/VIN	Net Horsepower @ rpm	Net Torque @ rpm (ft. lbs.)	Bore x Stroke (in.)	Compression Ratio	Oil Pressure @ rpm
2002	Sable	3.0	U	155@4900	185@3950	3.50x3.15	9.3:1	40-60@2500
		3.0	S	200@5750	200@4500	3.50x3.13	10.0:1	20-45@1500
	Taurus	3.0	U	155@4900	185@3950	3.50x3.15	9.3:1	40-60@2500
		3.0	S	200@5750	200@4500	3.50x3.13	10.0:1	20-45@1500
2003	Sable	3.0	U	155@4900	185@3950	3.50x3.15	9.3:1	40-60@2500
		3.0	S	200@5750	200@4500	3.50x3.13	10.0:1	20-45@1500
	Taurus	3.0	U	155@4900	185@3950	3.50x3.15	9.3:1	40-60@2500
		3.0	S	200@5750	200@4500	3.50x3.13	10.0:1	20-45@1500
2004	Sable	3.0	U	155@4900	185@3950	3.50x3.15	9.3:1	40-60@2500
		3.0	S	200@5750	200@4500	3.50x3.13	10.0:1	20-45@1500
	Taurus	3.0	U	155@4900	185@3950	3.50x3.15	9.3:1	40-60@2500
		3.0	S	200@5750	200@4500	3.50x3.13	10.0:1	20-45@1500
2005	Sable	3.0	U	153@5800	186@3250	3.50x3.10	9.4:1	40-60@2500
		3.0	S	201@5500	207@4500	3.50x3.13	10.0:1	11@1500
	Taurus	3.0	U	153@5800	186@3250	3.50x3.10	9.4:1	40-60@2500
		3.0	S	201@5500	207@4500	3.50x3.13	10.0:1	11@1500

SFI: Sequential Fuel Injection

06017-TAUR-C02

ENGINE TUNE-UP SPECIFICATIONS

Year	Engine Displacement Liters	Engine ID/VIN	Spark Plug Gap (in.)	Ignition Timing (deg.)	Fuel Pump (psi) ①	Idle Speed (rpm)	Valve Clearance	
							Intake	Exhaust
2002	3.0	U	0.042-0.046	10B	26-45	②	HYD	HYD
	3.0	S	0.052-0.056	10B	26-45	②	HYD	HYD
2003	3.0	U	0.042-0.046	10B	26-45	②	HYD	HYD
	3.0	S	0.052-0.056	10B	26-45	②	HYD	HYD
2004	3.0	U	0.042-0.046	10B	26-45	②	HYD	HYD
	3.0	S	0.052-0.056	10B	26-45	②	HYD	HYD
2005	3.0	U	0.042-0.046	10B	26-45	②	HYD	HYD
	3.0	S	0.052-0.056	10B	26-45	②	HYD	HYD

NOTE: The Vehicle Emission Control Information label often reflects specification changes made during production. The label figures must be used if they differ from those in this chart.

B: Before Top Dead Center

HYD: Hydraulic

① Fuel pressure with engine running, pressure regulator vacuum hose connected

② Refer to Vehicle Emission Control Information label

06017-TAUR-C03

3.0L (VIN U) engine
Firing order: 1–4–2–5–3–6
Distributorless ignition system

79223G25

FRONT OF VEHICLE

3.0L (VIN S) engine
Firing order: 1–4–2–5–3–6
Distributorless ignition system

79223G26

1. Generator
2. Drive belt tensioner
3. Power steering pump
4. A/C compressor
5. Crankshaft pulley
6. Idler pulley
7. Water pump
8. Drive belt

79224G24

Serpentine accessory drive belt routing—3.0L (VIN U) engine

1. Power steering pump
2. Water pump
3. A/C compressor
4. Drive belt
5. Crankshaft pulley
6. Drive belt tensioner
7. Generator

79224G25

Serpentine accessory drive belt routing—3.0L (VIN S) engine

CAPACITIES

Year	Model	Engine Displacement Liters	Engine ID/VIN	Engine Oil with Filter (qts.)	Transaxle (pts.) Auto. ①	Drive Axle (pts.)	Fuel Tank (gal.)	Cooling System (qts.)
2002	Sable	3.0	U	4.5	24.5	②	③	11.6
		3.0	S	5.8	27.0	②	③	10.5
	Taurus	3.0	U	4.5	24.5	②	③	11.6
		3.0	S	5.8	27.0	②	③	10.5
2003	Sable	3.0	U	4.5	24.5	②	③	11.6
		3.0	S	5.8	27.0	②	③	10.5
	Taurus	3.0	U	4.5	24.5	②	③	11.6
		3.0	S	5.8	27.0	②	③	10.5
2004	Sable	3.0	U	4.5	24.5	②	③	11.6
		3.0	S	5.8	27.0	②	③	10.5
	Taurus	3.0	U	4.5	24.5	②	③	11.6
		3.0	S	5.8	27.0	②	③	10.5
2005	Sable	3.0	U	4.5	26.8	②	18.0	11.6
		3.0	S	6.0	27.0	②	18.0	10.6
	Taurus	3.0	U	4.5	26.8	②	18.0	11.6
		3.0	S	6.0	27.0	②	18.0	10.6

NOTE: All capacities are approximate. Add fluid gradually and ensure a proper fluid level is obtained.
① Includes torque converter
② Included in transaxle capacity

06017-TAUR-C04

VALVE SPECIFICATIONS

Year	Engine Displacement Liters	Engine ID/VIN	Seat Angle (deg.)	Face Angle (deg.)	Spring Test Pressure (lbs. @ in.)	Spring Installed Height (in.)	Stem-to-Guide Clearance (in.)		Stem Diameter (in.)	
							Intake	Exhaust	Intake	Exhaust
2002	3.0	U	45	44	180@1.16	1.580	0.0010-0.0028	0.0015-0.0033	0.3126-0.3134	0.3121-0.3129
	3.0	S	44.75	45.5	153@1.18	1.570	0.0007-0.0027	0.0017-0.0037	0.2350-0.2358	0.2343-0.2350
2003	3.0	U	45	44	180@1.16	1.580	0.0010-0.0028	0.0015-0.0033	0.3126-0.3134	0.3121-0.3129
	3.0	S	44.75	45.5	153@1.18	1.570	0.0007-0.0027	0.0017-0.0037	0.2350-0.2358	0.2343-0.2350
2004	3.0	U	45	44	180@1.16	1.580	0.0010-0.0028	0.0015-0.0033	0.3126-0.3134	0.3121-0.3129
	3.0	S	44.75	45.5	153@1.18	1.570	0.0007-0.0027	0.0017-0.0037	0.2350-0.2358	0.2343-0.2350
2005	3.0	U	45	44	193@1.17	1.650-1.736	0.0010-0.0028	0.0015-0.0033	0.2744-0.2752	0.2740-0.2748
	3.0	S	44.75	45.5	156@1.18	1.570	0.0007-0.0027	0.0017-0.0037	0.2350-0.2358	0.2343-0.2350

06017-TAUR-C05

CRANKSHAFT AND CONNECTING ROD SPECIFICATIONS
All measurements are given in inches.

Year	Engine Displacement Liters	Engine ID/VIN	Crankshaft				Connecting Rod		
			Main Brg. Journal Dia.	Main Brg. Oil Clearance	Shaft End-play	Thrust on No.	Journal Diameter	Oil Clearance	Side Clearance
2002	3.0	U	2.5190-2.5198	0.0009-0.0027	0.004-0.0080	3	2.1253-2.1261	0.0009-0.0027	0.0060-0.0140
	3.0	S	2.4670-2.4790	0.0009-0.0018	0.004-0.0090	4	1.9670-1.9680	0.0010-0.0025	0.0039-0.0118
2003	3.0	U	2.5190-2.5198	0.0009-0.0027	0.004-0.0080	3	2.1253-2.1261	0.0009-0.0027	0.0060-0.0140
	3.0	S	2.4670-2.4790	0.0009-0.0018	0.004-0.0090	4	1.9670-1.9680	0.0010-0.0025	0.0039-0.0118
2004	3.0	U	2.5190-2.5198	0.0009-0.0027	0.004-0.0080	3	2.1253-2.1261	0.0009-0.0027	0.0060-0.0140
	3.0	S	2.4670-2.4790	0.0009-0.0018	0.004-0.0090	4	1.9670-1.9680	0.0010-0.0025	0.0039-0.0118
2005	3.0	U	2.5190-2.5198	0.0009-0.0027	0.004-0.0080	3	2.1253-2.1261	0.0009-0.0027	0.0060-0.0140
	3.0	S	2.4790-2.4800	0.0009-0.0018	0.0050-0.0010	4	1.9670-1.9680	0.0010-0.0025	0.0039-0.0118

06017-TAUR-C06

PISTON AND RING SPECIFICATIONS

All measurements are given in inches.

Year	Engine Displacement Liters	Engine ID/VIN	Piston Clearance	Ring Gap			Ring Side Clearance		
				Top Compression	Bottom Compression	Oil Control	Top Compression	Bottom Compression	Oil Control
2002	3.0	U	0.0014-0.0022	0.010-0.020	0.010-0.020	0.010-0.049	0.0012 0.0031	0.0012 0.0031	SNUG
	3.0	S	0.0005-0.0009	0.004-0.010	0.011-0.017	0.006-0.026	0.0015-0.0029	0.0015-0.0033	SNUG
2003	3.0	U	0.0014-0.0022	0.010-0.020	0.010-0.020	0.010-0.049	0.0012 0.0031	0.0012 0.0031	SNUG
	3.0	S	0.0005-0.0009	0.004-0.010	0.011-0.017	0.006-0.026	0.0015-0.0029	0.0015-0.0033	SNUG
2004	3.0	U	0.0014-0.0022	0.010-0.020	0.010-0.020	0.010-0.049	0.0012 0.0031	0.0012 0.0031	SNUG
	3.0	S	0.0005-0.0009	0.004-0.010	0.011-0.017	0.006-0.026	0.0015-0.0029	0.0015-0.0033	SNUG
2005	3.0	U	0.0012-0.0022	0.010-0.020	0.010-0.020	0.010-0.049	0.0602-0.0612	0.0602-0.0612	SNUG
	3.0	S	0.0005-0.0009	0.004-0.010	0.0005-0.017	0.006-0.026	0.0015-0.0029	0.0015-0.0033	SNUG

06017-TAUR-C07

TORQUE SPECIFICATIONS
All readings in ft. lbs.

Year	Engine Displacement Liters	Engine ID/VIN	Cylinder Head Bolts	Main Bearing Bolts	Rod Bearing Bolts	Crankshaft Damper Bolts	Flywheel Bolts	Manifold		Spark Plugs	Oil Pan Drain Plug
								Intake	Exhaust		
2002	3.0	U	①	56-62	23-28	93-121	54-64	②	15-18	7-15	10
	3.0	S	③	④	⑤	⑥	54-64	⑦	13-16	7-15	20
2003	3.0	U	①	56-62	23-28	93-121	54-64	②	15-18	7-15	10
	3.0	S	③	④	⑤	⑥	54-64	⑦	13-16	7-15	20
2004	3.0	U	①	56-62	23-28	93-121	54-64	②	15-18	7-15	10
	3.0	S	③	④	⑤	⑥	54-64	⑦	13-16	7-15	20
2005	3.0	U	①	59	26	107	59	⑧	⑨	11	10
	3.0	S	③	④	⑩	⑥	59	⑪	15	7-15	11

① Step 1: 35-39 ft. lbs.
   Step 2: Loosen one turn
   Step 3: 20-24 ft. lbs.
   Step 4: Rotate 85-95 degrees
   Step 5: Repeat Step 4

② Upper intake: 15-22 ft. lbs.
   Lower Intake:
   Step 1: 15-22 ft. lbs.
   Step 2: 20-23 ft. lbs.

③ Step 1: 30 ft. lbs.
   Step 2: Rotate 90 degrees
   Step 3: Loosen one turn
   Step 4: 30 ft. lbs.
   Step 5: Rotate 90 degrees
   Step 6: Repeat Step 5

④ Step 1: Cap bolts 1-8 (outer) 18 ft. lbs.
   Step 2: Cap bolts 9-16 (inner) 30 ft. lbs.
   Step 3: Rotate bolts 1-16, 90 degrees
   Step 4: Bolts 17-22; 18 ft. lbs.

⑤ Step 1: 30-33 ft. lbs.
   Step 2: Rotate 90-120 degrees

⑥ Step 1: 89 ft. lbs.
   Step 2: Loosen 360 degrees
   Step 3: Tighten to 37 ft. lbs.
   Step 4: Rotate 90 degrees

⑦ 71-106 inch lbs.

⑧ Upper intake:
   Step 1: Hand tighten all bolts
   Step 2: 89 inch lbs.
   Lower Intake:
   Step 1: 1 ft. lbs.
   Step 2: 24 ft. lbs.

⑨ Step 1: 8 ft. lbs.
   Step 2: 16 ft. lbs.

⑩ Step 1: 17 ft. lbs.
   Step 2: 32 ft. lbs.

⑪ Upper and lower intake:
   89 inch lbs.

06017-TAUR-C08

WHEEL ALIGNMENT

Year	Model		Caster Range (+/-Deg.)	Caster Preferred Setting (Deg.)	Camber Range (+/-Deg.)	Camber Preferred Setting (Deg.)	Toe-in (in.)
2002	Taurus Sedan	F	0.70	0	0.70	0	0.10 +/- 0.13
		R	—	—	0.70	0	0.18 +/- 0.13
	Taurus Wagon	F	0.70	0	0.70	0	0.10 +/- 0.13
		R	—	—	1.20	0	0.18 +/- 0.13
	Sable	F	0.70	0	0.70	0	0.20 +/- 0.25
		R	—	—	0.70	0	0.36 +/- 0.25
	Sable Wagon	F	0.70	0	0.70	0	0.20 +/- 0.25
		R	—	—	0.12	0	0.36 +/- 0.25
2003	Taurus Sedan	F	0.70	0	0.70	0	0.10 +/- 0.13
		R	—	—	0.70	0	0.18 +/- 0.13
	Taurus Wagon	F	0.70	0	0.70	0	0.10 +/- 0.13
		R	—	—	1.20	0	0.18 +/- 0.13
	Sable	F	0.70	0	0.70	0	0.20 +/- 0.25
		R	—	—	0.70	0	0.36 +/- 0.25
	Sable Wagon	F	0.70	0	0.70	0	0.20 +/- 0.25
		R	—	—	0.12	0	0.36 +/- 0.25
2004	Taurus Sedan	F	0.70	0	0.70	0	0.10 +/- 0.13
		R	—	—	0.70	0	0.18 +/- 0.13
	Taurus Wagon	F	0.70	0	0.70	0	0.10 +/- 0.13
		R	—	—	1.20	0	0.18 +/- 0.13
	Sable	F	0.70	0	0.70	0	0.20 +/- 0.25
		R	—	—	0.70	0	0.36 +/- 0.25
	Sable Wagon	F	0.70	0	0.70	0	0.20 +/- 0.25
		R	—	—	0.12	0	0.36 +/- 0.25
2005	Taurus Sedan	F	0.70	0	0.70	0	0.10 +/- 0.13
		R	—	—	0.70	0	0.18 +/- 0.13
	Taurus Wagon	F	0.70	0	0.70	0	0.10 +/- 0.13
		R	—	—	1.20	0	0.18 +/- 0.13
	Sable	F	0.70	0	0.70	0	0.20 +/- 0.25
		R	—	—	0.70	0	0.36 +/- 0.25
	Sable Wagon	F	0.70	0	0.70	0	0.20 +/- 0.25
		R	—	—	0.12	0	0.36 +/- 0.25

06017-TAUR-C09

TIRE, WHEEL AND BALL JOINT SPECIFICATIONS

Year	Model	OEM Tires Standard	OEM Tires Optional	Tire Pressure (psi) Front	Tire Pressure (psi) Rear	Wheel Size	Ball Joint Inspection	Lug Nut Torque (ft lbs)
2002	Sable GS, LS Sedan	P205/65R16	None	33	33	7J	0.030 in. ①	95
	Sable LS Premium Sedan	P215/60R16	None	33	33	7J	0.030 in. ①	95
	Sable Wagon	P215/60R16	None	31	31	7J	0.030 in. ①	95
	Taurus	P215/60R16	None	33	33	6.5-JJ	0.030 in. ①	95
2003	Sable GS, LS Sedan	P205/65R16	None	33	33	7J	0.030 in. ①	95
	Sable LS Premium Sedan	P215/60R16	None	33	33	7J	0.030 in. ①	95
	Sable Wagon	P215/60R16	None	31	31	7J	0.030 in. ①	95
	Taurus	P215/60R16	None	33	33	6.5-JJ	0.030 in. ①	95
2004	Sable GS, LS Sedan	P205/65R16	None	33	33	7J	0.030 in. ①	95
	Sable LS Premium Sedan	P215/60R16	None	33	33	7J	0.030 in. ①	95
	Sable Wagon	P215/60R16	None	31	31	7J	0.030 in. ①	95
	Taurus	P215/60R16	None	33	33	6.5-JJ	0.030 in. ①	95
2005	Sable GS, LS Sedan	P205/65R16	None	33	33	7J	0.030 in. ①	95
	Sable LS Premium Sedan	P215/60R16	None	33	33	7J	0.030 in. ①	95
	Sable Wagon	P215/60R16	None	31	31	7J	0.030 in. ①	95
	Taurus	P215/60R16	None	33	33	6.5-JJ	0.030 in. ①	95

OEM: Original Equipment Manufacturer

PSI: Pounds Per Square Inch

① Maximum radial tolerance in inches

06017-TAUR-C10

BRAKE SPECIFICATIONS
Ford Taurus, Mercury Sable
All measurements in inches unless noted

Year	Model		Brake Disc Original Thickness	Brake Disc Minimum Thickness	Brake Disc Maximum Run-out	Brake Drum Original Inside Diameter	Brake Drum Max. Wear Limit	Brake Drum Maximum Machine Diameter	Minimum Lining Thickness	Brake Caliper Bracket Bolts (ft. lbs.)	Brake Caliper Mounting Bolts (ft. lbs.)
2002	Sable	F	1.020	0.974	0.002	—	—	—	0.039	65-85	23-28
		R	0.550	0.500	0.004	8.85	0.05	8.92	0.039	65-87	23-25
	Taurus	F	1.020	0.974	0.002	—	—	—	0.039	65-85	23-28
		R	0.940	0.500	0.002	8.85	0.05	8.92	0.039	65-87	23-25
2003	Sable	F	1.020	0.974	0.002	—	—	—	0.039	65-85	23-28
		R	0.550	0.500	0.004	8.85	0.05	8.92	0.039	65-87	23-25
	Taurus	F	1.020	0.974	0.002	—	—	—	0.039	65-85	23-28
		R	0.940	0.500	0.002	8.85	0.05	8.92	0.039	65-87	23-25
2004	Sable	F	1.020	0.974	0.002	—	—	—	0.039	65-85	23-28
		R	0.550	0.500	0.004	8.85	0.05	8.92	0.039	65-87	23-25
	Taurus	F	1.020	0.974	0.002	—	—	—	0.039	65-85	23-28
		R	0.940	0.500	0.002	8.85	0.05	8.92	0.039	65-87	23-25
2005	Sable	F	1.020	0.974	0.002	—	—	—	0.039	76	26
		R	0.550	0.500	0.004	8.85	0.05	8.92	0.039	76	24
	Taurus	F	1.020	0.974	0.002	—	—	—	0.039	76	26
		R	0.940	0.500	0.002	8.85	0.05	8.92	0.039	76	24

NOTE: Follow specifications stamped on rotor or drum if figures differ from those in this chart.

F: Front

R: Rear

06017-TAUR-C11

SCHEDULED MAINTENANCE INTERVALS
Ford—Taurus & Mercury—Sable

TO BE SERVICED	TYPE OF SERVICE	VEHICLE MILEAGE INTERVAL (X1000)																			
		5	10	15	20	25	30	35	40	45	50	55	60	65	70	75	80	85	90	95	100
Engine oil & filter	R	✓	✓	✓	✓	✓	✓	✓	✓	✓	✓	✓	✓	✓	✓	✓	✓	✓	✓	✓	✓
Rotate tires	S/I	✓		✓		✓		✓		✓		✓		✓		✓		✓		✓	
Engine coolant protection, hoses & clamps	S/I			✓			✓			✓			✓			✓			✓		
Passenger compartment air filter	R				✓				✓				✓				✓				✓
Air cleaner filter	R						✓						✓						✓		
Automatic transaxle fluid & filter	R						✓						✓						✓		
Brake lines & connections	S/I						✓						✓						✓		
Exhaust heat shields	S/I						✓						✓						✓		
Front and rear disc brake pads & rotors	S/I						✓						✓						✓		
Accessory drive belt(s)	S/I												✓								
Engine coolant	R	At 5 years or 100,000 miles																			
Spark plugs (exc. 3.0L FF) ①	R																				✓
Spark plugs (3.0L FF)	R						✓						✓								
PCV valve (except 3.0L 4-valve)	R												✓								
PCV valve (3.0L 4-valve)	R																				✓

① Platinum tip spark plugs: change every 100,000 miles

R: Replace   S/I: Service and Inspect

FREQUENT OPERATION MAINTENANCE (SEVERE SERVICE)

If a vehicle is operated under any of the following conditions it is considered severe service:

- Extremely dusty areas.
- 50% or more of the vehicle operation is in 32°C (90°F) or higher temperatures, or constant operation in temperatures below 0°C (32°F).
- Prolonged idling (vehicle operation in stop and go traffic)..
- Frequent short running periods (engine does not warm to normal operating temperatures).
- Police, taxi, delivery usage or trailer towing usage.

Oil & oil filter: change every 3000 miles

Rotate tires at 6000 miles & every 9000 miles thereafter

Air cleaner element service or inspect every 15,000 miles

Automatic transaxle fluid & filter: change every 21,000 miles

06017-TAUR-C12

ENGINE REPAIR

➡ Disconnecting the negative battery cable on some vehicles may interfere with the functions of the on board computer system. The computer may undergo a relearning process once the negative battery cable is reconnected.

Alternator

REMOVAL

2002 3.0L (VIN U) Engine

1. Before servicing the vehicle, refer to the precautions section.
2. Remove or disconnect the following:
   • Negative battery cable
   • Accessory drive belt
   • Alternator wiring connectors
   • Alternator brace
   • Alternator

2003–05 3.0L (VIN U) Engine

1. Before servicing the vehicle, refer to the precautions section.
2. Disconnect the battery ground cable.
3. Release the accessory drive belt tension and remove the belt from the alternator pulley.
4. Remove the alternator B+ nut and the B+ cable.
5. Position the B+ cable cover aside.
6. Position the wiring harness aside.
7. Remove the bolts and the alternator.
8. Disconnect the voltage regulator electrical connector.

2002–04 3.0L (VIN S) Engine

1. Before servicing the vehicle, refer to the precautions section.
2. Remove or disconnect the following:
   • Negative battery cable
   • Right front wheel
   • Right fender splash shield
   • Accessory drive belt

➡ The crankshaft pulley retainer has left-hand threads.

25 Nm (18 lb-ft)   48 Nm (35 lb-ft)

9346KG03

Alternator bracket–3.0L (VIN U) engine

100 Nm (74 lb-ft)

9346KG04

Remove the crankshaft pulley to provide clearance—2002-04 3.0L (VIN S) engine

   • Crankshaft pulley
   • Alternator harness connectors
   • Alternator

2005 3.0L (VIN S) Engine

1. Before servicing the vehicle, refer to the precautions section.
2. Disconnect the negative battery cable.
3. Release the accessory drive belt tension and remove the belt from the alternator pulley.
4. Remove the power steering hose bracket nut from the alternator mounting bolt.

➡ The left side bolt will come out with the alternator.

5. Remove the upper alternator bolts.
6. Remove the right hand front wheel and tire assembly and front fender splash shield.
7. Remove the screws and the alternator splash shield.
8. Remove the torque converter inspection cover.
9. Install a flywheel holding tool.

➡ The crankshaft pulley bolt has a reverse thread. The crankshaft pulley and bolt are a molded composite and are one piece.

10. Remove the bolt and the crankshaft pulley.
11. Lower the alternator to allow access to the electrical connectors.
12. Disconnect the Crankshaft Position (CKP) sensor.
13. Release the Oxygen (O_2s) sensor connector locator from the bracket.
14. Remove the lower alternator bolt.
15. Disconnect the regulator electrical connector.
16. Remove the alternator from the vehicle and disconnect the B+ connector nut.

INSTALLATION

2002 3.0L (VIN U) Engine

➡ Do not tighten the alternator pivot and bracket bolts until the accessory drive belt is installed.

1. Install or connect the following:
   • Alternator
   • Alternator brace. Tighten the bolts to 76–97 inch lbs. (8–11 Nm), and the nut to 15–22 ft. lbs. (20–30 Nm).
   • Alternator wiring connectors
   • Accessory drive belt
2. Tighten the alternator pivot bolt to 30–40 ft. lbs. (40–55 Nm), and the bracket bolt to 15–22 ft. lbs. (20–30 Nm).
3. Install the negative battery cable. Start the engine and check for proper operation.

2003–05 3.0L (VIN U) Engine

1. Connect the voltage regulator electrical connector.
2. Install the alternator and tighten the bolts to 35 ft. lbs. (48 Nm).
3. Install the B+ cable and tighten the nut to 71 inch lbs. (8 Nm).
4. Install the accessory drive belt.

2002–04 3.0L (VIN S) Engine

1. Install or connect the following:
   • Alternator and tighten the fasteners to 18 ft. lbs. (25 Nm)
   • Alternator harness connectors

➡ The crankshaft pulley retainer has left-hand threads.

   • Crankshaft pulley and tighten to 74 ft. lbs. (100 Nm)
   • Accessory drive belt
   • Right fender splash shield
   • Right front wheel
   • Negative battery cable

48 Nm (35 lb-ft)

67197-TAUR-G01

Alternator and mounting bolts—2003–05 3.0L (VIN U) Engine

2. Start the engine and check for proper operation.

2005 3.0L (VIN S) Engine

1. Connect the B+ connector and tighten the nut to 71 inch lbs. (8 Nm).

2. Install the alternator and tighten the lower bolt to 18 ft. lbs. (25 Nm).

3. Connect the regulator electrical connector.

4. Connect the O2s sensor connector locator.

5. Connect the CKP sensor.

➡ The crankshaft pulley bolt has a reverse thread. The crankshaft pulley and bolt are a molded composite and are one piece.

6. Install the crankshaft pulley and tighten the bolt to 74 ft. lbs. (100 Nm).

7. Install the torque converter inspection cover.

8. Install the alternator splash shield.

9. Install the right hand front fender splash shield and wheel assembly.

10. Install the upper alternator bolts and tighten to 18 ft. lbs. (25 Nm).

11. Install the power steering hose bracket nut and tighten to 13 ft. lbs. (18 Nm).

12. Install the accessory drive belt.

13. Connect the negative battery cable.

Ignition Timing

ADJUSTMENT

The base ignition timing is set at 10 degrees Before Top Dead Center (BTDC) and is not adjustable.

Engine Assembly

REMOVAL & INSTALLATION

2002–04 Models

1. Before servicing the vehicle, refer to the precautions section.

2. Disconnect the negative battery cable.

3. Drain the engine oil.

4. Drain the cooling system.

5. Recover the A/C refrigerant.

6. On the 3.0L (VIN S) engine, remove or disconnect the following:
- Windshield wipers
- Cowl extension
- Emission vacuum control connector at the right side of the dash panel

- Hood
- Steering column pinch bolt
- Mass Airflow (MAF) sensor connector
- Intake Air Temperature (IAT) sensor
- Air cleaner assembly
- Idle Air Control (IAC) valve connector
- Throttle Position (TP) sensor connector
- Fuel lines
- Intake manifold vacuum hoses
- Ground straps
- Powertrain Control Module (PCM) connector
- Engine control sensor wiring harness connectors from the bracket located at the top of the transaxle
- Evaporative emission canister purge valve connector
- Crankcase vent hose
- Accelerator cable
- Cruise control actuator
- Shift cable and lever
- Secondary air injection pump relay connector
- Main emission vacuum control connector near the fan shroud
- Radiator hoses
- Heater hoses
- Transaxle oil cooler lines
- Power steering return hose
- Alternator wiring harness connectors
- A/C compressor lines
- A/C pressure cutoff switch, if equipped
- Front wheels
- Stabilizer bar links
- Lower ball joints
- Outer tie rod ends
- Halfshafts
- Radiator splash shield
- Heated Oxygen (HO2S) sensors
- Dual converter Y-pipe
- Power steering pressure switch connector
- Power steering pressure hose

7. Support the powertrain from below and remove the subframe bolts.

8. Raise the vehicle away from the powertrain.

9. Attach an engine hoist to the powertrain.

10. Remove the left, right, and rear powertrain support insulators and lift the powertrain away from the subframe.

11. Support the transaxle from below.

12. Remove or disconnect the following:
- Starter motor
- Torque converter
- Transaxle flange bolts

Throttle and cruise control cables—2002–04 3.0L (VIN U) engine shown

Throttle Position (TP) sensor and Intake Air Control (IAC) valve connectors—2002–04 3.0L (VIN U) engine shown

13. Separate the engine from the transaxle.

To install:

14. Install the engine to the transaxle. For 3.0L (VIN S) engines, tighten the flange bolts to 25–33 ft. lbs. (33–46 Nm). For all other engines, tighten the flange bolts to 30–44 ft. lbs. (40–60 Nm). Tighten the torque converter nuts to 20–34 ft. lbs. (27–46 Nm).

15. Install or connect the following:
- Starter motor and tighten the fasteners to 21 ft. lbs. (29 Nm)
- Powertrain support insulators. Tighten the bracket bolts to 65 ft. lbs. (88 Nm) and the subframe bolts to 90 ft. lbs. (122 Nm).

16. Lower the vehicle on to the powertrain assembly. Use 2 pieces of ¾ inch outside diameter pipe in the alignment holes behind the front subframe mounts to align the subframe to the body. Tighten the subframe mounting bolts to 57–76 ft. lbs. (77–103 Nm).

17. Install or connect the following:
- Power steering pressure hose
- Power steering pressure switch connector
- Dual converter Y-pipe
- HO2S sensors
- Radiator splash shield
- Halfshafts and tighten the hub retainer nuts to 170–202 ft. lbs. (230–275 Nm)

ROTUNDA
POWERTRAIN LIFT
014-00765

9346KG05

Support the powertrain when removing the subframe bolts

- Outer tie rod ends and tighten the nuts to 35–46 ft. lbs. (47–63 Nm)
- Lower ball joints and tighten the nuts to 50–68 ft. lbs. (68–92 Nm)
- Stabilizer bar links and tighten the nuts to 30–40 ft. lbs. (40–55 Nm)
- Front wheels
- A/C pressure cutoff switch, if equipped
- A/C compressor lines
- Alternator wiring harness connectors
- Power steering return hose
- Transaxle oil cooler lines
- Heater hoses
- Radiator hoses
- Main emission vacuum control connector near the fan shroud
- Secondary air injection pump relay connector
- Shift cable and lever
- Cruise control actuator
- Accelerator cable
- Crankcase vent hose
- Evaporative emission canister purge valve connector
- Engine control sensor wiring harness connectors from the bracket located at the top of the transaxle
- PCM connector

- Ground straps
- Intake manifold vacuum hoses
- Fuel lines
- IAC valve connector
- TP sensor connector
- Air cleaner assembly
- IAT sensor
- MAF sensor connector
- Steering column pinch bolt
- Hood

18. On the 3.0L (VIN S) engine, install or connect the following:
  - Windshield wipers
  - Cowl extension
  - Emission vacuum control connector at the right side of the dash panel

19. Fill the cooling system, fill the crankcase and recharge the A/C system.

20. Connect the negative battery cable.

21. Start the engine. Check for leaks and proper operation.

➡ **Whenever the vehicle's subframe is removed or lowered, the wheel alignment should be checked.**

2005 3.0L (VIN U) Models

1. Before servicing the vehicle, refer to the precautions section.

2. Disconnect the negative battery cable.

3. Drain the engine oil.

4. Drain the cooling system.

5. Recover the A/C refrigerant.

6. Remove the battery.

7. Remove the cowl vent screen and cowl extension.

8. Remove the halfshafts.

9. Remove the nuts and position the steering column input shaft coupling boot aside.

10. Remove the pinch bolt and slide the coupling off the steering gear input shaft.

11. Disconnect the tubes, loosen the clamps and remove the air cleaner outlet pipe.

12. Remove the pin-type retainers and the snow shield.

13. Disconnect the accelerator cable, speed control actuator cable and the throttle return spring from the throttle body.

14. Remove the bolts and position the accelerator cable bracket aside.

15. Disconnect the vacuum hose and the Evaporative Emissions (EVAP) return tube.

16. Remove the nut and disconnect the manual control lever cable.

17. Disconnect the manual control lever cable from the bracket and position aside.

18. Loosen the bolt and disconnect the 42-pin and Transmission Range (TR) sensor electrical connectors.

19. Disconnect the upper radiator hose and the heater hose from the thermostat housing.

20. Remove the nut and disconnect the ground strap electrical connector.

21. Disconnect the power steering return hose.

22. Disconnect the alternator electrical connectors and position the wire harness aside.

23. Disconnect the A/C suction tube from the accumulator drier.

24. Disconnect the fuel supply hose.

25. Disconnect the degas bottle hoses.

26. Remove the nut, the bolt and the engine roll restrictor brace.

27. Remove the bolts and the engine roll restrictor.

28. Disconnect the heater hose.

29. Remove the bolt and disconnect the ground electrical connectors.

30. Disconnect the EVAP canister purge valve electrical connector.

31. Loosen the bolt disconnect the Powertrain Control Module (PCM) electrical connector.

➡ **Support the exhaust with mechanic's wire.**

32. Remove the nuts and disconnect the three-way catalytic converter from the dual converter Y-pipe.

33. Remove the nuts and the muffler clamp.

34. Separate and remove the three-way catalytic converter from the muffler pipe.

35. Remove the nut and disconnect the A/C discharge tube.

36. Remove the radiator support bracket.

37. Disconnect the lower radiator hose from the radiator and degas bottle supply hose.

38. Disconnect the transmission oil cooler hose.

39. If equipped, disconnect the wire harness electrical connector.

40. Disconnect the auxiliary oil cooler assembly.

41. Disconnect the transmission oil cooler hose, if equipped.

42. Disconnect the power steering return hose.

43. Remove the oil pan drain plug and drain the engine oil. Install the drain plug when finished and torque to 10 ft. lbs. (13 Nm).

44. Remove the starter motor.

45. Remove the bolt and the engine inspection cover.

46. Remove the four torque converter nuts.

47. Remove the nuts and disconnect both stabilizer links from the stabilizer bar.

48. Remove the nuts and separate both tie-rod ends from the steering knuckles.

49. Position engine dolly/support tool 014-00765.

50. Remove the four front subframe-to-body bolts.

51. Using the engine dolly/support tool 014-00765, lower the powertrain and subframe assembly out of the vehicle.

52. Remove the nut and detach the Power Steering Pressure (PSP) line bracket.

53. Remove the nut and detach the PSP line bracket.

06017-TAUR-G01

Position engine dolly/support tool 014-00765—2005 3.0L (VIN U) models

54. Disconnect the PSP line from the power steering pump.

55. Disconnect the spark plug wires.

56. Disconnect the ignition coil electrical connector.

57. Disconnect the Output Shaft Speed (OSS) sensor electrical connector and position the wire harness aside.

58. Disconnect the Turbine Shaft Speed (TSS) sensor electrical connector.

59. Disconnect the transaxle harness electrical connector and position the wire harness aside.

60. Remove the bolt, the nuts and the right hand catalytic converter heat shield.

61. Install lifting eye tool 303-D030 to the left and right side of the engine.

62. Attach a hoist to the lifting hooks and support the engine.

63. Remove the two right hand engine support insulator-to-transmission bolts.

64. Remove the left hand engine support insulator-to-subframe nut.

65. Remove the right hand engine support insulator through bolt.

66. Disconnect the catalyst monitor sensor electrical connector.

67. Remove the bolts and the Y-pipe.

68. Disconnect the lower radiator hose from the coolant pump.

69. Remove the five transmission-to-engine bolts and one stud bolt.

70. Separate the engine from the transmission.

To install:

71. Position the engine on the subframe and transmission assembly.

72. Install the transmission-to-engine bolts and tighten to 37 ft. lbs. (50 Nm).

73. Connect the lower radiator hose to the coolant pump.

74. Install the right hand engine support insulator through bolt and tighten to 89 ft. lbs. (120 Nm).

75. Install the dual converter Y-pipe and the bolts and tighten to 30 ft. lbs. (40 Nm).

76. Connect the catalyst monitor sensor electrical connector.

77. Install the left hand engine support insulator-to-subframe nut and tighten to 66 ft. lbs. (90 Nm).

78. Install the two right hand engine support insulator-to-transmission bolts and tighten to 44 ft. lbs. (60 Nm).

79. Using the lifting eyes, secure the engine to the hoist.

80. Install the right hand catalytic converter heat shield and install the two nuts.

81. Position the ground strap and install the bolt and tighten to 15 ft. lbs. (20 Nm).

82. Position the 42-pin connector on the

06017-TAUR-G02

Install lifting eye tool 303-D030 to the right side of the engine—2005 3.0L (VIN U) models

06017-TAUR-G03

Install lifting eye tool 303-D030 to the left side of the engine—2005 3.0L (VIN U) models

bracket and connect the transaxle harness electrical connector.

83. Connect the TSS sensor electrical connector.

84. Connect the OSS electrical connector.

85. Install the ignition coil and the bolts.

86. Connect the ignition coil electrical connector.

87. Connect the spark plug wires to the ignition coil.

88. Connect the PSP tube to the power steering pump and tighten to 35 ft. lbs. (48 Nm).

89. Install the PSP tube bracket and the nut and tighten to 20 ft. lbs. (27 Nm).

90. Using the dolly/support tool, raise the engine, transmission and subframe into the vehicle.

91. Install the four front subframe-to-body bolts and tighten to 76 ft. lbs. (103 Nm).

92. Connect the tie-rod ends to the steering knuckles. Install the nuts and tighten to 41 ft. lbs. (55 Nm) and install new cotter pins.

93. Connect the stabilizer links to the stabilizer bar and tighten to 41 ft. lbs. (55 Nm).

94. Install the four torque converter nuts and tighten to 27 ft. lbs. (36 Nm).

95. Install the engine rear plate.

96. Install the starter motor.
97. Connect the auxiliary oil cooler assembly.
98. If equipped, connect the transmission oil cooler hose.
99. Connect the power steering return hose.
100. If equipped, connect the wiring harness electrical connector.
101. Connect the transmission oil cooler hose.
102. Connect the lower radiator hose to the radiator and the degas bottle supply hose.
103. Install the radiator support bracket.

➡ **Replace and lubricate the O-ring with PAG oil.**

104. Connect the A/C discharge tube and tighten to 89 inch lbs. (10 Nm).
105. Install a new gasket into the Y-pipe flange.
106. Connect the three-way converter to the dual converter Y-pipe and the muffler pipe and install the nuts and tighten to 30 ft. lbs. (40 Nm).
107. Install the exhaust clamp and tighten to 30 ft. lbs. (40 Nm).
108. Connect the PCM electrical connector.
109. Connect the EVAP canister purge valve electrical connector.
110. Connect the ground electrical connectors and install the bolt and tighten to 89 inch lbs. (10 Nm).
111. Connect the heater water hose.
112. Install the engine roll restrictor and the bolts and tighten to 35 ft. lbs. (48 Nm).
113. Install the engine roll restrictor brace and the nut and bolt and tighten to 35 ft. lbs. (48 Nm).
114. Connect the hoses to the degas bottle.
115. Connect the fuel tube spring lock coupling.
116. Connect the A/C suction tube to the accumulator drier.
117. Connect the alternator electrical connections.
118. Connect the power steering return hose.
119. Connect the ground strap electrical connector. and tighten to 18 ft. lbs. (25 Nm).
120. Connect the upper radiator hose and heater hose to the thermostat housing.
121. Connect the TR sensor and the 42-pin electrical connectors.
122. Install the manual control lever cable in the bracket.

123. Connect the manual control lever cable to the manual control lever and install the nut and tighten to 13 ft. lbs. (17 Nm).
124. Connect the vacuum hose and the EVAP return tube.
125. Position the accelerator cable and bracket and install the bolts and tighten to 13 ft. lbs. (17 Nm).
126. Connect the accelerator cable, speed control cable and throttle return spring.
127. Install the snow shield.
128. Position the air cleaner outlet pipe, connect the tubes and tighten the clamps.
129. Connect the steering column to the steering gear input shaft and install the bolt and tighten to 18 ft. lbs. (25 Nm).
130. Position the steering column input shaft coupling boot and install the nuts.
131. Install the halfshafts.
132. Install the cowl extension and cowl vent screen.
133. Install the battery and connect the negative battery cable.
134. Fill the engine with clean engine oil.
135. Fill and bleed the engine cooling system.
136. Charge the A/C system using approved equipment.
137. Fill and bleed the power steering system.
138. Start the vehicle and inspect the engine and cooling system for leaks.

2005 3.0L (VIN S) Models

1. Before servicing the vehicle, refer to the precautions section.
2. Disconnect the negative battery cable.
3. Drain the engine oil.
4. Drain the cooling system.
5. Recover the A/C refrigerant.
6. Remove the air cleaner and air cleaner outlet pipe.
7. Remove the cowl vent screen and cowl extension.
8. Remove the halfshafts.
9. Remove the pinch bolt and slide the coupling off the steering gear input shaft.
10. Remove the ground strap bolt.
11. Loosen the bolt and disconnect the Powertrain Control Module (PCM) electrical connector.
12. Disconnect the Evaporative Emission (EVAP) system hoses.
13. Remove the pin-type retainers and the accelerator cable splash shield.
14. Disconnect the accelerator cable and the cruise control cable.

15. Detach the cables from the bracket.
16. Disconnect the Transaxle Range (TR) sensor electrical connector.
17. Remove the nut and unclip the shifter cable from the bracket.
18. Loosen the bolt and disconnect the harness connector.
19. Disconnect the low pressure A/C tube spring lock coupling at the accumulator/drier.
20. Remove the heater hose bracket nut.
21. Disconnect the heater hoses.
22. Remove the heater hose bracket nut.
23. Disconnect the coolant recovery bottle hose.
24. Remove the nut and disconnect the starter motor ground cable.
25. Disconnect the upper radiator hose.
26. Disconnect the power steering fluid hose from the reservoir.
27. Remove the two left hand exhaust manifold-to-pipe nuts.
28. Remove the front splash shield.
29. Remove the inner fender splash shields.
30. Remove the starter cable cover.
31. Remove the nut and disconnect the cable from the starter.
32. Remove the stabilizer bar end link nuts.
33. Remove the tie-rod end cotter pin and nut.
34. Using the tool 211-001, separate the tie-rod end from the knuckle.
35. Remove the right hand exhaust manifold-to-pipe bolts.
36. Remove the nuts and the dual converter Y-pipe.
37. Disconnect the power steering and transaxle hoses from the cooler.
38. Disconnect the transaxle cooler hose.
39. Remove the nut and disconnect the A/C high pressure line at the condenser.
40. Remove four bolts and remove the radiator support bracket.
41. Disconnect the lower radiator hose and the reservoir supply hose.
42. If equipped, disconnect the engine block heater electrical connector.
43. If equipped, unclip the engine block heater harness.
44. Remove the torque converter inspection cover.

➡ **Index the flexplate and torque converter by marking one stud.**

45. Remove the four torque converter-to-flexplate nuts.
46. Using the dolly support tool 104-00765, support the subframe.
47. Remove the four subframe bolts and

lower the subframe engine/transaxle assembly.

48. Remove the coolant tube lower mounting bolt.

49. Remove the coolant tube mounting bolt and position the tube away from the cylinder head.

50. Remove the engine support brace.

51. Install lift eye tools 303-050.

52. Using the lift eye tools and a suitable engine hoist, support the engine and transaxle at the cylinder heads.

53. Disconnect the power steering pressure tube and remove the clamp bolt.

54. Remove the left hand engine mount-to-frame nut.

55. Remove the right hand engine mount-to-frame nut.

56. Remove the engine bracket-to-transaxle bolts.

57. Remove the bolts and separate the engine from transaxle.

58. Remove the oil pan, if necessary.

To install:

➡ **The studs from the torque converter must align with the engine flexplate holes to assemble the engine/transaxle. Install the indexed stud to the indexed hole in the flexplate.**

59. Using the lift hooks and a suitable engine hoist, support the engine and assemble with the transaxle.

60. Install the five engine-to-transaxle bolts and tighten to 30 ft. lbs. (40 Nm).

61. Install the right and left engine mount nut and tighten to 66 ft. lbs. (90 Nm).

62. Using the tool 211-D027, install new seals on the power steering hose fitting.

63. Connect the power steering pressure line and install the clamp bolt.

64. Install the engine support brace and tighten the bolts as illustrated.

65. Position the coolant tube to the head and install the coolant tube mounting bolt.

66. Install the coolant tube lower mounting bolt.

Tighten the engine support brace bolts as shown—2005 3.0L (VIN S) engine

67. Using the dolly, raise the engine/transaxle assembly into the vehicle and loosely install the four subframe bolts.

68. Align the subframe with the body. Install a 0.75 inch (19 mm) rod into the front left and right hand subframe-to-body alignment hole.

69. Hand-tighten the bolts.

70. Verify the subframe-to-body alignment at both of the alignment holes.

71. Tighten the four subframe bolts to 76 ft. lbs. (103 Nm).

➡ **Clean and degrease all sealing surfaces with metal surface cleaner. The oil pan must be installed and the bolts tightened within four minutes of the sealant application.**

72. Apply a 0.4 inch (10 mm) dot of silicone gasket and sealant to the front cover-to-cylinder block mating surface.

Apply a 0.4 inch (10 mm) dot of silicone gasket and sealant to the front cover-to-cylinder block mating surface—2005 3.0L (VIN S) engine

73. Install a new gasket, position the oil pan and loosely install the bolts and the studs in the sequence illustrated.

74. Install the oil pan to transaxle housing bolts and tighten to 30 ft. lbs. (40 Nm).

Loosely install the oil pan bolts and the studs in the sequence shown—2005 3.0L (VIN S) engine

Tighten the oil pan bolts and the studs in the sequence shown—2005 3.0L (VIN S) engine

75. Tighten the oil pan bolts and studs in sequence to 18 ft. lbs. (25 Nm).

76. Position the engine-to-transaxle bracket and install the nuts and the bolt. Tighten the bolts to 18 ft. lbs. (25 Nm) and the nuts to 89 inch lbs. (10 Nm).

77. Install the four torque converter-to-flexplate nuts and tighten to 27 ft. lbs. (37 Nm).

78. Install the torque converter inspection cover.

79. Clip the engine block heater harness to the engine block.

80. If equipped, connect the engine block heater electrical connector.

81. If equipped, connect the lower radiator hose and reservoir supply hose.

82. Install the radiator support bracket and bolts.

83. Connect the A/C high pressure tube to the condenser and install the nut.

84. Connect the transaxle cooler hose.

85. Connect the power steering and transaxle hoses to the cooler.

86. Position the dual converter Y-pipe and install the nuts and tighten to 30 ft. lbs. (40 Nm).

87. Install the right hand exhaust manifold-to-pipe bolts and tighten to 30 ft. lbs. (40 Nm).

88. Connect the tie-rod ends and tighten the nuts to 41 ft. lbs. (55 Nm). Install new cotter pins.

89. Install the sway bar end link nuts and tighten to 41 ft. lbs. (55 Nm).

90. Position the starter cable and install the nut.

91. Install the starter cable cover.

92. Install the inner fender splash shield.

93. Install the front splash shield.

94. Install the two left hand exhaust manifold-to-pipe nuts and tighten to 30 ft. lbs. (40 Nm).

95. Connect the power steering fluid hose to the reservoir.

96. Connect the upper radiator hose.

97. Position the starter motor ground cable and tighten the nut to 18 ft. lbs. (25 Nm).

98. Connect the coolant recovery bottle hose.

99. Connect the heater hoses.

100. Install the heater hose bracket nut.

101. Connect the low pressure A/C tube spring lock coupling at the accumulator/drier.

102. Connect the harness and tighten the bolt.

103. Clip the shifter cable to the bracket and install the nut.

104. Connect the TR sensor electrical connector.

105. Install the accelerator and cruise control cables in the bracket.

106. Connect the cables and the throttle return spring.

107. Install the accelerator cable splash shield.

108. Connect the EVAP system hoses.

109. Connect the PCM electrical connector and tighten the bolt to 53 inch lbs. (6 Nm).

110. Position the ground strap and install the bolt.

111. Install the steering shaft pinch bolt and tighten to 18 ft. lbs. (25 Nm).

112. Install the halfshafts.

113. Install the cowl vent screen.

114. Install the air cleaner and the air cleaner outlet pipe.

115. Evacuate, leak test, and charge the A/C system using approved equipment.

116. Fill and bleed the cooling system.

117. Fill and bleed the power steering system.

118. Fill the engine with clean engine oil.

119. Connect the negative battery cable.

Water Pump

REMOVAL & INSTALLATION

2002–04 3.0L (VIN U) Engine

1. Before servicing the vehicle, refer to the precautions section.
2. Drain the cooling system.
3. Remove or disconnect the following:
   - Negative battery cable
   - Accessory drive belt and tensioner
   - Water pump pulley
   - Heater hose
   - Engine control sensor wiring
   - Water pump

To install:

➡**The water pump bolts are of different lengths and must be installed in the correct locations.**

4. Install the water pump. Tighten bolts 1, 2, 3, 4, 5, 6, 7, 8, 9 and 15 to 15–22 ft. lbs. (20–30 Nm). Tighten bolts 10, 11, 12, 13 and 14 to 72–106 inch lbs. (8–12 Nm). Refer to the accompanying illustration for the bolt locations.
5. Install or connect the following:
   - Engine control sensor wiring
   - Heater hose
   - Water pump pulley and tighten the bolts to 15–22 ft. lbs. (20–30 Nm)
   - Accessory drive belt and tensioner
   - Negative battery cable
6. Fill the cooling system and check for leaks.

2005 3.0L (VIN U) Engine

1. Before servicing the vehicle, refer to the precautions section.
2. Disconnect the negative battery cable.
3. Drain the engine cooling system.
4. Loosen the water pump pulley.
5. Remove the accessory drive belt.
6. Remove the degas bottle.
7. Remove the alternator.
8. Remove the bolt and the nut, and position the roll restrictor bracket out of the way.
9. Remove the bolt and the accessory drive belt tensioner.
10. Remove the nut, the bolt and the support bracket.
11. Remove the bolts and the water pump pulley.
12. Disconnect the water pump inlet hose and the Crankshaft Position (CKP) sensor electrical connector.
13. Remove the bolts and the water pump, and clean the sealing surfaces.

To install:

14. Install the water pump. Tighten bolts in the sequence illustrated as follows:
   a. Bolts 1-7 to 18 ft. lbs. (25 Nm).
   b. Bolts 8-12 to 89 inch lbs. (10 Nm).
15. Connect the CKP sensor electrical connector and the water pump inlet hose.
16. Install the support bracket, the nut and the bolt and tighten to 18 ft. lbs. (25 Nm).
17. Install the accessory drive belt tensioner and the bolt and tighten to 18 ft. lbs. (25 Nm).
18. Install the water pump pulley and the bolts and tighten to 18 ft. lbs. (25 Nm).
19. Install the roll restrictor bracket, the bolt and the nut. Do not tighten at this time.
20. Install the alternator and the mounting bolt. Do not tighten at this time.
21. Install the alternator brace, the nut and the bolt. Tighten the bolts as illustrated.
22. Tighten the alternator lower mounting bolt to 18 ft. lbs. (25 Nm).
23. Tighten the roll restrictor bracket bolt and nut. Tighten the bolts as illustrated
24. Connect the alternator electrical connectors.
25. Install the accessory drive belt.
26. Install the degas bottle.
27. Connect the negative battery cable.
28. Fill the engine cooling system.
29. Start the engine and check for leaks.

2002–04 3.0L (VIN S) Engine

1. Before servicing the vehicle, refer to the precautions section.

Fasteners			Torque Specifications	
Fastener And Hole No.	Size	Fastener Application	Nm	Lb-Ft
1A	M8 x 1.25 x 43.5	F/C TO BLOCK	20-30	15-22
2A	M8 x 1.25 x 73	W/P & F/C TO BLOCK	20-30	15-22
3A	M8 x 1.25 x 104.3	W/P & F/C TO BLOCK	20-30	15-22
4A	M8 x 1.25 x 71.3	F/C TO BLOCK	20-30	15-22
5A	M8 x 1.25 x 71.3	W/P & F/C TO BLOCK	20-30	15-22
6A	M8 x 1.25 x 71.3	W/P & F/C TO BLOCK	20-30	15-22
7A	M8 x 1.25 x 104.3	W/P & F/C TO BLOCK	20-30	15-22
8A	M8 x 1.25 x 104.3	W/P & F/C TO BLOCK	20-30	15-22
9A	M8 x 1.25 x 52	F/C TO BLOCK	20-30	15-22
10B	M6 x 1 x 28.5	W/P TO F/C	8-12	71-106 (lb-in)
11B	M6 x 1 x 28.5	W/P TO F/C	8-12	71-106 (lb-in)
12B	M6 x 1 x 28.5	W/P TO F/C	8-12	71-106 (lb-in)
13B	M6 x 1 x 28.5	W/P TO F/C	8-12	71-106 (lb-in)
14B	M6 x 1 x 28.5	W/P TO F/C	8-12	71-106 (lb-in)
15A	M8 x 1.25 x 71.3	W/P & F/C TO BLOCK	20-30	15-22

W/P—Water Pump

F/C—Engine Front Cover

9300KG01

Exploded view of the water pump and timing front cover bolt locations—2002–04 3.0L (VIN U) engine

- Upper radiator hose
- Water pump belt
- Battery and tray
- Air cleaner assembly
- Radiator upper front tube bolt
- Radiator lower tube bolt
- Lower heater hose
- Splash shield

5. Fill the cooling system.
6. Start the engine and check for leaks.

2005 3.0L (VIN S) Engine

1. Before servicing the vehicle, refer to the precautions section.
2. Disconnect the negative battery cable.
3. Remove the air cleaner assembly.
4. Remove the battery and the battery tray.
5. Remove the water pump belt.
6. Drain and recycle the engine coolant.
7. Disconnect the heater hose from the bottom of the water pump.
8. Remove the radiator lower tube bolt.
9. Remove the radiator upper front tube bolt.
10. Disconnect the upper radiator hose, the heater hose and the thermostat housing hose.
11. Remove the radiator bypass hose assembly.
12. Remove the pin-type retainer, disconnect the Engine Coolant Temperature (ECT) sensor electrical connector and position the harness aside.
13. Remove the three bolts, and reposition the water pump to access the remaining hoses.
14. Loosen the clamps, disconnect the hoses and remove the water pump.

To install:
15. Connect the hoses to the water pump.
16. Position the water pump and install the mounting bolts. Tighten the bolts in two steps:
   a. Step 1: Tighten to 89 inch lbs. (10 Nm).
   b. Step 2: Tighten an additional 90 degrees.
17. Install the radiator bypass hose assembly.
18. Connect the ECT sensor electrical connector and attach the pin-type retainer.
19. Connect the upper radiator hose, the heater hose and thermostat housing hose.
20. Install the radiator upper front cooling tube bolt.
21. Install the lower cooling tube bolt.
22. Connect the heater hose to the bottom of the water pump.

06017-TAUR-G04

Install the water pump. Tighten bolts in the sequence illustrated—2005 3.0L (VIN U) models

2. Drain the cooling system.
3. Remove or disconnect the following:
   - Splash shield
   - Lower heater hose
   - Radiator lower tube bolt
   - Radiator upper front tube bolt
   - Air cleaner assembly
   - Battery and tray
   - Water pump belt
   - Upper radiator hose
   - Engine vent hose
   - Transaxle 10 pin connector
   - Radiator bypass hose assembly
   - Thermostat housing
   - A/C compressor brace, if equipped
   - Water pump mounting nuts
   - Connector hose
   - Water pump

To install:
4. Install or connect the following:
   - Water pump
   - Connector hose
   - Water pump mounting nuts and tighten them to 15–22 ft. lbs. (20–30 Nm)
   - A/C compressor brace, if equipped, and tighten the nuts to 15–22 ft. lbs. (20–30 Nm)
   - Thermostat housing
   - Radiator bypass hose assembly
   - Transaxle 10 pin connector
   - Engine vent hose

25 Nm (18 lb-ft)   48 Nm (35 lb-ft)

06017-TAUR-G05

Install the alternator brace, tighten the retainers as illustrated—2005 3.0L (VIN U) models

47 Nm (35 lb-ft)

25 Nm (18 lb-ft)

06017-TAUR-G06

Tighten the roll restrictor bracket bolt and nut as illustrated—2005 3.0L (VIN U) models

NUT
4 REQ'D
TIGHTEN TO
20-30 N·m
(15-22 LB-FT)

FRONT OF ENGINE

9346KG10

Water pump mounting—2002–04 3.0L (VIN S) engine

23. Install the water pump belt.
24. Install the battery and the battery tray.
25. Install the air cleaner assembly.
26. Fill and bleed the engine cooling system.
27. Connect the negative battery cable.
28. Start the engine and check for leaks.

Heater Core

REMOVAL & INSTALLATION

1. Disconnect the negative battery cable.

☀ CAUTION

After disconnecting the negative battery cable, wait for at least 1 minute for the SRS or air bag module to deplete its energy.

2. Place the front wheels in the straight-ahead position.
3. Lock the steering column.
4. Remove or disconnect the following:
   • SRS bolt covers from both sides of the steering wheel
   • SRS module-to-steering wheel bolts

• SRS module and disconnect the electrical connector
• Steering wheel bolt and discard it
5. Press the steering wheel from the steering column.
   • Lower steering column shaft bolt
   • 2 lower instrument panel cover-to-instrument panel screws and unsnap the lower cover from the instrument panel
6. Turn ignition switch to the RUN position
7. Insert a ⅛ in. (3mm) wire or pin punch in the lower steering column shroud hole, under the ignition switch; then, press on the pin while pulling out on the ignition switch lock cylinder and remove it from the steering column lock cylinder housing.
8. Remove or disconnect the following:
   • 3 steering column shroud screws and the shrouds
   • Shift control selector lever boot
   • Gearshift lever pin from the manual control lever and remove the lever
   • Wiring connector at the bottom of the steering column and remove the wiring from the column, (if equipped with an overdrive lockout switch on the manual control lever)
   • Electrical connectors

INSTRUMENT PANEL REINFORCEMENT

INSTRUMENT PANEL INSULATOR

FRONT OF VEHICLE

91170G05

Remove the instrument panel insulator from the instrument panel reinforcement—Taurus and Sable

Manual air conditioning/heater equipped vehicles disconnection points—Taurus and Sable

Item	Description
1	A/C Evaporator Housing
2	Blower Motor
3	Vacuum Hose Harness
4	Main Wiring

91170G06

EATC equipped vehicles disconnection points—Taurus and Sable

Item	Description
1	A/C Evaporator Housing
2	Blower Motor
3 & 4	Main Wiring
5	Automatic Temperature Control Sensor Hose and Elbow

91170G07

91170G08

On floor shift vehicles, remove the console center finish panel—Taurus and Sable

91170G09

On floor shift vehicles, remove the finish panel from around the integrated control panel—Taurus and Sable

91170G10

If equipped with autolamps, detach the light sensor amplifier—Taurus and Sable

FRONT OF VEHICLE

INSTRUMENT PANEL

INSTRUMENT PANEL UPPER FINISH PANEL

91170G12

Unsnap the upper finish panel from the instrument panel—Taurus and Sable

INSTRUMENT PANEL

RH SHOWN
LH SIMILAR

INSTRUMENT PANEL FINISH END PANEL

91170G11

Remove the instrument panel finish end panel—Taurus and Sable

- Multi-function switch screws and move it aside
- Shift indicator cable from the shifter tube
- Shifter indicator-to-column adjustment cable screw
- Interlock cable and actuator (if equipped with a column shift)
- 4 steering column-to-instrument panel bracket nuts and steering column
- Push pins and the lower instrument cover from the instrument panel reinforcement (at the passenger's side)
- Console finish panel

9. Under the steering column, 2 instrument panel brace screws, the courtesy lamp socket, the 2 Diagnostic Link Connector (DLC) screws and the instrument panel brace.

Item	Description
1	Wiring Assy Electrical Connector
2	Wiring Assy Electrical Connector
3	Bolt (Part of 14290 and 14A005)
4	Wiring Assy Electrical Connector
5	Instrument Panel
6	Parking Brake Control
A	Tighten to 4-6 N•m (36-53 Lb-In)

91170G13

Detach the electrical connectors—Taurus and Sable

Item	Description
1	Screw (3 Req'd)
2	Cowl Top Panel
3	Screw (4 Req'd)
4	Instrument Panel

Item	Description
A	Tighten to 7-9 N·m (62-79 Lb-In)
B	Tighten to 19-25 N·m (15-18 Lb-Ft)

91170G14

Instrument panel-to-cowl panel mounting—Taurus and Sable

A/C ELECTRONIC
BLEND DOOR
ACTUATOR

A/C EVAPORATOR HOUSING

91176G05

Remove the 4 blend door actuator retaining screws—Taurus and Sable

Item	Description
1	Metal Link (Part of 19B555)
2	Heater Core Cover
3	A / C Evaporator Housing
4	Spring (Part of 19B555)
5	Lever (Part of 19B555)
6	Locking Ramp (Part of Secondary A / C Air Temperature Control Door Shaft)
7	Secondary A / C Air Temperature Control Door Shaft (Part of 19B555)

Secondary air temperature door connections—Taurus and Sable

10. If not equipped with an Electronic Automatic Temperature Control (EATC), disconnect the electrical connectors and the vacuum harness from the evaporator housing and the blower motor.

11. If equipped with an Electronic Automatic Temperature Control (EATC), remove the sensor hose/elbow and disconnect the electrical harness connectors and the vacuum hose harness from the evaporator housing.

12. Insert the Radio Removing tools 415-001 into the integrated control panel faceplate; then, push the tools in approximately 1½ in. (38mm) to release the retaining clips. Spread the tools slightly and pull the integrated control panel from the instrument panel.

13. Remove or disconnect the following:
- Electrical connectors and the automatic temperature control sensor hose and elbow, if equipped
- 6 console center finish panel screws and the panel, (if equipped with a floor shift)
- 4 instrument panel finish panel screws and panel, located at the right side of the steering column, (if equipped with a column shift)
- Instrument cluster finish panel-to-instrument panel clips and pull the finish panel straight out
- 4 instrument cluster-to-instrument panel screws, the electrical connectors and the instrument cluster
- Upper finish panel (located at the top of the instrument panel)
- Light sensor amplifier connector, (if equipped with an autolamp)
- Instrument panel finish end panels on both sides of the instrument panel
- Front scuff plates and pull the door weatherstrip away from the instrument panel (on both sides of the instrument panel)
- 3 instrument panel-to-upper cowl top panel screws
- 2 screws at each end of the instrument panel
- Instrument panel electrical connectors and the parking brake switch
- Instrument panel

14. Drain the cooling system into a clean container for reuse

15. Remove or disconnect the following:
- Heater hoses from the heater core and plug the openings
- 4 air conditioning electronic blend door actuator-to-heater/air conditioning housing assembly screws and the actuator
- Metal cover, disengage the spring

91176G06

ALIGN FLATS ON STEERING WHEEL AND STEERING COLUMN SHAFT

VIEW A

TAURUS, SABLE (EXCEPT SHO)

VIEW A

1 Steering Wheel	10 Speed Control/Horn Wire Connector
2 Bolt	
3 Driver Side Air Bag Module	11 Speed Control Actuator Switch (Left Hand)
4 Screw (2 Req'd)	A Tighten to 34-46 N·m (26-33 Lb-Ft)
5 Steering Wheel Spoke Cover	
6 Steering Column Tube	B Tighten to 10-14 N·m (89-123 Lb-In)
7 Electrical Connector	
8 Air Bag Electrical Connector	
9 Speed Control Actuator Switch (Right Hand)	

93111G90

Exploded view of the SRS module and the steering wheel assembly—Taurus and Sable

from the heater core cover and from the lever

• Gently, depress the locking ramp and remove the lever from the secondary air temperature control door end

> **❄❄ WARNING**
>
> **Do not attempt to bend any part of the lever, for it is brittle and will break.**

16. Rotate the primary air conditioning air temperature control door shaft downward, swing the metal link counterclockwise and remove it from the pin.

17. Remove the 3 heater core cover-to-heater/air conditioning housing assembly screws, the cover and the seal.

18. Press on the heater core tubes and remove the heater core from the heater/air conditioning housing assembly.

To install:

19. Install or connect the following:

• Heater core to the heater/air conditioning housing assembly
• Heater core cover, the seal and the 3 heater core cover-to-heater/air conditioning housing assembly screws
• Lever to the secondary air conditioning air temperature control door end
• Spring to the lever, engage the spring to the heater core cover and install the metal cover
• Air conditioning electronic blend door actuator and the 4 actuator-to-heater/air conditioning housing assembly screws
• Heater hoses to the heater core
• Instrument panel with the help of an assistant
• Instrument panel electrical connectors and the parking brake switch
• 2 screws at each end of the instrument panel with the help of an assistant
• 3 instrument panel-to-upper cowl top panel screws
• Front scuff plates and the door weatherstrip (on both sides of the instrument panel)
• Instrument panel finish end panels (on both sides of the instrument panel)
• Electrical connector to the light sensor amplifier (if equipped with an autolamp)
• Upper finish panel (at the top of the instrument panel)

View of the integrated control panel—Taurus and Sable

- Instrument cluster, electrical connectors and the 4 instrument cluster-to-instrument panel screws
- Instrument cluster finish panel and engage the finish panel-to-instrument panel clips
- Instrument panel finish panel and 4 panel screws, located at the right side of the steering column (if equipped with a column shift)
- Console center finish panel and the 6 panel screws (if equipped with a floor shift)
- Electrical connectors and the automatic temperature control sensor hose and elbow, if equipped
- Integrated control panel faceplate

✷✷ WARNING

Do not use excessive force when installing the radio, the retaining clips can become damaged.

- Sensor hose/elbow, the electrical harness connectors and the vacuum hose harness to the evaporator housing (if equipped with an Electronic Automatic Temperature Control (EATC)
- Connect the electrical connectors and the vacuum harness to the evaporator housing and the blower motor (if not equipped with an EATC)
- Instrument panel brace and the 2 Diagnostic Link Connector (DLC) screws
- Courtesy lamp socket and install the 2 instrument panel brace screws
- Console finish panel (if equipped with a floor shift)
- Lower instrument cover to the instrument panel reinforcement (located on the passenger's side)
- Steering column and torque the 4 steering column-to-instrument panel bracket nuts to 10–13 ft. lbs. (13–17 Nm)
- Interlock cable and actuator (if equipped with a column shift)
- Shifter indicator-to-column adjustment cable screw
- Shift indicator cable to the shifter tube
- Multi-function switch and the screws
- Electrical connectors
- Wiring connector at the bottom of the steering column and install the wiring to the column (if equipped with an overdrive lockout switch on the manual control lever)

1	Screw (3 Req'd)	A	Tighten to 7-9 N·m (62-79 Lb-In)
2	Cowl Top Panel		
3	Screw (4 Req'd)	B	Tighten to 19-25 N·m (15-18 Lb-Ft)
4	Instrument Panel		

Exploded view of the instrument panel—Taurus and Sable

1 Metal Link	6 Locking Ramp (Part of Secondary A/C Air Temperature Control Door Shaft)
2 Heater Core Cover	
3 A/C Evaporator Housing	
4 Spring	7 Secondary A/C Air Temperature Control Door Shaft
5 Lever	

93111G93

View of the temperature control mechanism—Taurus and Sable

1 A/C Evaporator Housing	4 Heater Core Cover
2 Heater Core	5 Screw
3 Heater Core Cover Seal	

93111G94

Exploded view of the heater core—Taurus and Sable

- Gearshift lever and the lever pin to the manual control lever
- 3 steering column shroud screws and the shrouds
- Ignition switch to the steering column lock cylinder housing
- Lower instrument panel cover and the 2 lower cover-to-instrument panel screws
- Lower steering column shaft bolt and torque it to 17–20 ft. lbs. (22–26 Nm)
- Steering wheel to the steering column
- New steering wheel bolt and torque it to 26–33 ft. lbs. (34–46 Nm)
- SRS module and connect the electrical connector
- SRS module-to-steering wheel bolts and torque the bolts to 89–123 inch lbs. (10–14 Nm)
- SRS bolt covers
- Negative battery cable

20. Fill the cooling system.

21. Operate the engine to normal operating temperatures; then, check the climate control operation and check for leaks.

Cylinder Head

REMOVAL & INSTALLATION

2002–04 3.0L (VIN U) Engine

1. Before servicing the vehicle, refer to the precautions section.

2. Drain the cooling system.

3. Set the No. 1 cylinder to Top Dead Center (TDC) of the compression stroke.

4. Remove or disconnect the following:
- Negative battery cable
- Air cleaner outlet hose
- Vacuum hoses from the intake manifold
- Exhaust Gas Recirculation (EGR) valve
- EGR transducer
- EGR vacuum regulator solenoid
- Intake Air Temperature (IAT) sensor connector
- Throttle Position (TP) sensor connector
- Intake Air Control (IAC) valve connector
- Camshaft Position (CMP) sensor and housing
- Engine Coolant Temperature (ECT) sensor connector
- Accelerator cable
- Cruise control cable
- Fuel lines
- Upper intake manifold

- Fuel injector wiring harness
- Coolant hoses
- Ignition coil and bracket
- Spark plug wires
- Accessory drive belt and tensioner
- Alternator
- Power steering pump
- Oil dipstick tube
- Heater supply tube brackets
- Valve covers
- Rocker arms and pushrods. Keep the rocker arms and pushrods in order for installation.
- Lower intake manifold
- Spark plugs
- Exhaust manifolds
- Cylinder heads

To install:

➡ **The cylinder head bolts are a torque-to-yield design and cannot be reused.**

5. Install the cylinder heads with new gaskets.

6. Tighten the bolts in sequence as follows:

   a. Step 1: 36–39 ft. lbs. (47–53 Nm).

   b. Step 2: Loosen the bolts 1 complete turn.

   c. Step 3: Tighten the bolts to 20–24 ft. lbs. (27–33 Nm).

   d. Step 4: Tighten the bolts 85–95 degrees.

   e. Step 5: Tighten the bolts 85–95 degrees.

7. Install or connect the following:

- Exhaust manifolds
- Spark plugs
- Lower intake manifold
- Rocker arms and pushrods in their original locations. Tighten the rocker arm bolts to 24 ft. lbs. (32 Nm). Tighten each rocker arm bolt with the valve closed and the lifter on the camshaft lobe base circle. Rotate the crankshaft as necessary.
- Valve covers
- Heater supply tube brackets
- Oil dipstick tube and tighten the nut to 13 ft. lbs. (18 Nm).
- Power steering pump
- Alternator
- Accessory drive belt and tensioner
- Spark plug wires
- Ignition coil and bracket. Tighten the fasteners to 30–39 ft. lbs. (40–55 Nm).
- Coolant hoses
- Fuel injector wiring harness
- Upper intake manifold
- Fuel lines
- Cruise control cable
- Accelerator cable

Install the cylinder head gasket with the "UP" designation on gasket facing the cylinder head—2002–04 3.0L (VIN U) engines

Cylinder head torque sequence—2002–04 3.0L (VIN U) engine

- ECT sensor connector
- CMP sensor and housing
- IAC valve connector
- TP sensor connector
- IAT sensor connector
- EGR vacuum regulator solenoid
- EGR transducer
- EGR valve
- Vacuum hoses from the intake manifold

- Air cleaner outlet hose
- Negative battery cable

8. Fill the cooling system and check for leaks.

2005 3.0L (VIN U) Engine

RIGHT SIDE

1. Before servicing the vehicle, refer to the precautions section.

2. Remove the lower intake manifold.

TIGHTENING SEQUENCE — VALVE COVER

9300KG08

Valve cover tightening sequence—2002–04 3.0L (VIN U) engine

3. Disconnect the Crankshaft Position (CKP) sensor connector.

4. Disconnect the Heated Oxygen (HO₂S) sensor connector.

5. Remove the nut and disconnect the wiring harness ground connections.

6. Loosen the bolt and disconnect the Powertrain Control Module (PCM) electrical connector.

7. Disconnect the Evaporative Emissions (EVAP) canister purge valve electrical connector.

8. Disconnect the spark plug wires and position the wires aside.

9. Loosen the nut and disconnect the Exhaust Gas Recirculation (EGR) tube.

10. Remove the three bolts and the exhaust manifold heat shield.

11. Loosen the exhaust manifold-to-catalytic converter bolts.

12. Remove the six exhaust manifold-to-cylinder head bolts.

13. Disconnect the engine wire harness locator.

14. Remove the bolt and separate the catalytic converter heat shield from the cylinder head.

15. Remove the eight cylinder head bolts.

16. Remove the cylinder head.

17. Remove and discard the head gasket.

To install:

18. Using metal surface cleaner, clean the cylinder head sealing surfaces.

➡ **The "V" notch in the cylinder head gasket faces the front of the engine.**

19. Install a new cylinder head gasket.

20. Install the cylinder head.

21. Install the bolts. Tighten the bolts in five steps in the sequence shown.

  a. Step 1: Tighten to 37 ft. lbs. (50 Nm).

  b. Step 2: Loosen the bolts one full turn.

  c. Step 3: Tighten to 22 ft. lbs (30 Nm).

d. Step 4: Rotate each bolt 90 degrees.

e. Step 5: Rotate each bolt an additional 90 degrees.

22. Install the catalytic converter heat shield and wire harness ground connection.

23. Position the wiring harness and install the harness locator.

24. Install the exhaust manifold.

25. Tighten the exhaust manifold-to-catalytic converter bolts to 30 ft. lbs. (40 Nm).

26. Install the exhaust manifold heat shield.

27. Install the EGR tube and tighten to 37 ft. lbs. (50 Nm).

28. Connect the spark plug wires.

29. Connect the EVAP canister purge valve electrical connector.

30. Connect the PCM electrical connector.

31. Connect the wire harness ground connections.

32. Connect the HO₂S sensor electrical connector.

33. Connect the CKP electrical connector.

34. Install the lower intake manifold.

LEFT SIDE

1. Before servicing the vehicle, refer to the precautions section.

2. Remove the lower intake manifold.

3. Remove the accessory drive belt.

4. Remove the retainers and the engine anti-roll strut brace.

5. Disconnect the alternator electrical connections.

6. Remove the retainers and the alternator.

7. Remove the bolt and the drive belt idler pulley.

8. Remove the bolt and the drive belt tensioner pulley.

9. Remove the spark plug wires.

10. Remove the exhaust manifold.

11. Remove the power steering pump and bracket assembly as follows:.

  a. Disconnect the power steering return hose.

  b. Remove the bolt.

  c. Remove the two nuts.

12. Remove the eight cylinder head bolts.

13. Remove the cylinder heads.

14. Remove and discard the head gasket.

To install:

15. Using metal surface cleaner, clean the cylinder head sealing surfaces.

➡ **The "V" notch in the cylinder head gasket faces the front of the engine.**

16. Install a new cylinder head gasket.

17. Install the cylinder head.

18. Install the bolts. Tighten the bolts in five steps in the sequence shown.

  a. Step 1: Tighten to 37 ft. lbs. (50 Nm).

  b. Step 2: Loosen the bolts one full turn.

  c. Step 3: Tighten to 22 ft. lbs (30 Nm).

06017-TAUR-G07

Cylinder head torque sequence, right side show, left side similar—2005 3.0L (VIN U) models

d. Step 4: Rotate each bolt 90 degrees.

e. Step 5: Rotate each bolt an additional 90 degrees.

19. Install the cylinder head.

20. Install the bolts. Tighten the bolts in five steps in the sequence shown.

a. Step 1: Tighten to 37 ft. lbs. (50 Nm)

b. Step 2: Loosen the bolts one full turn.

c. Step 3: Tighten to 2 2ft. lbs. (30 Nm).

d. Step 4: Rotate each bolt 90 degrees.

e. Stage 5: Rotate each bolt an additional 90 degrees.

21. Install the power steering pump and bracket assembly as follows:

a. Install the bolt and the two nuts. Tighten to 35 ft. lbs. (48 Nm).

b. Connect the power steering return hose.

22. Install the exhaust manifold.

23. Install the spark plug wires.

24. Install the drive belt tensioner and bolt. Tighten to 35 ft. lbs. (48 Nm).

25. Install the drive belt idler pulley and bolt. Tighten to 35 ft. lbs. (48 Nm).

26. Install the alternator and the retainers.

27. Connect the alternator electrical connections.

28. Install the engine anti-roll strut brace and the retainers. Tighten to 35 ft. lbs. (48 Nm).

29. Install the accessory drive belt.

30. Install the lower intake manifold.

2002–04 3.0L (VIN S) Engine

RIGHT SIDE

1. Before servicing the vehicle, refer to the precautions section.

2. Remove the engine from the vehicle and mount it on an engine stand.

3. Remove or disconnect the following:

- Upper intake manifold
- Lower intake manifold
- Exhaust manifolds
- Valve covers
- Radiator hose tube and bracket
- Accessory drive belt
- Crankcase ventilation tube
- Heater hose bypass tube
- Power steering pump
- A/C compressor and bracket
- Water pump

➥The crankshaft accessory drive pulley shaft has left-hand threads. Rotate the pulley shaft clockwise to remove.

- Crankshaft pulley
- Oil pan
- Front cover

Cylinder head bolt removal sequence–right side head shown, left side similar—3.0L (VIN S) engine

Right side valve cover bolt removal sequence—3.0L (VIN S) engine

Left side valve cover bolt removal sequence—3.0L (VIN S) engine

Cylinder head torque sequence—3.0L (VIN S) engine

Left side valve cover bolt tightening sequence—3.0L (VIN S) engine

Right side valve cover bolt tightening sequence—3.0L (VIN S) engine

- Timing chains
- Camshafts
- Valve lash adjusters

➡**Keep the lash adjusters in order for installation.**

4. Remove the cylinder head bolts from the cylinder heads in sequence and remove the cylinder heads.

To install:

➡**The cylinder head bolts are a torque-to-yield design and cannot be reused.**

➡**Left and right cylinder head gaskets are not interchangeable.**

5. Install new gaskets and install the cylinder heads. Use new bolts and tighten as follows:
   a. Step 1: Tighten the bolts in sequence to 28–31 ft. lbs. (37–43 Nm).
   b. Step 2: Plus 85–95 degrees.
   c. Step 3: Loosen the bolts in reverse sequence 1 full turn.
   d. Step 4: Tighten the bolts in sequence to 28–31 ft. lbs. (37–43 Nm).
   e. Step 5: Tighten the bolts in sequence 85–95 degrees.
   f. Step 6: Tighten the bolts in sequence 85–95 degrees.
6. Install or connect the following:
   - Valve lash adjusters in their original locations
   - Camshafts
   - Timing chains
   - Front cover
   - Oil pan
   - Crankshaft pulley
   - Water pump
   - A/C compressor and bracket
   - Power steering pump
   - Heater hose bypass tube
   - Crankcase ventilation tube
   - Accessory drive belt
   - Radiator hose tube and bracket
   - Valve covers
   - Exhaust manifolds
   - Lower intake manifold
   - Upper intake manifold
   - Engine assembly into the vehicle.
7. Run the engine and check for leaks.

2005 3.0L (VIN S) Engine

LEFT SIDE

1. Before servicing the vehicle, refer to the precautions section.
2. Release the fuel system pressure.
3. Drain and recycle the engine coolant.
4. Remove the water pump bypass tube.
5. Remove the exhaust manifold-to-pipe nuts.
6. Remove the water pump.
7. Remove the lower intake manifold.
8. Remove the camshafts.
9. Install and engine support tool and lower the engine until it is aligned with the A/C compressor bracket, and finger-tighten the four bolts.
10. Remove the engine support tool.
11. Remove the camshaft followers.
12. Remove the hydraulic lash adjusters.
13. Remove the nuts and remove the exhaust manifold and discard the gasket.
14. Remove the dipstick tube.
15. Inspect and, if necessary, install a new O-ring seal.
16. Remove the left hand cylinder block drain plug.

➡**New cylinder head bolts must be installed. They are torque-to-yield designed and cannot be reused.**

17. Remove the bolts in the sequence shown and remove the cylinder head.
18. Discard the gasket and the bolts.

To install:

19. Position the cylinder head, gasket and install the bolts in the sequence shown in six steps:
   a. Step 1: Tighten to 30 ft. lbs. (40 Nm).
   b. Step 2: Tighten bolts 90 degrees.
   c. Step 3: Loosen one full turn.
   d. Step 4: Tighten to 30 ft. lbs. (40 Nm).
   e. Step 5: Tighten 90 degrees.
   f. Step 6: Tighten 90 degrees.
20. Install the cylinder block drain plug and tighten to 49 ft. lbs. (67 Nm).

➡**Lubricate the oil dipstick tube O-ring with clean engine oil.**

21. Install the oil dipstick tube and the nut.
22. Position a new gasket and install the exhaust manifold nuts.
23. Install the hydraulic lash adjusters. Lubricate the hydraulic lash adjusters with clean engine oil prior to installation.
24. Install the camshaft followers. Lubricate the camshaft followers with clean engine oil prior to installation
25. Install the engine support tool.
26. Remove the four A/C compressor bracket bolts.
27. Using the engine support tool, raise the engine away from the A/C compressor bracket.
28. Install the camshafts.
29. Install the lower intake manifold.
30. Install the water pump.
31. Install the exhaust manifold-to-pipe nuts.
32. Install the water pump bypass tube.
33. Run the engine and check for leaks.

RIGHT SIDE

1. Before servicing the vehicle, refer to the precautions section.
2. Release the fuel system pressure.
3. Remove the water pump bypass tube.
4. Remove the rear support brace.
5. Remove the lower intake manifold.
6. Remove the camshafts.
7. Install and engine support tool and lower the engine until it is aligned with the A/C compressor bracket, and finger-tighten the four bolts.
8. Remove the engine support tool.
9. Remove the exhaust manifold-to-pipe bolts.
10. Disconnect the Exhaust Gas Recirculation (EGR) tube from the exhaust manifold.
11. Remove the bolt and the ground wire.
12. Remove the camshaft followers.
13. Remove the hydraulic lash adjusters.
14. Remove the nuts and remove the exhaust manifold and discard the gasket.
15. Remove the right hand cylinder block drain plug.

➡**New cylinder head bolts must be installed. They are torque-to-yield designed and cannot be reused.**

16. Remove the bolts in the sequence shown and remove the cylinder head.
17. Discard the gasket and the bolts.

To install:

18. Position the cylinder head, gasket and install the bolts in the sequence shown in six steps:
   a. Step 1: Tighten to 30 ft. lbs. (40 Nm).
   b. Step 2: Tighten bolts 90 degrees.
   c. Step 3: Loosen one full turn.
   d. Step 4: Tighten to 30 ft. lbs. (40 Nm).
   e. Step 5: Tighten 90 degrees.
   f. Step 6: Tighten 90 degrees.
19. Install the cylinder block drain plug and tighten to 49 ft. lbs. (67 Nm).
20. Position a new gasket and install the exhaust manifold nuts.
21. Install the hydraulic lash adjusters. Lubricate the hydraulic lash adjusters with clean engine oil prior to installation.
22. Install the camshaft followers. Lubricate the camshaft followers with clean engine oil prior to installation
23. Install the ground wire and bolt.

24. Connect the EGR tube to the exhaust manifold. Tighten to 30 ft. lbs. (40 Nm).

25. Install the exhaust manifold-to-pipe bolts. Tighten to 30 ft. lbs. (40 Nm).

26. Install the engine support tool.

27. Remove the four A/C compressor bracket bolts.

28. Using the engine support tool, raise the engine away from the A/C compressor bracket.

29. Install the camshafts.

30. Install the lower intake manifold.

31. Install the rear support brace. Tighten to 35 ft. lbs. (47 Nm).

32. Install the water pump bypass tube.

33. Run the engine and check for leaks.

Rocker Arms/Shafts

REMOVAL & INSTALLATION

3.0L (VIN U) Engine

1. Before servicing the vehicle, refer to the precautions section.

2. Remove or disconnect the following:
- Negative battery cable
- Upper intake manifold
- Valve covers
- Rocker arms

➡**Keep the rocker arms in order for installation.**

1	Bolt (12 Req'd)
2	Rocker Arm Seat (12 Req'd)
3	Rocker Arm (12 Req'd)
4	Push Rod (12 Req'd)
5	Cylinder Head (2 Req'd)
6	2.15-4.69 mm (0.085-0.185 inch)
A	Tighten in Two Steps: 7-15 N·m (5-11 Lb-Ft) 26-38 N·m (19-28 Lb-Ft)

7922KG22

Exploded view of the rocker arms and related components—3.0L (VIN U) engines

To install:

➡**The rocker arm bolts are tightened with the valves closed. Rotate the crankshaft as necessary to position the lifter on the base circle of the camshaft lobe before tightening the corresponding rocker arm bolt.**

3. Install the rocker arms. The rocker arm bolts are tightened in two steps as follows:
   a. Step 1: 96 inch lbs. (11 Nm).
   b. Step 2: 24 ft. lbs. (32 Nm).

4. Install or connect the following:
- Valve covers
- Upper intake manifold
- Negative battery cable

5. Start the engine and check for proper operation.

3.0L (VIN S) Engine

1. Before servicing the vehicle, refer to the precautions section.

2. Remove or disconnect the following:
- Negative battery cable
- Upper intake manifold
- Valve covers

3. Rotate the crankshaft so that the camshaft lobe on the valve to be serviced is pointing directly away from the valve.

4. Install special tool Valve Spring Compressor 303-473.

5. Compress the valve spring and remove the rocker arm. Repeat for each arm to be removed.

➡**If the rocker arms are to be reused, ensure that they are installed in the same position that they were removed from.**

To install:

6. Compress the valve spring and install the rocker arm. Repeat for each arm to be installed.

7. Install or connect the following:
- Valve covers
- Upper intake manifold
- Negative battery cable

303-473

9346KG01

Valve spring compressor 303-473—3.0L (VIN S) engine

8. Start the engine and check for proper operation.

Intake Manifold

REMOVAL & INSTALLATION

2002–04 3.0L (VIN U) Engine

1. Before servicing the vehicle, refer to the precautions section.

2. Drain the cooling system.

3. Remove or disconnect the following:
- Negative battery cable
- Air cleaner and outlet tube
- Cruise control cable
- Accelerator cable and bracket
- Fuel lines
- Vacuum lines
- Exhaust Gas Recirculation (EGR) valve
- EGR transducer
- EGR vacuum solenoid
- Throttle Position (TP) sensor connector
- Intake Air Control (IAC) valve connector
- Camshaft Position (CMP) sensor connector
- Engine Coolant Temperature (ECT) sensor connector
- Coolant temperature indicator sender connector
- Coolant hoses
- Upper alternator bracket
- Engine control sensor wiring harness and bracket
- Upper intake manifold

➡**Note the position of the CMP sensor electrical connector. Installation requires that the connector be located in the same position.**

➡**Before removing the CMP sensor, position the engine at Top Dead Center (TDC) of the No. 1 cylinder compression stroke.**

4. Remove or disconnect the following:
- CMP sensor and housing
- Spark plug wires
- Ignition coil
- Valve covers
- No. 3 cylinder intake valve rocker arm and pushrod
- Lower intake manifold

To install:

5. Install the lower intake manifold. Use new gaskets and tighten the bolts in sequence as follows:
   a. Step 1: 15–22 ft. lbs. (20–30 Nm).
   b. Step 2: 19–24 ft. lbs. (26–32 Nm).

Item	Description
1	Stud Bolt
2	Bolt (2 Req'd)
3	Stud Bolt (2 Req'd)
4	Bolt
5	Upper Intake Manifold
6	Lower Intake Manifold
7	Intake Manifold Upper Gasket
A	Tighten to 20-30 N·m (15-22 Lb-Ft)

9300KG09

Exploded view of the upper intake manifold mounting—2002–04 3.0L (VIN U) engine

➡ **A special Synchro Positioning tool T95T-12200-A must be used when installing the CMP sensor housing.**

6. Attach the Synchro Position tool T95T-12200-A as follows:

a. Engage the CMP sensor housing vane into the radial slot of the tool.

b. Rotate the tool on the CMP sensor housing until the tool boss engages the notch in the CMP sensor housing.

c. Install the CMP sensor housing so the drive gear engagement occurs when the arrow on the locator tool is pointed about 75 degrees counterclockwise from the rear face of the cylinder block. This step will locate the CMP sensor electrical

connector in the same position as was noted on removal.

d. Install the hold-down clamp and tighten the bolt to 14–22 ft. lbs. (19–30 Nm).

7. Remove the Synchro Position tool.

8. Install or connect the following:
- CMP sensor
- No. 3 cylinder intake valve rocker arm and pushrod
- Valve covers
- Ignition coil and tighten the bolts to 30–40 ft. lbs. (40–55 Nm)
- Spark plug wires
- Upper intake manifold and tighten the bolts to 15–22 ft. lbs. (20–30 Nm)
- Engine control sensor wiring harness and bracket
- Upper alternator bracket
- Coolant hoses
- Coolant temperature indicator sender connector
- ECT sensor connector
- CMP sensor connector
- IAC valve connector
- TP sensor connector
- EGR vacuum solenoid
- EGR transducer
- EGR valve
- Vacuum lines
- Fuel lines
- Accelerator cable and bracket
- Cruise control cable
- Air cleaner and outlet tube
- Negative battery cable

9. Fill the cooling system.

10. Start the engine and check for leaks and proper operation.

2005 3.0L (VIN U) Engine

UPPER MANIFOLD

1. Before servicing the vehicle, refer to the precautions section.

7922KG17

Lower intake manifold torque sequence—2002–04 3.0L (VIN U) engine

2. Disconnect the negative battery cable.

3. Disconnect the tubes, loosen the clamps and remove the air cleaner outlet pipe.

4. Remove the snow shield.

5. Disconnect the accelerator cable, cruise control cable and throttle return spring from the throttle body.

6. Remove the bolts and position the accelerator cables and bracket aside.

7. Disconnect the Throttle Position (TP) sensor and the Intake Air Control (IAC) electrical connectors.

8. Detach the harness retainer from the throttle body stud bolt.

9. Disconnect the vacuum hose and the Evaporative Emissions (EVAP) return tube.

10. Remove the upper intake manifold support bracket bolt and nut.

11. Position the bracket and harness aside.

12. Loosen the nut and disconnect the Exhaust Gas Recirculation (EGR) tube from the EGR valve.

13. Disconnect the Positive Crankcase Ventilation (PCV) tube from the upper intake manifold.

14. Disconnect the fuel supply manifold pressure sensor vacuum tube.

15. Disconnect the PCV tube from the PCV valve and remove the tube.

16. Disconnect the EGR vacuum regulator solenoid and Intake Manifold Tuning valve (IMTV) electrical connectors.

17. Detach the pin-type harness retainer from the upper intake manifold.

18. Detach the spark plug wire holder from the valve cover stud and disconnect the radio interference capacitor electrical connector.

19. Remove the two nuts and position the EGR vacuum regulator solenoid and bracket aside.

20. Remove the bolts and the upper intake manifold.

To install:

➡**Clean all of the sealing surfaces. Inspect the upper intake manifold gaskets and install new gaskets as necessary.**

21. Install the upper intake manifold. Tighten bolts in 2 steps in the sequence shown:

    a. Step 1: hand-tighten the bolts.

    b. Step 2: Tighten the bolts to 89 inch lbs. (10 Nm).

22. Position the EGR vacuum regulator solenoid and bracket and install the nuts.

23. Attach the spark plug wire holder to

Upper intake manifold torque sequence—2005 3.0L (VIN U) engine

the valve cover stud and connect the radio interference capacitor electrical connector.

24. Connect the EGR vacuum regulator solenoid and IMRC electrical connectors.

25. Attach the pin-type harness retainer to the upper intake manifold.

26. Connect the PCV tube to the PCV valve.

27. Connect the fuel supply manifold pressure sensor vacuum tube.

28. Connect the PCV tube to the upper intake manifold.

29. Connect the EGR tube to the EGR valve and tighten the fitting to 41 ft. lbs. (55 Nm).

30. Position the upper intake manifold support bracket and install bolt and nut.

31. Connect the vacuum hose and the EVAP return tube.

32. Connect the TP sensor and the IAC electrical connectors.

33. Attach the harness retainer from the throttle body stud bolt.

34. Position the accelerator cable bracket and install the bolts.

35. Connect the accelerator cable, cruise control cable and throttle return spring to the throttle body.

36. Install the snow shield.

37. Install the air cleaner outlet pipe, connect the tubes and tighten the clamps.

38. Connect the negative battery cable.

LOWER MANIFOLD

1. Before servicing the vehicle, refer to the precautions section.

2. Remove the upper intake manifold.

3. Disconnect the fuel tube spring lock coupling.

4. Drain and recycle the engine coolant.

5. Remove the push rods.

6. Disconnect the Fuel Rail Pressure (FRP) sensor electrical and vacuum connections.

7. Disconnect the fuel injector electrical connectors.

8. Disconnect the heater hose and Engine Coolant Temperature (ECT) sender electrical connector.

9. Disconnect the ECT sensor electrical connector and the degas bottle hose.

10. Disconnect the upper radiator hose from the thermostat housing.

11. Remove the bolts and the lower intake manifold.

12. Remove the intake manifold gaskets and the end seals.

To install:

13. Using metal surface cleaner, clean all the sealing surfaces.

➡**If the lower intake manifold is not secured within four minutes, the sealant must be removed and the sealing area cleaned with metal surface cleaner. Allow to dry until there is no sign of wetness or four minutes, whichever is longer. Failure to follow this procedure can cause future oil leakage.**

Lower intake manifold torque sequence—2005 3.0L (VIN U) engine

14. Apply a 0.20–0.23 inch (5–6 mm) bead of silicone gasket and sealant at the four cylinder block-to-cylinder head seams.

15. Install the lower intake manifold gaskets and end seals.

16. Install the intake manifold. Tighten the bolts in two steps in the sequence shown:

　　a. Step 1: Tighten to 15 Nm (11 lb-ft).

　　b. Step 2: Tighten to 24 ft. lbs. (32 Nm.

17. Connect the upper radiator hose to the thermostat housing.

18. Connect the ECT sensor electrical connector and the degas bottle hose.

19. Connect the heater hose and the ECT sender electrical connector.

20. Connect the fuel injector electrical connectors.

21. Connect the FRP sensor electrical and vacuum connections.

22. Install the push rods.

23. Connect the fuel tube spring lock coupling.

24. Install the upper intake manifold.

25. Fill and bleed the engine cooling system

2002–04 3.0L (VIN S) Engine

1. Before servicing the vehicle, refer to the precautions section.

2. Disconnect or remove:
- Negative battery cable
- Windshield wipers and cowl top inner panels

● **REMOVE IN SEQUENCE SHOWN**

LOWER INTAKE MANIFOLD

9300KG11

Lower intake manifold bolt removal sequence—2002–04 3.0L (VIN S) engine

- Cruise control cable
- Accelerator cable and bracket
- Vacuum hoses
- Intake Air Control (IAC) valve supply hose
- IAC connector
- Throttle Position (TP) sensor connector
- Exhaust Gas Recirculation (EGR) valve
- Secondary air injection valve bracket
- Upper intake manifold
- Fuel lines
- Fuel injector wiring harness
- Intake Manifold Runner Control (IMRC) cable
- Spark plug wires
- Lower intake manifold

To install:

3. Install the lower intake manifold. Use new gaskets and tighten the bolts in sequence to 71–106 inch lbs. (8–12 Nm).

4. Install or connect the following:
- Spark plug wires
- IMRC cable
- Fuel injector wiring harness
- Fuel lines

5. Install the upper intake manifold. Use new gaskets and tighten the bolts in sequence to 71–106 inch lbs. (8–12 Nm).

6. Install or connect the following:
- Secondary air injection valve bracket
- EGR valve
- TP sensor connector
- IAC connector
- IAC valve supply hose
- Vacuum hoses
- Accelerator cable and bracket
- Cruise control cable
- Windshield wipers and cowl top inner panels
- Negative battery cable

7. Run the engine and check for leaks and proper engine operation.

2005 3.0L (VIN S) Engine

UPPER MANIFOLD

1. Before servicing the vehicle, refer to the precautions section.

2. Disconnect the negative battery cable.

3. Remove the air cleaner outlet pipe.

4. Remove the accelerator cable splash shield.

5. Disconnect the accelerator cable, the cruise control cable and the throttle return spring.

6. Remove the bolts and position the bracket and cables aside.

LOWER INTAKE MANIFOLD

7922KG16

Lower intake manifold torque sequence—2002–04 3.0L (VIN S) engine

7. Disconnect the Throttle Position (TP) sensor and the Idle Air Control (IAC) valve electrical connectors.

8. Detach the harness retainer from the throttle body stud bolt.

9. Disconnect the heated throttle body coolant tubes.

10. Disconnect the Positive Crankcase Ventilation (PCV) hose.

11. Disconnect the vacuum supply hose.

12. Disconnect the brake booster and Evaporative Emissions (EVAP) vacuum hoses.

13. Remove the Exhaust Gas Recirculation (EGR) valve as follows:

　　a. Disconnect the vacuum hose.

　　b. Remove the nuts and bolts.

　　c. Remove the EGR valve.

14. Remove the EGR Vacuum Regulator (EVR) valve as follows:

　　a. Disconnect the vacuum hose.

　　b. Disconnect the electrical connector.

　　c. Remove the bolts.

　　d. Remove the valve.

15. Remove the eight bolts and the upper intake manifold.

16. Remove and discard the gaskets.

17. Clean all mating surfaces.

To install:

18. Position new gaskets in the upper intake manifold.

19. Position the upper intake manifold and install the bolts in the sequence shown. Tighten to 89 inch lbs. (10 Nm).

20. Install the EGR vacuum regulator valve and tighten the retainers to 53 inch lbs. (6 Nm).

CAUTION: CAMSHAFTS MUST BE IN TIME WITH CRANKSHAFT PRIOR TO INSTALLATION OF ROCKER ARMS

Journal Cap

Intake Camshaft

Rocker Arm

Tappet

NOTE: VALVE TAPPET MUST BE PUMPED UP PRIOR TO INSTALLATION

Thrust Cap

Exhaust Camshaft

CAUTION: REMOVE CYLINDER HEAD JOURNAL THRUST CAPS FIRST. INSTALL CYLINDER HEAD JOURNAL THRUST CAPS LAST

CAUTION: ROTATE CRANKSHAFT KEYWAY TO 11 O'CLOCK POSITION PRIOR TO INSTALLATION OF LH CAMSHAFT AND ROCKER ARMS. ROTATE CRANKSHAFT KEYWAY CLOCKWISE TO 3 O'CLOCK POSITION PRIOR TO INSTALLATION OF RH CAMSHAFT AND ROCKER ARMS AND CAMSHAFT JOURNAL THRUST CAPS

7922KG10

Upper intake manifold torque sequence—2002–04 3.0L (VIN S) engine

10 Nm (89 lb-ft)

06017-TAUR-G14

Upper intake manifold torque sequence—2005 3.0L (VIN S) engine

06017-TAUR-G15

Lower intake manifold loosening sequence—2005 3.0L (VIN S) engine

10 Nm (89 lb-in)

06017-TAUR-G16

Lower intake manifold tightening sequence—2005 3.0L (VIN S) engine

21. Install the EGR valve. Tighten the bolts to 18 ft. lbs. (25 Nm) and the nuts to 89 inch lbs. (10 Nm).

22. Connect the brake booster and PCV vacuum hoses.

23. Connect the vacuum supply hose.

24. Connect the PCV valve hose.

25. Connect the TP and IAC valve electrical connectors.

26. Attach the harness retainer to the throttle body stud bolt.

27. Connect the heated throttle body coolant tubes.

28. Position the bracket and cables and install the bolts.

29. Connect the throttle return spring, accelerator cable and the cruise control cable.

30. Install the accelerator cable splash shield.

31. Install the air cleaner outlet pipe.

32. Connect the negative battery cable.

LOWER MANIFOLD

1. Before servicing the vehicle, refer to the precautions section.

2. Release the fuel system pressure.

3. Disconnect the fuel hose quick release coupling from the fuel supply manifold.

4. Remove the upper intake manifold.

5. Disconnect the fuel injection pressure and temperature sensor vacuum tube and the electrical connector.

6. Disconnect the fuel injector electrical connectors.

7. Remove the intake manifold bolts in the sequence shown, and remove the lower intake manifold.

8. Remove and discard the gaskets from the lower intake manifold.

To install:

➡️ **Clean and inspect all mating surfaces and install new gaskets in the lower intake manifold.**

9. Install the lower intake manifold and tighten the bolts in the sequence shown to 89 inch lbs. (10 Nm).

10. Connect the fuel injector electrical connectors.

11. Connect the fuel injection pressure and temperature sensor vacuum tube and the electrical connector.

12. Connect the fuel hose quick release coupling to the fuel supply manifold.

13. Install the upper intake manifold.

Exhaust Manifold

REMOVAL & INSTALLATION

3.0L (VIN U) Engine

RIGHT SIDE

1. Before servicing the vehicle, refer to the precautions section.

2. Remove or disconnect the following:
- Negative battery cable
- Cowl vent screen and extension
- Heated Oxygen (HO2S) sensor connector
- Exhaust Gas Recirculation (EGR) tube

- Exhaust manifold heat shield
- Catalytic converter
- Exhaust manifold

To install:

3. Install the exhaust manifold with a new gasket. Tighten the bolts in sequence as follows:
   a. Step 1: 89 inch lbs. (10 Nm)
   b. Step 2: 16 ft. lbs. (22 Nm)

4. Install or connect the following:
- Catalytic converter and tighten the bolts to 30 ft. lbs. (40 Nm)
- Exhaust manifold heat shield and tighten the bolts to 89 inch lbs. (10 Nm)
- EGR tube
- HO2S sensor connector
- Cowl vent screen and extension
- Negative battery cable

5. Start the engine and check for leaks.

LEFT SIDE

1. Before servicing the vehicle, refer to the precautions section.

2. Remove or disconnect the following:
- Negative battery cable
- Heated Oxygen (HO2S) sensor connector
- Oil dipstick tube
- Power steering pressure line
- Secondary air injection tube, if equipped
- Dual converter Y-pipe
- Exhaust manifold

To install:

3. Install the exhaust manifold with a new gasket. Tighten the bolts in sequence as follows:
   a. Step 1: 89 inch lbs. (10 Nm)
   b. Step 2: 16 ft. lbs. (22 Nm)

4. Install or connect the following:
- Dual converter Y-pipe and tighten the bolts to 30 ft. lbs. (40 Nm)
- Secondary air injection tube, if equipped
- Power steering pressure line
- Oil dipstick tube
- HO2S sensor connector
- Negative battery cable

5. Start the engine and check for leaks.

2002–04 3.0L (VIN S) Engine

1. Before servicing the vehicle, refer to the precautions section.

2. Remove or disconnect the following:
- Negative battery cable
- Upper intake manifold
- Right exhaust manifold heat shield
- Ignition coil
- Exhaust Gas Recirculation (EGR) tube
- Heated Oxygen (HO2S) sensors
- Secondary air injection tube
- Dual converter Y-pipe
- Exhaust manifolds

Right side Heated Oxygen (HO2S) sensor connector—3.0L (VIN U) engine

Left side Heated Oxygen (HO2S) sensor connector—3.0L (VIN U) engine

Right side exhaust manifold mounting bolt tightening sequence—2002–04 3.0L (VIN S) engine

Right exhaust manifold torque sequence—3.0L (VIN U) engine

Left exhaust manifold torque sequence—3.0L (VIN U) engine

Left side exhaust manifold mounting bolt tightening sequence—2002–04 3.0L (VIN S) engine

To install:

3. Install or connect the following:
- Exhaust manifolds. Use new gaskets and tighten the bolts in sequence to 13–16 ft. lbs. (18–22 Nm).
- Dual converter Y-pipe
- Secondary air injection tube
- HO$_2$S sensors
- EGR tube
- Ignition coil
- Right exhaust manifold heat shield
- Upper intake manifold
- Negative battery cable

4. Run the engine and check for exhaust leaks and proper operation.

2005 3.0L (VIN S) Engine

LEFT SIDE

1. Before servicing the vehicle, refer to the precautions section.
2. Remove the coolant tube bracket bolt.
3. Disconnect and unclip the Heated Oxygen (HO$_2$S) sensor electrical connector from the coolant tube.
4. Remove the dual converter Y-pipe.
5. Remove the front splash shield.
6. Remove the coolant tube lower bracket bolt.
7. Remove the nuts and the exhaust manifold.
8. Discard the gasket.

To install:

9. Installation is the reverse of removal. Tighten the manifold bolts in the sequence illustrated to 15 ft. lbs. (20 Nm).

RIGHT SIDE

1. Before servicing the vehicle, refer to the precautions section.
2. Disconnect the negative battery cable.
3. Remove the wiper mounting arm and pivot shaft.
4. Remove the air cleaner outlet pipe.
5. Remove the three retaining screws and the right hand cowl top inner panel.
6. Remove the two retaining nuts and

Left exhaust manifold tightening sequence—2005 3.0L (VIN S) engine

06017-TAUR-G17

06017-TAUR-G18

Right exhaust manifold loosening sequence—2005 3.0L (VIN S) engine

20 Nm (15 lb-ft)

06017-TAUR-G19

Right exhaust manifold tightening sequence—2005 3.0L (VIN S) engine

position the vacuum outlet manifold assembly aside.

7. Unseat the windshield wiper motor wiring harness grommet. Pull the wiring harness through the grommet.

8. Disconnect the cowl drain tube, remove the three retaining screws and the left hand cowl top inner panel.

9. Disconnect the differential pressure feedback Exhaust Gas Recirculation (EGR) sensor and the Heated Oxygen (HO$_2$S) sensor electrical connectors.

10. Remove the differential pressure feedback EGR sensor tube and bracket as follows:

   a. Detach the electrical connector pin-type retainer.

   b. Remove the bolts from the differential pressure feedback EGR bracket.

   c. Disconnect and remove the differential pressure feedback EGR and bracket from the vacuum tubes.

11. Remove the rear support brace.
12. Remove the steering gear heat shield.
13. Remove the EGR tube.
14. Remove the dual converter Y-pipe.
15. Remove the nuts in the sequence shown and remove the exhaust manifold.
16. Discard the gasket.

To install:

17. Position a new gasket and install the exhaust manifold nuts in the sequence shown. Tighten to 15 ft. lbs. (20 Nm).

18. Install the dual converter Y-pipe.

19. Install the EGR tube.
20. Install the steering gear heat shield.
21. Install the rear support brace.
Position the rear support brace and tighten the retainers to 35 ft. lbs. (47 Nm).
22. Install the differential pressure feedback EGR valve sensor and bracket.
23. Attach the HO$_2$S electrical connector.
24. Connect the differential pressure feedback EGR sensor and the HO$_2$S electrical connectors
25. Connect the negative battery cable.

Camshaft and Valve Lifters

REMOVAL & INSTALLATION

3.0L (VIN U) Engine

➡️ **If the rocker arms, pushrods or lifters are to be reused, they must be installed in the same positions they were removed from.**

1. Before servicing the vehicle, refer to the precautions section.
2. Remove the engine from the vehicle and mount it on an engine stand.
3. Remove or disconnect the following:
- Valve covers
- Rocker arms and pushrods
- Accessory drive belt and tensioner
- Alternator and brackets
- Intake manifold
- Valve lifter guide plate
- Valve lifters
- Crankshaft damper
- Oil pan
- Front cover
- Timing chain and gears
- Camshaft thrust plate
- Camshaft

To install:

➡️ **If replacing the camshaft, the valve lifters must also be replaced.**

4. Install or connect the following:
- Camshaft
- Camshaft thrust plate and tighten the bolts to 84 inch lbs. (10 Nm)
- Timing chain and gears. Align the timing marks on the sprockets.
- Front cover
- Oil pan
- Crankshaft damper
- Valve lifters
- Valve lifter guide plate
- Intake manifold
- Alternator and brackets
- Accessory drive belt and tensioner
- Rocker arms and pushrods
- Valve covers

Tighten the rocker arm nuts according to the position of the camshaft—3.0L (VIN U) engine

5. Install the engine assembly into the vehicle.

6. Start the engine and check for leaks and proper engine operation.

2002–04 3.0L (VIN S) Engine

1. Before servicing the vehicle, refer to the precautions section.

2. Remove the engine from the vehicle and mount it on an engine stand.

3. Remove or disconnect the following:
- Upper intake manifold
- Valve covers

➡ **If the rocker arms are to be reused, they must be kept in order so that they can be installed in their original positions.**

- Rocker arms
- Accessory drive belt
- Power steering pump
- Water pump
- A/C compressor and bracket

➡ **The crankshaft accessory drive pulley shaft has left-hand threads. Rotate the pulley shaft clockwise to remove.**

- Crankshaft pulley
- Oil pan
- Front cover
- Timing chains

➡ **The camshaft journal thrust caps must be removed first, before loosening the remaining camshaft journal cap bolts, to ensure that the camshaft journal thrust caps are not damaged.**

Exploded view of camshaft mounting—2002–04 3.0L (VIN S) engine

Camshaft journal bolt removal sequence—2002–04 3.0L (VIN S) engine

Camshaft journal bolt tightening sequence—2002–04 3.0L (VIN S) engine

4. Loosen the camshaft journal cap bolts in sequence and in several passes to allow the camshaft to raise off the cylinder head evenly.

5. Remove the journal caps and the camshafts.

➡If the hydraulic lifters are to be reused, they must be kept in order so that they can be installed in their original positions.

6. Remove the hydraulic lifters.

To install:

7. Rotate the crankshaft so the keyway is at the 11 o'clock position for installation of the camshafts.

8. Install or connect the following:
- Hydraulic lifters in their original positions
- Camshafts with the timing marks (marked RFF) on the back of the camshaft sprockets aligned

➡The camshaft journal caps and cylinder heads are numbered to ensure that they are assembled in their original positions.

➡Do not install the camshaft journal thrust caps until the rocker arms and timing chains have been installed and the camshaft journal caps are tightened into position.

- Camshaft journal caps in their original positions
- Timing chains

9. Tighten the camshaft journal cap bolts, in sequence, to 71–106 inch lbs. (8–12 Nm).

10. Install the thrust caps and tighten the bolts to 71–106 inch lbs. (8–12 Nm).

11. Install or connect the following:
- Front cover
- Oil pan
- Crankshaft pulley
- A/C compressor and bracket
- Water pump
- Power steering pump
- Accessory drive belt
- Rocker arms in their original positions
- Valve covers
- Upper intake manifold
- Engine assembly into the vehicle

12. Run the engine and check for leaks and proper operation.

2005 3.0L (VIN S) Engine

LEFT SIDE

1. Before servicing the vehicle, refer to the precautions section.

2. Remove the timing drive components.

3. Remove the water pump belt.

✳✳ CAUTION

The OEM pulley is pressed on 0.18 inch (4.74 mm) past flush of the camshaft. The service pulley gets pressed on flush to the end of the camshaft.

4. Using the tools illustrated, remove the water pump drive pulley and discard.

✳✳ CAUTION

In order to make sure of correct sealing, do not scratch the camshaft.

5. Using the tool illustrated, remove the camshaft oil seal and discard.

6. Remove the intake camshaft oil seal retainer and discard the press-in-place gasket.

Using the tools illustrated, remove the water pump drive pulley—2005 3.0L (VIN S) engine

Using the tools illustrated, remove the camshaft oil seal—2005 3.0L (VIN S) engine

Left camshaft cap loosening sequence—2005 3.0L (VIN S) engine

06017-TAUR-G22

12. Lubricate the bearing surfaces of the camshaft bearing caps with engine oil. Install the bearing caps and loosely install the bolts.

13. Lubricate the bearing surfaces of the camshaft bearing thrust caps with engine oil. Install the bearing thrust caps and loosely install the bolts.

> ✳✳ **CAUTION**
>
> The camshaft bearing caps are positional and must be installed in their original locations and orientations or engine damage can occur.

14. Verify that the camshaft bearing caps are in the correct locations.

15. Tighten the bolts in the sequence shown.

Left camshaft alignment—2005 3.0L (VIN S) engine

06017-TAUR-G23

Verify the left camshaft bearing caps are in their correct locations—2005 3.0L (VIN S) engine

06017-TAUR-G24

> ✳✳ **CAUTION**
>
> Record the camshaft bearing cap locations. The camshaft bearing caps are positional and must be installed in their original locations and orientations or engine damage can occur.

7. Loosen the bolts for the left camshaft thrust caps.

8. Loosen the left camshaft cap bolts in sequence to allow the camshafts to rise from the cylinder head and remove the caps.

9. Remove the camshafts.

To install:

➥ Be sure the camshaft bearing caps are installed in their original positions.

10. Lubricate camshafts with engine oil and carefully position the camshafts into the cylinder head.

11. Align the camshafts as illustrated

> ✳✳ **CAUTION**
>
> Do not install the camshaft journal thrust caps until all of the camshaft bearing caps have been installed, or damage to the thrust caps can occur.

8 Nm (71 lb-in)

06017-TAUR-G25

Left camshaft cap torque sequence—2005 3.0L (VIN S) engine

Using the tools illustrated positioned over the oil seal, turn to install the seal—2005 3.0L (VIN S) engine

Using the tools illustrated , install the water pump drive pulley—2005 3.0L (VIN S) engine

➡ **Clean and degrease the sealing surfaces with metal surface cleaner.**

16. Install a new press-in-place gasket and install the camshaft oil seal retainer. Tighten to 89 inch lbs. (10 Nm).

17. Lubricate the camshaft oil seal with engine oil.

18. Install the new camshaft oil seal.

19. Using the tools illustrated positioned over the oil seal, turn to install the seal.

> ⚡ **CAUTION**
>
> **The OEM pulley is pressed on 0.18 inch (4.74 mm) past flush of the end of the camshaft. The service pulley gets pressed on flush to the end of the camshaft.**

20. Using the tool illustrated, install the water pump drive pulley.

21. Install the water pump belt.

22. Install the timing drive components.

RIGHT SIDE

1. Before servicing the vehicle, refer to the precautions section.

2. Remove the timing drive components.

> ⚡ **CAUTION**
>
> **Record the camshaft bearing cap location. The camshaft bearing caps are positional and must be installed**

Right camshaft cap loosening sequence—2005 3.0L (VIN S) engine

in their original locations and orientations or engine damage can occur.

3. Loosen the right hand camshaft thrust cap bolts.

4. Loosen the right hand camshaft cap bolts in sequence to allow the camshafts to rise from the cylinder head and remove the caps.

5. Remove the camshafts.

To install:

➡ **Be sure the camshaft bearing caps are installed in their original positions.**

Right camshaft alignment—2005 3.0L (VIN S) engine

Right camshaft cap torque sequence—2005 3.0L (VIN S) engine

6. Lubricate camshafts with engine oil and carefully position the camshafts into the cylinder head.

7. Align the camshafts as illustrated

✳ CAUTION

Do not install the camshaft journal thrust caps until all of the camshaft bearing caps have been installed, or damage to the thrust caps can occur.

8. Lubricate the bearing surfaces of the camshaft bearing caps with engine oil. Install the bearing caps and loosely install the bolts.

9. Lubricate the bearing surfaces of the camshaft bearing thrust caps with engine oil. Install the bearing thrust caps and loosely install the bolts.

✳ CAUTION

The camshaft bearing caps are positional and must be installed in their original locations and orientations or engine damage can occur.

10. Verify that the camshaft bearing caps are in the correct locations.

11. Tighten the right camshaft bolts in the sequence shown to 89 inch lbs. (10 Nm).

12. Install the timing drive components

Valve Lash

ADJUSTMENT

3.0L (VIN U and S) Engines

The lash adjusters (valve tappets), are hydraulic and are not adjustable.

Starter Motor

REMOVAL & INSTALLATION

1. Before servicing the vehicle, refer to the precautions section.

2. Remove or disconnect the following:
- Negative battery cable
- Splash shield
- Starter electrical connectors
- Starter

To install:

3. Install or connect the following:
- Starter and tighten the bolts to 18 ft. lbs. (25 Nm)
- Starter electrical connectors and tighten the battery cable nut to 80–123 inch lbs. (9–14 Nm)

7 Nm (62 lb-in)

Remove the front splash shield for access

Starter motor—3.0L (VIN U) engine

Starter motor—3.0L (VIN S) engine

- Splash shield
- Negative battery cable

Oil Pan

REMOVAL & INSTALLATION

2002–04 3.0L (VIN U) Engine

1. Before servicing the vehicle, refer to the precautions section.

2. Drain the engine oil.

3. Remove or disconnect the following:
- Negative battery cable
- Oil dipstick tube
- Low oil level sensor, if equipped
- Starter motor and brace
- Heated Oxygen (HO_2S) sensor connectors
- Dual converter Y-pipe
- Engine rear plate
- Oil pan

To install:

4. Install the oil pan with new gaskets. Tighten the bolts as follows:

a. Step 1: Tighten the 4 corner bolts to 108 inch lbs. (12 Nm).

b. Step 2: Tighten the remaining bolts to 108 inch lbs. (12 Nm).

c. Step 3: Retighten all bolts to 108 inch lbs. (12 Nm).

5. Install or connect the following:
- Engine rear plate
- Dual converter Y-pipe
- HO_2S sensor connectors
- Starter motor and brace
- Low oil level sensor, if equipped
- Oil dipstick tube
- Negative battery cable

6. Fill the crankcase.

7. Start the engine and check for leaks.

Oil pan—3.0L (VIN U) engine

2005 3.0L (VIN U) Engine

1. Before servicing the vehicle, refer to the precautions section.
2. Disconnect the negative battery cable.
3. Remove the bolts and engine roll restrictor.
4. Remove the bolt, ground strap and nut.
5. Remove the nuts from the front exhaust manifold.
6. Remove the bolt from the rear upper manifold.
7. Remove the nut and disconnect the ground strap electrical connector.
8. Remove the nut and position the heat shield aside.
9. Loosen the rear lower exhaust manifold bolt and position the Y-pipe aside.
10. Remove the bolts and position the starter aside.
11. Remove the bolt and the engine rear plate.
12. Remove the oil pan drain plug and drain the engine oil.
13. Install the drain plug when finished and tighten to 10 ft. lbs. (13 Nm).
14. Remove the bolts and the oil pan.
15. Remove the oil pan gasket from the oil pan.
16. Using metal surface cleaner, clean the oil pan sealing surfaces.

To install:

17. Install the oil pan gasket onto the oil pan.

➡If the oil pan is not secured within four minutes, the sealant must be removed and the sealing area cleaned with metal surface cleaner. Allow to dry until there is no sign of wetness or four minutes, whichever is longer. Failure to follow this procedure can cause future oil leakage.

18. Apply a bead of silicone gasket and sealant in two places where the cylinder block meets the front cover and in two places where the rear main bearing cap meets the cylinder block.

➡Make sure the fasteners are clean and dry prior to installation.

19. Tighten the oil pan bolts in sequence to 9 ft. lbs. (12 Nm).
20. Install the engine rear plate and the bolt.
21. Position the starter and install the bolts.
22. Position the Y-pipe and tighten the rear lower exhaust manifold bolt to 35 ft. lbs. (48 Nm).
23. Position the heat shield and install the nut.

06017-TAUR-G31

Apply a bead of silicone gasket and sealant in two places where the cylinder block meets the front cover and in two places where the rear main bearing cap meets the cylinder block—2005 3.0L (VIN U) engine

06017-TAUR-G32

Oil pan bolt torque sequence—2005 3.0L (VIN U) engine

Oil Pan Removal Sequence

Item	Description
1	Upper Cylinder Block
2	Lower Cylinder Block (Part of 6010)
3	Oil Pan
4	Stud Bolt (5 Req'd)
5	Bolt (10 Req'd)
6	Oil Pan Gasket
7	Engine Front Cover

✱ LOCATION OF STUDS

● REMOVE BOLTS/STUDS IN SEQUENCE SHOWN

9300KG14

Oil pan bolt removal sequence—2002–04 3.0L (VIN S) engine

6. Start the engine and check for leaks and proper operation.

2005 3.0L (VIN S) Engine

1. Before servicing the vehicle, refer to the precautions section.
2. Disconnect the negative battery cable.

Apply a 0.4 inch (10 mm) dot of silicone gasket and sealant to the front cover-to-cylinder block mating surface—2005 3.0L (VIN S) engine

Oil pan bolt tightening sequence—2002–04 3.0L (VIN S) engine

24. Connect the ground strap electrical connector and install the nut.

25. Install the bolt in the rear upper manifold tighten to bolt to 35 ft. lbs. (48 Nm).

26. Install the nuts on the front exhaust manifold. Tighten to bolt to 30 ft. lbs. (40 Nm).

27. Install the nut, ground strap and bolt.

28. Install the engine roll restrictor and the bolts. Tighten to bolt to 35 ft. lbs. (48 Nm).

29. Connect the negative battery cable.

30. Fill the engine with clean engine oil.

31. Start the engine and check for oil leaks around the oil pan

2002–04 3.0L (VIN S) Engine

1. Before servicing the vehicle, refer to the precautions section.
2. Drain the engine oil.
3. Remove or disconnect the following:
   • Negative battery cable
   • Dual converter Y-pipe
   • Transaxle support bracket
   • Oil pan by following the bolt removal sequence shown

To install:

4. Install or connect the following:
   • Oil pan. Use a new gasket and tighten the bolts in sequence to 15–22 ft. lbs. (20–30 Nm).
   • Transaxle support bracket. Tighten the nuts to 71–106 inch lbs. (8–12 Nm) and the bolts to 15–22 ft. lbs. (20–30 Nm).
   • Dual converter Y-pipe
   • Negative battery cable
5. Fill the crankcase.

Loosely install the oil pan bolts and the studs in the sequence shown—2005 3.0L (VIN S) engine

Tighten the oil pan bolts and the studs in the sequence shown—2005 3.0L (VIN S) engine

3. Remove the right front inner wheel splash shield.

4. Remove the dual converter Y-pipe.

5. Secure the exhaust system and remove the jackstand.

6. Drain the engine oil. Install and tighten the drain plug to 20 ft. lbs. (27 Nm).

7. Remove and discard the oil filter.

8. Remove the bolts, nuts and the engine-to-transaxle bracket.

9. Remove the torque converter inspection cover.

10. Remove the 18 bolts and the oil pan.

To install:

11. Clean all sealing surfaces on the engine and the oil pan with metal surface cleaner.

Install a new gasket on the oil pan.

12. Apply a 0.4 inch (10 mm) dot of silicone gasket and sealant to the front cover-to-cylinder block mating surface.

13. Install a new gasket, position the oil pan and loosely install the bolts and the studs in the sequence illustrated.

14. Install the oil pan to transaxle housing bolts and tighten to 30 ft. lbs. (40 Nm).

15. Tighten the oil pan bolts and studs in sequence to 18 ft. lbs. (25 Nm).

16. Tighten the oil pan-to-transaxle bolts to 30 ft. lbs. (40 Nm).

17. Install the torque converter inspection cover.

18. Position the engine-to-transaxle bracket and install the nuts and the bolt. Tighten the bolts to 18 ft. lbs. (25 Nm) and the nuts to 89 inch lbs. (10 Nm).

19. Lubricate the oil filter O-ring with clean engine oil and install the new oil filter.

20. Position a jackstand and release the exhaust system.

21. Install the dual converter Y-pipe.

22. Install the front inner wheel splash shield.

23. Install the exhaust manifold-to-pipe nuts and tighten to 30 ft. lbs. (40 Nm).

24. Fill the crankcase with clean engine oil.

Oil Pump

REMOVAL & INSTALLATION

2002–04 3.0L (VIN U) Engine

1. Before servicing the vehicle, refer to the precautions section.

2. Drain the engine oil.

3. Remove or disconnect the following:
   - Negative battery cable
   - Oil dipstick tube
   - Low oil level sensor, if equipped
   - Starter motor and brace

Exploded view of the oil pump mounting—3.0L (VIN U) engine

Oil pump bolt removal sequence—2002–04 3.0L (VIN S) engine

- Heated Oxygen Sensor (HO$_2$S) connectors
- Dual converter Y-pipe
- Engine rear plate
- Oil pan
- Oil pump

4. Separate the intermediate shaft from the oil pump.

To install:

5. Insert the intermediate shaft into the oil pump until the snapring seats.

6. Install or connect the following:
- Oil pump and tighten the bolt to 30–40 ft. lbs. (40–55 Nm)
- Oil pan
- Engine rear plate
- Dual converter Y-pipe
- HO$_2$S sensor connectors
- Starter motor and brace
- Low oil level sensor, if equipped
- Oil dipstick tube
- Negative battery cable

7. Fill the crankcase.

8. Start the engine and check for leaks and proper operation.

2005 3.0L (VIN U) Engine

1. Before servicing the vehicle, refer to the precautions section.

2. Remove the oil pan.

3. Remove the bolts and the oil pump.

4. Clean all the sealing surfaces.

To install:

5. Installation is the reverse of removal, tighten the oil pump bolts to 35 ft. lbs. (48 Nm).

2002–04 3.0L (VIN S) Engine

1. Before servicing the vehicle, refer to the precautions section.

2. Remove the engine from the vehicle and mount it on an engine stand.

3. Remove or disconnect the following:
- Upper intake manifold
- Valve covers
- Accessory drive belt
- Power steering pump
- Alternator
- Water pump
- A/C compressor and bracket

➡The crankshaft accessory drive pulley shaft has left-hand threads. Rotate the pulley shaft clockwise to remove.

- Crankshaft pulley
- Oil pan
- Oil pump screen and tube
- Front cover
- Timing chains
- Crankshaft timing gears
- Oil pump

Item	Description
1	Cylinder Block
2	Oil Pan Baffle
3	Nut (5 Req'd)
4	Nut (2 Req'd)
5	Oil Pump Screen Cover and Tube
6	Oil Pump Inlet Tube Gasket
7	Oil Pump
A	Tighten to 14-25 Nm (11-18 Lb-Ft)
B	Tighten to 8-14 Nm (71-123 Lb-In)

9300KG16

Exploded view of the oil pump pickup tube and screen mounting—3.0L (VIN S) engine

7922KG43

Oil pump bolt tightening sequence—2002-04 3.0L (VIN S) engine

To install:

4. Install or connect the following:
- Oil pump and tighten the bolts in sequence to 71–106 inch lbs. (8–12 Nm)
- Crankshaft timing gears
- Timing chains
- Front cover
- Oil pump screen and tube. Tighten the bolts to 71–106 inch lbs. (8–12 Nm), and the nut to 15–22 ft. lbs. (20–30 Nm).
- Oil pan
- Crankshaft pulley

➡ **The air conditioning mounting bracket bolts are torque-to-yield bolts and must be replaced when removed.**

- A/C compressor and bracket
- Water pump
- Alternator
- Power steering pump
- Accessory drive belt
- Valve covers
- Upper intake manifold
- Engine in the vehicle.

5. Fill the crankcase.
6. Start the engine and check for leaks and proper operation.

2005 3.0L (VIN S) Engine

1. Before servicing the vehicle, refer to the precautions section.
2. Remove the oil pan.

Oil pump loosening sequence—2005 3.0L (VIN S) engine

Oil pump tightening sequence—2005 3.0L (VIN S) engine

3. Remove the nut, bolts and the oil pump screen and pick-up tube.
4. Remove and inspect the O-ring seal.
5. Remove the pump bolts in the sequence shown and remove the oil pump.

To install:

6. Position the oil pump and install the bolts in the sequence shown. Tighten to 89 inch lbs. (10 Nm).
7. Install a new O-ring seal.
8. Install the oil pump screen and pick-up tube. Tighten the nut in two steps:
   a. Step 1: Tighten to 89 inch lbs. (10 Nm).
   b. Step 2: Tighten an additional 45 degrees.
9. Install the oil pan.

Rear Main Seal

REMOVAL & INSTALLATION

3.0L (VIN U) Engine

1. Before servicing the vehicle, refer to the precautions section.
2. Remove the negative battery cable, transaxle and flexplate.
3. Use a sharp awl and punch a hole into the rear main seal metal surface between the seal lip and the cylinder block.

Jet Plug Remover 310-005—3.0L (VIN U) engine

Rear Seal Replacer 303-323—3.0L (VIN U) engine

4. Screw the threaded end of Jet Plug Remover 310-005 into the seal. Use the Jet Plug Remover to remove the rear main seal.

To install:

5. Position the rear oil seal on Rear Seal Replacer 303-323. Position the tool and seal on the rear of the engine. Tighten the bolts alternately to seat the rear main oil seal.

Rear Crankshaft Seal Remover T95P-6701-EH—2002–04 3.0L (VIN S) engine

FRONT OF ENGINE

CRANKSHAFT REAR OIL SEAL 6701

SEAL REPLACER T82L-6701-A

CRANKSHAFT 6303

REAR CRANKSEAL REPLACER ADAPTER T91P-6701-A

9346KG30

Rear Seal Replacer T82L-6701-A and Adapters T91P-6701-A—2002–04 3.0L (VIN S) engine

6. Install the flexplate and tighten the bolts to 54–64 ft. lbs. (73–87 Nm).

7. Install the transaxle and the negative battery cable.

8. Check all fluid levels and fill as needed.

9. Start the engine and check for leaks.

2002–04 3.0L (VIN S) Engine

1. Before servicing the vehicle, refer to the precautions section.

2. Attach an engine support fixture to the engine lifting eyes.

3. Remove or disconnect the following:

- Negative battery cable
- Transaxle
- Flexplate

4. Use Seal Remover T95P-6701-EH and a slide hammer to remove the rear crankshaft seal.

To install:

5. Install or connect the following:

- Rear main seal flush with the cylinder block surface. Use Rear Main Seal Replacer T82L-6701-A and Adapter Bolts T91P-6701-A.
- Flexplate and tighten the bolts to 54–64 ft. lbs. (73–87 Nm)

307-005

303-519

06017-TAUR-G35

Using the tools illustrated, remove the crankshaft rear oil seal—2005 3.0L (VIN S) engine

303-384

303-178

06017-TAUR-G36

Using the tools illustrated, install the crankshaft rear oil seal—2005 3.0L (VIN S) engine

- Transaxle
- Negative battery cable

6. Run the engine and check for leaks.

2005 3.0L (VIN S) Engine

1. Before servicing the vehicle, refer to the precautions section.

2. Remove the negative battery cable, transaxle and flexplate.

3. Using the tools illustrated, remove the crankshaft rear oil seal and discard.

To install:

➡**Clean all sealing surfaces with metal surface cleaner. Lubricate the crankshaft rear oil seal lips with clean engine oil.**

4. Using the tools illustrated, install the crankshaft rear oil seal.

5. Install the flexplate and tighten the bolts to 549 ft. lbs. (80 Nm)

6. Install the transaxle.

7. Connect the negative battery cable.

8. Run the engine and check for leaks.

Timing Chain, Sprockets and Front Cover and Seal

REMOVAL & INSTALLATION

2002–04 3.0L (VIN U) Engine

1. Before servicing the vehicle, refer to the precautions section.

2. Drain the engine oil and the cooling system.

3. Remove or disconnect the following:

- Negative battery cable
- Accessory drive belt
- Accessory drive belt tensioner and idler pulley
- Water pump pulley
- Crankshaft pulley
- Crankshaft Position (CKP) sensor
- Coolant hoses
- Dual converter Y-pipe
- Oil pan
- Front crankshaft seal
- Front cover

4. Rotate the crankshaft until the No. 1 piston is at Top Dead Center (TDC) and the timing sprocket marks are aligned.

5. Remove the timing chain and sprockets.

To install:

➡**The camshaft bolt has a drilled oil passage in it for timing chain lubrication. Prior to installation, clean the passage and be sure it is clear. Never replace the camshaft bolt with a standard bolt.**

Timing chain alignment marks—2002–04 3.0L (VIN U) engine

6. Install the timing chain and sprockets with the timing marks aligned. Tighten the camshaft sprocket bolt to 46 ft. lbs. (63 Nm).

7. Use a new gasket and install the front cover. Apply sealant to bolts 1, 2 and 3. Tighten bolts 1–10 to 19 ft. lbs. (25 Nm) and bolts 11–15 to 84 inch lbs. (10 Nm).

8. Install or connect the following:
- Front crankshaft seal
- Oil pan
- Dual converter Y-pipe
- Coolant hoses
- CKP sensor
- Crankshaft pulley and tighten the damper bolt to 107 ft. lbs. (145 Nm)
- Water pump pulley and tighten the bolts to 16 ft. lbs. (21 Nm)

Fastener And Hole No.	Fasteners		Torque Specifications	
	Size	Fastener Application	N·m	LB-FT
1A	M8 x 1.25 x 43.5	F/C TO BLOCK	20-30	15-22
2A	M8 x 1.25 x 43.5	F/C TO BLOCK	20-30	15-22
3A	M8 x 1.25 x 73	W/P & F/C TO BLOCK	20-30	15-22
4A	M8 x 1.25 x 104.3	W/P & F/C TO BLOCK	20-30	15-22
5A	M8 x 1.25 x 73	F/C TO BLOCK	20-30	15-22
6A	M8 x 1.25 x 73	W/P & F/C TO BLOCK	20-30	15-22
7A	M8 x 1.25 x 73	W/P & F/C TO BLOCK	20-30	15-22
8A	M8 x 1.25 x 104.3	W/P & F/C TO BLOCK	20-30	15-22
9A	M8 x 1.25 x 104.3	W/P & F/C TO BLOCK	20-30	15-22
10A	M8 x 1.25 x 52	F/C TO BLOCK	20-30	15-22
11B	M6 x 1 x 28.5	W/P TO F/C	8-12	71-106 (lb-in)
12B	M6 x 1 x 28.5	W/P TO F/C	8-12	71-106 (lb-in)
13B	M6 x 1 x 28.5	W/P TO F/C	8-12	71-106 (lb-in)
14B	M6 x 1 x 28.5	W/P TO F/C	8-12	71-106 (lb-in)
15B	M6 x 1 x 28.5	W/P TO F/C	8-12	71-106 (lb-in)

W/P—Water Pump
F/C—Engine Front Cover

Timing chain front cover bolt location and identification—2002–04 3.0L (VIN U) engine

- Accessory drive belt tensioner and idler pulley
- Accessory drive belt
- Negative battery cable
9. Fill the cooling system.
10. Fill the crankcase.
11. Start the engine and check for leaks and proper operation.

2005 3.0L (VIN U) Engine

1. Before servicing the vehicle, refer to the precautions section.
2. Disconnect the negative battery cable.
3. Drain and recycle the engine coolant.
4. Remove the engine anti-roll strut brace.
5. Remove the engine anti-roll strut.
6. Remove the coolant expansion tank.
7. Loosen the four water pump pulley bolts.
8. Remove the crankshaft damper.
9. Disconnect the lower radiator hose.
10. Remove the Crankshaft Position (CKP) sensor.
11. Remove the oil pan.
12. Remove the alternator.
13. Remove the accessory drive belt idler pulley.
14. Remove the accessory drive belt tensioner.
15. Remove the A/C compressor bracket.
16. Disconnect the heater hose.
17. Remove the water pump pulley.
18. Remove the engine front cover bolts, cover and the water pump as an assembly.
19. Remove and discard the front cover gasket.
20. Rotate the crankshaft and align the timing marks.
21. Remove the camshaft sprocket bolt.
22. Remove the timing chain, the camshaft sprocket and the crankshaft sprocket as an assembly.

25 Nm (18 lb-ft)

06017-TAUR-G38

Install the engine front cover and water pump assembly. Apply Pipe Sealant with Teflon® to the bolts illustrated—2005 3.0L (VIN U) engine

To install:
23. Align the timing marks.
24. Install the timing chain, the camshaft sprocket and the crankshaft sprocket as an assembly.
25. Install the camshaft sprocket bolt and tighten to 46 ft. lbs. (63 Nm).
26. Clean all sealing surfaces and position a new front cover gasket.
27. Install the engine front cover and water pump assembly. Apply Pipe Sealant with Teflon® to the bolts illustrated.
28. Install the water pump pulley and tighten the bolts to 18 ft. lbs. (25 Nm).
29. Connect the heater hose.
30. Install the A/C compressor bracket and tighten the bolts to 18 ft. lbs. (25 Nm).
31. Install the accessory drive belt tensioner and tighten the bolts to 35 ft. lbs. (48 Nm).
32. Install the accessory drive belt idler pulley and tighten the bolts to 35 ft. lbs. (48 Nm).
33. Install the alternator.
34. Install the CKP sensor.
35. Connect the lower radiator hose.
36. Install the oil pan.
37. Install the crankshaft damper.
38. Install the coolant expansion tank.

39. Install the engine anti-roll strut and tighten the bolts to 35 ft. lbs. (48 Nm).
40. Install the engine anti-roll strut brace and tighten the bolts to 35 ft. lbs. (48 Nm).
41. Connect the negative battery cable.
42. Fill and bleed the cooling system.

2002–04 3.0L (VIN S) Engine

1. Before servicing the vehicle, refer to the precautions section.
2. Remove the engine from the vehicle and mount it on an engine stand.
3. Remove or disconnect the following:
   - Upper intake manifold

9346KG31

Access the power steering pump bolts through the holes in the pulley—2002–04 3.0L (VIN S) engine

06017-TAUR-G37

Rotate the crankshaft and align the timing marks—2005 3.0L (VIN U) engine

- Valve covers
- Accessory drive belt
- Power steering pump
- Alternator
- Water pump
- A/C compressor and bracket

➡ **The crankshaft accessory drive pulley shaft has left-hand threads. Rotate the pulley shaft clockwise to remove.**

- Crankshaft pulley
- Crankshaft damper
- Oil pan
- Oil pump screen and tube
- Crankshaft Position (CKP) sensor connector
- Camshaft Position (CMP) sensor connector
- Front cover
- CKP sensor pulse ring. Note the keyway alignment for installation.

➡ **The Crankshaft Position Sensor (CKP) pulse ring is used on several different engines. Note the keyway alignment for installation.**

4. Rotate the crankshaft so that the keyway is at the 11 o'clock position to locate the crankshaft at Top Dead Center (TDC) for No. 1 cylinder.

5. Verify that the alignment arrows on the camshafts are aligned. If not, rotate the crankshaft 1 complete revolution and recheck.

6. Rotate the crankshaft so that the keyway is at the 3 o'clock position. This positions the right cylinder head camshafts to the neutral position.

7. Remove or disconnect the following:
- Right timing chain tensioner

➡ **The camshaft thrust caps must be removed before loosening the remaining camshaft journal cap bolts to ensure that the thrust caps are not damaged.**

- Camshaft thrust caps
- Camshaft journal caps. Loosen the bolts in sequence and in several passes to allow the camshaft to be raised from the cylinder head evenly.
- Rocker arms. Keep the rocker arms in order for installation.
- Right timing chain tensioner arm
- Right timing chain and crankshaft sprocket

8. Rotate the crankshaft 2 revolutions and locate the crankshaft keyway at the 11 o'clock position. This will position the left cylinder head camshafts to their neutral position.

9. Verify that the alignment arrows on the camshafts are aligned.

10. Remove the left cylinder head timing chain tensioner retaining bolts and the timing chain tensioner.

➡ **The camshaft thrust caps must be removed before loosening the remaining camshaft journal cap bolts to ensure that the thrust caps are not damaged.**

11. Remove or disconnect the following:
- Camshaft thrust caps
- Camshaft journal caps. Loosen the bolts in sequence and in several passes to allow the camshaft to be raised from the cylinder head evenly.
- Rocker arms. Keep the rocker arms in order for installation.
- Left timing chain tensioner arm
- Left timing chain and crankshaft sprocket

To install:

12. Prepare the timing chain tensioners for installation as follows:

a. Place the left chain tensioner in a vise.

1 RH Exhaust Camshaft	6 Timing Chain Tensioner
2 RH Intake Camshaft	7 Bolt (2 Req'd)
3 RH Timing Chain Crankshaft Sprocket	8 Timing Chain Tensioner Arm
4 RH Timing Chain	9 Timing Chain Guide
5 Bolt (2 Req'd)	10 RH Cylinder Head
	A Tighten to 20-30 N·m (15-22 Lb-Ft)

7922KG49

Exploded view of the right cylinder head timing chain and related components—2002–04 3.0L (VIN S) engine—left side similar

7922JG27

Using a thin prytool, release and hold the timing chain tensioner ratchet/pawl mechanism—2002–04 3.0L (VIN S) engine

7922JG28

Retain the piston with a 1.5mm wire or paperclip—2002–04 3.0L (VIN S) engine

b. Using a small prytool, release and hold the timing chain tensioner ratchet/pawl mechanism through the access hole in the timing chain tensioner.

c. Slowly compress the tensioner.

d. Lock the piston with a 1.5mm wire or paperclip.

e. Repeat for the right chain tensioner.

➡ **Be sure that the crankshaft keyway is still at the 11 o'clock position.**

13. Install or connect the following:
- Left timing chain and crankshaft sprocket. Align the colored links with the index marks on the camshaft and crankshaft sprockets.
- Left timing chain tensioner arm
- Left timing chain tensioner and tighten the retaining bolts to 15–22 ft. lbs. (20–30 Nm)
- Right timing chain and crankshaft sprocket. Align the colored links with the index marks on the camshaft and crankshaft sprockets.
- Right timing chain tensioner arm
- Right timing chain tensioner and tighten the retaining bolts to 15–22 ft. lbs. (20–30 Nm)

➡ **The crankshaft keyway must be in the 11 o'clock position to install the left cylinder head rocker arms.**

➡ **The camshaft journal caps and cylinder heads are numbered to ensure that they are assembled in their original positions.**

14. Install the left cylinder head rocker arms in their original positions.

➡ **Do not install the camshaft journal thrust caps until the other journal caps have been installed and tightened.**

15. Tighten the left camshaft journal caps in the order shown and in several passes to 71–106 inch lbs. (8–12 Nm).

16. Install the left camshaft journal thrust caps and tighten the bolts to 71–106 inch lbs. (8–12 Nm).

17. Remove the retaining wire from the left timing chain tensioner.

➡ **The crankshaft keyway must be in the 3 o'clock position to install the right cylinder head rocker arms.**

18. Rotate the crankshaft so that the keyway is in the 3 o'clock position.

➡ **The camshaft journal caps and cylinder heads are numbered to ensure that they are assembled in their original positions.**

Camshaft journal bolt tightening sequence—2002–04 3.0L (VIN S) engine

Be sure the timing marks are as shown after the chain has been installed—2002–04 3.0L (VIN S) engine

19. Install the right cylinder head rocker arms in their original positions.

➡ **Do not install the camshaft journal thrust caps until the other journal caps have been installed and tightened.**

20. Tighten the right camshaft journal caps in the order shown and in several passes to 71–106 inch lbs. (8–12 Nm).

21. Install the right camshaft journal thrust caps and tighten the bolts to 71–106 inch lbs. (8–12 Nm).

22. Remove the retaining wire from the right timing chain tensioner.

➡ **The Crankshaft Position Sensor (CKP) pulse ring is used on several different engines. The keyway position for the 3.0L (VIN S) engine will be marked 30, 30RFF or with an ORANGE stripe.**

23. Install the CKP sensor pulse ring. Align the keyway with the proper slot.

24. Replace the crankshaft seal in the front cover with a new one. Apply clean engine oil to the seal lip.

25. Apply silicone sealer to the 6 critical areas shown in View **A**, to the cylinder block to prevent oil seepage.

26. Place new front cover gaskets onto the dowel pins on the cylinder block and heads.

27. Place the front cover into position.

28. Install the 6 front cover retaining bolts and stud bolts where the silicone sealer was applied.

29. Tighten the bolts and stud bolts until the front cover contacts the cylinder block and heads, then turn the bolts and stud bolts an additional ¼ turn.

30. Install the remaining front cover retaining bolts and stud bolts.

31. Tighten all of the front cover retaining bolts and stud bolts in sequence to 15–22 ft. lbs. (20–30 Nm).

APPLY 3 mm (0.118 INCH)
BEAD OF SEALANT

**VIEW A
TYPICAL 6 PLACES**

7922JG29

To prevent oil leakage, apply sealant to the places indicated—2002–04 3.0L (VIN S) engine

➡**The air conditioning mounting bracket bolts are torque-to-yield bolts and must be replaced.**

32. Install or connect the following:
- CMP sensor connector
- CKP sensor connector
- Oil pump screen and tube
- Oil pan

33. Install the crankshaft damper and tighten the bolt as follows:

   a. Step 1: Tighten the bolt to 78–99 ft. lbs. (105–135 Nm).

   b. Step 2: Loosen the bolt one full turn.

   c. Step 3: Tighten the bolt to 35–39 ft. lbs. (47–53 Nm).

   d. Step 4: Tighten the bolt 85–95 degrees.

34. Install or connect the following:
- Crankshaft pulley and tighten counterclockwise to 70–77 ft. lbs. (95–105 Nm)
- A/C compressor and bracket
- Water pump
- Alternator
- Power steering pump and tighten the bolts to 18 ft. lbs. (25 Nm)
- Accessory drive belt
- Valve covers
- Upper intake manifold

- Engine assembly into the vehicle

35. Run the engine and check for leaks and proper operation.

2005 3.0L (VIN S) Engine

1. Before servicing the vehicle, refer to the precautions section.

2. Disconnect the negative battery cable.

3. Remove the valve covers.

4. Remove the power steering pump.

5. Remove the oil pan.

6. Remove the retaining nut and position the power steering pressure line and muffler out of the way.

7. Remove the upper retaining bolts from the alternator.

8. Remove the bolts, nuts and the engine-to-transaxle bracket.

9. Remove the torque converter inspection cover.

10. Remove the retainers and position the front rocker panel moulding aside.

11. Remove the rear portion of the front fender splash shield.

12. Remove the two screws and the alternator splash shield.

13. Install a Flywheel Holding Tool.

FRONT OF ENGINE

* LOCATION OF STUDS FOUR PLACES

● TIGHTEN BOLTS/STUDS IN SEQUENCE SHOWN FINGER-TIGHT

* STUD LOCATIONS

Front cover bolt torque sequence—2002–04 3.0L (VIN S) engine

7922KG52

06017-TAUR-G39

Using the tool illustrated, remove the crankshaft damper—2005 3.0L (VIN S) engine

➡ **The pulley bolt has a reverse thread.**

14. Remove the crankshaft damper pulley.

15. Remove the crankshaft damper bolt and washer.

16. Using the tool illustrated, remove the crankshaft damper.

17. Remove the Flywheel Holding Tool.

18. Disconnect the Crankshaft Position (CKP) sensor.

19. Remove the alternator bolt and position the alternator to gain access to the electrical connector and terminal nut.

20. Disconnect the alternator electrical connectors.

21. Remove the bolts and the A/C compressor to front cover bracket.

22. Disconnect the Camshaft Position (CMP) sensor

23. Remove the tie strap from the power steering hose.

24. Remove the retaining nut and position the A/C low-pressure hose out of the way.

25. Remove the bolts and position the A/C high-pressure hose on top of the belt tensioner.

26. Remove the bolt and remove the belt tensioner.

27. Disconnect the engine cooling fan electrical connector and unclip the wiring harness.

06017-TAUR-G40

Remove the ignition pulse wheel—2005 3.0L (VIN S) engine

06017-TAUR-G41

Rotate the crankshaft clockwise to position the crankshaft keyway in the 11 o'clock position and position the camshafts in the correct position—2005 3.0L (VIN S) engine

06017-TAUR-G42

Rotate the crankshaft clockwise 120 degrees to the 3 o'clock position to locate the right hand camshafts in the neutral position —2005 3.0L (VIN S) engine

06017-TAUR-G43

Verify that the right hand camshafts are in the neutral position—2005 3.0L (VIN S) engine

Remove the bolts, right hand timing chain guide and the timing chain—2005 3.0L (VIN S) engine

06017-TAUR-G44

Rotate the crankshaft clockwise 600 degrees (1⅔ turns) to position the crankshaft keyway in the 11 o'clock position—2005 3.0L (VIN S) engine

06017-TAUR-G45

Verify that the left hand camshafts are in the neutral position—2005 3.0L (VIN S) engine

06017-TAUR-G46

28. Remove the bolt and the engine cooling fan.

29. Install the lift eye hooks to support the engine.

30. Remove the two upper A/C compressor bolts.

31. Remove the two lower A/C compressor bolts and position the A/C compressor aside.

32. Remove the bolts from the A/C bracket.

33. Raise the engine away from the A/C bracket using an engine support tool.

34. Remove the bolts, studs and the engine front cover.

❈❈ CAUTION

This pulse wheel is used with several different engines. Install the pulse wheel with the keyway in the slot stamped "30" or "30 RFF" only (orange in color).

35. Remove the ignition pulse wheel.

36. Install the damper bolt.

37. Remove the spark plugs.

38. Rotate the crankshaft clockwise to position the crankshaft keyway in the 11 o'clock position and position the camshafts in the correct position. This will position the number one cylinder at Top Dead Center (TDC). Verify that the camshafts are correctly located. If not, rotate the crankshaft one additional turn and recheck.

39. Rotate the crankshaft clockwise 120 degrees to the 3 o'clock position to locate the right hand camshafts in the neutral position.

40. Verify that the right hand camshafts are in the neutral position.

41. Remove the right hand timing chain tensioner arm.

42. Remove the bolts, right hand timing chain guide and the timing chain.

43. Rotate the crankshaft clockwise 600 degrees (1-⅔ turns) to position the crankshaft keyway in the 11 o'clock position. This will position the left hand camshafts in the neutral position.

44. Verify that the left hand camshafts are in the neutral position.

45. Remove the left hand timing chain and tensioner arm.

46. Remove the left hand timing chain and timing chain guide.

47. Remove the damper bolt and the crankshaft sprockets.

To install:

❈❈ CAUTION

Failure to verify correct timing drive component alignment will result in severe engine damage.

➡**Install the sprockets with the timing marks facing outward.**

48. Install the crankshaft sprockets.

49. Position the chain tensioner in a soft-jawed vise.

50. Hold the chain tensioner ratchet lock mechanism away from the ratchet stem with a small pick.

✷✷ CAUTION

During tensioner compression, do not release the ratchet stem until the tensioner piston is fully bottomed in its bore or damage to the ratchet stem will result.

51. Slowly compress the timing chain tensioner.

52. Retain the tensioner piston with a 0.06 (1.5 mm) wire or paper clip.

53. If timing marks in the timing chains are not evident, use a permanent-type marker to mark the crankshaft and camshaft timing marks on the left and right hand timing chains as follows:

   a. Mark any link to use as the crankshaft timing mark (1).

   b. Starting with the crankshaft timing mark, count 29 links and mark the link (2).

   c. Continue counting to 42 and mark the link (3).

54. Position the left timing chain and guide and install the bolts.

55. Align the marks on the timing chain with the marks on the camshaft and crankshaft sprockets.

56. Install the left hand timing chain tensioner and tensioner arm. Tighten the bolts to 18 ft. lbs. (25 Nm).

57. Install the crankshaft damper bolt and rotate the crankshaft clockwise 120 degrees until the crankshaft keyway is in the 3 o'clock position.

58. Verify that the right hand camshafts are correctly positioned.

59. Position the right timing chain and chain guide and install the bolts. Tighten the bolts to 18 ft. lbs. (25 Nm).

60. Align the marks on the timing chain with the marks on the camshaft and crankshaft sprockets.

61. Install the right hand timing chain tensioner and tensioner arm. Tighten the bolts to 18 ft. lbs. (25 Nm).

62. Remove the right and left hand timing chain tensioner piston retaining wires.

63. Rotate the crankshaft counterclockwise 120 degrees to top dead center (TDC)

06017-TAUR-G47

Remove the left hand timing chain and timing chain guide—2005 3.0L (VIN S) engine

06017-TAUR-G48

If timing marks in the timing chains are not evident, use a permanent-type marker to mark the crankshaft and camshaft timing marks on the left and right hand timing chains—2005 3.0L (VIN S) engine

25 Nm (18 lb-ft)

06017-TAUR-G49

Align the marks on the left hand timing chain with the marks on the camshaft and crankshaft sprockets—2005 3.0L (VIN S) engine

06017-TAUR-G50

Install the crankshaft damper bolt and rotate the crankshaft clockwise 120 degrees until the crankshaft keyway is in the 3 o'clock position—2005 3.0L (VIN S) engine

06017-TAUR-G51

Verify that the right hand camshafts are correctly positioned—2005 3.0L (VIN S) engine

25 Nm (18 lb-ft)

06017-TAUR-G53

Align the marks on the right hand timing chain with the marks on the camshaft and crankshaft sprockets—2005 3.0L (VIN S) engine

❋❋ CAUTION

Failure to verify correct timing drive component alignment will result in severe engine damage.

64. Verify the timing drive component alignment with the following steps.

   a. There should be 12 chain links between the front camshaft timing marks.

   b. There should be 27 chain links between the camshaft and crankshaft timing marks.

   c. There should be 30 chain links between the camshaft and crankshaft timing marks.

65. Remove the crankshaft damper bolt.

66. Install the spark plugs.

❋❋ CAUTION

This pulse wheel is used with several different engines. Install the pulse wheel with the keyway in the slot stamped "30" or "30 RFF" only (orange in color).

67. Install the ignition pulse wheel.

❋❋ CAUTION

Do not use metal scrapers, wire brushes, power abrasive discs, or other means to clean the sealing surface. These tools cause scratches and gouges which make leak paths. Use a plastic scraping tool to remove all traces of sealant.

68. Clean all sealing surfaces with metal surface cleaner.

69. Install three new gaskets in the front cover.

➡ **The engine front cover must be installed and the bolts tightened within six minutes of applying the sealant.**

70. Apply a 0.24 inch (6 mm) diameter dot of gasket and sealer to the cylinder block to lower cylinder block and cylinder head mating surfaces.

➡ **Fasteners No. 1 and 6 are studs.**

71. Position the cover and install the bolts in the sequence shown. Tighten the bolts to 18 ft. lbs. (25 Nm).

72. Using the engine support tool, lower the engine back together with the A/C bracket.

73. Install the bolts and tighten the bolts in two steps:

Refer to the procedure to verify the timing drive component alignment—2005 3.0L (VIN S) engine

Apply a 0.24 inch (6 mm) diameter dot of gasket and sealer to the cylinder block to lower cylinder block and cylinder head mating surfaces—2005 3.0L (VIN S) engine

25 Nm (18 lb-ft)

Engine front cover torque sequence—2005 3.0L (VIN S) engine

a. Step 1: Tighten to 18 ft. lbs. (25 Nm).

b. Step 2: Tighten an additional 90 degrees.

74. Position the A/C compressor and install the two lower bolts. Tighten the bolts to 18 ft. lbs. (25 Nm).

75. Install the two upper A/C compressor bolts. Tighten the bolts to 18 ft. lbs. (25 Nm).

76. Remove the lifting hooks.

77. Install the engine cooling fan and install the bolt.

78. Connect the engine cooling fan electrical connector and clip the wiring harness.

79. Install the belt tensioner and install the bolt. Tighten the bolts to 18 ft. lbs. (25 Nm).

80. Install the A/C high-pressure hose and install the bolts.

81. Install the A/C low-pressure hose and install the retaining nut.

82. Install the tie strap to the power steering hose.

83. Install the CMP sensor

84. Position the A/C compressor to front

Apply silicone gasket and sealant to the keyway slot—2005 3.0L (VIN S) engine

Using the tool illustrated, install the crankshaft damper—2005 3.0L (VIN S) engine

cover bracket and install the bolts. Tighten the bolts to 18 ft. lbs. (25 Nm).

85. Install the alternator.

86. Install the CKP sensor.

87. Reposition the power steering line and muffler and install the retaining nut.

88. Install a flywheel holding tool.

➡ **Seal surfaces must be free of dirt and oil.**

89. Apply silicone gasket and sealant to the keyway slot.

➡ **Lubricate the outside diameter sealing surface of the crankshaft pulley with clean motor oil.**

90. Using the tool illustrated, install the crankshaft damper.

91. Install the damper bolt and washer and tighten in four steps:

a. Step 1: Tighten to 89 ft. lbs. (120 Nm).

b. Step 2: Loosen 360 degrees.

c. Step 3: Tighten to 37 ft. lbs. (50 Nm).

d. Step 4: Tighten 90 degrees.

➡ **The pulley has reverse threads.**

92. Install the crankshaft pulley. Tighten the bolt to 74 ft. lbs. (100 Nm).

93. Remove the Flywheel Holding Tool.

94. Install the alternator splash shield.

95. Install the rear portion of the front fender splash shield.

96. Install the front rocker panel moulding.

97. Install the torque converter inspection cover.

98. Install the engine-to-transaxle bracket and install the nuts and bolts. Tighten the bolts to 18 ft. lbs. (25 Nm).

99. Install the oil pan.

100. Install the valve covers.

101. Connect the negative battery cable.

102. Install the power steering pump.

103. Start the engine and check for leaks.

Piston and Ring

POSITIONING

Ring end-gap spacing—3.0L (VIN S and U) engines

Piston and connecting rod positioning—3.0L (VIN S and U) engines

FUEL SYSTEM

Fuel System Service Precautions

Safety is the most important factor when performing not only fuel system maintenance but any type of maintenance. Failure to conduct maintenance and repairs in a safe manner may result in serious personal injury or death. Maintenance and testing of the vehicle's fuel system components can be accomplished safely and effectively by adhering to the following rules and guidelines.

• To avoid the possibility of fire and personal injury, always disconnect the negative battery cable unless the repair or test procedure requires that battery voltage be applied.

• Always relieve the fuel system pressure prior to disconnecting any fuel system component (injector, fuel rail, pressure regulator, etc.), fitting or fuel line connection. Exercise extreme caution whenever relieving fuel system pressure, to avoid exposing skin, face and eyes to fuel spray. Please be advised that fuel under pressure may penetrate the skin or any part of the body that it contacts.

• Always place a shop towel or cloth around the fitting or connection prior to loosening to absorb any excess fuel due to spillage. Ensure that all fuel spillage (should it occur) is quickly removed from engine surfaces. Ensure that all fuel soaked cloths or towels are deposited into a suitable waste container.

• Always keep a dry chemical (Class B) fire extinguisher near the work area.

• Do not allow fuel spray or fuel vapors to come into contact with a spark or open flame.

• Always use a back-up wrench when loosening and tightening fuel line connection fittings. This will prevent unnecessary stress and torsion to fuel line piping. Always follow the proper torque specifications.

• Always replace worn fuel fitting O-rings with new. Do not substitute fuel hose or equivalent, where fuel pipe is installed.

Fuel System Pressure

RELIEVING

2002–04 Models

1. Before servicing the vehicle, refer to the precautions section.
2. Disconnect the negative battery cable.

3. Remove the fuel tank fill cap to relieve the pressure in the fuel tank.
4. Remove the cap from the Schrader valve located on the fuel supply manifold.
5. On gasoline engines, attach Fuel Pressure Gauge T80L-9974-A or equivalent, to the valve and drain the fuel through the drain tube into a suitable container.
6. On flex-fuel engines, connect Fuel Pressure Gauge T80L-9974-A or equivalent and Fuel Pressure Test Kit 134-R0035 or equivalent, to the Schrader valve. Drain the fuel through the drain tube into a suitable container.
7. After the fuel system pressure is relieved, remove the fuel pressure gauge and install the cap on the Schrader valve.
8. Install the fuel tank fill cap.
9. Connect the negative battery cable after system repairs are completed.

2005 Models

1. Before servicing the vehicle, refer to the precautions section.

➡**The fuel pump relay is located in the engine compartment fuse box.**

2. Remove the fuel pump relay.
3. Start the engine and allow it to idle until it stalls.
4. After the engine stalls, crank the engine for approximately 5 seconds to make sure the fuel injection supply manifold pressure has been released.
5. Turn the ignition switch to the OFF position.
6. When fuel system service is complete, install the fuel pump relay.

➡**It may take more than one key cycle to pressurize the fuel system.**

7. Cycle the ignition key and wait three seconds to pressurize the fuel system. Check for leaks before starting the engine.
8. Start the vehicle and check the fuel system for leaks.

Fuel Filter

REMOVAL & INSTALLATION

1. Before servicing the vehicle, refer to the precautions section.
2. Disconnect the negative battery cable.
3. Relieve the fuel system pressure.
4. Disconnect the fuel lines.
5. Loosen the filter retaining clamp and remove the fuel filter.

FUEL LINE DISCONNECT TOOL T90T-9550-B OR T90T-9550-C

7922KG55

Push connect fitting and removal tool

To install:

6. Install the fuel filter with the flow arrow facing the proper direction and tighten the filter retaining clamp.
7. Push the fuel lines on to the filter fittings until an audible click is heard.
8. Connect the negative battery cable.
9. Start the engine and check for fuel leaks and proper operation.

Fuel Pump

REMOVAL & INSTALLATION

2002–04 Models

1. Before servicing the vehicle, refer to the precautions section.
2. Relieve the fuel system pressure.
3. Drain the fuel tank.
4. Remove or disconnect the following:

• Negative battery cable
• Fuel lines
• Fuel pump module electrical connector
• Fuel tank
• Fuel pump module

To install:

5. Install the fuel pump module carefully to ensure the filter and hoses and float rod are not damaged. Use a new O-ring seal.
6. Align the fuel pump module and the fuel tank retainer and push the fuel pump module into the fuel tank retainer. When the fuel pump module is properly engaged, a definite click will be heard engaging 2 locking tabs on the outside of the fuel pump.
7. Install or connect the following:

• Fuel pump module locking ring
• Fuel tank
• Fuel pump module electrical connector
• Fuel lines
• Negative battery cable

8. Add a minimum of 10 gallons of clean fuel to the tank.
9. Start the engine and check for leaks.

1   Fuel Pump Locking Retainer
     Ring
2   Fuel Tank Pressure / Vacuum
     Transducer
3   Locking Tab
4   Fuel Tank
5   O-Ring Seal
6   Connector
7   Fuel Pump Module

7922KG58

Exploded view of the fuel pump module mounting

2005 Models

1. Before servicing the vehicle, refer to the precautions section.

2. Properly relieve the fuel system pressure.

3. Disconnect the negative battery cable.

4. Using suitable hose pinching pliers, clamp the fuel tank filler hose in the location illustrated.

5. Release the clamp to the fuel tank filler tube and remove. Disconnect the three quick release couplings from the fuel tank tubes.

6. Disconnect the fuel pump electrical connector.

7. Disconnect the fuel filter inlet tube.

06017-TAUR-G58

Using suitable hose pinching pliers, clamp the fuel tank filler hose in the location illustrated—2005 models

310-D006

06017-TAUR-G59

Using the tool illustrated, remove the fuel pump module retaining ring—2005 models

8. Position a suitable lifting device under the fuel tank.

9. Remove the two fuel tank support strap bolts (one shown).

10. Lower the fuel tank to access the fuel pump module while swinging the fuel tank support straps away from the fuel tank.

11. Clean the area around the fuel pump mounting flange.

12. Disconnect the fuel pump module electrical connector and the fuel pump module fuel outlet tube.

13. Using the tool illustrated, remove the fuel pump module retaining ring.

✳✳ CAUTION

The fuel pump module must be handled carefully to avoid damage to the float arm, filter and convolute hoses.

14. Turn the fuel pump module counterclockwise and the fuel pump module will release from the locking tabs.

15. Remove the fuel pump module from the fuel tank.

16. Drain the fuel from the fuel tank.

To install:

17. Installation is the reverse of removal.

18. Tighten the fuel tank strap bolts to 26 ft. lbs. (35 Nm).

19. Start the engine and check for leaks.

Fuel Injector

REMOVAL & INSTALLATION

3.0L (VIN U) Engine

1. Before servicing the vehicle, refer to the precautions section.

2. Relieve fuel system pressure.

3. Remove or disconnect the following:
   - Negative battery cable
   - Air cleaner outlet tube
   - Fuel lines
   - Upper intake manifold
   - Fuel injector electrical connectors
   - Fuel pressure regulator vacuum line
   - Fuel supply manifold with the injectors attached
   - Injectors from the supply manifold

To install:

4. Install or connect the following:
   - Fuel injectors. Use new O-ring seals.
   - Fuel supply manifold with the injectors attached and tighten the bolts to 89 inch lbs. (10 Nm)
   - Fuel pressure regulator vacuum line
   - Fuel injector electrical connectors

Fuel supply manifold—3.0L (VIN U) engine

Retaining bolts (1) and Fuel supply Manifold (2)—3.0L (VIN S) engine

- Upper intake manifold
- Fuel lines
- Air cleaner outlet tube
- Negative battery cable
5. Start the engine and check for leaks.

3.0L (VIN S) Engine

1. Before servicing the vehicle, refer to the precautions section.
2. Relieve fuel system pressure.
3. Remove or disconnect the following:
- Negative battery cable
- Air cleaner outlet tube
- Fuel lines
- Upper intake manifold
- Fuel injector electrical connectors
- Fuel pressure regulator vacuum line
- Fuel supply manifold
- Injectors from the lower intake manifold

To install:
4. Install or connect the following:
- Fuel injectors. Use new O-ring seals.
- Fuel supply manifold and tighten the bolts 89 inch lbs. (10 Nm)
- Fuel pressure regulator vacuum line
- Fuel injector electrical connectors
- Upper intake manifold
- Fuel lines
- Air cleaner outlet tube
- Negative battery cable
5. Start the engine and check for leaks.

DRIVE TRAIN

Transaxle

REMOVAL & INSTALLATION

2002–04 Models

1. Before servicing the vehicle, refer to the precautions section.
2. Attach a powertrain support to the engine lifting eyes.
3. Remove or disconnect the following:
- Battery and tray
- Air cleaner assembly
- Transaxle electrical connectors
- Shift cable
- Transaxle cooler lines
- Front wheels
- Stabilizer bar links
- Lower ball joints
- Axle halfshafts

- Heated Oxygen Sensor (HO2S) connectors
- Dual converter Y-pipe
- Starter
- Steering rack and pinion gear. Support the gear with safety wire.
- Powertrain support insulators
- Subframe
- Torque converter nuts
4. Support the transaxle with a transmission jack.
5. Remove the transaxle flange bolts and remove the transaxle.

To install:
6. Install the transaxle. For 3.0L (VIN S) engines, tighten the flange bolts to 25–33 ft. lbs. (33–46 Nm). For all other engines, tighten the flange bolts to 30–44 ft. lbs. (40–60 Nm).
7. Tighten the torque converter nuts to 20–34 ft. lbs. (27–46 Nm).
8. Install the subframe. Use 2 pieces of

¾ inch outside diameter pipe in the alignment holes behind the front subframe mounts to align the subframe to the body. Tighten the subframe mounting bolts to 57–76 ft. lbs. (77–103 Nm).
9. Install or connect the following:
- Powertrain support insulators. Tighten the bracket bolts to 65 ft. lbs. (88 Nm) and the subframe bolts to 90 ft. lbs. (122 Nm).
- Steering rack and pinion gear and tighten the nuts to 84–113 ft. lbs. (113–133 Nm)
- Starter and tighten the fasteners to 21 ft. lbs. (29 Nm)
- Dual converter Y-pipe
- HO2S sensor connectors
- Axle halfshafts and tighten the hub retainer nuts to 170–202 ft. lbs. (230–275 Nm)
- Lower ball joints and tighten the nuts to 51–67 ft. lbs. (68–92 Nm)

- Stabilizer bar links and tighten the nuts to 35–46 ft. lbs. (47–63 Nm)
- Front wheels
- Transaxle cooler lines
- Shift cable
- Transaxle electrical connectors
- Air cleaner assembly
- Battery and tray

10. Start the engine. Check for leaks and proper operation.

➡ **Whenever the vehicle subframe is removed or lowered, the wheel alignment should be checked.**

2005 Models

3.0L (VIN U) ENGINE

1. Before servicing the vehicle, refer to the precautions section.
2. Disconnect the battery cables, negative cable first.
3. Remove the battery.
4. Disconnect the Mass Air Flow (MAF) sensor and the two breather tubes.
5. Remove the engine air cleaner cover and tube assembly.
6. Remove the two bolts and the air filter assembly.
7. Remove the relay box and the battery tray.
8. Disconnect the electrical connector.
9. Remove the nut from the power steering line bracket.
10. Remove the electrical connector bracket.
11. Disconnect the Output Shaft Speed (OSS) sensor.
12. Disconnect the Turbine Shaft Speed (TSS) sensor.
13. Remove the manual control lever cable by removing the nut, then remove the shift actuator cable fitting from the shift cable and bracket.
14. Remove the ground cable and bracket.
15. Remove the cooler line retaining clips. Depress the tabs on the plastic insert retainer and remove cooler lines from fitting.
16. Remove the upper transmission bolts.
17. Install an engine support device such as the one illustrated.
18. Install the tools shown on the right (back) hand side of the engine.
19. Install the tools shown on the left (front) hand side of the engine.
20. Drain the transmission fluid by lowering the transaxle pan. Once the fluid has drained re-install the pan

Install an engine support device such as the one illustrated—2005 models

> ❋❋ **CAUTION**
>
> **If the steering gear is removed with the subframe, do not allow the steering wheel to rotate while the steering column intermediate is disconnected or damage to the clockspring can result. If there is evidence that the shaft has rotated, the clockspring must be recentered before the vehicle is driven.**

21. Remove the subframe assembly as follows:
   a. If equipped, remove the roll-restrictor bolt and nut.
   b. Remove the downstream catalytic converter and support the exhaust system to one side under the vehicle.
   c. If equipped, remove the five pin-type retainers from the chin spoiler.
   d. Remove the air dam.
   e. Remove the front tires and wheels.
   f. Remove the two lower ball joint nuts.
22. Disconnect the ball joint from the lower arm.
23. Push the lower arm downward until the ball joint is free of the arm.

> ❋❋ **CAUTION**
>
> **Use extreme care not to damage boot seal. Do not use power tools to tighten the nut, or bearing and seal damage can result. Install nut using the hex hold feature to prevent the stud from rotating.**

   a. Remove the two nuts and disconnect the front stabilizer bar links.

Install the tools shown on the right (back) hand side of the engine—2005 models

Install the tools shown on the left (front) hand side of the engine—2005 models

   b. Support the steering gear to prevent damage to the steering shaft and remove the steering gear nuts.
   c. Remove the four bolts holding the power steering line to the frame.
   d. Remove the right hand front and rear engine mount nuts.
   e. Remove the transmission insulator support bracket bolts from the frame.
24. Remove the pin-type retainers from the right hand subframe splash shield.

a. Using a suitable support table, support the subframe.

✳✳ CAUTION

Make sure all hoses and the steering gear assembly are clear of subframe to prevent damage to these components when lowering the subframe.

b. Remove the four subframe bolts and lower the subframe from the vehicle.

➡ **The use of mechanic's wire will aid in positioning the halfshaft out of the way.**

25. Remove the halfshafts from the transaxle and position out of the way.
26. Remove the starter motor.
27. Remove the transaxle housing cover.
28. Remove and discard the four torque converter to flexplate nuts.
29. Position a transmission jack under the transaxle.
30. Remove the transaxle retaining bolts.
31. Lower the transaxle assembly from the engine compartment.

To install:

➡ **Locate and align the orange dot adjacent to the stud on the torque converter and place it in the 6 o'clock position.**

32. Position the transaxle on a suitable transmission jack.
33. Apply multi-purpose grease to the torque converter pilot hub.
34. Raise the transmission into position and install the bolts. Tighten the bolts to 46 ft. lbs. (62 Nm).
35. Install new torque to flexplate nuts and tighten to 26 ft. lbs. (35 Nm).
36. Install the transmission housing cover.
37. Install the starter motor.
38. Install the halfshafts.

✳✳ CAUTION

If the steering gear was removed with the subframe and there is evidence that the steering column intermediate shaft has rotated, the clockspring must be recentered before the vehicle is driven or damage to the clockspring can result.

✳✳ WARNING

Incorrect centralization may result in premature component failure. If in doubt when centralizing the clockspring, repeat the centralizing proce-

dure. **Failure to follow this instruction may result in personal injury.**

✳✳ CAUTION

Make sure the road wheels are in the straight ahead position.

39. If a clockspring has rotated out of center, perform the following:
a. Centralize the clockspring.
b. Hold the clockspring outer housing (1) stationary.

✳✳ CAUTION

Overturning will destroy the clockspring. The internal ribbon wire acts as the stop and can be broken from its internal connection.

c. While turning the rotor (2) counterclockwise, carefully feel for the ribbon wire to run out of length and for a slight resistance. Stop turning at this point.
d. Turn the clockspring clockwise (3) approximately three turns. This is the center point of the clockspring. Do not allow the rotor to turn from this position.
40. Install the subframe assembly as follows:
a. Raise the subframe into the vehicle. Install the bolts and tighten to 76 ft. lbs. (103 Nm).

✳✳ CAUTION

Make sure all hoses and the steering gear assembly are clear of subframe to prevent damage to these components when lowering the subframe.

b. Install the splash shield.
c. Install the transmission insulator support bracket bolts to the frame and tighten to 65 ft. lbs. (88 Nm).
d. Install the right hand engine mount nuts and tighten to 66 ft. lbs. (90 Nm).
e. Install the four bolts holding the power steering line to the frame.
f. Install the steering gear nuts and tighten to 91 ft. lbs. (123 Nm).
g. Connect the front stabilizer bar links and tighten the nuts to 59 ft. lbs. (80 Nm).
h. Connect the ball joint to the lower arm. Use new nuts and tighten to 59 ft. lbs. (80 Nm).
i. Install the front wheels.
j. Install the air dam.
k. Install the downstream catalytic converter.
l. If equipped, install the roll-restric-

06017-TAUR-G63

Make sure the clockspring is properly centralized—2005 models

tor bolt and nut. Tighten to 46 ft. lbs. (62 Nm).
41. Remove the engine lifting brackets.
42. Remove the engine support tools.
43. Install the upper transmission bolts and tighten to 46 ft. lbs. (62 Nm).
44. Reconnect the transmission fluid cooler tubes.
45. Install the retaining clips.
46. Install the bracket and ground cable.
47. Verify that the digital Transmission Range (TR) sensor is correctly adjusted as follows:
a. Verify that the shift selector lever is in NEUTRAL.
b. Install the digital TR sensor and loosely install the bolts.
c. Using the tool illustrated, align the digital TR sensor and tighten the bolts.
d. Connect the connector.
48. Install the manual control lever cable onto the bracket and tighten the nut to 10 ft. lbs. (13 Nm).
49. Connect the TSS sensor.
50. Connect the OSS sensor.
51. Install the electrical connector bracket.
52. Install the power steering line bracket.
53. Reconnect the connector.
54. Install the battery tray.

307-351

06017-TAUR-G64

Using the tool illustrated, align the digital TR sensor and tighten the bolts—2005 models

55. Install the air filter assembly.

56. Install the engine air cleaner.

57. Reconnect the MAF sensor and the breather tubes.

58. Install the battery.

59. Connect the battery cables.

60. Fill the transaxle using clean automatic transmission fluid.

61. Start the engine.

62. Move the transaxle range selector lever through all positions.

63. Check the fluid level.

64. Check and adjust the wheel alignment.

2005 Models

3.0L (VIN S) ENGINE

1. Before servicing the vehicle, refer to the precautions section.

2. Disconnect the battery cables, negative cable first.

3. Remove the battery.

4. Disconnect the Mass Air Flow (MAF) sensor and the two breather tubes.

5. Remove the engine air cleaner cover and tube assembly.

6. Disconnect the breather tubes.

7. Remove the engine air cleaner.

8. Remove the air filter assembly.

9. Remove the battery tray.

10. Disconnect the connectors.

11. Remove the electrical connector and the wiring harness from the bracket.

12. Remove the manual control lever cable.

13. Disconnect the power steering sensor connector.

14. Disconnect the power steering harness and Output Shaft Speed (OSS) sensor.

15. Disconnect the wiring harness from the shift cable bracket.

16. Disconnect the Turbine Shaft Speed (TSS) sensor.

17. Remove the bolt from the coolant pipe bracket.

18. Remove the nut from the coolant pipe bracket.

19. Remove the left hand upper flange bolts.

20. Remove the cooler tube bolt for the cooling tube.

21. Install an engine lifting bracket.

22. Remove the right hand wiper, clips and upper cowl assembly.

23. Remove the right hand lower cowl assembly.

24. Remove the rear support brace.

25. Install the rear engine lifting bracket in the same location where the top bolt was removed from the rear support brace.

26. Install an engine support tool.

➡ It may be easier to access the nuts for the exhaust flange from the engine compartment.

27. Remove the front nuts from the front catalytic converter assembly.

28. Drain the transmission fluid by lowering the transaxle pan.

29. Disconnect the front heated Oxygen (HO_2S) sensor.

30. Disconnect the rear HO_2S sensor.

31. Remove the rear bolts from the front catalytic converter assembly.

32. Remove the nut from the rear catalytic converter assembly.

33. Remove the rear catalytic converter and cross over pipe assembly.

✳✳ CAUTION

If the steering gear is removed with the subframe, do not allow the steering wheel to rotate while the steering column intermediate is disconnected or damage to the clockspring can result. If there is evidence that the shaft has rotated, the clockspring must be recentered before the vehicle is driven.

34. Remove the subframe assembly as follows:

   a. If equipped, remove the roll-restrictor bolt and nut.

   b. Remove the downstream catalytic converter and support the exhaust system to one side under the vehicle.

   c. If equipped, remove the five pin-type retainers from the chin spoiler.

   d. Remove the air dam.

   e. Remove the front tires and wheels.

   f. Remove the two lower ball joint nuts.

35. Disconnect the ball joint from the lower arm.

36. Push the lower arm downward until the ball joint is free of the arm.

✳✳ CAUTION

Use extreme care not to damage boot seal. Do not use power tools to tighten the nut, or bearing and seal damage can result. Install nut using the hex hold feature to prevent the stud from rotating.

   a. Remove the two nuts and disconnect the front stabilizer bar links.

   b. Support the steering gear to prevent damage to the steering shaft and remove the steering gear nuts.

   c. Remove the four bolts holding the power steering line to the frame.

   d. Remove the right hand front and rear engine mount nuts.

   e. Remove the transmission insulator support bracket bolts from the frame.

➡ The use of mechanic's wire will aid in positioning the halfshaft out of the way.

37. Remove the halfshafts from the transaxle and position out of the way.

38. Remove the nut and the battery ground cable.

39. Remove the starter motor.

40. Remove the cooler line retaining clips.

41. Depress the tabs on the plastic insert retainer and remove the cooler lines from the fitting.

42. Remove the transmission housing cover.

43. Remove and discard the four torque converter to flexplate nuts.

44. Position a transmission jack under the transaxle.

45. Remove the transaxle retaining bolts.

46. Remove the support bracket.

47. Lower the transaxle assembly from the engine compartment.

To install:

➡ Locate and align the orange dot adjacent to the stud on the torque converter and place it in the 6 o'clock position.

48. Position the transaxle on a suitable transmission jack.

49. Apply multi-purpose grease to the torque converter pilot hub.

50. Raise the transmission into position and install the bolts. Tighten the bolts to 30 ft. lbs. (40 Nm).

51. Install the transaxle retaining bracket. Tighten the retainers as shown in the accompanying illustration.

52. Install new torque to flexplate nuts and tighten to 26 ft. lbs. (35 Nm).

53. Install the transmission housing cover.

Install the transaxle retaining bracket— 2005 3.0L (VIN S) models

54. Install the cooler lines.
55. Install the starter motor
56. Install the ground cable.
57. Install the halfshafts.

If the steering gear was removed with the subframe and there is evidence that the steering column intermediate shaft has rotated, the clockspring must be recentered before the vehicle is driven or damage to the clockspring can result.

Incorrect centralization may result in premature component failure. If in doubt when centralizing the clockspring, repeat the centralizing procedure. Failure to follow this instruction may result in personal injury.

Make sure the road wheels are in the straight ahead position.

58. If a clockspring has rotated out of center, perform the following:
    a. Centralize the clockspring.
    b. Hold the clockspring outer housing (1) stationary.

Overturning will destroy the clockspring. The internal ribbon wire acts as the stop and can be broken from its internal connection.

   c. While turning the rotor (2) counterclockwise, carefully feel for the ribbon wire to run out of length and for a slight resistance. Stop turning at this point.
   d. Turn the clockspring clockwise (3) approximately three turns. This is the center point of the clockspring. Do not allow the rotor to turn from this position.
59. Install the subframe assembly as follows:
    a. Raise the subframe into the vehicle. Install the bolts and tighten to 76 ft. lbs. (103 Nm).

Make sure all hoses and the steering gear assembly are clear of subframe to prevent damage to these components when lowering the subframe.

   b. Install the splash shield.
   c. Install the transmission insulator support bracket bolts to the frame and tighten to 65 ft. lbs. (88 Nm).
   d. Install the right hand engine mount nuts and tighten to 66 ft. lbs. (90 Nm).
   e. Install the four bolts holding the power steering line to the frame.
   f. Install the steering gear nuts and tighten to 91 ft. lbs. (123 Nm).
   g. Connect the front stabilizer bar links and tighten the nuts to 59 ft. lbs. (80 Nm).
   h. Connect the ball joint to the lower arm. Use new nuts and tighten to 59 ft. lbs. (80 Nm).
   i. Install the front wheels.
   j. Install the air dam.
   k. Install the downstream catalytic converter.
   l. If equipped, install the roll-restrictor bolt and nut. Tighten to 46 ft. lbs. (62 Nm).
60. Install the rear catalytic converter and cross over pipe assembly. Tighten the retainers to 30 ft. lbs. (40 Nm).
61. Install the nut for the rear catalytic converter assembly. Tighten the retainers to 30 ft. lbs. (40 Nm).
62. Install the bolts for the front catalytic converter assembly. Tighten the retainers to 30 ft. lbs. (40 Nm).
63. Connect the rear HO2S sensor.
64. Connect the front HO2S sensor.

➡**It may be easier to access the nuts from the engine compartment.**

65. Install the nuts for the front catalytic converter assembly. Tighten the retainers to 30 ft. lbs. (40 Nm).
66. Remove the engine support tool and lifting brackets.
67. Install the rear support brace. Tighten the retainers to 35 ft. lbs. (47 Nm)
68. Install the lower cowl assembly.
69. Install the wiper, upper cowl assembly and clips.
70. Install the nut for the cooling tube.
71. Install the left hand upper flange bolts. Tighten the retainers to 46 ft. lbs. (62 Nm)
72. Install the nut for the coolant pipe bracket.
73. Install the bolt for the coolant pipe bracket.
74. Connect the TSS sensor.
75. Connect the wiring harness to the shift cable bracket.
76. Connect the OSS sensor.
77. Connect the power steering sensor connector.
78. Verify that the digital Transmission

Using the tool illustrated, align the digital TR sensor and tighten the bolts—2005 models

06017-TAUR-G64

Range (TR) sensor is correctly adjusted as follows:
   a. Verify that the shift selector lever is in NEUTRAL.
   b. Install the digital TR sensor and loosely install the bolts.
   c. Using the tool illustrated, align the digital TR sensor and tighten the bolts.
   d. Connect the connector.
79. Install the manual control lever cable onto the bracket and tighten the nut to 18 ft. lbs. (25 Nm).
80. Install the electrical connector and wiring harness to the bracket.
81. Connect the connector.
82. Install the battery tray.
83. Install the air filter assembly.
84. Install the engine air cleaner.
85. Reconnect the breather tubes.
86. Reconnect the MAF sensor and the breather tubes.
87. Install the battery.
88. Connect the battery cables.
89. Fill the transaxle using clean automatic transmission fluid.
90. Start the engine.
91. Move the transaxle range selector lever through all positions.
92. Check the fluid level.
93. Check and adjust the wheel alignment.

Halfshaft

REMOVAL & INSTALLATION

2002–04 Models

1. Before servicing the vehicle, refer to the precautions section.
2. Remove or disconnect the following:
    • Front wheels
    • Lower ball joints
    • Outer tie rod ends
    • Stabilizer bar links
    • Height sensors, if equipped with air suspension

- Wheel speed sensors
- Steering knuckle from the halfshaft
- Halfshafts

To install:

➡ **Use new nuts, bolts and circlips.**

3. Install or connect the following:
- Halfshafts
- Steering knuckle to the halfshaft
- Hub retainer nut and tighten to 170–202 ft. lbs. (230–275 Nm)
- Wheel speed sensors
- Height sensors, if equipped with air suspension
- Stabilizer bar links and tighten the nuts to 57–75 ft. lbs. (77–103 Nm)
- Outer tie rod ends and tighten the nuts to 35–46 ft. lbs. (47–63 Nm)
- Lower ball joints and tighten the fasteners to 50–68 ft. lbs. (68–92 Nm)
- Front wheels

2005 Models

1. Before servicing the vehicle, refer to the precautions section.
2. The following are required when performing this procedure:
   a. A new front axle wheel hub retainer.
   b. A new nut to retain the knuckle to the strut.
   c. A new nut to retain the front suspension lower arm to the ball joint.
   d. A new circlip.

➡ **Do not reuse these parts during assembly. Their torque retention capability diminishes during removal.**

3. Remove the center hub cap from the wheel.
4. Loosen the axle nut.
5. Remove the front wheel and tire assembly.
6. Remove the axle nut and washer.
7. Remove the front brake anti-lock sensor harness from the clip.
8. Remove and position the front brake anti-lock sensor aside.
9. Remove and discard lower control arm to strut nut.
10. Remove the strut to knuckle nut and bolt. Discard the nut.
11. Press the ball joint stud until it is loose in the knuckle.

✳✳ WARNING

Wire the knuckle to the strut. This will prevent the knuckle from falling off of the strut while the knuckle and lower ball joint are separated. Failure to follow these instructions may result in personal injury.

➡ **Failure to reposition the knuckle on the strut body will prevent separation of the front suspension lower arm from the knuckle.**

12. Reposition the knuckle, and secure it to the strut.
13. Pull upward on the knuckle to raise it approximately 0.5 inch (12.7 mm) on the strut body. Wire the knuckle to the strut.
14. Disconnect the front suspension lower arm from the knuckle.

✳✳ CAUTION

Never use a hammer to separate the front wheel driveshaft joint from the wheel hub. Damage to the joint threads and internal components can result.

15. Press the front wheel driveshaft joint until it is loose in the wheel hub.

✳✳ CAUTION

Do not allow the halfshaft to hang unsupported. Damage to the CV joints and boots can result. Do not damage the front wheel driveshaft joint boot or the clamps.

16. Remove the front wheel driveshaft joint from the wheel hub.

✳✳ CAUTION

Do not damage the crank sensor when removing the right hand halfshaft assembly.

17. Separate the inboard CV joint housing assembly from the transaxle.

✳✳ CAUTION

Do not damage the differential seal.

18. Carefully remove the halfshaft assembly with both hands.
To install:
19. Lubricate the differential seal.
20. Use Ford Synthetic MERCON® Multi-Purpose Automatic Transmission Fluid E6AZ-19582-B or equivalent meeting Ford specification ESR-M2C163-A2.

✳✳ CAUTION

Install a new circlip every time you remove the halfshaft from the vehicle.

21. On the right hand side only, install a new circlip.
22. Start one end in the groove and work

the circlip over the shaft and into the groove. This will prevent the circlip from over-expanding.

✳✳ CAUTION

Do not damage the inboard CV joint housing assembly and the differential seal.

➡ **If necessary, use a non-metallic mallet to aid in seating the circlip in the differential side gear groove (right hand) or inboard CV joint housing (left). Tap only on the outboard front wheel driveshaft joint.**

23. Align the inboard CV joint housing splines with the differential side gear splines (right hand) or output shaft splines (left hand), and push the inboard CV joint housing assembly inward until the circlip seats in the differential side gear (right hand) or inboard CV joint housing (left hand).
24. Align the front wheel driveshaft joint splines and the wheel hub splines, and push the joint into the wheel hub as far as possible.
25. Connect the suspension lower arm to the knuckle.
26. Using the old nut, seat the front wheel driveshaft joint in the hub. Remove the old nut and discard it.
27. Install a new lower control arm nut and tighten to 59 ft. lbs. (80 Nm).
28. Remove the wire. Install the strut to knuckle bolt and nut tighten to 85 ft. lbs. (115 Nm).
29. Position the brake hose bracket and install the bolt.
30. Install the front brake anti-lock sensor.
31. Install the front brake anti-lock sensor harness in the clip.
32. Install the wheel and tire assembly.

✳✳ CAUTION

Install and tighten the new front axle wheel hub retainer to specification in a continuous rotation. Stopping the rotation during installation will cause the nylon lock to set incorrectly. This will cause incorrect torque readings while tightening the retainer, and lead to bearing failure. Install a new front axle wheel hub retainer every time after loosening the retainer or not installing the retainer to specification in a continuous rotation.

33. Tighten the new axle nut to 184 ft. lbs. (250 Nm) in a continuous rotation.

34. Check and, if necessary, fill the transaxle to the correct level.

35. Check and adjust the wheel alignment.

CV-Joint

REMOVAL & REPLACEMENT

Inner Tripod Joint

The inner CV-joint is serviced with the axle shaft as an assembly. The inner CV-joint boot can be serviced by removing the outer CV-joint.

Outer CV-Joint

1. Before servicing the vehicle, refer to the precautions section.
2. Place the halfshaft in a vise.
3. Remove the CV-joint boot clamps and slide the boot away from the joint.
4. Drive the CV-joint off the halfshaft with a brass drift and a hammer.

To install:

5. Replace the snapring.
6. Fill the CV-joint with fresh grease and slide the joint on to the halfshaft.
7. Use new clamps and install the CV-joint boot.

1. CV-Joint boot
2. CV-Joint

9306HG07

Removing the outer CV-joint

STEERING AND SUSPENSION

Air Bag

☀ CAUTION

Some vehicles are equipped with an air bag system. The system must be disarmed before performing service on, or around, system components, the steering column, instrument panel components, wiring and sensors. Failure to follow the safety precautions and the disarming procedure could result in accidental air bag deployment, possible injury and unnecessary system repairs.

PRECAUTIONS

Several precautions must be observed when handling the inflator module to avoid accidental deployment and possible personal injury.

• Never carry the inflator module by the wires or connector on the underside of the module.

• When carrying a live inflator module, hold securely with both hands, and ensure that the bag and trim cover are pointed away.

• Place the inflator module on a bench or other surface with the bag and trim cover facing up.

• With the inflator module on the bench, never place anything on or close to the module which may be thrown in the event of an accidental deployment.

DISARMING

1. Before servicing the vehicle, refer to the precautions section.

☀☀ WARNING

Always wear safety glasses when repairing an air bag supplemental restraint system (SRS) vehicle and when handling an air bag module. This will reduce the risk of injury in the event of an accidental deployment.

☀☀ WARNING

Never probe the connectors on the air bag module. Doing so can result in air bag deployment, which can result in personal injury.

☀☀ WARNING

The safety belt pretensioner is a pyrotechnic device. Always wear safety glasses when repairing an air bag equipped vehicle and when handling a safety belt buckle pretensioner or safety belt retractor pretensioner. Never probe a pretensioner electrical connector. Doing so could result in pretensioner or air bag deployment and could result in personal injury.

☀☀ WARNING

To reduce the risk of personal injury, do not use any memory saver devices.

➡ The air bag warning lamp illuminates when the RCM fuse is removed and the ignition switch is ON. This is normal operation and does not indicate a supplemental restraint system (SRS) fault.

➡ The SRS must be fully operational and free of faults before releasing the vehicle to the customer.

2. Turn all vehicle accessories OFF.
3. Turn the ignition switch to OFF.
4. At the Smart Junction Box (SJB), located below the left hand side of the instrument panel, remove the cover and the Restraints Control Module (RCM) fuse (10A) from the SJB.
5. Turn the ignition ON and visually monitor the air bag indicator for at least 30 seconds. The air bag indicator will remain lit continuously (no flashing) if the correct RCM fuse has been removed. If the air bag indicator does not remain lit continuously, remove the correct RCM fuse before proceeding
6. Turn the ignition switch to OFF.

☀☀ WARNING

To avoid accidental deployment and possible personal injury, the backup power supply must be depleted before repairing or replacing any front or side air bag supplemental restraint system (SRS) components and before servicing, replacing, adjusting or striking components near the front or side air bag sensors, such as doors, instrument panel, console, door latches, strikers, seats and hood latches.

➡ To deplete the backup power supply energy, disconnect the battery ground cable and wait at least one minute. Be sure to disconnect auxiliary batteries and power supplies (if equipped).

7. Disconnect the battery ground cable and wait at least one minute.

ARMING

1. Before servicing the vehicle, refer to the precautions section.

⁂ WARNING

The restraint system diagnostic tool is for restraint system service only. Remove from vehicle prior to road use. Failure to remove could result in injury and possible violation of vehicle safety standards.

2. Make sure all restraint system diagnostic tool(s) that may have been installed during the repair have been removed from the vehicle and all SRS components are connected.

3. Turn the ignition switch from OFF to ON.

4. Install the RCM fuse (10A) to the SJB and install the cover.

⁂ WARNING

Be sure that nobody is in the vehicle and that there is nothing blocking or set in front of any air bag module when the battery ground cable is connected.

5. Connect the battery ground cable.

6. Prove out the supplemental restraint system (SRS) as follows:

   a. Turn the ignition key from ON to OFF. Wait 10 seconds, then turn the key back to ON and visually monitor the air bag indicator with the air bag modules installed. The air bag indicator will light continuously for approximately six seconds and then turn off. If an air bag supplemental restraint system (SRS) fault is present, the air bag indicator will fail to light, remain lit continuously or flash.

   b. The flashing might not occur until approximately 30 seconds after the ignition switch has been turned from the OFF to the ON position. This is the time required for the restraints control module (RCM) to complete the testing of the SRS. If the air bag indicator is inoperative and a SRS fault exists, a chime will sound in a pattern of 5 sets of 5 beeps. If this occurs, the air bag indicator and any SRS fault discovered must be diagnosed and repaired.

   c. Clear all continuous DTCs from the restraints control module using a diagnostic tool.

Power Rack and Pinion Steering Gear

REMOVAL & INSTALLATION

2002–04 Models

1. Before servicing the vehicle, refer to the precautions section.

2. Remove or disconnect the following:
   - Negative battery cable
   - Intermediate steering shaft
   - Front wheels
   - Dual converter Y-pipe
   - Outer tie rod ends
   - Left stabilizer link
   - Heat shield
   - Power steering hose bracket
   - Auxiliary actuator connector, if equipped
   - Power steering pressure switch connector
   - Rear subframe bolts and lower the rear of the subframe about 4 inches
   - Steering gear mounting nuts, then move the steering gear to the left to access the power steering hoses
   - Power steering hoses
   - Steering gear through the left wheel opening

To install:

3. Install or connect the following:
   - Steering gear. Use new seals on the hydraulic fittings. Tighten the mounting nuts to 85–100 ft. lbs. (115–135 Nm).
   - Rear subframe bolts and tighten them to 57–76 ft. lbs. (77–103 Nm)
   - Power steering pressure switch connector
   - Auxiliary actuator connector, if equipped
   - Power steering hose bracket
   - Heat shield
   - Left stabilizer link
   - Outer tie rod ends and tighten the nuts to 35 ft. lbs. (48 Nm)
   - Dual converter Y-pipe
   - Front wheels
   - Intermediate steering shaft
   - Negative battery cable

4. Fill and bleed the power steering system. Check the system for leaks and proper operation.

5. Check the alignment and adjust as necessary.

Support the rear of the subframe with 2 tall jackstands while removing the steering gear

2005 Models

1. Before servicing the vehicle, refer to the precautions section.

2. Turn the steering wheel 1/4 turn to the right and turn the ignition switch to the OFF position.

3. Remove both front wheel and tire assemblies.

4. Loosen the tie-rod end jam nuts.

5. Remove and discard the cotter pins and nuts.

✳✳ CAUTION

Do not damage the tie rod end boot when installing the special tool.

6. Disconnect the tie-rod ends.

➡**Note the number of times the tie-rod ends turn for assembly reference.**

7. Remove the tie-rod ends.

8. Position the boot out of the way.

9. Remove the bolt and disconnect the intermediate shaft coupling. Discard the bolt.

10. Remove and discard the subframe nuts as shown.

11. Support the rear of the front subframe with jack stands.

12. Remove and discard the bolts. Lower the rear of the front subframe approximately 4 inches (102 mm).

13. On vehicles equipped with a 3.0L 4V DOHC engine, remove the bolts and the bracket with the heat shield attached.

14. On vehicles equipped with a 3.0L 2V MFI engine, remove the bolt and detach the bracket from the steering gear.

15. Loosen the fittings and disconnect the hoses. Discard the seals.

16. Remove the steering gear out of the left fender well.

17. Using the tool 211-D027, install new seals on the power steering hose fittings.

To install:

18. Install the steering gear.

19. Connect the hoses and tighten the fittings to 27 ft. lbs. (37 Nm) .

20. On vehicles equipped with a 3.0L 2V MFI engine, attach the bracket to the steering gear. Tighten the bolt to 8 ft. lbs. (11 Nm).

21. On vehicles equipped with a 3.0L 4V DOHC engine, install the bracket with the heat shield attached.

22. Guide the steering column intermediate shaft coupling on the steering gear input shaft as the subframe is raised into position.

23. Raise the rear of the front subframe approximately 4 inches (102 mm). Tighten the bolts to 85 ft. lbs. (115 Nm).

24. Remove the jack stands.

25. Install new subframe nuts as shown in the earlier illustration and tighten the bolts to 85 ft. lbs. (115 Nm).

26. Center, and then turn the steering gear input shaft 1/4 turn to the right to make sure the intermediate shaft coupling will fit correctly on the input shaft.

27. Connect the intermediate shaft coupling. Install a new bolt and tighten to 35 ft. lbs. (47 Nm).

28. Install the boot.

29. Install the tie-rod ends and tighten the same number of turns as noted during removal.

30. Connect the tie rod ends and using new nuts, tighten to 41 ft. lbs. (55 Nm). Install a new cotter pin.

31. Tighten the jam nuts to 41 ft. lbs. (55 Nm).

32. Install both front wheel and tire assemblies.

33. Fill and leak check the system.

34. Check and adjust front alignment as necessary.

Strut

REMOVAL & INSTALLATION

Front

2002–04 MODELS

1. Before servicing the vehicle, refer to the precautions section.

2. Remove or disconnect the following:
- Negative battery cable
- Front wheels
- Height sensor and wiring, if equipped with air suspension
- Brake hose bracket
- Disc brake caliper and rotor
- Wheel speed sensor and wiring
- Outer tie rod end
- Stabilizer link

06017-TAUR-G66

Remove and discard the subframe nuts as shown—2005 models

Front suspension component identification

Item	Description
1	Front Coil Spring
2	Front Shock Absorber
3	Tie Rod End
4	Wheel Hub

Item	Description
5	Front Wheel Knuckle
6	Front Suspension Lower Arm
7	Front Stabilizer Bar
8	Stabilizer Bar Link

7922KG61

Item	**Description**
1	Rear Stabilizer Bar
2	Rear Spring
3	Shock Absorber
4	Nut
5	Washer
6	Lower Suspension Arm (Rear)
7	Rear Wheel Spindle
8	Nut (4 Req'd)
9	Washer (2 Req'd)

Item	**Description**
10	Rear Suspension Tie Rod Bushing (4 Req'd)
11	Washer (2 Req'd)
12	Rear Suspension Tension Strut and Bushing (2 Req'd)
13	Rear Suspension Lower Arm (Front)
A	Tighten to 68-92 N·m (50-67 Lb-Ft)
B	Tighten to 46.7-63.3 N·m (35-46 Lb-Ft)

7922KG62

Rear suspension component identification—sedan models

- Steering knuckle pinch bolt
- Strut assembly

To install:

3. Install or connect the following:
- Strut assembly. Tighten the upper nuts to 22–29 ft. lbs. (30–40 Nm), and the knuckle pinch bolt to 73–97 ft. lbs. (98–132 Nm).
- Stabilizer link and tighten the nut to 55–75 ft. lbs. (75–101 Nm)
- Outer tie rod end and tighten the nut to 35 ft. lbs. (48 Nm)
- Wheel speed sensor and wiring
- Disc brake caliper and rotor and tighten the caliper anchor bracket bolts to 65–87 ft. lbs. (88–118 Nm)
- Brake hose bracket
- Height sensor and wiring, if equipped with air suspension
- Front wheels
- Negative battery cable

4. Road test the vehicle and check for proper operation.

2005 MODELS

1. Before servicing the vehicle, refer to the precautions section.

※※ WARNING

All vehicles are equipped with gas-pressurized front shock absorbers which will extend unassisted. Do not apply heat or flame to the front shock absorber during removal or disassembly. Failure to follow these instructions may result in personal injury.

※※ CAUTION

Suspension fasteners are critical parts because they affect performance of vital components and systems and their failure can result in major service expense. A new part with the same part number must be installed if installation is necessary. Do not use a new part of lesser quality or substitute design. Torque values must be used as specified during reassembly to make sure of correct retention of these parts.

➡Turn the ignition switch to position 1.

➡It is not necessary to remove the wheel hub and bearing or the disc brake dust shield from the knuckle.

2. Remove the wheel knuckle as follows:

※※ CAUTION

Discard the wheel hub retainer. It is a torque prevailing design and cannot be reused. If loosened, the retainer must be removed and a new one installed.

➡Make sure the steering wheel is in the unlocked position.

a. Remove the hub cap or wheel cover, the wheel hub retainer, and the washer. Discard the retainer.
b. Remove the front brake rotor.
c. Remove and discard the tie rod cotter pin and nut.

※※ CAUTION

Do not damage the tie-rod end boot when installing the removal tool 211-001.

➡Remove the adapter from the ball end of the removal tool. Apply a small amount of grease to the tie-rod end stud and the ball of the special tool.

d. Using the removal tools illustrated, disconnect the tie-rod end.
e. Remove the bolt and detach the ABS sensor, if equipped. Unclip the ABS sensor wire retainer and position the sensor out of the way.
f. Remove and discard the strut to knuckle nut.
g. Remove and discard the ball joint nut.
h. Using the tools illustrated, disconnect the ball joint from the lower arm.
i. Using a pry bar, push the lower arm down until the ball joint is free of the arm.

06017-TAUR-G67

Using the removal tools illustrated, disconnect the tie-rod end—2005 models

Using the tools illustrated, disconnect the ball joint from the lower arm—2005 models

06017-TAUR-G68

Using the tools illustrated, press the half-shaft from the wheel bearing and hub—2005 models

06017-TAUR-G69

✳✳ CAUTION

Do not allow the halfshaft to move outboard. Over-extension of the tripod CV joint can result in separation of internal parts, causing failure of the halfshaft.

➡**Hold the threaded rod while turning the nut.**

   j. Using the tools illustrated, press the halfshaft from the wheel bearing and hub. Wire the halfshaft in a level position.

   k. Remove the flag bolt and the knuckle.

✳✳ CAUTION

Do not use power tools to remove the nut. Damage to the boot or ball joint can result.

➡**To remove the nut, first loosen the nut, then use the hex holding feature to keep the stabilizer bar link ball joint from turning while removing the nut.**

   3. Remove the nut and disconnect the stabilizer bar link. Discard the nut.

   4. Remove the bolt and detach the brake hose bracket.

   5. Unclip the ABS sensor wire retainer and position the sensor out of the way.

   6. Lower the vehicle far enough to access the strut and spring assembly-to-body nuts.

   7. If necessary, detach the washer hose retainer.

   8. Remove the nuts and the strut and spring assembly. Discard the nuts.

   To install:

   NOTE: Make sure the spring is correctly seated in both spring seats

   9. Install the spring assembly and tighten the nuts to 26 ft. lbs. (35 Nm).

   10. If removed, attach the washer hose retainer.

   11. Install the ABS sensor.

   12. Attach the brake hose bracket.

➡**Use the hex holding feature to keep the stabilizer bar link ball joint from turning while installing the nut.**

   13. Connect the stabilizer bar link. Install a new nut and tighten to 66 ft. lbs. (90 Nm).

   14. Position the knuckle on the strut and install the flag bolt.

   15. Using the tools used during removal, press the halfshaft into the wheel bearing and hub.

   16. Using a pry bar, push the lower arm down until the ball joint can be installed.

   17. Install a new nut and tighten to 59 ft. lbs. (80 Nm).

   18. Install a new strut to knuckle nut and tighten to 85 ft. lbs. (115 Nm).

   19. Install the ABS sensor wire retainer and sensor, if equipped.

➡**Make sure the ABS sensor mounting hole does not have debris or corrosion build up, or false ABS activation can occur.**

➡**If the hole in the tie-rod end does not line up with the slots in the nut, continue tightening the nut. Do not loosen the nut to align the slots in the nut with the hole in the outer tie-rod end.**

   20. Connect the tie-rod end to the knuckle and install a new nut and cotter pin. Tighten the nut to 41 ft. lbs. (55 Nm).

   21. Install the brake rotor.

✳✳ CAUTION

The front axle wheel hub retainer must be tightened to specification immediately during installation. If the retainer is not tightened immediately, the nylon lock will set incorrectly, leading to incorrect torque readings and bearing failure. Any front wheel hub retainer that is not immediately tightened to specification or is loosened must be removed and a new retainer installed.

   22. Install the washer, the wheel hub retainer, tighten to 184 ft. lbs. (250 Nm). Install the hub cap or wheel cover.

Rear

2002–04 MODELS

   1. Before servicing the vehicle, refer to the precautions section.

   2. Remove or disconnect the following:

- Rear package tray trim panel
- Rear wheels
- Brake load sensor
- Brake hose
- Stabilizer bar bracket and link
- Tension strut
- Spindle pinch bolt
- Strut assembly

To install:

➡**Use new mounting nuts and bolts.**

   3. Install or connect the following:

- Strut assembly. Use a new pinch bolt and tighten to 50–67 ft. lbs. (68–92 Nm). Tighten the 3 upper mounting nuts to 19–25 ft. lbs. (25–34 Nm).
- Tension strut and tighten the nut to 35–46 ft. lbs. (68–92 Nm)
- Stabilizer bar link and tighten the nut to 60–81 inch lbs. (7–9 Nm)
- Stabilizer bar bracket and tighten the bolts to 25 –33 ft. lbs. (34–46 Nm)
- Brake hose
- Brake load sensor
- Rear wheels
- Rear package tray trim panel

2005 MODELS

   1. Before servicing the vehicle, refer to the precautions section.

   2. Measure the distance from the center of the hub to the lip of the fender with the vehicle in a level, static ground position (curb height).

   3. Remove the parcel tray.

   4. Loosen the upper strut nuts.

   5. Remove the wheel and tire assembly.

➡**It is not necessary to remove the stabilizer bar links from the stabilizer bar.**

   6. Position a jackstand at the front of the vehicle under the subframe.

   7. Detach the front muffler insulator from the muffler.

   8. Position jackstands under the rear suspension and raise the suspension until the distance between the center of the hub and the lip of the fender is equal to the measurement taken previously (curb height).

   9. Remove the nuts, washers, bushings,

and bolts. Detach the stabilizer bar assembly from the body and the links. Discard the nuts and bolts.

10. Lower the suspension and remove the jack stands.

11. Carefully remove the stabilizer bar assembly out of the right wheel well.

12. Remove the bolt and detach the rear wheel brake hose.

13. Remove the nut, washer and bushing. Detach the tension strut from the spindle and discard the nut.

❊ CAUTION

Care should be taken when removing the strut and spring assembly so the rear wheel brake hose is not stretched.

14. Remove and discard the pinch bolt. Detach the spindle from the strut and spring assembly. To aid in removal, use a large prytool to spread the pinch joint.

15. Partially lower the vehicle.

16. Remove the nuts and the strut and spring assembly and discard the nuts.

To install:

17. Position the strut and spring assembly and install new nuts but do not tighten the nuts at this time.

18. Partially raise the vehicle.

19. Position the spindle on the strut and install a new pinch bolt. Tighten to 59 ft. lbs. (80 Nm).

➡**Make sure the cupped side of the washer is facing away from the bushing.**

20. Position the tension strut in the spindle and install the bushing, the washer and a new nut. Tighten the nut to 41 ft. lbs. (55 Nm).

21. Position the rear wheel brake hose and install the bolt.

22. Install the stabilizer bar, if equipped. Tighten the bar to body bolts to 30 ft. lbs. (40 Nm) and the bar to link nuts to 71 inch lbs. (8 Nm).

23. Install the wheel and tire assembly.

24. Tighten the upper strut nuts to 22 ft. lbs. (30 Nm).

25. Install the parcel tray

Shock Absorber

REMOVAL & INSTALLATION

Wagons

1. Before servicing the vehicle, refer to the precautions section.

2. Remove the rear wheels and support the rear control arms on jackstands.

80 Nm (59 lb-ft)

9346KG34

Shock absorber lower mounting bolt—Wagon

3. Remove or disconnect the following:
- Rear compartment access panels
- Upper shock mounting nuts and insulators
- Lower shock mounting bolts
- Shock absorbers

To install:

➡**Use new mounting fasteners and insulators.**

4. Install or connect the following:
- Shock absorbers. Tighten the lower bolt to 50–68 ft. lbs. (68–92 Nm), and the upper nuts to 19–25 ft. lbs. (26–34 Nm).
- Rear compartment access panels
- Rear wheels

Coil Spring

REMOVAL & INSTALLATION

Struts

1. Before servicing the vehicle, refer to the precautions section.

2. Remove the strut from the vehicle.

3. Compress the coil spring using a suitable spring compressor until the spring comes away from the seat.

4. Remove the large center nut and slowly release the spring compressor.

To install:

5. Compress the spring and install it on the strut.

6. Install the lower washer and mounting bracket.

7. Install the upper washer and a new nut. Tighten the nut to 46 ft. lbs. (62 Nm).

8. Install the strut assembly in the vehicle.

Wagons with rear shock absorbers

1. Before servicing the vehicle, refer to the precautions section.

2. Remove or disconnect the following:
- Rear shock absorber
- Stabilizer bar link and bracket

Item	Description
1	Dust Boot (Part of 18124)
2	Nut (3 Req'd)
3	Washer
4	Nut
5	Front Shock Absorber Mounting Bracket
6	Washer
7	Front Suspension Bearing and Seal
8	Front Spring Insulator (Part of 18124)
9	Front Coil Spring
10	Front Shock Absorber
11	Jounce Bumper (Part of 18124)

7922KG63

Exploded view of the front strut and coil spring assembly

10 mm SIX POINT DEEPWELL SOCKET

OXYGEN SENSOR WRENCH

7922KG64

Hold the strut rod while loosening or tightening the nut

- Brake hose bracket
- Upper ball joint

3. Install spring keepers on the coil springs.

4. Slowly lower the lower control arm until the tension is relaxed on the coil spring. Remove the coil spring and the upper and lower spring insulators.

To install:

➡ **Use new mounting nuts and bolts.**

5. Install or connect the following:
- Coil spring with upper and lower spring insulators
- Upper ball joint and tighten the nut to 46 ft. lbs. (52 Nm)
- Brake hose bracket
- Stabilizer bar link and bracket
- Rear shock absorber

Lower Ball Joints

REMOVAL & INSTALLATION

The lower ball joint is an integral part of the steering knuckle. If the lower ball joint is found to be defective, the entire steering knuckle must be replaced.

Upper Control Arm

REMOVAL & INSTALLATION

Rear

2002–04 WAGON MODELS

1. Before servicing the vehicle, refer to the precautions section.
2. Support the lower control arm on a jackstand.
3. Remove or disconnect the following:
- Rear wheel
- Brake hose bracket
- Upper ball joint
- Upper control arm

To install:

➡ **Use new mounting nuts and bolts.**

4. Install or connect the following:
- Upper control arm. Tighten the ball joint nut to 50–67 ft. lbs. (68–92 Nm), then tighten the control arm mounting bolts to 73–97 ft. lbs. (98–132 Nm).
- Brake hose bracket
- Rear wheel
5. Check the wheel alignment and adjust as necessary.

2005 WAGON MODELS

1. Before servicing the vehicle, refer to the precautions section.

✳✳ CAUTION

Suspension fasteners are critical parts because they affect performance of vital components and systems and their failure can result in major service expense. A new part with the same part number, or an equivalent part, must be installed if installation is necessary. Do not use a new part of lesser quality or substitute design. Torque values must be used as specified to make sure of correct retention of these parts.

2. Remove the hub cap or wheel cover.
3. Measure the distance from the center of the hub to the lip of the fender with the vehicle in a level, static ground position (curb height).
4. Remove the wheel and tire assembly.
5. Position a jack stand under the lower control arm and raise the suspension slightly.
6. Remove the bolt and detach the shock absorber. Discard the bolt.
7. Remove the ball joint stud nut. Discard the nut.
8. Using the tools illustrated, separate the upper control arm from the ball joint.
9. Remove and discard the two flag nuts and bolts.
10. Remove the upper control arm.

To install:

11. Position the upper control arm and install two new bolts and flag nut but do not tighten the bolts at this time.
12. Attach the upper control arm and install a new nut. Tighten the nut to 59 ft. lbs. (80 Nm).
13. Using a jack stand, raise the suspension until the distance between the center of the hub and the lip of the fender is equal to the measurement taken in the removal procedure (curb height).
14. Tighten the two upper bolts to 85 ft. lbs. (115 Nm).
15. Attach the shock absorber. Install a new bolt and tighten to 59 ft. lbs. 80 Nm).
16. Lower the suspension and remove the jack stand.
17. Install the wheel and tire assembly.
18. Check and adjust the wheel alignment as necessary

CONTROL ARM BUSHING REPLACEMENT

The control arm bushings are serviced with the control arm as an assembly.

Lower Control Arm

REMOVAL & INSTALLATION

Front

1. Before servicing the vehicle, refer to the precautions section.
2. Remove or disconnect the following:
- Front wheel
- Wheel speed sensor wiring harness
- Lower ball joint
- Lower control arm

To install:

3. Install or connect the following:
- Lower control arm. Tighten the front bolt to 57–75 ft. lbs. (77–103 Nm) on 2002–04 models or 98 ft. lbs. (133 Nm) on 2005 models, and the rear bolt to 72–97 ft. lbs.

06017-TAUR-G70

Using the tools illustrated, separate the upper control arm from the ball joint—2005 wagon models

(98–132 Nm) on 2002–04 models or 85 ft. lbs. (115 Nm) on 2005 models .

- Lower ball joint. Use a new nut and tighten to 50–67 ft. lbs. (68–92 Nm) on 2002–04 models or 59 ft. lbs. (80 Nm) on 2005 models.
- Wheel speed sensor wiring harness
- Front wheel

4. Check the wheel alignment and adjust as necessary.

Rear

SEDAN

1. Before servicing the vehicle, refer to the precautions section.
2. Remove or disconnect the following:
   - Rear wheel
   - Parking brake cable
   - Proportioning valve
   - Lower control arm

To install:

➡Use new bolts and nuts.

➡The rear suspension lower control arms are marked BOTTOM on the lower edge. The flange edge of the right side rear suspension arm and bushing stamping must face the front of the vehicle. The other three must face the rear of the vehicle.

➡The rear suspension arms have two adjustment cams that fit inside the bushings at the arm-to-body attachment. Each adjustment cam is installed from the front on the rear suspension arm and bushing.

3. Install or connect the following:
   - Lower control arm and tighten the bolts to 50–67 ft. lbs. (68–92 Nm)
   - Proportioning valve
   - Parking brake cable
   - Rear wheel

4. Check the wheel alignment and adjust as necessary.

2002–04 WAGON

1. Before servicing the vehicle, refer to the precautions section.
2. Remove or disconnect the following:
   - Rear wheel
   - Coil spring
   - Wheel spindle
   - Lower control arm

To install:

➡Use new nuts and bolts.

➡Tighten the control arm fasteners with the vehicle weight resting on the wheels.

3. Install or connect the following:
   - Lower control arm
   - Wheel spindle
   - Coil spring
   - Rear wheel

4. Tighten the control arm-to-body bolt to 40–52 ft. lbs. (54–71 Nm), and the arm-to-spindle bolt to 50–67 ft. lbs. (68–92 Nm).

5. Check the wheel alignment and adjust as necessary.

2005 WAGON

1. Before servicing the vehicle, refer to the precautions section.
2. Remove the hub cap or wheel cover.
3. Measure the distance from the center of the hub to the lip of the fender with the vehicle in a level, static ground position (curb height).
4. Remove the wheel and tire assembly.
5. Detach the parking brake cable and, if equipped, the rear anti-lock brake sensor wiring harness.
6. To remove the front and rear lower control arms:
   a. Remove the nut, washer and flag bolt. Detach the lower control arm from the spindle. Discard the nut and flag bolt.
7. To remove the front lower control arm:
   a. Remove and discard the nut and flag bolt. Remove the front lower control arm and the adjusting cam.
8. To remove the rear lower control arm:
   a. Remove the nut, flag bolt, and rear lower control arm. Discard the nut and flag bolt.

To install:

➡When installing lower control arms, the offset must face up. The arms are stamped "BOTTOM" on the lower edge. The flanges of the rear lower control arm must face the front of the vehicle. The flanges on the front control arms must face the rear of the vehicle.

9. To install the rear lower control arm:
   a. Position the rear lower suspension arm and install a new flag bolt, and a new nut but do not tighten the nut at this time.
10. To install the front lower control arm:
   a. Position the front lower control arm and the adjusting cam, and install a new flag bolt and nut but do not tighten the nut at this time.
   b. Position the adjusting cam in the neutral position. Refer to the accompanying illustration for further details.

Item	Description
1	90° (maximum toe out [in])
2	Adjustment cam
3	90° (maximum toe in [out])
4	Notch pointing down (neutral position)

06017-TAUR-G71

Position the adjusting cam in the neutral position—2005 wagon models

11. To install the front and rear lower control arms:

a. Position the suspension arm and install a new flag bolt, the washer, and a new nut.

b. Using a jack stand, raise the suspension until the distance between the center of the hub and the lip of the fender is equal to the measurement taken in Removal procedure (curb height).

12. Tighten the control arm nuts on all arms to 59 ft. lbs. (80 Nm).

13. Lower the suspension and remove the jack stand.

14. On the front lower control arm, attach the rear brake anti-lock sensor wiring harness and the parking brake cable.

15. Install the wheel and tire assembly.

16. Check and adjust the wheel alignment as necessary

CONTROL ARM BUSHING REPLACEMENT

The control arm bushings are serviced with the control arm as an assembly.

Wheel Bearings

ADJUSTMENT

There is no adjustment for the front or rear wheel bearings due to the nature of their design. These bearings are permanently lubricated and require no periodic maintenance.

REMOVAL & INSTALLATION

Front

2002–04 MODELS

1. Before servicing the vehicle, refer to the precautions section.

2. Remove or disconnect the following:
- Front wheel
- Hub retainer nut
- Brake caliper and rotor
- Outer tie rod end
- Stabilizer bar link
- Wheel speed sensor
- Lower ball joint
- Steering knuckle
- Hub and bearing assembly

To install:

➡**Use new nuts, bolts, and split pins.**

➡**The knuckle must be clean enough to allow the wheel hub to be completely seated by hand. Do not press or draw the wheel hub into place.**

3. Install or connect the following:
- Hub and bearing assembly and

tighten the bolts to 61–78 ft. lbs. (83–107 Nm)
- Steering knuckle and tighten the pinch bolt to 72–97 ft. lbs. (98–132 Nm)
- Lower ball joint and tighten the nut to 50–67 ft. lbs. (68–92 Nm)
- Wheel speed sensor
- Stabilizer bar link and tighten the nut to 57–75 ft. lbs. (77–103 Nm)
- Outer tie rod end and tighten the nut to 35–46 ft. lbs. (47–63 Nm)
- Brake caliper and rotor and tighten the caliper anchor bracket bolts to 65–87 ft. lbs. (88–118 Nm)
- Hub retainer nut and tighten to 170–202 ft. lbs. (230–275 Nm)
- Front wheel

2005 MODELS

1. Before servicing the vehicle, refer to the precautions section.

✳✳ CAUTION

Discard the wheel hub retainer. It is a torque prevailing design and cannot be reused. If loosened, the retainer must be removed and a new one installed.

➡**Make sure the steering wheel is in the unlocked position.**

2. Remove the hub cap or wheel cover, the wheel hub retainer, and the washer. Discard the retainer.

3. Remove the front brake rotor.

a. Remove and discard the tie rod cotter pin and nut.

✳✳ CAUTION

Do not damage the tie-rod end boot when installing the removal tool 211-001.

06017-TAUR-G67

Using the removal tools illustrated, disconnect the tie-rod end—2005 models

06017-TAUR-G68

Using the tools illustrated, disconnect the ball joint from the lower arm—2005 models

➡**Remove the adapter from the ball end of the removal tool. Apply a small amount of grease to the tie-rod end stud and the ball of the special tool.**

4. Using the removal tools illustrated, disconnect the tie-rod end.

5. Remove the bolt and detach the ABS sensor, if equipped. Unclip the ABS sensor wire retainer and position the sensor out of the way.

6. Remove and discard the strut to knuckle nut.

7. Remove and discard the ball joint nut.

8. Using the tools illustrated, disconnect the ball joint from the lower arm.

9. Using a pry bar, push the lower arm down until the ball joint is free of the arm.

✳✳ CAUTION

Do not allow the halfshaft to move outboard. Over-extension of the tripod CV joint can result in separation of internal parts, causing failure of the halfshaft.

➡**Hold the threaded rod while turning the nut.**

10. Using the tools illustrated, press the halfshaft from the wheel bearing and hub. Wire the halfshaft in a level position.

06017-TAUR-G69

Using the tools illustrated, press the halfshaft from the wheel bearing and hub—2005 models

11. Remove the flag bolt and the steering knuckle.

✷✷ CAUTION

The wheel bearing and hub is not pressed into the knuckle. Do not use a slide hammer or strike the inner race of the bearing. Damage to the bearing can result.

➡️**If necessary, lubricate the knuckle and bearing with a rust inhibitor.**

12. Remove the bolts and the wheel hub and bearing. Discard the bolts.

✷✷ CAUTION

The knuckle bore must be clean enough to allow the wheel bearing and hub to seat completely by hand. Do not press or draw the wheel hub and bearing into place.

13. Clean and inspect the knuckle bearing bore. If the knuckle is cracked, install a new knuckle.

To install:

14. Lubricate the knuckle bearing bore with anti-seize lubricant

15. Install the wheel bearing and hub and tighten to 70 ft. lbs. (95 Nm).

16. Position the knuckle on the strut and install the flag bolt.

17. Using the tools used during removal, press the halfshaft into the wheel bearing and hub.

18. Using a pry bar, push the lower arm down until the ball joint can be installed.

19. Install a new nut and tighten to 59 ft. lbs. (80 Nm).

20. Install a new strut to knuckle nut and tighten to 85 ft. lbs. (115 Nm).

21. Install the ABS sensor wire retainer and sensor, if equipped.

➡️**Make sure the ABS sensor mounting hole does not have debris or corrosion build up, or false ABS activation can occur.**

➡️**If the hole in the tie-rod end does not line up with the slots in the nut, continue tightening the nut. Do not loosen the nut to align the slots in the nut with the hole in the outer tie-rod end.**

22. Connect the tie-rod end to the knuckle and install a new nut and cotter pin. Tighten the nut to 41 ft. lbs. (55 Nm).

23. Install the brake rotor.

✷✷ CAUTION

The front axle wheel hub retainer must be tightened to specification immediately during installation. If the retainer is not tightened immediately, the nylon lock will set incorrectly, leading to incorrect torque readings and bearing failure. Any front wheel hub retainer that is not immediately tightened to specification or is loosened must be removed and a new retainer installed.

24. Install the washer, the wheel hub retainer, tighten to 184 ft. lbs. (250 Nm). Install the hub cap or wheel cover.

Rear

1. Before servicing the vehicle, refer to the precautions section.

2. Remove or disconnect the following:

- Rear wheel
- Brake hose bracket
- Brake caliper and rotor, if equipped with rear disc brakes
- Brake drum, if equipped with rear drum brakes
- Hub and bearing assembly grease cap and retaining nut
- Hub and bearing assembly

To install:

➡️**Use new retaining nuts and grease caps**

3. Install or connect the following:

- Hub and bearing assembly and tighten the retaining nut to 188–254 ft. lbs. (255–345 Nm) on 2002–04 models, or 221 ft. Lbs. (300 Nm) on 2005 models
- Grease cap
- Brake caliper and rotor, if equipped with rear disc brakes
- Brake drum, if equipped with rear drum brakes
- Brake hose bracket
- Rear wheel

BRAKES

Brake Caliper

REMOVAL & INSTALLATION

Front

1. Before servicing the vehicle, refer to the precautions section.

2. Remove brake fluid from the brake master cylinder reservoir until the reservoir is ½ full.

3. Remove the wheel and tire assembly.

4. Mark the disc brake caliper to ensure that it is reinstalled in the correct location.

5. Remove the hollow bolt connecting the brake hose to the disc brake caliper and plug the brake hose. Discard the 2 copper sealing washers.

6. Remove the caliper locating pins and lift the caliper off the rotor using a rotating motion.

To install:

7. Retract the disc brake caliper piston fully in the piston bore, using an old brake pad or block of wood and a C-clamp.

➡️**Make sure the clip-on insulators are attached to the brake pads.**

8. Install the disc brake pads to the caliper. Make sure the brake pad insulators are correctly attached to the brake pad plate.

9. Position the disc brake caliper and pad assembly above the rotor and install it with a rotating motion. Make sure the inner and outer pads are properly positioned and the outer anti-rattle spring is properly positioned.

10. Lubricate the locating pins and the inside of the insulators with silicone grease. Torque the locating pins to 26 ft. lbs. (35 Nm).

11. Remove the plug and install the brake hose to the disc brake caliper. Use 2 new copper washers and torque the hollow bolt to 35 ft. lbs. (48 Nm).

12. Bleed the brake system, filling the master cylinder as required.

13. Install the wheel and tire assembly; torque the nuts to 85–104 ft. lbs. (115–142 Nm).

14. Pump the brake pedal several times to position the brake pads prior to moving the vehicle.

15. Road test the vehicle and check for proper brake system operation.

Rear

1. Before servicing the vehicle, refer to the precautions section.

2. Remove brake fluid from the brake master cylinder reservoir until the reservoir is ½ full.

3. Remove the wheel and tire assembly.

4. Remove the retaining bolt and disconnect the brake hose from the caliper

assembly. Discard the copper sealing washers.

5. Remove the retaining clip from the parking brake at the caliper. Disengage the parking brake cable end from the lever arm.

6. Lift the rear disc brake caliper away from the rear disc support bracket.

7. Remove the disc brake caliper locating pins and boots from the rear disc support bracket.

To install:

8. Using rear caliper piston adjuster tool T87P-2588-A, rotate the rear disc brake piston and adjuster clockwise until fully seated.

➡**Make sure one of the 2 slots in the rear disc brake piston and adjuster face is positioned so it will engage the nib on the disc brake pad.**

9. Apply silicone dielectric compound to the inside of the slider pin boots and the slider pins.

10. Position the slider pins and boots in the support bracket. Position the caliper assembly on the support bracket. Make sure the brake pads are installed correctly.

11. Remove the residue from the pin retainer threads and apply 1 drop of threadlock and sealer. Install the pin retainers and torque to 24 ft. lbs. (33 Nm).

12. Attach the cable end to the parking brake lever. Install the cable retaining clip on the caliper assembly.

13. Using new washers, connect the brake flex hose to the caliper. Torque the retaining bolt to 41 ft. lbs. (55 Nm).

14. Bleed the brake system, filling the master cylinder as required.

15. Install the wheel and tire assembly; torque the nuts to 85–104 ft. lbs. (115–142 Nm).

16. Pump the brake pedal several times to position the brake pads prior to moving the vehicle.

17. Road test the vehicle and check for proper brake system operation.

Disc Brake Pads

REMOVAL & INSTALLATION

Front

1. Before servicing the vehicle, refer to the precautions section.

2. Remove the master cylinder reservoir cap and check the fluid level in the reservoir. Remove brake fluid until the reservoir is ½ full. Discard the removed fluid.

3. On Continentals, turn the air suspension switch, located in the left side of the luggage compartment, to the **OFF** position.

4. Remove the wheel and tire assembly.

5. Remove the disc brake caliper locating pins. Lift the caliper assembly from the anchor plate and rotor using a rotating motion.

6. Suspend the caliper inside the fender housing with wire. Do not allow the caliper to hang from the brake hose.

7. Remove the inner and outer brake pads. Inspect the rotor braking surfaces for scoring and machine as necessary.

To install:

8. Use a C-clamp and an old brake pad or block of wood to seat the caliper piston in its bore.

9. Remove any rust buildup from the inside of the caliper in the brake pad contact area.

10. Install the inner pad in the caliper piston.

11. Install the outer pad onto the anchor plate. Make sure the clips are properly seated.

➡**Make sure the insulators are installed on the brake pads.**

12. Install the disc brake caliper onto the anchor plate.

13. Install caliper locating pins and torque to 26 ft. lbs. (35 Nm).

14. Install wheel and tire assembly.

15. Pump the brake pedal several times prior to moving the vehicle to position the brake pads to the rotor.

16. Refill the master cylinder reservoir as

93006G28

Front disc brake caliper, pads and related components

necessary, using only clean DOT 3 brake fluid from a closed container.

17. Road test the vehicle and check the brake system for proper operation.

Rear

1. Before servicing the vehicle, refer to the precautions section.

2. Remove the master cylinder reservoir cap and check the fluid level in the reservoir. Remove brake fluid until the reservoir is ½ full. Discard the removed fluid.

3. Remove the wheel and tire assembly.

4. Remove the screw retaining the brake hose bracket to the frame side rail.

5. Remove the retaining clip from the parking brake cable at the disc brake caliper. Remove the cable end from the parking brake lever.

6. Remove the upper disc brake caliper locating pin at the support bracket. Rotate the caliper away from the rotor.

7. Remove the disc brake pads.

8. Inspect the rotor braking surfaces for scoring and machine as necessary.

To install:

9. Using Rear Caliper Piston Adjuster T87P-2588-A, rotate the piston clockwise until it is fully seated. Make sure one of the slots in the piston face is positioned so it will engage the nib on the brake pad.

10. Install the brake pads in the support bracket. Rotate the caliper assembly over the rotor into position on the support

bracket. Make sure the brake pads are installed correctly.

11. Remove the residue from the rear brake pin retainer bolt threads and apply 1 drop of a suitable threadlock sealer. Install and torque the disc brake caliper locating pin to 24 ft. lbs. (33 Nm).

12. Attach the cable end to the parking brake lever. Install the cable retaining clip on the caliper assembly. Position the brake flex hose and bracket assembly to the side rail, and install the retaining screw. Torque to 11 ft. lbs. (16 Nm).

13. Install the wheel and tire assembly.

14. Pump the brake pedal several times prior to moving the vehicle, to position the brake pads to the rotor.

15. Refill the master cylinder reservoir if necessary, using only clean DOT 3 brake fluid from a closed container.

16. Road test the vehicle and check the brake system for proper operation.

Brake Drums

REMOVAL & INSTALLATION

1. Before servicing the vehicle, refer to the precautions section.
2. Remove the wheel and tire assembly.
3. Remove the brake drum.

➡If the brake drum cannot be removed easily, remove the brake tube-to-axle retention bracket and pry rubber plug from rear brake backing plate inspec-

tion hole. This will allow sufficient room for insertion of a screwdriver and brake tools to disengage brake shoe adjusting lever and back off the brake adjuster screw.

4. Inspect the drum for scoring and/or other wear. Machine or replace, as necessary.

To install:

5. Measure the brake drum inside diameter using a brake adjustment gauge.

6. Using the brake adjustment gauge, adjust the brake shoes to the same dimensions as the brake drum.

7. Position the brake drum over the brake shoes on the axle hub.

8. Install the wheel and tire assembly.

9. Pump the brake pedal several times to position the brake shoes and complete the adjustment.

10. Road test the vehicle and check for proper brake system operation.

Brake Shoes

REMOVAL & INSTALLATION

1. Before servicing the vehicle, refer to the precautions section.

2. Remove the wheel and tire assembly.

3. Remove the brake drum.

4. Remove the parking brake cable from the parking brake lever.

5. Remove the 2 brake shoe hold-down springs and pins.

Retracting the brake shoes to allow drum removal—Taurus and Sable

93006G39

1	Boot
2	Spring Expander
3	Rear Wheel Cylinder
4	Piston and Insert
5	Shoe Adjustment Access Hole
6	Wheel Cylinder Retaining Bolt (2 Req'd)
7	Brake Adjusting Hole Cover
8	Brake Shoe Hold-Down Spring Pin
9	Rear Brake Backing Plate
10	Trailing Shoe and Lining
11	Brake Shoe Hold-Down Spring
12	Brake Shoe Retracting Spring
13	Parking Brake Lever

14	Parking Brake Lever Pin (Inner)
15	Brake Shoe Adjusting Screw Socket
16	Brake Shoe Adjusting Lever
17	Parking Brake Lever Pin
18	Leading Shoe and Lining
19	Brake Shoe Adjusting Screw Spring
20	Cup
21	Washer
22	Brake Adjuster Screw
23	Washer
24	Parking Brake Lever Pin Retainer
25	Adjusting Pivot Nut
A	Tighten to 12-18 N·m (107-159 Lb-In)

Brake shoes and related components—Taurus and Sable

93006G56

6. Lift the brake shoes, springs and adjuster assembly off the backing plate and wheel cylinder assembly. When removing the assembly, be careful not to bend the adjusting lever.

7. Remove the retracting springs from the lower brake attachments and upper shoe-to-adjusting lever attachment points.

8. Remove the horseshoe retaining clip and spring washer and slide the lever off the parking brake lever pin on the trailing shoe. Discard the horseshoe clip.

To install:

9. Apply a light coating of disc brake caliper slide grease at the points where the brake shoes contact the backing plate.

10. Apply a thin coat of lubricant to the adjuster screw threads and socket end of the adjusting screw. Install the stainless steel washer over the socket end of the adjusting screw and install the socket. Turn the adjusting screw into the adjusting pivot nut to the limit of the threads and then back off ½ turn.

11. Assemble the parking brake lever to the trailing shoe by installing the spring washer and a new horseshoe retaining clip. Crimp the clip until it retains the lever to the shoe securely.

12. Position the trailing shoe on the backing plate and attach the rear parking brake cable.

13. Position the leading shoe on the backing plate and attach the lower brake shoe adjusting spring to the brake shoes.

14. Install the adjuster assembly in the slots on the brake shoes. The wide slot on the dual slotted end must fit into the leading shoe. The narrow slot on the dual slotted end fits into the shoe adjusting lever. The single slotted side of the adjuster assembly must fit into the slots on the trailing shoe and the rear parking brake cable bracket.

➡ **The adjuster socket blade is marked R for the right or L for the left brake assemblies. The adjuster blade must be installed with the letter R or L in the upright position, facing the wheel cylinder. Make sure the adjuster socket fits into the parking brake lever.**

15. Complete the installation by reversing the removal procedures.

16. Pump the brake pedal several times to position the brake shoes and finish the brake shoe adjustment.

17. Road test the vehicle and check the brake system for proper operation.

FORD

Thunderbird

16

SPECIFICATIONS AND MAINTENANCE CHARTS

ENGINE AND VEHICLE IDENTIFICATION

Code ①	Engine Liters (cc)	Cu. In.	Cyl.	Fuel Sys.	Type	Eng. Mfg.
A	3.9 (3947)	243	8	MFI	DOHC	Ford

MFI: Multi-Port Fuel Injection
DOHC: Double Overhead Camshaft
① 8th digit of the VIN
② 10th digit of the VIN

Model Year Code ②	Year
2	2002
3	2003
4	2004
5	2005

06017-TBRD-C01

GENERAL ENGINE SPECIFICATIONS

Year	Model	Engine Displacement Liters	Engine VIN	Net Horsepower @ rpm	Net Torque @ rpm (ft. lbs.)	Bore x Stroke (in.)	Compression Ratio	Oil Pressure @ rpm
2002	Thunderbird	3.9	A	252@6100	267@4300	3.38x3.34	10.6:1	NA
2003	Thunderbird	3.9	A	280@6000	286@4000	3.38x3.34	10.6:1	NA
2004	Thunderbird	3.9	A	280@6000	286@4000	3.38x3.34	10.6:1	61-73@400
2005	Thunderbird	3.9	A	280@6000	286@4000	3.39x3.35	10.75:1	61-73@400

NA: Not Available

06017-TBRD-C02

ENGINE TUNE-UP SPECIFICATIONS

Year	Engine Displacement Liters	Engine VIN	Spark Plugs Gap (in.)	Ignition Timing (deg.) ① MT	Ignition Timing (deg.) ① AT	Fuel Pump (psi)	Idle Speed (rpm) ① MT	Idle Speed (rpm) ① AT	Valve Clearance In.	Valve Clearance Ex.
2002	3.9	A	0.039-0.043	—	10-20B	43	—	650-750	0.007-0.009	0.009-0.011
2003	3.9	A	0.039-0.043	—	10-20B	43	—	650-750	0.007-0.009	0.009-0.011
2004	3.9	A	0.039-0.043	—	10-20B	43	—	650-750	0.007-0.009	0.013-0.015
2005	3.9	A	0.039-0.043	—	NA	NA	—	650-750	0.007-0.009	0.013-0.015

The underhood specifications sticker often reflects tune-up specification changes in production. Sticker figures must be used if they disagree with those in this chart.
① Controlled by the engine computer

06017-TBRD-C03

3.9L (VIN A) engine
Firing order: 1–5–4–2–6–3–7–8
Distributorless Ignition System (One coil on each plug)

1 Belt idler pulley—unflanged
2 Generator pulley
3 Water pump pulley
4 Drive belt tensioner
5 Power steering pump pulley
6 A/C clutch pulley
7 Crankshaft vibration damper

Serpentine accessory drive belt routing—2002–2003 3.9L (VIN A) engine

Serpentine accessory drive belt routing—2004–05 3.9L (VIN A) engine

CAPACITIES

Year	Model	Engine Displacement Liters	Engine VIN	Engine Oil with Filter (qts.)	Transmission (pts.)		Drive Axle Rear (pts.)	Fuel Tank (gal.)	Cooling System (qts.)
					Manual	Auto.			
2002	Thunderbird	3.9	A	NA	—	23.8	3.0	18.0	11.3
2003	Thunderbird	3.9	A	NA	—	23.8	3.0	18.0	11.3
2004	Thunderbird	3.9	A	6.9	—	23.8	2.6	18.0	11.9
2005	Thunderbird	3.9	A	6.9	—	23.8	2.6	18.0	11.9

N/A: Not Available

06017-TBRD-C04

VALVE SPECIFICATIONS

Year	Engine Displacement Liters	Engine VIN	Seat Angle (deg.)	Face Angle (deg.)	Spring Test Pressure (lbs. @ in.)	Spring Free Length (in.)	Stem-to-Guide Clearance (in.)		Stem Diameter (in.)	
							Intake	Exhaust	Intake	Exhaust
2002	3.9	A	NA	NA	NA	NA	NA	NA	NA	NA
2003	3.9	A	NA	NA	NA	NA	NA	NA	NA	NA
2004	3.9	A	①	45	NA	NA	0 0.001	0.001	0.196-0.197	0.195-0.196
2005	3.9	A	①	45	NA	NA	0 0.001	0.001	0.196-0.197	0.195-0.196

NA: Not Available

① Intake: 14 degrees
   Exhaust: 13 degrees

06017-TBRD-C05

CRANKSHAFT AND CONNECTING ROD SPECIFICATIONS
All measurements are given in inches.

Year	Engine Displacement Liters	Engine VIN	Crankshaft				Connecting Rod		
			Main Brg. Journal Dia.	Main Brg. Oil Clearance	Shaft End-play	Thrust on No.	Journal Diameter	Oil Clearance	Side Clearance
2002	3.9	A	NA	NA	NA	NA	NA	NA	NA
2003	3.9	A	NA	NA	NA	NA	NA	NA	NA
2004	3.9	A	NA	NA	NA	NA	NA	NA	NA
2005	3.9	A	NA	NA	NA	NA	NA	NA	NA

NA: Not Available

06017-TBRD-C06

PISTON AND RING SPECIFICATIONS
All measurements are given in inches.

Year	Engine Displacement Liters	Engine VIN	Piston Clearance	Ring Gap			Ring Side Clearance		
				Top Compression	Bottom Compression	Oil Control	Top Compression	Bottom Compression	Oil Control
2002	3.9	A	NA	NA	NA	NA	NA	NA	NA
2003	3.9	A	NA	NA	NA	NA	NA	NA	NA
2004	3.9	A	NA	NA	NA	NA	NA	NA	NA
2005	3.9	A	NA	NA	NA	NA	NA	NA	NA

NA: Not Available

06017-TBRD-C07

TORQUE SPECIFICATIONS
All readings in ft. lbs.

Year	Engine Displacement Liters	Engine VIN	Cylinder Head Bolts	Main Bearing Bolts	Rod Bearing Bolts	Crankshaft Damper Bolts	Flywheel Bolts	Manifold		Spark Plugs	Oil Pan Drain Plug
								Intake	Exhaust		
2002	3.9	A	①	NA	NA	②	③	18	18	19	17
2003	3.9	A	①	NA	NA	②	③	18	18	19	17
2004	3.9	A	①	NA	NA	②	③	18	18	19	17
2005	3.9	A	①	NA	NA	②	③	18	18	20	17

NA: Not Available

① Step 1: Tighten M10 bolts to 15 ft. lbs.
Step 2: Tighten M10 bolts to 26 ft. lbs.
Step 3: Tighten M10 bolts to 33 ft. lbs.
Step 4: Tighten M10 bolts plus 90 degrees
Step 5: Tighten M10 bolts plus 90 degrees
Step 6: Tighten M8 bolts to 15 ft. lbs.
Step 7: Tighten M8 bolts plus 90 degrees

② Step 1: 59 ft. lbs.
Step 2: Plus 80 degrees

③ Step 1: 11 ft. lbs.
Step 2: 81 ft. lbs.

06017-TBRD-C08

WHEEL ALIGNMENT

Year	Model		Caster		Camber		Toe-in (in.)
			Range (+/-Deg.)	Preferred Setting (Deg.)	Range (+/-Deg.)	Preferred Setting (Deg.)	
2002	Thunderbird	F	0.70	0	0.70	0	0.08 +/- 0.13
		R	—	—	0.75	0	0.13 +/- 0.13
2003	Thunderbird	F	0.70	0	0.70	0	0.08 +/- 0.13
		R	—	—	0.75	0	0.13 +/- 0.13
2004	Thunderbird	F	0.70	0	0.70	0	0.08 +/- 0.13
		R	—	—	0.75	0	0.13 +/- 0.13
2005	Thunderbird	F	0.70	0	0.70	0	0.08 +/- 0.13
		R	—	—	0.75	0	0.13 +/- 0.13

06017-TBRD-C09

TIRE, WHEEL AND BALL JOINT SPECIFICATIONS

Year	Model	OEM Tires		Tire Pressures (psi)		Wheel Size	Ball Joint Inspection	Wheel Lug Nut Torque
		Standard	Optional	Front	Rear			
2002	Thunderbird	P235/50VR17	—	30	30	7-J/7.5J	U ① L ① ②	100
2003	Thunderbird	P235/50VR17	—	30	30	7-J/7.5J	U ① L ① ②	100
2004	Thunderbird	P235/50VR17	—	30	30	7-J/7.5J	U ① L ① ②	100
2005	Thunderbird	P235/50VR17	—	30	30	7-J/7.5J	U ① L ① ②	100

OEM: Original Equipment Manufacturer

PSI: Pounds Per Square Inch

STD: Standard

OPT: Optional

L: Lower

U: Upper

① Replace if any measurable movement is found.

② Do not lift car. Inspect the boss into which the grease fitting is threaded. Replace if the boss is flush or receded below the surface of the ball joint.

06017-TBRD-C10

BRAKE SPECIFICATIONS
All measurements in inches unless noted

Year	Model		Brake Disc			Minimum Lining Thickness	Brake Caliper	
			Original Thickness	Minimum Thickness	Maximum Runout		Bracket Bolts (ft. lbs.)	Mounting Bolts (ft. lbs.)
2002	Thunderbird	F	1.180	1.120	0.004	0.079	76	26
		R	0.810	0.740	0.004	0.039	76	25
2003	Thunderbird	F	1.180	1.120	0.004	0.079	76	26
		R	0.810	0.740	0.004	0.039	76	25
2004	Thunderbird	F	1.180	1.120	0.004	0.079	76	26
		R	0.810	0.740	0.004	0.039	76	25
2005	Thunderbird	F	NA	1.120	NA	0.118	76	26
		R	NA	0.752	NA	0.118	76	26

06017-TBRD-C11

SCHEDULED MAINTENANCE INTERVALS
Ford—Thunderbird

TO BE SERVICED	TYPE OF SERVICE	VEHICLE MILEAGE INTERVAL (x1000)												
		5	10	15	20	25	30	35	40	45	50	55	60	65
Air cleaner filter	R			✓			✓			✓			✓	
Accessory drive belt	S/I												✓	
Automatic Transmission Fluid	R			✓			✓							
Brake system ①	S/I	✓	✓	✓	✓	✓	✓	✓	✓	✓	✓	✓	✓	✓
Cooling system hoses and clamps	S/I			✓			✓			✓			✓	
CV-joint boots & axle seals	S/I			✓			✓			✓			✓	
Engine coolant	R	Five years or 100,000 miles												
Engine oil & filter	R	Every 3 months or 3,000 miles												
Exterior Lights	S/I	Check monthly												
Fuel Filter	R			✓			✓			✓			✓	
PCV valve	S/I												✓	
Exhaust system & heat shields	S/I						✓						✓	
Parking brake system	S/I	Every 6 months												
Power steering fluid	S/I	Every 6 months												
Rotate tires	S/I	✓		✓		✓		✓		✓		✓		✓
Steering linkage	S/I	Every 3 months or 3,000 miles												
Spark plugs	R												✓	
Suspension components	S/I						✓						✓	

R: Replace S/I: Inspect and service, if necessary L: Lubricate A: Adjust C: Clean

① Inspect the reservoir fluid level, rotor and or drum, brake lines, hoses, calipers and or wheel cylinders

Special Operating Condition Requirements

When towing a trailer or using a camper or car-top carrier:

Change engine oil and install a new oil filter every 4,800 km (3,000 miles) or 3 months.

Change transfer case fluid every 96,000 km (60,000 miles).

Change manual transmission fluid as required.

Inspect and lubricate U-joints as required.

During extensive idling and/or low speed driving for long distances, as in heavy commercial use such as delivery, taxi, patrol car or livery:

Change engine oil and install a new oil filter, lube front lower control arm and steering linkage ball joints with zerk fittings (if equipped) every 4,800 km (3,000 miles) or 3 months.

Inspect brake system and check battery electrolyte level (Patrol cars) every 8,000 km (5,000 miles).

Install a new fuel filter every 24,000 km (15,000 miles).

Change automatic transmission fluid, lubricate 4x2 wheel bearings,

Install new spark plugs and change transfer case fluid every 96,000 km (60,000 miles).

Install a new cabin air filter as required.

When operating in dusty conditions such as unpaved or dusty roads:

Change engine oil and install a new oil filter every 4,800 km (3,000 miles) or 3 months.

Install a new fuel filter every 24,000 km (15,000 miles).

Change automatic transmission fluid every 48,000 km (30,000 miles).

Install a new engine air filter as required.

Install a new cabin air filter as required.

When operating in off-road conditions:

Change automatic transmission fluid every 48,000 km (30,000 miles).

Install a new cabin air filter as required.

Inspect and lubricate U-joints.

Inspect and lubricate steering linkage ball joints with zerk fittings.

ENGINE REPAIR

➡**Disconnecting the negative battery cable on some vehicles may interfere with the functions of the on board computer system. The computer may undergo a relearning process once the negative battery cable is reconnected.**

Alternator

REMOVAL

1. Before servicing the vehicle, refer to the Precautions Section.
2. Remove or disconnect the following:
   - Negative battery cable
   - Air intake tube
   - Accessory drive belt
   - Lower splash shield
   - Alternator mounting bolts
   - Alternator harness connectors
   - Alternator

INSTALLATION

1. Install the alternator harness connectors, then, install the alternator. Tighten the bolts as follows:
   a. Step 1: Tighten bolt No. 1 to 15 ft. lbs. (21 Nm)
   b. Step 2: Tighten nut No. 3 to 33 ft. lbs. (45 Nm)
2. Install or connect the following:
   - Lower splash shield
   - Accessory drive belt
   - Air intake tube
   - Negative battery cable

Ignition Timing

ADJUSTMENT

This vehicle is equipped with a Distributorless Ignition System (DIS). The ignition timing is not adjustable. It is controlled by the PCM.

Engine Assembly

REMOVAL & INSTALLATION

2002–2004 Models

1. Before servicing the vehicle, refer to the Precautions Section.
2. Secure the steering wheel in the straight-forward position.
3. Drain the cooling system.
4. Relieve the fuel system pressure.
5. Recover the A/C refrigerant.
6. Drain the engine oil.
7. Remove or disconnect the following:
   - Negative battery cable
   - Air cleaner inlet tube
   - Upper radiator shield
   - Upper radiator support brackets
   - A/C pressure switch connector
   - Power steering return line clip
   - Power steering reservoir
   - Vapor Management Valve (VMV) cover
   - VMV vacuum hose and canister purge hose
   - Main vacuum supply hose
   - Cowl vent screens
   - Cross vehicle support bar
   - Degas bottle hose
   - Accelerator cable
   - Cruise control cable
   - Ground strap
   - Fresh air filter and housing
   - Powertrain harness connectors at right strut tower
   - Fresh air filter panel
   - Main engine wiring harness connector
   - Main transmission wiring harness connector
   - Heater hoses at the water control valve. Note the locations for assembly.
   - Hydraulic cooling fan reservoir
   - Water control valve harness connector
   - Driveshaft
   - Front wheels
   - Inner splash shields
   - Wheel speed sensor connectors and harness clips
   - Brake calipers
   - Lower stabilizer bar links
   - Upper ball joints
   - Lower strut mount bolts

⑤ 8 Nm (71 lb-in)	
① 21 Nm (15 lb-ft)	⑧ 48 Nm (35 lb-ft)
③ 45 Nm (33 lb-ft)	

1 Generator upper bolt	4 Generator	7 Connector
2 Generator lower bolt	5 Generator B+ wire nut	8 Generator bracket bolts
3 Generator lower nut	6 Generator B+ wire	9 Generator bracket

67197-THUN-G01

Alternator mounting—3.9L engine

- Left, right and center splash shields
- A/C suction and discharge lines
- Shift cable and bracket
- Power steering line frame rail clip
- Rack and pinion harness connectors
- Steering shaft bolt and coupling
- Starter motor harness connectors and ground cable
- Alternator harness connectors
- Lower transmission flange bolts
- Torque converter
- Inner air deflector
- Engine block heater, if equipped

8. Support the engine, transmission, front and center crossmembers and the cooling system with a powertrain lift and transmission support bracket.

9. Support the rear of the vehicle with safety stands.

10. Remove or disconnect the following:
- Transmission crossmember bolts
- Front crossmember bolts
- Center crossmember bolts

11. Carefully lower the powertrain assembly from the vehicle.

12. Remove or disconnect the following:
- A/C compressor manifold and tube assembly
- Power steering pump return hose
- Hydraulic cooling fan return hose
- Lower radiator hose
- Upper radiator hoses
- Oil cooler hoses, if equipped
- Knock Sensor (KS) connector
- Heater hose
- Transmission cooler lines and bracket
- Power steering pressure line and bracket
- Hydraulic cooling fan pressure line and bracket

13. Attach a hoist to engine lifting brackets 303-050 and support the engine and transmission.

14. Remove the motor mount nuts and lift the powertrain out of the subframe.

15. Remove or disconnect the following:
- Wiring harness retainers
- Upper transmission flange bolts
- Transmission from the engine

To install:

16. Install or connect the following:
- Transmission to the engine. Tighten the upper flange bolts to 35 ft. lbs. (48 Nm).
- Wiring harness retainers. Tighten the nuts to 89 inch lbs. (10 Nm).
- Powertrain to the subframe. Tighten the mount nuts to 30 ft. lbs. (40 Nm).

- Hydraulic cooling fan pressure line and bracket
- Power steering pressure line and bracket
- Transmission cooler lines and bracket
- Heater hose
- Knock sensor connector
- Upper radiator hoses
- Lower radiator hose
- Hydraulic cooling fan return hose
- Power steering pump return hose
- A/C compressor manifold and tube assembly. Tighten the bolt to 15 ft. lbs. (21 Nm).
- Powertrain assembly. Tighten the front and center crossmember bolts to 76 ft. lbs. (103 Nm) and the transmission crossmember bolts to 30 ft. lbs. (40 Nm).
- Engine block heater, if equipped
- Inner air deflector
- Torque converter. Tighten the nuts to 28 ft. lbs. (38 Nm).
- Lower transmission flange bolts. Tighten the bolts to 35 ft. lbs. (47 Nm).
- Alternator harness connectors
- Starter motor harness connectors and ground cable
- Steering shaft bolt and coupling. Tighten the coupling pinch bolt to 26 ft. lbs. (35 Nm) and the shaft bolt to 22 ft. lbs. (30 Nm).
- Rack and pinion harness connectors
- Power steering line frame rail clip
- Shift cable and bracket
- A/C suction and discharge lines
- Left, right and center splash shields
- Lower strut mount bolts. Tighten the bolts to 129 ft. lbs. (175 Nm).
- Upper ball joints. Tighten the nuts to 66 ft. lbs. (90 Nm).
- Lower stabilizer bar links. Tighten the nuts to 41 ft. lbs. (55 Nm).
- Brake calipers
- Wheel speed sensor connectors and harness clips
- Driveshaft
- Inner splash shields
- Front wheels
- Water control valve harness connector
- Hydraulic cooling fan reservoir
- Heater hoses at the water control valve
- Main transmission wiring harness connector
- Main engine wiring harness connector
- Fresh air filter panel

- Powertrain harness connectors at right strut tower
- Fresh air filter and housing
- Ground strap
- Cruise control cable
- Accelerator cable
- Degas bottle hose
- Cross vehicle support bar. Tighten the bolts to 15 ft. lbs. (20 Nm).
- Cowl vent screens
- Main vacuum supply hose
- VMV vacuum hose and canister purge hose
- VMV cover
- Power steering reservoir
- Power steering return line clip
- A/C pressure switch connector
- Upper radiator support brackets
- Upper radiator shield
- Air cleaner inlet tube
- Negative battery cable

17. Fill the crankcase to the correct level.
18. Fill the cooling system.
19. Recharge the A/C system.
20. Start the engine and check for leaks.

2005 Models

1. Before servicing the vehicle, refer to the Precautions Section.

2. Secure the steering wheel in the straight-forward position.

3. Drain the cooling system.

4. Relieve the fuel system pressure.

5. Recover the A/C refrigerant.

6. Drain the engine oil.

7. Remove or disconnect the following:
- Negative battery cable
- Air intake assembly
- Fuel supply hose
- Upper radiator shield
- Two upper radiator support brackets
- Evaporative emissions (EVAP) hose from EVAP canister purge valve
- Vacuum hose
- Degas bottle hose
- Power steering reservoir mounting bolts
- Ground strap
- Engine wiring harness connectors on the backside of the strut tower
- Main wiring harness connectors
- Coolant valve electrical harness
- Cooling fan electrical connecter
- A/C pressure switch electrical connector
- Engine splash guards
- Inner air deflectors
- Catalytic converters

8. Matchmark the front driveshaft pinion flange.

9. Remove the 3 nuts, washers and bolts. Slide the driveshaft rearward and secure.

10. Remove or disconnect the following:
- Three heater hose quick release couplings
- A/C high pressure hose
- A/C low pressure hose
- Inner splash shields
- ABS electrical sensors
- Front brake calipers
- Stabilizer bar links from lower control arms
- Upper ball joints from wheel knuckles
- Two strut-to-control arm bolts and nuts
- Shift cable
- Shift cable bracket
- Variable Assist Power Steering (VAPS) actuator electrical connector
- Block heater
- Steering coupling
- Starter motor wiring
- Six lower transmission-to-engine bolts
- Torque converter

11. Support the rear of the vehicle with jack stands.

12. Support the engine, transmission, crossmembers, and the cooling system with a powertrain lift and a transmission support bracket.

13. Remove or disconnect the following:
- Four transmission crossmember mounting bolts
- Four front cross brace mounting bolts
- Front and center crossmember mounting bolts

14. Lower the powertrain assembly from the vehicle.

15. Remove or disconnect the following from the engine assembly:
- Upper and lower radiator hoses
- Heater hose
- Oil cooler hoses, if equipped
- Accessory drive belt
- A/C compressor and Power Steering Pressure (PSP) connectors
- A/C compressor and hang securely aside
- Power steering pump and reservoir and hang securely aside
- Transmission cooler tubes

16. Install Special Tool engine lifting eyelets 303-050.

17. Using an engine crane, support the engine and transmission assembly

18. Remove or disconnect the following the upper engine mount nuts.

Supporting the engine on the 014-00765 Powertrain Lift and 014-00341 Support Bracket.

19. Separate the engine/transmission assembly from the crossmember and cooling assembly.

20. Remove the retaining nuts and disconnect the wiring harness.

21. Remove the remaining transmission-to-engine mounting bolts.

To install:

22. Install or connect the following:
- Transmission-to-engine flange bolts and tighten to 35 ft. lbs (48 Nm)
- Wiring harness and tighten the retaining nuts to 89 inch lbs. (10 Nm)

23. Using an engine crane, move the engine and transmission assembly onto the crossmember and cooling assembly. Install the upper engine mount nuts and tighten to 37 ft. lbs. (50 Nm).

24. Install or connect the following to the engine assembly:
- Transmission cooler tubes and tighten to 26 ft. lbs. (35 Nm)
- Power steering pump and tighten to 18 ft. lbs. (25 Nm)
- A/C compressor and tighten bolts to 18 ft. lbs. (25 Nm)
- A/C compressor and PSP connectors
- Accessory drive belt
- Oil cooler hoses, if equipped
- Heater hose
- Upper and lower radiator hoses

25. Using a powertrain lift, raise the powertrain assembly into the vehicle.

Attaching the engine to a crane using 303-050 lifting eyelets.

26. Install or connect the following:
- Front and center crossmember bolts and tighten to 76 ft. lbs. (103 Nm)
- Front cross brace bolts and tighten to 46 ft. lbs. (63 Nm)
- Transmission crossmember bolts and tighten to 30 ft. lbs. (40 Nm)
- Torque converter and tighten nuts to 28 ft. lbs. (38 Nm)
- Six lower transmission-to-engine flange bolts and tighten to 35 ft. lbs. (48 Nm)
- Starter motor wiring
- Steering coupling
- Block heater
- VAPS actuator electrical connector
- Shift cable bracket and tighten to 89 inch lbs. (10 Nm)
- Shift cable
- Two strut-to-control arm bolts and tighten to 129 ft. lbs. (175 Nm)
- Upper ball joints to wheel knuckles
- Stabilizer bar links to lower control arms
- Front brake calipers
- ABS electrical sensors
- Inner splash shields
- A/C high and low pressure hoses
- Three heater hose quick release couplings
- Front driveshaft and tighten to nuts to 60 ft. lbs. (81 Nm)
- Catalytic converters
- Inner air deflector
- Engine splash guards
- A/C pressure switch electrical connector
- Cooling fan electrical connector
- Coolant valve electrical connector
- Main wiring harness electrical connectors and tighten 53 inch lbs. (6 Nm)
- Engine wiring harness connectors on the backside of the strut tower
- Ground strap and tighten to 89 inch lbs. (10 Nm)
- Power steering reservoir
- Degas bottle hose
- Vacuum hose
- EVAP hose
- Two upper radiator support brackets
- Upper radiator shield
- Fuel supply hose
- Air intake assembly
- Negative battery cable
27. Fill the engine with oil.
28. Recharge the A/C system.
29. Fill the cooling system to the correct level.
30. Start the engine and check for leaks.

Water Pump

REMOVAL & INSTALLATION

1. Before servicing the vehicle, refer to the Precautions Section.
2. Drain the cooling system.
3. Remove or disconnect the following:
- Thermostat housing
- Accessory drive belt
- Water pump pulley
- Water pump

To install:

4. Install or connect the following:
- Water pump, using a new gasket. Tighten the bolts to 71 inch lbs. (8 Nm) plus 90 degrees.
- Water pump pulley. Tighten the bolts to 89 inch lbs. (10 Nm) plus 45 degrees.
- Accessory drive belt
- Thermostat housing
5. Fill the cooling system.
6. Start the engine and check for leaks.

Heater Core

REMOVAL & INSTALLATION

1. Before servicing the vehicle, refer to the Precautions Section.
2. Drain the cooling system.
3. Recover the A/C refrigerant.
4. Remove or disconnect the following:
- Negative battery cable
- Heater hose assembly
- Cabin air filter plenum
- Thermostatic expansion valve manifold and tube assembly
- Powertrain Control Module (PCM) bracket bolt
- Driver's side air bag module
- Floor console and A/C duct
- Shift lever assembly
- Glove box assembly
- Left and right instrument panel insulators
- Left and right door sill scuff plates
- Left and right door weatherstrips
- Left and right A pillar lower trim panels
- Left and right windshield side garnish moldings

1 Coolant pump pulley bolts
2 Coolant pump pulley
3 Coolant pump bolts
4 Coolant pump
5 Coolant pump gasket

67197-THUN-G02

Water pump mounting—3.9L engine

- Instrument panel defroster opening grille assembly
- Instrument panel cowl top screws
- Instrument panel upper reinforcement bolts
- Left and right instrument panel side finish panels
- Hood release handle and cable
- Upper right bulkhead electrical connector
- Right instrument panel electrical connectors
- Passenger side tunnel electrical connector
- Steering column intermediate shaft
- Left junction box electrical connectors
- Left instrument panel electrical connectors
- Ignition shift interlock connector, if equipped
- 4 instrument panel tunnel brace bolts
- Left outer instrument panel cowl side cover and reinforcement bolt
- Left instrument panel cowl side bolt and nut
- Right instrument panel cowl side bolt and nut
- Instrument panel
- Evaporator housing electrical connector
- Cowl top attachment bolt
- Evaporator housing attachment bolt
- 3 evaporator nuts and washers in the engine compartment
- Rear seat floor ducts
- Evaporator core housing
- Evaporator core housing air inlet
- Bypass door harness connector
- Heater core

To install:

5. Install or connect the following:
   - Heater core
   - Bypass door harness connector
   - Evaporator core housing air inlet
   - Evaporator core housing
   - Rear seat floor ducts
   - 3 evaporator nuts and washers in the engine compartment
   - Evaporator housing attachment bolt
   - Cowl top attachment bolt
   - Evaporator housing electrical connector
   - Instrument panel
   - Instrument panel cowl top screws. Tighten to 27 inch lbs. (3 Nm).
   - Right instrument panel cowl side bolt and nut. Tighten the fasteners to 15 ft. lbs. (20 Nm).
   - Left instrument panel cowl side bolt

1 Heater hose clamp
2 Thermostatic expansion valve manifold bolt
3 PCM bracket bolt
4 Wire harness
5 Heater core and evaporator core housing bracket bolt
6 Heater core and evaporator core housing nut
7 Heater core and evaporator core housing
8 Thermostatic expansion valve O-ring

67197-THUN-G03

Heater core and evaporator core housing mounting—Thunderbird

9 Wire harness
10 Heater core and evaporator core housing screw
11 Heater core and evaporator core housing screw
12 Blower motor speed control screw
13 Blower motor speed control
14 Heater core tube seal
15 Heater core tube bracket screw
16 Heater core tube bracket
17 Heater core and evaporator core housing screw
18 Heater core and evaporator core housing cover
19 Heater core
20 Thermostatic expansion valve seal
21 Evaporator core
22 Thermostatic expansion valve

67197-THUN-G04

Exploded view of heater core—Thunderbird

and nut. Tighten the fasteners to 15 ft. lbs. (20 Nm).

- Left outer instrument panel cowl side cover and reinforcement bolt. Tighten the bolt to 15 ft. lbs. (20 Nm).
- Instrument panel upper reinforcement bolts. Tighten the bolts to 15 ft. lbs. (20 Nm).
- 4 instrument panel tunnel brace bolts. Tighten the bolts to 15 ft. lbs. (20 Nm).
- Ignition shift interlock connector, if equipped
- Left instrument panel electrical connectors
- Left junction box electrical connectors
- Steering column intermediate shaft. Tighten the pinch bolt to 26 ft. lbs. (35 Nm).
- Passenger side tunnel electrical connector
- Right instrument panel electrical connectors
- Upper right bulkhead electrical connector
- Hood release handle and cable
- Left and right instrument panel side finish panels
- Instrument panel defroster opening grille assembly
- Left and right windshield side garnish moldings
- Left and right A pillar lower trim panels
- Left and right door weatherstrips
- Left and right door sill scuff plates
- Left and right instrument panel insulators
- Shift lever assembly, if equipped with automatic transmission
- Floor console and A/C duct
- Driver's side air bag module
- Thermostatic expansion valve manifold and tube assembly
- Cabin air filter plenum
- Heater hose assembly
- Negative battery cable
6. Fill the cooling system.
7. Recharge the A/C system.
8. Start the engine and check for leaks.

Cylinder Head

REMOVAL & INSTALLATION

2002 Models

1. Before servicing the vehicle, refer to the Precautions Section.

2. Drain the cooling system.
3. Relieve the fuel system pressure.
4. Remove or disconnect the following:
   - Negative battery cable
   - Engine appearance cover
   - Intake manifold
   - Valve covers
   - Accessory drive belts
   - Front cover
   - Timing chains
   - Camshafts
   - Water outlet pipe
   - Cylinder head temperature sensor connector
   - Exhaust front pipes
   - Exhaust Gas Recirculation (EGR) tube

- Bolts and stud bolts at the rear of the cylinder heads
- Cylinder heads. Loosen the bolts in reverse of the tightening sequence.

To install:

5. Install the cylinder heads. Tighten the NEW bolts in sequence as follows:
   a. Step 1: M10 bolts to 15 ft. lbs. (20 Nm)
   b. Step 2: M10 bolts to 26 ft. lbs. (35 Nm)
   c. Step 3: M10 bolts to 33 ft. lbs. (45 Nm)
   d. Step 4: M10 bolts plus 90 degrees
   e. Step 5: M10 bolts plus 90 degrees
   f. Step 6: M8 bolts to 15 ft. lbs. (20 Nm)

67197-THUN-G05

Right cylinder head torque sequence—3.9L engine

67197-THUN-G06

Left cylinder head torque sequence—3.9L engine

g. Step 7: M8 bolts plus 90 degrees

6. Install or connect the following:
   - Bolts and stud bolts at the rear of the cylinder heads. Tighten the bolts to 37 ft. lbs. (50 Nm).
   - EGR tube
   - Exhaust front pipes
   - Cylinder head temperature sensor connector
   - Water outlet pipe
   - Camshafts
   - Timing chains
   - Front cover
   - Accessory drive belts
   - Valve covers
   - Intake manifold
   - Engine appearance cover
   - Negative battery cable

7. Fill the cooling system.

8. Start the engine and check for leaks.

To install:

9. Install the exhaust manifold with a new gasket onto the cylinder head. Tighten bolts to 18 ft. lbs. (25 Nm).

10. Install the cylinder head with a new gasket and tighten as follows:
   a. Step 1: M10 bolts to 15 ft. lbs. (20 Nm)
   b. Step 2: M10 bolts to 26 ft. lbs. (35 Nm)
   c. Step 3: M10 bolts to 33 ft. lbs. (45 Nm)
   d. Step 4: M10 bolts plus 90 degrees
   e. Step 5: M10 bolts plus 90 degrees
   f. Step 6: M8 bolts to 15 ft. lbs. (20 Nm)
   g. Step 7: M8 bolts plus 90 degrees

11. Install or connect the following:
   - Valves and shims
   - EGR valve
   - Catalytic converter to the exhaust manifolds
   - CHT sensor
   - Engine sound insulator
   - Water outlet pipe
   - Camshafts
   - Intake manifold

2003–2005 Models

1. Before servicing the vehicle, refer to the Precautions Section.

2. Remove the engine from the vehicle and mount on a suitable stand.

3. Remove or disconnect the following:
   - Oil filter
   - Oil cooler
   - Oil pressure and oil temperature sensors
   - Accessory drive belt
   - Idler pulley
   - Water pump pulley

Left valve cover torque sequence—3.9L Engine

4. Hold the crankshaft pulley with Special Tool 303-D055 and remove the main bolt.

5. Remove the crankshaft damper using Special Tool 303-009.

6. Remove or disconnect the following:
   - Front crankshaft seal
   - A/C compressor
   - Power steering pump
   - Ignition coil covers

   - Throttle position (TP) sensor
   - Fuel pressure sensor
   - Electronic throttle body electrical connector
   - Fuel rail temperature sensor
   - Oil level indicator tube
   - Vacuum harnesses
   - Evaporative emissions (EVAP) hose
   - Engine appearance cover
   - Engine wiring harness

Right valve cover torque sequence—3.9L Engine

- Ignition coils
- Crankcase vent tube
- EGR system module
- Coolant bypass housing
- Throttle body heater hoses
- Throttle body
- Intake manifold
- Right hand valve cover, removing bolts in sequence
- Left hand valve cover, removing bolts in sequence
- Exhaust manifold heat shields
- Right hand exhaust manifold
- Left hand exhaust manifold
- Front cover
- Variable Cam Timing (VCT) bush carriers
- Timing chain tensioners
- Primary timing chains
- VCT units
- Secondary timing chains
- Camshaft sprockets
- Crankshaft sprocket
- Camshafts
- Valves and valve shims
- Cylinder heads. Loosen the bolts in reverse of the tightening sequence.

To install:

7. Install the cylinder heads with a new gasket and tighten as follows:

   a. Step 1: M10 bolts to 15 ft. lbs. (20 Nm)

   b. Step 2: M10 bolts to 26 ft. lbs. (35 Nm)

   c. Step 3: M10 bolts to 33 ft. lbs. (45 Nm)

   d. Step 4: M10 bolts plus 90 degrees

   e. Step 5: M10 bolts plus 90 degrees

   f. Step 6: M8 bolts to 15 ft. lbs. (20 Nm)

   g. Step 7: M8 bolts plus 90 degrees

8. Install or connect the following:

- Valves and valve shims
- Camshafts
- Crankshaft sprocket
- Camshaft sprockets
- Secondary timing chains
- VCT units
- Primary timing chains
- Timing chain tensioners
- VCT bush carriers
- Front carrier
- Left hand exhaust manifold with a new gasket
- Right hand exhaust manifold with a new gasket
- Exhaust manifold heat shields
- Left and right valve covers. Tighten the bolts to using the sequence shown.
- Intake manifold with a new gasket
- Throttle body

06017-TBRD-G03

Holding crankshaft pulley with Special Tool 303-D055 to remove crankshaft bolt.

- Throttle body hoses
- Coolant bypass housing
- EGR system module
- Crankcase vent tube
- Ignition coils
- Engine wiring harness
- Engine appearance cover
- Oil level indicator tube
- Vacuum harnesses
- Evaporative emissions (EVAP) hose
- Throttle position (TP) sensor
- Fuel pressure sensor
- Electronic throttle body electrical connector
- Fuel rail temperature sensor

- Power steering pump
- Ignition coil covers
- A/C compressor
- Crankshaft front seal
- Crankshaft pulley using Special Tool 303-102
- Water pump pulley
- Idler pulley
- Accessory drive belt
- Oil pressure and oil temperature sensors
- Oil cooler
- Oil filter

9. Install the engine assembly into the vehicle.

06017-TBRD-G04

Removing crankshaft damper using Special tool 303-009.

06017-TBRD-G05

Install the crankshaft pulley using Special Tool 303-102.

10. Fill the cooling system to the correct level.
11. Fill the engine with oil
12. Start the engine and check for leaks.

Rocker Arms/Shafts

REMOVAL & INSTALLATION

The vehicles covered in this section are not equipped with rocker arms/shafts. The camshaft directly actuates the valves.

Intake Manifold

REMOVAL & INSTALLATION

1. Before servicing the vehicle, refer to the Precautions Section.
2. Drain the cooling system.
3. Relieve the fuel system pressure.
4. Remove or disconnect the following:
   - Negative battery cable
   - Air cleaner outlet tube
   - Cowl vent screen
   - Cross vehicle support bar
   - Fuel line
   - Exhaust Gas Recirculation (EGR) valve
   - Throttle Position (TP) sensor connector
   - Fuel pressure sensor connector and vacuum line
   - Throttle plate motor connector
   - Vapor management valve connector
   - Fuel rail temperature sensor connector
   - Accelerator cable
   - Cruise control cable
   - Intake manifold vacuum hoses
   - Idle Air Control (IAC) valve connector
   - Positive Crankcase Ventilation (PCV) tube
   - Knock Sensor (KS) connector
   - Camshaft Position (CMP) sensor connector
   - Evaporative Emissions (EVAP) canister purge valve line
   - Appearance covers and support brackets
   - Fuel rail
   - Fuel injector connectors
   - Throttle body coolant hoses
   - Delta Pressure Feedback Electronic (DPFE) system sensor
   - Intake manifold. Loosen the bolts

in reverse of the tightening sequence.

To install:

5. Install or connect the following:
   - Intake manifold. Tighten the bolts in sequence to 18 ft. lbs. (25 Nm) on 2002–2003 models, or 15 ft. lbs. (20 Nm) on 2004–2005 models.
   - Delta Pressure Feedback Electronic (DPFE) system sensor
   - Throttle body coolant hoses
   - Fuel injector connectors
   - Fuel rail
   - Appearance covers and support brackets
   - Evaporative Emissions (EVAP) canister purge valve line
   - Camshaft Position (CMP) sensor connector
   - Knock Sensor (KS) connector
   - Positive Crankcase Ventilation (PCV) tube
   - Idle Air Control (IAC) valve connector
   - Intake manifold vacuum hoses
   - Cruise control cable
   - Accelerator cable
   - Fuel rail temperature sensor connector
   - Vapor management valve connector
   - Throttle plate motor connector
   - Fuel pressure sensor connector and vacuum line
   - Throttle Position (TP) sensor connector

25 Nm (18 lb-ft)

9306TG06

Intake manifold torque sequence—2002–2003 3.9L engine

Intake manifold torque sequence—2004–2005 3.9L engine

- Exhaust Gas Recirculation (EGR) valve
- Fuel line
- Cross vehicle support bar. Tighten the bolts to 15 ft. lbs. (20 Nm).
- Cowl vent screen
- Air cleaner outlet tube
- Negative battery cable
6. Fill the cooling system.
7. Start the engine and check for leaks.

Exhaust Manifolds

REMOVAL & INSTALLATION

Left Side

1. Before servicing the vehicle, refer to the Precautions Section.
2. Remove or disconnect the following:
   - Negative battery cable
   - Power steering pump reservoir
   - Oil dipstick tube
   - Exhaust front pipes
   - Exhaust Gas Recirculation (EGR) tube
   - Heat shield
   - Exhaust manifold

To install:
3. Install or connect the following:
   - Exhaust manifold. Tighten the bolts in sequence to 18 ft. lbs. (25 Nm).
   - Heat shield
   - EGR tube
   - Exhaust front pipes

- Oil dipstick tube
- Power steering pump reservoir
- Negative battery cable
4. Start the engine and check for leaks.

Right Side

1. Before servicing the vehicle, refer to the Precautions Section.
2. Remove or disconnect the following:
   - Negative battery cable
   - Catalytic converters
   - Starter motor
   - Exhaust Gas Recirculation (EGR) tube
   - Lower splash shield
   - Heat shield
   - Exhaust manifold

To install:
3. Install or connect the following:
   - Exhaust manifold. Tighten the bolts in sequence to 18 ft. lbs. (25 Nm).
   - EGR tube and tighten the nuts to 15 ft. lbs. (20 Nm).
   - Heat shield
   - Lower splash shield
   - Starter motor
   - Catalytic converters
   - Negative battery cable
4. Start the engine and check for leaks.

Camshaft and Valve Lifters

REMOVAL & INSTALLATION

1. Before servicing the vehicle, refer to the Precautions Section.
2. Remove or disconnect the following:
   - Negative battery cable
   - Valve covers
   - Front cover
   - Timing chains

➡ **Keep all valve train components in order for assembly.**

- Camshaft journal bearing caps
- Camshafts
- Valve tappets and shims

To install:
3. Install or connect the following:
   - Valve tappets and shims
   - Camshafts
4. Install the camshaft journal bearing caps in their original positions. Tighten the bolts in sequence as follows:
   a. Step 1: Finger tight
   b. Step 2: 53 inch lbs. (6 Nm)
   c. Step 3: Plus 90 degrees
5. Install or connect the following:
   - Timing chains
   - Front cover
   - Valve covers

25 Nm (18 lb-ft)

Exhaust manifold torque sequence—3.9L engine

Camshaft journal bearing cap torque sequence—3.9L engine

- Negative battery cable
6. Start the engine and check for leaks.

Valve Lash

ADJUSTMENT

2002–2003 Models

1. Before servicing the vehicle, refer to the Precautions Section.
2. Remove or disconnect the following:
   - Negative battery cable
   - Engine appearance covers
   - Ignition coils
   - Valve covers
3. Measure the valve clearance while the camshaft lobe is pointed away from the valve shim. Rotate the crankshaft as necessary for each valve to be measured.
4. Valve clearance should be 0.007–0.009 in. (0.18–0.23mm) for intake valves or 0.009–0.011 in. (0.23–0.28mm) for exhaust valves.
5. If adjustment is necessary, compress the valves with the special tools and remove the shim with compressed air. Repeat for each valve to be adjusted.
6. Install or connect the following:
   - Valve covers

Valve adjustment tool base plate—2002–2003 3.9L engine

Valve adjustment tool attachment—2002–2003 3.9L engine

Remove the shims with compressed air—2002–2003 3.9L engine

- Ignition coils
- Engine appearance covers
- Negative battery cable

2004–2005 Models

1. Before servicing the vehicle, refer to the Precautions Section.
2. Remove or disconnect the following:
   - Negative battery cable
   - Engine appearance covers
   - Ignition coils
   - Valve covers
   - Spark plugs
3. Measure the valve clearance while the camshaft lobe is pointed away from the valve shim. Rotate the crankshaft as necessary for each valve to be measured.
4. Valve clearance should be 0.007–0.009 in. (0.18–0.23mm) for intake valves or 0.013–0.015 in. (0.33–0.38mm) for exhaust valves.
5. If adjustment is necessary, remove the timing chains and camshafts. Remove the shim and replace the shim with the correct shim to achieve the desired clearance. Repeat for each valve to be adjusted.

Measuring valve clearance—2004–2005 3.9L engine

Starter Motor

REMOVAL & INSTALLATION

1. Before servicing the vehicle, refer to the Precautions Section.
2. Remove or disconnect the following:
   - Negative battery cable
   - Starter motor wiring connectors
   - Starter motor

To install:
3. Install or connect the following:
   - Starter motor. Tighten the bolts to 18 ft. lbs. (25 Nm).
   - Starter motor wiring connectors
   - Negative battery cable

Oil Pan

REMOVAL & INSTALLATION

1. Before servicing the vehicle, refer to the Precautions Section.
2. Drain the engine oil.
3. Remove the engine crossmember
4. Remove the oil pan.

To install:
5. Install the oil pan. Tighten the bolts in sequence as follows:
   a. Step 1: 44 inch lbs. (5 Nm)
   b. Step 2: 108 inch lbs. (12 Nm)
6. Install the engine crossmember and tighten the bolts as follows:
   a. 2002–2004 models: 36 ft. lbs. (49 Nm)
   b. 2005 models: 46 ft. lbs. (63 Nm)
7. Fill the crankcase to the correct level.
8. Start the engine and check for leaks.

Oil Pump

REMOVAL & INSTALLATION

1. Before servicing the vehicle, refer to the Precautions Section.
2. Remove or disconnect the following:
   - Negative battery cable
   - Oil pan
   - Oil pump screen and pickup tube
   - Timing chains
   - Oil pump mounting bolts
   - Oil pump

To install:
3. Install or connect the following:
   - New gasket
   - Oil pump. Tighten the bolts to 53 inch lbs. (6 Nm) plus 90 degrees.
   - Timing chains and timing drive components
   - Oil pump screen and pickup tube
   - Oil pan
   - Negative battery cable

Rear Main Seal

REMOVAL & INSTALLATION

1. Before servicing the vehicle, refer to the precautions at the beginning of this section.

Oil pan torque sequence—3.9L engine

Removing the real seal using Special Tool 303-647 with adapter 204-208.

06017-TBRD-G06

2. Attach an engine support fixture to the engine lifting eyes.
3. Remove or disconnect the following:
   - Negative battery cable
   - Transmission
   - Flywheel
   - Rear crankshaft seal using Special Tool 303-647

To install:

4. Install or connect the following:
   - Rear main seal flush with the cylinder block surface using Special Tool 303-647
   - Flywheel

5. Tighten the flywheel bolts in sequence as follows:
   a. Step 1: 11 ft. lbs. (15 Nm)
   b. Step 2: 81 ft. lbs. (110 Nm)

6. Install or connect the following:
   - Transmission
   - Negative battery cable

7. Start the engine and check for leaks.

67197-THUN-G12

Flywheel torque sequence—3.9L engine

Timing Chain, Sprockets, Front Cover and Seal

REMOVAL & INSTALLATION

1. Before servicing the vehicle, refer to the Precautions Section.
2. Drain the cooling system.

3. Drain the engine oil.
4. Remove or disconnect the following:
   - Negative battery cable
   - Engine appearance cover and brackets
   - Valve covers
   - Engine cooling fan assembly
   - Accessory drive belt
   - Lower splash shield
   - Oil filter
   - Oil cooler
   - Crankshaft pulley and discard the bolt
   - Front crankshaft seal using Special Tool 303-409
   - A/C compressor
   - Lower radiator hose and pipe
   - Water pump
   - Heater hose
   - Power steering reservoir hose
   - Power Steering Pressure (PSP) switch connector
   - Power steering pump and bracket
   - Alternator and bracket
   - Front cover wiring harness clips
   - Front cover. Loosen the bolts in sequence.
   - Left and right side Variable Camshaft Timing (VCT) housings
   - Crankshaft Position (CKP) sensor

Front cover loosening sequence—3.9L engine

9306TG16

Front cover bolt removal sequence—3.9L engine

- Torque converter access panel

5. Rotate the crankshaft to 45 degrees After Top Dead Center (ATDC). The crankshaft keyway will be in the 6 o'clock position. Check that the camshaft lobes are facing upward. If not, rotate the crankshaft 1 full turn.

6. Install Crankshaft Holding Tool 303-645 in place of the CKP sensor.

Crankshaft Holding Tool—3.9L engine

21 RH VCT unit bolt	26 RH secondary timing chain tensioner bolts	30 LH VCT unit
22 RH exhaust camshaft sprocket bolt	27 RH secondary timing chain tensioner	31 LH secondary timing chain
23 RH VCT unit	28 LH VCT unit bolt	32 LH exhaust camshaft sprocket
24 RH secondary timing chain	29 LH exhaust camshaft sprocket bolt	33 LH secondary timing chain tensioner bolts
25 RH exhaust camshaft sprocket		34 LH secondary timing chain tensioner
		35 Crankshaft sprocket

Exploded view of timing chain sprocket assembly—3.9L engine

7. Install Camshaft Locking Tool 303-530 to the right bank camshafts.

8. Remove the right side exhaust camshaft sprocket and VCT bolts and discard.

9. Insert a drill rod into the right side timing chain tensioner to lock the piston.

10. Remove the timing chain tensioner and tensioner arm.

11. Remove the timing chain guide, right bank timing chain and camshaft timing sprockets.

Timing chain tensioner preparation—3.9L engine

Camshaft Locking Tool—3.9L engine

12. Remove the locking tool from the right bank camshafts and install it on the left bank camshafts.

13. Remove the left side exhaust camshaft sprocket and VCT bolts and discard.

14. Insert a drill rod into the left side timing chain tensioner to lock the piston.

15. Remove the timing chain tensioner and tensioner arm.

16. Remove the timing chain guide, left bank timing chain and camshaft timing sprockets.

To install:

17. Install the left bank secondary timing chain and camshaft timing sprockets as an assembly.

18. Hand tighten the NEW sprockets bolts.

19. Remove the locking pin from the tensioner.

20. Install the right bank secondary timing chain and camshaft timing sprockets as an assembly.

21. Hand tighten the NEW sprockets bolts.

22. Remove the locking pin from the tensioner.

23. To reset the timing chain tensioners, compress the tensioners and install a drill rod to lock the pistons.

> ❊❊ **CAUTION**
>
> **Ensure the camshaft holding tool is installed on the left camshaft.**

24. Place the primary timing chain on the left camshaft and crankshaft gears.

25. Install the left side timing chain guide. Torque the bolts to 97 inch lbs. (11 Nm).

26. Install the tensioner arm. Torque the bolts to 97 inch lbs. (11 Nm).

27. Install the timing chain tensioner and torque the bolts to 97 inch lbs. (11 Nm). Remove the drill rod.

28. Install special tool 303-532 onto the exhaust camshaft sprocket and apply 89 inch lbs. (10 Nm) of force in a counter-clockwise direction, then tighten the exhaust camshaft sprocket bolt in two stages: Step 1 to 26 ft. lbs. (35 Nm), step 2 and additional 90°.

29. Tighten the intake camshaft sprocket bolt in two stages: Step 1 to 26 ft. lbs. (35 Nm), step 2 an additional 90°.

30. Remove the locking tool from the left camshaft.

31. Install the locking tool on the right camshaft.

32. Place the primary timing chain on the right camshaft and crankshaft gears.

33. Install the right side timing chain guide. Torque the bolts to 97 inch lbs. (11 Nm).

34. Install the tensioner arm. Torque the bolts to 97 inch lbs. (11 Nm).

35. Install the timing chain tensioner and torque the bolts to 97 inch lbs. (11 Nm). Remove the drill rod.

36. Install special tool 303-532 onto the exhaust camshaft sprocket and apply 89 inch lbs. (10 Nm) of force in a counter-clockwise direction, then tighten the exhaust camshaft sprocket bolt in two stages: Step 1 to 26 ft. lbs. (35 Nm), step 2 and additional 90°.

37. Tighten the intake camshaft sprocket bolt in two stages: Step 1 to 26 ft. lbs. (35 Nm), step 2 an additional 90°.

Left side

Right side

67197-THUN-G15

Installing special tool to exhaust camshafts—3.9L engine

38. Remove the camshaft locking tool.

39. Install or connect the following:
- CKP sensor
- Torque converter access panel
- VCT housings. Tighten the bolts to 15 ft. lbs. (20 Nm), and the nut to 89 inch lbs. (10 Nm).

40. Apply silicone sealant to the areas indicated and install the front cover.

41. Tighten the front cover bolts in sequence as follows:
   a. Step 1: 44 inch lbs. (5 Nm)
   b. Step 2: 89 inch lbs. (10 Nm)

42. Install or connect the following:
- Front cover wiring harness clips
- Alternator and bracket
- Power steering pump and bracket
- PSP switch connector
- Power steering reservoir hose
- Front crankshaft seal using Special Tool 303-646

3 mm (0.12 in)

67197-THUN-G16

Apply silicone sealant to the areas indicated—3.9L engine

06017-TBRD-G07

Removing the front seal using Special Tool 303-409—3.9L engine

43. Install the crankshaft pulley and tighten the NEW bolt as follows:
    a. Step 1: 59 ft. lbs. (80 Nm)
    b. Step 2: Loosen the bolt 2 complete turns
    c. Step 3: 37 ft. lbs. (50 Nm)
    d. Step 4: Plus 90 degrees
44. Install or connect the following:
- Heater hose
- Lower radiator hose and pipe
- Water pump
- A/C compressor
- Oil cooler
- Oil filter
- Lower splash shield
- Accessory drive belt
- Engine cooling fan assembly
- Valve covers
- Engine appearance cover and brackets
- Negative battery cable

45. Fill the cooling system.
46. Fill the engine with clean engine oil.
47. Start the engine and check for leaks.

67197-THUN-G17

Front cover torque sequence—3.9L engine

303-646

06017-TBRD-G08

Installing the front crankshaft seal using Special Tool 303-646.

FUEL SYSTEM

Fuel System Pressure

RELIEVING

1. Before servicing the vehicle, refer to the Precautions Section.
2. Remove the fuel pump relay.
3. Start the engine and allow it to idle until it stalls.
4. Crank the engine to ensure the fuel injection supply manifold pressure has been released.
5. Turn the ignition to the **OFF** position.
6. Reinstall the fuel pump relay when service is complete.

Fuel Filter

REMOVAL & INSTALLATION

1. Before servicing the vehicle, refer to the Precautions Section.
2. Relieve the fuel system pressure.
3. Remove or disconnect the following:
   - Negative battery cable
   - Left front wheel rear splash shield
   - Fuel lines
   - Vapor line fittings
   - Fuel filter bracket cover
   - Fuel filter

To install:
4. Install or connect the following:
   - Fuel filter into the bracket making sure the flow direction is correct. Tighten the clamp to 15–25 inch lbs. (2–3 Nm).
   - Vapor line fittings
   - Fuel lines. Tighten the fittings to 22 ft. lbs. (30 Nm).
   - Splash shield
5. Start the engine and check for leaks.

Fuel Pump

REMOVAL & INSTALLATION

2002–2003 Models

1. Before servicing the vehicle, refer to the Precautions Section.
2. Relieve the fuel system pressure.
3. Drain the fuel tank.
4. Remove or disconnect the following:
   - Negative battery cable
   - Trunk liner
   - Trunk seal retainer

- Rear lamp assembly interior trim finisher
- Left and right side liners
- Fuel feed and return lines
- Fuel filler and vent hoses
- Fuel tank wiring connectors
- Fuel filler cap
- Fuel tank retaining straps
- Fuel tank
- Fuel pump module

To install:
5. Install or connect the following:
   - Fuel pump module
   - Fuel tank
   - Fuel tank retaining straps
   - Fuel filler cap
   - Fuel tank wiring connectors
   - Fuel filler and vent hoses

- Fuel feed and return lines
- Left and right side liners
- Rear lamp assembly interior trim finisher
- Trunk seal retainer
- Trunk liner
- Negative battery cable

6. Fill the fuel tank with at least 10 gallons (38L) of fuel.
7. Start the engine and check for leaks.

2004–2005 Models

1. Before servicing the vehicle, refer to the Precautions Section.
2. Relieve the fuel system pressure.
3. Drain the fuel tank.
4. Remove or disconnect the following:

18 Fuel transfer pump electrical connector	21 Fuel transfer pump lock ring	24 Fuel transfer pump gasket
19 Fuel delivery tube quick connect fitting	22 Fuel transfer pump	25 Fuel delivery module lock ring
20 Fuel delivery module electrical connector	23 Fuel crossover hose quick connect fittings	26 Fuel delivery module
		27 Fuel delivery module gasket

67197-THUN-G18

Fuel and transfer pump mounting—2004 Thunderbird

310-069

Using Special Tool 310-069 to remove the lock ring.

06017-TBRD-G09

- Driveshaft
- Right rear wheel well splash shield
- Fuel filler cap
- Fuel filler and vent hoses
- Fuel tube shield
- Fuel feed and return lines
- Under vehicle heat shield
- Support fuel tank with a jack
- Fuel tank retaining straps
- Vapor lines
- Fuel tank
- Fuel and transfer pump connectors
- Fuel and/or transfer pump using Special Tool 310-069

To install:

5. Install or connect the following:
   - Fuel and/or transfer pump module using Special Tool 310-069
   - Fuel and transfer pump connectors
   - Fuel tank
   - Vapor lines
   - Fuel tank retaining straps
   - Heat shield
   - Fuel feed and return lines
   - Fuel tube shield
   - Fuel filler and vent hoses
   - Fuel filler cap
   - Right rear wheel well splash shield
   - Driveshaft
   - Negative battery cable

6. Fill the fuel tank with at least 10 gallons (38L) of fuel.

7. Start the engine and check for leaks.

Fuel Rail And Injectors

REMOVAL & INSTALLATION

1. Before servicing the vehicle, refer to the Precautions Section.

2. Relieve fuel system pressure.

3. Remove or disconnect the following:
   - Negative battery cable
   - Engine appearance covers and mounting brackets
   - Fuel lines
   - Fuel pressure sensor connectors
   - Fuel rail
   - Fuel injector connectors
   - Fuel injectors

To install:

4. Install or connect the following:
   - Fuel injectors. Use new O-ring seals.
   - Fuel injector connectors
   - Fuel rail. Tighten the mounting bolts to 80 inch lbs. (9 Nm).
   - Fuel pressure sensor connectors
   - Fuel lines
   - Engine appearance covers and mounting brackets
   - Negative battery cable

5. Start the engine and check for leaks.

6 Nm (53 lb-in)

9 Nm (80 lb-in)

1	Engine appearance cover mounting bracket nuts	4	Fuel pressure sensor vacuum connector	7	Fuel tube retaining clip
2	Engine appearance cover mounting brackets	5	Fuel injector electrical connector	8	Fuel tube spring lock coupling
3	Fuel pressure sensor electrical connector	6	Fuel temperature sensor electrical connector	9	Fuel injection supply manifold bolts
				10	Fuel injection supply manifold and fuel injectors
				11	Fuel injectors

Fuel rail and injector assembly—3.9L engine

67197-THUN-G19

DRIVE TRAIN

Transmission Assembly

REMOVAL & INSTALLATION

1. Before servicing the vehicle, refer to the Precautions Section.
2. Install a support fixture to the engine lifting eyes.
3. Drain the transmission fluid.
4. Remove or disconnect the following:
   - Negative battery cable
   - Exhaust heat shields
   - Exhaust system
   - Driveshaft
5. Support the transmission with a suitable jack.
6. Remove or disconnect the following:
   - Rear transmission support
   - Shift selector cable
   - Transmission electrical connectors
   - Transmission oil cooler lines
   - Torque converter
   - Transmission mount
   - Transmission-to-engine bolts
   - Transmission

To install:

7. Install or connect the following:
   - Transmission to the engine. Tighten the bolts to 35 ft. lbs. (48 Nm).
   - Torque converter. Tighten the bolts to 35 ft. lbs. (48 Nm).
   - Transmission oil cooler lines
   - Transmission electrical connectors
   - Transmission mount. Tighten the mount screws to 41 ft. lbs. (55 Nm) and the center screw to 30 ft. lbs. (40 Nm).
   - Shift selector cable
   - Driveshaft. Tighten the bolts to 60 ft. lbs. (81 Nm).
   - Exhaust front pipes
   - Exhaust heat shields
   - Negative battery cable
8. Fill the transmission to the correct level with the proper fluid. Do not over-fill.
9. Start the engine and check for leaks.

81 Nm
(60 lb-ft)

67197-THUN-G20

Driveshaft flange bolts

Halfshaft

REMOVAL & INSTALLATION

1. Before servicing the vehicle, refer to the Precautions Section.
2. Remove or disconnect the following:
   - Negative battery cable
   - Rear wheel
   - Axle hub retaining nut
   - Brake caliper
   - Wheel speed sensor
   - Lower suspension arm bolt
   - Toe link bolt
3. Using Special Tool Remover 204-069, press the outer CV-joint until is it loose in the hub.
4. Raise the wheel knuckle and remove the CV-joint.
5. Using Special Tool 205-475, separate the inner CV-joint from the differential side gear.
6. Remove the halfshaft.

204-069

205-237

204-085

06017-TBRD-G11

Pressing the outer CV-Joint using Special Tool 204-069

205-475

06017-TBRD-G12

Using Special Tool 205-475 to separate the inner CV-joint from the differential.

To install:

7. Install or connect the following:
- Halfshaft inner joint to the differential side gear using a new axle circlip
- Halfshaft in the wheel hub by applying Loctite® 270 thread locking compound to the splines
- Lower suspension arm bolt. Tighten the nut to 111 ft. lbs. (150 Nm).
- Toe link bolt. Tighten the nut to 41 ft. lbs. (55 Nm).

- Wheel speed sensor
- Brake caliper
- Axle hub retaining nut. Tighten the new nut to 221 ft. lbs. (300 Nm).
- Rear wheel
- Negative battery cable

8. Check the wheel alignment and adjust as necessary.

➡️ **If the hub is removed for any reason, a new bearing assembly must be installed. Never attempt to re-use a bearing.**

CV-Joint

OVERHAUL

The CV-joints are serviced with the axle halfshaft as an assembly.

STEERING AND SUSPENSION

Air Bag

DEPOWERING

1. Turn all vehicle accessories OFF.
2. Turn the ignition switch to OFF.
3. Remove the right kick panel cover and remove the restraints control module fuse from the central junction box (CJB). For additional information, refer to the Owner's Manual.
4. Turn the ignition ON and visually monitor the air bag indicator for at least 30 seconds. The air bag indicator will remain lit continuously (no flashing) if the correct RCM fuse has been removed. If the air bag indicator does not remain lit continuously, remove the correct RCM fuse before proceeding.
5. Turn the ignition OFF.

✳️ CAUTION

To avoid accidental deployment and possible personal injury, the backup power supply must be depleted before repairing or replacing any front or side air bag supplemental restraint system (SRS) components and before servicing, replacing, adjusting or striking components near the front or side air bag sensors, such as doors, instrument panel, console, door latches, strikers, seats and hood latches.

➡️ **The side impact sensors are attached to the rocker panels, near the seats.**

6. To deplete the backup power supply energy, disconnect the battery ground cable and wait at least one minute. Be sure to disconnect auxiliary batteries and power supplies (if equipped).
7. Disconnect the battery ground cable (14301) and wait at least one minute.

REPOWERING

✳️ CAUTION

The restraint system diagnostic tool is for restraint system service only. Remove from vehicle prior to road use. Failure to remove could result in injury and possible violation of vehicle safety standards.

1. Make sure all restraint system diagnostic tool(s) that may have been installed during the repair have been removed from the vehicle and all SRS components are connected.
2. Turn the ignition switch from OFF to ON.
3. Install RCM fuse to the CJB and install the cover.

✳️ CAUTION

Be sure that nobody is in the vehicle and that there is nothing blocking or set in front of any air bag module when the battery ground cable is connected.

4. Connect the battery ground cable.
5. Prove out the supplemental restraint system (SRS) as follows:

   a. Turn the ignition key from ON to OFF. Wait 10 seconds, then turn the key back to ON and visually monitor the air bag indicator with the air bag modules installed. The air bag indicator will light continuously for approximately 6 seconds and then turn off. If an air bag SRS fault is present, the air bag indicator will:
   - fail to light.
   - remain lit continuously.
   - flash.

   b. The flashing might not occur until approximately 30 seconds after the ignition switch has been turned from the OFF to the ON position. This is the time required for the RCM to complete the testing of the SRS. If the air bag indicator is inoperative and a SRS fault exists, a chime will sound in a pattern of 5 sets of 5 beeps. If this occurs, the air bag indicator and any SRS fault discovered must be diagnosed and repaired.

6. Clear all continuous DTCs from the restraints control module using a scan tool.

Power Rack and Pinion Steering Gear

REMOVAL & INSTALLATION

✳️ WARNING

Do not allow the steering wheel to rotate when the intermediate shaft is disconnected, or damage to the clockspring can result. If it is suspected that the shaft has rotated, the clockspring must be removed and re-centered.

1. Before servicing the vehicle, refer to the Precautions Section.
2. Lock the steering wheel in the straight-ahead position.
3. Remove or disconnect the following:
- Negative battery cable
- Front wheels
- Outer tie rod ends
- Variable Assist Power Steering (VASP) connector
- Steering column intermediate shaft
- Power steering lines
- Steering rack and pinion gear

To install:

4. Install or connect the following:
- Steering rack and pinion gear. Tighten the steering gear-to-nuts to bolts for 2002–2004 models to 59

1 – 80 Nm (59 lb-ft)

80 Nm (59 lb-ft) – 2 [N]

4

3

1 Tie-rod end nuts

2 Steering gear-to-crossmember nuts

3 Steering gear-to-crossmember bolts

4 Steering gear

67197-THUN-G21

Steering gear mounting

ft. lbs. (80 Nm). Tighten 2005 models to 76 ft. lbs. (103 Nm).
- Power steering lines
- Outer tie rod ends. Tighten the nuts to 59 ft. lbs. (80 Nm).
- Steering column intermediate shaft. Tighten the bolts to 26 ft. lbs. (35 Nm).
- VASP connector
- Front wheels
- Negative battery cable

5. Fill the power steering fluid reservoir.

6. Start the engine and check for leaks.

7. Check the wheel alignment and adjust, as necessary.

Strut

REMOVAL & INSTALLATION

Front

1. Before servicing the vehicle, refer to the Precautions Section.

2. Remove or disconnect the following:
- Front wheel
- Stabilizer bar link
- Lower strut mounting bolt
- Upper strut mount cover and fasteners
- Strut and spring assembly

To install:

➡**Use new fasteners for assembly.**

3. Install or connect the following:

28 Nm (21 lb-ft) – 1

175 Nm (129 lb-ft) – 4

6

3

5

55 Nm (41 lb-ft) – 2

1 Shock absorber upper mount-to-body nuts

2 Stabilizer bar link-to-lower arm nut

3 Stabilizer bar link (detach only)

4 Shock absorber-to-lower arm nut

5 Shock absorber-to-lower arm bolt

6 Shock absorber and spring assembly

Front strut mounting

67197-THUN-G22

28 Nm (21 lb-ft) — 1

133 Nm (98 lb-ft) — 2

3

1 Nuts
2 Shock absorber-to-lower arm bolt
3 Shock absorber and spring assembly

67197-THUN-G23

Rear strut mounting

- Strut and spring assembly. Tighten the upper mount nuts to 21 ft. lbs. (28 Nm).
- Upper strut mount cover
- Lower strut mounting bolt. Tighten the bolts to 129 ft. lbs. (175 Nm).
- Stabilizer bar link. Tighten the nut to 41 ft. lbs. (55 Nm).
- Front wheel

Rear

1. Before servicing the vehicle, refer to the Precautions Section.
2. Remove or disconnect the following:
   - Trunk trim covers
   - Upper strut mount nuts
   - Lower strut mount bolt
   - Strut and spring assembly

To install:

➡ **Use new fasteners for assembly.**

3. Install or connect the following:
   - Strut and spring assembly. Tighten the lower bolt to 98 ft. lbs. (133 Nm) and the upper nuts to 21 ft. lbs. (28 Nm).
   - Trunk trim covers

Coil Spring

REMOVAL & INSTALLATION

1. Before servicing the vehicle, refer to the Precautions Section.
2. Remove the strut assembly from the vehicle.
3. Compress the coil spring and remove the piston rod nut.
4. Remove or disconnect the following:
   - Upper strut mount
   - Spring upper seat
   - Coil spring

To install:

5. Install or connect the following:
   - Coil spring
   - Spring upper seat
   - Upper strut mount. Tighten the piston rod nut to 37 ft. lbs. (50 Nm).
6. Remove the spring compressor and install the strut assembly to the vehicle.

Upper Ball Joint

REMOVAL & INSTALLATION

The upper ball joint is serviced with the upper control arm as an assembly.

Lower Ball Joint

REMOVAL & INSTALLATION

The lower ball joint is serviced with the lower control arm as an assembly.

Upper Control Arm

REMOVAL & INSTALLATION

1. Before servicing the vehicle, refer to the Precautions Section.
2. Remove or disconnect the following:
   - Front wheel
   - Strut and spring assembly
   - Upper ball joint and tapered washer
   - Inner control arm fasteners

➡ **To remove the upper arm on the left side, position the power steering reservoir aside. Disconnect the master cylinder electrical connector, remove master cylinder attaching nuts and position the master cylinder aside. This will allow access to the upper arm-to-body nut.**

   - Upper control arm

To install:

➡ **Use new fasteners for assembly.**

3. Install or connect the following:
   - Upper control arm. Tighten the inner fasteners to 35 ft. lbs. (48 Nm), with the weight of the vehicle resting on the suspension.
   - Upper ball joint and tapered washer. Tighten the nut to 66 ft. lbs. (90 Nm).
   - Master cylinder. Tighten the nut to 18 ft. lbs. (25 Nm).
   - Power steering reservoir. Tighten the nut to 35 ft. lbs. (48 Nm).
   - Strut and spring assembly
   - Front wheel

CONTROL ARM BUSHING REPLACEMENT

The control arm bushings are serviced with the control arm as an assembly.

Lower Control Arm

REMOVAL & INSTALLATION

1. Before servicing the vehicle, refer to the Precautions Section.
2. Remove or disconnect the following:
   - Front wheel
   - Splash shield

48 Nm (35 lb-ft)

17 - 175 Nm (129 lb-ft)

15 - 55 Nm (41 lb-ft)

175 Nm (129 lb-ft)

175 Nm (129 lb-ft)

12 Upper arm-to-body nuts
13 Upper arm-to-body bolts
14 Upper arm
15 Stabilizer bar link-to-lower arm nut

16 Stabilizer bar link (detach only)
17 Shock absorber-to-lower arm nut
18 Shock absorber-to-lower arm bolt

19 Lower arm-to-frame nuts
20 Lower arm-to-frame bolt
21 Lower arm-to-frame bolt
22 Lower arm RH/LH

67197-THUN-G24

Upper and lower control arms

- Stabilizer bar link
- Lower strut mounting bolt
- Lower ball joint
- Rack and pinion steering gear
- Inner control arm mounting bolts
- Lower control arm

To install:

➡ Use new fasteners for assembly.

3. Install or connect the following:
- Lower control arm. Tighten the inner mounting bolts to 129 ft. lbs. (175 Nm) with the weight of the vehicle resting on the suspension.
- Rack and pinion steering gear
- Lower ball joint. Tighten the nut to 148 ft. lbs. (200 Nm).
- Lower strut mounting bolt. Tighten the bolt to 129 ft. lbs. (175 Nm).
- Stabilizer bar link. Tighten the nut to 41 ft. lbs. (55 Nm).
- Splash shield
- Front wheel

4. Check the wheel alignment and adjust as necessary.

CONTROL ARM BUSHING REPLACEMENT

The control arm bushings are serviced with the control arm as an assembly.

Wheel Bearings

ADJUSTMENT

The wheel bearings are not adjustable.

REMOVAL & REPLACEMENT

Front

1. Before servicing the vehicle, refer to the Precautions Section.
2. Remove or disconnect the following:
- Front wheel
- Brake caliper and rotor
- Wheel speed sensor connector
- Hub and bearing assembly

➡ The hub and bearing assembly is not pressed into the knuckle. Do not use a slide hammer or press to remove the

hub and bearing assembly. Damage to the hub and bearing assembly may result.

To install:

➡ Do not remove the wheel speed sensor from the hub and bearing assembly unless it is being replaced. If installing a new hub and bearing assembly, a new wheel speed sensor must be installed.

➡ Use new fasteners for assembly.

3. Install or connect the following:
- Hub and bearing assembly. Tighten the bolts to 66 ft. lbs. (90 Nm).
- Wheel speed sensor connector
- Brake caliper and rotor
- Front wheel

Rear

1. Before servicing the vehicle, refer to the Precautions Section.
2. Remove or disconnect the following:
- Rear wheel
- Axle nut
- Wheel speed sensor
- Brake caliper and rotor
- Toe link bolt
- Lower arm bolt
- Hub, bearing and knuckle assembly
- Disc brake dust shield

3. Press the hub from the knuckle and bearing assembly.

4. Remove the snapring and press the bearing assembly out of the knuckle.

To install:

5. Press the bearing assembly into the knuckle.

6. Install the snapring and press the hub into the knuckle and bearing assembly.

7. Install or connect the following:
- Disc brake dust shield. Use aluminum rivets.
- Hub, bearing and knuckle assembly
- Lower arm bolt. Tighten to nut to 111 ft. lbs. (150 Nm).
- Toe link bolt. Tighten to nut to 41 ft. lbs. (55 Nm).
- Brake caliper and rotor
- Wheel speed sensor
- Axle nut. Tighten to nut to 302 ft. lbs. (410 Nm).
- Rear wheel

8. Check the wheel alignment and adjust as necessary.

BRAKES

Brake Caliper

REMOVAL & INSTALLATION

Front

1. Before servicing the vehicle, refer to the Precautions Section.
2. Remove or disconnect the following:
   - Front wheel
   - Brake fluid hose
   - Caliper mounting bolts
   - Brake caliper

To install:

3. Install or connect the following:
   - Brake caliper. Tighten the mounting bolts to 26 ft. lbs. (35 Nm).
   - Brake fluid hose. Use new copper washers and tighten the bolt to 35 ft. lbs. (47 Nm).
   - Front wheel
4. Bleed the brake system.
5. Before attempting to move the vehicle, pump the brake pedal to seat the pads against the rotors. Make sure the vehicle has a firm brake pedal. Check the level of the brake fluid and add DOT 3 or 4 brake fluid if necessary.

Rear

1. Before servicing the vehicle, refer to the Precautions Section.
2. Remove or disconnect the following:
   - Rear wheel
   - Parking brake cable
   - Caliper mounting bolts
   - Brake fluid hose
   - Brake caliper

To install:

3. Install or connect the following:
   - Brake fluid hose. Use new copper washers and tighten the bolt to 36 ft. lbs. (48 Nm).
   - Brake caliper. Tighten the mounting bolts to 25 ft. lbs. (33 Nm).
   - Parking brake cable
   - Rear wheel
4. Bleed the brake system.
5. Before attempting to move the vehicle, pump the brake pedal to seat the pads against the rotors. Make sure the vehicle has a firm brake pedal. Check the level of the brake fluid and add DOT 3 or 4 brake fluid if necessary.

1 Bolts	6 Bolts
2 Guide pin/boot	7 Brake caliper support bracket
3 Brake caliper assembly	8 Brake disc
4 Brake pads	9 Rivets
5 Slippers	10 Brake disc shield

67197-THUN-G25

Exploded view of front brake assembly

1 Guide pin assembly	5 Brake caliper support bracket
2 Brake caliper	6 Bolts
3 Brake pads	7 Brake disc
4 Slippers	8 Rivets
	9 Brake disc shield

67197-THUN-G26

Exploded view of rear brake assembly

Disc Brake Pads

REMOVAL & INSTALLATION

Front

1. Before servicing the vehicle, refer to the Precautions Section.

2. Remove the master cylinder reservoir cap and check the fluid level in the reservoir. Remove brake fluid until the reservoir is ½ full. Discard the removed fluid.

3. Remove the wheel and tire assembly.

4. Remove the disc brake caliper locating pins. Lift the caliper assembly from the anchor plate and rotor.

5. Suspend the caliper inside the fender housing with wire. Do not allow the caliper to hang from the brake hose.

6. Remove the inner and outer brake pads. Inspect the rotor braking surfaces for scoring and machine as necessary.

To install:

7. Use a C-clamp and an old brake pad or block of wood to seat the caliper piston in its bore.

8. Remove any rust buildup from the inside of the caliper in the brake pad contact area.

9. Install the inner pad in the caliper piston.

10. Install the outer pad onto the anchor plate. Make sure the clips are properly seated.

11. Install the disc brake caliper onto the anchor plate.

12. Install caliper locating pins and torque to 26 ft. lbs. (35 Nm).

13. Install wheel and tire assembly and torque lugs nuts to 100 ft. lb. (135 Nm).

14. Pump the brake pedal several times prior to moving the vehicle to position the brake pads to the rotor.

15. Refill the master cylinder reservoir as necessary, using only clean DOT 3 or 4 brake fluid from a closed container.

Rear

1. Before servicing the vehicle, refer to the Precautions Section.

2. Remove the master cylinder reservoir cap and check the fluid level in the reservoir. Remove brake fluid until the reservoir is ½ full. Discard the removed fluid.

3. Remove the wheel and tire assembly.

4. Remove the disc brake caliper locating pins at the support bracket.

5. Remove the caliper.

6. Remove the disc brake pads.

7. Inspect the rotor braking surfaces for scoring and machine as necessary.

To install:

8. Using Rear Caliper Piston Adjuster 206-026 or equivalent, rotate the piston clockwise until it is fully seated. Make sure one of the slots in the piston face is positioned so it will engage the nib on the brake pad.

9. Install the brake pads in the support bracket. Place the caliper assembly over the rotor into position on the support bracket. Make sure the brake pads are installed correctly.

10. Install and torque the disc brake caliper locating pin to 25 ft. lbs. (33 Nm).

11. Install the wheel and tire assembly and torque the lug nuts to 85–104 ft. lbs. (115–142 Nm).

12. Pump the brake pedal several times prior to moving the vehicle, to position the brake pads to the rotor.

13. Refill the master cylinder reservoir if necessary, using only clean DOT 3 or 4 brake fluid from a closed container.

14. Road test the vehicle and check the brake system for proper operation.

06017-TBRD-G10

Seating the brake piston using Special Tool 206-026—Rear caliper

MERCURY

17

Villager

SPECIFICATION AND MAINTENANCE CHARTS

VEHICLE AND ENGINE IDENTIFICATION CHART

			Engine Code						Model Year	
Code	Liters (cc)	Cu. In.	Cyl.	Fuel Sys.	Engine Type	Eng. Mfg.			Code	Year
T	3.3 (3275)	200	6	SEFI	SOHC	Nissan			1	2001
									2	2002

MFI: Multi-port Fuel Injection

SEFI: Sequential Multi-port Fuel Injection

67197-VILL-C01

GENERAL ENGINE SPECIFICATIONS

Year	Engine Displacement Liters (VIN)	Net Horsepower @ rpm	Net Torque @ rpm (ft. lbs.)	Bore x Stroke (in.)	Com-pression Ratio	Oil Pressure @ rpm
2001	3.3 (T)	195@4500	190@3800	3.60x3.27	8.9:1	60-65@2000
2002	3.3 (T)	195@4500	190@3800	3.60x3.27	8.9:1	60-65@2000

MFI: Multiport fuel injection

SEFI: Sequential Multi-port Fuel Injection

67197-VILL-C02

ENGINE TUNE-UP SPECIFICATIONS

			Ignition Timing (deg.)			Idle Speed (rpm)		Valve Clearance	
Year	Engine Displacement Liters (VIN)	Spark Plug Gap (in.)	MT	AT	Fuel Pump (psi) ①	MT	AT ②	In.	Ex.
2001	3.3 (T)	0.043	—	13-17B	34	—	650-750	HYD	HYD
2002	3.3 (T)	0.043	—	13-17B	34	—	650-750	HYD	HYD

NOTE: The Vehicle Emission Control Information label must be used if they differ from those in this chart.

B: Before top dead center

HYD: Hydraulic

① System pressure at idle with vacuum hose connected should increase to 43 psi when disconnected

② Transmission in Neutral

67197-VILL-C03

3.3L engines
Firing order: 1–2–3–4–5–6
Distributor rotation: Counterclockwise

Accessory V-belt routing—3.3L engines

CAPACITIES

Year	Model	Engine Displacement Liters (VIN)	Engine Oil with Filter (qts.)	Transmission (pts.) 4-Spd	5-Spd	Auto.	Drive Axle Front (pts.)	Rear (pts.)	Fuel Tank (gal.)	Cooling System (qts.) ②
2001	Villager	3.3 (T)	4.0	—	—	20.0	①	—	20	11.25
2002	Villager	3.3 (T)	4.0	—	—	20.0	①	—	20	11.25

NOTE: All capacities are approximate. Add fluid gradually and check to be sure a proper fluid level is obtained.

① Included in transaxle capacity

② Includes reservoir tank.

67197-VILL-C04

VALVE SPECIFICATIONS

Year	Engine Displacement Liters (VIN)	Seat Angle (deg.)	Face Angle (deg.)	Spring Test Pressure (lbs. @ in.)	Spring Installed Height (in.)	Stem-to-Guide Clearance (in.) Intake	Exhaust	Stem Diameter (in.) Intake	Exhaust
2001	3.3 (T)	45	45	①	②	0.0008-0.0021	0.0012-0.0019	0.2742-0.2748	0.3136-0.3138
2002	3.3 (T)	45	45	①	②	0.0008-0.0021	0.0012-0.0019	0.2742-0.2748	0.3136-0.3138

① Outer spring: 118@1.81
   Inner spring: 57.3@0.984

② Spring height measured unloaded
   Minimum length. outer spring: 2.016
   Minimum length. inner spring: 1.736

67197-VILL-C05

CRANKSHAFT AND CONNECTING ROD SPECIFICATIONS
All measurements are given in inches.

Year	Displacement Liters (VIN)	Engine			Connecting Rod		
		Main Brg. Oil Clearance	Shaft End-play	Thrust on No.	Journal Diameter	Oil Clearance	Side Clearance
2001	3.3 (T)	0.0011-0.0022	0.0020-0.0067	3	1.9667-1.9675	0.0006-0.0021	0.0079-0.00138
2002	3.3 (T)	0.0011-0.0022	0.0020-0.0067	3	1.9667-1.9675	0.0006-0.0021	0.0079-0.00138

67197-VILL-C06

PISTON AND RING SPECIFICATIONS
All measurements are given in inches.

Year	Engine Displacement Liters (VIN)	Piston Clearance	Ring Gap			Ring Side Clearance		
			Top Compression	Bottom Compression	Oil Control	Top Compression	Bottom Compression	Oil Control
2001	3.3 (T)	①	0.0083-0.0122	0.0197-0.0236	0.0079-0.0236	0.0016-0.0031	0.0012-0.0028	0.0006-0.0073
2002	3.3 (T)	①	0.0083-0.0122	0.0197-0.0236	0.0079-0.0236	0.0016-0.0031	0.0012-0.0028	0.0006-0.0073

① Journals 1, 2 and 6: 0.0010 - 0.0018 in.
  Journals 3 and 4: 0.0006 - 0.0010 in.
  Journal 5: 0.0012 - 0.0016 in.

67197-VILL-C07

TORQUE SPECIFICATIONS
All readings in ft. lbs.

Year	Engine Displacement Liters (VIN)	Cylinder Head Bolts	Main Bearing Bolts	Rod Bearing Bolts	Crankshaft Damper Bolts	Flywheel Bolts	Manifold		Spark Plugs	Oil Pan Drain Plug
							Intake	Exhaust		
2001	3.3 (T)	①	②	②	141-156	61-69	①	13-16	14-22	24
2002	3.3 (T)	①	②	②	141-156	61-69	①	13-16	14-22	24

① Intake manifold and cylinder heads are installed at the same time.
  Step 1: Cylinder head bolts to 22 ft. lbs.
  Step 2: Cylinder head bolts to 43 ft. lbs.
  Step 3: Loosen all bolts completely
  Step 4: Cylinder head bolts to 7 ft. lbs.
  Step 5: Intake manifold bolts to 2.9 (4Nm) ft. lbs.
  Step 6: Intake manifold bolts to 13 ft. lbs.
  Step 7: Intake manifold bolts to 14 ft. lbs.
  Step 8: Loosen all maifold bolts completely
  Step 9: Cylinder head bolts to 26 inch lbs.
  Step 10: Cylinder head bolts to 47 ft. lbs. Or, an additional 65 degrees
  Step 11: Cylinder head sub-bolts to 104 inch lbs.
  Step 12: Intake manifold bolts to 36 inch lbs.
  Step 13: Intake manifold bolts to 78 inch lbs.
  Step 14: Intake manifold bolts to 84 inch lbs.

② Step 1: 34-37 ft. lbs.
  Step 2: 67-74 ft. lbs.

67197-VILL-C08

TIRE, WHEEL AND BALL JOINT SPECIFICATIONS

Year	Model	OEM Tires		Tire Pressures (psi)		Wheel Size	Ball Joint Inspection	Lug Nut
		Standard	Optional	Front	Rear			
2001	Villager	P215/70R15	P225/60R16	30	30	Std: 5.5-JJ Opt: 6.5-JJ	①	80
2002	Villager	P215/70R15	P225/60R16	30	30	Std: 5.5-JJ Opt: 6.5-JJ	①	80

OEM: Original Equipment Manufacturer

PSI: Pounds Per Square Inch

① Replace if any measurable movement is found.

67197-VILL-C09

WHEEL ALIGNMENT

Year	Model		Caster Range (Deg.)	Caster Preferred Setting (Deg.)	Camber Range (Deg.)	Camber Preferred Setting (Deg.)	Toe-in (in.)
2001	Villager	F	0.75	2.75	0.75	-0.25	0.04 +/- 0.04
		R	—	—	0.75	-1.00	0.04 +/- 0.16
2002	Villager	F	0.75	2.75	0.75	-0.25	0.04 +/- 0.04
		R	—	—	0.75	-1.00	0.04 +/- 0.16

67197-VILL-C10

BRAKE SPECIFICATIONS
All measurements in inches unless noted

Year	Model		Brake Disc Original Thickness	Brake Disc Minimum Thickness	Brake Disc Maximum Runout	Brake Drum Diameter Original Inside Diameter	Brake Drum Diameter Max, Wear Limit	Brake Drum Diameter Maximum Machine Diameter	Minimum Lining Thickness	Brake Caliper Bracket-to-Hub Bolt (ft. lbs.)	Brake Caliper Mounting Pin or Bolt (ft. lbs.)
2001	Villager	F	1.002	0.940	0.0028	—	—	—	0.079	—	18-25
		R	—	—	—	9.84	9.90	9.86	0.079	—	—
2002	Villager	F	1.002	0.940	0.0028	—	—	—	0.079	—	18-25
		R	—	—	—	9.84	9.90	9.86	0.079	—	—

NOTE: Due to changes made during production, refer to the manufacturer's specifications if they differ from those in this chart

F: Front

R: Rear

67197-VILL-C11

SCHEDULED MAINTENANCE INTERVALS
MERCURY—VILLAGER

TO BE SERVICED	TYPE OF SERVICE	VEHICLE MILEAGE INTERVAL (x1000)												
		5	10	15	20	25	30	35	40	45	50	55	60	65
Engine oil & filter	R	✓	✓	✓	✓	✓	✓	✓	✓	✓	✓	✓	✓	✓
Rotate tires	S/I	✓		✓		✓		✓		✓		✓		✓
Engine coolant strength hoses & clamps	S/I			✓			✓			✓			✓	
Air cleaner filter	R						✓						✓	
Automatic transmission fluid & filter	R						✓						✓	
Engine coolant ①	R						✓						✓	
PCV valve	R												✓	
Spark plugs ②	R													
Drive belts ③	S/I			✓			✓			✓			✓	
Timing belt ④	S/I													
Exhaust system & heat shields	S/I			✓			✓			✓			✓	
Drive shaft boots	S/I			✓			✓			✓		✓		
Front & rear brake components	S/I	✓	✓	✓	✓	✓	✓	✓	✓	✓	✓	✓	✓	✓

R: Replace   S/I: Service or Inspect

① Engine coolant: change every 30,000 miles or 36 months.

② Replace every 105,000 miles

③ Inspect every 15,000 miles or 12 months and replace every 60,000 miles or 48 months.

④ Replace every 105,000 miles

FREQUENT OPERATION MAINTENANCE (SEVERE SERVICE)

If a vehicle is operated under any of the following conditions it is considered severe service:

- Extremely dusty areas.

- 50% or more of the vehicle operation is in 32°C (90°F) or higher temperatures, or constant operation in temperatures below 0°C (32°F).

- Prolonged idling (vehicle operation in stop and go traffic.

- Frequent short running periods (engine does not warm to normal operating temperatures).

- Police, taxi, delivery usage or trailer towing usage.

Engine oil & filter: replace every 3000 miles.

Rotate tires initially at 6000 miles and every 9000 miles thereafter.

Air cleaner filter: change every 15,000 miles.

Engine coolant strength, hoses & clamps: check every 15,000 miles.

Exhaust system: check every 15,000 miles.

Automatic transmission fluid & filter: change every 21,000 miles.

67197-VILL-C12

ENGINE REPAIR

Distributor

REMOVAL

1. Before servicing the vehicle, refer to the Precautions Section.
2. Remove or disconnect the following:
   - Negative battery cable
   - Distributor cap
   - Distributor wiring harness connector
3. Matchmark the rotor to the distributor housing and the distributor housing to the cylinder head.
4. Remove the distributor hold-down bolt and the distributor.

INSTALLATION

Timing Not Disturbed

1. Install or connect the following:
   - Distributor and align the matchmarks made during removal. Tighten the hold-down bolt to 10–12 ft. lbs. (14–17 Nm).
   - Distributor wiring harness connector

— Distributor drive gear
— Mark on shaft
— Mark on housing (protruding)
— Mark on housing (indented)

7924VG28

Distributor shaft alignment

— Rotor head position (No. 1 cylinder at TDC)

9308VG03

Distributor rotor alignment

DISTRIBUTOR GROUND
CONNECTOR

7924WG01

Disengage the distributor ground connector when removing the distributor

ROTOR POSITION
WHEN NO. 1
CYLINDER IS
AT TDC

7924WG02

Note the position of the rotor when the No. 1 piston is at TDC on the compression stroke

- Distributor cap
- Negative battery cable

2. Check the ignition timing and adjust, as necessary.

Timing Disturbed

1. Set the engine to Top Dead Center (TDC) of the compression stroke for the No. 1 cylinder.

2. Align the index mark on the distributor shaft with the protrusion on the distributor housing.

3. Install the distributor and check that the distributor rotor is aligned.

4. Install or connect the following:
- Distributor. Tighten the hold-down bolt to 10–12 ft. lbs. (14–17 Nm).
- Distributor cap
- Distributor harness connector

5. Check the ignition timing and adjust, as necessary.

Alternator

REMOVAL

1. Before servicing the vehicle, refer to the Precautions Section.

2. Remove or disconnect the following:
- Negative battery cable
- Idler adjusting bolt, loosen
- A/C belt
- Engine undercover
- Alternator electrical connectors and bracket
- Alternator mounting bolts
- Alternator belt
- Alternator

INSTALLATION

1. Install the components in the reverse order of removal. Tighten the fasteners to the following specifications:
   a. Alternator mounting bolts to 16–22 ft. lbs. (22–29 Nm).
   b. Alternator bracket bolt to 12–15 ft. lbs. (16–20 Nm).

Ignition Timing

ADJUSTMENT

1. Before servicing the vehicle, refer to the Precautions Section.

2. Check for trouble codes and make necessary repairs if needed.

3. Apply the parking brake and be sure that the vehicle is in PARK.

4. Start and run the engine until it reaches normal operating temperature.

5. Run the engine at about 2000 rpm for 2 minutes under no-load.

6. Turn off all electrical loads.

7. Disconnect the Throttle Position (TP) sensor electrical connector.

8. Be sure the engine speed is 700–800 rpm.

9. Rev the engine 2 or 3 times to 2,000–3,000 rpm and return the engine to idle speed.

10. Connect a timing light to the distributor end and check the ignition timing. Be sure that the timing pointer is pointing to the 15° BTDC mark on the crankshaft pulley.

➡**Each notch on the crankshaft pulley represents 5°.**

11. If the timing is not within the specification, loosen the distributor mounting bolt and adjust the distributor until the timing is at the proper specification.

12. Tighten the distributor mounting bolt to 10–12 ft. lbs., (14–17 Nm).

13. Stop the engine and connect the TP sensor.

Timing
indicator

7924WG03

Adjust the timing so the pointer on the engine indicates 15° before top dead center (3 notches from TDC) on the crankshaft pulley.

Engine Assembly

REMOVAL & INSTALLATION

1. Before servicing the vehicle, refer to the Precautions Section.

2. Properly relieve the fuel system pressure.

3. Drain the coolant and crankcase.

4. Remove or disconnect the following:
- Negative battery cable
- Front wheels
- All vacuum hoses, fuel lines, wires, harnesses and connectors that would interfere with engine removal
- Exhaust tube
- Ball joints
- Drive shafts

5. Recover the refrigerant from the A/C system
- A/C compressor manifold
- Power steering pump

6. Support the engine using a suitable lift.
- Left hand engine mount bolts
- Right hand engine mount
- Rear A/C refrigerant line bracket, if equipped
- Crossmember

7. Lower the engine and transaxle assembly and remove it from the vehicle.

To install:

8. Installation is the reverse of removal. Refer to the accompanying engine mounting illustration for all necessary torque specifications.

43 – 55
(4.4 – 5.6, 32 – 41)

43 – 55
(4.4 – 5.6, 32 – 41)

43 – 55
(4.4 – 5.6, 32 – 41)

41 – 52
(4.2 – 5.3,
30 – 38)

43 – 55
(4.4 – 5.6, 32 – 41)

41 – 52
(4.2 – 5.3,
30 – 38)

Slinger

Slinger

43 – 55
(4.4 – 5.6, 32 – 41)

22 – 29
(2.2 – 3.0, 16 – 22)

22 – 29
(2.2 – 3.0, 16 – 22)

64 – 74
(6.5 – 7.5,
47 – 54)

41 – 52
(4.2 – 5.3, 30 – 38)

78 – 88
(8.0 – 9.0,
58 – 65)

41 – 52 (4.2 – 5.3, 30 – 38)

41 – 52
(4.2 – 5.3, 30 – 38)

78 – 88
(8.0 – 9.0,
58 – 65)

78 – 88
(8.0 – 9.0,
58 – 65)

Center member

78 – 88
(8.0 – 9.0,
58 – 65)

Vehicle front

78 – 88
(8.0 – 9.0,
58 – 65)

78 – 88
(8.0 – 9.0, 58 – 65)

: N·m (kg-m, ft-lb)

9302WG01

Engine mounting components and specifications—3.3L engines

POWERTRAIN
LIFT WITH
TILTING PLATE

7924WG04

Carefully lower the engine/transaxle assembly from the vehicle.

Water Pump

REMOVAL & INSTALLATION

1. Before servicing the vehicle, refer to the Precautions Section.
2. Drain the coolant.
3. Remove or disconnect the following:
   - Negative battery cable
   - Radiator hoses and fan shroud
   - Drive belts
   - Water pump pulley using strap wrench 303–D055–(D85L–6000–A) to hold the pulley while removing the bolts
4. Remove the crankshaft pulley using the following procedure:
   a. Raise and safely support the vehicle.
   b. Remove the 5 right side inner engine and transmission splash shield bolts and 2 screws and remove the inner engine and transmission shield.

c. Remove the 4 right side outer engine and transmission splash shield bolts and 2 screws and remove the right side outer engine and transmission splash shields.
   d. Use a strap wrench to hold the crankshaft pulley while removing the crankshaft pulley bolt.
   e. Use a crankshaft damper remover to draw the crankshaft pulley off the front of the crankshaft.
5. Remove the 5 lower engine front cover bolts and take of the front cover.
6. Remove the 6 water pump bolts. Make note of the locations of the bolts since one should be a stud/bolt and must be returned to its original location. Remove the water pump.

To install:

7. Clean all parts well. The bolt threads should be cleaned of any old sealer or corrosion. Be sure the mating surfaces between the water pump and the engine block are cleaned of any old sealant. Apply a continuous bead of gasket maker type sealer approximately ⅛ inch (3mm) wide onto the

water pump and position the water pump on the engine block.
8. Install the 6 water pump bolts. Refer to any notes made at removal so the bolts can be returned to their original locations. Do not over-tighten the water pump bolts. Tighten the water pump bolts evenly to 12–15 ft. lbs. (16–21 Nm).
9. Position the water pump pulley on the water pump and install the 4 pulley bolts. Use a strap wrench to hold the pulley as the bolts are tightened to 12–15 ft. lbs. (16–21 Nm).
10. Install the front engine cover and the 5 lower front cover bolts. Tighten to 27–44 inch lbs. (3–5 Nm).
11. Install the crankshaft pulley using the following procedure:
    a. Install the crankshaft pulley and pulley bolt.
    b. Hold the pulley with a strap wrench. Tighten the crankshaft pulley bolt to 90–98 ft. lbs. (123–132 Nm).
    c. Install the inner and outer engine and transmission splash shields.
12. Install the drive belts.
13. Connect the negative battery cable.
14. Refill the cooling system.
15. Start the engine and check for leaks.

Heater Core

REMOVAL & INSTALLATION

Front System

1. Disconnect the negative battery cable.
2. Drain the cooling system into a clean container for reuse.
3. Remove or disconnect the following:
   - Heater hoses at the bulkhead and plug
   - Storage bin, then both side covers by the bin and the footlamp, if equipped
   - Control console bezel (1 screw in the center), then the ashtray assembly
   - Climate control console screws, pull the console rearward and detach the electrical connectors
   - 4 radio assembly screws and take the radio out of the vehicle
   - Floor duct and the right and left knee reinforcement plates
   - ABS control module.
4. The speed control module, keyless entry module (if equipped) and the passive restraint (air bag) module are all located behind the center console and can be removed after detaching the respective con-

WATER PUMP

WATER PUMP BOLTS (6)

STUD/BOLT

7924WG05

Water pump mounting. Note the location of the stud/bolt

nectors and removing the retaining nuts or screws.

※※ WARNING

The control modules are very sensitive to static electricity and can be damaged if exposed to static or stray electrical impulses.

5. Remove or disconnect the following:
   • Center air duct
   • 2 ground wire bolts, the U-bracket and the 2 console brackets
   • Glove box and lamp
   • Accelerator pedal and pedal stop
   • Floor air duct
   • Temperature blend sir door actuator and mode door actuator by unfastening the attaching bracket bolts and detaching the electrical connections
   • Center distribution duct
   • 4 evaporator/blower assembly screws, the 4 heater assembly screws and the heater assembly
   • Heater pipe plate from the assembly
   • Heater core retainer, disengage the shut-off valve control rod
   • Heater core from the assembly

To install:
6. Install or connect the following:
   • Heater core to the case, the retainer and pipe plate
   • Heater assembly in the vehicle and attach the 4 retaining screws
   • Center distribution duct, the blend air and mode door actuators
   • Foor air duct
   • Acelerator stop and pedal
   • Glove box and lamp, then the center console and U-brackets
   • Center air duct, the passive restraint, the keyless entry, the speed control and the ABS modules, as removed
   • Remaining center console components
   • Heater hoses to the heater core
7. Refill the cooling system.
8. Connect the negative battery cable.
9. Run the engine to normal operating temperatures; then, check the climate control operation and check for leaks.

Rear Auxiliary System

➡**The rear heater/air conditioning assembly must be removed as a complete unit in order to remove the heater core and/or evaporator core.**

1. Disconnect the negative battery cable.

Exploded view the front heater/air conditioning assembly

93113GC2

View the front heater core and heater housing assembly

2. Drain the cooling system into a clean container for reuse.

3. Discharge and recover the air conditioning system refrigerant.

4. Remove or disconnect the following:
- Heater hoses at the bulkhead and plug
- Center seats
- 2 left half seat belt lower anchor bolts
- Left rear cargo net retainers, if equipped
- Lift gate scuff plate and the 3 screws from the left rear quarter trim panel. Gently pry the rear seat remote control (if equipped) from the trim panel. Disconnect the remote control wiring connector and remove the rear radio control panel. Pull the top of the trim panel away from the body.
- Rear climate control panel wiring, if equipped
- Left front lap belt guide from the left quarter trim panel and pass the belt through the trim panel
- Trim panel from the vehicle
- Upper duct from the assembly (6 screws)
- Blower motor and resistor wiring
- Temperature blend and vent door actuator connectors

5. Raise and safely support the vehicle. Use the spring lock coupling tool to disconnect and plug the refrigerant line connections from beneath the vehicle.

6. Lower the vehicle.

7. Remove or disconnect the following:
- 4 heater/air conditioning assembly bolts and the assembly from the vehicle
- Heater core and/or evaporator core from the assembly

To install:

8. Install or connect the following:
- Heater core and/or evaporator core into the assembly and the 4 retaining bolts

9. Raise and safely support the vehicle.

10. Using new O-rings, reconnect the refrigerant lines to the evaporator.

11. Lower the vehicle.

12. Install or connect the following:
- All wiring connectors
- Upper air duct with the 6 screws
- Trim panel and pass the lap seat belt through the panel slot
- Rear climate control panel
- Rear radio and rear remote control
- Remaining trim panel and components

13. Refill the cooling system.

14. Connect the negative battery cable.

15. Evacuate and charge the air conditioning system.

16. Run the engine to normal operating temperatures; then, check the climate control operation and check for leaks.

Cylinder Head

REMOVAL & INSTALLATION

The factory specifies that the cylinder head bolts ARE NOT to be reused. Obtain the proper replacement parts before beginning this procedure. Check carefully that all bolts are removed before attempting to remove a cylinder head. A tab, part of the head, contains 1 lightly tightened head bolt that is external to the valve cover. Do not overlook this "hidden" bolt or the head will be damaged.

1. Before servicing the vehicle, refer to the Precautions Section.

2. Properly relieve the fuel system pressure.

3. Drain the coolant.

4. Remove or disconnect the following:
- Negative battery cable
- Air intake tube
- Timing belt
- Upper intake manifold (plenum)
- Fuel feed and return hoses from the fuel rail
- Fuel injector's electrical connections
- Fuel rail and injectors as an assembly
- Intake manifold (lower)
- Camshaft sprockets
- Rear timing belt cover
- Distributor
- Harness clamp from the right hand rocker cover
- Exhaust tube from the left hand manifold
- Left hand exhaust manifold from the right hand exhaust manifold
- Left hand manifold-to-bracket bolt
- A/C compressor, alternator and their brackets
- Rocker covers
- Cylinder head bolts in the sequence illustrated using several passes
- Cylinder head with the exhaust manifold and gasket. Discard the gasket.
- Exhaust manifold from the head

To install:

5. Clean all parts well.

6. Inspect the cylinder head for damage, cracks and leakage of water and oil. If necessary, replace the head. Check the head gasket surface for burrs and nicks. If the head is cracked, it must be replaced.

7. Install the exhaust manifold on the cylinder head.

8. Position a new head gasket and the cylinder head on the block. Examine the head bolt washers. Note that the washers have a chamfer or bevel on one side. The beveled side should face "up" when installed. Examine the new replacement head bolts. There are different lengths. The head bolts in positions 4, 7, 9 and 12 are 5.00 inches (127mm) long and the rest are 4.17 inches (106mm) long. Be sure the new cylinder head bolts are installed in the correct positions.

9. Tighten the new head bolts in the following sequence:
   a. First pass: cylinder head bolts to 22 ft. lbs. (29 Nm).
   b. Second pass: cylinder head bolts to 43 ft. lbs. (59 Nm).

For LH cylinder head   For RH cylinder head

No. 1   No. 3   No. 5

Engine front

Loosen in numerical order.

9348WG08

Remove the cylinder head bolts in the sequence shown—3.3L engines

Camshaft sprocket

9348WG10

Hold the camshaft sprocket while removing the sprocket retaining bolt—3.3L engines

RH side   No. 1   No. 3   No. 5

LH side

No. 2   No. 4   No. 6

Cylinder head bolt

9348WG11

Tighten the cylinder head bolts is sequence as shown—3.3L engines

c. Third pass: Loosen all of the cylinder head bolts completely.

d. Fourth pass: cylinder head bolts to 7 ft. lbs. (10 Nm).

e. Fifth pass: intake manifold bolts and nuts to 2.9 ft. lbs. (4 Nm).

f. Sixth pass: intake manifold bolts and nuts to 13 ft. lbs. (18 Nm).

g. Seventh pass: intake manifold bolts and nuts to 12–14 ft. lbs. (16–20 Nm).

h. Eight pass: Loosen all of the intake manifold bolts and nuts completely.

i. Ninth pass: cylinder head bolts to 22 ft. lbs. (29 Nm).

j. Tenth pass: cylinder head bolts to 40–47 ft. lbs. (54–64 Nm).

k. Eleventh pass: cylinder head sub-bolts to 6.7–8.7 ft. lbs. (9–12 Nm).

l. Twelfth pass: intake manifold bolts and nuts to 2.9 ft. lbs. (4 Nm).

m. Thirteenth pass: intake manifold bolts and nuts to 6.5 ft. lbs. (9 Nm).

n. Fourteenth pass: intake manifold bolts and nuts to 6–7 ft. lbs. (8–10 Nm).

- Rocker covers
- A/C compressor, alternator brackets
- A/C compressor and alternator
- Left hand manifold-to-bracket bolt
- Left hand exhaust manifold to the right hand exhaust manifold
- Exhaust tube to the left hand manifold
- Harness clamp to the right hand rocker cover
- Distributor
- Rear timing belt cover
- Camshaft sprockets
- Intake manifold (lower)
- Fuel rail and injectors as an assembly
- Fuel injector's electrical connections
- Fuel feed and return hoses to the fuel rail
- Upper intake manifold (plenum)
- Timing belt
- Air intake tube
- Negative battery cable

10. Fill the cooling system. An oil and filter change is recommended.

11. Start the vehicle and check for leaks. Check the ignition timing and adjust as required.

Rocker Arms/Shafts

REMOVAL & INSTALLATION

1. Before servicing the vehicle, refer to the Precautions Section.

Exhaust

RH cylinder head front   **LH cylinder head front**

Intake

1 - 3 (0.1 - 0.3, 9 - 26)

18 - 22 (1.8 - 2.2, 13 - 16)

LH rocker cover

Oil filler cap

Intake rocker shaft
Be sure to align cut portion to cylinder head bolt.

Gasket

Rocker arm

Hydraulic valve lifter

RH rocker cover

Bolt M6 with washer

Valve lifter guide
Cylinder head bolt

Washer*1

Exhaust rocker shaft

Inner spring seat

Valve collet
Valve spring retainer
Outer valve spring
Inner valve spring
Valve oil seal
Valve guide
Valve seat
Outer spring seat
Exhaust valve

Bolt
Cylinder head rear cover
Rear cover gasket

78 - 88 (8.0 - 9.0, 58 - 65)

Camshaft locate plate
LH cylinder head
Gasket*2

RH cylinder head assembly

Cylinder block

LH camshaft
Camshaft front oil seal

*1

Cylinder head side

*2 Cylinder head gasket identification

: N·m (kg-m, in-lb)

: N·m (kg-m, ft-lb)

: Lubricate with new engine oil

9302WG02

Rocker arm and shaft components

2. Remove or disconnect the following:
- Negative battery cable
- Upper intake manifold
- Valve covers
- Rocker arm and shaft assemblies
- Rocker arms from the shafts

➡**Keep all valvetrain components in order for assembly.**

To install:

3. Lubricate all contact points with clean engine oil and assemble the rocker

arms to the shafts in their original positions.

4. Install or connect the following:
- Rocker arm and shaft assemblies. Tighten the bolts to 13–16 ft. lbs. (18–22 Nm).
- Valve covers
- Upper intake manifold
- Negative battery cable

5. Start the engine and check for leaks.

Wire

7924WG09

Wire the lifters on top of the guide so they won't fall out when the guide is removed from the head

Intake Manifold

REMOVAL & INSTALLATION

1. Before servicing the vehicle, refer to the Precautions Section.
2. Drain the cooling system.
3. Relieve the fuel system pressure.
4. Remove or disconnect the following:
   - Negative battery cable
   - Air intake duct
   - Idle Air Control (IAC) valve connectors
   - Throttle Position (TP) sensor and switch connectors
   - Exhaust Gas Recirculation (EGR) solenoid valve connector
   - Evaporative Emissions (EVAP) canister vacuum and purge hoses
   - Water, heater and Positive Crankcase Ventilation (PCV) valve hoses
   - Vacuum hoses from the EVAP canister, brake cylinder, pressure regulator and EGR tube
   - Spark plug wires
   - Distributor cap
   - 3 left bank injector connectors
   - Thermal transmitter
   - Ground harness
   - Breather pipe
   - Upper manifold
   - Fuel feed and return lines from the fuel rail
   - Right injector harness connectors
   - Fuel rail and injectors
   - Coolant temperature switch harness connector
   - Water hose from the thermostat
   - Lower manifold bolts in the sequence illustrated.
   - Manifold gasket and discard

Tighten bolts in numerical order.

7924VG33

Intake manifold tightening sequence—3.3L engine

To install:

5. Install the lower intake manifold with a new gasket.
   a. Step 1: 35 inch lbs. (4 Nm)
   b. Step 2: 78 inch lbs. (9 Nm)
   c. Step 3: 70–84 inch lbs. (8–10 Nm)
6. Install or connect the following:
   - ECT sensor connector
   - Fuel supply manifold
   - Right bank injector connectors
   - Fuel lines
   - Upper intake manifold
   - Breather pipe
   - Upper intake manifold ground cable
   - Thermal transmitter
   - Left bank injector connectors
   - Distributor
   - Spark plug wires
   - EGR tube
   - Fuel pressure regulator vacuum hose
   - Brake booster vacuum hose
   - EVAP canister vacuum and purge hoses
   - PCV valve and hose
   - Heater hoses
   - Radiator hoses
   - EGR temperature sensor connector
   - EGR solenoid valve connector
   - Ignition coil and power transistor connectors
   - TP sensor and switch connectors
   - IAC valve connector
   - Cruise control cable
   - Accelerator cable
   - Air intake duct
   - Negative battery cable
7. Fill the cooling system.
8. Start the engine and check for leaks.

Exhaust Manifold

REMOVAL & INSTALLATION

Rear (Right-Hand) Exhaust Manifold

1. Before servicing the vehicle, refer to the Precautions Section.
2. Remove or disconnect the following:
   - Negative battery cable
   - Radiator overflow hose from the radiator
   - Radiator coolant-recovery reservoir off of the bracket
   - Reservoir
   - Air cleaner intake tube and the engine air intake resonator
   - 6 rear (right-hand) exhaust manifold crossover tube heat-shield bolts and the heat shields
   - 2 nuts and the 1 bolt securing the rear (right-hand) exhaust manifold tube to the front (left-hand) exhaust manifold. Discard the gasket.
   - Transmission fluid level indicator tube heat shield
3. Disengage the following electrical connectors:
   - Idle switch
   - Throttle Position (TP) sensor

Loosen bolts in numerical order.

7924VG32

Intake manifold loosening sequence—3.3L engine

- Exhaust Gas Recirculation (EGR) control solenoid

4. Raise and safely support the vehicle.

5. Remove or disconnect the following:

- EGR valve-to-back-pressure transducer valve tube nut and position it out of the way
- 2 EGR valve-to-exhaust manifold tube nuts and tube
- 6 rear exhaust manifold nuts in the reverse order of the tightening sequence

6. Safely lower the vehicle, remove the exhaust manifold and discard the exhaust manifold gasket.

To install:

7. Raise and safely support the vehicle.

8. Be sure that both the exhaust manifold and the cylinder head mating surfaces are clean of any old gasket material.

9. Install or connect the following:

- Rear (right-hand) exhaust manifold gasket onto the exhaust manifold mounting studs

10. Lower the vehicle safely.

- Rear (right-hand) exhaust manifold onto the studs
- 6 rear (right-hand) exhaust manifold nuts. Tighten the nuts in sequence to 13–16 ft. lbs. (18–22 Nm).
- EGR valve-to-exhaust manifold tube and tube nuts
- EGR valve-to-back-pressure transducer valve tube nut

11. Lower the vehicle carefully.

12. Reconnect the following electrical connectors:

- EGR solenoid
- TP sensor
- Idle switch
- Transmission fluid level indicator tube heat shield
- New gasket between the front (left-hand) exhaust manifold and the rear exhaust manifold crossover tube
- 2 nuts and the 1 bolt securing the rear (right-hand) exhaust manifold crossover tube to the front (left-hand) exhaust manifold. Tighten the rear exhaust manifold crossover tube-to-front (left-hand) exhaust manifold nuts and bolt.
- Rear (right-hand) exhaust manifold crossover tube heat shield with the 6 mounting bolts
- Rear (right-hand) exhaust manifold crossover tube bolts
- Air cleaner intake tube and the engine air intake resonator

RH exhaust

Front

LH exhaust

Front

Tighten in numerical order.

9348WG12

To avoid warping the exhaust manifolds, use this sequence when loosening the bolts—3.3L engine

- Radiator coolant recovery reservoir
- Radiator overflow hose to the radiator
- Negative battery cable

13. Start the engine and check for leaks and proper operation.

Front (Left-Hand) Exhaust Manifold

1. Before servicing the vehicle, refer to the Precautions Section.

2. Remove or disconnect the following:

- Negative battery cable and wait at least 90 seconds before performing any work. This allows time for the SRS or air bag system to deplete its back up energy supply.
- 2 nuts and the 1 bolt securing the front (left-hand) exhaust manifold to the rear (right-hand) exhaust manifold crossover tube. Discard the gasket.

3. Remove the transmission fluid level indicator tube heat shield.

- 6 front (left-hand) exhaust manifold nuts in 2 steps in the reverse order of the tightening sequence. Do not remove the 3 lower front (left-hand) exhaust manifold nuts.
- Front (left-hand) exhaust manifold-to-mounting bracket bolt

4. Raise and safely support the vehicle.

- Heated Oxygen Sensor (HO2S) electrical connector
- 3 front (left-hand) manifold-to-inlet pipe nuts
- Exhaust system flex tube bracket bolt
- Left-hand inner engine and transmission splash shield bolts and screws

- Left-hand inner engine and transmission splash shield
- 3 lower exhaust manifold nuts
- Front (left-hand) exhaust manifold and discard the exhaust manifold gasket

To install:

5. Be sure that both the exhaust manifold and the cylinder head mating surfaces are clean of any old gasket material.

6. Install or connect the following:

- New front exhaust manifold gasket in place
- Front (left-hand) exhaust manifold
- 3 lower exhaust manifold mounting nuts. Do not tighten the nuts at this time.
- Left-hand inner engine and transmission splash shield with their mounting bolts and screws
- Exhaust system flex tube bracket bolt
- 3 exhaust manifold-to-exhaust inlet pipe nuts
- HO2S electrical connector

7. Lower the vehicle.

- Front (left-hand) exhaust manifold-to-mounting bracket bolt
- 3 upper exhaust manifold mounting bolts and tighten all 6 exhaust manifold mounting bolts in sequence to 13–16 ft. lbs. (18–22 Nm)
- Transmission fluid level indicator tube heat shield
- 2 nuts and the 1 bolt securing the front (left-hand) exhaust manifold to the rear (right-hand) exhaust manifold crossover tube
- Negative battery cable

8. Start the engine, check for leaks and road test for proper operation.

Starter

REMOVAL & INSTALLATION

1. Before servicing the vehicle, refer to the Precautions Section.
2. Remove or disconnect the following:
   - Battery negative cable
   - Air cleaner
   - Nut attaching the positive cable to the starter
   - Positive cable from the starter
   - S-terminal connector
   - 2 starter bolts and the starter

To install:
3. Installation is the reverse of removal.
4. Tighten the starter bolts to 17–19 ft. lbs. (23–26 Nm) and the nut that attaches the positive battery cable to the starter to 87–104 inch lbs. (10–12 Nm).

Front Crankshaft Seal

REMOVAL & INSTALLATION

1. Before servicing the vehicle, refer to the Precautions Section.
2. Remove or disconnect the following:
   - Negative battery cable
   - Drive belts
   - Radiator hoses
   - Crankshaft pulley
   - Front cover
   - Timing belt
   - Crankshaft timing sprocket
   - Crankshaft seal using a suitable prytool

To install:
3. Install or connect the following:
   - Crankshaft seal using a driver and a hammer until its flush with the housing
   - Crankshaft timing sprocket
   - Timing belt

- Front cover and tighten the bolts to 26–43 inch lbs. (3–5 Nm)
- Crankshaft pulley and tighten the bolt to 141–156 ft. lbs. (191–211 Nm)
- Radiator hoses
- Drive belts
- Negative battery cable
4. Fill the cooling system, start the vehicle and check for leaks.

Camshaft And Valve Lifters

REMOVAL & INSTALLATION

1. Before servicing the vehicle, refer to the Precautions Section.
2. Drain the cooling system.
3. Remove or disconnect the following:
   - Negative battery cable
   - Upper intake manifold
   - Valve covers

➡**Keep all valvetrain components in order for assembly.**

   - Rocker arm and shaft assemblies
   - Valve lifter guide and valve lifters. Attach a wire to the top of the lifters so that they will not drop from the lifter guide.
   - Radiator
   - Accessory drive belts
   - Front cover
   - Timing belt
   - Camshaft sprockets
   - Camshaft seals
   - Rear timing cover
   - Distributor
   - Cylinder head rear covers
   - Camshaft locating plates
   - Camshafts

To install:
4. Install or connect the following:
   - Camshafts
   - Camshaft locating plates. Tighten the bolts to 58–65 ft. lbs. (78–88 Nm).

   - Cylinder head rear covers
   - Distributor
   - Rear timing cover
   - Camshaft seals
   - Camshaft sprockets. Tighten the bolts to 58–65 ft. lbs. (78–88 Nm).
   - Timing belt
   - Front cover
   - Accessory drive belts
   - Radiator
   - Valve lifter guide and valve lifters
   - Rocker arm and shaft assemblies. Tighten the bolts to 13–16 ft. lbs. (18–22 Nm).
   - Valve covers
   - Upper intake manifold
   - Negative battery cable
5. Fill the cooling system.
6. Start the engine and check for leaks.

Valve Lash

ADJUSTMENT

The engines covered in this section use hydraulic valve lifters that automatically adjust the valve lash. No periodic adjustment is needed.

Oil Pan

REMOVAL & INSTALLATION

1. Before servicing the vehicle, refer to the Precautions Section.
2. Drain the engine oil.
3. Remove or disconnect the following:
   - Negative battery cable
   - Front engine mount (support) insulator through-bolt
   - Rear engine mount (support) through-bolt
   - 2 rear refrigerant/heater pipe hold down bracket bolts
   - 4 crossmember (also called a transverse member) bolts, and remove the crossmember.
   - Exhaust inlet pipe
   - 4 rear transaxle-to-engine brace bolts and the 5 front transaxle-to-engine brace bolts
   - Front transaxle-to-engine brace
   - Low oil level sensor electrical connector
   - 18 oil pan bolts in the reverse order of the tightening sequence, working from the outside, towards the center bolts.
   - Oil pan and discard the seals

SUITABLE TOOL

7924WG13

Removal and installation of the front crankshaft seal—3.3L engines

To install:

4. Clean all parts well. Be sure that all old sealing material is removed from the oil pan and engine mating surfaces.

5. Position new oil pan seals. Apply Loctite® Ultra Gray 599 Silicone Sealer, or equivalent, to the ends of the oil pan seals.

6. Apply a bead of Loctite® Ultra Gray 599 Silicone Sealer or equivalent to the oil pan gasket rail inboard of the bolt holes.

7. Install or connect the following:
- Oil pan on the engine block. Tighten the 18 oil pan bolts in sequence, working from the inside, towards the outer bolts. Do not

TIGHTENING SEQUENCE

7924WG15

Tighten the 18 oil pan bolts in sequence, working from the inside, towards the outer bolts

Apply RTV silicone sealer to the seal ends and to the oil pan gasket rail

7924WG14

over-tighten. Tighten to 62–70 inch lbs. (7–8 Nm).
- Low oil level sensor electrical connector.
- Front and rear transaxle braces. Tighten all bolts to 22–30 ft. lbs. (30–40 Nm).
- Exhaust inlet pipe
- Crossmember and tighten the bolts to 58–65 ft. lbs. (78–88 Nm).
- Both engine support through-bolts and tighten to 58–65 ft. lbs. (78–88 Nm).

8. Remove the support jack from under the crankshaft pulley.

9. Lower the vehicle.

10. Fill the engine with the specified engine oil to the required level.

11. Connect the negative battery cable. Start the engine and check for leaks.

Oil Pump

REMOVAL & INSTALLATION

1. Before servicing the vehicle, refer to the Precautions Section.
2. Drain the engine oil.
3. Drain the cooling system.
4. Remove or disconnect the following:
- Negative battery cable
- Oil pan
5. After removing the oil pan, reinstall the crossmember and the mount bolts.

Exploded view of the oil pump assembly—3.3L engine

9348WG14

- Timing belt
- Crankshaft sprocket and timing belt plate
- Oil pump

To install:

6. Install or connect the following:
- Oil pump-to-body bolts to 52–69 inch lbs. (6–8 Nm)
- Timing belt plate and tighten the bolts to 52–69 inch lbs. (6–8 Nm)
- Crankshaft sprocket
- Timing belt
- Oil pan
- Negative battery cable
7. Fill the cooling system.
8. Fill the crankcase to the correct evel.
9. Start the engine and check for leaks.

Rear Main Seal

REMOVAL & INSTALLATION

3.3L Engine

1. Before servicing the vehicle, refer to the Precautions Section.
2. Disconnect the negative battery cable.
3. Remove the transaxle from the vehicle.
4. Remove the flexplate from the crankshaft.
5. Remove the rear oil seal retainer.

✳✳ WARNING

Do not scratch the seal bore of the oil seal retainer when removing the oil seal.

6. Remove the oil seal from the seal retainer.

To install:

7. Apply clean engine oil to the lip and outer surface of the new seal to aid during installation.
8. Install the seal in the retainer using a suitable seal driver.
9. Using a new gasket install the

retainer on the engine. Tighten the bolts to 52–61 inch lbs. (6–7 Nm).

10. Install the flexplate. Tighten the bolts to 61–69 ft. lbs. (83–93 Nm).

11. Install the transaxle and remaining components.

Timing Belt

REMOVAL & INSTALLATION

3.3L (VIN T) engines

On this vehicle, right side refers to the "rear" components (near the firewall) and left side refers to the "front" components (near the radiator).

1. If the timing belt is to be removed, it is good practice to turn the crankshaft until the engine is at Top Dead Center (TDC) of the No. 1 cylinder, compression stroke (firing position), before beginning work. This should align all timing marks and serve as a reference for all work that follows. After verifying that the engine is at TDC for the No. 1 cylinder, do not crank the engine or allow the crankshaft or camshaft sprockets to be turned otherwise engine timing will be lost.

2. Drain the cooling system.
3. Remove or disconnect the following:
- Negative battery cable
- Alternator drive belt, water pump and power steering pump belt and the air conditioning compressor belt, if equipped
- 3 air conditioning compressor drive belt idler pulley bolts and the idler pulley, if equipped with air conditioning
- Upper radiator hose bracket bolt
- Upper hose with the bracket from the vehicle
- Water bypass hose from between the thermostat housing and the lower water hose connection
- Main wiring harness from the upper engine front cover

- 8 upper engine front cover bolts and the upper cover
- Right side front wheel and tire assembly
- 4 right side engine and transmission splash shield bolts and 2 screws, and right side outer engine and transaxle splash shield

4. Use a strap wrench to hold the water pump pulley. Remove the 4 pulley bolts, and the water pump pulley.

5. Use a strap wrench to hold the crankshaft pulley. Remove the center pulley bolt, and the crankshaft pulley using a harmonic balancer (damper) puller to draw the pulley from the front of the crankshaft.
- 5 lower engine front cover bolts, then remove the lower engine front cover

6. Be sure that the timing marks between the crankshaft sprocket and the oil pump housing align.

7. If the timing belt is to be reused, mark an arrow on the belt indicating the direction of rotation. The directional arrow is necessary to ensure that the timing belt, if it to be reused, can reinstalled in the same direction.

8. Loosen the timing belt tensioner nut and slip the timing belt off of the sprockets.

9. If necessary, the camshaft sprockets can be removed. A special spanner tool is designed to hold the sprocket to keep it from turning while the center bolt is being loosened. Use care if using substitutes.

➡**The sprockets are not interchangeable.**

10. If necessary, the crankshaft sprocket can be removed. The outer timing belt guide (looks like a large washer) and the crankshaft sprocket simply pull off the front of the crankshaft.

➡**Be careful, there are 2 crankshaft keys. Use care not to loose them.**

To install:

11. Clean all parts well. If removed, inspect the crankshaft sprocket for warping or abnormal wear. Check the sprocket teeth for wear, deformation, chipping or other damage. Replace as necessary. Clean the sprocket mounting surface to ease installation. Install the key. Slip the sprocket onto the crankshaft. Tap it in place with a suitably-sized socket.

12. If removed, inspect the camshaft sprockets for damage and wear. Replace as required. The sprockets should be marked **L3** to designate the front, or left side camshaft and **R3** to designate the rear, or right side camshaft. Use care to install the

Exploded view of the oil seal, retainer and gasket—3.3L engine

7924WG17

Use a shop rag to clean the alignment marks for the timing belt— 3.3L (VIN T) engines

sprockets properly. A special spanner tool is designed to hold the sprocket to keep it from turning while the center bolt is being tightened. Use care if using a substitute. Tighten the camshaft sprocket center bolts to 61 ft. lbs. (83 Nm). Verify that the timing marks on the camshaft sprockets and the timing marks on the rear cover (called the seal plate) are aligned.

13. Use an Allen wrench to turn the timing belt tensioner clockwise until the belt tensioner spring is fully extended. Temporarily tighten the tensioner nut to 32–43 ft. lbs. (43–58 Nm).

14. If a new timing belt is to be installed, look for a printed arrow on the belt. Be sure the arrow is pointing away from the engine. If the original timing belt is to be reused, be sure that the directional arrow that was marked at disassembly is facing the correct direction.

15. A new Original Equipment Manufacture (OEM) timing belt should have 3 white timing marks on it that indicate the correct timing positions of the camshafts and the crankshaft. These marks are to help ensure that the engine is properly timed. When the engine is properly timed, each white timing mark on the timing belt will be aligned with the corresponding camshaft and crankshaft timing mark on the sprocket. Because the white timing marks are not evenly spaced, the technician needs to use care in installing the belt. There should be 40 timing belt teeth between the timing marks on the front and rear camshaft sprockets and 43 teeth between the timing mark on the front camshaft sprocket and the timing mark on the crankshaft sprocket.

16. Verify that the camshaft timing marks are aligned with the timing marks on the rear cover (seal plate) and that the crankshaft sprocket timing mark is aligned with the timing mark on the oil pump housing.

17. Install the timing belt starting at the crankshaft sprocket and moving around the camshaft sprockets following a counter-clockwise path. Do not allow any slack in the timing belt between the sprockets. After all of the timing marks are aligned with the timing belt installed, slip the timing belt onto the belt tensioner.

18. While holding the timing belt tensioner with an Allen wrench, loosen the tensioner nut. Allow the tensioner to put pressure on the timing belt. Use an Allen wrench to turn the timing belt tensioner 70–80 degrees clockwise and tighten the timing belt tensioner nut to 32–43 ft. lbs. (43–58 Nm).

❊❊ WARNING

If any binding is felt when adjusting the timing belt tension by turning the crankshaft, STOP turning the engine, because the pistons may be hitting the valves.

19. Rotate the crankshaft clockwise twice and align the No. 1 piston to TDC on the compression stroke (firing position).

20. Apply 22 lbs. (10kg) of force on the timing belt between the rear camshaft sprocket and the timing belt tensioner. An assistant may be needed. While holding the timing belt tensioner steady with an Allen wrench, loosen the timing belt tensioner nut. Remove the Allen wrench and adjust the timing belt tensioner using the following procedure:

   a. Install a 0.0138 in. (0.35mm) thick and 0.500 in. (12.7mm) wide feeler gauge where the timing belt just starts to go around the tensioner (approximately the 4 o'clock position, looking at the tensioner).

   b. Turn the crankshaft sprocket clockwise, which should force the feeler gauge between the timing belt and the tensioner, up to a position on the tensioner of about 1 o'clock.

   c. Tighten the timing belt tensioner nut to 61 ft. lbs. (83 Nm.

   d. Turn the crankshaft clockwise to

rotate the feeler gauge out from between the timing belt tensioner and the timing belt.

21. Rotate the crankshaft clockwise twice, and once again align the No. 1 piston to TDC on the compression stroke (firing position).

22. Apply 22 lbs. (10kg) of force on the timing belt between the front and rear camshaft sprockets. Measure the amount of belt deflection. Belt deflection should be between 0.51–0.59 in. (13–15mm). If belt deflection is out of specification, repeat Steps 29 through 33. If the timing belt deflection cannot be adjusted into specification, the timing belt will have to be replaced.

23. Install or connect the following:

- Lower engine front cover and the 5 lower cover bolts. Do not over tighten. Tighten to 27–44 inch lbs. (3–5 Nm).
- Outer timing belt guide next to the crankshaft sprocket with the dished side facing away from the cylinder block. Install the crankshaft pulley. Use a strap wrench to keep the crankshaft pulley from turning and tighten the center bolt to 148 ft. lbs. (201 Nm).
- Water pump pulley on the pump. Install the 4 bolts. Use a strap wrench to keep the water pump pulley from turning and tighten the 4 water pump pulley bolts to 89 inch lbs. (10 Nm).
- Right side outer engine and transaxle splash shield, and secure with the 4 bolts and 2 screws
- Right side front wheel. Tighten the lug nuts to 72–87 ft. lbs. (98–118 Nm).
- Upper engine timing belt front cover, and tighten the 8 bolts to 27–44 inch lbs. (3–5 Nm)
- Main wiring harness on the upper engine front cover
- Water bypass hose between the thermostat housing and water connection
- Upper radiator hose between the radiator and the water hose connection. Secure the hoses with clamps. Install the upper radiator hose bracket. Tighten the bracket bolt to 34–58 ft. lbs. (46–65 Nm).
- Air conditioning compressor drive belt idler pulley and install the 3 bolts. Tighten to 15 ft. lbs. (21 Nm), if equipped
- Alternator drive belt, the water pump and power steering pump drive belt and the air conditioning compressor drive belt, if equipped
- Negative battery cable

24. Fill the cooling system.

25. Start the engine and allow it to warm to operating temperature. Check and adjust the ignition timing. Road test to verify correct engine operation.

Piston and Ring Positioning

3.3L engines piston ring end-gap spacing

3.3L engines piston and connecting rod assembly positioning

3.3L engine piston positioning

FUEL SYSTEM

Fuel System Service Precautions

Safety is the most important factor when performing not only fuel system maintenance but any type of maintenance. Failure to conduct maintenance and repairs in a safe manner may result in serious personal injury or death. Maintenance and testing of the vehicle's fuel system components can be accomplished safely and effectively by adhering to the following rules and guidelines.

• To avoid the possibility of fire and personal injury, always disconnect the negative battery cable unless the repair or test procedure requires that battery voltage be applied.

• Always relieve the fuel system pressure prior to disconnecting any fuel system component (injector, fuel rail, pressure regulator, etc.), fitting or fuel line connection. Exercise extreme caution whenever relieving fuel system pressure, to avoid exposing skin, face and eyes to fuel spray. Please be advised that fuel under pressure may penetrate the skin or any part of the body that it contacts.

• Always place a shop towel or cloth around the fitting or connection prior to loosening to absorb any excess fuel due to spillage. Ensure that all fuel spillage (should it occur) is quickly removed from engine surfaces. Ensure that all fuel soaked cloths or towels are deposited into a suitable waste container.

• Always keep a dry chemical (Class B) fire extinguisher near the work area.

• Do not allow fuel spray or fuel vapors to come into contact with a spark or open flame.

• Always use a back-up wrench when loosening and tightening fuel line connection fittings. This will prevent unnecessary stress and torsion to fuel line piping. Always follow the proper torque specifications.

• Always replace worn fuel fitting O-rings with new. Do not substitute fuel hose or equivalent, where fuel pipe is installed.

Fuel System Pressure

RELIEVING

1. Before servicing the vehicle, refer to the Precautions Section.
2. Remove the left side engine compartment relay panel cover.
3. Locate and remove the fuel pump relay from the relay panel.
4. Start the engine.
5. Allow the engine to run until it stalls from fuel starvation. After the engine stalls, crank the engine over 2 more times to ensure all pressure has been released.
6. Turn the ignition switch to the **OFF** position and install the fuel pump relay.
7. Most service work that follows fuel pressure relief also requires that the negative battery cable (ground) be disconnected before service work begins. This also prevents accidental fuel pump energizing that could pressurize the system.

Fuel Filter

REMOVAL & INSTALLATION

In-Line—Except California

1. Before servicing the vehicle, refer to the Precautions Section.
2. Relieve the fuel system pressure using the recommended procedure.
3. Disconnect the negative battery cable.
4. Raise and safely support the vehicle.
5. Remove the fuel hose clamps.
6. Disconnect and plug the hoses to prevent leakage.
7. Remove the fuel filter from the bracket.

To install:

8. Install the fuel filter into the bracket with the arrow facing up, in the direction of the fuel travel to the engine.
9. Reconnect the fuel hoses.
10. Install and tighten the hose clamps. Verify that the clamps are properly tightened. System operating pressure is approximately 36 psi (248 kPa) and fuel will leak is connections are not properly made.
11. Lower the vehicle.
12. Reconnect the negative battery cable.
13. Check for leaks.

In-Line—California

1. Before servicing the vehicle, refer to the Precautions Section.

2. Relieve the fuel system pressure using the recommended procedure.

3. Disconnect the negative battery cable.

4. Raise and safely support the vehicle.

5. Remove the filter splash shield olts.

6. Disconnect the lines from each end of the filter and plug the hoses to prevent leakage.

7. Loosen the filter bracket nuts.

8. Remove the fuel filter from the bracket.

To install:

9. Install the fuel filter into the bracket with the arrow facing forward. Tighten the bracket bolts to 44 inch lbs. (5 Nm).

10. Reconnect the fuel hoses.

11. Lower the vehicle.

12. Reconnect the negative battery cable.

13. Check for leaks.

Fuel Pump

REMOVAL & INSTALLATION

1. Before servicing the vehicle, refer to the Precautions Section.

2. Properly relieve the fuel system pressure.

3. Disconnect the negative battery cable.

4. Raise and safely support the vehicle.

5. Remove the fuel tank as follows:
   a. Drain the fuel from the tank.
   b. Remove the filler protector.
   c. Disconnect the filler tube.
   d. Detach any electrical connectors related to the fuel pump and fuel level sending unit.
   e. Detach the fuel line quick connectors.
   f. Safely support the fuel tank.
   g. Remove the tank mounting straps, then lower the tank out of the vehicle.

6. Remove the 6 fuel pump bolts.

7. Lift the fuel pump out of the fuel tank. Use care. The fuel level sensor and fuel pump and bracket must be tipped to remove it from the fuel tank. Do not lift the fuel sensor and pump assembly straight out of the fuel tank or damage to the level sensor may occur.

8. Remove the 2 bolts attaching the level sensor to the fuel pump.

9. Remove the fuel pump level sensor and the gasket.

10. Discard the gasket.

11. Remove the fuel pump from the bracket.

To install:

12. Position the fuel level sensor on the fuel pump and bracket and install the 2 bolts.

13. Install a new level sensor gasket. Carefully install the level sensor and pump assembly.

14. Install the 6 fuel pump bolts. Do not over-tighten the bolts. Tighten the bolts to just 17–23 inch lbs. (2–3 Nm).

Fuel tank and related components

9302WG03

15. Install the fuel tank in the reverse order of removal, be sure to tighten the tank mounting straps to 20–26 ft. lbs. (27–35 Nm).

16. Lower the vehicle. Refill the fuel tank as required.

17. Connect the negative battery cable. Verify that the fuel pump relay has been properly installed. Start the engine and check for proper operation.

Fuel Injector

REMOVAL & INSTALLATION

1. Before servicing the vehicle, refer to the Precautions Section.

2. Disconnect the negative battery cable.

3. If removing a rear injector, remove the upper intake manifold.

4. Disengage the injector electrical connection.

➡ **When removing the fuel injectors, use a screwdriver head socket to remove the injector cap screws.**

5. Remove the injector cap screws and the cap.

6. Pull the injector from the fuel rail.

7. remove and discard the injector O-rings.

To install:

➡ **Use new insulators and O-ring seals for assembly.**

8. Install or connect the following:
- Fuel injectors with the rail and tighten the fasteners to 8–11 ft. lbs. (11–15 Nm)
- Fuel injector caps. Tighten the screws to 26–33 inch lbs. (3–4 Nm).
- Injector electrical connections.
- Intake manifold, if removed.
- Negative battery cable.

9. Start the vehicle and check for leaks.

DRIVE TRAIN

Automatic Transaxle Assembly

REMOVAL & INSTALLATION

1. Before servicing the vehicle, refer to the Precautions Section.

2. Remove or disconnect the following:
- Negative battery cable
- Battery and tray
- Resonator
- Terminal cord assembly harness connector
- Vacuum lines
- Starter motor
- Transaxle fluid from the unit
- Halfshafts
- Transaxle cooler hose and control cable
- Front exhaust manifold
- Crankshaft Position (CKP) sensor
- Engine gusset and torque converter undercover
- Bolts from the drive plate from the torque converter. Rotate the crankshaft to access all the bolts.

3. Support the transaxle with a suitable jack.
- Front mount
- Rear mount
- Bolts attaching the transaxle to the engine

4. Carefully separate the transaxle assembly from the engine assembly. Lower the assembly from the vehicle.

To install:

5. Be sure that the transaxle is secured firmly to the transaxle jack.

6. Carefully raise the transaxle into the vehicle and align the transaxle to the engine assembly, making sure that the alignment dowels are positioned properly.

7. Install or connect the remaining components in the reverse order of removal. Refer to the accompanying transaxle torque specification illustration for bolt locations and their specifications.

8. Connect the negative battery cable.

9. Fill the transaxle with the correct amount and type of fluid.

10. Start the engine.

11. Check for leaks and proper operation.

Halfshaft

REMOVAL & INSTALLATION

1. Before servicing the vehicle, refer to the Precautions Section.

2. Raise and safely support the vehicle.

3. Remove or disconnect the following:
- Wheel
- Fender splash shield
- Cotter pin, nut retainer, and the hub retainer washers from the front hub assembly
- Lower ball joint from the knuckle
- Sway bar from the lower control arm at the sway bar link nut

Bolt No.	Tightening torque N·m (kg-m, ft-lb)	ℓ mm (in)
1	39 - 49 (4.0 - 5.0, 29 - 36)	60 (2.36)
2	30 - 40 (3.1 - 4.1, 22 - 30)	25 (0.98)
3*	30 - 40 (3.1 - 4.1, 22 - 30)	25 (0.98)

*: TORX bolt

9348WG17

Transaxle torque specification and locations—3.3L engine

Removing the left side halfshaft by gently prying with 2 prybars to unseat the circlip

- Halfshaft and CV-joint from the wheel hub

4. Position a drain pan under the transaxle since some fluid may run out when the inner joint is disengaged from the transaxle.

5. A prybar is used to separate the inner CV-joint from the transaxle. Use great care that the prybar does not damage the transaxle case, differential oil seal, outer race or boot. If removing the left side halfshaft, position prybars on both sides of the outer race, between the outer race and the transaxle case. Gently pry outward to unseat the circlip.

6. When removing the right side halfshaft, it is not be necessary to remove the halfshaft bearing retainer bracket from the cylinder block. Remove the 3 bearing retainer bolts and pull the right side halfshaft CV-joint with the bearing retainer from the differential side gear.

7. Support the halfshafts and remove them from the vehicle. Use care not to damage the boots. Place the halfshafts on a flat, protected work area.

To install:

✳ CAUTION

Do not reuse the circlip used on the left side halfshaft.

8. To prevent over-expanding the circlip, install the circlip carefully, starting one end in the shaft groove, then working the circlip over the CV-joint splined end. Always use a new circlip. No circlip is used on the right side halfshaft.

9. Inspect the CV-joint boots. If service is required, replace the CV-joint boots.

10. Inspect the differential oil seals. If damaged, the factory recommends using a hook-type puller and slide hammer arrangement to remove the seals. A seal driver is used to install the replacement differential oil seals.

11. If installing the left side halfshaft and CV-joint assembly, position the CV-joint so the splines are aligned with the differential side gear splines, then push the halfshaft joint into the differential case. As the circlip

locks into the differential side gear groove, a click will be felt.

12. If installing the right side halfshaft and CV-joint assembly, simply push the CV-joint into the differential side gear. Position the bearing retainer onto the bearing retainer bracket that should still be on the cylinder block. Install the 3 bolts and tighten to 8–14 ft. lbs. (13–19 Nm).

13. Install or connect the following:
- Halfshaft
- Lower ball joint and tighten the lower ball joint stud nut to 52–63 ft. lbs. (71–86 Nm). Secure the nut with a new cotter pin.
- Sway bar link to the lower control arm and tighten the link nut to 12–16 ft. lbs. (16–22 Nm).
- Wheel outer bearing retainer, washer and axle nut. Tighten the hub nut to 174–231 ft. lbs. (235–314 Nm). Install the nut retainer and secure with a new cotter pin.
- Splash shield
- Wheel. Tighten the lug nuts to 72–87 ft. lbs. (98–118 Nm).

14. Lower the vehicle.

15. Check the transaxle fluid level.

16. Road test the vehicle to verify correct operation and no noise or vibration.

CV-Joint

OVERHAUL

Inner

1. Remove the boot bands.

2. Matchmark the slide joint housing and inner race, prior to separating the joint assembly.

3. Pry off the snapring and remove the ball cage, inner race and balls as a unit.

4. Remove the snapring and withdraw the boot.

To install:

➤**Cover the halfshaft serrations with tape, so as not to damage the boot.**

5. Thoroughly clean all parts in solvent and dry with compressed air. Check parts for evidence of damage and replace as necessary.

6. Install the boot and new boot band on the halfshaft.

7. Install a new inner snapring.

8. Install the ball cage, inner race and balls as a unit. Confirm that the matchmarks are aligned.

9. Install a new outer snapring.

10. Pack the CV-joint with 5.0–6.0 ounces (165–175 g) of grease.

Right side halfshaft bearing retainer bracket

Circular clip:
  Make sure circular clip is properly meshed with side gear (transaxle side) and joint assembly (wheel side), and will not come out.
Be careful not to damage boots. Use suitable protector or cloth during removal and installation.

Wheel side (Rzeppa joint)

Boot band ⊗

Joint assembly

Boot

Circular clip B ⊗

Drive shaft

Dynamic damper
(For M/T models: installed on right drive shaft)

Dynamic damper band ⊗

Boot

Snap ring A ⊗

Inner race

Ball

Boot band ⊗

Snap ring B ⊗

Cage

Snap ring C ⊗

Slide joint housing

Dust shield ⊗

Circular clip A ⊗

Left drive shaft

⊡ : N•m (kg-m, ft-lb)

⊡ 30 - 40 (3.1 - 4.1, 22 - 30)

⊡ 25 - 35 (2.6 - 3.6, 19 - 26)

⊡ 43 - 58 (4.4 - 5.9, 32 - 43)

Side joint housing with extension shaft

Snap ring ⊗

Dust shield

Support bearing

Support bearing retainer

Bracket

⊡ 13 - 19 (1.3 - 1.9, 9 - 14)

Snap ring D ⊗

Dust shield

Right drive shaft

Transaxle side (Double offset joint)

89617G09

Exploded view of the halfshafts and related components

89617G07

The inner CV-joint uses a large C-clip to retain the ball and cage assembly in the outer housing

89617G08

After the outer housing is removed, the ball and cage assembly can slide from the shaft by removing the C-clip

11. Ensure that the boot is properly installed on the halfshaft groove.
12. Set the boot so that it does not swell or deform when its length is 3.86 in. (98mm).
13. Lock the new boot bands securely.

Outer

The joint on the wheel side cannot be disassembled.
1. Prior to separating the joint assembly, matchmark the halfshaft and joint assembly.
2. Separate the joint using a slide hammer.

89617G02

Make sure to properly position the boot before tightening the boot clamps

3. Remove the boot bands and the boot.

To install:

4. Thoroughly clean all parts in solvent and dry with compressed air. Check parts for evidence of damage and replace as necessary.

➡**Cover the halfshaft serrations with tape, so as not to damage the boot.**

5. Install the boot and small boot band on the halfshaft.

6. Set the joint assembly onto the halfshaft and align the matchmarks.

7. Attach the joint assembly to the halfshaft by lightly tapping the serrated end with a plastic hammer.

➡**Using a metal hammer may damage the threads on the end of the joint.**

8. Pack the CV-joint with 4.76–5.11 ounces (135–145 g) of grease.

9. Ensure that the boot is properly installed on the halfshaft groove.

10. Set the boot so that it does not swell or deform when its length is 3.82 in. (97mm).

11. Lock the new boot bands securely.

Use vinyl tape and wrap the end of the shaft to protect the boot during installation

Use an old nut to protect the threads when tapping the outer CV-joint onto the shaft

STEERING AND SUSPENSION

Air Bag

PRECAUTIONS

Several precautions must be observed when handling the inflator module to avoid accidental deployment and possible personal injury.

- Never carry the inflator module by the wires or connector on the underside of the module.
- When carrying a live inflator module, hold securely with both hands, and ensure that the bag and trim cover are pointed away.
- Place the inflator module on a bench or other surface with the bag and trim cover facing up.
- With the inflator module on the bench, never place anything on or close to the module which may be thrown in the event of an accidental deployment.

DISARMING

❋❋ CAUTION

To avoid rendering the Supplemental Restraint System (SRS) inoperative, which could lead to personal injury or death in the event of a severe frontal collision, extreme caution must be taken when servicing the electrical related systems.

➡**All SRS electrical wiring harnesses and connectors are covered with YELLOW outer insulation. Do not use electrical test equipment on any circuit related to the SRS (air bag) sensors. When installing SRS components, always install with the arrow marks facing the front of the vehicle.**

Disarming

To disarm the Supplemental Restraint System (SRS) system turn the ignition switch to the **OFF** position. Then, disconnect the both battery cables starting with the negative cable first and wait at least 10 minutes after the cables are disconnected. Be sure to insulate the battery terminal ends.

Arming

To arm the Supplemental Restraint System (SRS) system turn the ignition switch to **OFF** position. Connect the both battery cables starting with the positive cable first.

➡The SRS or air bag system is equipped with a self-diagnostic operation. After turning the ignition key to the ON or START position, the AIR BAG warning lamp will illuminate for 7 seconds. After 7 seconds, the AIR BAG lamp will extinguish if no malfunction is detected. If the AIR BAG lamp does not extinguish after 7 seconds, check the SRS self-diagnostic system for a malfunction.

Power Rack and Pinion

REMOVAL & INSTALLATION

The power steering gear is held in position by 2 steering gear brackets and insulators. Note that the housing may move slightly when the steering wheel is turned. If the housing moves more than 0.080 inch (2mm), replace the steering gear insulators. If one or both of the brackets move, check the torque of the bracket bolts. The correct torque for these bolts is 54–72 ft. lbs. (73–97 Nm).

1. Before servicing the vehicle, refer to the Precautions Section.

Tighten the power steering rack mounting bolts in the sequence shown

2. Place a drain pan under the steering rack.

3. Remove or disconnect the following:
- Brake master cylinder remote reservoir bracket screws. Position the reservoir out of the way and secure with wire.
- Junction block/high pressure line from the steering rack. Position the junction block and line out of the way.
- Both front wheels
- Front sway bar
- Tie rod ends from the steering knuckles
- Lower steering column shaft clamp bolt
- Power steering fluid return hose and position out of the way.
- The steering rack clamp bracket bolts

4. Lower the steering rack from the vehicle.

To install:

5. Carefully slide the steering gear rack and pinion assembly in place from the left side of the vehicle. Position the input shaft so it is just below the lower steering column shaft clamp.

6. Raise the steering gear until the plastic aligning tab on the input shaft enters the clamp bolt gap on the lower column shaft. Do not install the clamp bolt yet.

7. Examine the steering gear brackets. They should be marked UP with arrows pointing to one end of the bracket. Be sure the brackets are installed correctly. Tighten the steering gear bracket bolts to 54–72 ft. lbs. (73–97Nm) in sequence, working counterclockwise from the number 1 bolt (upper right side).

8. Install or connect the following:
- Fluid return line to the steering gear
- Steering column shaft clamp bolt. Tighten the bolt to 17–22 ft lbs. (24–29 Nm). Install the dust cover.

- Tie rod ends
- Stabilizer bar
- Wheel. Tighten the lug nuts to 72–87 ft. lbs. (98–118 Nm).
- Junction block. Tighten the high-

pressure line to 11–18 ft. lbs. (15–25 Nm).
- Brake master cylinder reservoir

9. Check for leaks and proper operation.

MacPherson Strut

REMOVAL & INSTALLATION

1. Before servicing the vehicle, refer to the Precautions Section.

2. Disconnect the negative battery cable.

3. Matchmark the front strut upper mounting bracket and the chassis strut tower.

4. Raise and safely support the vehicle.

5. Remove the front wheel.

The strut is attached to the knuckle with 2 large bolts

Compress the coil spring in an approved spring compressor

When installing rubber parts, final tightening must be carried out under unladen condition* with tires on ground.

*: Fuel, radiator coolant and engine oil full. Spare tire, jack, hand tools and mats in designated positions.

39 – 54 (4.0 – 5.5, 29 – 40)

Strut cap

59 - 78 (6 - 8, 43 - 58)

Spacer

Strut insulator

Strut thrust bearing

Dust seal

Upper spring seat

Coil spring

Bound bumper urethane

Dust cover

Spring rubber seat

Bushing

Clamp

41 – 51 (4.2 – 5.2, 30 – 38)

Strut assembly

Stabilizer

41 – 51 (4.2 – 5.2, 30 – 38)

Connecting rod

62 - 70 (6.3 - 7.1, 46 - 51)

137 - 157 (14 - 16, 101 - 116)

Washer

Bushing

Knuckle

Baffle plate

Drive shaft

Plain washer

Gusset

Wheel bearing lock nut

118 – 147 (12 – 15, 87 – 108)

235 – 314 (24 – 32, 174 – 231)

Plain washer

Cotter pin

128 – 157 (13 – 16, 94 – 116)

Transverse link bushing

118 – 147 (12 – 15, 87 – 108)

Transverse link

Bolt assembly

Bushing

Cotter pin

Washer

71 – 86 (7.2 – 8.8, 52 – 64)

16 – 22 (1.6 – 2.2, 12 – 16)

Lower ball joint

74 - 88 (7.5 - 9, 54 - 65)

Front

: N·m (kg-m, ft-lb)

Coil spring and strut assembly

9302WG04

6. If equipped, remove the 2 front brake anti-lock sensor cable bracket bolts and position the anti-lock sensor cable out of the way.

7. Detach the brake tube from the strut.

8. Support the control arm.

9. Matchmark the knuckle to the strut so it can installed in the same position. This is important for the camber angle of the front wheel.

10. Remove the strut-to-steering knuckle bolts.

11. Support the strut and remove the 3 upper strut-to-chassis nuts. Remove the strut from the vehicle.

✳✳ WARNING

Never loosen the strut center nut until the spring is compressed or serious injury or vehicle damage may occur.

12. Place the strut and coil spring assembly in a suitable vise and remove the strut nut cover.

13. Slightly loosen, but **do not** remove the front strut nut.

If desired, use the following steps to remove the coil spring from the strut.

14. Using an approved coil spring compressor, compress the coil spring.

15. Remove the strut assembly top nut.

16. Remove the following components from the strut assembly:
- Upper mounting bracket
- Strut bearing
- The bearing seat.
- Upper coil spring seat and dust boot
- Coil spring

17. Slowly release the tension of the coil spring compressor and remove the coil spring from the compressor tool.

18. Remove the coil spring insulator and slide the jounce bumper off of the strut assembly.

To install:

19. Slide the jounce bumper onto the strut assembly and install the coil spring insulator.

20. Carefully compress the coil spring with an approved coil spring compressor.

21. Reinstall the following components to the strut assembly:
- Coil spring

➡**Install the coil spring to the strut assembly with the end of the spring in the lower coil spring seat indentation.**

- Upper coil spring seat and dust boot

- Bearing seat and the bearing
- Upper mounting bracket

22. Install and tighten the strut assembly nut and tighten the nut to 43–58 ft. lbs. (59–78 Nm).

23. Install the strut assembly onto the vehicle and tighten the following:
- Strut-to-body nuts: 29–40 ft. lbs. (39–54 Nm)
- Strut-to-knuckle bolts: 101–116 ft. lbs. (137–157 Nm)

24. Reattach the brake tube to the strut assembly.

25. Install and tighten the 2 front brake anti-lock sensor cable bracket bolts.

26. Reinstall the tire and wheel assembly.

27. Connect the negative battery cable and the adjustable strut electrical connectors, if equipped.

28. Check and/or adjust the wheel alignment.

Shock Absorber

REMOVAL & INSTALLATION

1. Before servicing the vehicle, refer to the Precautions Section.

2. Raise and safely support the vehicle.

3. Support the rear axle and slightly lower the vehicle enough to lessen tension on the shock absorber.

4. Remove the lower shock absorber retaining nut and washer.

5. Disconnect the lower end of the shock absorber from the mounting stud.

6. Remove the shock absorber upper end retaining nut and washer.

7. Remove the shock absorber from the vehicle.

To install:

8. Install the shock absorber onto the upper and lower mounting studs of the vehicle.

9. Install the washers and retaining nuts. Tighten the upper and lower retaining nuts to 22–30 ft. lbs. (30–41 Nm).

10. Lower the vehicle.

Lower Ball Joints

REMOVAL & INSTALLATION

To check if ball joint replacement is required, raise and safely support the vehicle clear of the floor and try to rock the wheel up and down. If any play is felt, have an assistant rock the wheel while observing the front suspension lower arm ball joint at the bottom of the steering knuckle. If any movement is seen, the ball joint should be replaced. If not, any wheel play indicates wheel bearing wear.

1. Before servicing the vehicle, refer to the Precautions Section.

2. Raise and safely support the vehicle.

3. Remove the tire and wheel.

4. Remove and discard the ball joint cotter pin. Loosen the ball joint attaching nut from the steering knuckle. Because of tight clearance, the nut likely cannot be removed until the ball joint stud is loosened and lowered slightly.

5. Strike the front knuckle with a hammer while pulling down on the lower control arm. There should now be enough clearance to allow removal of the ball joint stud nut. Separate the ball joint from the steering knuckle.

6. Remove the 3 bolts attaching the ball joint to the control arm.

7. Remove the ball joint from the control arm.

To install:

8. Install the ball joint to the control arm and install the attaching bolts.

9. Tighten the bolts to 54–65 ft. lbs. (74–88 Nm).

10. Install the ball joint into the steering

Cotter pin

Nut

Lower ball joint

7924WG26

Loosen the nut on the lower ball joint stud

knuckle, just enough to get the nut started on the stud. Then, push the ball joint stud fully in place. Tighten the nut to 52–63 ft. lbs. (71–86 Nm). Secure the nut with a new cotter pin.

11. Install the tire and wheel.
12. Lower the vehicle.
13. A front end alignment check is recommended.

Lower Control Arm

REMOVAL & INSTALLATION

1. Before servicing the vehicle, refer to the Precautions Section.
2. Remove the wheel.

3. Disconnect the ball joint.
4. Disconnect the stabilizer bar from the control arm.
5. Remove the 2 rear arm bolts and the mounting bracket.
6. Remove the lower arm nut.
7. Pull the rear of the arm down and gently pry the arm forward and off the gusset.
8. Installation is the reverse of removal. Observe the following torques:
   - Stabilizer bar-to-lower arm: 12–16 ft. lbs. (16–22 Nm)
   - Lower arm rear bolts: 87–108 ft. lbs. (118–147 Nm)
   - Lower arm nuts: 94–115 ft. lbs. (128–156 Nm)

- Ball stud nut: 56–80 ft. lbs. (76–109 Nm)

BUSHING REPLACEMENT

The bushings are press-fit types. Support the arm in a press, using the proper adapters. Ford tool numbers are: T93P-5493-A, T75L-1165-B and -DA.

Wheel Bearings

ADJUSTMENT

The wheel bearings are not adjustable. If the bearings become loose or make noise, they must be replaced using the following procedure.

Exploded view of the front suspension and drive axles

7924WG25

1. Cotter pin
2. Nut retainer
3. Insulator
4. Front axle wheel hub retainer
5. Front wheel outer bearing retainer washer

6. Wheel hub
7. Wheel hub bolt
8. Snap ring
9. Front wheel bearing
10. Front disc brake rotor shield
11. Front wheel knuckle

7924WG23

Exploded view of the knuckle, hub and bearing

REMOVAL & INSTALLATION

Front

1. Before servicing the vehicle, refer to the Precautions Section.

2. Raise and safely support the vehicle.

3. Remove the wheel and tire.

4. Remove the brake caliper assembly. DO NOT disconnect the brake hose. Hang the caliper on a piece of wire from a near by support such as the strut.

5. Remove the brake rotor.

6. Remove and discard the cotter pin from the end of the outboard CV-joint stub shaft. Remove the hub nut retainer, washer and the hub nut. There should be another washer under the hub nut that acts as a front wheel bearing outer bearing retainer.

7. Disengage the lower ball joint stud from the steering knuckle using the following procedure.

   a. Remove and discard the cotter pin from the front lower ball joint.

   b. Loosen the lower ball joint nut until it contacts the front halfshaft joint.

   c. Strike the front knuckle with a hammer while pulling down on the lower control arm until the ball joint stud separates from the knuckle.

   d. Remove the ball joint nut.

   e. Disengage the lower ball joint stud from the steering knuckle.

8. Disengage the outer tie rod end stud from the steering knuckle using the following procedure.

   a. Remove and discard the cotter pin from the outer tie rod end stud.

   b. Remove the outer tie rod end retaining nut.

   c. Use a tie rod end puller to carefully press the tie rod end from the steering knuckle.

9. Remove the front ABS sensor bolt.

10. Remove the 2 front strut-to-front knuckle nuts and remove the 2 bolts. Disengage the strut from the steering knuckle.

11. Use a 2-jaw puller to separate the front halfshaft outboard CV-joint stub shaft from the knuckle/bearing assembly.

12. Remove the front wheel hub, knuckle and wheel bearing assembly from the vehicle.

13. If the knuckle is being replaced with

Example of a puller set up to bear against the front wheel bearing inner race

a service part, change over the steering stop bolt and jam nut from the old knuckle to the replacement part.

14. To remove the front wheel bearing, jig up a puller to bear against the front wheel bearing inner race and pull the race from the hub/knuckle assembly.

15. Use a shop press to press out damaged wheel studs and also to press out the outer bearing race.

16. Use a shop press to press out the inner bearing race.

To install:

17. If the front wheel bearings were removed, assemble the ABS sensing ring, if removed and the disc brake dust shield under the steering knuckle. Use a shop press to push in new front wheel bearing inner and outer races. Support the knuckle and press the front wheel bearing into the knuckle and install the snapring retainer. Support the bearing assemblies and press the hub onto the knuckle and wheel bearing assembly.

18. Install the hub, knuckle and bearings as an assembly. Position the assembly on the halfshaft outer CV-joint stub axle end. Guide the knuckle into the front strut and install the 2 knuckle-to-strut bolts and nuts. Tighten the nuts to 83–91 ft. lbs. (113–123 Nm).

19. Install the ABS sensor bolt. Do not over-tighten. Tighten to just 16–21 inch lbs. (1.8–2.4 Nm).

20. Install the outer tie rod end to the steering knuckle. Tighten the nut to 22–29 ft. lbs. (29–39 Nm). If the cotter pin holes

1. Knuckle puller
2. Knuckle puller adapter
3. Step plate adapter
4. Front disc brake rotor shield
5. Front wheel knuckle

7924WG27

do not align, tighten the nut slightly until they do. Never loosen the nut to align the holes. Secure the nut with a new cotter pin.

21. Start the lower ball joint stud to the steering knuckle and partially install the nut, then push the ball joint stud fully in place. Tighten the ball joint stud nut to 52–63 ft. lbs. (71–86 Nm). Secure the nut with a new cotter pin.

22. Install the front wheel outer bearing retaining washer and the hub retainer nut. Tighten to 174–231 ft. lbs. (235–314 Nm). Install the nut retainer, insulator and a new cotter pin.

23. Install the front brake rotor and install the disc brake caliper.

24. If removed, install the steering stop bolt.

25. Install the tire and wheel assembly. Tighten the lug nuts to 72–87 ft. lbs. (98 to 118 Nm).

26. Lower the vehicle. Pump the brake pedal slowly to seat the front brake pads. Do not move the vehicle until a firm pedal is obtained.

27. A front end alignment is recommended.

Rear

1. Before servicing the vehicle, refer to the Precautions Section.
2. Raise and safely support the vehicle.
3. Remove the rear wheel(s).
4. Remove the brake drum.
5. Remove the grease cap for the hub.
6. Remove and discard the cotter pin.

7. Remove the wheel bearing nut and washer.

8. Remove the rear wheel hub and bearing assembly.

To install:

9. Install the rear wheel hub and bearing assembly onto the vehicle.

10. Install the rear wheel bearing washer and nut and tighten the bearing nut to 159–210 ft. lbs. (216–284 Nm). Install a new cotter pin.

11. Install the wheel hub grease cap. Install the brake drum.

12. Install the rear wheel(s) and lug nuts. Tighten the lug nuts, in a star sequence, to 72–87 ft. lbs. (98–118 Nm).

13. Lower the vehicle.

Spindle

Baffle plate

Wheel hub bearing

Wheel nut
98 – 118 (10 – 12, 72 – 87)

Washer

98 – 113
(10.0 – 11.5, 72 – 83)

Wheel bearing lock nut
216 – 284,
(22 – 29, 159 – 210)

Cotter pin

Hub cap

Front

: N·m (kg-m, ft-lb)

Rear hub assembly

BRAKES

Brake Caliper

REMOVAL & INSTALLATION

The front disc brake caliper slides on 2 stainless steel locating pins. The front disc brakes use a conventional pin slider-type front disc brake caliper with a 10.875 inch

(27.6cm) front disc rotor. The front disc brake caliper is attached to the front suspension with 2 Torx® head brake caliper bolts. Rubber insulators isolate the stainless steel locating pins from direct contact with the front disc brake caliper. The front disc brake calipers must be removed to replace the front brake pads.

1. Before servicing the vehicle, refer to the Precautions Section.

2. Remove or disconnect the following:
   • Wheel and tire

➡**If the brake caliper is being removed for brake pad replacement only, DO NOT disconnect the brake hose.**

BRAKE CALIPER
BOLTS (2 REQ'D)

93026G10

Caliper pin bolt removal

- 2 caliper pin bolts. Most applications will require a Torx® T-40 bit to remove the 2 brake caliper bolts.

3. If the brake caliper is being removed just for brake service, with the brake hose still attached to the caliper, use a length of wire to support the caliper from the front shock absorber. Do not let the caliper hang by the brake hose. If the caliper is being completely removed from the vehicle for overhaul, use care not to drip brake fluid on the paint.

➥If both calipers are being completely removed from the vehicle at the same time, mark them Left and Right so the calipers can be reinstalled to their original locations. The reason for this is that the bleeder screws must be positioned on the top of the front disc brake caliper when installed on the vehicle.

To install:

4. Clean all parts well. Use a C-clamp and a used brake pad to push the caliper piston fully in the piston bore. Inspect the caliper pins and clean any dirt and debris.

5. Install the caliper onto the rotor. Make sure the inboard and outboard brake pads are properly positioned.

6. Lubricate the stainless steel locating pins with a Silicone Dielectric Compound such as Ford DZAZ-19A331-A or equivalent silicone grease. Install the 2 caliper pin bolts and torque to 18–25 ft. lbs. (24–34 Nm).

7. If disconnected, install the brake hose using a new replacement copper washer, install the banjo bolt and torque to 12–14 ft. lbs. (17–20 Nm).

8. If the brake hose had been disconnected, bleed the brake system.

9. Install the wheel and tire.

10. Torque the lug nuts to 72–87 ft. lbs. (98–118 Nm).

11. Check the master cylinder reservoir and add fresh DOT 3 brake fluid as required.

12. Lower the vehicle. Pump the brake pedal slowly until a firm brake pedal is obtained, indicating that the brake pads are properly seated, before attempting to move the vehicle. Road-test and check for proper brake operation.

Disc Brake Pads

REMOVAL & INSTALLATION

Front

1. Before servicing the vehicle, refer to the Precautions Section.

2. Remove or disconnect the following:
   - Wheels
   - Bottom guide pin from the caliper and swing the caliper cylinder body upward; support the caliper with a wire
   - Brake pad retainers and the pads

To install:

3. Compress the piston of the disc brake caliper.

4. Install or connect the following:
   - Brake pads and caliper assembly. Torque the guide pin to 23–30 ft. lbs. (31–41 Nm).
   - Wheels

5. Apply the brakes a few times to seat the pads. Check the master cylinder and add fluid if necessary. Bleed the brakes, if necessary.

Rear

1. Before servicing the vehicle, refer to the Precautions Section.

➥Do not press the piston into the bore as performed on the front disc brakes. Due to the parking brake mechanism, the caliper piston must be turned into the bore using a special tool.

2. Remove or disconnect the following:
   - Rear wheels
3. Release the parking brake.
   - Parking brake cable bracket bolt
   - Pin bolts and lift off the caliper body
   - Pad springs and then remove the pads and shims

To install:

4. Clean the piston end of the caliper body and the area around the pin holes. Be careful not to get oil on the rotor.

5. Using the proper tool, carefully turn the piston clockwise back into the caliper

WEAR INDICATOR

OUTBOARD FRONT BRAKE
SHOE AND LINING

93026G33

Replacing the disc brake pads

Brake shoe adjustment may need to be loosened to remove the brake drum

body. Take care not to damage the piston boot.

6. Coat the pad contact area on the mounting support with a silicone based grease.

7. Install or connect the following:
- Pads, shims, and the pad springs. Always use new shims.
- Caliper body in the mounting support and tighten the pin bolts to 28–38 ft. lbs. (38–52 Nm)
- Wheels

8. Bleed the system if necessary.

Brake Drums

REMOVAL & INSTALLATION

1. Before servicing the vehicle, refer to the Precautions Section.

The rear drum brakes used on these vehicles are conventional expanding shoe-type with the brake shoe lining applied to the inside of the rotating drum. An incremental brake adjuster screw is designed to actuate whenever sufficient wear occurs.

2. Remove or disconnect the following:
- Wheel and tire
- Brake drum by pulling it from the wheel studs

3. If necessary for brake drum removal, pry off the access hole plug from the access hole. Insert a screwdriver and a brake adjustment tool. Press the screwdriver against the adjusting lever to disengage it from the adjuster. Loosen the adjuster using the brake adjusting tool.

To install:

4. Clean all parts well. It is good practice to inspect the wheel cylinder for leaks anytime the brake drum is removed. If a new replacement brake drum is being installed, inspect it for a protective coating on the machined inside braking surface. Remove any coating with suitable solvent.

5. Install or connect the following:
- Brake drum onto the wheel studs

6. In most all cases, manual brake adjustment IS NOT recommended. Adjustment is performed by driving the vehicle and applying the brakes.
- Tire and wheel and torque the fasteners to 72–87 ft. lbs. (98–118 Nm)

7. Adjust the rear brake shoes by sharply applying the brakes several times while driving the vehicle alternately forwards and backwards. Check the brake operation by making several stops while driving forward.

Brake Shoes

REMOVAL & INSTALLATION

1. Before servicing the vehicle, refer to the Precautions Section.

The rear drum brakes use an internal rear wheel cylinder with expanding shoes and lining that are applied against a rotating brake drum. An incremental brake adjuster screw is actuated whenever sufficient wear occurs. Brake adjustment takes place in forward or reverse braking but not with parking brake application.

2. Remove or disconnect the following:
- Wheels
- Brake drum
- Parking brake rear cable and conduit from the parking brake lever
- 2 brake shoe hold-down springs

Remove the upper retracting spring

Remove the lower retracting spring

1 Rear Wheel Cylinder
2 Dust Boot (2 Req'd)
3 Wheel Cylinder Piston (2 Req'd)
4 Cup (2 Req'd)
5 Wheel Cylinder Piston Cup Spring
6 Wheel Cylinder Housing
7 Brake Shoe Hold-Down Pin (2 Req'd)
8 Access Hole
9 Rear Brake Bleeder Screw
10 Access Hole Plug
11 Rear Wheel Cylinder Bolt (2 Req'd)
12 Rear Brake Backing Plate

13 Rear Brake Backing Plate Bolts (4 Req'd)
14 Parking Brake Lever Clip
15 Spring Washer
16 Secondary Brake Shoe and Lining
17 Brake Shoe Hold-Down Spring
18 Lower Retracting Spring
19 Parking Brake Lever
20 Parking Brake Lever Pin
21 Brake Shoe Adjusting Lever
22 Adjuster Lever Pin
23 Primary Brake Shoe and Lining
24 Upper Retracting Spring
29 Brake Adjuster Screw

93026G29

Rear drum brake assembly and related components

and the 2 brake shoe hold-down pins
• Upper retracting spring
• Lower retracting spring
• Brake adjuster screw
• Rear brake shoes and linings from the brake backing plate
• Parking brake lever clip and washer
• Parking brake lever from the secondary brake shoe and lining

To install:

3. Clean all parts well.

4. Inspect the wheel cylinder for signs of leaking. Service as required.

5. Inspect the retracting springs for heat damage, bends or damage to the coils or shank or loss of tension. A good retracting spring will make a full thud when dropped on a concrete floor. A heat-damaged retracting spring that has lost tension

will make a distinctive ringing sound when dropped on a concrete floor.

6. Check the brake backing plate for signs of scoring. The shoe contact points must be smooth and have a light coating of lithium grease. Verify that the brake lining thickness is between 0.059–0.232 in. (1.5–5.9mm). Failure to replace worn rear brake shoes will result in a scored drum.

7. Inspect the brake drum for scratches, scoring, bell mouth and out-of-round conditions. Remove minor scores on a brake drum with sandpaper. Do not refinish brake drums to remove scoring marks. A brake drum surface that is highly polished can cause the brakes to lock up. Remove polished surfaces with sandpaper or refinish the brake drum. Refinish a brake drum that is out-of-round enough to cause vehicle vibration or noise when braking. Remove

only enough surface metal to true-up the brake drum. Brake drum maximum inside diameter is shown on each drum. If the maximum inside diameter shown on the brake drum is exceeded through wear or refinishing, replace the brake drum. After a brake drum is refinished, wipe the refinished surface with a cloth soaked in clean denatured alcohol. If one brake drum is refinished, the brake drum on the opposite side of the vehicle should also be refinished to the same diameter. The standard inner brake drum diameter is 9.840 inches (250.0mm). Replace the brake drum if worn beyond 9.900 inches (251.5mm).

8. Install or connect the following:
• Parking brake lever to the secondary brake shoe and lining with a new parking brake lever clip
• Secondary (rear) shoe on the back-

BRAKE DRUM MICROMETER

MAXIMUM BRAKE DRUM DIAMETER

250.0MM

SPREAD OR COLLAPSED COILS

DAMAGED SHANK

BENT OR TWISTED SHANK

DISCOLORATION

93026G32

These size checks should be made and the retracting springs' condition checked. Replace questionable parts

ing plate and the brake shoe hold-down spring and pin
- Primary (front) shoe on the backing plate and the brake shoe hold-down spring and pin
- Parking brake rear cable and conduit to the parking brake lever
- Lower retracting spring to the rear brake shoes

9. Apply a light coat of high-quality grease to the threaded areas of the adjuster nut and adjuster socket. Turn the adjuster nut all the way down on the brake adjuster screw, then loosen the adjuster ½ turn. Install the adjuster screw in the slots on the rear brake shoes. The wider slot on the socket must fit in the slot on the primary (front) brake shoe. The slot on the adjuster nut end must fit into the slots in the secondary (rear) brake shoe and parking brake lever.

10. Install or connect the following:
- Brake shoe adjusting lever on the adjuster lever pin
- Upper retracting spring in the slot on the secondary shoe and in the slot on the brake shoe adjusting lever. The brake shoe adjusting lever should contact the brake adjuster screw.
- Brake drum
- Tire and wheel and torque the fasteners to 72–87 ft. lbs. (98–118 Nm).

➡ **In most all cases, manual brake adjustment IS NOT recommended. Adjustment is performed by driving the vehicle and applying the brakes.**

11. The rear brakes do not require adjustment when being serviced to obtain a firm brake pedal feel. To achieve a firm brake pedal after servicing the rear brakes, sharply apply the brake pedal several times while driving the vehicle alternately forwards and backwards. Check the brake operation by making several stops while driving forward. The self-adjusting mechanism will sufficiently adjust the rear brake shoes without any manual tightening at the brake shoe adjuster. If the rear brake shoes are manually adjusted, the additional action of the brake shoe adjuster can cause the brakes to become over-tightened and result in binding or overheated rear brakes.

FORD

Windstar

18

SPECIFICATIONS AND MAINTENANCE CHARTS

VEHICLE AND ENGINE IDENTIFICATION CHART

Code	Liters (cc)	Cu. In.	Cyl.	Fuel Sys.	Engine Type	Eng. Mfg.
			Engine Code			
4	3.8 (3802)	231	6	SEFI	OHV	Ford

SEFI: Sequential Multi-port Fuel Injection

Code	Year
	Model Year
1	2001
2	2002
3	2003

67197-WIND-C01

GENERAL ENGINE SPECIFICATIONS

Year	Engine Displacement Liters	Engine VIN	Net Horsepower @ rpm	Net Torque @ rpm (ft. lbs.)	Bore x Stroke (in.)	Com-pression Ratio	Oil Pressure @ rpm
2001	3.8	4	200@5000	225@3000	3.81x3.39	9.3:1	40-60@2500
2002	3.8	4	200@5000	225@3000	3.81x3.39	9.3:1	40-60@2500
2003	3.8	4	200@5000	225@3000	3.81x3.39	9.4:1	40-125@2500①

① Hot

67197-WIND-C02

GASOLINE ENGINE TUNE-UP SPECIFICATIONS

Year	Engine Displacement Liters	Engine VIN	Spark Plugs Gap (in.)	Ignition Timing (deg.) MT	Ignition Timing (deg.) AT	Fuel Pump (psi)	Idle Speed (rpm) MT	Idle Speed (rpm) AT	Valve Clearance In.	Valve Clearance Ex.
2001	3.8	4	0.052-0.056	—	①	②	—	①	HYD	HYD
2002	3.8	4	0.052-0.056	—	①	②	—	①	HYD	HYD
2003	3.8	4	0.052-0.056	—	①	②	—	①	HYD	HYD

NOTE: The Vehicle Emission Control Information label often reflects specification changes changes made during production.

The label figures must be used if they differ from those in this chart.

B: Before top dead center

HYD: Hydraulic

① Controlled by the Powertrain Control Module (PCM) and cannot be manually adjusted.

② Engine running: 28-45 psi

　　Key On, Engine Off (KOEO): 35-45 psi

67197-WIND-C03

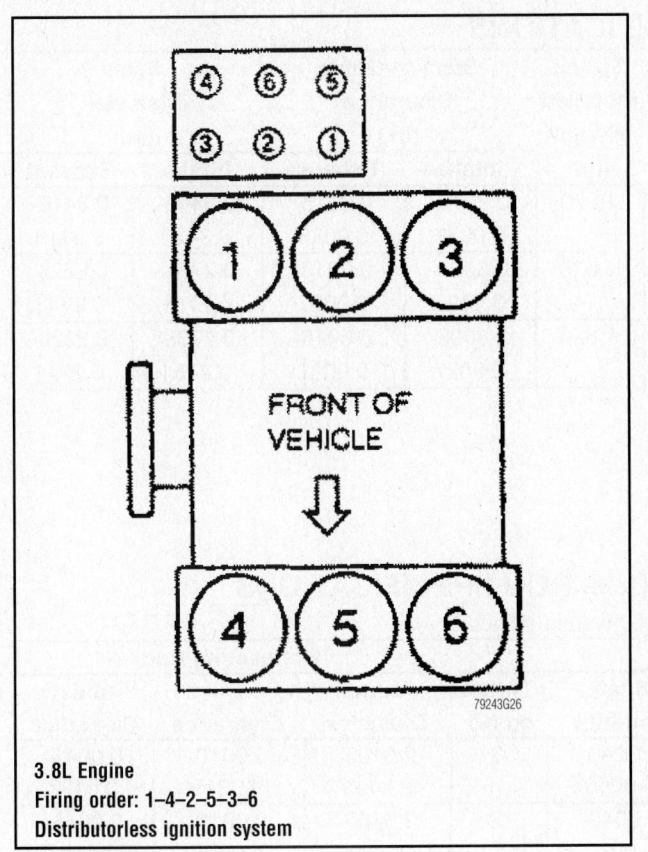

3.8L Engine
Firing order: 1–4–2–5–3–6
Distributorless ignition system

79243G26

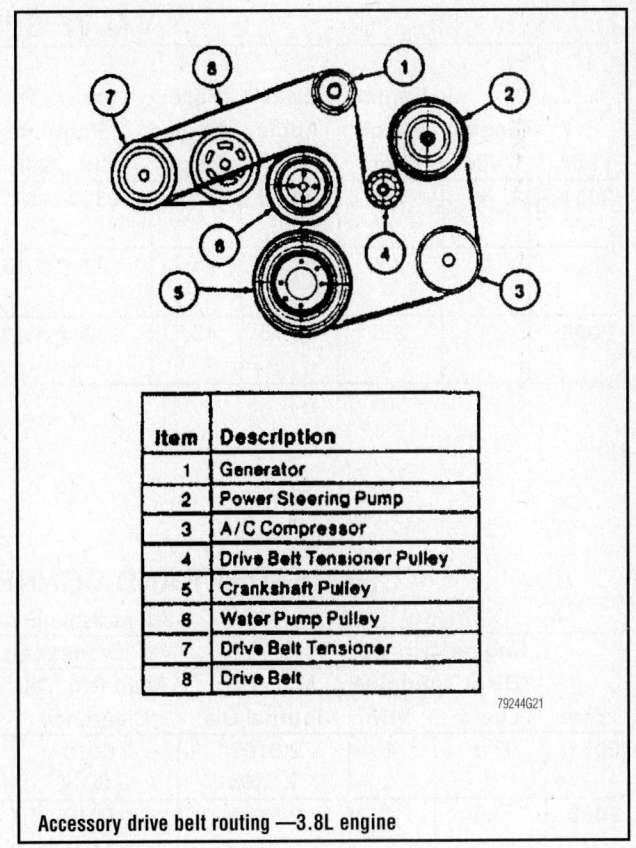

Item	Description
1	Generator
2	Power Steering Pump
3	A/C Compressor
4	Drive Belt Tensioner Pulley
5	Crankshaft Pulley
6	Water Pump Pulley
7	Drive Belt Tensioner
8	Drive Belt

79244G21

Accessory drive belt routing —3.8L engine

CAPACITIES

Year	Model	Engine Displacement Liters	Engine ID/VIN	Engine Oil with Filter (qts.)	Transmission (pts.) 4-Spd	5-Spd	Auto.	Drive Axle Front (pts.)	Rear (pts.)	Fuel Tank (gal.)	Cooling System (qts.)
2001	Windstar	3.8	4	5.0	—	—	24.5	①	—	26	16.0
2002	Windstar	3.8	4	5.0	—	—	24.5	①	—	26	16.0
2003	Windstar	3.8	4	5.0	—	—	24.5	①	—	26	16.0

NOTE: All capacities are approximate. Add fluid gradually and check to be sure a proper fluid level is obtained.

① Included in transaxle capacity

67197-WIND-C04

VALVE SPECIFICATIONS

Year	Engine VIN	Engine Displ. Liters	Seat Angle (deg.)	Face Angle (deg.)	Spring Test Pressure (lbs. @ in.)	Spring Installed Height (in.)	Stem-to-Guide Clearance (in.)		Stem Diameter (in.)	
							Intake	Exhaust	Intake	Exhaust
2001	4	3.8	44.75	45.8	198-220@1.18	1.970	0.0010-0.0028	0.0015-0.0033	0.3415-0.3423	0.3410-0.3418
2002	4	3.8	44.75	45.7	224@1.16	1.620	0.0008-0.0027	0.0018-0.0037	0.2738-0.2751	0.2728-0.2741
2003	4	3.8	44.75	45.7	224@1.16	1.620	0.0008-0.0027	0.0018-0.0037	0.2738-0.2751	0.2728-0.2741

67197-WIND-C05

CRANKSHAFT AND CONNECTING ROD SPECIFICATIONS

All measurements are given in inches.

Year	Engine Displ. Liters	Engine VIN	Crankshaft				Connecting Rod		
			Main Brg. Journal Dia.	Main Brg. Oil Clearance	Shaft End-play	Thrust on No.	Journal Diameter	Oil Clearance	Side Clearance
2001	3.8	4	2.5190-2.5198	0.0010-0.0014	0.0040-0.0080	3	2.3103-2.3111	0.0010-0.0015	0.0043-0.0192
2002	3.8	4	2.5190-2.5198	0.0010-0.0014	0.0040-0.0080	3	2.3103-2.3111	0.0010-0.0015	0.0043-0.0192
2003	3.8	4	2.5190-2.5198	0.0010-0.0014	0.0040-0.0080	3	2.3103-2.3111	0.0010-0.0015	0.0047-0.0192

67197-WIND-C06

PISTON AND RING SPECIFICATIONS

All measurements are given in inches.

Year	Engine Displ. Liters	Engine VIN	Piston Clearance	Ring Gap			Ring Side Clearance		
				Top Comp.	Bottom Comp.	Oil Control	Top Comp.	Bottom Comp.	Oil Control
2001	3.8	4	0.0007-0.0017	0.0098-0.0161	0.0150-0.0252	0.0059-0.0650	0.0012-0.0031	0.0012-0.0031	Snug
2002	3.8	4	0.0007-0.0017	0.0098-0.0161	0.0150-0.0252	0.0059-0.0650	0.0012-0.0031	0.0012-0.0031	Snug
2003	3.8	4	0.0007-0.0017	0.0067-0.0130	0.0118-0.0217	0.0059-0.0256	0.0012-0.0031	0.0012-0.0031	Snug

67197-WIND-C07

TORQUE SPECIFICATIONS

All readings in ft. lbs.

Year	Engine VIN	Engine Displ. Liters	Cylinder Head Bolts	Main Bearing Bolts	Rod Bearing Bolts	Crankshaft Damper Bolts	Flywheel Bolts	Manifold Intake	Manifold Exhaust	Spark Plugs	Oil Pan Drain Plug
2001	4	3.8	①	④	②	104-132	54-64	③	15-22	7-15	19
2002	4	3.8	①	④	②	118	59	③	18	11	19
2003	4	3.8	①	④	⑤	118	59	③	18	11	19

① See text for torque procedure

② Step 1: 30-34 ft. lbs.
   Step 2: Tighten an additional 90-120 degrees

③ Upper: 89 inch lbs.
   Lower: Step 1, 44 inch lbs.; step 2, 89 inch lbs.

④ Step 1: 37 ft. lbs.
   Step 2: plus 115-125 degrees

⑤ Step 1: 18 ft. lbs.
   Step 2: 33 ft. lbs.
   Step 3: an additional 105 degrees

67197-WIND-C08

WHEEL ALIGNMENT

Year	Model			Caster Range (+/-Deg.)	Caster Preferred Setting (Deg.)	Camber Range (+/-Deg.)	Camber Preferred Setting (Deg.)	Toe-in (in.)
2001	Windstar	Front	Left	0.75	+3.3	0.50	-0.44	-0.15+/-0.25
			Right	0.75	+3.8	0.50	-0.44	-0.15+/-0.25
		Rear		—	—	0.50	-0.10	0.06+/-0.34
2002	Windstar	Front	Left	0.75	+3.3	0.50	-0.44	-0.15+/-0.25
			Right	0.75	+3.8	0.50	-0.44	-0.15+/-0.25
		Rear		—	—	0.50	-0.10	0.06+/-0.34
2003	Windstar	Front	Left	0.75	+3.3	0.50	-0.44	-0.15+/-0.25
			Right	0.75	+3.8	0.50	-0.44	-0.15+/-0.25
		Rear		—	—	0.50	-0.10	0.06+/-0.34

67197-WIND-C09

TIRE, WHEEL AND BALL JOINT SPECIFICATIONS

Year	Model	OEM Tires Standard	OEM Tires Optional	Tire Pressures (psi) Front	Tire Pressures (psi) Rear	Wheel Size	Ball Joint Inspection	Lug Nut Torque (ft. lbs.)
2001	Windstar	P215/70R15	P205/70R15	①	①	NA	②	100
2002	Windstar	P215/70R15	P205/70R15	①	①	NA	②	100
2003	Windstar	P215/70R15	P205/70R15	①	①	NA	②	100

NA: Information not available

OEM: Original Equipment Manufacturer

PSI: Pounds Per Square Inch

① See placard on vehicle

② Replace if any perceptible movement is noticed

67197-WIND-C10

BRAKE SPECIFICATIONS
All measurements in inches unless noted

Year	Model	Brake Disc Original Thickness	Brake Disc Minimum Thickness	Brake Disc Maximum Runout	Brake Drum Diameter Original Inside Diameter	Brake Drum Diameter Max. Wear Limit	Brake Drum Diameter Maximum Machine Diameter	Minimum Lining Thickness	Brake Caliper Bracket-to-Hub Bolt (ft. lbs.)	Brake Caliper Mounting Pin or Bolt (ft. lbs.)
2001	All	1.020	0.974	0.003	9.84	9.90	9.90	①	85	25
2002	All	1.063	1.010	0.003	9.84	9.90	9.90	①	85	25
2003	All	1.063	1.010	0.003	9.84	9.90	9.90	①	85	25

NOTE: Due to changes made during production, refer to the manufacturer's specifications if they differ from those in this chart

① With drum brakes: 0.059
   With disc brakes: 0.125

67197-WIND-C11

SCHEDULED MAINTENANCE INTERVALS
2001-02 FORD—WINDSTAR

TO BE SERVICED	TYPE OF SERVICE	VEHICLE MILEAGE INTERVAL (x1000)												
		5	10	15	20	25	30	35	40	45	50	55	60	65
Engine oil & filter	R	✓	✓	✓	✓	✓	✓	✓	✓	✓	✓	✓	✓	✓
Rotate tires	S/I	✓		✓		✓		✓		✓		✓		✓
Engine coolant strength	S/I			✓			✓			✓			✓	
Air cleaner filter	R						✓						✓	
Automatic transmission fluid & filter	R						✓						✓	
Engine coolant ①	R						✓						✓	
PCV valve	R												✓	
Spark plugs ②	R						✓						✓	
Drive belts	S/I						✓						✓	
Exhaust system & heat shields	S/I						✓						✓	
Front & rear brakes	S/I						✓						✓	

R: Replace   S/I: Service or Inspect

① Engine coolant: change initially at 50,000 miles and every 30,000 miles thereafter.

② Spark plugs: replace every 100,000 miles.

FREQUENT OPERATION MAINTENANCE (SEVERE SERVICE)

If a vehicle is operated under any of the following conditions it is considered severe service:

- Extremely dusty areas.

- 50% or more of the vehicle operation is in 32°C (90°F) or higher temperatures, or constant operation in temperatures below 0°C (32°F).

- Prolonged idling (vehicle operation in stop and go traffic.

- Frequent short running periods (engine does not warm to normal operating temperatures).

- Police, taxi, delivery usage or trailer towing usage.

Engine oil & filter: replace every 3000 miles.

Rotate tires initially at 6000 miles and every 9000 miles thereafter.

Air cleaner filter: change every 15,000 miles.

Engine coolant strength, hoses & clamps: check every 15,000 miles.

Exhaust system: check every 15,000 miles.

Automatic transmission fluid & filter: change every 21,000 miles.

67197-WIND-C12

SCHEDULED MAINTENANCE INTERVALS
2003 FORD—WINDSTAR

TO BE SERVICED	TYPE OF SERVICE	5	10	15	20	25	30	35	40	45	50	55	60	65
		VEHICLE MILEAGE INTERVAL (x1000)												
Engine oil & filter	R	✓	✓	✓	✓	✓	✓	✓	✓	✓	✓	✓	✓	✓
Rotate tires	S/I	✓		✓		✓		✓		✓		✓		✓
Automatic transmission fluid	I			✓			✓			✓			✓	
Engine cooling system & hoses	I			✓			✓			✓			✓	
Steering linkage	I			✓			✓			✓			✓	
Ball joints	I			✓			✓			✓			✓	
Cabin filter	R			✓			✓			✓			✓	
Engine coolant (yellow)	R	every 100,000 miles												
Engine coolant (green)	R									✓				
Fuel filter	R						✓						✓	
Air cleaner filter	R						✓						✓	
Automatic transmission fluid & filter	R						✓						✓	
PCV valve	R	every 100,000 miles												
Spark plugs	R	every 100,000 miles												
Accessory drive belts	S/I	every 100,000 miles												
Exhaust system & heat shields	S/I						✓						✓	
Front & rear brakes	S/I			✓			✓			✓			✓	

R: Replace   S/I: Service or Inspect

② Spark plugs: replace every 100,000 miles.

Special Operating Condition Requirements

When towing a trailer or using a camper or car-top carrier:

Change engine oil and install a new oil filter every 4,800 km (3,000 miles), 3 months or 200 hours of engine operation (whichever occurs first).

Change transfer case fluid every 96,000 km (60,000 miles).

Change manual transmission fluid as required.

Inspect and lubricate U-joints as required.

During extensive idling and/or low speed driving for long distances, as in heavy commercial use such as delivery, taxi, patrol car or livery:

Change engine oil and install a new oil filter every 4,800 km (3,000 miles), 3 months or 200 hours of engine operation (whichever occurs first).

Lube front lower control arm and steering linkage ball joints with zerk fittings (if equipped) every 4,800 km (3,000 miles) or 3 months.

Inspect brake system and check battery electrolyte level (Patrol cars) every 8,000 km (5,000 miles).

Install a new fuel filter every 24,000 km (15,000 miles).

Change automatic transmission fluid, lubricate 4x2 wheel bearings, install new grease seals and adjust bearings every 48,000 km (30,000 miles). If equipped, change the in-line service installed transmission fluid filter.

Install new spark plugs and change transfer case fluid every 96,000 km (60,000 miles).

Install a new cabin air filter as required.

When operating in dusty conditions such as unpaved or dusty roads:

Change engine oil and install a new oil filter every 4,800 km (3,000 miles) or 3 months.

Install a new fuel filter every 24,000 km (15,000 miles).

Change automatic transmission fluid every 48,000 km (30,000 miles). If equipped, change the in-line service installed transmission fluid filter.

Change transfer case fluid every 96,000 km (60,000 miles).

Install a new engine air filter as required.

Install a new cabin air filter as required.

When operating in off-road conditions:

Change automatic transmission fluid every 48,000 km (30,000 miles). If equipped, change the in-line service installed transmission fluid filter.

Change transfer case fluid every 96,000 km (60,000 miles).

Install a new cabin air filter as required.

Inspect and lubricate U-joints.

Inspect and lubricate steering linkage ball joints with zerk fittings.

ENGINE REPAIR

Alternator

REMOVAL & INSTALLATION

1. Disconnect the negative battery cable.
2. Disconnect the alternator wiring harness.
3. Detach the alternator drive belt.
4. Remove the three alternator attaching bolts.
5. Remove the alternator.

To install:

6. Position the alternator on the engine.
7. Install the three alternator attaching bolts. Tighten the bolts 18 ft. lbs. (25 Nm).
8. Install and tension the alternator drive belt.
9. Connect the alternator wiring harness. Tighten the output terminal nut to 80–97 inch lbs. (9–11 Nm).
10. Connect the negative battery cable.

Engine Assembly

REMOVAL & INSTALLATION

1. Remove or disconnect the following:
   - Negative battery cable
   - Coolant
   - Refrigerant, into a refrigerant recovery station
   - Cowl top vent panel
   - Wiring from the alternator
   - Air cleaner assembly
   - Upper and lower radiator hoses from the engine

- Heater water hoses and secure to body
- Air conditioning discharge and suction hoses, then secure them to the engine. Cap the open lines to prevent moisture from entering the system.
- Accelerator cable and speed control cable from the throttle body lever
- Accelerator cable bracket from the throttle body
- Fuel supply and return lines from the fuel injection supply manifold
- All engine wiring harnesses from the engine and secure to the body
- All vacuum hoses from the engine
- Gear shift cable from the transaxle

➥**Damage to the steering column air bag wiring can result if the steering wheel is allowed to rotate freely. The wire is wound like a watch spring and can be over-tightened and break if the steering wheel is rotated too far in either direction.**

2. Lock the steering wheel with the wheels in the straight-ahead position by turning the ignition to the **OFF** position.
3. Remove or disconnect the following:
   - Wheels
   - Engine oil
   - Transaxle cooler lines at the transaxle. Secure the lines to the radiator.
   - Heated oxygen sensor wiring harness

➥**Do not allow the flex connector of the duel converter Y-pipe to hang unsup-**

ported or damage to the flex joint will result.

- Dual converter Y-pipe and support it from the body

➥**The routing of the battery ground cable to the cylinder block is critical. It should go between the transaxle and the bracket. Take note during disassembly.**

- Starter motor wires and secure out of the way
- Starter motor
- Engine rear plate and torque converter-to-flywheel nuts
- Power steering cooler lines
- Upper bolt from the sway bar links
- Dust boot from the steering rack pinion support
- Steering coupling pinch bolt from the steering column intermediate shaft at the steering gear
- Intermediate shaft from the steering gear
- Front stabilizer bar links
- Front suspension lower arms from the knuckles at the ball joint
- Tie rod ends from the knuckle
- Front axle wheel hub retainers from the halfshaft ends
- Halfshafts from the front wheel knuckle

4. Support the front subframe, engine and transaxle assembly.
5. Remove or disconnect the following:
   - The 4 retaining bolts, and lower the engine transaxle and front subframe from the vehicle

MOUNTING
BOLTS

89692G13

Alternator mounting—3.8L engine

- Power steering pressure hose from the power steering pump

6. Attach an engine hoist and lift the engine slightly.

7. Remove the engine support insulators.

8. Lift the engine and transaxle assembly from the front subframe.

9. Lower the engine and transaxle.

10. Support the transaxle on a level stationary surface and separate the engine from the transaxle.

To install:

11. If removed, install the transaxle on the engine.

12. Install the engine on the subframe.

13. Install the engine support insulators.

14. Connect the power steering hose to the pump.

15. Carefully raise the engine/transaxle assembly into the vehicle.

16. Install the 4 subframe bolts.

17. The remainder of the installation is the reverse of removal.

18. Please note the following torque specifications:

- Engine-to-transaxle: 37 ft. lbs. (50 Nm)
- Torque converter nuts: 20–34 ft. lbs. (27–46 Nm)

- Subframe-to-body bolts: 57–76 ft. lbs. (77–103 Nm)
- Steering coupling pinch bolt: 25–34 ft. lbs. (34–46 Nm)
- Dual converter Y-pipe-to-exhaust manifold: 25–34 ft. lbs. (34–46 Nm)
- Flex pipe retaining bolts: 25–34 ft. lbs. (34–46 Nm)
- Accelerator cable bracket retaining bolts: 71–106 inch lbs. (8–12 Nm)
- Front engine support insulator-to-subframe 50–68 ft. lbs. (68–92 Nm)
- Transmission insulator-to-subframe: 65–87 ft. lbs. (88–119 Nm)
- Rear engine and transaxle support insulator-to-subframe: 56–75 ft. lbs. (76–103 Nm)

19. Refill the engine, transaxle and cooling system with the correct amount of the appropriate fluids before starting the engine.

Water Pump

REMOVAL & INSTALLATION

1. Remove or disconnect the following:
- Negative battery cable

- Coolant
- Drive belts
- Lower radiator hose
- Lower nut on both front engine supports
- Alternator
- Power steering pressure line from the pump
- Power steering reservoir filler cap
- Water bypass hose and oil cooler hose from the heater water outlet tube
- Heater water outlet tube from the water pump
- Air conditioning bracket brace

2. Raise the engine approximately 2 inches (51mm) to provide necessary clearance for water pump removal.

3. Remove the water pump pulley.

4. Remove the drive belt tensioner from the power steering pump brace.

5. Remove the power steering pump brace and place the pump and brace aside in the engine compartment.

6. Remove the water pump.

To install:

7. Clean all gasket mating surfaces thoroughly.

WATER PUMP

CYLINDER BLOCK

ENGINE FRONT COVER

1. M8 x 1.25 x 98
2. M8 x 1.25 x 98
3. M8 x 1.25 x 131
4. M8 x 1.25 x 131
5. M8 x 1.25 x 25
6. M8 x 1.25 x 35
7. M8 x 1.25 x 35
8. M8 x 1.25 x 25
9. M8 x 1.25 x 61.5
10. M8 x 1.25 x 141
11. M8 x 1 x 131
12. M8 x 1 x 35
13. M8 x 1 x 35
14. M8 x 1 x 105
15. M8 x 1 x 20

Because of their varying lengths, be sure to install the water pump bolts in the correct bolt holes—3.8L engine

7924GG02

☀ WARNING

Be careful not to gouge the aluminum surfaces when scraping the old gasket material from the mating surfaces of the water pump and front cover.

8. Cover the threads of the No. 1 engine front cover stud with Teflon® tape.
9. Install or connect the following:
   - New water pump housing gasket on the water pump sealing surface using gasket sealant to hold the gasket in place
   - Water pump and tighten the bolts to 18 ft. lbs. (25 Nm)
   - Drive belt tensioner
   - Water pump pulley
   - Air conditioning bracket brace
   - Heater water outlet tube
   - Water bypass hose and oil cooler hose
   - Power steering reservoir filler cap
   - Power steering pressure line
   - Alternator
   - Lower nut on both front engine supports
   - Lower radiator hose
   - Drive belts
10. Fill and bleed the cooling system.
11. Connect the negative battery cable.

Heater Core

REMOVAL & INSTALLATION

Front System

1. Disconnect the negative battery cable.
2. Drain the cooling system into a clean container for reuse.

Item	Description
1	Outside Air Seal
2	A/C Air Inlet Duct
3	A/C Air Inlet Door Inner Seal
4	A/C Recirculating Air Duct
5	A/C Damper Door Shaft
6	Vacuum Control Motor
7	Blower Motor Wheel
8	Blower Motor
9	A/C Evaporator Housing
10	Rear Seat Airflow Duct

Item	Description
11	Heater Outlet Floor Duct
12	Heater Housing Core Plate
13	Heater Core
14	A/C Electronic Door Actuator Motor
15	Vacuum Control Motor
16	Shaft — Heater Air Damper Door
17	Windshield Defroster Door Shaft
18	Windshield Defroster Duct Connector

89696G06

Front heater housing assembly

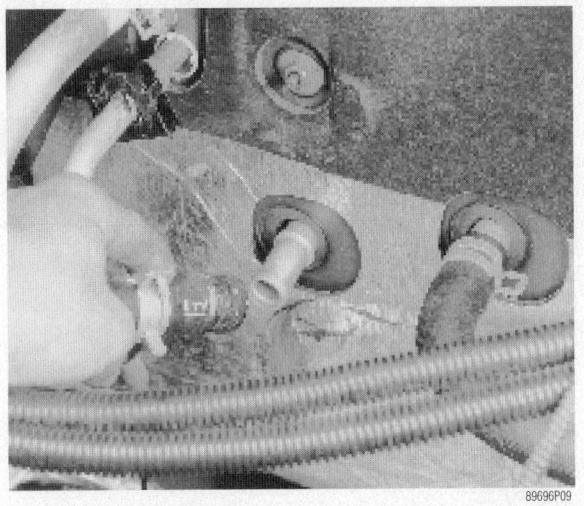

Be careful when removing the hoses from the heater core, as the tubes are easily damaged

The instrument panel support bracket must be removed to lower the heater core

The center instrument panel is attached using special plastic screws

The heater outlet floor duct is attached to the bottom of the housing core

The rear seat airflow duct carries air to the rear of the passenger compartment

The heater core is located in a box at the bottom of the housing core

A foam liner is used to cushion the heater core

89696P15

3. Remove the cowl top vent panel for clearance.

4. Disconnect the heater hoses from the heater core inlet and outlet tubes in the engine compartment.

5. Remove the cassette box/center instrument support trim.

6. Pull out and remove the ashtray cup holder by depressing the service lever.

7. Remove the floor/rear seat lower air duct.

8. Disconnect the keyless entry wiring harness, if equipped.

9. Remove the center instrument panel support brackets.

10. Disconnect the climate control vacuum harness connector.

11. Remove the heater floor/rear seat upper air duct.

12. Remove the heater core cover retaining screws and remove the heater core cover.

13. Remove the heater core.

14. Remove the seal from the heater core tubes and discard.

FRONT OF VEHICLE

Item	Description
1	U-Nut
2	Screw and Washer Assembly
3	Body Main Wiring
4	Screw and Washer Assembly

Item	Description
5	Nut Insert
6	Heater and A/C Assembly

89696G04

Rear heater and air conditioning assembly

To install:

15. Install the seal to the heater core tubes and discard.

16. Install the heater core.

17. Install the heater core cover and the heater core cover retaining screws.

18. Install the heater floor/rear seat upper air duct.

19. Connect the climate control vacuum harness connector.

20. Install the center instrument panel support brackets.

21. Connect the keyless entry wiring harness, if equipped.

22. Install the floor/rear seat lower air duct.

23. Install the ashtray cup holder by depressing the service lever.

24. Install the cassette box/center instrument support trim.

25. Connect the heater hoses to the heater core inlet and outlet tubes in the engine compartment.

26. Install the cowl top vent panel for clearance.

27. Refill the cooling system.

28. Connect the negative battery cable.

29. Run the engine to normal operating temperatures; then, check the climate control operation and check for leaks.

Rear Auxiliary System

1. Disconnect the negative battery cable.

2. Drain the cooling system into a clean container for reuse.

3. Discharge and recover the rear auxiliary air conditioning refrigerant.

4. Raise and support the vehicle safely.

5. Place a drain pan under the heater water hose connections for the rear heater assembly.

6. Drain and recycle the engine coolant from the auxiliary heater core and hoses.

7. Disconnect the 2 air conditioning spring lock couplings using a quick disconnect coupling tool.

8. Lower the vehicle.

9. Remove all rear passenger's seating.

10. Remove the 3 push pin retainers from the lower edge of the auxiliary heater and air conditioning service cover.

11. Lift the service cover outward and upward to remove.

12. Label and disconnect the electrical harnesses.

13. Label and disconnect the vacuum lines.

14. Remove the body side trim panels.

15. Disconnect and plug all air conditioning refrigerant lines.

➡ It is extremely important that the air conditioning lines be plugged to prevent the entry of dirt or moisture.

16. Remove the lower air conditioning recirculation air duct and heater extension air duct.

17. Remove the auxiliary heater and air conditioning assembly.

18. Remove the housing cover screws and the cover.

19. Remove the heater core and the heater core case seal.

To install:

20. Install the heater core and the heater core case seal.

21. Install the housing cover and screws.

22. Position the auxiliary heater and air conditioning assembly in the vehicle. Guide the heater water hoses through the floor grommet. Tighten the retaining screws to 16–23 inch lbs. (2–3 Nm).

23. Connect all air conditioning refrigerant lines.

24. Install the lower air conditioning recirculation air duct and heater extension air duct. Tighten the retaining screws to 16–23 inch lbs. (2–3 Nm).

25. Install the body side trim panels.

26. Connect the vacuum lines.

27. Connect the electrical harnesses.

FRONT OF VEHICLE

Item	Description
1	A/C Evaporator Housing Assy
2	Heater Core
3	Heater Housing Core Plate
4	Screw
5	Rear Seat Airflow Duct
6	Heater Water Hose

89696G05

Rear heater housing assembly

28. Install the auxiliary heater and air conditioning service cover and secure with 3 push pin retainers.

29. Install all rear passenger's seating.

30. Raise and support the vehicle safely.

31. Connect the heater water hoses to the heater core tubes.

32. Lower the vehicle.

33. Refill the cooling system.

34. Connect the negative battery cable.

35. Evacuate and charge the rear auxiliary air conditioning refrigerant.

36. Run the engine to normal operating temperatures; then, check the climate control operation and check for leaks.

Cylinder Head

REMOVAL & INSTALLATION

1. Remove or disconnect the following:
   - Negative battery cable
   - Coolant
   - Air cleaner assembly
   - Cowl top vent panel
   - Accessory drive belt

2. If the front cylinder head is being removed, perform the following:
   a. Remove the oil filler cap.
   b. Remove the air conditioning compressor mounting bracket and set the air conditioning compressor aside with the refrigerant lines still connected.
   c. Remove the power steering pump and bracket. Leave the power steering hoses connected and place the pump aside in the engine compartment.
   d. Remove the alternator and mounting bracket.

3. If the rear cylinder head is being removed, perform the following:
   a. Remove the accessory drive belt tensioner.
   b. Remove the PCV valve.
   c. Remove the power steering line bracket.
   d. Remove the tensioner bracket.
   e. Remove the coil pack assembly.

4. Remove or disconnect the following:
   - Upper intake manifold
   - Valve cover
   - Fuel system pressure
   - Fuel charging assembly
   - Lower intake manifold
   - Exhaust manifolds

➡**Pushrods must be installed in their original positions. Note pushrod location during removal.**

   - Pushrods
   - Cylinder head bolts

   - Cylinder head from the engine block and discard the gaskets

To install:

5. The cylinder head should be cleaned and inspected prior to installation.

6. Lightly oil all bolt and stud bolt threads before installation.

7. Clean all gasket mating surfaces thoroughly.

8. Position new head gaskets on the cylinder block, noting the **UP** position mark on the gasket face, using the dowels in the engine block for alignment. If the dowels are damaged, they must be replaced.

✳✳ WARNING

Always use new cylinder head bolts when installing cylinder head or damage to the engine may occur.

9. Position the cylinder head on the cylinder block.

10. Lubricate the cylinder head bolts with engine oil and install. Tighten the cylinder head bolts following the proper torque sequence as follows:
   - Step 1: 15 ft. lbs. (20 Nm)
   - Step 2: 29 ft. lbs. (40 Nm)
   - Step 3: 37 ft. lbs. (50 Nm)

➡**Do not loosen all of the cylinder head bolts at once. Only work on 1 bolt at a time or damage to the engine may occur.**

In sequence, loosen each cylinder head bolt 2–3 turns and retighten in 2 steps.
   - Long bolts, step 1: 29–37 ft. lbs. (40–50 Nm); step 2: tighten an additional 175–185 degrees.

   - Short bolts, step 1: 15–22 ft. lbs. (20–30 Nm); second 2: tighten an additional 175–185 degrees.

11. Dip each pushrod end in engine assembly lubricant. Install the pushrods in their original positions.

12. Lubricate all rocker arm components with engine assembly lubricant.

13. For the rocker arm being installed, rotate the engine clockwise until the valve tappet rests on the heel (base circle) of the camshaft lobe.

14. Install the rocker arms, seats and bolts and tighten to 44 inch lbs. (5 Nm).

15. Perform the previous 2 steps for each rocker arm.

16. After all rocker arms have been installed, final tighten all bolts to 22–29 ft. lbs. (30–40 Nm).

17. Install or connect the following:
   - Exhaust manifolds
   - Lower intake manifold
   - Fuel injection charging assembly
   - Valve covers with new gaskets. Tighten the bolts to 71–97 inch lbs. (8–11 Nm).
   - Upper intake manifold
   - Spark plugs and ignition wires

18. If the front cylinder head is being installed, perform the following:
   a. Install the alternator and mounting bracket.
   b. Install the power steering pump and bracket.
   c. Install the air conditioning compressor bracket.
   d. Install the oil filler cap.

TIGHTEN BOLTS IN SEQUENCE

7924GG04

Cylinder head bolt torque sequence—3.8L engine

19. If the rear cylinder head is being installed, perform the following:

    a. Install the coil pack assembly.

    b. Install the tensioner bracket.

    c. Install the power steering line bracket.

    d. Install the PCV valve.

    e. Install the accessory drive belt tensioner.

    f. Rotate the tensioner clockwise and install the accessory drive belt.

20. Install the cowl top vent panel.

21. Install the air cleaner assembly.

22. Fill and bleed the cooling system.

23. Connect the negative battery cable.

Rocker Arms

REMOVAL & INSTALLATION

1. Remove the valve cover.
2. Remove the rocker arm retaining bolt.

➡ **Rocker the arms should be installed in their original location during assembly.**

3. Remove the rocker arms. If more than 1 rocker arm is to be removed, identify each rocker arm location.

To install:

4. Lubricate the pushrods and rocker arms with engine assembly lubricant. Lubricate the retaining bolts with engine oil.

➡ **Prior to final tightening, the rocker arm seats must be fully seated into the cylinder head. The pushrods must be fully seated in the rocker arm and valve tappet sockets.**

5. Install the rocker arms into position with the pushrods and snug the retaining bolt.

ROCKER ARM SEAT AND BOLT MUST BE FULLY SEATED AFTER FINAL TORQUE

CLEARANCE SHOULD BE 2.25-4.79mm (0.09-0.19 INCH) WITH VALVE TAPPET FULLY COLLAPSED ON BASE CIRCLE OF CAMSHAFT AFTER ASSEMBLED.

7924GG17

When the lifter is fully collapsed and on the base circle of the cam, check for proper clearance between the tip of the valve and the rocker arm—3.8L engine

6. Rotate the crankshaft until the lifter for the rocker arm being installed, is on the base circle (heel) of the cam lobe.

7. Tighten the rocker arm retaining bolt to 44 inch lbs. (5 Nm)

8. Finally, tighten the bolt with the camshaft in any position to 20–28 ft. lbs. (26–38 Nm).

9. Install the valve cover.

Intake Manifold

REMOVAL & INSTALLATION

Upper Manifold

1. Remove or disconnect the following:
   - Air cleaner outlet tube
   - Accelerator cable and speed control actuator cable at the throttle body
   - Accelerator cable bracket and position it aside
   - Vacuum lines
   - Necessary electrical harnesses
   - Crankcase ventilation tube from the PCV valve
   - Throttle body
   - Idle air control valve
   - Intake manifold retaining bolts, noting their positions

➡ **Keep the intake manifold bolts in order, so they can be installed in their original positions.**

   - Upper intake manifold

To install:

2. Inspect the intake gasket to ensure the seals are completely installed in the manifold groove and the seals show no signs of damage.

3. Install or connect the following:
   - Intake manifold and tighten the bolts to 71–106 inch lbs. (8–12 Nm) in the sequence shown
   - Idle air control valve
   - Throttle body
   - Crankcase ventilation tube to the PCV valve
   - Electrical harnesses
   - Vacuum lines
   - Accelerator cable bracket and tighten the bolts to 71–106 inch lbs. (8–12 Nm)
   - Accelerator cable and speed control actuator cable at the throttle body
   - Air cleaner outlet tube

Lower Manifold

1. Remove or disconnect the following:
   - Coolant
   - Upper intake manifold
   - Water bypass hose from the heater water outlet tube
   - Bypass hose from the lower intake manifold
   - Fuel system pressure
   - Electrical wiring harnesses
   - Fuel injectors and fuel charging assembly
   - Vacuum motor and bracket assemblies
   - Valve assembly and linkage from the IMRC lever and bushing by using a prytool
   - Tube retaining bolts
   - Old bushing from the lever
   - EGR valve and adapter
   - Lower intake manifold retaining bolts

➡ **The lower intake manifold is sealed at each corner with sealer. To break the seal it may be necessary to pry on the front of the intake manifold with a prybar. If it is necessary, use care to prevent damage to the machined surfaces.**

   - Lower intake manifold

To install:

2. Thoroughly clean all gasket mating surfaces.

➡ **When using silicone rubber sealer, assembly must occur within 15 minutes after sealer application. After this time, the sealer may start to set up and its sealing effectiveness may be reduced.**

3. Install or connect the following:
   - New bushings into the IMRC levers
   - Apply a 3mm bead of RTV silicone sealer at each corner where the cylinder head joins the engine block
   - Front and rear intake manifold seals

Upper intake manifold bolt torque sequence—3.8L engine

Lower intake manifold and related components—3.8L engine

Lower intake manifold bolt torque sequence—3.8L engine

- Lower intake manifold into position on the cylinder block, using new gaskets
- Apply pipe sealant to the intake bolts and install in their original locations. Tighten in sequence to 44 inch lbs.; then to 89 inch lbs. (8–12 Nm).
- EGR valve and adapter
- IMRC vacuum motors, and tighten the retaining bolts to 71–106 inch lbs. (8–12 Nm)
- Fuel injectors and charging assembly. Tighten retaining bolts to 71–97 inch lbs. (8–11 Nm).
- Water bypass tube to the lower intake manifold. Tighten the retaining bolts to 71–97 inch lbs. (8–11 Nm).
- Water bypass tube hose to the outlet tube and tighten the hose clamp securely
- Electrical wiring harnesses
- Vacuum lines to the IMRC motors
- Upper radiator hose to water hose connection and tighten the hose clamp securely
- Upper intake manifold
4. Fill and bleed the cooling system.
5. Start the engine and check for leaks.

Exhaust Manifold

REMOVAL & INSTALLATION

➡Spray the exhaust system fasteners with penetrating lubricant before removing them to help prevent broken studs and bolts. The use of a 6-point socket is highly recommended when removing exhaust system fasteners.

✳✳ CAUTION

To prevent serious burns, allow the exhaust manifold to cool down before attempting to remove it.

Rear Manifold

1. Disconnect the negative battery cable.
2. Remove the cowl vent panel.
3. Remove the engine air cleaner and air cleaner outlet tube.
4. Disconnect the ignition wires from the rear cylinder head and ignition coil.
5. Remove the spark plugs from the rear cylinder head.
6. Raise and support the vehicle safely on jackstands.
7. Disconnect the dual converter Y-pipe from the exhaust manifold.

8. Lower the vehicle.

9. Remove the exhaust manifold.

To install:

10. Clean all gasket mating surfaces thoroughly.

➡A slight warpage in the exhaust manifold may cause a misalignment between the bolt holes in the cylinder head and exhaust manifold. Elongate the holes in the exhaust manifold as necessary to correct the misalignment. Do not elongate the pilot hole.

11. Install a new exhaust manifold gasket and the exhaust manifold on the cylinder head. Start 2 bolts to hold the manifold in position.

12. Install the remaining bolts. Tighten the bolts beginning from the center port and working outward to 15–22 ft. lbs. (20–30 Nm).

13. Raise and support the vehicle safely.

14. Connect the dual converter Y-pipe and tighten the bolts to 25–34 ft. lbs. (34–47 Nm).

15. Lower the vehicle.

16. Install the spark plugs in the rear cylinder head.

17. Connect the ignition wires.

18. Install the engine air cleaner and air cleaner outlet tube.

19. Install the cowl vent panel.

20. Connect the negative battery cable.

21. Start the engine and check for exhaust leaks.

Front Manifold

1. Disconnect the negative battery cable.

2. Remove the oil level indicator tube.

3. Label and disconnect the ignition wires from the front cylinder head.

4. Disconnect the EGR-to-exhaust manifold tube.

5. Raise and support the vehicle safely on jackstands.

6. Disconnect the dual converter Y-pipe from the exhaust manifold.

7. Lower the vehicle.

8. Remove the exhaust manifold.

To install:

9. Clean all gasket mating surfaces thoroughly.

➡A slight warpage in the exhaust manifold may cause a misalignment between the bolt holes in the cylinder head and exhaust manifold. Elongate the holes in the exhaust manifold as necessary to correct the misalignment. Do not elongate the pilot hole.

10. Install a new gasket and the exhaust manifold on the cylinder head. Start 2 bolts to hold the manifold in position.

11. Install the remaining bolts. Tighten the bolts, starting from the center port and working outward, to 15–22 ft. lbs. (20–30 Nm).

12. Raise and support the vehicle safely on jackstands.

13. Connect the dual converter Y-pipe and tighten the bolts to 25–34 ft. lbs. (34–47 Nm).

14. Lower the vehicle.

15. Connect the EGR to the exhaust manifold tube.

16. Connect the ignition wires.

17. Install the oil level indicator tube.

18. Connect the negative battery cable.

19. Start the engine and check for exhaust leaks.

Camshaft and Valve Lifters

REMOVAL & INSTALLATION

1. Rotate the crankshaft until the No. 1 piston is at the TDC on its compression stroke and the timing marks are aligned.

2. Remove or disconnect the following:

- Engine from the vehicle
- Valve covers
- Intake manifolds
- Pushrod
- Tappet guide plate
- Tappets
- Crankshaft pulley and damper
- Oil pan
- Engine front cover assembly

3. Check the camshaft end-play as follows:

a. Push the camshaft toward the rear of the engine and install a dial indicator, so the indicator point is on the camshaft sprocket attaching screw.

b. Zero the dial indicator. Position a small prybar between the camshaft sprocket or gear and block.

c. Pull the camshaft forward and release it. Camshaft end-play should be 0.001–0.006 in. (0.025–0.15mm).

d. If the camshaft end-play is not within specification, replace the thrust plate upon reassembly.

4. Remove or disconnect the following:

- Timing chain and sprockets
- Camshaft thrust plate

5. Carefully remove the camshaft by pulling it toward the front of the engine. Remove it slowly to avoid damaging the bearings, journals and lobes.

To install:

6. Clean and inspect all parts before installation.

7. Lubricate the camshaft lobes and journals with Molylube® or heavy engine oil.

8. Carefully install the camshaft.

➡If a new camshaft is being installed, recheck camshaft end-play.

9. Lubricate the engine thrust plate with engine assembly lubricant, then install the thrust plate. Tighten the retaining bolts to 71–124 inch lbs. (8–14 Nm).

10. Install the timing chain and sprockets.

➡Check the camshaft sprocket bolt for blockage of the drilled oil passages prior to installation, and clean if necessary.

11. Install or connect the following:

- Engine front cover
- Crankshaft damper and pulley
- Hydraulic tappets into their original bores
- Align the valve tappet flats and install the tappet guide plate with the word **UP** facing you
- Install the intake manifold assembly
- Lubricate and install the pushrods and rocker arms
- Install the oil pan
- Install the valve covers
- Install the engine assembly into the vehicle

Starter Motor

REMOVAL & INSTALLATION

1. Disconnect the negative battery cable.

2. Raise and support the vehicle safely.

➡When removing the hard shell connector at terminal "S", grasp the plastic shell. Do not pull on the wire.

3. Disconnect the starter electrical harness.

4. Remove the upper starter bolt.

5. Support the starter and remove the lower bolt.

6. Remove the starter from the vehicle.

To install:

7. Position the starter in the vehicle.

8. Install the upper and lower bolts. Tighten to 20 ft. lbs. (28 Nm).

9. Connect the starter electrical harness. Tighten the starter cable nut to 80–124 inch lbs. (9–14 Nm).

➡️When installing the hard shell connector, be careful to push it straight on and make sure it locks in position with a notable click or detent.

10. Lower the vehicle.
11. Connect the negative battery cable.

Oil Pan

✳✳ CAUTION

The EPA warns that prolonged contact with used engine oil may cause a number of skin disorders, including cancer! You should make every effort to minimize your exposure to used engine oil. Protective gloves should be worn when changing the oil. Wash your hands and any other exposed skin areas as soon as possible after exposure to used engine oil. Soap and water, or waterless hand cleaner, should be used.

REMOVAL & INSTALLATION

1. Disconnect the negative battery cable.
2. Raise and support the vehicle safely on jackstands.
3. Drain the engine oil.
4. Remove the oil filter.
5. Remove the dual converter Y-pipe assembly.
6. Remove the starter motor.
7. Remove the engine rear plate/converter housing cover.

8. Remove the retaining bolts and remove the oil pan.

To install:

9. Clean the gasket mating surfaces thoroughly.
10. Trial fit the oil pan to the cylinder block. Ensure that enough clearance has been provided to allow the oil pan to be installed without sealant being scraped off when pan is positioned under the engine.
11. Apply a bead of silicone sealer to the oil pan flange. Also apply a bead of sealer to the front cover/cylinder block joint and fill the grooves on both sides of the rear main seal cap.

➡️When using silicone rubber sealer, assembly must occur within 15 minutes after sealer application. After this time, the sealer may start to harden and its sealing effectiveness may be reduced.

12. Install the oil pan and secure to the block with the attaching screws. Tighten the screws to 80–106 inch lbs. (9–12 Nm). Torque the oil pan-to-transaxle bolts to 33 ft. lbs. (45 Nm).
13. Install a new oil filter.
14. Install the engine rear plate/converter housing cover.
15. Install the starter motor.
16. Install the Y-pipe converter assembly.
17. Lower the vehicle.
18. Fill the engine with the proper type and amount of clean oil.
19. Connect the negative battery cable.
20. Start the engine and check for leaks.

Oil Pump

REMOVAL & INSTALLATION

➡️The oil pump, oil pressure relief valve and drive intermediate shaft are contained in the front cover assembly.

1. Disconnect the negative battery cable.
2. If necessary for access, remove the oil filter.
3. Remove the oil pump and filter body-to-engine front cover retaining bolts, then remove the oil pump and filter body from the engine front cover.
4. Inspect the oil pump body seal, oil pump and filter body, and engine front cover for distortion. Replace damaged components as necessary.

To install:

5. Position the oil pump and filter body on the engine front cover, then install the retaining bolts.
6. Tighten the 4 large engine front cover retaining bolts to 17–23 ft. lbs. (23–31 Nm), then tighten the remaining retaining bolts to 71–97 inch lbs. (8–11 Nm).
7. If removed, install the oil filter.
8. Connect the negative battery cable.

➡️Check for proper engine oil pressure immediately after starting the engine. If engine oil pressure is not within specification a few seconds after starting the engine, stop the engine and determine the reason for the low oil pressure condition. Running an engine with low oil pressure may result in serious engine damage.

9. Start the engine and check for leaks.

1. Oil pan
2. Oil pan drain plug
3. End seal
4. Silicone gasket and sealant
5. Engine front cover
6. Guide pin
7. Bolt
8. Rear bearing cap
9. Cylinder block

SEALER MUST FILL SEAL GROOVE (AS SHOWN) TWO PLACES

VIEW A

NOTE: APPLY A ZIG-ZAG BEAD OF EXTRA SEALER ON JOINT OF FRONT COVER AND CYLINDER BLOCK IN TWO PLACES

Exploded view of the oil pan and related components. Apply silicone gasket sealant in the places shown—3.8L engine

7924GG19

OIL PUMP AND FILTER BODY

The oil pump assembly is mounted to the side of the 3.8L engine

Rear Main Seal

REMOVAL & INSTALLATION

1. Disconnect the negative battery cable.
2. Raise and support the vehicle safely on jackstands.
3. Remove the transaxle.
4. Remove the flywheel and the rear cover plate, if necessary.
5. Using a sharp awl, punch 1 hole into the crankshaft rear oil seal metal surface between the seal lip and the cylinder block.

✳✳ WARNING

Use caution when working near the crankshaft sealing surface. If the surface becomes damaged, an oil leak may occur.

6. Screw in the threaded end of a crankshaft rear seal replacer tool, then use the tool to remove the seal.

To install:

7. Inspect the crankshaft seal area for any damage that may cause the seal to leak. If damage is evident, service or replace the crankshaft as necessary.
8. Coat the crankshaft seal area and the seal lip with engine oil.
9. Using a crankshaft seal replacer tool, install the seal. Tighten the bolts of the seal installer tool evenly so the seal is straight and seats without misalignment.
10. Install the flywheel.
11. Install the rear cover plate, if necessary.
12. Install the transaxle, lower the vehicle and connect the battery.

Timing Chain, Sprockets, Front Cover and Seal

REMOVAL & INSTALLATION

1. Before servicing the vehicle, refer to the Precautions Section.

2. Remove or disconnect the following:
- Negative battery cable
- Coolant
- Air cleaner assembly and air intake duct
- Fan/clutch assembly and shroud
- Accessory drive belt idlers, drive belts and the water pump pulley
- Power steering pump bracket retaining bolts. Leaving the hoses connected, place the pump/bracket assembly aside in a position to prevent fluid from leaking out.
- Compressor front support bracket but leave the compressor in place
- Coolant bypass hose and heater hose at the water pump
- Upper radiator hose at the thermostat housing
- Coil wire from the distributor cap
- Cap with the secondary wires attached
- Distributor hold-down clamp and lift the distributor out of the front cover
- Crankshaft damper and pulley

➡️If the crankshaft pulley and vibration damper have to be separated, mark the damper and pulley so they may be reassembled in the same relative position. This is important as the damper and pulley are initially balanced as a unit. If the crankshaft damper is being replaced, check if the original damper has balance pins installed. If so, new balance pins must be installed on the new damper in the same position as the original damper. The crankshaft pulley, new or original, must also be installed in the same relative position as originally installed.

- Oil filter
- Lower radiator hose at the water pump
- Oil pan

➡️The front cover cannot be removed without lowering the oil pan.

- Front cover retaining bolts. It is not necessary to separate the water pump from the front cover.

➡️Do not overlook the cover retaining bolt located behind the oil filter adapter. The front cover will break if pried on, and all retaining bolts are not removed.

- Front cover and water pump as an assembly. Drive the crankshaft seal out of the front cover with a suitable seal driver. Remove and discard the cover gasket.

0.5-0.0mm (0.020-0.0 INCH) (DO NOT BOTTOM)

REAR FACE OF CYLINDER BLOCK
CRANKSHAFT REAR OIL SEAL
CRANKSHAFT
REAR CRANKSHAFT MAIN BEARING CAP

SECTION A
SEAL INSTALLED HEIGHT

SECTION A

CRANKSHAFT REAR OIL SEAL

CRANKSHAFT REAR SEAL REPLACER

The rear main seal must be installed with the proper tools to avoid damaging the seal or crankshaft

➡️The front cover contains the oil pump, water pump and crankshaft seal. **If a new front cover is to be installed, remove the water pump and oil pump from the old front cover and install them on the new cover along with a new crankshaft seal.**

- Camshaft bolt and washer from the end of the camshaft
- Distributor drive gear, camshaft sprocket, crankshaft sprocket and timing chain

➡️**If the crankshaft sprocket is difficult to remove, pry the sprocket off the shaft using a pair of large prybars positioned on both sides of the sprocket.**

To install:

3. Clean all gasket mating surfaces. If reusing the front cover, replace the front cover oil seal.

4. If removed, install the timing chain vibration damper. Tighten the mounting bolts to 71–123 inch lbs. (8–14 Nm).

5. Rotate the crankshaft to position the No. 1 piston at TDC and the crankshaft keyway at the 12 o' clock position.

6. Lubricate the timing chain with engine oil.

7. Install or connect the following:

- Camshaft sprocket, crankshaft sprocket and timing chain. Be sure the timing marks align.
- Distributor drive gear. Install the bolt and washer assembly on the end of the camshaft and tighten to 30–37 ft. lbs. (40–50 Nm).
- New crankshaft seal in the front cover and lubricate the seal lip with engine oil
- New gasket on the cylinder block and install the front cover using dowels for proper alignment. Install the front cover retaining bolts and tighten to 15–22 ft. lbs. (20–30 Nm).
- Oil pan
- Lower radiator hose
- Oil filter

8. Coat the crankshaft damper sealing surface with clean engine oil. Apply a small amount of silicone sealer to the crankshaft keyway.

9. Position the crankshaft pulley key in the crankshaft keyway and install the damper, using a suitable installation tool.

10. Install or connect the following:

- Damper washer and retaining bolt and tighten to 103–132 ft. lbs. (140–180 Nm).
- Crankshaft pulley and tighten the

POSITIONING OF TIMING MARKS AND KEYWAYS IN CAMSHAFT AND CRANKSHAFT SPROCKETS MUST BE IN LINE AS SHOWN WITH NO. 1 PISTON AT TOP DEAD CENTER FIRING

7922QG21

The timing marks should be facing each other, when the timing chain is installed correctly—3.8L engines

retaining bolts to 20–28 ft. lbs. (26–38 Nm).
- Coolant bypass hose
- Distributor with the rotor pointing at the No. 1 distributor cap tower
- Distributor cap and coil wire
- Upper radiator hose at the thermostat housing
- Heater hose
- Compressor and mounting brackets. Tighten the retaining bolts to 30–45 ft. lbs. (41–61 Nm).
- Power steering pump and mounting bracket. Tighten the retaining bolts to 30–45 ft. lbs. (41–61 Nm).
- Water pump pulley. Position the accessory drive belts over the pulleys.

- Fan/clutch assembly and fan shroud. Cross-tighten the fan/clutch assembly retaining bolts to 12–18 ft. lbs. (16–24 Nm).

11. Fill the crankcase with the proper type and quantity of engine oil. Fill and bleed the cooling system. Connect the negative battery cable.

12. Start the engine and check for leaks. Check the ignition timing and curb idle speed and adjust, as necessary.

Piston and Ring

POSITIONING

3.8L engines—piston and connecting rod assembly positioning

FUEL SYSTEM

Fuel System Service Precautions

Safety is the most important factor when performing not only fuel system maintenance, but any type of maintenance. Failure to conduct maintenance and repairs in a safe manner may result in serious personal injury or death. Work on a vehicle's fuel system components can be accomplished safely and effectively by adhering to the following rules and guidelines.

• To avoid the possibility of fire and personal injury, always disconnect the negative battery cable unless the repair or test procedure requires that battery voltage by applied.

• Always relieve the fuel system pressure prior to disconnecting any fuel system component (injector, fuel rail, pressure regulator, etc.) fitting or fuel line connection. Exercise extreme caution whenever relieving fuel system pressure, to avoid exposing skin, face and eyes to fuel spray. Please be advised that fuel under pressure may penetrate the skin or any part of the body that it contacts.

• Always place a shop towel or rag around the fitting or connection prior to loosening to absorb any excess fuel due to spillage. Ensure that all fuel spillage is quickly removed from engine surfaces. Ensure that all fuel-soaked cloths or towels are deposited into a flame-proof waste container with a lid.

• Always keep a dry chemical (Class B) fire extinguisher near the work area.

• Do not allow fuel spray or fuel vapors to come into contact with a light bulb, spark or open flame.

• Always use a second wrench when loosening or tightening fuel line connections fittings. This will prevent unnecessary stress and torsion to fuel piping. Always follow the proper torque specifications.

• Always replace worn fuel fitting O-rings with new ones. Do not substitute fuel hose where rigid pipe is installed.

Fuel System Pressure

RELIEVING

All Sequential Electronic Fuel Injection (SEFI) engines are equipped with a pressure relief valve located on the fuel supply manifold. Remove the fuel tank cap and attach a fuel pressure gauge to the valve to release the fuel pressure. Be sure to drain the fuel into a suitable container and to avoid gaso-

line spillage. If a pressure gauge is not available, disconnect the vacuum hose from the fuel pressure regulator and attach a hand-held vacuum pump. Apply about 25 in. Hg (84 kPa) of vacuum to the regulator to vent the fuel system pressure into the fuel tank through the fuel return hose. Note that this procedure will remove the fuel pressure from the lines, but not the fuel. Take precautions to avoid the risk of fire and use clean rags to soak up any spilled fuel when the lines are disconnected.

Fuel Filter

REMOVAL & INSTALLATION

Although the manufacturer does not specify a replacement interval for fuel filters, we at Chilton feel the fuel filter should be replaced every 30,000 miles (48,000 km) under normal conditions or 15,000 miles (24,000 km) under severe conditions. Those intervals are industry standards.

1. Relieve the fuel system pressure.
2. Raise and support the vehicle safely on jackstands.
3. Place a rag under the fuel filter to catch any residual fuel that may leak out when the filter is removed.
4. Remove the push-connect fittings at both ends of the fuel filter.
5. Install retainer clips in each fitting.
6. Note the flow arrow direction for installation reference.

7. Remove the fuel filter by pulling it from the bracket.

To install:

8. Install the fuel filter in its bracket, ensuring proper direction of flow as noted earlier.
9. Install push-connect fittings at both ends of the fuel filter.
10. Start the engine and check the filter connections for leaks by running the tip of your finger around each connection.
11. Turn the engine off and lower the vehicle.

Fuel Pump

REMOVAL & INSTALLATION

➡ **To gain access to the fuel pump, it is necessary to remove the fuel tank.**

1. Depressurize the fuel system and remove the fuel tank from the vehicle.
2. Remove any dirt that has accumulated around the fuel pump module attaching flange to prevent it from entering the tank during service.
3. Turn the fuel pump module locking ring counterclockwise using a locking ring removal tool or a brass drift, and remove the locking ring.
4. Remove the fuel pump module.
5. Remove the seal gasket and discard it.

To install:

6. Put a light coating of grease on a new seal ring to hold it in place during

In-tank electric fuel pump and related components

7924GG25

assembly. Install it in the fuel tank ring groove.

7. Insert the fuel pump module into the fuel tank, then secure it in place with the locking ring. Tighten the ring until secure.

8. Install the tank in the vehicle.

9. Install a minimum of 10 gallons (38L) of fuel and check for leaks.

10. Install a pressure gauge on the throttle body valve and turn the ignition **ON** for 3 seconds. Turn the key **OFF**, then repeat the key cycle 5 to 10 times until the pressure gauge shows at least 30 psi. (207 kPa).

11. Check for fuel leaks at the fittings.

12. Remove the pressure gauge.

13. Start the engine and check for fuel leaks.

Fuel Injectors

REMOVAL & INSTALLATION

1. Remove the upper intake manifold.

2. Remove the fuel injection supply manifold.

3. Carefully remove the fuel charging wiring harness connectors from the fuel injectors.

4. Pull fuel injector body up while gently rocking fuel injector from side to side.

To install:

5. Inspect fuel injector O-rings for signs of deterioration and replace as required.

➡**Never use silicone grease on fuel injectors.**

6. Lubricate O-rings with clean engine oil and install fuel injectors using a slight twisting motion.

7. Install fuel injection supply manifold.

8. Install the fuel charging wiring connectors.

9. Install the upper intake manifold.

DRIVE TRAIN

Transaxle Assembly

REMOVAL & INSTALLATION

1. Remove or disconnect the following:

- Battery and battery tray
- Hood and cowl vent
- Air cleaner assembly
- All transaxle electrical harnesses
- Transaxle shift cable from the lever by unsnapping the shift cable end from the lever ball stud
- Transaxle fluid cooler lines

➡**Leave the 2 lower engine-to-transaxle bolts in place to hold the transaxle secure against the engine block until a suitable jack can be placed under the transaxle to support it during removal.**

- Upper transaxle-to-engine bolts
- Engine electrical harness bracket
- Battery cable bracket

➡**Install engine lifting eyes to support the engine during transaxle removal.**

2. Install an engine support kit and suitably support the engine.

3. Raise and support the vehicle.

4. Remove or disconnect the following:

- Transaxle fluid
- Front wheels
- Halfshafts
- Bolts retaining the rear engine support to the transaxle
- Front subframe
- Speedometer cable from the vehicle speed sensor
- Starter
- Transaxle housing cover
- The 4 flexplate-to-converter nuts

5. Support the transaxle with a suitable

jack and remove the remaining transaxle-to-engine bolts.

6. Remove the engine bracket-to-transaxle bolts.

7. Separate the transaxle from the engine block by carefully moving the transaxle rearward until enough clearance exists to remove the transaxle from the engine compartment.

8. Slowly lower the transaxle from the engine compartment.

To install:

9. Place the transaxle on a suitable jack and position it in place.

10. Install or connect the following:

- Engine bracket-to-transaxle bolts, and tighten the bolts to 39–53 ft. lbs. (53–72 Nm)
- Lower transaxle-to-engine bolts, and tighten the bolts to 39–53 ft. lbs. (53–72 Nm)
- Flexplate-to-torque converter bolts, and tighten them to 20–34 ft. lbs. (27–46 Nm)
- Transaxle housing cover, and tighten the bolts to 80–106 inch lbs. (9–12 Nm)
- Starter motor, and connect the electrical harness
- Speedometer cable
- Front subframe
- The 4 bolts retaining the rear engine support, and tighten them to 39–53 ft. lbs. (53–72 Nm)
- Both halfshafts
- Front wheels and lower the vehicle

11. Remove the engine support kit.

12. Install or connect the following:

- Transaxle electrical harnesses
- Upper transaxle-to-engine bolts and tighten to 39–53 ft. lbs. (53–72 Nm)
- Fluid cooler-to-transaxle lines

- Transaxle shift cable to the manual lever ball stud
- Air cleaner assembly
- Cowl vent and hood
- Battery tray and battery

13. Fill the transaxle with proper amount of Mercon® fluid.

14. Connect the positive, then the negative battery cable.

Halfshafts

REMOVAL & INSTALLATION

➡**Do not begin this removal procedure unless a new wheel hub retainer nut, a new retainer circlip and a new lower ball joint-to-front wheel knuckle retaining bolt and nut are available. Once removed, these parts must not be reused during assembly. Their torque holding ability, or retention capability, is diminished during removal.**

1. Remove or disconnect the following:

- Front wheels
- Axle hub nut. Discard the nut.
- Ball joint-to-front wheel knuckle retaining nut. Drive the bolt out of the front wheel knuckle using a punch and hammer.
- Front brake anti-lock sensor and position it out of the way
- Ball joint from the knuckle

✷✷ WARNING

Use care to prevent damage to the CV-joint boot.

- Stabilizer bar link at the front stabilizer bar

➡**Make sure the CV-joint puller does not contact the transaxle shaft speed sensor. Damage to the sensor will result.**

Remove the halfshaft from the transaxle using a CV-joint puller, extension and impact slide hammer

The front hub adapter must be used to remove the hub without damage

2. Install a CV-joint puller between the inboard CV-joint and the transaxle case.

3. Install a CV-joint extension into the puller and hand-tighten.

4. Using a slide hammer, remove the driveshaft from the transaxle.

✳✳ WARNING

Do not allow the halfshaft to hang unsupported. Damage to the CV-joint may result. Do not wrap wire around the joint boot. Damage to the boot may result.

5. Support the end of the halfshaft assembly by suspending it from the chassis using a length of wire.

✳✳ WARNING

Never use a hammer to separate the outboard CV-joint from the wheel hub. Damage to the outboard CV joint threads and internal components may result.

6. Separate the outboard CV-joint from the wheel hub using a front hub remover/replacer. Make sure the hub remover adapter is fully threaded onto the hub stud.

✳✳ WARNING

Do not move vehicle without the outboard CV-joint properly installed as damage to the bearing may occur.

7. Remove the halfshaft assembly from the vehicle.
To install:

✳✳ WARNING

Do not reuse the retainer circlip. A new circlip must be installed each time the inboard CV-joint stub shaft is installed into the transaxle differential.

8. Install a new retainer circlip on the inboard CV-joint stub shaft by starting one end in the groove and working the retainer circlip over the inboard shaft housing end and into the groove. The will avoid over-expanding the circlip.

➡**A non-metallic mallet may be used to aid in seating the retainer circlip into the differential side gear groove. If a mallet is necessary, tap only on the outboard CV-joint shaft.**

9. Carefully align the splines of the inboard CV-joint stub shaft housing with

the splines in the differential. Exerting some force, push the inboard CV-joint stub shaft housing into the differential until the retainer circlip is felt to seat in the differential side gear. Use care to prevent damage to the inboard CV-joint stub shaft and transaxle seal.

10. Carefully align the splines of the outboard CV-joint with the splines in the wheel hub and push the shaft into the wheel hub as far as possible.

11. Temporarily fasten the front disc brake rotor to the wheel hub with washers and 2 lug nuts. Insert a steel rod into the front disc brake rotor and rotate clockwise to contact the front wheel knuckle to prevent the front disc brake rotor from turning when the nut is tightened.

➡️**A new front axle wheel hub retaining nut must be installed.**

12. Manually thread the front axle wheel hub retaining nut onto the outboard CV-joint stub shaft housing as far as possible.

➡️**A new bolt and nut must be used to connect the front suspension arm to the knuckle.**

13. Connect the front suspension lower arm to the front wheel knuckle. Tighten the nut and bolt to 40–55 ft. lbs. (54–75 Nm).

14. Install the front brake anti-lock sensor.

15. Connect the front stabilizer bark link and tighten to 35–45 ft. lbs. (47–65 Nm).

➡️**Do not use power or impact tools to tighten the hub nut.**

16. Tighten front axle wheel hub retaining nut to 157–212 ft. lbs. (213–288 Nm).

17. Install the front wheels and lower the vehicle.

18. Fill the transaxle to the proper level with Mercon® automatic transmission fluid.

CV-Joint

OVERHAUL

Inboard Joint

1. Remove the clamps.
2. Separate the front wheel driveshaft joint boot from the inboard CV joint housing.
3. Check the CV joint grease for contamination by rubbing it between two fingers. Any gritty feeling indicates contamination.

➡️**Other than the front wheel driveshaft joint boot, the interconnecting shaft and inboard CV joint housing are not repairable. Install a new assembly if worn/damaged.**

4. To install a new inboard CV joint housing assembly front wheel driveshaft joint boot, remove the front wheel driveshaft joint and boot.

5. Slide the front wheel driveshaft joint boot off the interconnecting shaft.

6. Install the front wheel driveshaft joint boot.

7. Position the boot into the small boot groove.

8. Using the special tool, install the small clamp.

9. Position the special tool on the clamp ear, and tighten the tool through bolt until the tool closes completely.

10. Install the front wheel driveshaft joint boot and the joint.

11. Fill the inboard CV joint housing with 475 grams (16.75 ounces) of grease. Spread the remaining grease evenly inside the front wheel driveshaft joint boot. Use constant velocity joint grease, only.

➡️**Remove all excess grease from the CV joint external surface and the front wheel driveshaft joint boot sealing surface.**

12. Seat the front wheel driveshaft joint boot in the inboard CV joint housing boot groove.

13. Set the halfshaft assembled length to specification. Halfshaft assembled length, left side: 612.6 mm (24.5 in); right side: 746.4 mm (29.85 in). Measure the entire assembly length. Push in or pull out on the inner joint as necessary to adjust the halfshaft assembled length to specification.

14. Hold the inner joint to prevent the assembled length from changing, and insert a small flat blade screwdriver between the boot and the joint to equalize the pressure.

➡️**Make sure the front wheel driveshaft joint boot seats in the groove.**

15. Install the clamps as tight as possible by hand.

16. Using the special tool, tighten the clamp.

17. Position the special tool on the clamp ear, and tighten the tool through bolt until the tool closes completely.

18. Move the CV joints through their full range of travel at various angles. The

joints must flex, extend and compress smoothly.

Outboard Joint

> ❋❋ **CAUTION**
>
> **Do not allow the vise jaws to contact the front wheel driveshaft joint boot or the clamp. Use a vise equipped with soft jaws or wood blocks.**

1. Clamp the halfshaft assembly in a vise.
2. Remove the clamps.
3. Separate the front wheel driveshaft joint boot from the front wheel driveshaft joint.
4. Check the CV joint grease for contamination by rubbing it between two fingers. Any gritty feeling indicates contamination.
5. If the grease is contaminated/additional disassembly is necessary, proceed as follows.

> ❋❋ **CAUTION**
>
> **Do not allow the front wheel driveshaft joint to fall.**

6. Using a brass drift and a hammer, separate the front wheel driveshaft joint from the interconnecting shaft.
7. Give a sharp tap to the inner bearing race to dislodge the circlip.

> ❋❋ **CAUTION**
>
> **Do not reuse the circlip.**

➡️**Install a new stop ring, located just below the circlip, only if the stop ring is damaged/worn, or additional halfshaft disassembly is necessary.**

8. Remove and discard the circlip. Remove the stop ring if damaged/worn, or additional halfshaft disassembly is necessary.
9. If necessary, remove the boot from the shaft.

> ❋❋ **CAUTION**
>
> **Do not remove the front brake anti-lock sensor indicator unless it is necessary to install a new one.**

10. Position the special tool on a press bed, and place the front wheel driveshaft joint on the special tool.
11. Press the damaged front brake anti-lock sensor indicator off of the front wheel driveshaft joint and discard it.

✳✳ CAUTION

Do not damage the front brake anti-lock sensor indicator. Tooth damage will affect brake performance.

12. If removed, position the special tool on a press bed, and place the new front brake anti-lock sensor indicator on the special tool.

13. Position the front wheel driveshaft joint in the special tool.

14. Place a steel plate across the front wheel driveshaft joint back face.

➡ **Installation is complete when the front wheel driveshaft joint bottoms out in the special tool.**

15. Press the front brake anti-lock sensor indicator on the front wheel driveshaft joint.

➡ **Make sure the front wheel driveshaft joint boot seats in the shaft boot groove.**

16. If removed, install the clamp and the front wheel driveshaft joint boot.

17. Install the clamps as tight as possible by hand.

18. Using the special tool, tighten the clamp.

19. Position the special tool on the clamp ear, and tighten the tool through bolt until the tool closes completely.

➡ **Make sure the stop ring seats in the groove.**

20. If removed, install the stop ring.

✳✳ CAUTION

Do not reuse the circlip.

✳✳ CAUTION

Do not over-expand or twist the circlip during installation.

21. Install a new circlip.

22. Start one end in the groove and work the circlip over the shaft and into the groove. This will avoid over-expanding the circlip.

23. Fill the front wheel driveshaft joint with 180 grams (6.3 ounces) of grease. Spread the remaining grease evenly inside

the front wheel driveshaft joint boot. Use constant velocity joint grease only!

➡ **The front wheel driveshaft joint has seated when the circlip locks in the groove cut in the inner race.**

24. Using a non-metallic hammer, tap the front wheel driveshaft joint onto the interconnecting shaft. Make sure the front wheel driveshaft joint has locked on the interconnecting shaft by attempting to pull the joint off the shaft.

➡ **Remove all excess grease from the front wheel driveshaft joint external surface and the front wheel driveshaft joint boot mating surface.**

25. Seat the front wheel driveshaft joint boot in the front wheel driveshaft joint boot groove.

26. Install the clamp as tight as possible by hand.

27. Using the special tool, tighten the clamp.

28. Position the special tool on the clamp ear, and tighten the tool through bolt until the tool closes completely.

STEERING AND SUSPENSION

Air Bag (Supplemental Restraint) System

The Supplemental Restraint System (SRS) is designed to work in conjunction with the standard 3-point safety belts to reduce injury in a head-on collision.

✳✳ CAUTION

The SRS can actually cause physical injury or death if the safety belts are not used, or if the manufacturer's warnings are not followed. The manufacturer's warnings can be found in your owner's manual, or, in some cases, on your sun visor.

The SRS is comprised of the following components:

- Driver's side air bag module
- Passenger's side air bag module
- Right-hand and left-hand primary crash front air bag sensors
- Air bag diagnostic monitor computer
- Electrical wiring

The SRS primary crash front air bag sensors are hard-wired to the air bag modules and determine when the air bags are deployed. During a frontal collision, the sensors quickly inflate the 2 air bags to reduce injury by cushioning the driver and front passenger from striking the dash-

board, windshield, steering wheel and any other hard surfaces. The air bag inflates so quickly (in a fraction of a second) that in most cases it is fully inflated before you actually start to move during a collision.

Since the SRS is a complicated and essentially important system, its components are constantly being tested by a diagnostic computer. The computer illuminates the air bag indicator light on the instrument cluster for approximately 6 seconds when the ignition switch is turned to the **RUN** position when the SRS is functioning properly. After being illuminated for the 6 seconds, the indicator light should then turn off.

If the air bag light does not illuminate at all, stays on continuously, or flashes at any time, a problem has been detected by the diagnostic computer.

✳✳ CAUTION

If at any time the air bag light indicates that the computer has noted a problem, immediately diagnose the problem. A faulty SRS can cause severe physical injury or death.

SERVICE PRECAUTIONS

Whenever working around, or on, the air bag supplemental restraint system, ALWAYS adhere to the following warnings and cautions.

- Always wear safety glasses when servicing an air bag vehicle and when handling an air bag module.
- Carry a live air bag module with the bag and trim cover facing away from your body, so that an accidental deployment of the air bag will have a small chance of personal injury.
- Place an air bag module on a table or other flat surface with the bag and trim cover pointing up.
- Wear gloves, a dust mask and safety glasses whenever handling a deployed air bag module. The air bag surface may contain traces of sodium hydroxide, a byproduct of the gas that inflates the air bag and which can cause skin irritation.
- Ensure to wash your hands with mild soap and water after handling a deployed air bag.
- All air bag modules with discolored or damaged cover trim must be replaced, not repainted.
- All component replacement and wiring service must be made with the negative and positive battery cables disconnected from the battery for a minimum of 1 minute prior to attempting service or replacement.
- NEVER probe the air bag electrical terminals. Doing so could result in air bag deployment, which can cause serious physical injury.

• If the vehicle is involved in a fender-bender that results in a damaged front bumper or grille, the air bag sensors should be inspected to ensure that they were not damaged.

• If at any time, the air bag light indicates that the computer has noted a problem, immediately diagnose the problem. A faulty SRS can cause severe physical injury or death.

DISARMING THE SYSTEM

1. Disconnect the negative battery cable from the battery.
2. Disconnect the positive battery cable from the battery.
3. Wait 1 minute. This time is required for the back-up power supply in the air bag diagnostic monitor to completely drain. The system is now disarmed.

ARMING THE SYSTEM

1. Connect the positive battery cable.
2. Connect the negative battery cable.
3. Stand outside the vehicle and carefully turn the ignition to the **RUN** position. Be sure that no part of your body is in front of the air bag module on the steering wheel, to prevent injury in case of an accidental air bag deployment.
4. Ensure the air bag indicator light turns off after approximately 6 seconds. If the light does not illuminate at all, does not turn off, or starts to flash, diagnose the problem. If the light does turn off after 6 seconds and does not flash, the SRS is working properly.

Rack and Pinion Steering Gear

REMOVAL & INSTALLATION

1. Remove or disconnect the following:
   • Front wheels
   • Tie rod end cotter pins and castle nuts
   • Tie rod ends from the knuckles
   • Front stabilizer bar
2. Position the dash opening weather seal for the steering column out of the way.
3. Remove or disconnect the following:
   • Pinch bolt retaining the steering column intermediate shaft coupling
   • Steering gear retaining nuts/bolts
   • Rear subframe bolts

➡**Use wire to support exhaust components unless you are removing them completely.**

4. Support the exhaust system flex tube and remove the flex tube-to-dual converter Y-pipe attachment.

Exploded view of the rack and pinion steering gear mounting on the front subframe of the vehicle

5. Lower the vehicle slightly until the rear subframe separates from the body approximately 4 inches (10cm).
6. Remove the heat shield band and fold the heat shield down.
7. Rotate the rack and pinion assembly to clear the bolts from the front subframe, and pull toward the driver's side of the vehicle.
8. Place a drain pan under the vehicle and disconnect the power steering lines.
9. Remove the rack and pinion assembly through the driver's side of the vehicle.

To install:

10. Install new Teflon® O-rings on the power steering line fittings.
11. Place the rack and pinion retaining bolts in the gear housing.
12. Install or connect the following:
   • Rack and pinion assembly through the driver's side of the vehicle
   • Power steering lines on the rack and pinion assembly
   • Rack and pinion assembly on the subframe
   • Strap on the heat shield
   • Tie rod ends to the knuckles. Tighten the castle nuts and install the cotter pins.
   • Stabilizer bar
   • Rack and pinion assembly retaining bolts, and tighten them to 85–99 ft. lbs. (115–135 Nm).
13. Raise the vehicle until the subframe contacts the body.
14. Install or connect the following:
   • Rear subframe retaining bolts, and

tighten them to 83–112 ft. lbs. (113–153 Nm).
   • Exhaust system flex tube-to-dual converter Y-pipe
   • Front wheels

15. Using a new pinch bolt, install the steering column intermediate shaft coupling on the rack input shaft. Tighten the pinch bolt to 25–33 ft. lbs. (34–46 Nm).
16. Position the steering column opening weather seal over the steering gear housing.
17. Lower the vehicle.
18. Fill the power steering oil reservoir.
19. Start the vehicle and check for leaks.
20. Check for proper wheel alignment and steering wheel position.

Struts

REMOVAL & INSTALLATION

✳✳ CAUTION

Suspension fasteners are critical parts because they affect performance of vital components and systems and their failure can result in major service expense. They must be exchanged with the same part number or an equivalent part if an exchange is necessary. Do not use an alternative part of lesser quality or substitute design. Torque values must be used as specified during reassembly to make sure these parts are correctly retained.

➡️**Make sure the steering wheel is in the unlocked position.**

1. Raise and support the vehicle.
2. Remove the wheel and tire assembly.
3. Remove and discard the front stabilizer bar link retaining nut.
4. Remove and discard the strut and spring retaining bolt.

✳✳ CAUTION

Do not allow the halfshaft to move outboard. Over-extension of the tripod CV joint can result in separation of internal parts, causing failure of the halfshaft.

5. Push down on the front wheel knuckle until the strut and spring assembly is free of the front wheel knuckle. Support the front wheel knuckle to prevent the front axle shaft from moving outboard.

6. Partially lower the vehicle.
7. Remove and discard the three strut and spring mounting bracket retaining nuts.
8. Remove the strut and spring assembly.

1 Front stabilizer bar
2 Front stabilizer bar bracket
3 Front stabilizer bar insulator
4 Tie-rod end
5 Front suspension lower arm

5 Front suspension lower arm
6 Front wheel hub (includes bearing)
7 Front stabilizer bar link
8 Front wheel knuckle
9 Front strut assy

10 Front spring insulator
11 Front coil spring
12 Front strut assy washer
13 Front suspension bearing and seal
14 Front strut assy mounting bracket

67197-WIND-G02

Front suspension components

✳ CAUTION

The strut assembly is under spring tension. Attempting to disassemble the strut without correctly compressing the spring can result in injury.

9. Using a suitable spring compressing tool, compress the coil spring to relieve spring tension on the spring seat and bearing assembly.

10. Position a suitable crows-foot wrench onto the strut shaft top retaining nut.

11. Place a 10-mm six point deep well socket on top of the strut shaft.

12. Remove the strut shaft top retaining nut while preventing the strut shaft from turning.

13. Remove the upper washer and strut mounting bracket.

✳ CAUTION

The spring seat and bearing assembly is press-fit onto the upper mount. The mount washers must be installed in the correct orientation.

14. Remove the lower washer, spring seat and bearing assembly.

15. Remove the spring compressing tool, coil spring and dust boot.

16. Remove the jounce bumper and lower spring insulator.

To install:

✳✳ CAUTION

Do not over tighten the lock nut; damage to the lock nut or piston rod can result.

➡ Be sure to check the spring insulator for damage before assembly.

➡ Make sure the correct assembly sequence and correct positioning of the seat and bearing assembly and the lower spring insulator are followed.

17. Install the lower spring insulator and jounce bumper.

18. Install the spring compressing tool, coil spring and dust boot.

19. Install the spring seat and bearing assembly and the lower washer.

20. Install the strut mounting bracket and upper washer.

21. Install a new strut and spring assembly top shaft retaining nut.

22. Remove the spring compressing tool.

23. Position the strut and spring assem-

bly into the vehicle and install new mounting bracket nuts.

24. Partially raise the vehicle.

25. Position the strut and spring assembly to the steering knuckle and install a new pinch bolt.

26. Position the stabilizer link to the strut and spring assembly and install a new nut.

27. Install the wheel and tire assembly.

28. Lower the vehicle.

Shock Absorbers

REMOVAL & INSTALLATION

1. Loosen the lug nuts on the rear wheels.

2. Raise and safely support the vehicle.

3. Remove the rear wheels.

4. Position a jack under the rear axle assembly and raise it slightly to put the suspension at normal ride height.

5. Remove the lower shock absorber bolt/nut and disconnect the shock from the rear axle.

6. Lower the rear axle slightly to help aid removal of the upper shock absorber bolt/nut.

7. Remove the shock absorber.

To install:

8. Attach the shock absorber to the upper mounting bracket and install a new retaining bolt/nut.

9. Slowly raise the rear axle assembly with a jack, and guide the lower shock absorber into the bracket on the rear axle assembly. Install a new retaining bolt/nut.

10. Raise the rear suspension to normal ride height and tighten the shock absorber retaining upper bolt to 76 ft. lbs. (103 Nm); the lower bolt to 59 ft. lbs. (80 Nm).

11. Install the wheels.

12. Lower the vehicle.

Coil Springs

REMOVAL & INSTALLATION

1. Raise and safely support the vehicle.

2. Remove the rear wheels.

1	J-Nut
2	Bolt M12-1.75 x 66 Hex Flanged Head
3	Rear Spring Insulator (Upper)
4	Rear Spring
5	Rear Spring Insulator (Lower)
6	Axle Assembly
7	Nut
8	Bolt
9	Shock Absorber

7924GG12

Exploded view of the rear coil spring and shock absorber mounting between the axle and chassis

➡The rear axle will need to be supported when the shock absorbers are removed.

3. Position an adjustable stand or jack under the rear axle.

4. Remove the shock absorber-to-rear axle nut and disconnect the shock from the rear axle.

5. Slowly lower the rear axle assembly until the rear spring can be removed.

6. Remove the rear spring.

To install:

7. Position the rear spring insulator on the rear axle assembly and press the insulator downward into place. Verify rear spring insulator is properly seated into correct position.

8. Slowly raise the rear axle assembly with a jack, and guide the upper rear spring insulator onto the upper spring seat on the underbody.

9. Position the shock absorber on the lower rear axle assembly and install new nuts and bolts. Tighten upper bolt to 76 ft. lbs. (103 Nm); the lower bolt to 59 ft. lbs. (80 Nm).

10. Install the wheels.

11. Lower the vehicle.

Lower Ball Joint

REMOVAL & INSTALLATION

The lower ball joint and seal are an integral part of the lower control arm assembly, and can not be replaced separately. If the lower ball joint or seal is found to be defective, the lower control arm must be replaced as an assembly.

Lower Control Arm

REMOVAL & INSTALLATION

➡Do not begin the removal procedure unless a new strut-to-lower arm nut, a new ball joint pinch bolt/nut and a new lower arm-to-front subframe bolt/nut are available.

1. Remove or disconnect the following:
- Wheels
- Lower arm strut nut and dished washer. Discard them.

➡Do not use a hammer to separate the ball joint from the front wheel hub and spindle.

2. Place the steering wheel in the unlocked position.

➡Do not allow the halfshaft to move outward. Over-extension of the halfshaft joint could result in separation of internal parts, causing failure of the halfshaft joint.

3. Remove or disconnect the following:
- Lower arm ball joint nut and pinch bolt. Discard them.
- Lower arm from the front wheel knuckle

➡Use extreme care to not damage the ball joint boot seal.

- Remove and discard the lower arm-to-front subframe bolt and nut
- Remove the lower arm from the vehicle

To install:

4. Install or connect the following:
- Lower arm strut into the lower arm rear strut bushing
- Lower arm into the front sub-frame bracket
- New lower arm-to-front subframe nut and bolt. While holding the lower arm horizontal, tighten to 98 ft. lbs. (133 Nm).
- Ball joint stud-to-wheel hub and

spindle, making sure that the ball stud groove is properly positioned.
- New lower arm ball joint pinch bolt and nut. Tighten to 46 ft. lbs. 63 Nm).

5. Clean the lower arm strut threads to remove dirt and contamination.

6. Install the dished washer with the dished side away from the lower arm rear strut bushing.

7. Install the front suspension lower arm strut-to-strut nut and tighten to 85 ft. lbs. (115 Nm).

8. Install the wheels.

9. Lower the vehicle.

LOWER ARM BUSHING REPLACEMENT

➡Prior to starting this procedure, ensure that new replacement fasteners are available.

1. Remove the lower arm from the vehicle.

2. Using a bushing remover, remove the bushing from the lower arm using a large C-clamp as a press.

To install:

3. Saturate the lower arm and new bushing in vegetable oil.

FRONT SUSPENSION LOWER ARM

FRONT SUSPENSION LOWER ARM REAR STRUT BUSHING

C-FRAME AND CLAMP ASSEMBLY T74P-3044-A1

BUSHING REMOVER T86P-5493-A5

89698G02

Removing the lower arm bushing using a C-clamp and the proper adapters

Installing the lower arm bushing using a C-clamp and the proper adapters

Item	Description
1	Front Suspension Lower Arm
2	Sleeve
3	Front Suspension Lower Arm Rear Strut Bushing
4	Bushing Driver
5	C-Frame and Clamp Assembly

89698G03

➡Use only vegetable oil. Any mineral or petroleum based oil or brake fluid will deteriorate the rubber.

4. Using a bushing driver, press the bushing into the lower arm using a C-clamp as a press.

5. Install the lower arm on the vehicle.

Hub and Wheel Bearing

ADJUSTMENT

The wheel bearings on the Ford Windstar are not adjustable. If the bearings become loose or make noise they must be replaced as an assembly.

REMOVAL & INSTALLATION

Front

❋❋ CAUTION

Suspension fasteners are critical parts because they affect performance of vital components and systems and their failure can result in major service expense. They must be exchanged with the same part number or an equivalent part if an exchange is necessary. Do not use an alternative part of lesser quality or substitute design. Torque values must be used as specified during reassembly to make sure these parts are correctly retained.

➡Make sure the steering wheel is in the unlocked position.

1. If equipped, remove the front wheel center cap.

❋❋ CAUTION

A new wheel hub retainer must be installed, if it is loosened or removed. Failure to install a new retainer after removal can result in loss of the front wheel or damage to the wheel hub and bearing assembly.

❋❋ CAUTION

The front axle wheel hub retainer must be tightened to specification immediately during installation. If the retainer is not tightened immediately, the nylon lock will set incorrectly, leading to incorrect torque readings and bearing failure. Any front wheel hub retainer that is not immediately tightened to specification or is loosened must be removed and a new retainer installed.

➡The Windstar uses a wheel hub and bearing assembly and is not repaired separately.

2. Remove and discard the wheel hub retainer.

3. Raise and support the vehicle.

4. Remove the wheel and tire assembly.

5. Remove the front brake disc.

6. Remove the ABS sensor retaining bolt, ABS sensor harness retaining bolt and the ABS sensor. Position out of the way.

7. Remove and discard the tie-rod end cotter pin and the castellated nut.

8. Using a puller, remove the tie-rod end.

9. Remove and discard the front stabilizer bar link nut and separate the front stabilizer bar link.

10. Remove and discard the ball joint pinch bolt and nut.

11. Using a pry bar between the front subframe and the front suspension lower arm, push down until the lower ball joint is free of the front wheel knuckle.

❋❋ CAUTION

Do not allow the halfshaft to move outboard. Over-extension of the tripod CV joint could result in separation of internal parts, causing failure of the halfshaft.

12. Using a hub puller, remove the front driveshaft and joint from the wheel hub.

➡The wheel hub is a slip-fit design and should not require a puller to remove.

13. Remove and discard the three wheel hub retaining bolts and remove the wheel hub.

➡Apply a small patch of Loctite®242 or equivalent meeting Ford specification WSK-M2G351-A5 to the last five front wheel driveshaft joint threads prior to installing the wheel hub retainer nut.

14. Using new fasteners, follow the removal procedure in reverse order. Observe the following torques:
- Hub retaining bolts: 85 ft. lbs. (115 Nm)
- Ball joint pinch bolt: 46 ft. lbs. (63 Nm)
- Stabilizer bar link nut: 66 ft. lbs. (90 Nm)
- Tie rod end nut: 41 ft. lbs. (55 Nm)
- Wheel hub retainer nut: 185 ft. lbs. (250 Nm)

Rear

1. Raise the vehicle on a hoist.
2. Remove the wheel and tire assembly.
3. Remove the brake drum.
4. Remove the hub cap grease seal.

1 Rear axle assy	6 Rear wheel spindle	11 Retainer and washer assy
2 Rear spring	7 Rear brake anti-lock sensor	12 Rear wheel gasket
3 Shock absorber	8 Bolt, rear wheel spindle-to-rear axle	13 Anti-lock brake sensor bracket
4 Trailing arm	9 Hub and bearing assy	14 Wheel studs
5 Nut, rear wheel spindle-to-rear axle	10 Hub grease cap	

67197-WIND-G01

Rear hub assembly

✳✳ WARNING

Discard the rear wheel hub retainer and washer assembly. It is a torque prevailing design and cannot be reused.

5. Remove the retainer and washer assembly.

6. Remove the wheel hub from the rear wheel spindle.

7. Thoroughly clean any grease from the surrounding surfaces. If a new hub assembly is being installed, remove the protective coating using carburetor degreaser.

8. To install, reverse the removal procedure. Use a new retainer and washer assembly. Torque the nut to 221 ft. lbs. (300 Nm).

BRAKES

Brake Caliper

REMOVAL & INSTALLATION

1. Raise and support the vehicle.
2. Remove the wheel and tire assembly.
3. Remove the disc brake caliper bolts.

✳✳ CAUTION

Brake fluid contains polyglycol ethers and polyglycols. Avoid contact with eyes. Wash hands thoroughly after handling. If brake fluid contacts eyes, flush eyes with running water for 15 minutes, get medical attention if irritation persists. If taken internally, drink water and induce vomiting. Get medical attention immediately.

✳✳ CAUTION

Brake fluid is harmful to painted and plastic surfaces. If brake fluid is spilled onto a painted or plastic surface, wash it with water immediately.

4. Remove the front brake hose bolt and discard the copper washers.
5. Remove the brake caliper.
6. Inspect the disc brake caliper. If leaks or damaged boots are found, disassembly is required.

To install:
7. Install the brake caliper.
8. Install the front brake hose bolt. Use new copper washers; connect the front brake hose. Torque the bolt to 35 ft. lbs. (48 Nm).

✳✳ CAUTION

Make sure guide pin boots are correctly seated or damage to guide pins can occur.

➡ Tighten the lower caliper bolt first.

9. Install the disc brake caliper.
10. Install the disc brake caliper bolts. Torque to 26 ft. lbs. (35 Nm).
11. Bleed the caliper.
12. Install the wheel and tire assembly.
13. Fill the brake master cylinder reservoir with clean High Performance DOT 3 Motor Vehicle Brake Fluid C6AZ-19542-AB or equivalent DOT 3 fluid meeting Ford specification ESA-M6C25-A. Install brake master cylinder filler cap.
14. Inspect the brake system operation.

Disc Brake Pads

REMOVAL & INSTALLATION

1. Remove the brake master cylinder filler cap. Check the brake fluid level in the brake master cylinder reservoir. Remove fluid until the brake master cylinder reservoir is half full.
2. Raise and support the vehicle.
3. Remove the wheel and tire assembly.

✳✳ WARNING

Install a new pad if worn to or past the specified thickness above the metal backing plate or rivets. Install new pads in complete axle sets.

4. Inspect the pads for wear and contamination.

✳✳ WARNING

Do not pry in caliper sight hole to retract pistons as this can damage the pistons and boots.

✳✳ WARNING

When removing the disc brake caliper, never allow it to hang from the brake hose. Provide a suitable support.

5. Remove two front disc brake caliper bolts.
6. Lift the disc brake caliper off the front disc brake caliper anchor plate.
7. Inspect the disc brake caliper. If leaks or damaged boots are found, disassembly is required.
8. Remove the front disc brake caliper anchor plate stainless steel slippers and discard.
9. Inspect the front disc brake anchor plate assembly. Check the guide pin boots for damage. Check the guide pins for binding and damage. Replace worn or damaged pins. Lube pins with Silicone Brake Caliper Grease and Dielectric Compound D7AZ-19A331-A or equivalent meeting Ford specification ESE-M1C171-A.

Pads and slippers

67197-WIND-G03

35 Nm (26 lb/ft) ②

35 Nm (26 lb/ft) ②

①

67197-WIND-G04

Caliper installation

To install:

> ⚙ **WARNING**
>
> **Do not allow grease, oil, brake fluid or other contaminants to contact the pad lining material. Do not install contaminated pads.**

➡ **Install all new hardware supplied with pad kit and spring kit.**

10. Install the new front disc brake caliper anchor plate stainless steel slippers. Make sure taps point away from brake disc as illustrated.

11. Install the pads.

➡ **Use a wood block or used pad to protect pistons and boots.**

12. Compress the caliper pistons.

> ⚙ **WARNING**
>
> **Use care not to damage the bleeder screw or brake disc shield.**

13. Install the disc brake caliper.

14. Install the disc brake caliper bolts; tighten the top bolt first.

15. Install the wheel and tire assembly.

16. Lower the vehicle.

17. Fill the brake master cylinder reservoir with clean High Performance DOT 3 Brake Fluid C6AZ-19542-AB or equivalent DOT 3 fluid meeting Ford specification ESA-M6C25-A. Install brake master cylinder filler cap.

18. Inspect brake operation.

Brake Drums

REMOVAL & INSTALLATION

1. Raise and safely support the vehicle.

2. Remove the wheel and tire assembly.

3. Remove the retainers holding the drum to the hub, if installed and discard.

4. Grasp the drum and remove.

5. If the drum will not slide off with light force, the brake shoes will need to be backed off. Remove the rubber plug on the backing plate and insert a screwdriver and a brake adjusting tool into the slot. Hold the adjuster lever away from the adjuster wheel with the screwdriver and back of the adjuster wheel with the brake adjusting tool.

6. Remove the brake drum. Inspect the drum for wear and/or damage. Machine or

ROTUNDA BRAKE DRUM MICROMETER

BRAKE DRUM

93026G14

Measuring the brake shoes and drum

replace as necessary. If machining, observe the maximum diameter specification.

To install:

7. If a new brake drum is being installed, remove the protective coating from the inner brake surface.

8. Use a suitable brake adjustment gauge to measure the inside diameter of the brake drum.

9. Adjust the brake shoes to match the inside diameter of the brake drum.

10. Slide the brake drum onto the hub. Make sure that the brake shoes are not tight to the brake drum.

11. Install the rubber plug in the access hole. Retainers do not need to be reused to hold the drum.

12. Install the wheel and tire assembly. Torque the lug nuts to 85–105 ft. lbs. (115–142 Nm).

13. Lower the vehicle.

14. Check and fill the brake master cylinder as required.

15. Road-test the vehicle and check for proper brake operation.

Brake Shoes

REMOVAL & INSTALLATION

1. Remove the brake drum.

2. Inspect the rear brake assembly for the following:

• The rear wheel cylinder for leakage.

1 Rear wheel cylinder	7 Brake backing plate (LH)	11 Brake shoe retracting spring	15 Brake shoe adjusting screw (RH)
2 Bleed screw	7 Brake backing plate (RH)	12 Parking brake lever pin (inner)	16 Brake adjusting lever (LH)
3 Bleeder dust cap	8 Shoe and lining		16 Brake adjusting lever (RH)
4 Wheel cylinder retaining screw (2 req'd)	9 Brake shoe hold-down spring	13 Parking brake lever pin retainer	17 Parking brake lever pin (outer)
5 Brake adjusting hole cover	10 Parking brake lever (LH)	14 Washer	18 Brake adjusting screw spring
6 Brake shoe hold-down spring pin	10 Parking brake lever (RH)	15 Brake shoe adjusting screw (LH)	

67197-WIND-G05

Rear drum brake assembly

- The rear brake shoes and linings for contamination, and install new as necessary.
- The springs for heat discoloration and install new as necessary.
- The adjusting lever contact with the brake adjuster screw.

3. Inspect the rear brake shoes and linings for minimum thickness, above the backing plate or rivets, and install new as necessary.

4. Remove the brake shoe adjusting screw spring.

5. Remove the brake adjuster lever.

6. Remove the brake adjuster screw.

7. Remove the brake shoe hold-down spring.

8. Remove the brake shoe and the retracting spring.

9. Remove the rear brake shoe hold-down spring.

10. Remove the rear brake shoes and linings.

11. Remove and discard the parking brake lever clip.

12. Remove the rear brake shoes and linings from the parking brake lever.

13. To install, reverse the removal procedure.

GLOSSARY

ABS: Anti-lock braking system. An electro-mechanical braking system which is designed to minimize or prevent wheel lock-up during braking.

ABSOLUTE PRESSURE: Atmospheric (barometric) pressure plus the pressure gauge reading.

ACCELERATOR PUMP: A small pump located in the carburetor that feeds fuel into the air/fuel mixture during acceleration.

ACCUMULATOR: A device that controls shift quality by cushioning the shock of hydraulic oil pressure being applied to a clutch or band.

ACTUATING MECHANISM: The mechanical output devices of a hydraulic system, for example, clutch pistons and band servos.

ACTUATOR: The output component of a hydraulic or electronic system.

ADVANCE: Setting the ignition timing so that spark occurs earlier before the piston reaches top dead center (TDC).

ADAPTIVE MEMORY (ADAPTIVE STRATEGY): The learning ability of the TCM or PCM to redefine its decision-making process to provide optimum shift quality.

AFTER TOP DEAD CENTER (ATDC): The point after the piston reaches the top of its travel on the compression stroke.

AIR BAG: Device on the inside of the car designed to inflate on impact of crash, protecting the occupants of the car.

AIR CHARGE TEMPERATURE (ACT) SENSOR: The temperature of the airflow into the engine is measured by an ACT sensor, usually located in the lower intake manifold or air cleaner.

AIR CLEANER: An assembly consisting of a housing, filter and any connecting ductwork. The filter element is made up of a porous paper, sometimes with a wire mesh screening, and is designed to prevent airborne particles from entering the engine through the carburetor or throttle body.

AIR INJECTION: One method of reducing harmful exhaust emissions by injecting air into each of the exhaust ports of an engine. The fresh air entering the hot exhaust manifold causes any remaining fuel to be burned before it can exit the tailpipe.

AIR PUMP: An emission control device that supplies fresh air to the exhaust manifold to aid in more completely burning exhaust gases.

AIR/FUEL RATIO: The ratio of air-to-gasoline by weight in the fuel mixture drawn into the engine.

ALDL (assembly line diagnostic link): Electrical connector for scanning ECM/PCM/TCM input and output devices.

ALIGNMENT RACK: A special drive-on vehicle lift apparatus/measuring device used to adjust a vehicle's toe, caster and camber angles.

ALL WHEEL DRIVE: Term used to describe a full time four wheel drive system or any other vehicle drive system that continuously delivers power to all four wheels. This system is found primarily on station wagon vehicles and SUVs not utilized for significant off road use.

ALTERNATING CURRENT (AC): Electric current that flows first in one direction, then in the opposite direction, continually reversing flow.

ALTERNATOR: A device which produces AC (alternating current) which is converted to DC (direct current) to charge the car battery.

AMMETER: An instrument, calibrated in amperes, used to measure the flow of an electrical current in a circuit. Ammeters are always connected in series with the circuit being tested.

AMPERAGE: The total amount of current (amperes) flowing in a circuit.

AMPLIFIER: A device used in an electrical circuit to increase the voltage of an output signal.

AMP/HR. RATING (BATTERY): Measurement of the ability of a battery to deliver a stated amount of current for a stated period of time. The higher the amp/hr. rating, the better the battery.

AMPERE: The rate of flow of electrical current present when one volt of electrical pressure is applied against one ohm of electrical resistance.

ANALOG COMPUTER: Any microprocessor that uses similar (analogous) electrical signals to make its calculations.

ANODIZED: A special coating applied to the surface of aluminum valves for extended service life.

ANTIFREEZE: A substance (ethylene or propylene glycol) added to the coolant to prevent freezing in cold weather.

ANTI-FOAM AGENTS: Minimize fluid foaming from the whipping action encountered in the converter and planetary action.

ANTI-WEAR AGENTS: Zinc agents that control wear on the gears, bushings, and thrust washers.

ANTI-LOCK BRAKING SYSTEM: A supplementary system to the base hydraulic system that prevents sustained lock-up of the wheels during braking as well as automatically controlling wheel slip.

ANTI-ROLL BAR: See stabilizer bar.

ARC: A flow of electricity through the air between two electrodes or contact points that produces a spark.

ARMATURE: A laminated, soft iron core wrapped by a wire that converts electrical energy to mechanical energy as in a motor or relay. When rotated in a magnetic field, it changes mechanical energy into electrical energy as in a generator.

ATDC: After Top Dead Center.

ATF: Automatic transmission fluid.

ATMOSPHERIC PRESSURE: The pressure on the Earth's surface caused by the weight of the air in the atmosphere. At sea level, this pressure is 14.7 psi at 32°F (101 kPa at 0°C).

ATOMIZATION: The breaking down of a liquid into a fine mist that can be suspended in air.

AUXILIARY ADD-ON COOLER: A supplemental transmission fluid cooling device that is installed in series with the heat exchanger (cooler), located inside the radiator, to provide additional support to cool the hot fluid leaving the torque converter.

AUXILIARY PRESSURE: An added fluid pressure that is introduced into a regulator or balanced valve system to control valve movement. The auxiliary pressure itself can be either a fixed or a variable value. (See balanced valve; regulator valve.)

AWD: All wheel drive.

AXIAL FORCE: A side or end thrust force acting in or along the same plane as the power flow.

AXIAL PLAY: Movement parallel to a shaft or bearing bore.

AXLE CAPACITY: The maximum load-carrying capacity of the axle itself, as specified by the manufacturer. This is usually a higher number than the GAWR.

AXLE RATIO: This is a number (3.07:1, 4.56:1, for example) expressing the ratio between driveshaft revolutions and wheel revolutions. A low numerical ratio allows the engine to work easier because it doesn't have to turn as fast. A high numerical ratio means that the engine has to turn more rpm's to move the wheels through the same number of turns.

BACKFIRE: The sudden combustion of gases in the intake or exhaust system that results in a loud explosion.

BACKLASH: The clearance or play between two parts, such as meshed gears.

BACKPRESSURE: Restrictions in the exhaust system that slow the exit of exhaust gases from the combustion chamber.

BAKELITE®: A heat resistant, plastic insulator material commonly used in printed circuit boards and transistorized components.

BALANCED VALVE: A valve that is positioned by opposing auxiliary hydraulic pressures and/or spring force. Examples include mainline regulator, throttle, and governor valves. (See regulator valve.)

BAND: A flexible ring of steel with an inner lining of friction material. When tightened around the outside of a drum, a planetary member is held stationary to the transmission/transaxle case.

BALL BEARING: A bearing made up of hardened inner and outer races between which hardened steel balls roll.

BALL JOINT: A ball and matching socket connecting suspension components (steering knuckle to lower control arms). It permits rotating movement in any direction between the components that are joined.

BARO (BAROMETRIC PRESSURE SENSOR): Measures the change in the intake manifold pressure caused by changes in altitude.

BAROMETRIC MANIFOLD ABSOLUTE PRESSURE (BMAP) SENSOR: Operates similarly to a conventional MAP sensor; reads intake manifold pressure and is also responsible for determining altitude and barometric pressure prior to engine operation.

BAROMETRIC PRESSURE: (See atmospheric pressure.)

BALLAST RESISTOR: A resistor in the primary ignition circuit that lowers voltage after the engine is started to reduce wear on ignition components.

BATTERY: A direct current electrical storage unit, consisting of the basic active materials of lead and sulfuric acid, which converts chemical energy into electrical energy. Used to provide current for the operation of the starter as well as other equipment, such as the radio, lighting, etc.

BEAD: The portion of a tire that holds it on the rim.

BEARING: A friction reducing, supportive device usually located between a stationary part and a moving part.

BEFORE TOP DEAD CENTER (BTDC): The point just before the piston reaches the top of its travel on the compression stroke.

BELTED TIRE: Tire construction similar to bias-ply tires, but using two or more layers of reinforced belts between body plies and the tread.

BEZEL: Piece of metal surrounding radio, headlights, gauges or similar components; sometimes used to hold the glass face of a gauge in the dash.

BIAS-PLY TIRE: Tire construction, using body ply reinforcing cords which run at alternating angles to the center line of the tread.

BI-METAL TEMPERATURE SENSOR: Any sensor or switch made of two dissimilar types of metal that bend when heated or cooled due to the different expansion rates of the alloys. These types of sensors usually function as an on/off switch.

BLOCK: See Engine Block.

BLOW-BY: Combustion gases, composed of water vapor and unburned fuel, that leak past the piston rings into the crankcase during normal engine operation. These gases are removed by the PCV system to prevent the buildup of harmful acids in the crankcase.

BOOK TIME: See Labor Time.

BOOK VALUE: The average value of a car, widely used to determine trade-in and resale value.

BOOST VALVE: Used at the base of the regulator valve to increase mainline pressure.

BORE: Diameter of a cylinder.

BRAKE CALIPER: The housing that fits over the brake disc. The caliper holds the brake pads, which are pressed against the discs by the caliper pistons when the brake pedal is depressed.

BRAKE HORSEPOWER (BHP): The actual horsepower available at the engine flywheel as measured by a dynamometer.

BRAKE FADE: Loss of braking power, usually caused by excessive heat after repeated brake applications.

BRAKE HORSEPOWER: Usable horsepower of an engine measured at the crankshaft.

BRAKE PAD: A brake shoe and lining assembly used with disc brakes.

BRAKE PROPORTIONING VALVE: A valve on the master cylinder which restricts hydraulic brake pressure to the wheels to a specified amount, preventing wheel lock-up.

BREAKAWAY: Often used by Chrysler to identify first-gear operation in D and 2 ranges. In these ranges, first-gear operation depends on a one-way roller clutch that holds on acceleration and releases (breaks away) on deceleration, resulting in a freewheeling coast-down condition.

BRAKE SHOE: The backing for the brake lining. The term is, however, usually applied to the assembly of the brake backing and lining.

BREAKER POINTS: A set of points inside the distributor, operated by a cam, which make and break the ignition circuit.

BRINNELLING: A wear pattern identified by a series of indentations at regular intervals. This condition is caused by a lack of lube, overload situations, and/or vibrations.

BTDC: Before Top Dead Center.

BUMP: Sudden and forceful apply of a clutch or band.

BUSHING: A liner, usually removable, for a bearing; an anti-friction liner used in place of a bearing.

CALIFORNIA ENGINE: An engine certified by the EPA for use in California only; conforms to more stringent emission regulations than Federal engine.

CALIPER: A hydraulically activated device in a disc brake system, which is mounted straddling the brake rotor (disc). The caliper contains at least one piston and two brake pads. Hydraulic pressure on the piston(s) forces the pads against the rotor.

CAPACITY: The quantity of electricity that can be delivered from a unit, as from a battery in ampere-hours, or output, as from a generator.

CAMBER: One of the factors of wheel alignment. Viewed from the front of the car, it is the inward or outward tilt of the wheel. The top of the tire will lean outward (positive camber) or inward (negative camber).

CAMSHAFT: A shaft in the engine on which are the lobes (cams) which operate the valves. The camshaft is driven by the crankshaft, via a belt, chain or gears, at one half the crankshaft speed.

CAPACITOR: A device which stores an electrical charge.

CARBON MONOXIDE (CO): A colorless, odorless gas given off as a normal byproduct of combustion. It is poisonous and extremely dangerous in confined areas, building up slowly to toxic levels without warning if adequate ventilation is not available.

CARBURETOR: A device, usually mounted on the intake manifold of an engine, which mixes the air and fuel in the proper proportion to allow even combustion.

CASTER: The forward or rearward tilt of an imaginary line drawn through the upper ball joint and the center of the wheel. Viewed from the sides, positive caster (forward tilt) lends directional stability, while negative caster (rearward tilt) produces instability.

CATALYTIC CONVERTER: A device installed in the exhaust system, like a muffler, that converts harmful byproducts of combustion into carbon dioxide and water vapor by means of a heat-producing chemical reaction.

CENTRIFUGAL ADVANCE: A mechanical method of advancing the spark timing by using flyweights in the distributor that react to centrifugal force generated by the distributor shaft rotation.

CENTRIFUGAL FORCE: The outward pull of a revolving object, away from the center of revolution. Centrifugal force increases with the speed of rotation.

CETANE RATING: A measure of the ignition value of diesel fuel. The higher the cetane rating, the better the fuel. Diesel fuel cetane rating is roughly comparable to gasoline octane rating.

CHECK VALVE: Any one-way valve installed to permit the flow of air, fuel or vacuum in one direction only.

CHOKE: The valve/plate that restricts the amount of air entering an engine on the induction stroke, thereby enriching the air/fuel ratio.

CHUGGLE: Bucking or jerking condition that may be engine related and may be most noticeable when converter clutch is engaged; similar to the feel of towing a trailer.

CIRCLIP: A split steel snapring that fits into a groove to hold various parts in place.

CIRCUIT BREAKER: A switch which protects an electrical circuit from overload by opening the circuit when the current flow exceeds a pre-determined level. Some circuit breakers must be reset manually, while most reset automatically.

CIRCUIT: Any unbroken path through which an electrical current can flow. Also used to describe fuel flow in some instances.

CIRCUIT, BYPASS: Another circuit in parallel with the major circuit through which power is diverted.

CIRCUIT, CLOSED: An electrical circuit in which there is no interruption of current flow.

CIRCUIT, GROUND: The non-insulated portion of a complete circuit used as a common potential point. In automotive circuits, the ground is composed of metal parts, such as the engine, body sheet metal, and frame and is usually a negative potential.

CIRCUIT, HOT: That portion of a circuit not at ground potential. The hot circuit is usually insulated and is connected to the positive side of the battery.

CIRCUIT, OPEN: A break or lack of contact in an electrical circuit, either intentional (switch) or unintentional (bad connection or broken wire).

CIRCUIT, PARALLEL: A circuit having two or more paths for current flow with common positive and negative tie points. The same voltage is applied to each load device or parallel branch.

CIRCUIT, SERIES: An electrical system in which separate parts are connected end to end, using one wire, to form a single path for current to flow.

CIRCUIT, SHORT: A circuit that is accidentally completed in an electrical path for which it was not intended.

CLAMPING (ISOLATION) DIODES: Diodes positioned in a circuit to prevent self-induction from damaging electronic components.

CLEARCOAT: A transparent layer which, when sprayed over a vehicle's paint job, adds gloss and depth as well as an additional protective coating to the finish.

CLUTCH: Part of the power train used to connect/disconnect power to the rear wheels.

CLUTCH, FLUID: The same as a fluid coupling. A fluid clutch or coupling performs the same function as a friction clutch by utilizing fluid friction and inertia as opposed to solid friction used by a friction clutch. (See fluid coupling.)

CLUTCH, FRICTION: A coupling device that provides a means of smooth and positive engagement and disengagement of engine torque to the vehicle powertrain. Transmission of power through the clutch is accomplished by bringing one or more rotating drive members into contact with complementing driven members.

COAST: Vehicle deceleration caused by engine braking conditions.

COEFFICIENT OF FRICTION: The amount of surface tension between two contacting surfaces; identified by a scientifically calculated number.

COIL: Part of the ignition system that boosts the relatively low voltage supplied by the car's electrical system to the high voltage required to fire the spark plugs.

COMBINATION MANIFOLD: An assembly which includes both the intake and exhaust manifolds in one casting.

COMBINATION VALVE: A device used in some fuel systems that routes fuel vapors to a charcoal storage canister instead of venting them into the atmosphere. The valve relieves fuel tank pressure and allows fresh air into the tank as the fuel level drops to prevent a vapor lock situation.

COMBUSTION CHAMBER: The part of the engine in the cylinder head where combustion takes place.

COMPOUND GEAR: A gear consisting of two or more simple gears with a common shaft.

COMPOUND PLANETARY: A gearset that has more than the three elements found in a simple gearset and is constructed by combining members of two planetary gearsets to create additional gear ratio possibilities.

COMPRESSION CHECK: A test involving removing each spark plug and inserting a gauge. When the engine is cranked, the gauge will record a pressure reading in the individual cylinder. General operating condition can be determined from a compression check.

COMPRESSION RATIO: The ratio of the volume between the piston and cylinder head when the piston is at the bottom of its stroke (bottom dead center) and when the piston is at the top of its stroke (top dead center).

COMPUTER: An electronic control module that correlates input data according to prearranged engineered instructions; used for the management of an actuator system or systems.

CONDENSER: An electrical device which acts to store an electrical charge, preventing voltage surges.

2. A radiator-like device in the air conditioning system in which refrigerant gas condenses into a liquid, giving off heat.

CONDUCTOR: Any material through which an electrical current can be transmitted easily.

CONNECTING ROD: The connecting link between the crankshaft and piston.

CONSTANT VELOCITY JOINT: Type of universal joint in a halfshaft assembly in which the output shaft turns at a constant angular velocity without variation, provided that the speed of the input shaft is constant.

CONTINUITY: Continuous or complete circuit. Can be checked with an ohmmeter.

CONTROL ARM: The upper or lower suspension components which are mounted on the frame and support the ball joints and steering knuckles.

CONVENTIONAL IGNITION: Ignition system which uses breaker points.

CONVERTER: (See torque converter.)

CONVERTER LOCKUP: The switching from hydrodynamic to direct mechanical drive, usually through the application of a friction element called the converter clutch.

COOLANT: Mixture of water and anti-freeze circulated through the engine to carry off heat produced by the engine.

CORROSION INHIBITOR: An inhibitor in ATF that prevents corrosion of bushings, thrust washers, and oil cooler brazed joints.

COUNTERSHAFT: An intermediate shaft which is rotated by a mainshaft and transmits, in turn, that rotation to a working part.

COUPLING PHASE: Occurs when the torque converter is operating at its greatest hydraulic efficiency. The speed differential between the impeller and the turbine is at its minimum. At this point, the stator freewheels, and there is no torque multiplication.

CRANKCASE: The lower part of an engine in which the crankshaft and related parts operate.

CRANKSHAFT: Engine component (connected to pistons by connecting rods) which converts the reciprocating (up and down) motion of pistons to rotary motion used to turn the driveshaft.

CURB WEIGHT: The weight of a vehicle without passengers or payload, but including all fluids (oil, gas, coolant, etc.) and other equipment specified as standard.

CURRENT: The flow (or rate) of electrons moving through a circuit. Current is measured in amperes (amp).

CURRENT FLOW CONVENTIONAL: Current flows through a circuit from the positive terminal of the source to the negative terminal (plus to minus).

CURRENT FLOW, ELECTRON: Current or electrons flow from the negative terminal of the source, through the circuit, to the positive terminal (minus to plus).

CV-JOINT: Constant velocity joint.

CYCLIC VIBRATIONS: The off-center movement of a rotating object that is affected by its initial balance, speed of rotation, and working angles.

CYLINDER BLOCK: See engine block.

CYLINDER HEAD: The detachable portion of the engine, usually fastened to the top of the cylinder block and containing all or most of the combustion chambers. On overhead valve engines, it contains the valves and their operating parts. On overhead cam engines, it contains the camshaft as well.

CYLINDER: In an engine, the round hole in the engine block in which the piston(s) ride.

DATA LINK CONNECTOR (DLC): Current acronym/term applied to the federally mandated, diagnostic junction connector that is used to monitor ECM/PC/TCM inputs, processing strategies, and outputs including diagnostic trouble codes (DTCs).

DEAD CENTER: The extreme top or bottom of the piston stroke.

DECELERATION BUMP: When referring to a torque converter clutch in the applied position, a sudden release of the accelerator pedal causes a forceful reversal of power through the drivetrain (engine braking), just prior to the apply plate actually being released.

DELAYED (LATE OR EXTENDED): Condition where shift is expected but does not occur for a period of time, for example, where clutch or band engagement does not occur as quickly as expected during part throttle or wide open throttle apply of accelerator or when manually downshifting to a lower range.

DETENT: A spring-loaded plunger, pin, ball, or pawl used as a holding device on a ratchet wheel or shaft. In automatic transmissions, a detent mechanism is used for locking the manual valve in place.

DETENT DOWNSHIFT: (See kickdown.)

DETERGENT: An additive in engine oil to improve its operating characteristics.

DETONATION: An unwanted explosion of the air/fuel mixture in the combustion chamber caused by excess heat and compression, advanced timing, or an overly lean mixture. Also referred to as "ping".

DEXRON®: A brand of automatic transmission fluid.

DIAGNOSTIC TROUBLE CODES (DTCs): A digital display from the control module memory that identifies the input, processor, or output device circuit that is related to the powertrain emission/driveability malfunction detected. Diagnostic trouble codes can be read by the MIL to flash any codes or by using a handheld scanner.

DIAPHRAGM: A thin, flexible wall separating two cavities, such as in a vacuum advance unit.

DIESELING: The engine continues to run after the car is shut off; caused by fuel continuing to be burned in the combustion chamber.

DIFFERENTIAL: A geared assembly which allows the transmission of motion between drive axles, giving one axle the ability to rotate faster than the other, as in cornering.

DIFFERENTIAL AREAS: When opposing faces of a spool valve are acted upon by the same pressure but their areas differ in size, the face with the larger area produces the differential force and valve movement. (See spool valve.)

DIFFERENTIAL FORCE: (See differential areas)

DIGITAL READOUT: A display of numbers or a combination of numbers and letters.

DIGITAL VOLT OHMMETER: An electronic diagnostic tool used to measure voltage, ohms and amps as well as several other functions, with the readings displayed on a digital screen in tenths, hundredths and thousandths.

DIODE: An electrical device that will allow current to flow in one direction only.

DIRECT CURRENT (DC): Electrical current that flows in one direction only.

DIRECT DRIVE: The gear ratio is 1:1, with no change occurring in the torque and speed input/output relationship.

DISC BRAKE: A hydraulic braking assembly consisting of a brake disc, or rotor, mounted on an axle shaft, and a caliper assembly containing, usually two brake pads which are activated by hydraulic pressure. The pads are forced against the sides of the disc, creating friction which slows the vehicle.

DISPERSANTS: Suspend dirt and prevent sludge buildup in a liquid, such as engine oil.

DOUBLE BUMP (DOUBLE FEEL): Two sudden and forceful applies of a clutch or band.

DISPLACEMENT: The total volume of air that is displaced by all pistons as the engine turns through one complete revolution.

DISTRIBUTOR: A mechanically driven device on an engine which is responsible for electrically firing the spark plug at a pre-determined point of the piston stroke.

DOHC: Double overhead camshaft.

DOUBLE OVERHEAD CAMSHAFT: The engine utilizes two camshafts mounted in one cylinder head. One camshaft operates the exhaust valves, while the other operates the intake valves.

DOWEL PIN: A pin, inserted in mating holes in two different parts allowing those parts to maintain a fixed relationship.

DRIVELINE: The drive connection between the transmission and the drive wheels.

DRIVE TRAIN: The components that transmit the flow of power from the engine to the wheels. The components include the clutch, transmission, driveshafts (or axle shafts in front wheel drive), U-joints and differential.

DRUM BRAKE: A braking system which consists of two brake shoes and one or two wheel cylinders, mounted on a fixed backing plate, and a brake drum, mounted on an axle, which revolves around the assembly.

DRY CHARGED BATTERY: Battery to which electrolyte is added when the battery is placed in service.

DVOM: Digital volt ohmmeter

DWELL: The rate, measured in degrees of shaft rotation, at which an electrical circuit cycles on and off.

DYNAMIC: An application in which there is rotating or reciprocating motion between the parts.

EARLY: Condition where shift occurs before vehicle has reached proper speed, which tends to labor engine after upshift.

EBCM: See Electronic Control Unit (ECU).

ECM: See Electronic Control Unit (ECU).

ECU: Electronic control unit.

ELECTRODE: Conductor (positive or negative) of electric current.

ELECTROLYSIS: A surface etching or bonding of current conducting transmission/transaxle components that may occur when grounding straps are missing or in poor condition.

ELECTROLYTE: A solution of water and sulfuric acid used to activate the battery. Electrolyte is extremely corrosive.

ELECTROMAGNET: A coil that produces a magnetic field when current flows through its windings.

ELECTROMAGNETIC INDUCTION: A method to create (generate) current flow through the use of magnetism.

ELECTROMAGNETISM: The effects surrounding the relationship between electricity and magnetism.

ELECTROMOTIVE FORCE (EMF): The force or pressure (voltage) that causes current movement in an electrical circuit.

ELECTRONIC CONTROL UNIT: A digital computer that controls engine (and sometimes transmission, brake or other vehicle system) functions based on data received from various sensors. Examples used by some manufacturers include Electronic Brake Control Module (EBCM), Engine Control Module (ECM), Powertrain Control Module (PCM) or Vehicle Control Module (VCM).

ELECTRONIC IGNITION: A system in which the timing and firing of the spark plugs is controlled by an electronic control unit, usually called a module. These systems have no points or condenser.

ELECTRONIC PRESSURE CONTROL (EPC) SOLENOID: A specially designed solenoid containing a spool valve and spring assembly to control fluid mainline pressure. A variable current flow, controlled by the ECM/PCM, varies the internal force of the solenoid on the spool valve and resulting mainline pressure. (See variable force solenoid.)

ELECTRONICS: Miniaturized electrical circuits utilizing semiconductors, solid-state devices, and printed circuits. Electronic circuits utilize small amounts of power.

ELECTRONIFICATION: The application of electronic circuitry to a mechanical device. Regarding automatic transmissions, electrification is incorporated into converter clutch lockup, shift scheduling, and line pressure control systems.

ELECTROSTATIC DISCHARGE (ESD): An unwanted, high-voltage electrical current released by an individual who has taken on a static charge of electricity. Electronic components can be easily damaged by ESD.

ELEMENT: A device within a hydrodynamic drive unit designed with a set of blades to direct fluid flow.

ENAMEL: Type of paint that dries to a smooth, glossy finish.

END BUMP (END FEEL OR SLIP BUMP): Firmer feel at end of shift when compared with feel at start of shift.

END-PLAY: The clearance/gap between two components that allows for expansion of the parts as they warm up, to prevent binding and to allow space for lubrication.

ENERGY: The ability or capacity to do work.

ENGINE: The primary motor or power apparatus of a vehicle, which converts liquid or gas fuel into mechanical energy.

ENGINE BLOCK: The basic engine casting containing the cylinders, the crankshaft main bearings, as well as machined surfaces for the mounting of other components such as the cylinder head, oil pan, transmission, etc.

ENGINE BRAKING: Use of engine to slow vehicle by manually downshifting during zero-throttle coast down.

ENGINE CONTROL MODULE (ECM): Manages the engine and incorporates output control over the torque converter clutch solenoid. (Note: Current designation for the ECM in late model vehicles is PCM.)

ENGINE COOLANT TEMPERATURE (ECT) SENSOR: Prevents converter clutch engagement with a cold engine; also used for shift timing and shift quality.

EP LUBRICANT: EP (extreme pressure) lubricants are specially formulated for use with gears involving heavy loads (transmissions, differentials, etc.).

ETHYL: A substance added to gasoline to improve its resistance to knock, by slowing down the rate of combustion.

ETHYLENE GLYCOL: The base substance of antifreeze.

EXHAUST MANIFOLD: A set of cast passages or pipes which conduct exhaust gases from the engine.

FAIL-SAFE (BACKUP) CONTROL: A substitute value used by the PCM/TCM to replace a faulty signal from an input sensor. The temporary value allows the vehicle to continue to be operated.

FAST IDLE: The speed of the engine when the choke is on. Fast idle speeds engine warm-up.

FEDERAL ENGINE: An engine certified by the EPA for use in any of the 49 states (except California).

FEEDBACK: A circuit malfunction whereby current can find another path to feed load devices.

FEELER GAUGE: A blade, usually metal, of precisely predetermined thickness, used to measure the clearance between two parts.

FILAMENT: The part of a bulb that glows; the filament creates high resistance to current flow and actually glows from the resulting heat.

FINAL DRIVE: An essential part of the axle drive assembly where final gear reduction takes place in the powertrain. In RWD applications and north-south FWD applications, it must also change the power flow direction to the axle shaft by ninety degrees. (Also see axle ratio.)

FIRING ORDER: The order in which combustion occurs in the cylinders of an engine. Also the order in which spark is distributed to the plugs by the distributor.

FIRM: A noticeable quick apply of a clutch or band that is considered normal with medium to heavy throttle shift; should not be confused with harsh or rough.

FLAME FRONT: The term used to describe certain aspects of the fuel explosion in the cylinders. The flame front should move in a controlled pattern across the cylinder, rather than simply exploding immediately.

FLARE (SLIPPING): A quick increase in engine rpm accompanied by momentary loss of torque; generally occurs during shift.

FLAT ENGINE: Engine design in which the pistons are horizontally opposed. Porsche, Subaru and some old VW are common examples of flat engines.

FLAT RATE: A dealership term referring to the amount of money paid to a technician for a repair or diagnostic service based on that particular service versus dealership's labor time (NOT based on the actual time the technician spent on the job).

FLAT SPOT: A point during acceleration when the engine seems to lose power for an instant.

FLOODING: The presence of too much fuel in the intake manifold and combustion chamber which prevents the air/fuel mixture from firing, thereby causing a no-start situation.

FLUID: A fluid can be either liquid or gas. In hydraulics, a liquid is used for transmitting force or motion.

FLUID COUPLING: The simplest form of hydrodynamic drive, the fluid coupling consists of two look-alike members with straight radial varies referred to as the impeller (pump) and the turbine. Input torque is always equal to the output torque.

FLUID DRIVE: Either a fluid coupling or a fluid torque converter. (See hydrodynamic drive units.)

FLUID TORQUE CONVERTER: A hydrodynamic drive that has the ability to act both as a torque multiplier and fluid coupling. (See hydrodynamic drive units; torque converter.)

FLUID VISCOSITY: The resistance of a liquid to flow. A cold fluid (oil) has greater viscosity and flows more slowly than a hot fluid (oil).

FLYWHEEL: A heavy disc of metal attached to the rear of the crankshaft. It smoothes the firing impulses of the engine and keeps the crankshaft turning during periods when no firing takes place. The starter also engages the flywheel to start the engine.

FOOT POUND (ft. lbs., lbs. ft. or sometimes, ft. lb.): The amount of energy or work needed to raise an item weighing one pound, a distance of one foot.

FREEZE PLUG: A plug in the engine block which will be pushed out if the coolant freezes. Sometimes called expansion plugs, they protect the block from cracking should the coolant freeze.

FRICTION: The resistance that occurs between contacting surfaces. This relationship is expressed by a ratio called the coefficient of friction (CL).

FRICTION, COEFFICIENT OF: The amount of surface tension between two contacting surfaces; expressed by a scientifically calculated number.

FRONT END ALIGNMENT: A service to set caster, camber and toe-in to the correct specifications. This will ensure that the car steers and handles properly and that the tires wear properly.

FRICTION MODIFIER: Changes the coefficient of friction of the fluid between the mating steel and composition clutch/band surfaces during the engagement process and allows for a certain amount of intentional slipping for a good "shift-feel".

FRONTAL AREA: The total frontal area of a vehicle exposed to air flow.

FUEL FILTER: A component of the fuel system containing a porous paper element used to prevent any impurities from entering the engine through the fuel system. It usually takes the form of a canister-like housing, mounted in-line with the fuel hose, located anywhere on a vehicle between the fuel tank and engine.

FUEL INJECTION: A system replacing the carburetor that sprays fuel into the cylinder through nozzles. The amount of fuel can be more precisely controlled with fuel injection.

FULL FLOATING AXLE: An axle in which the axle housing extends through the wheel giving bearing support on the outside of the housing. The front axle of a four-wheel drive vehicle is usually a full floating axle, as are the rear axles of many larger (1 ton and over) pick-ups and vans.

FULL-TIME FOUR-WHEEL DRIVE: A four-wheel drive system that continuously delivers power to all four wheels. A differential between the front and rear driveshafts permits variations in axle speeds to control gear wind-up without damage.

FULL THROTTLE DETENT DOWNSHIFT: A quick apply of accelerator pedal to its full travel, forcing a downshift.

FUSE: A protective device in a circuit which prevents circuit overload by breaking the circuit when a specific amperage is present. The device is constructed around a strip or wire of a lower amperage rating than the circuit it is designed to protect. When an amperage higher than that stamped on the fuse is present in the circuit, the strip or wire melts, opening the circuit.

FUSIBLE LINK: A piece of wire in a wiring harness that performs the same job as a fuse. If overloaded, the fusible link will melt and interrupt the circuit.

FWD: Front wheel drive.

GAWR: (Gross axle weight rating) the total maximum weight an axle is designed to carry.

GCW: (Gross combined weight) total combined weight of a tow vehicle and trailer.

GARAGE SHIFT: initial engagement feel of transmission, neutral to reverse or neutral to a forward drive.

GARAGE SHIFT FEEL: A quick check of the engagement quality and responsiveness of reverse and forward gears. This test is done with the vehicle stationary.

GEAR: A toothed mechanical device that acts as a rotating lever to transmit power or turning effort from one shaft to another. (See gear ratio.)

GEAR RATIO: A ratio expressing the number of turns a smaller gear will make to turn a larger gear through one revolution. The ratio is found by dividing the number of teeth on the smaller gear into the number of teeth on the larger gear.

GEARBOX: Transmission

GEAR REDUCTION: Torque is multiplied and speed decreased by the factor of the gear ratio. For example, a 3:1 gear ratio changes an input torque of 180 ft. lbs. and an input speed of 2700 rpm to 540 Ft. lbs. and 900 rpm, respectively. (No account is taken of frictional losses, which are always present.)

GEARTRAIN: A succession of intermeshing gears that form an assembly and provide for one or more torque changes as the power input is transmitted to the power output.

GEL COAT: A thin coat of plastic resin covering fiberglass body panels.

GENERATOR: A device which produces direct current (DC) necessary to charge the battery.

GOVERNOR: A device that senses vehicle speed and generates a hydraulic oil pressure. As vehicle speed increases, governor oil pressure rises.

GROUND CIRCUIT: (See circuit, ground.)

GROUND SIDE SWITCHING: The electrical/electronic circuit control switch is located after the circuit load.

GVWR: (Gross vehicle weight rating) total maximum weight a vehicle is designed to carry including the weight of the vehicle, passengers, equipment, gas, oil, etc.

HALOGEN: A special type of lamp known for its quality of brilliant white light. Originally used for fog lights and driving lights.

HARD CODES: DTCs that are present at the time of testing; also called continuous or current codes.

HARSH(ROUGH): An apply of a clutch or band that is more noticeable than a firm one; considered undesirable at any throttle position.

HEADER TANK: An expansion tank for the radiator coolant. It can be located remotely or built into the radiator.

HEAT RANGE: A term used to describe the ability of a spark plug to carry away heat. Plugs with longer nosed insulators take longer to carry heat off effectively.

HEAT RISER: A flapper in the exhaust manifold that is closed when the engine is cold, causing hot exhaust gases to heat the intake manifold providing better cold engine operation. A thermostatic spring opens the flapper when the engine warms up.

HEAVY THROTTLE: Approximately three-fourths of accelerator pedal travel.

HEMI: A name given an engine using hemispherical combustion chambers.

HERTZ (HZ): The international unit of frequency equal to one cycle per second (10,000 Hertz equals 10,000 cycles per second).

HIGH-IMPEDANCE DVOM (DIGITAL VOLT-OHMMETER): This styled device provides a built-in resistance value and is capable of limiting circuit current flow to safe milliamp levels.

HIGH RESISTANCE: Often refers to a circuit where there is an excessive amount of opposition to normal current flow.

HORSEPOWER: A measurement of the amount of work; one horsepower is the amount of work necessary to lift 33,000 lbs. one foot in one minute. Brake horsepower (bhp) is the horsepower delivered by an engine on a dynamometer. Net horsepower is the power remaining (measured at the flywheel of the engine) that can be used to turn the wheels after power is consumed through friction and running the engine accessories (water pump, alternator, air pump, fan etc.)

HOT CIRCUIT: (See circuit, hot; hot lead.)

HOT LEAD: A wire or conductor in the power side of the circuit. (See circuit, hot.)

HOT SIDE SWITCHING: The electrical/electronic circuit control switch is located before the circuit load.

HUB: The center part of a wheel or gear.

HUNTING (BUSYNESS): Repeating quick series of up-shifts and downshifts that causes noticeable change in engine rpm, for example, as in a 4-3-4 shift pattern.

HYDRAULICS: The use of liquid under pressure to transfer force of motion.

HYDROCARBON (HC): Any chemical compound made up of hydrogen and carbon. A major pollutant formed by the engine as a by-product of combustion.

HYDRODYNAMIC DRIVE UNITS: Devices that transmit power solely by the action of a kinetic fluid flow in a closed recirculating path. An impeller energizes the fluid and discharges the high-speed jet stream into the turbine for power output.

HYDROMETER: An instrument used to measure the specific gravity of a solution.

HYDROPLANING: A phenomenon of driving when water builds up under the tire tread, causing it to lose contact with the road. Slowing down will usually restore normal tire contact with the road.

HYPOID GEARSET: The drive pinion gear may be placed below or above the centerline of the driven gear; often used as a final drive gearset.

IDLE MIXTURE: The mixture of air and fuel (usually about 14:1) being fed to the cylinders. The idle mixture screw(s) are sometimes adjusted as part of a tune-up.

IDLER ARM: Component of the steering linkage which is a geometric duplicate of the steering gear arm. It supports the right side of the center steering link.

IMPELLER: Often called a pump, the impeller is the power input (drive) member of a hydrodynamic drive. As part of the torque converter cover, it acts as a centrifugal pump and puts the fluid in motion.

INCH POUND (inch lbs.; sometimes in. lb. or in. lbs.): One twelfth of a foot pound.

INDUCTANCE: The force that produces voltage when a conductor is passed through a magnetic field.

INDUCTION: A means of transferring electrical energy in the form of a magnetic field. Principle used in the ignition coil to increase voltage.

INITIAL FEEL: A distinct firmer feel at start of shift when compared with feel at finish of shift.

INJECTOR: A device which receives metered fuel under relatively low pressure and is activated to inject the fuel into the engine under relatively high pressure at a predetermined time.

INPUT: In an automatic transmission, the source of power from the engine is absorbed by the torque converter, which provides the power input into the transmission. The turbine drives the input(turbine)shaft.

INPUT SHAFT: The shaft to which torque is applied, usually carrying the driving gear or gears.

INTAKE MANIFOLD: A casting of passages or pipes used to conduct air or a fuel/air mixture to the cylinders.

INTERNAL GEAR: The ring-like outer gear of a planetary gearset with the gear teeth cut on the inside of the ring to provide a mesh with the planet pinions.

ISOLATION (CLAMPING) DIODES: Diodes positioned in a circuit to prevent self-induction from damaging electronic components.

IX ROTARY GEAR PUMP: Contains two rotating members, one shaped with internal gear teeth and the other with external gear teeth. As the gears separate, the fluid fills the gaps between gear teeth, is pulled across a crescent-shaped divider, and then is forced to flow through the outlet as the gears mesh.

IX ROTARY LOBE PUMP: Sometimes referred to as a gerotor type pump. Two rotating members, one shaped with internal lobes and the other with external lobes, separate and then mesh to cause fluid to flow.

JOURNAL: The bearing surface within which a shaft operates.

JUMPER CABLES: Two heavy duty wires with large alligator clips used to provide power from a charged battery to a discharged battery mounted in a vehicle.

JUMPSTART: Utilizing the sufficiently charged battery of one vehicle to start the engine of another vehicle with a discharged battery by the use of jumper cables.

KEY: A small block usually fitted in a notch between a shaft and a hub to prevent slippage of the two parts.

KICKDOWN: Detent downshift system; either linkage, cable, or electrically controlled.

KILO: A prefix used in the metric system to indicate one thousand.

KNOCK: Noise which results from the spontaneous ignition of a portion of the air-fuel mixture in the engine cylinder caused by overly advanced ignition timing or use of incorrectly low octane fuel for that engine.

KNOCK SENSOR: An input device that responds to spark knock, caused by over advanced ignition timing.

LABOR TIME: A specific amount of time required to perform a certain repair or diagnostic service as defined by a vehicle or after-market manufacturer.

LACQUER: A quick-drying automotive paint.

LATE: Shift that occurs when engine is at higher than normal rpm for given amount of throttle.

LIGHT-EMITTING DIODE (LED): A semiconductor diode that emits light as electrical current flows through it; used in some electronic display devices to emit a red or other color light.

LIGHT THROTTLE: Approximately one-fourth of accelerator pedal travel.

LIMITED SLIP: A type of differential which transfers driving force to the wheel with the best traction.

LIMP-IN MODE: Electrical shutdown of the transmission/transaxle output solenoids, allowing only forward and reverse gears that are hydraulically energized by the manual valve. This permits the vehicle to be driven to a service facility for repair.

LIP SEAL: Molded synthetic rubber seal designed with an outer sealing edge (lip) that points into the fluid containing area to be sealed. This type of seal is used where rotational and axial forces are present.

LITHIUM-BASE GREASE: Chassis and wheel bearing grease using lithium as a base. Not compatible with sodium-base grease.

LOAD DEVICE: A circuit's resistance that converts the electrical energy into light, sound, heat, or mechanical movement.

LOAD RANGE: Indicates the number of plies at which a tire is rated. Load range B equals four-ply rating; C equals six-ply rating; and, D equals an eight-ply rating.

LOAD TORQUE: The amount of output torque needed from the transmission/transaxle to overcome the vehicle load.

LOCKING HUBS: Accessories used on part-time four-wheel drive systems that allow the front wheels to be disengaged from the drive train when four-wheel drive is not being used. When four-wheel drive is desired, the hubs are engaged, locking the wheels to the drive train.

LOCKUP CONVERTER: A torque converter that operates hydraulically and mechanically. When an internal apply plate (lockup plate) clamps to the torque converter cover, hydraulic slippage is eliminated.

LOCK RING: See Circlip or Snapring

MAGNET: Any body with the property of attracting iron or steel.

MAGNETIC FIELD: The area surrounding the poles of a magnet that is affected by its attraction or repulsion forces.

MAIN LINE PRESSURE: Often called control pressure or line pressure, it refers to the pressure of the oil leaving the pump and is controlled by the pressure regulator valve.

MALFUNCTION INDICATOR LAMP (MIL): Previously known as a check engine light, the dash-mounted MIL illuminates and signals the driver that an emission or driveability problem with the powertrain has been detected by the ECM/PCM. When this occurs, at least one diagnostic trouble code (DTC) has been stored into the control module memory.

MANIFOLD ABSOLUTE PRESSURE (MAP) SENSOR: Reads the amount of air pressure (vacuum) in the engine's intake manifold system; its signal is used to analyze engine load conditions.

MANIFOLD VACUUM: Low pressure in an engine intake manifold formed just below the throttle plates. Manifold vacuum is highest at idle and drops under acceleration.

MANIFOLD: A casting of passages or set of pipes which connect the cylinders to an inlet or outlet source.

MANUAL LEVER POSITION SWITCH (MLPS): A mechanical switching unit that is typically mounted externally to the transmission/transaxle to inform the PCM/ECM which gear range the driver has selected.

MANUAL VALVE: Located inside the transmission/transaxle, it is directly connected to the driver's shift lever. The position of the manual valve determines which hydraulic circuits will be charged with oil pressure and the operating mode of the transmission.

MANUAL VALVE LEVER POSITION SENSOR (MVLPS): The input from this device tells the TCM what gear range was selected.

MASS AIR FLOW (MAF) SENSOR: Measures the airflow into the engine.

MASTER CYLINDER: The primary fluid pressurizing device in a hydraulic system. In automotive use, it is found in brake and hydraulic clutch systems and is pedal activated, either directly or, in a power brake system, through the power booster.

MacPherson STRUT: A suspension component combining a shock absorber and spring in one unit.

MEDIUM THROTTLE: Approximately one-half of accelerator pedal travel.

MEGA: A metric prefix indicating one million.

MEMBER: An independent component of a hydrodynamic unit such as an impeller, a stator, or a turbine. It may have one or more elements.

MERCON: A fluid developed by Ford Motor Company in 1988. It contains a friction modifier and closely resembles operating characteristics of Dexron.

METAL SEALING RINGS: Made from cast iron or aluminum, their primary application is with dynamic components involving pressure sealing circuits of rotating members. These rings are designed with either butt or hook lock end joints.

METER (ANALOG): A linear-style meter representing data as lengths; a needle-style instrument interfacing with logical numerical increments. This style of electrical meter uses relatively low impedance internal resistance and cannot be used for testing electronic circuitry.

METER (DIGITAL): Uses numbers as a direct readout to show values. Most meters of this style use high impedance internal resistance and must be used for testing low current electronic circuitry.

MICRO: A metric prefix indicating one-millionth (0.000001).

MILLI: A metric prefix indicating one-thousandth (0.001).

MINIMUM THROTTLE: The least amount of throttle opening required for upshift; normally close to zero throttle.

MISFIRE: Condition occurring when the fuel mixture in a cylinder fails to ignite, causing the engine to run roughly.

MODULE: Electronic control unit, amplifier or igniter of solid state or integrated design which controls the current flow in the ignition primary circuit based on input from the pick-up coil. When the module opens the primary circuit, high secondary voltage is induced in the coil.

MODULATED: In an electronic-hydraulic converter clutch system (or shift valve system), the term modulated refers to the pulsing of a solenoid, at a variable rate. This action controls the buildup of oil pressure in the hydraulic circuit to allow a controlled amount of clutch slippage.

MODULATED CONVERTER CLUTCH CONTROL (MCCC): A pulse width duty cycle valve that controls the converter lockup apply pressure and maximizes smoother transitions between lock and unlock conditions.

MODULATOR PRESSURE (THROTTLE PRESSURE): A hydraulic signal oil pressure relating to the amount of engine load, based on either the amount of throttle plate opening or engine vacuum.

MODULATOR VALVE: A regulator valve that is controlled by engine vacuum, providing a hydraulic pressure that varies in relation to engine torque. The hydraulic torque signal functions to delay the shift pattern and provide a line pressure boost. (See throttle valve.)

MOTOR: An electromagnetic device used to convert electrical energy into mechanical energy.

MULTIPLE-DISC CLUTCH: A grouping of steel and friction lined plates that, when compressed together by hydraulic pressure acting upon a piston, lock or unlock a planetary member.

MULTI-WEIGHT: Type of oil that provides adequate lubrication at both high and low temperatures.

needed to move one amp through a resistance of one ohm.

MUSHY: Same as soft; slow and drawn out clutch apply with very little shift feel.

MUTUAL INDUCTION: The generation of current from one wire circuit to another by movement of the magnetic field surrounding a current-carrying circuit as its ampere flow increases or decreases.

NEEDLE BEARING: A bearing which consists of a number (usually a large number) of long, thin rollers.

NITROGEN OXIDE (NOx): One of the three basic pollutants found in the exhaust emission of an internal combustion engine. The amount of NOx usually varies in an inverse proportion to the amount of HC and CO.

NONPOSITIVE SEALING: A sealing method that allows some minor leakage, which normally assists in lubrication.

O2 SENSOR: Located in the engine's exhaust system, it is an input device to the ECM/PCM for managing the fuel delivery and ignition system. A scanner can be used to observe the fluctuating voltage readings produced by an 02 sensor as the oxygen content of the exhaust is analyzed.

O-RING SEAL: Molded synthetic rubber seal designed with a circular cross-section. This type of seal is used primarily in static applications.

OBD II (ON-BOARD DIAGNOSTICS, SECOND GENERATION): Refers to the federal law mandating tighter control of 1996 and newer vehicle emissions, active monitoring of related devices, and standardization of terminology, data link connectors, and other technician concerns.

OCTANE RATING: A number, indicating the quality of gasoline based on its ability to resist knock. The higher the number, the better the quality. Higher compression engines require higher octane gas.

OEM: Original Equipment Manufactured. OEM equipment is that furnished standard by the manufacturer.

OFFSET: The distance between the vertical center of the wheel and the mounting surface at the lugs. Offset is positive if the center is outside the lug circle; negative offset puts the center line inside the lug circle.

OHM'S LAW: A law of electricity that states the relationship between voltage, current, and resistance. Volts = amperes x ohms

OHM: The unit used to measure the resistance of conductor-to-electrical flow. One ohm is the amount of resistance that limits current flow to one ampere in a circuit with one volt of pressure.

OHMMETER: An instrument used for measuring the resistance, in ohms, in an electrical circuit.

ONE-WAY CLUTCH: A mechanical clutch of roller or sprag design that resists torque or transmits power in one direction only. It is used to either hold or drive a planetary member.

ONE-WAY ROLLER CLUTCH: A mechanical device that transmits or holds torque in one direction only.

OPEN CIRCUIT: A break or lack of contact in an electrical circuit, either intentional (switch) or unintentional (bad connection or broken wire).

ORIFICE: Located in hydraulic oil circuits, it acts as a restriction. It slows down fluid flow to either create back pressure or delay pressure buildup downstream.

OSCILLOSCOPE: A piece of test equipment that shows electric impulses as a pattern on a screen. Engine performance can be analyzed by interpreting these patterns.

OUTPUT SHAFT: The shaft which transmits torque from a device, such as a transmission.

OUTPUT SPEED SENSOR (OSS): Identifies transmission/transaxle output shaft speed for shift timing and may be used to calculate TCC slip; often functions as the VSS (vehicle speed sensor).

OVERDRIVE: (1.) A device attached to or incorporated in a transmission/transaxle that allows the engine to turn less than one full revolution for every complete revolution of the wheels. The net effect is to reduce engine rpm, thereby using less fuel. A typical overdrive gear ratio would be .87:1, instead of the normal 1:1 in high gear. (2.) A gear assembly which produces more shaft revolutions than that transmitted to it.

OVERDRIVE PLANETARY GEARSET: A single planetary gearset designed to provide a direct drive and overdrive ratio. When coupled to a three-speed transmission/transaxle configuration, a four-speed/overdrive unit is present.

OVERHEAD CAMSHAFT (OHC): An engine configuration in which the camshaft is mounted on top of the cylinder head and operates the valve either directly or by means of rocker arms.

OVERHEAD VALVE (OHV): An engine configuration in which all of the valves are located in the cylinder head and the camshaft is located in the cylinder block. The camshaft operates the valves via lifters and pushrods.

OVERRUNCLUTCH: Another name for a one-way mechanical clutch. Applies to both roller and sprag designs.

OVERSTEER: The tendency of some vehicles, when steering into a turn, to over-respond or steer more than required, which could result in excessive slip of the rear wheels. Opposite of under-steer.

OXIDATION STABILIZERS: Absorb and dissipate heat. Automatic transmission fluid has high resistance to varnish and sludge buildup that occurs from excessive heat that is generated primarily in the torque converter. Local temperatures as high as 6000F (3150C) can occur at the clutch plates during engagement, and this heat must be absorbed and dissipated. If the fluid cannot withstand the heat, it burns or oxidizes, resulting in an almost immediate destruction of friction materials, clogged filter screen and hydraulic passages, and sticky valves.

OXIDES OF NITROGEN: See nitrogen oxide (NOx).

OXYGEN SENSOR: Used with a feedback system to sense the presence of oxygen in the exhaust gas and signal the computer which can use the voltage signal to determine engine operating efficiency and adjust the air/fuel ratio.

PARALLEL CIRCUIT: (See circuit, parallel.)

PARTS WASHER: A basin or tub, usually with a built-in pump mechanism and hose used for circulating chemical solvent for the purpose of cleaning greasy, oily and dirty components.

PART-TIME FOUR WHEEL DRIVE: A system that is normally in the two wheel drive mode and only runs in four-wheel drive when the system is manually engaged because more traction is desired. Two or four wheel drive is normally selected by a lever to engage the front axle, but if locking hubs are used, these must also be manually engaged in the Lock position. Otherwise, the front axle will not drive the front wheels.

PASSIVE RESTRAINT: Safety systems such as air bags or automatic seat belts which operate with no action required on the part of the driver or passenger. Mandated by Federal regulations on all vehicles sold in the U.S. after 1990.

PAYLOAD: The weight the vehicle is capable of carrying in addition to its own weight. Payload includes weight of the driver, passengers and cargo, but not coolant, fuel, lubricant, spare tire, etc.

PCM: Powertrain control module.

PCV VALVE: A valve usually located in the rocker cover that vents crankcase vapors back into the engine to be reburned.

PERCOLATION: A condition in which the fuel actually "boils," due to excessive heat. Percolation prevents proper atomization of the fuel causing rough running.

PICK-UP COIL: The coil in which voltage is induced in an electronic ignition.

PING: A metallic rattling sound produced by the engine during acceleration. It is usually due to incorrect ignition timing or a poor grade of gasoline.

PINION: The smaller of two gears. The rear axle pinion drives the ring gear which transmits motion to the axle shafts.

PINION GEAR: The smallest gear in a drive gear assembly.

PISTON: A disc or cup that fits in a cylinder bore and is free to move. In hydraulics, it provides the means of converting hydraulic pressure into a usable force. Examples of piston applications are found in servo, clutch, and accumulator units.

PISTON RING: An open-ended ring which fits into a groove on the outer diameter of the piston. Its chief function is to form a seal between the piston and cylinder wall. Most automotive pistons have three rings: two for compression sealing; one for oil sealing.

PITMAN ARM: A lever which transmits steering force from the steering gear to the steering linkage.

PLANET CARRIER: A basic member of a planetary gear assembly that carries the pinion gears.

PLANET PINIONS: Gears housed in a planet carrier that are in constant mesh with the sun gear and internal gear. Because they have their own independent rotating centers, the pinions are capable of rotating around the sun gear or the inside of the internal gear.

PLANETARY GEAR RATIO: The reduction or overdrive ratio developed by a planetary gearset.

PLANETARY GEARSET: In its simplest form, it is made up of a basic assembly group containing a sun gear, internal gear, and planet carrier. The gears are always in constant mesh and offer a wide range of gear ratio possibilities.

PLANETARY GEARSET (COMPOUND): Two planetary gearsets combined together.

PLANETARY GEARSET (SIMPLE): An assembly of gears in constant mesh consisting of a sun gear, several pinion gears mounted in a carrier, and a ring gear. It provides gear ratio and direction changes, in addition to a direct drive and a neutral.

PLY RATING: A. rating given a tire which indicates strength (but not necessarily actual plies). A two-ply/four-ply rating has only two plies, but the strength of a four-ply tire.

POLARITY: Indication (positive or negative) of the two poles of a battery.

PORT: An opening for fluid intake or exhaust.

POSITIVE SEALING: A sealing method that completely prevents leakage.

POTENTIAL: Electrical force measured in volts; sometimes used interchangeably with voltage.

POWER: The ability to do work per unit of time, as expressed in horsepower; one horsepower equals 33,000 ft. lbs. of work per minute, or 550 ft. lbs. of work per second.

POWER FLOW: The systematic flow or transmission of power through the gears, from the input shaft to the output shaft.

POWER-TO-WEIGHT RATIO: Ratio of horsepower to weight of car.

POWERTRAIN: See Drivetrain.

POWERTRAIN CONTROL MODULE (PCM): Current designation for the engine control module (ECM). In many cases, late model vehicle control units manage the engine as well as the transmission. In other settings, the PCM controls the engine and is interfaced with a TCM to control transmission functions.

Ppm: Parts per million; unit used to measure exhaust emissions.

PREIGNITION: Early ignition of fuel in the cylinder, sometimes due to glowing carbon deposits in the combustion chamber. Preignition can be damaging since combustion takes place prematurely.

PRELOAD: A predetermined load placed on a bearing during assembly or by adjustment.

PRESS FIT: The mating of two parts under pressure, due to the inner diameter of one being smaller than the outer diameter of the other, or vice versa; an interference fit.

PRESSURE: The amount of force exerted upon a surface area.

PRESSURE CONTROL SOLENOID (PCS): An output device that provides a boost oil pressure to the mainline regulator valve to control line pressure. Its operation is determined by the amount of current sent from the PCM.

PRESSURE GAUGE: An instrument used for measuring the fluid pressure in a hydraulic circuit.

PRESSURE REGULATOR VALVE: In automatic transmissions, its purpose is to regulate the pressure of the pump output and supply the basic fluid pressure necessary to operate the transmission. The regulated fluid pressure may be referred to as mainline pressure, line pressure, or control pressure.

PRESSURE SWITCH ASSEMBLY (PSA): Mounted inside the transmission, it is a grouping of oil pressure switches that inputs to the PCM when certain hydraulic passages are charged with oil pressure.

PRESSURE PLATE: A spring-loaded plate (part of the clutch) that transmits power to the driven (friction) plate when the clutch is engaged.

PRIMARY CIRCUIT: The low voltage side of the ignition system which consists of the ignition switch, ballast resistor or resistance wire, bypass, coil, electronic control unit and pick-up coil as well as the connecting wires and harnesses.

PROFILE: Term used for tire measurement (tire series), which is the ratio of tire height to tread width.

PROM (PROGRAMMABLE READ-ONLY MEMORY): The heart of the computer that compares input data and makes the engineered program or strategy decisions about when to trigger the appropriate output based on stored computer instructions.

PULSE GENERATOR: A two-wire pickup sensor used to produce a fluctuating electrical signal. This changing signal is read by the controller to determine the speed of the object and can be used to measure transmission/transaxle input speed, output speed, and vehicle speed.

PSI: Pounds per square inch; a measurement of pressure.

PULSE WIDTH DUTY CYCLE SOLENOID (PULSE WIDTH MODULATED SOLENOID): A computer-controlled solenoid that turns on and off at a variable rate producing a modulated oil pressure; often referred to as a pulse width modulated (PWM) solenoid. Employed in many electronic automatic transmissions and transaxles, these solenoids are used to manage shift control and converter clutch hydraulic circuits.

PUSHROD: A steel rod between the hydraulic valve lifter and the valve rocker arm in overhead valve (OHV) engines.

PUMP: A mechanical device designed to create fluid flow and pressure buildup in a hydraulic system.

QUARTER PANEL: General term used to refer to a rear fender. Quarter panel is the area from the rear door opening to the tail light area and from rear wheel well to the base of the trunk and roof-line.

RACE: The surface on the inner or outer ring of a bearing on which the balls, needles or rollers move.

RACK AND PINION: A type of automotive steering system using a pinion gear attached to the end of the steering shaft. The pinion meshes with a long rack attached to the steering linkage.

RADIAL TIRE: Tire design which uses body cords running at right angles to the center line of the tire. Two or more belts are used to give tread strength. Radials can be identified by their characteristic sidewall bulge.

RADIATOR: Part of the cooling system for a water-cooled engine, mounted in the front of the vehicle and connected to the engine with rubber hoses. Through the radiator, excess combustion heat is dissipated into the atmosphere through forced convection using a water and glycol based mixture that circulates through, and cools, the engine.

RANGE REFERENCE AND CLUTCH/BAND APPLY CHART: A guide that shows the application of clutches and bands for each gear, within the selector range positions. These charts are extremely useful for understanding how the unit operates and for diagnosing malfunctions.

RAVIGNEAUX GEARSET: A compound planetary gearset that features matched dual planetary pinions (sets of two) mounted in a single planet carrier. Two sun gears and one ring mesh with the carrier pinions.

REACTION MEMBER: The stationary planetary member, in a planetary gearset, that is grounded to the transmission/transaxle case through the use of friction and wedging devices known as bands, disc clutches, and one-way clutches.

REACTION PRESSURE: The fluid pressure that moves a spool valve against an opposing force or forces; the area on which the opposing force acts. The opposing force can be a spring or a combination of spring force and auxiliary hydraulic force.

REACTOR, TORQUE CONVERTER: The reaction member of a fluid torque converter, more commonly called a stator. (See stator.)

REAR MAIN OIL SEAL: A synthetic or rope-type seal that prevents oil from leaking out of the engine past the rear main crankshaft bearing.

RECIRCULATING BALL: Type of steering system in which recirculating steel balls occupy the area between the nut and worm wheel, causing a reduction in friction.

RECTIFIER: A device (used primarily in alternators) that permits electrical current to flow in one direction only.

REDUCTION: (See gear reduction.)

REGULATOR VALVE: A valve that changes the pressure of the oil in a hydraulic circuit as the oil passes through the valve by bleeding off (or exhausting) some of the volume of oil supplied to the valve.

REFRIGERANT 12 (R-12) or 134 (R-134): The generic name of the refrigerant used in automotive air conditioning systems.

REGULATOR: A device which maintains the amperage and/or voltage levels of a circuit at predetermined values.

RELAY: A switch which automatically opens and/or closes a circuit.

RELAY VALVE: A valve that directs flow and pressure. Relay valves simply connect or disconnect interrelated passages without restricting the fluid flow or changing the pressure.

RELIEF VALVE: A spring-loaded, pressure-operated valve that limits oil pressure buildup in a hydraulic circuit to a predetermined maximum value.

RELUCTOR: A wheel that rotates inside the distributor and triggers the release of voltage in an electronic ignition.

RESERVOIR: The storage area for fluid in a hydraulic system; often called a sump.

RESIN: A liquid plastic used in body work.

RESIDUAL MAGNETISM: The magnetic strength stored in a material after a magnetizing field has been removed.

RESISTANCE: The opposition to the flow of current through a circuit or electrical device, and is measured in ohms. Resistance is equal to the voltage divided by the amperage.

RESISTOR SPARK PLUG: A spark plug using a resistor to shorten the spark duration. This suppresses radio interference and lengthens plug life.

RESISTOR: A device, usually made of wire, which offers a preset amount of resistance in an electrical circuit.

RESULTANT FORCE: The single effective directional thrust of the fluid force on the turbine produced by the vortex and rotary forces acting in different planes.

RETARD: Set the ignition timing so that spark occurs later (fewer degrees before TDC).

RHEOSTAT: A device for regulating a current by means of a variable resistance.

RING GEAR: The name given to a ring-shaped gear attached to a differential case, or affixed to a flywheel or as part of a planetary gear set.

ROADLOAD: grade.

ROCKER ARM: A lever which rotates around a shaft pushing down (opening) the valve with an end when the other end is pushed up by the pushrod. Spring pressure will later close the valve.

ROCKER PANEL: The body panel below the doors between the wheel opening.

ROLLER BEARING: A bearing made up of hardened inner and outer races between which hardened steel rollers move.

ROLLER CLUTCH: A type of one-way clutch design using rollers and springs mounted within an inner and outer cam race assembly.

ROTARY FLOW: The path of the fluid trapped between the blades of the members as they revolve with the rotation of the torque converter cover (rotational inertia).

ROTOR: (1.) The disc-shaped part of a disc brake assembly, upon which the brake pads bear; also called, brake disc. (2.) The device mounted atop the distributor shaft, which passes current to the distributor cap tower contacts.

ROTARY ENGINE: See Wankel engine.

RPM: Revolutions per minute (usually indicates engine speed).

RTV: A gasket making compound that cures as it is exposed to the atmosphere. It is used between surfaces that are not perfectly machined to one another, leaving a slight gap that the RTV fills and in which it hardens. The letters RTV represent room temperature vulcanizing.

RUN-ON: Condition when the engine continues to run, even when the key is turned off. See dieseling.

SEALED BEAM: A automotive headlight. The lens, reflector and filament from a single unit.

SEATBELT INTERLOCK: A system whereby the car cannot be started unless the seatbelt is buckled.

SECONDARY CIRCUIT: The high voltage side of the ignition system, usually above 20,000 volts. The secondary includes the ignition coil, coil wire, distributor cap and rotor, spark plug wires and spark plugs.

SELF-INDUCTION: The generation of voltage in a current-carrying wire by changing the amount of current flowing within that wire.

SEMI-CONDUCTOR: A material (silicon or germanium) that is neither a good conductor nor an insulator; used in diodes and transistors.

SEMI-FLOATING AXLE: In this design, a wheel is attached to the axle shaft, which takes both drive and cornering loads. Almost all solid axle passenger cars and light trucks use this design.

SENDING UNIT: A mechanical, electrical, hydraulic or electromagnetic device which transmits information to a gauge.

SENSOR: Any device designed to measure engine operating conditions or ambient pressures and temperatures. Usually electronic in nature and designed to send a voltage signal to an on-board computer, some sensors may operate as a simple on/off switch or they may provide a variable voltage signal (like a potentiometer) as conditions or measured parameters change.

SERIES CIRCUIT: (See circuit, series.)

SERPENTINE BELT: An accessory drive belt, with small multiple v-ribs, routed around most or all of the engine-powered accessories such as the alternator and power steering pump. Usually both the front and the back side of the belt comes into contact with various pulleys.

SERVO: In an automatic transmission, it is a piston in a cylinder assembly that converts hydraulic pressure into mechanical force and movement; used for the application of the bands and clutches.

SHIFT BUSYNESS: When referring to a torque converter clutch, it is the frequent apply and release of the clutch plate due to uncommon driving conditions.

SHIFT VALVE: Classified as a relay valve, it triggers the automatic shift in response to a governor and a throttle signal by directing fluid to the appropriate band and clutch apply combination to cause the shift to occur.

SHIM: Spacers of precise, predetermined thickness used between parts to establish a proper working relationship.

SHIMMY: Vibration (sometimes violent) in the front end caused by misaligned front end, out of balance tires or worn suspension components.

SHORT CIRCUIT: An electrical malfunction where current takes the path of least resistance to ground (usually through damaged insulation). Current flow is excessive from low resistance resulting in a blown fuse.

SHUDDER: Repeated jerking or stick-slip sensation, similar to chuggle but more severe and rapid in nature, that may be most noticeable during certain ranges of vehicle speed; also used to define condition after converter clutch engagement.

SIMPSON GEARSET: A compound planetary gear train that integrates two simple planetary gearsets referred to as the front planetary and the rear planetary.

SINGLE OVERHEAD CAMSHAFT: See overhead camshaft.

SKIDPLATE: A metal plate attached to the underside of the body to protect the fuel tank, transfer case or other vulnerable parts from damage.

SLAVE CYLINDER: In automotive use, a device in the hydraulic clutch system which is activated by hydraulic force, disengaging the clutch.

SLIPPING: Noticeable increase in engine rpm without vehicle speed increase; usually occurs during or after initial clutch or band engagement.

SLUDGE: Thick, black deposits in engine formed from dirt, oil, water, etc. It is usually formed in engines when oil changes are neglected.

SNAP RING: A circular retaining clip used inside or outside a shaft or part to secure a shaft, such as a floating wrist pin.

SOFT: Slow, almost unnoticeable clutch apply with very little shift feel.

SOFTCODES: DTCs that have been set into the PCM memory but are not present at the time of testing; often referred to as history or intermittent codes.

SOHC: Single overhead camshaft.

SOLENOID: An electrically operated, magnetic switching device.

SPALLING: A wear pattern identified by metal chips flaking off the hardened surface. This condition is caused by foreign particles, overloading situations, and/or normal wear.

SPARK PLUG: A device screwed into the combustion chamber of a spark ignition engine. The basic construction is a conductive core inside of a ceramic insulator, mounted in an outer conductive base. An electrical charge from the spark plug wire travels along the conductive core and jumps a preset air gap to a grounding point or points at the end of the conductive base. The resultant spark ignites the fuel/air mixture in the combustion chamber.

SPECIFIC GRAVITY (BATTERY): The relative weight of liquid (battery electrolyte) as compared to the weight of an equal volume of water.

SPLINES: Ridges machined or cast onto the outer diameter of a shaft or inner diameter of a bore to enable parts to mate without rotation.

SPLIT TORQUE DRIVE: In a torque converter, it refers to parallel paths of torque transmission, one of which is mechanical and the other hydraulic.

SPONGY PEDAL: A soft or spongy feeling when the brake pedal is depressed. It is usually due to air in the brake lines.

SPOOLVALVE: A precision-machined, cylindrically shaped valve made up of lands and grooves. Depending on its position in the valve bore, various interconnecting hydraulic circuit passages are either opened or closed.

SPRAG CLUTCH: A type of one-way clutch design using cams or contoured-shaped sprags between inner and outer races. (See one-way clutch.)

SPRUNG WEIGHT: The weight of a car supported by the springs.

SQUARE-CUT SEAL: Molded synthetic rubber seal designed with a square- or rectangular-shaped cross-section. This type of seal is used for both dynamic and static applications.

SRS: Supplemental restraint system

STABILIZER (SWAY) BAR: A bar linking both sides of the suspension. It resists sway on turns by taking some of added load from one wheel and putting it on the other.

STAGE: The number of turbine sets separated by a stator. A turbine set may be made up of one or more turbine members. A three-element converter is classified as a single stage.

STALL: In fluid drive transmission/transaxle applications, stall refers to engine rpm with the transmission/transaxle engaged and the vehicle stationary; throttle valve can be in any position between closed and wide open.

STALL SPEED: In fluid drive transmission/transaxle applications, stall speed refers to the maximum engine rpm with the transmission/transaxle engaged and vehicle stationary, when the throttle valve is wide open. (See stall; stall test.)

STALL TEST: A procedure recommended by many manufacturers to help determine the integrity of an engine, the torque converter stator, and certain clutch and band combinations. With the shift lever in each of the forward and reverse positions and with the brakes firmly applied, the accelerator pedal is momentarily pressed to the wide open throttle (WOT) position. The engine rpm reading at full throttle can provide clues for diagnosing the condition of the items listed above.

STALL TORQUE: The maximum design or engineered torque ratio of a fluid torque converter, produced under stall speed conditions. (See stall speed.)

STARTER: A high-torque electric motor used for the purpose of starting the engine, typically through a high ratio geared drive connected to the flywheel ring gear.

STATIC: A sealing application in which the parts being sealed do not move in relation to each other.

STATOR (REACTOR): The reaction member of a fluid torque converter that changes the direction of the fluid as it leaves the turbine to enter the impeller vanes. During the torque multiplication phase, this action assists the impeller's rotary force and results in an increase in torque.

STEERING GEOMETRY: Combination of various angles of suspension components (caster, camber, toe-in); roughly equivalent to front end alignment.

STRAIGHT WEIGHT: Term designating motor oil as suitable for use within a narrow range of temperatures. Outside the narrow temperature range its flow characteristics will not adequately lubricate.

STROKE: The distance the piston travels from bottom dead center to top dead center.

SUBSTITUTION: Replacing one part suspected of a defect with a like part of known quality.

SUMP: The storage vessel or reservoir that provides a ready source of fluid to the pump. In an automatic transmission, the sump is the oil pan. All fluid eventually returns to the sump for recycling into the hydraulic system.

SUN GEAR: In a planetary gearset, it is the center gear that meshes with a cluster of planet pinions.

SUPERCHARGER: An air pump driven mechanically by the engine through belts, chains, shafts or gears from the crankshaft. Two general types of supercharger are the positive displacement and centrifugal type, which pump air in direct relationship to the speed of the engine.

SUPPLEMENTAL RESTRAINT SYSTEM: See air bag.

SURGE: Repeating engine-related feeling of acceleration and deceleration that is less intense than chuggle.

SWITCH: A device used to open, close, or redirect the current in an electrical circuit.

SYNCHROMESH: A manual transmission/transaxle that is equipped with devices (synchronizers) that match the gear speeds so that the transmission/transaxle can be downshifted without clashing gears.

SYNTHETIC OIL: Non-petroleum based oil.

TACHOMETER: A device used to measure the rotary speed of an engine, shaft, gear, etc., usually in rotations per minute.

TDC: Top dead center. The exact top of the piston's stroke.

TEFLON SEALING RINGS: Teflon is a soft, durable, plastic-like material that is resistant to heat and provides excellent sealing. These rings are designed with either scarf-cut joints or as one-piece rings. Teflon sealing rings have replaced many metal ring applications.

TERMINAL: A device attached to the end of a wire or cable to make an electrical connection.

TEST LIGHT, CIRCUIT-POWERED: Uses available circuit voltage to test circuit continuity.

TEST LIGHT, SELF-POWERED: Uses its own battery source to test circuit continuity.

THERMISTOR: A special resistor used to measure fluid temperature; it decreases its resistance with increases in temperature.

THERMOSTAT: A valve, located in the cooling system of an engine, which is closed when cold and opens gradually in response to engine heating, controlling the temperature of the coolant and rate of coolant flow.

THERMOSTATIC ELEMENT: A heat-sensitive, spring-type device that controls a drain port from the upper sump area to the lower sump. When the transaxle fluid reaches operating temperature, the port is closed and the upper sump fills, thus reducing the fluid level in the lower sump.

THROTTLE POSITION (TP) SENSOR: Reads the degree of throttle opening; its signal is used to analyze engine load conditions. The ECM/PCM decides to apply the TCC, or to disengage it for coast or load conditions that need a converter torque boost.

THROTTLE PRESSURE/MODULATOR PRESSURE: A hydraulic signal oil pressure relating to the amount of engine load, based on either the amount of throttle plate opening or engine vacuum.

THROTTLE VALVE: A regulating or balanced valve that is controlled mechanically by throttle linkage or engine vacuum. It sends a hydraulic signal to the shift valve body to control shift timing and shift quality. (See balanced valve; modulator valve.)

THROW-OUT BEARING: As the clutch pedal is depressed, the throwout bearing moves against the spring fingers of the pressure plate, forcing the pressure plate to disengage from the driven disc.

TIE ROD: A rod connecting the steering arms. Tie rods have threaded ends that are used to adjust toe-in.

TIE-UP: Condition where two opposing clutches are attempting to apply at same time, causing engine to labor with noticeable loss of engine rpm.

TIMING BELT: A square-toothed, reinforced rubber belt that is driven by the crankshaft and operates the camshaft.

TIMING CHAIN: A roller chain that is driven by the crankshaft and operates the camshaft.

TIRE ROTATION: Moving the tires from one position to another to make the tires wear evenly.

TOE-IN (OUT): A term comparing the extreme front and rear of the front tires. Closer together at the front is toe-in; farther apart at the front is toe-out.

TOP DEAD CENTER (TDC): The point at which the piston reaches the top of its travel on the compression stroke.

TORQUE: Measurement of turning or twisting force, expressed as foot-pounds or inch-pounds.

TORQUE CONVERTER: A turbine used to transmit power from a driving member to a driven member via hydraulic action, providing changes in drive ratio and torque. In automotive use, it links the driveplate at the rear of the engine to the automatic transmission.

TORQUE CONVERTER CLUTCH: The apply plate (lockup plate) assembly used for mechanical power flow through the converter.

TORQUE PHASE: Sometimes referred to as slip phase or stall phase, torque multiplication occurs when the turbine is turning at a slower speed than the impeller, and the stator is reactionary (stationary). This sequence generates a boost in output torque.

TORQUE RATING (STALL TORQUE): The maximum torque multiplication that occurs during stall conditions, with the engine at wide open throttle (WOT) and zero turbine speed.

TORQUE RATIO: An expression of the gear ratio factor on torque effect. A 3:1 gear ratio or 3:1 torque ratio increases the torque input by the ratio factor of 3. Input torque (100 ft. lbs.) x 3 = output torque (300 ft. lbs.)

TRACTION: The amount of usable tractive effort before the drive wheels slip on the road contact surface.

TORSION BAR SUSPENSION: Long rods of spring steel which take the place of springs. One end of the bar is anchored and the other arm (attached to the suspension) is free to twist. The bars' resistance to twisting causes springing action.

TRACK: Distance between the centers of the tires where they contact the ground.

TRACTION CONTROL: A control system that prevents the spinning of a vehicle's drive wheels when excess power is applied.

TRACTIVE EFFORT: The amount of force available to the drive wheels, to move the vehicle.

TRANSAXLE: A single housing containing the transmission and differential. Transaxles are usually found on front engine/front wheel drive or rear engine/rear wheel drive cars.

TRANSDUCER: A device that changes energy from one form to another. For example, a transducer in a microphone changes sound energy to electrical energy. In automotive air-conditioning controls used in automatic temperature systems, a transducer changes an electrical signal to a vacuum signal, which operates mechanical doors.

TRANSMISSION: A powertrain component designed to modify torque and speed developed by the engine; also provides direct drive, reverse, and neutral.

TRANSMISSION CONTROL MODULE (TCM): Manages transmission functions. These vary according to the manufacturer's product design but may include converter clutch operation, electronic shift scheduling, and mainline pressure.

TRANSMISSION FLUID TEMPERATURE (TFT) SENSOR: Originally called a transmission oil temperature (TOT) sensor, this input device to the ECM/PCM senses the fluid temperature and provides a resistance value. It operates on the thermistor principle.

TRANSMISSION INPUT SPEED (TIS) SENSOR: Measures turbine shaft (input shaft) rpm's and compares to engine rpm's to determine torque converter slip. When compared to the transmission output speed sensor or VSS, gear ratio and clutch engagement timing can be determined.

TRANSMISSION OIL TEMPERATURE (TOT) SENSOR: (See transmission fluid temperature (TFT) sensor.)

TRANSMISSION RANGE SELECTOR (TRS) SWITCH: Tells the module which gear shift position the driver has chosen.

TRANSFER CASE: A gearbox driven from the transmission that delivers power to both front and rear driveshafts in a four-wheel drive system. Transfer cases usually have a high and low range set of gears, used depending on how much pulling power is needed.

TRANSISTOR: A semi-conductor component which can be actuated by a small voltage to perform an electrical switching function.

TREAD WEAR INDICATOR: Bars molded into the tire at right angles to the tread that appear as horizontal bars when 1/16 in. of tread remains.

TREAD WEAR PATTERN: The pattern of wear on tires which can be "read" to diagnose problems in the front suspension.

TUNE-UP: A regular maintenance function, usually associated with the replacement and adjustment of parts and components in the electrical and fuel systems of a vehicle for the purpose of attaining optimum performance.

TURBINE: The output (driven) member of a fluid coupling or fluid torque converter. It is splined to the input (turbine) shaft of the transmission.

TURBOCHARGER: An exhaust driven pump which compresses intake air and forces it into the combustion chambers at higher than atmospheric pressures. The increased air pressure allows more fuel to be burned and results in increased horsepower being produced.

TURBULENCE: The interference of molecules of a fluid (or vapor) with each other in a fluid flow.

TYPE F: Transmission fluid developed and used by Ford Motor Company up to 1982. This fluid type provides a high coefficient of friction.

TYPE 7176: The preferred choice of transmission fluid for Chrysler automatic transmissions and transaxles. Developed in 1986, it closely resembles Dexron and Mercon. Type 7176 is the recommended service fill fluid for all Chrysler products utilizing a lockup torque converter dating back to 1978.

U-JOINT (UNIVERSAL JOINT): A flexible coupling in the drive train that allows the driveshafts or axle shafts to operate at different angles and still transmit rotary power.

UNDERSTEER: The tendency of a car to continue straight ahead while negotiating a turn.

UNIT BODY: Design in which the car body acts as the frame.

UNLEADED FUEL: Fuel which contains no lead (a common gasoline additive). The presence of lead in fuel will destroy the functioning elements of a catalytic converter, making it useless.

UNSPRUNG WEIGHT: The weight of car components not supported by the springs (wheels, tires, brakes, rear axle, control arms, etc.).

UPSHIFT: A shift that results in a decrease in torque ratio and an increase in speed.

VACUUM: A negative pressure; any pressure less than atmospheric pressure.

VACUUM ADVANCE: A device which advances the ignition timing in response to increased engine vacuum.

VACUUM GAUGE: An instrument used for measuring the existing vacuum in a vacuum circuit or chamber. The unit of measure is inches (of mercury in a barometer).

VACUUM MODULATOR: Generates a hydraulic oil pressure in response to the amount of engine vacuum.

VALVES: Devices that can open or close fluid passages in a hydraulic system and are used for directing fluid flow and controlling pressure.

VALVE BODY ASSEMBLY: The main hydraulic control assembly of the transmission/transaxle that contains numerous valves, check balls, and other components to control the distribution of pressurized oil throughout the transmission.

VALVE CLEARANCE: The measured gap between the end of the valve stem and the rocker arm, cam lobe or follower that activates the valve.

VALVE GUIDES: The guide through which the stem of the valve passes.

The guide is designed to keep the valve in proper alignment.

VALVE LASH (clearance): The operating clearance in the valve train.

VALVE TRAIN: The system that operates intake and exhaust valves, consisting of camshaft, valves and springs, lifters, pushrods and rocker arms.

VAPOR LOCK: Boiling of the fuel in the fuel lines due to excess heat. This will interfere with the flow of fuel in the lines and can completely stop the flow. Vapor lock normally only occurs in hot weather.

VARIABLE DISPLACEMENT (VARIABLE CAPACITY) VANE PUMP: Slipper-type vanes, mounted in a revolving rotor and contained within the bore of a movable slide, capture and then force fluid to flow. Movement of the slide to various positions changes the size of the vane chambers and the amount of fluid flow. **Note:** GM refers to this pump design as variable displacement, and Ford terms it variable capacity.

VARIABLE FORCE SOLENOID (VFS): Commonly referred to as the electronic pressure control (EPC) solenoid, it replaces the cable/linkage style of TV system control and is integrated with a spool valve and spring assembly to control pressure. A variable computer-controlled current flow varies the internal force of the solenoid on the spool valve and resulting control pressure.

VARIABLE ORIFICE THERMAL VALVE: Temperature-sensitive hydraulic oil control device that adjusts the size of a circuit path opening. By altering the size of the opening, the oil flow rate is adapted for cold to hot oil viscosity changes.

VARNISH: Term applied to the residue formed when gasoline gets old and stale.

VCM: See Electronic Control Unit (ECU).

VEHICLE SPEED SENSOR (VSS): Provides an electrical signal to the computer module, measuring vehicle speed, and affects the torque converter clutch engagement and release.

VESPEL SEALING RINGS: Hard plastic material that produces excellent sealing in dynamic settings. These rings are found in late versions of the 4T60 and in all 4T60-E and 4T80-E transaxles.

VISCOSITY: The ability of a fluid to flow. The lower the viscosity rating, the easier the fluid will flow. 10 weight motor oil will flow much easier than 40 weight motor oil.

VISCOSITY INDEX IMPROVERS: Keeps the viscosity nearly constant with changes in temperature. This is especially important at low temperatures, when the oil needs to be thin to aid in shifting and for cold-weather starting. Yet it must not be so thin that at high temperatures it will cause excessive hydraulic leakage so that pumps are unable to maintain the proper pressures.

VISCOUS CLUTCH: A specially designed torque converter clutch apply plate that, through the use of a silicon fluid, clamps smoothly and absorbs torsional vibrations.

VOLT: Unit used to measure the force or pressure of electricity. It is defined as the pressure needed to move one amp through the resistance of one ohm.

VOLTAGE: The electrical pressure that causes current to flow. Voltage is measured in volts (V).

VOLTAGE, APPLIED: The actual voltage read at a given point in a circuit. It equals the available voltage of the power supply minus the losses in the circuit up to that point.

VOLTAGE DROP: The voltage lost or used in a circuit by normal loads such as a motor or lamp or by abnormal loads such as a poor (high-resistance) lead or terminal connection.

VOLTAGE REGULATOR: A device that controls the current output of the alternator or generator.

VOLTMETER: An instrument used for measuring electrical force in units called volts. Voltmeters are always connected parallel with the circuit being tested.

VORTEX FLOW: The crosswise or circulatory flow of oil between the blades of the members caused by the centrifugal pumping action of the impeller.

WANKEL ENGINE: An engine which uses no pistons. In place of pistons, triangular-shaped rotors revolve in specially shaped housings.

WATER PUMP: A belt driven component of the cooling system that mounts on the engine, circulating the coolant under pressure.

WATT: The unit for measuring electrical power. One watt is the product of one ampere and one volt (watts equals amps times volts). Wattage is the horsepower of electricity (746 watts equal one horsepower).

WHEEL ALIGNMENT: Inclusive term to describe the front end geometry (caster, camber, toe-in/out).

WHEEL CYLINDER: Found in the automotive drum brake assembly, it is a device, actuated by hydraulic pressure, which, through internal pistons, pushes the brake shoes outward against the drums.

WHEEL WEIGHT: Small weights attached to the wheel to balance the wheel and tire assembly. Out-of-balance tires quickly wear out and also give erratic handling when installed on the front.

WHEELBASE: Distance between the center of front wheels and the center of rear wheels.

WIDE OPEN THROTTLE (WOT): Full travel of accelerator pedal.

WORK: The force exerted to move a mass or object. Work involves motion; if a force is exerted and no motion takes place, no work is done. Work per unit of time is called power. Work = force x distance = ft. lbs. 33,000 ft. lbs. in one minute = 1 horsepower

ZERO-THROTTLE COAST DOWN: A full release of accelerator pedal while vehicle is in motion and in drive range.

Commonly Used Abbreviations

2

2WD	Two Wheel Drive

4

4WD	Four Wheel Drive

A

A/C	Air Conditioning
ABDC	After Bottom Dead Center
ABS	Anti-lock Brakes
AC	Alternating Current
ACL	Air cleaner
ACT	Air Charge Temperature
AIR	Secondary Air Injection
ALCL	Assembly Line Communications Link
ALDL	Assembly Line Diagnostic Link
AT	Automatic Transaxle/Transmission
ATDC	After Top Dead Center
ATF	Automatic Transmission Fluid
ATS	Air Temperature Sensor
AWD	All Wheel Drive

B

BAP	Barometric Absolute Pressure
BARO	Barometric Pressure
BBDC	Before Bottom Dead Center
BCM	Body Control Module
BDC	Bottom Dead Center
BPT	Backpressure Transducer
BTDC	Before Top Dead Center
BVSV	Bimetallic Vacuum Switching Valve

C

CAC	Charge Air Cooler
CARB	California Air Resources Board
CAT	Catalytic Converter
CCC	Computer Command Control
CCCC	Computer Controlled Catalytic Converter
CCCI	Computer Controlled Coil Ignition
CCD	Computer Controlled Dwell
CDI	Capacitor Discharge Ignition
CEC	Computerized Engine Control
CFI	Continuous Fuel Injection
CIS	Continuous Injection System
CIS-E	Continuous Injection System - Electronic
CKP	Crankshaft Position
CL	Closed Loop
CMP	Camshaft Position
CPP	Clutch Pedal Position
CTOX	Continuous Trap Oxidizer System
CTP	Closed Throttle Position
CVC	Constant Vacuum Control
CYL	Cylinder

D

DBC	Dual Bed Catalyst
DC	Direct Current
DFI	Direct Fuel Injection
DIS	Distributorless Ignition System
DLC	Data Link Connector
DMM	Digital Multimeter
DOHC	Double Overhead Camshaft
DRB	Diagnostic Readout Box
DTC	Diagnostic Trouble Code
DTM	Diagnostic Test Mode
DVOM	Digital Volt/Ohmmeter

E

EBCM	Electronic Brake Control Module
ECM	Engine Control Module
ECT	Engine Coolant Temperature
ECU	Engine Control Unit or Electronic Control Unit
EDIS	Electronic Distributorless Ignition System
EEC	Electronic Engine Control
EEPROM	Electrically Erasable Programmable Read Only Memory
EFE	Early Fuel Evaporation
EGR	Exhaust Gas Recirculation
EGRT	Exhaust Gas Recirculation Temperature
EGRVC	EGR Valve Control
EPROM	Erasable Programmable Read Only Memory
EVAP	Evaporative Emissions
EVP	EGR Valve Position

F

FBC	Feedback Carburetor
FEEPROM	Flash Electrically Erasable Programmable Read Only Memory
FF	Flexible Fuel
FI	Fuel Injection
FT	Fuel Trim
FWD	Front Wheel Drive

G

GND	Ground

H

HAC	High Altitude Compensation
HEGO	Heated Exhaust Gas Oxygen sensor
HEI	High Energy Ignition
HO2 Sensor	Heated Oxygen Sensor

I

IAC	Idle Air Control
IAT	Intake Air Temperature
ICM	Ignition Control Module
IFI	Indirect Fuel Injection
IFS	Inertia Fuel Shutoff
ISC	Idle Speed Control
IVSV	Idle Vacuum Switching Valve

Commonly Used Abbreviations

K

KOEO	Key On, Engine Off
KOER	Key ON, Engine Running
KS	Knock Sensor

M

MAF	Mass Air Flow
MAP	Manifold Absolute Pressure
MAT	Manifold Air Temperature
MC	Mixture Control
MDP	Manifold Differential Pressure
MFI	Multiport Fuel Injection
MIL	Malfunction Indicator Lamp or Maintenance
MST	Manifold Surface Temperature
MVZ	Manifold Vacuum Zone

N

NVRAM	Nonvolatile Random Access Memory

O

O2 Sensor	Oxygen Sensor
OBD	On-Board Diagnostic
OC	Oxidation Catalyst
OHC	Overhead Camshaft
OL	Open Loop

P

P/S	Power Steering
PAIR	Pulsed Secondary Air Injection
PCM	Powertrain Control Module
PCS	Purge Control Solenoid
PCV	Positive Crankcase Ventilation
PIP	Profile Ignition Pick-up
PNP	Park/Neutral Position
PROM	Programmable Read Only Memory
PSP	Power Steering Pressure
PTO	Power Take-Off
PTOX	Periodic Trap Oxidizer System

R

RABS	Rear Anti-lock Brake System
RAM	Random Access Memory
ROM	Read Only Memory
RPM	Revolutions Per Minute
RWAL	Rear Wheel Anti-lock Brakes
RWD	Rear Wheel Drive

S

SBC	Single Bed Converter
SBEC	Single Board Engine Controller
SC	Supercharger
SCB	Supercharger Bypass
SFI	Sequential Multiport Fuel Injection
SIR	Supplemental Inflatible Restraint
SOHC	Single Overhead Camshaft
SPL	Smoke Puff Limiter
SPOUT	Spark Output
SRI	Service Reminder Indicator
SRS	Supplemental Restraint System
SRT	System Readiness Test
SSI	Solid State Ignition
ST	Scan Tool
STO	Self-Test Output

T

TAC	Thermostatic Air Clearner
TBI	Throttle Body Fuel Injection
TC	Turbocharger
TCC	Torque Converter Clutch
TCM	Transmission Control Module
TDC	Top Dead Center
TFI	Thick Film Ignition
TP	Throttle Position
TR Sensor	Transaxle/Transmission Range Sensor
TVV	Thermal Vacuum Valve
TWC	Three-way Catalytic Converter

V

VAF	Volume Air Flow, or Vane Air Flow
VAPS	Variable Assist Power Steering
VRV	Vacuum Regulator Valve
VSS	Vehicle Speed Sensor
VSV	Vacuum Switching Valve

W

WOT	Wide Open Throttle
WU-TWC	Warm Up Three-way Catalytic Converter

ENGLISH TO METRIC CONVERSION: TORQUE

To convert foot-pounds (ft. lbs.) to Newton-meters (Nm), multiply the number of ft. lbs. by 1.36

To convert Newton-meters (Nm) to foot-pounds (ft. lbs.), multiply the number of Nm by 0.7376

ft. lbs.	Nm	ft. lbs.	Nm	ft. lbs.	Nm	ft. lbs.	Nm
0.1	0.1	34	46.2	76	103.4	118	160.5
0.2	0.3	35	47.6	77	104.7	119	161.8
0.3	0.4	36	49.0	78	106.1	120	163.2
0.4	0.5	37	50.3	79	107.4	121	164.6
0.5	0.7	38	51.7	80	108.8	122	165.9
0.6	0.8	39	53.0	81	110.2	123	167.3
0.7	1.0	40	54.4	82	111.5	124	168.6
0.8	1.1	41	55.8	83	112.9	125	170.0
0.9	1.2	42	57.1	84	114.2	126	171.4
1	1.4	43	58.5	85	115.6	127	172.7
2	2.7	44	59.8	86	117.0	128	174.1
3	4.1	45	61.2	87	118.3	129	175.4
4	5.4	46	62.6	88	119.7	130	176.8
5	6.8	47	63.9	89	121.0	131	178.2
6	8.2	48	65.3	90	122.4	132	179.5
7	9.5	49	66.6	91	123.8	133	180.9
8	10.9	50	68.0	92	125.1	134	182.2
9	12.2	51	69.4	93	126.5	135	183.6
10	13.6	52	70.7	94	127.8	136	185.0
11	15.0	53	72.1	95	129.2	137	186.3
12	16.3	54	73.4	96	130.6	138	187.7
13	17.7	55	74.8	97	131.9	139	189.0
14	19.0	56	76.2	98	133.3	140	190.4
15	20.4	57	77.5	99	134.6	141	191.8
16	21.8	58	78.9	100	136.0	142	193.1
17	23.1	59	80.2	101	137.4	143	194.5
18	24.5	60	81.6	102	138.7	144	195.8
19	25.8	61	83.0	103	140.1	145	197.2
20	27.2	62	84.3	104	141.4	146	198.6
21	28.6	63	85.7	105	142.8	147	199.9
22	29.9	64	87.0	106	144.2	148	201.3
23	31.3	65	88.4	107	145.5	149	202.6
24	32.6	66	89.8	108	146.9	150	204.0
25	34.0	67	91.1	109	148.2	151	205.4
26	35.4	68	92.5	110	149.6	152	206.7
27	36.7	69	93.8	111	151.0	153	208.1
28	38.1	70	95.2	112	152.3	154	209.4
29	39.4	71	96.6	113	153.7	155	210.8
30	40.8	72	97.9	114	155.0	156	212.2
31	42.2	73	99.3	115	156.4	157	213.5
32	43.5	74	100.6	116	157.8	158	214.9
33	44.9	75	102.0	117	159.1	159	216.2

METRIC TO ENGLISH CONVERSION: TORQUE

To convert foot-pounds (ft. lbs.) to Newton-meters (Nm), multiply the number of ft. lbs. by 1.36
To convert Newton-meters (Nm) to foot-pounds (ft. lbs.), multiply the number of Nm by 0.7376

Nm	ft. lbs.	Nm	ft. lbs.	Nm	ft. lbs.	Nm	ft. lbs.	Nm	ft. lbs.
0.1	0.1	34	25.0	76	55.9	118	86.8	160	117.6
0.2	0.1	35	25.7	77	56.6	119	87.5	161	118.4
0.3	0.2	36	26.5	78	57.4	120	88.2	162	119.1
0.4	0.3	37	27.2	79	58.1	121	89.0	163	119.9
0.5	0.4	38	27.9	80	58.8	122	89.7	164	120.6
0.6	0.4	39	28.7	81	59.6	123	90.4	165	121.3
0.7	0.5	40	29.4	82	60.3	124	91.2	166	122.1
0.8	0.6	41	30.1	83	61.0	125	91.9	167	122.8
0.9	0.7	42	30.9	84	61.8	126	92.6	168	123.5
1	0.7	43	31.6	85	62.5	127	93.4	169	124.3
2	1.5	44	32.4	86	63.2	128	94.1	170	125.0
3	2.2	45	33.1	87	64.0	129	94.9	171	125.7
4	2.9	46	33.8	88	64.7	130	95.6	172	126.5
5	3.7	47	34.6	89	65.4	131	96.3	173	127.2
6	4.4	48	35.3	90	66.2	132	97.1	174	127.9
7	5.1	49	36.0	91	66.9	133	97.8	175	128.7
8	5.9	50	36.8	92	67.6	134	98.5	176	129.4
9	6.6	51	37.5	93	68.4	135	99.3	177	130.1
10	7.4	52	38.2	94	69.1	136	100.0	178	130.9
11	8.1	53	39.0	95	69.9	137	100.7	179	131.6
12	8.8	54	39.7	96	70.6	138	101.5	180	132.4
13	9.6	55	40.4	97	71.3	139	102.2	181	133.1
14	10.3	56	41.2	98	72.1	140	102.9	182	133.8
15	11.0	57	41.9	99	72.8	141	103.7	183	134.6
16	11.8	58	42.6	100	73.5	142	104.4	184	135.3
17	12.5	59	43.4	101	74.3	143	105.1	185	136.0
18	13.2	60	44.1	102	75.0	144	105.9	186	136.8
19	14.0	61	44.9	103	75.7	145	106.6	187	137.5
20	14.7	62	45.6	104	76.5	146	107.4	188	138.2
21	15.4	63	46.3	105	77.2	147	108.1	189	139.0
22	16.2	64	47.1	106	77.9	148	108.8	190	139.7
23	16.9	65	47.8	107	78.7	149	109.6	191	140.4
24	17.6	66	48.5	108	79.4	150	110.3	192	141.2
25	18.4	67	49.3	109	80.1	151	111.0	193	141.9
26	19.1	68	50.0	110	80.9	152	111.8	194	142.6
27	19.9	69	50.7	111	81.6	153	112.5	195	143.4
28	20.6	70	51.5	112	82.4	154	113.2	196	144.1
29	21.3	71	52.2	113	83.1	155	114.0	197	144.9
30	22.1	72	52.9	114	83.8	156	114.7	198	145.6
31	22.8	73	53.7	115	84.6	157	115.4	199	146.3
32	23.5	74	54.4	116	85.3	158	116.2	200	147.1
33	24.3	75	55.1	117	86.0	159	116.9	201	147.8

ENGLISH/METRIC CONVERSION: TEMPERATURE

To convert Fahrenheit (F°) to Celsius (C°), take F° temperature and subtract 32, multiply the result by 5 and divide the result by 9
To convert Celsius (C°) to Fahrenheit (F°), take C° temperature and multiply it by 9, divide the result by 5 and add 32

F°	C°	F°	C°	C°	F°	C°	F°
-40	-40.0	150	65.6	-38	-36.4	46	114.8
-35	-37.2	155	68.3	-36	-32.8	48	118.4
-30	-34.4	160	71.1	-34	-29.2	50	122
-25	-31.7	165	73.9	-32	-25.6	52	125.6
-20	-28.9	170	76.7	-30	-22	54	129.2
-15	-26.1	175	79.4	-28	-18.4	56	132.8
-10	-23.3	180	82.2	-26	-14.8	58	136.4
-5	-20.6	185	85.0	-24	-11.2	60	140
0	-17.8	190	87.8	-22	-7.6	62	143.6
1	-17.2	195	90.6	-20	-4	64	147.2
2	-16.7	200	93.3	-18	-0.4	66	150.8
3	-16.1	205	96.1	-16	3.2	68	154.4
4	-15.6	210	98.9	-14	6.8	70	158
5	-15.0	212	100.0	-12	10.4	72	161.6
10	-12.2	215	101.7	-10	14	74	165.2
15	-9.4	220	104.4	-8	17.6	76	168.8
20	-6.7	225	107.2	-6	21.2	78	172.4
25	-3.9	230	110.0	-4	24.8	80	176
30	-1.1	235	112.8	-2	28.4	82	179.6
35	1.7	240	115.6	0	32	84	183.2
40	4.4	245	118.3	2	35.6	86	186.8
45	7.2	250	121.1	4	39.2	88	190.4
50	10.0	255	123.9	6	42.8	90	194
55	12.8	260	126.7	8	46.4	92	197.6
60	15.6	265	129.4	10	50	94	201.2
65	18.3	270	132.2	12	53.6	96	204.8
70	21.1	275	135.0	14	57.2	98	208.4
75	23.9	280	137.8	16	60.8	100	212
80	26.7	285	140.6	18	64.4	102	215.6
85	29.4	290	143.3	20	68	104	219.2
90	32.2	295	146.1	22	71.6	106	222.8
95	35.0	300	148.9	24	75.2	108	226.4
100	37.8	305	151.7	26	78.8	110	230
105	40.6	310	154.4	28	82.4	112	233.6
110	43.3	315	157.2	30	86	114	237.2
115	46.1	320	160.0	32	89.6	116	240.8
120	48.9	325	162.8	34	93.2	118	244.4
125	51.7	330	165.6	36	96.8	120	248
130	54.4	335	168.3	38	100.4	122	251.6
135	57.2	340	171.1	40	104	124	255.2
140	60.0	345	173.9	42	107.6	126	258.8
145	62.8	350	176.7	44	111.2	128	262.4

LENGTH CONVERSION

To convert inches (in.) to millimeters (mm), multiply the number of inches by 25.4

To convert millimeters (mm) to inches (in.), multiply the number of millimeters by 0.04

Inches	Millimeters	Inches	Millimeters	Inches	Millimeters	Inches	Millimeters
0.0001	0.00254	0.005	0.1270	0.09	2.286	4	101.6
0.0002	0.00508	0.006	0.1524	0.1	2.54	5	127.0
0.0003	0.00762	0.007	0.1778	0.2	5.08	6	152.4
0.0004	0.01016	0.008	0.2032	0.3	7.62	7	177.8
0.0005	0.01270	0.009	0.2286	0.4	10.16	8	203.2
0.0006	0.01524	0.01	0.254	0.5	12.70	9	228.6
0.0007	0.01778	0.02	0.508	0.6	15.24	10	254.0
0.0008	0.02032	0.03	0.762	0.7	17.78	11	279.4
0.0009	0.02286	0.04	1.016	0.8	20.32	12	304.8
0.001	0.0254	0.05	1.270	0.9	22.86	13	330.2
0.002	0.0508	0.06	1.524	1	25.4	14	355.6
0.003	0.0762	0.07	1.778	2	50.8	15	381.0
0.004	0.1016	0.08	2.032	3	76.2	16	406.4

ENGLISH/METRIC CONVERSION: LENGTH

To convert inches (in.) to millimeters (mm), multiply the number of inches by 25.4
To convert millimeters (mm) to inches (in.), multiply the number of millimeters by 0.04

Inches		Millimeters	Inches		Millimeters	Inches		Millimeters
Fraction	Decimal	Decimal	Fraction	Decimal	Decimal	Fraction	Decimal	Decimal
1/64	0.016	0.397	11/32	0.344	8.731	11/16	0.688	17.463
1/32	0.031	0.794	23/64	0.359	9.128	45/64	0.703	17.859
3/64	0.047	1.191	3/8	0.375	9.525	23/32	0.719	18.256
1/16	0.063	1.588	25/64	0.391	9.922	47/64	0.734	18.653
5/64	0.078	1.984	13/32	0.406	10.319	3/4	0.750	19.050
3/32	0.094	2.381	27/64	0.422	10.716	49/64	0.766	19.447
7/64	0.109	2.778	7/16	0.438	11.113	25/32	0.781	19.844
1/8	0.125	3.175	29/64	0.453	11.509	51/64	0.797	20.241
9/64	0.141	3.572	15/32	0.469	11.906	13/16	0.813	20.638
5/32	0.156	3.969	31/64	0.484	12.303	53/64	0.828	21.034
11/64	0.172	4.366	1/2	0.500	12.700	27/32	0.844	21.431
3/16	0.188	4.763	33/64	0.516	13.097	55/64	0.859	21.828
13/64	0.203	5.159	17/32	0.531	13.494	7/8	0.875	22.225
7/32	0.219	5.556	35/64	0.547	13.891	57/64	0.891	22.622
15/64	0.234	5.953	9/16	0.563	14.288	29/32	0.906	23.019
1/4	0.250	6.350	37/64	0.578	14.684	59/64	0.922	23.416
17/64	0.266	6.747	19/32	0.594	15.081	15/16	0.938	23.813
9/32	0.281	7.144	39/64	0.609	15.478	61/64	0.953	24.209
19/64	0.297	7.541	5/8	0.625	15.875	31/32	0.969	24.606
5/16	0.313	7.938	41/64	0.641	16.272	63/64	0.984	25.003
21/64	0.328	8.334	21/32	0.656	16.669	1/1	1.000	25.400
			43/64	0.672	17.066			